P9-DXM-551

Television
Engineering
Handbook

Other McGraw-Hill Reference Books of Interest

Handbooks

Baumeister and Marks   STANDARD HANDBOOK FOR MECHANICAL ENGINEERS

Beeman   INDUSTRIAL POWER SYSTEMS HANDBOOK

Coombs   BASIC ELECTRONIC INSTRUMENT HANDBOOK

Coombs   PRINTED CIRCUITS HANDBOOK

Croft, Carr, and Watt   AMERICAN ELECTRICIANS' HANDBOOK

Fink and Beaty   STANDARD HANDBOOK FOR ELECTRICAL ENGINEERS

Fink and Christiansen   ELECTRONICS ENGINEERS' HANDBOOK

Giacoletto   ELECTRONIC DESIGNER'S HANDBOOK

Harper   HANDBOOK OF ELECTRONIC PACKAGING

Harper   HANDBOOK FOR MATERIALS AND PROCESSES FOR ELECTRONICS

Harper   HANDBOOK OF THICK FILM HYBRID MICROELECTRONICS

Harper   HANDBOOK OF WIRING, CABLING, AND INTERCONNECTING FOR ELECTRONICS

Hicks   STANDARD HANDBOOK OF ENGINEERING CALCULATIONS

Johnson and Jasik   ANTENNA ENGINEERING HANDBOOK

Juran   QUALITY CONTROL HANDBOOK

Kaufman and Seidman   HANDBOOK OF ELECTRONICS CALCULATIONS

Kaufman and Seidman   HANDBOOK FOR ELECTRONICS TECHNICIANS

Kurtz   HANDBOOK OF ENGINEERING ECONOMICS

Perry   ENGINEERING MANUAL

Stout   HANDBOOK OF MICROPROCESSOR DESIGN AND APPLICATIONS

Stout and Kaufman   HANDBOOK OF MICROCIRCUIT DESIGN AND APPLICATION

Stout and Kaufman   HANDBOOK OF OPERATIONAL AMPLIFIER DESIGN

Tuma   ENGINEERING MATHEMATICS HANDBOOK

Williams   DESIGNER'S HANDBOOK OF INTEGRATED CIRCUITS

Williams   ELECTRONIC FILTER DESIGN HANDBOOK

Encyclopedias

CONCISE ENCYCLOPEDIA OF SCIENCE AND TECHNOLOGY

ENCYCLOPEDIA OF ELECTRONICS AND COMPUTERS

ENCYCLOPEDIA OF ENGINEERING

Dictionaries

DICTIONARY OF SCIENTIFIC AND TECHNICAL TERMS

DICTIONARY OF COMPUTERS

DICTIONARY OF ELECTRICAL AND ELECTRONIC ENGINEERING

DICTIONARY OF ENGINEERING

Markus   ELECTRONICS DICTIONARY

Television
Engineering
Handbook

K. Blair Benson *Editor in Chief*

Engineering Consultant
Member, Audio Engineering Society
Senior and Life Member, Institute of
    Electrical and Electronics Engineers
Fellow and Life Member, Society of
    Motion Picture and Television Engineers

McGraw-Hill Book Company

New York   St. Louis   San Francisco   Auckland   Bogotá   Hamburg
London   Madrid   Mexico   Montreal   New Delhi
Panama   Paris   São Paulo   Singapore   Sydney   Tokyo   Toronto

Library of Congress Cataloging in Publication Data
Main entry under title:

Television engineering handbook.

    Includes index.
    1. Television—Handbooks, manuals, etc.
I. Benson, K. Blair.
TK6642.T437   1985   621.388   85-4293
ISBN 0-07-004779-0

Copyright © 1986 by McGraw-Hill, Inc. All rights reserved.
Printed in the United States of America. Except as permitted
under the United States Copyright Act of 1976, no part of this
publication may be reproduced or distributed in any form or by
any means, or stored in a data base or retrieval system, without
the prior written permission of the publisher.

    567890   DOC/DOC   987

ISBN 0-07-004779-0

The editors for this book were Harold B. Crawford and David
Fogarty, the designer was Mark E. Safran, and the production
supervisor was Teresa F. Leaden. It was set in Century School-
book by University Graphics, Inc.

Printed and bound by R. R. Donnelley & Sons Company.

To Dolly . . .

whose encouragement and patience
have been most valuable contributions
to the publication of this Handbook

Contents

Contributors

EDWARD W. ALLEN, *Federal Communications Commission (retired), Washington, D.C.* (Ch. 6)

WILLIAM F. BAILEY, *Hazeltine Corporation (retired), Little Neck, N.Y.* (Ch. 4)

ODED BEN-DOV, *RCA Corporation, Gibbsboro, N.J.* (coauthor Ch. 8)

K. BLAIR BENSON, *Engineering Consultant, Norwalk, Conn.* (Chs. 9, 17; coauthor Ch. 14)

H. NEAL BERTRAM, *Ampex Corporation, Redwood City, Cal.* (Ch. 15)

MICHAEL BETTS, *The Grass Valley Group, Grass Valley, Cal.* (Ch. 14)

JAMES E. BLECKSMITH, *The Grass Valley Group, Grass Valley, Cal.* (Ch. 14)

W. LYLE BREWER, *Eastman Kodak (retired), Rochester, N.Y.* (coauthor Chs. 1 and 3)

E. STANLEY BUSBY, *Ampex Corporation, Redwood City, Cal.* (Ch. 15)

RICHARD J. CALDWELL, *TVC Video Inc., New York, N.Y.* (author Ch. 20)

ROBERT A. CASTRIGNANO, *CBS Technology Center, Stamford, Conn.* (author Ch. 16)

RICHARD G. CLAPP, *Consultant, Havertown, Pa.; Philco-Ford (retired)* (coauthor Ch. 5)

HAROLD V. CLARK, *Ampex Corporation, Redwood City, Cal.* (Ch. 15)

ALISTAIR COCKBURN, *IBM Corporation, Zurich, Switzerland* (coauthor Ch. 19)

A. D. COPE, *David Sarnoff Research Center, RCA Corporation, Princeton, N.J.* (Ch. 11)

WILLIAM DANIEL, *Federal Communications Commission, Washington, D.C.* (author Ch. 6)

L. E. DONOVAN, *NAP Consumer Electronics Corporation, Knoxville, Tenn.* (Ch. 13)

CARL ELLISON, *Evans and Sutherland, Salt Lake City, Utah* (coauthor Ch. 19)

MICHAEL O. FELIX, *Ampex Corporation, Redwood City, Cal.* (Ch. 15)

DONALD G. FINK, *Somers, N.Y.; Director Emeritus, Institute of Electrical and Electronics Engineers;* (author Ch. 22)

JOSEPH F. FISHER, *Wynnewood, Pa.; Philco-Ford (retired)* (coauthor Chs. 2 and 5)

CHARLES P. GINSBURG, *Ampex Corporation, Redwood City, Cal.* (author Ch. 15)

BEVERLEY R. GOOCH, *Ampex Corporation, Redwood City, Cal.* (Ch. 15)

JOHN HARTNETT, *CBS Television Network, New York, N.Y.* (Ch. 14)

R. A. HEDLER, *Philips ECG, Seneca Falls, N.Y.* (coauthor Ch. 12)

DOUGLAS J. HENNESSY, *CBS Television Network, New York, N.Y.* (Ch. 14)

L. H. HOKE, JR., *NAP Consumer Electronics Corporation, Knoxville, Tenn.* (author Ch. 13)

ROBERT N. HURST, *RCA Corporation, Gibbsboro, N.J.* (Ch. 17)

ROBERT JULL, *The Grass Valley Group, Grass Valley, Cal.* (Ch. 14)

CHARLES A. KASE, *Satellite Systems Engineering, Bethesda, Md.* (Ch. 9)

KARL KINAST, *Consultant, Westwood, N.J.* (Ch. 14)

JOHN W. KING, *Ampex Corporation, Redwood City, Cal.* (Ch. 15)

J. D. KNOX, *NAP Consumer Electronics Corporation, Knoxville, Tenn.* (Ch. 13)

CHARLES J. KUCA, *The Grass Valley Group, Grass Valley, Cal.* (Ch. 14)

ANTHONY H. LIND, *RCA Corporation (retired), Camden, N.J.* (coauthor Ch. 17)

KENNETH G. LISK, *Eastman Kodak (retired), Rochester, N.Y.* (coauthor Ch. 17)

DONALD C. LIVINGSTON, *Sylvania Electric Products (retired), Bayside, N.Y.* (Ch. 4)

BERNARD D. LOUGHLIN, *Hazeltine Corporation, Greenlawn, N.Y.* (author Ch. 4)

RENVILLE H. McMANN, *CBS Inc., Stamford, Conn.* (Ch. 14)

WILLIAM McSWEENEY, *Ampex Corporation, Redwood City, Cal.* (Ch. 15)

D. STEVENS McVOY, *Coaxial Communications Corporation, Columbus, Ohio* (coauthor Ch. 9)

L. L. MANINGER, *Philips ECG, Seneca Falls, N.Y.* (coauthor Ch. 12)

D. E. MANNERS, *NAP Consumer Electronics Corporation, Knoxville, Tenn.* (Ch. 13)

JAMES MICHENER, *The Grass Valley Group, Grass Valley, Cal.* (Ch. 14)

W. G. MILLER, *NAP Consumer Electronics Corporation, Knoxville, Tenn.* (Ch. 13)

R. A. MOMBERGER, *Philips ECG, Seneca Falls, N.Y.* (coauthor Ch. 12)

ROBERT A. MORRIS, *Eastman Kodak (retired), Rochester, N.Y.* (coauthor Chs. 1 and 3)

ROBERT G. NEUHAUSER, *RCA Corporation, New Products Division, Lancaster, Pa.* (author Ch. 11)

REGINALD OLDERSHAW, *Ampex Corporation, Redwood City, Cal.* (Ch. 15)

R. J. PEFFER, *NAP Consumer Electronics Corporation, Knoxville, Tenn.* (Ch. 13)

ROBERT H. PERRY, *Ampex Corporation, Redwood City, Cal.* (Ch. 15)

KRISHNA PRABA, *RCA Corporation, Gibbsboro, N.J.* (coauthor Ch. 8)

DALTON H. PRITCHARD, *RCA Laboratories, David Sarnoff Research Center, Princeton, N.J.* (author Ch. 21)

WILBUR L. PRITCHARD, *Satellite Systems Engineering, Bethesda, Md.* (Ch. 9)

RAYMOND F. RAVIZZA, *Ampex Corporation, Redwood City, Cal.* (Ch. 15)

BRUCE RAYNER, *The Grass Valley Group, Grass Valley, Cal.* (coauthor Ch. 14)

FREDERICK M. REMLEY, *University of Michigan TV Center, Ann Arbor, Mich.* (coauthor Ch. 14)

J. D. ROBBINS, *Philips ECG, Seneca Falls, N.Y.* (coauthor Ch. 12)

ALAN R. ROBERTSON, *Division of Physics, National Research Council of Canada, Ottawa, Canada* (coauthor Ch. 2)

JOSEPH ROIZEN, *Telegen, Palo Alto, Cal.* (coauthor Ch. 14)

DENNIS M. RYAN, *Ampex Corporation, Redwood City, Cal.* (Ch. 15)

DONALD L. SAY, *Philips ECG, Seneca Falls, N.Y.* (coauthor Ch. 12)

MARK SCHUBIN, *Consultant, New York, N.Y.* (Ch. 14)

SOL SHERR, *Westland Electronics, Hartsdale, N.Y.* (author Ch. 10)

JOSEPH L. STERN, *Stern Telecommunications Corporation, New York, N.Y.* (coauthor Ch. 9)

ERNEST J. TARNAI, *Bell-Northern Research Ltd., Ottawa, Canada* (author Ch. 18)

DAVID E. TRYTKO, *Ampex Corporation, Redwood City, Cal.* (Ch. 15)

STEVEN WAGNER, *Ampex Corporation, Redwood City, Cal.* (Ch. 15)

JOHN T. WILNER, *Consultant, Boca Raton, Fla.* (author Ch. 7)

E. CARLTON WINCKLER, *Imero Fiorentino Associates, New York, N.Y.* (Ch. 14)

J. G. ZAHNEN, *NAP Consumer Electronics Corporation, Knoxville, Tenn.* (Ch. 13)

Foreword

Since 1957, when the first *Television Engineering Handbook* was published by McGraw-Hill, television engineering has been enriched by a host of new solutions to old problems. In the early 1950s, the content of television images could be stored only on photographic film. In 1956, the quadruplex method of video tape recording was demonstrated by Charles Ginsburg (a contributor to the new Handbook), and the subsequent development of magnetic recording has transformed program production and scheduling. Millions of home video cassette recorders have likewise transformed the viewing habits of the audience.

In those days video signals were used only in analog form. Digital methods were established in the computer industry, but the digital integrated circuit had not yet been invented. Several years passed before analog-to-digital converters could operate faster than 10 million bits per second, the rate required for digital video. Today, not only do the integrated circuits exist, but a worldwide standard for digitization of color video signals has been adopted. Digital signals are coming into use in studio and field operations prior to broadcast, and the digital video disc system for home entertainment is now available. Furthermore, digital video signal generation and processing have revolutionized the arts of visual effects and animation.

In 1957, the NTSC system of compatible color television had been on the air for four years, although it was not yet fully accepted by the public. By 1964 the logjam was broken as a result of compromises between the theoretically sound concepts of the NTSC system and the means for practical implementation adopted by industry.

At about the same time, two other major systems of color television, PAL and SECAM, made their appearance. With them came the need for line storage of the video signal. The subsequent development of large-scale integrated circuits made possible the digital storage of entire frames of video information. This, in turn, made possible the conversion of signals originating on one color standard to another standard, notably among NTSC, PAL, and SECAM systems. The cost of frame stores, initially in the tens of thousands of dollars, has been reduced so sharply that their use in domestic receivers is now confidently anticipated within a decade. Greatly improved performance of NTSC, PAL, and SECAM receivers will result. A first step in this direction was the introduction of comb filters in receivers, permitting full use of the luminance bandwidth normally limited in conventional receivers to avoid chrominance-luminance interference.

New methods of distributing televison signals, not treated in the first Handbook, are master-antenna and cable systems, satellite relay, video tapes, and video discs. Supplementary information is now transmitted during vertical blanking, as are reference signals for automatic control of receiver functions. Steady improvement in the quality of the received images has come from new camera-imaging tubes and devices, signal-enhancement systems, and the line-segment shadow-mask color tube. Most important has been the total replacement of vacuum tubes by transistors and integrated circuits throughout television equipment, with the exception of high-power transmitter output stages. Thus, the stability and durability of equipment have been so extended that the weakest links are now those involving mechanical elements, particularly manually operated receiver tuners, and these are being supplanted rapidly by voltage-controlled electronic components.

All this new technology has opened doors to greatly improved performance of the established systems, as well as wholly new approaches to advanced systems. These include methods of bandwidth compression, as well as the use of additional channel width in cable systems and direct satellite broadcasting to the home. The demonstration of 1125-line high-definition color images by the Japan Broadcasting Corporation in 1979 has offered the prospect of electronic motion-picture production without the use of film, except for release prints for theaters.

To cover these new developments in television engineering, as well as the underlying fundamentals, has required that the Handbook be entirely rewritten. To undertake this massive assignment, Editor in Chief K. Blair Benson assembled a team of seventy contributors, each a well-known specialist in his field. The new *Television Engineering Handbook* clearly fills the need for an authoritative and up-to-date compilation of the current state of television technology.

DONALD G. FINK

Editor in Chief, Television Engineering Handbook, 1957
Director Emeritus
Institute of Electrical and Electronic Engineers

Preface

Television technology involves, in broad terms, the conversion of visual images into electrical signals for transmission, distribution, and finally for restoration to a visual format on an electronic display. Television engineering is concerned with all aspects of the technology associated with the many intricate steps in this process from creative program production, with electronic processing and postproduction refinement, to picture reproduction. The intermediate transmission may be by analog or digital means, or a combination of the two, and may involve image or signal storage in a variety of media.

Television is a relatively new technology, as evidenced by the fact that less than half a century has passed since the invention of the first electronic camera tubes. In that period, television has evolved from a laboratory curiosity, popularized by public demonstrations, to a dominant form of worldwide entertainment and information distribution.

For example, the electronic image-storage camera, using a photosensitive vacuum tube invented by Zworykin in 1937, preceded by only a short four years the first regular, albeit experimental, television broadcasting operations by several organizations in both the United States and England. In the ensuing years, no less than seven basic camera pickup tubes were developed, some of which were put into practical use. These were followed more recently by several solid-state photosensitive devices which promise ultimately to dominate the field of monochrome and color camera design and application.

In an even shorter time, recording of television images on magnetic tape in the form of video signals has revolutionized the television program industry. This invention, combined with revolutionary digital video graphics and special effects, has spawned a new industry that is moving to supplant many creative functions heretofore performed, by necessity, on hard copy and reproduced on film and slides.

These rapid and continuing developments have dictated that in order to design, as well as operate, television equipment and systems, in addition to a working knowledge of present-day techniques and related hardware, it is essential that an engineer understand the basic principles of light, vision and information theory. Of equal importance is a familiarity with technical standards and good engineering practices accepted and adopted throughout the world.

The thorough coverage given in the Handbook to this broad range of subjects provides a comprehensive source of technical information and reference data for engineers engaged in design and development, as well as in maintenance and operation. Moreover, the tutorial dissertations serve as foundation for further study by engineers intending to embark on careers in television.

The topics may be categorized broadly as follows:

Fundamental concepts of television imagery and transmission

Signal generation and processing

Transmission

Reception

Picture reproduction

Reference data

In the introductory chapters, the basis for a better understanding of the many diverse engineering subjects dealt with in subsequent sections is provided by a detailed review of the principles of vision, photometry and optics, and their relationship to television engineering, followed by a theoretical discussion of color vision and the methods employed for graphical representation of luminance and color parameters.

Individual chapters are devoted to the allied fields of broadcasting, cable, and satellite distribution; signal and image storage on video tape, video disc and film; and digital techniques for signal processing and transmission, graphics generation and picture manipulation, and standards conversion. Although the emphasis is on the picture signal, sound-signal generation, transmission, and reception practices are covered where video and audio techniques are integrated or related.

Lastly, television system standards and recommended practices, adopted by industry agreement in the United States and in other areas throughout the world, are described and tabulated, followed by a full chapter devoted to a compilation of data, equations, and definitions not generally found in any single handbook on electronics or television engineering.

In an endeavor of this magnitude and scope it is necessary to call upon many experts and authorities in a wide diversity of engineering fields in order to provide an authoritative and complete treatise. It is only through their dedicated efforts that this publication has been possible. To all of the contributors, I offer my sincere appreciation and thanks.

K. BLAIR BENSON

Television
Engineering
Handbook

Light, Vision, and Photometry

W. Lyle Brewer
Eastman Kodak (retired)
Rochester, New York

Robert A. Morris
Eastman Kodak (retired)
Rochester, New York

NOTE: Parts of this chapter are adapted from D. G. Fink (ed.), *Television Engineering Handbook*, McGraw-Hill, New York, 1957. Used by permission.

1.1 LIGHT AND THE VISUAL MECHANISM

Vision results from stimulation of the eye by light and consequent interaction through connecting nerves with the brain. In physical terms, light constitutes a small section in the range of electromagnetic radiation, extending in wavelength from about 400 to 700 nanometers (nm) or billionths (10^{-9}) of a meter.

1.1.1 SOURCES OF ILLUMINATION. Light reaching an observer usually has been reflected from some object. The original source of such light, however, generally is radiated from molecules or atoms owing to internal changes. The various kinds of emissions depend upon the ways in which the atoms or molecules are supplied with energy to replace that which they radiate and upon their physical state, whether solid, liquid, or gaseous. The most common source of radiant energy is the thermal excitation of atoms in the solid or gaseous state.

1.1.2 THE SPECTRUM. When a beam of light travling in air falls upon a glass surface at an angle, it is refracted or bent. The amount of refraction depends upon the wavelength, its variation with wavelength being known as *dispersion*. Similarly when the beam, traveling in glass, emerges into air, it is refracted (with dispersion). A glass prism provides a refracting system of this type. Since different wavelengths are refracted by different amounts, an incident white beam is split up into several beams corresponding to the many wavelengths contained in the composite white beam. Thus is obtained the spectrum, in the simple manner which Newton first described. The gamut of colors was stated by him to consist of seven hues: violet, indigo, blue, green, yellow, orange, and red, arranged in that order. We know now that many more can be distinguished.

If a spectrum is allowed to fall upon a narrow slit arranged parallel to the edge of the prism, a narrow band of wavelengths passes through the slit. Obviously the narrower the slit, the narrower the band of wavelengths or the "sharper" is the spectral line. Also, more dispersion in the prism will cause a wider spectrum to be produced, and a narrower spectral line will be obtained for a given slit width.

It should be noted that purples are not included in the list of spectral colors. The purples belong to a special class of colors; they can be produced by mixing the light from two spectral lines, one in the red end of the spectrum, the other in the blue end. Purple (magenta is a more scientific name) is therefore referred to as a nonspectral color.

A plot of the power distribution of a source of light is indicative of the watts radiated at each wavelength per nanometer of wavelength. It is usual to refer to such a curve as an *energy distribution curve*.

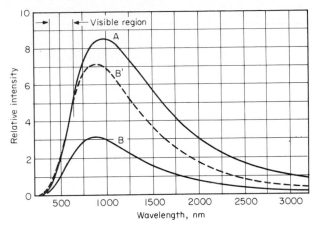

FIG. 1-1 Radiating characteristics of tungsten. Curve A: radiant flux from 1 cm^2 of a blackbody at 3000 K. Curve B: radiant flux from 1 cm^2 of tungsten at 3000 K. Curve B′: radiant flux from 2.27 cm^2 of tungsten at 3000 K (equal to curve A in visible region). (*From IES Lighting Handbook, Illuminating Engineering Society of North America, New York, 1981.*)

Individual narrow bands of wavelengths of light are seen as strongly colored. Increasingly broader bandwidths retain the appearance of color, but with decreasing purity, as if white light had been added to them. A very broad band extending generally throughout the visible spectrum is perceived as white light. Many white-light sources are of this type, such as the familiar tungsten-filament electric light bulb (see Fig. 1-1). Daylight also has a very broad band of radiation (see Fig. 1-2). The energy distributions shown in Figs. 1-1 and 1-2 are quite different and, if the corresponding sets of radiation were seen side by side, would be different in appearance. Either one, particularly if seen alone, however, would represent a very acceptable white. A sensation of white light can also be induced

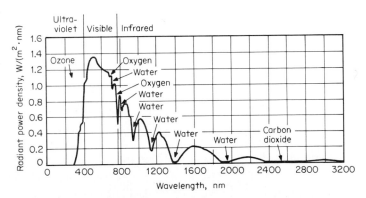

FIG. 1-2 Spectral distribution of solar radiant power density at sea level, showing the ozone, oxygen, water, and carbon dioxide absorption bands. (*From IES Lighting Handbook, Illuminating Engineering Society of North America, New York, 1981*).

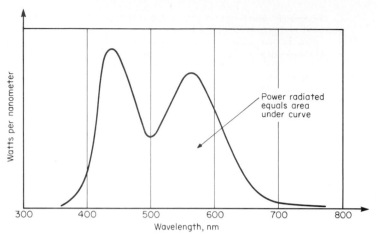

FIG. 1-3   Power distribution of a monochrome television picture-tube light source. (*From Fink.*[1])†

by light sources which do not have a uniform energy distribution. Among these is fluorescent lighting; this lighting exhibits sharp peaks of energy through the visible spectrum. Similarly, the light from a monochrome television picture tube is not uniform within the visible spectrum, generally exhibiting peaks in the yellow and blue regions of the spectrum; yet it appears as an acceptable white (see Fig. 1-3).

1.1.3 MONOCHROME AND COLOR VISION.   The color sensation associated with a light stimulus can be described in terms of three characteristics: hue, saturation, and brightness. The spectrum contains most of the principal hues: red, orange, yellow, green, blue, and violet. Additional hues are obtained from mixtures of red and blue light. These constitute the purples. Saturation pertains to the strength of the hue. Spectrum colors are highly saturated. White and grays have no hue and, therefore, have zero saturation. Pastel colors are of low or intermediate saturation. Brightness pertains to the intensity of the stimulation. If a stimulus has high intensity, regardless of its hue, it is said to be bright.

The psychophysical analogs of hue, saturation and brightness are dominant wavelength, excitation purity, and luminance, as shown in Table 1-1. By means of definitions and standard response functions which have received international acceptance through the International Commission on Illumination, the dominant wavelength, purity, and luminance of any stimulus of known spectral energy distribution may be determined by simple computations. Although roughly analogous to their psychophysical counterparts, the psychological attributes of hue, saturation, and brightness pertain to observer responses to light stimuli and are not subject to calculation. These sensation characteristics as applied to any given stimulus depend in part on other visual stimuli in the field of view and upon the immediately preceding stimulations.

Color sensations arise directly from the action of light on the eye. They are normally associated, however, with objects in the field of view from which the light comes. The objects themselves are therefore said to have color. Object colors may be described in terms of their hues and saturations, as is the case for light stimuli. The intensity aspect is usually referred to in terms of lightness, rather than brightness. The psychophysical

---

†Superscript numbers refer to References at end of chapter.

Table 1-1 Psychophysical and Psychological
Characteristics of Color

Psychophysical properties	Psychological properties
Dominant wavelength	Hue
Excitation purity	Saturation
Luminance	Brightness
Luminous transmittance ⎱ Luminous reflectance ⎰	Lightness

analogs of lightness are luminous reflectance for reflecting objects and luminous trans-mittance for transmitting objects.

At low levels of illumination objects may differ from one another in their lightness appearances but give rise to no sensation of hue or saturation. All objects appear to be of different shades of gray. Vision at low levels of illumination is called *scotopic* vision (see Table 1-2), as distinct from photopic vision which takes place at higher levels of illumination. Only the rods of the retina are involved in scotopic vision; the cones play no part. As the fovea centralis is free of rods, scotopic vision takes place outside the fovea. Visual acuity of scotopic vision is low as compared with photopic vision.

At high levels of illumination, where cone vision predominates, all vision is color vision. Reproducing systems such as black-and-white photography and monochrome television, however, cannot reproduce all three types of characteristics of colored objects. All images belong to the series of grays, differing only in their relative brightness.

The relative brightness of the reproduced image of any object depends primarily upon the luminance of the object as seen by the photographic or television camera. Depending upon the camera tube or the film, however, the dominant wavelength and purity of the light may also be of consequence. Some films, for example, have high sensitivity in the blue and ultraviolet spectral regions and low sensitivity in the green and red. Such film is not normally used for motion pictures; red and green objects are reproduced as if they were dark grays. Most films and television tubes now in use have sensitivity throughout the visible spectrum and, consequently, marked distortions in luminance as a function of dominant wavelength and purity are not encountered. Their spectral sensitivities sel-dom conform exactly to that of the human observer, however, so that some brightness distortions do exist.

1.1.4 VISUAL REQUIREMENTS FOR MONOCHROME AND COLOR TELEVI-SION. The objective in any type of visual reproduction system is to present to the viewer a combination of visual stimuli which can be readily interpreted as representing or having close association with a real viewing situation. It is by no means necessary that the light stimuli from the original scene be duplicated. There are certain characteristics in the reproduced image, however, which are necessary and others which are highly desir-able. Only a general qualitative discussion of such characteristics will be given here.

In monochrome television, images of objects are distinguished from one another and from their backgrounds because of luminance differences. In order that detail in the pic-ture be visible and that objects have clear, sharp edges, it is necessary that the television system be capable of rapid transitions from areas of one luminance to another. This degree of resolution need not match that possible in the eye itself, but too low an effective resolution results in pictures with fuzzy appearance and without fineness of detail.

Luminance range and the transfer characteristic associated with luminance repro-duction are likewise of importance in monochrome television. Objects seen as white usu-ally have minimum reflectances of the order of 80 percent. Black objects have reflec-tances of approximately 4 percent. This gives a luminance ratio of 20/1 in the range from white to black. To obtain the total luminance range in a scene, this reflectance range must be multiplied by the illumination range. In outdoor scenes the illumination ratio

Table 1-2 Relative Luminosity Values for Photopic and
Scotopic Vision

Wavelength, nm	Photopic vision	Scotopic vision
390	0.00012	0.0022
400	0.0004	0.0093
410	0.0012	0.0348
420	0.0040	0.0966
430	0.0116	0.1998
440	0.023	0.3281
450	0.038	0.4550
460	0.060	0.5670
470	0.091	0.6760
480	0.139	0.7930
490	0.208	0.9040
500	0.323	0.9820
510	0.503	0.9970
520	0.710	0.9350
530	0.862	0.8110
540	0.954	0.6500
550	0.995	0.4810
560	0.995	0.3288
570	0.952	0.2076
580	0.870	0.1212
590	0.757	0.0655
600	0.631	0.0332
610	0.503	0.0159
620	0.381	0.0074
630	0.265	0.0033
640	0.175	0.0015
650	0.107	0.0007
660	0.061	0.0003
670	0.032	0.0001
680	0.017	0.0001
690	0.0082	
700	0.0041	
710	0.0021	
720	0.00105	
730	0.00052	
740	0.00025	
750	0.00012	
760	0.00006	

between full sunlight and shadow may be as high as 100/1. The full luminance ranges involved with objects in such scenes cannot be reproduced in normal reproduction systems. Systems must be capable of handling illumination ratios of at least 2, however, and ratios as high as 4 or 5 would be desirable. This implies a luminance range on the output of the receiver of at least 40, with possible upper limits as high as 80 or 100.

Monochrome television transmits only luminance information, and the relative luminances of the images should correspond at least roughly to the relative luminances of the original objects. Red objects, for example, should not be reproduced markedly darker than objects of other hues but of the same luminance. Exact luminance reproduction, however, is by no means a necessity. Considerable distortion as a function of hue is entirely acceptable. Luminance reproduction is probably of primary consequence only if detail in some hues becomes lost.

Images in monochrome television are transmitted one point, or small area, at a time. The complete picture image is repeatedly scanned at frequent intervals. If the frequency of scan is not sufficiently high, the picture appears to flicker. At frequencies above the critical frequency no flicker is apparent. The critical frequency changes as a function of luminance, being higher for the higher luminance. The requirement for monochrome television is that the field frequency be above the critical frequency for the highest image luminances.

Images of objects in color television are distinguished from one another by luminance differences or by differences in hue or saturation. Exact reproduction in the image of the original scene differences is not necessary or even attainable (see Sec. 2.5.1). Nevertheless, some reasonable correspondence must prevail since the luminance gradation requirements for color are essentially the same as those for monochrome television.

1.2 LUMINOUS CONSIDERATIONS IN VISUAL RESPONSE CHARACTERISTICS

Vision is considered in terms of physical, psychophysical, and psychological quantities. The primary stimulus for vision is radiant energy. The study of this radiant energy in its various manifestations, including the effects on it of reflecting, refracting, and absorbing materials, is a study in physics. The response part of the visual process embodies the sensations and perceptions of seeing. Sensing and perceiving are mental operations and therefore belong to the field of psychology. Evaluation of radiant-energy stimuli in terms of observer responses they evoke is within the realm of psychophysics. Because observer-response sensations can be described only in terms of other sensations, psychophysical specifications of stimuli are made according to sensation equalities or differences.

1.2.1 **PHOTOMETRIC MEASUREMENTS.** Evaluation of a radiant-energy stimulus in terms of its brightness-producing capacity is a photometric measurement. An instrument for making such measurements is called a *photometer*. In visual photometers, which must be used in obtaining basic photometric measurements, the two stimuli to be compared are normally directed into small adjacent parts of a viewing field. The stimulus to be evaluated is presented in what is called the *test field*. The stimulus against which it is compared is presented in the *comparison field*. For most high-precision measurements the total size of the combined test and comparison fields is kept small, subtending about 2° at the eye. The area outside these fields is called the *surround*. Although the surround does not enter directly into the measurements, it has adaptation effects on the retina and thus affects the appearances of the test and comparison fields. It also influences the precision of measurement.

1.2.2 **LUMINOSITY CURVES.** A luminosity curve is a plot indicative of the relative brightnesses of spectrum colors of different wavelength or frequency. To a normal observer the brightest part of a spectrum consisting of equal amounts of radiant flux per unit wavelength interval is at about 555 nm. Luminosity curves are therefore commonly normalized to have a value of unity at 555 nm. If, at some other wavelength, twice as much radiant flux as at 555 nm is required to obtain brightness equality with radiant flux at 555 nm, the luminosity at this wavelength is 0.5. The luminosity at any wavelength λ is therefore defined as the ratio P_{555}/P_λ, where P_λ denotes the amount of radiant flux at the wavelength λ which is equal in brightness to a radiant flux of P_{555}.

The luminosity function which has been accepted as standard for photopic vision is given in Fig. 1-4. Tabulated values at 10 nm intervals are given in Table 1-2. This function was agreed upon by the International Commission on Illumination (CIE) in 1924. It is based upon a considerable amount of experimental work which had gone on for a number of years prior to that time. Chief reliance in arriving at this function was based on the step-by-step equality-of-brightness method. Flicker photometry provided additional data.

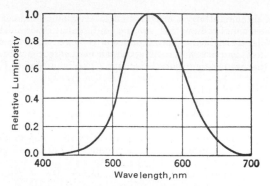

FIG. 1-4   Photopic luminosity function. (*From Fink.*[1])

In the scotopic range of intensities the luminosity function is somewhat different from that of the photopic range. The two curves are compared in Fig. 1-5. Values are listed in Table 1-2. The two curves are similar in shape, but there is a shift for the scotopic curve of about 40 nm to the shorter wavelengths.

Measurements of luminosity in the scotopic range are usually made by the threshold-of-vision method. A single stimulus in a dark surround is used. The stimulus is presented to the observer at each of a number of different intensities, ranging from well below the threshold to intensities sufficiently high to be definitely visible. Determinations are made of the amount of energy, at each chosen wavelength, which is reported visible by the observer a certain percentage of the time, say 50 percent. The reciprocal of this amount of energy determines the relative luminosity at this wavelength. The wavelength plot is normalized to have a maximum value of 1.00 to give the scotopic luminosity function.

In the intensity region between scotopic and photopic vision, called the Purkinje or mesopic region, the measured luminosity function takes on sets of values intermediate between those obtained for scotopic and photopic vision. Relative luminosities of colors within the mesopic region will therefore vary, depending upon the particular intensity level at which the viewing takes place. Reds tend to become darker in approaching scotopic levels; greens and blues tend to become relatively lighter.

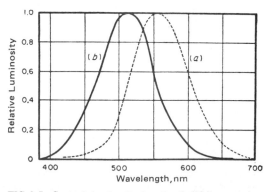

FIG. 1-5   Scotopic luminosity function (solid-line curve) as compared with photopic luminosity function (broken-line curve). (*From Fink.*[1])

1.2.3 LUMINANCE. Brightness is a term used to describe one of the characteristics of appearance of a source of radiant flux or of an object from which radiant flux is being reflected or transmitted. Brightness specifications of two or more sources of radiant flux should be indicative of their actual relative appearances. These appearances will depend in large part upon the viewing conditions, including the state of adaptation of the observer's eye.

Luminance, as indicated earlier, is a psychophysical analog of brightness. It is subject to physical determination, independent of particular viewing and adaptation conditions. Because it is an analog of brightness, however, it is defined in such a way as to relate as closely as possible to brightness.

The best established measure of the relative brightnesses of different spectral stimuli is the luminosity function. In evaluating the luminance of a source of radiant flux consisting of many wavelengths of light, therefore, the amounts of radiant flux at the different wavelengths are weighted by the luminosity function. This converts radiant flux to luminous flux. As used in photometry, the term *luminance* is applied only to extended sources of light, not to point sources (see Sec. 1.3.1). For a given amount (and quality) of radiant flux reaching the eye, brightness will vary inversely with the effective area of the source. Luminance is therefore expressed in terms of luminous flux per unit projected area of the source. The greater the concentration of flux in the angle of view of a source, the brighter it appears. Luminance is therefore expressed in terms of amounts of flux per unit solid angle or steradian. In summary, the luminance B of a source may be defined as

$$B = K_m \int \frac{V(\lambda)P(\lambda)}{\omega\alpha \cos\theta} \, d\lambda \qquad (1\text{-}1)$$

where V = value of luminosity function at wavelength λ

$\dfrac{P}{\omega\alpha \cos\theta}$ = radiant flux P at wavelength λ per steradian ω per projected area of source α $\cos\theta$

K_m = luminosity of radiant flux at wavelength (555 nm) of maximum efficiency lumens/watt

The numerical value of K_m depends upon the units of measure. If P is expressed in watts and B in candelas (cd)† per square centimeter, or stilbs, the defined value of K_m is 683 lumens per watt.

The normalizing constant K_m, in the equation for luminance B, is frequently dispensed with in studying relative luminances. Relative luminances are expressed in terms of luminance ratios. For two sources having luminances of B_1 and B_2, the ratio of luminances B_1/B_2 does not depend upon K_m.

In considering the relative luminances of various objects of a scene to be photographed or televised, it is convenient to normalize the luminance values so that the "white" in the region of principal illumination has a relative luminance value of 1.00. The relative luminance of any other object then becomes the ratio of its luminance to that of the white. This white is an object of highly diffusing surface with high and uniform reflectance throughout the visible spectrum. For purposes of computation it may be idealized to have 100 percent reflectance and perfect diffusion (see Sec. 1.3.3). A surface of pressed barium sulfate approaches this ideal.

1.2.4 LUMINANCE DISCRIMINATION. If an area of luminance B is viewed side by side with an area of luminance $B + \Delta B$, a value of ΔB may be established for which the brightnesses of the two areas are just noticeably different. The ratio of $\Delta B/B$ is known as Weber's fraction. The statement that this ratio is a constant, independent of B, is known as Weber's law.

Strictly speaking, the value of Weber's fraction is not independent of B. Furthermore, its value depends considerably on the viewer's state of adaptation. Values as determined

---

†For the definition of the candela, see Sec. 1.3.1.

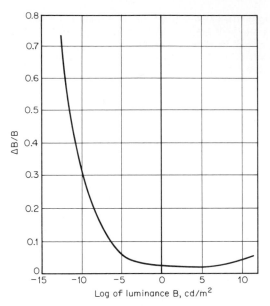

FIG. 1-6 Weber's fraction $\Delta B/B$ as a function of luminance B for a dark-field surround. (*From S. Hecht, The Visual Discrimination of Intensity and the Weber-Fechner Law, J. Gen Physiol., vol. 7, p. 241, 1924.*)

for a dark-field surround are shown in Fig. 1-6. It is seen that at very low intensities the value of $\Delta B/B$ is relatively large; that is, relatively large values of ΔB, as compared with B, are necessary for discrimination. A relatively constant value of roughly 0.02 is maintained through a brightness range of about 1 to 300 cd/m^2. The slight rise in the value of $\Delta B/B$ at high intensities as given in the graph may indicate lack of complete adaptation to the stimuli being compared.

The plot of $\Delta B/B$ as a function of B will change significantly if the comparisons between the two fields are made with something other than a dark surround. The greatest changes are for luminances below that of the adapting field. The loss of power of discrimination proceeds rapidly for luminances less by a factor of 10 than that of the adapting field. On the high-luminance side, adaptation is largely controlled by the comparison fields and is relatively independent of the adapting field.

Because of the luminance-discrimination relationship expressed by Weber's law, it is convenient to express relative luminances of areas from either photographic or television images in logarithmic units. Because $\Delta(\log B)$ is approximately equal to $\Delta B/B$, equal small changes in log B correspond reasonably well with equal numbers of brightness-discrimination steps.

1.2.5 PERCEPTION OF FINE DETAIL. Detail is seen in a picture image because of brightness differences between small adjacent areas in monochrome pictures or because of brightness, hue, or saturation differences in color pictures. Visibility of detail in a picture is important because it determines the extent to which small or distant objects of a scene are visible, and because of its relationship to the "sharpness" appearance of the edges of objects.

"Picture definition" is probably the most acceptable term for describing the general characteristic of "crispness," "sharpness," or image-detail visibility in a picture. Picture definition depends upon characteristics of the eye such as visual acuity and upon a variety of characteristics of the picture-image medium, including its resolving power, luminance range, contrast, and image edge gradients.

Visual acuity may be measured in terms of the visual angle subtended by the smallest detail in an object which is visible. One type of test object frequently employed is the Landolt ring. The ring, which has a segment cut from it, is shown in any one of four orientations, with the opening at the top or bottom or on the right or left side. The observer identifies the location of this opening. The visual angle subtended by the opening which can be properly located 50 percent of the time is a measure of visual acuity.

Test-object illuminance, contrast between the test object and its background, time of viewing, and other factors greatly affect visual-acuity measurements. Up to a visual distance of about 20 ft (6 m) acuity is partially a function of distance, because of changes in shape of the eye lens in focusing. Beyond 20 ft it remains relatively constant. Visual acuity is highest for foveal vision, dropping off rapidly for retinal areas outside the fovea.

A black line on a light background is visible if it has a visual angle no greater than 0.5 s. This is not, however, a true measure of visual acuity. For visual-acuity tests of the type described, normal vision, corresponding to a Snellen 20/20 rating, represents an angular discrimination of about 1 min. Separations between adjacent cones in the fovea and resolving-power limitations of the eye lens give theoretical visual-acuity values of about this same magnitude.

The extent to which a picture medium, such as a photographic or a television system, can reproduce fine detail is expressed in terms of resolving power or resolution. Resolution is a measure of the distance between two fine lines in the reproduced image which are visually distinct. The image is examined under the best possible conditions of viewing, including magnification.

Two types of test charts are commonly employed in determining resolving power, either a wedge of radial lines or groups of parallel lines at different pitches for each group. For either type of chart, the spaces between pairs of lines usually are made equal to the line widths.

Resolution in photography is usually expressed as the maximum number of lines (counting only the black ones or only the white ones) per millimeter which can be distinguished from one another. Measured values of resolving power depend upon a number of factors in addition to the photographic material itself. The most important of these probably are: (1) the density differences between the black and the white lines of the test chart photographed, (2) the sharpness of focus of the test-chart image during exposure, (3) the contrast to which the photographic image is developed, and (4) the composition of the developer. Sharpness of focus, in turn, depends upon the general quality of the focusing lens, image and object distances from the lens, and the part of the projected field in which the image lies. In determining the resolving power of a photographic negative or positive material, a test chart is generally employed which has a high-density difference, say 3.0, between the black-and-white lines. A high-quality lens is employed, the projected field is limited, and focusing is very critically adjusted. Under these conditions, ordinary black-and-white photographic materials generally have resolving powers in the range of 30 to 200 line-pairs per millimeter. Special photographic materials are available with resolving powers greater than 1000 line-pairs per millimeter.

Resolution in a television system is expressed in terms of the maximum number of lines, counting both black and white, which are discernible in viewing a televised test chart. The value of horizontal (vertical lines) or vertical (horizontal lines) resolution is the number of lines equal to the vertical dimension of the raster. Vertical resolution in a well-adjusted system equals the number of scanning lines, roughly 500. In normal broadcasting and reception practice, however, typical values of vertical resolution range from 350 to 400 lines. The theoretical limiting value for horizontal resolution (R_H) is given by

$$R_H = \frac{2(0.75)(\Delta f)}{30(525)} = 0.954 \times 10^{-4}\, \Delta f \qquad (1\text{-}2)$$

where Δf is the available bandwidth frequency in Hz. The constants 30 and 525 represent the frame and line frequencies, respectively. A factor of 2 is introduced because in one complete cycle both a black and a white line are obtainable. The factor of 0.75 is necessary because of the receiver aspect ratio; the picture height is three-fourths of the picture width. There is an additional reduction of about 15 percent (not included in the equation) in the theoretical value because of horizontal blanking time during which retrace takes place. A bandwidth of 4.25 MHz thus makes possible a maximum resolution of about 345 lines.

The appearance evaluation of a picture image in terms of the edge characteristics of objects is called *sharpness.* The more clearly defined the line which separates dark areas from lighter ones, the greater the sharpness of the picture. Sharpness is, of course, related to the transient curve in the image across an edge. The average gradient and the total density difference appear to be the most important characteristics. No physical measure has been devised, however, that in all cases will predict the sharpness (appearance) of an image.

Picture resolution and sharpness are to some extent interrelated but they are by no means perfectly correlated. Pictures ranked according to resolution measures may be ranked somewhat differently on the basis of sharpness. Both resolution and sharpness are related to the more general characteristic of picture definition. For pictures in which, under the particular viewing conditions, effective resolution is limited by the visual acuity of the eye rather than by picture resolution, sharpness is probably a very good indication of picture definition. If visual acuity is not the limiting factor, however, picture definition depends to an appreciable extent on both resolution and sharpness.

1.2.6 RESPONSES TO INTERMITTENT EXCITATION. The brightness sensation resulting from a single short flash of light is a function of the duration of the flash as well as of its intensity. For low-intensity flashes near the threshold of vision, stimuli of shorter duration than about ⅕ s are not seen at their full intensity. Their apparent intensities are very nearly proportional to the action times of the stimuli.

With increasing intensity of the stimulus, the time necessary for the resulting sensation to reach its maximum becomes shorter and shorter. A stimulus of 5 mL reaches its maximum apparent intensity in about ⅒ s; that of 1000 mL reaches it in less than ⅟₂₀ s. Also, for the higher intensities, there is a brightness overshooting effect. For stimulus times longer than that necessary to have the maximum effect, the apparent brightness of the flash is decreased. A 1000-mL flash of ⅟₂₀ s will appear to be almost twice as bright as a flash of the same intensity which continues for ⅕ s. These effects are substantially the same for colors of equal luminances, independent of their chromatic characteristics.

Intermittent excitations at low frequencies are seen as successive individual light flashes. With increase in frequency the flashes appear to merge into one another, giving a coarse, pulsating flicker effect. Further increases in frequency result in finer and finer pulsations until, at a sufficiently high frequency, the flicker effect disappears.

The lowest frequency at which flicker is not seen is called the *critical fusion frequency* or *critical frequency.* Over a wide range of stimuli luminances, critical fusion frequency is linearly related to the logarithm of luminance. This relationship is called the Ferry-Porter law. Critical frequencies for several different wavelengths of light are plotted as functions of retinal illumination (trolands) in Fig. 1-7. The second abscissa scale is in terms of luminance, assuming a pupillary diameter of about 3 mm. At low luminances critical frequencies differ for different wavelengths, being lowest for stimuli near the red end of the spectrum and highest for stimuli near the blue end. Above a retinal illumination of about 10 trolands (0.4 ft·L) critical frequency is independent of wavelength. This is in the critical frequency range above about 18 Hz.

Critical fusion frequency increases approximately logarithmically with increase in retinal area illuminated. It is higher for retinal areas outside the fovea than for those inside, although fatigue to flicker effects is rapid outside the fovea.

Intermittent stimulations sometimes result from rapid alternations between two color stimuli, rather than between one color stimulus and complete darkness. Critical fre-

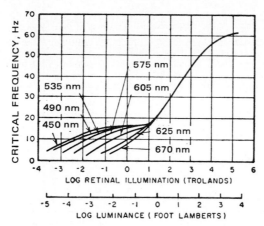

FIG. 1-7 Critical frequencies as functions of retinal illumination and luminance (1 ft·L $\cong$ 3.426 cd/m^2; 1 troland = retinal illuminance per square millimeter pupil area from surface with luminance of 1 cd/m^2). (*From S. Hecht, S. Shlaer, and E. L. Smith, Intermittent Light Stimulation and the Duplicity Theory of Vision, Cold Spring Harbor Symposia on Quantitative Biology, vol. 3, p. 241, 1935.*)

quency for such stimulations depends upon relative luminance and chromatic characteristics of the alternating stimuli. Critical frequency is lower for chromatic differences than for luminance differences. Flicker photometers are based upon this principle. Critical frequency also decreases as the difference in intensity between the two stimuli becomes smaller. Critical frequency depends to some extent upon the relative time amounts of the component stimuli and in the manner of changing from one to another. Contrary to what might be expected, smooth transitions such as of a sine-wave characteristic do not necessarily result in the lowest critical frequencies. Lower critical frequencies are sometimes obtained when the transitions are rather abrupt in one direction and slow in the opposite.

When intermittent stimuli are seen at frequencies above the critical frequency, the visual effect is that of a single stimulus which is the mean, integrated with respect to time, of the actual stimuli. This additive relationship for intermittent stimuli is known as the *Talbot-Plateau law.*

1.3 PHOTOMETRIC QUANTITIES

The study of visual response, it was noted in the preceding section, is facilitated by division of the subject into its physical, psychophysical, and psychological aspects. The subject of photometry is concerned with the psychophysical aspects: the evaluation of radiant energy in terms of equality or differences for the human observer. Specifically, photometry deals with the luminous aspects of radiant energy, or, in other words, its capacity to evoke the sensation of brightness.

1.3.1 LUMINANCE AND LUMINOUS INTENSITY.

By international agreement, the standard source for photometric measurements is a blackbody heated to the temperature at which platinum solidifies, 2042 K, and the luminance of the source is 60 candelas per square centimeter of projected area of the source. Before adoption of the interna-

Table 1-3 Conversion Factors for Luminance and Retinal Illumination Units

Multiply Quantity Expressed in Units of X by Conversion Factor to Obtain Quantity in Units of Y

X \ Y	Candelas per square centimeter	Candelas per square meter	Candelas per square inch	Candelas per square foot	Lamberts	Millilamberts	Footlamberts	Trolands[†‡]
Candelas per square centimeter	1	1×10^4	6.452	9.290×10^2	3.142	3.142×10^3	2.919×10^3	7.854×10^3
Candelas per square meter (nit)[§]	1×10^{-4}	1	6.452×10^{-4}	9.290×10^{-2}	3.142×10^{-4}	3.142×10^{-1}	2.919×10^{-1}	7.854×10^{-1}
Candelas per square inch	1.550×10^{-1}	1.550×10^3	1	1.440×10^2	4.869×10^{-1}	4.869×10^2	4.524×10^2	1.217×10^3
Candelas per square foot	1.076×10^{-3}	1.076×10	6.944×10^{-3}	1	3.382×10^{-3}	3.382	3.142	8.454
Lamberts	3.183×10^{-1}	3.183×10^3	2.054	2.957×10^2	1	1×10^3	9.290×10^2	2.5×10^3
Millilamberts	3.183×10^{-4}	3.183	2.054×10^{-3}	2.957×10^{-1}	1×10^{-3}	1	9.290×10^{-1}	2.500
Footlamberts	3.426×10^{-4}	3.426	2.210×10^{-3}	3.183×10^{-1}	1.076×10^{-3}	1.076	1	2.691
Trolands[‡]	1.273×10^{-4}	1.273	8.213×10^{-4}	1.183×10^{-1}	4.000×10^{-4}	4.000×10^{-1}	3.716×10^{-1}	1

[†] In converting luminance to trolands it is necessary to multiply the conversion factor by the square of the pupil
[‡] In converting trolands to luminance it is necessary to divide the diameter in millimeters.

[§] As recommended at Session XII in 1951 of the International Commission on Illumination, one nit equals one candela per square meter.

Source: From Fink.[1]

tional standard in 1948, the standard unit was luminous intensity, expressed in candles. The standard was originally specified in terms of a candle made of sperm wax but for many years had been maintained by primary and secondary standard tungsten lamps. The new standard source was found to have a luminance of 58.9 candles per square centimeter. The new unit, the candela, is therefore about 1.9 percent smaller in magnitude. Actually, at higher temperatures the differences are smaller owing to the methods by which the earlier substandards were prepared.

Luminance was defined in the preceding section as follows

$$B = K_m \int \frac{V(\lambda)P(\lambda)}{\omega\alpha\cos\theta}\,d\lambda \qquad (1\text{-}3)$$

where K_m = maximum luminous efficiency of radiation (683 lumens per watt)
   V = relative efficiency, or luminosity function
$P/(\omega\alpha\cos\theta)$ = radiant flux (P) per steradian (ω) per projected area of source ($\alpha\cos\theta$)

At first meeting, this appears to be an unnecessarily contrived definition but its usefulness, as was seen, lies in the fact that it relates most directly to the sensation of brightness, although there is no strict correspondence.

Other luminous quantities are similarly related to their physical counterparts, for example, luminous flux F is defined by: $F = K_m \int V(\lambda)P(\lambda)\,d\lambda$. When P is given in watts, F is given in lumens.

When the source is far enough away that it may be considered a point source, then the luminous intensity I in a given direction is

$$I = F/\omega$$

where F is measured in lumens and ω is the solid angle of the cone (in steradians) through which the energy is flowing. The luminous intensity unit is the candela (candle in the older terminology). Conversion factors for various luminance units are listed in Table 1-3. Luminance values for a few objects are given in Table 1-4.

1.3.2 ILLUMINANCE. In the discussion up to this point the photometric quantities have been descriptive of the luminous energy *emitted* by the source. When luminous flux reaches a surface, the surface is illuminated, and the illuminance E is given by $E = F/S$, where S is the area over which the luminous flux F is distributed. When F is expressed in lumens and S in square meters, the illuminance unit is lumens per square meter, or lux.

An element of area S of a sphere of radius r subtends an angle ω at the center of the sphere where $\omega = S/r^2$. For a source at the center of the sphere and r sufficiently large, the source, in effect, becomes a point source at the apex of a cone with S (considered small compared with r^2) as its base. The luminous intensity I for this source is given by $I = F/(S/r^2)$. It follows that $I = Fr^2/S$ and $I = Er^2$. Therefore, the illuminance E on a spherical surface element S from a point source is $E = I/r^2$. The illuminance thus varies

Table 1-4 Typical Luminance Values

	Luminance, ft·L
Sun at zenith	4.82×10^8
Perfectly reflecting, diffusing surface in sunlight	9.29×10^3
Moon, clear sky	2.23×10^3
Overcast sky	$9\text{–}20 \times 10^2$
Clear sky	$6\text{–}17.5 \times 10^2$
Motion-picture screen	10

Source: From Fink.[1]

Table 1-5   Conversion Factors for Illuminance Units

Multiply Quantity Expressed in Units of X by Conversion Factor to Obtain Quantity in Units of Y

X \ Y	Lux	Phot	Footcandle
Lux (meter-candle)			
lumens per square meter	1	1×10^{-4}	9.290×10^{-2}
Phot			
lumens per square centimeter	1×10^{4}	1	9.290×10^{2}
Footcandle			
lumens per square foot	1.076×10	1.076×10^{-3}	1

Source: From Fink.[1]

inversely as the square of the distance. This relationship is known as the *inverse-square law.*

As previously indicated, the unit for illuminance E may be taken as lumen per square meter or lux. It is also expressed in terms of the metercandle which denotes the illuminance produced on a surface 1 meter distant by a source having an intensity of 1 candela. Similarly, the footcandle is the illuminance produced by a source of 1 candela on a surface 1 ft distant and is equivalent to 1 lumen per square foot. Conversion factors for various illuminance units are given in Table 1-5.

The expression given above for illuminance, $E = I/r^2$, involves the solid angle S/r^2, so requires that the area S is normal to the direction of propagation of the energy. If the area S is situated so that its normal makes the angle θ with the direction of propagation, then the solid angle is given by $(S \cos \theta)/r^2$, as shown in Fig. 1-8. The illuminance E is given by $E = I \cos \theta/r^2$.

1.3.3 LAMBERT'S COSINE LAW.

Luminance was defined in Eq. (1-3) by its relationship to radiant flux because it is the fundamental unit for all photometric quantities. Luminance may also be defined as $B_\theta = I_\theta/(\alpha \cos \theta)$ where I_θ is the luminous intensity from a small element α of the area S at an angle of view θ, measured with respect to the normal of this element. For luminous intensities expressed in candelas (or candles), luminance may be expressed in units of candelas (or candles) per square centimeter.

A special case of interest arises if the intensity I_θ varies as the cosine of the angle of view, that is, $I_\theta = I \cos \theta$. This is known as *Lambert's cosine law.* In this instance $B = I/\alpha$ so that the luminance is independent of angle of view θ. Although no surfaces are known which meet this requirement of "complete diffusion" exactly, many materials conform reasonably well. Pressed barium sulfate is frequently used as a comparison standard for diffusely reflecting surfaces. Various milk-white glasses, known as opal glasses, are used to provide diffuse transmitting media.

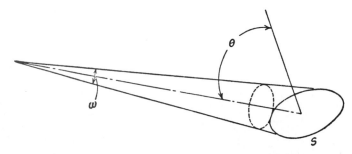

FIG. 1-8   Solid angle ω subtended by surface S with its normal at angle θ from the line of propagation. (*From Fink.*[1])

The luminous flux emitted per unit area F/a is called the luminous emittance. For a perfect diffuser whose luminance is 1 candela per square centimeter, the luminous emittance is π lumens per square centimeter. Or, if an ideal diffuser emits 1 lumen per square centimeter, its luminance is $1/\pi$ candelas per square centimeter. The unit of luminance equal to $1/\pi$ candelas per square centimeter is called the lambert. When the luminance is expressed in terms of $1/\pi$ candelas per square foot, the unit is called a footlambert.

The physical unit corresponding to luminous emittance is radiant emittance, measured in watts per square centimeter. Radiance, expressed in watts per steradian per square centimeter, corresponds to luminance.

1.3.4 MEASUREMENT OF PHOTOMETRIC QUANTITIES.

Of the photometric quantities luminous flux, intensity, luminance, and illuminance the last is perhaps most readily measured in practical situations. However, where the light source in question can be placed on a laboratory photometer bench, the intensity can be determined by calculation from the inverse-square law by comparison with a known standard.

Total luminous flux may be determined from luminous intensity measurements made at angular intervals of a few degrees over the entire area of distribution. It can also be found by inserting the source within an integrating sphere and comparing the flux received at a small area of the sphere wall with that obtained with a known source in the sphere.

In many practical situations the illuminance produced at a surface is of greatest interest. Visual and photoelectric photometers have been designed for such measurements. For most situations a photoelectric instrument is more convenient to use because it is portable and easily and quickly read. Because the spectral sensitivity of the cell differs from that of the eye, the instrument must not be used for sources differing in color from that for which the instrument was calibrated. For example, a meter calibrated for tungsten sources will usually require a correction factor if used for fluorescent or mercury-vapor lamps. Filters are also available to make the cell sensitivity conform more nearly to that of the eye.

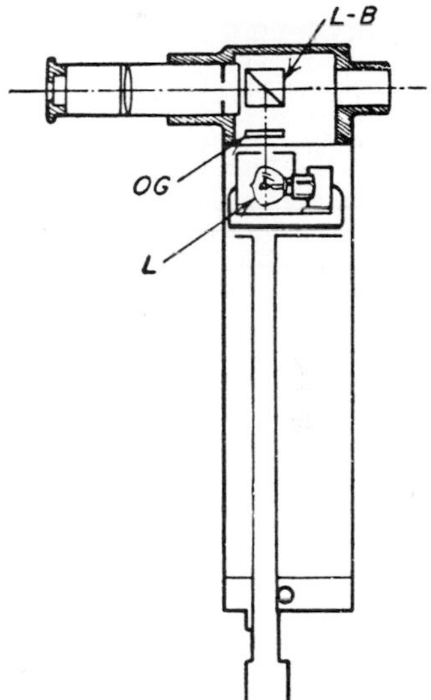

Visual measurements of illumination are generally more suitable where the light is colored. The general procedure is to convert the flux incident on the surface of interest to a luminance value which can be compared with the luminance of a surface within the photometer.

One visual instrument which is widely used is the Macbeth illuminometer. This is shown diagrammatically in Fig. 1-9. Light

FIG. 1-9 Macbeth illuminometer. (*From Fink.*[1])

from the test plate is incident on one face of the Lummer-Brodhun cube *LB*. Light from the opal-glass comparison surface *OG* is incident on the other face of the cube. The opal glass is illuminated by the lamp *L*. The distance of the lamp from *OG* can be varied by means of the rod to which the lamphouse is fixed. The illuminance may be read from the scale on the rod. In using the illuminometer at the surface to be measured, it is important to place the test plate so that it fills the field of view of the instrument, but so that no shadows are cast on it by the observer or instrument, and preferably, so that the viewing angle does not exceed 30°.

A visual instrument such as the Macbeth illuminometer can also be used to measure luminance by directing the instrument toward the luminous object rather than toward the test plate. A visual instrument designed specifically for luminance measurements is the Luckiesh-Taylor brightness meter.

1.3.5 RETINAL ILLUMINANCE. A psychophysical correlate of brightness is the measure of luminous flux incident on the retina, i.e., retinal illuminance. One unit designed to indicate retinal stimulation is the troland, formerly called a photon (not to be confused with the elementary quantum of radiant energy). It is defined (under restricted viewing conditions) as the visual stimulation produced by a luminance of 1 candela per square meter filling an entrance pupil of the eye whose area is 1 mm.[2] If luminance is measured in millilamberts and pupil diameter l in millimeters, then the retinal illuminance i is given approximately by

$$i = 2.5l^2B \tag{1-4}$$

For more accurate evaluation, the actual pupil area must be corrected to the effective pupil area, to take into account the fact that the brightness-producing efficiencies of light rays decrease as the rays enter the eye at increasing distances from the central region of the pupil (Stiles-Crawford effect). Variations in transmittance of the ocular media among individuals also prevent the complete specification of the visual stimulus on the retina.

1.4 RECEPTOR RESPONSE MEASUREMENTS

The eye, photographic film, and television cameras are receptors which respond to radiant energy. The television camera has a photoelectric response. Photons of energy absorbed by the photosensitive surface cause ejection of electrons from this surface. The resulting change in electrical potential in the surface gives rise to electrical signals either directly or through the scanning process.

The initial response of a photographic film is photoelectric. Photons absorbed by the silver halide grains cause the ejection of electrons with a consequent reduction of positive silver ions to silver atoms. Specks of atomic silver are thus formed on the silver halide grains. Conversion of this "latent image" into a visible one is accomplished by chemical development. Grains with the silver specks are reduced by the developer to silver; those without the silver specks remain as silver halide. Chemical reactions occurring simultaneously or subsequently to this primary development determine whether the final image will be negative or positive and whether it will be in color or black and white.

The direct response of the eye is either photoelectric, photochemical, or both. Absorption of light by the eye receptors causes neural impulses to the brain with a resulting sensation of seeing.

Each of these receptors, the eye, the photographic film, and the television camera, responds differentially to different wavelengths of light. Determination of these receptor responses for the photographic film and television camera provides a basis for correlating the reproduced image with that incident upon the receptor. Interpretation in terms of visual effects is made through a similar analysis of the eye response.

1.4.1. SPECTRAL RESPONSE MEASUREMENTS. In the photoelectric effect of releasing electrons from metals or other materials, light behaves as if it travels in discrete packets, or quanta. The energy of a single quantum, or photon, equals $h\nu$, where h is the Planck constant and ν is the frequency of the radiation. To release an electron, the photon must transfer sufficient energy to the electron to enable it to escape the potential-energy barrier of the material surface. For any material there is a minimum frequency, called the *threshold frequency,* of radiant energy which provides sufficient energy for an electron not already in an excited state to leave the material. Because of thermal excitation some electrons may be ejected at frequencies below the threshold frequency. The number of these, however, is usually quite small in comparison with those ejected at frequencies above the threshold frequency. It is because of the relationship between frequency and energy that ultraviolet light usually has a greater photoelectric effect than has visible light, and that visible light has a greater effect than infrared light.

For any given wavelength distribution of incident radiation, the number of electrons emitted from a photocathode is proportional to the intensity of the incident radiation. Photoelectric emission is therefore linear with irradiation. In practical applications in which there are space-charge effects, secondary emissions, or other complicating factors, this linear relationship does not always apply to the current actually collected.

Spectral response measurements are made by exposing the photosensitive surface to narrow-wavelength bands of light. The ratio of the emission current to the incident radiant power is a measure of the sensitivity for this wavelength region. A plot of this ratio as a function of wavelength gives the spectral response curve. If the photo-emissive device is a linear one, the intensity of the incident radiation in each spectral region may be taken at any convenient value without affecting the resulting curve.

If the electric output of the photoelectric device is not linear with intensity of illumination, the intensities of the spectral irradiations must be more carefully controlled. The electrical output for each spectral region should be the same. A plot of the reciprocal of the incident irradiance as a function of wavelength then gives the spectral response distribution. Response distributions are shown in Fig. 11-35.

Exposure and development of a piece of photographic film give a deposit of finely divided silver. This deposit is measured in terms of its transmittance or density (see Sec. 1.4.3). Photographic emulsions are generally not linear in their response characteristics. It is therefore necessary that the response evaluation be made with respect to the same transmittance (or density) value for each wavelength region. Density values of 1.0 (10 percent transmittance) are frequently chosen for this purpose. The spectral response characteristics of a film for these conditions of measurement are proportional to the reciprocals of the amounts of incident radiant energy necessary to produce a density of 1.0. The spectral response curves are given in Sec. 17.7.

If a different density level, say 0.2, were chosen for the determination of spectral response, a slightly different result might be obtained. Thus, there is no single curve that fully represents the spectral response characteristics of a film. For this reason it is important that a density level near the center of the density levels of primary interest for the film be used in arriving at the response characteristics. Also, the response characteristics as determined are in part a function of the conditions under which the film is exposed and developed. A change in time of exposure, for example, or a change in the developer may alter the spectral response characteristics.

Spectral response characteristics of color film are determined in much the same manner except that each component dye image of the system must be considered independently.

The eye is a very precise measuring device in judging the equality and nonequality of two stimuli if they are viewed side by side. It cannot be depended upon to give accurate results in ascertaining the amount of difference between two stimuli which are not alike. Therefore, in determining the spectral response characteristics of the eye, it is essential that measurements be made at equal response levels. For color response measurements, amounts of three primary stimuli are found which, in combination, identically match the fourth stimulus being evaluated. The relative amounts of the three primary stimuli necessary for the match are indicative of the response elicited by the test stimulus.

For spectral-luminance response measurements, the evaluations must be made at a common response level of equal brightness. The loss of precision associated with such measurements where chromatic differences exist is minimized by means of step-by-step comparisions or by means of flicker photometry. The brightness response characteristics, or luminosity function, of the spectrum colors are the reciprocals of the amounts of energy of these colors, all of which have the same brightness. Luminosity functions for photopic and scotopic vision were given in Figs. 1-4 and 1-5.

1.4.2 INTEGRATED RESPONSE CHARACTERISTICS.

Integrated response characteristics may be determined from a knowledge of the spectral response characteristics of a system. Consider first an ordinary vacuum phototube. Light reaching the photosensitive cathode has a certain radiant power distribution measured in, say, watts per unit wavelength. If the relative spectral response function for the phototube is $S(\lambda)$ and the incident spectral radiant power $P(\lambda)$, the response will be given by

$$R = k\!\int\! P(\lambda)S(\lambda)\,d\lambda \tag{1-5}$$

The factor k is determined by measuring some one response for a known incident power distribution $P(\lambda)$. For any other distribution $P(\lambda)$, R can then be calculated. The factor k can of course be incorporated into the spectral response distribution $S(\lambda)$. A direct integration then gives the desired result.

The distributions $P(\lambda)$ and $S(\lambda)$ are seldom known as analytical functions. If tabulated at discrete wavelengths, the integrated value is equal to the area under the product curve obtained by multiplying $P(\lambda)$ by $S(\lambda)$, wavelength by wavelength, through the spectrum.

1.4.3 TRANSMITTANCE.

Light incident upon an object is either reflected, transmitted, or absorbed. The transmittance of an object may be measured as illustrated in Fig. 1-10. Light from the source S passes through the object O and is collected at the receiver R. The spectral transmittance t_λ of the object is defined as

$$t_\lambda = \frac{P_\lambda}{P_{o\lambda}} \tag{1-6}$$

where P_λ is the radiance at wavelength λ reaching the receiver through the object, and $P_{o\lambda}$ is the radiance reaching the receiver with no object in the beam path.

The spectral density D_λ of the object at wavelength λ is defined as

$$D_\lambda = -\log t_\lambda = \log \frac{P_{o\lambda}}{P_\lambda} \tag{1-7}$$

An object with a transmittance of 1.00, or 100 percent, has a density of zero. One with a transmittance of 0.1, or 10 percent, has a density of 1.0.

Objects which transmit light also generally scatter the light to some extent. Consequently the transmittance measurement depends in part upon the geometrical conditions of measurements. In Fig. 1-10 the light incident upon the film is shown as a narrow collimated, or commonly called *specular,* beam. The receiver, in the form of an integrating sphere, is placed in contact with the object so that all the transmitted energy is collected. The transmittance measured in this fashion is called *diffuse transmittance.* The corresponding density value is called *diffuse density.* The same results are obtained if the incident light is made completely diffuse and only the specular component evaluated.

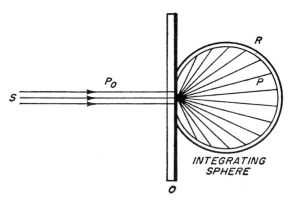

FIG. 1-10 Measurement of diffuse transmittance. (*From Fink.*[1])

If the incident beam is specular and only the specular component of the transmitted light is evaluated, the measurement is called *specular transmittance*. The corresponding density value is *specular density*. A smaller portion of the transmitted energy is collected in a specular measurement than in a diffuse measurement. The specular transmittance of an object is therefore always less than the diffuse transmittance unless the object does not scatter light, in which case the two transmittances are equal. Specular densities are therefore equal to or larger than diffuse densities. The ratio of the specular density to the diffuse density is a measure of the scatter of the object. It is defined as the Callier Q coefficient. For photographic silver images Q factors vary in the range of about 1.2 to 1.9. For color photographic images, Q factors are usually smaller than for silver images.

Transmittance measurements made with both the incident and collected beam diffuse are called *doubly diffuse* transmittances.

The integrated transmittance of an object depends upon the spectral radiant-flux distribution of the incident illumination and upon the spectral response characteristics of the receiver. Integrated transmittance T is defined as

$$T = \frac{\int P_t(\lambda) S(\lambda)\, d\lambda}{\int P_o(\lambda) S(\lambda)\, d\lambda} \tag{1-8a}$$

where $P_t(\lambda)$ is the radiant flux reaching the receiver through the sample, $P_o(\lambda)$ is the radiant flux which reaches the receiver with no sample in the beam path, and $S(\lambda)$ is the spectral response function of the receiver.

The radiant flux reaching the receiver is equal to the product $P_o(\lambda) t(\lambda)$, where $t(\lambda)$ is the transmittance function of the sample. Transmittance T is therefore equal to

$$T = \frac{\int P_o(\lambda) t(\lambda) S(\lambda)\, d\lambda}{\int P_o(\lambda) S(\lambda)\, d\lambda} \tag{1-8b}$$

1.4.4 REFLECTANCE. Reflectance of an object may be measured as illustrated in Fig. 1-11. Following reflection from the object, a portion of the light reaches the receiver. Spectral reflectance r_λ is defined as

$$r_\lambda = \frac{P_\lambda}{P_{o\lambda}} \tag{1-9}$$

where P_λ is the radiance at wavelength λ reaching the receiver from the object, and $P_{o\lambda}$ is the amount reaching the receiver when the sample object is replaced by a standard comparison object. Because of its high reflectance and diffusing properties, a surface of barium sulfate is frequently used as a standard. White paints also are available which have satisfactory reflectance characteristics.

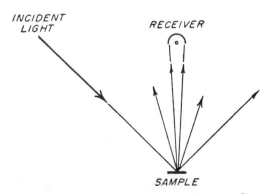

FIG. 1-11 Measurement of reflectance. (*From Fink.*[1])

The geometrical arrangement of the light source, sample, and receiver greatly influences reflectance measurements. The incident beam may be either specular or diffuse. If specular, it may be incident upon the object surface perpendicularly or at any angle up to nearly 90° from normal. Essentially all, or only a part, of the reflected light may be collected by the receiver. The effects of various combinations of these choices on the reflectance measurement will depend considerably upon the surface characteristics of the object being measured.

Integrated reflectance R is defined as

$$R = \frac{\int P_r(\lambda)S(\lambda)\,d\lambda}{\int P_o(\lambda)S(\lambda)\,d\lambda} \qquad (1\text{-}10a)$$

where $P_r(\lambda)$ is the radiant flux reaching the receiver by reflection from the sample, $S(\lambda)$ is the response of the receptor, and $P_o(\lambda)$ is radiant flux reaching the receiver when the sample is replaced by a standard which is completely reflecting and completely diffusing.

Integrated reflectances may be measured directly, provided that a light source of proper spectral distribution and a receptor with proper spectral response characteristics are available. The product curve obtained by a wavelength-by-wavelength multiplication of the light-source flux by the receptor response must equal $P_o(\lambda)S(\lambda)$. Filters placed in the beam can, of course, be used to help achieve the desired result.

Integrated reflectances may also be determined by numerical integration. In this case the spectral reflectance distribution of the object is generally measured. Designating this reflectance, as a function of wavelength λ, as $r(\lambda)$, then

$$R = \frac{\int r(\lambda)P_o(\lambda)S(\lambda)\,d\lambda}{\int P_o(\lambda)S(\lambda)\,d\lambda} \qquad (1\text{-}10b)$$

For any given illuminant and receptor whose spectral characteristics are known, R can then be determined.

1.4.5 PHOTOGRAPHIC FILM SPEED. The "speed" of a photographic film or television camera is a rating of the film or camera in terms of the minimum exposure necessary to obtain a high-quality picture. Speed is usually expressed as a number proportional to the inverse of some selected exposure value. Thus, the smaller the required exposure, the greater the speed and the "faster" the film.

For negative black-and-white materials film speed is defined in American National Standard PH2.5-1972 and International Standard ISO 6-1973. The procedure is illustrated in Fig. 1-12. The D versus log E curve shown is for a film that has been developed to the specified average gradient over a log E range of 1.3, originating at a density of 0.1

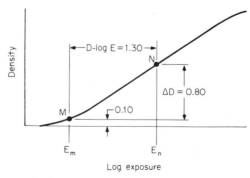

FIG. 1-12 D-log E curve illustrating the method of determining photographic speed according to the American Standard. (*From James.*[2])

above fog (point M) and extending to point N, at a density $0.80 \pm .05$ above the density at point M. Speed is computed at the point of exposure E_m by Eq. (1-11) when the exposure is expressed in units of meter-candle-seconds

$$S = \frac{0.8}{E_m} \tag{1-11}$$

Color-negative film speeds are computed by an analgous procedure described in American National Standard PH2.27-1965. Color-reversal film speed procedure is described in American National Standard PH2.21-1972.

1.4.6 CAMERA FILM EXPOSURE.   It is shown (see Sec. 3.1.6) that the illuminance†
I on a camera film is given approximately by

$$I \text{ (footcandles)} = 0.116 \left(\frac{B}{F^2} \right) \tag{1-12a}$$

where B is the scene luminance in foot-lamberts and F is the aperture ratio (f number). As indicated, I is expressed in footcandles. In photography the illuminance I is more frequently expressed in meter candles. With this change in units, the equation becomes

$$I = 1.25 \left(\frac{B}{F^2} \right) \tag{1-12b}$$

Photographic exposure E is defined as

$$E = It \tag{1-12c}$$

where I is the illuminance on the film in metercandles and t is the time of exposure in seconds. Combining the last two equations

$$E = 1.25 \left(\frac{Bt}{F^2} \right) \tag{1-12d}$$

The ISO (ASA‡) speed, S_x is defined as

$$S_x = \frac{k}{E_m} \tag{1-13a}$$

so it follows from Eq. (1-12d) that

$$S_x = \frac{KF^2}{tB_m} \tag{1-13b}$$

where B_m is the luminance in the scene corresponding to the exposure E_m. Assuming some average set of scene luminances, an exposure meter will give a reading proportional to B_m. Solving for t

$$t = \frac{K_g F^2}{S_x B_g} \tag{1-13c}$$

where B_g is the luminance reading obtained in the exposure meter and K_g is a constant. The constants differ for negative and reversal films and film format. The exposure index, or *film speed number,* is determined by taking into account film latitude, average lens transmittance, and luminance distribution of an average scene. These numbers are expressed in either International Standards Organization (ISO) or Deutsche Industrie Norm (DIN) units. ISO indices are expressed in arithmetic units, DIN indices in logarithmic units, both in intervals of one-third camera stop ($0.1 \log E$ or an exposure factor of 1.25). The relationship is shown in Table 1-6.

---

†The symbol I is used for illuminance in this section to avoid confusion with the symbol for exposure.
‡Superseded by ISO.

Table 1-6   Relationship of ISO and DIN Speeds

ISO	1000	640	400	250	200	160	125	100	80	64	50	40	32	25
DIN		31	29	27	25	24	23	22	21	20	19	18	17	16	15

1.4.7 TELEVISION CAMERA SPEEDS. The illumination on the photosensitive surface of a television camera depends essentially upon the same factors as does the illumination on the film in a photographic camera. Taking average sets of conditions, the illumination I in metercandles is given by

$$I = 1.25 \left(\frac{B}{F^2} \right) \tag{1-14a}$$

B is expressed in footlamberts and F is the aperture ratio of the camera lens. For conditions departing from these assumed "average conditions" the constant of proportionality, 1.25, would change but the form of the equation would be the same. If the total photosensitive area of the camera is A, the total luminous flux on this area is proportional to IA. If A is expressed in square centimeters, then the total luminous flux equals $IA \times 10^{-4}$. Letting S represent the number of microamperes per lumen, then the current i from the whole photosensitive surface, assuming that current is collected from all of it, is given by

$$i = IAS \times 10^{-4} \tag{1-14b}$$
$$= 1.25 \left(\frac{BAS}{F^2} \right) \times 10^{-4}$$

For nonstorage cameras only one element of the picture is being scanned at one time. If there are N elements in the picture, then the amount of current from the camera is less than that shown in Eq. (1-14b) by a factor of N, or

$$i = 1.25 \left(\frac{BAS}{NF^2} \right) \times 10^{-4} \tag{1-14c}$$

As the area a of each picture element equals the total picture area A divided by N, Eq. (1-14c) can also be written in the form

$$i = 1.25 \left(\frac{BaS}{F^2} \right) \times 10^{-4} \tag{1-14d}$$

Equations (1-14c) and (1-14d) would apply to television cameras such as the flying spot camera and the image dissector. They show that the output current is proportional to the scene luminance, the area (A) of the photosensitive surface, and inversely proportional to the number of picture elements and the square of the f number of the camera lens. Large photocathodes are seen to give high signals. On the other hand, large photocathodes require larger and more expensive lenses as well as heavier camera equipment.

Television camera speeds are generally thought of in terms of minimum signal-to-noise ratios. For a given value of i, which represents a picture of minimum accepted quality, Eqs. (1-14c) and (1-14d) provide a basis for establishing minimum scene luminances which will provide the required picture quality. These equations, however, do not indicate the current at the circuit point in which the signal-to-noise ratio is actually of importance. The most important source of noise in most cameras is in the first stages of video amplification. In a flying-spot scanner a photomultiplier is used so that some amplification precedes the regular amplifier. This makes possible smaller values of i, and of B, than given in Eqs. (1-14c) and (1-14d). The image-dissector tube makes use of electron multiplication within the camera. Such multiplication also makes possible lower values of i and of B.

In image-storage cameras, light incident upon each camera picture element throughout the complete frame-scanning interval results in a buildup of charge during that interval. The discharge takes place during the time that this element is scanned. Consequently, there is a theoretical multiplying factor equal to the number of scanned elements. For storage cameras, therefore,

$$i = 1.25 \left(\frac{BaSN}{F^2}\right) \times 10^{-4} = 1.25 \left(\frac{BAS}{F^2}\right) \times 10^{-4} \qquad (1\text{-}15)$$

This equation omits consideration of nonscanning portions of the frame interval, which would give a value of i slightly higher than that shown. Actually, however, the theoretical efficiency indicated by this equation is not fully attained because of the effects of secondary electrons and partial discharging of camera elements during the storage interval.

REFERENCES

1. D. G. Fink (ed.), *Television Engineering Handbook,* McGraw-Hill, New York, 1957.
2. T. H. James (ed.), *Theory of the Photographic Process,* 4th ed., Macmillan, New York, 1977.

BIBLIOGRAPHY

Boynton, R. M., *Human Color Vision,* Holt, New York, 1979.

Committee on Colorimetry, Optical Society of America, *The Science of Color,* New York, 1953.

Davson, H., *Physiology of the Eye,* 4th ed., Academic, New York, 1980.

Evans, R. M., W. T. Hanson, Jr., and W. L. Brewer, *Principles of Color Photography,* Wiley, New York, 1953.

Fink, D. G., *Television Engineering,* 2d ed., McGraw-Hill, New York, 1952.

Kingslake, R. (ed.), *Applied Optics and Optical Engineering,* vol. 1, Academic, New York, 1965.

Polysak, S. L., *The Retina,* University of Chicago Press, 1941.

Schade, O. H., "Electro-optical Characteristics of Television Systems," *RCA Rev.,* vol. 9, pp. 5–37, 245–286, 490–530, 653–686, 1948.

Wright, W. D., *Researches on Normal and Defective Colour Vision,* Mosby, St. Louis, 1947.

Wright, W. D., *The Measurement of Colour,* 4th ed., Adam Hilger, London, 1969.

Color Vision, Representation, and Reproduction

Alan R. Robertson

Division of Physics
National Research Council of Canada
Ottawa, Canada

Joseph F. Fisher

Wynnewood, Pennsylvania and
Philco-Ford (retired)

2.1 COLOR VISION AND COLOR MATCHING

2.1.1 COLOR STIMULI. The sensation of color is evoked by a physical stimulus consisting of electromagnetic radiation in the so-called *visible* spectrum. For most practical purposes, the visible spectrum comprises wavelengths between 380 and 780 nm, although often it is sufficient to consider only 400 to 700 nm.

The stimulus associated with a given object is defined by its spectral concentration of *radiance* $L_e(\lambda)$ and measured, e.g., in watts per steradian per square meter per nanometer ($W \cdot sr^{-1} \cdot m^{-2} \cdot nm^{-1}$) at each wavelength. (The subscript e is used to distinguish L_e from L, the symbol for luminance.) For a reflecting object, $L_e(\lambda)$ is given by

$$L_e(\lambda) = \frac{1}{\pi} E_e(\lambda) R(\lambda)$$

where $E_e(\lambda)$ is the *spectral irradiance* ($W \cdot m^{-2} \cdot nm^{-1}$) falling on the object and $R(\lambda)$ is the *spectral reflectance factor*. Sometimes, the spectral reflectance $\rho(\lambda)$ is used instead of $R(\lambda)$. Since $\rho(\lambda)$ is a measure of the total flux reflected by the object, whereas $R(\lambda)$ is a measure of the flux reflected in a specified direction (e.g., toward an eye or a television camera), the use of $\rho(\lambda)$ implies the assumption that the object reflects uniformly in all directions. For most objects this is approximately true, but for some, such as mirrors, it is not.

2.1.2 TRICHROMATIC THEORY. Full details of the mechanisms of color vision are not yet understood. However it is generally believed, on the basis of strong physiological evidence, that the first stage is the absorption of the stimulus by light-sensitive elements in the retina. These light-sensitive elements, known as *cones,* form three classes each having a different spectral sensitivity distribution. The exact spectral sensitivities are not known, but they are broad and overlap considerably. An estimate of the three classes of spectral sensitivity is given in Fig. 2-1.

It is clear from this *trichromacy* of color vision that many different physical stimuli can evoke the same sensation of color. All that is required for two stimuli to be equivalent is that they should each cause the same number of quanta to be absorbed by any given class of cone. In this case the neural impulses, and thus the color sensations, generated by the two stimuli will be the same. The visual system and the brain cannot differentiate between the two stimuli even though they are physically different. Such equivalent stimuli are known as *metamers* and the phenomenon as *metamerism.* Metamerism is fundamental to the science of colorimetery, and without it color television as we know it could not exist. The stimulus produced by a television receiver is almost always a metamer of the original object and not a physical (spectral) match.

2.1.3 COLOR MATCHING. Because of the phenomenon of trichromacy, it is possible to match any color stimulus by a mixture of three primary stimuli. There is no unique set of primaries; any three stimuli will suffice as long as none of them can be matched by a mixture of the other two. In certain cases it is not possible to match a given stimulus with positive amounts of each of the three primaries, but a match is always possible if the primaries may be used in a negative sense. This may seem to be a difficult experiment to make, but it is accomplished very simply by adding one or two of the primaries to the test stimulus in suitable amount until a match with the other primaries is achieved.

Experimental measurements in color matching are carried out with an instrument called a *colorimeter.* This device provides a split visual field and a viewing eyepiece. The two halves of the visual field are split by a line and are arranged so that the mixture of three primary stimuli appears in one-half of the field. The amounts of the three primaries can be controlled individually so that a wide range of colors can be produced in this half of the field. The other half of the field accepts light from the sample to be matched. The amounts of the primaries are adjusted until the two halves of the field match, and the amounts of the primaries are recorded. For those cases where negative values of one or more of the primaries are needed to secure a match, the instrument is arranged to trans-

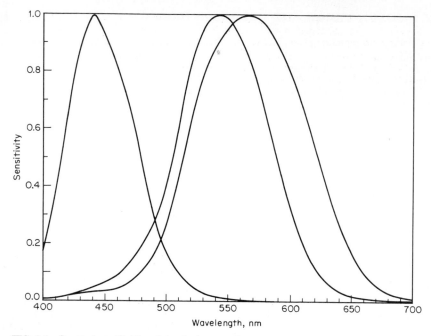

FIG. 2-1 Spectral sensitivities of the three types of cone in the human retina as estimated by Smith and Pokorny[1],† and tabulated by Boynton.[2] The curves have been normalized so that each is unity at its peak.

fer any of the primaries to the other half of the field. The amount of a primary inserted in this manner is recorded as negative.

The operation of color matching may be expressed by the match equation

$$\mathbf{C} \equiv R\mathbf{R} + G\mathbf{G} + B\mathbf{B} \tag{2-1}$$

which should be read as: "Stimulus $\mathbf{C}$ is matched by R units of primary stimulus $\mathbf{R}$ mixed with G units of primary stimulus $\mathbf{G}$ and B units of primary stimulus $\mathbf{B}$." The quantities R, G, and B are called *tristimulus values* and provide a convenient way of describing the stimulus $\mathbf{C}$. All the different physical stimuli that look the same as $\mathbf{C}$ will have the same three tristimulus values R, G, and B. [Throughout this chapter, primary stimuli will be denoted by boldface letters (usually $\mathbf{R}$, $\mathbf{G}$, and $\mathbf{B}$ or $\mathbf{X}$, $\mathbf{Y}$, and $\mathbf{Z}$) and the corresponding tristimulus values (the amounts of the primaries needed to match a given test stimulus) by italic letters R, G, and B or X, Y, and Z, respectively.]

The case in which a negative amount of one of the primaries is required is represented by the match equation

$$\mathbf{C} \equiv -R\mathbf{R} + G\mathbf{G} + B\mathbf{B}$$

assuming that it is the red primary that is required in a negative amount. The actual experiment, of course, is made in the form

$$\mathbf{C} + R\mathbf{R} \equiv G\mathbf{G} + B\mathbf{B}$$

---

†Superscript numbers refer to References at end of chapter.

The extent to which negative values of the primaries are required depends on the nature of the primaries, but no set of real physical primaries can eliminate the requirement entirely.

Experimental investigations have shown that in most practical situations color matches obey the algebraic rules of additivity and linearity. These rules, as they apply to colorimetry, are known as Grassmann's laws.[3] Suppose we have two stimuli defined by the match equations

$$\mathbf{C}_1 \equiv R_1\mathbf{R} + G_1\mathbf{G} + B_1\mathbf{B}$$

$$\mathbf{C}_2 \equiv R_2\mathbf{R} + G_2\mathbf{G} + B_2\mathbf{B}$$

Then if we add $\mathbf{C}_1$ to $\mathbf{C}_2$ in one half of a colorimeter field, and match the resultant mixture with the same three primaries in the other half of the field, the amounts of the primaries will be given by the sums of the amounts required in the individual equations given above; that is, the match equation will be

$$\mathbf{C}_1 + \mathbf{C}_2 \equiv (R_1 + R_2)\mathbf{R} + (G_1 + G_2)\mathbf{G} + (B_1 + B_2)\mathbf{B}$$

The symbols $\mathbf{R}$, $\mathbf{G}$, and $\mathbf{B}$ have been used above to signify that red, green, and blue will serve as a useful set of primaries. The meaning of these color names must be specified exactly before the colorimetric experiments have precise scientific meaning. Such specification may be given in terms of three relative spectral-power-distribution curves, one for each primary. Similarly, the amounts of each primary must be specified in terms of some unit, such as watts or lumens.

The concept of matching the color of a stimulus by a mixture of three primary stimuli is, of course, the basis of color television. The three primaries are the three colored phosphors, and the additive mixture is performed in the observer's eye because of the eye's inability to resolve the small phosphor dots or stripes one from another.

2.1.4 COLOR-MATCHING FUNCTIONS.

In general, a color stimulus is composed of a mixture of radiations of different wavelengths in the visible spectrum. An important consequence of Grassmann's laws is that if the tristimulus values R, G, and B of a monochromatic (i.e., single-wavelength) stimulus of unit radiance (say, 1 $W \cdot sr^{-1} \cdot m^{-2} \cdot nm^{-1}$) are known at each wavelength, the tristimulus values of any stimulus can be calculated by summation. Thus if the tristimulus values of the spectrum are denoted by $\bar{r}(\lambda)$, $\bar{g}(\lambda)$, and $\bar{b}(\lambda)$ per unit radiance, then the tristimulus values of a stimulus with a spectral concentration of radiance of $L_e(\lambda)$ are given by

$$R = \int_{380}^{780} L_e(\lambda)\bar{r}(\lambda)\ d\lambda$$

$$G = \int_{380}^{780} L_e(\lambda)\bar{g}(\lambda)\ d\lambda \qquad (2\text{-}2)$$

$$B = \int_{380}^{780} L_e(\lambda)\bar{b}(\lambda)\ d\lambda$$

If a set of primaries were to be selected and used with a colorimeter, all problems of color mixture could be set up on the colorimeter, and the appropriate matches could be made by an observer. This method has the disadvantage that any selected observer can be expected to have color vision which differs from the average vision of many observers. The color matches made might not be satisfactory to the majority of individuals with *normal* vision.

For this reason it is desirable to secure a set of universal data which would be prepared by averaging the results of color-matching experiments made by a number of individuals having normal vision. A spread of the readings taken in these color matches would indicate the variation to be expected among normal individuals. Averaging the results would give a reliable set of spectral tristimulus values.

Many experimenters have conducted such psychophysical experiments. The results

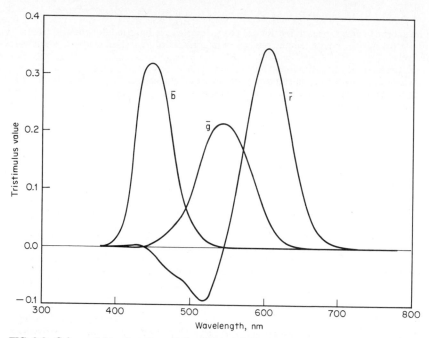

FIG. 2-2 Color-matching functions of the CIE standard observer based on matching stimuli of wavelengths 700.0, 546.1, and 435.8 nm, with units adjusted to be equal for a match to an equienergy stimulus.

are in good agreement and have been used as the basis for the standard data discussed in Sec. 2.3. In each series of experiments, monochromatic radiation at a succession of wavelengths was fed into a colorimeter and the primaries were adjusted to match. The amounts of each primary required were then plotted as a function of the wavelength of the monochromatic radiation being matched. In this way a set of three curves was obtained, one for each primary, showing how much of each primary was required to match each monochromatic radiation. The curves are now generally known as *color-matching functions,* although in older literature the terms *color-mixture functions* and *distribution coefficients* were sometimes used. The set of color-matching functions used by the CIE in 1931 as a basis for an international standard (Sec. 2.3) are shown in Fig. 2-2 in terms of a particular set of real primaries **R, G,** and **B.**

2.1.5 LUMINANCE RELATIONSHIPS. Luminances are, by their definition (Sec. 1.2.3), additive quantities. Thus, the luminance of a stimulus with tristimulus values R, G, and B is

$$L = L_R R + L_G G + L_B B$$

where L_R, L_G, and L_B are the luminances of unit amounts of the primaries.

In the special case of a monochromatic stimulus of unit radiance, the luminance is

$$L = L_R \bar{r}(\lambda) + L_G \bar{g}(\lambda) + L_B \bar{b}(\lambda)$$

This luminance is also given by $K_m V(\lambda)$ where $K_m = 683$ lm/w and $V(\lambda)$ is the spectral luminous efficiency function (see Sec. 1.2.3). It follows that $V(\lambda)$ must be a linear combination of the color-matching functions

$$K_m V(\lambda) = L_R \bar{r}(\lambda) + L_G \bar{g}(\lambda) + L_B \bar{b}(\lambda) \tag{2-3}$$

2.1.6 DEFECTIVE COLOR VISION. In the above discussion of color matching the term *normal* has been used deliberately to exclude those individuals (about 8 percent of males and 0.5 percent of females) whose color vision differs from that of the majority of the population. These people are usually called *color-blind,* although very few (about 0.003 percent of the total population) can see no color at all.

About 2.5 percent of males require only two primaries to make color matches. Most of these can distinguish yellows from blues but confuse reds and greens. The remaining 5.5 percent require three primaries, but their matches are different from the majority and their ability to detect small color differences is usually less.[4,5]

2.2 COLOR REPRESENTATION

2.2.1 COLOR APPEARANCE. Tristimulus values provide a very convenient way of measuring a stimulus. Any two stimuli with identical tristimulus values will appear identical in given viewing conditions. However, the actual appearance of the stimuli (whether they are red, blue, light, dark, etc.) depends on a large number of other factors such as the size of the stimuli, the nature of other stimuli in the field of view, and the nature of other stimuli viewed prior to the present ones. Color appearance cannot be predicted simply from the tristimulus values. Current knowledge of the human color-vision system is far from complete, and much remains to be learned before color appearance can be predicted adequately. However, the idea that the first stage is the absorption of radiation (light) by three classes of cone is accepted by most vision scientists and correlates well with the concept and experimental facts of tristimulus colorimetry. Further, tristimulus values, and quantities derived from them, do provide a useful and orderly way of representing color stimuli and illustrating the relationships between them. This is discussed further in Sec. 2.2.4.

It is possible to describe the appearance of a color stimulus in words based on people's perception of it. The trichromatic theory of color leads to the expectation that this perception will have three dimensions or attributes. Everyday experience confirms this. One set of terms for these three attributes is *hue, brightness,* and *colorfulness.*[6] *Hue* is the attribute according to which an area appears to be similar to one, or to proportions of two, of the perceived colors red, orange, yellow, green, blue, and purple. *Brightness* is the attribute according to which an area appears to be emitting, transmitting, or reflecting more or less light. *Colorfulness* is the attribute according to which an area appears to exhibit more or less chromatic color. Colorfulness is a new term, which is still somewhat controversial, and for most purposes it is better to use *saturation,* which is colorfulness judged in proportion to brightness. Thus *saturation* is the attribute according to which an area appears to exhibit a greater or smaller *proportion* of chromatic color. Consequently we usually describe a perceived color, especially that belonging to a self-emitting object, by its hue, brightness, and saturation.

For reflecting objects, two other attributes are often used. These are *lightness* and *perceived chroma,* which are brightness and colorfulness, respectively, judged in proportion to the brightness of a similarly illuminated area that appears to be white. Thus reflecting objects may be described by hue, lightness, and saturation or by hue, lightness, and perceived chroma. The perceptual color space formed by these attributes may be represented by a geometrical model as illustrated in Fig. 2-3. The achromatic colors (black, gray, white) are represented by points on the vertical axis with lightness increasing along this axis. All colors of the same lightness lie on the same horizontal plane. Within such a plane, the various hues are arranged in a circle with a gradual progression from red through orange, yellow, green, blue, purple, and back to red. Saturation and perceived chroma both increase from the center of the circle outward along a radius but in different ways depending on the lightness. All colors of the same saturation lie on a conical surface, whereas all colors of the same perceived chroma lie on a cylindrical surface.

If two colors have equal saturation but different lightness, the darker one will have

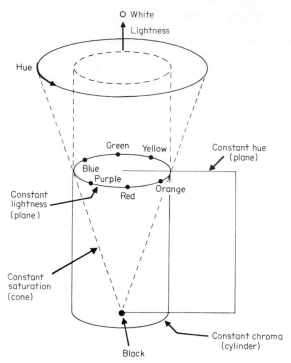

FIG. 2-3 Geometrical model of perceptual color space for reflecting objects.

less perceived chroma (because perceived chroma is judged relative to a white area). Conversely, if two colors have equal perceived chroma, the darker one will have greater saturation (because saturation is judged relative to the brightness or, in this case, lightness, of the color itself).[6]

2.2.2 MUNSELL SYSTEM.

It is possible to construct a sort of color atlas based on the above set of color attributes. One such atlas, devised by the artist A. H. Munsell, is known as the *Munsell system*. The names used for the attributes by Munsell are *hue, chroma,* and *value,*[7-9] corresponding, respectively, to the terms hue, perceived chroma, and lightness used above. In the Munsell system the colors are arranged in a circle in the order red (R), red-purple (RP), purple (P), purple-blue (PB), blue (B), blue-green (BG), green (G), green-yellow (GY), yellow (Y), and yellow-red (YR). This defines a hue circle having 10 hues. To give a finer hue division, each of the above 10 hue intervals is further subdivided into 10 parts. For example, there are 10 red subhues, which are referred to as $1\,R$ to $10\,R$.

The attribute of chroma is described by having hue circles of various radii. The greater the radius, the greater is the chroma.

Finally, the attribute of value is divided into 10 steps, from zero (perfect black or zero reflectance factor) to 10 (perfect white or 100 percent reflectance factor). At each value level there is a set of hue circles of different chromas, the lightness of all the colors in the set being equal.

A color is specified in the Munsell system by specifying in turn (1) hue, (2) value, and (3) chroma. Thus the color specified as 6 RP 4/8 has a red-purple hue of 6 RP, a value of

4, and a chroma of 8. Not all chromas or values can be duplicated with available pigments.

This arrangement of colors in steps of hue, value, and chroma was originally carried out by Munsell using his artistic eye as a judge of the correct classification. Later, a committee of the Optical Society of America[10] made extensive visual studies which resulted in slight modifications to Munsell's original arrangement. The committee's judgment is perpetuated in the form of a book of paper swatches colored with printer's ink and marked with the corresponding Munsell notation.[11] A set of chips arranged in the form of a color tree can also be obtained.

The relationship between Munsell notations and the CIE system of color specification (Sec. 2.3) was defined by the OSA committee in the form of tables and charts.[5,10]

2.2.3 OTHER COLOR-ORDER SYSTEMS.

In addition to the Munsell system, there are many other color atlases (also known as color-order systems.)[12] The three scales of the various systems, and the spacing of samples along the scales, are chosen by different criteria. In some the scales and spacing are determined by systematic mixture of dyes or pigments or by systematic variation of parameters in a printing process. In others they are based on the rules of additive mixture as in a tristimulus colorimeter. A third class of color-order systems (which includes the Munsell system) is based on visual perceptions. Within each class the exact rules by which the colors are ordered vary significantly from one system to another.

One rather special color-order system is the ISCC-NBS method of designating colors.[13] This system divides the color solid into 267 blocks, each one of which is given a color name such as moderate reddish brown, brilliant blue, dark gray. The boundaries of the blocks are defined in terms of the Munsell system.

2.2.4 COLOR TRIANGLE.

An alternative method of classifying and specifying colors is the color triangle. This method was originated by Newton and was used extensively by Maxwell. It is a method of representing the matching and mixing of stimuli and is derived from the tristimulus values discussed in Sec. 2.1.3.

The color triangle displays a given color stimulus in terms of the relative tristimulus values (i.e., relative amounts of three primaries needed to match it). Thus it displays only the quality of the stimulus and not its quantity. In one form, the stimulus is represented by a point chosen so that the perpendicular distances to each of the three sides are proportional to the tristimulus values. The triangle need not be equilateral, although the triangles used by Newton and Maxwell were of this type. The method is illustrated in Fig. 2-4.

This method of display is equivalent to the use of trilinear coordinates, which form a well-known coordinate system in analytical geometry. In this representation, the three primaries appear one at each of the three vertices of the triangle, because two of the trilinear coordinates vanish at each vertex.

Nowadays it is more common to use a right-angled triangle as shown in Fig. 2-5. Quantities r, g, and b, called *chromaticity coordinates,* are calculated by

$$r = \frac{R}{R + G + B}$$

$$g = \frac{G}{R + G + B} \tag{2-4}$$

$$b = \frac{B}{R + G + B}$$

They are plotted with r as abscissa and g as ordinate. Since $r + g + b = 1$ it is not necessary to plot b because it can be derived by $b = 1 - r - g$. This (r, g) diagram, and others like it, are known as *chromaticity diagrams.*

A chromaticity diagram is a specification of a color stimulus, not of appearance. A

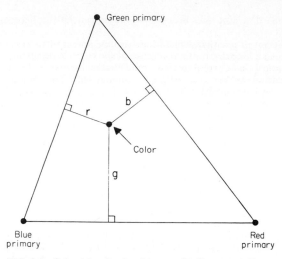

FIG. 2-4 Color triangle, showing use of trilinear coordinates.
The amounts of the three primaries needed to match the color
are proportional to r, g, and b.

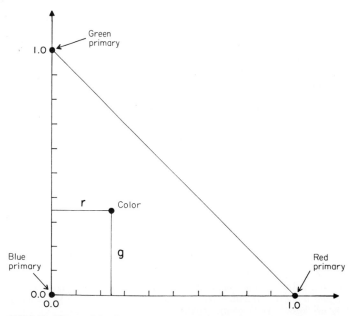

FIG 2-5 Chromaticity diagram. The amounts of the three primaries needed
to match the color are proportional to r, g, and b ($=1\text{-}r\text{-}g$).

particular chromaticity may have many different appearances depending on factors such as its luminance relative to a white reference in the field of view, and the observer's state of adaptation. For example, there is a wide range of chromaticities that will appear white to an observer who is adapted to them. Nevertheless, for a rigidly defined set of observing conditions it is sometimes useful to think of a chromaticity diagram in terms of appearance. Chromaticity coordinates are relative tristimulus values and thus correlate with hue and saturation but not with brightness. For most observing conditions, stimuli near the middle of the diagram produce color appearances of low saturation, and stimuli at the edges produce high saturations. The hues are arranged in the usual order of red, orange, yellow, green, blue, purple, and back to red. Conventionally, the diagram is arranged so that this progression is counterclockwise.

Center of Gravity Law. The chromaticity diagram is a very useful way of representing additive color mixture. Consider two stimuli C_1 and C_2

$$C_1 \equiv R_1\mathbf{R} + G_1\mathbf{G} + B_1\mathbf{B}$$

$$C_2 \equiv R_2\mathbf{R} + G_2\mathbf{G} + B_2\mathbf{B}$$

As explained in Sec. 2.1.3, the stimulus C_1 is matched by R_1 units of primary stimulus $\mathbf{R}$, mixed with G_1 units of primary stimulus $\mathbf{G}$, and B_1 units of primary stimulus $\mathbf{B}$. In a similar manner the stimulus C_2 is matched by R_2, G_2, and B_2 units of the same primaries. From Eq. (2-4), the chromaticity coordinates are

$$r_1 = \frac{R_1}{R_1 + G_1 + B_1} \qquad g_1 - \frac{G_1}{R_1 + G_1 + B_1} \qquad b_1 = \frac{R_1}{R_1 + G_1 + B_1}$$

and $R_1 + G_1 + B_1$ equals the total tristimulus value T_1. Then

$$r_1 = \frac{R_1}{T_1} \qquad g_1 = \frac{G_1}{T_1} \qquad b_1 = \frac{B_1}{T_1}$$

$$R_1 = r_1 T_1 \qquad G_1 = g_1 T_1 \qquad B_1 = b_1 T_1$$

In a similar manner

$$R_2 = r_2 T_2 \qquad G_2 = g_2 T_2 \qquad B_2 = b_2 T_2$$

In terms of chromaticity coordinates (r, g, b) the equations for C_1, and C_2 may be written as

$$C_1 = (r_1 T_1)\mathbf{R} + (g_1 T_1)\mathbf{G} + (b_1 T_1)\mathbf{B}$$

$$C_2 = (r_2 T_2)\mathbf{R} + (g_2 T_2)\mathbf{G} + (b_2 T_2)\mathbf{B}$$

Thus, by Grassmann's laws the stimulus C formed by mixing C_1 and C_2 is

$$C = R\mathbf{R} + G\mathbf{G} + B\mathbf{B}$$

where

$$R = r_1 T_1 + r_2 T_2$$
$$G = g_1 T_1 + g_2 T_2$$
$$B = b_1 T_1 + b_2 T_2$$

By application of Eq. (2-4), the chromaticity coordinates of the mixture are

$$r = \frac{r_1 T_1 + r_2 T_2}{T_1 + T_2}$$

$$g = \frac{g_1 T_1 + g_2 T_2}{T_1 + T_2}$$

$$b = \frac{b_1 T_1 + b_2 T_2}{T_1 + T_2}$$

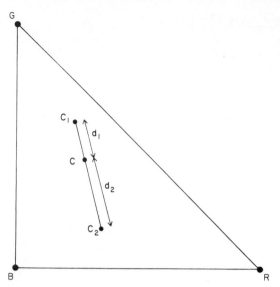

FIG. 2-6   Center of gravity law in chromaticity diagram. The additive mixture of color stimuli represented by C_1 and C_2 lies at C whose location on the straight line C_1C_2 is given by $d_1T_1 = d_2T_2$, where T_1 and T_2 are the total tristimulus values of the component stimuli.

The interpretation of this in the chromaticity diagram is simply that the mixture lies on the straight line joining the two components and divides it in the ratio T_2/T_1. This is illustrated in Fig. 2-6. It is known as the *center of gravity law* because of the obvious analogy with the center of gravity of weights T_1 and T_2 placed at the points representing $\mathbf{C}_1$ and $\mathbf{C}_2$.

Alychne.   The luminance of a stimulus with tristimulus values R, G, and B is (from Sec. 2.1.5)

$$L = L_R R + L_G G + L_B B$$

If we divide by $R + G + B$ and set $L = 0$, we have

$$0 = L_R r + L_G g + L_B b$$

which is the equation of a straight line in the chromaticity diagram. It is the line along which colors of zero luminance would lie if they could exist. The line is called the *alychne*.

The alychne is illustrated in Fig. 2-7, which is a chromaticity diagram based on monochromatic primaries of wavelengths 700.0, 546.1, and 435.8 nm with their units normalized so that equal amounts are required to match a stimulus in which the spectral concentration of radiant power per unit wavelength is constant throughout the visible spectrum (this stimulus is called the *equienergy stimulus*). The alychne lies wholly outside the triangle of primaries, as indeed it must, for no positive combination of real primaries can possibly have zero luminance.

Spectrum Locus.   Since all color stimuli are mixtures of radiant energy of different wavelengths, it is interesting to plot, in a chromaticity diagram, the points representing

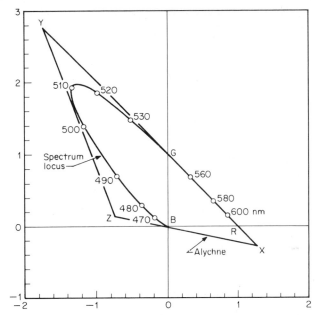

FIG. 2-7 Spectrum locus and alychne of the *CIE 1931 Standard Observer* plotted in a chromaticity diagram based on matching stimuli of wavelengths 700.0, 546.1, and 435.8 nm. The locations of the CIE primary stimuli *X,Y,* and *Z* are shown.

monochromatic stimuli (stimuli consisting of a single wavlength). These can be calculated from the color-matching functions (Sec. 2.1.4) by

$$r(\lambda) = \frac{\overline{r}(\lambda)}{\overline{r}(\lambda) + \overline{g}(\lambda) + \overline{b}(\lambda)}$$

$$g(\lambda) = \frac{\overline{g}(\lambda)}{\overline{r}(\lambda) + \overline{g}(\lambda) + \overline{b}(\lambda)}$$

When these spectral chromaticity coordinates are plotted (Fig. 2-7), they lie along a horseshoe-shaped curve called the *spectrum locus.* The extremities of the curve correspond to the extremities of the visible spectrum—approximately 380 nm for the blue end and 780 nm for the red end. The straight line joining the extremities is called the *purple boundary* and is the locus of the most saturated purples obtainable.

Since all color stimuli are combinations of spectral stimuli, it is apparent from the center of gravity law that all real color stimuli must lie on or inside the spectrum locus.

It is an experimental fact that the part of the spectrum locus lying between 560 and 780 nm is substantially a straight line. This means that broad-band colors in the yellow-orange-red region can give rise to colors of high saturation.

2.2.5 SUBJECTIVE AND OBJECTIVE QUANTITIES. It is important to distinguish clearly between perceptual (subjective) terms and psychophysical (objective) terms.

Perceptual terms relate to attributes of sensations of light and color. They indicate *subjective* magnitudes of visual responses. Examples are hue, saturation, brightness, and lightness.

Table 2-1   Perceptual Terms and Their Psychophysical Correlates

Perceptual (subjective)	Psychophysical (objective)
Hue	Dominant wavelength
Saturation	Excitation purity
Brightness	Luminance
Lightness	Luminous reflectance or luminous transmittance

Psychophysical terms relate to *objective* measures of physical variables which identify stimuli that produce equal visual responses in specified viewing conditions. Examples are tristimulus values, luminance, and chromaticity coordinates.

Psychophysical terms are usually chosen so that they correlate in an approximate way with particular perceptual terms. Examples of some of these correlations are given in Table 2-1. The perceptual terms in the table are defined in Sec. 2.2.1; the psychophysical correlates are defined in Secs. 1.2.3, 2.3.6, and 2.3.9.

2.3 CIE SYSTEM

2.3.1 SYSTEM PHILOSOPHY.   In 1931, the International Commission on Illumination (known by the initials CIE of its French name, Commission Internationale de l'Eclairage) defined a set of color-matching functions and a coordinate system that have remained the predominant international standard method of specifying color ever since.

The color-matching functions were based on experimental data from many observers measured by Wright[14] and Guild.[15] They are shown in Fig. 2-2. Wright and Guild used different sets of primaries, but the results were transformed to a single set, namely, monochromatic stimuli of wavelengths 700.0, 546.1, and 435.8 nm. The units of the stimuli were chosen so that equal amounts were needed to match an equienergy stimulus (constant radiant power per unit wavelength throughout the visible spectrum). Figure 2-7 shows the spectrum locus in the (r, g) chromaticity diagram based on these color-matching functions.

At the same time that it adopted these color-matching functions as a standard, the CIE introduced and standardized a new set of primaries involving some ingenious concepts. The set of real physical primaries of Figs. 2-2 and 2-7 were replaced by a new set of imaginary nonphysical primaries with special characteristics. These new primaries are referred to as **X**, **Y**, and **Z**, and the corresponding tristimulus values as X, Y, and Z. The chromaticity coordinates of **X**, **Y**, and **Z** in the **RGB** system are shown in Fig. 2-7. Primaries **X** and **Z** lie on the alychne (Sec. 2.2.4) and hence have zero luminance. All the luminance in a mixture of these three primaries is contributed by **Y**.

This convenient property depends only on the decision to locate **X** and **Z** on the alychne. It still leaves a wide choice of locations for all three primaries. The actual locations chosen by the CIE are illustrated in Fig. 2-7 and were based on the following additional considerations:

1. The spectrum locus lies entirely within the triangle **XYZ**. This means that negative amounts of the primaries are never needed to match real colors. The color-matching functions, $\overline{x}(\lambda)$, $\overline{y}(\lambda)$, and $\overline{z}(\lambda)$, shown in Fig. 2-8, are therefore all positive at all wavelengths. This reduced the chance of error in numerical computations in the pre-computer age.

2. The line $Z = 0$ (that is, the line from **X** to **Y**) lies along the straight portion of the spectrum locus. Thus Z is effectively zero for spectral colors with wavelengths greater than about 560 nm.

3. The line $X = 0$ (that is, the line from **Y** to **Z**) was chosen to minimize (approximately) the area of the **XYZ** triangle outside the spectrum locus. This choice led to a bimodal shape for the $\overline{x}(\lambda)$ color-matching function (Fig. 2-8) because the spectrum locus curves away from the line $X = 0$ at low wavelengths. A different choice of $X = 0$ (tangential to the spectrum locus at about 450 nm) would have eliminated the secondary lobe of $\overline{x}(\lambda)$ but would have pushed **Y** much further from the spectrum locus.

4. The units of **X**, **Y**, and **Z** were chosen so that the tristimulus values X, Y, and Z would be equal to each other for an equienergy stimulus.

This coordinate system and the set of color-matching functions that go with it are known as the *CIE 1931 Standard Observer*.

The color-matching data on which the *1931 Standard Observer* is based were obtained with a visual field subtending 2° at the eye. Because of the slight nonuniformities of the retina, color-matching functions for larger fields are slightly different. This prompted the CIE in 1964 to recommend a second *Standard Observer,* known as the *CIE 1964 Supplementary Standard Observer,* for use in colorimetric calculations when the field size is greater than 4°. Since the whole of a 19-in television screen viewed from 10 ft subtends only 9°, the *1964 Observer* has little relevance to television.

2.3.2 COLOR-MATCHING FUNCTIONS. The color-matching functions of the *CIE Standard Observer* are shown in Fig. 2-8 and are listed in Table 22-16. They are used, as described in the following sections, to calculate tristimulus specifications of color stimuli and to determine whether two physically different stimuli will match each other. Such calculated matches represent the results of the average of many observers, but may not represent an exact match for any single real observer. For most purposes, this restriction is not important; the match of an average observer is all that is required.

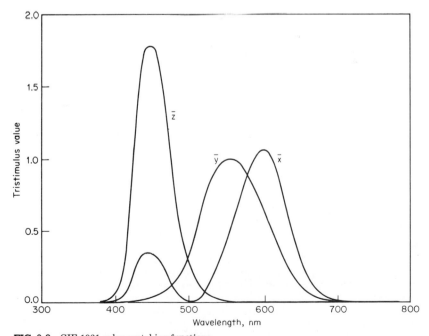

FIG. 2-8 CIE 1931 color-matching functions.

2.3.3 TRISTIMULUS VALUES AND CHROMATICITY COORDINATES. By exact analogy with the calculation of the tristimulus values R, G, B in Sec. 2.1.4, the tristimulus values X, Y, Z of a stimulus $L_e(\lambda)$ are calculated by

$$X = \int_{380}^{780} L_e(\lambda)\bar{x}(\lambda)\ d\lambda$$

$$Y = \int_{380}^{780} L_e(\lambda)\bar{y}(\lambda)\ d\lambda$$

$$Z = \int_{380}^{780} L_e(\lambda)\bar{z}(\lambda)\ d\lambda$$

The chromaticity coordinates x,y are then calculated by

$$x = \frac{X}{X + Y + Z}$$

$$y = \frac{Y}{X + Y + Z}$$

and plotted as rectangular coordinates to form the CIE 1931 Chromaticity Diagram (Fig. 2-9).

It is important to remember that the CIE Chromaticity Diagram is not intended to illustrate appearance. The CIE system tells only whether two stimuli match in color, not what they look like. Appearance depends on many factors not taken into account in the chromaticity diagram. Nevertheless it is often useful to know approximately where colors lie on the diagram. Figure 2-10 gives some color names for various parts of the diagram based on observations of self-luminous areas against a dark background.

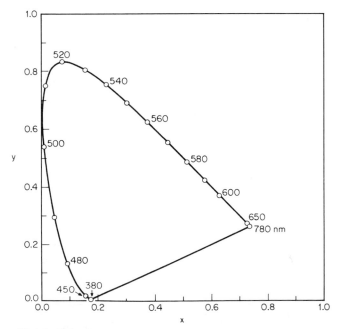

FIG. 2-9 CIE 1931 chromaticity diagram showing spectrum locus and wavelengths in nanometers.

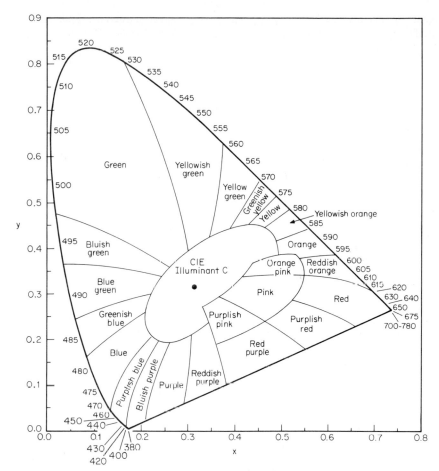

FIG. 2-10 CIE 1931 chromaticity diagram divided into regions corresponding to various color names derived from observations of self-luminous areas against a dark background. (*From K. L. Kelly,* "Color Designations for Lights," *J. Opt. Soc. Am., vol. 33, pp. 627–632, 1943.*)

2.3.4. CONVERSION BETWEEN TWO SYSTEMS OF PRIMARIES. To transform tristimulus specifications from one system of primaries to another, it is necessary and sufficient to know, in one system, the tristimulus values of the primaries of the other system. For example, consider two systems **R G B** (in which tristimulus values are represented by R, G, B and chromaticity coordinates by r, g) and **R′ G′ B′** (in which tristimulus values are represented by R', G', B' and chromaticity coordinates by r', g'). If one system is defined in terms of the other by the match equations

$$\mathbf{R'} \equiv a_{11}\mathbf{R} + a_{21}\mathbf{G} + a_{31}\mathbf{B}$$

$$\mathbf{G'} \equiv a_{12}\mathbf{R} + a_{22}\mathbf{G} + a_{32}\mathbf{B} \tag{2-5}$$

$$\mathbf{B'} \equiv a_{13}\mathbf{R} + a_{23}\mathbf{G} + a_{33}\mathbf{B}$$

then the following equations can be derived to relate the tristimulus values and chromaticity coordinates of a color stimulus measured in one system to those of the same color stimulus measured in the other system (details of the derivations are given in Sec. 2.6.1):

$$R = a_{11}R' + a_{12}G' + a_{13}B'$$
$$G = a_{21}R' + a_{22}G' + a_{23}B' \qquad (2\text{-}6)$$
$$B = a_{31}R' + a_{32}G' + a_{33}B'$$

$$R' = b_{11}R + b_{12}G + b_{13}B$$
$$G' = b_{21}R + b_{22}G + b_{23}B \qquad (2\text{-}7)$$
$$B' = b_{31}R + b_{32}G + b_{33}B$$

$$r = \frac{\alpha_{11}r' + \alpha_{12}g' + \alpha_{13}}{t}$$
$$g = \frac{\alpha_{21}r' + \alpha_{22}g' + \alpha_{23}}{t} \qquad (2\text{-}9)\dagger$$
$$t = \alpha_{31}r' + \alpha_{32}g' + \alpha_{33}$$

$$r' = \frac{\beta_{11}r + \beta_{12}g + \beta_{13}}{t'}$$
$$g' = \frac{\beta_{21}r + \beta_{22}g + \beta_{23}}{t'} \qquad (2\text{-}10)$$
$$t' = \beta_{31}r + \beta_{32}g + \beta_{33}$$

The coefficients b_{ij}, α_{ij}, and β_{ij} can all be calculated from the a_{ij}, as shown in Sec. 2.6.1. Equations (2-9) and (2-10) are of the form known as *projective transformations*. Such transformations have the property that they retain straight lines as straight lines. In other words, a straight line in the r, g diagram will transform to a straight line in the r', g' diagram. Another important property is that the center of gravity law continues to apply. This is apparent because the proof of the law (Sec. 2.2.4) could be made independently in either system.

The derivation of the CIE **XYZ** color-matching functions and chromaticity diagram (Figs. 2-8 and 2-9) from the corresponding data in the **RGB** system of Figs. 2-2 and 2-7 is an example of the use of the transformation equations given above. It is mainly of historical interest. A more important practical example, the relationship between the CIE system and a system based on television receiver phosphors, is discussed in Sec. 2.4.1.

2.3.5. LUMINANCE CONTRIBUTION OF PRIMARIES.

Because **X** and **Z** were chosen to be on the alychne, their luminances are zero. Thus all the luminance of a mixture of **X**, **Y**, and **Z** primaries is contributed by **Y**. This means that the Y tristimulus value is proportional to the luminance of the stimulus. In particular, if we consider monochromatic stimuli, the analog of Eq. (2-3) is

$$K_m V(\lambda) = L_X \bar{x}(\lambda) + L_Y \bar{y}(\lambda) + L_Z \bar{z}(\lambda) \qquad (2\text{-}11)$$

where L_X, L_Y, and L_Z are the luminances of unit amounts of **X**, and **Y**, and **Z**, respectively. L_X and L_Z are zero so that Eq. (2-11) reduces to

$$K_m V(\lambda) = L_Y \bar{y}(\lambda)$$

In fact the unit of **Y** was chosen to further simplify this to

$$V(\lambda) = \bar{y}(\lambda)$$

---

†See Sec. 2.6.1 for Eq. (2-8).

2.3.6 COLORIMETRIC COMPUTATIONS. The CIE tristimulus values of a stimulus $L_e(\lambda)$ are given by

$$X = k \int_\lambda L_e(\lambda)\bar{x}(\lambda)\, d\lambda$$

$$Y = k \int_\lambda L_e(\lambda)\bar{y}(\lambda)\, d\lambda$$

$$Z = k \int_\lambda L_e(\lambda)\bar{z}(\lambda)\, d\lambda$$

where the integrals are evaluated over the visible spectrum and k is a suitable normalizing constant whose value depends on the application.

If Y is to give the absolute value of a photometric quantity, the constant k must be set equal to K_m (see Sec. 1.2.3) and $L_e(\lambda)$ must be the spectral concentration of the radiometric quantity corresponding to the photometric quantity required. For example, if $L_e(\lambda)$ is the spectral radiance in watts per steradian per square meter per nanometer, Y will be the luminance in candelas per square meter.

For secondary light sources (reflecting or transmitting objects), $L_e(\lambda)$ is usually replaced by the relative spectral distribution of the stimulus, $R(\lambda)S(\lambda)$ or $T(\lambda)S(\lambda)$ where $R(\lambda)$ is the spectral reflectance factor, $T(\lambda)$ is the spectral transmittance factor, and $E(\lambda)$ is the relative spectral distribution of the illuminant. In this case, k is usually chosen so that $Y = 100$ for a perfect white-reflecting diffuser $[R(\lambda) = 1.0]$ or a clear transmitting object $[T(\lambda) = 1.0]$. Then (for the case of a reflecting object)

$$X = k \int_\lambda R(\lambda)E(\lambda)\bar{x}(\lambda)\, d\lambda$$

$$Y = k \int_\lambda R(\lambda)E(\lambda)\bar{y}(\lambda)\, d\lambda$$

$$Z = k \int_\lambda R(\lambda)E(\lambda)\bar{z}(\lambda)\, d\lambda$$

$$k = \frac{100}{\int_\lambda E(\lambda)\bar{y}(\lambda)\, d\lambda}$$

Thus, Y becomes the percent luminous reflectance or transmittance factor.

For television monitors and receivers, the constant k is sometimes chosen so that $Y = 100$ for *peak white*, the brightest white area obtainable on the particular monitor or receiver.

The usual method of evaluating these integrals is to divide the spectrum into narrow intervals, say 10 nm wide, adjacent to one another and to use the values of $R(\lambda)$, $E(\lambda)$, $\bar{x}(\lambda)$, $\bar{y}(\lambda)$, and $\bar{z}(\lambda)$ at the center of each interval in summations

$$X = k \,\Sigma\, R(\lambda)E(\lambda)\bar{x}(\lambda)$$
$$Y = k \,\Sigma\, R(\lambda)E(\lambda)\bar{y}(\lambda) \qquad (2\text{-}12)$$
$$Z = k \,\Sigma\, R(\lambda)E(\lambda)\bar{z}(\lambda)$$
$$k = \frac{100}{\Sigma\, E(\lambda)\bar{y}(\lambda)}$$

Evaluation of the sums in 10-nm steps from 400 to 700 nm is sufficient for many purposes. Values of $\bar{x}(\lambda)$, $\bar{y}(\lambda)$, and $\bar{z}(\lambda)$ for use in such computations are given in Table 22-16.

This method of evaluating X, Y, and Z is known as the *weighted ordinate method*. An alternative method, little used any more, is the *selected ordinate method*.[16] In this method the area under curves of $E(\lambda)\bar{x}(\lambda)$, $E(\lambda)\bar{y}(\lambda)$, and $E(\lambda)\bar{z}(\lambda)$ are divided, by ordinates, into equal areas, usually 30 in number for each curve. The integrals are evaluated simply by summing $R(\lambda)$ at the ordinates at the center of each area.

In either case, the chromaticity coordinates x and y can be calculated from the tristimulus values in the usual way

$$x = \frac{X}{X + Y + Z}$$

$$y = \frac{Y}{X + Y + Z}$$

2.3.7 STANDARD ILLUMINANTS. The CIE has recommended a number of standard illuminants $E(\lambda)$ for use in evaluating the tristimulus values of reflecting and transmitting objects by Eqs. (2-12). Originally, in 1931, it recommended three, known as A, B, and C. These illuminants are specified by tables of relative spectral distribution and were chosen so that they could be reproduced by real physical sources. (CIE terminology distinguishes between *illuminants*, which are tables of numbers, and *sources*, which are physical emitters of light.)

Source A is a tungsten filament lamp operating at a color temperature (see Sec. 2.3.8) of about 2856K. Its chromaticity coordinates are $x = 0.4476$ and $y = 0.4074$. It represents incandescent light.

Source B is source A with a composite filter made of two liquid filters of specified chemical composition.[17] The chromaticity coordinates of source B are $x = 0.3484$ and $y = 0.3516$. It represents noon sunlight.

Source C is also produced by source A with two liquid filters.[17] Its chromaticity coordinates are $x = 0.3101$ and $y = 0.3162$. It represents average daylight according to information available in 1931.

In 1971, the CIE[18] introduced a new series of standard illuminants which represent daylight more accurately than illuminants B and C. The improvement is particularly marked in the ultraviolet part of the spectrum which is important for fluorescent samples. The most important of the D illuminants is D_{65} (sometimes written D6500) which has chromaticity coordinates of $x = 0.3127$ and $y = 0.3290$. At present there is no recommended source D_{65} but there is a recommended method[19] for assessing how well a particular source conforms to D_{65} for colorimetric purposes.

The CIE[18] states that for general use in colorimetry, illuminants A and D_{65} should suffice.

The relative spectral power distributions of illuminants A, B, C, and D_{65} are given in Table 22-16 and in Fig. 2-11.

The hypothetical equienergy illuminant, for which the spectral concentration of power per unit wavelength is constant throughout the visible spectrum, is sometimes referred to as illuminant E. The units of the primaries of the CIE system were chosen so that the tristimulus values of illuminant E are all equal.

2.3.8 CORRELATED COLOR TEMPERATURE. Until the development of fluorescent lamps and other discharge lamps, all sources used for illumination were simply hot objects (e.g., the sun, a candle, or the filament of a tungsten lamp). The spectral power distribution of a heated black body is related to absolute temperature T by Planck's equation

$$E(\lambda) = c_1\lambda^{-5}(e^{c_2/\lambda T} - 1)^{-1} \qquad W \cdot m^{-3}$$

where $c_1 = 3.7415 \cdot 10^{-16}$ W·m^2 and $c_2 = 1.4388 \cdot 10^{-2}$ m·K. Because of this it became customary to describe the color of a source by the temperature of the equivalent Planckian radiator. This temperature is known as the *color temperature* of the source. The CIE chromaticity coordinates of Planckian radiators of various temperatures are shown in Fig. 2-12.

When new sources were introduced whose chromaticity did not fall on the Planckian locus, the term *correlated color temperature* was coined to denote the temperature of the Planckian radiator whose perceived color is closest to that of the source in question. The

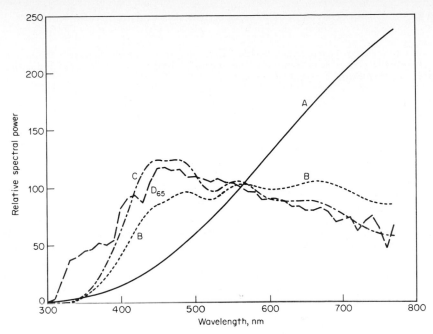

FIG. 2-11　Relative spectral power distributions of CIE standard illuminants A, B, C, and D$_{65}$.

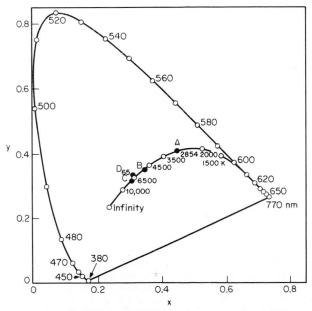

FIG. 2-12　CIE 1931 chromaticity diagram showing locus of Planckian radiators and chromaticities of CIE standard illuminants A, B, C, and D$_{65}$. (*From Judd and Wyszecki.*[3])

method currently recommended by the CIE to determine which Planckian radiator is perceptually closest is to determine the shortest distance in a chromaticity diagram obtained by plotting $\frac{3}{2} v'$ against u' where

$$u' = \frac{4x}{-2x + 12y + 3}$$

$$v' = \frac{9y}{-2x + 12y + 3}$$

(The significance of u' and v' is discussed in Sec. 2.5.5.) This shortest distance may be determined graphically[20] or by computation.[21]

The approximate correlated color temperatures of the CIE standard illuminants are

A: 2856 K,   B: 4900 K,   C: 6800 K, D_{65}: 6500 K

It is important to remember the limitations of the concept of correlated color temperature. It is a one-dimensional scale and gives no information about how far the source is from the Planckian locus. In most cases it is better to specify the chromaticity coordinates.

2.3.9 DOMINANT WAVELENGTH AND EXCITATION PURITY.

The *dominant wavelength* of a stimulus is defined as the wavelength of the monochromatic stimulus that must be mixed with a given achromatic stimulus to match the test stimulus. It can be found by drawing a straight line in the chromaticity diagram from the achromatic stimulus through the test stimulus to the point where it intersects the spectrum locus. Clearly, the dominant wavelength of the stimulus depends on the chromaticity of the achromatic stimulus. For reflecting objects this is usually taken to be the chromaticity of the illuminant. For television displays it is usually taken to be the nominal white of the display.

For some stimuli, the straight line from the achromatic point through the test chromaticity will strike the purple boundary rather than the spectrum locus. For these stimuli the line must be extended backwards from the achromatic point. The point where the extended line strikes the spectrum locus determines the *complementary wavelength* of such a stimulus.

The *excitation purity* of a stimulus is its distance from the achromatic point, expressed relative to the total distance from the achromatic point to the boundary stimulus (spectral or purple as the case may be).

The concepts of dominant wavelength, complementary wavelength, and excitation purity are illustrated in Fig. 2-13. Excitation purity p_e is calculated from whichever of the following formulas has the larger denominator and hence the smaller rounding error

$$p_e = \frac{x - x_n}{x_b - x_n} \qquad p_e = \frac{y - y_n}{y_b - y_n}$$

where x and y = chromaticity coordinates of test stimulus
x_n and y_n = chromaticity coordinates of achromatic stimulus
x_b and y_b = chromaticity coordinates of boundary stimulus

For those who are familiar with the colors of spectral stimuli, dominant (or complementary) wavelength and excitation purity are more suggestive of color appearance than are chromaticity coordinates. In very loose terms they correlate with hue and saturation, respectively.

2.3.10 GAMUT OF REPRODUCIBLE COLORS.

In a system which seeks to match or reproduce colors with a set of three primaries, only those colors can be reproduced that lie inside the triangle of primaries. Colors outside the triangle cannot be reproduced because they would require negative amounts of one or two of the primaries.

It is important in a color-reproducing system to have a triangle of primaries that is sufficiently large to permit a satisfactory gamut of colors to be reproduced. To illustrate the kinds of requirements that must be met, Fig. 2-14 shows the maximum color gamut for real surface colors and the triangle of typical modern color television receiver phos-

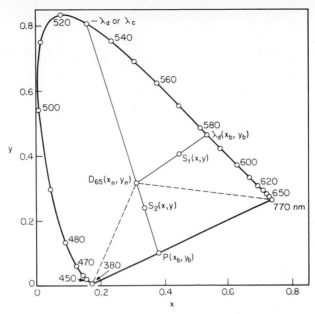

FIG. 2-13 CIE 1931 chromaticity diagram illustrating the determination of dominant wavelength λ_d and excitation purity p_e of a stimulus S_1 and complementary wavelength $-\lambda_d$ (or λ_c) and purity p_e of a stimulus S_2 relative to illuminant D_{65}. (*From Judd and Wyszecki.*[3])

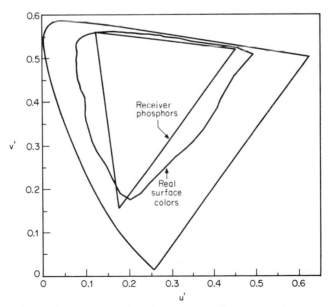

FIG. 2-14 The color triangle defined by a standardized set of color-television-receiver phosphors compared with the maximum real-color gamut on a u',v' chromaticity diagram. (*From M. R. Pointer, "The Gamut of Real Surface Colours," Color Res. Appl., vol. 5, pp. 145–155, 1980.*)

phors as standardized by the European Broadcasting Union (Sec. 2.4.5). These are shown in the CIE 1976 u', v' chromaticity diagram (Sec. 2.5.5) in which the perceptual spacing of colors is more uniform than in the x, y diagram. High-purity blue-green and purple colors cannot be reproduced by these phosphors, whereas the blue phosphor is actually of slightly higher purity than any real surface colors.

2.3.11 VECTOR REPRESENTATION.   In the preceding sections the representation of color has been reduced to two dimensions by eliminating consideration of quantity (luminance) and discussing only chromaticity. Since color requires three numbers to specify it fully, a three-dimensional representation can be made, taking the tristimulus values as vectors. For simplicity the discussion will be confined to CIE tristimulus values and to a rectangular framework of coordinate axes.

Tristimulus values have been treated as scalar quantities. If we now transform them into vector quantities by multiplying them by unit vectors $\mathbf{i}$, $\mathbf{j}$, and $\mathbf{k}$ in the x, y, and z directions, then X, Y, and Z will become vector quantities $\mathbf{i}X$, $\mathbf{j}Y$, and $\mathbf{k}Z$, and a color will be represented by three vectors $\mathbf{i}X$, $\mathbf{j}Y$, and $\mathbf{k}Z$ along the x, y, and z coordinate axes, respectively. Combining these vectors gives a single resultant vector $\mathbf{V}$, represented by the vector equation

$$\mathbf{V} = \mathbf{i}X + \mathbf{j}Y + \mathbf{k}Z$$

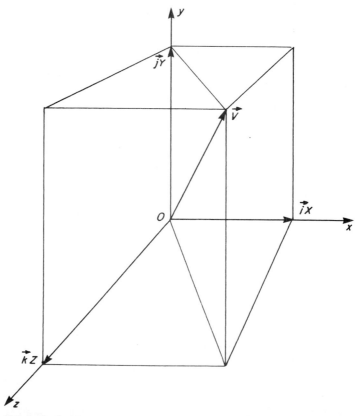

FIG. 2-15   Combination of vectors.

The resultant is obtained by the usual vector methods, which are illustrated in Fig. 2-15.

Some of the implications of this form of color representations are illustrated in Fig. 2-16. Here are drawn a vector **V** representing a color, a set of coordinate axes, and the plane $x + y + z = 1$ passing through the points $L(1,0,0)$, $M(0,1,0)$, and $N(0,0,1)$. The vector passes through the point Q in this plane, having coordinates x, y, and z. The point P is the projection of the point Q into the xy plane, and therefore has coordinates (x, y). From the geometry of the figure it can be seen that

$$\frac{X}{x} = \frac{Y}{y} = \frac{Z}{z} = \frac{X + Y + Z}{x + y + z}$$

Since Q is on the plane $x + y + z = 1$

$$\frac{X}{x} = \frac{Y}{y} = \frac{Z}{z} = X + Y + Z$$

Hence
$$x = \frac{X}{X + Y + Z} \qquad y = \frac{Y}{X + Y + Z}$$

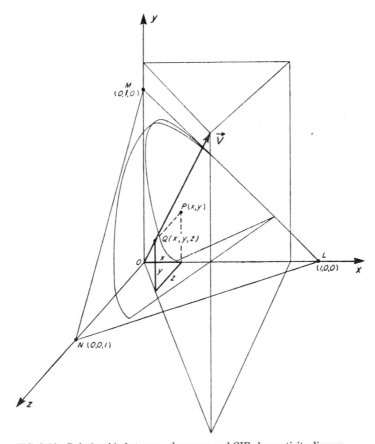

FIG. 2-16 Relationship between color space and CIE chromaticity diagram.

Thus x and y are the CIE chromaticity coordinates of the color. Hence the xy plane in color space represents the CIE chromaticity diagram, when the vectors have magnitudes equal to the CIE tristimulus values and the axes are rectangular. The triangle LMN is a Maxwell triangle for CIE primaries. The spectrum locus in the xy plane can be thought of as defining a cylinder with generators parallel to z; this cylinder intersects the plane of the triangle LMN in the spectrum locus for this Maxwell triangle.

2.4 TELEVISION COLORIMETRY

2.4.1 GENERAL PRINCIPLES.

A television display (receiver or monitor) may be regarded as a series of very small visual colorimeters. In each picture element a colorimetric match is made to an element of the original scene. The primaries are the red, green, and blue phosphors, and the mixing occurs inside the eye of the observer because the eye cannot resolve the individual phosphor dots since they are too closely spaced. The outputs (R, G, B) of the three phosphors may be regarded as tristimulus values. Their relation to the tristimulus values (X, Y, Z) of the CIE system is given by equations similar to Eqs. (2-6) and (2-7). The coefficients in these equations depend on the chromaticity coordinates of the phosphors and on the luminous outputs of each phosphor for unit electrical input. Usually the gains of each of the three channels are set so that equal electrical inputs to the three produce a standard displayed white such as CIE illuminant D_{65}.

A television camera must therefore produce, for each picture element, three electrical signals representative of the three tristimulus values (R, G, B) of the required display. It is clear from Eqs. (2-2) that to do this it must have three optical channels with spectral sensitivities equal to the color-matching functions $\bar{r}(\lambda)$, $\bar{g}(\lambda)$, and $\bar{b}(\lambda)$ corresponding to the three primaries of the display.

Thus, the information to be conveyed by the electronic circuits comprising the camera, transmitter, and receiver is the amount of each of the three primaries (phosphors) required to match the input color. This information is based on: (1) an agreement concerning the chromaticities of the three primaries to be used, (2) the representation of the amounts of these three primaries by electrical signals suitably related to them, and, usually, (3) the specification that the electrical signals shall be equal at some specified chromaticity. The electrical signal voltages are then representative of the tristimulus values of the original scene. They obey all the laws to which tristimulus values conform, including the property of being transformable to represent the amounts of primaries of other chromaticities than those for which the signals were originally composed. Such transformations can be arranged by forming three sets of linear combinations of the original signals.

Unfortunately the simple objective of producing an exact colorimetric match between each picture element in the display and the corresponding element of the original scene is difficult to fulfill and in any case may not achieve the ultimate objective of equality of *appearance* between the display and the original scene. There are several reasons. Among these are the following:

1. It may be difficult to achieve the luminance of the original scene through the use of a reproduction because of the limitation of the maximum luminance that can be generated by the reproducing system.

2. The adaptation of the eye may be different for the reproduction than it is for the original scene because the surrounding conditions are different.

3. Ambient light complicates viewing the reproduced picture and changes its effective contrast ratio.

4. The angle subtended by the reproduced picture may be different from that of the original scene.

While it is an oversimplification, it is often considered that adequate reproduction is achieved when the chromaticity is accurately reproduced, while the luminance is repro-

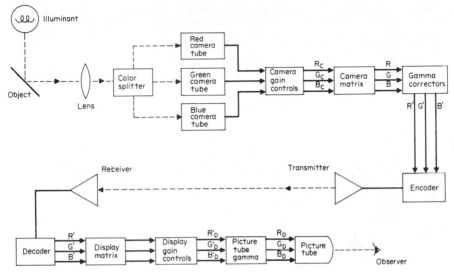

FIG. 2-17 Block diagram of simplified color television system.

duced proportionally to the luminance of the original scene. The adequacy of this approach is somewhat questionable (see Sec. 2.5.1), but it provides a starting point and enables the setting up of targets for system performance.

A block diagram of a simplified color television system is shown in Fig. 2-17. Light reflected (or transmitted) by an object is split into three so that a portion strikes each camera tube, producing outputs proportional to the tristimulus values R, G, and B. (In some cameras there is a matrix amplifier as explained in Sec. 2.4.7.) Gain controls are available so that the three signals can be made equal when the camera is viewing a standard white object. For transmission, the signals are encoded into three different signals and then decoded back to R, G, and B in the receiver. The nature and rationale of this encoding and decoding are explained in Chap. 4. The main purpose is to enable the signal to be transmitted within a limited bandwidth and to maintain compatibility with monochrome systems. The decoded R, G, and B signals are applied to the picture tube to excite the three phosphors. (Again, there may be a matrix circuit, often combined with the decoder; see Sec. 2.4.7.) Gain controls are provided so that equal inputs R, G, and B can be made to produce a standard white on the display.

As shown in Sec. 2.6.2, the following equations can be derived that relate R, G, and B to the CIE tristimulus values X, Y, and Z

$$X = a_{11}R + a_{12}G + a_{13}B$$
$$Y = a_{21}R + a_{22}G + a_{23}B \qquad (2\text{-}16)\dagger$$
$$Z = a_{31}R + a_{32}G + a_{33}B$$

The inverse equations are

$$R = b_{11}X + b_{12}Y + b_{13}Z$$
$$G = b_{21}X + b_{22}Y + b_{23}Z \qquad (2\text{-}17)$$
$$B = b_{31}X + b_{32}Y + b_{33}Z$$

---

†See Sec. 2.6.2 for Eqs. (2-13)–(2-18).

The coefficients a_{ij} and b_{ij} are functions of the CIE chromaticity coordinates of the phosphors $(x_r, y_r, z_r; x_g, y_g, z_g; x_b, y_b, z_b)$ and of the white (x_w, y_w, z_w) that is to be produced when $R = G = B$.

Gamma.   So far, a linear relationship has been assumed between corresponding electrical and optical quantities in both the camera and the receiver. In practice the "transfer function" is often not linear. For example, over the useful operating range of a typical color receiver, the light output of each phosphor follows a power-law relationship to the video voltage applied to the grid or cathode of the color cathode-ray tube (CRT). The light output (L) is proportional to the video-driving voltage (E_V) raised to the power γ $(L = KE_V^\gamma)$, where γ is typically about 2.5. This produces black compression and white expansion. Compensation for these three nonlinear transfer functions is accomplished by three electronic "gamma correctors" in the color camera video-processing amplifiers. Thus the three signals which are encoded, transmitted, and decoded (Fig. 2-17) are not in fact R, G, and B, but rather R', G', and B' given by

$$R' = R^{1/\gamma} \qquad G' = G^{1/\gamma} \qquad B' = B^{1/\gamma} \tag{2-19}$$

If the rest of the system is linear, application of these signals to the color picture tube causes light outputs which are linearly related to the R, G, and B tristimulus inputs to the color camera, so the correct reproduction is achieved.

2.4.2 NTSC STANDARDS.   In 1953 the National Television System Committee (NTSC) recommended a set of color television standards[22] which were later adopted by the Federal Communications Commission (FCC) for use in the United States. Subsequently, this system was chosen by many additional countries including Canada, Japan, and Mexico.

Other countries have chosen different systems. Sequentiel Couleur avec Mémoire (SECAM) is used by France and the USSR. Phase alternation line (PAL) is used by many European countries including Germany and the United Kingdom.[23] In this section, specifications will be given only for the NTSC system. Details of the other two systems will be found in Chap. 4 and in Ref. 23. From a colorimetrist's point of view, the differences are small. The differences are mainly in the encoding and decoding of the transmitted signals.

The NTSC system was designed for use with displays having three phosphors with the following CIE chromaticity coordinates

$$x_r = 0.67 \qquad y_r = 0.33$$

$$x_g = 0.21 \qquad y_g = 0.71$$

$$x_b = 0.14 \qquad y_b = 0.08$$

The R, G, and B signals were originally specified to be equal for illuminant C

$$x_w = 0.3101 \qquad y_w = 0.3162$$

although illuminant D$_{65}$ with

$$x_w = 0.3127 \qquad y_w = 0.3290$$

is now more generally used. The difference is not significant in practice and is a factor only when precise numerical calculations are required.

Substituting these values into Eqs. (2-14) and (2-15) of Sec. 2.6.2, we can derive numerical values for the coefficients in the transformation equations (Eqs. 2-16 of Secs. 2.4.1 and 2.6.2). For illuminant C

$$X = 0.607R + 0.174G + 0.200B$$

$$Y = 0.299R + 0.587G + 0.114B$$

$$Z = \qquad\qquad 0.066G + 1.116B$$

and for illuminant D_{65}

$$X = 0.588R + 0.179G + 0.183B$$
$$Y = 0.290R + 0.606G + 0.105B$$
$$Z = 0.068G + 1.021B$$

Inverting these equations for illuminant C

$$R = 1.910X - 0.532Y - 0.288Z$$
$$G = -0.985X + 1.999Y - 0.028Z$$
$$B = 0.058X - 0.118Y + 0.898Z$$

and for illuminant D_{65}

$$R = 1.971X - 0.549Y - 0.297Z$$
$$G = -0.954X + 1.936Y - 0.027Z \tag{2-20}$$
$$B = 0.064X - 0.129Y + 0.982Z$$

The required spectral sensitivities of the camera (for D_{65} as display white) are thus

$$\bar{r}(\lambda) = 1.971\bar{x}(\lambda) - 0.549\bar{y}(\lambda) - 0.297\bar{z}(\lambda)$$
$$\bar{g}(\lambda) = -0.954\bar{x}(\lambda) + 1.936\bar{y}(\lambda) - 0.027\bar{z}(\lambda)$$
$$\bar{b}(\lambda) = 0.064\bar{x}(\lambda) - 0.129\bar{y}(\lambda) + 0.982\bar{z}(\lambda)$$

These curves are shown as solid lines in Fig. 2-18. Note that the curves have negative lobes in addition to the main positive lobes. These negative lobes cannot be achieved

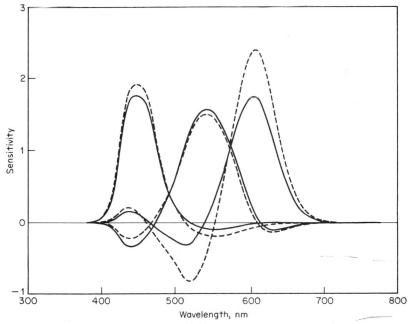

FIG. 2-18 Ideal camera spectral sensitivities for two sets of display primary colors. Full line, NTSC primaries. Dashed line, EBU primaries.

directly by the optical system shown in Fig. 2-17. Solutions to this problem are discussed in Sec. 2.4.7.

Contributions to Luminance. From Eqs. (2-21), Sec. 2.6.3, the relative contributions of the three NTSC phosphors to the luminance of a displayed stimulus with chromaticity coordinates x, y, and z are

Red:	$0.1806x - 0.0504y - 0.0273z$
Green:	$-0.1828x + 0.3710y - 0.0052z$
Blue:	$0.0021x - 0.0043y + 0.0325z$

Gamma. The NTSC specifications state that the transmitted signals should be designed for a display with $\gamma = 2.2$ [Eqs. (2-19)]. In practice, however, it is more realistic to assume $\gamma = 2.5$.[24]

2.4.3 DISPLAY WHITE. The NTSC signal specifications were designed so that equal signals $R = G = B$ would produce a display white of the chromaticity of illuminant C. For many years most home receivers (but not studio monitors) were set so that equal signals produced a much bluer white. The correlated color temperature was about 9300 K and the chromaticity was usually slightly on the green side of the Planckian locus. This was to achieve satisfactorily high brightness and to avoid excessive red/green current ratios with available phosphors. With modern phosphors, high brightness and red/green current equality can be achieved for a white at the chromaticity of D_{65} so that both monitors and receivers are now usually balanced to D_{65}. As D_{65} is very close to illuminant C, the color rendition is generally better than with the bluer balance of older receivers.

2.4.4 SCENE WHITE. When the original scene is illuminated by daylight (of which D_{65} is representative), it is clearly reasonable to aim to reproduce the chromaticities of each object exactly in the final display. However, many television pictures are taken in a studio with incandescent illumination of about 3000 K. In viewing the original scene, the eye adapts to a great extent so that most objects have very similar appearance in both daylight and incandescent light. In particular, whites appear white under both types of illumination. However, if the chromaticities were to be reproduced exactly, studio whites would appear much yellower than the outdoor whites. This is because the viewer's adaptation is controlled more by the ambient viewing illuminant than by the scene illuminant and therefore does not correct fully for the change of scene illuminant. Because of this, exact reproduction of chromaticities is not a good objective.

The ideal objective is that the reproduction should have exactly the same *appearance* as the original scene, but not enough is known about the chromatic adaptation of the human eye for us even to define what this means in terms of chromaticity. A simpler criterion is to aim to reproduce objects with the same chromaticity that they would have if the original scene were illuminated by D_{65}. This can be achieved by placing an optical filter (colored glass) in front of the camera with the spectral transmittance of the filter being equal to the ratio of the spectral power distributions of D_{65} and the actual studio illumination. As far as the camera is concerned, this has exactly the same effect as putting the same filter over every light source.

This solution has disadvantages because a different correction filter (in effect, a different set of camera sensitivities) is required for every scene illuminant. For example, every phase of daylight requires a special filter. In addition insertion of a filter increases light scattering and can slightly degrade the contrast, resolution, and signal-to-noise ratio.

An alternative method is to adjust the gain controls in the camera so that a white object produces equal signals in the three channels irrespective of the actual chromaticity of the illuminant. For colors other than white this solution does not produce exactly the same effect as a correction filter, but in practice it seems to be satisfactory. On some of the latest cameras, the operator focuses a white reference in the scene inside a cursor on the monitor, and a microprocessor in a camera does this white balancing operation automatically.

Often, the two methods are combined. A series of optical filters is available corresponding to common scene illuminants, and in addition there are gain controls for exact adjustment of $R = G = B$ for a reference white.

The main difficulty with these methods is that while, by themselves, scenes purporting to be either outdoors or indoors look acceptable, if the scene shifts from one to the other, the viewer subconsciously expects some change in color temperature. It is disturbing when no change occurs.[25] One suggestion for compensating for this difficulty is to correct somewhat less than completely for changes in illuminant color.[26]

2.4.5 PHOSPHOR CHROMATICITIES.
The phosphor chromaticities specified by the NTSC in 1953 were based on phosphors in common use for color television displays at that time. Since then, different phosphors have been introduced[27,28] mainly to increase the brightness of displays. These modern phosphors, especially the green ones, have different chromaticities so that the gamut of reproducible chromaticities has been reduced. However, because of the increased brightness, the overall effect on color rendition has been beneficial.

Two sets of modern phosphors are shown plotted in the CIE chromaticity diagram in Fig. 2-19. They are the set standardized by the European Broadcasting Union (EBU)[23,29] and a set used in many North American color monitors and being considered for adoption as a standard by the Society of Motion Picture and Television Engineers

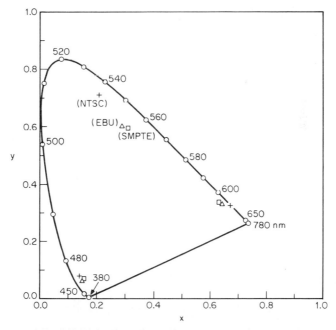

FIG. 2-19 CIE 1931 chromaticity diagram showing three sets of phosphors used in color television displays.

+ NTSC specification
△ EBU standard phosphors
□ SMPTE standard phosphors (proposed)

Table 2-2   Chromaticity Coordinates of Three Sets of Phosphors

	NTSC (+)†		EBU (Δ)†		SMPTE proposed (□)†	
	x	y	x	y	x	y
Red	0.67	0.33	0.64	0.33	0.635	0.340
Green	0.21	0.71	0.29	0.60	0.305	0.595
Blue	0.14	0.08	0.15	0.06	0.155	0.070

†Symbols correspond to Fig. 2.19.

(SMPTE).[30] The NTSC chromaticities are also shown. All three sets of chromaticity coordinates are listed in Table 2-2.

The electrical inputs required to produce accurate color rendition with the modern phosphors are, of course, not the same as the inputs required for the NTSC phosphors. Substituting the chromaticity coordinates of the EBU phosphors into Eqs. (2-14) and (2-15), and using D_{65} as the reference white, we obtain

$$X = 0.431R + 0.342G + 0.178B$$

$$Y = 0.222R + 0.707G + 0.071B$$

$$Z = 0.020R + 0.130G + 0.939B$$

Inverting these

$$R = \quad 3.062X - 1.393Y - 0.476Z$$

$$G = -0.969X + 1.876Y + 0.042Z \qquad (2\text{-}22)†$$

$$B = \quad 0.068X - 0.229Y + 1.069Z$$

where the coefficients are different from those in the NTSC equations (2-20).

This means that if signals which follow NTSC specifications are applied directly to display devices with modern phosphors, color errors will result. There are two ways to solve this problem.[31] One way, adopted by the EBU in Europe, is to require cameras to have spectral sensitivities appropriate to the modern phosphors. The ideal spectral sensitivities for such cameras are obtained by using the coefficients of Eq. (2-22)

$$\bar{r}(\lambda) = \quad 3.062\bar{x}(\lambda) - 1.393\bar{y}(\lambda) - 0.476\bar{z}(\lambda)$$

$$\bar{g}(\lambda) = -0.969\bar{x}(\lambda) + 1.876\bar{y}(\lambda) + 0.042\bar{z}(\lambda)$$

$$\bar{b}(\lambda) = \quad 0.068\bar{x}(\lambda) - 0.229\bar{y}(\lambda) + 1.069\bar{z}(\lambda)$$

These curves are shown together with the NTSC curves in Fig. 2-18.

The alternative approach, preferred in the United States,[31] is to incorporate circuitry in monitors and receivers to convert from NTSC signals to signals appropriate to the actual phosphors being used. This approach is discussed in Sec. 2.4.7.

2.4.6 CAMERA SPECTRAL SENSITIVITIES.   The ideal camera spectral taking characteristics are determined from the values of the chromaticity coordinates of the receiver primaries and of the reference white to which the receiver is adjusted [Eqs. (2-18)]. Over the years this white reference has varied from 9300 K to illuminant C and currently is D_{65}. The calculated ideal taking characteristics have negative lobes (Fig. 2-18). For this reason early cameras attempted to duplicate only the positive parts of the curves. In some cases, the sides of the curves were trimmed to offset the absence of the negative

---

†See Sec, 2.6.3 for Eqs. (2-21).

lobes, but these narrower curves resulted in loss of sensitivity and hence lower signal-to-noise ratio, so most manufacturers tended to prefer broader curves.[32]

With any set of all-positive curves, whether narrow or broad, significant color errors are generated. Thus it is preferable to try to incorporate the effect of negative lobes. One way to do this is to introduce a portion of the signal from one channel as a negative signal into another channel. For example, some of the blue signal could be subtracted from the green signal to duplicate the effect of one of the missing negative lobes of the green sensitivity curve.

In a generalization of this subtraction technique, known as matrixing, electronic circuitry in the camera transforms the three signals from the camera tubes into three new signals which are closer to the required NTSC R, G, B signals. Mathematically, the technique is equivalent to multiplication by a 3×3 matrix.

2.4.7 MATRIXING. There exist an infinite number of linear combinations of the ideal camera sensitivities

$$\bar{r}_c(\lambda) = c_{11}\bar{r}(\lambda) + c_{12}\bar{g}(\lambda) + c_{13}\bar{b}(\lambda)$$

$$\bar{g}_c(\lambda) = c_{21}\bar{r}(\lambda) + c_{22}\bar{g}(\lambda) + c_{23}\bar{b}(\lambda)$$

$$\bar{b}_c(\lambda) = c_{31}\bar{r}(\lambda) + c_{32}\bar{g}(\lambda) + c_{33}\bar{b}(\lambda)$$

Many of these are all-positive and can be produced by inserting appropriate colored filters in front of the three camera tubes. The signals generated by these camera tubes will be

$$R_c = c_{11}R + c_{12}G + c_{13}B$$

$$G_c = c_{21}R + c_{22}G + c_{23}B \qquad (2\text{-}23)$$

$$B_c = c_{31}R + c_{32}G + c_{33}B$$

Provided that the electrical signals are related linearly to the optical inputs (as they are with the Plumbicon and Saticon tubes but not with the Vidicon), an electrical matrix circuit can be introduced to transform R_c, G_c, B_c to R, G, B

$$R = d_{11}R_c + d_{12}G_c + d_{13}B_c$$

$$G = d_{21}R_c + d_{22}G_c + d_{23}B_c \qquad (2\text{-}24)$$

$$B = d_{31}R_c + d_{32}G_c + d_{33}B_c$$

where the matrix of d's is the inverse of the matrix of c's. Equations (2-23) and (2-24) are exactly analogous to Eqs. (2-6) and (2-7) in Sec. 2.3.4.

One point which must be considered is that when a matrix amplifier calling for negative values in some of the nine coefficients is used, the video signal-to-noise ratio is decreased. This is so because the noises from the individual camera tubes are not phase-related and they add in an rms fashion while the video signals are subtracted in a linear manner. The introduction of the Plumbicon television camera in the 1960s, with its greatly improved signal-to-noise ratio compared with the previously used image Orthicon and Vidicon tubes, made camera matrixing practical.[32] Nevertheless cameras are often designed so that the matrix can be switched out in low light levels to increase signal-to-noise ratio at the expense of color fidelity.

The coefficients of the matrix depend quite significantly on the phosphor chromaticities for which the camera is designed [compare Eqs. (2-20) and (2-22)]. European standards specify that the camera should be designed for modern phosphors, whereas North American standards assume that the receiver or monitor will expect signals appropriate to the NTSC phosphors.[31]

The coefficients of the R, G, B equations of the matrix units used by CBS in their color television studios in Hollywood, California, where the cameras use Plumbicon extended-red tubes, are given below. With a scene lighting of about 150 footcandles, (fc) (1614 lm/m^2) and a lens stop of f/4.5, the signal-to-noise ratio is excellent. The matrix is designed to compensate for the negative lobes of the ideal color camera spectral taking characteristics based on the use of NTSC display phosphors. Matrixing is done prior to gamma correction where the camera signals are linearly related to the R, G, B light inputs

$$R = 1.52R_c - 0.49G_c - 0.03B_c$$

$$G = 0.01R_c + 1.05G_c - 0.06B_c$$

$$B = -0.04R_c - 0.13G_c + 1.17B_c$$

The coefficients of R, G, and B each add up to unity, so the matrix amplifier does not affect the white balance.

Ideally, as discussed in Sec. 2.4.4, different camera sensitivities are required for different scene illuminants (e.g., studio incandescent or daylight). An approach to this ideal can be achieved by having two separate, switchable matrices, one for studio use and one for daylight use.[33,34]

One possible set of all-positive camera spectral sensitivities is the CIE color-matching functions. In the early 1950s a color-television CRT flying-spot film camera was designed and built with spectral taking characteristics (considering the spectral output of the flying-spot CRT, the colored filters used, and the response of the photomultiplier tubes) closely approximating the CIE $\bar{x}$, $\bar{y}$, $\bar{z}$ color-matching functions.[35] A matrix to convert from the signals proportional to the X, Y, Z tristimulus values to the R, G, B video voltages was incorporated.[36]

In practice, X, Y, Z cameras have not proved popular because other all-positive characteristics are easier to produce. In theory, there is some advantage to having one camera tube with the $\bar{y}(\lambda)$ sensitivity because this enables the luminance signal to be generated directly. However the inclusion of gamma correction in the encoding of the transmitted signals means that a separately generated luminance signal leads to chromaticity and luminance errors in the reproduced picture.[37]

Theoretically, matrixing can also be applied in the same way in color television display devices. If the input signals, R, G, B are appropriate to NTSC phosphors but the actual display has modern phosphors which require signals R_D, G_D, B_D, then the conversion can be made in the usual way

$$R_D = e_{11}R + e_{12}G + e_{13}B$$

$$G_D = e_{21}R + e_{22}G + e_{23}B \qquad (2\text{-}25)$$

$$B_D = e_{31}R + e_{32}G + e_{33}B$$

Unfortunately, since the transmitted signals are gamma-corrected to compensate for the power-law distortion of the color CRT, exact application of Eqs. (2-25) requires applying to NTSC-decoded signals a gamma corresponding to that of the display, corrective matrixing by Eqs. (2-25), then applying the inverse gamma correction to the matrixed signals before applying them to the actual display. This approach is complex and has not been used.

Direct application of the matrix of Eqs. (2-25) to the gamma-corrected signals improves the rendition of some colors[38] but causes some serious distortion of other colors, especially of bright-red saturated colors, which are reproduced with excess luminance, saturation and chroma noise.[39] Consequently, color monitors employing this type of correction are generally operated with the 3×3 video correcting matrix switched out. The matrices used in color TV monitors are generally of the 3×3 video matrix form, so that the monitor may be initially set up using an electronic color-bar-chart signal. The matrix is then switched in for viewing program material.

With a color TV receiver an equivalent or somewhat modified matrix may be obtained by decoding the composite color video signal at nonstandard phase angles and changing the gains in the $(R - Y)$, $(B - Y)$, and $(G - Y)$ video channels. A decoding system operating in this manner, which provides good color rendition for white, flesh tones, and grass and some distortion of other colors, is described by Parker.[40]

Another approach, described by Neal,[24] is to use a matrix which, applied to linear signals, would correct for only 80 percent of the difference between the NTSC and the actual phosphors. Applied to gamma-corrected signals, Neal reports that such a matrix gives better color fidelity than the "full" correction method.

A third method, applicable to color television receivers more than to monitors, described by Sasaki et al.,[41] uses a nonlinear method of decoding.

Whether matrixing is used or not, it is important to remember that the camera and the display device must be matched. A camera that provides signals intended for a display with modern phosphors will produce color errors in a display that is corrected on the expectation of receiving NTSC signals. Similarly, a camera designed for NTSC phosphors will give errors in a display with modern phosphors and no correction.

2.4.8 MEASUREMENT OF COLOR DISPLAYS. The chromaticity and luminance of an area of a color display device such as a monitor or receiver may be measured in several ways.

The most fundamental way is a complete spectroradiometric measurement followed by computation using tables of color-matching functions (Sec. 2.3.6). Several portable spectroradiometers with built-in computers are available on the market. Another method, somewhat faster but less accurate, is to use a photoelectric colorimeter. These devices have spectral sensitivities approximately equal to the CIE color-matching functions and thus provide direct readings of tristimulus values.

For setting up the reference white, it is often simplest to use a split-field visual comparator and to adjust the display device until it matches the reference field (usually D_{65}) of the comparator. However, because there is usually a large spectral difference (i.e., large metamerism) between the display and the reference, different observers will often make different settings by this method. Thus settings by one observer, or a group of

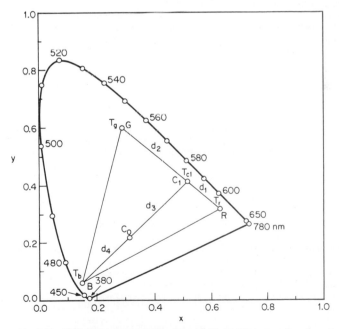

FIG. 2-20 CIE 1931 chromaticity diagram illustrating use of center of gravity law $(T_r d_1 = T_g d_2, T_{c1} = T_r + T_g, T_{c1} d_3 = T_b d_4)$.

observers, with normal color vision are often used just to provide a reference point for subsequent photoelectric measurements.[42]

An alternative method of determining the luminance and chromaticity coordinates of any area of a display is to measure the output of each phosphor separately and combine them using the center of gravity law (Sec. 2.2.4) in which the total tristimulus output of each phosphor is considered as an equivalent weight located at the chromaticity coordinates of the phosphor.

Consider the CIE chromaticity diagram shown in Fig. 2-20 to be a uniform flat surface positioned in a horizontal plane. For the case illustrated, the center of gravity of the three weights (T_r, T_g, T_b), or the balance point, will be at the point C_0. This point determines the chromaticity of the mixture color. The luminance of the color C_0 will be the linear sum of the luminance outputs of the red, green, and blue phosphors. The chromaticity coordinates of the display primaries may be obtained from the manufacturer. The total tristimulus output of one phosphor may be determined by turning off the other two CRT guns, measuring the luminance of the specified area, and dividing this value by the y chromaticity coordinate of the energized phosphor. This procedure is then repeated for the other two phosphors. From these data the color resulting from given excitations of the three phosphors may be calculated as follows (Fig. 2-20)

$$\text{Chromaticity coordinates of red phosphor} = x_r, y_r$$

$$\text{Chromaticity coordinates of green phosphor} = x_g, y_g$$

$$\text{Chromaticity coordinates of blue phosphor} = x_b, y_b$$

$$\text{Luminance of red phosphor} = Y_r$$

$$\text{Luminance of green phosphor} = Y_g$$

$$\text{Luminance of blue phosphor} = Y_b$$

$$\text{Total tristimulus value of red phosphor} = X_r + Y_r + Z_r = \frac{Y_r}{y_r} = T_r$$

$$\text{Total tristimulus value of green phosphor} = X_g + Y_g + Z_g = \frac{Y_g}{y_g} = T_g$$

$$\text{Total tristimulus value of blue phosphor} = X_b + Y_b + Z_b = \frac{Y_b}{y_b} = T_b$$

Consider T_r as a weight located at the chromaticity coordinates of the red phosphor and T_g as a weight located at the chromaticity coordinates of the green phosphor. The location of the chromaticity coordinates of color C_1 (blue gun of color CRT turned off) can be determined by taking moments along line RG to determine the center of gravity of weights T_r and T_g

$$T_r \times d_1 = T_g \times d_2$$

The total tristimulus value of C_1 is equal to $T_r + T_g = T_{C1}$. Taking moments along line C_1B will locate the chromaticity coordinates of the mixture color C_0

$$T_{C1} \times d_3 = T_b \times d_4$$

The luminance of the color C_0 is equal to $Y_r + Y_g + Y_b$.

2.5 ASSESSMENT OF COLOR REPRODUCTION

2.5.1 FUNDAMENTAL SYSTEM REQUIREMENTS. Benson[38] has identified many factors that cause bad color rendition in color television. To assess the effect of these factors, we need to define system objectives and have a way of measuring departures from the objectives.

Hunt[26] has distinguished between several different objectives for color reproduction, as noted below.

1. *Spectral color reproduction* is the exact reproduction of the spectral power distributions of the original stimuli. Clearly this is not possible in a television system with three primaries.

2. *Exact color reproduction* is the exact reproduction of tristimulus values. The reproduction is then a metameric match to the original. Exact color reproduction will result in equality of appearance only if the viewing conditions for the picture and the original scene are identical. These conditions include the angular subtense of the picture, the luminance and chromaticity of the surround, and glare. In practice, exact color reproduction often cannot be achieved because of limitations to the maximum luminance that can be produced on a color monitor or receiver.

3. *Colorimetric color reproduction* is a variant of exact color reproduction in which the tristimulus values are proportional to those in the original scene. In other words, the chromaticity coordinates are reproduced exactly, but the luminances are all reduced by a constant factor. Traditionally, color television systems have been designed and evaluated for colorimetric color reproduction. If the original and the reproduced reference whites have the same chromaticity, if the viewing conditions are the same, and if the system has an overall gamma of unity, colorimetric color reproduction is indeed a useful criterion. However, these conditions often do not hold and then colorimetric color reproduction is inadequate.[43]

4. *Equivalent color reproduction* is the reproduction of the original color appearance. This might be considered as the ultimate objective but cannot be achieved because of the limited luminance that can be generated.

5. *Corresponding color reproduction* is a compromise in which colors in the reproduction have the same appearance as the colors in the original would have had if they had been illuminated to produce the same average luminance level and the same reference white chromaticity as that of the reproduction. For most purposes, corresponding color reproduction is the most suitable objective of a color television system.

6. *Preferred color reproduction.* It is sometimes argued that even corresponding color reproduction is not the ultimate aim for color television but that account should be taken of the fact that people prefer some colors to be different from their actual appearance. For example, sun-tanned skin color is preferred to average real skin color, and sky is preferred bluer and foliage greener than they really are.

Even if we accept corresponding color reproduction as the target, it is important to remember that some colors are more important than others. Thus flesh tones must be acceptable—not obviously reddish, greenish, purplish, or otherwise incorrectly rendered. Similarly the sky must be blue and the clouds white, within the viewer's range of acceptance. Similar conditions apply to other well-known colors of common experience. For this reason, special circuits have been designed[44,45] which attempt to ensure that flesh tones are correctly reproduced, often at the expense of accurate rendition of other colors.

2.5.2 CHROMATIC ADAPTATION AND WHITE BALANCE. With properly adjusted cameras and displays, whites and neutral grays are reproduced with the chromaticity of D_{65}. Tests have shown that such whites (and grays) appear satisfactory in home viewing situations even if the ambient light is of quite different color temperature. The only problems occur when the white balance is slightly different from one camera to the next or when the scene shifts from studio to daylight or vice versa. In the first case, unwanted shifts of the displayed white occur, whereas in the other, no shift occurs even though the viewer subconsciously expects a shift.[25]

By always reproducing a white surface with the same chromaticity, the system is mimicking the human visual system which adapts so that white surfaces always appear the same whatever the chromaticity of the illuminant (at least within the range of common light sources). The effect on other colors, however, is more complicated. In television cameras the white adjustment is usually made by gain controls on the R, G, and B channels. This is similar to the von Kries model of human chromatic adaptation although the R, G, and B primaries of the model are not the same as the television primaries. It is known that the von Kries model does not accurately account for the appearance of colors after chromatic adaptation,[46] so it follows that simple gain adjustments in a television camera are not the ideal adjustments. Nevertheless they seem to work fairly well in practice, and the viewer does not object to the fact, for example, that the relative increase in the luminances of reddish objects in tungsten light is lost.[47]

2.5.3 OVERALL GAMMA REQUIREMENTS.

Colorimetric color reproduction requires that the overall gamma of the system, including the camera, the display, and any gamma-adjusting electronics, should be unity. This simple criterion is the one most often used in the design of television color rendition. However, the more sophisticated criterion of corresponding color reproduction takes into account the effect of the viewing conditions. In particular, several authors have shown that the luminance of the surround is important.[48-50] For example, a dim surround requires a gamma of about 1.2, and a dark surround requires a gamma of about 1.5 for optimum color reproduction.[51] DeMarsh[43] also suggested that an offset of the black level is important. This is often done in television cameras where the black level is set to 7.5 ± 2.5 IRE units (of the 100 IRE units between blanking level and peak white level) for the darkest object in the scene.

2.5.4 PERCEPTIBILITY OF COLOR DIFFERENCES.

The CIE 1931 chromaticity diagram does not map chromaticity on a uniform-perceptibility basis. A just-perceptible change of chromaticity is not represented by the same distance in different parts of the diagram. Many investigators have explored the manner in which perceptibility varies over the diagram. The most often quoted study is that of MacAdam,[52] who determined a set of ellipses (Fig. 2-21) which are contours of equal perceptibility about a given color.

From this and many other similar studies[53] it is apparent, for example, that large distances represent relatively small perceptible changes in the green sector of the diagram. In the blue region, much smaller changes in the chromaticity coordinates are readily perceived.

2.5.5 UNIFORM COLOR SPACES.

Several suggestions have been made for projective transformations of the CIE x, y chromaticity diagram to a diagram in which perceptual spacing is more uniform. In 1960 the CIE provisionally recommended the use of one such Uniform Chromaticity Scale (UCS) diagram formed by plotting v against u in rectangular coordinates, where

$$u = \frac{4x}{-2x + 12y + 3}$$

$$v = \frac{6y}{-2x + 12y + 3}$$

This diagram was called the CIE 1960 UCS diagram.

Subsequently, in 1964, the CIE provisionally recommended an extension of this diagram to three dimensions, based on a proposal by Wyszecki.[54] The recommended rectangular coordinates U^*, V^*, and W^* were defined by

$$W^* = 25Y^{1/3} - 17$$

$$U^* = 13W^*(u - u_n)$$

$$V^* = 13W^*(v - v_n)$$

where u_n and v_n are the u, v values of the reference white such as illuminant A or D_{65}. The distance between two points U^*, V^*, W^* and $U^* + \Delta U^*$, $V^* + \Delta V^*$, $W^* + \Delta W^*$ defined a measure ΔE_{CIE64} for the perceptual size of the color difference between the corresponding stimuli

$$\Delta E_{CIE64} = [(\Delta U^*)^2 + (\Delta V^*)^2 + (\Delta W^*)^2]^{1/2}$$

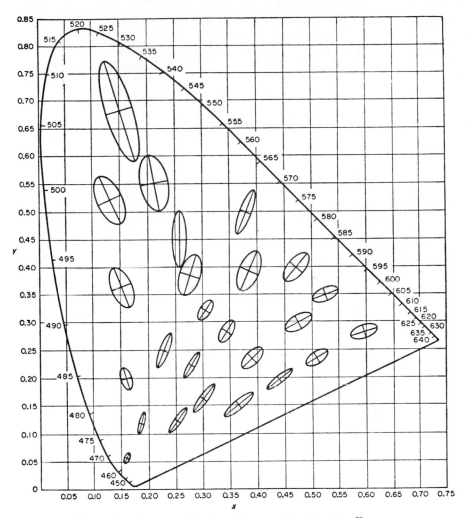

FIG. 2-21 Ellipses of equally perceptible color differences. (*From MacAdam.*[52])

Following the provisional recommendation of the $U^*V^*W^*$ formula, Eastwood[55] and others found that an expansion of the $v-$ (and hence the V^*-) scale by 50 percent improved considerably the agreement with visual judgments. This led the CIE to develop a new formula[56,57]

$$L^* = 116 \left(\frac{Y}{Y_n}\right)^{1/3} - 16$$

$$u^* = 13L^*(u' - u'_n)$$

$$v^* = 13L^*(v' - v'_n)$$

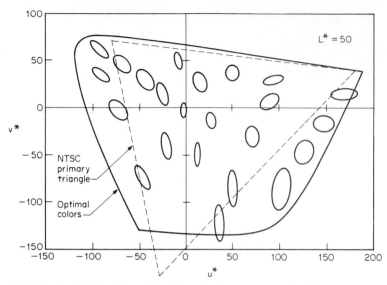

FIG 2-22   MacAdam's ellipses plotted in a u^*, v^* cross section of the CIE LUV uniform color space. The triangle formed by the NTSC primary colors is shown as a dashed line.

where

$$u' = \frac{4x}{-2x + 12y + 3}$$

$$v' = \frac{9y}{-2x + 12y + 3}$$

and where Y_n, u_n, and v_n are the values of Y, u, and v for the reference white. With $Y_n = 100$, L^* is almost exactly the same as W^*.

For values of $Y/Y_n \leq 0.008856$, the formula for L^* should be replaced by

$$L^* = 903.3 \left(\frac{Y}{Y_n}\right)$$

The uniform color space formed by plotting L^*, u^*, and v^* as rectangular coordinates is called the CIE 1976 $L^*u^*v^*$ space (abbreviated CIELUV). The associated color-difference formula

$$\Delta E_{uv}^* = [(\Delta L^*)^2 + (\Delta u^*)^2 + (\Delta v^*)^2]^{1/2} \tag{2-26}$$

is called the CIE 1976 L*u*v* (CIELUV) color-difference formula.

If v' is plotted against u', a new uniform chromaticity diagram, called the CIE 1976 UCS diagram, is obtained. When MacAdam's ellipses are replotted in this diagram or in a u^*, v^* diagram (Fig. 2-22), it is clear that the perceptual uniformity is much better than in the CIE 1931 diagram (Fig. 2-21), though it is still not perfect. The CIE recommendation is an interim one, pending the development of a space and formula giving substantially better correlation with visual judgments.

In 1976 the CIE also recommended an alternative uniform color space and formula.[56,57] These are known as the CIE 1976 L*a*b* space and formula, respectively (CIELAB for short). The perceptual uniformity is about equally as good (or bad) as that of CIELUV. The defining equations are

$$L^* = 116 f\left(\frac{Y}{Y_n}\right) - 16$$

$$a^* = 500 \left[f\left(\frac{X}{X_n}\right) - f\left(\frac{Y}{Y_n}\right) \right]$$

$$b^* = 200 \left[f\left(\frac{Y}{Y_n}\right) - f\left(\frac{Z}{Z^n}\right) \right]$$

where $\qquad f(q) = q^{1/3} \qquad$ for $q > 0.008856$

$$f(q) = 7.787q + \frac{16}{116} \qquad \text{for } q \leq 0.008856$$

and where X_n, Y_n, and Z_n are the values of X, Y, and Z for the reference white. L^* is exactly the same as L^* in the CIELUV formula. The CIELAB formula is less useful than the CIELUV formula in color television because it does not have an associated chromaticity diagram in which additive mixtures of stimuli lie on the straight line joining the components and in which the center of gravity law applies, as it does in the u', v' diagram.

The coordinates of the CIELUV formula (or the CIELAB formula) can be used to define approximate correlates of the perceptual terms lightness, hue, perceived chroma, and saturation. The correlate of lightness is L^*, which is thus called CIE 1976 lightness. The correlate of hue is CIE 1976 u, v hue angle

$$h_{uv} = \arctan \frac{v^*}{u^*}$$

The correlate of perceived chroma is CIE 1976 u, v chroma

$$C_{uv}^* = [(u^*)^2 + (v^*)^2]^{1/2}$$

The correlate of saturation is CIE 1976 u, v saturation

$$s_{u,v} = 13[(u' - u_n')^2 + (v' - v_n')^2]^{1/2}$$

These new quantities correlate better with the corresponding perceptual quantities than do the ones listed in Table 2-1.

The total color difference ΔE_{uv}^* is sometimes written as the sum of lightness, chroma, and hue differences

$$\Delta E_{uv}^* = [(\Delta L^*)^2 + (\Delta C_{uv}^*)^2 + (\Delta H_{uv}^*)^2]^{1/2} \qquad (2\text{-}27)$$

where ΔH_{uv}^*, the CIE 1976 u, v hue difference, is defined to make Eqs. (2-26) and (2-27) equivalent

$$\Delta H_{uv}^* = [(\Delta E_{uv}^*)^2 - (\Delta L^*)^2 - (\Delta C_{uv}^*)^2]^{1/2}$$

It has been found that different circumstances (e.g., side-by-side comparison of textiles, side-by-side comparison of paints, comparison of color television displays) may require different weights for the ΔL^*, ΔC_{uv}^*, and ΔH_{uv}^* components of Eq. (2-27). For color television a weight of 0.25 for the lightness component has been suggested by several authors.[26,31,58,59] With this weight, Eqs. (2-26) and (2-27) become

$$(\Delta E_{uv}^*)' = [(0.25\Delta L^*)^2 + (\Delta u^*)^2 + (\Delta v^*)^2]^{1/2}$$

$$(\Delta E_{uv}^*)' = [(0.25\Delta L^*)^2 + (\Delta C_{uv}^*)^2 + (\Delta H_{uv}^*)^2]^{1/2}$$

On the basis of previous work using the $U^*V^*W^*$ formula,[31] errors of less than 5 to 8 units of $(\Delta E_{uv}^*)'$ are likely to give very acceptable color rendition of most colors, although critical colors such as flesh tones may require $(\Delta E_{uv}^*)' < 5$ units.

2.5.6 COLOR-RENDERING INDICES.

A good index of the fidelity of color rendering in a television system may be found by calculating the color differences (for example, using the CIELUV formula) between the original stimulus and the reproduction of a number of standard test objects. This is analogous to the CIE general color-rendering index for light sources[60,61] in which the average color difference for eight test objects is calculated and subtracted from 100.

However, for color television the calculations are complicated by the fact that the whole system of camera, encoder, transmission, decoder, and display device must be considered. Even if all these components are considered to be standard, the calculation of a color-rendering index for light sources to be used in television is much more involved than if the sources are to be used for general illumination.[62]

A further complication is that colorimetric color reproduction is not really an adequate criterion. Proper evaluation must take into account such factors as the influence of the surround on brightness response.[63]

A number of different sets of test objects have been used to assess color rendition in television. Many of these are available to other investigators.[28,64−66]

2.6 DERIVATION OF EQUATIONS

Section 2.6 is an appendix in which equations quoted in earlier sections are derived in detail.

2.6.1 CONVERSION BETWEEN TWO SYSTEMS OF PRIMARIES.
This section explains the details of the derivation of the transformation equations given in Sec. 2.3.4. The starting point is the relationship between two sets of primaries

$$R' \equiv a_{11}\mathbf{R} + a_{21}\mathbf{G} + a_{31}\mathbf{B}$$
$$G' \equiv a_{12}\mathbf{R} + a_{22}\mathbf{G} + a_{32}\mathbf{B} \qquad (2\text{-}5)$$
$$B' \equiv a_{13}\mathbf{R} + a_{23}\mathbf{G} + a_{33}\mathbf{B}$$

This is a match equation showing how the $\mathbf{R'G'B'}$ primaries can be matched by a mixture of the $\mathbf{RGB}$ primaries. The coefficients a_{ij} are the tristimulus values of $\mathbf{R'}$, $\mathbf{G'}$, and $\mathbf{B'}$ in the $\mathbf{RGB}$ system. From the a_{ij} we can derive relationships between the tristimulus values and chromaticity coordinates of any stimulus in one system and the tristimulus values and chromaticity coordinates of the same stimulus in the other system.

The match equation of any color $\mathbf{C}$ in the $\mathbf{R'G'B'}$ system is [Eq. (2-1), Sec. 2.1.3]

$$\mathbf{C} \equiv R'\mathbf{R'} + G'\mathbf{G'} + B'\mathbf{B'}$$

Substituting Eq. (2-5) into this

$$\mathbf{C} \equiv R'(a_{11}\mathbf{R} + a_{21}\mathbf{G} + a_{31}\mathbf{B}) + G'(a_{12}\mathbf{R} + a_{22}\mathbf{G} + a_{32}\mathbf{B}) + B'(a_{13}\mathbf{R} + a_{23}\mathbf{G} + a_{33}\mathbf{B})$$

Rearranging the right-hand side

$$\mathbf{C} \equiv (a_{11}R' + a_{12}G' + a_{13}B')\mathbf{R} + (a_{21}R' + a_{22}G' + a_{23}B')\mathbf{G} + (a_{31}R' + a_{32}G' + a_{33}B')\mathbf{B}$$

Thus the tristimulus values in the $\mathbf{RGB}$ system are

$$R = a_{11}R' + a_{12}G' + a_{13}B'$$
$$G = a_{21}R' + a_{22}G' + a_{23}B' \qquad (2\text{-}6)$$
$$B = a_{31}R' + a_{32}G' + a_{33}B'$$

Note that the matrix of a's in Eqs. (2-6) has its rows and columns transposed compared with Eqs. (2-5).

Equations (2-6) can be solved to find R', G', and B' in terms of R, G, and B

$$R' = b_{11}R + b_{12}G + b_{13}B$$
$$G' = b_{21}R + b_{22}G + b_{23}B \qquad (2\text{-}7)$$
$$B' = b_{31}R + b_{32}G + b_{33}B$$

where

$$b_{11} = \frac{a_{22}a_{33} - a_{23}a_{32}}{D} \qquad b_{12} = \frac{a_{13}a_{32} - a_{12}a_{33}}{D} \qquad b_{13} = \frac{a_{12}a_{23} - a_{13}a_{22}}{D}$$

$$b_{21} = \frac{a_{23}a_{31} - a_{21}a_{33}}{D} \qquad b_{22} = \frac{a_{11}a_{33} - a_{13}a_{31}}{D} \qquad b_{23} = \frac{a_{13}a_{21} - a_{11}a_{23}}{D}$$

$$b_{31} = \frac{a_{21}a_{32} - a_{22}a_{31}}{D} \qquad b_{32} = \frac{a_{12}a_{31} - a_{11}a_{32}}{D} \qquad b_{33} = \frac{a_{11}a_{22} - a_{12}a_{21}}{D}$$

and

$$D = a_{11}(a_{22}a_{33} - a_{23}a_{32}) + a_{12}(a_{23}a_{31} - a_{21}a_{33}) + a_{13}(a_{21}a_{32} - a_{31})$$

Note that in matrix formulation, Eqs. (2-5) to (2-7) can be written simply as

$$\begin{bmatrix} R' \\ G' \\ B' \end{bmatrix} = \begin{bmatrix} a_{11}\ a_{12}\ a_{13} \\ a_{21}\ a_{22}\ a_{23} \\ a_{31}\ a_{32}\ a_{33} \end{bmatrix}^{T} \begin{bmatrix} R \\ G \\ B \end{bmatrix}$$

$$\begin{bmatrix} R \\ G \\ B \end{bmatrix} = \begin{bmatrix} a_{11}\ a_{12}\ a_{13} \\ a_{21}\ a_{22}\ a_{23} \\ a_{31}\ a_{32}\ a_{33} \end{bmatrix} \begin{bmatrix} R' \\ G' \\ B' \end{bmatrix}$$

and

$$\begin{bmatrix} R' \\ G' \\ B' \end{bmatrix} = \begin{bmatrix} a_{11}\ a_{12}\ a_{13} \\ a_{21}\ a_{22}\ a_{23} \\ a_{31}\ a_{32}\ a_{33} \end{bmatrix}^{-1} \begin{bmatrix} R \\ G \\ B \end{bmatrix}$$

The chromaticity coordinates of **C** in the **RGB** system are

$$r = \frac{R}{T}$$

$$g = \frac{G}{T}$$

where

$$T = R + G + B.$$

Substituting Eqs. (2-6)

$$r = \frac{a_{11}R' + a_{12}G' + a_{13}B'}{T}$$

$$g = \frac{a_{21}R' + a_{22}G' + a_{23}B'}{T}$$

$$T = (a_{11} + a_{21} + a_{31})R' + (a_{12} + a_{22} + a_{32})G' + (a_{13} + a_{23} + a_{33})B'$$

Dividing the numerators and denominators by $(R' + G' + B')$

$$r = \frac{a_{11}r' + a_{12}g' + a_{13}b'}{t}$$

$$g = \frac{a_{21}r' + a_{22}g' + a_{23}b'}{t} \tag{2-8}$$

$$t = (a_{11} + a_{21} + a_{31})r' + (a_{12} + a_{22} + a_{32})g' + (a_{13} + a_{23} + a_{33})b'$$

where r', g', and b' are the chromaticity coordinates of **C** in the **R'G'B'** system, given by

$$r' = \frac{R'}{R' + G' + B'}$$

$$g' = \frac{G'}{R' + G' + B'}$$

$$b' = \frac{B'}{R' + G' + B'}$$

Substituting $b' = 1 - r' - g'$ into Eqs. (2-8) yields

$$r = \frac{\alpha_{11}r' + \alpha_{12}g' + \alpha_{13}}{t}$$

$$g = \frac{\alpha_{21}r' + \alpha_{22}g' + \alpha_{23}}{t} \qquad (2\text{-}9)$$

$$t = \alpha_{31}r' + \alpha_{32}g' + \alpha_{33}$$

where

$$\alpha_{11} = a_{11} - a_{13} \qquad \alpha_{12} = a_{12} - a_{13} \qquad \alpha_{13} = a_{13}$$

$$\alpha_{21} = a_{21} - a_{23} \qquad \alpha_{22} = a_{22} - a_{23} \qquad \alpha_{23} = a_{23}$$

$$\alpha_{31} = a_{11} + a_{21} + a_{31} - a_{13} - a_{23} - a_{33}$$

$$\alpha_{32} = a_{12} + a_{22} + a_{32} - a_{13} - a_{23} - a_{33}$$

$$\alpha_{33} = a_{13} + a_{23} + a_{33}$$

The inverse of Eqs. (2-9) is

$$r' = \frac{\beta_{11}r + \beta_{12}g + \beta_{13}}{t'}$$

$$g' = \frac{\beta_{21}r + \beta_{22}g + \beta_{23}}{t'} \qquad (2\text{-}10)$$

$$t' = \beta_{31}r + \beta_{32}g + \beta_{33}$$

where the β_{ij} are related to the b_{ij} of Eqs. (2-7) in exactly the same way that the α_{ij} are related to the a_{ij}.

2.6.2 RELATION BETWEEN DISPLAY PHOSPHOR CHROMATICITIES AND CAMERA SPECTRAL SENSITIVITIES.

This section explains the details of the equations given in Sec. 2.4.1 which relate the three spectral sensitivities of a television camera to the chromaticity coordinates $(x_r, y_r, z_r; x_g, y_g, z_g; x_b, y_b, z_b)$ of the three phosphors in the display and the chromaticity coordinates (x_w, y_w, z_w) of the white which is produced when the three input signals R, G, and B are equal.

It is assumed that the chromaticity of each phosphor remains unchanged when its degree of excitation is changed. It is also assumed that the light output from each phosphor is related linearly to the R, G, B input signals. This second condition will hold if the gamma correction applied in the camera exactly compensates for the gamma of the display device, and if there are no other nonlinearities in the system.

With these assumptions, the total CIE tristimulus values, $T = X + Y + Z$, produced by inputs of R, G and B are

$$T_r = K_r R \qquad T_g = K_g G \qquad T_b = K_b B \qquad (2\text{-}13)$$

where K_r, K_g, and K_b are constants that can be set by gain controls in the three channels. The tristimulus values of the displayed color are the sums of the tristimulus values of the individual phosphors

$$X = x_r T_r + x_g T_g + x_b T_b$$

$$Y = y_r T_r + y_g T_g + y_b T_b$$

$$Z = z_r T_r + z_g T_g + z_b T_b$$

Substituting Eqs. (2-13) yields

$$X = x_r K_r R + x_g K_g G + x_b K_b B$$
$$Y = y_r K_r R + y_g K_g G + y_b K_b B \qquad (2\text{-}14)$$
$$Z = z_r K_r R + z_g K_g G + z_b K_b B$$

The values of K_r, K_g, and K_b can be set so that equal input signals, $R = G = B$, produce a white with chromaticity coordinates x_w, y_w, z_w and luminance $Y_w = 1$. (This arbitrary choice of $Y_w = 1$ is not important because the aim of the system is to reproduce relative rather than absolute luminances.) For this white color, Eqs. (2-14) become

$$X_w = \frac{x_w}{y_w} = x_r K_r + x_g K_g + x_b K_b$$

$$Y_w = 1 = y_r K_r + y_g K_g + y_b K_b$$

$$Z_w = \frac{z_w}{y_w} = z_r K_r + z_g K_g + z_b K_b$$

Solving these for K_r, K_g, K_b yields

$$K_r = \frac{x_w(y_g z_b - y_b z_g) + y_w(x_b z_g - x_g z_b) + z_w(x_g y_b - x_b y_g)}{y_w D_1}$$

$$K_g = \frac{x_w(y_b z_r - y_r z_b) + y_w(x_r z_b - x_b z_r) + z_w(x_b y_r - x_r y_b)}{y_w D_1}$$

$$K_b = \frac{x_w(y_r z_g - y_g z_r) + y_w(x_g z_r - x_r z_g) + z_w(x_r y_g - x_g y_r)}{y_w D_1} \qquad (2\text{-}15)$$

$$D_1 = x_r(y_g z_b - y_b z_g) + y_r(x_b z_g - x_g z_b) + z_r(x_g y_b - x_b y_g)$$

These values can be substituted in Eq. (2-14), which can be rewritten as

$$X = a_{11} R + a_{12} G + a_{13} B$$
$$Y = a_{21} R + a_{22} G + a_{23} B \qquad (2\text{-}16)$$
$$Z = a_{31} R + a_{32} G + a_{33} B$$

where the a_{ij} are now functions of x_r, y_r, z_r; x_g, y_g, z_g; x_b, y_b, z_b; x_w, y_w, z_w

$$a_{11} = x_r K_r \qquad a_{12} = x_g K_g \qquad a_{13} = x_b K_b$$
$$a_{21} = y_r K_r \qquad a_{22} = y_g K_g \qquad a_{23} = y_b K_b$$
$$a_{31} = z_r K_r \qquad a_{32} = z_g K_g \qquad a_{33} = z_b K_b$$

In fact, since $z = 1 - x - y$, the a_{ij} could be written as functions of x_r, y_r; x_g, y_g; x_b, y_b; x_w, y_w only.

Equations (2-16) can be inverted to give R, G, and B in terms of X, Y and Z

$$R = b_{11} X + b_{12} Y + b_{13} Z$$
$$G = b_{21} X + b_{22} Y + b_{23} Z \qquad (2\text{-}17)$$
$$B = b_{31} X + b_{32} Y + b_{33} Z$$

where the b_{ij} are related to the a_{ij} in the usual way. Thus

$$b_{11} = \frac{K_g K_b(y_g z_b - y_b z_g)}{D_2} \qquad b_{12} = \frac{K_g K_b(x_b z_g - x_g z_b)}{D_2} \qquad b_{13} = \frac{K_g K_b(x_g y_b - x_b y_g)}{D_2}$$

$$b_{21} = \frac{K_r K_b(y_b z_r - y_r z_b)}{D_2} \qquad b_{22} = \frac{K_r K_b(x_r z_b - x_b z_r)}{D_2} \qquad b_{23} = \frac{K_r K_b(x_b y_r - x_r y_b)}{D_2}$$

$$b_{31} = \frac{K_r K_g (y_r z_g - y_g z_r)}{D_2} \qquad b_{32} = \frac{K_r K_g (x_g z_r - x_r z_g)}{D_2} \qquad b_{33} = \frac{K_r K_g (x_r y_g - x_g y_r)}{D_2}$$

$$D_2 = K_r K_g K_b [x_r(y_g z_b - y_b z_g) + x_g(y_b z_r - y_r z_b) + x_b(y_r z_g - y_g z_r)]$$

For monochromatic stimuli, Eqs. (2-17) become

$$\bar{r}(\lambda) = b_{11}\bar{x}(\lambda) + b_{12}\bar{y}(\lambda) + b_{13}\bar{z}(\lambda)$$

$$\bar{g}(\lambda) = b_{21}\bar{x}(\lambda) + b_{22}\bar{y}(\lambda) + b_{23}\bar{z}(\lambda) \qquad (2\text{-}18)$$

$$\bar{b}(\lambda) = b_{31}\bar{x}(\lambda) + b_{32}\bar{y}(\lambda) + b_{33}\bar{z}(\lambda)$$

where $\bar{r}(\lambda)$, $\bar{g}(\lambda)$, and $\bar{b}(\lambda)$ are the required camera spectral sensitivities and $\bar{x}(\lambda)$, $\bar{y}(\lambda)$, and $\bar{z}(\lambda)$ are the CIE color-matching functions.

Note that all these equations can be written very simply in matrix notation.[27,51] For example, Eqs. (2-14) become

$$\begin{bmatrix} X \\ Y \\ Z \end{bmatrix} = \begin{bmatrix} x_r & x_g & x_b \\ y_r & y_g & y_b \\ z_r & z_g & z_b \end{bmatrix} \begin{bmatrix} K_r & 0 & 0 \\ 0 & K_g & 0 \\ 0 & 0 & K_b \end{bmatrix} \begin{bmatrix} R \\ G \\ B \end{bmatrix}$$

Equations (2-15) become

$$\begin{bmatrix} K_r \\ K_g \\ K_b \end{bmatrix} = \frac{1}{y_w} \begin{bmatrix} x_r & x_g & x_b \\ y_r & y_g & y_b \\ z_r & z_g & z_b \end{bmatrix}^{-1} \begin{bmatrix} x_w \\ y_w \\ z_w \end{bmatrix}$$

and Eqs. (2-17) become

$$\begin{bmatrix} R \\ G \\ B \end{bmatrix} = \begin{bmatrix} K_r & 0 & 0 \\ 0 & K_g & 0 \\ 0 & 0 & K_b \end{bmatrix}^{-1} \begin{bmatrix} x_r & x_g & x_b \\ y_r & y_g & y_b \\ z_r & z_g & z_b \end{bmatrix}^{-1} \begin{bmatrix} X \\ Y \\ Z \end{bmatrix}$$

2.6.3 RELATIVE CONTRIBUTION OF PRIMARIES TO LUMINANCE.

It is sometimes useful to know the relative contribution of each primary to the luminance of a given reproduced stimulus. This is easy to calculate from Eqs. (2-16) and (2-17) of Sec. 2.6.2. If the chromaticity coordinates are (x, y, z), Eqs. (2-17) become

$$R = T(b_{11}x + b_{12}y + b_{13}z)$$

$$G = T(b_{21}x + b_{22}y + b_{23}z)$$

$$B = T(b_{31}x + b_{32}y + b_{33}z)$$

because $X = xT$, $Y = yT$, and $Z = zT$, where T is the total tristimulus value $X + Y + Z$. From Eqs. (2-16), the luminance of the stimulus is

$$Y = a_{21}R + a_{22}G + a_{23}B$$

Thus the relative contributions of the three primaries are

Red: $a_{21}(b_{11}x + b_{12}y + b_{13}z)$

Green: $a_{22}(b_{21}x + b_{22}y + b_{23}z)$ $\qquad (2\text{-}21)$

Blue: $a_{23}(b_{31}x + b_{32}y + b_{33}z)$

It can easily be shown that, in terms of the chromaticity coordinates of the phosphors, these relative contributions become

Red: $y_r[(y_gz_b - y_bz_g)x + (x_bz_g - x_gz_b)y + (x_gy_b - x_by_g)z]$

Green: $y_g[(y_bz_r - y_rz_b)x + (x_rz_b - x_bz_r)y + (x_by_r - x_ry_b)z]$

Blue: $y_b[(y_rz_g - y_gz_r)x + (x_gz_r - x_rz_g)y + (x_ry_g - x_gy_r)z]$

It is not surprising that these expressions are independent of x_w and y_w because x_w and y_w are used simply to adjust the relative size of the units of **R**, **G**, and **B**. Clearly, the size of the units cannot affect the contributions of **R**, **G**, and **B** to the luminance of a mixture.

REFERENCES

1. V. C. Smith and J. Pokorny, "Spectral Sensitivity of the Foveal Cone Pigments between 400 and 500 nm," *Vision Res.* **15**: 161–171 (1975).

2. R. M. Boynton, *Human Color Vision,* Holt, New York, 1979, p. 404.

3. D. B. Judd and G. Wyszecki, *Color in Business, Science, and Industry,* 3d ed., Wiley, New York, 1975, pp. 44–45.

4. R. M. Boynton, op. cit.

5. G. Wyszecki and W. S. Stiles, *Color Science,* 2d ed., Wiley, New York, 1982.

6. R. W. G. Hunt, "Colour Terminology," *Color Res. Appl.* **3**: 79–87 (1978).

7. D. Nickerson, "History of the Munsell Color System, Company and Foundation, I," *Color Res. Appl.* **1**: 7–10 (1976).

8. D. Nickerson, "History of the Munsell Color System, Company and Foundation, II. Its Scientific Application," *Color Res. Appl.* **1**: 69–77 (1976).

9. D. Nickerson, "History of the Munsell Color System, Company and Foundation, III," *Color Res. Appl.* **1**: 121–130 (1976).

10. S. M. Newhall, D. Nickerson, and D. B. Judd, "Final Report of the OSA Subcommittee on the Spacing of the Munsell Colors," *J. Opt. Soc. Am.* **33**: 385–418 (1943).

11. *Munsell Book of Color.* Munsell Color Co., 2441 No. Calvert Street, Baltimore MD 21218.

12. Judd and Wyszecki, op. cit., 244–274.

13. K. L. Kelly and D. B. Judd, *Color-Universal Language and Dictionary of Names,* National Bureau of Standards, Washington, SP 440 (1976).

14. W. D. Wright, "A Redetermination of the Trichromatic Coefficients of the Spectral Colours," *Trans. Opt. Soc.* **30**: 141–164 (1928–1929).

15. J. Guild, "The Colorimetric Properties of the Spectrum," *Phil. Trans. Roy. Soc. A* **230**: 149–187 (1931).

16. Judd and Wyszecki, op. cit., p. 149.

17. Judd and Wyszecki, op. cit., p. 118.

18. "Colorimetry," Publication no. 15, Commission Internationale de l'Eclairage, Paris, 1971.

19. "A Method for Assessing the Quality of Daylight Simulators for Colorimetry," Publication no. 51, Commission Internationale de l'Eclairage, Paris, 1982.

20. K. L. Kelly, "Lines of Constant Correlated Color Temperature Based on MacAdam's (u, v) Uniform Chromaticity Transformation of the CIE Diagram," *J. Opt. Soc. Am.* **53**: 999–1002 (1963).

21. A. R. Robertson, "Computation of Correlated Color Temperature and Distribution Temperature," *J. Opt. Soc. Am.* **58**: 1528–1535 (1968).

22. D. G. Fink, *Color Television Standards,* McGraw-Hill, New York, 1955.

23. D. H. Pritchard and J. J. Gibson, "Worldwide Color Television Standards—Similarities and Differences," *SMPTE J.* **89**: 111–120 (1980).

24. C. B. Neal, "Computing Colorimetric Errors of a Color Television Display System," *IEEE Trans. CE* **21**: 63–73 (1975).

25. R. Brodeur, K. R. Field, and D. H. McRae, "Measurement of Color Rendition in Color Television," in M. Pearson (ed.), *Proc. ISCC Conf. Optimum Reproduction of Color,* Williamsburg, Va., 1971, Graphic Arts Research Center, Rochester, N.Y., 1971.

26. R. W. G. Hunt, *The Reproduction of Colour,* 3d ed., Fountain Press, England, 1975.

27. C. B. Neal, "Television Colorimetry for Receiver Engineers," *IEEE Trans. BTR* **19**: 149–162 (1973).

28. R. F. D. Corley, "Test Materials for the Alignment of Telecine Colorimetry," *J. SMPTE* **90**: 1064–1071 (1981).

29. C. B. Wood and W. N. Sproson, "The Choice of Primary Colours for Colour Television," BBC Eng. Rep. No. 85, 19–36 (1971).

30. C. W. Rhodes, SMPTE Television Video Technology Committee, "Report on Standardization of Monitor Colorimetry by SMPTE," *J. SMPTE* **91**: 1201–1202 (1982).

31. L. E. DeMarsh, "Colorimetric Standards in US Color Television," *J. SMPTE* **83**: 1–5 (1974).

32. A. H. Jones, "Optimum Color Analysis Characteristics and Matrices for Color Television Cameras with Three Receptors," *J. SMPTE* **77**: 108–115 (1968).

33. W. N. Sproson, "3 x 3 Matrices for Use with Plumbicon Colour Cameras Using Three Tubes," BBC Research Rep. PH-15 (1967).

34. A. N. Heightman, "A High-Performance Automatic Color Camera," *IEEE Trans. BC* **18**: 1–7 (1972).

35. R. C. Moore, J. F. Fisher, and J. B. Chatten, "Measurement and Control of the Color Characteristics of a Flying-Spot Color Signal Generator," *Proc. IRE* **41**: 730–733 (1953).

36. J. F. Fisher, "Generation of NTSC Color Signals," *Proc. IRE* **41**: 338–343 (1953).

37. I. C. Abrahams, "Analysis of Color Errors in Color Television Cameras," *J. SMPTE* **72**: 595–601 (1963).

38. K. B. Benson, "Report on Sources of Variability in Color Reproduction as Viewed on the Home Television Receiver," *IEEE Trans. BTR* **19**: 269–275 (1973).

39. L. E. DeMarsh, "Progress Report of the SMPTE Working Group on Television Studio/Field Camera Colorimetry," *J. SMPTE* **84**: 1–2 (1975).

40. N. W. Parker, "An Analysis of the Necessary Decoder Corrections for Color Receiver Operation with Non-Standard Receiver Primaries," *IEEE Trans. BTR* **12**: 23–32 (1966).

41. R. Sasaki, Y. Nagaoka, and T. Tomimoto, "Non-linear Chrominance Signal Decoding," *IEEE Trans. CE* **23**: 335–341 (1977).

42. "Setting Chromaticity and Luminance of White for Color Television Monitors Using Shadow Mask Picture Tubes," SMPTE Recommended Practice 71-1977.

43. L. E. DeMarsh, "Color Rendition in Television," *IEEE Trans. CE* **23**: 149–157 (1977).

44. R. Ekstrand, "A Flesh-Tone Correction Circuit," *IEEE Trans. BTR* **17**: 182–189 (1971).

45. L. A. Harwood, "A Chrominance Demodulator IC with Dynamic Flesh Correction," *IEEE Trans. CE* **22**: 111–118 (1976).

46. C. J. Bartleson, "Predicting Corresponding Colors with Changes in Adaptation," *Color Res. Appl.* **4**: 143–155 (1979).

47. R. W. G. Hunt, op. cit., p. 439.
48. C. J. Bartleson and E. J. Breneman, "Brightness Reproduction in the Photographic Process," *Photog. Sci. Eng.* **11**: 254–262 (1967).
49. S. B. Novick, "Tone Reproduction from Colour Telecine Systems," *Br. Kin. Sound TV* **51**: 342–347 (1969).
50. L. E. DeMarsh, "Color Reproduction in Color Television," in M. Pearson (ed.), op. cit., pp. 69–97.
51. S. Herman, "The Design of Television Color Rendition," *J. SMPTE* **84**: 267–273 (1975).
52. D. L. MacAdam, "Visual Sensitivities to Color Differences in Daylight," *J. Opt. Soc. Am.* **32**: 247–274 (1942).
53. A. R. Robertson, "Colour Differences," *Die Farbe* **29**: 273–296 (1981).
54. G. Wyszecki, "Proposal for a New Color-Difference Formula," *J. Opt. Soc. Am.* **53**: 1318–1319 (1963).
55. D. Eastwood, "A Simple Modification to Improve the Perceptual Uniformity of the CIE 1964 U*V*W* Colour Space," *Die Farbe* **24**: 97–108 (1975).
56. *Uniform Color Spaces—Color Difference Equations—Psychometric Color Terms,* Publication No. 15, Supplement No. 2, Commission Internationale de l'Eclairage, Paris, 1978.
57. A. R. Robertson, "The CIE 1976 Color-Difference Formulae," *Color Res. Appl.* **2**: 7–11 (1977).
58. L. E. DeMarsh and J. E. Pinney, "Studies of Some Colorimetric Problems in Color Television," *J. SMPTE* **79**: 338–342 (1970).
59. A. R. Robertson, "Color Error Formulas," *J. SMPTE* **89**: 947 (1980).
60. J. E. Kaufman (ed.), *IES Lighting Handbook—1981 Reference Volume,* Illuminating Engineering Society of North America, New York, 1981.
61. *Method of Measuring and Specifying Colour Rendering Properties of Light Sources,* Publication No. 13.2, Commission Internationale de l'Eclairage, Paris, (1974).
62. J. J. Opstelten and L. B. Beijer, "Specification of Colour Rendering Properties of Light Sources for Colour Television," *Ltg. Res. Tech.* **8**: 89–102 (1976).
63. L. E. DeMarsh, "Evaluation of Color Rendering in Film and Television," *J. SMPTE* **86**: 624–625 (1977).
64. G. B. Townsend, "Coloured Fabrics for Use in Colour Television Test Scenes," *TV Soc. J.* **10**: 208–212 (1963).
65. C. S. McCamy, H. Marcus, and J. G. Davidson, "A Color-Rendition Chart," *J. Appl. Photogr. Eng.* **2**: 95–99 (1976).
66. E. W. Taylor and S. J. Lent, "BBC Test Card No. 61 (Flesh Tone Reference): Colorimetric and Other Optical Considerations," *J. SMPTE* **87**: 76–78 (1978).

BIBLIOGRAPHY

Basic Colorimetry

Boynton, R. M., *Human Color Vision,* Holt, New York, 1979.
"Colorimetry," Publication no. 15, Commission Internationale de l'Eclairage, Paris, 1971.
Judd, D. B., and G. Wyszecki, *Color in Business, Science, and Industry,* 3d ed., Wiley, New York, 1975.
Kaufman, J. E. (ed.), *IES Lighting Handbook—1981 Reference Volume,* Illuminating Engineering Society of North America, New York, 1981.

Wright, W. D., *The Measurement of Colour,* 4th ed., Adam Hilger, London, 1969.

Wyszecki, G., and W. S. Stiles, *Color Science,* 2d ed., Wiley, New York, 1982.

Television Colorimetry

Bingley, F. J., "Colorimetry in Color Television—Pt. I," *Proc. IRE* **41**: 838–851 (1953).

Bingley, F. J., "Colorimetry in Color Television—Pts. II and III," *Proc. IRE* **42**: 48–57 (1954).

Bingley, F. J., "The Application of Projective Geometry to the Theory of Color Mixture," *Proc. IRE* **36**: 709–723 (1948).

DeMarsh, L. E., "Colorimetric Standards in US Color Television," *J. SMPTE* **83**: 1–5 (1974).

Epstein, D. W., "Colorimetric Analysis of RCA Color Television System," *RCA Review* **14**: 227–258 (1953).

Fink, D. G., *Color Television Standards,* McGraw-Hill, New York, 1955.

Herman, S., *"The Design of Television Color Rendition," J. SMPTE* **84**: 267–273 (1975).

Hunt, R. W. G., *The Reproduction of Colour,* 3d ed., Fountain Press, England, 1975.

Neal, C. B., "Television Colorimetry for Receiver Engineers," *IEEE Trans. BTR* **19**: 149–162 (1973).

Pearson, M. (ed.), *Proc. ISCC Conf. on Optimum Reproduction of Color,* Williamsburg, Va., 1971, Graphic Arts Research Center, Rochester, N.Y., 1971.

Pritchard, D. H., "US Color Television Fundamentals—A Review," *IEEE Trans. CE* **23**: 467–478 (1977).

Sproson, W. N., *Colour Science in Television and Display Systems,* Adam Hilger, Bristol, England, 1983.

Wentworth, J. W., *Color Television Engineering,* McGraw-Hill, New York (1955).

Wintringham, W. T., "Color Television and Colorimetry," *Proc. IRE* **39**: 1135–1172 (1951).

Optical Components and Systems

W. Lyle Brewer
Eastman Kodak (retired)
Rochester, New York

Robert A. Morris
Eastman Kodak (retired)
Rochester, New York

NOTE: Parts of this chapter are adapted from D. G. Fink (ed.), *Television Engineering Handbook*, McGraw-Hill, New York, 1957. Used by permission.

3.1 GEOMETRIC OPTICS FUNDAMENTALS

Geometric optics is the branch of optics that deals with image formation using geometric methods. It is based on two postulates: that light travels in straight lines in a homogeneous medium, and that two rays may intersect without affecting the subsequent path of either. The fundamental laws of geometric optics may be developed from general principles such as Maxwell's electromagnetic equations or Fermat's principle of least time. However, the laws of reflection and refraction can also be determined in a very simple way by means of Huygen's principle. This states that every point of a wave front may be considered as a source of small waves which spread out in all directions from their centers to form the new wave front along their envelope.

3.1.1 LAWS OF REFLECTION AND REFRACTION. The laws of reflection and refraction may be stated as follows:

Law of Reflection. The angle of the reflected ray is equal to the angle of the incident ray.

Law of Refraction. A ray entering a medium in which the velocity of light is different is refracted so that

$$n \sin i = n' \sin r \qquad (3\text{-}1a)$$

where i is the angle of incidence, r is the angle of refraction, and n and n' are the indexes of refraction of the two media.

A ray is an imaginary line normal to the wave front. The angle the advancing ray forms with the line normal to the surface in question is the angle of incidence and is equal to the angle the wave front forms with the surface. The index of refraction is the ratio of the velocity of light c in a vacuum to the velocity v in the medium

$$n = \frac{c}{v} \quad n' = \frac{c}{v'} \qquad (3\text{-}1b)$$

For air, the velocity is generally considered equal to the velocity *in vacuo* so $n = 1.0$ and the equation may be simplified to

$$\frac{\sin i}{\sin r} = n' \qquad (3\text{-}1c)$$

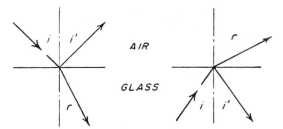

FIG. 3-1 Refraction and reflection at air-glass surface: (*left*) beam incident upon glass from air, (*right*) upon air from glass. (*From Fink.*[1])†

When a ray passes from a medium of smaller index into one of larger index, as from air to glass, the angle of refraction is less than the angle of incidence, and the ray is bent toward the normal. In passing from glass to air, the ray is bent away from the normal (see Fig. 3-1). The incident ray, reflected ray, refracted ray, and the normal to the surface at the point of incidence all lie in the same plane.

A ray passing from a medium of higher index to one of lower index may be totally internally reflected. From the relation

$$n \sin i \doteq n' \sin r$$

the value of sin r is always greater than sin i when n is greater than n'. The maximum value for sin r is unity ($r = 90°$) and occurs for some value of i, called the *critical angle,* which is determined by the refractive indexes of the two media. For a water-air surface, $n/n' = 1.33$ and the critical angle is 48.5°. When the angle of incidence exceeds the critical angle, the ray is not refracted into the medium of lower index but is totally reflected (see Fig. 3-2). For angles smaller than the critical angle the rays are partially reflected.

Application of the sine law to two parallel surfaces, such as a glass plate, shows that the ray emerges parallel to the entering ray, but is displaced. The most important applications of the laws of reflection and refraction are to the formation of images by means of spherical surfaces, i.e., mirrors and lenses.

3.1.2 REFRACTION AT A SPHERICAL SURFACE IN THIN LENSES.

It can be shown by tracing a ray through a single refracting surface that

$$\frac{n}{s} + \frac{n'}{s'} = \frac{n' - n}{R} \tag{3-2a}$$

where s = object distance to refracting surface
s' = image distance to refracting surface
R = radius of curvature of surface
n = index of refraction of object medium
n' = index of refraction of image medium

A ray traversing two refractive surfaces, as in a lens in air, has a path whose image distance and object distance are found by applying Eq. (3-2a) to each of the two surfaces. For a lens whose thickness may be considered negligible relative to the image distance

$$\frac{1}{s} + \frac{1}{s'} = (n - 1)\left(\frac{1}{R_1} - \frac{1}{R_2}\right) \tag{3-2b}$$

where n is the index of refraction of the lens and R_1 and R_2 are the radii of curvature of the two surfaces. The right side of Eq. (3-2b) contains quantities which are characteristic

---

†Superscript numbers refer to References at end of chapter.

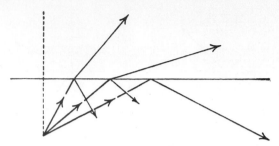

FIG. 3-2 When the angle of refraction exceeds the critical angle, the ray is totally reflected. (*From Fink.*[1])

of the lens and is called the power of the lens. The reciprocal of this expression is called the focal length f

$$\frac{1}{f} = (n - 1)\left(\frac{1}{R_1} - \frac{1}{R_2}\right) \tag{3-2c}$$

For thin lenses in air the object distance, image distance, and focal length are related as follows

$$\frac{1}{s} + \frac{1}{s'} = \frac{1}{f} \tag{3-2d}$$

Certain conventions of algebraic sign must be observed in the use of this and the other equations above. The conventions used here may be summarized as follows:

1. Draw all figures with the light incident on the reflecting or refracting surface from the left.
2. Consider the object distance s positive where the object lies at the left of the vertex. The vertex is the intersection of the reflecting or refracting surface with the axis through the center of curvature of the surface.
3. Consider the image distance s' positive when the image lies at the right of the vertex.
4. Consider radii of curvature positive when the center of curvature lies at the right of the vertex.
5. Consider angles positive when the slope of the ray with respect to the axis is positive.
6. Consider dimensions, such as image height, positive when measured upward from the axis.

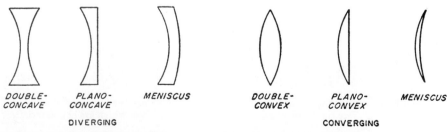

FIG. 3-3 Various forms of simple converging and diverging lenses. (*From Fink.*[1])

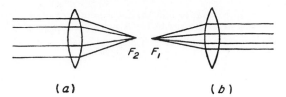

FIG. 3-4 (*a*) Parallel rays incident upon the lens pass through the second focal point; (*b*) rays passing through the first focal point incident on the lens emerge parallel. (*From Fink.*[1])

Briefly, after observing conventions 1 and 2, the others follow the rules of coordinate geometry with the vertex as origin.

From Eq. (3-2*c*) and the sign conventions it is apparent that the sign of the focal length may be negative or positive. For a lens in air, and parallel incident rays, the focal length is positive when the transmitted rays converge and negative when they diverge. Cross sections of simple converging and diverging lenses are shown in Fig. 3-3.

There are two focal points of a lens, located on the lens axis. All incident rays parallel to the lens axis are refracted to pass through the second focal point; all incident rays from the first focal point emerge parallel to the lens axis (see Fig. 3-4). For a thin lens, the distances from the two focal points to the lens are equal and denote the focal length.

The magnification m provided by a lens is defined as the ratio of the image height to the object height

$$m = \frac{y'}{y} \tag{3-3a}$$

From the similar triangles ABC and CDE (see Fig. 3-5)

$$m = \frac{y'}{y} = \frac{s'}{s} \tag{3-3b}$$

3.1.3 REFLECTION AT A SPHERICAL SURFACE.

By considering reflection as a special case of refraction, many of the equations above apply to reflection by a spherical mirror if the convention is adopted that $n' = -n$. Applying this to Eq. (3.2*a*) yields

$$\frac{1}{s} - \frac{1}{s'} = -\frac{2}{R} \tag{3-4a}$$

The focal point is the axial point which is imaged at infinity by the mirror, and its distance from the mirror is the focal length. Hence, substituting in Eq. (3.26*a*),

$$f = -\frac{R}{2} \tag{3-4b}$$

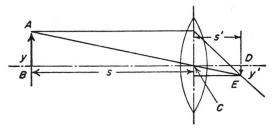

FIG. 3-5 Magnification for a simple lens. (*From Fink.*[1])

For a concave mirror, R is negative and the focal point lies at the left of the mirror, following the convention of signs described in Sec. 3.1.2. For any spherical mirror

$$\frac{1}{s} - \frac{1}{s'} = \frac{1}{f} \tag{3-4c}$$

and

$$m = \frac{y'}{y} = \frac{s'}{s} \tag{3-4d}$$

Such mirrors are subject to spherical aberration as in lenses, i.e., failure of the centrally reflected rays to converge at the same axial point as the rays reflected from the mirror edge. Aspherical surfaces formed by a paraboloid of revolution have the property that rays from infinity incident on the surface are all imaged at the same point on the axis. Thus, for the focal point and infinity, spherical aberration is eliminated. This is a useful device in projectors and searchlights where the light source is placed at the focal point to secure a beam of nearly parallel rays.

Spherical aberration in mirrors can also be eliminated by inserting lenses before the mirror. The Schmidt corrector is an aspherical lens, with one surface convex in the central region and concave in the outer region. The other surface is plane. A Schmidt system of spherical mirror and corrector plate can be made with a very high relative aperture; $f/0.6$ is a typical value. Because of the efficiency and low cost of these systems compared with projection lens systems, they have been used to obtain enlarged images from television picture tubes. (See Sec. 3.5.5.)

Another type of corrector for a spherical mirror is a meniscus lens having no aspherical surfaces, known as a Maksutov corrector. The spherical surfaces of the meniscus lens are more easily made by machine methods.

3.1.4 THICK (COMPOUND) LENSES.

The equations of Sec. 3.1.2 apply to thin lenses. When the thickness of the lens cannot be ignored, measurements must be made from reference points other than the lens surface, i.e., from the focal points, which have already been defined, or from the principal points. The principal points are located as follows. Consider ray OA (Fig. 3-6) proceeding from the object parallel to the lens axis. This will be refracted to pass through the focal point F'. The ray OB which passes through the focal point F will emerge along DI parallel to the lens axis. If OA and $F'I$ are extended, their point of intersection lies in the second principal plane. The point H' where this plane intersects the axis is called the second principal point. Similarly, the intersection of OF and DI extended lies in the first principal plane, and H is the first principal point. The distances FH and $F'H'$ are the first and second focal lengths, respectively. When the index of the medium on both sides of the lens is the same, as for a lens in air, the first and second focal lengths are equal.

If the direction of the light ray is reversed, i.e., the object is placed at the image position, the ray retraces its path and the image is formed at the former object position. Any

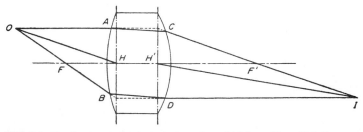

FIG. 3-6   Focal points and principal planes for a thick lens. (*From Fink.*[1])

two corresponding object and image points are said to be conjugate to each other, and hence are conjugate points.

Equation (3-2d) continues to hold for the thick lens, but s and s' are measured from their respective principal points as is the focal length f. The object distance, image distance, and focal length are related in another form, known as the Newtonian form of the lens equation. If x is the distance of the object from its focal point, and x' the image distance from its focal point, then

$$xx' = f^2 \tag{3-5}$$

3.1.5 LENS ABERRATIONS.

Up to this point optical images have been considered to be faithful reproductions of the object. The equations given have been derived from the general expressions for the refraction of a ray at a spherical surface when the angle between the ray and the axis is small so that $\sin \theta = \theta$. This approximation is known as first-order theory. The departures of the actual image from the predictions of first-order theory are called *aberrations*. In 1855 Ludwig von Seidel extended the first-order theory by including the third-order terms of the expanded sine function. The third-order theory contains five terms to be applied to the first-order theory. When no aberrations are present, and monochromatic light is passed through the optical system, the sum of the five terms is zero. Thus von Seidel's sums provide a logical classification for the five monochromatic aberrations. In addition, two forms of chromatic aberration can occur because of variation of index with wavelength. The five monochromatic aberrations are spherical aberration, coma, astigmatism, curvature of field, and distortion. These will be described briefly.

Spherical aberration is illustrated in Fig. 3-7. The rays near the axis converge at a point farther from the lens than the rays incident near the edge. By stopping down such a lens the spherical aberration can be reduced. In general, spherical aberration can be minimized if the deviation of the rays is equally divided between the front and rear surfaces of the lens. In a system of two or more lenses spherical aberration can be eliminated by making the contribution of the negative elements equal and opposite to that of the positive elements.

Coma is an aberration affecting the rays from points off the lens axis. It is similar to spherical aberration in that rays passing through the center and outer zones of the lens fail to image at the same point. Rays from a point off the axis which pass through the central zone of the lens form a point image. Rays passing through the outer zones form overlapping circles of increasing diameter below the point image, forming a cometlike image from which the aberration derives its name. It can be eliminated for a given object and image distance in a single lens by proper choice of radii of curvature.

Spherical aberration was described as the failure of rays from an axial point to form a point image in the direction along the axis. Coma relates to failure of the rays from an off-axis point to converge at the same point in the plane perpendicular to the axis. Astigmatism contains aspects of both. It resembles coma in that the off-axis points are affected, but, like spherical aberration, results from spreading of the image in a direction along the axis. The rays from a point converge on the other side of the lens to form a line

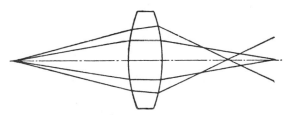

FIG. 3-7   Spherical aberration. (*From Fink.*[1])

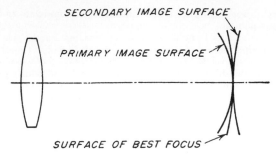

FIG. 3-8 Surfaces of best focus, illustrating lens astigmatism. (*From Fink.*[1])

image, actually the axis of a degenerate ellipse; continuing, the rays join with other rays to form a circle and then at a still further distance form a second line image crossed perpendicularly to the first. The best focus occurs when a circular image is formed. The locus of inner line images, the primary images, is a surface of revolution about the lens axis, called the *primary image surface.* The locus of outer line images forms the secondary image surface. The locus of circles of least confusion forms the surface of best focus. As shown in Fig. 3-8, these surfaces are tangent to one another at the lens axis. The failure of the primary and secondary image surfaces to coincide is called *astigmatism.* The surface of best focus is usually not a plane but a curved surface. This is the fourth type of aberration, known as *curvature of field.* It is not possible to eliminate both astigmatism and curvature of field in a single lens.

All rays passing through a lens from the center to the edge should result in equal magnification of the image. Distortion of the image occurs when the magnification varies with axial distance. If the magnification increases with axial distance, the effect is known as *pincushion distortion,* and the opposite effect is known as *barrel distortion* (see Fig. 3-9).

The five types of aberration described above can occur in uncorrected lenses even though light of a single wavelength forms the image. When the image is formed by the light from different regions of the spectrum, two types of chromatic aberration can occur, axial or longitudinal chromatism, and lateral chromatism. The first results from convergence of rays of different wavelength at different points along the axis, i.e., the lens focal length varies with wavelength. Since the magnification depends upon the focal length, the images are also of different size, producing lateral chromatism. In many instances lenses are corrected so that the focal points coincide for two or three colors, thus eliminating longitudinal chromatism. However, unless the focal lengths are also made to coincide, the images will be of slightly different size. This defect results in color fringing in

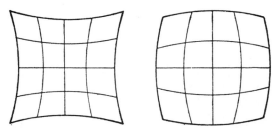

FIG. 3-9 Pincushion and barrel distortion in the image of a lens system. (*From Fink.*[1])

the outer portions of the field. Lenses for critical color photography, as in high-quality cameras, enlargers, and photoengravers' cameras, must be well corrected for lateral chromatism.

3.1.6 PHOTOMETRY OF LENSES.

It was mentioned earlier that the light passed by a lens varies with the area of the aperture stop and hence inversely with the square of the f-number. A simple approximate relationship exists between the luminance B of an extended surface and the illuminance E' of the image formed at the focal point; i.e., the object surface is presumed at infinity

$$E' = \frac{\pi BT}{4F^2} \qquad (3\text{-}6a)$$

where T is the lens transmittance as limited by losses due to reflection and absorption, and F is the f-number. When B is given in candles per square foot, E' is given in lumens per square foot (footcandles). For near objects the expression must be modified to

$$E' = \frac{\pi B f^2 T}{4 s'^2 F^2} \qquad (3\text{-}6b)$$

in which f is the focal length and s' the image distance. For images off the axis the illumination decreases approximately by the fourth power of cos θ where θ is the angle the image point and axis subtend at the lens. An additional factor H is frequently applied to account for limiting of the image illuminance by vignetting from the lens barrel. The complete expression then becomes

$$E' = \frac{\pi B f^2}{4 s'^2 F^2} HT \cos^4 \theta \qquad (3\text{-}6c)$$

The factor π is eliminated if E' is expressed in footcandles and B in footlamberts. Using these units and an object distance of $40f$, average displacement of image elements from the axis of 15 deg, $T = 0.70$, $H = 0.80$, Eq. (3-6c) becomes

$$E' = 0.116 \frac{B}{F^2} \qquad (3\text{-}6d)$$

Strictly speaking, since in many photographic and television applications the response is not limited to that of the eye, the relationship should be expressed in units of radiant energy, watts per square centimeter for E', and watts per steradian per square centimeter for B. In this case the factor π is not eliminated.

3.1.7 LENS STOPS.

It is obvious in the case of a simple lens that the rim of the lens forms the limiting boundary for rays transmitted by the lens. The introduction of smaller apertures before or after the lens can further limit the bundle of transmitted rays. This is done to eliminate unwanted rays which would produce distortions, to control the quantity of light transmitted, or to control the field of view. An aperture which controls the quantity of light transmitted, as the iris diaphragm in a camera, is called an *aperture stop* (see Fig. 3-10). An aperture which controls the field of view is called a *field stop*. The image of the aperture stop, projected into the object space, is called the *entrance pupil of the lens system*. The image of the aperture stop in the image space is called the *exit pupil*.

The relative aperture of a lens, usually called the f-number, is the ratio of the focal length to the effective lens diameter. A lens of $f/3.5$ has a focal length 3.5 times its effective diameter. In photographic objectives the lens stop may be reduced from its maximum, rated value to a limiting value, usually $f/22$. Since the focal length remains constant, the effective area of the lens, and hence the amount of light transmitted, varies inversely as the square of the f-number. Thus a lens set at $f/8$ passes nearly twice as much light as the same lens set at $f/11$.

The angular field of view of a lens system is the limiting angle for objects in front of the lens which are visible in the image plane. If the total angular field is 2θ, the half-

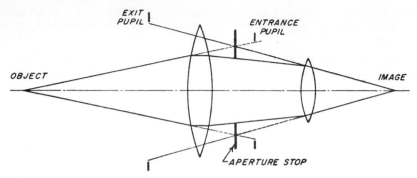

FIG. 3-10   Aperture stop, entrance pupil, and exit pupil for a lens system. (*From Fink.*[1])

angle θ can be determined when the lens is focused on distant objects by means of Eq. (3-7)

$$\tan \theta = \frac{W}{2f} \tag{3-7}$$

The quantity W is the maximum dimension of the image plane, and f is the focal length of the lens.

In addition to determining the amount of light transmitted by the lens, the size of the aperture stop also determines the range of distances which appear uniformly sharp for a given focus setting of the lens. Equation (3-2d) gives the relationship between focal length, object distance, and image distance for sharp focus. However, objects at distances between $s + d_1$ and $s - d_2$ will also appear acceptably sharp where

$$d_1 = \frac{s}{mf/aF - 1} \qquad d_2 = \frac{s}{mf/aF + 1} \tag{3-8a}$$

F denotes the f-number, m is the magnification, and a is the diameter of the circle of confusion expressed in the same units as the focal length, f. The circle of confusion is the unsharp image of a point. The range of image distances corresponding to the depth of field $d_1 + d_2$ is called the *depth of focus*.

From Eq. (3-8a) it may be inferred that all points from s to infinity will be in focus when $mf/aF = 1$ or $1/m = f/aF$. For a thin lens the magnification is

$$m = \frac{f}{s - f} \tag{3-8b}$$

Thus, when the lens is focused for a distance s, where

$$s - f = \frac{f^2}{aF} \tag{3-8c}$$

all objects from s to infinity will appear sharp. The distance $s - f$ is called the *hyperfocal distance* for the lens at aperture F. Similarly, it follows from Eq. (3-8a) that objects on the near side will appear in focus to a point midway between the lens and the object plane in focus at s.

The acceptable diameter of the circle of confusion varies with the application. In the most critical work the angle subtended at the eye should not exceed 2 min arc. For a viewing distance of 10 in (254 mm), the diameter of the circle of confusion in the reproduction should, therefore, not exceed 0.006 in (0.152 mm). In television the width of two scanning lines in the standard 525-line image, for a viewing distance of five times the picture height, is about 2.8 min arc.

The lower limit of the circle of confusion is reached when a point is imaged by the lens as a diffraction pattern of dark and light rings formed by interference of light waves (see Sec. 3.4.2). The ability of the lens to distinguish two points by forming two distinguishable diffraction patterns is called the *resolving power* of the lens. In some optical instruments, such as telescopes or microscopes, this lower limit of resolving power is realized. In lenses which must cover broader fields, such as those used for photographic or television objectives, the resolving power is limited by other factors such as aberrations of the system, and the term is used in a broader sense to describe the limit of the lens to record fine detail. The diameter a_o of the diffraction circle is determined by the diameter of the lens stop as follows

$$a_o = 2.4\lambda F \qquad (3\text{-}9)$$

For light of wavelength 550 nm (0.000055 cm) and $f/4.5$ lens aperture the diameter of the diffraction circle is 0.00059 cm.

3.2 LENS SYSTEMS

A combination of lenses may be treated as a thick lens (see Fig. 3-11). Consider two thin lenses of focal lengths f_1 and f_2 separated by a distance d. The second principal plane is found as for a single thick lens. The focal length of the combination f is the distance from the focal point to the principal plane. It is related to the focal lengths of the two thin lenses as follows

$$\frac{1}{f} = \frac{1}{f_1} + \frac{1}{f_2} - \frac{d}{f_1 f_2} \qquad (3\text{-}10)$$

3.2.1 FIXED FOCAL LENGTH. A TELEPHOTO LENS SYSTEM. Equation (3-10) may be applied to calculate the focal length of a typical telephoto lens system shown simplified in Fig. 3-12. If the positive lens has a focal length of $+20$ cm, the negative lens a focal length of -20 cm, and they are separated by a distance of 10 cm, Eq. (3-10) shows that the focal length of the system is 40 cm. The system has a long focal length, but the rear-element-to-film distance is half the focal length.

3.2.2 VARIABLE FOCAL LENGTH. If the distance d in Eq. (3-10) is variable, then the focal length is variable. So-called zoom lenses can produce satisfactory images over a wide range of focal lengths, for example, 25 mm (wide angle) to 250 mm (telephoto) for a motion picture or television camera.

3.2.3 ANAMORPHIC LENSES. Lenses are shaped to have spherical properties, i.e., uniform properties about the *center* of the lens, or cylindrical properties, i.e., uniform about the horizontal or vertical *axis* of the lens. An anamorphic lens is designed to pro-

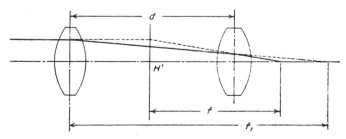

FIG. 3-11 Lens system treated as a single thick lens. (*From Fink.*[1])

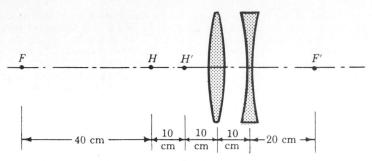

FIG. 3-12 Principle of the telephoto lens.

duce different magnification of the image in the horizontal and vertical axes. Anamorphic lenses are described in terms of their aspect ratio, the ratio of width to height of the screen image. The historical motion picture screen dimensions are 4 units wide to 3 units high, or an aspect ratio of 1.33:1. In 1953 the Cinemascope system was introduced, with an aspect ratio of 2.35:1 at the screen. Other aspect ratios are 1.65:1 and 1.85:1. The television screen ratio is the historical standard 1.33:1. When wide-screen motion picture prints are televised, accommodations are necessary, and some information, as in titles, is usually lost (see Chap. 17).

3.2.4 OPTICAL COMPENSATORS.

In most motion-picture projectors the motion of the film is intermittent. Small claws in the film sprocket holes pull the film frame into position and hold it there while light is flashed through it onto the screen. During an interval when the light beam is interrupted, the claws engage sprocket holes corresponding to the next film image and pull it into position.

There are optical methods, however, by which the image from a continuously moving film frame can be made stationary. Devices for accomplishing this result are called *optical compensators*. There are some possible advantages for such a system in ordinary motion-picture projection, but such projectors have never been widely used in the United States. They have certain advantages which make them particularly well suited for film utilization in color television. A few of the basic principles involved in optical compensation are:

1. *Moving Lenses.* If a lens is moved in such a way that the center of the picture frame, the center of the lens, and the center of the projection screen are always on a straight line, the image will be stationary. In passing from one frame to the next the lens must either be returned to the starting point, or a succession of lenses must follow successive frames. Because of the relatively large physical dimensions and masses of lenses, proper control of the motion is difficult to maintain. Cylindrical lenses can be more easily arranged to accomplish the desired result than can spherical ones, but cylindrical lenses with the desired optical properties are difficult to design.

2. *Moving Mirrors.* Assume a beam of light incident upon a mirror at 45° and reflected onto a screen, as in Fig. 3-13. The beam path to the screen will then be at right angles to the original beam. If the beam then moves through an angle θ from the original direction and the mirror moves through an angle $\theta/2$, the reflected beam will remain superimposed. Distortions in the screen image produced by such a mirror system are large unless the angle through which the mirror is allowed to move remains small. Relatively long beam paths (long focal-length lenses) are therefore used. Usually a number of mirrors are placed on a rotating wheel. Each mirror follows a single frame of the film. Tilting of each individual mirror is accomplished by cam mechanisms.

3. *Moving Prisms.* A beam of light passing through a piece of glass with plane parallel sides is undeviated in direction but laterally displaced according to the angle of inci-

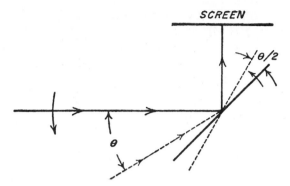

FIG. 3-13 Moving-mirror optical compensation. (*From Fink.*[1])

dence. This is illustrated in Fig. 3-14a. For small changes in angle, the amount of lateral displacement is proportional to the change. Use of such a device in a continuous projector is illustrated in Fig. 3-14b. As the film moves downward at A, the glass prism rotates counterclockwise. The rate of rotation is adjusted so that the emergent beam remains fixed. Successive prisms are made to follow successive film frames.

In practical applications a many-sided polygon is used as the moving prism. The faces of the polygon are of the same size as a film frame and, as the polygon rotates, successive faces of the polygon come in contact with successive frames. Light enters through the film and into one side of the polygon. The central part of the polygon is cut away, providing for additional optics necessary in correcting for film shrinkage. Passing through this central region, the light enters the inner part of the polygon ring on the opposite side, and out through the flat polygon surface. Opposite pairs of polygon faces operate, in effect, as the outer faces of a single prism such as shown in Fig. 3-14.

3.3 COLOR BEAM-SPLITTING SYSTEMS

Color television recording and reproduction require spatial separation of the red, green, and blue light. These beams must also be filtered to eliminate spurious color signals. The design goal in color beam-splitting systems is to reflect *all* the light of one primary color and to transmit *all* remaining visible radiation. Dichroic mirrors and prisms are used with supplemental trimming filters. The most efficient systems utilize dichroic mirrors.

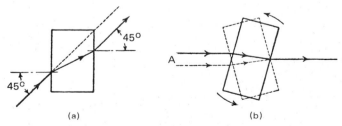

FIG. 3-14 Moving-prism optical compensation. (*From Fink.*[1])

3.3.1 DICHROIC MIRRORS.

Dichroic mirrors are made by coating glass with alternate layers of two materials having high and low indices of refraction. The material must

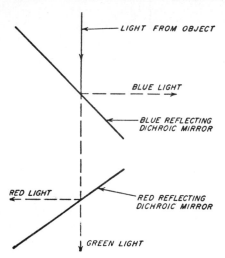

have a thickness of one-quarter wavelength at the center of the band to be reflected. Figure 3-15 shows a typical mirror arrangement. The blue light is reflected by the first mirror, and the red and green light is transmitted. The red light is reflected by the second mirror, and the green light is passed. The curves of Fig. 3-16 show typical transmittance versus wavelength. The blue reflecting mirror transmits about 90 percent of the green and red light, and the red reflecting mirror transmits nearly 90 percent of the blue and green light.

FIG. 3-15 Arrangement of dichroic-mirror beam-splitting system. (*From Fink.*[1])

3.3.2 DICHROIC PRISMS.

Figure 3-2 showed that when the angle of incidence of a light ray exceeds the critical angle the ray is totally reflected. The critical angle for an air-glass surface is 42° for a typical index of refraction for glass of 1.50. Hence a 45-45-90° glass prism offers a totally reflecting surface. Other designs permit partial reflection and refraction. Coatings at the prism surface, as for dichroic mirrors, will selectively pass or reflect different colors.

3.3.3 SPECTRAL TRIM FILTERS.

Figure 3-16 shows that a typical dichroic mirror does not abruptly change spectral reflection at some specific wavelength. Instead, there is a gradual transition over a wide band. This gradual transition must be eliminated to maintain purity of the red, green, and blue color signals. The spectral reflectance trans-

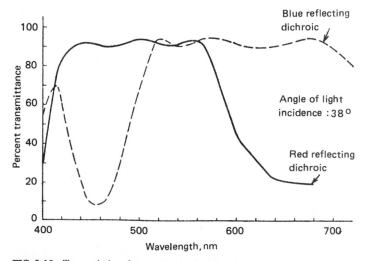

FIG. 3-16 Transmission characteristics of typical dichroic mirrors.

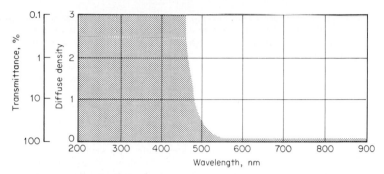

FIG. 3-17   Spectral trim filter (yellow). (*Courtesy of Eastman Kodak Company, 1981.*)

mittance bands are trimmed by inserting filters having abrupt divisions between high and low transmittance. These filters are glass, plastic, or gelatin containing light-absorbing substances. Wratten filters,[2] an example of the latter, are available in a wide variety of abrupt bandpasses from yellow to red, and minus green (magenta).

Another type of filter useful in beam splitters is the neutral density filter. It absorbs equally, or nearly so, all wavelengths in the visible spectrum. These filters are available in different densities so that color beams may be balanced for equal signal output.

Curves for a spectral trim filter (yellow) and a neutral density filter ($D = 1.0$) are shown in Figs. 3-17 and 3-18.

3.4 PHYSICAL OPTICS FUNDAMENTALS

Section 3.1 dealt with those optical problems which could be handled using geometric methods, considering the propagation of light in terms of rays. In this section such phenomena as interference, diffraction, and polarization are discussed in terms of light as wave motion.

3.4.1 INTERFERENCE EFFECTS.

Huygen's principle was mentioned as forming the basis for the laws of reflection and refraction. To this concept should now be added the principle of superposition which states that the resultant effect of the superposition of two or more waves at a point may be found by adding the instantaneous displacements which would be produced at the point by the individual waves if each were present alone.

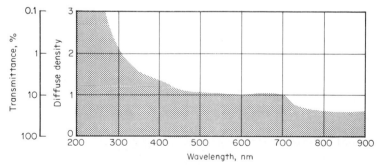

FIG. 3-18   Neutral density filter, visible spectrum. (*Courtesy of Eastman Kodak Company, 1981.*)

If the wave path is thought of as a sinuous path consisting of alternate crests and troughs, the maximum height of the crests or depth of the troughs is called the *maximum amplitude*. Starting at zero amplitude and progressing through a crest back to zero and through a trough to zero constitutes one cycle, and the distance traveled through the medium is one wavelength. The number of cycles per unit time is the frequency. The distance traveled per unit time is the velocity. Therefore, the velocity v is the product of the wavelength λ and the frequency ν

$$v = \nu\lambda \qquad (3\text{-}11)$$

If two waves meet in such a manner that the crests reinforce each other to produce the maximum possible amplitude, the waves are said to be in phase, but if the crest meets the trough to produce the minimum possible amplitude, they are said to be 180° out of phase. Thus, the phase expresses the distance between the crests of the waves. If two waves of the same amplitude, traveling in the same or opposite directions, are 180° out of phase, they are said to completely interfere and no disturbance is noted. If the phase, or amplitude, or frequency of the waves is not the same, then the waves can reinforce at certain points and destroy at other points to produce an interference pattern. A commonly observed example of interference is the array of colors seen in a thin film of oil on a wet pavement. Light waves are reflected from the front and rear surfaces of the film. When the thickness of the film is an odd number of quarter wavelengths, the light of that wavelength reflected from the front and back surfaces of the film reinforces itself and light is strongly reflected (assuming normal incidence). For an even number of quarter wavelengths, destructive interference occurs. Thus, from one area of film only blue-green light may be reflected while from another area of different thickness red light may be observed.

If, instead of a thin film of oil, two reflecting surfaces, such as two glass surfaces, are placed together but not in complete optical contact, an interference pattern is formed by light from the front and rear surfaces. Frequently the pattern takes the form of concentric rings, called *Newton's rings*.

Interference patterns are useful in grinding optical surfaces. The new surface may be tested by bringing it in contact with a surface of known curvature and noting the shape and separation of the fringes. By repeating the test at intervals, the new surface may be gradually worked to the desired precision.

3.4.2 DIFFRACTION EFFECTS. If an obstacle such as a slit or straight edge is placed in a beam of light, according to Huygen's principle each point along the slit becomes a source for new wavelets. It can be shown that as these wavelets fan out beyond the obstacle they tend to reinforce or destroy each other in various regions, forming an interference pattern. As the wavelets fan out beyond the obstacle, the light "bends around" it, producing light areas in regions which would be dark if the light traveled only in straight lines. The effects produced by blocking part of a wave front to form interference patterns are called *diffraction effects*. The phenomenon has already been mentioned briefly in discussing the resolving power of lenses (see Sec. 3.1.7). If a wave front is incident on a circular opening such as a lens aperture, the diffraction pattern consists of a bright central disk surrounded by alternate dark and bright rings. The angle α formed at the lens by the diffraction circle is dependent upon the diameter of the lens opening D as follows

$$\alpha = \frac{2.4\lambda}{D} \qquad (3\text{-}12)$$

An important device utilizing diffraction is the diffraction grating. This consists, essentially, of a very large number of parallel slits of the same width spaced at regular intervals. Light passing through the slits is diffracted to form interference patterns. The waves will reinforce to form a maximum when

$$\sin\theta = \frac{n\lambda}{d} \qquad (3\text{-}13)$$

where θ = angle of deviation from direction of incident light
  d = distance between successive grating slits
  n = integer denoting order of the maximum

Some light will pass directly through the grating. This is called the *zero order*. The first maximum (assuming monochromatic light) lies beyond the zero order and is called the *first order*. The next maximum is the second order and so on. If white light is incident on the grating, the zero order is a white image followed by a first-order spectrum, second-order, etc. By proper ruling of the grating lines, a large proportion of the incident light may be directed into one of the first-order spectra. These gratings are used in many spectral-analysis instruments because of their high efficiency.

3.4.3 POLARIZATION EFFECTS.

Another optical phenomenon which should be briefly mentioned is polarization. Since light is a series of electromagnetic waves, each wave can be separated into its electric (E) and magnetic (H) vectors vibrating in planes at right angles to each other. A series of electromagnetic waves will have E vectors, for example, vibrating in all possible planes perpendicular to the direction of travel. By means of reflection, double refraction, or scattering, the waves can be sorted into two resultant components with their E vectors at right angles to each other. Each ray is said to be plane-polarized, i.e., made up of waves vibrating in a single plane. If two rays with waves of equal amplitude are brought together, they can form elliptically, plane, or circularly polarized light depending upon whether the phase difference between the vibrating waves lies between 0 and $\pi/2$ for elliptical polarization, is at 0 or π for plane polarization, or is at $\pi/2$ for circular polarization.

The angle of incidence at which light reflected from a polished surface will be completely polarized is given by the equation known as *Brewster's law*

$$\tan \theta = \frac{n'}{n} \tag{3-14}$$

where n' and n are the indexes of refraction of the two media. For glass and air, $n' = 1.5$ and $n = 1$, the polarizing angle is 56°. Of the natural light incident at the polarizing angle about 7.5 percent is reflected and is polarized with its vibration plane perpendicular to the plane of incidence. The rest of the light is transmitted and consists of a mixture of the light with vibration plane parallel to the plane of incidence and the balance of the perpendicular component. By passing the mixture through successive sheets of glass stacked in a pile, more of the perpendicular component is removed at each reflection and the transmitted fraction consists of the parallel component.

The velocity of a light wave through many transparent crystalline materials is not the same in all directions. Since the ratio of the velocity of light in the medium to the velocity in a vacuum is the index of refraction, these materials have more than one index of refraction. When oriented in one position with respect to the direction of the incident ray, the crystal behaves normally and that direction is called the *optic axis* of the crystal. A ray incident on the crystal to form an angle with the optic axis is broken into two rays, one of which obeys the ordinary laws of refraction and is called the *ordinary ray*. The second ray is called the *extraordinary ray*. The two rays are plane-polarized in mutually perpendicular planes. By eliminating one of the rays, such doubly refracting materials can be used to obtain plane-polarized light. In some materials one of the components is much more strongly absorbed than the other. Crystals of iodoquinine sulfate are an example. The parallel orientation of layers of such crystals in plastic has been used to form polarizing filters.

In 1875 Kerr discovered that some liquids become doubly refracting when an electric field is applied. The Kerr effect makes it possible to control the transmission of light by an electric field. A Kerr cell consists of a transparent cell containing a liquid such as nitrobenzene. The cell is placed between crossed polarizers. When an electric field is applied, light is transmitted and is cut off when the field is removed.

3.5 VIDEO PROJECTION SYSTEMS

Two systems of large-screen television employing diffraction principles have been devised. These are the Eidophor system and the General Electric large-screen system. Both systems employ a conventional light source, such as a high-intensity xenon lamp, and modulate the incident light on the screen to form the raster. Of historical interest is the Scophony system, also utilizing diffraction effects. Refractive and reflective systems of conventional types are also briefly described.

3.5.1 EIDOPHOR[3] LIGHT MODULATOR. Figure 3-19 is a diagram of one color element of an Eidophor projector.[4] The rays from the lamp (1) pass through a system of lenses (2) and mirror bars (3) and an oil layer (4) coated on a spherical mirror surface (5). The bars and lenses are aligned so that no light is passed when the liquid surface is undeformed. When the control layer is deformed, part of the light is diffracted and passes between the mirror bars, and a bright spot is then visible on the screen, the brightness dependent on the depth of the deformation. The deformations are produced by bombarding the layer with an electron beam.[7] The electric charge on the thin oil film deforms the surface, which returns to normal as the charge decays. The electron beam scans the picture area of the liquid at constant intensity and variable speed. The variable

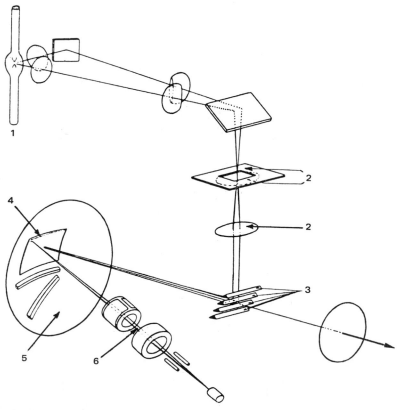

FIG. 3-19 Schematic diagram of Eidophor projector. (*Courtesy of Eidophor, Ltd., Regensdorf, Zurich, Switzerland.*)

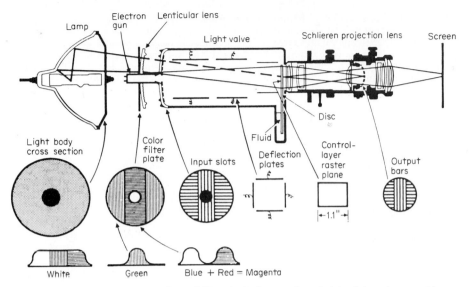

FIG. 3-20 Schematic diagram of General Electric single-gun color television light-valve assembly.

speed is produced by an alternating voltage of constant frequency and variable amplitude superimposed on the line sweep voltage. The frequency of this alternating voltage determines the size of the picture raster elements; its amplitude determines their relative brightnesses.

3.5.2 GENERAL ELECTRIC LARGE-SCREEN SYSTEM. This system was described by W. E. Glenn[5] and W. E. Good.[6,7] A deformable liquid surface modified by an electron gun produces diffraction patterns which are imaged by means of an external light source and Schlieren projection lens upon a screen. See Fig. 3-20. Color filters, input slots, and output bars position the color signals to create three simultaneous and superimposed red, green, and blue images from the same electron beam.

3.5.3 SCOPHONY LIGHT MODULATOR. The Scophony system, which is of historical interest, also utilized the diffraction principle. This system employs a liquid cell containing a piezoelectric quartz crystal as the light modulator. An HF signal applied to the piezoelectric crystal creates compression waves in the liquid which produce striations with a higher index of refraction. The cell, therefore, acts like a diffraction grating. The cell is situated in the optical system as shown in Fig. 3-21. A stop at A obscures the direct

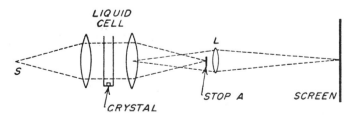

FIG. 3-21 Schematic illustration of Scophony projection system.

image of the source. When the crystal is excited, diffraction images of the source are formed on either side of the stop A. Lens L collects the light in these side images and forms an image of the cell on a screen. The illumination of the screen image can be made approximately linear to the excitation voltage. If this voltage is modulated with picture signals, the intensity of the line image at any instant will be modulated along the length of the image, since the waves are traveling uniformly across the cell. Thus the image of the cell is stationary, but the image of the modulated waves moves with uniform velocity across the screen. Two rotating mirror drums are placed between the lens and the screen. The first compensates the moving image and renders it stationary on the screen. The second drum is synchronized to present the successive lines of the picture. For a 405-line image the first mirror drum must rotate at 30,750 r/min.

3.5.4 REFRACTIVE PROJECTION SYSTEMS.

A large-screen television image, i.e., larger than that attainable by direct viewing of a picture tube, can be achieved by a trinoscope color display where the images from three cathode-ray tubes are enlarged by projection and in optical superposition form an image on a reflecting or translucent screen. Each tube provides one primary color image, red, green, or blue. The image is enlarged by lenses or spherical mirrors (see Sec. 3.5.5). A separate lens is used for each tube. To achieve acceptable screen brightness, the lenses must have high relative aperture, $f/1.0$ or better, which introduces problems of image aberration. Various geometric configurations of lenses and mirrors are designed for maximum screen brightness versus cathode-ray voltage.

Television receivers employing this system are used in homes and small public rooms. Kits and systems are also available for enlarging a conventional picture-tube image by means of a single lens to a larger screen size for home viewing.

3.5.5 REFLECTIVE PROJECTION SYSTEMS.

As mentioned in Sec. 3.1.3, spherical mirrors with Schmidt correctors can be made with a very high relative aperture, for example, $f/0.6$. Figure 3-22 shows a schematic Schmidt system. The aspherical corrector can be molded from plastic material. A small high-brightness cathode-ray tube emitting red, green, or blue light is positioned at the focal point of the mirror. The three images are registered on a reflecting or translucent screen, usually with first surface mirrors interposed to make the projection path more compact.[8] Television receivers employing this system are also used in homes and small public rooms.

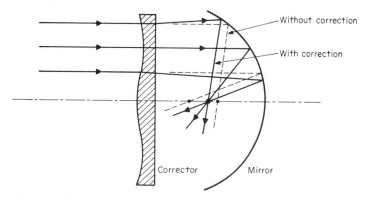

FIG. 3-22 The Schmidt corrector for minimizing spherical aberration of a mirror. Dashed lines show paths of rays without aspherical corrector.

REFERENCES

1. D. G. Fink (ed.), *Television Engineering Handbook,* McGraw-Hill, New York, 1957.
2. "Kodak Filters for Scientific and Technical Uses," Eastman Kodak Co., Rochester, N.Y.
3. E. Bauman, "The Fischer Large-Screen Projection System," *J. SMPTE,* vol. 60, p. 351, 1953.
4. Gretag AG, "What You May Want to Know about the Technique of Eidophor," Regensdorf, Switzerland.
5. W. E. Glenn, "Principles of Simultaneous Color Projection Using Fluid Deformation," *J. SMPTE,* vol. 79, p. 788, 1970.
6. W. E. Good, "A New Approach to Color Television Display and Color Selection Using a Sealed Light Valve," *Proc. National Electronics Conf.,* vol. 24, 1968.
7. W. E. Good, "Recent Advances in the Single-Gun Color Television Light-Valve Projector," vol. 59, Soc. Photo-Optical Instrumentation Engrs., 1975.
8. T. Poorter and F. W. deVrijer, "The Projection of Color Television Pictures," *J. SMPTE,* vol. 68, p. 141, 1959.

BIBLIOGRAPHY

Hardy, A. C., and F. H. Perrin, *The Principles of Optics,* McGraw-Hill, New York, 1932.
Kingslake, Rudolf (ed.), *Applied Optics and Optical Engineering,* vol. I, chap. 6, Academic, New York, 1965.
Sears, F. W., *Principles of Physics, III, Optics,* Addison-Wesley, Cambridge, Mass., 1946.
Williams, Charles S., and Becklund, Orville A., *Optics: A Short Course for Engineers and Scientists,* Wiley Interscience, New York, 1972.

Monochrome and Color Visual Information Transmission

Bernard D. Loughlin

Hazeltine Corporation
Greenlawn, New York

With contributions by

Donald C. Livingston

Sylvania Electric Products (retired)
Bayside, New York

William F. Bailey

Hazeltine Corporation (retired)
Little Neck, New York

NOTE: Parts of this chapter are adapted from D. G. Fink (ed.), *Television Engineering Handbook,* McGraw-Hill, New York, 1957. Chapters 8 and 9. Used by permission. These chapters were prepared by: Donald C. Livingston, then with Sylvania Electric Products, Bayside, New York; William F. Bailey, then with Hazeltine Corporation, Little Neck, New York; the author of this present chapter.

This chapter is concerned with the nature of the video information carried by television signals and with the changes in information content which occur in the normal course of generating and transmitting these signals and using them to operate a receiver.

4.1 VIDEO INFORMATION CONCEPTS

4.1.1 FUNDAMENTAL DEFINITIONS AND CLASSIFICATION OF INFORMATION.

For the purpose of this handbook, video information may be defined as data conveying a description of a picture which can be displayed by an appropriate picture-reproducing system through the use of television signals.[1-8]†

When the information under discussion is that possessed by the electrical signals occurring at a given location in a television system, the character of the information can be interpreted in terms of the picture which would be displayed by some reference reproducer when driven by the given signals. For purposes of this handbook, the reference reproducer is here defined as a television signal–processing and display system which correctly displays the full information content of any set of signals which have been properly formed and have been properly inserted into the signal-processing circuits of the system. All the signal-processing circuits are assumed to be free of distortion, noise, and interference. The only nonlinear element in the reference reproducer is taken to be its picture display device, whose photoelectric transfer characteristics are specified to be the same as those of the program director's studio monitor.

In monochrome systems and compatible color systems, the information in the associated pictures and signals can be broadly classified as:

1. *Monochrome Information:* the information carried by the monochrome signal in both monochrome systems and compatible color systems

2. *Coloring Information:* the information carried by all signals other than the monochrome signal in a compatible color television system

Monochrome information by itself describes a monochrome picture without reference to its chromaticity. Monochrome and coloring information together describe a color picture.

When it is desired to discuss the types of information carried by the signals in the primary-color signal channels of a color television transmitter or receiver, the following classifications are pertinent:

1. *Red-primary Information:* the information carried by a red-primary signal

2. *Green-primary Information:* the information carried by a green-primary signal

3. *Blue-primary Information:* the information carried by a blue-primary signal

If a picture or a scene is under discussion without reference to any electrical signals which may be associated with it, a useful classification for the types of information contained in it is:

1. *Luminance Information:* the information which describes the distribution of luminance levels in the picture.

2. *Chromaticity Information:* the information which describes the distribution of chromaticity values in the picture if it is a color picture

---

†Superscript numbers refer to References at end of chapter.

There are two methods for classifying the information in a monochrome picture: The information in the picture associated with a monochrome television system can be called *monochrome information* in general and *luminance information* in particular. It is monochrome information because it is brought to the picture by a monochrome signal; it is luminance information since it describes the distribution of luminance levels in the picture.

In the same way, the information content of a color picture produced by a compatible color television system can be described in at least three ways:

1. It can be said to consist of a combination of monochrome information and coloring information since it is brought to the picture by the monochrome signal and by an auxiliary coloring signal (the chrominance signal).

2. It can also be said to consist of a combination of luminance and chromaticity information since it describes the distribution of luminance levels and chromaticity values in the picture.

3. Finally, it can be said to consist of red-, green-, and blue-primary information on analogous grounds.

The information content of video signals can be described in terms of the picture which they would elicit from a reference reproducer. Various systematic procedures for carrying out such descriptions can be developed through use of the following: (1) luminance level, (2) monochrome level, (3) chromaticity value, (4) raster coordinates, (5) point image, (6) line image, and (7) picture element

The *luminance level* at a given point in a picture is the ratio of the luminance at that point to the maximum luminance occurring anywhere in the picture during some appropriate interval of time. Its value can lie anywhere between zero and unity. The reciprocal of the lowest luminance level in the scene is called the *contrast range* of that scene.

The *monochrome level* at a given point in a color television picture produced through use of the transmission signals of a compatible color television system is the luminance level in the monochrome picture which appears in the absence of the chrominance signal.

The *chromaticity value* at a given point in a color television picture is the quantitative specification of the chromaticity at that point in terms of a suitable set of chromaticity coordinates.

The *raster coordinates*[9] of a given point in a picture are a pair of numbers specifying the position of that point with respect to a suitable set of reference axes. The horizontal and vertical axes may be chosen along the upper edge and left-hand edge of the raster, respectively, and a distance equal to the raster height may be designated as unity along both axes. Thus, raster coordinates of a point in a raster with a 4/3 aspect ratio would range between zero and unity for the vertical coordinate and between zero and 1.33 for the horizontal one.

A *point image* is the pattern of light formed on a television display in response to light received by a television camera from a single object point in its field of view. The position of any given point image on a reproducing raster is determined by the position of the corresponding object point together with whatever geometric distortions (pincushioning, etc.) may exist in the raster. Its position is also influenced by the line structure in the scanning raster unless its size is large in comparison with the distance between adjacent scanning lines, as happens when the object point lies appreciably outside the camera field of focus.

A *line image* is the pattern of light formed on a television display in response to light received by a television camera from a uniformly illuminated ideally narrow line in its field of view. The position of any given line image on a reproducing raster is determined by the position of the corresponding object line in a manner analogous to that for a point image. A line image oriented perpendicular to the scanning path may be called a *transverse* line image, and one parallel to this path may be called a *longitudinal* line image.

A *picture element*[10] is the smallest area of a television picture capable of being delineated by an electrical signal passed through the system or part thereof. The number of picture elements (pixels) in a complete picture, and their geometric characteristics of vertical height and horizontal width, provide information on the total amount of detail which the raster can display and on the sharpness of that detail, respectively.

For a more complete understanding of the concept, the following ad hoc specification developed by the IEEE in considering the definition of a picture element may prove to be useful. The picture element is rectangular in shape, its boundaries being parallel and perpendicular to the scanning lines. Its center lies on the path traversed by the center of the scanning spot. Its dimensions are defined through the use of luminance variations which occur at the center of the element when line images are moved over it in prescribed ways. Its *length*, denoting the dimension parallel to the scanning path, is defined as the greatest distance interval which a transverse line image can travel along the scanning path under the conditions (1) that the luminance level at the center of the element shall reach its highest possible value when the line image is at some point between its extreme positions, and (2) that the luminance level shall be a fraction $2/\pi$ of this value when the line image is at either end point in its range.

For the picture element *width*, which is the dimension perpendicular to the scanning path, an essentially analogous specification applies. In this case, the line element is oriented parallel to the scanning path, and it is moved over the element center in a direction perpendicular to that path. The one difference to be noted is that the width of the element is arbitrarily constrained to be not less than the raster pitch distance, since this distance represents a minimum spacing below which spurious resolution can occur because of the discrete line structure in the raster itself if the reproducing aperture is small.

The description of the information content of a picture through use of the above quantities may be carried out in varying degrees of exactness. In the case of a monochrome picture, the method used would be expected to lead to the specification of definite numerical values for the displayed luminance level at points having given raster coordinates. If the monochrome picture concerned is that resulting from withdrawal of the chrominance signal from a color display during a color transmission, the observed luminance levels represent monochrome levels of the color picture. In the case of a color picture, the method would lead to the specification of numerical values for both luminance levels and chromaticity values at points having given raster coordinates.

The most direct rigorous approach involves specification of numerical values at *all* points in the picture. This requires associating the pertinent colorimetric values with the numerical values of explicit mathematical functions of the raster coordinates. In general, these are continuous functions because of the continuous nature of the distribution functions representing typical aperture transmittances.

It is meaningful to specify only as much information as can be recognized to consist of separate and independent elements. To carry out this type of description, the raster area can be subdivided into picture elements, the total number of elements being determined by the spectrum characteristics of the monochrome signal and by the percentage of the total frame scanning time which is consigned to blanking intervals. The video signal waveform for one frame period can be regarded as a portion of a repetitive wave corresponding to a stationary picture, and its spectrum then turns out to have a finite number of discrete frequency components (see Chap. 5). The total information capacity of this spectrum is twice the number of components, since each component can be adjusted in both amplitude and phase.

The waveform for a single frame has the same total information capacity, and the waveform for any fractional part of a frame has a corresponding fraction of this total information capacity. Consequently, the portion of the waveform used for active scanning of the *raster* during one frame period has an information capacity equal to twice the number of frequency components in the spectrum for the signal multiplied by the number representing the fractional portion of the frame period used in active scanning. This information capacity is a number which may be interpreted as the number of picture elements in the raster area. The elements may be regarded as being uniformly distributed over the raster in the sense that their separations along the scanning lines represent distances traversed by the scanning spot in equal times.

4.1.2 PICTURE-ELEMENT CHARACTERISTICS. An array of picture elements can be considered to be distributed over the television raster. Since the information content of a signal can be described in terms of the picture which the signal might elicit from a reference reproducer, it follows that picture elements can be associated with signals as well as with pictures.

The specification for the dimensions of the picture element leads to a particularly simple result in a television system wherein the camera scanning aperture is a mathematical point, all signal channels have sharp cutoffs at some frequency below which transmission is distortionless, and all transducers including the picture display device are linear. It is clear that this signal itself has the form of an ideal impulse function. Its frequency expression is given by

$$E_0(t) = \frac{1}{\pi} \int_0^\infty \cos 2\pi ft \; df \tag{4-1}$$

when its amplitude has been chosen so that

$$\int_{-\infty}^\infty E(t) \; dt = 1 \tag{4-2}$$

f and t here denote frequency and time, respectively. After band limiting, the frequency expansion of the resulting signal becomes

$$E(t) = \frac{1}{\pi} \int_0^{f_c} a(f) \cos \left[2\pi ft - \alpha(f) \right] df \tag{4-3}$$

in which $a(f)$ and $\alpha(f)$ are the attenuation and phase shift, respectively, of the channel at frequency f, and f_c is the frequency above which no appreciable transmission occurs.

The luminance distribution across the associated transverse line image of a signal in this system has the form of the impulse response of the signal channel, which, in turn, is a curve of the form $\sin x/x$. Specifically, it is given by Eq. (4-3) with $a(f)$ constant and $\alpha(f)$ equal to zero. This integrates to

$$E(t) = \left(\frac{f_c}{\pi} \right) \frac{\sin 2\pi f_c t}{2\pi f_c t} \tag{4-4}$$

where f_c is the cutoff frequency of the channel.

It is now to be noted that the function $E(t)$ has its maximum value at $t = 0$ and extends over the time range $t = \pm 1/4f_c$. The time interval $\Delta t = 1/2f_c$ is then representative of the maximum distance through which the line image can be moved along the scanning path so that the luminance at a given point on the raster reaches the value $E(0)$ at an internal point in the range and is down to $2E(0)/\pi$ at either end point. Specifically, this distance is the distance traversed by the scanning spot in time Δt. The length of the picture element is defined to be this distance.

The number of picture elements in the raster is twice the maximum number of frequency components which the given signal can possess, multiplied by the factor representing the fractional portion of the total frame scanning interval used in active raster scanning. The elements are specified to be distributed uniformly over the raster, the distance between centers of adjacent elements on a given scanning path being fixed. That this distance is precisely equal to the picture-element length in the television system under discussion may be shown as follows.

The spectrum for a signal representing a stationary picture is shown in Chap. 5 to contain only frequencies which are integral multiples of the picture repetition frequency f. Thus, if the channel cutoff frequency is f_c, the signal can contain not more than f_c/f frequency components. This corresponds to $2f_c/f$ information elements, and these are to be conceived as distributed uniformly throughout the raster scanning and blanking intervals. Picture elements are simply those information elements which occur during the raster scanning intervals. The time interval between successive elements of either type is evidently given by the frame scanning interval $1/f$ divided by the total number $2f_c/f$ of information elements occurring in that time. This turns out to be $1/2f_c$, which is precisely the time interval Δt found just previously to be indicative of the picture-element length. It follows that in the system under discussion, adjacent picture elements on a scanning path are in contact with each other, although they do not overlap. It is in order to bring about this particular result that the picture-element length is defined in terms of luminance ratios of $2/\pi$.

The definition for picture-element width leads to a similar result. In the system under discussion, displacing a longitudinal line image from exact coincidence with a given scanning path results in an abrupt drop of the displayed luminance from the value characterizing the line image to zero. Hence, the added condition appended to the definition becomes applicable and prescribes that the picture-element width in this case is equal to the raster pitch distance.

It is thus seen that in a television system having a mathematical-point scanning aperture in the camera, having sharp-cutoff signal transmission channels with distortionless passbands, and having no nonlinear transducers, the picture elements exactly fill the raster area without overlapping.

Each such picture element may be called an *ultimate picture element* for the given television system. It has the smallest possible dimensions which the picture element in a linear system with given cutoff frequency can possess. Enlarging the camera scanning aperture from a point to a configuration with finite dimensions immediately brings about an increase in the picture-element length; a considerably enlarged aperture also causes the picture-element width to increase so that overlap occurs both longitudinally and transversely. Replacement of the sharp-cutoff frequency characteristic of the signal channel by a gradual one with the same cutoff frequency changes the impulse response from that in Eq. (4-4) to the more general form in Eq. (4-3). The latter invariably yields a greater interval between points for which $E(t)$ has a value equal to a fraction $2/\pi$ of its maximum value than does the former. Lowering the cutoff frequency of the channel causes the total number of picture elements to decrease and causes each element to become longer.

In a nonlinear system, the actual picture element can be smaller than the ultimate element. This can occur, for example, when the picture display device has a gamma exponent greater than unity. This is significant since the gamma exponent is usually found to lie in the range $\gamma = 2.2 \pm 0.2$ for conventional cathode-ray picture tubes. In practice, the picture-element length is nevertheless usually larger than that of the ultimate element because of the effects of camera aperture size and the gradual cutoff in the frequency characteristic of the signal channel. The width, however, is ordinarily that of the ultimate element.

The total number of ultimate picture elements in the raster of a given television system is the product of the number $2f_c/f$ of information elements in the frame interval and the percentage of the frame interval consumed in active scanning of the raster. Thus, if a percentage p_H of each line and a percentage p_V of the total number of lines are used in active scanning of the raster, then there are $2p_Hp_Vf_c/f$ picture elements in the raster. Moreover, the number of picture elements on each line of the raster is given by this number divided by the number of lines in the raster. The number of lines in a frame is f_H/f, where f_H is the line scanning frequency, and the number of lines in the raster is therefore p_Vf_H/f. Consequently, there are $2p_Hf_c/f_H$ picture elements on each raster line.

The length of a single ultimate picture element in terms of raster coordinates is the raster width divided by the number of elements on a scanning line. The raster width is ⅔ when the aspect ratio is 4/3, so the length of an ultimate picture element is $2f_H/3p_Hf_c$. Similarly for one of the HDTV systems under consideration, the raster width is ⅗, so the length of an ultimate picture element is $5f_H/6p_Hf_c$. A list of picture-element statistics for various television systems is given in Table 4-1, which is based on data from the various signal transmission standards.

It is to be noted that although the picture element may indicate how much detail the television system can carry and how sharply each element of detail will be reproduced, it does not indicate directly the extent to which this detail will be resolved. To fill this need, it is convenient to introduce the concept of a *picture resolution element*. By a picture resolution element is meant a geometrical figure which, like the picture element, is

Table 4-1 Data on Picture Elements in Various Television Systems

Quantity	CCIR standard					
	M U.S. NTSC	A U.K.‡ MONO	B/G CCIR PAL	L France SECAM	E France‡ MONO	HDTV§
Information elements per frame	280,000	240,000	400,000	480,000	800,000	1,300,000
Picture elements per raster	210,000	180,000	300,000	360,000	590,000	1,000,000
Picture elements per line	440	480	520	620	790	990
Picture-element length†	30	28	26	21	17	17
Picture-element width†	21	27	17	17	13	10

†Based on raster height = 10,000 units.
‡Discontinued services.
§Based on provisional standards.

rectangular in form and has its sides respectively parallel and perpendicular to the scanning paths. Its dimension parallel to the scanning path may be called its *length* and may be defined to be the width of each stripe in the finest pattern of alternating light and dark transverse stripes of uniform width which is resolved by the signal. It measures the horizontal resolution of the signal. Similarly, its dimension perpendicular to the scanning path may be called its *width* and may be defined as the width of each stripe in the finest pattern of alternating light and dark longitudinal stripes of uniform width which is fully resolved by the signal when one edge of at least one stripe in the group coincides with a scanning path.

Unlike the picture element, the total number of resolution elements which may be placed in the raster is not determined by the cutoff frequency of the signal channel but by the specification that the complete raster area is to be subdivided into a system of resolution elements with their adjacent edges in contact so that every point in the raster is in one and only one resolution element or on a boundary edge between adjacent elements. It follows in consequence of this specification that no television signal can contain as many resolution elements as it does picture elements. This is evident both from the fact that the picture element generally overlaps its neighbor whereas the resolution element does not and from the fact that the resolution element can never be smaller than the picture element. The latter fact follows from the experimental observation on which the *Kell factor* is based. The Kell factor is the number obtained by dividing the raster pitch distance by the above-specified width of the picture resolution element. Various writers have reported different numerical values for this factor as follows:

Source	Kell factor
Kell, Bedford, and Trainer (1934)[2]	0.64
Mertz and Gray (1934)[9]	0.53
Wheeler and Loughren (1938)[3]	0.71
Wilson (1938)[5]	0.82
Kell, Bedford, and Fredendall (1940)[6]	0.85
Baldwin (1940)[11]	0.70

The variations from one source to another are probably attributable to differences in scanning-spot geometry in the picture display systems used by different observers. For the present discussion, a value of 0.70 is used.

The total amount of detail which can be resolved simultaneously by a given signal is specified by the ratio of the raster area to the resolution-element area. The *vertical resolution* is measured by the ratio of the raster dimension across the scanning path to the resolution-element width. This number is the maximum number of uniform longitudinal stripes which can be fully resolved by the signal when the edge of at least one stripe coincides with a scanning path. The *horizontal resolution* is measured by the ratio of the raster dimension across (*not* parallel to) the scanning path to the resolution-element length. The resolution-element length would correspond to half a cycle at the channel cutoff frequency if the channel had a sharp cutoff characteristic and if the camera scanning aperture was a mathematical point. Under these conditions, the length of the picture element would also have this value. The resolution-element width, however, would be equal to the raster pitch distance divided by the Kell factor, or to 1.4 times the raster pitch distance. At the same time, the picture-element width would be exactly equal to the raster pitch distance. Just as the picture element under this special condition may be called the ultimate picture element, the resolution element under this same condition can be called the *ultimate resolution element*. Its dimensions indicate the maximum resolving power of which the given television system is capable.

When the signal-channel cutoff frequency is unchanged although cutoff is made gradual rather than sharp, both picture element and resolution element become longer. Their widths remain unchanged as long as the camera scanning aperture remains a mathematical point. When the camera aperture is enlarged beyond a diameter comparable with the raster pitch distance, both picture-element and resolution-element width become greater than their ultimate values, and it cannot be tacitly assumed that the ratio of their widths remains equal to the Kell factor.

Actual dimensions of the resolution element in a given television system cannot be calculated on theoretical grounds but must be established by subjective experimental observation. Statistics on the ultimate resolution element, however, are readily calculated on the basis of the specifications given above. They are presented for various television systems in Table 4-2, a value of 0.70 being assumed for the Kell factor.

4.1.3 COLOR ANALYSIS AND REPRESENTATION. The analysis of the color information into luminance and chrominance coordinates and the choice of video signals representative of these quantities are discussed in this section. The formation of a single composite color television signal from the basic video-frequency component signals and the utilization of this information are covered in subsequent sections.

It is not known at present that accuracy of colorimetric reproduction is either nec-

Table 4-2 Data on Resolution Elements in Various Television Systems

Quantity	CCIR standard					
	M U.S. NTSC	A U.K.† MONO	B/G CCIR PAL	L France SECAM	E France† MONO	HDTV‡
Resolution elements per raster	150,000	130,000	210,000	250,000	410,000	730,000
Vertical resolution	340	260	400	400	530	740
Horizontal resolution	330	360	390	470	590	740

†Discontinued services.
‡Based on provisional standards.

essary or desirable. It is believed, however, that colorimetric distortions should not occur in the transmission process. The philosophy followed in the transmission of chrominance values is that the object of the transmission of these signals is to reproduce on the picture screen of a receiver the same picture that exists on the screen of a reference monitor at the transmitter. If distortions of the information are employed to produce pleasing pictures or other desired effects, they should be introduced in the channel between the pickup of the original scene and the reference monitor, and not in the subsequent transmission path.†

Color Coordinates.   A set of color coordinates which has been found very useful for color television work is to specify the color in terms of luminance, dominant wavelength, and excitation purity (purity).

These are psychophysical terms and represent quantities which may be measured in a quantitative fashion.[12,13] There is also a concept to classify color in terms of a perception, and in this concept the color may be described by brightness, hue, and saturation.[14,15] By brightness is meant that attribute of a perception by which a color is classified as equivalent to a white, a light gray, a darker gray, etc. Hue is the attribute of a perception by which color is classified as red, or yellow, or green, or the like. Saturation is the attribute of a perception which classifies the color as described by the terms strong or intense as contrasted with faint or pastel. These terms which describe perception of a color are not subject to quantitative measurement, but they do have a correlation in a general way with the psychophysical quantities luminance, dominant wavelength, and purity, mentioned earlier.[16]

The signal in monochrome television is one which varies the brightness point by point in the reproduced picture. Thus, if a photometrically accurate reproduction of the scene is desired, this signal should control the reproduced luminance to make it proportional to the original, or scene, luminance. The addition to a monochrome signal of signals representative of dominant wavelength and purity adds the necessary factors for the transmission of complete color information.

Dominant wavelength and purity are a way of denoting chromaticity equivalent to the stating of the CIE chromaticity coordinates x and y. Figure 4-1 shows a CIE chromaticity diagram on which colors are mapped in terms of their chromaticity coordinates x and y. The luminance quantity has been completely suppressed in this map. The outer horseshoe curve represents spectral colors and the horseshoe is closed by the straight line representing saturated nonspectral purples. All real colors fall within this closed area. Certain points are designated on the curve and the numbers give the corresponding wavelength in nanometers of the monochromatic radiant energy.

The point C is the chromaticity of the white light from standard source C. Using C as a reference point, radial lines may be drawn to the boundary curve, and these lines are the loci of colors having constant dominant wavelength. Thus all colors lying on the line between points C and 580 nm would have the dominant wavelength of 580 nm when standard source C is considered as the reference white. The dominant wavelength of purples is generally stated by giving the complementary wavelength, that is, the wavelength at the intersection of a locus extended through point C until it intersects the curved spectral locus.

Purity is shown by loci which surround the point C at some fixed ratio of the total distance from C to the spectral locus and closing line. The curve labeled purity = 50 percent is the locus of the points halfway along any line of constant dominant wavelength from the white point C to the boundary.

Thus, loci of constant dominant wavelength and purity are a set of coordinates which map out the chromaticity space, and which bear a general correlation to the subjective quantities hue and saturation, respectively.

---

†However, in spite of this principle, it has become rather usual for receivers to include certain forms of colorimetric distortion to, on the average, make more pleasing pictures, particularly from transmissions which may have been undesirably distorted, or which may emanate from source material of questionable quality. (See Sec. 4.6.4.)

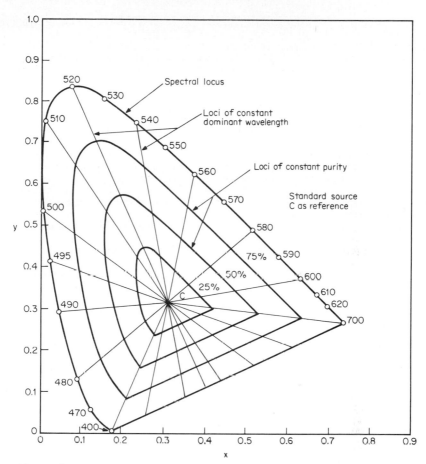

FIG. 4-1 CIE chromaticity diagram showing loci of dominant wavelength and purity. (*From Fink.*[17])

Definitions of Chrominance Quantities. All the picture reproducers known at present synthesize the reproduced picture point by point by generating light of three different colors. These reproducing lights are called *primaries*. The most useful gamut of reproduced colors results from using primaries having the hues red, green, and blue.[18] Current practice in the adjustment of monitors and receivers is to balance for D6500 white, the coordinates of which are $x = 0.3127$ and $y = 0.3290$ (see Sec. 2.4.2). Similarly, cameras normally resolve the light of each point in the scene into three components from which the proper signals to control three reproducing primary lights are formed. The three primary signals so produced are E_R, E_G, and E_B. For colorimetrically exact reproduction, the photoelectric transducers should have a proper set of spectral sensitivity curves, that is, they should be related by a linear transformation to the color-mixture curves for some set of three primary colors.

For the transmission of color in the NTSC system (standard in the United States) the basic primary-color signals are denoted E_R, E_G, and E_B and these correspond to signals which produce a suitable picture on a monitor having the reference primary colors defined by their chromaticities as follows:[19,20]

	Chromaticity	
Primary	x	y
Red	0.67	0.33
Green	0.21	0.71
Blue	0.14	0.08

The convention is also followed that on reproduction of the reference chromaticity, standard source C, the three color signals E_R, E_G, and E_B are all equal.

All the broadcast color television standards now in use employ essentially a simultaneous system in which the three pieces of information are all continuously present.† This method has the advantage that the information to be transmitted need not be the primary signals E_R, E_G, and E_B, but may be other signals derived from them. By this means it is possible to take advantage of certain of the characteristics of the human visual process to reduce the total information transmitted.

The formation of the luminance signal E_Y is the first step in the process of choosing the transmitted signal to minimize the transmitted information. The luminance signal E_Y is the only one which is transmitted if the coloring or chrominance signals are zero. In this case the picture should be reproduced with the chromaticity of standard source C ($x = 0.310$, $y = 0.316$). Many studio control-room color monitors, and present-day color receivers, are being balanced for white to D6500, the chromaticity coordinates of which are $x = 0.3127$ and $y = 0.3290$ (see Sec. 2.4.2). For the reference primaries cited above, the luminance signal should have the composition

$$E_Y = 0.30E_R + 0.59E_G + 0.11E_B \qquad (4\text{-}5)$$

Note that the coefficients of the primary signals in this equation equal unity. Thus, when reproducing standard source C, the condition

$$E_Y = E_R = E_G = E_B \qquad (4\text{-}6)$$

is obtained. The coefficients of the primary signals in the luminance signal, Eq. (4-5), represent the luminance contributions of the individual primaries to the total luminance when reproducing standard source C.

The coloring information is transmitted in two additional signals also derived from the primary signals. These additional signals are called *chrominance* or *color-difference signals* since they are representative of the difference in the reproduced chromaticity from some reference chromaticity (standard source C in this case).[21] Color-difference signals of the composition $E_R - E_Y$, $E_G - E_Y$, and $E_B - E_Y$ are used in conjunction with the luminance signal as follows. By the operation of adding the red color-difference signal to the luminance signal, the red signal E_R can be obtained as

$$(E_R - E_Y) + E_Y = E_R \qquad (4\text{-}7)$$

The color-difference signal disappears on the reference chromaticity since, as stated before, for this chromaticity

$$E_Y = E_R = E_G = E_B \qquad (4\text{-}6)$$

The three color-difference signals are not independent, since only two are required, in conjunction with the luminance signal, to transmit all the information regarding the three primaries. Thus one can be derived from the other signals as[22]

$$E_G - E_Y = -[0.51(E_R - E_Y) + 0.19(E_B - E_Y)] \qquad (4\text{-}8)$$

---

†In SECAM (which uses rapid-line sequential transmission of color-difference signals) delay lines are used in the receiver producing simultaneous signals for application to the color image reproducer.

By means of transformations well known in colorimetry it is possible to transform information in the form of the three signals for the reference primaries E_R, E_G, and E_B to other signals for any other three primaries.[18,23] One such set is for the nonphysical primaries standardized by the CIE, X, Y, Z. In this case the luminance signal is E_Y and the color-difference signals have the form $E_X - E_Y$ and $E_Z - E_Y$ (or suitable modification) so that the color-difference signals disappear on the reference chromaticity.

The color-difference signals of the type $E_R - E_Y$ and $E_B - E_Y$ are not the only form in which this information can be transmitted. For the NTSC signal it is desirable to form color-difference or chrominance signals for transmission purposes in which each signal is constituted partially by $E_R - E_Y$ and $E_B - E_Y$. These derived chrominance signals have been given the symbols E_I and E_Q and their compositions are

$$E_I = 0.74(E_R - E_Y) - 0.27(E_B - E_Y) \qquad (4\text{-}9)$$

$$E_Q = 0.48(E_R - E_Y) + 0.41(E_B - E_Y) \qquad (4\text{-}10)$$

The term *chrominance* is defined as the difference between a color and a reference color when both are at the same luminance. This may be visualized somewhat more clearly by considering a three-dimensional representation of color as shown in Fig. 4-2. The coordinates shown are the CIE tristimulus values X, Y, and Z. As Y is the luminance, any plane parallel to the XZ plane is one of constant luminance. Point W is the reference color and point A is a color having the same luminance. The vector WA is therefore chrominance, by definition, and it is necessarily two-dimensional since it requires two quantities, such as a pair of rectangular coordinates or a magnitude and angle, to define it.

It is possible to design a color television receiver and correspondingly set the proportions of the two transmitted color-difference signals so that they do not affect the reproduced luminance. Thus the luminance signal E_Y exerts the control over the reproduced luminance. This is termed the *constant-luminance principle*.[21,24] With this method of operation, the color-difference signals may be called chrominance signals since they do

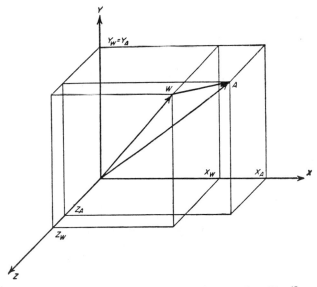

FIG. 4-2 Illustration of chrominance in color space. (*From Fink.*[17])

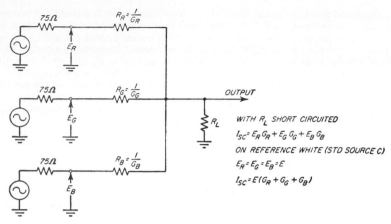

FIG. 4-3   Simple resistive matrix. (*From Fink.*[17])

not affect the reproduced luminance but do control the reproduced color, which would be the reference chromaticity (standard source C) in their absence. The use of chrominance components in the composite color signal results in a system of high compatibility as the chrominance signals are transmitted only in the degree necessary, and disappear on the reference color. The proportioning for constant-luminance operation results in a minimum of added subjective noise in a color picture since the color-difference or chrominance channel does not contribute luminance noise to the picture.†

Transformation of Chrominance Coordinates or Matrixing.   The signals generated by a camera or other pickup device are generally suitable to reproduce a picture, using the reference reproducing primaries. The signals which are desired, to optimize the transmitted information, are the luminance signal E_Y and two chrominance signals. The chrominance signals may be of the composition $E_R - E_Y$ and $E_B - E_Y$ or they may be of the composition E_I and E_Q.

From the equations given previously it is seen that all the derived signals are expressed by linear algebraic equations. The operation of producing the derived signals is termed *matrixing* and is readily performed by simple circuits.

A reliable way to form the luminance signal is as shown in Fig. 4-3.[25] Here the three primary signals are available at a low impedance, and separate resistors R_R, R_G, and R_B are used to control the relative currents flowing into the common load R_L. The relative current from each input is independent of R_L but is governed by the input voltage and the series resistance to the junction at the common load. This arrangement exhibits a high degree of stability and is simple, thus it is widely used. There is an attenuation of the level through this matrix, but the loss is not serious, and with low-impedance sources, the cross coupling between the sources is usually low enough to be negligible.

Amplifiers are frequently used in a matrix where one of the component signals has the opposite sign from the others.[25,26] In this case, phase-reversing amplifiers can be used to produce a signal of the opposite sign, which is then added to the other component signals to produce the resultant.

---

†Usual receiver practice results in *first-order* constant-luminance operation such that these statements apply very well for the less-saturated (pastel) colors. On highly saturated colors, the picture-tube nonlinearity (gamma) results in some luminance noise contribution from the chrominance channel.

One requirement which must be satisfied, for any matrix to function properly, is that all the component signals should be simultaneous, that is, they should represent information in the various channels corresponding to the same picture element. The NTSC standard is that the several components should correspond to each other in time with an error of less than 0.05 μs.

Bandwidths Required for Chrominance Components.   One reason for transforming the information contained in the primary signals to information of the type in luminance and chrominance signals is that this is an efficient way to minimize the required amount of transmitted information for a picture of good quality.

Experiments with a simultaneous color television system using the mixed-highs principle have shown that at normal viewing distances an observer does not perceive the color in the fine-detail portions of the picture.[27] Tests conducted for Panel 11 of the NTSC indicated that if the color bandwidth were about 1 MHz for each primary and the fine detail corresponding to video frequencies in the band from 1 to 4 MHz were reproduced at the correct luminance but in monochrome, satisfactory pictures were produced.[28] The results of a number of tests of this type are shown in Fig. 4-4 which shows for the case of the crossover frequency between color and monochrome of 1.0 MHz that out of 77 observations using 11 observers, about 46 observations showed no perceptible defect in the picture, 22 additional observations showed perceptible defects but satisfactory pictures, 8 additional observations indicated increased defects giving marginal pictures, and 2 additional observations indicated unsatisfactory pictures, as compared with the reference pictures produced with signals of 4-MHz bandwidth for each primary.

The picture quality in the example just given for the case of mixed-highs transmission would be exactly duplicated if the information were transmitted by means of a luminance signal having the bandwidth 0 to 4 MHz and two chrominance signals each having the bandwidth 0 to 1 MHz. Further studies of the properties of the human eye, in regard to its ability to distinguish color, show that while the normal eye has three-color vision for large areas, and is only achromatically responsive to the luminance component in very small areas, there is an intervening region in which objects subtend an angle of about 10

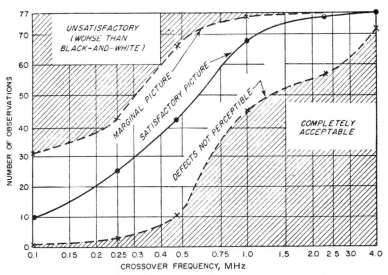

FIG. 4-4   Results of tests of determine necessary color bandwidth in simultaneous color television. (*From Fink.*[17])

to 25 min arc at the eye in which the eye appears to have only two-color vision.[29-31] The type of two-color vision exhibited by the eye in this case is called *tritanopia* and in general it may be said that when the vision is tritanopic the observer confuses purples, blues, grays, and greenish yellow. However, the observer usually exhibits good acuity for red, orange, and cyan.

Subjective tests have been made of color television picture quality in which this last-described characteristic of the eye was also exploited.[32] In these tests the luminance information was transmitted by a normal luminance signal of about 4 MHz bandwidth, and two chrominance signals were used. One of these, having a bandwidth of about 1.5 MHz, carried the information regarding the difference in the reproduced color from white in the orange or cyan direction. The other chrominance signal, which had a bandwidth of about 0.5 MHz, carried the information of the deviation of the reproduced color from white in the direction of green or its complement, magenta. This arrangement produced pictures of about the same quality as those produced in the mixed-high experiments described earlier.† In the arrangment just described, since all three signals are present for frequencies 0.5 MHz, detail in the picture corresponding to this frequency range is reproduced in three colors. For detail lying in the frequency range from 0.5 to about 1.5 MHz, there are only two signals present; thus the detail is reproduced in only two colors. For frequencies higher than 1.5 MHz, only the luminance component is present and this detail is therefore reproduced only in brightness. The wider-band chrominance signal, carrying the information for color changes in the orange or cyan direction, is denoted as the E_I chrominance signal. Its composition in terms of the red and blue color-difference signals is

$$E_I = 0.74(E_R - E_Y) - 0.27(E_B - E_Y) \qquad (4\text{-}11)$$

By substituting the value for the E_Y signal into the above equation the composition of the E_I signal in terms of the original E_R, E_G, and E_B primary signals may be found. It is

$$E_I = 0.60E_R - 0.28E_G - 0.32E_B \qquad (4\text{-}12)$$

The narrow-band signal used here is denoted by the symbol E_Q and it supplies the information on color changes in the green or magenta direction. Its composition in terms of the red and blue color-difference signals is

$$E_Q = 0.48(E_R - E_Y) + 0.41(E_B - E_Y) \qquad (4\text{-}13)$$

Similarly E_Q may be defined in terms of the primary signals as

$$E_Q = 0.21E_R - 0.52E_G + 0.31E_B \qquad (4\text{-}14)$$

Figure 4-5 shows the passbands versus frequency of the signal components E_Y, E_I, and E_Q, transmitted when this method of producing a color picture is used. The passbands of the signals controlling the red-, green-, and blue-reproducing primaries of the picture reproducer are also shown and each of these passbands is divided into three regions showing the contributions to the red-, green-, and blue-primary signals at the receiver which are made by the three transmitted signals E_Y, E_I, and E_Q.

The advantages of transmitting the information by means of the chrominance signals E_I and E_Q are more fully explained in Sec. 4.2.2, treating the coding of the three component signals in a single compatible signal. Pickup devices at the transmitter generally produce primary camera signals E_R, E_G, and E_B, having bandwidths of at least 4 MHz.

---

†Only the NTSC system, which required a maximum of packing of information, uses the arrangement of three different bandwidths for encoded transmission. PAL and SECAM systems are equal-bandwidth chrominance channels and correspondingly use $R - Y$ and $B - Y$ specifications (instead of I and Q) for simplicity. Nevertheless, it is interesting to note that all broadcast color television systems have a summation for the two chrominance-channel bandwidths of approximately 50 to 60 percent of the luminance channel. Similarly, the newly evaluated data for proposed HDTV systems resulted in 20 MHz for Y, 7 MHz for the wider $R - Y$ channel, and 5 MHz for the narrower $B - Y$ channel. All these are in general agreement with the early *mixed-highs* test results given in Fig. 4-4. (For further recent data also see Ref. 130.)

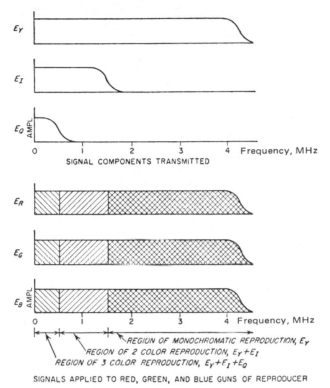

E_Y

E_I

E_Q

AMPL

0   1   2   3   4   Frequency, MHz

SIGNAL COMPONENTS TRANSMITTED

E_R

E_G

E_B AMPL

0   1   2   3   4   Frequency, MHz

REGION OF MONOCHROMATIC REPRODUCTION, E_Y
REGION OF 2 COLOR REPRODUCTION, $E_Y + E_I$
REGION OF 3 COLOR REPRODUCTION, $E_Y + F_I + E_Q$

SIGNALS APPLIED TO RED, GREEN, AND BLUE GUNS OF REPRODUCER

FIG. 4-5 Bandwidths of transmitted signals and composition of reconstituted primary signals when using E_Y, E_I, and E_Q. (*From Fink.*[17])

The usual practice in producing the three signals to be transmitted, namely the luminance signal E_Y and the two chrominance signals E_I and E_Q, is that these are matrixed without any band limitation to form three wideband signals. Subsequent to the matrixing, the luminance signal is normally used without band limitation. The E_I signal is limited to about 1.5 MHz and the E_Q signal to about 0.5 MHz. The process of band limiting the three signals to different bandwidths usually results in their suffering different amounts of time delay through the filters, the least delay being suffered by the luminance signal, and greatest delay being suffered by the E_Q signal. The delays are equalized to that of the greatest one by delay networks in the channels handling the luminance signal E_Y and E_I chrominance signal so that after traversing the filters and delay networks the three signals are synchronous. The delay networks normally used are either wide-band delay devices or lumped-element circuits having constant-amplitude transmission over the band, with the correct amount of uniform time delay.

As a result of limiting the bandwidth of the chrominance signals to a smaller value than that for the luminance signal, transitions in the level of the chrominance signals take a longer time to occur than does a transition in the luminance signal. Thus, when the camera-scanning spot scans across a boundary from one colored area to another, the transitions in the chrominance signals start earlier and last longer than does the transition of the luminance signal. If the time delays in the channels are properly equalized, all the transitions pass through their half-amplitude points at the same time. This effect results in a relatively sharp transition in luminance with a somewhat smeared transition

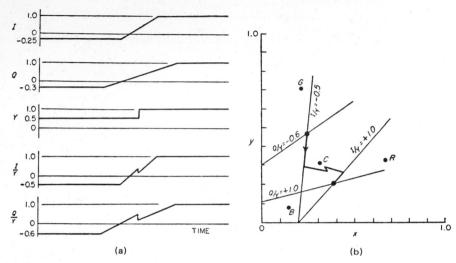

FIG. 4-6   Signal transitions and locus of reproduced chromaticity change when using luminance and chrominance signals. (*From Fink.*[17])

of color centered on the luminance change.[33] For the case where the transmitted signal components are E_Y, E_I, and E_Q, and considering a transition from a cyan to a magenta color, the changes in signal levels with time are shown schematically in Fig. 4-6. The steepness of these changes is of course governed by the bandwidth available for each of the signals. In Fig. 4-6 there is shown a CIE chromaticity diagram on which is traced the path representing the transition in reproduced chromaticity for the change under consideration. The transition starts with a change in chromaticity due to the Q signal. After this change is well under way, the I signal also begins to change, which results in the path being modified. At the halfway point in the transition, the luminance signal changes abruptly and this puts the little jog in the path as the luminance signal instantaneously changes the saturation. Following the completion of the luminance signal change, the I-signal transition goes to completion; then the Q signal finally completes its transition and the chromaticity is stable at the final point.

Gamma Correction.   For the perception of adjacent areas of differing luminance the human eye exhibits a logarithmic characteristic. That is, the sensation tends to be proportional to the logarithm of the luminance difference. This relation may be derived from the Weber-Fechner law which states that the threshold of perception of the difference in luminance between two adjacent areas is characterized over a considerable range by a constant ratio of the difference in luminances to the greater of the two.[34] This may be written

$$\Delta S = \frac{\Delta B}{B} \tag{4-15}$$

where ΔS = threshold sensation change
  ΔB = difference in luminance
  B = greater of luminances

It is found that the fraction $\Delta B/B$ is essentially constant over the ranges of luminance in which television pictures are produced.

An important aspect of the fact that the visual process follows this law relates to the perception of noise in the received picture. Assume that some noise is added to the signal in the transmission path between the transmitter and the receiver. The noise tends to have a constant rms value independent of the modulation level of the signal. In a receiver in which the reproduced brightness is a linear function of the modulation level it can be shown by application of the Weber-Fechner law that the noise will have greatest visibility in the darkest parts of the picture and least visibility in the brightest parts. The noise visibility smoothly increases as the parts of the picture become darker. If the receiver has the characteristic that the reproduced luminance is an exponential function of the modulation level, then it is found that the sensitivity for perceiving added random noise is uniform regardless of the reproduced brightness. Thus, the perception of the added random noise in the signal channel at various brightness levels is governed by the transfer characteristics of the receiver.

At the present time, there is no definite standard on the linearity characteristic which a receiver should have, and therefore none on the linearity characteristic which a transmitter should have, so that the overall system produces satisfactory pictures. The problem is complex, since there are many additional factors which affect the results such as the effects of signal-level changes, ambient illumination at the viewing position, and the exactness of the bias voltage adjustment on the reproducer. A good compromise for monochrome practice appears to be to use a receiver whose reproduced luminance varies in accordance with the input modulating signal raised to some power generally lying in the range of 2 or 3, and to use approximately the inverse power law for modifying the signal at the program source.[35] It is desirable that the color television system operation be performed with a power-law receiver having an exponent in the range of 2 or 3, as this will provide operation tending to have substantially a uniform subjective signal-to-noise ratio for noise added in the path.

At the present time, television pictures are almost universally reproduced by cathode-ray tubes in which the reproduced brightness is varied by modulating the intensity of the electron beam. With the usual space-charge-limited emission, and with modulation of the beam current by one of the electrode potentials, picture tubes generally exhibit a power-law characteristic of the type expressed by the equation

$$B = KE^n \qquad (4\text{-}16)$$

where B is the reproduced brightness, K is a constant, E represents the grid-drive voltage above cutoff, and n is an exponent lying in the range of 2 to 3.

For monochrome television the picture tube itself can normally be used to provide the desired nonlinearity at the receiver to optimize the subjective signal-to-noise ratio in the reproduced picture. This makes for a simple receiver since the nonlinearity is inherent in the picture tube, and all the other amplifying or translating stages of the receiver can be arranged to be linear or approximately so. The adjustment of only one critical bias level, that for the picture tube, is required. In this case the program source should have a substantially inverse characteristic to that of the nominal receiver, over the required contrast-ratio range, in order to reproduce pictures with acceptable tone-reproduction values.

The practice usually followed for gamma correcting the transmitted color signal is as follows:[20] It is assumed that the primary colors in the picture-reproducing tube have the reference chromaticities and that the reproduced light from each primary varies according to the 2.2 power of the grid-driving signal in volts above cutoff. Gamma-corrected signals, denoted by the symbols E_R', E_G', and E_B' to distinguish them from linear signals E_R, E_G, and E_B, are produced in the R, G, and B color-camera video-processing amplifiers located at the program source. There is no definitive standard on what the gamma-corrected signals E_R', E_G', and E_B' consist of, other than that they be suitable for a color picture tube with the reference primaries and a transfer gradient of 2.2.

One method which has had wide use up to the present time for producing gamma-corrected signals is to give them the characteristic

$$E_R' = E_R^{1/\gamma} \qquad (4\text{-}17)$$

with no particular value specified for γ (gamma). The signals E_B' and E_G' are formed in the same manner, using the same value of gamma. Values of gamma used range from about 1.4 to about 2.5.

The normal practice is to use the gamma-corrected signals E'_R, E'_G, and E'_B in place of the linear primary signals to make up the luminance and chrominance signals which are used to form the composite signal for transmission. In the deliberations of the NTSC, and also in the FCC standards for the transmission of color, it is conventional to denote any signal formed of gamma-corrected components by the use of the prime. Thus the luminance signal E'_Y and the color-difference signals E'_I and E'_Q denote gamma-corrected signals made up from E'_R, E'_G, and E'_B.

Chrominance Axes on the CIE Chromaticity Diagram. It is possible to plot on the CIE chromaticity diagram the loci of the chrominance axes.[23,33,37] The plot then shows the relation of the reproduced chromaticity to the level of the three components of the transmitted signal, that is, the luminance signal and the two chrominance signals. For a constant reproduced chromaticity, the chrominance signal is proportional to the luminance signal being transmitted. Thus the plots are in terms of the ratio of each chrominance-signal component to the luminance signal, as the plots show only chromaticity and do not convey any information as to the actual luminance. Figure 4-7 shows, for the case of a linear reproducer and linear signals E_Y, $E_R - E_Y$, and $E_B - E_Y$, the mapping of the signal-amplitude ratios on the chromaticity diagram. Figure 4-8 shows the mapping of the ratios $(E'_R - E'_Y)/E'_Y$ and $(E'_B - E'_Y)/E'_Y$ on the chromaticity diagram for the case of a reproducer having the characteristics

$$R = K_1 E_1^{2.2} \tag{4-18}$$

$$G = K_2 E_2^{2.2} \tag{4-19}$$

$$B = K_3 E_3^{2.2} \tag{4-20}$$

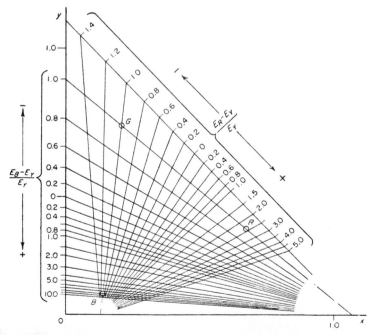

FIG. 4-7 Map of reproduced chromaticity as a function of normalized chrominance signals $(E_R - E_Y)/E_Y$ and $(E_B - E_Y)/E_Y$ (linear case). (*From Fink.*[17])

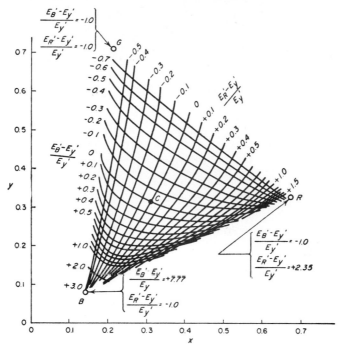

FIG. 4-8 Map of reproduced chromaticity as a function of normalized chrominance signals $(E'_R - E'_Y)/E'_Y$ and $(E'_B - E'_Y)/E'_Y$ (nonlinear case). (*From Fink.*[17])

where R = red light
      G = green light, etc.
E_1, E_2, E_3 = grid-drive signals on red, green, and blue tubes, respectively, in volts above cutoff
K_1, K_2, K_3 = constants

The transmitted gamma-corrected signals are E'_Y, $E'_R - E'_Y$, and $E'_B - E'_Y$ where these signals are formed from gamma-corrected primary signals of the characteristic

$$E'_R = E_R^{1/2.2} = KR^{1/2.2}, \text{ etc.} \tag{4-21}$$

where R = red light to be reproduced and E_R = signal voltage proportional to red light.

Plots of the same sort of information are shown when using the signals E_Y, E_I, and E_Q in Fig. 4-9 and E'_Y, E'_I, and E'_Q in Fig. 4-10. It may be seen that the general result of using gamma-corrected signals of the forms described on the color reproduction is to make little change in the appearance of the maps near white, although the chromaticity changes at a more rapid rate with chrominance signal than for the linear case. However, the maps are crowded in the regions of the reproducing primaries, in such a fashion that reproduced chromaticity changes are in general reduced for a given signal change when compared with the linear case.

Reproduced Luminance on the CIE Chrominance Diagram. Another colorimetric map of value is that of reproduced luminance versus reproduced color. In a linear system, such as represented by Figs. 4-7 and 4-9, the relative reproduced luminance would be unity and independent of chromaticity. However, as described before, in the nonlinear

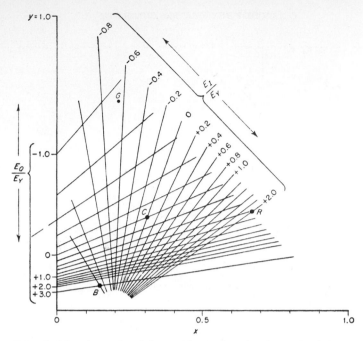

FIG. 4-9 Map of reproduced chromaticity as a function of normalized chrominance signals E'_I/E'_Y and E'_Q/E'_Y (linear case). (*From Fink.*[17])

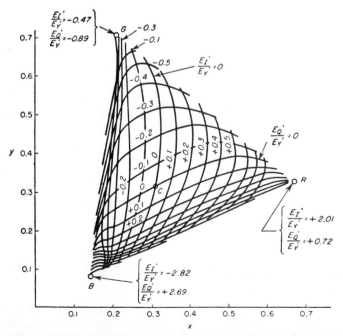

FIG. 4-10 Map of reproduced chromaticity as a function of normalized chrominance signals E'_I/E'_Y and E'_Q/E'_Y (nonlinear case). (*From Fink.*[17])

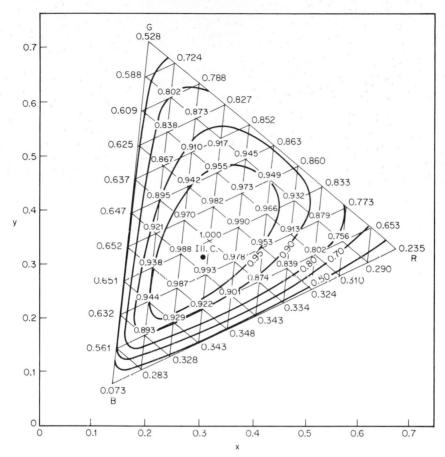

FIG. 4-11 Map of fractional part of total reproduced luminance produced by luminance signal, as a function of chromaticity (nonlinear case). $E_M = E'_Y$, $E_{RD} = E_R^{1/\gamma} - E'_Y$, $E_{BD} = E_B^{1/\gamma} - E'_Y$, $E'_Y = 0.299 E_R^{1/\gamma} + 0.587 E_G^{1/\gamma} + 0.114 E_B^{1/\gamma}$, $\gamma = 2.2$.. *(From Fink.[17])*

case, some luminance is contributed by the chrominance signals. Figure 4-11 shows a map (for the nonlinear-receiver case) of fractional luminance provided by the luminance channel—and thus 1 minus the value is the fractional luminance provided through the chrominance channel. The preponderance of luminance through the luminance channel is indicated, as well as the effect of *constant-luminance failure* upon highly saturated colors.

4.2 COMPOSITE VIDEO INFORMATION SIGNALS

4.2.1 COMPOSITE MONOCHROME SIGNALS. For many years it has been *standard* to generate a composite monochrome signal containing video (picture) information portions with interspersed horizontal and vertical blanking periods, at approximately black level of the picture signal, to accommodate horizontal and vertical retrace periods of display scanning. In addition, synchronizing pulses are added in the *blacker-than-black* signal region to permit synchronization of receiver scanning. While the precise details of such blanking periods and synchronization pulses vary from one television standard to another, the general formats are quite similar.

In this section the main concern is with composite video information signals, that is, with the forms of composite signals used during the picture portions of the signal in order to encode the color information.

4.2.2 COLOR-ENCODING PRINCIPLES

Compatibility Considerations.   To build up a color television broadcasting service in a country having an existing monochrome broadcasting service, it is economically desirable for the color system to be compatible, that is, one which provides color pictures on color receivers and at the same time provides monochrome pictures of adequate quality on existing unmodified monochrome receivers. Also in a country (such as the United States) where frequency spectrum is at a premium, it is desirable to fit the compatible color system completely within the channel occupied by the existing monochrome service. To accomplish this (for the United States standards) a method of coding of chrominance values is needed which permits transmitting the 6 MHz of information required for a satisfactory simultaneous color picture in the available 4-MHz video channel, in a compatible manner. This compact packaging of information can be accomplished using the band-shared color-systems specifications developed by the NTSC and approved by the FCC.[19,20] Even in cases where the compact packaging of information is not required,† because frequency spectrum is not at a premium and/or because compatibility is not a consideration, many features of the NTSC chrominance coding method are of interest since they are based upon fundamental properties of the eye.[24,27,31,38,39]

Band Sharing Using a Low-Visibility Subcarrier: Additional information can be transmitted in an existing monochrome channel because of certain characteristics of the system, which result from scanning. The spectral distribution of the video signal contains energy concentrated near harmonics of line-scanning frequency (see Chap. 5). This results from the fact that the signal, while not identical, is quite similar on successive scanning lines. Figure 4-12 illustrates an expanded portion of the frequency spectrum of a typical video signal. Additional sideband components exist at multiples of 60 Hz from each line-scanning harmonic, and these represent the variation of the picture in the vertical direction. Substantially no energy exists halfway between the line-frequency harmonics, that is, at odd harmonics of one-half line frequency. These blank spaces in the frequency spectrum suggest that an additional signal may be transmitted which fits into these gaps. A comb-shaped filter might be used to separate the two sets of signals from each other.[40-42]

The comb-shaped filter mentioned is effectively obtained in a display system, due to the scanning process. In other words, in addition to the fact that substantially no energy exists at odd harmonics of one-half line frequency in a normal video signal, it is also found that if a signal is injected at this frequency (shown dotted in Fig. 4-12) it will have low visibility in the reproduced image. As illustrated by Fig. 4-13, even harmonics of one-half line frequency (that is, harmonics of line frequency) have the same phase on successive scanning lines and in successive frames. However, odd harmonics of one-half line frequency have opposite phase on successive scanning lines and in successive frames, requiring four fields ($\frac{1}{15}$ s) to repeat. The raster patterns produced by these frequencies are illustrated in Fig. 4-14. The even harmonics line up to produce vertical bars which have a relatively high visibility. On the other hand, the odd harmonics produce a checkerboard pattern in any one field, and on any two successive fields, and, in addition, pro-

---

†Many countries, particularly those with approximately another megahertz of channel bandwidth available, were not forced to the extremely compact packaging of NTSC, and chose a somewhat different set of compromises as is pointed out in Sec. 4.2.3.

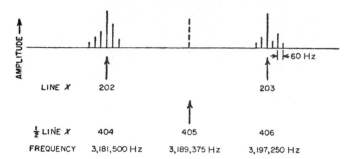

FIG. 4-12   Energy concentration in television signal. (*From Fink.*[17])

duce an opposite polarity of checkerboard pattern during the next frame. This is a pattern of relatively low visibility with a tendency for cancellation to occur between successive frames. Thus, odd harmonics of one-half line frequency are referred to as low-visibility frequencies and even harmonics of one-half line frequency as high-visibility frequencies. When viewing stationary images, the difference in just-perceptible signal levels can be as much as 20 dB between signals at low-visibility and high-visibility frequencies. This illustrates the effective comb-filter action obtained in the display.[22,43,44]

The relative visibility of an added signal is inversely related to its frequency since the resulting "dot pattern" has finer structure at higher frequencies. Also, the energy concentration in a normal video signal is less at higher video frequencies. Thus, if additional information is to be transmitted by a low-visibility subcarrier, the subcarrier frequency should be as high as is practical.

If a video signal (containing energy concentrated near harmonics of line-scanning frequency) is used to modulate a low-visibility subcarrier, then the sidebands due to modulation are also of low visibility. Figure 4-15 illustrates how this interleaving of a subcarrier and its sidebands can be used to transmit coloring information. The spectrum at the top of Fig. 4-15 represents the monochrome signal; the second spectrum illustrates a

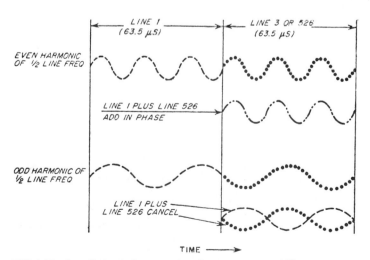

FIG. 4-13   Cancellation in frequency interleaving. (*From Fink.*[17])

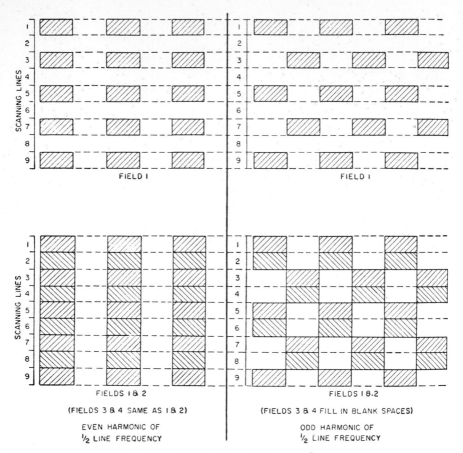

FIG. 4-14 High-visibility and low-visibility patterns. (*From Fink.*[17])

reduced-bandwidth video signal representing coloring or chrominance information. The third spectrum shows the result of modulating an interleaved subcarrier by the chrominance component; the spectrum diagram at the bottom illustrates the interleaving of the modulated subcarrier (and its sidebands) with the HF components of the monochrome signal.[22,43-46]

Other low-visibility patterns can be obtained. For example, if a frequency is considered which is removed from a line-scanning harmonic by only 30 Hz, it will produce a pattern which has substantially vertical lines in one field but produce a set of interleaved dots during the next field, such as illustrated in Fig. 4-16. This pattern repeats in ⅟₃₀ s as compared with the ⅟₁₅ s required for an odd harmonic of one-half line frequency. This 30-Hz checkerboard pattern is satisfactory as long as no motion exists (including motion of the eye). However, a small amount of motion which makes the apparent frequency of the pattern (as far as the eye is concerned) shift by only 30 Hz will produce a high-visibility pattern of vertical lines. Thus the odd harmonic of one-half line-scanning frequency pattern appears to be more satisfactory than the 30-Hz checkerboard pattern, when motion is considered.

Another low-visibility pattern, which has been considered, uses an odd harmonic of

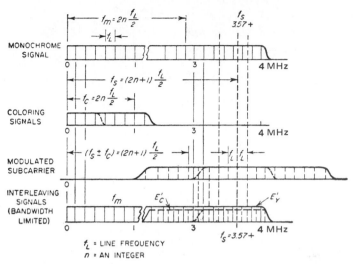

FIG. 4-15 Band-sharing of chrominance information. (*From Fink.*[17])

one-half line frequency with its phase shifted by 90° on every other field.[21,47] This 90° phase shift eliminates the predominant upward crawl of the odd harmonic of one-half line-frequency pattern, as is illustrated by Fig. 4-17. While some benefit is obtained from this pattern, it was not approved for transmission standards because of receiver complications resulting from the noncontinuous frequency.

When a high subcarrier frequency is used, the subcarrier-sound beat produced by various nonlinear effects in the system can become important. Due to the relatively low frequency of this beat it can, in certain cases, be more visible than the direct subcarrier energy. To reduce the visibility of this subcarrier-sound beat, the exact subcarrier frequency can be so chosen, with respect to the picture-to-sound spacing, that a low-visibility pattern is produced by the subcarrier-sound beat note. With frequency modulation of the sound carrier, an optimum low-visibility pattern is not obtained at all times. Experimental observations indicate that an average improvement of 4 to 6 dB can be obtained (for the United States standards) if the beat frequency is of low visibility under conditions of no sound modulation. If the picture-to-sound spacing is to remain unchanged (at 4.5 MHz in the United States standards), a slight change in horizontal scanning frequency is necessary to obtain both the subcarrier and the subcarrier-sound beat-note frequencies at odd multiples of one-half line-scanning frequency.[48]

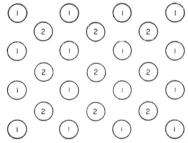

FIG. 4-16 30-Hz checkerboard pattern. Harmonic of line frequency ±30 Hz. Each dot represents peak of sine wave, and number represents field number. (*From Fink.*[17])

Color Coding for Minimum Subcarrier Visibility. In the band-sharing arrangement, a form of crosstalk occurs between the channels because subcarrier energy exists in the monochrome channel, and because HF monochrome components exist in the subcarrier channel. Due to the odd harmonic of one-half line-frequency relationship between

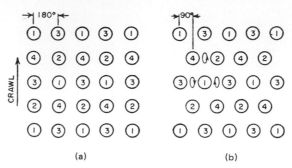

(a)   (b)

FIG. 4-17   Elimination of crawl by 90° shift.

the interleaved components, these crosstalk components should be of low visibility, but due to imperfect cancellation they are not of zero visibility. Factors contributing to the imperfect cancellation include the nonlinear characteristic of electron guns and the incomplete averaging over two frames by the display and the eye. Because of this incomplete cancellation it has been found desirable to use forms of color coding which additionally reduce the visibility of crosstalk components resulting from band sharing.[21]

The visibility of the subcarrier energy in the monochrome channel directly affects compatibility, and also affects the usable monochrome-signal information in a color receiver. The average value of the transmitted subcarrier energy can be substantially reduced by using a color-coding system in which the subcarrier and all its sideband components vanish when transmitting white and neutral-gray colors. The substantial reduction in transmitted subcarrier energy results because of the preponderance of pastel and gray colors in ordinary scenes.

The requirement that the subcarrier energy disappear on white and neutral-gray calls for the transmission of so-called color-difference signals on suppressed subcarriers. Thus, the subcarrier energy should convey the manner in which the color signals differ from each other (or from the monochrome signal, which is a weighted average of the color signals)—this difference disappearing on white when the red, green, and blue signals are equal to each other (and to the monochrome signal).[21] In other words, the transmitted subcarrier energy should be proportional to chrominance signals, chrominance being defined as the difference between a colored area and a neutral-gray area of the same luminance.

The transmission of chrominance information requires the transmission of two independent quantities which in combination with the monochrome signal give the necessary three independent quantities required to transmit color. These two independent chrominance components might be transmitted by modulating two low-visibility subcarriers at different subcarrier frequencies.[43,45] Such a two-subcarrier frequency system has a problem of spurious beats between the subcarriers, which would turn out to be an LF beat of high visibility, because the difference between two odd numbers is an even number. Additionally, when adequate guard bands are provided to permit frequency selection of the individual modulated subcarrier signals, one of the subcarrier frequencies is low enough to increase its visibility substantially. Thus, it has been found to be preferable to transmit the two chrominance components on the same subcarrier frequency by modulating a pair of subcarrier signals which are in quadrature with each other. The resulting modulated subcarrier signal may be alternately thought of as a pair of modulated quadrature subcarrier components, or as one subcarrier which is modulated in both amplitude and phase. Each of these points of view is found to be useful, depending upon the problem at hand.

The method of color coding the chrominance information on one subcarrier frequency can be directly compared with a conical three-dimensional representation of color space, as shown in Fig. 4-18. Using the polar coordinate concept (that is, considering the subcarrier is modulated in amplitude and phase), the amplitude of the subcarrier is a measure of the saturation of the color and the phase of the subcarrier is a measure of the hue of the color. The monochrome signal can be considered to represent brightness, and thus corresponds to the vertical direction on the diagram, while the subcarrier phase represents the hue of the desired color and the subcarrier amplitude represents saturation, which corresponds to a motion away from the white or achromatic colors in the direction of the hue corresponding to the subcarrier phase.

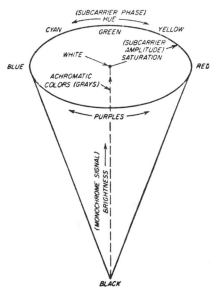

Color coding of the two chrominance components on one subcarrier frequency implies the use of some form of synchronous detection to recover the desired information. The operation of synchronous detectors is described in Sec. 4.3.2.

A further reduction in the visibility of subcarrier energy in the monochrome channel can be obtained by using as high a subcarrier frequency as is possible, consistent with other limitations of the signal-design problem. Since the chrominance information is conveniently analyzed in terms of a wideband component and a narrowband component (as discussed in Sec. 4.1.3), a higher subcarrier frequency can

FIG. 4-18 Color coding compared with color space. (*From Fink.*[17])

be used if vestigial sideband modulation is employed for the wideband component. The subcarrier frequency is chosen just high enough to permit double-sideband transmission of the narrowband component. The wideband component has its lower-modulation frequency components transmitted by double-sideband modulation and its higher-modulation frequency components transmitted by single-sideband modulation to the LF side of the subcarrier.[49,50] This type of operation is illustrated in Fig. 4-19.

An additional increase in subcarrier frequency has been considered, by employing a system which would permit vestigial sideband modulation of both chrominance components on a single subcarrier frequency. This system employed *color phase alternation* (CPA, also called *oscillating color sequence*) which permitted visual cancellation of the crosstalk products resulting from vestigial sideband modulation of both chrominance components.[21,22]† In the arrangement used in early NTSC field tests, the cancellation occurred between successive fields, and thus occurred between successive scanning lines in space on the display (because of interlaced scanning). However, the use of a substantially higher subcarrier frequency (3.9 MHz in the field tests) produced interference in the sound channel of certain receivers. In addition, the possibility of 30-Hz flicker (in large areas and on edges) when using high-brightness displays operating with incomplete constant-luminance performance discouraged the further use of CPA.[50] Furthermore, extensive field tests indicated that the compatibility obtainable with a subcarrier frequency of approximately 3.6 Hz was almost the same as that produced with a subcarrier frequency of 3.9 Hz when all factors in the system were considered.

---

†Color phase alternation will be discussed further in Sec. 4.2.5 on PAL.

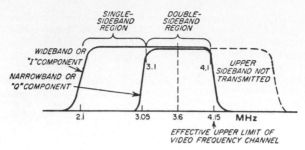

FIG. 4-19 Vestigial sideband transmission of wideband chrominance component. (*From Fink.*[17])

Another item affecting the visibility of the subcarrier energy in the monochrome channel is the relative registry of the subcarrier and monochrome information. Since there is substantial correlation between the subcarrier and monochrome information (in a color system) the subcarrier energy is less visible if it fits within (that is, is registered with) the monochrome signal, as displayed. Thus, for optimum compatibility a reasonable degree of time coincidence should exist between the desired monochrome signal and the undesired subcarrier energy as displayed on an average monochrome receiver. The time delay between the monochrome and subcarrier energy as transmitted is standardized substantially to accomplish this result.[51]

The items just discussed for color coding to produce minimum subcarrier visibility can be summarized as:

1. Arrange the system so that the subcarrier and all its sidebands vanish when transmitting white and neutral-gray colors.

2. Use a single-subcarrier frequency to transmit both components of the chrominance signal.

3. Use vestigial sideband transmission of the wideband chrominance component so that a high subcarrier frequency can be used.

4. Use suitable delay equalization of the monochrome and subcarrier signals so that substantial registry of the signals results in a monochrome receiver.

The block diagram of a band-shared simultaneous system designed in accordance with the above principles is shown in Fig. 4-20. The camera converts the red, green, and blue components of light into electrical voltages, shown as E'_R, E'_G, and E'_B. These voltages are applied to a matrix unit to produce a monochrome signal E'_Y and wideband and narrowband chrominance components E'_I and E'_Q, respectively. The chrominance components are filtered to the appropriate bandwidths and then used to modulate quadrature subcarrier signals. The modulated quadrature subcarrier signals are then added to the monochrome signal to produce the color-picture signal.

In the receiver the color-picture signal is applied to the three guns of a tricolor reproducer to produce a monochrome picture. The modulated subcarrier components are selected in the bandpass filter and applied to a set of quadrature synchronous demodulators. The output components of the demodulators, after suitable filtering, correspond respectively to the wideband and narrowband chrominance components. By matrixing, a set of color-difference signals is obtained which is combined with the monochrome signal to give the desired set of red, green, and blue output signals. (While not shown in the simplified block diagram, some form of delay equalization is used in the wider-bandwidth channels.)

It is instructive to compare the band-shared simultaneous system described above

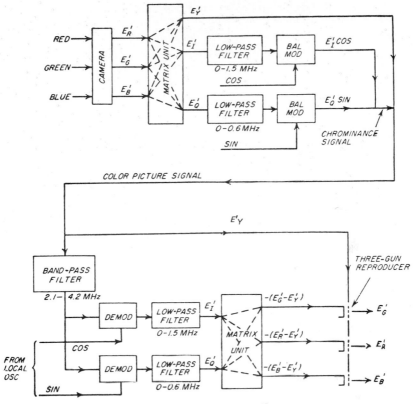

FIG. 4-20   Band-shared simultaneous system. (*From Fink.*[17])

with a "dot-sequential" system using mixed highs such as proposed by RCA and demonstrated to the FCC in 1949.[52,53] While the equipment employed was of a substantially different form, the actual transmitted dot-sequential signal, using mixed highs, was a form of band-shared simultaneous color signal, having a wideband monochrome signal and a narrowband color signal, which interleaved between the HF components of the monochrome signal. The receiving equipment then demonstrated was truly dot-sequential in nature, and the signal was sampled into short pulses which were directly displayed, producing color pictures which were full of sampling dots. On the other hand, the equipment illustrated in Fig. 4-20 is of a form using a shunt-monochrome channel arranged to preserve the substantially simultaneous nature of the system.[21,54,55]

Color Coding for Maximum Visual Signal-to-Noise Ratio (The Constant-Luminance Principle).   The method of color coding should be such that a maximum visual signal-to-noise ratio is provided by the band-shared subcarrier channel. This permits maximum protection against thermal noise and external interference signals, thus providing a maximum service range for a given transmitted subcarrier energy. (Alternatively, an improvement in compatibility might be obtained by transmitting less subcarrier energy.) Additionally, the intersignal interference due to band sharing should be visually minimized—specifically, "cross color" which results due to the presence of the HF monochrome-signal components in the chrominance channel, and which produces a crawling beat-note pattern on edges. The desired minimizing in visibility of such interference signals can be accomplished when it is recognized that the eye is considerably more sensitive to luminance fluctuations than to chromaticity fluctuations, both in time and in space.[21]

An experiment which shows that the eye has more sensitivity to spatial and temporal variations in luminance than to variations in chromaticity is illustrated in Fig. 4-21. Two well-registered rasters are produced on a color display, the electron guns biased to a moderate beam current, and a certain amount of random noise produced on one raster, say a red raster. By means of a phase inverter and attenuator, oppositely phased noise can be applied in an adjustable amount to another raster, say a green raster. It will be found that a minimum in annoyance value of the noise is obtained in the composite image for a particular setting of the attenuator. The waveforms illustrate the conditions existing when the attenuator is adjusted for minimum annoyance value of the noise. Under this condition the signal voltage to the green picture tube is about half of that applied to the red picture tube. Since the contribution of luminance of the green tube is about twice that of the red tube, this results in luminance variations which are substantially equal but of opposite polarity on the two tubes, so that the sum of the two luminances remains substantially constant. To evaluate the subjective improvement, like noise of reduced amplitude can be applied to the two rasters and the amplitude adjusted to produce the same subjective effect as that obtained with oppositely phased noise. Experiments of this type have indicated that an attenuation of 6 to 8 dB is required for the like-noise signal to produce similar subjective effects to those produced with oppositely phased noise.[24]

The above experiment leads to the constant-luminance principle: "to permit optimum color coding, signals in the band-shared color subcarrier channel should not affect the luminance of the reproduced picture." Practicing of the constant-luminance principle is fundamentally a receiver-design problem. Since in a receiver using this principle the reproduced luminance is controlled by the monochrome signal, the transmitted monochrome signal should be representative of luminance in order to obtain correct color reproduction. The dot-sequential system using mixed-highs did not operate in accordance with the constant-luminance principle.[21]

Because of the substantial difference in luminance efficiency of representative reproducing primaries (the relative luminances of the NTSC primaries when reproducing illuminant C are: 0.59, 0.30, and 0.11 for green, red, and blue, respectively), practice of the constant-luminance principle in a receiver implies an asymmetrical set of channels for applying the subcarrier information to the respective reproducing primaries. (The dot-sequential system used a symmetrical set of channels.) Realizing that the subcarrier information contains two independent quadrature components, each component individually must be applied to the display primaries in a suitable asymmetrical manner to give

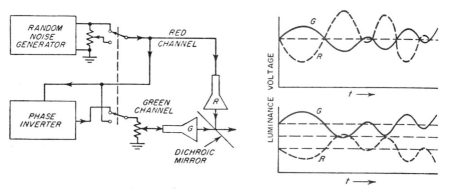

FIG. 4-21 Luminance versus chrominance noise. (*From Fink.*[17])

constant-luminance operation. For example, the E_I' component of the chrominance signal is applied to the green, red, and blue reproducing primaries with relative intensities of -0.28, $+0.96$, and -1.10, respectively. The net luminance effects produced by this E_I' signal are proportional to $-0.28 \times 0.59 = -0.17$ for the green primary, $+0.96 \times 0.30 = +0.29$ for the red primary, and $-1.10 \times 0.11 = -0.12$ for the blue primary. This gives a total luminance effect of zero. Thus, the signal gains are suitably asymmetrical so that when the luminance efficiency of each primary is considered, the overall effect is to produce no change in luminance, but to produce a change in the chromaticity. Consideration of the relative intensities with which the E_Q' signal is applied to the various primaries shows a similar constant-luminance operation but with the chromaticity changes along a different axis. Thus, the E_I' and E_Q' signals can produce chromaticity changes, but, to a first-order effect (see next paragraph), the reproduced luminance is constant being determined by the luminance signal.

The above description of constant-luminance operation has been on the basis of a linear-slope system, that is, one in which the light output is linearly related to the voltages applied to the display. Due to the nonlinearity of electron guns employed in present-day displays, the equivalence of linear-slope-system operation does not exist over the complete color gamut, when employing current receiver practices. However, an approach to linear-slope operation is obtained for pastel colors, so that the relations cited above are approximately correct when reproducing such pastel colors.

If the chrominance channel of the receiver employs no additional nonlinear operations, beyond that resulting from the nonlinear characteristic of the electron guns, then the simple linear-slope case of constant-luminance operation cited above does not apply on highly saturated colors. The normal power-law characteristic of electron guns results in a greater increase and a smaller decrease in luminance from the respective primaries, compared with the linear-slope case. This is illustrated by the characteristic curves shown in Fig. 4-22. Thus, if the chrominance channel of a receiver has gains proportioned in accordance with the linear-slope case, then (with current picture tubes) the subcarrier will always cause an increase in net luminance—this increase in luminance being negligibly small on pastel colors, but becoming rather significant on highly saturated colors.[56,57] A plot of the reproduced luminance as a function of the chrominance signal (for a receiver designed for the NTSC signal and utilizing a reproducer having a gamma of 2.2 directly associated with each primary color) is illustrated in Fig. 4-23.[50]

It will be noted from Fig. 4-23 that the contours of reproduced luminance are somewhat elliptical in shape. Actually, if the value of gamma is assumed to be exactly 2.0, it is possible to proportion the various chrominance channel gains in the receiver so that

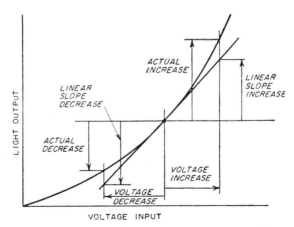

FIG. 4-22 Effect of nonlinear electron guns. (*From Fink.*[17])

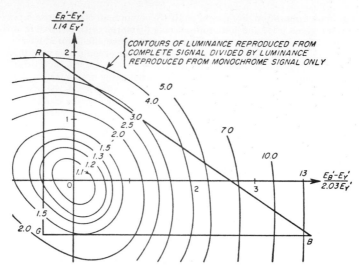

FIG. 4-23 Reproduced luminance versus chrominance signal. (*From Fink.*[17])

the luminance contours are exactly circles.[58] The appropriate chrominance signal to give the desired color reproduction in such a receiver has been called a *circular subcarrier*. With the circular subcarrier the luminance changes, in reproduced image, depend only upon the normalized chrominance-signal amplitude and are independent of the chrominance-signal phase.[24,57] However, the standardized chrominance signal was chosen, in preference to the circular subcarrier, because it was felt that the quadrature relationship between $E'_R - E'_Y$ and $E'_B - E'_Y$ resulted in simplifications which are more important than perfect circularity.

4.2.3 COLOR ENCODING SYSTEMS. The color principles outlined in Sec. 4.2.2 were first applied and widely used as a complete color television system with the development of the NTSC signal standards in the United States. The more important of these details will be described in Sec. 4.2.4. In applying these same principles throughout the world, various compromises were considered to be more important than others. Thus, during the late 1950s and early 1960s, different sets of these principles were tested, studied, and finally combined to produce various color television standards. The most notable of these additional standards are PAL and SECAM, to be described in Secs. 4.2.5 and 4.2.6.

The three combinations just mentioned (NTSC, PAL, and SECAM) all have resolution capabilities of the same general order of magnitude. However, recent exploratory work is taking place on high-definition television (HDTV) systems with a clear order-of-magnitude greater resolution than the earlier systems. This is described in Sec. 4.2.7.

4.2.4 NTSC SYSTEM. The general specifications for the NTSC-FCC color-signal specifications state that "the color picture signal shall correspond to a luminance component and a simultaneous pair of chrominance components transmitted as a-m sidebands of a pair of suppressed subcarriers in quadrature having the common frequency of +3.579545 MHz."[20] Thus, in terms of the basic principles outlined in Sec. 4.2.2, the standardized signal employs the constant-luminance principle (since the monochrome component is stated as being a luminance component), and employs a band-sharing operation with the color coding designed for minimum subcarrier visibility—using suppressed subcarrier transmission, with a single subcarrier frequency to transmit the two chrominance components. The specification further states that the horizontal scanning frequency shall be $\frac{2}{455}$ times the color subcarrier frequency, thus providing a low-visibility pattern for the subcarrier. The exact number used for the subcarrier frequency additionally produced a sound-color beat of low visibility.

The standardized color picture signal is specified as having the composition given by the following set of equations

$$E_M = E_Y' + [E_Q' \sin (\omega t + 33 \text{ deg}) + E_I' \cos (\omega t + 33 \text{ deg})] \qquad (4\text{-}22)$$

$$E_Q' = 0.41(E_B' - E_Y') + 0.48(E_R' - E_Y') \qquad (4\text{-}23)$$

$$E_I' = -0.27(E_B' - E_Y') + 0.74(E_R' - E_Y') \qquad (4\text{-}24)$$

$$E_Y' = 0.30E_R' + 0.59E_G' + 0.11E_B' \qquad (4\text{-}25)$$

The portion of the expression between the brackets represents the chrominance signal. In these expressions the symbols are defined as follows:

E_M is the total video voltage corresponding to the scanning of a particular picture element.

E_Y' is the gamma-corrected voltage of the monochrome portion of the color picture signal corresponding to a given picture element.

E_R', E_G', and E_B' are the gamma-corrected voltages corresponding to red, green, and blue signals during the scanning of the given picture element.

E_Q' and E_I' are the amplitudes of two orthogonal components of the chrominance signal corresponding, respectively, to narrowband and wideband axes. (The significance of the term *gamma-corrected voltages* is discussed in more detail below.) The quantities E_Y', E_Q', and E_I' can be considered as *transmission-primary signals;* that is, E_Y' is a wideband luminance-primary signal, E_Q' a coarse chrominance-primary signal, and E_I' a fine chrominance-primary signal.[59]

The equivalent bandwidths assigned (prior to modulation) to the chrominance components E_Q' and E_I' are given by

$Q -$ *channel bandwidth*

at 400 kHz less than 2 dB down

at 500 kHz less than 6 dB down

at 600 kHz at least 6 dB down

$I -$ *channel bandwidth*

at 1.3 MHz less than 2 dB down

at 3.6 MHz at least 20 dB down

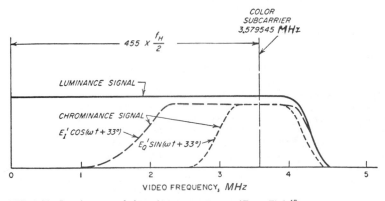

FIG. 4-24 Luminance and chrominance spectrums. (*From Fink.*[17])

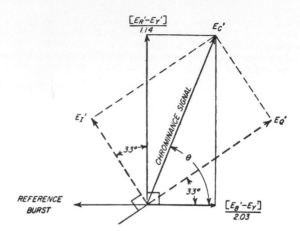

FIG. 4-25 Vector diagram of chrominance signal. (*From Fink.*[17])

The effective bandwidth for the E_Q' component is about 550 kHz while the bandwidth for the E_I' component is of the order of 1.5 MHz (as transmitted). The resulting chrominance signal channels (after modulation) compared with the luminance signal channel, and the manner in which the respective spectrums interleave, are illustrated by Fig. 4-24.

For the standardized color picture signal it is stated that E_Y', E_Q', E_I', and the components of these signals, shall match each other in time to 0.05 μs. Thus, in spite of the substantial difference in bandwidth between the various components, all the components are delay-equalized to agree substantially in time in the color picture signal. However, as noted in more detail in Sec. 4.3.3, a delay equalizer is specified within the transmitter unit which upsets this time equality between the monochrome and chrominance components to compensate for the inequality in time of transmission for these components through a receiver.[51]

For modulation frequencies below 500 kHz both the E_I' and E_Q' components are present, so the transmitted chrominance signal can vary in both amplitude and phase. Over this frequency range the chrominance signal has, therefore, two degrees of freedom and it can be represented by other components than E_I' and E_Q'. One convenient set of components for representing the chrominance signal includes the $E_R' - E_Y'$ and $E_B' - E_Y'$ components. For these lower color-difference frequencies (below 500 kHz) the signal can be conveniently represented by the alternative equation

$$E_M = E_Y' + \left\{ \frac{1}{1.14} \left[\frac{1}{1.78} (E_B' - E_Y') \sin \omega t + (E_R' - E_Y') \cos \omega t \right] \right\} \qquad (4\text{-}26)$$

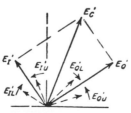

DOUBLE-SIDEBAND REGION   SINGLE-SIDEBAND REGION

FIG. 4-26 Sideband representation of chrominance signal. (*From Fink.*[17])

Vector Representation. Over the double-sideband region (about ± 500 kHz around the subcarrier frequency) the chrominance signal can be represented as pure AM quadrature components. This can be illustrated by a vector diagram in which the amplitude of the vectors varies in accordance with the stated chrominance modulation. Such a vector diagram is shown in Fig. 4-25. The chrominance signal E'_C can be considered to be made up of the components E'_I and E'_Q, or alternatively, the components $(E'_R - F'_Y)/1.14$ and $(E'_B - E'_Y)/2.03$. As shown by the vector diagram, each of these sets is a quadrature set of components, and they differ from each other by the angle 33°. [The vector diagram further illustrates the phase of the reference burst which is along the $-(E'_B - E'_Y)$ axis.] This angle of 33° was chosen so that the wideband and narrowband chrominance signals (E'_I and E'_Q) are conveniently obtainable by quadrature demodulation and at the same time the color-difference signals ($E'_R - E'_Y$) and ($E'_B - E'_Y$) are conveniently obtainable by quadrature demodulation.

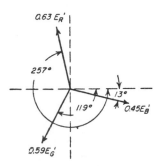

FIG. 4-27 Alternative vector representation of chrominance signal. (*From Fink.*[17])

To represent the signal more completely, a vector diagram illustrating the sideband components can be drawn as shown in Fig. 4-26. Over the double-sideband region the chrominance components E'_I and E'_Q can be considered to consist of upper- and lower-sideband components E'_{IU}, E'_{IL}, E'_{QU}, and E'_{QL}, respectively. However, over the single-sideband region only the lower-sideband component E'_{IL} exists.[60]

Over the double-sideband region the chrominance signal can also be represented by the combination of three vectors respectively proportional to the three primary-color signals E'_R, E'_G, and E'_B.[61] This alternative vector representation, shown in Fig. 4-27 is useful in calculating the chrominance signal obtained when transmitting saturated color bars.

Gamma Correction The gamma-correction specifications are intended to be somewhat flexible in nature. The signal specifications state that the gamma-corrected voltages E'_R, E'_G, and E'_B are suitable for a color picture tube having primary colors with the following chromaticities in the CIE system of specification:

	x	y
Red (R)	0.67	0.33
Green (G)	0.21	0.71
Blue (B)	0.14	0.08

and having a transfer gradient (gamma exponent) of 2.2 associated with each primary color. Further, the specifications state that the voltages E'_R, E'_G, and E'_B may be respectively of the form $E_R^{1/\gamma}$, $E_G^{1/\gamma}$, and $E_B^{1/\gamma}$, although other forms may be used with advances in the state of the art. Thus, the standards are not restricted to direct gamma correction of the primary-color signals, but imply a reasonable degree of compatibility of the transmitted signal for a color receiver designed in accordance with present gamma practice. (Note that the specifications do not state that the exact intended scene is obtained on the display defined above).

To obtain reasonably pleasing chromatic and brightness reproduction, some compensation for the nonlinear characteristic of currently employed electron guns should be used. In accordance with present practice, this nonlinear gamma-correction characteristic is included in each of the color channels handling the red, green, and blue signals at the program source. This type of gamma correction, which is the form specifically mentioned as being permissible in the standards, results in some receiver simplifications when employing currently available picture tubes.

Even if picture-tube characteristics were linear, it would probably be desirable to include a nonlinear operation at the receiver, with a corresponding inverse nonlinear operation at the program source, to improve the visual signal-to-noise ratio. The reason for such a nonlinear operation at the receiver results predominantly because of the logarithmic-type characteristic of the eye, such as is expressed by the Weber-Fechner law. For monochrome reception the nonlinear characteristic of typical picture tubes is in the right direction, and is of a suitable power law so that a substantial portion of the benefit to be gained is obtained automatically from the picture tube.[34]

On the other hand, the present gamma-correction practice used in color transmission does not appear to produce optimum over-all results in terms of signal-to-noise ratio and transmitted information. The present practice operation results in an increasing proportion of the luminance information being reproduced from chrominance-channel signals, as the color saturation is increased.[56] This produces a loss of fine-luminance detail of saturated colors (due to the restricted bandwidth of the chrominance channels), and an increase in visual noise on saturated colors (due to the luminance reproduction from the relatively noisy chrominance channel).[50,62,63] Systems have been devised (but not thoroughly field-tested) to overcome these effects, by using a nonlinear operation at the color receiver and by transmitting slightly different signals, so that all luminance information can be reproduced from the luminance channel.[58,64] The possibility of obtaining substantial benefits in the future, by changing the exact form of gamma correction led the NTSC to recommend a flexible set of standards regarding gamma correction. Although to date no significant use has been made of this flexibility in production designs, the continuing trend toward higher-quality reproduction, and the advent of low-cost integrated circuits, promise to result in use of the technique.

Maps of Chrominance Signal for Current Gamma-Correction Practice. For a specified form of gamma correction, maps can be plotted which show the interrelation of the chrominance-signal phase and amplitude with chromaticity, expressed according to the CIE form of color coordinates. Here it is the normalized chrominance signal (chrominance signal/luminance signal) which is of interest since chromaticity is directly related to the ratio of the chrominance to luminance signal, in the NTSC system. Consideration will be limited to current gamma-correction practice in which the color signals E'_R, E'_G, and E'_B are respectively of the form $E_R^{1/\gamma}$, $E_G^{1/\gamma}$, and $E_B^{1/\gamma}$ and where the voltages E_R, E_G, and E_B are assumed to be linearly related to the respective amounts of the standardized reproducing primaries which are present in the intended scene. Maps resulting from other forms of gamma correction are available in the literature.[65]

Figure 4-28 shows a map of the normalized chrominance-signal phase and amplitude plotted on the CIE diagram for the intended chromaticity, assuming current gamma-correction practice for a display having a gamma exponent of 2.2. Figure 4-29 shows the reverse type of map in which the CIE coordinates are plotted on the normalized chrominance-signal plane. Figure 4-23 shows contours of the luminance factor (representing the total luminance divided by the luminance transmitted in the luminance channel) plotted on the normalized chrominance-signal plane.[65]

4.2.5 PAL SYSTEM. Phase alternation line-rate, known by the acronym PAL, uses a principle which was studied[113] and field tested by the NTSC before adoption of the current 525-line NTSC color-television standards. In 1966, it became the basis for the development of 625-line color-system standards for most of Europe,† making use of new technology (inexpensive delay lines) to provide a more *rugged* system.[114,116]

PAL uses the principle of color phase alternation (CPA)[113] in which the color-phase sequence of the chrominance signal is periodically reversed so that phase errors represent opposite hue errors during the successive periods. Different rates for sequence reversal can be used with different convenient nomenclature—PAL standing for phase reversal after each scanning line and PAF for reversal after each scanning field.[115]

---

†PAL-M, a 525-line PAL system, is used in Brazil.

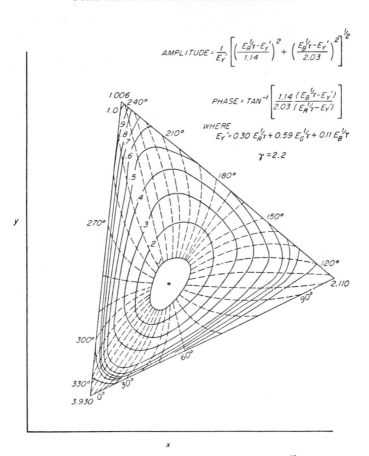

$$AMPLITUDE = \frac{1}{E_Y'} \left[\left(\frac{E_R^{\frac{1}{\gamma}} - E_Y'}{1.14} \right)^2 + \left(\frac{E_B^{\frac{1}{\gamma}} - E_Y'}{2.03} \right)^2 \right]^{\frac{1}{2}}$$

$$PHASE = TAN^{-1} \left[\frac{1.14 \, (E_B^{\frac{1}{\gamma}} - E_Y')}{2.03 \, (E_R^{\frac{1}{\gamma}} - E_Y')} \right]$$

WHERE
$$E_Y' = 0.30 \, E_R^{\frac{1}{\gamma}} + 0.59 \, E_G^{\frac{1}{\gamma}} + 0.11 \, E_B^{\frac{1}{\gamma}}$$

$$\gamma = 2.2$$

FIG. 4-28 Map of normalized chrominance signal. (*From Fink.*[17])

How CPA Works. The operation of CPA can be illustrated by reference to the signal vectors for PAL shown in Fig. 4-30. On alternate lines the chrominance signal has normal quadrature modulation, similar to NTSC; then in intervening periods the phase of $R - Y$ is reversed so that the phase-versus-hue sequence is also reversed.

Consider a chrominance signal level change, for example, going from a large amplitude to a reduced amplitude. If the chrominance passband is not flat or if significant phase distortion exists, the resulting chrominance transient, during the *even* lines, may be as illustrated by the dotted curve on the left-hand vector diagram. This shows a clockwise perturbation which corresponds to a bluish tint. However, during the *odd* lines, the resulting phase perturbation will still be in the clockwise direction, but this now corresponds to a reddish tint. If these oppositely colored disturbances are nearby in space or time, they may be averaged either visually or electrically.

Thus, it can be seen from this example that CPA works to an advantage by averaging not only transient phase errors on edges as above, but also large-area phase errors, such as might result from incorrect reference subcarrier phase, or from differential-phase distortion versus luminance level.

Visual Averaging. With CPA, visual averaging of color errors can be used. However, when the errors are very large, spurious effects may also be obtained with such visual

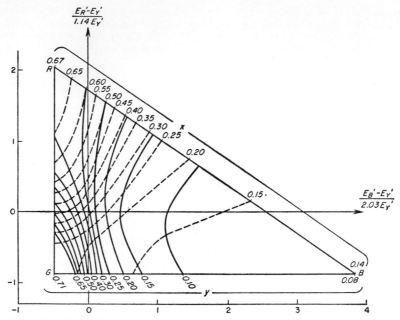

FIG. 4-29 CIE coordinates on normalized chrominance-signal plane (x = dotted lines, y = solid lines). (*From Fink.*[17])

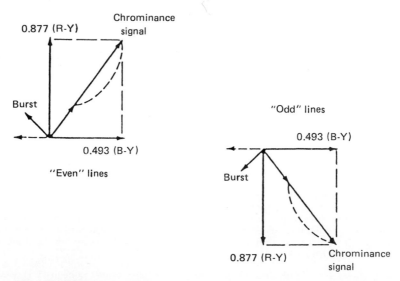

FIG. 4-30 PAL vectors, with transient illustrations.

averaging. With PAL, a line-crawling *venetian blind* (sometimes referred to as Hanover bars) pattern results with visual averaging (which may have a luminance component if constant luminance versus phase is not perfect), and this can be quite visible when phase errors are large. The alternate-line pattern crawls because of the odd-line interlace scanning, requiring four fields (two frames) to repeat.

Electrical or Delay-Line Averaging. With PAL, since the oppositely phased sequences are one scanning line apart, a delay line of delay equal to one scanning line can be used to provide electrical averaging of phase errors. This substantially eliminates the spurious line patterns cited above at the price of reduced vertical chrominance resolution.

Subcarrier Frequency. The chrominance energy distribution for the PAL signal is illustrated in Figure 4-31. For the $B - Y$ component, the energy basically bunches at the color subcarrier frequency and at frequencies removed therefrom by harmonics of line-scanning frequency. However, for the alternating $R - Y$ component, the energy bunches at sideband frequencies which are removed from the chrominance frequency by odd harmonics of one-half line-scanning frequency. Therefore, if the color subcarrier frequency is chosen with one-half line offset as in NTSC (i.e., an odd harmonic of one-half line-scanning frequency), the $R - Y$ energy will be bunched at harmonics of line-scanning frequency. In other words, the $B - Y$ signal would be of low visibility, but the $R - Y$ signal would produce a high-visibility subcarrier pattern. For this reason, PAL uses one-quarter line offset instead, to produce somewhat reduced visibility of both the stationary and alternating components. To further reduce the subcarrier visibility (and thereby improve compatibility), additionally a frame-frequency offset is used beyond the one-quarter line offset.

CPA Sync. The PAL system uses a color burst whose phase is oscillated by $\pm 45°$ between successive scanning lines in order to provide CPA synchronization information. In reference to Fig. 4-30, the normal NTSC burst phase is illustrated here by the dashed vector, and the burst phase for PAL is 45° toward the $R - Y$ vector and thereby permits identification of the even from the odd scanning lines.

A normal, long-time constant integrating color sync circuit will come to stable phase condition at the average phase of the alternating burst (shown dotted). Then a signal can be taken from the phase detector of the color sync APC loop providing the alternating line pulses for CPA sync.

Simple PAL Receivers. This name is used to refer to PAL receivers employing visual averaging. Such receivers can be substantially like NTSC receivers except for the need to "undo" the CPA and to properly synchronize this function. This receiver function can

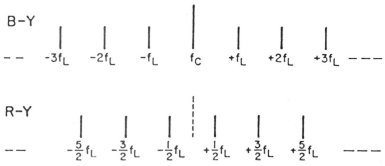

FIG. 4-31 Chrominance sidebands in PAL.

be performed in a variety of ways.[113] For example, the subcarrier phase to the $R - Y$ demodulator can be flipped by 180° on successive scanning lines, or the chrominance signal to the $R - Y$ demodulator can be flipped, or the $R - Y$ video signal can be phase-inverted in alternate lines.

Delay-Line PAL Receivers. This terminology is used to refer to PAL receivers employing electrical averaging. Figure 4-32 illustrates how this electrical averaging is performed by the equivalent of a comb filter.[114] Here the summing and subtracting functions contain either gain or attenuation so that the signal from the output of the delay line is combined with a signal of comparable amplitude from the input of the delay line. If the delay line has exactly one scanning line of delay, the output from the summing function will correspond to frequencies which are harmonics of line-scanning frequency, while the output from the difference function will correspond to frequencies which are odd harmonics of one-half scanning-line frequency. Note that the peak spacings from either output are equal to line-scanning frequency, but each output has nulls where the other output has peaks.

Considering the chrominance sideband distribution in PAL, it can be seen that one output can be arranged to correspond to the $B - Y$ signal and the other output to the $R - Y$ signal. Of course, each output signal will actually be the average signal for two scanning lines. When quarter-line offset is used for the subcarrier frequency, a fine adjustment of the delay line will permit moving the peaks and valleys to the desired frequencies. Of course, even after using this electrical averaging by a comb filter in order to select one chrominance component from the other, it is still necessary to unflip the alternating component.

Books and Technical Papers on PAL. Several references exist on the various system studies made in Europe and the system details decided upon. In March 1966, the issue of worldwide color television standards was explored at the IEEE International Convention and reported in a special volume of *Transactions, BTR*.[118] A paper by Dr. W. Brüch on PAL is included in this volume.[114] Also of particular interest is a tutorial article by Dr. R. Theile reviewing the various European proposals for subcarrier modulation.[116]

Excellent detailed discussions are given on NTSC, PAL, and SECAM in a two-volume series by Carnt and Townsend,[117] the second volume of this series being devoted to PAL and SECAM.

A concise article on similarities and differences between various worldwide color television standards was published in February 1980 in the *SMPTE Journal*.[120] This material neatly summarizes the many recommendations made regarding color television by the International Radio Consultative Committee (CCIR) over the past 15 to 20 years.

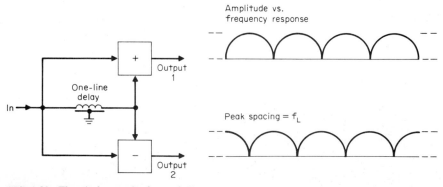

FIG. 4-32 Electrical averaging by comb filter.

Comparison of Delay-Line PAL and NTSC.   The following summary covers some of the differences between an NTSC system and a PAL system using delay-line (electrical) averaging in the receiver.

PAL is both more *rugged* and more complicated than NTSC; more rugged because the system includes first-order correction for analog chrominance-channel passband ambiguities and phasing errors; more complicated because it includes the added functions of CPA and its synchronization.

In order to provide compatibility for monochrome reception by reducing the color subcarrier visibility to an acceptable level, PAL requires less of an overlap between chrominance and luminance spectra than NTSC. Thus, a higher subcarrier frequency and channel bandwidth are used.

PAL has sacrificed some chrominance resolution through the delay-line averaging, which makes the chrominance spot approximately three times greater in the vertical direction because averaging of two successive lines in time, in an interlaced raster, averages lines which spatially are separated by two lines.

Nevertheless, PAL is a skillfully executed compromise between total color image-resolution capability and color fidelity which, at the expense of loss in the time-bandwidth product, can provide better performance than NTSC in the presence of certain types of transmission channel errors.

On the other hand, the future error-free digital channels, although requiring an even greater bandwidth product, will negate the need for the exigencies of the CPA. However, for analog systems the principles of Sec. 4.2.2 remain axiomatic.

4.2.6 SECAM SYSTEM.   SECAM, another compromise combination of the principles of Sec. 4.2.2, uses the equivalent of band-limiting of the color-difference signals in the vertical direction as well as the horizontal direction. This is done using line sequential transmission of the two-color difference signals $R - Y$ and $B - Y$ plus memory (delay line) at the receiver (one horizontal line period long) to have both signals present simultaneously for reconstruction of R, G, and B display signals. This overcomes the *phasing problems of NTSC* where the two-color difference components are simultaneously present at quadrature phases of the color subcarrier.

Since the late 1950s, a number of versions of SECAM have been proposed, and the form generally being referred to today uses frequency modulation of line-sequential color subcarriers. Since the unmodulated subcarriers are present in the absence of color (white or neutral grays), and are frequency-modulated, when color information is being transmitted (and, therefore, vary in degree of low visibility), the subcarrier signal visibility is significantly greater than for PAL or NTSC. Thus, the monochrome compatibility of SECAM basically is inferior to PAL or NTSC.

This shortcoming is offset partially by the higher color-subcarrier frequency, compared with NTSC, and by passing the frequency-modulated subcarrier through a filter with an inverted *bell-shaped* characteristic. This processing reduces, but does not eliminate, the transmitted subcarrier upon neutral white and grays.

Some additional features included in SECAM are preemphasis of the color-difference signals, a different *resting* frequency for the undeviated $R - Y$ and $B - Y$ subcarrier frequencies, and special identification signals for line-switching sequence.

SECAM versus PAL.   SECAM was developed earlier than PAL during a period of great concern for the effect of transmission vagaries on the color subcarrier signal and resultant color fidelity. Thus, among the factors contributing to the promulgation of the SECAM system, the major considerations were to permit the use of poor-quality terrestrial network interconnection circuits and early design monochrome video tape recorders. The subsequent technological advances in these fields have negated the need for safeguards (with their accompanying limitations in spectrum utilization) provided by SECAM and indicate the inadvisability of adopting standards tailored to meet the technical limitations of the time.

4.2.7 HDTV SYSTEMS.   Since 1968, the Japan Broadcasting Corporation (NHK) has been carrying out research into a high-definition wide-screen television system.[121–127] Starting in 1978, transmission tests of this HDTV system have been carried out via broadcast satellite. This has provided a new opportunity to look again at color television without the previous critical constraints of compatibility. Thus, scanning and encoding optimization is open for consideration.

Aspect Ratio and Viewing Distance.   The dimension of increased resolution, together with the visual impact of wide-screen movies, makes it apparent that the quantities of aspect ratio and normal viewing distance need to be reconsidered. The NHK studies conclude that for television images of 1000 to 1500 lines resolution, the optimum aspect ratio is 5/3 (instead of 4/3) with viewing distances of up to about three times picture height.

Number of Scanning Lines and Luminance Bandwidth.   Many factors bear upon the optimum choice here, such as cost, complexity, signal-to-noise ratio, bandwidth, and equipment state-of-the-art. The current studies are using 1125 lines, with 20-MHz luminance bandwidth.

Line-Interlace Ratio and Field Repetition Frequency.   Further tests have shown 2/1 interlace to be optimum and also show 60-Hz field rate desirable to permit bright flicker-free images.

Chrominance Axes and Bandwidths.   Further NHK tests confirm that the three-bandwidth system (similar to NTSC) is optimum except that a somewhat different compromise has been selected. It has been stated "recent studies show $R - Y$ and $B - Y$ to be quite appropriate for the wideband and narrow bandwidth axes." For this selection of axes, the bandwidths being used experimentally are 7 and 5.5 MHz, respectively.

Color Encoding Using a Subcarrier.   This arrangement uses an NTSC-like, or PAL-like, subcarrier at 24 MHz with half-line offset (HLO) like NTSC, but color phase alternation at line rate like PAL (the combination being referred to as HLO-PAL). The subcarrier is high enough in frequency so that no overlap exists between the Y signal and the higher-visibility narrow-band subcarrier energy. The wider bandwidth chrominance energy on the lower side does overlap the Y signal, but it is interleaved with a low-visibility pattern due to half-line offset.

Color Encoding Using $Y - C$ Separate FM Transmission.   This arrangement has seen use in broadcast satellite transmission. To overcome the triangular noise problem at a subcarrier frequency in an FM transmission net, the luminance and chrominance are transmitted on separate FM carriers (no overlap). In this case, with separate Y and C signals, and no compatibility question, it is also feasible to use simple line-sequential transmission of the chrominance signals.

4.3 COMPOSITE COLOR SIGNAL TRANSMISSION

This section is directed toward the problems and requirements placed upon transmission systems which need to handle composite color signals (luminance and chrominance) with a suitable minimum of distortion and crosstalk between the components of the composite signal.

4.3.1 CHROMINANCE TRANSFER CHARACTERISTICS.   When considering the chrominance transfer characteristic (the transfer characteristic between program source chrominance and receiver-image chrominance), a decision of where to measure the program source chrominance must be made. If exact colorimetric reproduction were a desirable characteristic, then the scene of interest would be the original scene in front of the cameras. There are many reasons why exact colorimetric reproduction does not appear to be desirable. These include such items as limited contrast range of the receiver display, difference in color adaptation of the home viewer as compared with a viewer at the original scene, and freedom of artistic license for the program director. Additionally, experience indicates that an overall system gamma somewhat greater than unity is desired, which tends to increase the contrast and color saturation of the scene. Thus, the scene of interest when considering chrominance transfer characteristics is not the scene before the camera (or the film being used in the case of film scanning) but instead it is the scene intended by the program director, as viewed on the studio monitor.

Considering that the chrominance transfer characteristic of interest is between the program source monitor-intended scene and the receiver-reproduced image, this should presumably be a direct one-to-one characteristic. The maps given in Sec. 4.2.3 are useful in estimating the chrominance error resulting from imperfect translation of the chrominance information. The map shown in Fig. 4-28 not only represents the amplitude and phase of the normalized chrominance signal versus chromaticity of the intended scene, but also ideally represents the chromaticity of the reproduced image versus amplitude and phase of the received normalized chrominance signal. Thus, the chromatic error which results due to an error in either the phase or amplitude of the normalized chrominance signal can be found from this map. The effect of a 50 percent increase in chrominance-signal amplitude or of a $\pm 20°$ error in phase† is illustrated in Fig. 4-33 for the case of several intended scene chromaticities.

From the map of Fig. 4-28, it is seen that the phase of the chrominance signal relates predominantly to the hue of the scene and the amplitude of the normalized chrominance signal relates to the saturation of the scene. However, as high-purity colors are being transmitted, the chrominance amplitude produces less effect upon the saturation, and, as illustrated by Fig. 4-23, it produces a greater effect upon the luminance. Also, the phase of the chrominance signal has a reduced effect upon hue near high-purity primary colors, and an increased effect upon hue near high-purity complementary colors—these effects being indicated respectively by the convergence and divergence of the radii near the outer regions of the diagram. Thus, the chromaticity varies in a nonlinear manner with changes in the chrominance-signal amplitude and phase, this being due to the present gamma-correction practice.[50] The relative importance of chromaticity changes (for large areas of color) in various parts of the diagram can be roughly estimated by a comparison with the map showing the relation between Munsell and CIE designations, as shown in Fig. 4-34. The Munsell system has been set up with subjective uniformity as an aim.[66] The maps do not agree too well, particularly with regard to the nonlinear chromaticity variations cited above.

While the maps referred to above show the relative importance of amplitude and phase errors in reproducing various colors (in large areas), they do not give a set of working tolerances for amplitude and phase errors. Figure 4-35 shows curves of the just-noticeable errors in phase and amplitude of the chrominance signal, as measured by a group of Hazeltine engineers viewing selected critical scenes. The figure also shows the curves for maximum tolerable errors in phase and amplitude of the chrominance signal, obtained under the same test conditions. These curves indicate that approximately 50 percent of the observations gave a just-noticeable phase error of $\pm 5°$† and a corresponding chrominance amplitude error of ± 1.5 dB. Approximately 90 percent of the observations indicated a maximum tolerable phase error of $\pm 15°$† and a corresponding chrominance amplitude error of ± 2.5 dB.

---

†Specifically for NTSC.

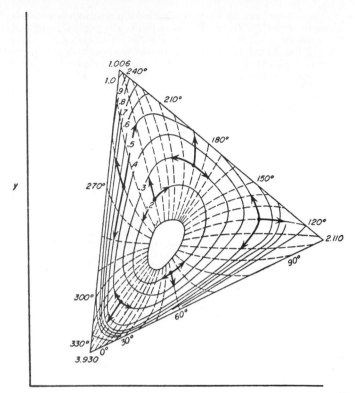

FIG. 4-33   Chromatic errors due to chrominance-signal errors. (*From Fink.*[17])

These data are indicative of the overall tolerances which can be placed upon the transfer of chrominance phase information (which is particularly related to the accuracy of color synchronization) and upon the transfer of chrominance amplitude information (which is particularly related to the uniformity of the frequency response of the system). However, it should be noted that the tolerances cited may be rather conservative since the data were taken with critical viewers and selected critical scenes. Other workers in the field have obtained data which show larger tolerances.[67]

To transfer the chrominance-signal phase information between the program source and the receiver, a color-phase reference signal is transmitted. The color-phase reference signal consists of a short burst of approximately nine cycles of the subcarrier at a reference phase [the phase of $-(E_B' - E_Y')$] during the blanking period, immediately following each horizontal synchronizing pulse. This color synchronizing signal must be translated with good phase accuracy, since any phase error in this signal will directly change the hue of all colors. The manner in which certain forms of system nonlinearities can affect the phase of the chrominance signal and the color synchronizing signal is discussed in Sec. 4.3.4.

When considering tolerances on the transmitted signal, in order to ensure adequate chrominance transfer characteristics, the items of primary importance are not the exact angles and exact color composition of the E_I' and E_Q' signal, but rather the exact amplitude and phase of the complete chrominance signal (with respect to the burst) when transmitting certain reference colors. (Of course, in most color coders, this is indirectly

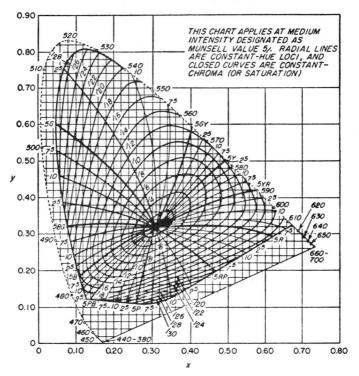

THIS CHART APPLIES AT MEDIUM
INTENSITY DESIGNATED AS
MUNSELL VALUE 5/. RADIAL LINES
ARE CONSTANT-HUE LOCI, AND
CLOSED CURVES ARE CONSTANT-
CHROMA (OR SATURATION)

FIG. 4-34 Relation between Munsell and CIE designations. (*From Fink.*[17])

determined by the phase and color composition of the E_I' and E_Q' signals.) As an attempt to tie down the chrominance transfer characteristics, the specifications do not give tolerances on the E_I' and E_Q' signals but instead state that the angles of the subcarrier (measured with respect to the burst phase) when reproducing saturated primaries and their complements (at 75 percent of full amplitude) shall be within $\pm10°$, and their amplitudes shall be within ±20 percent of the values specified by the equations. However, there is some indication of doubt that this system tolerance is tight enough since it is further stated that closer tolerances may prove to be practical and desirable with advances in the state of the art.[20]

4.3.2 RECOVERY OF CHROMINANCE INFORMATION. To recover the two independent types of chrominance information from the chrominance signal, synchronous detection is employed at the receiver. A balanced-diode synchronous detector using this technique is shown in Fig. 4-36. The chrominance subcarrier input signal E_1, and the locally generated reference signal, are applied to an envelope detector, the output of which is proportional to the modulation of the color subcarrier along the reference axis. By using diodes oppositely poled, and with the reference signal of the same phase applied to both in combination with opposite-phase color signals, the modulation signals from the subcarrier are recovered without an offset voltage from the rectified reference signal. The output signal contains components at twice the frequency of the locally generated reference signal. These components are eliminated in the low-pass filter shown in the output circuit.

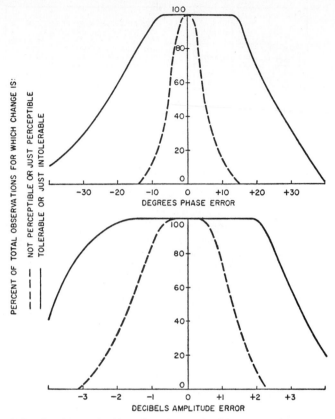

FIG. 4-35 Just-noticeable and maximum tolerable errors of chrominance signals. (*From Fink.*[17])

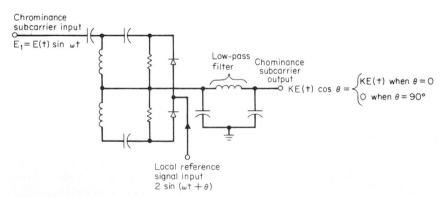

FIG. 4-36 Principle of synchronous detection.

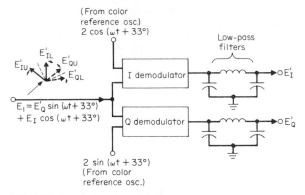

FIG. 4-37   Quadrature demodulation.

Figure 4-37 shows how a set of synchronous detectors can be used for independent detection of quadrature components of the chrominance signal. Assuming that the input chrominance signal is modulated by E_Q' at an angle of 33° from a sine component, then an output of E_Q' will be obtained in the lower demodulator. In the upper demodulator none of this component will appear in its output, since its reference subcarrier signal is in quadrature with this component. Correspondingly, if the input signal is assumed to be modulated by E_I' at an angle of 33° from a cosine term, then the output of the upper demodulator will be proportional to E_I', and again none of this component will occur in the lower demodulator output.

This quadrature demodulation operation can also be thought of in terms of the output due to the upper and lower sidebands of the input signal. Using this point of view, it will be seen that in the lower demodulator the outputs due to the upper- and lower-sideband components of the E_Q' signal add, while the outputs due to the upper and lower E_I' components cancel each other, thus giving no output due to the E_I' component. The correspondingly converse situation exists in the upper demodulator. Thus over the region of uniform double-sideband transmission the two components E_Q' and E_I' can be recovered independent of each other.

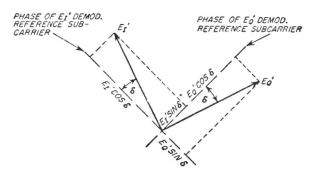

$$E_I' \ DEMOD. \ OUTPUT = E_I' \ COS \ \delta - E_Q' \ SIN \ \delta$$
$$E_Q' \ DEMOD. \ OUTPUT = E_Q' \ COS \ \delta + E_I' \ SIN \ \delta$$

FIG. 4-38   Effect of reference subcarrier phase error. (*From Fink.*[17])

If the phase of the reference subcarrier applied to a synchronous detector is in error, then both the desired and some of the undesired chrominance signals will be obtained in the output of the particular detector. As shown by the vector diagram and equations of Fig. 4-38, the desired term reduces slowly (according to cos δ) but the crosstalk from the undesired term enters in rather rapidly (according to sin δ) with an error in phase of the reference subcarrier signal.[21] When the phases of the reference subcarrier signals are in error in the same direction, as applied to the set of quadrature demodulators (such as due to an error in phase in the color synchronizing channel), the effect of the undesired crosstalk terms is to modify the hue of the reproduced image, thus producing the same effect discussed in Sec. 4.3.1.

Besides using product demodulators, synchronous detection action can also be obtained by using an exalted carrier reception employing a conventional envelope detector.[69,70] As illustrated in Fig. 4-39, if a large amplitude of reference subcarrier is added to the received chrominance signal, then the amplitude variations of the composite signal correspond mainly to the chrominance component which is in phase with the injected reference subcarrier. The major effect of the quadrature chrominance component is to change the phase of the composite signal. Since the output of an envelope detector does not respond to phase but only to amplitude of the applied signal, the output will correspond substantially only to the desired chrominance component. It will be noted, however, that the quadrature term does produce a small effect upon the amplitude when the ratio of the injected subcarrier to the chrominance signal is not very large. When using a linear envelope detector, the peak-to-peak value of resulting undesired crosstalk can be kept to less than 5 percent of the desired output produced along the quadrature axis, by maintaining the ratio of the injected subcarrier amplitude to the chrominance signal amplitude at a value greater than 5 to 1 (5 percent crosstalk produces the equivalent of a maximum phase error of 3°). A similar type of crosstalk can exist in a product demodulator due to a nonlinear characteristic of the first control, which results in some direct square-law rectification of the applied input chrominance signal.[68]

As illustrated by the vector diagram on Fig. 4-37, output signals corresponding to E_I' and E_Q' without crosstalk can be obtained over a frequency band corresponding to uniform double-sideband transmission, that is, for modulation frequencies up to about 500 kHz. However, over the single-sideband region the E_I' signal occurs in the output of both demodulators at one-half amplitude (since only one sideband is present), and with correct phase in the E_I' demodulator output, but shifted in phase by 90° in the E_Q' demodulator output.

If the output of the E_I' demodulator is boosted by 6 dB above 500 kHz and the output of the E_Q' demodulator is limited in bandwidth to about 500 kHz, then a set of E_I' and E_Q' signals is obtained free of crosstalk, as illustrated by the curves in Fig. 4-40. Thus, the E_Q' channel of the receiver should be limited in bandwidth to about 500 kHz to elim-

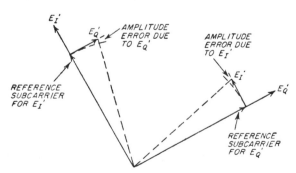

FIG. 4-39   Vector relations in exalted carrier detection. (*From Fink.*[17])

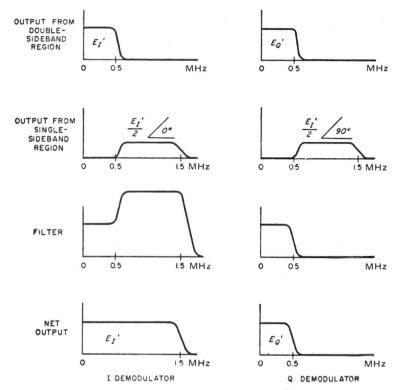

FIG. 4-40 Receiver chrominance passbands for crosstalk elimination. (*From Fink.*[17])

inate crosstalk of the higher-modulation frequency E_I' components into the E_Q' channel; and, if the higher-frequency components of the E_I' signal are desired at correct amplitude, a 6-dB boost should be included for these frequencies (which are transmitted by a single sideband). This bandwidth shaping to obtain signals free of crosstalk can be accomplished by video filters after the demodulators, as indicated by Fig. 4-40, or by suitable bandpass filters employed ahead of the demodulators. In either case, the different shapes of passbands used in the two channels generally result in different signal delays, thus calling for some delay equalization of the two channels, such as by using a delay line in the wider-bandwidth channel.

Receiver decoders have been designed to use the same bandwidth for all components of the chrominance signal information, and this has been called *equiband operation*. Equiband operation completely free of crosstalk voltages is obtainable by limiting the bandwidth to about 500 kHz. However, it has been experimentally found that some benefit in sharpness can be obtained, before the crosstalk becomes objectionable, by using bandwidths of the order of 700 to 800 kHz. Equiband operation is attractive as a means for simplifying receiver decoder design, since it permits a direct recovery of desired color-difference signals, such as the $E_B' - E_Y'$ and $E_R' - E_Y'$ signals, without the necessity of the matrix unit employed when demodulating for E_I' and E_Q' signals. Additionally, the problem of delay equalization of the various chrominance channels is eliminated since all chrominance channels are of the same bandwidth.

In certain forms of one-gun displays the chrominance information is recovered by a synchronous detector action occurring within the picture tube. In these displays the cathode-ray beam is permitted to impinge upon the various phosphors for certain periods of time, and this action is equivalent to a gating or sampling circuit which connects the input signal (corresponding to the cathode-ray beam current) to the particular phosphor during repetitive intervals of time. This periodic sampling action contains an effect which is equivalent to synchronous detection action at the sampling frequency. In these cases where the chrominance information is effectively recovered by a synchronous detection action at the phosphor screen it is convenient to practice the equivalent of equiband operation.

The equivalent synchronous detection action of electronic gating or sampling circuits can also be used to recover the chrominance signal components. By suitable design of the circuits, the output signal levels can be made to be substantially independent of the characteristics of the electronic elements (vacuum tubes, germanium diodes, or transistors). When using gate circuits, this desirable stability generally requires operation at a high signal level and results in no gain through the circuit.

4.3.3 REQUIREMENTS ON FREQUENCY RESPONSE OF TRANSMISSION SYSTEM (NTSC).

The ratio of the chrominance signal to the luminance signal determines the color saturation. To maintain the desired degree of saturation, therefore, the frequency response of the transmission system should be substantially flat. Specifically, the saturation of large-area colors will be determined by the response at the subcarrier frequency compared with the response at low frequencies. Since the overall system tolerance on the chrominance level is only on the order of a few decibels (see Fig. 4-35), each part of the system should be designed to have substantially identical transmission at low frequencies and at subcarrier frequency—unless specific compensation is intended.

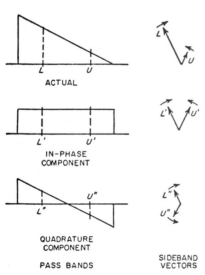

ACTUAL

IN-PHASE COMPONENT

QUADRATURE COMPONENT

PASS BANDS

SIDEBAND VECTORS

FIG. 4-41   Effect of sloping frequency response. (*From Fink.*[17])

As indicated by Figs. 4-37 and 4-40, uniform double-sideband transmission for ± 500 kHz about the subcarrier frequency is required to obtain E'_I and E'_Q signals free of crosstalk. If the frequency response of the overall transmission system is nonuniform around the subcarrier frequency, then crosstalk is present and produces color-boundary effects.[71] As illustrated by Fig. 4-41 a uniformly sloping response characteristic can be considered as the equivalent of the desired flat passband characteristic, plus a passband in which corresponding sideband phases are reversed, thus producing crosstalk into the quadrature channel.[60] For the case of a uniformly sloping amplitude characteristic with no phase distortion, the quadrature crosstalk component is proportional to the first derivative of the modulating signal.[21] Correspondingly, if the amplitude characteristic is flat but a small curvature (equivalent to a square-law term) exists in the phase characteristic, the quadrature crosstalk component is proportional to the second derivative of the modulating signal.

Thus, uniform amplitude and phase characteristics over the double-sideband region about the subcarrier frequency are required† in order to prevent quadrature crosstalk (and corresponding color errors) at color boundaries. Uniform double-sideband trans-

---

†Unless the color system uses CPA (as PAL does) to provide a first-order correction for such effects.

mission of the chrominance signal requires particular attention to the transmission of the upper sidebands, because of the relatively small frequency interval (about 920 kHz) between the radiated chrominance subcarrier and the sound carrier frequency. To assure adequately flat transmission characteristics for the transmitter, the specifications state that the amplitude of the radiated signal (for sine-wave modulation) shall not vary by more than ± 2 dB between the modulating frequencies of 2.1 and 4.18 MHz.

When the desired flat passband characteristic is obtained in a receiver, a relatively sharp-cutoff characteristic on the HF side results because of the relative nearness of the sound signal, and the large amount of attenuation required at this sound carrier frequency. When such a sharp-cutoff characteristic is obtained with simple minimum phase-shift networks, a substantial phase distortion is produced near the upper end of the passband, that is, in the vicinity of the chrominance signal. The resulting phase characteristic delays the chrominance signal compared with the luminance signal, and in addition, may produce quadrature crosstalk between the chrominance components. To compensate for these effects, delay precompensation is employed at the transmitter.[20,51,72] The standardized transmitter delay characteristic and its tolerances are indicated in Fig. 4-42. This is the characteristic which is believed to compensate for the nonlinear phase-versus-frequency characteristic of the IF channel of an average color receiver.

In the early NTSC color field tests, color phase alteration (CPA) was employed to permit visual cancellation of various forms of crosstalk between the chrominance components. By using the opposite color-phase sequence during successive fields, the crosstalk between chrominance components, due to a nonuniform frequency response or due to purposeful vestigial sideband transmission, produced opposite chromatic effects during successive fields.[21,22] Thus, visual cancellation of these crosstalk products resulted, due to the averaging by the eye of the chromaticity of adjacent lines in space on the display. The use of field-rate CPA was discontinued by NTSC,[50] for the reasons indicated in Sec. 4.2.2. However, more than a decade later, the technology of low-cost delay lines had developed, permitting use of electrical averaging with line-rate CPA. This technique became the basis for the PAL color television system, in which quadrature crosstalk is canceled.

4.3.4 REQUIREMENTS ON LINEARITY OF TRANSMISSION SYSTEM.

In the portions of the transmission system which handle the chrominance and luminance information as a composite color signal, the input-output characteristic must be linear to pre-

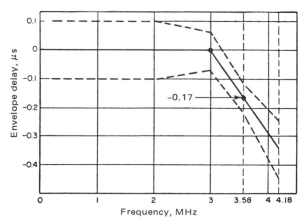

FIG. 4-42 Standardized transmitter delay characteristics. (*From Fink.*[17])

vent crosstalk between the two signals. A curvature in the input-output characteristic permits the luminance signal to vary the effective gain of the channel for the chrominance signal. Also a curved characteristic may permit rectification of the chrominance signal, thus producing an LF term which modifies the luminance signal. Since in most scenes the chrominance signal is smaller than the luminance signal, the former type of crosstalk is more usual than the latter.

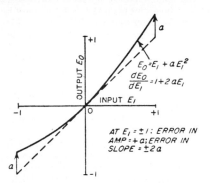

FIG. 4-43 Nonlinear input-output characteristics. (*From Fink.*[17])

The gain for a small chrominance signal superimposed upon a luminance signal is represented by a slope of the input-output characteristic measured at the operating point established by the luminance signal. If the input-output characteristic has a gradual change in slope, such as one represented by a small square-law term, then the change in chrominance gain can be simply evaluated and compared with the percent distortion produced by the same amplifier when handling sine waves. Referring to Fig. 4-43, if the input-output characteristic deviates from the linear case by a square-law term having a value of 5 percent at maximum amplitude, then the slopes at the corresponding maximum amplitude points will deviate by 10 percent from the original slope. This produces a total 20 percent change in chrominance signal gain, as the luminance signal deviates to the two extremes. But, if the same amplifier were used to handle sine waves, it would produce only 2½ percent of second harmonic distortion. Thus, the degree of linearity required to handle a composite color signal correctly is substantially greater than that required for a monochrome signal, and comparable with that required for a high-quality audio system.

The effect cited above results in the chrominance signal amplitude being modulated by the luminance signal. This same differential-gain change occurring in a feedback amplifier can result in the chrominance signal phase being varied by the luminance signal.[73] A typical example of a circuit in which this effect can occur is shown in Fig. 4-44, which is an amplifier employing an emitter resistor of substantial magnitude. Due to stray emitter-to-ground capacitance, a partial bypass results and a substantial phase shift at subcarrier frequency can occur, which has a value that is determined by the mag-

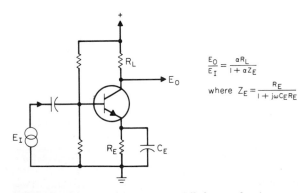

FIG. 4-44 Phase variation due to partially bypassed emitter.

nitude of the α of the transistor. When the luminance signal is sufficiently large to vary the gain of the amplifier, it also modulates the phase of the chrominance signal.

Another source of phase modulation of the chrominance signal by the luminance signal can be produced in FM relay links. In this case, if the phase characteristic versus frequency of the IF amplifiers is not linear, then a variation in envelope delay at various frequencies in the IF channel is produced. Thus, as the luminance signal deviates the carrier to various frequencies in the IF channel, the delay and therefore the phase of the chrominance signal are varied.[74] This effect has been adequately controlled in certain FM links by using an IF channel-phase equalizer combined with a signal predistorter which reduces the deviation produced by the LF components of the luminance signal by three or four times.

Tape recorders use FM signals and thus are subject to the problem just discussed with regard to FM-relay links. As with FM-relay links, proper circuit design and precorrection permits this problem to be controlled. PAL has first-order correction for such chrominance phase-modulation versus luminance.

In both transmitters and network facilities, nonlinear effects can occur at signal levels near black or sync and also near white. It is common to employ clamping during the back porch to establish black level, and also to employ a nonlinear operation for the sync signals to reshape them. In addition, transmitters often have a nonlinear modulation characteristic in the vicinity of white level due to many causes. A careful check of transmitter and network facilities is therefore desirable under conditions which are reasonably representative of normal program material.[73]

When electronically generated color-bar signals, having an amplitude of 100 percent for the primary and complementary colors are used for this purpose, unduly pessimistic answers may result. To avoid this, it is standard to check the performance[95,105] with color bars having an amplitude of 75 percent of maximum system amplitude. Under this condition all the errors, whether due to coder inaccuracies, nonuniform frequency response, or the effects of nonlinearities, must fall within the specified tolerances stated at the end of Sec. 4.3.1.[20] It should be noted that phase errors produced in the color burst are particularly important since they vary the apparent phase (and thus hue) of all colors.

In receivers, additional sources of nonlinearity can occur. For example, if the second detector operates at a low signal level so that square-law detection results, nonlinearity is produced. Beyond this, even if linear envelope detection is employed, a certain degree of nonlinearity results due to envelope detection of a single-sideband signal,[128] since the chrominance signal is transmitted wholly as single-sideband modulation of the picture carrier, in accordance with United States standards.[75]

In the receiver, the electron gun of the normal picture tube represents a highly nonlinear element. The effect of this nonlinear element upon the intended color signals is normally compensated for by the gamma correction currently employed at the program source. However, if the chrominance signal is permitted to come directly through the luminance channel, and is so applied to the picture tube, a rectification component is produced due to the nonlinearity of the electron gun. This rectification component results in an apparent increase in the luminance signal, and thus reduces the color saturation of the reproduced image.[21] This effect is normally eliminated by adequate trapping of the chrominance signal in the luminance channel of the receiver.

4.3.5 DISTURBANCES IN CHROMINANCE TRANSMISSION.

Thermal noise can produce two distinctive types of disturbance in chrominance transmission. Due to the relatively narrow bandwidth normally employed in the color synchronizing circuits (generally of the order of several hundred hertz) thermal noise in this channel produces gyrations in color phase which last for many scanning lines. Thus thermal noise in the color synchronizing channel produces wide horizontal streaks of incorrect chromaticity in the picture. Due to incomplete constant-luminance operation, these streaks of fluctuating chromaticity are frequently accompanied by fluctuating luminance streaks. By employing a narrow bandwidth in the color synchronizing channel, these fluctuations can be kept below the perceptible level.[76]

Thermal noise coming in through the chrominance demodulation channel produces disturbances in the form of colored specks. This is much like the effect of thermal noise in a monochrome receiver, except that here the specks take on various chromaticities, depending upon the instantaneous phase of the noise. Also the specks are somewhat coarser, due to the narrower bandwidth of the chrominance channel. Here, again, due to the present gamma-correction practice, incomplete constant-luminance operation results, and on saturated colors the noise produces a significant luminance fluctuation, which can be more annoying than the resulting fluctuation in chromaticity. When the relative effects of thermal noise in the chrominance and luminance channels are evaluated, another item to be considered is that only one sideband, resulting from modulation of the picture carrier by the chrominance signal, is transmitted. This impairs the LF signal-to-noise ratio of the chrominance channel compared to the luminance channel, since the latter employs double-sideband transmission for the LF components.

An HF CW signal which gets into the chrominance channel of the receiver heterodynes with the locally generated subcarrier, in the synchronous demodulators, to produce an LF beat. Thus HF tweets, such as those due to IF harmonics or to oscillator radiation which may be relatively unimportant in monochrome reception, may become more objectionable in color reception due to the coarser pattern produced in the chrominance channel.[21] Here again, the degree of visibility varies with color saturation due to the variation in constant-luminance operation.

The band-shared nature of the signal causes the HF luminance-signal components to produce a form of disturbance in chrominance transmission. This crosstalk of the fine-detail luminance components into the chrominance channel has been called "cross color." It appears as a low-visibility crawling pattern on edges or in areas of repeating fine detail. Due to the relative increase in luminance-signal energy at the lower frequencies, some increase in visibility of cross color results when the chrominance-channel bandwidth is increased. In particular, wideband utilization of the E'_I component with single sideband boost results in more cross color than equiband operation (described in Sec. 4.3.2). While specially selected scenes can produce objectionable cross-color disturbances, the effect on normal color scenes is usually unobjectionable.

When a color receiver is used to view a monochrome transmission, the subcarrier oscillator is not locked to an odd harmonic of one-half the line-scanning frequency. Thus, as the line-scanning frequency drifts back and forth within its permissible monochrome tolerance, the patterns due to cross color may become of high visibility instead of low visibility, as in color transmissions. Observations on high-quality monochrome transmissions indicate that the chrominance channel of a color receiver should either be turned off or reduced in gain when viewing monochrome transmissions. Thus it appears desirable for color receivers to use a "color killer," which recognizes the absence of the color-burst signal and reduces the gain of the chrominance channel.

Another form of disturbance, resulting from chrominance transmission, is due to the chrominance signal existing in the luminance channel of the receiver. As discussed in Sec. 4.2.2, this produces a low-visibility crawling-dot pattern in the image. However, subcarrier-frequency trapping in the luminance channel is generally employed to eliminate the rectification effects cited in Sec. 4.3.3. When this is done, the main dot patterns remaining are on colored edges, due to the sidebands of the chrominance signal which are not completely eliminated by the trap. In certain forms of displays a moiré pattern can be produced if subcarrier-frequency energy is applied to the display. In particular, displays using a phosphor dot array can produce an LF moiré by the heterodyning of the subcarrier-frequency energy with the dot pattern. With such displays trapping of the chrominance signal from the luminance channel to eliminate the LF beat pattern may be more important than the problems of a dot pattern or rectification previously cited.

4.4 DERIVATION OF COMPOSITE COLOR SIGNAL

The purpose of this section is to describe the functional steps involved in the derivation of a composite color signal. The specific illustrations will be for an NTSC composite color signal.

4.4.1 PRIMARY-COLOR SIGNALS FROM COLOR CAMERA. The purpose of the color camera is to examine the scene, to analyze the light values, and to transform them into electrical signals representative of the red-, green-, and blue-light values of the scene. The camera should deliver signals of such characteristics to be suitable to control three reproducing primaries to produce the intended scene on the station monitor. The subsequent coding and decoding operations on the signal should be such that the intended scene, as displayed on the monitor, is duplicated at a receiver using the reference-reproducing primaries. By *duplication* is meant that the color is reproduced without brightness distortion and with the same chromaticity. This is not meant to imply that the intended scene is a faithful reproduction of the original scene; in fact, faithful reproduction is frequently not desired.

The so-called taking characteristics, which are the relations between the output current in each channel and the wavelengths of the radiant energy supplied to the camera, govern the ability of the camera signals to provide the intended scene upon the monitor. The desired taking characteristics or spectral response curves for the three camera channels can be specified in terms of a set of reproducing primaries. For a given set of reproducing primaries R, G, and B, colorimetrically correct reproduction of the color is possible if, in the camera, one of the channels has a spectral sensitivity characteristic of the same shape as the color-mixture curve $\bar{r}$, another channel has a spectral sensitivity characteristic similar in shape to the color-mixture curve $\bar{g}$, and a third channel has a sensitivity characteristic similar in shape to the third color-mixture curve $\bar{b}$.[18] Such a set of taking characteristics is known as a "proper" set, since for the average observer it will reproduce the scene chromaticity exactly at the reproducer.

If the desired camera output signals are to be correct to control a real set of primaries R, G, and B, it will be found that each of the taking characteristics should have a negative response over certain portions of the visible spectrum. It is possible to use camera taking responses of a "proper" type in which only positive responses are necessary, by producing signals from the camera which are suitable for nonrealizable primaries. As an example, camera sensitivities having the shape $\bar{x}$, $\bar{y}$, and $\bar{z}$ corresponding to the CIE tristimulus values have only positive response functions and would therefore produce proper signals to excite the nonrealizable CIE primaries X, Y, and Z.[23] If the camera operates in a linear fashion, that is, the output signal is proportional to the scene luminance, it is possible, in accordance with the theory of color transformation, to matrix the three camera signals suitable for one set of primaries to three other signals suitable for a different set of reproducing primaries.

In the past the usual practice for direct-pickup color cameras has been to design them to produce, without matrixing, signals for a reproducer having the reference primaries and to make the taking characteristics duplicate as closely as possible only the positive portions of the color-mixture curves for these primaries.[77,78] The introduction of camera tubes and preamplifiers with greatly improved signal-to-noise ratios, compared with the previously used image orthicons and vidicons, has made matrixing for the negative lobes of the ideal taking characteristics practical (see Sec. 2.4.7). The three channel gains are so adjusted in practice that equal signals are produced by the camera for that chromaticity in the scene which is to be reproduced as the reference chromaticity, standard source C, in the intended scene on the monitor. Figure 4-45 shows a representative set of camera taking characteristics.

A similar problem for the adjustment of the taking characteristics of a flying-spot scanner for transparency or motion-picture-film reproduction exists, except that here the quality of the light source is constant, being that generated by the flying-spot cathode-ray tube. The usual practice with flying-spot scanners has been to use taking characteristics which approximate the positive portions of the color-mixture curves for the reference-reproducing primaries, although some scanners have been built in which the taking characteristics were close approximations of the CIE tristimulus values $\bar{x}$, $\bar{y}$, and $\bar{z}$.[79-81]

In practice the proper taking characteristics are obtained by the interposition in the light path for each channel of an optical color filter having the necessary transmission characteristics. The relation to be satisfied here, for the red channel, as an example, is

$$A_R(\lambda) = S(\lambda)F_1(\lambda)F_2(\lambda) \tag{4-27}$$

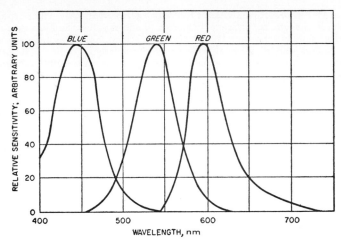

FIG. 4-45 Representative set of camera taking sensitivities. (*From Fink.*[17])

where the quantities are functions of the wavelength of light as follows:

$A_R(\lambda)$ = desired characteristic for light of equal energy per wavelength interval, amperes per watt

$S(\lambda)$ = photocell sensitivity, amperes per watt

$F_1(\lambda)$ = transmission of all optical elements except color filter

$F_2(\lambda)$ = transmission of color filter

This equation should be satisfied as closely as possible for all wavelengths over the visual region for positive values of A_R.

With cameras using image-orthicon pickup tubes or flying-spot scanners using cathode-ray-tube light sources, it is usual to blank off the electron beam during the horizontal and vertical return intervals, so the camera output signal during these intervals is representative of black level. For color operation, the cameras are operated in their linear range, so the average level of the signal, measured from the black level, is proportional to the average brightness of the scene being transmitted. If the level corresponding to black is not lost in the subsequent operations on the signals, the dc component of the signal is retained and constancy of the level corresponding to black can be reestablished at any subsequent stage by clamping it to a fixed value.[82,83]

All the frequency components generated in the camera from about 30 Hz to about 4.5 MHz should be faithfully transmitted in both amplitude and phase to the color coder. The amplifiers need not be of a type which transmits direct current, as the dc component can be reestablished as discussed above. The amplifiers for the camera signals are generally sufficiently wide, usually 6 or 8 MHz so that the amplitude response and time delay can be made constant to within about $\pm\frac{1}{2}$ dB and $\pm 0.05\mu s$, respectively, over the required 4-Mhz band.

4.4.2 MATRIXING OPERATIONS. The output signals from the camera are generally operated on by a gamma corrector as a first step, and the gamma-corrected primary-color signals are then used as the input for the matrixing operations. The production of the luminance signal E'_Y is accomplished by proper addition of the three primary signals according to the equation

$$E'_Y = 0.30E'_R + 0.59E'_G + 0.11\,E'_B \qquad (4\text{-}28)$$

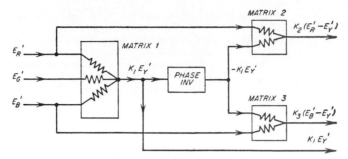

FIG. 4-46   Schematic matrix to form signals E_Y', $E_R' - E_Y'$, $E_B' - E_Y'$. (*From* *Fink.*[17])

The primes denote gamma-corrected signals. A simple matrix to perform this operation is shown in Fig. 4-3.[25] To produce the set of color signals comprising E_Y', $E_R' - E_Y'$, and $E_B' - E_Y'$ the basic operations are as shown in Fig. 4-46. Matrix unit 1 produces the luminance signal as previously described. In one channel the E_Y' signal is reversed in polarity by means of the phase inverter and the $-E_Y'$ signal is added to E_R' and E_B' in matrix units 2 and 3, respectively. This type of matrix design is satisfactory if the generator impedances supplying the three input signals E_R', E_G', and E_B' are low compared with the combining resistors in the matrix elements, so that each input signal is not contaminated with spurious signals coupled back into it from other inputs.

To produce chrominance signals in accordance with the FCC Recommended Practices the signals E_I' and E_Q' should be formed. E_I' and E_Q' may be defined either in terms of $E_R' - E_Y'$ and $E_B' - E_Y'$ or in terms of primary signals E_R', E_G', and E_B', and they may be synthesized according to either of these definitions. A widely used method is the latter, and this is shown as a portion of Fig. 4-48.

Flying-spot-scanner types of pickups have been built in which the output signals were good approximations to E_X, E_Y, and E_Z, where the subscripts X, Y, and Z denote that these signals were representative of the CIE tristimulus values. With signals of this type there is a problem concerned with gamma correction since if the signals E_X, E_Y, and E_Z are directly gamma-corrected, the use of some form of nonlinear matrix is required to obtain E_Y', E_I', and E_Q'. The practice with this type of camera therefore had been to matrix the signals to the form E_R, E_G, and E_B before gamma correction. The operation to be performed by this matrix is defined by the equations[79]

$$E_R = 1.91E_X - 0.53E_Y - 0.29E_Z \tag{4-29}$$

$$E_G = -0.98E_X + 2.0E_Y - 0.03E_Z \tag{4-30}$$

$$E_B = 0.06E_X - 0.12E_Y + 0.90E_Z \tag{4-31}$$

If it were desired to produce signals of the form E_Y, $E_X - E_Y$, and $E_Z - E_Y$, it is obvious that this last matrixing operation would not be necessary and the chrominance signals could be formed by similar matrixing operations as already described.

The reproduction, by a color television system, of pictures originating from motion-picture film or transparencies has considerable importance. Certain factors such as the brightness-reproduction range of film, which is greater than that for television, and the absorption characteristics of the dyes used in the film are involved in the quality of the reproduction. Studies of this problem have indicated the reproduced picture quality may be improved by the use of masking techniques, the purpose of which is to minimize some of the distortions generally encountered in reproducing pictures from film.[85-87] The dyes used in most of the commercially available color films depart in some degree from the ideal characteristics, in that they exhibit abnormal absorption of light in some regions of the spectrum. This abnormal absorption may be thought of as spurious cross coupling between the colors.

If the dye characteristics of the film being televised are known, it is possible to provide electrical cross coupling, in the signal channel following the camera, of an inverse characteristic to that exhibited by the film to neutralize the undesired dye absorptions. A method of accomplishing this for one signal channel is shown in Fig. 4-47. The output signal from this type of masking amplifier is given by the equation[86]

$$E''_G = K_1 E_G + [K_2 E_G + K_3 E_R + K_4 E_B] \qquad (4\text{-}32)$$

where $K_2 + K_3 + K_4 = 0$.

The signals within the brackets represent the modification to the original signal by the masking operation, and the modification is equal to zero when

$$E_G = E_R = E_B \qquad (4\text{-}33)$$

The added masking signal has the form of a color-difference signal which disappears on the reference chromaticity, standard source C, and thus does not modify the reproduction of white or near white. Similar circuits may be used to provide masked signals E''_R and E''_B. The constants K_1, K_2, K_3, and K_4, etc., in the above equations, take on a set of values which are governed by the particular type of film being televised. Calculations may be made from the film characteristics to provide values for the K's, but in practice the final values are determined by trial for optimum reproduction. The masking operation as defined by the above equation provides a first-order linear correction for the dye deficiency, and is therefore not exact, since a logarithmic form of correction should be used. If the masking operation is performed subsequent to gamma correction, a better approximation to the logarithmic operation is obtained.

4.4.3 FILTERING OPERATIONS PRIOR TO MODULATION.

To produce a composite color television signal meeting the standards of the FCC, the video frequency chrominance signals should be limited in bandwidth according to the following specifications:[19,20] The bandwidth of the channel handling the E'_I video chrominance signal shall have an attenuation of less than 2 dB at 1.3 MHz and an attenuation of more than 20 dB at 3.6 MHz. The bandwidth of the channel handling the E'_Q video frequency-chrominance signal shall have a bandwidth with less than 2 dB attenuation at 400 kHz, less than 6 dB attenuation at 500 kHz, and at least 6 dB attenuation at 600 kHz.

There are several ways of forming the chrominance-modulated subcarrier signal in conformity with these specifications. Figure 4-48 shows, in block-diagram form, the elements of a color coder or colorplexer in which the E'_I and E'_Q signals are formed directly from the gamma-corrected primary signals and then are band-limited in accordance with the specifications given above. To optimize the transient response in the two chrominance channels and minimize the overshoot, the filters for providing the band limitation are generally designed to have moderate roll-off characteristics. Filter configurations

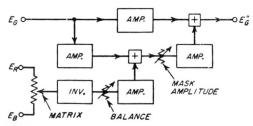

FIG. 4-47 Block diagram of a masking amplifier. (*From Fink.*[17])

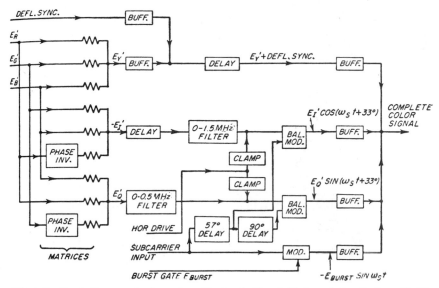

FIG. 4-48 Block diagram of color coder with signals E_I' and E_Q' formed directly from E_R', E_G', and E_B'. The symbol E_{BURST} denotes the envelope of the color burst. (*From Fink.*[17])

having linear phase characteristics to at least the 6-dB attenuation point are desired, and phase-correcting types of filters using bridging impedances are often employed.[26]

Figure 4-49 shows, in block-diagram form, a color coder in which the method of forming the two chrominance-signal components is somewhat different.[88] In this case the chrominance signals E_I' and E_Q' are produced from the primary signals. After band limitation of E_Q' and delay equalization, so that E_Q' and E_I' are synchronous, these signals are supplied to two matrices to form chrominance signals of the composition $E_R' - E_Y'$ and $E_B' - E_Y'$. These latter signals are then each band-limited to 1.5 MHz and are denoted by $(E_R' - E_Y')_1$ and $(E_B' - E_Y')_1$ to indicate that, although the compositions correspond to $E_R' - E_Y'$ and $E_B' - E_Y'$, the signals have been effectively band-limited along the E_I' and E_Q' axes. Since one of the balanced modulators handles the $E_B' - E_Y'$ signal, it is also used to produce the color burst [phase $-(E_B' - E_Y')$] and thus reduce the chance of phase error due to circuit drift between the burst and the modulated chrominance signal corresponding to a color. Since, in the chrominance channel, signals with the composition E_I', E_Q', $E_R' - E_Y'$, and $E_B' - E_Y'$ are produced and may be switched off individually, this coder is somewhat more flexible than that of Fig. 4-48. A circuit is included to blank off the $E_R' - E_Y'$ modulator during horizontal retrace to assure that during transmission of the color burst there is no spurious subcarrier signal to cause burst-phase error.

4.4.4 CHROMINANCE-SUBCARRIER GENERATOR. The chrominance information is carried in the composite picture signal as the sidebands produced by the process of modulating the color subcarrier signal by the two video frequency-chrominance components. The CW signal at the chrominance-subcarrier frequency is generated at the program source by a suitable oscillator. The standards specified by the FCC are that the color subcarrier frequency shall be 3.579545 MHz ± 10 Hz and the maximum rate of change of chrominance-subcarrier frequency shall not exceed $\frac{1}{10}$ Hz/s.[19,20] To meet these requirements, it is mandatory to use a high-quality oscillator with crystal control. The crystal is usually mounted in an oven so that it operates in a stabilized ambient temperature to minimize frequency drift. The tolerances on frequency and on the rate of change

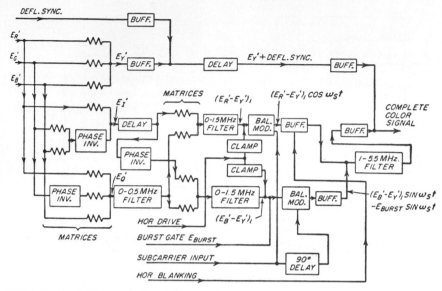

FIG. 4-49   Block diagram of color coder using signals $E_R' - E_Y'$ and $E_B' - E_Y'$ to form chrominance subcarrier signal. (*From Fink.*[17])

of frequency were given these limits to simplify the problem of decoding the composite color signal at the receiver. The chrominance-subcarrier frequency must be reestablished as a CW signal at the receiver, and the tight tolerance on transmitted frequency simplifies this problem at the receiver. The small rate of change of frequency allows circuits of high noise immunity (i.e., narrow noise bandwidth) to be used at the receiver without the production of chrominance noise due to random phase modulation of the chrominance-subcarrier frequency.

Since the modulated chrominance information is band-shared with the luminance information, as discussed in Sec. 4.2, the chrominance-subcarrier frequency should be precisely related to the horizontal deflection frequency. The FCC standards state that the horizontal scanning frequency shall be $\frac{2}{455}$ times the frequency of the chrominance subcarrier. This relation must be established at the program source and it is usually accomplished by using the chrominance-subcarrier frequency as a reference to control the frequency of horizontal deflection and, therefore, also the vertical deflection.[46] The control of the horizontal scanning rate is effected by a divider chain between the chrominance-subcarrier-frequency generator and the normal blanking and sync-signal generator. The blanking and sync-signal generator generally has an oscillator running at twice the horizontal scanning frequency, or about 31.5 kHz. This frequency must be $\frac{4}{455}$ times the chrominance-subcarrier frequency. This relationship is maintained precisely both in frequency and phase by a chain of dividers with one multiplier which divides the chrominance-subcarrier frequency by 455 and multiplies it by 4, to give the nominal frequency 31.5 kHz (more exactly 31.468+ kHz) which is then injected into the normal blanking and sync-signal generator to synchronize it. The number 455 has the factors 5, 7, and 13. A divider chain having these factors and also a multiplying operation of 4 times is used as shown in Fig. 4-50 to lock the horizontal deflection frequency to the chrominance-subcarrier frequency.[89] The frequency dividers may be of any of the standard types, the regenerative divider and locked oscillator being used most frequently for this purpose.†

---

†Modern-day dividers make use of digital ICs.

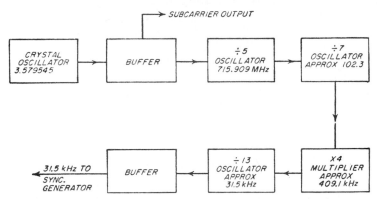

FIG. 4-50 Block diagram of divider chain from 3.57 + MHz/s to 31.5 kHz/s. (*From Fink.*[17])

Sometimes an LC stabilized multivibrator has been used for the lowest frequency division in the chain.

The color-synchronizing signal which is used to reestablish the chrominance-subcarrier-frequency reference at the receiver consists of a burst of 8 cycles (minimum) of the chrominance-subcarrier frequency following each horizontal synchronizing pulse. This color-synchronizing signal is called the *color burst* and the phase of the subcarrier during the burst is standardized as being that of $-(E'_B - E'_Y)$. It is generated at the program source by supplying a burst keying pulse to a gate circuit or modulator whose other input is a continuous chrominance-subcarrier-frequency signal having the correct phase to be used for the burst.[89] The output of the gate circuit thus consists simply of short pulses of chrominance-subcarrier frequency.

The burst-keying pulse is produced by a circuit comprising a delay network which triggers an auxiliary-pulse generator. The delay network assures that the initiation of the burst will occur at a specified time interval after the leading edge of the horizontal synchronizing signals. The width of the pulse generated by the auxiliary-pulse generator is adjusted to that necessary to produce the specified burst duration. The burst signal is not transmitted during the nine-line portion of the vertical retrace interval composed of the equalizing pulses and the broad pulses, and it is necessary therefore to provide a circuit to disable it during this interval.

4.4.5 CHROMINANCE-SUBCARRIER MODULATION. The two chrominance-signal components of the composite color-picture signal are each composed of only the sidebands of the process of modulating a suitably phased chrominance-subcarrier-frequency signal by one of the video chrominance signals. Two of these modulation processes occur, one for the E'_I chrominance signal, the other for the E'_Q chrominance signal. The chrominance-subcarrier-frequency signals supplied to the two modulators differ in phase by 90°. It is supplied to the E'_Q modulator 90° later in phase than to the E'_I modulator. The modulation is normally performed in a modulator of the balanced type so that neither the input video signal nor the supplied chrominance-subcarrier-frequency signal appears in the output. The only useful output signal is composed of the sidebands of the modulation process.

For the E'_I modulator, of the modulation process, signal E'_I consists of many frequency components at harmonics of the field- and line-scanning frequency and it may be expressed mathematically (considering only the line-frequency harmonics for simplicity) as follows

$$E'_I = \sum_{n=0}^{n=k} E'_{In} \cos n\omega_H t \qquad (4\text{-}34)$$

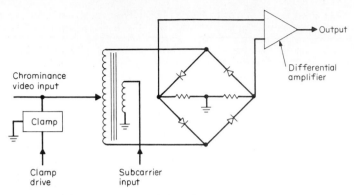

FIG. 4-51 Schematic circuit of a balanced modulator.

where ω_H is the angular frequency of horizontal deflection. This is the modulating signal supplied to the balanced modulator. The chrominance-subcarrier-frequency signal supplied to the balanced modulator is $\cos(\omega_s t + 33°)$. The output of the I modulator thus is seen to be

$$E'_I \cos(\omega_s t + 33°) = \sum_{n=0}^{n=k} E'_{I\,n} \cos n\omega_H t \cos(\omega_s t + 33°)$$

$$= \frac{1}{2} \sum_{n=0}^{n=k} E'_{I\,n} \{\cos[(\omega_s + n\omega_H)t + 33°] \qquad (4\text{-}35)$$

$$+ \cos[(\omega_s - n\omega_H)t + 33°]\}$$

which consists of an upper sideband $(\omega_s + n\omega_H)\ \underline{/33°}$ and a lower sideband $(\omega_s - n\omega_H)$ $\underline{/33°}$ disposed symmetrically about the suppressed carrier $\omega_s\ \underline{/33°}$, up to the kth harmonic line of frequency as limited by the bandwidth of the video frequency-chrominance signal.

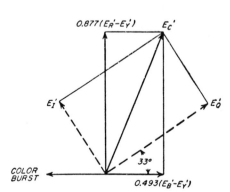

FIG. 4-52 Vector diagram of chrominance signal. (*From Fink.*[17])

The modulation process for the Q signal is the same except that the reference chrominance-subcarrier-frequency signal is delayed by 90° and is therefore $\sin(\omega_s t + 33°)$.

There are several forms of balanced modulator, one of which is shown in Fig. 4-51. In this circuit, the chrominance video signal is fed to both sides of a balanced four-diode bridge. The signal is clamped to a fixed potential during horizontal blanking to establish a zero-chrominance level. By means of a balanced transformer, a continuous sine-wave subcarrier signal at the appropriate phase (I or Q) is applied in opposite polarities to the same terminals of the bridge. With the bridge properly balanced for the video and subcarrier input signals, the subcarrier is modulated by the video by the alternating conduction of the diodes, and only the sideband products appear at the output. The complete chrominance signal is formed by combining the outputs of two such modulators. Thus, the subcarrier chrominance signal consists of the modulation products produced by the suppressed-carrier modulation of two components of the reference subcarrier in quadrature. A vector diagram representing the chrominance signal is shown in Fig. 4-52. The axes on which the E'_I sidebands and the E'_Q sidebands are produced are, as shown in this diagram, in quadrature. The actual chrominance signal is the vector sum of the E'_I and E'_Q signals. Thus the subcarrier is modulated by two quadrature components having independent values depending on the chrominance signals E'_I and E'_Q, and the resultant signal is one which varies both in amplitude and in phase. When the video-frequency signals E'_I or E'_Q change polarity and have video-voltage excursions below blanking level, the phase of the chrominance output signal from that modulator is shifted 180°. As a result, the chrominance signal E'_c can be at any phase between zero and 360°. The axes of the color difference signals $E'_R - E'_Y$ and $E'_B - E'_Y$ are also shown and since both sidebands of each chrominance signal are necessary for the diagram to be correct, the diagram holds for video chrominance frequencies below about 500 kHz and either representation is correct. For reference the color-sync burst phase is shown; it leads the E'_I axis by 57°. The phase of the complete chrominance signal relative to that of the color burst varies with the hue, while the amplitude of the resultant chrominance signal relative to the E'_Y signal varies with the saturation. Thus the subcarrier may be said to carry hue information in its phase and saturation information in its normalized amplitude.

4.4.6 OPERATIONS ON THE CHROMINANCE SIGNAL AFTER MODULATION.

When balanced modulators are used to produce the two components of the subcarrier-chrominance signal, in general only the desired sidebands are produced at a useful level. However, the modulators are not perfect, with the result that some spurious signals are generated in the modulation process. For example, it is possible to have disturbing amounts of the modulating video signal, the second harmonic of the carrier frequency, and higher-order modulation products as undesirable signals also present along with the desired chrominance signals. These are attenuated in low-pass or bandpass filters following the modulators.

It is usual to combine the two chrominance-signal components into a single signal which is passed through a bandpass filter having a transmission band extending from about 2 to about 5 MHz or a low-pass filter which transmits signals to about 5 MHz.[25,26] This filter is wide enough to transmit both sidebands of both the E'_I and the E'_Q chrominance signals. This filter should be uniform in amplitude transmission to within about ±¼ dB and its time delay over the band to 6-dB attenuation points should be constant to within about ±0.01 μs. The upper sideband of the E'_I signal, lying in the frequency range above 4.1 MHz, will normally be attenuated in the RF transmitter when the composite color signal is used to modulate the final picture carrier for radiation.

The useful outputs of the encoder (the luminance signal E'_Y, the subcarrier-chrominance signal, the deflection-synchronizing signal, and the color-synchronizing signal) are normally added together to form the single composite color-picture signal. This is shown in Figs. 4-48 and 4-49. Any additional filters in the path handling the composite color-picture signal should be flat in amplitude response and constant in time delay over the band containing the useful signal up to about 5 MHz. One practice is to use a linear-phase type of filter which is flat in both phase and amplitude response to about 8 MHz; this produces no noticeable degradation of the signal.[26]

4.5 DECODING OF A COMPOSITE COLOR SIGNAL

This section describes the basic methods for decoding a composite color signal in order to make a color image on either three-gun simultaneous displays or single-gun sequential displays. The description is in terms of an NTSC signal, but for the additional or different functions needed for PAL or SECAM, see Secs. 4.2.5 and 4.2.6, respectively.

4.5.1 COLOR-DIFFERENCE SIGNAL DEMODULATION.

Video signals corresponding to the E_I' and E_Q' components of chrominance can be derived, free of crosstalk, by using a set of quadrature synchronous detectors, followed by suitable filtering, as described in Sec. 4.3.2. However, in general, these signals are not of the desired form to operate a color display. Usually color-difference chrominance signals of the form $E_B' - E_Y'$, $E_R' - E_Y'$, and $E_G' - E_Y'$ are required. The desired signals can be obtained from the E_I' and E_Q' components by a simple matrix operation expressed by the following matrix equations

$$E_B' - E_Y' = 1.70E_Q' - 1.10E_I' \tag{4-36}$$

$$E_R' - E_Y' = 0.62E_Q' + 0.96E_I' \tag{4-37}$$

$$E_G' - E_Y' = -0.65E_Q' - 0.28E_I' \tag{4-38}$$

To perform such a matrix operation correctly, the E_I' and E_Q' signals should be in time coincidence. As mentioned in Sec. 4.3.2 this generally requires the use of some added delay in the wideband (E_I) channel. Following such delay equalization, the above matrix operation might be performed by using a pair of phase splitters, to obtain plus and minus E_I' and E_Q' signals, and then combining these signals through suitable resistive networks.[92-94] Such an arrangement is illustrated in Fig. 4-53.

When equiband operation is considered, the delay-equalization problem is simplified and demodulation along any convenient set of axes may be employed. For example, a set of quadrature demodulators may be used to obtain directly the desired color-difference signals $E_B' - E_Y'$ and $E_R' - E_Y'$. If desired a third demodulator can be employed to obtain directly the remaining color-difference signal $E_G' - E_Y'$. The relative gain and angle of demodulation required for the various color-difference signals are shown in Fig. 4-54. On the other hand, the desired color-difference signal $E_G' - E_Y'$ may be obtained by a simple video matrix circuit, using a resistive adding network plus a phase inverter.[95] Such an arrangement is illustrated by the block diagram on Fig. 4-55, and operates according to the following matrix equation

$$E_G' - E_Y' = -0.51(E_R' - E_Y') - 0.19(E_B' - E_Y') \tag{4-39}$$

Primary-color signals of the form E_R', E_G', and E_B' are required to operate a three-gun color display. These primary-color signals may be obtained by combining the respective color-difference signals with the luminance signal. This combining function may occur in external adding circuits, so that the primary-color signals E_R', E_G', and E_B' are applied respectively to one control electrode of each gun of the display. On the other hand, these signals may be effectively combined in the picture tube such as by applying a negative-polarity luminance signal $-E_Y'$ to each of the cathodes and by applying the appropriate color-difference signals $E_R' - E_Y'$, $E_G' - E_Y'$, and $E_B' - E_Y'$ to the respective control grids of the picture-tube electron guns. The former operation may be called *R-G-B operation* and the latter *color-difference operation*.

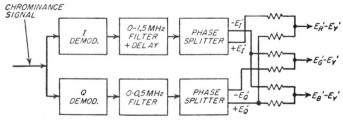

FIG. 4-53 Matrix for obtaining color-difference signals from E_I' and E_Q' signals. (*From Fink.*[17])

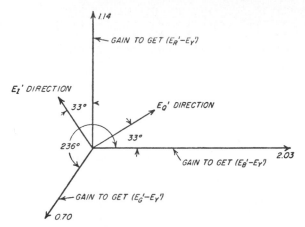

FIG. 4-54 Demodulation characteristics for color-difference signals. (*From Fink.*[17])

4.5.2 APPLICATION OF COLOR SIGNALS TO A THREE-GUN DISPLAY. For correct colorimetric presentation the color signals applied to a three-gun display should include their correct respective dc components, which may be substantially different in the three color signals. The correct dc components might be established by using a suitable dc restorer for each of the various color signals. Another possible solution is to use dc coupling in the luminance channel, from the second detector through to the picture tube, plus dc coupling in the chrominance channel from the demodulators through to the picture tube. However, with arrangements employing both matrixing and amplification operations following detection, dc coupled circuits may not be particularly attractive due to the large number of stages through which the dc coupling must be maintained.

When using the R-G-B operation, so that primary-color signals are applied directly to each gun of the display, simple dc restorers (which conduct on the deflection sync pulses) have been employed in each color channel. The design of these dc restorers generally represents a suitable compromise between good dc insertion (such as obtained by efficient peak detection) and good noise performance (such as obtained by average detection).

When dc insertion is employed with color-difference operation, the dc insertion action should operate on both the luminance and color-difference components, because the various signals can have individual dc components. Direct-current insertion on the color-difference signals separately would require "clamps," since the instantaneous values as well as the dc components of these signals can have either polarity. Therefore, it is convenient to operate the dc restorers on the individual primary-color signals, such as by connecting them essentially in parallel with the grid and cathode of each gun of the picture tube. In this arrangement, illustrated in Fig. 4-56, the luminance and color-difference signals are effectively added in each dc restorer (just as in each electron gun) to give the desired primary-color signals.

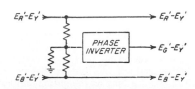

FIG. 4-55 Matrix for $E_G' - E_Y'$ signal. (*From Fink.*[17])

Correct colorimetric presentation additionally requires that the display be operated with the correct transfer characteristic for each of the three electron guns. This requires a correct setting of bias (background control) for each of the three electron guns. Monochrome experience indicates that it is

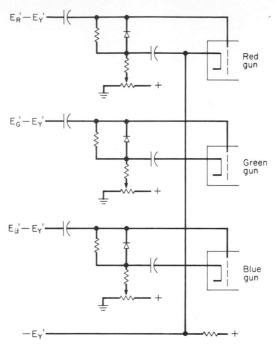

FIG. 4-56 DC restorer system for color-difference operation.

desirable to have a background or brightness control which the viewer can adjust to suit the particular conditions. For color receivers it appears desirable to have a common brightness control for the three guns which simultaneously adjusts the operating conditions of the three guns in an achromatic manner—that is, so that the color of a gray scale is not affected. The modern technique for accomplishing this is to disable the vertical deflection by means of a service switch and, with no video signal applied, to adjust each of the three CRT screen voltages to visible cutoff. Next, with the receiver operating normally and chrominance off, the red, blue, and green video gains are adjusted for a white balance, ideally at D6500.

4.5.3 SAMPLING OF COLOR SIGNALS FOR A ONE-GUN SEQUENTIAL DISPLAY.
One way of utilizing the composite color signal to operate a one-gun sequential display is to decode the composite color signal to obtain three simultaneous color signals, and then to sample these signals sequentially at the correct timing required by the particular type of color switching employed with the display.[96] Such signal processing by decoding plus gating is illustrated by the block diagram of Fig. 4-57. By such an arrangement, any desired color sequence may be employed by using suitable signals to actuate the gate circuits. Additionally, the color switching could conceivably operate at any desired rate, as long as the gating and switching are synchronous. However, in certain arrangements it is convenient to operate the color switching at the input chrominance-subcarrier frequency (3.6 MHz).

While the above arrangement is straightforward, it can have a gray-scale tracking problem substantially similar to that of three-gun displays, except here it is three samplers or gates that must be tracked. Also, correct colorimetric operation requires that the three simultaneous signals, which are sampled, contain their individual dc components.

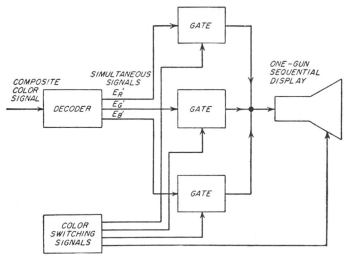

FIG. 4-57 Signal processing by decoding plus gating. (*From Fink.*[17])

Again, this calls for either dc coupling or three dc restorers at the inputs of the samplers or gates.

The gray-scale tracking problem can be reduced by using a shunted-monochrome arrangement in which the luminance signal is directly applied to the display, via a shunt path, and only the color-difference signals are gated.[96] Since the color-difference signals are zero on a neutral scale, linearity of tracking the gates is unimportant, in this case, for obtaining a good colorless gray scale. However, the three gates must be so balanced with respect to each other that a net output of zero is obtained on a neutral gray scale.

Beyond the tracking problem cited above, correct colorimetric presentation also requires correct transfer characteristic in the picture tube. This calls for a single background control, to set the bias on the electron gun, and a single dc restorer (or dc coupling) to provide the proper dc component for the composite signal applied to the electron gun of the display.

The problem of unequal phosphor efficiencies also exists in one-gun displays. In the arrangement shown in Fig. 4-57 this might be taken care of by increasing the gain in the appropriate gated channel. However, in general, it has been found desirable effectively to modify the phosphor efficiencies by other means, such as by diluting the higher efficiency phosphors when manufacturing the picture tube, or by using a suitable color filter in front of the picture tube.

4.5.4 DIRECT PROCESSING OF COLOR SIGNAL FOR A ONE-GUN SEQUENTIAL DISPLAY. In the arrangement shown in Fig. 4-57 the signal applied to the one-gun display can be considered as being a coded signal, which is then decoded within the display. The NTSC color-picture signal is also a coded signal, using a code chosen for optimum transmission through the present channels. However, this NTSC signal code may not be the particular signal code desired by the form of one-gun display being considered. In this case, for optimum results, the receiver should convert from one form of signal code to the other form, and the portion of the receiver performing the conversion or translation of code might be called a *translator unit*. In general, this translator-unit function can be performed by a more straightforward technique than decoding to simultaneous color signals or than gating or sampling the color signals, as shown in Sec. 4.5.3. To evaluate the desired translator-unit functions, the desired display coded signal can be expressed in terms similar to the NTSC signal code. In other words, the color composition of the various harmonic components of the display-coded signal can be calculated and then compared with the color composition of the various components of the NTSC signal.

For the study of translator units it is convenient to classify sequential display operation into two groups, namely, continuous (repeating) color and reversing color-sequence operation. Continuous color-sequence operation means that the beam of the display device can successively excite the color phosphors in simple repeating color sequence such as G-R-B-G-R-B . . . , and so forth. Such operation has been called *dot-sequential* operation.[97] Reversing color-sequence operation differs in that the beam of the display can successively excite the color phosphors in a color sequence which reverses in order before the entire sequence of colors is repeated, such as B-G-R-R-G-B . . . , and so forth. Figure 4-58 illustrates how these two forms of operation can be obtained with a certain type of phosphor-strip display. As illustrated, a third harmonic gating provides a continuous color sequence, and a sixth harmonic gating (or no added gating signal) provides a reversing color sequence, with this particular form of display.[98,99]

A "first-order" solution to the desired signal code for continuous color-sequence operation can be obtained by assuming that samples of the coded signal at 120° intervals produce the respective primary-color signals E'_G, E'_R, and E'_B. This is a dot-sequential signal which has a monochrome component of

$$E'_M = \tfrac{1}{3}E'_G + \tfrac{1}{3}E'_R + \tfrac{1}{3}E'_B \tag{4-40}$$

and has a modulated subcarrier component which can be described as a symmetrical set of three vector components having respective amplitudes of $\tfrac{2}{3}E'_G$, $\tfrac{2}{3}E'_R$, and $\tfrac{2}{3}E'_B$.[21,100] This subcarrier signal can also be described by a set of $E'_R - E'_Y$ and $E'_B - E'_Y$ components for convenient comparison with the NTSC signal as shown in Fig. 4-59.

The NTSC monochrome signal E'_Y can be modified to the monochrome component E'_M by the addition of a signal having the composition of

$$E'_M - E'_Y = -0.25E'_G + 0.03E'_R + 0.22E'_B \tag{4-41}$$

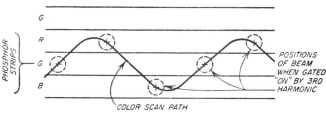

GATING FOR CONTINUOUS COLOR-SEQUENCE

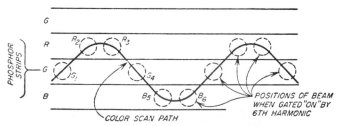

GATING FOR REVERSING COLOR-SEQUENCE

FIG. 4-58 Continuous versus reversing color-sequence operation. (*From Fink.*[17])

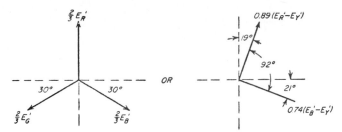

FIG. 4-59   Vector diagram of symmetrical subcarrier. (*From Fink.*[17])

This signal has the same form as the color-difference signals which modulate the NTSC chrominance signal; specifically, $E'_M - E'_Y$ goes to zero when transmitting reference white. Signals of this form can be obtained by synchronous detection of the chrominance signal at a certain phase and with a certain gain; and the color-difference signal $E'_M - E'_Y$ can be obtained by synchronous detection at 19° leading phase with respect to $E'_B - E'_Y$ and with a gain of 0.58 relative to the luminance channel.

The color composition of the chrominance signal can be modified to the symmetrical form by the addition of a reversed-phase-sequence chrominance signal, that is, a signal in which the colors versus phase occur in opposite sequence. For example, if a signal in which the $E'_R - E'_Y$ component lags the $E'_B - E'_Y$ component is added to the NTSC signal, in which the $E'_R - E'_Y$ leads $E'_B - E'_Y$, then the amplitudes and phases of the respective components will be modified in opposite directions, so that the resulting signal has modified amplitudes and phases versus color. This subcarrier-modification operation can be conveniently performed in a modulator amplifier in which the input chrominance signal is heterodyned with a second-harmonic reference subcarrier (7.2 MHz) to produce the desired additional reversed-phase-sequence component. Such a subcarrier modifier circuit has also been called an *elliptical-gain* amplifier.[101] The vector diagram for this desired processing is shown in Fig. 4-60, which indicates that the reversed-phase-sequence component should have a relative amplitude of 0.19 and that an angle of $-4.9°$ should exist between the corresponding $E'_B - E'_Y$ components. This gives a signal which is only 0.8 times the desired subcarrier-signal amplitude, and thus the subcarrier-modifier channel should have a gain of 1.25 relative to the shunt-monochrome channel.

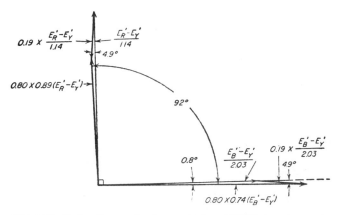

FIG. 4-60   Vector diagram showing processing of NTSC chrominance signal to a symmetrical subcarrier. (*From Fink*[17]).

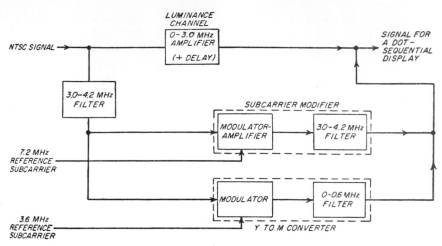

FIG. 4-61   Processing of NTSC signal for continuous color-sequence operation. (*From Fink.*[17])

Combining the $E_Y' - E_M'$ converter and the subcarrier modifier just described gives a complete arrangement for processing or translating the NTSC signal code to the signal code desired for continuous color-sequence display operation. Such an arrangement is illustrated by the block diagram in Fig. 4-61, for the case of a display operating at a color-switching frequency equal to the NTSC subcarrier frequency.[100] If the display operates at a color-switching frequency different from the NTSC subcarrier frequency, then the appropriate frequency conversion can be performed before the modified subcarrier signal is applied to the display.

A "first-order" solution to the signal code desired for reversing color-sequence operation can be obtained by assuming that samples of the coded signal at 60° intervals produce the respective primary-color signals E_B', E_G', E_R', E_R', E_G', and E_B'.[100] Analysis of such a "six-sample" signal shows that the monochrome component has the same equally weighted value as for the continuous color-sequence case. The analysis further shows that the desired fundamental frequency term is a constant phase component, having an amplitude proportional to $E_R' - E_B'$, and that additionally a second harmonic term exists, which is also a constant phase term, having an amplitude porportional to $E_G' - \frac{1}{2}E_R' - \frac{1}{2}E_B'$.

In the reversing color-sequence case the desired fundamental and second-harmonic subcarrier components each have a constant phase but are variable in amplitude. The function of effectively selecting one subcarrier axis and effectively eliminating the quadrature axis (to produce a constant phase signal) can be performed by circuits similar to the subcarrier modifier previously cited. If the modulator-amplifier operates as a narrow-angle modulator, or sampler, then the reversed-sequence subcarrier produced by heterodyning with the second harmonic will have the same amplitude as the translated original subcarrier. The combination of these two equal-amplitude but opposite-sequence subcarriers will add along one axis and cancel along a quadrature axis, thus acting as an "axis selector." This axis selector or "synchronous filter" can be considered as a narrow-angle sample which connects the input signal to the output circuit twice per cycle, for short intervals of time, and thereby ignores the quadrature component which is passing through zero at these instances.

The desired $E_R' - E_B'$ and $E_G' - \frac{1}{2}E_R' - \frac{1}{2}E_B'$ components can be obtained from the chrominance signals by axis selection along the particular phases shown in the vector diagram of Fig. 4-62. By combining a set of axis selectors with a frequency converter (to produce the desired second-harmonic term) and an E_Y' to E_M' converter, a complete translator unit can be made to translate the NTSC signal code to the code desired for a revers-

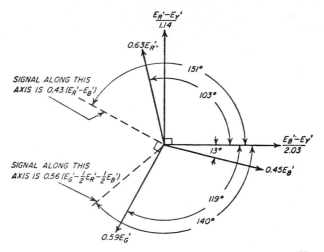

FIG. 4-62 Components for a reversing color-sequence display. (*From Fink.*[17])

ing color-sequence display.[100] Such an arrangement is shown in the block diagram of Fig. 4-63.

In the direct-processing arrangements, described in this section, no critical balance between circuit elements is required to obtain gray-scale tracking. Additionally, when using a display with effectively balanced phosphors, the desired transfer characteristics

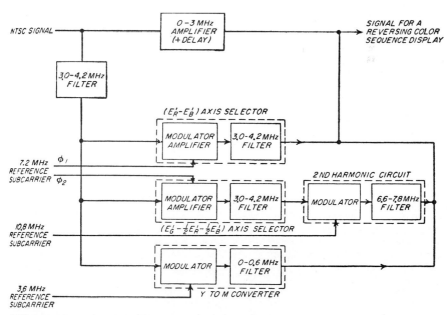

FIG. 4-63 Processing of NTSC signal for reversing color-sequence operation. (*From Fink.*[17])

to each phosphor can be obtained using a single background control to set the electron-gun bias, and a single dc restorer (or dc coupling) for the display coded signal.

4.6 DECODING IMPROVEMENTS

Beyond the basic decoding arrangements described in Sec. 4.5, there have been, in the last quarter-century since the first *Television Engineering Handbook* (D. G. Fink, ed.), a series of improvements in decoding of the composite color signal. These combine to provide modern-day color television receivers of good quality and stability. These improvements will be described in the context of an NTSC system, but most of them have more or less applicability to PAL and SECAM. Toward the end of this recent quarter-century period, great strides have been made in integrated-circuit decoders which can provide many special features, with great stability, linearity, and low cost.[140,145]

4.6.1 AUTOMATIC FINE TUNING. One of the early and very valuable receiver improvements was automatic fine tuning (AFT) which is an automatic frequency control system to help ensure that the luminance, chrominance, and sound signals are properly placed on the IF response. This helps to minimize the 920-kHz chrominance-sound beat by having the sound in the trap; ensure a more uniform ratio of chrominance to luminance at the picture detector; and improve the luminance and chrominance transient response of the receiver.

Statistically, AFT appears to be a distinct advantage. There are, however, isolated cases where manual tuning can improve the picture. These cases can occur owing to multipath conditions, antenna response, or RF tuner responses which upset the luminance-to-chrominance ratio so that some compensation by modifying the frequency placement of these signals on the IF response results in a net visual improvement.[128,132] Therefore, receivers normally include a manual AFT-defeat switch.

4.6.2 AUTOMATIC CHROMINANCE CONTROL. Automatic chrominance control (ACC) is another valuable tool to improve picture quality by maintaining the chrominance signal level range approximately constant, at least as measured by the color-burst amplitude. Assuming that the luminance signal range is maintained constant through a proper AGC system, and further assuming that the burst adequately represents the chrominance information, then ACC can assist in maintaining a constant color saturation even though the luminance and chrominance signals change compared with each other for various reasons. However, as will be discussed in Sec. 4.6.6, the FCC signal specifications do not ensure that the burst adequately represents the chrominance information. Therefore, some receivers are making use of the vertical-interval reference (VIR) signal (see Sec. 4.6.6).[132,146,147]

4.6.3 NEW DISPLAY PRIMARIES. Early in this recent quarter-century period, new display primaries came into general use which provided both brighter pictures and proper color balance on white with approximately equal drive. These primaries had a more efficient and more saturated red primary (rare earth) and a green primary which was less saturated (and correspondingly brighter). Use of these new primaries with a standard NTSC decoding matrix resulted in incorrect color. A first-order correction toward producing the intended picture can be made with a linear change in the matrix.[133] A more complete correction can be obtained by making a nonlinear matrix.[134] Most receivers use only the linear correction.

4.6.4 CHANGES TO REDUCE NEED FOR CONSUMER ADJUSTMENTS. An early study (1964) showed that when using correct encoding and decoding, there are enough other variable items to make color controls needed.[131] These other variable items include viewer preference for flesh tones, ambient illumination, differences in taking responses of cameras, and differences in individual color responses. With encoder and decoder incorrect adjustments, system amplitude and phase errors, and incorrect camera adjustments compounded on the above items, the viewer, during the early part of the past quarter-century period, found it necessary to regularly adjust the receiver controls in order to get pleasing color pictures.

Gradually, during this period, a substantial improvement in uniformity in the transmitted signal for skin color was obtained. Also, at the same time, the commercial impetus to sell acceptable color receivers forced receiver designers to *make more pleasing color images* in spite of the system variabilities, such as in the transmitted signal, and to make receiver designs such that control adjustments were less necessary.

The first step made by receiver designers was to modify the decoding matrix so colors near skin tones were *pulled closer* to skin tone. When this is done in a linear matrix, it is like reducing the gain for the Q signal. Skin tones generally correspond to a $+I$ signal and a Q signal near zero. Thus, with reduced Q signal gain, signals intended to be skin tones, but having an error, will be reproduced at a color closer to an expected skin color. But, with such reduced Q gain matrix, colors off the I axis are reproduced at reduced saturation. In particular, saturated reds and greens will have less saturation. Such reduced color gamut receivers have been widely sold in the United States. However, as uniformity of the transmitted signal continued to improve, and more use is made of the VIR signal in receivers, the color gamut of new receivers can be increased.[132]

Another approach, instead of effectively reducing the Q signal gain, is to distort the color gamut so that only signals near skin tones are pulled closer to such colors. A dynamic flesh correction circuit can maintain flesh color without noticeably affecting the three primaries. One such arrangement preserves the original colors in the $-I$ half-plane while signals in the $+I$ half-plane are shifted toward the $+I$ axis. Such nonlinear matrix arrangements have been shown both to make more pleasing *pictures* on the average and to make consumer adjustments of controls less necessary.[138-140]

4.6.5 USE OF COMB FILTERS. The comb filter is a type of transversal filter which can be used to reduce spurious effects due to band-sharing at some expense in fine-detail vertical resolution.[141] During the development of NTSC, comb filters in receivers were not considered to be economically feasible. However, since then, developments for PAL, plus more recent developments with charge-coupled devices (CCDs), have made the cost more reasonable. Additionally, in recent years, improvements in cameras and the increased use of both horizontal and vertical aperture equalization have made the spurious effects of band-sharing for the color subcarrier signal more noticeable.[142-144]

An arrangement similar to that used to separate the $R - Y$ and $B - Y$ components in PAL can be used to separate the high-frequency energy of the luminance and chrominance signals in NTSC. Such a comb filter used to select the chrominance signal will substantially cancel the cross-color component from the luminance signal. Furthermore, a comb filter used in the upper end of the luminance channel will filter out the chrominance signal and thus eliminate the dot crawl on edges, permitting the usable luminance passband to be increased.

By appropriate selection of passbands and appropriate reinsertion of the combed luminance from the chrominance channel, it is possible to obtain vertical detail enhancement of the lower frequencies (0 to 1 MHz).[145] This feature, when used with appropriate nonlinear processing, can produce pictures with both observable sharpness improvement and substantial reduction in spurious patterns due to band-sharing. Such circuits are being used in higher-quality NTSC receivers.

4.6.6 USE OF VIR SIGNAL AT RECEIVER. The specification for the NTSC signal overall is a highly practical system, but in certain aspects is somewhat incomplete in signal-component relationship. Perhaps the most outstanding tolerance problem is the tie of burst amplitude to chrominance amplitude range—a tie which is important if burst amplitude is to be used as a reference for chrominance, as it is in automatic chroma correction circuits. This problem was not considered by the NTSC, since the effects introduced by complex systems were not foreseen. The buildup of tolerances is such that if broadcast signals were to make full use of the permissible tolerances, then ACC circuits would have a very limited utility. Fortunately, the broadcasting industry exercises a much tighter control so that statistically automatic chroma correction circuits do have utility.

As described in Sec. 4.7.7, the VIR signal provides an improved reference for black and chrominance amplitude and phase, the chrominance reference being within the black-to-white region, and specifically on a luminance level corresponding closely to usual flesh luminance. While the VIR signal was intended mainly as a distribution network tool, it is also available at the receiver, and can be used to provide an improved chrominance amplitude and phase reference. Many receivers now are using the VIR.[146,147]

4.7 TRANSMISSION SYSTEM VERIFICATION

Special electrically generated test signals have proved to be very useful in testing chrominance-channel transmission characteristics, and for checking the performance of encoders and decoders. Special color test charts for use in front of color cameras, and for use in flying-spot scanners, have been devised, but, in general, these appear to be more useful for checking the performance of the camera or flying-spot scanner than for checking the performance of the chrominance channel of the system. Standard-frequency sweeping techniques, such as used in monochrome practice, are useful for testing the frequency response characteristics of various individual chrominance channels and of the channels handling the composite color signal. This section discusses the additional special test signals which have been found to be particularly useful for color work.

4.7.1 COLOR-BAR TEST SIGNAL.   The color-bar test pattern is in wide use.[26,102,103] This pattern provides a sequence of vertical bars in the picture area showing the saturated primaries and their complements, as well as black-and-white. The color-bar pattern generator provides a sequence of flat-top pulses to the green-, red-, and blue-signal inputs of the coding equipment. By suitable overlap of the pulses in certain portions of the raster, and nonoverlap in other portions, the three complementary saturated colors as well as the three saturated primary colors are produced. The exact order of colors in the bar pattern can take on several forms. That used in the industry standard (Fig. 21-46), together with the primary-color voltages required to make these patterns, are shown in Fig. 4-64. The form of pattern shown in Fig. 4-64 is attractive since it provides a monochrome component which appears as a continuous step wedge (but not with all steps having equal value).

To prevent overload of the transmitter, which would not be representative of operation with normal pictorial material, the amplitude of the saturated color bars is reduced to 75 percent of the full amplitude.

4.7.2 SOME SPECIAL TEST SIGNALS.   A neutral-gray step-wedge signal is particularly useful for checking the subcarrier balance of encoders and also for checking the gray-scale balance on color displays. Such a signal is generated by applying a monochrome step-wedge signal in equal amplitude to the three color channels of the encoder. When such a special step-wedge signal is not available, the monochrome color balance of a display might be checked using the monochrome component of the color-bar patterns mentioned above, and either turning off the subcarrier channel at the encoder, or turning off the chrominance channel at the receiver.

Flat fields of a saturated color are particularly useful for checking the decoding and color-purity adjustments of a display, particularly of one-gun sequential displays. Such a flat-field pattern can be generated by applying a blanking signal, in adjustable amplitude, to the three color channels of a color coder. Alternatively, the blanking signal might be used directly to generate a flat monochrome signal and to key in a subcarrier signal, which is adjustable in amplitude and phase.

While sweep-frequency generators can be used to check the frequency response of a system, it is sometimes desirable to obtain an approximate answer, with simple equip-

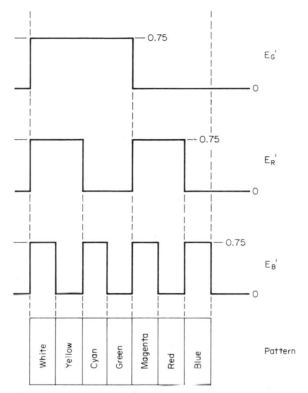

FIG. 4-64 Waveforms of color-bar patterns.

ment, and using a signal which permits normal operation of black-level clamps. Such a test signal uses the format illustrated in Fig. 21-15. It contains components which approximate the horizontal sync and blanking signals, and then, during the picture time, generates a series of discrete bursts of different frequency.[104,106] By observing the relative amplitude of these bursts, at any point in the system, an approximate idea is obtained of the frequency response to that particular point.

The effects of nonlinearities in the system can be conveniently checked using a waveform such as shown in Fig. 4-65. This includes a constant-amplitude component at subcarrier frequency, which is superimposed upon a luminance signal consisting either of a step wedge, as shown in the figure, or possibly a sine wave.[75,104–106] Using such a signal, any modulation of the chrominance signal produced by the luminance signal, while passing through nonlinear circuits, can be readily measured.

4.7.3 AMPLITUDE MEASUREMENTS. Measurements of signal amplitudes at various points in the system are conveniently made using a wide-band oscilloscope. If the relative amplitude of the luminance and chrominance signals is to be measured, then the frequency response of the oscilloscope at subcarrier frequency should be equal to the response at low frequencies. Additionally, if the oscilloscope is to be used for relative delay measurements of the luminance and chrominance signals, then the phase characteristic versus frequency should be linear up to, and somewhat beyond, the subcarrier frequency. These requirements are generally met when using wide-band oscilloscopes of the "10-MHz" category.

FIG. 4-65   Test signal for checking linearity. (*From Fink.*[17])

Observation of the composite signal, on a wide-band oscilloscope, when transmitting color bars, is particularly useful for checking the relative amplitude of the luminance and chrominance signals. Referring to Fig. 21-46, it will be noted that for the correct relative amplitudes of luminance and chrominance signals the subcarrier excursions during the green bar come just tangent to the black level. Additionally, the subcarrier excursions during the yellow and cyan bar are tangent to the peak-white level, when 75 percent amplitude color bars and a full-amplitude white bar are used. Furthermore, an oscilloscope observation that no subcarrier exists during the black and white bars is a useful check of the chrominance-channel balance.

For certain types of measurements it may be desirable to view substantially only the luminance component, or substantially only the chrominance component. This may be accomplished by inserting the appropriate filters ahead of the oscilloscope.[75,103,106] For example, a low-pass filter having an upper cutoff of approximately 2 MHz (with substantial attenuation at subcarrier frequency) permits viewing of the luminance signal only. On the other hand, use of a bandpass or high-pass filter having a lower cutoff of approximately 2 MHz permits viewing of mainly the chrominance components. The low-pass filter can be used to check rectification of the chrominance component by observing the luminance signal and turning on and off the chrominance signal. A change in pattern indicates that rectification of the chrominance signal is taking place. Correspondingly, the high-pass filter can be used to observe whether or not the luminance is modulating the chrominance signal. Here again, the luminance signal can be turned on and off, and a change in the observed chrominance signal represents the modulation effect due to the luminance signal. Alternatively, the step-wedge pattern, with superimposed constant-amplitude subcarrier, as mentioned in Sec. 4.7.2 can be used. As observed through the high-pass filter this should be just a constant-amplitude CW signal, and variations in amplitude of this signal represent modulation by the luminance signal.

4.7.4 PHASE MEASUREMENTS.   Accurate measurements of the chrominance signal involve not only relative amplitude measurements, which can be made by a wide-band oscilloscope, but also phase measurements.[103,107] While there are numerous arrangements for measuring phase, a device which has proved to be most useful for color television measurements is a so-called vectorimeter or vectorscope.[105,108–110] The basic arrangement

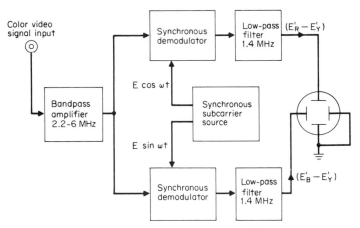

FIG. 4-66 Block diagram of a vectorscope. (*From Fink.*[17])

of a vectorscope is shown in Fig. 4-66. A bandpass amplifier is used to select the chrominance signal, and this signal is applied to a pair of synchronous quadrature demodulators. The outputs of these demodulators, after filtering, are applied to the horizontal and vertical deflection systems of an oscilloscope. Assuming a quadrature relation for the synchronous demodulators, accurate 90° relations between the horizontal and vertical deflection of the oscilloscope, and equal gains for the horizontal and vertical channels, the deflection of the spot then represents the head of a vector diagram of the applied chrominance information.

Several additional self-checking features can be added to vectorscope to improve its usefulness.[108] If the bandpass amplifier is keyed off during certain intervals of time, a spot is obtained on the cathode-ray-tube display which accurately represents zero subcarrier. Additionally, during another period of time, the input chrominance signal can be turned off and a signal which is about 10 kHz away from the subcarrier frequency can be injected. This draws a circular pattern on the display, and circularity of this pattern is a measure of the accuracy of quadrature adjustments and equality of channel gains.

The vectorscope can be used to show the vector diagram produced by a color bar signal, and by means of a suitably marked template or cursor, laid on the face of the tube, the signal can be checked to determine whether or not it is within specified tolerances. Also, by observing the pattern, it is generally possible to tell quickly what item of the color encoder is out of adjustment, when the signal is not correct.

The vectorscope can also be used to observe directly the effects of nonlinearities which result in modulation of the chrominance signal by the luminance signal. If a constant-amplitude chrominance signal is superimposed upon a varying-luminance signal then, under perfect conditions, the vectorscope should show a single spot. Variations in amplitude or phase of the chrominance signal, caused by the luminance signal, can be directly observed on the vectorscope. Similar effects can also be observed using the color-bar pattern by noting whether certain colors change in phase or amplitude when the luminance component of the signal is turned on and off. These types of measurements are useful not only for checking the color coding, the transmitter, and the network facilities, but also to check the performance of various circuits in a color receiver.

Measurements of color-synchronizing performance can also be made using a vectorscope. For example, the phase of the locally generated reference signal in a receiver can be measured with respect to the original subcarrier signal. Then, as the received signal is reduced in amplitude, gyrations in the color-sync phase, due to thermal noise, can be noted, and the signal level to produce a predetermined phase variation can be measured.

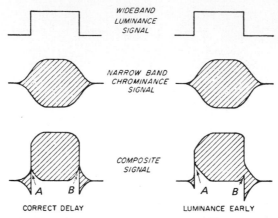

FIG. 4-67   Color-coder delay test. (*From Fink.*[17])

4.7.5 DELAY MEASUREMENTS. Another item of importance in testing the accuracy of chrominance transmission is the relative delay of the chrominance information compared with the luminance information. The correctness of this delay in the color encoder can be checked using a good wide-band oscilloscope and observing the pattern when transmitting a suitable single color bar. Since the chrominance signal has less bandwidth than the luminance signal, a characteristic type of pattern is produced during the starting and stopping transients, and symmetry of these two transients is a measure of the accuracy of relative delay. Figure 4-67 illustrates the type of pattern which is seen on the oscilloscope, and the equality of the areas marked A and B is a measure of the equality of delay of the two signals. The delay of E_Q' component compared with the E_Y' component can be measured by turning off the E_I' channel of the encoder, and correspondingly E_I' versus E_Y' delay can be checked by turning off the E_Q' channel.

Using a color coder which has been accurately delay-equalized, and transmitting a special test pattern, the relative delays in a decoder can be readily measured. A convenient test pattern is made up of two vertical bars which do not overlap. Both bars use no blue signal but have suitable amplitudes of green and red signals so that one bar pro-

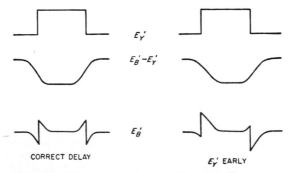

FIG. 4-68   Color-decoder delay test. (*From Fink.*[17])

duces a $+E_I'$ component, with no E_Q' component, and the other bar produces a $-E_Q'$ component, with no E_I' component. If the output of the blue channel of the decoder is observed, it should have zero value at all times except during transients. During both bars the zero output is obtained by a cancellation between the luminance component and the chrominance component, and during one bar the chrominance component is an E_I' component, and during the other bar it is an E_Q' component. When the delay of the respective chrominance components and the luminance components is correct, then the starting and stopping transients of the bars, observed in the blue channel, are symmetrical, such as illustrated in Fig. 4-68.

Relative-delay measurements of various channels can also be made by determining the phase slope versus frequency of the various channels. Envelope curve tracers and phase curve tracers which can be used for such measurements have been described in the literature.[111,112]

4.7.6 CONTINUOUS MONITORING BY VERTICAL-INTERVAL TEST SIGNALS.

The tremendous expansion of distribution networks for composite color signals through various media has made periodic and possible continuous testing of the transmission system an important function. Land lines, microwave links, tape recorders, satellite links, cable television, video disc, etc., all handle composite color television signals in some form, and all are subject to some of the problems discussed in Sec. 4.3. Broadcasters, common-carrier and cable-television operators, for example, have found continuous testing to be desirable. The most widely used method employs vertical-interval test signals (VITS) which are inserted on a scanning line near the end of the vertical interval. VITS can test or verify transmission system amplitude, phase, and delay characteristics using sine-squared baseband and modulated pulses. Also, a bar pattern may be included on one vertical-interval line. The Recommended Practices for these VITS formats[150] are shown in Figs. 21-15 to 21-17. These signals normally are treated as part of the vertical blanking period and thus may be removed in processing amplifiers which strip and reestablish sync and blanking.

4.7.7 IMPROVED COLOR REFERENCE BY VIR SIGNAL.

Extensive experience in distribution networks (and including receivers) has indicated that an improved reference for the composite signal, beyond that provided by the color-burst amplitude and phase, could be advantageous. The color signal primarily is in the sync-signal region, both in time and in amplitude. Thus, certain purposeful nonlinearities, such as those produced by stabilizing amplifiers in distribution systems, and unintended nonlinearities, particularly in broadcast transmitters, may mishandle the color burst—in extreme cases reinserting it at an incorrect amplitude and phase for a reference relative to the program chrominance. After much study, the Broadcast Television Systems Committee of the Electronic Industries Association recommended a specific VIR signal to be transmitted on line 19.[151,152] Figure 21-18 shows the format of this signal. FCC approval for such transmission in the United States, made in November 1974, allocated line 19 exclusively to the VIR signal. The signal is now widely used on the major United States and Canadian television networks.

The VIR signal is intended to be representative of the program picture information and, accordingly, is on a late line in the vertical interval so it can be treated like picture, and regarding amplitude is in the picture black-to-white region, additionally so it is treated like picture. The VIR signal contains reference levels for black, gray (50 percent), and chrominance and amplitude at reference phase, the latter being superimposed on a luminance level corresponding close to typical skin-tone luminance.

The VIR signal was formatted initially as a reference in distribution networks for use in resetting black level, white-to-black gain, burst amplitude, and phase to correspond to program chrominance.

While intended primarily as a networking tool, the VIR can also be used as an improved color reference in receivers to permit a more complete reference than that of the burst.[146,147]

REFERENCES

1. E. W. Engstrom, "Study of Television Image Characteristics," *Proc. IRE,* vol. 21, pp. 1631–1651, 1933.
2. R. D. Kell, A. V. Bedford, and M. A. Trainer, "Experimental Television System—the Transmitter," *Proc. IRE,* vol. 22, pp. 1246–1265, 1934.
3. H. A. Wheeler, and A. V. Loughren, "Fine Structure of Television Images," *Proc. IRE,* vol. 26, pp. 540–575, 1938.
4. A. V. Bedford, "Figure of Merit for Television Performance," *RCA Rev.,* vol. 3, pp. 36–44, 1938.
5. J. C. Wilson, "Channel Width and Resolving Power in Television Systems," *J. Television Soc.,* vol. 2, part 2, pp. 397–420, 1938.
6. R. D. Kell, A. V. Bedford, and G. L. Fredendall, "Determination of Optimum Number of Lines in Television System," *RCA Rev.,* vol. 5, pp. 8–30, 1940.
7. O. H. Schade, "Electro-Optical Characteristics of Television Systems—Part II," *RCA Rev.,* vol. 9, pp. 245–286, 1948.
8. O. H. Schade, *New System of Measuring and Specifying Image Definition, Optical Image Evaluation. Natl. Bur. Standards Circ.* 526, pp. 231–258, Apr. 29, 1954.
9. Pierre Mertz and F. Gray, "Theory of Scanning and Its Relation to Characteristics of Transmitted Signal in Telephotography and Television," *Bell System Tech. J.,* vol. 13, pp. 464–515, 1934.
10. M. W. Baldwin, "Subjective Sharpness of Simulated Television Images," *Proc. IRE,* vol. 28, pp. 458–468, 1940.
11. *IEEE Standard Dictionary of Electrical & Electronics Terms,* IEEE Std. 100–1977, 2d ed., p. 494.
12. A. C. Hardy, *Handbook of Colorimetry,* pp. 11 and 59, Technology Press, Massachusetts Institute of Technology, Cambridge, Mass., 1936.
13. R. M. Evans, *An Introduction to Color,* p. 209, Wiley, New York, 1948.
14. Committee on Colorimetry, Optical Society of America, *The Science of Color,* pp. 42 and 101, Thomas Y. Crowell, New York, 1953.
15. D. B. Judd, "Introduction to Color," *Symposium on Color,* p. 2, American Society for Testing Materials, Philadelphia, 1941.
16. R. M. Evans, W. T. Hanson, Jr., and W. L. Brewer, *Principles of Color Photography,* pp. 42 and 45, Wiley, New York, 1953.
17. D. G. Fink (ed.), *Television Engineering Handbook,* McGraw-Hill, New York, 1957.
18. W. T. Wintringham, "Color Television and Colorimetry," *Proc. IRE,* vol. 39, pp. 1135–1172, October, 1951.
19. Federal Communications Commission, Technical Standards Amended to Incorporate Color, FCC Public Notice No. 53-1663, Mimeo 98948, U.S. Government Printing Office, Washington, 1953.
20. NTSC, "NTSC Signal Specifications," *Proc. IRE,* vol. 42, pp. 17–19, January, 1954.
21. B. D. Loughlin, "Recent Improvements in Band-Shared Simultaneous Color Television Systems," *Proc. IRE,* vol. 39, pp. 1264–1279, October 1951.
22. C. J. Hirsch, W. F. Bailey, and B. D. Loughlin, "Principles of NTSC Compatible Color Television," *Electronics,* vol. 25, pp. 88–95, February 1952.
23. F. J. Bingley, "Colorimetry in Color Television," *Proc. IRE,* vol. 41, pp. 838–851, July, 1953.
24. W. F. Bailey, "The Constant Luminance Principle in NTSC Color Television," *Proc. IRE,* vol. 42, pp. 60–66, January 1954.

25. R. P. Burr, "Transmitting Terminal Apparatus for NTSC Color Television, Communication and Electronics," *J. Am. Inst. Elec. Engrs.*, pp. 26–32, March 1953.

26. E. E. Gloystein and A. H. Turner, "The Colorplexer—A Device for Multiplexing Color Television Signals in Accordance with the NTSC Signal Specifications," *Proc. IRE*, vol. 42, pp. 204–212, January 1954.

27. A. V. Bedford, "Mixed Highs in Color Television," *Proc. IRE*, vol. 38, pp. 1003–1009, September 1950.

28. K. McIlwain, "Requisite Color Bandwidth for Simultaneous Color-Television Systems," *Proc. IRE*, vol. 40, pp. 909–912, August 1952.

29. E. N. Willmer and W. D. Wright, "Colour Sensitivity of the Fovea Centralis," *Nature*, vol. 156, pp. 119–121, July 28, 1945.

30. W. E. Middleton and M. C. Holmes, "The Apparent Colors of Surfaces of Small Subtense—A Preliminary Report," *J. Opt. Soc. Amer.*, vol. 39, pp. 582–592, July, 1949.

31. R. D. Kell and A. C. Schroeder, "Optimum Utilization of the Radio Frequency Channel for Color Television," *RCA Rev.*, vol. XIV, pp. 133–143, June 1953.

32. NTSC, "Some Supplementary References Cited in the National Television System Committee Reports," NTSC, pp. 13–422, Radio-Electronics-Television Manufacturers Association, New York, 1953 (see documents NTSC-P13-286, NTSC-P13-287, and NTSC-P13-288, which are parts of this reference).

33. F. J. Bingley, "Colorimetry in Color Television—Part III," *Proc. IRE*, vol. 42, pp. 51–57, January 1954.

34. P. Mertz, "Perception of Television Random Noise," *J. SMPTE*, vol. 54, pp. 8–34, January 1950.

35. B. M. Oliver, "Tone Rendition in Television," *Proc. IRE*, vol. 38, pp. 1288–1300, November 1950.

36. RCA, Petition of Radio Corp. of America and National Broadcasting Co., Inc., pp. 376–377, June 25, 1953.

37. F. J. Bingley, "Colorimetry in Color Television—Part II," *Proc. IRE*, vol. 42, pp. 48–51, January 1954.

38. E. W. Engstrom, "Basic Concepts and Evolution of Color Television," *Proc. IRE*, vol. 42, pp. 7–9, January 1954.

39. A. V. Loughren, "Psychophysical and Electrical Foundations of Color Television," *Proc. IRE*, vol. 42, pp. 9–11, January 1954.

40. F. Gray, U.S. Patent 1769920, filed April 30, 1929, issued July 8, 1930.

41. P. Mertz and F. Gray, "A Theory of Scanning and Its Relation to the Characteristics of the Transmitted Signal in Telegraphy and Television," *Bell System Tech. J.*, vol. 13, pp. 464–515, July 1934.

42. P. Mertz, "Television—The Scanning Process," *Proc. IRE*, vol. 29, pp. 529–537, October 1941.

43. R. B. Dome, "Frequency-Interlace Color Television," *Electronics*, vol. 23, pp. 70–75, September 1950.

44. A. V. Loughren and C. J. Hirsch, "Comparative Analysis of Color TV Systems," *Electronics*, vol. 24, pp. 92–96, February 1951.

45. R. B. Dome, "Spectrum Utilization in Color Television," *Proc. IRE*, vol. 39, pp. 1323–1331, October 1951.

46. I. C. Abrahams, "The Frequency Interleaving Principle in the NTSC Standards," *Proc. IRE*, vol. 42, pp. 81–83, January 1954.

47. D. G. Fink, *Television Engineering*, pp. 506–508, McGraw-Hill, New York, 1952.

48. I. C. Abrahams, "Choice of Chrominance Subcarrier Frequency in the NTSC Standards," *Proc. IRE,* vol. 42, pp. 79–80, January 1954.

49. G. H. Brown, "The Choice of Axes and Bandwidths for the Chrominance Signals in NTSC Color Television," *Proc. IRE,* vol. 42, pp. 58–59, January 1954.

50. P. W. Howells, "Transients in Color Television," *Proc. IRE,* vol. 42, pp. 212–220, January 1954.

51. R. C. Palmer, "System Delay Characteristics in NTSC Color Television," *Proc. IRE,* vol. 42, pp. 92–95, January 1954.

52. RCA, "Six-Megacycle Compatible High-Definition Color Television System," *RCA Rev.,* vol. X, pp. 504–524, December 1949.

53. W. P. Boothroyd, "Dot Systems of Color Television," part 1, *Electronics,* vol. 22, pp. 88–92, December 1949; part 2, *Electronics,* vol. 23, pp. 96–99, January 1950.

54. D. G. Fink, "Alternative Approaches to Color Television," *Proc. IRE,* vol. 39, pp. 1124–1134, October 1951.

55. D. G. Fink, *Television Engineering,* pp. 529–549, McGraw-Hill, New York, 1952.

56. S. Applebaum, "Gamma Correction in Constant Luminance Color Television Systems," *Proc. IRE,* vol. 40, pp. 1185–1195, October 1952.

57. D. C. Livingston, "Colorimetric Analysis of the NTSC Color Television System," *Proc. IRE,* vol. 42, pp. 138–150, January 1954.

58. C. J. Hirsch, "Proposal for Modifications of the Complete Color Signal to Provide Improved Constant Luminance Transmission," NTSC-P13-284.

59. P. W. Howells, "The Concept of Transmission Primaries in Color Television," *Proc. IRE,* vol. 42, pp. 134–138, January 1954.

60. W. F. Bailey and C. J. Hirsch, "Quadrature Cross Talk in NTSC Color Television," *Proc. IRE,* vol. 42, pp. 84–90, January 1954.

61. G. H. Brown, "Mathematical Formulations of the NTSC Color Television Signal," *Proc. IRE,* vol. 42, pp. 66–71, January 1954.

62. J. B. Chatten, "Transition Effects in Compatible Color Television," *Proc. IRE,* vol. 42, pp. 221–228, January 1954.

63. D. C. Livingston, "Reproduction of Luminance Detail by NTSC Color Television Systems," *Proc. IRE,* vol. 42, pp. 228–234, January 1954.

64. C. H. Jones, "Effects of Noise on NTSC Color Standards," *Proc. Natl. Electronics Conf.,* pp. 185–200, 1952.

65. F. J. Bingley, "Transfer Characteristics in NTSC Color Television," *Proc. IRE,* vol. 42, pp. 71–78, January 1954.

66. D. B. Judd, *Color in Business, Science, and Industry,* pp. 231–253, Wiley, New York, 1952.

67. H. Weiss, "Significance of Some Receiver Errors to Color Reproduction," *Proc. IRE,* vol. 42, pp. 1380–1388, September 1954.

68. D. C. Livingston, "Theory of Synchronous Demodulator as Used in NTSC Color Television Receiver," *Proc. IRE,* vol. 42, pp. 284–287, January 1954.

69. D. H. Pritchard and R. U. Rhodes, "Color Television Signal Receiver Demodulators," *RCA Rev.,* vol. XIV, pp. 205–226, June 1953.

70. K. Schlesinger, "TV Color Detectors Use Pulsed-Envelope Method," *Electronics,* vol. 27, pp. 142–145, March 1954.

71. J. S. S. Kerr, "Transient Response in Color Carrier Channel with Vestigial Side Band Transmission," *IRE Convention Record,* part 4, pp. 18–23, 1953.

72. G. L. Fredendall, "Delay Equalization in Color Television," *Proc. IRE,* vol. 42, pp. 258–262, January 1954.

73. T. M. Gluyas, Jr., "Television Transmitter Considerations in Color Broadcasting," *RCA Rev.,* vol. XV, pp. 312–334, September 1954.

74. J. R. Roe, "Transmission of Color over Inter-City Television Networks," *Proc. IRE,* vol. 42, pp. 270–273, January 1954.

75. G. L. Fredendall and W. C. Morrison, "Effect of Transmitter Characteristics on NTSC Color Television Signals," *Proc. IRE,* vol. 42, pp. 95–105, January 1954.

76. D. Richman, "Color-Carrier Reference Phase Synchronization Accuracy in NTSC Color Television," *Proc. IRE,* vol. 54, pp. 106–133, January 1954.

77. D. W. Epstein, "Colorimetric Analysis of RCA Color Television System," *RCA Rev.,* vol. XIV, pp. 227–258, June 1953.

78. R. G. Neuhauser, A. A. Rotow, and F. S. Vieth, "Image Orthicons for Color Cameras," *Proc. IRE,* vol. 42, pp. 161–165, January 1954.

79. J. F. Fisher, "Generation of NTSC Color Signals," *Proc. IRE,* vol. 41, pp. 338–343, March 1953.

80. W. E. Tucker and F. J. Jauda, "RCA Color Slide Camera, TK-4A," *Broadcast News,* pp. 52–57, January–February 1954.

81. J. H. Haines, "Color Characteristics of a Television Film Scanner," *IRE Convention Record,* part 7, pp. 100–104, 1954.

82. D. G. Fink: *Television Engineering,* pp. 216–222, McGraw-Hill, New York, 1952.

83. J. H. Roe, "New Television Field-Pickup Equipment Employing the Image Orthicon," *Proc. IRE,* vol. 35, pp. 1532–1546, December 1947.

84. W. R. Feingold, "Matrix Networks for Color TV," *Proc. IRE,* vol. 42, pp. 201–203, January 1954.

85. W. L. Brewer, J. H. Ladd, and J. E. Pinney, "Brightness Modification Proposals for Televising Color Film," *Proc. IRE,* vol. 42, pp. 174–191, January 1954.

86. R. P. Burr, "The Use of Electronic Masking in Color Television," *Proc. IRE,* vol. 42, pp. 192–200, January 1954.

87. J. H. Ladd and W. L. Brewer, "Photographic Simulation of Color Television Brightness Modifications," *IRE Convention Record,* part 7, pp. 110–113, 1954.

88. This is a modification of the coder described in reference 13 and represents the color coders developed for use in the Hazeltine Laboratories.

89. A. H. Lind, "RCA Color Sync Generator Equipment," *Broadcast News,* pp. 50–51, January–February 1954.

90. H. S. Black, *Modulation Theory,* pp. 141–166, Van Nostrand, New York, 1953.

91. R. S. Caruthers, "Copper Oxide Modulators in Carrier Telephone Systems," *Bell System Tech. J.,* vol. 18, pp. 315–337, April 1939.

92. M. H. Kronenberg and E. S. White, "Design Techniques for Color Television Receivers," *Electronics,* vol. 27, pp. 136–143, February 1954.

93. C. Masucci, J. J. Insalaco, and R. Zitta, "A Laboratory Receiver for Study of the NTSC Color Television System," *Proc. IRE,* vol. 42, pp. 334–343, January 1954.

94. Wm. Quinn, Jr., "Methods of Matrixing in NTSC Color Television Receiver," *IRE Convention Record,* part 4, pp. 167–172, 1953.

95. K. E. Farr, "Compatible Color TV Receiver," *Electronics,* vol. 26, pp. 98–104, January 1953.

96. J. D. Gow and R. Dorr, "Compatible Color Picture Presentation with the Single Gun Tricolor Chromatron," *Proc. IRE,* vol. 42, pp. 308–315, January 1954.

97. R. R. Law, "A One-Gun Shadow-Mask Color Kinescope," *Proc. IRE,* vol. 39, pp. 1194–1201, October 1951.

98. P. K. Weimer and N. Rynn, "A 45-Degree Reflection-Type Color Kinescope," *Proc. IRE,* vol. 39, pp. 1201–1211, October 1951.

99. R. Dressler, "PDF Chromatron—Single or Multi-Gun Tri-Color Cathode-Ray Tube," *Proc. IRE,* vol. 41, pp. 851–858, July 1953.

100. B. D. Loughlin, "Processing of the NTSC Color Signal for One-Gun Sequential Color Displays," *Proc. IRE,* vol. 42, pp. 299–308, January 1954.

101. S. K. Altes and A. P. Stern, "Single-Gun Picture Tubes in NTSC Color Television," *IRE Convention Record,* part 7, pp. 46–52, 1954.

102. R. P. Burr, W. R. Stone, and R. O. Noyer, "Picture Generator for Color Television," *Electronics,* vol. 24, pp. 116–120, August 1951.

103. A. C. Luther, Jr., "Methods of Verifying Adherence to the NTSC Color Signal Specifications," *Proc. IRE,* vol. 42, pp. 235–240, January 1954.

104. W. C. Morrison, K. Karstad, and W. L. Behrend, "Test Instruments for Color Television," *Proc. IRE,* vol. 42, pp. 247–258, January 1954.

105. J. F. Fisher, "Alignment of a Monochrome TV Transmitter for Broadcasting NTSC Color Signals," *Proc. IRE,* vol. 42, pp. 263–270, January 1954.

106. J. A. Bauer, "A Versatile Approach to the Measurement of Amplitude Distortion in Color Television," *Proc. IRE,* vol. 42, pp. 240–246, January 1954.

107. A. P. Stern, "Phase Measurements at Subcarrier Frequency in Color Television," *IRE Convention Record,* part 4, pp. 57–60, 1953.

108. C. E. Page, "Monitoring System for NTSC Color Television Signals," *IRE Convention Record,* part 4, pp. 61–65, 1954.

109. K. Schlesinger and L. W. Nero, "Phase Indicator for Color Television," *Electronics,* vol. 25, pp. 112–114, October 1952.

110. W. L. Firestone and R. A. Richardson, "Simplified Vectorscope Measures Phase," *Electronics,* vol. 26, pp. 180–182, September 1953.

111. A. C. Schroeder, "A Sweep Method for Measuring Envelope Delay," *RCA Industry Service Laboratory Bulletin* 883, October 1952.

112. B. D. Loughlin, "A Phase-Curve Tracer for Television," *Proc. IRE,* vol. 29, p. 107, March 1941.

113. B. D. Loughlin, "Recent Improvements in Band-Shared Simultaneous Color Television Systems," pt. II, "Color Television Systems with Oscillating Color Sequence," *Proc. IRE,* vol. 39, pp. 1273–1279, October 1951. Description of Color Phase Alternation (CPA), also called Oscillating Color Sequence (OCS).

114. W. Bruch, "The PAL Colour TV Transmission System," *IEEE Trans. BTR,* vol. BTR-12, pp. 87–96, May 1966.

115. B. D. Loughlin, "The PAL Color Television System," *IEEE Trans. BTR,* vol. BTR-12, pp. 153–158, July 1966.

116. R. Theile, "Principles of Compatible Color Signal Transmission with Special Reference to the Various European Proposals for the Subcarrier Modulation Technique," *IEEE Trans. BTR,* vol. BTR-12, pp. 27–37, May 1966.

117. P. S. Carnt and G. B. Townsend, *Colour Television,* vol. 1 NTSC, vol. 2 PAL and SECAM, Iliff Books (Wireless World), London, 1961 and 1969.

118. "The Issue of Worldwide Color Television Standards," papers and discussions presented at IEEE International Convention, New York, March 22, 1966, *IEEE Trans. BTR,* vol. BTR-12, May 1966. A compilation of over a dozen good technical papers on the subject.

119. Panel Discussion of Color Television Systems, Chicago Spring Conference, June 1966, *IEEE Trans. BTR,* vol. BTR-12, pp. 147–166, July 1966.

120. D. H. Pritchard and J. J. Gibson, "Worldwide Color Television Standards—Similarities and Differences," *J. SMPTE,* vol. 89, pp. 111–120, February 1980. Also, see addendum, pp. 948–949, December 1980.

121. T. Fujio, "A Study of High-Definition TV System in the Future," *IEEE Trans. Broadcast.,* vol. BC-24, no. 4, pp. 92–100, December 1978.

122. T. Fujio et al., "High-Definition Television System—Signal Standard and Transmission," *J. SMPTE,* vol. 89, pp. 579–584, August 1980.

123. T. Fujio, "A Universal Weighted Power Function of Television Noise and Its Application to High-Definition TV System Design," *J. SMPTE,* vol. 89, pp. 663–669, September 1980.

124. T. Fujio, "High-Definition Wide-Screen Television System for the Future—Present State of the Study of HD-TV Systems in Japan," *IEEE Trans. Broadcast.,* vol. BC-26, no. 4, pp. 113–124, December 1980.

125. T. Fujio, "The NHK High-Resolution Wide-Screen Television System, Television Technology in the 80's," *J. SMPTE,* vol. 90, pp. 166–176, 1981. A collection of papers on television production and postproduction technology, presented during the 15th Annual SMPTE Television Conference, San Francisco, Feb. 6–7, 1981.

126. K. Hayashi, "Research and Development on High Definition Television in Japan," *J. SMPTE,* vol. 90, pp. 178–186, March 1981.

127. T. Kubo, "Development of High-Definition TV Displays," *IEEE Trans. Broadcast.,* vol. BC-28, no. 2, pp. 51–64, June 1982.

128. B. D. Loughlin, "Color Signal Distortions in Envelope Type of Second Detectors," *IRE Trans. BTR,* vol. BTR-3, pp. 81–93, October 1957; also *IEEE Trans. Consumer Electronics,* vol. CE-28, pp. 19–31, February 1982.

129. C. B. Neal and S. K. Goyal, "Frequency- and Amplitude-Dependent Phase Effects in Television Broadcast Systems," *IEEE Trans. Consumer Electronics,* vol. CE-23, no. 3, pp. 234–247, August 1977.

130. H. Isono et al., "Subjective Evaluation of Apparent Reduction of Chromatic Blur Depending on Luminance Signals," *IEEE Trans. Broadcast.,* vol. BC-24, no. 4, pp. 107–115, December 1978.

131. C. J. Hirsch, "A Study of the Need for Color Controls on Color TV Receivers in a Color TV System Operating Perfectly," *IEEE Trans. BTR,* vol. BTR-10, no. 3, pp. 71–96, November 1964.

132. B. D. Loughlin, "Techniques for Improving Broadcast Color Television," *IEEE Intercon 1973 Convention Record* 39/1, H-208, pp. 1–5, March 1973.

133. N. W. Parker, "An Analysis of the Necessary Decoder Corrections for Color Receiver Operation with Nonstandard Receiver Primaries," *IEEE Trans. BTR,* vol. BTR-12, 1, pp. 23–32, 1966.

134. R. Sasaki et al., "Non-Linear Chrominance Signal Decoding," *IEEE Trans. Consumer Electronics,* vol. CE-23, pp. 335–341, August 1977.

135. S. P. Ronzheimer, "A New Approach to the DC-Restoration Problem," *IRE Trans. BTR,* vol. BTR-8, pp. 39–55, November 1962.

136. R. Citta, "Frequency and Phase Lock Loop," *IEEE Trans. Consumer Electronics,* vol. CE-23, no. 3, pp. 358–365, August 1977.

137. L. A. Harwood, "A New Chroma-Processing IC Using Sample-and-Hold Techniques," *IEEE Trans. BTR,* vol. BTR-19, pp. 136–141, May 1973.

138. L. A. Harwood, "A Chrominance Demodulator IC With Dynamic Flesh Correction," *IEEE Trans. Consumer Electronics,* vol. 22, pp. 111–118, February 1976.

139. L. A. Harwood, "An Integrated One-Chip Processor for Color TV Receivers," *IEEE Trans. Consumer Electronics,* vol. CE-23, no. 3, pp. 300–310, August 1977.

140. L. A. Harwood and E. J. Witmann, "Integrated NTSC Chrominance/Luminance Processor," *IEEE Trans. Consumer Electronics,* vol. CE-26, pp. 693–706, November 1980.

141. D. H. Pritchard et al., "A High Performance Television Receiver Experiment," *IEEE Trans. BTR,* vol. BTR-18, pp. 82–90, May 1972.

142. S. Barton, "A Practical Charge-Coupled Device Filter for the Separation of Luminance and Chrominance Signals in a Television Receiver," *IEEE Trans. Consumer Electronics,* vol. CE-23, no. 3, pp. 342–357, August 1977.

143. R. Turner, "Some Thoughts on Using Comb Filters in the Broadcast Television Transmitter and at the Receiver," *IEEE Trans. Consumer Electronics,* vol. CE-23, no. 3, pp. 248–257, August 1977.

144. W. A. Lagoni, "A Base-Band Comb Filter for Consumer Television Receivers," *IEEE Trans. Consumer Electronics,* vol. CE-26, pp. 94–99, February 1980.

145. D. H. Pritchard, "A CCD Comb Filter for Color TV Receiver Picture Enhancement," *RCA Rev.,* vol. 41, pp. 3–28, March 1980.

146. S. K. Kim, "VIR II System," *IEEE Trans. Consumer Electronics,* vol. CE-24, no. 3, pp. 200–208, August 1978.

147. I. Nishimura et al., "A Newly Developed Correction System of Transmission Distortion with Vertical Interval Reference Signal," *IEEE Trans. Consumer Electronics,* vol. CE-25, pp. 71–81, February 1979.

148. W. Ciciora et al., "A Tutorial on Ghost Cancelling in Television Systems," *IEEE Trans. Consumer Electronics,* vol. CE-25, pp. 9–44, February 1979.

149. N. W. Parker, "The Cost of Using PAL or SECAM and Possible Improvements in NTSC Receivers," *IEEE Trans. BTR,* vol. BTR-12, no. 3, pp. 159–161, July 1966.

150. R. A. O'Connor, "Current Usage of Vertical Interval Test Signals in Television Broadcasting," *IEEE Trans. Consumer Electronics,* vol. CE-22, pp. 220–229, August 1976.

151. C. B. Neal, "A Brief History of the VIR: Vertical Interval Reference Signal," *Communications/Engineering Digest,* vol. 2, no. 1, pp. 20–25, January 1976.

152. C. B. Neal, "Improving Television Color Uniformity Through Use of the VIR Signal," *IEEE Trans. Consumer Electronics,* vol. CE-22, pp. 230–237, August 1976.

Waveforms and Spectra of Composite Video Signals

Joseph F. Fisher

Wynnewood, Pennsylvania and
Philco-Ford (retired)

Richard G. Clapp

Consultant
Havertown, Pennsylvania and
Philco-Ford (retired)

NOTE: Parts of this chapter are adapted from D. G. Fink (ed.), *Television Engineering Handbook*, McGraw-Hill, New York, 1957. Used by permission. The source material in Chaps. 7 and 10 was prepared by Daniel E. Harnett, then Engineer, General Electric Company; John B. Chatten, then Project Engineer, Philco Corporation Research Division; Richard G. Clapp, then Executive Engineer, Philco Corporation Research Division; and Donald G. Fink, then Director of Research, Philco Corporation Research Division.

5.1 SIGNIFICANCE OF THE COMPOSITE VIDEO WAVEFORM AND SPECTRUM

The composite video signal waveform is the amplitude-versus-time representation of the signal produced jointly by the television camera, the camera-control unit, the sync generator, and, for color operation, the color encoder. The sync generator produces the mixed blanking, the mixed sync, and the color subcarrier. For color operation the color encoder packages the processed red, green, and blue signals into a composite color signal. The time and amplitude relationships of these portions of the composite signal are discussed in Secs. 5.2 and 5.3 and are shown in Figs. 5-4 and 5-5 of this chapter.

The excellence of television transmission is inherent in the composite video signal as presented to the picture tube and synchronizing circuits of the receiver. The distortions and disturbances present in the signal at these points affect in a major way the resolution, tonal gradation, color fidelity, contrast, and raster stability exhibited by the image. To maximize these aspects of image quality, the system must be designed in terms of the video waveform, taking account of the amplitude levels, rates of change, noise levels, amplitude compression, and other distortions encountered under typical receiving conditions.

The *video spectrum* is the corresponding *amplitude-versus-frequency* representation of the signal. The distribution of energy in the spectrum is affected by the rates of change and periodicities of the waveform (notably the line- and field-scanning rates). Those aspects of system design having to do with bandwidth and frequency interleaving are most conveniently understood by reference to the spectra of typical video signals.

The present chapter describes waveforms and spectra arising from the scanning of typical pictures, test charts, and stationary images which are generally available to system designers and operators. These include the spectra of the synchronizing signals, flat fields in monochrome and color, a color-bar chart, a monochrome test chart having resolution wedges, and a typical color picture.

The spectra were experimentally measured by the method described in Sec. 5.11. Computation of the spectra is presented, in the flat-field cases, as a check on the experimental method.

5.2 SYNCHRONIZATION WAVEFORMS AND FUNCTIONS

5.2.1 DEFINITION OF SYNCHRONIZATION AND APPLICATION IN TELEVISION.

In monochrome television transmissions, two synchronization signals are used to control deflection. In color transmissions, a third synchronizing signal provides the color-coding information. The production, transmission, and use of these signals constitute the synchronizing functions encountered in television.

The synchronizing problems considered in this chapter apply to the 525-line NTSC color television system used in the United States, Canada, Mexico, and Japan. Specifications for the PAL and SECAM color television systems are included in Chaps. 4, 21, and 22 and in an article by Pritchard and Gibson.[1]†

5.2.2 PERFORMANCE REQUIREMENTS FOR SATISFACTORY RECEPTION.

In each of the three synchronizing functions the receiver operation must follow precisely the control provided in the transmitted signal, and the transmitted synchronizing frequencies must be correspondingly exact. Their phases must be held within close tolerances. The changes in phase from one scan to the next must be so small that they do not degrade the picture, and the relation of the receiver scanning information to the picture information must be correct. In deflection synchronization the steady-state displacement in phase must be small, not only to permit correct centering of the picture on the raster, but also to avoid the appearance of video modulation on the retrace ("foldover").

A precise phase relationship between the transmitted chrominance information and the synchronized CW color subcarrier regenerated in the color television monitor or receiver must be maintained. NTSC color television receivers and monitors have a tint or hue control by means of which the phase of the CW color subcarriers applied to the color demodulators may be adjusted to fine-tune the reproduced color. Once set, this adjustment should remain stable and locked to the same phase of the transmitted color burst.

The change of phase between successive vertical scans must be so small that the error in timing does not impair the interlace. The vertical difference in line spacing should be less than 10 percent, since this has been found to give the subjective impression of good

---

†Superscript numbers refer to References at end of chapter.

interlace. Complete loss of interlace occurs for a vertical timing error of one half-line in alternate fields.

If the start of one field relative to the next is advanced by approximately 3.0 μs, an impairment of interlace of one-tenth the line separation occurs. Since the vertical field is 16,667 μs, this means that the timing of the vertical synchronization must be correct within approximately $\frac{1}{5000}$ of the vertical field interval.

The accuracy requirement for the horizontal scan depends on the range of video frequencies transmitted. The highest video frequency is about 4.2 MHz, and the highest frequency actually used by the receiver may be less than this. If the timing on alternate lines is shifted by as much as $\frac{1}{2}$ cycle, the detail in a fine vertical line would be destroyed. Using 4.2 MHz as the upper limit, this occurs if successive lines are mistimed by 0.119 μs. Since the line period is 63½ μs, this means that the horizontal accuracy must be approximately 1 part in 500. The fact that the vertical timing cycle is 10 times more critical than the horizontal timing illustrates the great sensitivity of the interlaced picture to vertical displacement.

In color transmissions, errors in hue can be caused by the improper phase of the receiver CW subcarrier voltage which is synchronized to the transmitted color burst. In a few instances an error of 2° can be detected. In most cases an error of 10 to 15° must be present before the degradation in picture quality is sufficient to be objectionable. This tolerance question has been studied by Weiss.[2]

Since it is known that an error of about 2° can be observed in the picture under critical viewing conditions, this value is customarily taken as a permissible tolerance for the color television receiver subcarrier-synchronizing system.

Since the period of the subcarrier is 0.28 μs, an error of 2° represents a timing error of 0.0015 μs. In terms of time interval this is by far the most precise of the three synchronizing tolerances. In terms of the fraction of the cycle, which is a better indication of the difficulty encountered in maintaining accuracy, the subcarrier-synchronizing problem is the easiest of the three.

Any timing changes introduced in the receiver must be small enough so that they can be corrected by the receiver adjustments. Once set, the variations should not exceed the values given in Table 5-1.

5.2.3 FREQUENCY RELATIONSHIPS.   To secure a stable, interference-free picture, the relations between the several synchronizing frequencies must be properly chosen. After detection the frequencies are:

1. Vertical scanning frequency
2. Horizontal scanning frequency
3. Chrominance-subcarrier frequency
4. Difference frequency between sound and picture carrier

The choice of the approximate numbers for each of these frequencies is dependent upon the general considerations outlined in Chap. 22. These lead, for example, to the choice of about 3.6 MHz as the proper chrominance-subcarrier frequency. Frequencies

Table 5-1   Synchronization Periods and Tolerances

Synchronizing signal	Nominal frequency	Period, μs	Tolerance, μs	Approx. fraction of period
Vertical	60 Hz	16,667	3.0	1/5,000
Horizontal	15,750 Hz	63.5	0.119	1/500
Chrominance	3.58 MHz	0.28	0.0015	1/180

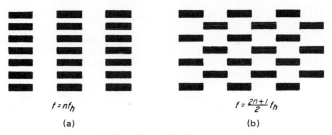

$$f = nf_h \qquad\qquad f = \frac{2n+1}{2} f_h$$

(a)   (b)

FIG. 5-1   Pattern visibility. (*From Fink.*[3])

in the upper part of the video-signal range are more or less visible, depending on their harmonic relationship to the line frequency. As shown in Fig. 5-1, a frequency which is an exact harmonic of the line-scanning rate produces a noticeable vertical grid whereas signals of frequency halfway between harmonics of the line-scanning rate give a much less visible pattern. Tests have shown that an interfering signal at an odd harmonic of half the line-scanning rate may have approximately five times the voltage of a signal which is at a harmonic of the line-scanning rate, for equal visibility. After the nominal values of scanning frequencies and chrominance subcarrier have been chosen, the exact values are selected to take advantage of the least visible frequency ratios.

The chrominance subcarrier and sound carrier may cause the following interference in the picture:

1. The chrominance-subcarrier frequency (approximately 3.6 MHz)

2. Sound carrier (4.5 MHz)

3. Beat between sound and chrominance (approximately 0.9 MHz)

The color-subcarrier frequency used is 3.579545 MHz (± 10 Hz) which is the 455th harmonic of one half the line-scanning frequency. As shown in Fig. 5-2, the synchronizing generator uses a crystal-controlled oscillator operating at the color subcarrier frequency. The number 455 can be factored ($455 = 5 \times 7 \times 13$), and these countdowns are

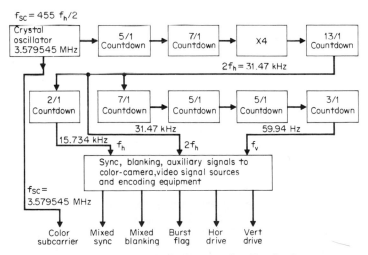

FIG. 5-2   Derivation of frequencies in the composite video signal.

Table 5-2   Synchronizing Frequencies and Ratios for Color Signals

	Actual	Nominal
Vertical scanning frequency, f_v	59.94 Hz	60 Hz
Horizontal scanning frequency, f_h	15,734.264 Hz	15,750 Hz
Chrominance subcarrier, f_{sc}	3.579545 MHz	3.58 MHz
Sound-carrier–picture-carrier difference, f_s	4.5 MHz	4.5 MHz

Ratios

f_h/f_v 525/2
f_{sc}/f_h 455/2
f_s/f_h . . . . 572/2 or 286/1
$(f_s - f_{sc})/f_h$ 117/2

used along with a 4:1 multiplier to derive one of the required timing signals ($2f_h$). The other timing signals f_h, f_v are derived as shown in Fig. 5-2. The actual and nominal frequencies are shown in Table 5-2.

Having determined the relation between the horizontal frequency and the chrominance-subcarrier frequency, the next frequency to be considered is the beat between the sound carrier and the chrominance subcarrier. This beat is also arranged to be an odd harmonic of half the line frequency.

Since the location of the sound carrier relative to the picture carrier is determined by the sum of the chrominance subcarrier and the chrominance-sound beat frequency, each of which is an odd harmonic of half-line frequency, the sound-picture carrier separation must be an even harmonic of half-line frequency, which is one of the more visible frequencies. This is not considered objectionable because selective circuits in the receiver are provided to remove sound interference from the picture.

The exact synchronizing frequencies for color transmissions are shown in Table 5-2.

5.2.4   TOLERANCE ON FREQUENCIES.   The frequency ratios specified in Table 5-2 are exact except for those which include the sound frequency. Since the sound carrier is frequency-modulated, the ratio is given for the average frequency. To maintain exact relationships between the other frequencies, one of them must be chosen as the master and the others obtained by frequency division or multiplication. The reference frequency chosen is the chrominance subcarrier.

One of the methods used in receivers to select the chrominance subcarrier involves a crystal filter circuit. Such a circuit responds with the necessary phase accuracy over a very limited range of frequencies. The frequency f_{sc} is therefore specified to as close a value as can be readily held at the video signal source. This was chosen as plus or minus 10 Hz. If the frequency were to vary within these tolerance limits at a fast rate, the 2° phase-error limit would be exceeded. The frequency stability is therefore specified as tight as can be held with reasonable expense at the video signal source. The value chosen is ⅒ Hz per second.

Once the subcarrier frequency with its tolerance is chosen, the specified ratios determine the tolerance on the transmitted horizontal and vertical scanning frequencies. The synchronizing system cannot be synchronized to the power line, since commercial power does not meet the frequency-stability requirements.

5.3   CHARACTERISTICS OF THE COMPOSITE SIGNAL AND METHODS OF ENCODING SYNCHRONIZING INFORMATION

5.3.1   POLARITY OF MODULATION.   The polarity of modulation is negative. An increase in the radiated signal amplitude corresponds to a decrease in picture brightness.

5.3.2 DIVISION OF AMPLITUDE RANGE. One portion of the amplitude range is selected to transmit the brightness information. With negative modulation the minimum radiated signal is selected as the brightest to be transmitted. A higher level is chosen as the value of black. A small amplitude range between the black level and a higher level, the blanking level, is used as a guard interval. This is the setup. It facilitates the separation of video modulation from synchronizing and also avoids distortion in the blacks from the circuits—limiters or clamps—which maintain the blanking level. The amplitude range between this blanking level and the maximum level to be transmitted is available for the transmission of synchronizing information. The signals within this synchronizing range are not visible.

The division of amplitude between luminance modulation and synchronizing is a constant percentage of the maximum of the radiated signal. At the receiver input, the synchronizing signal peaks received from a distant transmitter may be considerably lower than the white-signal level from a local transmitter. Since the amplitude range over which the synchronizing circuits operate satisfactorily is much less than the range of input signals at the antenna, automatic gain control (AGC) circuits are used to keep the output voltage nearly constant. The variation in signal at the receiver input between weak and strong signals may be 10^5 whereas the variation in video signal level, measured at the synchronizing separator input, usually should not exceed 10.

5.3.3 SIGNAL FREQUENCY. The frequency interval assigned for the transmission of video modulation extends from 0 to 4.2 MHz. The luminance and chrominance signals occupy the entire band, leaving no unoccupied frequency range which would be suitable for the transmission of the synchronizing information. The synchronizing signals must therefore be transmitted within the video passband.

Since the synchronizing signals and the video modulation are transmitted in a common frequency band, their time relationships are easily maintained, whereas if they were sent over separate channels the relative timing might be seriously in error. The five signals to be transmitted (vertical, horizontal, and chrominance synchronizing signals, and luminance and chrominance modulation signals) must be coded in such a way that each can be used in the receiver without interference from the others.

For the synchronizing signals the phase must be transmitted. If the time at which the signal goes through a reference phase, such as crossing the zero axis in a positive direction, is known for each cycle, the frequency is determined. For the two deflection synchronizing signals, the information is transmitted by including sufficient harmonics so that the sum produces an abrupt change in signal level at the reference time. The first 60 harmonics of the deflection frequencies are transmitted by video frequencies below 1 MHz. Thus only the lower frequency portion of the video passband need be used for the transmission of deflection sync. Since the chrominance signal is near the upper frequency edge of the band, the fundamental is the only component which can be transmitted and a different procedure must be used.

The luminance modulation differs from the synchronizing signals in that the amplitude must be transmitted as well as the correct frequency and phase. In order to minimize the effect of noise or other interference on the luminance modulation, the range available for luminance modulation is made as great as possible.

The chrominance signal, whose maximum peak-to-peak amplitude is of the same order as the luminance modulation, must in certain circumstances extend into both the synchronizing and video modulation-amplitude ranges. The overlap in the modulation ranges has been found to introduce no difficulty.

The frequency differences are such that the chrominance signal and the deflection synchronizing signals can be separated by selective filters. The interference in the chrominance channel from the luminance signal and in the luminance channel from the chrominance signal are both reduced by the selection of a least-visible frequency for the chrominance subcarrier. The maximum chrominance amplitude is fixed by the signal specification at a level which causes no objectionable interference.

Many color television receivers of the latest design use comb filters to minimize the crosstalk of luminance signals into the chrominance channel and vice versa. By use of these techniques, it is possible to increase the bandwidth of the luminance channel in the color television receiver and provide higher horizontal resolution without objectionable interference.

5.3.4 METHODS OF ENCODING SYNCHRONIZING SIGNAL.

The essential information to be transmitted is the exact phase of each of the three signals. Referring to the *horizontal deflection signal,* the desired information is a short timing pulse once per line which is properly timed with respect to the video modulation. To transmit the greatest amount of information, the pulse should extend over the full amplitude range assigned for the synchronizing signal. The duration of each pulse must be great enough so that it will not be lost if the high frequencies in the video band are attenuated. If the pulse were made so short that a large portion of its energy existed in the upper end of the video passband, one of the first effects noted as the set is tuned away from the sound carrier would be the loss of horizontal synchronism. The duration chosen is such so that synchronism can be maintained even though the receiver is appreciably mistuned.

The peak amplitude of the pulse varies with the strength of the received signal. In order to select the synchronizing portion, and only the synchronizing portion, it is necessary that the duration be great enough to permit satisfactory clamping operation. The duration of the maximum signal level must be sufficient for the level to be determined by simple means. These requirements lead to a pulse duration of 7 or 8 percent of the horizontal period. This length of pulse has also been found to be a good compromise between a longer pulse which would give some improvement in the presence of noise and the use of the available time for other purposes, i.e., level stabilization and the transmission of the color subcarrier burst.

A synchronizing signal of rectangular shape transmits the required information even though its amplitude is modified by nonlinear circuits.

If the signal suffers compression in the receiver, and is applied to a synchronizing separator which passes only the peaks of the remaining signal, complete synchronizing information will be obtained, even though only a small fraction of the synchronizing amplitude is used. This operation is illustrated in Fig. 5-3.

The *vertical synchronizing signal* is included within the same amplitude range. It also takes advantage of the rectangular shape. The vertical pulse is, however, considerably longer than the horizontal pulse to enable the receiver to distinguish between the two signals by simple means. A pulse about 35 times the length of the horizontal pulse is used. The two pulses are shown in Fig. 5-4.

The ratio of the vertical and horizontal signal frequencies is necessarily an odd multiple of half-line frequency. This makes the relative timing of the pulses different in successive fields. The differences are illustrated in Fig. 5-4. The basic signals to be trans-

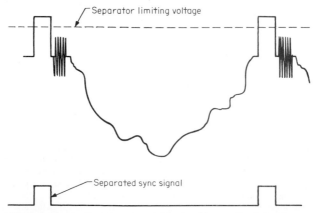

FIG. 5-3   Separation of synchronizing signal by amplitude level.

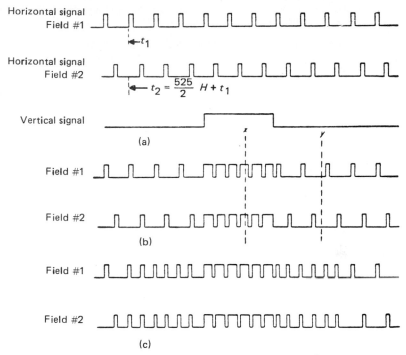

Horizontal signal
Field #1

$\leftarrow t_1$

Horizontal signal
Field #2

$\leftarrow t_2 = \dfrac{525}{2} H + t_1$

Vertical signal

(a)

Field #1

Field #2

(b)

Field #1

Field #2

(c)

FIG. 5-4   Composition of deflection synchronizing signals. (*From Fink.*[3])

mitted are shown in the three lines at the top of the figure labeled (*a*). The combined form for the two fields appears in the two lines labeled (*b*). The interruptions in the vertical pulse supply the horizontal frequency information through the vertical interval. The double-frequency interruptions ("equalizing pulses") make the vertical pulse identical on successive scans. The waveform in Fig. 5-4*b* contains sufficient information to operate both deflection synchronizing circuits.

To avoid the possibility of interference from the horizontal pulses, the vertical circuit must respond slowly; that is, its output must depend on the information received during several horizontal periods. If, for example, the vertical retrace starts at point *x* (Fig. 5-4*b*), the signals immediately preceding are different on the successive field scans. In one case the last horizontal pulse is one half-line period preceding the broad pulse, and in the next scan it will precede by a full-line period. Similarly the interval between point *y*, the end of the vertical retrace, and the preceding horizontal pulse will differ by a half-line interval for the two scans. This difficulty is avoided by doubling the horizontal pulse frequency for a short period preceding and following the broad pulses. The duration of the double-frequency pulses is reduced 50 percent to avoid any disturbances in the clamping circuits which set the signal level. The resulting signal is shown in Fig. 5-4*c*.

The third synchronizing signal is the *color-synchronizing burst* which gives the reference phase for the chrominance subcarrier. The synchronizing information is contained in a short burst of the subcarrier frequency which follows the horizontal pulse and is completed before the start of the picture modulation. The chrominance burst violates the principle of amplitude separation between synchronizing and video information. The burst extends in each range, half of it being in the synchronizing amplitude range and half of it in the video range.

The burst can be completely separated from the deflection synchronizing signal by frequency-selective filters, since the burst frequency is near the upper-frequency limit of the passband, and the deflection synchronizing signals are transmitted in the lower half of the passband.

A selective filter for the chrominance burst will, of course, pass the chrominance signals transmitted during the picture interval. These occur at any phase and so would interfere with the proper synchronism. The difficulty is avoided by gating the chrominance synchronizing channel. Since the burst follows the horizontal synchronizing pulse by a definite short interval, the horizontal flyback pulse can be used to gate the burst, completely separating it from the chrominance information.

5.4 DETAILED DESCRIPTION OF SYNCHRONIZING SIGNAL

The synchronizing waveform shown in Fig. 5-5 is from FCC Rules and Regulations (Part 73.699, Fig. 6) and was in effect in 1983. The waveform is specified in terms of the modulating signal which must be applied to an ideal transmitter. Such a transmitter is one with perfectly linear modulation characteristics and with the specified vestigial sideband attenuation and envelope delay.

The envelope of the signal radiated is identical with this waveform for all the components lower than 0.75 MHz. The higher-frequency components are reduced in amplitude to half that shown, by the removal of the lower-frequency sideband.

When the signal of Fig. 5-5 is used to modulate a transmitter with the ideal characteristics, the radiated envelope will be that shown except that the HF components will be reduced in amplitude and advanced in phase with respect to the lower-frequency components. The effect would be to reduce the burst by 6 dB and to reduce the interval following the horizontal pulse, shown in the drawing as $0.006H$. After this signal has been amplified by a receiver having the specified attenuation and delay, the envelope will be identical to that shown in Fig. 5-5.

The division of amplitude range between the video modulation and the synchronizing signal should be such that relatively simple circuits will be able to hold satisfactory synchronism. That is, the picture should be stably synchronized when the signal is so weak that it is difficult to tell from visual inspection of the raster whether or not the picture is present. The synchronizing signal amplitude equal to one-quarter of the peak carrier amplitude was chosen for monochrome television and was continued for color transmission. Experience has shown that the value is an appropriate one.

The following considerations led to the specification of the duration of each portion of the signal: Referring to the detailed drawing of the horizontal pulse (5), the interval preceding the pulse is used to stabilize the conditions at the blanking level. If the video modulation were continued until the time of start of the horizontal pulse, the timing of the pulse would be affected by the modulation content. This would be particularly serious in a narrow-band receiver.

The duration of the pulse, as explained in Sec. 5.3.4, is 7½ percent of the horizontal period. The total period at the blanking level provided for receiver retrace is $0.165H$ minimum or $0.145H$ after the start of the horizontal pulse. The time specified for the horizontal retrace is a compromise between two conflicting requirements. The duration of the video modulation interval should be as large as possible since the amount of information which can be transmitted is directly proportional to its duration. The retrace time should be as long as possible to simplify the receiver scanning requirements. The blanking interval duration must be sufficient to include the complete receiver retrace. The 17.5 percent duration was selected as a compromise between simple receiver retrace requirements and the maximum time for the transmission of video modulation.

The chrominance synchronizing burst must be received without interference from the chrominance information present during video modulation. A guard interval of $0.02H$ is provided. The difference in envelope delay for the low frequencies and the burst frequency will advance the burst in the radiated signal with respect to the horizontal pulse

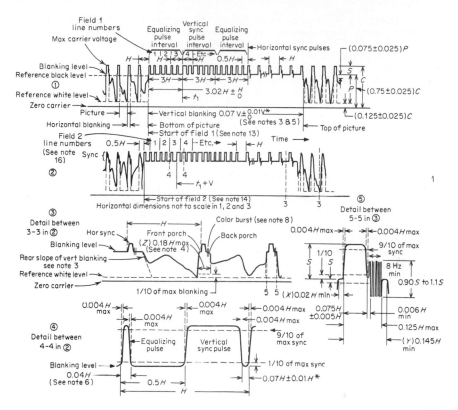

FIG. 5-5 Television synchronizing waveform for color transmission. *(FCC 1983.) Notes:* (1) H = time from start of one line to start of next line. (2) V = time from start of one field to start of next field. (3) Leading and trailing edges of vertical blanking should be complete in less than $0.1H$. (4) Leading and trailing slopes of horizontal blanking must be steep enough to preserve minimum and maximum values of $(x + y)$ and (z) under all conditions of picture content. *(5) Dimensions marked with an asterisk indicate that tolerances are permitted only for long time variations and not for successive cycles. (6) Equalizing pulse duration shall be between 0.45 and 0.55 of the horizontal synchronizing pulse duration. (7) Color burst follows each horizontal pulse but is omitted following the equalizing pulses and during the broad vertical pulses. (8) Color bursts to be omitted during monochrome transmission. (9) The burst frequency shall be 3.579545 MHz. The tolerance on the frequency shall be ± 10 Hz with a maximum rate of change not to exceed 0.1 Hz/s (10) The horizontal scanning frequency shall be $\frac{2}{455}$ times the burst frequency. (11) The dimensions specified for the burst determine the times of starting and stopping the burst but not its phase. The color burst consists of amplitude modulation of a continuous sine wave. (12) Dimension P represents the peak excursion of the luminance signal from blanking level but does not include the chrominance signal. Dimension S is the synchronizing pulse amplitude above blanking level. Dimension C is the peak carrier amplitude. (13) Start of field 1 is defined by a whole line between first equalizing pulse and preceding H sync pulses. (14) Start of field 2 is defined by a half line between the first equalizing pulse and the preceding H sync pulses. (15) Field 1 line numbers start with the first equalizing pulse in field 1. (16) Field 2 line numbers start with second equalizing pulse in field 2. (17) Refer to text for further explanations and tolerances. (18) During color transmissions, the chrominance component of the picture signal may penetrate the synchronizing region and the color burst penetrates the picture region.

by approximately 0.17 μs, or approximately 0.003H. A guard interval is therefore needed preceding the burst. The remaining interval. 0.039H minimum, is available for the transmission of the burst. This corresponds to 8⅓ cycles at 3.58 MHz. The specifications call for a minimum of 8 cycles.

In November 1977 the Electronic Industries Association (EIA) issued Tentative Standard RS170A, "Color Television Studio Picture Line Amplifier Output." This tentative standard on synchronization waveforms was in effect in June 1983 and is shown in Fig. 21-41.

To facilitate measurements, video voltage levels are expressed in IRE units, where sync level equals 40 IRE units, blanking level 0 IRE units, and reference-white level 100 IRE units. When applied to a linear television transmitter, these levels correspond to 100 percent carrier, 75 percent carrier, and 12.5 percent carrier, respectively, which are the FCC specifications.

Critical pulse measurements during the horizontal-blanking interval and preceding the first portion of the vertical sync pulse are expressed in microseconds.

One innovation in the drawing is the showing of four color fields of a color signal. Field 1 is defined by a particular subcarrier to horizontal-sync timing relationship; all other fields then follow accordingly.

FCC Rules specify the parameters of the radiated signal. It has been recommended that the target values and tolerances shown in EIA Drawing RS-170A be maintained in television programs as they leave the video signal source to other broadcasters or to transmitter plants. In practice, maintaining the various timing values within the tolerances specified will facilitate compliance with the FCC Rules.

Compliance with these EIA-recommended values and tolerances is voluntary. However, it is expected that with the adoption of these values as industry standards for analog television signal sources, most broadcasters will begin to use these recommended values and that equipment manufacturers will design new equipment so that eventually all video signal sources will be able to follow these specifications.

5.5 WAVEFORM AND SPECTRUM FOR FLAT FIELDS

5.5.1 FLAT MONOCHROME FIELD, LINE-SCAN WAVEFORM.   Figure 5-6 shows the waveform arising from the lines of a medium-gray monochrome flat field not including the vertical blanking interval. The complete waveform extending over a full frame period consists of approximately 490 line scans (see Fig. 5-5), interspersed with two vertical blanking and sync periods. The line horizontal period is 63.5 μs; the field (vertical) period is 1/59.4 s or 16,835 μs.

The line-scan waveform may be separated into two parts (Fig. 5-7). The first in Fig. 5-7a is a trapezoidal wave having a broad portion due to the luminance signal during active scanning (about 52.4 μs) and a narrow pedestal during horizontal blanking duration (about 11.1 μs); the amplitude of this portion extends from blanking to a medium-

FIG. 5-6   Video waveform corresponding to a medium-gray monochrome flat field (successive line scans and horizontal sync signals).

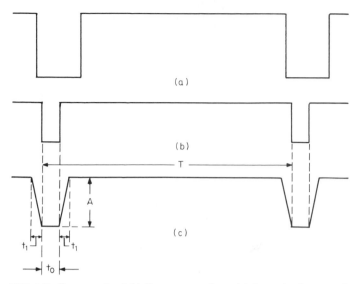

FIG. 5-7 Components of the line-scan waveform: (a) due to luminance and horizontal blanking; (b) due to horizontal sync signals; (c) idealized trapezoidal waveform.

gray luminance level of 64 IRE units. The second portion in Fig. 5-7b is the horizontal-sync pulse, a trapezoidal waveform (about 4.7 μs wide at the base and 4.1 μs at the peak) located off-center on the horizontal-blanking pedestal and having an amplitude of 40 IRE units.

5.5.2 LINE-SCAN SPECTRA.   The amplitude spectra of these waveforms, considered separately, can be derived from Fourier analysis, using the formula for symmetrical trapezoidal pulses given in Chap. 22, Table 22-8. The luminance and sync waves have the same period (1/15, 734 s) and consequently their spectra consist of a dc component and harmonics at multiples of their fundamental frequency (15,734 Hz). However, the waves differ in amplitude, duty cycle, and times of rise and fall. On this account the distribution of energy among the harmonics differs in the two waves.

The amplitude of the nth harmonic in a trapezoidal wave is given by

$$C_n = \frac{2A(t_o + t_1)}{T} \frac{\sin (\pi n t_1/T)}{\pi n t_1/T} \frac{\sin [\pi n(t_o + t_1)/T]}{\pi n(t_o + t_1)/T} \tag{5-1}$$

where the symbols have the significance shown in Fig. 5-7c. Computation is facilitated by reference to the values of $\sin x/x$ given in Chap. 22, Table 22-9. The spectra of the two waves, computed by Eq. (5-1), are given in Figs. 5-8 and 5-9.

It will be noticed that the amplitude of the harmonics fall off rapidly with increasing frequency, and that particular harmonics are very small or absent altogether (near the "zeros" of the spectrum) at frequencies determined jointly by the times of rise and fall t_1 and the duty cycle $(t_0 + t_1)/T$ of the respective waves.

The spectra of the two waves can be combined by vector (phasor) addition of harmonics of like frequency to obtain the overall spectrum of the line period; that is, the spectrum of the monochrome flat field less vertical blanking and sync pulses. Vector addition requires knowledge of the relative phase angles of each pair of harmonics being added.

For the FCC-standard signal, with typical front and back porches of 1.3 and 4.6 μs duration, respectively, the sync pulses are off-center by 1.65 μs, which is a phase angle

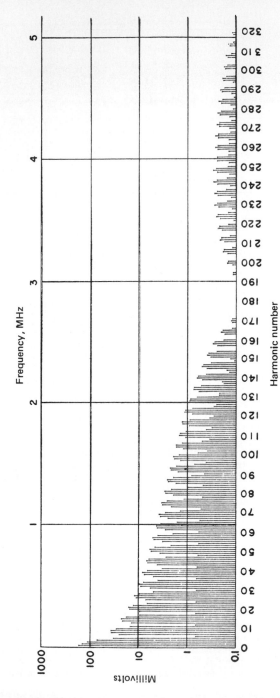

FIG. 5-8 Computed spectrum of luminance and horizontal blanking. No vertical sync or vertical blanking. (*From Fink.*[3])

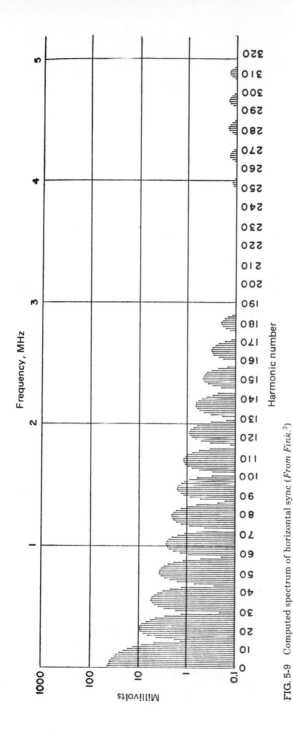

FIG. 5-9 Computed spectrum of horizontal sync (*From Fink.*[3])

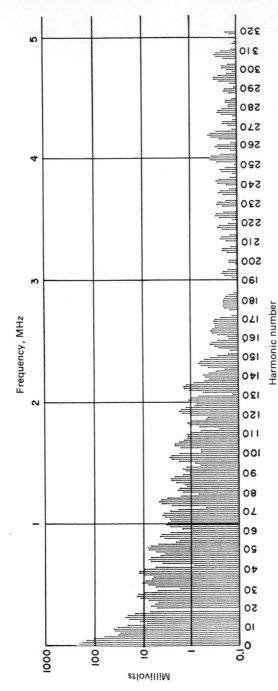

FIG. 5-10 Computed spectrum of the monochrome flat field with the horizontal synchronizing pulses having a 1.3-μs front porch and a 4.6-μs back porch. No vertical sync or vertical blanking. (*From Fink.*[3])

of $2\pi n(1.65/63.5)$ rad at the nth harmonic. When the harmonic pairs are added at these respective phase angles, the combined spectrum of Fig. 5-10 results.

It is evident from the foregoing discussion that the energy distribution in the spectrum is affected by any time shift between the two portions of the wave. Since the durations of front and back porches are open to substantial tolerances in practice, it must be emphasized that Fig. 5-10 is illustrative of but one of many typical conditions. Figure 5-11 shows the corresponding experimentally measured spectrum of a medium-gray flat field, using the method outlined in Sec. 5.11.

It will be noticed that the differences between the measured spectrum in Fig. 5-11 and the computed spectrum in Fig. 5-10 become progressively greater as the frequency or harmonic number increases. Very small differences between the waveforms used for the computation and for the measurement, in the rise and fall times of the trapezoids, and in the number of microseconds by which the sync pulses are off-center easily account for these differences at high frequencies. The same comment applies to all the computed and measured spectra in this chapter.

5.5.3 FLAT-FIELD MONOCHROME SIGNAL, INCLUDING VERTICAL SIGNALS.

The remaining portion of the spectrum of the monochrome flat-field signal, that produced by vertical blanking and sync, is not so readily computed in view of the multiplicity of rising and falling edges, at various intervals, associated with the equalizing pulses and the vertical serrations. However, the contribution of the vertical sync and equalizing pulses to the spectrum energy distribution is small, in view of their low fundamental frequency and low duty cycle.

The more important effect is that of vertical blanking, which in effect separates the line-scanning periods into groups of about 245 each, every $\frac{1}{60}$ s, with no luminance signal during about 18 line periods between each group. The harmonic components of the line-scan spectrum may thus be thought of as carrier waves, each with a 59.94-Hz modulation envelope. The harmonic components are thus amplitude-modulated, and two sets of symmetrically disposed sidebands occur at intervals of 59.94 Hz (Fig. 5-12).

The energy distribution of each set of sidebands depends on the modulation envelope mentioned above, which in the case of the medium-gray flat field can be taken as a rectangular wave, of duty cycle approximately $\frac{18}{263} = 0.07$. The spectrum of such a wave develops its first zero at the 14th side band, that is, at the frequency separation of $14 \times 60 = 840$ Hz each side of the line-scan harmonic. Energy further removed is generally of low amplitude.

Consequently, the flat-field spectrum including vertical blanking is more diffuse than that of an uninterrupted succession of line scans, but there are large regions between the line-scan harmonics which are free of energy of significant amplitude. This is equally true of any signal representing subject matter without variation in luminance or chrominance from top to bottom of the picture, such as certain gray-scale-bar charts and color-bar charts. When top-to-bottom variations are present, the spectrum is more diffuse.

5.5.4 WAVEFORM FOR FLAT COLOR FIELD.

The flat-field monochrome signal discussed in the previous section is transformed to a flat-field color signal by the addition of the chrominance signal at 3.579545 MHz and color sync burst. A typical line scan, not including vertical sync and blanking for the entire frame interval of $\frac{1}{30}$ s, for a flat color field appears in Fig. 5-13.

The chrominance signal may be considered as a phase-modulated carrier which is amplitude-modulated by two trapezoidal modulation envelopes. One of the envelopes occurs during the active line scan, about 53 μs, and has an amplitude depending on the color saturation and luminance of the flat field; the other occurs during the color burst (about 2.5 μs), and its amplitude is specified by the FCC standards to be 40 IRE units, averaged around the back porch of the horizontal-blanking level.

5.5.5 WAVEFORM AND SPECTRUM FOR YELLOW FLAT FIELD.

Since the chrominance frequency is an odd multiple of one-half the line-scanning frequency (see Sec. 5.2.3), the spectrum components of the chrominance signal are precisely interleaved

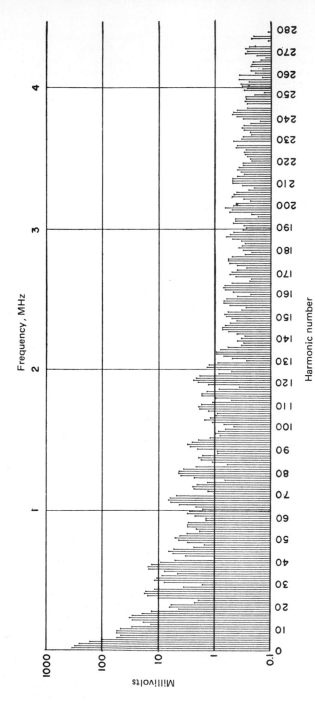

FIG. 5-11 Measured spectrum of monochrome flat field corresponding to computed spectrum of Fig. 5-10. No vertical sync or vertical blanking. (*From Fink.*[3])

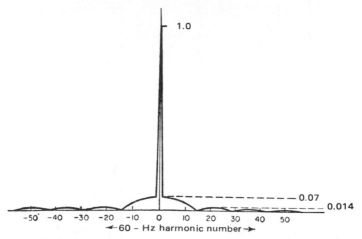

FIG. 5-12 Because of the interruption of line scans by vertical blanking, corresponding to amplitude modulation of spectrum components, each 15.734-kHz line-scan harmonic is surrounded by sidebands spaced at intervals of 59.94 Hz. (This figure shows the envelope of the sideband amplitudes. The line-scan harmonic amplitude is shown as 1.0 in this figure.) (*From Fink.*[3])

among the components of the luminance spectrum. Consequently, the flat-field chrominance spectrum may be computed separately and its components interspersed in the luminance spectrum (Fig. 5-10) to obtain the overall computed spectrum for the flat color field. In so doing it must be recalled that Fig. 5-10 is the spectrum of an uninterrupted succession of luminance line scans (without vertical blanking or vertical sync), and that vertical blanking has the effect of broadening each line harmonic. The flat-field chrominance signal is similarly interrupted during vertical blanking and its spectrum components are broadened thereby in about the same degree as the luminance components.

Computation of a typical flat-color-field spectrum is considerably simplified if the phase difference between the chrominance signal and the burst is taken as zero. This removes the phase modulation and requires the computation merely of an AM spectrum. The image produced by such a signal is a flat yellow field having a slight greenish tinge (see Fig. 5-17).

On this basis, we find that the chrominance signal consists of symmetrically disposed sidebands centered around the suppressed 3.579-MHz color subcarrier corresponding to the modulation envelopes occurring during the active horizontal scan and sync intervals (Fig. 5-13). The amplitudes of the subcarrier sidebands correspond to the spectrum of the trape-

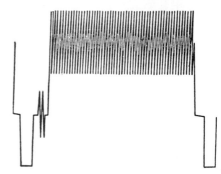

FIG. 5-13 Addition of chrominance information to transform a monochrome flat field to a color flat field. Line scan of colored flat field, containing luminance and chrominance components, horizontal sync pulse, and burst. No vertical blanking or vertical sync is included. (*From Fink.*[3])

zoidal envelopes, which in turn are computed by vector addition of two spectra, one for each envelope, taking the typical relative time delay between them (Fig. 5-13) as 2.6 μs, or a phase shift of $2\pi n(2.6/63.5)$ rad at the nth harmonic.

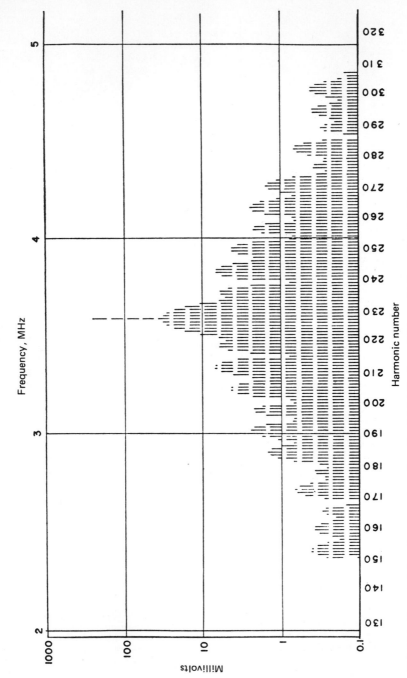

FIG. 5-14 Computed chrominance spectrum for yellow flat field having a slight greenish tinge without vertical sync or vertical blanking. (*From Fink.*[3])

The resulting computed chrominance spectrum is shown in Fig. 5-14. The combined computed spectrum of luminance and chrominance appears in Fig. 5-15. The experimentally measured spectrum, using the method described in Sec. 5.11 and a signal conforming to Fig. 5-13, is shown in Fig. 5-16.

The difference between the computed and measured chrominance spectra is due in large part to the filtering employed in the encoder in the measured case to confine the chrominance signal bandwidth, which affects the rates of rise and fall of the chrominance signal envelopes. It is also caused in part by having idealized the modulating envelope for the chrominance signal slightly, in order to facilitate computations.

5.5.6 FINE STRUCTURE OF A STANDARD COMPOSITE COLOR VIDEO SIGNAL.

The components in the clusters surrounding each line-scan harmonic are described above as being separated by 60 Hz, since these are most prominent. Actually, additional minor components separated 30 Hz from the 60-Hz components are also always present and, in addition, chroma components may, in some cases, be interleaved with these 30-Hz components.

The presence of components separated at 30 Hz arises from the fact that successive pairs of field scans, including the effect of the vertical sync signals, are never exactly alike due to line interlace and the slight differences in the positions of the equalizing pulses and vertical serrations, relative to the start of vertical blanking. Similar differences may exist when a top-to-bottom luminance variation in the raster falls differently on adjacent lines.

The 15-Hz harmonics arise from the fact that the entire chrominance signal consists of sidebands of the color subcarrier, which is, itself, an odd harmonic of 15 Hz. The modulation envelope of the color subcarrier repeats itself exactly every $\frac{1}{30}$ s, so that all the sidebands are separated in frequency from the subcarrier by multiples of 30 Hz. This results in all of them falling on odd harmonics of 15 Hz. In any event, the components of chrominance are interleaved with the corresponding 30-Hz, 60-Hz, and 15.7-kHz components of the luminance spectrum (see Sec. 5.9).

Motion of the image, involving a change of the vertical luminance or chrominance distribution of any substantial portion of the image, has a similar broadening effect (see Sec. 5.10).

Experimental examination of the 60-Hz fine structure of the video spectrum is difficult in view of the extremely high resolution required. A spectrum analyzer having a resolution of about 800 Hz was used in measuring the spectra presented here. Such an instrument is just capable of indicating the 1600-Hz broadening of the line harmonics of a flat-field spectrum, due to vertical blanking. For this reason little change is noticed in the measured spectra when vertical blanking is removed.

5.6. WAVEFORM AND SPECTRUM OF A COLOR-BAR-CHART SIGNAL

The color-bar chart used in this analysis consists of vertical bars of different luminance, but in addition they display different hues and saturations. A typical chart of this type contains nine bars; the primary colors red, green, and blue; the complementary colors yellow, cyan, and magenta; and the zero-saturation colors white, gray, and black. The colors may be arranged in any convenient order; one order of particular interest is (from left to right) red, yellow, green, cyan, gray, magenta, blue, white, and black. This order is one in which only one primary color is changed at the boundary between color bars. All three are changed at the boundaries of the white bar.

The primary color waveforms for each line scan are shown in Fig. 5-17. The phase of the chrominance signal producing each bar, relative to the color-burst phase, is that implied by the FCC standards. The luminance and chrominance levels and phases of such a bar chart are given in the table in Fig. 5-17. They produce the line-scan waveform

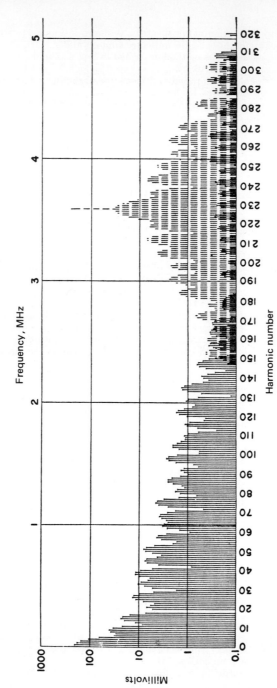

FIG. 5-15 Computed luminance-chrominance spectrum of yellow flat field having a slight greenish tinge without vertical sync or vertical blanking. (*From Fink.*[3])

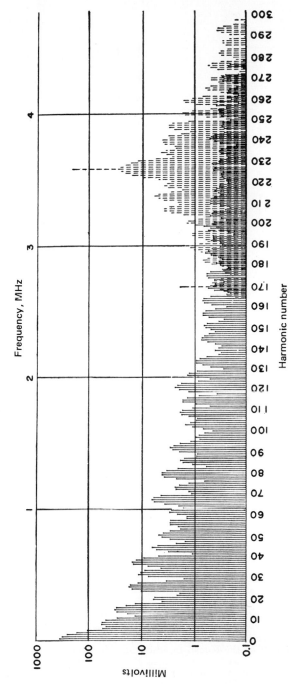

FIG. 5-16 Measured luminance-chrominance spectrum of yellow flat field, corresponding to Fig. 5-15. (*From Fink.*[3])

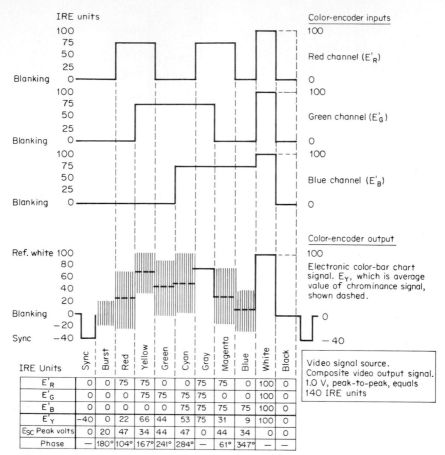

IRE Units	Sync	Burst	Red	Yellow	Green	Cyan	Gray	Magenta	Blue	White	Black
E'_R	0	0	75	75	0	0	75	75	0	100	0
E'_G	0	0	0	75	75	75	75	0	0	100	0
E'_B	0	0	0	0	0	75	75	75	75	100	0
E'_Y	-40	0	22	66	44	53	75	31	9	100	0
E_{SC} Peak volts	0	20	47	34	44	47	0	44	34	0	0
Phase	—	180°	104°	167°	241°	284°	—	61°	347°	—	—

FIG. 5-17   Makeup of the color-bar-chart signal. Color encoder input waveforms, line-scan waveform, and tabulation for each bar of the amplitude of each color encoder input, the amplitude of the luminance component, and the amplitude and phase of the chrominance component.

shown in Fig. 5-17. This bar chart is similar to, but not identical with, the EIA Standard RS-189-A for saturated color bars at 75 percent full amplitude, with 7½ percent setup referred to in Chap. 21 (see also Fisher[4] and Bailey and Loughlin[5]). The measured luminance spectrum of the color-bar-chart signal is shown as the solid lines in Fig. 5-18.

The individual line-harmonic amplitudes, at low and medium luminance frequencies, vary in amplitude over a rather wide range. This range of amplitude is almost independent of video frequency.

The chrominance spectrum displays a number of features owing to the fact that the color subcarrier is modulated both in amplitude (by horizontal blanking and the burst envelope) and in phase (assuming in succession the six phase angles shown in Fig. 5-17, plus reference phase during the burst). In addition, the chrominance signal is absent altogether (amplitude-modulated to zero) during the gray, white, and black bars.

In measuring the chrominance spectrum, the indicating instrument responds to the average amplitude of each chrominance-sideband cluster. These components are sepa-

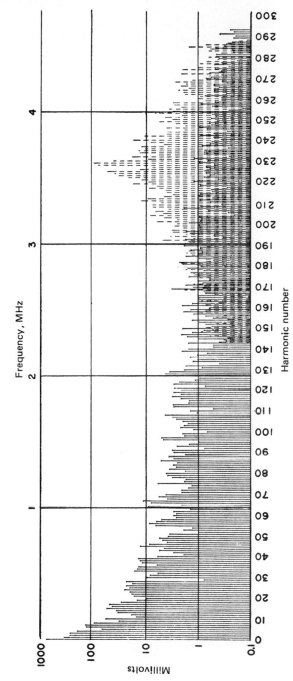

FIG. 5-18 Measured luminance-chrominance spectrum of color-bar-chart signal. (*From Fink.*[3])

rated from the color subcarrier at multiples of 15.734 kHz and their individual amplitudes are determined by the combination of amplitude and phase modulation of the subcarrier corresponding to the hue and saturation of the individual bars.

The measured average chrominance-component amplitudes are distributed as shown by the dashed lines in Fig. 5-18. Their distribution about the subcarrier is generally symmetrical, but there is no correlation between the amplitudes of individual sideband pairs. The measured distribution of the amplitudes further than several hundred kHz from the subcarrier is determined in part by the constants of the chrominance-signal filters.

5.7 WAVEFORM AND SPECTRUM OF MONOCHROME-RESOLUTION TEST CHART

The next signal in order of complexity is typified by the resolution wedges of monochrome test charts. These wedges consist of converging groups of narrow lines, used in measuring vertical and horizontal resolution. Such charts also contain other elements,

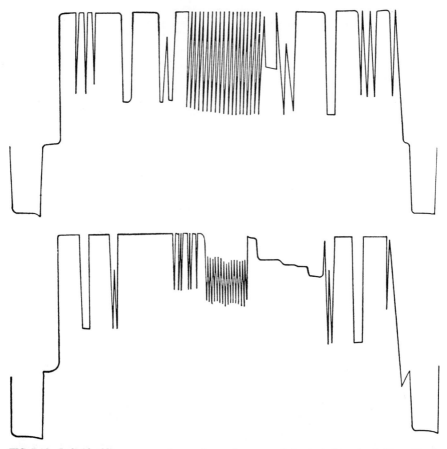

FIG. 5-19   Individual line-scan waveforms of monochrome resolution test-chart signal. (*From Fink.*[3])

such as circles to indicate linearity of scanning and adjacent areas of different luminance to indicate tonal gradation.

Two important features distinguish such test charts from the monochrome flat field and bar chart previously discussed: they display variations in luminance from top to bottom of the picture and, in particular the transitions of luminance on successive line scans are not identical. These variations and transitions are of such complexity as to render impractical the computation of the spectrum of the signal.

The spectrum of the resolution-chart signal is found to differ from those of the monochrome flat field and color bar chart in two particulars. First, the space between line harmonics is more completely occupied by signal energy (more diffused). Second, the diffusion of the spectrum increases with increasing video frequency. The waveforms of selected line scans of a typical monochrome test-chart signal, as seen on a television line-selector waveform monitor, are shown in Fig. 5-19. The high-frequency sinusoids shown at the centers of these waveforms correspond to the luminance variations in scanning through two different portions of the vertical wedge of the monochrome television test chart. The measured spectrum of the signal is shown in outline form in Fig. 5-20.

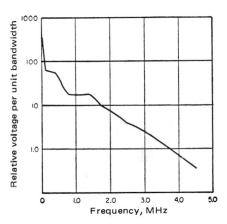

FIG. 5-20 Outline of spectrum of resolution-chart signal. (*From Fink.*[3])

The fine structure of 15.734-kHz harmonics of the test-chart spectrum is shown in Figs. 5-21 and 5-22, at low and high video frequencies. At very low frequencies it will be seen that the line harmonics are discrete, that is, most of the space between harmonics is unoccupied. At higher frequencies the spectrum is more diffuse.

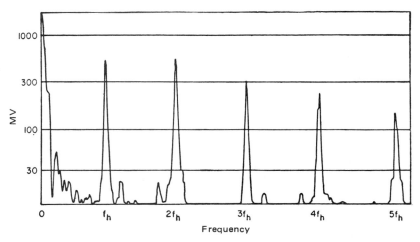

FIG. 5-21   Fine structure of 15.734-kHz harmonics of resolution-chart spectrum in vicinity of direct current. (*From Fink.*[3])

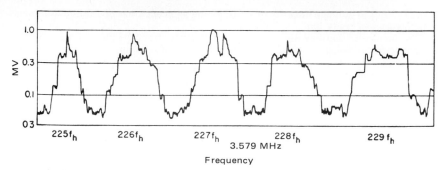

FIG. 5-22   Fine structure of 15.734-kHz harmonics of resolution-chart spectrum in vicinity of 3.58 MHz. (*From Fink.*[3])

5.8   WAVEFORMS AND SPECTRA OF STATIONARY COLORED IMAGES

The test signals most nearly representative of program material are those provided by scanning photographic images (test slides or films) in monochrome and color. Such images occasionally display sharp boundaries and striations of the types previously discussed. More usually, however, the geometry of the image is difficult to describe in such terms. Rather, program images display detail, contrast, hues, and saturations peculiar to the subject matter. The angle subtended by the image at the camera has an important bearing on the detail structure; close-up shots usually have less essential detail than medium and long shots. Although such generalizations are of interest, it has proved essential, in checking the overall performance of a television system, to use a series of test slides and films covering a wide range of subject matter.

A set of such slides, provided by the Eastman Kodak Co. for the use of the National Television System Committee, was widely used in color-system testing. Signals produced by some of these slides have been subjected to spectrum measurements. A representative case is "Tulip Garden" (NTSC Slide 13), a long shot containing fine detail in various colors. A selected line scan from this slide is shown in Fig. 5-23.

The overall luminance and chrominance spectra resulting from scanning this slide are

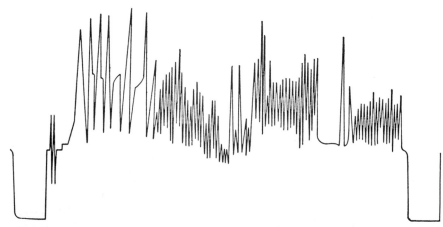

FIG. 5-23   Individual line scan from NTSC test-slide signal "Tulip Garden." (*From Fink.*[3])

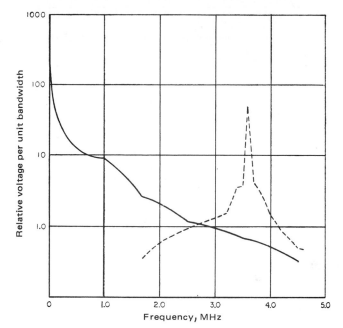

FIG. 5-24   Outline of luminance-chrominance spectra for NTSC test-slide signal, "Tulip Garden." (*From Fink.*[3])

shown in outline form in Fig. 5-24. The fine structures of the spectra (Figs. 5-25 to 5-27a) also illustrate the HF content in the luminance signal spectrum corresponding to this detailed slide.

Certain facts concerning the luminance and chrominance spectra of typical program material are evident from these measurements. First, the luminance-signal components decrease in amplitude at a rate of about 6 dB per octave in the range from 0.0157 to 0.75 MHz, 10 dB per octave from 0.75 to 4 MHz, and 12 dB per octave at higher frequencies. The chrominance components decrease at about 7 dB per octave in the range from 15.734 to 400 kHz from the color subcarrier, within which the chrominance signal filters have little effect. At higher frequencies, the attenuation is influenced by the filters, and is about 5 dB per octave from 400 kHz to 1.5 MHz on the side toward the picture carrier, and 9 dB per octave from 400 kHz to 1.5 MHz on the side away from the picture carrier.

Second, the amplitude of the luminance signal in the vicinity of the color subcarrier frequency (3.58 MHz) is very small, typically −40 dB relative to the 100-kHz luminance amplitude. The range of luminance-signal amplitudes encompassed by the chrominance signal includes −37 dB at 3.0 MHz and −42 dB at 4.2 MHz. The chrominance-signal sideband distribution, as measured under the long-term-average conditions of the experimental procedure, displays a high degree of symmetry about the subcarrier frequency, out to ±400 kHz.

5.9 INTERLEAVING OF LUMINANCE AND CHROMINANCE COMPONENTS

The distribution of luminance and chrominance components of the color-signal spectra follows the interleaving principle; that is, the maxima of the chrominance spectra are

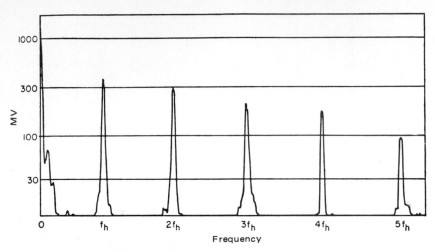

FIG. 5-25   Fine structure (15.734-kHz harmonics) of test-slide spectrum, "Tulip Garden," in the vicinity of direct current. (*From Fink.*[3])

located precisely midway between the adjacent maxima of the luminance spectra. The precision of interlacing arises from the fact that the line-scanning and field-scanning frequencies are derived by frequency division from the source of the color subcarrier, as shown in Fig. 5-2. The line-scanning frequency is thereby caused to be $\frac{2}{455}$ the color-subcarrier frequency. The field-scanning frequency, in a 525-line image, is $\frac{2}{525}$ the line frequency. It follows that the components of luminance and chrominance spaced at 15.734 kHz are interleaved, and this is equally true of the respective components spaced at 30 Hz.

In considering the spectra of the previous sections, it should be noted that 15-kHz interleaving suffices to separate luminance and chrominance information only in the simple cases of the flat fields and color bar charts, since there is no overlap of the 15-kHz components in these cases. In the more complex signals (resolution test chart and photographic slide images), the diffusion of the 15-kHz components of luminance at frequencies higher than about 2.5 MHz, is such as to cause an apparent overlap of luminance and chrominance signals (Fig. 5-27). There is, however, no *actual* overlap, since

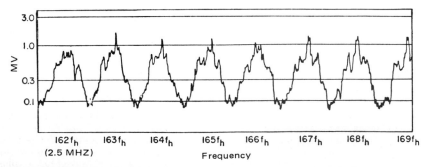

FIG. 5-26   Fine structure (15.734-kHz harmonics) of test-slide spectrum, "Tulip Garden," in the vicinity of 2.5 MHz (luminance only). (*From Fink.*[3])

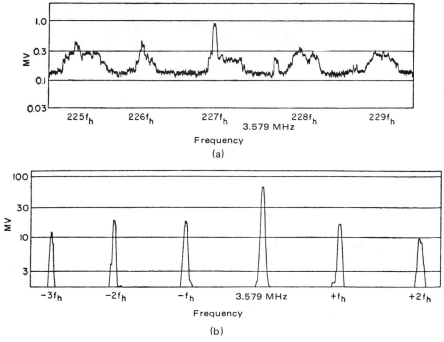

FIG. 5-27   Fine structure of test-slide spectrum, "Tulip Garden," (*a*) 15.734-kHz harmonics in the vicinity of 3.579 MHz (luminance only). (*From Fink.*[3]) (*b*) Odd harmonics of half-line frequency in the vicinity of 3.579 MHz (chrominance only).

the diffused luminance spectrum is made up of 30-Hz components which are interleaved with the 30-Hz components of the chrominance spectrum.

This fact would be revealed by a spectrum analyzer of sufficiently high resolution to show the 30-Hz components individually.

5.9.1 TIME-DOMAIN EXAMPLE OF THE ADVANTAGE OF FREQUENCY INTER-LEAVING.

The waveforms in Figs. 5-28 and 5-29 illustrate the mechanism by which frequency interleaving helps to reduce the effects of luminance crosstalk in the chrominance channel. These are waveforms recorded in a color receiver which is fed a color-bar-chart signal without the chroma information. In the absence of luminance crosstalk, there should be no signals in the chroma channel of the receiver except those corresponding to the color burst. Actually, luminance crosstalk appears in the chrominance channel and manifests itself primarily as perturbations coincident with sharp transitions in the luminance signal. The color-bar-chart signal chosen for this illustration is characterized by sharp transitions of large amplitude and thus emphasizes luminance-into-chrominance crosstalk.

Figure 5-28*a* is the color-bar-chart signal without the chrominance signal, but with the color-burst signal, as it would appear at the output of the second detector in a color receiver. This is the same signal as in Fig. 5-17 with chrominance removed. Figure 5-28*b* is the corresponding signal applied to the blue gun from the $B - Y$ color-difference amplifier. The large negative pulse is the horizontal blanking signal applied to the color-difference amplifiers; the small pulses represent the luminance-into-chrominance crosstalk, coincident with the luminance signal transitions.

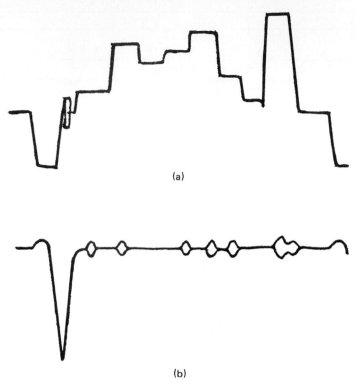

(a)

(b)

FIG. 5-28   Line-scan waveforms in receiver: (a) composite color-bar-chart signal with chroma signal removed; (b) corresponding signal applied to blue gun. (*From Fink.*[3])

Figure 5-29 shows expanded views of the waveforms of successive lines corresponding to a single one of the transitions in Fig. 5-28. Each of these expanded waveforms was observed on an oscilloscope whose horizontal sweep was triggered with horizontal sync pulses, and in each case some continuous color-carrier-frequency sinusoid was deliberately included in the signal. This appears in two polarities on the expanded waveforms due to the fact that the color carrier is chosen as an odd harmonic of one-half line and frame frequency. Figure 5.29a is a section of the composite signal as it would appear at the output of the receiver second detector. Figure 5-29b is the signal corresponding to the same transition as it appears at the output of the receiver chrominance filter. It is seen that the perturbation caused by the luminance signal appears in only one polarity whereas the color carrier appears in opposite polarities on alternate lines and frames. Figure 5.29c is the output of the $B - Y$ demodulator which is the result of mixing the signal of Fig. 5.29b with the color carrier and recovering the video-frequency component. The resultant perturbation is of opposite polarity on alternate lines and frames because of the interleaved nature of luminance components and color carrier.

These waveforms illustrate three characteristics that reduce the visibility of luminance-into-chrominance crosstalk. First, the crosstalk is largely confined to a small area about transitions, where it tends to be obscured by other transition effects. Secondly, it is of small amplitude with respect to the normal signal components. Thirdly, it is of opposite polarity on alternate lines and frames so that the integrated (average) color error of two adjacent lines in a field, or of a single line in two frames, is zero. The nature of the human visual process is such as to make this integration.

The preceding statement assumes that the display-tube brightness is linearly related to signal amplitude. While this is not true in general, it is a good approximation in the case discussed here, where the luminance-into-chrominance crosstalk components are small in amplitude with respect to the other components of the display drive signals. The line-to-line integration of color errors loses some of its advantage on horizontal and diagonal transitions in the picture material, but the frame-to-frame integration still applies in these cases. The frame-to-frame integration of color errors due to luminance-into-chrominance crosstalk can be disturbed by motion in the scene, but the human eye is less critical to these effects when associated with objects in motion.

5.10 EFFECT OF MOTION AND BACKGROUND LUMINANCE CHANGES

The spectra and waveforms presented in this chapter arise from stationary rasters, charts, and images and contain no ac component lower than 30 Hz. In fact, the differences between successive field scans of such images are so small (as previously noted) that the signal repeats itself in all essential particulars at the field-scanning rate; accordingly, the spectra can be

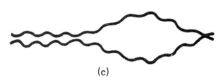

FIG. 5-29 Magnified portions of line-scan waveforms in receiver: (a) single transition of waveform 5.28; (b) corresponding input to $B - Y$ demodulator (output of chroma bandpass filter); (c) corresponding output of $B - Y$ demodulator. (From Fink.[3])

described in terms of 60-Hz components clustered about the dc component and the 15-kHz line harmonics.

When the image is in motion, successive field and frame scans are not alike and the additional periodicities associated with the motion must be taken into account. This is also true when the outlines and the boundaries of the image are stationary but the luminance or chrominance of any part of the image changes with time. When such time variations are revealed to the eye by differences in successive frame scans, the period of the change is lower than 30 Hz.

When such changes affect the background luminance of the scene, they are reflected in the dc component of the spectrum. On this account, changes in background luminance can be conveyed without reference to the ac spectrum.

When motion is present, the effect on the spectrum depends on the size of the moving area, its direction of motion relative to the scanning lines, and the velocity of motion. A simple case of image motion amenable to analysis consists of a dark-gray raster on which a narrow horizontal white bar appears, the bar moving up and down continuously between top and bottom of the raster at a constant velocity of, say, one picture height per second. In such an image the dc component of luminance remains constant.

Successive field scans consist of groups of pulses generated as the scanning spot encounters the moving bar. Between successive field scans the bar moves only ⅟₆₀ the picture height. Consequently, adjacent field scans differ only by a small amount. The signal repeats itself at intervals of 2 s, that is, its fundamental frequency is 0.5 Hz. The spectrum of such a signal consists of components separated from one another by 0.5 Hz.

It does not follow that each 0.5 Hz harmonic is of significant amplitude. Rather these components are grouped around the 60-Hz harmonics as sidebands. This conclusion follows from the fact that the 120 field scans which occur during the full period of the motion are in fact phase modulated by a triangular-shaped "envelope."

In this moving image, then, the spectrum consists of major concentrations of energy spaced at 15-kHz intervals, due to line scanning. These are in turn surrounded by clusters of 30-Hz harmonics due to frame scanning. These are in turn surrounded by very closely spaced "motion" harmonics due to amplitude or phase modulation of the field scans, arising from the image motion. Whenever the motion is regularly repeated, a line spectrum occurs, but it is always more diffuse than that of the same image when stationary.

The spacing of the "motion" harmonics depends on the period of the motion. In violent motion, say when the bar moves from top to bottom of the raster in slightly less than ⅟₆₀ s, the spectrum becomes almost completely diffused. Fortunately the eye is not critical of the detail, contrast, or color exhibited by such rapidly moving objects, so the luminance-chrominance crosstalk then present is not readily identified.

5.11 EXPERIMENTAL TECHNIQUES OF SPECTRUM ANALYSIS

The experimentally determined frequency spectra illustrated in this chapter were obtained with the equipment shown in block form in Fig. 5-30. The spectrum analyzer consists of the RF and IF sections of a standard communication receiver with two modifications. A variable-speed motor drive is coupled to the tuner to produce continuous observation of portions of the spectra. A magnetostrictive IF filter is inserted in the receiver to reduce the aperture of the spectrum analyzer. The overall frequency response of the communications receiver and filter is as shown in Fig. 5-31. This is essentially the characteristic of the magnetostrictive filter.

The signals studied were those arising from stationary images, and consequently the spectra consisted of discrete frequency components separated in frequency by the frame frequency, 30 Hz, or in the case of color signals, by half the frame frequency, 15 Hz. The

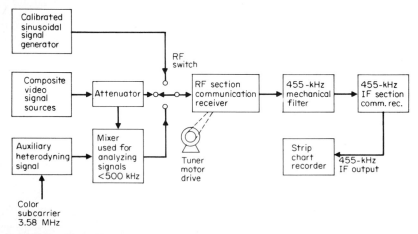

FIG. 5-30   Block diagram of spectrum-analysis equipment.

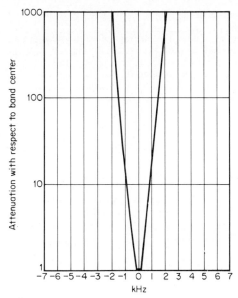

FIG. 5-31 Frequency aperture of spectrum analyzer. (*From Fink.*[3])

aperture of the filter and the inertia of the strip-chart recorder were such that the recording represents a true sum of the frequency components within the passband of the spectrum analyzer. Thus, although this equipment is incapable of resolving the finest details of the spectra, it correctly resolves the gross characteristics of spectrum detail between harmonics of line frequency.

The experimentally determined data can be divided into three categories: the surveys of the complete video signal from 0 to 4.25 MHz typified by Figs. 5-11, 5-16, and 5-18; more detailed studies of restricted portions of the video signal as typified by Figs. 5-21, 5-22, 5-25, and 5-27; and outlines of the spectrum from 0 to 4.5 MHz typified by Figs. 5-20 and 5-24.

The first category was restricted to flat-field and bar-chart signals having their energy clustered about harmonics of line frequency and, in the cases of color transmission, odd harmonics of half-line frequency. This resulted from the fact that each of these signals was periodic at line rate, with the exception of the vertical blanking interval. The overall surveys were restricted to samplings at harmonics of the line frequency and odd harmonics of half-line frequency. The sampled points shown in Figs. 5-11, 5-16, and 5-18 correspond to the peak values of the detailed spectra. To determine the amplitudes of the low-frequency (0 to 550 kHz) line-scan harmonics of the luminance signal with this equipment, it was necessary to heterodyne the luminance signal with an auxiliary signal of constant frequency and to translate the measured values back to LF range. For convenience, the auxiliary heterodyning signal was derived from the color-subcarrier oscillator. (See Fig. 5-30.)

For the second category of data, the detailed sections of spectra, it was not necessary to restrict the type of signal. This made possible a study of complex signals such as those corresponding to monoscope and slides. Figures 5-22 and 5-25 are the strip-chart recordings made as the spectrum analyzer was continuously driven through a number of line-frequency harmonics in different parts of the video spectrum. The speed with which the tuner was driven was adjusted to be sufficiently slow so that the response time of the recording pen would not impose any significant limitation on the reproduced spectra.

Thus, the detailed spectra are accurate within the limitations of the spectrum-analyzer passband characteristic.

For the third category of data, the outlines of complete spectra, the bandwidth of the spectrum analyzer was made as wide as possible for the communication receiver, or about 10 kHz. It was driven continuously through the spectrum at a rate much more rapid than that used for the detailed sections of spectra, but still slow enough that the response time of the recorder did not impose any limitation in reading the true rms value of the portion of the spectrum passing through the analyzer at the given instant. The curves of Figs. 5-20 and 5-24 represent the average rms value of the spectrum over several successive line-frequency harmonics near each frequency.

In each measurement, the signals consisted of signal plus noise. The noise component was negligible, except in those measurements of color-slide luminance components in the vicinity of 2.5 and 3.58 MHz.

REFERENCES

1. D. H. Pritchard and J. J. Gibson, "Worldwide Color Television Standards—Similarities and Differences," *J. SMPTE,* vol. 89, no. 2, February 1980.

2. Harold Weiss, "Significance of Some Receiver Errors to Color Reproduction," *Proc. IRE,* vol. 42, p. 1380, September 1954.

3. D. G. Fink (ed.), *Television Engineering Handbook,* McGraw-Hill, New York, 1957.

4. J. F. Fisher, "Color Terminal Equipment," in ibid., pp. 17-81–17-87.

5. William F. Bailey and Bernard D. Loughlin, "Transmission of Chrominance Values," in ibid., pp. 9-58 to 9-60.

BIBLIOGRAPHY

Chatten, John B., Richard G. Clapp, and Donald G. Fink, "Composite Video Signals, Waveforms and Spectra," in Donald G. Fink (ed.), *Television Engineering Handbook,* McGraw-Hill, New York, 1957, chap. 10.

Harnett, Daniel E., "Synchronization of Scanning and Color Coding," in ibid., chap. 7.

Mertz, P., "Television—The Scanning Process," *Proc. IRE,* vol. 29, p. 529, October 1941.

Mertz, P., and W. Gray, "Theory of Scanning and Its Relation to the Transmitted Signal in Telephotography and Television," *Bell System Tech. J.,* vol. 13, p. 464, July 1934.

Schade, O. H., "Electro-optical Characteristics of Television Systems: Introduction; Part I—Characteristics of Vision and Visual Systems," *RCA Rev.,* vol. 9, no. 1, p. 5, March 1948; "Part II—Electro-optical Specifications for Television Systems," *RCA Rev.,* vol. 9, no. 2, p. 245, June 1948; "Part III—Electro-optical Characteristics of Camera Systems," *RCA Rev.,* vol. 9, no. 3, p. 490, September 1948; "Part IV—Correlation and Evaluation of Electro-optical Characteristics of Imaging Systems," *RCA Rev.,* vol. 9, no. 4, p. 653, December 1948.

Wendt, K. R., "The Television D-C Component," *RCA Rev.,* vol. IX, no. 1, pp. 85–111, March 1948.

Propagation

William Daniel

Federal Communications Commission
Washington, D.C.

with contributions by

Edward W. Allen

Federal Communications Commission (retired)
Washington, D.C.

NOTE: Parts of this chapter by Edward W. Allen are adapted from D. G. Fink (ed.), *Television Engineering Handbook*, McGraw-Hill, New York, 1957. Used by permission.

6.1 INTRODUCTION

The portion of the electromagnetic spectrum currently used for radio transmissions lies between approximately 10 kHz and 18 GHz. Future systems, both terrestrial and satellite, will use even higher frequencies. The influence on audio waves of the medium through which they propagate is frequency-dependent. The lower frequencies are greatly influenced by the characteristics of the earth's surface and the ionosphere, while the highest frequencies are greatly affected by the atmosphere and especially rain. There are no clear-cut boundaries between frequency ranges but instead considerable overlap in propagation modes and effects of the path medium. In this chapter, theoretical models for propagation in free space over a plane earth and over a smooth spherical earth are presented, as well as the effects of the ionosphere, the atmosphere, hills, buildings, and trees. In addition, statistical methods for calculating television coverage and interference are presented.

In the United States those frequencies allocated for television-related use include the following:

54–72 MHz:	*TV channels 2–4*
76–88 MHz:	*TV channels 5–6*
174–216 MHz:	*TV channels 7–13*
470–806 MHz:	*TV channels 14–69*
0.9 – 12.2 GHz:	*Nonexclusive TV terrestrial and satellite ancillary services*
12.2–12.7 GHz:	*Direct satellite broadcasting*
12.7–40 GHz:	*Direct satellite broadcasting*

6.2 PROPAGATION IN FREE SPACE

For simplicity and ease of explanation, propagation in space and under certain conditions involving simple geometry, in which the wave fronts remain coherent, may be treated as ray propagation. It should be kept in mind that this assumption may not hold in the presence of obstructions, surface roughness, and other conditions which may be encountered in practice.

For the simplest case of propagation in space, namely that of uniform radiation in all directions from a point source, or isotropic radiator, it is useful to consider the analogy to a point source of light. The radiant energy passes with uniform intensity through all portions of an imaginary spherical surface located at a radius r from the source. The area of such a surface is $4\pi r^2$, and the power flow per unit area $W = P_t/4\pi r^2$, where P_t is the total power radiated by the source. In the engineering of broadcasting and of some other radio services, it is conventional to measure the intensity of radiation in terms of the strength of the electric field E_o rather than in terms of power density W. The power density is equal to the square of the field strength divided by the impedance of the medium, so for free space $W = E_o^2/120\pi$, and $P_t = 4\pi r^2 E_o^2/120\pi$, or

$$P_t = \frac{r^2 E_o^2}{30} \qquad (6\text{-}1)$$

where P_t is in watts radiated, W is in watts per square meter, E_o is the free space field in volts per meter, and r is the radius in meters. A more conventional and useful form of this equation, which applies also to antennas other than isotropic radiators, is

$$E_o = \frac{\sqrt{30 g_t P_t}}{r} \qquad (6\text{-}2a)$$

where g_t is the power gain of the antenna in the pertinent direction compared to an isotropic radiator.

An isotropic antenna is useful as a reference for specifying the radiation patterns for more complex antennas but does not in fact exist. The simplest forms of practical antennas are the electric doublet and the magnetic doublet, the former a straight conductor which is short compared with the wavelength and the latter a conducting loop of short radius compared with the wavelength. For the doublet radiator the gain is 1.5 and the field strength in the equatorial plane is

$$E_o = \frac{\sqrt{45 P_t}}{r} \qquad (6\text{-}2b)$$

For a half-wave dipole, namely, a straight conductor one-half wave in length, the power gain is 1.64 and

$$E_o = \frac{7\sqrt{P_t}}{r} \qquad (6\text{-}2c)$$

From the above formulas it can be seen that for free space: (1) the radiation intensity in watts per square meter is proportional to the radiated power and inversely proportional to the square of the radius or distance from the radiator; (2) the electric field strength is proportional to the square root of the radiated power and inversely proportional to the distance from the radiator.

6.2.1 TRANSMISSION LOSS BETWEEN ANTENNAS IN FREE SPACE[1]† The maximum useful power P_r that can be delivered to a matched receiver is given by

$$P_r = \left(\frac{E\lambda}{2\pi}\right)^2 \frac{g_r}{120} \qquad \text{W} \qquad (6\text{-}3)$$

where E = received field strength, volts per meter
λ = wavelength, meters, $300/F$
F = frequency, megacycles per second
g_r = receiving antenna power gain over an isotropic radiator

---

†Superscript numbers refer to References at end of chapter.

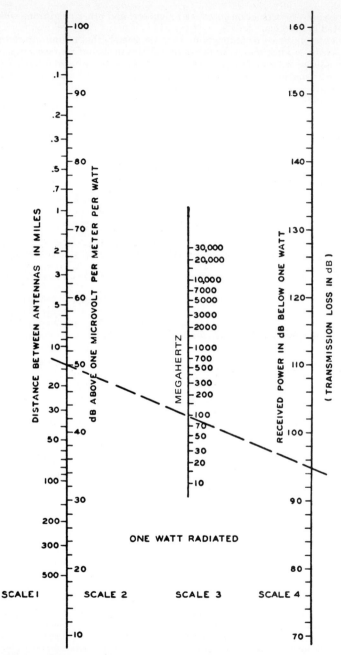

FIG. 6-1 Free-space field intensity and received power between half-wave dipoles. (*From Fink.*[2])

6.4

This relation between received power and the received field strength is shown by scales 2, 3, and 4 in Fig. 6-1 for a half-wave dipole. For example, the maximum useful power at 100 MHz that can be delivered by a half-wave dipole in a field of 50 dB above 1 μV per meter is 95 dB below 1 W. A general relation for the ratio of the received power to the radiated power obtained from Eqs. (6-2a) and (6-3) is

$$\frac{P_r}{P_t} = \left(\frac{\lambda}{4\pi r}\right)^2 g_t g_r \left(\frac{E}{E_o}\right)^2 \tag{6-4a}$$

When both antennas are half-wave dipoles, the power-transfer ratio is

$$\frac{P_r}{P_t} = \left(\frac{1.64\lambda}{4\pi r}\right)^2 \left(\frac{E}{E_o}\right)^2 = \left(\frac{0.13\lambda}{r}\right)^2 \left(\frac{E}{E_o}\right)^2 \tag{6-4b}$$

and is shown on scales 1 to 4 of Fig. 6-3. For free-space transmission $E/E_o = 1$.

When the antennas are horns, paraboloids, or multielement arrays, a more convenient expression for the ratio of the received powr to the radiated power is given by

$$\frac{P_r}{P_t} = \frac{B_t B_r}{(\lambda r)^2} \left(\frac{E}{E_o}\right)^2 \tag{6-4c}$$

where B_t and B_r are the effective areas of the transmitting and receiving antennas, respectively. This relation is obtained from Eq. (6-4a) by substituting $g = 4\pi B/\lambda^2$, and is shown in Fig. 6-2 for free-space transmission when $B_t = B_r$. For example, the free-space loss at 4000 MHz between two antennas of 10 ft^2 (0.93 m^2) effective area is about 72 dB for a distance of 30 mi (48 km).

6.3 PROPAGATION OVER PLANE EARTH[3,4]

The presence of the ground modifies the generation and the propagation of the radio waves so that the received field strength is ordinarily different than would be expected in free space. The ground acts as a partial reflector and as a partial absorber, and both of these properties affect the distribution of energy in the region above the earth.

6.3.1 **FIELD STRENGTHS OVER PLANE EARTH.** The geometry of the simple case of propagation between two antennas each placed several wavelengths above a plane earth is shown in Fig. 6-3. For isotropic antennas, for simple magnetic-doublet antennas with vertical polarization, or for simple electric-doublet antennas with horizontal polarization the resultant received field is[4,5]

$$E = \frac{E_o d}{r_1} + \frac{E_o dR e^{j\Delta}}{r_2}$$
$$= E_o (\cos \theta_1 + R \cos \theta_2 e^{j\Delta}) \tag{6-5a}$$

For simple magnetic-doublet antennas with horizontal polarization or electric-doublet antennas with vertical polarization at both transmitter and receiver, it is necessary to correct for the cosine radiation and absorption patterns in the plane of propagation. The received field is

$$E = E_o (\cos^3 \theta_1 + R \cos^3 \theta_2 e^{j\Delta}) \tag{6-5b}$$

where E_o is the free-space field at distance d in the equatorial plane of the doublet; R is the complex reflection coefficient of the earth; $j = \sqrt{-1}$; $e^{j\Delta} = \cos \Delta + j \sin \Delta$; and Δ is the phase difference between the direct wave received over path r_1 and the ground-reflected wave received over path r_2, which is due to the difference in path lengths.

For distances such that θ is small and the differences between d and r_1 and r_2 may be neglected, Eqs. (6-5a) and (6-5b) become

$$E = E_o(1 + Re^{j\Delta}) \tag{6-6}$$

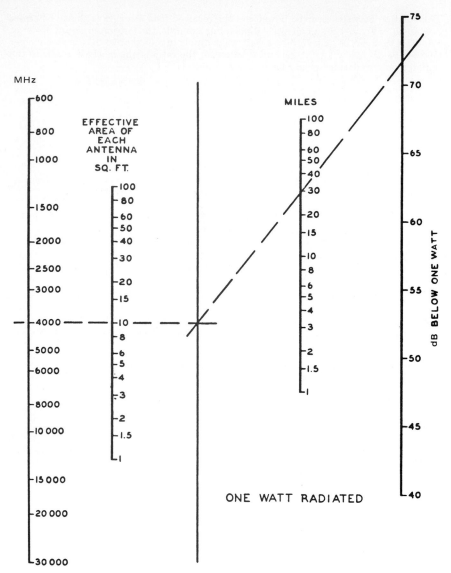

FIG. 6-2 Received power in free space between two antennas of equal effective areas. (*From Fink.*[2])

When the angle θ is very small, R is approximately equal to -1. For the case of two antennas, one or both of which may be relatively close to the earth, a surface-wave term must be added[3,6] and Eq. (6-6) becomes

$$E = E_o[1 + Re^{j\Delta} + (1 - R)Ae^{j\Delta}] \qquad (6\text{-}7a)$$

The quantity A is the surface-wave attenuation factor which depends upon the frequency, ground constants, and type of polarization. It is never greater than unity and decreases with increasing distance and frequency, as indicated by the following approximate equation.[1]

$$A \simeq \frac{-1}{1 + j(2\pi d/\lambda)(\sin\theta + z)^2} \tag{6-7b}$$

This approximate expression is sufficiently accurate as long as $A < 0.1$, and it gives the magnitude of A within about 2 dB for all values of A. However, as A approaches unity, the error in phase approaches 180°. More accurate values are given by Norton[3] where, in his nomenclature, $A = f(P,B)e^{i\phi}$.

The equation (6-7a) for the absolute value of field strength has been developed from the successive consideration of the various components which make up the ground wave, but the following equivalent expressions may be found more convenient for rapid calculation

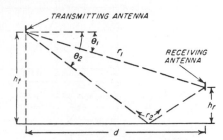

FIG. 6-3 Ray paths for antennas above plane earth. (*From Fink.*[2])

$$E = E_o \left[2\sin\frac{\Delta}{2} + j[(1 + R) + (1 - R)A]e^{j\Delta/2} \right] \tag{6-8a}$$

When the distance d between antennas is greater than about five times the sum of the two antenna heights h_t and h_r, the phase difference angle Δ is equal to $4\pi h_t h_r/\lambda d$ rad. Also when the angle Δ is greater than about 0.5 rad, the terms inside the brackets, which include the surface wave, are usually negligible, and a sufficiently accurate expression is given by

$$E = E_o \left(2\sin\frac{2\pi h_t h_r}{\lambda d} \right) \tag{6-8b}$$

In this case the principal effect of the ground is to produce interference fringes or lobes, so that the field strength oscillates about the free-space field as the distance between antennas or the height of either antenna is varied.

When the angle Δ is less than about 0.5 rad, there is a region in which the surface wave may be important but not controlling. In this region $\sin\Delta/2$ is approximately equal to $\Delta/2$ and

$$E = E_o \frac{4\pi h_t' h_r'}{\lambda d} \tag{6-8c}$$

In this equation $h' = h + jh_o$, where h is the actual antenna height and $h_o = \lambda/2\pi z$ has been designated as the minimum effective antenna height. The magnitude of the minimum effective height h_o is shown in Fig. 6-4 for seawater and for "good" and "poor" soil. "Good" soil corresponds roughly to clay, loam, marsh, or swamp, while "poor" soil means rocky or sandy ground.[1]

The surface wave is controlling for antenna heights less than the minimum effective height, and in this region the received field or power is not affected appreciably by changes in the antenna height. For antenna heights that are greater than the minimum effective height, the received field or power is increased approximately 6 dB every time the antenna height is doubled, until free-space transmission is reached. It is ordinarily sufficiently accurate to assume that h' is equal to the actual antenna height or the minimum effective antenna height, whichever is the larger.

When translated into terms of antenna heights in feet, distance in miles, effective power in kilowatts radiated from a half-wave dipole, and frequency F in megacycles per second, Eq. (6-8c) becomes the following very useful formula for the rapid calculation of

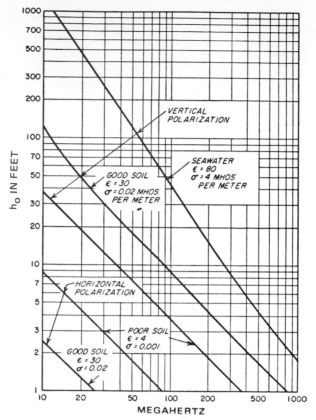

FIG. 6-4 Minimum effective antenna height. (*From Fink.*[2])

approximate values of field strength for purposes of prediction or for comparison with measured values.

$$E \simeq F \frac{h_t' h_r' \sqrt{P_t}}{3d^2} \tag{6-8d}$$

6.3.2 TRANSMISSION LOSS BETWEEN ANTENNAS OVER PLANE EARTH. The ratio of the received power to the radiated power for transmission over plane earth is obtained by substituting Eq. (6-8c) into (6-4a), resulting in

$$\frac{P_r}{P_t} = \left(\frac{\lambda}{4\pi d}\right)^2 g_t g_r \left(\frac{4\pi h_t' h_r'}{\lambda d}\right)^2 = \left(\frac{h_t' h_r'}{d^2}\right)^2 g_t g_r \tag{6-9}$$

This relation is independent of frequency, and is shown on Fig. 6-5 for half-wave dipoles ($g_t = g_r = 1.64$). A line through the two scales of antenna height determines a point on the unlabeled scale between them, and a second line through this point and the distance scale determines the received power for 1 W radiated. When the received field strength is desired, the power indicated on Fig. 6-5 can be transferred to scale 4 of Fig. 6-1, and a line through the frequency on scale 3 indicates the received field strength on scale 2. The results shown on Fig. 6-5 are valid as long as the value of received power indicated

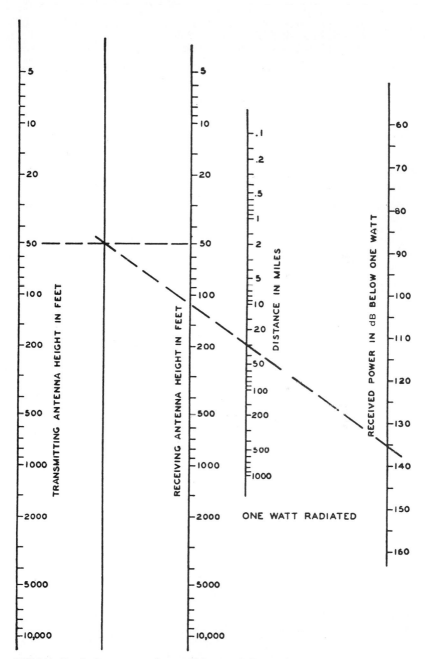

FIG. 6-5 Received power over plane earth between half-wave dipoles. *Notes:* (1) This chart is not valid when the indicated received power is greater than the free space power shown in Fig. 6-1. (2) Use the actual antenna height or the minimum effective height shown in Fig. 6-4, whichever is the larger. (*From Fink.*[2])

is lower than that shown on Fig. 6-3 for free-space transmission. When this condition is not met, it means that the angle Δ is too large for Eq. (6-8c) to be accurate and that the received field strength or power oscillates around the free-space value as indicated by Eq. (6-8b).[1]

6.4 PROPAGATION OVER SMOOTH SPHERICAL EARTH

6.4.1 PROPAGATION WITHIN THE LINE OF SIGHT.

The curvature of the earth has three effects on the propagation of radio waves at points within the line of sight. First, the reflection coefficient of the ground-reflected wave differs for the curved surface of the earth from that for a plane surface. This effect is of little importance, however, under the circumstances normally encountered in practice. Second, since the ground-reflected wave is reflected against the curved surface of the earth, its energy diverges more than would be indicated by the inverse distance-squared law, and the ground-reflected wave must be multiplied by a divergence factor D. Finally, the heights of the transmitting and receiving antennas h_t' and h_r', above the plane which is tangent to the surface of the earth at the point of reflection of the ground-reflected wave, are less than the antenna heights h_t and h_r, above the surface of the earth, as shown in Fig. 6-6.

Under these conditions Eq. (6-6), which applies to larger distances within the line of sight and to antennas of sufficient height that the surface component may be neglected, becomes

$$E = E_o(1 + DR'e^{j\Delta}) \tag{6-10}$$

Similar substitutions of the values which correspond in Figs. 6-3 and 6-6 may be made in Eqs. (6-7a) through (6-9). However, under practical conditions, it is generally satis-

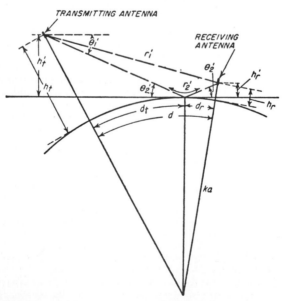

FIG. 6-6   Ray paths for antennas above spherical earth. (*From Fink.*[2])

factory to use the plane-earth formulas for the purpose of calculating smooth-earth values. An exception to this is usually made in the preparation of standard reference curves, which are generally calculated by the use of the more exact formulas.[1,4-9]

6.4.2 PROPAGATION BEYOND THE LINE OF SIGHT. Radio waves are bent around the earth by the phenomenon of diffraction, with the ease of bending decreasing as the frequency increases. Diffraction is a fundamental property of wave motion, and in optics it is the correction to apply to geometrical optics (ray theory) to obtain the more accurate wave optics. In wave optics, each point on the wave front is considered to act as a radiating source. When the wave front is coherent or undisturbed, the resultant is a progression of the front in a direction perpendicular thereto, along a path which constitutes the ray. When the front is disturbed, the resultant front may be changed in both magnitude and direction with resulting attenuation and bending of the ray. Thus all shadows are somewhat "fuzzy" on the edges and the transition from "light" to "dark" areas is gradual, rather than infinitely sharp.

The effect of diffraction around the earth's curvature is to make possible transmission beyond the line of sight, with somewhat greater loss than is incurred in free space or over plane earth. The magnitude of this loss increases as either the distance or the frequency is increased and it depends to some extent on the antenna height.

The calculation of the field strength to be expected at any particular point in space beyond the line of sight around a spherical earth is rather complex, so that individual calculations are seldom made. Rather, nomograms or families of curves are usually prepared for general application to large numbers of cases. The original wave equations of Van der Pol and Bremmer[6] have been modified by Burrows[7] and by Norton[3,5] so as to make them more readily usable and particularly adaptable to the production of families of curves. Such curves have been prepared. These curves have not been included herein, in view of the large number of curves which are required to satisfy the possible variations in frequency, electrical characteristics of the earth, polarization, and antenna height. Also, the values of field strength indicated by smooth-earth curves are subject to considerable modification under actual conditions found in practice. For VHF and UHF broadcast purposes, the smooth-earth curves have been to a great extent superseded by curves modified to reflect average conditions of terrain.

Figure 6-7 is a nomogram to determine the additional loss caused by the curvature of the earth.[1] This loss must be added to the free-space loss found from Fig. 6-1. A scale is included to provide for the effect of changes in the effective radius of the earth, caused by atmospheric refraction. Figure 6-7 gives the loss relative to free space as a function of three distances; d_1 is the distance to the horizon from the lower antenna, d_2 is the distance to the horizon from the higher antenna, and d_3 is the distance between the horizons. The total distance between antennas is $d = d_1 + d_2 + d_3$.

The horizon distances d_1 and d_2 for the respective antenna heights h_1 and h_2 and for any assumed value of the earth's radius factor k can be determined from Fig. 6-8.[1]

6.5 EFFECTS OF HILLS, BUILDINGS, VEGETATION, AND THE ATMOSPHERE

The preceding discussion assumes that the earth is a perfectly smooth sphere with a uniform or a simple atmosphere, for which condition calculations of expected field strengths or transmission losses can be computed for the regions within the line of sight and regions well beyond the line of sight, and interpolations can be made for intermediate distances. The presence of hills, buildings, and trees has such complex effects on propagation that it is impossible to compute in detail the field strengths to be expected at discrete points in the immediate vicinity of such obstructions or even the median values over very small areas. However, by the examination of the earth profile over the path of propagation and by the use of certain simplifying assumptions, predictions which are

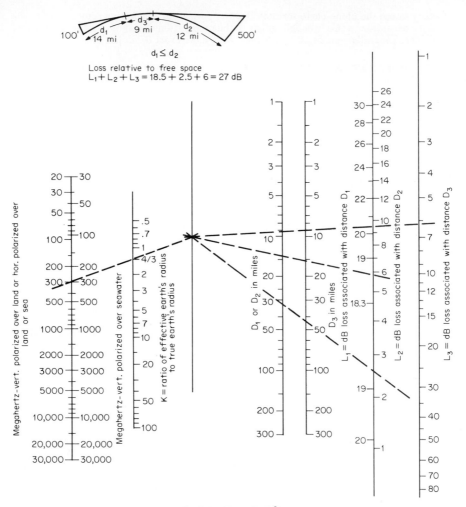

FIG. 6-7   Loss beyond line of sight in decibels. (*From Fink.*[2])

more accurate than smooth-earth calculations can be made of the median values to be expected over areas representative of the gross features of terrain.

6.5.1 EFFECTS OF HILLS. The profile of the earth between the transmitting and receiving points is taken from available topographic maps† and is plotted on a special chart which provides for average air refraction by the use of a four-thirds earth radius, as shown in Fig. 6-9. The vertical scale is greatly exaggerated for convenience in displaying significant angles and path differences. Under these conditions vertical dimensions are measured along vertical parallel lines rather than along radii normal to the curved surface, and the propagation paths appear as straight lines. The field to be expected at a low receiving antenna at A from a high transmitting antenna at B can be predicted by

---

†Available from the U.S. Geological Survey, Dept. of the Interior, Washington, D.C.

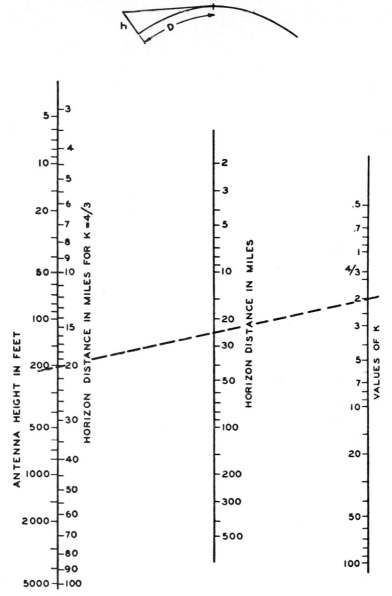

FIG. 6-8 Distance to horizon. (*From Fink.*[2])

plane-earth methods, by drawing a tangent to the profile at the point at which reflection appears to occur with equal incident and reflection angles. The heights of the transmitting and receiving antennas above the tangent are used in conjunction with Fig. 6-5 to compute the transmission loss, or with Eq. (6-8d) to compute the field strength. A similar procedure can be used for more distantly spaced high antennas when the line of sight does not clear the profile by at least the first Fresnel zone.[10]

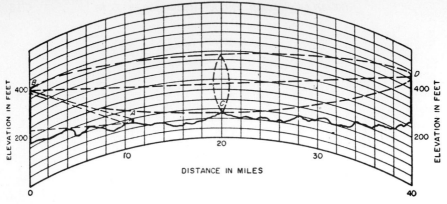

FIG. 6-9 Ray paths for antennas over rough terrain. (*From Fink.*[2])

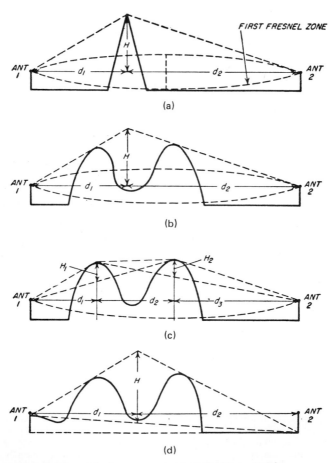

FIG. 6-10 Ray paths for antennas behind hills. (*From Fink.*[2])

Propagation over a sharp ridge, or over a hill when both the transmitting and receiving antenna locations are distant from the hill, may be treated as diffraction over a knife edge, shown schematically in Fig. 6-10a.[1,9-14] The height of the obstruction H is measured from the line joining the centers of the two antennas to the top of the ridge. As shown in Fig. 6-11, the shadow loss approaches 6 dB as H approaches 0, grazing incidence, and it increases with increasing positive values of H. When the direct ray clears the obstruction, H is negative, and the shadow loss approaches 0 dB in an oscillatory manner as the clearance is increased. Thus, a substantial clearance is required over line-of-sight paths in order to obtain free-space transmission. There is an optimum clearance, called the first Fresnel-zone clearance, for which the transmission is theoretically 1.2 dB

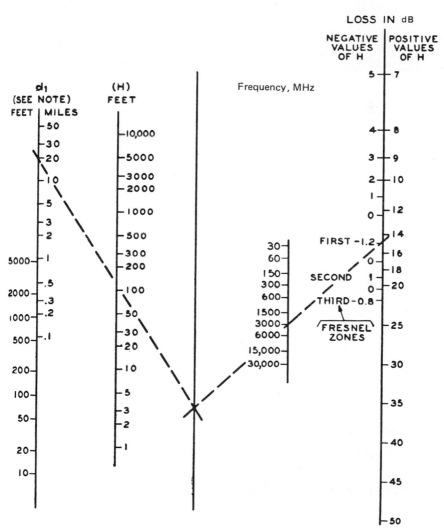

FIG. 6-11 Shadow loss relative to free space. *Note:* When accuracy greater than ±1.5 dB is required, values on the d_1 scale should be $d_1[\sqrt{2}/(1 + d_1/d_2)]$. (*From Fink.*[2])

better than in free space. Physically, this clearance is of such magnitude that the phase shift along a line from the antenna to the top of the obstruction and from there to the second antenna is about one-half wavelength greater than the phase shift of the direct path between antennas.

The locations of the first three Fresnel zones are indicated on the right-hand scale on Fig. 6-11, and by means of this chart the required clearances can be obtained. At 3000 MHz, for example, the direct ray should clear all obstructions in the center of a 40 mi (64 km) path by about 120 ft (36 m) to obtain full first-zone clearance, as shown at C in Fig. 6-9. The corresponding clearance for a ridge 100 ft (30 m) in front of either antenna is 4 ft (1.2 m). The locus of all points which satisfy this condition for all distances is an ellipsoid of revolution with foci at the two antennas.

When there are two or more knife-edge obstructions or hills between the transmitting and receiving antennas, an equivalent knife edge may be presented by drawing a line from each antenna through the top of the peak that blocks the line of sight, as in Fig. 6-10b.

Alternatively, the transmission loss may be computed by adding the losses incurred when passing over each of the successive hills, as in Fig. 6-10c. The height H_1 measured from the top of hill 1 to the line connecting antenna 1 and the top of hill 2. Similarly, H_2 is measured from the top of hill 2 to the line connecting antenna 2 and the top of hill 1. The nomogram in Fig. 6-11 is used for calculating the losses for terrain conditions represented by Fig. 6-10a to c.

The above procedure applies to conditions for which the earth-reflected wave can be neglected, such as the presence of rough earth, trees, or structures at locations along the profile at points where earth reflection would otherwise take place at the frequency under consideration; or where first Fresnel-zone clearance is obtained in the foreground of each antenna and the geometry is such that reflected components do not contribute to the field within the first Fresnel zone above the obstruction. If conditions are favorable to earth reflection, the base line of the diffraction triangle should not be drawn through the antennas, but through the points of earth reflection, as in Fig. 6-10d. H is measured vertically from this base line to the top of the hill, while d_1 and d_2 are measured to the antennas as before. In this case Fig. 6-12 is used to estimate the shadow loss to be added to the plane-earth attenuation.[1]

Under conditions where the earth-reflected components reinforce the direct components at the transmitting and receiving antenna locations paths may be found for which the transmission loss over an obstacle is less than the loss over spherical earth. This effect may be useful in establishing VHF relay circuits where line-of-sight operation is not practical. Little utility may be expected for mobile or broadcast services.[14]

An alternative method for predicting the median value for all measurements in a completely shadowed area is as follows:[15] (1) The roughness of the terrain is assumed to be represented by height H, shown on the profile at the top of Fig. 6-13. (2) This height is the difference in elevation between the bottom of the valley and the elevation necessary to obtain line of sight with the transmitting antenna. (3) The difference between the measured value of field intensity and the value to be expected over plane earth is computed for each point of measurement within the shadowed area. (4) The median value for each of several such locations is plotted as a function of $\sqrt{H/\lambda}$.

These empirical relationships are summarized in the nomogram shown in Fig. 6-13. The scales on the right-hand line indicate the median value of shadow loss, compared with plane-earth values, and the difference in shadow loss to be expected between the median and the 90 percent values. For example, with variations in terrain of 500 ft (150 m), the estimated median shadow loss at 4500 MHz is about 20 dB and the shadow loss exceeded in 90 percent of the possible locations is about $20 + 15 = 35$ dB. This analysis is based on large-scale variations in field intensity, and does not include the standing-wave effects which sometimes cause the field intensity to vary considerably in a matter of a few feet.

6.5.2 EFFECTS OF BUILDINGS.   Built-up areas have little effect on radio transmission at frequencies below a few megacycles, since the size of any obstruction is usually

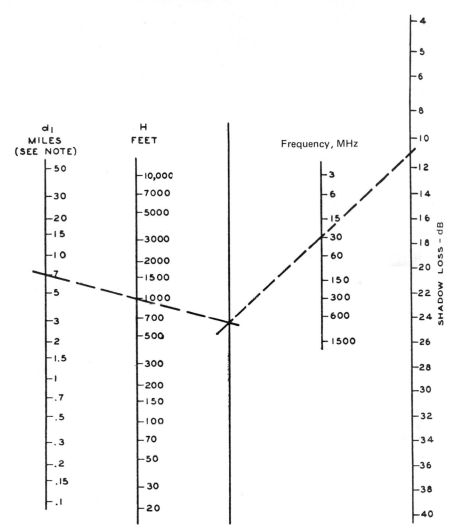

FIG. 6-12 Shadow loss relative to plane earth. *Note:* When accuracy greater than ±1.5 dB is required, values on the d_1 scale should be $d_1[\sqrt{2}/(1 + d_1/d_2)]$. (*From Fink.*[2])

small compared with the wavelength, and the shadows caused by steel buildings and bridges are not noticeable except immediately behind these obstructions. However, at 30 MHz and above, the absorption of a radio wave in going through an obstruction and the shadow loss in going over it are not negligible, and both types of losses tend to increase as the frequency increases. The attenuation through a brick wall, for example, may vary from 2 to 5 dB at 30 MHz and from 10 to 40 dB at 3000 MHz, depending on whether the wall is dry or wet. Consequently, most buildings are rather opaque at frequencies of the order of thousands of megacycles.

For radio-relay purposes, it is the usual practice to select clear sites; but where this is not feasible the expected fields behind large buildings may be predicted by the pre-

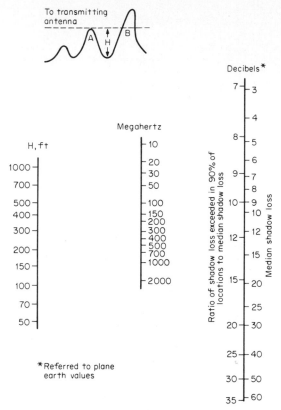

FIG. 6-13 Estimated distribution of shadow loss for random locations (referred to plane-earth values). (*From Fink.*[2])

ceding diffraction methods. In the engineering of mobile- and broadcast-radio systems it has not been found practical in general to relate measurements made in built-up areas to the particular geometry of buildings, so that it is conventional to treat them statistically. However, measurements have been divided according to general categories into which buildings may readily be classified, namely, the tall buildings typical of the centers of cities on the one hand, and typical two-story residential areas on the other.

Buildings are more transparent to radio waves than the solid earth, and there is ordinarily much more backscatter in the city than in the open country. Both of these factors tend to reduce the shadow losses caused by the buildings. On the other hand, the angles of diffraction over or around the buildings are usually greater than for natural terrain, and this factor tends to increase the loss resulting from the presence of buildings. The quantitative data on the effects of buildings indicate that in the range of 40 to 450 MHz there is no significant change with frequency, or at least the variation with frequency is somewhat less than the square-root relationship noted in the case of hills. The median field strength at street level for random locations in Manhattan (New York City) is about 25 dB below the corresponding plane-earth value. The corresponding values for the 10 percent and 90 percent points are about −15 and −35 dB, respectively.[1,15]

Measurements in congested residential areas indicate somewhat less attenuation than among large buildings. In the Report of the Ad Hoc Committee[16] measurements between

4 and 10 mi (6 and 16 km) from the transmitter, which include some large building areas and some open areas but are made up principally of residential areas, show median values of 4 to 6 dB below plane earth for frequencies below 100 MHz, and about 10 dB for frequencies near 200 MHz. More recent measurements at 850 MHz[17] show values of 15 to 26 dB below free space, which appear to be about 10 to 15 dB below plane earth. These measurements were not random, however, but were the maximum values of field between 10 and 30 ft (3 and 9 m) above the earth. The average effects for measurements taken in built-up areas have been accounted for in the preparation of the modified propagation curves of Figs. 6-17 and 6-18.

6.5.3 EFFECTS OF TREES AND OTHER VEGETATION.

When an antenna is surrounded by moderately thick trees and below treetop level, the average loss at 30 MHz resulting from the trees is usually 2 or 3 dB for vertical polarization and negligible with horizontal polarization. However, large and rapid variations in the received field strength may exist within a small area, resulting from the standing-wave pattern set up by reflections from trees located at a distance of as much as 100 ft (30 m) or more from the antenna. Consequently, several nearby locations should be investigated for best results. At 100 MHz the average loss from surrounding trees may be 5 to 10 dB for vertical polarization and 2 or 3 dB for horizontal polarization. The tree losses continue to increase as the frequency increases, and above 300 to 500 MHz they tend to be independent of the type of polarization. Above 1000 MHz trees that are thick enough to block vision present an almost solid obstruction, and the diffraction loss over or around these obstructions can be obtained from Fig. 6-11 or 6-12.[1,9]

There is a pronounced seasonal effect in the case of deciduous trees, with less shadowing and absorption in the winter months when the leaves have fallen. However, when the path of travel through the trees is sufficiently long that it is obscured, losses of the above magnitudes may be incurred, and the principal mode of propagation may be by diffraction over the trees.

When the antenna is raised above trees and other forms of vegetation the prediction of field strengths again depends upon the proper estimation of the height of the antenna above the areas of reflection and of the applicable reflection coefficients. For growth of fairly uniform height and for angles near grazing incidence, reflection coefficients will approach -1 at frequencies near 30 MHz. As indicated by Rayleigh's criterion of roughness, the apparent roughness for given conditions of geometry increases with frequency so that near 1000 MHz even such low and relatively uniform growth as farm crops or tall grass may have reflection coefficients of about -0.3 for small angles of reflection.[17]

The distribution of losses in the immediate vicinity of trees does not follow normal probability law but is more accurately represented by Rayleigh's law, which is the distribution of the sum of a large number of equal vectors having random phases. This distribution is shown by the graph R of Fig. 6-22.

6.5.4 EFFECTS OF THE LOWER ATMOSPHERE, OR TROPOSPHERE.

Radio waves propagating through the lower atmosphere, or troposphere, are subject to absorption, scattering, and bending. Absorption is negligible in the VHF-UHF frequency range but becomes very significant at frequencies above 10 GHz. The index of refraction of the atmosphere, n, is slightly greater than 1 and varies with temperature, pressure, and water vapor pressure, and therefore with height, climate, and local meteorological conditions. An exponential model showing a decrease with height to 37 to 43 mi (60 to 70 km) is generally accepted.[18,19] For this model variation of n is approximately linear for the first kilometer above the surface in which most of the effect on radio waves traveling horizontally occurs. For average conditions the effect of the atmosphere can be included in the expression of earth diffraction around the smooth earth without discarding the useful concept of straight-line propagation by multiplying the actual earth's radius by k to obtain an effective earth's radius, where

$$k = \frac{1}{1 + a(dn/dh)}$$

(6-11)

where a is the actual radius of the earth and dn/dh is the rate of change of the refractive index with height. By use of average annual values of the refractive index gradient, k is found to be ⅔ for temperate climates.

Stratification and Ducts. Due to climatological and weather processes such as subsidence, advection, and surface heating and radiative cooling, the lower atmosphere tends to be stratified in layers with contrasting refractivity gradients.[20] For convenience in evaluating the effect of this stratification, *radio refractivity N* is defined as $N = (n - 1) \times 10^6$ and can be derived from

$$N = 77.6\frac{P}{T} + 3.73 \times 10^5\frac{e}{T^2} \qquad (6\text{-}12)$$

where P = atmosphere pressure, mbar
T = absolute temperature, K
e = water vapor pressure, mbar

When the gradient of N is equal to -39 N-units per kilometer, normal propagation takes place, corresponding to the effective earth's radius ka, where $k = $ ⅔.

When dN/dh is less than -39 N-units per kilometer, subrefraction occurs and the radio wave is bent strongly downward.

When dN/dh is less than -157 N-units per kilometer, the radio energy may be bent downward sufficiently to be reflected from the earth, after which the ray is again bent toward the earth, and so on. The radio energy thus is trapped in a duct or waveguide. The wave also may be trapped between two elevated layers, in which case energy is not lost at the ground reflection points and even greater enhancement occurs. Radio waves thus trapped or ducted can produce fields exceeding those for free-space propagation since spread of energy in the vertical direction is eliminated as opposed to the free-space case, where the energy spreads out in two directions orthogonal to the direction of propagation. Ducting is responsible for abnormally high fields beyond the radio horizon. These enhanced fields occur for significant periods of time on overwater paths in areas where meteorological conditions are favorable. Such conditions exist for significant periods of time and over significant horizontal extent in the coastal areas of southern California and around the Gulf of Mexico. Over land the effect is less pronounced because surface features of the earth tend to limit the horizontal dimension of ducting layers.[20]

Tropospheric Scatter. The most consistent long-term mode of propagation beyond the radio horizon is that of scattering by small-scale fluctuations in the refractive index due to turbulence. Energy is scattered from multitudinous irregularities in the common volume which consists of that portion of troposphere visible to both transmitting and receiving site. There are some empirical data that show a correlation between the variations in the field beyond the horizon and ΔN, the difference between the reflectivity on the ground and at a height of 1 km.[21] Procedures have been developed for calculating scatter fields for beyond the horizon radio relay systems as a function of frequency and distance.[22,23] These procedures, however, require detailed knowledge of path configuration and climate. Investigation of this is continuing, and values of parameters are frequently updated. The effect of scatter propagation is incorporated in the statistical evaluation of propagation considered in Sec. 6.6, where the attenuation of fields beyond the diffraction zone is based on empirical data and shows a linear decrease with distance of approximately 0.2 dB/mi (0.1 dB/km) for the VHF-UHF frequency band.

Atmospheric Fading. Variations in the received field strengths around the median values are caused by changes in atmospheric conditions. Field strengths tend to be higher in summer than in winter, and higher at night than during the day, for paths over land beyond the line of sight. As a first approximation, the distribution of long-term variations in field strength in decibels follows a normal probability law, as described in Sec. 6.6.

Measurements indicate that the fading range reaches a maximum somewhat beyond the horizon and then decreases slowly with distance out to several hundred miles. Also the fading range at the distance of maximum fading increases with frequency, while at the greater distances where the fading range decreases, the range is also less dependent on frequency. Thus the slope of the graph N must be adjusted for both distance and frequency. This behavior does not lend itself to treatment as a function of the earth's radius factor k, since calculations based on the same range of k produce families of curves in which the fading range increases systematically with increasing distance and with increasing frequency. Methods for the statistical treatment of fading are described in Sec. 6.6.

6.5.5 EFFECTS OF THE UPPER ATMOSPHERE, OR IONOSPHERE. Four principal recognized layers or regions in the ionosphere are the E layer, the $F1$ layer, and the $F2$ layer, centered at heights of about 100, 200, and 300 km, respectively, and the D region, which is less clearly defined but lies below the E layer. These *regular* layers are produced by radiation from the sun, so that the ion density, and hence the frequency of the radio waves which can be reflected thereby, is higher in the day than at night. The characteristics of the layers are different for different geographic locations and the geographic effects are not the same for all layers. The characteristics also differ with the seasons and with the intensity of the sun's radiation, as evidenced by the sunspot numbers, and the differences are generally more pronounced upon the $F2$ than upon the $F1$ and E layers. There are also certain random effects which are associated with solar and magnetic disturbances. Other effects which occur at or just below the E layer have been established as being caused by meteors.[24]

The greatest potential for television interference by way of the ionosophere is from Sporadic E which consists of occasional patches of intense ionization occurring 62 to 75 mi (100 to 120 km) above the earth's surface and apparently formed by the interaction of winds in the neutral atmosphere with the earth's magnetic field. Sporadic E ionization can reflect VHF signals back to earth at levels capable of causing interference to television reception for periods lasting from 1 h or more and in some cases totaling more than 100 h per year. In the United States, VHF Sporadic E propagation occurs a greater percentage of the time in the southern half of the country and during the May to August period.[25]

Information regarding potential interference to VHF services from all propagation by way of the ionosphere is available from the CCIR.[26]

6.6 STATISTICAL EVALUATION OF PROPAGATION

In previous sections a partial statistical description has been given of the separate effects of terrain and of the variation of field strengths with time. Methods have been given for the prediction of the median field strengths to be expected for areas of size comparable with the gross features of the terrain, to which correction factors may be applied for the presence of buildings and vegetation and within which the fields may be described in terms of the strengths which are expected to be exceeded at a given percentage of locations. Alternatively, the distribution of field strengths as a function of the percentage of locations may be regarded as the probability, in percent, that a given field strength will be exceeded at a particular location within the area in question.

For the purpose of formulating a national plan for the allotment of television channels, it was deemed impractical by the FCC and their industry advisors to consider in detail even the gross feature of terrain. Therefore, a statistical approach was adopted to prepare applicable propagation curves reflecting the median values extracted from all available data.

6.6.1 FIELD-STRENGTH PREDICTION PROCEDURES. A family of curves was prepared and a general planning approach developed as part of the work of a government-industry ad hoc committee[16] and adopted by the FCC in 1952.[27] Subsequently these curves were revised a number of times as more experience was gained and more empirical data became available. The development of the curves, now contained in the

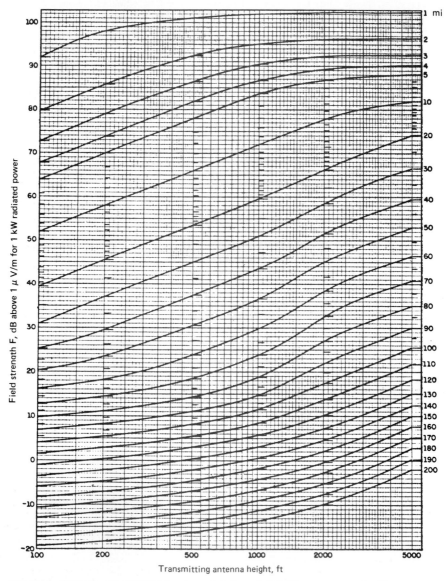

FIG. 6-14 Field strength $F(50,50)$ for channels 2 through 6. In Figs. 6-14–6-16 the $F(50,50)$ designation gives the estimated field strength exceeded at 50 percent of the potential receiver locations for at least 50 percent of the time [receiving antenna height 30 ft (9 m)]. *(FCC.)*

FCC Rules and Regulations,[28] is described in FCC Report R6602. These curves are shown in Figs. 6-14 through 6-19.

Figures 6-14 through 6-16 show field strength in decibels relative to one microvolt per meter, dB (μV/m), for one kilowatt of effective radiated power (ERP) expected to be exceeded at 50 percent of receiving locations for 50 percent of the time, for antenna heights from 100 to 5000 ft (30 to 1500 m) for the low VHF, high VHF, and UHF bands.

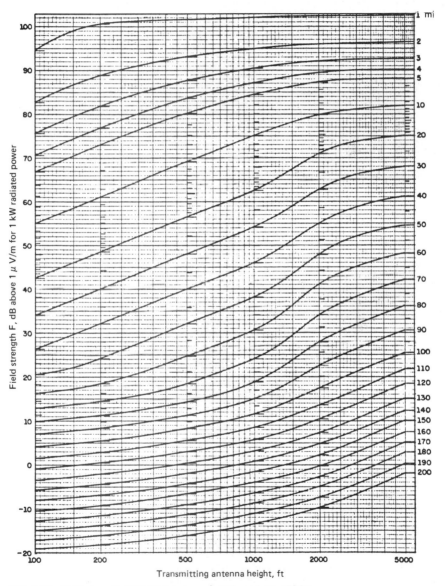

FIG. 6-15 Field strength $F(50,50)$ for channels 7 through 13. *(FCC.)*

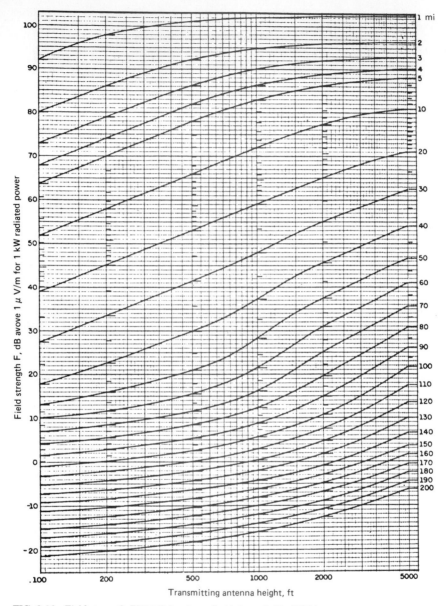

FIG. 6-16   Field strength $F(50,50)$ for channels 14 through 69. *(FCC.)*

These field strengths are referred to as $F(50,50)$ values. The field strengths are based on an effective radiated power of 1 kW from a half-wave dipole in free space, which produces an unattenuated field strength at 1 mi of 102.8 dB above 1 μV/m [102.8 dB (1 μV/m)]. The antenna height to be used with these charts in any particular case is the height of the center of the radiating element above the average height of the terrain profile

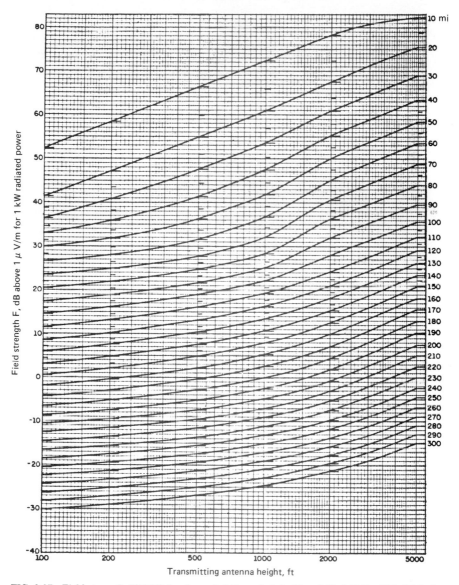

FIG. 6-17 Field strength $F(50,10)$ for channels 2 through 6. In Figs. 6-17–6-19 the $F(50,10)$ designation gives the estimated field strength exceeded at 50 percent of the potential receiver locations for at least 10 percent of the time [receiving antenna height 30 ft (9 m)]. *(FCC.)*

between 2 and 10 mi (3 to 16 km) from the transmitter along the desired radial. Figures 6-17 through 6-19 show the corresponding $F(50,10)$ field-strength curves, i.e., field strength exceeded for 50 percent of the locations and 10 percent of the time. These families of curves may be used to estimate the service provided by television stations in accordance with procedures described below.

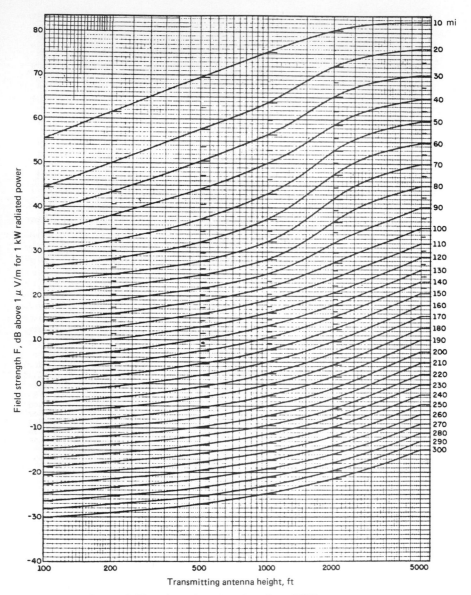

FIG. 6-18 Field strength $F(50,10)$ for channels 7 through 13. *(FCC.)*

6.6.2 MINIMUM FIELD-STRENGTH REQUIREMENTS. The minimum field strength for an acceptable grade of service was determined by the ad hoc committee as follows. The minimum signal into a typical receiver needed to overcome the combined thermal, receiver, and ambient noise power in the 4-MHz video bandwidth was determined. This was found by statistical sampling of receivers and viewers to be 30 dB. The power-flux density at the receiver needed to produce this receiver input was found by

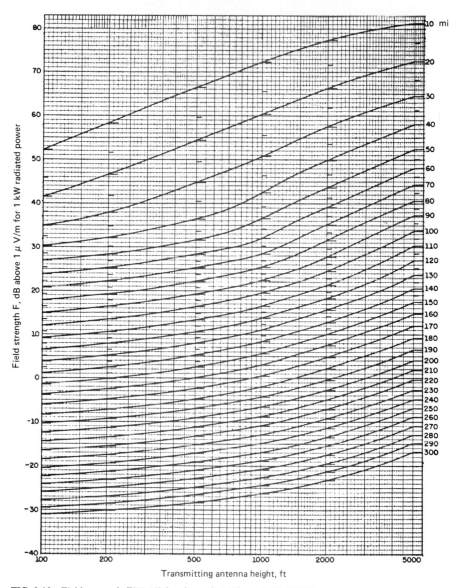

FIG. 6-19   Field strength $F(50,10)$ for channels 14 through 69. *(FCC.)*

considering the effective area of the antenna ($A_{\text{eff}} = \lambda^2/4\pi$) and losses in the transmission line. The power-flux density was converted to steady-state field strength ($E = \sqrt{377p}$). A time-fading factor was added to ensure service for 90 percent of the time.

Originally two service contours were defined. *Grade B* service was intended to provide service to rural areas for at least 50 percent of the locations for 90 percent of the time, and *grade A* service was intended to provide service in an urban environment to at least

Table 6-1 Service Contours

Service Contour	Probability of service	F(50,50) in dBμV/m		
		Low VHF	High VHF	UHF
Principal community	90% locations for 90% time	74	77	80
Grade A	70% locations for 90% time	68	71	74
Grade B	50% locations for 90% time	47	56	64

70 percent of the locations for 90 percent of the time. Later a third, the *principal-city-service* contour, was added to provide service to the community of licensee for at least 90 percent of the locations for 90 percent or more of the time. The minimum field strengths for these contours for the three television bands are shown in Table 6-1. Detailed explanations of the derivation of these values appear in the literature.[29-31]

6.6.3 PROPAGATION-PREDICTION PROCEDURES. The field-strength curves in Figs. 6-14 through 6-19 show the relationship between field strength, distance, and transmitting antenna height. If any two of these quantities are known, the third can be found. For example, the field strength at a given distance may be found by locating the intersection of that curve with the antenna height from the abscissa, adding the field strength from the left-hand scale to the correct number of decibels for the difference in power above the reference level of 1000 W, or 10 log (P/1000). To find the service range of a station, subtract the power in decibels above 1 kW (dBk) from the minimum field strength required, at that decibel level on the left-hand scale, locate the intersection with the desired antenna height, and read the distance from the right-hand scale.

6.6.4 LOCATION AND TIME FACTORS. In general, field-strength variation with *location* has been found to follow a log-normal distribution and a log-normal distribution with *time* between the 10 and 90 percentiles. The field strength for any percentage of *location* and *time* can be found using the following equation

$$F(L,T) = F(50,50) + k(T)\sigma_T + k(L)\sigma_L \qquad (6\text{-}13)$$

where $F(50,50)$ = median field-strength level (field-strength level exceeded for 50 percent of the locations and 50 percent of the time)
$\quad k(T)\sigma_T$ = time-fading factor
$\qquad k(T)$ = standard variate for normal distribution for T percent
$\qquad\quad \sigma_T$ = standard deviation of field strength varying with time
$\quad k(L)\sigma_L$ = location variability factor
$\qquad k(L)$ = standard variate for normal distribution for L percent
$\qquad\quad \sigma_L$ = standard deviation of field strength varying with location

$F(L,T)$ is defined as the field strength available for L percent of the locations in a given area for T percent of the time. $F(50,50)$ can be read from the charts in Figs. 16-14 through 6-19. The values for $k(T)$ and $k(L)$ can be found from Fig. 6-20 or from published statistical tables[32] and

$$\sigma_T = \frac{F(50,10) - F(50,50)}{1.282} \qquad (6\text{-}14)$$

$$\sigma_L = 4.74 \log f - 1.45 \qquad (6\text{-}15)$$

where f is the frequency in megahertz.

Example. Given a channel 2 station with an effective radiated power (ERP) of 100 kW (20 dBk) and an antenna height of 1000 ft (300 m) above average terrain. What percentage of locations will receive a field-strength level of 61 dBμ or greater for 90 percent of the time in an area 25 mi (40 km) from the station? The field strength for L percent of the locations and 90 percent of the time, in a given area, is equal to or greater than 61 dB (μV/m), expressed below in the notation of Eq. (6-13).

$$F(L,T) = F(L,90) = 61 \text{ dB } (\mu\text{V/m})$$

$$\text{ERP} = 20 \text{ dBk}$$

$$F(50,50) = 55 \text{ dB } (\mu\text{V/m}) \qquad \text{from Fig. 6-14}$$

$$F(50,10) = 56.8 \text{ dB } (\mu\text{V/m}) \qquad \text{from Fig. 6-17}$$

$$\sigma_L = 4.74 \log 55.25 - 1.45 = 6.8 \qquad \text{from Eq. (6-15)}$$

$$\sigma_T = \frac{F(50,10) - F(50,50)}{1.282} = 1.4 \qquad \text{from Eq. (6-14)}$$

$$k(T) = k(90) = -1.28 \qquad \text{from Fig. 6-20†}$$

$$F(L,T) = F(50,50) + k(T)\sigma_T + k(L)\sigma_L \qquad \text{from Eq. (6-13)}$$

$$F(L,90) = F(50,50) + k(90)\sigma_T + k(L)\sigma_L + 20 \text{ (dBk ERP)} = 61$$

Rearranging terms and solving for $k(L)$

$$k(L) = -1.79$$

From Fig. 6-20

$$L = 96.3 \text{ percent}$$

6.6.5 PREDICTION OF SERVICE IN THE PRESENCE OF INTERFERENCE.

The percentage of receiving locations, or the probability in percent L, at any given distance

---

†More precise values may be found in statistical tables.[32]

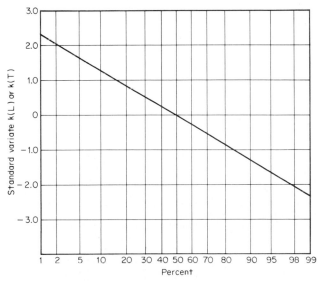

FIG. 6-20 Standard variate for normal (N) distribution.

from a desired station and an undesired station (for which an acceptable ration A, in decibels of desired-to-undesired signals is exceeded for at least T percent of the time at the receiver input) may be determined from the following equation

$$\sqrt{2}\, k(L)\sigma_L = A + P_u - P_d + g_u - g_d + F_u\,(50,50) - F_d\,(50,10) + k(T)\,\sqrt{\sigma_{Td}^2 + \sigma_{Tu}^2}$$

The subscript d denotes values applicable to the desired signal, and the subscript u denotes values applicable to the undesired signal. The effective radiated powers of the desired and undesired stations P_d and P_u are expressed in decibels above 1 kW radiated from a half-wave dipole. g_u and g_d are, respectively, the gains of the receiving antenna in the directions of the undesired and desired transmitters. $F_d(50, 50)$ and $F_u(50, 10)$ are the indicated field strengths from the desired and undesired transmitters taken from the appropriate curves of Figs. 6-14 through 6-19. σ_{Td} and σ_{Tu} are the standard deviations for time-varying desired and undesired field strengths, respectively. σ_L is the standard deviation of the field strength as it varies with location from Eq. (6-15). σ_T for $L = 90$ percent can can be determined from Eq. (6-14). For any other percentage of time, σ_T can be found from the same equation by substituting $k(T)$ from Fig. 6-20 for 1.28 in the equation. Once $k(L)$ has been calculated, the percentage of receiving locations, or the probability in percent, L, can be read from Fig. 6-20.

The time variability applicable to two fading signals combines as the square roots of the individual factors. When fading of the undesired signal is three or more times the fading of the desired, the latter fading may be neglected with negligible error. Thus, the charts for the desired signal $F(50, 50)$ and the $F(50, 10)$ charts for the undesired, with appropriate corrections for the effective radiated powers of the two stations and the directional patterns of the receiving antennas, can be used to determine directly the iso-service contours for 90 percent of the time and 50 percent of the locations.

The approximate method, in which the fading of the desired signal is neglected, may be expressed in the following formula

$$\sqrt{2}\, k(L)\sigma_L = A + P_u - P_d + g_u - g_d + F_u(50,10) - F_d(50,50)$$

This equation permits the approximate method to be applied to the case where it is desired to locate the contour at which an acceptable ratio is exceeded for a percentage of the locations other than 50 percent.

6.7 SELECTION OF STATION SITES

The sites for the antennas of permanent relay stations and for television broadcast stations should be selected carefully, as the success of the operation depends to a great extent upon the care and foresight used in selecting the sites. In previous sections, quantitative information has been given which will assist in estimating the transmission loss incurred over television relay links under specified conditions and in estimating the probable service obtained from a television broadcast station. The purpose of this section is to give a few guides which engineers have found to be of assistance in selecting sites which will yield optimum results for the area in question.

6.7.1 SELECTION OF SITES FOR RELAY STATIONS   A large amount of preliminary work is necessary in laying out a relay system. For microwave systems it is usual that adjacent sites have a clear line-of-sight path between their antennas. To determine station locations, the best available topographical maps should be used. These, however, usually will not be adequate for final station location, particularly since for some areas no maps are available, and for others existing maps may be found to be inaccurate. In some cases, aerial surveys have been used. Before finally selecting a site, radio transmission tests should be made with the adjacent sites. For these tests, portable towers with parabolic antennas and transmitters mounted on carriages which can be moved up and down the towers are used.

If the intervening terrain is rough and the intermediate clearances satisfactory, no large change in received signal will be found as the antennas are raised and lowered on the towers, until the antennas are lowered so far that clearance of the first Fresnel zone no longer exists.

In some cases, however, substantial earth reflection may be encountered, and this will be evidenced by substantial variations in received signal strength as the antennas travel up and down. It is desirable that transmission be largely confined to a single ray arriving from the distant station. If one or more additional rays are present, due to reflection from the earth at some intervening point or points, the resultant signal will be that due to the combination of rays and will depend on their relative phase relations. Sites showing such earth reflection are not desirable, as substantial amounts of fading may be expected at times when changes in the atmosphere cause the amount of bending of the waves to change with consequent changes in phase relation of the arriving rays. If such variations are observed in these path tests, the intervening terrain should be inspected with a view to determining whether by moving one or both sites a short distance the reflecting earth surface can be avoided.

It is usually difficult to recognize the areas which are responsible for earth reflection, but a few guides can be given to assist in inspection at the site. If the suspected area is fairly flat, areas of a size equivalent to an ellipse capable of reflecting the wave front over the first Fresnel zone should be inspected. Smaller areas are capable of supporting a reflection, but in general the strength of the reflected component will be decreased. When the intervening area is rolling or irregular, the determination of the location and size of the responsible area is still more difficult. It will also be necessary to decide whether the surface roughness is too great to support reflection at the frequency of interest. For this purpose, Rayleigh's criterion of roughness is used. The surface is considered to be smooth if $h \sin \theta < \lambda/8$, where h is the average height of the features of roughness, θ is the angle of incidence of the wave to the reflecting surface, and λ is the wavelength expressed in the same units as h.

In one case where transmission was to take place over extensive salt flats which are smooth and of high conductivity, it was not possible to avoid earth reflection by any reasonable change in the station locations. The fading due to such reflection was minimized in this case by employing very low antennas at one end of the section and high mountaintop antennas at the adjacent station. With this arrangement, the earth-reflection point was close to one of the stations thereby minimizing the change in phase relations between the direct and reflected rays during periods of varying transmission conditions.

6.7.2 SELECTION OF SITES FOR BROADCAST STATIONS. Sites for the antennas of broadcast stations should be so chosen that at least first Fresnel-zone clearance is obtained over all near obstructions in the directions of the areas to be served. Thus hills with gentle slopes or with foothills which prevent such clearance should be avoided. Not only will the field strengths be reduced in the shadows of foothills and along the slope of the hill, due to the low height of the antenna above the effective plane of reflection, but also nonuniform fields and ghosts may occur in distant areas which are within the line of sight of the transmitting antenna. Similarly, sites in the midst of tall buildings should be avoided unless the antenna can be placed well above them.

If relief maps of the proposed site are available or can be made, small grain-of-wheat lamps placed at the antenna location will assist in locating shadowed areas. Both theory and experience indicate that the radio shadows are of lesser length than the optical shadows.

Profiles, taken from topographic maps, should be drawn for at least eight radials from the antenna site, and for any additional radials which from inspection appear to present particular problems, in the manner shown in Fig. 6-9. Estimates of the areas of service should be made, both by the methods provided by the rules of the FCC,[28] described in Sec. 6.6, and the more detailed methods of Sec. 6.5.

In doubtful areas, actual measurements should be made over these radials, either from existing transmitters at or near the chosen site which have frequencies near the chosen frequency, or from test transmitters installed for the purpose. The field strength of the visual carrier should be measured with a voltmeter capable of indicating accurately the peak amplitude of the synchronizing signal using procedures outlined in Sec. 63.685 of the FCC Rules and Regulations.[28] These procedures generally follow those recommended by the Television Allocations Study Organization.[29]

REFERENCES

1. K. Bullington, "Radio Propagation at Frequencies above 30 Mc," *Proc. IRE,* p. 1122, October 1947.
2. D. G. Fink (ed.), *Television Engineering Handbook,* McGraw-Hill, New York, 1957.
3. T. L. Eckersley, "Ultra-Short-Wave Refraction and Diffraction," *J. Inst. Elec. Engrs.,* p. 286, March 1937.
4. K. A. Norton, "Ground Wave Intensity over a Finitely Conducting Spherical Earth," *Proc. IRE,* p. 623, December 1941.
5. K. A. Norton, "The Propagation of Radio Waves over a Finitely Conducting Spherical Earth," *Phil. Mag.,* June 1938.
6. Balth van der Pol and H. Bremmer, "The Diffraction of Electromagnetic Waves from an Electrical Point Source Round a Finitely Conducting Sphere, with Applications to Radiotelegraphy and to Theory of the Rainbow," pt. 1, *Phil. Mag.,* July, 1937; pt. 2, *Phil. Mag.,* November 1937.
7. C. R. Burrows and M. C. Gray, "The Effect of the Earth's Curvature on Groundwave Propagation," *Proc. IRE,* p. 16, January 1941.
8. "The Propagation of Radio Waves through the Standard Atmosphere," Summary Technical Report of the Committee on Propagation, vol. 3, National Defense Research Council, Washington, 1946, published by Academic Press, New York.
9. "Radio Wave Propagation," Summary Technical Report of the Committee on Propagation of the National Defense Research Committee, Academic Press, New York, 1949.
10. E. W. de Lisle, "Computations of VHF and UHF Propagation for Radio Relay Applications," RCA, Report by International Division, New York.
11. H. Selvidge, "Diffraction Measurements at Ultra High Frequencies," *Proc. IRE,* p. 10, January 1941.
12. J. S. McPetrie and L. H. Ford, "An Experimental Investigation on the Propagation of Radio Waves over Bare Ridges in the Wavelength Range 10 cm to 10 m," *J. Inst. Elec. Engrs.,* pt. 3, vol. 93, p. 527, 1946.
13. E. C. S. Megaw, "Some Effects of Obstacles on the Propagation of Very Short Radio Waves," *J. Inst. Elec. Engrs.,* pt. 3, vol. 95, no. 34, p. 97, March 1948.
14. F. H. Dickson, J. J. Egli, J. W. Herbstreit, and G. S. Wickizer, "Large Reductions of VHF Transmission Loss and Fading by the Presence of a Mountain Obstacle in Beyond-Line-of-Sight Paths," *Proc. IRE,* vol. 41, no. 8, p. 96, August 1953.
15. K. Bullington, "Radio Propagation Variations at VHF and UHF," *Proc. IRE,* p. 27, January 1950.
16. "Report of the Ad Hoc Committee, Federal Communications Commission," vol. 1, May 1949; vol. 2, July 1950.
17. J. Epstein and D. Peterson, "An Experimental Study of Wave Propagation at 850 Mc." *Proc. IRE,* p. 595, May 1953.
18. "Documents of the XVth Plenary Assembly," CCIR Report 563, vol. 5, Geneva, 1982.
19. B. R. Bean and E. J. Dutton, "Radio Meteorology," National Bureau of Standards Monograph 92, March 1, 1966.
20. H. T. Dougherty and E. J. Dutton, "The Role of Elevated Ducting for Radio Service and Interference Fields," NTIA Report 81-69, March 1981.
21. "Documents of the XVth Plenary Assembly," CCIR Report 881, vol. 5, Geneva, 1982.
22. "Documents of the XVth Plenary Assembly," CCIR Report 238, vol. 5, Geneva, 1982.
23. A. G. Longley and P. L. Rice, "Prediction of Tropospheric Radio Transmission over Irregular Terrain—A Computer Method," ESSA (Environmental Science Services

Administration), U.S. Dept. of Commerce, Report ERL (Environment Research Laboratories) 79-ITS 67, July 1968.

24. National Bureau of Standards Circular 462, "Ionospheric Radio Propagation," June 1948.

25. E. E. Smith and E. W. Davis, "Wind-induced Ions Thwart TV Reception," *IEEE Spectrum,* February 1981, pp. 52–55.

26. "Documents of the XVth Plenary Assembly," CCIR Report 159, vol. 6, Geneva, 1982.

27. Federal Communications Commission, "Sixth Report and Order," Docket Nos. 8736, 8975, 8976, 9175, April 11, 1952.

28. Federal Communications Commission, Rules and Regulations, Section 73.685.

29. "Engineering Aspects of Television Allocations," TASO (Television Allocations Study Organization) Report to the FCC, March 1, 1977.

30. G. S. Kalagian, "A Review of the Planning Factors for VHF Television Service," FCC Report RStt-01, March 1, 1977.

31. R. A. O'Connor, "Understanding Television's Grade-A and Grade-B Service Contours," *IEEE Trans. Broadcast.,* vol. BC-14, no. 4, December 1968.

32. B. W. Lindgren and D. W. McElrath, *Introduction to Probability and Statistics,* Macmillan, New York, 1959.

33. "Wide Band Systems Operating in the VHF, UHF, and SHF Bands," Report of C.C.I.R. Seventh Plenary Assembly, London, 1953.

Transmitters

John T. Wilner

Consultant
Boca Raton, Florida

7.1 INTRODUCTION

Television transmitters are classified in terms of operating band, power level, final-tube type, cooling method, translators, and low power.[1,†] The visual transmitter refers to that portion of the television transmitter which accepts the video signal, modulates a radio frequency (RF) carrier, and amplifies the signal to feed the antenna system. The aural transmitter is that portion of the television transmitter that accepts the audio signal, FM-modulates an RF carrier, and amplifies the signal to feed the antenna system. In the United States, the ratio of aural power permitted by FCC Rules and Regulations may be between 10 and 20 percent of the peak signal power.

The differences between low VHF, high VHF, and UHF, as far as the transmitter design is concerned, are mainly in the up-converter and the output power level. Low VHF are channels 2 to 6, and high VHF are channels 7 to 13. UHF are channels 14 to 83, except for channels 70 to 83, presently assigned to mobile radio service. Present-day practice is to use an exciter operating at or near the intermediate frequency (IF) followed by an up-converter, which translates the IF signal to the desired channel frequency. It is also common practice to perform most of the various distortion corrections at IF, although some may be done at baseband and RF.

Power and height limitations in the United States of television antennas are given in Table 7-1.

7.1.1 TRANSMITTER CLASSIFICATION.

Power Output.   Output power refers to the peak power of the visual transmitter. The FCC-licensed power is the transmitter output power, less the feedline losses times the power gain of the antenna. Low VHF power output, sufficient to produce 100 kW, can be 35 kW with an antenna gain of 4, to 10 kW with an antenna gain of 12. High VHF power output can be 50 kW with an antenna gain of 8, to 30 kW with an antenna gain of 12, to produce maximum power. UHF output power, in order to achieve 5000 kW, could use a 220-kW transmitter with an antenna gain of 25. Not all UHF stations can afford to install a 220-kW transmitter, and most operate considerably below the maximum. The above examples are appropriate, since transmission line losses vary according to frequency and length and will affect the final ERP.

Modulation Level.   Transmitters classified by level are of the IF type. Some of the reasons for using IF modulation are ease of correcting distortions, ease in vestigial sideband shaping, and economic advantages to the manufacturer, the latter since only one type of exciter is required for the VHF band, and only one basic design is required for the UHF band.

---

†Superscript numbers refer to References at end of chapter.

Table 7-1   Power and Height Limitations

Band	Channels	ERP†, kW	Height, ft (m)	Zone
Low VHF	2–6	100	1000 (305)	1
			2000 (610)	2, 3
High VHF	7–13	316	1000 (305)	1
			2000 (610)	2, 3
UHF	14–69	5000	2000 (610)	1, 2, 3

†Effective radiated power.

Source: Federal Aviation Administration.

Final-Stage Tube Type.   Tetrodes generally are used for VHF transmitters and for lower-level UHF transmitters below 5 kW, and klystrons are used for UHF above about 10 kW. The choice is dictated by the tube that will best operate at the required frequency, taking into account initial cost, operating cost per hour, and the cost of the auxiliary equipment.

Final-Tube Cooling.   Cooling can be by water, air, vapor, or a combination of these. Again, operating frequency, operating costs, tube life, noise, and efficiency determine the final selection.

Translators.   Low-power stations, or translators, rebroadcast another station's programs on another frequency, usually in the grade B service area of the mother station. They do not originate their own programs, and their power output is limited to 100 W VHF and 100 W UHF. (See FCC Rules and Regulations, Par. 74.701, for other restrictions.)

Low-Power Stations.   These are new and, in effect, are translators, except that they originate programs and can be assigned to any channel meeting the FCC Rules. The power limitations are 100 W VHF and 1000 W UHF with the stipulation that they do not interfere with other television stations even if it requires reducing their power output to a few watts. See Sec. 7.6 for further details.

7.1.2 TRANSMITTER SYSTEM DESCRIPTION.[2]   See Fig. 7-1.

Visual.   The input signal is fed first to a video processing amplifier (Proc Amp) and equalizer unit. The Proc Amp performs the following functions:

Video input amplification

Receiver delay equalization

Video clamping and hum cancellation

Phase-delay compensation

White clipping

Sync regeneration

The processed video is converted to amplitude modulation of an IF frequency, either on a carrier generated by a crystal oscillator, or by a frequency synthesizer. The modulated IF is band-shaped in a vestigial sideband filter, typically a surface-acoustic-wave (SAW) filter. No envelope-delay correction is required because of its inherent stable characteristics. Envelope-delay compensation may, however, be required for other parts of the transmitter.

The modulated visual-IF signal is processed for distortion corrections and subsequent RF transmission in the following stages:

Power amplifier linearity correction

Power amplifier group delay correction

Driver amplifier linearity correction

Incidental phase modulation correction

RF up-conversion

Channel bandpass filtering

The IF power amplifier and the final power amplifiers then raise the power level to the station license requirement. An RF sample, obtained from a directional coupler installed at the output of the transmitter, is used in an automatic power-level control circuit located in the RF processor section. A color-notch filter is used to provide the

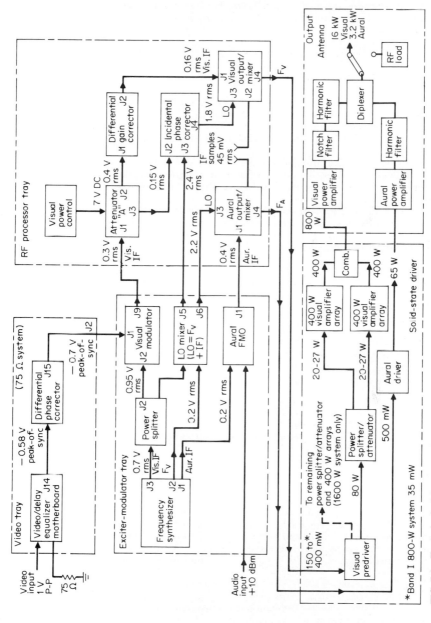

FIG. 7-1 Typical television transmitter. *(RCA Corp.)*

7.5

necessary additional attenuation of the lower sideband at 3.58 MHz, and a harmonic filter is used to attenuate out-of-band radiation as required by the FCC. Additional sample couplers provide signals for the following monitoring facilities:

VSWR metering

VSWR protection

RF for the demodulator

RF for the spectrum analyzer and sideband units

Aural. The audio signal is preemphasized to the 75-μs standard and used to frequency-modulate the IF carrier generated in the oscillator section. The modulated signal is then up-converted to the final RF-carrier frequency and passed to the intermediate power amplifier (IPA) and power amplifier (PA) stages. A harmonic filter attenuates out-of-band radiation.

The visual and aural output signals are fed to a hybrid diplexer where the two signals are combined to feed the antenna. For those installations which require dual-antenna feedlines, a hybrid combiner with quadrature-phased outputs is used.

7.1.3 TRANSMITTER PERFORMANCE CHARACTERISTICS. Many shortcomings in the received picture and sound, resulting from reflections and terrain limitations, cannot be controlled by the broadcaster. However, other problems, such as poor resolution, color fringing, hue shift, luminance shift, and poor audio quality, can be prevented if proper test equipment is available and if the transmitter has been designed to provide correction for such problems. Among the critical performance parameters causing these types of degradations are incidental phase modulation, differential phase and gain, frequency response, envelope delay, quadrature distortion, luminance-to-chrominance shift, audio response, and distortion. The following paragraphs describe these distortions.

There are two broad classifications of signal distortions in television systems, linear and nonlinear. Linear distortions occur which are independent of picture level, while nonlinear distortions vary with the amplitude of the picture signal.

Linear distortions usually occur in systems with incorrect frequency response and are divided into four general categories:

1. Short-time distortions affecting horizontal sharpness
2. Line-time distortion causing horizontal streaking
3. Field-time distortion causing errors in the low-frequency response and vertical shading in the picture
4. Long-time distortions in the direct current to extremely low-frequency range causing flicker and slow changes in brightness of the picture

Incidental phase-modulation distortion occurs when the quadrature sideband components of the vestigial sideband televsion system are unequal, causing the instantaneous resultant of the two sideband vectors around the carrier change in phase, and appearing as though the picture and color subcarriers are being phase-modulated. In other words, the normal AM picture and color subcarriers appear to be FM signals. This is undesirable because it results in noise in the audio reproduction and color distortion in the picture.

Nonlinear distortions occur with changes in the average picture level (APL). Luminance nonlinear distortion occurs when the gain of the system varies as the signal changes from black to white. Differential gain distortion is the change in the amplitude of the chrominance signal as the amplitude of the luminance changes from black to white. Differential phase distortion is the change in the color subcarrier phase as the luminance signal changes in amplitude.

IF Group Delay.   Group delay is the distortion that results when a group of related frequencies differ in arrival time when passing through a circuit. Examples are distortion due to the restrictions in frequency response of the vestigial sideband filter, the aural sound notch, the color-notch filter, and the average receiver response.

Low-Frequency Delay.   The characteristic LF delay has a significant effect upon the luminance signal, while high-frequency delay primarily affects the chrominance signal. The 2-T pulse is representative of the energy content of the typical character generator waveform. As can be seen in Fig. 7-2, most of the energy of the 2-T pulse is distributed below 2 MHz.

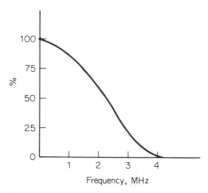

FIG. 7-2   Spectral distribution of a 2-T sine-squared pulse. *(Harris Corp.)*

The 12.5-T pulse is an excellent test signal to measure chrominance-to-luminance timing error because its energy spectral distribution is bunched in relatively narrow bands at very low frequencies as well as around the color subcarrier frequencies. The use of this signal detects differences in the luminance and chrominance phase distortion, but not between other frequency groups. See Fig. 7-3 for the spectral energy distribution of the 12.5-T pulse.

Quadrature Distortion.   This is the apparent phase shift of the visual carrier when energy exists unequally in the two sidebands. A simple test for quadrature distortion is to observe the transmitter output when a 2-T pulse modulates the transmitter and an envelope detector is used. When the input and output pulses at normal modulation are compared with the input and output pulses at 50 percent modulation, there should be no change in the amplitude and the wave shape if no quadrature distortion is present. See Fig. 7-4 which explains how this type of distortion occurs.

Differential Phase and Incidental Phase Modulation Distortions.   Differential phase is the change in phase of the color subcarrier for a change in the brightness level. Incidental phase modulation is the undesirable phase modulation of the visual carrier and its color subcarrier relative to the visual carrier being amplitude-modulated. In most

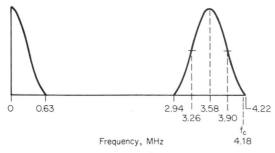

FIG. 7-3   Spectral distribution of a modulated 12.5-T sine-squared pulse. *(Harris Corp.)*

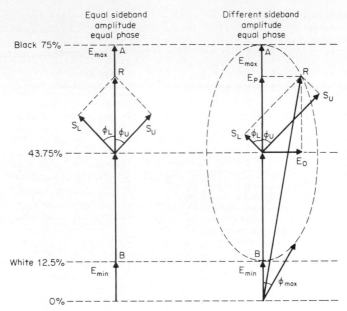

FIG. 7-4 Quadrature distortion caused by unequal upper and lower sidebands around carrier. *(Harris Corp.)*

transmitters differential phase modulation usually is less than incidental phase modulation. The most common observation of incidental phase modulation is the audio buzz from sync and bright-level picture content such as character generators. This occurs because aural detection requires a constant difference between the picture and sound carriers to produce the proper aural intermodulation carrier. If the picture carrier is phase-modulated as the carrier is normally amplitude-modulated, a changing separation between the two carriers will result in an unwanted frequency modulation or buzz.

Earlier test equipment designs cannot separate differential from incidental phase distortion. Presently several demodulators are available which can demodulate the RF signal by envelope or synchronous modes. Modern television transmitters have incidental phase modulation correction circuits to minimize this modulation problem.

Differential Gain.   Differential gain distortion occurs when the amplitude of the chrominance signal changes with respect to the amplitude of the luminance signal. This causes a change in hue saturation as the luminance varies between black level and white level.

Audio Response and Distortion.   Little can be said about the audio response and distortion in today's transmitters, except that these parameters have improved dramatically in the past few years. It was not uncommon to barely comply with the FCC response and distortion requirements in older transmitters. In the new transmitters it is uncommon to measure more than 1 percent amplitude-frequency response deviation and more than 1 percent distortion. It is to be noted that this refers only to the transmitter and does not include the studio-to-transmitter links (STL).

7.2 ELEMENTS OF A TRANSMITTER

While each manufacturer has its own proprietary design philosophies, nevertheless the final products generally have a commonality that is the result of many years of design development to meet current state-of-the-art performance.

7.2.1 GROUP-DELAY EQUALIZER. The purpose of the group-delay equalizer (Fig. 7-5) is to introduce video delays to correct the notch diplexer, the tuned amplifier, and the receiver group delays. Each delay equalizer contains up to five all-pass delay networks, each to correct a specific frequency band. These networks are constant-impedance devices, which allow composite delay curves to be made by cascading sections, resulting in an overall delay curve which complements the transmitter's delay characteristics. Figure 7-5 shows an all-pass network which is made up of several such units. Such networks can independently correct phase and amplitude problems without undue interactions.

7.2.2 VIDEO CLAMPER. The video clamper is required to hold the back porch level constant, regardless of the changes in the average picture level. A sample of the input signal is sent to the clamper circuit where the clamp-pulse generator produces a pulse that is coincident with the back porch of the video signal. A signal from the output of the video board, after removal of the color subcarrier, is applied to a tip-of-sync detector, causing the output level to be clamped, and then sent to a transistor switch that is open between pulses and closed during pulses. The bias developed during the closure of the switch holds until the next switch closure, thus keeping the level constant at all times. In the event of loss of video, the amplifier is automatically set to blanking level to prevent overdrive to the solid-state amplifiers.

7.2.3 WHITE-CLIP CIRCUIT. The white-clip circuit is used to prevent the negative-going video modulation signal going below 12.5 percent of reference white. If the video is below 12.5 percent, overmodulation and a high incidental phase modulation occur, and audio buzz may be encountered in home receivers.

7.2.4 DIFFERENTIAL PHASE CORRECTOR. The differential phase corrector circuit shown in Fig. 7-6, by means of diode gating at selected video-waveform levels, forms signals of opposite phase delays which, when added to the input video signal, will cancel subsequent unwanted differential-phase distortion. In a typical system design, this is accomplished by using seven such correction circuits.

The 10-kΩ differential phase adjustment potentiometer will produce a bias of up to 6.2 V on the zener diodes, CR1 and CR2. Depending upon this bias value, the zener diodes will conduct over a portion of the video-signal amplitude level, and a correction signal will be added to the output through the 7-pF capacitor and 2.2-kΩ resistor.

An inverted input signal, produced by an amplifier not shown on the diagram, is added to the output via a diode, CR3. This diode can be inserted into the mounting

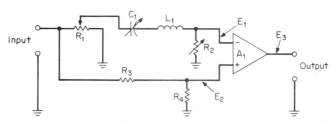

FIG. 7-5 Simplified diagram of an active all pass network. (*Harris Corp.*)

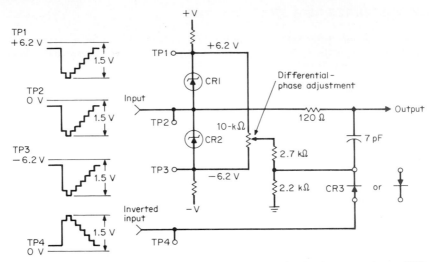

FIG. 7-6 Simplified diagram of differential phase corrector showing input waveforms. *(RCA Corp.)*

circuit by means of a fuse-clip type of mounting with the polarity in either direction, as required, during the adjustment for differential phase correction.

Differential phase correction is performed on a precorrection basis to make up for the distortion in subsequent intermediate and final power amplifiers (IPA and PA) of a transmitter. Since, in the example described, there are seven correctors all fed from a low-impedance driver, and since there is a selection of diode polarity for correction, a total of 14° correction is possible.

7.2.5 VIDEO MODULATOR. The video modulator receives a precorrected video signal from the video processor and a CW signal from the IF synthesizer. The resulting AM signal is produced with double sidebands. An SAW filter removes one sideband, resulting in a single sideband signal which conforms with FCC specifications.

7.2.6 SAW FILTER. The SAW filter[3] provides wave shaping of the IF-modulated signal in a superior manner, compared with the vestigial sideband filter used heretofore. The SAW filter requires no adjustments, does not have to be envelope-delay-compensated, and is stable with respect to temperature and time. SAW filters are made on a substrate about 2 mm thick and generally use a piezoelectric crystal such as lithium tantalate, although other crystalline structures are also used successfully. The dimensions are determined by the frequency response characteristics of the crystal. The unit has exceptionally steep skirts at the band edges and is obtainable with up to 60-dB attenuation of the out-of-band frequencies. The insertion loss is about 25 dB, making it necessary to use carefully designed low-noise amplifiers to maintain a satisfactory signal-to-noise ratio. A color-notch filter is required at the output of the transmitter because the imperfect linearity of the solid-state and power amplifiers introduces some unwanted modulation products.

7.2.7 FREQUENCY SYNTHESIZER. The frequency synthesizer (Fig. 7-7) used to generate the aural and visual IF carriers makes use of a phase-locked loop (PLL) to accurately control the carriers. A single unit can cover the entire VHF or UHF channel ranges. Thus, any single television channel frequency can be obtained from the synthesizer by merely strapping the proper jumpers on the frequency-dividing circuits.

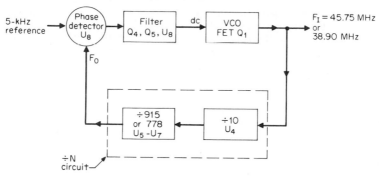

FIG. 7-7 Visual IF-synthesizer block diagram. *(RCA Corp.)*

The 5-kHz signal from the local oscillator is fed to the phase detector where it is compared with a sample of the final IF carrier which has been divided down to 5 kHz. If the two signals differ in phase or frequency, an error voltage is generated and is applied to the voltage-controlled oscillator, causing it to correct in the direction required for lock. The output of the synthesizer at visual IF is sent to the modulator. A second visual output plus the visual carrier is used for incidental phase correction and up-conversion, and a similar signal is sent to the aural up-converter. In the event that precise frequency is desired for more precise frequency stability when there is a possibility of on-channel interference, the local oscillator is replaced with a precise frequency generator.

7.2.8 LINEARITY CORRECTOR. The linearity corrector predistorts the modulated IF signal to compensate for the transmitter differential-gain distortion. It is accomplished by compressing or stretching the nonlinear portions of the transfer characteristic using several differentially biased diodes in the resistive loads of an IF-signal amplifier, much as is done in a gamma-correction video amplifier.

7.2.9 EXCITER. All the preceding units, the visual modulator, SAW filter, frequency synthesizer, linearity corrector, and differential phase corrector, grouped together comprise the visual exciter (Fig. 7-1). A local-oscillator circuit produces additional signals whose outputs are the sum of the IF and the final carrier frequencies and are used in the up-converter. In the aural exciter a preemphasized audio signal modulates a CW IF carrier and is passed on to the aural up-converter.

7.2.10 UP-CONVERTER. The modulated IF signal (Fig. 7-1) is mixed with the signal from the local oscillator which is the sum of the IF carrier and the final carrier. A sum and difference frequencies are produced in the mixing process, and the final carrier difference signal is extracted by the use of a bandpass filter. The visual output of the mixer (up-converter) at the final carrier frequency is sent to the visual solid-state driver, and the aural signal is sent to the aural solid-state driver. Pin-diode attenuators are used to automatically adjust for proper level inputs to the solid-state drivers in order to prevent overmodulation and protection.

7.2.11 INCIDENTAL PHASE MODULATION CORRECTOR. The incidental phase corrector cancels the incidental phase modulation distortion of the visual carrier.

A sample of the modulated visual RF carrier is detected to develop a detector video signal which contains all the delay and differential phase precorrections. This video signal is filtered and then used to phase-modulate the local oscillator feed to the visual up-converter. The modulation is equal and opposite in phase with the undesired incidental phase modulation (ICPM). Once correction is obtained, envelope and synchronous demodulators should produce essentially the same output waveforms.

7.2.12 SOLID-STATE DRIVERS.   The solid-state driver is the broadband amplifier which covers the entire low or high VHF channels, without the need for tuning adjustments. Present VHF technology economically limits the RF power output of such devices to about 2000 W. In UHF, where high-gain klystrons are used for the final amplifier, the driver does not have to be greater than about 20 W. About 800-W RF drive is required for a low-band 16- to 20-kW transmitter, and about 1600 W RF drive is required for a high-band 35- to 50-kW transmitter.

To develop 800 W in a solid-state driver, it is necessary to use a number of smaller 200-W amplifiers connected in parallel by hybrid combiners. The inputs are also connected together by hybrids, but here they are used as splitters. Since the power level at the output of the RF processor (up-converter) is in the order of about 10 W, an intermediate amplifier is used to produce the required power for the paralleled units. When higher drive is required, more 200-W units are used. At the output of the driver, a bandpass filter allows only the required modulated carrier and its sidebands to pass.

Solid-state drivers are cooled in one make of transmitter by a heat pipe in which a closed container absorbs heat from the transistors through a wick in a highly conductive liquid. The liquid transfers the heat to external cooling fins. Fault detectors elsewhere in the transmitter automatically adjust the transistors to prevent overload, or to control the output power to safe operating levels.

7.2.13 POWER AMPLIFIER.   The power amplifier raises the output RF power to the transmitted power level. For VHF, the screen-grid tetrode has become the standard output tube because of its high output power, its tolerance to temporary abnormal conditions, and its relatively high plate efficiency. For UHF, the klystron is the standard output tube, although tetrode transmitters of 10 kW or less can be used.

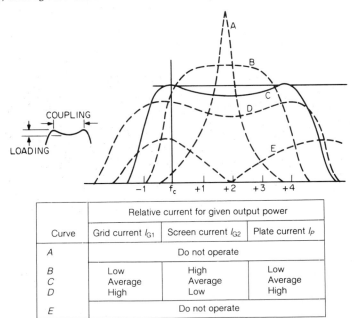

Curve	Relative current for given output power		
	Grid current I_{G1}	Screen current I_{G2}	Plate current I_P
A	Do not operate		
B	Low	High	Low
C	Average	Average	Average
D	High	Low	High
E	Do not operate		

FIG. 7-8 Tetrode-plate tuning adjustments. *(Harris Corp.)* Curves: *A*, high VSWR; *B*, narrow band, more coupling needed and less loading (optimum for aural centered around aural carrier); *C*, optimum (visual); *D*, overcouple (plate dissipation may be exceeded); *E*, high VSWR. *Note:* In general, *coupling* affects frequency excursion between saddle peaks; *loading* affects depth of saddle.

The VHF tetrode is available for television service in power ranges of 500 W to 50kW. It is operated class B to obtain better efficiency and is available in coaxial construction to keep the stray capacitance and inductance to low values. Class B amplifiers, when operated in tuned circuits, provide linear performance because of the fly-wheel effect of the resonance circuit. This allows a single output tube to be used instead of two in push-pull. The bias of the linear amplifier must be so chosen that the transfer characteristic at low modulation levels is as close as possible to that at higher modulation levels. Even so, some nonlinearity is produced requiring differential-gain correction. The impedance that the tube must work into is considerably lower than that of the plate. Thus, it is evident that the capacitive reactance (in kilovolt amperes) should be kept as low as possible.

Tuning a visual television transmitter output stage requires a basic knowledge of how the tube acts under various conditions. The following description is applicable to tetrodes and is general in nature. More specific information will be found in the manufacturers' transmitter instruction manuals.

1. Connect a sideband analyzer to the tetrode input and observe the reflected signal via the directional coupler. The drive power should be reduced to 25 percent, and the visual input tuning and match control should be adjusted for minimum reflected signal over the 5-MHz frequency span.
2. Adjust the plate tuning to near the carrier frequency.
3. Adjust the secondary tuning to about 4 MHz above the carrier frequency.
4. Adjust the coupling control for the waveform C in Fig. 7-8.
5. Adjust the loading control to achieve an optimum waveform as indicated in the figure.
6. Repeat steps 4 and 5 until the waveform is correct.
7. Verify that input match, plate current, grid current, and screen current are all correct; if not, make adjustments where needed and redo all the steps so far.
8. Confirm that operating parameters are within 10 percent of factory test data sheets.
9. Aural tuning is similar except that the aural cavity does not need a secondary tuning adjustment and the tube will be operating at class C conditions with narrower bandwidth.

After all conditions have been met, slowly raise the drive until required power output has been achieved, checking the operating voltages, currents, power levels, and voltage standing-wave ratios (VSWRs) constantly. Then make enough tests to see that all the distortions are within specifications.

Checklist. When all the operating goals have been met, all or most of the following criteria will have been met:

1. Lowest-necessary plate voltage
2. Lowest-necessary filament voltage
3. Just-sufficient bandwidth
4. Moderate screen current
5. Just-sufficient headroom
6. Low input VSWR
7. Lowest plate dissipation
8. Lowest sync compression
9. Just-sufficient zero signal plate current
10. High tube gain
11. Low tube-seal temperatures

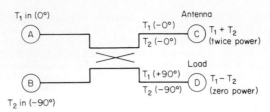

FIG. 7-9   Power combiner. *(Micro Communications, Inc.)*

7.2.14 RF POWER COMBINERS.   The RF combiner[4] (Fig. 7-9) is used to combine two RF power sources, such as transmitters or amplifiers, and present a constant impedance match to each source. One RF unit must be phase-delayed by 90° in order to cancel out the inherent quadrature splitting by the combiner. The combiner is a hybrid device that accepts one RF source and splits it equally into two parts, one arriving at output port C with 0° phase and the other delayed by 90° at port D. A second RF source connected to input port B, but with a phase delay of 90°, will also split in two, but that part arriving at output port C will now be in phase with source 1 and the other will cancel. If output port C were connected to a load, it would absorb all the power from both transmitters. Connecting output port D to an RF load would take care of any power resulting from slight differences in amplitude or phase between the two sources. If one RF source would disappear, such as a transmitter failure, then half of the remaining transmitter output would be absorbed by the second load and the other half would go to the first load. Note that in this case the power to the first load would be one-fourth of both. If one-half of the power is desired, the combiner would have to be bypassed via a patch or switch panel.

7.2.15 COLOR-NOTCH FILTER.   The color-notch filter is used to further attenuate the lower sideband color subcarrier energy in order that the transmitter can meet -42-dB FCC requirements at $f_c = 3.58$ MHz. The color-notch filter is placed across the transmitter output feedline. The filter consists of coax or waveguide stub series tuned to $f_c = 3.58$ MHz below the picture carrier. The Q is high enough so that energy around the lower sideband is not materially affected, but provides high attenuation at $f_c = 3.58$ MHz.

7.2.16 DIRECTIONAL COUPLERS.   The directional coupler[5] (Fig. 7-10) is used to couple signals from an RF feedline for the purpose of obtaining samples of the output in

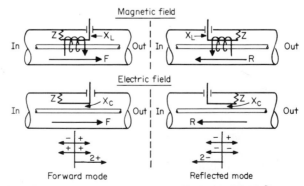

FIG. 7-10   Directional coupler. *(From Terman and Pettit.[5])*

order to operate alarms, demodulators, frequency monitors, forward and reverse power monitoring, forward and reverse VSWR indicators, automatic power-level controls, and such.[5] It does so by responding only to either the forward or reflected wave with discriminations of better than 30 dB. The unit consists of a small loop which is source-terminated and fed to an output connector. If a demodulated output is required, a diode is inserted in the output lead. The loop is part of a carefully designed mechanical device that is coupled to the inner conductor of an RF feedline at precise calibrated depths and can be rotated at that depth. The principle of operation depends on the fact that two components of induced loop currents are produced, one by the electric field and the other by the magnetic field.

Since the reflective voltage is in phase with the forward voltage, and the reflective current is 180° out of phase with the forward current, a means is available to differentiate between the forward and the reflective waves. The electric field produces two currents in the loop, one flowing toward the source termination and the other toward the output, but with opposite polarities. The magnetic field also produces two voltages, but they are of the same polarity and add. The polarity of the voltages due to the magnetic field depends upon the physical placement of the loop with respect to the induced magnetic field. Thus, for a forward wave, the electric and the magnetic fields add, while they subtract for a reflective wave. By reversing the loop 180°, the coupler becomes sensitive only to the reflective wave.

For installation of the directional coupler, determine the amount of sample voltage that is required. Assume that 1 V is needed, that 30 kW is in the line, and that the line has an impedance of 50 Ω. Therefore

$$E = (30,000 \times 50)^{1/2}$$

$$= 1225 \text{ V}$$

$$= 20 \log_{10} 1225 = 62 \text{ dB}$$

Using the manufacturer's data for 62-dB coupling, and the optimum angle for the loop, set the coupler to these values. Adjust the meter for 100 percent. To measure reflected power or VSWR, reverse the coupler 180° and read the reflected value on the meter.

7.2.17 DIPLEXERS. The notch diplexer[6] (Fig. 7-11) combines the aural and visual transmitter signals in order to feed a common antenna system and to provide a constant impedance load to each transmitter.

The notch diplexer is made up of two hybrid combiners and two sets of reject cavities. In Fig. 7-11 the signals inputted at A_1 and A_2 will have a 90° relationship

$$A_1 = 0.7 \text{ (at } 0°) \qquad A_2 = 0.7 \text{ (at } 90°)$$

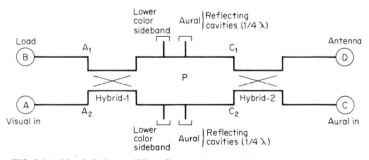

FIG. 7-11 Notch diplexer. *(Micro Communications, Inc.)*

Since the cavities are tuned outside the visual band, the signals passing through from A_1 to C_1 and from A_2 to C_2 will neither change amplitudes nor undergo a phase change. The signals at C_1 and C_2 will split as follows

$$C_1 \text{ (at C)} = 0.7 \text{ (at } 0°)$$
$$C_2 \text{ (at C)} = 0.7 \text{ (at } 180°) \left.\right\} \text{ cancel}$$

$$C_1 \text{ (at D)} = 0.7 \text{ (at } 90°)$$
$$C_2 \text{ (at D)} = 0.7 \text{ (at } 90°) \left.\right\} \text{ add}$$

As can be seen, the signals will cancel at C and power will go directly from A to D with no bandwidth restrictions. The signal C is split in the hybrids

$$C_2 = 0.7 \text{ (at } 90°),$$
$$C_1 = 0.7 \text{ (at } 0°).$$

Since the aural cavities are resonant, a low impedance is shunted across the line at point P. The signals reverse

$$C_2 = 0.7 \text{ (at } 270°)$$
$$C_1 = 0.7 \text{ (at } 180°)$$

$$C_1 \text{ (at C)} = 0.7 \text{ (at } 180°)$$
$$C_2 \text{ (at C)} = 0.7 \text{ (at } 360°) \left.\right\} \text{ cancel}$$

$$C_1 \text{ (at D)} = 0.7 \text{ (at } 270°)$$
$$C_2 \text{ (at D)} = 0.7 \text{ (at } 270°) \left.\right\} \text{ add}$$

Therefore, all the power of *transmitter 1* passes from C to D.

7.3 MAINTENANCE OF THE VISUAL TRANSMITTER[7]

Today's sophisticated transmitters require comparable monitoring and measuring equipment. Fortunately, available measuring equipment has kept pace with the steady improvements in the picture and sound qualities of modern television transmitters. This section describes some of the measurements required to properly maintain and monitor the television transmitter performance.[7]

After each description the applicable FCC specification,[8] when available, is given together with tighter figures from an average of a number of FCC station proof-of-performance tests and manufacturers' published specifications, and are given as performance goals.[9]

All measurements are made on the transmitter with an input of 1 V peak sync input feeding a 75-Ω impedance. All RF monitor samples are taken from a directional coupler in the transmission line at the point of measurement. The directional coupler output shall be 2 V rms at an impedance of 50 Ω. Some special monitors may require different RF inputs, and in these cases the appropriate instruction manual should govern.

7.3.1 **RF OUTPUT POWER.** The average power output is measured with a standard load, preferably a self-calibrated water load. If an in-line load is used, care should be taken to see that it has been carefully calibrated. The load shall be connected after any harmonic and color-notch filters and the power diplexer. During the measurement, the transmitter shall be modulated with precisely 25 percent normal sync and blanking corresponding to 75 percent of the zero carrier to peak sync amplitude. The water flow should be maintained at a constant rate, and the test continued until the output temperature stops rising and has stabilized. The peak power is calculated by multiplying the average power reading by a factor of 1.68.

The average power dissipated in the load is

$$(T_{out} - T_{in}) \text{ °C} \times 0.264 \times \text{LPM}$$

where T_{out} = temperature of output water
T_{in} = temperature of inlet water, °C
0.264 = water specific heat factor
LPM = water flow, L/min

The transmitter should be equipped with a true-reading meter which indicates peak incident voltage, current, or power and should have a means of output power adjustment. The control shall permit operation from 100 percent down to 80 percent and from 100 percent up to 110 percent of the rated power output. For UHF stations desiring maximum beam efficiency of klystrons, little or no head room should be used, and the transmitter should be calibrated only at 100 and 80 percent.

7.3.2 CARRIER REFERENCE-WHITE LEVEL. A sample of the transmitter output should be connected to the demodulator in the envelope-detection mode. The zero-carrier reference and the peak sync pulse levels should be displayed on the waveform monitor (WFM) and adjusted to 120 IRE units and −40 IRE units, respectively. White level should be set at 12.5 percent, with the zero carrier reference to peak sync on the calibrated WFM display at 100 IRE units.

7.3.3 CARRIER BLANKING LEVEL. The blanking level shall be set at 75 percent of the peak voltage for any fixed picture content. With variations of the APL from 50 to 10 percent and from 50 to 90 percent the carrier blanking level should not change by more than

 *FCC: 75% ± 2½%   Goal: 75% ± 2%*

7.3.4 VARIATION OF THE OUTPUT. A sample of the transmitter output should be demodulated and the waveform displayed on the WFM. The amplitudes of the highest and the lowest sync peak shall not exceed

 *FCC: no specification   Goal: 5%*

7.3.5 REGULATION OF THE OUTPUT. With a white picture at the input to the transmitter, a sample of the output is fed to the demodulator. The demodulator should have its automatic gain control (AGC) turned off and be in the envelope-detection mode. The waveform monitor should be set to 120 IRE units for zero carrier and at −40 IRE units for sync peaks. The input should then be changed to all black, and the change in peak to reference zero measured and expressed in percent. This change shall not exceed

 *FCC: no specification   Goal: 3%*

7.3.6 MODULATION CAPABILITY. With a signal composed of normal sync and blanking plus a ramp, verify that the sync peaks are at −40 IRE units and zero carrier reference at 120 IRE units. Increase the ramp amplitude of the above signal to the transmitter input until the limits of differential gain and phase are reached. Modulation capability shall be at least:

 *FCC: 12.5%   Goal: 3%*

7.3.7 CARRIER-FREQUENCY ACCURACY. The frequency of the carriers shall be measured using samples of the modulated or unmodulated carriers depending upon the type of frequency monitor used. The preferred method is to reference to the National

Bureau of Standards (NBS) 60-kHz standard from station WWVB. The frequency stability shall be as follows:

	Visual	Aural
FCC requirements	± 1000 Hz	Visual + 4.5 MHz ± 1000 Hz
Operating goals	± 200 Hz	Visual + 4.5 MHz ± 200 Hz

7.3.8 AMPLITUDE-FREQUENCY RESPONSE. A spectrum analyzer with a sideband analyzer, or other frequency-selective voltmeter, can be used to measure the transmitter amplitude-frequency response. The transmitter should be operating into a standard load. The sideband analyzer generates a composite video signal which is a constant amplitude variable frequency set to sweep beyond the passband of the transmitter. The axis of the sine wave is set at 50 percent of the normal peak-to-peak signal, and the amplitude of the swept signal is set at ± 15 IRE units. The output then is displayed on the analyzer cathode-ray oscilloscope (CRO). Limits shall not exceed:

FCC: 200 kHz reference, + 6 dB ± 2 dB from 3.58 MHz

2.10 to 4.1 MHz, 0 dB ± 2 dB from 3.58 MHz

4.18 MHz not more than 0 dB −4 dB from 3.38 MHz

The above holds for a diode demodulator.

Goal: Using sideband analyzer, all frequencies from

0.5 MHz to 4.1 MHz ± 1.0 dB

4.1 MHz to 4.18 MHz, VHF − 2.0 dB

4.1 MHz to 4.18 MHz, UHF − 4.0 dB

7.3.9 UPPER AND LOWER SIDEBAND ATTENUATION. The upper sidebands between 4.75 and 7.75 MHz and the lower sidebands of 1.25 and 4.25 MHz should be attenuated at least 20 dB below the visual carrier, using 200 kHz as the reference. In addition, the 3.58 MHz lower sideband should be attenuated at least 42 dB.

The input signal for the above measurements should consist of a sine-wave sweep normal sync and blanking plus a variable sinusoidal-swept signal spaced between the blanking pulses. The axis of the sine waves shall be maintained at a level of 50 percent of the sync peak voltage. The amplitude of the upper sideband of 200 kHz is measured and used as a reference. The modulating frequency should be varied over the range of 200 kHz to 8 MHz, and each discrete frequency recorded. In lieu of discrete measurements, a photograph of the attenuation characteristic as seen on the WFM may be used. Two pictures should be taken, one with a vertical scale of 10 dB/cm and another with 2 db/cm.

7.3.10 ENVELOPE DELAY VERSUS FREQUENCY. A sine wave introduced at the input to the transmitter should produce a radiated signal having an envelope delay of:

Frequency, MHz	Delay
0.05–3.0	0 ns
3.58	−170 ns
3.0–4.18	Linearly

The measurement should be made with the transmitter operating into a standard load and should be made on the transmitter output signal detected by a vestigial sideband (VSB) demodulator with known delay characteristics, or on separate sideband sig-

Table 7-2   Envelope-Delay Tolerances

Frequency band, MHz	FCC, ns	Goal, ns
0.05–2.1	± 100	± 50
2.1–3.0	± 70	± 35
3.0–3.58	± 50	± 25
3.58–4.18	± 100	± 60

nals as detected by a synchronous sweep receiver (Table 7-2).[10] The maximum excursion of the modulating signal shall not exceed 25 percent of the reference-white to black signal limits.

7.3.11 LINEAR-WAVEFORM DISTORTION.   The transmitter input should be fed a composite video signal consisting of sync pulses and a 12.5-T modulated pulse-and-bar signal. If a synchronous demodulator is used, the sound notch should be switched out. The relative gain change is the difference between the amplitude of the 12.5-T pulse and the reference white-level bar.

   FCC: no specification   *Goal: ± 3 IRE units*

7.3.12 VISUAL SIGNAL-TO-NOISE.   The signal to weighted random noise is the ratio of the modulation, not counting the sync, over a spectrum from 30 Hz to 4.2 MHz from blanking to reference white to the weighted rms noise amplitude measured at the output of the demodulator

$$S/N = 20 \log \text{(video amplitude)}/\text{(rms noise)}$$

   FCC: no specification   *Goal: 55 dB*

7.3.13 FREQUENCY RESPONSE VERSUS BRIGHTNESS.   This measurement will show how the frequency response might vary with changes in the luminance signal. The same setup as that used for the measurement described in Sec. 7.3.8 on amplitude-frequency response will be used, except that the frequency response will be made using the 45, 25, and 65 percent axes. The amplitude of the swept signal in each case will be held to ± 12.5 percent of modulating voltage between reference black to reference white. The variation of the frequency response at 25 and 65 percent shall be compared with the frequency response of the reference 45 percent:

   FCC: no specification   *Goal: ±I dB*

7.3.14 LOW-FREQUENCY LINEARITY.   Low-frequency, or luminance, nonlinearity is the measure of gain variation of the system for a luminance signal as a function of average picture level and instantaneous luminance level. The transmitter shall be fed a composite signal of sync and five equal stairsteps. The demodulated signal is passed through a differentiating network. The stairsteps are transformed into a train of five pulses. By comparing the amplitudes of the pulses, the ratio of the largest to the smallest is obtained.

   FCC: no specification   *Goal: 10% between luminance levels of black and white*

7.3.15 DIFFERENTIAL GAIN.   For measurement use composite signal consisting of sync pulses and a low-frequency stairstep, modulated with a 3.58-MHz sine wave, the latter at an amplitude of 20 percent blanking to reference white, to modulate the transmitter. The demodulator in the synchronous mode passes the transmitter output signal through a high-pass filter. The resulting signal is observed on the WFM. Any deviation from flat response, expressed in percent, is the differential gain.

   FCC: no specification   *Goal: 5%*

Quadrature distortion occurs when an envelope detector is used to demodulate the upper chrominance sideband after the lower sideband has been filtered out. This distortion produces apparent differential gain, reduction in the subcarrier amplitude as the luminance component of the test signal approaches white. For this reason the differential gain is measured with a synchronous detector.

FCC: no specification   Goal: 5%

7.3.16 DIFFERENTIAL PHASE. The transmitter should be modulated with the same input signal as that used for differential gain measurements. The transmitter output should be detected with the demodulator in either the envelope or the synchronous mode.[11] Where the transmitter suffers from ICPM distortion by the luminance signal, the measured differential phase will differ between the envelope and the synchronous detection modes. When synchronously demodulated, differential phase due to ICPM will not occur. Measurement of ICPM requires an instrument such as a vectorscope designed to measure differential phase of the color subcarrier. It is desirable to make this measurement with the sound trap out to minimize burst envelope distortions.

FCC: no specification   Goal: ±3°

7.3.17 INCIDENTAL PHASE MODULATION. The transmitter should be fed a signal having sync, color burst, and a stairstep with a constant-phase 3.58-MHz sine-wave signal phased with the color burst. The p-p amplitude of the 3.58-MHz signal shall be 20 percent of the blanking to reference white (100 IRE). The quadrature component of the synchronous detector may be used to display the luminance ICPM characteristic on a WFM with one axis synchronized at the line rate and the other axis calibrated in degrees of RF phase shift. Color subcarrier is removed from the test signal to facilitate measurement of the luminance characteristic.

FCC: no specification   Goal: 7°

7.4 MAINTENANCE OF THE AURAL TRANSMITTER

The aural transmitter converts baseband audio and ancillary signals to a frequency-modulated output signal with the characteristics outlined as follows.

Transmitter Input. The input impedance over the range of frequencies from 30 to 15 kHz should be 600 Ω ±10 percent referenced to 400 Hz. Modulation of 100 percent will be ±25 kHz RF swing of the aural carrier.

RF Monitoring Connection. The RF monitoring connection shall be a directional coupler set for 2 V rms minimum when the transmitter is fed to a standard 50-Ω resistive load.

7.4.1 MODULATION CAPABILITY. The modulation capability of the transmitter is the maximum frequency deviation of which it is capable without generating objectionable products. The modulation capability shall not be less than ±40 kHz. This measurement is usually made by instruments capable of measuring frequency deviation directly. An absolute method makes use of the fact that the carrier goes to zero at specific ratios of RF carrier deviation to modulating frequency. Thus, the carrier goes to zero at ratios of 2.405, 5.520, 8.654, 11.792, 14.931, 18.071, 21.212, and 24.353.

FCC: ±40 kHz   Goal: ±50 kHz

7.4.2 CARRIER-FREQUENCY STABILITY. The carrier-frequency stability is the ability of the aural transmitter to maintain its assigned frequency at 4.5 MHz higher than the visual carrier. The aural carrier frequency shall be such that it stays within ± 1000 Hz of the sum of 4.5 MHz and its assigned visual frequency.

FCC: 1000 Hz   Goal: 500 Hz

The preferred method is reference to NBS 60-kHz transmission from station WWVB.

7.4.3 AMPLITUDE-FREQUENCY RESPONSE.

The audio frequency response is the ratio of input voltages expressed in decibels required to obtain a constant frequency deviation over a specified range of frequencies referenced to 400 Hz. The maximum departure of the audio frequency response from the standard 75-μs preemphasis curve over the range of 30 to 15,000 Hz should not exceed:

FCC: −4 dB at 50 Hz to −3 dB at 100 Hz
  −3 dB at 100 Hz to 7500 Hz
  −3 dB at 7500 Hz to −5 dB at 15,000 Hz

Measurements should be made with 100 percent modulation at

30, 50, 100, 400, 1000, 5000, 7500, 15,000 Hz

and for 25 and 50 percent modulation at

50, 100, 400, 1000, 5000 Hz

FCC: as above   Goal: ±1 dB from 75-μs preemphasis curve

7.4.4 FM NOISE LEVEL.

The FM noise level should be measured by sampling the RF output by an rms device under conditions of no modulation and 100 percent modulation. At no modulation, the input should be terminated in a 600-Ω load. The deemphasis circuit should be left in.

FCC: −55 dB   Goal: −58 dB

7.4.5 AM NOISE LEVEL.

The AM noise level of the aural carrier is the ratio, in decibels, of the rms value of the residual amplitude modulation component in the band of frequencies of 30 to 15 kHz of the carrier envelope to the rms value of the carrier in the absence of the input signal. Measurements may be made by using a peak-reading linear AM detector connected to the output directional coupler of the aural transmitter. The dc component and the rms ac component are measured. The dc measurement is multiplied by 0.707 to obtain the rms value of the unmodulated carrier, and this value is set for the 100 percent reading of the noise meter. The ac component (noise) is read on the linear noise meter. The ratio is the signal to noise.

FCC: −50 dB   Goal: 50 dB

7.4.6 AMPLITUDE NONLINEARITY DISTORTION.

This is the ratio of the rms output voltage to the total harmonics produced by the sinusoidal input signal. The FCC limits on this type of distortion are listed in Table 7-3.

A 75-μs preemphasis network in the transmitter and a 75-μs deemphasis network in the demodulator are used. The audio output of the aural demodulator is connected to a 30-kHz low-pass filter. The audio input to the transmitter should have less than 0.5 percent distortion. The amplitude of the audio signal supplied to the transmitter should be adjusted to keep the carrier deviation as read on the modulation meter constant at the modulation level being measured. No deemphasis network is used in the deviation meter.

FCC: as above   Goal: 2% at 100% modulation

Table 7-3 Amplitude Distortion

Frequency band	Distortion, %	
	100% Modulation	25 and 50% Modulation
50–100 Hz	3.5	3.5
100–5000 Hz	2.5	2.5
5000–7500 Hz	2.5	No specification
7500–15,000 Hz	3.0	No specification

7.5 TYPICAL TELEVISION TRANSMITTERS

7.5.1 REPRESENTATIVE DESIGNS. This section describes several transmitters that have been selected to bring out design philosophies, outstanding features, and operational conveniencies (Table 7-4). Of the transmitters listed, Harris, NEC, Philips, and RCA manufacture a full line of VHF and UHF transmitters, Comark and Townsend manufacture a full line of UHF transmitters, and Acrodyne and Emcee manufacture a full line of low-power transmitters. Other manufacturers supply television transmitters in the United States. However, those listed are typical and a cross section of design approaches.

7.5.2 TERMINOLOGY. A solid-state transmitter identification signifies that the transmitter uses only solid-state components up to about 1500 W output. Above that limit tubes are used in the output stages. Tetrodes are used from 5 to 50 kW in UHF and up to 10 kW in UHF, and klystrons are used in UHF up to 220 kW.

All television transmitters today use low-level modulation, usually at the IF stages. VHF transmitters are air-cooled, while UHF transmitters are water- or vapor-cooled above 10 kW and can be air-cooled below that level.

UHF transmitters are divided into two general classifications, i.e., one using integral-cavity klystrons and the other external-cavity klystrons. Integral klystrons are available in four- and five-cavity designs, while external-cavity klystrons are usually of the four-cavity design. This is discussed in greater detail in Sec. 7.8.

Single-tube transmitters in which the aural and the visual signals are both amplified in a single tube generally are not used above 5 kW because of the nonlinearity shortcomings.

Transmitters producing power levels greater than that shown above are achieved for UHF by paralleling output finals, by using higher-power tetrodes, or for VHF by paralleling lower-powered transmitters for VHF.

Table 7-4 Typical Transmitter Designs

Manufacturer	Band	Visual power, kW	Model	Cooling
Acrodyne	UHF	1.0	TT-3400U	Air
Comark	UHF	30.0	TTU-30	Water
Emcee	VHF	0.01	TTU-ID	Air
Harris	UHF	55.0	TVE-55	Vapor
NEC	UHF	10.0	710-H	Air
Philips	VHF	17.5	LDM-1211/06	Air
RCA	VHF	16.0	TTG-16L	Air
Townsend	UHF	10.0	TA-10	Water

7.5.3 LOW-POWER 1-kW UHF ACRODYNE MODEL TT-3400U.

This transmitter is an example of a design for low-power and translator use. It features a single tube in the output and has a wide bandwidth exciter which covers the entire UHF band without requiring tuning on channels 14 through 69. Low-level IF modulation is used. The visual output is a vestigial sideband signal and internally diplexed.

Description. The transmitter produces carriers of 1000 W visual and 100 W aural. The final amplifier is driven by a 30-W solid-state exciter which produces an IF output that is envelope-delay and phase-compensated. The aural carrier is phase-locked with the visual carrier. A built-in aural tone generator is available for use in maintenance.

The up-converter uses an oven-stabilized crystal oscillator to drive a frequency multiplier chain to produce the final carrier frequency. This is followed by an RF amplifier to raise the output drive level to 3 W.

A 30-W driver, made up of four low-power amplifiers, is used to drive the final output stage. Any single-stage failure decreases the output power by 6 dB. The tuned cavity uses a Thompson CSF TH-347 tetrode operating in class AB1 and is followed by an output notch filter, to reduce the out-of-band intermodulation products to a low level, and a low-pass filter to provide a high degree of harmonic rejection.

The transmitter incorporates a control system for automatic recycling, fault monitoring, VSWR protection, and provision for remote control.

7.5.4 30-kW UHF COMARK MODEL TTU-30

Main Features. This transmitter is an example of a four-cavity, external-klystron, water-cooled transmitter which is compact enough to allow installation directly against a wall. All maintenance can be done from the front of the transmitter. All parts are nonproprietary. Glycol can be added to the coolant to prevent freezing, and the heat exchangers and power supplies may be placed out of doors. Each klystron has its own power supply and heat exchangers for redundancy. Either klystron can be used for visual or aural service since both klystron assemblies are the same. For maximum efficiency, however, the aural klystron should be narrow-banded for normal operation, and in an emergency the aural klystron can be used for multiplexed operation. (Visual requires broadband cavities. In this case the cavities can be preset to previously determined wideband settings.)

Description. The exciter is used to produce a 20-W peak visual output and a 20-W average aural power output. It consists of the IF modulator, the up-converter, and its power amplifier. The equipment can supply separate aural and visual drive signals to either of the two klystrons, as well as providing a combined (multiplexed) aural and visual signal to either klystron in an emergency.

A modulator, Philips Model PM-5580, is used to convert the baseband audio and video to the 45.75-MHz visual and the 41.25-MHz aural frequencies. Peak visual output is $+13$ dBm, and the average aural power is $+3$ dBm.

The modulated outputs are routed through separate amplifiers to the up-converter. Here the carriers are converted to the final transmitted carrier frequencies. A dual-output local oscillator provides the required mixing frequencies, ensuring proper relationship between the two carriers. An IF incidental phase-modulation corrector permits correction for klystron-induced time delays when the modulating-anode (mod-anode) pulser is used. The *mod-anode pulser* is a device which "pulses in" the upper 50 percent of the sync pulses to improve the klystron efficiency. This will be treated in more detail in Sec. 7.8.3. The aural and visual signals are then fed into a remote-controlled transfer switch and do not require envelope delay compensation. If the signals are combined, both up-converters are fed the same combined signal. If separate signals are desired, the transfer switch will return the up-converters to the conventional configuration.

A solid-state driver, used to raise the output of the modulator to 20-W peak-visual power, uses a combination of integrated microcircuits and broadband hybrid-combined amplifiers to achieve linear gain and stable frequency response over the entire UHF band without the need for tuning adjustments. In combined aural/visual service at a power output of 15 W, the third-order intermodulation levels are lower than 56 dB.

A mod-anode pulser, available as an option, improves the klystron beam efficiency of the visual amplifier from its normal 40 to 45 percent to 50 to 55 percent with no output couplers.

7.5.5 LOW-POWER 10-W VHF EMCEG MODEL TTU-ID. The Emcee Model TTU-ID transmitter is an example of a low-power transmitters which has been designed for either a stand-alone audio and video baseband modulated transmitter or translator service. The transmitter will accept a standard television signal and convert this signal to either a VHF 10-W output or with the addition of a suitable UHF RF amplifier, a 1-, 10-, or 100-W UHF output.

Description. The IF section is the control center of the transmitter. The IF signal from the modulator is amplified and processed. The automatic-on circuit places the transmitter on the air when the proper audio and video signals are present. If the modulator-output power fluctuates, or if the gain of the power amplifier changes with temperature, detuning, or other drift in characteristics, the automatic circuits will compensate for these variations. The IF signal is distortion-corrected, and an automatic-gain-controlled attenuator regulates the power delivered to the up-converter, thus controlling the power output of the transmitter. The modulator processes the audio and video baseband signals to provide 41.25- and 45.75-MHz IF signals. The modulator has the capability of accepting an off-air signal for translator operation.

7.5.6 55-kW UHF HARRIS MODEL TVE-55. This transmitter is the improved version of the BT-55U transmitter which is representative of a five-cavity, integral-klystron, vapor-cooled design. It features relatively simple, high-gain klystrons, use of aural and visual output couplers, more efficient heat exchanger, optional mod-anode pulser, and an improved quadrature phase corrector.

Description. Video is amplified and clamped, then modulates a CW IF carrier at 36.75 MHz. Audio is preemphasized and then frequency-modulates a CW carrier at 32.25 MHz. A sample of each exciter's output is divided down and compared with the first local oscillator, and any error voltage derived is used to continually correct the frequency of the oscillator.

The visual modulated output is band-shaped by means of a SAW filter, amplified and linearity-corrected, and passed to the up-converter. The aural modulated output is sent directly to its up-converter.

A second local oscillator generates two CW carriers, one at 36.75 MHz plus the final carrier and the second at 32.25 MHz plus the final carrier. Two resultant signals are produced, one a sum and the other a difference frequency in the mixing process. A bandpass filter selects the difference signals and is amplified to 1-W p-p for the visual drive and 0.5 W for the aural drive for the klystrons.

Adjustable-output couplers and a mod-anode pulser can be used to increase the beam efficiencies of the klystrons. The visual klystron is adjusted for saturation at 90 percent of rated power. The coupler is then adjusted in incremental steps by varying the drive power while observing the output on the power meter and the detected output waveform from the envelope detector. This continues until the increase in saturated power levels off. A reduction of 25 kW with the visual coupler and 28 kW with the mod-anode pulser has been achieved at channel 26. As a result of these efficiency increases, the heat exchanger blower motor has been replaced by a 6-HP smaller motor.

Other improvements made are the introduction of a new SAW filter, which features lower ripple in the video IF signal and the elimination of the receiver discrete equalizer, and a new quadrature phase corrector, which incorporates linearity correctors in both the quadrature and in-phase signals. Up to 10° of phase modulation distortion can be corrected.

A 62 percent beam efficiency of the visual transmitter has been achieved at channel 26, with all performance specifications more than meeting the manufacturer's published data.

7.5.7 10-kW UHF NEC MODEL 710-H. This transmitter is representative of an air-cooled, external, four-cavity klystron transmitter. The company manufactures UHF television transmitters in 5-, 10-, 20-, 30-, and 40-kW ranges. Above 40 kW, suitable klytstron amplifiers are combined to produce 60 and 80 kW.

Features. It features a high-efficiency klystron with beam efficiency of 45 percent at peak sync power. Since all power supplies are contained in the cabinets, the transmitter can be placed against the wall. Other features are use of avalanche-diode rectifiers, SAW filters, pedestal AGC, and safety interlocks.

Description. This transmitter uses the HPA-3696 IF television modulator. The modulator accepts the 1-V video signal through a differential amplifier to minimize low-frequency distortions and correct for differential phase and gain, optional quadrature correction, receiver and transmitter equalization, and clamped. The processed video then modulates the IF carrier and then is band-shaped by the SAW filter. The aural input is preemphasized and then modulates the aural IF carrier. Next the modulated signals go to the amplitude and phase correctors and then to an AGC circuit for automatic gain control. The local-oscillator frequencies are then used to up-convert the IF signals to the correct carrier frequencies, then passed through a passband filter to remove the unwanted sideband products, and finally sent to the final klystron amplifiers. Before combining the two signals in the diplexer, harmonic filters remove the carrier harmonics. The diplexer output is then routed through a patch panel (or optional RF switch) and then to the antenna.

7.5.8 17.5-kW VHF HIGH-BAND PHILIPS MODEL LDM 1211/06. This transmitter is representative of a European design and is the building block for higher-power transmitters, by either changing the output tube, or paralleling amplifiers, or complete transmitters.

Features. The transmitter is all solid-state except for two ceramic tubes in the visual and one ceramic tube in the aural. It can be operated either in the automatic, manual, or remote-control modes. The transmitter can be installed with the cabinets against the wall for space conservation, has automatic power regulation, and has a key-controlled safety interlock system.

Description. The transmitter incorporates an IF-modulated solid-state exciter, an RF mixer linear amplifier, and the necessary power supplies. The modulator accepts the incoming audio and video signals and produces a signal which is corrected for delay and bandwidth. Only clamping and receiver group delay correction is performed at baseband video, all other corrections being done at IF frequencies.

The visual IF frequency is generated in a temperature-stabilized oscillator and supplies the aural and video modulators. Phase nonlinearities in the transmitter resulting from the roll-offs in the sideband filter and in the subsequent stages including the sound notch in the diplexer are compensated in the envelope delay corrector. Both envelope delay and amplitude distortion can be independently corrected. Differential phase at color subcarrier and incidental phase modulation of the visual carrier are minimized in this section. The automatic power regulator acts on the signal at this point to keep the visual power level at blanking constant.

The audio signal is preemphasized and then modulates the aural carrier. Both the aural and visual signals are fed at a 1-mW level to the mixer where both signals are up-converted to their assigned carrier frequencies, and then amplified to a 50-mW level. Filters then attenuate out-of-channel emissions, and the power level is brought up to a 1-W level. A frequency synthesizer may be substituted for the local oscillator.

Two solid-state amplifiers raise the level of the signals to about 10 W to feed the output stages. The visual signal is amplified in the tube driver using a YL 1540 IPA stage followed by a YL 1430 (tetrode amplifiers) final. The aural carrier is amplified directly by a YL 1540 vacuum tube. All three tubes are ceramic tetrodes with high gain and good linearity. Following the finals are the harmonic filters, the color-notch filter, the directional couplers, and, optionally, the RF switching system to the out or the dummy load.

7.5.9 16-kW VHF LOW BAND RCA MODEL TTG-16L (FIG. 7-1). This transmitter represents the state-of-the-art transmitter for the RCA VHF line. The transmitter is available for power levels from 16 to 60 kW for low band and from 17.5 to 100 kW for high band.

Features.   It is entirely solid-state except for the aural and visual finals, a frequency synthesizer covers the entire VHF band by jumper selectable straps, a SAW filter is used for band shaping, no transmitter tuning is required except in the finals, and space-saving construction allows a compact installation.

Description.   The transmitter is housed in three cubicles. The harmonic filters and the color-notch filter are located on the transmission lines external to the cabinets.

The same exciter/modulator is used for the entire line of transmitters. The video processor unit contains the input amplifier, delay equalization, video clamper, differential-phase corrector, and the white clipper. This processed video signal then modulates the IF carrier which is generated by the frequency synthesizer. This signal is referenced to the local oscillator or a precise frequency standard. The synthesizer also generates an IF signal for the aural FM exciter.

The aural exciter generates an FM signal which is phase locked to the reference generator and shared with the visual frequency synthesizer. A second input to the aural exciter is available for an optional telemetry circuit for transmitter remote control.

The frequency synthesizer uses one crystal, which operates in conjunction with a phase-locked loop. Any television channel may be selected merely by changing jumpers. Since both the aural and the visual frequencies are obtained from a single crystal, the visual and the aural separation is maintained to with ±5 Hz.

The RF processor performs the following functions: automatic power control, correction of nonlinear distortions, up-conversion to assigned carrier frequencies, correction for incidental phase modulation, protection for deterioration of VSWR conditions, and power amplification of the aural and visual signals for proper operation of drive to the solid-state IPA stages.

The solid-state drivers are combinations of lower-level amplifiers which raise the drive level to the finals to 800 W peak visual and 85 W average aural. The heat generated by the transistor amplifiers is safely conducted to external cooling fins by means of a heat pipe. This device operates on the principle of a wick in a fluid, contained in a metal pipe, causing movement of the fluid from the warmer parts to the external cooling fins without the use of any mechanical moving parts.

The power amplifier uses the only tubes in the transmitter. The visual amplifier uses RCA 8976 and the aural uses RCA 8977 ceramic tetrodes. Only 10 tuning adjustments are needed to align the entire transmitter, six in the visual final—input, input matching, plate tuning, and output coupling—and four in the aural final—input, input matching, plate, and output coupling. The filament is heated by direct current, and a single direct-drive blower cools the output stages.

A solid-state logic system controls the complete automatic sequences such as start-up, shutdown, overload protection, and primary-power-failure protection. In the event of a power failure, a logic memory is maintained by a built-in chargeable battery system.

7.5.10 10-kW UHF VISUAL-KLYSTRON/AURAL-TETRODE TOWNSEND MODEL TA-10.   The full line of UHF transmitters ranges from 10 to 220 kW·s and is available in water- or vapor-cooled, integral and external-cavity klystrons. The Model TA-10 represents a novel approach to reducing costs by substituting a tetrode in the aural final stage instead of using a more costly klystron.

Features. It features back-to-the-wall installation, plug-in vacuum contactors, separate power supplies, tetrode aural final, mod-anode separate power supply bias instead of resistor bleeder bias, and an improved mod-anode pulser.

Description. A solid-state IF exciter and a SAW filter are used. This section will not cover the details of the visual transmitter since it has already been essentially covered in other manufacturers' models. However the tetrode tube aural amplifier is the first such used in a klystron transmitter. Elimination of the aural klystron can result in substantial capital savings since a second magnet assembly with its power supply and an aural heat exchanger are not required. Counterbalancing these savings are the higher drive requirements for the tetrode and an additional blower for the tetrode. The replacement cost of the tetrode is substantially less than that of a klystron. In summary, the operating and the capital costs are reduced.

Townsend uses a mod-anode pulser with a fiberoptic circuit which isolates the controls from the high-voltage components. The two power supplies to which the mod anode is switched during normal and elevated sync operation use a separate mod-anode bias power supply for control of the beam perveance and hence result in increased efficiency. The latter eliminates the rather crude method of setting proper beam current by changing taps on the high-voltage bleeder resistor string.

7.6. LOW-POWER TELEVISION TRANSMITTERS

Low power (LP) television transmitters were described in Sec. 7.5. This section discusses the method of determining how the low-power television station must comply with FCC Rules. On July 19, 1982, the FCC adopted regulations which are both complex and detailed. Only the technical aspects of these regulations, not operational requirements, are discussed herewith.

The initial and general requirements to be met are listed in Table 7-5.

If the proposed channel is less than the distance from the full-power station, it may be authorized if the low-power station meets or exceeds the signal strengths listed in Table 7-6 for the grade B contour of the protected full-power station.

If the criteria in Table 7-6 are met, the low-power applicant must also ensure that the intended location is not in excess of the distances shown in Table 7-7.

A low-power station may have a maximum of 10 W VHF and 1000 W UHF, except for areas near the Canadian and Mexican borders (see FCC Rules and Regulations for these areas). If the applicant proposes a circular-polarized antenna, the peak power may be doubled. No heights are specified for the antenna. However, the applicant must cal-

Table 7-5 LP Station General Requirements

Full-power stations	Distance from LP station, mi (km)
VHF co-channel, no offset	210 (338)
VHF co-channel, offset	150 (241)
VHF ± 1 channel	90 (145)
UHF co-channel, no offset	210 (338)
UHF co-channel, offset	150 (241)
UHF ± 1 channel	75 (121)
UHF ± 2, 3, 4, 5 channels	20 (32)
UHF ± channels	69 (110)
UHF −14 channels	70 (113)
UHF −15 channels	75 (121)

Source: Federal Communications Commission.

Table 7-6 LP-Station Power Level

Full-power station	LP/FPS, dB
Co-channel, no offset	−45
Co-channel, offset	−28
VHF + 1 channel	6
VHF − 1 channel	12
UHF ± 1 channel	15
UHF − 14 channels	23
UHF − 15 channels	6

Source: Federal Communications Commission.

culate the height above average terrain as specified in Par. 73.684 of the FCC Rules and Regulations.

The low-power station must protect the full-power station to its grade B contour for the following levels;

Channels 2–6	*Grade B contour: 47 dB*
Channels 7–13	*Grade B contour: 56 dB*
Channels 14–69	*Grade B contour: 64 dB*

The low-power station also must not interfere with other low-power or translator stations for the following levels:

Channels 2–6	*Protected to their 62-dB contour*
Channels 7–13	*Protected to their 68-dB contour*
Channels 14–69	*Protected to their 74-dB contour*

Protection for existing stations is determined as follows. The ratio of the protected signal at its grade B contour to the signal that would exist from the proposed station at that point is called the desired-to-undesired (D/U) ratio (measured in decibels). By use of the propagation curves in Par. 73.699, the distance of the protected signal to its grade B contour is found. Knowing this, the distance of the low-power station can be determined, since this distance must comply with the above guidelines. Again, turning to the propagation curves, the ERP and the antenna height can be determined for the proposed station. If either station uses a directional antenna, calculations of the desired to undesired signal strengths must be made at 0° (direct line between the stations), ±15°, and ±30°. All D/U ratios must be satisfied. Here the ingenuity of the design engineer comes into play. In order to fit in a low-power station, the engineer must resort to the use of directional antennas, choice of antenna, heights, transmitter power output, frequency offset, and natural terrain shielding. The last item must be demonstrated convincingly.

Table 7-7 LP-Station Maximum Spacing

Full-power station	Distance to low-power site
Co-channel VHF or UHF	Beyond the grade B contour
VHF or UHF ± 1 channel	Beyond the grade B contour
UHF −14 or −15 channels	Beyond the grade B contour
UHF + 7 channels	At least 61 mi
UHF ± 2, 3, 4, 5 channels	At least 20 mi

Source: Federal Communications Commission.

7.7 MICROWAVE SYSTEMS

A microwave system used in television broadcasting is the point-to-point transmission of a television signal by means of an RF carrier in the 2-, 7-, or 13-GHz frequency spectrum. The RF power of the transmitter is in the order of a fraction of a watt up to about 15 W. The relatively high gain of the antenna provides a high enough ERP from a relatively low power transmitter to produce excellent transmission qualities.

A microwave beam behaves much like a light beam insofar as atmospheric effects are concerned, and is subject to a number of transmission factors which must be considered in the design of the system.

7.7.1 TERRAIN OBSTRUCTIONS.

A microwave beam follows a straight line in azimuth until intercepted by structures, natural or made by humans, along its path. In the atmosphere the beam follows a slightly curved path in the vertical plane, owing to the variations in the dielectric with respect to elevation. The amount of refraction varies with time owing to changes in temperature, pressure, and humidity, all of which affect the dielectric. As the beam just grazes the obstruction, the receiving antenna will incur a loss in signal which can range from up to 6 dB for knife-edge and tree refraction to 20 dB for smooth-surface refraction.

7.7.2 FRESNEL-ZONE CLEARANCE (FIG. 7-12).

As the microwave beam passes close to an object, some energy is reflected by the object, causing a variation in the received signal level. The microwave frequency, the distance from the center of the microwave beam, and the distance to the ends of the microwave path determine where the nulls and the peaks of the signal will occur. The type of obstruction and its shape determine the magnitude of the variation.

The distance from the microwave beam center, as it passes over an obstruction, to the obstruction is measured in units of Fresnel zones, taking into account both frequency and distance. The first Fresnel zone (F_1) is the surface of points along the path in which the total distance between the endpoints is exactly ½ wavelength longer than the direct

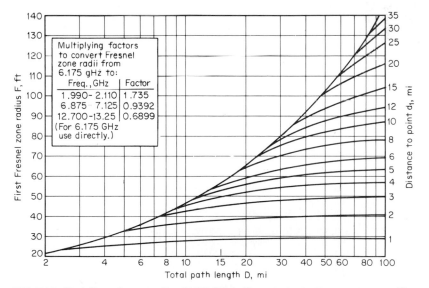

FIG. 7-12 First Fresnel zone radius (6.175 GHz). [$D = d_1 + d_2$, $F = 72.1(d_1 d_2/fD)^{1/2}$.] *(GTE-Lenkurt.)*

end-to-end path. It is important to note that clearance of at least $0.6F_1$ of the first Fresnel zone is required to maintain minimum free-space loss. When less than $0.6F_1$ clearance is present, the microwave beam is considered to be obstructed. When the object is at or outside the first Fresnel zone, the distance from the object to the center of the microwave beam is measured in actual Fresnel zones.

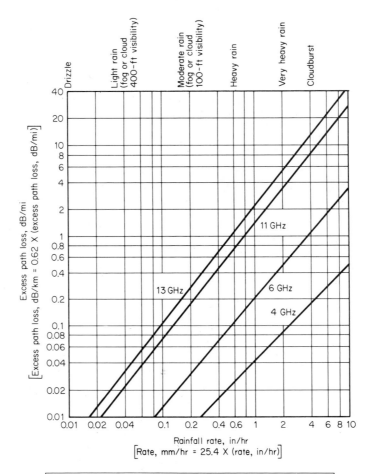

Attenuation due to atmospheric absorption				
Path length mi (km)	2–6 GHz	8 GHz in dB	12 GHz	14 GHz
20 (32)	0.20	0.26	0.38	0.48
40 (64)	0.40	0.52	0.76	0.96
60 (96)	0.60	0.78	1.14	1.44
80 (128)	0.80	1.04	1.52	1.92
100 (160)	1.00	1.30	1.90	2.40

FIG. 7-13 Rain attenuation versus rainfall rate. *(GTE-Lenkurt.)*

Table 7-8 Microwave Feedline Losses

	Rectangular waveguide	Loss/100 ft (30 m)	Elliptical waveguide	Loss/100, ft (30 m)
4 GHz	WR-229	0.85	EW-37	0.85
6 GHz	WR-137	2.00	EW-59	1.75
8 GHz	WR-112	2.70	EW-71	2.50
13 GHz	WR-75	4.5	EW-122	4.50

Source: Lenkurt Corp. (GTE Communication Systems Corp.)

Microwave paths with strong, shallow reflections can be used to advantage so that the odd Fresnel zone reflections can be combined to obtain increases in signal strength.

Fresnel zone clearance, earth's bulge, and microwave beam refraction are combined in clearance criteria for path design. For heavy routes, clearance should be at least $0.3F_1$ at $K = \frac{2}{3}$, and $1.0F_1$ at $K = \frac{4}{3}$, whichever is greater. In very difficult propagation areas, grazing at $K = \frac{1}{2}$ may be added to the criteria to ensure adequate clearance. For light routes, use $0.6F_1$ at $K = 1.0$ plus 10 ft (3 m).

7.7.3 EFFECTS OF RAIN, FOG, AND ATMOSPHERE (FIG. 7.13). At frequencies below 6 to 8 GHz, rain, fog, and atmosphere attenuation are not thought sufficient to warrant special consideration. However, above 6 to 8 GHz they cause a loss in signal strength, and variations in the instantaneous intensity can cause serious fading.

7.7.4 MICROWAVE-FEEDLINE SYSTEM. Three waveguide systems are commonly used above about 2 GHz. These are rectangular, elliptical, and circular. The first two are easier to construct and install but have somewhat higher losses. The elliptical system is flexible, allowing ease in installation. The circular one has the advantage of dual-mode operation both in frequency and in polarization, but unless it is designed properly assembly can be complicated because of bending and termination complexities. Table 7-8 shows several rectangular and elliptical types with manufacturers' type numbers, frequencies, and losses.

7.7.5 ANTENNAS. The basic microwave antenna for television use is the parabolic dish. Alternatively, horn antennas can be used where high front-to-back ratios are required, but they are more expensive. Reflectors have the problem of possible interference with satellite communications because the transmitting periscope type of antenna could radiate unwanted energy toward the satellite. The periscope antenna makes use of a parabolic dish looking upward to a rectangular plane reflector oriented to the proper azimuth.

Table 7-9 Microwave Antenna Gains

Plane polarized diameter, ft (m)	Gain relative to isotropic, dB					
	2 GHz	4 GHz	6 GHz	7 GHz	8 GHz	13 GHz
4 (1.2)	26.5		35.2	35.9	37.0	41.3
6 (1.8)	29.0	35.0	38.7	39.4	40.6	44.8
8 (2.4)	31.5	37.3	41.1	41.9	43.1	47.3
10 (3.0)		40.8	44.6	45.5	46.7	
15 (4.6)		42.6	46.0	46.9	48.7	

Source: Lenkurt Corp. (GTE Communication Systems Corp.)

In choosing an antenna, besides the required gain, consideration must be given to bandwidth, side lobes, off-axis radiation, polarization, and impedance match (return loss). Parabolic antennas usually have efficiencies of 55 percent.

7.7.6 SYSTEM-NOISE OBJECTIVES. A 59-dB S/N ratio (p-p video/rms noise, the latter weighted for the bandwidth in use) is considered satisfactory, while an S/N ratio of 33 dB or lower is considered unusable with an unacceptable noise level.

7.7.7 MICROWAVE TOWERS. Microwave towers may be either self-supporting or guyed. Because of the narrow beam widths some sort of antitwist mechanism must be incorporated to avoid misalignment with variations in wind loading. EIA standard RS-222-C lists recommended specifications for towers in different zones in the United States. For example, one of these calls for 30 lb/ft^2 [1437 pascals (Pa)] for zone A, 40 lb/ft^2 (1916 Pa) for zone B, and 50 lb/ft^2 (2395 Pa) for zone C.

Wind-loading increases as the square of the wind velocity and may be calculated from

$$P \text{ (lb/ft}^2) = K \text{ (wind factor)} \times \text{velocity}^2$$

EIA recommends a wind factor of 0.004. By use of this figure, the following values are obtained

$$20 \text{ lb/ft}^2 \text{ (958 Pa): 71.0 mi/h (31.7 m/s)}$$

$$30 \text{ lb/ft}^2 \text{ (1437 Pa): 86.0 mi/h (38.4 m/s)}$$

$$40 \text{ lb/ft}^2 \text{ (1916 Pa): 100.0 mi/h (44.7 m/s)}$$

$$50 \text{ lb/ft}^2 \text{ (2395 Pa): 112.5 mi/h (50.3 m/s)}$$

The maximum twist and sway of the tower should not exceed that value which attenuates the received microwave signal by more than 10 dB.

7.7.8 USEFUL MICROWAVE FORMULAS

Free-Space Loss

$$A = 96.6 + 20 \log_{10} F + 20 \log_{10} D$$

where A is in decibels, F is in gigahertz, and D is in miles.

First Fresnel Zone Clearance

$$F^1 = \frac{72.1 \; (D^1 + D^2)}{F \times D}$$

where F^1 is in feet, D^1 is the distance of site 1 in miles to the obstruction, D^2 is the distance from the obstruction in miles to site 2, F is the frequency in gHz, and D is the distance in miles between sites 1 and 2.

Parabolic Antenna Gain

$$G = 20 \log_{10} D + 20 \log_{10} F + 7.5$$

where G is in decibels relative to an isotropic source, D is the diameter of the dish, and F is in gigahertz.

3 dB of the Antenna Beam Width

$$\phi = \frac{70}{F \times \text{diameter}}$$

where ϕ is in degress, F in gigahertz, diameter in feet.

Earth's Curvature Bulge

$$h = \frac{d^1 \times d^2}{15 \times K}$$

where h is in feet, d^1 is the distance from the obstruction to site 1, d^2 is the distance from the obstruction to site 2, both in miles, and K is the exaggerated earth's radius ranging from flat ($K = 0$) to $K = \frac{4}{3}$.

Recommended Clearance Criteria. For highest reliability, Fresnel zone clearance should be at least $F = 1$ at $K = \frac{4}{3}$, to $F = 0.3$ at $K = \frac{2}{3}$, whichever gives the greatest clearance. Clearance must include earth's bulge, Fresnel height, trees, or other obstructions.

7.7.9 EXAMPLE OF A MICROWAVE PATH CALCULATION (FIG. 7-14).

1. Determine distance and bearing between sites 1 and 2 [measured from a United States geodetic map or by calculation using the FCC Rules and Regulations, Par. 73.611 (5), Tables 1 to 3].

2. Calculate path loss using above formula.

3. Check Fresnel clearance.

4. Check annual rainfall averages and atmospheric attenuation.

5. Determine transmitter (TX) and receiver (RX) antenna gains.

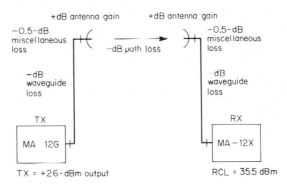

TX output	+ 26 dBm
TX WG, 100 ft (30 m) of PE-122	− 4.2 dB
Miscellaneous TX system loss	− 0.5 dB
TX antenna gain, 4-ft (121.9 cm) diameter	+ 41.6 dB
Path Loss, 10 m (16 km)	−138.9 dB
RX antenna gain, 6-ft (182.9 cm) diameter	+ 45.1 dB
Miscellaneous RX system loss	− 0.5 dB
RX W/G, 50 ft (15 m) of PA-122	− 2.1 dB
Field allowance	− 2.0 dB
RCL (net of above)	− 35.5 dBm
RX threshold	− 79 dBm
Fade margin (difference between clear-sky RCL and RX threshold)	+ 43.5 dB

FIG. 7-14 Example of simplified fade margin calculations. Practice is to add about 2-dB loss for field allowance or antenna misalignment. (*Microwave Associates Communications, Inc.*)

6. Calculate waveguide losses.

7. Calculate miscellaneous losses (including random losses and waveguide connection losses).

8. Determine microwave transmit power needed.

9. Obtain receiver threshold sensitivity.

10. Make final calculations.

7.7.10 SUMMARY. In this short discourse it is not possible to cover all the factors that can influence performance of a microwave system. Variations from normal wave propagation can occur, depending upon changes in the vertical dielectric. Such changes can result in greater or lesser bending of the beam. For further information see the reference list at the end of this chapter.[12]

7.8 NEW DEVELOPMENTS IN UHF

UHF transmitters are required to produce more RF power than VHF counterparts in order to overcome the greater losses in free-space propagation and in the receiver antenna system and in receiver tuners. Unfortunately, the generation of such power entails a substantial increase in capital and operating cost. With the cost of primary energy increasing dramatically, improvements in UHF transmitter efficiency have become an increasingly more important consideration.[13]

Several areas in which improvements have been made or are being investigated are the following:[14]

1. Design higher-efficiency klystrons[15,16]

2. Use of aural and visual couplers

3. Klystron operation with reduced perveance

4. Mod-anode sync pulsing

5. Variable perveance operation

6. Sold-state exciters

7. Linearity correctors with greater range

8. Reclamation of waste heat

9. Lower-loss feedline components

10. Aural operation of 10 percent

Another development, the depressed-collector klystron,[17] offers a further improvement in efficiency, but it is still in the development stage. Since such a tube has not been successfully demonstrated, no further comment will be made in this chapter.

The klystron is the most suitable output tube to develop the required high power for UHF transmitters because of its high gain, long life, and stable characteristics. However, until recently, when primary power costs were low, the tube was operated very inefficiently for several reasons. First, color transmission demanded linear transfer characteristics, and the linearity correctors available had very limited range. Hence, the tube operated with a high perveance, resulting in a high collector current. Second, the tube-type IPAs were limited in the amount of linear drive they could produce, and the klystron itself was not efficient because of the internal drift-tube design.

Figure 7-15 shows the elements of the klystron. The klystron is made up of a heater, cathode, and, at the far end, a collector which absorbs the modulated electron flow. Along this path the electrons traverse a modulating anode whose bias is used to control the amount of electrons that pass through the cavities on their way to the collector. The cavities offer a tuned resonance, and velocity modulates the electron beam. The collector

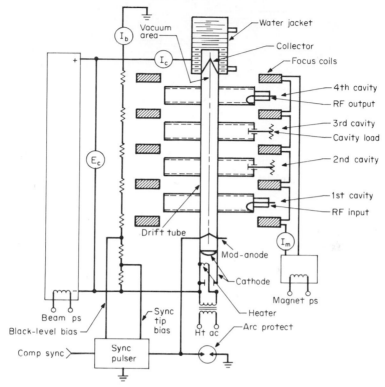

FIG. 7-15 Typical elements of a four-cavity klystron. (© *Public Broadcasting Service 1980.*)

is at ground potential while the cathode is at a high negative potential. The cavities can be either integral to the tube or added to the klystron assembly, in which case the klystron is of the external type. In the integral type all parts are in the vacuum of the tube, and because of this the input and output couplings are fixed at an average position so that the coupling will cover the entire band of frequencies over which the tube can operate. In the external klystron the cavities are not in the vacuum of the tube and thus can be adjusted for optimum performance on any channel.[18]

7.8.1 AURAL AND VISUAL COUPLERS (FIG. 7-16). As noted above, one of the advantages of the external-type klystron is the ability to optimize the input and output couplings for high efficiency. An adjustable coupler has been made available for integral klystrons which allows operation with reduced perveance. However, operating with reduced perveance increases the resistance to the beam, making it necessary to raise the loaded Q of the cavity to provide better matching to the load. The design of the coupler consists of adding a stub section across the output line which is approximately ⅜ wavelength from the coupling loop inside the output cavity. The stub is adjustable over ¼ wavelength so that either a capacitance or an inductive susceptance is seen by the tube. The ⅜-wavelength separation converts the susceptance to a conductance so that there is no pulling of the cavity resonance as the stub length is changed. The stub is adjustable so that it presents a capacitive susceptance when it is less than ¼ wavelength and an inductive susceptance when it is more than ¼ wavelength. Measurements with and with-

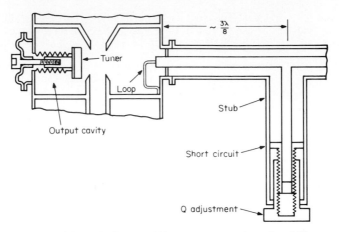

FIG. 7-16 Schematic diagram of klystron output cavity and variable coupler (transformer). *(Varian Associates, Inc.)*

out the coupler have shown about an 8 to 10 percent improvement. The coupler for the aural klystron is of the fixed type, and no adjustments are required. However, it is good practice to adjust the aural klystron on low voltage so that during initial coupling misadjustments cannot damage the tube.

7.8.2 LOW-PERVEANCE OPERATION OF A KLYSTRON (FIG. 7-17). With the availability of solid-state exciters, greater-range linearity correctors, and improved klystrons, it is now possible to operate the tube with a longer, but more nonlinear, characteristic by biasing the modulating anode electrode to reduce the beam current.[19] The beam voltage must be simultaneously increased. The result is a longer, less-linear transfer characteristic. Although the black level operates at a higher level requiring greater RF drive and correction, solid-state drivers can provide the additional drive and correction. Beam-perveance reduction procedures are successful with both external and integral klystrons, but the latter require external output couplers to optimize the RF power output. Efficiencies of between 45 and 50 percent have been achieved with external klystrons without additional output couplers, and almost as much with the internal klystrons

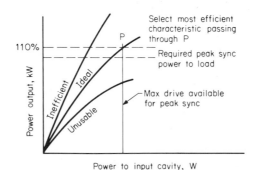

FIG. 7-17 Adjustments for perveance. (© *Public Broadcasting Service 1980.*)

fitted with output couplers. These efficiencies should be compared with high-perveance operation with efficiencies of between 25 and 35 percent. Further improvement in efficiencies can be obtained by the addition of mod-anode pulsing which is explained below.

7.8.3 MOD-ANODE PULSERS. The mod-anode used in a klystron allows the tube to be biased by a dc voltage usually obtained from a resistor divider connected across the high-voltage supply. Biasing toward cathode increases the beam current, while biasing toward ground (collector) decreases the beam current. In the composite signal waveform the sync occupies 25 percent of the amplitude and the video 75 percent, and the tip of sync represents the peak power of the signal while the blanking (black) represents 56 percent of the peak power. If the black could be operated at 100 percent and the sync pulsed in, like a radar transmitter, the peak sync would then be at a power level of 178 percent and the efficiency would increase accordingly. Unfortunately, the color burst extends 50 percent into the sync region, and any attempt to pulse the complete sync would distort the color-burst reference phase. Therefore, pulsing the sync would have to be done at the 12.5 percent level above black in order to preserve the color-burst waveform. Several years ago a proposal for pulsing in the sync was made by the Public Broadcasting Service, and RCA produced a mod-anode pulser shortly thereafter.[20] The RCA mod-anode pulser incorporated a high-voltage switch which raised the voltage on the mod-anode only during the sync peaks and switched back to normal for the remainder of the waveform. This required additional drive so that the black level could be raised, and thus improve efficiency. Greater nonlinearity resulted but this was handled by linearity correctors. This technique was first used on an integral four-cavity klystron operated at 30 kW. An improvement of about 20 percent in beam efficiency was obtained.

7.8.4 VARIABLE-PERVEANCE KLYSTRONS. The operation of a UHF transmitter using a method of variable-beam perveance was announced by Pye TVT Limited of England. In the above paragraph it was shown that the addition of a mod-anode sync pulser could improve the beam efficiency by about 20 percent at sync tip power output. In order to achieve still greater efficiencies, Pye fitted a normal external-cavity klystron with an additional beam control electrode in the gun assembly. This electrode, when driven by a negative-going narrow-band video signal of a few hundred volts peak, proportionately controlled the beam current to be adequate, but not more than necessary, at all points of the complete modulation cycle from peak white to peak sync. The klystron operates in a quasi class AB condition rather than the normal class A condition. This results in a reduction in beam current in the order of 35 percent, compared with the same tube adjusted for maximum efficiency but without application of the beam control modulation.

This technique has been applied to an external-cavity, low-perveance, high-efficiency klystron normally adjusted for a 45 percent beam efficiency. An overall beam efficiency of 65 to 70 percent has been achieved without the use of output couplers or mod-anode sync pulsing. This is the highest reported klystron efficiency which still met FCC broadcasting performance specifications. The same tube can be used in aural service by applying an appropriate voltage to the new grid so the tube operates as if no control electrode were in the circuit.

Description. A video signal, derived from the drive circuits after the first video clamp and sync stabilization module, is fed to the video precorrection and optic assembly where the signal employed to modulate the normal visual carrier and that fed to the new control electrode are both processed. The normal modulation channel has amplitude and frequency correction to compensate for the changes in response caused by the variable beam loading of the klystron cavities.

The signal fed to the beam control modulator is timed so that the modulation applied to the control electrode coincides correctly with that of the RF carrier fed to the input cavity. It is then passed through a low-pass filter to limit the bandwidth to somewhat less than the double sideband part of the vestigial single sideband television signal. The use of a narrow bandwidth control signal reduces the power that the modulating amplifier must produce to drive the control beam electrode capacitance. The processed signal is passed to the modulating amplifier by a fiber-optic guide system which uses pulse-width-modulated, infrared light. In the optic receiver the signal is clipped to remove the pulse-shape degradations caused by the optic link and is passed through a low-pass filter from which the signal is recovered. The modulating amplifier, which is solid-state, produces a negative signal of about 800 V and is connected directly to the beam-control electrode. Overall negative feedback is provided to maintain good linearity. The tube is able to provide full power required for the sync. As the video signal increases toward peak white, the control voltage becomes more negative and the beam power is thus reduced.

7.8.5 RECLAMATION OF WASTE HEAT. In spite of welcome improvements in klystron efficiency, the typical UHF transmitter still consumes about 90 percent of its input power in klystron dissipation. An examination of two transmitters, one operating in 1977 at a 55-kW level with external four-cavity klystrons and the second transmitter with the same power level and external four-cavity klystrons, but fully modified in 1981 with a mod-anode pulser and operation at reduced perveance, provided the data on improved overall transmitter efficiency shown in Table 7-10.

Thus it can be seen that the energy dissipated in the heat exchangers due to the klystrons is still a substantial source of possible further power savings. There are a number of stations, both VHF and UHF, that make use of such energy sources. In the case of air-cooled transmitters, the ducts carrying the waste heat to the outside have bypass vanes, thermostatically controlled, to route the air to the inside of the building. In the case of water-cooled transmitters, the majority of which are UHF, the pipes carrying the hot water from the klystrons are routed through space heaters, also thermostatically controlled, to parts of the building which require heat.

The main problem with the above system is the absence of heat when the transmitter is shut down, this being when the building needs the heat the most, both for the comfort of the occupants and to keep the water-cooled equipment from freezing.

One of the UHF stations of Connecticut Public Television is in the process of designing a heat-exchange system in which a large 2500-gal (9462-liter) insulated tank will be installed so that the energy normally dissipated in the regular heat exchanger will travel through a second heat exchanger through a coil installed within the large tank. The water used to cool the klystrons has a mixture of glycol to prevent freezing since the pumps and primary heat exchangers are installed out-of-doors. The water in the 2500-gal tank is untreated since the water will remain inside the building at all times. Calculations show that to heat the 1600-ft^2 (149-m^2) building, about 50,000 Btu/h ($52,750 \times 10^3$ J/h) will be required. Since the storage tank will keep 750,000 Btu on a 40°F (4°C) heat rise, it is expected that the system will maintain proper building temperature for upward of 10 h. If more heat is required, as in the case of unusual weather conditions, the regular building furnace will cut in automatically.

Table 7-10 Efficiency Improvement

Item	Unmodified	Modified
Aural and visual beam†	230.6 kW†	103.9 kW‡
Total fixed losses†	41.1 kW†	22.8 kW§
Total transmitter input†	276.3 kW†	126.7 kW¶
Beam kilowatts to total kilowatts	83.5%	82%

†From WMAH Mississippi Public Broadcasting, Dec. 16, 1971.
‡Aural reduced from 20 to 10%.
§Heat exchanger reduced in size.
¶Pulser installed, 60% beam efficiency.

REFERENCES

1. John Shipley, "Tell Me about TV Transmitters," NAB Report, Mar. 21, 1976, RCA Corp., Camden, N.J.
2. "TT 1000B Series TV Transmitters," Specifications Publication, RCA Corp., Camden, N.J.
3. David R. Hertling, "Surface Acoustic-wave Filters," Engineering Report, Harris Corp., Quincy, Ill.
4. Thomas J. Vaughan, "Combiners and Diplexers," RF Planning Guide, Micro Communications, Inc., Manchester, N.H.
5. F. E. Terman and J. M. Pettit, "Directional Couplers," in *Electronic Measurements,* 2d ed., McGraw-Hill, New York, 1952, pp. 57–64.
6. Thomas J. Vaughan, "Waveguide, Diplexer Losses," RF Planning Guide, Micro Communications, Inc., Manchester, N.H.
7. "Television Operational Measurements, Video and RF for NTSC Systems," Publication AX-3323-1, Tektronix, Inc., February 1979.
8. FCC Rules and Regulations, Section VIII, Parts 73 and 74.
9. "Television Operational Standards for TV Transmitters," Interim Report PN-1432, Electronic Industries Association, Sept. 30, 1981.
10. L. J. Stanger, "Active IF Group-Delay Correction of TV Transmitters," Engineering Report 2.5M-775, Harris Corp., Quincy, Ill.
11. A. G. Utzen Daele and A. H. Butt, "Envelope and Synchronous Demodulation of VSB Signals," Harris Corp., Quincy, Ill.
12. Robert F. White, "Engineering Considerations for Microwave Communication Systems," GTE/Lenkurt, Inc., Phoenix, Ariz., June 1970.
13. C. Chapin Cutter, "New Opportunities for UHF TV Transmitters," Federal Communications Commission Report, February 1980.
14. John T. Wilner and Thomas B. Keller, "UHF Transmitter Improvements," Public Broadcasting Service, Report no. E-8012, July 30, 1980.
15. R. D. Symons, "Klystrons for UHF Television," IEEE Broadcast Symposium, Washington, November 1982.
16. "Internal and External-Cavity Klystrons," Publications 4511, March 1982, and 2967, February 1973, Varian Associates, Inc.
17. R. D. Symons, "Depressed-Collector Klystrons," NAB Report, Dallas, April 1979, Varian Associates, Inc., Palo Alto, Cal.
18. "Aligning Klystrons for Optimum Efficiency," British Broadcasting Company, Engineering Report no. 8/80, Dec. 10, 1980, Addendum 1, Dec. 17, 1980, Addendum 2, March 11, 1981, London.
19. R. M. Unetch and R. W. Zborowski, RCA Corp., and R. D. Symons, Varian Associates, Inc., "Perveance Reduction and Efficiency Enhancement of Intregal-cavity Klystrons," April 1981.
20. John Bullock and Robert C. Schmidt, "System for Reducing Power Required for UHF Transmitters," NAB Report, March 1976.

Transmitting Antennas

Oded Ben-Dov
Krishna Praba

RCA Corporation
Gibbsboro, New Jersey

8.1 BASIC RADIATING SYSTEMS

Broadcasting is accomplished by the emission of coherent electromagnetic waves in free space from a single or group of radiating-antenna elements, which are excited by modulated electric currents. Although, by definition, the radiated energy is composed of mutually dependent magnetic and electric vector fields, conventional practice in television engineering is to measure and specify radiation characteristics in terms of the electric field.[1-3],†

The field vectors may be polarized, or oriented, linearly, horizontally, vertically, or circularly. Linear polarization is used for some types of radio transmission. Television broadcasting has used horizontal polarization for the majority of the system standards in the world‡ since its inception.

More recently, the advantages of circular polarization, described in subsequent sections, have resulted in an increase in the use of this form of transmission, particularly for VHF channels.

Both horizontal and circular polarization designs are suitable for tower-top or tower-face installations. The latter option is dictated generally by the existence of previously installed tower-top antennas. On the other hand, in metropolitan areas where several antennas must be located on the same structure, either a *stacking* or a *candelabra-type* arrangement is feasible. For example, in New York a stacking of antennas, first on the Empire State Building and presently on the World Trade Center, has permitted the installation of 10 channels (Fig. 8-1). In Chicago, atop the John Hancock Center, antennas for six channels are stacked on twin towers (Fig. 8-2). Alternatively, on Mt. Sutro in San Francisco the antennas for eight channels are mounted on a candelabra assembly (Fig. 8-3).

Another solution to the multichannel location problem, where space or structural limitations prevail, is to diplex two stations on the same antenna. This approach, while eco-

---

†Superscript numbers refer to References at end of chapter.
‡Vertical polarization is used for the first broadcasting service established in the United Kingdom in 1937 on 405-line monochrome standards. This service will be discontinued in the late 1980s.

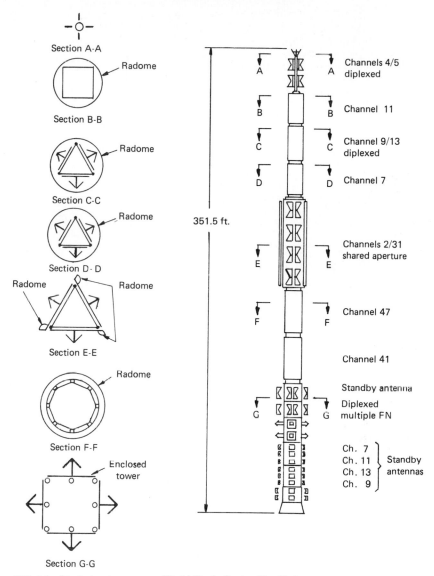

FIG. 8-1 Stacked antenna array, World Trade Center, New York.

nomical from the installation aspect, results in transmission degradation because of the diplexing circuitry and antenna-pattern and impedance broadbanding required.

8.1.1 VHF ANTENNAS FOR TOWER-TOP INSTALLATION. The typical television broadcast antenna is a broadband radiator operating over a bandwidth of several megahertz with an efficiency of over 95 percent. Reflections in the transmission between the transmitter and antenna must be small enough to introduce negligible picture degradation. Furthermore, the gain and pattern characteristics must be designed to achieve the desired coverage within acceptable tolerances, and operationally with a minimum of

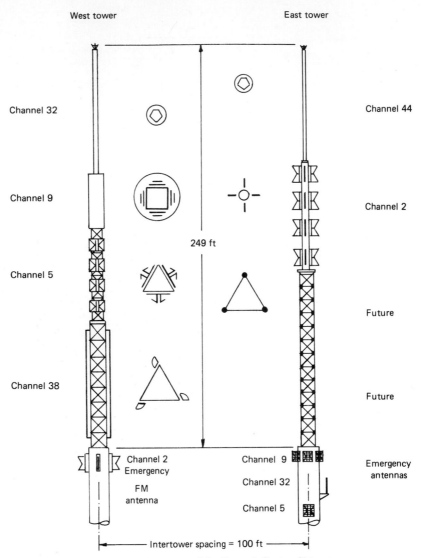

West tower

East tower

Channel 32

Channel 44

Channel 9

Channel 2

249 ft

Channel 5

Future

Channel 38

Future

Channel 2
Emergency

Channel 9

Emergency
antennas

FM
antenna

Channel 32

Channel 5

Intertower spacing = 100 ft

FIG. 8-2   Twin tower antenna array atop John Hancock Center, Chicago.

maintenance. Tower-top, pole-type antennas designed to meet these requirements can be classified into two categories: (1) resonant dipoles and slots, and (2) multiwavelength traveling-wave elements.[4]

Turnstile Configuration.   The main consideration in the top-mounted antennas is the achievement of excellent omnidirectional azimuthal fields with minimum of windload. The earliest and most popular resonant antenna for the VHF applications is the "turnstile," which is made up of four batwing-shaped elements mounted on a vertical pole in a manner resembling a turnstile. The four "batwings" are, in effect, two dipoles which are fed in quadrature phase. The azimuthal-field pattern is a function of the diam-

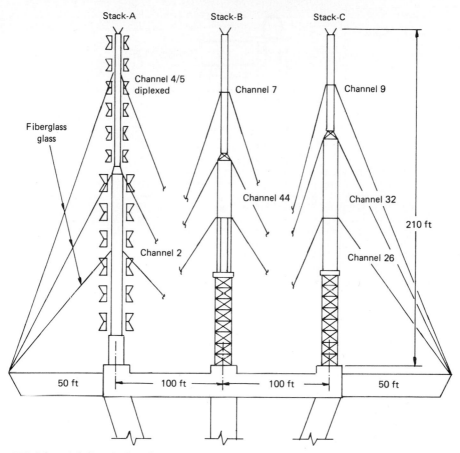

FIG. 8-3 100-ft (30 m) triangular "candelabra" of antennas, Mt. Sutro, San Francisco.

eter of the support mast and is within a range of 10 to 15 percent from the average value. The antenna is made up of several layers, usually six for channels 2 to 6 and twice that number for channels 7 to 13. This antenna is suitable for horizontal polarization only. It is unsuitable for side-mounting, except for standby applications where coverage degradation can be tolerated.

Multislot Radiator.   A multislot antenna has been developed for channels 7 to 13 which consists of an array of axial slots on the outer surface of a coaxial transmission line. The slots are excited by an exponentially decaying traveling wave inside the slotted pole. The omnidirectional azimuthal pattern deviation is less than 5 percent from the average circle. The antenna is generally about fifteen wavelengths long. A schematic of such an antenna and its principle of slot excitation are shown in Fig. 8-4.

Circular Polarization.   For circular polarization, both resonant and traveling-wave antennas have been developed. The traveling-wave antenna is essentially a side-fire helical antenna supported on a pole. A suitable number of such helices around the pole provide excellent azimuthal pattern circularity. This type of antenna is especially suited for application with channels 7 to 13 since only 3 percent pattern and impedance bandwidth are required. For channels 2 to 6, circular polarization applications where the

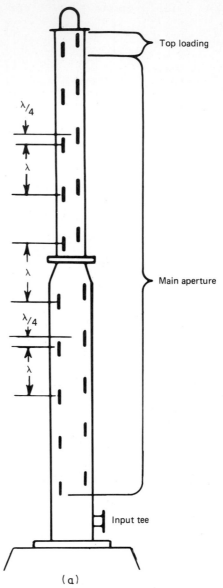

(a)

FIG. 8-4 (*a*) Schematic of multislot traveling-wave antenna.
(*b*) Principle of slot excitation to produce omnidirectional pattern.

bandwidth is approximately 10 percent, resonant dipoles around a support pole are a preferred configuration.

Transmission Lines. Table 8-1 lists the required output from a transmitter for various lengths of transmission line runs to the antenna. The FCC restrictions on the peak effective radiated power (ERP) of 100 kW for channels 2 to 6 and 316 kW for channels

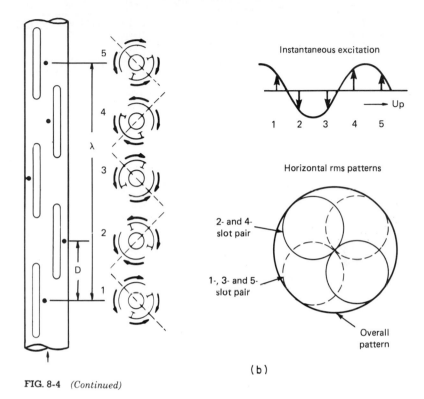

Instantaneous excitation

Horizontal rms patterns

2- and 4-slot pair

1-, 3- and 5-slot pair

Overall pattern

(b)

FIG. 8-4   *(Continued)*

7 to 13 are used. For circular-polarization applications, twice the values shown in Table 8-1 are permissible.

8.1.2 UHF ANTENNAS FOR TOWER-TOP INSTALLATION.   The slotted-cylinder antenna, commonly referred to as the *pylon* antenna, is the most popular top-mounted antenna for UHF applications. Horizontally polarized radiation is achieved using axial resonant slots on a cylinder to generate circumferential current around the outer surface

Table 8-1   Transmitter Output in Kilowatts

Line		Channels 2–6, ERP = 100 kW		Channels 7–13, ERP = 316 kW	
Size, in (cm)	Length, ft (m)	Antenna gain = 3	Antenna gain = 6	Antenna gain = 6	Antenna gain = 12
	500 (152)	36.9	18.5	62.1	31.0
3⅛ (8)	1000 (305)	40.9	20.5	73.2	36.6
50 Ω	1500 (457)	45.3	22.7	86.3	43.2
	2000 (610)	50.2	25.1	101.7	50.9
	500	35.0	17.5	56.7	28.5
6⅛ (15.6)	1000	36.7	18.9	61.6	30.8
75 Ω	1500	38.5	19.3	66.5	33.3
	2000	40.5	20.3	71.9	36.0

of the cylinder. An excellent omnidirectional azimuthal pattern is achieved by exciting four columns of slots around the circumference of the cylinder which is a structurally rigid coaxial transmission line.

The slots along the pole are spaced approximately one wavelength per layer and a suitable number of layers used to achieve the desired gain. Typical gains range from 20 to 40.

By varying the number of slots around the periphery of the cylinder, directional azimuthal patterns are achieved. It has been found that the use of fins along the edges of the slot provide some control over the horizontal pattern.

The ability to shape the azimuthal field of the slotted cylinder is somewhat restricted. Instead, arrays of panels have been utilized such as the zigzag antenna.[5] In this antenna the vertical component of the current along the zigzag wire is mostly cancelled out, and the antenna can be considered as an array of dipoles. Several such panels can be mounted around a polygonal periphery, and the azimuthal pattern can be shaped by the proper selection of the feed currents to the various panels.

For UHF antennas with a gain of 30, the required transmitter output for various transmission line runs is given in Table 8-2, assuming maximum effective radiated power from the antenna to be 1000 kW.

8.1.3 VHF ANTENNAS FOR TOWER-FACE INSTALLATION. One of the most popular panel antennas used for tower face applications in the United States is the so-called butterfly[6] or batwing panel antenna developed from the turnstile radiator. This type of radiator is suitable for the entire range of VHF applications. Recent innovations include modification of the shape of the turnstile-type wings to rhombus or diamond shape. Another popular version is the multiple dipole panel antenna used in most installations outside the United States. For circularly polarized applications,[7,8] two crossed dipoles or a pair of horizontal and vertical dipoles are used. A variety of cavity-backed crossed-dipole radiators are also utilized for circular polarization transmission.

The azimuthal pattern of each panel antenna is unidirectional, and three or four such panels are mounted on the sides of a triangular or square tower to achieve omnidirectional azimuthal patterns. The circularity of the azimuthal pattern is a function of the support tower size.[9]

Calculated circularities of these antennas are shown in Table 8-3 for idealized panels. The panels can be fed in-phase, with each one centered on the face of the tower, or fed in rotating phase with proper mechanical offset on the tower face. In the latter case, the input impedance match is far better.

Directionalization of the azimuthal pattern is realized by proper distribution of the feed currents to the individual panels in the same layer. Stacking of the layers provides gains comparable with those of top-mounted antennas.

Table 8-2   Transmitter Output in Kilowatts

Line, (75 Ω)		Antenna gain = 30			
Size, in (cm)	Length, ft (m)	Channels 14–26	Channels 27–40	Channels 41–54	Channels 55–70
6⅛ (15.6)	500 (152)	38.0	38.4	38.8	39.2
	1000 (305)	43.2	44.7	45.4	49.2
	1500 (457)	49.4	51.0	52.7	54.4
	2000 (610)	56.2	58.6	61.3	64.0
8³⁄₁₆ (21)	500	36.8	37.0	37.3	Use waveguide†
	1000	40.5	41.1	41.7	
	1500	44.7	45.7	46.6	
	2000	49.3	50.7	52.1	

†Use of waveguide recommended above channel 52.

Table 8-3   Circularities of Panel Antennas†

Shape	Tower-face size, ft (m)	Circularity, ± dB	
		Channels 2–6	Channels 7–13
Triangular	5 (1.5)	0.9	1.8
Triangular	6 (1.8)	1.0	2.0
Triangular	7 (2.1)	1.1	2.3
Triangular	10 (3.0)	1.3	3.0
Triangular	4 (1.2)	0.5	1.6
Square	5 (1.5)	0.6	1.9
Square	6 (1.8)	0.7	2.4
Square	7 (2.1)	0.8	2.7
Square	10 (3.0)	1.2	3.2

†Add up to ± 0.3 dB for horizontally polarized panels and ± 0.6 dB for circularly polarized panels. These values are required to account for tolerances and realizable phase patterns of practical hardware assemblies.

The main drawbacks of panel antennas are high wind-load, complex feed system inside the antenna, and the restriction on the size of the tower face in order to achieve smooth omnidirectional patterns. However, they provide an acceptable solution for vertical stacking of several antennas or where ease of installation is paramount.

8.1.4   UHF ANTENNAS FOR TOWER-FACE INSTALLATION.   Utilization of panel antennas in a manner similar to those for VHF applications is not always possible for the UHF channels. The high gains, which are in the range of 20 to 40 compared with those of 6 to 12 for the VHF, will require far more panels with the associated cumbersome branch feed system. It is also very difficult to mount a large number of panels on all the sides of a tower, the cross section of which must be restricted to achieve a good omnidirectional azimuthal pattern.

The zigzag panel described earlier has been found to be applicable for special omnidirectional and directional situations. For special directional azimuthal patterns, such as a cardioid shape, the pylon antenna can be side-mounted on one of the tower legs.

The use of tangential firing panels around the periphery of a large tower has resulted in practical antenna systems for the UHF applications. Zigzag panels or dipole panels are stacked vertically at each of the corners of the tower and oriented such that the main beam is along the normal to the radius through the center of the tower. The resultant azimuthal pattern is acceptable for horizontal polarization transmission.

8.1.5   VERTICALLY STACKED MULTIPLE ANTENNAS.   In metropolitan areas where there are several stations competing for the same audience, usually there will be only one preferred location for all the transmitting antennas.[10,11] A straightforward approach to the problem is stacking the antennas one on top of the other. One of the earliest installations of stacked antennas was on the Empire State Building in New York. Recently it has been replaced by another vertically stacked arrangement at the World Trade Center. A schematic of the World Trade Center stack is shown in Fig. 8-1. Since the heights of the centers of radiation decrease from the antenna at the top to that at the lowest level, there is a preference to be at a higher level. Thus, the final arrangement depends on both technical and commercial constraints. For relatively uniform coverage, technical considerations usually dictate that the top of the mast be reserved for the higher channels and the bottom of the mast for the low channels. However, because of contractual stipulations, this is not always possible, as Fig. 8-1 clearly shows.

Vertical stacking provides the least amount of interference among the antennas. However, since the antennas in the lower levels are panels on the support tower faces, which tend to be larger because the level is lower, the azimuthal pattern characteristics may be less than optimum. Further, the overall height constraint may compromise the desirable gain for each channel.

8.1.6 CANDELABRAS. In order to provide the same, as well as the highest, center of radiation to more than one station, the antennas are arranged on platform in a *candelabra* style. There are many *tee bar* arrangements in which there are two stacks of antennas. A triangular or even a square candelabra is utilized in some cases. A schematic of the Mt. Sutro, San Francisco, three-stack candelabra[12-14] is shown in Fig. 8-3. When several antennas are located within the same aperture, the radiated signal from one antenna is partly reflected and partly reradiated by the opposing antenna or antennas. Since the interference signal is received with some time delay, with respect to the primary signal, it can introduce picture distortion and radiation-pattern deterioration. The choice of opposing antennas and the interantenna spacing and orientation of the antennas determine the trade-offs among the in-place performance characteristics. These characteristics are azimuthal pattern, video bandwidth response, echoes, and differential windsway effect on the color subcarrier and isolation among channels. Some of these will be treated in more detail in Secs. 8.3.9 through 8.3.11.

8.1.7 MULTIPLE-TOWER INSTALLATIONS. The *candelabra* multiple-antenna system[15] requires cooperation of all the broadcasters in an area and considerable planning prior to erection. In many cases new channels are licensed after the first installation and addition of the antenna on an existing tower is not possible. Consequently, location on a nearby new tower is the only solution. In some cases, the sheer size of the candelabra may make it more expensive than multiple towers. The antenna farms around Philadelphia and Miami are good examples of several towers located in the same area.

The case of multiple towers, the creation of long-delayed *echoes* is of major concern, if economically or practically towers cannot be located close enough to result in echoes that either (a) cannot be resolved by the television system, or (b) will not affect color subcarrier phase and amplitude. This would dictate an impractical spacing of under 100 ft (30 m). Circularly polarized antennas provide an advantage since the reflected signal sense of polarization rotation is usually reversed, and the effect of echoes can be reduced with a proper circularly polarized receiving antenna.[16,17]

The twin stack of antennas atop the John Hancock Center in Chicago, shown in Fig. 8-2, is illustrative of a multiple-tower system.[18]

8.1.8 MULTIPLEXING OF CHANNELS. Another technique of accommodating more than one channel in the same antenna location is by combining the signals from these stations and feeding them to the same antenna for radiation. Broadband antennas are designed for such applications. The antenna characteristics must be broadband in more than terms of input impedance. Pattern bandwidth and, in the case of circularly polarized antennas, axial-ratio bandwidth are equally important. Generally, it is more difficult to broadband antennas if the required bandwidth will be in excess of 20 percent of the design-center frequency. Another of the problems of multiplexing is that the antenna must be designed for the peak voltage breakdown, which is proportional to the square of the number of channels. A third problem is the resolution of all technical commercial and legal responsibilities that arise from joint usage of the same antenna.

Multiplexed antennas typically are designed with integral redundance. The antenna is split into two halves with two input feedlines. This provides protection from failure due to the potential reduction in the high-power breakdown safety margin. This added hardware complexity may, in fact, lower the long-term reliability of the antenna. The Mt. Sutro multiple antenna installation (Fig. 8-3) is a successful paradigm of vertical stacking, candelabra, and diplexing design.

8.1.9 KEY CONSIDERATIONS IN SYSTEM DESIGN. Antenna system design is always an iterative process of configuration analysis, subject to some well-defined and many ill-defined constraints. Each iteration basically is a study with a sequence of binary decisions (see diagram of decision process and checklist of constraints on next page). The number of antennas at the same location increases the complexity of such a study. The

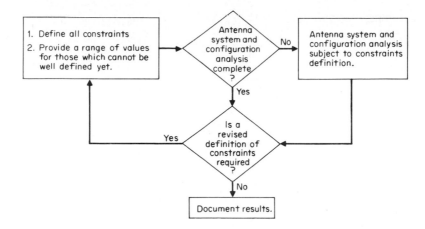

main reason for this is that most constraints must be viewed from three angles: engineering, legal, and commercial. Following is a checklist of items that are key considerations in system design:

1. Coverage and picture quality
2. Transmitter power
3. Antenna mechanical aperture
4. Channel multiplexing desirability or necessity
5. System maintenance requirements for broadcaster and service organization
6. Performance deterioration due to ice, winds, and earthquakes
7. Initial investment
8. Project implementation schedule
9. Building code requirements
10. Aesthetic desires of the broadcaster and community
11. Radiation hazards protection
12. Environmental protection
13. Beacon height
14. Accessibility of beacon for maintenance

It is worth reemphasizing that these considerations are interrelated and cover more than isolated technical issues.

8.2 INTERPRETATION OF FCC RULES AND REGULATIONS

This section discusses the requirements for television-broadcasting antenna performance in the United States stipulated in the FCC Rules and Regulations, Secs. 17 and 23. Although limited specifically to the United States, because of the worldwide similarity of broadcasting standards much of the information and explanations are applicable in other countries.†

———————

†For worldwide standards, see Secs. 21.2 and 21.4.

8.2.1 EFFECTIVE RADIATED POWER.

The *effective radiated power (ERP)* of an antenna is

$$\text{ERP} = P_{in} \times \text{gain}$$

The input power to the antenna is the transmitter output power minus the losses in the interconnection hardware between the transmitter output and the antenna input. This

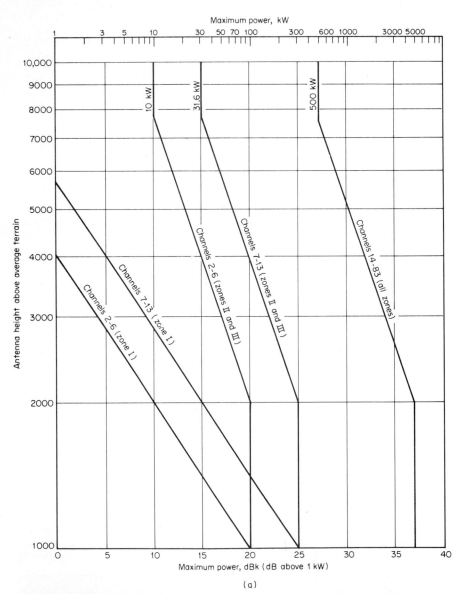

(a)

FIG. 8-5　Allowable effective radiated power for all zones: (*a*) maximum power versus antenna height; (*b*) FCC zones assignments.

hardware consists of the transmission lines and the filtering, switching, and combining complex discussed in Chap. 9.

The *gain* of an antenna denotes its ability to concentrate the radiated energy toward the desired direction rather than dispersing it uniformly in all directions. The property of gain and its analytical definition will be discussed later in Sec. 8.3.3. At this point it is important to note that higher gain does not necessarily imply a more effective antenna. It does mean a lower transmitter power output to achieve the allowable ERP.

The visual ERP, which must not exceed the FCC specifications, depends on the television channel frequency, the geographical zone, and the height of the antenna-radiation center above average terrain. The FCC Rules with respect to maximum ERP are shown in Fig. 8-5. These rules are for *either* horizontal or vertical polarization of the transmitted field. Thus, the total permissible ERP for circularly polarized transmission is doubled.

The visual ERP must not be below 100 W (−10 dBk) in the azimuthal plane, although there is no minimum specification for the antenna height above average terrain.

The FCC-licensed ERP is based on average-gain calculation for omnidirectional antennas and peak-gain calculation for antennas designed for a directional pattern. Details on this are in Sec. 8.2.4.

The aural ERP must not be less than 10 percent or more than 20 percent of the visual ERP.

8.2.2 COVERAGE. Knowledge of the ERP, the antenna height above average terrain, and the FCC $F(50,50)$ propagation curves is sufficient to estimate the field-strength contours over the desired coverage area.

Figures 8-6 through 8-10 show the estimated field strength for the *maximum* ERP for all zones and all channels for variations in the center of radiation from 500 to 4000 ft (152 to 1219 m). On each curve, the depression angle versus distance from the antenna location is also indicated. A more detailed calculation of depression angle versus distance for various antenna heights is shown in Sec. 8.4.1. The relationship between depression

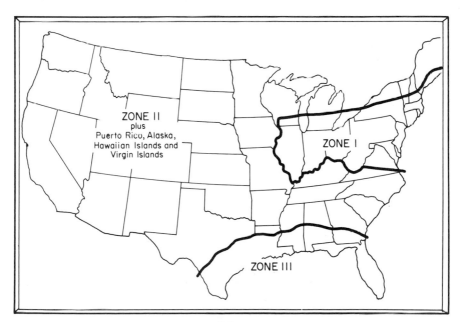

(b)

FIG. 8-5 *(Continued)*

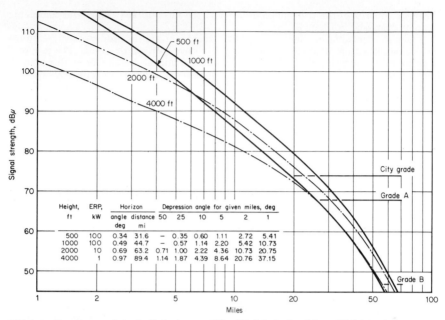

FIG. 8-6 Signal strength in decibels above 1 μV/m (dBμ) [calculated from FCC (50,50) data]. Zone 1, channels 2 to 6.

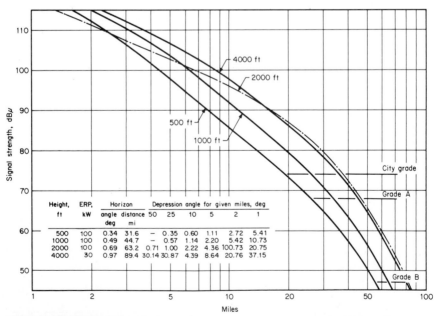

FIG. 8-7 Signal strength in decibels above 1 μV/m [calculated from FCC (50,50) data]. Zones 2 and 3, channels 2 to 6.

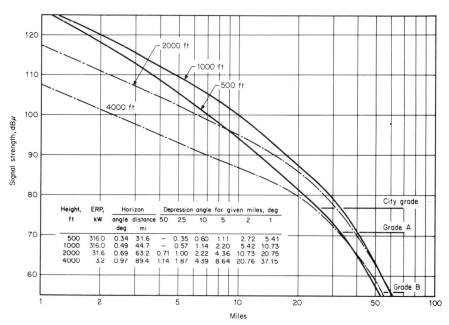

FIG. 8-8 Signal strength in decibels above 1 μV/m [calculated from FCC (50,50) data]. Zone 1, channels 7 to 13.

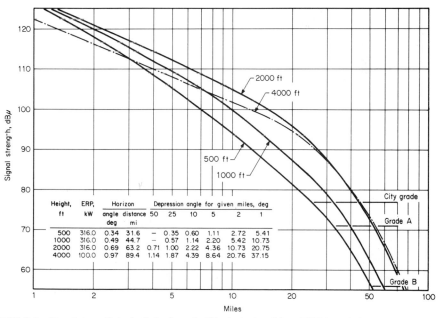

FIG. 8-9 Signal strength in decibels above 1 μV/m [calculated from FCC (50,50) data]. Zones 2 and 3, channels 7 to 13.

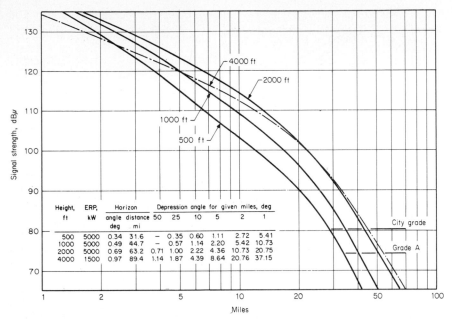

The figure contains an embedded data table:

Height, ft	ERP, kW	Horizon angle deg	Horizon distance mi	Depression angle for given miles, deg 50	25	10	5	2	1
500	5000	0.34	31.6	–	0.35	0.60	1.11	2.72	5.41
1000	5000	0.49	44.7	–	0.57	1.14	2.20	5.42	10.73
2000	5000	0.69	63.2	0.71	1.00	2.22	4.36	10.73	20.75
4000	1500	0.97	89.4	1.14	1.87	4.39	8.64	20.76	37.15

FIG. 8-10 Signal strength in decibels above 1 μV/m [calculated from FCC (50,50) data]. Zones 1 to 3, channels 14 to 70.

angle with respect to the horizontal plane and the distance to a point in the service area has been defined by the FCC on the basis of an idealized spherical earth with a radius of 5280 mi (8497 km) ($\frac{4}{3}$ times the earth's radius) to account for diffraction effects due to the earth's atmosphere. On this basis, the depression angle of the horizon can be calculated from the formula

$$A_h = 0.0153 \sqrt{H} \qquad \text{deg}$$

and distance from the formula

$$D_h = 5280 \times 0.0174533 \, A_h \qquad \text{mi}$$

where H is the antenna height above average terrain in feet.

There are three critical field strength contours which must be analyzed once the ERP and H are known. They are specified in millivolts per meter or in decibels above one microvolt (dBμ). The FCC specifications for these field strengths are listed in Table 8-4.

As stated previously, Figs. 8-6 to 8-10 show the estimated field strength for the max-

Table 8-4† FCC Field-Strength Specifications

	Channels 2–6	Channels 7–13	Channels 14–83
City grade	74 dBμ 5 mV/m	77 dBμ 7 mV/m	80 dBμ 10 mV/m
Grade A	68 dBμ 2.5 mV/m	71 dBμ 3.5 mV/m	74 dBμ 5 mV/m
Grade B	47 dBμ 0.22 mV/m	56 dBμ 0.63 mV/m	64 dBμ 1.6 mV/m

†These values are also indicated in Figs. 8-6 to 8-10.

imum ERP. Their usefulness mainly is in determining the maximum-expected grade contours for various antenna heights. The field-strength contours of a given antenna depend on the shape of the elevation and azimuthal radiation patterns of the given antenna.

For example, consider an omnidirectional channel 10 antenna at a height of 1000 ft

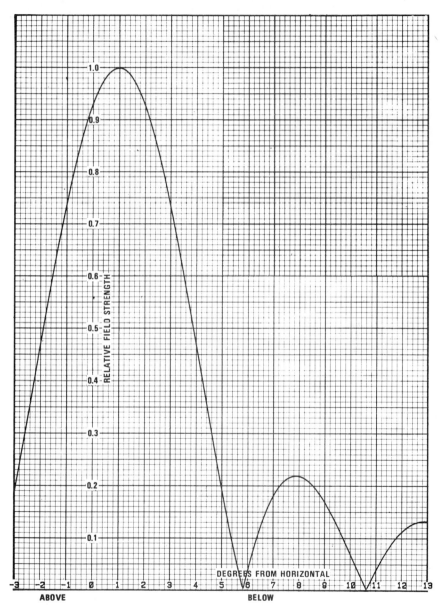

FIG. 8-11 Elevation pattern of a 12-bay channel 10 antenna.

(305 m) above average terrain and an ERP of 316 kW. The elevation pattern of the antenna is shown in Fig. 8-11. When this elevation pattern is applied to the 1000-ft antenna height curve in Fig. 8-8, the resulting field strength versus miles is shown in Fig. 8-12. Note that in some locations the field strength dropped to zero owing to the nulls in the elevation pattern. In Sec. 8.3.4 it is shown how these nulls can be *filled* if they fall in areas where coverage is desired.

Estimated field strengths may vary substantially from measured values due to actual terrain and the limited data used to generate the *F*(50,50) curves. In particular, no propagation curves are presently available for the vertical polarization component. Nonetheless, these curves have proved useful and are required for calculating the *filing* data.

8.2.3 POLARIZATION. Polarization is the orientation of the electric field as radiated from the transmitting antenna. When the orientation is parallel to the ground in the radiation direction-of-interest, it is defined as *horizontal polarization*. When the direction of the radiated electric field is perpendicular to the ground, it is defined as *vertical polarization*. These two states are shown in Fig. 8-13. Therefore, a simple dipole arbitrarily can be oriented for either vertical polarization or any tilted polarization between these two states.

There are numerous advantages and disadvantages to the transmission of either horizontal or vertical polarization. These are discussed further in Sec. 8.3.2.

If a simple dipole is rotated at the picture carrier frequency, it will produce circular polarization, since the orientation of the radiated electric field will be rotating either clockwise or counterclockwise during propagation. This is shown in Fig. 8-13. Instead of rotating the antenna, the excitation of equal longitudinal and circumferential currents in phase quadrature will produce circular polarization. Since any state of polarization can be achieved by judicious choice of vertical currents, horizontal currents, and their

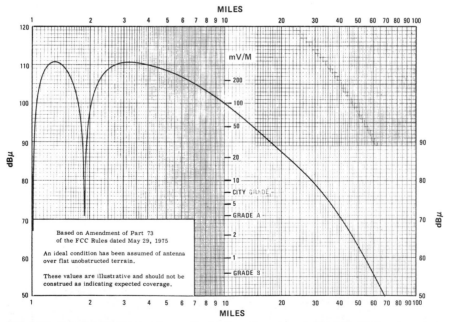

FIG. 8-12 Calculated field strength vs. miles for the pattern shown in Fig. 8-11. [12-bay channel 10 antenna; beam tilt, 1.0; effective radiated power, 316 kW; height, 100 ft (30 m)].

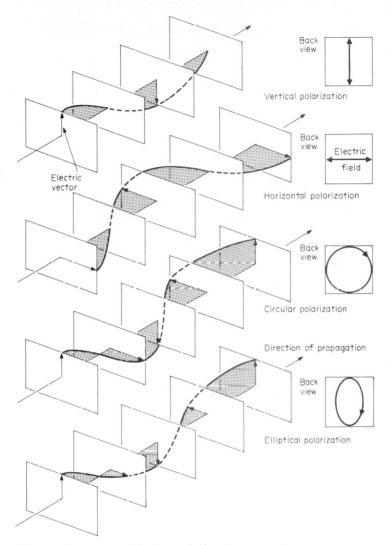

Back view

Vertical polarization

Back view

Electric field

Horizontal polarization

Back view

Circular polarization

Direction of propagation

Back view

Elliptical polarization

Electric vector

FIG. 8-13   Polarizations of the electric field of the transmitted wave.

phase difference, it follows that the reverse is also true. That is, any state of polarization can be described in terms of its vertical and horizontal phase and amplitude components.

Perfectly uniform and symmetrical circular polarization in every direction is not possible in practice. Circular polarization is a special case of the general elliptical polarization which characterizes practical antennas. Other special cases occur when the polarization ellipse degenerates into linear polarization of arbitrary orientation.

As yet, the FCC has not set any specifications on the permissible deviations from ideally perfect circular polarization.[19,20] Nevertheless, there are three FCC requirements for circular polarization transmission of television broadcasting.

1. The sense of rotation should be right hand (RHCP). This means that, as viewed from behind the antenna and in the direction of propagation, the electric field will rotate in a clockwise fashion.

2. The maximum ERP of the vertical polarization component shall not exceed the maximum ERP of the horizontal polarization component for both directional and omnidirectional antennas.

3. For directional antennas, the ERP of the vertical polarization component shall not exceed the ERP of the horizontal polarization component in *any* direction.

8.2.4 DIRECTIONAL ANTENNAS.   According to the FCC rules, an antenna which is *intentionally* designed or altered to produce a noncircular azimuthal radiation pattern is a directional antenna. When an antenna is filed with the FCC as directional, certain rules must be met. These rules are:

1. Minimum to maximum ratio of the field in the azimuthal plane shall not exceed − 10 dB for channels 2 to 13 and −15 dB for channels 14 to 83 with ERP of more than 1 kW.

2. The gain of a directional antenna is peak gain rather than the average (rms) gain of an omnidirectional antenna. The difference is discussed in Sec. 8.3.3.

3. If the directional antenna is also designed for circular polarization transmission, the ERP of the vertical polarization component must not exceed the ERP of the horizontal polarization component in any direction.

There are a variety of reasons for designing directional antennas. In some instances there are interference protection requirements. In other instances, the broadcaster may desire to improve service by diverting useful energy from an unpopulated area toward population centers. Generally speaking, directional antennas are more expensive than omnidirectional antennas.

While the FCC has not set minimum requirements for the azimuthal pattern of omnidirectional antennas, a maximum to minimum field deviation of 6 dB is considered acceptable. The deviations of most omnidirectional antennas are within ± 3 dB.

8.3 ELECTRICAL CONSIDERATIONS

The antenna system designer must focus on three key electrical considerations:

1. Adequate power handling capability. This is the subject of Sec. 8.3.1.

2. Adequate signal strength over the coverage area. This is analyzed in Secs. 8.3.2 to 8.3.5 and in 8.3.9 for antenna farms.

3. Distortion-free picture. This issue is discussed for all antennas in Secs. 8.3.6 to 8.3.8 and for multiple-antenna installation in Secs. 8.3.10 and 8.3.11.

There is sufficient information in these sections to allow the system designer to write specifications and to evaluate available hardware. The transformation of the shaped vertical pattern to field-strength curves requires the usage of Figs. 8-6 to 8-10.

8.3.1 POWER AND VOLTAGE RATING.   Television antennas are conservatively rated assuming a continuous black level. The nomenclature used in rating the antenna is "peak of sync TV power + 20% (or 10%) aural." The equivalent heating (average) power is 0.8 of the power rating if 20 percent aural power is used and 0.7 if 10 percent aural power is used. The equivalent heating power value is arrived at as follows:

	Carrier levels, %		Fraction of time, %	Average power, %
	Voltage	Power		
Sync	100	100	8	8
Blanking	75	56	92	52
Visual black-signal power				60
Aural power (percent of sync power)				20 (or 10)
Total transmitted power (percent peak-of-sync)				80 (or 70)

Thus an antenna power rating increases by 14 percent when the aural output power is reduced from 20 to 10 percent.

In the design of feed systems, the transmission lines must be derated from the manufacturer's catalog values (based on VSWR = 1.0) to allow for the expected VSWR under extraordinary circumstances such as ice and mechanical damage. The derating factor is

$$\left(\frac{1}{1 + |\Gamma|}\right)^2 = \left(\frac{\text{VSWR} + 1}{2\text{VSWR}}\right)^2 \tag{8-1}$$

where $|\Gamma|$ = expected reflection coefficient due to ice, etc. This derating factor is in addition to the derating required due to the normally existing VSWR in the antenna's feed system.

The manufacturer's power rating for feed system components is based on a fixed ambient temperature. This temperature is typically 40°C (104°F). If the expected ambient temperature is higher than the quoted value, a good rule of thumb is to lower the rating by the same percentage. Hence, the television power rating (including 20 percent aural) of the feed system is given by

$$P_{\text{TV}} \approx \frac{1}{0.8} P_{T/L} \left(\frac{T_{T/L}}{T}\right) \left(\frac{\text{VSWR} + 1}{2\text{VSWR}}\right)^2 \tag{8-2}$$

where $P_{T/L}$ = quoted average power for transmission line components with VSWR = 1.0

$T/T_{T/L}$ = ratio of expected to quoted ambient temperature
VSWR = worst-possible expected VSWR

Television antennas must also be designed to withstand voltage breakdown due to high instantaneous peak power both inside the feed system and on the external surface of the antenna. Improper air gaps or sharp edges on the antenna structure and insufficient safety factor could lead to arcing and blooming. The potential problem due to instantaneous peak power is aggravated when multiplexing two or more stations on the same antenna. In the latter case, if all stations have the same input power, the maximum possible instantaneous peak power is proportional to the number of the stations squared as derived below.

For a single channel, the maximum instantaneous voltage can occur when the visual and aural peak voltages are in phase. Thus, with 20 percent aural, the worst-case peak voltage is

$$V_{\text{peak}} = \sqrt{2Z_0 P_{\text{PS}}} + \sqrt{0.4 Z_0 P_{\text{PS}}} = 2.047 \sqrt{Z_0 P_{\text{PS}}} \tag{8-3}$$

where P_{PS} = peak of sync input power, Z_0 = characteristic impedance, and the equivalent peak power is

$$P_{\text{peak}} = \frac{V_{\text{peak}}^2}{2Z_0} = 2.047 P_{\text{PS}} \tag{8-4}$$

For N stations multiplexed on the same antenna, the equivalent peak voltage is

$$\frac{1}{\sqrt{2Z_0}} \, V_{\text{peak}} = 1.431\sqrt{P_{\text{PS}}} + 1.431\sqrt{P_{\text{PS}}} + \cdots = 1.431N\sqrt{P_{\text{PS}}} \qquad (8\text{-}5)$$

and the equivalent peak power is

$$P_{\text{peak}} = 2.047N^2 P_{\text{PS}} \qquad (8\text{-}6)$$

Experience has shown that the design peak power and the calculated peak power should be related by a certain safety factor. This safety factor is made of two multipliers. The first value, typically 3, is for the surfaces of pressurized components. This factor accounts for errors due to calculation and fabrication and/or design tolerances. The second value, typically 3, accounts for humidity, salt, and pollutants on the external surfaces. The required peak power capability is

$$\text{Safe peak power} = 2.047 \times F_s \times N^2 \times P_{\text{PS}} \qquad (8\text{-}7)$$

where $F_s = 3$ for internal pressurized surfaces and $F_s = 9$ for internal external surfaces.

8.3.2 POLARIZATION. The polarization of the electric field of television antennas was limited in the United States to horizontal polarization during the first 30 years of broadcasting. But in other parts of the world, both vertical and horizontal polarizations were allowed primarily to reduce co-channel and adjacent channel interference.[21]

The U.S. FCC modified the rules in the early seventies to include circularly polarized transmission for television broadcasting. By mid-1982, 50 television stations worldwide replaced their horizontally polarized antennas with circularly polarized antennas.

Although well-documented scientific tests comparing horizontally and circularly polarized transmissions are not available, the available experience validated by theoretical analysis suggests the continued growth of circularly polarized transmission for television broadcasting.

The expected advantages of circular polarization broadcasting are:

1. Improved coverage without affecting service grade contours. First, the signal-to-noise ratio is increased for sets with adjustable indoor dipole (rabbit ears) and monopole antennas. Second, the signal strength at the receiver is relatively constant when the receiving antenna height is changed or when the frequency is changed across the channel. The same applies to weather-beaten rooftop antennas.

2. Potential 3-dB improvement in signal-to-noise ratio. Since an equal amount of ERP is allowed in the vertical vector, this is possible with a circularly polarized (CP) receiving antenna. This will be of significance in the fringe area of coverage.

3. Improved picture clarity with circularly polarized receiving antennas. Structures close to the transmitting and receiving antennas can produce reflections. Long-delayed reflections cause *ghosting*. Short-delayed reflections reduce edge sharpness and degrade color purity. With a circularly polarized receiving antenna, most of the first-order reflections are filtered out, thus improving picture clarity.

Items (2) and (3) could provide significant advantage to cable systems and large apartment buildings with outdoor CP receiving antennas.

It should be pointed out that the investment required in a circularly polarized transmission facility is approximately twice that required for horizontal polarization, mostly because of the doubling of transmitter power. While doubling of antenna gain instead of transmitter power is possible, the coverage of high-gain antennas with narrow beamwidth may not be adequate within a 2- to 3-mi (3- to 5-km) radius of the support tower unless a proper shaping of the elevation pattern is specified and achievable.

Practical antennas do not transmit ideally perfect circular polarization in all directions. The figure of merit of CP antennas is the *axial ratio*. The axial ratio is not the ratio of the vertical to horizontal polarization field strength. The latter is called *polarization ratio*. Practical antennas produce elliptical polarization; that is, the magnitude of the propagating electric field prescribes an ellipse when viewed from either behind or in front of the wave. Every elliptically polarized wave can be broken into two circularly polarized waves of different magnitudes and sense of rotation. This is shown in Fig. 8-14. For television broadcasting, usually a right-hand (clockwise) rotation is specified when viewed from behind the outgoing wave, in the direction of propagation.

Referring to Fig. 8-14, the axial ratio of the elliptically polarized wave can be defined in terms of the axes of the polarization ellipse or in terms of the right-hand and left-hand components. Denoting the axial ratio as R, then

$$R = \frac{E_1}{E_2} = \frac{E_R + E_L}{E_R - E_L} \tag{8-8}$$

When evaluating the transfer of energy between two CP antennas, the important performance factors are the power-transfer coefficient and the rejection ratio of the unwanted signals (echoes) to the desired signal. Both factors can be analyzed using the coupling-coefficient factor between two antennas arbitrarily polarized.

For two antennas whose axial ratios are R_1 and R_2, the coupling coefficient is

$$f = \frac{1}{2}\left[1 \pm \frac{4R_1 R_2}{(1 + R_1^2)(1 + R_2^2)} + \frac{(1 - R_1^2)(1 - R_2^2)}{(1 + R_1^2)(1 + R_2^2)}\cos(2\alpha)\right] \tag{8-9}$$

where α = the angle between the major axes of the individual ellipses of the antennas.

The plus sign is used if the two antennas have the same sense of rotation (either both

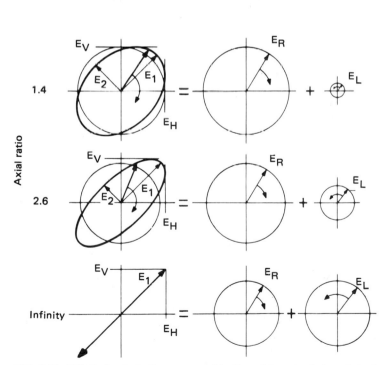

FIG. 8-14 Elliptical polarization as a combination of two circularly polarized signals.

right hand or left hand). The minus sign is used if the antennas have opposite senses of polarization.

Two special cases are of importance when coupling between two elliptically polarized antennas is considered. The first is when the major axes of the two ellipses are aligned ($\alpha = 0$). The second case is when the major axes are perpendicular to each other ($\alpha = \pm \pi/2$).

Case 1. $\alpha = 0$. The major axes of the polarization ellipses are aligned.

$$\text{Maximum power transfer} = f = \frac{(1 \pm R_1 R_2)^2}{(1 + R_1^2)(1 + R_2^2)} \tag{8-10}$$

$$\text{Minimum power rejection ratio} = \frac{f_-}{f_+} = \frac{(1 - R_1 R_2)^2}{(1 + R_1 R_2)^2} \tag{8-11}$$

Case 2. $\alpha = \pm \pi/2$. The major axes of the two polarization ellipses are perpendicular.

$$\text{Maximum power transfer} = f = \frac{(R_1 \pm R_2)^2}{(1 + R_1^2)(1 + R_2^2)} \tag{8-12}$$

$$\text{Minimum power rejection ratio} = \frac{f_-}{f_+} = \frac{(R_1 - R_2)^2}{(R_1 + R_2)^2} \tag{8-13}$$

The ability to reject unwanted reflections is of particular importance in television transmission. Since in many cases the first-order reflections have undergone a reversal of the sense of rotation of the polarization ellipse, the minimum rejection ratio is given by f_-/f_+. These are shown graphically in Figs. 8-15 and 8-16. The word *minimum* is used since other factors such as additional attenuation for the reflection path have not been considered.

8.3.3 GAIN.

The antenna gain is a figure of merit that describes the antenna's ability to radiate the input power within specified sectors in the directions of interest. Broadcast antenna gains are defined relative to the peak gain of a half-wavelength-long dipole.

The gain is one of the most critical figures of merit of the television broadcast

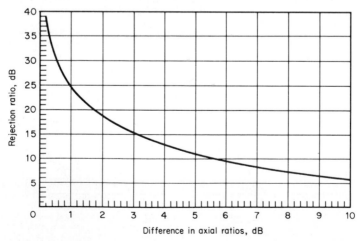

FIG. 8-15 Rejection ratio between two circularly polarized antennas whose major axes of polarization ellipses are perpendicular.

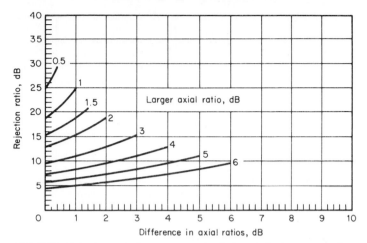

FIG. 8-16 Rejection ratio between two circularly polarized antennas whose major axes of polarization ellipses are parallel.

antenna. It determines the transmitter power required to achieve a given ERP (see Sec. 8.2.1). It is related to the beamwidth of the elevation pattern which in turn affects the coverage and sets limits on the allowable windsway. It is related to height (windload) of the antenna and to the noncircularity of the azimuthal pattern.

The gain of any antenna can be estimated quickly from its height and knowledge of its azimuthal pattern. It can be calculated precisely from measurements of the radiation patterns. It can be also measured directly, although this is rarely done for a variety of practical reasons. The gain of television antennas is always specified relative to a half-wavelength dipole. This practice differs from that used in nonbroadcast antennas.

Broadcast antenna gains are specified by elevation (vertical) gain, azimuthal (horizontal) gain, and total gain. The total gain is specified at either its peak or average (rms) value. In the United States, the FCC allows average values for omnidirectional antennas but requires the peak-gain specification for directional antennas. For circularly polarized antennas, the aforementioned terms also are specified separately for the vertically and horizontally polarized fields.

For an explanation of elevation gain, consider the superimposed elevation patterns in Fig. 8-17. The elevation pattern with the narrower beamwidth is obtained by distributing the total input power equally among 10 vertical dipoles stacked vertically 1 wavelength apart. The wider-beamwidth, lower peak-amplitude elevation pattern is obtained when the same input power is fed into a single vertical dipole. Note that at depression angles below 5°, the lower-gain antenna transmits a stronger signal. In the direction of the peak of the elevation patterns, the gain of the 10 dipoles relative to the single dipole is

$$g_e = \left(\frac{1.0}{0.316}\right)^2 = 10 \tag{8-14}$$

As can be seen from Fig. 8-17, the elevation gain is proportional to the vertical aperture of the antenna.

The theoretical upper limit of the elevation gain for practical antennas is given by

$$G_E = 1.22 \, \eta \, \frac{A}{\lambda} \tag{8-15}$$

where η is the feed system efficiency and A/λ is the vertical aperture of the antenna in wavelengths of the operating television channel. In practice, the elevation gain varies from $\eta \times 0.85 \, A/\lambda$ to $\eta \times 1.1 A/\lambda$.

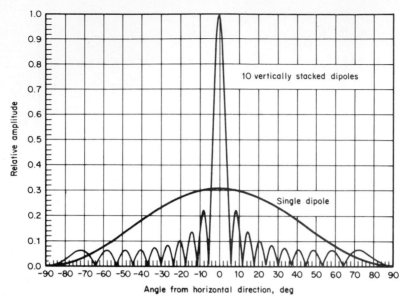

FIG. 8-17　Elevation pattern gain of two antennas of different aperture.

Azimuthal-Gain Principles. The concept of azimuthal gain can be understood by considering again two antennas. Antenna A is the same single vertical dipole as before, while antenna B is made of two single vertical dipoles spaced horizontally ¼ wavelength apart. The same input power is applied to either antenna, but in the case of antenna B the two dipoles split the power equally but in phase quadrature. As can be seen from Fig. 8-18, the directional antenna B has an azimuthal gain relative to the omnidirectional antenna A given by

$$g_a = \left(\frac{1.0}{0.8}\right)^2 = 1.56 \tag{8-16}$$

The total gain is specified as

$$G = g_e g_a \tag{8-17}$$

or

$$G = 10 \log g_e + 10 \log g_a \quad \text{dB} \tag{8-18}$$

The factoring of the total antenna gain into elevation and azimuthal components is valid only if the elevation pattern is the same in all azimuthal directions. This generally is the case for broadcast antennas.

Omnidirectional Antennas. In computing the gain of omnidirectional antennas, in the United States the FCC permits the total peak gain of the antenna to be reduced to its value as shown in Fig. 8-19. In that case, the total gain of the antenna becomes

$$G = g_e \times g_a \times (\text{rms})^2 \tag{8-19}$$

This is very advantageous to the broadcaster, since the allowable ERP is based on the rms-gain figure rather than peak gain. For the example shown in Fig. 8-19, the allowable ERP of 100 kW is met on average, while in some directions it is exceeded by 1.3 dB and in other directions it is reduced by 3.0 dB. Had the peak value been used to specify

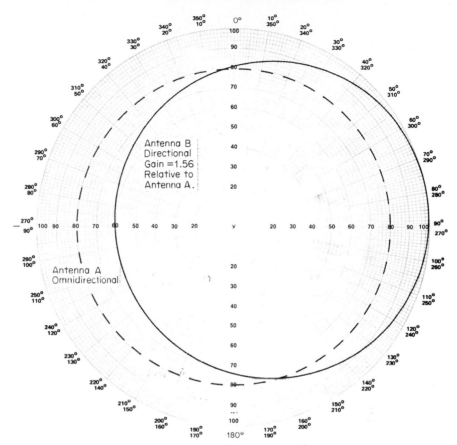

FIG. 8-18 Illustrating the concept of azimuthal gain of directional antennas.

the gain of the antenna in Fig. 8-19, the ERP at the minima of the azimuthal pattern would be down to 37 kW.

Circular Polarization. To clarify the factors contributing to the gain figure for circular polarized antennas, the derivation of the applicable expressions for the gain of each polarization and the total gain follow.

The total gain of a mathematical model of an antenna can be described as the sum of the individual gains of the field polarized in the vertical plane and the field polarized in the horizontal plane regardless of their ratio.

Starting with the standard definition of antenna gain G

$$G = 4\pi\eta \frac{|E_v|^2 + |E_h|^2}{1.64 \iint |E_v|^2 + |E_h|^2 \sin\theta \; d\theta \; d\phi} \tag{8-20}$$

where $|E_v|$, $|E_h|$ are the magnitudes of the electric fields of the vertical polarization and the horizontal polarization, respectively, in the direction of interest. This direction usually is the peak of the main beam. Now define

$$G_h = 4\pi\eta \frac{|E_h|^2}{1.64 \iint |E_h|^2 \sin\theta \; d\theta \; d\phi} \tag{8-21}$$

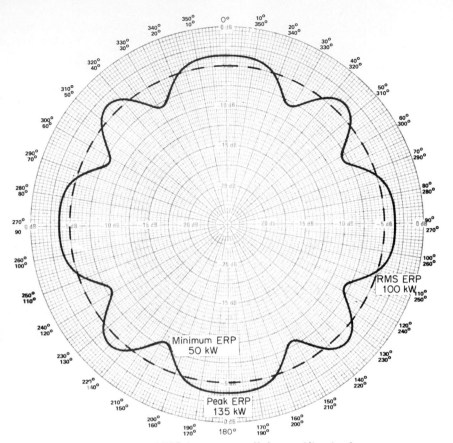

FIG. 8-19   Illustrating the actual ERP for an antenna filed as omnidirectional.

as the total (azimuthal and elevation) gain for the horizontal polarization component in the absence of vertical polarization, and let

$$G_v = 4\pi\eta \, \frac{|E_h|^2}{1.64 \iint |E_h|^2 \sin\theta \, d\theta \, d\phi} \tag{8-22}$$

be the total gain for the vertical polarization component in the absence of horizontal polarization. Then

$$\frac{G}{G_h} = \frac{1 + |E_v/E_h|^2}{1 + (G_h/G_v)\,|E_v/E_h|^2} = \frac{1 + |P|^2}{1 + (G_h/G_v)\,|P|^2} \tag{8-23}$$

and

$$\frac{G}{G_v} = \frac{1 + |E_h/E_v|^2}{1 + (G_v/G_h)\,|E_h/E_v|^2} = \frac{1 + |P|^2}{|P|^2 + G_v/G_h} \tag{8-24}$$

where $|P|$ is the magnitude of the polarization ratio. When the last two expressions are added and rearranged, the total gain G of the antenna is obtained as

$$G = \left[\, 1 + |P|^2 \,\right] \left[\, \frac{1}{(1/G_h) + |P|^2/G_v} \,\right] \tag{8-25}$$

The total gain can be broken into two components whose ratio is $|P|^2$.

$$\text{Gain of horizontal polarization} = \frac{G_h}{1 + (G_h/G_v)\,|P|^2} \qquad (8\text{-}26)$$

$$\text{Gain of vertical polarization} = \frac{G_v}{1 + (G_v/G_h)/|P|^2} \qquad (8\text{-}27)$$

The last three expressions specify completely any antenna provided G_h, G_v, and $|P|$ are known. From the definitions it can be seen that the first two can be obtained from measured-pattern integration and the magnitude of the polarization ratio is either known or can be measured.

When the antenna is designed for horizontal polarization, $|P| = 0$ and $G = G_h$. For circular polarization, $|P| = 1$ in all directions, $G_h = G_v$, and the gain of each polarization is half of the total antenna gain.

8.3.4 ELEVATION-PATTERN SHAPING.[22-25]

The elevation pattern of a vertically stacked antenna array of radiators can be computed from the *illumination* or *input currents* to each radiator of the array and the elevation pattern of the single radiator. Mutual coupling effects generally can be ignored when the spacing between the adjacent radiators is approximately a wavelength, a standard practice in most broadcast antenna designs. The elevation pattern $E(\theta)$ of an antenna consisting of N vertically stacked radiators, as a function of the depression angle θ, is given by

$$E(\theta) = \sum_{i=1}^{N} A_i P_i\,(\theta)\,\exp\,(j\phi)\,\exp\,(j\,\frac{2\pi}{\lambda}\,d_i \sin\,\theta) \qquad (8\text{-}28)$$

where $P_i(\theta)$ = vertical pattern of ith panel
$\quad A_i$ = amplitude of current in ith panel
$\quad \theta_i$ = phase of current in ith panel
$\quad d_i$ = location of ith panel

In television applications, only the normalized magnitude of the pattern is of interest.

For an array consisting of N identical radiators, spaced uniformly apart (d), and carrying identical currents, the magnitude of the elevation, or vertical, radiation pattern is given by

$$|E(\theta)| = \left| \frac{\sin\,[(N\pi/\lambda)\,d \sin\,\theta]}{\sin\,[(\pi d/\lambda) \sin\,\theta]} \right|\,|p(\theta)| \qquad (8\text{-}29)$$

where the first part of the expression on the right is commonly termed the *array factor*. The *elevation pattern* of a panel antenna comprising six radiators, each 6 wavelengths long, is given in Fig. 8-20. The elevation-pattern power gain g_e of such an antenna can be determined by integrating the power pattern and is given by the expression

$$g_e \simeq \frac{Nd}{\lambda} \qquad (8\text{-}30)$$

Thus, for an antenna with N half-wave dipoles spaced 1 wavelength apart, the gain is essentially equal to the number of radiators N. In practice, slightly higher gain can be achieved by the use of radiators which have a gain greater than that of a half-wave dipole.

The *array factor* becomes zero whenever the numerator becomes zero, except at $\theta = 0$ when its value equals 1. The nulls of the pattern can be easily determined from the equation

$$\frac{N\pi d}{\lambda}\,\sin\,\theta_m = m \qquad (8\text{-}31)$$

or
$$\theta_m = \sin^{-1}\frac{m}{g_e} \qquad (8\text{-}32)$$

where $m = 1, 2, 3 \ldots$ refers to the null number and θ_m = depression angle at which a null is expected in radians.

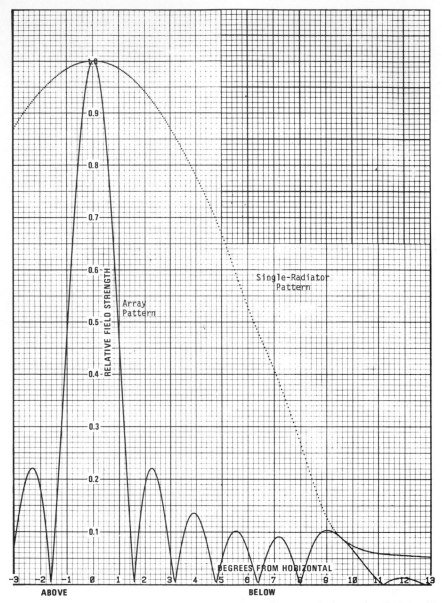

FIG. 8-20 Elevation pattern of an antenna array of six radiators, each six wavelengths long. All radiators feed equally and in-phase.

The approximate beamwidth corresponding to 70.7 percent value of the field (50 percent of power), or 3 dB below the maximum, can be determined from the **array factor** and is given by the expression

$$\text{Min beamwidth} = \frac{50.8}{N(d/\lambda)} = \frac{50.8}{g_e} \quad \text{deg} \qquad (8\text{-}33)$$

It is interesting to note that the gain-beamwidth product is essentially a constant. The gain-beamwidth product of practical antennas varies from 50.8 to 68, depending on the final shaping of the elevation pattern.

In television broadcasting, the desired gain of an omnidirectional antenna generally is determined by the maximum-allowable effective radiated power and the transmitter power output to the antenna. Provision for adequate signal strength in the coverage area of interest requires synthesis of the antenna pattern to ensure that the nulls of the pattern are filled to an acceptable value. In addition, in order to improve the transmission efficiency, the main beam is tilted down to reduce the amount of unusable radiated power directed above the horizontal.

Although the previous discussion has been concerned with arrays of discrete elements, the concept can be generalized to include antennas which are many wavelengths long in aperture and have a continuous illumination. In such cases, the classic techniques of null filling by judicious antenna aperture current distribution, such as cosine-shaped or exponential illuminations, are employed. The elevation patterns for several distributions of aperture illuminations are shown in Figs. 8-21 through 8-25.

Elevation-Pattern Derivation. The signal strength at any distance from an antenna is directly related to the maximum ERP of the antenna and the value of the antenna elevation pattern toward that depression angle (see Sec. 8.2.2, "Coverage"). For a signal strength of 100 mV/m, which is considered adequate and assuming the FCC (50,50) propagation curves, one can derive the desired elevation pattern of an antenna. Typical 100 V/m curves are shown in Figs. 8-26 to 8-28 for low-band VHF (United States channels 2 to 6), high-band VHF (United States channels 7 to 13), and UHF band (United States channels 14 to 20), respectively.

These curves provide a basis for antenna selection and as such require further discussion. Only the region of the curve for which the maximum relative field is unity and the depression angle corresponding to 1 mi or 16° is shown in the above curves. Any antenna pattern which lies to the right of these curves will thus provide a signal strength in excess of 100 mV/m, as determined from the U.S. FCC data. Since the relative field pattern can never exceed unity, the signal strength will be lower than 100 mV/m for depression angles between the horizon and the values corresponding to unity in the above curves. The angle to the horizon is to the left of the curve. The distance from the antenna for any depression angle can be determined from Table 8-5, for any antenna height. Hence an antenna can be selected with a gain such that the beamwidth of the antenna's elevation pattern is wide enough to reach over the region to the right. If the antenna beamwidth is narrow, the beam tilt can be increased to reach over the right of the curve.

Beam Tilting. Any specified tilting of the beam below the horizontal is easily achieved by progressive phase shifting of the currents in each panel in an amount equal to $-2(\pi d_i/\lambda) \sin \theta_T$, where θ_T is the required amount of beam tilt in degrees. Minor corrections, necessary to account for the panel pattern, can be determined easily, by iteration.

In some cases, a progressive phase shifting of individual radiators may not be cost effective, and several sections of the array may have the illuminating current of the same phase, thus reducing the number of different phasing lengths. The correct value of the phase angle for each section can be computed.

When the specified beam tilt is comparable with the beam width of the elevation pattern, the steering of the array reduces the gain from the equivalent array without any beam tilt. To improve the antenna gain, mechanical tilt of the individual panels, as well as the entire antenna, is resorted to in some cases. However, mechanically tilting the entire antenna results in variable beam tilt around the azimuth.

Null-Fill. Another criterion is the null-fill requirement. If the antenna is near the center of the coverage area, depending on the minimum gain, the nulls in the coverage area must be filled in to provide a pattern that lies above the 100-mV/m curve for that particular height. For low-gain antennas, this problem is not severe, especially when the

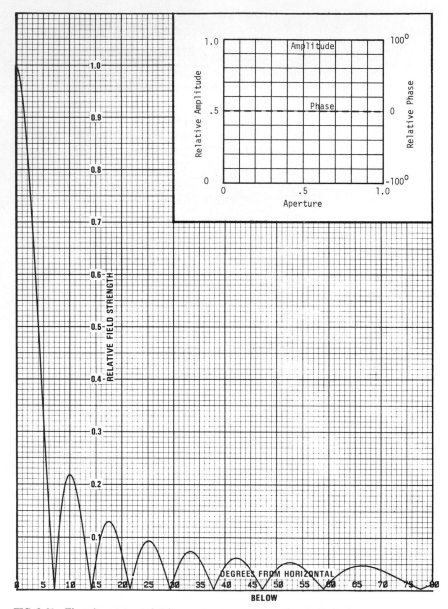

FIG. 8-21   Elevation pattern of eight-wavelength-aperture uniform distribution.

antenna height is lower than 2000 ft (610 m) and only the first null has to be filled. But in the case of UHF antennas, with gains greater than 20, the nulls occur close to the main beam and at least two nulls must be filled.

When the nulls of a pattern are filled, the gain of the antenna is reduced. A typical gain loss of 1 to 2 dB generally is encountered.

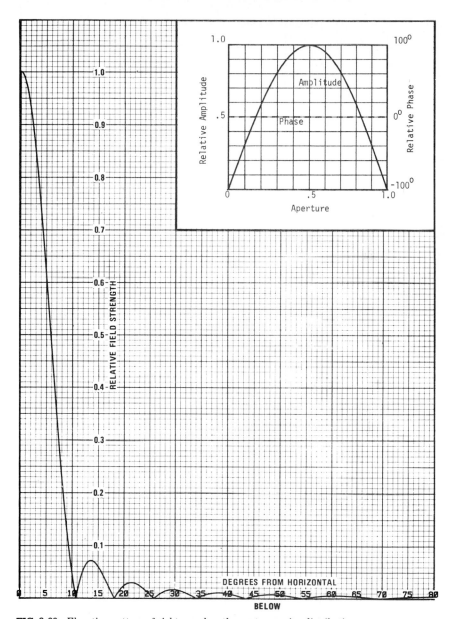

FIG. 8-22   Elevation pattern of eight-wavelength-aperture cosine distribution.

The variables for pattern null-filling are the *spacing* of the radiators and the *amplitudes and phases* of the feed currents. The spacings generally are chosen to provide the minimum number of radiators necessary to achieve the required gain. Hence, the only variables are the $2(N-1)$ relative amplitudes and phases.

The distance from the antenna to the null can be approximated if the *gain* and the

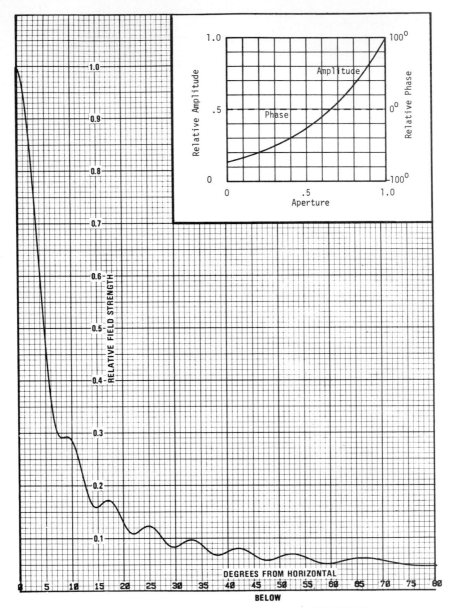

FIG. 8-23   Elevation pattern of eight-wavelength-aperture exponential distribution.

height of the antenna above average terrain are known. Since the distance at any depression angle θ can be approximated as

$$D = 0.0109 \frac{H}{\theta} \qquad (8\text{-}34)$$

with H = antenna height in feet and the depression angle of the mth null is

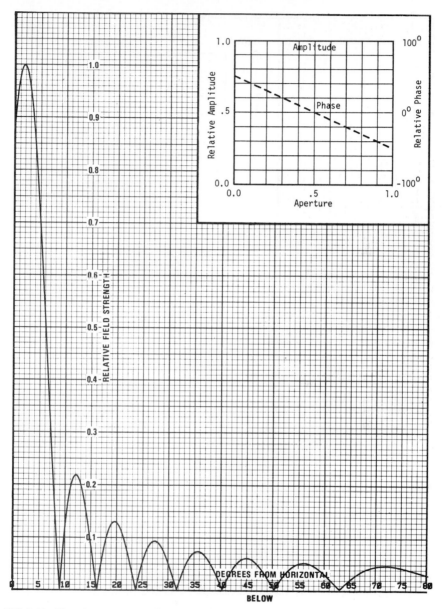

FIG. 8-24 Elevation pattern of eight-wavelength-aperture linear phase distribution.

$$\theta_m = 57.3 \sin^{-1}\left(\frac{m}{g_e}\right) \qquad g_e = \text{elevation power gain of antenna} \qquad (8\text{-}35)$$

then

$$D_m = 0.00019\,\frac{Hg_e}{m}\;\text{miles} \qquad \text{for}\; \sin\frac{m}{g_e} \approx \frac{m}{g_e} \qquad (8\text{-}36)$$

is the distance from the antenna to the mth hull.

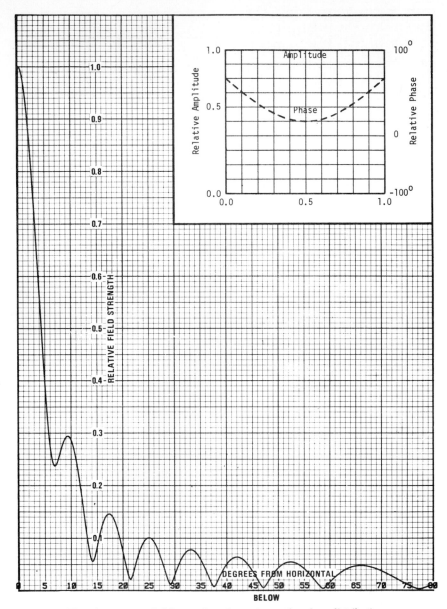

FIG. 8-25 Elevation pattern of eight-wavelength-aperture cosine phase distribution.

A simple method of null-filling is by power division among the vertically stacked radiators. For example, a 70:30 power division between each half of the array produces a 13 percent fill of the first null. Power division usually is employed where only the first null is to be filled. A typical pattern is shown in Fig. 8-29.

Another approach to null-filling by changing only the phases of the currents is very

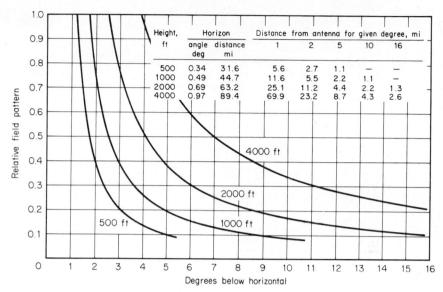

FIG. 8-26 Relative field for 100 mV/m ERP = 100 kW channels 2 to 6 [based on U.S. FCC (50,50) data].

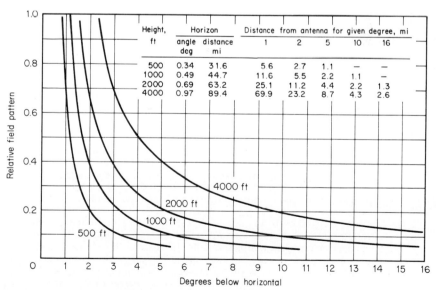

FIG. 8-27 Relative field for 100 mV/m ERP = 316 kW channels 7 to 13 [based on U.S. FCC (50,50) data].

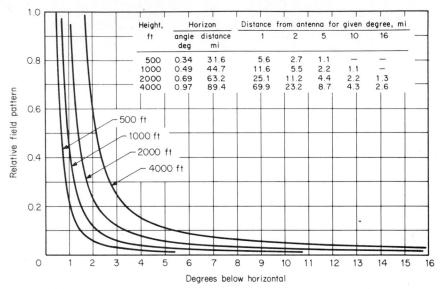

Height, ft	Horizon angle deg	Horizon distance mi	Distance from antenna for given degree, mi				
			1	2	5	10	16
500	0.34	31.6	5.6	2.7	1.1	—	—
1000	0.49	44.7	11.6	5.5	2.2	1.1	—
2000	0.69	63.2	25.1	11.2	4.4	2.2	1.3
4000	0.97	89.4	69.9	23.2	8.7	4.3	2.6

FIG. 8-28   Relative field for 100 mV/m ERP = 5000 kW channels 14 to 70 [based on U.S. FCC (50,50) data].

useful since the input power rating of the antenna is maximized when the magnitude of the current to each radiator in the array is adjusted to its maximum value. For example, if the bottom and the top layers of an N-layer array differ in phase by θ from the rest, the first $(N/2) - 1$ nulls are filled, as shown in Fig. 8-30.

In practice, the elevation pattern is synthesized, taking into consideration all the design constraints, such as power-rating of the individual radiators and the restrictions imposed on the feed system due to space, access, etc. Beam tilting is achieved by progressive or discrete phasing of sections of the antenna. The pattern in Fig. 8-31 illustrates the final design of the array pattern of a high-gain antenna, determined by a computer-aided iteration technique.

8.3.5 AZIMUTHAL-PATTERN SHAPING.

For omnidirectional antennas, a circular azimuthal pattern is desired. However, in practice, the typical circularity may differ from the ideal by ± 2 dB. The omnidirectional pattern is formed by the use of several radiators arranged within a circle having the smallest-possible diameter. If a single radiator pattern can be idealized for a sector, several such radiators can produce truly circular patterns (see Fig. 8-32). Practical single-radiator-element patterns do not have sharp transitions, and the resultant azimuthal pattern is not a perfect circle. Furthermore, the interradiator spacing becomes important, since for a given azimuth the signals from all the radiators add vectorially. The resultant vector, which depends on the space-phase difference of the individual vectors, varies with azimuth, and the circularity deteriorates with increased spacing.

In the foregoing discussion, it was assumed that the radiators around the circular periphery were fed in-phase. Similar omnidirectional patterns can be obtained with radial-firing panel antennas when the panels differ in phase uniformly around the azimuth wherein the total phase change is a multiple of 360°. The panels then are offset mechanically from the center lines as shown in Fig. 8-33. The offset is approximately 0.19 wavelength for *triangular-support* towers and 0.18 wavelength for *square-support* towers. The essential advantage of such a phase rotation is that, when the feedlines from all the radiators are combined into a common junction box, the first-order reflections

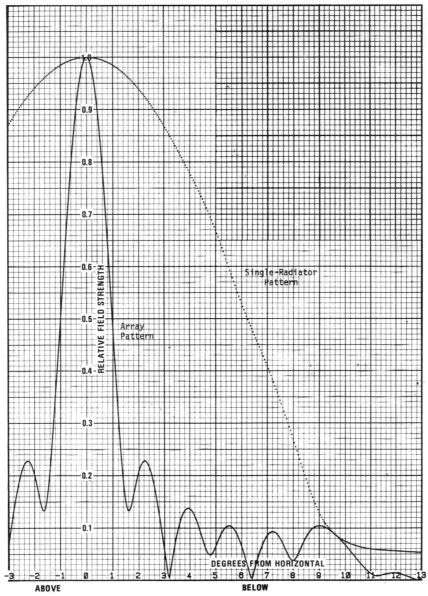

FIG. 8-29 Elevation pattern of an antenna array of six radiators, each six wavelengths long. Null-fill by power division (70 percent to top three and 30 percent to lower three).

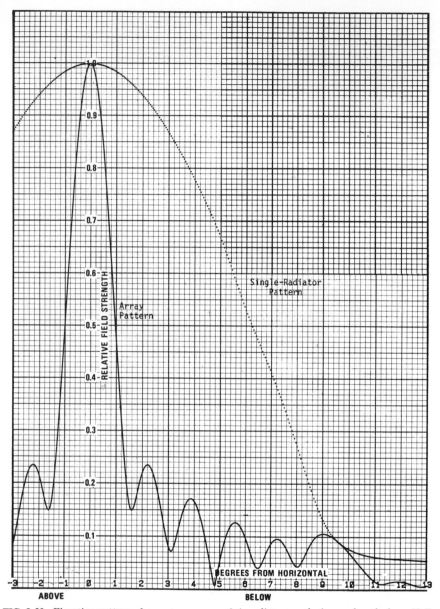

FIG. 8-30 Elevation pattern of an antenna array of six radiators, each six wavelengths long. Null-fill by phasing (top and bottom, 30°; others, 0°).

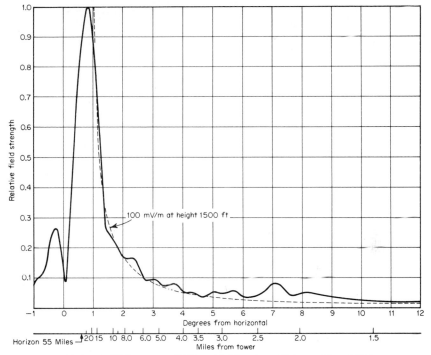

100 mV/m at height 1500 ft

Horizon 55 Miles

FIG. 8-31 High-gain UHF antenna (ERP = 5 MW; gain = 60, beam tilt = 0.75°).

from the panels tend to cancel out at the input to the junction box, resulting in a considerable improvement in the input impedance.

The azimuthal pattern of a panel antenna is a cosine function of the azimuthal angle in the front of the radiator, and its back lobe is small. The half-voltage width of the frontal main lobe ranges from 90° for square-tower applications to 120° for triangular towers. The panels are affixed on the tower faces. Generally, a 4-ft-wide tower face is small enough for United States channels 7 to 13. For channels 2 to 6, the tower-face size could be as large as 10 ft (3 m). See Table 8.3 for typical circularities.

In all the above cases, the circular omnidirectional pattern of panel antennas is achieved by aligning the main beam of the panels along the radials. This is the *radial-fire* mode. Another technique utilized in the case of towers with a large cross section (in wavelengths) is the *tangential-fire* mode. The panels are mounted in a skewed fashion around a triangular or square tower, as shown in Fig. 8-34. The main beam of the panel is directed along the tangent of the circumscribed circle as indicated in Fig. 8-34. The optimum interpanel spacing is an integer number of wavelengths when the panels are fed in-phase. When a rotating-phase feed is employed, correction is introduced by modifying the offset, as, for example, by adding ⅓ wavelength when the rotating phase is at 120°. The table of Fig. 8-34 provides the theoretical circularities for ideal elements. Optimization is achieved in practice by model measurements to account for the back lobes and the effect of the tower members.

A measured pattern of such a tangential-fire array is shown in Fig. 8-35.

In the case of directional antennas, the desired pattern is obtained by choosing the proper input-power division to the panels of each layer, adjusting the phase of the input signal, and/or mechanically tilting the panels. In practice, the azimuthal-pattern synthesis of panel antennas is done by superposition of the azimuthal phase and amplitude

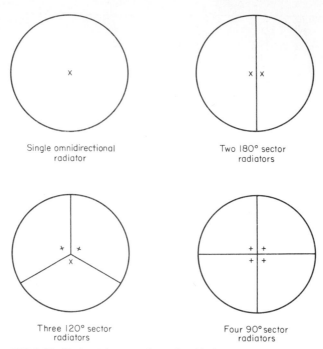

FIG. 8-32   Horizontal pattern formation, ideal case.

patterns of each radiator, while adjusting the amplitudes and phases of the feed currents until a suitable configuration of the panels on the tower yields the desired azimuthal pattern.

The turnstile antenna (see Sec. 8.1.1) is a top-mounted omnidirectional pole antenna. Utilizing a phase rotation of four 90° steps, the crossed pairs of radiators act as two dipoles in quadrature, resulting in a fairly smooth pattern. The circularity improves with decreasing diameter of the turnstile support pole.

The turnstile antenna can be directionalized. As an example, a peanut-shaped azimuthal pattern can be synthesized, either by power division between the pairs of radiators or by introducing proper phasing between the pairs. The pattern obtained by phasing the pairs of radiators by 70° instead of the 90° used for an omnidirectional pattern is shown in Fig. 8-36.

A directional pattern of a panel antenna with unequal power division is shown in Fig. 8-37. The panels are offset to compensate for the rotating phase employed to improve the input impedance of the antenna.

8.3.6 VOLTAGE STANDING-WAVE RATIO AND IMPEDANCE.   The transmission line connecting the transmitter to the antenna is never fully transparent to the incoming or *incident* wave. Due to imperfections in manufacture and installation, some of the power in the incident wave can be reflected at a number of points in the line. Additional reflection occurs at the line-to-antenna interface, since the antenna per se presents an imperfect match to the incident wave. The reflections set up a *reflected wave* which combines with the incident wave to form a *standing* wave inside the line. The characteristic of the standing-wave pattern is periodic maximums and minimums of the voltage and current along the line. The ratio of the maximum to minimum at any plane is called the *voltage standing-wave ratio* (VSWR). Since the VSWR is varying along the transmission line, the reference plane for the VSWR measurement must be defined. When the reference plane is at the gas stop input in the transmitter room, the measured value is *system*

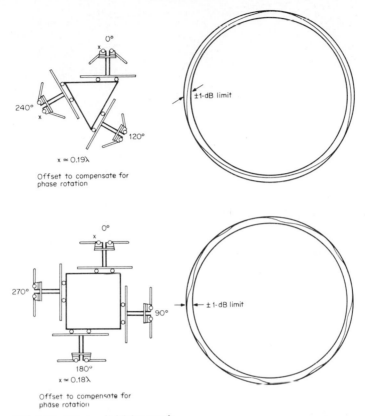

FIG. 8-33 Offset radial firing panels.

VSWR. When the reference plane is at the input to the antenna on the tower, the measured value is *antenna VSWR*. The system VSWR differs from the antenna VSWR owing to the introduction of standing waves by the transmission line. When two sources of standing waves S_1 and S_2 exist, then

$$\text{Maximum of VSWR} = S_1 S_2$$

$$\text{Minimum of VSWR} = \frac{S_2}{S_1} \quad \text{for } S_1 < S_2$$

More generally, the expected system VSWR for n such reflections is

$$\text{Maximum VSWR} = S_1 S_2 S_3 \cdots S_n$$

$$\text{Minimum VSWR} = \frac{S_n}{S_1 S_2 \cdots S_{n-1}} \quad \text{for } S_1 < S_2 \cdots S_n$$

If the calculated minimum VSWR is less than 1.00, then the minimum VSWR = 1.00.

As an example, consider an antenna with an input VSWR of 1.05 at visual and 1.10 at aural carrier frequencies. If the transmission line per se has a VSWR of 1.05 at the visual and 1.02 at aural carriers, the *system* VSWR will be between

$$1.000 = \frac{1.05}{1.05} \leq S_{\text{vis}} \leq 1.05 \times 1.05 = 1.103 \tag{8-37}$$

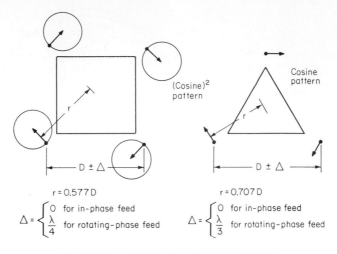

r = 0.577 D

$$\Delta = \begin{cases} 0 & \text{for in-phase feed} \\ \dfrac{\lambda}{4} & \text{for rotating-phase feed} \end{cases}$$

r = 0.707 D

$$\Delta = \begin{cases} 0 & \text{for in-phase feed} \\ \dfrac{\lambda}{3} & \text{for rotating-phase feed} \end{cases}$$

Square configuration		Triangular configuration	
D/λ	Circularity, $\pm$ dB	D/λ	Circularity, $\pm$ dB
1	0.09	1	0.61
2	0.33	2	0.70
3	0.63	3	0.98
4	0.93	4	1.23
5	1.24	5	1.53
6	1.50	6	1.83

FIG. 8-34 Tangential-fire mode. *Note:* With back lobes and tower reflections, the circularities tend to be worse by about 2 dB more, especially for large tower sizes.

$$1.078 = \frac{1.10}{1.02} \leq S_{\text{aural}} \leq 1.10 \times 1.02 = 1.122 \qquad (8\text{-}38)$$

The VSWR due to any reflection is defined as

$$S = \frac{1 + |\Gamma|}{1 - |\Gamma|} \qquad (8\text{-}39)$$

where $|\Gamma|$ is the magnitude of the reflection coefficient at that frequency.

For example, if 2.5 percent of the incident voltage is reflected at the visual carrier frequency, the VSWR at that frequency is

$$S = \frac{1 + 0.025}{1 - 0.025} = 1.05$$

A high value of VSWR is undesirable since it contributes to

1. Visible ghosts if the source of the VSWR is more than 250 ft (76 m) away from the transmitter
2. Short-term echoes ($< 0.1\text{-}\mu\text{s}$ delay)
3. Aural distortion
4. Reduction of the transmission line efficiency

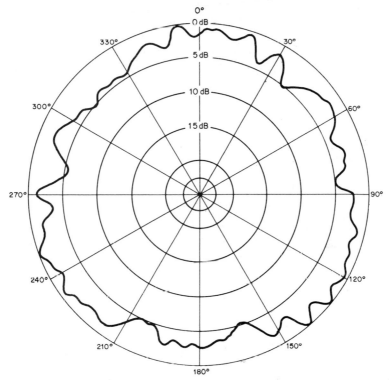

FIG. 8-35 Measured azimuthal pattern of a "tangential-fire" array of three panels fed in-phase around a triangular tower with $D = 7.13$ wavelengths.

Of all the undesirable effects of the system VSWR, the first, a visible ghost due to input VSWR, is the most critical. The further the antenna is from the transmitter, the higher the subjective impairment of the picture, due to the reflection at the input of the antenna, will be. This effect is illustrated in Fig. 8-40.

Antenna input specification in terms of VSWR is not an effective figure of merit to obtain the best picture quality. For example, Fig. 8-38 shows a comparative performance of two antennas. Antenna A has a maximum VSWR of 1.08 across the channel and a VSWR of 1.06 at picture (pix) carrier. Antenna B has a maximum VSWR of 1.13 and a VSWR of 1.01 at pix. It is hard to tell anything about the relative picture impairment of these two antennas by inspecting the VSWR alone. However, the reflection of a 2-T sine-squared pulse by each antenna is also shown in the same figure. It can be seen that antenna A, with maximum VSWR of 1.08, produces a reflection of more than 3 percent. This reflection results in a ghost which could be perceptible if the transmission line to the antenna is at least 600 ft long as shown in Fig. 8-40. Antenna B with a maximum VSWR of 1.13 produces only 1 percent reflection for the same pulse.

The pulse response of an antenna mounted on top of a tower can be measured if the transmission line is sufficiently long and "clean" to resolve the incident from the reflected pulse. If the line is not sufficiently long or if knowledge of the antenna's pulse response is required prior to installation, a calculation is possible. Pulse response cannot be calculated from the VSWR data alone since it contains no information with respect to the phase of the reflection at each frequency across the channel. Since the impedance representation contains the amplitude and phase of the reflection coefficient, the pulse response can be calculated if the impedance is known. The impedance representation is

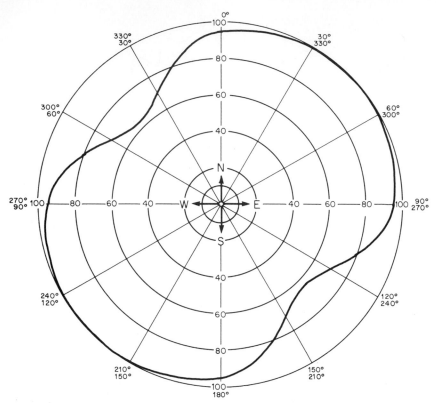

FIG. 8-36 Directionalizing the superturnstile antenna by phasing.

typically done on a Smith chart which is shown in Fig. 8-38 for antennas A and B. Some attempts at relating various shapes of VSWR curves to subjective picture quality have been made. A good rule of thumb is to minimize the VSWR in the region from -0.25 to $+0.75$ MHz of the visual carrier frequencies to a level below 1.05 and not to exceed the level of 1.08 at color subcarrier.

8.3.7 VSWR OPTIMIZATION TECHNIQUES. As noted in previous sections, the shape of the antenna VSWR across the channel spectrum must be optimized to minimize the subjective picture impairment. Frequently it may be desirable to perform the same VSWR optimization on the transmission line, so that the entire system appears transparent to the incoming wave. The optimization of the transmission line VSWR is a relatively time-consuming and expensive task, since it requires laying out the entire length of line on the ground and slugging it at various points. *Slugging* describes a technique of placing a metallic sleeve of a certain diameter and length which is soldered over the inner conductor of a coaxial transmission line. In some instances it is more convenient to use a section of line with movable capacitive probes instead of a slug. This section is usually called variously a *variable transformer, fine tuner,* or *impedance matcher.*

The single-slug technique is the simplest approach to VSWR minimization. At any frequency, if the VSWR in the line is known, the relationship between the slug length in wavelengths and its characteristic impedance is given by

$$\frac{L}{\lambda} = \frac{1}{2\pi} \tan^{-1} \left[\frac{S - 1}{\sqrt{(S - R^2)\left[(1/R^2) - S\right]}} \right] \tag{8-40}$$

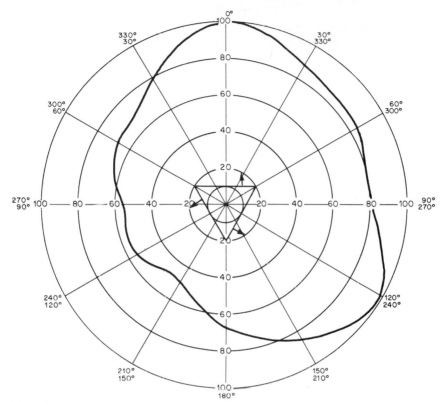

FIG. 8-37 Directional panel antenna by P/D and offsetting.

where S = existing VSWR

$$R = \frac{\text{slug characteristic impedance}}{\text{line characteristic impedance}}$$

L/λ = length of slug

A graphic representation of this expression is given in Fig. 8-39. The effect of the fringe capacitance, due to the ends of the slug, is not included in the design chart because it is negligible for all channels. Once the characteristic impedance of the slug is known, its diameter is determined from

$$Z_c = 138 \log_{10} \frac{D}{d} \qquad (8\text{-}41)$$

where D = the outside diameter of the slug conductor and d = the outside diameter of the inner conductor. The slug thus constructed is slowly slid over the inner conductor until the VSWR disappears. This occurs within a line length of ½ wavelength.

There are two shortcomings in the single-slug technique:

1. Access to the inner conductor is required.

2. The technique is not applicable if the VSWR at two frequencies must be minimized.

The first shortcoming can be eliminated by using the variable transformer mentioned earlier. While it is more expensive, slug machining and installation sliding adjustment

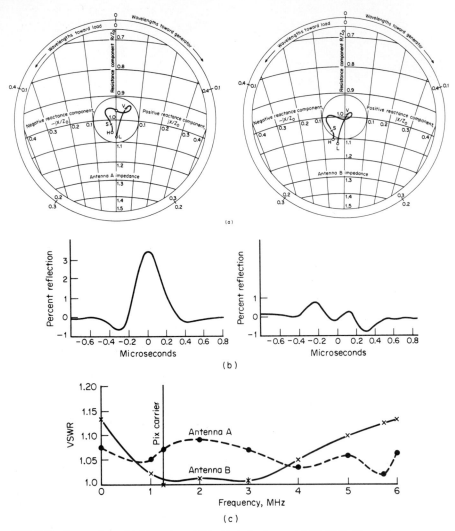

FIG. 8-38   Comparison of (a) impedance, (b) reflected pulse, and (c) VSWR of two antennas.

are avoided. The second shortcoming can be overcome with the double-slug technique, or more conveniently with two variable transformers.

To perform a diplexed-frequencies VSWR minimization, a first variable transformer is tuned such that the VSWR at both frequencies f_1 and f_2 is simultaneously equal and minimum. A third frequency which will be denoted f_0 is then assigned to that frequency which is closest to the midrange between f_1 and f_2 and simultaneously measures a VSWR of 1.00. If f_0 cannot be found, the initial adjustment of the variable transformer, as described above, was not performed adequately. A second variable transformer then is inserted down the line (toward the transmitter) at a distance of

$$L = \frac{0.25\lambda_0 f_0}{f_2 - f_1} \tag{8-42}$$

and is adjusted until the VSWR at f_1 and f_2 is minimum.

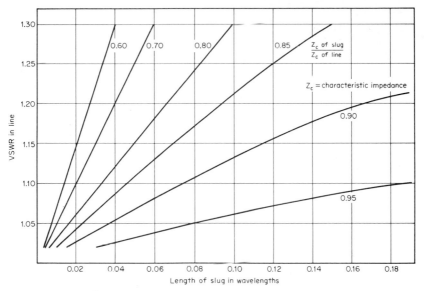

FIG. 8-39 Single-slug matching of coaxial line.

The single- and double-slug techniques may have an undesirable effect if not properly done. The slugs should be placed as near as possible to the source of the undesirable VSWR. Failure to do so could lead to higher VSWRs at other frequencies across the bandwidth of importance. Thus, if both the system and the antenna VSWRs are high at the visual carrier, slugging at the transmitter room will lower the system VSWR but will not eliminate the undesirable echo. Another effect of slugging is the potential alteration of the power and voltage rating. This is of particular importance if the undesirable VSWR is high. Generally, the slugging should be limited to situations where the VSWR does not exceed 1.25.

8.3.8 LONG-DELAYED ECHOES. Long-delayed echoes are defined here as reflections which are delayed at least 0.25 μs with respect to the main signal. This is equivalent to a round-trip path between two points separated by 125 ft (38 m). For shorter delays it is difficult to subjectively resolve the echo as a *ghost*.

The subjective impairment of picture quality by long-delayed echoes depends on both the magnitude and the delay of the reflections much like the K factor.[26–29] The K factor, however, remains constant for delays beyond 1 μs. Subjective tests have shown that the impairment is somewhat worse for delays greater than 1 μs for the same magnitude of echo. This is shown in Fig. 8-40 where the 2 and 3 percent K factors are superposed on the echo-perceptibility curves. Generally speaking, an echo of 2 percent is just perceptible provided it is delayed at least 1 μs. In other words, the signal-to-echo path difference is 984 ft (300 m). On a 19-in (48-cm) television screen, such an echo would be displaced 0.29 in (0.74 cm), assuming no overscan.

There are two primary sources of echoes in broadcast antenna systems. The first is the echo which originates within the transmission line run to the antenna input. This echo could be a result of multiple discontinuities in the transmission line and improper match between the antenna input and the transmission line. Since the phase velocity of the electromagnetic wave is

$$V = 983.6 \text{ ft/μs } (300 \text{ m/μs})$$

the distance between any discontinuity on the transmission line and the transmitter is

$$d \text{ (ft)} = 492 \times \text{delay (μs)}$$

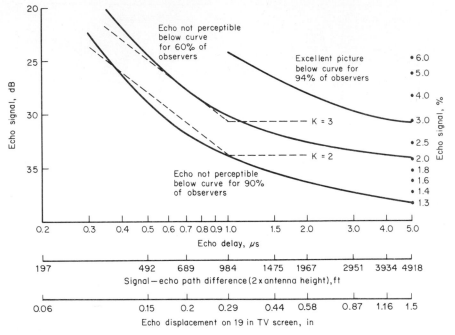

FIG. 8-40   Subjective echo visibility.

assuming the delay is measured on the radiated signal. The same distance can be deduced from measuring the displacement of the echo on the picture tube screen as follows

$$d \text{ (ft)} = 25{,}770 \frac{\text{displacement of echo (in)}}{\text{width of screen (in)}} \tag{8-43}$$

The second primary source of echoes is the proximity effect of nearby interfering structures. These structures could be other antennas and their support towers. In most instances only reflections from the surfaces of the interfering structures need be considered. However, if the interfering structure is a co-channel or adjacent-channel antenna, some of the intercepted signal may enter the feed system and thereafter be reradiated as an additional ghost to that produced by the reflection from the external surfaces.

In considering the echoes due to proximity effects, two geometries are of interest. These geometries are shown in Fig. 8-41. The geometry on the left is useful for the analysis of echoes around the azimuth of multiple-antenna installations. The geometry on the right is useful for the analysis of echoes due to specular reflection where the angle of incidence is equal to the angle of reflection.

Where more than one reflection exists and the reflections are not separable in time, a good rule of thumb for the equivalent echo amplitude is

$$K = (K_1^{3/2} + K_2^{3/2} + \cdots)^{2/3} \tag{8-44}$$

As noted earlier, if multiple antennas are located less than 125 ft apart, the echoes due to reflections and scattering alone are too close to be resolved by 525-line and 625-line systems. However, a visible ghost from such a system is possible if a portion of the intercepted signal enters the feed system and is reradiated by the interfering antenna with a greater time delay.[30,31] It is of interest to note that for a high-definition television (HDTV) system, close-in echoes may be highly visible.

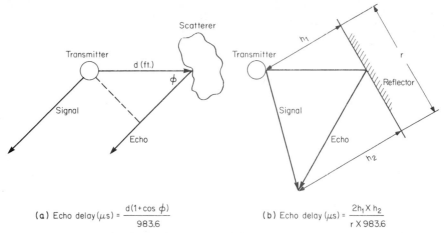

(a) Echo delay $(\mu s) = \dfrac{d(1+\cos \phi)}{983.6}$

(b) Echo delay $(\mu s) = \dfrac{2h_1 \times h_2}{r \times 983.6}$

FIG. 8-41 Geometries for delay-echo analysis. (a) Receiver far from reflector. (b) Receiver near reflector.

8.3.9 AZIMUTHAL-PATTERN DISTORTION IN MULTIPLE-ANTENNA INSTAL-LATIONS.

In a stacked arrangement, the cross sections of the support structure of the antenna on the lower levels are large in order to support the antenna above. Hence, the circularity of the azimuthal pattern of the lower antennas will not be as uniform as for the upper antennas where the support structure is slimmer.

In the case of a candelabra arrangement, the centers of radiation of most antennas are approximately equal. However, the radiated signal from each antenna is scattered by the other opposing antennas and, owing to the reflected signal, there results a deterioration of azimuthal-pattern circularity and video-signal performance criteria. When the proximity of one antenna to others is equal to its height or less, the shape of its elevation pattern at the interfering antennas is essentially the same as its aperture illumination. Consequently the reflections of significance are from the sections of the interfering structures parallel to the aperture of the radiating antenna. In this case, a two-dimensional analysis of the scatter pattern can be utilized to estimate the reflected signal and its effect on the free-space azimuthal pattern.

When the opposing structure is truly cylindrical, as for example a pylon antenna or a steel cylinder supporting an antenna above, a theoretical two-dimensional analysis can be used to compute the proximity effects of the opposing structures. The scatter geometry is shown in Fig. 8-42. The scatter from a cylinder is well known, and the total signal toward any azimuth angle θ is given by the expression

$$E(\phi)_{\text{total}} = E(\phi)_{\text{primary}} + E(\phi_1)_{\text{primary}} \frac{g(\phi)}{\sqrt{d/\lambda}} \exp \left\{ j \frac{2\pi d}{\lambda} [1 - \cos(\phi - \phi_1)] \right\} \quad (8\text{-}45)$$

where $g(\phi)$ is the reflection coefficient from a cylinder and is a function of its cross section and $E(\phi_1)$ is the incident signal toward the scattering cylinder.

The magnitude of the reflection coefficient, $g(\phi)$ from a cylinder of 0.3 wavelength in radius, is shown in Fig. 8-43 for both horizontally and vertically polarized signals.

Owing to the physical separation between the transmitting antenna and the reflecting structure, the primary and reflected signals add in-phase in some azimuthal directions and out-of-phase in others. Thus, the overall azimuthal pattern is serrated. The minimum-to-maximum level of serrations in any azimuthal direction is a function of the cross section of the opposing structure. It also is directly proportional to the incident signal on the reflecting structure and inversely proportional to the square root of the spacing, expressed in wavelengths. Computed in-place azimuthal patterns in the presence of a

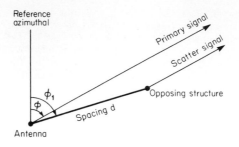

Signal received = primary + scatter

$$\left[\text{use Eq. } (8\text{-}45) \quad \text{for } \frac{d}{\lambda} > 2\left(\frac{2a}{\lambda}\right)^2\right]$$

FIG. 8-42 Geometry of scatter.

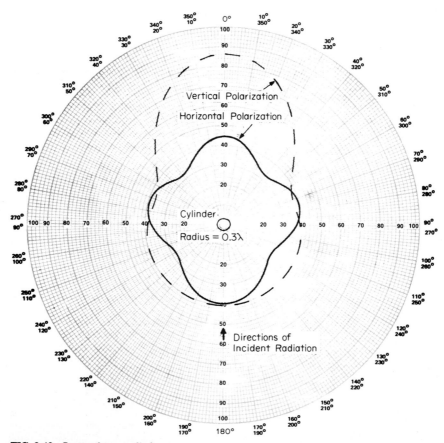

FIG. 8-43 Scatter from a cylinder.

cylinder of 0.6 wavelength diameter are shown in Figs. 8-44 and 8-45 for both polarizations. The relative value of the reflection is shown in the center of each figure. It is obtained by dividing the value of the reflection shown in Fig. 8-43 by the square root of the spacing. The in-place pattern is computed taking into account the reduced magnitude of the primary azimuthal pattern toward the cylinder. For circularly polarized antennas mounted as a candelabra, both the azimuthal pattern and the axial ratio are distorted.

Furthermore, the reflections due to the vertical polarization component are higher than those due to the horizontal polarization component. Consequently, candelabras for circularly polarized antennas require larger spacing or fewer antennas. Figures 8-46 and 8-47 show the azimuthal pattern distortion of a channel-4 circularly polarized antenna 35 ft (11 m) away from a channel-10 circularly polarized antenna, for horizontal and vertical polarizations, respectively.

The calculated in-place pattern is based on ideal assumptions, and the exact locations of the nulls and peaks of the in-place pattern cannot always be determined accurately prior to installation. However, the outer and inner envelopes of the pattern provide a reasonable means for estimating the amount of deterioration that can result in the pattern circularity. For larger cross sections of the opposing cylinder, the scatter pattern

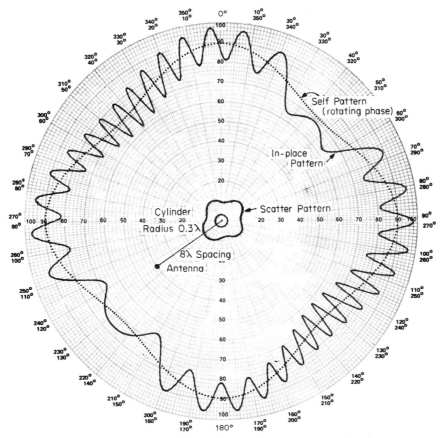

FIG. 8-44 In-place azimuthal pattern of an antenna (horizontal polarization) in the presence of a cylinder.

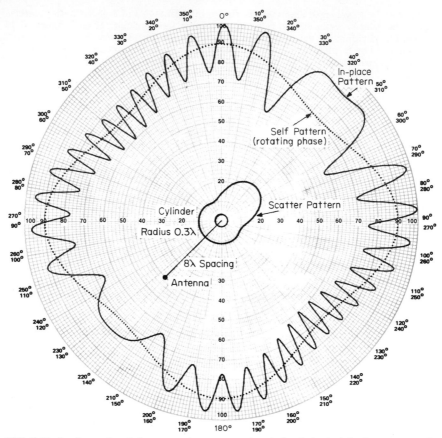

FIG. 8-45 In-place azimuthal pattern of an antenna (vertical polarization) in the presence of a cylinder.

can be approximated by a uniformly constant value around a major portion of the azimuth and a larger value over the shadow region. The former is very close to the rms value of the scatter serration. The maximum value occurs toward the shadow region except for the very small diameter of the opposing structure. The variation of the rms and peak values of $g(\phi)$, the reflection coefficient, is shown in Fig. 8-48. The rms value of the reflection coefficient can be utilized to estimate the in-place circularity from an obstructing cylinder, over most of the periphery, and the peak value can be used to judge the signal toward the shadow region. The example below illustrates the technique.

Example. With the primary pattern circularity $= \pm 1.0$ dB (horizontal polarization), the diameter of interfering cylinder $= 1.0$ wavelength, and the distance to the interfering cylinder $= 9.0$ wavelengths, then from Fig. 8-48 the rms value of $g(\phi) = 0.5$ for horizontal polarization. Hence, rms reflections from cylinder $= 0.5/\sqrt{9} = 0.167$ of incident signal. Thus the envelope of the pattern is ± 16.7 percent $= \pm 1.5$ dB and the estimate of in-place circularity $\pm 1.0 + 1.5 = \pm 2.5$ dB.

In the shadow region, the circularity will be worse. It can be seen that the antennas are generally oriented such that the value of the pattern toward the opposing antenna is as small as possible, so as to reduce the reflected signal proportionately.

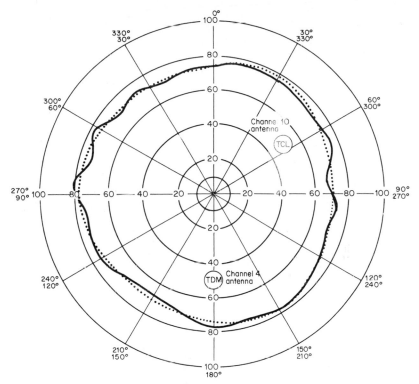

FIG. 8-46 In-place azimuthal pattern of a TDM antenna in the presence of a TCL antenna, 35 ft (11 m) away from it (horizontal polarization). Self pattern is dotted curve.

When the cross section of the reflecting structure is triangular or square, it can be approximated by a solid cylinder of equal perimeter and the in-place azimuthal pattern determined theoretically. In some cases, a scaled-model study is a faster, less expensive, and more reliable approach to determine proximity effects of very complex structures. When the separation exceeds approximately 125 ft, 0.25-μs pulses are resolvable, as noted previously, and ghosts may be apparent.

In the foregoing discussions, the only effect of the opposing antenna as a passive reflector was considered. If the opposing antenna has a sufficiently large impedance bandwidth, a portion of the incident energy will penetrate the feed system of the antenna and reradiate after internal reflections. The ghost or echo problem for these types of reflections is of more concern than the pattern-deterioration effect. The problem can be minimized by the use of absorption or rejection filters in the transmission system of the opposing antenna.

8.3.10 FREQUENCY RESPONSE OF AZIMUTHAL PATTERNS IN MULTIPLE-ANTENNA INSTALLATIONS. The azimuthal pattern of an antenna is essentially independent of the channel frequency spectrum, since the interradiator spacing between adjacent radiators in the same plane is comparable with a wavelength. However, in a multiple-antenna system, if the interfering structures are removed by several wavelengths, the reflected signal from these structures can change in-phase considerably in some azimuthal directions. This results in variation of the primary azimuthal pattern within the spectrum of the channel. Since the aural signal is frequency-modulated, the

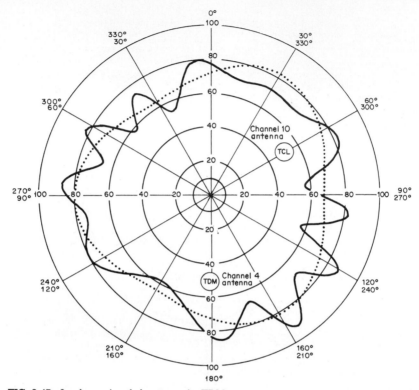

FIG. 8-47 In-place azimuthal pattern of a TDM antenna in the presence of a TCL antenna, 35 ft from it (vertical polarization). Self pattern is dotted curve.

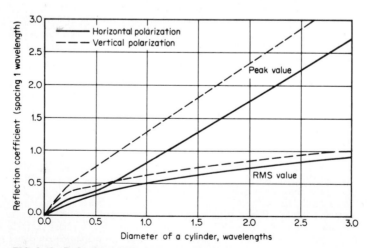

FIG. 8-48 Reflection coefficient from a cylinder.

video distortion analysis is not applicable to it and only the variation in the azimuthal pattern across the video band is of interest here. Most television sets are equipped with automatic chroma gain control (ACC) circuitry to adjust the luminance-to-chrominance gain ratio, and thus reduce the severity of the transmission variation.

The geometry of the reflection problem is shown in Fig. 8-41. The phase difference between the main and reflected signals is given by the equation below

$$\text{Phase difference } \phi = \frac{2\pi d}{\lambda} (1 - \cos \phi) \qquad (8\text{-}46)$$

where ϕ = azimuthal angle
d = spacing
λ = wavelength

or
$$\theta = \frac{2\pi d}{c} (1 - \cos \phi)f \qquad (8\text{-}47)$$

where c = velocity of light and f = frequency. When the effect of frequency variation is derived by differentiating the phase with respect to frequency, it is interesting to note

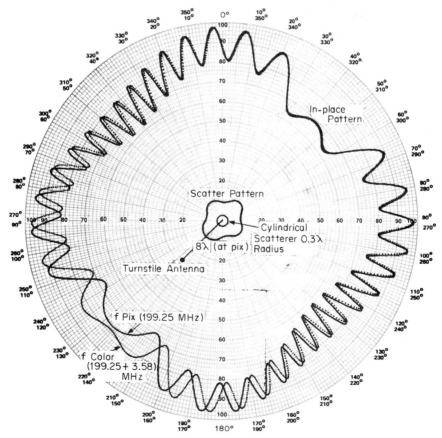

FIG. 8-49 Frequency response of an antenna's in-place azimuthal pattern for horizontal polarization.

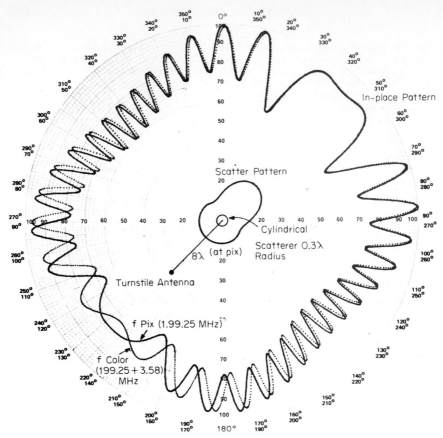

FIG. 8-50 Frequency response of an antenna's in-place azimuthal pattern for vertical polarization.

that the change in the phase of the reflected signal for a specified bandwidth, say, 4.5 MHz (visual carrier to color subcarrier), is proportional to the spacing d between the two structures and independent of the channel. The change is $165(1 - \cos \phi)$ deg for every 100 ft of spacing, and is in excess of 180° for azimuthal angles beyond $\phi = \pm 95°$. In those regions, the reflected signal adds with the main signal in one portion of the channel band and subtracts from it at another frequency in the channel, resulting in video amplitude variation across the channel. Where there are more than two structures not in-line, the video variation occurs around the entire azimuthal circle.

The reduction in the spacing between antennas reduces the phase variation, but the magnitude of the reflection increases inversely with the square root of the spacing. Such an increase leads to deterioration of the azimuthal pattern circularity at the picture-carrier frequency. Thus, a design of a multiple-antenna system requires a compromise of the azimuthal pattern circularity deterioration, without substantially increasing the frequency-response variation.

A typical azimuthal pattern of a horizontally polarized component of an antenna in the presence of an opposing cylindrical structure is shown in Fig. 8-49 at picture-carrier and color subcarrier frequencies. The azimuthal pattern of the vertical-polarization component of the same geometry is shown in Fig. 8-50. The video variation is determined for any azimuthal angle by varying the frequency across the band from picture to color. For two antennas, the video variation is negligible in the direction of the interfering antenna and is maximum in directions away from it. When there are more than two antennas, computation or model measurement can provide the amount of video visual signal variation toward any azimuthal angle. The maximum video variation around the azimuth is used as a figure of merit when optimizing the interantenna spacing of a multiple-antenna system.[32-34]

When the antennas in a multiple system sway in the wind, the spacing between them can change if the deflections are not synchronous. Owing to the resulting phase variations, the magnitude of the received signal can increase at picture-frequency and decrease at color, or vice versa. If automatic gain correction across the channel band is not sufficient, wind flutter may be noticeable. Thus, more stringent mechanical design may be warranted than is the case for a single stack. The automatic chroma-control (ACC) circuitry capability to adjust the luminance-to-chrominance-gain ratio automatically in most modern television receivers has minimized this flutter problem. The response time in the receiver is much faster than the period of mechanical sway of the antennas.

8.3.11 ECHOES IN MULTIPLE-TOWER SYSTEMS. In multiple-tower systems, at least two towers are utilized to mount the antennas for the same coverage area. The discussions on multiple-antenna systems in the previous sections are applicable when the towers are located within 100 ft of each other. The in-place azimuthal patterns can be determined and the echoes are not perceptible, since the delay is less than 0.2 μs if the reradiation due to coupling into the feed system of the interfering antennas is negligible.

When towers are located at spacings greater than 100 ft (30 m), both the problem of azimuthal pattern deterioration and the magnitude of the ghost have to be considered. The magnitude of the reflection from an opposing structure decreases as the spacing increases, but not linearly. For example, if the spacing is quadrupled, the magnitude of the reflection is reduced by only one-half. When the antennas are located more than several hundred feet from each other, the magnitude of the reflection is small enough to be ignored, as far as the pattern deterioration is concerned. However, a reflection of even 3 percent is noticeable to a critical viewer (see Sec. 8.3.8). Thus, large separations, usually more than a thousand feet, are necessary to reduce the echo visibility.

In the previous analysis for the illumination of the interfering structure by the

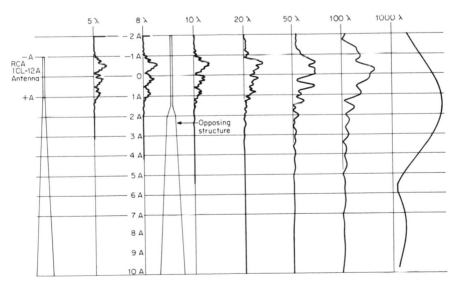

FIG. 8-51 Illustration of opposing tower illumination for various spacings. Spacing is in wavelengths between the TCL antenna and the opposing structure.

antenna, it was assumed that only the portion of the interfering structure in the aperture of the antenna was of importance. This is true for separation distances comparable with the antenna aperture. However, as the separation distance from the antenna increases, the elevation pattern in any vertical plane changes its shape from a *near-field* to a *far-field* pattern. As the elevation pattern changes, more of the opposing structure is illuminated by the primary signal. This effect of distance from the antenna on the elevation pattern is illustrated in Fig. 8-51. Note that the elevation pattern shown is plotted against the height of the opposing structure, rather than the elevation angle.

In practice, optimization of a multiple-tower system may require both a model study and a computer simulation. The characteristics of the reflections of the particular tower in question can best be determined by model measurements and the effect of spacing determined by integration of the induced illumination on the interfering structure.

8.4 REFERENCE DATA

8.4.1 DEPRESSION ANGLE VERSUS ANTENNA HEIGHT AND DISTANCE FROM ANTENNA. See Table 8-5.

8.4.2 VSWR REFLECTION COEFFICIENT, AND RETURN LOSS CONVERSION TABLE. See Table 8-6.

8.4.3 VOLTAGE RATIO, POWER RATIO, AND DECIBELS CONVERSION TABLE. See Table 8-7.

Table 8-5   Distance in Miles vs. Depression Angle for Various Antenna Heights†

Height, ft	Horizon mi	Horizon deg	0.1	0.2	0.3	0.4	0.5	0.6	0.7	0.8	0.9	1.0	1.2	1.4	1.6	1.8	2.0
100	14.14	0.15	...	6.61	3.89	2.82	2.22	1.84	1.57	1.37	1.21	1.09	0.91				
200	20.00	0.22	...	...	8.56	5.90	4.57	3.74	3.18	2.76	2.45	2.20	1.82	1.56	1.36	1.21	1.09
300	24.49	0.27	...	...	14.83	9.32	7.05	5.72	4.83	4.19	3.70	3.31	2.75	2.35	2.05	1.82	1.63
400	28.28	0.31	...	...	...	13.22	9.70	7.78	6.53	5.64	4.97	4.45	3.68	3.14	2.74	2.43	2.18
500	31.62	0.34	...	...	...	17.92	12.56	9.94	8.28	7.13	6.26	5.60	4.62	3.93	3.43	3.04	2.73
600	34.64	0.38	...	...	...	24.26	15.69	12.20	10.09	8.64	7.58	6.76	5.56	4.74	4.13	3.66	3.28
700	37.42	0.41	...	...	...	...	19.19	14.58	11.96	10.20	8.92	7.94	6.52	5.54	4.83	4.27	3.84
800	40.00	0.43	...	...	...	...	23.21	17.12	13.90	11.79	10.28	9.13	7.49	6.36	5.53	4.89	4.39
900	42.43	0.46	...	...	...	...	28.10	19.83	15.91	13.43	11.67	10.35	8.46	7.17	6.23	5.52	4.95
1000	44.72	0.49	...	...	...	...	34.98	22.78	18.02	15.11	13.09	11.58	9.44	8.00	6.94	6.14	5.51
1100	46.90	0.51	...	...	...	...	...	26.01	20.22	16.84	14.54	12.83	10.44	8.83	7.66	6.77	6.07
1200	48.99	0.53	...	...	...	...	...	29.65	22.54	18.63	16.01	14.10	11.44	9.66	8.37	7.40	6.63
1300	50.99	0.55	...	...	...	...	...	33.91	24.99	20.48	17.52	15.39	12.46	10.50	9.10	8.03	7.19
1400	52.91	0.57	...	...	...	...	...	39.25	27.61	22.39	19.07	16.70	13.48	11.35	9.82	8.66	7.76
1500	54.77	0.59	...	...	...	...	...	47.72	30.43	24.37	20.66	18.04	14.51	12.20	10.55	9.30	8.32
1600	56.57	0.61	...	...	...	...	...	...	33.50	26.44	22.28	19.40	15.56	13.06	11.28	9.94	8.89
1700	58.31	0.63	...	...	...	...	...	...	36.91	28.61	23.95	20.79	16.62	13.93	12.02	10.58	9.46
1800	60.00	0.65	...	...	...	...	...	...	40.81	30.88	25.68	22.21	17.69	14.80	12.76	11.23	10.04
1900	61.64	0.67	...	...	...	...	...	...	45.50	33.28	27.45	23.65	18.77	15.68	13.50	11.88	10.61
2000	63.24	0.69	...	...	...	...	...	...	51.80	35.84	29.28	25.13	19.87	16.56	14.25	12.53	11.19
2500	70.71	0.77	...	...	...	...	...	...	...	52.86	39.59	33.05	25.56	21.10	18.06	15.82	14.10
3000	77.46	0.84	...	...	...	...	...	...	...	53.28	42.22	31.65	25.83	21.98	19.19	17.06
3500	83.66	0.91	...	...	...	...	...	...	...	...	53.51	38.27	30.80	26.03	22.64	20.08
4000	89.44	0.97	...	...	...	...	...	...	...	...	69.94	45.54	36.03	30.22	26.17	23.15
4500	94.86	1.03	...	...	...	...	...	...	...	...	...	53.75	41.57	34.56	29.79	26.28
5000	99.99	1.09	...	...	...	...	...	...	...	...	...	63.35	47.49	39.08	33.52	29.47

†Distances beyond the horizon and below 1 mi (1.6 km) are not indicated.

Table 8-6 Conversion Chart for Reflection Coefficient, VSWR, and Return Loss

Reflection coefficient, %	VSWR	Return loss, dB	Reflection coefficient, %	VSWR	Return loss, dB
0.5	1.010	46.0	3.0	1.062	30.5
0.6	1.012	44.4	3.5	1.073	29.1
0.7	1.014	43.1	4.0	1.083	28.0
0.8	1.016	41.9	4.5	1.094	26.9
0.9	1.018	40.9	5.0	1.105	26.0
1.0	1.020	40.0	5.5	1.116	25.2
1.1	1.022	39.2	6.0	1.128	24.4
1.2	1.024	38.4	6.5	1.139	23.8
1.3	1.026	37.7	7.0	1.151	23.1
1.4	1.028	37.1	7.5	1.162	22.5
1.5	1.031	36.5	8.0	1.174	21.9
1.6	1.033	35.9	8.5	1.186	21.4
1.7	1.035	34.4	9.0	1.198	20.9
1.8	1.037	34.9	9.5	1.210	20.5
1.9	1.039	34.4	10	1.222	20.0
2.0	1.041	34.0	11	1.247	19.2
2.1	1.043	33.6	12	1.272	18.4
2.2	1.045	33.2	13	1.299	17.7
2.3	1.047	32.8	14	1.326	17.1
2.4	1.049	32.4	15	1.353	16.5
2.5	1.051	32.0	16	1.381	15.9
2.6	1.053	31.7	17	1.410	15.4
2.7	1.055	31.4	18	1.439	14.9
2.8	1.058	31.1	19	1.469	14.4
2.9	1.060	30.8	20	1.500	14.0

angle, deg

2.5	3.0	3.5	4.0	4.5	5.0	6.0	7.0	8.0	9.0	10.0	12.0	14.0	16.0	18.0	20.0	25.0
0.87																
1.31	1.09	0.93														
1.74	1.45	1.24	1.09	0.96												
2.18	1.81	1.55	1.36	1.21	1.08	0.98										
2.62	2.18	1.86	1.63	1.45	1.30	1.08	0.93									
3.06	2.54	2.17	1.90	1.69	1.52	1.26	1.08	0.94								
3.50	2.91	2.49	2.17	1.93	1.74	1.44	1.24	1.08	0.96							
3.94	3.27	2.80	2.45	2.17	1.95	1.62	1.39	1.21	1.08	0.97						
4.38	3.64	3.11	2.72	2.41	2.17	1.80	1.54	1.35	1.20	1.07	0.89					
4.82	4.00	3.42	2.99	2.66	2.39	1.99	1.70	1.48	1.32	1.18	0.98					
5.27	4.37	3.74	3.26	2.90	2.61	2.17	1.85	1.62	1.44	1.29	1.07	0.91				
5.71	4.74	4.05	3.54	3.14	2.82	2.35	2.01	1.75	1.56	1.40	1.16	0.99				
6.16	5.11	4.36	3.81	3.38	3.04	2.53	2.16	1.89	1.68	1.51	1.25	1.06	0.92			
6.60	5.48	4.68	4.09	3.63	3.26	2.71	2.32	2.02	1.80	1.61	1.34	1.14	0.99			
7.05	5.84	4.99	4.36	3.87	3.48	2.89	2.47	2.16	1.92	1.72	1.43	1.22	1.06	0.93		
7.50	6.21	5.31	4.63	4.11	3.70	3.07	2.63	2.29	2.04	1.83	1.52	1.29	1.12	0.99		
7.94	6.58	5.62	4.91	4.35	3.91	3.25	2.78	2.43	2.15	1.94	1.61	1.37	1.19	1.05	0.94	
8.39	6.95	5.94	5.18	4.60	4.13	3.43	2.94	2.56	2.28	2.04	1.69	1.44	1.26	1.11	0.99	
8.85	7.32	6.25	5.46	4.84	4.35	3.62	3.09	2.70	2.39	2.15	1.78	1.52	1.32	1.17	1.04	0.81
11.11	9.19	7.84	6.83	6.06	5.44	4.52	3.87	3.38	2.99	2.69	2.23	1.90	1.65	1.46	1.30	1.02
13.40	11.06	9.43	8.22	7.28	6.54	5.43	4.64	4.05	3.59	3.23	2.68	2.28	1.98	1.75	1.56	1.22
15.72	12.95	11.03	9.60	8.51	7.64	6.34	5.42	4.73	4.20	3.77	3.12	2.66	2.31	2.04	1.82	1.42
18.06	14.85	12.63	11.00	9.74	8.74	7.26	6.20	5.41	4.80	4.31	3.57	3.04	2.64	2.33	2.08	1.63
20.43	16.77	14.25	12.40	10.97	9.85	8.17	6.98	6.09	5.40	4.85	4.02	3.42	2.98	2.63	2.34	1.83
22.82	18.70	15.87	13.80	12.21	10.95	9.08	7.76	6.77	6.00	5.39	4.46	3.80	3.31	2.92	2.60	2.03

Table 8-7 Conversion Chart for Voltage Ratio, Power Ratio, and Decibels

Voltage ratio	Power ratio	− dB +	Voltage ratio	Power ratio
1.000	1.000	0	1.000	1.000
0.989	0.977	0.1	1.012	1.023
0.977	0.955	0.2	1.023	1.047
0.966	0.933	0.3	1.035	1.072
0.955	0.912	0.4	1.047	1.096
0.944	0.891	0.5	1.059	1.122
0.933	0.871	0.6	1.072	1.148
0.923	0.851	0.7	1.084	1.175
0.912	0.832	0.8	1.096	1.202
0.902	0.813	0.9	1.109	1.230
0.891	0.794	1.0	1.122	1.259
0.881	0.776	1.1	1.135	1.288
0.871	0.759	1.2	1.148	1.318
0.861	0.741	1.3	1.161	1.349
0.851	0.734	1.4	1.175	1.380
0.841	0.708	1.5	1.189	1.413
0.832	0.692	1.6	1.202	1.445
0.822	0.676	1.7	1.216	1.479
0.813	0.661	1.8	1.230	1.514
0.804	0.646	1.9	1.245	1.549
0.794	0.631	2.0	1.259	1.585
0.785	0.617	2.1	1.274	1.622
0.776	0.603	2.2	1.288	1.660
0.767	0.589	2.3	1.303	1.698
0.759	0.575	2.4	1.318	1.738
0.750	0.562	2.5	1.334	1.778
0.741	0.550	2.6	1.349	1.820
0.733	0.537	2.7	1.365	1.862
0.724	0.525	2.8	1.380	1.905
0.716	0.513	2.9	1.396	1.950
0.708	0.501	3.0	1.413	1.995
0.700	0.490	3.1	1.429	2.042
0.692	0.479	3.2	1.445	2.089
0.684	0.468	3.3	1.462	2.138
0.676	0.457	3.4	1.479	2.188

Voltage ratio	Power ratio	− dB +	Voltage ratio	Power ratio
0.447	0.200	7.0	2.239	5.012
0.442	0.195	7.1	2.265	5.129
0.437	0.191	7.2	2.291	5.248
0.432	0.186	7.3	2.317	5.370
0.427	0.182	7.4	2.344	5.495
0.422	0.178	7.5	2.371	5.623
0.417	0.174	7.6	2.399	5.754
0.412	0.170	7.7	2.427	5.888
0.407	0.166	7.8	2.455	6.026
0.403	0.162	7.9	2.483	6.166
0.398	0.159	8.0	2.512	6.310
0.394	0.155	8.1	2.541	6.457
0.389	0.151	8.2	2.570	6.607
0.385	0.148	8.3	2.600	6.761
0.380	0.145	8.4	2.630	6.918
0.376	0.141	8.5	2.661	7.079
0.372	0.138	8.6	2.692	7.244
0.367	0.135	8.7	2.723	7.413
0.363	0.132	8.8	2.754	7.586
0.359	0.129	8.9	2.786	7.762
0.355	0.126	9.0	2.818	7.943
0.351	0.123	9.1	2.851	8.128
0.347	0.120	9.2	2.884	8.318
0.343	0.118	9.3	2.917	8.511
0.339	0.115	9.4	2.951	8.710
0.335	0.112	9.5	2.985	8.913
0.331	0.110	9.6	3.020	9.120
0.327	0.105	9.7	3.055	9.333
0.324	0.105	9.8	3.090	9.550
0.320	0.102	9.9	3.126	9.772
0.316	0.100	10.0	3.162	10.000
0.313	0.0977	10.1	3.199	10.23
0.309	0.0955	10.2	3.236	10.47
0.306	0.0933	10.3	3.273	10.72
0.302	0.0912	10.4	3.311	10.96

Voltage ratio	Power ratio	− dB +	Voltage ratio	Power ratio
0.200	0.0398	14.0	5.012	25.12
0.197	0.0389	14.1	5.070	25.70
0.195	0.0380	14.2	5.129	26.30
0.193	0.0372	14.3	5.188	26.92
0.191	0.0363	14.4	5.248	27.54
0.188	0.0355	14.5	5.309	28.18
0.186	0.0347	14.6	5.370	28.84
0.184	0.0339	14.7	5.433	29.51
0.182	0.0331	14.8	5.495	30.20
0.180	0.0324	14.9	5.559	30.90
0.178	0.0316	15.0	5.623	31.62
0.176	0.0309	15.1	5.689	32.36
0.174	0.0302	15.2	5.754	33.11
0.172	0.0295	15.3	5.821	33.88
0.170	0.0288	15.4	5.888	34.67
0.168	0.0282	15.5	5.957	35.48
0.166	0.0275	15.6	6.026	36.31
0.164	0.0269	15.7	6.095	37.15
0.162	0.0263	15.8	6.166	38.02
0.160	0.0257	15.9	6.237	38.90
0.159	0.0251	16.0	6.310	39.81
0.157	0.0246	16.1	6.383	40.74
0.155	0.0240	16.2	6.457	41.69
0.153	0.0234	16.3	6.531	42.66
0.151	0.0229	16.4	6.607	43.65
0.150	0.0224	16.5	6.683	44.67
0.148	0.0219	16.6	6.761	45.71
0.146	0.0214	16.7	6.839	46.77
0.145	0.0209	16.8	6.918	47.86
0.143	0.0204	16.9	6.998	48.98
0.141	0.0200	17.0	7.079	50.12
0.140	0.0195	17.1	7.161	51.29
0.138	0.0191	17.2	7.244	52.48
0.137	0.0186	17.3	7.328	53.70
0.135	0.0182	17.4	7.413	54.95

		x		
0.668	0.447	3.5	1.496	2.239
0.661	0.437	3.6	1.514	2.291
0.653	0.427	3.7	1.531	2.344
0.646	0.417	3.8	1.549	2.399
0.638	0.407	3.9	1.567	2.455
0.631	0.398	4.0	1.585	2.512
0.624	0.389	4.1	1.603	2.570
0.617	0.380	4.2	1.622	2.630
0.610	0.372	4.3	1.641	2.692
0.603	0.363	4.4	1.660	2.754
0.596	0.355	4.5	1.679	2.818
0.589	0.347	4.6	1.698	2.884
0.582	0.339	4.7	1.718	2.951
0.575	0.331	4.8	1.738	3.020
0.569	0.324	4.9	1.758	3.090
0.562	0.316	5.0	1.778	3.162
0.556	0.309	5.1	1.799	3.236
0.550	0.302	5.2	1.820	3.311
0.543	0.295	5.3	1.841	3.388
0.537	0.288	5.4	1.862	3.467
0.531	0.282	5.5	1.884	3.548
0.525	0.275	5.6	1.905	3.631
0.519	0.269	5.7	1.928	3.715
0.513	0.263	5.8	1.950	3.802
0.507	0.257	5.9	1.972	3.890
0.501	0.251	6.0	1.995	3.981
0.496	0.246	6.1	2.018	4.074
0.490	0.240	6.2	2.042	4.169
0.484	0.234	6.3	2.065	4.266
0.479	0.229	6.4	2.089	4.365
0.473	0.224	6.5	2.113	4.467
0.468	0.219	6.6	2.138	4.571
0.462	0.214	6.7	2.163	4.677
0.457	0.209	6.8	2.188	4.786
0.452	0.204	6.9	2.213	4.898
0.299	0.0891	10.5	3.350	11.22
0.295	0.0871	10.6	3.388	11.48
0.292	0.0851	10.7	3.428	11.75
0.288	0.0832	10.8	3.467	12.02
0.285	0.0813	10.9	3.508	12.30
0.282	0.0794	11.0	3.548	12.59
0.279	0.0776	11.1	3.589	12.88
0.275	0.0759	11.2	3.631	13.18
0.272	0.0741	11.3	3.673	13.49
0.269	0.0724	11.4	3.715	13.80
0.266	0.0708	11.5	3.758	14.13
0.263	0.0692	11.6	3.802	14.45
0.260	0.0676	11.7	3.846	14.79
0.257	0.0661	11.8	3.890	15.14
0.254	0.0646	11.9	3.936	15.49
0.251	0.0631	12.0	3.981	15.85
0.248	0.0617	12.1	4.027	16.22
0.246	0.0603	12.2	4.074	16.60
0.243	0.0589	12.3	4.121	16.98
0.240	0.0575	12.4	4.169	17.38
0.237	0.0562	12.5	4.217	17.78
0.234	0.0550	12.6	4.266	18.20
0.232	0.0537	12.7	4.315	18.62
0.229	0.0525	12.8	4.365	19.05
0.227	0.0513	12.9	4.416	19.50
0.224	0.0501	13.0	4.467	19.95
0.221	0.0490	13.1	4.519	20.42
0.219	0.0479	13.2	4.571	20.89
0.216	0.0468	13.3	4.624	21.38
0.214	0.0457	13.4	4.677	21.88
0.211	0.0447	13.5	4.732	22.39
0.209	0.0437	13.6	4.785	22.91
0.207	0.0427	13.7	4.842	23.44
0.204	0.0417	13.8	4.898	23.99
0.202	0.0407	13.9	4.955	24.55
0.133	0.0178	17.5	7.499	56.23
0.132	0.0174	17.6	7.586	57.54
0.130	0.0170	17.7	7.674	58.88
0.129	0.0166	17.8	7.762	60.26
0.127	0.0162	17.9	7.852	61.66
0.126	0.0159	18.0	7.943	63.10
0.125	0.0155	18.1	8.035	64.57
0.123	0.0151	18.2	8.128	66.07
0.122	0.0148	18.3	8.222	67.61
0.120	0.0145	18.4	8.318	69.18
0.119	0.0141	18.5	8.414	70.79
0.118	0.0138	18.6	8.511	72.44
0.116	0.0135	18.7	8.610	74.13
0.115	0.0132	18.8	8.710	75.86
0.114	0.0129	18.9	8.811	77.62
0.112	0.0126	19.0	8.913	79.43
0.111	0.0123	19.1	9.016	81.28
0.110	0.0120	19.2	9.120	83.18
0.108	0.0118	19.3	9.226	85.11
0.107	0.0115	19.4	9.333	87.10
0.106	0.0112	19.5	9.441	89.13
0.105	0.0110	19.6	9.550	91.20
0.104	0.0107	19.7	9.661	93.33
0.102	0.0105	19.8	9.772	95.50
0.100	0.0102	19.9	9.886	97.72
0.100	0.0100	20.0	10.000	100.00
0.089	0.008	21.0	11.22	125.90
0.079	0.006	22.0	12.59	158.49
0.071	0.005	23.0	14.13	199.53
0.063	0.004	24.0	15.85	251.19
0.056	0.003	25.0	17.78	316.23
0.050	0.003	26.0	19.95	398.11
0.045	0.002	27.0	22.39	501.19
0.040	0.002	28.0	25.14	630.96
0.035	0.0001	29.0	28.18	794.33
0.032	0.0001	30.0	31.62	1000.0

REFERENCES

General

1. J. D. Kraus, *Antennas,* McGraw-Hill, New York, 1950.
2. T. Moreno, *Microwave Transmission Design Data,* Dover, New York.
3. R. C. Johnson and H. Jasik, *Antenna Engineering Handbook,* 2d ed., McGraw-Hill, New York, 1984.

Broadcast Antennas

4. M. S. Siukola, "The Traveling Wave VHF Television Transmitting Antenna," *IRE Trans. Broadcasting,* vol. BTR-3, no. 2, October 1957, pp. 49–58.
5. R. N. Clark and N. A. L. Davidson, "The V–Z Panel as a Side Mounted Antenna," *IEEE Trans. Broadcasting,* vol. BC-13, no. 1, January 1967, pp. 3–136.
6. D. A. Brawn and B. F. Kellom, "Butterfly VHF Panel Antenna," *RCA Broadcast News,* vol. 138, March 1968, pp. 8–12.
7. R. E. Fisk and J. A. Donovan, "A New CP Antenna for Television Broadcast Service," *IEEE Trans. Broadcasting,* vol. BC-22, no. 3, September 1976, pp. 91–96.
8. G. G. DeVito and L. Manis, "Improved Dipole Panel for Circular Polarization," *IEEE Trans. Broadcasting,* vol. BC-28, no. 2, June 1982, pp. 65–72.
9. J. Perini, "Improvement of Pattern Circularity of Panel Antenna Mounted on Large Towers," *IEEE Trans. Broadcasting,* vol. BC-14, no. 1, March 1968, pp. 33–40.

Multiple-Antenna System

10. "Predicting Characteristics of Multiple Antenna Arrays," *RCA Broadcast News,* vol. 97, October 1957, pp. 63–68.
11. P. C. J. Hill, "Measurements of Reradiation from Lattice Masts at VHF," *Proc. IEEE,* vol. III, no. 12, December 1964, pp. 1957–1968.
12. "Bay Area TV Viewers Turn to Sutro Tower," *RCA Broadcast News,* vol. 150, August 1973, pp. 18–31.
13. H. H. Wescott, "A Closer Look at the Sutro Tower Antenna Systems," *RCA Broadcast News,* vol. 152, February 1944, pp. 35–41.
14. M. R. Johns and M. A. Ralston, "The First Candelabra for Circularly Polarized Broadcast Antennas," *IEEE Trans. Broadcasting,* vol. BC-27, no. 4, December 1981, pp. 77–82.
15. "WBAL, WJZ and WMAR Build World's First Three-Antenna Candelabra," *RCA Broadcast News,* vol. 106, December 1959, pp. 30–35.
16. P. Knight, "Reradiation from Masts and Similar Objects at Radio Frequencies," *Proc. IEEE,* vol. 114, January 1967, pp. 30–42.
17. M. S. Siukola, "Size and Performance Trade Off Characteristics of Horizontally and Circularly Polarized TV Antennas," *IEEE Trans. Broadcasting,* vol. BC-23, no. 1, March 1976.
18. "Big John Blankets Chicago," *RCA Broadcast News,* vol. 141, March 1969, pp. 8–13.

Circular Polarization

19. O. Ben-Dov, "Measurement of Circularly Polarized Broadcast Antennas," *IEEE Trans. Broadcasting,* vol. BC-19, no. 1, March 1972, pp. 28–32.

20. S. J. Dudzinsky, Jr., "Polarization Discrimination for Satellite Communications," *Proc. IEEE*, vol. 57, no. 12, December 1969, pp. 2179–2180.

21. N. Furnes and K. N. Stokke, "Reflection Problems in Mountainous Areas: Tests with Circular Polarization for Television and VHF/FM Broadcasting in Norway," *E.B.U. Review*, Technical Part, no. 184, December 1980, pp. 266–271.

Vertical Pattern Shaping

22. P. C. J. Hill, "Methods for Shaping Vertical Pattern of VHF and UHF Transmitting Aerials," *Proc. IEEE*, vol. 116, no. 8, August 1969, pp. 1325–1337.

23. G. DeVito, "Considerations on Antennas with no Null Radiation Pattern and Pre-established Maximum-Minimum Shifts in the Vertical Plane," *Alta Frequenza*, vol. XXXVIII, no. 6, 1969.

24. J. Perini and M. H. Ideslis, "Radiation Pattern Synthesis for Broadcast Antennas," *IEEE Trans. Broadcasting*, vol. BC-18, no. 3, September 1972, p. 53.

25. K. Praba, "Computer-aided Design of Vertical Patterns for TV Antenna Arrays," *RCA Engineer*, January–February 1973, vol. 18-4.

References on Echoes

26. A. M. Lessman, "The Subjective Effect of Echoes in 525-Line Monochrome and NTSC Color Television and the Resulting Echo Time Weighting," *J. SMPTE*, vol. 81, December 1972.

27. P. Mertz, "Influence of Echoes on Television Transmission," *J. SMPTE*, vol. 60, May 1953.

28. J. W. Allnatt and R. D. Prosser, "Subjective Quality of Television Pictures Impaired by Long Delayed Echoes," *Proc. IEEE*, vol. 112, no. 3, March 1965.

29. A. D. Fowler and H. N. Christopher, "Effective Sum of Multiple Echoes in Television," *J. SMPTE*, vol. 58, June 1952.

Antenna Performance Measurements and Evaluation

30. M. S. Siukola, "TV Antenna Performance Evaluation with RF Pulse Techniques," *IEEE Trans. Broadcasting*, vol. BC-16, no. 3, September 1970.

31. J. Perini, "Echo Performance of TV Transmitting Systems," *IEEE Trans. Broadcasting*, vol. BC-16, no. 3, September 1970.

32. D. J. Whythe, "Specification of the Impedance of Transmitting Aerials for Monochrome and Color Television Signals," Tech. Rep. E-115, BBC, London, 1968.

33. K. Praba, "R. F. Pulse Measurement Techniques and Picture Quality," *IEEE Trans. Broadcasting*, vol. BC-23, no. 1, March 1976, pp. 12–17.

34. D. W. Sargent, "A New Technique for Measuring FM and TV Antenna Systems," *IEEE Trans. Broadcasting*, vol. BC-27, no. 4, December 1981.

Cable and Satellite Home Distribution Systems

D. Stevens McVoy
Coaxial Communications Corporation
Columbus, Ohio

Joseph L. Stern
Stern Telecommunications Corporation
New York, New York

with contributions by

K. Blair Benson
Engineering Consultant
Norwalk, Connecticut

Charles A. Kase
Satellite Systems Engineering
Bethesda, Maryland

Wilbur L. Pritchard
Satellite Systems Engineering
Bethesda, Maryland

9.1 OVERVIEW OF CATV SYSTEMS

9.1.1 DEFINITION.   Cable-television systems, also known as community-antenna television (CATV) systems, use coaxial cable to distribute television, audio, and data signals to homes or establishments subscribing to the service. CATV systems with bidirectional capability can also transmit signals from various points within the cable network to the central originating point (head end).

Networks which serve primarily residential subscribers are referred to as *subscriber networks,* while networks which serve business, commercial, educational, and governmental users are called *institutional networks.*

9.1.2 PROGRAM SOURCES. Television picture and sound program material may originate from local or distant broadcasting stations picked up over the air or relayed by microwave or satellite. A number of advertiser-supported networks serving CATV exclusively are distributed by satellite, as are several pay television services. In addition, programs may be originated at the facilities of the CATV system, utilizing studio facilities, video tape playback equipment, or computerized alphanumeric character generators for continuous automated service.

9.1.3 EVOLUTION OF CATV. CATV began in rural areas in the early 1950s as a means of bringing television service to areas with no broadcast stations. These systems typically had a five-channel capability and carried only the three major television networks. During the 1960s, CATV moved into areas served by broadcast stations but without a full complement of network stations. Channel capacity was increased to 12 or 20 and carried almost exclusively broadcast television signals. By the mid-1970s, satellite distribution of pay television programming made CATV viable in urban areas with a good selection of over-the-air television. The systems offered a greater variety of television, including independent stations from distant cities and pay television.

Systems being built in the 1980s typically have a 50- to 100-channel capacity and are capable of bidirectional transmission, thus allowing interactive television programming, information retrieval, home monitoring, and point-to-point data transmission. Many modern systems include an institutional network in addition to one or two subscriber network cables.

By 1976, cable served 12 million subscribers (17 percent of all United States television households), 565,000 of which also subscribed to pay television service. By the middle of 1983 CATV systems served a total of almost 32 million subscribers, which represented 38 percent of all television households in the United States, and pay television subscriptions over CATV totaled almost 19 million. It is projected that by 1990, 59 million homes will subscribe to CATV, or 62 percent of all United States television households. See Table 9-1.

9.2 ELEMENTS OF A CATV SYSTEM

9.2.1 SUBSCRIBER NETWORK. A typical CATV system comprises four main elements: a *head end,* the central originating point of all signals carried, where signals are received and processed; a *trunk system,* the main artery carrying signals through a community; a *distribution system,* which is bridged from the trunk system and carries signals into individual neighborhoods for distribution to subscribers; and *subscriber drops,* individual lines into subscribers' television sets fed from taps in the distribution system (see Fig. 9-1).

Within the subscriber's home, the drop may terminate directly into the television receiver in the case of systems carrying only 12 channels, or into a converter where more than 12 channels are utilized. For pay television service, signals are often carried in scrambled form, requiring a descrambler in the subscriber's home. In some systems, converters and descramblers are addressable from the head end, allowing control of which channels a subscriber is authorized to receive from the head end.

While the main purpose of CATV is to deliver high-quality television signals to subscribers, another function, interactive communications, has been developed to allow the subscriber to interact with the program source and to request various types of information (videotext). Interactive systems can also provide monitoring capability for such services as home security and pay-per-view pay television. With these services, additional equipment is required within the subscriber's home or establishment. For monitoring

Table 9-1 Growth of CATV in the United States

Pay and basic cable growth†

Year	Basic subscribers	% of TVHHs	Pay subscribers	% of TVHHs
1990	58,900,000	62	43,700,000	46
1989	54,988,000	59	40,076,000	43
1988	52,098,000	57	36,560,000	40
1987	49,280,000	55	33,152,000	37
1986	45,656,000	52	29,852,000	34
1985	40,467,000	47	25,830,000	30
1984	35,448,000	42	21,944,000	26
1983	30,636,000	37	18,216,000	22
1982	26,016,000	32	14,634,000	18
1981	22,596,000	28	11,804,000	14
1980	18,672,000	24	7,780,000	10
1979	16,023,000	21	5,341,000	7
1978	14,155,000	19	2,980,000	4
1977	13,194,000	18	1,466,000	2
1976	12,094,000	17	565,000	1

Cable barometer‡	
Television households (TVHH)	83,900,000
Homes passed by cable (HP)	51,019,700
Percent of TVHH passed by cable	60.8%
Basic cable subscribers	28,933,300
Percent of TVHH with basic cable	35.7%
Percent of HP with basic cable	56.7%
Pay cable subs	16,800,000
Percent of TVHH with pay cable	20.2%
Percent of HP with pay cable	32.9%
Percent of basic subs with pay cable	58.1%

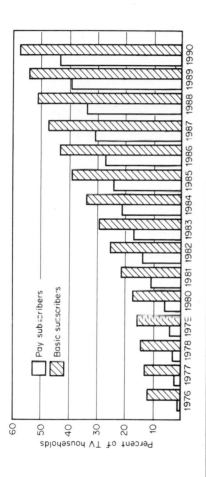

†Pay and basic subscriber counts for 1982–1990 are projections.
‡Figures are as of May 31, 1983.

Source: *Cablevision*, August 1983. From ICR, Cable Information Service, a Titsch Communications division.

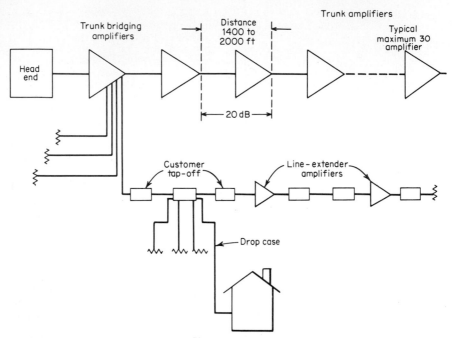

FIG. 9-1 Typical CATV network. [*D. G. Fink and D. Christiansen (eds.), Electronics Engineers' Handbook, 2nd ed., McGraw-Hill, New York, 1982.*]

services, a home terminal is required. For information retrieval, a videotext decoder is needed, and for data transmission, cable modems must be provided.

9.2.2 INSTITUTIONAL NETWORKS.   Many newer CATV systems include institutional networks. Generally, these networks share the same head end as the subscriber network and follow the same route. Institutional networks, however, usually have no feeder system, with all users being fed directly from the trunk.

9.2.3 SUPERTRUNKS.   In networks serving large areas, supertrunks are often used to tie together trunk systems. Supertrunks use large-diameter [¾-in (2-cm) or greater] coaxial cable with high-quality amplifiers and battery standby power. Often FM transmission is used for video signals to reduce signal degradation.

9.2.4 CONSTRUCTION.   CATV distribution systems are usually built on leased space on utility poles owned by the telephone or power companies. In some cases, such as in downtown areas or residential areas with underground utilities, cable systems are installed underground, either in conduit or directly buried, depending on local building codes and soil conditions.

9.2.5 CHANNEL CAPACITY.   Newer CATV systems are capable of transmitting signals in the range of 5 to 400 MHz, though the majority of existing systems have an upper-frequency limit of 300 MHz. Generally, systems are designed for bidirectional transmission, with the spectrum divided between the two directions. In subscriber networks, 5 to 35 MHz is utilized for upstream transmission (toward the head end) and 50 to 400 MHz utilized for downstream transmission (toward the subscriber), since a greater number of video channels is desired from the head end to subscribers. In institutional networks, however, it is desirable to have an equal number of channels in each direction, so upstream transmission is usually from 5 to 150 MHz and downstream transmission utilizes the frequencies from 200 MHz up.

As hybrid amplifiers, the main element in CATV amplifiers, improve, the upper-frequency limit of CATV systems is being increased. At present, 500-MHz amplifiers have been developed, and it is likely that the upper-frequency maximum will be even higher in the near future. Other elements of the CATV network, such as passive line splitters, cable, and converters, are already available with capabilities of 500 MHz and higher.

With a single cable, a maximum of 54 video channels can be carried in the downstream direction on a 400-MHz system (36 on a 300-MHz system). However, FCC Rules and Regulations which prohibit CATV systems from using frequencies in the aircraft navigation and communications bands often make several channels unusable. As a result, the actual number of channels carried on most 400-MHz systems is approximately 50 (32 on a 300-MHz system). Many systems are now being built with a dual-cable subscriber network. With this approach, the channel capacity is doubled.

In addition to the video channel capacity of CATV systems, there is space for the FM radio band (88 to 108 MHz), various control signals, and digital data transmission. In the upstream direction, a theoretical capacity for five video channels exists. However, few cable systems utilize more than one video channel in the upstream direction. Most uses for the upstream spectrum are digital.

The channels carried between 54 and 88 MHz are referred to as *low band* (channels 2 to 6), and those between 120 and 174 MHz are called *Midband* (channels A to I or 14 to 22). The frequencies from 174 to 216 MHz are called *high band* (channels 7 to 13), and those above 216 MHz are referred to as *superband, hyperband,* and *ultraband,* (channels J and up or 23 and up). See Table 13-2 for a typical CATV channel allocation and identification.

Institutional networks generally allocate 20 or 25 channels in the downstream direction and 15 to 20 channels in the upstream direction. Since most institutional networks are carrying only experimental services, the final channel configurations have not yet evolved. It is likely, however, that a good portion of the spectrum on institutional networks will be dedicated to digital transmission.

9.3 HEAD END

The *head end* of a CATV system is the origination point for all signals carried on the system. Signals are received off the air, from satellites, and from terrestrial microwave systems. In addition, many signals are originated at the head end. Signals are processed and then combined for transmission over the cable system. In bidirectional systems the head end also serves as the collection point for all signals originating within the subscriber and institutional networks.

The major elements of the head end are the antenna system, signal-processing equipment, pilot carrier generators, combining networks, and equipment for bidirectional and interactive services. See Fig. 9-2 for a typical CATV head end.

A CATV antenna system includes a tower and antennas for reception of local and distant stations. For distant signals, tall towers with high-gain directional receiving antennas are utilized to provide sufficient gain to pick up the desired signal and provide discrimination against unwanted adjacent channel, co-channel, or reflected signals. Antennas are located in an area of low ambient electrical noise where it is possible to receive the desired television channels with a minimum of interference and at a sufficient level to obtain a high-quality signal. For weak signals, low-noise preamplifiers are used near the pickup antennas. Strong adjacent channels are attenuated through the use of bandpass and bandstop filters.

Satellite earth receiving stations, or television receive only (TVRO) stations, are used by most CATV systems. Earth stations are located to minimize interference from terres-

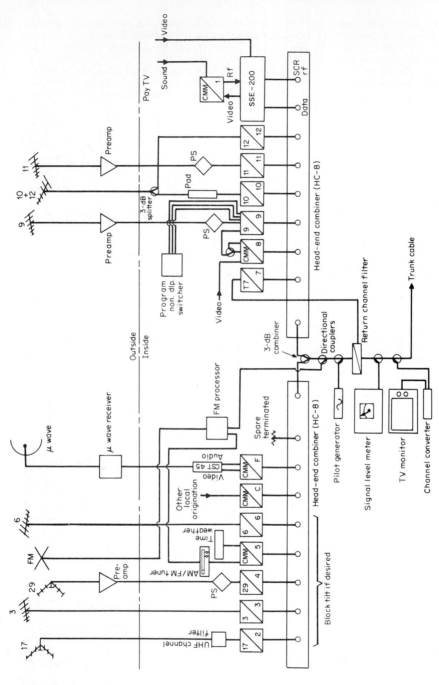

FIG. 9-2 Typical CATV head end. [*D. G. Fink and D. Christiansen (eds.), Electronics Engineers' Handbook, 2d ed., McGraw-Hill, New York, 1982.*]

trial microwave transmissions, which share the same 4-GHz spectrum, and to allow an unobstructed line-of-sight path to the desired satellites. Earth stations consist of a receiving antenna [usually a parabolic reflector 13 to 23 ft (4 to 7 m) in diameter], low-noise preamplifiers at the focal point, and a waveguide to transmit the signals to receivers.

Signal processing is performed at the head end for the following reasons:

1. To regulate the signal-to-noise ratio at the highest practical value
2. To control the output level of the signal to a close tolerance automatically
3. To reduce the aural carrier level of television signals to avoid interference with adjacent cable channels
4. To suppress undesired out-of-band signals

Processing is used also to convert the received signals to a different cable channel and to convert UHF signals to VHF cable channels. Processing of off-air television signals is done either with a heterodyne processor or a demodulator-modulator pair. Signals received by satellite are processed by a receiver and then placed on an appropriate vacant cable channel by a modulator. Similarly, locally originated signals are converted to a cable channel with a modulator.

Pilot-carrier generators provide precise reference levels for the proper operation of the trunk system. Generally, two reference pilots are provided, one near each end of the cable spectrum. Combining networks are used to group all signals from individual processors and modulators into a single output for connection to the CATV network.

In bidirectional systems, a computer system is located at the head end. The configuration of the computer varies with the type of service to be offered and can range from a small microprocessor and single display terminal to multiprocessor minicomputers with many peripherals. Such computers control the flow of data to and from terminals located within the CATV network. See Fig. 9-3 for a typical bidirectional head end.

Interactive services require one or more data receivers located at the head end. Polling of home terminals is controlled by data transmitters. Polling and data collection are controlled by the computer system. Where CATV networks are used for point-to-point data transmission, *modems* are supplied at each end location, and *RF-turnaround converters* are used to redirect incoming upstream signals back downstream.

Modern CATV systems utilize computerized switching systems to program one or more video channels from multiple-program sources. In addition, computer-controlled alphanumeric character generators are used to program automated information channels.

Institutional networks in CATV systems require switching, processing, and turnaround equipment at the head end. Video, data, and audio signals, originating within the network, must be routed back out over either the institutional or subscriber network. One method of accomplishing this is to demodulate the signals to baseband, route them through a switching network, and then remodulate them onto the desired network using demodulators and modulators. Another method is to convert the signals to a common intermediate frequency, route them through a radio-frequency switching network, and then up-convert them to the desired frequency. This method utilizes heterodyne processors.

In larger systems, different areas of the network may be tied together with a central head end utilizing supertrunks or multichannel microwave. In this method, called the *hub system,* supertrunks are high-quality trunks often utilizing FM transmission of video signals or feed-forward amplifier techniques to reduce distortion. Multichannel microwave transmission may use either amplitude or frequency modulation.

Supertrunks or multichannel microwave systems are also used where the pickup point for distant over-the-air stations is located away from the central head end of the system, and to interconnect CATV systems owned by different operators in the same geographic area.

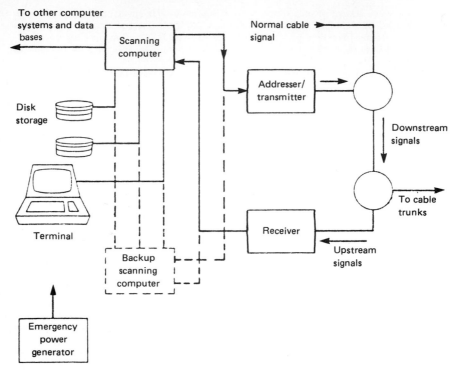

FIG. 9-3 Typical bidirectional head end. *(T. F. Baldwin and D. S. McVoy, Cable Communications,* © *1983. Reprinted by permission of Prentice-Hall, Inc., Englewood Cliffs, N.J.)*

9.3.1 HETERODYNE PROCESSORS. The heterodyne processor, a simplified block diagram of which is shown in Fig. 9-4, is the most common type of head-end processor currently in use. A typical specification is shown in Table 9-2. The processor heterodynes the incoming signal to an intermediate frequency, where the aural and visual carriers are separated. Next the signals are independently amplified, filtered, and level-controlled. The signals are then recombined and heterodyned to the desired output channel. The following functions are performed:

1. Amplification
2. Rejection of adjacent channels (filtering)
3. Automatic level control of visual and aural carriers
4. Setting of desired level ratios between visual and aural carriers
5. Channel conversion (if the channel to be applied to the CATV system differs from the received channel)

The third and fourth functions are required because most television receivers are not designed to operate with adjacent channels, and the ratio of aural to visual signals transmitted over the air would cause adjacent-channel interference in subscribers' television receivers if not changed for operation over CATV networks. A visual-to-aural carrier ratio of 15 to 16 db is typical in CATV systems as a compromise between intolerable adjacent-sound-carrier interference and poor sound quality.

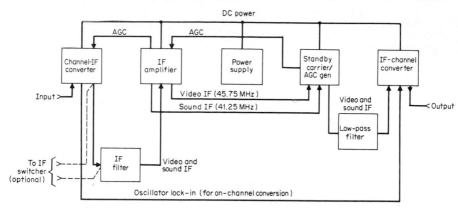

FIG. 9-4 Block diagram of typical heterodyne-type head-end processor. (*Jerold Manufacturing Co., model CHC.*)

Channel conversion (function 5, above) provides a change in the received signal to a transmission channel optimized for application to the cable system. Processors are generally modular, so that by using appropriate input and output modules any input channel can be accepted, processed, and translated to any other location in the 5- to 450-MHz spectrum. Conversion may be necessary when a local broadcast station is carried on a CATV network. If the local station is carried on its original channel, direct pickup of the station over the air within the subscriber's television receiver would interfere with the signal carried over the CATV system. Similarly, when a CATV system receives off the air two stations from different locations broadcasting on the same channel, one must be moved to a different channel on the CATV system. Furthermore, when a UHF channel is received off the air, it must be converted to a channel in the CATV spectrum.

Table 9-2 Heterodyne Processor Specifications (Scientific Atlanta Model 6150)

Characteristic	Specification
Overall sensitivity	100-μV (-20-dBm V) input for $+60$-dBm V output
Input frequency range	Any television channel
Input level dynamic range	-20 to $+30$ dBm V ($+25$ dBm V UHF) with adjacent channels at the same level
Input impedance and voltage standing-wave ratio (VSWR)	75Ω; VSWR less than 1.35:1 over entire dynamic range for 6-MHz channel of interest
Output impedance and VSWR	75Ω; VSWR less than 1.25:1 over 6-MHz channel of interest

Noise figure
   Low band
   High band
   UHF
	
	6 dB at maximum gain
	7 dB at maximum gain
	9 dB at maximum gain

Frequency response (video and color signal)	.5 dB from 0.75 MHz below video carrier to 4.2 MHz above video carrier
Envelope delay from picture carrier to color subcarrier	Flat, ± 25 ns

Table 9-2 Heterodyne Processor Specifications (Scientific Atlanta Model 6150) (*Continued*)

Characteristic	Specification
AGC type	Keyed, sync tip referenced, noise immune
AGC regulation	±0.5-dB maximum output variation with input level variation with input level variation from −20 to +30 dBmV
Adjacent channel rejection	60 dB minimum
Sound limiting	10 dB (minimum) with −25-dBmV input
Frequency	
Picture carrier	45.75 MHz
Sound carrier	41.25 MHz
Standby carrier	
Delay on	10s ±3s (can be internally programmed for instant-on operation)
Delay off	Less than 1s
Standby carrier mode	Unmodulated carrier
	Carrier modulated internally with 15.75 kHz, 0 to 90%
	Carrier modulated with external video, 0 to 90%
	Carrier modulated with external 4.5-MHz signal
Test points front panel	Input and output RF levels 20 dB ±1 dB below actual levels
Frequency accuracy of crystal oscillators	±0.0025% at 77°F (25°C)
Frequency stability of crystal oscillators	
VHF	±0.0025% from +20°F (−7°C) to 120°F (49°C)
UHF	±0.001% from +20 to +120°F
Maximum output levels	
Video carrier	+60 dBmV (less than 0.5 dB of sync compression
Sound carrier	+45 dBmV
Spurious outputs	Greater than 60 dB below video carrier output level from 5 to 300 MHz with video carrier set at +60 dBmV and sound set at +45 dBmV
Output level range for video and substitution carriers	Continuously variable from +45 to +60 dBmV
Output level range for sound carrier	Continuously variable from 10 to 20 dB below picture carrier
Differential gain	±0.5 dB maximum
Differential phase	±1.0° maximum
Output frequency range	Any standard sub-low, VHF, midband, superband, or hyperband channel
External video signal required for 90% modulation of subcarrier	0.5V p-p minimum, sync negative
External 4.5-MHz signal required to produce a sound signal 15 dB below subcarrier level	0.1V p-p minimum

The visual-signal intermediate-frequency passband of a typical heterodyne processor is shown in Fig. 9-5. Note that this curve is not like that of a television set, where the visual carrier is set at a point 6 db down on the response curve. This is because the heterodyne processor is designed to reproduce the received signal with a minimum of differential phase and amplitude, and this is best accomplished with a flat passband. This is one of the reasons that the heterodyne processor has better differential phase characteristics than the modulator-demodulator pair.

Heterodyne processors also contain standby carrier generators, which automatically switch on when the incoming signal disappears owing to the broadcast station going off the air or to an antenna or preamplifier malfunction. This is necessary to maintain a constant visual carrier level on CATV systems which use a television channel as a pilot carrier for trunk automatic level control.

9.3.2 DEMODULATOR-MODULATOR PAIR. In a demodulator-modulator pair, the demodulator is basically a high-quality television receiver, with baseband video and audio output, whereas the modulator is essentially a low-power television transmitter. The demodulator-modulator pair provides increased selectivity, better level control, and more flexibility in switching of input and output signal paths compared with the heterodyne processor or the single-channel amplifier (see Sec. 9.3.3).

The demodulator (Fig. 9-6) consists of an input amplifier, local oscillator and mixer to down-convert the incoming signal to an intermediate frequency, an IF amplifier section, and a video detector. The aural subcarrier is processed by a 4.5-MHz amplifier and demodulated by a discriminator. The input down-conversion is accomplished with either a high-quality conventional television tuner, where the demodulator is tunable, or with a crystal-controlled local oscillator, if the demodulator is fixed-tuned. The input downconverter has high gain, a good noise figure, and a wide range of AGC.

The IF amplifier is very similar to that in a high-quality television receiver. A typical

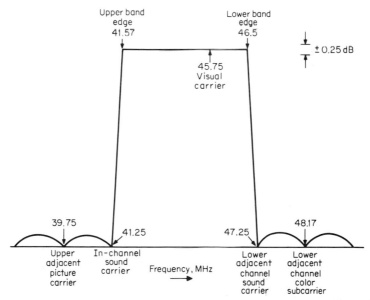

FIG. 9-5 Typical idealized video IF response curve of heterodyne processor. Note that the visual carrier is located on the flat-top portion of the curve to permit improved phase response. [*D. G. Fink and D. Christiansen (eds.), Electronics Engineers' Handbook, 2d ed., McGraw-Hill, New York, 1982.*]

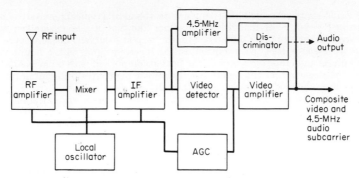

FIG. 9-6   Block diagram of demodulator portion of a demodulator-modulator pair. [*D. G. Fink and D. Christiansen (eds.), Electronics Engineers' Handbook, 2d ed., McGraw-Hill, New York, 1982.*]

IF response curve is shown in Fig. 9-7. Note the point at which the visual carrier is located on the passband. The 4.5-MHz intercarrier sound is taken off before video detection, amplified, and limited to remove video components.

The demodulator must be carefully designed to minimize phase and amplitude distortion. This can be done by linearizing the detector, but quadrature distortion is inherent in a system using an envelope detector on a vestigial-sideband signal and can be corrected for only by video processing. Typical demodulator specifications are shown in Table 9-3.

In the modulator (Fig. 9-8) the composite input is applied to the separation section, where the video is separated from the sound-subcarrier and the video fed to a video

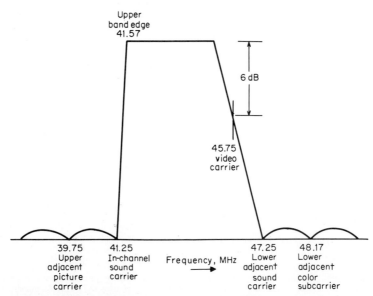

FIG. 9-7   Idealized video IF-response curve of a demodulator. The visual carrier is located 6 dB below the maximum response. [*D. G. Fink and D. Christiansen (eds.), Electronics Engineers' Handbook, 2d ed., McGraw-Hill, New York, 1982.*]

Table 9-3 Demodulator Specifications (Scientific Atlanta Model 6250)

Characteristic	Specification
Overall sensitivity	100-μV (-20-dBmV) input for 1.0-V peak-to-peak video output
Input frequency range	Any standard VHF or UHF television channel, also sub-low, midband, and superband
Input converter local oscillator	
Frequency accuracy of crystal oscillator	$\pm 0.0025\%$ at 77°F
Frequency stability of crystal oscillator	
VHF	$\pm 0.0025\%$ from $+20$ to 120°F
UHF (crystal oven)	$\pm 0.001\%$ from $+20$ to 120°F
Input level dynamic range (with equal level adjacent channels)	
VHF	-20 to $+30$ dBmV
UHF	-20 to $+25$ dBmV
Input impedance and VSWR	75Ω (50 Ω available) VSWR $<1.35{:}1$; VSWR specification applicable to channel of interest over entire dynamic range of input
Noise figure (at maximum gain)	
Low band	6 dB
High band	7 dB
UHF	9 dB
AGC type	Selectable-keyed, sync tip referenced, noise immune, peak detecting, manual gain control
AGC regulation	± 0.5-dB maximum output variation for signal variation over full input level dynamic range
Image rejection	
VHF	>60 dB
UHF	>50 dB
IF rejection	>60 dB
Adjacent channel rejection	60 dB, minimum
IF frequencies	
Video	45.75 MHz
Audio	41.25 MHz
Optional IF switching	Either composite IF or audio IF
Video amplitude/frequency characteristic	10 Hz to 4.18 MHz ± 0.5 dB
Envelope delay	Complement of standard transmitter group delay predistortion 0.0 to 3.6 MHz ± 40 ns $3.58 +170$ ns ± 20 ns
Differential gain	$\pm 2.5\%$ (envelope detector) $\pm 2\%$ (synchronous detector)
Differential phase	$\pm 1°$ (envelope detector) $\pm 0.5°$ (synchronous detector)

Table 9-3 Demodulator Specifications (Scientific Atlanta Model 6250) (*Continued*)

Characteristic	Specification
Chrominance-to-luminance cross talk	3% (envelope detector) unmeasurable (synchronous detector); above values are percentage of blanking-to-white-level luminance shift as chrominance signal varies from 0 to 100 IRE units; 87.5% modulation depth
Zero chop position, length	Start-time adjustable from three to six lines after vertical sync pulse; width adjustable one to three lines
Video output level	Variable 0 to 1.5 V peak-to-peak at each of two outputs
Video output impedance	75Ω (30-dB minimum return loss)
Squelch	On, delayed approximately 10 s; off, instantaneous; adjustable threshold
Optional carrier loss relay	Provides capability for remote indication of loss of input signal
Sound output levels	
Audio	+11.5-dBm maximum across 600-Ω balanced load
Aural subcarrier	0.2-V peak-to-peak maximum across 37.5 Ω (levels adjustable down to 0)
Audio frequency response	±0.5 dB, 30 Hz to 15 kHz with deemphasis
Deemphasis time constant	75 μs
Audio harmonic distortion	1% maximum at any frequency, maximum deviation and output level
Audio monitor	Standard phone jack with level control— will drive most headphones
Output level controls	20-dB minimum range on all level controls

amplifier. From this point the video signal is processed, mixed with a carrier oscillator to obtain the desired output frequency, amplified to obtain the necessary power level, and filtered to remove any undesired products. Following the RF amplifier is the vestigial-sideband filter required to remove most of one sideband to allow adjacent-channel operation. The characteristics of this filter are designed to meet the same requirements as that of a television transmitter. Following the filter, the carrier-level adjustment occurs.

The audio signal is handled in one of two ways. If the video input is composite, the audio subcarrier is separated from the video signal, filtered, amplified, and limited, then mixed with the visual carrier. If the audio and video signals are fed to the modulator separately, the modulator generates a frequency-modulated subcarrier 4.5 MHz above the visual carrier frequency and combines it with the visual carrier.

Most modulator designs generate the visual and aural carriers at the standard television-receiver intermediate frequencies (45.75 and 41.25 MHz). An up-converter then places the signal on the desired CATV channel.

In the modulation process, a high percentage of modulation is desirable for high signal-to-noise ratio, but this produces differential gain and phase on the color subcarrier. The vestigial-sideband filter must be optimized to minimize phase distortion near the

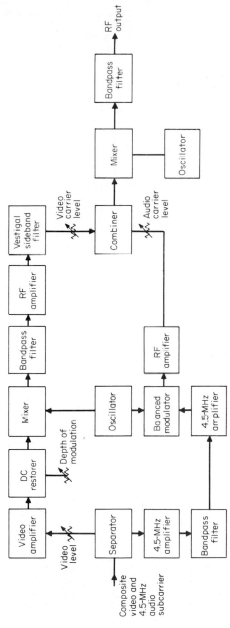

FIG. 9-8 Modulator block diagram. [*D. G. Fink and D. Christiansen (eds.), Electronics Engineers' Handbook, 2d ed., McGraw-Hill, New York, 1982.*]

Table 9-4 Modulator Specifications (Scientific Atlanta Model 6350)

Characteristic	Specification
Video	
Input type	Composite video, sync negative
Input level	0.5-V p-p minimum for 87.5% depth of modulation
Input impedance	75-Ω, unbalanced (return loss 30-dB min) or high-impedance loop-through
Frequency response	±0.5 dB from 10 Hz to 4.2 MHz
Differential gain	±0.18 dB maximum at 87.5% modulation
Differential phase	±0.5° maximum at 87.5% modulation
Hum and noise	60 dB down with respect to 90% modulation
Tilt	1% maximum on 60Hz, 50% square wave
Audio input	

Input level	Normally set for −10 to +10 dBm; internally switchable to two other ranges:
	    +5 to +25 dBm
	    −25 to −5 dBm

Input impedance	600-Ω balanced, may be field-modified for high-impedance bridging input, may be used with unbalance sources
Common-mode hum rejection	40 dB
Frequency response	±0.5 dB from 30 Hz to 15 kHz
Carrier shift with modulation	±100 Hz or less
Harmonic distortion	0.5% maximum, 30 Hz to 15 kHz at 25 kHz deviation
FM hum and noise	60 dB down with respect to 25 kHz deviation
Intercarrier frequency tolerance	Within ±500 Hz of being 4.5 MHz above video carrier (−7 to +50°C)
4.5-MHz FM input	
Input level	10 to 200 mV rms
Input impedance	75-Ω unbalanced (normally supplied as composite picture and aural subcarrier input)
RF	
Output frequency	Any standard VHF channel, sub-low, midband, superband; transposed video-sound channels available; incremental and harmonically related coherent outputs are also available, as are nonstandard output frequencies
Output impedance	75-Ω unbalanced (VSWR 1.35:1)
Output level	+40 to +60 dBmV continuously variable
Spurious outputs	60 dB below video carrier with video carrier at +60 dBmV and sound carrier at +45 dBmV
Frequency tolerance stability	+8 kHz VHF midband; ±10 kHz superband
Group delay response	Meets FCC predistortion requirements for color transmission
Vestigial-sideband response	−20 dB at channel edge; −40 dB at adjacent picture and sound carrier frequencies and all frequencies farther removed from channel

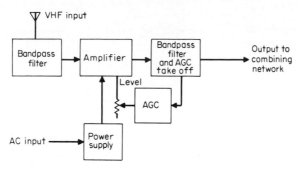

FIG. 9-9 Block diagram of single-channel amplifier (strip amplifier) with AGC. [*D. G. Fink and D. Christiansen (eds.), Electronics
Engineers' Handbook, 2d ed., McGraw-Hill, New York, 1982.*]

picture carrier. The compensation required for the demodulator-modulator pair is usually provided in a video equalizer in the modulator. Typical modulator specifications are
shown in Table 9-4.

The modulator portion of a demodulator-modulator pair is in common use in CATV
systems to place locally generated signals and signals received through microwave or satellite receivers on the cable.

9.3.3 SINGLE-CHANNEL AMPLIFIER.

Single-channel amplifiers, or strip ampli-
fiers, are the simplest head-end processors. They amplify one channel and have bandpass
and bandstop filters to reject other channels. In simplest form they consist of a filter,
amplifier, and power supply. More elaborate designs include the above functions plus
automatic gain control and, in some cases, independent control of aural and visual carrier
levels. A representative type is shown in block diagram form in Fig. 9-9. Typical specifications are shown in Table 9-5.

Single-channel amplifiers are used where the desired signal levels are fairly high and
the undesired signals low or absent. They do not offer the selectivity of the more complicated heterodyne and demodulator-modulator processors. They also generally lack
such features as independent control of aural and visual carrier levels and the ratio
between them, independent AGC of these carriers, or limiting for the aural carrier. Their
use is restricted to 12 channels because they cannot convert input channels to different
output channels, and they are difficult to use in adjacent-channel applications because
of their limited selectivity.

Single-channel amplifiers provide an inexpensive means of processing signals for master antenna systems, very small CATV systems, and specialized RF distribution systems
used in business communications.

Table 9-5 Typical Single-Channel Amplifier Specifications

Minimum gain	51 dB (channels 2 to 13 and FM)
Maximum output for 0.5-dB gain compression	+66 dBm (2V)
AGC range	40 dB
AGC capability	±0.5 dB for 40-dB input change
Minimum input for TASO† "excellent" picture	0 dBm
Bandpass	6 MHz, ±0.5 dB for TV
Skirt selectivity	25 dB down ±9 MHz from midchannel

†Television Allocations Study Organization.

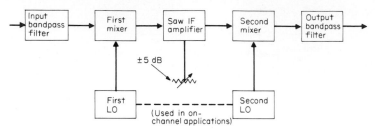

FIG. 9-10   FM-radio heterodyne-processor block diagram.

9.3.4 HEAD-END EQUIPMENT FOR NONVIDEO SERVICES. CATV systems have the capability for carriage and distribution of signals other than television. This includes aural radio services, data transmission, interactive signaling and specialized audio/video transmissions.

FM radio service is commonly provided on CATV systems in the frequency band between 88 and 108 MHz, duplicating the over-the-air FM band. Heterodyne processors are usually utilized to place the FM radio signal in the CATV cable spectrum, at a point where it will not be interfered with by direct pickup of over-the-air signals. Figure 9-10 shows a block diagram of the FM-signal heterodyne processor.

Data transmission on a CATV network is accomplished through the use of modems designed for the particular bandwidth, frequency, and data transmission speed required. Many CATV systems will dedicate a portion of the spectrum on the subscriber network to data transmission or install a separate institutional network for data transmission. These systems are usually bidirectional to provide for interactive data transmission and interactive signaling.

Modems are available for use on CATV systems to carry data at speeds of up to 1.544 Mbits/s. Modems are also available for carriage of voice signals on cable. Table 9-6 shows the specifications for a data modem presently in use on cable systems.

Although most modems in use in CATV systems today use discreet frequencies for each point-to-point link, various time division and contention architectures have been proposed, and some equipment is becoming available.

Table 9-6   RF-Data Modem, E-COM Model TRM-202 Specifications

Data rates	50–9600 bits/s
Data formats	Binary-serial, asynchronous or synchronous; RS-232-c 20-mA current loop
Clock	DTE/DCE
RF frequencies	5–250 MHz, crystal controlled
Modulation	Three-level FSK† with alternate-space inversion
Bandwidth	100 kHz (-40 dB)
Channel spacing	200 kHz, typical
Receive level	-20 to $+10$ dBmV,
Output level	$+40$ dBmV, adjustable
Size	1.75 in H $\times$ 8.125 in W $\times$ 15 in D (45 mm $\times$ 206 mm $\times$ 382 mm)

†FSK = frequency shift keying.

Source: Courtesy E-COM Products Division, AM Cable TV Industries, Inc.

9.4 TRUNK SYSTEM

The trunk system (see Fig. 9-11) is designed for bulk transportation of a multiplicity of channels. The trunk system may connect a number of distribution points or may inter-connect widely spaced sections of a cable system or more than one cable system. When a trunk system is used to carry a multiplicity of channels over very long distances without any intermediate distribution, it is commonly called a *super trunk*.

The coaxial cable used in trunk systems is made with a solid aluminum shield, a gas-injected polyethylene dielectric insulator, and a copper or copper-clad aluminum center conductor. Underground cables also have an outer polyethylene jacket with a flooding compound underneath to seal small punctures which might occur during or after instal-lation. Trunk cables are usually ¾ to 1 in (2 to 2.5 cm) in diameter and typically have a loss of about 1 dB per 100 ft (30 m) at 400 MHz.

9.4.1 AMPLIFICATION.
Amplifiers with equalizers are utilized to overcome the losses in the coaxial cable used in the trunk system. From the output of a given repeater ampli-fier, through the span of coaxial cable and the equalizer, to the output of the next repeater amplifier, unity gain is required so that the same signal level is maintained on all channels at the output of each trunk unit.

As indicated in Fig. 9-11, repeater amplifiers are spaced from 1400 to 2000 ft (427 to 610 m), depending on the diameter of the coaxial cable. This represents an electrical loss of about 20 dB at the highest frequency carried. Systems with trunk-amplifier cascades of up to 64 amplifiers are possible, depending on the number of channels the system carries, the performance specifications chosen for the system, the modulation scheme utilized, and the distortion-corrrection techniques used. Table 9-7 gives representative figures of distortion versus the number of amplifiers cascaded.

9.4.2 LONG-DISTANCE TRUNKING TECHNIQUES.
For long-distance trunking with minimum distortion, frequency-modulation techniques are usually utilized. While frequency modulation requires a larger portion of the available bandwidth than ampli-

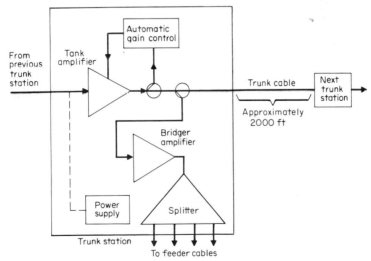

FIG. 9-11   Trunk system. *(T. F. Baldwin and D. S. McVoy, Cable Communications,* © *1983. Reprinted by permission of Prentice-Hall, Inc., Englewood Cliffs, N.J.)*

Table 9-7   Distortion versus Number of Amplifiers Cascaded in a Typical CATV System

No. of amplifiers in cascade	Cross modulation, dB	Second order, dB	S/N ratio, dB
1	−96	−86	60
2	−90	−81.5	57
4	−84	−77	54
8	−78	−72.5	51
16	−72	−68	48
32	−66	−63.5	45
64	−60	−59	42
128	−54	−54.5	39

†Dashed line indicates lower limit of acceptable system performance.

tude modulation, the resultant distortion in the delivered signal is considerably lower, and it thus permits trunking for longer distances.

Bridging amplifiers are used to feed signals from the trunk system to the distribution system, en route to the subscriber drops. A wide-band directional coupler is used to select a portion of the signal from the trunk amplifier to be fed to the bridger amplifier. The bridging amplifier acts as a buffer, isolating the distribution system from the trunk system while providing the signal level required to drive the distribution lines.

9.5 DISTRIBUTION SYSTEM

As shown in Fig. 9-12, up to four distribution lines are fed from a bridging station. Distribution lines are routed through the subscriber area. Amplifiers and tap-off devices are

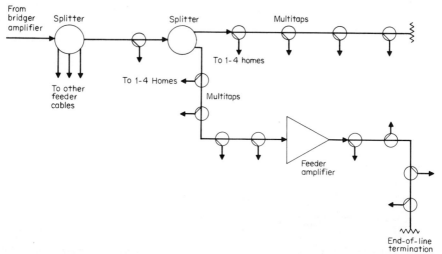

FIG. 9-12   Distribution and feeder system. *(T. F. Baldwin and D. S. McVoy, Cable Communications, © 1983. Reprinted by permission of Prentice-Hall, Inc., Englewood Cliffs, N.J.)*

provided to meet the needs of the subscriber density of the particular area being served. The cable used in distribution systems is identical to that used in trunks, except that the diameter is somewhat less, ½ or ⅝ in (1.3 or 1.6 cm), and the loss is higher, typically 1.5 db per 100 ft (30 m) at 400 MHz.

As the signal proceeds along the distribution line, the attentuation of the coaxial cable and the insertion loss of tap devices reduce its level to a point where small line-extender amplifiers may be required. These inexpensive booster amplifiers have a gain of 25 to 30 dB. Generally, no more than two are cascaded, since they operate at relatively high output levels and create considerable cross-modulation and intermodulation distortion. In addition, the length of distribution systems is limited because distribution amplifiers usually do not have automatic level-control circuitry to compensate for variations in cable attentuation due to temperature changes. This attenuation change creates significant level changes at the ends of the distribution system which adversely affect picture quality.

9.5.1 MULTITAPS.

A commonly-used tap-off device is the *directional-coupler multitap*, which allows two, four, or eight subscribers to be served by one unit. Individual taps, called *pressure taps*, fastened directly to the distribution cable were once common in cable systems but are now utilized only in extremely small systems or special-application systems, owing to the undesirable signal reflections produced by their mechanical distortion of the coaxial cable. Figure 9-13 shows the design of a typical multiple-output tap.

The multitap removes the appropriate amount of energy from the distribution cable with a directional coupler and splits this energy into multiple paths, each proceeding from this tap location via a subscriber drop line into the subscriber home. Multitaps are selected to produce a signal level adequate to provide a good signal-to-noise ratio at the subscriber's receiver, but not so high as to create intermodulation in the television set. Multitaps introduce attenuation into the distribution system which is related to the amount of signal tapped off to the subscribers. The tap portion of the multitap is of the directional type, resulting in a much higher attenuation from the tap back into the cable downstream. This minimizes the possibility of interference originating in the drops reaching subscribers served by other multitaps and attenuates reflected signals in the distribution cable which might otherwise cause visible ghosts on subscribers' receivers.

The splitter portion of the multitap unit is of the hybrid design to introduce substantial isolation from reflections or interference coming from the home of a subscriber, and

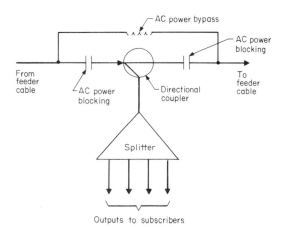

FIG. 9-13 Multitap block diagram.

to prevent such interference from affecting the picture quality of other subscribers connected to the same multitap.

9.5.2 ADDRESSABLE SUBSCRIBER TAPS. Since the tap is the final service point immediately prior to the subscriber location, this is the point where the channels authorized to individual subscribers can be controlled. An addressable tap can be used in place of a standard multitap to control access to basic cable service or individual channels from a central location. The addressable tap consists of a small receiver and address recognition circuitry which control a diode switch to enable or disable the entire spectrum to a drop, and notch filters, jamming oscillators, and/or scrambling devices to control individual channels or groups of channels.

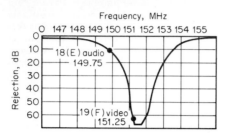

FIG. 9-14 Notch-filter, negative-trap method for rejection of premium service channel 19(F) to nonauthorized subscribers. Authorized channel 18(E) audio attenuated only 10 dB.(*Vitek Corp.*)

9.5.3 TIER AND CHANNEL CONTROL. Control of service tiers and channels at the subscriber's location is also effected through the use of filters which can be added to the output of the multitap. These can take the form of notch filters which will remove a particular channel or a group of channels from service. Positive filters can also be used to remove an interfering carrier which would otherwise cause a particular channel to be presented in a scrambled mode. Figure 9-14 shows the characteristics of typical notch filters utilized for control of premium program service in cable systems.

9.6 TRUNK AND DISTRIBUTION AMPLIFIERS

Since the trunk line is the main artery of the CATV system, the trunk amplifiers must provide minimum degradation to the system. A typical trunk amplifier station includes plug-in modules for trunk amplification, bridging, automatic level control, and splitting signal to feed multiple distribution cables. Amplifiers in CATV systems are powered by multiplexing 30- or 60-V alternating current on the coaxial cable, and trunk stations contain power supplies which supply dc voltage for operation of the above modules. Figure 9-11 illustrates a typical trunk station.

Specifications of a typical trunk amplifier are presented in Table 9-8. Figure 9-15

Table 9-8 Trunk and Bridge Amplifier, Jerrold Starline X-1000-450 Specifications

Performance parameters		Notes†	AGC/ASC specifications
Passband (forward)	MHz	1	50–450 MHz
Response flatness:			
Trunk amplifier	±dB	2	0.4
Bridger or distribution amplifier			0.75
Minimum full gain:			
Trunk amplifier	dB	3	26
Bridger or distribution amplifier			45

Table 9-8 Trunk and Bridge Amplifier, Jerrold Starline X-1000-450 Specifications (*Continued*)

Performance parameters		Notes†	AGC/ASC specifications
Gain control range:		4	
Trunk amplifier	dB		0–8
Bridger or distribution amplifier			0–9
Slope control range at 450 MHz			
Trunk amplifier	dB		2–8
Bridger or distribution amplifier			2–8
Control accuracy	±dB	5, 6	0.5
Control pilots			
ASC: Tuned to channel			4
Tunable to channels			2 or 3
Operating range min/max	dB		21/29
AGC: Tuned to channel			36
Tunable to channels			35 or 37
Operating range min/max	dB		27/35
Return loss at 75-Ω impedance	dB		16
Noise Figure	dB	7	7.5
Rec oper levels w/equalizers:	dBmV		
Trunk in			9
Trunk out			25/31
Bridger or dist amp out			39/47
Distortion char at rec oper level (60 channels)			
Composite triple beat	−dB	8,9	
Trunk			87
Bridger or distribution amplifier			57
Cross modulation	−dB	8	
Trunk			85
Bridger or distribution amplifier		8,9,10	57
Second order	dB		
Trunk			85
Bridger or distribution amplifier			69
Hum modulation by 60-Hz line	−dB		70
Chroma delay	ns		
CH-2			6
CH-3			4
T-7			8
T-10			6
Other channels			2

† *Notes:* (1)For return amplifier specifications refer to the appropriate return amplifier. (2) Single station measured at recommended operating gain, with slope compensated by cable. Trunk system flatness is $N/10 + 1$ typical at balance temperature. (3) Measured from trunk input to trunk output or single bridger output at 450 MHz without equalizer. (4) Range is referenced from minimum full gain. (5) AGC/ASC trunk amplifier output level accuracy, at pilots, for -3 dB $+ 4$ dB of input level change at 50 MHz and 450 MHz. (6) Manual trunk amplifier output level accuracy, at 450 MHz, from ambient over the range of -40 to $+140°$F (-40 to $+60°$C) for 18-dB cable spans. (7) Measured in station without equalizer at maximum gain, flat slope. (8) Standard channels, flat, per NCTA test methods, or sloped where indicated by dual levels. (9) Measured with CW signals and spectrum analyzer. (10) Any combination of channels 2, 20 (G), and 13.

Source: Jerrold Division, General Instrument Corp.

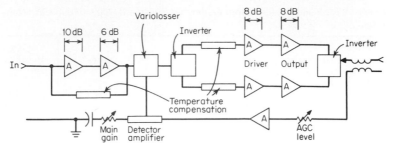

FIG. 9-15 Functional block diagram of a trunk-AGC amplifier and its gain distribution; A = amplifier. [*D. G. Fink and D. Christiansen (eds.), Electronics Engineers' Handbook, 2d ed., McGraw-Hill, New York, 1982.*]

shows the block diagram of a typical trunk amplifier with automatic level control (ALC) and the distribution of gain. The first stage is designed for low-noise figure and the remainder for low cross modulation. Modern trunk amplifiers are built using hybrid circuits containing both integrated circuits and discreet components in a single package. Such hybrid modules are available with a variety of gain and performance characteristics.

Because coaxial cable changes attenuation with temperature, trunk amplifiers must be designed to compensate for such changes and correct their output level to a predetermined standard. This is accomplished through automatic level control circuitry. Typically, a CATV system uses two frequencies as system pilots. One is at the lower end of the spectrum, and the other near the upper end. In some cases separate carriers are placed on the system at the head end, and in other cases existing video channels are used as pilots. The ALC circuits sample the trunk amplifier output on these two pilots and feed control voltages back to diode attenuators at an intermediate point in the trunk amplifier. The upper pilot adjusts the gain of the amplifier, while the lower pilot adjusts the slope of an equalizer. In this way the entire spectrum is maintained at a constant output level.

A second form of trunk amplifier using feed-forward distortion correction is also available and permits carriage of a larger number of channels for equivalent gain than the standard trunk amplifier. Feed-forward amplifiers are considerably more expensive than standard trunk amplifiers, and are used mostly for supertrunk and unusually long trunk lines. Table 9-9 outlines the performance specifications of a feed-forward amplifier.

Bridger modules are provided in trunk stations where connection from the trunk to the distribution system is required. These bridger amplifiers tap off a portion of the energy in the trunk system and provide a buffer with multiple outputs to feed distribution lines. Table 9-8 shows the specifications of a typical bridger module.

Line-extender stations, like trunk stations, usually contain a plug-in amplifier module and power supply. The specifications for the line-extender amplifier utilized to extend the length of the distribution plant to serve individual subscriber locations is shown in Table 9-10 along with a functional block diagram in Fig. 9-16.

9.6.1 CROSS MODULATION. The maximum output level allowable in CATV system amplifiers is almost always determined by cross modulation of the picture signals. Cross modulation is most likely to occur in the output stage, where levels are high. The cross-modulation distortion products at the output of a typical amplifier (Fig. 9-15) are 96 dB below the desired signal at an output level of +32 dBmV. Note from Fig. 9-15 that the gains of the third and fourth stages each are 8 dB. With an amplifier output level of 32 dBmV, the output level of the third stage therefore would be

$$P_3 = P_4 - G_4 = +32 - 8 = +24 \text{ dBmV}$$

Table 9-9 Feed-Forward Amplifier Specifications, Century III 4100 Series
Amplifier—400 MHz

Response	
Frequency range, MHz	45–400
Flatness, dB, at operating gain	+0.25
Return loss, dB, all ports	16
Typical operating levels, dBmV	
54–88 MHz	29
318–400 MHz	35
Gain	
Full-gain minimum, dB	29
Operating gain, dB	26
Gain control range, dB	8
Dynamic range	
Distortion at typical operating levels	
Cross modulation, dB (52 channels)	−94
Composite triple beat, dB (52 channels)	−101
Intermodulation, dB (2 channels)	−92
Hum modulation, dB	−66
Noise figure, dB	10.0
Rated output	
At −77 dB cross modulation, dBmV	+43.5
At −70 dB composite triple beat, dBmV	+50.5
At −80 dB second order, dBmV at	
channel 2	+47.0
AGC operation	
Low pilot	Channels 3, 4, and 5; 73 or 107.9 MHz
High pilot	Channels R or W
Output stability, dB	±0.5

Figure 9-17 illustrates the relationship between cross-modulation distortion and amplifier output level. (Note that the distortion products at the output of a typical amplifier are 96 dB below the desired signal at an output level of +32 dBmV.) As shown, a 1-dB decrease in output level produces a corresponding 2-dB decrease in the cross-modulation distortion. By use of this relationship, the output cross-modulation distortion of the third stage would be −112 dB. It can be seen that the contribution by the third stage to the total cross-modulation distortion is only 1.27 dB and that the output stage, where signal levels are high, principally determines the amplifier cross-modulation distortion. Thus, it follows that the contribution of the first and second stages to the total amplifier cross modulation is insignificant.

Cross modulation increases by 6 dB each time the number of amplifiers in cascade doubles, and also increases 6 dB each time the number of channels carried is doubled.

As noted above, a typical single-trunk amplifier introduces a cross-modulation component of approximately −96 dB relative to the normal visual carrier output level. It has been determined by subjective tests that at about −60 dB cross modulation becomes objectionable to CATV subscribers. Thus, approximately 64 amplifiers can be cascaded before cross modulation becomes objectionable.

9.6.2 FREQUENCY RESPONSE. The most important design consideration in CATV amplifiers is the frequency response. A frequency response flat within ±0.25 dB over 40 to 450 MHz is required of an amplifier carrying 50 or more 6-MHz channels to permit a cascade of 20 or more amplifiers. To meet this requirement, special attention must be paid not only to the high-frequency parameters of the transistors and associated components but to good high-frequency layout and packaging techniques as well.

Table 9-10 Line-Extender Amplifier, Jerrold JLE Series Specifications

Performance parameters	Notes†	Specifications	
		JLE-6-450-2W	JLE-7-450-2W
Passbands		52-450 MHz 5-32 MHz	
Response flatness		±0.75 dB, all models	
Min. full gain, at 450 MHz with 9-dB slope	2.7	41 dB	33 dB
Typical operating gain	3	37 dB	29 dB
Manual gain control range		6 dB min., all models	
Manual slope control range	6	0 to 9 dB referenced to 450 MHz; additional compensation for attenuation vs. frequency by optional equalizer Models SEE‡, and vs. temperature by optional compensator Models CTF-‡	
Noise figure; with JXP-O and without equalizer, at 450 MHz minimum full gain referenced to highest channel		10 dB Ref. to Ch. 61	9 dB Ref. to Ch. 61
Distortion characteristics with 9-dB slope without equalizer, with typical output level of:	4, 5	47 dBmV	47 dBmV
2d order beats, Channels 2, 20 (G), 13		72 dB	72 dB
Composite triple beat, for 60 channels		59 dB	59 dB
Maximum chroma delay Forward		Ch 2: 7 ns; Ch. 3: 4 ns; Ch. 4: 2 ns; all other forward channels less than 1 ns	
Return		Ch. T7 and T8: 4 ns; Ch. T9: 7.5 ns; Ch. T10: 15 ns	
Hum modulation, by 60 Hz source		60 dB or better, all models	
Operating ambient temperature range	1	−40° to +60°C all models	
Terminal match, at 75-Ω impedance, input and output		16 dB minimum return loss	

†Notes: (1) Specifications apply to module installed in housing. (2) Minimum full gain specification is referenced to use without optional models SEE-* or optional temperature compensator models CTF-*; for relevant insertion losses see pertinent catalog sheets. (3) Typical operating gain is referenced to use with equalizer and temperature compensator and with factory-inserted JXP-O. (4) When JLE-6 or JLE-7 station is used with 40 or more channels, we recommend a phase-locked head end. (5) Phase lock rated output established at 3dBmV below subjective level of barely perceptible distortion. (6) 9 dB slope implies channel 2 level operated 9 db below channel 61. (7) Composite triple beat is measured with CW signals and spectrum analyzer at phase lock rated output.

Source: Jerrold Division, General Instrument Division.

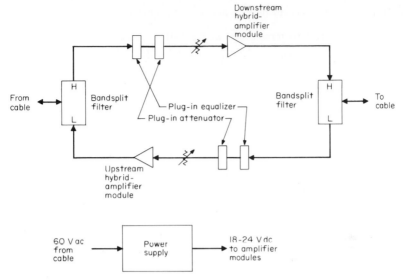

FIG. 9-16 Bidirectional line-extender block diagram.

The circuit shown in Fig. 9-18 is representative of amplifiers designed to achieve flat response. By properly designing the feedback network, comprising C_1, R_1, and L_1, sufficient negative feedback can be used to maintain a nearly constant output over a wide frequency range. The collector transformer T_2 and the splitting transformers T_1 and T_3 play an important part in the amplifier's performance. In a representative amplifier of this type, transformer T_2 is bifilar-wound on a ferrite core and presents an essentially constant 75-Ω impedance over the entire frequency range. Transformers T_1 and T_3 are similar in construction but have the additional function of providing the required 180°

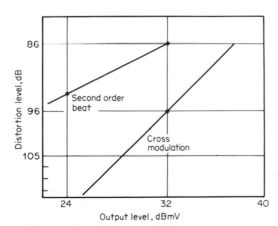

FIG. 9-17 Cross-modulation and second-order beat distortion vs. output level of a trunk amplifier. [*D. G. Fink and D. Christiansen (eds.), Electronics Engineers' Handbook, 2d ed., McGraw-Hill, New York, 1982.*]

phase shift for the push-pull pair Q_1 and Q_2, while maintaining a 75-Ω input-output impedance.

The sweep-response patterns shown in Fig. 9-19 illustrate the wide bandwidth and amplitude linearity necessary to allow cascading of trunk amplifiers. The response irregularity indicated in this figure is, by itself, of little concern, but if this small irregularity (perhaps 0.2 dB) occurs at the same frequency in each amplifier, it becomes a response "signature" and accumulates to a magnitude of 6.4 dB at the end of 32 amplifiers. The degree of signature in a high-quality CATV trunk amplifier is typically no more than 0.1 dB.

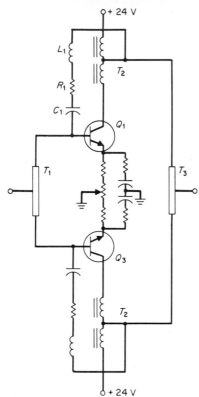

FIG. 9-18 High-performance push-pull wideband amplifier stage for a trunk amplifier. [*D. G. Fink and D. Christiansen (eds.), Electronics Engineers' Handbook, 2d ed., McGraw-Hill, New York, 1982.*]

9.6.3 SECOND-ORDER DISTORTION.

Originally, CATV systems carried only 12 channels, spaced so that the second harmonic of any channel would not fall on any other channel. As systems were expanded to carry more channels covering more than one octave, the second-order-beat distortion characteristics of an amplifier became important to CATV equipment manufacturers. A number of approaches have been used to reduce amplifier second-order distortion, but the most successful is the push-pull configuration. The circuit illustrated in Fig. 9-18 is representative of the 75-Ω push-pull building-block approach. Each push-pull stage is of this basic configuration and differs essentially in component values only. The splitting transformers T_1 and T_3 are the key to the operation of the push-pull amplifier, since the maximum second-order cancellation occurs when the proper 180° phase relationship is maintained over the full amplifier bandwidth. Additionally, it is necessary that the gain be equal in both push-pull halves and that the individual transistors be optimized for maximum second-order linearity. Second-order distortion increases approximately 1.5 dB for each 1 dB in amplifier output level. Table 9-7 illustrates how second-order distortion increases with amplifier cascade. A typical single amplifier introduces a second-order distortion of approximately −86 dB relative to the visual carriers. Subjective tests have shown that second-order beats greater than −55 dB are objectionable. It can be seen that approximately 64 amplifiers can be cascaded before that level is reached.

9.6.4 NOISE FIGURE.

A typical trunk amplifier hybrid module has a noise figure of approximately 7 dB. However, equalizers, bandsplit filters, and other components which precede the module add to the overall station noise figure, which is usually approximately 10 dB.

A single amplifier produces a signal-to-noise ratio of approximately 60 dB with a typical signal input level. Subjective tests have shown that a signal-to-noise ratio of approximately 43 dB is acceptable to most CATV subscribers. Table 9-7 shows that noise

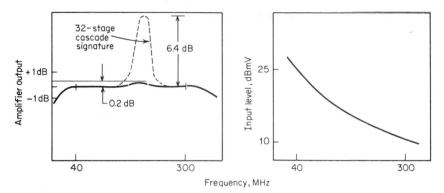

FIG. 9-19 Input and output levels of the amplifier shown in Fig. 9-18 as functions of frequency over the band 40 to 300 MHz. The dashed curve shows the cumulative effect of the 0.2-dB hump when the signal is passed through 32 such stages in cascade. [*D. G. Fink and D. Christiansen (eds.), Electronics Engineers' Handbook, 2d ed., McGraw-Hill, New York, 1982.*]

increases 3 dB each time the number of amplifiers in cascade is doubled, and a cascade of 32 amplifiers would produce a signal-to-noise ratio of 45 dB, a typical CATV design specification.

9.6.5 HUM MODULATION. Powering of CATV amplifiers through the coaxial cable results in hum modulation of the radio frequency signals carried in the network. Several factors cause this hum modulation, including power-supply filtering and saturation of inductors used to bypass ac current around active and passive devices within the system. A typical specification for hum modulation in a CATV system is −40 dB relative to the visual carriers, a level which subjective tests have shown to be acceptable to subscribers.

9.7 SUBSCRIBER-PREMISE EQUIPMENT

The output of the tap device feeds a 75-Ω coaxial cable drop line into the home. Typically, this cable is about ¼ in (0.6 cm) in diameter and is constructed of an outer polyethylene jacket, a shield made of aluminum foil surrounded by braid, a polyethylene insulator, and a copper center conductor. Loss at 400 MHz is typically 6 db per 100 ft (30 m).

9.7.1 ISOLATION. Between the subscriber's television receiver and the distribution cable isolation is provided for the purposes of both safety and signal purity. The multitap utilizes blocking capacitors to prevent amplifier line powering energy from reaching the subscriber drop cable. As the drop cable enters the subscriber's premises, a grounding connection is provided in accordance with local regulations and/or utility requirements. The grounding connection protects the television receiver from power surges, such as those caused by lightning discharges, and protects the subscriber from the possibility of shock due to voltages which might otherwise be present on the shield of the drop cable. The signal from the unbalanced 75-Ω coaxial cable is converted to a 300-Ω balanced output for connection to the subscriber's television set with an unbalanced-to-balanced matching transformer. Where high ambient signal levels exist, excellent balance is required on the 300-Ω side of this transformer. Transformers are available with very good balance plus a form of Faraday shield, which helps to minimize direct pickup of over-the-air signals. The matching transformer also provides one further level of isolation to the carriage of energy along the sheath of the cable.

9.8 CONVERTERS AND DESCRAMBLERS

In order to allow reception of more than the 12 VHF channels on a standard television set, most CATV systems provide a converter at the subscriber's receiver. In addition, to secure pay television services from unauthorized viewing, signal scrambling is provided on some channels, and descramblers are required at the receiver location.

9.8.1 CONVERTERS.

The simplest and least expensive means of conversion is provided by a block converter. In one type, a group of channels, typically midband or superband channels, is converted to VHF channels 7 to 13. In another, the entire CATV spectrum is up-converted to the UHF band, where the subscriber may receive the CATV channels on a standard receiver. Block converters are generally used when a small number of additional channels are desired at a low cost.

Block converters consist of an input bandpass filter, a mixer, a local oscillator, an output bandpass filter, and sometimes an output amplifier. See Fig. 9-20 for a block diagram of a block converter.

Most converters in use today are tunable. See Fig. 9-21 for a block diagram. Incoming signals are first routed through an input bandpass amplifier (sometimes coupled with a buffer amplifier), then converted to a UHF intermediate frequency utilizing a tunable local oscillator and mixer. The signal is then put through a bandpass filter, and next down-converted to a single VHF channel with a fixed-frequency local oscillator and mixer. The output channel is filtered and fed to the subscriber's receiver. Tunable converters are available in many configurations, including models which are located on top of the television set, models with the channel selection switch connected to the converter by a thin cord, and models with channel selection done with a small hand-held remote-control unit which communicates with the converter by infrared light. Tunable converters are used extensively in the cable industry as a means of providing 30 or more channels to subscribers.

The earliest tunable converters utilized switched inductors and capacitors to select channels. Later versions utilized varactors, controlled by a voltage developed by a series of potentiometers selected by the channel switch. The most recent tunable converters utilize frequency synthesizers controlled by microprocessors for the first local oscillator.

The least-expensive tunable converters have a fine-tuning control on one of the local oscillators to allow the subscriber to compensate for converter frequency drift. More sophisticated converters utilize automatic frequency control circuitry to control drift. In these converters, the output signal is compared with the standard output frequency through a discriminator, and a control voltage is generated, which adjusts the frequency of one of the local oscillators through a varactor.

The most recent converters with microprocessor-controlled frequency synthesizers are very frequency-stable and require no additional frequency control, since they are referenced to a crystal.

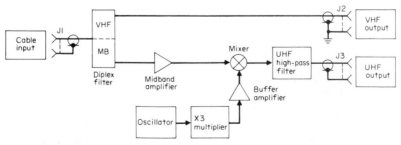

FIG. 9-20   Block-converter block diagram. *(Sylvania.)*

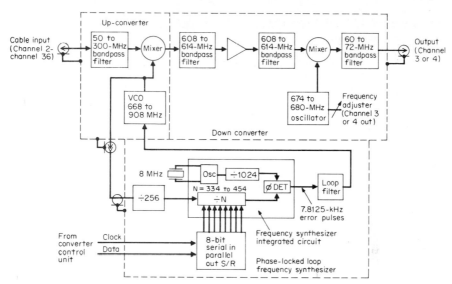

FIG. 9-21 Tunable-converter block diagram. *(Sylvania.)*

9.8.2 SCRAMBLING.

Descrambling of premium programming is done in many ways. The simplest is the trap. In the positive-trap method, one or more interfering carriers are inserted in the television channel at the head end. Authorized subscribers are given filters which attenuate the interfering signals with minimal distortion of the desired video signal. Though the positive-trap method is very inexpensive, it is relatively easy to defeat, since all that is needed to receive a clear picture is a notch filter of the correct frequency, a device easily constructed by someone with minimal technical knowledge.

The negative-trap method utilizes a sharp-notch filter on the visual carrier frequency of the channel. These filters are installed outside the homes of those subscribers who are not authorized to receive the pay-television channel. This method probably is the most secure security system in use today, since defeating the trap requires access to the CATV plant located outside the home and often on a utility pole. However, control of more than one channel is difficult, since a separate trap is usually required for each channel. In addition, when only a small percentage of subscribers desire the pay-television channel, the cost of the negative trap method becomes prohibitive, since a trap must be installed on the drop of every subscriber not authorized to receive that channel.

One common scrambling system, of which there are many versions, operates by suppressing the vertical and horizontal synchronizing signals as the channels are transmitted from the head end. A pilot signal, containing the timing information for the synchronizing signal, is transmitted either on a separate carrier or within the video or audio signals of the television channel. Within the descrambler, which is located within the subscriber's home, the pilot signal is used to add the synchronizing signals back to the video signal. Although this type of scrambling is relatively easy to defeat, it is the predominant method in use today. Recent improvements in the technique have made the security of the system greater through the use of random delays between the pilot signal and the desired location of the synchronizing signal, and through transmitting the pilot signal in two or more parts on both the audio and video carriers. See Fig. 9-22 for a block diagram.

Another method of scrambling, called *baseband scrambling,* utilizes inversion of the polarity of individual lines of the picture on a random basis. The code to instruct the descrambler which lines to restore to normal polarity is transmitted within the video

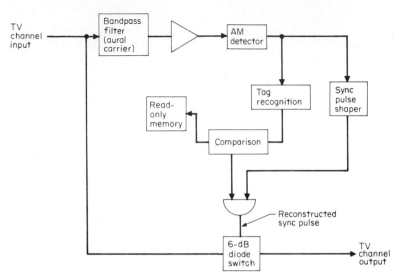

FIG. 9-22 Sync-suppression descrambler.

signal of that channel. This type of descrambler requires demodulation of the television signal to baseband and subsequent remodulation to a channel. As a result, it is considerably more expensive than other methods but is more difficult for unauthorized users to decode. In addition, control of the television set audio volume level is possible from the converter remote control, and videotext decoders can be more easily incorporated.

In most cases, the descrambler and converter are located in the same housing. Recently, addressable converters have become common in CATV systems. These converters combine the tunable converter and a descrambler with an RF receiver and address-recognition circuitry. Each converter may be addressed and controlled by a computer at the head end. Individual channels may be turned off and on, or the entire unit may be disabled from the head end. In this way a subscriber may be authorized for new services or disconnected from them without a CATV service person visiting the subscriber's home.

Addressable converters use either an addressing carrier in the midband or incorporate the addressing commands within the vertical interval of each video channel. At the head end, a computer system is required to store information on which channels each subscriber is authorized to receive and to continually address the converters within the network. See Fig. 9-23 for a block diagram.

Signal security remains a significant problem for the CATV industry, and new technology is emerging in an attempt to provide solutions. Off-premise devices, located outside the subscriber's home or establishment, control which channels are allowed into the home under control from the head end. Since only authorized channels are allowed into the subscriber's home, unauthorized viewing is difficult. Two types of off-premise devices are available. The off-premise converter performs all the functions of an indoor converter, sending only one channel at a time down the drop to the subscriber. The off-premise addressable tap jams or removes those channels which a subscriber is not authorized to view under control of the head end. A converter or television set which can tune mid- and superband CATV channels is needed, but no descrambler is required.

With the off-premise devices, no scrambling is required, since security is provided by preventing nonauthorized signals from entering the subscriber's home. Off-premise equipment is new and for the most part unproven, and significant technical and cost problems remain unsolved.

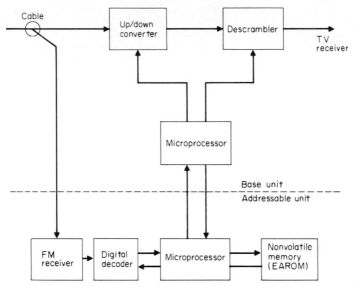

FIG. 9-23 Addressable-converter block diagram. *(Scientific Atlanta.)*

Manufacturers are presently developing more sophisticated scrambling systems, utilizing many techniques. Digital processing of video and audio, using complicated encryption methods, may emerge in the near future as a solution to the signal security problem.

9.9 BIDIRECTIONAL SYSTEMS

Bidirectional transmission over CATV can be accomplished by the use of two cables, one carrying signals from the head end (downstream transmission) and one carrying signals to the head end (upstream transmission). Few such systems have been built because of the high construction cost. A more common method utilizes frequency division on a single cable.

9.9.1 **FREQUENCY DIVISION.** Most single-cable bidirectional networks transmit signals upstream from 5 to 35 MHz and downstream from 50 MHz to the upper frequency limit of the system. This division is referred to as a *sub-split*. In some networks, particularly institutional networks, a mid-split arrangement is used, with upstream transmission from 5 to approximately 150 MHz and downstream transmission from 200 MHz to the upper limit of the network.

To accomplish frequency division, each repeater location is made up of two amplifiers with bandsplitting filters at the input and output. Both amplifiers are located in the same housing and share a common power supply. See Fig. 9-16 for a block diagram of a bidirectional amplifier.

9.9.2 **SYSTEM DESIGN.** CATV networks utilize a tree architecture, with a large number of branches consisting of many miles of cable plant and dozens of upstream amplifiers feeding into a single point at the head end. As a result, the cumulative thermal noise created by the upstream amplifiers makes reliable transmission, especially of amplitude-modulated video, difficult. In addition, bidirectional systems are subject to signal ingress in the upstream direction. Since many over-the-air transmitters exist in the 5- to 35-MHz spectrum, often located near the CATV plant and utilizing high-power outputs, it is difficult to prevent significant ingress into the CATV network. Ingress interferes with reliable upstream transmission, and, in the case of point-to-point data transmission or home security service, reliability is of extreme importance. In an attempt to solve the ingress and noise accumulation problems, many CATV systems segment the return plant by switching the upstream signal at the bridger locations on the trunk using a *code-operated switch* or *bridger switch*. The code-operated switch consists of band-splitting filters to route downstream signals through with no interruption. Upstream signals are routed through a switch controlled by a receiver which is in turn controlled by a computer at the head end. Each code-operated switch can be individually addressed, allowing upstream signals from selected areas of the CATV network to enter the trunk. See Fig. 9-24 for a block diagram of a code-operated switch.

Code-operated switches successfully control ingress but are difficult to use with continuous signals, such as point-to-point data services. Some code-operated switch designs switch only a portion of the upstream spectrum, allowing continuous services to operate in an unswitched portion of the spectrum.

9.9.3 TERMINAL EQUIPMENT.

Terminal equipment for bidirectional systems, at present, is in use in only a few CATV systems, since most bidirectional services are still experimental. Three basic types of terminals are used: transmitter, transponder, and contention.

The transmitter-type terminal consists of a transmitter which is continually transmitting data from the subscriber's home. Within a bridger area, each transmitter is on a discrete frequency, spaced 15 or 20 kHz apart. The flow of signals into the trunk is controlled at the bridger by use of a code-operated switch, so that only one group of signals enters the trunk at a time. At the head end, a scanning receiver extracts the data from

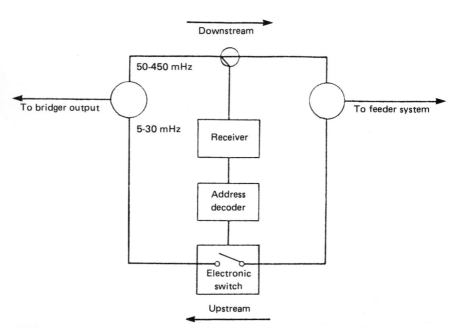

FIG. 9-24   Code-operated switch. *(T. F. Baldwin and D. S. McVoy, Cable Communications, © 1983. Reprinted by permission of Prentice-Hall, Inc., Englewood Cliffs, N.J.)*

each carrier sequentially, or a bank of individual fixed-tuned receivers gathers the data from all transmitters within a bridger area simultaneously. The data transmission rate from a terminal is relatively slow (2 to 3 kbits/s), but since most bidirectional services require little information from the home, the time to poll a large CATV system can be as low as a few seconds. Transmitter terminals are simple and inexpensive, making them suitable for such services as home security and pay-per-view pay television, but since they are limited to monitoring functions and are not addressed, they may not be suitable for more sophisticated services. See Fig. 9-25 for a block diagram of a transmitter-type terminal.

Transponder terminals consist of a receiver, address-recognition circuitry, and a transmitter. Each terminal is polled by the computer at the head end through an addressing channel. When a terminal recognizes its address, it turns on its transmitter and sends back data to a receiver at the head end. All transponder terminals within a CATV system utilize the same channel for addressing, and a relatively high data rate is used (256 kbits/s is typical). A bandwidth of approximately 2 MHz is required in the downstream spectrum, usually in the midband. In the upstream direction, relatively high data transmission rates are used. In some systems, terminals are frequency-agile in the upstream direction, to allow moving the transmitter carrier away from signal ingress which might otherwise cause unreliable data. Depending on the application, bidirectional systems with transponder-type terminals can be polled in as little as 1 to 2 s.

The contention-type terminal also consists of a receiver, address-recognition circuitry, and a transmitter. When a terminal has data it wishes to transmit to the head end, it first monitors the addressing channel to sense that no other terminal is transmitting before sending out its data. The data are received at the head end and echoed back over the addressing channel. The terminal then monitors the channel to verify that its message was received accurately. See Fig. 9-26 for a block diagram of a transponder and contention terminal. Because contention terminals transmit only when new data are presented to them, it is impossible to know if a terminal is still operational if it is idle for a long period of time. Contention terminals may be polled periodically to assure that they are still operational.

Both the transponder and contention terminals can be used with a code-operated switch. With the transponder, the computer that controls the addressing of home ter-

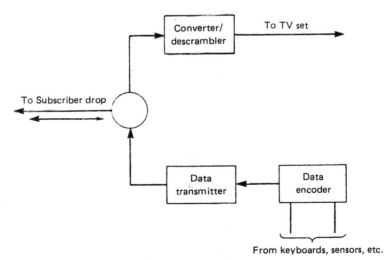

FIG. 9-25 Transmitter-type terminal. *(T. F. Baldwin and D. S. McVoy, Cable Communications, © 1983. Reprinted by permission of Prentice-Hall, Inc., Englewood Cliffs, N.J.)*

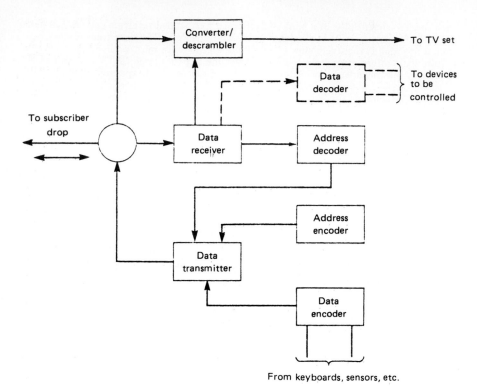

FIG. 9-26 Transponder-type terminal. *(T. F. Baldwin and D. S. McVoy, Cable Communications, ©*
1983. Reprinted by permission of Prentice-Hall, Inc., Englewood Cliffs, N.J.)

minal units also addresses the appropriate code-operated switch prior to addressing the
home terminal. With the contention terminal, code-operated swtiches are used in the
normally closed mode. When signal ingress is detected from one segment of the network,
that area can be switched off to prevent interference with the rest of the system.

9.10 REGIONAL INTERCONNECTION

As cable systems have developed and extended throughout communities, interconnection between these systems has been found to be a cost-effective method of sharing services. In many cases a single program origination, satellite reception, and/or over-the-air
reception location can provide service to eight or ten separate cable systems. The individual cable systems can operate as intermediate hub locations for the overall system,
integrating their own local programs with the interconnected signals.

Regional interconnection is also advantageous for the distribution of regional advertising as well as for the interconnection of areawide institutions and businesses, tying
together a multiplicity of cable systems as well as interconnecting with common-carrier
gateway services.

The regional-interconnection-system concept has some of the features proposed for
New York Communication Teleport operations in the consolidation of service and con-

servation of spectrum. Spectrum conservation is extremely important when scarce terrestrial microwave frequencies are utilized for interconnection.

A common approach to a microwave interconnection is the utilization of amplitude-modulated links (AML) which provide for the transmission of up to 50 television channels on a single microwave system. Through the utilization of highly efficient transmitting and receiving antennas, these frequencies may be reused to provide for up to 100 channel interconnections. Interconnection is also possible through the use of coaxial cable or fiber-optic supertrunking.

9.11 TECHNICAL STANDARDS

Cable-television systems in the United States operate under the authorization of the Federal Communications Commission when these systems carry off-air broadcast signals. It is not possible to construct a cable television system that does not carry FCC-licensed broadcast transmissions and to operate without the jurisdiction of the Federal Communications Commission.† State and local regulations also govern the operation of cable television systems without regard to whether FCC-licensed programs are carried on the system. Most CATV systems operate under a franchise, granted by a municipality or other local government. Franchise contractuality binds the CATV operator to meet financial, technical, and programming standards that are often substantially stricter than FCC regulations, which are generally minimal standards for system performance.

9.12 DIRECT-BROADCAST SATELLITES (DBS)

9.12.1 **SYSTEM CONCEPTS.** In the mid-1940s direct broadcast of television signals to home receivers was envisioned by Arthur C. Clark, a pioneer in the development of satellite communication concepts. He proposed the utilization of communication satellites in geostationary orbits to relay microwave signals from stationary ground transmitters to home terminals feeding conventional television receivers. Present developments and actions by government regulatory bodies in the United States and other countries promise a service of 30 or more channels by a number of broadcasting and communication organizations in the K_u band. The services will include signals on both conventional 525- and 625-line video standards and HDTV standards using a substantially greater number of lines and wider video baseband.

Transmitters with 200-W output capability will transmit signals over a large area of the earth to 3-ft (1-m) antennas on each viewer's premises from 22,500 mi (36,202 km) in space. Thus, the broad concepts are established for a multichannel service to augment VHF and UHF broadcast and wideband cable services. Remaining to be resolved are the satellite transmission and home-terminal requirements.

9.12.2 PERFORMANCE CONSIDERATIONS‡

Carrier-to-Noise Ratio. The performance of any communications link, irrespective of its bandwidth, is determined by the carrier-to-noise ratio. In the case of the broadcast satellite, that ratio is determined by the effective isotropic radiated power (EIRP) of the

---

†See FCC Rules and Regulations, Secs. 76.601 to 76.617 (Subpart K, Technical Standards) and Sec. 76.5 (Subpart A, General Definitions). Rules and Regulations regarding power levels of certain frequencies used in certain areas by the FAA and DOD are contained in Secs. 76.610, 76.611, and 76.613. Prior FCC authorization is required for use of some of the frequencies listed therein.

‡Sections 9.12.2 to 9.12.4 including Figs. 9-27 to 9-32 excerpted from C. A. Kase and W. L. Pritchard, "Getting Set for Direct-Broadcast Satellites," *IEEE Spectrum*, vol. 18, no. 8, 1981, pp. 22–28. © 1981 IEEE.

satellite; frequency; home terminal figure of merit (G/T); and degradations due to rain. For a fixed G/T, improving the system performance requires increasing the EIRP (Fig. 9-27).

These curves show that the carrier-to-noise ratio depends on the strength of the transmitter power and on the size of the reception area. Increasing the transmitter power can be costly, whereas limiting the coverage reduces the service and the income from it. An optimum consumer television picture would require a carrier-to-noise ratio of 90 dBHz.

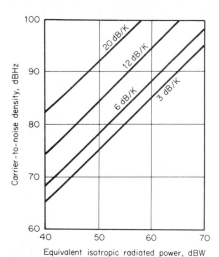

FIG. 9-27   Carrier-to-noise ratio is determined by the spacecraft's effective radiated power (EIRP) and the home receiver figure of merit (G/T). For a fixed receiver quality, carrier-to-noise ratio can be improved by increasing EIRP. This may be accomplished by either increasing satellite transmitter power or reducing coverage area (increasing antenna gain). These curves assume an operating frequency of 12 GHz. For a typical receiver noise figure of 3.0 dB, the antenna diameters would range from 0.36 m for the 3-dB/K case to 2.5 m for the 20-dB/K case.

Picture quality is also basically related to the ratio, so a way of improving the picture is to use higher figures of merit for the home terminal. This can be achieved with a larger antenna that has a higher gain or with a more expensive receiver with lower noise-temperature characteristics.

Noise, Antenna, and Carrier Relationships.   The relationship between the carrier-to-noise ratio and home-terminal antenna size is independent of carrier frequency. It is independent because the antenna gain of the home terminal depends on the square of the frequency in exactly the same way as the free space loss. Therefore in the equation for transmission power the two variables cancel each other, all other things being equal. On the other hand, for a given power level in a satellite and a given coverage area, the move to higher carrier frequencies requires larger antennas to achieve the same picture quality because of atmospheric and rain losses. Thus, to maintain picture quality with atmospheric losses, high-power signals have to be used in the satellite to accommodate the smaller home-receiver antennas.

FM-carrier transmission is necessary in order to conserve satellite-transmitter power and maintain an acceptable signal-to-noise ratio in the received television picture. A basic tradeoff between carrier-to-noise ratio and the available radio frequency bandwidth per channel is crucial to determining the modulation scheme. If a weighted video signal-to-noise ratio is used, the desired picture quality can be plotted against the available RF bandwidth (Fig. 9-28). For an excellent consumer TV image, the range for signal-to-noise ratio is 45 to 50 dB. For an acceptable picture it can be as low as 40 dB. If an FM carrier-to-noise ratio of 10 dB can be met, tremendous improvements in picture quality are possible with only small increases in power by going to broader bandwidths. To transmit an optimum consumer television picture with its required carrier-to-noise ratio of 90 dBHZ, and to cover all United States time zones, a 200-W transmitting tube must be used. In turn, the home terminal should have a figure of merit of 12 dB/K. The desired bandwidths for such a scheme, even with a margin for rain, are probably above 20 MHz. At present, potential operators are considering 27-MHz bandwidths.

9.12.3   HOME TERMINAL VARIABLES.

Picture-Quality Perception by Viewers.   To deliver a quality picture to the home terminal, two random variables must be addressed: the subjective perception of picture

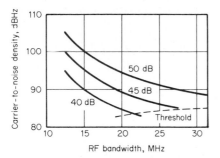

FIG. 9-28 When the TV signal-to-noise ratio is used as a parameter, the requirement for carrier-to-noise ratio decreases as bandwidth increases. In FM systems, the bandwidth can be increased and that ratio decreased up to the point where FM threshold is reached. In a practical system, where rain attenuation must be accommodated, a good design would allow a margin of several decibels above the threshold point for clear-weather performance.

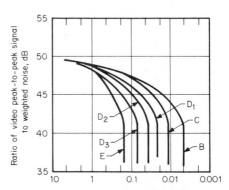

FIG. 9-29 In the 12-GHz band, video performance will degrade during rain. This example shows performance for a system designed to have a signal-to-noise ratio of 50 dB during clear weather. Climate types range from cold and dry areas such as Maine (B) to tropical areas such as Florida and Louisiana (E).

quality and attenuation due to rain. Both ultimately determine how successfully the system performs.

Studies of subjective evaluations of picture quality have been made worldwide, with the results hinging on video signal-to-noise ratio. An average ratio of peak-to-peak signal to weighted noise level of 40 dB is acceptable to a majority of viewers.

Cable systems are typically designed for reception from satellites with picture signal-to-noise ratios of 50 dB, although much lower values are perfectly satisfactory for a home terminal. The smallest terminals and the worst conditions of high rain and low angles of elevation decrease the minimum ratio to the 30-dB level, a very poor-quality picture.

Rain Attenuation. Attenuation due to rain is a more difficult problem. Both the rate and volume of rain vary extensively around the United States. A rain attenuation model, drawn up by the National Aeronautics and Space Administration for the 12-GHz broadcast band, summarizes the different climates in the United States. The hardest to deal with is that of the southeast, particularly in Louisiana and Florida.

To illustrate how the climate curves reflect the video signal-to-noise ratio, they are plotted against percentage of time for each climate type (Fig. 9-29). The picture quality in the worst climate would be usable until the carrier-to-noise ratio dropped below the FM threshold. This would happen about 0.2 percent of the time in the United States. This amounts to almost 18 h a year—enough to prove frustrating to viewers, who would inevitably have to acquire larger and more expensive antennas.

A critical characteristic for receiving quality pictures from a direct-broadcast satellite is the figure of merit (G/T) at the home terminal. At a particular frequency, this figure simply depends on the size of the antenna and on the system noise temperature of the receiver. The system noise temperature, for its part, is affected by rain. Rain not only weakens the signal but also increases the antenna temperature. This temperature must be added to the low-noise amplifier temperature to set the system noise temperature. Variation of the antenna temperature is particularly distressing because the system temperature margin that can account for a high G/T is obliterated just when the margin is needed the most. System temperature in clear weather and antenna size are thus not perfectly interchangeable.

For the same ratio, a larger antenna is better than a lower clear-weather system temperature, because the increased gain of a larger antenna is not affected by rain at all.

A plot of the figure of merit versus antenna diameter for different low-noise amplifier

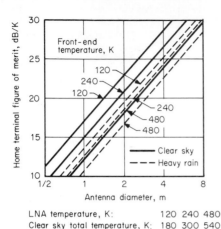

LNA temperature, K: 120 240 480
Clear sky total temperature, K: 180 300 540
Heavy rain total temperature, K: 410 530 770

FIG. 9-30 Receiving station G/T is a function of antenna diameter, receiver temperature, and antenna temperature. In heavy rain, the antenna temperature increases from a clear-weather value of perhaps 50K to a maximum value of 290K (ambient).

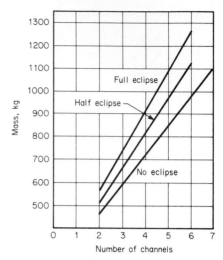

FIG. 9-31 A satellite's mass increases with the number of channels and with eclipse capability. The number of channels determines the solar power requirements and consequently a great part of the mass. Eclipse capability requires heavy batteries and also additional solar power to recharge batteries between eclipses.

temperatures in both clear weather and heavy rain yields the results in Fig. 9-30. A 5.9-ft (1.8 m) antenna with a 480-K temperature and a 3.3-ft (1-m) antenna with a 120-K temperature both give the same clear-weather figure of merit: 17 dB/K. However, in heavy rain the larger antenna G/T deteriorates to only 16 dB, whereas the smaller antenna system drops all the way to 13 dB.

9.12.4 SATELLITE DESIGN. A direct relationship between the total transmitter power required and the mass in orbit of the satellite should be viewed in terms of the total system cost. The power and mass of the satellite, in turn, determine the number of channels that the satellite is able to accommodate. A typical satellite mass can be related to the number of channels available for coverage of an area that is approximately the size of a United States time zone, as can be seen in Fig. 9-31.

The sun's eclipse has a direct bearing on the behavior of a mass in space and is used in the figure as a parameter.

The relationships between cost and mass are rather complicated, involving such considerations as transmitter power, efficiency, weight and redundancy, battery charging, structure, attitude control, antennas, and station-keeping propellant. Satellite designers, in addition, have to take into consideration the satellite's operation under various positions of the sun's eclipse.When the sun is close to the equator during the spring and autumn, the satellite will enter the shadow of the earth once a day, and it will be necessary to carry batteries to supply whatever power the satellite needs when its solar cells are inactive. The eclipse period is a maximum of about 72 min, and the season in which any eclipse takes place lasts twice a year for exactly 6 weeks before and 6 weeks after the equinoxes.

The batteries must have electronic circuitry to control their charging and discharging and extra solar cells to charge them during the 22 h or so between daily eclipses. This added weight of batteries, solar cells, and electronics requires extra structure and station-keeping propellant so the satellite will stay in position at its prescribed longitude and in the plane of the equator. This increases the total mass of the satellite in orbit and the requirements for the perigee and apogee kick engines and the launch vehicle itself. In other words, ensuring operation under differing eclipse conditions means cumulative cost increases.

Although some batteries will always be necessary for housekeeping during the eclipse, the greatest fraction of the battery power is required to operate the transmitter. The transmitting power requirement depends on the local clock time at the onset of an eclipse. If eclipse transmitting power could be reduced or eliminated, the saving in space-craft costs could run into the tens of millions of dollars.

The peak of an eclipse occurs at apparent solar midnight, but the time on a viewer's clock can be considerably earlier. At the longest eclipse, the onset is some 36 min before the peak. There is a further difference of about 8 min at the equinox between apparent solar time and mean clock time. This still presupposes that the viewer is immediately below the satellite on the central meridian of his or her time zone. Since the viewer can be considerably to the east or west of the subsatellite point, a discrepancy of several hours is possible between the apparent solar time of the peak eclipse at the sub-satellite point and the time on the viewer's clock. The local time of the onset of the longest eclipse at the equinox can be calculated as a function of the relative longitude between the viewer and the satellite.

For example, an eclipse might start at 10:30 or 11:00 P.M., when the requirements for television service are extremely high. Since this is still prime time, the satellite could be moved, say, 25° to the west of the covered area, so that to the viewer the onset of the eclipse would be delayed until 1 A.M. Little or no television service would have to be transmitted during the eclipse. The effects of reducing eclipse capabilities on satellite mass and costs are shown in Fig. 9-32.

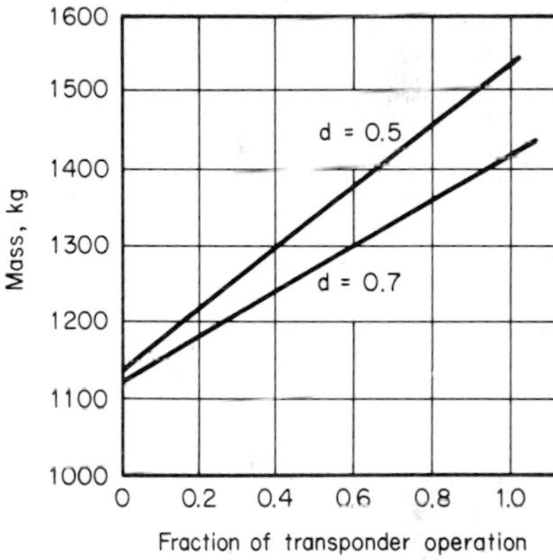

There is a drawback to this maneuver, however. Although the eclipse situation becomes more and more favorable as the satellite is moved to the west of the covered area, rain attenuation becomes more acute. The lower angles of elevation to the satellite cause longer rain paths. At very low angles of elevation, the signal is blocked from neighboring buildings, hills, and trees. Thus to cover a particular service area, the satellite should be "parked" in a sector of the geostationary arc that offers the best compromise between the eclipse service and elevation angle.

For a given United States time zone, this sector of the orbit is approximately from 25 to 40° west of the time zone. Compromises like this must be made for both fixed and broadcast satellites operating in the K_u band over the United States and the rest of the Western Hemisphere.

FIG. 9-32 As eclipse capability increases, so does the mass required for batteries and primary power to recharge them. A greater depth of discharge (d) reduces battery mass but not primary power requirements.

Electron Optics, Deflection, and Color Registration

Sol Sherr

Westland Electronics
Hartsdale, New York

10.1 PRODUCTION AND CONTROL OF ELECTRON BEAMS

10.1.1 ELECTRON GUNS. The electron gun in a CRT may be defined as "the electrode structure which produces and may control, focus, and deflect an electron beam."[1]† Similarly, a definition is proposed by the author for a cathode-ray device as "any device which uses the electron beam to cause a change in, or affect some characteristics of, one or more other elements of the device."[2] The electron gun therefore is basic to the structure and operation of any cathode-ray device, and specifically to those of most interest to television engineers, namely, the camera and picture tube which first convert the visual scene into electrical signals and then convert them back into visual images. Although this section is not intended as a detailed and comprehensive discussion of the design and operation of electron guns, which are covered elsewhere, both in this volume and other publications,[3-3c] it is necessary to present certain simplified aspects of electron guns in order to be able to cover the other topics included in this chapter.

Triode Gun. In its simplest schematic form, an electron gun may be represented by the diagram in Fig. 10-1, which shows a triode gun in cross section. The electrons are emitted by the cathode which has been heated by the filament to a temperature sufficiently high to emit electrons. However, this stream of electrons issues from the cathode as a cloud, and it is necessary to accelerate, focus, deflect, and otherwise control this electron cloud so that it becomes a beam and can be made to strike a phosphor at the proper location and with the desired beam cross section.

The laws of motion for an electron in a uniform electrostatic field are obtained from Newton's second law. When this is solved, the velocity of the electron is given by

$$v = \left(\frac{2eV}{m}\right)^{1/2} \tag{10-1}$$

where $e = 1.6 \times 10^{-19}$ C
    $m = 9.1 \times 10^{-28}$ g
    $V = -Ex$ (potential through which electron has fallen)

Substituting these values in Eq. (10-1) and using practical units throughout results in

$$v = 5.93 \times 10^5 V^{1/2} \quad \text{m/s} \tag{10-2}$$

as the expression for the velocity of the electron. Then if the electron velocity is at the angle Θ to the potential gradient in a uniform field, the motion of the electron is given by

$$y = -\frac{Ex^2}{4V_0 \sin^2 \Theta} + \frac{x}{\tan \Theta} \tag{10-3}$$

---

†Superscript numbers refer to references at end of chapter.

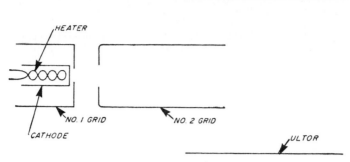

FIG. 10-1   Triode electron gun. (*After Fink.*[4])

where V_0 = potential at initial velocity. This equation defines a parabola, and the electron trajectory is shown in Fig. 10-2, in which y_m = maximum height, x_m = x displacement at maximum height, and α = slope of the curve. These three parameters are given in turn by

$$y_m = \frac{V_0 \cos^2 \Theta}{E} \tag{10-4}$$

$$x_m = \frac{2 V_0 \sin \Theta \cos \Theta}{E} \tag{10-5}$$

$$\tan \alpha = \frac{Ex}{2 V_0 \sin^2 \Theta} + \frac{1}{\tan \Theta} \tag{10-6}$$

Similarly, for a uniform magnetic field, the radius of the electron path when the electron enters the field at right angles is expressed by

$$R = \frac{3.38 \times 10^{-6} V^{1/2}}{B_m} \quad \text{m} \tag{10-7}$$

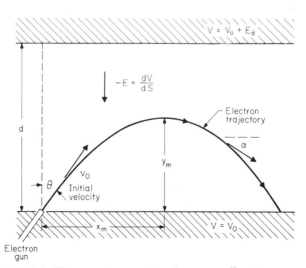

FIG. 10-2   Electron trajectory. (*After Spangenberg.*[10])

and
$$R = \frac{3.38 \times 10^{-6} V^{1/2} \sin \Theta}{B_m} \quad \text{m} \tag{10-8}$$

where B_m = magnetic-flux density and Θ = angle at which the electron enters the field. The path of the electron is a helix, with the pitch given by

$$P = \frac{21.1 \times 10^{-6} V^{1/2} \cos \Theta}{B_m} \quad \text{m} \tag{10-9}$$

Equation (10-6) is used to develop the equation for electrostatic deflection, and Eq. (10-9) that for magnetic focusing. These equations are correct, in a strict sense, only for uniform fields, and when the fields are not uniform, or when both fields are present, the differential equations are much more complex and are not easily solved. However, in general the equations given for uniform fields of one type are satisfactory for most purposes.

Tetrode Gun. In addition, there are some designs that have a fourth electrode, resulting in a tetrode gun, but we limit the discussion to the triode gun, which is sufficiently representative to serve our purposes. Nevertheless, for information purposes, a diagram of a tetrode gun is shown in Fig. 10-3.[4] Its main advantage is that the additional electrode causes the beam to converge more than for the triode gun.[5]

10.1.2 ELECTROSTATIC FOCUS. The general principles involved in focusing the electron beam can be best understood by initially examining optical lenses and then establishing the parallelism between them and the electrical focusing techniques. The exact forms of various types of such lenses are discussed in detail in Chap. 12. Thus the discussion here is limited to the physical basis for the focusing action and illustrated only by simple structures.

Electrostatic Lenses. The optical analogy may be established by examining the illustration shown in Fig. 10-4, which is a simplified diagram of an electrostatic lens. An electron emitted at zero velocity enters the V_1 region and moves in that region at a constant velocity, since that region has a constant potential. The velocity of the electron in that region is defined by Eq. (10-2) for the straight-line component, with V_1 replacing V in Eq. (10-2). After passing through the surface into the V_2 region, the velocity changes to a new value determined by V_2. The only component of the velocity that is changed is the one normal to the surface, so that

$$v_t = v_1 \sin I_1 \tag{10-10}$$

and
$$v_1 \sin I_1 = v_2 \sin I_2 \tag{10-11}$$

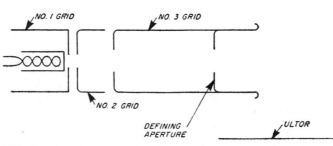

FIG. 10-3   Tetrode electron gun. (*After Fink.*[4])

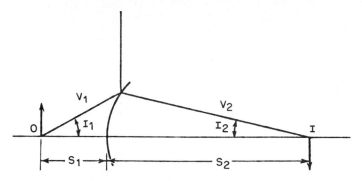

FIG. 10-4   Electron optics. (*After Sherr.*[2])

Snell's law, also known as the law of refraction, has the form

$$N_1 \sin I_1 = N_2 \sin I_2 \qquad (10\text{-}12)$$

where N_1 = index of refraction for first medium
  N_2 = index of refraction for second medium
  I_1 = angle of incident ray with surface normal
  I_2 = angle of refracted ray with surface normal

The parallelism between the optical and the electrostatic lens is apparent if we make the appropriate substitutions. This becomes even clearer if we make the substitutions for the velocity from Eq. (10-1) leading to

$$V_1 \sin I_1 = V_2 \sin I_2 \qquad (10\text{-}13)$$

$$\frac{\sin I_1}{\sin I_2} = \frac{V_2}{V_1} \qquad (10\text{-}14)$$

whereas for Snell's law

$$\frac{\sin I_1}{\sin I_2} = \frac{N_2}{N_1} \qquad (10\text{-}15)$$

making the analogy complete. The magnification of the electrostatic lens is given by

$$m = \frac{(V_1/V_2)^{1/2} \, S_2}{S_1} \qquad (10\text{-}16)$$

with the symbols as defined in Fig. 10-4. Similarly, for the thin, unipotential lens, where V_1 equals V_2, and referring to Fig. 10-5,

$$m = \frac{h_2}{h_1} = \frac{f_2}{X_1} = -\frac{X_2}{f_2} \qquad (10\text{-}17)$$

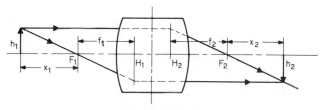

FIG. 10-5   Thin unipotential lens. (*After Sherr.*[2])

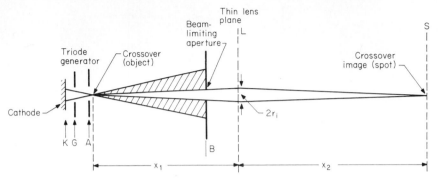

FIG. 10-6   Electron beam shape. (*After Moss.*[3])

The shape of the beam under these conditions is shown in Fig. 10-6, and if the potential at the screen is the same as the potential at the anode, then the crossover is imaged at the screen with the magnification given by

$$m = \frac{x_2}{x_1} \tag{10-18}$$

The magnification can be controlled by changing this ratio, which in turn changes the size of the spot, and is one way to control the quality of the focus. Although the actual lens may not be "thin," and in general is more complicated than is shown in Fig. 10-6, this leads to much more difficult calculations and is not essential to an understanding of the operation of electrostatic focus. Therefore, this more complex aspect is not dealt with here. For more detail the reader is referred to other publications.[6] However, it is clear from the discussion and Eqs. (10-16) and (10-18) that the size of the spot can be controlled by changing the ratio of V_1 to V_2 or of x_2 to x_1. Since x_1 and x_2 are established by the design of the CRT, the voltage ratio is the one available to the circuit designer who wishes to control the size or focus of the spot, and it is by this means that focusing is achieved for CRTs using electrostatic focus.

Aberrations.   Before leaving electrostatic lenses, a few of the more common types of aberrations that are associated with these lenses should be mentioned. There are five important types, termed *astigmatism, coma, curvature of field, distortion of field,* and *spherical aberration*[7] for monochromatic systems, to which should be added *chromatic aberration* for color systems. To begin with chromatic aberration, this can be understood by reference to Fig. 10-7. This is due to the effect that is analogous to the effect in geometrical optics which results in light of different wavelengths having different focal lengths, as is illustrated in Fig. 10-7. In electrostatic lenses, electrons with different velocities will have different focal points, as shown in Fig. 10-7. However, since electron velocity is different only insofar as the electrons leave the cathode with different emission velocities, the effect is generally small at the accelerating potentials that are used, and the error is usually not significant. Coma is more important but applies only to images and objects not on the axis of the lens system. This is illustrated by Fig. 10-8, which shows that circles are imaged in the distorted form shown. Coma may be reduced by using less of the lens center, but this reduces the amount of beam current and therefore luminance available, so that it may not be a desirable approach.

A more important aberration is that termed astigmatism. This results because objects that are off the axis lines toward the axis have different focal lengths than lines that are perpendicular to them. Again, this effect is well known in geometrical optics and is shown in Fig. 10-9. It can be seen from Fig. 10-9 that compromises must be made when focusing the entire image. Changing the focusing voltage changes the portion of the image that is sharply focused, while the rest of the image may be blurry.

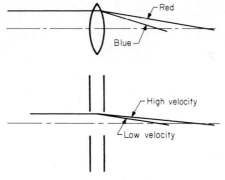

FIG. 10-7 Chromatic aberration. (*After Spangenberg.*[10])

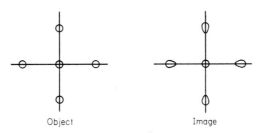

Object Image

FIG. 10-8 Coma. (*After Spangenberg.*[10])

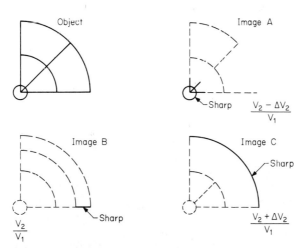

FIG. 10-9 Astigmatism. (*After Spangenberg.*[10])

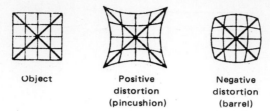

Object | Positive distortion (pincushion) | Negative distortion (barrel)

FIG. 10-10 Pincushion and barrel distortion. (*After Spangenberg.*[10])

Curvature of field is usually associated with astigmatism but is a more noticeable effect. This is due to the image lying on a curved surface for an object that is in a plane at right angles to the axis. The result is that concentric circles can be adjusted in the image plane so that only one radial distance is sharp. Thus, if the center is focused, the outside will be unfocused, or the opposite if the outer circle is focused.

A very important aberration is that termed distortion of field, which results from variations in linear magnification with the radial distance. These are the well-known pincushion and barrel distortions, with the former resulting from an increase in magnification and the latter from a decrease. They are pictured in Fig. 10-10 and are familiar to all users of CRTs.

Finally, there is spherical aberration where parallel rays entering the lens system have different focal lengths depending on the radial distance of the ray from the center of the lens. This effect is shown in Fig. 10-11 and is perhaps the most serious of all the aberrations. It can be seen from Fig. 10-11 that the focal length becomes smaller as the dis-

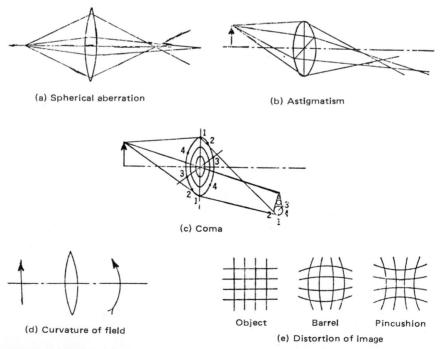

(a) Spherical aberration

(b) Astigmatism

(c) Coma

(d) Curvature of field

(e) Distortion of image

Object | Barrel | Pincushion

FIG. 10-11   Nature of the five third-order aberrations. (*After Zworykin.*[7])

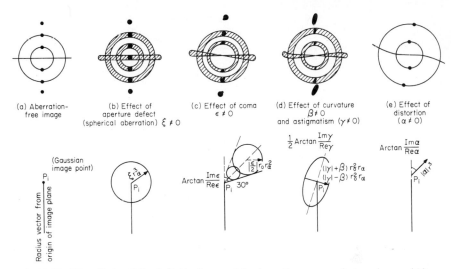

FIG. 10-12 The effects of the individual geometric aberrations on an electron image. (*After Zworykin.*[9])

tance increases. This is known as *positive* spherical aberration and is always found when electron lenses are used. The focal length decreases slowly at first and then much more rapidly as the radial distance increases. Figure 10-12 contains an image that illustrates this effect, as well as the others. This type of aberration is always positive in electron lenses, and it cannot be eliminated by adding a lens with equal negative spherical aberration as is the case with optical systems. However, it is possible to somewhat reduce the effect by using dual-cylinder lenses with a high-potential inner cylinder and a lower-potential outer cylinder.

10.1.3 MAGNETIC FOCUS. It is well known that electron beams can be focused with magnetic fields as well as with electrostatic fields, but the analogy with optical systems is not as apparent. When electrons leave a point on a source with the principal component of the velocity parallel to the axis of a long magnetic field, they travel in helical paths and come to a focus at a further point along the axis. The helical paths have essentially the same pitch, which is given by Eq. (10-9). From this equation it is clear that the pitch is relatively insensitive to Θ for small angles, and the electrons will return to their original relative positions at some distance P on the magnetic path that is parallel to the electron beam. Thus, spreading of the beam is avoided, but there is no reduction in the initial beam diameter. Focusing is achieved by changing the current in the focusing coil until the best spot size is achieved. The manner in which the focus coil is placed around the neck of the CRT is shown in Fig. 10-13. The focal length of such a coil is given by[8]

$$f = \frac{48.6 \ Vd}{N^2 I^2} \tag{10-19}$$

where f = focal length
  d = diameter of wire loop
  NI = current in ampere-turns
  V = potential of region

and the image rotation is expressed by

$$\Theta = \frac{0.19 \ NI}{V^{1/2}} \tag{10-20}$$

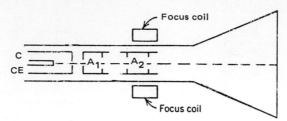

FIG. 10-13   Magnetic focusing. (*After Sherr.*[2])

Equation (10-19) may be used for a short coil with a mean diameter of d that has N turns and a current equal to I. This equation is plotted as a nomograph in Fig. 10-14[9] and may be used for an iron-encased coil with a small NI and d as the diameter of the pose pieces. In addition, a form factor correction should be added to the numerator of Eq. (10-19) for completeness, and this form factor is between 1.00 and 1.25.

Aberrations and Distortions.   As is the case for electrostatic lenses, magnetic lenses are also subject to the same aberrations, to which should be added another distortion that is associated with the rotation of the image. This distortion is called *spiral distortion*

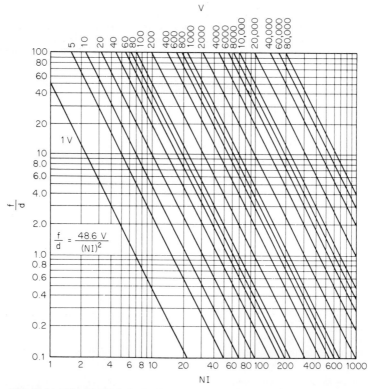

FIG. 10-14   The focal length of a short coil as a function of ampere-turns and beam voltage. (*After Pender.*[8])

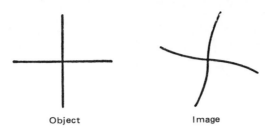

Object   Image

FIG. 10-15   Spiral distortion in magnetic-lens images. (*After Spangenberg.*[10])

and is illustrated in Fig. 10-15. Spiral distortion results from different parts of the image being rotated different amounts as a function of their radial position. The effect is reduced by using very small apertures or essentially eliminated by having a pair of lenses that rotate in different directions. There is also the possibility of distortion due to stray fields or ripple in the current driving the focus coil. The current ripple causes a point to become a blurred spot, whereas the stray fields cause a point to elongate to a line. Both of these effects can be minimized by careful design of the current source and the focus coil, respectively.

Applications.   Magnetic focusing is rarely used in television displays, since the majority of CRTs are designed with electrostatic focus elements. Therefore, it is unlikely that there will be much occasion to consider the characteristics of magnetic lenses when designing a television type of display. However, when the best resolution is desired, magnetic focus is the approach that can achieve this resolution. Thus, it is well to understand the operation of this type.

Design Principles.   The more general expression for the image rotation is given by

$$\theta = \frac{1}{2V^{1/2}} \left(\frac{e}{2m} \right)^{1/2} \int_{z_1}^{z_2} B_0 \, dz \tag{10-21}$$

which becomes

$$\theta = \frac{1.48 \times 10^5}{V^{1/2}} \int_{z_1}^{z_2} B_0 \, dz \tag{10-22}$$

when the usual substitutions are made for e and m. For the specific case of the short coil, Eq. (10-20) may be derived from Eq. (10-22). Similarly, the focal length is found from

$$\frac{1}{f_1} = \frac{-1}{f_2} = \frac{2.2 \times 10^{10}}{V} \int_{z_1}^{z_2} B_0{}^2 dz \tag{10-23}$$

where z_1 and z_2 are points to the left and right, respectively, of any significant variations in the field and B_0 is the axial component of flux density, expressed in webers per square meter for all three equations.

When the integral is evaluated, this leads to Eq. (10-19) for the focal length, which completes the derivations of the two equations necessary to understand the operation of magnetic lenses. Much more complicated expressions are needed to describe the operation of combined electrostatic and magnetic fields, but they are of little interest to the designer who must work with a set of given parameters, so we do not present those expressions here and refer the interested reader to the references for further information.[9]

Mechanical Configurations.   The gun structures and the location of the focus coil are shown in Fig. 10-13 and are simpler than those needed for electrostatic focusing. Only the cathode, control electrode, and accelerating electrode are needed, with the focus coil

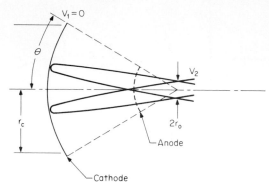

FIG. 10-16 Idealized cathode with spherical field. (*After Spangenberg.*[10])

usually located externally on the neck of the CRT. A constant current source must be provided for the magnetic focus coil, which can be varied in order to control the focal point as given by Eq. (10-19). The ripple from this current source must be kept small to minimize the blurring effect noted above.

10.1.4 OTHER FOCUSING EFFECTS

Beam Crossover.   The beam crossover is used as the object whose image appears on the screen of the CRT. Therefore, the location and size of the crossover are very important in determining the minimum spot size attainable by means of the focusing techniques described in Secs. 10.1.2 and 10.1.3. While the exact values are difficult to determine, a good approximation can be achieved by assuming a spherical field in the vicinity of the cathode. This idealized arrangement is shown in Fig. 10-16, and the radius of the crossover is given by[10]

$$r_0 = \frac{2r_c}{\sin 2\Theta \, (V_2/V_e)^{1/2}}$$ (10-24)

where r_0 = crossover radius
$\quad\;\; r_c$ = cathode radius
$\quad\;\; V_2$ = crossover potential
$\quad\;\; V_e$ = voltage equivalent of emission velocity
$\quad\;\; \Theta$ = cathode half-angle viewed from crossover

It is clear from Eq. (10-24) that the crossover radius will differ for different velocities of emission. Therefore, since the electrons will be emitted at all possible velocities, an average value of V_e must be used. In addition, the equation is valid only for small values of Θ, that is, less than 20°. A further description of crossover effects may be found in Sec. 12.3. It should be noted here that these effects, while of interest to the designer of a CRT system, are controlled by the tube design and are not subject to any further control once the CRT has been chosen.

Focus Masks.   Another effect which is controlled by the CRT design, but is of interest in determining the focusing characteristics, is that introduced by the focus mask. This applies only to tubes that use some kind of mask. The purpose of the focus mask is to reduce the amount of the beam that is intercepted by the mask. Three-beam and single-beam focus-mask tubes have been designed to accomplish this end by increasing the size of the aperture, and then creating an electron lens at the aperture that imaged the electron beam cross section in the deflection plane on the phosphor screen. The manner in which this is accomplished is shown in Fig. 10-17, and for perfect imaging the spot

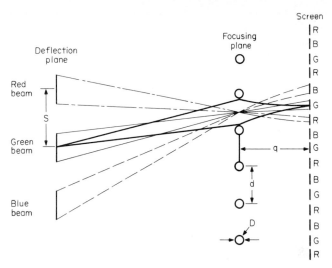

FIG. 10-17   Principle of the focus mask. (*After Morell.*[11])

size on the screen will be no larger than the pinhole at the center of the aperture.[11] A similar principle was applied to the three-beam focus-grill tube. It has also been used for three-beam focus-mask tubes. For this latter type, the spot diameter is reduced from that attained in a standard shadow-mask tube by the factor $\frac{1}{2}[3-(V_s/V_m)]$,[2] and the apertures can be increased in diameter for equal projected size according to[12]

$$\frac{T}{T_{sm}} = \frac{4}{[3 - (V_s/V_m)^{1/2}]^2}$$

(10-25)

where T = focus-mask transmission
   T_{sm} = shadow-mask transmission
   V_m = focus-mask and gun-anode potential
   V_s = screen potential

The increase in luminance is then given by

$$\frac{L}{L_{sm}} = \left[\frac{T_{sm}}{T(3 - 2(T_{sm}/T)^{1/2})} \right] - 1$$

(10-26)

for the focus-mask tube, and

$$\frac{L}{L_{sm}} = \left[\frac{T_{sm}}{T(2 - T_{sm}/T)^2} \right] - 1$$

(10-27)

for the focus-grill tube, where L = luminance of focus-mask or focus-grill tube, L_{sm} = luminance of shadow-mask tube, and the other terms are as before. These relationships are plotted in Fig. 10-18, and it can be seen that for the focus-mask tube there is no gain in luminance until the transmission ratio exceeds 4, whereas for the focus-grill tube a ratio of 2.6 is sufficient. This requires the same ratio to exist for V_s/V_m, leading to the requirement for a high transmission auxiliary mask or the operation of the anode at the screen voltage. Both techniques have been used, and an improvement of 1.8 has been achieved. Other techniques have also been used with similar results, but unfortunately the reduction in contrast ratio due to secondary and back-scattered electrons has reduced the improvement, although in theory this loss could be eliminated by the use of three-beam double-grill tubes. Unfortunately, as noted in Sec. 12.2.1, manufacturing difficulties in producing focus-grill tubes have resulted in such designs no longer being

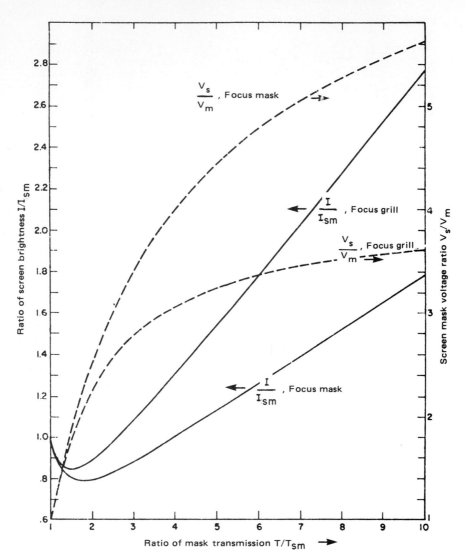

FIG. 10-18 Ratio of screen intensities of focus-mask or grill tubes to screen intensity of shadow-mask tubes. (*After Morell.*[11])

built. In addition, improvements in luminance and contrast ratios of color CRTs have made the use of focus-mask structures unnecessary, and they are not presently in use. However, they remain as possible techniques for improving the performance of color tubes, and may return as manufacturing techniques are developed which make them less difficult to construct and as the needs of nonentertainment television color-data display demand such improvement. The resurgence of high-resolution color, after it was neglected for close to 20 years, is an example of such an event.

10.2. MAGNETIC DEFLECTION

10.2.1 **FUNDAMENTALS.**[12] The basic, simplified magnetic-deflection equation is derived from Eq. (10-7) and Fig. 10-19. Assuming that the magnetic field is constant within the area delineated by the dots, the electron beam will follow the arc of a circle whose radius is given by Eq. (10-8). The angle at which the beam leaves the magnetic field is then given by

$$\sin \alpha = 2.97 \times 10^5 \frac{lB_m}{(V)^{1/2}} \qquad (10\text{-}28)$$

The deflection is related to this angle by

$$y_d = l \tan \alpha \qquad (10\text{-}29)$$

since $\tan \alpha$ is approximately equal to y_d/l because the center of deflection is at the center of the field. Substituting in Eq. (10-28) from Eq. (10-28) results in

$$y_d = 2.97 \times 10^5 \frac{l^2 B_m}{(V)^{1/2}} \qquad (10\text{-}30)$$

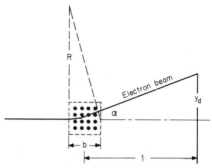

FIG. 10-19 Magnetic deflection. (*After Spangenberg.*[10])

assuming that $\sin \alpha$, $\tan \alpha$ and α are all equivalent, which holds for small values of α. For cases where this equivalence does not hold, the deflection is not directly proportional to the current since $\sin \alpha$ rather than $\tan \alpha$ is proportional to the current. This leads to the need for corrective circuitry to compensate for the error. This is covered in more detail later (Sec. 10.2.2).

Another derivation of the magnetic deflection equation may be made by using the somewhat more elaborate diagram shown in Fig. 10-20 and noting that the angle α is approximately equal to the angle β. Therefore

$$\frac{d_m}{r} \approx \frac{d_b}{Z_m} \qquad (10\text{-}31)$$

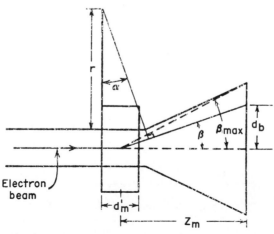

FIG. 10-20 Magnetic-deflection CRT. (*After Spangenberg.*[10])

for r, and using the relationship given for R in Eq. (10-7) results in

$$d_b = \frac{d_m Z_m B_m}{3.38 \times 10^{-6} V^{1/2}} \qquad (10\text{-}32)$$

with d_b in meters for B in webers per square meter, and in centimeters if B is in gauss and the factor 10^{-6} in the denominator is dropped. In addition, the deflection angle may be expressed by

$$\sin \alpha = \frac{(k_m L_y)^{1/2} I_y}{(2 V_c)^{1/2}} \qquad (10\text{-}33)$$

where $L =$ yoke inductance,
$\quad I_y =$ current through the yoke, A
$\quad V_c =$ CRT anode voltage
$\quad k_m =$ yoke sensitivity factor

The significance of the yoke sensitivity factor is discussed later in Sec. 10.3.

10.2.2 FLAT-FACE DISTORTION.[13]

As noted above, flat-face distortion results from the difference between the radius of curvature of the deflected beam and the actual radius of the display surface, as illustrated in Fig. 10-21 for the general case. The ratio of the true deflection to the deflection on an ideal surface that has its deflection center at the center of curvature is given by

$$\frac{d_a}{d_i} = \frac{R_a \sin \Theta_a}{R_i \sin \Theta_i} \qquad (10\text{-}34)$$

We may express the deflection angle in terms of the current through the yoke by substituting for B_m in Eq. (10-28) and using a constant to replace all the other terms. This results in

$$\sin \Theta = KI \qquad (10\text{-}35)$$

and it is clear that only in the ideal case will the deflection be directly proportional to the current. If we then apply the general case to the specific one of the flat-faced screen, as illustrated in Fig. 10-22, Eq. (10-35) becomes

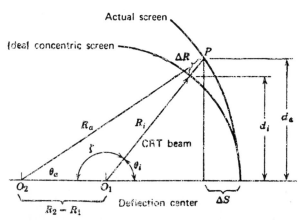

FIG. 10-21 Linearity distortion due to CRT screen curvature. (*After Popodi in Sherr.*[18])

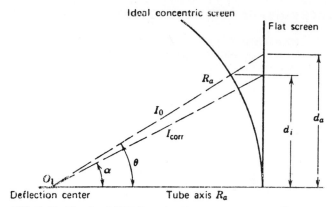

FIG. 10-22 Flat-faced CRT linearity distortion. (*After Sherr.*[18])

$$\frac{d_a}{d_i} = \frac{R_a \tan \Theta}{R_a \sin \Theta} \qquad (10\text{-}36)$$

which may be simplified to

$$\frac{d_a}{d_i} = \frac{1}{\cos \Theta} \qquad (10\text{-}37)$$

If Eq. (10-37) is rewritten as

$$\frac{d_a}{d_i} = \frac{1}{(1 - \sin^2 \Theta)^{1/2}} \qquad (10\text{-}38)$$

this may be further reduced by using Eq. (10-35) to give

$$\frac{d_a}{d_i} = \frac{1}{(1 - \sin^2 \Theta)^{1/2}} \qquad (10\text{-}38)$$

and the denominator may be expanded giving

$$d_a = d_i \left(1 + \frac{K^2 I^2}{2} \right) \qquad (10\text{-}40)$$

if all but the first two terms of the expansion are neglected. This error can be compensated for dynamically by introducing special circuitry, which is described in Sec. 10.5.1.

10.2.3 DEFLECTION DEFOCUSING.[14]   Another effect due to deflection of the electron beam is the defocusing that occurs when the beam is moved from the center to some other location on the CRT screen. The changes in focus are due to the edges of the beam entering the magnetic field at different angles because of convergence. The edges of the beam will be deflected by the same radius of curvature and leave the field with a new convergence angle. Then the change in spot size as a function of the deflection angle is given by

$$r_s = r_0 \frac{(d_b^2 + Z^2)^{1/2}}{Z - 1} \qquad (10\text{-}41)$$

with the terms defined in Fig. 10-18. This effect can be reduced by the use of dynamic-focusing circuits that are described in Sec. 10.5.2.

10.3 ELECTROMAGNETIC DEFLECTION YOKES

10.3.1 YOKE INDUCTANCE. In order to understand the operation and function of the deflection yoke, it is necessary to examine its structure in more detail than has been done in the previous sections. Some of the characteristics of the magnetic field are covered in Sec. 10.1.1 and defined by Eqs. (10-7) and (10-8). It is the purpose of this section to apply those general considerations to the specific requirements of the magnetic deflection yoke. To do this, we begin with the definition of field strength, given by

$$HD_a = ni \tag{10-42}$$

where H = field strength, H
D_a = yoke diameter, m
n = number of turns
i = current through yoke, A

The length of the field is expressed by[15]

$$l = r \sin \Theta \tag{10-43}$$

where l = length of uniform magnetic field
r = radius of electron path
Θ = deflection angle

The radius r can be obtained in terms of H and V by rewriting Eq. (10-1) to include the self-inductance factor u_0. Then

$$ni = 2.69 \frac{(\sin \Theta) D_a}{l(V)^{1/2}} \tag{10-44}$$

with u_0 equal to 4×10^{-9} H/m and the other terms as defined in Eqs. 10-1, 10-43, and 10-44. Then combining Eq. (10-42) with the expression for self-inductance

$$L = \frac{n\phi}{i} \tag{10-45}$$

where the flux $\phi = u_0 H D_n l_f$, and D_n is the diameter of the inside of the yoke that fits around the neck of the tube, we get[16]

$$L = \frac{1.26n^2 l_f D_n}{D_a} \times 10^{-8} \quad \text{H} \tag{10-46}$$

Equation (10-46) may now be substituted for L_y in Eq. (10-33) to arrive at the complete expression for the deflection angle in terms of the yoke parameters as well as the voltage and current involved. Finally, the yoke sensitivity factor in Eq. (10-33), also called the *yoke energy constant,* can be used to estimate the maximum inductance for a specific application.[17] First, the dimensions of the yoke are set by the maximum deflection angle and the outside diameter of the CRT neck. These establish the maximum length and minimum inside diameter of the yoke. Given these parameters, and the accelerating voltage of the CRT, the relationship between the yoke inductance and the current is given by

$$k = \frac{LI^2}{2} \tag{10-47}$$

where $k = k_m$ = yoke energy constant for a given deflection angle and accelerating voltage, μs
L = yoke inductance, μH
I = yoke current to deflect the beam through the half-angle and the selected accelerating voltage, A

Then the maximum voltage induced across the yoke is found from

$$e = L\frac{di}{dt} \tag{10-48}$$

where e = induced potential, V
$\quad L$ = yoke inductance, H
$\quad di/dt$ = rate of change of current, A/s

Further, if the maximum applied voltage V_m is used for e, and $(2k/L)^{1/2}$ divided by the minimum time t to deflect the beam through the angle is substituted for di/dt, then

$$L_m = \frac{V^2 T^2}{2k} \tag{10-49}$$

where V = maximum allowable induced voltage, V
$\quad T$ = time to deflect the beam through half-angle, μs
$\quad k$ = yoke energy constant for half-angle deflection at specified acceleration voltage, μJ
$\quad L_m$ = maximum allowable yoke inductance, μH

Similarly, the equation for the resonant mode, assuming that the deflection current is a sine wave, becomes

$$L_m = \frac{2V^2 T^2}{\pi^2 k} \tag{10-50}$$

Using Eq. (10-49) or (10-50), it is now possible to determine the required current from

$$I = \left(\frac{2k}{L}\right)^{1/2} \tag{10-51}$$

where I is the current required for half-angle deflection, and the other terms are as above. Another factor which determines the upper limit for the yoke is the retrace time which must be less than the approximately 10 μs allowed. The natural retrace time for a yoke may be found from

$$T_r = \pi(L_y C_y)^{1/2} \leq 10 \ \mu s \tag{10-52}$$

where L_y = yoke inductance, H
$\quad C_y$ = yoke capacitance, F
$\quad T_r$ = retrace time, s

Using a shunt capacitance value of 330 pF leads to a maximum yoke inductance of 30 mH. This applies to the horizontal-deflection yoke, and the vertical-deflection yoke can be larger because of the longer retrace time of about 2 ms, leading to yoke inductances as high as 1 H.

10.3.2 **OTHER USEFUL FORMULAS.**[18,19]   There are several other formulas that are useful in determining the yoke current and inductance required when they are different from the ones given in the yoke data sheet. First, for a specific yoke, with a constant anode voltage, the deflection current required for deflection varies as the sine of the deflection angle. Therefore

$$\frac{I_2}{I_1} = \frac{\sin \Theta_2}{\sin \Theta_1} \tag{10-53}$$

and

$$I_2 = I_1 \frac{\sin \Theta_2}{\sin \Theta_1} \tag{10-54}$$

where I_1 = given current
$\quad I_2$ = new current

Θ_1 = given deflection angle
Θ_2 = new deflection angle

Equation (10-54) may be used to find the current required for a deflection angle different from the deflection angle in the data sheet. Similarly, the yoke inductance may be related to a different current using

$$\frac{I_2}{I_1} = \left(\frac{L_1}{L_2}\right)^{1/2} \tag{10-55}$$

where L_1 = given inductance and L_2 = other inductance for similar type of yoke.

Next, if the deflection yoke is given and the deflection angle is constant, then the deflection current will vary as the square root of the anode voltage, or

$$\frac{I_2}{I_1} = \left(\frac{V_2}{V_1}\right)^{1/2} \tag{10-56}$$

where V_1 = given anode voltage and V_2 = different anode voltage.

Finally, the step response of the yoke can be found from[20]

$$I = \frac{E}{R(1 - e^{-Rt/L})} \tag{10-57}$$

where E = voltage across yoke
R = yoke resistance
t = settling time
I = current through yoke

If R/L is much smaller than 1, this reduces to

$$I = \frac{ERt}{RL} \tag{10-58}$$

or

$$t = \frac{IL}{E} \tag{10-59}$$

which is the commonly used equation for settling time and is accurate to about 1 percent for settling time to 99 percent of the final value, but may be in error by as much as 25 percent for settling time to 99.9 percent of the final value, or if R/L is not small enough. However, it is usually adequate for most calculations and is in general use.

All these equations have been combined in a slide rule offered by one manufacturer of yokes and may be obtained from that manufacturer.[21] Alternatively, with the availability of electronic calculators it is probably just as easy to calculate the required values by substituting numbers in the equations and solving for the results. In either case, the use of these equations makes it simple to design with any of a number of yokes that are offered by yoke manufacturers. All the proportionalities may be combined in a single equation as

$$I_2 = I_1 \left(\frac{L_1}{L_2}\right)^{1/2} \left(\frac{E_2}{E_1}\right)^{1/2} \frac{\sin \Theta_2}{\sin \Theta_1} \tag{10-60}$$

10.3.3 **YOKE SELECTION PROCEDURE.** A yoke selection procedure has been developed by the Electronic Industries Association and is contained in a publication issued by that organization.[22] It is summarized here for convenience, and the completed procedure may be found by consulting that publication. Briefly, then, the steps are as follows:

1. Select a CRT and determine the maximum deflection angle.
2. Determine yoke ID dimensions from the CRT neck size.

3. Find the yoke energy constant from the yoke manufacturer, based on the ID and deflection angle.

4. Establish the anode voltage.

5. Determine the half-angle deflection for the two axes.

6. Calculate the energy constants.

7. Set the minimum time for the half-angle deflection.

8. Find the maximum allowable induced voltage.

9. Calculate the maximum allowable yoke inductance from Eq. (10-49).

By following this procedure, it is possible to arrive at an optimum selection of a yoke for a given set of requirements.

Table 10-1a to c provides reference data on the mechanical and electrical characteristics of commonly used 90 and 110° tubes and accompanying magnetic deflection systems.

10.3.4 DEFLECTION OF MULTIPLE BEAMS.

The most common forms of color CRTs used in television are the shadow mask in both the *delta* and *in-line* versions, and the Trinitron. All these tube types are described in detail in Chap. 12. For an explanation of principles, the multiple-beam arrangements of only the delta and in-line guns are described. These are shown in Figs. 10-23 and 10-24.[22] The main problems that occur when the three beams are deflected by a common deflection system are associated with the spot distortions that occur in single-beam tubes. However, the effect is intensified by the need for the three beams to cross over and combine as a spot on the shadow mask. The two effects that are most significant are curvature of the field and astigmatism. These distortions result in a misconvergence of the beam as is illustrated in Fig. 10-25 for the delta gun when astigmatism is considered and in Fig. 10-26 for astigmatism and coma. The misconvergence that occurs in the four corners of the raster is shown in Fig. 10-27, with the result that the color rendition will not be true, particularly at the edges of the raster. This can be partially compensated for by introducing quadripole fields that cause the beams to be twisted and restore the equilateral nature of the triangle. The shapes that these fields may take are illustrated in Fig. 10-28 along with the currents required to produce these fields. Another technique for producing the fields is to place magnets around the neck of the CRT or to use internal pole pieces that are magnetized by external coils. Such a field is shown in Fig. 10-29 for a four-pole field, and a six-pole field is also used. These techniques have been used with some success but require a rather cumbersome adjustment procedure to be effective. However, their adequacy is attested to by the success of the delta-gun shadow-mask tube in commercial television. The situation is less satisfactory for data display, where misregistration results in color errors that cannot be ignored, especially at the edges of the display.

In-Line Gun. The situation is somewhat less severe in the case of the in-line gun. Here, the three beams need only be converged into a vertical line rather than the round spot required by the delta gun. In the latter case, the quasi-anastigmatic deflection systems cause misconvergence because the almost uniform deflection fields that are generated result in misconvergence since the beams are overconverged when deflected, although they are statically converged at the center.[23] For the in-line gun, a precision self-converging system has been devised where the yoke is designed to operate with one specific tube. This self-converging yoke causes the beams to diverge horizontally in the yoke, resulting in non-uniform fields that counteract the overconvergence. The shape of the fields for horizontal and vertical deflection are shown in Figs. 10-30 and 10-31 for the two cases. The horizontal yoke generates a pincushion field, while the vertical yoke generates a barrel-shaped field to accomplish these ends. However, complete self-convergence without the need for compensating adjustments is possible only with narrow-angle tubes, and for the 110° tubes it is necessary to add some convergence adjustments, although much less than is required for the delta gun. The drawback is that the yoke

Table 10-1a Mechanical Characteristics

Type	Viewable-screen dimensions				Tube dimensions			
	Diagonal, in/mm	Width, in/mm	Height, in/mm	Area, in/mm	Tube length, in/mm	Neck length, in/mm	Neck diameter, in/mm	Beam offset in deflection plane (s), in/mm
RCA 25-V 90° system	24.658 / 626.3	20.776 / 527.7	15.582 / 395.8	315 / 2032	21.430 / 544.3	6.693 / 170.0	1.438 / 36.5	0.210 / 5.33
A66-140X wide-neck, 24-V 110° system	24.323 / 617.8	20.394 / 518	15.354 / 390	306 / 1973	17.248 / 438.1	6.169 / 156.7	1.437 / 36.5	0.216 / 5.48
RCA A67-150X narrow-neck PST 25-V 110° system	24.658 / 626.3	20.776 / 527.7	15.582 / 395.8	315 / 2032	17.009 / 431.3	5.968 / 151.59	1.146 / 29.1	0.148 / 3.76

Table 10-1b Circuit Parameters of 110° Deflection Systems

Color tube	Neck diameter, in (mm)	Yoke	Type	Yoke impedances				Peak deflection currents Center to edge scan at V_a = 25 kV		Peak deflection power	
				Horiz. coils		Vert. coils				Stored energy in horiz. coils $W = \tfrac{1}{2} L_H I_H^2$, mJ	Power dissipated in vert. coils $P = I_V^2 R_V$, W
				L_H, mH	R_H, gV	L_V, mH	R_V, gV	I_H, A	I_V, A		
RCA 25-V 90° system	1.438 (36.5)	XD4099-D2	Saddle	0.45	0.35	45.0	35.0	3.31	0.31	2.5	3.4
25-V 110° narrow-neck saddle yoke	1.15 (29)	XD4269-A4	Saddle	0.20	0.13	24.5	13.6	7.0	0.6	4.9	4.9
A66-140X wide-neck 24-V 110° system	1.438 (36.5)	AT1060	Dynamic saddle	4.42†	3.51†	25.4	15.0	1.58	0.6	5.6	5.1
				1.10‡	0.92‡	25.4	15.0	2.16	0.6	5.6	5.1
RCA A67-150X narrow-neck PST 25-V 110° system	1.15 (29)	XD4422-J12	PST	1.25†	1.75†	0.95	1.47	2.97	2.8	5.5	11.3
				0.31‡	0.44‡	0.95	1.47	5.94	2.8	5.5	11.3

†Series connected.
‡Parallel connected.

Table 10-1c Performance of 110° Color Systems

| Color tube | Deflection systems | | | | | Beam landing pattern | | | | | | System performance | | | |
| | Neck dia., (mm) | Yoke | Type | Difference current drive required | Pullback to equal neck shadow, in/mm | Average of poorest trio distortion ratios§ | | | | Intrinsic edge mask transmission T, % | Convergence | | Focus | |
						E-W	S-N	SSE-SSW	NNE-NNW		Quality	Complexity (number of controls)	Quality	Dynamic focusing required
RCA 25-V, 90° system	1.438 (36.5)	XD4099-D2	Saddle	No	$\frac{0.55}{14}$	_1.21_	0.86	1.16	1.19	100	Reference	Reference (12)	Reference	No
25-V 110° narrow-neck saddle yoke†	1.15 (29)	SD4269-A4†	Saddle	No	$\frac{0.39}{10}$	_1.33_	.083	1.19	1.18	80	Not optimized	Comparable to 90° ref. (12)	Better than 90° ref.	No
A66-140X wide-neck 24-V 110° System	1.438 (36.5)	AT1060	Dynamic saddle	Yes	$\frac{0.32}{8}$	1.6	0.89	1.08	_1.21_	100	Corners inferior to 90° ref.	More complex than 90° ref. (12 + 4)	Slightly better than 90° ref.	Yes
RCA A67-150X narrow-neck PST 25-V 110° system‡	1.15 (29)	XD4422-J12	PST	No	$\frac{0.28}{7}$	_1.21_	0.84	1.20	1.20	100	Comparable to 90° ref.	Comparable to 90° ref. (12)	Significantly better than 90° ref.	No

†Deflection yoke design for the 18-V 110° color tube; illustrates the loss of light output by use of a static saddle yoke on a 25-V 110° picture tube.
‡Precision Static Toroid (PST) deflection yoke designed for RCA narrow-neck 110° system.
§Measured ¾ in from edge of screen. Mask transmission-limiting trio distortion ratios are underlined.

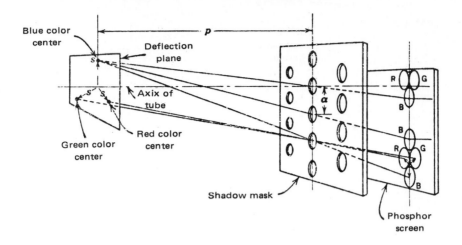

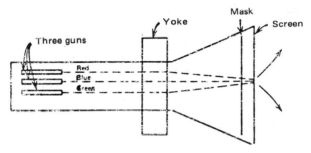

FIG. 10-23   Shadow-mask color CRT. (*After Law in Sherr.*[18])

must be tailored to the specific tube with which it is used. A quadripole winding is added to the yoke which generates poles at the 45° diagonals and is driven by one of the scanning frequencies. The results have been quite successful, as is apparent from examining a television display using an in-line gun with this type of yoke and comparing it with the delta gun that is as well adjusted as is possible. There are still a few adjustments needed for the in-line gun, but they are much simpler to accomplish. Further details on the design of the yoke may be found in Ref. 23.

10.3.5 DEFLECTION AMPLIFIER DESIGN CONSIDERATIONS.[24] In order to establish the requirements for the horizontal and vertical deflection amplifiers, it is advisable to describe the general characteristics of magnetic deflection amplifiers and arrive at a set of equations that characterize this type of amplifier. Of prime interest is the response of this amplifier to the two most prevalent types of signals, the sawtooth used for the horizontal and vertical linear deflection periods, and the step signal resulting from the flyback portion of the deflection signal. To this end, we may examine the general diagram of such a deflection amplifier as shown in Fig. 10-32. It should be noted that the yoke is represented as a separate circuit element, and contribution of the yoke to the amplifier response is of prime significance. The feedback resistor is another important circuit element since it contributes to both the small-signal and large-signal response of the amplifier. The various elements are represented by their Laplace transforms, and

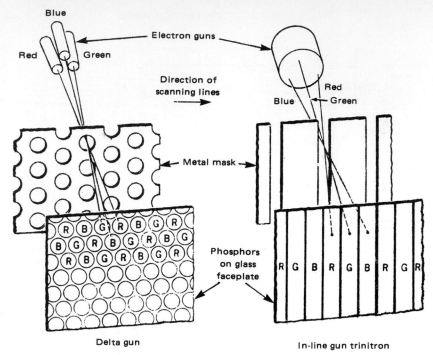

Blue

Red   Green

Electron guns

Direction of
scanning lines →

Red

Blue   Green

Metal mask →

R B G R B G
B G R B G R
R B G R B G

Phosphors
on glass
faceplate

R G B R G B R G R

Delta gun

In-line gun trinitron

FIG. 10-24 Delta gun compared with in-line gun Trinitron. (*After Herold in Sherr.*[14])

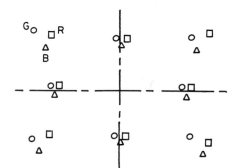

FIG. 10-25 Astigmatism errors. (*After Hutter.*[23])

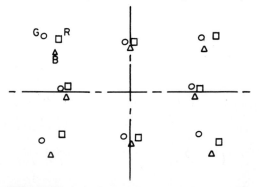

FIG. 10-26 Astigmatism and coma errors. (*After Hutter.*[23])

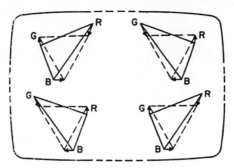

FIG. 10-27 Misconvergence in the four corners of the raster. (*After Hutter.*[23])

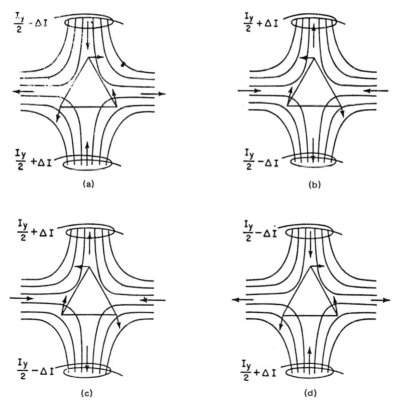

FIG. 10-28 Field configurations suitable for correcting misconvergence. (*After Hutter.*[23])

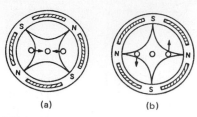 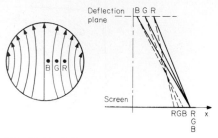

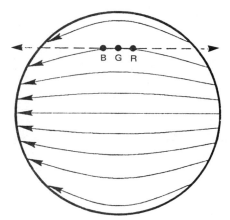

(a)   (b)

FIG. 10-29   Four-pole field. (*After Hutter.*[23])

FIG. 10-30   Self-converging horizontal deflection field. (*After Barkow.*[31])

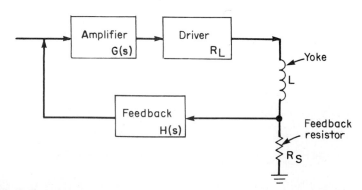

FIG. 10-31   Self-converging vertical deflection field. (*After Barkow.*[31])

FIG. 10-32   Block diagram of CRT magnetic-deflection amplifier. (*After Sherr.*[18])

10.28

$$C(s) = \frac{EmA}{s[A + (1+m)(1+sT_a)(1+sT_y)]} \qquad (10\text{-}61)$$

where E = driving function
$\quad m = R_f/R_i$ (feedback factor)
$\quad A$ = amplifier gain
$\quad T_a$ = amplifier time constant
$\quad T_y$ = yoke time constant

is the expression for the small-signal response. This may be solved in terms of the natural frequency (W_n), the damping factor (δ), and a third term (ψ), to give

$$C(t) = \frac{EmA}{1 + m + A}\left\{1 + \frac{1}{(1 - \delta^2)^{1/2}} e^{-\delta w_n t} \sin\left[W_n(1 - \delta^2)^{1/2}t - \psi\right]\right\} \qquad (10\text{-}62)$$

where

$$W_n = \left[\frac{1 + m + A}{(1 + m)T_aT_y}\right]^{1/2}$$

$$\delta = \frac{T_a + T_y}{2(t_aT_y)^{1/2}}\left(\frac{1 + m}{1 + m + A}\right)^{1/2}$$

$$\psi = \tan^{-1}\frac{(1 - \delta^2)^{1/2}}{-\delta}$$

Similarly, the response to a ramp or sawtooth is given by

$$C(s) = \frac{EmA}{t_0(1 + m + A)}\left[s^2\frac{(1+m)T_aT_y}{1+m+A} + s\frac{(1+m)(T_a + T_y)}{1+m+A} + 1\right] \qquad (10\text{-}63a)$$

and the solution is

$$C(t) = \frac{E_1mA}{t_0(1 + m + A)}\left\{t - \frac{2\delta}{W_n} + \frac{1}{W_n(1 - \delta^2)^{1/2}} e^{\delta w_n t} \sin\left[W_n(1 - \delta^y)^{1/9}t - \psi_2\right]\right\}$$

$$(10\text{-}63b)$$

By the use of these expressions it is possible to determine the response of any system to either a small signal or a sawtooth and find the natural frequency of the combination.

The large-signal response is different in nature, since the amplifier is overloaded, and it is only the yoke with the sum of its resistance and the series resistor that make up the time constant. In this case Eq. (10-57) or (10-59) applies, and the settling time may be calculated from either or both of these equations. Therefore, all these equations must be used to fully characterize the response of the deflection amplifier-yoke combination.

A few other equations of interest should be noted before leaving the analysis of the general deflection amplifier. First, Eq. (10-63) for the ramp response is obtained by using the Laplace for a ramp which is

$$E(s) = \frac{E}{s^2t_0} \qquad (10\text{-}64)$$

Next are two expressions for the band width of the amplifier and the amplifier-yoke combination. The amplifier may be treated as a simple RC, for which the bandwidth is related to the rise time by the well-known

$$T_r = \frac{0.35}{f_c} \qquad (10\text{-}65)$$

where T_r = rise time in microseconds from the 10 to the 90 percent point and f_c = 3-dB bandwith in MHz. When combined with the yoke in a feedback arrangement, it becomes a second-order system for which the equation is

$$T'_r = \frac{0.45}{f_c} \tag{10-66}$$

where T'_r is the rise time of the second-order system. These equations are useful in comparing the bandwidth requirements of the horizontal and vertical deflection amplifiers, in particular when we look at the characteristics of electrostatic deflection amplifiers. It is true that the latter are rarely used in television systems, but they should not be neglected since it is possible to attain greater bandwidths with such systems than with magnetic deflection amplifiers, and there are conditions where they may be the better choice.[25]

10.4 ELECTROSTATIC DEFLECTION

10.4.1 FUNDAMENTALS[26] Although magnetic deflection is the most common form used in television, building a system using an electrostatic deflection CRT is entirely feasible. This has been done by at least one manufacturer and results in a monitor that has certain advantages over a magnetic deflection unit. Beginning with Eq. (10-6), and the deflection arrangement shown in Fig. 10-33, the equations for electrostatic deflection may be derived. Referring to Figs. 10-2 and 10-33 and assuming that the electron enters the deflection field between the deflection plates at right angles, that is, θ equals 90°, then Eq. 10-6 may be rewritten as

$$\tan \alpha = \frac{V_d \, d_l}{2d_p V_b} \tag{10-67}$$

where V_d = voltage between deflection plates
 d_l = length of plates
 d_p = distance between plates
 V_b = beam voltage

Then since, tan α is approximately equal to y_d/l

$$y_d = \frac{l d_l V_d}{2d_p V_b} \tag{10-68}$$

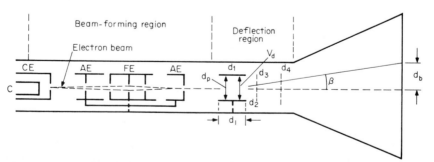

FIG. 10-33 Electrostatic deflection and focus CRT. C, cathode; CE, control electrode; AE, accelerating electrode; FE, focus electrode; d_1, d_2, vertical deflection plates; d_3, d_4, horizontal deflection plates; d_p, distance between plates; d_l, length of plates; β, beam deflection angle; d_b, beam deflection. (*After Sherr.*[2])

Equation (10-68) holds for parallel plates and neglects fringe effects. For nonparallel plates, the gradient is

$$\frac{dv}{dy} = \frac{V_d}{a_1 + (a_2 - a_1)\, X/d_l} \tag{10-69}$$

where a_1 = plate separation at entering end
 a_2 = plate separation at leaving end
 X = distance for beam in field of plates

The deflection equation then becomes

$$y = \frac{d_l V_d}{2 V_b a_1} \frac{\ln (a_2/a_1)}{(a_2/a_1) - 1} \tag{10-70}$$

Equation (10-70) becomes Eq. (10-68) when $a_1 = a_2$.

10.4.2 ACCELERATION VOLTAGE EFFECTS.

Examination of Eq. (10-87) demonstrates that the deflection distance is directly proportional to the deflection voltage V_d and inversely proportional to the acceleration or beam voltage V_b. This leads to the requirement that the deflection voltage must be increased a proportional amount when the beam voltage is increased. This may be compared with Eq. (10-33) for magnetic deflection, where the deflection is proportional to the square root of the beam voltage. This leads to two considerations when electrostatic and magnetic deflection are compared. First, it can be shown that the beam spread as a function of beam voltage (in kilovolts) and current (in milliamperes) is given by[27]

$$K^{1/2} = \frac{32.3 r_0 V^{3/4}}{I^{1/2}} \tag{10-71}$$

This expression is represented by the nomograph shown in Fig. 10-34 where r_0 = crossover spot size and z = position on the beam. From this it is clear that the beam size is affected by the beam voltage, and the higher the beam voltage, the smaller the spot. However increasing the beam voltage increases the required deflection voltage by a proportional amount, so that magnetic deflection can use larger beam voltages and therefore attain smaller spot sizes.

A second effect of beam voltage is that on light output from the phosphor, or luminance. Phosphor luminance is given by the empirical expression[38]

$$L = A f(p) V^n \tag{10-72}$$

where L = phosphor luminance
 $f(p)$ = function of current
 V = accelerating voltage
 A = constant
 n = 1.5 to 2

The phosphor output or luminance may also be expressed by

$$L = \frac{k_b I_b V_b^n}{A} \tag{10-73}$$

where k_b is a proportionality factor termed *luminous efficiency* and ranging from 5 to 62 for the various phosphors, I_b and V_b are the beam voltage and current, and A is the area of the phosphor surface. The proportionality factor is given in terms of lumens per watt and may be found in various references.[28] Equation (10-73) clearly demonstrates the effect of the beam voltage on the light output and illustrates the desirability of maintaining the beam voltage at the highest value consistent with the other requirements such as deflection sensitivity and focus voltage. Equation (10-73) is approximately correct over the linear region but does not hold when phosphor saturation occurs, so that it is not possible to increase phosphor luminance merely by raising the beam voltage.

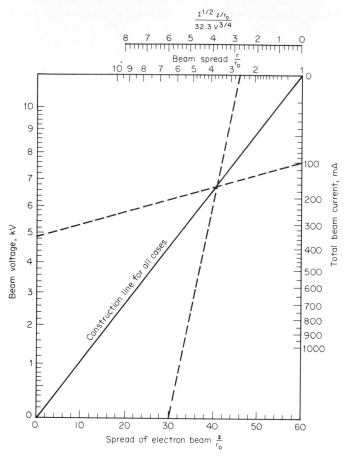

FIG. 10-34 Beam spread nomograph. (*After Spangenberg.*[10])

10.4.3 COMPARISON OF ELECTROSTATIC AND MAGNETIC DEFLECTION.

The effect of beam voltage on phosphor luminance described in the previous section also holds for magnetic deflection, and it is appropriate at this point to compare the characteristics of electrostatic and magnetic deflection CRTs. This can be done conveniently by examining Table 10-2a and *b*, which lists various parameters of the two types. The first difference that should be noted is the longer bottle required for the electrostatic type. This becomes more significant as the wide deflection angle magnetic deflection tubes become most common, and the longer length of the electrostatic types impose packaging limitations on the assembly which contains the tube. Related to the increased length are the narrower deflection angle available in the electrostatic types and the higher focus voltage required, as well as the need for a post-accelerator voltage. Of greatest significance, however, are the higher luminance and smaller spot size attainable with the electromagnetic deflection type. The one characteristic which is not shown in Table 10-2 is the faster deflection speed possible with the electrostatic deflection tube, which can be as low as 1μs, compared with the 10 μs that is possible with the electromagnetic deflection tube. This parameter is not included in the table because it is influ-

Table 10-2 CRT Monitors

Parameter	Value
(a) Magnetic Deflection	
Deflection settling time	10 μs
Small-signal bandwidth	2 MHz
Video bandwidth	15–30 MHz
Linear writing speed	1 μs/cm
Resolution	>1000 (shrinking raster) TV lines
Luminance	300 nits
Spot size, mm	0.25 (53-cm CRT)
Viewing area	13 in $\times$ 13 in (21-in CRT)
	[33 cm $\times$ 33 cm (53-cm CRT)]
Accelerator voltage	10 kV
Phosphor	Various
Refresh data	30–60 Hz
Power consumption	250 W
(b) Electrostatic Deflection	
Deflection settling time	<1 μs to one spot diameter
Small-signal bandwidth	5 MHz
Video bandwidth	25 MHz
Linear writing speed	25 cm/μs
Resolution	17 lines/cm (shrinking raster)
Luminance, nits	150–500
Spot size, mm	0.25–0.38
Viewing area	11 in high $\times$ 13 in wide
	(28 cm high $\times$ 38 cm wide)
Acceleration voltage	28 kV
Phosphor	P-31
Power consumption	130–140 W

enced by the choice of deflection amplifier and may be higher or lower, depending on what kind of amplifier is used. However, with use of the customary types of amplifiers, this 10/1 advantage is not unusual. In addition, the power consumption of the amplifier is much lower for the electrostatic deflection unit, which is another point in its favor. Finally, the wide use of magnetic deflection CRTs for television has made tubes of this type cheaper and more readily available than the other, so that this factor must also be considered when making a choice between the two.

To summarize this comparison, there are advantages and disadvantages for each type, and the choice of one over the other must be made on the basis of examining the requirements of the display system and selecting the one that meets these requirements best.

10.5 DISTORTION-CORRECTING CIRCUITS

10.5.1 FLAT-FACE DISTORTION. The theoretical basis for flat-face distortion[29] is covered in Sec. 10.2.2 and is defined by Eq. (10-34). We now consider how this distortion may be minimized by the use of special correction circuits. This departure from linearity is illustrated in Fig. 10-22, and the correction equation is Eq. (10-40). Linearity correction can be achieved by using a circuit that compensates for the second term of Eq. (10-

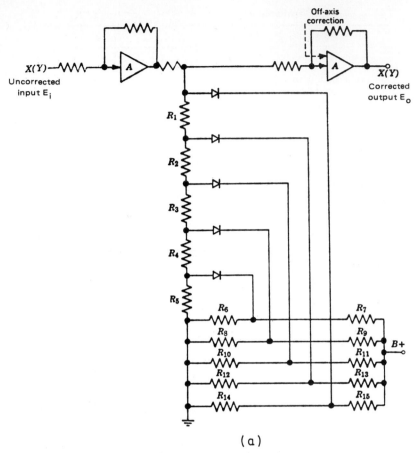

(a)

FIG. 10-35 (a) Linearity correction network; (b) plot of E_o versus E_i where $E_o = KI^3$; (c) linearity correction block diagram. (*After Sherr.*[18])

40), rewritten by substituting KI for d_i, resulting in

$$d_a = KI + \frac{K^3 I^3}{2} \tag{10-74}$$

Since $d_i = KI$, then the second term in Eq. (10-74) must be subtracted to achieve linearity, with the same type of correction applied to both the X and the Y axes. This is done by obtaining the current for each axis from its yoke winding and using a circuit of the kind shown in Fig. 10-35a. The cubic term is generated by means of a piecewise approximation, using as many diodes as necessary to achieve the desired accuracy of correction. Ten segments are usually sufficient, although only five are shown in Fig. 10-35a for simplicity. The circuit operates by biasing the diodes to different voltages so that they will conduct only when the output of the first summing amplifier exceeds these voltages. The diode currents are then summed in the output amplifier, and the output voltage is given by

$$E_o = \frac{E_i R_f}{R_p} \tag{10-75}$$

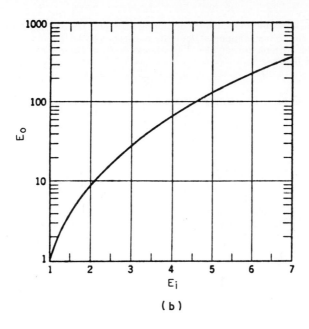

(b)

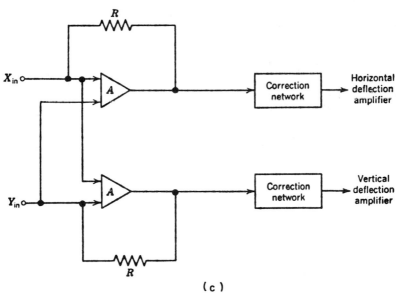

(c)

FIG. 10-35 *(Continued)*

where R_p is the equivalent value of the summing resistors. The similarity to a digital-to-analog (D/A) converter is apparent, but the resistors are not binary weighted, and E_i depends on the current in the yoke. The values for R_n in the resistor network may be determined by plotting

$$E_o = KI^3 \qquad (10\text{-}76)$$

on semilog paper as is done in Fig. 10-35b and then constructing a piecewise approximation, reading the bias levels from the horizontal axis where a new segment begins, and deriving the summing resistor values from the slope of the segment. A fairly accurate result may be attained by choosing R_1 to R_5, with each increasing by a factor of 2 from the previous one so that the resistors are defined by $R_n/2^n$, where n takes on the values from zero to the maximum number of segments.

The network shown in Fig. 10-35a corrects only for on-axis nonlinearity. However, when both X and Y deflection signals are present, they will affect each other and it is necessary to cross couple the two inputs as is shown in Fig. 10-35c that illustrates the manner in which the signal from one axis is added to that from the other axis.

Figure 10-36 illustrates the effect of off-axis deflection, and it is obvious that another correction is required, which is given by

$$\frac{\sin \alpha_2}{\sin \alpha_1} = \frac{(\sin \alpha_0^2 - 1)^{1/2}}{(\sin^2 \alpha_0 + \sin^2 \beta_0 + 1)^{1/2}} \qquad (10\text{-}77)$$

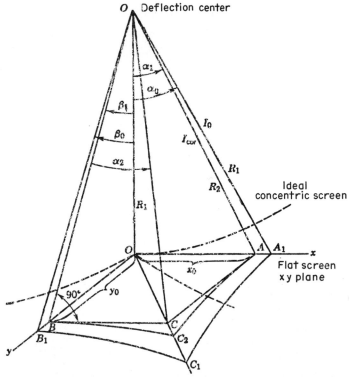

FIG. 10-36 Simultaneous horizontal and vertical deflection for flat-faced tube. (*After Popodi in Sherr.*[18])

and the total correction is expressed by

$$\frac{I_{cor}}{I_0} = \frac{1}{(\sin^2 \alpha_0 + \sin^2 \beta_0 + 1)^{1/2}} \tag{10-78}$$

The correction given by Eq. 10-78 applies to the X axis only, but it is symmetrical for both axes and applies to the Y axis as well. In Eq. 10-77 and 10-78, I_0 is the uncorrected signal, I_{cor} is the corrected signal, α_0 is the uncorrected angle in X, and β_0 is the uncorrected angle in Y. Equation 10-78 is difficult to use to obtain specific values for the resistors in the network, and it is usually sufficient and simpler to use a network similar to that shown in Fig. 10-35a, with final correction achieved by empirical means. One approach that has been suggested is to observe spot movement for full-scale variation of the signal on one axis and a succession of values on the other.[30]

10.5.2 DYNAMIC FOCUSING.
The effect of deflection on the focus of the electron beam is described in Sec. 10.2.3, with the change in spot size given by Eq. (10-41). It is possible to minimize this effect by means of the correction circuit shown in Fig. 10-37. It operates on the basis that the diameter of the electron beam is affected by the deflection distance on the face of the CRT, given by

$$r_s = Kd^2 \tag{10-79}$$

where r_s = change in spot size
d = deflection distance
K = constant

The deflection from the center d consists of the X and Y components, and it is given by

$$d^2 = X^2 + Y^2 \tag{10-80}$$

The correction signal requires that the X and Y terms be generated and then summed. These square functions may be produced by piecewise approximations using a diode network similar to that used for the cubic function described in the previous section. In the circuit for generating the dynamic focus correction signal shown in Fig. 10-37, the diode breakpoints and the values of the summing resistors may be determined by plotting the correction signal as shown in Fig. 10-38 and then converting it into the number of segments required, which is usually not more than five. In the case illustrated in Fig. 10-37, all resistors except the first may have the same value, which simplifies the circuit, and if more exact function generation is desired, it may be achieved by determining the exact values for each resistor more precisely and adding more segments.

Once the X^2 and Y^2 functions have been attained, they are summed in the output summing amplifier and applied to the focusing circuit. It is possible, by the use of this correction circuit, to maintain a spot size ratio of better than 1.5 over 35° of deflection instead of a variation of 3 to 1 which is not uncommon without correction. Thus, the use of dynamic focus correction is a necessary part of any well-designed deflection system.

10.5.3 PINCUSHION CORRECTION AND CONVERGENCE.
Pincushion and barrel distortions are briefly mentioned in Sec. 10.1.2 and are illustrated in Fig. 10-10. Pincushion correction is used in all wide-angle deflection systems and may be achieved either by using a special yoke design that has controlled field distortion or by predistorting the deflection current and applying it to a separate pincushion transformer which connects the correction current to the vertical yoke. The type of correction required is illustrated in Fig. 10-39a where the top lines bow down and the bottom lines bow up with the center line straight. Figure 10-39b shows the corresponding correction to the raster, which is achieved by introducing a parabolic correction to the vertical deflection signal. The vertical deflection is modulated by the parabolic correction signal at the horizontal deflection rate and is then combined with the vertical deflection signal to produce the corrected vertical deflection signal, as shown in the block diagram given in Fig. 10-40.

Another form of pincushion correction is that used for side correction. In this case, the distortion to be corrected is that occurring in the vertical dimension, also shown in

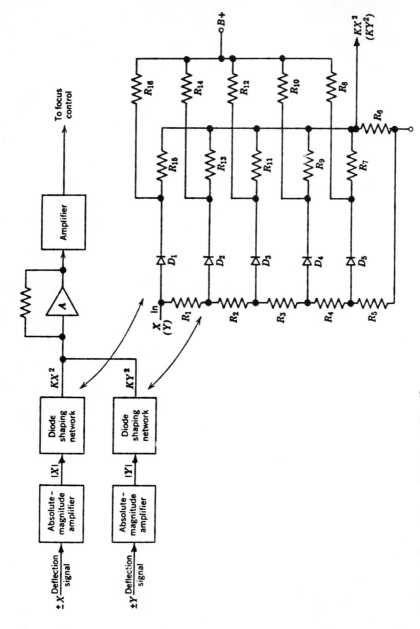

FIG. 10-37 Dynamic focus circuit. (*After Sherr.*[18])

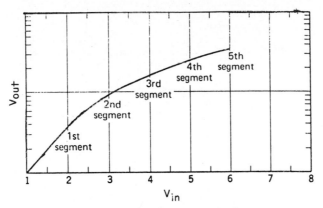

FIG. 10-38 Plot of correction signal. (*After Sherr.*[18])

Fig. 10-39*a*, with the form of the required correction illustrated in Fig. 10-39*b*. In this case, the vertical deflection current is used to modulate the horizontal deflection current at the vertical scanning frequency. To accomplish this, the vertical signal is modified by an RC network to produce the desired correction wave shape, which is then applied through a *pincushion transformer* to the horizontal yoke.

10.5.4 DYNAMIC CONVERGENCE. The last correction that should be mentioned is dynamic convergence, an essential requirement for all three-gun CRTs before the advent of the new in-line tubes which use predistortion in the deflection yokes for this purpose and thus much simplify the convergence process. However, it is useful to understand the effects that cause misconvergence when the beam is swept through angles of 70° and more. These occur as the result of two effects: (1) beam parallax, where the beams are off axis when they arrive at the deflection yoke, and (2) beam tilt, where the beams contain a component of radial velocity due to being converged before deflection. The effect of beam parallax is shown in Fig. 10-41 where, with the three guns arranged as illustrated, the three sweeps take on rhombic patterns. The effect of beam tilt is shown in Fig. 10-42, and from the triangle *A, D, Q* it follows that

$$\sin (\alpha + \beta) = K(i_d + i_0) \tag{10-81}$$

where K = constant
i_d = deflection current
i_0 = direct current = $\sin \beta / K$

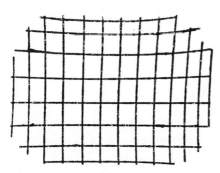 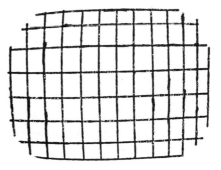

FIG. 10-39 Pincushion correction. (*After Fink.*[4])

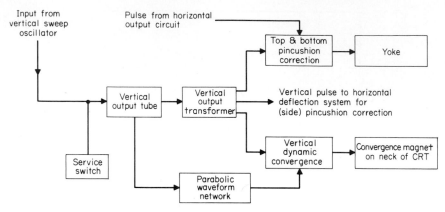

FIG. 10-40   Color vertical output block diagram.

The dc term is not found in Eq. (10-28) and causes an initial convergence of the beam prior to its entering the yoke. This dc term may be either positive or negative, depending on the sign and amplitude of the convergence angle β, and displaces the beams as is shown in Fig. 44a and b, for horizontal and vertical deflections. It is evident from Fig. 10-43 that the green beam is leading, and the red trailing the blue beam, with the resultant shifts illustrated by the patterns shown in Fig. 10-44a and b. Correction may be achieved by generating parabolic waveforms of the type shown in Fig. 10-45. The waveforms may then be applied to the yokes to achieve dynamic convergence.

Alternatively, a special convergence yoke may be included. In either case, this discussion is primarily of historical interest, since the in-line guns and predistorted yokes have essentially eliminated the need for separate dynamic-convergence circuits or yokes.[31]

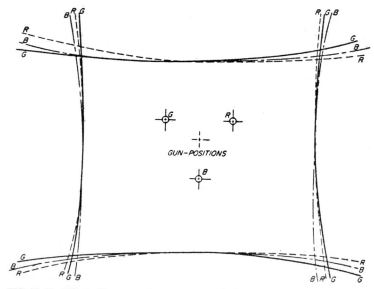

FIG. 10-41   Effect of beam parallax. (*After Fink.*[4])

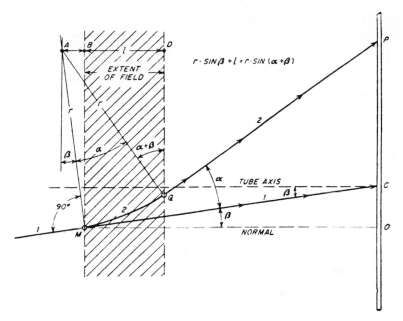

$$r \cdot \sin \beta + l = r \cdot \sin (\alpha + \beta)$$

FIG. 10-42 Effect of beam tilt. (*After Fink.*[4])

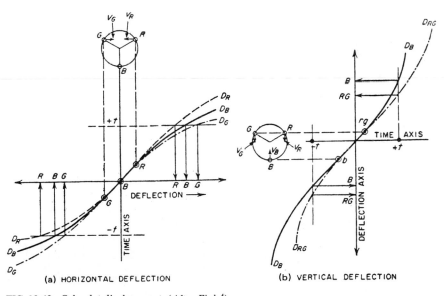

(a) HORIZONTAL DEFLECTION

(b) VERTICAL DEFLECTION

FIG. 10-43 Color dot displacement. (*After Fink.*[4])

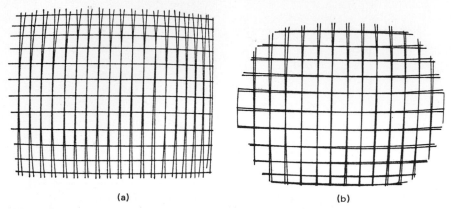

(a)   (b)

FIG. 10-44 (*a*) Horizontal shift of vertical green bars. (*b*) Vertical shift of horizontal blue bars. (*After Fink.*[4])

In-Line Systems. The modern in-line display systems eliminate the need for the special circuits and convergence yokes by providing nonuniform fields to overcome the basic overconvergence of in-line beams that are statically converged at the center. These beams, when deflected, will cross over before they reach the screen, as is shown in Fig. 10-46. By using a self-converging yoke, nonuniform fields are generated that balance out the overconvergence by causing the beams to diverge horizontally inside the yoke so that the horizontal yoke generates a pincushion-shaped field with its intensity increasing as the horizontal distance from the axis increases. The overconverged beams that are deflected by this field are shown in Fig. 10-30. Similarly, the vertical yoke creates a barrel-shaped field which diverges the beams horizontally as is illustrated by Fig. 10-31. Thus, the self-converging yoke causes almost perfect vertical-line focus along the axis because of the horizontal-negative and vertical-positive isotropic astigmatism. In addi-

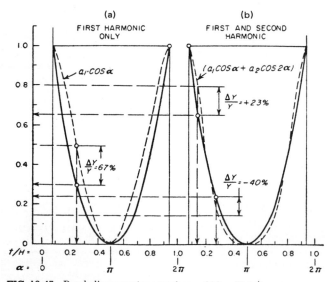

FIG. 10-45 Parabolic correction waveforms. (*After Fink.*[4])

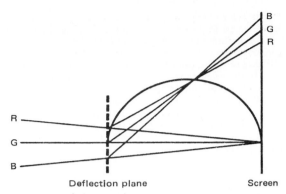

FIG. 10-46 Image field showing beam crossover. (*After Barkow.*[31])

tion, the anisotropic astigmatism is removed and the coma due to yoke is eliminated in the gun. However, this self-convergence without any dynamic convergence is effective only in small-angle (90°) systems. For 110° systems there is a small systematic convergence error which can be overcome only by limiting it to the horizontal and using only one scanning frequency for correction. This may be achieved by either of the two techniques shown in Fig. 10-47. In either system the horizontal lines converge over the whole raster and the vertical lines converge along the horizontal axis. Convergence is achieved by means of quadripole windings on the yoke, that are energized by one scanning frequency. The in-line system with self-convergence along the horizontal axis results in better convergence and requires only vertical frequency correction current, which is less expensive and results in less deflection defocusing. The net result is improved convergence with only two preset dynamic controls. At the same time, the yoke is more compact and sensitive, and the CRT is shorter by 10 mm. Therefore, this approach is used on all in-line guns, with demonstrably improved results over the older types, and it does not require the multiple adjustments and magnets used in the older system.

The self-converging television display system has been the subject of a patent[32] and is described in more detail there. In addition, there is an interesting technique, developed by Tektronix,[33] that uses additional phosphor strips to produce illumination into the rear of the CRT. This output, which is detected by a photomultiplier tube placed in the side of the CRT, is used to determine the misconvergence, and the resultant electrical signals correct the error, reducing color fringing and the loss of definition to negligible amounts. The use of these techniques has virtually eliminated misconvergence as a significant factor in color CRTs, although it may arise again as higher and higher resolution color tubes are developed.

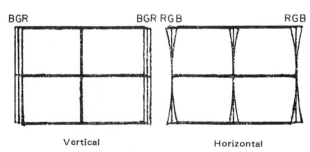

FIG. 10-47 Self-converging in-line systems. (*After Barkow.*[31])

In addition to the color CRTs that use more or less standard gun structures and deflection systems, there are also several approaches that incorporate a flat CRT but are not constructed or driven in a matrix fashion so that they fall into this section rather than the one on matrix systems.[34,34a] In one of them, the gun is placed parallel to the screen[34] as was the case in the Aiken tube,[35] and electrostatic deflection plates are used to deflect the beam. In the other type, magnetic deflection is used for horizontal deflection and electrostatic deflection for vertical deflection. The same considerations apply for these deflection systems as do for the deflection amplifiers covered in the previous sections, except that corrections must be applied to reduce the trapezoidal patterns that result. In the first this is achieved by reducing the screen height, and in the second by applying correction waveforms to the scanning signals. Both tubes use single guns and do not produce color displays.

10.6 MATRIX SYSTEMS[36]

10.6.1 FLAT-PANEL DISPLAYS. Flat-panel displays have become increasingly common and offer alternatives to the CRT when the smaller volumes and thin front-to-back dimensions are desired. They are briefly described in Chap. 12, but only in relation to their use as large-screen displays. They are by no means restricted to this type of application and have been used as the display devices for small-screen television systems as well.[37,37a] The most successful technologies for this purpose are those built around the gas discharge panels,[38,39] thin-film electroluminescence,[40,41] and liquid crystal displays[42,43] although only the last one has been actually incorporated in a receiver that is on the market.[36] None of these technologies offers color displays as yet, but development is continuing, and we may find the flat-panel displays to be the viewing device of the future. A variety of special waveforms are required to properly activate the display elements, but they all use some variety of matrix addressing.

10.6.2 MATRIX ADDRESSING.[44] The basic technique used in addressing and driving matrix flat-panel displays is the one known as *matrix addressing*. It differs considerably from the deflection techniques used for CRT displays, with the difference due to the unique nature of the construction of the panels. Instead of the electron beams found in the CRT that may be deflected by means of magnetic or electrostatic fields, the flat panel is made up of an assembly of discrete emitting or transmitting elements that must be individually selected and driven to achieve the desired results. Although it is in theory possible to make direct connections to each element and select the proper combinations by driving each one by means of these direct connections, in practice this rapidly becomes impractical, since for a standard 525-line display this means that more than 250,000 connections must be made. Clearly, this requirement is unreasonable from both an economic and physical point of view, and the expedient that is used is to resort to the technique termed *matrix addressing*.

Matrix addressing signifies that the elements on the panel are arranged in groups or rows and columns, as is illustrated in Fig. 10-48 for a 4 × 4 matrix, with each element in a row or column connected to all other elements in that row or column as shown in the figure. Given this type of connection, it is then possible to select the drive signals so that when only a row or a column is driven, no elements in that row or column will be activated, but when both the row and column containing a selected element are driven, the sum of the two signals will be sufficient to activate that element. The reduction in the number of connections required from the discrete element approach is quite large, since the number of lines is reduced from one per element to the sum of the number of rows and columns in the panel. For example, in the 4 × 4 array shown in Fig. 10-48, the reduction is from 16 to 8, which is not very large, but for a 500 × 500 array, the reduction is from 250,000 to 1000, which is certainly significant. In general

$$T = XY \qquad (10\text{-}82)$$

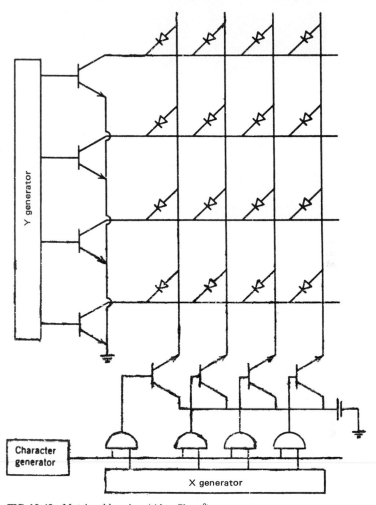

FIG. 10-48   Matrix addressing. (*After Sherr.*[2])

and
$$C = X + Y \tag{10-83}$$

where T = number of addressable elements
  X = number of rows
  Y = number of columns
  C = number of connections

and the advantage of the matrix addressing technique is quite apparent. However, along with the advantage come certain requirements that must be placed on the display elements so that this addressing technique is feasible, the most important of which is the need for a sharp discontinuity between the signal level below which the element will not be actuated and the level above which it will be fully on. This type of response is pictured in Fig. 10-49 for both the ideal case, where the element switches from off to on for an infinitesimal change in signal level, and the more practical case, also shown in Fig. 10-49, where the switching occurs over some change in the signal input. This necessitates

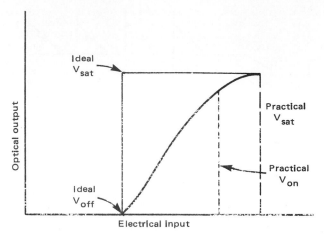

FIG. 10-49   Comparison of ideal and practical response of electrooptical device. (*After Sherr.*[2])

that some level, such as half the on requirement, be applied to either the row or the
column, with the further proviso that this half value be less than the amount needed to
turn the element even partially on. Then, if the same signal level is applied to the column
or row, the element should turn at least partially on. This is not too difficult to achieve
when the element at least approximates the ideal case, but in the majority of situations
twice the maximum allowable off voltage is not sufficient to place the element fully in
the on condition, with the resultant loss in either light output or transmission capability.
As a result, the contrast ratio is less than the best that can be achieved. The criteria for
best operation are

$$V_{\text{off}} \simeq V_{\text{on}}$$
$$2V_{\text{off}} = V_{\text{sat}} \quad \text{(saturation)}$$

(10-84)

but this situation is rarely fully achieved in practical terms. As a result, the phenomenon
known as crosstalk occurs, where elements that should be off are partially on, or else
elements that should be fully on are not driven into saturation. A number of techniques
have been evolved to overcome this problem, such as an additional layer in the dc gas
discharge panel which provides a means for scanning the panel while turning only the
desired element[38] or adding a switching device with the desired characteristics at the
intersections of the rows and columns[45] and these have achieved considerable success in
attaining desired performance.

Another deficiency that occurs when matrix addressing a flat-panel display is the loss
of contrast ratio that results from the element being on only during the time that it is
addressed. This is the same phenomenon that is encountered in the CRT, where the
beam moves from point to point, and the contrast ratio is lowered by the duty cycle as a
result of the average luminance being given by

$$L_a = nL_p$$

(10-85)

where L_a = average luminance
L_p = peak luminance
n = duty factor (on-time/frame time)

This is not too serious in the case of the CRT where the peak luminance is very high, but
for the matrix panels the peak luminances are much lower and limit the number of elements that can be sequentially addressed before the average luminance drops below that
required to maintain an adequate contrast ratio. The same limitation also applies to

nonemissive display elements such as liquid crystals, where the peak contrast ratio is the determining factor. The expedients that have been devised to overcome this difficulty range from incorporating an inherent memory in the panel as in the gas discharge types,[38,46] adding switching elements that can remain on after the signal has been removed,[39] and driving multiple elements in parallel. All these expedients increase the complexity of either the panel or the electronics, and it remains to be determined what the most successful approach will be, both as to the panel technology and the matrix addressing. However, regardless of these difficulties, several television sets have incorporated liquid crystal flat-panel displays, and we may anticipate that others will be developed in the future. For the present, the liquid crystal display is the only one that is being used in a commercially available unit, with matrix addressing as the means for selecting and driving the elements, using parallel addressing to overcome the contrast-ratio problem.

REFERENCES

1. *IEEE Standard Dictionary of Electrical and Electronics Terms,* 2d ed., Wiley, New York, 1977.

2. S. Sherr, *Electronic Displays,* Wiley, New York, 1979, pp. 69–70.

3. Hilary Moss, *Narrow Angle Electron Guns and Cathode Ray Tubes,* Academic, New York, 1968.

3a. K. R. Spangenberg, *Vacuum Tubes,* McGraw-Hill, New York, 1948, p. 100.

3b. Ibid., p. 101.

3c. S. Sherr, *Fundamentals of Display System Design,* Wiley, New York, 1970, pp. 63–64.

4. Donald Fink (ed.), *Television Engineering Handbook,* McGraw-Hill, New York, 1957, p. 5-4.

5. H. Moss, op. cit., pp. 156–158.

6. H. Moss, op. cit.

7. V. K. Zworykin and G. Morton, *Television,* 2d ed., Wiley, New York, 1954, pp. 159–167.

8. H. Pender and K. McIlwain (eds.), *Electrical Engineers Handbook,* Wiley, New York, 1950, pp. 14-61.

9. Ibid.

10. K. R. Spangenberg, op. cit, pp. 419–420.

11. A. M. Morell et al., "Color Television Picture Tubes", in *Advances in Image Pickup and Display,* vol. 1, B. Kazan (ed.), Academic, New York, 1974, p. 136.

11a. Ibid., pp. 145–146.

12. S. Sherr, *Electronic Displays,* p. 79.

13. Ibid., pp. 368–370.

14. Ibid., pp. 103–106, 120.

15. D. Fink, op. cit., p. 6-2.

16. Ibid., p. 6-4.

17. "Yoke Selection Guide," Syntronic Instruments, Inc.

18. S. Sherr, *Fundamentals of Display System Design,* pp. 334–335.

19. Syntronic Instruments, Inc., op. cit.

20. S. Sherr, *Fundamentals of Display System Design,* pp. 334–335.

21. "CRT Display Computer Instructions," Constantine Engineering Laboratories Co., Mahwah, N.J.

22. "TEPAC Engineering Bulletin," no. 22, Electronic Industries Association, Washington, 1979.

23. Rudolph G. E. Hutter, "The Deflection of Electron Beams," in *Advances in Image Pickup and Display,* B. Kazan (ed.), vol. 1, Academic, New York, 1974, pp. 212–215.

24. S. Sherr, *Fundamentals of Display System Design,* pp. 319–341.

25. S. Sherr, *Electronic Displays,* p. 424.

26. S. Sherr, *Electronic Displays,* pp. 55–57, 360–362.

27. S. Sherr, *Fundamentals of Display System Design,* pp. 58–59.

28. Ibid., pp. 67–68.

29. Ibid., pp. 370–374.

30. Ibid., p. 374.

31. W. H. Barkow and J. Gross, "The RCA Large Screen 110° Precision In-Line System," ST-5015, RCA Entertainment, Lancaster, Pa.

32. Josef Gross et al., Self-Converging Television Display System, Patent 4,376,924, March 15, 1983.

33. "Automatic Convergence 'Paints' Unprecedented Color Display," *Electronic Products,* May 12, 1983, pp. 17–18.

34. Clive Sinclair, "Small Flat Cathode Ray Tube," *SID Digest,* 1981, pp. 138–139.

34a. S. Sherr, *Electronic Displays,* op cit., pp. 505–540.

35. W. R. Aiken, "A Thin Cathode Ray Tube," *Proc. IRE,* vol. 45, no. 12, December 1957, pp. 1599–1604.

36. M. Takeda, et al., "Practical Application Technologies of Thin-Film Electroluminescent Panels," *Proc. SID,* vol. 22, no. 1, 1981, pp. 57–62.

37. T. Yamazaki et al., "A Liquid Crystal TV Display Panel Incorporating Drivers," *SID Proceedings,* vol. 23, no 4, 1982, pp. 223–226.

37a. A. G. Fischer et al., "Mass-Producible TFT-LC Modules for Square-Meter-Sized or for Pocket-Sized Flat TV Panels," *Conference Record 1982 Int. Display Research Conf.,* October 1982, pp. 161–165.

38. R. Colz et al., "Gas Discharge Panels with Internal Line Sequencing," in *Advances in Image Pickup and Display,* B. Kazan (ed.), vol. 3, Academic, New York, 1977, pp. 83–107.

39. P. Pleshko, "AC Plasma Display Technology Overview," *SID Proceedings,* vol. 20, no. 3, 1979, pp. 127–130.

40. J. E. Bernard et al., "Mechanism of Thin Film Electroluminescence," *Conference Record,* op. cit., pp. 20–24.

41. H. Kozawaguchi et al., "Low-Voltage ac Thin-Film Electroluminescent Devices," *SID Proceedings,* vol. 23, no. 3, 1982, pp. 181–186.

42. S. Sherr, *Fundamentals of Display System Design,* pp. 221–232.

43. T. Umeda et al., "Liquid Crystal Display Panel Using Plastic Substrates," *SID Proceedings,* vol. 23, no. 4, pp. 227–232.

44. S. Sherr, *Fundamentals of Display System Design,* pp. 250–254.

45. A. L. Lakatos, "Promise and Challenge of Thin-Film Silicon Approaches to Active Matrices," *Conference Record,* op cit., pp. 146–151.

46. G. Holz et al., "A 2000 Character Self-Scan[R] Memory Plasma Display," *SID Digest,* 1983, pp. 130–131.

Photosensitive Camera Tubes and Devices

Robert G. Neuhauser

Manager, Broadcast Tubes Marketing and Application Engineering
RCA Corporation
New Products Division
Lancaster, Pennsylvania
Fellow of Society of Motion Picture and Television Engineers

with contributions by

A. D. Cope

David Sarnoff Research Center
RCA Corporation
Princeton, New Jersey

11.1 PHOTOSENSITIVITY

A photosensitive camera tube is the light-sensitive device utilized in a television camera to develop the video signal. The image of the scene being televised is focused upon the light-sensitive member of this tube.

The energy present in the photons of light generates free electrons. In the case of a photoemitting light-sensitive surface, they are emitted from the surface into the vacuum of the tube. When a photoconducting material is used, the electrons are freed in the bulk of the material to carry current through it.

The function of the camera tube is to interrogate the electron charges that are developed in this process at successive points of the optical image, in synchronism with the television raster scan.

11.1.1 PHOTOEMITTERS.

Photoemitters are alloys of metals (usually the alkali metals) that form a semiconductor. They have both a low band gap and a low electron affinity. The sum of the band gap and the electron affinity is sometimes called the *work function,* i.e., the minimum amount of energy that an electron from the material must receive to cause it to be emitted from the surface of the material into a vacuum. (See Fig. 11-1.) The energy involved is expressed in electronvolts. In metals there are many free electrons, and so energy from the photon does not have to be expended to free electrons from valance states to conduction states; i.e., the band gap is negligible. (See Fig. 11-2.)

The energy for photoemission of an electron must come from the energy of a single photon of light. (The probability that any electron will receive energy from two photons is extremely small.) The energy of photons of visible light ranges from 3.11 to 1.77 eV, corresponding to wavelengths of 400 and 700 nm, respectively, the limits of perception of visible light by the average person.

The energy of light in a photon of any wavelength is expressed as

$$E = h\nu \tag{11-1}$$

where h = Planck's constant (6.625×10^{-34} J·s)
  ν = frequency of radiation, Hz
  E = energy, J

and
$$\nu = \frac{c}{\lambda}$$

where c = velocity of light, 3.0×10^8 m/s, and λ = wavelength in meters

or
$$E = h\frac{c}{\lambda}$$

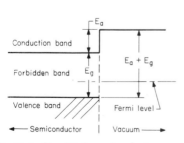

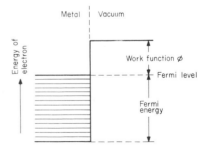

FIG. 11-1 Simplified semiconductor energy-band model (E_a, electron affinity; E_g, band gap). *(RCA Corp.)*

FIG. 11-2 Energy model for a metal showing the relationship of the work function and the Fermi level. *(RCA Corp.)*

For a wavelength of 700 nm (700×10^{-9}) (the red threshold of visible light)

$$E = \frac{(6.625 \times 10^{-34})\,(3 \times 10^{8})}{700 \times 10^{-9}} = 2.840 \times 10^{-19}\ \text{J} \qquad (11\text{-}2)$$

Expressed in electronvolts

$$E = \frac{J}{J/eV}$$

where $J/eV = 1.602 \times 10^{-19}$, which gives

$$E = \frac{2.840 \times 10^{-19}}{1.602 \times 10^{-19}} = 1.773\ \text{eV} \qquad (11\text{-}3)$$

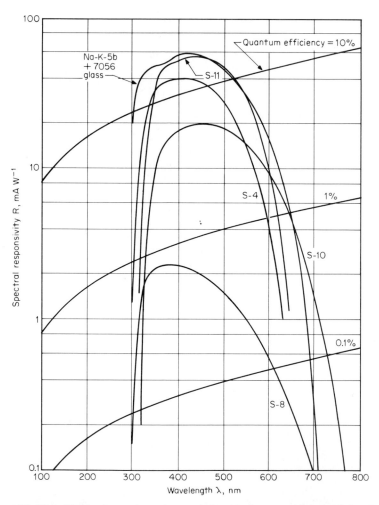

FIG. 11-3 Visible photoemitter characteristics. Absolute spectral responsivity of various photocathodes useful for the detection of visible light. *(RCA Corp.)*

An electronvolt is the energy of one electron when it has been accelerated through an electric field of 1 V.

When a photon of light is absorbed in a photoemitter, its energy is transferred to an electron in the valance band. This energy allows the electron to be liberated from the valance band, and it enters the conduction band. If it still has enough energy to overcome the electron affinity barrier, it may be liberated or expelled into the vacuum, and it becomes a *photoelectron*. The efficiency of the best photoemitters is about 0.3 or 30 percent. This refers to the average number of electrons emitted per photon of incident light at the wavelength of maximum photosensitivity.

Photoemitters are generally composed of complex alloys of the various metals, particularly the alkali metals or their oxides. These materials include various combinations of sodium, potassium, cesium lithium, rubidium, antimony, and silver.

In general, photoemitters have reasonably good electrical conductivity. If an electrical connection is made to the photoemitting material, it can supply large quantities of emitted electrons in response to large input fluxes of photons. The spectral and efficiency characteristics of photoemitters useful in television camera tubes are shown in Fig. 11-3.

11.1.2 PHOTOCONDUCTORS.

Photoconductors are semiconducting materials designed to absorb light and utilize the energy of photons to raise electrons from the valance band to the conduction band. Thus they serve as charge carriers to increase electrical conduction through the material.

This is in contrast to the photoemitters, where the electrons are emitted from the surface of the material into a vacuum. Instead here the photons of light absorbed in a photoconductor serve to change the electrical conductivity of the material. The semiconductor band structure of a photoconductor can be represented almost exactly as the photoemitter. (See Fig. 11-1.)

The energy of the incident photon that is absorbed by the electrons in the valence band of the photoconductor need only be enough to allow the electron to overcome the band-gap barrier and enter the conduction band. These free electrons contribute electrical conductivity to an otherwise high-resistance material.

Photoconductors for camera tubes usually have very high dark resistance. Specific resistivity is in the order of 10^{12} $\Omega \cdot$ cm. Low-resistance photoconductors can be used if the material is formed into special structures. For detailed descriptions of photoconductors utilized in camera tubes and their relative performance characteristics see Sec. 11.7.

11.2 PHOTOELECTRIC-INDUCED TELEVISION SIGNAL GENERATION

Any practical camera tube designed for operation in a camera that must operate in normally available light levels is of the storage type. This means that the light from each point on a scene generates an electrical charge corresponding to the brightness of that portion of the scene. The device integrates and stores that charge in a two-dimensional array during the interval between successive scans of that portion of the scene. This greatly increases the effective sensitivity. If this storage did not occur, all the potential information developed during the interval between successive scans would not be utilized and therefore lost.

The function, then, of the photosensitive portion of the camera tube is to absorb photons of light. Next the device must generate, integrate, and store a two-dimensional pattern of charges. This stored pattern is then interrogated to develop a television signal whose waveform corresponds in amplitude to the scene brightness at each point in succession as the television scanning process proceeds.

11.2.1 PHOTOEMISSION-INDUCED CHARGE IMAGES.

If a photoemitter is to generate and store an electrical image, the image charges that are developed will be positive, since electrons will be lost from the surface by the photoemission process.

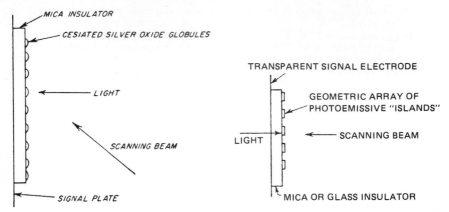

FIG. 11-4 Iconoscope mosaic. (*From Fink.*[1,†])   **FIG. 11-5** CPS emitron mosaic. (*From Fink.*[1])

If storage of the image is to take place on the photoemitter itself, the photoemitter must have sufficient electrical resistivity to store the charge without significant leakage from one element of the picture to adjacent elements during the storage time of a single television field (16 to 20 ms). Photoemitters are generally good electrical conductors and therefore cannot store an image charge on a continuous sheet or film of photoemitter material.

Early camera tubes surmounted this problem by having the photoemissive material either in discrete particles or in a discrete geometric array of photoemitting patches. This presented construction difficulties, problems of electrical stability, and difficulties of scanning with an electron beam.

One type of tube, the iconoscope, used a mosaic consisting of small particles of photoemitting material on an insulating substrate (Fig. 11-4). Light absorbed by the photosensitive silver-oxygen-cesium photoemitter emitted electrons to a surrounding collector, and the isolated globules of the mosaic charged up positively in proportion to the light on each.

Another approach was to use a thin, semitransparent two-dimensional array of square patches of light-sensitive material. These were placed on a transparent insulator and transparent signal electrode (Fig. 11-5). This structure functioned in the same way as the mosaic shown in Fig. 11-4, except that the light entered from the side opposite the photoemitter. This greatly simplified the scanning process. It was made possible by the development of an exceedingly thin photoemitter so that electrons generated near one side of the material could travel through it and escape from the vacuum interface surface.

11.2.2 SECONDARY-EMISSION-INDUCED CHARGE IMAGES.

A second method of developing a charge image by a photoemitter is to use the principle of secondary emission. Here the electrons are accelerated away from the photoemitter surface, which is a continuous film on the inside of the tube faceplate. The individual streams of electrons are focused on an insulating material that is also a good secondary emitter (Fig. 11-6).

The electrons can be focused by either magnetic or electrostatic fields, but it must be done with sufficient precision that each stream of electrons striking the insulator develops a charge that has the same spatial relationship on the insulator as the source of electrons on the photoemitter.

The secondary emission process is identical to the photoemission process, except that the incident energy is provided by electrons having energy of many hundreds of electronvolts rather than only the several electronvolts of energy available in visible-light

---

† Superscript numbers refer to References at end of chapter.

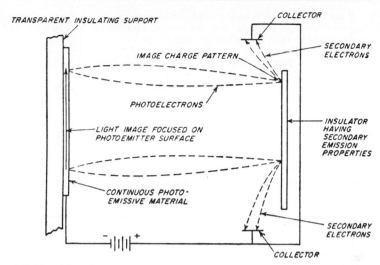

FIG. 11-6 Transfer of potential images by secondary emission. (*From Fink.*[1])

photons. The primary electrons penetrate into the material and transfer a portion of their energy to each electron with which they interact. If this energy is sufficient to free the electrons from the valance band and overcome the band gap (Fig. 11-1) and the electron can migrate to the surface of the material with enough energy to overcome the electron affinity of the surface, that electron can escape from the material as a secondary-emitted electron.

The charge pattern that is developed on the secondary-emitter insulator surface is positive if more secondary electrons are emitted than are incident on it from the stream of primary photoelectrons. The insulating surface must have high enough resistance to store the electron charges and prevent lateral leakage in the interval between successive scans of each portion of the image.

Secondary emission will take place at the surface of any material if the energy of the impinging electrons is high enough. Most materials have a secondary-electron-emitting characteristic similar to that illustrated in Fig. 11-7. At low energies, the bombarding electrons cause less than one secondary electron to be emitted for every primary incident electron, and a negative voltage will be built up on the insulator. When the energy of the primary electron is high enough so that more than one electron (on the average) is liberated, the charge developed on the surface will be positive. It is desirable to have positive-image charges developed by a television camera tube since positive charges can be more readily interrogated by an electron beam than negative charge patterns.

The process of generating charge images in a television camera tube can result in amplification. The gain at the secondary emitting surface is the secondary emission ratio minus 1.

The voltage chosen for acceleration of the photoelectrons determines the amount of gain. The gain reaches a maximum at a particular voltage for each material, and then decreases as the energy of the incident electrons is increased. This is caused by the electrons penetrating deep into the material where they lose their momentum by transferring energy to electrons in the

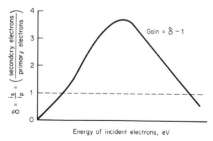

FIG. 11-7 Secondary-emission ratio of a typical surface.

material. These excited electrons generated deep within the material have a much lower probability of escape than electrons that are excited near the surface. In some materials the secondary emission ratio actually decreases below 1 at very high energies of the primary electrons.

Unlike a photon of light that transfers all its energy to a single electron as it is annihilated, a primary electron transfers some momentum to each electron with which it interacts and continues to do so till the electron is captured within the structure of the material. There is a wide spectrum in the energy of the secondary electrons emitted from the surface of a secondary emitter, as shown in Fig. 11-8. Many electrons have a low energy, in the range of several electronvolts. A few have much higher energies. The right-hand peak of the curve represents electrons that have an energy equal to that of the incident electron beam. In all probability these are primary electrons that are reflected from the surface without interacting with any electrons in the structure. Because of the wide range of velocities of the secondary electrons, it is difficult to image these electrons to another surface for further storage or amplification.

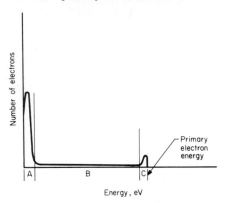

FIG. 11-8   Secondary-electron energy distribution.

11.2.3 ELECTRON-BOMBARDMENT-INDUCED CONDUCTIVITY.

Electron-bombardment-induced conductivity is a natural extension of both secondary-emission-generated image charge and photoconductivity technology.

Here the insulating material of the secondary-emission-generated charge storage target is not used. In its place there is a material that exhibits high electron-bombardment-induced conductivity. This material is placed at the plane where the electrons generated by photoemission are brought to focus corresponding to the point on the photoemitter from which they were emitted (Fig. 11-9). The electrical conductivity is produced by the primary electrons penetrating the material and releasing electrons from the valance band into the conduction band. Typically, the photogenerated electrons from the photoemitter are accelerated to an energy of several thousand electronvolts before they strike the target.

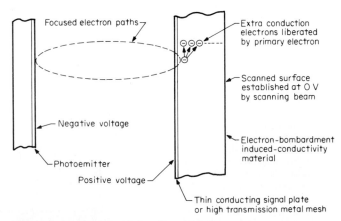

FIG. 11-9   Bombardment-induced conductivity.

Unlike photon-induced conductivity (where the photons are annihilated by transferring all their energy to a single electron), this process is not limited to one free electron per incident photon. Here the fast-moving electrons continue losing energy as they hit other electrons, until they are no longer able to excite any more into conductivity.

This process is capable of extremely high amplification since very high-energy primary electrons are needed to penetrate the signal plate and deep into the material. This energy is available for transfer to numerous electrons. Gains in the thousands of electrons can be achieved in practical camera tubes using this method of signal generation.

The secondary-electron bombardment-induced conductivity method to generate a stored-charge image is the same one used by photoconductive-sensitive camera tubes (see Sec. 11.2.4).

Materials having a narrow band gap (needed for photoconductors) are not necessary for electron-bombardment-induced conductivity. This is because the energy of the bombarding electrons can be in the range of many thousands of volts. Thus, large portions of the energy of the bombarding electrons can be transferred to the electrons of the material. These excited electrons can then either jump into the conducting band or, if there is enough energy left over, some can interact with other electrons in the valance band, which are in turn pushed into the conducting band. The gain of this process (conducting electrons/primary electrons) can be varied by varying the velocity of the primary electrons.

A widely used bombardment-induced conductivity target is silicon. This silicon target is virtually identical to the one used as a light-sensitive photoconductor (see Sec. 11.4.4). The charge image is generated from the electrons (and holes) that are freed as charge carriers in the silicon and stored as charges on the diode array.

11.2.4 PHOTOCONDUCTIVE-GENERATED CHARGE IMAGES.
A photoconductor can be used to develop a charge image which is then stored on its surface.

To utilize the property of photoconductivity, an electric field must be impressed across the photoconductor. The usual way to do this is to deposit photoconductive material on an electrically conductive and transparent substrate (Fig. 11-10). In order to develop a positive charge pattern easily interrogated by an electron beam, a positive voltage is applied to the signal electrode. A more negative voltage is established on the opposite side by an electron beam (see Sec. 11.2.5).

To be effective as a charge pattern generator and storage medium, the surface on which the charges are to be stored must have high resistivity. This prevents the charges from being lost laterally during storage. The bulk resistance of the material must also be high enough to prevent loss of the voltage across the thickness of the layer during intervals between successive scans. The equivalent electrical circuit of a photoconductor used in a camera tube is shown in Fig. 11-11.

The voltage E that is applied across the photoconductor is called the *target voltage*.

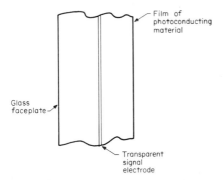

FIG. 11-10 Cross section of a camera tube target using a photoconductor.

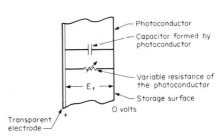

FIG. 11-11 Equivalent electrical circuit of a photoconductor.

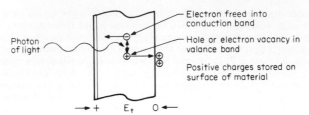

FIG. 11-12 Photon-generated electron-hole pair liberated in a photoconductor, producing stored positive charges on the photoconductor surface.

When light is absorbed in the photoconductor, the effective resistance decreases and more current flows through the resistor. This resistor discharges the capacitor, and a more positive voltage appears on the storage surface where light is absorbed than where there is little or none. The positive voltages that are built up on this surface are in proportion to the illumination at each portion of the image which is focused on the plane of the photoconductor.

Another way of understanding this process is to consider the photoconductor in terms of the dynamics of energy bands of the material and photon energy (discussed in Sec. 11.1.2) as visualized in Fig. 11-12. Under the influence of the electric field within the material, photons liberate electrons and holes. The electrons are free to move toward the positive electrode in the conduction band, and holes or electron vacancies are free to move toward the negative surface. The positive charges are stored on the storage surface and constitute the charge image that is then used when generating the television picture signal.

Several necessary properties of the photoconductive material are implied by the discussion of the method of operation.

The first is a band gap narrow enough that all photons in the spectrum in which the device is to be sensitive have enough energy to boost electrons across this band gap.

Second, the material must optically absorb the light that is to be detected. The resistance of the material must be high enough on the storage surface that the charge integrity is maintained at each point between successive scans; it is preferable that the resistance in the dark be high enough so that very little signal is generated in the absence of light.

Third, there must be few trapping states in the photoconductor that can slow up or delay the transit of charge carriers from their point of excitation to their destination.

Fourth, the carriers should have good mobility through the material so they can be swept through before they recombine with charge carriers of the opposite polarity.

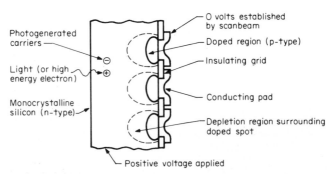

FIG. 11-13 Cross section of a silicon diode array photoconductive target.

Though high-resistance materials are best, low-resistance materials that are good photoconductors in other respects can be used in a television camera tube. Special structures, such as silicon diode arrays, are used in some tubes. Figure 11-13 illustrates such a structure. Silicon has good electrical conductivity and can serve as its own contact electrode. By reverse biasing the diode formed by the p-doped dots in n-type silicon, the resistance between the adjacent diodes, and between the diodes and the substrate, can be made high. A high electric field exists in the depletion region around each diode dot. Since the diodes are reverse biased, very little dark current flows through the depletion region.

When light (or high-energy electrons) enters the silicon, charge carriers are generated. The material is n-type and has poor mobility for p-type carriers. It is processed in such a way that the hole mobility and lifetime are great enough for most of the p-carriers to survive long enough to diffuse through the silicon and enter the depletion region. Here they are accelerated to the diode doped region and establish a positive charge on the contact pad.

11.2.5 GENERATION OF VIDEO SIGNALS BY SCANNING.

In a storage-type television camera tube, the signal is developed by an electron beam that scans the stored charge images that are produced by photosensitivity. The assembly on which these charges are stored is called the *target*. At the present time all camera tubes use a low-velocity scan beam.

11.2.6 LOW-VELOCITY SCANNING.

Low-velocity scan does not pertain to the speed with which the beam progresses across the picture area of the target in the television scanning process. Instead, the term indicates that the electrons in the beam are moving slowly as they approach the stored charges of the charge image.

The purpose of the scanning electron beam is to deposit electrons on the positively charged areas of the stored charge image that correspond in voltage to the brightness of the scene at that point. The use of a low-velocity beam assures that most of the electrons of the beam will land on the stored charge until its voltage drops to near zero and that very few secondary electrons will be emitted from the surface. If the kinetic energy of the beam electrons is high when they land on the storage surface, secondary electrons may be emitted from the surface, defeating the positive-charge neutralizing process.

The element of the storage target being interrogated by the scanning beam (Fig. 11-14) develops a positive voltage in the interval between successive scans of that spot. The magnitude of the voltage depends on the capacitance C of the element and the quantity of positive charge being developed

$$C_t V = Q$$

The scanning beam deposits electrons on this element and almost instantaneously drives the voltage down to zero in interval BC in Fig. 11-15. It then continues to the next portion of the scan image, where the process is continued. The point that was just neutralized will start to charge positively again along the curve CD if light is still present at that portion of the image. At time t_2 when that element is scanned again and the process will be repeated, that portion of the target capacitance will be recharged to its original voltage.

A problem in terminology often develops when describing this process. The stored charge image is a positive voltage with respect to the reference, which is the cathode of the electron beam. However,

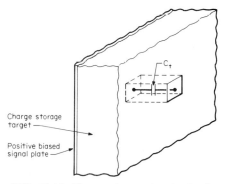

FIG. 11-14 Charge storage target showing capacitance of individual areas storing an image charge.

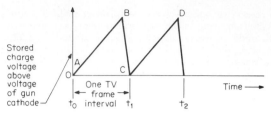

FIG. 11-15 Discharging and recharging of the capacitance of an elemental area on a storage-type target.

the positive-image charge voltage is actually a result of *discharging* the target capacitance. The electron beam is usually said to be "discharging" this positive voltage and restoring it to zero volt. In actuality it is *recharging* the target capacitance.

A television signal may be developed by this process in two ways. The first is by amplifying the current that flows in the signal plate electrode of the target. The second is by capturing the return beam that is unused in the stored-image charge neutralization process.

The method of signal generation using the signal plate current can be visualized in Fig. 11-16. Figure 11-16a illustrates what takes place in the time interval between scans of this point on the target of a photoconductive tube. The photo-induced current is flowing through the resistance R_p and the target capacitance C_t. No net charge is being added to the target assembly, nor does any of this current flow into or out of the signal plate at this time. As the capacitor C_t is being partially discharged, it becomes more positive on the scanned surface. In Fig. 11-16b the scanning beam deposits electrons on the surface. This flow of electron current recharges the capacitor, and an equal current flows through the signal plate to the amplifier. In the absence of light, little or no buildup of voltage occurs on the scan surface; consequently no current is developed when the beam scans that spot. The current that flows out of the signal electrode always flows in the same direction and is zero in the absence of light. This is extremely important since there is always a zero dc component in the video-signal current when no beam is landing on the target, and the absolute level of the current flowing out of the signal plate relates directly to the voltage of the charge pattern that is being scanned by the electron beam.

A low-velocity beam is achieved by designing the surface which is scanned so that the secondary emission ratio of the scanned surface is less than 1 at the highest voltage applied to the target. Even if the beam is accelerated and formed at very high voltages

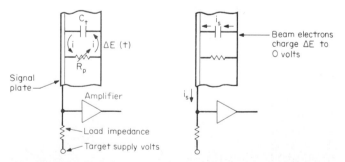

FIG. 11-16 (a) Currents flowing in a photoconductive target elemental area while the target is integrating a charge ΔE because of light-induced change in the photoconductor resistance. (b) Current flow when a low-velocity electron beam is interrogating the element.

and producing high beam velocities in the electron gun, it will be decelerated as it approaches the target surface. Maximum velocity (expressed in electronvolts) the beam will have as it lands on the scanned target is the voltage of the scanned surface with respect to the cathode of the electron gun from which the electrons are emitted. If the secondary emission of the surface is less than 1, more electrons will be deposited than emitted, and the surface will be charged in a negative direction until the voltage is nearly equal to the voltage on the emitting cathode of the electron gun. When this condition is reached, no more electrons will be deposited, and the unused electrons of the beam will be accelerated back toward the positive gun electrodes.

In a photoemissive target, or when a charge is being developed on a storage target by the process of secondary emission, current flows in both directions into the signal plate. There is an inflow of electron current caused by the photoelectrons or secondary electrons, and there is an outflow of electrons caused by the scanning beam recharging the stored-image charge pattern. See Fig. 11-17. This presents problems in establishing a proper dc or black-level reference, especially if the scene illumination is not constant. The signal current, however, is developed by the beam current as illustrated in Fig. 11-16.

11.2.7 RETURN-BEAM SIGNAL GENERATION.

When a low-velocity scanning beam is used in a camera tube, it may be convenient to utilize the return beam. The return beam is the portion of the beam that does not land on the target during the process of scanning the stored charge image on the target. The electrons that are not used to recharge the capacitance of the target are slowed to a stop in front of the target and then returned toward the electron gun, following roughly the same path that the scanning beam took as it passed through the gun. See Fig. 11-18. This return beam has information that is equal and opposite in polarity to the signal current flowing through the target.

This return beam can be steered by electron optical means and captured on an electrode, where it is channeled as a current to an external amplifier. Or it can be amplified within the device before it is collected in an output electrode to be channeled to the external amplifier.

This return beam has additional noise in addition to any noise in the signal component of the return beam. This noise is contributed by the random noise in any excess beam current beyond that required to neutralize the charge image on the target.

The return beam actually has two components. One is a reflected component; the other, a scattered component. The reflected electrons are reflected mirrorlike from the surface; the other electrons are scattered by interacting with the charge pattern on the target.

The *reflected* electron beam has a maximum amount of beam reflected from dark portions of an image. Consequently there is a lot of noise in the dark portion of the image, but very little in the light.

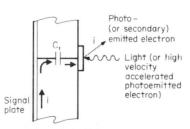

FIG. 11-17 Current flow in target structure during integration interval (photoemissive or secondary emitting storage target).

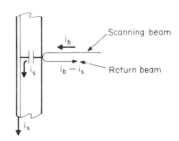

FIG. 11-18 Return-beam and scanning-beam current relationships.

The opposite occurs with the scattered beam. Here the maximum number of electrons are scattered by the positive charges on the target, and very little scattering takes place on the small charges representing black. As a consequence there is little noise in this component of the return beam in the dark, and maximum noise is present at highlights where its visibility is lessened.

The scattered and reflected portions of the return electron beam can be separated by unique operation and separation methods, and the most desirable portion of the return beam can be used to constitute the video signal. (See Image Iconoscope in Sec. 11.3 and Ref. 2.)

11.2.8 HIGH-VELOCITY SCANNING.

High-velocity scanning takes place when the beam lands on the target with very high energy and maintains the scanned surface at high voltage by the process of secondary emission. More secondary electrons are emitted than the number of electrons on the primary scanning beam. As a consequence the target surface charges positively when bombarded by the electron beam. The voltage increases as secondary electrons are emitted and then collected by a more positive electrode. When the surface reaches, or slightly exceeds, the voltage of the collector electrode, secondary electrons leaving the target are repelled by the collector and fall back on the target. When they land, they unavoidably reduce the positive stored image charges there. A video signal can be generated in this high-velocity scan process, as illustrated in Fig. 11-19.

There is a net change in charge of the elements as they are scanned by the beam. The areas that were charged positively by photoemission have a lesser change of charge when scanned than do the unilluminated areas. This change is accompanied by a varying charging current flowing through the capacitor which is formed by photosensitive elements spaced from the signal plate by a dielectric insulator. These currents flow through the deflection plate, and in and out through the video amplifier, and constitute the video signal.

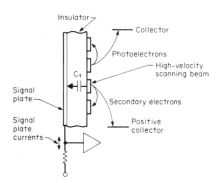

If one has difficulty determining the resulting magnitude of the video signal from this description, it is quite understandable. The camera tubes using this process had the same difficulty in producing a signal that was a faithful representation of the absolute, or even the relative, amount of light at each portion of the scene. Instead of all the emitted photoelectrons going to the collector, some fell back on unpredictable portions of the storage area. Secondary electrons followed similar unpredictable paths, and the output signal consequently lacked fidelity to the scene brightness at each point. However, these devices are capable of good resolution, since the charges do not migrate laterally across the storage surface.

FIG. 11-19 Flow of currents in and from a high-velocity scanned target.

11.3 EVOLUTION AND DEVELOPMENT OF TELEVISION CAMERA TUBES

11.3.1 NONSTORAGE TUBES.

The early television camera tubes all used photoemitters as light detectors. The first one was a nonstorage device, the *image dissector*. It was reasonably uncomplicated. The operating principles are illustrated in Fig. 11-20. A photocathode film (photoemitter) was formed on the inside of the faceplate. The elec-

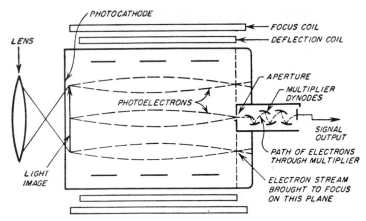

FIG. 11-20 Operating principles of the image dissector. (*From Fink.*[1])

trons emitted by light were accelerated by a uniform electric field. They were then focused on a plane at the opposite end of the field by a uniform axial magnetic field produced by a solenoid focusing coil. A small aperture followed by a secondary electron multiplier structure was positioned at the center of the plane of focus. Focused streams of electrons from the photocathode were deflected across the aperture by horizontal and vertical magnetic deflection coils. Those electrons that passed through the aperture were multiplied by a factor of several thousand by the electron multiplier. The electrons not passing through the aperture were lost.

There were several disadvantages to this tube, the most significant of which was the high light level required to achieve an adequate signal-to noise ratio. A large amount of information was lost because only the photoelectrons emitted from each point of the image at the time the point was being interrogated could be used.

11.3.2 STORAGE TUBES

Iconoscope. The next camera tube to evolve was a storage type. The *iconoscope* (Fig. 11-21) tube had a mosaic photoemitter consisting of isolated granules of cesium silver oxygen on an insulating mica substrate, with a conducting signal plate on the opposite side. The light image was focused on the same side on which the charge image of the mosaic was scanned. This tube utilized a high-velocity beam that scanned the surface at an angle of approximately 45°. It was much more sensitive than the image dissector since it stored the charges developed by the photoemission process between successive scans. The output signal was developed as capacitively coupled signal currents flowing through the signal plate. These currents were directly amplified by an external low-noise amplifier (see Sec. 11.2.8, High-Velocity Scanning).

The iconoscope has good resolution but suffered from imprecise signal levels. These were caused by both photoelectrons and secondary electrons raining back randomly on the storage surface, thus discharging some of the stored image charges. The tube required illumination levels on the photosensitive mosaic of the order of 5 to 10 fc (50 to 100 lux).

Image Iconoscope. A further evolution was the *image iconoscope* (Fig. 11-22). This tube had an image section with an efficient continuous-film semitransparent photocathode on the inside of the faceplate. The streams of light-emitted electrons from this surface were focused on the plane of the storage plate (target) by the axial magnetic-focusing field. This field was produced by focusing coils in the same manner as in the image

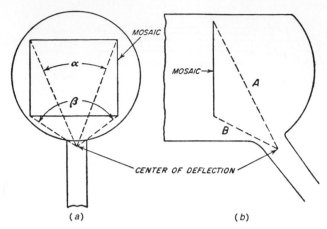

FIG. 11-21 Iconoscope deflection geometry (a) front view and (b) side view, where α and β are horizontal deflection angles of the scanning beam, A at the top, and B at the bottom, of the raster. (*From Fink.*[1])

dissector. The electrons were accelerated by a uniform electrical field and struck the storage target with an energy of several hundred electronvolts. They formed a positive image charge on the insulating surface of the target by the process of secondary emission, producing a gain in level in the process (see Sec. 11.2.2). The video signal was developed by

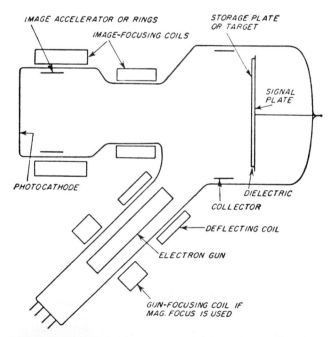

FIG. 11-22 Schematic diagram of the image iconoscope and its associated components. (*From Fink.*[1])

scanning the storage target with a high-velocity scanning beam at an angle of about 45°, in exactly the same manner as in the iconoscope.

Both of these devices had a unique virtue not present in the image dissector. That tube had a linear relationship between input light and output signal. But in the iconoscope and image iconoscope the output signal versus input light was less than linear. In other words, the highlights were compressed and the signals more closely matched the nonlinear input-voltage output-brightness characteristics of the picture display tubes. This produced more natural tones in the picture.

Orthicon and CPS Emitron.   The next development was a family of tubes called *orthicons* and *CPS emitrons.* These tubes used a target with a mosaic-type photoemitter on one side of an insulating support membrane and a transparent signal plate on the other side. The target was scanned by a low-velocity electron beam.

The name orthicon was derived from the fact that the beam landed orthogonal to the target in the low-velocity scanning process, whereas the CPS emitron takes its name from the process of cathode-potential-stabilized target scanning.

The orthicon photoemitter was composed of isolated grains of light-sensitive material deposited on the insulator, while the CPS emitron used a precise mosaic of squares of a semitransparent photoemitter. This structure was formed by evaporating the base material of the photoemitter through a fine mesh. Both of these tubes produced high-resolution pictures with precise gray scales.

The signal from these tubes was taken from the signal plate and fed directly to a video amplifier (see Sec. 11.2.6).

A cross-sectional view of a CPS emitron (Fig. 11-23) illustrates the configuration of these tubes. It is extremely important that in low-velocity scan the beam land perpendicular to the target and alignment coils were developed to correct for any off-axis alignment of the electron beam. In addition, fine and precise high-transmission mesh structures were developed to produce a uniform deceleration field at the target.

This class of tube had two drawbacks. The first was instability caused by highlights. If an abnormally bright highlight raised the charge on the mosaic photoemitter more than a few volts during the storage interval between scans, the scanning-beam electrons would land with sufficient velocity to emit more secondary electrons than the number landing from the beam. The surface would then charge positively in a runaway manner and progressively wipe out the picture across the target. The tube would then have to be shut down and restarted to restore a normal picture.

Secondly, photoemission from the target caused an electron current to flow into the signal plate as electrons were emitted from the mosaic elements. If the illumination came from fluorescent lights, for example, this produced a low-frequency ac signal in the signal plate, which was amplified by the video amplifier. This unwanted signal was added to the video signal currents which were induced in this circuit when the scanning beam deposited electrons on the positively charged mosaic elements, and restored them to the cathode potential of the electron-gun cathode.

These tubes used a second-electron optical feature. The beam went through multiple nodes of focus as it progressed from the gun to the target. This aided the deflected scanning beam to land perpendicular to the target and cut down on the amount of magnetic focus field and deflecting power required.

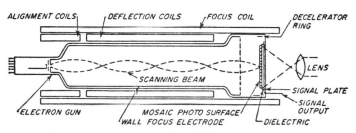

FIG. 11-23   Diagram of CPS emitron. (*From Fink.*[1])

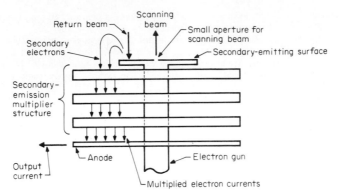

FIG. 11-24   Electron gun and return beam secondary emission multiplier.

Multiplier Orthicon.   The next stage in the evolution was the multiplier orthicon. This tube used the return beam to produce the video signal. (See Sec. 11.2.7.)

The return beam in this type of electron-optical system returns more or less along the same path as the outgoing scanning beam. It is amplitude modulated by the extraction of electrons in the target scanning process. This return beam is directed into an electron multiplier structure, which is situated as shown in Fig. 11-24. The secondary electrons emitted from the front surface of the gun are directed into the first dynode of the multiplier structure. They are amplified approximately 1000 times before being collected on the anode, where the signal is then fed into an external amplifier. This method of amplification is essentially noiseless and raises the amplitude of the signal above the noise of the first amplifier stage.

Return beam amplification improves the signal-to-noise ratio of the output signal, compared with the signal-to-noise ratio achieved when the target readout signal is fed directly into an amplifier. There is a difference in the character of the noise. The noise energy in the return beam signal is constant at all frequencies. On the other hand, the noise in the spectrum of the directly amplified target signal is proportional to the frequency. (See the discussion of video amplifiers in Sec. 11.5.6.)

Image Orthicon.   This tube of advanced design evolved from the multiplier orthicon with the development of a two-sided target structure. This solved the problems of target instability and allowed the use of a more efficient continuous-film semitransparent photocathode. This tube also eliminated the problem of ac photocurrents produced by fluorescent lights.

The image orthicon consists of three separate sections, each of which operates differently from the others. They are termed the *image section,* the *scanning section,* and the *multiplier section.* (See Fig. 11-25.) The action of the image section is considered first.

The image to be televised is focused on the transparent photocathode on the inside of the faceplate of the tube. This faceplate is maintained at a negative voltage of approximately 450 V. The electrons are emitted from this photosurface at each point in proportion to the illumination at that point and are accelerated toward the target-mesh assembly, which is maintained at nearly zero potential. The G_6, or image accelerator electrode, serves by rapidly accelerating the electrons to prevent them from being deflected by extraneous magnetic fields, and serves to maintain geometric symmetry of the image created on the target mesh assembly. The electrons emitted from each point of the photocathode are brought into sharp focus at a corresponding position on the target by the action of the electrostatic field of the electrodes of the image section, and the axial magnetic field produced by the external focus coil. The target mesh asesmbly consists of an

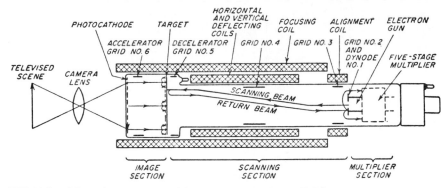

FIG. 11-25 Schematic arrangement of the image orthicon. (*From Fink.*[1])

extremely fine mesh closely spaced to a thin glass membrane. Most of the photoelectrons pass through this mesh and strike the glass target, causing secondary electrons to be emitted. These secondary electrons are collected by the target mesh, which is maintained at a slightly positive potential with respect to the target. This action produces a positive charge on the target glass and also produces an amplification of charge that is equal to the secondary emission ratio minus 1. The charge image on the glass is stored until neutralized by the scanning beam.

The scanning section consists of an electron gun which produces a high-resolution beam. This beam scans the target of the image section under the influence of the transverse magnetic field produced by the deflection coils located along the neck of the tube. The target consists of a thin membrane of semiconducting glass. The resistivity of this glass is of particular importance, since it must be low enough that the charge on the mesh side can migrate through the glass in the interval between successive scans of the tube, yet high enough that the lateral leakage of the stored charge of one element to the adjacent one is negligible. The electron gun consists of a thermionic cathode which is maintained at ground potential, a control grid, and an accelerating grid with a limiting aperture (G_2) that forms an extremely sharp and collimated electron stream. This beam travels through the tube at a velocity determined by the G_4 or wall electrode voltage. The electrons within this beam go through several loops of focus and are brought to sharp focus on the target after passing through the decelerating field of the G_5 electrode. They land with nearly zero velocity on the gun side of the target. The electron beam scans the surface of this glass and deposits sufficient electrons on the charged areas to neutralize the charge on the target under the beam, and drive it down to the potential of the cathode of the electron gun. When this condition is reached, no more electrons can be deposited, and the remainder of the electrons return toward the electron gun. Figure 11-26 illustrates the action of the target-mesh assembly and the discharge of the target charge by the electron beam.

The scanning beam is caused to return substantially along the same path for its return trip as for the scanning trip. This return beam has been amplitude modulated by the loss of electrons to the charges on the target, and constitutes the video signal information that is ultimately taken from the tube. The return beam scans a small area of the surface of the first dynode (which is also grid No. 2), the surface of which has a high secondary emission ratio. The secondaries, emitted when the beam strikes the first dynode, are attracted toward the multiplier stages by the influence of the G_3 electrode and the field of the second dynode. These multipliers are made in the shape of a pinwheel, and as such channel the electrons efficiently through successive multiplier stages to the anode. The total current gain in this multiplier is between 500 and 1000.

The optical input may use conventional components because the useful area of the photocathode is 1.6 in (41 mm) in diameter. This corresponds to the same area as used

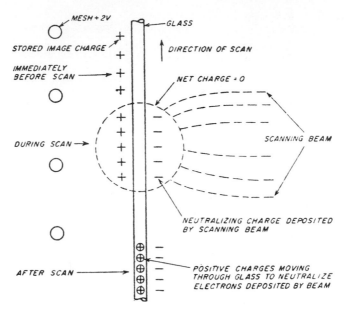

FIG. 11-26 Illustration of the target charge and discharge cycle of the image orthicon. (*From Fink.*[1])

for double-frame 35-mm film. The photocathode can be composed of a variety of different semitransparent alkali metal-based photocathode surfaces. It is deposited directly on the inside of the faceplate of the tube.

The signal output of the image orthicon is in the order of 10 μA peak to peak The signal developed across the load resistance of the anode is black negative and white positive.

The output impedance of the multiplier is very high in its resistive component and is essentially a constant-current generator. The shunt capacity to ground of the multiplier anode is approximately 12 pF. The signal contains fairly accurate black-level information during retrace and may be coupled directly to the preamplifier with a capacitor, since the anode potential is normally operated at a high voltage.

The signal-to-noise ratio of the signal developed by the image orthicon is determined by the shot noise in the scanning beam. The signal-to-noise ratio, therefore, varies as the square root of the beam current necessary to discharge the target charge. The signal-to-noise ratio of the image orthicon tube is in the range of 36-dB peak signal to rms noise. The noise energy distribution is flat, having equal components in all portions of the video frequency band utilized. Maximum noise is in the black portion of the signal.

The basic light transfer characteristic is shown in Fig. 11-27. The knee of the curve represents the point where the target charges up to the potential of the mesh and is stabilized by secondary-emission mechanics to this potential. When this point is reached, the secondaries from the target are no longer collected by the mesh, but are free to be redistributed over the adjacent areas. This characteristic is important because it limits highlight signals and prevents the camera tube from becoming unstable in the presence of extreme highlights.

This is a rough interpretation of the stabilizing action of the image orthicon target-mesh assembly. The action is modified by the nature of the velocity distribution of the secondary electrons emitted by the target when bombarded by the photoelectrons. Secondary electrons with low velocities are collected by the mesh, or fall on adjacent areas

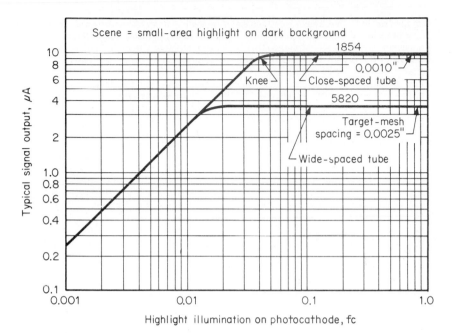

FIG. 11-27 Transfer characteristic of wide-spaced and close-spaced image orthicon tubes. (1 fc = 10.76 lm/m².) (*From Fink.*[1])

of the target and discharge these areas toward black. This enhances the apparent contrast and resolution of the picture. Secondary electrons with high velocities travel further from their point of origin and have sufficient energy to produce a positive spurious image charge.

The light transfer characteristic of the image orthicon is complex, considering these scattered secondary electrons, particularly when operated with the highlights substantially over the knee of the curve. The signal output of any portion of a scene is modified by the influence of the adjacent areas when operated in this manner. Operation with the highlights over the knee has some advantages. It produces control of the average transfer characteristic. Operation with twice the amount of light necessary to raise the highlights of the scene to the knee produces a signal that is nearly complementary to the characteristics of the kinescope and produces normal and pleasing tone rendition. The average gamma characteristic follows a 0.5 to 0.6 power law in this case.

Use of image orthicons in color cameras presents a different problem, requiring a predictable gamma or light transfer characteristic. A constant-gamma characteristic can be corrected electrically by compression or stretch circuitry. Random electron redistribution cannot be tolerated, since it will differ from color to color, dependent upon the amount of highlight energy in each color band of the scene.

Therefore the image orthicon in a color camera must be operated so that accurate colors can be reproduced. For this reason a high-capacity (close-spaced) target mesh structure tube is used, to extend the linear range of the signal-transfer characteristic.

Resolution of the tube is primarily limited by crosstalk of deflection fields into the image section. These fields cause a slight transverse motion of the photoelectron streams from the photocathode and degrade the fine-detail resolution. External shielding of the image section from these deflection fields must be provided for both the line and field frequencies. Resolution is also affected by the operating temperature.

Image Isocon. The image isocon is an advanced version of the image orthicon. It is designed to produce a better signal-to-noise ratio and to operate at lower light levels than the image orthicon. It accomplishes this by a more efficient use of the video information contained in the return beam. Unlike the image orthicon, it has practically no noise in the black portions of the signal.

The design of the image isocon is based on the discovery that the return scanning beam has two components, a reflected component and a scattered component, as illustrated by Fig. 11-28. The reflected electrons are the ones that do not land on the target, but are instead reflected mirrorlike from the target when they do not actually land on the target. When a charge is present on the target, a large number of electrons interact with this charge. Those that do not actually land are scattered by the electric field produced by the highlights of the stored image. If the beam is *misaligned* so that it approaches the target at a slight angle to the perpendicular, the reflected electrons will return along a slightly different path from that of the symmetrically scattered electrons.

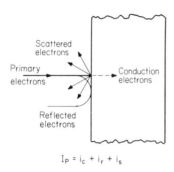

$$I_P = i_c + i_r + i_s$$

FIG. 11-28 Beam-current distribution of target of both image orthicon and image isocon. *(RCA Corp.)*

Figure 11-29 illustrates the paths of these different beam components and how they are separated in the return process. The return reflected beam is steered through the same aperture through which the primary beam passed. The scattered electrons strike the first dynode of the internal electron-multiplier section of the tube and are used as the video signal. The highest video-output current signal is in the highlights, while practically zero current is produced in the low lights. The polarity of the signal is therefore opposite to that of the image orthicon signal. Figure 11-30 illustrates an image isocon tube, its internal components, and the external deflecting, focusing, and aligning circuits required to operate this device.

The isocon is designed to handle a wide range of scene content. The knee characteristic limits the signal range of highlights, and the low noise at low light levels of a scene allows a wide contrast of information to be produced in the video signal. This tube is particularly useful in systems where very low light information must be seen in the presence of high-contrast highlights. It is used for outside nighttime surveillance in the presence of lights in the field of view, and for x-ray inspection systems where there are areas of very high and very low x-ray absorption in the field of view. It is made as 3- and 2-in-diameter bulbs with image facemats of 36.6 by 27.4 mm (45.7 mm diagonal) and 28.8 by 21.6 mm (36 mm diagonal), respectively.

Image isocon performance can be extended further to low light levels by coupling an image intensifier stage to the input of the tube. This is usually done by building the tube with a fiber-optic input window and butting an image intensifier tube with a fiber-optic output window to the image isocon fiber-optic faceplate.

Silicon Intensifier Tube. The silicon intensifier tube (SIT) is used where performance is expected at extremely low light levels that vary from very low light to daylight (see Fig. 11-31). The tube operates on the principle of electron bombardment–induced conductivity. This produces a high gain in the number of electrons stored at the target (see Sec. 11.2.3).

Typically the gain of the device can be varied from 10 to 1600. The gain at the target varies linearly with the voltage across the intensifier section (Fig. 11-32). This tube can produce a signal in medium light levels that is limited by the noise in the electrons generated by the photons interacting with the photoemitter. At lower light levels, noise in the signal is limited by the amplifier input noise, and at higher light levels by the loss of

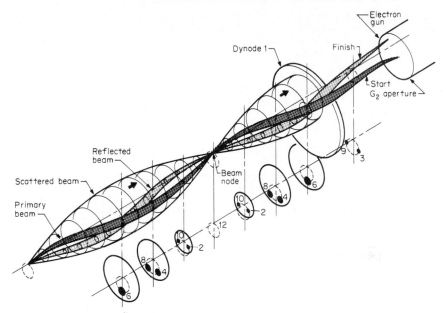

FIG. 11-29 Idealized trajectories of image-isocon electron beams. *(RCA Corp.)*

electrons in the gain-controlling buffer layer on the target structure (Fig. 11-33). Because of the high gain, the tubes are subject to rapid deterioration if they are overexposed, requiring fast-acting control and protection circuits in the camera.

The scanning section of the tube operates identically to a silicon-diode vidicon-type tube (see Sec. 11.1.2). Resolution in this type is limited by the diode structure of the target and reduced by the image section of the tube. The high gain at the target raises the signal level there by the gain factor. This greatly reduces the lag when operating at very low light levels. Recent versions of the SIT tube have reduced blooming because of an altered target structure, and target doping to prevent lateral migration of charge when

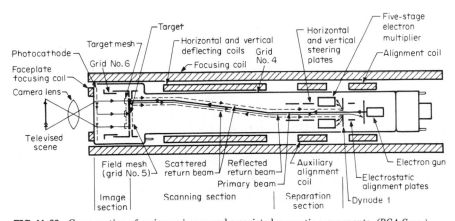

FIG. 11-30 Cross section of an image isocon and associated magnetic components. *(RCA Corp.)*

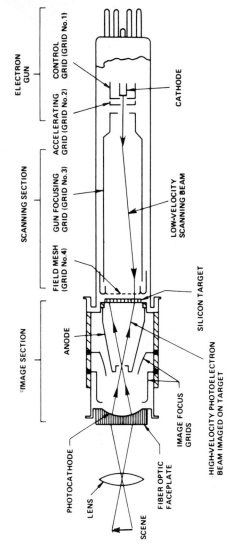

FIG. 11-31 Light enters the SIT tube through a fiber-optic faceplate, which transfers the flat-scene image onto the curved photocathode. The light then travels through the focusing grids and strikes the target, which is a matrix of over 1800 silicon diodes per inch (per 2.5 cm). The image is typically stored there and read out by the scanning beam every $\frac{1}{30}$ s. *(RCA Corp. and SMPTE.)*

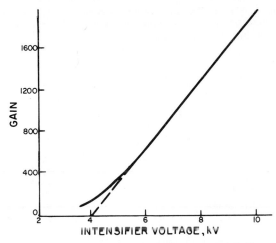

FIG. 11-32 Electron gain (solid line) is an essentially linear function of the intensifier voltage. An energy-absorbing "buffer layer" requires keeping the voltage above 3 kV, where the intensifier section performs best. Deviation from linearity (the dashed line) is caused by high-energy photoelectrons penetrating the buffer layer. *(RCA Corp. and SMPTE.)*

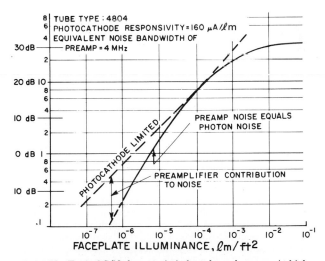

FIG. 11-33 Typical S/N characteristic for tube and camera. At higher illumination levels, intensifier voltage decreases to limit gain. This lowers the number of primary photoelectrons striking the target, so the S/N flattens out from the photocathode-limited line. *(RCA Corp.)*

11.25

the target becomes overloaded. The tube is generally made with a scanning section in the 25-mm diameter (1-in) vidicon size. Special designs with larger image formats are made when higher resolution is required.

11.4 VIDICON-TYPE CAMERA TUBES

The vidicon tube employing antimony trisulfide as the photoconductor was the first successful photoconductive television camera tube. The name *vidicon* has been adopted as a generic classification for photoconductive camera tubes. Various other photoconductive camera tubes that have been developed have been identified by trade names or the type of photoconductor that is utilized. All these tubes operate with a target readout signal and use low-velocity scanning with an electron beam. (See Sec. 11.2.6.)

The deflecting and focusing of the electron beam vary in many of these different tubes. Most tubes utilize magnetic focusing and magnetic deflection. Other types have electrostatic focus and magnetic deflection, or magnetic focus and electrostatic deflection, and a few use electrostatic focus and deflection.

All use an electron gun producing an electron beam, derived from a thermionic cathode, that emerges from a small aperture. Various versions of these guns have been designed to produce specific resolution values, low-impedance beams, or beams that can switch to a high-current erase mode during different portions of the scanning cycle.

The tubes are commonly ½ to 1½ in (13 to 38 mm) in nominal diameter, although special 2- and 4½-in-diameter (51- and 114-mm) tubes have been made for special high-resolution systems. These large tubes utilize the return-beam signal and have a built-in electron multiplier system for signal amplification.

All photoconductor-type vidicons, except the silicon diode tube, utilize a continuous-film structureless photoconductor deposited on a transparent signal electrode. The silicon diode photoconductor is a wafer of silicon on which an array of diodes with contact pads are fabricated.

11.4.1 ANTIMONY TRISULFIDE PHOTOCONDUCTOR.
Antimony trisulfide photoconductors consist of alternating layers of porous and solid antimony trisulfide. The thickness of the material is approximately 1 to 2 μm. The operating target voltage can vary between a few volts and 100 V. Part of this variation in voltage is necessary to accommodate manufacturing variations. The rest of the range is used to control the sensitivity and/or dark current of the tube.

Dark current is fairly high on this type of tube. In a 25-mm-diameter tube it can vary between 1 and 2 nA at a low target voltage to 100 nA at a high target voltage. The dark current increases approximately as the cube of the target voltage, and the sensitivity varies as its square. At a high target voltage the tube is quite sensitive and laggy, and the dark current approaches the signal current level. At low target voltages the reverse takes place. Typical operating dark currents are in the 10- to 30-nA region. When tubes are set up to operate with similar dark currents, the sensitivities will be nearly identical. However the operating target voltages under these conditions will not necessarily be the same on all tubes. The adjustment of target voltage is therefore best established by setting for a desired value of dark current.

Dark current is a function of temperature. It approximately doubles for every 8° C increase in temperature. This is significant when operating in the high-sensitivity mode.

Antimony trisulfide is sensitive through the entire visible spectrum and has some sensitivity in the infrared spectrum. (See Fig. 11-34.) The signal output of antimony trisulfide tubes as a function of the light level is not linear. This signal varies at ~0.65 power of the input light. (See Fig. 11-35.) This is a desirable feature since the progressive compression of higher signal currents partially compensates for the nonlinearity of a typical picture tube, which compresses blacks and stretches whites. Consequently the overall picture is brighter and has a more natural tonal rendition than that produced by a video signal which has a linear characteristic.

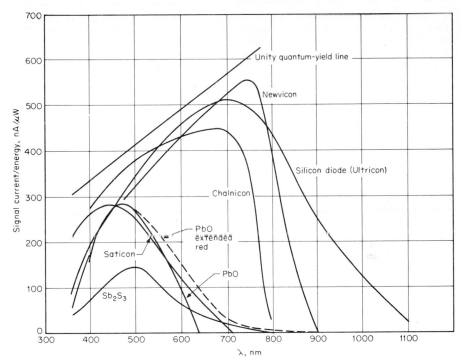

FIG. 11-34 Absolute spectral response curves of various camera tube photoconductors.

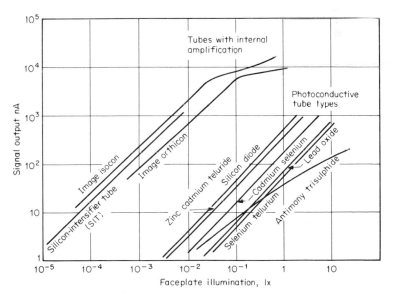

FIG. 11-35 Light-transfer characteristics of typical camera tubes.

Resolution of an antimony trisulfide photoconductor is not greatly degraded by the photoconductor, since it has high resistivity and is very thin. In the first place, scattered light cannot progress far laterally through this thin layer. Second, the photoconductor is dark in appearance. Therefore it absorbs most of the incident light and does not allow it to scatter laterally through the photoconductor, where it otherwise might contribute to lower resolution.

The low reflectance of the photoconductor also helps maintain the contrast of the image, since very little is reflected from the faceplate or from other portions of the optical system, which could scatter back onto the photoconductor.

Antimony trisulfide tubes are used primarily in industrial surveillance closed-circuit television systems and in telecine cameras in broadcast service. They are used where variable sensitivity of the tube is desired and where low cost is an objective. They are used in telecine cameras where the tubes can be operated with high light levels and at low dark current, which results in minimum lag.

11.4.2 LEAD OXIDE PHOTOCONDUCTOR.

One of a group of photoconductors called *heterojunction* photoconductors, lead oxide photoconductor tubes, are variously called Plumbicons, Vistacons, Leddicons, or Hi-sensicons by different manufacturers. The lead oxide photoconductor is a porous vapor-grown microcrystalline layer of lead monoxide. The material is processed during the vapor growth so that it is approximately i-type; i.e., it has nearly equal hole and electron mobility. It is vapor-grown on an n-type transparent signal electrode (tin or indium oxide). The vacuum interface surface is treated so it has p-type conductivity. The cross section of this 10- to 20-μm-thick layer is illustrated in Fig. 11-36.

In operation, the ni and ip junctions are reverse biased by a positive voltage applied to the signal electrode and negative or zero volts established by the beam on the scanned surface, hence the categorization as a heterojunction structure. Reverse biasing greatly reduces the dark current, since holes from the signal plate are prevented from entering the photoconductor, and the p-type material on the scanned surface discourages accep-

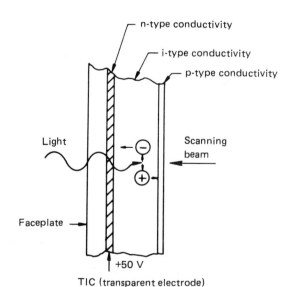

TIC (transparent electrode)

FIG. 11-36   Cross-sectional view of lead oxide photoconductor illustrating generation of output signal. *(RCA Corp.)*

tance of electrons from the scanning beam. Photons of light are absorbed throughout the bulk of the material, which has a band gap of 1.9 eV. Electron and hole-carrier pairs are generated from photons having more than this energy. Most of the applied target voltage appears across the bulk of this *i*-type layer. This field pulls electrons to the signal plate and holes to the scanned side, where they produce the positive-image charge pattern that is interrogated by the electron beam. Dark current from this photoconductor is less than 1 nA/cm^2 at the typical target voltage of 45 V. The 1.9-eV band gap is too wide to detect all visible light (700 to 650 nm). The spectral response is negligible from the 650- to 700-nm end of the visible spectrum. Red sensitivity and color fidelity suffer as a result, particularly in color cameras. Doped versions of this photoconductor, called *extended red,* produce additional red sensitivity, and extend the red sensitivity throughout the entire visible spectrum and into the near infrared (see Fig. 11-35).

The layer thickness is chosen as a compromise between sensitivity (particularly the red), lag, and resolution. Lead oxide does not highly absorb red and green light. Greater thickness increases the absorbed light and the red and green light sensitivity. The material is microcrystalline and therefore scatters light that is not yet absorbed. A thin surface has higher resolution than a thick layer, since scattered light cannot spread as far laterally. Blue light is absorbed strongly. Resolution of an image of blue light is not limited by the photoconductor thickness but is progressively degraded for light that is more weakly absorbed. Resolution for red light is lower than for green.

Because of the high reflection and dispersion and the low absorption of visible light, particularly in the red-green portion of the spectrum, severe halation can result, which reduces contrast and fidelity (see Fig. 11-37). A partial solution to this problem, used in most lead oxide tubes, is an antihalation glass faceplate button. This reduces halation within the tube. Light dispersed within the cone angle 2 reflects back into the optical system (Fig. 11-38). Here it can impair the contrast, as any other surface-reflecting element of the optical system will do.

Thicker layers have less storage capacitance. Since lag is proportional to the storage capacitance, thick layers produce the lowest lag. The doped extended-red version of the lead oxide photoconductor has higher red and green resolution than the undoped, because the doping material (generally sulfur) increases the absorption of the longer-wavelength light. The trade-off with this doping is higher lag and highlight image reception produced by trapping states in the material. These parameters are varied in different versions of these tubes to achieve the desired performance objectives.

The principal usage of lead oxide photoconductor tubes is in three-tube broadcast color cameras and in television intensifier systems used in conjunction with medical and industrial x-ray units.

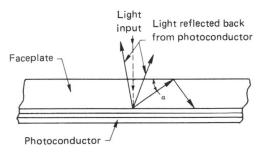

FIG. 11-37 Light paths without antihalation button. *Note:* Reflected light striking the faceplate at an angle less than the angle α (the total internal reflection angle) is reflected downward toward the photoconductor. *(RCA Corp.)*

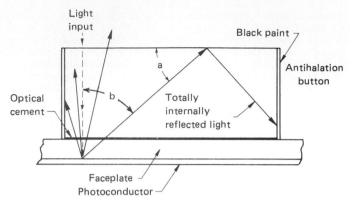

FIG. 11-38 Light paths with antihalation button. *Note:* Button thickness is designed so that all totally reflected light strikes the sides of the button and is absorbed by the black paint on the diffusing side of the glass. *(RCA Corp.)*

11.4.3 SELENIUM PHOTOCONDUCTOR. Used in Saticons, it is a glassy or amorphous selenium-based photoconductor of the heterojunction type. It incorporates arsenic to inhibit crystallization of the glassy material and tellurium to produce adequate red response. Its physical structure is illustrated in cross section in Fig. 11-39. The tellurium doping is located close to, but not on, the signal electrode. The antimony trisulfide coating on the scanned side acts primarily to reduce secondary emission of electrons from the scanning beam. The *p*-type conductivity discourages unwanted beam electrons from entering the photoconductor on that surface.

The electronic structure is illustrated in Fig. 11-40. The reverse-biased junction on the light input side prevents holes from entering the photoconductor from that side. The *p*-type nature of the bulk of the material and the thin antimony trisulfide prevent excess electrons from entering from the other side. As a consequence the dark current is less

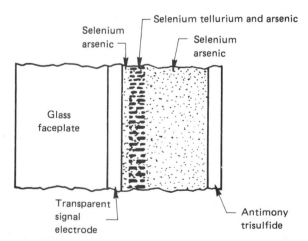

FIG. 11-39 Cross section of Saticon tube photoconductor and faceplate. *(RCA Corp. and SMPTE.)*

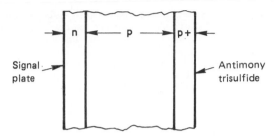

FIG. 11-40 Cross section of Saticon tube photoconductor showing electrical conductivity characteristics. *(RCA Corp. and SMPTE.)*

than 0.5 nA/cm^2. Selenium is highly absorptive for blue and green light. It has a band gap of 2.0 eV, which effectively limits sensitivity to wavelengths above 620 nm. Tellurium increases the absorption of red light to wavelengths as long as 700 nm, giving the photoconductor response throughout the visible spectrum. Photons of light are absorbed either in the selenium arsenic directly behind the signal plate or in the tellurium-doped region. Pairs of charge carriers are liberated in this region, and the holes are pulled through the p-type layer to the scanned side, where they generate a stored positive-image charge. Electrons having low mobility in the p-type material are able to traverse the short distance to the positive signal plate. The photoconductive conversion takes place in the first few dozen nanometers of this layer. The remainder of the layer acts primarily as a charge-transport layer. Because the material has high resistivity (10^{13} $\Omega \cdot$ cm) and the field through the photoconductive layer is high (125 kV/cm), these charges travel without lateral diffusion, maintaining high resolution for all wavelengths of light. The material is glassy (noncrystalline). As a consequence light is not dispersed in the photoconductor. These factors preserve resolution and make the resolution of the photoconductor independent of the thickness or of the size of the useful photoconductive area of the tube. Because the photon conversion takes place in the front portion of the layer, sensitivity is independent of the photoconductor thickness. These factors make the photoconductor more versatile. Its parameters can be varied to produce changes in storage capacitance, to reduce lag, or be suitable in small-size tubes. Tubes with Saticon photoconductors have low flare. The high absorption of light by the photoconductor results in little light being dispersed or reflected back from the photoconductor. This helps to preserve picture contrast and tonal fidelity. It is particularly important when high-contrast images are encountered and in three-tube color cameras where brightness distortions become color distortions.

Advanced Saticon-type photoconductors designated types II and III are designed to reduce the memory of specular highlights and reduce any tendency for burn-in of stationary images.

Saticon tubes are used primarily in three-tube broadcast color cameras for studio, field production, and electronic news gathering, and in telecine cameras. Saticon tubes are also widely used in single-tube color cameras having color-filter systems of one form or another to perform the color separation functions, and in medical x-ray television systems.

11.4.4 SILICON-DIODE PHOTOCONDUCTIVE TARGET. The silicon-diode structure (Sec. 11.1.2) is a versatile photoconductor. It has the highest sensitivity of any camera tube photoconductor in the visible portion of the spectrum and extends its sensitivity well into the infrared region. Silicon-diode photoconductor tubes are made by various manufacturers under such names as Sivicor, Ultricor, Tivicor, and Silicon Vidicon. Dark current is in the range of 5 to 10 nA/cm^2. Resolution is limited by the diode structure, which usually has a square array of about 700 diodes per centimeter in either direction.

The recommended target voltage is in the range of 8 to 15 V. The storage capacitance of such a target is about 3500 pF/cm², making it one of the higher-capacitance photoconductors. This has an influence on low-level signal lag. The silicon diode target is also immune to image burn. In fact, it can be focused directly on the sun for short intervals with no permanent damage. With a 25-mm-diameter (1-in) tube, images of highlights tend to bloom when the target capacitance is completely discharged at a peak signal current in the range of a 1300-nA signal. Reduced blooming versions produced by variations in the structure of the target minimize this tendency in some tube models. The silicon target structure usually is self-supporting. It is positioned behind the faceplate rather than being deposited onto it.

The signal output, up to the saturation point, is linear; i.e., it has a gamma of 1.0. Special versions are made with targets that have enhanced infrared sensitivity to wavelengths of 1100 nm. Ultraviolet sensitivity to 330 nm can be obtained if a faceplate having good ultraviolet transmission is used.

Silicon diode tubes are most prominently utilized in closed-circuit monochrome systems where low-light sensitivity, infrared sensitivity, or resistance to highlight and other image burns is desired.

11.4.5 CADMIUM SELENIDE PHOTOCONDUCTOR.

This photoconductor, used in the Chalnicon, consists of a microcrystalline layer of cadmium selenide (CdSe), an n-type photoconductor. Accordingly, it does not have blocking-type contact at the n-type signal plate, and dark current is not as low as in the heterojunction types of photoconductors. Versions of the Chalnicon are made by doping with cadmium telluride to increase the red and infrared sensitivity of the cadmium selenide. Cadmium selenide has a band gap of 1.7 eV, which limits the red response to wavelengths shorter than 730 nm, unless doped. Cadmium selenide layers are thin (1 μm) and consequently have high storage capacitance.

Latest versions have a composite structure with an added layer of arsenic selenide (As_2Se_3) in a porous-type structure on the scanned side. This increases the thickness and reduces the capacitance, thus reducing the lag. Tubes with this type of photoconductor are called Chalnicon-FR.

Sensitivity in the visible spectrum is very high and approaches unity quantum yield throughout most of the visible spectrum. The infrared-sensitive version has somewhat lower sensitivity in the visible spectrum but extends in sensitivity to beyond 800 nm (Fig. 11-35).

Target voltage for the Chalnicon tube varies between 20 and 50 V for different tubes and must be individually adjusted. Low target voltage produces excessive image retention. High target voltage produces spots and graininess. The required target voltage varies with temperature and is controlled by a temperature-sensitive control circuit. Dark current is in the 1- to 2-nA/cm² range for normal operating temperatures.

Lag and dark current are higher than in the selenium and lead oxide tubes, while the sensitivity is much higher. Resolution is about the same as in the Saticon. Some use is made of these tubes in three-tube color cameras, but their primary use is in surveillance-type cameras.

11.4.6 ZINC SELENIDE PHOTOCONDUCTOR.

The Newvicon uses this heterojunction type of photoconductor. It consists of an amorphous or glassy interface-layer of zinc selenide, which is deposited on the transparent signal plate, and an additional amorphous layer consisting of a complex mixture of zinc cadmium and

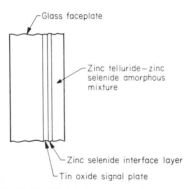

FIG. 11-41 Newvicon photoconductor cross-section schematic.

tellurium (Fig. 11-41). The zinc selenide layer provides the proper substrate for the photosensitive portion of the photoconductor, is n-type in conductivity, and is transparent to visible light. The zinc selenide–zinc telluride glossy layer is the light-sensitive portion of the photoconductor. It has p-type conductivity.

When the photoconductive target is reverse biased by application of a positive voltage to the signal plate, the dark current is in the order of 7 nA /cm^2.

This dark current is higher than most other heterojunction camera tube photoconductors. The photoconductor operates best with a target voltage of 10 to 25 V across the layer, which is individually adjusted for each tube and for different temperatures to achieve a condition of minimum burn-in of images.

The Newvicon photoconductor is highly absorbing in the visible spectrum. As a consequence, it has very little reflectance. Flare caused by reflected light from the photoconductor is extremely low. Resolution of tubes using the Newvicon photoconductor is high. It is not primarily limited by the photoconductor because of its high optical absorbtion, high resistance to lateral leakage on the charge storage surface, and the negligible scattering of light in the thin, amorphous layer. It is also resistant to blooming or spreading of specular highlight images and has a low memory of specular highlights.

The Newvicon tubes are utilized primarily in surveillance, medical, and industrial x-ray intensification systems and other industrial-type cameras where high sensitivity is desired. The Newvicon photoconductor is also widely used in stripe color filter-type tubes for inexpensive single-tube VTR color cameras.

11.5 INTERFACE WITH THE CAMERA

The principal function of the camera is to operate the camera tube and process the video signal developed by it. The interface controls and connections between the tube and the camera are important and are discussed separately. Listed below, and illustrated in Fig. 11-42, they pertain primarily to photoconductor-type camera tubes:

Optical input
Dynamic focusing
Blanking the beam
Alignment coils
Video output
Deflection coils and deflection circuits
Magnetic shielding

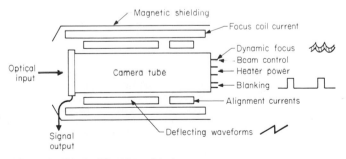

FIG. 11-42 Principal interface with the camera.

Table 11-1 Recommended Image Sizes of Various Photoconductive Camera Tubes

Nominal tube diameter		Image size (3 × 4 aspect ratio)					
		Width		Height		Diagonal	
mm	in	mm	in	mm	in	mm	in
38	1.5	20.32	0.80	15.24	0.60	25.4	1.0
25	1.0	12.7	0.50	9.5	0.375	15.9	0.625
18	0.66	8.8	0.346	6.6	0.260	11.0	0.432
13	0.5	6.5	0.256	4.9	0.192	8.13	0.320
30	1.2	16.9	0.667	12.3	0.500	21.1	0.834

11.5.1 OPTICAL INPUT. The image is focused on the photoconductive material deposited on the inside surface of the tube's faceplate. The exception is the silicon diode target tube, where the silicon wafer is generally mounted separately behind the faceplate. The faceplate is 1 to 3 mm (0.04 to 0.12 in) thick (specified in data sheets by each manufacturer) and has an index of refraction of approximately 1.5. The faceplate thickness and refractive index must be considered when designing the mechanical positioning of the tube relative to the lens and other portions of the optical system.

When the photoconductor reflects a substantial portion of the light incident on it, an antihalation button is needed. This reduces loss of contrast caused by light being reflected back to the photoconductor from the front surface of the glass. Figure 11-37 shows that when this diffused reflected light strikes the interface between faceplate and air at an angle less than α, all the light is reflected back on the photoconductor and produces an unwanted flare or halo around the highlight. The antihalation faceplate button (Fig. 11-38) moves this interface farther from the photoconductor so that the reflected light from the photoconductor is reflected back to the wall of the button, where a black coating absorbs it. When reflected light from the photoconductor strikes the faceplate-air interface at less than the angle α, a small percentage is reflected back from the interface and contributes to poor contrast.

Each class or size of camera tube utilizes a prescribed image size for optimum performance. These sizes are shown in Table 11-1. The *visible* area of the photoconductor is much larger than the values cited. When a larger area than specified is utilized, problems of geometric distortion, poor focus, nonuniform signal amplitude, and blemishes will probably be encountered as progressively more of the area is utilized. It is important to position the tube in the camera so that the longitudinal axis of the tube is precisely on the axis of the optical system. It is not good practice to utilize the additional sensitive area of the tube to accommodate for off-axis positioning of the tube in the camera.

11.5.2 OPERATING VOLTAGES. All voltages discussed in this section refer to the voltage between the cathode and the various other tube electrodes. G_4 and G_3 voltages in magnetically focused tubes control the transit time of electrons through the tube. They establish the transit time so as to bring electrons that start out from the gun with a radial component of velocity back to the axis of the beam after they have completed one 180° circle about the axial magnetic-focus lines of force. These electrodes therefore operate as a beam-focus control. A form of collimating lens is formed between the G_3 and G_4 electrodes. The shape of this lens is determined by the difference in voltage between the two electrodes. The purpose of the collimating lens is to assist in correcting the trajectory of the deflected electron beam so that the beam will land perpendicular to the faceplate. The strength and/or shape of this lens must be determined empirically for each focus-coil and deflection-coil combination. It is also a function of the axial positioning of the tube with respect to the coil. The strength and shape of this lens are controlled primarily by the ratio of the G_4 to G_3 voltage. G_4 (mesh) must *always* be higher (in voltage) than G_3. If not, any positive ions generated by the electron beam striking gas molecules in the tube will be accelerated to a diffuse spot on the photoconductor, causing permanent damage.

If the G_4 voltage is fixed more positive than the G_3, it creates a field that suppresses secondary electrons emitted from the mesh by the scanning beam. It prevents them from being trapped in orbits around the magnetic field lines. Trapped electrons can contribute a space charge that tends to disperse and defocus the scanning beam. Early vidicon tubes did not have separate mesh and wall electrodes. These were electrically and mechanically connected.

Beam Focus.   Once the G_4/G_3 voltage ratio is established, the best way to achieve beam focus by varying the tube voltages is to vary the supply voltage to a fixed voltage divided circuit that always provides the proper voltage ratio. Currents drawn by the tube electrodes in normal operation are in the order of 10 μA, even when the beam is set to handle extreme highlights. Focus stability requires good stability of these voltages. A variation of 1 or 2 V can produce noticeable degradation of resolution.

The G_4, or mesh electrode, and to a lesser extent the focus electrodes, must be bypassed to the video input amplifier ground. This minimizes pickup of deflection transients or video signals that can be capacitively coupled into the signal electrode through the capacitance of these electrodes to the target.

Resolution will be higher if the G_3 and G_4 are operated at high voltages. This also requires using higher magnetic focusing fields and focusing power. G_4 voltages below 300 V are generally unacceptable. Voltages of about 1000 V do not contribute much to improved resolution and can reduce tube life. High G_4 voltages do produce a stronger field between the G_4 mesh and the target and reduce beam bending (localized image distortion). This high field reduces the tendency for the scanning beam to be bent laterally as it approaches highly positive image charges on the scanned target or toward the positive unscanned areas around the edge of the scanned areas. In an *electrostatically focused tube* the G_5 and G_6 electrodes provide a similar function of beam collimation or deflected scanning beam trajectory correction. High G_6 voltages also reduce beam-bending effects at the target.

In electrostatically focused tubes, focus is achieved by an electrostatic lens or lenses placed midway in the electron optical system. Beam focus is achieved by varying the dc voltage in the appropriate focus electrodes, which in turn controls the strength of these lenses (see Fig. 11-43). Once the proper voltages for G_5 and G_6 and focus are established, and if the electrodes are all supplied by a voltage divided circuit from a common power supply, good beam focus is maintained over a wide variation in the high-voltage supply voltage.

G_2 voltage is used to provide an accelerating voltage for electrons from the cathodes of the tube. This voltage is not critical for beam focus, but it does determine the range of voltages for G_1 beam control and to a lesser extent determines the beam impedance and, hence, the lag of the tube. This voltage is usually fixed at the voltage recommended by the tube manufacturer. Currents into this electrode can be as high as 1 to 2 mA, although normally they are in the 200-μA region. The G_2 voltage is normally in the range of 300 V.

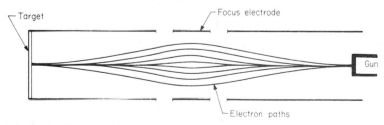

FIG. 11-43   Electrostatic focus.

G_1 voltage is used to control the amount of beam current. This is a tube setup control that must be adjusted for each tube in each camera. Sufficient negative voltage range (below the cathode voltage) should be provided in controls to allow the beam to be cut off completely. If the G_2 voltage is high, a more negative G_1 voltage range must be provided for proper beam control. Normally about -100 V is provided for G_1 beam control.

Constant beam is often desired. One method of achieving this is to utilize a circuit that senses the current in the G_2 circuit and provides an automatic adjustment of the G_1 voltage to provide the proper constant G_2 current.

Optimum performance for lag and resolution can be obtained if the beam is set as low as possible to handle the signal level being developed. When specular highlights are encountered, high values of beam current are needed. Some cameras utilize circuitry that can "instantaneously" provide sufficient beam to handle the extra signal level. These circuits sample the video signal and feed a video-rate signal to the G_1 to automatically increase the beam current. These circuits are variously called CTS, ABO, DBC, or other acronyms that describe their function.

Target voltage is a positive dc level applied to the target in order to establish the proper electric field across the photoconductor. The scanned side of the photoconductor is established at the voltage of the thermionic cathode of the electron gun by the low-velocity scanning process of the electron beam. Target voltage is therefore the difference in voltage between the target electrode and the cathode. Each type of tube uses a different target voltage, ranging from 8 to 75 V.

If the target voltage is applied directly to the target, rather than to the cathode, a coupling capacitor of adequate size and voltage rating must be provided between the target and the video amplifier. This adds to the shunt capacitance of the tube output and degrades the signal-to-noise ratio. Many cameras connect the target directly to the amplifier at approximately zero *chassis* voltage and apply the target voltage by biasing the cathode negatively. The actual target voltage is the absolute voltage difference between these two electrodes. Target currents are in the 1- to 2-μA maximum range, and cathode currents are in the 1- to 2-mA maximum range.

The target voltage for most tubes is fixed for any one tube or tube design. The exception is the antimony trisulfide photoconductors. Sensitivity of this type varies drastically with target voltage. This factor is used to advantage in some cameras where sensitivity is adjusted either manually or automatically by varying the target voltage. A very high resistance in series with the target supply (when voltage is applied to the target) provides a degree of automatic sensitivity control to accommodate the light conditions of different scenes.

Heater voltage can be either alternating or direct current. The manufacturer's recommendation should be observed. Direct current is preferable, to avoid any possibility of ac hum getting into the video signal. Voltages higher or lower than normal will shorten tube life. Regulation of ± 5 percent is adequate. The voltage rating between heater and cathode should be observed, particularly when a dc voltage is applied to the cathode to achieve the proper target voltage.

11.5.3 DYNAMIC FOCUSING. Dynamic focusing is the process of correcting the focus of the beam to accommodate defocusing caused by deflection of the beam. The deflected beam travels a longer distance than the undeflected beam in its trip from the cathode to the target. Hence, it comes to focus slightly behind the target. Focus can be restored by applying an ac voltage wave form to the G_3 focus electrode. This voltage is parabolic in shape and increases in voltage as the beam is deflected on each side of center. Both vertically and horizontally synchronous parabolic waveforms are needed to produce proper focus uniformity. The peak-to-peak voltage required for focus modulation is in the range of 10 to 20 V.

11.5.4 BEAM BLANKING. It is necessary to prevent the beam from landing on the target during both vertical and horizontal retrace. Blanking prevents erasure of the stored charge during retrace and also provides a black-level reference signal during each retrace interval. Blanking can be accomplished by applying a negative blanking signal to the G_1 sufficient to cut off the beam. An alternative is to apply a positive blanking signal to the cathode sufficient to raise its voltage above the few volts potential of the image charges on the target. The blanking voltage pulse should be as long as the retrace time of the deflecting circuit and narrower than the final picture blanking. A spurious signal which is developed at the very edge of the scan, as defined by the start and stop of the camera tube blanking pulses, should be blanked off on both sides and top and bottom from the final video signal by the wider system blanking pulse. To accomplish this vertically, it is desirable to generate a vertical system blanking pulse that starts several scan lines before the camera vertical retrace and the camera vertical-blanking pulse. A few microseconds delay of the video signal in the video amplifiers usually provides this guard time at the right side of the picture in the horizontal direction. (See Fig. 11-44.)

When the beam is prevented from landing on the target during retrace, the signal output is zero. The signal level during this retrace interval can be used to provide an absolute black-level signal reference and can be the signal level used for clamping or dc restoration purposes.

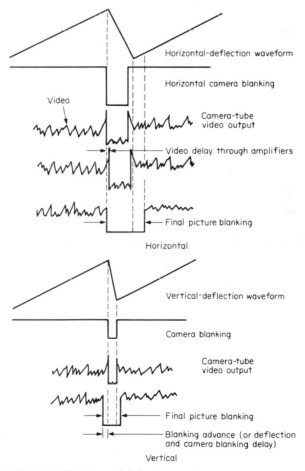

FIG. 11-44 Blanking and deflection waveform timing.

When both blanking and *target* voltage are applied to the cathode of a tube, the target voltage is the absolute dc voltage of the cathode during the active scan time with respect to the target.

11.5.5 BEAM TRAJECTORY CONTROL

Alignment Coils.   Alignment coils are usually employed to correct the beam trajectory so that the undeflected beam lands perpendicular to the target in the center of the picture. One symptom of nonperpendicularity is translation of the center of the image, as focus is varied. A useful technique to properly adjust alignment is to rock the beam focus manually or electrically. This can be done by applying a frame-scan rate square wave to G_3. The alignment coil current (or field) is then adjusted so that the image rotates about the center of the picture.

Pairs of orthogonal alignment coils (or adjustable magnets) are positioned over G_2 of the tube and are used to provide the magnetic field. The consequences of improper alignment can be poor lag, nonuniform signal output, nonuniform focus, and geometric distortion or bowing of lines across the picture.

Focus coils that have a uniform field strength extending from the target to the G_2 aperture usually do not require alignment coils. Shorter focus-coil systems suffer in performance if alignment is not provided.

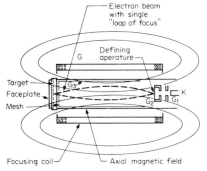

FIG. 11-45 Magnetically focused vidicon, tube system. *(RCA Corp. and SMPTE.)*

Magnetic-Focusing Coils.   Magnetic-focusing coils are used on many vidicon and other types of camera tubes to provide beam focus. The focusing coil is a solenoid surrounding the tube producing a reasonably uniform axial focus field (Fig. 11-45). The electrons emitted from the gun that move parallel to the focus field are unaffected by it. Electrons with a radial component of velocity cross the magnetic field lines, complete a circle, and arrive back at the same magnetic field line at a time proportional to the magnetic field

$$T = \frac{3.56 \times 10^{-7}}{B} \qquad (11\text{-}4)$$

where B = magnetic field in gauss. Focus of the beam is achieved by adjusting the voltages on the tube electrodes (and hence the beam velocity), so that the transit time of the beam through the tube is equal to the time of rotation of the radial directed electrons (Figs. 11-46 and 11-47). In most vidicon camera tube systems a single loop of focus is utilized. Increasing the magnetic field by a factor of 2, or decreasing the accelerating

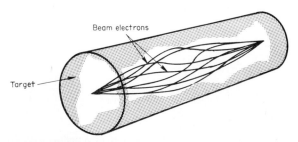

FIG. 11-46 Magnetic focusing.

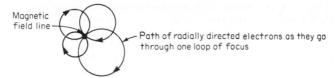

FIG. 11-47   Axial view of the path followed by beam electrons having a radial component of velocity.

electron field by a factor of 4, will result in the beam going through two loops of focus instead of one. This method of operation generally increases the focus error caused by electrons having different thermal velocities. Multiple focus loops are used in some specialized camera tube systems, where this design is useful in obtaining better collimation of the beam at the target. By judicious use of multiple loops of focus, aberrations caused by the deflection process can be minimized or canceled.

Care must be taken in the design of both the focusing and deflecting coils so that the focusing coil is not excited by the deflecting coils at a resonant harmonic frequency of the scanning frequencies. If this happens, localized focusing and geometric distortion will occur.

The shape of the focusing-coil field can influence the system resolution and resolution uniformity. A strong magnetic field over the target and a weak field over the electron gun will demagnify the beam and improve the center resolution, usually at the expense of corner resolution. Some of this degraded corner resolution can be recovered by the use of focus modulation voltage waveforms applied to the focusing electrodes of the tube.

The shape of the magnetic field at the front of the tube has an influence on the collimation of the beam. The deflected beam has both a radial and a tangential component of velocity as the beam approaches the faceplate (Fig. 11-48). The collimating lens formed by the G_3 and G_4 can correct for the radial component of velocity. A flared mag-

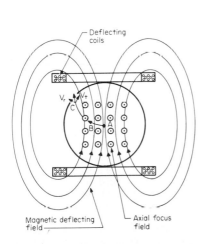

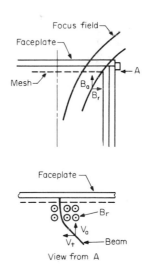

FIG. 11-48   Front view of deflected beam trajectory in a magnetically focused, magnetically deflected vidicon showing the radial (v_r) and tangential (v_t) component of beam velocity as it approaches the target. *(RCA Corp. and SMPTE.)*

FIG. 11-49 v_r compensation using radial component of magnetic-focusing field. *(RCA Corp. and SMPTE.)*

netic focus field (having radial components) can compensate for tangential components of beam velocity and bend the beam so that it approaches the target perpendicularly (Fig. 11-49). For this reason the axial position of the focusing coil with respect to the deflecting coil and the tube is critical It must be carefully determined and properly maintained for any tube, focusing-coil, and deflecting-coil system.

Focus current must be regulated very accurately to maintain focus, more accurately than the voltage on the tube electrode. The length of the focus *loop* in a tube has the following relationship to the voltages and focusing field

$$L = K \frac{\sqrt{V}}{H}$$

where L = loop length
V = voltage on focusing electrode
H = magnetic field, G

11.5.6 VIDEO OUTPUT.

Video output is taken from the target. In most tubes contact is made through the target ring. This ring serves the dual function of providing the seal between the faceplate and the bulb and providing electrical contact through the tube envelope to the target signal electrode. The shunt capacitance of this circuit to ground (including the amplifier input and wiring capacitance) controls the signal-to-noise ratio of the video signal

$$\frac{\text{S}}{\text{N}} = K_1 \frac{i_s}{\sqrt{\Delta f + K_2\,(AF)^3 C^2}} \tag{11-5}$$

where i_s = peak signal output
f = bandwidth, Hz
C = total shunt capacitance of tube and first amplifier stage
S/N = peak signal to rms noise

Negligible noise is produced by the signal current from the tube itself, compared with the noise in the input amplifier. In the signal current rms noise is $\sqrt{2ei\,\Delta f}$. Δf is modified by tube resolution, since the output signal does not have a flat frequency response.

The output impedance of the target is extremely high and can be considered as a constant-current generator. The shunt capacitance to ground can be as low as 2 pF for the low-output capacitance versions, and as high as 12 pF for the larger tubes. These values are increased when the tube is mounted in a coil assembly.

The high-frequency components of the signal are shunted to ground through the shunt capacitance of the target and amplifier input circuit. The signal can be processed in an amplifier to restore the system frequency response drop-off and phase shift produced by the progressive loss of high-frequency information caused by the shunt capacitance. When this is done, the flat-spectrum noise of the input stage noise is boosted in the high frequencies (above a few hundred kilohertz) and produces a signal dominated by high-frequency noise. The spectrum of this noise is called a *triangular-noise spectrum*, continuously rising to the frequency cutoff of the amplifier.

Where high signal-to-noise is a major factor in camera performance, low-output capacitance tubes are used to reduce the shunt capacitance of the target circuit. Figure 11-50 illustrates the components of capacitance of a conventional tube. Figure 11-51 illustrates the structure and the shunt-capacitance factors of this design.

The target electrode is a good antenna for high frequencies, and, therefore, must be inside a well-shielded enclosure that is bypassed to the input ground of the first amplifier stage. This enclosure includes the internal shielding of the focusing- and deflecting-coil system and the G_4 (mesh) electrode of the camera tube. This electrode must be well bypassed to the same ground system as the remainder of the target-shielding enclosure. Deflecting-coil shield grounds should not run through this same circuit.

Signal currents range from several nanoamperes to 1 or 2 μA. For broadcast television, and other reasonably critical systems, peak signal currents of 150 to 600 nA are utilized for normal-sensitivity operation. For high-sensitivity operation in limited light condi-

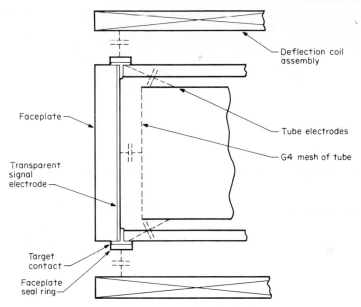

FIG. 11-50 Output-capacitance sources of standard tube.

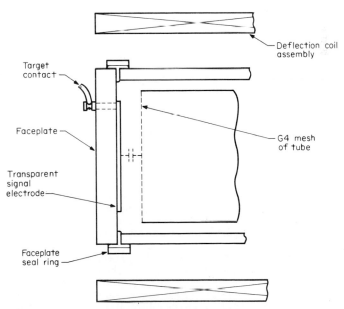

FIG. 11-51 Output-capacitance sources of low-capacitance tube.

tions, signal currents of one-tenth these values are often utilized. The polarity of the signal highlights is negative.

The signal during the interval of retrace across the target is zero. This provides a reference level on which black-level sensing or dc restoration can be performed.

11.5.7 DEFLECTING COILS AND CIRCUITS.
Two sets of magnetic deflecting coils produce a pair of transverse deflecting fields. Both of these sets of coils produce fields that have the same position along the axis of the tube, which implies that coils are usually concentric. Each coil of a pair is wrapped around the tube nearly 180° (Fig. 11-52). Impedances (inductance) are kept low (about 1 μH) for the horizontal deflecting coil to reduce the peak voltage during retrace and to keep the resonant frequency high so that the retrace time (determined by the resonant frequency) is short.

The coils are shielded from the tube electrodes by internal Faraday shields that do not buck the magnetic deflecting fields, but instead reduce the electrostatic pickup that can be coupled from the coils to the tube electrodes and, in turn, into the signal plate of the target.

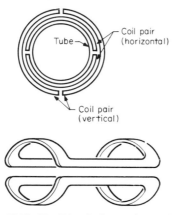

In electrostatically deflected tubes, the direction of deflection is perpendicular to the magnetic-field lines. In magnetic-focus tubes, the beam is initially deflected perpendicular to the magnetic field of the coils. This causes it to traverse the focusing field, which in turn causes it to take a spiral path. Thus, the actual deflection path at the target occurs nearly parallel to the deflection coil field.

The deflecting-current waveform is essentially a linear sawtooth for both vertical and horizontal deflection coils. The G_3 or focus electrodes are made of high-resistance material to reduce the loading on the deflecting coil and to minimize induced circulation currents which buck the deflection field.

The unavoidable bucking field produces some nonlinearity of scan, particularly at the horizontal-deflection fre-

FIG. 11-52 Traditional shape of a pair of deflecting coils.

quency. This necessitates putting slight compensating distortions in the scanning-current waveforms to achieve good linearity of scan. Tubes made by different manufacturers often require individual adjustment of horizontal linearity to compensate for the loading or eddy-current bucking effects.

Timing of waveforms for deflecting and blanking is important. The typical timing is shown in Fig. 11-44.

11.5.8 MAGNETIC SHIELDING.
Magnetic shielding is generally required to prevent deflection, or misaligning, of the beams by the earth's magnetic field and other stray magnetic fields. Ideally, the tube and deflection coils should be completely enclosed in magnetic shielding. An input port must, of course, be provided for the optical input. Adequate magnetic shielding is particularly important for color cameras where external magnetic fields can produce misregistration of images.

11.5.9 ANTI-COMET-TAIL TUBE.
Versions of lead oxide tubes called ACT or HOP by their manufacturers are designed to erase signals generated by excessive highlights during the retrace interval of each horizontal line. This erasure minimizes the *comet tail* produced by brightly moving objects going across the field of view. The gun is designed so that pulsing an extra electrode between the control grid and the small beam-defining aperture in the gun will focus the beam crossover directly on this aperture. As a result,

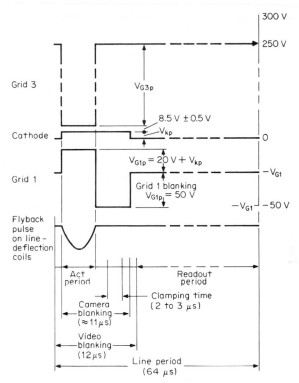

FIG. 11-53 Voltage and pulse timing for ACT- or HOP-type guns. (© *1983 Philips Export B.V.*)

a very high beam current can be produced during this pulse interval which takes place during horizontal scan retrace. During this interval, the cathode is pulsed several volts more positive than during the scan readout interval. Consequently, it will land on the target only in places where the image charge exceeds this additional pulse voltage. This pulse voltage is adjusted in operation so that the excessive beam erases charge signals which occur just above the normal highlight signal level.

The typical pulse voltages and timing required to operate this type of tube are shown in Fig. 11-53. In this case, the voltage on the extra gun electrode controls the position of the beam crossover. The G_1 voltage is pulsed positively during the beam retrace interval to produce a very high beam current. It is cut off during the remainder of the blanking cycle to produce a video signal black-level reference for video clamping purposes. The camera video amplifier used with these tubes must be capable of handling a peak signal current at least 300 times normal highlight level without overload to accommodate the large signal pulse occurring during retrace. Tube life is maximized if this circuitry is activated and utilized only when specular highlights occur in the picture.

11.6 CAMERA TUBE PERFORMANCE CHARACTERISTICS

Sensitivity, resolution, and lag are the three most important camera tube performance characteristics.

11.6.1 SENSITIVITY AND OUTPUT. Photoconductive-tube sensitivity is straightforward. Relative sensitivity can be determined by inspection of the absolute spectral response curves of the various photoconductors (Fig. 11-34). None of these photoconductors produces greater than unity quantum yield, i.e., more than one electron of signal current per incident photon of light. In broad photographic terms, the photoconductors on 25-mm-diameter (1-in) tubes used for broadcast television service have a sensitivity equivalent to film that has an ISO exposure index (EI) of 300 when used with tungsten illumination. Sensitivity is lower when utilized in a color camera that splits the light three ways to three different tubes, or when other color filters are introduced.

In a camera tube, unlike film, the equivalent ISO sensitivity changes as a function of the image size. If a smaller image size is used, fewer lumens of light will fall on the image for the same scene illumination and lens opening. Therefore, a lower signal output current, in direct proportion to the reduction of the scanned area, will result. A lower lens f-number, or larger aperture, must be utilized to increase the signal level to the expected level. At first glance it might seem that the smaller tubes are less sensitive. This is not necessarily the case, because of the geometry of the optics. When the optics are analyzed and a comparison is made on the basis of the same *angle of view* and the same *depth of focus* for the small and for the large image format lens, it turns out that the smaller image format lens will have the same diameter of opening as the large one; i.e., it collects the same number of lumens of light. The f-number of the small image format lens will be smaller, but the system performance factors, depth of focus, angle of view, and signal output from the tubes will be identical to those of the large image format system.

Light-transfer characteristics (Fig. 11-35) are more useful in relating scene illumination and signal output from a camera tube. By industry convention, light-sensitive devices are evaluated with tungsten illumination with a color temperature of 2856 K. This is typical of normal tungsten-bulb illumination. (Studio lighting is usually a higher color temperature of 3200 K, and sunlight is $\sim$ 5900 K.)

Scene illumination and the illumination of the faceplate of a tube in a simple optical system are related in the following manner

$$I_s = 4I_{pc}F^2/TR \tag{11-6}$$

where I_{pc} = illumination of photo surface
F = lens f-number
T = lens transmission
R = scene highlight reflectance

T of a lens is typically 0.8. The simplified equation can be expressed as

$$I_{pc} = I_sR/5F^2 \tag{11-7}$$

Signal Output. From the light transfer characteristics, a signal output can be determined if the faceplate illumination I_{pc} is known. Light transfer characteristics are traditionally published as dc signal output measurements made with uniform illumination. Peak-signal currents will be higher because part of the time the scanning beam is blanked off, and all the signal is developed during the active scan time. Peak-signal output will be the indicated signal output from the light transfer characteristics divided by the percentage of unblanked scanning time

$$I_s = \frac{I_s \text{ indicated}}{\% \text{ unblanked camera scan time}} \tag{11-8}$$

Equivalent ISO Exposure Index. Camera tube sensitivity can be related to film exposure indexes (ISO ratings) by the use of Fig. 11-54. In the television system, the *exposure time* is fixed by the television scan rate. A comparison of photographic to television reproduction of moving images reproduced by a low-lag camera tube shows that the most realistic time interval for a television signal corresponds closely to a photographic exposure of 1/60 s. Consequently, if the tube specification recommends a certain faceplate illumination in footcandles, the television camera will require the same lens

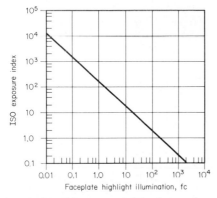

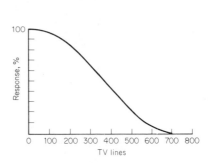

FIG. 11-54 ISO exposure index values vs. faceplate illumination for camera tubes operated on United States broadcast scanning standards (equiv. shutter speed 1/60 s). *(RCA Corp. and SMPTE.)*

FIG. 11-55 Typical camera tube amplitude response curve.

opening and light level as a photographic camera operating at a shutter speed of 1/60 s would with film of the ISO rating determined by Fig. 11-54.

11.6.2 RESOLUTION. Resolution is the ability of a tube to faithfully reproduce fine-detail information and transitions between dark and light parts of an image.

The use of *limiting resolution* as a measure of a tube's resolution fidelity is usually subjective and therefore given to error. However, an even more compelling reason for abandoning the process of using limiting resolution for evaluation of a camera tube is the fact that the adjustment of beam and lens focus for maximum limiting resolution is different from the point-of-focus adjustment required for maximum subjective sharpness of the television picture.

Consequently, resolution of camera tubes is commonly expressed as amplitude response or *depth of modulation*. The data are presented either as a number defining the response of the tube to a square-wave test pattern consisting of black and white bars of a certain size, or as a curve showing the response from coarse to very fine test-bar patterns (Fig. 11-55). The size of the test pattern bars is expressed in a television line number, which is the number of *black and white* bars of equal width that will occupy the vertical dimension of the television picture. Actual response curves to a square-wave pattern are often referred to as the *contrast transfer function* (CTF). This can be contrasted to lens and film technology which utilizes test patterns calibrated in line pairs (one black and one white) per unit length across the actual image. The response data usually are converted to *modulation transfer function* (MTF), which is the response to a sine-wave test pattern. The resolution performance of a system can be predicted from the product of the MTF of each component.[3] For this purpose it is necessary to convert camera tube CTF data to MTF data, using the following relationship

$$MTF_n = \frac{\pi}{4}\left(CTF_n + \frac{CTF_{3n}}{3} - \frac{CTF_{5n}}{5} + \frac{CTF_{7n}}{7} - \cdots\right) \qquad (11\text{-}9)$$

MTF_n = MTF at television line number n and CTF_n = CTF at the corresponding television line number. MTF data for lenses must be converted to television lines per picture height of the image size used in the particular case.

In television technology, several shortcuts are often used to express camera tube resolution. The most common is the response at 400 television lines. This represents a frequency of 5 MHz for vertically oriented bars in both 525- and 625-line systems. This value is used because it corresponds roughly to the bandwidth of the television channel.

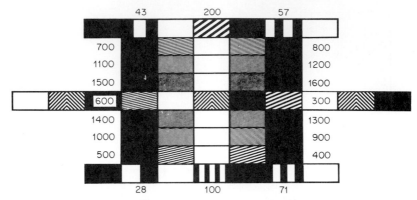

FIG. 11-56   Center section of P200 test chart. *(RCA Corp. and SMPTE.)*

It represents the point where the response of camera tubes is usually dropping off rapidly and is therefore a sensitive single-number comparision figure. The video frequency of any vertically oriented pattern in 525- and 625-MHz line systems can be determined from the relationship 1 MHz = 80 television lines.

Resolution Measurement Methods. Tube resolution cannot be measured independently of the deflecting- and focusing-coil systems, but must be measured in the coils in an operating camera, whose design can influence the resolution of the tube.

One method of determining resolution involves measurement (by means of an oscilloscope) of the response of a camera tube to a line pattern calibrated in television lines per picture height. However, this measurement should be made with the aid of a test method that is independent of observer judgment and test equipment variation. For vertically oriented test bars the measurements vary with the frequency response of the amplifier, since finer-pitched test lines produce higher-frequency signals. The bandwidth must be significantly greater than the frequency being measured, so that sufficient harmonics of the fundamental frequency can be present to reproduce the square-wave signal.

The technique of testing a high-impedance video preamplifier for flat response is a difficult one, and one subject to many pitfalls that produce errors. As a result, differences in measurement by factors of 2 to 1 can be experienced between test equipments in various laboratories. One way to avoid frequency-response errors is to use a special test pattern designed to measure resolution data independent of the amplifier frequency response. One test pattern, the RCA P200, shown in Fig. 11-56, has been adopted by several television organizations and laboratories for evaluating camera tube resolution. This test pattern is designed with line patterns that are rotated from their usual vertical orientation so that the beam traverses them at a slower rate (Fig. 11-57). The signal output from these line patterns is, then, produced at a lower frequency. Since the frequency of the video signal varies as the sine of the angle from the horizontal, the maximum and minimum signal outputs, as the beam traverses the light and dark bars of the test-pattern image, are essentially independent of the direction of scan across the test pattern. This fact allows a design of test patterns that produces two important results: First, the bandwidth of the required video amplifier system is reduced; second, each group of lines of the bar pattern can be inclined at an angle such that each set of bars representing different television line numbers produces the same fundamental video frequency. This latter feature eliminates the amplifier frequency response as a factor in the measurements.

In the center of the pattern is a chevron-shaped pattern of 400 television line bars. The lines are oriented 90° from one another, and both sets of lines are oriented 45° from the vertical. This portion of the pattern is utilized when focusing the beam, and to detect

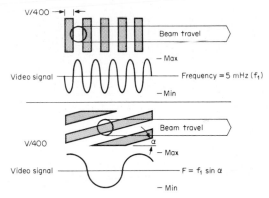

FIG. 11-57 The principles of the P200 slant-line test chart showing how the measurements can be made at a lower video frequency and how all line number measurements above a low line number, such as 100 television lines, can be made at the same video frequency. *(RCA Corp. and SMPTE.)*

astigmatism in the beam. If the amplitude of the signals from these two patterns is not equal, the beam has some astigmatism (ellipticity) that will favor the resolution of information in one direction and reduce it in others. The beam focus should be adjusted to produce as nearly equal signals as possible from the two parts of this chevron, to maximize the response to the two orthogonal blocks of lines.

Each block of lines is bounded left and right by a black and white block having the same transmission characteristics as the black and white line of the line patterns. These blocks produce 0 and 100 percent reference signals that eliminate tube signal uniformity and test-pattern illumination nonuniformity from the measurements. Chevrons of 400 television lines are located also at strategic positions around the pattern to allow measurement of any astigmatism.

The test pattern is designed for use with a line-selector oscilloscope equipped with provisions for displaying on the picture monitor the location of the line being examined. The desired peak signal level and proper beam setting must be established first. With the selected line positioned so that it passes through the center chevron of the test pattern, the optical and electrical beam-focus adjustments are set for maximum, and preferably equal, signals from the two chevrons. The selected line is then moved so that it passes through the appropriate television line-number block. The amplitude of this signal is measured as a percentage of the peak black-to-white signal of the adjacent blocks.

A more sophisticated way of measuring resolution takes into account the fact that the effective beam is not always symmetrical, and thus produces different resolutions along different axes of the picture. The RCA P300 test chart produces data defining the actual beam shape, thus permitting computation of the equivalent resolving characteristics of the beam. Figure 11-58 shows the central portion of this test pattern. Data to reconstruct the effective beam shape are shown in Fig. 11-59.

System and Component Resolution. The resolution of the camera tube and its coil system can be divorced from the lens used to generate the data and the characteristics of the test pattern if the MTF of the lens and the CTF of the test pattern are known. Actual tube coil system-resolution data can be determined in the following way:

1. Measure CTF of the camera tube in the system.
2. Divide the data obtained in step 1 by measured CTF of the test pattern.

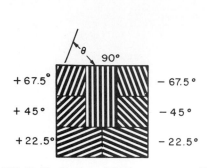

FIG. 11-58 Portion of P300 test pattern, 400 television lines slanted at various angles. *(RCA Corp. and SMPTE.)*

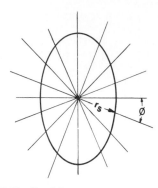

FIG. 11-59 Resolving aperture geometry determined from slanted test patterns (P300). *(RCA Corp. and SMPTE.)*

3. Convert the results of step 2 to MTF.

4. Divide the results of step 3 by the comparable lens MTF data.

5. Convert the results of step 4 to CTF. This will be the tube CTF

$$\text{CTF} = \frac{4}{\pi}\left(\text{MTF}_n - \frac{\text{MTF}_{3n}}{3} + \frac{\text{MTF}_{5n}}{5} - \frac{\text{MTF}_{7n}}{7} + \cdots\right) \qquad (11\text{-}10)$$

(Some tube manufacturers publish resolution data for the tube/coil combinations in this manner.)

Factors Affecting Tube Resolution. In photoconductive camera tubes, resolution is determined primarily by the photoconductor, the scanning beam size and shape, the tube size, and the number of scan lines specified by the system standards.

If the photoconductor is nondispersive of light and has good optical absorption, and little lateral leakage caused by a conductive layer of the scanned surface, the photoconductor will not limit the resolution for even very small size tubes. The scan-line pitch of a 525-line system on the popular 18-mm-diameter (0.66-in) tube is 13.2 μm. The thickness of photoconductors ranges from 1.5 μm for antimony trisulfide through 4 to 5 μm for Saticon photoconductors to 12 to 13 μm for lead oxide tubes. Even if light is dispersed by the antimony trisulfide photoconductor, it cannot scatter any appreciable distance because of the thinness of the material. At the other end of the range, lead oxide is about as thick as the scan line pitch, allowing greater dispersion and resolution loss. In silicon diode tubes the resolution is ultimately limited by the number of diodes in the diode array.

The electron beam has the major influence on camera tube resolution. Very small size beams can be produced, but they may be limited in current capabilities. Operation of the tube elements, particularly the mesh electrode and electrodes that control the beam velocity at high voltage, will produce higher-resolution beams. The use of excess beam in the setup of the tube can also degrade resolution.

The scanning-line standards determine the upper resolution of a system in the vertical direction. System standards initially recognized the fact that the vertical resolution is limited to 70 percent of the number of scan lines (commonly called the *Kell* factor). Resolution is highest in a diagonal direction in a bandwidth-limited system.

The beam of a camera tube can be too small for the television system in which it is used. If the effective beam size is smaller than the scan line pitch, areas between lines will not be fully scanned. Low-frequency flicker in some areas of the pictures can result, and moving objects may be followed by a succession of light and dark images, making it appear that the scene is being illuminated by stroboscopic lights. This effect is often called *sternwave*.

The deflecting and focusing coil systems used with the camera tube also have an important influence on resolution. The design or type of these coils is usually specified when camera tube resolution data are presented, since the quality and the design of these components have a major influence on the beam size and shape.

Adjustment procedures, such as alignment and beam focusing, cannot be ignored if best resolution performance is expected.

11.6.3 LAG. Lag in a television camera tube is a measure of the rate of decay of the video signal when the illumination is changed abruptly or cut off. It is influenced by carrier-trapping effects called photoconductive lag, target storage capacitance, scanning-beam impedance, the signal level being utilized, bias-lighting level, and the amount of beam current.

Lag Terminology and Measurements. The photoconductor is a capacitor that stores the charge carriers generated by light at each point on the image and integrates these charges during the interval between successive scans of each point of the image. Immediately after the light is applied, no charge is developed, and the full signal level at each point will be built up only at the end of 1/30 or 1/25 s, which represents the frame repetition rate. When light is removed, the signal output does not drop immediately to zero, because a charge is stored on each *element* of the charge image in proportion to the integrated illumination of each point since it was last scanned. In a perfect camera tube the signal should drop off linearly during the time interval of one complete frame (two fields in an interlaced system) after the light is removed.

It has been a convention to measure the signal level during the first field after a complete frame (two interlaced fields) is scanned following removal of light. The complete lag curve commencing after the first frame interval is a better measure of lag when comparing camera tubes. This curve detects long-term persistence caused by photoconductor trapping effects and high scanning-beam impedance, and portrays the long-tail lag effects that are particularly noticeable to the eye.

Photoconductor Trapping Effects. Trapping effects in a photoconductor can produce long-lasting image retention or lag (Fig. 11-60). In the tubes illustrated, lag is dependent upon the color of light. Lag is higher for blue light in the undoped lead oxide and lower for green and red light. Lag of the doped extended-red tubes (shown here) is higher than for the undoped lead oxide types when red illumination is used. Lead-oxide photoconductors have lower storage capacitance than the same-size selenium photoconductor, but the lead oxide has more trapping effects, which can produce the characteristic *long-tail* red or blue characteristic of the pictures produced by cameras equipped with these tubes.

When a photoconductor has lag that is controlled by photoelectron-trapping effects, the lag is a variable quantity. If the exposure to light is short, the lag will be low and determined primarily by the photoconductor capacitance-beam time-constant (RC) characteristics. If the exposure is longer, the lag will be higher and caused by trapping effects. Figure 11-61 illustrates the typical lag characteristics of a photoconductor which has negligible trapping effects. The tests that produced the curves were made using different periods of exposure before the light was removed. The exposure time ranged between 2 and 256 fields (4.27 s).

Figure 11-62 illustrates the lag characteristics of typical lead oxide photocon-

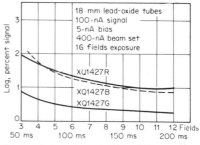

FIG. 11-60 Lag characteristics of 18-mm-diameter (0.7-in) lead oxide photoconductor tubes. *(RCA Corp. and SMPTE.)*

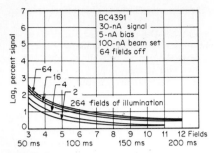

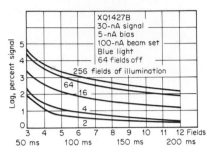

FIG. 11-61   A family of decay lag curves produced by a typical Saticon tube as a function of the exposure time before the light is removed. Low trapping (photoconductor lag) produces very little change in lag as a function of exposure time. This is also independent of the color of light. *(RCA Corp. and SMPTE.)*

FIG. 11-62   Decay lag characteristics of a blue 18-mm (0.7-in) lead oxide photoconductor as a function of exposure time (using blue light). *(RCA Corp. and SMPTE.)*

ductor tubes. In these tubes the lag is partially controlled by trapping and/or doping effects. The measured and the subjective lags are much higher where an area is exposed for an interval longer than two fields.

11.6.4   LAG-REDUCTION TECHNIQUES.

Lag can be progressively reduced by employing several methods. The effects are cumulative, as illustrated by Fig. 11-63. This curve shows the improvements in lag of a 25-mm-diameter Saticon photoconductor tube.

Beam Impedance and Photoconductor Capacitance.   When the lag of a photoconductor is not limited by photoconductor-trapping (or transit-time) effects, it is controlled by the storage capacitance of the photoconductor and the effective resistance of the scanning beam at low signal levels. The signal decays with an RC time constant determined by these factors. The photoconductor storage capacitance is the capacitance formed by the two sides of the photolayer, with the photoconductor itself being the dielectric (Fig. 11-64).

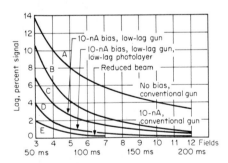

FIG. 11-63   Progressive improvement in lag of 25-mm-diameter (1-in) Saticon tubes operating with 50-nA signal current by utilizing bias light, a low-impedance (low-lag) gun, a low storage capacitance photoconductor, and beam reduction made possible by automatic highlight overload compensation circuitry (bias: 10 nA). *(RCA Corp. and SMPTE.)*

The scanning-beam resistance results from beam electrons having a range of velocities in the direction perpendicular to the photoconductor. The scanning-beam electrons approach the photoconductor with nearly zero velocity perpendicular to the surface. When the beam drives the surface to zero, or cathode voltage, the excess beam electrons return again toward the gun. Those electrons in the scanning beam with the highest velocity will land on the photoconductor last and drive the surface slightly more negative than will others with lower velocity. On successive scans of that point, if it remains unilluminated, the higher-velocity electrons will drive the scanned area even more negative and produce a small amount of signal current each time until an equilibrium voltage is reached. This situation produces a residual or *lag* signal.

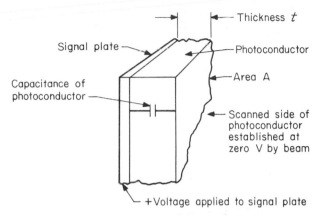

FIG. 11-64 The photoconductor as a storage capacitor. *(RCA Corp. and SMPTE.)*

A resistance value can be assigned to the electron beam. A plot of collector voltage versus current collected from the beam is shown in Fig. 11-65. At a negative collector voltage, no current will be collected. At a high collector voltage, all electrons will be collected, as in the case of a high positive charge image on the photoconductor. The transition curve between these conditions represents beam resistance: $R = (\Delta I / \Delta E)^{-1}$.

R in series with the photoconductor C produces the effective RC time constant that determines the shape of the decay curve. Reducing either C, R, or both will decrease the lag.

Reduction in Capacitance. The viable options available to decrease the capacitance are: (1) Decrease the scanned area, and (2) increase the thickness or decrease the dielectric constant of the photoconductor. Decreasing the scanned area allows the production of a smaller tube. Increasing the thickness of the photolayer decreases the capacitance. Antimony trisulfide and lead oxide photoconductors are made porous to some extent to reduce the effective dielectric constant.

Beam Resistance. The beam resistance can be altered by the design of the electron gun. Rossmalen pointed out that the mutual repulsions of electrons in the high-density region occurring in the beam crossover portion of the beam path in a conventional electron gun (Fig. 11-66) can cause a spread in the velocities of the electrons. B. H. Vine showed that a diode-type gun could limit the axial spread of electron velocities and contribute to a lower beam impedance. Careful design of a conventional triode electron gun can minimize the charge density in the crossover and reduce lag, as illustrated in Fig. 11-63, curves B and C.

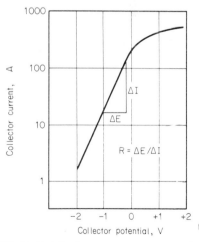

FIG. 11-65 The beam acceptance curve of a low-velocity electron beam used to establish the effective resistance of the beam. *(RCA Corp. and SMPTE.)*

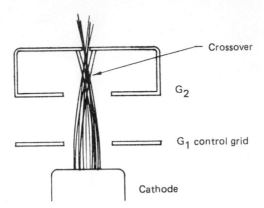

FIG. 11-66   Cross section of a conventional triode gun showing the high electron density crossover region that can contribute to high beam resistance. *(RCA Corp. and SMPTE.)*

Operating Conditions and Beam Impedance.   The beam impedance is a function of the beam current. When excess beam current is used, the effective beam resistance is increased, and lag increases. Figure 11-67 shows the progressive decrease in lag as the beam current is decreased. This improvement is possible in cameras designed to use highlight-sensing circuits that provide extra beam current to instantaneously handle brighter-than-normal scene highlights when they occur.

Bias Lighting.   Bias lighting is perhaps the most effective method of reducing lag. With bias lighting, a small amount of uniform illumination is applied on the photoconductor. The amount of bias light usually is enough to develop a uniform signal current from 5 to 10 nA.

The charge voltage developed by the bias light raises the voltage on the scanned side of the photoconductor in the absence of scene light to a level that is near the effective velocity spread of electrons within the beam.

The beam lands more fully on the lowest charges developed by light from the scene image. Under these conditions the beam has a low resistance, and the lag is substantially

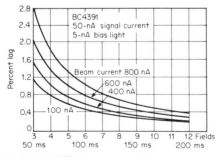

FIG. 11-67   The progressive reduction in lag by reducing the amount of beam current. *(RCA Corp. and SMPTE.)*

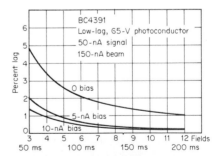

FIG. 11-68   Progressive improvement in lag from an 18-mm (0.66-in) Saticon tube as a function of bias light level. *(RCA Corp. and SMPTE.)*

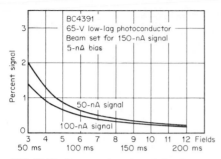

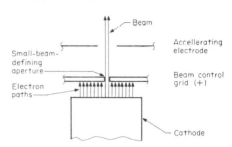

FIG. 11-69   Lag as a function of signal current in an 18-mm camera tube. *(RCA Corp. and SMPTE.)*

FIG. 11-70   Diode-gun configuration.

reduced. Bias light does not reduce the contrast, since the added dc signal level is canceled in the signal-processing amplifier. Figure 11-68 shows the progressive improvements in lag as bias lighting is increased. Lag is also a function of the signal current, as illustrated by Fig. 11-69. This fact must be taken into consideration when applying bias light to a camera in which the signal current for white light is different in the different channels.

Lag is never reduced to zero. In a color camera the amount of bias light in the three channels should be roughly in inverse proportion to the operating signal current level in each channel for best lag performance.

Diode Gun Tubes.   Diode gun tubes are camera tubes with a gun structure designed so that the G_1 control grid is run positive with respect to the thermionic cathode. This prevents a space charge from altering the velocity spread of the electrons. The configuration is different for different designs, but all have in common an electron beam that flows nominally parallel to the axis without developing a high current-density crossover region. The intent is to minimize beam impedance and produce a beam with as narrow a divergence angle as possible, to maintain good resolution of the focused beam.

A diode gun allows design-change flexibility. In lead oxide tubes the reduced beam impedance permits a thinner, and hence higher-capacitance, photolayer without increasing lag. This thinner photoconductor layer has higher resolution than a thick layer. When used with a photoconductor such as a Saticon layer, whose resolution does not depend on thickness, slightly higher resolution is produced by the small beam divergence angle from the diode gun.

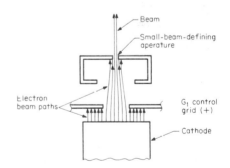

FIG. 11-71   Diode and triode beam resistance or equivalent temperature determined from collector voltage-current curves.

A diode gun has the structure and beam electron trajectory as shown in Fig. 11-70. The beam resistance and the equivalent beam temperature of conventional triode and diode guns are shown in Fig. 11-71.

In some designs of this type of gun, *diode gun* may be a misnomer. In actuality, it is a triode gun operated in a positive G_1 mode. The scanning beam in this gun is formed at a small aperture in a third electrode (Fig. 11-72).

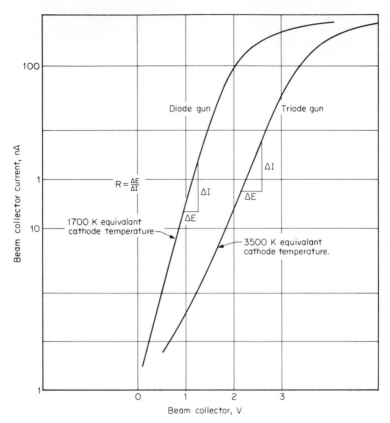

FIG. 11-72 Positive-grid triode, sometimes called a diode gun.

11.7 SINGLE-TUBE COLOR CAMERA SYSTEMS

Most low-cost home use cameras, and many professional color cameras, utilize a single camera tube to produce the complete color television signal. Nearly all these tubes incorporate an array of fine color stripes from which the color information is derived. These tubes are vidicon types; i.e., they use a photoconductor to detect the light and use a target readout.

There are two categories of these tubes. One has a single output, and the color signals are derived by retrieving color information from high-frequency carrier signals present in the output signals. The other utilizes multiple output signals from multiple discrete-signal plate structures inside the tube. Most conventional camera tubes having adequate resolution and color response can be utilized in a single-tube color camera by imaging the scene onto an appropriate color striped-filter array and reimaging this image to the faceplate of the tube. The tube then acts as a single-output, single-color tube.

Some camera designs have also been made in which a single tube produces all the color information and a second tube produces the luminance and detail information of the composite color signal.

11.7.1 SINGLE-OUTPUT-SIGNAL TUBES.

A typical stripe color system will use two different sets of stripes from which three-color information can be derived. One example is the structure shown in Fig. 11-73. Yellow (Y) stripes are utilized to produce blue information, and cyan stripes (C) are used to detect red color information. The sets of stripes are inclined from one another so that there are two different color carrier frequencies developed as the beam scans across these periodic structures. When blue light is present in the scene, no blue light will pass through the yellow filter lines, but it will pass through the open areas between the yellow lines, including that portion covered by the cyan filter. This will produce a blue carrier signal in the output. Similarly, red light will be absorbed by the cyan filter and will be transmitted through the space between the cyan filters (including the portion of the yellow filter in this space), producing a red carrier signal in the output. The frequency domain of the output signal is shown on Fig. 11-74. The information below the color carrier frequencies is extracted to form the white or luminance signal (Y), and the two-color carriers are detected separately as shown in Fig. 11-75. The amplitude-modulated color carriers are then detected to produce separate red and blue signals. Subtracting (matrixing) these two signals from the luminance signal will produce a green signal, as illustrated in Fig. 11-76.

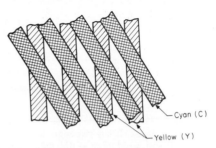

FIG. 11-73   Dual sets of color stripe filters.

The beam size of the camera tube does not have to be small enough to resolve each line with 100 percent contrast to produce a color picture. Low resolution of either the beam or the photoconductor will result in low color-signal levels and poor color-noise performance, but will not necessarily result in poor colorimetry. Loss of resolution during operation will produce a fail-to-green mode since luminance information will remain constant while red and blue information will drop.

A variation of the stripe-filter design can produce a single-carrier system that contains red and blue information which can be extracted later. The filters are configured as shown in Fig. 11-77. They are angled so that at each successive scanning line, the carriers produced by the lines slanted clockwise advance in phase by 90° and the ones slanted counterclockwise drop back by 90° (Fig. 11-77, lines A and D). If the line 1 information is stored in a $1H$ delay and line 3 is delayed for 90° of the carrier frequency, the red information on line 3 compared with line 1 (delayed) will be exactly 180° out of phase (A and E). The blue information on line 1 (delayed $1H$) compared with line 3 (C and F)

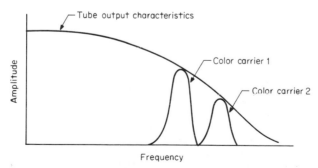

FIG. 11-74   Frequency characteristics of a two-carrier single-tube color camera tube.

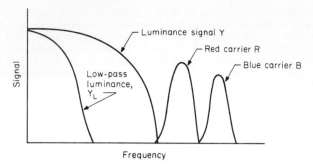

FIG. 11-75 Signals extracted from the color tube output.

will be *in* phase. If the two signals are added, red information will cancel and blue signals will add, producing a blue signal carrier. If the two signals are subtracted, the blue signals will cancel and the red signals will add to produce a blue carrier. These two carriers can then be detected to produce the blue and red signals (Fig. 11-78).

Other systems proposed, but not widely used, employ parallel vertical stripe filters that consist of triplets of different colors. These systems operate on a system of synchronous detection. This means they can recognize the position of the beam relative to the stripes so that the detector can sequentially interrogate the signal being developed by the beam over a particular stripe and divert that signal into an appropriate color channel. Recognition of the position of the beam is determined from an additional black or white reference stripe that can be detected (to time the synchronous detector). These systems suffer from color fidelity if the beam is not as narrow as the color stripe. Timing of the synchronous detector is uncertain if the reference signal becomes ambiguous in the presence of other video information.

Basic Design of Single-Color Tube. High-resolution photoconductors such as antimony trisulfide, Saticon, Newvicon, or Chalnicon are used because of their inherent resolution and color response. The electron optical systems utilized are of superior quality for resolution and resolution uniformity, since color level and color uniformity depend upon high resolution and resolution uniformity, respectively. The filters are incorporated in the tube between the faceplate and the signal plate (or plates), and a barrier layer produces a smooth substrate for the signal plate and the photoconductor plate (Fig. 11-79).

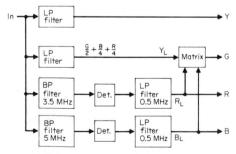

FIG. 11-76 System block diagram of a two-carrier system. *(RCA Corp.)*

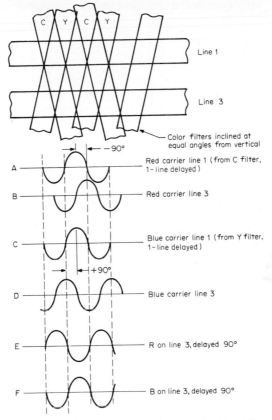

FIG. 11-77 Configuration and phase relationship of the interleaved color carriers in the single-carrier system.

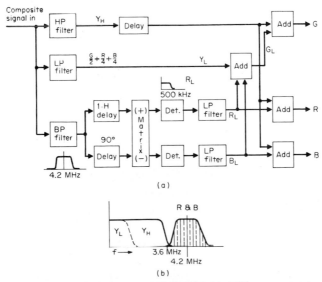

FIG. 11-78 Single-carrier system: (a) decoder employing 1-H delay comb filter and (b) typical frequency spectrum. (RCA Corp.)

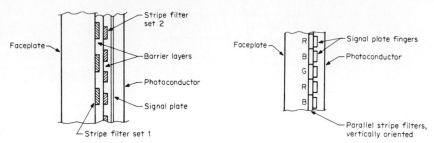

FIG. 11-79 Cross section of typical multistripe color tube target construction.

FIG. 11-80 Cross section of trielectrode color vidicon.

11.7.2 MULTIPLE-OUTPUT-SIGNAL TUBES. Two single-tube color vidicons have been developed using multiple-output signals. The *trielectrode vidicon* has three outputs from three independent sets of signal plates, each associated with a set of appropriate color filters. The *Trinicon* has two separate output terminals from the target.

The trielectrode vidicon has a filter structure consisting of vertically oriented triplets of red, blue, and green filters. Each filter stripe is backed up by a transparent signal plate. All the red, blue, and green signal plate stripes are tied together to their respective output pins protruding through the faceplate (Figs. 11-80 and 11-81). Even if the beam is larger than the stripes, the signal produced at each signal stripe will relate only to the color of light coming into the photoconductor through its associate color filter. This preserves color fidelity and prevents loss of color information in noise at low light levels. High-frequency detail information is cross coupled into all channels because of the high interelectrode capacitance.

The Trinicon tube has vertically oriented color stripes and is designed and operated to produce an unambiguous phase-reference signal that can be used to synchronously detect the color signals from the output signal.

The Trinicon tube has two separate outputs that are connected to two sets of interleaved signal plate fingers. Vertical triplets of red, blue, and green filter stripes are positioned in front of these fingers as shown on Fig. 11-82. There are three filter stripes for

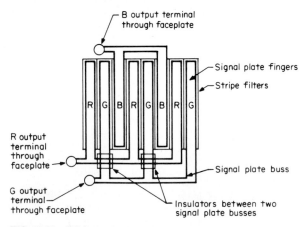

FIG. 11-81 Trielectrode vidicon structure schematic.

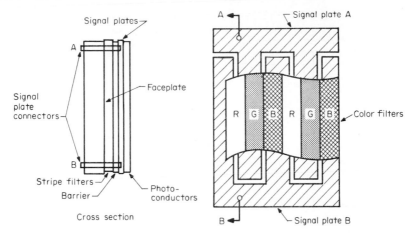

FIG. 11-82 Cross section and plan view of Trinicon showing relationship of stripe filters to the signal plate fingers.

each pair of interleaved fingers. The filters are located so that one color filter stripe straddles the gap between two fingers and the other gap coincides with the junction between the other two color filter stripes. During operation, the voltage between the two fingers is changed on every other scan line so that alternately one set of fingers and then the other is more positive.

The signal output from the photoconductor which is behind the signal plate fingers and filters that are more positive will be higher than the signal produced behind the low-voltage fingers. This is illustrated in Fig. 11-83. It is shown that in line 1, among other differences, the red signal is different on one side of the filter than on the other. This situation is reversed on line $n + 1$ since the voltages on the signal-plate fingers are reversed.

The signals are processed as shown in Fig. 11-84. The outputs from both signal plates are coupled into the same amplifier through a small transformer. The signal is delayed for exactly one horizontal scan-line interval. The active and the delayed scan line are then added, as well as subtracted (Fig. 11-84). The sum of the two produces the proper video output corresponding to the light and color filter transmission. When the signals are subtracted, the video signals cancel and a properly phased index signal is generated that is then used in the synchronous detector to key the sequential color signals into the proper color channel.

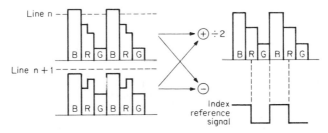

FIG. 11-83 Alternate-line signal output from Trinicon tube.

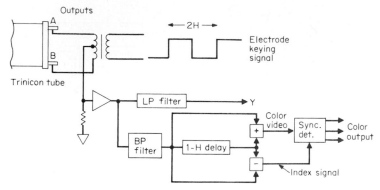

FIG. 11-84   Trinicon tube and signal-processing block diagram.

11.8 SOLID-STATE IMAGER DEVELOPMENT

The technological developments for fabricating and manufacturing solid-state imaging arrays with the objective of achieving performance which is competitive with that of camera tubes was started 20 years ago. The earliest devices to employ self-scanning were arrays of thin-film transistors (TFTs).[4] Since then the extensive activity directed toward the perfection of large-scale integrated circuits using silicon technology has provided a rapidly expanding base of manufacturing technology and facilities which can be applied to imagers. In the United States, the complexity and high production cost of solid-state sensor arrays initially limited their use to applications employing single-line scanners and small-area arrays. The limited number of large-scale arrays produced by Fairchild, RCA, and Texas Instruments were most widely used in scientific applications where radiometric and geometric accuracy in a video output was of particular importance.

With the rapidly expanding markets for closed-circuit television systems and home video recorders, several Japanese manufacturers have made an intensive effort to develop a single-sensor solid-state color camera, comparable in performance with counterparts using camera tubes.

11.8.1 EARLY IMAGER DEVICES.   The first solid-state imagers consisted of two-dimensional arrays of photosensing diodes. The operating potentials were provided by x and y bus bars, and the signal from an individual sensor was sampled by activating on-chip switching circuits as shown schematically in Fig. 11-85. An imager of this type having a checkerboard of color filters is used in a VCR color camera introduced by Hitachi in 1981.[5]

Various design concepts were developed for an on-chip output stage to detect the signal from successive pixels as required for a video output. At this time, the most successful means for eliminating the characteristic fixed-pattern noise of the x-y addressed devices is that included in the Hitachi camera.

A more complex type of x-y addressed imager called a *charge injection device* (CID) is being marketed by the General Electric Co.[6] The sensing element of this architecture consists of an isolated pair of MOS capacitors. As the integrated photocharge is shifted from one capacitor to the other, its amplitude can be sensed by one of the bus bar lines. This nondestructive readout provides a unique capability for multiple readout of one frame of data. Erasure is achieved by dumping the charge into the substrate.

11.8.2 IMPROVEMENTS IN SIGNAL-TO-NOISE RATIO.   The introduction of internal charge transfer concepts such as the bucket brigade[7] and the charge-coupled device[8] (CCD) opened the way to the development of area sensors with improved signal-

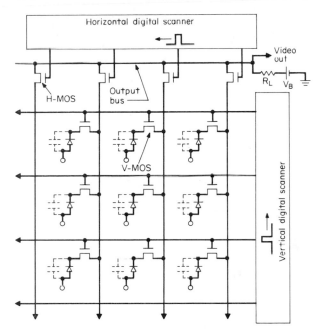

FIG. 11-85 A schematic diagram of an *x-y* addressed imager with an MOS switching transistor at each pixel.

to-noise performance at low light levels. This results from the capability to move the packets of photocharge from the sensor site to the output stage by controlled transfer through the silicon substrate. The effective output capacitance can be less than 1 pF, providing an rms noise per pixel of 50 electrons or less from an imager having a full pixel charge of 2×10^5 electrons. This was accomplished with an off-chip signal processor employing correlated double sampling that eliminates the reset noise of the charge-sensing floating diffusion and the $1/f$ noise of the output transistors.

11.8.3 CCD STRUCTURES

Frame-Transfer Structure. There are two basic structures for CCD imagers. The *frame-transfer* structure shown in Fig. 11-86a is one. The photocharge is generated by the illumination incident on the image register. At the end of the exposure interval (⅟₆₀ s for United States standards) this entire charge pattern is transferred in parallel through vertical CCD columns into the storage register, thereby freeing the image register to begin the integration of the next field. During this same time interval the charge packets in the storage register are transferred one row at a time into the output register through which they are transferred serially to the output stage. When all rows of the storage register have been read out, this register is ready to accept the next parallel transfer from the image register, and the entire operating cycle repeats.

Interline-Transfer Structure. Shown in Fig. 11-86b is an interline-transfer structure, the second basic structure for CCD imagers, in which columns of photosensor elements alternate with CCD transfer registers. During the optical exposure, charge is generated in the photosites. At the completion of the exposure/integration period all packets are shifted into the neighboring stage of the adjacent CCD column. These charge transfer

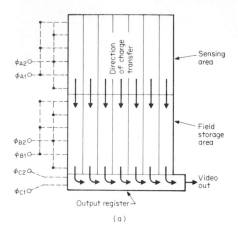

(a)

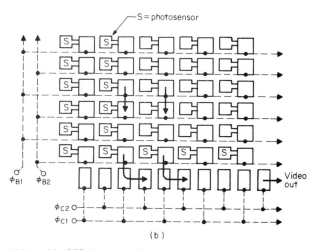

(b)

FIG. 11-86   CCD imager architectures: (a) frame transfer structure; (b) interline transfer structure.

columns are the equivalent of the storage register of the frame transfer structure. After all rows of signal charge have been shifted into the output register and transferred to the output stage, the operating sequence is repeated.

During the early stage of CCD-imager development, when the transferred charges were moved along the surface of the silicon substrate adjacent to the gate structure, there was a significant amount of charge left behind as the result of surface state trapping. The smaller number of transfers, required by the interline-transfer structure was an attractive feature. The subsequent development of buried channel structures in which the movement of charge takes place in an n-doped channel formed on a p-doped substrate eliminates surface trapping losses and provides a significant increase in charge transfer efficiency.

Because the charge-transfer columns of the interline-transfer structure must be cov-

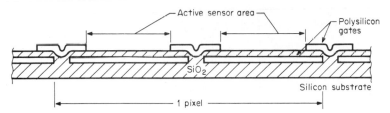

FIG. 11-87   A cross section of a typical multilevel polysilicon gate structure used in CCDs.

ered by an opaque mask, only a part of the area illuminated by the light is photosensitive. This reduces the area utilization of the incident photon flux to the order of 30 to 50 percent. Similarly, frame transfer imager structures which are illuminated on the structured side waste incident radiation through absorption in the multilevel gate structure such as that diagrammed in Fig. 11-87. Only those photons which are absorbed in the substrate are capable of generating photocharge. A variety of gate structures with transparent windows have been devised to increase device sensitivity.

11.8.4   NEW DEVELOPMENTS.   The introduction by RCA of rear-illuminated thin substrate devices has made available frame-transfer imagers with full-area utilization of the incident light. By assuring that the depletion region extends across the full thickness of the 10-μm-thick substrate, quantum efficiency comparable with the silicon-target vidicon is obtained. Figure 11-88 shows the reported quantum efficiency as a function of wavelength for several different CCD imagers.

The latest generation of solid-state imagers has 10- to 20-μm wide pixels which are 15 to 30 μm long. Horizontally there are between 300 and 500 elements and, with interlace, 500 elements vertically. Because the modulation transfer function remains high out to the Nyquist limit, this number of pixels gives acceptable dynamic image detail for most applications. The illumination-to-signal transfer function is close to linear over the full dynamic range which typically exceeds 10^3. Not unexpectedly, these new and complex structures have certain undesirable characteristics not encountered in camera tubes. While these new devices are free of image lag, even at low light levels, they are subject to image blooming when exposed to a light overload. This has been overcome by adding complexity to the pixel structure.

There is an additional problem of transfer smear which is the result of spurious charge generated in high light regions of the scene contaminating the signal from low-light regions. Since this effect is limited to the direction of charge transfer, it tends to produce low-contrast vertical bands in the video output. Again, this defect has been reduced or eliminated by adding further complexity to either the device architecture or the camera.

By taking advantage of the continuously improving quality of silicon wafers and the better understanding of processing constraints which must be respected, there has been continuous improvement in the yield of cosmetically acceptable devices over the past 10 years. The density of both high dark current and low sensitivity pixels is directly related to the processing cycle used to fabricate the device. A high density of crystalline anomalies which are potential defect sites is present in the best quality silicon wafers. The number which become electrically active is controlled by the wafer processing both before and during device manufacture.

In this brief review of solid-state imager development all discussion of device physics has been omitted. Listed in the references are several publications dealing in detail with this aspect of an ongoing technological development.[2,6,9,10] A 1983 review of the reported performance for the large number of architectures which have been devised and fabricated either as laboratory or manufactured products is given in Ref. 11.

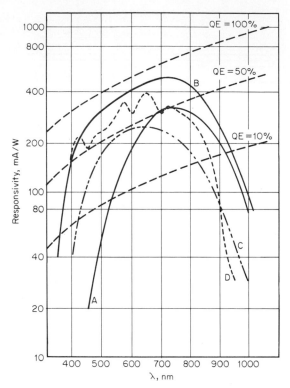

FIG. 11-88 Responsivity-vs.-wavelength curves for image sensors employing MOS-capacitor sensing elements. Curve *A*: front-illuminated CCD imager with single-level polysilicon gates. Curve *B*: rear-illuminated thinned, glass-laminated CCD imager. Curve *C*: front-illuminated virtual-phase CCD imager. Curve *D*: front-illuminated CCD imager with transparent second level gates. Dashed lines indicate, at different wavelengths, the varying relationship between responsivity and quantum conversion efficiency (QE) at levels of 10, 50, and 100%.

REFERENCES

1. D. G. Fink (ed.), *Television Engineering Handbook,* McGraw-Hill, New York, 1957.

2. D. Barbe and S. Campara, "Imaging Arrays Using the Charge-Coupled Concept," in *Advances in Image Pickup and Display,* vol. 3, B. Kazan (ed.), Academic, New York, 1977, pp. 172–296.

3. L. D. Miller, "A New Method of Specifying the Resolution Power of Television Camera Tubes Using the RCA P30C Test Chart," *J. SMPTE,* vol. 89, 1980, pp. 249–256.

4. P. Weimer, "Image Sensors for Solid State Cameras," in *Adv. Electron. Electron Phys.,* vol. 37, 1975, pp. 181–262.

5. S. Ohba, et al., "MOS Area Sensor: Pt. II—Low Noise MOS Area Sensor with Anti-blooming Photodiodes," *IEEE Trans. Electron Devices,* vol. ED-27, 1980, pp. 1682–1687.

6. G. Michon and H. Burke, "CID Image Sensing," in *Topics in Applied Physics,* vol. 28, D. Barbe (ed.), Springer-Verlag, Berlin, 1979, chap. 2, pp. 5–24.

7. F. Sangster and K. Teer, "Bucket-Brigade Electronics," *IEEE J. Solid-State Circuits,* vol. SC-4, 1969, pp. 131–136.

8. W. Boyle and G. Smith, "Charge Coupled Devices—A New Approach to MIS Structures," *IEEE Spectrum,* July 1971, pp. 18–27.

9. C. Séquin and M. Tomsett, "Charge Coupled Devices," *Adv. Electron. Electron Phys.,* suppl. 8, 1975.

10. W. Kosonocky and J. Carnes, "Basic Concepts of CCDs," *RCA Rev.,* vol. 36, 1975, pp. 566–593.

11. P. Weimer and D. Cope, "Image Sensors for Television and Related Applications," in *Advances in Image Pickup and Display,* vol. 6, B. Kazan (ed.), Academic, 1983, pp. 177–252.

BIBLIOGRAPHY

Photosensitivity

Engstrom, R. W., "Absolute Spectral Response Characteristics of Photosensitive Devices," *RCA Rev.,* vol. 21, 1960, pp. 184–190.

Kazan, B., and M. Knoll, *"Electronic Image Storage,"* Academic, New York, 1968.

McConnell, R. A., "Video Storage by Secondary Emission from Simple Mosaic," *Proc IRE,* vol. 35, November 1947, pp. 58–64.

RCA Electro-Optics Handbook, no. EOH-11, RCA Solid State Division, Sommerville, N.J., 1974.

RCA Photomultiplier Handbook, no. PMT62, RCA Solid State Division Electro-Optics, Lancaster, Pa., 1980.

Rose, A., *Vision—Human and Electronics,* Plenum, New York, 1973.

Sommer, A. H., *"Photoemissive Materials,"* Wiley, New York, 1968.

Wright, D. A., *"Semiconductors,"* Methuen, New York, 1955.

Zworykin, V. K., and E. G. Romberg, *Photoelectricity and Its Application,* Wiley, New York, 1949.

Photoelectric-Induced Television Signal Generation

Alexander, J. W. F., and R. B. Burtt, "Bombardment Induced Conductivity Targets for Image Orthicons," *Adv. Electron Physics,* vol. 16, 1962, pp. 247–248.

Boerio, A. H., R. R. Beyer, and G. W. Goetze, "The SEC Target," *Adv. Electron. Electron Phys.,* vol. 22A, 1960, pp. 229–234.

Bube, R. H., *"Photoconductivity in Solids,"* Wiley, New York, 1960.

Filby, R. S., S. B. Mende, and N. D. Tridley, "A Television Camera Tube Using Low Density Potassium Chloride Target," *Adv. Electron. Electron Phys.,* vol. 22A, 1966, pp. 273–290.

Iams, H., and A. Rose, "Television Pickup with Cathode Ray Scanning Beam," *Proc. IRE,* vol. 25, 1937, pp. 1048–1070.

Isozuki, Y., "2 Inch Return Beam Saticon—A High Resolution Camera Tube," *J. SMPTE,* vol. 28, no. 8, M1978, pp. 489–493.

Rose, A., "An Outline of Some Photoconductive Image Processes," *RCA Rev.,* vol. 12, 1951, p. 362.

Rose, A., and H. Iams, "Television Pickup Tubes Using Low Velocity Beam Scanning," *Proc. IRE*, vol. 27, 1939, pp. 517–555.

Schade, O. H., "Electron Optics and Signal Readout of High Definition Return Beam Vidicon Cameras, *RCA Rev.*, vol. 31, 1970, p. 60.

Simon, R. E., and B. F. Williams, "Secondary-Electron Emission," *IEEE Trans., Nuclear Science*, vol. NS15, 1968, pp. 166–170.

Sommer, A. H., "Relationship between Photoelectronic and Secondary Emission; with Special Reference to Ay-O-Cs (S-I) Photocathode," *J. Appl. Phys.*, vol. 42, 1971, pp. 567–569.

Zworykin, V. K., "The Iconoscope," *Proc. IRE*, vol. 22, 1934, p. 16.

Evolution and Development of Camera Tubes

Cope, J. E., L. W. Germany, and R. Theile, "Improvements in Design and Operation of Image Iconoscope Type Camera Tubes," *J. Br. IRE*, vol. 12, 1952, pp. 139–144.

Farnsworth, P. J., "Television by Electron Image Scanning," *J. Franklin Institute*, vol. 218, 1934, p. 411.

Gibbons, D. J., "The Tri-Alkali Stabilized CPS Emitron, a New Television Camera Tube of High Sensitivity," *Adv. Electron. Phys.*, vol 2, 1960, pp. 203–218.

Iams, H., G. A. Norton, and V. K. Zworykin, "The Image Iconoscope," *Proc. IRE*, vol. 27, 1934, pp. 541–547.

Janes, R. B., R. E. Johnson, and R. S. Moore, "Development and Performance of Television Camera Tubes," *RCA Rev.*, vol. 10, 1949, pp. 191–223.

Kazan, B., and M. Knoll, *Electronic Image Storage*, Academic, New York, 1968.

Knoll, M., and F. Schroter, "Electronic Picture Transmission by Insulating or Semiconductor Layers," *Physik*, vol. K38, 1932, pp. 330–333.

McKee, J. D., and H. K. Lubszynski, "EMI Cathode-Ray Television Transmission Tubes," *J. Inst. Elect. Engr.*, vol. 84, 1939, pp. 468–482.

Rose, A., and H. Iams, "The Orthicon, a Television Pickup Tube," *RCA Rev.*, vol. 4, 1939, pp. 186–199.

Schagen, P., H. Bruining, and J. C. Franken, "The Image Iconoscope, a Camera Tube for Television," *Philips Tech. Rev.*, vol. 13, 1961, pp. 119–133.

Weimer, P. K., "A Historical Review of the Development of Television Pickup Devices, 1930–1976," *IEEE Trans. Electron Devices*, vol. ED-23, no. 7, 1976.

Zworykin, V. K., G. A. Morton, and L. E. Flory, "Theory and Performance of the Iconoscope," *Proc. IRE*, vol. 25, 1937, pp. 1071–1092.

Zworykin, V. K., and G. A. Morton, *Television*, 2d ed., Wiley, New York, 1954.

Image Orthicon Tubes

Banks, G. B., K. Frank, and E. E. Hendry, "Image Orthicon Camera Tube," *J. Television Soc.*, vol. 7, 1953, pp. 92–104.

Janes, R. B., R. E. Johnson, and R. R. Handel, "A New Image Orthicon," *RCA Rev.*, vol. 10, 1944, pp. 586–592.

Janes, R. B., and A. A. Rotow, "Light Transfer Characteristics of Image Orthicons," *RCA Rev.*, vol. 11, 1950, pp. 364–376.

Morton, G. A., and J. E. Ruedy, "The Low Light Level Performance of the Intensifier Orthicon," *Adv. Electron. Electron Physics*, vol. 12, 1960, pp. 183–195.

Neuhauser, R. G., "Picture Characteristics of Image Orthicon and Vidicon Camera Tubes," *J. SMPTE*, vol. 70, 1961, pp. 696–698.

Neuhauser, R. G., A. A. Rotow, and F. S. Veith, "Image Orthicons for Color Cameras," *Proc. IRE,* vol. 42, 1954, pp. 161–165.

Rose, A., P. K. Weimer, and H. B. Law, "The Image Orthicon, a Sensitive Television Pickup Tube," *Proc. IRE,* vol. 34, 1946, pp. 424–432.

Rotow, A. A., and F. D. Marschka, "Image Intensifier Orthicon," *Proc. Image Intensifier Symposium,* Ft. Belvoir, Va., October, 1958.

Vine, B. H., "Analysis of Noise in the Image Orthicon," *J. SMPTE,* vol. 70, 1961, pp. 432–435.

Image Iconoscope and Image Isocon

Cope, A. D., "The Isocon, a Low Noise Wide Signal Range Camera Tube," *Proc. Image Intensifier Symposium,* Ft. Belvoir, Va., October, 1961.

VanAsselt, R. L., "The Image Isocon as a Studio, X-Ray and Low Light Camera Tube," *NEC Proc.,* 1968.

Weimer, P. K., "The Image Iconoscope, an Experimental Television Pickup Tube Based on Scattering at Low Velocity Electrons," *RCA Rev.,* vol. 10, 1944, pp. 366–386.

Low-Light Tubes and Devices

Engstrom, R. W., and R. L. Rodgers III, "Camera Tubes for Night Vision," *Optical Spectra,* February, 1971, pp. 26–30.

Robinson, G. A., "The Silicon Intensifier Target Tube—Seeing in the Dark," *J. SMPTE,* vol. 86, 1977, pp. 414–418.

Vidicon-Type Camera Tubes

Crowell, M. W., et al., "The Silicon Diode Array Target," *Bell System Tech. J.,* vol. 6, 1967, p. 491.

De Haan, E. F., F. M. Klassen, "The Plumbicon, a Camera Tube with Photoconductive Lead Oxide Layers," *J. SMPTE,* vol. 37, 1964.

De Haan, E. R., F. M. Klassen, and P. P. Schampers, "An Experimental Plumbicon Camera Tube with Increased Sensitivity to Red Light," *Philips Tech. Rev.,* vol. 26, 1965, pp. 49–51.

De Haan, E. F., A. VanderDrift, and P. P. M. Schampers, "The Plumbicon, a New Television Camera Tube," *Philips Tech. Rev.,* vol. 25, 1964, pp. 130–180.

Forgue, S. V., R. R. Goodrich, and A. D. Cope, "Properties of Some Photoconductors, Principally Antimony Trisulphide," *RCA Rev.,* vol. 12, 1951, pp. 335–349.

Fujiwara, S., et al., "High Sensitivity Camera Tube—Newvicon," *Natl. Tech. Rep. (Japan),* vol. 25, no. 2, April 1974, pp. 286–297.

Goto, N., "Saticon, a New Photoconductive Camera Tube with Se-As-Te Target," *IEEE Trans. Electron Devices,* vol. ED-21, 1971, p. 662.

Goto, N., Y. Isozaki, and K. Shidara, "New Photoconductive Camera Tube, Saticon," *NHK Laboratory,* Note N, 171, NHK Research Lab., Tokyo, 1973.

Heijne, L. J., P. Schagen, and H. Bruining, "An Experimental Photoconductive Camera Tube for Television," *Philips Tech. Rev.,* vol. 16, 1954, pp. 23–25.

Heine, L. J., "Physical Principles of Photoconductivity," *Philips Tech. Rev.,* vol. 25, 1963, pp. 120–131.

Neuhauser, R. G., "The Silicon Target Vidicon," *J. SMPTE,* vol. 86, 1977, pp. 408–413.

Neuhauser, R. G., "The Saticon Color Television Camera Tube," *J. SMPTE,* vol. 87, 1978, pp. 147–152.

Neuhauser, R. G., "Vidicon for Film Pickup," *J. SMPTE,* vol. 62, 1954, pp. 142–152.

Neuhauser, R. G., "Camera Tubes for Color Broadcast Service," *J. SMPTE,* vol. 65, 1956, pp. 626–642.

Rodgers, R. L., III, "Beam Scanned Silicon Target for Camera Tubes," *IEEE Intercom,* March 1973.

Schichiri, S., "18 mm Electrostatic Focusing Camera Tube PE20PE14," *Toshiba Rev.,* vol. 117, September, October 1978.

Shimizu, K., and Y. Kiuchi., "Characteristics of the New Vidicon Type Camera Tube Using Cd Se as a Target Material," *Japan J. Appl. Phys.,* vol. 6, 1967, pp. 1089–1095.

Vine, B. H., R. B. Janes, and F. S. Veith, "Performance of the Vidicon, a Small Developmental Television Camera Tube," *RCA Rev.,* vol. 13, 1952, pp. 3–10.

Weimer, P. K., and A. D. Cope, "Photoconductivity in Amorphous Selenium," *RCA Rev.,* vol. 12, 1951, pp. 314–334.

Weimer, P. K., S. V. Forgue, and R. R. Goodrich, "The Vidicon Photoconductive Camera Tube," *Electronics,* vol. 23, May 1950, pp. 70–73.

Yoshida, O., "Recent Chalnicon Developments," *Proc. 7th Symp. Electronic Imaging Devices,* September 1978, Imperial College, London.

Yoshida, O., and K. Shimizo, "New Camera Tubes for Color Television," *J. SMPTE,* 1975, vol. 84.

Interface with the Camera

Kuehne, J. E., and R. G. Neuhauser, "An Electrostatically Focused Vidicon," *J. SMPTE,* vol. 71, 1962, pp. 772–775.

Kumbuta, Y., et al., "A New Magnetic-Focus Static Deflection System for Single Tube Color Camera," *IEEE Trans. Consumer Electronics,* vol. CE24, no. 1, 1978, pp. 114–119.

Miller, L. D., and B. H. Vine, "Improved Developmental One Inch Vidicon for Television Cameras," *J. SMPTE,* vol. 67, 1955, pp. 154–156.

Neuhauser, R. G., "Developments in Electro Optics Produce Two New Lines of Vidicon Tubes," *RCA Engineer,* no. 10, 1964, pp. 16–20.

Neuhauser, R. G., and L. D. Miller, "Beam Landing Errors and Signal Output Uniformity of Vidicons," *J. SMPTE,* vol. 67, 1958, pp. 149–153.

Sadashige, K. "A Study of Noise in Television Camera Preamplifiers," *J. SMPTE,* vol. 73, 1964, pp. 202–206.

Schlesinger, K., and R. A. Wagner, "A Mixed-Field Type of Vidicon," *IEEE Trans. Electron Devices,* vol. ED-14, 1967, pp. 163–170.

VanRoosmalen, J. H. T., "Adjustable Saturation in a Pickup Tube with Linear Light Transfer Characteristics," *4th Symp. Photoelectronic Image Devices,* Imperial College London, 1965.

Performance Characteristics

Clark Jones, R., "On the Defective Quantum Efficiency of Television Camera Tubes," *J. SMPTE,* vol. 68, 1959, pp. 462–466.

Neuhauser, R. G., "Sensitivity and Motion Capturing Ability of Television Camera Tubes," *J. SMPTE,* vol. 68, 1959, pp. 455–461.

Resolution

Miller, L. D., "A New Method of Specifying the Resolution Power of Television Camera Tubes Using the RCA P30C Test Chart," *J. SMPTE,* vol. 89, 1980, pp. 249–256.

Neuhauser, R. G., "Measuring Camera Tube Resolution with the RCA P200 Test Chart," *J. SMPTE,* vol. 89, 1980, pp. 249–256.

RCA Electro Optics Handbook, no. EOH-11, RCA Solid State Div., Sommerville, N.J.

Schade, O. H., "Image Graduation Graininess and Sharpness in Television and Motion Picture Systems," *J. SMPTE,* pt. 1, vol. 56, 1951, pp. 132–177; pt. II, vol. 58, 1952, pp. 181–222; pt. III, vol. 61, 1953, pp. 97–164; pt. IV, vol. 64, 1956, pp. 543–617.

Schade, O. H., "An Evaluation of Photographic Image Quality and Resolving Power," *J. SMPTE,* vol. 73, 1964, pp. 81–119.

Lag Reduction

Ehata, S., "Low-Lag Gun Saticon H8397, Its Design and Characteristics," *Hitachi Rev.,* vol. 27, no. 3, 1978.

Franken, A. A. J., "Lag and Light Bias in Lead Oxide Tubes," *Broadcast Systems and Operations,* May 1979, p. 164.

Franken, A. A. J., "Miniature Plumbicon Tube for Portable Cameras," *Electronic Components and Applications,* vol. 3, no. 3, N. V. Philips, Eindhover, Netherlands, 1981.

Meltzer, B., and P. L. Holmes, *British Applied Physics,* vol. 9, 1958, pp. 139–148.

Neuhauser, R. G., "Lag Reduction and Lag Characteristics of Television Camera Tube Signals," *J. SMPTE, Television Technology, 80's,* 1981.

VanRoosmalen, J. H. T., "A New Concept for Television Camera Tubes," *Philips Technical Rev.,* vol. 39, no. 8, 1980, pp. 201–209.

Single-Tube Color Camera Systems

Nishimura, S., and V. Tomii, "Single Tube Color Camera Using Tri-Electrode Vidicon," *Natl. Tech. Rept. (Japan),* vol. 22, no. 5, October 1976, pp. 591–600.

Nobutoki, S., "A Color Separation Filter Integrated Vidicon for Frequency Multiplex System Single Pickup Color Television Camera," *IEEE Trans. Electron Devices,* vol. ED-18, no. 11, 1971.

Pritchard, D. H., and G. L. Fredendall, "Stripe Color Encoded Single Tube Color-Television Camera Systems, Filter Colorimetry for Single Tube Color Camera," *RCA Rev.,* no. 34 1973, pp. 217–275.

See, L., "New Single Tube Eng. Camera Enters the Marketplace," *Broadcast Communications,* vol. 4, no. 12, 1981.

Weimer, P. K., "A Developmental Tri-Color Vidicon Having a Multiple-Electrode Target," *IRE Trans. Electron Devices,* vol. ED-7, 1960, pp. 147–153.

Solid State Imager Development

Barbe, D., and S. Campara, "Imaging Arrays Using the Charge-Coupled Concept," in *Advances in Image Pickup and Display,* vol. 3, B. Kazan (ed.), Academic, New York, 1977, pp. 172–296

Boyle, W., and G. Smith, "Charge Coupled Devices—A New Approach to MIS Structures," *IEEE Spectrum,* July 1971, pp. 18–27.

Kosonocky, W., and J. Carnes, "Basic Concepts of CCDs," *RCA Rev.,* vol. 36, 1975, pp. 566–593.

Kosonocky, W., H. Elabd, and M. Cantella, "Infrared Imaging Systems Using CCD's," *Microstructure Science and Engineering,* vol. 4, N. Einspruch (ed.), Academic, New York, 1982, chap. 6, pp. 227–295.

Michon, G., and H. Burke, "CID Image Sensing," in *Topics in Applied Physics,* D. Barbe (ed.), Springer-Verlag, Berlin, 1979, vol. 28, ch. 2, pp. 5–24.

Ohba, S., "MOS Area Sensor: pt. II—Low Noise MOS Area Sensor with Anti-blooming Photodiodes," *IEEE Trans. Electron Devices,* vol. ED-27, 1980, pp. 1682–1687.

Sangster, F., and K. Teer, "Bucket-Brigade Electronics," *IEEE J. Solid-State Circuits,* vol. SC-4, 1969, pp. 131–136.

Séquin, C., and M. Tomsett, "Charge Coupled Devices," *Adv. Electron. Electron Phys.,* suppl. 8, 1975.

Weimer, P. "Image Sensors for Solid State Cameras," *Adv. Electron. Electron Phys.,* vol. 37, 1975, pp. 181–262.

Weimer, P., and D. Cope, "Image Sensors for Television and Related Applications," in *Advances in Image Pickup and Display,* vol. 6, B. Kazan (ed.), Academic, N.Y., 1983, pp. 177–252.

Monochrome and Color Image-Display Devices

Donald L. Say
R. A. Hedler
L. L. Maninger
R. A. Momberger
J. D. Robbins
Philips ECG
Seneca Falls, New York

12.1 VIDEO DISPLAY DEVICES

Video displays using present–day technology can be classified into six areas as follows:

1. Cathode-ray tube, direct view, color
2. Large-area displays projected from a cathode-ray-tube source, color
3. Large-area displays from discreet elements, monochrome or color
4. Cathode-ray tube, direct view, monochrome
5. Matrixed flat displays, monochrome
6. Flat devices, modified electron-beam type, monochrome

In the past 25 years remarkable improvements and simplifications have taken place in shadow-mask cathode-ray-tube (CRT) technology. With these improvements the shadow-mask tube continues to be the dominant video display device. It will be described in some detail in Secs. 12.2 and 12.3.

Large-area displays, both cathode-ray-tube projection and discrete elements, are described in Sec. 12.4. Monochrome cathode-ray tubes along with matrixed flat displays and modified electron-beam flat displays are described in Sec. 12.5. Phosphors, brightness, and contrast are covered in Sec. 12.6 and in Sec. 12.7 a table of the more widely used color CRTs is presented along with a description of the new worldwide designation system for CRTs.

Throughout the chapter techniques to achieve the higher resolution needed for high-definition satellite and cable television systems of the future are noted.

12.2 PRINCIPLES OF COLOR PICTURE-TUBE DESIGN

12.2.1 COLOR SELECTION. Many schemes for cathode-ray-tube color selection were investigated early in color television history.[1,2]† Most were abandoned; five of the more important approaches are shown in Fig. 12-1.

Three-beam grill focusing (Fig. 12-1a) offers the advantage of very high brightness,[3,4] i.e., three beams with very little mask intercept. One-beam grill focusing (Fig. 12-1b) offers less brightness, with the simplicity of single-beam operation.[5,6] Both of the grill focusing schemes, however, suffer from one fundamental problem; i.e., secondary electrons generated at the grill and screen are all attracted without color selection to the high voltage of the screen where they reduce the contrast of the display. A further problem with grill focusing involves the screening process. That is, with the absence of an optical shadowing device, such as a shadow mask, screening requires the use of printing masters, and interchangeable parts—a formidable task.[7] For these reasons focus-grill tubes are no longer being built, even though much development effort was spent by a number of companies on focus-grill designs.

A large-hole focus mask (Fig. 12-1c,) with the mask at a fraction of screen potential, will also focus the beamlets to permit increased mask transmission and screen brightness, but again at the expense of reduced contrast.

Figure 12-1d depicts an "index" color selection arrangement. In this scheme a periodic index signal generated by the beam, from either secondary electrons or a fast-response light-emitting phosphor, makes it possible to monitor the position of a single beam and control which screen color is being excited at any instant. Much of the early work in color followed this approach.[8–10] Two major problems existed. First, high-speed

---

†Superscript numbers refer to References at end of chapter.

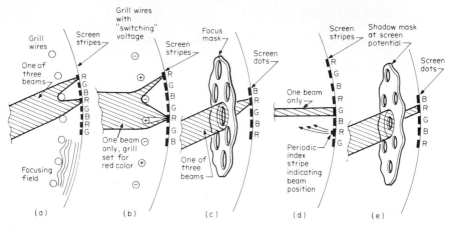

FIG. 12-1 Color selection schemes (plan view). (*a*) Three-beam focus grill. Wires at 30 to 40 percent of screen voltage. (*b*) One-beam focus grill. Wires at 30 to 40 percent of screen voltage, with 300- to 500-V color "switching" voltage between alternate wires. (*c*) Focus mask. Three beams. Mask at 30 to 40 percent of screen voltage. (*d*) Index scheme. Electron beam smaller than phosphor stripe. Periodic index signal (secondary electrons or light) indicating beam position. (*e*) Shadow mask at screen voltage. No focusing. Three beams. Can consist of holes and dots as shown, or slots and stripes. *Note:* In (*a*), (*c*), and (*e*) two additional beams approach at other angles to excite the other two colors.

extensive circuitry was needed to monitor the position of the beam and properly control it. Second, the beam itself must be no wider than a phosphor stripe, even when deflected over the full area of the picture tube. With the advent of integrated circuits and solid-state photo detectors, the first problem has become much simpler. The problem of beam width, however, persists. A recent report[11] indicates that beam width feasibility has been demonstrated in an envelope as large as 30-in (76-cm) diagonal.

A more important reason for the emergence of the shadow-mask design, with mask and screen at the same potential (Fig. 12-1*e*,) as the dominant color selection scheme is the significant improvement in brightness and contrast of color screens that has been accomplished in recent years. These improvements are discussed in Sec. 12-6. The higher brightness of Fig. 12-1*a* to *c*, with the resulting complications, is no longer needed.

12.2.2 GEOMETRY OF SHADOW-MASK TUBES.

Figure 12-2 illustrates the shadow-mask geometry at face center for in-line guns and a shadow mask of round holes (or of vertical slots).[12–14] For round holes, the screen is as shown in detail view; for slots, the screen is continuous vertical stripes (Fig. 12-3). The three guns and their undeflected beams lie in the horizontal plane. The beams are shown here converged at mask surface. The beams may overlap more than one hole and the holes are encountered only as they happen to fall in the scan line. By convention, a beam in the figure is represented by a single straight line projected backward at the incident angle from an aperture to an apparent "center of deflection" located in the "deflection plane."

In Fig. 12-2, the points B', G', and R', lying in the deflection plane, represent such apparent centers of deflections for blue, green, and red beams striking an aperture under study. (These deflection centers move with varying deflection angles.) Extending the rays forward to the facepanel denotes the printing location for the respective colored dots (or stripes) of a tricolor group. Thus, *centers of deflection* become *color centers* with a spacing S in the deflection plane. The distance S projects in the ratio Q/P as the dot spacing within the trio. The diagram also shows how the mask hole horizontal pitch b projects as screen horizontal pitch in the ratio L/P. The same ratio applies for projection of mask vertical pitch a. The *Q-space* (mask to panel spacing) is optimized to obtain the largest

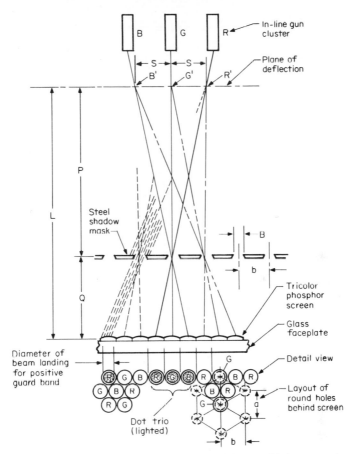

FIG. 12-2 Shadow mask with in-line guns. Mask has round holes as per view (or vertical slots with stripe sequence as per top row of dots).

possible theoretical dots without overlap. At panel center, the ideal screen geometry is then a mosaic of equally spaced dots (or stripes).

The stripe screen of Fig. 12-3 with the in-line gun geometry of Fig. 12-2 plan view is used in most of today's television tubes. One variation of this stripe (or line) screen uses a cylindrical faceplate with a vertically tensioned grill shadow mask without tie bars.[15] Prior to the stripe screen, the standard construction was a tri-dot screen with delta gun cluster as in Fig. 12-4.

12.2.3 POSITIVE AND NEGATIVE GUARD BAND. Chapter 10 discussed color purity and beam registration with target phosphor dots. One feature for aiding purity is *guard band,* where the lighted area is smaller than the theoretical tangency condition.[16,17] In Fig. 12-2, the leftmost red phosphor exemplifies positive guard band; the lighted area is smaller than the actual phosphor segment, accomplished by mask hole diameter design. Figure 12-3, on the other hand, shows *negative guard band* (NGB) (or window-limited) construction for stripe screens. Vertical black stripes about 0.1 mm (0.004 in) wide separate the phosphor stripes forming "windows" to be lighted by a beam wider than the window opening by about 0.1 mm. Figure 12-4 shows NGB construction of a tri-dot screen. Black matrix advantages in tube performance are discussed in Sec. 12.6.

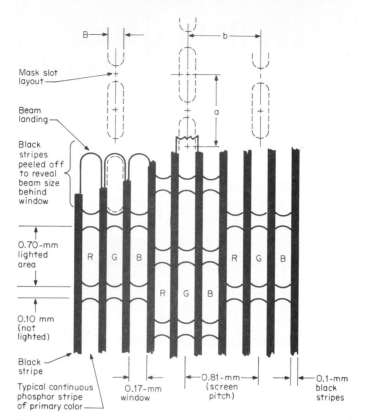

FIG. 12-3 Negative guard band stripe screen with slot mask and in-line guns. Mask slots concentric with green beam landings.

12.2.4 DETAILS OF SHADOW MASKS. The shadow mask is a 0.13-mm-thick low-carbon sheet steel chemically etched[18] to the desired pattern of apertures in the flat form using photoresist techniques. The photographic masters are usually made by a precision laser plotter. The completed flat mask is then press-formed to a contour more or less concentric to the faceplate (or panel) except that mask-to-panel distance (Q-spacing) is modified locally to achieve optimum nesting of screen triplets.

Round-Hole Masks. The array layout for a typical round-hole shadow mask is shown in Figs. 12-2 and 12-4. The round holes, numbering about 400,000, are placed at the vertices and centers of iterated regular hexagons (long axes vertical).[19] If b is the spacing between hole centers in a horizontal row and a is the spacing between holes in a vertical column, then $b/a = \sqrt{3} = 1.732$ is the ratio of horizontal pitch to vertical pitch. This ratio applies both to mask holes and screen triads. When references are made to mask or screen pitch without stating direction, the meaning is the closest center-line spacing between like elements—i.e., vertical pitch for tri-dot screens, horizontal pitch for stripe screens. As shown in Fig. 12-2, each aperture is tapered to present a more sharply defined limiting aperture plane to an angled incident beam, thus increasing transmission efficiency and preventing color desaturation due to electrons being scattered by sidewalls of the apertures. Apertures are graded radially to smaller diameters at screen edge because that is where the trio configuration and beam quality are less ideal and registry is more

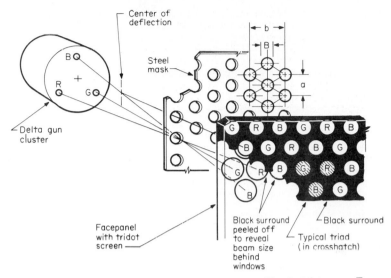

FIG. 12-4 Delta gun, round-hole mask, and negative guard band tri-dot screen. Taper on mask holes not shown.

critical. At the tube center, shadow-mask transmission is typically 16 percent for non-matrix construction and 22 percent for a *black matrix* shadow-mask construction.

The beam triad pattern may become distorted near the screen edges, resulting in poorer packing factors (geometric nesting) for the related phosphor dots. For *delta-gun* systems the beam triad triangles compress radially at all screen edges due to foreshortening. For in-line gun dot screens, on the other hand, the beam triad is three circles in a horizontal line; using compass points to designate axes and areas, there is no foreshortening at N and S, while at E and W, the mask-panel spacing can be increased enough to restore nesting quality. However, near the screen corners, there will be some rotation of the line trio, thereby demanding smaller holes, dots, and beam landings and resulting in less efficient nesting. For the NE corner the rotation direction would be clockwise.

Slot Masks. The relationship between a and b, horizontal and vertical pitch, can be as chosen by the designer. Only the horizontal pitch will affect Q-space. The factors affecting choice of horizontal pitch are resolution, practical mask-panel spacing, attainable slot widths, and manufacturability of masks and screens. The main considerations relating to choice of vertical pitch and tie bars (or bridges) are avoidance of moiré pattern, strength of tie bars, and transmission efficiency. Vertical screen stripes may be regarded as lined-up oblong dots that have been merged vertically. As a result, there is no color purity or registration requirement in the vertical direction; this simplifies the design and manufacture of tubes and receivers.

12.2.5 RESOLUTION AND MOIRÉ. *Resolution* is a measure of the definition or sharpness of detail in television image.[20] Resolution may be measured vertically and horizontally—often expressed in the equivalent number of television lines. The layout and center-line spacings of the mask apertures are designed to provide enough horizontal and vertical lines in the pattern to ensure that they are not the limiting factor in resolution, compared with the number of raster lines. In addition, the number of lines running horizontally in the layout pattern is chosen to avoid moiré fringes due to a "beat" with the scan lines. (This latter criterion is not applicable to the cylindrical grill structure of continuous slits.)

For round-hole masks, the effective number of pattern lines running horizontally, allowing for the staggered pattern, is typically about 2.25 times the number of picture horizontal scan lines; and there are about 440,000 dot trios on the screen compared with about 200,000 pixels in the raster. This applies to tube sizes 19-in (48-cm) through 25-in (63-cm) screen diagonal. For smaller tubes, the pattern is made relatively coarser to avoid excessively small apertures and to avoid moiré pattern in the display. The hexagonal pattern used to lay out the centers of round holes is favorable for increasing the number of pattern lines, reducing moiré likelihood, and for "nesting" of screen dots. The zigzag or staggering of alternate columns effectively increases the number of columns and rows as long as the pattern is significantly finer than the raster scan.

For slot apertures, the spacings between columns and resulting number of columns are chosen for resolution and cosmetic appearance. The tie bars or bridges, only about 0.13 mm (0.005 in) broad, are placed according to moiré and strength considerations and are not thought significant for resolution. Some specific examples of structure fineness are as follows from EIA-registered values for the 25VFLP22 90° color tube, with delta gun and tri-dot NGB black matrix screen.

$$\text{Nominal screen diagonal} = 25 \text{ in (63 cm)}$$

$$h = 396 \text{ mm, screen height}$$

$$w = 528 \text{ mm, screen width}$$

$$a = 0.74 \text{ mm, vertical spacing of trio centers}$$

Using the geometry of the hexagon pattern (but without correcting h or w for panel curvature)

$$N_V = \frac{2h}{a} = 1070 \text{ horizontal rows of trios in pattern}$$

$$N_H = \frac{2w}{\sqrt{3}a} = 820 \text{ vertical columns in pattern}$$

$$N = \frac{(2)\,hw}{(\sqrt{3})a^2} = 441{,}417 \text{ dot trios (published value of 439,000)}$$

The values of N_V and N_H calculated above are not actual vertical and horizontal resolution measurements, but instead are the pattern structure values in comparison with the raster. The mask vertical pitch a_m for the above structure would be about 0.70 mm, and the center-hole diameter would be about 0.5 a_m = 0.35 mm, which is smaller than the width of a television scan line. A scan line would then be expected to encompass at least a part of an aperture in both odd and even columns. Also, assuming no more than 475 visible scan lines in a 525-line picture

$$\frac{1070}{475} \left[\frac{\text{horizontal lines}}{\text{scan lines}} \right] = 2.25$$

An example for an in-line gun, NGB stripe screen is 90° color tube type 25VGDP22 which has registered data as follows.

$$\text{Nominal screen diagonal} = 25 \text{ in (63 cm)}$$

$$w = 528 \text{ mm, screen width}$$

$$b = 0.82 \text{ mm, horizontal center-line spacing of green stripes}$$

For the number of columns, we have

$$N_H = \frac{w}{b} = 640 \text{ vertical columns in pattern}$$

This number is smaller than the comparable figure for the round-hole example, but it is still adequate compared with the nominal 525- or 625-line television raster. The vertical resolution of a slot mask is the inherent value for the actual signal under test.

12.2.6 MASK-PANEL ASSEMBLY AND TEMPERATURE COMPENSATION.

The printing of screen patterns requires that the mask assembly be removed and reinserted four or five times without loss of registry. In the most common mask suspension system, the interior skirt of the glass faceplate has three or four protruding metal studs which engage springclips welded to the mask frame. Since the shadow mask intercepts more than 75 percent of the beam current and undergoes thermal expansion, misregistry would occur if thermal correction were not provided. This temperature compensation is accomplished by automatically shifting the mask slightly closer to the screen surface by means of thermal expansion of a bimetal structure incorporated into each spring support[22] or by means of a designed lever action resulting from the transverse expansion.

12.2.7 SCREENING OF FACEPLATE.

Three fields of phosphor color segments are sequentially printed on a faceplate by photoresist techniques using its own shadow mask as a master and a different lighthouse setup for each color. Each lighthouse (Fig. 12-5) has a mercury lamp for exposure and corrective optical lenses such that light rays passing through the mask land on the photoresist in a simulation of the deflection electron beam. The light intensity across the panel is controlled by means such as a graded density pattern on a glass sheet. It black matrix is used, it is applied prior to phosphor screening using photoresist methods.

The above exposure methods apply to biaxially stretched domed shadow masks; their stretch-forming variability prohibits both interchangeability and the use of a dedicated corrective mask master for exposure instead of lenses. (However, the cylindrical grill mask, which is not stretch-formed, is more amenable to modification of the above methods using special procedures.)

The screened faceplate is aluminized (described in Sec. 12.6) and then low-temperature fritted to a glass funnel using fixturing which essentially duplicates that used during lighthousing. A conductive graphite coating covers the interior wall of the funnel and also some of the exterior surface according to a prescribed pattern.

12.2.8 MAGNETIC SHIELDING.

An external or internal magnetic metal shield compensates for the effect of the earth's magnetic field and stray magnetic fields. The low-carbon steel shadow mask itself contributes to the shielding. The shield does not extend into the yoke field since this would cause *loading*. External shields must also be well clear of the anode button area. With set orientation in chosen compass direction, the shield and tube must be thoroughly *degaussed* to gain full benefit of purity.[23] This treatment not only removes any residual magnetization of the shield and shadow mask, but also induces therein a residual static magnetic field which bucks out the ambient magnetic field. A suitable degaussing coil for 120-V alternating current, 60 Hz, uses 425 turns of No. 20 enameled wire on a 305-mm-diameter form 19 mm wide. The coil should be passed

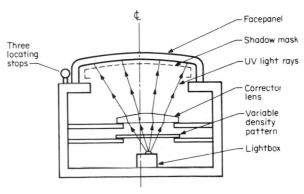

FIG. 12-5 Exposure lighthouse.

over and around the screen end of tube for a few seconds, slowly withdrawn to a point at least 2 m (6.6 ft) from the tube, and disconnected (or the coil voltage gradually reduced to zero).

Television sets offer automatic degaussing by permanently mounted degaussing coils which are briefly activated at turn-on or turn-off. These are particularly useful in portable television sets.[24]

12.2.9 CLASSIFICATION OF TUBES BY CHARACTERISTICS. Tubes may be classified in terms of bulb parameters and screen/gun geometry.

1. *Tube Size.* Conventionally, this is the screen diagonal dimension in rounded inch dimensions. This number is included in tube-type numbers. In some older conventions tube size was the bulb diagonal.
2. *Neck Diameter* (OD). Guns, yoke, neck hardware, and socketing are affected by this dimension (given in millimeters). Currently neck sizes are 36.5, 29, and 22.5 mm (1.4, 1.1, and 0.9 in).
3. *Deflection Angle.* Calculated from the rated full-screen diagonal and glassware drawings, using the yoke reference plane as an assumed center of deflection. Angles in use are 90, 100, and 110°. Higher deflection angles enable shorter tubes but entail other tradeoffs.
4. *Other Characteristics.* Some of these are *gun type* (delta or in-line); *screen structure* (stripes or dots); *black matrix* (NGB) or *nonmatrix; implosion protection;* and panel percent *transmission.*

12.2.10 ARC PROTECTION. Because of the high operating voltages applied to a CRT, internal arcing is possible. Arcs or flashovers can cause tube damage and/or circuit component failures and can give auditory or visual annoyance to viewers.[25] Corrective or protective methods include:

1. Internal cleanliness. No sharp edges or points on electrodes.
2. Subjecting the finished tube to a programmed high-voltage conditioning process called *spot knocking.*
3. Spark gaps on the tube base, socket, and diodes, capacitors, chokes, and resistors as external protection.
4. Internal resistive coating between bulb anode button and gun anode.[26] Special high resistance internal coatings in the lower neck region (but above the gun).
5. Discrete internal resistors such as incorporated into getter wand or electrode connections.[27]
6. Special arc shields in gun.

12.2.11 INTEGRAL IMPLOSION PROTECTION. The evacuated picture-tube bulb has severe forces imposed on the faceplate by atmospheric pressure. The glassmakers use finite element computer modeling and demanding stress tests to avoid weaknesses in glass design. However, the tube face in a receiver set is exposed to the possibility of an accidental blow. To protect the viewer from possible harm, stringent safety test specifications have been established by the appropriate agencies.[28] The various implosion protection systems[29] designed to meet these tests include:

1. *Laminated Safety Panel.* A glass safety panel formed and bonded to the tube faceplate with a transparent resin. All indices of refraction must be identical.
2. *Kimcode.* (Kimball Controlled Devacuation[29]). A rim band of two pieces of stamped sheet metal affixed around the skirt of the faceplate and secured by a tension band.

3. *Tension Band.* A metal tension band around the skirt of the faceplate, secured by a clip (or welds).

4. *Shelbond.* A resin-filled steel shell affixed around the skirt of the faceplate.

5. *Prewelded Tension Bands.* Tension bands expanded by heating to enable installation around the panel skirt periphery.

Mounting ears may be included in all the above systems except the laminated safety panel. With these implosion systems, breakage of the faceplate causes the tube to devacuate quietly and safely without a violent implosion.

12.2.12 X-RADIATION.
The shadow-mask color tube operates with extremely high anode voltages (typically 20 to 30 kV), and the possibility of x-rays must be considered for the safety of tube technicians and home viewers.

Color tubes are made from glasses which have been specially designed for x-ray-absorbing characteristics, and the glassmaker closely controls glass thickness and x-ray absorption. Also, the power levels at which the tube is operated must be controlled. Tube-type data sheets show the relevant absolute maximum ratings (for voltages and currents) which will keep x-radiation below 0.5 mR/h throughout the useful life of the tube, using recommended test procedures.[30]

12.2.13 IMPROVEMENTS IN SHADOW-MASK TUBES FOR HIGH-RESOLUTION AND COLOR DATA DISPLAY.
This trend was noted in Sec. 12.1. Shadow-mask tubes are also used in a variety of applications such as video games, home receivers operating as computer terminals, and commercial monitors. For the shadow-mask assembly, this means finer pitch, smaller-diameter apertures and screen dots, smaller Q-space, and generally thinner mask material. Etching of mask holes becomes much more demanding when diameters are smaller than material thickness. The black matrix tri-dot system has the mask-manufacturing advantage that mask aperture diameters are larger than in comparable nonmatrix tri-dot tubes or in comparable matrix (or nonmatrix) stripe tubes.

Table 12-1 shows comparative resolution capabilities of three 19-in (48-cm) visible (19V) constructions (with round-hole masks) as screen pitch is reduced to increase resolution. The value N_r, a comparative measure of resolution, assumes that resolution is proportional to the square root of the number of trios in a square area with sides equal to screen height. To achieve higher resolution than the third design (or for smaller tube sizes), it would be necessary to use even thinner mask material and smaller holes. The Q-space also reduces proportionally to pitch, becoming critically small for manufacturing.

Proposals for high-definition television (HDTV) of about 1100 lines include changing the aspect ratio from the 4:3 domestic standard to 5:3 or even 2:1 for greater visual impact.[31] Some experimental tubes have been made for this format but glassware is not generally available for production tubes.

Table 12-1 Comparative Resolution of Shadow-Mask Designs (Round-Hole Mask, Tri-Dot Screen)†

19V (48-cm) NGB tube type	Flat shadow mask			Screen vert. pitch, mm	N_T, trios in screen	$N_r = \sqrt{NT}/1.33$
	Mask material, mm	Vert. pitch, mm	Center hole diam., mm			
Conventional	0.15	0.56	0.27	0.60	400,000	500 lines
Monitor	0.15	0.40	0.19	0.43	800,000	775
High resolution	0.13	0.30	0.15	0.32	1,400,000	1025

†Based on 404- × 303-mm screen; aspect ratio = 1.33.

12.3 ELECTRON GUNS FOR SHADOW-MASK COLOR TUBES

The general electrode configuration for the shadow-mask color electron gun can be seen in Fig. 12-6. A beam-forming region consists of the cathode, grid-1, and grid-2 electrodes; a prefocus lens consisting of the grid-2 and lower grid-3 electrode, and a main lens where the grids 3 and 4 create a focusing field. In more complicated lens systems, additional elements follow the grid 4.

12.3.1 CATHODE CURRENT VERSUS DRIVE VOLTAGE.

The cathode current I_k, under conditions of space-charge-limited emission at the cathode surface, can be calculated for grid drive by the empirically derived relation[32,33]

$$I_k = K V_D^{3.0} V_c^{-1.5} \ ua$$

where K = *modulation constant* V_D-positive-going drive signal applied to negative grid
1 (grid drive), V
V_c = grid-1 spot cutoff voltage

This formula holds until V_D reaches approximately one-half of V_c. Above that point the formula is revised to

$$I_k = K V_D^{3.5} V_c^{-2} \ ua$$

The modulation constant varies between 3.0 and 4.5, depending on geometry in the cathode grid-1/grid-2 region of the electron gun.

When a negative-going signal is applied to a positive cathode for cathode drive, the grid-2 accelerating voltage varies with respect to cathode and becomes a factor in the above equation. In this case the formula must be modified to include this grid-2 effect. Grid-1 spot-cutoff voltage in the formula is replaced with cathode spot cutoff, and the entire right side of the equation is multiplied by the added factor

$$1 + \left[\frac{\text{cathode drive signal}}{\text{grid-2 accelerating voltage}} \right]^{1.5}$$

This added factor is valid as long as grid-2 voltage does not reach unusually small values, on the order of cathode drive signal, or less. Further, any grid-3 field penetrating through grid 2 will cause the effective accelerating voltage to be higher than the measured grid-2 voltage, requiring some adjustment in the grid-2 value in the formula.

12.3.2 ELECTRON GUN CLASSIFICATION.

Electron guns for color tubes can be classified according to the main lens configuration. They include unipotential, bipotential, tripotential, and *"hybrid"* lenses.

Unipotential Lens. The unipotential gun is discussed in Sec. 14.5.2 as it applies to monochrome. In color, this type of gun is rarely used except for small screen sizes. It suffers from a tendency toward arcing at high anode voltage, and relatively large low-current spots.

Bipotential Lens. The bipotential lens is the most commonly used gun in shadow-mask color tubes.[34-37] The arrangement of gun electrodes is shown in Fig. 12-6 along with a computer-generated plot of equipotential lines and electron trajectories. (See Sec. 12.5.2 for a description of the computer model.) The main lens of the gun is formed in the gap between grid 3 and grid 4. When grid 3 operates at 18 to 22 percent of the grid-4 voltage, the lens is referred to as a *low bipotential* configuration, often called *LoBi* for short. With grid 3 at 26 to 30 percent of grid 4 the lens is referred to as a *high bipotential* or *HiBi* configuration.

The LoBi has the advantages of a short grid 3 and shorter overall length, with parts assembly generally less critical than the HiBi. However, with its shorter object distance (grid-3 length), it suffers from somewhat larger spot size than the HiBi. The HiBi, on the

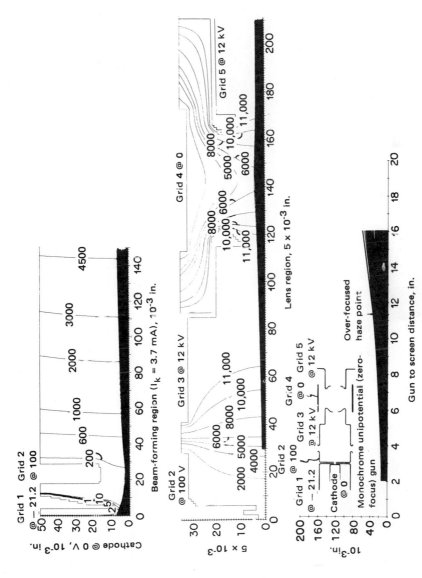

FIG. 12-6 Bipotential electron gun electrode configuration shown with the corresponding field and beam plots.

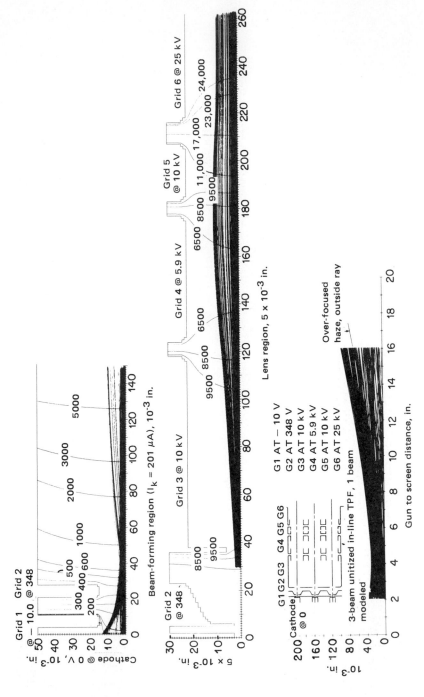

FIG. 12-7 Tripotential electron gun electrode configuration shown with the corresponding field and beam plots.

other hand, with a longer grid-3 object distance has improved spot size and resolution. Focus voltage supply for the LoBi can be less expensive.

As seen from the computer plot Fig. 12-6, the bipotential beam starts with a crossover of the beam near the cathode, rising to a maximum diameter in the lens. After a double bending action in the lens region the rays depart in a convergent attitude toward the screen of the tube.

Tripotential lens. Further improvement in focus characteristic has been achieved with a *tripotential lens*.[38] The electrode arrangement is shown in the computer model of Fig. 12-7. The lens region has more than one gap and requires two focus supplies, one at 40 percent and the other at 24 percent of anode. With this refinement the lens has lowered spherical aberration. Together with a longer object distance (grids 3 to 5), the resulting spot size at the screen is smaller than the bipotential designs. The tripotential gun is longer, needs two focus supplies, and requires a special base to deliver the high focus voltage through the stem of the tube.

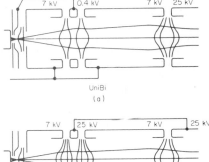

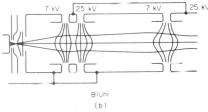

FIG. 12-8 Electrode arrangement of the (a) UniBi gun; (b) BiUni gun.

Hybrid Lenses. Another recent approach to improved focus has involved combinations of unipotential and bipotential lenses in series.[36] The more common of these are known as *UniBi* and *BiUni*.

The UniBi structure (sometimes called *quadripotential* focus) has the HiBi main lens gap with the addition of an earlier unipotential type of lens structure to collimate the beam. In Fig. 12-8a the grid-2 voltage is tied to a grid 4 inserted in the object region of the gun, causing the beam bundle to collimate to a smaller diameter in the main lens. With this added focusing, the gun is slightly shorter than a bipotential gun having an equal focus voltage.

The BiUni structure (Fig. 12-8b) achieves a similar beam collimation by tying the added element to the anode, rather than grid 2. Again the gun structure must be shorter because of the added focusing early in the gun. With three high-gradient gaps, arcing can be a problem in the BiUni.

Both of the hybrid designs are shorter than equivalent bipotential. Whether the shorter structure and improved beam collimation will justify the problems of added parts count and parts alignment remains to be seen.

Trinitron.† In this design the three beams are focused by the use of a single large main focus lens of unipotential design.[34,39,40] Figure 12-9 shows the three in-line beams mechanically tilted to pass through the center of the lens, then reconverged toward a common point on the screen. The design, with a large-focus lens, is claimed to have favorable spot size. The gun is somewhat longer than other color guns; also the mechanical structuring of the device requires unusual care and accuracy. As yet, other manufacturers have not adopted this approach.

12.3.3 GUN ARRANGEMENTS, DELTA VERSUS IN-LINE. Prior to 1970 most color gun clusters used a delta arrangement. Since that time the design trend has been toward the three guns in-line on the horizontal axis of the tube. These two arrangements are shown in Fig. 12-10, with the *in-line* shown for both 36- and 29-mm-neck diameters.

---

†A trademark of the Sony Corporation.

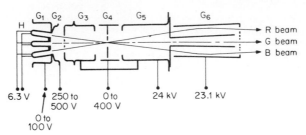

FIG. 12-9 Electrode arrangement of the Trinitron gun. (*From Morrell et al.*[34] *Used by permission.*)

Delta Gun. The electron optical performance of a delta gun is superior to in-line by virtue of a larger lens diameter, resulting from more efficient use of the available area inside the neck. Larger neck diameters will of course improve lens diameter, but at the expense of higher deflection power and more difficult convergence problems over the face of the tube.[35,41,42] Lens diameters used in delta guns are typically 12 mm (0.5 in) for a 51-mm (2-in) neck OD, 9 mm (0.4 in) for a 36-mm (1.4-in) neck, and 7 mm (0.3 in) for a 29-mm (1.1-in) neck. Delta guns use individual cylinders to form the electron lenses, and the three guns are tilted toward a common point on the screen. The individual cylinders are subject to random errors in position that can cause misconvergence of the three unde-flected spots. Further, carefully tailored current waveforms are applied to magnetic pole pieces on the separate guns to dynamically converge the three beams over the full face of the tube.

In-Line Guns. These guns enjoy one major advantage that more than offsets their less efficient use of available neck area. That is, with three beams on the horizontal axis, a deflection yoke can be built that will maintain dynamic convergence over the full face of the tube, without the need for correcting waveforms (see Chap. 10). With this major simplification in needed circuitry almost all guns now being built for commercial tele-vision are of the in-line design. In-line guns are being built for 36-mm, 29-mm, and more recently for 22.5-mm neck diameters. Typical lens diameters for these three guns are 7.5, 5.5, and 3.5 mm, respectively.

12.3.4 UNITIZED CONSTRUCTION. Most in-line guns presently being built have the three apertures or lens diameters formed from a single piece of metal. Hence, the name *unitized*. With this arrangement the apertures and lens diameters are accurately fixed with respect to each other so that beam landings at the screen are more predictable than with the older *cylinder* guns. Also, self-converging deflection yokes need the more accurate positioning of the beams in the yoke area. The only trio of gun elements not electrically tied together are the cathodes. Therefore, all varying voltages controlling luminance and color must be applied to the cathodes only. It should be noted, too, that the three beams in unitized guns travel parallel until they reach the final lens gap. Here, an offset, or tilted, lens on the two outboard beams bends the two outer beams toward the center beam at the screen. Any change in the strength of this final lens gap, such as a focus voltage adjustment, will cause a slight change in the undeflected convergence pattern at the screen.

12.3.5 ASYMMETRICAL BEAM-FORMING REGION. With the self-converging deflection yokes mentioned above, a unique overfocusing feature occurs on the vertical center line of the spot.[43] This causes *haze tails* to appear above and below the spot around the periphery of the screen. This haze degrades picture contrast. To avoid the haze tails, it is necessary to generate an electron beam having a horizontally elongated cross section (oval shape) in the yoke region. This horizontal elongation can be generated with either a vertically slotted grid-1 aperture or a horizontally slotted grid-2 aperture.[44–46] Figure 12-11 shows photographs of pulsed spots,[47] both undeflected and

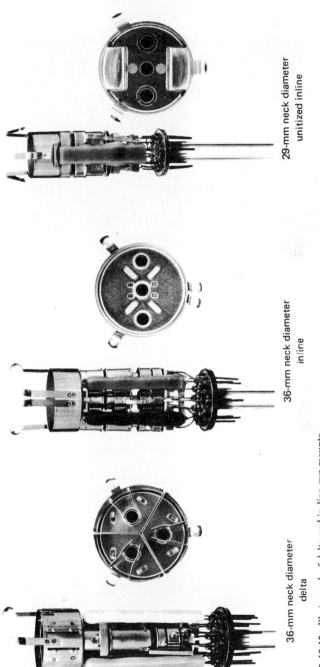

36-mm neck diameter
delta

36-mm neck diameter
in line

29-mm neck diameter
unitized inline

FIG. 12-10   Photograph of delta and in-line gun mounts.

12.17

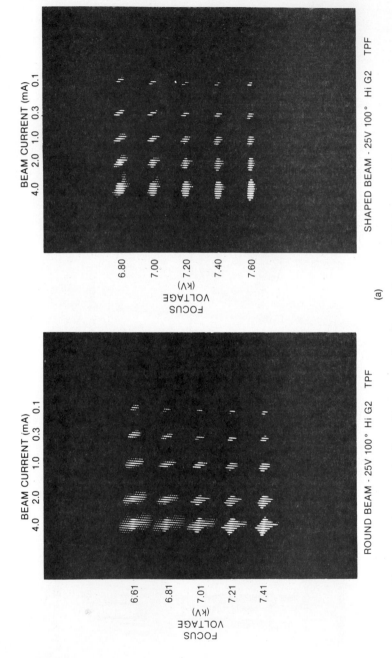

FIG. 12-11 Photos comparing conventional round beam versus shaped beam pulse spots: (*a*) deflected top left corner (self-converging yokes); (*b*) undeflected.

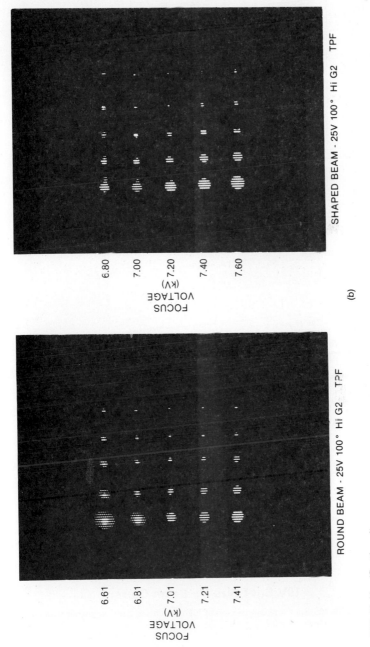

SHAPED BEAM - 25V 100° Hi G2   TPF

(b)

ROUND BEAM - 25V 100° Hi G2   TPF

FIG. 12-11 (*Continued*)

12.19

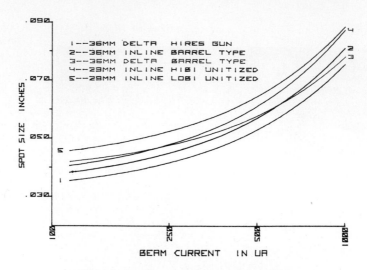

13V Spot Size Vs Beam Current

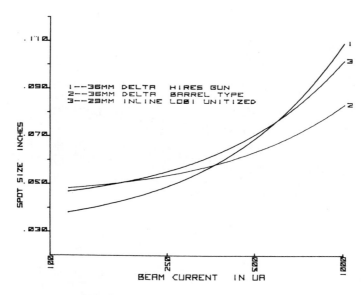

19V Spot Size Vs Beam Current

FIG. 12-12   Spot-size comparison of high-resolution guns compared with commercial television guns.

deflected through a range of beam currents and focus voltages. Note the absence of haze tails above and below the spot in the top right photograph compared with the top left photograph. This is an electron beam deflected to the top left corner of the tube, in one case a round beam, in the other case an oval-shaped beam.

12.3.6 OVERLAPPING LENS. A recent development in color electron guns involves the use of *overlapping* lenses.[48] In this approach the effective diameter of *in-line* lenses is increased to the point that the outboard lenses overlap the center lens. It is then necessary to carefully tailor interfacing partitions at the plane of overlap. If it is done with sufficient care and repeatability, it could result in improved resolution with no change in neck diameter.

12.3.7 ELECTRON GUNS FOR HIGH-RESOLUTION AND DATA DISPLAY. Improved versions of both the delta and in-line guns noted above are used in high-resolution data display applications. In both cases the guns are adjusted for the lower beam current and higher resolution needed in data display. The advantages and disadvantages noted earlier for delta and in-line guns apply here also. For instance, the use of a delta-type cylinder, or *barrel*-type gun, needs as many as 20 carefully tailored convergence waveforms to obtain near-perfect convergence over the full face of the tube. The in-line guns, with self-converging yokes, avoid the need for these extensive waveforms, at the expense of slightly larger spots, particularly in the corners where the overfocused haze tails can cause problems.

Figure 12-12a and b compares spot sizes, up to 1 mA of beam current, for high-resolution (hi-res) designs compared with commercial receiver-type designs, both delta and in-line in 13- and 19-in vertical (13V and 19V) applications. Note the marked improvement in spot size at currents below 500 μA for the high-resolution designs.[41]

12.4 LARGE-SCREEN DISPLAYS

12.4.1 PROJECTION SYSTEMS. Projection television systems provide video displays having much greater area than can be generated on cathode-ray picture tubes. Optical magnification is employed to throw an expanded image on a passive viewing surface which may have a diagonal dimension of 60 in (1.5 m) or more [CRTs are generally restricted to less than 39-in (1-m) diagonal].

A primary factor limiting CRT size is glass strength. In order to withstand atmospheric pressure on evacuated envelopes, CRT weight increases exponentially with linear dimension (such as viewable diagonal) as shown by Table 12.2.

Projection techniques have been employed from the earliest days of television to magnify images produced by Nipkow discs and 3-in (76-mm) CRTs on *large* rear-view frosted screens.[49]

Now that direct-view devices approach a 30-in (1-m) diagonal, projection methods serve to expand images to full life-size or greater.

Table 12-2 Color CRT Weight Versus Screen Diagonal Dimension

Screen diagonal dimension, cm	CRT weight, kg
33 (13V)	5.5
48 (19V)	12
64 (25V)	23
76 (30V)	40

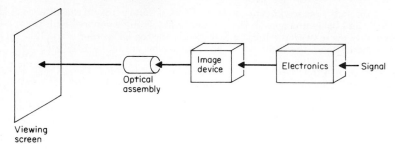

FIG. 12-13   Constituent parts of a projection television receiver.

The basic constituents of a projection receiver are (see Fig. 12-13):

1. Viewing screen
2. Optical elements
3. Image source
4. Electronics

The major differences of projection systems from direct-view systems are embodied in the first three areas, while the electronics assembly is essentially the same as for direct-view systems.

Display Requirements.   To be acceptable, a projection system must approach or equal the performance of direct-view systems in terms of brightness, contrast, and resolution. Whereas the first two parameters may be compromised to some extent, large displays must excel in resolution due to the tendency of viewers to be positioned less than the normal relative distance of three to eight times the picture height from the viewing surface. Table 12-3 indicates performance levels achieved by direct-view television receivers and conventional film theater equipment.

Evaluation of overall projection system brightness *B*, as a function of its optical components, can be calculated using Eq. (12-1)

$$B = \frac{L_G GTR^M D}{4 W_G (f/N)^2 (1 + m)^2} \qquad (12\text{-}1)$$

where L_G = luminance of green source (CRT, etc.)
   G = screen gain
   T = lens transmission
   R = mirror reflectance

Table 12-3   Performance levels

	Luminous output (brightness), nits (ft-L)	Contrast ratio at ambient illum., fc	Resolution (TVL)
Television receiver	200–400 (60–120)	30:1 at 5	275
Theater (film projector)	55 + 14 (16 + 4) 55 − 21 (16 − 6)†	100:1‡ at 0.1	4800

†United States Standard (PH 22.124-1961); see Ref. 50, p. 291.
‡Limited by lens flare.

M = number of mirrors
D = dichroic efficiency
W_G = green contribution to desired white output, %
f/N = lens f-number
m = magnification

For systems in which dichroics or mirrors are not employed, those terms drop out.

Screens. Two basic categories of viewing screen are employed for projection television (Fig. 12-14).

1. Front projection: Image is viewed from the same side of the screen as that on which it is projected.
2. Rear projection: Image is viewed from the opposite side of the screen as that on which it is projected.

The former depends upon reflectivity to provide a bright image, the latter requires high transmission to achieve that characteristic. In either case, screen size influences display brightness inversely as follows

$$B = L/A \qquad (12\text{-}2)$$

where B = apparent brightness, cd/m^2
L = projector light output, lm
A = screen-viewing area, m^2

Thus, for a given projector luminance output, viewed brightness varies in proportion to the reciprocal of the square of any screen linear dimension (width, height, or diagonal).

To improve apparent brightness, directional characteristics are designed into viewing screens. This is termed *screen gain G*, and Eq. (12-2) becomes

$$B = G \times L/A \qquad (12\text{-}3)$$

Gain is expressed as screen brightness relative to a lambertian surface. Table 12-4 lists some typical front-projection screens with their associated gains.

Screen contrast is a function of the manner in which ambient illumination is treated. Figure 12-14 illustrates that a highly reflective screen (as in front projection) reflects ambient illumination as well as the projected illumination (image). The reflected light thus tends to dilute contrast, although highly directional screens diminish this effect. A rear-projection screen depends upon high transmission for brightness but can capitalize on low reflectance to improve contrast. A scheme for achieving this is equivalent to the black matrix utilized by tricolor CRTs as illustrated in Fig. 12-15.

The technique focuses projected light through lenticular lens segments onto strips of the viewing surface allowing intervening areas to be coated with a black (nonreflective) material. The lenticular segments and black stripes normally are oriented in the vertical

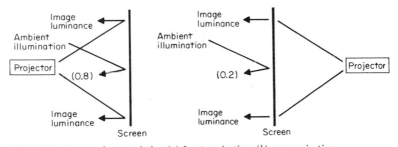

FIG. 12-14 Screen characteristics: (*a*) front projection; (*b*) rear projection.

Table 12-4   Screen Gain

Screen type	Gain
Lambertian (flat-white paint magnesium oxide)	0.85–0.90
White semigloss	1.5
White pearlescent	1.5–2.5
Aluminized	1–12
Lenticular	1.5–2
Beaded	1.5–3
Kodak Ektalite	10–15
Scotch-light	Up to 200

Source: Luxenberg and Kuehn,[50] pp. 306, 307.

dimension to broaden the horizontal viewing angle. The overall result is a screen which transmits most of the light (typically 60 percent) incident from the rear while absorbing a large percentage of the light (typically 90 percent) incident from the viewing side, thus providing high contrast.

Rear-projection screens usually employ extra elements, including diffusers and directional correctors, to maximize brightness and contrast in the viewing area.

As with direct-view CRT screens, resolution can be affected by screen construction. This is not usually a problem with front-projection screens, although granularity or lenticular patterns can limit image detail. In general, any screen element, such as the matrix arrangement described above, which quantizes the image (breaks it into discrete segments) limits resolution. For 525- and 625-line broadcast television, this factor does not provide the limiting aperture. High-resolution applications, however, may require attention to this parameter.

Projection Optics.   Several of the more frequently employed optical configurations for projecting full-color television images are diagrammed schematically in Fig. 12-16.

Tricolor CRT and trinescope designs minimize image registration (convergence) problems, which must be addressed in multilens systems. In the multilens configurations outboard tube rasters must be keystoned to adjust for optical-path length differences (see Fig. 12-17). This may be done electronically or mechanically[55] for the angled format of Fig. 12-16b. A variety of folded optical-path schemes have been employed to reduce the volume in which a required "throw" distance is contained, particularly in rear-projection receivers.

Variables to be evaluated in choosing among the many schemes include source luminance, source area, image magnification, optical-path transmission, light collection effi-

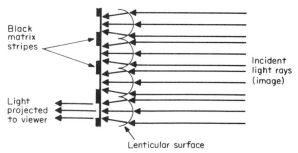

FIG. 12-15   High-contrast rear-projection screen.

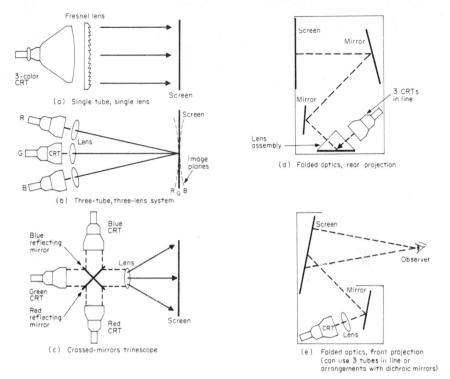

FIG. 12-16   Optical configurations.

ciency (of lens), and, of course, cost, weight, and complexity of components and corrective circuitry.

A critical factor in rendering a projection system cost-effective is the lens package. It must possess good luminance collection efficiency (small f-number), high transmission, good modulation transfer function (MTF), light weight, and low cost. Table 12-5 compares characteristics of some available lens complements.[64]

The total light incident upon a projection screen is equal to the total light emerging from the projection optical system, neglecting losses in the intervening medium. The distribution of this light generally is not uniform. Its intensity is less at screen edges than at the center in most projection systems as a result of light ray obliquity through the lens ($\cos^4 \theta$ law) and vignetting effects (Ref. 50, p. 292) as illustrated by Fig. 12-18. Luminance decrease to 80 percent from center to edge of a projected display is considered excellent performance (Ref. 50, pp. 185, 292; Ref. 51). Low-cost television projection lenses such as used in Fig. 12-18 allow corner luminance to drop to 35 percent (Ref. 52, p. 49). Light output from a lens is determined by collection efficiency and transmittance as well as source luminance. Typical figures for these characteristics are 15 to 20 percent[53] and 75 to 85 percent (Ref. 52, p. 49), respectively. The former is partially a function of the light source, and the figure given is typical for a lambertian source (CRT) and lens having a half-field angle of approximately 25° (Ref. 52, p. 48).

Optical distortions are important to image geometry and resolution. Geometry is generally corrected electronically, both for pin cushion and barrel effects and keystoning which results from the fact that the three image sources are not coaxially disposed in the common in-line array (see Fig. 12-17).

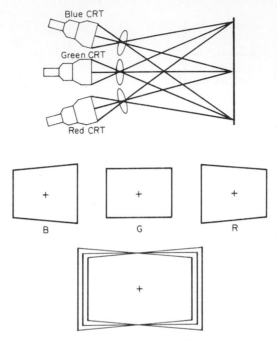

FIG. 12-17   Three-tube in-line array. Outboard tubes (red and blue) and optical assemblies tilted so that axes intersect axis of green tube at screen center. Red and blue rasters show trapezoidal (keystone) distortion and resulting misconvergence when they are superimposed on the green raster.

Resolution, however, is affected by lens astigmation, coma, spherical aberration, and chromatic aberration. The first three factors are dependent upon the excellence of the lens, but chromatic aberration can be minimized by using line (monochromatic) or very narrow band emitters for each of the three image sources. Because a specific lens design possesses different magnification for each of the three primary colors (index of refraction varies with wavelength), throw distance for each must also be adjusted independently to attain perfect registration.

In determining final-display luminance, transmission, reflectance, and scattering by additional optical elements such as dichroic filters, optical-path folding mirrors, or corrective lenses must also be accounted for [Eq. (12-1)]. Dichroics exhibit light attenua-

Table 12-5   Projection Lens Performance

Lens	Aperture	Image diagonal, mm	Focal length, mm	Magnification	Response at 300 TVL, %
Refractive, glass	$f/1.6$	196	170	8	33
Refractive, acrylic	$f/1.0$	127	127	10	13
Schmidt	$f/0.7$	76	87	30	15
Fresnel	$f/1.7$	305	300	5	6

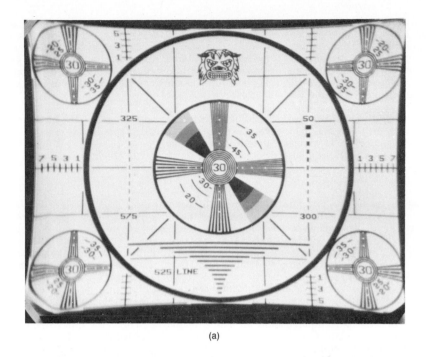

(a)

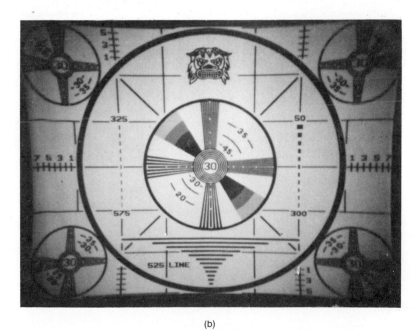

(b)

FIG. 12-18 Projection lens light ray obliquity: (*a*) source image; (*b*) projected image.

tions of 10 to 30 percent, and mirrors can reduce light transmission by as much as 5 percent each (front-surface mirrors exhibit minimum absorption and scattering but are susceptible to damage during cleaning). Contrast is also affected by the number and nature of optical elements employed in the projection system. Each optical interface generates internal reflections and scattering which dilute contrast and reduce MTF amplitude response. Optical coatings may be utilized to minimize these effects, but their contribution must be balanced against their cost.

Image Devices.   The two most utilized devices for creating images to be optically projected are CRTs and light valves. Each has been created in a multitude of variations, the most successful of which are described below.

Projection CRTs have historically ranged in size from 1 in (2.5 cm) to 13 in (33 cm) diagonal (diameter for round envelope types). Since screen power

$$W_S = I_b \times E_{a2} \tag{12-4}$$

must increase in proportion to the square of the magnification ratio, it is clear that faceplate dissipation for CRTs used in projection systems must be extremely high. Electrical-to-luminance conversion efficiency for television phosphors is on the order of 15 percent.[53] A 50-in (1.3-m) diagonal screen at 60 ft·L requires a 5-in (12.7-cm) CRT to emit 6000 ft·L, exclusive of system optical losses, resulting in a faceplate dissipation of approximately 20 W in a 3-in (7.6-cm) by 4-in (10.2-cm) raster. A practical limitation for ambient air-cooled glass envelopes (to minimize thermal breakage) is 1 mW/mm² or 7.74 W for this size display. Accommodation of this incompatibility must be achieved in the form of improved phosphor efficiency or reduced strain on the envelope via cooling. Since phosphor development is a mature science, maximum benefits are found in the latter course of action with liquid cooling assemblies employed to equalize differential strain on the CRT faceplate. Such implementations produce an added benefit through reduction of phosphor thermal quenching and thereby supply up to 25 percent more luminance output than is attainable in an uncooled device at equal screen dissipation.[56]

A liquid-cooled CRT assembly as shown in Fig. 12-19 depends upon a large heat sink (part of receiver chassis) to carry away and dissipate a substantial portion of the heat generated in the CRTs. Very large screen projectors using such assemblies commonly operate CRTs at four to five times their rated thermal capacities. Economic constraints mitigate against the added cost of cooling assemblies, however, and methods to improve

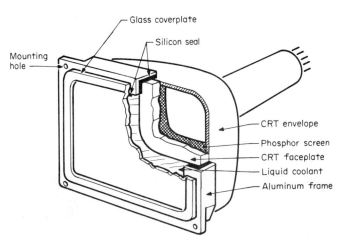

FIG. 12-19   Liquid-cooled projection CRT.

Table 12-6 Paths to Increased Luminance Output

Parameter	Increased anode voltage	Increased beam current
Spot diameter	$\propto I/E_a$	$\propto I_b$
Phosphor aging[54]	No effect	I_b
X-radiation[57]	$\propto E_a^n$†	$\propto I_b$
Glass browning	$\propto E_a^n$†	$\propto I_b$
Flashover, stray, emission	$\propto E_a$	No effect
Density quenching	No effect	$\propto I_b$

†n = 20 at 30 kV; 16 at 36 kV.

phosphor conversion efficiency and optical coupling/transmission efficiencies continue to be investigated.

Concomitant to high power are high voltage and/or high beam current. Each has benefits and penalties, as noted by the cause-effect relations listed in Table 12-6.

Resolution, dependent on spot diameter, is improved by increased anode voltage and reduced beam current. For a 525- or 625-line display, spot diameter should be 0.006 in (0.16 mm) on the 3- by 4-in (7.6- by 10.2-cm) raster previously described (Ref. 49, pp. 1–10). Higher resolving-power displays require yet smaller spot diameters, but a practical maximum anode-voltage limit is 30 to 32 kV when x-radiation,[56a,57] arcing, and stray emission are considered. Exceeding that value requires special shielding and CRT processing.

Figure 12-20 depicts a CRT design which utilizes the advantageous large relative aperture and low surface reflectance losses of cassegrainian mirror (Schmidt) optics. This design employs a relatively small metal-backed phosphor target (screen), and magnetic focusing provides the necessarily small spot diameter. Despite the very low numerical aperture (approximately $f/0.7$), physical size constraints and central area light blockage by the screen result in an overall luminance efficiency of about 65 percent that of a CRT coupled to $f/1.0$ refractive optics in a similarly sized system.[53]

Light valves may be defined as devices which, like film projectors, employ a fixed light source modulated by an optical-valve intervening source and projection optics. Table 12-7 lists some of the many light valve technologies which have been successfully demonstrated. The most successful type commercially has been the Eidophor/Schlierenoptic configuration which can produce light outputs of 7000 lm versus 200 to 500 lm for CRT-

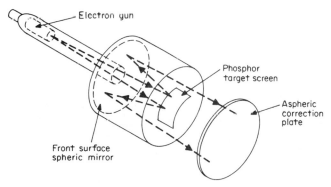

FIG. 12-20 Projection CRT with integral Schmidt optics.

Table 12-7   Light Valve Technologies

Light valve technology	Reference
Oil film Eidophor/Schlieren	50, 58, 59
Pockels effect (EDP, KD_2P, ADP)	50
Pin matrix	50
Thermoplastic	50, 60
Suspension	50
Liquid crystal	65

based projectors (Ref. 8, p. 48) (Fig. 12-21). These devices are utilized primarily in commercial applications (theaters, stock market displays, etc.) owing to their relatively high cost.

Laser Projectors. Two approaches to laser projection displays have been implemented on a limited scale. The first employs three optical laser light sources whose coherent beams are modulated electrooptically and deflected by electromechanical means to generate a raster display on a projection screen as shown in Fig. 12-22.[63]

The second category employs an electron-beam pumped monocrystalline screen in a CRT to create a small 1-in (2.5-cm) raster.[61] The laser screen image is then projected by conventional optics per Fig. 12-23. This technology promises three optical benefits: very high luminance in the image plane, highly directional luminance output for efficient optical coupling, and compact, lightweight, inexpensive projection optics.

12.4.2 DISCRETE-ELEMENT LARGE-SCREEN DISPLAYS. The essential characteristic defining discrete-element display devices is division of the viewing surface (volume for three-dimensional displays) into separate segments which are individually controlled to generate an image. Each element therefore embodies a dedicated controlling switch or valve as opposed to raster display devices which employ one (or a very small number of) control device(s) for activation of all display elements. Two categories of discrete-element display have been used for large-screen monochrome applications (see Chap. 10).

Luminescent Panels. The prevalent technology of large-screen luminescent panels employs plasma excitation of phosphors overlying each display cell. The plasma is formed in low-pressure gas which may be excited by either ac or dc fields. Figure 12-24 depicts a color television dc plasma panel. Used principally in business displays, such panels have been made up to a 5-ft (1.5-m) diagonal dimension with 1000 cells per side (10^6 cells total). Color rendition requires red/green/blue trios to be excited preferentially

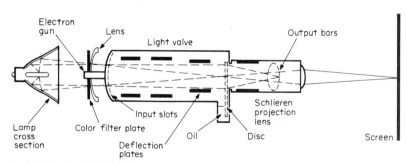

FIG. 12-21   Eidophor light valve.

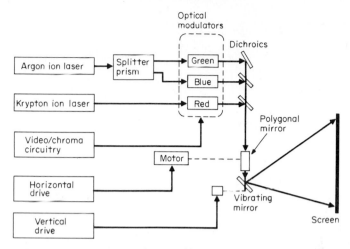

FIG. 12-22 Schematic block diagram of laser projection television receiver.

or proportionally just as in a color CRT, but this triples the number of actual cells; i.e., three cells comprise one picture element.

Flood Beam CRT. Very large displays (up to 20 m diagonal) have been employed for mass audiences in stadiums to provide special coverage of sports events. These consist of flood beam CRT arrays in which each device fulfills the function of a single phosphor dot in a delta-gun shadow-mask CRT. Thus each display element consists of a trio of flood beam tubes, one red, one green, one blue, each with a 1 to 6 in (2.5 to 15 cm) diameter. A fully NTSC capable display would require in excess of 400,000 such tubes (147,000 of each color) to be individually addressed. Practical implementations have employed less than 40,000, thus requiring substantially less drive complexity.[62]

Matrix Addressing. Provision for individual control of each element of a discrete element display implies a proliferation of electrical connections and, therefore, rapidly expanding cost and complexity. These factors are minimized in discrete-element displays through matrixing techniques wherein a display element must receive two or more gating signals before it is activated. Geometric arrangements of the matrixing connectors guar-

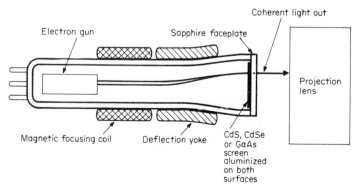

FIG. 12-23 Laser-screen projection CRT.

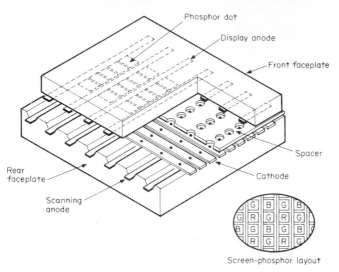

Phosphor dot

Display anode

Front faceplate

Spacer

Rear faceplate

Cathode

Scanning anode

Screen-phosphor layout

FIG. 12-24   Dc plasma panel for color television.

antee that only the appropriate elements are activated at a given time. A simplified version of such a scheme is illustrated in Fig. 12-25.

12.5 MONOCHROME VIDEO DISPLAY DEVICES

Monochrome video displays, of less importance than color in recent years, continue to be dominated by the cathode-ray tube. Its electron optics and spot size will be described. In addition, the long-time goal of a flat matrixed display, with its formidable problems, will be summarized. Finally, recent innovations in thin, flat CRTs and their significance will be discussed. Monochrome phosphors, brightness, and contrast are covered in Sec. 12.6. Deflection characteristics are detailed in Chap. 10.

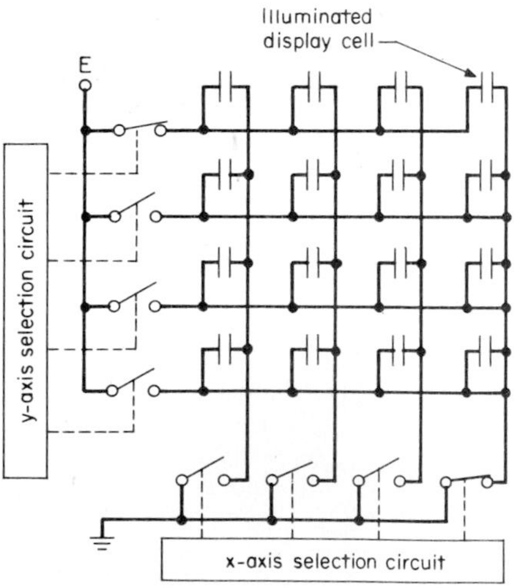

Illuminated display cell

y-axis selection circuit

x-axis selection circuit

FIG. 12-25   Matrix addressing.

12.5.1 MONOCHROME CATHODE-RAY TUBES. Monochrome picture tubes were the workhorse of the television industry from its start in the 1940s until the emergence of color as the dominant technology in the mid-1960s. The progression of monochrome envelope sizes from 10-in (25-cm) diagonal with 50° deflection up to a 27-in (69-cm) diagonal with 90° deflection is covered in the first *Television Engineering Handbook*.[66] Of interest is the reverse trend in the years from 1965 to the present toward smaller screen sizes, higher deflection angles, and smaller neck diameters. The need for low cost in monochrome video displays is today an overriding consideration. As a result screen sizes of 12-in (30-cm) diagonal and smaller, and neck diameters of 0.8 in (20 mm), are typical in monochrome picture tubes in present use.

12.5.2 MONOCHROME ELECTRON GUNS.

With the need for low cost at an acceptable performance level, monochrome electron guns usually use a low focus voltage unipotential design like that shown in Fig. 12-26. In this design a focus voltage at, or near, ground potential is a cost-saving feature. In the figure a cross section of the gun is shown on a matrix of points with the gun elements, equipotential lines, and electron beam displayed on a computer-generated plot. The availability of high-speed computer programs[67,68] with large memories makes possible a rigorous analysis of undeflected electron beams from cathode to screen with all the factors that affect spot size. The gun model shown in Fig. 12-26 is divided into three sections, (1) beam-forming region, (2) lens region, and (3) gun to screen, each shown on an appropriate scale. In each case only one-half of the gun is shown, with the other half a mirror image.

From this model the analogy to light optics may be seen. The waist of the beam, in the beam-forming region, is imaged by the main lens to the screen of the tube. The contribution to final spot size of this imaged crossover may be calculated from the well-known magnification equation

$$\text{Image size} = \text{crossover size} \times \left(\frac{V_o}{V_i}\right)\frac{1}{2} \times \frac{\text{image distance}}{\text{object distance}} \tag{12-5}$$

where V_o is the potential at the crossover, V_i is the potential at the screen, and object and image distances are measured from the main lens. In the computer analysis a *virtual* crossover diameter is obtained by projecting rays backward in straight lines from the main lens to form a crossover with slightly different size and location from the actual crossover. The virtual crossover can then be assigned the grid-3 potential in Eq. (12-5).

The size of the crossover itself results from (1) thermal velocity of electrons leaving the cathode,[09] (2) imperfect focusing of the rays in the beam-forming region, and (3) electron self-repulsion in the beam-forming region. These three factors are taken into account in the computer model.

The total spot size at the screen consists of (1) the magnified crossover previously discussed, (2) a spherical aberration contribution of the main lens,[70] and (3) electron self-repulsion from crossover to screen. These three factors are modeled in the computer solution. It should be noted that the three factors are related in a complex manner, so that if calculated separately and added arithmetically, they will total somewhat more than the true computed spot size.

The low-focus unipotential gun modeled above has the advantage of not only a low-cost focus supply, but a small beam diameter in the lens and yoke region, resulting from rapid acceleration in the grid-2/grid-3 region. With this small beam bundle the tube suffers less from deflection defocusing than designs having larger beam bundles.

Disadvantages of the monochrome gun include a relatively large undeflected spot, for two reasons. First, V_o in Eq. (12-5) is at the relatively high value of screen potential and, second, that part of spot size related to thermal velocities at the cathode surface, according to a well-known rule of electron optics,[69] varies inversely with beam bundle diameter in the main lens region. A further disadvantage of this design is the presence of the high screen voltage in the bottom end of the gun where flashover may occur at anode potentials above, say, 16 kV, unless special precautions are taken.

For higher-resolution applications, such as occur in certain data display applications, and in wide bandwidth television applications, better quality monochrome guns will be needed. The gun types discussed in Sec. 12.3 offer better quality lensing with larger beam bundles in the lens region. With these larger beam bundles it is likely that corrective fields at scan rates for focus and astigmatism[71] will be needed in high-resolution monochrome applications.

Spot Size Measurement. The measurement of spot size at the screen of an operating tube can be accomplished by the following methods:

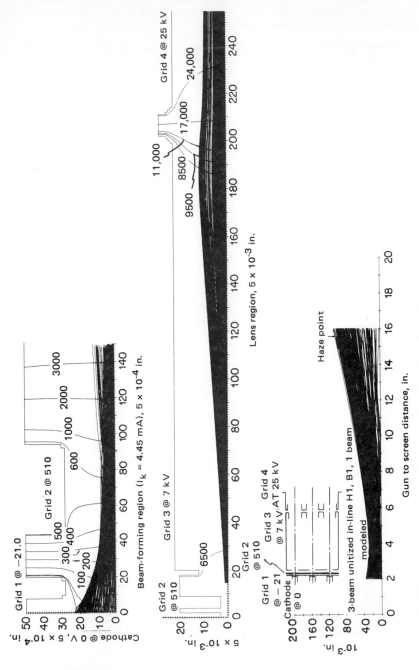

FIG. 12-26 Monochrome electron gun electrode configuration shown along with corresponding field and beam plots.

1. *Shrinking raster*[72] in which the vertical raster size is lowered until individual scan lines just merge together. Resulting line width can be readily calculated.

2. *Scanning slit* or probe in which a small light-sensing slit or probe is scanned across either a single line or a pulsed spot, giving a line profile or spot profile.

3. A *two-slit* method in which the spacing between two slits further serves to calibrate the resulting double-spot profile.

4. *Spot photography*[73] in which a pulsed spot is photographed at a range of current and focus settings on a single piece of film. This is effective at spot sizes down to approximately 0.02 in (0.5 mm) (see Fig. 12-11).

Methods 2 and 3 have the further advantage that their resulting spot profiles may be analyzed, by a Fourier transform, into a modulation transfer function (MTF), giving a spatial response characteristic[72,74] that may be readily included in a calculated MTF for an entire system.

The following approximate rules of thumb refer to cathode-ray-tube spot size.

1. Spot size varies inversely with neck diameter (lens diameter). Many factors are involved. A first-power relationship is approximate.

2. With increasing anode voltage beam current can be increased in direct proportion to anode voltage, for a given spot size.

3. Spot size varies directly with gun-to-screen distance—a function of image distance-deflection angle.

12.5.3 MONOCHROME MATRIXED DISPLAYS.

"Flat wall" video displays addressed by a matrixed system have long been a shining hope of the display industry. Many technologies have been proposed and most have gone through excursions of high hope followed by pessimism. The more notable of these technologies are: (1) electroluminescence, (2) gas discharge, (3) liquid crystal, and, (4) light-emitting diodes. A detailed description of these four technologies is beyond the scope of this chapter. Three references are suggested as a start for the interested researcher.[75-77]

Two major problem areas in all matrixed video displays are, first, the 10^5 to 10^6 picture elements to be addressed and, second, the uniformity needed among the outputs of these elements.

The growth of large-scale integrated-circuit technology has brought most of the circuit-related problems within reach of a solution. The one overriding problem that remains is the 10^5 to 10^6 picture elements that need a uniformity and dependability competitive with a single CRT electron beam swept over the same number of points. This requirement of uniformity and freedom from flaws is formidable. Whether it can be solved effectively remains to be seen.

Two matrixed technologies that appear most favorable at present are gas discharge and thin-film electroluminescence. The information shown in Table 12-8 summarizes the performance of ac and dc versions of these two technologies.

Light-emitting diodes, in arrays of 10^5 to 10^6 elements, are expensive and dissipate considerable power. They are not presently in the running.

Liquid crystal displays, on the other hand, are less expensive and use little power. As passive devices they modulate reflected or transmitted light from another source. Like most passive displays, liquid crystals have a relatively slow response time that may "smear" a moving image. Two successful video displays using small experimental liquid crystal devices have recently been reported.[79,80] Whether liquid crystals can, in production, meet the requirements for competitive uniformity and fast response time remains to be seen.

12.5.4 FLAT CRTs.

A significant comment on small, flat video displays is the fact that two devices which are reportedly nearing commercial production are, in fact, not matrixed displays. Both are modifications of a monochrome CRT.[81,82] The well known

Table 12-8 Performance of Developmental Television Matrix Displays

Display	Resolution (no. horizontal lines × elements per line)	Average luminance, ft·L	Contrast	Power, W	Color	Video mod. voltage, V	Size	Eff., lm/W	Line address time, µs	Duty cycle
DC plasma	40 × 100	50–60	10:1	....	Monochrome (orange)	150	5 × 12.5 cm	....	420/field	1/40
DC plasma	100 × 100	25	20:1	≈100	Monochrome (orange)	150	12.5 × 12.5 cm	....	....	1/100
Selfscan plasma	212 × 77	8	40:1	....	Monochrome (orange)	40	2.4 × 6.3 in	0.1	60	1/500
Selfscan plasma	222 × 77	25	30:1	25	Monochrome (orange)	40	2.4 × 6.3 in	0.2	125	1/267
Selfscan plasma	222 × 77	15	25:1	30–50	Monochrome (orange)	40	2.4 × 6.3 in	....	125	1/250
DC plasma	212 × 282	25	40:1	10 panel 80 circuitry	Monochrome (orange)	160	10.5 × 14.0 cm	0.5	60	1/250
AC-EL Zn(S,Se):Cu,Br	80 × 80	11	10:1	....	5100 Å	450 @ 70 kHz	10 × 7.5 cm	....	180	1/100
AC-EL Zn(S,Se):Cu,Br	200 × 250	3	....	....	5100 Å	450 @ 70 kHz	20 × 25 cm	....	....	....
DC-EL ZnS(Cu,Al)	230 × 230	1	4:1	100	5850 Å	300	7.8 × 10.4 in	....	60	1/250
DC-EL ZnS(Cu,Al)	225 × 225	10	20:1	150	5850 Å	200	7.8 × 10.4 in	....	120	1/250
EL ZnS:Mn	81 × 108	60	50:1	3	5800 Å	130	3.6 × 4.8 cm	0.3	100	1/80
DC plasma-color	212 × 94 color triads	6 (peak)	20:1	15 (panel only)	Color	230	10.5 × 14.0 cm	0.05	60	1/250
DC plasma-color (self-scanning type)	120 × 53 color triads	5	8:1	14	Color	....	12 × 16 cm	−0.05	120	1/120

Source: After Van Raalte,[78] by permission.

Kaiser-Aiken flat tube of the 1950s,[83] with the electron gun parallel to the plane of the screen, serves as the basis for the newer designs. As shown in Fig. 12-27, a combination of magnetic and electrostatic deflection is used for the best space and power conservation.

Extensive literature attests to efforts that have gone into larger flat versions of vacuum devices with the axis of the gun in the plane of display.[84,85] Here an overriding consideration is the mechanical structuring of a large, flat vacuum device, where the atmospheric force is sizable, without interfering with the resolution of the display. At present none of these devices is known to be nearing production.

12.6 PHOSPHOR AND SCREEN CHARACTERISTICS

Many materials, naturally occurring and synthetic, organic and inorganic, have the ability to give off light. In the field of television we are concerned only with crystalline inorganic solids that are stable under cathode-ray tube fabricating and operating conditions. These materials are generally powders having average particle sizes in the range of 5 to 15 μm. Figure 12-28 shows some typical particle size distributions. Because of defects and irregularities in the crystal lattice structure, these materials have the ability to absorb incident energy, in the case of a television tube high-energy electrons or cathode rays, and convert this energy into visible light. This process involves the transfer of energy from the electron beam to electrons in the phosphor crystal. The phosphor electrons are thereby excited or raised to levels higher than the ground state. Light is emitted when the electrons return to the more stable states. Figure 12-29 is a representation of these changes in energy levels.

Phosphors are composed of a host crystal which comprises the bulk of the material and one or more activators which may be present in amounts from parts per million to a few mole percent. Either the host or activator can determine the luminescent properties of a phosphor system. For example, in the zinc sulfide/cadmium sulfide:silver (ZnS:Ag/CdS:Ag) system, the emitted color ranges from blue at zero cadmium through green, to yellow, and into red as the cadmium content is increased. Contrasted to this is the yttrium oxysulfide (Y_2O_2S) system which is orange to red when activated with europium and white to green when activated with terbium.

In the commercial preparation of phosphors the highly purified host and the required amount of activator are intimately mixed, normally with a flux, such as an alkali or alkaline earth halide or phosphate, which supplies a low-temperature melting phase. The flux controls the particle development and aids in the diffusion of the activator into the lattice. This mixture is then fired at high temperature, 1472 to 2192°F (800 to 1200°C) on a prescribed schedule in order to develop the desired physical and luminescent properties. In some cases the firing is carried on under specific controlled atmospheres. After firing, the resultant cake is broken up, residual soluble materials are removed by washing, and any required coatings are applied. For further descriptions of phosphor mechanism and preparation see Refs. 86 and 87.

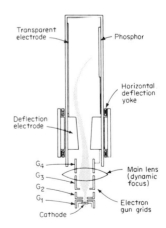

FIG. 12-27 Sony flat CRT. *(Reprinted from Electronics, Feb. 10, 1982. Copyright © McGraw-Hill, Inc. All rights reserved.)*

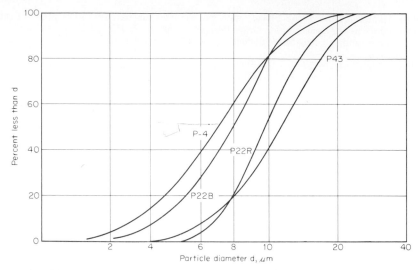

FIG. 12-28   Particle size distribution of representative phosphors.

12.6.1 LUMINESCENT PROPERTIES

Efficiency and Luminosity.   Table 12-9 presents important characteristics of the most common television phosphors. A more complete listing of registered phosphors is given in Ref. 88. Standard methods of measuring luminescent properties are described in Ref. 89.

Absolute phosphor efficiency is measured as the ratio of total absolute energy emitted to the total excitation energy applied. Since we are dealing with visual displays when evaluating or comparing picture tubes, it is more meaningful to measure luminescence in footlamberts using a system the response of which matches the eye. In addition to the use of a suitable detector, a number of other parameters must be controlled if meaningful measurements are to be obtained. Most important of these is the total energy to the screen, i.e., anode voltage, cathode current, and raster size. The raster should be synchronized and be linear.

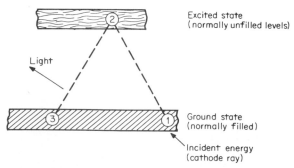

FIG.   **12-29** Energy   transitions   of   electrons   leading   to luminescence.

Table 12-9 Typical Characteristics of Some Common Phosphors†

EIA no.	Worldwide designation	Use	Composition	Relative efficiency	Typical CIE coordinates				Decay
					x	y	u'	v'	
P-4	WW	Black and white television	$ZnS:Ag +$ $Zn_{(1-x)}Cd_xS:Cu,Al$	100	0.270	0.300	0.178	0.446	Medium short
P-1	GJ	Projection green	$ZnSiO_4:Mn$	130	0.218	0.712	0.079	0.577	Medium
P-43	GY	Projection green	$Gd_2O_2S:Tb$	155	0.333	0.556	0.148	0.556	Medium
P-22R	X	Red direct-view projection	$Y_2O_3:Eu$ $Y_2O_2S:Eu$	65	0.640 0.625	0.340 0.340	0.441 0.429	0.528 0.525	Medium short
P-22G	X	Green direct-view	$Zn_{(1-x)}Cd_xS:Cu,Al$ $ZnS:Cu,Al$	180	0.340 0.285	0.595 0.600	0.144 0.119	0.566 0.561	Medium
P-22B	X	Blue direct-view projection	$ZnS:Ag$	25	0.150	0.065	0.172	0.168	Medium short

†Values are nominal; they may change with measurement methods and source of phosphor.

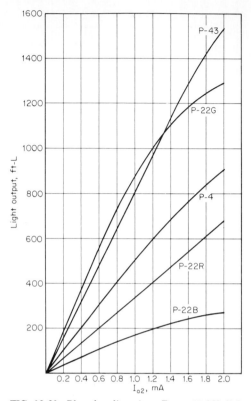

FIG. 12-30 Phosphor linearity—E_{a2} = 25 kV; light output response to increasing current. (1 ft·L = 3.426 cd/m^2.)

Phosphor Linearity. Under normal picture-tube operating conditions the luminescence of phosphors is essentially proportional to the beam current applied. However, when high beam currents are employed, as in projection sets, some phosphors saturate and depart from linear behavior, in which case the brightness response is less than proportional to the current applied. The same effect can be observed in direct-view color television in areas of high light brightness and with electron guns having small spot size. If linearity of the three primary phosphors is not closely matched, noticeable shift in white field color can result in highlight areas. Figure 12-30 shows some typical light output/beam current curves.

Thermal Quenching. Another mechanism which may result in loss of phosphor efficiency is thermal quenching. In most phosphors the energy transitions become less efficient as screen temperature increases. This phenomenon can be quite pronounced in projection systems where the high power loading can be responsible for a large increase in screen temperature. Figure 12-31 shows some examples of thermal quenching. Quenching is a transient condition, and screens return to normal efficiency after being cooled.

Screen Burn and Aging. Aging is a nonreversible loss in phosphor efficiency caused by permanent damage to the crystal lattice. The susceptibility to burning varies among phosphor types, with higher-melting materials such as silicates generally being more resistant to burning than materials with lower heats of formation such as fluorides.

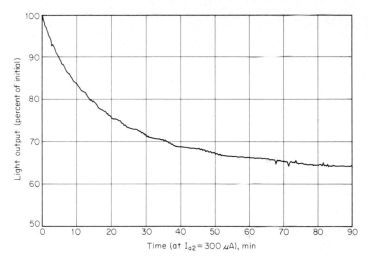

FIG. 12-31 Loss in phosphor efficiency as screen heats at high-current operation—rare-earth green.

Within given types the efficiency loss is proportional to the product of the beam current and the time it is applied. The term *coulomb aging* is often applied to the phenomenon. Screen burn is a discoloration or change in body color and can be visible on both excited and unexcited screens. In a picture tube differentiating between phosphor screen burn and glass solarization is difficult. References 90 and 91 review screen burn and aging.

Chromaticity. The fundamental method of measuring color is to disperse the radiant energy into its component parts using a prism or grating device such as a spectroradiometer. The resulting spectral energy distribution (SED) curve is a plot of relative energy as a function of wavelength. Figure 12-32 shows a typical composite curve for a

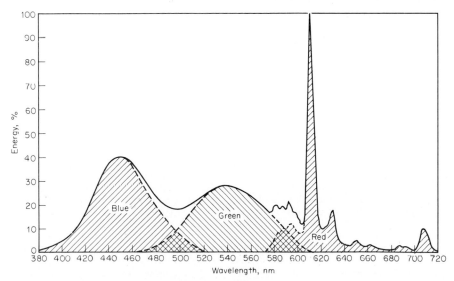

FIG. 12-32 Typical spectral energy distribution (SED) color primaries at equal current density.

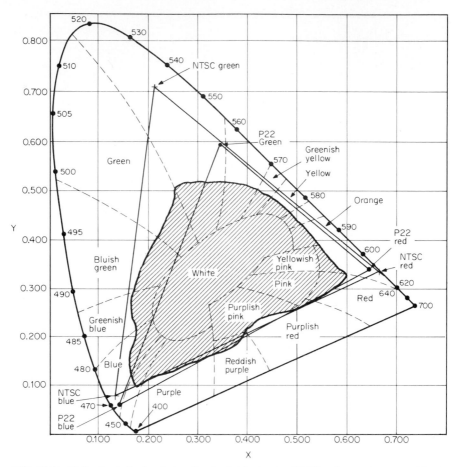

FIG. 12-33 Kelly chart of color designation for lights showing NTSC and current commercial primary phosphors and locus of dyes, paints, and pigments.

color tube at equal current density and the SED curves of the primary phosphors composing the screen. This figure also illustrates the two basic types of phosphor emission curves. The green and blue are *band* emitters, while the red is a *line* emitter.

The SED curve completely identifies the color of a light source, and colors can be matched or reproduced by exactly matching their SED curves. This is not an easy task in practice, and materials having very different SED curves can be perceived as the same colors. The most common method of color matching is based on the three-component, "standard observer" system developed by the International Commission on Illumination (CIE). The development of this system is discussed in Refs. 92 and 93. In this system all colors can be designated in terms of two coordinates. Figure 12-33 shows such a representation. The curved perimeter of this chart designates the locus of pure color points. Within the chart, if any two points are picked as primary colors, all colors lying on a line connecting those points can be achieved by blending suitable amounts of those primaries. Furthermore, if any three primaries are chosen, all colors lying within the resulting triangle can be reproduced by suitable blends of the primaries. This blend may be the actual physical mixing of two phosphors as in P4 or may be done electrically as in a color picture tube.

The NTSC specified primaries for color television are shown in Fig. 12-33, as well as the location of a set of coordinates for typical P22 color phosphors. Although the difference in the range of reproducible colors seems fairly large, viewing tests have shown that there is no significant objection to the reduced range of color rendition. As a further reference also included in Fig. 12-33 is the locus of colors currently achievable with dyes, inks, and pigments. This is discussed in Ref. 94.

Figure 12-34 shows an alternative CIE diagram in which color coordinates are expressed in terms of u' and v'. These values are derived from x and y by the following

$$u' = \frac{4x}{-2x + 12y + 3} \qquad v' = \frac{9y}{-2x + 12y + 3}$$

The advantage of the u', v' system is that equal differential changes in u', v' relate more directly to observable color differences than do changes in x and y. The CIE diagrams in Figs. 12-33 and 12-34 are adapted from Ref. 88.

Persistence.   The luminescence of all phosphors can be divided into two parts: fluorescence, in which the emission continues essentially only so long as the excitation is continued, and phosphorescence, in which the emission continues after the excitation is

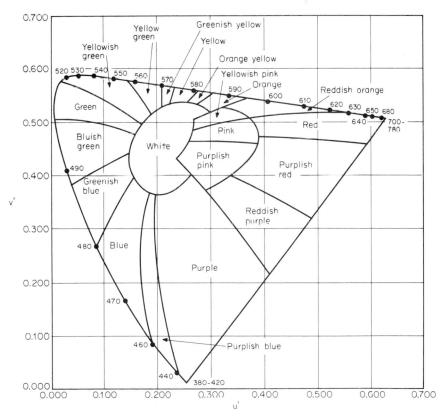

FIG. 12-34   Kelly chart of color designation for lights, u', v' coordinates.

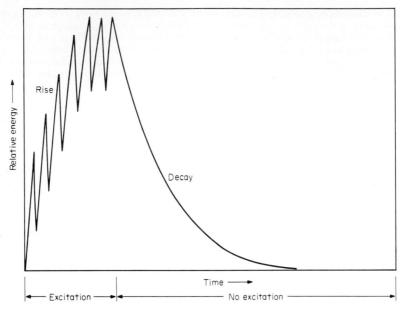

FIG. 12-35   Typical rise time–decay curve.

removed. Persistence is a measure of how much phosphorescence a phosphor exhibits. It is generally reported as the time required, after excitation is removed, for the light output to decrease to a specified level. Figure 12-35 shows a typical buildup and decay curve. The shape of the persistence curve for most phosphors is dependent on the excitation conditions and method used in measurement. The method employed in measurement should relate to the final application of the display device. For special uses persistence values for decay to 10 percent of the initial brightness may range from nanoseconds to several seconds. For television typical values range from 20 μs to a few milliseconds. The main criterion is that the decay is fast enough so that there is no objectionable smearing of moving objects due to afterglow. At standard NTSC refresh rates flicker due to too-rapid phosphor decay is not a serious problem.[95]

12.6.2 SCREEN CHARACTERISTICS

White Field Brightness.   Figure 12-36 shows the increase in color picture tube white field brightness (WFB) over the last two decades. This chart indicates nearly a fivefold increase in brightness during this time period. While some of this increase is due to improvements in phosphor efficiency, especially the introduction of the rare earth phosphors, many other factors have contributed to this improvement.

Significant increases in WFB have occurred owing to changes in chromaticity and color balance among the primary phosphors. As mentioned earlier, the primaries currently used do not correspond to the original NTSC primaries, but the marginal loss in color rendition is more than offset by the significant increase in WFB, owing to the more efficient balance of electron-gun current ratios.[96]

All CRT screens are now metal backed or aluminized. This thin, 150- to 300-nm film serves a number of purposes. Acting as a mirror, it reflects light from the phosphor which would have otherwise been lost to the interior of the tube. The film provides an electrically conductive path which eliminates screen charging and removes the need for phosphors to have good secondary emission levels. The aluminum film also serves to protect the phosphor screen from ion bombardment which would result in screen burn and loss in screen efficiency. Some of the increases in WFB relate to improvements in the lacquering and aluminizing processes.

Improvements in electron gun design and tube design have led to the ability to operate tubes at higher anode voltages and currents, and this has added to the increase in WFB.

The *black matrix* process which has been developed during this period has also added significantly to WFB. This process will be discussed in more detail under Contrast below. The combination of all these increases has led to the development of color picture tubes which no longer need to be viewed in a semidarkened room, but can be viewed comfortably under high ambient light levels.

Contrast.   Contrast ratio is normally defined as the ratio of the excited screen brightness to the level of brightness in the unexcited screen area. Unexcited screen brightness is composed of several factors. Since all common phosphors tend to be white-bodied or very light in body color, they reflect ambient light very efficiently. The reflected ambient light is a major component of the unexcited screen brightness. In addition to this, scattered electrons excite the "dark" areas, and the crystal nature of the phosphors tends to scatter light from the excited area into the nominally dark area.[97]

Historically contrast has been enhanced by placing neutral density filters between the phosphor screen and the viewer. Currently color picture tube faceplates are available in transmissions ranging from about 50 to 90 percent. Light emitted from the phosphor screen passes through this faceplate once, while ambient light is attenuated twice, so that

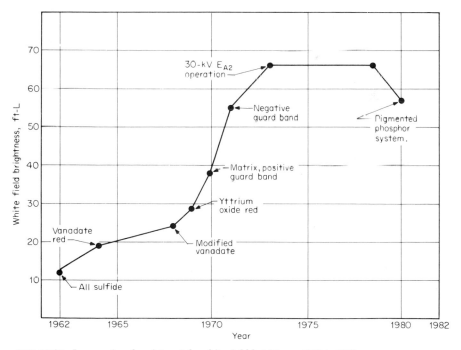

FIG. 12-36   Increase in color picture tube white-field brightness, 1962 to 1982.

Table 12-10 Typical Color Picture Tubes†

Tube type	Screen diag., cm‡	Defl. angle, deg.	Neck OD,‡ mm	Guns	Base/basing	Implosion protection	Panel % trans./ screen
				EIA types			
13VBFP22	33	90	29	I LB	B12-262/13D	TBE	85 NGB
13VBRP22	33	90	29	I LB	B12-262/13D	TB	52 NGB
13VCAP22	33	90	29	I LB	B12-262/13M	TBE	86 NGB
13VBMP22	33	100	29	I TPF	B9-271/13K	KE	86 NGB
15VAYP22	38	90	29	I HB	B8-274/13L	TBE	85 NGB
17VBTP22	43	90	29	I HB	B8-274/13L	KE	86 NGB
19VLWP22	48	90	29	I HB	B10-276/13N	KE	52 NGB
19VLYP22	48	90	29	I LB	B12-262/13D	TBE	85 NGB
19VMHP22	48	90	29	I HB	B10-276/13N	KE	86 NGB
19VMTP22	48	90	29	I LB	B12-262/13M	TBE	85 NGB
19VNDP22	48	90	29	I HB	B10-276/13N	TBE	85 NGB
19VMEP22	48	100	29	I TPF	B9-271/13K	TBE	86 NGB
21VBHP22	53	90	36	D LB	B12-244/14BE	K	52 NGB
25VFLP22	63	90	36	D LB	B12-244/14BE	KE	85 NGB
25VGDP22	63	90	29	I HB	B8-274/13L	KE	85 NGB
25VFQP22	63	100	29	I HB	B10-276/13N	KE	85 NGB
25VFTP22	63	100	29	I TPF	B9-272/13K	KE	85 NGB
25VGSP22	63	100	29	I TPF	B9-271/13K	K	85 NGB
25VHEP22	63	100	29	I TPF	B9-271/13K	K	52 NGB
25VHLP22	63	100	29	I TPF	B9-271/13K	KE	85 NGB
25VGRP22	63	110	29	I HB	B8-274/13L	KE	85 NGB
				Non-EIA types			
A42-592X	38	90	29	I HB	B10-277/13S	ShB	68 PGB
A51-592X	48	90	29	I HB	B10-277/13S	ShB	64 PGB
510-UTB22	48	90	29	I HB	Bi0-277/13S	KE	85 NGB
510-YWB22	48	90	23	I QPF	B8-288/-	KE	85 NGB
17-in Trinitron	44	90	29	Trin	B8-218/-	KE	70 NGB
19-in Trinitron	49	100	31	Trin	B12-262/-	KE	69 NGB
26-in Trinitron	67	100	31	Trin	. . . .	KE	66 NGB

†Key to abbreviations

D: Delta gun (dots)   TPF: Tripotential   K: Kimcode
I: In-line gun (stripes)   QPF: Quadripotential   TB: Tension band
LB: Low potential   NGB: Negative guard band   E: Mounting ears
HB: High potential   PGB: Positive guard band (nonmatrix)   ShB: Shelbond, ears

‡1 cm = 0.3937 in; 1 mm = 0.03937 in.

contrast is significantly improved. In some current tube designs further improvement is being made by matching the absortivity of the glass more closely to the emission of the phosphor.[98]

While the use of neutral-density filters has improved contrast at the expense of tube brightness, the development of the black matrix/black surround process for color picture tubes has produced an increase in both light output and contrast at the same time. In this process a layer of light-absorbing graphite is deposited so as to surround the phosphor dots or stripes.[99] By thus reducing screen reflectivity, contrast was significantly improved. In the first version of this process, the so-called beam-limited or positive guard band process, a relatively small amount of graphite was applied to the screen, and the exposed phosphor area was larger than that defined by the aperture mask holes. The more current version, window limited or negative guard band, employs matrix windows, or phosphor areas, smaller than the electron beam. The result is that more graphite remains on the faceplate and contrast is further improved. This increase in screen contrast has permitted a significant reduction in glass tint, which has led to a substantial increase in tube brightness while contrast has been maintained at acceptable levels.

Some further increase in contrast has been achieved recently through the technique of encapsulating the phosphor particles within a shell of pigment. These pigments have normally been applied to the red and blue phosphors. These pigments do not significantly absorb the emitted light of the phosphor, while absorbing other frequencies of ambient light. Since it is difficult to find pigments that exactly match the phosphor's emission, there is some loss in brightness in order to gain this contrast improvement.

In addition to the loss of picture quality due to reflections from the phosphor screen surface, picture viewability can be further degraded by reflections from the front surface of the tube face. Specular reflections of well-defined light sources are especially objectionable. They may be controlled in several ways. Converting the faceplate to a diffusing surface through grinding or etching reduces these specular reflections. Diffusing coatings may also be applied. Coatings which reduce specular reflections by means of optical interference are also available.[100]

Screen resolution or image sharpness is generally a function of electron gun design but may be influenced to some extent by phosphor particle size distribution and panel surface roughness. Image sharpness may be assessed by MTF. MTF is the sine-wave response of the system under test and is defined as the ratio of the modulation at each spatial frequency present in the final image normalized to the zero frequency modulation size. A screen's MTF indicates its ability to reproduce image detail. Resolution can be determined by the high-frequency cutoff of the MTF curve.[101]

12.7 COLOR PICTURE TUBES, TYPICAL DATA

Table 12-10 lists some typical color television tubes with descriptive data. The nonexhaustive list includes representative sizes and constructions based on specification sheets of the early 1980s (up to April 1982). As an example, type 13VBFP22 of the then-current EIA system is explained as follows: 13V means viewable screen diagonal in inches, BF are serially assigned letters for the screen potential, and P22 indicates a tricolor phosphor registered with the EIA Tube Engineering Panel Advisory Council (TEPAC).

12.7.1 **WORLDWIDE TYPE DESIGNATION SYSTEM FOR TELEVISION PICTURE TUBES AND MONITOR TUBES.** This system, developed jointly by the Electronic Industries Association (EIA) in the United States, EIA of Japan, and Pro-Electron of Europe, applies to all new registrations after April 1982. An example type assignment is A63AAAOOX for a color television picture tube with a 24-in (63-cm) screen diagonal. Details on the six-symbol system are explained in EIA Publication TEP-106 and are summarized in Table 12-11. The third symbol (three letters) defines tubes having specific mechanical and electrical characteristics.

Table 12-11 Worldwide Type Designation System for Television Picture Tubes and Monitor Tubes

Tube type category	First symbol (application code)	Second symbol (screen diagonal, mm)	Third symbol (family code)	Fourth symbol (member of family)	Fifth symbol (phosphor)	Sixth symbol (only for tubes with integral components)
Color television	A	Two digits	(Three letters assigned in alphabetic sequence to all four categories: AAA,AAB AAC . . . ABA, ABB)	00 to 99	Letter X	(If applicable, two-digit integer from 01 to 99)
Color monitor	M	Two digits		00 to 99	X, or other single letter except I,O,W	
Monochrome television	A	Two digits		0 to 9	WW	
Monochrome monitor	M	Two digits		0 to 9	WW, or two letters, except I or O	

REFERENCES

Principles of Color Picture Tube Design

1. D. G. Fink (ed.), *Television Engineering Handbook,* McGraw-Hill, New York, 1957, chap. 5, pp. 91–105.

2. J. Gow and R. Door, "Compatible Color Picture Presentation with the Single-Gun Tri Color Chromatron," *Proc. IRE,* vol. 42, no. 1, January 1954, pp. 308–314.

3. C. Carpenter et al., "An Analysis of Focusing and Deflection in the Post-Deflection-Focus Color Kinescope," *IRE Trans. Electron Devices,* vol. 2, 1955, pp. 1–7.

4. E. Herold, "A History of Color TV Displays," *Proc. IEEE,* vol. 64, no. 9, September, 1976, pp. 1331–1337.

5. R. Dressler, "The PDF Chromatron—A Single or Multi-Gun CRT," *Proc. IRE,* vol. 41, no. 7, July 1953, p. 853.

6. A. Morrell et al., *Color Television Picture Tubes,* Academic, New York, 1974, pp. 135–154.

7. K. Palac, Method for Manufacturing a Color CRT Using Mask and Screen Masters, U.S. Patent 3,989,524, 1976.

8. R. Clapp et al. "A New Beam Indexing Color Television Display System," *Proc. IRE,* vol. 44, no. 9, September 1956, pp. 1108–1114.

9. J. Swartz, "Beam Index Tube Technology," *SID Proceedings,* vol. 20, no. 2, 1979, p. 45.

10. J. Hasker, "Astigmatic Electron Gun for the Beam Indexing Color TV Display," *IEEE Trans. Electron Devices,* vol. ED-18, no. 9, September 1971, p. 703.

11. A. Ohkoshi et al., "A New 30V″ Beam Index Color Cathode Ray Tube," *IEEE Trans. Consumer Electronics,* vol. CE-27, August 1981, p. 433.

12. R. Barbin and R. Hughes, "New Color Picture Tube System for Portable TV Receivers," *IEEE Trans. Broadcast TV Receivers,* vol. BTR-18, no. 3, August 1972, 193–200.

13. W. Rublack, In-Line Plural Beam CRT with an Aspherical Mask, U.S. Patent 3,435,668, 1969.

14. W. Flechsig, CRT for the Production of Multicolored Pictures on a Luminescent Screen, French Patent 866,065, 1939.

15. S. Yoshida et al., "The Trinitron—A New Color Tube," *IEEE Trans. Consumer Electronics,* vol. CE-28, no. 1, February 1982, 56–64.

16. A. Morrell et al., *Color TV Picture Tubes,* Academic, New York, 1974, p. 119.

17. J. Fiore and S. Kaplin, "A Second Generation Color Tube Providing More Than Twice the Brightness and Improved Contrast," *IEEE Trans. Consumer Electronics,* vol. CE-28, no. 1, February 1982, pp. 65–73.

18. N. Mears, "Method and Apparatus for Producing Perforated Metal Webs," U.S. Patent 2,762,149, 1956.

19. H. Law, "A Three-Gun Shadowmask Color Kinescope," *Proc. IRE,* vol. 39, October 1951, pp. 1186–1194.

20. "CRTs, Glossary of Terms and Definitions," Publication TEP92, Electronic Industries Association, Washington, 1975.

21. J. Robbins and D. Mackey, "Moire Pattern in Color TV," *IEEE Trans. Consumer Electronics,* vol. CE-28, no. 1, February 1982, pp. 44–55.

22. R. Godfrey et al., "Development of the Permachrome Color Picture Tube," *IEEE Trans. Broadcast TV Receivers,* vol. BTR-14, no. 1, 1968, pp. 8–11.

23. "Degaussing TV Tubes," (author unlisted), *Philips Tech. Rev.* vol. 29, December 1968, p. 368.

24. R. Blaha, "Degaussing Circuits for Color TV Receivers," *IEEE Trans. Broadcast TV Receivers,* vol. BTR-18, no. 1, February 1972, pp. 7–10.

25. J. Schwartz and M. Fogelson, "Recent Developments in Arc Suppression for Picture Tubes," *IEEE Trans. Consumer Electronics,* vol. CE-25, no. 1, February 1979, pp. 82–90.

26. J. Gerritson, "Soft Flash Picture Tubes," *IEEE Trans. Consumer Electronics,* vol. CE-24, no. 4, November 1978, pp. 560–561.

27. Y. Kobari et al., "A Novel Arc Suppression Technique for CRTs," *IEEE Trans. Consumer Electronics,* vol. CE-26, no. 3, August 1980, pp. 446–450.

28. Underwriters Laboratory Report UL492.8, Jan. 25, 1974.

29. W. Dickenson, "Monochrome Picture Tubes—Status Report," *IEEE Trans. Broadcast TV Receivers,*" vol. BTR-13, no. 3, 1967, pp. 46–48.

30. "Recommended Practice for Measurement of X-Radiation from Direct View TV Picture Tubes," Publication TEP 164, Electronics Industries Association, Washington, 1981.

31. N. Mokhoff "A Step Toward Perfect Resolution," *IEEE Spectrum,* vol. 18, no. 7, July 1981, pp. 56–58.

Electron Guns for Shadow-Mask Color Tubes

32. H. Moss, *Narrow Angle Electron Guns and Cathode Ray Tubes,* Academic, New York, 1968, pp. 15–26.

33. F. Oess, "CRT Considerations for Raster Dot Alpha Numeric Presentations," *Proc. SID,* vol. 20, no. 2, second quarter, 1979, pp. 81–88.

34. A. Morrell et al., *Color Television Picture Tubes,* Academic, New York, 1974, pp. 91–98.

35. A. Morrell, "Color Picture Tube Design Trends," *Proc. SID,* vol. 22, no. 1, 1981, pp. 3–9.

36. C. Davis and D. Say, "High Performance Guns for Color TV—A Comparison of Recent Designs," *IEEE Trans. Consumer Electronics,* vol. CE-25, August 1979.

37. D. Say, "The High Voltage Bipotential Approach to Enhanced Color Tube Performance," *IEEE Trans. Consumer Electronics,* vol. CE-24, no. 1, February 1978, p. 75.

38. A. Blacker et al., "A New Form of Extended Field Lens for Use in Color Television Picture Tube Guns," *IEEE Trans. Consumer Electronics,* August 1966, pp. 238–246.

39. S. Yoshida et al., "25-V Inch 114-Degree Trinitron Color Picture Tube and Associated New Development," *Trans. BTR,* August 1974, pp. 193–200.

40. S. Yoshida et al., "A Wide Deflection Angle (114°) Trinitron Color Picture Tube," *IEEE Trans. Electron Devices,* vol. 19, no. 4, 1973, pp. 231–238.

41. A. Johnson, "Color Tubes for Data Display—A System Study," Philips ECG, Electronic Tube Division.

42. W. Masterson and R. Barbin, "Designing Out the Problems of Wide-Angle Color TV Tube," *Electronics,* April 26, 1971, pp. 60–63.

43. B. Lucchesi and M. Carpenter, "Pictures of Deflected Electron Spots from a Computer," *IEEE Trans. Consumer Electronics,* vol. CE-25, no. 4, 1979, pp. 468–474.

44. H. Chen and R. Hughes, "A High Performance Color CRT Gun with an Asymmet-

rical Beam Forming Region," *IEEE Trans. Consumer Electronics,* vol. CE-26, August 1980, pp. 459–465.

45. CRT Control Grid Having Orthogonal Openings on Opposite Sides, U.S. Patent 4,242,613, Dec. 30, 1980.

46. Electron Gun with Astigmatic Flare—Reducing Beam Forming Region, U.S. Patent 4,234,814, Nov. 18, 1980.

47. D. Say, "Picture Tube Spot Analysis Using Direct Photography," *IEEE Trans. Consumer Electronics,* vol. CE-23, February 1977, pp. 32–37.

48. K. Hoskoshi et al., "A New Approach to a High Performance Electron Gun Design for Color Picture Tubes," *1980 IEEE Chicago Spring Conf. Consumer Electronics.*

Large-Screen Displays

49. D. Fink, *Television Engineering Handbook,* McGraw-Hill, New York, 1957, pp. 15–42.

50. H. Luxenberg and R. Kuehn, *Display Systems Engineering,* McGraw-Hill, New York, 1968.

51. S. Sherr, *Fundamentals of Display System Design,* Wiley-Interscience, 1970.

52. R. Howe and B. Welham, "Developments in Plastic Optics for Projection Television Systems," *IEEE Trans.,* vol. CE-26, no. 1, February 1980, pp. 44–53.

53. S. McKechnie, Philips Laboratories (NA) report, 1981, unpublished.

54. A. Pfahnl, "Aging of Electronic Phosphors in Cathode Ray Tubes," *Advances in Electron Tube Techniques,* Pergamon, New York, pp. 204–208.

55. K. Schiecke, "Projection Television: Correcting Distortions," *IEEE Spectrum,* vol. 18, no. 11, November 1981, pp. 40 45.

56. M. Kikuchi et al., "A New Coolant-Sealed CRT for Projection Color TV," *IEEE Trans.,* vol. CE-27, no. 3, August 1981, pp. 478–485.

56a. "X-Radiation Measurement Procedures for Projection Tubes," TEPAC Publication 102, Electronic Industries Association, Washington.

57. S. Wang et al., "Spectral and Spatial Distribution of X-Rays from Color Television Receivers," *Proc. Conf. Detection and Measurement of X-radiation from Color Television Receivers,* Washington, March 28–29, 1968, pp. 53–72.

58. A. Robertson, "Projection Television—1 Review of Practice," *Wireless World,* vol. 82, no. 1489, September 1976, pp. 47–52.

59. W. Good, "Projection Television," *IEEE Trans.,* vol. CE-21, no. 3, August 1975, pp. 206–212.

60. A. I. Lakatos and R. F. Bergen, "Projection Display Using an Amorphous-Se-Type Ruticon Light Valve," *IEEE Trans. Electron Devices,* vol. ED-24, no. 7, July 1977, pp. 930–934.

61. A. Nasibov et al., "Electron-Beam Tube with a Laser Screen," *Sov. J. Quant. Electron.* vol. 4, no. 3, September 1974, pp. 296–300.

62. K. Kurahashi et al., "An Outdoor Screen Color Display System," *SID Int. Symp. Digest 7,* Technical Papers, vol. XII, pp. 132–133, April 1981.

63. T. Taneda et al., "A 1125-Scanning Line Laser Color-TV Display," *SID 1973 Symp. Digest Technical Papers,* vol. IV, May 1973, pp. 86–87.

64. N. Itah et al., "New Color Video Projection System with Glass Optics and Three Primary Color Tubes for Consumer Use," *IEEE Trans. Consumer Electronics,* vol. CE-25, no. 4, August 1979, pp. 497–503.

65. J. Grinberg et al., "Photoactivated Birefringent Liquid-Crystal Light Valve for Color Symbology Display," *IEEE Trans. Electron Devices*, vol. ED-22, no. 9, September 1975, pp. 775–783.

Monochrome Video Display Devices

66. D. Fink, *Television Engineering Handbook*, McGraw-Hill, New York, 1957, pp. 5-81–5-85.

67. J. Boers, "Computer Simulation of Space Charge Flows," Rome Air Development Command RADC-TR-68-175, University of Michigan, 1968.

68. R. True, "Space Charge Limited Beam Forming Systems Analyzed by the Method of Self-Consistent Fields with Solution of Poisson's Equation on a Deformable Relaxation Mesh," Ph.D. thesis, University of Connecticut, Storrs, 1968.

69. D. Langmuir, "Limitations of Cathode Ray Tubes," *Proc. IRE*, vol. 25, 1937, pp. 977–991.

70. D. Say, "The High Voltage BiPotential Approach to Enhanced Color Tube Performance," *IEEE Trans. Consumer Electronics*, vol. CE-24, no. 1, 1978.

71. L. Nix, "Spot Growth Reduction in Bright, Wide Deflection Angle CRT's," *SID Proc.*, vol. 21, no. 4, 1980, p. 315.

72. S. Sherr, *Electronic Displays*, Wiley, New York, 1979, pp. 602–609.

73. D. Say, "Picture Tube Spot Analysis Using Direct Photography," *IEEE Trans. Consumer Electronics*, vol. CE-23, February 1977, pp. 32–37.

74. R. Donofrio, "Image Sharpness of a Color Picture Tube by MTF Techniques," *IEEE Trans. Broadcast TV Receivers*, vol. BTR-18, no. 1, February 1972, pp. 1–6.

75. S. Sherr, *Electronic Displays*, Wiley, New York, 1979, Chap. 3.

76. *SID Proc.* vol. 17, no. 1, first quarter, 1976.

77. I. Chang, "Recent Advances in Display Technologies," *Proc. SID*, vol. 21, no. 2, 1980, p. 45.

78. J. Van Raalte, "Matrix TV Displays: Systems and Circuit Problems," *Proc. SID*, vol. 17, no. 1, 1976, p. 10.

79. T. Hosokawa et al., "Dichroic Guest-Host Active Matrix Video Display," *SID '81 Digest*, p. 114.

80. K. Kaneko et al., "Liquid Crystal TV Display," *Proc. SID*, vol. 19, no. 2, 1978, p. 49.

81. C. Sinclair, "Small Flat Cathode Ray Tube," *SID Digest*, 1981, p. 138.

82. C. Cohen, "Sony's Pocket TV Slims Down CRT Technology," *Electronics*, Feb. 10, 1982, p. 81.

83. J. A. Aiken, "A Thin Cathode Ray Tube," *Proc. IRE*, vol. 45, 1957, p. 1599.

84. Thomas L. Credelle et al., "Cathodoluminescent Flat Panel TV Using Electron Beam Guides," *SID Int. Symp. Digest*, 1980, p. 26.

85. T. Stanley, Flat Cathode Ray Tube, U.S. Patent 4,031,427; J. Schwartz, Electron Beam Cathodoluminescent Panel Display, U.S. Patent 4,137,486; T. Credelle, Modular Flat Display Device with Beam Convergence, U.S. Patent 4,131,823.

Phosphor and Screen Characteristics

86. H. Leverenz, *An Introduction to Luminescence of Solids*, Dover, New York, 1968.

87. P. Goldberg, *Luminescence of Inorganic Solids*, Academic, New York, 1966.

88. "Optical Characteristics of Cathode Ray Tube Screens," TEPAC Publication 116, Electronic Industries Association, Washington, 1980.

89. "Optical Characteristics of Cathode Ray Tube Screens," JEDEC Publication 16C, Electron Tube Council, Washington, 1971.

90. A. Pfahnl, "Aging of Electronic Phosphors in Cathode Ray Tubes," in *Advances in Electron Tube Techniques*, Pergamon, New York, 1961, pp. 204–208.

91. R. Donofrio, "Low Current Density Aging," *Proc. Electrochemical Society*, May 12, 1981.

92. D. Judd and G. Wyszecki, *Color in Business, Science and Industry*, Wiley, New York, 1975.

93. G. Wyszecki and W. Stiles, *Color Science*, Wiley, New York, 1967.

94. W. Wintringham, "Color Television and Colorimetry," *Proc. IRE*, vol. 39, 1951, pp. 1135–1172.

95. T. Rychlewski and R. Vogel, "Phosphor Persistence in Color Television Screens," *Electrochemical Technology*, vol. 4, no. 1–2, January-February 1966, pp. 9–12.

96. R. Donofrio, "Color in Color T.V.—A Phosphor Approach," *Color Engineering*, February 1971, pp. 11–14.

97. R. Vogel, "Contrast Measurements in Color T.V. Tubes," *IEEE Conf.* Chicago, 1970.

98. A. V. Gallaro and R. A. Hedler, Process for Forming a Color CRT Screen Structure Having Optical Filter Therein, U.S. Patent 3,884,694, 1973.

99. J. Fiore and S. Kaplan, "A Second Generation Color Tube Providing More than Twice the Brightness and Improved Contrast," *Spring Conf. Broadcast and Television Receivers*, IEEE, June 1969.

100. R. Hunter, *The Measurement of Appearance*, Wiley, New York, 1975.

101. R. Donofrio, "Image Sharpness of a Color Picture Tube by Modulation Transfer Techniques," *IEEE Tran. Broadcast Television Receivers*, vol. BTR-18, no. 1, February 1972, p. 16.

Receivers

L. H. Hoke, Jr.

NAP Consumer Electronics Corporation
Knoxville, Tennessee

with contributions by

L. E. Donovan

J. D. Knox

D. E. Manners

W. G. Miller

R. J. Peffer

J. G. Zahnen

NAP Consumer Electronics Corporation
Knoxville, Tennessee

13.1 FUNCTIONAL CONFIGURATION OF MONOCHROME RECEIVERS

13.1.1 GENERAL CONSIDERATIONS. Television receivers perform the functions of receiving the appropriate signal being broadcast through the air or via cable and of displaying it for viewing. Broadcast channels in the United States are 6 MHz wide and have been allocated for television service as given in Table 13-1. Cable system channels are similar to broadcast and are discussed in greater detail in Sec. 13.4.1

Television receivers typically operate in an environment having desired signal levels at the antenna terminals as low as 10 to 20 μV (sensitivity level) and upward to several hundred millivolts. For the purpose of station allocation, the Federal Communications Commission (FCC) has established two standard signal levels or classes of service. *Grade A* is considered to be an urban area near the transmitting tower. *Grade B* is considered to be the suburban and rural area, several miles from the transmitting antenna (fringe reception.) The standard field-strength values assigned to the outer edge of these service areas by the FCC, as well as the resultant dipole antenna terminal voltage into a *matched receiver load,* as calculated by the formula below, are shown in Table 13-1 (derivation is covered in Sec. 13.9.10)

$$e = E \frac{96.68}{\sqrt{f_1 \times f_2}} \tag{13-1}$$

where e = terminal voltage, μV, 300 Ω
E = field, μV/m
f_1 and f_2 = band-edge frequencies, MHz

Many sizes and form factors of monochrome receivers are available. The personal portable class includes pocket-sized or hand-held models having picture sizes of 2 to 4 in (5 to 10 cm) (diagonal) powered by four AA cells (6 V) to nine-C cells (12 V).[1]† Picture display types in this rapidly growing class are liquid crystal, unique flat cathode-ray tube, and the more conventional CRTs. The table-model class covers receivers having picture dimensions of 10 to 12 in (diagonal). These are available in models operated by 12-V batteries, as well as models intended for 120-V ac power line operation. Monochrome monitors intended for home personal computers and video games also fit in this category. Console monochrome receivers have picture sizes of 15 to 21 in (38 to 53 cm) (diagonal) and are ac-line powered.

The functional block diagram of a conventional monochrome receiver is shown in Fig. 13-1. The dashed lines indicate optional connections which are applicable to various circuit implementation schemes as discussed in Secs. 13.11 and 13.12.

---

†Superscript numbers refer to References at end of chapter.

Table 13-1   Television Service

Band and channels	Frequency, MHz	City grade		Grade A		Grade B	
		μV/m	μV	μV/m	μV	μV/m	μV
VHF 2–6	54–88 MHz	5,010	7030	2510	3520	224	314
VHF 7–13	174–216 MHz	7,080	3550	3550	1770	631	315
UHF 14–69	470–806 MHz	10,000	1570	5010	787	1580	248
UHF 70–83‡	806–890 MHz	10,000	1570	5010	571	1580	180

‡Receiver coverage of channels 70 to 83 has been on a voluntary basis since July 1982. This frequency band was reallocated by the FCC to land mobile use in 1975 with the provision that existing transmitters could continue indefinitely.

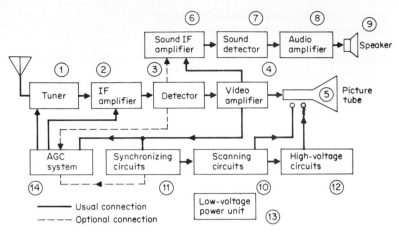

FIG. 13-1   Fundamental block diagram, monochrome receiver.

13.1.2 RF AND IF AMPLIFICATION   The purpose of the tuner and IF amplifier sections labeled ① and ② in Fig. 13-1 is to select the desired signal available at the antenna and exclude all other signals, and to amplify it to a level such that demodulation of the information on the carrier can take place. The tuner section, shown in Fig. 13-2, consists of a VHF tuner which can be tuned selectively to any of the 6-MHz-wide channels in the 54- to 88-MHz and 174- to 216-MHz bands, and a UHF tuner which can tune to any of the channels in the 470- to 890-MHz band.

VHF Tuner.   The VHF tuner contains a tunable RF stage which provides gain of 15 to 20 dB and passband selectivity of 10 MHz at the 3-dB points, and the lowest noise figure possible consistent with cost, bandwidth, and dynamic range. The mixer stage combines the RF signal with the output of the tunable local oscillator to yield the intermediate frequency (IF) signals of 45.75 MHz for picture carrier and 41.25 MHz for sound carrier. The local oscillator operates at a frequency which is always 45.75 MHz above the desired incoming picture frequency. Choice of these frequencies was made originally to minimize the interference of one television receiver with another by always having the local oscillator fundamental above each VHF television band. Further, the image response produced by a high-VHF station and the local oscillator would also fall above the low-VHF band. The images which might fall in the high-VHF band would be from lower-powered nontelevision transmitters well above the VHF band. In the UHF band, no such gap between groups of channels exists. Therefore, the FCC UHF-channel allocation plan avoids assigning transmitters with a 15-channel separation in the same area.

UHF Tuner.   The UHF tuner contains a preselective tunable filter followed by a diode mixer. The tunable local-oscillator fundamental also operates at 45.75 MHz above the desired UHF picture carrier, thus producing an IF output signal having picture carrier of 45.75 MHz and a 41.25-MHz sound carrier. Since the diode mixer and input circuitry result in a conversion loss instead of a gain, the IF output of the UHF tuner is typically fed to the VHF mixer, which operates as an IF amplifier in this configuration. Choice between UHF and VHF is made by selectively turning on or off the power supply to the UHF tuner and VHF RF stage.

Both *mechanically* and *electronically controlled* tuners are utilized in monochrome and color receivers. The mechanically controlled VHF tuner has a 13-position rotary switch with wafers or sections which switch various tuning coils into the RF, mixer, and local-oscillator circuits. The channel-1 position, no longer required for VHF channels, establishes the configuration for UHF reception. Mechanical UHF tuners utilize rotating

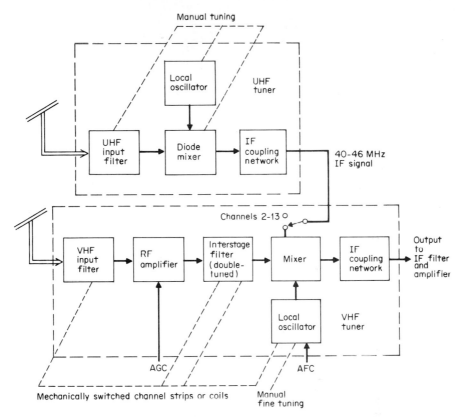

FIG. 13-2 Typical mechanical-tuner configuration.

capacitor plates to tune stripline inductors for the preselector and oscillator circuits. Seventy-channel discrete selection, necessary to meet the FCC *Comparability of Tuning Regulations*,[2] has been achieved by attaching mechanical detent mechanisms to these same tuners.

Varactor Tuner.   Varactor tuners utilize the change in capacitance by voltage characteristic of a varicap diode. One diode is used in each tank circuit. Additional gain is required to compensate for the lower Q and noise-figure degradation; therefore, an RF stage is typically included in the varactor UHF tuner, making it functionally similar to a VHF tuner (Fig. 13-3). Varactor tuners have no moving parts or mechanisms and, therefore, have about one-fourth to one-third the volume of their mechanical counterparts. The current trend in varactor tuner design is to incorporate both the UHF and VHF tuners on the same printed-circuit board, in the same enclosure, with a common antenna input connection.

Tuning Systems.   The purpose of the tuning system is to set the tuners to the desired channel. In mechanical tuners, this obviously is the rotary switch or capacitor shaft and knob on the dial. In electronically tuned systems, the dc tuning voltage can be supplied from the wiper arm of a potentiometer control connected to a voltage source, or for discrete selection of channels, a multiple of this configuration with each control set to a

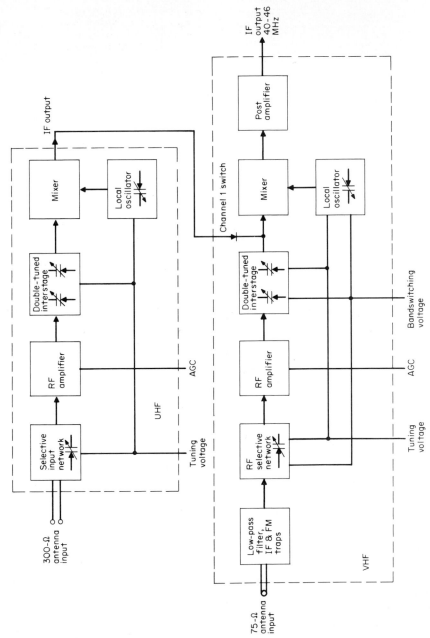

FIG. 13-3 Typical varactor-tuner configuration.

13.8

specific channel voltage and the appropriate one connected to the tuner by a selector switch as shown in Fig. 13-4. In digital systems, the tuning voltage can be read as a digital word from one of n memory locations, then, after conversion from digital to analog, sent to the tuner. Currently, the most popular scheme employs a microprocessor-controlled phase-locked loop in which a channel number request from a keyboard is converted to a

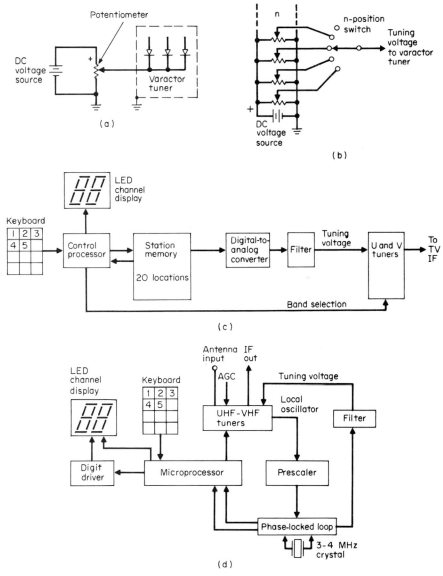

FIG. 13-4   Varactor-tuning systems: (a) simple potentiometer control varactor-tuning system; (b) multiple of potentiometers provides n-channel selection; (c) simplified memory-tuning for 20-station selection; (c) microprocessor phase-locked loop.

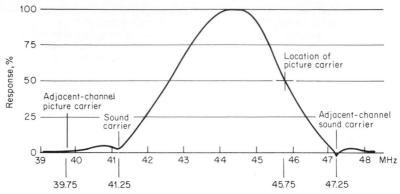

FIG. 13-5 Typical overall IF response.

medium-frequency square-wave signal in the range 400 kHz to 3.6 MHz.[3,4] This then is compared with the local oscillator frequency which has been processed by a divide-by-256 prescaler. The error signal generated by the difference of these two frequencies is filtered and used to correct the tuning voltage being supplied to the tuner.

The *intermediate frequency section* is tuned to a specific frequency band and provides the sharpest rejection selectivity for adjacent-channel carriers. Typically, the upper adjacent-channel picture carrier and lower adjacent-sound carrier must be attenuated 40 and 50 dB, respectively, to eliminate visible disturbances in the picture. The IF passband is also tailored to compensate for the vestigial sideband character of the transmitted television signal (Nyquist slope) (Fig. 13-5). Amplification needed to increase the signal level from the mixer output (200 μV) to the detector or demodulator stage input (2 V) represents a gain of 80 dB (Fig. 13-6).

Automatic gain control (AGC) is a closed feedback loop used with all tuner and IF sections to prevent overload in the IF and mixer stages from a strong signal at the antenna. This is accomplished by monitoring the signal level at the detector and adjusting the gain of the RF stage and one or more IF stages to keep a constant carrier level at the detector. Best signal-to-noise ratio for signals from threshold sensitivity (10 to 20 μV) to medium level (1 mV) is obtained if only the IF stages are gain-reduced while the RF gain is held at its maximum value. For signals greater than this level, both RF and IF are gain-reduced.

Automatic frequency control (AFC), another closed feedback loop, monitors the frequency of the video carrier in the IF section and develops an error signal whenever the carrier is above or below 45.75 MHz. This dc error signal is then sent to the tuner to shift the local oscillator in the direction to diminish the error.

13.1.3 VIDEO SIGNAL DEMODULATION AND AUDIO SIGNAL SEPARATION.

The function of the video demodulator, ③ of Fig. 13-1, is to recover the picture information which has been placed on the video carrier as vestigial-sideband amplitude

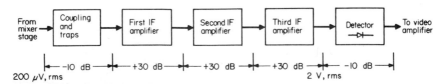

FIG. 13-6 Typical IF-amplifier strip block diagram and gain distribution.

modulation. The detector receives its signal from the IF-amplifier section and sends the demodulated output (direct current to 4.5 MHz) to the video amplifier.

Undesired beat products can be generated during the process of demodulation. One of the more common is produced by the combination of the 41.25-MHz sound signal and the 42.17-MHz chroma subcarrier, resulting in a 920-kHz spurious signal in the video output. Nonlinear demodulator circuits, such as the conventional diode envelope detector and the simple transistor detector, require that both the sound and chroma IF signals be greatly attenuated ahead of the detector. Synchronous and balanced detectors, more commonly used in color receivers because of their linear amplitude characteristic, produce considerably lower levels of spurious output and, therefore, can be operated with a wider bandwidth signal at the input.

Intercarrier Sound Separation. In this system, shown in Fig. 13-7, both the picture and sound carriers of the desired channel are amplified by the IF section. However, the sound carrier amplification is, typically, about 10 to 20 dB less than the picture carrier. The two carriers are mixed in the video-detector, and the resulting 4.5-MHz difference-frequency signal, containing sound-carrier frequency modulation, appears in the detector output. In monochrome designs, which use a single-diode or transistor-envelope detector, 10- to 20-dB attenuation of the lower-frequency portion of the IF response (43.25 to 41.25 MHz, including color subcarrier) is usually provided to reduce the 920-kHz sound-chroma beat in the picture signal. If a synchronous demodulator or balanced-diode detector with a more linear characteristic is used, considerably less attenuation of the lower frequencies is needed.

In a diode-detector system, to achieve improved video and sound performance, with lower beat-pattern and intercarrier *buzz* levels, it has been customary to greatly attenuate sound (>40 dB) just ahead of the video deomodulator. A portion of the 41.25- and 45.75-MHz carriers is taken from the second or third IF stage and mixed in a separate diode detector to produce 4.5 MHz (Fig. 13-8). This signal is then processed in a separate sound IF amplifier.

The primary advantage of the intercarrier system is that, because of the fact that the frequency difference between the carriers is established accurately by the transmitter, the frequency and detection of the resulting sound carrier is independent of receiver tuning. Consequently, the effects of local-oscillator drift and microphonics are minimized, and tuning of the receiver has only a minor effect on the sound performance, except under very weak signal or adverse noise conditions.

Nevertheless, good performance with an intercarrier detection system can be obtained only if certain prerequisites are met. The transmitter picture carrier must not be overmodulated in the white direction (zero carrier). Transmission standards call for the maintenance of carrier level on peak-white modulation at 12.5 percent of maximum picture carrier level (Sec. 21.5.1). Such overmodulation causes a temporary interruption of the picture carrier and, hence, the intercarrier beat signal. The resulting error in the audio modulation may result in a 60-Hz buzz of distorted sound.

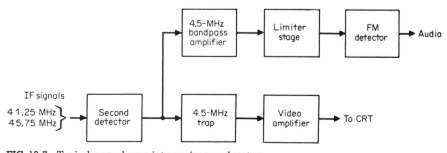

FIG. 13-7 Typical monochrome intercarrier-sound system.

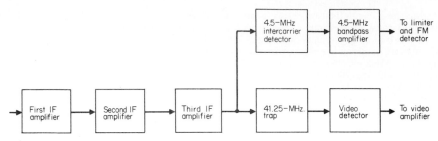

FIG. 13-8   Intercarrier-sound takeoff ahead of detector.

Split-Carrier Sound.   To eliminate the sound and picture beat problems and to achieve high-fidelity sound, the original split-sound system has recently been revived.[5,6] The 41.25-MHz carrier at the output of the tuner is bandpassed to a separate IF amplifier system while a sound trap is placed in the video IF chain, as shown in Fig. 13-9. The

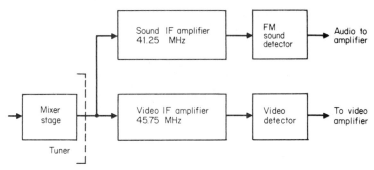

FIG. 13-9   Split-carrier sound system.

system, in theory, avoids the sound buzz caused by video modulation in a common demodulator. In practice, however, it is subject to the *incidental carrier phase modulation* (ICPM) created in the mixer stage.[7] ICPM can also occur in the low-cost modulator supplied in a typical home video cassette recorder or video disc player.

Quasi-Parallel Sound.   A more effective solution for high-fidelity sound appears to be the quasi-parallel system, a hybrid of split and intercarrier (Fig. 13-10). Parallel paths for sound and picture carriers are created at the output of the tuner. For the sound channel, a portion of video and sound carriers is passed through a *double-humped* bandpass which has a rejection trap at the color subcarrier frequency. The object is to ensure that sound and picture carriers are both on flat response curves so that little or no ICPM can occur. At this point a separate diode detector creates the 4.5-MHz intercarrier signal which is then processed conventionally. As in the split system, the 41.25-MHz signal in the video IF channel is attenuated 50 dB or greater to eliminate the sound-color beat products. This approach has become feasible only since the advent of integrated circuits and surface-wave filters.[8]

13.1.4   VIDEO AMPLIFICATION.   The last block, ④ of Fig. 13-1, in the picture signal processing chain is the video amplifier. This circuit amplifies the output signal from the video detector (1 to 2 V peak-to-peak) to a signal large enough to drive the cathode-ray

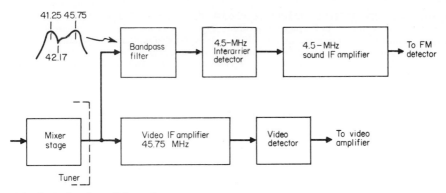

FIG. 13-10   Quasi-parallel sound system.

tube or other display device. Typical monochrome CRTs in the 5- to 12-in (13- to 30-cm) class require 20 to 40 V of video drive. Figure 13-11 shows a typical video amplifier block diagram.

Picture Controls.   In addition to driving the CRT, the video output stage usually provides the receiver-operator control functions. The brightness control typically shifts dc bias to the CRT to raise or lower the brightness level of all parts of the picture. The contrast control changes the peak-to-peak video drive level to the picture tube, either by attenuator action between the output of the video stage, or more commonly in larger screen sizes, by changing the ac gain of the video stage.

Blanking.   During the sync portion of the video signal, the CRT beam is moving rapidly from right to left (horizontal retrace of ≈12 μs) and from bottom of the picture to the top (vertical retrace ≈1.4 ms). To avoid displaying the beam path during these periods, it is necessary that the CRT beam be biased off. This is accomplished by coupling both vertical and horizontal retrace (flyback) pulses into the video amplifier or to the control grid of the CRT.

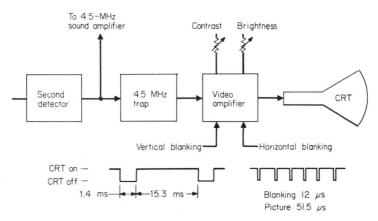

FIG. 13-11   Video amplifier.

13.1.5 SOUND SIGNAL AMPLIFICATION AND DEMODULATION. The first major penetration of multistage integrated circuits in television receivers was in the sound channel, beginning in the late 1960s. Although discrete transistor stage designs are still found, the majority of television receivers, including monochrome, utilize one or two ICs and 15 to 25 external resistors, capacitors, and coils to implement the complete sound channel. A functional diagram is shown in Fig. 13-7.

Bandpass Filtering. After the 4.5-MHz sound IF is generated in the video detector or by a separate intercarrier detector, it is necessary to remove as much of the video (especially sync) as is possible. Ahead of an IC amplifier a single- or double-tuned *LC* network having a sharp low-frequency cutoff characteristic is frequently used. A more recent alternative consists of a nonadjustable ceramic resonator-type filter. In a multistage transistorized design, tuned *LC* interstage coupling networks provide excellent selectivity.

Sound IF Amplifier and Limiter. Since the sound IF signal is frequency-modulated, a great deal of extraneous video amplitude-modulation can be removed by amplifying and limiting prior to the FM detector. The amount of gain and the limiting threshold will depend upon the level of input and the type of detector to be used. For example, video detectors typically produce a higher amplitude of intercarrier signal than that generated by a mixing diode in an early IF stage. In the former, a gain of 30 to 40 dB may suffice, while the latter will require at least 60 dB.

FM Demodulators. Following the limiter stage, the FM 4.5-MHz intercarrier sound signal is processed by an FM detector. The resulting output is the audio signal which is then sent to the audio amplifier and speaker.

Several types of FM detectors are used in television receivers. The classic ratio-detector has been preferred in discrete designs because of its simplicity and inherent AM rejection capability. In an IC format, the choice consists of the balanced coincidence or quadrature detector, the differential peak detector, and the phase-lock loop detector. A more detailed description of these is found in Sec. 13.12.

13.1.6 SCANNING SIGNALS AND HIGH-VOLTAGE GENERATION. In order to display a recognizable picture, the information in the video signal must be displayed on a planar surface in exactly the same position relative to all the information transmitted from the scene pickup device. Conventionally, CRTs have been the display device; however, with the more recent mosaic- or matrix-driven displays, the process is similar. Three major components make up the scanning function, i.e., row addressing (horizontal scan or sweep), column addressing (vertical scan), and master synchronization. For simplification, the further discussion will consider, primarily, the cathode-ray tube display techniques.

Picture scan, as shown in Fig. 13-12, is from left to right and top to bottom, consisting of 525 horizontal lines per frame and 30 frames per second. Each frame is divided into two alternating fields. Referring to Fig. 13-12 beginning at the upper left, the lines 21, 22, 23, . . . up to the first half of line 263 make up the first field. The beam then travels back to the picture top during the vertical interval period. This is not an instantaneous bottom-to-top jump, but actually requires the length of time to scan 20 horizontal lines (left-right diagonal trajectory to the top of the picture). These lines are numbered 264 through 283.

The second field then begins with the second half of line 283 and continues with 284 · · · to 525, then back up to the upper left corner, using up another 20 lines in the process, to begin line 21 of field 1 once more.

Sync Separator. The synchronizing signals, which are needed to lock the picture display beam to the camera or other source, are transmitted at a higher modulation level than the normal video information. The sync separator extracts these synchronizing pulses from the composite video signal and sends them to the horizontal and vertical scan waveform generators. The waveform of these sync pulses is shown in Fig. 13-13.

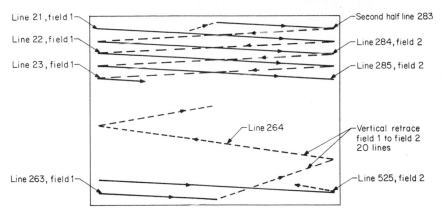

FIG. 13-12 Interlaced-scanning pattern (raster).

The vertical sync information consists of a group of six pulses which last for a total duration of three horizontal lines, approximately 190 μs, and is transmitted during the vertical interval. This block can be extracted from the higher-frequency horizontal pulses by low-pass filtering. A detailed description of the shape and timing of the synchronizing signal is contained in Sec. 21.3.4.

Vertical Scanning.   The vertical waveform is used to synchronize an oscillator having a frequency of approximately 60 Hz. The output of this stage is a sawtooth-shaped current waveform to the magnetic deflection yoke, or voltage sawtooth to the electrostatic deflection plates of the CRT as described in Chap. 10.

Horizontal Scanning.   The horizontal scanning waveform is generated by an oscillator and power output stage operating at a frequency of approximately 15,734 Hz. In a manner similar to that in the vertical, the horizontal stage sends a ramp, or sawtooth

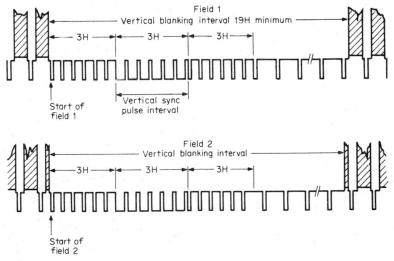

FIG. 13-13 Synchronizing waveforms.

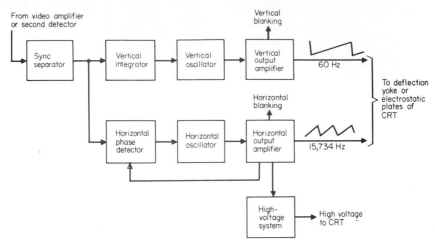

FIG. 13-14   Picture-scanning section.

current, to the deflection yoke (or voltage to deflection plates) to produce the linear left-to-right scan. A block diagram of the picture-scanning section of a television receiver is shown in Fig. 13-14.

High Voltage.   Previous discussions on cathode-ray tubes have indicated the requirement for a voltage up to 20 kV for direct-view monochrome, and up to 30 kV for color receivers. This voltage is easily obtainable from the horizontal output section by utilizing the high peak voltage produced by the stored energy in the deflection yoke at the end of each horizontal scan line. By transformer action, this voltage is stepped-up to the required level and converted to direct current by rectifying and filtering.

13.1.7 MONOCHROME PICTURE TUBES AND DISPLAY DEVICES.   The typical display device for a monochrome picture is the cathode-ray tube (CRT). These devices are available in sizes from 1-in (2.5-cm) diagonal, for use as video casette recorder (VCR) camera viewfinders, up to, typically, 19 in (48 cm). The electron gun of the CRT produces a beam of electrons which are targeted on the phosphor coating covering the inside of the front glass surface or screen. The beam intensity is modulated by the signal developed by the video amplifier. The beam is scanned or swept across the CRT screen by the magnetic deflection yoke (or electrostatic plates inside the tube) using signals developed by the horizontal and vertical amplifiers. A more detailed discussion can be found in Sec. 12.5.

Flat television picture displays as small as 2 in (51 mm) have been developed which consist of a liquid crystal display containing a P-channel MOS switching matrix of 240 horizontal by 220 vertical picture elements.[9] The active display area measures 1.25 by 1.65 in (31.8 by 41.8 mm). The gates of all transistors in a given horizontal row are connected to a common bus (Fig. 13-15). Similarly, the drains of all transistors in a given vertical column are connected to a common bus. Vertical scan (row addressing) is produced by sequentially driving the gate buses from a 220-stage shift register.

During horizontal scan, video information is placed on each column bus. This is somewhat more difficult because of the large stray capacitance (≈ 60 pF) and crossunder resistances associated with each drain bus. A given line of video is broken into 240 pieces and stored in 240 sample-and-hold (S/H) stages all of which drive their respective drain bus lines simultaneously, thus creating a line sequential display. The information on a drain line is, therefore, changed only once for each horizontal period (63.5 μs). The 265 hori-

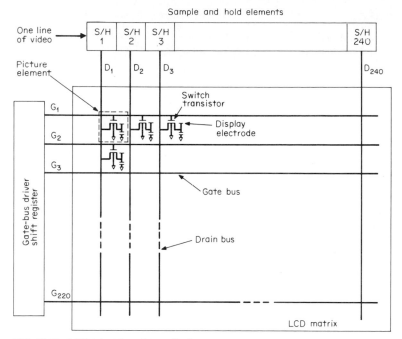

FIG. 13-15   LCD television picture display.

zontal lines per field are accommodated by 220 rows by eliminating 11 lines from top and bottom. Further, each field is *scanned* on the same rows of picture elements, resulting in only 220 lines per frame instead of the conventional $2x$ lines-per-field interlaced display.

13.2 FUNCTIONAL CONFIGURATION OF COLOR RECEIVERS

13.2.1   GENERAL CONSIDERATIONS.   Color television receivers are dimensionally larger than monochrome receivers of an equivalent picture size, primarily because the tri-gun CRT is larger and more complex than its monochrome counterpart. With the recent increased demand for second and third sets in the household, the trend toward smaller sets has grown. The 19-in (48-cm) screen size (measured diagonally) has become dominant in the second-set market, with 13-in (33-cm) and smaller sizes providing the bulk of the third-set market. Personal-sized, battery-powered color sets having 5- and 6-in (13- and 15-cm) pictures have recently become available. Although primarily intended for entertainment use from off-the-air cable or VCR sources, the color set now has found commercial use as a monitor or display for camera chains, computers, and videotex.

Color Signal.   The color television RF carrier modulation signal consists of a luminance signal, similar to that for monochrome transmission, a suppressed-carrier color signal, and an FM sound signal. As outlined in Chap. 4 (Sec. 4.2.4), the luminance signal Y is formed by adding the signals $0.30R$, $0.59G$, and $0.11B$, where R, G, and B represent equal primary color voltages referenced to white light. The color signal is a modulation of the picture carrier by a 3.579-MHz subcarrier. The subcarrier is modulated in a suppressed-carrier circuit, by color information which generates sidebands extending from 2.5 to 4.3 MHz.

The color subcarrier is formed by the addition of two subcarrier frequency signals which are phase-displaced by 90°. The chrominance modulations of the two subcarrier signals are independent and are proportioned as[10]

$$I = -0.60R + 0.28G + 0.32B$$

$$Q = 0.21R + 0.31B - 0.52G$$

Note that the amplitude of each subcarrier becomes zero on white, i.e., when $R = G = B$. In the receiver, red, blue, and green video channel voltages, corresponding to the initial camera signals, are derived by adding luminance and chrominance signals as follows

$$R = Y + 0.96I + 0.62Q$$

$$G = Y - 0.28I + 0.65Q$$

$$B = Y - 1.10I + 1.70Q$$

The color signal includes a phase-reference component, termed the *color burst,* consisting of nine cycles of the subcarrier frequency located on the back porch of each horizontal blanking interval.

Block Diagram.   Figure 13.16 shows the fundamental block diagram of a television receiver utilizing a simultaneous, three-gun picture-tube display. The function of several of the blocks is similar to those of the monochrome receivers (Sec. 13.1). Functions exclusive to a color receiver are shown with heavier outline. The discussion in the remainder of this section will concentrate on the simultaneous-display system.

Color Bandpass Amplifier.   The composite video signal is sent to the color bandpass amplifier which separates the color subcarrier and burst from the video signal. A gain control on this amplifier serves as a chroma control for manual adjustment of the ratio of luminance to color signal, i.e., color saturation in the displayed picture.

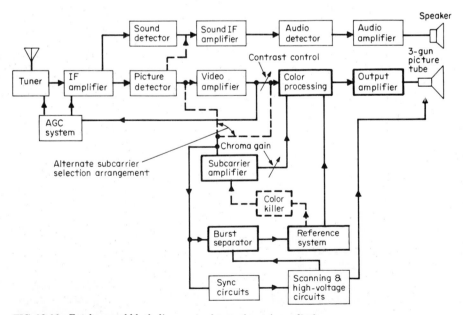

FIG. 13-16 Fundamental block diagram—color-receiver tri-gun display.

Burst Separation and Reference System. The color processing of the color subcarrier requires that a reference signal of the subcarrier frequency be generated in the receiver. This is accomplished by first selecting and amplifying the color burst signal from the chroma subcarrier signal (burst separator) and then by phase-locking a local reference oscillator having the same frequency as the burst. The oscillator output serves as a continuous phase reference for color processing.

Color Processor. Synchronous detectors are used to demodulate the color subcarrier, producing either I and Q or $R - Y$ and $B - Y$ color difference signals. Carrier drive of reference phase for the demodulator circuitry is obtained from the reference circuit. Accurate control of the reference phase is required for accurate demodulation, which in turn determines color fidelity. The third color-difference signal $G - Y$ is created by matrixing the $R - Y$ and $B - Y$ signals.

Output Stage. The color processor output signals must be amplified up to 20 or 30 times to have sufficient amplitude to drive a three-gun CRT. Several techniques for luminance and color-difference signal matrixing to provide R, G, and B drives have been used.

1. CRT matrixing utilizes the three color-difference signals on the three grids and the luminance signal applied simultaneously to the three cathodes. This was a conventional system used with vacuum tubes and delta gun CRTs. Four output amplifiers were required.

2. Newer in-line CRTs have a common grid-1 barrel for all three beams.[11] This requires that the matrixing be performed prior to the CRT. The *RGB* signals are applied to the three cathodes. The two choices for forming *RGB* are to matrix in the output stages or matrix within the color demodulators.

13.2.2 REQUIREMENTS FOR COLOR SIGNAL RECEPTION. The additional information contained in a color transmission requires greater complexity in the receiver as compared with a monochrome set. Design constraints and more critical circuitry add further to a color receiver's complexity. The considerations in several areas of circuit design are outlined below.

1. *IF bandwidth* must be extended on the high-frequency video side to accommodate the color subcarrier modulation sidebands which extend to an IF frequency of 41.5 MHz, as shown in Fig. 13-17. The IF amplitude versus frequency response must be

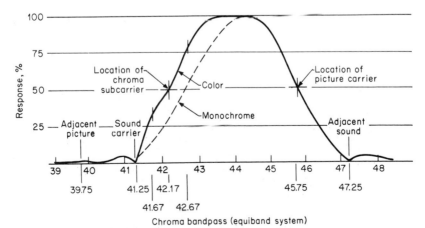

FIG. 13-17 Overall IF bandwidth for color reception.

stable and not change as a function of input signal level (AGC), in order to maintain a uniform level of color saturation.

2. *More accurate tuning* of the received signal must be accomplished to preserve the location of carriers on the tuner-IF passband response. Deviation from their prescribed positions will alter the ratio of chrominance to luminance. Further, a change in transient response characteristic of the luminance signal will result as the amplitude and phase relationships of the higher- and lower-frequency components within the luminance spectrum change with respect to the low frequencies. Nearly all color receivers utilize automatic fine tuning (AFT) or some form of automatic frequency control (AFC) to maintain proper tuning.

3. The *color subcarrier* presence as a second signal requires greater freedom from intermodulation, detection, and overload. Beat frequency signals in the video bandpass, or modulation of the luminance or chrominance carriers, resulting from nonlinear operation, cannot be removed by subsequent filtering or signal processing. This requires adequate dynamic range in the tuner and IF amplifiers and proper application of AGC to avoid excessive signal levels or overloading.

4. *Envelope delay* between the wide-band luminance and narrow-band chrominance signals must be equalized so that the displayed picture will have the correct geometric match between color luminance transitions.

5. *The more complex display device* requires three or four input signals instead of one. Further, the CRT requires a higher level of drive voltages and currents than a similarly sized monochrome display. This translates into a substantial increase in power consumption and heat dissipation.

6. *Additional circuitry,* plus the need for more power by the display device, results in a heavier power supply with significantly better regulation for both line and load changes than a monochrome receiver.

13.2.3 CHROMA SIGNAL SEPARATION, SYNCHRONIZATION, AND DEMODULATION.
The chroma subcarrier amplifier performs the bandpass function of separating the subcarrier sidebands, which extend from 2.2 to 4.2 MHz, from the luminance signal. The overall amplitude response from the tuner to the subcarrier demodulator inputs should be uniform throughout this frequency range to prevent incidental phase-modulation of the two quadrature-phased, amplitude-modulated color-difference subcarriers. Typically, the chroma sidebands are usually positioned on the high-frequency slope of the IF response; the subcarrier amplifier must, therefore, supply the required amplitude (and phase) equalization, usually by means of a low-Q, LC tank circuit tuned to the upper sideband region. These complementary responses are shown in Fig. 13-18.

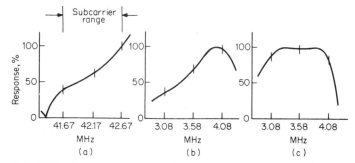

FIG. 13-18 IF and subcarrier-amplifier complementary responses: (*a*) IF response; (*b*) subcarrier amplifier response; and (*c*) overall chroma subcarrier response.

Gain control of the chroma level to achieve proper balance between color and luminance amplitude is handled in this stage in two ways.

1. Manual control by the set operator.

2. Automatic chroma control (ACC) consists of a feedback system which detects burst amplitude and supplies a dc control voltage to the subcarrier amplifier to keep burst at a constant level. This system tends to compensate for variations in scenes, transmission, propagation, and receiver fine tuning. In addition, chroma limiting can be used to prevent oversaturation of bright colors, usually shades of red. This circuit approach simply prevents the maximum amplitude of the subcarrier from exceeding a predetermined value, while having no effect on lower-amplitude chroma.

Burst Separator. To recover the color information contained in the subcarrier modulation, it is necessary to generate a local reference signal of subcarrier frequency and reference phase. To accomplish this, the reference burst of subcarrier present during the horizontal blanking interval is selected by the burst gate and applied to the reference system.

The burst separator is usually a gated amplifier which is driven into conduction during the burst period by the horizontal frequency pulses from the scanning circuit. It is important that the separator pass the complete burst signal without suppressing either its initial or final cycles. Incomplete gating results in the introduction of a quadrature error signal, causing inaccurate reference information.

Reference Circuit. The local reference signal required in a receiver is obtained by controlling the frequency and phase of a locally generated CW carrier signal by the transmitted burst. The most popular reference system is the phase-locked loop or phase-controlled oscillator type. Although more complex than earlier systems such as a crystal filter ringing amplifier and an injection-locked oscillator, the phase-locked loop (PLL) type system offers best noise immunity and lowest phase error, typically less than 2° within the pull-in range.[12] When implemented within an integrated circuit, the performance benefits far outweigh the cost.

Demodulation of Subcarriers. The color subcarrier signal must be demodulated to obtain the chrominance information. A form of synchronous detection must be used since two independent chroma signals are to be derived by virtue of the quadrature-phase relationship of their subcarriers. A typical circuit arrangement of a wide-band demodulation system is shown in Fig. 13-19. A synchronous demodulator is driven with the color subcarrier signal, and with a reference signal whose phase has been adjusted relative to the phase of the subcarrier to demodulate one of the chrominance signals. Figure 13-20 shows the phasor arrangement of the demodulators for this wide-band system. The Q-channel demodulator output circuit is bandwidth-limited by a low-pass filter approximately 600 kHz wide, to provide rejection of unwanted components resulting from high-frequency crosstalk from the I channel.

Crosstalk is due to phase modulation of the I subcarrier in the single-sideband frequency range of the receiver amplifiers. Another demodulator for the I channel is similarly driven with the subcarrier, and with a reference signal phased to be in quadrature with the drive for the Q demodulator. This arrangement provides the other chrominance signal with components over a bandwidth of up to 1.25 MHz. The wider-band I-channel chrominance signal must be delayed to compensate for the greater delay of the narrow-band Q channel.

Equiband Demodulation. Although the wide-band I and Q system has potential for producing the theoretically ideal picture, the complexity has prevented its use in all but a few receiver models, and in professional monitors, over the years since introduction of the first commercial receiver designs. Instead, designers have opted for the equiband system having a bandwidth of approximately 800 kHz. The need for dissimilar filters and delay equalization in the chrominance channels is thereby eliminated. The resulting

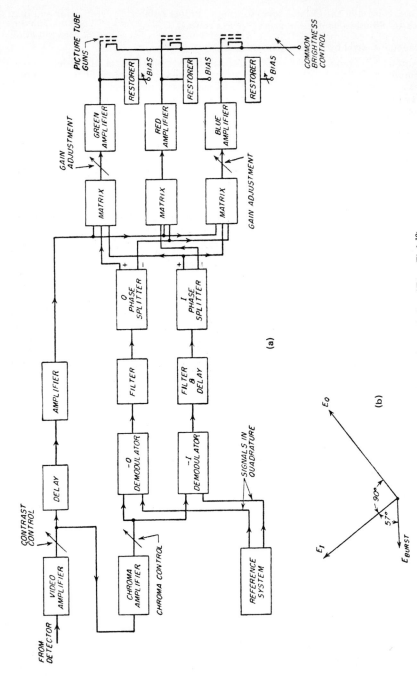

FIG. 13-19 (a) I-Q demodulator system; (b) reference-demodulator phase relation. (*From Fink.*[10])

block diagram is shown in Fig. 13-21. Demodulation is usually done on the $R - Y$ and $B - Y$ axes of Fig. 13-21, although in some current systems the I and Q demodulation phases have been retained. In Fig. 13-22, the I vector lies in the area of flesh-tone colors; by varying its recovered amplitude, a form of dynamic flesh-tone correction can be achieved.[13]

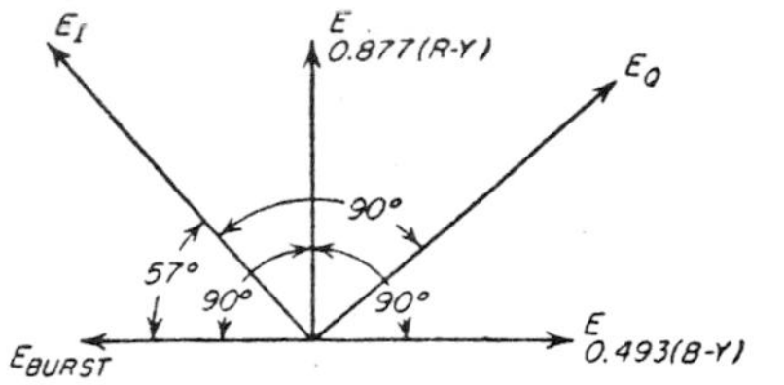

13.2.4 COLOR SIGNAL PROCESSING AND DERIVATION OF COLOR PRIMARIES.

To recombine the luminance and chrominance portions of the signal into the required red, green, and blue signals for the color picture tube, the I and Q signals of a wide-band system are matrixed to provide color-difference signals. This may be accomplished by deriving positive and negative drives from phase splitters and combining the components according to the equations

$$R - Y = 0.96I + 0.62Q \tag{13-2}$$

$$G - Y = 0.28I + 0.65Q \tag{13-3}$$

$$B - Y = 1.10I + 1.70Q \tag{13-4}$$

In an equiband system of Fig. 13-23, the $(R - Y)$ and $(B - Y)$ chrominance signals are matrixed to produce $(G - Y)$ according to the following equation

$$G - Y = 0.51(R - Y) - 0.19(B - Y) \tag{13-5}$$

Luminance Addition. The process of adding the luminance signal Y to the color-difference signals can be accomplished in several ways. In older receivers utilizing delta

FIG. 13-20 Phasor diagram of color subcarrier. (*From Fink.*[10])

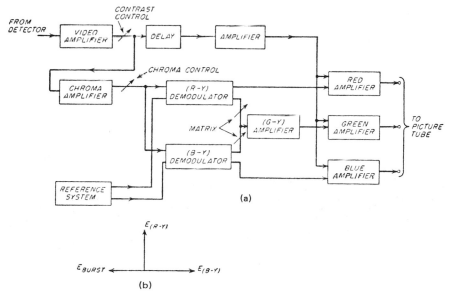

FIG. 13-21 (*a*) Equiband color-difference demodulator system; (*b*) reference-demodulator phase relations. (*From Fink.*[10])

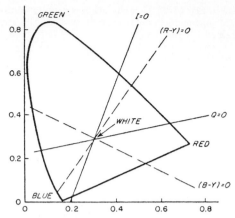

FIG. 13-22 Chrominance axes on chromaticity diagram. (*From Fink.*[10])

tri-gun CRTs and vacuum-tube output stages, the addition was performed in the CRT itself by driving the three color-difference signals to the three number-1 grids and the luminance signal to the three cathodes in parallel as shown in Fig. 13-23. This scheme required four output stages. Drive attenuation and dc bias controls were required on each cathode and grid-1 in order to match the operating point and transconductance of the

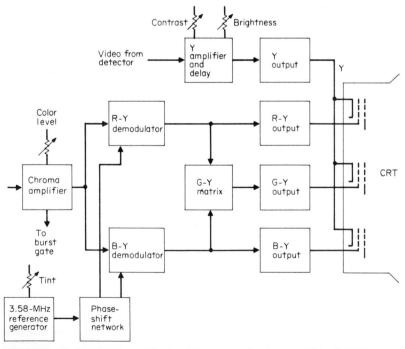

FIG. 13-23 Equiband system with color-difference and luminance addition in CRT guns.

three guns, and thus achieve a uniform white-point temperature or white balance for all luminance shades from black to white.

Newer color CRTs having the in line, unitized gun arrangement as described in Sec. 12.3.3 have a common grid-1 for all three beams. These tubes require fully matrixed color drives R, G, and B to the three cathodes. One technique commonly used in transistorized designs is to add the luminance and color-difference signals in the three output stages as shown in the block diagram in Fig. 13-24. Typically, color-difference signals are fed to the three transistor bases independently, while the luminance signal drives the emitters in parallel.

A third technique for combining the color-difference and luminance signals is to perform resistive addition in a low-level stage. This scheme is generally used in integrated-circuit formats and is located after the color-difference demodulators and $G - Y$ matrix. Emitter follower isolation and drive stages supply the IC output pins. Three relatively simple discrete video stages then drive R, G, and B color signals to the CRT cathodes.

13.2.5 SCANNING AND REGISTRATION.   The most popular display device for color picture reproduction is the three-gun shadow-mask picture tube shown in Fig. 10-23. Several configurations of this tube exist and are described in detail in Sec. 12.2.2. The picture is produced by scanning the three beams, in unison, back and forth across the face of the CRT, from left to right and top to bottom, as described in Sec. 13.1.6 for the single-beam monochrome tube.

Color Purity.   Electron optical components associated with a tri-gun tube must be aligned to obtain color purity and convergency. *Color purity* is the term used to express the condition of uniform saturation of primary colors over the tube screen. It can be best obtained when the electron beam from each gun travels to the screen substantially along the light paths of the original photographic processing of the screen (See Fig. 12-5): That is, when converged at the center of the screen, the three electron beams should pass through the color centers of the tube. This ensures proper angle of arrival of beams at the screen and proper shadowing of the screen by the aperture mask as shown in Figs. 10-24 and in greater detail in Figs. 12-2 and 12-4.

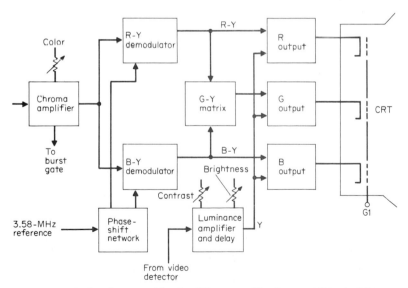

FIG. 13-24   Equiband system with color-difference and luminance addition in *RGB*-output stages.

Further, deflection of the beams should approximate theoretical deflection about the color centers to maintain purity over the screen area. To align the beams to pass through the color centers when undeflected, the tri-gun assembly is mechanically oriented and aligned during assembly of the tube. For final adjustment, a magnetic field of variable strength, polarity, and orientation is used to produce a small radial deflection of all three beams simultaneously.

To approximate deflection of the beams about the color centers, the deflection yoke is positioned along the axis of the tube so as to have its apparent center of deflection in the plane of the color centers of the tube. However, the center of deflection of a yoke moves axially backward with increasing deflection current, and the position adjustment is a compromise. A clear guard space is usually provided between phosphor dots on the screen to permit small errors in the angle of beam arrival at the screen without accompanying desaturation. Stray magnetic fields, including the vertical component of the earth's magnetic field, are suppressed by a magnetic shield around the tube. The horizontal component of the earth's field should be neutralized by an electromagnetic field from a coil around the tube face.

Convergence.   When the three beams are deflected during scanning, it is necessary that they always converge on the same point of the screen. This requirement may be considered in two parts. First, the undeflected beams must converge at the center of the screen. This action is independent of the deflection yoke. It is obtained partly by an initial mechanical alignment of the guns and partly by independent radial bending of the beams by static magnetic fields generated in adjustable "convergence magnets." Second, convergence must be maintained with deflection. Since the mid-1970s, tube and yoke designs which have magnetic fields capable of maintaining convergence over the entire face of the tube have been available. These tubes have utilized the newer in-line gun arrangement and slotted-mask construction.

Colorimetry.   The operating characteristics of the picture-tube gun assembly must be adjusted for correct colorimetry. Balance of gun and phosphor characteristics for each color channel relative to the others is particularly important in the reproduction of a neutral gray scale (white balance). The grid voltage-beam current characteristic of each gun is dependent upon details of gun assembly. Furthermore, the screen efficiencies for each color are unequal and may vary among production lots. Operating parameters for one channel having the lowest phosphor efficiency are set up first to obtain maximum usable brightness consistent with good focus. To normalize the channels, adjustments of first-anode voltage, grid bias, and video drive amplitude for two channels relative to the third are usually provided. This procedure provides moderate control of the composition of the gray scale.

13.3 DESIGN AND PERFORMANCE SPECIFICATIONS

13.3.1 **DESIGN AND TEST STANDARDS.**   Numerous documents on test techniques and recommended design goals and practices have been established to aid the receiver designer. Organizations which have published these are the Institute of Electrical and Electronic Engineers (IEEE), Electronics Industry Association (EIA), Joint Electron Devices Engineering Council (JEDEC), and Society of Motion Picture and Television Engineers (SMPTE). Worldwide design information has been published by the International Electrotechnical Commission (IEC), Technical Centre of the European Broadcast Union (EBU), and EIA-Japan (EIAJ). Three of those most widely used by receiver designers are the following:

IEEE Standard 190.   "Methods of Testing Monochrome Television Receivers" (1960).

IEC Publication 107. "Recommended Methods of Measurement on Receivers for Television Broadcast Transmission," Parts 1 and 2.

Japan Industrial Standard (JIS) C6101. "Methods of Testing Television Receivers."

13.3.2 GOVERNMENT AGENCY REGULATIONS. Television receivers throughout the world are subject to rules and regulations of governmental agencies. In the United States the agencies are the Federal Communications Commission (FCC) and the Department of Health and Human Services (DHHS), and in Canada, the Canadian Standards Association (CSA) and the Department of Communications (DOC). The pertinent documents are as follows:

1. *U.S. FCC Regulations.* Part 15, Radio Frequency Devices, Subpart C and Part 2, Subpart J specify: (*a*) the electromagnetic radiation emitted by a television receiver from its antenna terminals, chassis, or power line cord, and (*b*) the noise-figure and peak-power sensitivity of the receiver on the UHF channels. (Refer also to Pars. 21.1.3 and 21.5.3–6.)
2. *U.S. Department of Health and Human Services, Bureau of Radiological Health.* Document 21, Code of Federal Regulations, Subchapter J, pertains to the x-radiation emission from a television receiver which employs a high voltage on the anode of the picture tube.
3. *Canadian CSA Standard C22.2.* Supplement 1 deals with electrical shock hazard and flammability of electronic devices.
4. *Canadian DOC General Radio Regulation.* Part II, Chaps. 134 through 138, pertains to the requirements for electromagnetic radiation and RF characteristics, which are similar to those in FCC Regulations.

13.3.3 INDUSTRY STANDARDS AND RECOMMENDATIONS. In support of these federal regulations, the IEEE and EIA have developed standard techniques for measuring parameters called for in the legal documents. Several of these are:

IEEE Standard 187. *Open Field Methods of Measurement of Spurious Radiation from FM and TV Broadcast Receivers.*

IEEE Standard 213. *Measurement of Conducted Interference Output to the Power Line from FM and TV Broadcast Receiver in the Frequency Range of 300 KHz to 25 MHz.*

IEC Publication 106 and 106A. *Radiated Interference.*

EIA Standard RS-378. *Measurement of Spurious Radiation from FM and TV Broadcast Receivers in the Frequency Range 100 to 1000 MHz, using the EIA Laurel Broadband Antenna.*

EIA Consumer Products Engineering Bulletin CPEB 4-A. *Standard Form for Reporting Measurements of TV and FM Broadcast Receivers in Compliance with FCC Part 15 Rules.*

EIA CPEB3. *Measurement Instrumentation for X-radiation from Television Receivers.*

13.3.4 VOLUNTARY SAFETY COMPLIANCE. In the United States the Underwriters Laboratories, Inc. (UL), a private industry-supported testing firm, provides a service in the areas of electrical shock hazard, internal combustion from electrical fault or overstress, and CRT implosion hazard. UL is recognized throughout the United States as a reputable and competent standards and certifying organization in the many areas of safety. UL Document No. 1410 is applicable to television receivers and video products. Other documents are applicable to cabinets, printed wiring boards, and other component parts. In Canada, the CSA provides a similar function with their activities supported by government funding.

13.3.5 DESIGN GUIDELINES.   A very numerous collection of "good engineering practice" guidelines and standards has been published by agencies and committees listed in Secs. 13.3.1 and 13.3.2. A longer listing and description are contained in Secs. 21.5.4 through 21.5.7.

13.4 CABLE TELEVISION AND SATELLITE DISTRIBUTION

13.4.1 CABLE CHANNEL IDENTIFICATION PLAN.   The spectrum utilization for television broadcast and cable systems is shown in Fig. 13-25. Frequency allocations for cable systems are shown in Table 13-2. The channel-numbering plan shown here has recently been developed by the EIA/NCTA Joint Committee on Receiver Compatibility and has been published by both organizations as an Engineering Standard.[14]

Standard frequencies refer to cable systems which transmit on the standard off-air frequencies for channels 2 to 6 and 7 to 13. Supplemental channels are in 6-MHz increments, counting down from channel 7 (175.25 MHz) to 91.25 MHz and upward from channel 13 (211.25 MHz).

The harmonic related carriers (HRC) channeling plan refers to cable systems having a coherent head end and visual carriers located at multiples of 6 MHz starting at 54 MHz. All visual carriers in the system are phase locked to a 6-MHz master oscillator. This ensures that all the carriers are harmonically related to 6 MHz and no matter what shift occurs to the master oscillator, all carriers maintain the same relative frequency separation. The second- and third-order intermodulation products resulting from any two carriers, therefore, always fall exactly on the visual carrier frequencies, and their undesirable effects on television pictures will be reduced or eliminated. When compared with the broadcast or standard plan, HRC channels are frequency displaced by -1.25 MHz on all standard and cable supplementary channels (midband, etc.) except channels 5 and 6, where the displacement is $+0.75$ MHz.

For the *interval related carriers* (IRC) system, the cable channels are on frequencies starting at 55.25 MHz with increments of 6 MHz (6 N + 1.25 MHz). These channels are the same as standard frequencies except for channels between 67.25 and 91.25 MHz.

13.4.2 CABLE COMPATIBLE TELEVISION RECEIVER.   EIA TV Systems Bulletin No. 2, *Cable Compatible Television Receiver and Cable System Technical Standards* (March 1975), specified desirable and standard characteristics for a cable compatible receiver. Many advances in the cable industry since that time have made the EIA document obsolete. These issues have been updated by the EIA/NCTA Joint Engineering Committee.

By Canadian regulation, two types of receivers exist: the standard receiver and the cable compatible receiver,[15] the latter having additional requirements as listed below.

Minimum selection of 18 preset channels, *with all others in the VHF, mid- and superband being accessible by mechanical means.*

Fine tuning or AFC capability *of ± 0.55 MHz on VHF channels and -1.31 MHz on mid- and superband channels.*

Noise figure *not to exceed 10 dB, unless double conversion is utilized; noise figure can then be up to 13 dB.*

No noticeable co-channel interference *when receiver is in 100 mV/m radiated field from a broadcast station with receiver 75-Ω cable input adjusted to 1 mV.*

No receiver overload *for signal input up to 5 mV.*

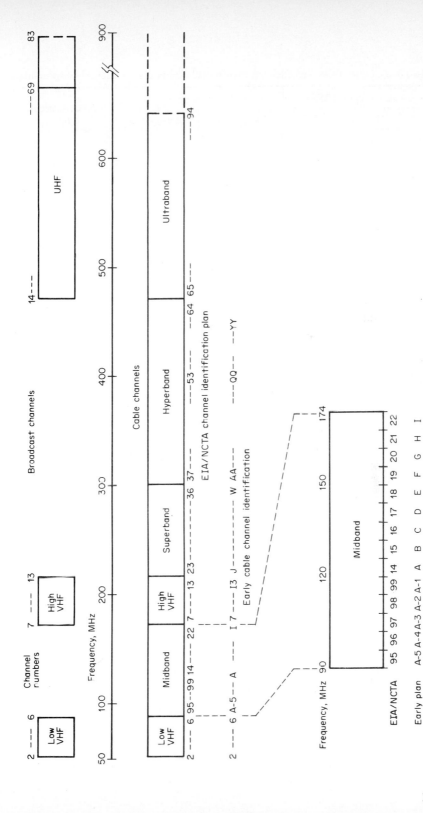

FIG. 13-25 Frequency spectrum and channel designation for broadcast and cable television.

13.29

Table 13-2 Cable Television Channel Identification Plan (by Frequency Assignments)

Pix carrier frequency, MHz			Band name	Channel designation	Historical reference‡
Std.	HRC	IRC			
55.25	54.00	55.25	↑	2	
61.25	60.00	61.25		3	
67.25	66.00	67.25		4	
†	72.00	73.25	Low VHF	1	4+, A-8
77.25	78.00	79.25	↓	5	A-7 (HRC, IRC)
83.25	84.00	85.25	——	6	A-6 (HRC, IRC)
91.25	90.00	91.25	↑	95	A-5
97.25	96.00	97.25		96	A-4
103.25	102.00	103.25		97	A-3
109.25	108.00	109.25		98	A-2
115.25	114.00	115.25		99	A-1
121.25	120.00	121.25	Midband	14	A
·	·	·		·	·
·	·	·			·
·	·	·		·	
169.25	168.00	169.25	↓	22	I
175.25	174.00	175.25	↑	7	
·	·	·		·	·
·	·	·	High VHF	·	·
·	·	·		·	·
211.25	210.00	211.25	↓	13	
217.25	216.00	217.25	↑	23	J
·	·	·		·	·
·	·	·	Superband	·	·
295.25	294.00	295.25	↓ ——	36	W
301.25	300.00	301.25	↑	37	AA
·	·	·		·	·
·	·	·		·	·
325.25	324.00	325.25		41	EE
·	·	·		·	·
·	·	·	Hyperband	·	·
·	·	·		·	
397.25	396.00	397.25		53	QQ
·	·	·		·	·
·	·	·			·
463.25	462.00	463.25	↓	64	·
469.25	468.00	469.25	↑	65	
·	·	·		·	·
·	·	·		·	·
·	·	·	Ultraband	·	·
493.25	492.00	493.25		69	
·	·	·		·	·
·	·	·		·	·
547.25	546.00	547.25		78	·
·	·	·		·	·
·	·	·		·	·
·	·	·		·	

Table 13-2 Cable Television Channel Identification Plan (by Frequency Assignments) (*continued*)

| Pix carrier frequency, MHz | | | | Channel | Historical |
Std.	HRC	IRC	Band name	designation	reference‡
595.25	594.00	595.25	Ultraband	86	·
·	·	·		·	·
·	·	·		·	·
·	·	·		·	·
643.25	642.00	643.25	↓	94	·

†Undesignated.

‡Identification used prior to adoption of EIA/NCTA channel designation system.

Source: EIA Interim Standard no. 6, Electronic Industries Association, Washington, 1983.

Local oscillator signal *or any other signal generated within the receiver shall not exceed the limits shown in Table 13-3. These regulations also apply to cable converters.*

13.4.3 SATELLITE CHANNELS. The two bands allocated for television satellite broadcast are the C band (4 to 6 GHz) and the K_u band (12 to 17 GHz). The down-link spectrum for the C-band satellite system consists of 24 channels, each having 40-MHz bandwidth, in the 3.7- to 4.2-GHz-frequency band. Each channel overlaps its adjacent channel by 20 MHz. Even-numbered channels are horizontally polarized, and the odd-numbered are vertically polarized to give a minimum of 30-dB separation. Satellite transponder power output is 5 to 10 W. This band is intended for commercial applications.

Direct-broadcast satellite (DBS) television program service to homes in the K_u band was established by the 1977 World Administrative Radio Conference.[16] The down link consists of a band of frequencies between 12.2 and 12.7 GHz. Channel allocation and technical specifications for the western hemisphere (Region 2 of WARC) was established at the 1983 Western Hemisphere Regional Administrative Radio Conference. As of 1983, the FCC had approved applications from numerous companies to launch and establish DBS service for standard broadcast, premium pay-television programming, and high-definition television (HDTV) in the 1983 to 1985 period. Bandwidths for these services will range from 16 to 27 MHz per channel. Transponder output power will be 100 to 400 W.

The ground antenna for satellite reception usually consists of a parabolic reflector having a diameter of 3 to 5 m for the C band and 1 m for the DBS band. A low-noise amplifier (LNA) having gain of approximately 40 dB and a noise figure of less than 1.5 dB (120 K) is placed at the antenna reflector feed point. Double-conversion receivers typically are used to achieve the required gain and selectivity. The output of this type of receiver consists of conventionally modulated video and sound carriers on one of the low VHF channels (usually three or four) which is fed to the antenna terminals of a standard VHF/UHF television receiver.

Table 13-3 Canadian Regulation for Local Oscillator or Spurious Signal at the 75-Ω Receiver Terminals

Frequency range, MHz	Maximum level, dBmV
5–54	−50
54–300	−26
300–1000	−10

Future designs proposed will provide wide-band R-B-G outputs to feed decoded video signals to either conventional or high-definition television (HDTV) receivers.

13.5 RECEIVER NOISE AND SENSITIVITY

13.5.1 TYPES OF NOISE.
The types of noise encountered in electronic circuits are thermal noise, shot noise, and flicker noise.

Thermal noise comes from the thermal agitation of electrons within a resistance. Thermal noise sets a lower limit on the noise present in a circuit. The open-circuit noise voltage or short-circuit noise current produced by a resistor is shown in Fig. 13-26. The

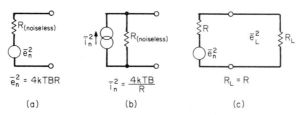

FIG. 13-26 Thermal-noise source: (*a*) voltage source, (*b*) current source, and (*c*) matched noise source, where $\bar{e}_n^2$ = mean square open-circuit noise voltage across resistor R, k = Boltzmann constant ($= 1.38 \times 10^{-23}$ J/K), T = temperature (K), B = bandwidth (Hz), and R = resistance (Ω).

maximum thermal noise power which can be delivered to a load occurs when the value of the load R_L connected across the output terminals is equal to R. In this case, the power delivered is given by

$$P_n = \frac{e_n^2}{4R} = kTB \qquad \text{W} \tag{13-6}$$

and is referred to as *available noise power*. Thermal noise has a power density which is constant with frequency.

Shot noise is associated with current flow across a potential barrier, such as a semiconductor junction or electrons leaving the cathode of a vacuum tube, or photoelectrons emitted from the light-sensitive surface of a photomultiplier tube. Shot noise is usually described in terms of a current value as shown in Fig. 13-27. Like thermal noise, shot noise has a power density which is constant with frequency.

Flicker noise, also called $1/f$ noise because of its inverse frequency characteristic, is generally associated with an imperfect contact between two materials or the generation-recombination of charge carriers in semiconductor devices. An equivalent circuit for this type of noise is given in Fig. 13-28.

13.5.2 NOISE SPECIFICATIONS.
Several ways to specify the noise characteristics of a circuit element or network are given below.

The *noise factor F* is a quantity which compares the noise performance of a device or circuit with the performance of a noiseless device or circuit

$$F = \frac{\text{noise power output of actual device (circuit)}}{\text{noise power output of ideal device (circuit)}}$$

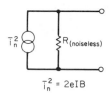

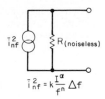

$$\overline{i}_n^2 = 2eIB$$

$$\overline{i}_{nf}^2 = k\frac{I^a}{f^n}\Delta f$$

FIG. 13-27 Shot-noise-source equivalent circuit where $\overline{i}_n^2$ = mean square short-circuit noise current, e = electron charge (= 1.6×10^{-19} C), I = direct current through R (A), R = resistance (Ω), and B = bandwidth (Hz).

FIG. 13-28 Flicker-noise-source equivalent circuit where $\overline{i}_{nf}^2$ = mean square short-circuit flicker noise current, R = resistance (Ω), I = direct current (A), f = frequency (Hz), Δf = frequency interval, and k, α, n = empirical constants depending on physical properties of device.

An alternative and frequently used definition for the noise factor compares the signal-to-noise power ratio at the input with that at the output at a standard temperature of 290 K

$$F = \frac{S_{in}/N_{in}}{S_{out}/N_{out}} \tag{13-7}$$

The noise factor expressed in decibels is called the *noise figure* (NF) and is given by

$$\text{NF (db)} = 10 \log F \tag{13-8}$$

The noise factor for a system of two or more networks in cascade, as shown in Fig. 13-29, is given by

$$F_{tot} = F_a + \frac{F_b - 1}{G_a} \qquad \text{for two networks} \tag{13-9}$$

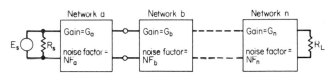

FIG. 13-29 Networks in cascade.

and

$$F_{tot} = F_a + \frac{F_b - 1}{G_a} + \frac{F_c - 1}{G_a G_b} + \cdots + \frac{F_n - 1}{G_a G_b G_c \cdots G_{n-1}} \tag{13-10}$$

for n networks.

The *noise temperature* of a network is defined as the average noise power available divided by Boltzmann constant (1.38×10^{-23} J/K) and bandwidth B (Hz)

$$T = \frac{P_{nav}}{kB} \tag{13-11}$$

At a single input-output frequency in a two-part network, the effective input noise temperature is given by

$$T_e = T_0(F - 1) \tag{13-12}$$

where $T_0 = 290$ K.

The *excess noise level* is defined as the equivalent excess available noise power density contributed by the system relative to the available thermal noise power density contributed by the termination which is $kT_0 = 4 \times 10^{-21}$ W/Hz and is given by

$$E = \frac{T_e}{T_0}$$

For a cascade system, the effective noise temperature is given by

$$T_{e(\text{tot})} = T_{ea} + \frac{T_{eb}}{G_a} + \frac{T_{ec}}{G_a G_b} + \cdots \frac{T_{en}}{G_a G_b \cdots G_{n-1}} \qquad (13\text{-}13)$$

If the first network of a two-stage cascade system consists of an attenuator using pure resistive elements whose gain $G_a = 1/L$ is less than 1, then the noise factor and noise temperature can be shown to be[17]

$$F = 1 + \frac{T_{ea}}{T_0}(L - 1) \qquad (13\text{-}14)$$

and $$T_{ea} = (L - 1)T_0 \qquad (13\text{-}15)$$

where L is the loss of the attenuator.

If the attenuator has noise temperature $T_{ea} = T_0$, the usual case for a television receiver, then $F = L$ and the noise figure of the system is

$$F = LF_b \qquad (13\text{-}16)$$

Therefore, an antenna transmission line having an insertion loss of 3 dB will have a noise figure of 3 dB.

$V_n - I_n$ is an additional set of parameters which are frequently used to specify the noise contribution of a network or device. Here the device is modeled as an ideal device or network to which the noise generators V_n and I_n are connected as shown in Fig. 13-30. For this model the noise factor can be expressed as[18]

$$F = 1 + \frac{1}{4kTB}\left(\frac{V_n^2}{R_s} + I_n^2 R_s\right) \qquad (13\text{-}17)$$

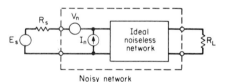

Noisy network

FIG. 13-30   Noisy network modeled with noise-voltage and noise-current sources.

13.5.3 NOISE FACTOR FOR TYPICAL RF AMPLIFIER DEVICES.

Two semiconductor amplifier devices which are commonly used in the RF stage of television receivers are the bipolar transistor and MOSFET. The general noise factor expression for a bipolar transistor in either the common-base or common-emitter connection, neglecting the $1/f$ low-frequency effects, is given by[19]

$$F = 1 + \frac{r_b'}{R_s} + \frac{r_e}{2R_s} + \frac{R_s + r_e + r_b'}{2r_e R_s \beta_0}\left[1 + \left(\frac{f}{f_a}\right)^2(1 + \beta_0)\right] \qquad (13\text{-}18)$$

where r_b' = transistor base-spreading resistance
  R_s = source resistance
  r_e = intrinsic emitter resistance
  β_0 = dc value of common-emitter current gain
  f_a = alpha cutoff frequency

This equation has a $+6$ dB per octave slope with increasing frequency beginning at a frequency of $f_a\sqrt{1 - a_0}$.

The optimum source resistance for lowest noise factor is given by

$$R_{s\ \text{opt}} \cong \left[(r_e + r_b')^2 + \frac{\beta_0 r_e(2r_b' + r_e)}{1 + (f/f_a)^2(1 + \beta_0)} \right]^{1/2} \qquad (13\text{-}19)$$

Noise in a MOSFET used as a high-frequency amplifier device is generated by the following device mechanisms:

1. Thermal channel noise current generated by the motion of carriers in the resistive channel between the source and drain. This noise has a "white" spectrum.

2. Induced gate noise arising from the capacitive coupling of the gate to the channel. This noise current becomes significant at high frequencies.

Other sources, such as gate shot noise and generation-recombination noise, both of which have a $1/f$ characteristic, have little or no effect above the midfrequency region for the device (several hundred kilohertz). An equivalent circuit of the active noise sources is shown in Fig. 13-31. The noise factor for a MOSFET at high frequencies is given by[20]

$$F = 1 + \frac{1}{g_m' R_s}\left[0.67 + 0.12\left(\frac{\omega C_{\text{in}}}{g_m'}\right)^2 + 0.3(\omega C_{\text{in}} R_s)^2 \right] \qquad (13\text{-}20)$$

where g_m' = saturation value of g_m (which equals y_{12})
ωC_{in} = imaginary part of y_{11}
R_s = source resistance

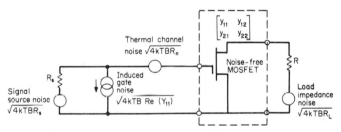

FIG. 13-31   High-frequency noise model of MOSFET.

The final term represents the induced gate noise contribution. For small values of R_s, this is less than the other terms. As R_s increases, F begins to decrease, reaches a minimum, and then increases with further increase in R_s.

The dual-gate MOSFET can be analyzed as a cascode connection of two devices, the lower one as common source with the upper one connected as common gate. The noise factor in terms of device parameters has been given by[21]

$$F = 1 + \frac{(\omega C_{\text{in}})^2}{2g_s g_{m1}[1 + (C_{\text{in}}/2g_{m1})^2]} + \frac{1}{g_s}\left\{ g_s + \frac{(\omega C_{\text{in}})^2}{2g_{m1}[1 + (C_{\text{in}}/2g_{m1})^2]} \right\}$$
$$\times \frac{2}{3g_{m1}}\left[1 + \left(\frac{L_1}{L_2}\right)^2\left(\frac{\omega C_t}{g_{m2}}\right)^2 \right] \qquad (13\text{-}21)$$

where L_1 and L_2 = gate lengths of transistors 1 and 2
C_t = total capacitance consisting of drain 1 to source 1, gate 2 to source 2, and drain 1–source 2 junction to substrate
C_{in} = gate 1 input capacitance

13.5.4 MEASUREMENT OF TELEVISION RECEIVER NOISE FIGURE. The pre-detection method of noise-figure measurement has been recommended by the Advisory Committee to the FCC as being the most accurate and practical method at the present state-of-the-art.[22] In this technique, as shown in Fig. 13-32, a gas-discharge tube or hot-cold noise source is connected to the receiver antenna terminal. The noise power level in the receiver IF is sensed at some point and sent to the Y-factor meter. The ratio of the output power when the noise source is hot and when the source is cold is Y, and the noise factor can be determined by the equation below

$$F = \frac{(T_2/T_0) - (YT_1/T_0)}{Y - 1} + 1 \tag{13-22}$$

where T_1 and T_2 represent the cold and hot temperature of the source and T_0 is ambient (290K). If the cold state (T_1) is room temperature (T_0), then the equation simplifies to

$$F = \frac{(T_2/T_0) - 1}{Y - 1} = \frac{E_2}{Y - 1} \tag{13-23}$$

E_2 is called the *excess noise ratio* of the noise source when hot and is specified for each noise source by the manufacturer. The noise figure in decibels is given by

$$\text{NF (dB)} = 10 \log_{10} (E_2) - 10 \log_{10} (Y - 1) \tag{13-24}$$

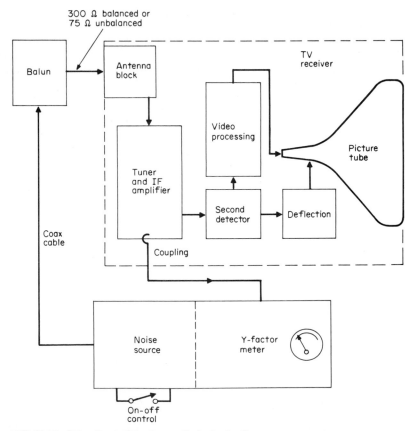

FIG. 13-32 Setup for predetection method of noise-figure measurement.

An alternative method of measuring noise factor, described in IRE 190,[23] which measures the output of the television receiver second detector (postdetection) has been found to give results which are unreliable and nonrepeatable. The problem arises from the nonlinearity of the second detector. Attempts to prebias the detector by using a small amount of injected carrier have not proved acceptable.

Typical television receivers have noise-figure values on VHF channels of 4 to 8 dB and 7 to 12 dB on UHF channels.

13.5.5 RECEIVER SIGNAL-TO-NOISE RATIO. The signal-to-noise ratio S/N can be computed from the definition of noise factor given in Sec. 13.5.1 and Fig. 13-29

$$F = \frac{\text{signal in}}{kTB} \Big/ \frac{\text{signal out}}{\text{noise out}} \tag{13-25}$$

The formula for receiver signal-to-noise ratio becomes

$$\text{S/N output (dB)} = S_{\text{in}} \text{ (dBm)} - \text{NF (dB)} - kTB \text{ (dBm)} \tag{13-26}$$

where the signal and noise levels are stated in decibels compared with 1 mW. For example, a receiver having an input power of -93 dBm (6 mV across the 75-Ω input impedance) and a 6-dB noise figure (noise factor $F = 2.0$) would yield a 10-dB signal-to-noise ratio at the video detector.

This calculation is expanded in Sec. 13.9.11 to cover the entire receiving system which includes the antenna, preamplifier (if used), and transmission line.

13.5.6. PEAK PICTURE SENSITIVITY. IRE-190 defines peak picture sensitivity[24] as "the lowest input signal which results in standard picture test output when the receiver is tuned for maximum picture output." It measures the maximum system gain by means of a frequency-selective voltmeter at the CRT and as such does not measure the noise which is also present. A typical receiver design will have sufficient gain such that the noise output at the CRT will equal the standard picture level (S/N = 0 dB). If additional gain is built into the system, it may cause the AGC to prematurely reduce the gain of the IF amplifiers and RF stage, thereby causing a desensitizing of the receiver and/or an inferior signal-to-noise ratio for medium level signals.

13.5.7 TELEVISION RECEIVER PICTURE GRADES. In the late 1950s the Television Allocations Study Organization (TASO) developed a six-level subjective picture quality scale based upon the amount of noise and interference which an observer could see in a picture.[25] Later experimentation applied signal-to-noise levels to these observer grades[26] (see Table 13-4).

Table 13-4 Picture Quality Grades

TASO grade	Picture quality	S/N, dB
1	Excellent	>41
2	Fine	33–41
3	Passable	28–33
4	Marginal	23–28
5	Inferior	17–23
6	Unusable	<17

13.5.8 DETERMINATION OF SERVICE GRADES AND CONTOURS. One of the factors which the FCC uses in predicting the coverage of a television station is a set of assumptions about television receiving equipment and the receiving environment. These are commonly referred to as the *television planning factors* and are composed of the following elements:[27-29]

1. Inherent thermal noise at the receiver input, $P = kTB$; for a 300-Ω antenna, the noise voltage becomes $+7$ dB above 1 μV

2. Television receiver noise figure

3. Signal-to-noise ratio required for a satisfactory picture, usually a TASO grade 3 (S/N $\approx$ 30 dB)

4. Receiving antenna dipole factor, which accounts for decreasing efficiency of receiving antennas as frequency increases: $K = 20 \log_{10} (96.68/F_{\text{MHz}})$ for 300-Ω antenna

5. Receiving antenna gain

6. Signal attenuation of the lead-in or transmission line

7. A factor that accounts for the radio noise present in urban areas

8. A time probability factor, delta T, to account for signal fading

9. A terrain or location probability factor to account for terrain irregularities

The following service areas have been defined by the FCC; the derivation of field strength values is given in Table 13-5.

Grade A is that area in which the level of field strength 30 ft (9 m) above the ground is sufficient to provide an *acceptable* quality picture to a receiving installation typical of suburban areas. This signal is provided to the best 70 percent of receiving locations at least 90 percent of the time.

The *grade A contour* is the geographic boundary within which the median field strength is equal to or better than the grade A value.

Grade B is that area in which the level of field strength 30 ft (9 m) above the ground is sufficient to produce an "acceptable" picture to a receiving installation considered typical of outlying areas, the best 50 percent of the locations for at least 90 percent of the time. The grade B contour defines the outer boundary to the grade B service area.

City grade contour is the geographic boundary of an area having a median field strength which is 6 dB higher than the grade A contour. The area within this boundary usually has a picture quality which is one TASO grade better than the grade A area; i.e., passable improves to fine.

Indoor Antenna Contour. This designation was developed during the UHF comparability studies[27] and relates to an area in which antenna height is designated as 6 ft (1.8 m) above ground instead of 30 ft.

13.6 PICTURE DISPLAY REQUIREMENTS

The objective of commercial receiver design is to serve the maximum number of installations at an optimum product cost. Since a wide range of field conditions exists, not all receiver characteristics are significant at all locations. The receiver designer has some flexibility in determining the relative importance of the various performance characteristics. Definitions and measurement techniques for several of these parameters are contained in standards discussed in Sec. 13.3.1.

13.6.1 **BRIGHTNESS.** Adequate display brightness to yield a pleasing picture over a wide range of ambient lighting conditions is required for a monochrome as well as color television receiver. Table 13-6 gives values for large-area brightness levels for several popular screen sizes. It is important that the picture stay in focus at these high levels and that all other picture characteristics do not deteriorate. Small-area brightness is the measure of how well a receiver can maintain focus on small portions of the picture. A

Table 13-5 Derivation of Median Field Strength for Various Service Grade Contours

Factors	Units	Low VHF 2–6			High VHF 7–13			UHF 14–83		
		B	A	Indoor	B	A	Indoor	B	A	Indoor
Receiver noise level	dB/μV	7	7	7	7	7	7	7	7	7
Receiver noise figure	dB	6	12	6	7	12	7	15	15	12
Required signal-to-noise	dB	30	30	30	30	30	30	30	30	30
Dipole factor†	dB	−3	−3	−3	6	6	6	16	16	16
Receiver antenna gain†	dB	−6	0	4	−6	0	3	−13	−8	0
Line loss	dB	1	1	0	2	2	0	5	5	0
Delta T	dB	6	3	3	5	3	3	4	3	6
Delta L	dB		4	4		4	4		6	6
To overcome urban noise	dB		14	14		7	7		0	0
Building penetration	dB			17			20			24
Required median field	db/μV/m	47	68	82	56	71	87	64	74	102

†Since this is a table of losses, gains appear as a negative number.

Source: P. B. Gieseler et al., "Comparability for UHF Television: Final Report," Federal Communications Commission, Office of Plans and Policies, Washington, September 1980.

Table 13-6   Large-Area Brightness Values†

Screen size, in (cm)	Type	Brightness, ft·L‡
5 (13)	Monochrome	40–60
12 (30)	Monochrome	50–70
19 (48)	Monochrome	50–70
13 (33)	Color	80–120
19 (48)	Color	80–120
25 (64)	Color	70–100

†Signal is typical monoscope circle test pattern modulating the RF carrier 87.5% (tip of sync to white reference) with black at 7.5%. Color tube faceplate transmission is 85%.
‡For conversion to metric SI unit see Table 22-10.

small white square on a full black background is the pattern for this test. Under these conditions the high voltage usually does not fall nearly as far owing to supply loading as it does for large areas. Small-area brightness is very dependent upon picture-tube defocusing as a function of beam current.

13.6.2   CONTRAST.   The maximum contrast ratio available in the television picture is usually determined by ambient light falling on the screen. Numerous techniques to improve contrast ratio under high ambient conditions have been used in television receivers, glass faceplates having lower light transparency, black paint material surrounding the phosphor dot triad or stripes, and light-absorbing phosphors, to name a few.

Three circuit requirements which impact on contrast are:

1. *Background (brightness) control* must have sufficient range to provide adequate picture-tube bias for all types of picture content at all contrast settings, within the range of the picture-tube beam cutoff tolerance.
2. *The video amplifier* (R-G-B amplifiers for color set) should be capable of driving the picture tube to the full extent permitted by the electron gun design.
3. *Automatic beam control* or beam-limiting circuitry must not interact in such a way that high drive conditions produce a washed-out (low-contrast) picture.

13.6.3   BLACK-LEVEL STABILITY.   Black-level reference must be maintained in luminance and color output channels to produce accurate gray scale and color saturation. Failure to maintain a reference of 80 percent or higher will result in washed-out pictures and loss of detail due to incorrect contrast ratio. Black-level reference is usually maintained in the circuit design by a form of dc coupling or clamping in the video (R-G-B) output stages.

13.6.4   TRANSFER CHARACTERISTIC.   This term, applied to a receiver, refers to the gradation of the gray tones of the image. The desired objective is that the transmission and reception systems provide a linear transfer of tonal values. At the program source, compensation for the nonlinear voltage characteristic of the receiver picture tube (gamma correction) has been added. If the receiver gain characteristic from RF to video output stage is linear, then the picture-tube scene will be a faithful reproduction of the original. Stages in the receiver which must receive special attention to ensure linearity and wide dynamic range are the baseband demodulators (both luminance and chrominance) and output stage to the picture tube (luminance and/or R-G-B).

13.6.5   SHARPNESS.   Sharpness is determined by the ability of the circuitry and display device to reproduce a transition from a dark area to an adjacent light area or vice versa. These transitions can be enhanced by use of high-frequency peaking which produces preshoots and overshoots. If used, care must be taken that the preshoots and overshoots are of nearly equal amplitude and that they do not exceed the main transition by more than 10 percent. Excessive peaking and ringing may deteriorate fine detail in the picture.

A second part of picture sharpness is resolution, the ability to display fine detail. This parameter is optimized when the phase response is linear throughout the entire receiver for the total bandwidth of the luminance channel. Subjective evaluation of resolution can be made by using an EIA (formerly RETMA) or SMPTE resolution chart display. Observed horizontal resolution values of 250 to 275 lines are typical for receivers having overall bandwidth of 3 MHz. When a comb filter is employed to permit use of the full-luminance bandwidth, values of 320 to 330 lines can be seen. Vertical resolution is derived from the 525-line system. The only requirement necessary in order to achieve the typical value of 340 lines is that the scan have proper interlace.

13.6.6 GEOMETRY.   Proper picture geometry is important to perceived picture quality. Geometry is usually measured by using a circle test pattern and cross-hatch pattern. Several factors which contribute to overall geometry and typical values achieved in production designs are listed below.

Aspect ratio	*+5%*
Overscan	*5–10%*
Vertical linearity error	*5–8%*
Horizontal linearity	*5–8%*
Parallelogram	*90° ±1° at four corners*
Pincushion (barrel)	*±1–2%*

13.6.7 COLOR SATURATION.   This is the ability of the receiver to achieve a light output corresponding to the color point of the phosphor on the screen. Good performance requires approximately 90 percent saturation at the maximum brightness of the display.

13.6.8 COLOR UNIFORMITY (PURITY).   This is the ability of the display to reproduce a uniform hue over its entire area. The eye can detect hue differences near saturation resulting from as little as 5 electrical degrees phase change, particularly when the error is visible on a side-by-side basis. Hence, the tolerance on uniformity is small, particularly on red scenes where visual sensitivity is most acute. White point, the ability of the receiver to display a pure white field corresponding to a white or monochrome transmission, is a subset requirement of color uniformity. White point requires that the drive circuitry as well as the display tube be electrically balanced.

13.6.9 COLOR FIDELITY.   The requirement for color fidelity in reproduction is the most difficult to evaluate numerically. Four types of errors are the most objectionable.

1. Shifts of white point with output level.

2. Errors in flesh-tone rendition. This requirement is critical because of the subjective response to small errors in the hue range. The incorporation of various flesh-tone correction circuits can be beneficial in reducing this problem.

3. Errors of excessive saturation. This defect shows as a readily identified, unnatural appearance of subject matter. Chroma peak limiting circuits relieve this situation substantially.

4. Hue accuracy, generally measured with the bar chart display, although not usually as objectionable as the first three, can still be a source of complaint to a viewer, particularly if the scene has recognizable colors such as green grass (not blue) and blue sky. Hue accuracy in some receiver designs is sacrificed to achieve accurate flesh-tone rendition.

13.6.10 DELAY EQUALIZATION. Luminance and chrominance information are processed by different circuits in a receiver resulting in differential delay of the two signals. This time delay must be equalized in the luminance channel by use of a delay network so that both the luminance and color information are written on the display device with good registration.

13.6.11 SYNCHRONIZING STABILITY. Strong-, medium-, and weak-signal conditions must all be considered. Strong-signal conditions include signals generated by auxiliary equipment such as VCRs, video disc players, and home computers. Nonstandard sync format is generated under certain modes of operation, e.g., still frame (pause), fast forward, and fast backward (scan) on VCRs. Receiver design characteristics should be chosen to eliminate the vertical roll or jitter and horizontal jitter of the top several scan lines (flag waving) under these conditions.

Protection against impulse and ignition noise disturbance of the synchronizing circuits becomes important at medium-signal levels. These disturbances are capable of disabling sync-separator and AGC circuits and of driving IF and video amplifiers into cutoff and saturation regions. Protection includes use of short time constants in IF coupling networks, amplitude clipping and noise inversion in video amplifier and sync-separator circuits, filtering and gating the AGC circuits, and appropriate dynamic range in video amplifiers.[30]

The ability of the synchronizing circuits (particularly the horizontal and chroma reference) to function in the presence of thermal noise is important in weak-signal reception. The horizontal AFC circuitry must be immune to noise-induced phase modulation of the sync signal. Failure of the chroma reference circuits to remain in phase with the reference burst signal may result in either a static hue shift or streaking of the chroma information due to intermittent failure of the reference generator.

13.6.12 SPURIOUS RADIATION. Spurious radiation or self-generated radiation can be emitted by the receiver local oscillator, chroma reference oscillator, sound limiter stages, video detector (harmonics of IF), video and R-G-B drive circuits, deflection output stages, and power supplies. These radiations may affect the receiver emitting them or may affect other receivers in the near vicinity. Acceptable limits and methods of measurement are given in the references listed in Sec. 13.3.

13.6.13 SELECTIVITY. Rejection of an undesired adjacent channel is an important design characteristic on cable systems having moderately strong signal levels[31,32] as well as for medium- and weak-signal over-the-air broadcasts. Typical color receivers provide at least 50 dB of rejection to the adjacent-channel sound and picture IF carriers (47.25 and 39.75 MHz, respectively) relative to the desired picture carrier.

13.6.14 INTERFERENCE SUSCEPTIBILITY. The television receiver must be designed to reject interference from many sources in addition to undesired adjacent-channel broadcasts. The more common sources are listed below:

1. *Image frequency.* There is typically 60-dB rejection to image frequencies for VHF and 45 dB for UHF.

2. *IF rejection* of greater than 60 dB is usually provided in the tuner input circuit. Police communications and private paging services operate in the IF frequency band.

3. *Citizens band (CB) radio* on frequencies from 26.965 through 27.405 MHz is harmonically related to channels 2, 5, 6, 9, and 13. Based upon the CB power level and proximity to home television receivers, these harmonics can be generated in the television receiver tuner.[33-36]

4. *Broadcast and amateur-radio transmitters* on frequencies from 540 kHz to 30 MHz are being discussed by several industry committees and the FCC. Voluntary guidelines similar to those for CB are being developed.

5. *Medium- and high-powered FM transmitters,* especially in the frequency band 88 to

92 MHz, create interference to channel 6 reception by means of cross modulation of the channel 6 sound carrier and FM carrier onto the picture carrier, and intermodulation between the FM carrier, channel 6 picture, and local oscillator.[37] FM band filters typically are built into the VHF tuner to eliminate the problems which might otherwise be caused by FM stations in the upper end of the band. These filters, if moved too close to the low end of the band, can have a deleterious effect on the channel-6 sound-carrier and chroma-subcarrier information.

In addition to the undesired sources of antenna input discussed above, the receiver can be influenced by direct electromagnetic radiation into the chassis and cabinet internal wiring. Evidence suggests that television receivers and business electronic equipment should be designed for immunity to radiation fields up to 1 V/m.[38] Evaluation of receiver susceptibility can be made by placing the receiver in a transverse electromagnetic (TEM) test cell. A cell which will accommodate table model receivers has outside dimensions of 2 m (6.5 ft) square by 5m (16 ft) long. Frequency-range capability of this size TEM test cell is 10 kHz to 350 MHz.[39]

13.7 POWER-LINE VOLTAGE CONSIDERATIONS

13.7.1 VARIATIONS. Three nonstandard characteristics of the power line which impact upon the design of line-powered television receivers are voltage deviations from the normal value, interruptions, and transients. Voltage deviations considered here are the changes in rms value which have duration of several seconds to several hours. A supply voltage lower than normal, sometimes called a *brownout*, usually results from overload on the power generation plant or distribution system. Values as low as 95 to 100 V rms are typical in certain localities. Internal regulation in the receiver power supply will generally handle these situations, with perhaps a slight reduction in scan width and height, while the set still maintains usable picture and sound. Voltage values of 20 V above the nominal 120 V rms must also be handled by the receiver with no apparent loss in performance or reliability.

13.7.2 POWER SOURCE INTERRUPTION. Interruption of the power source, referred to as *power crash*, is defined as a sudden drastic decrease in the rms value for a short period of time, milliseconds to seconds or minutes. The longer interruptions will obviously shut down the receiver. When power returns, receiver recovery should be natural and uneventful. Shorter interruptions may affect digital circuitry (control processors, channel memories, etc.) much more than analog circuitry. A receiver should be immune to power crashes which have a duration of up to 100 ms.

13.7.3 TRANSIENTS. Transients on the power line are caused by two significant types of sources. The first is load switching within the residence where the transient is observed. Sources include oil-burner igniters, fan, pump, and refrigerator motors, fluorescent-lamp ballasts, and food mixers. Surge values of 400 to 800 V in some homes and 1200 to 2500 V in others have been recorded.[40] The second type consists of externally produced disturbances, for example, lightning strikes. Here values of 800 to 1200 V are typical on a 120-V line. A few peaks in the 4000- and 5600-V range have been recorded.

A surge transient condition can be created within the receiver itself if a power transformer is utilized between the ac line and the dc power supply. Energizing the transformer primary at the peak of the supply voltage will produce an oscillatory transient with amplitude of up to twice the normal peak secondary voltage. In a similar manner, deenergizing the primary side when the input current peak is at its maximum value can produce an oscillatory surge of up to 10 times the normal secondary peak voltage if driving a high-impedance load.

13.7.4 TRANSIENT PROTECTION. The television receiver can be protected from both internally and externally generated transients by placing metal oxide varistors in the supply circuit. Complete design information can be found in Ref. 41.

Standard test procedures have been developed by both IEEE and Underwriters Laboratories.[42,43] The IEEE *surge withstand capability test,* primarily intended for equipment reliability evaluation, calls for an oscillatory wave of frequency 1.0 to 1.5 MHz and peak value of 2.5 to 3 kV for a duration of 6 μs to be impressed on the power line feeding the appliance under test. The UL test utilizes a 0.01-μF capacitor charged to 10 kV to dump surges into the power line on a 5-s interval. This test is primarily aimed at determining the operator safety of the television set after exposure to power transients.

13.8 MODERN CIRCUIT DESIGN TRENDS

The block diagrams and system design philosophy of the television receiver described in Secs. 13.1 and 13.2 have changed only slightly over the past 30 years. Some concepts discarded earlier have recently been reintroduced in light of changing technology and a reevaluation of performance requirements. The most evolutionary aspect of the television receiver has been the actual circuit implementation of these classic concepts.

In the 1960s vacuum tubes were gradually replaced by the newer solid-state transistor and diode technology. At first, only small-signal circuits (tuner, IF, video, and chroma processing) were transistorized. Designers stayed with tubes in the power circuits (deflection and signal outputs to speaker and CRT), thereby creating a hybrid design. By the late 1960s to early 1970s, even the power stages became transistorized, bringing in the era of full solid-state designs. Power consumption decreased drastically, up to 50 percent, although the total number of electrical components increased slightly. Physical circuit partitioning matched the conceptual block diagram on a 1:1 basis.

In the mid-1960s IC amplifiers began to be used in color television receivers in applications such as AFC amplifiers and burst reference oscillators. By the early 1970s, ICs became available which contained many of the active and passive components of several of the blocks of the conventional block diagram. New circuit approaches were under development which would yield devices having a much greater number of internal transistors, diodes, and resistors but requiring fewer external components to perform the circuit function. One of these circuit areas was the sound IF and detector, which by the late 1970s expanded to include an electronically controlled volume circuit, audio amplifier, and output stage (1 to 5 W). Another early IC was the chroma demodulator which then expanded to include the reference regeneration system, and by the late 1970s encompassed all the circuitry in the chroma and luminance processing sections, from video detector to low-level *RGB* output with only a few additional discrete components.

As the technology has progressed from simple single-stage integration to large-scale integration, the number of ICs in a television receiver increased from the one introductory unit to five and as high as ten in receivers having electronic digital tuning systems. The present trend is toward fewer ICs with higher levels of integration (more circuit functions) in each device. A low-cost monochrome receiver chassis utilizing two ICs to perform all small-signal functions, excluding tuner, was described by Lunn and McGinn in 1981.[44] A block diagram is shown in Fig. 13-33. A year later, Yoshitomi and colleagues described a color chassis which incorporated two large-scale integrated circuits (LSIC) (Fig. 13-34) to perform the small-signal functions with a considerable reduction in total electrical components.[45]

As more circuitry areas are included in a single IC, the correspondence between the classic block diagram and the physical implementation becomes more obscure. Although the block diagram still provides a sound basis for analysis and functional understanding, the shrinking size of the chassis electrical layout with only a few ICs and a handful of discrete parts must be regarded from the system viewpoint. Beneficial or detrimental interactions which might exist between subsystems may now become the overriding design consideration rather than performance optimization of an individual part of the

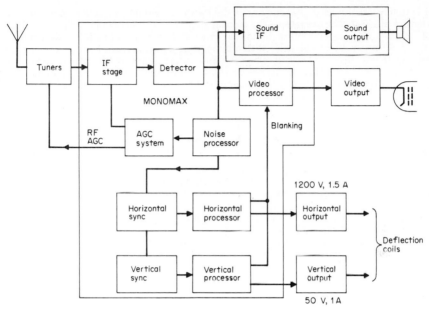

FIG. 13-33 Monochrome receiver having two integrated circuits. *(From G. Lunn and M. McGinn, "Monomax—An Approach to the One-Chip TV," IEEE Trans. Consum. Electron., vol. CE-27, no. 3. © 1981 IEEE.)*

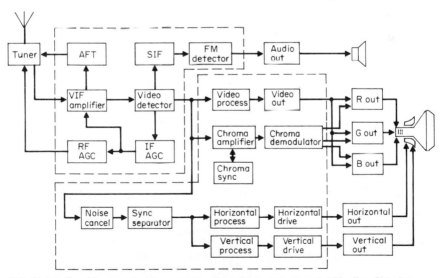

FIG. 13-34 Two-IC color television receiver. *(From M. Yoshitomi et al., "The Two-Chip Integration for Color Television Receivers," IEEE Trans. Consum. Electron., vol. CE-28, no. 3. © 1982 IEEE.)*

circuit. The remainder of this section will deal with individual and fundamental circuits, whether integrated or discrete, in the manner shown in the earlier block diagrams (Figs. 13-1 and 13-16). Any attempt to discuss the interrelationship of circuits and design techniques within an IC can be dealt with only on a generalized basis because the ongoing state-of-the-art brings its own obsolescence. Instead, emphasis will be placed upon those areas which still use primarily discrete circuitry and those areas in which solutions to old problems have been effected by the clever use of the much larger quantity of solid-state components that are available in an IC.

13.9 RECEIVER ANTENNA SYSTEMS

The antenna system is one of the most important circuits in television reception. Signal-to-noise ratio of both picture and sound information is determined primarily by the ability of the antenna system to capture the transmitted field and pass this signal to the receiver.

In an urban area the antenna system typically consists of a set-top antenna (rabbit ears for VHF plus loops for UHF) connected to the receiver terminals with a short length of 300-Ω transmission line. In suburban and rural areas (grade B areas) roof-mounted or chimney-mounted antennas provide adequate reception. For the deep fringe areas, usually at or beyond the grade B contour, a roof-mounted antenna combined with a signal preamplifier is the usual choice. Several choices of 300-Ω twin-lead and 75-Ω shielded coaxial cable are available for the transmission line from the antenna to the receiver. The characteristics of various antennas, transmission lines, and system configurations will be covered throughout this section.

13.9.1 BASIC CHARACTERISTICS.
Antennas have several characteristics which are used to define their ability to receive the radiated field.[46]

Beamwidth in a principal plane of the radiation is defined by the angular width of the pattern at a level which is 3 dB down from the beam maximum.

Polarization of an antenna is defined in terms of the orientation of the electric field vector in the direction of maximum radiation. For example, a vertical dipole above ground is considered to be vertically polarized.

Gain of an antenna is a basic property which is frequently used as a figure of merit. Gain is closely associated with directivity, which in turn is dependent upon the radiation pattern of the antenna. High values of gain are associated with narrow beamwidths. Gain can be calculated only for very simple shapes. It is usually determined by measuring the antenna's performance compared with a standard reference dipole.

Effective area of an antenna is directly related to its gain and is an important parameter in the calculation of energy collected by a receiving antenna in a radiated field. Gain and effective area are related by the formula $A = G\lambda^2/4\pi$ where λ is the wavelength of the E-field radiation.

Input impedance of an antenna is a parameter which must be known in order to extract the maximum received energy from the antenna. All the available power will be transferred to the receiving load only when the load impedance presents a conjugate match to the antenna impedance. For any other match conditions, the transferred power will be smaller by the mismatch loss factor. Like gain, input impedance can be calculated for only very simple shapes and is usually determined by actual measurement.

Radiation resistance is the ratio of the power radiated by an antenna to the square of the current driving its terminals. For short, simple antennas it is closely related to the real part of the input impedance.

Bandwidth describes the frequency band in which the antenna is effective, i.e., has gain, directivity, etc. Unlike the other parameters, no unique definition exists for bandwidth. It is related, however, to the antenna Q in the classic sense and is the function of the stored or reactive energy to the radiated (absorbed) energy.

The bandwidth requirements for television broadcast reception are quite variable.

The single-channel reception requirement is 6, the channel width. The fractional band width b/f varies from 0.11 for channel 2 to 0.0075 for channel 69. In many installations the antenna design has been selected to cover several channels, e.g., low VHF, channels 2 to 6, with a fractional bandwidth of 0.49.

13.9.2 DIPOLE ANTENNA. This antenna is the simplest to construct and analyze and has been the basis for many other derivative designs. It consists of two elements (rods or wires) each of length $l = \lambda/4$ placed in a line and connected at their center to a parallel-wire transmission line (Fig. 13-35). For the half-wave ($\lambda/2$) dipole, the magnitude of the E field, which is an indication of the reception directivity, is given by the equation below and Fig. 13-36.

$$E_\theta = \frac{60I}{r}\left(\frac{\cos \pi/2 \cos \theta}{\sin \theta}\right) \qquad (13\text{-}27)$$

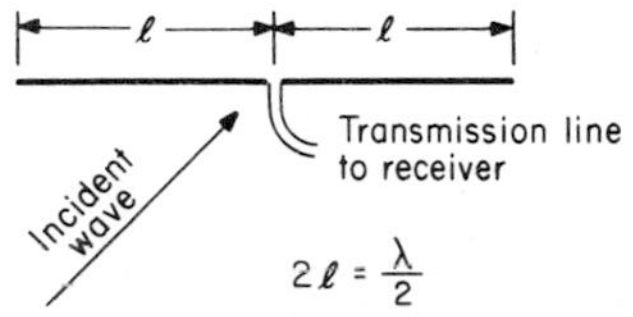

FIG. 13-35 Half-wave dipole antenna.

The half-power (3-dB) beam width is 78°. The radiation resistance is 73 Ω. An impedance transformation is therefore required to minimize VSWR and maximize power transfer when using 300-Ω balanced transmission line. A 75-Ω balanced-to-unbalanced (balun) transformation is used with 75-Ω shielded cable.

If the total length of the dipole antenna becomes much less than $\lambda/2$, the characteristics become those of a "short dipole" for which the radiation resistance is

$$R_{\text{rad}} = 20\pi^2\left(\frac{L}{\lambda}\right)^2 \qquad (13\text{-}28)$$

where $L = 2l$. The E-field pattern becomes more circular, as shown in Fig. 13-36, with the beamwidth increasing to 90°. The input impedance becomes highly reactive which results in an increasing Q and narrower frequency bandwidth. The typical set-top rabbit-ears antenna operates somewhere between a half-wave dipole and a short dipole, especially on the lower VHF channels. For example, on channel 2 an antenna having a length

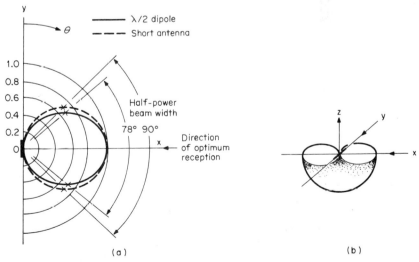

(a)   (b)

FIG. 13-36 (a) Normalized E-field pattern of $\lambda/2$ and short dipole antennas. (b) Three-dimensional view of radiation pattern.

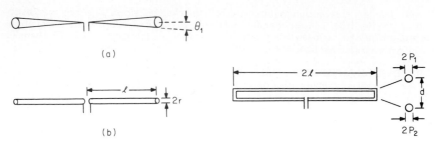

(a)

(b)

FIG. 13-37　(a) Conical dipole; (b) cylindrical dipole.

FIG. 13-38　Folded dipole.

of 27 in (68.5 cm) ($L = \lambda/8$) would show a radiation resistance of only 3 Ω and a capacitive reactance of several hundred ohms.[47]

The frequency bandwidth of a dipole antenna can be increased by increasing the diameter of the elements, for example, by using cylinders and cones instead of wire. As the diameter of the conductor is increased, the capacitance per unit length increases and the inductance per unit length decreases. Since the radiation resistance is affected only slightly, the decreased L/C ratio causes the Q of the antenna, as given below, to decrease

$$Q = \frac{\pi Z_0}{4R_{rad}} \tag{13-29}$$

where $Z_0 = \sqrt{L/C}$. The characteristic impedance for a conical antenna (Fig. 13-37a) is given by Jordan[48] as $Z_0 = 120 \ln (\cot \theta_1/2)$, where θ_1 is the cone angle as shown in Fig. 13-37a. Usually θ_1 is chosen so that the characteristic impedance of the conical dipole

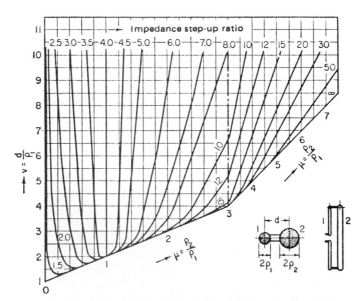

FIG. 13-39　Step-up transformation chart for a folded dipole. (*From H. Jasik, Antenna Engineering Handbook, McGraw-Hill, New York, 1961, chap. 3.*)

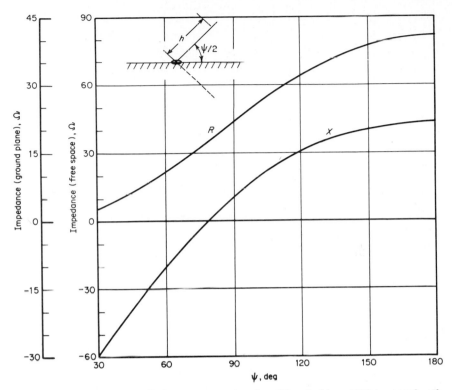

FIG. 13-40 Impedance of a V-dipole antenna as a function of V angle. $h/a = 1000$, h = arm length $= \lambda/4$, a = wire radius. *(Courtesy of J. E. Jones, Wright Patterson AFB, Ohio.) (From D. G. Fink and D. Christiansen, Electronics Engineers' Handbook, McGraw-Hill, New York, 1982, chap. 18.)*

matches as closely as possible the characteristic impedance of the transmission line which it feeds. The characteristic impedance for a cylindrical dipole defined in Fig. 13-37b is $Z_0 = 120 \ln [(2l/r) - 1]$. This broadbanding technique is widely utilized in UHF indoor and outdoor antennas where cylinders, cones, planar bow ties, wire-outlined bow ties, and other fat shapes are typical antenna forms.[49]

The *folded dipole* is another example of a broadbanded antenna derived from the basic dipole (Fig. 13-38). By using rods of different diameter for the half-wave element compared with the quarter-wave elements and by varying the spacing between the parallel elements, one can achieve impedance transformation ratios from 2 to 10 or more, as shown in Fig. 13-39. The typical folded dipole, utilizing elements of equal diameter, has a radiation resistance which is 290 Ω, four times that of the standard half-wave dipole. Broadbanding of nearly a 2-to-1 frequency range is brought about by the unique configuration in which the dipole elements connected to the transmission line act as matching stubs for the tightly coupled $\lambda/2$ radiating element.[50]

V Dipole. Another variation of the dipole antenna, the V dipole, results when the two legs are bent into a V. Again, the similarity between this configuration and the typical set-top antenna can be visualized. Figure 13-40 gives the input impedance versus angle of the V. The effect of bending the dipole is to tune it, giving another practical method in addition to adjusting its length. With increasing tilt angle, the antenna develops a cross-polarization component which may improve its reception capability in the

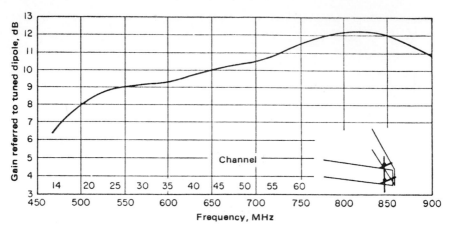

FIG. 13-41   Measured gain of a stacked V antenna. *(From H. Jasik, Antenna Engineering Handbook, McGraw-Hill, New York, 1961, chap. 24.)*

home application. The V antenna also finds application as a UHF outdoor antenna where the selection of element length of 2.1 λ at channel 14 (4.2 λ at channel 83) and an angle between the elements of 50° results in an antenna gain of 8 to 12 dB over the entire band (Fig. 13-41).[51]

The *fan-type dipole* is also based upon the V antenna. As shown in Fig. 13-42, the fan consists of two or more dipoles connected in parallel at their driving point and spread at the ends, thus giving it a further broadband characteristic similar to a fat antenna. This particular design[52] has been optimized for television VHF reception by tilting the dipoles forward by 33° to reduce the beam splitting, i.e., improve the gain, on the high VHF channels, a frequency range in which the antenna is longer than λ/2.

A ground plane or flat reflecting sheet placed behind a dipole antenna and spaced from it a distance of λ/16 to λ/4 can increase the gain in the primary direction by a factor of approximately two. The E-field pattern becomes more like a ball in the third dimension than the torus-shaped pattern which surrounds the dipole in free space. This design is frequently used for UHF antennas, e.g., a bow-tie or cylindrical dipole with reflector. A further improvement in gain and directivity of the half-wave dipole can be made by using a corner reflector (Fig. 13-43). For this configuration gain is a function of the reflector angle α, as well as the distance between the reflector and dipole elements (Fig. 13-44). The terminal radiation resistance of the dipole element is shown in Fig. 13-45, and the typical radiation pattern is shown in Fig. 13-46.[53] As with the other dipole-derived antennas, the bandwidth of this antenna can be increased by using one of the fat dipole shapes. Wind resistance of the ground plane can be reduced by replacing the sheet with metal screening or by a series of parallel conductive rods, spaced approximately 0.1 λ apart from each other.

The *quarter-wave monopole* above a ground plane is another antenna derived from the elementary dipole (Fig. 13-47). This antenna is supplied with many personal portable television sets. Although the monopole is theoretically intended to receive vertically polarized E-field waves, the practical monopole on a receiver has a pivot or ball joint to allow it to be tilted from vertical to a nearly horizontal position. The curves of Fig. 13-40 can be used to determine the impedance by decreasing the values on the y axis (impedance scale) by a factor of 2. The theoretical monopole has radiation resistance of 37 Ω, which is half the value for the dipole in free space.

The loop antenna configuration (Fig. 13-48) is frequently used as an indoor UHF

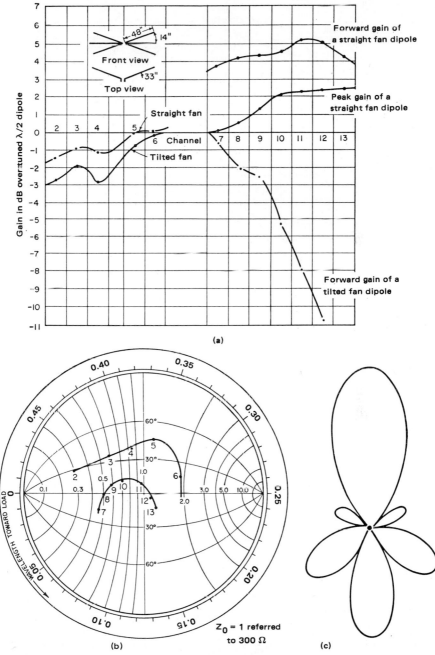

FIG. 13-42 (a) Gain of a fan dipole. (b) Impedance characteristics of a fan dipole. (c) Field pattern of a fan dipole at channel 10. (From H. Jasik, Antenna Engineering Handbook, McGraw-Hill, New York, 1961, chap. 24.)

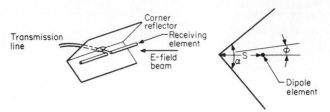

FIG. 13-43 Corner-reflector antenna.

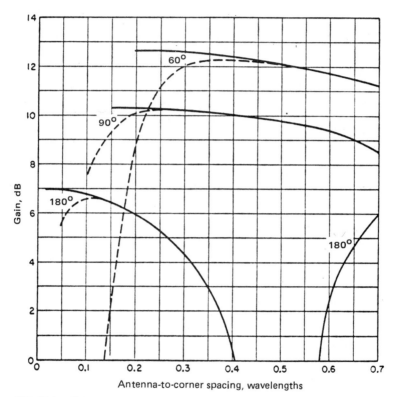

FIG. 13-44 Gain of corner-reflector antennas over a ½-wavelength antenna in free space with the same power input as a function of the antenna-to-corner spacing. Gain is in the direction $\phi = 0$ and is shown for zero loss resistance (solid curves) and for an assumed loss resistance of 1 Ω ($R_{1L} = 1$ Ω) (dashed curves). *(From J. D. Kraus, Antennas, McGraw-Hill, New York, 1950, chap. 12.)*

antenna. The equation for radiation resistance for a small current loop, if analyzed as a magnetic dipole, is given by

$$R_{\text{rad}} = 320\pi^6\left(\frac{a}{\lambda}\right)^4 \tag{13-30}$$

where a = the radius of the loop; $a \ll \lambda$. Calculations for a typical single-conductor loop used as a set-top antenna in the UHF band are given in Table 13-7. An analysis of this

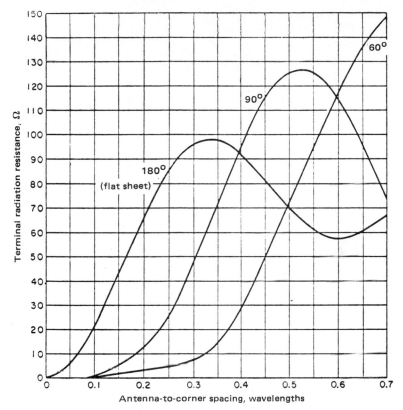

FIG. 13-45 Terminal radiation resistance of driven ½-wavelength antenna as a function of the antenna-to-corner spacing for corner reflectors of three corner angles. *(From J. D. Kraus, Antennas, McGraw-Hill, New York, 1950, chap. 12.)*

same loop as a form of wire antenna approximated with a 12-sided polygon having the same length as the loop[54] yields the input impedance results shown in Fig. 13-49. This approximation shows the antenna resonance at the upper end of the UHF band. In reality, the proximity of the television set (lossy ground plane) as well as the loop connection terminal capacitance will add capacitive reactance, thereby lowering the frequency of resonance toward midband.

Multielement arrays can be used to achieve higher gain and directivity. These find widest application in the fringe signal region, near or beyond the grade B signal contour. One design that has been popular for single-channel reception has been the Yagi-Uda array (Fig. 13-50) which consists of a $\lambda/2$ driven dipole, a reflector element ($l > \lambda/2$), and 1 to n director elements ($l < \lambda/2$) spaced approximately $0.25\,\lambda$ from each other. Gain and beam width values for the typical designs are given in Table 13-8. Typically, the bandwidth is quite narrow, e.g., one or two channels. This can be overcome by making the directors slightly shorter than theoretical and the reflector slightly longer, with only a slight loss in gain. A second technique involves adding much shorter elements in groups between the standard directors, such as interleaving television VHF high-band elements and low-band elements.[55] Gain characteristics are shown in Figs. 13-51 through 13-53.

The *log-periodic design* represents the widest-bandwidth, high-gain antenna in use

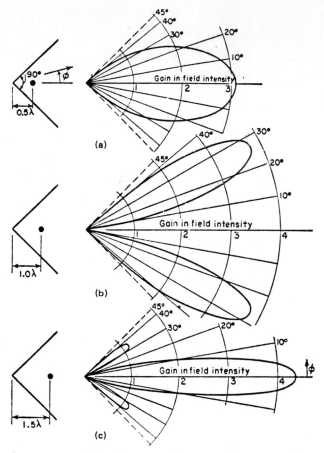

FIG. 13-46 Calculated patterns of square corner-reflector antennas with antenna-to-corner spacings of: (*a*) 0.5 wavelength; (*b*) 1.0 wavelength; (*c*) 1.5 wavelengths. Patterns give gain relative to ½-wavelength antenna in free space with same power input. (*From J. D. Kraus, Antennas, McGraw-Hill, New York, 1950, chap. 12.*)

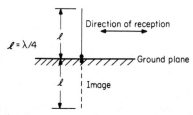

FIG. 13-47 Vertical monopole above ground plane.

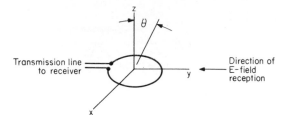

FIG. 13-48 Loop antenna.

Table 13-7 Radiation Resistance for a Single-Conductor Loop Having Diameter of 17.5 cm and Wire Thickness of 0.2 cm

Channel	Frequency, MHz	R_{rad}, Ω
14	473	108
42	641	367
69	803	954
83	887	1342

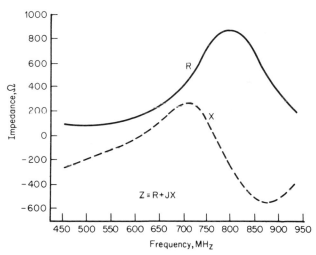

FIG. 13-49 Impedance of a 6.9-in (17.5-cm) loop in the UHF band.

today. The design basis for this type is that the electrical properties, i.e., input impedance, directivity, gain, must repeat periodically with the logarithm of the frequency

$$\tau = \frac{R_n + 1}{R_n} = \frac{l_{n+1}}{l_n} \tag{13-31}$$

where τ = geometric ratio

    R = distance from vertex to each element

    l = length of each element (Fig. 13-54)

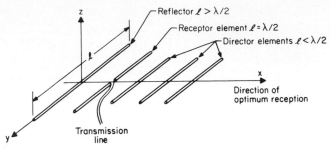

FIG. 13-50   Yagi-Uda array.

Table 13-8   Typical Characteristics of Single-Channel Yagi-Uda Arrays

No. of elements	Gain over $\lambda/2$ element, dB	Beamwidth, deg
2	3–4	65
3	6–8	55
4	7–10	50
5	9–11	45
9	12–14	37
15	14–16	30

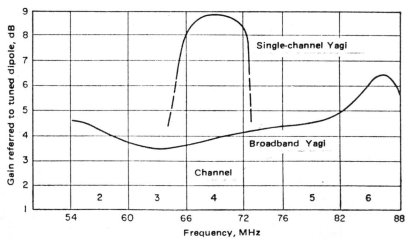

FIG. 13-51   Measured gain of five-element Yagi: (*a*) single-channel Yagi; (*b*) broadband Yagi. (*From H. Jasik, Antenna Engineering Handbook, McGraw-Hill, New York, 1961, chap. 24.*)

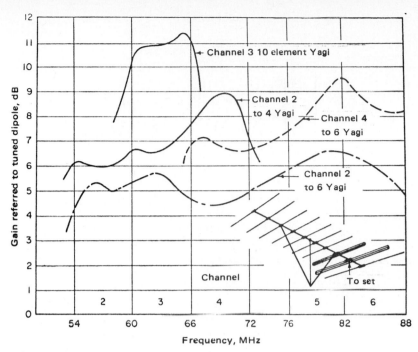

FIG. 13-52 Measuerd gain of three twin-driven 10-element Yagis and a single-channel 10-element Yagi. *(From H. Jasik, Antenna Engineering Handbook, McGraw-Hill, New York, 1961, chap. 24.)*

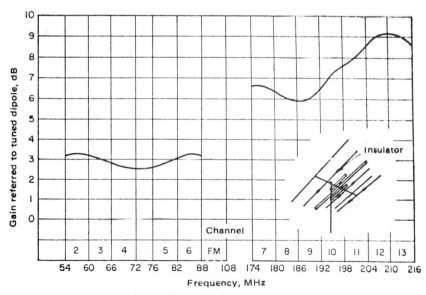

FIG. 13-53 Measured gain of an all VHF-channel Yagi antenna. *(From H. Jasik, Antenna Engineering Handbook, McGraw-Hill, New York, 1961, chap. 24.)*

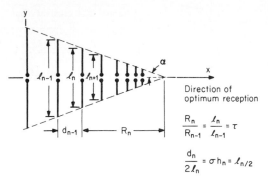

FIG. 13-54   Log-periodic dipole array.

Angles as well as repetitive measurements become the controlling parameters in the design procedure.

The *Vee* log-periodic (LP) configuration (Fig. 13-55) is most often used. This provides a high front-to-back ratio compared with the planar LP antenna which has equal front and back responses. Frequency bandwidths of greater than 20/1 are easily achieved if the theoretical design is closely approximated.[56]

13.9.3 MEASUREMENTS ON COMMERCIALLY AVAILABLE ANTENNAS.

Georgia Institute of Technology made a cost and performance evaluation of 55 antennas as part of the FCC program on UHF comparability. This work has been reported by Free.[57] The characteristics of several of these antennas are given in Tables 13-9 through 13-11. The following conclusions were reported:

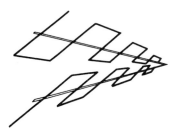

FIG. 13-55   Log-periodic VEE.

1. "UHF only" antennas provide better performance than UHF-VHF combination antennas.

2. The average gain for all UHF antennas is approximately 8 dB, compared with 6 dB for all UHF-VHF combination antennas.

3. The average gain for all VHF antennas is approximately 6 dB, compared with 8 dB for UHF antennas.

4. Considering both overall performance and cost, the four-bay bow-tie antenna at an approximate cost of $10.00 is considered the best antenna choice for UHF reception.

13.9.4 TRANSMISSION LINES.

The basic transmission line equations are derived for distributed parameters R (series resistance), G (shunt conductance), L (series inductance), and C (shunt capacitance). All are defined per unit length of the line. Useful relationships are given in Table 13-12. Relationships between the voltage standing-wave ratio (VSWR) and wave transmission and reflection are given in Table 13-13. The nomograph of Fig. 13-56 provides a numerical solution for these equations. The Smith chart (Fig. 13-57) can be used to graphically determine the input impedance of a transmission line of a known length with a known resistive or reactive load.

Coaxial shielded cable and several types of parallel line or balanced twin lead are commonly used between the antenna and receiver. Figure 13-58 shows the insertion loss for five types of line. Typically, the VSWR for the shielded coaxial lines is in the range of 1.3 to 1.5 in the VHF band and 1.1 to 1.2 in the UHF band. The unshielded twin lead also has low VSWR (1.2 to 1.5) in the VHF band, but increases to 2.0 in the UHF band.

Table 13-9 Characteristics of Indoor Antennas

Channel no. (freq. MHz)	Gain, dB	Beamwidth, deg	Front-to-back, dB	ML-to-SL, dB	VSWR
Type: UHF loop					
14 (473)	−1.5	100.8	2.0	2.0	5.0
24 (533)	0.5	81.6	−0.1	−0.1	2.0
34 (593)	3.0	81.6	0.2	0.2	1.2
44 (653)	3.0	81.0	0.2	0.2	1.0
54 (713)	−0.5	78.6	0.2	0.2	1.8
64 (773)	−3.0	74.4	1.3	1.3	3.6
74 (833)	0.0	121.2	3.1	3.1	4.7
83 (887)	−0.5	64.8	2.8	2.8	4.0
Type: UHF two-bay bow tie with screen					
14 (473)	−4.0	54.5	9.1	9.1	5.3
24 (533)	3.1	57.0	8.6	8.6	2.9
34 (593)	7.6	53.5	10.3	10.3	<1.4
44 (653)	7.6	50.0	11.8	11.8	<1.7
54 (713)	7.7	49.0	9.9	9.9	2.0
64 (773)	0.8	48.0	9.5	9.5	2.1
74 (833)	3.2	49.0	10.5	10.5	3.2
83 (887)	5.8	47.5	9.7	9.7	2.2
Type: VHF rabbit ears/UHF concentric loops					
14 (473)	−4.5	15.0	−1.1	−1.1	4.0
24 (533)	−0.5	70.8	−0.3	−0.3	1.2
34 (593)	−4.3	58.8	−1.8	−1.8	2.0
44 (653)	0.0	96.0	−2.6	−2.6	1.8
54 (713)	−5.5	78.0	−1.2	−1.2	1.2
64 (773)	−6.0	58.8	4.3	4.3	1.3
74 (833)	0.0	96.0	4.1	4.1	2.0
83 (887)	3.0	67.2	1.9	1.9	>10.0

Source: Ref. 57.

Shielded twin-lead VSWR has values up to 2.5 in the VHF band and up to 3 at UHF frequencies.

Although the twin-lead transmission line has lowest insertion loss under ideal conditions, the presence of metal and moisture, as well as outdoor aging, will cause the electrical properties to deteriorate, and, in fact, insertion loss may even exceed that of coaxial lines. Coaxial cable is preferred in many installations because it can be routed along the antenna mast adjacent to metallic structural members and directly through a hole in the house wall with no need for spacers, which are required with twin lead. A study by Free et al.[57] showed that the insertion loss of flat twin lead can increase from 5 to 10 dB [50-ft (15-m) section] when a section is placed in close proximity to metal. The tubular foam-filled types increased only 1 to 2 dB under the same conditions, while coaxial line showed no change. In a wetness test, the flat twin lead again increased by 5 to 10 dB with a change of 1 to 5 dB for the foam-filled tubular or oval lead. The greatest increases occurred in the UHF band for all types of twin lead.

13.9.5 BALUN TRANSFORMERS. These transformers are a type of matching network which provides transition between the balanced and unbalanced sections of the

Table 13-10　Characteristics of Combined VHF-UHF Antennas

Channel no. (freq., mHz)	Gain, dB	Beamwidth, deg	Front-to-back, dB	ML-to-SL, dB	VSWR
Type: VHF LP/UHF Yagi with CR					
4 (69)	5.3	62.5	18.6	18.6	2.1
9 (189)	8.2	32.5	16.3	8.3	<1.6
14 (473)	3.0	25.2	4.5	4.5	1.5
24 (533)	0.0	50.4	11.4	11.4	1.4
34 (593)	0.5	48.0	13.1	13.1	2.1
44 (653)	0.5	38.4	12.2	12.2	2.3
54 (713)	−5.0	33.6	17.4	17.4	2.0
64 (773)	0.5	30.0	12.6	7.1	2.4
74 (833)	6.0	36.0	0.0	−0.2	4.1
83 (887)	0.0				1.8
Type: VHF LP "V"/UHF Yagi with CR					
4 (69)	1.2	75.0	8.6	8.6	2.0
9 (189)	6.5	28.5	12.1	6.2	<1.6
14 (473)	4.4	20.4	−1.1	−1.1	3.2
24 (533)	7.5	24.0	10.1	7.5	1.7
34 (593)	7.3	21.6	8.5	8.5	1.8
44 (653)	10.0	42.0	12.9	12.9	<1.7
54 (713)	10.5	32.0	15.9	10.9	2.7
64 (773)	8.3	25.2	8.8	4.4	5.0
74 (833)	0.0	32.4	−2.0	−3.9	4.5
83 (887)	1.0	52.8	3.4	3.2	5.0
Type: VHF LP/UHF LP					
4 (69)	4.3	69.0	10.3	10.3	1.8
9 (189)	6.5	39.0	12.0	12.0	<1.6
14 (473)	8.5	49.2	5.5	2.0	2.5
24 (533)	3.5	50.4	16.7	16.7	2.2
34 (593)	3.5	58.8	17.7	17.7	<1.4
44 (653)	2.0	58.8	22.0	22.0	1.9
54 (713)	1.5	43.2	10.7	10.7	3.5
64 (773)	10.0	51.6	11.4	11.4	<1.9
74 (833)	9.8	44.4	10.3	10.3	2.4
83 (887)	−1.0				2.7

Source:　Ref. 57.

antenna system, e.g., match balanced antenna terminals to unbalanced shielded cable or shielded cable to balanced receiver input. In receivers these networks typically consist of several turns of 18 to 30 AWG insulated wire, bifilar-wound on a ferrite toroidal core of ⅜- to ½-in (9.5- to 12.7-mm) OD and ⅛- to ³⁄₁₆-in (3.2-to 4.8-mm) thickness. Impedance transformations of 1/4 and 1/1 can easily be obtained by the choice of winding format and interconnection between the balun wires and the transmission lines and/or loads as shown in Fig. 13-59.

Grossner[58] shows that the two wires of the bifilar winding are equivalent to a transmission line in the shape of a helix. Therefore, the analysis and synthesis can be accomplished by using transmission line theory. The characteristic impedance Z_0 for parallel

Table 13-11 Characteristics of Outdoor UHF Antennas

Channel no. (freq., MHz)	Gain, dB	Beamwidth, deg	Front-to-back, dB	ML-to-SL, dB	VSWR
Type: UHF four-bay bow-tie with screen					
14 (473)	9.0	66.0	10.4	10.4	3.5
24 (533)	8.5	55.2	13.5	13.5	2.6
34 (593)	10.3	54.0	15.4	15.4	1.6
44 (653)	13.5	51.6	16.1	16.1	1.1
54 (713)	11.5	48.0	14.9	14.9	1.8
64 (773)	16.0	43.2	13.0	13.0	1.9
74 (833)	11.0	35.4	10.6	10.6	1.3
83 (887)	5.0	39.6	13.2	11.2	1.6
Type: UHF single bow-tie with corner reflector					
14 (473)	4.0	48.0	4.8	4.8	1.8
24 (533)	6.5	61.2	14.2	14.2	2.6
34 (593)	5.8	64.2	13.4	13.4	2.1
44 (653)	8.0	60.0	13.9	13.9	1.7
54 (713)	8.0	57.0	14.7	14.7	1.1
64 (773)	12.0	52.8	14.0	14.0	1.3
74 (833)	10.3	48.0	14.6	14.6	2.5
83 (887)	4.0	50.4	11.7	11.7	1.7
Type: UHF, log-periodic					
14 (473)	4.8	57.0	17.5	17.5	1.8
24 (533)	2.5	58.0	17.3	17.3	2.5
34 (593)	7.2	56.0	14.8	14.8	2.2
44 (653)	7.4	55.0	15.1	15.1	<1.7
54 (713)	6.5	60.0	13.4	13.4	2.4
64 (773)	2.4	58.0	4.4	4.4	3.5
74 (833)	3.0	51.0	15.9	15.9	3.0
83 (887)	6.8	52.5	15.4	15.4	2.0

Source: Ref. 57.

wires is given by

$$Z_0 = \sqrt{\frac{L}{C}} = \sqrt{R_s R_L} = 120 \sqrt{\frac{\mu_r}{\epsilon_r}} \cosh^{-1}\frac{D}{d} \qquad (13\text{-}32)$$

where L = line inductance, per unit length
C = line capacitance, per unit length
R_s = source resistance, Ω
R_L = load resistance, Ω
μ_r = relative permeability = 1
E_r = relative permittivity, dielectric constant of line
D = wire spacing, center to center
d = wire diameter

A transformer to match 75 to 300 Ω for television reception requires a D/d ratio of 1.88 for bifilar parallel wire ($\epsilon_r \cong 1$). This requirement can easily be met with flat two-wire plastic-molded cord. The length of the wires in the coil is related to the upper transmis-

Table 13-12 Summary of Transmission-Line Equations

Quantity	General line expression†	Ideal line expression†
Propagation constant	$\gamma = \alpha + j\beta = \sqrt{(R + j\omega L)(G + j\omega C)}$	$\gamma = j\omega\sqrt{LC}$
Phase constant β	$\text{Im } \gamma$	$\beta = \omega\sqrt{LC} = 2\pi/\Lambda$
Attenuation constant α	$\text{Re } \gamma$	0
Impedance characteristic	$Z_0 = \sqrt{\dfrac{R + j\omega L}{G + j\omega C}}$	$Z_0 = \sqrt{\dfrac{L}{C}}$
Input	$Z_{-l} = Z_0 \dfrac{Z_r + Z_0 \tanh \gamma l}{Z_0 + Z_r \tanh \gamma l}$	$Z_{-l} = Z_0 \dfrac{Z_r + jZ_0 \tan \beta l}{Z_0 + jZ_r \tan \beta l}$
Of short-circuited line, $Z_r = 0$	$Z_\alpha = Z_0 \tanh \gamma l$	$Z_\alpha = jZ_0 \tan \beta l$
Of open-circuited line, $Z_r = \infty$	$Z_\alpha = Z_0 \coth \gamma l$	$Z_\alpha = -jZ_0 \cot \beta l$
Of line an odd number of quarter wavelengths long	$Z = Z_0 \dfrac{Z_r + Z_0 \coth \alpha l}{Z_0 + Z_r \coth \alpha l}$	$Z = \dfrac{Z_0^2}{Z_r}$
Of line an integral number of half wavelengths long	$Z = Z_0 \dfrac{Z_r + Z_0 \tanh \alpha l}{Z_0 + Z_r \tanh \alpha l}$	$Z = Z_r$
Voltage along line	$V_{-l} = V_i(1 + \Gamma_0 e^{-2\lambda l})$	$V_{-l} = V_i(1 + \Gamma_0 e^{-2+\beta l})$
Current along line	$I_{-l} = I_i(1 - \Gamma_0 e^{-2\gamma l})$	$I_{-l} = I_i(1 - \Gamma_0 e^{-2j\beta l})$
Voltage reflection coefficient	$\Gamma = \dfrac{Z_r - Z_0}{Z_r + Z_0}$	$\Gamma = \dfrac{Z_r - Z_0}{Z_r + Z_0}$

†l = length of transmission line.

Source: From J. Feinstein, "Passive Microwave Components," in D. Fink and D. Christiansen (eds.), *Electronic Engineers Handbook*, McGraw-Hill, New York, 1982, chap. 9.

Table 13-13 Some Miscellaneous Relations in Low-Loss Transmission Lines

Equation	Explanation
$r = \dfrac{1 + |\Gamma|}{1 - |\Gamma|}$	r = VSWR
$|\Gamma| = \dfrac{r - 1}{r + 1}$	$|\Gamma|$ = magnitude of reflection coefficient
$\Gamma = \dfrac{R - Z_0}{R + Z_0}$	Γ = reflection coefficient (real) at a point in a line where impedance is real (R)
$r = \dfrac{R}{Z_0}$	$R > Z_0$ (at voltage maximum)
$r = \dfrac{Z_0}{R}$	$R < Z_0$ (at voltage minimum)
$\dfrac{P_t}{P_i} = |\Gamma|^2 = \left(\dfrac{r - 1}{r + 1}\right)^2$	P_r = reflected power P_i = incident power
$\dfrac{P_t}{P_i} = 1 - |\Gamma|^2 = \dfrac{4r}{(r + 1)^2}$	P_t = transmitted power
$\dfrac{P_b}{P_m} = \dfrac{1}{r}$	P_b = net power transmitted to load at onset of breakdown in a line where VSWR = r exists P_m = same when line is matched, $r = 1$
$\dfrac{\alpha_t}{\alpha_m} = \dfrac{1 + \Gamma^2}{1 - \Gamma^2} = \dfrac{r^2 + 1}{2r}$	α_m = attenuation constant when $r = 1$, matched line α_t = attenuation constant allowing for increased ohmic loss caused by standing waves
$r_{\max} = r_1 r_2$	$r_{\max}$ = maximum VSWR when r_1 and r_2 combine in worst phase
$r_{\min} = \dfrac{r_2}{r_1} \quad r_2 > r_1$	$r_{\min}$ = minimum VSWR when r_1 and r_2 are in best phase
$|\Gamma| = \dfrac{|X|}{\sqrt{X^2 + 4}}$	Relations for a normalized reactance X in series with resistance Z_0
$|X| = \dfrac{r - 1}{\sqrt{r}}$	
$|\Gamma| = \dfrac{|B|}{\sqrt{B^2 + 4}}$	Relations for a normalized susceptance B in shunt with admittance Y_0
$|B| = \dfrac{r - 1}{\sqrt{r}}$	

Source: From J. Feinstein, "Passive Microwave Components," in D. Fink and D. Christiansen (eds.), *Electronic Engineers' Handbook*, McGraw-Hill, New York, 1982, chap. 9.

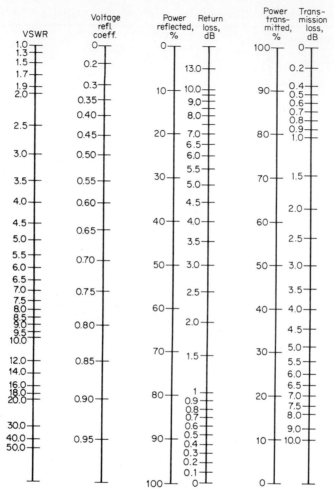

FIG. 13-56 Nomograph for transmission and deflection of power at high-voltage standing-wave ratios (VSWR). *(From J. Feinstein, "Passive Microwave Components," in D. Fink and D. Christiansen (eds.), Electronic Engineers' Handbook, McGraw-Hill, New York, 1982, chap. 9.)*

sion frequency by

$$l < \frac{\lambda_2}{4} = \frac{v}{4f_2} \qquad (13\text{-}33)$$

where λ_2 and f_2 = wavelength and upper transmission frequency and v = the phase velocity of the line.

Data published by Free et al.[57] indicate that television receiver system baluns typically have an insertion loss of less than 1 dB in the VHF band and from 0.7 dB at the low end of the UHF band to 2.0 dB at the high end. VSWR values of 1.4 to 2.0 are typical.

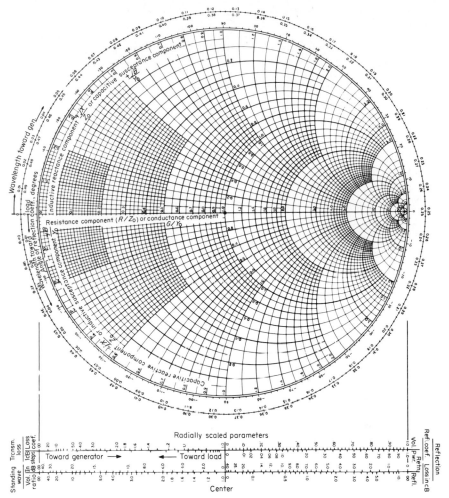

FIG. 13-57 The Smith chart.

13.9.6 BAND SEPARATORS. Splitters, as they are often called, are frequently used to separate the signals on a common lead-in and route them to the separate VHF and UHF input terminals on the television receiver. The splitter consists of a passive *LC* network in a conventional second-order high-pass, low-pass filter arrangement (Fig. 13-60). The splitter can be designed for a totally 300-Ω balanced system or a combination of 75- and 300-Ω input and output. For example, a popular configuration has a 75-Ω input from the lead-in, a 75-Ω output to the receiver VHF terminals, and 300-Ω output (includes balun) to the receiver UHF terminals. Splitter insertion loss values are typically in the range of 1.2 to 2.5 dB. VSWR values of 1.5 to 4.0 are common.[57]

13.9.7 RF TRAPS. Television reception impairment caused by strong, undesired out-of-channel sources can sometimes be best remedied by filtering or trapping at the RF

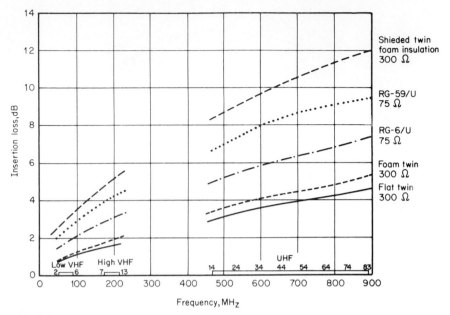

FIG. 13-58   Insertion loss for common transmission lines.

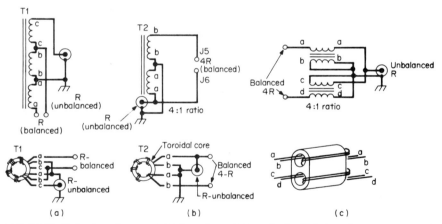

FIG. 13-59   Balun transformers: (*a*) balanced to unbalanced, 1/1; (*b*) balanced to unbalanced, 4/1; (*c*) balanced to unbalanced, on symmetrical balun core, 4/1, [(*a*) and (*b*) *from Radio Amateur's Handbook, American Radio Relay League, Newington, Conn. 1983.*].

input terminals of the television receiver. Commercially available high-pass filters designed for 300- and 75-Ω systems can be installed at the receiver input to reject those signals which have radiation frequencies below the VHF television band. These sources include amateur and CB communications transmitters, diathermy, and RF heating generators. For interference sources in the television frequency range (54 to 890 MHz), for example, FM broadcast stations and VHF amateur transmitters, a quarter-wave, half-

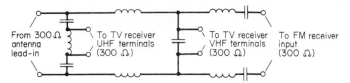

FIG. 13-60   VHF/UHF/FM bandsplitter.

wave, or tuned stub section of transmission line can be used as a trap. This stub is attached to the receiver RF input terminals and cut to a length which will cause it to behave like a short circuit at the exact frequency of the undesired radiation. Details are shown in Fig. 13-61.

13.9.8 ACTIVE ANTENNA SYSTEMS. Such systems consist of a preamplifier integrated with an antenna structure. When a full-sized antenna is used, this combination will provide a signal level which has been increased by the gain factor of the amplifier over a similar passive antenna. The technique for calculating the improvement in noise figure and signal level to the receiver from this combination is described later in this section.

The use of a preamplifier with a much smaller antenna structure ($L < \lambda/2$) can also provide satisfactory reception to sites at which the space for a full-sized VHF antenna is not available. An all-band (VHF/UHF) structure having a diameter of 21 in (53 cm) and thickness of 7 in (18 cm) has been described by Gibson and Wilson.[59] This unit uses a form of broadband, resistive-loaded loop (Fig. 13-62) having a channel 2 gain of -19 dB and a broadband amplifier having a gain of approximately 22 dB across the VHF band. For UHF a broadbanded seven-element Yagi-type antenna provides $+3$- to $+5$-dB gain referenced to a dipole and, therefore, needs no preamplifier. This unit is described as providing adequate performance for color television reception at distances of up to 35 mi (56 km) from the transmitter.

13.9.9 PREAMPLIFIERS. Five major parameters determine their system performance.

1. Input and output impedance (300 or 75 Ω or a combination of both) which must be selected to match the antenna and transmission line characteristics.
2. Gain: Defined as power output to the transmission line divided by power available from the antenna source.
3. Noise figure: See Secs. 13.5.2 and 13.5.5.
4. VSWR caused by mismatch between the source (antenna) and preamplifier input.
5. Distortion level or maximum undesired signal level capability, frequently stated as third-order intercept point of the amplifier.

In an amplifier both cross-modulation and intermodulation distortion can result from the presence of two or more signals. Two equations can be used to relate the third-order intercept level and the power level of the undesired signals to the distortion products[60]

$$P_{mn} = \frac{P_n^2 P_f}{P_{3I}^2} \qquad \text{for } m + n = 3 \tag{13-34}$$

$$P_u = \frac{P_{3I}}{4(m/m' - \frac{1}{2})} \tag{13-35}$$

where P_{mn} = intermod output power at frequencies $\pm m\omega_1 \pm n\omega_2$ (Fig. 13-63)
      P_n = power of undesired carrier nearest in frequency to desired carrier
      P_f = power of undesired carrier farthest in frequency from desired carrier

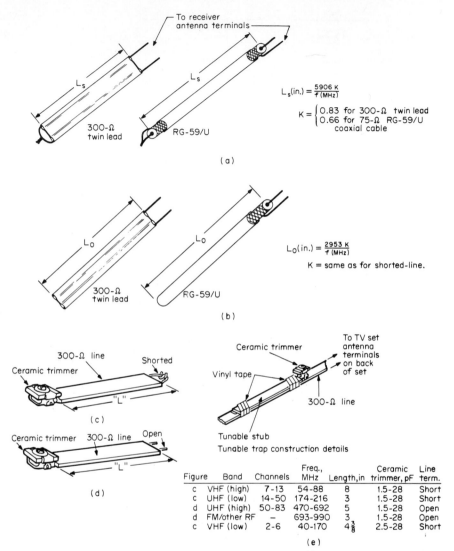

$$L_s(\text{in.}) = \frac{5906\ K}{f\ (\text{MHz})}$$

$$K = \begin{cases} 0.83 \text{ for } 300\text{-}\Omega \text{ twin lead} \\ 0.66 \text{ for } 75\text{-}\Omega \text{ RG-59/U} \\ \quad\ \text{coaxial cable} \end{cases}$$

$$L_0(\text{in.}) = \frac{2953\ K}{f\ (\text{MHz})}$$

K = same as for shorted-line.

Figure	Band	Channels	Freq., MHz	Length, in	Ceramic trimmer, pF	Line term.
c	VHF (high)	7-13	54-88	8	1.5-28	Short
c	UHF (low)	14-50	174-216	3	1.5-28	Short
d	UHF (high)	50-83	470-692	5	1.5-28	Open
d	FM/other RF	–	693-990	3	1.5-28	Open
c	VHF (low)	2-6	40-170	$4\frac{3}{8}$	2.5-28	Short

(e)

FIG. 13-61 Transmission-line stub traps. (*a*) Half-wave shorted-end stub, (*b*) quarter-wave open-end stub, (*c*) VHF shorted-line tunable-stub trap, (*d*) UHF open-line tunable-stub trap, and (*e*) tunable-stub trap construction and installation details. *(From Electronic Technician's Interference Handbook, Television Interference, Consumer Electronics Group, Electronic Industries Association, Washington, November 1977.)*

P_{3I} = third-order intercept power point of amplifier
P_u = power of undesired signal causing cross modulation to desired signal
m/m' = ratio of modulation index of undesired signal to that induced in desired signal

P_{mn} is frequently expressed as a ratio of the desired carrier; e.g., an undesired signal level at -45 dB with respect to the desired carrier P_D will produce a "just perceptible"

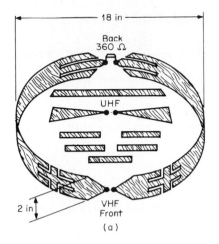

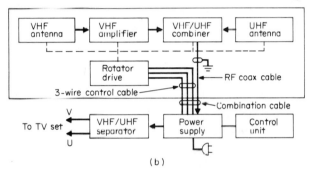

FIG. 13-62 Miniature active antenna: (*a*) basic antenna elements; (*b*) block diagram of the system. *(From J. Gibson and R. Wilson, "The Mini-State—A Small Television Antenna," IEEE Trans. Consum. Electron., vol. CE-22, no. 2. © 1976 IEEE.)*

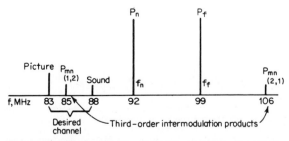

FIG. 13-63 Intermodulation products created by two strong FM stations.

beat interference pattern in the television picture

$$P_{mn} = P_D P_{R/D} \tag{13-36}$$

where P_D = power of desired carrier and $P_{R/D}$ = ratio of undesired resultant to desired signal level.

Converting all values in Eq. (13-34) to decibels referred to one milliwatt and decibels, and rearranging terms to solve for input signal power level, yields

$$2P_n + P_f = P_D + P_{R/D} + 2P_{3I} \tag{13-37}$$

If the assumption is made that both interfering carriers are of equal power, then Eq. (13-37) becomes

$$P_1 = \frac{P_D + P_{R/D} + 2P_{3I}}{3} \tag{13-38}$$

from which

$$3P_1 = 2P_n + P_f \tag{13-39}$$

Equation (13-37) or a combination of Eqs. (13-38) and (13-39) can be used to solve for the maximum internal interfering signals which the amplifier can handle.

For cross modulation [Eq. (13-35)] the value of m/m' may be determined from the assumption of 90 percent modulation of the undesired carrier ($m = 0.9$) and 1 percent induced modulation on the desired carrier ($m' = 0.01$) for *just perceptible* interference in the picture. This gives a value of 19.5 dB for ($m/m' - \frac{1}{2}$). Converting Eq. (13-35) to decibels referred to one milliwatt of input level yields

$$P_2 = P_{3I} - 6 - 19.5 \tag{13-40}$$

P_1, that is, P_n and P_f, can represent the carrier signal with either an AM or FM modulation, for example, an FM station, land mobile, or another television station. P_2 has been derived as though it were only an AM carrier; however, it, too, can represent the carrier level of an FM station to a close approximation. FM band traps are included in most preamplifiers to reduce this problem. It is important to note that intermodulation [Eqs. (13-37) to (13-39)] is dependent upon the desired signal level, while cross modulation [Eq. (13-40)] is not.

Table 13-14 gives the mean values of the data on five typical VHF-UHF preamplifiers derived from the evaluation data developed by Free and staff.[57]

Table 13-14   Mean Values for Five VHF-UHF Preamplifiers Measured in the Georgia Institute of Technology Study, 1981

	Channel						
	4	9	14	34	54	74	83
Power gain, dB	19.1	19.1	13.9	18.0	15.1	10.7	8.6
Noise figure, dB	6.6	6.9	6.9	6.5	6.2	7.9	8.9
VSWR, dB	2.4	2.0	3.6	3.2	2.3	4.2	2.6
Third-order intercept,† dBm	24.2	17.2	13.5	15.2	16.3	14.1	13.9
−45-dB intermodulation,‡ dBm	−23.3	−27.8	−29.5	−28.5	−25.9	−24.3	−23.1
−45-dB cross modulation, dBm	−20.0	−26.2	−29.2	−27.8	−23.9	−21.6	−19.8

†Subtract amplifier gain (dB) from these values to refer them to input level.
‡Based upon a desired level $P_D = -35$ dBm.

Source: Ref. 57.

13.9.10 DIPOLE FACTOR. For an antenna immersed in a radiating field, the power available at the antenna terminals is given by the product of the power density (watts per square meter) of the linearly polarized wave and the effective area of the antenna

$$P_r = \frac{E_2}{120\pi} \frac{g\lambda^2}{4\pi} \qquad (13\text{-}41)$$

where E = field strength,
 g = gain of antenna compared with isotropic radiator
 λ = wavelength, m

For a half-wave dipole ($g = 1.64$) the available power in terms of frequency is

$$P_r \text{ (W)} = [E^2 \text{ (V/m)}] \frac{31.16}{f^2 \text{ (MHz)}} \qquad (13\text{-}42)$$

For a 300-Ω system, the voltage at the antenna terminals is given by

$$e = E \frac{96.68}{f \text{ (MHz)}} \qquad (13\text{-}43)$$

If E is given in microvolts per meter, e will be in microvolts. The dipole factor (dB) is defined as 20 log 96.68/f and accounts for the decreasing ability of a reference dipole antenna to convert radiated field intensity into voltage at its terminals as the frequency increases.

13.9.11 ANTENNA SYSTEM CALCULATIONS. These calculations can provide information on the signal-to-noise ratio to be expected in a television receiver for various antenna system configurations. Both passive and active systems will be considered.

   Example 1. The signal-to-noise ratio in the receiver for the simple antenna and transmission line shown in Fig. 13-64 can be calculated from the equation

$$\text{S/N} = \frac{P_r}{kTB} \frac{1}{F_s} \qquad (13\text{-}44)$$

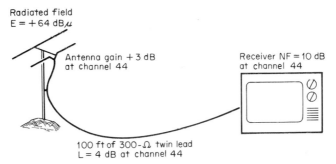

FIG. 13-64 Simple 300-Ω antenna system (Example 1).

where the system noise factor F_s consists of the transmission line loss L and the receiver noise factor F, and k is the Boltzmann constant. Converting the equation to decibels yields

$$\text{S/N (dB)} = E + K + G - F_s - N_i \qquad (13\text{-}45)$$

where E = field strength dBμ (referred to 1 μV per meter)
 K = dipole factor
 G = antenna gain compared with dipole

$$F_s = L \ (4 \ \text{dB}) + F \ (10 \ \text{dB}) = 14 \ \text{dB}$$
$$N_i = kTB, \ 1.6 \times 10^{-14} \ \text{W}, \ +7 \ \text{dB}\mu$$

$$\text{S/N} = 64 \ \text{dB}\mu + (-16) + 3 - 14 - 7 = 30 \ \text{dB} \qquad (13\text{-}46)$$

which is TASO grade 3, "passable."

Example 2. The 300-Ω twin lead of Example 1 is replaced with 75 Ω RG-59/Ω plus the appropriate 300- to 75-Ω balun at the antenna and the corresponding 75- to 300-Ω balun at the receiver as shown in Fig. 13-65. The new value for the system noise figure

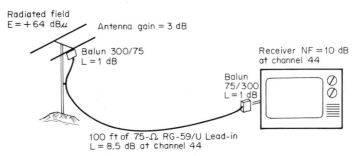

FIG. 13-65 75-Ω antenna system (Example 2).

F_s must include the insertion loss of each balun as well as that for the transmission line and the receiver noise figure

$$F_s = \text{lead-in loss} + \text{balun loss} + \text{receiver NF}$$

$$= 8.5 + 2 + 10 \qquad (13\text{-}47)$$

$$= 20.5 \ \text{dB}$$

$$\text{S/N} = 64 + (-16) + 3 - 20.5 - 7 = 23.5 \ \text{dB}$$

This represents a degradation to TASO grade 4, "marginal."

Example 3. As shown in Fig 13-66, this system is similar to that of Example 2 with a preamplifier at the antenna. Although this calculation can be done using voltages (300-Ω system) as above, we will use power levels which are independent of impedance.

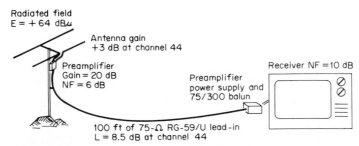

FIG. 13-66 Antenna system with preamplifier (Example 3).

STEP 1. Calculate the power available at the antennas

$$P_r \text{ (mW)} = [E^2 \text{ (V/m)}] \frac{31.16 \times 10^3}{f^2 \text{ (MHz)}}$$

$$E \text{ (dB}_{V/m}) = +64 \text{ dB}\mu - 120 = -56 \text{ dB}_{V/m} \tag{13-48}$$

$$P_r \text{ (dBm)} = E \text{ (dB}_{V/m}) + 10 \log 31.16 - 20 \log f + 30$$

$$= -56 + 14.9 - 56.3 + 30 = -67.4 \text{ dBm}$$

STEP 2. Calculate the system noise figure F_s

Preamplifier power gain = 20 dB $G_a = 100$

Preamplifier noise figure = 6.0 dB $F_a = 4.0$

Transmission line loss = 8.5 dB

Receiver noise figure = 10 dB

$$8.5 + 10 = 18.5 \text{ dB} \qquad F_b = 70.8$$

$$F_s = F_a + \frac{F_b - 1}{G_a} = 4.0 + \frac{70.8 - 1}{100} = 4.7 \tag{13-49}$$

$$F_s \text{ (dB)} = 10 \log 4.7 = 6.7 \text{ dB}$$

STEP 3. Calculate noise at the input N_i

$$N_i = kTB = (1.38 \times 10^{-23} \text{ J/K})(290 \text{ K})(4 \times 10^6 \text{ Hz})$$

$$= 1.6 \times 10^{-14} \text{ W} \tag{13-50}$$

$$= -108 \text{ dBm}$$

STEP 4. Calculate signal-to-noise ratio at receiver

$$S/N \text{ (dB)} = P_r - F_s - N_i \tag{13-51}$$

$$= -67.4 - 6.7 - (-108) = 34 \text{ dB}$$

This signal-to-noise corresponds to a TASO grade 2, "fine." Use of the preamplifier gives an improvement of 10.5 dB (2 TASO grades) over example 2 and an improvement of 4 dB compared with Example 1.

13.10 TUNERS

13.10.1 GENERAL CONSIDERATIONS.   As the input stage to the television receiver, the tuner provides the primary role of selecting the desired signal and excluding all others. General characteristics of tuners which are of considerable design importance in providing optimum reception are discussed below.

Selectivity.   The tuner typically provides a 10-MHz passband for the 6-MHz channel. In color receivers this passband is usually achieved by four tuned circuits: a single-tuned preselector located between the antenna input and the RF stage, a double-tuned interstage network between the RF and mixer stages, and a single-tuned coupling circuit at the output of the mixer. The first three are frequency-selective to the desired channel by varying their inductance, capacitance, or both. The mixer output is always tuned to the IF frequency of approximately 44 MHz. Small-screen monochrome receivers typically contain only a single-tuned circuit in the interstage.

The selectivity provided by the RF stage is to reduce those signals which are several channels removed in frequency from the desired channel, e.g., image, as well as out-of-band signals, thereby eliminating or reducing interference problems. The input section of VHF band tuners usually contains a high-pass filter and trap section to reject signals having frequencies lower than 54 MHz, for example, standard broadcast, amateur, and citizen's band transmissions. A two- or three-pole FM trap is also included to reduce severity of those problems involving FM stations. A partial list of potential interference problems in the VHF band is given in Table 13-15.

In the UHF band, which is continuous from 470 to 806 MHz, the receiver must protect against six major interference mechanisms. These are listed in Table 13-16 and have been given the name "UHF taboos," stemming from the 1952 UHF station allocation study to determine channel separation and mileage restraints.[61]

Spurious Responses.   Transfer characteristic nonlinearities internal to the television receiver are a mechanism for producing interference to picture and sound. Spurious responses can be grouped into three classes: cross modulation, intermodulation, and overload. The first two result primarily from third-order nonlinearity in the amplifiers, mixers, and demodulators. Overload results from excessively large signal levels which can cause the bias of a stage to shift from the intended linear operating range to a nonlinear region having significant second, third, and higher orders of nonlinearity. Television transmitter co-channel and CB harmonic radiation are not considered in this discussion since these interfering signals exist in nature within the frequency range of the desired channel, and therefore receiver selectivity has no effect.

The transfer function of an active stage can be represented mathematically as a power series having real coefficients

$$\text{Ouptut} = A_0 + A_1 X + A_2 X^2 + \cdots + A_n X^n \tag{13-52}$$

where X = the total input to the given stage. For RF and IF amplifier stages, the input consists of all signals passed by the selectivity networks. A mixer stage contains these, plus the local oscillator excitation. The input X can therefore be expressed as

$$X = P \cos \omega_p t + S \cos \omega_s t + O \cos \omega_0 t + U_1 \cos \omega_{u_1} t + \cdots + U_n \cos \omega_{u_n} t \tag{13-53}$$

where P, S, and O represent the desired picture carrier, sound carrier, and local oscillator, respectively, and $U_1 \cdots U_n$ represent undesired interfering signals present at the input. Substitution of the input equation into the transfer series and solving for those terms having a third-power relationship yields results having the general form of

$$K_1 A_3 \cos (3\omega_a)t \tag{13-54}$$

$$K_2 A_3 \cos (2\omega_a \pm \omega_b)t \tag{13-55}$$

$$K_3 A_3 \cos (\omega_a \pm \omega_b \pm \omega_c)t \tag{13-56}$$

where K = a constant determined by the trigonometric expansion of ω^2 and ω^3 terms. ω_a, ω_b, and ω_c represent any of the combinations of the signals making up the input. A

Table 13-15   Potential VHF-Interference Problems

Desired channel	Interfering signals	Mechanism
5	Ch. 11 picture	2 × ch. 5 osc. − ch. 11 pix = IF
6	Ch. 13 picture	2 × ch. 6 osc. − ch. 13 pix = IF
7 and 8	Ch. 5, FM (98–108 MHz)	Ch. 5 pix + FM = ch. 7 and 8
2–6	Ch. 5, FM (97–99 MHz)	2 × (FM − ch. 5) = IF
7–13	FM (88–108 MHz)	2 × FM = ch. 7–13
6	FM (89–92 MHz)	Ch. 6 pix + FM − ch. 6 osc. = IF
2	6M amateur (52–54 MHz)	2 × ch. 2 pix − 6M = ch. 2
2	CB (27 MHz)	2 × CB = ch. 2
5 and 6	CB (27 MHz)	3 × CB = ch. 5 and 6

Table 13-16 Potential UHF Interference Taboos

Type	Interfering channels	Example: receiver tuned to ch. 30
IF beat	N ± 7, ± 8	22, 23, 37, 38
Intermodulation	N ± 2, ±3, ±4, ±5	25–28, 32–35
Adjacent channel	N + 1, − 1	29, 31
Local oscillator	N ± 7	23, 37
Sound image	N + ⅙ (2 × 41.25)	44
Picture image	N + ⅙ (2 × 45.75)	45

potential intermodulation spurious response or interference to the picture or sound results when the aggregate of the cosine functions in Eqs. (13-54) and (13-56) are equal to the desired channel or receiver IF frequency.

Cross modulation can be expressed by substituting into Eq. (13-52) the representation of the input shown below

$$X = V_1 \cos \omega_1 t + V_2(1 + M_2 \cos \omega_{m2} t) \cos \omega_2 t \qquad (13\text{-}57)$$

where $V_1 \cos \omega_1 t$ = desired signal, unmodulated for simplicity of calculation, and $V_2 \cos \omega_2 t$ = undesired signal containing modulation $M_2 \cdot \cos \omega_{m2} t$.

For small amounts of cross modulation the modulation transferred to the desired carrier is given by

$$M_i \simeq \frac{3A_3}{A_1} M_2 V_2^2 \qquad (13\text{-}58)$$

Voltage standing-wave-ratio (VSWR) is a parameter which indicates the degree of match between the antenna system and the receiver input. Low values of VSWR (1.0 to 1.5) can be achieved if the input circuit, which includes high-pass filters, traps, and selective tuned circuits, can be made to match the characteristic impedance of the transmission line within the television passband.

Noise figure goes hand-in-hand with VSWR in transforming the radiated signal into a useful picture as was shown in the examples at the end of Sec. 13.9.11. The resultant value of noise figure in a tuner is dependent upon the insertion loss between the antenna system (source) and the impedance reflected to the input of the RF amplifier device. Insertion loss is a function of Q ratio and power match. These relationships will be covered in the latter part of this section. Values of source resistance for lowest noise figure, as a function of operating bias, is given on manufacturers' data sheets for RF devices. Some typical values are listed in Table 13-17.

Table 13-17 Typical Values of Noise Figure for RF-Amplifier Devices

Type	VHF (200 MHz)	UHF (800 MHz)
Bipolar transistor	3 dB	
Si dual-gate MOSFET	2 dB	4 dB
GaAs MESFET	......	2 dB

Gain of a tuner, as a minimum, should be sufficient to minimize the contribution to the receiver noise factor of the thermal noise generated by the IF filter and amplifier. From Eq. (13-9) in Sec. 13.5.2 for the noise factor of a two-stage cascode

$$G_A = \frac{F_b - 1}{F_{\text{tot}} - F_a} \qquad (13\text{-}59)$$

where G_a = tuner power gain
F_b = IF section noise factor
$F_{tot} - F_a$ = excess noise contributed by IF section

A typical design criterion is to have the IF contribute less than 0.5 dB (1.122 noise factor)

$$F_{tot} - F_a = 1.22$$

Therefore
$$G_a \, \min = \frac{F_b - 1}{1.122} \tag{13-60}$$

A similar calculation gives the required gain of the RF stage needed to reduce the contribution of the mixer. Note that G_a includes the insertion loss of the interstage network.

Gain reduction of the RF stage is intended to maintain the signal level in the tuner at a constant level, below that at which spurious response products become noticeable in the picture (approximately 50 dB below picture carrier level). Typically the RF stage is operated at full gain for signals below -50 dBm, the point at which the noise background in the picture ceases to be noticed. Above -50 dBm the RF stage gain reduces nearly linearly with input level increase. For example, an input level of 0 dBm (0.55 V, 300 Ω) requires a gain-reduction capability of 50 dB. (See Table 13-18.)

Oscillator radiation from the tuner can cause interference to UHF channels and to other services located between low and high VHF and above the high VHF band. Radiation comes primarily from three sources: directly out the antenna input connection, from leads connecting the tuner to other parts of the chassis (supply, AGC, tuning voltage,

Table 13-18   Gain-Reduction Capability of RF Devices

Type	VHF, dB	UHF, dB
Bipolar transistor		
Common emitter	30	
Common base	50	
JFET cascode	36	
Dual-gate MOSFET	50	35
MESFET	....	35

etc.), and direct radiation from the mechanical structure, be it a metal box or other shielded configuration. Oscillator energy on the antenna terminals (see Sec. 13.4.1) may arrive there by direct coupling within the tuner structure or by a path from the mixer through the interstage network and RF stage to the input. It is always advisable to keep the local oscillator power as low as possible, consistent with acceptable mixer conversion gain, spurious response generation, and oscillator stability.

13.10.2 CHANNEL SELECTION—MECHANICAL SWITCH AND ROTARY TUNERS. *VHF tuners* have a rotary shaft which switches a different set of three or four coils or coil taps into the circuit at each VHF channel position (2 to 13). These factory-aligned coils are in the following circuits: RF preselector (input), single-tuned (monochrome) or double-tuned transitionally coupled, RF-mixer interstage, and local oscillator. Values of inductance and capacitance are selected to yield nearly equal bandwidth and gain on all VHF channels, consistent with noise figure, VSWR, and amplifier gain stability.

In the channel 1 position (VHF) the RF stage and local oscillator are disabled, and the mixer stage becomes the IF amplifier having bandpass centered at 44 MHz for the output of the companion UHF tuner. Supply voltage is applied to the UHF tuner only when the VHF tuner is in the channel 1 position.

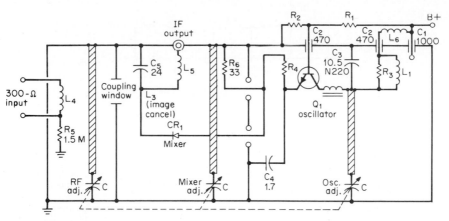

FIG. 13-67 UHF rotary tuner. *(Courtesy of NAP Consumer Electronics Company.)*

UHF tuners have a shaft which, when rotated, moves the plates of air-dielectric variable capacitors in three series-resonant tuned circuits, i.e., double-tuned preselector and local oscillator (Fig. 13-67). The inductor for each tuned circuit consists of a rigid conductive metal strip fastened at one end to the tuner enclosure (ground) and connected at the other end to the stationary plate of the variable capacitor whose rotary plates are grounded. The surface of the metal bar may be polished or plated to reduce its surface resistivity to reduce skin-effect loss. Separation between the tuned circuits is provided by two internal partitions which divide the tuner enclosure into three compartments. A small window in the partition between the two RF preselector tanks provides transitional coupling between the two strips. Design of each of the tuned circuits can be done by treating it as a conventional series LC circuit or by considering it as a distributed transmission line in an enclosed box having a square cross section. The characteristic impedance is given by[62]

$$Z_0 = 138\epsilon^{-1/2} \log_{10}\left(\frac{D}{d} + 6.48 - 2.34A\right) \tag{13-61}$$

where ϵ = dielectric constant for air = 1
  D = height and width of box
  d = diameter of conductor
  $A = [1 + 0.405 \, (d/D)^4]/[1 - 0.405 \, (d/D)^4]$ \hfill (13-62)

For example, if $d/D = 0.025$, $Z_0 \approx 134 \, \Omega$. The electrical and mechanical parameters can be determined by using the equations for a short transmission line considered as having a capacitor at one end and a short circuit at the other. Both quarter- and half-wave resonant lines have been used in tuner design. Tuners utilizing microstrip lines on substrates of alumina and epoxy-glass laminate board have also been built.[63,64]

13.10.3 INPUT FILTERS. A considerable reduction can be made in the number of spurious responses which can be generated with a tuner if the frequency band at the RF input is restricted. Common practice for VHF tuners is to place at the input a high-pass filtering network which greatly attenuates those frequencies below 54 MHz and traps specific frequencies, for example, 27-MHz CB, 44-MHz IF-band, or 88 to 108 MHz-FM band. Circuit configurations based upon several half-sections of constant-K and m-derived image-parameter calculations[62] are shown in Figs. 13-68 and 13-69.

13.10.4 BANDPASS LC TUNED CIRCUIT DESIGN (FIG. 13-70a). The equation for the response of a parallel *single-tuned* circuit as a function of frequency deviation from

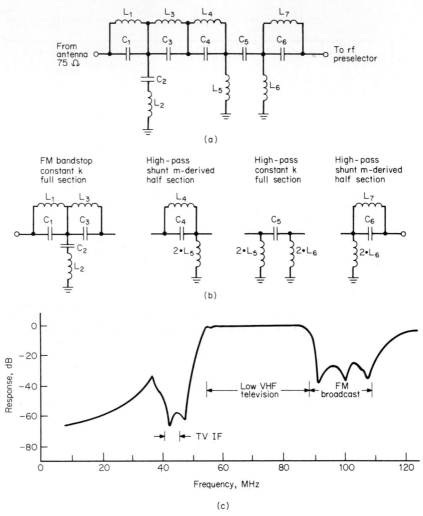

FIG. 13-68 Tuner input network having FM bandstop filter: (*a*) schematic; (*b*) filter analysis sections; (*c*) response.

the resonant point is given by the equation below and is plotted in Fig. 13-70*b*

$$\left| \frac{E}{E_0} \right| = \sqrt{1 + (FQ_L)^2} \tag{13-63}$$

and

$$F = \frac{f}{f_0} - \frac{f_0}{f} \tag{13-64}$$

where $|E/E_0|$ = ratio of voltage at any frequency compared with the voltage at resonance

f = frequency of interest
f_0 = resonant frequency of LC circuit
Q_L = loaded Q of circuit

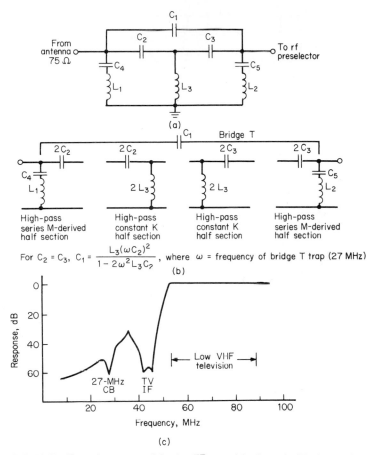

FIG. 13-69 Tuner input network having CB trap: (*a*) schematic; (*b*) filter analysis; (*c*) response.

Insertion loss (IL) at resonance can be thought of as having the two components mismatch loss (between source and load) and Q loss (ratio of unloaded to loaded Q of circuit). The equation is given below, and the curves are plotted in Fig. 13-70*c* and *d*

$$ IL = 10 \log \left[\frac{1}{(1 - Q_L/Q_U)^2} \right] + 10 \log \left[\frac{4R_sR_L}{(R_s + R_L)^2} \right] \qquad (13\text{-}65) $$

where Q_L and Q_U = loaded and unloaded Q of circuit, and R_s and R_L = resistive component of source and load.

Double-tuned response for air-coupled coils (Fig. 13-71*a*) is given by the equation below (modified from Valley and Wallman[65]) and plotted in Fig. 13-71*b*

$$ \frac{|V_2|_f}{|V_2|_{f_0}} = \frac{1 + K^2}{\{[1 + K^2 - (FQ_L)^2]^2 + (FQ)^2[(Q_{Lp} + Q_{Ls})/Q_L]\}^{1/2}} \qquad (13\text{-}66) $$

where $K = k/k_c$ (the coefficient of coupling), $F = f/f_0 - f_0/f$ $\quad Q_L = \sqrt{Q_{Lp}Q_{Ls}}$ and Q_{Lp} and Q_{Ls} are the loaded Q of the primary and secondary.

Other coupling configurations, e.g., high or low side L or C and node coupling, give

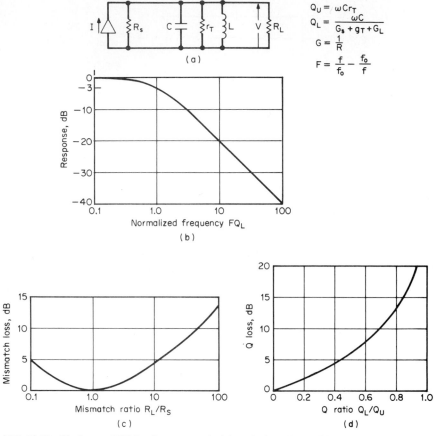

FIG. 13-70 Single-tuned LC-bandpass network: (*a*) equivalent circuit; (*b*) amplitude response; (*c*) mismatch loss; (*d*) Q loss.

identical results for small values of the normalized frequency parameter ($FQ_L < 10$) but deviate appreciably outside that range (Ref. 62, Chap. 9).

Insertion loss (IL) for a double-tuned network can be broken into the three components of primary Q loss, secondary Q loss, and coupling loss, as given below. These components are plotted in Fig. 13-71*c* and *d*

$$\text{IL} = 10 \log \left[\frac{1}{1 - Q_{Lp}/Q_{Up}} \right] + 10 \log \left[\frac{1}{1 - Q_{Ls}/Q_{Us}} \right] + 10 \log \left[\frac{1 + K^2}{2K} \right]^2 \quad (13\text{-}67)$$

UHF image rejection must be provided by the preselector tuned circuits. Derivation of the relationship between rejection at the image frequency and insertion loss at the desired bandpass frequency has been given by Cohn[66]

$$L_0 \text{ (dB)} = \frac{4.343n \text{ antilog } [L'_s + 6.02)/20n]}{\omega_s Q_U} \quad (13\text{-}68)$$

where L_0 = insertion loss at midband
      n = number of equal-element filter sections
      L'_s = loss (dB) at image frequency
      ω_s = fractional image bandwidth $(\omega_{i2} - \omega_{i1})/\omega_0$

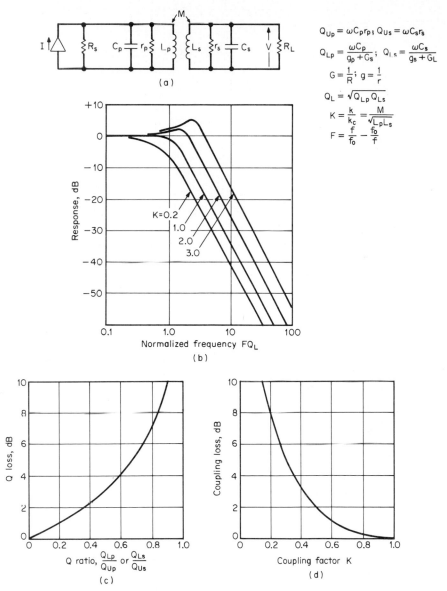

$$Q_{Up} = \omega C_p r_p; \quad Q_{Us} = \omega C_s r_s$$

$$Q_{Lp} = \frac{\omega C_p}{g_p + G_s}; \quad Q_{Ls} = \frac{\omega C_s}{g_s + G_L}$$

$$G = \frac{1}{R}; \quad g = \frac{1}{r}$$

$$Q_L = \sqrt{Q_{Lp} Q_{Ls}}$$

$$K = \frac{k}{k_c} = \frac{M}{\sqrt{L_p L_s}}$$

$$F = \frac{f}{f_o} - \frac{f_o}{f}$$

FIG. 13-71 Double-tuned LC bandpass network: (a) equivalent circuit; (b) amplitude response; (c) Q loss; (d) coupling loss.

Perlow[67] extended this result in terms of the parameters of the UHF resonant cavities. Figure 13-72 is a plot of these relationships.

A technique sometimes used to achieve an additional 6 to 10 dB of image rejection involves a feed-forward technique of coupling the opposite phase of the input signal (image) to the output of the preselector circuit to achieve a degree of cancellation.[68] A simple coupling loop in series with the mixer diode performs this function.

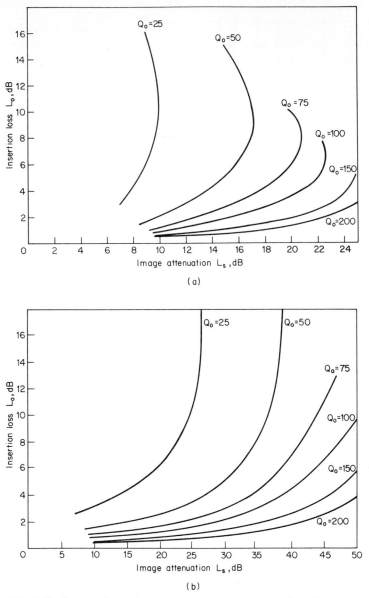

FIG. 13-72 Insertion loss and image rejection: (a) one-section filter; (b) two-section filter. *(From S. Perlow, "Noise Performance Factors in Television Tuners," RCA Rev., vol. 37, March, 1976, p. 119.)*

13.10.5 VARACTOR TUNER DESIGN CONSIDERATIONS. The equivalent circuit of the voltage-controlled capacitance of a reverse-biased diode which performs as the tuning element is shown in Fig. 13-73. The relationship between voltage and capacitance is given by

$$C_j = \frac{\epsilon k}{(V + \phi)^\gamma} \qquad (13\text{-}69)$$

where C_j = junction capacitance
ϵ = dielectric constant of basic material
k = lumped constant
V = applied voltage
ϕ = contact potential of junction, approximately 0.7 V for silicon
γ = capacitance exponential, depends upon doping profile and geometry of junction

This equation can also be expressed as

$$C = \frac{C_0}{(1 + V/\phi)^\gamma} \qquad (13\text{-}70)$$

where C_0 = capacitance at reference (zero volts).

VHF varactor tuner configuration has the same block diagram (Fig. 13-3) and stages as its mechanical counterpart (Sec. 13.10.2). The technique of varying capacitance instead of varying inductances to change channels is the major difference.

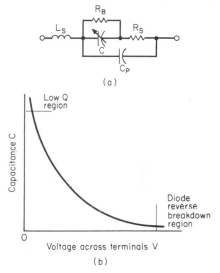

(a)

(b)

FIG. 13-73 Varactor diode characteristics: (a) equivalent circuit; (b) C vs. V curve.

Band switching from the low VHF band to high VHF is accomplished by short-circuiting a part of the tuning coil in each resonant tank, thereby reducing its inductance. This allows the full varactor capacitance range to be used to tune two limited-frequency ranges (57 to 85 MHz and 177 to 213 MHz) instead of one large range. The short circuit is provided by a diode having low series resistance in the forward conduction direction and low capacitance in the reverse-bias condition. Diode parameters are shown in Table 13-19. Figure 13-74 shows a typical RF, interstage, and oscillator inductor arrangement and the diode circuitry used to alter the frequency from low VHF to high VHF.

Table 13-19 Band-Switching Diode Characteristics

Parameter	Type		
	1	2	3
Series resistance R_s, Ω	0.6	0.7	0.4
Forward current, mA	2.0	5.0	10
Capacitance C, pF	2.0	1.0	0.5
Reverse bias, V	15	20	20

UHF varactor tuner configuration typically contains a single-tuned preselector tank, an RF stage, and double-tuned interstage network between the RF and mixer, much like the VHF tuner. The RF stage in the UHF tuner is needed to compensate for the higher insertion losses of the varactor tuned resonant circuits.

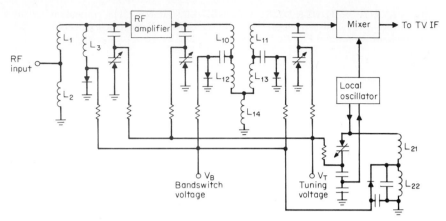

FIG. 13-74 VHF-tuner band-switching and tuning circuits. $+V_B$ = high VHF (active tuning inductors = L_3 in parallel with L_1 and L_2, L_{10}, L_{11}, L_{21}). $-V_B$ = low VHF (active tuning inductors = $L_1 + L_2$, $L_{10} + L_{12}$, $L_{11} + L_{13}$, L_{14}, $L_{21} + L_{22}$).

Tuned-circuit design, when varactor diodes are used, follows much the same philosophy as that for switched coils or air-dielectric rotary capacitors. Design parameters of C_{max}, C_{min}, and the voltage range which can be applied to the diode can be found from data sheets. Transmission line analysis as described for UHF rotary tuners (Sec. 13.10.2) is applicable to UHF varactor tuners.

When compared with a lumped LC circuit, however, the varactor-tuned circuit suffers several shortcomings described by several authors.[69,70] Some of these are more dominant at UHF frequencies than at VHF.

1. Lower Q caused by series resistance R_s results in higher insertion loss and higher noise figure in an RF stage and greater frequency drift with temperature, especially in the local oscillator. Series resistance has been shown to have a voltage-sensitive component. Figure 13-75 compares the voltage sensitivity of silicon diodes with those made of gallium arsenide. The influence of series resistance on noise figure and image rejection in a UHF tuner having a dual-gate MOSFET RF stage is shown in Fig. 13-76.

2. The second derivative of the capacitance versus voltage characteristic d^2C/dV^2 contributes a static frequency shift in the local oscillator because of the larger signal level in that circuit compared with the RF stage networks. This must be compensated by additional fixed parallel capacitance across the RF and interstage circuits to achieve tracking. The third derivative d^3C/dV^2 contributes a third-order nonlinearity which can lead to noticeable cross modulation and intermodulation if extremely large signals are present.

3. With capacitive frequency tuning, a series-resonant circuit rather than parallel type is more appropriate if constant insertion loss and bandwidth are to be maintained across the band. Both impedance and Q increase with increasing frequency in a variable C series-tuned circuit.

4. The temperature coefficient of a varactor diode results from both the dielectric coefficient of the base material (silicon = 30 ppm/°C) and the contact potential of the junction (-2mV/°C).

5. The series inductance L_s, although quite small, reduces the actual capacitance seen at the diode terminals. In a parallel circuit this can lead to reduced Q and mistrack-

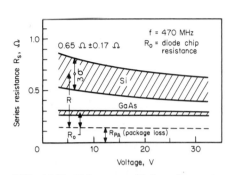

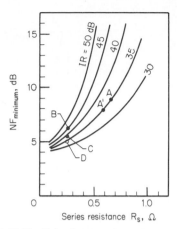

FIG. 13-75 Voltage sensitivity of varactor diodes. *(From T. Hara et al., "Resistance of GaAs Diodes for TV Tuners," IEEE Trans. Consum. Electron., vol. CE-26, no. 4, 1980.)*

FIG. 13-76 Noise figure, series resistance, and image rejection for varactor tuner having single-tuned preselector and double-tuned interstage. *(From T. Hara et al., "TV Tuners Using Low Loss and Low Distortion GaAs Varactor Diodes," IEEE Trans., CE-26, no. 2, May 1980.)*

ing. In a series-tuned circuit, this series inductance is of no consequence because it becomes part of the total circuit inductance.

Tracking. For optimum performance the RF stage and interstage circuits must tune to exactly the same frequency as the tuner is swept across the band. Simultaneously, the local oscillator must be operating at a frequency exactly 44 MHz higher. Diodes matched to ± 1 or ± 2 percent can provide adequate tracking of the RF and interstage circuits if each has the same inductance value and the same amount of stray capacitance. The oscillator circuit, however, requires L and C values which differ from those of the other circuits in order to cover necessary frequency range.

The ratio of varactor capacitance values which tune the RF stage to the lower and upper ends of the band is given by

$$\frac{C_{\max}}{C_{\min}} = \gamma = \left(\frac{f_{\max}}{f_{\min}}\right)^2 \tag{13-71}$$

The additional capacitance C_1 which can be placed across the diode can be calculated from

$$\frac{C_1 + C_a}{C_1 + C_b} = \gamma \tag{13-72}$$

from which

$$C_1 = \frac{C_a - \gamma C_b}{\gamma - 1} \tag{13-73}$$

C_a and C_b are values found on a C versus V plot such as in Fig. 13-77a. Table 13-20 shows the relationships for the three sections of the television broadcast band. The value of RF inductance is given by

$$L_1 = \frac{1}{4\pi^2 f_{\min}^2 (C_1 + C_a)} \tag{13-74}$$

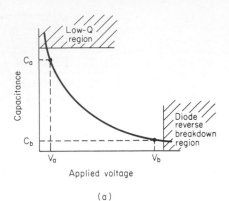

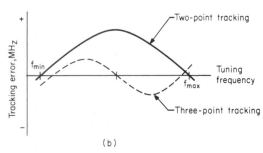

FIG. 13-77 (*a*) Varactor diode C vs. V characteristic; (*b*) tracking error between RF and oscillator circuits.

Table 13-20 Calculation of Shunt Capacitance for RF-tuned Circuits

Band	f_{min}, MHz	f_{max}, MHz	γ	C_1
Low VHF	57	85	2.2	$0.82\,C_a - 1.82\,C_b$
High VHF	177	213	1.45	$2.22\,C_a - 3.22\,C_b$
UHF	473	803	2.88	$0.53\,C_a - 1.53\,C_b$

The relationship for the local oscillator tank capacitance is similar to the RF stage, but frequencies are different

$$C_2 = \frac{C_a + \delta C_b}{\delta - 1} \tag{13-75}$$

where C_2 = additional capacitance to be placed across oscillator varactor diode
C_a and C_b = same value as for RF networks
δ = ratio of C_{max} to C_{min}

The values found from using Tables 13-20 and 13-21 will result in tracking at the two end points of the band (Fig. 13-77*b*). The value of diode capacitance which gives the maximum error is given by Hopkins[71]

$$C_D = \frac{C_2(L_2)^{1/3} - C_1(L_1)^{1/3}}{(L_1)^{1/3} - (L_2)^{1/3}} \tag{13-76}$$

Table 13-21 Calculation of Shunt Capacitance for
Oscillator-Tuned Circuit

Band	Min, MHz	Max, MHz	δ	C_2
V Low	101	129	1.63	1.59 C_a–2.59 C_b
V High	221	257	1.35	2.48 C_a–3.84 C_b
U 14-69	517	847	2.68	.594 C_a–1.59 C_b

where C_D = diode capacitance
    L_1 = RF inductance
    L_2 = oscillator inductance
    C_1 = RF nondiode capacitance
    C_2 = oscillator nondiode capacitance

Midpoint mistracking can be decreased by selecting the tracking points slightly shifted inward from the band edges; or a three-point tracking circuit, originally developed for AM broadcast receivers having equal section capacitors, can be used.[72]

13.10.6 MINIMIZING SPURIOUS RESPONSES. The generation of spurious responses within a television receiver can be reduced by appropriate selection of devices for amplifiers and mixers and setting their operating bias points to yield minimum third-order terms in the transfer characteristic. Figure 13-78 shows that the dual-gate MOS-FET has a clear advantage over other devices for RF stage usage. Figure 13-79, from Weaver,[73] illustrates the graphical technique for determining the proper bias line for the MOSFET. In this example, a compromise solution was chosen in order to have a more reasonable value of drain current (6 mA) rather than the 20-mA line which would have resulted from the more optimum third-order bias point.

The mixer is the most critical stage for spurious response generation. The application of a dual-gate MOSFET device in a VHF tuner has shown a 10- to 20-dB improvement over bipolar devices, especially for reduction of channel 6 color beat, an in-band beat produced by the intermodulation of channel 6 picture and sound carriers[74,75]

$$\text{Picture 6 + sound 6 — local oscillator = IF spur}$$
$$83.25 + 87.75 \quad - 129 \qquad = 42.0 \text{ MHz}$$

Another popular VHF mixer consists of two silicon transistors connected in the common-emitter, common-base cascode configurations. This circuit gives up to 6 dB greater gain and requires a lower level of oscillator injection than the MOSFET. Linearity is not quite as good as for the MOSFET.

Double-balanced mixer circuits which have the characteristic of canceling all even-order products have been built into *IC* format. One utilizes four silicon transistors in a bridge arrangement to achieve a 25 percent intermodulation index compared with an unbalanced circuit.[76] Another has achieved a conversion gain of 35 dB and low spurious generation with a double-balanced differential amplifier arrangement.[77] Various mixer configurations are shown in Fig. 13-80.

UHF tuner spurious responses primarily originate in the nonlinearity of the varactor tuning diodes and the mixer. Tuning diodes should be selected which have a smooth C versus V curve with a small value for d^3C/dV^3 over the range of operation. Further, the lower the value of diode series resistance, the more optimally the tank circuits will operate, providing additional rejection to signals at frequencies well out of the passband.

The bias point for the mixer hot-carrier diode is somewhat critical for minimum non-linearity. Constant-voltage forward-bias rather than constant-current bias has been found to be more effective.[75] Schottky-barrier diodes have much lower forward-biased knee (0.3 V) than silicon diodes and therefore require little if any bias.

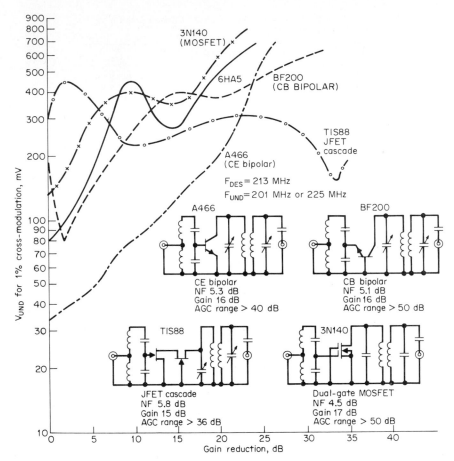

FIG. 13-78 Cross-modulation characteristics of RF devices and circuits. *(From T. H. Moon, "A High Performance VHF Solid-State TV Tuner," IEEE Trans. Broadcast Telev. Receivers, vol. BTR-15, no. 2. © 1969 IEEE.)*

13.10.7 VARACTOR CONTROL SYSTEMS.   Nearly all television varactor tuners are controlled by one of two systems, i.e., voltage synthesis type or frequency synthesis type.

The simplest voltage synthesis technique consists of an array of potentiometers connected between a stable voltage supply and ground. The current-source equivalent has also been used. A manually operated switch, either rotary or pushbutton latching type, connects the appropriate potentiometer to the tuner. Precise adjustment of each potentiometer presets its voltage output to tune one of the desired channels. Band-switching (low VHF, high VHF, UHF) is accomplished by a three-position auxiliary switch associated with each potentiometer. Automatic fine tuning or frequency control (AFT, AFC) can be implemented by adding the frequency control voltage to the reference voltage. Typical potentiometer tuning systems consist of six to ten positions in Europe and Japan. Systems having 12 to 20 positions are common in the United States.

Digital voltage synthesis involves storing the tuner control information, in the form of one digital word per channel, in a memory which is frequently referred to as a *station memory*. Word lengths of 14 to 17 bits have been used, with typical composition being

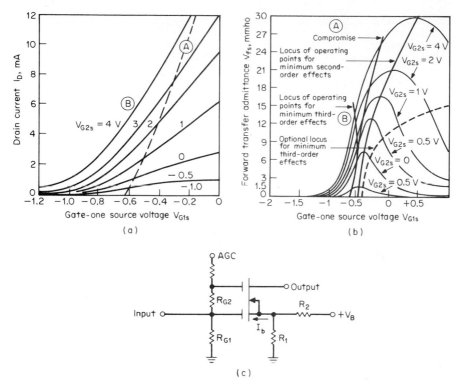

FIG. 13-79 Biasing a MOSFET for minimum spurious response generation. (*a*) Transfer characteristic. (*b*) Transfer admittance characteristic. (*c*) Bias circuit for AGC operation. *A*, compromise locus of operating points for RF AGC operation which yields nearly optimum gain with simple bias scheme and only minor degradation in third-order linearity; *B*, optimum point for mixer operation with fixed bias. (*From S. Weaver, "TV Design Considerations Using High-Gain Dual Mosfets," IEEE Trans. Broadcast Telev. Receivers, May 1973, p. 87.*)

12 to 14 bits for tuning voltage and the remaining bits for band-switching and AFC. The size of the memory determines the number of channels to be made available. Although 20-channel systems are popular, a system handling all the broadcast and midband cable channels in a 100-word by 14-bit electrically alterable read-only memory (EAROM) has been marketed.[81]

Upon human activation of the control keyboard, the system control, in the form of either custom logic circuitry or microprocessor with software control, will illuminate the proper segments of the LED channel number display and will address or extract the word from the corresponding memory location. The tuning voltage portion will be sent to a digital-to-analog converter, the output of which is combined with the receiver AFC voltage to put the varactor tuner on the proper frequency. The remainder of the data stream is analyzed by the microprocessor to determine the band-switching status.

Frequency synthesis utilizes a phase-locked loop (PLL) or frequency-locked loop (FLL) to control the tuner (Fig. 13.81). Human input to the keyboard, the desired television channel number, is converted by the microprocessor to a square wave (1- to 4-MHz range) representing the frequency of the desired channel (scaled down). This square wave is then compared in a phase or frequency comparator with the local oscillator signal which has been processed by a ÷64 or ÷256 prescaler to put it into the same frequency range as the reference. Prescaler ICs have frequency capability to greater than 1 GHz and are usually located in or near the tuner box. The output of the phase comparator consists of a train of error pulses which are integrated by an RC filter to produce a low-level dc voltage. This voltage is then amplified and applied to the tuning varactors. The keyboard also interprets the keyboard entry number and outputs the appropriate band-switching voltage to the tuner.

The frequency-locked loop, in its simplest form, utilizes a counter which has been

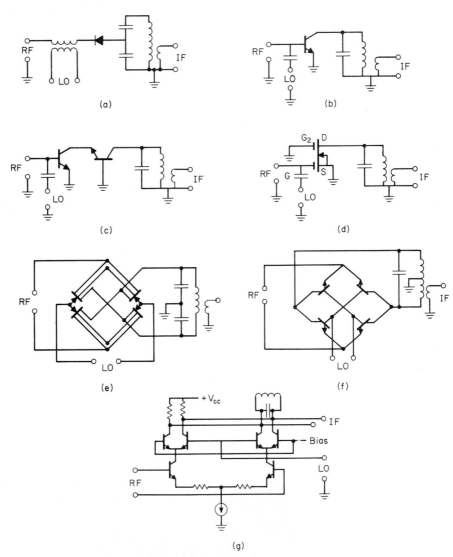

FIG. 13-80   Various mixer configurations: (a) diode; (b) common-emitter bipolar; (c) bipolar cascade, CE-CB;[78,79] (d) dual-gate MOSFET;[73] (e) dual-balanced MESFETS or JFETS;[78,79] (f) double-balanced bipolar;[76] (g) bipolar differential balance circuit.[80]

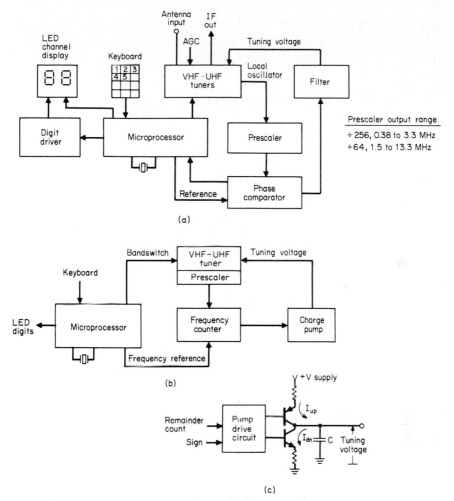

FIG. 13-81 Frequency synthesizer tuning system: (*a*) phase-locked loop; (*b*) frequency-locked loop; (*c*) charge pump circuit.

preloaded by the microprocessor with a number representing the desired channel frequency.[82] The prescaled local oscillator pulses then decrement the counter during a fixed time interval. At the end of the interval, the remaining count is read out to an up-down charge pump which changes the charge on a capacitor. If the count is positive, charge is pumped to the capacitor; if negative, indicating a count past zero, charge is withdrawn from the capacitor. The resultant voltage is applied to the varactor diodes in the tuner. This system corrects for frequency drift only when drift occurs, as compared with the continuous correcting action of the PLL system.

Although the FLL and PLL systems are designed to tune to the precise broadcast frequency, offset frequency capability to accommodate HRC cable channels (Sec. 13.4) or MATV systems can be incorporated. One technique tunes to the standard frequency and then increments the local oscillator in a ±50-kHz step search routine until AFC crossover occurs.[83] Since this technique is under software control (read-only-memory, ROM), a fixed offset can be built in for certain applications. Another system accommodates manual fine tuning by increasing or decreasing the reference frequency in steps of 0.027 percent by use of a programmable divider.[4]

13.11 COMPOSITE-SIGNAL IF AMPLIFIERS AND DETECTORS

13.11.1 RECOMMENDED FREQUENCIES. The important factors to be considered in the choice of intermediate frequencies for a television receiver include the following:

1. Interference from other services
2. Interference from local oscillators of other television receivers and local-oscillator harmonics
3. Spurious responses from images and IF harmonics

Figure 13-82 is a chart which shows the relationship of IF harmonics and images to the desired signal as a function of the intermediate frequency. This chart indicates the soundness of the choice of 45.75 MHz for IF picture carrier. The chart reveals that no images fall within the desired VHF band, except that the channel 6 image just touches the channel 7 passband. All channels are clear of picture harmonics except that the fourth harmonic of IF picture carrier falls near the channel 8 picture carrier. This can cause a serious beat, and the IF designer must check its existence carefully in any new design, especially if the receiver is intended for operation with a set-top antenna. Local-oscillator frequencies do not interfere with another receiver with the same IF frequency

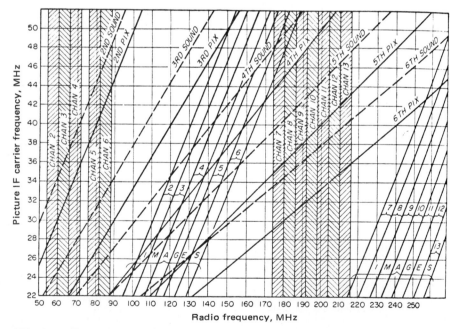

FIG. 13-82 Harmonics of sound and picture IF carriers. *(From Fink, Ref. 10, Chap. 16.)*

on any channel or any channel image. The frequency allocations for UHF stations have been based on a receiver IF frequency of 45.75 MHz so that no station falls on another station's image frequency.

One serious difficulty with the standard intermediate frequency is that it lies in a police communications band where transmitters may radiate significant power. Careful shielding of the input section of the IF amplifier may be necessary to eliminate this problem.

13.11.2 DISCRETE STAGES WITH *LC* COUPLING NETWORKS.

This approach has been the classic technique for many years. First implemented with tubes and later bipolar and MOSFET transistors, the art has changed little.

Gain Requirements. The gain of the tuner and IF must be of such value as to produce a 1- or 2-V video signal at the second detector when the tuner RF input voltage is at or near the sensitivity level. The typical IF design consists of a three-stage chain. Figure 13-83 shows block-by-block gain values for bipolar and MOSFET discrete IFs. The conventional diode detector is used with each design. Each gain stage is separated by a bandpass *LC* interstage network.

Bipolar Transistor Gain Stages. The maximum available gain of an active two-port under matched conditions, as shown in Fig. 13-84 , is given by[84]

$$G_{T(\max)} = \frac{|y_{21}|}{4 \operatorname{Re} y_{11} \operatorname{Re} y_{22}} \tag{13-77}$$

Bipolar transistors have sufficient forward gain (y_{21}) and feedback admittance (y_{12}) in the common-emitter configuration at the IF frequency to cause the amplifier stage to become unstable for a certain range of values of source and load. This precludes the use of the design philosophy of matching input and output for maximum power transfer. Stern[85] and Linvill[86] have both developed techniques for stabilizing the amplifier stages which use these devices. In Stern's approach the source and load are mismatched to the transistor input and output by an equal amount to achieve stability. The gain of the stage as defined in Fig. 13-85 is given by

$$G_T = \frac{4 |y_{21}| G_L G_S}{|(y_{11} + Y_s)(y_{22} + Y_L) - y_{12}y_{21}|^2} \tag{13-78}$$

For stable stage design the following relationship, as defined by Stern, must be met

$$K = \frac{2(G_s + g_{11})(G_L + g_{22})}{y_{12}y_{21}(1 + \cos \theta)} \tag{13-79}$$

where

$$\theta = \tan^{-1} \frac{\operatorname{Im} (y_{12}y_{21})}{\operatorname{Re} (y_{12}y_{21})} \tag{13-80}$$

For stable alignability of the several *LC* networks of an *n*-stage cascade amplifier and for acceptable interchangeability of transistor devices, the value of *K* should be between 2 and 10. The larger the value of *K*, the better the stability margin and the lower the transducer gain (Ref. 84). Linvill's approach matches the input and places all the mismatch loss at the output to achieve nearly the same results. Although these equations appear cumbersome to use, they have led to empirical design techniques for both single-tuned and double-tuned amplifiers.[87,88]

MOSFET Transistor Gain Stages. Because of its extremely low value of capacitance from drain to gate 1, the MOSFET in common-source configuration gives higher stable gain, more AGC range, and less passband skewing. These factors, plus high-input impedance, lead to a simpler design procedure for FETs than for bipolar transistors. The design approach described by Weaver[73,89] has the load, consisting of the interstage bandpass network, matched to the output admittance of the MOSFET to maximize power

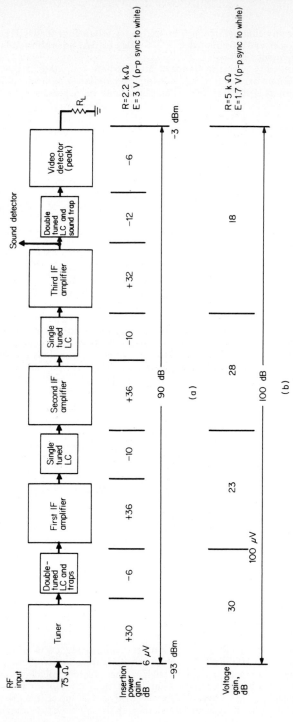

FIG. 13-83 Gain distribution for typical receiver RF and IF section: (*a*) bipolar IF amplifiers; (*b*) MOSFET IF amplifiers.

transfer. The tuned interstage is loaded to achieve the desired Q. It is then direct-coupled to the high impedance of the following stage (Fig. 13-86).

The stage gain equation is similar to that for a vacuum-tube amplifier

$$A_v = \frac{V_{\text{out}}}{V_{\text{in}}} = Y_{fs}R_L \tag{13-81}$$

where Y_{fs} is the forward transfer admittance, common-source configuration. Typical values of $Y_{fs} = 25$ mmho and $R_L = 3$ kΩ will give a voltage gain of 75 (37.5 dB). Typical stage gains of 27 dB, including interstage tuned-circuit losses, are required for the television IF application.

Amplitude Response. The ideal amplitude response of intercarrier-sound monochrome and color receivers is shown in Fig. 13-87. Notice that the color IF is wider and shows greater attenuation of the self-sound carrier. There are at present two techniques for dealing with the problem of separating the three signals in a color television discrete IF:

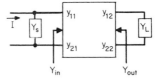

FIG. 13-84 Two-port network having matched conditions: $Y_s = Y_{\text{in}}$; $Y_{\text{out}} = Y_L$.

1. Separate the sound signal from the IF amplifier in the last IF stage and place additional sound attenuation just ahead of the second detector, as described in Sec. 13.1.3 and shown in Fig. 13-8. The detector now provides the luminance and chrominance baseband signals only.

2. Separate the sound and chroma signals from the IF amplifier prior to the third IF stage. Pass these signals together through one common branch amplifier stage and detect them. The detector following the main IF chain provides only baseband luminance information.

The first method requires that the bandwidth of the IF chain be sufficiently wide to pass the chrominance sidebands to the final detector. The sound carrier must be held to an amplitude of 50 dB or more below the luminance carrier at the luminance-chrominance diode detector to minimize the 920-kHz intermodulation beat between the chrominance and sound signals. The signal level at the sound detector should be -26 dB, as has been common practice in intercarrier monochrome receivers. There is no need to attenuate the chrominance carrier ahead of the sound detector, because the 920-kHz beat formed in the intercarrier sound detector cannot pass through the sound IF system.

Variations in the design of the passband to the luminance-chrominance detector are possible. If the chrominance amplifier after the detector has an amplitude response which is peaked at about 4 to 4.1 MHz, the curve of this amplifier may be made to compensate for a drooping curve of the main IF amplifier in the region of 3.6 to 4.1 MHz. The latter amplifier may then be made somewhat narrower in bandwidth. The wide bandwidth shown in Fig. 13-87, combined with the high attenuation of the sound carrier,

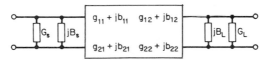

FIG. 13-85 Single-frequency description of an active two-port. *(From M. Ghausi, Principles and Design of Linear Active Circuits, McGraw-Hill, New York, 1965.)*

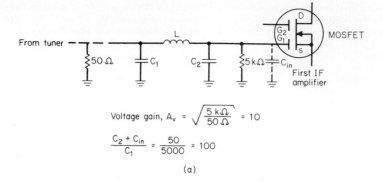

$$\text{Voltage gain, } A_v = \sqrt{\frac{5 \text{ k}\Omega}{50 \Omega}} = 10$$

$$\frac{C_2 + C_{in}}{C_1} = \frac{50}{5000} = 100$$

(a)

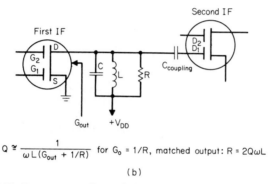

$$Q \cong \frac{1}{\omega L (G_{out} + 1/R)} \quad \text{for } G_o = 1/R, \text{ matched output: } R = 2Q\omega L$$

(b)

FIG. 13-86 Interstage coupling techniques used with MOSFET IF amplifiers: (a) tuner to IF voltage step-up. (b) interstage network designed for maximum power transfer from drain of MOSFET.

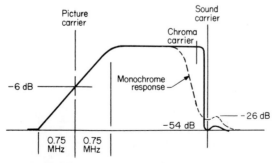

FIG. 13-87 Ideal IF amplitude response, color and monochrome receivers.

can be achieved by using a non-minimum-phase-shift trap, which will be discussed later in this section. A color receiver using this type IF configuration must be tuned precisely to place the sound carrier exactly on the frequency of the sound traps to minimize 920-kHz beat.

The second method mentioned above offers the greater degree in tuning with less critical trap frequency stability. The response at the luminance in the chrominance region must be attenuated considerably more than for the system described above. The higher-frequency video components will also be attenuated, thus decreasing the picture resolution. For this reason, this configuration has not been used in recent years.

The ideal response has been modified in practice to achieve certain specific characteristics learned from field experience. Most monochrome and some color receivers have been designed to have IF amplitude response characteristics which change with AGC bias. At low RF signal levels the response shifts as shown by the dashed line of Fig. 13-88. Higher sensitivity is provided to both picture and sound, and maximum response to

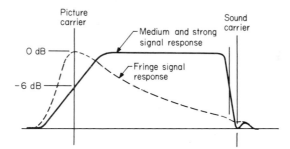

FIG. 13-88   IF response shift with AGC.

the low-frequency picture components is achieved at the tuning point for best sound reception. The picture quality at this point is poor, but the noise which is present obscures fine detail. Under these conditions, the viewer is interested only in reasonable contrast and readability of the picture.

The television receiver's adjacent channel attenuation requirements have been reevaluated in light of the growing number of community antenna television (CATV) systems. At present, it is estimated that some 30 percent of the homes in the United States are served by cable. The number is higher for Canada. The study done by Schwarz indicated that the nearest carrier of the adjacent channel was not the only offender in creating beat products.[32] In fact, the CATV systems operate sound at -15 dB below picture carrier as compared with -3 to -10 dB for broadcast stations. The proposed frequency response is shown in Fig. 13-89. This work was based upon the analysis of the beat visibility threshold plotted for various beat frequencies in Fig. 13-90.

The amplitude response of Fig. 13-87 can be approximated with discrete networks which can be synthesized by using pole-zero theory. The amplitude response is given by network function

$$N_{(s)} = \frac{P_{(s)}}{Q_{(s)}} = \frac{a_n S^n + a_{n-1} S^{n-1} + \cdots + a_0}{b_m S^m + b_{m-1} S^{m-1} + \cdots + b_0} \tag{13-82}$$

where $S = \sigma + j\omega$, the coefficients a_i and b_i are real, and the polynomials can be factored as

$$N_{(s)} = \frac{(S - z_1)(S - z_2) \cdots (S - z_n)}{(S - P_1)(S - P_2) \cdots (S - P_n)} \tag{13-83}$$

The terms $(S - z_i)$ represent the locations of the zeros of the network and can be plotted in the complex plane. Similarly, terms $(S - P_i)$ represent the poles of the network and

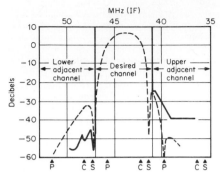

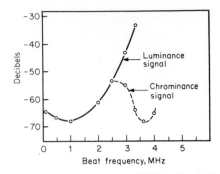

FIG. 13-89 Receiver frequency response, RF and IF: (dashed curve) actual response, (solid curve) required attenuation in adjacent channels to guarantee interference-free operation under "worst-case" conditions. *(From H. Schwarz, "Selectivity Requirements to Eliminate Adjacent Channel Interference," IEEE Trans. vol. CE-22, no. 4, p. 342. © 1976 IEEE.)*

FIG. 13-90 Beat frequency visibility threshold as a function of beat frequency. Threshold is defined as rms signal ratio of interfering frequency level, rms/peak picture carrier level, at input of second detector. *(From H. Schwarz, "Selectivity Requirements to Eliminate Adjacent Channel Interference," IEEE Trans., vol. CE-22, no. 4, p. 341. © 1976 IEEE.)*

can also be plotted. By use of a computer program or manual graphic technique, the designer can determine the amplitude response which will result from given pole-zero locations. Each pole and zero, or combination thereof, can be synthesized as an *RLC* circuit. Numerous texts covering this topic are available; for example, Ghausi,[84] Chap. 15, covers the topic in detail.

A plot of the poles and zeros for a three-stage discrete IF and the resultant amplitude response is shown in Fig. 13-91. The movement of the pole controlled by AGC is also shown. The details on each pole and zero are given in Table 13-22.

Table 13-23 gives design parameters for several *RLC* networks which are commonly used in television receivers as discrete IF *RLC* interstage coupling. Response curves and insertion loss charts contained in Sec. 13.10.4 are also applicable.

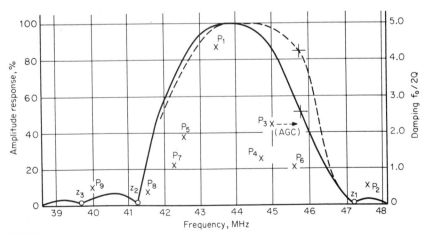

FIG. 13-91 Response and pole-zero locations for three-stage IF circuit.

Table 13-22 Poles and Zeros for a Discrete Three-Stage Color Television IF

Description	f_0	Q	$f_0/2Q$
P_1 Mixer output, broadband	43.5	5	4.35
P_2 47.25 trap pole (Z_1)	47.5	47	0.5
P_3 First IF input	45.0	10	2.25
P_4 First-second IF interstage	44.8	16	1.40
P_5 Second-third IF interstage	42.4	11	1.93
P_6 Third IF output, double-tuned P_7	45.65	19	1.2
P_7 Detector, double-tuned P_6	42.35	18	1.2
P_8 41.25 trap pole	41.5	52	0.4
P_9 39.75 trap pole	40.00	40	0.5
Z_1 Adjacent sound trap	47.25	236	0.1
Z_2 Self–sound trap	41.25	206	0.1
Z_3 Adjacent picture trap	39.75	100	0.2

Phase Response and Envelope Delay. The importance of correct amplitude response is obvious; the effect of variations in phase response is more subtle. Basically, the receiver is required to reproduce rapid changes from black to white or from white to black with as little distortion as possible. This means that, from a design standpoint, the important characteristic of the receiver is its response to a step function. Phase distortion may be even more objectionable than amplitude distortion.

Fundamentally, the absolute time delay in a television receiver is immaterial. The important condition is that the delay imparted to all modulation frequencies shall be as constant as possible. The envelope delay introduced by a network to a steady modulated signal may be expressed as

$$\tau = \frac{\Delta\phi}{2\pi f_m} \tag{13-84}$$

where τ = envelope delay, μs
  $\Delta\phi$ = shift of sidebands with respect to carrier, rad
  f_m = modulating frequency, MHz

It is apparent from the above equation that, for a network to introduce the same delay to all modulating frequencies, the phase difference between carrier and sidebands must be proportional to the modulating frequency, that is, that the phase characteristic of the network be as linear as possible. As long as the phase curve is linear, the value of the slope is not important. The average slope represents the total envelope delay.

Television picture signals are transmitted by vestigial sideband. This is unfortunate from the viewpoint of the receiver phase response. The picture IF amplifier must provide the proper bandwidth and the major portion of the selectivity. This implies the use of networks with sharp cutoff, which tend to exhibit highly nonlinear phase curves in the region near cutoff. In the vestigial sideband system the picture carrier is near the cutoff frequency of the IF amplifier and is thus in a region of phase nonlinearity. Since this fundamental design difficulty is well recognized, transmitters normally introduce some compensating distortion in the modulation waveform to compensate for receiver distortions. The designer must use great care in choosing a pole diagram which has as linear a phase response as possible, especially in the region around the picture carrier.

The phase slope in the area of chrominance carrier also has special considerations. The NTSC defined a group delay specification for transmitters which would compensate for the known receiver circuit roll-off in the chrominance region (Fig. 13-92). The IF group delay should not complement this transmitter predistortion. For proper transient response, the delay complement should be provided by the sum of the group delays of the IF, the 4.5-MHz trap, and the baseband pole which compensates for the sloping IF-amplitude response in the chrominance region.[90]

Table 13-23 RLC Interstage Bandpass and Trap Networks (Narrow-band Approximation)

Circuit	S-plane diagram	Equations	Comments
1. Parallel R L C bandpass		$$S_P = -\frac{\omega_0}{2Q} + j\omega_0$$ $$\omega_0 = \frac{1}{\sqrt{LC}}$$ $$\sigma = \frac{\omega_0}{2Q} = \frac{1}{2RC}$$ $$Q = \omega_0 CR = \frac{R}{\omega_0 L}$$	Response and insertion loss curves: Fig. 13-70, Sec. 13.10.4.
2. Series R L C trap		$$S_z = -\frac{\omega_0}{2Q} + j\omega_0$$ $$\omega_0 = \frac{1}{\sqrt{LC}}$$ $$\sigma = \frac{\omega_0}{2Q} = \frac{R}{2L}$$ $$Q = \frac{\omega_0 L}{R} = \frac{1}{\omega_0 CR}$$	

3. Double-tuned bandpass air coupled		$S_{z1},\ S_{P2} = -\dfrac{\omega_c}{2Q} + j\omega_0\left(1 \pm \dfrac{K}{2}\right)$ $\omega_0 = \dfrac{1}{\sqrt{L_1 C_1}} = \dfrac{1}{\sqrt{L_2 C_2}} \qquad \sigma = \dfrac{\omega_0}{2Q}$ $K = \dfrac{M}{\sqrt{L_1 L_2}}$ $Q_1 = \dfrac{R_1}{\omega_0 L_1} = \omega_0 R_1 C_1$ $Q_2 = \dfrac{R_2}{\omega_0 L_2} = \omega_0 R_2 C_2$	1. Both poles move when M is varied. 2. R_1 and R_2 include the source and load resistance, respectively. 3. Response and insertion loss curves: Fig. 13-71, Sec. 13.10.4.
4. Double-tuned, low–side L coupled		$S_{P1} = -\dfrac{\omega_0}{2Q} + j\omega_0$ $S_{P2} = -\dfrac{\omega_0}{2Q} + j\omega_0\left(1 - \dfrac{1}{\sqrt{1+K}}\right)$ $K = \dfrac{L_M}{[(L_1 + L_M)(L_2 + L_M)]^{1/2}}$ $= \omega_0^2 L_M (C_1 C_2)^{1/2}$ $\approx \dfrac{L_M}{\sqrt{L_1 L_2}}$ $Q_1 \approx \dfrac{R_1}{\omega_0(L_1 + L_M)} \qquad Q_2 \approx \dfrac{R_2}{\omega_0(L_2 + L_M)}$	1. Low-frequency pole moves with variations in L_M. 2. R_1 and R_2 include the source and load resistance, respectively.

Table 13-23   RLC Interstage Bandpass and Trap Networks (Narrow-Band Approximation) (Continued)

Circuit	S-plane diagram	Equations	Comments
5. Double-tuned, low-side C coupled		$S_{P_1} = -\dfrac{\omega_0}{2Q} + j\omega_0\left(1 + \dfrac{1}{\sqrt{1+K}}\right)$ $S_{P_2} = -\dfrac{\omega_0}{2Q} + j\omega_0$ $K = -\dfrac{C_1 C_2}{[(C_1+C_M)(C_2+C_M)]^{1/2}} \approx -\dfrac{\sqrt{C_1 C_2}}{C_M}$ $= -\dfrac{1}{\omega_0^2 C_M(L_1 L_2)}$ $Q_1 = R_1\omega_0\left(\dfrac{C_1 C_M}{C_1+C_M}\right)$ $Q_2 = R_2\omega_0\left(\dfrac{C_1 C_M}{C_1+C_M}\right)$	1. High-frequency pole moves with variations in C_M. 2. R_1 and R_2 include the source and load resistance, respectively.
6. Bridged−T trap		$S_Z = 0 + j\omega_0$ $\omega_0 \cong \sqrt{L\dfrac{C^2}{2C}}$ $C_1 = C_2$ $R = 4R_L = \dfrac{4\omega_0 L}{Q}$	1. R_L represents the loss in the inductor if placed in a simple series-resonant circuit (no. 2). 2. More detailed design procedure is contained in Weaver.[89]

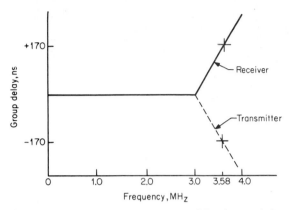

FIG. 13-92   NTSC complementary group delay characteristics.

Tuner-IF Link Coupling.   With mechanically switched tuners it has been necessary to place the tuner behind the customer control panel and connect it to the IF section, located on the chassis, with a length of 50- to 75-Ω shielded cable. Applied to a circuit, this cable is a portion of the low-side coupling between the two half sections of a double-tuned network, one tuned circuit in the tuner (mixer output) and the other in the IF section (first amplifier input) as shown in Fig. 13-93. Both synchronously tuned undercoupled ($k/k_c \simeq 0.8$) and overcoupled staggered pair arrangements have been used. In addition, both low-side C and low-side L (Table 13-23) have been used. Low-side C has

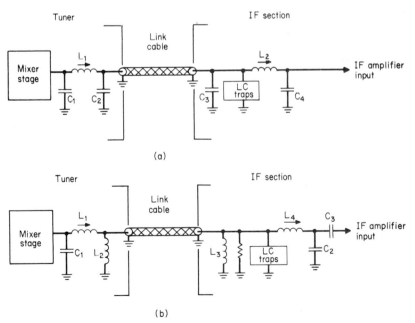

FIG. 13-93   Tuner to IF section link couplings: (*a*) low-side capacitive; (*b*) low-side inductive.

a low-pass configuration which attenuates the local-oscillator and mixer harmonic currents prior to the IF amplifier. This can be a definite advantage in controlling local-oscillator radiation, as well as reducing spurious response generation in the IF section. Lowside L gives better termination to the link cable and therefore reduces cable loss in the interstage.

13.11.3 BLOCK FILTER—BLOCK GAIN IF SECTION. The development of integrated-circuit gain blocks in the late 1960s spurred the need for block filtering. One or two ICs could now supply all the required gain and demodulation with only an input port and one interstage available for bandpass and trap networks. Mertes, in 1972, reported on a predistortion transitional gaussian design which used 10 sections and required each section to be shielded from the others.[91] Although the performance results were excellent, the cost appeared to be prohibitive for the wide range of consumer television receivers. By the mid-1970s the state-of-the-art of surface acoustic wave filters had progressed to the point of producing commercially acceptable devices.

Surface Acoustic Wave Filter. Recent practice has been to implement the IF filtering and gain in the configuration of a block filter followed by a gain-block IC amplifier. A surface acoustic wave (SAW) filter can provide the entire passband shape and adjacent channel attenuation required by a television receiver.[92] A typical amplitude response and group delay characteristics are shown in Fig. 13-94. The sound carrier (41.25 MHz) attenuation of the SAW filter has been designed to operate with a synchronous video detector. The LC discrete response shows 60-dB attenuation of sound carrier which is required for suppression of the 920-kHz sound-chroma beat when using a diode detector. In SAW technology it is possible to make wider adjacent channel traps. This improves their performance and, in part, makes allowance for the temperature coefficient of the substrate materials (Table 13-24). Desirable characteristics for a substrate material are:

1. High value of surface wave coupling k^2 to minimize the inherent loss within the device. The lower the k^2, the higher the impedance and Q, making the filter characteristics more critical to peripheral components.

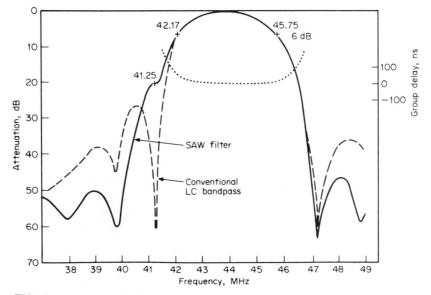

FIG. 13-94 Surface acoustic wave (SAW) IF filter response.

Table 13-24 Characteristics of Substrate Materials

Material	$LiNbO_3$	$LiNbO_3$	$LiTaO_3$	$Bi_{12}SiO_{20}$
Cut and propagation plane	y, z	$128°y, x$	$x - 112°, y$	$\langle 110\rangle, 100$
Coupling, k^2	0.048	0.055	0.008	0.018
Wave velocity, V_s, m/s	3488	3875	3295	1700
Wavelength at 45 MHz, m	75	83	73	37
Temperature coefficient, ppm/°C	95	75	18	118
Trap drift per 10°C, kHz	50	40	12	62
Data source, reference	93	99	100	101

2. Short wavelength to minimize the physical dimensions of the device, giving a greater number of devices on a given size wafer.

3. Low value of temperature coefficient to yield stable realizable bandpass and trap responses.

The schematic diagram of a SAW filter IF circuit is shown in Fig. 13-95. The filter typically has an insertion loss of 15 to 20 dB and therefore requires a preamplifier to maintain satisfactory signal-to-noise ratio in the receiver.

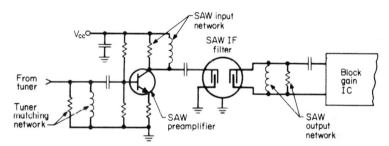

FIG. 13-95   Schematic of block filter–block gain configuration.

The SAW filter consists of a piezoelectric substrate measuring 4 to 8 mm (0.16 to 0.31 in) on a side and 0.4 mm (0.016 in) thick, upon which has been deposited a pattern of aluminum interdigitated fingers, typically 50 to 500 nm thick and 10 to 20 μm in width,

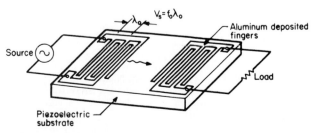

FIG. 13-96   SAW filter.

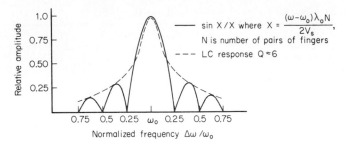

FIG. 13-97 Response of uniform interdigitated SAW filter. [*From A. DeVries et al., "Characteristics of Surface-Wave Integratable Filters (SWIFS)," IEEE Trans. vol. BTR-17, no. 1, p. 16.*]

as shown in Fig. 13-96. Typical substrate materials for television IF applications are lithium niobate ($LiNbO_3$) and lithium tantalate ($LiTaO_3$), although quartz and other materials have been researched.[93] When one set of fingers is driven by an electrical voltage, an acoustic wave moves across the surface to the other set of fingers connected to the load. The transfer frequency response appears as sin x/x (Fig. 13-97). The synchronous frequency f_0 at which the maximum surface energy is generated is given by

$$f_0 = \frac{V_s}{\lambda_0}$$

where V_s = the surface wave velocity and λ_0 = the periodicity of the comb. This finger pattern is called the *uniform interdigitated* (IDT) *pattern*. The transfer function equation is given by Hartmann[94] as

$$H(\omega) = 2k\sqrt{C_s f_0} N \frac{\sin [N\pi(\omega - \omega_0)/\omega_0]}{[N\pi(\omega - \omega_0)]/\omega_0} \exp\left(\frac{-j\omega N}{2f_0}\right) \qquad (13\text{-}85)$$

where k = electromechanical coupling
C_s = static capacitance per finger pair
N = number of finger pairs in pattern
$\omega_0 \equiv 2\pi f_0$ = resonant frequency defined by finger spacing

A modification to the design which gives more optimum television bandpass and trap response consists of changing the length of the fingers or apodizing (Fig. 13-98). This is

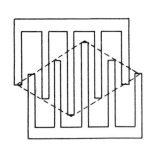

FIG. 13-98 Apodized IDT pattern.

equivalent to parallel-connecting several transducers which have slightly different resonant frequencies and bandwidths.[96-97] Other modifications consist of varying aperture spacings, distance between transducers, and use of passive multistrip coupler patterns in the space between the transducers.

The SAW filter transducers are non-minimum-phase networks; therefore the designer has freedom to independently tailor both the amplitude and phase response. Group delay is held flat to within 50 ns across the passbands in most designs. This is a value which is as good as the best discrete-stage designs and better than most. Nishimura, however, used this independence to produce a SAW filter which has greatly differing group delay for the video carrier and sidebands compared with the chroma sidebands (Fig. 13-99), thereby eliminating the need for a separate 450-ns delay line in the video channel of the receiver.[98]

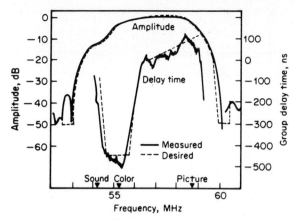

FIG. 13-99 SAW filter having group-delay compensation for chroma channel (Japanese frequencies). *(From Nishimura et al., "Delay-Line-Less Video Signal Processing Circuit," IEEE Trans., vol. CE 26-3, no. 3, p. 384, © 1980 IEEE.)*

Yamada and Uematsu[8] developed a SAW filter having a single-input transducer and two output transducers, one for picture (luminance and chrominance), the other for sound. This filter is used in a quasi-parallel IF system (Sec. 13.1.3). Frequency-response curves of the SAW filter outputs are shown in Fig. 13-100.

A SAW filter is a traveling-wave device, and reflections on the surface and within the crystal substrate will cause interferences with the desired wave. Ripples and echos in the response will result (Fig. 13-101). Triple transit echo (TTE) occurs when the output transducer retransmits the wave back to the input, which then transmits it again to the output. Each return trip is attenuated by the natural loss of the transducers and coupling factor k of the substrate. TTE must be suppressed by greater than 50 dB or it will be seen in the television picture as a delayed ghost of the intended image. One technique for reducing the effects of TTE is to split the transducer fingers and to mismatch the input impedance by driving from a lower value of source resistance. Considering the SAW filter as a two-port network (Fig. 13-102) leads to the following equations for power loss and TTE level[100]

$$\text{Power loss} = \frac{2G_g G_{\text{in}}}{(G_g + G_{\text{in}})^2} \frac{2G_{\text{out}} G_L}{(G_{\text{out}} + G_L)^2} \frac{Y_{12}}{G_{\text{in}} G_{\text{out}}} \tag{13-86}$$

$$\text{TTE level} = \frac{G_{\text{in}}^2}{(G_g + G_{\text{in}})^2} \frac{G_{\text{out}}^2}{(G_{\text{out}} + G_L)^2} \frac{Y_{12}}{G_{\text{in}} G_{\text{out}}} \tag{13-87}$$

where G_g = source conductance
$\quad G_L$ = load conductance
$\quad G_{\text{in}}$ = SAW input conductance (tuned by coil)
$\quad G_{\text{out}}$ = SAW output conductance (tuned by coil)

The third term of both equations is independent of source and load parameters, and its value may be treated as 1 when internal losses are negligible. A chart relating insertion loss, TTE, and input and output conductance ratios is shown in Fig. 13-102b.

Other reflection problems, such as bulk reflection and substrate edge reflections, can be reduced by use of resin absorbers along the substrate edges and by cutting the edges on the diagonal to the path of the main wave.

It is important that the incidental paths between external input and output components be kept to a minimum in order to fully realize the adjacent channel trap attenua-

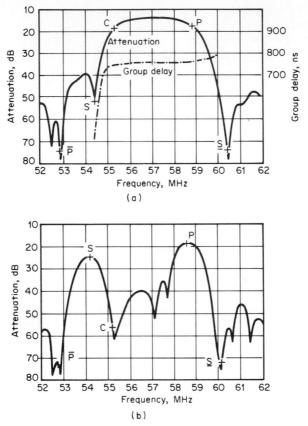

FIG. 13-100 SAW IF filter which has separate outputs for picture carrier and sound carrier: (*a*) frequency response of picture filter; (*b*) frequency response of sound filter. (*From Yamada and Uematsu, "New Color TV with Composite SAW IF Filter Separating Sound and Picture Signals," IEEE Trans. vol. CE-28, no. 3, p. 193. © 1982 IEEE.*)

tion capabilities of the SAW filter. This can be accomplished by proper selection of grounding points and printed-circuit board copper layout.

Noise Figure of SAW IF Filter Circuits. The degradation to receiver noise figure can be calculated by using the equations as given in Sec. 13.5.2 for noise figure of a cascade system

$$\text{Receiver NF (dB)} = 10 \log \left[F_T + \frac{(1/\alpha) - 1}{G_T} + \frac{F_{\text{IF}} - 1}{G_t \cdot \alpha} \right]$$

$$= 10 \log \left[F_T + \frac{(F_{\text{IF}}/\alpha) - 1}{G_T} \right]$$

(13-88)

where F_T = tuner noise factor
G_T = gain of tuner

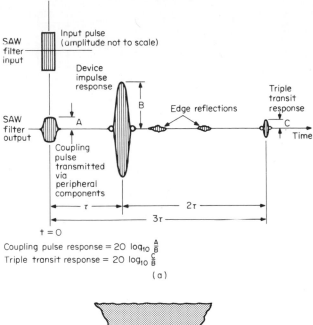

Coupling pulse response = $20 \log_{10} \frac{A}{B}$

Triple transit response = $20 \log_{10} \frac{C}{B}$

(a)

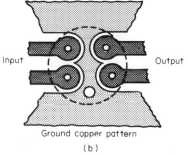

Ground copper pattern

(b)

FIG. 13-101 Spurious echoes at SAW filter output: (*a*) time relationships; (*b*) printed-circuit layout to reduce coupling pulse direct breakthrough.

α = attenuation of SAW filter circuit

F_{IF} = noise factor of IF amplifier

Since the gain of the amplifier and the noise factors of both the tuner and IF stage change with gain reduction (increasing signal level), the calculation should be done at several levels to determine the full influence on receiver signal-to-noise ratio. Figure 13-103 shows a plot of calculations for a typical application. The result of adding a low-noise preamplifier ahead of the SAW filter is also shown. This amplifier has been designed to operate at fixed bias with excellent linearity.[102]

In making these calculations, one obtains signal-to-noise ratio results which are slightly better than reality. Nobeyama et al.[103] have modified the noise-factor equations to account for the video carrier and chrominance subcarrier being at −6 dB on the IF response curve. The IF block amplifier also has a broadband frequency response which

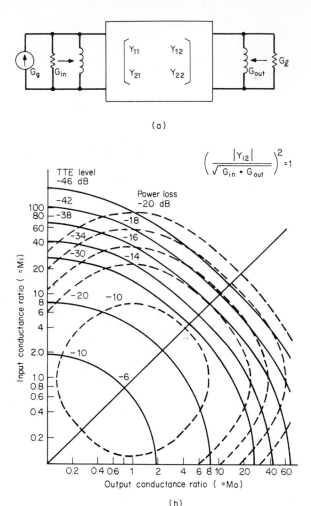

FIG. 13-102 (*a*) Two-port electrical representation of SAW filter. (*b*) Power loss and triple transit echo (TTE). *(From S. Takahashi et al., "SAW IF Filter on LiTaO₃ for Color TV Receivers," IEEE Trans., vol. CE-24, no. 3, p. 343. © 1970 IEEE.)*

allows double-sideband spectrum noise (picture carrier ± 3 MHz) to appear at the second detector. Their modified equations for the noise factor of the luminance and chrominance channels appear below

$$F_y = 3.73 \times F_T + 6.4 \frac{F_{IF} - 1}{G_T(1/L_F)} \tag{13-89}$$

$$F_c = F_T + 8 \frac{F_{IF} - 1}{G_T(1/L_F)} \tag{13-90}$$

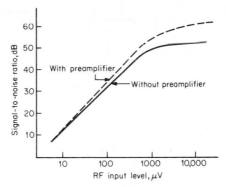

FIG. 13-103   Receiver signal-to-noise ratio for SAW Filter IF.

where F_y = equivalent noise factor for luminance channel
   F_c = equivalent noise factor for chrominance channel
   F_T = noise factor of tuner
   F_{IF} = noise factor of IF amplifier
   L_F = SAW filter insertion loss
   G_T = tuner gain

Integrated-circuit gain blocks have the same basic requirements as discrete-stage IF amplifiers, namely, high gain, low noise, low distortion, and large linear gain control range with high stability under all operating conditions. The differential amplifier, which is basic to many linear integrated-circuit gain stages, meets these objectives. One configuration has a gain of nearly 20 dB and a gain control range of 24 dB. A direct-coupled cascade of three stages yields an overall gain of 57 dB and a gain control range of 64 dB. The gain control system internal to this IC begins to gain-reduce the third stage at an IC input level of 100 μV of IF carrier. With increasing input signal level the third-stage gain reduces to 0 dB, followed by the second stage to a similar level, and followed then by the first. By this means, a noise figure of 7 dB is held constant over an IF input signal level range of 40 dB. The need for a preamplifier ahead of the SAW filter becomes much less of a necessity when an IF amplifier having constant noise figure, such as this one, is used.

The high gain and small physical size of the integrated-circuit IF amplifier require that the designer pay strict attention to the printed-circuit copper layout in order to achieve a design which is stable under all operating conditions. Ground paths must be designed for low impedance and be properly placed, usually with a segment separating input peripheral circuit components and output circuitry. Manufacturers' data sheets and applications notes generally give suggested circuit board layout tips.

A desirable characteristic of discrete stage IF amplifiers is the ability to pole-shift at weak signal levels to improve the receiver gain and picture readability (Sec. 13.11.2 and Fig. 13-88). This has been achieved in an IF IC gain block by placing an external series-tuned *LC* network in the emitter circuit of the third IF stage.[105]

13.11.4 Video Signal Demodulators.   The function of the video demodulator is to recover the picture information which has been placed on the video carrier as vestigial sideband amplitude modulation. The detector receives its signal from the IF amplifier section and sends the demodulated output (direct current to 4.5 MHz) to the video amplifier.

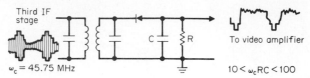

FIG. 13-104   Envelope detector.

Envelope Detector.   Of the several types of demodulators, the simplest type is the envelope detector, consisting of a rectifying diode driving a parallel *RC* network (Fig. 13-104). This circuit has a significant conversion insertion loss and, therefore, must be driven by the IF amplifier with signal levels of several volts peak-to-peak in order to recover 1 to 2 V of video. The inherent nonlinearity of this circuit along with its large input signal leads to several design problems and receiver performance deficiencies.

1. With full bandwidth signals present (41.25^- to 45.75^+ MHz), undesirable beat products will occur between the color subcarrier (42.17 MHz), sound carrier, and strong video frequency components, for example, 920-kHz picture beat interference and sound buzz. In monochrome receivers, the sound carrier is attenuated by greater than 20 dB and the color subcarrier is attenuated from 10 to 15 dB to alleviate these effects.

2. Quadrature distortion caused by the (vestigial) sideband nature of the signal and the receiver Nyquist slope can cause a luminance shift toward black of up to 10 percent as well as asymmetric transient response.[90,105] This situation can also be helped by attenuating the color subcarrier by 6 dB or more with respect to the top of the IF response.

3. The fourth harmonic of the video IF carrier, caused by detector action, can radiate directly from the chassis to the antenna terminals and interfere with VHF channel 8 (180 to 186 MHz).

Even with these deficiencies, the diode envelope detector has been used in a great many monochrome and color receivers dating from vacuum-tube designs through the era of discrete transistor designs.

Transistor Detector.   A transistor biased near collector current cutoff and driven with a modulated carrier having amplitude greater than the bias level (Fig. 13-105) provides a demodulator which can be designed to have a gain of 15 to 20 dB greater than the diode detector. Since this requires less gain in the IF section, the third IF stage in

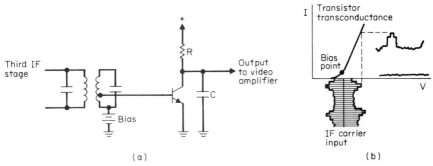

FIG. 13-105   Transistor detector: (*a*) schematic; (*b*) detection characteristic.

some designs has been completely eliminated. Typically, this circuit suffers from the same deficiencies as the diode envelope detector.

Synchronous Detector. A third type is the synchronous demodulator or balanced multiplier in which the modulated carrier is sampled by a pure unmodulated carrier of the same frequency (45.75 MHz). In the quasi-synchronous version the reference carrier is derived by passing a portion of the IF signal (modulated carrier) through a separate tuned high-Q limiting amplifier which removes the amplitude modulation.[106]

Another version, the phase-locked loop (PLL) type, uses a similarly stripped carrier to phase-lock a local oscillator operating at 45.75 MHz.[104] This approach gives the ideal reference waveform, hence, the most accurate recovery of the original modulating waveform. The major advantages to synchronous demodulation are:

1. Higher gain than diode detector.
2. Low-level input considerably reduces undesired beat generation.
3. Low-level switching reduces IF harmonics by >20 dB.
4. Little or no quadrature distortion (Fig. 13-106), depending upon the purity (lack of residual phase modulation) of the reference carrier.[107]
5. Circuit topology which can be easily implemented in IC format.

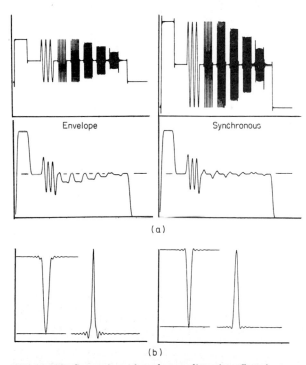

FIG. 13-106 Comparison of quadrature distortion effects in envelope and synchronous detectors: (*a*) axis shift; (*b*) inverted and normal 2T pulse response. *(From C. B. Neal and S. Goyal, "Frequency and Amplitude Phase Effects in Television Broadcast Systems," IEEE Trans., vol. CE-23, no. 3, August 1977, p. 241.)*

An improved form of diode demodulator recently described uses a balanced full-wave configuration with a like element in a feedback loop.[44] Excellent linearity and low-beat product equivalent to the synchronous demodulator types have been achieved. The balanced diode configuration has a simpler circuit configuration and needs no adjustable tuning elements.

13.11.5 AUTOMATIC GAIN CONTROL (AGC).   Each amplifier, mixer, and detector stage in the television receiver has a set of optimum operating conditions at which the following occur:

1. Input level exceeds the internal noise by a chosen factor.

2. Input level does not overload the device thereby causing bias shift.

3. Bias operates each device at its optimal linearity point, i.e., lowest third-order product for amplifiers and mixers, highest second order for mixers and detectors or a reasonable compromise thereof.

4. Spurious responses in the output current are low (< 50 dB) compared with desired signal.

The function of the automatic gain control system is to maintain the signal levels in these stages at the optimum value over a large range of receiver input signal levels. The control voltage which operates this system is usually derived from the video detector or first video stage. A block diagram is shown in Fig. 13-107.

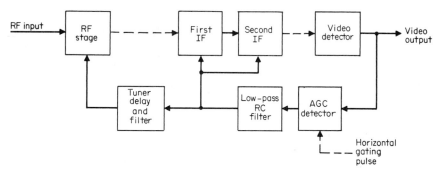

FIG. 13-107   Receiver AGC system.

Average AGC operates on the principle of keeping the carrier level constant in the IF and is influenced by the changing video modulation levels of the carrier. This type is used only in low-cost receivers.

Peak or sync clamp AGC compares the video sync tip level with a fixed dc value. If the amplitude of the tip exceeds the reference level, a control voltage is applied to the RF and IF stages to reduce their gain, thereby restoring sync tip to the reference level.

Keyed or gated AGC is similar to sync clamp AGC. The sync tip is compared with a dc reference in a stage which is active only during the period when sync is transmitted. The horizontal flyback pulse operates the key or gate for the stage. Noise immunity is considerably enhanced over the other two systems.

AGC Delay.   For best receiver signal-to-noise ratio the tuner RF stage is operated at maximum gain for RF signal levels of threshold level up to approximately 1 mV. In discrete amplifier chains, the AGC system begins to reduce the gain of the second IF proportionately as the RF signal level increases from just above the sensitivity level to the second stage limit of gain reduction (20 to 25 dB). For increasing signals, the first IF stage gain is reduced. Finally, above the tuner delay point, 1 mV, the tuner gain is

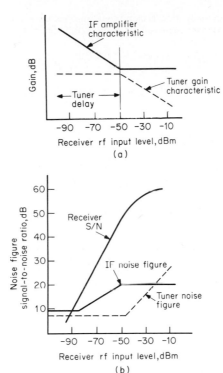

FIG. 13-108 (a) Gain control with input level.
(b) Noise figure of RF and IF stages with gain
control and resultant receiver signal-to-noise
ratio.

reduced. A plot illustrating this activity is shown in Fig. 13-108a. Figure 13-108b shows
the noise figure of each stage and the overall receiver signal-to-noise ratio.

System Analysis. The interconnection of gain stages (RF, mixer, and IF), detector,
and low-pass filter constitutes a feedback system which can be analyzed by classic means
by using piecewise analysis over incremental sections of the range. Over the total range

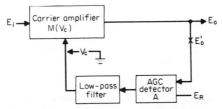

FIG. 13-109 AGC system block diagram for
analysis of loop gain. *(From T. Mills and H.
Suzuki, "Design Concepts for Low-Cost Transistor AGC Systems," IEEE Trans., vol. BTR-14,
no. 3, p. 43. © 1968 IEEE.)*

it can be shown that the loop gain is nonlinear, increasing with increasing signal level. Analysis of Fig. 13-109 leads to the following equations for AGC dc loop gain derived by Mills and Suzuki.[108]

$$\frac{dE_{\text{out}}}{dE_{\text{in}}} = \frac{M(V_c)}{1 + K} \qquad K = 0.115 E_o A \frac{dG}{dV_c} \tag{13-91}$$

where G = gain of carrier amplifier, dB, 20 log (M)
     V_c = control or AGC voltage
     K = loop gain
dG/dV_c = slope of amplifier gain control characteristic

This may also be expressed as

$$\frac{dE_{\text{out}}/E_{\text{out}}}{dE_{\text{in}}/E_{\text{in}}} = \frac{1}{1 + K} \tag{13-92}$$

Loop gain K should be large to maintain good regulation of the second detector output. As loop gain increases, however, the stability of the system will decrease owing to the phase shift caused by the filter network and poles at the sampling frequency in the bypass networks.

A second consideration in establishing loop gain is impulse noise. The ability of a receiver to reject impulse noise is inversely proportional to its loop gain. Excessive impulse noise can saturate the detector and reduce the RF-IF gain, thereby causing the picture to lose contrast or the picture to be lost completely. Two techniques help to reduce this problem.

1. Bandwidth limiting the video to the AGC detector
2. Gating the AGC detector with horizontal pulse and closing the detector during scan time to prevent false inputs

A good compromise between video output regulation and noise immunity will result for dc loop gain factors in the range of 20 to 50.

The filter network and filter time constants play an important role in AGC operation. The filter network performs the function of removing the 15.750-kHz horizontal sync pulses and the equalizing pulses which occur in blocks at the beginning and end of vertical sync interval. The effective rate of these blocks is 60 Hz. The frequency response of the typical loop is given by[109]

$$\frac{E_{\text{out}}(s)}{E_{\text{in}}(s)} = \frac{[1/(1 + K)](1 + \tau s)}{1 + [\tau/(1 + K)]s} \tag{13-93}$$

AGC closed-loop cutoff frequency is defined as

$$f_{cc} = \frac{1 + K}{2\pi} \tag{13-94}$$

Filter time constants must be chosen carefully to eliminate or minimize the AGC-related receiver problems listed below.

Airplane flutter can occur when an airplane flying overhead reflects signal to the antenna which causes rapid cancellation and reinforcement of the main signal path. This problem establishes a lower limit on the response speed of the system. If the time constants are too long, especially that of the control voltage to the RF stage, the gain will not change rapidly enough to track the fluctuating resultant signal. The result will be a flutter or pulsating of the picture. In the laboratory, an airplane flutter signal can be simulated by adding an RF signal which has been phase-modulated at an audio rate to the original RF source. Modulation rates of 50 to 150 Hz and peak-to-trough amplitudes of up to 2 to 1 are representative of field conditions.

Vertical sync pulse sag results from the AGC system speed being fast enough to follow the change in sync pulses during the vertical interval period. Gain first increases during the initial equalizing pulses, then decreases during the wide vertical pulse, then increases

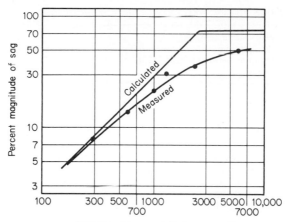

FIG. 13-110   Relationship between sag during vertical interval and AGC loop cut-off frequency. *(From H. Nabeyama and G. Miyazaki, "An AGC System Design Based upon the dc Restoration and the Dynamic Characteristics," IEEE Trans., vol. BTR-16, no. 4, p. 334. © 1970 IEEE.)*

again during the final equalizing pulses. A sag of 64 percent occurs in the AGC voltage, based upon the duty cycle, during this period.[100] This sag may be suppressed by reducing the frequency response of the loop as shown in Fig. 13-110. Sag waveform in the control voltage will result in the recovery of a distorted sync pulse at the video detector. With excessive sag this can lead to interlace problems, and loss of sync pulses at the sync separator which may result in excessive vertical jitter and bounce in the picture. In a gated AGC detector, sync sag can be made less of a problem than with a peak or average AGC system.

Lock-out during channel switching is also caused by excessive AGC system speed and the interaction with the horizontal AFC system. This can result in as much as a 2/1 decrease in pull-in range for the horizontal AFC system. If a gated AGC system is used, the AGC detector and the horizontal AFC loop are dependent upon each other for proper operation. Improper timing of the horizontal pulses could mean improper gating and incorrect level for the AFC voltage which further compounds the problem by allowing excessive or insufficient sync at the sync separator. This, naturally, will further upset the operation of the horizontal AFC loop.

13.11.6 AUTOMATIC FREQUENCY CONTROL (AFC).   Also called automatic fine tuning (AFT), this circuit senses the frequency of the picture carrier in the IF section and sends a correction voltage to the local oscillator, in the tuner section, if the picture carrier is not on the frequency of 45.75 MHz.

Typical AFT systems consist of a frequency discriminator prior to the video detector, a low-pass filter, and varactor diode controlling the local oscillator. The frequency discriminator in discrete transistor sets has been the Foster-Seely type with circuit components adjusted for wide-band operation centered at 45.75 MHz. Typically a small amount of unbalance is built into the circuit to compensate for the vestigial sideband components of the carrier. (See Table 13-25 for characteristics of AFC closed loops.)

An early application of integrated circuits in television receivers was the AFT detector block as a single IC needing only two adjustable coils as external components. More recent designs have included the AFC circuit in the form of a synchronous demodulator on the same IC die as the other functions of the IF section.

Table 13-25 AFC Closed-Loop Characteristics

Pull-in range	± 750 kHz
Hold-in range	± 1.5 MHz
Frequency error for ±500 kHz offset	< 50 kHz

13.12 SOUND CHANNEL

13.12.1 SOUND CARRIER SEPARATION SYSTEMS. The sound transmission is a frequency-modulated signal having a maximum deviation of ±25 kHz (100 percent modulation) and capable of providing an audio bandwidth of 50 to 15,000 Hz. Sound carrier is transmitted at a frequency 4.5 MHz above the RF picture carrier.

The *intercarrier sound system* passes the IF picture and sound carriers (45.75 and 41.25 MHz, respectively) through a detector (nonlinear stage) to create the intermodulation beat of 4.5 MHz. The intercarrier sound signal is then amplified, limited, and FM-demodulated to recover the audio. Block diagrams of intercarrier sound systems are shown in Figs. 13-7 and 13-8. In a discrete component format, the intercarrier detector is typically a simple diode detector feeding a 4.5-MHz resonant network. If an IC IF system is used (Sec. 13.11.3), the sound and picture IF signals are carried all the way to the video detector where one output port of the balanced synchronous demodulator supplies both the 4.5-MHz sound carrier and the composite baseband video. Schematics are shown in Figs. 13-111 and 13-112.

The coupling network between the intercarrier detector and the sound IF amplifier usually has the form of a half-section high-pass filter which is resonant at 4.5 MHz. This form gives greater attenuation to the video and sync pulses in the frequency range from 4.5 MHz to direct current (carrier), thereby reducing buzz in the recovered audio, especially under the conditions of low picture carrier. An alternative implementation uses a piezoelectric ceramic filter which is designed to have a bandpass characteristic at 4.5 MHz and needs no in-circuit adjustment.[110]

Buzz results when video-related phase-modulated components of the visual carrier (ICPM) are transferred to the sound channel. The generation of ICPM can be transmitter-related or receiver-related; however, the transfer to the sound channel takes place in the receiver. This can occur in the mixer or the detection circuit. Here a synchronous detector represents very little improvement over an envelope diode detector unless a narrow-band filter is used in the reference channel.[7]

Split-carrier sound processes the IF picture and sound carriers as shown in Fig. 13-9 and described in Sec. 13.1.3.

Quasi-parallel sound utilizes a special filter such as the SAW filter of Fig. 13-100 to eliminate the Nyquist slope in the sound-detection channel, thereby eliminating a major source of ICPM generation in the receiver. The block diagram of this system is shown in Fig. 13-10.

13.12.2 INTEGRATED-CIRCUIT IMPLEMENTATION. Nearly all sound channels in present-day television receivers are designed as a one- or two-IC configuration. The single IC contains the functions of sound IF amplifier-limiter, FM detector, volume control, and audio output. Two-chip systems usually have the functions of amplifier-limiter, detector, and volume control in one IC and the audio amplifier and output in the second.

13.12.3 AMPLIFIER-LIMITER SECTIONS. These consist of from three to eight direct-coupled, balanced, differential-amplifier stages. Although somewhat similar in design to the amplifier stages in the video-IF IC, the sound IF stages have no variable gain control (AGC). As signal level increases, the signal in the last stage begins to symmetrically limit (simultaneously clip the positive and negative peaks of the sine wave). Bias networks are stiff so that no shifting of the bias voltage occurs. Increased signal level forces the earlier stages successively into limiting. Limiting action removes much of the amplitude-modulated video and sync components from the sound IF signal. Limiting

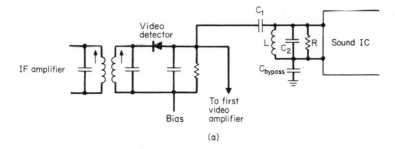

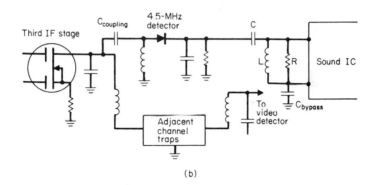

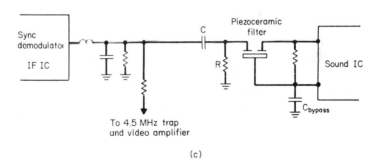

FIG. 13-111 Intercarrier sound detector and filter circuits: (*a*) monochrome receiver; (*b*) FET color IF section; (*c*) IC IF color receiver.

threshold for an amplifier-limiter section typically occurs at a 4.5-MHz input level of 30 to 100 μV.

13.12.4 FM DETECTORS. Three types of detector circuits are used in ICs for the demodulation of the FM sound carrier. These are quadrature detector, balanced-peak detector, and the phase-locked loop detector.

Quadrature detector, also called gated coincidence detector and analog multiplier,[111] measures the instantaneous phase shift across a reactive circuit as the carrier frequency shifts. Referring to the block diagram shown in Fig. 13-113 at center frequency (zero deviation) the LC phase network gives a 90° phase shift to V_2 compared with V_1. As the carrier deviates, the phase shift changes proportionately to the amount of carrier devia-

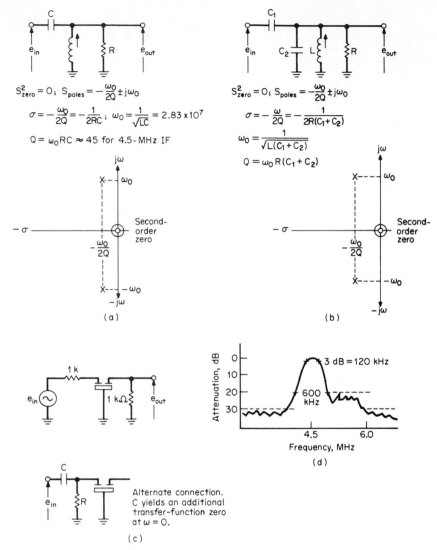

FIG. 13-112   Analysis of intercarrier sound bandpass networks: (*a*) and (*b*) *RLC* networks; (*c*) piezoceramic filter schematics; (*d*) amplitude response of piezoceramic filter.

tion and direction.[112] A piezoceramic discriminator can be used in place of the *LC* phase-shift network.

    The square-wave signal, provided by the limiter, and sinusoidal signal from the phase-shift network are compared in the gated detector (Fig. 13-113). The resultant current pulses at the output collector have a width equal to the phase difference ϕ of the signals, i.e., the angle difference between the zero crossings of both waveforms.

    Residual amplitude modulation gives rise to asymmetries in the limited waveform V_1. This problem can be alleviated by sampling twice per period in a full-wave or double-

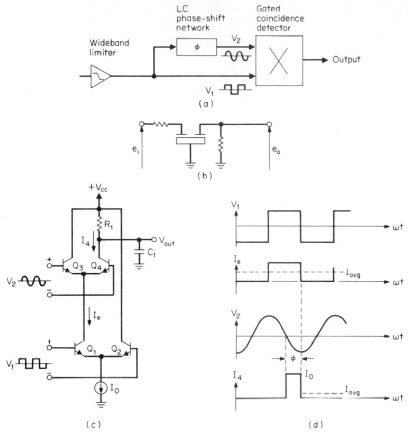

FIG. 13-113 FM quadrature detector: (a) block diagram: (b) piezoceramic discriminator; (c) single-balanced coincidence detector; (d) detector waveforms.

balanced detector circuit, which cancels the error in the output caused by the waveform asymmetry.

Features which make this circuit attractive as an FM detector are

1. Low-cost, easily adjustable LC tuning circuit as phase-shifting element.

2. Coincidence detector circuitry directly transforms phase shift into amplitude without regard to the amplitude characteristics of the reactive phase-shift circuit.

3. Coincidence detector exhibits complete linearity between the phase deviation and the resulting output signal.

The *balanced peak detector* or differential peak detector, described by Peterson,[113] utilizes two peak or envelope detectors, a differential amplifier, and a frequency-selective circuit or piezoceramic discriminator as shown in Fig. 13-114. Also shown are the voltages at each detector input and the resultant difference signal.

The differential peak detector operates at a lower voltage level and does not require square-wave switching pulses. Therefore it creates far lower harmonic radiation than the

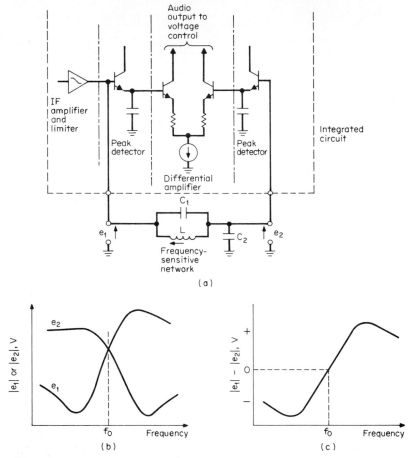

FIG. 13-114 Differential peak detector: (*a*) schematic; (*b*) voltages at the frequency-sensitive network; (*c*) resultant FM discriminator characteristic. (*From R. Peterson, "High-Performance Integrated Circuits for High-Gain FM-IF Systems," IEEE Trans. Broadcast Telev. Receivers, vol. BTR-16, no. 4, p. 257. © 1970 IEEE.*)

quadrature detector. In some designs a low-pass filter has been placed between the limiter and peak detector to further reduce harmonic radiation and increase AM rejection.[114]

The *phase-locked-loop detector* requires no frequency-selective *LC* network to accomplish demodulation.[115] In this system, shown in Fig. 13-115, the voltage-controlled oscillator (VCO) is phase-locked by the feedback loop into following the deviation of the incoming FM signal. The low-frequency error voltage which forces the VCO to track is indeed the demodulated output.

Only external capacitors and resistors are required for the integrated-circuit implementation of the PLL detector.[116] C_o determines the frequency of the VCO, thereby establishing the center frequency of the demodulator. Components C_1 and R_1 form the external low-pass loop filter which determines the interference rejection (capture) characteristics of the system.

13.12.5 VOLUME AND TONE CONTROLS.

The theory of dc-operated controls is based upon the current-splitting characteristic of a differential amplifier. As a differen-

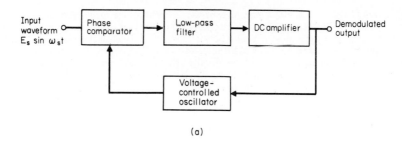

(a)

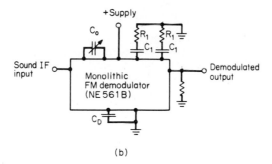

(b)

Time constant of low-pass filter, $\tau = C_1(R_1 + R_0) \approx C_1 R_0$ ($R_0 = 6\ k\Omega$)

De-emphasis time constant, $\tau_D = C_D R_D$ ($R_D = 8\ k\Omega$) $C_D \geq 200\ pF$

For 4.5-MHz TV application, $C_O \approx 65\ pF$, $0 < R_1 < 50\ \Omega$, $C_1 \approx 2000\ pF$

FIG. 13-115 PLL FM detector: (*a*) block diagram; (*b*) application to television sound detector. *(From A. Grebene, "An Integrated Frequency-Selective AM/FM Demodulator," IEEE Trans. Broadcast Telev. Receivers, vol. B-17, no. 2, p. 71. © 1971 IEEE.)*

tial amplifier is unbalanced, the signal current shifts to the unused side, away from the load. Attenuation ranges of 80 to 90 dB have been achieved with these circuits.[114,117]

The tone control provides up to 6 dB per octave of high-frequency attenuation.

13.12.6 AUDIO AMPLIFIERS. The audio amplifier contained in a sound system IC can provide output power in the 2- to 5-W range. Speaker impedances are usually 16Ω. Sufficient audio gain is provided to drive full-rated output with a 30 percent modulated input signal.

Audio power output requirements for small-screen monochrome receivers range from 100 mW to 1 W. These applications use a conventional discrete-component design consisting of a two-transistor, push-pull, class B output stage driven by a single transistor amplifier.

13.13 VIDEO AND CHROMA PROCESSING

13.13.1 MONOCHROME VIDEO AMPLIFICATION REQUIREMENTS. A range of video signals of 1 to 3 V at the second detector has become standard for many practical reasons, including optimum gain distribution between RF, IF, and video sections and distribution of signal levels so that video detection and sync separation may be effectively performed. The video amplifier gain and output level are designed to drive the picture tube with this input level. Sufficient reserve is provided to allow for low percentage modulation and signal strengths below the AGC threshold.

Picture Controls. A video gain or *contrast control* and a *brightness* or *background control* are provided to allow the viewer to select the contrast ratio and overall brightness level which produce the most pleasing picture for a variety of scene material, transmission characteristics, and ambient lighting conditions. The contrast control usually provides a 4/1 gain change. This is accomplished either by attenuator action between the output of the video stage and the CRT or by changing the ac gain of the video stage by means of an ac-coupled variable resistor in the emitter circuit. The brightness control shifts the dc bias level on the CRT to raise or lower the video signal with respect to the CRT beam cutoff-voltage level.

Ac and dc Coupling. For perfect picture transmission and reproduction, it is necessary that all shades of gray are demodulated and reproduced accurately by the display device. This implies that the dc level developed by the video demodulator, in response to the various levels of video carrier, must be carried to the picture tube. Direct coupling or dc restoration is often used, especially in color receivers where color saturation is directly dependent upon luminance level.

Many low-cost monochrome designs utilize only ac coupling with no regard for the dc information. This eases the high-voltage power supply design as well as simplifying the video circuitry. These sets will produce a picture in which the average value of luminance remains nearly constant. For example, a night scene having a level of 15 to 20 IRE units and no peak-white excursions will tend to brighten toward the luminance level of the typical daytime scene (50 IRE units). Likewise a full-raster white scene with few black excursions will tend to darken to the average luminance level. More deluxe monochrome receivers reduce this condition by use of partial dc coupling in which a high-resistance path exists between the second detector and the CRT. This path usually has a gain of one-half to one-fourth that of the ac-signal path.

The *transfer characteristic* of the amplifier can be expressed in terms of the exponent *gamma* (γ) of the input voltage which, together with a gain constant K, determines the output voltage. The expression $e_{out} = Ke_{in}^{\gamma}$ has been adopted from the conventional technique for expressing the brightness transfer characteristic of the complete television system. By this convention, greater than unity produces expansion of white relative to black; this is the distortion introduced by the cathode-ray tube. Because the cathode-ray-tube nonlinearity is often compensated for by transmitter predistortion (gamma correction), it is desirable to have the video amplifier as linear as economically feasible. If amplifier distortion cannot be reduced to a negligible amount, subjective evaluation indicates the preferred nonlinearity to be that produced by γ less than unity. This is based on the observation that a system with γ greater than unity produces a subjective sensation best described as "harsh."

A "staircase" test signal provides a convenient dynamic check for transfer distortion. The signal is divided into equal voltage steps to facilitate direct interpretation of oscilloscope observations.

The *transient response* of the video amplifier is controlled by its amplitude and phase characteristics. The *low-frequency transient response,* including the effects of dc restoration, if used, is measured in terms of distortion to the vertical blanking pulse. Faithful reproduction requires that the change in amplitude over the pulse duration, usually a decrease from initial value called *sag* or *tilt,* be less than 5 percent. In general, there is no direct and simple relationship between the sag and the lower 3-dB cutoff frequency. However lowering the 3-dB cutoff frequency will reduce the tilt.

Equations for low-frequency tilt due to various interstage network components in a common-emitter stage (Fig. 13-116) are

Output coupling circuit:

$$\% \text{ tilt/s} \approx -\frac{100}{(R_c + R_L)C_2} \tag{13-95}$$

Input coupling circuit:

$$\% \text{ tilt/s} \approx -\frac{100}{C_1(R_s + R_1)} \tag{13-96}$$

where $R_1 = R_B R_I/(R_B + R_I)$.

Emitter bypass circuit:

$$\% \text{ tilt/s} \approx -\frac{100B_o}{(R_I + R_s')C_E} \tag{13-97}$$

where $R_s' = R_s R_B/(R_s + R_B)$. These equations apply only for small values of tilt, which require the assumption that the circuit time constants are much longer than the pulse duration (Ref. 84, p. 515).

Low-Frequency Response Requirements. The effect of inadequate low-frequency response appears in the picture as vertical shading. If the response is so poor as to cause a substantial droop of the top of the vertical blanking pulse, then incomplete blanking of retrace lines may occur.

It is not necessary or desirable to extend the low-frequency response to achieve essentially perfect LF square-wave reproduction. First, the effect of tilt produced by imperfect LF response is modified if dc restoration is employed. Direct-current restorers, particularly the fast-acting variety, substantially reduce tilt, and their effect must be considered in specifying the overall response. Second, extended LF response makes the system more susceptible to instability and low-frequency interference. Current coupling through a common power-supply impedance can produce the low-frequency oscillation known as "motorboating." Motorboating is not usually a problem in television receiver video amplifiers since they seldom employ the number of stages required to produce regenerative feedback, but in multistage amplifiers the tendency toward motorboating is reduced as the LF response is reduced.

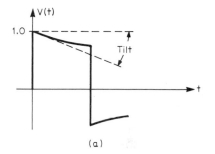

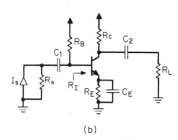

FIG. 13-116 Video stage low-frequency response: (a) square-wave output showing tilt; (b) RC time constant circuits in common-emitter stage which affect low-frequency response.

A more commonly encountered problem is the effect of airplane flutter and "line bop." Although a fast-acting AGC can substantially reduce the effects of low-frequency amplitude variations produced by airplane reflections, the effect is so annoying visually as to warrant a sacrifice in excellence of LF response to bring about further reduction. A transient inline-voltage amplitude, commonly called a line bop, can also produce an annoying brightness transient which can similarly be reduced through a sacrifice of LF response. Special circuit precautions against line bop include the longest possible power supply time constant, bypassing the picture tube electrodes to the supply instead of ground, and the use of special coupling networks to attenuate the response sharply below the LF cutoff frequency. The overall receiver response is usually an empirically determined compromise.

The *high-frequency transient response* is usually expressed as the amplifier response to an ideal input voltage or current step. This response is described in the following terms and is shown in Fig. 13-117.

Rise time τ_R *is the time required for the output pulse to rise from 10 to 90 percent of its final (steady-state) value.*

Overshoot *is the amplitude by which the transient rise exceeds its final value, expressed as a percentage of the final value.*

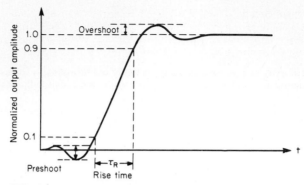

FIG. 13-117   Response to a step input.

Preshoot *is the amplitude by which the transient oscillatory output waveform exceeds its initial value.*

Smear *is an abnormally slow rise as the output wave approaches its final value.*

Ringing *is an oscillatory approach to the final value.*

In practice, rise times of 0.1 to 0.2 µs are typical. Overshoot, smear, and ringing amplitude are usually held to 5 percent of the final value, and ringing is restricted to one complete cycle.

13.13.2 PICTURE-TUBE DRIVE REQUIREMENTS.

Cathode-ray tubes have two primary elements for controlling beam current, i.e., grid-1 and cathode. In low-cost monochrome receivers cathode drive is preferred for two reasons.

First, g_m, the ratio of beam current change to input voltage change, is greater for cathode drive than for grid drive. The value of beam current produced by a given grid-drive voltage is given by the equation

$$I_{bG} = \frac{K \cdot V_{dG}^3}{V_{cG}^{3/2}} \qquad \mu\text{A} \qquad (13\text{-}98)$$

where I_{bG} = beam current, grid drive
V_{dG} = drive voltage at grid, positive-going from cutoff
V_{cG} = cutoff voltage at grid, measured from reference cathode voltage to a negative value which is sufficient to cause visual extinction of undeflected spot in low ambient light level
K = perveance constant, $\mu\text{A/V}^{3/2}$ ($K \approx 3$)

For the same tube, the equation for beam current produced by cathode drive is

$$I_{bK} = K \frac{[(1 + 1/\mu)V_{dK}]^3}{[V_{cK} + (V_{dK}/\mu)]^{3/2}} \qquad (13\text{-}99)$$

where K = same perveance constant as in grid-drive equation
V_{dK} = drive voltage at cathode
V_{cK} = cutoff voltage at cathode, measured from reference grid voltage to beam cutoff value (positive polarity)
μ = amplification factor

$$\mu = \frac{V_{G2G1} - V_{cK}}{V_{cK}} \qquad (13\text{-}100)$$

where V_{G2G1} = grid-2 voltage relative to grid-1 in volts. These equations are plotted in Fig. 13-118. A comparison shows that if the grid-2 voltage is adjusted to produce the identical value of beam current at zero-bias for each drive configuration, then cathode-drive will require less voltage change to reach cutoff.

Second, cathode drive requires a lower total operating voltage between the cathode, grid-1, and grid-2 than does grid drive to produce the same beam current range, as depicted in Fig. 13-119. This eases considerably the power supply design requirements and cost.

Monochrome picture-tube characteristics and voltage requirements are given in Table 13-26.

13.13.3 MONOCHROME VIDEO AMPLIFIERS.

These amplifiers are typically based upon the one- or two-transistor configurations shown in Fig. 120a. Choice is usually dictated by the type and impedance level of the video detector. A diode envelope detector having an output of 1 to 2 V (sync to white) and a load of 3 to 5 kΩ requires the two-transistor configuration. An IC detector which typically supplies an output of 2 to 3 V from an internal low-impedance video amplifier can adequately drive the one-transistor stage.

The video stage output requirement can be determined by selecting the CRT drive requirement (black to white, i.e., cutoff to zero bias) and multiplying by 1.6 to allow for the additional voltage swing necessary to prevent video stage clipping on sync tips. For a CRT requiring a 40-V drive, the video stage must be capable of at least a 64-V swing. An analysis of the transient response of this stage may indicate a slightly higher voltage requirement. Video stage gain is in the range of 25 to 60, with 40 being a typical value for many monochrome sets. Characteristics of transistors used in video stage applications are given in Table 13-27.

The typical transistor video amplifier stage utilizes one or a combination of several types of high-frequency broadbanding techniques in order to achieve the desired bandwidth and transient response. Derivation and discussion of these techniques are contained in numerous texts on transistor circuit analysis.

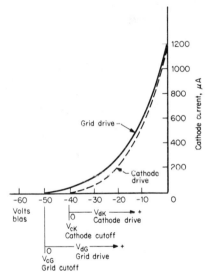

FIG. 13-118 Grid vs. cathode drive conditions for a CRT. Grid drive V_{G2K} = 250 V; cathode drive V_{G2G1} = 250 V.

Two techniques, i.e., *series feedback* via an emitter RC network and *shunt peaking* with an inductor in series with the collector load resistor, are most often used in monochrome video amplifiers. The common-emitter amplifier equivalent circuit which contains the hybrid-pi transistor representation (Fig. 13-120b) shows the collector feedback capacitance $C_{b'c}$ which leads to a complex relationship between output and input. By substituting the unilateralized transistor equivalent circuit (Ref. 84, p. 213), the problem can be simplified into a more straightforward circuit having input and output networks as shown in Fig. 13-120c and d.

The relationship between input current and current flowing from the source $g_m V$ in the output side can be found by considering only the input side network. In general this

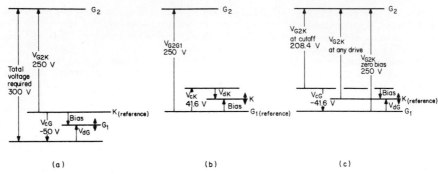

FIG. 13-119 Potential diagrams showing relationship between grid drive, cathode drive, and the total power supply requirement for each: (a) grid drive; (b) cathode drive; (c) cathode drive using grid-drive parameters.

gives an equation of the form[118]

$$A_i = \frac{g_m V}{I_s}$$

$$= \frac{g_m}{C_{b'e}} \frac{s + 1/R_e C_e}{s^2 + s[\omega_\beta + \omega_2 + \omega_3 + (1+K)\omega_1 + \cdots + (1+K)\omega_1\omega_3 + \omega_\beta(\omega_2 + \omega_3)]}$$

where $\omega_1 = \dfrac{1}{Z_e C_{b'e}}$ $\omega_2 = \dfrac{1}{Z_e C'}$ $\omega_3 = \dfrac{1}{R'_s C'}$ $K = g_m Z_e$

$$\omega_\beta = \omega_t/1 + \beta \qquad (-3\text{-dB point of current gain})$$

$$Z_e = \frac{1}{C_e[s + (1/R_e C_e)]}$$

Table 13-26 Monochrome Television Picture Tubes, Typical Values

Size, diagonal, in (cm)	Type number	Deflection, deg	Anode voltage, kV	Grid 2 V_{G2G1}†	Grid 4 V_{G3G1}†	Video drive (V_{CK} to $V_{GK} = 0$,† V)
19 (48)	19GTP4	114	16	50	0	50
19	19VGSP4	114	16	135	0	50
16 (41)	16VCSP4	114	16	135	0	50
16	16VBQP4	114	16	135	0	50
12 (30)	12VCUP4	90	12.5	100	0	40
12	12BWP4	90	12	3.5	0	40
12	12CXP4	110	12	100	0	50
9 (23)	23OAUB4	90	9	100	0	45
9	9VADP4	90	12	100	0	45
5 (13)	14OAJB4	55	8	300	100	35

†*Notes:* 1. All voltages listed for cathode-drive service are positive with respect to G1.
2. V_{CK} = voltage for cathode current cutoff.
V_{GK} = Voltage between grid-1 and cathode. $V_{GK} = 0$ is zero-bias, i.e., maximum beam current.

Table 13-27 Video Output Transistors, Monochrome and Color

Type	Collector breakdown voltage		Power dissipation P total, W		h_{FE}, min.	f_t, MHz, min.	$C_{re}(C_{cb})$, pF, max.
	V_{CBO}	V_{CEO}	Amb.†	HS.‡			
BF422	250	250	0.8	.....	50	60	1.6
BF457	160	160	1.2	6	26	90	3.5
BF458	250	250	1.2	6	26	90	3.5
MPSA42	300	300	0.6	1.5	40	50	3.0
BF869	300	300	1.6	5	50	60	2.0
D4ON	300	500	2	10	40	60	3.5

†Transistor operating with no heat sink in 25°C ambient air.
‡Transistor operating with large heat sink (HS).

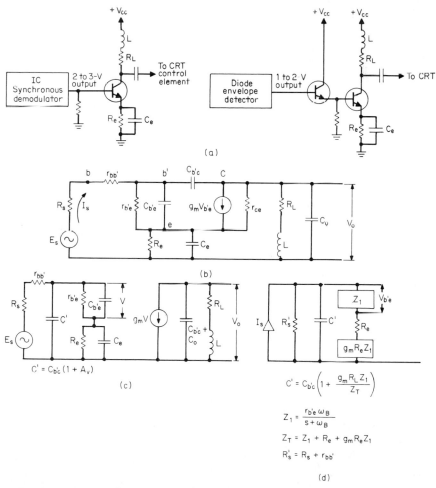

FIG. 13-120 (a) Typical monochrome video amplifiers. Equivalent circuit of video amplifier having emitter peaking: (b) general circuit; (c) simplified by use of unilateralized transistor; (d) further simplified input side of circuit (Ref. 118).

If the assumption is made that $R'_s \ll 1/\omega C'$ (a low-impedance source is typical for this stage), then the equation reduces to

$$\frac{g_m V}{I_s} = \frac{g_m}{C_{b'e}} \frac{s + A}{s^2 + sB + D}$$

where

$$A = \frac{1}{R_e C_e} \qquad B = \omega_\beta \left(1 + \frac{r_{b'e}}{R'_s} \right) + \frac{1}{R_e C_e} \left(1 + \frac{R_e}{R'_s} \right)$$

$$D = \frac{\omega_\beta}{R_e C_e} \left[\frac{r_{b'e} + R'_s + R_e (1 + \beta)}{R'_s} \right]$$

Setting $B = A + D/A$ causes the zero of the current-gain equation to be placed on top of one of the poles in the pole-zero diagram. This yields the relationship $1/R_e C_e = \omega_t$ (the radian frequency at which current gain has fallen to 1). The remaining pole, positioned as given below, will control the frequency response of the input circuit or the relationship of output current to input current for the stage. The 3-dB cutoff frequency is given by

$$\omega_{3 \text{ dB}} = \frac{D}{A} = \omega_\beta \left[\frac{r_{b'e} + R'_s + R_e(1 + \beta)}{R'_s} \right] \qquad (13\text{-}101)$$

The low-frequency current gain of this stage is

$$A_i \text{ (midband)} = \frac{\beta R'_s}{r_{b'e} + R'_s + R_e(1 + \beta)} \qquad (13\text{-}102)$$

and the midband voltage gain is given by

$$A_v \text{ (midband)} = \frac{R_L}{R_e + r_{b'e}} \qquad (13\text{-}103)$$

The output side of Fig. 13-120a shows the shunt peaked circuit consisting of load resistor R_L in series with the peaking coil L. Capacitor C_o consists of the capacitance of the CRT drive element (usually cathode), socket, lead, transistor collector to base, collector to emitter, and heat tab or heat sink. The relationship between output voltage V_o and current $g_m V$ produced by the current source is given by

$$\frac{V_o}{g_m V} = \frac{1}{C_T} \frac{s + R/L}{s^2 + s(R_L/L) + (1/L \, C_T)} = K \frac{s + Z_1}{s^2 + as + b} \qquad (13\text{-}104)$$

where $C_T = C_{b'c} + C_o$, which leads to a zero and two complex conjugate poles if $4/LC_T > (R_L/L)^2$.

One can now consider adjusting the circuit parameters to position these poles to achieve prescribed amplitude or transient response.

Maximum flat magnitude response will result when the coefficients conform to the equation $a^2 - 2b = (b/Z_1)^2$. This gives a value for the inductance $L = 0.414 C_T R_L^2$.

The *transient response* of the network to a step input is shown in Fig. 13-121 for various values of m defined by the equation

$$m = \frac{L}{R_L^2 C_T} \qquad (13\text{-}105)$$

Compared with the uncompensated case ($m = 0$ or $L = 0$), a value of $m = 0.42$ gives a rise time improvement of 1.84 and a bandwidth increase of 1.72 with 3 percent overshoot.

More complex circuits having a combination of *series, shunt,* and *tuned-shunt peaking* are shown in Fig. 13-122 and in Table 13-28. For the series type, the highest ratio of end-circuit capacitance for which the circuit is generally useful is 1/2. For this condition and $m_2 = 0.4$, the rise time improvement is 2.5 percent with 6 percent overshoot. It is more usual to employ the combination of series inductance and damping resistor which will give the greatest rise time improvement for the desired amplitude of overshoot or ringing and a given capacitance ratio.

Practical Video Stage Design. One can begin the solution of a video stage by selecting the value of collector load resistor which equals the capacitative reactance of the

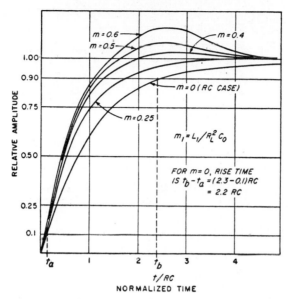

FIG. 13-121 Transient response of shunt-peaked amplifier. (*From Fink.*[10])

output side capacitance C_T at the upper cutoff frequency

$$R_L = \frac{1}{C_T} \tag{13-106}$$

A second consideration is that the collector-circuit time constant should be less than that of the emitter circuit. If it were longer, in the event of a rapid turn-off of the base-emitter circuit, the only discharge path for collector voltage would be through the load resistor, resulting in diagonal clipping. Engel[119] suggests a ratio of 1/3 for maximum worst-case rise time to collector time constant. The collector load resistor is therefore

$$R_L \; (\text{k}\Omega) = \frac{\tau_R \; (\text{ns})}{1.3 C_T \; (\text{pF})} \tag{13-107}$$

If the power-dissipation ratings of both collector resistor and transistor are satisfied, the next step is to determine the value of emitter resistor which satisfies the midband gain requirement

$$R_e = \frac{R_L}{A_V} \tag{13-108}$$

The emitter peaking capacitor can be selected by the following equation to produce a maximum flat magnitude response: $C_e = 1/R_e \omega_t$.

The final step involves selecting the collector circuit peaking configuration and calculating the component values from the equations of Fig. 13-122. Somewhat different design procedures have been covered elsewhere in the literature.[119-121]

Contrast Control. Considerations of unwanted signal pickup, stray capacitance, and circuit stability generally lead to incorporating the contrast control in the low impedance emitter circuit rather than the base or collector circuits. A typical circuit is shown in Fig. 13-123. Alternating-current coupling is preferred for ease of design, although a dc-coupled contrast control could have lower cost, provided the shifting dc bias point does not place a hardship on the power supply voltage and device dissipation requirements.

In Fig. 13-154 the emitter resistor has been split into a fixed resistor and a potenti-

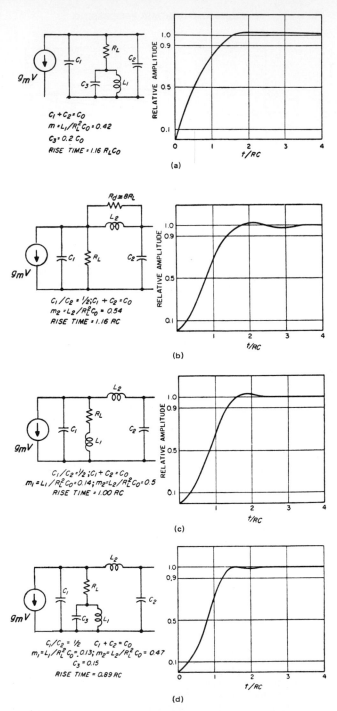

FIG. 13-122 Peaking circuits for video amplifiers: (*a*) tuned shunt peaking (Doba's network); (*b*) series peaking; (*c*) series-shunt peaking; (*d*) series-tuned shunt peaking (Dietzold's network). (*Adapted from Fink.*[10])

Table 13-28 Comparative Performance of Peaking Circuits

Circuit	m_1	m_2	C_1/C_2	t_0/t_1	f_1/f_0	C_3	R_d
Shunt (Fig. 13-121)	0.42	....	...	1.85	1.72		
Tuned-shunt (Fig. 13-122a)	0.43	....	...	1.89	1.83	$0.21C_0$	
Series (Fig. 13-122b)	....	0.54	½	1.90	2.07	......	$7.8R_L$
Series-shunt (Fig. 13.122c)	0.14	0.51	½	2.21	2.28		
Tuned-shunt series (Fig. 13.122d)	0.13	0.47	½	2.47	2.48	$0.15C_2$	

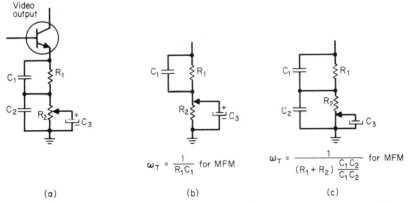

$$\omega_T = \frac{1}{R_1C_1} \text{ for MFM}$$

$$\omega_T = \frac{1}{(R_1 + R_2)\frac{C_1C_2}{C_1C_2}} \text{ for MFM}$$

(a)   (b)   (c)

FIG. 13-123 (a) Contrast control network in emitter circuit. (b) Equivalent circuit at maximum contrast (max. gain). (c) Minimum contrast.

ometer to yield a variable ac-voltage gain ratio of 4/1, that is, 0.5 to 2.0 times the nominal gain requirement. Capacitors C_1 and C_2 are then selected to give the desired transient response at various settings of the control, especially the upper gain range, where video transients are most noticeable.

Brightness control for small monochrome receivers is usually in the video stage collector circuit. Figure 13-124a shows a CRT luminance drive circuit in which changes in the background or black level (brightness) are achieved by changing the bias level of the picture tube by means of the potentiometer connected to a bias supply. An alternative configuration shown in Fig. 13-124b changes the bias voltage at the grid of the CRT.

Direct Coupling. The dc component of the video signal is coupled to the picture tube via resistor R in Fig. 13-124a. The dc coupling factor, expressed as a percentage of the ac gain, is given by the resistor attenuation formula

$$\text{DC coupling factor } (\%) = \frac{R_2}{R_1 + R_2} \times 100 \tag{13-109}$$

DC Restorer. The circuit of Fig. 13-125 can be used between the video output transistor and the CRT to reestablish a dc reference related to the video transmission. The discharge time constant must be long enough to prevent sag in the average video level between horizontal charging pulses; therefore

$$\frac{E_d}{R_L + R_d} t = \frac{e_i - E_d}{R_k'} T \quad \text{or} \quad (R_k' + R_L) \gg (R_L + R_d) \tag{13-110}$$

where the circuit parameters are defined in Fig. 13-125.

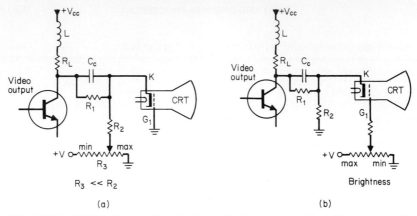

FIG. 13-124 CRT luminance drive circuit: (*a*) brightness control in cathode circuit; (*b*) brightness control in CRT grid circuit.

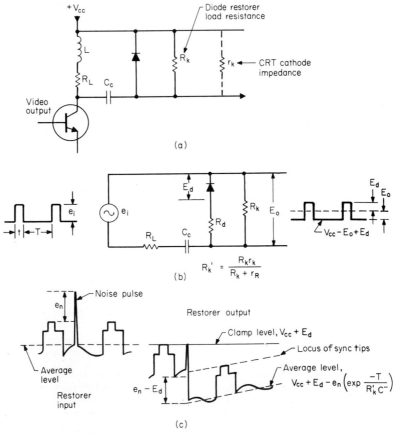

FIG. 13-125 (*a*) Simple dc restorer circuit. (*b*) Equivalent circuit. (*c*) Effect of impulse noise.

This simple circuit lacks noise immunity. A positive-going noise spike of equal or greater amplitude than sync tip will cause a false charge, thereby upsetting the bias level. This can be corrected by gating the diode into conduction during sync time and blocking it off during scan, in a similar manner to that used in the AGC system (Sec. 13.11.5).

Blanking. Horizontal and vertical retrace waveforms are introduced into the video chain to cut off the CRT during the blanking intervals. Typically, the vertical waveform is injected into the video amplifier interstage, while the horizontal is coupled directly to the CRT grid bias circuit. Injection of both signals into the emitter circuit is also frequently used.

13.13.4 COLOR RECEIVER LUMINANCE CHANNEL.

Suppression of chroma subcarrier is necessary to reduce the objectionable dot crawl in and around colored parts of the picture as well as reduce the distortion of luminance levels due to the nonlinear transfer characteristic of CRT electron guns. Traditionally a simple high-Q LC trap, centered around the color subcarrier, has been used for rejection, but this necessitates a trade-off between luminance channel bandwidth and the stop band for the chroma sidebands. Luminance channel comb filtering (Sec. 13.13.11) largely avoids this compromise and is one reason why it is becoming widely used in present receivers.

The luminance channel also provides the time delay required to correct the time delay registration with the color difference signals which normally incur delays in the range of from 300 to 1000 ns in their relatively narrow-bandwidth filters.

While delay circuits having substantially flat amplitude and group delay out to the highest baseband frequency of interest can and have been used, this is not necessarily required nor desirable for cost-effective overall design. Since the other links in the chain, i.e., tuner, IF, traps at 4.5 and 3.58 MHz, CRT driver stage, and peaker may all contribute significant linear distortion individually, it is frequently advantageous to allow these distortions to occur and use the *delay block* as an overall group delay and/or amplitude equalizer.

For example, the emitter-peaking network used to produce an overshoot in the step response (Fig. 13-126) individually distorts the group delay. If placed in tandem with a

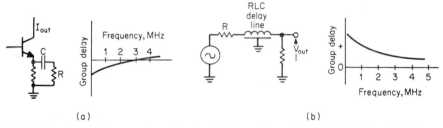

(a)   (b)

FIG. 13-126   Group-delay complementing circuits: (*a*) emitter-peaking circuit; (*b*) video *RLC* delayline.

delay network which can be made to possess the complementary group delay response, also shown in Fig. 13-159, the complete system can be made reasonably flat.

Amplitude Response Shaping (Peaking, Aperture Correction).   Although it is well known that, for "distortionless" transmission, a linear system must possess both uniform amplitude and group delay responses over the frequency band of interest, the limitations of a finite bandwidth lead to noticeably slower rise and fall times rendering edges less sharp or distinct. By intentionally distorting the receiver amplitude response and boosting the relative response to the mid and upper baseband frequencies to varying degrees, both faster rise and fall times can be developed along with enhanced fine detail. If carried too far, however, objectionable *outlining* can occur, especially to those transients in the white direction. Furthermore, the visibility of background noise is increased.

For several reasons, including possible variations in transient response of the transmitted signal, distortion due to multipath, antennas, receiver tolerances, signal-to-noise ratio, and viewer preference, it is difficult to define a fixed response at the receiver that is optimum under all conditions. Therefore it is useful to make the amplitude response variable (Fig. 13-127) so it can be controlled to best suit the individual situation. Over

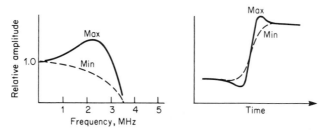

FIG. 13-127   Variable-amplitude peaking: (*a*) amplitude response; (*b*) time-domain response.

the range of adjustment, it is assumed that the overall group delay shall remain reasonably flat across the video band. The exact shape of the amplitude response is directly related to the desired time domain response (height and width of preshoot, overshoot, and ringing) and chroma subcarrier sideband suppression.

A constant group delay variable-peaking system, sometimes called an *aperture corrector* or *edge enhancer,* is shown in Fig. 13-128.[122] The system can be represented by

$$V_p = V_{in}e^{-j\omega t}\left[\frac{1}{2} - \frac{1}{4}\left(e^{j\omega T} + e^{-j\omega T}\right)\right] \quad\quad (13\text{-}111)$$

$$= V_{in}\frac{e^{-j\omega t}}{2}\left(1 - \cos \omega T\right)$$

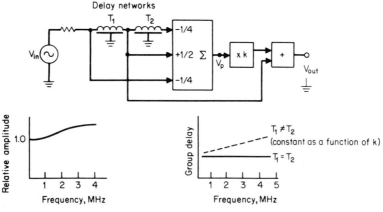

FIG.   **13-128**   Constant-delay   variable-peaking   system   which   produces   edge enhancement.

$$\frac{V_{\text{out}}}{V_{\text{in}}} = e^{-j\omega t}\left[1 + \frac{k}{2}(1 - \cos \omega T)\right] \tag{13-112}$$

The group delay is constant ($= T$) and is independent of frequency and sharpness-control setting k. It is not necessary for the T delay sections to be phase-linear to ensure group delay independence for control variation, only that their phase responses match.

An alternative approach, which achieves nearly the same results, combines a properly phased *second derivative* with the incoming signal followed by a low-pass filter (Fig. 13-129).[123]

An example of a *luminance-channel filter* providing subcarrier rejection, peaking, and

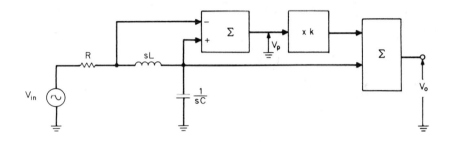

$$V_p = \frac{-s^2\,LC}{s^2\,LC + sRC + 1}$$

$$\frac{V_{\text{out}}}{V_{\text{in}}} = \frac{1 - k\,s^2\,LC}{s^2\,LC + sCR + 1} = L(s) \bullet \left[1 - k\,s^2\,LC\right]$$

$$\frac{V_{\text{out}}}{V_{\text{in}}}(j\omega) = L(j\omega)\left[1 + k\omega^2\,LC\right]$$

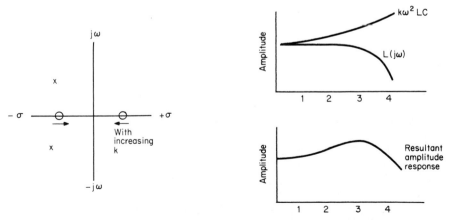

FIG. 13-129 Constant-delay peaking system using second-derivative edge enhancement.

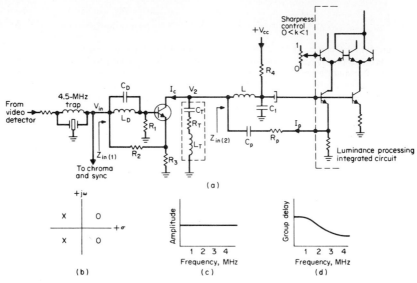

FIG. 13-130 Luminance-channel filter and peaking network: (*a*) system schematic; (*b*) pole-zero plot; (*c*) amplitude; and (*d*) group-delay characteristics of delay-equalizer network. (*Courtesy of NAP Consumer Electronics Co.*)

delay equalization is shown in Fig. 13-130. The transfer function of the *delay equalizer* is

$$\frac{I_c}{V_{in}}(s) = \frac{1}{R_3} \frac{R_1 (R_1/R_3) - Z}{R_1 + Z} \qquad Z = \frac{sL_D}{s^2 L_D C_D + 1} \qquad (13\text{-}113)$$

which will be an all-pass function for $R_2 = R_3$. Furthermore, its input impedance given by

$$Z_{in(1)} = R_1 \frac{R_1 + Z}{R_2 + Z} \qquad (13\text{-}114)$$

can be made to have a constant value at all frequencies of $R_1 = R_2$.

The transfer function of the peaking path is given by

$$\frac{I_p R_4}{V_2}(s) = \frac{-s^2 L C_p}{s^2 L C + s(L/R_4) + 1} \frac{sCR + 1}{sC_p R_p + 1} \qquad (13\text{-}115)$$

while that of the direct unpeaked signal is

$$\frac{1}{s^2 L C + s(L/R_4) + 1} \qquad (13\text{-}116)$$

Therefore if $R_4 C_1 = R_p C_p$, the peaking signal will have the same phase response as the unpeaked signal, and the video enhancement will be provided. Scaling the peaking signal by k will yield variable peaking. The resultant peaking system transfer function will then be

$$\frac{I_c R}{V_{in}} = \frac{1 - (ks^2 L C_p)}{s^2 L C + s(L/R_4) + 1} \qquad \text{for } k \geq 0 \qquad (13\text{-}117)$$

The impedance looking into the peaker circuit will be

$$Z_{in(2)} = R_4 \frac{s^2 L C + s(L/R_4) + 1}{s^2 L C_p + sCR_4 + 1} \qquad (13\text{-}118)$$

If $C = C_p$ and $L/R = R_4C_1$, then $Z_{in} = R$, and the 3.58-MHz trap will be loaded by only a pure resistive component. The exact choice of parameters must be selected to produce characteristics which suitably complement the overall tuner-IF-video output chain.

Nonlinear Video Compression. Because the peaked signal later operates on the non-linear CRT gun characteristics, large white preshoots and overshoots can contribute to excessive beam currents, which can cause CRT spot defocusing. To alleviate this, circuits have been developed which compress large excursions of the peaking component in the white direction.[122] One choice is shown in Fig. 13-131. For best operation it is desirable that the signal being processed have symmetrical transient response in the time domain, i.e., equal preshoot and overshoot. A low-passed second derivative is applied as a differential signal across the bases of the two transistors. For large-signal excursion the peaks are compressed by the transfer function of the amplifier.

Noise Reduction. Low-level, high-frequency noise in the luminance channel can be removed by a technique called *coring*. One technique involves nonlinearly operating on the peaking or edge-enhancement signal, discussed earlier in this section. The peaking signal is passed through an amplifier having the transfer characteristic shown in Fig. 13-132. When this modified peaking signal is added to the direct video, the large transitions will be enhanced, but the small ones (noise) will not be, giving the illusion that the picture sharpness has been increased while the noise has been decreased.

13.13.5 CHROMA SUBCARRIER PROCESSING

Chroma Bandpass. In the equiband chroma system, typical of all consumer receivers, the chroma amplifier must be preceded by a bandpass filter network which complements the chroma sideband response produced by the tuner and IF (Sec. 13.2.3 and Fig. 13-18). In addition, frequencies below 3 MHz must be attenuated to reduce not only possible video cross-color disturbances but also crosstalk caused by the lower-frequency I channel chroma information.

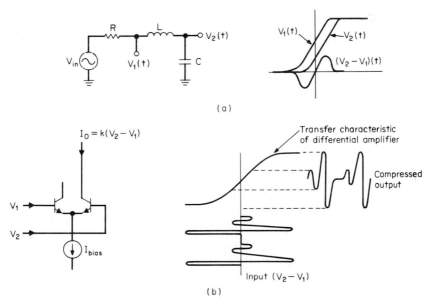

FIG. 13-131 Peak-white compression circuit: (*a*) circuit for generating second derivative; (*b*) differential amplifier processes second derivative.

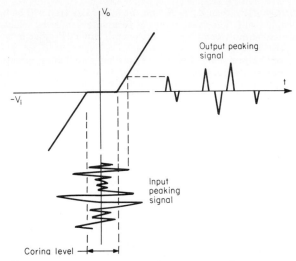

FIG. 13-132 Transfer characteristic of noise-reduction coring circuit.

A third requirement is that the filter have a gentle transition from passband to stop band in order to impart a minimum amount of group delay in the chroma signal which then must be compensated by additional group delay circuitry in the luminance channel. The fourth-order high-pass filter (Fig. 13-133) is a practical realization of these requirements.

As described in Sec. 13.2.3, this stage also serves as the chroma gain control stage. The usual implementation in an IC consists of a differential amplifier having the chroma signal applied to the current source. The gain control dc voltage is applied to one side of the differential pair to divert the signal current away from the output side. This type of stage was discussed in Sec. 13.12.5.

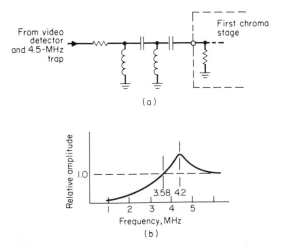

FIG. 13-133 Luminance-suppression filter in chrominance channel: (*a*) schematic; (*b*) amplitude response.

13.13.6 BURST SEPARATION

Gating Requirements. Complete separation of the color synchronizing burst from video requires time gating. The gate requirements are largely determined by the horizontal sync and burst specifications illustrated in Fig. 13-134. It is essential that all video information be excluded. It is also desirable that both the leading and trailing edges of burst be passed so that the complementary phase errors introduced at these points by quadrature distortion average to zero. Widening the gate pulse to minimize the required timing accuracy has negligible effect on the noise performance of the reference system and may be beneficial in the presence of echoes. The ≈ 2-μs spacing between trailing edges of burst and horizontal blanking determines the total permissible timing variation. Noise modulation of the gate timing should not permit noise excursions to encroach upon the burst since the resulting cross modulation will have the effect of increasing the noise power delivered to the reference system.

Gating-signal Generation. The gate pulse generator must provide both steady-state phase accuracy and reasonable noise immunity. The horizontal flyback pulse has been widely used for burst gating since it is derived from the horizontal-scan oscillator system which meets the noise immunity requirements and, with appropriate design, can approximate the steady-state requirements. A further improvement in steady-state phase accuracy can be achieved by deriving the gating pulse directly from the trailing edge of the horizontal sync pulse. This technique is utilized in several current chroma system integrated circuits.

The *burst gate* in conventional discrete component circuits has the form of a conventional amplifier which is biased into linear conduction only during the presence of the gating pulse. In IC format, the complete chroma signal is often made available at one input of the automatic phase control (APC) burst-reference phase detector. The gating pulse then enables the phase detector to function only during the presence of burst.

13.13.7 SUBCARRIER REFERENCE SYNCHRONIZATION

General Requirements. The color subcarrier reference system converts the synchronizing bursts to a continuous carrier of identical frequency and close phase tolerance. Theoretically, the long-term and repetitive phase inaccuracies should be restricted to the same value, approximately $\pm 5°$. Practically, if transmission variations considerably in excess of this value are encountered, and if an operator control of phase ("hue control")

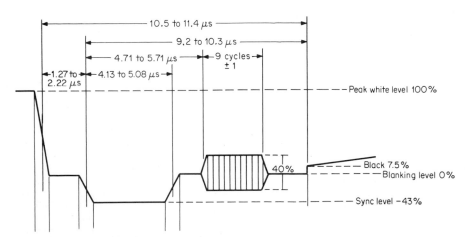

FIG. 13-134 Horizontal-blanking interval specification.

is provided, the long-term accuracy need not be so great. Somewhat greater instantaneous inaccuracies can be tolerated in the presence of thermal noise so that an rms phase error specification of 5 to 10° at a signal-to-noise ratio of unity may be regarded as typical.

Reference Systems. Three types of reference synchronization systems have been used. These are automatic phase control (APC) of a voltage-controlled oscillator (VCO), injection lock of a crystal, and ringing of a crystal filter. Table 13-29 gives significant parameters for each system.[124] Best performance can be achieved by the APC loop.[125] In typical application the *figure of merit* can be made much smaller (better) for the APC loop than for the other systems by making the factor $(1/y) + m$ have a value considerably less than 1, even as small as 0.1. The APC loop system currently is used in all color television receivers. The parts count for each type of system, at one time much higher for the APC system, is no longer a consideration, especially in the IC implementation where the oscillator and phase detector are integrated and only the resistors and capacitors of the filter network and oscillator crystal are external.

The *APC circuit* is a phase-actuated feedback system consisting of three functional components: a phase detector, a low-pass filter, and a dc voltage-controlled oscillator (Fig. 13-135). The characteristics of these three units define both the dynamic and static loop characteristics and hence the overall system performance (see Table 13-30). The following analysis is based on the technique of Richman.[126]

The phase detector generates a dc output E whose polarity and amplitude are proportional to the direction and magnitude of the relative phase difference $d\phi$ between the oscillator and synchronizing (burst) signals. The dc loop gain of the system f_c, assuming unity dc transfer for the filter, is the product

$$f_c \text{ (Hz)} = \mu\beta$$

where μ = the sensitivity of the phase detector, $dE/d\phi$, and β = the sensitivity of the VCO, df/dE. The loop gain f_c has the dimensions of frequency and represents the maximum possible frequency difference from which the system could pull in. The magnitude of the dc loop gain determines the steady-state phase error ϕ_{ss}, which will result from a given frequency difference Δf between the burst and oscillator signals as given by

$$\phi_{ss} = \frac{\Delta f}{f_c} \tag{13-119}$$

The dynamic performance of the APC system is determined by the effects of the low-pass filter on the dc loop gain. The compromise between noise performance and pull-in characteristics can be handled effectively by a filter circuit having a double time constant (Fig. 13-135).

Noise power bandwidth f_{nn} is the effective bandwidth of a rectangular filter which passes the same amount of noise as the loop filter

$$f_{nn} = \frac{\pi}{2}(mf_c + f_2) \tag{13-120}$$

The principal practical design limitation is the time required for frequency stabilization (pull-in time) when the detuning is appreciably greater than the noise bandwidth. For values of detuning less than half the maximum pull in, the minimum pull-in time T_f is given by

$$T_f \approx \frac{4(\Delta f)^2}{f_{nn}^{\ 3}} \tag{13-121}$$

If the pull-in time is to be visually acceptable, it must not exceed 0.5 s.

The *voltage-controlled oscillator* (VCO) is an IC implementation which requires only an external crystal and simple phase-shift network.[127] The oscillator can be shifted ± 45° by varying the phase-control voltage. This leads to symmetrical pull-in and hold-in ranges.

Table 13-29 Comparison of Color Synchronizing Systems

Item	APC	Injection lock	Ringing
Static phase error $\Delta\phi$, rad	$\sin^{-1}\dfrac{\Delta f}{\mu\beta} \simeq \dfrac{\Delta f}{\mu\beta}$	$\sin^{-1}\dfrac{E_2}{E_1}\dfrac{\Delta f}{f_0}2Q \simeq \dfrac{E_2}{E_1}\dfrac{\Delta f}{f_0}2Q$	$\tan^{-1}\dfrac{\Delta f}{f_0}2Q \simeq \dfrac{\Delta f}{f_0}2Q$
Loop gain† $G_l = \dfrac{\Delta f}{\Delta\phi}$ Hz/rad	$\mu\beta$	$\dfrac{f_0}{2Q}\dfrac{E_1}{E_2}$	$\dfrac{f_0}{2Q}$
Noise bandwidth f_{NN}, Hz	$\dfrac{\pi G_l}{2}\left(\dfrac{1}{y}+m\right)$	$\dfrac{\pi}{4}\dfrac{E_1}{E_2}\dfrac{f_0}{Q}$	$\dfrac{\pi}{2}\dfrac{f_0}{Q}$
Figure of merit $\Delta\phi\cdot f_{NN}$ Hz·rad	$\dfrac{\pi}{2}\Delta f\left(\dfrac{1}{y}+m\right)$	$\dfrac{\pi}{2}\Delta f$	$\pi\Delta f$
Parameter	μ: sensitivity of phase detector β: sensitivity of VCO m: AC loop gain y: $\dfrac{\text{filter time constant}}{\text{loop time constant}}$	f_0: subcarrier frequency Q: Q of crystal oscillator E_1: voltage of impressed burst signal E_2: voltage induced at oscillator input	f_0: subcarrier frequency Q: Q of cryustal filter

†Δf = frequency difference.

Source: From K. Mohri, M. Shimano, and T. Kuroyanagi, "Chroma Systems Trend in the Past and Future and Latest Chroma IC with Versatile Flexibility," *IEEE Trans. Consum. Electron.*, vol. CE-24, no. 1, February 1978, p. 82.

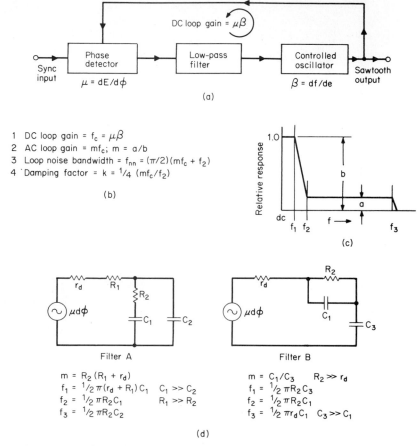

FIG. 13-135 APC system: (*a*) block diagram; (*b*) basic relationships; (*c*) simplified filter response; and (*d*) equivalent circuits.

Table 13-30 Performance Parameters of Typical APC Color Reference System

DC loop gain, $f_c = \mu\beta$	30–50 Hz/deg
Pull-in range	± 500 Hz
Static phase error, ϕ_{ss}	2–5 deg
Noise bandwidth, f_{nn}	100–150 Hz

The *phase detector* is another adaptation of the balanced multiplier or synchronous detector described earlier (Secs. 13.11.4 and 13.12.4).[111]

13.13.8 CHROMA DEMODULATION. The chroma signal can be considered to be made up of two amplitude-modulated carriers having a quadrature phase relationship. Each of these carriers can be individually recovered by use of a synchronous detector or balanced multiplier[111,128,129] which performs the function of time domain multiplication of the chroma subcarrier with the locally generated reference signal.

In Fig. 13-136 the chroma subcarrier is represented by $I(t) \cos \omega_c t + Q(t) \sin \omega_c t$, where $I(t)$ and $Q(t)$ represent the I and Q modulation. The reference to the demodulator is that phase which will demodulate only the I signal. The output contains no Q signal.

Demodulation products of 7.16 MHz in the output current may contribute to an optical interference moiré pattern in the picture. This is related to the line geometry of the shadow mask. The 7.16-MHz output can also result in excessively high line-terminal radiation from the receiver (Sec. 13.18.7). A first-order low-pass filter with cutoff of 1 to 2 MHz usually provides sufficient attenuation. In extreme cases an LC trap may be needed.

Choice of Demodulation Axis. Over the double sideband region (± 500 kHz around the subcarrier frequency) the chrominance signal can be demodulated as pure quadrature AM signals along either the I and Q axis or $R - Y$ and $B - Y$ axis. The latter leads to a simpler matrix for obtaining the color drive signals R, G, and B.

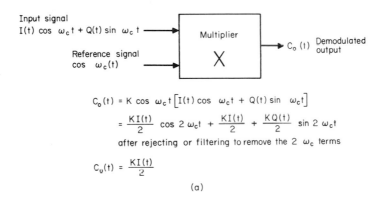

Input signal
$I(t) \cos \omega_c t + Q(t) \sin \omega_c t$ ──→ Multiplier ──→ $C_o (t)$ Demodulated output

Reference signal
$\cos \omega_c (t)$ ──→ X

$$C_o(t) = K \cos \omega_c t \left[I(t) \cos \omega_c t + Q(t) \sin \omega_c t \right]$$

$$= \frac{KI(t)}{2} \cos 2\omega_c t + \frac{KI(t)}{2} + \frac{KQ(t)}{2} \sin 2\omega_c t$$

after rejecting or filtering to remove the $2\omega_c$ terms

$$C_o(t) = \frac{KI(t)}{2}$$

(a)

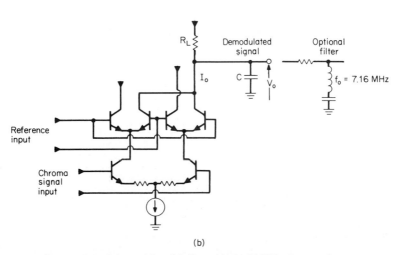

R_L Demodulated signal   Optional filter

I_o C V_o $f_0 = 7.16$ MHz

Reference input

Chroma signal input

(b)

FIG. 13-136 Chroma demodulator: (*a*) multiplier analysis; (*b*) IC implementation.

Table 13-31 Chroma Demodulator Gains and Angles for Two Different
Color Standards

		White-point standard, phosphor type	
		9300 K, NTSC	D65, 1972 #2
$R - Y$	Gain	1.14	2.547
	Normalized	1.0	1.0
	Angle	90	87.67
$G - Y$	Gain	0.7025	0.7314
	Normalized	0.6162	0.2872
	Angle	235.7	244.2
$B - Y$	Gain	2.03	2.174
	Normalized	1.781	0.8539
	Angle	0	−0.38

Source: From C. B. Neal, "Television Colorimetry for Receiver Engineers," IEEE
Trans. *Broadcast Telev. Receivers*, vol. BTR-19, no. 3, August 1973, p. 160.

Current practice has moved away from the classic demodulation angles for two main
reasons. First, receiver picture tube phosphors have been modified to yield greater light
output and can no longer produce the original NTSC primary colors. Currently, the chromaticity coordinates of the primary colors as well as the *RGB* current ratios to produce
white balance vary from one CRT manufacturer to another. Second, the white-point
setup has, over the years, moved from the cold 9300 K of monochrome tubes to the
warmer illuminant C and D65 which produce more vivid colors, more representative of
those which can be seen in natural sunlight. Several authors have developed techniques
for determining the decoding angles and matrix constants to optimize a given set of chromaticity coordinates, fleshtones, green grass, CRT gamma values, or whatever the design
engineer selects.[130−132] Table 13-31 gives an example of the changes which have occurred.

13.13.9 *RGB* MATRIXING AND CRT DRIVE. Since *RGB* primary color signals driving the display are required as the end output, it is necessary to combine or "matrix" the
demodulated color-difference signals with the luminance signal. Several circuit configurations can be used to accomplish this.

Color-Difference Drive Matrixing. In this technique $R - Y, G - Y$, and $B - Y$ are
applied to respective control grids of the CRT while luminance is applied to all three
cathodes; the CRT thereby matrixes the primary colors. It has the advantage that gray
scale is not a function of linearity matching between the three channels since at any level
of gray the color-difference driver stages are at the same dc level. Also, since the luminance driver is common, any dc drift shows up only as a brightness shift. Luminance
channel frequency response uniformity is ensured by the common driver. It does, however, require significantly greater drive on G1 than on the cathodes (Sec. 13.13.2). For
example, to calculate the effective color difference drive to produce a 100-V color bar
pattern, +89 V is needed for blue and −89 V for yellow, therefore requiring a total $B -
Y$ swing of 178 V. However, because of lower G1 sensitivity compared with the cathode,
somewhat more than 178 V will be needed. Although gray-scale balance can be achieved
for a given CRT by adjusting the relative luminance drives, it is necessary to provide
additional gain ratio adjustment for the color-difference signals to produce color fidelity.
Furthermore, the unequal transfer characteristic associated with cathode versus grid
drive produces color errors in highly saturated areas.

RGB drive, wherein *RGB* signals are applied to respective cathodes and G1 is dc-
biased, requires less drive and has none of the potential color fidelity errors of the former
system. *RGB* drive places higher demands on the drive amplifiers for linearity, frequency
response, and dc stability, plus requiring a matrixing network in the amplifier
chain.[133−135]

Color-differences drive and CRT matrix provided a suitable low-cost solution for vacuum-tube receiver designs. The priority of maintaining adequate gray-scale tracking with inexpensive circuitry was deemed more important than the color fidelity requirements.

With the adoption of transistor output amplifiers and IC demodulators in the early 1970s, the aforementioned problems diminished and low-level *RGB* matrixing was commonly adopted.

Low-level RGB matrixing and CRT drive are universally used today, especially with the newer CRTs which have unitized guns in which the common G1 and G2 requires differential cathode bias adjustments and drive adjustments to give gray-scale tracking. In this technique, *RGB* signals are matrixed at a level of a few volts and then amplified to a higher level (100 to 200 V) suitable for CRT cathode drive.

The design considerations are basically the same as discussed in Sec. 13.13.3 for monochrome receivers but with additional attention to certain parameters. Direct-current stability, frequency response, and linearity, of the three stages, even if somewhat less than ideal, should be reasonably well matched to ensure overall gray-scale tracking. Bias and gain adjustments should be independent in their operation, rather than interdependent, and should minimally affect those characteristics listed above.

Figure 13-137 illustrates a simple example of one of the three CRT drivers. If the

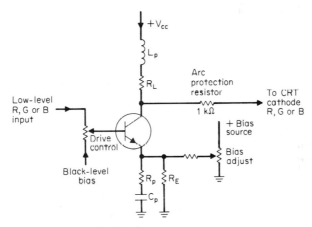

FIG. 13-137 Simplified *R*, *G*, or *B* output stage.

amplifier black-level bias voltage equals the black level from the *RGB* decoder, drive adjustment will not change the amplifier black-level output voltage level or affect CRT cutoff. Furthermore, if $R_B \gg R_E$, drive level will be independent of bias setting. Note also that frequency response–determining networks are configured to be unaffected by adjustments.

For improved dc stability and frequency response the cascode circuit shown in Fig. 13-138 can be used.

Frequently the shunt peaking coil can be made common to all three channels, since differences between the channels are predominantly color-difference signals of relatively narrow bandwidth. Although the frequency responses could be compensated to provide widest possible bandwidth, this is usually not necessary when the frequency response of preceding low-level luminance processing (especially the peaking stage) is factored in. One exception in which output stage bandwidth must be increased to its maximum is in an application, television receiver or video monitor, where direct *RGB* inputs are provided for auxiliary services, such as computers and teletext.

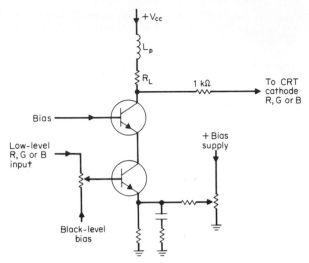

FIG. 13-138 Cascode *RGB* output has improved frequency response.

13.13.10 PICTURE CONTROLS

Contrast Control.   In present-day receivers, contrast control usually regulates a dc-controlled amplifier in the luminance chain. This amplifier is also the control point for automatic beam current limiting (ABL), automatic contrast, and adjustment of the picture to compensate for ambient light variations. When the dc control voltage is connected to both the luminance control stage and the dc-controlled chroma amplifier, the ratio of chroma level to luminance can be held to unity, thereby providing proper saturated color levels for all settings of the contrast control. This configuration is commonly referred to as a *picture control*.[122,136] Tracking errors should be less than approximately 1.0 dB for acceptable operation.

The luminance gain control function typically introduces some variable dc offset over the adjustment range, which is normally resolved by black-level restoration in subsequent stages. In systems in which the black-level is established at a level which differs from blanking, i.e., back porch of sync, additional means are needed to compensate for the dc shift.

Saturation or Color Level.   To compensate for system tolerances as well as individual viewer preference, it is typical to dc-control an attenuator or amplifier in the common chroma channel, preceding the demodulators. A range of from substantially zero to 200 to 300 percent of that required for fully saturated color bars is usually provided.

Hue Control (Tint).   This is provided to permit viewer correction for possible static phase errors in the regenerated subcarrier reference or other parts of the system chain. Traditionally a range of ±45° has been considered adequate. However, since most frequently the control is adjusted near the midpoint of its range, it has been found desirable to reduce the control sensitivity. In some designs the range has been reduced to ±15°, thereby giving the illusion of having an automatic control system for hue.

Brightness (black-level) control compensates for possible black-level setup variations in received signals, receiver variations, and viewer preference. A viewer-controlled variable dc level is added to the receiver's luminance channel, usually at the dc reference for the black-level clamp circuit.

Sharpness control allows the viewer to select the amount of edge sharpening and compromise the amount of background noise or grainy picture desired. Circuit descriptions are contained in Sec. 13.13.4.

13.13.11 COMB-FILTER PROCESSING. The frequency spectrum of a typical NTSC composite video signal is shown in Fig. 13-139. A filter having amplitude response like teeth of a comb, i.e., 100 percent transmission for desired frequencies of a given channel and substantially zero transmission for the undesired interleaved signal spectrum, can separate chroma and luminance components from the composite signal. Such a filter can be easily made, in principle, by delaying the composite video signal one horizontal scan period (63.555 μs in NTSC-M) and adding or subtracting to the undelayed composite video signal (Fig. 13-140).

The output of the *sum channel* will have frequencies at f (horizontal), and all integral multiples thereof reinforce in phase, while those interleaved frequencies will be out of phase and will cancel. This can be used as the luminance path. The *difference channel* will have integral frequency multiples cancel while the interleaved ones will reinforce. This channel can serve as the chrominance channel. For the filter in Fig. 13-140 the luminance channel response is defined by[137]

$$\frac{V_L}{V_{\text{in}}}(j\omega) = 1 + e^{-j\omega T} \tag{13-122}$$

$$\left|\frac{V_L}{V_{\text{in}}}(j\omega)\right| = \sqrt{(1 + \cos \omega T)^2 + \sin^2 \omega T} \tag{13-123}$$

$$= \left|2 \cos \frac{\omega T}{2}\right|$$

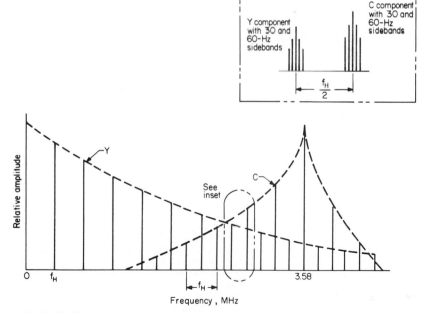

FIG. 13-139 Frequency spectrum of NTSC system showing interleaving of signals.

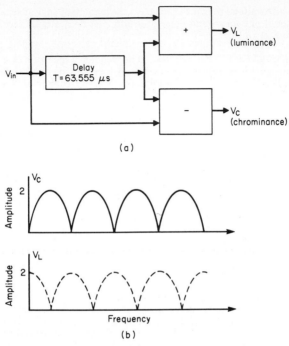

FIG. 13-140 Simplified $1H$-delay comb filter: (a) block diagram; (b) outputs V_L and V_C.

Similarly, the chroma response is defined by

$$\frac{V_C}{V_{in}}(j\omega) = 1 - e^{-j\omega T} \tag{13-124}$$

$$\left|\frac{V_C}{V_{in}}(j\omega)\right| = \left|2\sin\frac{\omega T}{2}\right| \tag{13-125}$$

This particular configuration using one delayed signal has been called a $1H$ comb filter and is characterized by an amplitude response in the shape of rectified sinusoids. A $2H$ comb filter as shown in Fig. 13-141 has the amplitude response given by[138]

$$\frac{V_L}{V_{in}} = e^{-j\omega T}(1 - \cos\omega T) \tag{13-126}$$

While such filters can separate interleaved chroma and luminance spectra, difficulties arise when television pictures do not conform exactly with the idealized spectrum. If vertical correlation is not present from one line to the next, a comb filter circuit will introduce spurious information from the earlier line. Considering that the chroma spectrum is not present below 2.0 MHz, combing has no effect below this frequency for chroma-luminance separation. In fact, it is necessary to delete such combing action below 2.0 MHz to prevent the loss in resolution that would otherwise be experienced on line-to-line picture changes. An additional problem results when there are line-to-line chroma changes. These will be apparent as residual subcarrier in the luminance, and care must be taken in controlling the relative luminance channel amplitude response in the chroma region for acceptable overall performance.

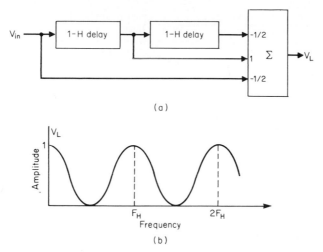

(a)

(b)

FIG. 13-141 $2H$-delay-line comb filter: (a) block diagram; (b) output V_L.

In addition to superior chroma-luminance separation, the comb filter provides more complete rejection of chroma subcarrier from the luminance channel with less compromise in the bandwidth and resolution of the luminance channel. The signal-to-noise ratio of both channels can be improved by 3 dB. Better rejection of cross color (high-frequency luminance crosstalk into chroma) also results.

There are at present two types of comb filter devices in common use in color television receivers. They are the glass delay line and the charge-coupled device (CCD).

Ultrasonic Glass Delay Line.   Figure 13-142 shows a typical glass delay line. Table 13-32 gives typical electrical characteristics. Being essentially a bandpass device, the external resistors and inductors tune with the capacitance of the piezoelectric transducers.

A commonly employed comb filter chrominance-luminance circuit is shown in Fig. 13-143. Transistor amplifier Q_1 drives the 1H-delay-line input with sufficient amplitude to overcome the insertion loss of the device. The voltage V_L at the wiper of R_5 consists of the sum of the 1H-delayed-signal output plus the original input signal. R_5 and L_1 are usually adjusted for a null at or near the color subcarrier frequency.

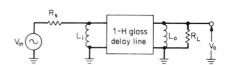

FIG. 13-142   Glass ultrasonic delay-line test circuit.

The amplitude response of the glass delay can conveniently be made zero for frequencies less than 2.0 MHz. This eliminates much of the degradation of vertical resolution which would otherwise occur with full baseband filtering. The fact that the comb filter bandpass response is added to the full bandwidth video response gives rise to some frequency response distortion in the luminance channel. This must be taken into consideration in the design of the overall system.

The chroma signal is formed by taking the combined luminance V_L, inverting it in transistor Q_2, and adding to the original composite signal available from resistor R_4. Further chroma bandpass filtering is usually employed to give additional suppression to any uncombed frequencies resulting from the bandwidth limitations of the delay line.

Table 13-32   Characteristics of Typical
Ultrasonic Glass Delay Lines

Insertion loss	8–20 dB
Phase delay	
at 3.579545 MHz	63.555 μs $\pm$ 5 ns
3.1–4.1 MHz	63.555 μs $\pm$ 10 ns
Amplitude response	
−3 dB points	2.8–4.4 MHz
−20 dB points	2.0–5.2 MHz
Spurious responses	−26 dB max.

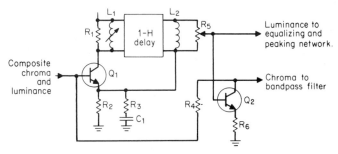

FIG. 13-143   1H-delay comb filter circuit for chrominance and luminance
separation.

CCD.   In this approach the 1H-delay function is provided by a charge-coupled device
which is inherently full bandwidth and whose time delay, independent of frequency, is
given by

$$T = \frac{\text{no. of elements}}{f_{\text{clock}}} \qquad (13\text{-}127)$$

The number of elements is fixed and the clock frequency is locked to a harmonic of the
chroma subcarrier, resulting in a very stable delay characteristic.

Since the full baseband spectrum is combined, vertical detail is lost. The missing
information can, however, be restored by taking the low-frequency output of the difference channel and adding it back into the combined luminance signal. If a greater amount
of low-frequency signal is added than that which was lost, the process will result in an
enhancement of the vertical detail.[137,139] An example of a CCD comb filter system is
shown in Fig. 13-144.

13.13.12 SCAN-VELOCITY MODULATION (SVM).   Selective modulation of the constant horizontal scan velocity can improve the final spatial brightness distribution on
small picture details. For example, the effect of adding the first derivative of a white
pulse luminance signal to the horizontal deflection is shown in Fig. 13-145. The width of
the white pulses can be made appreciably narrower, partly compensating for the finite
video bandwidth and picture tube spot size under high drive conditions.

Figure 13-146 shows a block diagram of a receiver with SVM using auxiliary deflection
coils.[140] Electrostatic deflection has also been used in commercial receivers.[141]

In developing an SVM design, it is important, first, to ensure that the time delays
through the luminance and SVM channels match to a tolerance of less than 30 ns. Second, the amount of modulation must be controlled in order to prevent reversal of the
beam scan direction on very large transient peaks.

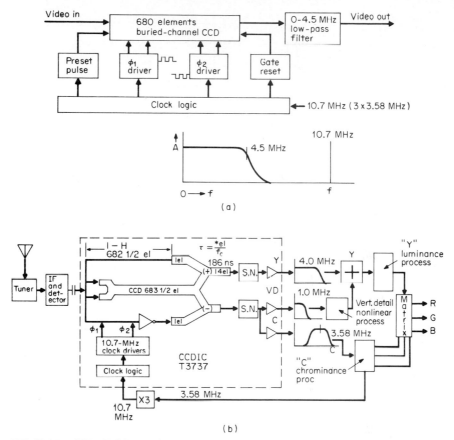

FIG. 13-144 CCD 1*H*-delay comb filter system: (*a*) basic CCD operation (delay = number of elements/f_c); (*b*) system diagram. (*From D. Pritchard, "A Comb Filter for Color TV Receiver Picture Enhancement," RCA Rev., vol. 41, March 1980, p. 18.*)

13.13.13 AUTOMATIC CIRCUITS

Automatic Chroma Control (ACC). The relative level of chroma subcarrier in the incoming signal is highly sensitive to transmission path disorders, thereby introducing objectionable variations in saturation. These can be observed between one received signal and another or over a period of time on the same channel unless some adaptive correction is built into the system. The color-burst reference, transmitted at 40 IRE units peak-to-peak, is representative of the same path distortions and is normally used as a reference for automatic gain controlling the chroma channel. A balanced peak detector or synchronous detector, having good noise rejection characteristics, detects the burst level and provides the control signal to the chroma gain-controlled stage.

Color Killer. Allowing the receiver chroma channel to operate during reception of a monochrome signal will result in unnecessary cross color and colored noise, made worse by the ACC increasing the chroma amplifier gain to the maximum. Most receivers, therefore, cut off the chroma channel transmission when the received burst level goes below

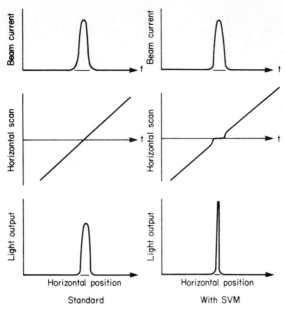

FIG. 13-145 A comparison of the effect of SVM on spotlight output.

approximately 5 to 7 percent. Hysteresis has been used to minimize the flutter or threshold problem with varying signal levels.

Chroma-limiting. Burst-referenced ACC systems perform adequately when receiving correctly modulated signals with appropriate burst-to-chroma ratio. Occasionally, however, burst level may not bear correct relation to its accompanying chroma signal,

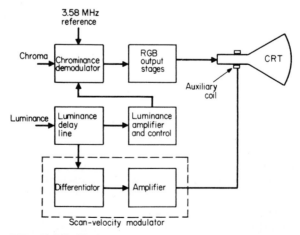

FIG. 13-146 Block diagram of scan-velocity modulation implementation.

leading to incorrectly saturated color levels. It has been determined that most viewers are more critical of excessive color levels than insufficient ones. Experience has shown that when peak chroma amplitude exceeds burst by greater than 2/1, limiting of the chroma signal is helpful. This threshold nearly corresponds to the amplitude of a 75 percent modulated color bar chart (2.2/1). At this level, negligible distortion is introduced into correctly modulated signals. Only those which are nonstandard are affected significantly. The output of the peak chroma detector is also sent to the chroma gain-controlled stage.

Automatic Black-Level. Systems using the back porch of the horizontal blanking interval as a black-level reference are satisfactory only to the extent that this level correlates with picture black-level, which frequently is not the case. Some improvement in the receiver picture characteristics can be achieved by using the picture content itself to derive a reference. For example, the blackest excursion of the video (luminance) is restored to a prescribed level defined by the viewer *brightness* control (Fig. 13-147).[142] It is necessary to limit the amount of this correction so that video scenes with no large amplitude black excursions will not be objectionably distorted.

Auto-contrast. To ease the high-voltage power supply requirements and prevent harsh overdrive of the picture tube on large white scenes, it is desirable to prevent the average beam current from exceeding some prescribed limit. In the past, execution of this problem has been handled by a feedback loop which shifted the brightness or background in a black-going direction. Present practice normally reduces contrast, i.e., video drive, in the interest of preserving black-level stability.

A second feedback loop or AGC system can be placed in the luminance channel to compensate for less than fully modulated signals. This system tends to maintain more uniform contrast or picture black-to-white ratio. Again, there is a subjective limit to the gain which can be designed into the system. When considering the possible picture distortion which will occur on a fade-to-black, for example, it is clear that the pull-up capability of the system should not exceed 2 or 3 dB.

Automatic Sharpness. Sharpness variations which exist from station to station due to receiving antennas, receiver tolerances, transmitted program material, etc., can be reduced by an adaptive feedback loop driving the sharpness or peaking circuit (Fig. 13-148). These systems typically utilize a high-pass filter and detector to determine the amplitude of the picture high-frequency content.[143] A second input to the loop may be the tuner AGC voltage which will ensure that the peaking is reduced to nominal or lower (smear) when the receiver is operating under fringe signal conditions.

Automatic Tint. One major objective in color television receiver design is to minimize the incidence of flesh-tone reproduction with incorrect hue. Automatic hue-correcting systems can be categorized into two classes.

Static flesh-tone correction is achieved by selecting the chroma demodulating angles and gain ratios to desensitize the resultant color-difference vector in the flesh-tone region ($+I$ axis). The demodulation parameters remain fixed, but the effective Q axis gain is reduced. This has the disadvantage of distorting hues in all four quadrants.

Dynamic flesh-tone corrective systems can adaptively confine correction to the region within several degrees of the positive I axis, leaving all other hues relatively unaffected (Fig. 13-149). This is typically accomplished by detecting the phase of the incoming chroma signal and modulating the phase angle of the demodulator reference signal to result in effective phase-shift of 10 to 15° toward the I axis for a chroma vector which lies within 30° of the I axis.[144] This approach produces no amplitude change in the chroma. In fact, for chroma saturation greater than 70 percent the system is defeated on the theory that the color is not a flesh tone.

A simplification in circuitry can be achieved if the effective correction area is increased to the entire positive-I 180° sector.[145] A conventional phase detector can be utilized and the maximum correction of approximately 20° will occur for chroma signals

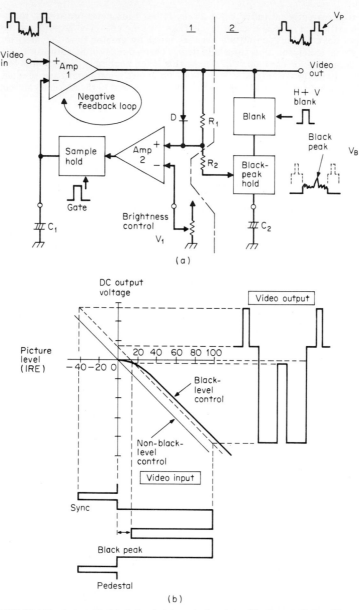

FIG. 13-147 Automatic black-level: (*a*) control system; (*b*) characteristics. (*From M. Yoshitomi, H. Yamashita, K. Kojima, and H. Takano, "The Two-Chip Integration for Color Television Receivers," IEEE Trans., vol. CE-28, no. 3, p. 189. © 1982 IEEE.*)

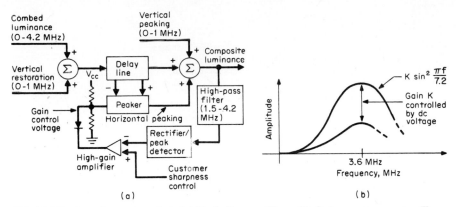

FIG. 13-148 Auto-sharpness system: (a) block diagram; (b) amplitude-frequency response. (*From W. Harlan, "An Automatic Sharpness Control," IEEE Trans., vol. CE-27, no. 2, p. 311. © 1981 IEEE.*)

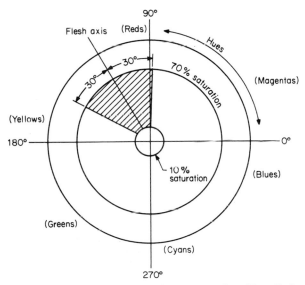

FIG. 13-149 Area of effective flesh-tone correction. (*From T. S. Rzeszewski, "A Novel Automatic Hue Control System," IEEE Trans. Consum. Electron., Vol. CE-21, no. 2. © 1975 IEEE.*)

having phase of $\pm 45°$ from the I axis. Signals with phase greater or less than 45° will have increasingly lower correction values (Fig. 13-150).

Vertical-interval reference (VIR) signal, as shown in Fig. 21-18, provides references for black level, luminance, and, in addition to burst, a reference for chroma amplitude and phase.[146] While originally developed to aid broadcasters, it has been employed in television receivers[147,148] to correct for saturation and hue errors resulting from transmitter or path distortion errors.

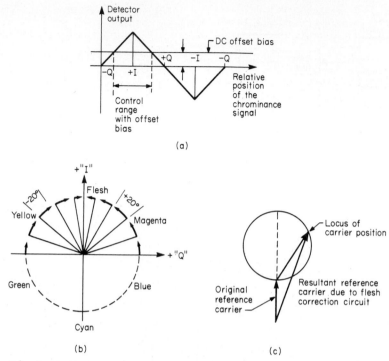

FIG. 13-150 Flesh-tone correction system characteristics: (*a*) phase detector response; (*b*) rotation of color vectors due to flesh-tone correction; (*c*) locus of the reference carrier modified by the flesh-tone correction circuit. (*From L. Harwood, "A Chrominance Demodulator IC with Dynamic Flesh Correction," IEEE Trans., vol. CE-22, no. 1, p. 112. © 1976 IEEE.*)

13.13.14 VIDEO ACCESSORY INTERFACE

Composite Video. The accepted standard for video accessories such as cameras, video tape recorders, and video disc players is 1 V peak-to-peak into 75 Ω ac-coupled, with sync negative. At the present time phono plugs are the most common form of connector in use in the United States.

Power line isolation can be provided by a fully *cold chassis* or an *iso-hot* type (Sec. 13.18.1). Alternatively, transformer coupling of the video signals can be used with a hot chassis. However, this necessitates more extensive frequency response equalization.

The frequency response at the output of the video detector usually is not flat. Signals to be taken from the receiver to accessories or from accessories to the receiver therefore must be compensated to match the peculiar characteristics of the receiver video and chroma channels.

Function switching can be accomplished with either mechanical switches or electronic switches. Electronic switches have the advantage of being placed directly in the circuit area of the chassis with only a dc control lead, rather than shielded cables, connecting to the user control panel. Both CMOS and bipolar device switches have been used in consumer receivers.

RGB Signal Input. No single standard exists, at present, for direct-drive nonencoded format. Generally, black is negative-going, but levels range from 0.7 V black-to-white to TTL levels (5 V). Composite sync can exist either as a separate line or be contained in one of the three channels, usually green.

Power-line isolation and switching considerations are similar to those for composite video.

The bandwidth of the *RGB* output amplifiers (CRT drivers) generally requires improvement over what is considered adequate for displaying restricted bandwidth NTSC-encoded signals. In addition, high-frequency crosstalk, usually not a problem with narrow-bandwidth color-difference signals, will deserve added attention.

A further consideration in displaying *RGB* signals depends upon the need for the picture controls to be active in this mode. A simple two-state primary-color signal may not require controlling, whereas a full analog *RGB* signal probably will need viewer preference control.

13.14. SCANNING SYNCHRONIZATION

13.14.1 SYNC SEPARATOR.

The scan-synchronizing signal, consisting of the horizontal and vertical sync pulses (Sec. 13.1.6), is removed from the composite video signal by means of a sync-separator circuit. The classic approach has been to ac-couple the video signal to an overdriven transistor amplifier (Fig. 13-151), which is biased off for signal levels smaller than V_{co} and saturates for signal levels greater than V_{sat}, a range of approximately 0.1 V. Analysis of the equivalent circuits also shown in Fig. 13-151, leads to the value for sync separator slicing level V_s as given below. This analysis is based upon the concept of equilibrium in the capacitor charge and discharge during sync time (charge) versus nonsync time (discharge) and follows the clamping theorem given by Millman and Taub[149]

$$\frac{A_f}{A_r} = \frac{R_s + R_f}{R_s + R_b} \qquad \text{assuming } R_b \gg R_d \qquad (13\text{-}128)$$

where A_f, the area of the voltage curve during forward conduction, is given by

$$A_f = V_s T_h \qquad (13\text{-}129)$$

and A_r, the area during the discharge of C, with transistor reverse-biased, is given by

$$A_r = (0.286 \text{ V} - V_s)(T_b + T_s + T_f) + 0.714 \text{ V}(T_s) \qquad (13\text{-}130)$$

From which the slice level V_s can be found

$$V_s = \frac{0.714 \text{ V} [T_s/(T_b + T_s + T_f)] + 0.286 \text{ V}}{1 + [T_h/(T_b + T_s + T_f)](R_s + R_b)/(R_s + R_d)} \qquad (13\text{-}131)$$

In the NTSC system the sync-pulse widths have the following nominal values:

$$T_h = 4.7 \text{ }\mu\text{s} \qquad T_b = 3.0 \text{ }\mu\text{s}$$
$$T_s = 54.3 \text{ }\mu\text{s} \qquad T_f = 1.5 \text{ }\mu\text{s}$$

Therefore, for full white scene (12.5 percent modulation level)

$$V_s = \frac{0.946 \text{ V}}{1 + 0.080 [(R_s + R_b)/(R_s + R_d)]} \qquad (13\text{-}132)$$

Frequently the approximation $R_b \gg R_s$ is made to simplify the calculations.

The values for R_s, R_b, and R_d must be selected such that the level V_s stays as near the middle of the sync pulse as possible with the full range of video content, i.e., from full-white scene to black scene. For a conservative reserve, the calculation frequently is made with black-reference level at the blanking level. The following example demonstrates the shift in slicing level for the simple sync-separator circuit.

Given: $R_s = 1 \text{ k}\Omega$
$R_d = 1 \text{ k}\Omega$

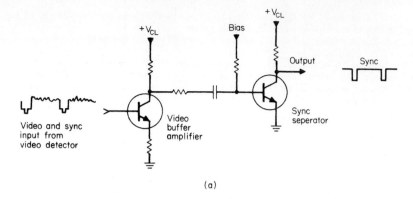

(a)

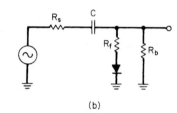

(b)

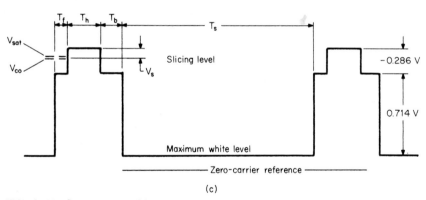

(c)

FIG. 13-151 Sync separator: (a) typical circuit; (b) equivalent circuit; (c) sync waveform.

R_b = 100 kΩ
V_1 = 4.9 V, white to sync tip
V_2 = 1.4 V, blanking to sync tip
V_s (white scene) = 0.927 V or 66% clamp level

For a blanking to sync-tip transmission, the equation for V_s becomes

$$V_s = \frac{V_2}{1 + (R_s/R_d) + (R_b/R_d)\,[T_h/(T_b + T_s + T_f)]}$$

$$= 0.14 \text{ V or } 10\% \text{ clamp level}$$

(13-133)

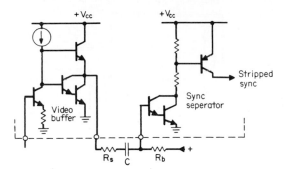

FIG. 13-152 IC sync separator with Darlington input stage.

A common sync-separator implementation in IC format uses a Darlington stage as shown in Fig. 13-152. This type reflects less load back to the video stage and requires a smaller, less costly capacitor.

Another type is the *common base clipper* circuit shown in Fig. 13-153, which requires negative-going sync from the source, i.e., second detector or video amplifier. Analysis of

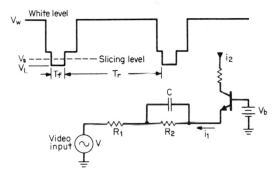

FIG. 13-153 Common-base sync separator has slice level which is independent of video white-level.

this circuit shows that the clamp level is independent of the white-video level.[150] The slicing level in this case is given by

$$V_s = \frac{V_L + (V_b + V_{be})(t_r R_1)/(t_f R_2)}{1 + (t_r R_1)/(t_f R_2)} \tag{13-134}$$

The schematic of Fig. 13-154 shows an IC implementation which utilizes an operational amplifier and feedback loop to set a proportionally constant slicing level between sync tip and pedestal level, independent of sync amplitude. Capacitor C_1 charges to the sync tip level V_T. Transistor Q_1 conducts on the tips, thereby canceling them in the signal sent to C_2, which therefore charges to the value of the pedestal level V_b. The resistor divide consisting of R_1 and R_2 controls the slice level according to the formula[151]

$$V_s = \frac{R_2}{(T_s/T_h)R_1 + R_2} \tag{13-135}$$

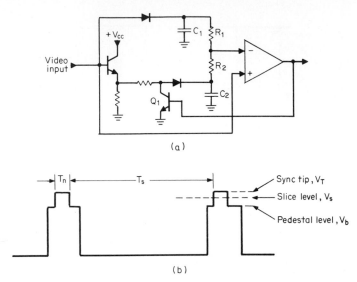

(a)

(b)

FIG. 13-154   IC sync circuit which maintains constant slicing level: (*a*) circuit diagram; (*b*) sync waveform.

13.14.2 IMPULSE NOISE SUPPRESSION.   The simplest type of suppression which will improve the basic circuit in a noise environment of human origin consists of a parallel resistor and capacitor in series with the sync charging capacitor (Fig. 13-155). Capacitor C_n is selected between one-tenth and one-fiftieth the value of C_s. In the presence of a large, narrow noise pulse, the capacitors will charge to voltages inversely proportional to their capacitance. The much larger voltage on C_n will discharge rapidly through R_n, thereby unblocking the input and restoring the separator to handle the next sync pulse. Variations of this simple circuit which include diode clamps and switches have been developed to speed up the noise performance while not permitting excessive tilt in the sync output during the vertical block.[152]

More complex solutions to the noise problem include the *noise canceler* and *noise gate* (Fig. 13-156). The canceler monitors the video signal between the second detector and the sync separator. A noise spike, which exceeds the sync or pedestal level, is inverted and added to the video after the isolation resistor R. This action cancels that part of the noise pulse which would otherwise produce an output from the sync separator.

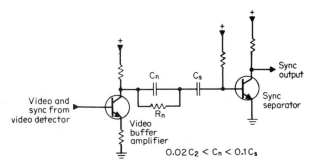

FIG. 13-155   Sync separator with simple noise protection.

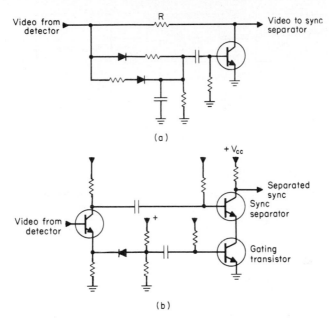

FIG. 13-156 Noise-immune circuits: (a) noise canceler; (b) noise gate.

For proper operation the canceler circuit must track the sync-tip amplitude from strong RF signal to fringe level. The noise gate, in a similar manner, recognizes a noise pulse of large amplitude and prevents the sync separator from conducting either by applying reverse bias or by opening a transistor switch in the emitter circuit.[152]

13.14.3 HORIZONTAL APC LOOP. Horizontal scan synchronization is accomplished by means of an APC loop, with theory and characteristics similar to those used in the chroma reference system (Sec. 13.13.7). Input signals to the horizontal APC loop are sync pulses from the sync separator and horizontal flyback pulses which are integrated to form a sawtooth waveform. The phase detector compares the phase (time coincidence) of these two waveforms and sends a dc-coupled low-pass-filtered error signal to the voltage-controlled oscillator, to cause the frequency to shift in the direction of minimal phase error between the two input signals.

The characteristics of the APC loop of Fig. 13-157 are given by the following equations[153]

$$\text{Loop gain } f_c \text{ (Hz)} = \frac{\omega_c}{2\pi} \tag{13-136}$$

$$\text{Noise bandwidth } f_{nn} \text{ (Hz)} = \frac{1 + \omega_c x^2 T}{4xT} \tag{13-137}$$

$$\text{Loop frequency } \omega_n \text{ (rad)} = \sqrt{\frac{\omega_c}{(1 + x)T}} \tag{13-138}$$

$$\text{Damping factor } K \approx \frac{x^2 T \omega_c}{4} \tag{13-139}$$

when $K = 1$, the loop is critically damped
$K > 1$, the loop is overdamped
$K < 1$, the loop is underdamped and a period of oscillatory ringing will result

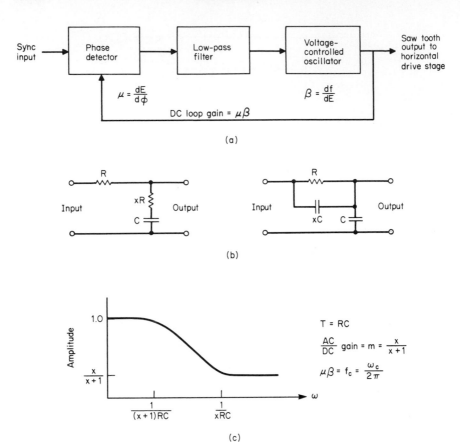

FIG. 13-157 Horizontal APC loop: (a) loop diagram; (b) typical filter configurations; (c) filter response plot. (*From Giles.*[153])

The APC loop-recovery time from a step transient is given by

$$T_r = \frac{2.303}{(1 + xT\omega_c)/[2(1 + x)T]} \tag{13-140}$$

Maximum pull-in range $\Delta f_{max} = \frac{2\phi_s f_c \sqrt{m}}{3 + m} \tag{13-141}$

where ϕ_s = the width of the stable operating region of the loop and $m = x/(x + 1)$.

For various operating conditions the following choice of parameters has been found to be nearly optimum: fringe-signal performance, $f_{nn} = 500$ Hz, $K \approx 1$; video-cassette recorder operation, $f_{nn} = 1000$ Hz, $K \approx 18$. The values of these parameters are typical of many current designs and indicate that no single set of values is entirely adequate for both applications. Numerous manufacturers are currently incorporating, within sync and horizontal ICs, an automatic means to switch the time constant as conditions warrant. Typically, a noise-sensing detector will put the circuit in the slow mode whenever noise is present in the video signal. When no noise can be detected, the circuit switches to the fast mode to reduce the random horizontal shifting of the top several lines of scan when the receiver is operating from a VCR source.

The recovery of the APC loop to a step transient input involves the parameters of the natural resonant frequency, as well as the amount of overshoot permitted as the correct phase has been reached. Both of these characteristics can be evaluated by use of a jitter generator which creates a time base error between alternate fields.[154,155] The resultant sync error and system dynamics can be seen in the picture display of the receiver shown in Fig. 13-158. This type of disturbance occurs when the two or three playback heads of a consumer helical-scan VCR switch tracks. It will also occur when the receiver is operated from a signal which does not have horizontal slices in the vertical sync block.

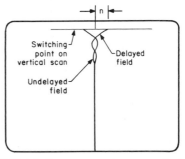

FIG. 13-158 Pattern on receiver screen displaying vertical line with the even and odd fields offset by "n" μs.

Phase detector circuitry in the form of classic double-diode bridge detectors, in either the balanced or unbalanced configuration, are still found in low-cost receivers. Integrated circuits commonly utilize a gated differential amplifier (Fig. 13-159) in which the common feed from the current source is driven by the sync pulse, and the sawtooth waveform derived from the flyback pulse is applied to one side of the differential pair.

Voltage-Controlled Oscillator (VCO). An LC tuned Hartley oscillator is typically used in discrete component designs. Frequency of oscillation can be shifted by changing the dc bias to the base of the oscillator transistor. This changes the transistor conduction level and the stored charge on the capacitor which couples the top of the tank to the transistor base. A schematic of this arrangement is shown in Fig. 13-160a.

Relaxation oscillators of the form in Fig. 13-160b are preferred in IC designs.[153] These circuits charge a capacitor C_t from a current source R_t until the voltage across the capacitor reaches the IC trigger level V_2. At this point the internal circuitry switches modes and discharges capacitor C_t at a known rate until the capacitor reaches the low threshold V_1. The IC then resets and begins to charge C_t. The charge rate is given by

$$T_c = R_t C_t \ln \frac{V_{cc} - V_1}{V_{cc} - V_2} \tag{13-142}$$

The discharge time T_d, if the condition $R_t \gg R_4$ is assumed, is given by

$$T_d = R_4 C_t \ln \frac{V_2}{V_1} \tag{13-143}$$

The frequency is given by

$$F_0 = \frac{1}{\text{charge time} + \text{discharge time}} \tag{13-144}$$

$$= \frac{1}{R_t C_t \ln [f_1(R_1, R_2, R_3)] + R_4 C_t \ln [f_2(R_1, R_2, R_3)]}$$

The frequency contains no dependence upon the supply voltage V_{cc}. The oscillator control sensitivity is given by

$$\beta = \frac{1}{C_t(V_2 - V_1)} \tag{13-145}$$

This is a general relationship applicable to many relaxation oscillators which switch between two selected voltages.

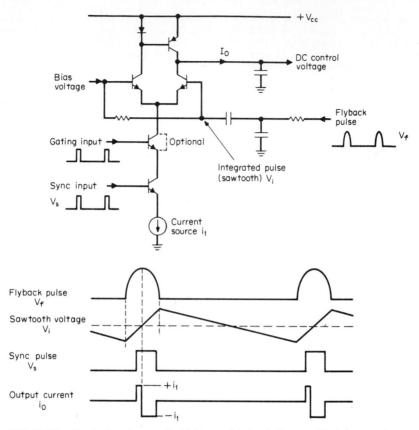

FIG. 13-159 Gated phase detector for IC format: (*a*) circuit diagram; (*b*) timing waveforms.

13.14.4 VERTICAL INTEGRATION AND OSCILLATOR. Vertical synchronizing information can be recovered from the output of the sync separator by the technique of integration. The classic two-section *RC* integrator provides a smooth, ramp waveform which corresponds to the vertical-sync block as shown in Fig. 13-14. The ramp is then sent to a relaxation or blocking oscillator, which is operating with a period slightly longer than the vertical frame period. The upper part of the ramp will trigger the oscillator into conduction to achieve vertical synchronization. The oscillator will then reset and wait for the threshold level of the next ramp. The *vertical-hold* potentiometer controls the free-running frequency or period of the vertical oscillator.

One modification to the integrator design is to reduce the integration (speed up the time constants) and provide for some differentiation of the waveform prior to applying it to the vertical oscillator (Fig. 13-161). Although degrading the noise performance of the system, this technique provides a more certain and repeatable trigger level than the full integrator, thereby leading to an improvement in interlace over a larger portion of the hold-in range. A second benefit is to provide a more stable vertical lock when receiving signals which have distorted vertical-sync waveforms. These can be generated by video-cassette recorders in nonstandard playback modes, such as fast/slow forward, still, and reverse. Prerecorded tapes having antipiracy nonstandard sync waveforms also contribute to the problem.

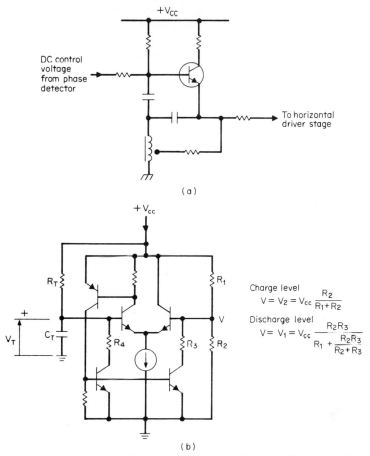

FIG. 13-160   Voltage-controlled horizontal-sweep oscillators: (*a*) Hartley oscillator; (*b*) relaxation oscillator.

13.14.5 VERTICAL-COUNTDOWN SYNCHRONIZING SYSTEMS.

A vertical-scan system using digital logic elements can be based upon the frequency relationship of 525/2 which exists between horizontal and vertical scan. Such a system can be considered to be phase-locked for both horizontal scan and vertical scan which will result in improved noise immunity and picture stability. Vertical sync is derived from a pulse train having a frequency of twice the horizontal rate. The sync is therefore precisely timed for both even and odd fields, resulting in excellent interlace. This system needs no hold control.

The block diagram of one design is shown in Fig. 13-162.[156] The 31.5-kHz clock input is converted to horizontal drive pulses by a divide-by-2 flip-flop. A two-mode counter, set for 525 counts for standard interlaced signals and 541 for noninterlaced signals, produces the vertical output pulse. Noninterlaced signals are produced by a variety of VCR cameras, games, and picture generators used by television service people. The choice of 541 allows the counter to continue until the arrival of vertical sync from the composite video waveform. This actual vertical sync then resets the counter.

Other systems make use of clock frequencies of 16 and 32 times the horizontal frequency.[157,158] Low-cost crystal or ceramic resonator-controlled oscillators can be built to

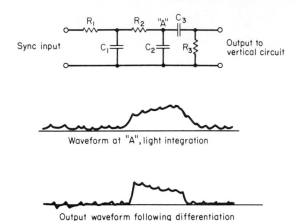

Waveform at "A", light integration

Output waveform following differentiation

FIG. 13-161   Modified integrator followed by differentiator.

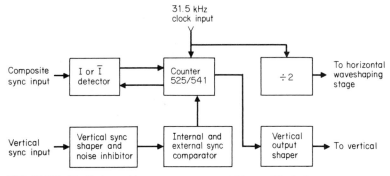

FIG. 13-162   Vertical-countdown sync processor which uses $2f_H$ clock.

operate at these frequencies. As in the first system, dual-mode operation is necessary in order to handle standard interlaced and nonstandard sync waveforms. Exact 525/2 countdown is used with interlaced signals. For noninterlaced signals, the systems usually operate in a free-running mode with injection of the video-derived sync pulse causing lock.

A critical characteristic, and probably the most complex portion of any countdown system, is the circuitry which properly adjusts for nonstandard sync waveforms. These can be simple 525 noninterlaced fields, distorted vertical-sync blocks, blocks having no horizontal serrations, fields with excessive or insufficient lines, and a combination of the above. As discussed in Sec. 13.14.4, home VCRs operating in the "tricks" modes can present unique sync waveforms. An example of the logic flowchart for this decision making is shown in Fig. 13-163.

13.15   VERTICAL SCANNING

13.15.1   CLASS A DESIGN.   Early vertical-drive circuits were the class A type, transformer coupled to the yoke as shown in Fig. 13-164. With the advent of solid-state components, class A was short-lived for the following reasons:

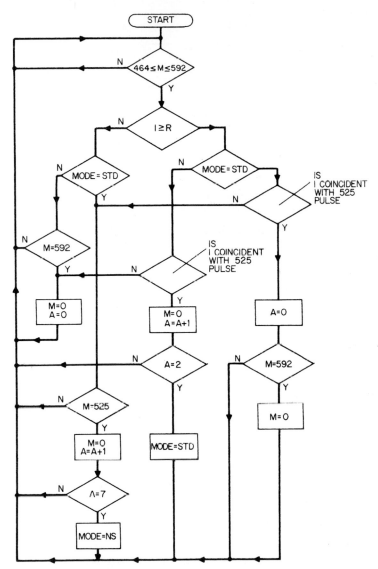

FIG. 13-163 Vertical-logic flowchart, where M is the main counter state, A the auxiliary counter state, I the input, R the reference level, and where STD is standard and NS nonstandard. *(From S. Steckler, A. Balaban, and R. Fernsler, "Horizontal, Vertical and Regulator Control IC," IEEE Trans., vol. CE-28, no. 2, p. 152. © 1982 IEEE.)*

1. An expensive coupling transformer was required.
2. Power dissipation was high.
3. A linearity control was required.
4. A thermistor was needed to compensate for yoke resistance changes with temperature, including receiver warm-up.

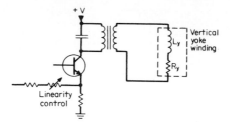

FIG. 13-164   Class A vertical output stage.

13.15.2 CLASS B DESIGN.   Class B vertical circuits consist of an audio amplifier with current feedback.[159] This approach maintains linearity without the need for an adjustable linearity control. Yoke-impedance changes caused by temperature changes will not also affect the yoke current; thus the thermistor is not required. The current-sensing resistor must, of course, be temperature stable. An amplifier which uses a single *npn* and a single *pnp* in the form of a complementary output stage is sketched in Fig. 13-165.

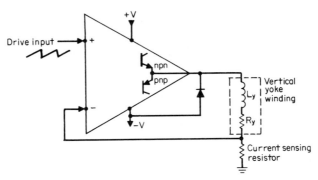

FIG. 13-165   Class B vertical-output stage.

Quasi-complementary, Darlington outputs and other common audio-output stage configurations can be used.

Establishing proper dc bias for the output stages is quite critical. Too little quiescent current will result in crossover distortion which will put a faint horizontal white line in the center of the picture even though the distortion may not be detectable on an oscilloscope presentation of the yoke current waveform. Too much quiescent current results in excessive power dissipation in the output transistors.

13.15.3 WAVEFORMS AND POWER REQUIREMENTS.   The energy required in the magnetic field is given by the equation

$$W = 2.78 \ DE \left(\sin^3 \frac{\theta}{4} \right) \left(\cos \frac{\theta}{4} \right) \tag{13-146}$$

where D = CRT neck diameter, i.e., electron beam path diameter
E = CRT anode voltage
θ = peak-to-peak deflection angle

In actual practice, the yoke is not 100 percent efficient. Accurate energy requirements can be calculated after making measurements on an operating system

$$W = \tfrac{1}{2}LI_{pk}^2 \qquad (13\text{-}147)$$

where L = yoke inductance and I_{pk} = peak yoke current required for full deflection.

The yoke inductance and resistance can be changed to suit a particular circuit configuration, supply voltage, etc., but the value of $\tfrac{1}{2}LI^2$ must be met in order to achieve full scan. During the trace or scan period the yoke voltage and current waveforms are nearly a ramp as shown in Fig. 13-166

$$i(t) = -I_{pk} + \frac{2I_{pk}}{T_t}t \qquad for\ 0 < t < T_t \qquad (13\text{-}148)$$

$$V(t) = \left(-I_{pk} + \frac{2I_{pk}}{T_t}t\right)R_{yoke} + \frac{2I_{pk}}{T_t}L_{yoke} \qquad (13\text{-}149)$$

$$V_{pk+} = I_{pk}R_{yoke} + \frac{2I_{pk}}{T_t}L_{yoke} \qquad (13\text{-}150)$$

$$V_{pk-} = -I_{pk}R_{yoke} + \frac{2I_{pk}}{T_t}L_{yoke} \qquad (13\text{-}151)$$

where L_{yoke} and R_{yoke} include impedances which are in series with the yoke.

In actual practice, the somewhat flat faceplate of the CRT requires that the slope of the yoke current ramp decrease slightly near the start and finish as shown in Fig. 13-167. This characteristic improves the geometrical linearity of the picture near the top and bottom. The shape of the waveform gives rise to the name "S" correction.

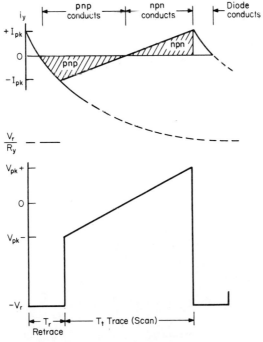

FIG. 13-166 Class B output waveforms.

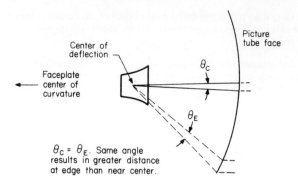

$\theta_C = \theta_E$. Same angle results in greater distance at edge than near center.

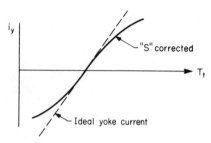

FIG. 13-167 "S" correction.

The waveforms during retrace for the class B circuit are given in Fig. 13-166. During retrace, the yoke current cannot change instantaneously; therefore the amplifier goes into saturation until the yoke current starts to ramp again at the start of trace. The result is an exponential current waveform

$$\text{Yoke voltage } v = -V_r$$

where V_r = the voltage during retrace

$$\text{Yoke current } i(t) = I_{pk} - \left(I_{pk} + \frac{V_r}{R} \right)(1 - \epsilon^{-t\,L_y/R_y}) \tag{13-152}$$

The required retrace voltage is then given by

$$V_r = R_y \left[\frac{2I_{pk}}{1 - \epsilon^{-T_rL_y/R_y}} - I_{pk} \right]$$

The power in the output transistors can be calculated by multiplying the voltage and current on each device and averaging over the complete cycle.[160]

$$P_{npn} = \frac{T_t}{T_t + T_r} \frac{I_{pk}}{4} \left(V_+ - \frac{5(V_{pk+} + V_{pk-})}{6} \right) \tag{13-153}$$

$$P_{pnp} = \frac{T_t}{T_t + T_r} \frac{I_{pk}}{4} \left(V_- + \frac{1}{6} V_{pk} - \frac{5}{6} V_{pk-} \right) + \frac{T_r}{T_t + T_r} \frac{I_{pk} V_{ce(sat)}}{2} \tag{13-154}$$

13.15.4 DC AND AC YOKE COUPLING. The circuit of Fig. 13-165, as described in the preceding paragraphs, is direct-coupled to the yoke and requires two power supplies,

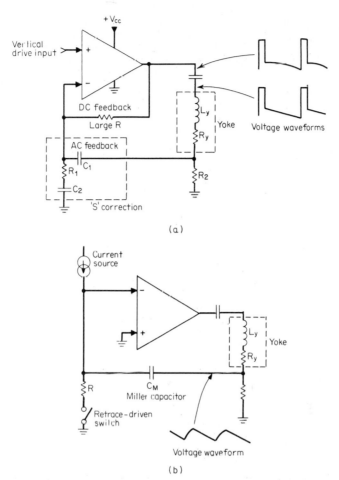

FIG. 13-168 (*a*) Class B capacitor-coupled circuit. (*b*) Miller-integrator vertical system.

each with opposite polarity. The direct current to the yoke can be adjusted to achieve vertical centering. The amplifier dc-bias point must be stable to avoid vertical shift in the picture. A failure in the output circuit can result in excessive direct current through the yoke. If this current is high enough, the electron beam can be deflected into the CRT neck which can result in localized heating of the glass and breakage in the neck area, if allowed to continue in excess of several seconds.

Just as in audio amplifiers, a capacitor can be used to couple the yoke. This avoids the neck-breakage problems associated with dc-coupled circuits. Only one power supply is required in this case. The dc-bias point is no longer as critical, since the direct current is not coupled to the yoke. Raster-centering adjustment must be provided by some other method. The power-dissipation equations shown earlier must be changed to account for the voltage on the coupling capacitor. Figure 13-168a shows a capacitor-coupled circuit. The input waveform has the ramp inverted compared with the previous circuit. The coupling capacitor C and voltage feedback resistor R provide the wave shaping to accomplish the required S correction.

13.15.5 MILLER CAPACITOR INTEGRATION.   The vertical stage of this system is shown in Fig. 13-168*b*. The current source in the input results in a ramp of voltage across the Miller capacitor C_m and the current sense resistor. The switch is closed during retrace to discharge C_m. The dc voltage across the Miller capacitor is determined for the most part by the closed-time of the retrace switch. The dc-bias voltage on the output is usually not as well defined as with the previous circuits. This variation must be provided for in the supply voltage.[161]

13.15.6 TOTEM POLE OUTPUT.   This configuration (Fig. 13-169) is designed to allow the vertical circuit to operate from the main power supply of the receiver, that is, 110 to 130 V, thereby saving the cost of the separate low-voltage supply required by the previous configurations. The need for a vertical driver transistor can also be eliminated

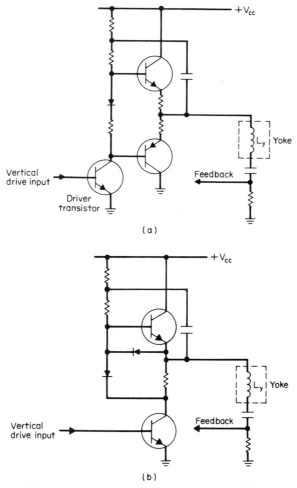

FIG. 13-169   Output-stage configurations: (*a*) typical class B output stage; (*b*) totem-pole stage, while using only two transistors, requires higher supply voltage.

by the totem pole arrangement. Good linearity is more difficult to obtain, and the outputs must be biased to a higher level of conduction in order to avoid crossover distortion. The increased bias current will result in increased power dissipation.

13.15.7 RETRACE POWER REDUCTION. The voltage required to accomplish retrace results in a substantial portion of the power dissipation in the output devices. The supply voltage to the amplifier and corresponding power dissipation can be reduced by using a "retrace-switch" or "flyback-generator" circuit to provide additional supply voltage during retrace.[162] One version of a retrace switch is given in Fig. 13-170. During

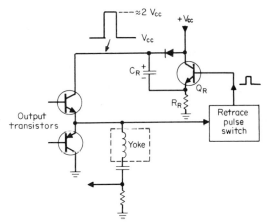

FIG. 13-170   Retrace switch.

trace time the capacitor is charged to a voltage near the supply voltage. As retrace begins, the voltage across the yoke goes positive, thus forcing Q_R into saturation. This places the cathode of C_R at the supply potential and the anode at a level of 1.5 to 2 times the supply, depending upon the values of R_R and C_R.

13.15.8 INTEGRATED-CIRCUIT VERTICAL SYSTEMS. These systems are becoming commonplace in color television chassis. In one implementation the IC contains the vertical-sync processor, oscillator, and driver stage. The power output stage consists of external discrete components. A second implementation requires drive-pulse input and contains the driver, class B output stage, and bias and protection circuits as well as retrace switch. This circuit can drive repetitive yoke current peaks of 1.5 A from a 25-V supply.

13.16 HORIZONTAL SCAN SYSTEM

13.16.1 BASIC FUNCTIONS. The horizontal scan system has two primary functions. It provides a modified sawtooth-shaped current to the horizontal-yoke coils to cause the electron beam to travel horizontally across the face of the CRT. It also provides drive to the high-voltage or *flyback* transformer to create the voltage needed for the CRT anode. Frequently low-voltage supplies are also derived from the flyback transformer (Sec. 13.18.4). The major components of the horizontal-scan section consist of a driver stage, horizontal output device (either bipolar transistor or SCR), yoke-current damper diode, retrace capacitor, yoke coil, and flyback transformer.

13.16.2 SCAN-CIRCUIT OPERATION. During the scan or trace interval $t_1 - t_3$ in Fig. 13-171e, the deflection yoke may be considered a pure inductance L_y, with a dc voltage ($V_{cs} = E_{in}$) impressed across it. This creates a sawtooth waveform of current given by

$$i_y(t) = -I_{y(pk)} + \frac{1}{L_y} \int_{t_1}^{t} V_y \, dt = -I_{y(pk)} + \frac{E_{in}(t - t_1)}{L_y} \tag{13-155}$$

This current flows through the damper diode D_d during the first half of scan ($t_1 - t_2$) (Fig. 13-171b). After i_y reverses direction at t_2, it flows through the horizontal output transistor (HOT) collector with the equivalent circuit now modified as shown in Fig. 13-171c. This sawtooth-current waveform deflects the electron beam across the face of the picture tube. A similarly shaped current flows through the primary winding L_p of the high-voltage transformer. Other current in this winding results from the reflection of secondary loads.

At the beginning of the retrace intervals t_3, the equivalent circuit of Fig. 13-171d is applicable. The inductors L_p and L_y contain stored energy equal to $\frac{1}{2}LI^2$ which is transferred to the retrace tuning capacitor C_k, thereby causing a half sine wave of voltage having frequency

$$f_0 = \frac{1}{2\pi} \sqrt{L_{eq} C_t} \tag{13-156}$$

where

$$L_{eq} = \frac{L_p L_y}{L_p + L_y} \tag{13-157}$$

and C_t contains the tuning capacitance and all winding and stray capacitance reflected to the flyback primary. This high-energy pulse appears on the transistor collector and is

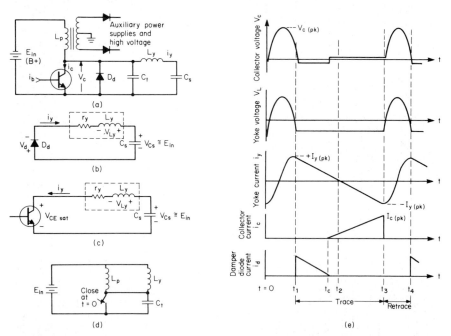

FIG. 13-171 (*a*) Equivalent circuits and waveforms of basic horizontal-scan circuit; (*b*) first half of scan; (*c*) second half of scan; (*d*) retrace interval; (*e*) waveform-timing diagrams.

stepped up, via the flyback transformer, to become the high voltage for the picture-tube anode. The collector pulse amplitude is given by

$$V_{c(pk)} = \left(1.79 + \frac{\pi}{2}\frac{T_t}{T_r}\right) E_{in}$$

where T_t and T_r = the trace and retrace periods, respectively. This value will be reduced by approximately 20 percent if third-harmonic tuning is employed. Finally, at t_4, the damper diode conducts and conditions return to those at t_1.

13.16.3 YOKE AND TRANSFORMER DESIGN CONSIDERATIONS. The peak current I_y required to scan the CRT can be approximated by

$$I_{y(pk)} = 0.053 \left[\frac{VD\,(\sin\theta/4)^3\cos\theta/4}{L_y}\right]^{1/2}$$

where V = anode voltage, kV
  D = yoke-shell diameter, cm; a good approximation is twice the diameter of CRT neck
  θ = deflection angle, deg

Variations in yoke sensitivity, anode voltage, capacitance C_s, and supply E_{in} contribute to the requirement for having the raster overscan the front viewing area of the CRT to prevent a black border under worst-case conditions. Nominal values with a regulated power supply are between 6 and 10 percent. With an unregulated supply the overscan requirement goes somewhat higher to accommodate both extremes of low and high power-line voltage.[163] If an adjustable-width control is provided, the overscan can be reduced safely to approximately 4 percent.

The output transistor collector current is composed of the yoke current, current for the primary inductance of the flyback, and current of the secondary- and high-voltage loads reflected to the primary. With early transistor devices, a high ratio of L_p to L_y was favored, for example, 4/1, to minimize the peak collector-current and to lessen variations in retrace time due to the wider tolerance of L_p caused by earlier construction techniques. Most current designs use a lower ratio of L_p to L_y, approximately 3/2. This comes about because newer transistors can withstand higher current levels and L_p can be controlled to within ± 5 percent. In addition, transformer designs, especially those driving a diode voltage multiplier or having diode-integrated high-voltage sections (Sec. 13.17.4), are not as critically tuned to a harmonic of the flyback frequency. Consequently, the peak collector current may have twice the peak value of the yoke current.

13.16.4 OUTPUT TRANSISTOR DRIVE CONSIDERATIONS. The horizontal-output transistor must have a special construction in order to handle the high peak-currents and voltages with a fast turnoff time. This also requires careful design of the base-current driver waveforms.

High peak voltage requires that the transistor have a thick, high-resistivity collector region. Collector dissipation at high collector currents can be minimized by driving the transistor into hard saturation with ample base current. Thus, the base-collector junction becomes forward biased and a surplus of minority carriers flows into the collector region. This yields a very low value of saturation voltage $V_{ce(sat)}$ at the high current level. At the end of scan, therefore, this large surplus of minority carriers must be removed before the collector current will decrease. Figure 13-172 shows the transistor collector and base waveforms. Figure 13-172c shows that the collector current I_c turnoff at t_5, initiated by the rapidly falling base drive current i_b, is delayed by the storage time, $t_s = t_3 - t_5$. The base drive current actually becomes negative in order to conduct the excess charge out of the transistor. If the reverse base drive is insufficient, the device will turn off slowly, resulting in high dissipation, i.e., the product of i_c and v_c during this period. Base circuits shown in Fig. 13-173 can effectively increase the drive by a controlled amount.[164,165] If reverse drive is excessive, however, some of the surplus carriers in the collector may be temporarily trapped, resulting in tailing of the collector current,[166] also

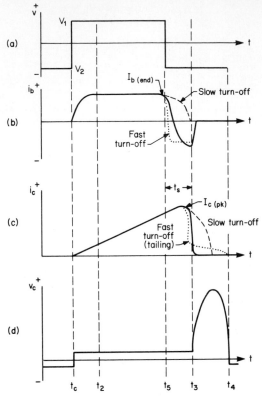

FIG. 13-172 Horizontal-output transistor current and voltage waveforms: (a) idealized base drive transformer voltage; (b) base current; (c) collector current; (d) collector voltage.

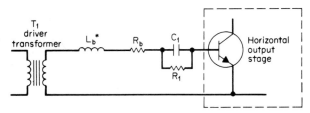

*L_b may be incorporated into the leakage inductance of T_1

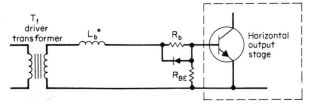

FIG. 13-173 Base-drive circuits to reduce collector turnoff dissipation.

a high-dissipation condition. The optimum values of $I_{b(end)}$, V_2, and L_b depend upon the peak collector current, upon the transistor parameters (and tolerances), especially H_{FE}, and upon the transistor operating temperature.

13.16.5 PICTURE GEOMETRY WAVEFORM CORRECTION. a linear sawtooth deflection current will not yield a linear reproduction of the original image on the CRT screen (Fig. 13-174).

S Correction. An angular change of $\Delta\theta$ in the angle θ_2 at the edge of the faceplate will move the beam a greater distance across the screen than will the same change in θ_1 near the center of the screen. An improvement in edge-to-center linearity can be accomplished by tuning the capacitor C_s and yoke inductance L_y to approximately 5 kHz.[167] This will impart a slight sinusoidal current to i_y and the voltage across C_s during scan, as illustrated in Fig. 13-175. This compresses the edges of the image and slightly increases the peak-to-peak current amplitude.

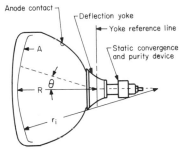

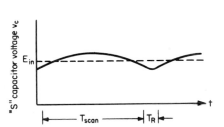

FIG. 13-174 CRT dimensions which produce geometric errors in pictures, where A is the inner surface of the plate, R the radius of deflection, θ the angle of deflected electron beam, and r_i the radius of inner faceplate surface.

FIG. 13-175 S-shaping waveform.

Right-side compression results from dc resistance of the yoke winding. Referring back to Fig. 13-171, the voltage across the yoke inductance at the left edge of scan (Fig. 13-171b) is given by

$$v_{Ly}(\text{left}) = E_{in} + r_y I_{y(pk)} + V_d \qquad (13\text{-}158)$$

At the right edge of scan the voltage is given by

$$v_{Ly}(\text{right}) = E_{in} - r_y{}_{(pk)} - V_{ce(sat)} \qquad (13\text{-}159)$$

To compensate for the difference in these values, a *linearity coil* is often added in series with the yoke. This coil consists of a small inductor with a dc magnetic bias. When the yoke current flows in the direction to produce right-side scan, the coil saturates, becoming a very small inductor. On the left side, the current opposes the magnetic bias, thereby increasing the inductance and compressing the left edge to be geometrically equivalent to the right.

Pincushion Correction. The geometric analysis used above indicates that the corners of the raster will naturally stretch, compared with the edges at the cardinal points. There are three common methods for correcting this error. The *pincushion-correcting transformer* is a saturable reactor with one winding in series with the horizontal windings of the deflection yoke. The other winding is fed with a parabolic waveform developed from the vertical-scan current. At the middle of the raster, the parabolic current is in the direction to saturate the reactor, thereby reducing its inductive effect. Toward the top and bottom of the raster, the parabolic waveform increases the inductance, thereby decreasing the horizontal-scan width to the same value as at the east–west points. The technique employed by the saturable reactor is similar to that of the linearity coil, described above.

The pin-corrected yoke has a unique winding distribution which causes underscan in the corners as compared with the centerline. Usually a 5 to 10 percent loss in deflection sensitivity results. In many recent designs this technique has been combined with picture tube and yoke integration to result in a self-converging pin-corrected system.[168,169]

The *diode-modulator* circuit (Fig. 13-176) can also provide east–west correction with

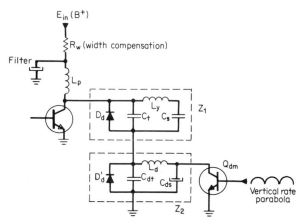

FIG. 13-176 Diode-modulator circuit. *(Courtesy of NAP Consumer Electronics Co.)*

several advantages. When the product of L_yC_t equals that of L_dC_{dt}, the currents and voltages in the top circuit Z_1 will be proportional to those in the bottom circuit Z_2. Impressing a vertical-rate parabolic waveform on transistor Q_{dm} will cause the voltage across C_{ds} to decrease while that on C_s will increase according to the equation $E_{in} = V_{Cs} + V_{Cds}$. Varying V_{Cs} controls the peak yoke current (Sec. 13.16.1)

$$I_{y(pk)} = \frac{V_{Cs}T_t}{2L_y} \qquad (13\text{-}160)$$

The result will be increased scan width at the E–W cardinal points of the picture and proportionally reduced width toward the corners.

One advantage of this circuit over the pin transformer is that width can be changed without affecting the retrace time; therefore the high-voltage pulse remains unchanged. Second, raster-width adjustment can be accomplished simply by changing the dc-bias level on transistor Q_{dm}.

13.16.6 RASTER-WIDTH REGULATION.   The equation which relates yoke energy to beam deflection[170]

$$W = \tfrac{1}{2}L_yI_{y(pk)}^2 = k_1E_aD\theta^{2.5} \qquad (13\text{-}161)$$

can be rearranged into the form

$$\theta = k_2\frac{I_{y(pk)}^{0.8}}{E_a^{0.4}} \approx k_2\frac{I_{y(pk)}}{\sqrt{E_a}} \qquad (13\text{-}162)$$

where θ = horizontal deflection angle

k_1 and k_2 = constants including yoke-to-beam sensitivity and dimensional conversion factors

E_a = CRT anode voltage, kV

This equation shows that the deflection angle (raster width) will change approximately as an inverse function of the square root of the high voltage. The impedance or regulation of the high-voltage power supply determines the relationship between CRT anode voltage and anode (or cathode) current

$$E_a \cong E_{a0} - \frac{I_a}{R_s}$$

where E_{a0} = value of anode voltage (kV) at zero beam current, $I_a = 0$

I_a = average anode or beam current, mA

R_s = supply impedance, MΩ (this quantity varies somewhat over range of beam current because of flyback transformer leakage inductance and tuning)

As average beam current increases, anode voltage will decrease, leading to a corresponding increase in width.

Numerous solutions exist for compensating for width change. The simplest consists of a small-value resistor in series with the supply voltage.[171] As CRT beam current increases, thereby drawing more current from the supply, the resistor will reduce the *effective* supply E_{in} which will decrease the yoke current $I_{y(pk)}$. The series resistor also limits the peak collector current in the event of a CRT gun arc.

A second solution utilizes the diode modulator described in Sec. 13.16.5 (Fig. 13-176). A current which is inversely proportional to the average value of beam current is fed to the drive transistor Q_{dm}. This modifies the voltage across the capacitors C_{ds} and C_s to produce the desired width compensation.

13.17 HIGH-VOLTAGE GENERATION

13.17.1 **GENERAL.** High voltage in the range 8 to 16 kV is required to supply the anode of monochrome picture tubes. Color tubes have anode requirements in the range 20 to 30 kV and focus voltage requirements of 3 to 12 kV, as shown in Table 13-33. The horizontal flyback transformer is the common element in the current methods of generating high voltage. The three variations of this design which are most popularly used are the flyback with half-wave rectifier, flyback driving a voltage multiplier, and the voltage-adding integrated flyback.

A simplified horizontal-scan circuit is shown in Fig. 13-177. The voltage at the top of the high-voltage (HV) winding consists of a series of pulses delivered during retrace from the stored energy in the yoke field (Sec. 13.16.2). The yoke voltage pulses are then multiplied by the turns ratio of the HV winding to primary winding. The peak voltage across the primary during retrace is given by

$$V_{p(pk)} = E_{in}0.8\left(1.79 + 1.57\,\frac{T_{trace}}{T_{retrace}}\right) \qquad (13\text{-}163)$$

where E_{in} = supply voltage (B$^+$)

0.8 = accounts for pulse shape factor with third harmonic tuning

T_{trace} = trace period, $\approx 52.0\ \mu s$

$T_{retrace}$ = retrace period, $\approx 11.5\ \mu s$

The amplitude of the pulses at the top of the HV winding, assuming ideal transformer action, is therefore

$$V_{t(pk)} = V_{p(pk)}\,\frac{N_t}{N_p} \qquad (13\text{-}164)$$

where N_t and N_p = the number of turns in the primary and tertiary windings.

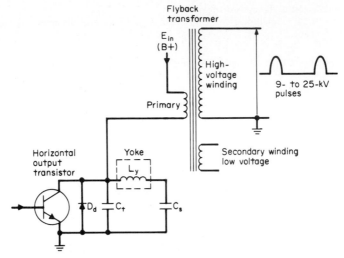

FIG. 13-177   Simplified horizontal-scan circuit.

13.17.2 FLYBACK WITH HALF-WAVE RECTIFIER.

The most common HV supply for small-screen monochrome and color television receivers uses the direct half-wave rectifier circuit shown in Fig. 13-178. The pulses at the top of the HV winding are rectified by the single diode, or composite of several diodes in series. The charge is then stored in the capacitance of the anode region of the picture tube. Large-voltage step up is required from the primary to the HV winding. This results in a large value of leakage inductance for the HV winding which decreases its efficiency as a step-up transformer.

Harmonic tuning of the HV winding improves the efficiency by making the total inductance and the distributed capacitance of the winding plus the CRT anode capacitance resonate at an odd harmonic of the flyback pulse frequency (Sec. 13.16.2). For the single-rectifier circuit, usually the third harmonic resonance is most easily implemented by proper choice of winding configuration which results in appropriate leakage inductance and distributed capacitance values. This will result in HV pulse waveforms, as shown in Fig. 13-179, which will give an improvement in the HV supply regulation (internal impedance)[172,173] as well as a reduction in the amplitude of ringing in the current and voltage waveforms at the start of scan.

13.17.3 VOLTAGE-MULTIPLIER CIRCUITS.

A voltage-multiplier circuit consists of a combination of diodes and capacitors connected in such a way that the dc output voltage is greater than the peak amplitude of the input pulse. The circuit diagram and equations for both the even and odd number-of-diodes multipliers[174] are shown in Fig. 13-180. For television receiver application the ratio of input-pulse voltages is established by the trace and retrace periods. For example

$$V_1 = \frac{52}{11.5} V_2 \tag{13-165}$$

and

$$V_1 + V_2 = V_1 \left(1 + \frac{11.5}{52}\right) = 1.22 \, V_1 \tag{13-166}$$

The *tripler* has been the most popular HV system for color television receivers requiring anode voltages of 25 to 30 kV in the 1970 period. Both even and odd configurations have been used by various manufacturers. The considerably reduced pulse voltage required from the HV winding has resulted in a flyback transformer with a more tightly

Table 13-33 Typical Anode and Focus Voltage Requirements for Color Picture Tubes

Screen size, in (cm)†	Representative type	Deflection angle, deg	Anode		Focus range, kV	Gun type	Neck diameter, mm (in)
			Voltage, kV	Current, mA			
9 (23)	9VACP22	90	20	0.6	2.7–4.5	LoBi	29 (1.14)
9	9VBAP22	90	20	0.6	5.2–6.0	HiBi	29
13 (33)	13VBFP22	90	25	1.2	4.7–5.5	LoBi	29
15 (38)	15VAEP22	90	25	1.2	4.2–5.0	LoBi	29
19 (48)	19VMFP22	90	27.5	1.5	7.3–8.2	HiBi	29
25 (64)	25VGEP22	100	30	1.8	12, 8‡	TPF	29
25	A66AAMOOX	100	30	1.8	8.0–8.9	CFF/HiBi	29

†Diagonal measurement.
‡Tripotential focus (TPF) requires two separate focus voltages.

coupled HV winding. Fifth-harmonic tuning which produces pulse waveforms as shown in Fig. 13-181 has been favored over third.[175] This combination has resulted in HV-supply regulation approaching 1.5 MΩ, approximately a 2/1 improvement compared with the third-harmonic half-wave rectifier system.[172]

13.17.4 INTEGRATED FLYBACKS.

Most current medium- and large-screen color receivers utilize an integrated flyback transformer in which the HV winding is segmented into three or four parallel-wound sections. These sections are series-connected with a diode between adjacent segments as shown in Fig. 13-182. These diodes are physically mounted as part of the HV section. The transformer is then encapsulated in high-voltage polyester or epoxy.[176]

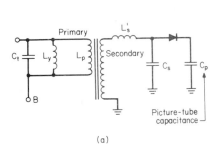

FIG. 13-178 Half-wave rectifier circuit, where C_A is the anode (dag) capacitance and i_b the CRT beam current.

Two HV-winding construction configurations are being used. One, the *layer* or *solenoid-wound* type, has very tight coupling to the primary and operates well with no deliberate harmonic tuning. Each winding (layer) must be designed to have balanced voltage and capacitance with respect to the primary. The second, *bobbin* or *segmented-winding* design, has high leakage inductance and usually requires tuning to an odd harmonic, e.g., the ninth.[177] Regulation of this construction is not quite as good as the solenoid-wound unit.

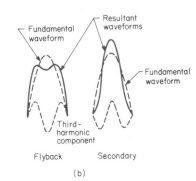

FIG. 13-179 Harmonic tuning: (a) simplified model of flyback transformer showing leakage inductance L_s and distributed capacitance C_s; (b) third-harmonic waveforms.

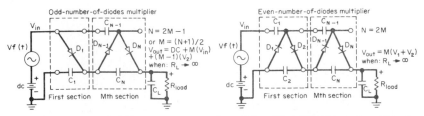

FIG. 13-180 Voltage-multiplier circuits where $V_{in} = DC + (V_1 + V_2)f(t)$ and where V_1 is the plus portion of $Vf(t)$ and V_2 the minus portion of $Vf(t)$. *(From M. Buechel, "High Voltage Multipliers in TV Receivers," IEEE Trans., vol. BTR-16, no. 1, p. 35. © 1970 IEEE.)*

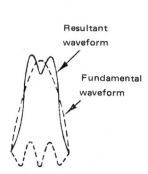

FIG. 13-181 Waveform on HV winding for fifth-harmonic tuning.

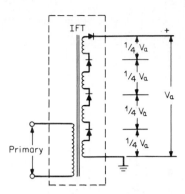

FIG. 13-182 Integrated-flyback transformer schematic diagram.

13.17.5 FOCUS VOLTAGE SUPPLIES. Table 13-33 gives the voltages required by the focus electrode of present-day picture tubes. These values are dependent upon electron gun construction and anode voltage as shown in Table 13-34. Typical arrangements for providing these voltages from the high-voltage systems discussed in the previous three sections are shown in Fig. 13-183.

Table 13-34 Focus Voltage Requirements for Various Color CRT Guns

Gun type	G_2, V	F_1/V_a, %	F_2/V_a, %
LoBi	200–600	16.8–22	
HiBi	200–600	26–35	
TPF†—a	200–800	22–26	40
TPF†—b	600–2000	22–26	40

†TPF = tripotential focus.

13.17.6 HIGH-VOLTAGE BLEEDER RESISTOR. Connected from the anode supply to ground, the HV bleeder can serve three purposes: (1) Provide focus voltage source, (2) improve HV regulation, and (3) provide discharge path for HV charge stored on the CRT anode *dag* coating. The resistance value of the bleeder is usually selected to provide the fixed load on the HV system which will flatten the regulation characteristic between 0 and 100 μA of CRT beam current (Fig. 13-184). Other parameters to be considered are the power dissipation capabilities of the bleeder, the focus voltage required, and the range and voltage-breakdown value for the adjustable resistor R_1.

13.17.7 HIGH-VOLTAGE PROTECTION CIRCUIT. Strict limits for the generation of x-rays are placed upon color television receivers by the federal government (Sec. 13.3.2). Excessive x-ray generation is caused by the product of beam current and anode voltage exceeding the maximum allowed levels for a given picture tube. An overvoltage-protection circuit is usually built into large-screen color receivers which have zero beam-current HV of 25 kV or more. Occasionally an overcurrent-protection circuit will be included to ensure that the limits cannot be exceeded by either parameter.

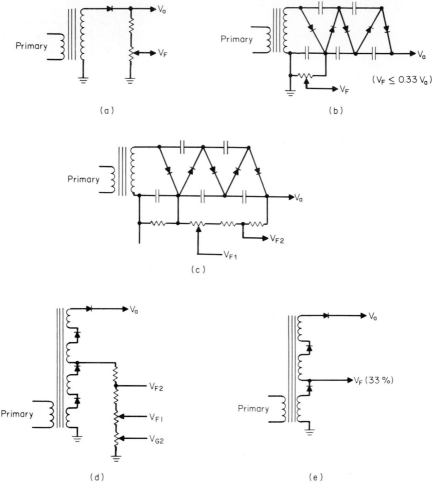

FIG. 13-183 CRT focus-voltage supplies. (*a*) Simple focus-bleeder circuit; (*b*) focus voltage derived from first section of multiplier; (*c*) bleeder network supplies both voltages for TPF CRT; (*d*) TPF voltages derived from IFT; (*e*) HiBi focus voltage derived from three-section IFT.

    These circuits must be relatively simple and have demonstrated high reliability in order to be acceptable. In most designs the overvoltage circuit monitors the voltage on one of the flyback low-voltage windings. If the HV increases, so too will the low voltage. At the overvoltage threshold value the protection circuit will activate and shut down the horizontal-driver stage or by some other technique will disable or lower the high voltage. In a similar manner, the overcurrent-detector stage monitors the current in the HV winding, approximately the CRT beam current, and also operates on the horizontal driver if the beam current is excessive. The time constants of these detectors are made to be of the order of 1 to 2 s to avoid accidental tripping when video content changes, channel is changed, or the set is turned on or off.

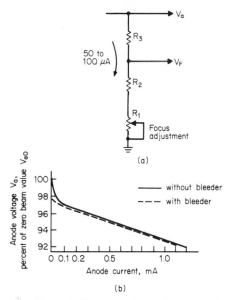

FIG. 13-184 Effect of bleeder resistor on HV regulation: (*a*) bleeder circuit; (*b*) high-voltage regulation characteristic.

13.18 POWER SUPPLIES

13.18.1 POWER LINE ISOLATION. Modern television receivers have one of the following types of power line connection configurations (Fig. 13-185)

1. *Cold chassis:* Chassis is completely isolated from the power line.
2. *Hot chassis:* Common chassis ground is connected to one side of the ac power line.
3. *Iso-hot chassis:* Hybrid of cold and hot wherein some of the chassis ground returns are common with one side of the ac power line and others are isolated from the power line.

In the *cold chassis,* the ac line connects to a power transformer primary. The secondary winding then supplies the ac waveforms needed by the various receiver power supplies. The separation between primary and secondary windings of the transformer serves as the sole isolation between the chassis circuitry and the power line.

Until recently, a type of power transformer known as voltage regulating or ferroresonant constant-voltage transformer was used as both isolation and regulating system.[178] This system was reliable and had excellent regulating characteristics, especially during periods of very low ac-line voltage or brownout (Sec. 13.7.1). The cost, physical size, weight, and dissipation, however, led to its being replaced by active-device, electronic designs.

Hot-chassis receivers have the ac-line cord polarization connected such that the common-signal and internal chassis grounds are connected to the low or ground side of the ac-power line. The hazard with the system occurs when the ac network is incorrectly wired or the ground is not properly connected. The receiver chassis "common" points then may be connected to the hot side of the ac line. Although the receiver itself will

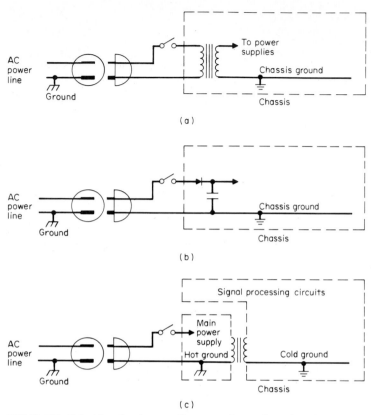

FIG. 13-185 Television-chassis power-line connection configurations: (*a*) cold; (*b*) hot; (*c*) iso-hot.

perform well under these conditions, externally grounded auxiliary equipment cannot be safely connected to the receiver. For this reason, all user-accessible parts of a hot-chassis receiver are electrically isolated from the chassis ground for operator safety. An isolation network, *capristor,* consisting of a high-resistance parallel RC network, is placed in series with each side of the 300-Ω antenna input terminals. Further, servicing of a hot-chassis receiver should be done only after the set has been connected to the ac line through an isolation transformer. The hot chassis has been popular, especially in monochrome and portable color receivers, because of its low cost and lighter weight, since no isolating transformer is required.

The *iso-hot chassis* has been developed over the past few years to provide a well-isolated transformerless system which meets the demand for connecting accessory units to the basic television receiver. These accessories include video games, video recorders, disc players, computers, and external speakers. Generally some form of switched-mode regulator is used for the main supply to the horizontal-scan system. This power supply, as well as the horizontal-output stage, is connected directly to the ac power-line ground in a manner similar to the hot chassis. The remainder of the circuitry, including tuner, IF, signal processing, and vertical, are all referenced to a separate cold chassis ground which is isolated from the main power supply and ac line. Power for these circuits is derived from a secondary winding on the horizontal flyback transformer (Sec. 13.18.4) or a secondary winding on the switched-mode supply inductor. Isolation between the hot and cold portions of the chassis is usually provided by several components, e.g., low-power transformers and opto-couplers.

Two special requirements are peculiar to the iso-hot chassis. First, all components involved in the isolation interface must meet the Underwriters Laboratory requirements for such isolation (Ref. 43, Sec. 23). This involves voltage breakdown limits as well as physical dimensions between the two sides of the circuit.

Second, a chassis, having scan circuit–derived power supplies, is not self-starting. A means, at least momentary, must be provided to start the horizontal oscillator. Frequently this start-up voltage is supplied by a small transformer operating from the ac line or from the switched-mode regulator. After the start-up has occurred, the transformer output can be open-circuited to reduce the receiver's total power consumption.

13.18.2 POWER INPUT RECTIFIER CIRCUITS.

Four types of ac to dc rectifying circuits are utilized in television receivers. These are shown in Fig. 13-186. The output voltage for capacitor-input filters is dependent upon the values of source and load resistance, and filter capacitor.[179] These relationships are shown graphically in Fig. 13-187 (a) through (c). Other important parameters in the design of silicon diode rectifier circuits are the following.

Peak inverse voltage (PIV): *The reverse voltage across the terminals of the rectifying device, including line transients.*

Surge current time (I^2t): *Given by*

$$I^2t = \left(\frac{\text{one-cycle surge current}}{2}\right)^2 (16.67 \times 10^{-3}) \qquad (13\text{-}167)$$

Power loss (watts): *Given by*

$$P = (V_{dc}I_{dc}) + (I^2_{rms}R_{dyn}) \qquad (13\text{-}168)$$

where $V_{dc} \simeq 0.4$ to 0.9 V
$\quad I_{dc} \simeq I_{avg}$
$\quad R_{dyn} -$ dynamic resistance of diode over current range

Half wave with capacitor input: *The type used with many small-screen, low-power receivers in the hot- or iso-hot-chassis configuration.*

Full wave with capacitor input: *Requires a center-tapped transformer and is therefore applicable to the cold-chassis configuration.*

Full-wave bridge with capacitor input: *Can be utilized in a cold, hot, or iso-hot configuration. This supply is most popular in large-screen color receivers.*

Direct-coupled doubler with capacitor input: *Frequently used in a hot or iso-hot chassis for small-screen table-model color receivers. The cascade configuration is preferred over the conventional configuration because it has a common input and output terminal (ground), although it requires C_2 to be rated at twice the voltage of the capacitors in the conventional circuit.*

13.18.3 REGULATORS.

Once the ac-input voltage is rectified and filtered, it must be regulated in order to stabilize the dc-voltage level from the normal variations of line and load. In general, the main horizontal dc supply of modern television receivers is well regulated. Many of the low-voltage supplies for other parts of the receiver are then derived from this supply, either directly or by transformer action (Sec. 13.18.4) and, as a result, are sufficiently regulated.

The *series-pass* regulator circuit, shown in Fig. 13-188, utilizes a power transistor operating in the linear mode. A voltage reference circuit senses both source and load variations and adjusts the current into the base to maintain a constant value of output voltage.

A practical application of this circuit operating as the regulator of the main dc-volt-

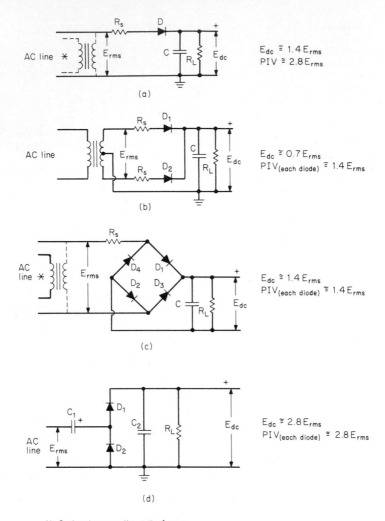

$E_{dc} \cong 1.4 E_{rms}$
$PIV \cong 2.8 E_{rms}$

(a)

$E_{dc} \cong 0.7 E_{rms}$
$PIV_{(each\ diode)} \cong 1.4 E_{rms}$

(b)

$E_{dc} \cong 1.4 E_{rms}$
$PIV_{(each\ diode)} \cong 1.4 E_{rms}$

(c)

$E_{dc} \cong 2.8 E_{rms}$
$PIV_{(each\ diode)} \cong 2.8 E_{rms}$

(d)

$*$ Optional power line transformer

FIG. 13-186 Power-rectifier circuits: (a) half wave; (b) full wave; (c) full-wave bridge; (d) cascade doubler.

age supply ($+120$ V) in a color television receiver is shown in Fig. 13-188b. A reference amplifier, Q3, reduces the effects of load change on the reference-voltage circuit consisting of R4 and zener diode D1. A potentiometer control (HV adjust) provides for adjustment of the supply voltage to set up the proper CRT anode voltage. Transistor Q2 supplies the necessary current to the base of series-pass transistor Q1. In the event that the supply output becomes short-circuited to ground, resistor R7 and diode D2 supply sufficient current to Q1 to keep it in a low power-saturated state. The dissipation then occurs in resistor R1.

Series-pass regulators operate on the principle of voltage dropping which dissipates

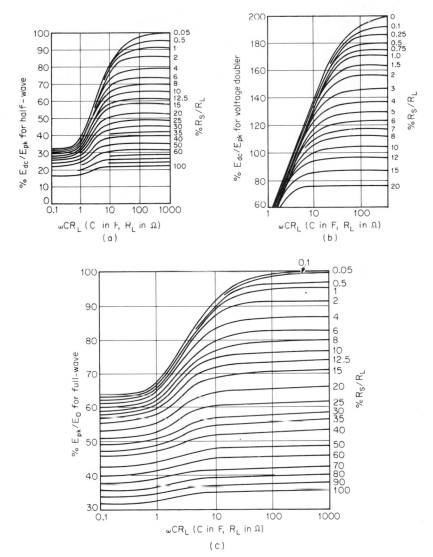

FIG. 13-187 Relationship of applied ac peak voltage to dc output voltage in capacitor-input circuits: (*a*) half wave; (*b*) doubler; (*c*) full wave.

power. This usually necessitates a large heat sink and power transistor. The general equation, derived from Fig. 13-188, is

$$P_d = (V_{\text{out}} - V_{\text{in}})I_{\text{out}} \frac{\beta_o}{1 + \beta_o} + \left(V_{\text{BE}} \frac{I_{\text{out}}}{1 + \beta_o}\right) \qquad (13\text{-}169)$$

where P_d = power dissipated by the series-pass transistor, V_{out}, V_{in}, and I_{out} are defined in Fig. 13-188, and β_o = the transistor dc current gain.

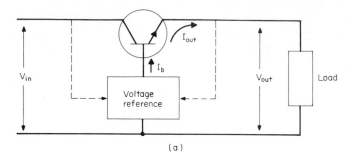

(a)

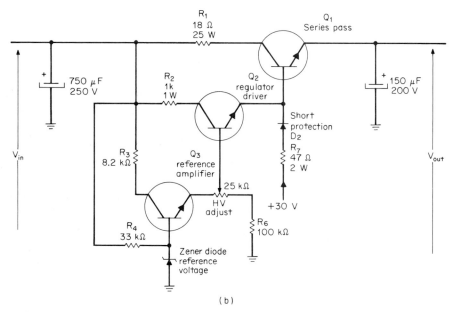

(b)

FIG. 13-188 Series-pass regulator: (*a*) basic circuit; (*b*) practical implementation in a color receiver. (*Courtesy of NAP Consumer Electronics Co.*)

The *phase-control regulator* operates on the principle of varying the conduction angle of the ac-rectifying element in the power supply to control the level of recovered dc voltage and current being supplied to the load. The usual rectifying element is a silicon-controlled rectifier (SCR) operating as a half-wave rectifier. The SCR turn-on point is referenced to the positive peak of the input ac sine wave and is controlled by a trigger-pulse generator and reference voltage or element (Fig. 13-189). The SCR ceases conduction on the negative-going zero crossing and will not fire again until the next ac input positive half-cycle and the coincidence of the trigger pulse at the SCR gate. The trigger pulse results when the charge voltage of an *RC* series network reaches the threshold voltage of a unijunction-type transistor, which then fires to create energy pulse at the SCR gate. Either the threshold level or the charge rate can be made proportional to the load voltage to provide regulation.[180,181] The phase-control regulator is locked to the 60-Hz power-line frequency.

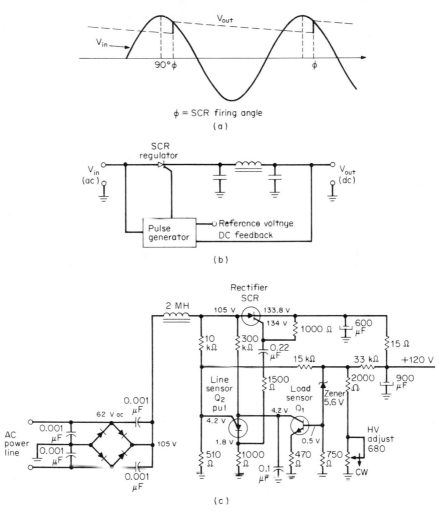

FIG. 13-189   SCR phase-controlled regulator: (*a*) input-output waveforms; (*b*) basic circuit; (*c*) practical circuit. *(Courtesy of NAP Consumer Electronics Co.)*

The *switched-mode regulator,* as the name implies, operates the control device as a switch, i.e., either fully on (saturated) or fully off, with the duty cycle (ratio of "on" time to "off" time) providing the regulation of the output-voltage level. The main advantage of this approach is that the control device has minimal power loss (dissipation) since it is in a low-dissipation mode for nearly the entire duty cycle. The typical control-switch devices used in a switched-mode circuit are bipolar transistors, field-effect transistors (FETs), or silicon-controlled rectifiers (SCRs).

Three primary configurations are shown in Fig. 13-190. These offer several additional advantages compared with a series-pass regulator.

1. Capability to produce an output voltage which is higher than the input voltage

2. Capability to produce either a positive or negative output voltage from a positive input voltage

3. Capability to produce an ac output voltage from a dc input

Energy is stored in inductor L while switch S1, the control device, is closed. When S1 is opened, diode D1 provides a path for the current I_L to continue to flow. The energy in the circuit is transferred to the load during the time when S1 is open. In practice, inductor L is often a transformer primary with the dc-isolated secondary providing an output voltage and current which can be used as alternating current, or rectified and filtered into direct current. A practical arrangement of this circuit, referred to as the *parallel* type of switched-mode regulator, is shown in Fig. 13-191.[182] The production of CRT anode voltage by the flyback transformer falls into this category (Sec. 13.17.2).

In applications of commercial equipment, switched-mode regulators typically operate in the frequency range of 20 to 100 kHz. Operating frequency is sometimes allowed to shift as required to meet system demands. When used in television receivers, the operating frequency is locked to the horizontal-line frequency of 15,734 kHz, and the on-off duty cycle is varied within that frequency. This resembles a pulse-width modulation technique. Horizontal-line frequency is convenient because it is readily available and any switching transient interference that may appear in the video display will be stationary on the raster. Such undesirable signal pickup may move left or right with a change in the regulator duty cycle, e.g., line voltage change, but will not drift through the picture.

Design information on switched-mode power supplies is given in Refs. 183 through 186.

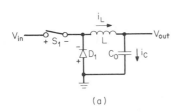

(a)

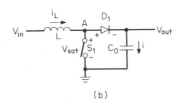

(b)

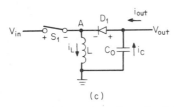

(c)

FIG. 13-190 Basic configurations of switched-mode regulator: (*a*) step-down; (*b*) step-up; (*c*) inverting.

13.18.4 AUXILIARY POWER SUPPLIES. In addition to the main horizontal scan power supply, television receivers require power sources for the other functions, among them the following:

1. Tuner and tuning system
2. Signal-processing circuitry
3. Video and chroma
4. Audio system
5. Vertical scan
6. CRT filament
7. CRT anode and focus voltages
8. CRT G$_2$ (screen) control

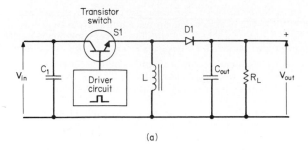

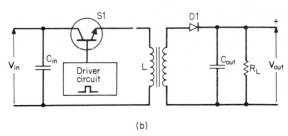

FIG. 13-191   Parellel switched-mode regulator: (a) conventional design; (b) adapted to provide power-line isolation.

Most present-day receivers utilize the flyback transformer, switched by the horizontal system, to provide the required auxiliary power supplies. Since the waveform and duty cycle of a winding on the flyback transformer are approximately as shown by Fig. 13-192, one can see that by proper winding direction and grounding either end of the winding, four possible types of dc power supplies can be created.

1. *Positive-retrace rectified* supply formed by rectifying $9V_1$ peaks from time A to B yields a relatively high voltage and low current-loading (due to low duty cycle of the source). Typically, this type is used to power the video and chroma output circuits.

2. *Positive-scan rectified* supply for lower-voltage, higher-current requirements such as vertical output, signal processing levels, and audio. This is formed by reversing the direction of the winding such that a mirror image of the waveform of Fig. 13-192 results and rectifying the V_1 portion from time B to C.

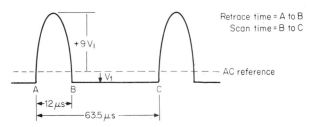

FIG. 13-192   Auxiliary supply-voltage waveform from flyback transformer.

3. The *negative-polarity-retrace rectified* supply uses the reversed winding as in type 2 above and rectifies the negative-going retrace pulses from time *A* to *B*. This yields a relatively high voltage negative level with low current-loading capacity such as some tuner systems might require for a negative reference level or bandswitching voltage.

4. The *negative-polarity-scan rectified* supply conducts on the *B* to *C* portion of the $-V_1$ level with winding polarity as shown in Fig. 13-192, to produce a negative supply of relatively low voltage, but capable of higher current drain.

"Scan rectified" supplies are operated at a duty cycle of approximately 80 percent and are thus better able to furnish higher current loads. Also, the diodes used in such supplies must be capable of blocking voltages that are nine to ten times larger than the level they are producing. Diodes having fast-recovery characteristics are used to keep the power dissipation at a minimum during the turn-off interval due to the presence of this high reverse voltage.

A typical receiver system containing the various auxiliary power supplies derived from flyback transformer windings is shown in Fig. 13-193. Transistor Q 452 switches the

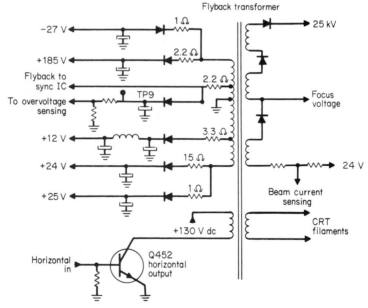

FIG. 13-193 Auxiliary supplies derived from horizontal-output transformer. *(Courtesy of NAP Consumer Electronics Co.)*

primary winding at a horizontal-frequency rate. The +12-V, + 24-V, +25-V, and −27-V supplies are *scan-rectified*. The +185-V, overvoltage sensing, focus voltage, and 25-kV anode voltage are derived by *retrace-rectified* supplies. The CRT filament is used directly in its ac mode.

Flyback-generated supplies provide a convenient means for isolation between different ground systems as needed for iso-hot chassis.

13.18.5 AC-DC SUPPLIES FOR PORTABLE RECEIVERS. Small-screen portable monochrome receivers operate from either the ac power line or a nominal 12-V battery supply. The main voltage in the chassis (+11 V) is supplied either directly from the battery or from a four-diode bridge rectifier and series-pass transistor regulator. The power transformer gives cold-chassis isolation from the power line.

13.18.6 TYPICAL COLOR TELEVISION POWER SUPPLY.   Figure 13-194 shows the block diagram of a television receiver power supply system. In this system, which oper-

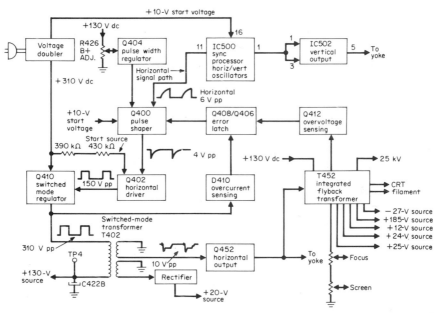

FIG. 13-194   Color receiver power supply using a switched-mode regulator system. *(Courtesy NAP Consumer Electronics Co.)*

ates at horizontal frequency, some cost savings have been achieved by combining functions that are normally separate networks. For example, the horizontal-driver transformer operation is accomplished by a winding on the switch mode transformer, T402. Likewise, another winding on this transformer provides +20-V dc power for use by the sound system.

The flyback transformer, T 452, itself part of the horizontal output switched-mode subsystem, provides several power levels required by means of both scan and retrace-derived supplies. The CRT filament is simply driven by an ac waveform from a winding on T 452 which has an RMS value of approximately 6.3 V. Another type of television switched-mode regulator system features an SCR as the switching device. An SCR is reset each horizontal cycle via a pulse from a winding on the flyback transformer. This pulse forces the SCR anode to fall well below the cathode level during horizontal retrace

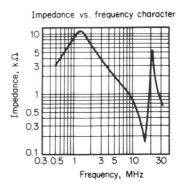

FIG. 13-195   Impedance characteristic of typical RFI line-terminal choke wound on a ferrite core.

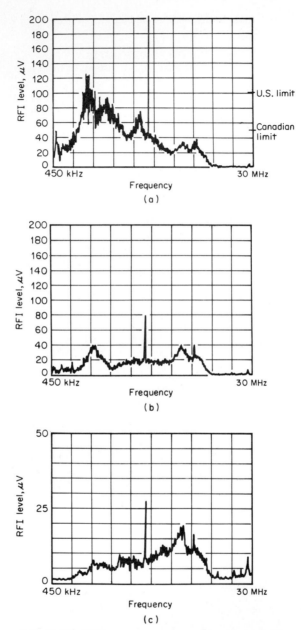

FIG. 13-196 RFI-spectrum signature for color television receiver: (*a*) no line filter or other major RFI components installed; (*b*) wire jumper bridging two ground-foil areas on printed-circuit board; (*c*) all RFI-reduction components installed.

time, thus unlatching it. The SCR turn on occurs during the next horizontal scan period. The turn-on point will shift laterally in time with changes in the raw dc level due to either line variations or load requirement changes.

13.18.7 LINE-TERMINAL RADIATION AND CURES.

Any switching circuit, high-frequency oscillator, or large-signal amplifier in a television receiver may generate radio-frequency interference (RFI) which can appear on the power line and interfere with the proper operation of other electronic appliances connected to that line. Consequently, federal regulations and measurement standards have been established for line terminal radiation (Sec. 13.3).

The most common scheme for removing the RFI energy from the power line involves use of a *common-mode* dual-winding series line RFI choke and an *across-the-line* capacitor. The capacitor may be on either side of the choke and is typically 0.1 to 0.3 μF. The choke is typically wound as a gapless-transformer configuration, with 1 mH or less per leg, connected as shown to avoid being saturated by the input current. The unbiased choke impedance may measure several kilohms at the frequency of interest (Fig. 13-195). Larger inductance values which are more useful at lower frequencies, such as the AM broadcast band, have less impedance at higher frequencies owing to interwinding capacitance.

The most common sources of RFI are the horizontal output and driver stages, with the HV transformer and driver transformer acting as radiating antennas. The regulator inductor and driver transformer of switched-mode power-supply output and driver stages, if used, also are likely sources of RFI. These stages are usually bypassed with a damping network and/or a ceramic capacitor mounted as close as possible to the switching transistor to help alleviate the problem.[187,188]

Other sources of RFI are the video and chroma or *RGB* output amplifiers, typically radiating the 2-MHz and 3.58-MHz signals and harmonics of 3.58 MHz, especially via the lead wires to the CRT socket. The amount of 3.58 MHz generated in the chroma circuit must be minimized, and traps for the harmonics of 3.58 MHz may be added to these output amplifiers. Mounting the output amplifiers on a printed-circuit board directly attached to the CRT socket decreases the problem.

Other common sources of line-conducted RFI include the power-supply rectifiers, vertical output (especially the *retrace-switch* type), the synchronization and oscillator circuits, any digital circuitry, any lead wire going to the CRT, and long leads from scan-derived power supplies. Solutions consist of placing ceramic capacitors or ferrite leads on the wires near the RFI source.

Sources may be discovered by observing a tuned voltmeter or spectrum analyzer and moving a loop of the line cord near the suspected source. The ground-return patterns for the circuits mentioned above (especially higher-current circuits) are particularly important in avoiding RFI in the power line. As a final precaution, the chassis layout should geometrically separate the power-input area from the major sources of RFI as much as possible.[189]

Typical Receiver Line-conducted Radiation. In Fig. 13-196 the RFI spectrum signature of a typical 19-in color television receiver is shown. In Fig. 13-196a the receiver has no line filter or other major RFI-reduction components. In Fig. 13-196b a wire jumper has been added between two ground foil areas on the printed-circuit board. Addition of the line choke yields the signature shown in Fig. 13-196c.

REFERENCES

1. C. Cohen, "Sony's Pocket TV Slims Down CRT Technology," *Electronics,* vol. 55, no. 3, Feb. 10, 1982, p. 81.
2. FCC Regulations, 47 CFR, 15.65, Washington.
3. E. G. Breeze, J. Rothstein, P. Alfke, and H. Hamaui, "A Frequency Synthesizer for

Television Receivers," *IEEE Trans. Broadcast Telev. Receivers,* vol. BTR-20, no. 4, Nov. 1974, p. 259.

4. A. N. Borg, "Synthesized Tuning Comes to Black and White TV," *IEEE Trans. Consum. Electron.,* vol. CE-28, no. 3, August 1982, p. 373.

5. Service Manual for VTX-1000R Component TV Tuner, Sony Corporation, Tokyo, Japan, 1981.

6. L. W. Ocker, J. R. Simanton, and C. R. Wegner, "TV Multichannel Sound—Savior or a New Sacred Cow," *IEEE Trans. Consum. Electron.,* vol. CE-27, no. 3, August 1981, p. 361.

7. P. Fockens and C. G. Eilers, "Intercarrier Buzz Phenomena Analysis and Cures," *IEEE Trans Consum. Electron.,* Vol. CE-27, no. 3, August 1981, p. 381.

8. J. Yamada and M. Uematsu, "New Color TV Receiver with Composite SAW IF Filter Separating the Sound and Picture Signals," *IEEE Trans. Consum. Electron.,* vol. CE-28, no. 3, August 1981, p. 192.

9. N. Kokado et al., "A Pocketable Liquid—Crystal Television Receiver," *IEEE Trans. Consum. Electron.,* vol. CE-27, no. 3, August 1981, p. 462.

10. D. G. Fink (ed.), *Television Engineering Handbook,* McGraw-Hill, New York, 1957, p. 15–31.

11. A. H. Morrell, H. B. Law, E. G. Ramberg, and E. W. Herold, *Color Television Picture Tubes,* Academic, New York, 1974, p. 98.

12. J. L. Rennick, "An IC Approach to the Subcarrier Regeneration Problem," *IEEE Trans. Broadcast Telev. Receivers,* vol. BTR-15, no. 2, July 1969, p. 226.

13. L. A. Harwood, "A Chrominance Demodulator IC with Dynamic Flesh Correction," *IEEE Trans. Consum. Electron.,* vol. CE-22, no. 1, February 1976, p. 112.

14. *Recommended Cable Television Channel Identification Plan, Interim Standard No. 6 (CP),* Electronic Industries Association, Washington, May 1983.

15. *General Radio Regulations,* Pt. II, Secs. 136 and 137, Department of Communications, Ottawa, Canada, October 1978.

16. *ITU Radio Regulations,* "Final Acts, World Administrative Radio Conference (WARC-79)," Geneva, 1979, Appendix 30.

17. A. B. Carlson, *Communication Systems,* McGraw-Hill, New York, 1975.

18. H. W. Ott, *Noise Reduction Techniques in Electronic Systems,* Wiley-Interscience, New York, 1976, p. 224.

19. E. G. Nielsen, "Behavior of Noise Figure in Junction Transistors," *Proc. IRE,* vol. 45, July 1957, pp. 957–963.

20. R. S. Cobbold, *Theory and Applications of Field-Effect Transistors,* Wiley-Interscience, New York 1970, p. 352.

21. S. Weaver, "TV Design Considerations Using High-Gain Dual-Gate MOSFETS," *IEEE Trans. Broadcast Telev. Receivers,* vol. BTR-19, no. 2., May 1973, p. 91.

22. "Accuracy and Tolerance in Measurements of Television Receiver Noise Figures," *Final Report of Subcommittee B of the Advisory Committee,* submitted to FCC, December 1979, p. 13.

23. IRE Standards on Television 60-IRE-17.S1, *Proc. IRE,* vol. 48, June 1960 (IRE 190, p. 1138).

24. Ibid., pp. 1124 and 1137.

25. "Engineering Aspects of Television Allocations," Television Allocations Study Organization (TASO), report to the FCC, March 1959.

26. H. Fine, "A Further Analysis of TASO Panel 6 Data on Signal to Interference Ratios and Their Application to Description of Television Service," FCC/OCE Report T.R.R., no. 5.1.2, April 1, 1960.

27. P. B. Gieseler et al., "Comparability for UHF Television: Final Report," Staff Report of Office of Plans and Policy, FCC, September 1980, pp. 243–254.

28. G. S. Kalagian, "A Review of the Technical Planning Factors for VHF Television Service," FCC/OCE Report RS-77-01, Washington, March 1, 1977.

29. R. A. O'Connor, "Understanding Television's Grade A and Grade B Service Contours," *IEEE Trans. Broadcasting,* vol. BC-14, no. 4, December 1968.

30. G. F. Rogers, "Methods of Measurement of Impulse Noise Susceptibility of TV Receivers," *IEEE Trans. Broadcast Telev. Receivers,* vol. BTR-18, no. 3, August 1972, p. 143.

31. W. L. Hand, "Television Receiver Requirements for CATV Systems," *IEEE Trans. Broadcast Telev. Receivers,* vol. BTR-18, no. 3, August 1972, p. 183.

32. H. G. Schwarz, "Selectivity Requirements to Eliminate Adjacent Channel Interference on Multi-Channel CATV Systems," *IEEE Trans. Consum. Electron.,* vol. CE-22, no. 4, November 1976, p. 339.

33. W. L. Hand, "Personal Use Radio (CB) and Its Effect on TV Reception," *IEEE Trans. Consum. Electron.,* vol. CE-23, no. 1, February 1977, p. 78.

34. R. H. Beeman, "Citizens Band Expansion and the Consumer Environment," *IEEE Trans. Consum. Electron.,* vol. CE-24, no. 1, February 1978, p. 16.

35. R. M. Smith, "The Extent and Nature of TV Reception Difficulties Associated with CB Radio Transmission," *IEEE Trans. Consum. Electron.,* vol. CE-24, no. 1, February 1978, p. 6.

36. *Susceptibility of TV Tuners to Harmonically Generated Interference, pt. I, Performance Guideline—Rejection of CB Interference by TV Receivers,* Consumer Products Engineering Bulletin, no. 8-A, EIA, Washington, February 1980.

37. L. H. Hoke, "Characteristics of Television Receivers on Ch. 6," NAB Engineering Conference, April 1983.

38. *Recommended Practice on Procedures for Control of System Electromagnetic Compatibility,* ANSI C63.12, American National Standards Committee for C63, IEEE, New York, 1977.

39. W. P. Kruse, "A TEM RF Test Cell for Consumer Products Emission and Susceptibility Measurements," *IEEE Trans. Consum. Electron.,* vol. CE-25, no. 2, May 1979, p. 198.

40. F. D. Martzloff and G. J. Hahn, "Surge Voltages in Residential and Industrial Power Circuits," *IEEE Trans. Power Apparatus Systems,* vol. PAS-89, no. 6, July/August 1970, p. 1052.

41. D. C. Kay, *Transient Voltage Suppression Manual,* General Electric Co., Semiconductor Products Dept., Syracuse, N.Y., 1976.

42. *IEEE Guide for Surge Withstand Capability (SWC) Tests,* ANSI C37.90a-1974/IEEE Std. 472-1974, IEEE, New York, 1974.

43. *Television Receivers and Video Products UL 1410,* Section 71, Underwriters Laboratories, Inc., New York, 1981.

44. G. Lunn and M. McGinn, "'Monomax'—An Approach to the One-Chip TV," *IEEE Trans. Consum. Electron.,* vol. CE-27, no. 3, August 1981, p. 284.

45. M. Yoshitomi et al., "The Two-Chip Integration for Color Television Receivers," *IEEE Trans. Consum. Electron.,* vol. CE-28, no. 3, August 1982, p. 184.

46. H. Jasik, "Properties of Antennas and Fundamentals of Antennas," *Antenna Engineering Handbook,* McGraw-Hill, New York, 1961, chaps. 1 and 2.

47. R. S. Elliot, *Antenna Theory and Design,* Prentice-Hall, Englewood Cliffs, N.J., 1981, p. 64.

48. E. C. Jordan, *Electromagnetic Waves and Radiating Systems,* Prentice-Hall, Englewood Cliffs, N.J., 1950, pp. 457–461.

49. Y. T. Lo, "TV Receiving Antennas," in H. Jasik (ed.), *Antenna Engineering Handbook,* McGraw-Hill, New York, 1961, pp. 24–25.

50. Jordan, op. cit., pp. 534–537.

51. Lo, op. cit., p. 24–26.

52. Lo, op. cit., p. 24–4.

53. J. D. Kraus, *Antennas,* McGraw-Hill, New York, 1950, pp. 328–336.

54. M. C. Bailey and W. F. Croswell, "Antennas," in D. G. Fink and D. Christiansen (eds.), *Electronics Engineers' Handbook,* McGraw-Hill, New York, 1982, p. 18-20.

55. Lo, op. cit., pp. 24-19–24-23.

56. R. L. Carrel, "The Design of Logarithmically-Periodic Dipole Antennas," *IRE Natl. Conv. Record,* 1961, pt. 1, pp. 61–75.

57. W. R. Free, J. A. Woody, and J. K. Daher, *Program to Improve UHF Television Reception,* prepared for UHF Comparability Task Force, Office of Plans and Policy, Contract FCC-0315, Georgia Institute of Technology, Atlanta, September 1980.

58. N. Grossner, *Transformers for Electronic Circuits,* 2d ed., McGraw-Hill, New York, 1983, pp. 344-358.

59. J. Gibson and R. Wilson, "The Mini-State—A Small Television Antenna," *IEEE Trans. Consum. Electron.,* vol. CE-22, no. 2., May 1976, p. 159.

60. H. B. Goldberg, "Predict Intermodulation Distortion," *Electronic Design,* vol. 10, May 10, 1970, pp. 76–78.

61. L. Middlekamp, "UHF Taboos—History and Development," *IEEE Trans. Consum. Electron.,* vol. CE-24, no. 4, November 1978, p. 514.

62. *Reference Data for Radio Engineers,* 6th ed., ITT/Howard W. Sams & Co., Indianapolis, 1977.

63. K. Torii et al., "Varactor Tuned UHF Tuner Using Microstrip Lines," *IEEE Trans. Consum. Electron.,* vol. CE-23, no. 1, February 1977, p. 44.

64. M. Saito et al., "UHF TV Tuner Using PC Board with Suspended Striplines," *IEEE Trans. Consum. Electron.,* vol. CE-24, no. 4, November 1978, p. 553.

65. G. Valley and H. Wallman, *Vacuum Tube Amplifiers,* McGraw-Hill, New York, 1948, p. 203.

66. S. Cohn, "Dissipation Loss in Multiple-Coupled-Resonant Filters," *Proc. IRE,* vol. 47, August 1959, p. 1342.

67. Perlow, S., "Noise Performance Factors in Television Tuners," *RCA Rev.,* vol. 37, no. 1, March 1976, p. 119.

68. S. Perlow and B. Bossard, "Feed-Forward Filter Technique," *Southwest IEEE Conf. Record,* April 1966.

69. S. Hilliker, "Avoiding Pitfalls in Varactor Circuit Designs," *IEEE Trans. Consum. Electron.,* vol. CE-22, no. 3, August 1976, p. 195.

70. T. Hara et al., "Resistance of GaAs Diodes for TV Tuners," *IEEE Trans. Consum. Electron.,* vol. CE-26, no. 4, November 1980, p. 729.

71. J. Hopkins, "Design Considerations of Printed Circuit VHF TV Tuners," *IEEE Trans. Broadcast Telev. Receivers,* vol. BTR-18, no. 1, February 1972, p. 18.

72. R. Landee, D. Davis, and A. Albrecht, *Electronic Designers' Handbook,* McGraw-Hill, New York, 1957, p. 7–85.

73. S. Weaver, "TV Design Considerations Using High-Gain Dual Gate MOSFETS," *IEEE Trans. Broadcast Telev. Receivers,* vol. BTR-19, no. 2, May 1973, p. 87.

74. H. Gasquet, "The Channel 6 Color Beat—What Can Be Done About It," *IEEE Trans. Broadcast Telev. Receivers,* vol. BTR-18, no. 1, February 1972, p. 11.

75. G. Carter, "New VHF and UHF Varactor Tuners," *IEEE Trans. Consum. Electron.*, vol. CE-22, no. 3, August 1976, p. 205.

76. R. Taeuber and J. Fenk, "VHF Tuner IC for Use in Television Receivers and CATV Receivers," *IEEE Trans. Consum. Electron.*, vol. CE-28, no. 4, November 1982, p. 508.

77. K. Torii et al., "Monolithic Integrated VHF TV Tuner," *IEEE Trans. Consum. Electron.*, vol. CE-26, no. 2, May 1980, p. 180.

78. D. Ash, "High Performance TV Receiver," *IEEE Trans. Consum. Electron.*, vol. CE-24, no. 1, February 1978, p. 39.

79. A. Evans (ed.), *Designing with Field-Effect Transistors* (Siliconix, Inc.), McGraw-Hill, New York, 1981.

80. K. Torii et al., "Monolithic Integrated VHF TV Tuner," *IEEE Trans. Consum. Electron.*, vol. CE-26, no. 2, May 1980, p. 180.

81. D. Heuer, "Microprocessor Controlled Memory Tuning System," *IEEE Trans. Consum. Electron.*, vol. CE-25, no. 4, August 1979, p. 677.

82. F. Sutaria and D. Vazirani, "Modern Techniques of FLL Tuning," *IEEE Trans. Consum. Electron.*, vol. CE-28, no. 4, November 1982, pp. 536–540.

83. K-H. Seidler, "Integrating Analog Control and Digital Tuning for TV on One Chip," *IEEE Trans. Consum. Electron.*, vol. CE-28, no. 4, November 1982, p. 541.

84. M. Ghausi, *Principles and Design of Linear Active Circuits*, McGraw-Hill, New York, 1965, p. 458.

85. A. Stern, "Stability and Power Gain of Tuned Transistor Amplifiers," *Proc. IRE*, vol. 45, March 1956, pp. 335–337.

86. J. Linvill and J. Gibbons, *Transistors and Active Circuits*, McGraw-Hill, New York, 1961, chaps. 11 and 18.

87. W. Lancaster, "Analysis of Transistor Tuned Amplifiers Using Y Parameters," Advanced Development Report 63-067, Philco Corporation, Philadelphia, 1963.

88. K. Redmond, "Rapid Design Procedures for Double Tuned IF Amplifiers," *IEEE Trans. Broadcast Telev. Receivers*, vol. BTR-9, no. 3, November 1963, pp. 52–82.

89. S. Weaver, "Dual-Gate MOSFETS in TV IF Amplifiers," *IEEE Trans. Broadcast Telev. Receivers*, vol. BTR-16, no. 2, May 1970, pp. 96–111.

90. C. B. Neal and S. Goyal, "Frequency and Amplitude-Dependent Phase Effects in Television Broadcast Systems," *IEEE Trans. Consum. Electron.*, vol. CE-23, no. 3, August 1977, pp. 234–247.

91. I. Mertes, "Application of Bandpass Filters to Color TV IF Amplifiers," *IEEE Trans. Broadcast Telev. Receivers*, vol. BTR-18, no. 3, August 1972, p. 151.

92. A. DeVries and R. Adler, "Case History of a Surface-Wave TV IF Filter for Color Television Receivers," *Proc. IEEE*, vol. 64, no. 5, May 1976, p. 671.

93. A. Slobodnik, "Surface Acoustic Waves and SAW Materials," *Proc. IEEE*, vol. 64, no. 5, May 1976, p. 581.

94. C. Hartmann, D. Bell, and R. Rosenfeld, "Impulse Model Design of Acoustic Surface-Wave Filters," *IEEE Trans. Microwave Theory Technology*, vol. MTT-21, April 1973, pp. 162–175.

95. R. Tancrell and M. Holland, "Acoustic Surface-Wave Filters," *Proc. IEEE*, vol. 59, no. 3, March 1971, p. 393.

96. D. Parker, R. Pratt, F. Smith, and R. Stevens, "Acoustic Surface-Wave Bandpass Filters," *Philips Technical Rev.*, vol. 36, no. 2, February 1976, p. 29.

97. J. Rypkema, A. DeVries, and F. Banach, "Engineering Aspects of the Application of Surface-Wave Filters in Television," *IEEE Trans. Consum. Electron.* vol. CE-21, no. 2, May 1975, p. 105.

98. I. Nishimura et al., "A Delay-Line-Less Video Signal Processing Circuit Using a New Surface Acoustic Wave Filter for Television Receivers," *IEEE Trans. Consum. Electron.* vol. CE 26, no. 3, August 1980, p. 376.

99. K. Shibayama et al., "Optimum Cut for Rotated Y-Cut LiNbO$_3$ Crystal Used as the Substrate of Surface-Acoustic Wave Filters," *Proc. IEEE*, vol. 64, no. 5, May 1976, p. 595.

100. S. Takahashi et al., "SAW IF Filter on LiTaO$_3$ for Color TV Receiver," *IEEE Trans. Consum. Electron.*, vol. CE-24, no. 3, August 1978, p. 337.

101. D. Parker, R. Pratt, and R. Stevens, "A Television IF Acoustic Surface-Wave Filter on Bismuth Silicon Oxide," *Proc. IEEE*, vol. 64, no. 5, May 1976, p. 677.

102. "Single-Transistor Preamplifier for Surface-Wave IF Filters," *Electronic Components and Applications*, N. V. Philips/Mullard, Ltd., vol. 3, no. 3, May 1981, p. 135.

103. H. Nobeyama et al., "New Color Television Receiver with SAW IF Filter and One-Chip PIF IC," *IEEE Trans. Consum. Electron.*, vol. CE-25, no. 1, February 1979, p. 54.

104. M. Long, M. Wilcox, and R. Baker, "An Integrated Video IF Amplifer and PLL Detection System," *IEEE Trans. Consum. Electron.*, vol. CE-28, no. 3, August 1982, p. 168.

105. L. Bluestein, "Envelope Detection of Vestigial Sideband Signals with Application to Television Reception," *IEEE Trans. Consum. Electron.*, vol. CE-21, no. 4, November 1975, p. 369.

106. G. Lunn, "A Monolithic Wideband Synchronous Video Detector for Color TV," *IEEE Trans. Broadcast Telev. Receivers*, vol. BTR-15, no. 2, July 1969, p. 159.

107. T. Gluyas, "Television Demodulator Standards," *IEEE Trans. Consum. Electron.*, vol. CE-23, no. 3, August 1977, p. 222.

108. T. Mills and H. Suzuki, "Design Concepts for Low-Cost Transistor AGC Systems," *IEEE Trans. Broadcast Telev. Receivers*, vol. BTR-14, no. 3, October 1968, p. 38.

109. H. Nabe-yama and G. Miyazaki, "An AGC System Design Based upon the DC Restoration and the Dynamic Characteristics," *IEEE Trans. Broadcast Telev. Receivers*, vol. BTR-16, no. 4, p. 329.

110. *Ceramic Filter Application Manual*, Cat. No. P11E, Murata Mfg. Co., Kyoto, Japan, 1982.

111. A. Bilotti "Applications of a Monolithic Analog Multiplier," *IEEE J. Solid-State Circuits*, vol. SC-3, no. 4, December 1968, p. 373.

112. A. Bilotti and R. Pepper, "A Monolithic Limiter and Balanced Discriminator for FM and TV Receivers," *Proc. National Electronics Conf.*, vol. 23, October 1967, p. 489.

113. R. Peterson "High-Performance Integrated Circuits for High-Gain FM-IF Systems," *IEEE Trans. Broadcast Telev. Receivers*, vol. BTR-16, no. 4, November 1970, p. 257.

114. P. Menniti and B. Murari, "An Integrated-Circuit Sound for Television Receivers," *IEEE Trans. Consum. Electron.*, vol. CE-21, no. 1, February 1975, p. 74.

115. A. Grebene and H. Camenzind, "Frequency-Selective Integrated Circuits Using Phase-Lock Techniques," *IEEE J. Solid-State Circuits*, vol. SC-4, no. 4, August 1969, p. 216.

116. A. Grebene, "An Integrated Frequency-Selective AM/FM Demodulator," *IEEE Trans. Broadcast Telev. Receivers*, vol. BTR-17, no. 2, p. 71.

117. B. Cooke, "A 2.5 Watt Monolithic TV Sound System with Refined DC Volume Control," *IEEE Trans. Broadcast Telev. Receivers*, vol. BTR-20, no. 3, August 1974, p. 206.

118. M. Joyce and K. Clark, *Transist Circuit Analysis,* Addison-Wesley, Reading, Mass., 1961, p. 255.

119. C. Engel, "Considerations in the Design of Transistor Luminance Output Amplifiers," *IEEE Trans. Broadcast Telev. Receivers,* vol. BTR-14, no. 2, p. 34.

120. D. Bray, "Semiconductor Video Amplifiers for Monochrome and Color Television Receivers," *IEEE Trans. Broadcast Telev. Receivers,* vol. BTR-12, no. 4, p. 20.

121. *Solid-State Television Video Amplifiers,* Applications Notes 165 and 171, Motorola Semiconductor Products, Inc., Phoenix, Ariz., September, 1967.

122. J. Bingham, M. Norman, R. Shanley, and B. Yorkanis, "A New Low-Level Luminance Processing System," *IEEE Trans. Consum. Electron.,* vol. CE-22, no. 2, May 1976, pp. 135–148.

123. K. Skinner and W. Cocke, "A Versatile Low-Level Luminance IC for TV," *IEEE Trans. Consum. Electron.,* vol. CE-24, August 1978, pp. 169–175.

124. K. Mohri, M. Shimano, and T. Kuroyanagi, "Chroma Systems Trend in the Past and Future and Latest Chroma IC with Versatile Flexibility," *IEEE Trans. Consum. Electron.,* vol. CE-24, no. 1, February 1978, pp. 81–88.

125. N. Doyle, "A Comparison of Solid State Subcarrier Oscillators for Color TV Receivers," *IEEE Trans. Broadcast Telev. Receivers,* vol. BTR-16, no. 1, pp. 37–42.

126. D. Richman, "Color Carrier Reference Phase Accuracy in NTSC Color Television," *Proc. IRE,* vol. 42, no. 1, January 1954, p. 106.

127. M. Shimano et al., "One-Chip Chroma IC for Color TV Receiver," *IEEE Trans. Consum. Electron.,* vol. CE-21, no. 2, May 1975, pp. 164–170.

128. L. Blaser, D. Bray, and J. Rennick, "A Monolithic Chroma Demodulator Integrated Circuit for Color Television Receivers," *IEEE Trans. Consum. Electron.,* vol. CE-28, no. 1, February 1982, p. 84; also *IEEE Trans. Broadcast Telev. Receivers,* vol. BTR-14, no. 3, 1968.

129. F. Hilbert, G. Cecchin, and J. Feit, "Stereo and Color Signal Demodulation with Integrated Circuit Techniques," *IEEE Trans. Broadcast Telev. Receivers,* vol. BTR-14, no. 3, October 1968, pp. 58–73.

130. W. Bretl, "Colorimetric Performance of Non-Standard Television Receiver with Customer Adjustment," *IEEE Trans. Consum. Electron.,* vol. CE-25, no. 1, February 1979, pp. 100–110.

131. C. B. Neal, "Computing Colorimetric Errors of a Color Television Display System," *IEEE Trans. Consum. Electron.,* vol. CE-21, no. 1, February 1975, pp. 63–73.

132. C. B. Neal, "Television Colorimetry for Receiver Engineers," *IEEE Trans. Broadcast Telev. Receivers,* vol. BTR-19, no. 3, August 1973, pp. 149–162.

133. E. Surowiec, "Analysis of a Large-Screen RGB System," *IEEE Trans. Broadcast Telev. Receivers,* vol. BTR-18, no. 4, November 1972, pp. 222–233.

134. R. Thielking, "A Cost Optimized, Heatsinkless RGB Amplifier," *IEEE Trans. Broadcast Telev. Receivers,* vol. BTR-18, no. 2, May 1972, pp. 104–108.

135. D. Poppy, "A Semiconductor Video Output Amplifier for a Red-Blue-Green Large Screen Television Receiver," *IEEE Trans. Broadcast Telev. Receivers,* vol. BTR-15, no. 2, July 1969, pp. 167–170.

136. G. Cecchin, "One-Knob Picture Control," *IEEE Trans. Broadcast Telev. Receivers,* vol. BTR-20, no. 3, May 1976, p. 179.

137. S. Barton, "A Practical Charge-Coupled Device Filter for the Separation of Luminance and Chrominance Signals in a Television Receiver," *IEEE Trans Consum. Electron.,* vol. CE-23, no. 3, August 1977, pp. 342–357.

138. R. Turner, "Some Thoughts on Using Comb Filters in the Broadcast Television Transmitter and at the Receiver," *IEEE Trans. Consum. Electron.,* vol. CE-23, no. 3, August 1977, pp. 248–257.

139. W. Lagoni, D. Pritchard, and J. Fuhrer, "A Baseband Comb Filter for Consumer Television Receivers," *IEEE Trans. Consum. Electron.*, vol. CE-26, no. 1, February 1980, pp. 94–99.

140. G. Haenen and H. Simons, "Scan-Velocity Modulation Increases TV Picture Sharpness," *Electronic Components and Applications*, vol. 4, no. 1, November 1981, N. V. Philips, Eindhoven, The Netherlands, pp. 38–41.

141. S. Yoshida, A. Ohkoshi, and K. Shinkai, "Achievement of High Picture Quality in Color CRTs with the Beam-Scan Velocity Modulation Method," *IEEE Trans. Consum. Electron.*, vol. CE-23, no. 3, August 1977, pp. 366–374.

142. M. Yoshitomi et al., "The Two-Chip Integration for Color Television Receivers," *IEEE Trans. Consum. Electron.*, vol. CE-28, no. 3, August 1982, pp. 184–191.

143. W. Harlan, "An Automatic Sharpness Control," *IEEE Trans. Consum. Electron.*, vol. CE-27, no. 3, August 1981, pp. 311–314.

144. T. Rzeszewski, "A Novel Automatic Hue Control System," *IEEE Trans. Consum. Electron.*, vol. CE-21, no. 2, May 1975, pp. 155–163.

145. L. Harwood, "A Chrominance Demodulator IC with Dynamic Flesh Correction," *IEEE Trans. Consum. Electron.*, vol. CE-22, no. 1, February 1976, pp. 112–118.

146. C. B. Neal, "Improved Color Television Uniformity through Use of VIR Signal," *IEEE Trans. Consum. Electron.*, vol. CE-22, no. 3, August 1976, pp. 230–237.

147. S. Kim, "VIR II System," *IEEE Trans. Consum. Electron.*, vol. CE-24, no. 3, August 1978, pp. 200–208.

148. S. Barton and B. Sadler, "A New LSI Integrated Circuit for Line Recognition and VIR Signal Processing in Television Receivers," *IEEE Trans. Consum. Electron.*, vol. CE-24, no. 3, August 1978, pp. 191–199.

149. J. Millman and H. Taub, *Pulse, Digital and Switching Waveforms,* McGraw-Hill, New York, 1965, pp. 262–278.

150. J. Sakamoto, *Application Note for Color Television Deflection, IC LA7800,* Tokyo Sanyo Electric Co., Semiconductor Division, Japan, November 1980.

151. A. Cense et al., *Synchronization and Stabilization with TDA 2571 and 2581,* Elcoma Lab Report, N. V. Philips, Eindhoven, The Netherlands, August 1979.

152. M. Giles, *Television Sync Separator Design,* Consumer Engineering Report 105, Motorola Semiconductor Products, Inc. Phoenix, Ariz., October 1972.

153. M. Giles, *Horizontal APC/AFC Loops,* Consumer Engineering Report, CER-110, 1973, and Semiconductor Application Note, AN-921, 1984, Motorola Semiconductor Products, Inc.

154. W. Gruen, "Test Generator for Horizontal Scanning AFC Systems," *IRE Trans. PG-Broadcast Telev. Receivers,* January 1953, pp. 36–43.

155. Y. Shiraishi et al., "The Jitter of Television Signals," *IEEE Trans. Consum. Electron.*, vol. CE-25, no. I, February 1979, pp. 1–8.

156. D. Rhee, "Systems Approach to Deflection Drive with IC," *IEEE Trans. Broadcast Telev. Receivers,* vol. BTR-19, no. 3, August 1973, pp. 188–193.

157. S. Cox, M. Hendrickson, and R. Merrell, "Digital Vertical Sync System," *IEEE Trans. Consum. Electron.*, vol. CE-23, no. 3, August 1977, pp. 311–326.

158. S. Steckler, A. Balaban, and R. Fernsler, "Horizontal, Vertical and Regulator Control IC," *IEEE Trans. Consum. Electron.*, vol. CE-28, no. 2, May 1982.

159. D. Rhee, "New Transformerless Vertical Deflection Systems," *IEEE Trans. Broadcast Telev. Receivers,* vol. BTR-16, no. 3, August 1970, pp. 173–177.

160. L. Donovan and O. Kolody, *Complementary Vertical Deflection—Two Approaches,* App. Note 200.63, General Electric Co., Semiconductor Products Dept., Auburn, N.Y., June 1972.

161. J. McDonald, "A Transistorized TV Vertical Deflection Circuit," *Television Receivers, Related Circuits and Devices*, RCA, 1966.

162. A. Romano and L. Venutti, "A Monolithic Integrated Circuit for Vertical Deflection in Television Receivers," *IEEE Trans. Consum. Electron.*, vol. CE-21, February 1975, pp. 85–94.

163. C. F. Wheatley, "An AC/DC Line-Operated Transistorized TV Receiver," *IEEE Trans. Broadcast Telev. Receivers*, vol. BTR-13, no. 1, April 1967, p. 21.

164. M. Maytum and A. Lear, "Driver-Circuit Design Considerations for High Voltage Line-Scan Transistors," *IEEE Trans. Broadcast Telev. Receivers*, vol. BTR-19, no. 2, pp. 127–135.

165. W. Hetterscheid, "Turn-on and Turn-off of High Voltage Switching Transistors," *Electronic Applications Bulletin*, N. V. Philips, vol. 33, no. 2 October 1975, p. 59.

166. R. Walker and R. Yu, "Horizontal Output Transistor Base Circuit Design," *IEEE Trans. Broadcast Telev. Receivers*, vol. BTR-20, no. 3, August 1974, pp. 185–192.

167. F. Bate, "'S' Correction," *Wireless World*, May 1964, pp. 219–222.

168. T. Maruyama, I. Niitsu, and K. Osakabe, "Pincushion Distortion-Free Self-Convergence Deflection System," *IEEE Trans. Consum. Electron.*, vol. CE-25, no. 4, August 1979, pp. 491–496.

169. A. Morrell, "An Overview of the COTY –29 Tube System, an Improved Generation of Color Picture Tubes," *IEEE Trans. Consum. Electron.*, vol. CE-28, no. 3, pp. 290–296.

170. C. F. Wheatley, "Design Considerations for Transistorized Television Deflection Circuits," *Solid-State Design/Communications and Data Equipment*, October and November, 1964, pp. 39–44.

171. C. F. Wheatley, "Destructive Circuit Malfunctions and Corrective Techniques in Horizontal Deflection," *IEEE Trans. Broadcast Telev. Receivers*, vol. BTR-11, no. 2, July 1965, p. 111.

172. R. G. Woodhead, "The Influence on EHT Regulation of the Harmonic Content in the Retrace Voltage of a TV Horizontal Output Stage," *IEEE Trans. Consum. Electron.*, vol. CE-21, no. 1, February 1975.

173. E. M. Cherry, "Third Harmonic Tuning of EHT Transformers," Institute of Electrical Engineers, London, March 1961, p. 227.

174. M. Buechel, "High Voltage Multipliers in TV Receivers," *IEEE Trans. Broadcast Telev. Receivers*, vol. BTR-16, no. 1, February 1970, pp. 32–36.

175. J. Hornberger, and W. Robinson, "A Transistorized Fifth Harmonically Tuned Horizontal-Deflection Circuit for Large-Screen Color TV," *IEEE Trans. Broadcast Telev. Receivers*, vol. BTR-16, no. 1, pp. 43–49.

176. M. Buechel and D. Waltz, "Design Considerations for HV Silicon Rectifiers Integrated into Television Flyback Transformers," *IEEE Trans. Consum. Electron.*, vol. CE-23, no. 1, February 1977, pp. 101–106.

177. R. Takeuchi, H. Sawada, and K. Fujita, "Multi-Stage-Singular Flyback Transformer," *IEEE Trans. Consum. Electron.*, vol. CE-23, no. 1, February 1977, pp. 107–113.

178. W. Lucarz, "An Old Solution to a New Problem," *Electronic Products Magazine*, Sept. 17, 1973.

179. O. Schade, "Analysis of Rectifier Operation, *Proc. IRE*, vol. 31, July 1943, pp. 341–361.

180. J. Hicks and A. Hopengarten, "Low-Cost SCR Line-Connected DC Regulator," *IEEE Trans. Broadcast Telev. Receivers*, vol. BTR-19, no. 4, November 1973, p. 239.

181. W. Babcock, *SCR-Regulated Power Supply*, Application Note ST-6191, RCA Solid State Div., Somerville, N.J., June 1973.

182. W. Hetterscheid, and G. vanSchaik, "Power Supply System for Colour Television Receivers," *IEEE Trans. Broadcast Telev. Receivers,* vol. BTR-16, no. 3, August 1970, p. 203.

183. R. Widlar, *Designing Switching Regulators,* Applications Note AN-2, National Semiconductor, Inc., Sunnyvale, Calif., November 1981.

184. R. Middlebrook, and S. Cuk, "Advances in Switch Mode Power Conversion," *Robotics Age,* Winter 1979, pp. 6–19, and Summer 1980, pp. 28–41.

185. *Designing Transformers and Chokes with Ferrite Cores for Switched-Mode Power,* Applications Note F602, Ferroxcube, Inc., 1975.

186. P. Wilson, *Design of Isolated Switching Mode Power Supplies for Colour Televisions,* Application Note B176, Texas Instruments, Inc., 1975.

187. R. Farber, "Control of Spurious Radiation from FM and Television Receivers," *Electrical Engineer,* vol. 76, May 1957, pp. 414–416.

188. A. Intrator, "Design Considerations in the Reduction of Sweep Interference from Television Receivers," *IEEE Trans. Broadcast Telev. Receivers,* vol. BTR-2, April 1956, pp. 1–5.

189. D. White, *Handbook on Electromagnetic Interference and Compatibility,* vol. III, Don White Consultants, Inc., Germantown, Md., 1972.

Broadcast Production Equipment, Systems, and Services

K. Blair Benson

Engineering Consultant
Norwalk, Connecticut

Bruce Rayner

The Grass Valley Group
Grass Valley, California

Frederick M. Remley

University of Michigan TV Center
Ann Arbor, Michigan

Joseph Roizen

Telegen
Palo Alto, California

with contributions by

Michael Betts

The Grass Valley Group
Grass Valley, California

James E. Blecksmith

The Grass Valley Group
Grass Valley, California

John Hartnett
CBS Television Network
New York, New York

Douglas J. Hennessy
CBS Television Network
New York, New York

Robert Jull
The Grass Valley Group
Grass Valley, California

Karl Kinast
Consultant,
Westwood, New Jersey

Charles J. Kuca
The Grass Valley Group
Grass Valley, California

Renville H. McMann
CBS Inc.
Stamford, Connecticut

James Michener
The Grass Valley Group
Grass Valley, California

Mark Schubin
Consultant
New York, New York

E. Carlton Winckler
Imero Fiorentino Associates
New York, New York

14.1 SYNC GENERATION AND DISTRIBUTION

by James Michener

14.1.1 SYNC GENERATOR FUNCTIONS AND APPLICATIONS.   When any video
sources are combined, either through electronic video tape editing or video mixing, all
sources must be timed to scan the images in exact synchronism. Timing requirements

became more critical when encoded color signals were routed through a television facility. The mixing of such signals requires not only that each video source be synchronized vertically and horizontally, but also that the color subcarrier phase of the reference burst be synchronized between sources. With the advent of color-frame editing, the synchronization requirement became even more stringent because both horizontal sync and subcarrier must be precisely timed and must maintain a constant subcarrier-to-horizontal (ScH) phase relationship. This ensures proper operation of the color-frame editor used with video tape machines.

The function and design of sync generators have changed over the years to respond to these new requirements and to simplify facility wiring. Sync generators are typically employed in one of three different applications: as master sync generators, as slave sync generators, or as source-oriented sync generators.

Master Sync Generators.   The master sync generator is just that, the master clock for an entire video facility. This generator determines the absolute frequency stability of the facility and should therefore be very stable and absolutely reliable. Any problems with the master clock will ripple throughout the facility. Often two sync generators and an automatic changeover switch are employed for redundancy. The master generator often uses an atomic-based oscillator as the master clock to precisely determine subcarrier frequency. This generator can generate standard television pulses or any single-line reference pulses, as described in Sec. 14.1.3. The pulses and reference signals are often distributed to other source equipment.

Many pieces of source equipment employ an internal sync generator. This provides regeneration of all the required television pulses at the proper time for internal use and permits the output video time to be set arbitrarily with respect to the reference to simplify system input timing requirements.

Slave Sync Generators.   A slave sync generator is like a master in that it derives standard television pulses and reference signals that can be distributed to source equipment; however, this generator is usually locked (slaved) to a master. The slave can be locked precisely to the master but typically provides outputs that are advanced or retarded in time compared with the master. A slave generator is extremely useful when a single piece of source equipment, or a group of source equipment, has excessive delay due either to internal electronic delays or propagation delays caused by interconnecting cables. The timing pulses to this equipment can be advanced with respect to the master so that video from the source equipment will be timed with video from the master generator.

Source-oriented Sync Generators.   Source-oriented sync generators use an alternative approach of delaying all the composite video from, or all reference signals to, source equipment timed from the master sync generator. This approach provides more flexibility but can be costly because of the equipment required and, in the case of delayed video, may degrade signal quality.

14.1.2 CONCEPTS IN SYSTEM TIMING AND DISTRIBUTION.   Timing pulses can be distributed to source equipment by several methods. The method of distribution used is often a good indicator of the age of the equipment. Older monochrome source equipment used the full five television pulses to the equipment. These so called five-line standard pulses, composite sync, horizontal drive, vertical drive, vertical blanking, and horizontal blanking, are discussed in detail in Sec. 14.1.3. A limited amount of color equipment was produced that used all five pulses, plus burst flag and continuous wave (CW) subcarrier. Second-generation color equipment, and first-generation solid-state equipment, used three-line distribution: composite sync, composite blanking, and subcarrier. Modern equipment, which uses integrated-circuit technology, often includes self-contained sync generators and requires only a single reference pulse for operation. The de facto standard for single-line distribution is a composite color-black signal. There are some difficulties associated with black-burst distribution, however. Each time a generator is to be locked, sync and subcarrier must be carefully separated from video in order to maintain an exact timing relationship. An alternate single-line distribution signal used to couple master and slave generators together is encoded subcarrier. This signal is continuous subcarrier which has two cycles of subcarrier inverted during every color frame to precisely color-frame lock the slave generator.

Zero-time Concept.   The first concept in timing a video facility is often referred to as a *zero-time point.* It is a location within a system where all the composite video sources are exactly synchronized with one another. This location can be either at the input bus of the switcher or at the input to the video routing or distribution system of the facility. *Zero time* means zero-time difference between sync and subcarrier of any two video inputs. If the zero-time point is at the switcher input, then dissolves or effects can be performed without horizontal picture shifts or chroma hue changes. To achieve a zero-time point from a single sync generator for the facility, the sum of the delays of the reference signals from the sync generator through the source equipment and back to the zero-time point must be equal for all paths. The delay elements include: (1) the delay of timing pulses from the generator to the video sources; (2) the delay in any pulse or video distribution amplifier; (3) the electronic delay within the video source, and; (4) the time delay of the composite video to the zero-time point, including any delay in additional distribution amplifiers. Compensation can be accomplished by making the delay of each path from the generator to the zero-time point equal to the delay of the longest path. The delay can take the form of either: (1) delaying the pulses to the source equipment and rotating subcarrier phase; (2) delaying the reference pulse (normally color black) to the source equipment; or (3) delaying the composite video signal from the source equipment to the zero-time point.

To facilitate timing, there are several distribution amplifiers (DAs) which are specifically designed for these tasks. (A more complete discussion of distribution amplifiers is included in Sec. 14.1.4.) A DA which is useful for retiming subcarrier is a regenerative subcarrier DA. This type of DA receives continuous-wave subcarrier frequency and provides a constant level of subcarrier at the output, regardless of input level, and allows a complete 360° rotation of phase. A DA which is useful for delaying pulses is a regenerative pulse DA. This type of DA receives pulses, strips the pulse at the 50 percent amplitude point, delays each edge by an equal amount, and then regenerates the proper amplitude negative-going pulse. A DA which is useful for delaying either the complete composite video signal, color black, or any of the single-line reference signals is a video delaying DA. This type of DA includes a normal video distribution amplifier with a high-quality broadband, low group delay–generating delay line.

In designing a video facility, it is very useful to inventory the source equipment to be timed and note not only what synchronizing signals are required, but also the cable delay expected along with the electronic delay associated with both the source equipment and any distribution amplifiers in the path to or from the equipment. Many newer devices have built-in sync generators and therefore already provide a horizontal and subcarrier retiming range; this should also be noted in the inventory. With this information the amount of delay required to bring each piece of source equipment into time can be calculated.

Delay within each path can be increased by using one of the three distribution amplifiers previously discussed, by using excess coaxial cable, or by using a slave sync generator.

Zero Timing.   One method of timing a facility is called *zero timing,* implying that there is a sync generator with sufficient retiming range at each piece of source equipment so that no other cable, pulse, or video delaying is required. With older equipment this often requires adding a slave sync generator. Newer equipment often contains sufficient retiming capabilities. This method does provide a large degree of flexibility; however, because of the number of generators required, it may not be a cost-effective solution.

Area Timing.   A modification of the zero timing concept is *area timing.* A slave sync generator is used to time a smaller subsection of source equipment, such as a telecine, videotape room, a production suite, or perhaps an edit bay. This area could be timed in a traditional manner, and the timing of the entire area could be adjusted to match the timing of the master sync generator. Using a slave generator has the added benefit that it may be placed in a system where an area could be run completely independently of the master generator. An example of this is an editing control room or suite, where most of the time the equipment runs as an independent studio. However, on occasions the equipment may be used in, say, the production studio. Within a broadcasting station, it may be desirable to operate the production facility nonsynchronously. This ensures that the production facility is unaffected by possible signal interruptions or hits which may occur if the on-air operation locks the master generator to off-premise station-identification slides over a network program.

Reentry Timing Problems. Timing becomes complicated when video reentry is required. An example is where video sources are to be timed into two switchers and the output of one switcher must also appear as a source of the second switcher which is in time with all the sources. The delay of the first switcher must be inserted in the video path of the second switcher in order to accomplish this goal completely. This is not only costly, but the video delays may impair video quality through the second switcher.

In a large, multiple-switcher facility it may be difficult to time the entire operation, especially where studios were built at differing times with equipment of different vintages. A complete change of the sync and timing distribution systems would be costly. There are often times when an expensive piece of equipment such as a digital video effects unit, or a flying-spot scanning telecine, must be utilized in more than one studio. One solution is to receive the signal as a nonsynchronous source and time it into the system through a digital frame synchronizer. Another method is to use a routing switcher and select reference pulses from one of several studios to time the machine. If the reference pulses are sent in the proper time and selected to control the device by the sync assignment router, then the source will be timed for that destination.

Sync Assignment Concepts. Another concept in timing involves the generation of several reference signals which represent different delay timing, one possibly for each studio. Through an assignment router the reference signal is sent to the source equipment which is being assigned to that studio. The router facilities have several preset timings to lock up the sync generators. In terms of complexity of timing, this approach is the most difficult. The total path delay from the output of the sync assignment system through each source and back to the studio must be matched. Such sync assignment routers usually require a limited number of inputs and many outputs. The size of the matrix may exceed the capacity of most low-cost video routers that are available.

Another approach with the highest degree of flexibility is to use a slave sync generator that can remember several preset timing offsets from the master generator. Then the exact timing of each slave generator can be stored for each destination, and the generator is merely commanded to switch to the appropriate preset.

Feedback Sync Systems. Still another concept is the feedback sync generator, in which the timing of the sync generator is controlled by a signal from a unit that is comparing the output of the source and the facility reference, and an error signal is fed back to the slave sync generator. This system does require a reverse router to feed the error signal back to the generator from the many comparators, but it has the advantage that no timing is required of the slave.

For any of the assignment systems, a protocol must be established so that two production control rooms cannot try to utilize the same piece of equipment at the same time. An indication should appear in all other studios showing that the source is not available; a switcher lockout is also desirable to avoid accidental selection of the nonsynchronous source.

Other Timing Considerations. The preceding considerations are a survey of timing methods presently being employed by broadcasters. Cost and reliability are two elements which were not considered in the survey but are of upmost importance. Reliability is easiest to address; generally, the fewer components used, the less the chance of a failure. It is much more difficult to draw conclusions about costs. Operational features often override cost considerations in system design; nevertheless, a few recommendations can be made for a cost-effective design. Timing signals should be delayed before delaying the video. A slave sync generator is desirable whenever timing advance is required, or whenever approximately three or more pieces of equipment need equally delayed pulses. Nonreentry designs are the most cost effective.

One device which assists in maintaining facility timing is an *isophase* distribution amplifier. Signals can be timed to a zero time point and then distributed through the use of this type of intelligent distribution amplifier. An isophase DA is one which takes reference subcarrier and then automatically adjusts the absolute video delay through the amplifier to align the burst reference subcarrier with the applied reference subcarrier. This scheme then is useful in canceling out the long-term variations of timing. Because of system functions as a variable-length delay line, an isophase amplifier does not alter the sync-to-subcarrier phase relationship of the source. There are several factors to consider if an isophase amplifier is to be used. The DA has a considerable delay (120 ns nominal), and if isophase DAs are going to be used with normal distribution amplifiers, video to the isophase amplifiers must be advanced by this amount to have the output of all DAs properly timed. This complicates the problem of incorporating isophase amplifiers within an existing, already timed facility. If a new facility is being planned, however, using the isophase DA as a fan out from the zero-timed point is a very attractive feature since the isophase amplifiers maintain all signals properly subcarrier timed with the zero-time point. The cost savings in not having to regularly make drift adjustments and trim the timing of video sources more than equal the investment.

14.1.3 SYNCHRONIZING PULSES AND COMPOSITE VIDEO. There are many types and formats of signals which deliver timing information to the various pieces of video-signal source equipment. They range from multiple, single-function pulses to composite color-black signals distributed on a single line.

Pulse Definitions and Composite Video Relationships. A television synchronizing pulse which is distributed between a sync generator and source or test equipment is negative-going at a standard peak-to-peak amplitude of either 4 V for NTSC systems or 2 V for PAL systems† (see Secs. 4.2.4, 4.2.5, and 21.4.4). The rise time of all pulses is recommended to have a sine-squared shape (or elevated cosine shape) with a rise time equal to the specified rise time of the corresponding picture element. This is because most source equipment that uses these pulses to generate composite video does not regenerate or shape them, the pulse being used simply as the input to a multiplier. Therefore, if the equipment uses a linear multiplier, the resulting video will have the proper rise times. Linear gain relationship between pulse input and video output is relatively uncommon; however, it is common to find pulses distributed with nearly infinite rise times. Because little or no processing of pulses is done on the input to most source equipment, the pulse timing position and width directly correspond to their corresponding composite video definitions. For NTSC, the composite video is defined by the Electronics Industries Association (EIA) document number RS-170 revision A for NTSC 525/60. This definition is restated by the International Radio Consultative Committee (CCIR) (international standards) specifications for NTSC system M. The European Broadcasting Union (EBU) has also adopted the CCIR standards recommendations for PAL (625/50) systems.

Composite Sync. Composite sync contains both horizontal and vertical synchronization information. If composite sync were to be added with the proper ratio to a noncomposite video signal, the results would be a normal composite video signal. The rise times and duration of composite sync are identical to those of the negative-going sync signal of composite video. In composite sync, a downward transition occurs every hori-

---

†NTSC: National Television System Committee; PAL: Phase Alternation by Line.

zontal period; this is referred to as the *leading edge of horizontal sync.* Measurement of the exact timing point is made at the half-amplitude level of the leading edge of sync. Within a composite video signal this transition is the timing reference point for measurement of all pulse timing data.

Vertical Drive.   Vertical drive (VD) is used to trigger the vertical scan and internal vertical blanking in some video source equipment. Vertical drive does not directly correspond to the composite video signal and is generally not used directly to generate a composite video signal. The leading edge of vertical drive is coincident with the end of the active picture within a field. The trailing edge of vertical drive is coincident with the leading edge of sync on the first line following vertical interval.

Horizontal Drive.   Horizontal drive (HD) is used to trigger the horizontal scan and internal horizontal blanking in some video source equipment. Horizontal drive does not directly correspond to the composite video signal. The leading edge of horizontal drive is coincident with the end of the active picture on a line, or the leading edge of horizontal blanking. The trailing edge of horizontal drive is coincident with the trailing edge of composite sync. During the vertical interval the horizontal-drive signal continues undisturbed.

Horizontal Blanking.   Horizontal blanking (H BLANK) is used to blank the source output during the horizontal rescan time. It occurs on every line except during vertical interval, begins 1.5 μs before sync, and terminates 9.2 to 10.6 μs later. This pulse is used to gate the video, and its rise time may be transferred directly to the video signal.

Vertical Blanking.   Vertical blanking (V BLANK) is used to blank the source output during the vertical rescan time. The pulse occurs every vertical interval, beginning with the start of horizontal blanking of the last line of active video before the preequalizing pulses and ends at the trailing edge of horizontal blanking of the first line of the active picture. The minimum blanking time is 19 lines, and at present Federal Communications Commission (FCC) regulations permit only 21 lines of blanking. This pulse is used to gate the video, and its rise time may be transferred directly to the video signal.

Composite Blanking.   Composite blanking (C BLANK) can be thought of as the result of a logic ORing of vertical and horizontal blanking, such that whenever the signal is low, the picture is blanked.

Burst Flag.   Burst flag (BF) is a pulse which is used to gate the reference subcarrier onto each active picture line. In PAL, burst flag also contains the "meander" burst-blanking sequence (see Fig. 21-21) during the vertical interval. In NTSC the pulse begins 5.3 μs after sync and has a duration of 2.41 μs. The rise time of the BF pulse may be transferred directly to the composite video waveform.

Color Black.   Color black is a composite video color-black signal. The signal should contain no video information, hum, or other distortions and must have the proper ScH phase relationship, sync and burst amplitude, and rise times. In NTSC the signal should contain the proper set up level.

Color Frame Reference Pulse.   Color frame reference pulse is a one-line-long 9-bit coding pulse used to identify field one of the four-field color sequence in NTSC or field one of the eight-field PAL color sequence. It begins at the leading edge of horizontal blanking of line 11 and ends one line later (see Fig. 21.19c).

Encoded Subcarrier.   Encoded subcarrier is a 1-V peak-to-peak signal of continuous subcarrier where, coincident with the start of each color-frame pulse, two cycles of subcarrier are inverted. This phase-reversal timing flag resets both horizontal and vertical counters.

Pulse and Composite Video Tolerances. Table 14-1 shows pulse timing specifications and tolerances for the various sync signal components, and Table 14-2 shows frequency and intersync signal relationships.

Table 14-1   Pulse Timing†

Pulse name	NTSC M ± tol, μs	PAL B ± tol, μs
Horizontal sync	4.7 ± 0.1	4.7 ± 0.1
Equalizing pulse	2.3 ± 0.1	2.35 ± 0.1
Vertical serration	4.7 ± 0.1	4.7 ± 0.1
Burst start	5.3	5.6 ± 0.1
Burst end	7.82	7.85
Front porch	1.5 + 0.7 − 0.1	1.55 ± 0.25
Back porch	9.2 + 1.1 − 0.1	10.5 ± 0.7

†All times measured with the leading edge of sync as the datum.

Subcarrier-to-Horizontal Phasing. The ScH phase relationship has been established for the NTSC system by the EIA TR4.4 subcommittee to revise RS170A (see Sec. 5.4 and Fig. 21.41). THe EBU has also drafted a statement defining a preferred ScH phase for PAL recordings. The current definitions are given below.

1. *NTSC ScH phase definition.*

> *The extrapolation of the reference subcarrier burst should intersect the 50 percent point of the leading edge of sync at the zero crossing of subcarrier. Furthermore field one of the four field color sequence is identified in that on all even numbered lines, the extrapolated subcarrier will be observed rising at the leading edge of sync.*

A tolerance at the time of writing has not been given for ScH phasing. Any error exceeding ± 40° is considered as having an ambiguous ScH phase. Therefore recordings should clearly have ScH phase errors of less than 40°. A timing tolerance of ± 10° is a good long-term goal; however ± 30° is sufficient.

2. *PAL ScH phase definition.* The CCIR standards for PAL specifically state that phase of subcarrier has no specific relationship to the horizontal sync pulse. In 1979 the EBU drafted a statement first defining a preferred ScH phase relationship for PAL recordings. It states:

> *The subcarrier to line-sync (ScH) phase is defined as the phase of the Eu' component of color burst extrapolated to the half-amplitude point of the leading edge of the synchronizing pulse of Line 1 of Field 1. In the definition of Field*

Table 14-2   Frequency and Intersync Signal Relationships

Pulse name	NTSC system M	PAL system B
Subcarrier frequency	3,579,545 ± 10 Hz	4,433,618.75 Hz
Horizontal frequency	15,734.2657 Hz	15,625 Hz
Vertical frequency	59.94 Hz	50 Hz
Lines per frame	525	625
H to Sc relationship	$H = \dfrac{2 \times Sc}{455}$	$H = \dfrac{4 \times Sc}{1135 + (4/625)}$
Sc to H relationship	$Sc = 455 \times H/2$	$Sc = (283.75) \times H + 25$

1 of the eight PAL fields, the EBU has adopted the value of zero degrees for this central value of the Sc-H phase.

The EBU gave a target tolerance of $\pm 20°$ in the statement.

14.1.4 SYNC-TO-SUBCARRIER PHASING. As outlined in 14.1.3, in NTSC and PAL systems with four and eight fields, respectively, are required before the sequence for both sync and subcarrier repeats. This parameter of video is of no significance in television viewing; it is of great significance during video tape editing.

Color-Frame Editing. A typical example of an NTSC edit where the field sequence recorded on the tape with both playback machines locked to the edit-room sync generator is illustrated below:

Edit

Video A　　　｜　　　Video B

Field #　1 2 3 4 1 2 3 4 1 2｜1 2 3 4 1 2 3 4 1 2 3 4 1 2 3

At the edit point during playback, since the tape machine must remain locked to the edit-room generator, and since the subcarrier must match the reference subcarrier in phase, the recording tape machine is forced to do one of two things: either shift the horizontal phase by one-half cycle of subcarrier timing (140 ns), which is the normal playback mode of operation, or unlock the servo system and slide the tape one frame to realign the tape video to the same color frame as the edit room. This is operation of the machine in a color-frame playback mode. If the 140-ns shift occurs, and if the edit occurred at a time when there is a portion of the B video which is identical to the A video, a very noticeable shift will occur. Also, since the output processor of the tape machine is adding new sync and new blanking, sync appears to remain stationary, and the picture moves either right or left. The tape machine may be inserting a new horizontal blanking interval in the video, and when the picture shifts, a portion of active picture may be blanked, causing a widening of blanking and a narrowing of the active picture.

In cases where a program undergoes an extensive amount of editing and where color-frame editing is not followed, if the tape machine blanking width is set for the maximum standard (10.6 μs), the growth of blanking can be serious enough to exceed FCC tolerances. For example, after five edits there is a possibility of as much as a 1-μs increase in blanking and an equivalent horizontal picture shift.

The only solution to this problem is to ensure that all video on the tape has a consistent and stable sync-to-subcarrier phase relationship. For the sake of uniformity and interchangeability of tapes among machines, RS170A states the preferred sync-to-subcarrier relationship (see Fig. 21-41). If all video which can be recorded is timed and properly ScH phased, and if the tape machine is referenced to a stable and properly phased source, the tape machine color-frame editor will provide contiguous color-frame timing.

A video processor ahead of the tape machine can alternate color-frame editing problems. The processor at the output of a switcher, or at the input to a tape machine, can be operated with external sync and subcarrier to ensure consistent ScH phased signals to the tape machines. If the processor is fed mistimed sources, it can still shift the picture and widen blanking before recording; however, the processor will prevent any picture shift during playback.

An operating practice is outlined in the EIA RS170A specifications which prevents blanking growth when improperly timed sources are passed through an externally referenced processor. The practice includes recommended blanking widths at various locations in a facility. It uses sources with narrow blanking and gates wide blanking in just before transmission to the consumer. Running narrow blanking internal to the facility ensures that there will be ample picture available to meet FCC specifications for over-the-air broadcast even if several non-color-frame edits, or passes of mistimed video through an externally referenced processor, occur.

PAL editing has the particular problem that color-frame editing of video can be achieved only when the two video sources are aligned at eight fields, or ⅕-s intervals. For exacting vocal editing and video effects, this restriction is prohibitive. If a color-frame edit would result in awkward audio track editing under such circumstances, a decision has to be made as to which is most objectionable, a verbal editing flaw or a shift in horizontal position. When this creative decision has been made, the repositioning of the horizontal sync should be accomplished at video between tape playback and recorded with an externally referenced processor. The new tape that is generated will again have a consistent ScH phased sequential frame relationship.

ScH Phase Measurement. An early method of timing a zero-time point of a switcher input without regard to ScH phase was to align the subcarrier from each source to be precisely in time and then adjust horizontal so that they all match. To time with regard to ScH phase adds the constraint on horizontal timing that all sources must be exactly timed and also aligned to the proper ScH. This can be easily accomplished if it is observable. ScH phase can be easily viewed in NTSC systems with a procedure outlined in Sec. 14.1.5. For PAL, with the inclusion of the 25-Hz offset between sync and subcarrier, observation without additional circuitry is nearly impossible. In either system a simple to operate and read ScH phase-monitoring device is desirable.

In measuring ScH phase there are two common problems which can cause confusion. First is sync-to-subcarrier time-base error. Since subcarrier rarely jitters in phase, in effect this is a measure of the short-term jitter of the horizontal sync. Many devices in a system can inject instabilities in the sync, including any sync generator (whether a master, source, or slave), sync regenerative processors, and regenerative pulse distribution amplifiers. The second problem is normal transmission distortions. Group delay at low video frequencies is very common on long coaxial runs, and even though a modulated 20-T pulse may be acceptable, low-frequency group delay can alter the ScH phase consid-

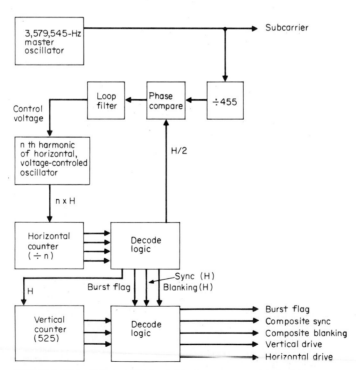

FIG. 14-1   Free-running NTSC 525-line sync-generator block diagram.

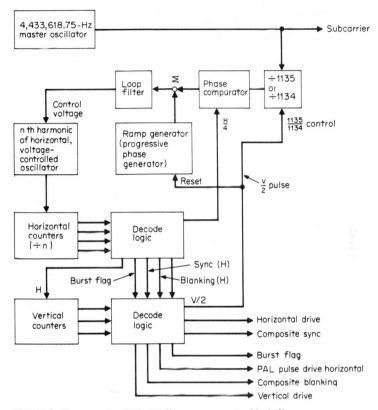

FIG. 14-2 Free-running PAL 625-line sync-generator block diagram.

erably. Smear, a low-frequency response problem, can also cause video to run into the sync portion of the video waveform and cause picture-dependent sync time-base errors.

14.1.5 SYNC GENERATOR DESIGN. The controlling portion of all modern sync generators is similar in concept. A clock feeds an LSI circuit which contains the counters and decoders needed to generate all the standard synchronizing signals. The clock frequency is chosen to be a harmonic of the sync rate so that the edges of the clock pulse can be used to form the edges of the television waveforms. For NTSC systems the 910th harmonic of horizontal (14.32 MHz) makes an ideal clock frequency, since it is also four times the color subcarrier frequency, which ensures a phase lock between subcarrier and horizontal. LSI, due to power density, is generally restricted to metal-oxide semiconductor (MOS) technologies, and MOS devices are generally slow. Thus a 14-MHz clock frequency is not practical, and a lower harmonic of horizontal (such as 320) is often used. In such systems, sync and subcarrier must be locked together by using a phase-locked loop. In PAL, with the inclusion of the 25-Hz offset, there is no practical single-frequency clock which is harmonically related to both sync and subcarrier. A phase-locked loop must therefore be utilized in PAL.

Free-running Generator Design. Figures 14-1 and 14-2 show typical block diagrams of NTSC and PAL free-running sync generators. PAL requires generation of an offset between sync and subcarrier. This is often done by summing a phase-advancing term into the loop and changing the counter by one count every $V/2$ which is coincident with the reset of the phase-advancing ramp. The comparison between sync and subcarrier is often done at an $H/4$ rate, utilizing a divide-by-1135/1136 counter.

In all generators the frequency of subcarrier determines the exact pulse repetition rate. The stability of subcarrier is critical only in an on-air situation. The television receiver locks its subcarrier oscillator to the master sync-pulse generator. If that generator is off frequency, it is possible that many viewers' sets will be unable to lock color to that station.

In NTSC, subcarrier is often derived from an atomic clock, or a precision standard which operates at 5 MHz. The clock is then locked to a subcarrier oscillator by first dividing the 5 MHz by 88 and then dividing the 3.58 MHz by 63; the results are compared and the error signal is used to control the frequency of the 3.58-MHz oscillator. The resulting frequency is precisely 3,579,545.45454545 . . . Hz. In many PAL countries, the subcarrier frequency is used as a reference for the visual transmitter carrier. With this system, the carrier frequency of the transmitters can be precisely set to the exact offset frequency to minimize co-channel interference.

The most accurate oscillator is an atomic-based one. However, there are two types of crystal oscillators which are also used widely in sync generators. A crystal oven oscillator is excellent where a high degree of accuracy is required at minimal cost. A crystal changes frequency as a function of temperature depending on the angular cut of the crystal in the lattice. In a crystal oven oscillator the temperature is elevated to the highest expected operating temperature and regulated at that temperature with a heating element. A crystal can be cut so that it exhibits a zero temperature coefficient such that the oscillator frequency does not change at temperatures near the operating temperature of the oven. The result is a high stable oscillator. Crystal oven oscillators typically operate at 60 to 70°C and require considerable power to operate the heating element. In an enclosed, dirt-free chassis, this may raise the temperature of other components and reduce reliability.

An alternative design is called a *temperature-compensated oscillator*. A crystal is specified with a known drift in frequency with respect to temperature. The temperature of the crystal is sensed with a thermistor or p-n junction which is used to control the frequency to minimize drift. The result is an oscillator whose frequency stability with temperature meets broadcast specifications. Most sync generators also allow referencing the generator to an externally supplied subcarrier.

Oscillator aging is primarily due to the transfer of mass to and from the crystal. This transfer is reduced at low drive levels and at low temperatures. It is most important that the crystal be kept clean and contained in either a cold-weld or glass enclosure.

Locking Generator Design.   There are many different schemes for locking a sync generator to another source, frequently called *gen-locking* or *sync-locking*. In addition to maintaining frequency synchronization, these generators are often required to lock at a different phase or time from the incoming video. There are two basic concepts for gen-locking. One utilizes independent locking of sync and subcarrier, and the other involves dependent locking where the sync generator is gen-locked but the sync-to-subcarrier relationship of the free-run generator is preserved. A conceptual block diagram of each gen-locking scheme is shown in Fig. 14-3. In either system, separated sync, vertical information, and a continuous burst-locked subcarrier are required. In the independent-locking generator, the sync circuits of the generator are locked to incoming video separated sync during gen-lock. In the dependent-locking generator, a pulse is generated to *crash-lock* the master generator during gen-lock, and the clock for the locked generator is locked to the reference subcarrier. The dependent method permits gen-locking to non-phased composite color signals and also permits exact horizontal locking. However, in the gen-lock mode, the sync-to-subcarrier phase relationship is determined by the incoming signal.

In the dependent locking generator, the generator will always have the same ScH phase; however, when gen-locked, there may be a quarter subcarrier cycle of horizontal phase error if the generator is locking to an ambiguous ScH phased signal. Some advantages of this approach are assurance of an ScH phased master sync generator, and locking

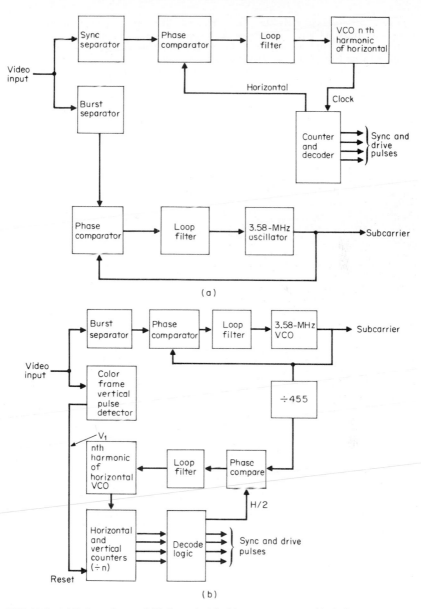

FIG. 14-3 (*a*) Independent and (*b*) dependent locking sync-generator block diagram.

with subcarrier alone (by removing the reset pulses) which is totally immune to changes in sync owing to noise and sync-only time-base error. It also eliminates the generator following nonsynchronous cuts or loss of video. This is important only when playing or recording a tape while gen-locked to an external source.

The effects on sync-only time-base error are illustrated dramatically in Fig. 14-4. Notice the apparent blurring of the sync waveform. Each photo shows the display of a dual-trace oscilloscope; the top trace shows the composite video signal to which the generator is being locked. This video is an off-air signal from a station with a 35-dB S/N ratio. The lower trace is the time-base error at the 50 percent point of the leading edge of sync. The vertical scale is 20 ns/cm. The shutter speed was 1s, and so the illustration represents many fields as a sample. Figure 14-5 shows the sync time base of the off-air signal with no processing. Notice the interruption in time-base error at the vertical interval and the effects of random noise. Figure 14-6 shows the sync time-base error which results from an independent horizontal and subcarrier locking generator with a moder-

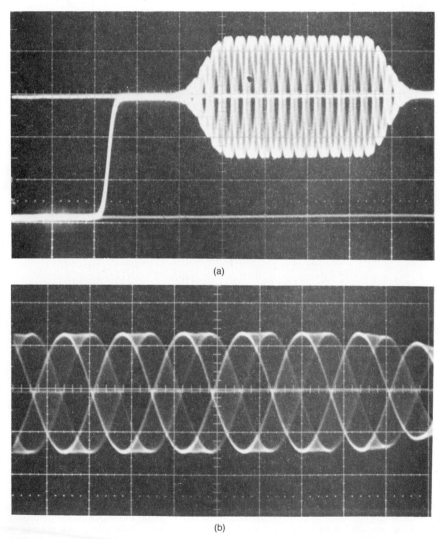

(a)

(b)

FIG. 14-4   (a) Sync-to-Sc time base error waveform relationship. In (b) the sweep is expanded to show the phase displacement.

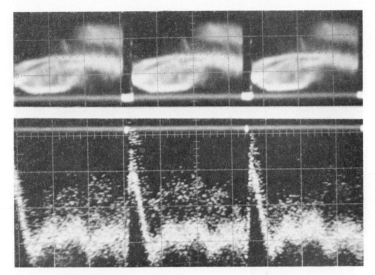

FIG. 14-5 Sync time-base error without processing of an off-air signal. *Upper:* Video signal; *lower:* error signal.

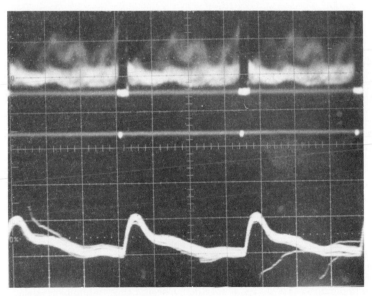

FIG. 14-6 Sync time-base error with an independent-locking sync generator. *Upper:* Off-air signal; *lower:* error signal.

ately slow-speed analog phase-locked loop for horizontal sync. Note the integration of the sync time-base error. This generator is able to follow the time-base hit in the vertical interval. Also note that the response to such a hit creates no ringing or overshoot in the time-base error, indicating that the loop is critically damped. Figure 14-7 shows the sync time-base error from a dependent-locking sync generator. The short-term time-base

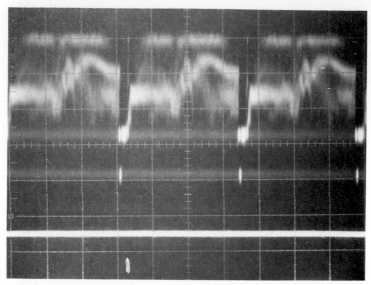

FIG. 14-7 Sync time-base error with a dependent-locking sync generator. *Upper:* off-air signal; *lower:* error signal.

error is based upon the stability of the subcarrier-locking loop, which can have a very narrow bandwidth. The residual time-base error is essentially zero.

Another iteration of the independent-locking sync generator is one in which the horizontal phase-locked loop is implemented with digital instead of analog circuitry. With either the digital or analog locking technique the result is similar except for the horizontal step in the digital system. The time-base error may show a pronounced one-unit step as it attempts to follow noise and time-base error on the input. The speed of the horizontal loop can be slowed to the point where lock time becomes excessive and therefore performs a longer time integration of time-base error. This problem occurs only when the sync generator is being operated with program material and must slew between nonsynchronous sources.

Both systems require separation of sync and locking of a continuous subcarrier to the reference burst. Detection of the precise location of the sync edge is less accurate in the presence of video and noise. Even an ideal 1 percent error in pickoff for a PAL signal, whose sync rise time is 250 ns, is 2.5 ns, or about 3.5° at subcarrier frequency. Thus, to keep an independent-locking generator properly locked with ScH phase within the +20° EBU range, the 50 percent pickoff point must be determined within 5 percent. Equipment that specifies color black as a reference signal will perform differently depending on the amplitude of rise time of the applied color-black signal. If a composite video with actual video is applied, the unit may fail to function properly.

Accurate detection of subcarrier from a video signal is also complicated by several factors. For example, the subcarrier generator must be locked to a small burst whose location in practice is not always as specified. Chrominance-to-luminance delay within a signal path can also easily displace the burst. Many burst modulators generate incidental phase modulation; therefore, the phase of burst at the edges of the envelope may be inaccurate. What is required are circuits which locate and sample the burst phase near the middle of the envelope. Most equipment that specifies color black as an input merely samples burst phase at a certain time after the sync pulse and is inadequate for use on picture signal video.

The composite-video dc axis of a picture signal may shift or bounce with varying average picture level (APL), whereas most color-black inputs will not. To operate properly when color-black inputs are switched without retiming, all color-black signals should have proper and equal amplitudes and rise times, along with proper burst position and minimal incidental phase modulation throughout the burst period.

When ScH phasing is important, such as in an editing operation and in a multiple sync generator facility, one solution to problems arising from color-black distribution between slave generators is using *encoded* subcarrier. Encoded subcarrier is the color subcarrier with two cycles inverted in phase once every four fields. The first positive subcarrier zero crossing after the inversion defines the time to reset the slave generators. With such a system, there are no distortions which can disrupt the precise locking of the slave generators. Furthermore, all generators are then operating with precise ScH phased outputs and can be exactly locked, and color-frame locked.

ScH Phase Design Considerations.   The requirements of ScH phasing include some subtle design constraints. There are many sync generators which, when placed in free-run mode, will typically achieve a horizontal-to-subcarrier frequency lock in one of two phasing conditions, either at the adjusted ScH phase or 90° from that phasing. If such a generator is set so that its output has a proper ScH phase, and either the power is removed or the generator is switched to a gen-lock mode and returned to a free-run state, there is equal probability that the generator may be either ScH phased or ScH misphased by 90°.

Sync generator outputs typically provide redundant information. For example, a generator may output sync, subcarrier, and color-black signals. Color black contains information representing both sync and subcarrier. The intersignal relationships are not critical unless two generators are used in a sync generator changeover system. In this case the output timing information from both generators must match precisely. As an example, assume that two generators have no phase difference in burst phase of the color-black signal but have an 8-ns time difference in the leading edge of sync between the outputs of each generator. Further assume that the generators were operating two cameras, one which uses sync and subcarrier and the other which uses color black. If both cameras were set up on one generator so that video was received at a zero time point in the proper ScH phase, and then the generator changeover were activated, the camera fed by color black would still be properly ScH phased, while the camera fed with sync and subcarrier would show a 10-ns ScH phase error. This error is the direct result of a horizontal phase shift of 8 ns. This problem becomes more complex when a generator outputs several signals containing redundant information which are used differently by various video source and processing equipment.

In a television facility which has a large number of slave sync generators, it is very common to find a chain of sync generators, one locking the next in a serial sequence. The last, or independent-locking sync generator, may exhibit both horizontal and subcarrier problems associated with accumulated error. If five generators are connected in series, each with 30 ns of horizontal phase error (output referred to input), the fifth generator could exhibit as much as 150 ns of error and appear properly ScH phased, but in reality would be phased to the opposite color frame as the first generator.

At the time of publication there remains an unresolved problem regarding color-frame identification. A composite video signal contains color-frame information to allow color frame 1 to be detected. Because of the complexity of accurately identifying color frame 1, proposals have been made for incorporating the four-field NTSC color sequence, or eight-field PAL color sequence, as an identification tag within the video for color frame 1. One proposal in the United States is for an identification based on a subcarrier burst in the breezeway before the leading edge of sync on one line every four fields; however, this approach may require adding a sync processing unit to each source to force alignment to the color frame tag. The EBU recommends that color-black signals include a white flag on line 7 of field 1. The RS170A committee at present has rejected such proposals as adding redundant information.

Equipment such as video tape machines must still determine color-frame information from the incoming video based on sync and subcarrier for sources that do not conform to any new standards that may be recommended. ScH phase meters are now available which operate from standard signals and provide accurate color-frame identification and comparison between sources without the need to specially tag the video in a typical studio.

External Locking Design Considerations.   The features designed into sync generators vary as a function of their applications. All sync generators should, of course, be stable, with respect to both frequency and time-base error. The required output pulses must also be provided. If a sync generator is to be used for retiming, it should have more than the required retiming range. Generally in such applications the generator is being used to advance pulses in order to overcome system delays. The most important timing parameter is the amount of advance available. Again, if a generator is used for timing, it is operating as a slave from a master generator. It should therefore provide stable and accurate color-frame locking.

If a sync generator is to be used as a master generator for a facility, it is very desirable to be able to gen-lock to any composite video signal. Gen-locking as an operating practice is slowly being superseded by digital frame synchronizers. Gen-locking, however, is still a good backup in case of failure of the more costly synchronizer, and it does not add any video delay. A generator designed to gen-lock to a color-black or black-burst reference signal will generally have problems locking to a noisy or rapidly changing APL source. In a facility where it is important to maintain ScH phasing even when gen-locked, it is desirable for the sync generator to maintain proper ScH phasing and to have minimum horizontal time-base error. This implies the use of a dependent-gen-locking system. Master sync generators should be lockable to a precision external subcarrier source in the free-run mode. The generator should provide indications of the mode it is operating in. Also, in the master sync generator location, it is highly desirable to utilize two sync generators and a changeover switch. The changeover switch should contain enough intelligence to properly detect generator faults and switch to a working unit if a failure occurs. If power is removed from the changeover unit, signals should continue to pass. This implies that the changeover unit is a passive, multipole switch. The changeover unit should not be so sensitive that it reacts to momentary signal variations or degradation. It should be simple and as reliable as, or more reliable than, the generators.

If the master generator is often gen-locked, another useful feature is *protected gen-lock.* Upon initial locking, the gen-locked generator is phased to the incoming video. Then from that point on, the generator remains locked only via subcarrier. If any disturbance occurs to the sync signals of the incoming signal, the gen-lock generator will be totally unaffected. The transition from gen-lock to free-run mode should not cause a glitch in horizontal sync. To appreciate the advantage of such an arrangement, assume that a station superimposes an identification over the network logo just before the local break. To do this, the local station must be gen-locked to the network. The problem occurs when the network does a switch while the local station is running a video tape spot. The sync generator would normally try to follow the network, causing the tape machine to relock its servo system. With protected gen-lock, however, the master generator would switch to the free-run mode without introducing a glitch.

Source-oriented sync generators should again have sufficient retiming range to accommodate even the largest studio. Timing controls should be back in the racks, for example, at the camera control unit. Many generators even provide for multiple lines of advance and delay and include a vertical-phase adjustment.

Most modern generators allow selection of vertical-blanking lines. This is important so that vertical information may be passed through equipment that may use the blanking output from the generator. This feature has become more in demand with the increased use of the vertical interval for such applications as closed captions and teletext. Most generators also provide independent adjustments for setting horizontal blanking widths. This allows sources to have the widest possible blanking widths to allow for buildup due to non-color-frame or non-ScH-phased edits. It also allows other generators to be set with wider blanking for use at the facility output or before broadcast transmission.

There are many features being offered for special-purpose applications. A generator used for electronic field production (EFP) may have several independently timeable color-black outputs. This allows the cameras for each shoot to be timed from one piece of equipment. Another feature includes automatic feedback. By monitoring the camera output compared with reference black, an error signal can be generated which controls the timing of each color-black output. This eliminates the timing requirements for a simple EFP shoot.

If a sync generator is to be used as a test generator in a maintenance operation, other than providing sufficient quantity and quality of test signals, it is useful to be able to unlock subcarrier from horizontal in order to view the burst envelope rise time. This feature is not required in PAL.

14.1.6 SYNC GENERATOR MEASUREMENTS. There are a number of television measurements which are unique to sync generators. Although most are included with information regarding the measurement of the composite video-signal waveform, their purpose and application to this class of television equipment are discussed in detail.

Pulse Width. Pulse width measurements are made per IEEE standards, measured at the 50 percent point of the pulse. Rise times of pulses are measured between the 10 and 90 percent levels. There is a problem in performing this measurement for the NTSC burst envelope since the subcarrier is locked to horizontal and inverts phase every line. This inhibits viewing the rise-time limitation imposed by the burst modulator. This measurement can be effectively made only when subcarrier and horizontal sync can be unlocked from one another. Many test-signal generators include this feature.

Stability. Stability is an implied specification, in that it is usually unspecified but is required by equipment users. There are two elements of stability, short term and long term. The short-term stability of sync-to-subcarrier time-base error is usually caused by phase noise within the phase-locked loop that locks the horizontal pulse train to subcarrier. This phase jitter can be observed by several methods. The most accurate is to loosely lock a stable crystal oscillator to a harmonic of horizontal sync. If the time constant of the phase-locked loop is long in comparison with a field, that is, 2s, the output of the phase detector, when calibrated, will give an excellent picture of the short-term phase stability of the horizontal phase-locked loop. The photos in Fig. 14-4 were taken of such a device.

A second approach, assuming that the subcarrier has low phase noise, is to observe the changes in subcarrier-to-sync (ScH) phase. This can be done in NTSC by triggering a scope on the 50 percent point of the leading edge of sync and observing the burst envelope. If there are not two clean sine waves which intersect at the zero crossings, then there is some ScH time-base error. Unfortunately this does not quantify the problem, nor show whether it is field or randomly related. Using this method in PAL becomes difficult owing to insertion of the 25-Hz sine-wave variation between horizontal and subcarrier. In PAL there is a greater possibility for problems due to the various methods of generating the 25-Hz offset lock.

Also included in short-term stability considerations is gen-lock stability. In independent-locking generators, this includes the lock stability of both the horizontal and subcarrier circuits. A short-term problem is associated with the response of a generator to a rapid change of horizontal or subcarrier phase. All phase-locked loops should be greater than critically damped. If the loops are underdamped, a generator will appear to swing when the gen-lock source takes a minor *hit* in phase. The problem of underdamped sync generator loops is aggravated when a number of generators are placed in a system where one is locked to another in series. If each horizontal loop rings, the cumulative results may be intolerable.

Long-term frequency stability is also important. A properly designed, temperature-compensated crystal oscillator (TCXO) is preferred for best long-term stability. For a high degree of absolute stability, an atomic clock is by far the best. Another consideration in long-term stability is sync-to-subcarrier phase stability. Many generators can be set to give a proper ScH output but drift considerably with aging and temperature changes over the long term. Also very common, when operating in the free-run mode, is failure of the generator to come up in the same ScH phase after a power loss. Many generators do not properly reset all the necessary counters to ensure correct ScH phase after an interruption of its counting sequence.

Long-term stability includes all temperature effects. A variety of problems can and do arise with respect to temperature. Most common are ScH phase drift, gen-lock phase-drift of either subcarrier or horizontal, and increase in sync-to-subcarrier time-base error.

Many of these measurements are not directly specified by sync generator manufacturers; however, they are necessary requirements.

Sync-to-subcarrier phase measurements, along with sync-to-subcarrier time-base error measurements, can be made with specific equipment designed to measure these parameters, including ScH phase.

14.2 VIDEO SYSTEM DESIGN: AN OVERVIEW

by Charles J. Kuca

Television is a major force in modern society. Over the years, it has evolved from a technical curiosity to a universally accepted communications medium. Along the way, its applications have grown beyond mere entertainment. Today, television is used to probe, to inform, and to educate. Modern technology has fueled this growth. Solid-state electronics, magnetic tape recording, and space communications have made program generation, storage, and transmission practical and widely available.

14.2.1 EVOLUTION OF VIDEO SYSTEM DESIGN.   In the early days of commercial television broadcasting, video systems were relatively simple. The typical station consisted of a single, stand-alone studio feeding a transmitter. Even in the multistudio environment of the network center, individual studios were self-contained entities.

There were relatively few signal sources available, and most of them were cameras located within the studio itself. The telecine was just another camera, and video tape recorders did not yet exist. The concept of a pool of sources shared by several studios was unknown at the time. Since the early systems were monochrome, facility timing was considerably less precise than is required today for color.

Only the simplest of production techniques were used. Transitions were cuts or mixes. Special effects—including keys—would not appear for several years. Advanced techniques, such as digital picture manipulation, could be accomplished only in the established medium of motion picture film.

There were only a few manufacturers of video equipment at the time, and their product lines were not very broad. Peculiarities in equipment interconnections from manufacturer to manufacturer made it difficult to build a system using several brands of equipment. It was much easier to purchase a fully engineered and integrated system from one supplier.

As the number of stations grew, the demand for equipment increased. New manufacturers appeared, and equipment-interconnect standards were developed. Many of the new companies used the newly emerging solid-state technology to produce designs that were more stable, reliable, and compact and less costly than their vacuum-tube predecessors.

During the same time period, video tape recorders (VTRs) became available. The VTR had considerable impact on system design. Before it appeared, the typical system had to deal with only one (live) program production at a time. With the advent of video tape, systems suddenly had to be capable of simultaneous off-line, as well as live, production. The subsequent emergence of electronic editing further encumbered the video system with additional program assembly and postproduction tasks.

Following the integration of VTRs into the video system, a universal conversion from monochrome to color occurred in North America. This rendered much of the existing equipment obsolete and necessitated total redesign of most facilities. Unprecedented tolerances on signal timing precision and stability confronted the system designer.

Electronic news gathering (ENG) equipment, 1-in (2.5-cm) VTRs, and digital picture processors (time-base correctors, frame synchronizers, and digital-effects units) have more recently added to the complexity of the video system. The production techniques made possible by this technology place considerable demands on the capability of the system.

14.2.2 CONTEMPORARY VIDEO SYSTEMS. Modern video facilities are divided into various operating areas such as production studio(s), VTR/telecine pool, news and ENG, postproduction, and master control. With the possible exception of postproduction (which is often totally self-contained), there is a great deal of interdependence among operating areas. This contributes much to the modern system-design problems.

The solution would be relatively trivial if program video and audio were the only signals involved. However, timing, control, tally, communications, and other ancillary signals are also typically required for proper operation. To further complicate matters, the relationships between subsystems is ever-changing. For example, a pool VTR which is assigned to record a studio production one hour may be assigned to play back a prerecorded program through master control the next hour. An effective system design should accommodate such resource sharing and ensure that the machine is properly timed and controllable from the desired operating area.

The video industry is no longer dominated by one or two suppliers of turn-key systems. There are hundreds of companies producing video equipment. Nearly all of them follow the prescribed standards for video signals and video interconnections. But standards for control, timing, and other ancillary signals are not as well defined. The challenge of making it all work together is the task of the video system designer.

14.2.3 GENERALIZED APPROACH TO SYSTEMS DESIGN. It is very easy to become preoccupied with equipment and technology and lose sight of the function of the system. This is especially true of an industry as exciting as television. Therefore, it is essential to approach the design task objectively and systematically.

Needs Analysis. The most fundamental question that must be asked is, "How is the system going to be used?" It can by very tempting to bypass this step and move directly to equipment selection, perhaps asking the question along the way. However, a less than optimum design is the likely result, and interesting alternatives will not have been explored.

Many factors must be considered in order to analyze system design needs completely. What are the functions to be performed? What is the desired level of performance for each function? How flexible must the system be? What level of operator skill is required?

When looking at the functions that the system must perform, it is advisable to note the frequency of usage. Heavy usage may warrant an entire sybsystem dedicated to that function, while occasional usage may be accommodated by a subsystem that is used primarily for another function, but not fully utilized. As an example, heavy postproduction is best handled by a dedicated system, while occasional postproduction jobs can be easily handled by a production studio and pool VTRs.

Most types of video equipment are available in various performance (and price) levels. Therefore, it is important to determine the level of performance required in each part of the system. Electronic news gathering would never have become a reality if broadcasters had insisted on studio-quality pictures.

System flexibility is an area which demands careful consideration, since it can be costly. In designing a multistudio facility, for example, it may be tempting to provide for timed reentries through several studios into master control. Before investing in the pulse-timing system and/or frame synchronizers necessary to accomplish the task, it would be wise to determine whether timed studio reentries are really required to meet the production demands.

The level of operator skill is a factor which is sometimes overlooked in video system design. It is very easy for the system designer to erroneously assume that the personnel who will be operating the equipment are thoroughly familiar with the layout of the system. It is much safer to design the system so that all routine operations (including activation of backup subsystems) can be performed by persons with average skills and without referring to system diagrams, patch panel run sheets, or other documentation.

Future needs, as well as present, should be considered. Design, procurement, and construction will take time. Therefore, the designer should project what will be required of the system at the time of completion. Going one step further, the designer should attempt to anticipate future needs and factor the necessary expansion capability into the design.

The designer must also be aware of impending changes in technology. In times of rapid change, it is possible to design a system that will be obsolete before it is built. Most of the time, however, changes will occur more slowly. A thorough analysis of needs should consider the system's adaptability to new technologies.

Financial Considerations. Large projects are usually administered by means of a budget. Ideally, the budget should be driven by the needs analysis. For this to occur, the system designer must consider the financial implications and return on investment (ROI) of the proposed design.

The most important financial consideration is ROI. High-cost items should be scrutinized carefully to ensure that they are fully utilized and contribute significantly to profits. For example, an expensive digital video-effects unit might be justified on the basis of its ability to improve the quality of a local news program, thereby increasing the program's ratings and revenue. However, if the same equipment can be used to garner more commercial production contracts (perhaps at premium rates), the financial justification is strengthened even further.

System design and final equipment choice can have tremendous impact on operating costs. Staffing and maintenance costs are two areas which demand special attention.

Risk factors must also be considered. There is a fair amount of risk associated with dealing with new technologies and new manufacturers. In most cases, the potential benefit greatly outweighs the risk; however, to ensure success of the project, risk areas should be identified and necessary contingencies planned.

Other Considerations. Many questions will arise as the design begins to take shape. Some of these do not relate as much to the function of the system as they do to its support such as space, air conditioning, and power requirements.

In new plants, sufficient space should be provided for the proposed design plus future expansion. In existing plants, new construction may be required to provide the necessary space. If space is limited and no new construction is possible, modification of the system design and careful equipment selection may be required to enable the system to meet its functional objectives.

Adequate air conditioning must be provided in all equipment areas. To determine air-conditioning requirements, heat load should be estimated for each area and a reserve factor added. Backup should be provided for all air-conditioning systems, since temperature rise would be excessive in the event of failure.

In addition to loading, the power distribution system should take into account serviceability and reliability. It is good practice to provide a separate circuit for each equipment rack, and if the rack contains equipment with dual power supplies, two separate circuits should be provided. Power conditioning (surge suppression and regulation) is not required for most video equipment but may be needed for computer systems used in conjunction with video systems.

Physical layout of the system should take foot traffic and work flow into account. For example, film and tape storage areas should be located close to the VTR/telecine area.

Equipment should be arranged to minimize long video cable runs. Where long runs cannot be avoided, plan to terminate the cable with a high-quality distribution amplifier with differential input, clamping, and cable loss equalization.

Equipment layout should also take maintenance requirements into account. Equipment must be accessible, and if it must be serviced in place (as opposed to on a workbench), adequate space must be provided for test equipment access.

14.2.4 PLANNING. Translating the functional requirements of the system into a detailed design will require several iterations. It is best to start with a simple one-line drawing of the system, showing each operating area and the relationships between areas. This should help to identify basic problems. For example, if a 24-input production switcher must have access to 35 sources, an input preselector is required. The second iteration should resolve any of the problems identified on the first. Before long, a level of detail will be reached where a one-line drawing will no longer suffice. At this point, it would be advisable to produce a master work sheet of sources and destinations.

The work sheet should be arranged on a grid, with source names along a horizontal axis and destination names along a vertical axis. X's or other marks should be made at the source-destination intersections which represent utilization possibilities. The work sheet should then be expanded to reflect audio, control, communications, and timing requirements. Color coding may be used to consolidate all requirements onto one work sheet, or separate sheets may be made for each category. The work sheet will provide a complete and concise picture of all signal distribution requirements in the facility. It will reveal areas where signals may be hard-wired and other areas where some type of switching is required. Flexibility requirements should help to determine the size and type of distribution switcher(s) needed.

Timing and other ancillary signals should be considered next. Various approaches are possible, and detailed discussions of these appear in Secs. 14.1.2 and 14.6.2.

Once the system design is complete, the marked-up drawings and work sheet should be used to produce a complete set of final system drawings. If reasonable care is taken in their preparation, these drawings can serve as permanent system documentation. Wire numbers, rack numbers, and a uniform system of equipment designations should be used.

14.2.5 EXECUTION. To minimize conflicts and wasted time, equipment ordering and system installation should be managed using program evaluation and review technique (PERT), critical path method (CPM), or an equivalent scheduling process. Regardless of the method employed, the schedule must be monitored carefully and problems resolved in a timely fashion.

Areas requiring special attention include changes in vendor lead times, interdependency of various phases of the project (including construction and electrical work), and, if applicable, maintenance of existing services.

14.2.6 SUMMARY. Modern video systems are extremely complex. They are highly integrated with their ancillary and support systems. The design process begins with a thorough needs analysis. Successful execution requires careful planning, scheduling, and monitoring. The remaining sections in this chapter provide detailed descriptions of various elements of the video system. Applications information of interest to the systems designer is included in each section.

14.3 VIDEO DISTRIBUTION

by James Michener and Robert Jull

Video signals are distributed by video distribution amplifiers (DAs). These are wide-bandwidth amplifiers designed to drive the low-impedance, unbalanced (75-Ω nominal) coaxial cables used in television facilities. Video DAs typically provide bridging (high-impedance) inputs with two paralleled input connectors to allow the input signal to be *looped through* to additional equipment and ultimately terminated in 75 Ω. They also normally provide multiple, isolated 75-Ω source-terminated outputs to drive distribution cables for one or more destinations.

There are three basic types of DAs, designated by their intended use as video DAs, subcarrier DAs, and pulse DAs. Video DAs are designed to accommodate standard 1-V (nominal) peak-to-peak (p-p) composite or 0.7 V (nominal) p-p noncomposite video signals. Some video DAs may be used interchangeably as subcarrier DAs because they have sufficient headroom to accommodate the 2-V p-p (nominal) terminated or 4-V p-p unterminated subcarrier signal levels. Pulse DAs typically accommodate up to 4-V p-p (nominal) pulse signal inputs and may provide shaped output pulses having controlled rise and fall times.

Subcarrier and pulse DAs may be further categorized as being linear or regenerative. Linear DAs linearly amplify the signal, whereas regenerative DAs replace the original sync, burst, and blanking portion of the signal with a regenerated version.

14.3.1 VIDEO DISTRIBUTION SYSTEM DEVELOPMENT.

Standard transmission line practice requires that a low-impedance, shielded coaxial cable be used to transmit wide-band video signals in order to minimize overall signal loss, frequency response roll-off, and hum or noise pickup. To optimize transmission line performance, it is also important to ensure that the driving source and terminating impedances match the characteristic transmission line impedance. Although these practices were generally observed in early television installations, the need to route a single video output to several locations, and the lack of adequate distribution amplifiers, caused some problems.

The most serious problem resulted when a piece of equipment needed to be removed for servicing or when a faulty cable or connection occurred in the series (looped-through) signal path. This would interrupt the signal to all destinations downstream of the break, and the resulting open termination would cause twice the video signal level to be applied to all destinations upstream of the break. Serious frequency-response problems would also occur on the upstream equipment because of the unterminated line. If such an interruption occurred on a transmitter-feed line, the result could be very serious.

A secondary problem was caused by the loop-through connections themselves. Each loop-through input tended to load and degrade the video signal. Since the equipment which received the looped-through signals was not generally designed for a high return loss (minimal loading), each loop-through connection tended to generate a step impedance on the transmission line and cause ripples in the frequency response. The solution to these problems was to use distribution amplifiers having high input and low-impedance outputs to route the video signals to several destinations. At first, despite the series connection and signal degradation problems, loop-through connections continued to be used, mainly because the vacuum-tube DA designs that were available tended to be bulky, costly, and unreliable.

The solid-state solution increased the use of DAs dramatically, with today's television installations often using hundreds of DAs. The typical modern solid-state DA is reliable and inexpensive, provides six or more outputs, dissipates only a few watts of power, and requires only two rack-units of space for eight DAs.

In addition to the basic loop-through inputs and multiple, isolated outputs, distribution amplifiers often provide capabilities such as differential input, level control, short-cable equalization, long-cable equalization, clamping, dc restoration, signal delay adjustment, and automatic delay adjustment to satisfy a variety of DA applications. Figure 14-8 is a simplified block diagram of a typical distribution amplifier.

14.3.2 DISTRIBUTION AMPLIFIER FEATURES AND TRADE-OFFS.

Differential Inputs. Differential input circuits provide balanced input impedances with respect to ground and amplify only the difference between the signals. Thus, *common-mode* (in-phase) signals, such as power-line hum caused by equipment ground loops and other noise and interference components commonly encountered on long cable runs, are automatically canceled.

Most modern DAs now provide differential inputs in order to avoid hum and noise

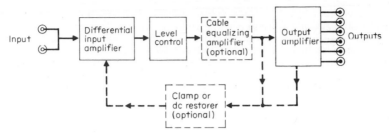

FIG. 14-8 Distribution-amplifier simplified block diagram.

pickup problems. It is not uncommon in typical installations to encounter several volts of common-mode hum on a 1-V p-p video signal. (In one installation, nearly 100 V of common-mode hum was present on a DA input owing to a power-line ground loop between two buildings.) In some installations clamp-on ammeters have been attached to coaxial cables and have measured several amps of ground loop current passing through the shield. In severe cases, power-line ground loops have been known to melt video cables. In installations where such conditions may exist, DA input specifications should be examined carefully to avoid damage to the DA and distortion of the video signal.

Probably the most serious problem that occurs in specifying DA inputs results from using a nondifferential input in a noisy studio environment. Unfortunately, nondifferential-input DAs will amplify any hum and noise that are present as shield currents, in addition to amplifying the video signal. If a clamping DA is used to remove the hum and noise already added by a nondifferential-input DA, further irreversible distortion is added to the signal (even the most carefully designed clamping circuits add some distortion). The problem is that once hum and noise become additive (are amplified by a nondifferential-input DA), they are nearly impossible to remove without causing further problems. The solution is to always specify a differential-input DA for noisy signal environments to minimize video hum and noise in the first place, and to use a clamping DA to remove hum only when absolutely necessary. Generally, if there is any question about how clean the electromagnetic environment of the installation is, the cost of specifying a differential-input DA is so small that it is worthwhile insurance. Most modern DAs provide differential inputs. In any case, installations should be designed using good engineering practices. Single-point grounding of video equipment and coaxial cables with a separate, clean video grounding system helps to eliminate unwanted hum and noise at the source that might otherwise cause hum and noise currents in coaxial cable shields.

Level Control. Although DAs normally operate at the standard video levels of 1 or 0.7 V p-p, most DAs provide some form of selectable and/or variable level control to compensate for nonstandard input levels or line losses. The gain control typically has an adjustment range of from −2 to +3 dB and is normally accessible from the front panel.

Cable Equalization. Distribution amplifiers usually provide provision for cable loss equalization. The distributed capacitance of long cable runs degrades frequency response, particularly at the higher frequencies. Consequently, many DAs will accommodate optional plug-in equalizer networks which compensate for distributed high-frequency roll-off effects by selectively boosting the high-frequency gain of the amplifier. Typically, cables are equalized when their lengths exceed 50 to 100 ft (depending on cable type) where high-frequency losses begin to be excessive. Standard equalizing DAs will compensate for up to 1000 ft (304 m) of Belden 8281/9231, Western Electric WE724, or equivalent cable.

It is important to realize that distribution amplifiers are designed to equalize video cables and not RF cables. If proper video cables are not used, correct equalization may not be possible. RF cables are used for applications such as CATV where low loss at VHF or UHF frequencies is required. These cables often use a silver-coated steel center conductor (for strength and efficient RF transmission) and an aluminum shield which exhibit severe low-frequency roll-off characteristics below 1 MHz and are nearly impractical to equalize for video. It should be noted that the amount of cable loss specified for these cables applies to operation at RF frequencies only. Video transmission cables should have a full copper shield and a copper center conductor to provide a low dc-loop resistance (the series resistance of the shield and center conductor). This type of cable, for example, Belden 8281/9231, Western Electric WE724, or equivalent, ensures proper response across the video band.

Both fixed and variable equalizers are generally available for use with DAs. Fixed equalizers compensate for a fixed cable length and are typically specified in 50-ft (15.2-m) factory-adjusted increments. An advantage of fixed equalizers is that the parameters are generally calculated by computers and checked by sophisticated equipment at the factory, and so it is not necessary to be concerned with the complex equalization parameters that are being compensated for. (Proper equalization involves more than simple frequency-response correction.) In applications where fixed equalizers can be used, selection simply involves choosing the value for the related cable length. Another advantage of fixed equalization is that there is no variable-equalization control that must be adjusted with test signals and measuring equipment. A factor often ignored with variable equalizers is that varying the equalization changes both the frequency response and subcarrier timing or delay through the DA (often considerably, depending on cable length). A disadvantage of fixed equalizers is that cable length must be known before installation and must not be subject to change. If this is not the case, a variable equalizer should be specified.

Variable-equalizer DAs typically allow rapid delay adjustment, usually with a single front-card edge control. They are often used for mobile operations and situations where setup with unknown cable lengths is required. As an example, typical variable-equalizer DAs are available for lengths of 50 to 100 ft (15 to 30 m), 100 to 250 ft (30 to 76 m), 250 to 500 ft (76 to 152 m), and 500 to 1000 ft (152 to 304 m). Custom equalizers are sometimes available for nonstandard cable lengths or types. Variable equalizers are only a first-order approximation of frequency-response correction versus distance and will therefore compensate accurately over only a limited range close to the cable length specified. For example, a variable equalizer for 1000 ft of cable will compensate correctly at 1000 ft but will become increasingly inaccurate on either side of 1000 ft. (A 1000-ft variable equalizer should not be expected to compensate for cables from 0 to 1000 ft, although a 50-ft variable equalizer may provide a fairly accurate compensation over the entire 0 to 50-ft range.) Variable equalizers from most manufacturers will compensate adequately from 75 percent to more than 100 percent of their specified range. The variable equalization control varies the cable response and also significantly affects the subcarrier delay (and ScH phase) through the DA. Therefore a variation in the timing to all destinations that the DA services will be encountered. If a variable equalizer is specified, it will usually be necessary to measure subcarrier delay through the DA after the cable response has been equalized to ensure proper system timing.

Video cables are usually equalized at the receiving end of the cable. For very long cables it is imperative that the cable be equalized at the receiving end. For example, 1000 ft of RG/59 75-Ω cable exhibits about 10 dB of loss at 8 MHz, and would require a DA gain of 3 at 8 MHz for proper compensation. The output stages of most DAs do not have sufficient headroom or slew rate (ability to follow rapid input signal changes) to provide this amount of high-frequency boost for preequalization. Equalizing at both ends of the cable provides only a negligible signal-to-noise improvement. Consequently common practice is to equalize cables at the receiving end. An exception to this rule might occur if a DA was used to feed several sources that were the same distance away. In this case equalizing at the transmit end would be an advantage since it would require only one equalizer. If this is done, the amount of cable loss at the maximum frequency of interest should not exceed about 2 dB since most DAs will not accommodate more high-frequency boost than this. (This technique will generally work at 100 ft but not at 1000 ft.)

Long-Cable Equalization. Special long-cable-equalizing DAs will compensate for up to 3000 ft (914 m) of Belden 8281/9231, WE724, or equivalent cable. Long-cable DAs often provide some form of lightning protection. Long-cable DAs may employ as many as three specially designed cascaded equalizer stages (depending on cable length) to provide the correct equalization at up to 3000 ft (914 m). It is important to note that because cable loss is not linear with distance, three 1000-ft equalizing DAs placed in series will not properly equalize 3000 ft of cable. When an equalizing DA is selected, the total distance to be equalized must be specified.

Fiber-optics offer a unique alternative to resolving the long-cable-equalization problem. For example, equalizing a mile of RG/59 cable, which exhibits about 45 dB of loss at subcarrier frequency, would not be very practical. A fiber-optic communications system, such as the Grass Valley Group Wavelink, provides a 10-MHz broadcast quality video or a 6-MHz video plus audio FM transmission link which does not need to be equalized, is totally unaffected by electromagnetic interference and hum pickup, and is highly cost effective for long cable runs. The cost of suitable fiber cable is less than half the cost of coaxial cable (and is still decreasing). Thus, the additional cost of terminal equipment is often more than offset by the lower cost of fiber compared with coaxial cable alone.

Clamping. Clamping is a video-processing operation that provides a line-by-line correction of the video blanking or sync tip level to a fixed dc reference voltage. The primary application of clamping amplifiers is to: (1) reduce additive low-frequency noise and hum, and (2) minimize dc-level bounce upstream of the video switching system when switching between synchronous video sources. Clamping also increases the dynamic range of amplifiers by reducing the swing in peak levels with APL changes. Clamping DAs can usually accept either composite video inputs (self-driven mode) or noncomposite inputs (external sync mode). Clamping is also used for restoring the dc component of the video signal prior to processing circuitry such as clipping, blanking insertion, and gamma correction.

All clamp circuits inherently introduce a certain amount of distortion in the video signal at the point of clamping. One effective clamping DA design, for example, uses both feedback and feedforward clamping circuitry which operates only on the sync tip to prevent distortion during the vertical interval (see Fig. 14.9). In the ac-coupled mode, the integrator senses any average dc component on the output signal and feeds an error current back to the input amplifier to maintain less than 20-mV average dc level at a terminated output. In the clamped mode, the composite video signal at the equalizing-amplifier output is low-pass filtered, amplified, and inverted by IC4 to provide a 2-V p-p sync signal. A sync separator clamps the sync tip and separates the sync at a constant pickoff level. A seven-pole low-pass Bessel filter filters and delays the signal for proper sampling. The sample-and-hold samples and stores the sync voltage in a holding capacitor until the next sync pulse is sampled. The sample-and-hold voltage is amplified and fed forward to the input of the output amplifier to maintain sync at a relatively constant level. In the clamped mode, any hum voltage present at the output passes through the slew rate circuit, the mode switch, and the integrator and is fed back to the input amplifier for cancellation. The slew rate switch selects the rate at which the system can follow rapid variations of the input signal. In the fast slew rate mode, this type of clamping DA can provide 50- or 60-Hz hum rejection of 45 dB. Alternatively, the slow slew rate mode can be used to minimize the quantification effects of high-frequency noise due to clamping, with only a small reduction in hum rejection.

DC Restoration. DC restoration, as opposed to clamping, provides a slow return of the blanking or sync tip level to a fixed dc reference voltage. Because of their long time constants, dc restorers will not respond to rapid changes such as noise, hum, or sync pulses and cannot make the rapid corrections possible with clamping DAs. Nevertheless, they are useful in restoring the dc component of the video without introducing the distortion caused by clamping.

The dc restorer is usually used as an anti-bounce circuit to reduce the effects of mul-

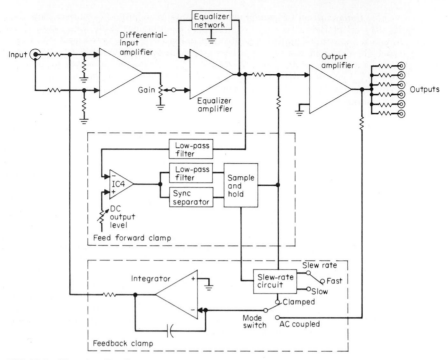

FIG. 14-9 Clamping distribution amplifier block diagram.

tiple ac couplings in large systems without introducing the noise-translation characteristics of line-by-line, or so-called hard, clamps. DC restorers can operate on either composite or, when a sync-gating signal is provided, noncomposite signals.

Delay Adjustment. Some distribution amplifiers provide for signal timing or delay adjustment over a 60- to 500-ns range by a combination of switchable steps and a continuously variable vernier. This allows system signal delays to be matched for a variety of video applications.

Automatic Delay Adjustment. A recent development in delay distribution amplifiers is the introduction of automatic delay capability. Even though stable delay DAs employing high-quality delay lines are now available, it is still possible for overall system delay to shift by as much as ±15 ns owing to daily temperature effects. Generally, an automatic-delay DA is one whose delay automatically varies as the function of an error signal derived by monitoring the difference in timing between reference subcarrier and burst.

An automatic-delay DA will adjust the timing of source signals to ensure precise color-phase timing at the input to a video switcher or processing facility, to eliminate undesirable hue shifts. Generally, a range of ±20° (NTSC) or ±24° (PAL) of automatic video delay correction is provided with less than 1° (NTSC) or 1.5° (PAL) of phase error.

Since the amplifier varies the total video delay (shifts the luminance, chrominance, sync, and burst simultaneously) to compensate for overall system timing errors, it does not affect the critical ScH phase relationship.

14.3.3 DA DESIGN AND SPECIFICATIONS. Good engineering practice for video system design dictates that distribution amplifier specifications be an order of magnitude better than system requirements. This rigorous demand results from the fact that a video signal may be routed through 10 or more DAs in a single installation. A typical network program may pass through literally hundreds of DAs from the originating point in a camera to a local transmitter. Signal degradation or attenuation can result from system interconnections, such as uncompensated loop-through connections, and reflections from improper sending and receiving end terminations.

Measured Characteristics. The manner in which DA performance parameters are measured should reflect actual, dynamic operating conditions. For instance, differential gain change through the amplifier should be measured during a 10 to 90 percent APL bounce test. Similarly, a low-frequency bounce test should be used to ensure proper amplifier ac coupling. Other important parameters often not specified by manufacturers are cable frequency-response degeneration and clamping distortion, which was previously discussed.

To achieve the high-quality specifications required, DAs typically employ carefully designed feedback amplifiers for good linearity and gain stability. Input stages use differential transistors rather than op amps to maintain good low-noise performance and high amplifier slew rate. All critical gain determining resistors have typically 1 percent or better tolerance. Distribution amplifiers also use GHz-frequency transistors to achieve the frequency performance needed to provide extremely low-output impedance and output isolation across the video band. Complementary output stages are used to provide optimum output transistor matching and linearity, and are carefully biased to run close to class B operation to ensure low power consumption. (When used throughout an installation, this can result in significant power savings.) These design techniques ensure the type of DA specifications shown in Table 14-3.

14.4. VIDEO PROCESSING

by James Michener and Robert Jull

Video processing amplifiers, or *proc amps,* are high-performance video amplifiers which regenerate the sync, blanking, and subcarrier portions of the video signal. Video processors are typically used to improve noisy video signals, to provide stable and standard video inputs to video tape recorders (VTRs), studio systems, network distribution services, and transmitters. In addition, special proc amps are used for more complex processing applications such as image enhancement, gamma correction, and automatic white- and black-level control. Figure 14-10 is a simplified block diagram of a typical video processing system.

14.4.1 VIDEO PROCESSING SYSTEM DEVELOPMENT

Early Designs. During the early monochrome television years, a primary concern was the problem of variations in the sync level that occurred when switching between video sources. Since the peak power of the visual transmitter was determined by the peak sync amplitude, source signals with higher than normal sync amplitude could cause overload of the transmitter, and protection circuits would shut down the transmitter. In order to avoid this type of service interruption, the first video processing amplifiers were used as sync (amplitude) stabilizing amplifiers, or *stab amps.* Both clipping- and regenerative-type stab amps are still in use today in many applications. The early sync stab amp designs were essentially a threshold-sensitive *stretch-and-clip* type of sync processor. A sync amplifier, followed by a sync clipper, was used to maintain a fixed sync output amplitude. A threshold control was used to adjust the sync gain.

Table 14-3 Typical Video DA Specifications

Differential distortion		
Output 1 V p-p:	Phase	0.1°
	gain	0.15%
Output 1.4 V p-p:	Phase	0.25°
	gain	0.3%
Electrical length (nominal)		28 ns
Delay match		±0.4 ns (adjustable +5 to −3 ns)
Frequency response		
10 Hz–5 MHz		±0.05 dB
5–8 MHz		+0.05 dB, −0.1 dB
10 MHz		+0 dB, −0.3 dB
Cable equalizers		To 8 MHz for up to 1000 ft of Belden type 8281
Frequency response error		±0.1 dB/dB cable loss
T pulse to bar		±1%
Common-mode rejection		
60 Hz		>60 dB
5 MHz		>25 dB
Common-mode range		±4 V about chassis ground
Gain		
Total range		−2–+11 dB
Internal		0-, +3-, +6-, +9-dB steps
Front panel		±0.5 or ±2 dB, internally selectable
Outputs		6 BNC connectors
Delay match		±0.2 ns
Output to output isolation		>40 dB to 5 MHz
Output return loss		>40 dB to 5 MHz
Output impedance		750 ±0.5%
Amplifier to amplifier isolation		>60 dB to 5 MHz
Input return loss		>55 dB to 100 kHz, >40 dB to 5 MHz
Field tilt		Composite field square-wave 0.5%
Line tilt		<0.5%
Chrominance/luminance delay		<10 ns
Chrominance/luminance amplitude error		<0.05 dB
Hum		>60 dB p-p below 1 V p-p
Output noise		>70 dB rms below 1 V p-p (10 MHz BW unweighted)
DC on output		<20 mV in ac-coupled mode
DC restorer		
50% APL video at 1 V p-p		<50 mV dc
Blanking level		Holds within ±25 mV of 50% APL value for 10–90% APL input; may be set to ground, ±25 mV, for the entire APL range
Line voltage required (switch selectable)		95–125, 190–250 V, 50–60 Hz; 85–111, 170–222 V, 50–60 Hz; or 105–139, 210–278 V, 50–60Hz
Input power required		2 W per amplifier plus 2 W per each power supply
Temperature range for specification		0–50°C

Source: Courtesy of the Grass Valley Group, Inc.

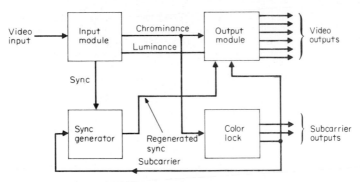

FIG. 14-10   Video processor simplified block diagram.

Current Designs.   The stretch-and-clip systems were succeeded by regenerative video processor designs which removed the entire blanking interval (including sync, color burst, and setup) from the incoming video signal and replaced it with an accurate, adjustable blanking interval from a local sync generator that was locked to the incoming signal. The main advantage of this approach is that incorrect amplitude, or missing, sync or burst is removed automatically and accurate sync and burst reinserted; in fact, the entire vertical interval may be reinserted, if necessary.

In summary, video processing amplifier designs presently provide the following features: (1) regeneration and amplitude control of sync, burst, and blanking, with accurate clamping of the blanking level, (2) color-black generation and automatic switching between station generator signals, (3) subcarrier-to-horizontal (ScH) phasing, (4) adjustable blanking width, soft and/or hard white-level clipping, (5) high slew rate capability, fast sync-lockup time, (6) phase-linear separation of chroma and luminance, (7) deletion of vertical interval noise, (8) common-mode rejection, (9) cable equalization capability, (10) vertical interval reference (VIR) correction capability, and (11) remote-control capability.

14.4.2 VIDEO PROCESSOR FEATURES AND APPLICATIONS

Improving Noisy Video Signals.   Probably the most critical application of video processors is in improving noisy video signals. (Some video processors are designed to operate only in clean-signal environments and therefore do not improve noisy signals.) Under these conditions it is necessary to separate and regenerate sync reliably and to remove (clip) any luminance noise spikes that go below the video blanking level. For example, noise spikes can occur during video time owing to microwave signal fade-out, VTR dropouts, or spurious random transients. If these negative excursions are not removed, downstream equipment may malfunction by processing noise pulses as sync pulses.

Other video processor requirements are: (1) regeneration of subcarrier (burst), control of burst level, and correction of burst axis offset errors, (2) accurate detection and clamping of the original video blanking level and adjustment of the pedestal setup portion of the dc picture axis, (3) accurate clipping of the video at the maximum white level to limit the signal amplitude to downstream equipment such as VTRs or television transmitters, (4) removal of noise pulses that occur during the vertical interval, (5) automatic gain control (AGC) correction of video, chroma, burst, and setup level, and (6) generation of controlled rise time pulse outputs for sync, blanking, V and H drive, burst flag (PAL), color-frame identification, and PAL identification pulses.

A desirable optional feature is provision for signal outputs with controlled rise times. Another desirable proc amp option is an external reference which eliminates horizontal picture movement due to timing errors at the input to a studio switching system.

Processing Switcher and VTR Signals. Another important application of video processors is in improving switching system output signals or signals into a VTR. (Typically the output of a switcher is processed before being routed to a VTR.) In order to perform properly in this application, the processor must be able to maintain a constant sync amplitude between different video sources that are selected by the switcher. This is particularly important for VTRs, since many base their video-AGC action on sync amplitude. The processor must also be able to maintain a consistent color-burst amplitude for different video input signals. This is also very important for VTRs, since tape machines often reference video head equalization to burst level; if the burst level is allowed to change as video sources are switched, the overall VTR frequency response could be affected.

The processor must also maintain uniform sync and subcarrier timing for multiple video inputs to minimize VTR time-base errors. A typical amplifier with this feature inserts new sync, blanking, and subcarrier and allows manual or automatic switching between source and station sync, based on programmable nonsynchronization timing windows. This allows automatic switching to station sync in case a nonsynchronous source is selected. In addition to normal sync clipping, the external reference is also useful for clipping video excursions below the blanking level to ensure that the dynamic signal range of VTRs is not exceeded. In some master control switching systems a video AGC mode of operation may be desirable to maintain picture brightnesss without manual attention.

Processing the Signal at a Transmitter. In order to perform adequately in a transmitter application, the video processor must provide all the functions required for processing noisy signals since transmitter signals may be subject to fade-out and resulting noise problems from microwave studio-to-transmitter links (STLs) or terrestrial conditions. The processor must also maintain constant amplitude sync, blanking, and peak video and must clip negative-going luminance spikes to prevent overload of the transmitter. White-level hard clipping is also required to limit peak luminance to 100 IRE units and prevent modulation of the transmitter through zero carrier. To ensure that peak luminance is not exceeded for all luminance and chroma conditions, it is important that the processor separately process the luminance and chroma signals.

Processors are also used as sync generators at the transmitter to accurately lock test equipment to the transmitter signal for off-air testing. For transmitter use, a processor should provide the capability of increasing setup. The luminance clip level must also be adjustable to satisfy FCC requirements of 7.5 IRE units in the United States and 0 in PAL and SECAM countries. The processor should be capable of establishing FCC blanking levels and widths. (Throughout the signal path it is often desirable to maximize the picture-signal horizontal duration by operating all facilities at narrow blanking and allowing the video processor at the transmitter to insert the correct blanking width before transmission.)

14.4.3 VIDEO PROCESSOR TECHNICAL DESCRIPTION. Figure 14-11 shows the block diagram of a typical video processor. The system consists of four plug-in modules: an input module, a color lock module, a sync generator module, and an output module. (Optional modules are also available for additional functions.)

Input Module. The input module (Fig. 14-12) provides common-mode rejection and equalizes the video signal to be processed, controls the system video gain, clamps the video using feedforward and feedback clamps, separates the chroma, luminance, and sync signals, generates a wide-burst gate and clamp pulses for the output module, and interfaces the local and remote input function controls.

The most critical function provided by the input stage is clamping of the video blanking level. Since clamp stages inherently add some distortion, it is imperative that the clamp be designed to identify the true blanking level of the original signal (often in the presence of noise and distortion) without generating appreciable distortion of its own.

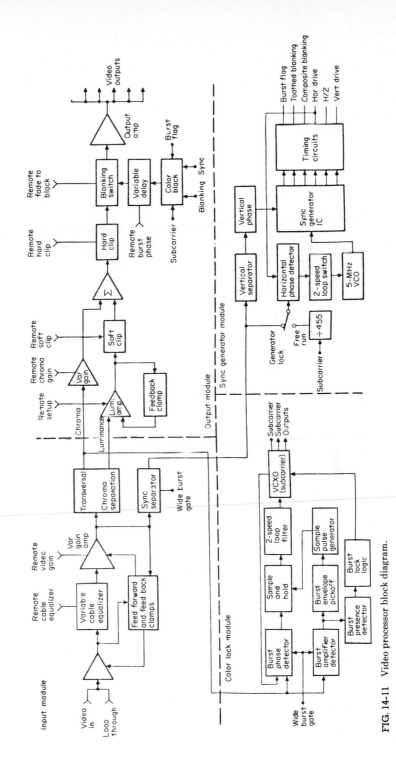

FIG. 14-11 Video processor block diagram.

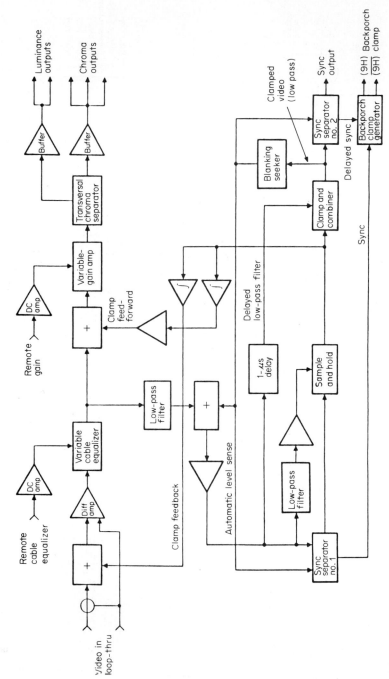

FIG. 14-12 Video processor input-section block diagram.

Two types of distortion that can obscure the true blanking level of the signal are amplifier slew rate limiting and burst distortion.

An amplifier that is slew rate limited will not follow the subcarrier signal properly, generating a large second harmonic of burst component which will cause a shift in burst dc axis. If the processor clamps only during the back-porch period, because of this axis shift, it will clamp on the incorrect blanking level. This can cause several IRE units of blanking-level error. This is prevented by sampling blanking during the vertical interval (which has no burst) and comparing it with horizontal line time burst (which contains the long-term error signal corresponding to burst offset error) to automatically correct the clamp point for the lines having burst.

One of the simplest clamping schemes used in some video processors is to clamp each horizontal line to a dc reference level. Unfortunately, since line-by-line clamps maintain any error in level for the following line, the clamping level is modulated by the sum of any noise that is present during the clamping period. An example of this type of clamping that is subject to the greatest error is a very high speed clamp that functions only during burst. On alternate lines this type of clamp will sample opposite-phase burst, causing a very noticeable line-by-line blanking level shift called *piano keying*. Once this type of level distortion is added, removing it is extremely difficult.

To avoid the line-by-line blanking distortion problems, it is essential to provide minimum gain at the 15,750-Hz line rate (to eliminate piano keying) and maximum gain at the 60-Hz field rate to maintain good noise rejection. This is accomplished by using a combination of feedback and low-frequency feedforward-clamping† techniques.

Another important function provided by the input module is sync separation. The processor must be able to accurately identify the 50 percent point of sync (usually amid noise) and lock the sync generator module to it. This is accomplished by slicing the sync signal, after a 1 μs delay, at the 50 percent level, which is determined on a line-by-line basis before the delay. Thus, variations due to hum are reduced by a factor of 63.5/1 (63.5 versus 1 μs). This ensures that the sync separator is not affected by line-by-line variations in hum or noise. Sync separators that store and separate sync by simply slicing the 50 percent level of sync, determined by an average of many lines, are highly susceptible to hum since hum can shift the 50 percent level of sync during each horizontal line. Another critical factor, often ignored in simple line-by-line sync separators, is that because of the sine-squared shape of sync pulses, when the sync pickoff level varies owing to hum or line-by-line variations, it also varies the horizontal timing of the recovered sync signal which shifts ScH phase and affects overall facility timing.

Separation of the chroma and luminance information is also an important function of the input module. Not all video processors separate the two signals; however, to provide accurate luminance (maximum white level) clipping, the two signals must be separated and later recombined without generating amplitude or group delay disturbances across the video band. A transversal filter provides a phase-linear separation of the chroma and luminance information, which are recombined in the output module.

Color Lock Module. The color lock module (Fig. 14-13) regenerates the color-burst signal supplied from the input module by generating a color-subcarrier reference signal which is phase locked to the incoming burst. Burst-detector circuitry and burst-lock logic determine whether the burst needs to be regenerated. The subcarrier reference oscillator operates in a free-run mode if burst is temporarily lost during operation. The regenerated subcarrier is fed the sync generator and output modules.

Video from a high-pass filter in the input module is fed to the burst-phase detector and burst-amplitude detector. These detectors are keyed on during back porch by the wide-burst-gate generator. The burst-phase detector compares the phase of subcarrier produced by the voltage-controlled crystal oscillator (VCXO) with the input burst. The burst-phase error is then filtered by a low-pass filter and detected by the burst-phase sample-and-hold circuit.

---

*A clamping method patented by the Grass Valley Group.

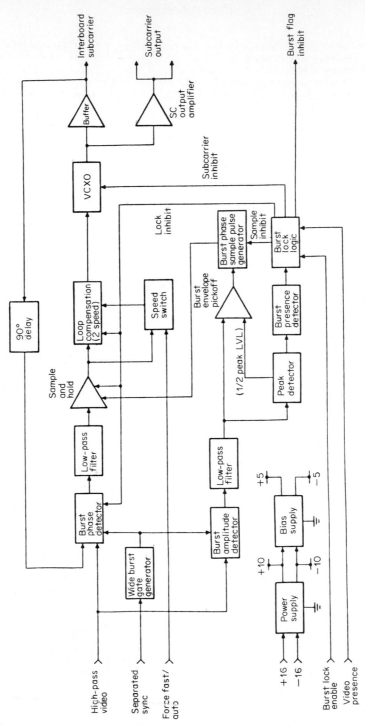

FIG. 14-13 Color lock section block diagram.

The sample pulse for the burst-phase sample-and-hold circuit is derived from input burst by the burst-phase sample pulse generator. The position of the sample pulse will track the position of burst relative to sync.

The phase-error signal is coupled to the VCXO by a two-speed loop-compensation filter. The fast speed allows fast lockup time and the ability to follow rapid changes in subcarrier phase. The slower speed allows precise phase control after the high-speed filter brings the signal within lock range. The loop speed is controlled by the speed switch, which senses an out-of-lock condition.

There are two subcarrier output amplifiers. One provides subcarrier to the other modules and the 90° delayed signal to the burst-phase detector.

The output of the keyed burst-amplitude detector is coupled through a low-pass filter which has a longer delay than the corresponding filter following the burst-phase detector. The resulting burst-envelope pulse is fed to a peak detector, which passes the pulse at its half-amplitude point, and to a burst-envelope pickoff voltage comparator. The pulse from the burst-envelope pickoff comparator is shortened in the burst-phase sample pulse generator. The resulting sample pulse is included in relation to the correct time within the position of the burst-phase envelope at the sample-and-hold section.

The burst-amplitude peak detector output is fed to the burst-presence detector. If input burst is less than approximately -8 dB (from 140 mV), the detector applies a *no burst present* signal to the burst-lock logic section.

The burst-lock logic section switches subcarrier off or on and causes it to lock or free run depending upon the condition of the burst presence, burst-lock enable, and video presence logic inputs.

Sync-Generator Module. The sync-generator module (Fig. 14-14) normally regenerates the incoming sync signal supplied by the sync-separator section of the input module. The module provides video (V sync) presence detection, vertical- and horizontal-phase adjustment, generation of sync pulses from a 5-MHz clock, vertical-interval test signal (VITS) line deletion selection, horizontal blanking width adjustment, and burst-flag timing adjustment. The sync generator module produces all pulses required by the video processor and accessory functions available with the processor. It also serves as the pulse generator when used as a sync generator system. Either separated sync from the input video or external reference sync can be used to phase-lock the module. Operation proceeds as follows.

1. *Sync preconditioning circuitry* receives the separated input sync or external reference-sync signal and applies it to a noise window, a 210-ns delay, an Sc-clocking section, an Sc-clocking switch, and a fine-H ϕ (phase) section. In the absence of input video, an output free-run/lock switch automatically switches to the H/2 signal provided by the free-run ScH ϕ section (derived from reference 3.58 MHz and the divide-by-455 section).

   The noise window section ignores the vertical sync component and increases system noise immunity by ignoring all input pulses except those occurring during or close to the horizontal-sync period. The sync window width is selectable (from 0.5 to 20 μs), and the system automatically switches from wide-window mode while acquiring lock (or when the time-base error is execessive) to narrow-window mode during sync lock.

   The 210-ns delay section compensates for the delay of input sync through the 2 $\times$ Sc-clocking section to prevent changes in horizontal phase when the system switches between Sc clocking and nonclocking modes.

   The 2 $\times$ Sc-clocking section includes frequency doubling and squaring circuits which clock (synchronize) the incoming horizontal sync with twice the subcarrier frequency (provided by the 2 $\times$ 3.58-MHz section). The 2 $\times$ Sc-clocking section is disabled by a burst enable signal when in monochrome mode.

   The preconditioned horizontal sync signal is selected by the Sc-clocking switch and applied to the fine-H ϕ section. The fine-H ϕ section allows adjustment of the H

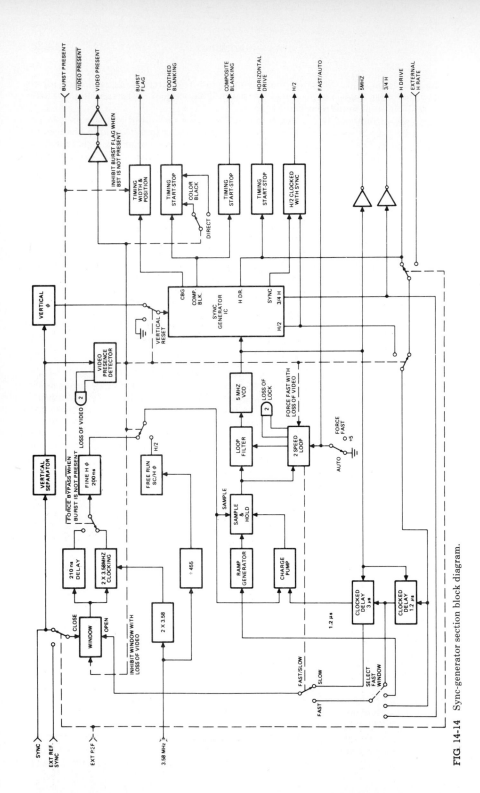

FIG 14-14 Sync-generator section block diagram.

phase relative to subcarrier. The output of the fine-H ϕ section is applied to the lock/free-run switch along with the $H/2$ ScH ϕ from the free-run ScH ϕ section. If video is not present, the switch selects $H/2$ which is derived by dividing the subcarrier from the divide-by-455 section.

2. *Horizontal sync phase-locked loop circuitry* receives the output of the sync preconditioning circuitry and applies it a charge pump (phase coincidence detector), horizontal ramp generator, sample and hold, loop filter, two-speed loop section, 5-MHz VCO, sync generator IC, and a 1.2- and 3-μs clocked delay section (which feeds the other input of the charge pump to close the phase-locked loop).

To prevent vertical roll on television sets, sync generators must be able to operate in both fast and slow sync lock modes. The *fast mode* is used when switching between video sources to ensure fast lock-up time; the *slow mode* is used during normal sync-lock operation to integrate time-base fluctuation from random noise. The video processor employs a two-speed phase-locked loop and a sample-and-hold phase-detector in parallel with a coincidence phase detector to provide this dual capability.

The ramp generator section generates and buffers a ramp waveform for the sample-and-hold phase detector section.

The charge pump (phase coincidence detector) compares trigger pulses from the incoming (preconditioned) H sync with the phase-locked loop feedback (from the clocked delay sections) to keep the loop locked to the sync frequency.

The sample-and-hold section samples the ramp voltage from the ramp generator during the preconditioned H sample pulse and holds the sampled voltage until the next sync pulse arrives and a new sample is taken. The sample-and-hold buffers the sample voltage and applies it to the loop filter and two-speed loop sections.

The loop filter contains a two-speed integrator circuit whose integration time constant is selected by a switch signal from the two-speed loop section. The two-speed loop section full-wave peak-detects and level-compares the sample-and-hold output voltage to determine if the dynamic phase error of the phase-locked loop exceeds approximately one microsecond. If this is the case, the two-speed loop section places the loop filter (phase-locked loop) in the fast (sync acquisition) mode, places the fast/slow window switch in the fast mode, and turns on the loss-of-lock indicator.

The 5-MHz VCO section contains an LC oscillator whose frequency is determined by current from a constant-current source and a voltage-controlled phase shifter which receives the error voltage from the loop filter block. The phase shifter output is applied to the LC oscillator current source to form a closed-loop VCO system. The buffered LC oscillator output is squared up by a clock driver circuit which drives the sync generator IC and also provides a 5-MHz feedback signal to the clocked delay blocks.

The sync generator IC is a complementary metal-oxide semiconductor (CMOS) LSI sync generator that provides all the timing pulses for the processor, $H/2$ pulses for the 1.2-μs clocked delay section, and $H/2$ plus ¾-H pulses for the fast-window-select switch. Pulses such as vertical drive, composite sync, window, vertical blanking, 15V, 20H, V1, and 2H are simply buffered before being used by other sections of the module or by optional equipment. Other pulses, such as burst flag, toothed blanking, composite blanking, horizontal drive, and $H/2$, which are gated and/or adjustable, are first processed by timing start-stop or clock circuits consisting of one-shots and gating logic. The toothed blanking timing start-stop section includes a counter with manually programmable switches plus gating logic which allows the vertical blanking of the processed video to be defeated during any line from 10 to 21.

The clocked-delay sections contain horizontal delay counters that delay the timing of the horizontal reference feedback pulse to the charge pump (phase coincidence detector) which has the effect of advancing the ramp generator start pulse by about

1.4 μs so that the sample pulse occurs approximately in the center of the ramp. Since the delay is programmable, it allows horizontal phase adjustment in increments of 200 ns.

3. *Vertical-sync circuitry* includes a vertical-sync separator section, a vertical-ϕ detector block, and a video presence detector section. The vertical-sync separator includes a dual-slope integrator and comparator which integrates and squares up the vertical-sync interval pulses, plus a ¾-*H* retriggerable one-shot and divide-by-6 counter which separates the vertical-sync interval by counting whether six vertical-sync serrations occur within a 3½-line period. If, owing to noise, six vertical pulses have not been counted, the sync generator IC will continue to lock to incoming horizontal sync until vertical sync is detected or until the video presence detector causes it to free run.

The separator circuitry also includes vertical-ϕ (reset) circuitry which delays the reset pulse to the sync generator IC by the correct number of lines to ensure that the processed vertical sync has the same phase as the input video.

The video presence detector section includes a retriggerable one-shot and a divide-by-4 counter, which counts four consecutive vertical-sync pulses to detect the presence of video or approximately eight fields of missing vertical-sync pulses to indicate that video is not present.

Output Module.   The output module (Fig. 14-15) interfaces the local and remote controls and combines the regenerated subcarrier signal (from the color lock module) and blanking, sync, and burst flag signals (from the sync generator module) with the chrominance and luminance signals (from the input module). This provides the desired video output(s), each with regenerated blanking, sync, and burst. Some of the major functions provided by the output module circuitry include the following: (1) setup adjustment, (2) feedback clamping, (3) soft black and white clipping, (4) chroma gain control, (5) hard black and white clipping, (6) burst regeneration and gain control, (7) burst 90° phase shifting (PAL only), (8) sync regeneration and level control, (9) local color-black signal generation, (10) burst phase control (for maintaining ScH phasing), (11) blanking switching, (12) fade-to-black control, and (13) video output amplification.

The luminance-sum amplifier receives the separated luminance signal from the input module and sums it with the variable setup voltage from the four-quad multiplier section, plus the feedback clamp error voltage from integrators Nos. 1 and 2, to provide a sync-negative output signal with the blanking level clamped rigidly to 0 V.

The four-quadrant multiplier section is keyed on during picture time (nonblanking time) and processes the local and/or remote setup control voltages to provide a variable setup current having sine-squared edges into the sum amplifier section.

The feedback-clamp circuitry consists of two samplers and two integrators which provide a clamping-error voltage to the sum amplifier. The first sampler is keyed on during back-porch time every line except during the vertical serration pulses (9*H*). The clamp pulses are derived from the incoming video signal on the input module. The second sampler is keyed on during back porch only during 9H. This provides a sample of the video signal that is free from burst axis shifts since there is no ᵇurst present during 9H. The second sampler feeds a fast integrator which uses the second integrator as a reference voltage.

The soft clipping section consists of a soft-black (SB) clipper, a soft-white (SW) clipper, and a clip control network. The soft-black clipper provides a forced setup on the signal by clipping any luminance below a preset value. The soft-white clipper removes any luminance above a given value. The clip control network processes the local and remote control voltages, such as soft clip, normal/direct, and forced setup, and provides a weighted current to the soft-black and -white clippers to generate the desired clip voltages to the chroma sum amplifier.

The chroma variable gain amplifier is a feedback amplifier which processes the separated chroma input signal from the input module and varies the chroma gain between

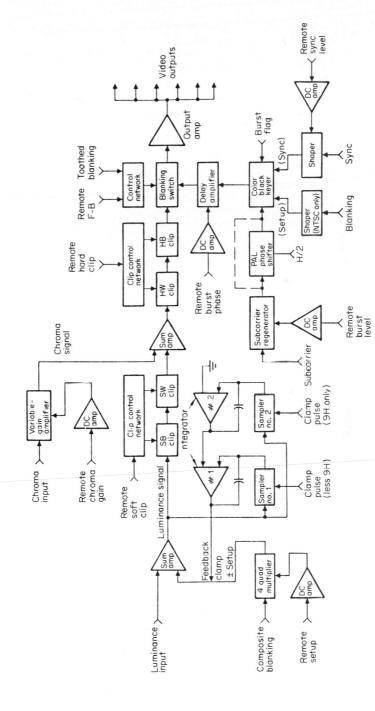

FIG. 14-15 Output section block diagram.

two preset limits. It also receives a remote and/or local chroma-gain signal from the chroma controls to allow chroma gain adjustment.

The chroma-sum amplifier is a feedback amplifier which combines the chroma and clipped luminance signals. The two signals are summed as currents at the inputs of the amplifier.

The hard-clipping circuitry consists of a hard-white clipper, a hard-black clipper, and a clip-control network. The hard clipper removes any chrominance or luminance signal component which goes above or below the respective white and black clip points. The clip thresholds are determined by operational amplifiers in a similar manner to the soft clippers. The hard clippers can never overlap the action of the soft clippers due to diode gating of the white and black clipping circuits. Operational amplifiers are also used to determine the minimum separation of the soft and hard clippers. The clip-control network provides the remote control input for the hard clippers.

When operating in the normal (regenerative) mode, the blanking-switch section replaces the incoming sync and burst with locally generated sync and burst. This is accomplished by switching between program video (from the hard-black clipper section) and the locally generated color-black signal (from the delay-amplifier section) at blanking times.

The subcarrier-regenerator section contains a differential amplifier which switches at the zero crossings of incoming subcarrier. The amplitude of the resulting square wave signal is a function of the burst level control which is processed by the DC amplifier section. The square wave is low-pass filtered to remove harmonics of the color subcarrier and applied to the color-black keyer (NTSC systems) or to the burst-phase shifter (PAL systems).

The PAL burst-phase-shifter section (not used for NTSC) consists of a burst phase shifter and a color-black keyer which produces a 90° phase shift of burst at an $H/2$ rate to satisfy the burst phase-alteration-by-line (PAL) requirement.

The blanking shaper section (not used for PAL) processes and buffers blanking pulses from the sync generator module to provide setup blanking level with sine-squared pulse edges to the color-black keyer section.

The sync-shaper circuitry produces sine-squared edges on the sync signal (similar to the blanking shaper section) which is applied to the color-black keyer section. The sync signal amplitude is adjusted by a local and/or remote sync level control. (The remote sync level control voltage is processed by the dc-amplifier section.)

Burst-flag (PAL only) pulses from the sync-generator module are given sine-squared edges in a similar manner to the blanking and sync shapers before applied to the color-black keyer section.

The color-black keyer section combines the processed subcarrier, blanking setup (PAL only), sync, and burst flag (PAL only) signals and applies the composite signal to the delay amplifier section.

The burst phase delay amplifier is a variable delay network used to maintain the sync-to-subcarrier (ScH) phase relationship produced in the color-black keyer. It also adds the remote burst phase control voltage from the remote burst phase dc-amplifier section. The ScH timed output is applied to the blanking switch for combining with the luminance and chroma signals.

The toothed blanking control network adds the remote fade to black (F-B) and toothed blanking control inputs to the regenerated composite signal controlled by the blanking switch section. Toothed composite blanking from the sync generator module is converted to sine-squared edges during blanking transitions. In the direct mode, the program signal is passed through the blanking switch, thus retaining original sync and burst.

When fade-to-black operation is selected, the control network switches the processor output to a locally generated color-black signal. Level comparators are used to sense the black-lever position.

The output amplifier is a unity gain, noninverting feedback amplifier capable of driving six 75-Ω loads. It provides the regenerated output signals for the system and may be operated in either the ac- or dc-coupled mode.

14.5 SWITCHING SYSTEMS FOR SIGNAL ROUTING AND DISTRIBUTION

by Charles J. Kuca

Routing switchers may differ considerably in configuration to suit a variety of applications. They range from simple, single-bus input selectors to complex, master-grid systems with many inputs and outputs. They may switch program and monitoring video and audio signals, intercommunication signals, control and data signals, or any combination thereof. Control of the switching function may be centralized or distributed, performed locally or remotely, and may follow any number of different types of operational logic.

To clarify the descriptions that follow, some basic terms are included with their definitions:

Source.   *Any device which outputs a (video/audio) signal.*

Destination.   *Any device which accepts a (video/audio) signal as an input.*

Switching system.   *An assemby of hardware whose function is to establish a signal path between one of several sources and one or more destinations.*

Crosspoint.   *A circuit element which acts as a switch, allowing information to pass only in the* ON *state.*

Matrix.   *An array of crosspoints represented schematically on horizontal and vertical axes, with inputs along one axis and outputs along the other. This term is often used in reference to a complete switching system but may also be used to describe a single subsection or collection of subsections (usually on circuit cards) within a complete system.*

Switching bus.   *A group of crosspoints along a common output row.*

14.5.1 SWITCHING-BUS ROUTING SWITCHERS.   This is the simplest type of routing switcher. It may be nothing more than a passive device consisting of a mechanically interlocked multiple-pushbutton switch assembly. However, most broadcast applications demand advanced features (e.g., remote control, vertical interval switching) and performance (e.g., high input return loss, low crosstalk) which passive devices cannot provide. Therefore, the trend is toward the use of modern solid-state designs.

Applications.   Single-bus switchers are generally used where it is necessary to select one of a relatively small number of dedicated sources at a given destination. For example:

1. Studio camera match monitoring
2. Transmitter signal path monitoring (video in, modulator out, IPA diode out, PA diode out, demod out)
3. Edit-bay record-VTR input selector
4. Sync generator gen-lock source selector
5. Studio camera viewfinder input selector

The relative simplicity of a single-bus routing switcher, when compared with a master control or production switcher, for example, equates directly to a high mean time before failure (MTBF) and a low mean time to repair (MTTR). Therefore, it is common practice to use them as *emergency bypass* switchers on the output of master control. A limited number of often-used sources (network, studio outputs, and VCRs) appear, along with the output of the master control switcher, as inputs to the bypass switcher. The output of the bypass switcher, in turn, feeds the STL. Under normal circumstances, the output of the master-control switcher is selected on the bypass switcher. Should the master control switcher fail, the bypass switcher is used to keep the station on the air.

Single-bus routing switchers are frequently used to extend the input capabilities of a production or master-control switcher. Large numbers of sources are common in modern television facilities, and quite often a substantial percentage of them are needed in studio control rooms or master control. Since the number of inputs on production and master control switchers is generally limited to a maximum of about 20, it is not uncommon to find that a single unit simply does not have enough. An excellent solution to this problem is to conncect the most often used sources directly, reserving one input for a single-bus routing switcher *preselector*. This provides a quantity of additional sources, equal to the number of inputs on the routing switcher, available to the production or master-control switcher. Of course, it is necessary to account for the delay through the preselector and take appropriate steps to preserve system timing.

Single-Bus Routing Switcher Design. The functional circuit blocks used in most single-bus routing switchers are very similar to those employed in larger multiple-bus designs. Thus, the single-bus system is a good starting point for a close look at basic routing switcher design and operation. Figure 14-16 is a functional block diagram of a 10-input, single-bus switcher. Only the video section is shown; the functional blocks employed in the audio section are similar.

The *input amplifier* serves several purposes. The most important one is to provide a stable, predictable termination for the source feeding it. *Termination* does not necessarily mean a 75-Ω load, since routing switchers (especially the single-bus type) may have looping inputs. Regardless of input impedance, the input characteristics should not change when the input is selected or deselected. Neither should they change appreciably when electrical power is switched on or off. Typically, the input amplifier is designed with inherently high input impedance. A passive-component input network (perhaps as simple as a resistive termination) provides the necessary predictable input characteristics.

Quite often, single-bus switchers are used as building blocks to make a multiple-bus switcher. This is usually done by looping cables through the inputs of several single-bus units. While this will work, performance is compromised, mainly because of the effects of uncompensated looping on system input return loss.

Input return loss is an important factor in any video system component. Input return loss is the ratio of the power delivered by the signal source to the power reflected back to the source, expressed in decibels. A high-input return-loss figure is desirable, because the reflections caused by mismatch can affect video-system transient and frequency response. The best way to ensure a high-input return-loss figure is to make the input impedance of the component purely resistive and exactly equal to the impedance of the source. This assumes, of course, an *ideal* cable impedance between them.

Even the best grades of coaxial cable are less than perfect, to some degree. More important, they become even less perfect (as transmission lines) if they are looped through several pieces of equipment. For this reason, this practice should be used with caution. The best system performance will always be realized in systems where looping is not used. When looping cannot be avoided, the cables should be kept as short as possible, and equipment should be selected which has been properly designed to compensate for looping inputs. Unless the equipment includes an input compensation network, it will probably introduce a significant impedance discontinuity when attached to the middle of a section of transmission line.

In addition to providing the proper characteristics for input-signal termination, the input amplifier may provide input-signal conditioning. This includes gain trim, input cable loss equalization, and clamping or dc restoration. The input amplifier's output stage is matched to the type of crosspoint and bus structure which follows; this usually means a low output impedance and some sort of dc offset.

The *crosspoint* and *bus* act as a multiposition switch to allow the operator to select one of many sources. The crosspoint must therefore behave as much like a perfect switch as possible; that is, it must exhibit zero transmission loss in the ON state and infinite transmission loss in the OFF state. With most solid-state devices, low transmission loss in the ON state is usually not a problem. High OFF-state loss can be a problem, especially at high video frequencies; therefore, a T configuration is often employed.

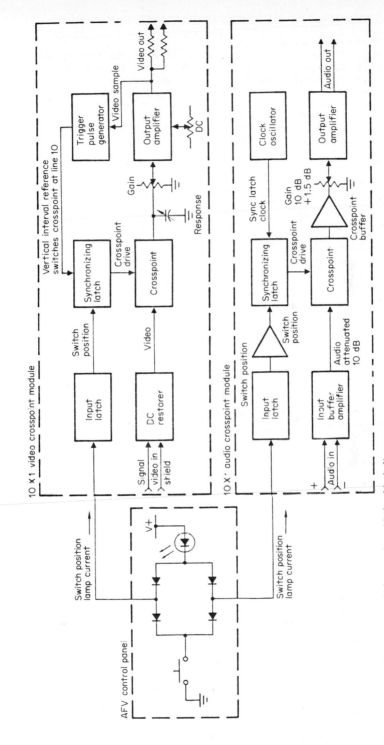

FIG. 14-16 Ten-input single-bus routing-switcher block diagram.

14.47

A T crosspoint consists of three solid-state switch elements. Two are connected in series and are placed in the signal path. The third is connected between the junction of the other two and ground. When the crosspoint is selected, the two series elements are turned on and the shunt element to ground is turned off. The signal is able to flow with little or no attenuation. When the crosspoint is not selected, the series elements are turned off and the shunt element is turned on. Thus, the signal path is opened and any residual signal leakage through the first series element is diverted to ground.

Crosspoint and bus design influence a number of important routing-switcher performance parameters. Crosstalk is one of these. At the level of an individual crosspoint, the goal is to reduce the signal leakage through an unselected crosspoint to a minimum. The same result is desired at the system level, and the measure of success at achieving it is the system crosstalk specification. Improvements in solid-state crosspoint design have produced crosspoints with extremely low signal leakage in the OFF state. Even so, as the number of inputs on routing switchers increases and buses become wider, crosstalk becomes more problematic. Most *wide-bus* switchers address the problem by dividing the bus into groups of inputs and using a secondary crosspoint to select the desired group. With this technique, the level of crosstalk is confined primarily to the signal leakage through the unselected crosspoints in the selected group.

Many solid-state crosspoint designs that exhibit nearly ideal transmission characteristics are less than ideal in regard to switching transients. Moderate transients can usually be tolerated in a vertical-interval video switcher, since the disturbance is hidden from the viewer, or removed by subsequent signal processing. In an audio switcher, however, this is not the case. Even small-amplitude disturbances are noticeable, especially with wide dynamic ranges in program material. Therefore, for audio switching, it is common practice to employ crosspoint designs which are optimized for low switching noise. Alternatively, a noisy switch may be masked using a circuit which momentarily reduces system gain during the switching interval.

The *output amplifier* serves primarily to buffer the output of the bus to the following circuitry and system. Usually, a gain-trim adjustment is provided. However, since a routing switcher is normally regarded as a unity gain device, the output gain trim has only enough range to compensate for minor system variations.

The output amplifier may also include provisions for output cable-loss equalization. This is similar to the equalization function which may be included in the input amplifier. There is one important difference that should be noted, however. Equalization is accomplished in both cases by employing an amplifier stage with a transfer characteristic that is the complement of the transfer characteristic of the cable. Postequalization (equalizing high-frequency losses at the receiving end of a cable) of fairly extreme losses requires an equalizer stage with high gain at high frequencies in the input amplifier. Once the signal has been equalized, it passes through the rest of the input amplifier, and indeed the system, as a *flat* response signal.

Conversely, preequalization (equaling high-frequency losses at the sending end of a cable) places rigorous demands upon the stages following the equalization stage—including the stage which must drive the cable. Attempting to preequalize excessively long cable runs will result in slew rate limiting, probably in the final output stage. Slew rate limiting is the inability of an amplifier to handle large voltage swings at high rates of change (high frequencies). The problem will be most severe in the output stage because high-voltage swings into a low-impedance load dictate current swings beyond the capabilities of the typical video-output driver. Preequalization should therefore be applied with caution.

Crosspoint Control.   Control schemes vary in complexity depending on the number of crosspoints being controlled and the desired control functions. Once again, the single-bus case provides a good model, since control is limited in terms of both crosspoints and functions. Generally speaking, the control of single-bus routing switchers is restricted to the selection and tally of a relatively small number of crosspoints in one dimension (a single switching bus).

If the number of crosspoints is 10 or less, wire-per-crosspoint control is usually employed. A control line is provided for each crosspoint, and shunting any line to ground causes the corresponding crosspoint to be selected. Crosspoint status (confirmation of actual occurrence of the switch) is provided by the switcher control system in the form of a sustained ground on the control line after the temporary ground is removed.

Simplicity is the primary advantage of wire-per-crosspoint control. Control panels may be constructed using momentary-action normally open pushbuttons with LED or incandescent indicators. A properly executed wire-per-crosspoint control system should accommodate a wide variety of inputs for selection, while providing a status signal acceptable to a wide variety of devices. A good design will support control/status by external logic as well as switches and indicators.

Since it is possible to ground more than one line at a time, wire-per-crosspoint control systems must include logic to inhibit simultaneous selections. Priority encoding is usually used, resulting in selection of the highest-numbered input.

When the number of crosspoints being controlled exceeds 10, it becomess attractive to use some form of coded-control selection which reduces the number of wires between the switcher and control panel. Typically, binary or binary-coded decimal (BCD) coding is employed. In addition to a sufficient number of data lines, a *strobe* or *take* line is required. Occasionally an *acknowledge* or *confirm* line will also be included to signal completion of a valid crosspoint selection.

Various serial data selection schemes provide yet another means of crosspoint control. The use of serial data permits nearly unlimited control possibilities. Serial control is rarely used in single-bus switchers since extra flexibility is not needed and the additional hardware would add unnecessary cost to the design. Serial control is quite common, however, in large multiple-bus switchers.

Regardless of the method used to control crosspoint selection, a means of timing the switch point is required. Nearly all modern video switchers switch during the vertical interval. Since the inputs to routing switchers are often nonsynchronous with respect to each other, the question arises as to which vertical interval should be used. A popular solution, especially on single-bus switchers, is to strip sync from the output of the switcher and use it to phase-lock a trigger-pulse generator operating near the vertical rate. The switch will then be timed to the vertical interval of the last-selected input. If there was no signal present at the last-selected input, the switch will occur randomly.

14.5.2 MULTIPLE-BUS ROUTING SWITCHERS.

It is natural to think of a multiple-bus routing switcher as a group of single-bus switchers with common (looped) inputs. This approach can be, and in fact is, used, but there are a number of reasons why it is a makeshift one at best. The most serious technical limitation, the effects of looping on performance, has already been discussed. The difficulty in implementing full-matrix xy control is yet another. But the most compelling reason for avoiding this approach is its inefficiency, both in terms of cost and routing density. A truly efficient and cost-effective multiple-bus routing switcher is a difficult design exercise involving a number of trade-offs such as reliability versus packaging efficiency, price versus performance, and flexibility versus complexity.

Applications.   Multiple-bus routing switchers are generally used when it is necessary to make a fairly large number of sources (usually 20 or more) available to multiple destinations. In many instances, a *master-grid* concept is employed, where virtually all sources in the facility appear as inputs to a large matrix with many switching buses. Alternatively, smaller matrices may be employed, each performing a specific function, such as VTR input selection, production switcher input preselection, and production switcher iso-bus switching.

The master-grid concept provides the greatest flexibility, since all sources are available at each destination. System timing is greatly simplified, since the length of program paths is predictable. Efficiency is sacrificed, however, because some of the crosspoints (the ones representing unused or seldom-used source-to-destination connections) are not required.

The alternative approach, employing a number of *functional-area matrices,* eliminates the efficiency problem since each matirx is fed only by the required sources. Thus, there are fewer redundant crosspoints. Flexibility is sacrificed, however, because the seldom-used sources are not available, except by patching them temporarily. Preservation of system timing is also difficult, unless strict rules are observed regarding program paths through multiple matrices.

Both approaches can be appropriate, but the decision as to which is best for a given facility can be determined only after a careful study of present and future needs, operating philosophy, and budget.

Multiple-Bus Routing-Switcher Design.   The most fundamental consideration in the design of a multiple-bus routing switcher is system architecture—how the system is put together. There are two broad architectural classifications: output-oriented and matrix-oriented. In an output-oriented system, the crosspoints and output amplifier for a single bus are confined to a single circuit card (or group of cards). In a matrix-oriented system, a single circuit card contains crosspoints (and perhaps output amplifiers) for several switching buses.

The chief advantages of an output-oriented architecture are reliability and serviceability. If a component should fail, diagnosis is simple and repairs may be made without disturbing other paths. This can be extremely important, especially in large master-grid systems. Output-oriented systems require more interconnect cabling, however, and are not as dense (in terms of the number of crosspoints-per-unit-volume) as matrix-oriented systems. An example is a 64 by 64 video-switching system constructed from 64 by 1 circuit cards, each with input amplifier and control card. Multiples of these 64 by 16 building blocks are combined to produce 64 by N systems.

Inputs are distributed (rather than looped) to each building block using *fanout amplifiers*. Sixteen of these amplifiers are packaged in a frame that is 7 in (18 cm) high [4 rack units (RU)]. This particular fanout-amplifier module provides signal conditioning in the form of gain trim, dc restoration, and input-cable loss equalization.

Output expansion of the system is accomplished by adding 64 by 16 building blocks. Input expansion is somewhat more involved and requires an additional building block—a *secondary crosspoint*. This system employs an 8 by 1 module for secondary switching. These modules are packaged 16 to a frame with from two to eight of the 64 by 16 switching frames, to produce N by 16 building blocks (where N = 128, 192, 256, 320, 384, 448, or 512). Figure 14-17 is a block diagram of a signal path through the system just described.

The first multiple-bus routing switchers were mostly output-oriented designs. They also accommodated relatively small numbers of inputs and outputs. As input/output numbers increased, the need for higher packaging density became greater. This led to the development of the matrix-oriented architecture.

Matrix-oriented structures are generally constructed using relatively small input groups (10 or less). This is in contrast to the relatively large input group common with output-oriented systems. Any switching system is most efficient when the number of inputs is an exact multiple of the basic input group. For example, the output-oriented system just described is optimized for input sizes of 64, 128, 192, and so on; it is not as efficient at 40 or 80 inputs. A smaller input group (8 or 10 inputs, for example) is required to produce optimum efficiency over a wide range of input requirements.

In a typical 64 by 64 video switching system based on a matrix-oriented architecture, the basic building block in this system is an 8 by 16 circuit module. This module is used with another (an 8× output amplifier module) to provide switching systems up to 128 by 128 in size.

This system is exactly twice as dense as the output-oriented system described earlier, that is, 64 by 32 versus 64 by 16 in 10½ in (6 RU) of rack space. The increased density results from including more crosspoints per module and by input bussing (as well as output bussing) at the module level. It should be noted that, while matrix-oriented architectures inherently are more dense than output-oriented architectures, the dramatic (2-to-1) increase in density is made possible by using specialized integrated circuits in the matrix-oriented example.

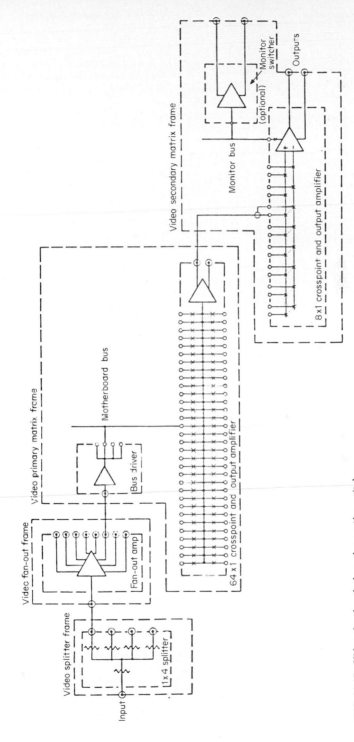

FIG. 14-17 Video-signal path through a routing switcher.

14.51

In the basic 8 by 16 circuit module, high-impedance (loop through) video inputs are buffered and dc restored by a single integrated circuit. The output of this IC drives buses on the circuit card. The switching function is performed by another special IC, a (1 by 1) crosspoint. In addition to the video switch itself, the crosspoint IC provides compensation circuits and the crosspoint-latch logic.

In addition to the eight primary crosspoints on each bus, there is a ninth crosspoint on the output of the bus. This crosspoint performs exactly the same function as the secondary crosspoints on the output-oriented system—input expansion of the primary input group. The secondary crosspoints for a single output bus are all on one module in the output-oriented system; they are distributed in the matrix-oriented system.

All the secondary crosspoints on a given output are connected to a common trace on the motherboard in the matrix frame. This trace is, in turn, connected to an output amplifier module. Each output amplifier module contains eight amplifiers. Thus, every group of 16 output buses requires two output amplifier modules.

Nearly all matrix-oriented routing switchers accomplish output expansion by looping the input signals through multiple crosspoint modules; fanout amplifiers are rarely used. This is consistent with the basic goal of a matrix-oriented design—efficiency, but seems to contradict the comments earlier in this section about looping. In fact, looping is always a compromise, but careful attention is given in most matrix-oriented systems to providing the proper input compensation and keeping the connections between frames as short as possible. Thus, reasonable system input return loss is maintained.

Crosspoint Control.   It should be fairly obvious that controlling a multiple-bus routing switcher is a more complicated task than controlling a single-bus routing switcher. This is true not only because of the increased number of crosspoints, but also because a unique crosspoint in a single-bus switcher can be specified by a single number (x, the input number). In a multiple-bus switcher, two numbers must be used (x, the input number, and y, the output-bus number).

Operationally, a multiple-bus routing switcher is generally controlled in one of two ways: on a dedicated bus-by-bus basis, or full-matrix xy control. Quite often, both methods are employed. For example, a given system may include several buses for VTR input selection with control panels for these buses located at the VTRs. At the same time, an xy panel would be provided at a central maintenance/monitor location. Under normal circumstances, the VTR operator would control his or her own bus. In some cases, several buses are controlled using a single control panel.

Although xy control is used primarily for maintenance purposes, it does have other applications. An xy control panel can be used for *emergency* control of a bus whose dedicated control panel has failed. Small routing switchers which are being used as patch panel replacements may use a single xy panel as the primary means of control. Regardless of how they are used, xy control panels should be applied with caution, since an inadvertent operator error has the potential of causing a selection to be made on a bus other than the desired one.

Routing-switcher control-system technology has become quite sophisticated. Most contemporary designs are microprocessor based. The choices as to basic control system architecture closely parallel those of basic signal system architecture. Instead of the choices being output-oriented versus matrix-oriented, they are distributed versus centralized. Choice criteria are similar in both cases and primarily relate to the importance of reliability and serviceability.

A distributed control system is comparable to an output-oriented signal system. Isolated failures in a distributed control system will affect only a small portion of the system. If the control system is properly designed, faults can be traced and repaired with minimum impact on the total system.

A control fault in a routing switcher with a centralized control system can be castastrophic. Such a fault might, at the very least, inhibit control of the entire system. At the very worst, a faulty control system might cause random spurious crosspoint selctions to occur. To minimize the effects of control system faults on system operation, prudent designers provide on-line diagnostic systems, backup control modules, and changeover mechanisms.

Regardless of whether their control systems are distributed or centralized, most microprocessor-controlled routing switchers communicate with external control facilities and status displays using serial data. This includes communication between matrix and control panels. With serial data, it is possible to control very large matrices using a single pair of wires (sometimes two pairs are used, one for *transmit* data and one for *receive* data). The use of a serial-data format also opens the door to a virtually unlimited repertoire of commands beyond simple switching and tally. Some examples are crosspoint and/or bus protect, multilevel switching, and salvo switching, to cite a few.

Switchers with distributed control systems generally provide a dedicated serial-data channel for each bus. If the system supports xy control, a separate xy port is provided with data paths to all the pieces of the distributed control system.

In a switcher with a centralized control system xy control is almost inherent. Typically, all control panels are connected to a common (time-division multiplexed) data channel and, by definition, have the capability of controlling all buses. While this is somewhat beneficial, care must be exercised in programming each control panel (this is usually done with dipswitches) to control the proper bus.

14.5.3 CONTROL PANELS. Routing switcher manufacturers' literature typically includes a dazzling array of types of control panels. Selecting a panel for any given application can be greatly simplified if three primary criteria are considered: function, form, and features.

What control function is required? There are three possibilities: single bus, multibus, and xy. If the routing switcher has multiple levels (e.g., video, audio 1, audio 2, time code), the question arises whether independent control of each level is necessary or whether all switching operations will be married.

There are numerous ways to enter and display data. Data entry methods include individual pushbuttons, key-pad pushbutton arrays, and thumb-wheel switches. Data display methods include indicator lamps (incandescent or LED) and multiple-segment numeric and alphanumeric displays.

Button-per-Source. The button-per-source control panel consists of one or more rows of individual pushbuttons. Each button is equipped with an illuminating indicator to show the operator which source is active. The unit shown in the figure is designed to accept film legends with the name of the source in each button. This enables the operator to select sources rapidly. Typical applications for this type of panel include production switcher iso-buses and camera match monitor selector. There are, of course, physical limitations to the number of buttons that can be mounted on a reasonable-sized control panel; button-per-source panels are generally constrained to deal with a maximum of 30 or so sources. If the switching system has more inputs than this, the control panel must provide some means for designating which subset of sources it controls.

Thumb wheels. The thumb-wheel control panel consists of a multidigit thumb-wheel switch, a take button, and a multidigit numeric display. The major limitation of thumb wheels is their slow data entry speed. Their main advantage is that they do not require much space.

The most popular type of general-purpose control panel for routing switchers is the *key pad*. Key pads provide access to a large number of sources without requiring much space. A key pad is *natural* for data entry, a characteristic no doubt attributable to the widespread use of electronic calculators and pushbutton telephones.

Mechanical Design. Physical dimensions and form are extremely important since the plethora of equipment jammed into operating areas allows little excess space. Mounting methods should also be carefully considered. Most panels are rack-mountable, but if there is no rack space in the operator's console, the control panel must be capable of being custom-mounted. As human engineering and esthetics become more important in television facility design, the extent to which the control panel harmonizes with other equipment must be given more consideration.

Operating Functions. Features should not be overlooked. Microprocessor-based control panels are capable of many useful functions. Intelligent panels can be programmed to function with source names (for example, "VTR12" or "STU32") instead of abstract numbers. Smart panels can send commands to the matrix to prevent other panels in the system from changing the selected source.

Status Display Devices. Control panels are almost always status display devices as well. But there are a number of devices whose sole function is to display status.

Using a routing switcher as a production switcher input preselector solves one problem (limited inputs on the production switcher) very nicely. But at the same time, it creates another: How does the director and/or technical director keep track of sources? One solution involves fitting the production switcher with status readouts which serve, in effect, as *dynamic* button legends.

If preview monitors are provided for the preselectable production switcher inputs, it is advisable to identify the source associated with each monitor. This may be done by keying information into the monitor video or by means of an external display.

A system designed and built by the British Broadcasting Corporation (BBC) and used in their main transmission center in London is quite unique. Each status readout is connected to a data bus carrying routing switcher status. The readout assembly's logic is programmed to extract the number of currently selected input from this data stream, convert it to a source-name mnemonic, and display it on the dot-matrix display. Also connected to the data bus is a typewriter-like keyboard which can send commands to the display units, instructing them to append *tags* to the source. For example, a name can be added to identify a particular operator of a camera or VTR. All tags sent out on the bus are known to all display devices, so anyone selecting a source will see the source and operator's name displayed.

Tally Functions. From an operational viewpoint, it is essential that the operator of a program source such as a camera or VTR know when the source is on air (in use). The control subsystem which provides this information is called the *source tally system.*

Source tally is built in to most production switchers, where it serves primarily to activate the tally lamps on studio cameras. On air is a well-defined condition in a production switcher; a source is on air when a path exists between it and the program output of the switcher. Tally in a routing switcher is yet another matter because the on-air condition is not as clearly defined.

The first step toward defining it is to designate a bus (or buses) where tally is required. Having done so, a source should be considered on air if it is connected to the output of the designated bus (or buses). For example, a routing switcher bus which feeds the transmitter would be a likely candidate for tally; any source selected on that bus would most certainly be on air. A station or network feeding multiple transmitters or transmission circuits would undoubtedly want a tally indication at the source(s) feeding any of the outgoing lines.

The most obvious way of handling tally in a routing switcher is with a relay matrix slaved to a level of the switcher. With such an approach, a separate and distinct closure is provided for each and every crosspoint, affording the system designer tremendous flexibility. However, relay matrices are expensive, and the flexibility they offer may be of questionable value.

A logic-based source-tally system is an inexpensive alternative to the full-relay matrix. Inputs to a logic-based source-tally system are *bus enables* (one ON-OFF signal per switching bus) and switcher status. Outputs are relay closures (one per routing switcher source input). A closure will occur on any relay corresponding to a source that is selected on an *enabled* bus. Multiple closures are possible if more than one bus is enabled and different sources are selected on the enabled buses.

Enables may be hardwired; it would be appropriate to permanently hardwire the enables for outgoing transmission circuits. However, the flexibility of a logic-based source-tally system is enhanced considerably if enables can be altered externally. External enables are particularly useful where the on-air path makes multiple passes through the routing switcher and perhaps through external equipment as well.

Consider the example in Fig. 14-18. A VTR is selected on one bus of a routing switcher whose output is connected to the input of a production switcher. The output of the production switcher reenters the routing switcher and is selected on a bus which feeds the station transmitter. The on-air tally is propagated back through this complex program path to the source VTR by external enables.

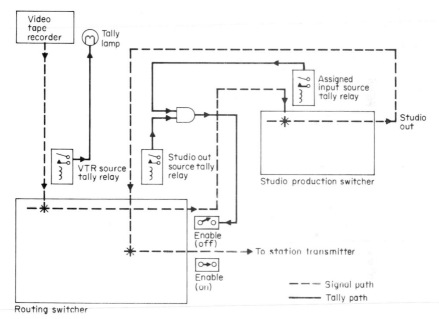

FIG. 14-18   A typical application of source tally.

The enable for the bus feeding the transmitter is hardwired. When the production-switcher output is selected on this bus, the corresponding relay closure is used to enable the routing-switcher bus feeding the production switcher. Thus, when the VTR is selected on this bus, the VTR's tally relay will energize. Note that the intermediate closure (the one corresponding to the production switcher output) could be used as a *studio on-air* tally.

In facilities that require multiple levels of tally, *live on-air* and *record on-air*, for example, separate tally systems are necessary. It is necessary to provide an enable to only the correct tally system at the ultimate destination in the program path. Enables to a record-tally system could be provided by the record relays on the VTRs.

Operational Philosophies.   The primary function of a routing switcher, as the name implies, is to establish a *route,* or connection path, between pairs of terminals (i.e., between input and output). Usually, control is associated with destinations. Occasionally, however, control is associated with the source. Routing switchers with source-oriented control are often called *assignment switchers* because sources *assign* themselves to a destination or destinations.

Assignment logic is often employed in facilities where signal routing, machine control, timing, and perhaps intercom are all handled by a single integrated system. The addition of these functions, especially machine control and timing, complicates the rules by which the system must operate. Source-oriented control serves to simplify the operational logic.

In a routing switcher with destination-oriented control, selection of sources is unrestricted; any destination may select any source. However, any given destination can select only one source at a time. In other words, each output bus is interlocked with respect to sources, but the output buses are not interlocked with respect to each other. This form of operational logic works well in a signal (e.g., video/audio) routing system because most routing switchers are also distribution switchers; that is, an input can be fed to multiple outputs.

However, when machine control or timing is considered, paths through the matrix are reversed, and conventional interlock logic breaks down. From a signal (e.g., video or audio) point of view, the VTR is a *source*. From a machine control or timing point of view, the same VTR is a *destination*. Distribution of the VTR's output signals (audio/video) to multiple destinations is legitimate; control and timing of the VTR from multiple locations is not. A destination cannot be fed by multiple sources.

One way to prevent such conflicts is with an *xy interlock* system. An *xy* interlock prevents multiple destinations from selecting the same source. In a system with integrated machine control and/or timing, *xy* interlock prevents control and timing conflicts. To be effective, *xy* interlock must be automatic; in other words, once a destination selects a particular source, if another destination attempts to select the same source, it will get a busy signal. To ensure availability of nonbusy sources, a release mechanism (analogous to putting a telephone back on the hook) is required. Selection of input zero (black/silent) is the usual procedure.

Automatic *xy* interlock is one way of resolving conflicts in an assignment system. Placing control at the source and putting a human operator in the loop makes it possible to avoid *xy* interlocks entirely. Instead, the operator is given the option of assigning signal output, machine control, and timing independently. The operator may assign a machine's signal outputs to multiple destinations for viewing and listening but must guard against assigning machine control and/or timing to multiple locations.

14.5.4 ALTERNATIVES TO RECTANGULAR CROSSPOINT ARRAYS. Our discussion of multiple-output routing switchers has assumed a rectangular matrix in which the total number of crosspoints is the product of inputs times outputs. It should be obvious that as these dimensions increase, the total number of crosspoints, and hence the *cost* of the matrix, increases geometrically. This phenomenon has stimulated a great deal of interest in various methods for crosspoint reduction.

The telecommunications industry has been responsible for most of the research in this area. And little wonder; the dimensions of most telephone exchanges are staggering when compared with even the largest of broadcast routing switchers. As an example, a 100 by 100 routing switcher implemented as a rectangular matrix contains 10,000 crosspoints. A 10,000 by 10,000 telephone switch would require 100 million crosspoints!

Most of the telephone companies' early research was concerned with the basic concept of using a number of small matrices interconnected in such a way as to provide the necessary number of inputs and outputs. This technology reached its peak in the 1950s when Charles Clos of Bell Laboratories published a paper entitled "A Study of Non-Blocking Switching Networks."[1]† Clos described a network structure composed of multiple stages or submatrices. The crosspoint savings over a rectangular matrix are significant: 43 percent savings at 100 by 100 and over 94 percent savings at 10,000 by 10,000.

The main reason the Clos design has failed to replace the more conventional rectangular matrix in broadcast applications is because broadcast routing switchers are *distribution* switchers as well. The Clos matrix is a telephone exchange designed to connect one telephone instrument to another; any connection which is made is always from a

---

†Superscript numbers refer to references at end of chapter.

single input to a single output. The inherent nonblocking characteristic of the Clos matrix is predicated on this sort of operation. (Note: *Blocking* is a condition in which the control logic is unable to make a connection between an input and an output.)

When the distribution requirement is considered, the simple connection algorithm described by Clos is no longer valid and must be replaced by a more complex set of rules. The basic strategy is to avoid making connections in a manner which leads to blocking while recognizing that statistically, blocking is inevitable. Successful broadcast implementations of the Clos design have overcome this problem by using smart control systems which are capable of reassigning paths through the matrix, allowing blockages to be cleared.

There is another method (also commonly employed in the telecommunications industry) which can be applied in the design of large broadcast routing switchers: time division multiplexing (TDM). TDM does not merely reduce crosspoints; it eliminates them!

TDM can be better understood if one considers a conventional matrix to be a *space division* system. The crosspoints occupy a divided space (the matrix), and selections are made by moving around in that space and activating crosspoints. In a TDM system, input signals are digitized and clocked into time slots on a high-speed parallel data bus. Outputs need only "grab" the data in the time slot corresponding to the desired input and convert it back into analog form. See Fig. 14-19.

TDM has been used in the telecommunications industry for some time, chiefly for handling voice grade (3-kHz bandwidth) communications. The technology has not existed until recently to apply this technique to wider-bandwidth signals in a commercially viable product.

A prototype TDM system, designed by The Grass Valley Group, accommodates wideband, program-quality (20-kHz bandwidth) audio signals. A system with 128 inputs and 128 outputs occupies only 34 in (20 RU) of rack space (plus power supplies). For com-

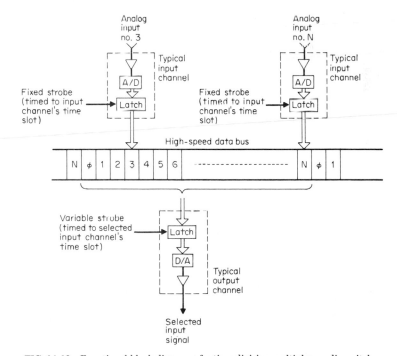

FIG. 14-19   Functional block diagram of a time division multiplex, audio switcher.

parison, a rectangular audio matrix of the same dimensions using high-density packaging techniques would require 84 in (48 RU). In addition to the significant improvement in space required, the TDM system also provides better performance, particularly with regard to crosstalk and signal-to-noise ratio.

Digital technology has not yet advanced far enough to permit conjecture as to when a TDM video switcher might be feasible. Extrapolating the design to video suggests data buses operating near 3 gHz, so it is unlikely to occur in the foreseeable future.

14.5.5 SUMMARY. Routing switcher systems can be characterized in many different ways: size, architecture, and control philosophy, to name just a few. Selecting a particular system requires a careful analysis of the application. Factors to consider include performance and reliability, flexibility, expansion capability, operating features, and control requirements.

Alternatives to the classic rectangular crosspoint array have begun to emerge and should be considered where they offer a significant advantage in terms of performance and/or cost.

14.6 VIDEO PRODUCTION

by Michael Betts, James E. Blecksmith, and Bruce Rayner

The portion of the studio video system that handles video production includes equipment that switches, modifies, or creates video for the purpose of enhancing program material supplied by cameras, VTRs, or other outside sources. Typical video production devices and their interfaces are shown in the video system block diagram in Fig. 14-20.

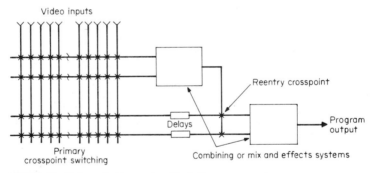

FIG. 14-20 Video switching-system block diagram.

Video production has been heavily impacted by recent developments in digital control and digital video. While the basic functions of studios have remained the same, several new devices have changed video production considerably. These devices, video switching systems, switcher memory systems, and digital video effects systems, will be discussed in this section. Video tape editors also play an important role in video production, and the expanding interface between switchers, editors, and other production devices is also discussed.

14.6.1 BASIC FUNCTIONS. A video production switching system *(switcher)* is one of the primary devices used to produce a television program. Production switchers are essential to all live operations and most postproduction situations. The main function of a production switcher is to either switch or cut between two video sources, or combine them in a variey of ways. The principal methods used to combine video are (1) mixing, (2) wiping, and (3) keying.

A production switcher consists of two main sections: (1) an input selection matrix which provides the input switcher functions, and (2) video mixing amplifiers, usually called the *mix and effects (mix/effects) system* which provides the combining functions. Improvements in production switchers have been quite rapid in recent years. Early systems could only switch or mix video sources using an input selection matrix and a simple mixing amplifier controlled by a lever arm. This selected the proportions of the two video signals being combined and added the result to form the output signal. In Europe, video switchers were patterned after audio mixers where the amplitude of each input was controlled by a separate control (knob), allowing two or more signals to be combined simultaneously. This approach is called a *knob-a-channel* system.

As technological advances provided high-speed switching transistors and integrated circuits, wipe and keying functions were added to production switchers, usually in the form of separate two-input combining amplifiers in addition to the mixing amplifier. A selection of different wipe patterns was provided to allow the operator to produce a variety of wipe effects between one picture and another. These units also provided the ability to key or switch from one picture to another under control of a video or external input signal. Keying is typically used for adding captions or titles to the picture.

Currently produced switchers usually provide all the above functions of mixing, wiping, and keying, in a single mix/effects (M/E) amplifier. Controls for the functions are usually grouped together to allow wipe patterns to be used to mask out parts of keys, while mixing in and out of the resultant combination. The largest and most sophisticated production switchers usually have two or three mix/effects systems, each with four to five input rows that can produce multilevel keys and utilize a large number of wipe patterns.

The configurations of production switchers vary in a number of ways to accommodate production requirements. The design variations generally encountered are as follows.

Number of Inputs. Usually between 8 and 32 video inputs can be accommodated. The number is determined by the requirements of the application, such as mobile, studio, production, and postproduction. Mobile use often requires a small panel size and fewer inputs than a studio or postproduction switcher. When a very large number of sources are available, a preassignment or routing switcher may be employed to preselect the sources, such as video tape or film, to limit the inputs to the production switcher's input matrix.

Number of M/E systems. The number of M/E systems varies considerably and depends largely on the complexity of the production, the number of effects needed simultaneously, and the extent of postproduction required. Mobile use often has less need for multiple-effects production than live and postproduction. In live operations more than one M/E system is useful to enable one to be used while a second is being set up. Usually M/E systems can be cascaded or reentered one into another to allow production of complex source combinations.

M/E Functions. Most mix/effects systems provide several basic functions which include:

1. *Mix:* An additive combination of two video sources (usually summing to unity).

2. *Mix transition:* A change from one video source to another using a mix operation (duration greater than one field).

3. *Wipe:* A switch occurring during the active video at specific points on the raster to produce a pattern between two video sources (switching is usually controlled by an internal waveform generator).

4. *Wipe transition:* A change from one video source to another using a wipe operation (duration greater than one field).

5. *Key:* A switch occurring during active video between two video sources controlled by a video or video-related signal.

M/E Features.   A great variety of operating features are associated with these modes. Some examples are listed below:

1. *Mixing:*

   Manual and auto transitions

   Nonadditive mixing (ability to combine two pictures so that only the brightest elements of each is present in the result)

2. *Wiping:*

   Number of patterns

   Rotational patterns

   Bordering modes, hard or soft, and colored edges

   Direction modes

   Modulation modes

   Positioner modes

   Aspect change

   Multiple patterns

3. *Keying:*

   Title keys (or luminance keys)-

      Source selection

      Invert mode

      Insert selection (video or color fill)

      Bordering selection

   Chroma keys-

      Source selection

      Key shadow features

      Key separate or hold modes

      Encoded chroma keys

Often these features are provided as add-on options to enable a certain degree of customizing to be selected by the user. Other common options are a quad split system, a final or downstream keying (DSK) system, and additional switching rows for utility use. Each of the features will be described later in more detail.

The availability of each of these features to the operator is largely dependent on the design of the control panel and its logic. To get the most effective use from a system, the control panel layout must provide a logical arrangement of controls for the operator. Grouping controls by functions, providing easily identifiable repeating patterns for similar functions (each effects system should look like the others), and making often-used functions such as keys, wipes, and mixes easy to find help make a switcher easy to use.

Effects Memory Systems.   With the increased capabilities of larger switchers, the number of operating changes that are required between operations has increased. This has led to the introduction of mix/effects memory systems, usually microprocessor controlled, that enable the operator to store settings of the panel for later recall. This allows the operator to go from one setting to another quickly and to utilize the full capability of the production switcher. The memory systems usually operate independently on each mix/effects system, providing the ability to recall and preview the control settings before taking them to the switcher's output.

Two general types of effects memory systems are in use:

1. *Learn mode programming:* The operator sets up the desired effect in the normal manner and then learns (enters) the effect into user memory register. A more permanent storage medium, such as flexible magnetic disc storage, is often provided as well.
2. *User programming:* The operator uses a keyboard to enter source and mode selections, for example, into memory locations, and can recall these on the switcher to preview and air the result.

The memory system for switcher control also makes it easier to interface with editing computers for postproduction editing by providing access to the full switcher operations, and the editing computers are easier for the operator to change than the editor listing for a particular editing sequence. The ability to interface to editing computers is usually by means of a serial or parallel interface between two microcomputer systems. The serial mode often provides access to other peripheral devices used to feed production switchers, such as still stores and character generators. This can provide full interaction with editing computers.

Additional Switcher Features.   Besides the basic M/E functions, most switchers provide additional and sometimes optional features.

1. *Separate mix systems.* These are quite useful in providing frequent mix and cut operations between two separate rows of buttons usually labeled *program* and *preset.* Sources are selected on the preset row and transferred to program via the mix lever arm or *take* button. This allows the M/E systems to be available for more complex functions.
2. *Title or downstream keying.* An additional title keying system (or downstream keyer) may be provided to enable titles to be keyed in after other functions have been performed. This function is often employed at the final output of the switcher (DSK) and typically incorporates a separate or master fade-to-black control.
3. *Quad split.* This is an additional feature providing the ability of displaying four different sources (pictures) in a four-quadrant display. The position of the quadrant dividing sections and colored quadrant borders can usually be changed.
4. *Tally systems.* An essential feature of all switchers is a tally system that provides output contacts to provide routing of camera and VTR tally signals to the on-air sources. Any path through the switchers, either direct, key, or key source, that contributes to the on-air or output picture should provide a tally indication for that source. Often the control panel lamp brightness levels are also used to indicate the on-air sources at any time as a further aid to the operator.
5. *Auxiliary switching.* Video input, M/E outputs, and program output are usually at different timing relationships owing to the internal delays in the production switcher. If a selectable timed feed is required for monitoring, VTR input, or digital video system source selection, timing has to be performed externally or provided as an internal addition to the switcher. Auxiliary buses provide delays and adjustments to enable such timed selections between sources, M/E outputs, and program to be made. Some switchers provide some auxiliary output capability internal to the switcher; however, if a large number of outputs are required, an entire timed auxiliary matrix may be needed.
6. *Digital video effects.* An effects system (Fig. 14-21) is often used in conjunction with a production switcher to provide picture manipulation such as compression and expansion. These systems often receive their input selections from ancillary switching systems in the production switcher, such as from the ouput of a particular mix/effects system. The output, usually one field later in time, is returned to the switcher's input matrix as an input, for selection as a direct or key source, to be used for combining

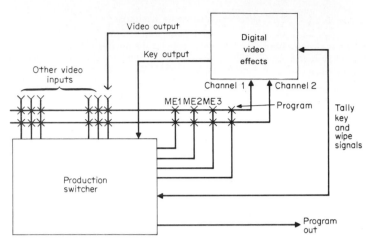

FIG. 14-21   Digital video effects system block diagram.

with other video signals. This provides feedback effects such as the repeating images in a hall of mirrors or the more usual compressed picture or picture slide modes.

14.6.2 SYSTEM APPLICATIONS. Production switchers have both internal and external timing requirements. The internal requirements will be covered later in this section. The main external requirement is for input video and pulse timing relationships to be met.

External Timing.   Section 14.1.2 discussed the timing of a system with reference to the production switcher as a time zero reference point. Since the production switcher must switch or combine video signals, coincident timing is essential to avoid color shifts or picture jumps that would occur if timing differences existed. Most production switchers are equipped with synchronization detectors on each input channel to inhibit mix/effects operations if vertical sync is asynchronous and horizontal sync is outside a prescribed window (on the order of ±500 ns from the local timing). This can usually be overridden if necessary to enable a source which is synchronous but slightly mistimed to be used (a b/w title or caption generator can be keyed in where horizontal position is usually not critical, for instance).

Some controls may be provided in the switcher to allow fine adjustment of sync and blanking signals. If this is not possible, external timing is important for these reference signals as well, since they are used to generate internal reference black and background matte signals. Phase adjustment of 360° at subcarrier frequency is usually provided to allow color phasing among inputs.

A typical switcher has coincident input video timing, and ideally key-signal timing as well, while the reference signals are usually in advance of video by a few hundred nanoseconds to allow the black and background that are generated to be correctly timed at the input matrix. The total video delay through the switcher is typically 200 ns for a single M/E switcher and up to 600 ns for a large system. Typical switcher internal delays are shown in Fig. 14-22. They consist of the input matrix delay plus a delay equal to that of each mix/effects system. Since each path must be equal from input to output, the shortest path must be delayed to equal the longest path, the longest path being equivalent to that through the video matrix and all M/E systems in cascade.

Care in the installation and design of the input matrix and internal connections to the M/E amplifiers and delay circuits is essential to minimize path differences and reduce the number of timing adjustments.

The timing relationship of the output signals of a production switcher is less impor-

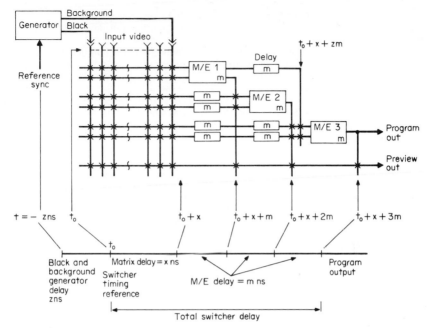

FIG. 14-22 Typical switcher internal-timing relationships.

tant, unless the program output is to be integrated with other time-related areas of a system. In systems where outputs of M/E systems are required to drive other equipment, such as routing switchers, time coincidence is desirable to avoid the need for delay amplifiers. Preview outputs usually do not need to be timed since they normally feed only preview picture monitors.

Live Studio Operation. Depending on the specific production format, the number of graphics captions, wipes, and digital effects required may vary enormously. Since a live studio must often be multipurpose, input requirements and monitoring must allow for many types of dramatic or variety productions, news programs, or commercials. In the example, three output monitors are used for the program, preset, and preview outputs. These allow current (PGM), preset event, and effects signals to be viewed and give the operator time to prepare a future event on the preview system while current and next events are always available. Input monitors (often b/w) can usually be smaller and provide instant observation of input signals for direct cuts during live switching operations. Some situations require monitors on the M/E systems as a permanent feature, especially if they are used for feeding a record videotape machine or other destination.

Remote Operation. Applications in an outside remote vehicle are either live, when used for sports or news events, or postproduction, when used to gather material for production at a later date. The main remote operation differences are that fewer inputs are usually needed and the switcher typically has less M/Es, primarily owing to less complex requirements but partly to space limitations. Since remote trucks vary greatly in size, small vehicles often have eight-input, 1-M/E switchers, while a large network vehicle may have the same switcher as in a studio.

Postproduction Operation. A postproduction switcher differs from a studio switcher primarily in its capability to be interfaced to a videotape editor. The specific needs of a postproduction facility depend primarily on whether the control room will be used: (1)

for on-line editing, (2) for off-line editing, (3) for both on- and off-line editing, (4) for on-air production, (5) with live cameras, or (6) with complex multilevel postproduction effects and titling.

There is also a trade-off between the size of the switcher (cost, number of mix/effects systems) and the time required for production, since complex effects that can be produced in a single pass on a large switcher can also be produced using multiple generations and a smaller switcher. The *average* and probably the most used postproduction switcher that represents a good compromise of features and cost might be a single-effects switcher with a switcher memory system, a separate mix row, and capability for two- to three-level keys.

Most of the newer switchers that can be used effectively for postproduction work are designed with a serial remote control, primarily for the editor interface. This feature allows control of practically all switcher functions and has facilitated design of editing systems with almost unlimited capabilities.

By comparison, remote control functions for switchers that were interfaced with an editing system by a parallel, one-wire-per-function approach, were generally limited to a few crosspoints on one mix/effects, wipes, and limited key functions. The development of more sophisticated editing techniques has provided total control of the editing functions of both machines and switcher using a standard or modified American Standard Code for Information Interchange (ASCII) data transmission code and compatible keyboard. This allowed a large number of switcher functions to be easily controlled directly or by programming into each edit event.

Even *on-the-fly* editing (switching manually during the edit process) can be done conveniently from the newer edit keyboards. The only need to operate the switcher manually is to preset the keys or effects to be used by the editor. Requirements for editors have changed, however, since using the newer switcher effects for postproduction requires controlling a multitude of real-time effects, previously not used or done off-line, such as digital video effects and graphic effects.

These effects are dynamic, with motion characteristics that can be synchronized by the switcher memory system to modify other switcher effects, or triggered to coincide with an edit event. While an effect can be taped and edited into the master to provide repeatability and accuracy, a better approach is to program the effect as a repeatable sequence on the switcher. This allows a higher degree of artistic expression by readjusting the sequence timing to match the program director's conception. The resulting production is completed without additional takes, and fewer tape generations are required.

The availability of more switcher features and digital effects for postproduction has resulted in the development of editing equipment with new capabilities. For example, the sequence start, memory, and recall can be controlled, as well as housekeeping chores such as preloading an initial setup and transferring switcher memory data between switching and editing equipment.

Thus, the editing philosophy of total control from the editing keyboard has been maintained, at the expense of an increase in operating complexity and operating burden on editing personnel. There is therefore a need for examination of the operator-machine interface to identify a better way for an operator to be in creative control of a number of diversified pieces of equipment in a postproduction environment. One solution is for the operator to edit using the switcher control panel as the primary creative control input. This is possible with switchers that can send operating status information back to the editing equipment. Thus an operator can operate the switcher in real time while the editing equipment records the operator's motions on the edit list. Using this approach, the operator can replay the sequence and continue to modify the video using the switcher control panel.

Whatever the technique, the art of postproduction editing is heavily dependent on the level of technology of both the editing equipment and the switcher, and the interface between them. Editing techniques and problems are dealt with in greater depth in Chap. 20.

Audio Switching.   In both live and remote applications, audio and video switching are usually handled separately.

Some production switchers incorporate audio or audio-follow-video (AFV) switching, which is primarily used in postproduction. Audio switching can be controlled by a separate audio control panel or by the selections on the video switcher panel. Audio switching architecture is typically a mix of video-type crosspoint controls operating on audio signals. It has the advantage of being easily adaptable to editor control. Its disadvantage is that only limited production is possible; thus audio production functions in this application are usually limited to those necessary to handle standard editor commands. More complex audio production can still be edited at the same time as video if the audio is handled by a separate mixing system. If extensive audio *sweetening* is required in postproduction, video can be edited first and audio can be remixed at a later time.

Because of the increased time and expense required for separate audio-video editing sessions, a new generation of audio-follow-video switchers with increased production capability is now available, usually as an option to the video production switcher. Features like memory control for audio function and auto transitions under memory control for multiple audio sources make it practical to do more audio production along with the video editing process. Even so, an audio console is usually present in an editing suite to handle more complex production.

The audio-follow-video switcher can be incorporated into the system either upstream or downstream of the audio console (Fig. 14-23). The audio inputs can be paralleled to both devices. Depending on the type of operation, the AFV switcher can also be fed from the console output or can feed an input to the audio board. Another possibility for AFV switcher interconnection is shown in Fig. 14-24. The AFV inputs are taken from the audio console preamp outputs, allowing the audio board faders and equalization to be used as preset conditions to the AFV switcher.

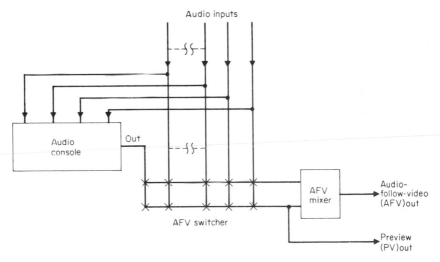

FIG. 14-23   Basic audio-follow-video (AFV) switcher configuration.

14.6.3 TECHNICAL DESCRIPTION

System Integration.   In an overall block diagram for a television studio the production switcher may be considered the reference for system alignment. In other words, by monitoring the switcher program output, much of the final system alignment can be accomplished. This places a number of important requirements on system alignment.

A switcher should, by its specification, be transparent to video through all paths. The timing, response, and gain must therefore be the same from any input to the main output

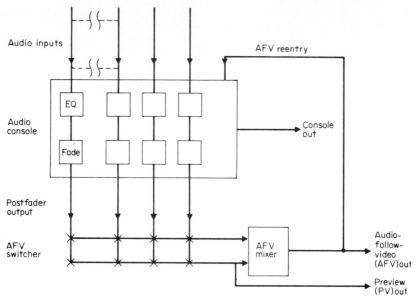

FIG. 14-24 AFV configuration using console equalization and level controls.

within close tolerances: (1) within 3° of phase difference at subcarrier frequency, (2) within 2 percent of gain difference, and (3) within 0.2 dB of frequency response difference over a bandwidth well in excess of the transmission system standards.

Ideally, all video inputs should arrive as nominal signals, correct for gain and phase and free of distortion. Owing to cable lengths, equipment variations, amplifier drift, etc., accurate timing is the most difficult specification to achieve and maintain, and most switchers therefore include a vernier timing control at the input. Except for this adjustment, the switcher should not be used to compensate for distorted input video and sync, or downstream equipment errors so that source signals can be patched to different inputs without system alignment changes.

When a system is planned, equalization and timing requirements should be taken into account as well as overall system delays around a switcher. Figure 14-22 shows inputs and outputs for a typical switcher as well as the timing requirements referenced to the video inputs. The requirements stated for advanced RGB signals are typical of the compensation for delay in camera encoders. The delay in the chroma keyer can be trimmed over a wide range to accommodate system requirements.

Input Video. The input circuitry of the switcher is generally high impedance, allowing for compensated loop-through or terminated feeds. Terminated feeds are preferable, though proper input-stage design with low return loss (40 dB typically) will permit short loop-through connections to an auxiliary switcher or other device without introducing signal degradation due to reflections. On high-quality switchers, a differential input removes common-mode noise, and a dc restorer accurately sets all video inputs to the same dc blanking reference. Input video is normally composite; however, noncomposite sources typically can be used, with sync and the FCC-specified 7½ percent fixed setup (NTSC only) added at the switcher output as required.

Crosspoint Matrix. The crosspoint matrix can be source- or destination-oriented, depending on system board design. Input orientation has the advantage that boards can be serviced without affecting the operation of other inputs, while destination-oriented systems provide some advantages in system control.

The matrix size is determined by input-to-output size, and many variations of number of crosspoints per circuit board can be accommodated. The main criterion, however, is to provide identical path lengths from all inputs to all outputs to maintain minimum phase variation as signals are routed through the system. Most systems use a motherboard or large circuit board for interconnecting the crosspoint boards to keep path lengths constant while distributing the signals to the crosspoint system. Two common techniques are: (1) to use contacts on the underside of the board which mate directly to the motherboard, and (2) to use three edges of the circuit board to supply input and output connections using zero-insertion-force connections for the sides of the boards.

Another important requirement of the crosspoint system is to provide signal switching with minimum disturbance at the switch point. The switching point is typically during the first full line after the vertical pulse period (line 10), although no standard has been established for this. Since the video has already been dc restored at the input buffer, there should be very little difference between the dc level of the signals being switched. Any difference will cause a dc shift at this point in the system. Clamping of this dc level is performed later to remove this error. Newer systems employ transient-suppression circuits to virtually eliminate this error.

The crosspoints per se usually consist of complementary transistors that are either reverse biased to block the video signal or forward biased to pass the signal. The main criterion of the crosspoint is to prevent the input video from leaking through to the output when the crosspoint is turned off. A common method is to use a control line that exhibits a low impedance to ground to turn the crosspoint off. This reverse biases two series transistors to present a very high impedance to the input signals. The result is high impedance at the input, a low impedance to virtual ground, followed by another high impedance to the video output bus as shown in Fig. 14-25.

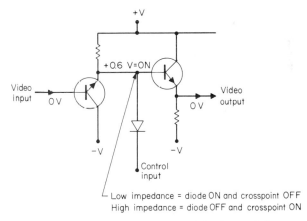

FIG. 14-25 Typical video crosspoint, where LO indicates diode on and crosspoint off and HI indicates diode off and crosspoint on.

Control of the crosspoints can be single wire or coded in a variety of ways to minimize cabling for each input bus. The control system is dealt with later in more detail; however, some specific requirements of the crosspoint system are listed below.

1. *Crosspoint priority.* Only one crosspoint can be on at one time. This is usually implemented using a priority system. An example might be that if two buttons were pressed

simultaneously, only the first source button pressed would be latched (excluding the other).

2. *Vertical interval switching.* To prevent picture disturbances and other problems, the video switch must take place only at the vertical interval switch point and usually requires that the pushbutton be held down during this time.

3. Crosspoint reset. Selecting any crosspoint must reset any other previous selection. This is automatic in a coded system, but requires separate circuitry in a wire-per-crosspoint system.

   Mix/Effects System.   The outputs from the crosspoint buses are fed either directly or via delay and reentry amplifiers to the mix/effects amplifier. A typical M/E amplifier block diagram is shown in Fig. 14-26. The delay system is necessary in some paths to provide timing coincidence between direct video bus outputs and video outputs from another mix/effects system. Crosspoint selection is then provided between the two paths as an extension of the direct or primary bus. The delay system is a bulk delay, typically 50 to 100 ns, and is adjustable to match the delay of the reentered video from the mix/effects system. Delay-reentry crosspoint requirements are similar to those of the primary matrix.

1. *Inputs.* The number of video buses feeding a mixing system can be from two to four, depending on the system; two inputs are common in smaller or older systems. The two main inputs are used for basic mix and wipe operations and are usually referred to as A and B buses. In systems with more inputs, the additional sources are used for adding keys over the A or B buses. These additional buses are usually associated with external key input buses or camera red, green, and blue switching for use in chroma key operations. The requirement for all video and external keying signals is that they should be time coincident at the combining point in the mix/effects amplifier.

2. *M/E amplifiers.* As mix/effects systems vary considerably, we will consider a conventional two-input system first and then consider systems with three or four inputs. The main feature of a mix/effects system shown in Fig. 14-26 is the mixing or combining

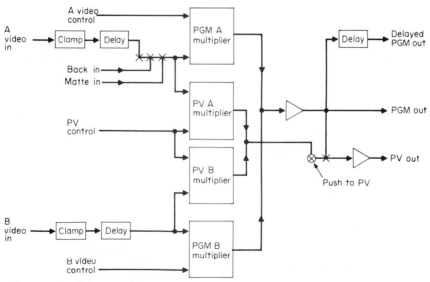

FIG. 14-26   Mix/effects amplifier-section block diagram.

arrangement. This consists of two multipliers that can select various combinations of A and B video to be summed to form the mix/effects video output. Input clamping and previewing features are also provided with additional switching for black and matte signals in the A channel. The selected and sync-tip clamped video is applied to the mix/effects amplifier clamps. Here it is back-porch clamped by a self-driven pulse generator to produce video with a stable blanking level and, therefore, a stable black level for the multiplier circuits (self-driven clamps are used to provide correct clamping even if a nonsync signal is applied). The A channel has an additional two crosspoints providing access to color black and the internal color-matte signals for specific operations.

The two multipliers which control the video amplitude provide linear control so that, as the control voltage changes, a corresponding proportion of the video is produced at the output of the multipliers. For example, when mixing from B to A with the control voltage reduced by 25 percent, the resultant video would consist of 75 percent B video and 25 percent A video, the sum of the two multipliers always producing 100 percent over the range of the control voltage. For key and wipe operations, the only difference is in the control signals which are generated by the wipe and key circuitry (discussed below). In some switchers separate control of A and B video is possible to give A and B sums which are less than 100 percent, with control limiting used to prevent summing to greater than 100 percent. In true additive mixing with no control limiting, care must be taken to ensure that the two video signals do not sum to greater than 100 percent. Linear multipliers are used to provide good cross fading from A to B video so that the two signals sum to unity; this is best observed when mixing with the same signal on both A and B buses.

The mixing operation referred to affects the active (picture) video only. During the horizontal and vertical blanking period, the control signals can perform one of two functions: (1) gate sync and burst from one of the two video channels, or (2) mix sync and burst during the transition. In the first case, sync and burst are switched at the end of the transition when a limit is reached. Sync and burst presence detector logic provides selection of the correct sync and burst sources when using one color and one monochrome source, and inhibits operations when sources are nonsynchronous. If the video is slightly out of time, a horizontal shift would occur at the lever arm limit when sync and burst are switched. In mix mode, a horizontal timing error would create a step in the sync edge which could cause serious problems such as horizontal jitter, if allowed to appear on the output of the system.

The color black input to the A channel is selected when the control system uses separate control of A and B video and reduces both channels to zero. In this position, neither A nor B video appears on the output, and so logic selects the color-black input and turns the A channel full on to provide full-amplitude color-black at the output. The color-matte signal is used for two basic functions: (1) to provide color fill for borders on wipe patterns, and (2) to colorize title or other key modes of operation. (Selection of these modes and the appropriate control signals is produced by the key and wipe system discussed subsequently.)

Mix/effects amplifiers also provide previewing capabilities. In the example, a monitoring mode provides an output of the program video to enable the mix/effects output to be examined before taking to air. An additional selectable mode provides access to a preview multiplier system that can combine the same video that feeds the main output. This allows an operator to look ahead at a combination of video sources to check key modes, border colors, etc., before using them on the mix/effects output. This function is available while the mix/effects system is selected on the switcher output and utilizes additional control signals to preview the same combination that will be seen when the transition is actually made. Switchers use this system to provide a push-to-preview capability which also overrides the preview bus selection while a specific mix/effects title-clip control is held down. This allows instant monitoring of titles prior to on-air selection.

Systems that provide additional input sources are typically used as keying inputs to enable the A and B buses in our example to provide mixing or wiping operations behind a key. This third input can usually be controlled by the operator independent of the A and B buses, but can be tied to the main control system to provide wiping of A and B video while simultaneously bringing on the title key signals. For systems with four inputs, a combination of two keys such as a title key and chroma key can be used over the background A or B video signals. Control of these larger mix/effects systems must be carefully designed to allow simple and logical operator control while providing complex combinations of video sources.

Key and Wipe Systems. Both *key* and *wipe* functions are similar operations since they control switching from one source to another at specific points during the active video. Keying, however, is related directly to a video signal, whereas a wipe is controlled by a nonvideo signal such as a ramp or triangle waveform.

1. *Keying.* A basic keying operation is illustrated in Fig. 14-27 using a simplified scanning system and three lines as examples of a monochrome signal. These are fed to the key system where the video is compared with an operator-selectable clip level. This produces the switching or key signal which is used to feed the mix/effects system. In this example, wherever the incoming monochrome signal is above the clip level (white), the key output is used to switch the A video to the output. Wherever the incoming monochrome signal is low, the mix/effects system is switched back to the B video. This is called a *luminance key* where the level (amplitude) of the incoming monochrome signal is used to control the switch. If this signal is applied to the A input as well, the result would be a white fill from the level (amplitude) of the incoming monochrome signal. This is referred to as a *self key* where only two signals are used for sources and one provides the key signal. Key operation is often described as the key signal cutting a hole (shape of the letter T) in the B video, with the hole being

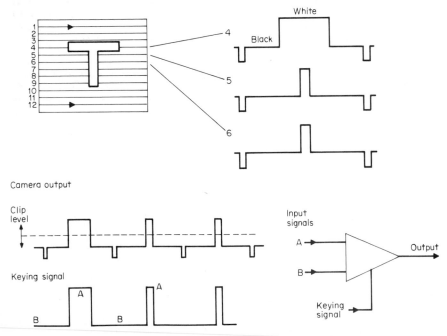

FIG. 14-27 Video-keying processing.

filled by the A or matte video. Some variations of the basic luminance key mode are listed below:

2. *Key invert* is a mode that inverts the key signal so that the positions of the A and B video are reversed and the B video appears in the key hole. This is typically used to produce a normal key when the source is black letters on white background.

3. *External key* is a control that allows selection of either the internal key (self key) or an external key input to be used to derive the key signal.

4. *Key borders* are quite often incorporated into the mix/effects system to enhance a basic luminance key by adding a border or outline effect. Three examples are:

   Border: provides a symmetrical border around the key.

   Shadow: produces an offset border usually to the right and below the key signal.

   Outline: replaces the original key with a signal which is an outline of the original key.

5. *Matte fill* is usually a standard feature that provides a color fill with adjustable luminance, chrominance, and hue levels to colorize a key signal. When used with bordering modes (that can also be controlled), matted titles produce signals that will contrast with any surrounding picture content.

6. *Key follow* is the ability of a keying signal selector to follow the appropriate fill video. Many character and graphics systems provide a key signal and video fill for production switchers. It is usually desirable to select the appropriate key for the particular fill being used without having to make both selections on the panel.

7. *Chroma key* is a key that differs from a luminance key in that it is derived from the color components of the picture rather than brightness components. Thus, keys are possible from blue, green, orange, etc. For best results, the red, green, and blue signals direct from the camera should be used; however, where these are not available (such as VTR signal), an encoded system can be used to decode the video back to component level. Encoded keys, however, contain encoding and decoding errors, as well as additional noise.

   A basic chroma key is shown in Fig. 14-28 where talent standing in front of a blue wall is keyed into another picture. The chroma key system selects only blue signals to produce the key output. Although any color can be used, green and blue are typical. Best results are obtained with high saturation colors, while the talent and other parts

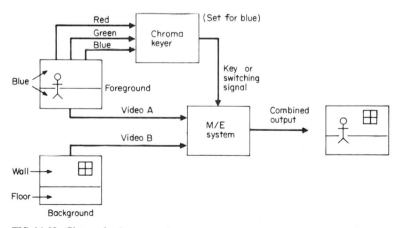

FIG. 14-28   Chroma-keying processing.

of the set must be different colors to prevent being keyed as well. In the example, both camera pictures must be aligned to enable the talent to move in correct relation to the inserted scene. A blue object in the foreground scene allows the talent to move behind the background or disappear "into" the background scene.

In some systems, shadows from the talent on the blue set can be reinserted in the background scene by use of a shadow function. This uses luminance variations to reduce the intensity of the background scene depending on the level of the shadow signal. The clip level is also used in the chroma keys to select the range of colors giving a key output. In chroma keying, a gain control is also often used to allow some control over the edge transition. Low gain allows the rise time of the key signal to be controlled by the incoming *RGB* sources to provide a soft effect or mix at the edge of the key. High gain provides edges but is more susceptible to noise.

Key-follow modes are usual for chroma key so that *RGB* signals are switched together with the foreground camera selection, although separation of the camera and key signals is usually provided for keying in different foreground signals to provide unusual effects.

8. *Wipe systems* can be used either to make a transition or as an effect to divide the A and B video sources, such as a corner wipe. Wipes differ from keys in that the switch point is related to waveform such as a ramp locked to horizontal or vertical scanning rates and is generated in the switcher itself.

A simple wipe system would consist of a few basic wipe patterns and be produced from ramp waveforms compared with the control voltage. A vertical wipe would be produced by comparing a horizontal ramp with a control voltage (Fig. 14-29). As the control voltage moves positive from zero, the comparison point varies along the ramp. With the control voltage midway at $+0.5$ V, the output waveform switches in the middle of the active line. This waveform then produces an A to B transition midway along every line so that part of the B picture appears on the left and part of the A picture on the right.

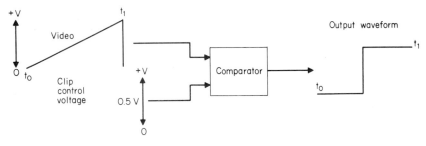

FIG. 14-29   Basic wipe-signal generation.

More complex wipe systems generate additional waveforms for multiple patterns, provide rotational effects, and allow wipe pattern border, modulation, and position modifications to be made. Simple patterns can be easily produced by logical combinations of different waveforms. However, the production of more complex patterns is better handled by a microprocessor. While the power of the digital control is necessary, and all pattern waveforms can be created digitally, a combination of analog and digital techniques often provides better performance.

Three main functional areas in a pattern generator must be optimized to produce the desired visual result. The requirements of a typical pattern generator are listed below:

a. Waveform generation. Generation of basic waveforms that are locked at horizontal (H) or vertical (V) rates to the program video. These can be generated by either analog or digital techniques.

b. Pattern selection. A wide variety of simple and complex wipe patterns are generated by processing the basic waveforms. Inversion, adding, borders, controlling level as a function of lever arm position or a sine-cosine algorithm that produces pattern rotation are typically best handled by a microprocessor.

c. Video-key signal. The linear waveform that produces soft or hard pattern or border edges can be produced conventionally by comparing the wipe waveform with the lever arm (size) control voltage, or it can be produced completely digitally by taking digital pattern information and multiplying it with digital lever arm information. Both techniques have been used. Considering cost and results, the analog technique is preferred. Cost is incurred in the digital approach because high bit rates are required to produce acceptably smooth visual transitions.

A typical modern production switcher uses a dedicated microprocessor for pattern generation and uses a combination of the design options described above:

a. Basic vertical waveforms are generated by a 12-bit D/A converter from a clock locked to vertical rate. Horizontal waveforms are produced by an analog integrator.

b. Pattern waveforms are produced using a dedicated microprocessor to select waveforms for the desired single or multiple pattern, hard or soft, bordered or rotating effects. In a combination of analog and digital techniques, selected waveforms are controlled by digitized lever arm and control signals using multiplying D/A converters. The processor also provides the sine-cosine functions that modulate the basic H- and V-ramp waveforms, providing the rotation effect which can be applied to certain patterns. The pattern generator is described in more detail below.

The main waveform generator (Fig. 14-30) produces a simple H- and V-ramp output for single patterns. It produces a multiple-ramp signal or triangle for any multiple pattern. The greater the number of ramps, the greater the number of patterns. Sinusoidal or triangular modulation can be applied to the ramp generators, or an external modulation signal can be used. Control of the amplitude and frequency of the internal signals is also provided. Positioning of the horizontal and vertical ramps is controlled to allow patterns to be positioned anywhere on or off screen. Additional positioner modes allow wipes to start from anywhere on screen and to be completed within the control voltage range. The resultant single or multiple ramps are applied to the rotation generator where they are modified by cosine and sine multiplying factors to pro-

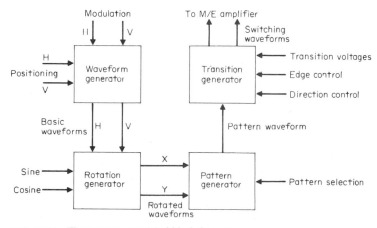

FIG. 14-30 Wipe-system simplified block diagram.

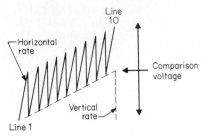

FIG. 14-31   Diagonal-wipe signal generation.

duce *rotary wipes*. A simple example of a rotating wipe can be used to explain how a diagonal pattern is produced.

The diagonal transition in Fig. 14-31 is produced by adding the horizontal waveform to the vertical waveform so that the resultant waveform, when applied to the transition generator, produces a switching signal that moves to the left progressively every line. When the amplitudes of the H and V ramps are equal, a diagonal at 45° is produced. To produce this effect from the rotation circuitry, two multipliers and a summing amplifier are required as shown in Fig. 14-32. The multiplying constants of sine and cosine for 45° are 0.7, and so the sum of $0.7V + 0.7H$ is the sum of 70 percent of both H and V ramps. For different angles of rotation, the H and V ramps are added with varying amplitudes. At 0° and 90°, sine or cosine is 0 or 1 and only H or V ramps are produced. By using both positive and negative values of sine and cosine, amplitudes and polarities of H and V ramps can be summed to produce any angle of wipe. The Y output has sine and cosine reversed, and so it is always 90° from the X component. This is needed for cross- and square-type patterns.

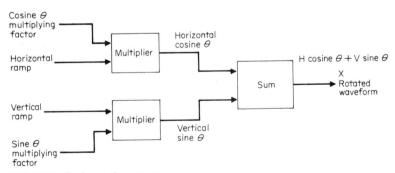

FIG. 14-32   Basic rotation system.

The combination of H and V ramps used to produce the X and Y waveforms allows any of the patterns made from them to be rotated. The specific pattern produced is determined by the pattern generator circuitry selected by the pattern select pushbutton. Patterns produced from the basic H and V ramps can be categorized into three groups:

a. *Parallel-edge patterns* are patterns whose edges are parallel to the H and V axis before any rotation is used. This includes basic H and V wipes, corner wipes, and square and cross patterns. These are produced by combining the H and V ramps in a nonadditive or gating mode.

b. *Nonparallel patterns* are patterns that have straight edges but are not parallel with the H and V axis, such as any diagonal, diamond, or star. These are produced by summing the H and V ramp components.

c. *Circular patterns* are produced from parabolic or squared functions of the H and V ramps. Triangular waveforms can be added to produce the ellipse and heart shapes.

The resultant waveform produced by the pattern generator is applied to the transition generator where it is used to produce the final wipe control signal. On the transition generator, two comparisons are made to produce the control waveforms for both edges of the border (Fig. 14-33.) The control voltage applied is modified by the border width to generate two waveforms which control the A-to-border and border-to-B transitions. Gain control of the comparator is used to change edge softness. High gain is used to provide fast transitions for hard edges and is usually filtered to prevent generation of transitions faster than those allowable in NTSC signal bandwidths. Low gain provides slow rise times for soft edges. The start and end of the soft transition should be gradual to give the best visual effect. Wipe direction can be changed by selecting waveform polarity before the waveform is gated by the blanking signal to determine sync and burst operation.

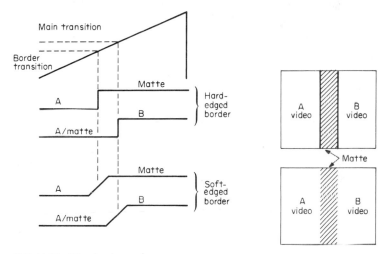

FIG. 14-33 Wipe border modes.

Additional wipe circuitry can be used to generate *clock wipes.* These consist of two basic functions to produce a moving arm (normal rotating transition) which is partly masked out by secondary switching waveforms to produce the stationary arm. As the moving arm rotates, additional sections of the waveform are blanked to produce the final clock-wipe effect.

Another type of wipe system is used to produce random *matrix wipes* such as those producing random square effects (Fig. 14-34). These are usually preprogrammed into a read-only memory (ROM) that is addressed by counters running at multiples of horizontal and vertical rates. These produce an output data word for specific sections per line and groups of lines on the screen. By using different ROM data, the resultant pattern can be changed for different effects. The data words produced are converted to analog signals and applied to the transition generator where they are compared with the control voltages.

Most wipe systems can also produce a preset or pattern-limited effect utilizing the same wipe system already described, but using modified wipe control voltages to pro-

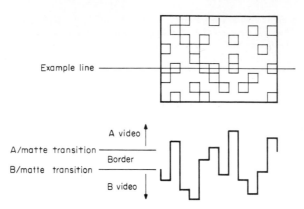

FIG. 14-34   Random-matrix wipe.

duce the limited-wipe effect. In conjunction with the mix function, mixes to preset patterns can also be produced. Most M/E systems combine mix, wipe, and key functions together to produce combinations of these, such as using a pattern to limit the area where a title key is present. The control of a M/E system reflects its complexity. A well-designed control system allows the switcher to be used effectively and efficiently.

Control Systems.   The control system in a switcher includes the control panel, the logic that controls the video crosspoints and mix/effects mode selections, and the tally system. The type of control system employed depends on the complexity of the switcher. Generally, information is sent from the control panel to the switcher electronics where it allows switcher function or energizes a video crosspoint. This can be done with direct, one-wire-per-function systems or by a form of multiplexing to reduce the number of control wires. The control system logic interprets panel commands and gates the control information sent to the actual crosspoints to ensure that video switching occurs in the vertical interval.

As switcher size increases, the logic interaction required between mix/effects systems, preview systems, and keyers increases considerably. After a point, microprocessors become a viable alternative to discrete logic, since they allow software to more readily solve the complexities of implementing logic control functions. The advantages of microprocessors must always be weighed against the possibility of a microprocessor failure that could shut down all functions that it controlled. As discussed below, using microprocessors in switcher control applications typically provides advantages that far outweigh the limitations:

1. *Memory functions.* Microprocessors provide the ability to memorize mix/effects set-ups which can be stored for later recall when needed. This provides assistance for operators to switch quickly from one complex condition to another when necessary, for fast-moving situations. Memory systems also allow sequences of events and simultaneous control changes to be performed readily by the switcher.

2. *Peripheral interfacing.* Another useful benefit of microprocessor-based systems is their ability to be easily interfaced to external devices. One such peripheral device is a tape or disc system for semipermanent storage of switcher setup conditions or conditions needed for use on a frequent basis, such as show openings. Editing computers can also be interfaced to M/E microprocessors to allow external control of all production switcher functions. Other external devices that can be interfaced to a switcher are still-store devices and character generator or graphics systems, which can be controlled by either the editor computer or the switcher memory.

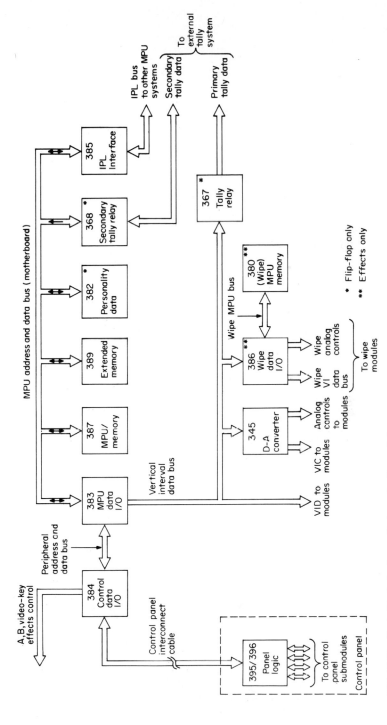

FIG. 14-35 Control logic for three-effects production switcher.

3. *System diagnostics.* Although this list is not comprehensive, another feature that can be mentioned is the ability to perform routine setup operations quickly, as well as self-diagnosis or external monitoring of system functions. This can be performed either by internal or external diagnostic systems.

Control Functions. The description of the control functions that follows applies to a typical, large-size production switcher.

Starting with the control panel, button pushes can be encoded in hardware or by a microprocessor into parallel or serial data that can be sent to the switcher electronics. Lamps are usually energized by local latches driven by encoded data from the switcher, providing feedback to the operator that a switch did take place. Analog functions may be sent directly to the switcher or converted to digital format before being sent. There is an advantage to sending digital information when the data will be further processed digitally or when it will be digitally stored by the switcher memory system.

Good design practices can reduce the chance of system failure due to failure of a single component. Typical production switchers now use multiple processors, allowing each effects system to operate independently. In addition, panel pushbutton data are encoded by hardware logic and, in case of microprocessor failure, are patched through directly to the video crosspoints, to provide fail-safe crosspoint control.

Within the processor-based switcher, data from the control panel are distributed to various functional groups in each effects bank: crosspoint control logic, key and wipe logic, and memory system. Although the exact routing and control mechanism depend on the switcher design, Fig. 14-35 shows the control logic for a three-effects production switcher.

In this example of a large switcher there are as many as eight microprocessors, each performing a specific function. Because of the many interrelated functions, it is ncecessary that these processors communicate with each other. A parallel data bus serves this purpose, interconnecting the master processor and the three effects systems to allow interchange of status, tally, and other information. While the design concept is complex, a digitally controlled switcher with an interprocessor bus has many advantages. Design changes can usually be implemented in software. New or optional features can be added by a single interface to the bus. The optional features (external remote control, digital video effects, memory functions, etc.) can interchange information with any part of the switcher via the bus and therefore can appear to be fully integrated operationally.

Remote Control. The interprocessor bus also provides an excellent method for supplying remote control of the switcher. Switchers using hardwired logic typically require hardwired remote control functions, which usually result in a minimum of functions available for remote control. With a software-based control system as described, remote control can be accomplished by providing another processor that interprets remote control information at a data port and loads it onto the interprocessor bus.

Figure 14-36 is an example of a serial interface system that can be used to interface editors or other peripherals, or to perform maintenance functions. The ability to communicate through the port to most functions in the switcher provides a variety of external control functions:

1. Under editor control, the switcher reacts to commands to set crosspoints, start transitions, recall preset memory registers, load memory registers. The switcher can send handshakes, status, and the contents of memory registers.

2. When communicating with a peripheral device, the switcher interrogates a character generator; for instance, to learn its status, the switcher can later send the same information to return the peripheral to the same status.

3. For maintenance or setup purposes, an external computer can be connected to set up nominal paths for alignment and perform total function checks.

Audio. Audio systems used with video switchers have an architecture similar to that of the video path. Typically, sources are selected on two buses which are then fed to a

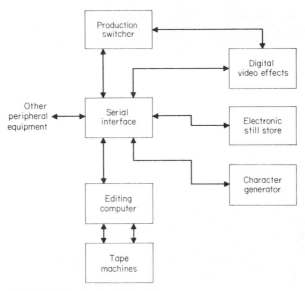

FIG. 14-36 Typical serial interface system.

crossfade mixer. Variations on this structure might include a second mix bus or an audio under/over bus.

A control panel may not be required in an AFV-only system. The control logic is similar to that found in the video portion of the switcher.

The audio crosspoints may be relays, FETs, or other solid-state devices. A solid-state design is preferred for an audio switcher under editor control where precise timing and repeatability are desirable. Solid-state designs are more prone to dc offsets which can result in switching clicks unless close design tolerances are maintained.

The mixer is usually a dual voltage controlled amplifier, with a crosstracking characteristic that yields approximately constant loudness as two sources of program material are crossfaded. Mixing circuitry typically used can be FETs, operational transconductance amplifiers (OTAs), or variations of the OTA, using a form of feedback in the control or audio path to reduce distortion and improve control conformance to a desired linearity curve.

Other functional blocks in the audio system are conventional, consisting of buffer amplifiers and line-level output amplifiers as required.

14.6.4 SMPTE STANDARD DIGITAL CONTROL INTERFACE. The serial remote-control port plays an important role in linking the switcher with other intelligent devices in a studio application. Because of the large number of intelligent devices in a studio that could be linked together for purposes such as editing, machine control, or other special purposes, a standard digital control bus has been proposed by the Society of Motion Picture and Television Engineers (SMPTE). The proposed standard consists of three parts, as described below.

Mechanical Standard. A nine-pin female D connector is used on each piece of equipment. A number of devices can be connected to the bus; however, one of them must be the controller. Communication is by separate send and receive balanced pairs. Data are sent over the bus at levels compatible with the EIA RS422 standard at a rate of 38.4 kbits and at distances up to 4000 ft (1219 m).

Bus Protocol. The bus is controlled by a few standard control characters and an intializing break character. In operation, all devices respond to the break character and only those devices addressed respond to further data. A special mode allows two devices to escape to their own protocol and communicate for any period of time, independent of bus protocol. This feature makes it possible for a wide variety of equipment in a studio application to make use of a standard bus for communications without burdening one device with the more complex protocol requirements of another.

Standard Messages. By standardizing on a number of specific messages that may be sent on the bus, such as motion commands for video tape recorders, certain standard control functions can be accomplished in a system without the need for customized interface software for different manufacturers. Overall, the standard is well thought out and very versatile. It provides for over 200 discrete tributary addresses and can handle devices of widely varying complexity on the same bus.

It would not be practical, though, to use one bus as a universal data bus for all control requirements in a studio. Several different buses would likely be used for different groups of devices, such as video tape editor machine control, studio machine control, switcher-editor control, and switcher-peripheral control. In the case of the switcher-editor interface, only two devices are interconnected by the bus. Standardization is still important, though. It allows for easier interface between different equipment manufacturers, the possibility of standardized test equipment, and a telecommunications bus that can be universally understood.

Other production devices that may use a SMPTE control bus include switcher memory disc drives and switcher peripheral devices such as character generators, still-store devices, graphics systems, and digital video effects systems.

14.6.5 SPECIFICATIONS. Production equipment specifications can be divided into two categories—those that determine the required video performance for proper system operation and those that determine the utility or performance for the user. Present equipment is nearly transparent to video, as far as visible distortion is concerned. Tight specifications are important primarily when the whole system is considered, the whole system being a local closed-circuit operation or a nationwide network involving hundreds of DAs and dozens of switches between the source and the home viewer.

Specifications that affect the visible performance of the equipment should be understood and considered when choosing equipment for a system. A few of the visible specifications for switchers are described below.

Mix/Effects Characteristics. Mix tracking error of less than 0.5 percent assures that identical portions of the picture will not visibly change level during a mix.

Production switchers should add minimal distortion to the video path as with other on-line video equipment. Besides the specifications normally included for video equipment, additional parameters are also important as is the path used for these measurements. Production switchers offer a selection of paths from any input to the main program output. The specifications are normally given for the *longest path.* The longest path is the path to which all other paths are delayed to be made equal and is usually the path that cascades all M/E systems to the output. This adds the largest number of gain-setting components in series and usually causes the most distortion.

Input Characteristics. These characteristics are typically standardized for 1 V peak-to-peak of composite video into 75Ω, and most systems provide for noncomposite inputs as well. Return loss and 60-Hz rejection are also specified for differential input systems. Specifications for subcarrier and reference pulse inputs are usually less important and can often tolerate unterminated operation.

Output Characteristics. These characteristics usually relate to program output and are specified at 1 V peak-to-peak into 75Ω with return loss and output dc mentioned as well. More than one output is available on most systems, and isolation between outputs is an important consideration. These specifications do not necessarily apply to all outputs of preview and M/E systems, since these are sometimes designed solely for monitoring functions.

Video Path. The longest path is used for these measurements with the control lever arms on a limit (A or B), and generally cover the following:

1. *Frequency response.* Often flat over most of the video spectrum within ∓0.2 dB of reference with a gradual roll-off as frequency increases.

2. *Waveform distortion.* Differential gain, differential phase, and K ratings of test waveforms are normally in the 0.5 to 1.0 percent range, or 0.5 to 1.0°, depending on the specific rating.

3. *Path differences.* Gain and phase are defined as a range of percentages or degrees to show path tolerances through various M/E and delay combinations.

4. *Stability.* An important characteristic of a switcher is stability of the parameters such as gain and dc levels. These are almost more important than the specifications for gain and dc, since a short-term change is usually more noticeable than an overall long-term error.

14.6.6 DIGITAL VIDEO EFFECTS.   Digital video effects or *digital effects* are created by digitizing the video effects signal so that it can be stored, retrieved, and manipulated in a digital format. Digital effects include all video effects that can be produced by moving the picture on the screen, changing its size, or giving it motion. To produce these effects, the individual picture elements must appear on the screen at locations different than in the original picture. This can be accomplished only by writing the picture into a portion of the memory that represents the normal raster, and then reading the processed picture from the memory at a later time. A digital video effects device can be a stand alone unit that adds special effects to one or more video signals, or it can be an integrated unit that is part of the production switcher, much the same as wipe and key special effects are integrated.

Features.   Similar digital effects are possible whether or not digital effects are an integrated part of a switcher, although more creative flexibility can be achieved with an integrated system. Some common terms that describe these effects are:

Compression

Expansion

Reversal

Splitting

Decay

Freeze

Resolution or data-bit reduction

Perspective

Rotation

A digital-effects unit can be separated, for purposes of discussion, into two parts—the digital-effects processor and the digital-effects controller. A typical digital-effects system consists of the control panel, controller electronics to generate the required control signals to define the effect, and the digital video processor which uses the control signals to control the video signal in the digital domain producing the desired effect.

The digital-effects processor digitizes the analog television signal and performs all the high-speed processing that is required for digital manipulation of the picture (compression, decay, rotation, etc.). The digital-effects controller typically provides the human interface between the switcher and the digital-effects processor. In some cases, the controller also provides the system integration of the digital effects. The effects controller normally sends the data required to control the effect (effects mode, size, rotation, etc.) to the effects processor, once per field to control the digital effect. The digital-effects processor and the effects controller may or may not be physically separated; however the operations are quite different. Thus, the effects processor is described in Chap. 19, while the effects controller is described here.

System Considerations.   Digital effects can be used in a wide variety of operational environments. For example, on-air and post-production are clearly very different environments that both use digital effects. The following is a list of system considerations whose relative importance will depend on the operational environment.

1. *Video input source selection.* How are the sources selected? Are they selected by a separate routing switcher or is the selection interpreted with the effects controller? How many input sources are required?

2. *Digital effect video and key output.* Normally, the digital-effects output is connected to the input of a production switcher to be combined with other video sources as part of the final picture. To accomplish this, a key signal must also be supplied by the digital-effects processor and must be of a form that is usable by the switcher. Many digital-effects processors provide the capability of keying the transformed picture onto a matte background. Without a production switcher, the final picture would be limited to the matte background.

3. *Control delegation.* To save costs, it is sometimes desirable to share one digital-effects processor among several studios or postproduction suites with the ability to delegate controls to one studio or suite when needed.

4. *Video tape editor control.* In postproduction applications it is necessary to control digital effects from a video tape editor. This control is normally provided through a serial interface.

Stand-Alone Digital-Effects System.   A controller for a stand-alone system provides the human interface to the digital-effects processor and would not normally include interfaces to other devices. There are two basic types of human interface to the controller—a menu-directed interface and a dedicated interface.

In the menu-directed interface, the human interface inputs are *soft* keys, i.e., switches and analog controls whose functions change depending on which mode or *menu* is selected. The human interface output may include a CRT display which would also change with each menu selected. The advantage of the menu system is that many control functions can be accommodated with a small number of switches, potentiometers, and display devices. Its disadvantage is that it may take some time (i.e., several menu selections) to get access to any particular operation.

The dedicated interface provides an input device (pushbutton switch, potentiometer, etc.) and a display indicator for each operation. The advantages and disadvantages are the opposite of the menu system. Quick access to all functions is provided at the cost of a large number of control devices and the accompanying large control panel.

Most systems are a combination of the two basic types. The mix of menu-directed and dedicated interface functions must be evaluated on the basis of what would work best in the particular operational environment.

Integrated Digital-Effects System.   An integrated system with its digital data and control paths (video paths are not shown) is shown in Fig. 14.37. The major components of the system are:

1. *Mix/effects system.* Each mix/effects system and the flip-flop system of the switcher are controlled.

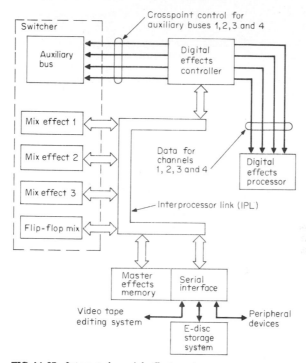

FIG. 14-37   Integrated special-effects system.

2. *Digital effects unit.* This component consists of the digital-effects processor and the digital-effects controller.

3. *Master effects memory.* This component learns (stores) and recalls all the control parameters of the switcher and digital-effects unit.

4. *Serial interface adapter.* This provides communication with external equipment such as a video tape editor, disc storage system, and other peripheral devices.

5. *Auxiliary video crosspoint bus.* This component selects the video input to the digital-effects unit. The selection can be made either manually or by the digital-effects unit.

These components communicate with each other via a high-speed data bus called the *inter processor link* (IPL). The IPL is a significant factor in the integration of the digital effects unit, the switcher, and peripheral devices.

The integrated system provides the following capabilities in addition to those found in a stand-alone system:

1. *Video input sources* can be selected by the digital-effects units, which allows changing the input source to become part of the overall effects, for example, changing the input source when the picture goes to zero size or changing the input to produce a sequence of frozen pictures each from a different source.

2. *Video tape editors* can control both the switcher and the digital-effects unit from one serial port.

3. *Effects memory functions* can include both the switcher and digital-effects unit. (The effects memory stores the condition of all the pushbutton and analog controls in

memory where it can be recalled as required to accurately repeat the digital-effects operation.)

4. *Additional functions.* By using the video output from the digital-effects unit as an input to the switcher system, additional functions can be performed, for example, on-air tally and improved key or wipe-tracking operations. (*Key* or *wipe tracking* is the ability to make the compressed picture fit into a key or wipe shape and follow or track that shape as it moves.)

Technical Description—Effects Controller.   Digital-effects controllers are micropro-cessor- or computer-based units. Most of their capabilities are implemented in software and vary widely from one manufacturer to another. The following description is based on an integrated effects controller whose architecture is shown in Fig. 14-38. The hard-ware architecture is described first, followed by a general description of the software modules.

The system hardware includes an 8086 microprocessor with 60 kilobytes of program-mable read-only memory (PROM) for program storage and 16 kilobytes of random access memory (RAM) for data storage. The microprocessor input and output information con-sists of the following: (1) bidirectional control panel data, including pushbutton and ana-log control input information and indicator and numeric display output (status) infor-mation; (2) output data to the digital-effects processor; (3) crosspoint select data to control the auxiliary video bus; (4) key and wipe signals used for the size- and effect-follow operations; (5) bidirectional IPL data used for communication with the switcher, serial interface adapter, and master effects memory. The serial interface adapter allows communication with video tape editors and disc storage systems.

The software consists of three types of routines: initialization routines, interrupt ser-vice routines, and processing routines.

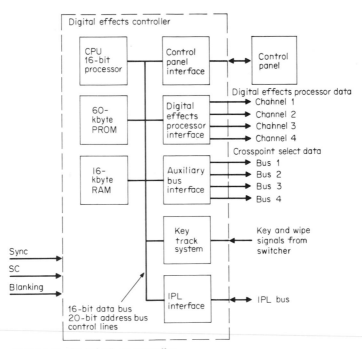

FIG. 14-38 Digital-effects controller.

The initialization routines are called only when the program is reset. A reset occurs on power up or when a reset pushbutton switch is pressed. Initialization routines perform functions such as (1) clearing the memory, (2) initializing the I/O ports, and (3) setting the interrupt vector table.

Interrupt service routines provide input or output of data with little or no processing. The size follow (key tracking) interrupt routine is an exception to this rule. It does considerable processing in order to eliminate an additional field of delay between key or wipe signals and the digital-effects picture.

There are basically two types of interrupts—those that are synchronous with video sync and those that are asynchronous. The synchronous interrupts are:

1. *Size follow:* occurs on line 1 of every field
2. *Aux bus:* occurs on line 4 of the field in which the crosspoint changes
3. *Effects processor output:* occurs on line 10 of every field
4. *Analog input:* occurs for every analog input (that is, 13 times each field)

The asynchronous interrupts are:

1. Pushbuttons
2. Lamp outputs
3. IPL input and output

The processing routines are scheduled by the base line scheduler program. All the processors run exclusive of one another; i.e., one processor will not interrupt another processor. Each processor runs to completion before another can start. The processors are interrupted by interrupt service routines; however, with the exception of size follow, they input or output data without processing it.

With this structure, the effects generator processor will not use partially processed data such as data from the pushbutton or analog processors.

Each software function/module is briefly described below:

1. *Initialization.* The initialization routine sets up the hardware and software for the effects controller. This includes segment registers, stack pointers, I/O port, RAM, interrupt table, lamp output port, displays, and control panel.

2. *Base line scheduler.* The base line scheduler routine schedules the base processing routines, which are pushbuttons, analog signals, effects generator, effects memory, interprocessor link, and lamps.

3. *Pushbutton interrupt.* The pushbutton input routine is an interrupt service routine. Depressed buttons are identified and saved for later processing. At the end of every control panel scan, the button status is grouped into three sections. These are buttons pushed, buttons released, and buttons held down. The pushbutton processor and lamp processor are then scheduled.

4. *Lamp processor and interrupt.* The lamp processor routine takes data from the pushbutton processor program and prepares it for output to the lamps. It is scheduled by the pushbutton input routine at the end of scan. The lamp output routine is called by the lamp interrupt and gives the control panel one group of lamp information for each interrupt.

5. *Effects memory.* The effects memory (E-MEM) processor is scheduled each time an effects memory operation is performed. The processor defines the type of effects memory function, inhibits illegal operations, and performs the necessary transfers. Four main operations, learn, recall, swap, and last X, divide the program into functional parts. The resulting operations are modifications of channel data information during recall and modifications of E-MEM register information during a learn operation.

6. *Interprocessor link (IPL).* The IPL program contains the necessary routines to communicate with other services external to the effects controller over the IPL data link. Messages are created, encoded, and transmitted as well as received, decoded, and interpreted by these routines. Send and receive routines are interrupt driven, whereas the encoder, decoder, and message service routines are scheduled processors.

7. *Pushbutton processor.* The pushbutton processor accepts data from the pushbutton interrupt routine and makes the logical decisions necessary for proper control-panel operation. This provides modified channel data information which is then interpreted by the effects generator for output to the digital processor and modified lamp data information which is used by the lamp output routine to update the control panel indicators.

8. *Effects generator.* The effects generator produces all the data that are sent to the digital-effects processor to control up to four channels of video. The effects-generator input is the data produced by the pushbutton processor and the analog processor. It is called by the baseline scheduler after the lever arm, aspect, positioner horizontal, and positioner vertical analog parameters are processed.

9. *Analog module.* The analog module receives the digitized data from the analog interrupt routine, processes it, and stores it in the channel data structure(s) determined by the control panel status and the nature of the analog variable.

   The digitized analog data are received in the analog interrupt routines and stored in a first-in, first-out buffer for later processing. The first interrupt occurs after line 10 and is followed by the additional interrupts for a total of 14. Each interrupt supplies one 12-bit analog variable. At each interrupt the analog processor is scheduled.

   The analog processor is called by the base line scheduler. It processes each analog variable until the FIFO buffer is empty.

10. *Size follow (key tracking).* The size follow module contains the interrupt and processing routines needed for the size follow and effect follow functions.

    The size follow processor controls the entry into, and the exit from size follow or effect follow functions and does processing that is not time critical.

    The interrupt routines read the key tracker left, right, top, and bottom values, and calculate the digital-effects processor addresses needed to fit the key or wipe area.

11. *Sequence input.* The sequence input module provides operating functions for the auxiliary bus.

Specifications. Digital video effects are difficult to specify in concrete terms since the objective of an effects device is to create new and different effects. However, experienced operators can determine if the effects are easily repeatable, easy to set up, and produce acceptable picture quality. Certain qualities of the effects controller are subtle and should be verified, such as whether screen movement appears smooth, whether controls have good resolution, and whether automatic transitions start and stop smoothly.

Video quality is a function of the analog and digital video path through the digital video processor. Refer to Sec. 14.2 for a more detailed discussion of digital-effects video quality.

14.7 COLOR CAMERAS

by K. Blair Benson, Karl Kinast, and Renville H. McMann

14.7.1 **EVOLUTION OF THE THREE-TUBE CAMERA.** The first commercially available color cameras for the FCC-approved 525-line NTSC broadcasting system in 1952 utilized three image-orthicon pickup tubes and vacuum-tube electronic circuitry.

The bulky packaging of the camera head, and the instability of both the pickup tubes and the accompanying circuits, significantly impeded the widespread use of color for broadcasting. Consequently, the majority of programming from studios and field locations continued in black and white, with color limited to one-time specials and sports and a few regularly scheduled programs with recurring sets and relatively simple production requirements.

Pickup Tubes and Optics. It was not until the introduction of three-tube, and later the four-tube, highly stable all-transistorized Plumbicon cameras in 1960 and 1961, respectively, that a rapid replacement of black-and-white cameras with color cameras using Plumbicon and similar lead oxide pickup tubes was undertaken in the United States by network broadcasters, to be followed shortly by affiliated and independent stations.

Concurrent with the development of the Plumbicon was the introduction of prism optics and zoom lenses. Cameras employing these features established the design concepts for present-day cameras.

During the past 30 years, much has been done to make the color camera smaller, lighter, more stable and rugged, with improved sensitivity, resolution, and colorimetry. This has been accomplished through major advances in the design of lenses, light-splitting optics, pickup tubes, integrated and microprocessor circuitry, and signal multiplexing on interconnecting cables. These developments have resulted from contributions by many organizations throughout the world in addition to those in the United States, most notably Holland, England, Germany, and France in Europe and Japan in Asia.[2]

Differences in System Standards.[3] Although there are three basic television system standards in the world, this factor has not impeded camera development and manufacture of universal designs, since all cameras are identical in that a format of color-primary signals, generally red, green, and blue, is derived for processing and encoding in accordance with the particular standard in use.

The major requirement for interchangeability among standards is the accommodation of 525- or 625-line raster-scanning rates. The differences in video bandwidths are of no significance since, in order to ensure negligible high-frequency transient distortions, the baseband cutoff usually is well beyond the system restrictions.

In summary, present-day cameras are suitable for use on all system standards with a minimal change in scanning rates and with the appropriate ancillary synchronizing-pulse generator and encoding equipment. Accordingly, this section covers designs of broadcast cameras intended for the 525-line NTSC system, with appropriate references to differences dictated by 625-line PAL and SECAM requirements.

14.7.2 BROADCAST APPLICATIONS. The designs are broadly categorized by the operational applications as follows: (1) studios, (2) news gathering, and (3) field production. In these various operational environments, a commonality of design fundamentals will be found to exist, with the major variations related to optical and mechanical features and packaging. In all cases, broadcast-quality picture signals are provided.

Broadcast quality is a term used throughout the industry, without a clear definition of its meaning. There are many types of cameras built specifically for industrial, military, cable, and commercial broadcast users. In general, broadcast quality does not define the video signal quality, since in a literal interpretation the definition implies conformity to FCC specifications concerning the vertical and horizontal blanking widths of the radiated signal (see Secs. 21.3 and 21.5.1). Instead, broadcast quality may be defined as the quality picture obtainable consistent with the circumstances surrounding the event to be televised, and that which will satisfy viewer requirements. In other words, entertainment programming, under carefully controlled conditions, warrants and demands the highest quality. On the other hand, at the other end of the scale, fast-breaking news events and sports under less-than-ideal operating conditions can suffer considerable degradation and still be acceptable. The latter is typified by the increased use of extremely compact cameras using three ½-inch pickup tubes, and more recently a single three-color tube. Nevertheless, the gap between the most expensive studio camera and the least expensive consumer camera, at one time a formidable one, has been significantly reduced with innovative technical achievements and mass production techniques.

14.7.3 BASIC DESIGN. The basic elements of a color television camera, from the three primary-color signals produced by the pickup tubes to the encoded, composite video-signal output, are shown in Fig. 14-39. Light from the scene being televised is imaged on the targets of the pickup tubes by an objective lens, either of a fixed focal-length variety or a variable focal-length zoom type. The primary-color images are derived by means of a dichroic-mirror or split-cube color-separation optical system. In reference to Fig. 14-39, the relatively low-level video-output signals from the pickup tubes are amplified by preamplifiers and corrected in amplitude-frequency response characteristic for losses in the pickup-tube coupling circuits. At this point parabola and sawtooth-correction signals are added to the video signal to compensate for shading errors in the pickup tubes and spurious flare light in the optical system. This is followed by line-by-line blanking-level clamping, image enhancement for aperture losses, and gamma correction for the linear transfer-characteristic of the pickup tube.

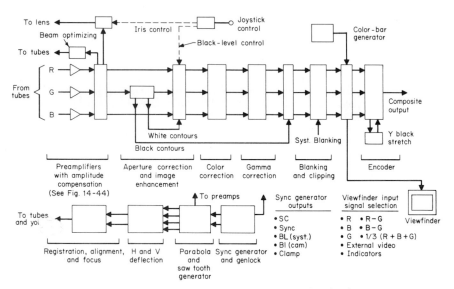

FIG. 14-39 Block diagram of a typical color camera showing major functional components.

The lens and camera pickup tube contribute to a loss in resolution at the higher spatial frequencies, in other words, both horizontally and vertically. Theoretically, the loss should follow a zero phase-shift amplitude-frequency, or $(\sin X)/x$, characteristic. Practically, the summation of response roll-offs does not precisely follow this simple law. Therefore, the green signal from the preamplifier is processed by a combination aperture corrector and image enhancer in the green channel. The aperture-corrected green signal, mixed with the *white* "contours out-of-green" enhancement components, is fed onto the color-correction, or masking, amplifier. The *black* enhancement signal, in order not to degrade the low-light signal-to-noise ratio, is mixed with green video after color and gamma correction.

All three color signals are color corrected by linear matrixes and gamma corrected, the latter to compensate for the essentially linear transfer characteristic of the Plumbicon, or Saticon, pickup tubes. Since the gamma correction is applied before insertion of system blanking and clipping, the nonlinear amplitude characteristic is referenced to the camera blanking level by a horizontal-rate clamping pulse.

The corrected and processed color signals are mixed with *system blanking* and clipped to produce three color signals for encoding to a composite 525- or 625-line signal. A technique to improve the rendition of shadow detail in low-key scenes, and to accommodate the extreme contrasts encountered under daylight illumination, additional adjustable gamma correction is available in the luminance (Y)-channel of the encoder. This has the advantage of producing no effect on color balance throughout its range.

The viewfinder has a number of signals selectively available for camera operation and for checking of camera alignment. These are:

- Red, blue, or green channels
- −Green with +red or +blue for registration checks
- Mixed red, blue, and green for normal operation
- External video for cue or special effects
- Indicators of iris setting and video level

In order to permit a maximum of flexibility for either studio or electronic news gathering and electronic field production (ENG/EFP) operation, the latter where the camera must be fully self-sufficient, a sync generator is included in the camera head which can be locked to its own internal crystal oscillator or to an external video (or composite sync) signal. The internal generator provides the following signals and drive pulses for the video processing and scanning circuits:

- Subcarrier
- Composite horizontal and vertical sync
- System blanking
- Camera blanking
- Horizontal-rate clamp pulses

The sync and subcarrier are mixed in the encoder with noncomposite video to produce a composite signal. For studio and field systems designed to add sync in the video switcher, the sync may be deleted, providing a noncomposite signal containing video, subcarrier burst, and composite blanking.

Blanking is added to video after gamma correction and clipped at the desired setup level. This may be at 7½ percent in accordance with FCC Rules and Regulations, or at zero with the required setup added in the encoder or the production switching system. For PAL and SECAM 625-line systems, no setup is used.

Operating a camera at zero setup results in an accurate adherence to black level, since the blanking waveform level is more easily discernible than a 7½ percent line on an oscilloscope graticule.

Associated with the sync generator are the line and field-rate video correction-signal generators, and pickup-tube scanning and focus coil drives.

14.7.4 FUNCTIONAL COMPONENTS.   Section 14.7.3 described the major components of a typical three-tube studio or field color camera. This section is devoted to a detailed description of each of these components and the design considerations. The discussion is limited to three-channel cameras, since the present-day level of technology obviates the need for the use of a fourth channel to produce sharp images free from registration errors.

Optical Color Separation.   A color camera must produce video signals which complement the characteristics of the three-phosphor standard additive display tube. For both live and film cameras it is now common to use a high-efficiency dichroic light splitter to divide the optical image into three images of red, blue, and green. The spectral characteristics of the three paths of such a camera are shown in Fig. 14-40.

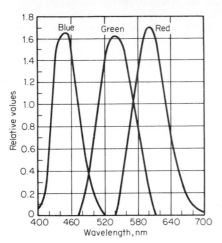

FIG. 14-40   Practical *RGB* taking sensitivities. (*From Fink and Christiansen.*[4])

The light splitting is accomplished either by a prism (Fig. 14-41) or by a relay lens and dichroic system (Fig. 14-42). The prism has the advantages of small size and high optical efficiency but disadvantage in that the three tubes are not parallel to each other and are thus more susceptible to misregistration produced by external magnetic fields. A more serious problem is that of obtaining a uniform bias light on the face of the tubes. Bias light producing 2 to 10 percent of the signal current is used in most modern cameras to reduce lag effects. Nonuniformity of the bias light can produce color shading in dark areas of the picture. Careful optical design can minimize this problem, and most new designs now use the prism splitter.

The relay optical system shown in Fig. 14-42 is one or two *f*-stops slower than the prism, but it has several compensating advantages. First, bias-light requirements are very low. Second, since an aerial image is produced in the optical system, a mask can be included around the edge of the picture which will produce a true optical black in the video signal. This black level can be used as a clamp reference point to establish true signal-black, keeping glare, bias-light, and dark current changes from upsetting the color balance of the picture. The aerial image is also a convenient point in which to insert points of light via fiberoptics, to permit automatic horizontal and vertical centering and size adjustment of the pickup-tube scanning circuits.

In both the prism and relay light splitters the colorimetry is, to some degree, a function of the polarization of the light entering the lens. One solution is to include a quarter-wave plate ahead of the splitter so that light of either horizontal or vertical polarization is converted to elliptical polarization. The splitter then sees the same depolarized light no matter what the polarization of the scene illumination. High-quality cameras currently are being produced using both prisms and relay optics.

Preamplifier.   Photoconductive tubes have an inherently high internal signal-to-noise ratio, the advantages of which cannot be fully realized because of (1) the low level of the target-signal current, and (2) the combined shunt capacitance of the target and output circuit. Since the capacitive impedance is less than 10 kΩ, coupling via a load resistor of under 500 Ω would provide a full-bandwidth signal requiring relatively little high-frequency equalization. However, the resultant low-amplitude signal would be masked by the flat-spectrum *thermal (Johnson) noise* of the resistor and, more significantly, shot noise of the preamplifier. Visually, the low-frequency components of a flat-noise spectrum are extremely disturbing.

The solution generally followed with vacuum-tube preamplifiers has been to use a high-value load resistor, in the order of 50 to 100 kΩ, in order to provide a signal level well above the preamplifier noise. The higher-value resistor, though, results in a higher thermal noise level and falloff in signal level with increasing frequency because of the shunt capacitance. The high-frequency loss can be equalized by a "high-peaker" amplifier at the expense of a similar increase in noise level.

The resultant signal, with a uniform amplitude-frequency characteristic, has a redistributed noise spectrum that is low in amplitude, below 0.5 MHz, but increases 6 dB per

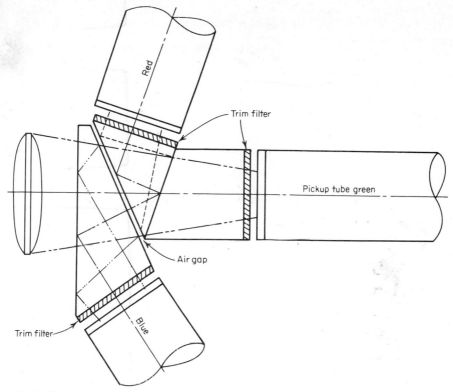

FIG. 14-41 Prism optical system of three-tube camera. (*From Fink and Christiansen.*[4])

octave at higher frequencies. This so-called *triangular noise* spectrum is visually less noticeable than *flat-channel* noise and will provide a better figure of *weighted* signal-to-noise. The high-frequency distribution of the noise explains the concern for noise deterioration when pictures are to be enhanced by boosting their high-frequency content. It is also important to note that this is virtually the only source of noise in a well-designed camera chain.[4]

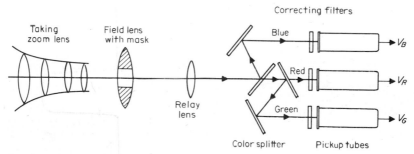

FIG. 14-42 Color separation by relay lens and dichroic filters. (*From Fink and Christiansen.*[4])

Considerable effort has been applied to improve the signal-to-noise ratio by reducing capacitance to a minimum,[5] and increasing the load resistor to as high as 1 MΩ, without excessive high-frequency loss, by use of a feedback configuration. In addition, availability of FETs with their high input impedance has eliminated many of the shortcomings of vacuum tubes. A convenient graph of signal-to-noise versus pickup-tube capacity and the formula for calculating the S/N is shown in Fig. 14-43.

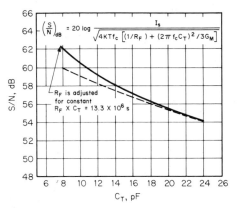

$$\left(\frac{S}{N}\right)_{dB} = 20 \log \frac{I_s}{\sqrt{4KTf_c\left[(1/R_F) + (2\pi f_c C_T)^2/3G_M\right]}}$$

R_F is adjusted for constant $R_F \times C_T = 13.3 \times 10^6$ s

FIG. 14-43 S/N performance vs. total pickup-tube capacitance. *(Courtesy of S. L. Bendell and SMPTE.)*

For example, an FET cascode preamplifier has no Miller-capacitance loading of its input, and its shunt capacitance is reduced to a minimum if the input stage and the 1-MΩ load resistor are located within the yoke housing in direct contact with the target. The high-peaker feedback equalization is fed to the input through the load resistor (see Fig. 14-44). The signal at this point is a voltage which is directly related to the target

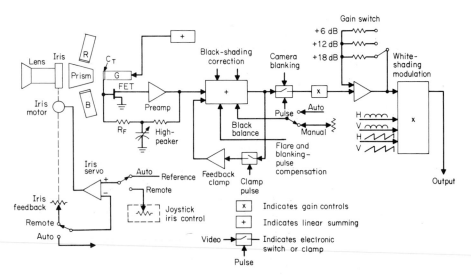

FIG. 14-44 Block diagram of typical color camera preamplifier with amplitude compensation.

signal current by a factor of 10^6. In other words, a 100-mV signal corresponds to 100 mA of target current, thus providing a test-signal injection point of known sensitivity and a test point.

Aperture Correction and Image Enhancement. The lens, optical beam-splitting system, and pickup tube, in total, contribute to a loss in resolution at higher spatial frequencies, both horizontally and vertically. These elements exhibit a (sin x)/x(phaseless) type of loss which, in practice, while valid for each, can produce an overall response curve that does not follow a simple law. Aperture correction and image enhancement therefore are used in all broadcast-quality cameras to improve the subjective picture quality. The horizontal aperture corrector is adjusted to restore the depth of modulation at 400 or 500 lines to that obtained at approximately 50 lines. Transversal delay lines and second-derivative types of corrector are frequently used since they exhibit high-frequency boost without phase shift, thus complementing the (sin x)/x rolloff, i.e., boosting the high frequencies without introducing ringing.

Once the camera response is flat to 400 lines, an additional correction is applied to increase the depth of modulation in the range of 250 to 300 lines, both vertically and horizontally. This additional correction, known as image enhancement, usually takes the form of a transversal filter (Fig. 14-45). The delay elements T_1 and T_2 are one picture element (or approximately 125 ns) long for horizontal correction and 63.5 μs for vertical correction. Such a corrector produces a correction signal with symmetrical overshoots around transitions in the picture. If overdone, this correction produces an unnatural image, characteristic of the early image orthicons which outlined midfrequency detail. Image enhancement must therefore be used very sparingly if a natural appearance is to be maintained.

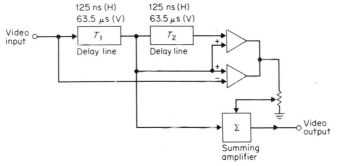

FIG. 14-45 Image enhancement using a transversal filter. (*From Fink and Christiansen.*[4])

Subjectively the eye is most sensitive to detail in the mid-gray-scale range of the picture. It is therefore beneficial to modulate the aperture and image-enhancement detail correction signals as a function of brightness. In this manner distracting noise in the dark areas of the picture can be eliminated, and the viewer is not aware of the accompanying loss of detail in the shadows.

Coring is sometimes used to slice out the mid-amplitude range of the detail signal so that noise and low-level detail signals are removed. This process not only removes noise but also prevents the performers' skin from appearing too rough, while at the same time permitting the highlights in the eyes to pass unattenuated.

Color Correction (Masking). The ideal camera colorimetric taking characteristics for the NTSC color system require negative lobes which cannot be generated optically in the beam splitter of pickup tubes because this would require negative light, nonexistent in nature. However, negative lobes can be produced electrically in the camera signal processing by a matrix operation called *masking.* One version of this technique balances the matrix for equal red, blue, and green signals, i.e., white, and generates correction signals only when chrominance is present. If the matrix is made polarity-sensitive (Fig. 14-46) it is possible to correct individual hues independently to match the NTSC vectors exactly, or alternatively to match camera tubes to dichroic filters and one camera to another.

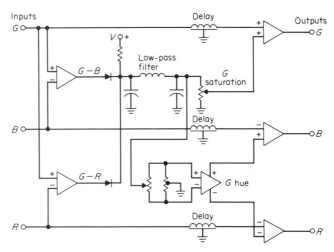

FIG. 14-46   Masking matrix with polarity sensing (matrix shown for G correction only). (*From Fink and Christiansen.*[4])

In the camera shown in Fig. 14-39, the color correction is performed before gamma on the linear signals from the red, blue, and green color tubes. Generally the adjustments are preset to establish the correction necessary for camera match. However, this "fixed matrix" is often made operationally adjustable. In fact, in some studio cameras a more elaborate "nonlinear matrix" is added with conveniently adjustable "paint pots" for operational purposes. The three primary and three complementary colors are available for independent adjustment, and the settings stored digitally for use in difficult staging and lighting situations.

Gamma Correction.   In order to provide a pleasing tonal and color rendition on color receivers and monitors, camera gamma† should be the inverse of the picture tube, or approximately 0.4. This is slightly lower than that specified in the FCC Rules and Regulations, Subpart E of Part 73, which states, "the gamma-corrected voltages . . . are suitable for a color picture tube . . . having a transfer gradient (gamma exponent) of 2.2 associated with each primary color." Since the tubes used in studio and field broadcasting have a linear characteristic, for a gain exponent of 0.4 to 0.5, the individual red, blue, and green signals must be amplified nonlinearly. The resulting slope, or amplification, is greater in the blacks, thus increasing black noise visibly by 12 dB or more.

Active or passive diode function generators are used to shape the signal, and can be made to have a level-dependent frequency response falloff in the black region to reduce black noise.

---

†Television system gamma: Exponent of the power law representing the transfer curve, as opposed to photographic film gamma which is the slope of the straight-line portion of the Hurter and Driffield curve.

A common technique for producing the wide-band nonlinear video signal is shown in Fig. 14-47. Operation depends on the fact that the voltage across a diode is approximately proportional to the square root of the current through it. By providing two balanced paths, a linear signal can also be produced. The setting of potentiometer R produces a mixture of linear and nonlinear signal which can be used to match exactly the slight differences in gamma of the red, blue, and green channels. The high gain of such a gamma circuit for signals near black makes it very susceptible to temperature drift. The use of a feedback-clamp circuit operating on the picture black level, after the gamma diode, reduces such drift to a negligible value.

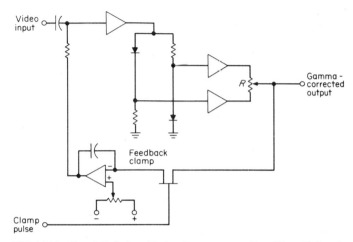

FIG. 14-47   Use of diode for wide-band gamma correction. (*From Fink and Christiansen.*[4])

Alternatively, the nonlinear function may be produced by several diodes biased to key in different values of load resistance at different video levels. If the diode-conduction amplitude characteristic between on and off is over a moderate difference in video levels, a smooth and exceedingly stable gamma curve can be obtained. In all such circuits, a means to replace the gamma corrector output with a linear signal of the same reference-white level is provided (gamma-off).

The first use of gamma correction for artistic, rather than technical, purposes in camera design has been to, in effect, replace the gamma-off switch with a control so that the contributions of the linear and 0.4 gamma-correction signals may be mixed in varying amounts to provide a smooth, variable adjustment. The tracking among channels, when using ganged controls, has not been found to be precise enough to maintain a high degree of color balance accuracy. Therefore, selection of any of several preset gamma-correction curves is more common and operationally convenient.

An extension of the classic gamma-correction curves is a *black stretch* applied to the luminance signal in the encoder, resulting in a decrease in low-light color saturation, not noticeable to the viewer, in return for greatly enhanced shadow and low-light detail.

Conversely, white compression can be applied to the white region of the video signal, at the expense of an increase in color saturation. The improvement in contrast-handling capability of the camera more than outweighs the color saturation errors.

Black-level Control.   The three color signals, after gamma correction, contain horizontal and vertical scan blanking signals inserted in the pickup tubes, with no adjustment having been made for the level of peak black signals to blanking level. Furthermore, the width of the blanking signals is narrower than the system blanking, and the vertical interval does not have the required equalizing and serrated vertical-sync pulses. This processing is accomplished in the next stage, where system blanking is added and the combined signals clipped to establish black level.

The *clipping level,* relative to peak blacks and camera blanking, is controlled by the camera-control operator through the black-level control, either on the joystick or the accompanying control panel. The control is applied to an amplifier stage which is clamped to camera blanking and direct coupled to the blanking adder and clipper stage.

This control, which sets the level of the blackest portion of the video signal, usually has a large range in clipping level. This provides the operator with a wide latitude of as much as 50 IRE units in setting black level, including clipping of blacks if desired for artistic reasons or camera matching.

14.7.5 STUDIO CAMERAS. There are several important distinctions that characterize studio cameras. These distinctions are, however, becoming less significant with the advent of more sophisticated and adaptable electronic field production (EFP) cameras. Nevertheless, the difference in size and cost are the result of these major distinctions:

- The highest attainable picture quality

- Multicamera program, rather than single-camera production, designs

- Full complement of features and accessories for broadcast production

The first television camera, or so-called camera chain, consisted of a camera head connected via a multiconductor cable to a rack-mounted camera control unit (CCU). This was a misnomer, because whatever could not be fitted in the camera head, for mechanical reasons or lack of remote-control technology, was relegated to the CCU. In other words, the CCU is the central connection point for all the auxiliary panels that are part of the camera system and its external monitors, one of which is the operating control panel. Here most of the operating controls and features, including iris, black level, gamma curve, and colorimetry, are manipulated by control room personnel. Their adjustments are guided by color-picture and waveform monitors. A high-performance picture monitor is provided as an integral part of the camera, frequently on a swiveled mount, for use as a "viewfinder" by the camera operator.

The *viewfinder* utilizes a picture tube, usually as large as 7 in (18 cm), with 15-kV accelerating potential in order to provide a high-resolution and high-brightness display, the latter being necessary to display a full-contrast range under high ambient illumination. To facilitate focusing, resolution is enhanced further by a switchable "peaking" control.

In the interest of ergonomics and the conservation of control room panel space, another small control panel is provided for each camera, so that three or four may be mounted adjacently in a single compact control desk. Each panel has a joystick control for iris (video level) and master black, with individual black level and gain controls for the three color channels. The controls located heretofore on panels in the CCU are usually grouped on a larger *master control panel.* Thus, the CCU no longer serves as an operational control panel and instead serves as a camera-signal processing unit, containing the majority of the electronic circuitry and the junction for all cable connections within the camera and to the studio system. In the more sophisticated systems this master control panel may be supplanted by a control multiplexer, by which all other cameras in the system can be controlled by a single master control unit.

This is possible with digital memory for the storage of function selection and control-voltage magnitude within each individual CCU, so that the status quo is maintained after master control is transferred to another camera. In this case, the control functions fall into either of two categories: (1) subsidiary operation controls and (2) setup controls. The subsidiary operation controls allow selection of filters and precalibrated controls for scene color temperature compensation, overall gain step increase, several choices of preset gamma and black stretch, and the degree of picture enhancement (contours).

Test Signals. The standard test signal complement includes test signal substitution (color bars, or sawtooth), and diascope for test patterns if supplied as part of or built into the optical system, and electronic lens capping. This, together with modern remote control and signal sensing technology, has made auto black and auto white balance standard, as well as automatic registration centering. With microprocessors and more memory, this technique has lately been extended to completely automated registration, relegating analog controls to basic setup adjustments on the circuit boards. More recently, using microprocessor technology, complete digital adjustment and memory have been developed to perform test-point check and adjustment operation, and to convert the readings to a video test display. To date, net effectiveness in regard to maintenance economics remains to be proven.

Digital Setup—Triaxial-Cable Cameras. A studio camera may have over 100 potentiometers which must be adjusted after changing tubes. Approximately 20 of these may need adjustment on a daily basis to maintain optimum performance. To make many routine adjustments as accessible as possible, studio cameras until recently were designed so that much of the signal processing occurred at the camera control unit (CCU) rather than at the camera head. In this case a CCU-to-camera-head cable might contain five or more individual cables and 60 to 80 individual wires. Also, maintaining the connectors in such a bulky, heavy camera cable in the field can be a major problem.

Cameras are now available which transmit all the necessary signals and power over a single triaxial cable using digital-multiplexed commands and video subcarriers (Fig. 14-48). When such cameras became available, it was realized that if a microprocessor was used to generate the command signals, suitable software could be used to set up the camera without human intervention. A test-pattern diascope projector is included in the camera lens so that the microprocessor can control special setup patterns without the need for stagehands and lighting technicians. The potentiometers are replaced by digital-to-analog converters, which are far more reliable and not subject to misadjustment from vibration or mishaps in the field.

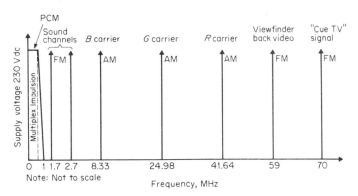

FIG. 14-48 Single triaxial-cable signals for color cameras. (*From Fink and Christiansen.*[4])

Operating Control Requirements. Regardless of degree of automation, proof of performance, if not manual setup, requires that the master panel have switches for selection of monitors, signals, and monitoring modes. These may be laterally passed comparison combinations for matching or standardizing cameras to each other, or one of the traditional modes; i.e., on the waveform monitor the final encoded output signal can be replaced by *R, G,* or *B* signals in an overlay for side-by-side sequential display, for the comparison of shading and gamma tracking, or singly for determination of overall modulation transfer function (resolution). On the viewfinder or monochrome monitor the luminance signal is replaceable by an inverted green signal, summed with either R or B to observe degree of registration.

Performance Requirements.   A studio camera need not be automated, or have the flexibility of triax camera cabling, but it is expected to have, above all, uncompromised quality at its output. This requires at least three pickup tubes, 1 in (2.5 cm) or larger, with aperture correction to provide a substantial modulation transfer function improvement, and full control of shading, enhancement, and colorimetry that at least matches other cameras in the same studio. Direct or wide-band multiplexed and demodulated red, green, and blue outputs for chroma key are required, with preferably two high-quality program audio outputs, if it is a floor camera. Stability and accuracy of registration, based on factory matching of deflection components and good thermal design, are more important than mode of registration. Finally the camera and its manipulation should be quiet so as not to intrude on audio quality.

14.7.6   ELECTRONIC NEWS GATHERING (ENG).   ENG cameras, at this time, are in a state of transition from tube designs using the popular ⅔-in (1.7-cm) pickup to a more compact design using three ½-in (1.3-cm) tubes, and for some applications a single 1-in (2.5-cm) three-color striped tube. The first ENG-dedicated cameras were a packaging of all the basic design features described in Sec. 14.3.3 in a self-contained shoulder-mounted camera with extreme attention being paid to portability and the reduction of power consumption.

Partly in the interest of miniaturization, and partly to simplify camera operation, external controls were eliminated wherever internal trimpots would suffice, with only essential operating controls, such as gain boost, retained and with operation of these simplified by the use of step switches. The knobless regimen requires black level be preset and, only if time permits, to be corrected before a recording or on-air transmission.

Display of signal level (peak white) in the viewfinder, available but not essential in studio work, is mandatory in the technically unsupervised application of ENG camera work. Any other ancillary indicators, such as audio level, must occupy field of view on and surrounding the small screen of the viewfinders.

Power conversion from "snap-on" rechargeable batteries is done via switch-mode power supply (SMPS) circuitry, with no intervening linear regulators. Acting as a variable-ratio dc to dc transformer, the SMPS converts battery power efficiently, so that the current drain on the battery is actually lower when the battery is in its highest voltage (freshly charged) state. The traditional ± 15- or ± 12-V internal circuit voltages required in early designs have been supplanted by ± 5-V circuitry in ENG cameras, the minimum required for operational (OP) amplifiers and logic circuitry.

Automatic white-balance control and adjustment data are stored in CMOS digital form, decoded by CMOS digital-to-analog (D/A) converter chips, and held for several days in memory by voltage from an internal "keep-alive" battery. Many circuits are integrated into dedicated LSI chips, and amplifier-gain and modulated-gain circuits are in the form of proprietary hybrid chips. Initial cost and maintenance spares costs are reduced by using the same type of chip in a dozen or more different circuits in a single camera. On the other hand, encoder and sync-generator chips are more expensive because only one is used per camera, and they require the largest scale of integration. The manufacturing technique of trimming these chips by lasers is being used increasingly in order to eliminate the majority of trimpots.

More recently, the addition of a ½- or ¼-in (1.3- or 0.6-cm) video cassette recorder (VCR) within the body of the camera results in an ENG camera that, for the first time, provides an electronic packaging to equal that of 8-mm motion picture film. In fact, it is called a *video-recording-camera* (VRC) by some users. Furthermore, a high-quality signal, significantly superior to conventional VCRs, is achieved by encoding and recording in a non-NTSC separate-luminance and multiplexed-chrominance format. Consequently, in order to utilize the signal for broadcast, an adapter containing a standard encoder must be added (or a circuit board substituted). This permits the use of a standard recorder, or transmitting via a microwave link for program integration or broadcast. A similar adaptation makes most ENG cameras remotely controllable in an EFP configuration.

14.7.7 ELECTRONIC FIELD PRODUCTION (EFP) CAMERAS. Depending on the mode and size of the base station (large van, vehicular mounted, or portable), the complexity and sophistication of an EFP camera can range from a full-blown studio camera with multiplexed triax cable downward to an upgraded version of a high-quality ENG camera. The underlying determinator of mode, size, and quality is the nature and frequency of assignments that are anticipated, coupled with the economics of return on investment.

Historically, field production operated from large remote or outside broadcast (OB) vans that were, in essence, mobile broadcast control rooms with camera-control and switching equipment originally intended for studio use. The cameras deployed were actual studio cameras, or "portable" cameras that were adapted to operate with a standard CCU. This practice continues with 1-in studio/field cameras having integral triax capability and special outfitting to cope with weather, radio-frequency interference (RFI), and very long cable runs. Although these complex setups can be used effectively for electronic field production, the identification EFP has become associated with portable cameras whose mobility and features are easily adaptable with modular accessories to accommodate the requirements of different day-to-day assignments.

Compactness and low power consumption are gained by downscaling lens, prism, and electron-optics to the ⅔-in (1.7-cm) Plumbicon or Saticon tube at the cost of less-than-proportionate, and not subjectively significant, loss in modulation transfer function (MTF) and signal-to-noise ratio. The MTF is restorable to 100 percent at 5 MHz by means of aperture correction, or "out-of-band" contours, in combination with standard "in-band" contour correction and the use of diode-gun tubes. Additional aperture correction reduces signal-to-noise ratio. However, low-output-capacitance tubes and selected field-effect transistors (FETs) can be used to reduce noise at the first preamplifier stage. Thus, a figure of 56 dB (on 525 lines) signal-to-noise ratio, with contours and other corrections off, is comparable to previous generations of studio cameras.

As stated previously, basic ENG cameras are often converted to EFP work. In any case, the essential difference between ENG and EFP is supervision of the output signal to maximize whatever quality the camera is capable of producing. This takes the form of a base station equipped with studio-type monitoring and remote-control equipment.

Master black, color balance, iris, gain, black stretch, and centering are controlled via the cable, which powers the camera. Cables can be up to 1000 f (300 m) in length, which necessitates gen-lock with advance timing adjustment and video cable compensation. Studio-type ancillary circuits (signaling, audio, and intercom) are easily accommodated, with the exception of wide-band chroma key. Further cable extension is possible with triax-adaptor boxes. Mechanically, an EFP camera must be adaptable to larger lenses than used for ENG, and mountable on a field tripod. In general, until recently EFP cameras have been adaptations rather than dedicated designs. The increased use of EFP to supplant film in many production operations has resulted in a new generation of cameras designed specifically to meet the cinematographer's needs.

14.8 CONTROL ROOM DESIGN AND LAYOUT

by Frederick M. Remley

A comprehensive definition of television control rooms should include all spaces that are allocated to control of the creative and technical phases of television production. By this definition, video control rooms, audio control rooms, master control rooms, continuity booths, and tape-editing suites would all be included. Each of these spaces has a specific function to perform in the final television production. In addition, some aspects of the design will vary depending on the type of facility being considered. For instance, the studio control room requirements of a general-purpose production center will differ from those of a control room intended for the nightly news broadcasts of a local television station. By the same token, an audio control area dedicated to serve a large, multipurpose studio will have requirements different from the needs of an audio booth assigned to a production or master-control operation.

It is important to identify carefully the primary use of a new control room before beginning the design of the room. A variety of design factors are affected by the planned use for the space. Since few architectural designers have experience in control room layout, the responsibility for careful definition of form and function usually lies with the future owner, the owner's engineering staff, or a consultant selected by the owner or architect.

Although skilled consultants may be found to assist in control room design, the owner or owner's technical experts must, in all cases, define carefully the specifications for any space that is to serve a control room function; both architects and consultants will require such guidance. Some relevant factors are listed in the following paragraphs.

14.8.1 OPERATING-PERSONNEL REQUIREMENTS

Accommodations for Personnel.   This factor is primarily determined by the specific, planned function of the new control room. If it is a studio control room, very likely a director and one or more technical persons, for example, the audio operator and perhaps the switcher operator, will require seating and monitor-viewing space. It may be that additional persons must also be accommodated. Perhaps these will include an assistant director, a technical director, or a video operator—or some combination of these people. It may even be necessary, in the case of a large production control room, to accommodate an announcer, a lighting supervisor, and several clients, and—for sporting events—advisors and statisticians.

Director-Personnel Relationships.   The arrangement of a control room must take into account the communication and interaction needs of the persons occupying it. Even though the television facility will probably be equipped with a complete intercommunication system, visual and spoken cues may also be necessary within a single control area. Hand signals remain an important part of the technique of television production, and visual contact with the director may be important for some control room occupants such as the audio operator or the technical director.

Because of conflicting requirements for listening to program-related audio, the audio operator may be placed in a separate booth adjoining the television studio production control room. By this means, the audio operator will be in close physical proximity to the production area but will not be exposed to the hubbub of the video production process. The result will usually be better audio—more accurate judgment of sound levels, microphone balance, etc. In addition, the other control room personnel will not be distracted by high loudness levels of the audio-monitoring system since the audio operator can make the monitor adjustments independently of those involved in the video part of the production. Other control room personnel who might be considered for separate booth locations include, of course, an announcer and perhaps a lighting control director in the case of a studio control room.

Spatial Relationships in Control Room Production and Technical Areas.   Examples of special control room space relationships can include a requirement for direct visual contact with a studio through soundproof windows, or for immediate walking access to a studio or VTR room. Careful consideration of space relationships and of requirements for adjacency of key personnel and equipment will result in reduced loss of valuable production time by permitting immediate, direct coordination.

For example, present television station operating practice often requires that the master control room operator be responsible for operation of video tape playback systems, plus slide and film chain operation, together with the associated audio and video switching requirements. Accordingly, the operator in such a facility will require easy physical access to the areas housing the video tape equipment and the telecine equipment. Direct sight lines between the master control switching position and these equipment areas may also be required. These requirements may be met by providing windows between master control and the ancillary equipment areas. In many cases, the problem has been solved by placing the operator and all the tape and film equipment together in a single large room. It should be noted that this last solution can result in high acoustic noise levels, and, as a result, it may be difficult for the operators to evaluate and control the loudness and quality of program audio.

14.8.2 ARCHITECTURAL CONSIDERATIONS

Noise Control. Among the main constraints on architectural design of control room areas are the problems of noise control. Noise from the control room loudspeakers must not intrude into the studio or announce booth. Noise from the projectors and video tape recorders should be prevented from entering the control room. Automobile, aircraft, office, and other external noises are to be avoided.

Most architectural techniques for noise control apply as well to the problem of television facility design as they do to the design of more conventional buildings. Sound (noise) transmission can be controlled only through isolation or attenuation of noise sources. For example, noises originating in rotating machinery can best be controlled by massive steel or concrete mounting pads and specifically designed resilient vibration isolation mountings. Airborne sound transmission can be controlled by massive walls, by isolation of rooms with double-wall construction containing air spaces, and by careful sealing of door openings with gaskets and thresholds. All these techniques raise the cost of control room (and studio) construction, but each approach is well proven and can be successful in improving the usability of critical television production spaces.

Room Illumination. Since most control rooms will house television monitors, viewing conditions that make picture quality judgment most accurate are greatly to be desired. SMPTE RP-71[6] describes the conditions necessary for critical viewing of television monitors, and the principles described therein can be taken into account in the design of the control room illumination system. Control of the light level and of the distribution of illumination within the control room will be necessary. It may be desirable to use lighting fixtures that permit quite precise spotlighting of the important work areas within the room, while at the same time protecting the video monitors from glare, reflections, and direct illumination by stray light. It may be necessary, also, to control stray light reaching the monitor screens through windows opening into studios, other control areas or corridors. Usually, this can be accomplished by using tinted glass or the aluminized plastic film material sometimes employed to reduce direct sunlight through windows facing the outside in office buildings.

Air Conditioning. Most control room areas will require conditioned air. The design of a suitable air-handling system, one that will provide properly controlled air volumes at the correct temperature and humidity and as silently as possible, is beyond the scope of this handbook. However, it is vitally important to discuss this aspect of facility design with both the architect and the mechanical engineer assigned to the project. Because of the requirements placed upon ventilation systems when they are properly specified for television production areas, it should be expected that the cost of these systems will exceed the costs encountered in conventional building design. Good design in air handling for control rooms calls for large distribution ducts properly lined on the inside for noise attenuation and insulated on the outside for trouble-free operation during times of high humidity. Location of the air-handling equipment at a suitably calculated distance from the critical areas to be served is important in order to move noise sources, such as fans and compressors, as far away as practical. Low air velocities in the ducts and at the outlet grills or anemostats will reduce hiss. Proper air distribution must be achieved within the control room for the comfort of personnel.

Good practice dictates that maximum outlet (jet) velocities of less than 500 ft/min (150 m/min) are required and, to avoid chilling breezes and rushing sounds in microphones, a maximum velocity of 100 ft/min (30 m/min) at 5 ft (1.5 m) above the floor is also desirable. Typically, modern radio and television control room ambient noise levels should be in the vicinity of NC 30–35 (42–47 dBA). This figure[7] should take into account both the noise caused by the ventilation system and the noise caused by the equipment located in the room.

The ventilation system designer must also consider the heat gain produced by the control room electronic equipment that, if great enough, may require a heat-exhaust system separate from the main air-handling system. Although modern television equipment generates much less heat than earlier apparatus, units such as video tape recorders, color monitors, and camera power supplies generate enough heat to demand special consideration. Rooms housing these items may well require their own air-conditioning zone with individual thermostat control, since the heat load changes drastically when the equipment is turned off or on.

In addition to the factors already noted, good building design will, in some climates, require humidity control. Most commonly, addition of moisture to the air in the winter months is required, but, in some locations, reduction of excessive summertime humidity is also necessary in critical areas such as video tape recording rooms and storage vaults for films and tapes. Both types of humidity control can be costly and will certainly require special consideration in the design of the air-handling equipment. Such requirements should be brought to the attention of the design team as early as possible.

It is important that the air supplied by the ventilation system be as clean as possible, especially in the rooms that house video recording and telecine equipment. In some instances this may be accomplished by using 10-μm, or finer, filtering systems at the air-handling fan or fans. Such filtering systems use elements known as *high-efficiency particulate* (HEP) filters that take the form of specially constructed baglike paper or fabric filter elements mounted in metal wire frames. The high dirt-removing efficiency of these filters is accompanied by an increase in air resistance on the inlet side of the air-handling supply fan, and this must be taken into account in the specification of the fan system. In the case of large systems, it is very likely that a return-air fan will be required, in addition to the usual supply-air fan. The maintenance of HEP filter systems can be rather expensive in dusty environments, since the cost of the filter elements is high and the labor required to install replacement elements is significant. Accordingly, it may be a better choice to use an electrostatic dust precipitation system in such cases. Electrostatic systems provide excellent dust removal and can be cleaned by maintenance personnel as a routine matter. As might be expected, however, the initial cost of electrostatic precipitation equipment is much higher than the cost of an HEP filter system. Again, the mechanical engineer responsible for the project should be consulted early in the design process, and the proposals studied carefully to ensure that the air supplied to the control areas will be clean enough for modern-day equipment requirements. The engineer may recommend HEP or electrostatic filtering for only part of the facility and may suggest conventional disposable glass fiber filters for use in the air-handling units serving noncritical areas. Such a solution may be the best compromise and should be carefully considered.

Control Room Acoustics.   Given that the ambient noise level of a control room used for audio monitoring must be satisfactorily low, it is also necessary that the room not exhibit reverberation or other acoustic defects that might impair the assessment of sound quality by the director or the technicians responsible for audio.[7] Reverberation may be reduced by use of conventional good design techniques. Ceilings may be treated with acoustic tile—perforated fiberboard panels either applied to hard surfaces with adhesive or comprising lay-in panels mounted in metal and suspended from ceiling frameworks. In the latter case, it is important to consider two facts: first, sound absorption is greatly improved if battens of 2-in-thick paper-backed glass fiber material are laid on top of the individual panels and, second, all walls surrounding the control room's suspended-ceiling area must be full-structure height. In other words, even if the lay-in paneling is suspended some distance down from the structural ceiling, the walls must extend from the floor to the primary building structure.

In addition to ceiling absorption, improvement in room sound quality can be achieved by applying absorbent materials to the walls. These materials may range from commercially available perforated hollow metal panels, about 1-in thick and filled with glass fiber material, to high-pile carpeting fastened to the walls with adhesive. The carpeting solution is quite popular; carpeting by its nature provides a durable and attractive wall covering. In fact, specially manufactured carpeting is available for this purpose. As an example of such an application, many control room designers have found that covering the wall opposite the main loudspeaker used in a control room helps to prevent reverberation from originating behind persons in the room.

It is important to know that the carpeting chosen for this purpose must meet building code requirements. In general, building codes require that carpeting used for wall covering be certified to pass more stringent flammability tests than those used for floor covering. All reputable carpet manufacturers indicate in their sample books and catalogs the flammability tests that their products have passed; it is important to check these data and to determine the rating necessary to meet local codes before the use of carpeting on walls is attempted.

In some control rooms it may be feasible to provide carpeted floors. Again, the use of carpeting can improve the reverberation characteristics of the room. Often this is not possible, however, since carpeting makes chair movement in a busy control room more difficult.

Probably the most common use of carpeted control room floors is found in video tape recorder equipment rooms and editing suites. In these cases, another important factor must be taken into account—the potential danger that the carpeting will generate static electricity. Static electricity can cause dangerous voltages that can destroy solid-state electronic devices, and sometimes static discharges will confuse digital computers and computer-based editing equipment. Accordingly, it is important to choose floor carpeting with this fact in mind. A type of carpeting designed to be electrically conductive (either through the interweaving of metal wires or through the choice of special fibers) must be used in most control room areas. This carpeting is often called *computer room carpet* in the trade. It is usually installed with special carpet adhesive, without padding, directly onto the structural floor. The type of backing supplied with the carpet must permit this method of installation. Such an installation permits easier movement of wheeled equipment into and out of the control area.

Control Room Windows. Other factors to be considered in the acoustic design of control rooms are the sizes and types of windows to be provided for visual contact with the studio or with adjacent control areas. A decision, often a good one, may even be made to use no windows at all. Generally, windows into studios are obscured much of the time with scenery or studio draperies. In addition, the rather large windows that are necessary for vision into a studio are costly, provide opportunity for sound to leak from the control room into the studio, and often adversely affect the acoustics of the control room itself. Direct vision into the studio is seldom a necessity, since experienced directors use the television cameras to show them what they need to see, both in directing the talent and in judging the overall studio situation. If it is decided to install windows, then they must be designed according to proper acoustical and noise-control practices. Excellent designs for noise-reducing windows may be found in Rettinger.[7]

Noise Control. The fundamental principle of reducing noise transmission is to use massive structures. Studio designers have, in difficult situations, even gone so far as to use walls covered with heavy lead sheets to reduce the transmission of sound from outside sources. Generally, television studios are constructed with poured concrete or concrete block walls. It is good practice, when block construction is used, to require the contractor to fill the core holes in the blocks with dry sand as each course is laid, since the sand significantly increases the mass of the wall. In addition, parallel walls should be avoided wherever possible to reduce acoustic reflections that always occur between parallel surfaces.

It is not always practical to use concrete block construction in control room design, because of either cost factors or structural requirements. Good walls can be made using more conventional stud-and-wallboard construction, although it is often necessary to use two thicknesses of wallboard on each side of staggered studs to achieve sufficient sound isolation between adjacent rooms. Most texts on architectural acoustics describe this method of building walls for achieving reduced sound transmission.

A wall is no better than the seals of the openings that penetrate it. Windows, doors, and cable entry points must be carefully engineered. Solid wood doors, or even specially designed acoustic doors, may be required between noisy areas and control rooms. Such doors must be properly gasketed to provide good closure seals. Windows should be double- or triple-glazed and must be checked carefully during construction to make certain that no leaks are covered over during the finishing process.[7]

A carefully sound-isolated room can be rendered imperfect if the air ducts supplying it are improperly designed. Use of *sound traps* (commercially available duct-noise attenuators) and very careful layout of ductwork are necessary if sound leaks between adjacent areas are to be avoided. Mechanical contractors will often object to the resulting duct layout as being wasteful of materials; the consultant and the owner should stand firm, however, since sound leaks passing through ductwork are very difficult to repair after the conclusion of construction.[7]

14.9 PROGRAM AUDIO AND COMMUNICATIONS SYSTEMS

by Mark Schubin

Television is defined as: "The transmission and reception of transient visual images, generally with accompanying sound." Similarly, television engineering tends to be concerned more with the aspects relating to pictures than sounds. In part, this may be due to the high cost of video equipment compared with audio. On the other hand, television has been described also as "radio with pictures." This reverse emphasis has some support in the fact that the loss of the sound portion of a program more often will result in viewers tuning to another channel than will the loss of the picture portion. With simultaneous stereo transmissions by FM radio stations accompanying television broadcasts (simulcasts) occurring regularly and with stereo video cassettes, video discs, and even television broadcasts growing in popularity, television audio production is becoming still more important.

14.9.1 MICROPHONES, STANDS, AND BOOMS.   Microphones can be divided into two broad categories based on their directional characteristics. Omnidirectional microphones will, with the exception of high sound frequencies for which the size of the microphone becomes significant relative to the wavelength of the sound, pick up sounds equally well from all directions (Fig. 14-49a). The major use of omnidirectional microphones in television has been in lavalier microphones—tiny microphones [often less than 1 cm (0.4 in) in diameter or width] that are attached to the clothing of a speaker. In live television production (such as news programming) or when it will be difficult to reshoot a scene due to microphone failure, it is common to use two such microphones on a single clothing pin or clip. Omnidirectional microphones are also hand-held by news reporters and are sometimes used for music recording or dramatic productions.

Directional microphones are purposely designed not to pick up sounds equally from all directions. One means of obtaining directivity is by the surrounding surface. *Pressure-zone microphones* are omnidirectional types attached to large, flat surfaces, such as walls or floors, so as to provide a hemispherical sensitivity characteristic. They are most often used to capture dramatic or musical performances. *Cardioid microphones* pick up sounds best in one direction (referred to as the *front*), less well to the sides, and worst to the rear (Fig. 14-49*b*). They are suited for use on talk shows, where the presence of a desk microphone is not considered objectionable, or hand-held by singers. Cardioid microphones are also available as lavaliers for use in noisy environments.

Figure-eight microphones (Fig. 14-49*d*) are named for their directional patterns—good to both front and rear and poor to the sides. They are rarely used in television production, except in recording musical performances. *Hyper-cardioid* microphones (Fig. 14-49*c*) offer less sound pickup at the sides and rear than do cardioid microphones. However, they pick up more sound directly to the rear than do cardioid microphones. These microphones are often used on "fish poles" (hand-held poles employed to suspend a microphone above a camera's shooting area) because they are lighter than other highly directional types and because their rear pickup, aimed toward the ceiling, will not detract from the sounds they are aimed at.

The most directional microphones fall into two categories, called *shotguns* and *parabolics*. Shotgun microphones are so-named because they are long and round, like the barrel of a shotgun. They are the most directional of microphones that offer relatively flat frequency response (relatively equal pickup of all frequencies of sound). They are very often used at the ends of microphone booms (similar to fish poles, but balanced on a column attached to a wheeled stand, and with mechanical arrangements that allow swiveling the microphone and moving it in all three axes of space) in studio productions. They are used also attached to pistol grips or directly to cameras for field production. Parabolic microphones use transducers placed at the focus of a parabolic reflector to achieve a directional characteristic. Since they react very differently to different frequencies, they are rarely used outside of sports productions for long-range voice pickups.

A number of microphones can be switched among omnidirectional, cardioid, hyper-cardioid, and figure-eight patterns. Some, using four transducers in a tetrahedral configuration, can be electronically reconfigured to even more possibilities.

Many microphones require either a charging voltage or actual powering for operation. Charging voltages, when required, are usually supplied by a battery internal to the microphone. Powering is usually supplied from a mixing console or tape recorder, although separate power supplies are always available, and occasionally, a microphone will contain its own battery power supply. The most common forms of powering require either 48 V between both signal lines and a ground reference (phantom powering) or 12 V between the two signal lines (T, AB, reverse T, or BA powering). Transformers in a microphone's signal path will not allow either form of powering to reach the microphone.

Wireless Microphones.   Although any microphone can be used with a wireless transmission system, usually only lavalier microphones or hand-held cardioids are. Transmitters are always battery-powered and are designed to be worn, either in a pocket or in a holster or belt designed for the purpose, except in those cases where they are internal to a hand-held microphone. Both VHF and UHF wireless-microphone transmitters are available for use in accordance with FCC licensing requirements, generally with frequency modulation. Some also use a form of dynamic loudness-range compression. Antennas are usually simply trailing wires, often concealed in clothing. Body moisture, however, can degrade an antenna's function.

Receivers may be either battery- or mains-powered, depending on whether they are designed for field or studio use. Elaborate antennas, even manually aimed parabolic reflectors, are used to feed the receivers. *Diversity receivers* allow for transmission problems without loss of audio signal. Some diversity receivers select between two or more antenna inputs for the one with the highest radio-frequency level. Others use two antennas and two receiving sections and then select the one with the highest audio-frequency level. It is common to use more than one wireless microphone in a production, and frequencies must be selected to avoid mutual interference.

Microphone Stands.   Besides booms and fish poles, there are a wide variety of microphone stands available for television production needs. It is usually advisable to use a stand that includes shock isolation mounting to prevent sounds from being mechanically

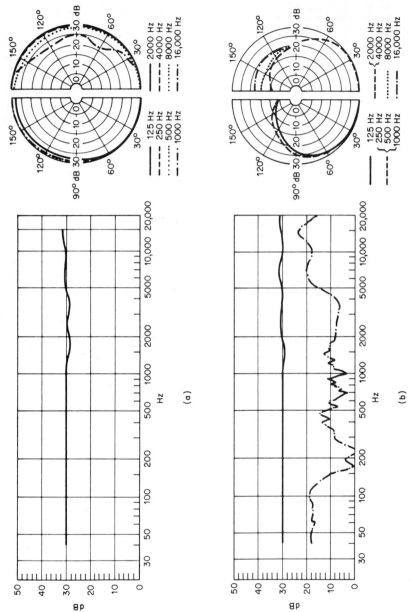

FIG. 14-49 Frequency-response characteristics and directivity patterns for typical omnidirectional and directional microphones: (*a*) omnidirectional; (*b*) cardioid; (*c*) hypercardioid; (*d*) figure eight.

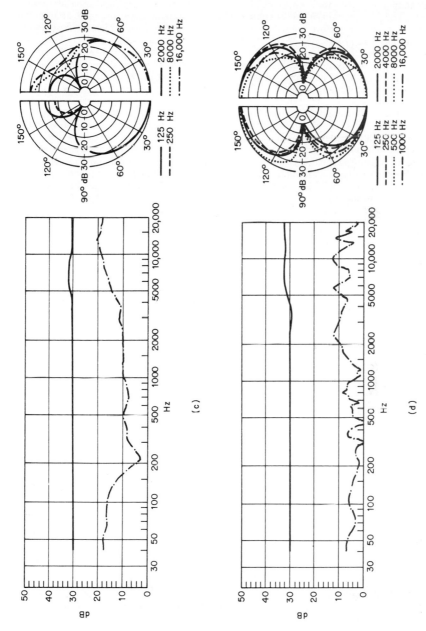

FIG. 14-49 *(Continued)*

14.107

transmitted along the stand to the microphone. The first length of signal cable from the microphone is often sheathed in cloth for the same reason. A number of microphones come apart into very small "capsules" (containing the transducer) and separate preamplifiers. These microphones can be connected to very narrow tubes to create an unobtrusive desk or podium pickup point. They can also be hung by thin cables when necessary.

Boom and fish pole operators must be careful to avoid creating a shadow in the picture as a result of being positioned between the lighting and the performers. Therefore, booms are usually equipped with picture monitors. It is also useful for boom and fish pole operators to hear the amplified sound of their microphones and instructions from the audio mixer via headphones.

Microphone Phasing.   When multiple microphones are used in a production, it is important to keep them all *phased* the same way (putting out a positive voltage when confronted with a positive pressure wave). Devices called *phase poppers* are often used to ascertain microphone phase. The transmitting popper emits a sequence of popping sounds, all beginning with a positive pressure wave. The receiving device, either connected electronically to a microphone or other audio device or placed in front of a loudspeaker, indicates, usually by means of lamps, whether the phase is correct or incorrect. In stereo production, phasing is particularly important, since stereo signals cannot be recombined into a monophonic signal unless phase is correct.

14.9.2 MIXERS AND CONSOLES.   Microphones emit signals at levels so low that they must be amplified prior to recording or transmission. Although some recorders designed for news gathering contain built-in microphone amplifiers and can be used without a microphone mixer, in most other cases a mixer is the first element of the audio processing chain to which a microphone is connected. Therefore, a great deal of the design of a mixer is devoted to amplifying and processing the microphone signals without introducing distortion or noise.

Many mixers provide a differential input balanced with respect to ground. This input is often through a transformer but more and more frequently uses direct connection to a differential-input amplifier. Either technique reduces *common-mode noise* (longitudinal) that may have been introduced in the microphone cable. This is defined as electrostatic or electromagnetic noise which appears equally and in phase from each signal conductor to ground. Microphone powering, if offered, must be introduced prior to this differential input, which may need to be isolated from the powering source by capacitors. Of course, in a multi-input mixer, power for each microphone must be isolated from that for the other microphones. Some mixers offer different forms of powering for different microphones. On some designs, however, phantom powering is always present on all microphone inputs, a situation to be kept in mind when using other than phantom-powered microphones.

It is often necessary to use more than one mixing console for a single group of microphones. For example, a music performance might require a public-address (PA) mixer for the audience, a stage mixer for the performers' monitor speakers, a mixer devoted to recording the performance for a phonograph record, a mixer for a radio broadcast, and a television mixer. While some very large mixing consoles can deal with all those mixes simultaneously, more often one of two methods will be used to distribute signals to all the mixers. The first is microphone splitting, usually using a transformer with one or more secondary windings to feed secondary mixes. Phantom power can pass from the primary console to the microphone since both are on the same side of the transformer. Sometimes such splitters include amplification stages. The second technique is to distribute an amplified signal for each microphone directly from the mixing console. Such signals are referred to as *direct outputs*.

The input stage of a mixing console will often include signal phase reversal capability and an attenuator or preamplifier gain control to compensate for microphones of varying sensitivity or placement. After amplification, the signal is often introduced to an *equalizer* (EQ) section, with the capability to affect its frequency response. Even field mixers usually include at least a low-frequency cutoff filter to allow wind noise to be reduced.

Complex mixing consoles offer "shelving" (positive or negative gain of all frequencies above or below a certain point) or "peak/dip" (positive or negative gain based around a single frequency) equalization in three or more overlapping frequency bands, with variable amplitude, frequency, and bandwidth. Such manipulation helps make different types of microphones and rooms sound the same and can compensate for problems such as excessive sibilance.

The microphone's signal, either before or after equalization, is available to a single mixing amplifier, in the case of a rudimentary field mixer, or to a large number of mixing amplifiers in a complex mixing console. In either case, the level to be mixed to the output signal from a particular microphone is controlled by a variable control called a *fader,* which may be adjusted in either a rotary or linear fashion. In some mixing consoles, in lieu of a mechanical fader, a voltage-controlled attenuator or voltage-controlled amplifier (VCA) is used, controlled by a dc voltage from a potentiometer or other source. By using a single adjustment to control several VCAs, microphones can be grouped into sections to be controlled together. Some mixing consoles also offer control of "subgroups" and "mastergroups" (non-VCA-equipped mixing consoles can also offer sub-master faders, but these require an additional stage of amplification and attenuation for each level of sub-mastering).

In large mixing consoles, there may be several mixing buses feeding a variety of outputs. A console designed specifically for multitrack audio recording will have at least 16 such buses (if not 24 or more), with the faders controlling the feeds to different tracks of a multitrack recorder, and with secondary faders providing a stereo or mono monitoring feed to amplifiers and speakers. In a console designed for broadcasting, the roles of the primary and secondary faders would be reversed. Some consoles can be switched between recording and broadcasting configurations. Those designed for use with multitrack audio recorders also include switching facilities for monitoring or mixing recorded tracks. There is usually a control for each input channel that controls the relative level fed to odd-numbered or even-numbered mixing buses. Since this control can be used to move sounds from left to right in much the way that a camera can move images, it is referred to as a "panning" control or "pan pot."

When previously recorded tracks or multitrack recordings are remixed, particularly if the original tracks were recorded on video tape, the process is often referred to as "sweetening." When multiple tracks are mixed to a stereo pair or a single monophonic track, the process is referred to as a "mixdown."

In a large mixing console there are usually a number of auxiliary mixing buses. In a console designed exclusively for broadcasting, these auxiliary buses (like those in a radio broadcasting console) might be referred to as *audition* and *cue* and would be used primarily to listen to possible inputs without putting them on the air. In studio consoles, they will more often be referred to as "echo sends" or "reverb sends" and "foldback," "cue," or "Q" buses. There is usually more than one of each type of bus.

Unlike the auxiliary buses on broadcasting consoles, these auxiliary buses are designed to be used while the same input signals (microphone or otherwise) are being recorded or broadcast. Therefore, each bus has its own fader for each input signal, and, often, inputs can be selected with or without an EQ and with or without the attenuation of the main fader for that input channel. *Echo send* feeds a signal to an echo chamber or artificial reverberation unit (*echo return* adds the reverberant signal to one or more main output buses). Foldback or cue buses feed to the studio floor loudspeakers or to a performer's headphones or earphones.

The auxiliary buses most closely approaching the broadcast audition and cue buses are called *solo* and *PFL* (or prefade listen). Depending on the configuration of a particular mixing console, one of these will usually substitute the sound of a single input channel for whatever was previously heard on the monitoring speakers, while the other might allow several channels to be heard at once. Different configurations will also allow listening to channels with or without EQ and before or after the attenuation of the main fader.

14.9.3 LEVEL MONITORING, CONTROL, AND EQUALIZATION.

A variety of different level-monitoring systems are now in use on audio mixing consoles. By far the most common of these is the VU meter, a signal-averaging meter standardized in ANSI C16.5, although few meters actually meet all aspects of the standard. Another popular meter, more prevalent in Europe than the VU meter, is the peak program meter (PPM), a meter that responds to shorter peaks than does the VU meter and that takes longer to return to a no-signal reading. It is also common to find a variety of nonmeter level indicators, most using light-emitting diodes, some approximating the characteristics of VU meters, some those of PPMs, some neither. It is also common to find light-emitting diodes used as amplifier peak indicators at various points in a console. These usually illuminate at a somewhat lower level than will cause distortion to be introduced into an audio signal, allowing an operator to take steps to reduce levels.

Besides the equalization and level control facilities found in a large mixing console, it is common to use "outboard" signal-processing equipment, such as the artificial reverberation units suggested by the echo send buses. A wide variety of artificial reverberation units exists, ranging from springs and plates connected to transducers to digital delay systems with characteristics programmed to match both real (cathedral, concert hall, etc.) and artificial reverberations.

Outboard equalizers are available in types similar to those used in consoles, as well as sharp notch filters for attenuating noises, as "graphic" filters that can vary gain at all octave or one-third octave frequencies in the audio band simultaneously, and as "comb" filters (often created with digital delays) to attenuate harmonics of a certain frequency (as might be required to attenuate a buzzing caused by a lamp dimmer).

Limiters and compressors reduce dynamic range (the level between the loudest and softest parts of a signal), while expanders and gates increase it. Used in conjunction with equalizers, delays, and the like, these processing tools can be used for special effects, as can devices such as *phasers* and *flangers* (that introduce phase distortion) and *harmonizers* (that change pitch).

Automated consoles are frequently used in mixdowns and sweetening sessions. These will accurately reproduce a particular mixing arrangement (including, on some consoles, equalization settings), allowing only certain changes to be made (and recorded) each time.

14.9.4 AUDIENCE REACTION AND SOUND SYSTEMS.

One of the most difficult tasks in television audio is the simultaneous recording or broadcasting of a program while it is being amplified for a live audience. The task is even more difficult when sounds from the audience must be heard.

Fortunately, large audio mixing consoles can easily create what is referred to as a "mix-minus signal" (a signal that includes a mixture of all the input channels except those being used for certain microphones). The loudspeakers facing the audience would thus be fed a mix-minus-audience signal (while the performers might be fed a mix-minus-performers signal). The use of shotgun microphones on booms rapidly positioned to a speaker in the audience reduces the need for high gain in the audience microphone channels and allows some of the speaker's signal to be fed back to the audience loudspeakers for the rest of the audience to hear.

When a performer enters the audience area to speak directly to audience members, it is best to use two microphones, one a lavalier on the performer, and the other a hand-held microphone in the performer's hand. Experienced performers can make do with a single, hand-held microphone. The use of wireless microphones will enhance the mobility of the performer.

14.10 COMMUNICATIONS

by John Hartnett and Douglas J. Hennessy

The relative importance of communications in the complex operation of television broadcasting is rarely appreciated. Ordinarily it is something that is taken for granted, since the greatest technical emphasis is placed on the primary program systems, audio and video. Nevertheless, a successful broadcast operation is based on split-second timing requiring the precise coordination of numerous activities. The production and technical personnel (whose number may be a hundred or more for a single production) performing these activities must be able to communicate effectually with each other in order to ensure against costly disruptions in programming.

14.10.1 DESIGN CRITERIA. An "effectual" communications system is difficult to define precisely. Instead, it is perhaps best to list some criteria upon which a communications system should be designed:

1. *Immediacy.* There must be no delay in the transmission of a message from source to destination.

2. *Priority.* No message should have priority over any other, and messages arriving simultaneously at a destination should be superimposed.

3. *Distribution.* A source should be able to transmit to one or a number of destinations (simultaneously), as desired.

4. *Privacy.* A message intended for a specific destination should not reach any other.

5. *Uniform volume levels.* An individual should control only the receive level, and the range of receive control should never permit the listening level to be brought to a point of inaudibility. The transmit level should remain fixed, and variations in individual speaking levels should be compensated for by limiting circuitry in the originator's microphone amplifier.

6. *Interference.* All spurious signals, either electrical or acoustical, present in the communication system must be eliminated or minimized.

7. *Feedback.* Electrical or acoustical feedback cannot be tolerated in the communications system.

8. *Crosstalk.* Communication signals must never be allowed to appear in the program material. This indicates elimination of electrical and acoustical crosstalk between communications and program systems.

9. *Communication limitations.* An individual should not be overcommunicated, that is, should have only those communications facilities necessary for the proper performance of his or her job function.

10. *Human engineering.* Great emphasis should be placed on the human-engineering design of a system, so that it is simple to operate (a special consideration for nontechnical personnel), easy to maintain, and in accord with the decor and style of the facilities.

Many of these criteria are simply good engineering practices. They are worth noting, however, since they are sometimes overlooked in communication system design.

Ancillary Telephone Systems. It should be noted that besides the communication functions discussed in this section (intercom, interphone, and cue systems), the telephone system is an additional valuable tool for communicating.

There are two types of telephone systems (besides business phones): *technical private line* (TPL) serving technical personnel and associated areas, and *production private line* (PPL) serving production personnel and associated areas. Both are generally two-digit systems and are of the "no busy" or conference type. This means that a circuit in use may be dialed by another party trying to reach either of the two parties on the circuit in question. These systems, while efficient and convenient, do have two faults:

1. An undesirable time delay is incurred, since the individual calling must acquire the number to dial, dial it, and then wait for the second party to answer.

2. A person may call only one location at a time.

14.10.2 **INTERCOMMUNICATIONS.** Provision for two-way, point-to-point communications is probably best achieved in a broadcast facility by the intercom system. It is a system employing an amplified microphone-to-loudspeaker path. The path is established by the originator who routes the signal, via a switching matrix of some type, to one or more destinations. Such a transmission should be, for all practical purposes, instantaneous.

There are a great number of intercom systems commercially available. They range in price anywhere from a few dollars to several hundreds of thousands of dollars, depending upon their quality and complexity. Some of these, however, have a number of faults which make them undesirable or unusable in television broadcasting or production service. They are as follows:

1. Intelligibility. The inability to reproduce a message intelligibly may result from shortcomings such as poor frequency response, excessive distortion, or susceptibility of the system to electrical noise pickup.

2. Reliability. Unreliability and requirement for a great deal of maintenance to keep the systems operational.

3. Flexibility. Flexibility that would permit modifications or expansion of a system to fit future requirements.

4. Design and operating simplification. Poor design, with little thought given to human engineering, thus making the systems difficult to install and operate.

5. Muting. The most common defect is the absence of a proper arrangement for muting. Some systems will provide only overall muting; that is, when an individual originates a transmission, all incoming messages to him or her are muted or dimmed for the period of the transmission. During this time, short though it might be, critical messages from others could be missed or seriously delayed.

It is not always possible to purchase a system without the aforementioned faults, and it may be necessary for broadcasters to design and construct all or most of it themselves.

The *functional diagram* of a simplified broadcast intercom system is shown in Fig. 14-50. Besides the four local stations shown in the diagram, there is also an external intercom line (11) from another area with a similar matrix.

The microphone (1) is normally an omnidirectional, low-impedance, dynamic type. In some areas a noise-canceling or close-talking type is used having the advantage of reducing background noise (which can become rather high in a control room) from the message information. It has, however, the disadvantage of requiring the person using it to speak very close to the microphone. It is not unusual to learn that production personnel find it objectionable to be confined by such a speaking requirement on a communications microphone.

The microphone preamplifier (2) brings the microphone output to line level. The preamplifier should have a limiting characteristic which allows for the level variations in different voices (people who tend to shout or crowd the microphone) and keeps the output level of the amplifier from causing excessive distortion.

The first routing element is the transmit crosspoint (4). It is normally open, and when closed provides a bridging path through a monitor amplifier to loudspeaker "T" (5), thus enabling "R" to talk to "T."

Control of the crosspoint is normally provided by a contact closure of a momentary switch. A similar path is provided through crosspoint (6) for "S" to talk to "T."

A complementary crosspoint (7) enables "T" to talk to "R" and "S." A logic interlock is provided (8) between these complementary crosspoints such that if a crosspoint is activated, its complement is locked out to prevent acoustic feedback. For the paths just mentioned, if either (4) or (6) is activated, crosspoint (7) cannot be activated. Likewise, if (7) is activated, both (4) and (6) are locked out. Note, however, that "U" (whose message may be of paramount importance) can still transmit to "T" or "R" and "S." Similar interlocks are provided for the paths between "R" and "S" and "U" (9) and between "T" and "U" (10). Such a system of interlocked complementary crosspoints is usually referred to as *selective muting*.

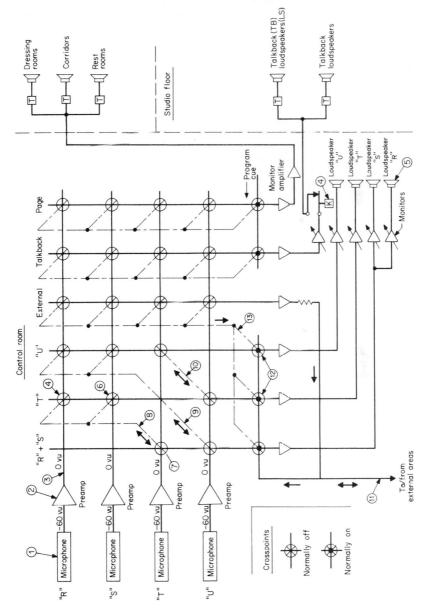

FIG. 14-50 Intercom-system block diagram.

14.113

The situation shown in Fig. 14-50 for stations "R" and "S" is quite common. These are two stations sharing a common receive circuit. This is done when two stations are located close to each other and do not require discrete listening positions (as in the case of the stations at a production console).

The two-way intercom line (11) external to the matrix would normally route to another studio, an equipment center, master control, outside studio, video tape machine, telecine chain, etc. The line might reach these areas, particularly video tape machines and film chains, via an assignment switcher or a patch-cross. The circuit at the other end could be similar to the matrix of Fig. 14-50. The transmit signals from the external area are received through normally on crosspoints (12). These are turned off by interlock (13) when any station in the control room transmits to the external areas.

Talkback and paging systems shall be considered here as part of the intercom system. Even though they are only one-way circuits, the crosspoints for them are normally contained within the intercom matrix. Both of these systems are illustrated in Fig. 14-50.

The talkback circuit is a transmit-only circuit to overhead loudspeakers in the studio area. It is used for communications from the control room to talent and other personnel on the stage. Return communications are usually accomplished by using an open program microphone or a special communications link supplied for a stage manager. The talkback system is unique in that it is the one portion of the communications system that is interlocked with the audio console. This interlock (14) prevents feedback but, more important, prevents the possibility of communications being heard on stage during a taping session or live show.

The paging circuit is simply what the name implies: a method for the stage manager or select personnel in the control room to page or call talent from outside areas to the stage floor or control room.

14.10.3 INTERPHONE.   A conference-type communication system that allows hands-free operation is a necessary part of the overall broadcast facilities. The director must be able to communicate with personnel on the studio floor, particularly camera operators, while a show is in progress. These conversations cannot be permitted to interfere with the show in any way (i.e., distracting talent or being picked up by program microphones on the stage floor). This rules out the use of a microphone-loudspeaker system such as an intercom. Intercom is also undesirable for this purpose since it is selective and would not permit hands-free operation.

The system used to fulfill this communication requirement is interphone. It allows a conference connection between technical and production personnel on stage and in the control room and provides isolation of portions of the system from the main conference bus, permitting private conversation (e.g., the isolation of a camera by the video operator for the purpose of working on a technical problem without interfering with the director's conversation on the main bus).

Figure 14-51 is a diagram of a simplified interphone system which employs standard telephone-type headsets (and occasionally handsets), a two-wire common power supply circuit, retardation coils† for isolation, and an interphone terminal unit.

This diagram illustrates the ability of stations to isolate themselves from the main (director) conference bus and the need for the retardation coil to provide isolation from the common power supply. Director microphone reinforcement is also shown in Fig. 14-51. This is a full-time feed from the output of the director's intercom microphone amplifier, which permits the director to give instructions without having to wear a headset.

The interphone terminal unit must be able to maintain a balance between the receive level and the sidetone‡ level in a headset (or handset). The level of sidetone should be the same as that of the receive level. If the sidetone is lower than the receive level, an

---

†The purpose of the retardation coil is to prevent crosstalk between portions of the system that have been isolated from each other. The coil allows passage of direct current, but inhibits the audio signal from returning to the common power-supply point.

‡Sidetone is that portion of the signal in a headset receiver that is originated in the transmitter of that headset.

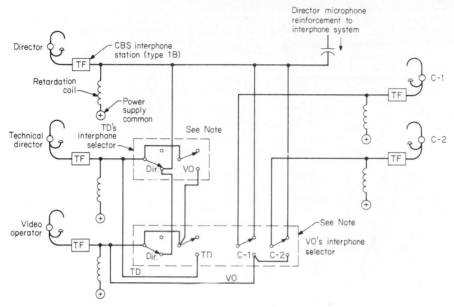

FIG. 14-51 Interphone system. Circuit is shown with all stations on director's-conference bus. *Note:* Selector switch is mechanically interlocked for positive lockout (i.e., only one position can be selected at a time).

individual on stage tends to speak louder, and his or her voice could be picked up by a nearby program microphone. If the sidetone level is higher than the receive level, the tendency is to speak more softly, thus making the transmission inaudible to others on the conference bus.

14.10.4 CUE. Many production and technical personnel, not located in the control room, are required to hear instructions from the control room (the director in particular) or program audio in order to perform their duties properly. This requirement is satisfied by a system of cueing circuits.

There are three basic cue signals which, in various combinations, provide the cueing requirements. They are:

1. *Director cue.* An amplified mix of the director, associate director, and technical director microphones.
2. *Program cue.* A bridged feed of the program output of the audio console.
3. *Audio operator cue.* The amplified output of the audio operator's microphone.

These cue signals are used to produce the following circuits:

1. *Director cue.* As previously defined.
2. *Regular cue.* Used to feed a double headset, with director cue on one side and program cue on the other.
3. *Boom cue.* Another double headset feed, with director cue on one side and program cue on the other. It differs from regular cue in that the program cue side can be interrupted by the audio operator and audio operator cue put in its place.
4. *RF cue.* Director cue is fed to a radio frequency transmitter and is picked up on small receivers carried by personnel on the stage floor. This is used by those who must be

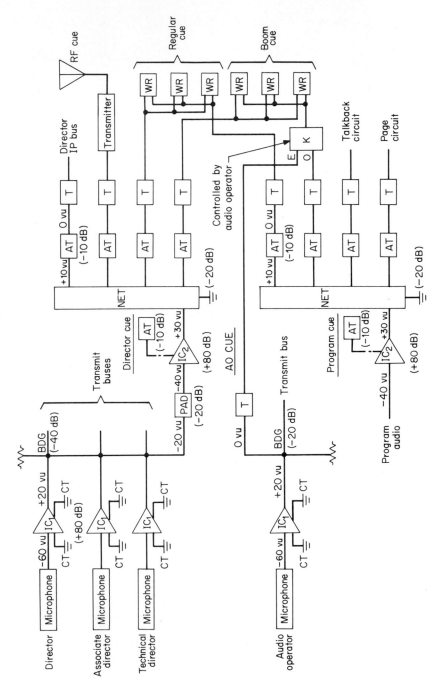

FIG. 14-52 Cue-distribution system.

14.116

able to move freely about the stage without the restraint of a cabled headset (used mainly by stage managers).

5. *Orchestra cue.* This is the same as boom cue with the exception of a separate interrupt control circuit for the audio operator.

6. *Camera cue.* This is a circuit provided to the cameras for situations where the camera operator finds it necessary to wear a double, rather than the usual single, interphone headset. Either director cue or program cue can be placed on this circuit to feed the second earpiece.

Figure 14-52 is a simplified illustration of how some of the above cue circuits are derived.

14.10.5 PROGRAM CUE INTERRUPT (INTERRUPTED FOLDBACK). Program cue interrupt (PI), also referred to as interrupted foldback (IFB), is the primary means of communication from the control room to talent on the stage floor, particularly during the actual production of a show. It is a one-way system, and the only return path is via a live program microphone. The talent generally wear small, inconspicuous earpieces and receive a program feed when no message is being sent. Designated personnel in the control room have a set of keys enabling them to "interrupt" the program feed to any individual on the stage floor and to substitute their own amplified microphone in place of program cue when a message needs to be sent. In actual practice, the program cue to each individual usually consists of a so-called mix-minus feed, that is, all the program mix except for the individual's own microphone. A simplified diagram of a PI system is shown in Fig. 14-53.

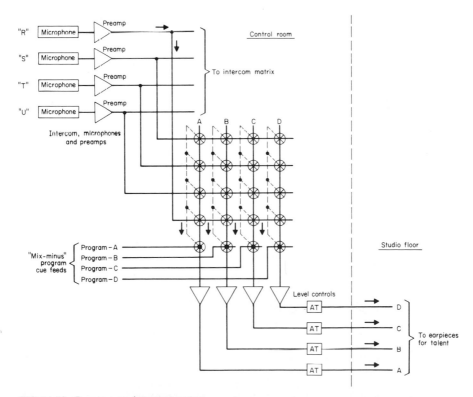

FIG. 14-53 Program cue interrupt system.

14.10.6 STUDIO COMMUNICATIONS REQUIREMENTS

Control Room. The working personnel in the control room during the production of a show should be provided with all the communications facilities they require to perform their job functions. This area is a command post, normally staffed by a director, assistant director, technical director, audio operator, video operator, and lighting director. Besides their telephone circuits (TPL, PPL, and business phones) they are provided with intercom and interphone, and in the case of the audio operator, video operator, and lighting director, a loudspeaker-level feed of director cue.

The areas to which these people must communicate are not the same for all of them. The director and assistant director, for example, have no need to communicate with certain technical areas such as camera control, maintenance, and equipment center. The technical personnel in the control room (technical director, audio operator, and video operator), on the other hand, have no requirements for the paging circuit. These differences in intercom requirements are also true in the interphone facilities. The facilities should generally not be planned around a particular show or type of show, and enough flexibility should be designed into the systems to allow for any modifications because of variations in show format.

Stage. Communications are supplied to the stage through audio-video (AV) boxes strategically situated around the perimeter of the stage area. The AV boxes contain all the cue circuits previously mentioned, as well as director bus and lighting interphone, talkback loudspeaker feeds, and orchestra, boom, and floor manager intercom circuits. Camera communications are provided via the camera connections located in the AV boxes.

Associated Areas. Production of a television show often requires facilities other than those ordinarily found in the studio. These outside areas, when used by a studio, must be able to communicate with the control room of that studio. In a small broadcast facility where the number of studios is limited to one or perhaps two at most, and the video tape machines, film chains, and other facilities in the plant are limited as well, the provision of communication is a relatively simple task. In a large plant, however, the situation is much more complex, where numerous studios and a multitude of external facilities must be able to communicate with the production or coordinating studio. An assignment switcher frequently is used, much as a switchboard is used for telephone service, to route the video tape machine or film chain's audio, video, control, and communications circuits to the studio wherein they will be part of the show facilities. This eliminates the great number of circuits that would otherwise have to be installed between the machines and studios, as well as a large and very complex communications system.

14.10.7 STATE-OF-THE-ART COMMUNICATIONS SYSTEMS.
Recent developments in the area of solid-state circuits have provided a new generation of equipment and components for the design of communications systems. These systems are smaller, as well as more economical to construct and maintain.

Integrated-circuit and microprocessor techniques permit multiplexing of communication-signal and control circuits, thus eliminating the need for installing many multiconductor cables, an unwieldy and undesirable part of an installation. They also provide new and more efficient means to route and process communications signals and control.

14.11 STAGING AND LIGHTING

by E. Carlton Winckler

14.11.1 OBJECTIVES AND FUNCTIONS

A Definition of Lighting. Lighting is the creative utilization of controlled illumination in television production to achieve artistic visual reproduction. Light sources and fixtures currently available provide the means to apply, with little difficulty, a wide range of light levels anywhere in a television scene. On the other hand, using light for specific artistic objectives presents more complex problems. The obvious functions of television lighting are to accent, model, separate, and illuminate, but the less-familiar functions, such as attracting and holding the viewer's attention to the principal element in a picture composition, are of at least equal importance. By lighting an area slightly brighter than its surround, the lighting designer attracts the viewer's eyes, moving or holding them at will. This is creating a *center of interest,* a major requisite for effective, creative lighting.

Production Coordination. Lighting requires total coordination among all areas of picture making, including both the technical operations and production staging segments. However skillfully applied, lighting cannot correct all visual problems. Although improperly staged scenes can be improved by judicious use of lighting, the time, effort, and tedious adjustments required may be disproportionate to simpler methods of correcting the basic source of picture shortcomings, namely, the backgrounds, wearing apparel, makeup, or positions of the performers.

Separation. To achieve the desirable separation of action and surrounds, more is required than aiming various luminaires at one object or another and dividing controls. One prime requirement is spatial separation of the two elements—preferably by at least 6 ft—so that the subject light illuminating the background will be less intense. Another technique is to use focusable luminaires, in order to permit concentration of the light on the action areas and away from the surrounds. A third approach is to adjust the relative intensities of the light sources, either mechanically or electrically.

While dimmers are most desirable to adjust intensity precisely for balance among individual luminaires or entire areas, nets, wire screen, or repositioning of the sources also can achieve approximations of the needed levels.

Illumination. All too often the objectives and functions of controlled lighting are ignored, and instead mere *illumination*—excess uncoordinated light from every possible direction to provide correct camera exposure—is applied. While pictures under mere illumination are possible, the visual effect has no meaningful composition. This type of illumination frequently is referred to as a *news shot.*

14.11.2 LIGHT INTENSITY. The use of a large number of intensely bright luminaires, necessary in the 1940s and early 1950s for studios equipped with image-orthicon cameras, is no longer essential for present-day cameras. The modern camera is capable of producing excellent pictures with very low levels of illumination. On the other hand, light levels in our everyday lives have taken an opposite turn toward very bright interior illumination. Consequently television action areas having much less than about 100 fc (1076 lm/m^2) of light are a depressing and abnormal environment for the talent. Consequently, the lighting levels should be coordinated with the director for aesthetic consideration and the video operators for camera requirements. With properly arranged lighting at the agreed-upon level, the camera can be exposed to reproduce all tonal values of the subject composition accurately.

Excessive light is a very common cause of camera exposure problems and resultant picture degradations. In practice, excessive light intensity causes the lighter-colored, more-reflective areas in a composition to appear brighter than normal while the darker areas which reflect smaller amounts of the light are minimally affected. The camera is exposed for the bright areas, and, therefore, the darker, less-reflective portions are pushed toward black and all detail is lost, producing an unpleasant picture with highlights and unintelligible, compressed shadow areas.

Measuring Light Intensity. In determining the light intensity, the lighting director (LD) relies on an averaging light meter employing a photoelectric cell (PEC) to indicate the light intensity in footcandles. There are two types of targets available, one a flat light-sensing surface, and the other a spherical, or "golfball," form. The latter is cumulative in action by collecting and averaging light over an acceptance area covering nearly

180°. This type of exposure meter is useful in determining film exposure. However, such information is not required for television cameras, since the waveform and picture monitors permit very accurate exposure measurements and judgments. The flat target meter is used generally for television lighting because of its greater directivity and its convenience for measurement of the output of single luminaires, its primary application. *Depth of field* refers to the range of distance from the camera lens wherein elements of the scene are in sharp focus, compared with other portions which are discernible but slightly to grossly out of focus. By controlling the *depth of field,* the *center of interest* can be accentuated, and elements of lesser importance de-emphasized. Since the depth of field is a function of the lens focal length and aperture setting, the light level is the controlling factor. Higher light levels require a smaller aperture, or *f*-stop setting, for proper camera exposure and result in a greater depth of field.

Consequently, with a fixed scene light level, if artistic considerations dictate a reduction in depth of field, this can be achieved only by an increase in lens aperture (lower *f*-stop) and the addition of neutral-density light-attention filtering in the camera.

14.11.3 THREE-POINT LIGHTING. The basic approach to television, or film, lighting is the three-point technique. In other words, light from three directions is directed toward the scene and subjects. This simple approach provides accent, modeling, and separation. (See Fig. 14-54*a*).

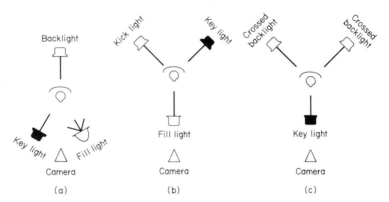

FIG. 14-54 (*a*) Classic three-point, (*b*) three-point variation, (*c*) three-point variation.

The three-point nomenclature refers to the following:

- *Key light*
- *Fill light*
- *Backlight*

These are essentially "people (talent) lights" and must be considered in conjunction with the set lights (scenery or surround lighting) as well.

The *key light* contributes the dominant front illumination for the subject at each camera position and is usually provided by a spotlight for control of the coverage area. The *key* is described as "the apparent source of directional illumination in the composition"—light from an unseen window, a lamp, the rising sun, or the primary source of scene light. The key is the brightest light in the scene composition and as such determines the relative intensity of the other sources and the camera exposure. While generally a front light, if staging requirements dictate, the key may be positioned to project from any direction. However, if it is from the back of the set, it usually should be directed from a ten or two o'clock position, in order to model the subject effectively. (See Fig. 14-54*b*.)

The position of the key light has a significant bearing on the overall lighting effect. A dead-center position directly over the camera (Fig. 54*c*) is very effective in the case of one subject shot with one camera, since all details of the face or facial expressions will be clearly illuminated. A position 10 to 30° off-center, in relation to the camera position, may also be effective, because it results in modeling and accenting the features, with another source of lesser intensity than the key providing softening of shadows and avoiding any loss of facial detail. In these instances the height of the key source is equally as important as the horizontal location. A sharp downward angle from a support which is too high will result in elongated, unattractive facial shadows and dark eye sockets. If the key is placed too low, the subject may be blinded and will tend to squint and be unable to read prompting devices, or see any communication signals from floor personnel.

Key light is ideally provided by a Fresnel-lens spotlight for maximum shadow and spill-light control. Alternatively, an open-face spotlight may be used although the unwanted spill light will require careful masking.

A floodlight is a less desirable key-light source. Thus, its use is most often restricted to compositions where space permits illumination of the subject without the flood's unfocusable light striking surrounds too brightly.

One further note on key lights. The use of dimmer control for these "people lights" requires careful application. Reducing intensity by more than one or two dimmer points will cause the light to take on a red tinge which may adversely effect the color of skin tones. When more reduction is desired, nets or screen wire, which reduce intensity without color distortion, are used.

Fill light, intended to soften shadows resulting from the stronger key source, may be over the complete scene or confined to the same single subject as the key. The fill may be a floodlight or a spotlight, depending upon the size of the area of control desired. For floodlight sources, it is advisable to apply a diffusion filter, either Mylar polyester film frost, spun glass, or etched glass, to soften shadows. This is necessary because an open floodlight source usually has a distinct hot spot, resulting in a fairly sharp shadow pattern. An ideal fill source of the floodlight type is the *softlight,* in which the light source is shielded so that only indirect light from the reflector reaches the subject. It is to be noted for planning purposes that indirect light fixtures are inefficient. Consequently, units of higher wattage will be required to provide adequate light levels than with direct sources.

Backlight is designed to rim the subject with light from the rear, or toward the camera position to provide the important separation illusion of subject and background. Usually this source is a spotlight, but a floodlight, very carefully masked to prevent lens spill, can be used.

More attention should be devoted to backlighting than is usually accorded this important function. The separation objective can be achieved with light of surprisingly low intensity, whereas a backlight of high brightness poses a major composition problem, causing the top of the subject, horizontal surfaces, or shoulders to become the bright center of interest rather than the face.

The position of the backlight is quite flexible and need not be restricted to its generally accepted place directly opposite the camera. In fact, it is usually more effective when placed a bit to one side to model the side as well as the top of the head and shoulders. In this position there are several advantages since the entire subject, head to floor, is rimmed with light and separation is much improved. Even better is the use of two backlights, one at each of the usual ten and two o'clock positions. Between the two sources a 90° angle with the subject at the apex should result but is not mandatory. This arrangement models and rims both sides of the head and figure. Another highly valuable contribution of the crossed-backlight approach is the provision for any subject to turn to speak to a neighbor without facing into darkness, but rather into light. Facing into light adds strength and importance to the subject's delivery. A still further advantage of the double backlight results from adjustment of one side to be brighter than the other to provide either an accent or even an upstage key (Fig. 14-54*c*).

Variations in three-point lighting with the use of backlights at ten and two o'clock and a single, center key for the subject may be highly effective. Variations of intensity among the three luminaires often can provide interesting configurations. Should it be decided to make either the ten or the two o'clock luminaire a back key light, the front source may be a floodlight or several flood sources—always providing, of course, that the subject being illuminated is far enough away from the background that the unfocused light drops off in intensity enough to avoid the background becoming brighter than the face.

14.11.4 LUMINAIRE PLACEMENT. Positioning the luminaires required to provide the normal key, fill, and backlight is a matter for serious consideration since it determines not only effective illumination of areas to be seen by the camera, but also the location of the shadow resulting from every light source.

A backlight projecting at too low an angle can introduce lens flare in the camera as well as being ineffective in rimming the subject. Placed too high, and projecting at a very steep downward angle, the light will spill over the forehead and highlight the nose and clothing detail in the front of the subject. Somewhere in between there is an ideal angle at which the source will rim and model in a satisfactory manner.

A key light projecting at a steep downward angle leaves eye sockets dark and elongates nose, chin, and facial contour shadows to a disturbing degree. If too low, the flat angle casts a shadow on elements in the background. The position in between the two extremes provides normal eye-socket appearance and short shadows of the facial features, with unobtrusive shadows on background elements.

Set Lights. The set lights, or background lighting, are normally decorative in nature. Designed to provide a pleasant surround, often in colored light, or to provide an indication of time or place as the composition requires, this lighting may be accomplished with floodlights, accent spotlighting, or a combination of the two. Prime considerations are a lesser brightness than foreground faces and control by separate circuitry. Use of the same luminaires lighting the talent should be avoided whenever possible to retain meaningful control of overall brightness and separation.

Discussions of the basic three-point plus set lights technique usually describe a one- or two-person static scene, while a large proportion of television production involves movement and extensive action.

Lighting for Movement. In the explanation of three-point lighting, *point* connotes *direction,* and each point may include several luminaires. Therefore, if the action area is more extensive, the original three-point concept may be expanded to cover a large area by adjusting barndoors on the spotlight units to avoid extensively overlapping beams that would result in confused shadow patterns. The more usual approach provides pools of "three-point" at the location of each important action, with the crosses between these "pools" covered by either fill light alone, fill and backlights covering the cross area, or an extra key light making a path to be followed between pools. It is of no concern if the areas between the pools may be of somewhat lower intensity, since this is a normal scenic transition which occurs in natural surroundings.

14.11.5 VARIATIONS FROM THREE-POINT LIGHTING. The basic three-point lighting is an effective and safe approach to most lighting projects and is used with or without variations by professionals as well as beginners. However, three-point is not a fixed rule by any means, and the creation of an effective composition, or the satisfactory lighting of varied subjects, may lead the designer into the use of other configurations.

One-Point. A single subject in front of a limbo background (a background having no form or definition in blacks, grays, or a single color) may well look best with a single key light and no other sources. Such a key may be head on, or perhaps from an off-center position to model the face slightly from one side, with shadow detail softly apparent in the other side. This is one-point lighting. In the same creative vein the single source may come from the ten or the two o'clock position with strong shadow detail for a stark dramatic effect or a night scene.

Two-point lighting presents many interesting possibilities in creative composition. The *opposition theory* of source placement is one effective approach to two-point. The sources are positioned directly opposite each other—for example the key is placed at eight o'clock and the backlight at two o'clock (camera at six o'clock). This will light, model, and separate one subject from the background, providing an effective picture in many instances. Applied to two subjects facing each other, the two sources in opposition serve as a key for one, backlight for the other, with the reverse result from the second source, thereby keying, modeling, and providing good separation. Where conditions prevent the opposition placement, one source for a key and the other at as different an angle as possible usually results in a reasonably acceptable composition, certainly better than front light alone.

Four-point lighting, wherein the sources are placed at the four corners of a rectangular area, offers an extremely effective approach to many compositions. Ideal for panel discussions, choral groups, orchestras, crowds, or almost any type of group action, each person or object is *keyed, filled, backlit,* and allowed to turn into light to speak. The effectiveness of the four-point approach is often enhanced by the addition of soft-fill light from the "front" or camera side. While this would appear to be an ideal arrangement for nearly any subject matter from one person to a crowd, it is not ideal for many compositions and must be used with care.

14.11.6 CONTROLLED LIGHT.

The mention of controlled light has appeared several times in this discourse, and explanatory comment should be offered. The basis of controlled light is, of course, inherent in the luminaire or light source providing the illumination. There are two generic types of luminaires, *spotlights* (focusable sources) and *floodlights* (unfocused sources), with many variations of form in each category.

Floodlights. Floodlights, aptly described by their name, respond in a very limited manner to control techniques. Their value lies in their broad coverage, soft shadow patterns, and sharp drop-off in field intensity as the subject moves away from the light source.

Spotlights. Spotlights do essentially what their name describes—project a confined area of light—and are capable of providing a defined field of intense light at a considerable distance from the luminaire. The output beam of a spot is "organized" by a lens or contoured reflector, permitting the beam to be condensed (spotted down) or spread out (flooded) for area coverage control within prescribed dimensions. In a flooded or semiflooded mode the spotlight may be shaped with barndoors or cutters, patterned with cutout screens, or shaped in a number of ways. This effective masking is not readily achieved with a unit at spot focus. The Fresnel lens, and to a less effective degree the contoured reflector, are designed for efficient performance at flood or semiflood positions. Attempting to shape the tight spot beam, which is not fully organized, results in a very unsatisfactory cutoff of indefinite form and may adversely affect intensity of the light.

No Lighting Formula. Perhaps the most common deterrents to good lighting are the popular, preconceived "rules" for intensity ratios between key and backlight, key and fill light, etc., taking one angle as 1.0 and another as 1.5 for a ratio of 1/1.5, or 2/1 or 3/1. Such ratios are thoroughly impractical and meaningless in effective lighting. In lighting, the reflectance values of compositional elements determine the intensity of light each requires to assure its proper importance in a scene, the separation or modeling.

For example, a subject with black hair and black clothing against a dark background may require a high level of backlight to make separation apparent—whereas a subject with pale blonde or white hair would require only a fraction of backlight intensity for good separation. In both of these instances the key light would be about the same.

It is important that any formula approach be discarded in forming light compositions. Substitute instead imaginative judgments as to the effectiveness of subjects for the television camera and screen.

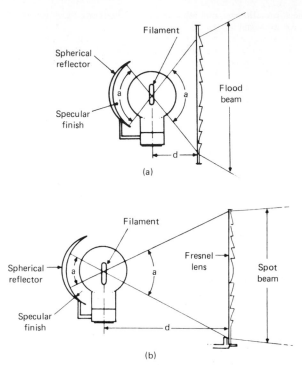

FIG. 14-55 Optical system of standard Fresnel spotlight: (*a*) full flood beam; (*b*) spot beam. *(From M. Forman, American Cinematographer Manual, Gardiner/Fulmer Lithograph. © 1973 A.S.C. Holding Corp.)*

14.11.7 LIGHTING EQUIPMENT.   In planning any lighting project, the planner must exercise considerable self-discipline to avoid the composition visualized being compromised by the availability and placement of suitable luminaires. Having the right hardware for each requirement is ideal, but in practice available equipment may dictate a compromise. Thus, although each type of luminaire is designed for a specific job, with some ingenuity other types can be made to do. While substitutes will not perform as effectively as the correct unit and require greater effort to achieve the desired result, it is better to compromise with equipment requirements and expend the necessary additional effort than to compromise on lighting composition.

To provide a basis for selection of lighting equipment, a listing of the basic types of light sources, their characteristics, and general applications in television lighting follows.

Fresnel-lens spotlights provide a soft-edge beam with an even field of light over a range of flood (Fig. 14-55*a*) to spot (Fig. 14-55*b*). In medium to flood focus, their most efficient mode, the beam can be shaped quite accurately with barndoors (Fig. 14-56), or a foreground frame, known

FIG. 14-56 Fresnel spotlight with eight-way barndoors. Side doors have collapsible *wings* to conform with setting of top and bottom doors.

as a cuke or gobo. Available in a wide range of lens diameters for control of beam spread and wattages for intensity requirements [(3 in)/(150 W) to (24 in)/(20,000 W)], they are used wherever their attributes meet specific lighting needs—including key light, fill, backlight, modeling, accent, and set lights in studio or on major remote (out-of-studio) originations.

Ellipsoidal-framing spolights project an efficient sharp-edge beam which may be framed with built-in shutters, shaped with built-in iris, or used with stamped-metal patterns and especially prepared glass slides for decorative effects on backgrounds. Valuable for sharp accents or framed areas of all types, they are available in several sizes and beam widths. The larger sizes are often used as follow spots. The extreme sharpness of the beam edges can be softened through adjustment of the lens barrels, but they never reach the soft-edge quality of the Fresnel. *Open-face spotlights* depend upon the configuration of the reflector to form the beam into a spot. These highly efficient units are designed for remote work or the lighting of large areas where a certain amount of inherent spill light merely adds to the illumination. Their use in studio work is by no means ruled out, but it is restricted somewhat by limited beam control. They have a tendency toward a secondary beam of spill light around the main beam. This secondary beam is not easy to mask off of unwanted areas, although masking can, of course, be done. They are available in a wide range of sizes and intensities and usually are of light weight for easy handling and mounting.

Sealed-beam units, with a large variety of housings, generally called *par heads,* are worthy of consideration for many purposes, especially in remote work. Available in several lens diameters, wattages, and beam widths, these efficient lamps project an elliptical beam, rather than the more conventional round field. The housings should, therefore, provide for turning the bulb to properly orient the field with the projection angle. These sources are of light weight. Tubular housings extending a foot or more in front of the lamp face are advantageous in masking off the inherent spill and providing holders for gels, screens, or diffusion far enough from the lens face for some heat dispersal to take place.

Floodlights (Nonfocusable Light).   Most floodlight luminaires require added diffusion because the light source often produces a hot spot capable of causing clearly defined shadows, whereas the objective of such a unit is a nearly shadowless field. Small floodlights, with which this category abounds, are closer to a pinpoint of light and thus produce distinct shadows. Small flood luminaires require a diffusion screen bigger than the unit itself to enlarge the source and soften the shadow detail.

Floodlights in general consist of a light source and a matte-surface reflector. General classes are scoops, broadsides, striplights, and, of course, softlights. Softlights are the ideal floodlight since they are almost entirely shadowless. Consisting of a large reflector illuminated by light sources concealed behind baffles, the output is indirect light (Fig. 14-57). Indirect light is not efficient—a consideration to be borne in mind when planning the needed intensity. Softlights come in many sizes and forms from 500 to 8000 W.

Hand-held Luminaires.   Many small luminaires, usable on battery packs or line power, are available for remote pickups, interviews, or hard-news stories. Usually these are focusable from flood to a reasonable spot. They are positioned in the same manner as other luminaires and react in a similar manner; extreme care must be exercised to avoid getting so close to subjects that exposure problems can destroy the picture. Again, careful consideration must be given to the use of diffusion filters where shadow detail may cause serious composition problems.

Accessories.   These items can be as important on any lighting project as the luminaires themselves. This is equally true in either studio or remote work. Prime accessories are barndoors, without which no spotlight is complete. Barndoors come in a variety of types and sizes—twoway, fourway, and eightway—the latter having the smaller pair of doors made with collapsible "wings" to accommodate themselves to the setting of the larger doors for a cleaner masking job.

Stands, grip stands, extension hangers, color frames (a convenient holder for any

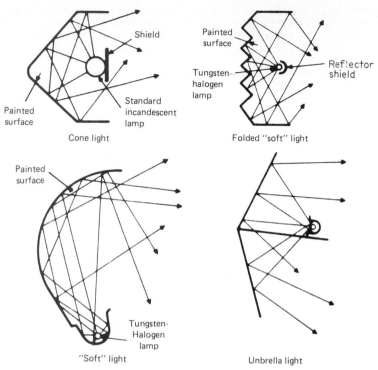

FIG. 14-57   Optical system of various *softlights*.

inserted material), screen wire, aluminum foil, diffusion material in a variety of densities, daylight blue gel, and 85A filters to correct color temperature are essential elements for a good complement of accessories.

14.11.8 COLOR TEMPERATURE.   For the purpose of lighting, color temperature may be defined as a visual comparison of the shift in hue of white light toward red or blue. Color temperature is measured in kelvin (K) units, which are on the Celsius scale. Zero kelvin is equal to −273°C.

The color-temperature scale is based upon the visual appearance of a standard *black-body* heated to a given temperature in degrees kelvin. Low color temperatures of light are toward red, and high toward blue. Owing to the flexibility of color balance in television cameras, color temperature is much less of a restraint for lighting than is the case in the other photographic media. It has been found that variations of ±250 K from the temperature of camera setup are quite acceptable, giving the television designer considerable freedom in lighting sources and the use of dimmers. However, some restrictions must be observed. The television camera is set up to operate at a specific color temperature, and changes in the color temperature of lighting beyond the 250-K range require camera rebalance. For example, set for interior light at 3200 K, the camera cannot move into daylight at 5400 K without readjustment or the addition of color-correction filters. Color-temperature variations have the most noticeable effect on recognizable subject matter such as commercial products. Most incandescent (including "quartz") sources radiate light at 3200 K. This is considered the nominal value for interior white light. Sources such as HMI, CID, or carbon arcs radiate at 5400 K, approximating daylight. Incandescent units may be adapted to match daylight with diachroic filters or *Tuf-Blue*

50 gels. Daylight may be adapted to match interior lights with diachroic filters or 85A gels. Such corrections result in a substantial intensity loss of as much as 40 percent, thus necessitating additional luminaires to maintain the desired illumination level.

14.11.9 COLORED LIGHT.   The use of colored light presents no major problem for the television camera, although the reproduction of some colors may provide the lighting designer with a problem. Some colors of light in the same manner as some colors of scenic materials are slightly distorted by the picture monitor or receiver balance to a white balance of 6500 K or higher. Blue intensifies slightly, with the result that a bluish red or lavender may appear entirely blue. There is also the minor problem of moonlight effects when pale blue light is used exclusively in their production. The "moonlight" may appear excessively blue and quite unnatural unless a proper, desaturated blue has been selected.

14.11.10 DIMMERS.   The ability to decrease or increase the intensities of luminaires quickly from an off-set location is a great time-saver in any production work. Even more important, dimmers are invaluable in the setup and control of light balance. Previous paragraphs have emphasized the importance of light balance in both artistic composition and camera operations. In addition, operating within the limits imposed by acceptable color temperature differences, dimmers can provide the desired adjustments quickly and effectively. Therefore, the inclusion of dimmers in permanent or portable setups is essential. *Silicon-controlled-rectifier* (SCR) *dimmers,* with their electronic-gate action to efficiently control the flow of current, are used in virtually all modern studio installations. These units use no power when the dimmer is "down." The power losses are low and proportional to the lamp current and brightness. They are most economical to operate, and, having no moving parts, maintenance is minimal. Being small and controllable through low voltage, SCRs readily lend themselves to being located at a distance from the studio-action areas, with only the control console in the operating areas. The use of computer control permits a variety of presets, including intensity controls, and regrouping of control circuits at will for easy operation. The favored and most flexible designs feature a single dimmer for each luminaire. An alternative, less expensive, and less complex method of grouping several luminaires in a larger-capacity SCR is workable and used in some installations, although with some loss of flexibility in regrouping and control. At a control console, operating station computer or tape memories activate the control switching and intensity presets, recalling them and activating the previously established cues, at the touch of a button. Many control consoles provide a complete readout display of current and upcoming cues.

Prior to the availability of the SCR a variety of different dimmers were used, such as Thyratrons, saturable reactors, and Autrastats. Many Autrastat installations continue in operation, and, aside from cumbersome bulk and high power demands, they continue to work well. For the most part, other types have been discarded, and new installations all use SCRs.

Patching.   Many dimmer installations of minimal complexity are accompanied by a cross-patch panel wherein a cord feeding a studio outlet may be plugged into a jack field of dimmer outputs (with multiple jacks for large-capacity dimmers). Such cross-patching devices provide breakers for each circuit as a protection and as a means of avoiding "hot patching." Using the cross-patch system permits any luminaire or group of luminaires to be connected to any dimmer control. In manual operation, having related areas controlled by adjacent dimmer handles presents many operating conveniences. In other instances patching permits grouping of related luminaires on a single control.

Cost-Effective Design.   Dimmers and the necessary control devices, consoles, and installations involve a substantial expense which may be a significant portion of the capital cost for a studio. Therefore, it is essential that the complexity of the facilities not exceed the needs of the types of productions envisioned for the facility (or for the single production on a remote operation). If the system saves production time and meets artistic demands, this cost is not difficult to justify. In other words, the ability to control the intensity of one or more lights instantly, to soften a facial shadow, avoid a bright background (causing the face of a major "talent" to go dark), or to provide an exciting visual change in form or center of interest, together with the time and delay saved in accomplishing it, can more than justify the economies of the dimmer system.

14.11.11 THE LIGHT PLOT. A professional lighting designer always starts work with the making of a light plot (Fig. 14-58). It may be either a rough floor plan with luminaire

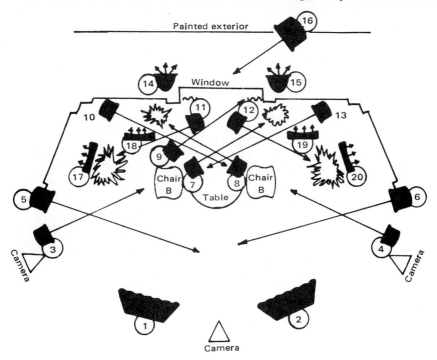

FIG. 14-58 Typical *light plot* for simple restaurant scene. Action: Two people seated while waiters enter and exit past the three cameras.

① ②	4000-W softlights (may not use all circuits) for blending (fill or base light).
③ ④	1000-W Fresnels focused low on chairs A and B and floor behind chairs.
⑤ ⑥	2000-W Fresnels focused high across foreground to supply highlights to waiters.
⑦ ⑧	1000-W Fresnels focused on upstage greens and into upstage corners of sets; these will show in the backgrounds of close-ups.
⑨	1000-W Fresnel into window to blend the harsh effect of 16.
⑩ ⑬	1000-W Fresnels as keys to chairs A and B. Bottom doors up off tablecloth, top doors down to avoid cameras on opposite sides.
⑪ ⑫	1000-W Fresnels spotted on large greens, dimmed to supply pleasing background for over-the-shoulder close-ups.
⑭ ⑮	1000-W scoops for painted exterior—may be dimmed or colored for time of day.
⑯	5000-W Fresnel through window—may be colored for *time of day* or broken up with a leaf pattern.
⑰ ⑱ ⑲ ⑳	500-W broads with light diffusion to control wall brightness—focus on top part of walls—may add small Fresnels if walls have large pictures or for a moodier look. Both may be used, or none.

locations noted during setup or a plan to scale superimposed on a ground plan of the studio settings or of remote locations. A light plot is an outline of the project to be accomplished and later a record of the setup for reference and possible future use. Aside from the obvious advantages of a plot to the efficiency of setup, the operational phase of a production continues the benefits through efficiency in operation. The identification of each luminaire and its control circuit indicated on the plot allows the lighting designer to make adjustments of intensity or to add or delete existing units in a setup, if the need arises, and to do so without delay.

14.11.12 REMOTE OR LOCATION SHOOTING. With the advent of small cameras, portable recorders, microwave links, and satellite relays, much of television production has moved out of the studio, or involves a combination of studio and outside, or remote activity. Lighting for a "remote" or a "location" follows the same principles outlined earlier in this section but presents a number of special considerations to the lighting designer. A normal aspect of any remote is that conditions at the site exist and are not entirely under the control of the television crew.

Interiors with Daylight Windows. On an interior remote there are two things to consider, the intrusion of daylight and existing fluorescent light. Daylight through a window requires some corrective action since cameras set up for daylight will render uncorrected incandescent light as being extremely red, while cameras set for 3200-K sources will render daylight as excessively blue. As noted earlier, there are small versions of sources operating at 5400 K which may be used to light the room for daylight balance, or the 3200-K sources can be corrected with diachroic filters or Tuf-Blue 50 gel to match daylight quality. A second approach calls for correcting the daylight window with 85A gels on the window itself, converting the daylight to approximately 3200 K. It must be noted that most window correction materials are not optically clear, and any shots through this material may be limited in definition, unless recently available gels are employed.

Fluorescent Light. Fluorescent light fixtures are present at most locations and are so much a recognizable part of the architecture that turning them off would result in an unrecognizable locale. Fluorescent light has an irregular color spectrum and in most instances photographs green. Normal practice is to accent these sources and use their light output without corrective measures, then lighting the action areas with regular 3200-K luminaires as if no other light sources were present. Subjects or areas illuminated in this way provide satisfactory color rendition. For example, in a fluorescent-lit location, lighting the action area in the foreground, a few elements of decor a bit further back, or even a small area of wall in the background, with normal-level incandescent sources appears quite satisfactory on camera, creating the illusion that the whole area has been lit to television requirements.

Fresnel spotlights and other types of so-called standard equipment perform extremely well and with their usual effectiveness on remote operations, but there is a tendency to prefer smaller and lighter-weight equipment for easy handling even though some control of beam is sacrificed. There are also extremely high-powered luminaires such as HMI, CSI, plus single and multiple sealed-beam housings, or even arc lights to cover large areas.

Daylight Remotes. No remote may reasonably be regarded as an easy project, but it is probable that the most difficult of all, from the viewpoint of a lighting designer, are the ones intended for shooting entirely in daylight. Daylight, or as many producers insist, "natural light," is probably all very well if the site is in the open plains of Kansas or some parts of Arizona where the full dome of horizon sky does indeed provide perfectly balanced light with natural fill light for sun shadows. In populated areas natural light tends to come straight down and to be inhibited by shadows of trees, buildings, or mountains, and thus is far from ideal. The shadow cast downward by a bright sky or slanting sun can be most unkind to faces—shadows of trees or buildings cause dark areas of high (blue) color temperature. On top of that, light intensity changes because of drifitng clouds or other atmospheric disturbances. As a measure of control the experienced designers use auxiliary artificial light to soften the shadows, fill the darker periods, and generally smooth out the facial areas and important color areas. High-powered HMI, CID, or arc lights are the most suitable luminaires for this task.

Reflectors are of great value in remote work. Available in many forms (white or silver, matte or specular), reflectors can redirect sunlight to fill or model almost any subject, live or architectural. Care must be exercised in using the specular side of a reflector on a live subject, however, since the glare it produces is comparable with direct sunlight. In practice the specular reflector is most often used to model or accent, or to redirect sun to a matte reflector for more practical diffuse illumination. The brightness of the matte unit, incidentally, can be controlled somewhat by misdirecting its reflection so that the less effective portion actually illuminates the subject. This is possible owing to the extreme diffusion of the reflected light which has no visible edge boundaries such as a specular reflector or spotlight. If at all possible live subjects should not be required to face into the sun for photography. Few people can do this without squinting. Rather, try to place the subject so that the sun serves as a backlight or with a matte reflector providing the front illumination. It is not imperative to have a professional reflector board for matte use, because a white card, a white bed sheet, or even a white truck or white wall does the trick very well. Should background considerations of the locale force the talent to face into the sun, some bobbinet or any netting material or thin translucent material can be stretched off-camera between subject and sun to soften the glare. Even cut tree branches or similar obstructional devices can serve this purpose and aid the talent in appearing relaxed.

Backgrounds on Remotes. Whenever conditions at a remote will permit, extensive thought devoted to the background or surrounds for any pickup will pay handsome dividends in picture quality. A background which can accept the same light striking the subject without appearing brighter than, or even as bright as, the face will make the subject more important, hold color values, and allow facial expressions to be visible. These same advantages, and many additional ones, will be achieved if the subject can possibly be removed from close proximity to any background, anytime, anywhere, allowing a picture of a real dimensional subject to appear. This application of space is a secret ingredient of both good pictures and good lighting.

14.11.13 ENG REMOTES. There are many remotes, especially of the news, interview, or documentary type, where either hand-held lights or two spotlights or floodlights on stands provide the main illumination. There is a tendency to use one light on each side of the camera on the assumption that this is the best that can be expected under the circumstances. More often than not, this is an excuse for using the easiest approach, rather than an actual condition. One source at one side of the camera focused diagonally across the field of camera vision, with the other unit placed out of camera range, and focused more or less toward the other unit, will key, model, and backlight a subject or two subjects placed in the center, while providing separation from a background not directly illuminated by either beam. This is the "opposition" placement of light sources and is often quite effective and usually provides a much better picture than flat lighting.

A final point should be mentioned. When placing light units for a remote, projecting the source from more than one direction always results in a better picture than having all illumination proceed from a single front direction. The multidirectional light models, separates, and permits features to show more clearly. Even projecting a single light source from a slight angle is better than a head-on beam for modeling and separation.

14.11.14 SINGLE- OR MULTIPLE-CAMERA STAGING. For maximum effectiveness of lighting it would be desirable to use only one camera to televise a scene. The practice has been used effectively in motion-picture productions. The advantage is that, with a single camera angle for each shot, lighting can be balanced easily for optimum picture quality. The use of multiple cameras dictates the need to compromise. The choice usually is to favor a center camera. As a result, the other cameras at different angles may see quite different light composition. Intercutting or assemble-editing the output of several camera angles often results in a disturbing mismatch. This is one of the contributing factors characterizing the *film look* versus the *tape look*.

In urging the lighting practitioner to strive relentlessly for the utmost quality in a composition, in spite of obstacles to such achievement, it would be a glaring omission if it was not pointed out that the acme of lighting quality can be achieved only when the composition is shot by one camera.

14.12 BROADCAST INFORMATION SERVICES

by Joseph Roizen

14.12.1 SCOPE OF SERVICES. Ancillary information services using existing television broadcast channels are in use in many countries around the world. The services identified by the generic terms *teletext, viewdata,* or *videotex* provide one-way or interactive information transfer. The systems also have proprietary trade names that define a particular technology associated with a specific country, or identify a public service provided by a television programming or broadcasting organization.

Teletext is the one-way information service that can be transmitted by a television station using cyclical digital data inserted in the vertical blanking interval (VBI). This teletext information does not interfere with the normal program material being broadcast at the same time. A color-television receiver equipped with an external or integral teletext decoder will convert this digital data into full-screen alphanumeric or graphic displays that are called "pages" of an "electronic magazine."

Teletext services, which originated in the United Kingdom, are called Ceefax by the BBC and Oracle by the IBA. In France the service is named Antiope†, and in the United States two of the major networks call their teletext services Extravision (CBS) and Tempo (NBC).

Viewdata or *videotex* is a two-way information service system using a color television receiver as a display device. However, it usually includes an entry keyboard and a return link, via a phone line, to the data base, which provides the interactive feature. The viewer can call up a wide variety of information, limited only by the size and scope of the data bases that can be accessed.

Viewdata service in Great Britain is called Prestel, in France Teletel, in Canada Telidon, in the United States Viewtron, and in Japan Captain.

Teletext services can provide a viewer with a wide variety of useful information, which can be program-oriented or completely independent. Current teletext pages, which comprise the electronic magazines being broadcast, contain such topics as news headlines, weather forecasts, sports results, classified ads, program logs, traffic conditions, air line schedules, and many more. The viewer uses a remote-control key pad to select normal television images, teletext pages or a combination of both. In the teletext mode, an index page provides the key to selecting the desired information, and when the proper page number is entered on the key pad, the page will appear on the television screen after an average wait of 5 to 10 s.

Teletext can also provide additional services to certain groups of viewers. For the hearing-impaired, teletext supplies flexible closed-captioning in multiple colors and several languages, or it can even cater to the variable comprehension and reading rates among different viewers. A conceal/reveal mode, which can be used for educational or game applications, allows the viewer to select a page of text containing questions, to which the answers are supplied when the *reveal* button on the key pad is used.

The *alarm* mode permits the viewer to watch normal television but to be notified of any important news break or other significant event by an insert into the picture that lists the teletext page to call up to get the full information.

---

†Acquisition Numerique et Televisualisation d'Images Organisees en Pages d'Ecriture.

Teletext can produce graphics, alphanumerics, and even color pictures in various sizes, fonts, differing resolutions, and a wide range of colors. Flashing, scrolling, and some animation are also possible.

14.12.2 TELETEXT SYSTEM DEVELOPMENT. Four major teletext systems have been developed by various national research laboratories in Great Britain, France, Japan, and Canada. The first country to launch a teletext service was England, where both the BBC and the IBA started experimental teletext broadcasts in 1972. At the beginning the BBC's Ceefax and the IBA's Oracle teletext transmissions were incompatible with each other. However, in cooperation with BREMA†, a common format for the digital data, in the VBI, was achieved, and today there is a single specification for all teletext in the United Kingdom. The same decoder in a normal British PAL receiver can display either the Ceefax or Oracle pages, depending upon which channel the television set is tuned to at any given time.

The French Antiope system was next to appear, developed by the CCETT,‡ a joint-research group formed by French television and telephone services. Antiope was part of an overall national telecommunications plan to integrate teletext, viewdata, and other electronic services into a large network spanning the country. This plan included the replacement of printed phone directories with small-screen, interactive terminals and the establishment of full-channel, teletext information services in the spectrum formerly occupied by the discontinued 819-line, monochrome television service.

The NHK Research Laboratories in Japan also developed a teletext system, tailored to the unique pictographic form of their written (Kanji) language. This system is known as Captain and encompasses both teletext and viewdata capabilities. As in the United Kingdom, teletext development work was also carried on by an independent commercial broadcaster and publisher, Asahi Broadcasting, in Osaka.

In Canada, the Department of Communications, a branch of the federal government in Ottawa, created the Telidon system, which also is a teletext and viewdata combination. Telidon offered a different approach to graphics which permitted color images of higher resolution and of variable gray scales than the earlier systems. This added the element of pictorial teletext to the range of growing capabilities.

Although no teletext systems had been developed in the United States by 1980, when it became apparent that there was a growing need for such systems for experimental and pilot broadcast tests, several of the already developed 525-line television systems were modified to operate on United States standards. The most widely used American teletext system, known as North American Broadcast Teletext Specification (NABTS), combines some of the technical features of Antiope and Telidon into a hybrid format optimized for 525-line, 60-field NTSC use.

14.12.3 CEEFAX/ORACLE (UNITED KINGDOM). Teletext services in the United Kingdom are provided by the national (BBC1, BBC2) and IBA commercial public television networks. Both operate on the same technical standards, using a fixed format, synchronous with the horizontal-line frequency, and a transmission rate of 6.9375 Mbits/s (444 times the line frequency).

Pages of alphanumerics consist of 40 characters per row and 24 rows per page, with a special top row called the *page header.*

The teletext data lines in the vertical-blanking interval that were used initially in 1974 consisted of lines 17, 18 in field one, and 330, 331 in field two. In 1981, two more lines per field were brought into use, bringing the total to four lines per field. This was done primarily to reduce access time to available pages. For example, a 100-page magazine has an average access time of 7.5 s per page. However, up to five key pages, containing such useful information as the main index, are cycled more frequently and appear almost instantaneously.

---

†British Radio and Electronic Manufacturers Association.
‡Centre Commun d'Etudes de Television et Telecommunications.

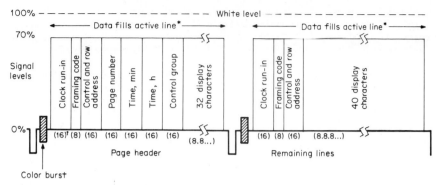

FIG. 14-59 Ceefax/Oracle data organization.

Data Organization.   The elements of the data organization, shown in Fig. 14-59, are tabulated and described below.

Page header (first television line)	
Clock run-in	Eight binary 1–0 sequences to synchronize the receiver data clock
Framing code	A particular eight-bit word (11100100) for data line recognition and word group synchronization
Control and row address	Sixteen bits containing the hundreds digit of the *page number,* the row number (both with *Hamming* code[2] error protection bits), header suppression bits, and other control bits
Page number	Sixteen bits containing the last two digits of the page number (with Hamming protection bits)
Time code	Thirty-two bits for nominal transmission time in hours and minutes (with Hamming protection bits)
Control group	Sixteen bits (including Hamming protection bits for various control functions, such as header suppression)
Display group	Thirty-two 8-bit words to define display (each with odd-parity bit)
Remaining rows	
Clock run-in	Similar to page header
Framing code	Similar to page header
Control and row address	Similar to page header
Display group	Forty 8-bit words to define display (each with odd-parity bit), which may be alphanumeric codes (ISO 7), graphic-combination codes, or instruction codes (when the following line is to be in a specified *color, flashing,* or *boxed*)

British teletext has also pioneered in the use of this medium for special services to the deaf or hearing-impaired viewers. In addition to normal, closed captioning on regular program material, they have developed a system for simultaneous subtitling on live transmissions. The captioning operator uses a Palantype keyboard, which resembles a court stenographer's phonetic typewriter. The Palantype system is interfaced with a computer, which produces the appropriate caption video signals that are superimposed. In the United States President Reagan's inaugural address and in England the royal wedding were both handled this way.

Other special services include a breakthrough mode for announcing important news events, flashing capabilities for emphasis, a conceal/reveal mode for question and answer sequences, and commercial messages on teletext pages, which may or may not be program related. Graphics are produced in the alpha-mosaic format which results in a six-element, relatively coarse image. This is considered adequate for large headlines or rough outline maps, etc. However, fine-line structures cannot be accommodated.

Enhanced Teletext.   A number of enhancements to the basic teletext system in the United Kingdom have been developed to improve future services, while retaining reverse compatibility with existing *level 1* decoders already in use. The enhanced teletext system uses a technique known as alpha-geometric coding, in which the picture elements are smaller than the six-element mosaic method. At *level 2,* the smoother graphics are the result of augmenting the rectangular outlines of such a six-element mosaic with alternative characters containing sloping boundaries. The diagonal lines formed between corners of the original element tend to give the resultant graphic a more rounded look. Level 2 also permits overwriting of any available character, in a space previously used by a control or displayed character, a feature which applies to graphics, as well. Pastel shades of color, in addition to saturated primaries and secondaries, are also possible.

Level 3 uses a technique known as dynamically redefinable character sets (DRCSs), and this extends the range of graphic symbols to an almost limitless potential. DRCS can specifically define the foreground and background picture elements or pixels from which the characters are created on the television screen, thus producing a wider range of alphanumerics and graphics.

Level 4 employs *alpha-geometrics* to extend the capabilities of teletext even further. It can be used for fine-line geometric drawings or even handwritten renditions. It can give continuous range for up to 16 colors and produce higher-resolution graphics than ever before. Alpha-geometric instructions are similar to computer programs and can therefore become a subset for a telesoftware language. Increased animation of the teletext image components is also a potential of level 4.

Level 5 is the optimum version of enhanced United Kingdom teletext, which allows presentation of alpha-photographic images that have full gray scale and complete color capability. The quality of these teletext pages can equal or exceed that of 525- and 625-line color television images. Picture teletext was demonstrated by the BBC in early 1982, and telesoftware trials have also been conducted by both United Kingdom networks, in conjunction with secondary schools.

Teletext in the United Kingdom is widely used by home viewers, travel agencies, air line companies, hotels, and others who have a need or desire to access timely information relating to their activities. The channel 4 television service, which began regular broadcasting in 1982, also uses a Ceefax/Oracle format teletext system, with its own unique page content, but meeting compatible BREMA television signal specifications.

14.12.4   ANTIOPE (FRANCE).   Antiope is the teletext system developed in France by the CCETT under the joint sponsorship of TDF† and the French PTT‡. Antiope teletext services are part of a nationwide telecommunications plan involving compatible videotex systems for interactive applications. Antiope transmissions are now available on the

---

†Telediffusion de France.
‡Postes Telecommunications et Telephone.

Table 14-4 Antiope Specifications for French SECAM (CCIR System L) Transmission

Data	Coding	Information
Packet header		Synchronization
		Procedural and prefix data
Clock run-in	16 bits	Sampling synchronization
Framing code	6 bits	Reference for byte synchronization
Data link		
Identifier	24 bits	Data-stream link identification
Continuity index	8 bits	Numbers of successive packets on a data-stream link
Format filler	8 bits	Normalizes variability among packets by matching number of useful bytes in a block
Data block	184 bits	Variable source data, packeted according to source rate and network availability

three French television networks: TF1, Antennes 2, and FR3, and are planned for a fourth network that is to replace the terminated 819-line monochrome service.

Antiope uses a variable-format asynchronous coding method, not locked to the horizontal-line frequency. The bit rate used in France is 6.203 Mbits/s (397 times the line frequency). Antiope also uses 24 rows of 40 characters each to display the alphanumerics on the television screen. A twenty-fifth row contains the page-header information with its control bits, and it may be displayed or suppressed. The technical specifications for the Antiope system used in France are listed in Table 14-4 and Fig. 14-60.

The Antiope system, as it is used in France, has two character sets, including an international alphabet (ISO 646) with Latin letters and appropriate accent marks. The secondary set allows rudimentary drawings, and additional optional sets can produce non-Latin alphabets, such as Cyrillic or Arabic. Antiope's display modes include alphanumerics only, or overlaid on television images, captioning for foreign languages or the hearing-impaired, a *reveal* mode, and an *alarm* mode. Higher-resolution graphics are achieved by DRCS and by the use of alpha-geometric coding with appropriate decoders.

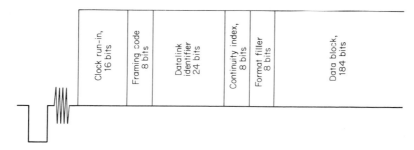

FIG. 14-60 Data sequence in horizontal line for French SECAM (CCIR System L) transmission of ANTIOPE teletext.

Antiope may be easily expanded to cover more than a few lines in the vertical blanking interval, and in France it is used in its full-channel capacity to transmit very large amounts of data, with rapid access time to individual blocks of information.

Antiope also has defined levels from zero to six, most of which have been demonstrated at least at prototype levels. These include finer graphics, pastel colors, animation, and photographic images via teletext. Antiope is compatible with the Canadian Telidon system and with the presentation level protocol (PLP) of the American Telephone and Telegraph (AT&T) videodata system. As a result, it has been incorporated into a hybrid system being tested in the United States, NABTS, or more recently the North American Presentation Level Protocol Syntax (NAPLPS).

In France, it is estimated that 250,000 to 300,000 people watch the Antiope electronic magazine on the national network, Antennes 2, during its morning broadcasts, before regular programming begins. At the start of 1983, Antiope was transmitting 2000 pages of national information and 600 pages of regional material. The topics covered include news, weather, stock market results, and multilingual captioning.

The four French television networks plan considerable expansion of the Antiope services in 1984 and 1985. Page count will go up to 5000, and eventually 40,000 using full-field transmission; 50 percent will be for professional use, and 50 percent for the general public.

14.12.5 JAPANESE TELETEXT.   Development of a national teletext system in Japan was carried out by the NHK Technical Research Laboratories, and on-air experimental transmissions have been going on since 1978. In March of 1981, the Radio Technical Council of the Japanese Ministry of Posts and Telecommunications recommended technical standards for teletext, and shortly thereafter a government-approved system was in operation.

The major problem confronted by the designers of Japanese teletext is the ideographic form of their written language. Unlike the relatively small number of characters in written European languages, Japanese uses thousands of Chinese Kanji characters and large numbers of Japanese Kana syllabics. To display these, a memory device with a very large capacity would be needed in the home receiver, thus making it prohibitively expensive.

To overcome some of these limitations, Japanese teletext utilizes a system called *photographic coding,* which can transmit and display ideographic letters, graphics, and a Latin alphabet. Text and graphics are transmitted as separate dots and are not encoded; however, those signals needed for control functions and display attributes are encoded. At the receiver, a frame memory is used instead of the typical character generator used in other teletext systems.

In the Japanese teletext system, text and graphics are disassembled into dots by a scanning technique. The matrix consists of 248 by 204 dots for luminance, and 31 by 17 color blocks. These are transmitted in the form of binary nonreturn to zero (NRZ) pulses. While this system is slower in its data transfer function than the western systems, it is well suited for the complex ideographs and fine-line graphics needed in Japan. It is also very rugged and virtually immune to random noise, electrical transients, co-channel interference, and ghosting. Japan also has a videotex system known as CAPTAIN†, which is dedicated, at present, to interactive phone-line operation.

Japanese teletext is transmitted in data packets, at a bit rate of 5.727 Mbits/s, and multiplexed on blank lines in the VBI. The VBI lines used are from 16 to 21 in field 1 and 279 to 284 in field 2; eventually this may be extended to lines 10 and 273. Since Japan uses CCIR System-M color television transmission standards, the bit rate is the same as that proposed in the United States for NABTS NTSC standards.

The display modes for Japanese teletext, similar to those of British or French systems, include text and graphics over the entire screen, superimposed text, and subtitling. In addition, there is a *vertical-scroll mode* where a full-screen text is rolled upward, and a *horizontal-scroll mode* where one line of text is moved from right to left.

Recent Japanese teletext receivers also have two convenience features. The first is an eight-page memory, which can be programmed to store pages the viewer most likely wants to see, thus providing almost instant access. The second is a small printer that provides monochrome hard copy of the teletext pages.

---

†Character Pattern Telephone Access Information Network.

Japanese teletext is quite different in its data transmission characteristics. These technical details are shown in Fig. 14-61, which illustrates the data line and its various data packets. Five types of data packets are used to transmit pages; they are the page control packet (PCP), the color code packet (CCP), the pattern data packet (PDP), the horizontal data packet (HDP), and the program index packet (PIP). The combination of all these provides the control of page display, coloring, flashing, and conceal functions, and two-axis scrolling. A single page of memory in the decoder requires 55 kbits, including the color code memory.

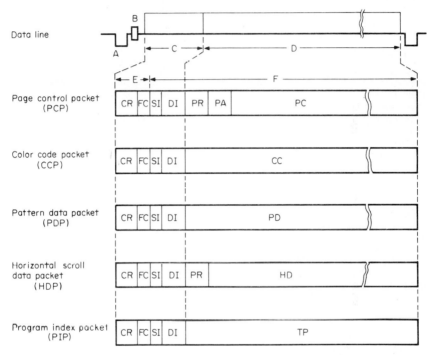

FIG. 14-61 Data packets for Japanese teletext. [*From Radio Technical Council, annual report of fiscal 1980, Recommendation "Teletext System (Pattern Transmission)," March 1981, (Dec. 11/ J-21) (in Japanese).*]

A : line-sync signal
B : color burst
C : prefix (3 bytes)
D : data block (31 bytes, 248 bits)
E : synchronization (3 bytes)
F : data packet (34 bytes)
CR : clock run-in (bit sync) (2 bytes)
FC : framing code (byte sync) (1 byte)
SI : service identification and interrupt control
DI : data packet identification
PR : program number
PA : page number
PC : page control codes
CC : color codes for 31 color blocks in one row
PD : pattern data signals for page, subtitling, and vertical-scroll displays
HD : pattern data signals for horizontal-scroll display (27 bytes)
TP : program numbers currently transmitted

Japanese teletext, while officially established in its present form, is still undergoing intensive study for future extensions. For example, in late 1982 an enhanced system using the Moji code was being tested in Tokyo.

The Japanese have also been very innovative in enhancing the display techniques of their teletext material, both on the television receiver screen and on hard-copy printers built into the set. In the first case, it is possible to program the teletext decoder to capture eight selected pages of maximum interest to the viewer, so that access to these pages is almost instantaneous, after the first page is displayed. In addition, low-cost, monochrome, hard-copy printers, using paper rolls about 4 in wide, take only a few seconds to provide a finished print of the data on the television screen. For the more sophisticated viewdata applications, full-color printers are also available that render good renditions of high-resolution color graphics.

14.12.6 TELIDON (CANADA). The Telidon system, developed by the Department of Communications in Ottawa, is the most recent entry into the videotex field. Like Antiope, it is a combined teletext and viewdata system using an asynchronous variable format for its basic operating principle. Its major difference is its abiility to store and display high-resolution graphics through a code system known as picture description instructions (PDIs). Telidon, with an alpha-geometric display, produces much finer graphics than those systems using alpha-mosaics. The PDI codes are capable of operating at different resolution levels, depending upon the sophistication of the decoder in the display terminal. Graphic images may vary from a 60 by 80 matrix of 4800 picture elements (pixels), with a character-only terminal, to as much as a 960 by 1280 matrix of 1,228,800 pixels, on a high-resolution business color terminal.

Since Canada uses the same NTSC color television standards as the United States (CCIR System M), Telidon is compatible for application throughout North America. Because its basic alphanumeric form is so similar to Antiope, a proposal to combine the French and Canadian teletext systems to form a hybrid 525-line, 60-field standard was adopted in May of 1981. An important feature of the Telidon system is an alpha-photographic mode that can transmit and display still photographic images, with color and gray scale equal to a normal broadcast-television image.

14.12.7 EIA/NAB/FCC (UNITED STATES). Teletext in the United States has been a subject of considerable concern to a variety of technical and public policy agencies dealing with television signal formats. The FCC in November of 1978 asked for demonstrations of existing teletext systems, and subsequently declared that they were in favor of the principle that teletext services should be provided by United States broadcasters. The FCC authorized a number of such on-air tests, which included stations in Salt Lake City (KSL), St. Louis (KMOX), Los Angeles (KNXT, KCET, KNBR), and others.

The FCC also suggested that an all-industry committee under the auspices of the Electronic Industries Association be formed to investigate the various systems being proposed and to recommend a set of technical specifications for a national teletext service in the United States. Such a committee was formed and included members from system proponents, television receiver manufacturers, television broadcast networks, and other interested parties. About 40 members met regularly, on an almost monthly basis, to produce a set of documents comparing the specifications and operating characteristics of the British, Canadian, and French teletext systems. While no compromise seemed possible between the synchronous and asynchronous systems, the committee did arrive at a recommendation for a single hybrid system, which included features from Antiope, Telidon, and the AT&T proposal for the PLP. This teletext system was called NABTS, and was supported by CBS, NBC, Westinghouse, and others.

However, a number of other industry groups were concerned that their interests were perhaps not being met by the EIA teletext committee, and additional committees or ad hoc teletext advisory groups were formed by the National Association of Broadcasters (NAB), the National Cable Television Association (NCTA), and the Videotex Industries Association (VIA).

Notwithstanding the very comprehensive work done by the EIA group, and its recommendation of a single teletext standard in the NABTS format, the FCC in mid-1983 decided to adopt a "marketplace" attitude toward teletext, allowing each broadcaster to transmit teletext signals in whatever format it chose. Two major teletext systems emerged, and adherents to both of these began providing public teletext services. The NABTS system became a national teletext service on the CBS and NBC networks in the second quarter of 1983. The World System Teletext patterned after the British Ceefax and Oracle, was adopted by Taft Broadcasting and began operating in Cincinnati in the summer of 1983.

Teletext in the United States is now a multistandard operation, with incompatible formats being transmitted by adherents to NABTS and World System Teletext. Both systems are capable of a variety of levels of upgradability, from simple alpha-mosaic graphics, through alpha-geometric, and full-picture teletext. Decoders for both systems are also in the initial stages of going from stand-alone, set-top types to fully integrated circuit boards within the receiver.

14.12.8 VERTICAL-INTERVAL DATA TRANSMISSION. Teletext uses the technique of transmitting digital data during selected empty lines of the vertical blanking interval of a standard television signal. The digital data contain alphanumeric or graphic information which the teletext decoder in the home receiver converts into a series of visual displays called pages of an electronic magazine.

The transmission of these digital data does not interfere with the broadcasting of regular program material, and the viewer can use a key pad to select either normal television images, teletext pages, or a combination of both.

These teletext pages usually contain a wide variety of user-oriented information, such as news headlines, weather, sports news, classified ads, program details, captioning, and the like. Teletext services first started in the United Kingdom in the early 1970s and have now spread to many countries all over the world. France, Japan, and Canada have also developed teletext systems which are now in use in those countries. Some of these systems are the basis for a United States system that is now undergoing field testing in major cities in North America.

While basic teletext services are limited to a few lines in the vertical interval, it is possible to have partial or full-channel teletext transmissions, in which a dedicated television channel can carry a huge amount of data. It is also possible to use a teletext-equipped color television receiver to decode similar digital data sent over a phone line or relayed by a radio carrier. These television data services are known as viewdata or videotex.

In the case of the phone-line videotex system, a modem is used to couple the line to the television set (or terminal), and the system is considered "interactive" because the viewer can request information via an entry keyboard attached to the television receiver.

Line Selection. The television waveform of a standard NTSC, 525-line, 60-field signal contains 12 unused lines between the end of the postequalizing pulses and the beginning of active picture time in each field. Some of these lines have already been allocated for specific use; they are as follows:

1. Multiburst   line 17, field 1
2. Color bars   line 17, field 2
3. Composite test   line 18, field ½
4. Vertical-interval reference   line 19, field ½
5. Closed captioning   line 20, field ½

It is anticipated that in the future some of these signals on lines 17 through 20 may be rearranged, and the potential teletext transmission lines can be between the start of line 10 and the end of line 21; however, the initially allocated teletext lines are 15, 16, 17, and 18. Field trials using digital data on lines 10 through 14 have demonstrated that older television receivers, with inadequate vertical-interval blanking, show the digital data as bright dots on the screen. As these older sets are phased out of service, the earlier VBI lines will become useful for teletext transmission.

Summarizing, the current FCC-authorized teletext services occupy four lines per field, five lines after vertical-sync and equalizing pulses, thus leaving an adequate guard space between the end of the vertical-synchronizing pulses and the start of teletext to avoid visibility on receivers during scanning retrace.

Multiplexing. The digital data, which represent the pages of text and graphics to be transmitted, must be inserted into the allocated lines of the vertical blanking interval. They must also be sequenced in a cyclic fashion so that each page will be available some time during the cycle, for the decoder to gate it into the receiver memory on command. The device which inserts the digital data into the video composite signal is called a *multiplexer.* The multiplexer may be set up to accept data pages from a variety of sources or magazines, and each of these magazines may have a different weighting factor ascribed to it, so that the sequence of pages going to the multiplexer may be accessible in varying amounts of time. As an example, topical, high-impact news pages may cycle twice as often as pages carrying listings of local movie offerings.

Access time to a particular page of multiplexed teletext data is affected by three factors: the bit rate of the transmission, the number of VBI lines allocated, and the weighting factor applied by the information provider. Certain kinds of teletext data, like program-related closed captions, are given immediate priority through the Multiplexer, because they must appear in synchronism with the action on the screen. Since closed captions use only a small bit rate, the delay imposed on other teletext pages is very small.

The multiplexer accepts the single- or multiple-sourced digital-data signals and gates them into the appropriate lines in each television field.

Bit Rates. The data bit rates used in broadcast teletext vary over a range dictated by the transmission channel bandwidth and the desired reliability of the system. A lower bit rate assures better error protection but increases the access time to a page; conversely, a higher bit rate reduces access time but increases error potential. The actual transmission bit frequency selected is a compromise between these two limits and is usually related mathematically to the horizontal-line scanning rate. Bit rates tested in the United States have ranged from 3.0 to 7.0 Mbits/s, and the bit rate currently proposed is 5.727 Mbits/s.

Data Coding. Teletext information is made up of a sequence of codes that are transmitted over the air. These codes meet ISO 2022 standards and are in binary form containing a seven-bit byte followed by an eighth bit for odd parity checking. There are three major groups of such codes whose functions can be described as follows:

1. to display text alphanumerics or images
2. to describe control and layout functions
3. to describe character and graphic attributes

In general teletext alphanumeric coding is based on ASCII 3.4 and includes 96 different code configurations to cover a primary set of characters. In addition, there are code sets to cover mosaic and geometric graphics, as well as a DRCS for higher-resolution purposes.

A different set of codes is employed for the PDI level of teletext, where more detailed images are created. This PDI set has six simple geometric graphics such as line, arc, and polygon, and eight codes that establish color, texture, blinking, and other attributes.

Data codes are described in rectangular tables which define the binary sequence for each character or graphic segment. Subsequent codes cover the control functions such as normal or reverse video, text size, color selection, texture, fill in, scrolling, and flashing. Different teletext systems use different data coding structures, and it is in this area that standardization becomes the most important element, if compatibility of receiving and display devices is to be achieved.

14.12.9 CURRENT PRACTICES AND STANDARDS. There are two basic teletext standards in use in the United States today, and they are known as the North American Broadcast Teletext Specification (NABTS) and World System Teletext. The broadcasters who use each system are free to encourage viewers to buy decoders that receive only their teletext transmission, since no multistandard decoder exists at present. Notwithstanding the strong recommendations by the EIA Teletext Committee, and several other television industry groups, that a single teletext standard should be adopted, efforts to agree on one teletext system for the United States, and other North American countries, have failed.

Teletext services in the United States that went public after 1982 operate on these two different standards and are described below.

North American Broadcast Teletext Specification. NABTS is the hybrid system evolved by the EIA teletext group using the French Antiope and Canadian Telidon systems as the base, with modifications and additions to make them operable on the NTSC, 525-line, 60-field television system. World System Teletext consists of similar NTSC-oriented modifications to the British teletext system based on the Ceefax/Oracle operating principles. NABTS was supported by CBS, NBC, Westinghouse, and several other major television broadcasting entities; World System Teletext was supported by Taft, Keycom, Bonneville, and a few others. Public teletext services being implemented in 1983 included NABTS broadcasts over CBS and NBC affiliates in New York, Los Angeles, and San Francisco, and World System Teletext transmissions in Chicago and Cincinnati.

NABTS is an asynchronous teletext transmission system derived from the Antiope coding language and the Didon transmission system developed in France by the CCETT. It is also compatible with the Telidon broadcast teletext system developed in Canada and the presentation level protocol for interactive videotex developed by AT&T in the United States. NABTS is a seven-layer system with forward and reverse compatibility for future enhancement. Both videotex and teletext services make use of the same presentation layer protocol data syntax and semantics but may have different protocols at other layers. The seven layers may be viewed in two major groupings. Layers 1 to 4 concern the transference of data, while layers 5 to 7 concern how the data are processed and used. The architectural structure of the seven layers of NABTS is as follows:

1. *Physical:* The physical layer provides mechanical, electrical and procedural functions in order to establish, maintain, and release physical connections.

2. *Data link:* the data link layer provides a data transmission link across one or several physical connections. Error corrections, sequencing, and flow control are performed in order to maintain data integrity.

3. *Network:* The network layer provides routing, switching, and network access considerations in order to make invisible to the transport layer how underlying transmission resources are utilized.

4. *Transport:* The transport layer provides an end-to-end transparent virtual data circuit over one or several tandem network transmission facilities.

5. *Session:* The session layer provides the means to establish a session connection and to support the orderly exchange of data and other related control functions for a particular communication service.

6. *Presentation:* The presentation layer provides the means to represent information in a data-coding format in a way that preserves its meaning. The detailed coding formats for the scheme described in this document provide the basis of a presentation layer protocol data syntax for videotex, teletext, and related applications.

7. *Application:* The application layer is the highest layer in the reference model, and the protocols of this layer provide the actual service sought by the end user. As an

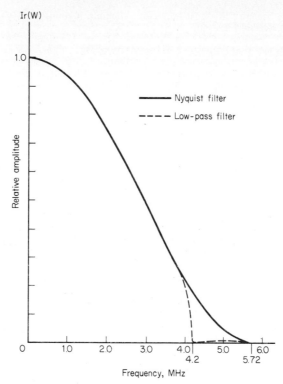

FIG. 14-62   Data spectrum after 100 percent roll-off Nyquist filter plus low-pass filter.

example, the information retrieval service commands of a videotex application form part of the application layer.

NABTS is designed to operate with CCIR, System M television signals using 525-line, 60-field scanning, and NTSC color coding. If the teletext signal is carried in the vertical blanking interval, lines 10 through 21 on both fields can be used. However, current usage is between lines 15 and 18. Lines 10 through 14 could be used in the future when the performance of retrace blanking on home receivers is improved, and lines 19 through 21 may carry teletext data if present test and captioning signals are shifted or multiplexed. NABTS may also be used for full field data transmission where no picture information is being sent. In this case, the full television channel is allocated for teletext purposes only.
NABTS technical specs are as follows:

1. *Transmission bit frequency.* The transmission bit frequency shall be 5.727272 Mbits/s ± 16 bits/s, which is the 364 multiple of the horizontal line scanning rate.

2. *Color television transmission.* When the data signal is being inserted into a color television transmission, the transmission bit frequency shall be 8/5 of the color subcarrier frequency (3.579545 MHz $\pm$ 10 Hz) and shall be phase-locked to the color subcarrier. The maximum rate of change of the transmission bit frequency shall be 0.16 bits/s.

3. *Monochrome television transmission.* When the data signal is being inserted into a monochrome television transmission (with no burst), the transmission bit frequency shall be 5.727272 Mbits/s ± 16 bits/s and phase-locked to the horizontal line scanning rate (364 times).

4. *Data modulation type.* The amplitude-modulated data shall use NRZ binary code.

5. *Data-pulse shape.* After shaping, the spectrum of the NRZ data at the output of the transmitter is described by the curve in Fig. 14-62. This spectral content is determined by a Nyquist filter with a roll-off of 100 percent followed by a phase-corrected low-pass filter with a cutoff frequency of 4.2 MHz. The impulse response of the combined filters is shown in Fig. 14-63.

6. *Data timing.* The start of the data signal shall be 10.5 ±0.34 μs from the half-amplitude point of the negative-going edge of horizontal sync to the half-amplitude point of the first transition from logical 0 to logical 1 of the clock synchronization.

7. *Data amplitude.* The nominal data amplitude, as seen in Fig. 14-64 shall be 70 ±2 IRE for a logical 1 and 0 ±2 IRE for a logical 0. By definition, the logical 0 is at blanking level and, therefore, the data signal should not contain any pedestal or setup. This section has specified a nominal data amplitude; however, the data waveform will contain overshoots, and so the peak-to-peak data amplitude will exceed the nominal data amplitude. Therefore, the peak-to-peak data amplitude shall not exceed 86 IRE units. In addition, the positive overshoots shall not exceed +6 IRE above nominal logical 1 level, and the negative overshoots shall not exceed −6 IRE below blanking level.

8. *Data line.* The data line consists of a string of bits with a maximum length of 288 bits (36 eight-bit bytes). All bytes have b1 (least significant bit) transmitted first. The data line is divided into two parts: the synchronization sequence and the data packet (see Fig. 14-65).

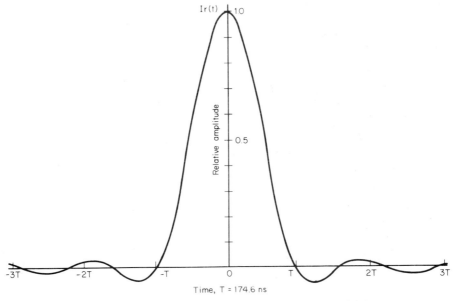

FIG. 14-63 Impulse response of 100 percent roll-off Nyquist filter plus low-pass filter.

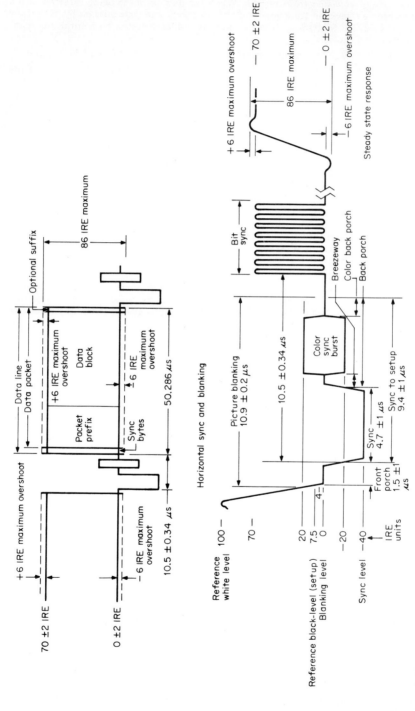

FIG. 14-64 Data amplitude.

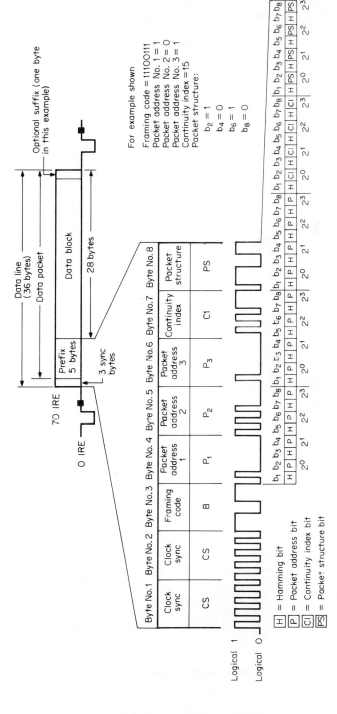

FIG. 14-65 Data line.

14.145

The coding sets for NABTS start with a simple seven bit in use table, followed by ASCII alphanumerics that include simple accents, punctuation marks, and other commonly used symbols (Fig. 14-66). A supplementary graphics character code adds many foreign language letters (Cyrillic, Greek, etc.), common fractions, arrows, diagonals, financial symbols, and the like (Fig. 14-67). The PDI and control sets of codes provide a wide range of teletext display manipulation so as to give the end user a most readable, or most understandable, electronic page under a wide variety of circumstances (Fig. 14-68).

The Teletext Display.   An NABTS display consists of 40 characters per horizontal row, and 20 rows of screen information on a CRT screen using a 4/3 aspect ratio. The display uses 200 scan lines and should fall in the safe title area outlined in SMPTE Recommended Practices.

b_7	0	0	0	0	1	1	1	1
b_6	0	0	1	1	0	0	1	1
b_5	0	1	0	1	0	1	0	1
$b_4 b_3 b_2 b_1$	**0**	**1**	**2**	**3**	**4**	**5**	**6**	**7**
0 0 0 0 **0**			SPACE	0	@	P	`	p
0 0 0 1 **1**			!	1	A	Q	a	q
0 0 1 0 **2**			"	2	B	R	b	r
0 0 1 1 **3**			#	3	C	S	c	s
0 1 0 0 **4**			$	4	D	T	d	t
0 1 0 1 **5**			%	5	E	U	e	u
0 1 1 0 **6**			&	6	F	V	f	v
0 1 1 1 **7**			'	7	G	W	g	w
1 0 0 0 **8**			(	8	H	X	h	x
1 0 0 1 **9**			)	9	I	Y	i	y
1 0 1 0 **10**			*	:	J	Z	j	z
1 0 1 1 **11**			+	;	K	[	k	{
1 1 0 0 **12**			,	<	L	\	l	|
1 1 0 1 **13**			−	=	M	]	m	}
1 1 1 0 **14**			.	>	N	^	n	_
1 1 1 1 **15**			/	?	O	_	o	∎

FIG. 14-66   ASCII alphanumerics.

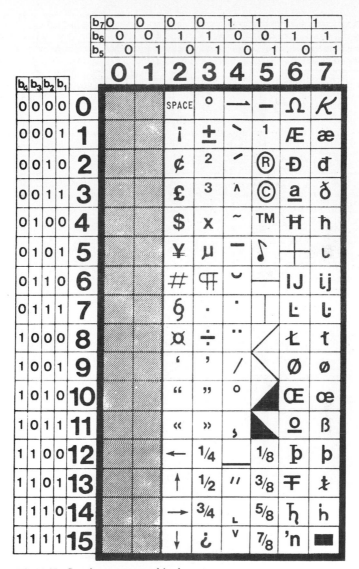

FIG. 14-67 Supplementary graphic characters.

There is an optional service row that is made up of two groups of 20 characters each. Twenty characters are sent by the teletext broadcaster; the remaining 20 characters are reserved for decoder interaction with the viewer. The optional service row may be displayed in a location chosen by the decoder, i.e., above or below the displayed page. When the optional service row is displayed, a total of 210 scanning lines is used. Most display devices will have 210 scanning lines available. If not, the optional service row may be displayed separately (on a remote-control panel, for instance), or, it will be recalled, the presentation layer allows for the redefining of the text size so as to accommodate display devices which utilize fewer than 210 scanning lines but with no guarantee of legibility.

b_7	0	0	0	0	1	1	1	1
b_6	0	0	1	1	0	0	1	1
b_5	0	1	0	1	0	1	0	1
$b_4\,b_3\,b_2\,b_1$	0	1	2	3	4	5	6	7
0 0 0 0 — 0					DEF MACRO	PROTECT		
0 0 0 1 — 1					DEFP MACRO	EDC_1		
0 0 1 0 — 2					DEFT MACRO	EDC_2		
0 0 1 1 — 3					DEF DRCS	EDC_3		
0 1 0 0 — 4					DEF TEXTURE	EDC_4		
0 1 0 1 — 5					END	WORD WRAP ON		
0 1 1 0 — 6					REPEAT	WORD WRAP OFF		
0 1 1 1 — 7					REPEAT TO EOL	SCROLL ON		
1 0 0 0 — 8					REVERSE VIDEO	SCROLL OFF		
1 0 0 1 — 9					NORMAL VIDEO	UNDER LINE START		
1 0 1 0 — 10					SMALL TEXT	UNDER LINE STOP		
1 0 1 1 — 11					MED TEXT	FLASH CURSOR		
1 1 0 0 — 12					NORMAL TEXT	STEADY CURSOR		
1 1 0 1 — 13					DOUBLE HEIGHT	CURSOR OFF		
1 1 1 0 — 14					BLINK START	BLINK STOP		
1 1 1 1 — 15					DOUBLE SIZE	UNPRO-TECT		

FIG. 14-68 Control set C1.

The basic teletext display is in the three primary and three secondary colors at saturation levels, with black and white added. Characters may appear at normal size, or may be double height or double size. Depending on the sophistication of the NABTS format used (alpha-mosaic, alpha-geometric, PDI, etc.), the graphics can be of low, medium, or high resolution, with smoothed edges and intermediate colors. Scrolling, underlining, and variable-rate flashing can be implemented by the transmitted codes, while the home viewer can select such optional services as closed captioning (on captioned programs), a reveal mode for question and answer teletext transmissions, or a breakthrough (bulletin) mode for important news events.

Teletext receivers can also have multiplepage memories with user-programmed cycles to store the most desirable page sequences and thus reduce access time, as well as hardcopy printers that provide a permanent record of the teletext page or magazine.

World System Teletext. World System Teletext (WST) is based on the British-developed Ceefax/Oracle teletext system in use in the United Kingdom and several other countries. Modified to operate on the NTSC line and field rates, WST can employ the same unused lines in the vertical blanking interval as NABTS (10 through 21), with the same limitations imposed as those stated for NABTS.

WST can also transmit teletext on all active picture lines (full field), if that is desired, and in general has the same attributes of data display as NABTS at comparable service levels. The fundamental difference is that WST uses a synchronous signal which is referred to as a defined (fixed) format relative to the video signal horizontal scan rate. The system can take advantage of a reference pulse for a decoder-associated adaptive equalizing filter, to minimize the effect of ghosting and multipath, and provision is made for visual emergency messages taking precedence over the teletext transmission.

WST proponents claim that this system is basically simpler, more rugged, more widely used, and less expensive that the NABTS system. WST also has multilevel standards that come in five layers or levels of operation that are defined as follows:

1. *Level one.* This level uses an alpha-mosaic system to display a basic character set of a selected Latin-based alphabet together with standard graphics. The display attributes include six colors, black and white, double height, and captioning. The display format is 40 characters per row and 24 rows per screenful. It is compatible with more advanced levels and can be upgraded in the future.

2. *Level two.* This level adds a full character set of all Latin-based alphabets including multilanguage text, accents, and other special symbols. This level is also known as Basic Polyglot-C and permits enhanced graphics and text. Its parallel attributes permit color changes within the same line and improved graphics where they are needed. There is a reverse compatibility with level one, in that an older decoder will still display the same data with some limits on accenting and colors.

3. *Level three.* This level accommodates other alphabets (non-Latin) and higher-definition graphics. It is also compatible with the West German Deutsche Bundespost standard for videotex. Level three employs the DRCS technique for fine definition of shapes and images. Sixteen colors, with a potential of 4096 different shades, can be used for the high-definition graphics to produce fine-line drawings, logos, graphics, music, and maps. Languages covered include Cyrillic, Greek, Japanese, and Chinese, and video images from a television camera can be digitized.

4. *Level four.* This level depends on alpha-geometric coding to produce very-high-resolution graphics. Geometric shapes can be generated in great detail by a special terminal, and computer-generated graphics can be mixed with text as required. Complex weather maps can be easily drawn using "chain coding" of the contours and outline. Extra intelligence in the television set is required to interpret the codes and produce the high-resolution graphics. Level four also allows for a Telesoftware service, where computer programs are transmitted via teletext for home reception and use.

5. *Level five.* Known as the Alpha-Photographic Transmission system, level five will permit the reproduction of full-color still pictures by the teletext decoder equipped to handle this level. Pictures and text can be mixed, thus producing teletext pages with images of the product being advertised or accurate reproductions of individual signatures for identification purposes. Less sophisticated teletext receivers will display the text and simpler pictorial graphics. Called Picture Teletext, this service is not expected until after 1985. WST is compatible with both videotex and home computer technologies and has operated that way in the United Kingdom for some years.

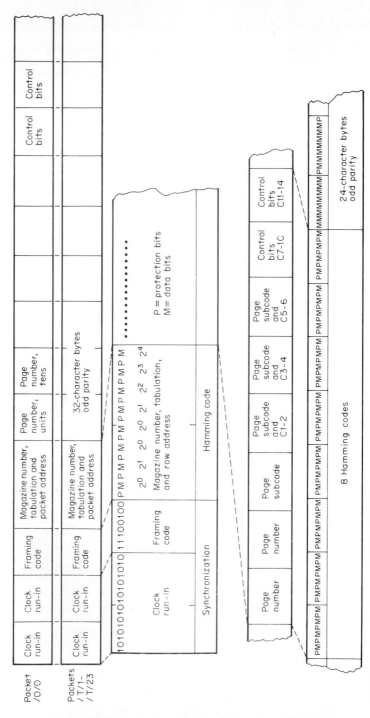

FIG. 14-69 Format of packets X/0/0 to X/T/23.

A petition for rule-making by the United Kingdom Teletext Industry Group was made to the FCC on March 26, 1981. The technical specifications proposed for WST are as follows:

a. *Transmission bit frequency.* The data bit rate for transmission shall be 5.727272 Mbits/s which is 364 times the line frequency and 8/5 times the color subcarrier frequency.

b. *Data format.* The data shall be in the form of a data packet consisting of 37 eight-bit bytes arranged into a prefix and a data block. The prefix shall consist of five bytes.

c. *Data modulation type.* The data signal is coded in an NRZ format.

d. *Signal levels for 0 and 1.* Black level ± 2 percent for 0; 70(± 6) percent of the difference between black level and peak white for 1.

e. *Spectrum of data pulse.* Skew symmetrical about 0.5 times bit rate, substantially zero at 4.2 MHz.

f. *Data timing reference point.* Peak level of penultimate 1 of clock run-in.

g. *Data line content.* 290 bits as 37 bytes of eight bits each.

h. *Data identification.* Clock run-in and framing code in appropriate time slot as defined in Fig. 14-69.

Many of the characteristics of WST are similar to those described for NABTS, including such features as conceal/reveal, flashing, boxing, double height, bulletin, and captioning. The smoothing of block mosaics utilizes a special smoothed mosaic graphics character set which minimizes the boxy appearance of the alpha-mosaic basic graphics.

The major difference in the display of WST and NABTS lies in the number of character rows shown on the screen. WST displays 24 rows as compared with 20, or in special circumstances 21, that appear on NABTS.

British videotex has also demonstrated a transcoding technique which makes WST compatible with line 21 captioning now provided on some American television stations for viewers with a hearing impairment. The low data rate captioning signal, normally transmitted on the last line of the vertical blanking interval, is also encoded on the WST signals on lines 15 and 16, thus providing a parallel subtitling service for WST-equipped home receivers.

14.12.10 TELETEXT NETWORKS. A teletext system starts with the studio hardware and software capable of creating the graphic pages that the viewer will see. In order to do this, the studio requires a frame-creation terminal, a computer storage device capable of holding a large number of teletext frames, a cycling system that scans the available pages at some predetermined rate, and an insertion device (multiplexer) that places the teletext data in the television lines assigned to carry it.

Editors working with the frame-creation terminal use alphanumeric keyboards and graphic-bit pads to enter the data needed to create a teletext page. Entry can also be made by a video camera, whose analog output is converted into the digital format of the teletext system. A typical teletext control room will have several editing consoles that connect to the storage and multiplexing devices.

Teletext signals may also be networked to affiliate stations who may retransmit the signal just as it is, or may insert material of local interest with, or in place of, parts of the network teletext transmission. Cable companies carrying network or local station broadcasts may also retransmit the teletext signal to their viewers, or may delete or replace it as they see fit. The FCC has been petitioned by the major networks to make the carriage of teletext signals mandatory for cable companies, if they are transmitted together by the originating television broadcaster, but the FCC has not instituted such a rule.

REFERENCES

1. Charles Clos, "A Study of Non Blocking Switching Networks," *Bell Systems Tech. J.,* March 1953.
2. K. B. Benson, "A Brief History of Television Camera Tubes," *SMPTE J.,* vol. 90, no. 8, Aug. 1981, pp. 708–712.
3. D. H. Pritchard and J. J. Gibson, "World-Wide Color Television Standards—Similarities and Difference," *SMPTE J.,* vol. 89, no. 2, Feb. 1980, pp. 111–120.
4. D. G. Fink and D. Christiansen (eds.), *Electronics Engineers' Handbook,* 2d ed., McGraw-Hill, New York, 1982, p. 20–30.
5. Sidney L. Bendel and C. A. Johnson, "Matching the Performance of a New Pickup Tube to the TK-47 Camera," *SMPTE J.,* vol. 89, no. 11, Nov. 1980, pp. 838–841.
6. SMPTE Recommended Practice, "Setting Chromaticity and Luminance for Color Television Monitors," *SMPTE J.,* vol. 86, no. 6, June 1977, p. 442.
7. M. Rettinger, *Acoustic Design and Noise Control,* Chemical Publishing Co., New York, 1973.

BIBLIOGRAPHY

American National Standard, "Electrical and Mechanical Characteristics for Digital Control Interface," *SMPTE J.,* June 1984.

EBU Review, "Disturbances Occurring at Edits on PAL 625 Line Video Tapes," Tech. No. 172, December 1978.

Grass Valley Group, "Establishing and Maintaining SC/H Phase," *Grass Valley Group Tech. Bull.,* no. MP-T-01, 1980.

Grass Valley Group, "Three Stage Switching Matrices in the Broadcast Environment," *Grass Valley Group Tech. Bull.,* no. RS-G-01, 1982.

Proposed recommended practice "Supervisory Protocol for Digital Control Interface," *SMPTE J.,* vol. 91, no. 9, Sept. 1981, p. 894.

Rayner, Bruce, "Application of Switcher Intelligent Interfaces to Video Tape Editing," *SMPTE J.,* vol. 88, Oct. 1979, p. 715.

Roizen, Joseph, "Teletext in the USA," *SMPTE J.,* vol. 90, no. 6, July 1980, pp. 602–610.

Van Dale, J. W., "Is SECAM the only remedy to overcome PAL editing problems?" Netherlands Broadcasting Systems, 1977.

Weaver, L. E., "Television Video Transmission Measurements," Marconi Instruments, Ltd., February 1979.

Wurtzel, Alan, *Television Production,* 2d ed., McGraw-Hill, New York, 1983, chapters 5 and 6.

Video Tape Recording

Charles P. Ginsburg

Ampex Corporation
Redwood City, California

with contributions by

H. Neal Bertram
E. Stanley Busby
Harold V. Clark
Michael O. Felix
Beverley R. Gooch
John W. King
William McSweeney
Reginald Oldershaw
Robert H. Perry
Raymond F. Ravizza
Dennis M. Ryan
David E. Trytko
Steven Wagner

Ampex Corporation
Redwood City, California

15.1 BASIC PRINCIPLES OF MAGNETIC RECORDING

by Charles P. Ginsburg and Beverley R. Gooch

15.1.1 OVERVIEW. The video tape recorders that first went on the air in 1956 were notable in their ability to store programs for later release with a picture quality not available from kinescope recording. Nevertheless, the recorders then were extremely simple in comparison with today's machines. Even in this age of high technology, the progress that has taken place in video tape recording is quite remarkable and may be attributed to two factors. One was the evolution in the field of electronics: transistor technology; integrated circuits and then large-scale integrated circuits, first in digital and then in linear devices; digital signal processing, and microprocessors. The second was the combination of foresight and creativity applied by engineers involved in improvements in the art after 1956 (and in user operations as well) in anticipating the need for improved or new capabilities, and bringing them to pass.

Most of the advances made in the field could be classified either as further development or as innovation of methods and techniques for meeting new requirements. Neither of the categories predominated over the other in milestones. For example, the high-band standard for video tape recording, which was introduced in the mid-1960s, and to which essentially all video tape recorders in broadcast use were eventually converted, definitely fell into the first group, but was without any doubt a breakthrough of the highest order. It should be noted that the values given in Sec. 15.4.2 for carrier frequency and modulating frequency are based on the assumption of the use of the high-band standard.

The several-step evolution of electronic editing, described in Sec. 15.9, falls into the second category but has had an impact on the television broadcast and postproduction industries as great as any other development or set of developments.

The initial conception and development of electronic time-base correction, covered in Sec. 15.5, is difficult to assign to either of the categories but was an essential step toward eliminating geometrical errors, both step-function and continuous, in off-tape pictures, and in eventually being able to recover from tape high-quality color recordings in concurrence with NTSC-color stipulations.

The multitude of subsets in the technology of video tape recording today makes adequate treatment in an engineering handbook by a single individual quite unlikely. Therefore, contributions to this chapter were obtained from more than a dozen experts, each a recognized authority in a particular speciality.

The modern magnetic tape recorder represents the application of highly developed scientific technologies, the result of many innovations and refinements since its invention by Valdemar Poulsen in 1898. Today many technical and business disciplines depend on the magnetic recorder in one form or another as an information storage device. The advancements in recording media, heads, and signal-processing techniques have made it possible to achieve packing densities that rival or exceed most other information-storage systems.

Magnetic recording is basically a moving-medium information storage device, which requires a means to transport the medium at a constant velocity past the record and reproduce heads. To achieve the necessary bandwidth for video recording, the heads are rotated at a relatively high velocity with respect to the tape, and a series of narrow tracks is recorded sequentially.

Fundamental Record and Reproduce Process.   The basic elements of a magnetic tape recorder are shown diagramatically by Fig. 15-1. A magnetic tape is moved in the direction indicated by a *tape-drive device* or *transport*. The magnetic coating of the tape contacts the magnetic heads in a prescribed sequence, starting with the erase head and ending with the reproduce head.

The erase head demagnetizes the tape coating by exposing the magnetic particles to a high-frequency field that is several times greater in strength than the coercivity of the particles. As the tape is drawn past the erase head, the erasing field gradually decays, leaving the magnetic coating in a demagnetized state.

The tape then moves into contact with the record head, which consists of a ring-shaped core made of a relatively high-permeability material, and having a nonmagnetic gap. A magnetic field fringes from the gap, varying in accordance with the magnitude of the current signal flowing in the head coil. With low-level signals the field is small, and some magnetic particles in the tape coating will be forced into alignment with the field. As the signal field is increased, a larger number of particles will become oriented in the direction of the recording field. As the tape is moved past the record gap, the magnetic coating acquires a net surface magnetization having both magnitude and direction. This magnetization is a function of the recording field at the instant the tape leaves the *recording zone,* a small region in the vicinity of the trailing edge of the gap.

The magnetization of the fundamental record system just described is not necessarily linear with respect to the head current. Linear magnetization can be achieved by adding a high-frequency ac bias current to the signal current. Audio recorders use such a scheme to linearize the tape and reduce the distortion. In video recorders, the signal information in the form of a frequency-modulated carrier is recorded directly, without ac bias.

When the tape approaches the nonmagnetic gap of the reproduce head, the flux Φ from the magnetized particles is forced to travel through the high-permeability core to link the signal windings and produce an output voltage. The output volgage is proportional to $d\Phi/dt$, the rate of change of the induced flux, and therefore will rise at the rate

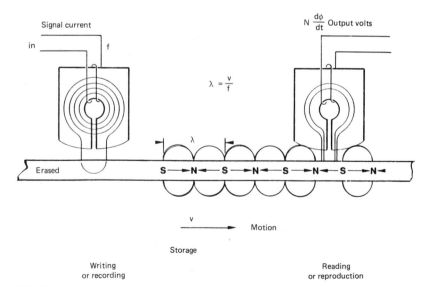

FIG. 15-1   Fundamental recording and reproduction process.

of 6 dB per octave until a wavelength is reached where the gap and spacing losses begin to reduce the head output. These effects are discussed in detail in the following section.

Wavelength of the Recorded Signal. The physical distance that one cycle of the recorded signal occupies along the tape is called the *wavelength,* which is directly proportional to the relative velocity between the head and the tape, and inversely proportional to the frequency of the recorded signal, and may be expressed as

$$\lambda = \frac{v}{f} \tag{15-1}$$

where λ = wavelength, in
v = velocity, in/s
f = frequency, Hz

Linear Packing Density. The linear packing density is the number of flux reversals per unit length along the recording medium. Since there are two flux reversals, or bits, per cycle, the linear packing density may be expressed as

$$\frac{\text{bits}}{\text{in}} = 2\left(\frac{1}{\lambda}\right) = \frac{2}{\lambda} \tag{15-2}$$

Area Packing Density. The area packing density is the number of bits per unit area and is, therefore, equal to the number of recorded tracks per inch times the linear packing density, or

$$\frac{\text{bits}}{\text{in}^2} = \left(\frac{2}{\lambda}\right)\frac{\text{tracks}}{\text{in}} \tag{15-3}$$

Table 15-1 compares the linear packing density and the shortest recorded wavelength of various types of magnetic recorders.

Table 15-1 High- and Medium-Density Recording Applications

Recorder	Tape speed, in/s (m/s)	Maximum frequency	Minimum wavelength μin (μm)	Linear packing density, bits/in (bits/cm)
High-density recording applications				
Quadruplex video	1500 (38.1)	15 MHz	100 (2.5)	20,000 (7874)
Type-C video	1000 (25.4)	15 MHz	70 (1.778)	30,000 (11,811)
Consumer video	220 (5.59)	7 MHz	30 (0.762)	64,000 (25,197)
Instrumentation	1–120 (0.254–3.05)	2 MHz	60 (1.524)	33,000 (12,992)
Audio cassette	1⅞ (0.0476)	20 kHz	94 (2.3875)	21,300 (8386)
Medium-density recording applications				
Professional audio	15–30 (0.38–0.76)	20 kHz	0.750 (0.019)	2700 (1063)
Computer tape	45 (1.1)	36 kHz	1.25 mils (32)	1600 (630)
Computer disc	1000 (25)	4.5 MHz	222 (5.64)	9000 (3543)

Directions of Recorded Magnetization. When the magnetization is oriented in the direction of relative motion between the head and tape, the process is referred to as *longitudinal recording.* If the magnetization is aligned perpendicular to the surface of the tape, it is called *vertical* or *perpendicular* recording.

Transverse recording exists when the magnetization is oriented at right angles to the direction of relative head-to-tape motion. From these definitions, longitudinal magnetization patterns are produced by both rotary- and stationary-head recorders. Therefore, stationary-head recorders should be referred to as such, rather than as longitudinal recorders.

15.1.2 RECORDING PROCESS

by H. Neal Bertram

Saturation Recording. The recording process consists of applying a temporally changing signal voltage to a record head as the tape is drawn by. The magnetic field which results from the energized head records a magnetization pattern which spatially approximates the voltage waveform. In saturation or direct recording the signal consists of polarity changes with modulated transition times or zero crossings. Strict linear replication of this signal is not required since the information to be recovered depends only on a knowledge of when the polarity transitions occur. Examples are digital recording, where the transitions are synchronized with a bit time interval and occur at bit positions depending upon the coded pattern, and FM video recording, where a modulated sine wave is applied so that the transitions occur not regularly but according to the signal information contained in the modulation.

The essential process in direct recording therefore is the writing of a transition or polarity change of magnetization. In Fig. 15-2 the resulting magnetization from a step change in head voltage is shown. In saturation recording the spatial variation of magnetization will not be a perfect replica of the time variation of signal voltage. Even if the

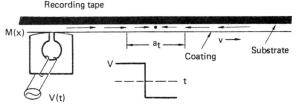

FIG. 15-2 Resultant magnetization from a step change in head voltage.

head field change is perfectly abrupt, the magnetization will gradually change from one polarity to another. In Fig. 15-2 this is indicated by a gradual change in vector lengths; the notation a_t denotes an estimate of the distance along tape over which the magnetization reverses. The nonzero distance between polarity changes of magnetization is due to the finite-loop slope at the coercivity, combined with the gradual decrease of the head field away from gap center. This process is illustrated in Figs. 15-3 and 15-4. In Fig. 15-3 the M-H remanence loop is shown for a well-oriented tape sample. The magnetization M is the remanence magnetization which results from the application of a field H which is subsequently removed. If the tape is saturated in one direction, for example, $-M$, and a positive field is applied, then the magnetization will start to switch toward the positive direction when the field is close to the remanent coercivity H_c^r. Since the slope of this M-H loop is not infinitely steep for fields near H_c^r, the switching will take place gradually. H_1 denotes the field which switches 25 percent of the particles to leave the magnetization at $-M/2$; H_c^r is, in fact, the 50 percent reversing field which leaves $M = 0$. H_2 denotes the 75 percent switching field that leaves the magnetization halfway to positive satura-

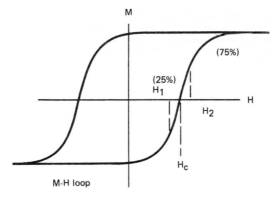

FIG. 15-3 *M-H* hysteresis loop for a typical tape.

tion $(+M/2)$. During recording a finite transition width will occur, depending on how H_1 and H_2 are spatially separated. In Fig. 15-4 three contours of recording field are plotted for the three fields H_1, H_c^r, and H_2. In plots of head fields, larger fields are closer to the surface of the head and toward the gap center. Thus, along the midplane of the tape the field magnitudes H_2, H_c^r, H_1 are in decreasing order away from the gap centerline. Therefore, if the tape is initially magnetized negatively, a positively energized head (H_0) will switch the magnetization according to Fig. 15-3 following the spatial change of the fields. This yields a finite transition width a_t. The transition width can be narrowed by using tape with a steeper loop gradient making H_1 closer to H_2 in magnitude (a narrower spread in switching fields) and decreasing the head-to-tape spacing which moves H_1 and H_2 closer together spatially (a larger head field gradient) as indicated in Fig. 15-4. In addition, spatial changes in magnetization cause demagnetization fields in the tape which further broaden the transition. The demagnetization broadening may be reduced by increasing the tape coercivity. The physics of this process may be seen in the calculation of the transition width a_t by Williams and Comstock for thin media.[1]† As the following section shows, a large transition width reduces the short-wavelength output and broadens the isolated voltage pulse.

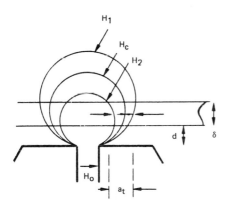

FIG. 15-4 Head-field contours showing recording zone in tape.

In saturation recording the signal current is held fixed for all wavelengths. The current level is set to optimize the short-wavelength output, and complete saturation of the tape does not occur. In Fig. 15-5 reproduce voltage versus input current is shown for two different wavelengths in square-wave recording for video tape (Ampex 196). If true saturation were to occur, the curves would increase initially with current as the tape is recorded, and then level, representing a magnetization saturated to full remanence and

---

†Superscript numbers refer to References at end of chapter.

recorded fully through the tape thickness of 200 μin (5 μm). However, at short wavelengths these curves are peaked, and the current which yields the maximum output represents recording only a very small distance into the tape. For video recording on a type C format machine optimized at 10 MHz ($\lambda \approx 100$ μin), this is a record depth of approximately 50 μin. For a video home system (VHS) cassette recorder (see Table 15-7, SMPTE type G) optimized at 30-μin wavelength, only 20 μin of the surface layers is recorded under optimum conditions. A mechanism for this optimum can be seen by considering the change in transition with record current. As the current is raised, the point of recording shifts continuously downstream from the gap center. The transition width depends upon the head-field gradient at the recording point. This field derivative, $H_2 - H_1$, divided by the separation between them, increases with distance along the head surface, as shown in Fig. 15-4, reaching a maximum near the gap edge and thereafter decreasing. Since the reproduce voltage increases with decreasing transition width, a maximum voltage will occur as the current is increased. From the reproduce expression discussed in Sec. 15.1.3, it is evident that this peaking becomes more pronounced as the wavelength is reduced.

A form of linearity known as *linear superposition* is found in saturation recording. For constant-current recording (strictly, *constant-field amplitude*) the reproduce voltage from a complicated pattern can be shown to closely resemble the linear superposition of isolated transition voltage pulses, according to the timing and polarity change of the series of transitions. The lack of complete linear superposition is believed due to demagnetization fields. This accompanies large head-to-medium separations, as in rigid-disc applications, where the increase in the demagnetization fields can cause significant nonlinearities.

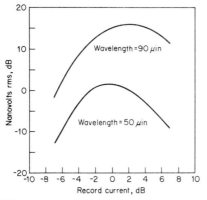

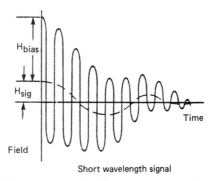

FIG. 15-5 Reproduce voltage versus record current for typical video tape.

FIG. 15-6 Signal-field history for ac-bias recording.

Bias Recording.   In some applications, predominantly audio recording, strict linearity is required between the reproduced voltage and the input signal. This may be achieved by superimposing a high-frequency, large-amplitude bias current on the signal (Fig. 15-6). The physical process is called *anhysteresis*.[2] The bias field supplies the energy to switch the particles while the resulting remanent magnetization is a balance between the signal field and the interparticle magnetization interactions. In Fig. 15-7 a comparison is shown between the magnetic sensitivity of ac bias recording and direct recording. Alternating-current bias or anhysteresis results in an extremly linear characteristic with a sensitivity an order of magnitude greater than that for unbiased recording. In typical audio applications the bias current is somewhat greater than that of the signal in direct recording. The signal current is approximately an order of magnitude less than the bias current and is set to maintain the harmonic distortion below 1 to 2 percent. For complete anhysteresis there should be many field reversals as the tape passes the recording point where the bias field equals the tape coercivity. This is achieved if the bias wavelength is substantially less than the record-gap length. In fact, to avoid reproducing the bias signal itself, the bias wavelength is usually less than the reproduce-gap length.

As in direct recording, a current optimization occurs, but in bias recording it is with respect to the bias current. In Fig. 15-8, reproduce voltage is shown versus bias at short and long wavelengths. At long wavelengths, the optimum bias occurs approximately when the bias field has recorded through to the back of the medium. This is often taken to be the usable bias current since close to this optimization a minimum in the distortion occurs. For shorter-wavelength machines (such as audio cassettes) the bias is chosen as a compromise between short-wavelength bias optimization (SWBO) and long-wavelength bias optimization (LWBO).

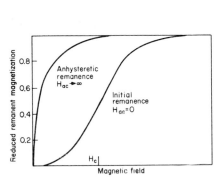

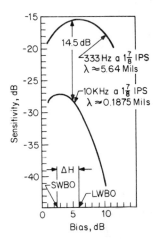

FIG. 15-7 Comparison of sensitivities of ac-bias and direct recording.

FIG. 15-8 Low-level output sensitivity versus bias for long and short audio cassette wavelengths.

Particle Orientation. The previous discussion applies generally to all types of recording tape; the most common is that composed of uniaxial elliptical particles oriented *longitudinally* in the direction of head-tape motion. However, *isotropic* tape does exist and is composed of particles of cubic (threefold) symmetry which exhibit high remanences in all directions. In addition, it is conceptually possible to *vertically* orient the grains to result in a tape isotropic in the plane, but capable of recording signals perpendicular to the surface. These last two would be advantageous for transverse recording since the difficult process of orienting elliptical particles along the tape cross direction could be avoided.

During the tape-coating process, elliptical particles will naturally orient along the tape-coating direction. A field applied during coating improves the orientation even further. It is extremely difficult to orient these particles vertically since the hydroscopic coating forces overwhelm the magnetic force from a vertical-orienting field. Thus, the only success has been with systems that *inherently* yield vertical orientation. As an example, barium-ferrite platelets have been successfully coated to yield perpendicular media since the magnetization anisotropy axis is perpendicular to the plane of the particles. To date one other vertical medium has been made by sputtering CoCr on either a tape substrate or a rigid disk.

Sufficient experimental comparisons or theoretical analyses have not yet been performed to properly compare the performance of various particle orientations. However, with regard to contact recording on tape composed of individual particles, it appears that vertical or isotropic media may record larger short-wavelength signals than longitudinally oriented tape.[3] The reason is that at the surface of a recording head the fields are primarily vertical, and it is the surface magnetization which dominates the short-wavelength signal. For media with large magnetizations, such as the vertically oriented cobalt chromium or thin planar cobalt nickel metallic films on tape or rigid disc, the performances seem comparable.[4] Both of these high-magnetization media are superior to conventional oxide media.

Erasure.   The writing of new information on previously recorded media requires that the previous information be completely removed. Erasure requirements, in terms of the previous-signal to new-signal ratio, vary from -30 dB for digital systems to as much as -90 dB for professional analog audio. Video recorders require about -60 dB.

Erasure is the ac-bias or anhysteretic process with zero-signal field. If a reel of tape is placed in a large ac field which is slowly reduced so that many field cycles occur when the field is near the coercivity, then complete erasure is easily obtained. In addition, the largest amplitude of the ac-erasing field must be sufficient to reverse at least 99.9 percent of the particles (for -60 dB), and in practice that field is about three times the coercivity.

Most tape recorders utilize erase heads to remove old information before recordng new data. Similar to bias recording, the requirement of the erase frequency is that the wavelength be much less than the erase-head gap to provide sufficient reversals of the particles. However, one important problem occurs with an erase head. As the erase current increases, the erasure level does not continuously increase as more of the M-H–loop *tail* is switched. There is an erasure plateau of about -40 dB for erase-gap lengths of 1 to 2 mils (0.0254 to 0.0508 mm) and tape thicknesses of 200 to 400 μin (5.08 to 10.16 μm). This leveling is believed to be due to the phenomenon of rerecording.[5] As the tape passes the erase head, the field from the portion yet to be erased (entering the gap region) acts as a signal for the bias-erase field to record a residual signal at the recording zone on the far side of the gap. This effect is seen only at long wavelengths where the field is significantly high. The problem is eliminated with double erasure by using a double-gap erase head. The erasure level may be decreased by decreasing the ratio of the tape thickness (or recording depth) to erase-gap length; this reduces the rerecording field.

15.1.3 REPRODUCTION PROCESS

Signal Spectrum.   In the previous section the process of writing magnetic information on tape was described. Here the reproduction of a recorded signal will be described in terms of the spectrum of a recorded square wave. A square wave of record current is assumed, resulting in a magnetization pattern of alternating polarity of fundamental wavelength λ. Each transition of magnetization is taken to be arctangent in shape with a transition width a_t and saturation magnetization M_r (Fig. 15-2). The mathematical form is $M(x) = (2/\pi)M_r \tan^{-1}(x/a_t)$. The magnetization pattern is assumed to be invariant with depth in both magnitude and direction. The direction can be at any angle, from longitudinal to vertical, since orientation yields only a constant phase shift which amplitude spectra will not show. When Fig. 15-1 is viewed, as the recorded tape passes the head, the external fields or flux due to the spatially varying magnetization pattern enter and circulate through the highly permeable reproduce head. The flux changes with time as the tape moves past the record head, giving a reproduce voltage. This voltage may be expressed as

$$V = \frac{4}{\pi} NW\varepsilon\upsilon\mu_0 M_r \frac{2\pi d_c}{\lambda} \frac{[1 - \exp(-2\pi d_c/\lambda)]}{2\pi d_c/\lambda} \exp\left[\frac{-2\pi(a + a_t)}{\lambda}\right] \frac{\sin 1.11\pi g}{1.11\pi g/\lambda} \qquad (15\text{-}4)$$

where N = no. of reproduce head turns
  W = trackwidth
  ε = reproduce head efficiency
  v = head-to-tape relative speed
  μ_0 = permeability of free space = $4\pi \times 10^{-7} H/M$
  M_r = tape remnant magnetization
  d_c = depth of recording
  λ = wavelength
  a = head-to-tape spacing
  a_t = recorded transition width
  g = reproduce gap length

The voltage is the fundamental component of wavelength λ. This expression may be separated into various terms which relate to different physical effects. The first terms $(4/\pi)NW\varepsilon v\mu_0 M_r$ are the calibration constants. The term $4/\pi$ is the amplitude of the fundamental of a square wave. N, W, ε, and v are, respectively, the number of turns of the reproduce head, the reproduce track width, the head efficiency, and the head-to-tape relative velocity. The term μ_0 is the permeability of free space ($4\pi \times 10^{-7} H/m$). M_r is the remanent magnetization. The voltage may be evaluated using metric (MKSA) units for these constants. The subsequent terms in Eq. (15-4) depend on length ratios only, and so the dimension chosen is not critical. A convenient working relation is that the rms voltage in nanovolts per turn per width of track (in mils) per unit efficiency per speed (in in/s) is approximately $M_r/16$ times the other factors. Here M_r is in gauss. As an example, the maximum voltage for a video tape (Ampex 196) on a type C format VTR is evaluated. In that case $N = 6$, $W = 5$ mils (0.127 mm), $v = 1000$ in/s, $\varepsilon \approx 0.8$, $M_r \sim 1100$ G. Then the reproduce spectral voltage constant is

$$\frac{NW\varepsilon v M_r n V \text{ (rms)}}{16}$$

or approximately 1.65 mV. Measured voltages are a factor of 5 to 10 less than 1.65 mV; primarily, this is due to other factors in Eq. (15-4) which are discussed next.

Reference to Fig. 15-9 will help in the discussion of the wavelength-dependent fac-

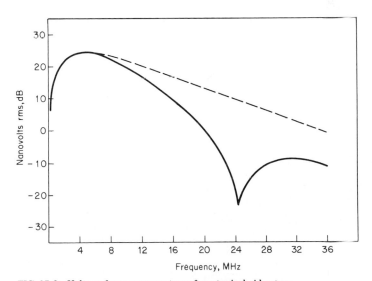

FIG. 15-9 Voltage-frequency spectrum for a typical video tape.

tors. This is a plot of the spectrum of video tape (Ampex 196) using a conventional type C format head with a gap of 35 μin (0.889 μm). In this experiment the current was set before the spectral sweep to a value which maximized the shortest measured wavelength [λ = 60 μin (1.5 μm)]. The output from the reproduce head was amplified by a flat amplifier; no pre- or postequalization was used. Over the frequency range measured, the head losses were minimal, and so the head efficiency was constant with frequency. Thus, the shape of the spectrum represents the wavelength-dependent factors in Eq. (15-4). The term $2\pi d_c/\lambda$ in Eq. (15-4) represents the head differentiation of the flux and varies linearly with frequency. In Fig. 15-9 this is the initial 6-dB/octave rise at low frequencies (curved on a log-linear plot).

Thickness Loss. The recorder does not reproduce direct current—only temporal changes in flux yield voltage for a conventional reproduce head. If this were the only factor, the voltage would continue to rise at 6 dB/octave indefinitely. The term

$$\frac{1 - \exp\,(-2\pi d_c/\lambda)}{2\pi d_c/\lambda}$$

gives the first limiting factor to that rise, i.e., the *thickness loss.* As Fig. 15-9 indicates, this term causes the spectrum to level. The reason is that the reproduce head senses only wavelength components which are approximately one-third of a wavelength, $\lambda/3$, of the coating surface. Thus, as the wavelength is shortened below about three times the recording depth, $3d_c$, the reproduce process senses increasingly less of the recorded layer. The combination of the increase in voltage due to differentiation and the decrease due to the restricted sensing depth, balances to give a constant voltage.

Spacing Loss. The next term $\exp\,[-2\pi(a\,+\,a_t)/\lambda]$ is the spacing loss which causes a decrease in the output with increasing frequency. This effect is physically the same as the thickness loss term since that term arose from the head not sensing the far layers of the recording. The exponential spacing loss, plotted log linear, is a straight line equal to $54.6(a\,+\,a_t)/\lambda$. As the frequency is increased, the level response is reduced linearly with slope $a\,+\,a_t$. The actual head-to-tape spacing is a, and a_t is the transition width discussed in the previous section. The transition width enters the reproduce voltage spectrum as an effective spacing. In fact, from Fig. 15-9, $a\,+\,a_t$ from the data is approximately 15 μin. Since the data are for contact recording, the measured slope must be due primarily to the width of the recorded transition. The terms discussed so far involve only tape parameters (a, d_c, a_t). The term $\sin\,(1.11\pi g/\lambda)/(1.11\pi g/\lambda)$ is the gap loss which is the primary head geometric effect; g is the gap length of the reproduce wavelength. As the wavelength approaches the gap length, the flux is shunted across the gap, and, at a wavelength of λ = 1.11g, there is a null in the output. The data (dashed curve) in Fig. 15-9 have been corrected for this factor in order to clearly exhibit the spacing loss term; the solid curve includes the gap loss term as measured. After the first null at λ = 1.11g, the spectra rise and peak before decreasing to the second null and so on according to the $(\sin x)/x$ function. The figure shows that the response of a recording channel is bounded between d_c (λ = ∞) and the first gap null (λ = 1.11g). A practical channel which would not invoke excessive equalization would be limited to an upper frequency or shortest wavelength of about twice the gap length. The lower frequency or largest wavelength would be limited by the decrease of the spectrum to zero at d_c which occurs at a wavelength about an order of magnitude larger than the recording depth. For the video example shown in Fig. 15-9, a practical channel would cover the wavelength range from 400 μin to 60 (10.16 to 1.5 μm) or, at 1000 in/s (25.40 m/s) relative speed, frequencies from 2.5 to 15 MHz. This spectrum also applies approximately to bias recording where M_r is replaced by the signal field (H_{sig}) times the anhysteretic susceptibility.[6]
There are additional reproduce phenomena which affect the spectral shape. Three are discussed below.

Azimuth Loss. The azimuth loss occurs if the reproduce head is rotated with respect to the recording phase fronts of the original recording. In Fig. 15-10 a reproduce head is

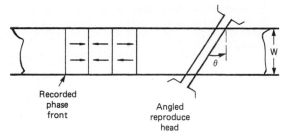

FIG. 15-10   Azimuth misalignment geometry.

shown angled at an angle θ with respect to the original recording. This rotation causes the signal to be reproduced at different times across the track. This continuous phase shift across the track causes an additional loss of the form

$$\frac{\sin\left[(\pi W \tan \theta)/\lambda\right]}{(\pi W \tan \theta)/\lambda}$$

This loss can easily be severe, since for a video recorder with $W = 5$ mils (0.127 mm) and $\lambda = 100$ μin (2.54 μm), a 4-dB loss occurs for a value of only 3.3 min of arc! Dual-head machines have an azimuth-alignment adjustment. Even machines which use the same head for record and reproduce, where theoretically this loss should not occur, may show this effect as a random fluctuation due to tape stretching during play.

Head-Length Loss.   When the wavelength of recording lengthens to approach the length of the reproduce head or longitudinal distance of head-to-medium contact, undulations in the response will occur. At these wavelengths the flux can return partially through the air to the tape and not enclose the windings. For certain conditions a flux reinforcement can occur to raise the level above that given in Fig. 15-9. This effect is strongest with sharp 90° head edges. Rounded corners soften the undulations, and, in fact, a head with a circular shape shows no undulations.[7] It is desirable to choose a contour which removes the undulations, since they are hard to equalize and cause phase shift. A theoretical discussion of the undulations due to a rectangular head is given by Lindholm,[8] who also includes interaction effects when the length of the head is reduced to the order of the gap length (thin film head).

Gap-Irregularity Effect.   Another example of a reproduce loss is that due to a gap which is not straight. *Gap irregularity* refers to the gap length being constant across the head but the gap position varying in a statistical manner. If the distribution is gaussian with scatter of spread, σ, then the loss can be shown to be

$$\text{Scatter loss} = -\frac{170}{\lambda^2} (\sigma_{\text{rec}}^2 + \sigma_{\text{play}}^2) \qquad \text{dB} \qquad (15\text{-}5)$$

where noncorrelated scatter for the reproduce and record head is included.[9] For example, if $\sigma_{\text{rec}} = \sigma_{\text{play}} = 6$ μin (0.15 μm) and $\lambda = 60$ μin (1.5 μm), the loss is 4 dB. For a common record-reproduce head this loss will not occur except owing to fluctuations due to tape stretching.

Undesired Signal Effects.   There are two primary factors which lead to extra signal components that will deteriorate the picture quality. One is noise which arises from the granularity of the tape and the random fluctuations in bulk properties, such as pigment loading surface and smoothness. The other is signal interference which arises primarily from crosstalk between adjacent channels.

Noise can be separated into *additive* and *modulation* components. Additive noise arises from the granularity or particulate nature of the media and is always present independent of the signal level. As the tape passes the reproduce head, the flux from each particle threads the head (Fig. 15-11), yielding a voltage pulse. These pulses occur randomly since the particles are distributed randomly in the tape. The noise power spectrum has been calculated on the basics of this phenomenon, excluding interparticle correlation.[10] In addition, signal-to-noise ratios for video recorders have been calculated on the basis of signal spectra given in Sec. 15.1.3 and the particulate noise power spectrum.

The derivation for the signal-to-noise ratio involves parabolic weighting of the noise power to include FM triangulation effects.[11] The result is

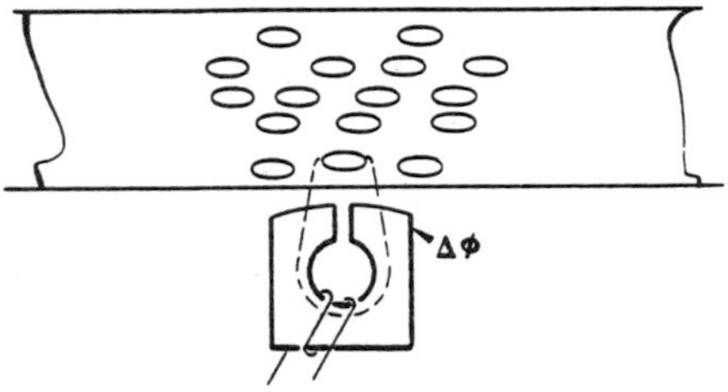

$$S/N = \frac{3nWV^2m^2}{8\pi f_c f_s} \qquad (15\text{-}6)$$

where n = particle density
W = track width
V = speed
m = modulation index
f_c = carrier frequency
f_s = video bandwidth

FIG. 15-11 Flux contribution from each particle as a source of noise.

For example, the S/N of a transverse scan quadruplex video tape recorder, broadcast quadruplex recording (Ampex AVR1) with W = 10 mils (0.254 mm), V = 1500 in/s (38 m/s), f_c = 9 MHz, f_s = 4.6 MHz, $m \approx$ 0.75, and n = 10^{14} particles/cm^3 yields

$$S/N \approx 56 \text{ dB (p-p/rms)}$$

S/N varies as the track width W, such that halving the track width reduces S/N by only 3 dB. The term n is the number of particles per unit volume: the use of smaller particles in the tape will increase S/N. For fixed-frequency operation, higher speeds will improve the S/N. However, for shorter wavelengths (f_c/v, f_s/v) the S/N will decrease. Almost all S/N expressions, independent of the particular application, vary linearly with track width and inversely as the square of the shortest wavelength. Thus, in designing for higher densities, it is better to narrow the track width rather than to shorten the wavelength.

The calculated number is higher than measured, and one contributing factor may be additional noise due to modulation effects. Modulation noise is proportional to the signal level and arises from processing variations of media density and surface smoothness. The former affects the saturation magnetization and therefore the signal [M_r in Eq. (15-4)], and the latter modulates the reproduce spacing loss [e^{-ka} in Eq. (15-4)]. All sources of modulation noise contribute to the *broadband*, or *luminance, noise*. *Chroma noise* is seen as a fluctuating disturbance in highly saturated areas of the picture and can be due primarily to fluctuations in head-to-tape spacing. As discussed in Sec. 15.4.2, the reproduce voltage of an FM video recorder is *straight-line equalized*. Variations which change the ratio of lower to upper sideband voltages proportionally, corresponding to a change only in the slope of the straight-line equalization, will not introduce noise. However, head-to-tape spacing fluctuations, due to tape roughness, cause nonlinear changes (by changes in the exponential spacing loss e^{-ka}) and will introduce chroma noise in the form of a visible *flicker*.

Side-reading Crosstalk Signal Effects. The most common signal interference effect is that due to side reading of adjacent channels and yields crosstalk in the reproduce channel. As shown in Fig. 15-12, as the video head is reproducing one track, the flux generated by a recorded adjacent track will circulate through the nearby high-permeability reproduce head. The on-track signal will thereby have a small additional signal due to the adjacent tracks. The phase of these crosstalk signals will be somewhat random

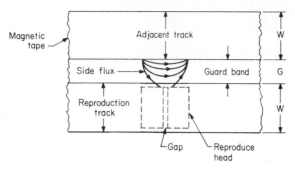

FIG. 15-12 Side-reading flux schematic.

since signals are not recorded coherently from track to track. Therefore the side-recording terms will resemble a noise signal and cause errors in, for example, reading of the zero signal crossings in video FM recording. This phenomenon has been computed by Lindholm.[12] An approximate formula for this side-reading ratio (SRR) may be written as

$$\text{SRR} = e^{-kG}(1 - e^{-kW})/\text{RW} \qquad (15\text{-}7)$$

where G = guard band
$\quad\ \ W$ = track width
$\quad\ \ k$ = wave number ($= 2\pi/\lambda$)

This is the ratio of adjacent side-reading voltage to the on-track signal. It is doubled from Ref. 12, since the worst case of two adjacent tracks adding in phase is considered. This formula yields a side-reading ratio which continually decreases as the wavelength is shortened. The expression is reasonably accurate for wavelengths not significantly longer than 10 times the guard band. Side-reading can be significant, as an example from high-density digital recording shows. For a 1000-tracks/in, 20-kbit/in system, where W = 0.060 in, G = 0.003 inches, and λ = 0.002 in (a long wavelength in the coded-information equation), (15-7) yields about -26 dB. For shorter wavelengths, the side-reading reduces rapidly, mainly because of the side-spacing loss e^{-kG}. This side-reading is a long-wavelength phenomenon and is highly dependent on the guard band.

15.1.4 PROPERTIES OF MAGNETIC MATERIALS

by Beverley R. Gooch

The performance of a magnetic tape recorder depends heavily on the properties of the magnetic materials used to make the recording heads and tapes. Today's magnetic materials are the product of sophisticated metallurgy and advanced manufacturing techniques, which in large measure are responsible for the advancement of the magnetic recording technology.

Magnetic materials are classified as either magnetically *hard* or magnetically *soft*. Both types are used in magnetic tape recorders.

The hard magnetic materials are so-called because of their ability to retain magnetism after being exposed to a magnetic field. The measure of this property is called *remanence*. These materials may be further characterized by high coercivity and low permeability. *Coercivity* is the resistance of the material to being magnetized or demagnetized. *Permeability* is a measure of the magnetic conductivity relative to air.

In magnetic recording, hard magnetic materials are used chiefly in the manufacturing of recording tape and other related media. Some examples are gamma ferric oxide, iron oxide, and chromium dioxide. Hard materials are also used to make permanent magnets for use in loudspeakers, electric motors, and other applications.

Table 15-2 Properties of Soft and Hard Magnetic Materials

Material	M_s, G	$B_s = 4\pi M_s$, G	H_c, O	B_r, G	μ (dc) initial	Resistivity, $\Omega \cdot$cm	Thermal expn.	Curie temp.	Vickers hardness
Soft magnetic materials									
Iron Fe	1700	21,362	1	......	20,000				
Hi-Mu 80 80% Ni, 20% Fe	661	8,300	0.02	......	50,000	65	12.9×10^{-6} cm/(cm°C)	733 °C	127
Alfesil (Sendust) 85% Fe, 6% Al, 9% Si	796	10,000	0.06	......	10,000	90	11.3×10^{-6} cm/(cm°C)	773 °C	496
Mn Zn, Hot-pressed ferrite	358	4,500	0.02–0.2	≈900	2000–5000	10^4	$10\text{–}15 \times 10^{-6}$ cm/(cm°C)	100–300 °C	650–750
Ni Zn, Hot-pressed ferrite	238	3,000	0.15–3	≈1800	100–2000	10^{10}	$7\text{–}9 \times 10^{-6}$ cm/(cm°C)	150–200 °C	700–750
Hard magnetic materials					*Squareness ratio*				
γ-Ferric oxide	400	5,026	300–350	1300†	0.75†				
Chromium dioxide	470	6,000	300–700	1600†	0.9†				
Metal particles	800	10,000	1000	3500†	0.8†				

†Value typical for finished tape.

On the other hand, soft magnetic materials such as Alfesil, hot-pressed ferrite, and Permalloy exhibit low coercivity, low remanence, and relatively high permeability. These materials are used to make cores for magnetic heads.

Ferromagnetic materials have permeabilities much greater than unity and show a strong magnetic effect. Ferromagnetism is exhibited mostly by metallic elements such as iron, cobalt, nickel, and magnetic metals which are alloys of these elements. With the exception of ferrites,[13,14] most magnetic materials used in tape recorders are ferromagnetic.

Paramagnetic substances have permeabilities which lie between 1.000 and 1.001. These materials do not show hysteresis, and their permeabilities are independent of field strength. Some examples of paramagnetic materials are sodium, potassium, oxygen, platinum, and ferromagnetic metals above the Curie temperature.[13]

Diamagnetic materials have a relative permeability slightly less than 1. Many of the metals and most nonmetals are diamagnetic.[13]

Magnetic anisotropy is the term applied to magnetic materials that exhibit preferred directions of magnetization. These preferred and nonpreferred directions are referred to as the *easy* and *hard* axes of magnetization, respectively. The higher the magnetic anisotropy, the harder it is to change the magnetization away from the preferred direction. In most polycrystalline materials, the crystals are randomly oriented and are magnetically isotropic. Single crystal ferrites and magnetic particles used in tape coating are examples of magnetic materials that are anisotropic.[13,15]

Table 15-2 shows properties of materials commonly used in magnetic heads and tapes. Throughout this section CGS units are used. Conversion factors to change to MKSA (or SI) units are given in Table 15-3.

Basic Theory of Magnetism.   Magnetism results from two sources: orbital motion of electrons around the nucleus and the spinning of the electrons on their own axes (see Fig. 15-13). Both the orbital and spin motions contribute to the *magnetic moment* of the atom, although in most magnetic substances almost all the magnetic moment is due to the spin motion. As the electron spins on its axis, the charge on its surface moves in a circular pattern. This moving charge, in turn, produces a current that creates a magnetic field. This phenomenon occurs in all substances. However, the electrons of the atoms in nonmagnetic materials occur in pairs with the spins in opposite directions, balancing each other and rendering the atom magnetically neutral. The atoms can produce the external effect of a magnet only when the electron spins are unbalanced.

The iron atom, for example, has 26 electrons in rotation around its nucleus (see Fig.

Table 15-3   Conversion Factors from CGS to MKSA or SI Units

Parameter	CGS units†	Multiply by	To obtain MKSA or SI units†
Flux Φ	Maxwell	10^{-8}	Webers (Wb)
Flux density B	Gauss	10^{-4}	Webers/meter^2 = 1 tesla (T)
Magnetization M	Gauss (1 gauss = 1 emu/cm^3)	10^3	Ampere turns/meter (At/m)
Permeability μ_0 of free space	1	$4\pi \times 10^{-7}$	Henry/meter (H/m)
Magnetomotive force F	Gilbert	$\dfrac{1}{0.4\pi}$	Ampere turns (At)
Field H (magnetomotive force per unit length)	Oersted	$\dfrac{10^3}{4\pi}$	Ampere turns/meter (At/m)

†Unit system abbreviations: CGS (centimeter-gram-second), MKSA (meter-kilogram-second-ampere), and SI (International System).

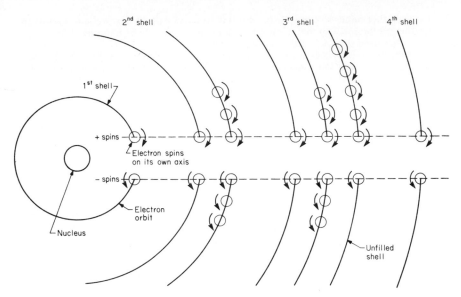

FIG. 15-13   Schematic diagram of iron atom.

15-13). These orbiting electrons occur in regions called *shells*. According to quantum theory, the maximum number of electrons that can exist in each shell is $2N^2$, where N is the number of the shell. Starting from the nucleus, the first, second, third, and fourth shells could have a maximum number of 2, 8, 18, and 32, respectively. The maximum number of electrons in each shell may not be reached before the next shell begins to form. The iron atom actually has two electrons in the first shell, the second has eight, the third and fourth shells have fourteen and two, respectively. The plus and minus signs show the direction of the electron spins. The electron spins in the first, second, and fourth shells balance each other, and produce no magnetic effect. It is the third shell that is of particular interest in the iron atom. In this shell there are five electrons with positive spins and one with a negative spin, which gives the atom a net magnetic effect.

The thermal agitation energy, even at low temperatures, would prevent the atomic magnets from being aligned sufficiently to produce a magnetic effect. However, powerful forces hold the electron spins in tight parallel alignment against the disordering effect of thermal energy. These forces are called *exchange forces.*

The parallel alignment of the electron spins, due to the exchange forces, occurs over large regions containing a great number of atoms. These regions are called *domains*. Each domain is magnetized to saturation by the aligned electron spins. Since this magnetization occurs with no external field applied, it is referrred to as *spontaneous magnetization*. When the magnetic material is in the demagnetized state, the direction of the magnetization of the saturated domains is distributed in a random order, bringing the net magnetization of the material to zero. The domains are separated from each other by partitions called *Bloch walls*.[13,15] The domain wall pattern is determined by the strains within the material and its composition.

In soft magnetic materials the magnetization takes place by the displacement of the domain walls.[13-15] The wall movement is not continuous but occurs in discrete steps called *Barkhausen steps* or *jumps* that are related to imperfections or inclusions in the crystalline structure of the material.

The particles used in magnetic tape coating are so small that Bloch walls do not form. They behave as single-domain particles which are spontaneously magnetized to saturation. Irreversible magnetization is achieved only through irreversible rotation of the individual particle magnetizations.[16,17]

Curie Point. The Curie point is the temperature at which the thermal agitation energy overcomes the exchange forces. The spontaneous magnetization disappears and the material is rendered nonmagnetic. This process is reversible; when the temperature is lowered below the Curie point, the spontaneous magnetization returns and the material is again magnetic. Figure 15-14 shows the effect of temperature on the permeability of a typical ferrite.

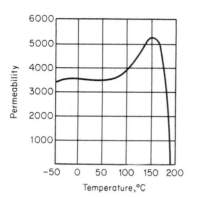

FIG. 15-14 Effect of temperature on permeability of typical ferrite.

Magnetic Induction. When a current I is connected to a solenoid coil of N turns, a magnetic field H is created which has direction as well as strength, and is defined by

$$H = \frac{0.4\pi NI}{l} \tag{15-8}$$

where H is in oersteds, l is the length of the solenoid in centimeters, and I is in amperes. As a result of the field H, flux lines are produced in the surrounding space (see Fig. 15-15). The flux lines form closed loops which flow from one end of the solenoid coil, into the air, and reenter the coil at the opposite end. The measure of the intensity or the concentration of the flux lines per unit area is called the *flux density,* or the *induction B.*

Figure 15-15a shows that with no magnetic material present in the solenoid coil the flux density B is relatively low and is equal to the applied field H. When a piece of magnetic material is placed in the solenoid coil, the flux density is increased (see Fig. 15-15b). This results from the magnetic moments[13,18] of the electron spins aligning themselves with the applied field H, causing the magnetic material to become a magnet. The sum of the magnetic moments per unit volume is the magnetization M. The magnetization of a material creates magnetic fields. Inside the material these fields are called *demagnetization fields* because they oppose the magnetization. Outside the material, they are called *stray* or *fringing fields*. The net field acting on the material is the vectorial sum of the demagnetization field and the applied field. The flux density is the net field plus the magnetization M, that is

$$B = H + 4\pi M \tag{15-9}$$

where H = net field and M and B are in gauss

Initial Magnetization B-H Curve. The relationship of the induced flux density B and the net field H of soft magnetic material is typically described by the initial B-H magnetization curves and the B-H hysteresis loop.

Figure 15-16a shows the initial magnetization curve of a typical soft magnetic material. This curve is obtained by starting with a toroid ring in the demagnetized state and plotting the flux density B against the field H. The demagnetization field in a toroid ring is zero; the net field is therefore equal to the applied field. The slope of the initial magnetization curve is the permeability μ, defined by

$$\mu = \frac{B}{H} \tag{15-10}$$

In CGS units the permeability is a dimensionless ratio and represents the increase in flux density relative to air caused by the presence of the magnetic material. The perme-

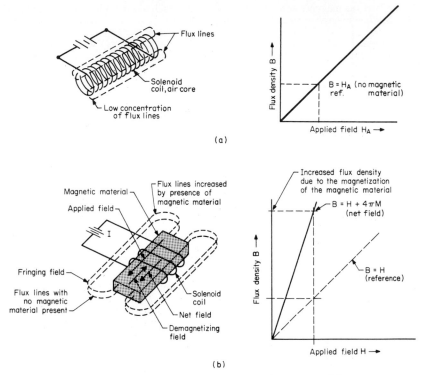

FIG. 15-15   Increase in flux with (*a*) air core and (*b*) magnetic core material.

ability can also be defined in terms of the magnetization M as

$$\mu = 1 + \frac{4\pi M}{H} \tag{15-11}$$

Starting at the origin, the curve has a finite slope which is the initial permeability. As the field H is increased, the slope becomes steeper. This is the maximum *permeability region.* The value of the maximum permeability is determined with a straight line of the steepest slope that passes through the origin and also contacts the magnetization curve. Finally, as H is further increased, a point is reached on the initial B-H curve where the magnetization approaches a finite limit indicated by the dotted line. At this point the magnetization of the material does not increase with further increases in the field. This is the saturation flux density B_s which is equal to the spontaneous magnetization of the magnetic material. After the material has reached saturation, the slope of the B-H curve changes and the flux density B continues to rise indefinitely at the rate of B equals to H_A as if the magnetic material were not present. Figure 15-16*b* shows a plot of the permeability as a function of the field.

Hysteresis Loop. If the H field is decreased after the initial magnetization curve reaches the saturated state, it is found that the induction does not follow the same initial curve back to the origin but traces a curve called the *hysteresis loop* which is shown by Fig. 15-17. As the magnetization is gradually decreased from the saturation point C, it follows along the lines CD and reaches a finite value B_r the *remanence,* which is the flux density remaining after removal of the applied field. In order to reduce the remanence to zero, a negative field, the coercive force H_c, must be applied. The curve from D to E

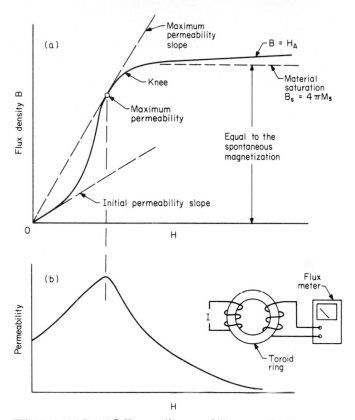

FIG. 15-16 (a) Initial B-H curve; (b) permeability versus H field.

is the demagnetization curve. As H is further increased in the negative direction, the magnetization will proceed from E to F, and the material will eventually become saturated in the opposite direction. If at this point the field is again reversed to the positive direction, the magnetization will trace the line F, G, C and the hysteresis loop is completed.

Hysteresis Losses.   The area of the hysteresis loop is the energy necessary to magnetize a magnetic substance. This energy is expended as heat. The loop area is a measure of the heat energy expended per cycle, per unit volume, and is called the *hysteresis loss*

$$W_h = \frac{A}{4\pi} \qquad \text{ergs/(cm}^3 \cdot \text{cycle)} \qquad (15\text{-}12)$$

A practical expression for power loss P in watts is given by

$$P = \frac{fal}{4\pi} \times A \times 10^{-7} \qquad (15\text{-}13)$$

where A = area of the loop, gauss-oersteds
    f = frequency, Hz
    a = cross-sectional area of core, cm^2
    l = magnetic path length, cm

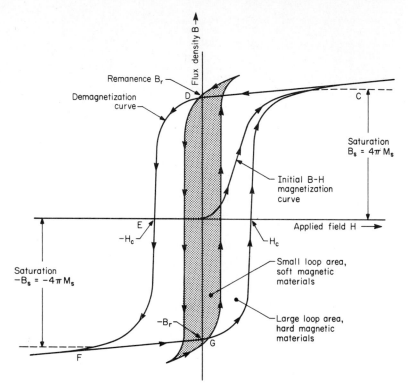

FIG. 15-17 *B-H* loops for hard and soft materials.

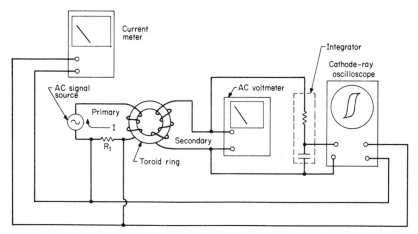

FIG. 15-18 Measurement of permeability and *B-H* loop for soft materials.

Figure 15-17 shows a comparison between the hysteresis loops for hard and soft magnetic materials. As indicated by the difference in the areas of the loops, more energy is required to magnetize the hard magnetic materials.

Permeability Measurements.   A simple method for measuring the low-frequency permeability (below 10,000 Hz) of soft magnetic materials used in magnetic head cores is illustrated by Fig. 15-18. The effect of core losses and the measurement of high-frequency permeability are discussed in Sec. 15.1.5 on head design.

Figure 15-18 shows a toroid ring with a thickness T and a mean length l wound with a primary and secondary coil. The ring typically is composed of a number of laminations of the same thickness that is to be used in the head. Because ferrite head cores are not laminated structures, the toroid ring typically is made from a solid piece of the material to be evaluated.

An alternating current I is applied to the primary coil while the secondary coil is connected to an ac voltmeter. The ac field thus created produces a time changing flux in the secondary coil, which in turn produces a voltage that is proportional to the rate of change of the flux. By means of an integrating circuit, a hysteresis loop can also be produced on an oscilloscope screen.

If the ac field is varied in amplitude from a low to a high value, or vice versa, a family of minor hysteresis loops is traced as shown by Fig. 15-19. The peak value of the flux density and the ac field corresponds to a line through the tips of these minor loops which

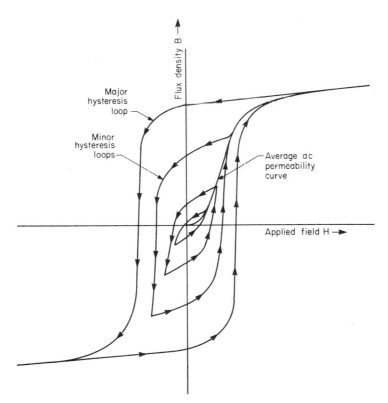

FIG. 15-19   Minor hysteresis loops.

closely approximates the initial magnetization curve. The slope of this curve is the ac permeability. The initial and maximum permeabilities are defined in the same manner as previously indicated.

The ac permeability may be determined by the following procedure.

1. Select the flux density that corresponds to the initial permeability on the magnetization curve, which for most magnetic head materials is approximately 40 G.
2. Calculate the peak ac output voltage E

$$E = 2\pi BANf \times 10^{-8} \tag{15-14}$$

where B = flux density, G
N = number of turns
f = frequency, Hz
A = cross-sectional area of toroid ring, cm^2

3. Adjust the current applied to the primary coil to give the voltage calculated in Step 2.
4. Measure the current applied in Step 3 and determine the H field

$$H = \frac{0.4\pi NI}{l} \tag{15-15}$$

where N = turns
I = peak current, A
l = mean length of toroid ring, cm

5. Now calculate μ as

$$\mu = \frac{B}{H} = \frac{40\text{ G}}{H}$$

Because the flux levels in magnetic heads are relatively low, the initial permeability is of prime interest. However, ac permeability at any flux density may be determined by the foregoing procedure.

Initial M-H Curve and M-H Hysteresis Loop. The initial *M-H* curve and *M-H* hysteresis loops are plots of the magnetization M versus the net field H and are typically used to describe the intrinsic properties of hard magnetic materials such as those used in recording media. An initial *M-H* curve is shown in Fig. 15-20. The slope of the *M-H* curve is the *susceptibility x* and is defined by

$$x = \frac{M}{H} \tag{15-16}$$

The permeability may be related to the susceptibility by

$$\mu = 1 + 4\pi\chi \tag{15-17}$$

When the saturation magnetization M_s is reached, the *M-H* curve approaches a finite limit and does not increase indefinitely as in the case of the *B-H* curve.

If, at the saturation point of the initial *M-H* curve, the applied field is made to follow the same sequence as previously outlined for the *B-H* loop, an *M-H* hysteresis loop will be traced (see Fig. 15-21).

The ratio of the remanent magnetization M_r to the saturation magnetization M_s is called the *squareness ratio* and is an important parameter in evaluating the magnetic orientation of the particles in magnetic tape. The squareness ratio is 1.0 for perfectly oriented particles. More practical values for oriented particles range from 0.7 to 0.9. Randomly oriented particles are approximately 0.5.

The vibrating sample magnetometer is used for measuring magnetic properties of recording media materials. A schematic representation of such a device is shown in Fig.

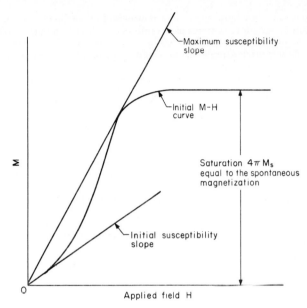

FIG. 15-20 Initial *M-H* curve.

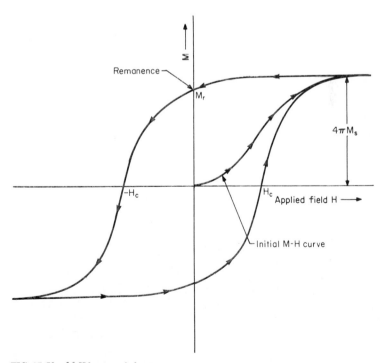

FIG. 15-21 *M-H* hysteresis loop.

15.25

15-22. The tape sample is vibrated in the magnetic field of the electromagnets, and the magnetic moment is measured in terms of the induced voltage in a pair of coils in which the exciting field from the electromagnets has been balanced out. In addition to the magnetic moment, the vibrating sample magnetometer can be used to measure such properties as coercivity and remanence.[16,19,20]

Hysteresis loops for magnetic media are usually obtained by an air core B-H meter as shown in Fig. 15-23, in which the field supplied by a solenoid coil is driven from a 60-Hz source. The tape sample is placed inside a small search coil connected in series with an identical coil to balance out the applied field. The signal from the search coil is amplified and integrated to obtain a voltage that is proportional to the flux from the tape sample.[16,19]

Demagnetization.   If a short bar of magnetic material is magnetized by an applied field H, poles are created at each end. These poles in turn create a magnetic field in the opposite direction to the applied field. This opposition field is called the *demagnetizations field* H_d (see Fig. 15-24). The net field H acting on the bar is

$$H = H_A - H_d \qquad (15\text{-}18)$$

The demagnetizing field H_d is dependent on the shape of the magnetic object and the magnetization M.[13,15]

The demagnetization field is zero in a ring core with no air gap. However, when an air gap is cut, creating poles at the gap-confronting surfaces, the resulting demagnetization field shears the hysteresis loop from the original position. This effect is shown by Fig. 15-25.

To bring a magnetic substance to a demagnetized state, a field that is equal to the

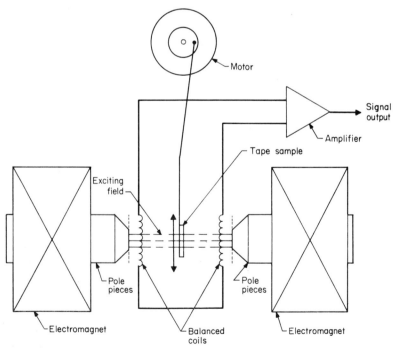

FIG. 15-22   Vibrating sample magnetometer.

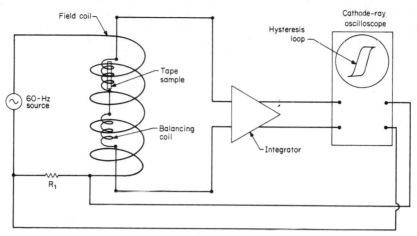

FIG. 15-23 Loop tracer.

coercive force H_c must be applied. However, upon removal of this field, the residual flux density will rise to a value B_1 as illustrated by Fig. 15-26. It is possible to reduce this residual flux density to zero by increasing the demagnetization field to a value greater than H_c and then decreasing it to zero as shown by the dashed lines. This technique requires a knowledge of the magnetic history of the material.

A more effective method to completely demagnetize a magnetic material is demagnetization by reversals. In this method the material is first saturated by an ac field, then cycled through a series of diminishing field reversals as shown by Fig. 15-27. The magnetic material will be left in a demagnetized state when zero field is reached regardless of its magnetic history. This technique is used to bulk erase magnetic tape and other recording

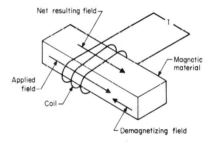

FIG. 15-24 Demagnetization field.

media, by exposing it to a strong ac field and then slowly removing the magnetic media from the field.

15.1.5 HEAD DESIGN.

The ring head was invented in 1935 by E. Schiller of Germany. In the intervening years, magnetic heads have become a highly developed technology and are used in a wide variety of applications for entertainment and data storage.

The heads on a magnetic tape recorder are the means by which the electrical signals are recorded on and reproduced from a magnetic tape. The record head is a transducer that changes the electrical energy from the signal system into a magnetic field that is emitted from a physical gap in the head. The field impresses a magnetic pattern on the tape proportional to the electrical signal. The reproduce head, on the other hand, is a transducer that collects the flux from the tape across a physical gap and changes it into an electrical signal that is proportional to the recorded flux.

Ring-type magnetic heads are composed of two highly permeable magnetic cores, with a nonmagnetic gap spacer and a coil to which the signal information is connected. Today most of the magnetic cores for video heads are made from either polycrystalline hot-pressed ferrite or single-crystal ferrite. However, for some video applications Alfesil

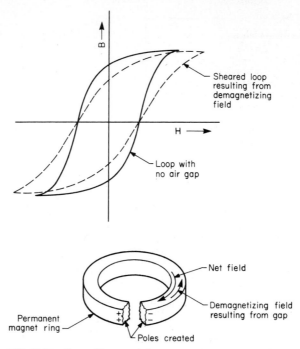

FIG. 15-25   Sheared hysteresis loop.

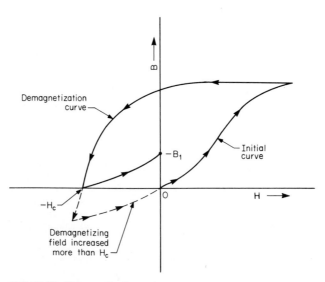

FIG. 15-26   Demagnetization curve.

(Senldust) is used, which is a hard-cast alloy composed of aluminum, iron, and silicon. Audio head cores typically use Permalloy-type materials.

The primary objective in magnetic head design is to achieve the highest possible head efficiency, which for a playback head is the ratio of the core flux to the total available tape flux. The record-head efficiency is the proportion of flux generated by the coil that fringes the gap and results in a useful field to magnetize the tape. Additionally, an important design requirement regarding the record head is that the gap fields do not exceed the saturation induction of the gap pole tips.

The critical design considerations that dictate the performance of magnetic heads are track width, gap length, gap depth, core geometry (e.g., path length), and the magnetic properties of the head core materials. Each of these parameters must be selected in accordance with the design criteria of the magnetic tape recorders, and at the same time, maintain the head efficiency as high as possible. Often such requirements as short-wavelength resolution, head life, and high record fields required to drive high-coercivity media conflict, and some design trade-offs are necessary.

The performance of magnetic heads depends heavily on maintaining accurate mechanical dimensions. To meet the short-wavelength demands of today's video recorder, the gap edges must be well defined and straight, and so the gap-confronting surfaces must be lapped to a high degree of flatness. The gaps in ferrite video heads are created by using sputtering techniques to deposit a thin film of glass or silicon dioxide

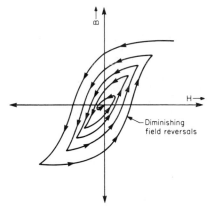

FIG. 15-27   Demagnetization by reversals.

(SiO_2) on the gap-confronting surfaces, and then they are glass-bonded to prevent edge erosion. The sides of small video tracks (less than 0.003 in, or 0.076 mm) are typically supported with glass or other means to provide adequate protection against the stresses involved when running in contact with tape at high velocities.

While the discussions in this section are directed primarily to video heads, the basic principles and design considerations presented apply to all ring-type heads.

Thin-film integrated heads are in increasing demand for computer disc applications. However, for video and audio recorders they are of limited interest and will not be addressed. The reader is referred to the many papers on the subject.

There are a wide variety of magnetic-head designs and engineering considerations. The information presented here is intended to provide an understanding of only the essential characteristics of ring-type magnetic heads.

Basic Structure.   The basic elements of a typical video head core are shown in Fig. 15-28. The core material is either a magnetic metal such as Alfesil (Sendust) or a ferrite. Ferrites are used in most present-day video heads because of their superior magnetic properties and long life. To create a nonmagnetic gap of controlled dimensions, the core is constructed in two half sections. To facilitate this construction, the gap spacer usually runs the entire length of the core. The gap-confronting surfaces are lapped to a high finish to ensure a sharp, well-defined edge. A winding aperture is cut either in one half or in each half to allow the signal coil to be wound on the core. The coil aperture is located directly beneath the front gap area. The front gap area is determined by the gap depth and the track width.

The rear gap area comprises the remainder of the core and is made much larger than the front gap. The tape contact surface is shaped to conform to the requirements of the scanner to provide intimate contact with the tape.

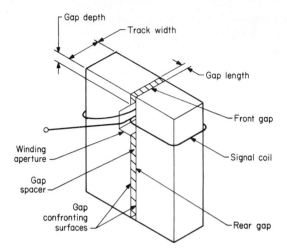

FIG. 15-28   Typical video head core.

Basic Record-Head Magnetic Circuits.   The effect of the head-core parameters and how they may be optimized to achieve maximum head efficiency can be more easily understood by considering simple electric circuits that are analogous to the magnetic circuits. Figure 15-29a is a plan view of the head core shown in Fig. 15-28, when connected as a record head. Figure 15-29b shows an electrical analog of the magnetic circuit, in which the gap and core reluctances, and the magnetomotive force (mmf) are represented by resistances and voltage (emf), respectively.

When a current I is applied to the coil of N turns, a magnetomotive force (mmf) is produced, with a magnitude proportional to the current and the number of turns (mmf in cgs units is $0.4\pi NI$). As a result of this magnetomotive force, a flux Φ flows in the magnetic core. This flux is analogous to a current in the electrical analog, and thus the relationship that exists among magnetomotive force, flux, and reluctance of magnetic

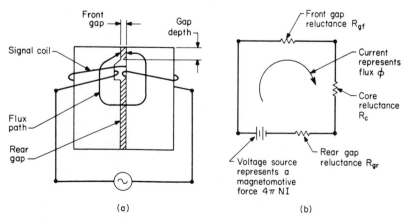

FIG. 15-29   (a) Plan view of record head; (b) electrical analog of record head.

circuits is analogous to voltage, current, and resistance in electric circuits. The basic equation for magnetic circuits can be written as

$$F = 0.4\pi NI = \varphi R \tag{15-19}$$

where φ = flux
R = reluctance
F = magnetomotive force

Reluctance R is the magnetic resistance and is related to the core geometry and permeability in the same manner as resistance is related to the physical dimensions and conductivity of a wire. Thus, to make the analogy, the wire resistance is given by

$$R = \frac{1}{\gamma} \frac{l}{A} \tag{15-20}$$

where l = length of wire
A = wire cross-sectional area
γ = conductance

Therefore, the corresponding magnetic resistance or reluctance is defined by

$$R = \frac{1}{\mu} \frac{l_c}{A_c} \tag{15-21}$$

where l_c = mean path length of magnetic core, cm
A_c = cross-sectional area of magnetic core, cm^2
μ = permeability of core material

The reluctance for magnetic head cores may be defined more specifically in terms of core and gap dimensions as

$$R_c = \frac{l_c}{\mu_c A_c} \tag{15-22}$$

where R_c = reluctance of core
l_c = mean magnetic path length
A_c = core-cross sectional area
μ = core permeability (as the result of core losses, the core reluctance is frequently dependent—core losses will be discussed in more detail later).

and

$$R_g = \frac{l_g}{\mu_0 A_g} \tag{15-23}$$

where R_g = reluctance of head gap
l_g = gap length
A_g = cross-sectional area of gap-confronting surfaces, cm^2
μ_0 = permeability of air, equal to 1 in cgs units

The cross-sectional areas of a magnetic head core are usually not constant. It therefore becomes necessary to calculate the reluctance of small segments of the core and sum the reluctance values of each segment. Thus, the total core reluctance becomes

$$R_{ct} = R_{c1} + R_{c2} + R_{c3} + \cdots + R_{cn} \tag{15-24}$$

where R_{ct} = the total core reluctance and $R_{c1} + \cdots + R_{cn}$ = the reluctance for each segment.

In reference again to the electrical analog (Fig. 15-29b), each reluctance (resistance) has a potential drop that is equal to the value of reluctance times the flux. By Kirchhoff's law, the potential drop across each series reluctance must equal the total mmf (voltage). The total mmf F_t is expressed thus

$$F_t = \Phi R_{gf} + \Phi R_{gr} + \Phi R_c \tag{15-25}$$

The total mmf may also be written as

$$0.4\pi NI = \Phi R_{gf} + \Phi R_{gr} + \Phi R_c$$

By rearranging the equation, the flux Φ may be defined as

$$\Phi = \frac{0.4\pi NI}{R_{gf} + R_{gr} + R_c} \qquad (15\text{-}26)$$

The condition for high head efficiency exists when most of the flux generated by the coil appears across the gap, or a large voltage drop across R_{gf} and a small drop across R_c and R_{gr}. The head efficiency ε therefore, is the ratio of the gap reluctance to the total core reluctance, which can be stated as follows

$$\varepsilon = \frac{\Phi R_{gf}}{F_t} = \frac{\Phi R_{gf}}{\Phi R_t} = \frac{R_{gf}}{R_{gf} + R_{gr} + R_c} \times 100\% \qquad (15\text{-}27)$$

Basic Reproduce-Head Magnetic Circuit. Figure 15-30a shows the electrical analog of a playback core, which is similar to the record head shown in Fig. 15-29. Now, however, the flux source Φ is the available flux from a magnetized tape, instead of a current flowing in the record coil. When the flux source Φ is connected across the gap front reluctance R_{gf}, it enters the core and splits into two paths. One path is across the front gap reluctance R_{gf} and the other is through each half of head core reluctances $R_c/2$ and the rear gap reluctance R_{gr}. The tape flux Φ that flows through the head core links the coil and thereby produces an output voltage. However, the tape flux that is shunted by the front gap reluctance R_{gf} does not link the coil, and no contribution to the output voltage is made. The available flux from the tape that actually links the coil is determined by the ratio of the gap reluctance R_{gf} to the total reluctance R_t. Therefore, the reproduce flux efficiency ε can be written as

$$\varepsilon = \frac{R_{gf}}{R_t} = \frac{R_{gf}}{R_{gf} + R_{gr} + R_c} \times 100\% \qquad (15\text{-}28)$$

SHUNT RELUCTANCE. The stray flux that fringes from the nonmagnetic front gap tends to enlarge the physical gap area. This effect can be seen in Fig. 15-30c and the electrical analog shown by Fig. 15-30b. The reluctance value R_g, as given by the physical dimensions Eq. (15-23), is shunted by additional reluctances, resulting from the stray flux, which are represented by R_{s1} and R_{s2}. R_{s1} is the combined shunt reluctance of the top and sides of the front gap region. R_{s2} is the shunt reluctance of the area beneath the front gap. The net effect of these shunt reluctances is to reduce the physical gap reluctance, which results in less potential drop across the front gap relative to the core, thereby reducing the head efficiency.

The shunt reluctances R_{s1} and R_{s2} may be estimated by the following approximations, which are derived from the work of Unger and Fritzsch.[21]

$$R_{s1} = \frac{\pi}{[2(\text{gD}) + \text{TW } l_n][\pi \, l_s/l_g]} \qquad (15\text{-}29)$$

$$R_{s2} = \frac{\varphi}{\text{TW } l_n[\varphi \, L/l_g]} \qquad (15\text{-}30)$$

where gD = gap depth
     TW = track width
     l_s = one-half of core length (see Fig. 15-31)
     l_g = gap length
     L = length shown in Fig. 15-31
     φ = total aperture angle, rad

All linear dimensions are in centimeters.

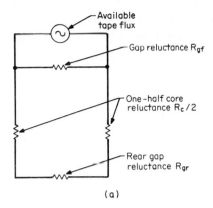

(a)

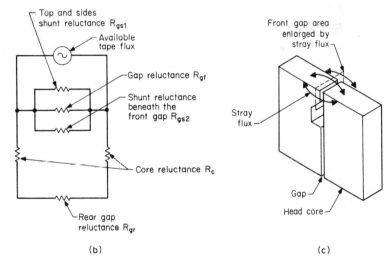

(b)   (c)

FIG. 15-30 (a) Electrical analog of playback core; (b) effect of shunt reluctance; (c) stray-flux effect.

The total shunt reluctance R_{st} is obtained by

$$R_{st} = \frac{R_{s1}R_{s2}}{R_{s1} + R_{s2}} \tag{15-31}$$

The effective front gap reluctance R_{ge} can be determined by the following

$$R_{ge} = \frac{R_g R_{st}}{R_g + R_{st}} \tag{15-32}$$

To minimize the shunting reluctance R_{s2} beneath the front gap, the aperture angle (Fig. 15-31) should be as large as possible. The optimum angle would be 90° relative to the gap line. However, if the aperture angle is made excessively large, a high stress point is produced and the pole tip area is subject to damage. This is especially true with ferrite

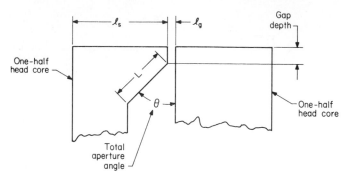

FIG. 15-31 Aperture angle.

video heads that operate at gap depths below 0.002 in (0.05 mm). Typical angles range from 45 to 60° relative to the gap line.

The shunt reluctance of the rear gap region has a minimal effect on the rear gap reluctance and may be neglected for most video head cores.

HIGH-FREQUENCY CORE LOSSES. The effect of core losses on video heads is of particular significance, for as the operating frequency of the head increases, the permeability of the core materials decreases relative to the low-frequency value, because of the effects of eddy currents and other residual core losses.

In low resistivity materials such as the magnetic metal alloys the reduction in permeability is primarily due to eddy currents. These losses force the flux to flow around the periphery of the core in the same manner as the high-frequency skin effects in electrical conductors. Thus, the effective flux conducting area is reduced and the core reluctance is increased. The effect on the core permeability at high frequencies is illustrated with the aid of Fig. 15-32 and the skin-depth equation for electrical conductors, which is

$$\delta = \frac{1}{2\pi} \sqrt{\frac{p}{\mu_{\mathrm{dc}} f}} \tag{15-33}$$

where δ = skin depth
p = resistivity
f = frequency
μ_{dc} = dc permeability

In reference to Fig. 15-32, the permeability is constant from dc up to a frequency where the skin depth is equal to magnetic core thickness T. Past this the permeability of the core is reduced by the skin effects at the rate of 3 dB per octave. The high-frequency core reluctance may be found by calculating the effective cross-sectional area at the frequency of interest and substituting the result for the low-frequency cross sectional area Ac in Eq. (15-22). However, this method is not very applicable to ferrite materials where the high-frequency losses are primarily due to the spin-relaxation effects,[19] and the eddy currents are significantly reduced owing to the relatively high resistivity of these materials. A more convenient way to quantify all losses for both metal and ferrite magnetic material is to treat the core permeability as a complex function which may be expressed as

$$\mu_f = \mu' - j\mu'' \tag{15-34}$$

where μ_f = value of permeability as result of core losses
μ' = real (magnetization) component
μ'' = imaginary (loss) component

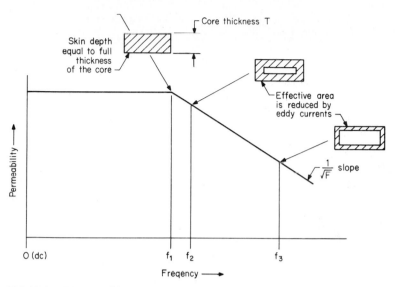

FIG. 15-32 Effect of eddy currents on permeability.

The core reluctance may now be defined in terms of the complex permeability. The basic reluctance relation

$$R_c = \frac{1}{\mu_c} \frac{l_c}{A_c} \tag{15-35}$$

becomes the frequency-dependent, complex relation given by

$$R_c = \left[\frac{1}{\mu' - j\mu''} \right] \frac{l_c}{A_c} \tag{15-36}$$

and the resulting real and imaginary components of the core reluctance can be shown as

$$R_c' = \frac{l_c \mu'}{A_c [(\mu')^2 + (\mu'')^2]} \tag{15-37}$$

$$R_c'' = \frac{l_c \mu''}{A_c [(\mu')^2 + (\mu'')^2]} \tag{15-38}$$

As previously stated, R_c' and R_c'' must be calculated for small segments of each cross section (A_c) and summed to obtain the total core reluctance R_{ct}' and R_{ct}''.

The gap reluctances R_{ge} and R_{gr} are not frequency dependent but purely resistive, and therefore are added as scalar quantities to the real part of the core reluctance. Thus, the total magnetic circuit reluctance, or magnetic impedance, at any specified frequency is

$$R_f = \sqrt{(R_{ge} + R_{gr} + R_{ct}')^2 + (R_{ct}'')^2} \tag{15-39}$$

where R_f = complex core reluctance at any frequency
  R_{ge} = effective front-gap reluctance
  R_{gr} = rear-gap reluctance
  R_{ct}' = total real component of core reluctance
  R_{ct}'' = total imaginary component of core reluctance

To obtain the head efficiency at any frequency, the complex core reluctance R_f can be substituted in the basic head efficiency Eq. (15-27); thus

$$\varepsilon = \frac{R_{ge}}{R_f} = \frac{R_{ge}}{\sqrt{(R_{ge} + R_{gr} + R_{ct}')^2 + (R_{ct}'')^2}} \times 100\% \qquad (15\text{-}40)$$

where ε is the head efficiency.

COMPLEX PERMEABILITY MEASUREMENT. To establish values for R_c' and R_c'', the values for μ' and μ'' of the complex permeability must first be determined. Some material manufacturers specify complex permeability; however, the design requirements for most video heads are so specialized that the complex permeability data must be obtained by the designer. This may be accomplished by winding a toroidal ring made frequently from the head core material (Fig. 15-33a) with N turns. The toroid is then treated as an impedance Z with a cross-sectional A_c and a mean length l_c, which can be represented by the equivalent circuit shown by Fig. 15-33b consisting of an inductance L_s, a loss resistance R_1 (in the dotted area), and the dc resistance of the coil R_{dc}. To minimize leakage losses, the size of the toroid dimension must be kept small (0.250-in OD, 0.125-in ID). A relatively large wire is used to reduce the dc resistance of the coil. The impedance Z and the phase angle can be measured with a vector impedance meter. The real and imaginary complex permeabilities can then be determined as

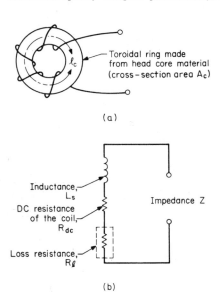

Toroidal ring made from head core material (cross-section area A_c)

(a)

Inductance, L_s

DC resistance of the coil, R_{dc}

Loss resistance, R_ℓ

Impedance Z

(b)

FIG. 15-33 (a) Top view of toroid; (b) equivalent circuit of toroid.

$$\mu' = \frac{\sin \varphi Z l_c}{\omega N^2 A_c} \times 10^8 \qquad (15\text{-}41)$$

$$\mu'' = \frac{(\cos \varphi Z - R_{\mathrm{dc}}) l_c}{\omega N^2 A c} \times 10^8 \qquad (15\text{-}42)$$

where $\omega = 2\pi f$, f is in Hertz, l_c is in centimeters, and A_c is in centimeters squared.

Plots of μ' and μ'' with respect to frequency for typical manganese zinc hot-pressed ferrite and Alfesil (Sendust) core materials are shown by Figs. 15-34 and 15-35, respectively.

Reproduce Heads. Head efficiency is of prime concern in a video reproduce head, since it determines the output voltage at all wavelengths. (The entire reproduce head function is calculated in Sec. 15.1.3.) Since video heads are not required to operate at wavelengths much longer than 0.003 in (0.076 mm), the contribution from flux outside the gap region is minimal, and therefore the long-wavelength contour effects (see Sec. 15.1.3) are of little consequence. Because the reproduce flux levels are very low, neither pole tip nor core saturation is of concern.

CORE DESIGN CONSIDERATIONS. To minimize the core reluctance, the magnetic path length l_c must be short and the cross-sectional area A_c large, compatible with obtaining the required number of turns on the coil (see Fig. 15-36). In video head cores a large percentage of the flux flows close to the periphery of the coil aperture, and therefore the path length l_c is largely determined by the coil aperture (window) dimension. The head core is typically constructed in two half sections, with the gap running the entire length of the core. To minimize the effect of the rear-gap reluctance, the confronting surface areas of the rear gap and front gap should be in the ratio of 10/1. Additionally, the core

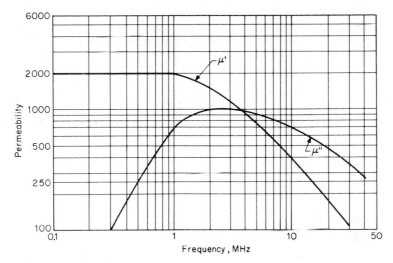

FIG. 15-34 Permeability versus frequency for hot-pressed ferrite.

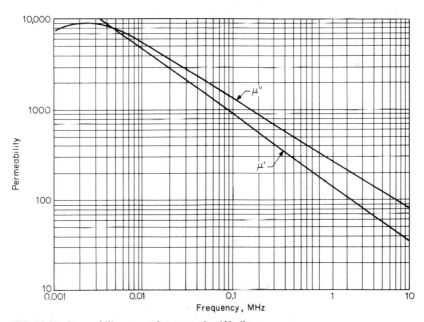

FIG. 15-35 Permeability versus frequency for Alfesil.

material should be selected for minimum losses (low μ'') and maximum permeability (high μ') for the required frequency range of the head.

GAP LENGTH.   In order to maintain the gap reluctance as high as possible, the effective gap length is usually determined by the maximum gap loss (see Sec. 15.1.3) that can

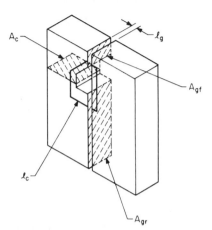

be tolerated by the recorder, at the shortest wavelength to be reproduced. The general rule for most applications is to select the effective gap length equal to one-half the shortest wavelength to be reproduced. Owing to mechanical imperfections in the gap-confronting surfaces, the effective gap is somewhat longer than the thickness of the physical spacer. The effective gap lengths for ferrite and for metal-core video heads are approximately 5 and 10 percent longer than the physical gap, respectively.

HEAD EFFICIENCY CALCULATION.   In the following example, the procedure for calculating the head efficiency and inductance of a typical video reproduce head shown in Fig. 15-36 is given.

FIG. 15-36   Physical parameters of video core.

1. For this example, the track width TW and gap length l_g for a type-C NTSC recorder are taken as 0.005 in (0.0127 cm) and 35 μin (89 $\times$ 10^{-6} cm), respectively.

2. A good estimate of the magnetic path length is approximately 0.010 in (0.0254 cm) larger than the periphery of the coil aperture, which for this example is equal to 0.318 cm.

3. The various cross-sectional areas for a typical video core are

$$A_c = 0.0016 \text{ cm}^2$$

$$A_{gf} = 0.000064 \text{ cm}^2$$

$$A_{gr} = 0.003 \text{ cm}^2$$

4. The front- and rear-gap reluctances R_{gf} and R_{gr} may be determined from Eq. (15-23)

$$R_{gf} = \frac{89 \times 10^{-6}}{0.000064} = 1.39$$

$$R_{gr} = \frac{89 \times 10^{-6}}{0.003} = 0.030$$

5. The front-gap shunt reluctances may be estimated by Eqs. (15-29) and (15-30). Substituting the appropriate core parameters, we have

$$R_{s1} = \frac{\pi}{[2(0.0051) + 0.0127l]N[\pi(0.1524)/(89 \times 10^{-6})]} = 15.97$$

$$R_{s2} = \frac{0.524}{0.0127lN[0.524(0.031)/(89 \times 10^{-6})]} = 8$$

6. Calculate the total shunt reluctance from Eq. (15-31)

$$R_{st} = \frac{(8)(15.97)}{8 + 15.97} = 5.28$$

7. The effective front-gap reluctance is obtained by using Eq. (15-32)

$$R_{ge} = \frac{(1.39)(5.28)}{1.39 + 5.28} = 1.1$$

8. The values for μ' and μ'' are selected at the operating frequency (9.5 MHz) from the curve shown by Fig. 15-41.

9. The real and imaginary parts of the core reluctance are calculated by using Eqs. (15-37) and (15-38)

$$R_c' = \frac{0.318 \times 400}{0.0016[(400)^2 + (700)^2]} = 0.122$$

$$R_c'' = \frac{0.318 \times 700}{0.0016[(400)^2 + (700)^2]} = 0.214$$

10. The magnetic impedance is calculated by Eq. (15-39)

$$R_f = \sqrt{(1.1 + 0.030 + 0.122)^2 + (0.214)^2} = 1.27$$

11. The head efficiency may now be determined from Eq. (15-40)

$$\varepsilon = \frac{1.1}{1.27} \, 100\% = 86.6\%$$

INDUCTANCE. As in the case of head efficiency, the head inductance is frequency dependent (core losses). The inductance L_f at any frequency may be obtained by substituting the magnetic impedance in the following expression

$$L_f = \frac{0.4\pi N^2 \times 10^{-8}}{R_f} = \frac{0.4\pi N^2 \times 10^{-8}}{\sqrt{(R_{ge} + R_{gr}' + R_{ct}')^2 + (R_{ct}'')^2}} \qquad (15\text{-}43)$$

where L_f = the inductance in H, and N = the number of turns. In the example considered

$$L_f = \frac{0.4\pi(6)^2 \times 10^{-8}}{1.27} = 3.56 \times 10^{-7} = 0.356 \; \mu H$$

GAP DEPTH. Assuming a video head core with a given track width (dependent on the tape format), in which the core reluctance has been reduced to the minimum possible, and the gap length established as above, the gap depth then becomes the controlling dimension with regard to efficiency and head life. Under these conditions the gap depth is usually established by the signal-to-noise requirements of the recorder. Because of the lower wear rate of ferrite core material, the gap depth can be reduced to relatively small dimensions and still retain a reasonably high head life. Gap depth on video heads ranges between 0.00075 in (0.02 mm) and 0.003 in (0.076 mm), depending on the application.

Record Heads. The basic design objective for a record head is that the gap field be less than the saturation induction of the head pole tip material and at the same time provide the necessary recording field above the gap to magnetize the tape at the required recording depth. Many video tape recorders use the same head for recording and reproduce. In such cases, a compromise must be struck between the optimum performance of each function. However, most professional recorders use separate heads, and both the record and reproduce functions may be optimized.

HEAD FIELD. Figure 15-37 shows a record head in which a signal current I is applied to a coil of N turns. As a result of this current, a gap field H_g is created across the head pole tips and is related to the gap length l_g, amp turns NI, and head efficiency ε by

$$H_g = \frac{0.4\pi NI}{l_g} \varepsilon \qquad (15\text{-}44)$$

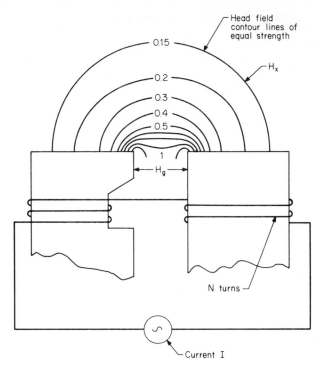

FIG. 15-37   Recording field.

where H_g is in oersteds, I is in amperes, and l_g is in centimeters. The gap field H_g produces a recording field H_x above the surface of the head. This field may be thought of as a bubble that emanates from the head gap to magnetize the tape. The intensity or size of the bubble relative to H_g is a function of the ratio of the gap length l_g to the recording depth y, and is expressed by the Karlquist equation for the maximum longitudinal field above the gap center line

$$H_x = \frac{H_g}{\pi} 2 \tan^{-1} \frac{l_g}{y} \tag{15-45}$$

Figure 15-38 shows a plot of the equation in which H_g is normalized to 1.

CRITICAL-RECORDING FIELD.   In order to magnetize the tape, the recording field H_x must be equal to or exceed a certain critical value, which is approximately equal to the coercivity H_c of the tape at the required recording depth y. Irreversible magnetization results when the tape passes through the *record zone*, the region within the field contour line $H_x = H_c$ (Fig. 15-39).

RECORDING DEPTH.   The recording depth is usually established at $\lambda/3$, within which 90 percent of the available reproduce flux occurs. Typical recording depths for video recorders range from 20 to 50 μin (0.51 to 1.27 μm).

POLE-TIP SATURATION.   Pole tip saturation occurs if the gap field approaches the saturation induction B_s of the pole tip material before the required recording depth is reached. The gap edges saturate first because of the high field concentration at these points. If the record current is increased in an attempt to reach the required recording depth, the entire pole tip may become saturated. As a result of this saturation, the permeability of the pole tips will be reduced, which tends to increase the record gap

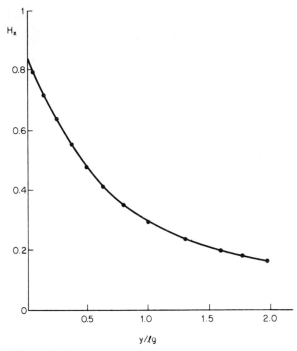

FIG. 15-38 Longitudinal field versus recording depth to gap length ratio.

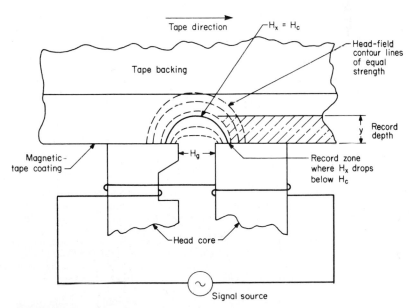

FIG. 15-39 Recording zone.

length. The increased gap length reduces the gap field H_g, producing a self-limiting effect on the recording field H_x and preventing the required recording depth y from being reached. These saturation effects lower the gap edge definition, reducing the head field gradient and resulting in poor short-wavelength resolution. The effects of pole tip saturation can be avoided if the gap field H_g is approximately one-half the saturation induction B_s of the pole tip material. [By definition, when $\mu = 1$ (cgs), as is the case with a nonmagnetic gap, then the induction B is equal to the gap field H_g.]

RECORD GAP CALCULATIONS. The procedure to determine the record gap length will be outlined by the following example.

1. The value for B_s for a typical manganese zinc ferrite is approximately 4500 G. Using the factor one-half, $H_g = 2250$.

2. The coercivity of a typical video tape is 650 Oe.

3. In this example the recording depth y is selected for a wavelength corresponding to the gray level carrier frequency (9.5 MHz) of a type C video recorder, which is where the record current is optimized. Thus, record depth is

$$y = \frac{v}{3f} = \frac{1000}{3(9.5 \times 10^6 \text{ Hz})} = 35 \times 10^{-6} \text{ in } (0.889 \ \mu\text{m}).$$

4. The record gap l_g can now be determined by solving Eq. (15-45) (Karlquist) for l_g. Thus

$$l_g = 2y \tan \left[\frac{\pi H_x}{2H_g} \right] \tag{15-46}$$

5. Substituting the parameter selected in steps 1 to 3

$$l_g = 2(35 \times 10^{-6}) \tan \left[\pi \frac{650}{2(2250)} \right] \approx 34 \times 10^{-6} = 34 \ \mu\text{in}$$

6. The amp turns to produce the gap field H_g can now be computed by rearranging Eq. (15-44)

$$NI = \frac{l_g H_g}{0.4\pi\varepsilon} \tag{15-47}$$

The head efficiency of the record head is determined in the same manner as the reproduce head. Thus, using the same core geometry as in Fig. 15-36 and the record gap length just calculated, the record efficiency is found to be 86 percent. Substituting the appropriate values in Eq. (15-47), the amp turns are

$$NI = \frac{2.54(34 \times 10^{-6})2250}{0.4\pi(0.86)} = 0.179 \text{ A} \tag{15-48}$$

The preceding calculations for both record and reproduce head parameters provide reasonable estimates, which are adequate for the initial design; final design is ultimately determined empirically.

RECORD GAP DEPTH. If for the record core the track width is determined by the format, the core reluctance minimized, and the gap length established as in the foregoing example, the gap depth again is the controlling dimension with respect to head efficiency and life.

OTHER CONSIDERATIONS. To optimize the recording resolution at short wavelengths, it is necessary that the recording-gap edges be sharp and straight. Additional improvement with regard to short-wavelength resolution may be obtained by reducing the record gap length and thereby improving the head field gradients. However, the gap length must not be reduced to the point that pole saturation occurs.

As the tape coercivity H_c is increased, the gap field H_g required to provide the critical recording field H_x must increase a proportionate amount. Thus, to avoid pole-tip saturation effects with high-coercivity tape, it becomes necessary to operate at reduced recording depths.

Video Head Construction

METAL VIDEO HEADS. It was generally recognized that ferrites, because of their superior electrical properties at high frequencies, were better suited to video applications than the metal core materials. However, high porosity and other mechanical characteristics made it impossible to fabricate ferrite video heads that did not show rapid gap erosion under the abrasive action of the tape.

It was not until the glass-bonding and high-density ferrite technology had been perfected in the late 1960s and early 1970s that ferrite video heads could be made to withstand the abrasive effects of tape contact. In the early days of video recording, the use of ferrites was limited to erase heads where gap resolution was of little importance.

The early video heads were confined to metal core materials, chiefly Alfenol and Sendust (Alfesil). Alfenol was being investigated by the Office of Naval Ordinance Laboratories, and was made available for commercial applications in thin-sheet form in the early 1950s. Alfenol is 16 percent aluminum and 84 percent iron, and its wear rate is approximately one-fourth that of Permalloy. Extreme brittleness and critical heat treating were its biggest drawbacks. Sendust (Alfesil) is the most widely used metal core material for video heads. Developed in Japan in the 1950s, its approximate composition is 10 percent silicon, 5 percent aluminum, and 85 percent iron. However, it was not available in the United States until the 1960s. It is a vacuum-melted alloy, approximately twice as hard as Alfenol, and much less critical to heat treatment. Both Alfenol and Sendust have relatively low resistivity, resulting in high eddy current losses and low head efficiency at video frequencies. However, novel designs to circumvent the losses of metal core materials have made their use possible for video head applications, some of which will be discussed briefly.

COMPOSITE HEADS. A novel head design (Fig. 15-40) that utilized the advantages of both metal and ferrite was proposed by Kornei at Brush Instruments in 1954. This was the *composite* head, so-named because of the use of two different materials to make one magnetic core.

Although many constructional and manufacturing refinements have been made in the intervening years, the basic design remains unchanged and is the backbone of modern wide-band fixed-gap heads. Figure 15-40 shows metal pole tips made from Alfenol, Sendust, or Permalloy, bonded to a ferrite core to form composite core halves. The core halves are joined with a nonmagnetic gap in the center to form the complete head core. The metal pole tip provides high gap definition, and the ferrite reduces the core losses at high frequencies, resulting in increased head efficiency. To minimize high-frequency losses, the amount of metal in the pole tips is reduced to the absolute minimum, compatible with sound structural integrity. In some designs, the pole tips are also laminated to further reduce the losses.

The construction of these heads requires that the mating surfaces between the metal and ferrite be lapped to a high finish and held in intimate contact to ensure a low magnetic reluctance.

One of the first video heads was a composite head designed by Pfost in 1955 and shown schematically in Fig. 15-41. The design consisted of two confronting Alfenol metal pole tips with a gap spacer between them and held in intimate contact with one side of a slotted ferrite ring core, with a coil

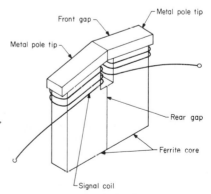

FIG. 15-40 Composite head.

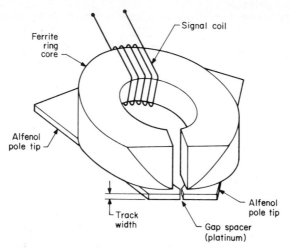

FIG. 15-41   Early composite video head.

connected to the signal source. The pole tips were oriented to bridge the slot in the ferrite. The slot provided a high-reluctance gap in the ferrite so that the flux path was completed through the Alfenol pole tips. The various components of the magnetic core were held in the required relationship to one another by high-precision mechanical parts which were assembled with screws and epoxy. A head of this type was used on the first quadruplex video tape recorder.

Another type of composite video head, developed by Pfost in 1959, is shown in Fig. 15-42. In this head Sendust was used as the core material and brazed together instead of being mechanically clamped. The advantage of this design over the preceding was a more rigid structure, thereby producing better-quality and longer-lasting gaps. This head was constructed by brazing two Alfesil core blocks approximately 0.3 in (0.762 cm) long, in which a winding aperture had been cut. The brazed block was sliced into the individual head cores, and a ferrite core member was placed on the Alfesil core to form a composite

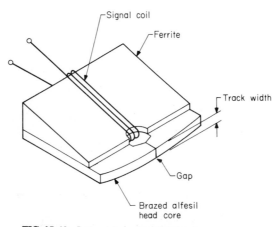

FIG. 15-42   Later composite video head.

head. A coil was wound around both the Alfesil and the ferrite cores, to complete the head. This type of structure was used in video recorders until the mid-1960s.

NONCOMPOSITE ALFESIL HEADS.   Chupity, in 1962, found that the effects of high-frequency losses could be substantially reduced by using a composite brazed Alfesil core, in which the coils were placed through a very small winding aperture near the front of the head. This design brought about a large reduction in the magnetic path length, which lowered the core reluctance and resulted in a much-improved head efficiency at high frequencies. Additionally, by eliminating the ferrite member, the construction was simplified, resulting in lower cost. The design is still widely used for quadruplex video recorders. Figure 15-43 shows a flow chart of the construction steps of a typical metal video head.

In order to facilitate batch fabrication, the head core is constructed from an Alfesil core block 0.3 in (0.762 cm) to 0.5 in (1.27 cm) long in which the required profile section has been cut by special grinding techniques. Carbide grinding wheels, along with very slow feed rates, have been found to give the best results. To obtain the gap definition that is necessary for optimum short-wavelength performance, the gap-confronting surfaces must be lapped flat to within 3 μin (0.076 μm). Flatness is typically measured by optical interference techniques, using monochromatic light and an optical flat. The lapping must be gentle and have low material removal rates to avoid damage to the gap-confronting surfaces. The effect of such damage is reduction of the permeability of the gap-confronting surfaces, in turn producing a gap length longer than the physical spacer.

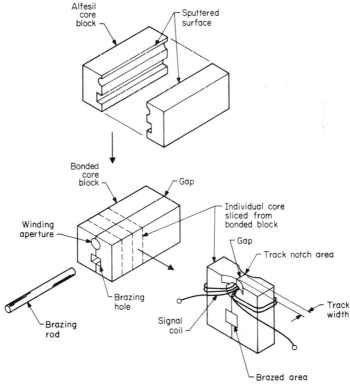

FIG. 15-43   Alfesil head construction.

After lapping, the gap-confronting surfaces are vacuum-deposited or sputtered with the required gap material, which is typically silicon dioxide (SiO_2), aluminum oxide (Al_2O_3), or chrome. The core blocks are then bonded by holding them in a fixture so that the gap-confronting surfaces are in intimate contact. Brazing rods coated with a flux are inserted into the brazing hole in the core blocks. The fixture and blocks are heated to the melting temperature of the brazing rod and bonded. With this technique, bonding takes place only within the brazing hole. The gap spacer is maintained by the mechanical clamping forces of the core blocks.

After the bonding operation, the bonded core blocks are sliced into the individual head cores, as shown. If the track widths are relatively large, approximately 0.008 in (0.020 cm) or over, the entire core area can be lapped to the track width dimensions while still maintaining sufficient mechanical strength. However, when the core thickness is reduced to below 0.008 in (0.02 cm), the strength, which is proportional to the cube root of the thickness, is substantially lowered. Moreover, the core reluctance increases and poor head efficiency results. To avoid these problems, the cores are notched to the required track width dimensions in the front only. The track width notch extends from the tape contact area to the aperture angle just beneath the front gap. This technique provides high mechanical strength and low reluctance.

HOT-PRESSED POLYCRYSTALLINE FERRITES. Ferrite materials have found wide use in video heads during the last decade, largely owing to development of the hot-pressing techniques in the late 1960s. This technology made it possible to produce high-density polycrystalline ferrite on a large scale. The properties of these magnetic ceramics that make them particularly desirable for video head work are hardness, high resistivity (low losses), high density, and relatively high permeability at video frequencies.

The two types of hot-pressed ferrite most commonly used for magnetic heads are referred to simply as manganese zinc or nickel zinc. Of these two types, manganese zinc ferrite is used more widely for video heads. This is primarily owing to its higher saturation induction, which makes it better suited for recording on high-coercivity tapes. Nickel zinc ferrites typically have higher resistivity and lower permeability. The permeability of nickel zinc is, however, more constant over a wider frequency range. Apart from the low-saturation induction, the lower permeability results in reduced head efficiency relative to heads made with manganese zinc materials. Hence, the nickel zinc ferrites find little use in video heads.

Hot-pressed ferrites are made from powders of either manganese and zinc (MnO + ZnO) or nickel and zinc (NiO + ZnO) mixed with an iron oxide (Fe_2O_3). After many intervening steps to process the powder, the final sintering (hot pressing) takes place in which high pressures (5000 lb/in^2) and heat (approximately 1300°C) are applied, either in one direction or isostatically.

Very small amounts (5 ppm) of oxygen can have a drastic effect on the magnetic properties of manganese zinc ferrites, and therefore the sintering must be carried out in an oxygen-free atmosphere. On the other hand, the magnetic properties of nickel zinc materials are not sensitive to oxygen and may be sintered in air. Temperature, pressure, and time control the final grain size. Typical grain sizes range from 0.0005 in (0.0127 mm) to 0.010 in (0.254 mm).

SINGLE-CRYSTAL FERRITES. Owing to the lack of porosity, single-crystal ferrite materials offer some advantages in achieving gaps with highly defined edges. The drawback of these materials is that the magnetic and wear characteristics can vary as much as 2 to 1, depending on the crystalographic orientation. For consistent results, strict attention must be given to the manufacturing process to ensure that the same crystalographic orientation is maintained from head to head. Single-crystal video heads have found wide use in consumer video tape recorders in recent years.

Table 15-4 shows the physical and magnetic properties of some typical materials used for video heads.

Ferrite Video Head Fabrication. Present-day ferrite material allows highly defined gap edges to be produced. However, owing to the brittle nature of ferrites, the gap edges, if not adequately supported by the gap spacer material, will rapidly erode and often structurally fail when exposed to the abrasive action of the tape. A gap spacer placed in

Table 15-4 Properties of Typical Core Materials

Material	Permeability μ	Coercivity H_c, Oe	Saturation induction, G	Resistivity ρ	Curie temp., °C	Coef. of expan., 10^{-6} cm/ (cm·°C)
Alfesil	See Fig. 15-35	0.06	10,000	90 $\mu\Omega$·cm	500	18
Hot-pressed ferrite Mn Zn	See Fig. 15-34	0.02–0.15	4000–6000	Approx. 10^5 $\mu\Omega$·cm	90–300	9.5–11.5
Single-crystal ferrite	Approx. 300 500, at 5 MHz	0.05	4000–5000	Approx. 10^5 $\mu\Omega$·cm	140–250	9–11

physical contact with the gap-confronting surfaces, as in the Alfesil head, does not provide the ferrite material with the needed structural support and protection against gap erosion. The problem of gap edge erosion was solved by the development of glass-bonding techniques. In this process a vitreous glass that approximately matches the physical properties of the ferrite materials is fused to the gap-confronting surfaces, forming a monolithic structure that resists the erosive forces of the tape. In addition to providing protection for the gap edges, the glass also acts as a precision nonmagnetic spacer that controls the gap length dimension. Glass is used because it is inherently compatible with the ferrite; moreover, good adhesion, durability, and wear properties that match the ferrite can be obtained. Most of the glasses used in ferrite heads are proprietary compositions that have been developed for optimum compatibility with a given ferrite material.

There are many ferrite-head fabrication techniques in use today, mostly proprietary. The following descriptions of the fabrication of ferrite video heads are intended only to give an understanding of the processes involved.

WIDE-TRACK HEADS. Two ferrite core blocks, shown in Fig. 15-44, are lapped to a flatness of 3 μin. In lapping ferrite material, the same precautions should be taken as for metal heads, to avoid reducing the permeability of the gap surfaces. Owing to the hardness of ferrites, diamond lapping techniques are usually employed.

In the next step, the gap-confronting surfaces are *sputtered* with a thin film of glass that is equal in thickness to one-half of the desired gap length dimension. *Sputtering* is a *momentum transfer deposition process,* in which an electrical discharge is set up between two plates in the presence of a low-pressure inert gas, such as argon. The ionized gas atoms are accelerated by the high electric field of the glass target. These ions strike the target and release their kinetic energy, knocking off glass atoms, which in turn deposit themselves on the ferrite core block below. The thickness of the sputtered glass is controlled by the time, the partial pressure of the argon, and the power of the electric field.

The sputtered blocks are clamped in a fixture so that the sputtered surfaces face each other. The temperature is then elevated in an oxygen-free atmosphere to a point where the glass molecules migrate together and the core blocks are thus fused along the entire gap line. To complete the head, the bonded block is cut into separate cores, and the track width is notched. A signal coil is then wound through the winding aperture. The core is mounted on a support means and contoured to the final gap depth in the *head scanner* (see Sec. 15.6).

Differential wearing between the gap glass and the ferrite occurs if the wear rates of both materials are not well matched. As a result of differential wear, the gap edges will become unsupported and erosion will take place, thus degrading the short-wavelength resolution.

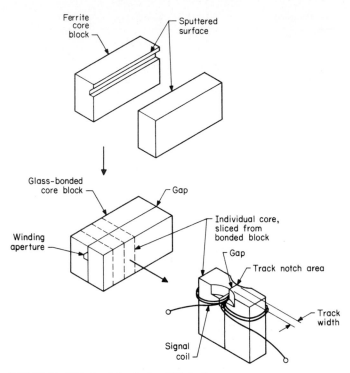

FIG. 15-44 Wide-track ferrite head fabrication.

An alternative method of glass bonding is illustrated by Fig. 15-45, which shows a pair of ferrite core blocks ready to be bonded. The gap-confronting surfaces are separated so that a free air space is created between them, which has the same dimensions as the desired gap length. Many techniques have been employed to create the air space to the precision required. Examples are vacuum-depositing, or sputtering, small spacer strips on the gap-confronting surface; precision metal foils; and cleaved mica shims. Glass in the form of rods, sheets, or powders is placed in close proximity to the air space. The blocks and glass are then raised in temperature until the viscosity of the glass is low enough to be drawn into the air space by capillary action. Adhesion takes place between the gap-confronting surface and the glass, bonding the core blocks. A drawback to this process is that at small gap lengths [smaller than 50 μin (1.27 μm)] the glass does not always fill the air space and *inclusions* result, giving rise to poor manufacturing yields. This technique has found wide use in the fabrication of computer disc heads, where the gap lengths are usually in excess of 50 μin (1.27 μm).

NARROW-TRACK HEADS. The track-notching technique described previously is adequate to produce relatively wide-track video heads. However, when the tracks are reduced to 0.003 in (0.076 mm) or smaller, the notched area becomes very fragile and prone to breakage. To overcome this problem, the notched areas are filled with glass to support the track width on each side. The support glass is fused to the ferrite, greatly increasing the structural integrity of the track notch area. Additionally, the glass pockets isolate the track and reduce the edge clipping. Using the technique allows the manufacture of extremely small tracks [below 0.001 in (0.025 mm)]. Most home video recorders use heads of this type. An example of the basic fabrication steps of a glass-notched video head is illustrated by Fig. 15-46.

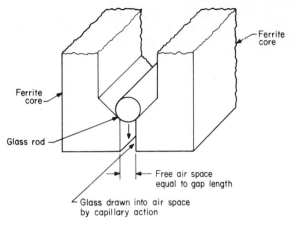

FIG. 15-45 Capillary bonding technique for ferrite heads.

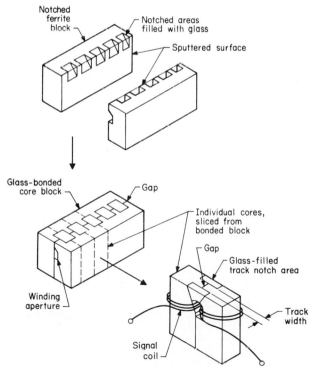

FIG. 15-46 Narrow-track ferrite head fabrication.

The track width is defined by the land area between a series of small notches that are cut along the length of the ferrite core block. Next, the notches are filled with glass, and the gap-confronting surfaces of the core block are lapped and sputtered with glass. The individual head cores are produced by cutting along the center of the glass-filled areas. The heads are completed in the manner outlined earlier.

15.2 MAGNETIC TAPE

by Robert H. Perry

Magnetic tape includes a multiplicity of products used for magnetic recording, all consisting of a magnetizable medium on a flexible substrate. Because of the great variety of machine types and recording formats in use and being developed, magnetic tape is designed and produced with widely different magnetic media, widths, thicknesses, lengths, and other properties optimized for each application. Media are used in either strip form in reels, cartridges, cassettes, and cards or in discs of different diameters. Similar technologies are used to produce all these products. The industry is a high-technology one, and the highest standards of quality and precision are maintained in the selection of raw materials, in the product design, and ultimately in the manufacture, testing, and packaging of the products. This chapter is a brief description of magnetic tape and how it is made.

15.2.1 CONSTRUCTION. Magnetic tape consists of three components: a magnetic film or coating supported by a flexible substrate, or base film, which in many applications is coated on the back by a nonmagnetic coating (Fig. 15-47).

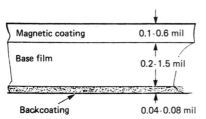

FIG. 15-47 Magnetic-tape base-film and coating thickness.

Backcoatings are used primarily in the most demanding tapes, such as professional and some consumer video, professional audio, instrumentation, and data products, where special winding and handling characteristics are required. This coating contains a conductive pigment, usually carbon black, which reduces buildup of static charge and therefore minimizes the accumulation of dirt and debris on the tape, factors which can cause drop-outs, or loss of signal with attendant loss of stored information. The backcoating also provides better frictional characteristics than raw base film does, and air is more easily eliminated from adjacent layers during winding. This reduces the tendency of the tape to cinch or form *pop strands,* and there is less likelihood of uneven stacking, edge damage, and creasing of the tape.

The *base film* is an integral and significant part of the whole tape system and is largely responsible for its mechanical strength and stability. Other factors such as stiffness and surface smoothness have a profound influence on tape performance in many applications, and base films having the proper characteristics for a given application must be carefully selected.

The principal substance used in the great preponderance of magnetic tapes today is poly(ethyleneterephthalate), or simply *polyester,* abbreviated PET. PET has an excellent combination of properties including chemical stability and mechanical properties, such as tensile strength, elongation and modulus, tear resistance, availability, and cost. Some typical properties are shown in Table 15-5.

Many different types and grades of PET are on the market for both magnetic tape-related and unrelated applications. In all cases mechanical strength in the plastic film is achieved during its manufacture by a process of biaxial, and sometimes uniaxial, orientation of polymer chains in the hot film after extrusion of the melt. Biaxial orientation

Table 15-5 Physical Properties of Poly(ethyleneterephthalate) Base†

Property	Balanced	Tensilized
Tensile strength, lb/in²	25,000	40,000
N/m²	172.38×10^6	275.8×10^6
Force to elongate 5%, lb/in²	14,000	22,000
N/m²	96.53×10^6	151.69×10^6
Elastic modulus, lb/in²	550,000	1,100,000
N/m²	3.79×10^9	7.58×10^9
Elongation, %	130	40
Thermal coefficient of linear expansion, per °C	1.7×10^{-5}	1.7×10^{-5}
Shrinkage at 100°C% (per 30 min interval)	0.4	2.5

†Measurements in machine direction.

is achieved by stretching in both the machine and the transverse directions, and the resulting film has a balance of properties in the two directions. Balanced film is adequate for many magnetic tape applications, especially those employing gauge thicknesses greater than 0.5 mil (0.0127 mm). In thinner gauges greater resistance to stretching is needed, and PET is used which is *tensilized,* i.e., oriented by drawing additionally in the machine direction.

Base films for magnetic tape range in thickness from about 0.2 to 1.5 mils (3 mils for flexible disks). They are employed by tape manufacturers in widths ranging from 12 to 60 in (0.3 to 1.5 m) and in lengths up to 15,000 ft (4572 m). The base-film manufacturer must ensure that the base film has the right balance of surface smoothness for recording performance and roughness for runability in the coating and processing steps. Small-particle-size, inorganic additives are incorporated in the PET to provide slip properties in film that would be otherwise unmanageable. These surface asperities must be critically controlled, especially for short-wavelength recording applications, since the base-film surface-roughness profile can be carried to a degree through the magnetic coating and reflected in the tape-surface roughness. An asperity of 10 μin, for example, in a typical 100-μin-wavelength video recording can result in a loss of signal of 5.5 dB due to head-to-tape separation, as seen from the Wallace formula

$$\text{Spacing loss (dB)} = 54.6\ d/\lambda$$

where d = the head-tape spacing and λ = the wavelength.

15.2.2 MAGNETIC COATING. There are two types of magnetic coatings used in magnetic tape. Most of them use magnetic particles bound in a matrix of organic, polymeric binder that is applied to the substrate from a dispersion in solvents. To a limited extent, other types are made by vapor deposition of thin films of metal alloys, and this is an area in which much development work is in progress.

Most magnetic coatings contain a single layer, although a few tapes are made with dual-layer magnetic coatings having different coercivities and are designed to have flat response over a range of frequencies. Magnetic-tape performance is a function of both the formulation of ingredients in the coating and the process by which the coating is applied and processed. The most important component in the formulation is the magnetic material.

Magnetic Materials. A wide variety of single-domain magnetic particles is used having different properties depending on the electrical requirements of each tape application. Retentivities range from about 1000 to 3000 G, and coercivities range from about 300 to 1500 Oe. Size and shape are important, since they relate to how well the particles pack in the coating; the signal-to-noise ratio achievable is proportional to the number of particles per unit volume in the coating. The length of the particles is about 8 to 40 μin (0.2 to 1.0 μm), and they are acicular with aspect ratios of 5/1 to 10/1. Acicularity makes the particles magnetically anisotropic, and thus it governs magnetic properties not inherent in the material. In general, magnetic pigments are loaded to as high a level as possible commensurate with retention of desirable physical properties and avoidance of shedding. The limiting factor is the amount of pigment the binder can retain without loss of cohesion and, hence, durability.

There are four types of magnetic particles used in magnetic tape: γ-ferric oxide, doped iron oxides, chromium dioxide, and metallic particles that usually consist of elemental iron, cobalt, and/or nickel. γ-Ferric oxide has been by far the most widely used material (Hc 300 to 360 Oe) and is useful for many of the lower-energy applications in which the ultimate in recording density or short-wavelength recording capability is not required. The sequence of steps used in the commercial production of γ-ferric oxide is as follows: precipitation of seeds of α-FeOOH (goethite) from solutions of scrap iron dissolved in sulfuric acid, or from copperas (ferrous sulfate obtained as a by-product from titanium dioxide manufacture); growth of more goethite on the seeds; dehydration to α-Fe_2O_3 (hematite); reduction to Fe_3O_4 (magnetite); and oxidation to γ-ferric oxide (maghemite). An improved γ-ferric oxide is produced starting with ferrous chloride rather than ferrous sulfate and precipitating γ-FeOOH (lepidocrocite) rather than α-FeOOH in the initial step.

Cobalt doping of iron oxide affords particles with higher coercivities (500 to 1200 Oe). The older process involves precipitation of cobaltous salts with alkali in the presence of yellow iron oxide (α-FeOOH), dehydration, reduction to cobalt-doped magnetite, and oxidation to cobalt-doped magnetite containing varying amounts of FeO. The resulting particles have cobalt ions within the lattice of the oxide, and they exhibit a marked magnetocrystalline anisotropy. This gives rise both to a strong temperature dependence of the coercivity and to magnetostrictive effects, which can cause problems of greatly increased printthrough, increased noise, and loss of output from stress on the tape through head contact. Somewhat improved stability has been achieved by using other additives, such as zinc, manganese, or nickel, with cobalt.

In recent years, epitaxial cobalt-doped particles have been developed which have largely overcome these problems because cobalt ion adsorption is limited to the surface of the oxide. Epitaxial particles have superseded lattice-doped particles in most applications.

Chromium dioxide provides a range of coercivities similar to that of cobalt-doped iron oxide (450 to 650 Oe) and possesses a slightly higher saturation magnetization, that is, 80 to 85 emu/g compared with 70 to 75 emu/g for γ-ferric oxide. It has uniformly good shape and high acicularity and lacks voids and dendrites, factors which undoubtedly account for the excellent rheological properties of coating mixes made with it. Its low Curie temperature (128°C) has been exploited in thermal contact duplication, a process which was largely developed in the late 1960s but because of problems in obtaining high-quality duplicates has not been commercialized. New machine designs have resolved the earlier problems, and commercial, thermal contact duplication of chromium dioxide tape at 50 to 100 times real time appears to be on the horizon.

A disadvantage of chromium dioxide is its abrasiveness, which can cause excessive head wear. Also, it is chemically less stable than iron oxide, and under conditions of high temperature and humidity it can degrade to nonmagnetic chromium compounds, resulting in loss of output of the tape. Chromium dioxide and cobalt-doped iron oxide yield tapes having 5 to 7 dB higher S/N ratio than those made from γ-ferric oxide.

The presence of metallic particles results in tapes which have a 10 to 12-dB higher S/N ratio than those made from γ-ferric oxides because of their having much higher saturation magnetization (150 to 200 emu/g), retentivity (2000 to 3000 G), coercivity (1000 to 1500 Oe), and smaller particle size. These factors, together with a square shape of the hysteresis loop, permit recording at shorter wavelengths with less self-erasure. Thus, recordings can be made at slower speeds without sacrifice in dynamic range, and higher bit-packing densities are possible.

Metallic particles are made by several different kinds of processes, the more important commercial ones being reduction of iron oxide with hydrogen, and chemical reduction of aqueous ferrous salt solutions with borohydrides. Metallic particles are more difficult to disperse than iron oxides because of their smaller size and higher remanence, and they are highly reactive. Processes such as partial oxidation of the surface or treatment with chromium compounds are used in their preparation to stabilize them for handling during tape manufacture. The corresponding tapes are more stable, but their susceptibility to corrosion at an elevated temperature and humidity is a disadvantage.

Magnetic-tape manufacturers are also developing and beginning to introduce tapes consisting of thin films (100 to 150 nm) of metal alloys deposited on the substrate under vacuum or by sputtering. The retentivity of these tapes (1.2×10^4 G) is almost an order of magnitude higher than that of γ-ferric oxides, with a corresponding increase in recording density.

In other areas, research is being devoted to very small, isotropic particles, which have aspect ratios of 1/1 to 2/1, because of advantages which can be taken of magnetization vectors in more than one direction, i.e., vertical as well as longitudinal recording, and because of the increased number of particles that can be packed in a coating per unit volume. New particles having the shape of rice grains are used in some new tapes recently introduced and will undoubtedly find increased application in the next few years.

Binders.   Binders must be capable of holding the magnetic pigment together in a flexible film which adheres to the base film with a high degree of toughness and chemical stability, and with thermoplastic properties enabling the pigmented film to be compacted to give smooth surfaces. It should also be soluble in suitable solvents. These requirements are not met by many substances available today for producing current state-of-the-art magnetic tapes.

Polyurethanes, either used as such or prepared in situ, represent the most important class of polymers for this purpose because of their affinity for pigments, their toughness and abrasion resistance, and their availability in soluble forms. Of the two types in use, poly(esterurethanes) are preferred over poly(etherurethanes) because of their superior mechanical properties in tape. Some physical properties of a typical poly(esterurethane) are shown in Table 15-6.

Other polymers may be used alone or in combination with one or two other polymers to obtain the desired properties. Although a great many types are claimed in the patent literature, the other most important polymers include poly(vinyl chloride-co-vinyl acetate/vinyl alcohol), poly(vinylidene dichloride-co-acrylonitrile), polyesters, cellulose nitrate, and phenoxy resin.

Most modern magnetic-tape coatings are cross-linked with isocyanates to provide durability. Isocyanate-curing chemistry is rather complex and difficult to control, and for this reason the industry is researching an emerging new technology involving curing with electron-beam radiation. A whole new field of binders is being developed for this purpose which polymerize extremely rapidly to high polymers in a much more controllable fashion.

Dispersants are surface-active agents which aid in the separation of magnetic particles, a process necessary for achieving the desired electrical performance of the tape (see

Table 15-6 Physical Properties of a Poly(esterurethane)

Tensile strength, lb/in^2	8000
N/m^2	55.16×10^6
Stress at 100% elongation, lb/in^2	300
N/m^2	2.07×10^6
Ultimate elongation, %	450
Glass transition temperature, °C	12
Hardness, shore A	76
Density, g/cm^3	1.17
Viscosity at 15% solids/tetrahydrofuron, cP	800
Pa·s	0.8

Dispersion, below). They facilitate separation of charges on the particles and stabilize particle separation. Common dispersants are lecithin, organic esters of phosphric acid, quaternary ammonium compounds, fatty acids, and sulfosuccinates.

Conductive materials are often added to tape formulations to reduce electrostatic charge buildup on tape as it is run on machines. Conductive carbon blacks are commonly used to reduce the resistivity of tape by about four to six orders of magnitude, from 10^{10} Ω/square or higher.

Lubricants are necessary to prevent stiction of the tape as it comes in contact with the record or playback head. A great many different materials are effective as lubricants, including silicones; fatty acids, esters, and amides; hydrocarbon oils; triglycerides; perfluoroalkyl polyethers; and related materials, often from natural products. Lubricants may be either incorporated in the tape coating formulation or added topically at the end of the tape process.

Miscellaneous Additives. Small amounts of other materials are included in many tape products to achieve special properties. For example, fine-particle alumina, chromia, or silica is often added to prevent debris obtained during use of the tape from accumulating on the heads and clogging them. This is not normally a requirement in tapes containing chromium dioxide as a magnetic pigment. Other additives include fungicides, which are used in certain limited applications.

Solvent choice is determined by chemical inertness, binder solubility and mix rheology, evaporation rate, availability, toxicity, ease of recovery, and cost. The most commonly used solvents for magnetic tape processes are tetrahydrofuran, methyl ethyl ketone, cyclohexanone, methyl isobutyl ketone, and toluene. Many common types of coating defects can be avoided by the combinations of solvents to provide differential evaporation rates from the coating during the drying process. Finished tape normally has very low levels of residual solvent.

15.2.3 MANUFACTURING PROCESS. The following sequence of steps is employed in manufacturing the tape: mix preparation; dispersion, or milling; coating; drying; surface finishing; slitting; rewind and/or assembly; testing; and packaging.

Dispersion. The magnetic particles must be deagglomerated without reducing the size of individual particles. This step is accomplished by agitating the combined ingredients as a wet mix in one of several types of mills, such as pebble, steel ball, sand, or Sweco, which produce high shear between agglomerates. Milling efficiency in a given system is controlled by mix solids content, viscosity, mix-to-media ratio, and temperature. The end point is reached when visual examination of a drawdown sample under magnification shows the absence of agglomerates or that it meets a predetermined standard of dispersion quality. Another method is to mill until a maximum in the derivative of the *B-H* loop is attained. Some commercial dispersion testers are available based on dc noise measurements.

Coating. The coater is perhaps the most critical processing step in the entire operation. There are trade-offs between advantages and disadvantages among the different types of coating methods used, principal among which are *reverse roll* (Fig. 15-48b) and *gravure* (Fig. 15-48c). *Reverse roll* is the most widely used, general-purpose method. *Gravure* is especially suited for very thin coatings (0.2 mil or less). *Knife coating* (Fig. 15-48a), one of the oldest methods, is gradually disappearing with the advent of thin coatings on thin films and high-speed, precision coatings generally. *Extrusion* and *curtain* coating are becoming increasingly important by affording high-quality coatings at high speeds. Coaters vary in width from 12 to 60 in as do the base films, and operate at speeds of approximately 250 to 1000 ft/min.

Orientation. Maximum signal-to-noise ratio is obtained when the magnetic particles are aligned, before drying, to the maximum extent possible in the direction of the intended recording. Accordingly, immediately after the wet-coating mix is applied, the web is passed through the field of an orienting magnet having a field strength (500 to

2000 G) optimized for the particular magnetic particle being used. Most tapes are longitudinally oriented, although some are oriented transversely to some degree for quadruplex rotary-head video recorders, in which the recording is in the transverse direction. The coater itself exerts shearing forces on the mix and thus often imparts some longitudinal orientation in the particles even in this stage.

Drying. The web is next passed through an oven containing circulating forced hot air. Many modern oven designs use air bearings at web-turnaround points to avoid rubbing between plastic and metal surfaces, and to minimize the formation of abrasion products, which can cause drop-outs. Once the coating is dried, the magnetic particles are no longer free to move. During the eventual recording process, only magnetization vectors, or aligned spins of electrons within the molecular species of the particle domains, rotate.

Surface Finishing. Surface finishing is generally required to produce an extremely smooth surface to maximize head-to-tape contact, an absolute necessity for short-wavelength recording. This is accomplished by calendering the tape, or passing the web one or more times through a *nip,* or line of contact, between a highly polished metal roll and a plastic or cellulosic compliant roll. This compaction process also reduces voids in the coating and increases the magnetic pigment volume concentration and in turn the retentivity of the tape.

Slitting. The web is slit into strands of the desired width, from 150 mils to 3 in (3.8 mm to 7.6 cm). Tolerances in width variation are about ±1.0 mil for most tapes except consumer ½-in video, which must be slit to within ±0.4 mil. Edge weave, or width waviness *(country laning)* over extended length, should not vary more than about 1 to 2 mils in 1-in video tapes and 0.4 mil in video cassette tape. Tape must be free from jagged edges and debris. Additional tape cleaning processes are sometimes used to ensure that loosely held dropout contributors are effectively removed.

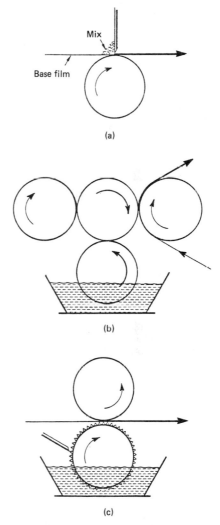

FIG. 15-48 (*a*) Knife coating; (*b*) reverse-roll coating; (*c*) gravure coating.

Testing. Sophisticated magnetic tape manufacturers test every component of tape in every step in the process, from individual raw materials through packaging. The most exacting specifications are set forth and followed. Electrical tests, including those for dropouts, are especially stringent, and in professional audio, instrumentation, video, and computer tapes, each reel of tape is tested, in some cases end to end, before shipment. In addition, warehouse audits are performed to ensure maintenance of quality.

Assembly and Packaging. Tape is assembled in various formats but mainly in reels, pancakes, cassettes, and cartridges of different sizes. A description of processes by which these components are made and assembled is beyond the scope of this chapter. However, the same standards of precision, cleanliness, and quality exist in these areas as in tape making per se, and final assemblies of tape components and packages are all performed in ultra-clean-room environments.

15.3 RECORDING FORMATS AND SYSTEMS

by Charles P. Ginsburg

In the early 1950s, when efforts to record television programs on magnetic tape began in earnest, methods for solving the problem could be classified as those which used stationary video heads, pulling the tape past them at a speed between 100 and 360 in/s, and those which used rotating video heads. Many different kinds of stationary-head methods were investigated. Some were straightforward, single-video-track techniques; some used frequency or time-division multiplexing; and at least one used one type of signal-handling method for the lower video frequencies and another for the higher ones. At the time of this writing, no stationary-head approach to recording television programs has met with commercial success.

Numerous types of rotating-head video tape recorders have been investigated, but only three are described; one because of its historical significance, another because it dominated the broadcast scene for two decades, and the third because versions of it had taken over markets ranging from consumer to broadcast by the late 1970s.

15.3.1 ARCUATE SWEEP The *arcuate sweep* configuration used four heads mounted so that their tips protruded about 0.003 in (0.076 mm) above the plane surface of a disc rotating at 240 r/s. With a rotational velocity of 1700 in/s (43 m/s) the heads recorded arcuate tracks across a 2-in-wide tape, guided to be in contact with the heads as they described descending arcs. With a reel-to-reel tape speed of 30 in/s, the recorded tracks were distorted slightly from circular shape. Audio and control tracks were recorded longitudinally along the edges of the tape surface in a conventional manner by stationary heads. The arcuate sweep technique was chosen initially to avoid the effect of the impact of heads coming into contact with the edge of the tape at right angles. However, timing errors unique to the arcuate method brought about a change in the scanning method in late 1954.

15.3.2 QUADRUPLEX TRANSVERSE SCAN. The 2-in quadruplex scan configuration, which followed the arcuate sweep developments, was introduced at the annual convention of the National Association of Broadcasters in 1956 and dominated the broadcast industry for the next two decades. The transverse scan is accomplished by four heads mounted on a 2.064-in-diameter (5.2-cm) headwheel with their tips protuding about 0.002 in beyond the periphery of the wheel. With a rotational rate of 240 r/s for 525-line television systems, the head writing speed is 1550 in/s. These values are 250 r/s and 1600 in/s for 625-line systems. The video tracks are 0.10-in wide with a *guard band* between tracks of 0.005 in. The tape speed is 15 in/s.

In recording, after the tape passes the rotating recording heads, stationary heads erase portions of the tape edges for the recording of auxiliary information which includes control track, program audio, and cue or time code. The same video and auxiliary heads are used for both record and playback. The quadruplex format is shown in Fig. 15-49a.

15.3.3 HELICAL SCAN. A profusion of *helical-wrap* recorders have made their appearance since 1961. They all have in common the characteristic that the tape is wrapped about the *scanning assembly* in a helix, resulting in a recorded track that describes a straight, slanted line across the tape. Helical recorders may be classified as *full wrap* and as *half wrap*. Alternatively, they may be classified as *field-per-scan* or as

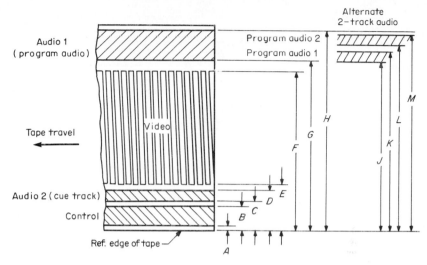

Dimensions	in		mm		Dimensions	in		mm	
	Min.	Max.	Min.	Max.		Min.	Max.	Min.	Max.
A	0.000	0.004	0.00	0.10	G	1.921	1.930	48.79	49.02
B	0.040	0.049	1.02	1.24	H	1.988	1.996	50.50	50.70
C	0.058	0.062	1.47	1.57	J	1.920	1.928	48.77	48.97
D	0.078	0.085	1.98	2.16	K	1.945	1.951	49.40	49.56
E	0.087	0.094	2.21	2.39	L	1.965	1.971	49.91	50.06
F	1.902	1.914	48.31	48.62	M	1.988	1.996	50.50	50.70

FIG. 15-49a Quadruplex 2-in tape format. [*D. G. Fink and D. Christiansen (eds.), Electronics Engineers' Handbook, 2d ed., McGraw-Hill, New York, 1982.*]

segmented field (Fig. 15-82). The field-per-scan machines write, or read, one complete field with each head pass. The segmented-field machines require several head passes, generally five or six, depending on the television system on which the machine is operating, to write or read one field. Since 1978, the broadcast version of the 1-in segmented-field helical video tape recorders have been manufactured in accordance with the standard for the type B format (Fig 15-49*b*), and the 1-in field-per-scan machines have conformed to the standard for the type C format (Fig. 15-49*c*). These standards were established by the Society of Motion Picture and Television Engineers (SMPTE) to make possible the interchange of tapes recorded on machines of a given television system standard made by different manufacturers.

Although helical recorders found wide application in industrial and commercial markets in the second half of the decade of the 1960s and beyond, and in the consumer market beginning in the early 1970s, their use in broadcasting was extremely limited. The problem was one of excessive time-base instability for on-air use. This was finally solved after the introduction of *digital time-base correction* in 1973, making it possible to incorporate far wider windows in time-base correctors than was practical in analog correctors. The excessive timing errors, characteristic of most helical recorders, were now under control.

Next, the development of a method for servoing the video heads themselves to permit slow-, stop-, and reverse-motion effects in some designs of helical recorders, coupled with a substantial decrease in operating costs through a reduction in tape costs by a factor of

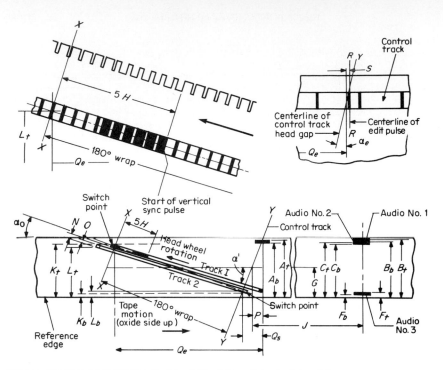

Dimensions†	mm Min.	mm Max.	mm Ref.	in Min.	in Max.	in Ref.
A_b Control track	23.55	26.65		0.9272	0.9311	
A_t Control track	23.95	24.06		0.9429	0.9472	
B_b Audio 1 track	24.35	24.45		0.9587	0.9626	
B_t Audio 1 track	25.15	25.26		0.9902	0.9945	
C_b Audio 2 track	22.35	22.45		0.8799	0.8839	
C_t Audio 2 track	23.15	23.26		0.9114	0.9157	
F_b Audio 3 track	0.195	0.205		0.00768	0.00807	
F_t Audio 3 track	0.990	1.01		0.03898	0.0398	
G Center of video tape			12.70			0.5000
J Position of audio heads	232.0	233.0		9.134	9.137	
K_b Full video width	1.18			0.0465		
K_t Full video width		22.19			0.8736	
L_b Video width (180°)	1.82			0.0717		
L_t Video width (180°)		21.55			0.8484	
N Video track pitch			0.200			0.00787
O Video track width	0.155	0.165		0.00610	0.00650	
P Position of control head	2.84	2.88		0.1118	0.1134	
Q_b Switch point video track 2	82.096	82.121		3.23213	3.23311	
Q_t Switch point video track 1	5.523	5.533		0.21744	0.21783	
S Distance between control-track head gap and center edit pulse at 180° switch point			0.040			0.00157
α^0 Scanning angle		14.434°				
α^1 Video track angle (525/60)		14.288°				

†b, t are the dimensions from the reference edge to the bottom and top of the record, respectively.

FIG. 15-49b SMPTE type B segmented 1-in tape format. [*D. G. Fink and D. Christiansen (eds.), Electronics Engineers' Handbook, 2d ed., McGraw-Hill, New York, 1982.*]

two-thirds compared with the transverse scan recorders, finally brought to an end the dominance of the quadruplex machines. The reduction in the amount of tape used per unit of time was the result of continuous improvements in the quality of both heads and tape. When the quality of signals played back from quadruplex recorders became so good that the track width, or even the writing speed, could have been reduced, the standard for those machines was so firmly entrenched that a change in either was not considered practical by most users. However, the improvements were exploited in the design of the helical recorders. In 1982, the manufacture of quadruplex machines was less than 2 percent of the total number of videotape recorders turned out for the broadcast market.

15.3.4 CASSETTE RECORDERS

Type E ¾-in Format. Introduced in 1972 by the Sony Corporation as the U-matic recorder, by use of the *color-under* technique for recording of color subcarrier, acceptable picture quality is achieved at the relatively slow tape-running speed of 3.752 in/s (95.3 mm/s). This format has found widespread use in industrial, CATV, and broadcast applications. The recorded track dimensions are shown in Fig. 15-50 and the salient characteristics in Table 15-7.

Home-Video ½-in Formats. By the use of a tape-running speed substantially lower than the type E, 1.57 in/s for beta and 1.3 in/s for VHS, these formats permit the recording of a full-length feature film on a single cassette. Thus, they are being used in rapidly increasing numbers in the home. In addition, by the use of a different, color-component signal, recording format, the basic mechanical design is finding application in broadcast and cable electronic news-gathering operations. The characteristics of the two types currently most popular are listed in Table 15-7.

Table 15-7 Characteristics of Video Cassette Recorders†

	U-matic	Beta	VHS
SMPTE type	E	G	H
Tape width, in	0.75	0.5	0.5
mm	19	12.65	12.65
Head drum diam, in	4.33	2.95	2.44
mm	110	75	62
Video track width, mils	3.35	2.3/1.15/0.77	2.3/1.15
μm	85	58.5/29.2/1.95	58.5/29.2
Head-to-tape speed, in/s	410	276	229
m/s	10.4	7.0	5.8
Audio track width, mils	32	41	41
mm	0.8	1.05	1.0
Tape speed, in/s	3.75	1.57/0.79/0.52	1.3/0.66/0.441
cm/s	9.53	4/2/1.33	3.34/1.67/1.12
FM carrier, MHz, sync	3.8	3.5	3.4
White	5.4	4.8	4.4
Chrominance subcarrier, kHz	688	688	629
Cassette or reel size, in	7.3 × 4.8 × 1.2	6.1 × 3.8 × 1	6.4 × 6.1 × 1
mm	186 × 123 × 31	156 × 96 × 25	162 × 104 × 25
Playing time,‡ min	60	60/120/180	20/240
Tape thickness,‡ mils	1.1	0.8	0.8
μm	27	20	20

†Tape-wrap angle is 180° on all models.
‡Thinner tapes permit longer playing time. The longest time available in the VHS system beginning in 1979 was 6 h of recording.

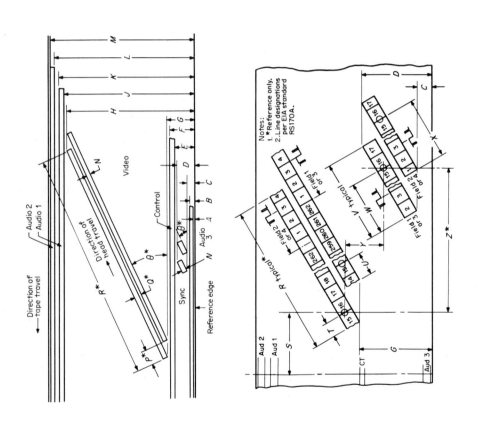

	Dimensions	mm Minimum	mm Maximum	in Minimum	in Maximum
A	Audio 3 lower edge	0.000	0.200	0.00000	0.00787
B	Audio 3 upper edge	0.775	1.025	0.03051	0.04035
C	Sync track lower edge	1.385	1.445	0.05453	0.05689
D	Sync track upper edge	2.680	2.740	0.10551	0.10787
E	Control tract lower edge	2.870	3.130	0.11299	0.12323
F	Control track upper edge	3.430	3.770	0.13504	0.14843
G	Video track lower edge	3.860	3.920	0.15197	0.15433
H	Video track upper edge	22.355	22.475	0.88012	0.88484
J	Audio 1 lower edge	22.700	22.900	0.89370	0.90157
K	Audio 1 upper edge	23.475	23.725	0.92421	0.93406
L	Audio 2 lower edge	24.275	24.525	0.95571	0.96555
M	Audio 2 upper edge	25.100	25.300	0.98819	0.99606
N	Video and sync track width	0.125	0.135	0.00492	0.00531
P	Video offset	4.067 (2.5H) ref.		0.16012 nom.	
Q	Video track pitch	0.1823 ref.		0.007177 nom.	
R	Video track length	410.764 (252.5H) ref.		16.17181 nom.	
S	Control track head distance	101.60	102.40	4.0000	4.0315
T	Vertical phase odd field	1.220 (0.75H)	2.030 (1.25H)	0.04803	0.07992
U	Vertical phase even field	2.030 (1.25H)	2.850 (1.75H)	0.07992	0.11220
V	Sync track length	25.620 (15.75H)	26.420 (16.25H)	1.00866	1.04016
W	Vertical phase odd sync field	22.360 (13.75H)	23.170 (14.25H)	0.88031	0.91220
X	Vertical phase even sync field	23.170 (14.25H)	28.980 (14.75H)	0.91220	0.94409
Y†	Vertical head offset	1.529 nom.		0.06020 nom.	
Z†	Horizontal head offset	35.350 nom.			
θ	Track angle	2°34' ref.		1.39173 nom.	

†Reference value only.

FIG. 15-49c SMPTE type C field-per-scan 1-in tape format. [*D. G. Fink and D. Christiansen (eds.), Electronics Engineers' Handbook, 2d ed., McGraw-Hill, New York, 1982.*]

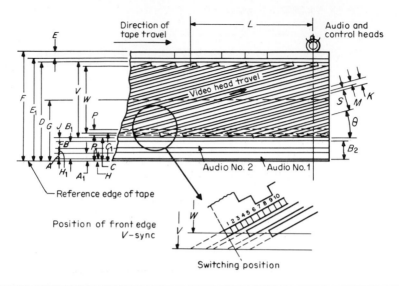

Dimensions		mm	in
A	Audio 1 width	0.80 ± 0.05	0.0315 ± 0.0020
A_1	Audio 1 reference	1.00 nom.	0.0394 nom.
B	Audio 2 width	0.80 ± 0.05	0.0315 ± 0.0020
B_1	Audio 2 reference	2.50 nom.	0.0984 nom.
B_2	Audio track total width	2.30 ± 0.08	0.0906 ± 0.0031
C	Video area lower limit	2.70 min.	0.1063 min.
C_1	Video effective area lower limit	3.05 min.	0.1201 min.
D	Video area upper limit	18.20 max.	0.7165 max.
E	Control track width	0.60 nom.	0.0236 nom.
E_1	Control track reference	18.40 + 0.28 − 0.18	0.7244 + 0.7244 − 0.0071
F	Tape width	19.00 ± 0.03	0.7480 ± 0.0012
G	Video trace center from reference edge	10.45 ± 0.05	0.4114 ± 0.0020
H	Audio guard band to tape edge	0.2 ± 0.1	0.008 ± 0.004
H_1	Audio-to-audio guard band	0.7 nom.	0.028 nom.
J	Audio-to-video guard band	0.2 nom.	0.008 nom.
K	Video track pitch (calculated)	0.137 nom.	0.00539 nom.
L	Audio and control head position from end of 180° scan	74.0 nom.	2.913 nom.
M	Video track width	0.085 ± 0.007	0.00335 ± 0.00028
P†	Address track width	0.50 ± 0.05	0.0197 ± 0.0020
P_1	Address track lower limit	2.90 ± 0.15	0.1142 ± 0.0059
S	Video guard band width	0.052 nom.	0.00205 nom.
Y	Video width	15.5 nom.	0.610 nom.
W	Video effective width	14.8 nom.	0.583 nom.
θ	Video track angle, moving tape	4°57′ 33.2″	
	stationary tape	4°54′ 49.1″	

†For reference value only.

FIG. 15-50 SMPTE type E ¾-in cassette format. [*D. G. Fink and D. Christiansen (eds.), Electronics Engineers' Handbook, 2d ed., McGraw-Hill, New York, 1982.*]

15.4 SIGNAL SYSTEMS AND PROCESSING

by Michael O. Felix and William McSweeney

15.4.1 THE NEED FOR A MODULATION SYSTEM. As shown in Sec. 15.1 any magnetic recorder has a bandpass characteristic with a null at zero frequency and a rapidly falling characteristic at high frequencies. Since the spectrum of a television signal extends from zero frequency to 5 or 6 MHz, it is impractical to record it without the use of a modulation system. Three forms of modulation can be considered: pulse, amplitude, and frequency (or its related phase). Pulse modulation systems involve high bit rates, and only until recently has their use been technologically feasible.

The modulation system selected preferably should provide a demodulated noise spectrum which complements that of the response of the eye. The eye is sensitive to large-area (low-frequency) disturbances (Fig. 15-51); it is much less sensitive as the frequency increases, except where the effects of noise close to the color subcarrier are considered because of the demodulation to base band.

At first analysis, amplitude modulation appears usable. By placing the carrier on the upper band edge (Fig. 15-52), a single-sideband system of adequate bandwidth results. However, such systems are excessively sensitive to head-to-tape separation loss. In accordance with the playback head-to-tape separation loss formula of 54.6 $[(a + a_t)/\lambda]$ dB described in Sec. 15.1.3, with a carrier of 9 MHz and a head-to-tape speed of 25 m/s, the playback signal decreases 1 dB for every 2.0 μin (0.05 μm) of separation. Tape-surface roughness alone is of this order, since much smoother tape will cling to any polished surface, such as a tape guide. Thus, the carrier-to-noise ratio of an AM system is poor, and since the noise power is essentially constant with frequency, the demodulated signal contains large amounts of the low-frequency noise to which the eye is so sensitive.

Variations in losses due to head-to-tape spacing produce no timing jitter. Consider an isolated magnetized point on a tape (Fig. 15-53); whether the tape is close to the head or far from it, the maximum flux links the head at precisely the same moment. Therefore an FM system using the same carrier frequency is unaffected by head-to-tape spacing until the carrier-to-noise ratio falls below the FM signal-limiting threshold. In addition,

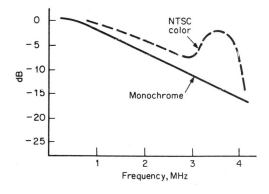

FIG. 15-51 Sensitivity of the eye to noise.

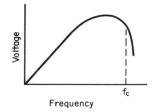

FIG. 15-52 Carrier on upper band edge.

the noise triangulation characteristic of an FM system, in which the noise increases with increasing frequency in the demodulated baseband, results in a S/N ratio variation with modulation frequency that closely complements the sensitivity of the eye. To reduce high-frequency noise, which is particularly important in color systems because of the subcarrier demodulation process, the video signal is preemphasized before being fed to the frequency modulator; typical values provide an 8-dB boost at frequencies above 1 MHz.

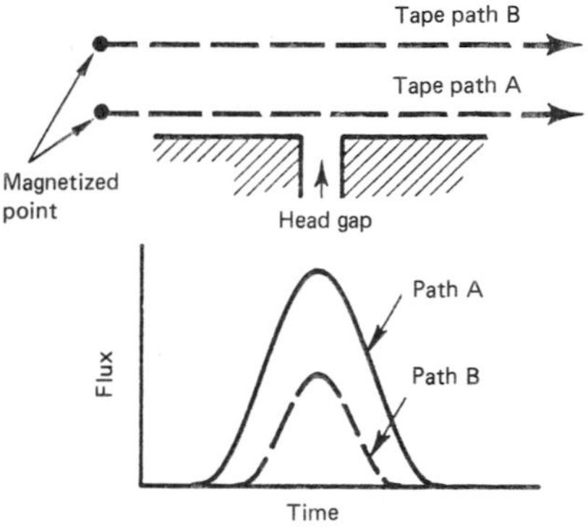

15.4.2 THE FM SYSTEM. A frequency-modulated signal can be analyzed either using a quasi-stationary approach or a sideband approach. To use the latter, consider the FM signal, $v = V \sin (\omega t + \beta \sin mt)$, where ω is the carrier frequency, m the modulation frequency, and β the peak phase deviation in radians. The peak-frequency deviation is then $m\beta$ Hz. Bessel showed that this could be split into an infinite number of sidebands, spaced m Hz apart above and below the carrier. Tables of Bessel functions for a given β give values for J_o (the carrier amplitude), J_1 (the amplitude of the first sidebands on $\omega + m$ and $\omega - m$); J_2 (the amplitude of the second sidebands on $\omega + 2m$ and $\omega - 2m$), and so on.

FIG. 15-53 Independence of timing from head-tape spacing.

The FM systems used for video tape recording differ fundamentally from those found in transmission systems in two respects: (1) the carrier frequency is not appreciably higher than the highest modulating frequency; and (2) the deviation ratio $\beta = \Delta f / f_m$ is much lower. For example, the highest modulation frequency is around 5 MHz, and carrier frequencies are in the 8- to 10-MHz range. In the following discussions an 8-MHz carrier modulated by a 4.4-MHz sine wave (approximately the frequency of the PAL color subcarrier) will be the standard example, since this presents the most severe performance requirements. A peak deviation of 2.2 MHz, giving a modulation index of 0.5, will be assumed.

The spectrum of this wave is shown in Fig. 15-54a. The power in the carrier, with modulation, is less than in the unmodulated case, as the total power in the carrier plus its sidebands is constant, as it is for all frequency-modulation systems. By use of the unmodulated carrier as reference level, the modulated carrier is at -0.6 dB, the first sidebands at 3.6 and 12.4 MHz are at -12 dB, and the second sidebands at -0.8 and 16.8 MHz are at -30 dB.

Negative Frequencies and Unavoidable Distortion. It is the significance and effect of these sidebands on *negative frequencies* that are unique to VTR FM systems. In com-

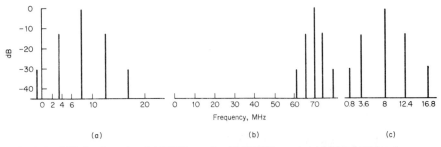

FIG. 15-54 FM signal spectra: (a) 8-MHz carrier; (b) 70-MHz carrier; (c) folded sideband.

parison, consider a *wide-band* FM system in which the carrier frequency at 70 MHz is modulated by the same 4.4-MHz signal with the same 2.2-MHz deviation (Fig. 15-54b). The sixteenth lower sideband, the first to appear on a negative frequency, is at an approximate level of −460 dB and may be neglected.

If the signal corresponding to Fig. 15-54a is fed to a spectrum analyzer or to a sharply tuned 0.8-MHz filter, a real component will be found, as shown in Fig. 15-54c. The analyzer display is shown in Fig. 15-55 and the 0.8-MHz *folded sideband* is spaced 7.2 MHz below the carrier. A demodulator will produce an unwanted output at this frequency that is not harmonically related to the wanted 4.4 MHz.

One source of unavoidable distortion comes from the folded sidebands. The vector diagram corresponding to Fig. 15-54 is shown in Fig. 15-55. The wanted output comes from the sum of the two first sidebands, each of amplitude 0.25. The peak deviation is therefore 0.5 rad, or (0.5 × 4.4) = 2.2 MHz. The folded sideband (Fig. 15-55b) has an amplitude of 0.03 (it has no corresponding upper sideband) and so produces a peak deviation of 0.03 rad or (0.03 × 7.2) = 0.22 MHz. The unwanted output from a frequency demodulator will be down in the ratio of 2.2/0.22, or 20 dB.

Since the FM system will be subjected both to FM and AM noise (the latter, for example, from head-to-tape spacing), it is essential that FM demodulators include limiters to eliminate AM components. Limiters in general introduce odd carrier harmonics,[22] and these introduce further amounts of unavoidable distortion, on the same frequencies as those from folded sidebands but generally at higher levels.

Unavoidable distortion can be kept at an acceptable level only by the choice of three parameters: the highest modulation frequency, the modulation index, and the carrier frequency. At a high carrier frequency, the folded sidebands are negligible; as the carrier frequency is decreased, lower-order sidebands are folded, and the resulting spurious outputs also decrease in frequency until they enter the wanted passband.

Avoidable Distortion. The predominant source of avoidable distortion comes from even-order carrier harmonics.

Consider a carrier which is slowly changing once per second between 8 and 9 MHz. Its second harmonic will change once per second between 16 and 18 MHz; the modulation frequency is unchanged but the deviation is doubled.

Suppose the FM signal is fed through an amplifier which introduces 1 percent of second-harmonic distortion. The unmodulated second-harmonic carrier on 16 MHz would then have a level of −40 dB. However, the deviation ratio of the second harmonic is twice that of the fundamental, and so is 1.00, with $J_0(\beta)$ equal to 0.77 (−2 dB) and $J_1(\beta)$ equal to 0.44 (−7 dB). The resulting spectrum (Fig. 15-56) has a 16-MHz carrier at −42 dB, with its first sidebands on 11.6 and 20.4 MHz at −47 dB.

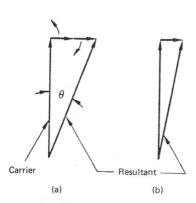

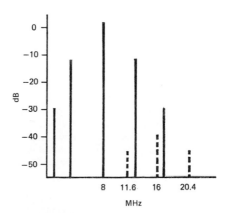

FIG. 15-55 FM vectors: (a) first-order sidebands; (b) folded sideband.

FIG. 15-56 FM spectra—1% second-harmonic distortion.

The 11.6-MHz sideband is spaced 3.6 MHz from the 8-MHz carrier, and upon demodulation produces a spurious 3.6-MHz output. (Note that as this is in the wanted passband, it cannot be removed by filtering.) The level of the 11.6-MHz component is (47 − 12) = 35 dB below each of the wanted first sidebands, or 41 dB below the sum of them; the spurious phase modulation is therefore 41 dB below the wanted, and the output of a frequency demodulator is changed by the ratio of the unwanted to the wanted frequency, or −2 dB. The spurious output is therefore down 43 dB.

Note that a 1 percent second harmonic has produced approximately a 1 percent spurious output. The RF circuits of VTRs, both record and playback, therefore must introduce very low levels of second-harmonic distortion.

Other possible sources of even-order harmonics are direct current in the head windings, dc magnetic fields in the area of head-to-tape contact, and ac fields synchronous with the head scanning rate.

Straight-Line Equalization. Conventional FM systems generally use bandpass responses in which all significant sidebands are amplified equally. However, it can be shown[23] that any response in which the voltage gain varies linearly with frequency does not distort the FM signal; all except the flat response do introduce unwanted amplitude modulation, which can later be removed by a limiter. An intuitive explanation can be seen from the vector diagrams of an FM signal at the time of peak phase shift (Fig. 15-57a) and zero phase shift (Fig. 15-57b). In the first case, each pair of upper and lower sidebands is rotated 90° more than the previous pair; in the second, odd pairs cancel and even pairs add. The resultant vector has the same length in each case, as required to give a constant-amplitude FM signal.

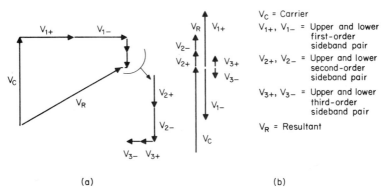

V_C = Carrier

V_{1+}, V_{1-} = Upper and lower first-order sideband pair

V_{2+}, V_{2-} = Upper and lower second-order sideband pair

V_{3+}, V_{3-} = Upper and lower third-order sideband pair

V_R = Resultant

(a)   (b)

FIG. 15-57 Vector diagrams of FM signal: (a) peak phase shift condition; (b) zero phase shift condition.

Suppose this signal is fed through a *straight-line* equalizer (Fig. 15-58) and that the gain at carrier frequency is unity. Then the increase in gain on each lower sideband is precisely balanced by the decrease on the upper sideband, and the resultant vector in Fig. 15-57a is unchanged. However, in Fig. 15-57b the odd-order sideband pairs no longer cancel, and the length of the resultant vector increases at one time and decreases 180° later. Thus, the amplitude is changed, and spurious AM at the modulating frequency has been introduced.

Mathematical analysis shows that the linear response may pass through zero amplitude (Fig. 15-58, point A), providing that the instantaneous carrier frequency does not pass through this frequency. This is an obvious practical limitation, since at such times the signal would have zero amplitude. Note that the restriction says nothing about sideband frequencies which always exist on both sides of point A, and may even occur at A without distortion.

The analysis also says that the characteristic passes through zero frequency, as shown dotted. This is not realizable and may be considered another facet of the folded sideband problem.

In practical systems, the deviation ratio is chosen so that the amplitude of sidebands above frequency A and below zero is small. No response above A is provided, which introduces a small but acceptable level of distortion.

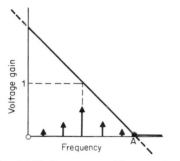

FIG. 15-58 Spectrum with straight-line equalizer.

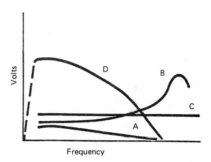

FIG. 15-59 Sources of noise: (a) tape noise; (b) head noise; (c) electronic noise; (d) signal.

Noise in VTRs. Noise in VTRs comes from three sources (Fig. 15-59): (1) tape noise, (2) head noise, and (3) electronic noise. Tape noise is approximately uniform in flux, and so signal and tape noise are approximately constant with frequency (both being equally affected by spacing and head losses).

Head noise increases with frequency owing to increasing core and copper losses. It shows a peak at head resonance which is typically above the wanted band (some early VTRs put it in the band and included electronic compensation to cancel its effects). Electronic noise is approximately flat. The resultant signal plus total noise, therefore, decreases rapidly with frequency. Since in the recorded signal each lower sideband always has an upper sideband of equal amplitude, a playback equalizer that emphasizes the lower at the expense of the upper will give a better S/N ratio. A straight-line equalizer in which point A (Fig. 15-58) is moved to the lowest possible frequency gives the highest S/N ratio. Frequency A is generally chosen just above the frequency of the highest first sideband; it typically gives an improvement of 10 dB over a flat system while introducing less than 1 percent distortion.

15.4.3 RECORD ELECTRONICS

Modulator. Figure 15-60 shows a typical single-head record system, such as found in type C format equipment. The source video is terminated at the input to a variable-gain amplifier. Gain adjustment provides the correct blanking-to-peak-white deviation. The input video is ac-coupled, and a feedback clamp is used to stabilize the back-porch dc level. The clamp has a fast time-constant of about 10 television lines, to minimize the effects of hum or dc shift on the input signal. Preemphasis is then applied as demanded by the standard on which the equipment is operating. A low-pass filter after the modulator to eliminate out-of-band frequencies, or a peak white clip circuit before the modulator, can be used to protect against overdeviation. Since the FM modulator is required to deviate a high percentage of its center frequency, that is, $2.2/8 = 27.5$ percent in the example given in Sec. 15.3.2, a multivibrator oscillator generally is chosen. The impor-

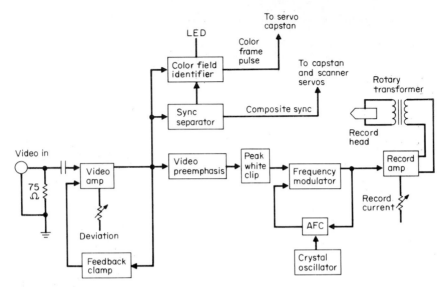

FIG. 15-60   Single-head record system.

tant design criteria are modulation linearity and minimum even-order harmonics. Linearity is an inverse function of the inactive time of the oscillator and can be minimized by fast switching and minimum recharge times. However, some compensation for non-linear effects is often required. An oscillator operating at twice the desired frequency, followed by an appropriate divider, can be used to achieve excellent symmetry.

In an alternative beat-frequency approach, two high-frequency reactance oscillators, offset from each other by the blanking-level frequency, for example, 108 and 100 MHz, are deviated in opposite directions. Linearity is good in this case since the deviation is a low percentage of the center frequency, and cancellation of square-law modulation non-linearities takes place. The resultant modulated signal differs from the multivibrator output in that it is sinusoidal.

Automatic Frequency Control.   The modulator back-porch frequency can be stabilized by an automatic frequency control loop, as shown in Fig. 15-60. This frequency is compared with a standard, such as a crystal oscillator, and suitable correction applied.

Record Amplifier.   The modulated signal is applied to the record head through a record amplifier. This provides a current drive with amplitude adjustable to give the maximum RF playback level during the setup procedure. Adjustment is made to over-saturation of the tape at carrier frequency. Above this saturation level, short-wavelength demagnetization will occur. The amplifier can be a *current-switching* type for the square-wave signal, but must be linear for the *sine wave* method to avoid generating harmonics. The resulting frequency spectrum on tape is the same for both techniques, because the tape acts as a limiter to the sine wave, introducing odd harmonics just as in the case of the current switching method. Levels of even-order harmonics must be kept below −40 dB and are necessary to keep *moiré patterns*, i.e., the avoidable distortion products discussed in Sec. 15.3.2, to acceptable levels in playback.

Record equalization is employed to produce either a constant current over the frequency spectrum for Alfesil heads or a desired roll-off with increasing frequency for ferrite heads.

Automatic modes of record current optimization involve a combination of record and playback at low tape speeds. In this manner, the playback level can be monitored continuously as the current drive is varied.

The output of the record amplifier is normally routed to the rotating head in a balanced manner through a rotary transformer in the scanner assembly.

Color Field Identification. Information derived from the record video signal is used for the control track. Vertical sync and an identification pulse for the first field in the sequence of four in NTSC (eight in the PAL system) are required. In playback, the machine is *color-frame synchronized* by locking color field no. 1 of the control track to color field no. 1 of reference video.

When input signals meet the RS170A specification (or the EBU equivalent) for subcarrier-to-horizontal-sync (ScH) phase, detection can be consistent. Arbitrary decisions are necessary if this phase relationship is within 90° of a subcarrier cycle away from the center value of the specification. A fixed amount of hysteresis, for example, $\pm 45°$, applied after the initial decision-making process, can prevent frame jumping of the identification pulse due to ScH phase drift or detector ambiguity in this region.

In playback, the color-field information should be ignored after initial servo lock to avoid capstan reframing.

A calibrated detection system can indicate the ScH phase relationship of input video by metering or other monitoring, such as a light-emitting diode (LED) as shown in Fig. 15-60.

15.4.4 REPRODUCE ELECTRONICS. A block diagram for a complete single-head playback system is shown in Fig. 15-61.

Preamplifiers. Figure 15-62 shows frequency responses for different values of preamplifier input impedance. Head capacitance C_h is assumed to be low compared with preamplifier input capacitance C_{in}. The high-input impedance approach has the disadvantage that resonance cancellation is required. This involves frequency response and Q adjustments, which must be altered as the head wears or is replaced. A negative-feedback, low-input impedance preamplifier overcomes this problem with a small sacrifice in noise figure, and fixed compensation for the -6 dB per octave response characteristic is easily applied later.

The mid-impedance approach is achieved with partial negative feedback and requires no resonance compensation if head parameters do not vary widely. Use of a step-up transformer can lower the *knee frequency* of the feedback configurations, i.e., the frequency at which $Z_h \approx R_{in}$.

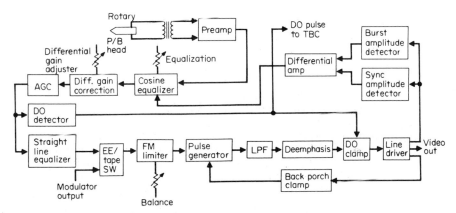

FIG. 15-61   Single-head playback system.

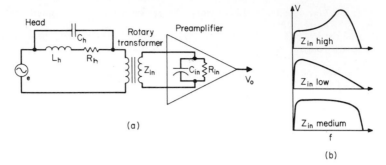

FIG. 15-62 Response as function of preamplifier input impedance: (*a*) equivalent circuits for head, transformer, and preamplifier; (*b*) frequency response for different values of input impedance.

The turns ratio is optimized for best S/N across the spectrum. Signal-to-noise requirements dictate the choice of amplifiers used. Inherently, field-effect transistors (FET) have an advantage over bipolar in handling the complex source impedance of the head. In most systems, however, tape noise is dominant, and a bipolar differential amplifier, for example, can give adequate S/N while excelling when isolation requirements are stringent, as in simultaneous record/playback modes.

Equalization. To correct the losses in the record/playback process, a network with constant group delay and high-frequency boost is ideal. This equalization boost restores to flat the frequency response of the carrier and its important sidebands, and is made variable to allow correction of variations in heads and tape. An aperture corrector (cosine equalizer) has the required characteristics. In the implementation shown in Fig. 15-63*a* the input signal 15-63*b* to the source-terminated delay line is combined in an adjustable amount with the output signal from the delay line. Exact compensation within the range shown in Fig. 15-64 for the head-tape frequency response is made by the adjustment of *K*. This avoids differential-gain and frequency-response problems, which occur when the output of the equalizer is not flat for the deviated carrier and its sidebands.

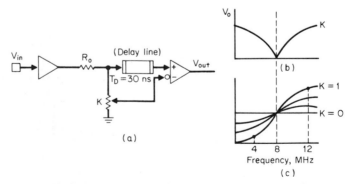

FIG. 15-63 Aperture correction by cosine equalizer—single delay line method: (*a*) circuit; (*b*) response at input to delay line; (*c*) range of compensation.

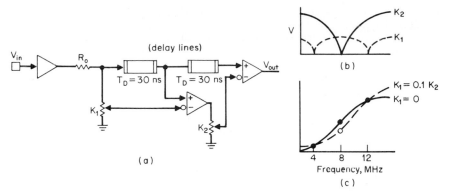

FIG. 15-64   Aperture correction by cosine equalizer—two delay line method: (*a*) circuit; (*b*) response at inputs to delay lines; (*c*) range of compensation.

Ferrite heads have less response curvature at low frequencies than Alfesil. Matching can be achieved with use of a second delay line connected as shown in Fig. 15-64. Here, the low-frequency curvature is controlled by the amplitude of K_1 and the amount of boost by K_2. A differential-gain adjustment can also be provided to compensate for any errors. This operates on the carrier, altering its amplitude response as it deviates from frequencies corresponding to black and white. Finally, to improve the S/N of the demodulated signal, a straight-line equalization network as described in Sec. 15.3.2 precedes the limiter stages. This should have constant group delay while attenuating the higher frequencies.

Drop-out Detection.   As shown in Fig. 15-61, an automatic-gain-control (AGC) circuit stabilizes the RF amplitude level prior to drop-out detection. The detector is designed to sense a drop in signal amplitude, greater than 16 dB, for example, and lasting slightly longer than the period of the lowest carrier frequency. Some amount of hysteresis, perhaps 4 dB, is then applied to this threshold level to avoid a marginal decision point. The drop-out pulse generated in this fashion is used to clamp the output video to blanking level, and to gate in compensation in the time-base corrector, (see Sec. 15.5.4, Drop-Out Compensation).

Demodulation.   Sufficient limiting is required to remove all amplitude modulation. This enables demodulation of varying signal levels down to the theoretical S/N threshold point. Balance adjustments are included in the limiter strip to minimize even-order harmonics. A pulse-counter type discriminator is the most suitable method for achieving good linearity over the wide carrier deviations involved. Here, a pulse is formed at each crossover of the limited RF and integrated by a low-pass filter. Filtering requirements are made easier since this doubles the fundamental carrier frequencies, which are then rejected by a low-pass filter with high stop-band attenuation. Any group-delay difference across the video passband is corrected by appropriate phase-equalizing networks.

The filter is followed by an amplifier which applies video deemphasis and correction of any passband losses. Drop-out clamping is followed by a video line driver providing video to a feedback back-porch clamp, autochroma circuitry, and the machine output.

Autochroma.   By amplitude detecting the sync and burst of the output video and comparing their levels, a difference voltage is generated. This is used to control the response of the cosine equalizer, making the burst level equal to the sync level. When monochrome status is detected, this control is switched to the manual mode. In segmented formats, response changes as the head scans its track are more noticeable than on one-field-per-track systems. Complex line-by-line compensation is needed to correct the burst amplitude across the scan. This is used in combination with head-by-head correction, depending on system noise immunity and reaction time.

15.5 TIME-BASE CORRECTION

15.5.1 HISTORY OF TIME-BASE CORRECTION

by E. Stanley Busby

Electromechanical.   In very early quadruplex VTRs, the headwheel rotated at a multiple of the power-line frequency. The timing of vertical sync was unrelated to that of other video sources in the studio. A video switch between a VTR and another source caused a vertical-scanning resynchronization in receivers, usually resulting in a disturbing picture roll.

The first level of synchronization of VTRs with the studio synchronizing generator was accomplished through the use of two servomechanisms:

1. A capstan servo, which positioned a recorded track containing a vertical-sync pulse under the headwheel at reference vertical time
2. A headwheel servo which adjusted the phase of the headwheel rotation until playback and reference vertical pulses were coincident

Both of these servos are discussed in greater detail in Sec. 15.7.

The departure from ideal synchronism was on the order of a few microseconds. Only slowly changing perturbations were corrected, however, owing to the low sampling rate of 60 or 50 Hz.

Mispositioning of the tape guide, which forces the tape against the headwheel, was responsible for a particularly objectionable time disturbance called *skewing*, described in Sec. 15.6.1. A tape-guide servo was devised which compared the interval between the last horizontal sync pulse in a head pass and the first one of the next pass, and compared it with the average interval, supplied by a phase-locked oscillator. The error operated a small motor which positioned the tape guide for minimum error.

The remaining time-base error had three principal spectra:

1. Slowly changing components caused mainly by variation in tape friction
2. Components at the head-pass rate, stemming from errors of the headwheel and tape-guide geometry
3. Very rapidly changing components caused by imperfect ball bearings in the headwheel motor

The horizontal-scanning generators in receivers of the day were phase-locked oscillators to improve noise immunity. They could follow a slowly changing phase but not a rapidly varying one.

Early Color Correction.   The stability required of the color subcarrier is orders of magnitude greater than an electromechanical servo can deliver. The first color playback from quadruplex machines stabilized the chrominance component only, leaving the luminance uncorrected.

In one implementation, the chroma was filtered from the composite playback signal, and an oscillator was phase-locked to each color burst. This oscillator was used to demodulate the chroma to color difference signals, which were then remodulated using a stable subcarrier, and mixed with the uncorrected luminance (see Fig. 15-65).

In another implementation, an oscillator synchronized with playback burst was used to translate the chroma component to a higher frequency well out-of-band, where it was

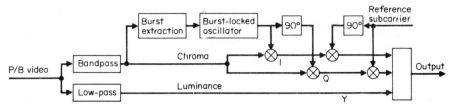

FIG. 15-65 Signal flow in early color corrector.

filtered and returned to the video spectrum by heterodyning with a stable frequency derived from reference color burst.

Both these systems were called *heterodyne color,* and produced viewable images. The relationship between horizontal scanning and the subcarrier frequency was not maintained, and artifacts of this deficiency could be seen at vertical edges of colored objects.

Full Monochrome Synchronization. An improvement was made to the mechanical servomechanism which further reduced time-base error, discussed in more detail in Sec. 15.7.

Once vertical synchronism was attained, the system changed modes to compare playback and reference horizontal pulses. The wider bandwidth resulting from the higher sampling rate reduced residual low-rate time-base error to less than 1 μs.

With full synchronization, it was possible for the first time to mix a VTR playback signal with other video sources, without an interruption in synchronizing-signal timing.

Electronic Time-Base Correction. Shortly after this level of performance was reached, a voltage-variable video delay line was developed to treat short-term errors. It had two principal modes:

1. For VTRs not equipped with the new servo mentioned above, playback horizontal-sync pulses were compared with a phase-locked oscillator having a time constant similar to that used in receivers. The error signal was used to adjust the delay line over its 2.5- to 3.5-μs range.
2. For systems with the new servo, playback horizontal pulses were directly compared with reference horizontal. A low-bandwidth feedback path was provided for the servo system which caused it to adjust the average arrival time of horizontal sync such that the electronic corrector operated in the center of its delay range.

In this mode, all residual time-base errors were reduced to less than 30 ns. This was adequate for monochrome, but not for color.

Full Color Correction. Yet another electronic corrector was added to the signal-processing path, comparing playback and reference color bursts, and manipulating a similar but shorter delay line. The residual error was thereby reduced to about 4 ns, limited primarily by the signal-to-noise ratio of playback burst.

Correction of Second-Order (Velocity) Errors. Since color burst and sync pulses are available only at horizontal intervals, time-base errors are detectable only then, and represent the accumulated error during one scan line. In the presence of a changing time error, color was correct at the outset of a horizontal scan, but became progressively worse as the line proceeded, reaching a maximum at the right edge of the picture. For the most part, these errors were a function of the headwheel and tape guide geometry, and repeated at the head pass rate. The error magnitude was proportional to the rate of change of time-base error.

A second-order corrector was devised which, for each of the horizontal line periods (16 or 17 television lines) scanned during a head pass, measured and integrated the error steps at each horizontal boundary to improve noise immunity. These stored voltages were used to generate a linear ramp function which was furnished to the color corrector to effect a progressive correction throughout the line.

The residual error, in the end, was either random or greater than second-order. The processing sequence of an entire analog system is shown in Fig. 15-66.

A Hybrid System. As either a forerunner or outgrowth of digital time-base correctors, a system was developed in which the playback video was sampled at an adequately high rate, but not quantized. The analog samples were entered in a charge-coupled analog delay line, and passed through it at the sample rate. The delay through the line was inversely proportional to the clocking frequency, which was varied to make the output uniform with time.

The dynamic range of the charge-coupled device was limited, and the approach was applied in systems which, for other reasons, had a low signal-to-noise ratio.

15.5.2 DIGITAL TIME-BASE CORRECTORS.

The decrease in cost of digital circuits, especially memory devices, made digital time-base correctors (TBCs) an attractive alternative to analog systems. This was especially so for correctors used with helical-scan VTRs, whose time error accumulated for an entire field, and which therefore required a correction range of tens of microseconds. Large-range analog systems required a long, expensive, analog delay. Long digital delays involve only additional memory.

As in any digital video system, the input signal must be dc-restored, bandwidth-limited, sampled, and quantized. In early digital TBCs, analog-to-digital conversion was accomplished with discrete circuits, using two sequential four-bit measurements. This method was soon replaced by an integrated circuit *one-look* flash converter, used in modern units.

A digital time-base corrector is characterized by sampling and writing into memory

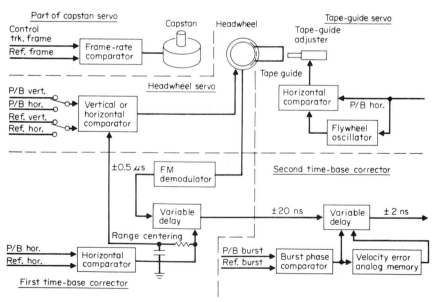

FIG. 15-66 Processing block diagram of analog system.

at a frequency f_1, while recovering and outputting samples at another frequency f_2. Of f_1 and f_2, one is derived from the time-unstable playback, and the other from a stable reference. The average frequency of f_1 equals f_2. The sampling frequency is typically a multiple of the color subcarrier, usually the third or fourth.

If the unstable frequency is derived from playback burst by setting its phase at each burst time, the system is a line-by-line corrector. Instantaneous discontinuities in timing, as might be produced in quadruplex VTRs at head-switch time, or as the result of an edit, will be corrected by such a system, but velocity errors will not.

If the unstable frequency is derived from the playback signal using a phase-locked oscillator with a fairly long time constant, instantaneous discontinuities are uncorrected, but some reduction of velocity error takes place. This method is frequently used with helical-scan recorders which are relatively free of sharp time discontinuities in the active part of the picture. It also has the advantage of being more immune to noise and occasional missing pulses.

In professional systems, the stable frequency is always derived from the studio synchronizing generator, so that the output video may be mixed with other video sources. In lower-cost systems, the stable frequency may be derived from the playback signal, using a phase-locked oscillator with a long time constant. Professional TBCs often offer this mode as a switchable option, for use in processing the playback of VTRs which have no servo system.

Otherwise, the playback timing of the VTR is adjusted so that the average delay through the TBC is half the total available, maximizing the correction range. Delay lengths range from one horizontal line time to several lines. Helical-scan VTRs which offer slow, fast, and reverse motion accumulate large time errors, and the longer the delay in the TBC, the wider is the range of speeds over which the VTR can perform these special effects. Details of the interaction between variable-speed playback and the TBC appear in Sec. 15.5.5.

Figure 15-67 illustrates the major components of a digital TBC. A phase-locked oscillator operating at three or four times F_{sc} is locked to playback video using a fairly long

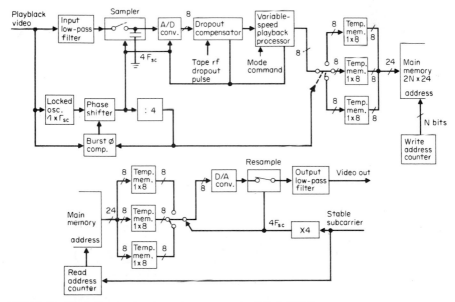

FIG. 15-67 Block diagram showing major components of digital TBC.

time constant. At the arrival of each color-sync burst, its output is phase adjusted such that when divided by 3 or 4, the result is synchronous with playback burst. This phase is maintained throughout the horizontal scan.

The band-limited playback video is sampled and converted to an eight-bit unsigned binary quantity. It is typical to temporarily store three or four samples in high-speed memory to form a 24- or 32-bit word, which is then strobed into the main memory once each cycle at F_{sc}, to extend the read/write time.

The choice of the number of samples stored temporarily is a function of the main memory bandwidth and is independent of the sampling frequency. The multiplexing into the main memory must, however, match the demultiplexing at its output.

The output clock is the same multiple of a stable subcarrier. Its output, suitably divided, is used to recall words from the main memory. The word is then commutated at the sampling rate to obtain the original stream of samples which then are applied to a digital-to-analog (D/A) converter. The D/A converter is followed by a resampler to remove transients, and by a low-pass filter to remove any out-of-band energy. The response of the output filter is sometimes shaped to compensate for the $(\sin x)/x$ loss associated with the sampling process.

15.5.3 VELOCITY ERROR COMPENSATION.

by David E. Trytko

Off-tape video signals contain many instabilities that must be eliminated to reproduce a high-quality color picture. While the line-by-line correction provides a significant improvement in the overall signal stability, it nonetheless fails to completely eliminate all time-base errors. Those that remain, caused by geometric errors, tension variations, or atmospheric conditions, are termed *velocity errors*. They are the result of the difference in the effective head-to-tape speed between record and playback, and are characterized by a progressive chroma phase shift throughout each horizontal line. Therefore, although the line-by-line correction eliminates the errors that exist at the beginning of each horizontal line, the velocity errors generated throughout the line remain.

The constant change in magnitude of line-by-line error measurements indicates that the overall time-base error profile is more complicated than a simple zero-order approximation (see Fig. 15-68). A zero-order estimation is the profile provided by the line-by-line process. The visual result of using this approximation alone is a video signal whose instabilities (and therefore chroma impurities) increase throughout the duration of the horizontal line. Clearly, a more accurate error approximation is required for high-quality video reproduction.

As seen from Fig. 15-68, the zero-order approximation provided by the line-by-line process can be significantly improved through linear interpolation. That is, a first-order approximation can be constructed by assuming that the rate of change (slope) throughout the line is constant. Consider Section AB in Fig. 15-68; a linear interpolation is performed to distribute the step changes in the line-by-line error samples across the horizontal line. Note that the slope M is constant and requires advance knowledge of the line-by-line error of line 3 at the beginning of line 2. This condition implies a noncasual process which can be performed with some type of video delay of at least one horizontal line.

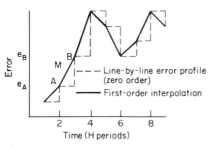

FIG. 15-68　First-order interpolation of velocity error.

More radical errors, such as impact errors, require an even more detailed error profile before they can be substantially eliminated. These types of errors primarily are the result of tape deformation caused by the impact of the rotating video heads as they make initial contact with the tape. This kind of disturbance creates a condition where the rate of change of error throughout the line can no longer be considered constant. Thus, the presence of *acceleration error* becomes apparent and may well require at least a second-order profile to be substantially reduced.

Manufacturers use various methods of eliminating these higher-order velocity errors. In general, however, the basic method of correction incorporates a sampled system of line-by-line error measurements that are in some way interpolated into a more accurate and detailed error profile.

15.5.4 DROPOUT COMPENSATION.

by Steven Wagner

In developing new videotape and videotape recorders, much effort has been made to reduce the frequency and severity of tape dropouts. Nevertheless there is still some dropout activity in practical videotape recording systems. Additionally, dropouts increase rapidly as tape becomes worn and contaminated with dust and other matter.

To minimize the visibility of residual dropouts, the technique of dropout compensation must be used. *Dropout compensation* is a method in which those portions of a playback video signal containing dropouts are first detected, then replaced with a video signal which is free of dropouts. The dropout-free replacement video should be a good estimate of the missing video for the masking process to be effective.

Dropout Characteristics. Dropouts in videotape recording have statistical characteristics which greatly affect dropout compensator design. When they occur, dropouts are almost always much shorter than one scan line in duration, and the time between dropouts is usually many scan lines, even in areas of severe dropout activity. Typically, dropouts might occur once every few seconds and last for a small fraction of a scan line.

Basic Previous-Line Dropout Compensation. Effective dropout compensation (DOC) relies on a good choice of substitute video when playback dropouts occur. It was realized in the early 1960s that the video contents of adjacent scan lines of a field are usually similar (ignoring for the moment the fact that subcarrier phase will not be identical). When a dropout occurs in a given line, then a section of the previous line can be substituted for the section of the line affected by the dropout (Fig. 15-69). Since the two lines will tend to be similar, the substitution is nearly invisible. This method, known as

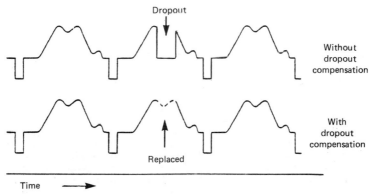

FIG. 15-69 Waveform showing without DOC and with DOC.

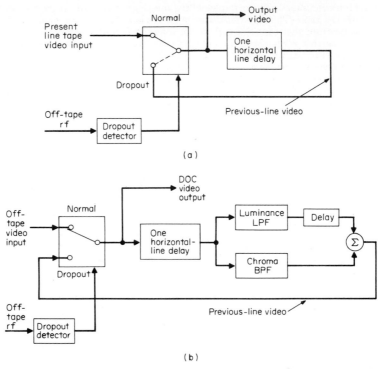

FIG. 15-70 (*a*) Monochrome previous-line DOC; (*b*) NTSC-color previous-line DOC.

previous-line dropout compensation, is the most common form of dropout compensation in use today.

Figure 15-70*a* is a block diagram of a simple monochrome DOC. The key element here is a one-horizontal-line delay that has incoming video as its input. The output of the delay thus is video from the previous input scan line of the input signal. When the input video contains a dropout, the output is switched from the input to the delayed signal and input video is replaced with dropout-free video from the previous line.

When dropouts occur that are one or more scan lines in length, *recirculation* is used. This is accomplished by having the output continue to be fed the delayed signal. Thus, the delay line also receives one-line-previous video at its input. This prevents any dropouts from ever entering the delay line so that its output will always be dropout-free. The delay line will always contain the most recent video that contains no dropouts even if it is from many lines previous.

Full-Color Dropout Compensation. All DOCs rely on nearby lines for replacement video in the case of a dropout. This replacement is straightforward for monochrome but is more complicated for color television systems because the color encoding structure is not identical for each horizontal line. Direct substitution video from an adjacent line has chrominance that is incorrectly encoded for the incoming line and would result in grossly wrong color if used in place of a dropout. There are several solutions to this problem. The simplest is to take substitute video from a recent scan line that has the same color encoding structure as the line containing the dropout.

In the NTSC 525-line system, every other scan line has the same encoded-chroma subcarrier phase. Thus, if video contains a dropout, substitute video from two lines previous can be used directly in place of the dropout and will give correct color. The main disadvantage is that the substitute video is spatially further away from the dropout and as a consequence will be statistically less similar to the missing video. This method has been used commercially in the past and is judged to be marginally acceptable, compared with newer methods. In the PAL system the color encoding structure repeats after four lines; hence, substitutional video must be taken from four lines away to have the same chroma structure as the video containing the dropout, and this is much too distant spatially for adequate performance.

A more satisfactory approach uses previous-line video for substitution and processes it to make the color encoding structure the same as the line containing the dropout. This technique is used in NTSC for the DOC shown in Fig. 15-70b. As in the simple DOC described above, tape playback video first passes into the one-line delay. Previous-line video at the output of the line delay is then separated into luminance and chrominance channels by luminance low-pass and chrominance bandpass filters, respectively. In NTSC, chrominance on a given line has chroma subcarrier opposite in phase from that of an adjacent line of the same field. Thus, the output of the chroma-bandpass filter is inverted to yield the same encoded subcarrier phase as the incoming off-tape video. Since only the color-encoding structure changes from line to line, the luminance from the previous line is substituted directly.

The PAL-625-line system is somewhat more complicated in that the color subcarrier of the adjacent line is shifted in phase by 90° and the V color-difference signal is opposite in polarity. One common method of PAL dropout compensation is shown in the DOC of Fig. 15-71. As in NTSC, the previous-line video at the delay line output first is separated into luminance and chrominance by low-pass/bandpass filtering. The chroma signal at the bandpass filter output is further decoded into color difference signals U and V using a decode subcarrier which matches that of the previous-line video. U and V, being independent of the color structure, can be reencoded into chroma of any subcarrier phase. Here, the encoded subcarrier phase is the same as that of the off-tape video, giving previous-line chroma identical in structure to that of the off-tape video. Adding previous-line luminance to the reencoded chroma gives a composite color-video signal derived from the previous line that can be substituted directly for dropouts in the off-tape video, and will give proper color.

Further Improved DOC Methods.   The full color DOCs just described have one weak point. Since the composite video must be separated by low-pass and bandpass filters, there will necessarily be some loss in signal fidelity when a dropout occurs. In particular, the luminance bandwidth will be reduced, and high-frequency luminance will be sepa-

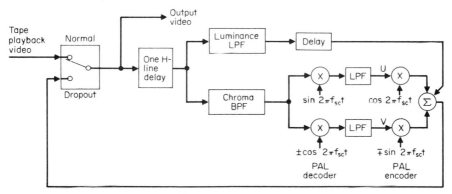

FIG. 15-71 PAL-color previous-line DOC.

rated as chrominance by the bandpass filter and will be processed incorrectly. Both of these defects can be noticeable for picture material of high luminance detail, especially if the low-pass and bandpass filters must be simple low-order designs for cost or space reasons.

A technique presently in wide use, shown in Fig. 15-72, has an additional one-line delay in the output of the chroma BPF in Fig. 15-70 to effectively form a comb filter-based luminance/chrominance separator. A comb filter is a device which can be used to average video from adjacent lines in such a way that luminance is passed through with full bandwidth, if the picture content is similar from line to line. In a comb filter-based DOC for the NTSC 525-line system, as with DOCs already discussed, the off-tape video passes first into a one-horizontal-line delay. The one-line delayed video from the previous scan line is then separated into luminance and chrominance using conventional low-pass and bandpass filters. Luminance is passed straight to the output with no further processing, while the separated chroma is passed through an additional one-line delay.

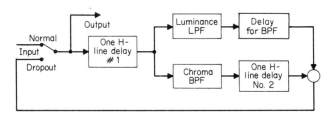

FIG. 15-72   Comb-filter-based DOC.

The *dropout* video thus consists of luminance from the previous line and chroma from two lines earlier. Since the chroma structure in NTSC is the same every two lines, this two-line-previous chroma has the same chroma phase as incoming video and will give correct color when used in substitution.

In this system, the low-frequency luminance passes through the low-pass filter, and high-frequency luminance near chroma passes through the bandpass filter. If it is assumed that the luminance is similar from one line to the next, then the luminance signal at the output of the chroma delay will be the same as its input. Thus, for luminance, the outputs of the low-pass and bandpass filters are added directly together, and if the filters are complementary, the resulting frequency response for luminance will be flat. This differs from the one-line-delay-based DOCs that necessarily reduce luminance bandwidth. The property of flat luminance response depends on the luminance being similar from line to line, which is generally the case, and on the low-pass and bandpass filters being approximately complementary. The sharpness of the filters does not affect the flat luminance property, and so simple filters may be used with little visible degradation.

One seeming disadvantage of the method is that chroma is derived from video two lines removed from the line containing the dropout, and hence will be less similar than previous line video. Fortunately, the eye is less sensitive to mispositioned chroma than to mispositioned luminance, and the trade-off here is a good one: flat full-resolution luminance derived from the previous line, and chroma from two lines previous.

DOC Circuit Implementation.   The type of circuitry, analog or digital, used to implement the chosen DOC system can greatly affect such factors as cost, size, complexity, stability, and reliability. While workable analog DOCs can and have been built in the past, recent and new designs are digital. While digital designs can be more costly, complex, and space-consuming than equivalent analog ones, the digital system inherently is stable and predictable, and requires no calibration or adjustment. Digital integrated circuits are continuously being developed that cost less, do more, and use less power than their predecessors.

The DOC function is a feature of modern time-base correctors, which are mainly digital, making it especially convenient to implement the DOC in a digital form. As shown in Fig. 15-67, the DOC circuit follows the analog-to-digital converter, where digital video data are first available in the time-base corrector (TBC). The output of the DOC then passes to the main memory, or to other TBC enhancements such as variable-speed processing, which is described in the next section.

15.5.5 VARIABLE-SPEED PLAYBACK PROCESSING.

by Steven Wagner

Modern professional helical-scan videotape recorders, besides providing excellent performance in standard playback use, also can be used to vary the tape playback speed from zero to greater than normal speed in the forward direction. In addition, they can be operated in a reverse slow-motion mode, always maintaining a disturbance-free broadcast quality signal. This is made possible by both the system of automatic scan tracking, described in Sec. 15.7.5, and by a unique processing technique in the TBC. Automatic scan tracking, by keeping the playback head centered on the recorded track over a wide range of tape speeds, eliminates the severe cross-tracking noise which occurs in an unequipped helical VTR at incorrect tape speeds. However, the signal must be processed by the time-base corrector to make it suitable for broadcast or recording.

Helical-Scan Playback Signal Characteristics in Variable-Speed Mode. The playback signal of a helical-scan VTR in the slow-motion mode has several properties with which the TBC must cope:

1. Depending on tape speed, the playback head will skip or repeat a track (one field on a type C format) on an intermittent basis to stay on the track center. This means that the off-tape video no longer conforms to the color-frame sequence (four fields in NTSC, eight in PAL) of reference sync.

2. When the playback head jumps a track, there will be a step change in time-base error of several lines, occurring in the vertical interval, because of the track geometry on tape.

3. There will be a constant frequency error in the playback signal of up to about 1 percent, related to the longitudinal component of the head-to-tape speed.

TBC Operation in Variable-Speed Mode. Properties 2 and 3 above can be handled readily by the structure of modern time-base correctors. The step time-base error is removed simply by shifting the TBCs main memory vertical reset pulse in the amount and direction of the playback-video step time-base error. The VTR normally sends special control signals to the TBC when the step error occurs, as well as the amount and direction of the error. At first it would seem that property 3 would present a severe problem: If the video contains a constant frequency error, then the time-base error between the playback signal and the reference sync is continually increasing, requiring an infinite time window in the TBC. However, the scanner rotation is locked in frequency to the reference vertical sync in all modes, including variable play, and thus the time-base error accumulates only over one field and then resets to zero in the vertical-blanking interval. If the frequency error is +1 percent, as would be the case in forward two times normal speed for an NTSC type C helical-scan VTR, then the time-base error which builds up over a field will be 2 percent of 262.5, or slightly more than five scanning lines. This sets a lower boundary of six on the number of lines the TBC memory must have to allow variable playback to two times normal speed.

TBC Variable-Speed Video Processing.   Property 1 above, the skipping or repeating of fields which occurs in the variable-speed mode, is the most difficult to handle in the TBC. When and how often this field jumping occurs depends only on playback variable tape speed relative to normal. In other words, the TBC output field type must match that of the reference sync. Therefore in the variable-speed mode the TBC must be able to convert or modify any given playback field into any other type field as dictated by the station reference sync. Conversion without side effects is made difficult by substantial difference in the structure of the fields.

1. Because of the system of interlace, adjacent fields are offset vertically by one-half scan line.

2. The color encoding structure is not identical on corresponding lines of different fields.

In NTSC there are two unique color frames, each having two interlaced fields. The fields are normally indicated as fields 1 and 2 of color-frame A and fields 3 and 4 of color-frame B. Fields 1 and 3 differ in that the encoding chroma subcarrier phase will be offset 180° for a given line on field 1 compared with the corresponding line on field 3. The same is true when fields 2 and 4 are compared. In PAL, the sync-to-subcarrier relationship and the V axis phase give four unique frames (eight unique fields). As indicated before, the TBC must convert any off-tape field to the field type indicated by the reference. The details of this process are discussed in the following sections.

Basic Chroma Processing of the Playback Signal.   To have correct color playback in the variable-speed mode, the TBC contains circuits to process playback video such that its chroma structure matches that of the reference sync at all times. There are several methods of chroma processing in use that vary in performance and complexity. The most common method uses standard low-pass and bandpass filters to separate luminance and chrominance, respectively. Chroma is decoded, then reencoded as necessary to match the chroma structure required by the TBC reference. A typical system is shown in Fig. 15-73. The off-tape video passes into a bandpass filter centered at the frequency of the subcarrier with a bandwidth corresponding to that of typical chroma. The off-tape subcarrier and a suitable decoder are used to decode the output of this filter to baseband color-difference signals. These are then reencoded using reference subcarrier. In this way the TBC output video always has correct color in variable-speed modes independent of playback-field type. Since the luminance need not be affected by the chroma processor, the output of the luminance low-pass filter is simply added back to the output.

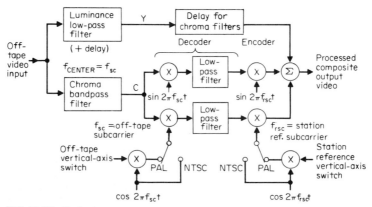

FIG. 15-73   Basic chroma processor.

Comb Filter for Improved Chroma Processing. The low-pass bandpass chroma processor of the type just discussed has one major drawback: It necessarily band-limits luminance to adequately separate chroma, causing loss in horizontal picture resolution in the processed signal. By adding a chroma comb filter in cascade with the chroma bandpass filter, the horizontal luminance bandwidth of the processed signal can largely be preserved. Figure 15-74 shows a comb filter-based chroma processor for NTSC. Off-tape video first enters the comb filter. The comb filter subtracts the composite video of one scan line from that of the adjacent field scan line. The luminance portion of off-tape composite video, if it is similar from line to line, will largely be blocked by the comb filter and instead will take the upper path directly to the output with no resolution loss. This is in contrast to the simpler circuit of Fig. 15-73 where the composite video must be low-pass filtered to a bandwidth less than subcarrier frequency to adequately separate luminance and chroma.

The output of the comb filter is mainly chroma with some residual luminance which is not line repetitive. The signal is further bandpass filtered at subcarrier frequency to remove most of the residual luminance. Because the comb filter adds full contributions from two lines, the chroma will be twice normal amplitude.

When the switch in Fig. 15-74 is in the *normal* position, the output is simply the unprocessed input with a short delay. This position is used when off-tape and reference-video fields have chroma which matches on corresponding lines. When the switch is in the *invert* position, the inverted, twice amplitude chroma $(-2C)$ at the bandpass filter output is added to the delayed composite signal $(Y + C)$ giving an output of $Y - C$; that is, the output has chroma which is opposite in phase from that of the input. The *invert* position is used when the chroma phase is opposite on corresponding lines of the off-tape and reference fields. One drawback of the simple comb filter is that it causes some loss of diagonal luminance resolution and a one scan-line vertical offset of luminance and chroma.

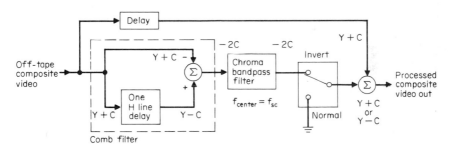

FIG. 15-74 Comb-filter-based chroma processor (NTSC).

Field-Interlace Conversion. With the chroma corrected by one of the techniques described, the problem of dealing with the interlace structure remains. A broadcastable picture can be obtained with no correction of the interlace at all. The off-tape field is used directly as the output field whether or not it is of the same type. The main impairment from this is a vertical bounce of the playback picture of one frame line at a rate depending on playback speed, being most disturbing at speeds slightly off normal. By processing the playback signal, it is possible to remove most of this impairment. The processing consists of the use of a vertical interpolator.

As noted before, odd and even fields differ only in that they are offset vertically by one raster line (one half-field line). An odd field can be made into an even one by shifting it vertically by one frame line. This can be done by vertical filtering or interpolation. One simple vertical interpolator does this shift by averaging adjacent lines of a field with equal weight. The spatial center of this average falls halfway between the scan lines or directly on top of the other field, thus forming an estimate of it. The averaging process introduces some loss in vertical resolution which can be quite noticeable. Higher-order vertical filters involving more lines can improve this at the cost of more complexity.

Variable-Speed Processor in Digital Time-Base Correctors. As shown in Fig. 15-67, the variable-speed processing circuit is placed before the main memory and following the DOC. In this way, the processor receives dropout-free video data, preventing the comb filter from spreading dropouts across two lines due to its line-averaging property.

15.5.6 COLOR-UNDER TIME-BASE CORRECTION

by David E. Trytko

Unique Characteristics of Color-Under Systems. Tape machines employing color-under techniques of recording are capable of providing a video signal which can be monitored directly without the use of a time-base corrector. To provide this ability, it is necessary that the chrominance information be retrieved in a manner which provides a color subcarrier sufficiently stable to lock a local oscillator. This creates an output color subcarrier which is not coherent with horizontal sync. Therefore, the video output of these tape machines cannot maintain an accurate sync-to-burst relationship and cannot be considered in accordance with FCC Rules and Regulations. Although a high-quality picture can be obtained by viewing this signal on a color monitor, time-base correction is required to reestablish a coherent sync-to-burst relationship. The type of time-base correction required for color-under systems therefore is different from that of a direct-color system. It is important to recognize the distinction between the color-under signal and the direct-color signal.

In a direct-color signal, the same time-base error exists in both the luminance and chrominance components. The burst-to-sync relationship therefore remains coherent, and the time-base error of the composite signal can be characterized by a single error profile. Since both components (luminance and chrominance) of the video signal possess the same instability, measuring and eliminating the instability in one component will eliminate the instability in the other. Typically, the error information of the composite signal is determined from the chrominance information; the variation of burst zero crossings is generally accepted as the most accurate representation of signal instability.

In a color-under system, the heterodyne process substantially reduces the noticeable time-base error from the chrominance component. This creates a video signal which is made up of relatively stable chrominance encoded on relatively unstable luminance. As a result, each component of the video signal now contains its own time-base error profile. The time-base corrector, therefore, must stabilize two noncoherent and uniquely unstable components into one coherent composite-video signal.

TBC Methods for Color-Under Systems. The most straightforward method of accomplishing time-base correction of color-under signals is to treat the luminance and chrominance components individually (see Fig. 15-75). After separating the two components, independent error measurements can be made and used to correct the respec-

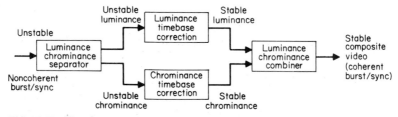

FIG. 15-75   Time-base correction of color-under signals.

tive instabilities. Once the time-base errors have been removed, a stable and coherent composite signal can then be reconstructed.

A second and more cost-effective method of time-base correcting a color-under signal involves the reconstruction of the noncoherent color-under signal into a composite signal which is essentially coherent (see Fig. 15-76). Apart from circuit complexity, the advantage of this method is that it easily facilitates color-under processing using time-base corrector configurations whose primary function is to stabilize direct-color signals. The key to accomplishing this second method lies in the recognition that the instability on the tape machine output color subcarrier is small. This condition is guaranteed by the requirement that the video signal be viewable on a color monitor without requiring time-base correction. To do this, the output color burst must be capable of phase-locking a crystal oscillator that can be used to demodulate the chrominance information. Therefore, the time-base corrector can be made capable of performing this demodulation just like any color monitor.

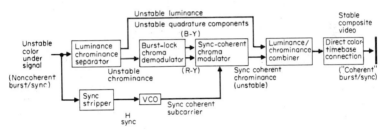

FIG. 15-76 Interactive time-base correction of color-under signals (one-wire operation).

At the same time, the time-base corrector can lock its voltage-controlled oscillator (VCO) input clock to the incoming horizontal sync, which has the same time-base error profile as the luminance component. A sync-coherent color subcarrier can then be derived from this input clock and used to remodulate the demodulated color quadrature components. The time-base error profile of this resulting color subcarrier now resembles that of the incoming luminance information. By combining these two components, an unstable but essentially coherent composite video signal is created. This method, therefore, has effectively simulated a direct-color signal from a color-under signal.

When reviewing the above system operation, it becomes apparent that separate encoding processing in the color-under machine and in the time-base corrector is not required. The decoder in the color-under machine utilizes a relatively stable subcarrier reference to be able to color-lock a monitor. However, this stable encoding process is of no use to the time-base corrector. In fact, it requires the time-base corrector to perform both a demodulation and remodulation process in order to effectively encode the chrominance information using the sync-coherent subcarrier.

If it is known that the output of the color-under machine is to be fed into a time-base corrector before being viewed on a color monitor, then the requirement of *stable* chrominance at the tape machine output is no longer necessary. As long as the chrominance is stable at the output of the time-base corrector, the stability of chrominance at the output of the color-under machine is insignificant. The encoder in the machine, therefore, need not be referenced to a stable subcarrier. Specifically, there is no reason why the on-board encoder of the color-under machine cannot be referenced directly to the sync-coherent subcarrier signal developed by the time-base corrector. By providing the sync-coherent subcarrier as an output, the time-base corrector can stabilize the signal without having to decode and reencode the chrominance information. This method, shown in Fig. 15-77 more faithfully processes the chrominance, and a higher-quality picture is obtained.

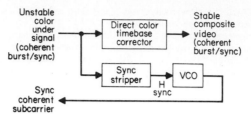

FIG. 15-77 Time-base correction of color-under signals by direct-color TBC.

This improved method of chrominance processing is referred to as *two-wire* operation in reference to the additional signal feed required between the color-under machine and the time-base corrector. The standard operation previously detailed is often described as *one-wire* operation.

15.5.7 SYNCHRONIZERS

by E. Stanley Busby

A synchronizer may be thought of as a time-base corrector having a correction range of one field or one frame. Furthermore, it cannot be assumed that the synchronizer has any influence over the average timing of the input video. The rate at which video is clocked in may be permanently different from the output rate. The purpose of a synchronizer is to accept inputs from uncontrolled sources such as unservoed VTR playbacks, remote sources arriving by microwave, and satellite feeds, and so delay them that the output remains entirely synchronous with local signal sources.

In the case of data entering the buffer faster than they are being withdrawn, inevitably the buffer becomes full. In the case of slower input than output, a condition is reached in which even though there are stored data, data are being read out very shortly after having been written. In the first case, one field or one frame of video is not output at all; i.e., the accumulated difference between input and output rates is discarded a field or frame at a time. In the second case, it becomes necessary to repeat the field or frame just stored, while some more data are accumulated.

Frame synchronizers sometimes discard or repeat in one-frame increments, since it is simple to implement. The jerkiness of motion at the time of adjustment is sometimes visible, depending on the picture content. This effect can be lessened by adjusting only one field at a time.

Whenever a field is repeated, or one omitted, the normal odd-even progression is disturbed, and it becomes necessary to synthesize a field that by all appearances preserves the sequence. Means to do this are discussed in detail in the preceding sections on dropout compensation and variable-speed playback.

When a field synchronizer is serving the dual functions of synchronizer and time-base corrector, at around the point of decision to repeat or discard a field the simultaneous presence of a time-base error can cause multiple conflicting decisions and disturb the picture. Hysteresis is usually applied to the decision circuitry in this case to minimize the effect.

A frame synchronizer, lacking any input at all, will present the last stable and full frame stored, offering *freeze frame* when done deliberately. A field synchronizer can do this too, but with reduced vertical resolution and with the feature of fewer interfield artifacts arising from motion in the picture at the instant of capture. This feature is sometimes used at the end of a tape playback when a few more seconds of video are needed and a still frame is an appropriate filler.

15.6 VIDEO HEAD ASSEMBLIES AND SCANNERS

15.6.1 QUADRUPLEX HEAD ASSEMBLIES

by John W. King

Description and Functions.   The video headwheel panel records and reproduces the video and control track portions of the quadruplex format. In addition, it generates tachometer signals that are used to servo the headwheel motor, to provide video switching information, and to generate the major component of the control track signal.

The headwheel panel is a compact screw-secured subassembly that plugs into the tape transport. This permits quick and easy removal, and replacement by a spare assembly, when the panel is to be returned to the factory for refurbishing.

Each headwheel panel has its own high-speed motor to which the headwheel is directly mounted. Mounted at the opposite end of the shaft are the magnetic and optical tachometers. The motor is equipped with hydrostatic air bearings. Air at 50 lb/in^2 (344,700 N/m^2) is supplied by a compressor located in the console. Air is introduced into front and rear bearing journals through small orifices around the circumference of the bearing, and is exhausted out the ends against the headwheel and the magnetic tachometer disc that serve as thrust bearings.

The headwheel consists of a stainless-steel disc about $\frac{1}{4}$-in (0.6-cm) thick and 2 in (5 cm) in diameter. On its face are mounted four transducer assemblies spaced 90°. Each transducer assembly is made up of a shoe to which a tip of high-permeability, wear-resistant material is mounted. The tip contains the magnetic recording gap and windings and is the only part that contacts the tape. Each transducer assembly must be adjusted radially and circumferentially (this pertains to the head quadrature adjustment) to a high degree of accuracy. A tapered, eccentric pin extending into a hole in the headwheel provides the radial adjustment. Quadrature (circumferential) adjustment is accomplished by means of segments which are attached to the face of the drum between the transducers. The segments are slotted to provide thin sections at each end. One end serves as a spring that is compressed when the unit is assembled, so that it exerts a force against the adjacent transducer which pushes it against the segment on the opposite side. At the other end of the segment, the slot is spanned by a threaded hole which accepts a tapered screw. By loosening the transducer hold-down screw and adjusting the tapered screw, the transducer can be made to rotate around the tapered pin, moving the tip around the circumference of the headwheel. After adjustment, the hold-down screw is tightened. Since spring pressure always exists against the transducers, backlash is eliminated, and very accurate adjustments can be made. Properly adjusted, the tips should project above the rim of the headwheel about 0.003 in (0.08 mm) when new, and should be separated by exactly 90°.

The vacuum guide plays an important role in controlling the head-to-tape interface. When vacuum is applied to the two slots located on either side of the guide center line, the tape is drawn in snugly against the radiused surface of the guide. A relief slot, located in the plane of the rotating tips, allows space for the tips to stretch the tape past the surface of the guide, ensuring that intimate contact is maintained between tips and tape. The guide is movable, being switched between the operating and retracted positions by the guide solenoid. In the retracted position, the tape is pulled clear of the tips, which eliminates unnecessary wear during shuttle and stand-by modes. The operating position of the guide is adjustable through a narrow range, up or down and in or out, about the center of rotation of the headwheel.

The rotary transformer that conducts video signals to and from the transducers is made up of four pairs of ferrite cores, each pair constituting an isolated transformer associated with its separate transducer. The rotating cores have a single-turn winding connected to the transducers and rotate in close proximity to matching stationary cores. Spacing between the rotor and stator cores is adjusted by a screw in the end of the transformer housing which also serves to ground the entire rotating assembly. Spacing between rotor and stator cores is adjusted to about 0.001 in.

The control-track head, which records and reproduces the control track and frame pulses at the bottom edge of the tape, is placed as close as practical to the headwheel to minimize the effect of tape stretch on tracking.

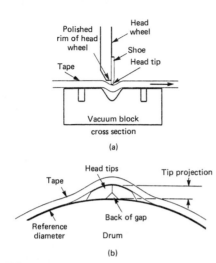

FIG. 15-78 (*a*) Cross section of vacuum guide; (*b*) head-tips–tape-engagement detail.

Figure 15-78 is a cross section of the vacuum guide showing the correct relationship of the headwheel with the guide and the tape in the operating position. Ideally, the center of rotation of the headwheel and the center of curvature of the vacuum guide should be coincident. Dimensions are chosen such that the guide radius exceeds the headwheel radius by 0.0014 in (0.035 mm) the nominal thickness of the tape. The tips stretch the tape locally into the center relief groove, providing the intimate head-to-tape contact necessary to reproduce the required 100-μ in (2.65 μm) wavelengths. The distance that the tips protrude past the unstretched surface of the tape is termed *tip penetration.* It can be seen that the rim of the headwheel just grazes the surface of the tape and serves to damp out oscillations that would otherwise be set up in the tape, causing erratic head-to-tape contact.

The same video signal is fed to all four transducers through its associated transformer during the record mode, and each transducer records across the full width of the tape. Some VTRs gate the record current to prevent disturbing audio tracks during video-only recording. Subsequent erasure of portions of the video signal clears the tape for recording the longitudinal tracks. Some overlap remains, meaning that a small amount of information recorded by the transducer leaving the tape is duplicated by the following transducer as it enters the tape. During the reproduce mode, switching occurs during the redundant interval so that no information is lost. Switching takes place during the horizontal blanking interval, thus allowing the switching transients to be eliminated by subsequent signal-processing circuitry.

Picture Geometry. The geometry of the reproduced picture is sensitive to the positional relationship between the head and the guide. A recording made under the ideal conditions of Fig. 15-78 must be reproduced under identical conditions if geometric errors are to be avoided. Differences in position of 50 μin (1.33 μm) or even less will result in noticeable time-base errors in the reproduced picture. For this reason, provisions are made for making fine mechanical adjustments in the operating position of the guide. While it is true that electronic time-base correctors will compensate for a wide range of time-displacement errors, good operating practice dictates that adjustments be made to produce the best possible uncorrected picture. It is interesting to note that the position of the vacuum guide need not be changed as the tip radius is reduced by wear but remains the same throughout the life of the head tip. The amount of stretch that occurs in the tape due to penetration decreases as the tips wear down, decreasing the length of the tape traversed by the tip for a given angular displacement. The tip, however, having a smaller radius, travels at reduced tangential velocity so that equal amounts of information are scanned during the same angular displacement. This self-compensating feature permits establishing a unique position for the guide.

Additional geometric errors result from incorrect quadrature adjustment as well as imperfections in the vacuum guide.

Evolution. A review of the evolution of the headwheel assembly from its beginnings to its present state of sophistication may help to understand some of the problems inherent in the quadruplex recording process. It should be remembered that when the first quadruplex recorders were placed in operation, time-base correctors had not been invented, and so mechanical precision was essential to produce error-free pictures. To make a recording and reproduce it with the same headwheel assembly is a relatively simple process. As long as the machine is adjusted such that each transducer reproduces its own recorded track, mechanical displacement errors introduced in the record mode are cancelled during the reproduce mode. The requirement for interchangeability among different headwheel assemblies is a different story and requires high precision in the manufacturing process.

Early versions of the headwheel had no provisions for adjusting quadrature, and so it was virtually impossible to reproduce recorded tapes with a headwheel other than that which recorded it. During those early days it was not unusual for headwheel panels to be stored with their recorded tapes if future airing of the program was contemplated. Also, headwheel panels were shipped to other locations with the recorded programs for satisfactory playback. This was a costly and cumbersome procedure and acceptable only as a stop-gap measure. Two solutions were developed to circumvent the quadrature problem. A means of adjusting the 90° relationship among the transducers after final assembly was devised. The method made use of tapered screws and was similar to that used on present-day assemblies. By making a recording and reproducing it so that each transducer reproduced the information recorded by one of the other transducers, it was possible to deduce from the picture which transducer had to be moved, and in which direction. The headwheel was stopped and the appropriate tapered screw was adjusted. Another recording was made and the process repeated until quadrature errors were reduced to acceptable limits under all combinations.

Another method made use of delay lines in the record channels. Nominal time delays were introduced into each channel and a recording was made. If the recording was played back with the machine adjusted so that each transducer played a track recorded by another transducer, then the amount of positive or negative delay adjustment required to make a recording with minimal quadrature error could be estimated. Using these methods, acceptable interchangeable monochrome pictures could be produced.

The first headwheel panels used hysteresis synchronous motors with ball bearings. The ball bearings proved to be a major source of trouble. Even the slightest imperfection in races or balls caused a nonsynchronous time-displacement error which traveled up or down through the picture. The phenomenon, known as *waterfall,* was eliminated with the introduction of the air bearing in 1963, which represented a significant step forward in the state of the art.

The first assemblies used slip rings to conduct video signals to and from the rotating transducers. Dirty slip rings and worn brushes were a common cause of transient noise. Considerable maintenance was required to keep slip rings and brushes in good condition. The rotary transformer introduced in 1962 proved to be the solution to the slip ring problem and produced yet another major improvement in picture quality and machine reliability.

Soon after video tape recording emerged from the laboratory and became a reality, the SMPTE assumed an active role in establishing industry standards. Dimensions and tolerances necessary to ensure interchangeability were published as SMPTE recommended practices. A standard reference tape was fabricated and developed that allowed all headwheel panels to be adjusted to conform to published standards, and thus provide interchangeability.

15.6.2 HELICAL SCANNERS

by Dennis M. Ryan

Helical-scan recording provides a practical means of achieving the high head-to-tape speeds necessary for videotape recording and playback. Instead of forming the tape into a concave shape to mate with transversely rotating heads, as in the quadruplex system, the tape is wrapped convexly around the surface of a cylindrical drum, containing either a centrally located rotating head panel or a rotating upper half to which the head tips are mounted. By inclining the tape so that it wraps the drum in a helical fashion, the tracks are laid diagonally across the tape. The angle of inclination, or *helix angle,* determines the width of tape occupied by the video tracks, and the tape motion establishes the spacing, or pitch, between adjacent tracks.

Types of Helical Configurations. There are two major categories of helical-scan recorders: field-per-scan types and segmented-scan types. In field-per-scan systems, one video field (262.5 horizontal lines in NTSC or 312.5 lines in PAL or SECAM) is recorded on each track. Segmented-scan systems, such as quadruplex recorders, record only a part of a field on each track, usually a half or less.

Within these categories are also found two common types of tape-wrap configurations: full-wrap (or near-360° wrap) and half-wrap (or near-180° wrap) systems. In practice, the wrap angle of a full-wrap scanner is slightly less than 360° to allow for tape threading and the placement of entry and exit guide pins. That of a half-wrap scanner is slightly greater than 180° to provide some overlap, or redundancy, between the entering and exiting heads. Figure 15-79 shows the scanner configurations mentioned.

In a full-wrap, field-per-scan system the head panel, or upper-drum half, rotates at field rate (60 Hz for 525-line systems, 50 Hz for 625). There is a short dropout, or loss of information, which occurs while the video tip traverses the gap between the tape-drum tangency lines. This dropout can be timed to occur during the normally unused portion of the vertical blanking interval. In the 1-in type C format, an optional sync channel is provided to eliminate this loss of information by use of another head tip located below and in front of the video head, which is switched on only during the drop-out period, resulting in a short *sync track* below and ahead of the corresponding video track.

In a half-wrap, field-per-scan system the upper-drum half rotates at one-half of field rate (30 Hz for 525-line systems, 25 Hz for 625-line systems), such that one field is

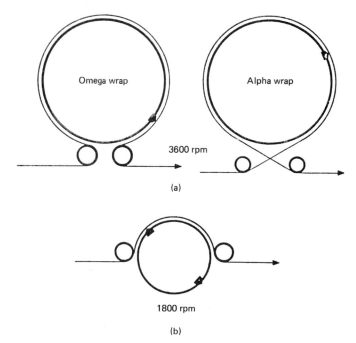

FIG. 15-79 Typical helical layouts: (*a*) full wrap, field per scan; (*b*) half wrap field per scan or segmented.

recorded during 180° of rotation. A second video head, located diametrically opposite the first, records the next field. The signal is switched between the heads during the brief period when both are in contact with the tape, such that no drop-out occurs.

Since the writing speed is directly proportional to the drum diameter and the rotational speed, a half-wrap drum must be twice the diameter of a full-wrap drum for a given writing speed. This is the principal advantage of the full-wrap approach. The half-wrap, on the other hand, is easier to thread and is used almost universally in cassette-loaded systems.

Segmented-scan systems have the flexibility of using any integral number of segments or tracks to compose one video field. If the number of scans per second is chosen as a common multiple of 60 and 50 Hz (as in the type B format, which is a half-wrap, 150-r/s, or 300-tracks/s, system), the drum speed remains the same in both 525 and 625 versions, and there is no difference in writing speed. Another basic advantage of segmented-scan systems is that small-diameter scanners can be used with no penalty in writing speed, enhancing portability. The major disadvantages include the danger of banding in the picture, as in the quadruplex system, and the inability to produce slow-motion, stop-motion, and reverse-play pictures without complicated electronic buffering.

Geometry and Format.   The direction of helix of the tape around the scanner may be either left-hand or right-hand, using terminology borrowed from screw-thread convention. In a right-hand helix, such as the type C format, the scanner rotates in a clockwise direction, viewed from above, and the tape rises in a counterclockwise spiral. In a left-hand helix, the scanner rotates counterclockwise and the tape rises in a clockwise spiral. Thus, for reasons to be discussed later regarding air film, the head always scans from the bottom of the tape toward the top; i.e., it begins the scan where the tape is highest on the drum and ends the scan where the tape is lowest.

The direction of tape motion relative to scanner rotation depends on other design parameters, which will be discussed subsequently. In the type C format (Fig. 15-80) the tape moves opposite to the scanner, thus adding the component of longitudinal tape motion to that of the head and generating a track angle shallower than the helix angle. In most auto-threaded cassette recorder formats, the tape is moved in the same direction as the scanner surface to avoid problems in threading and unthreading tape with the drum rotating. Here, the tape-motion component subtracts from the head motion, and the track angle is steeper than the helix angle.

Figure 15-81 shows the configuration of the tracks recorded by (a) a right-hand helix with tape motion opposite to the head, and (b) a left-hand helix with tape motion in the

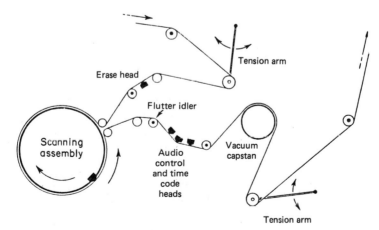

FIG. 15-80   Advanced type C format transport layout.

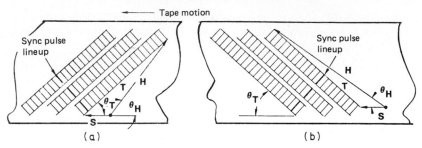

FIG. 15-81   Alternative track configurations.

same direction as the head, viewed from the magnetic oxide side of the tape. All but a few video tracks have been omitted for clarity. Line H represents the path of the head tip during one scan and illustrates the plane of the head motion. The edge of the tape is inclined to this plane by the helix angle ϑ_H. Line S represents the tape motion during one scan, that is, the tape speed divided by the number of scans per second (the field frequency for a field-per-scan system). Line T is the vector sum of these two motions and represents the track itself, inclined to the tape edge by the track angle ϑ_T.

This same figure helps to illustrate another important criterion for helical formats, known as *sync-pulse lineup*. The divisions shown on the track line T represent horizontal video lines, each track containing 262.5 lines in NTSC, 312.5 lines in PAL or SECAM, per field. Because of overlap or drop-out, each track may contain a greater or lesser number of lines, and its active length will be increased or decreased accordingly.

Note that the recorded information is physically positioned so that the horizontal-sync pulses are aligned with those on adjacent tracks. This ensures minimum phase jump in H-sync pulse timing between tracks during still frame operation, and provides the best possible slow-motion picture without special accessories. This condition also eases the design of special accessories that provide full-quality slow- and stop-motion playback or picture-in-shuttle features. Because of the odd number of lines in a frame, the offset between adjacent tracks must equal an integer plus one-half number of horizontal lines such as 2.5 as shown in the illustration.

Owing to this sync-pulse line-up constraint, the field-per-scan formats have a limited choice of parameters, compared with the segmented-scan formats. Consider, for example, the following list of video parameters:

1. Sync-pulse lineup number
2. Video rise (the width of tape occupied by the video tracks)
3. Writing speed
4. Drum diameter
5. Tape speed
6. Track pitch
7. Track angle
8. Helix angle

The sync-pulse lineup number is perhaps more fundamental than the others because it provides a quantifying constraint on other parameters, such as track angle. Selection of any two other parameters in addition to this one will determine the rest, and it is important that the choice of the three be prudent so that systems optimization of all the parameters will be achieved.

Typically, video rise will be established according to tape width and the requirement for longitudinal tracks, i.e., audio, control, and time code. Writing speed, and thus drum diameter, will be established commensurate with required video quality and transport-size limitations, and, as noted above, the sync-pulse lineup number.

Figure 15-82 shows the path of the tape before and following its lines of tangency with the drum. As can be seen, the center line of tape lies in diverging planes that are perpendicular to the tape and inclined at the helix angle θ_H, above and below the plane of head rotation. The tape will remain in these planes of travel as long as it is wrapped about cylindrical guiding elements whose axes are perpendicular to the tape center line. In some helical recorders, the supply and takeup tape reels, and all other tape-path components, lie in these planes. However, it is more common to redirect the tape onto parallel planes, or onto a common plane as in cassette-loaded systems with coplanar reels. This can be accomplished by twisted tape runs between inclined cylindrical guideposts, by helically wrapped pins, conical posts, or some combination of these, arranged such that the center line of the tape remains undeflected from its natural path. In doing so, it is also common to incline the axis of the scanner such that the plane of the cassette or tape reels is parallel to the plane of the deck, or top plate.

Tape Guiding.   Accurate tape guiding around the scanner drum is accomplished with some form of guiding element fixed to the stationary lower drum half, in conjunction with adjustable entry and exit guide pins. These guide pins may be mounted to the scanner assembly, mounted adjacent to the scanner, or even be incorporated into the tape auto-threading mechanism. In any case, they are an integral part of the scanner guiding system.

Most modern helical-scan systems use a continuous guiding surface on the lower drum half. An example of a noncontinuous system is the 1-in type A format developed by Ampex, which was superseded by the type C format in the mid-1970s. In this system the scanner guide is a cylindrical, ceramic pin mounted perpendicular to the drum surface at the midpoint of the scan, which, being eccentrically mounted, can be rotated to adjust its position up or down. Both the upper- and lower-drum halves are the same diameter.

To align this system, it was necessary to use a standard tape made with the proper track curvature and the proper track positioning. Because of the difficulty in specifying and measuring the track geometry, plus the ability to misadjust the guiding and thus produce noninterchangeable recordings, it was superseded by the similar type C format, using a continuous edge guide.

All standard helical formats intended for multimanufacturer, universal interchange use specify video tracks that are straight within very close limits, commensurate with the track width and the size of the track-to-track guard bands. These are achieved by registering the lower edge of the tape against a guiding surface that may be either a helically profiled step machined into the lower drum itself or a separate band that is wrapped around the drum in a helical path and attached with screws at close intervals. In the case of the machined step, the scanner design may also provide a means of adjusting the bearings which support the shaft of the rotating head panel or upper drum, to eliminate runout with respect to the drum of the helical step. In the case of the separate band, the setting fixture may be referenced to the rotational axis of the shaft for the same reason.

The entry and exit guide pins have little effect on steering the tape against the edge guide beyond the first and last quadrants of wrap. Obviously, then, these guide pins have a greater influence on guiding tape in half-wrap systems than they do in full-wrap systems. Full-wrap systems generally require some other means to bias the tape down against the edge guide during the remaining portion of the wrap. Some helical recorders with stationary upper drums use spring-loaded ceramic pushers against the upper tape edge to achieve this purpose. For those systems requiring a rotating upper drum, such an approach would be impractical. In the type C system, the diameter of the upper drum is made slightly larger than the lower by 0.0016 in (0.04 mm). This subtle difference is enough to cause the tape to slide down and register against the edge guide.

The wrap angle of the tape around the entry and exit guide pins must be sufficient to provide enough column strength to the tape so that guide adjustments will have the desired steering effect around the scanner. In half-wrap systems the tape-path compo-

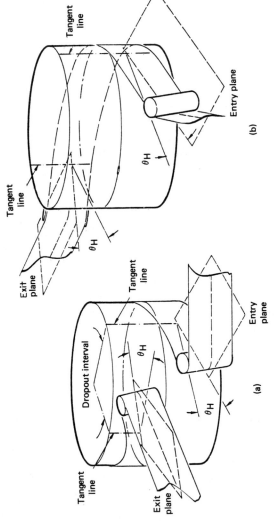

FIG. 15-82 Tape paths: (*a*) full wrap (omega); (*b*) half wrap.

nents adjacent to the scanner can be positioned so as to optimize this wrap angle. For full-wrap systems, the proximity of the two guide pins dictates a minimum wrap angle near 90° each, with a corresponding increase in tension buildup around the pin due to rubbing friction. In a typical type C design, this might total 6 to 8 oz of tape tension increase around the guide pins, while the increase around the drum itself is only 1 or 2 oz. This is not necessarily a problem, except in the case of portable recorders requiring low power consumption, and thus lower drive-motor torques, or in the case of high-performance studio recorders which require lower tension buildups for bidirectional operation and rapid direction changes. In these cases, high-precision roller guides or forced-air lubricated guides help reduce guide pin friction.

To effect tape guiding, the guide pins are usually adjustable for both elevation and tilt in the plane of the tape between the guide pin and the drum. In addition, the guide-pin-to-drum spacing and the angular orientation around the drum may be controlled to establish the length of the scan, or conversely, the length of the dropout interval, plus the location of the drop-out interval. In addition, this spacing and orientation establish the location of the drop-out interval, that is, the physical location of the ends of the track with respect to the tachometer-phased video information.

The guide pins may embody other usable features, such as being slightly conical or tilted toward the drum to locally increase tape tension and thus improve head-to-tape contact near the ends of the scan. They may contain other features to attenuate the tape disturbance effects associated with the head tips entering and leaving the tape.

Air Film.   With a rotating upper drum, a significant air film is generated between the drum surface and the tape, minimizing tension buildup and stick-slip friction. For this reason, all such helical formats scan from the bottom of the tape to the top, meaning that the tape is almost entirely on the upper rotating drum at the start of the scan and that the direction of head movement, and thus of drum rotation, acts to pump air between the tape and the drum. At this point the air-film thickness is at maximum, on the order of 0.001 in (0.03 mm). The film collapses to near zero at the end of the scan, where the tape is almost entirely on the stationary lower drum and the film is generated by tape motion alone. Variation in film thickness is nonlinear because of the helical wrap, the effect of the gap between the drum halves, the pressure distribution under the tape, and other geometric factors. Since the tape speed is relatively low compared with the drum surface speed, the direction of tape motion has little effect on air film and is usually determined to meet other constraints, or in some cases may be bidirectional, such as in the reverse-play mode. In the type C format, the tape moves opposite to the drum rotation in the normal record/play modes, which has the advantage of combining the outgoing or high-tension end of the tape with the point of maximum air film, where the head-tip penetration is the lowest.

Grooving of the rotating drum surface can serve two distinct purposes. Deep grooves in the vicinity of the plane of the head tips have the effect of collapsing the air film locally to improve head-to-tape contact. Shallow grooves are sometimes cut near the center of the rotating drum to channel the air and reduce the problem of wringing or adhesion if the tape is tensioned around the drum prior to startup.

Number of Heads.   The number of rotating heads used in a helical-scan system can vary from one (record/play) to three (separate record, play, and edit erase) heads per channel in a full-wrap system, to twice that number, mounted in diametrically opposite pairs, in a half-wrap system. The type C system, being a full-wrap two-channel system, uses six heads (see Fig. 15-83). The sync-channel head is activated just prior to the drop-out interval and remains so only during that interval.

Most other direct-recording video systems are single-channel type. Recent developments in upgrading half inch half-wrap helical systems have resulted in two-channel formats using separate luminance and chrominance component signals recorded on parallel tracks.

On some half-wrap formats, the azimuth of the diametrically opposite head pairs are radically different to minimize the effect of one head infringing on an adjacent track during playback. This *azimuth recording* permits the reduction or elimination of inter-track guard bands for maximum utilization of the tape.

Scanner Construction. Aluminum is the preferred material for scanner drums, because of its light weight, machinability, and dimensional stability. It has the additional advantage of a coefficient of thermal expansion very near that of polyester-based videotape, thus minimizing time-base errors when playing back tapes over a wide temperature range. Drums have been successfully produced from wrought, forged, die-cast, and gravity-cast processes.

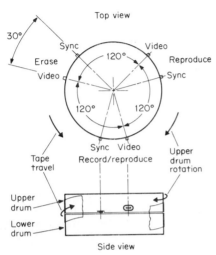

FIG. 15-83 SMPTE type-C head layout. [*D. G. Fink and D. Christiansen (eds.), Electronics Engineers' Handbook, 2d ed., McGraw-Hill, New York, 1982.*]

Despite the advantages of design flexibility, low tooling investment, and good density and uniformity, turning drums from bar stock is usually avoided for two reasons. One is the high cost of machining. The other is the problem of surface condition due to grain orientation within the bar stock, which can adversely affect friction characteristics and air film at the tape interface.

Forged drums are common, particularly in the consumer recorder market, and have the advantages of high density and low waste. Disadvantages include high tooling costs and possibly the grain structure problems alluded to above. Die-casting is attractive because of minimum machining and permits more complex details to be cast in. Material cost is low, but tooling cost is high. The choice of alloys is restricted, and the biggest problem is the control of microporosity and segregation of alloying elements on the drum surface. Gravity-cast drums can be made more uniform, with lower porosity and from wider choice of alloys. Parts costs are higher than die castings, and parts require more machining.

The tape-contacting surface of the drum must be finished to a smooth and highly controlled condition. Some manufacturers send drums out for use in the as-machined condition, while others prefer an anodizing treatment to reduce wear and increase longevity. Both the machining and finishing processes involve proprietary techniques developed through each manufacturer's experience and testing.

On rotating upper drums, the head tips protrude through windows in the drum face or through notches in the lower rim. The amount of *protrusion,* or *tip projection,* is carefully controlled to about 0.0032 in (0.08 mm). The head tips are mounted to carriers or shoes that, in some designs, are factory adjusted and permanently fixed to the upper drum. In other designs, the tip adjustments are made within the shoe, which in turn is very accurately positioned with respect to the drum so that the shoe may be individually removed and replaced in the field.

In addition to tip projection, the azimuth of the gap, head elevation, and angular location of each tip must be accurately controlled. Azimuth is normally specified as perpendicular to the plane of head motion, except for *azimuth recording,* mentioned previously. Because of tape motion, the azimuth is not actually perpendicular to the track itself.

Head elevation establishes the track location with respect to the lower reference edge of the tape via the edge guide on the lower drum. In the type C format the elevation of the record and playback heads are the same since the drum is rephased (120°) to interchange these heads between record and playback modes. The edit erase head is positioned 120° ahead of, and one-third of a track pitch higher than, the record head to follow the same path when both heads are functioning simultaneously. There is a similar relationship between the three sync-channel heads which are located on a lower plane. Thus, the tracks occupy a band beneath the video area, and the heads are switched off during the time when these head tips physically overlap the video track area.

The angular relationship between record, play, and edit-erase head tips is critical, with tolerances of about $\pm 0°5'$ of arc. The angular relationship between head tips which are *hot switched,* such as sync-to-video in type C, or opposed tips in half-wrap systems, is much more critical since errors will result in step-errors in video timing. The same is particularly true for head switching in segmented-scan systems. Here the angular tolerance will be about $\pm 0°0'15''$, which amounts to 0.2 μs.

The shaping and contouring of the head tips involves a series of manufacturing processes designed to obtain the proper gap depth, the proper *footprint* on the tape for optimum contact, and a suitable radius of the edges so that no demagnetization occurs owing to high contact pressure. During the first few hours of use the tips will wear and continue to develop a contour which conforms to the particular video tape magnetic material and base characteristics. This process will repeat whenever another type of tape is substituted. To obtain best performance, it is always recommended that one type of tape be consistently used on a given scanner.

Signals to and from the video heads are brought out through rotary transformers, or on low-cost recorders, through slip-ring contacts. It is quite common to install playback preamplifier electronics and solid-state head switching relays within the rotating upper drum with strict attention to the mounting of the components. As these are subjected to dynamic loads approaching 900 g, there is a need for careful dynamic balancing.

Rotary transformers are generally the type using flat ferrite-disc cores with windings fitted into shallow annular grooves. The gap between the rotor and mating stator is about 0.001 in (0.03 mm). In some designs, the highly polished faces of the rotor and stator cores are lightly spring-loaded together and rely on the air film generated by rotation to separate them. Most type C scanners have six sets of rotary transformers, one for each video and sync head. Type C units not equipped for either the sync channel or the video confidence options may contain only two transformers, one for edit erase and one switchable for either record or playback.

Slip-ring and brush assemblies, even if not used for head signals, are commonly used to conduct power and control signals to rotating preamplifiers or to playback-head servos for automatic scan tracking. These may use a variety of exotic materials and alloys in both the ring and brush elements to improve wear characteristics, reduce noise, and provide low contact resistance. As a general rule, the ring diameter is kept to the minimum to reduce surface speed, while still being large enough to provide mechanical rigidity and to accommodate the internal wiring of each ring. Brushes may either be wire-wiper type or spring-loaded plunger type, similar in principle to electric motor commutator brushes.

A common feature of helical scanners is a tachometer, usually a disc attached to the shaft below the drum interfacing with a stationary transducer. Optical encoding is typical, using slits in an opaque disc or bars printed photographically on a translucent disc, to interrupt a light beam focused on a photosensor. Other types include capacitive or magnetic coupling techniques.

The tachometer is accurately phased to the location of the video record head so that the angular position of the head tip can be detected and servoed during both record and playback. On some units the tachometer is also used to provide head-switching information, as between the sync and video tracks of type C. The number of tach pulses per revolution can vary from one pulse to several hundred, the higher-resolution systems being common in portable recorders for improved scanner servo control during high angular inertial disturbances, and also in more sophisticated studio recorders.

High resolution optical tachometers are commonly used in conjunction with a matching stationary grating to shutter the light beam, or with some means of collimating the light source to prevent stray light from reducing the sensitivity of the system.

Most modern helical-scan systems use a dedicated dc motor to drive the scanner, although in some systems the input torque is provided through a belt drive shared by other loads, with the scanner speed being regulated by a magnetic-particle brake. On some compact recorders the scanner motor is detached and coupled through a polyester drive belt. However, in most professional recorders the motor is integral with the scanner assembly. Both brushless and brush-commutated dc motors are common, the pancake or flat-rotor type being popular because of compactness, low cogging characteristics, and ease of mounting and maintenance. Commutation of brushless motors may be incorporated into the scanner tachometer disc.

The rotating upper drum or head panel is coupled to the motor and tachometer through a shaft, usually stainless steel, whose high torsional stiffness places the torsional resonant frequencies well above any commutating frequencies or other periodic system disturbances. A tubular shaft may be used as a conduit for wire from the slip rings and/or rotary transformers to the head tips or preamplifier assembly. Ball bearings are commonly used, typically selected American Bearing Manufacturers Association class 7 type, preloaded to reduce radial and axial play.

15.7 SERVO SYSTEMS

15.7.1 OVERVIEW

by Harold V. Clark

The components of video tape recorders referred to as the *servos* or *servo systems* are synonymously defined as *feedback control systems*.

Video tape recorders, unlike audio or other stationary-head tape recorders, have a rotating scanner, or headwheel, to achieve the high head-to-tape speed necessary for video bandwidths. It is then necessary to have at least one feedback control to coordinate the longitudinal tape motion with the rotating heads.

The servos used on videotape recorders may control rotary motion, as for example a rotating headwheel or capstan, or linear motion, such as the vacuum tape guide position on a quadruplex recorder. The variable controlled may be either velocity, position, or a combination of both. Rotating components such as capstans or scanners are often operated in a *phase-locked* mode, which from an analytical standpoint is analogous to a *position feedback loop*.

If the motor is driven by a current source, as is usually the case with a capstan motor, *motor torque* is the output quantity that is directly controlled by a voltage input to the motor drive amplifier. Any torque more than the constant or viscous friction loads on the motor will produce acceleration proportional to the excess torque. The angular velocity is the integral with respect to time of the angular acceleration. The shaft angle (phase) at any time is the integral with respect to time of the angular velocity. Thus, the phase output of the unit is the double integral with respect to time of the voltage input to the motor drive amplifier. This produces 180° phase shift of the feedback signal, so feedback is no longer inverse, but instead positive. The loop is therefore inherently unstable and will oscillate unless some phase-correcting components are added to it.

On the other hand, a loop that feeds back velocity information only, without regard to phase, will nearly always be inherently stable without additional phase-correcting circuits. A velocity loop may be added around the motor, within the phase-locked loop, to provide the phase correction necessary to stabilize the phase-locked loop. This combination has been used in several typical VTR servos.

The early quadruplex transverse scan recorders generally had three feedback control, or servo, systems—a headwheel servo, a capstan servo, and a guide position servo. If an attempt were made to list these in the order of their importance to the function of reproducing a video signal from tape, the capstan servo would rate first place, followed by the headwheel servo, and then the guide position servo. A reproducible, although not standard, recording could be made with no control of the headwheel or capstan other than turning at a constant speed, and with the tape guide in a fixed position. To achieve even a limited playback, however, either the capstan or the headwheel must be under feedback control with reference to the other to assure that the transverse video tracks coincide in time of passing the guide shoe with the rotational passage of a video transducer on the headwheel.

In the interest of standardization and interchangeability, all tapes are recorded as uniformly as possible. In the record mode, therefore, the servos are dedicated to achieving a standard recording. The headwheel or scanner servo causes the start of a video field to be recorded at a precise position on the video tracks. The capstan servo causes the longitudinal tape speed to be accurately controlled, so that the video tracks will be spaced to the standard dimension. Video track length is controlled on a quadruplex machine by accurately adjusting the vacuum tape guide to the standard position. On a helical machine the track length is controlled by accurately controlling tape tension in the record mode.

In the playback mode, the servo functions are similar, but generally oriented to synchronize the reproduced video to studio sync and to compensate for any departure of the recording from the center-line values of various parameters. One example is the action of the guide position servo in compensating for any error in positioning of the guide during either record or playback. Another is the function of the capstan servo in compensating for slight differences between the recording machine capstan diameter and the playback machine capstan diameter.

15.7.2 COMPONENTS

by Harold V. Clark

Motors.   The earlier models of Ampex quadruplex recorders used a hysteresis synchronous motor for the headwheel drive. This type of motor operating in the synchronous mode required a variable-frequency power amplifier to drive it. Since the motor shaft was phase locked to within the torque lag angle of the driving signal frequency, the parameters that had to be controlled were the frequency and phase of the driving signal. The motors were two-pole, three-phase designs, requiring 115-V, 240-Hz, three-phase drive power for 60-fields/s television standards, or 115-V, 250-Hz, three-phase power for 50-fields/s standards.

The drive frequency originated from a single-phase voltage-controlled oscillator. Since the system operated over a narrow range of frequencies, simple *RC* phase-shift circuits were used to generate two signals 90° out of phase. This two-phase signal was amplified by two linear power amplifiers to the required power level. To convert to three-phase power, the output transformers were connected in a Scott-T connection for two-to three-phase conversion.

Later models used a three-phase power switching drive in which the power output transistors operated in a saturated switching mode for greater efficiency. The signal originated in a voltage-controlled oscillator operating at six times the output frequency. A three-stage ring counter, triggered by the oscillator, generated the three-phase switching waveforms for the power output stages.

Early RCA models used the same type of three-phase motor, but operated in a subsynchronous voltage-controlled mode. In this mode the motor output torque was proportional to the applied voltage as long as the motor ran below synchronous speed. See Fig. 15-84. The servo feedback controlled the applied voltage amplitude to

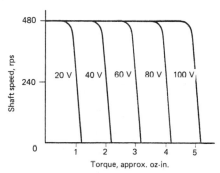

FIG. 15-84 Hysteresis motor operating below synchronous speed.

control the torque and consequently the speed and phase of the motor. In this way the headwheel was phase-locked to four or five times the reference vertical-sync frequency.

Later Ampex and RCA models used brushless dc types of headwheel motors. These motors have simpler, more easily controlled characteristics and are generally better suited to semiconductor drive circuits.

Early models also used hysteresis synchronous motors for the capstan drive. A two-phase dual-winding design provided operation on 15 or 7.5 in/s. One typical motor was a dual 6-pole/12-pole design. In either case the drive power was nominally 115 V, two-phase, 60 Hz (or 62.5 Hz for 625/50 television standards) supplied from a two-phase motor-drive amplifier.

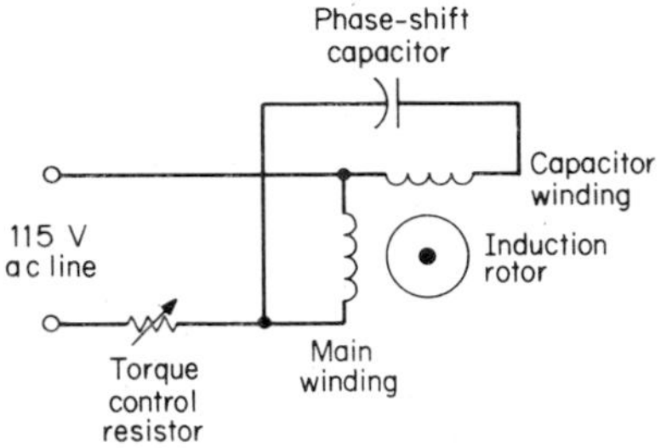

The reel-drive motors in the early machines were 50/60 Hz, 115-V ac torque motors. The designs were high-slip, capacitor induction motors. One winding was designed to operate with a series phase-shifting capacitor to generate the required phase-shifted drive. Torque was adjusted by merely raising or lowering the applied voltage with a variable series resistor. See Fig. 15-85.

FIG. 15-85 AC torque-motor reel-drive circuit.

Later models generally used dc torque motors of one type or another. The Ampex AVR-1 used a series-wound motor with a silicon-controlled rectifier (SCR) drive circuit. The Ampex ACR-25 used a shunt-wound motor with an SCR drive. Still later machines have generally used permanent-magnet-dc or brushless-dc motors.

Tachometers. Tachometers are employed on various rotating parts of VTRs to obtain a measure of the rotation—either velocity or phase. Optical or magnetic types generally are used. The first machine used only a simple photoelectric-cell pickup on the headwheel shaft. It consisted of a reflecting ring painted with nonreflecting black paint over 180° of its rotating area, an incandescent light source illuminating the ring, and a photoelectric cell to receive the light from the reflecting surface. The signal output approximated a square waveform and was used for various purposes within the VTR. One use was for controlling the commutation of the four playback channels in the head switcher. The servo loop used the signal for phase and velocity information. Since the information was obtained only once per revolution, the feedback-loop bandwidth was limited to only one-sixth to one-tenth of that information rate.

For higher-bandwidth servos, multiple-point tachometers are used. Optical tachometer discs with hundreds or even thousands of lines per revolution are used on higher-bandwidth capstan-servo loops. A high degree of accuracy, both in the manufacturing of the disc and its mounting on the capstan shaft, is necessary for low-flutter performance.

Magnetic-pickup devices are also widely used as tachometers. Usually some form of variable-reluctance device is used. Information rates range from one or a few pulses per revolution to hundreds of cycles per revolution on a *gear-tooth* type of device. The latter sums the magnetic-reluctance effect from all the points around the circumference simultaneously and thereby averages out any tooth-to-tooth inaccuracies. Considerable precision is still required in centering and elimination of wobble to avoid once-around disturbances.

15.7.3 HEADWHEEL SERVOS

by Harold V. Clark

Record Mode. The headwheel servo of a quadruplex machine in the record mode is designed to cause the vertical-sync pulses of the recorded video to be written in a location on the tape as specified by the industry standards for the television signal being recorded (see Fig. 15-86).

The basic headwheel servo functions for the record mode are illustrated in Fig. 15-87. The 60- or 50-Hz reference pulse is derived from the vertical-sync signal of the video signal being recorded. A timing point that is easily extracted is the trailing edge of the first wide pulse in the field synchronizing pulse sequence. Its timing accuracy will be that of the tolerance on the sync-pulse width, and will generally be consistent within 1 µs. The headwheel tachometer once-per-revolution signal is phase-compared with the extracted vertical pulse once each fourth (or fifth) revolution of the headwheel to accurately position the time of writing vertical sync on the tape to within the specified standard. The phase comparison of signals with a 4/1 or 5/1 frequency ratio is easily accomplished with a sample-and-hold circuit in which the field-rate pulse samples every fourth or fifth cycle of a ramp signal generated from the once-around tachometer.

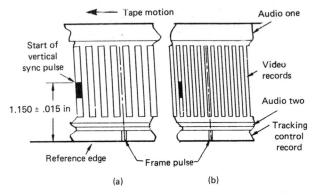

FIG. 15-86   The 525/60 recorded-track format.

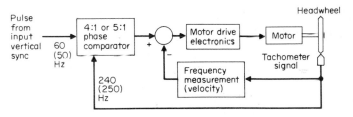

FIG. 15-87   Record-mode headwheel servo block diagram.

Although the standards allow a tolerance of about 20 µs peak-to-peak on the timing accuracy, good engineering practice dictates keeping the error as small as possible. A tolerance no greater than 3 or 4 µs peak-to-peak, including fixed offsets due to mechanical adjustment tolerance, is readily attainable.

Playback Mode.   In the playback mode, the headwheel servo determines the timing of the reproduced video signal. The basic function is similar to the record mode, except that the reference signal is usually local-studio sync instead of sync decoded from input video. The reproduced video signal is timed so that the final video output from the TBC is in time with other studio video signals at the video switcher. The demodulator video output will need to be earlier than the studio reference sync by an amount that will keep the TBC operating about the center of its dynamic range.

Another playback servo mode, used on some machines, involves *horizontal lock*. This mode compares the timing of horizontal sync pulses of the demodulated video signal with studio or other reference horizontal sync pulses. Because the feedback information rate is over 15 kHz, as opposed to 60 or 50 Hz with the other modes of operation, a much higher feedback-loop bandwidth can be obtained. This results in a correspondingly smaller playback timing error of the demodulated video signal before electronic time-base correction. Since the early TBCs had only 1-μs correction range, this mode was necessary to keep the timing error within the range of the TBC.

Later TBCs have much greater correction range, easing the requirement for tight servo control except for the case of electronic editing of tapes that may be played on a machine with a 1-μs TBC range. Because such use is likely, electronic editing of quadruplex tapes is nearly always done using horizontal lock to keep the edit-point-timing transient below 1 μs. A complete line-for-line and subcarrier-phase match is needed to make a perfect edit, and so the servo lock-up sequence involves first framing the playback to within one horizontal line of a correct line-for-line match of playback and reference video, including subcarrier phase, and then proceeding to lock on horizontal sync. Feedback-loop bandwidths to 100 Hz or more are typical with horizontal lock, giving timing accuracies to within 0.2 μs. With field-rate sampling, 5- or 6-Hz bandwidth would be typical, and timing accuracies are reduced correspondingly.

15.7.4 SERVOS FOR HELICAL RECORDERS

by Harold V. Clark

Record Mode. Helical scan VTRs have similar requirements to those of quadruplex machines. A type C recorder in the record mode compares scanner once-around tachometer phase to input-video vertical sync. In this way vertical sync and its associated blanking is positioned with respect to the scanning drop-out region that occurs as a result of the omega-wrap format. See Bibliography.

Playback Mode. Helical equipment in the playback mode, similar to quadruplex compares a tachometer once-around signal with a studio vertical-sync signal to time the final video output from the TBC with other studio video signals at the video switcher. The demodulated video output will require a timing advance equal to the average TBC total delay.

Another scanner playback-timing requirement occurs as a result of the use of separate record and playback heads. The playback head is displaced 120° from the record head, so the scanner must be phased 120° differently between the record and the playback modes.

Helical scanners can be servoed with off-tape horizontal sync in a horizontal-lock mode. This was actually done on some early helical recorders, but modern type C machines do not use the method because the wide dynamic range of TBCs now available makes it unnecessary. The result is that the timing requirements on the servo are less stringent and reliability is increased.

15.7.5 CAPSTAN SERVOS

by Reginald Oldershaw

Control Track. The purpose of the control track is to provide a record of the rotational position of the scanner, in the case of the helical recorder (or the headwheel, in the case of a quadruplex VTR) as the tape passes the locus of rotation of the video heads during the recording process. At the same time, the reel-to-reel tape motion is servolocked to the scanner (or headwheel) by the capstan. In playback, a tracking control is used to adjust the angular position of the capstan motor so that the rotating heads are centered on the previously recorded video tracks to obtain the maximum playback voltage and the maximum S/N ratio.

The control-track signal is recorded near the lower edge of the tape by a fixed head.

The record format showing the control-track position on tape for helical type C 1-in tape, and quad 2-in tape, is shown in Fig. 15-88.

In playback a longitudinal head placed a fixed distance from the scanner reads back the control-track signal. One cycle of control track corresponds to one cycle of scanner rotation, and an odd/even frame identifier is added. In type C helical 1-in recorders, one cycle of scanner rotation is one field. For quadruplex recorders, one cycle of headwheel rotation is one-quarter field for 60-Hz systems and one-fifth field for 50-Hz systems. The record waveforms are shown in Fig. 15-89.

Motor Assembly. At less than 50 in-oz, the torque requirements of the capstan motor are modest. The tape path is designed to have nearly zero-tension difference across the capstan hub to minimize creep between hub and tape. The torque requirements are composed of the accelerative torque needed plus the frictional torque lost in the motor bearings. An important requirement of the motor assembly is uniform angular rotation

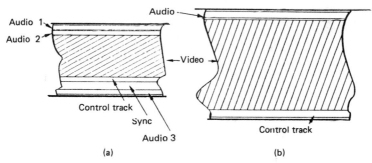

FIG. 15-88   Control-track recorded formats: (*a*) type C; (*b*) quadruplex.

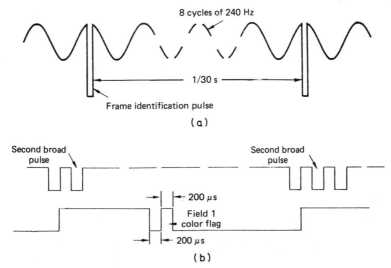

FIG. 15-89   (*a*) Quadruplex control-track record waveform. (*b*) Timing of frame—identification pulse.

at play speed. To achieve this, a motor having minimum brush/commutator or magnetic cogging is essential, and so a skewed armature motor or a multicommutator printed-circuit motor is used. To help still further, a small flywheel having an inertia of only 0.1 to 0.2 in-oz/s^2 and a high-resolution optical tachometer are added to the capstan shaft.

The tachometer design is chosen such that at play speed the output is a multiple of the scanner rotation and so can be phase-locked to an oscillator that in turn is phase-locked to the scanner. For example, in the Ampex VTR models AVR1, AVR3, and ACR25, the tachometer frequency is 3840 Hz, 16 times the scanner frequency of 240 Hz. This forms the inner loop of the capstan servo that, having this high carrier frequency, usually has a bandwidth of 100 to 200 Hz. The high-resolution tachometer is also useful for maintaining adequate feedback information for slow-speed search.

The capstan motor assemblies may include a pinch roller that holds the tape against the capstan hub and is engaged by means of a solenoid during the play and record modes. Capstans using pinch rollers have small shafts, under 0.5 in (1.3 cm), and do not control the tape movement in high-speed modes. In shuttle or search modes, the pinch roller is disengaged from the tape, and a low-inertia tape idler having an optical tachometer meters the tape movement for displaying tape-time and search-to-cue-point information. Since the capstan is used only in the play mode at a single speed, a flywheel is used having an inertia of 0.2 to 0.3 in-oz/s^2.

Capstan assemblies that do not have a pinch roller have a larger hub diameter [up to 2 in (5 cm)] and use either a thin neoprene coating on the hub to prevent tape slip or a perforated surface with an interior vacuum manifold to hold the tape on the capstan surface. Where a pinch roller is not used, the angle of wrap required around the hub is more than 180°. The capstan can be used to control both play/record modes and shuttle/search modes, information for the timer and search-to-cue circuits being provided by an optical tachometer on the capstan shaft.

Analog Capstan Servos. In older video tape recorders in which the servos were completely analog, the playback loop was typically configured as shown in Fig. 15-90. The inner velocity loop had a bandwidth of 100 to 200 Hz. The capstan tachometer signal, after processing, was applied to a frequency discriminator and compared with a reference signal that was 16 times the head-tachometer frequency. Errors in capstan speed were amplified and applied as a correcting signal through the motor-drive amplifier (MDA) to the capstan motor. This type of loop provided excellent damping and reduced commutator and magnetic cogging.

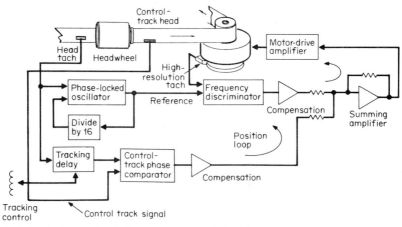

FIG. 15-90   Quadruplex analog capstan-servo block diagram.

An outer-position loop has a much lower bandwidth and provides the phasing correction necessary to keep the video tips on the center of the recorded tracks. In a quad recorder operating on 60-Hz standards, the control-track signal and headwheel rotational rate are 240 Hz and 240 r/s, respectively, and the bandwidth of this loop is 10 to 15 Hz. For 1-in helical recorders, the control track signal and scanner rotational rate are 60 Hz and 60 r/s, with a loop bandwidth of approximately 5 Hz. Errors in phase between the input signals are amplified and summed with the velocity error to provide the necessary correction to the capstan motor.

In record, the control-track phase-comparator is switched out and a capstan-tachometer phase-comparator switched in, using the same input signals as the velocity-loop frequency discriminator. This is to ensure that the capstan remains phased to better than one tachometer line of the high-resolution tachometer.

Digital Capstan Servos. In more recent video tape recorders, the extensive use of digital integrated circuits permits a capstan-servo design that is more compact and has less drift with time and temperature.

Figure 15-91 shows the manner in which this is accomplished using the same motor-drive amplifier (MDA) quad recorder example as in Fig. 15-90. The loop compensation and MDA remain analog functions, but a tachometer-locked loop replaces the velocity loop of the analog example. Once again, this loop is a high-bandwidth loop and has a phase comparator whose output is proportional to the difference in phase between a reference signal derived from the head tachometer and the capstan tachometer signal. When the input signals to the phase comparator exceed one cycle of the capstan tachometer, the output is forced to the limit voltage. This is the case during lock-up and does result in overshoot, since full MDA power is applied until the capstan tachometer frequency reaches the reference frequency, the dynamic range of the phase comparator being only 180° of the tachometer rate. On new servos using microprocessors, this limited dynamic range is extended to ± 8 or ± 64 cycles of tachometer, and is discussed in Sec. 15.7.10.

Figure 15-92 shows one type of digital-phase comparator using a *shift-left, shift-right* register in which only the first three stages are used. When tachometer and reference are equal in frequency, and when the two signals arrive in an alternating fashion, the middle section of the register toggles back and forth with a *mark/space* ratio equal to the dif-

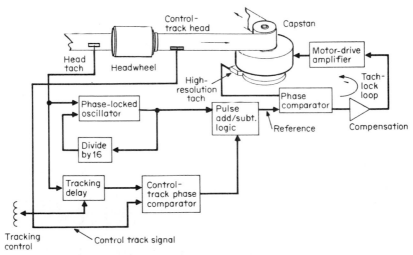

FIG. 15-91   Quadruplex digital capstan-servo block diagram.

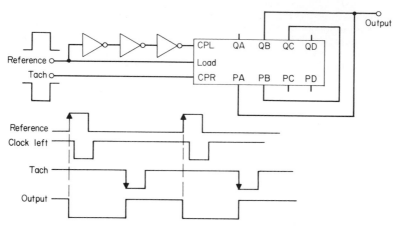

FIG. 15-92   Digital phase-comparator block diagram and waveforms.

ference between the two signals. If the tachometer frequency is lower than the reference frequency, then more *left-clock* pulses arrive, and the output from the middle stage of the register remains high. If the tachometer frequency goes above the reference frequency, more *right-clock* pulses arrive, and the middle stage of the register remains low.

Shifting the phase of the capstan to place the video head tips on the center of the recorded video tracks is done by shifting the phase of the reference to the tachometer lock loop. This requires simply adding pulses to the reference clock stream to shift the phase in one direction.

In record the *add/subt* logic is disabled, and the capstan is once again locked to the headwheel. For the larger shifts in phase required when matching off tape television frames to studio reference television frames, the reference signal to the phase comparator is switched to a fixed clock that is offset by the required framing-speed offset, and then switched back once a frame match has been achieved.

Every effort is made to obtain the maximum gain and bandwidth from the available information rate. For this purpose, the filtering of the clocking signals (capstan tachometer, reference clock, control track, etc.) is done by a *sample-and-hold* method on a ramp instead of integrating over, perhaps, 10 to 50 cycles of tachometer signal, since this adds excessive delay to the loop. Figure 15-93 shows the method. Typically, the output of a digital servo is a flip-flop being alternatively toggled from tachometer and reference. One-shot No. 1 is fired by the reference edge and forms a delay to clamp the capacitor C_1 to ground. At the end of the delay period the capacitor is charged through a constant current generator to form a ramp. One-shot No. 2 forms a sample pulse for a sample-and-hold, and the sample is held on C_2. The sample pulse also turns off the current generator during the sample time to further reduce the carrier component. The output signal is proportional to the difference in phase between reference and tachometer signal.

15.7.6 REEL SERVOS

by Reginald Oldershaw

Constant-Torque Method.   The reel motors are the highest power-consuming components of the video tape recorder. An important part of their design is to minimize noise, since the RF signal off tape is in the order of tens of microvolts, whereas the switching spikes on the motors can be tens of volts when changing modes.

The reel motors can operate in either a constant-torque or a constant-tension mode.

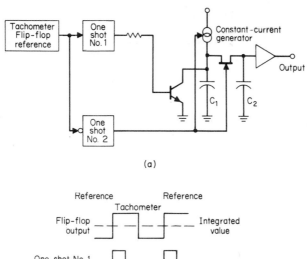

(a)

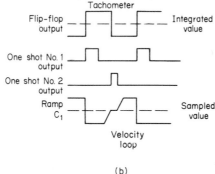

(b)

FIG. 15-93   Removal of clocking signals by clamping: (a) circuit diagram; (b) waveforms.

In early machines, constant torque was provided by means of a resistor connected in series with the motor from a voltage source much higher than the motor back-emf in order to provide a constant current through the motor. The torque provided was $K_t \times I_m$, where K_t = motor-torque constant and I_m = motor current.

On a 2-in (5-cm) quad recorder with a 12-in-diameter (30-cm) reel, the torque required is 90 to 120 in-oz for 15 to 20 oz of tension. On a 1-in helical recorder the required torque is approximately 60 to 90 in-oz for the same reel size.

For fast-wind modes, forward or reverse, the capstan pinch roller is disengaged, and the motor currents in the supply and take-up motors are unbalanced, to control the tape direction. This requires switching the large high-wattage resistors that are in series with the motors, which has been done with relays or electronic switches. The advantage of this method is that small reel motors are required (less than 150 oz-in), and because of the limited available torque, the system is very gentle with tape and makes a uniform tape pack in rewind, an important consideration for long-term or archival storage.

A disadvantage is that this is not a closed-loop servo system and as the pack diameter changes, so does the tension across the video head. For example, with a tach-diameter change of 3/1 the tension might vary from 7 to 20 oz. A further drawback of this simple system is the limited acceleration of the reels due to insufficient torque. This makes the video tape recorder unsuitable for the new automatic editing systems, in which prompt access to each cue point is significant. In addition, the high-voltage and high-wattage resistors needed do not lend themselves to the new compact helical-machine designs.

Constant-Tension Method. Long vacuum columns (2 to 3 ft) to isolate the read/write portion of the transport from the supply and take-up reels have been in wide use in the computer industry for many years. This technique first appeared in a video recorder in the Ampex AVR-1, with vacuum columns less than 12 in long, and was used subsequently on the cassette video recorder Ampex ACR-25 with vacuum columns of 6-in length.

In the *constant-tension* method, the tension across the video and audio heads remains constant. It is independent of reel torque and pack size, and requires a closed-loop servo that varies the motor torque to maintain the required tension across the video head as the pack size varies. In machines using vacuum columns, the tension is determined by the cross-sectional area of the column and the applied vacuum. The columns have ample tape clearance forming a high-loss system for the vacuum supply, but ensuring that tape slitting errors do not cause sticking. Excellent isolation of reel-torque variations from the video head is provided at all speeds and acceleration. A closed-loop servo between column sensors and reel motor keeps the tape loop in the center of the column. (See Fig. 15-94.)

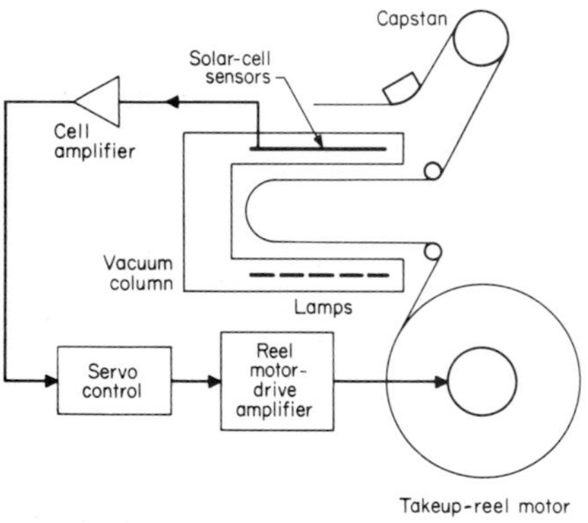

All tape motion is controlled by the capstan. As the capstan pulls tape out of the column, more of the photocell sensor strip is exposed to light from the lamps in the opposite wall. The change in cell current is amplified, a correcting signal is sent to the reel drive circuitry to replace tape in the column from the tape reel, and so the tape is kept centered in the column.

Another constant-tension method has been developed in recent years which eliminates the need for a large vacuum pump in the machine. This is the low-mass tension arm which rotates through a small arc, usually less than 90°. The tape tension is determined by the spring force on the arm when in the center of its range. Isolation of reel torque from the video head during acceleration is not as good by this method since displacement of the arm causes small changes in the spring force. A closed-loop servo between tension arm and reel motor keeps the arm in the center of its range (see Fig. 15-95), and all tape motion is controlled by the capstan. The closed-loop system operates in the same manner as that of the vacuum columns using a sensor whose output varies with change of tension arm shaft angle.

FIG. 15-94 Closed-loop servo using vacuum column.

A variety of sensors has been used for this purpose. One frequently used is the cadmium sulfide photo-pot in which a voltage is applied across the ends of the pot network and an illuminated slit of light photo-couples a portion of the voltage out (Fig. 15-96) as the illuminated slit rotates the voltage swings from zero to the fully applied voltage. Linearity is good, and thermal drift is insignificant since the output is a ratio of the applied voltage. The response time of the cadmium sulfide element is inclined to be slow.

Other methods used on video tape recorders include a rotating Hall sensor in which a magnet rotates over a Hall effect sensor. Both response time and linearity are good in this method.

A magnetic E core assembly also has been used, excited at 100 kHz. The end limbs are wound in opposition, and the *I* portion of the transformer is attached to the rotating tension arm. When the arm is in the center of range, the output signal from the center limb of the transformer core has equal coupling from both end limbs and the signals cancel, but as the *I* portion of the core rotates, the output signal unbalances, and by means of a synchronous detector will rectify to a dc signal proportional to the angular displacement. Response time and drift are good, but linearity is poor over large angular swings.

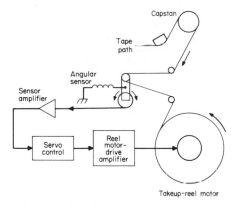

FIG. 15-95   Closed-loop servo using tension arm.

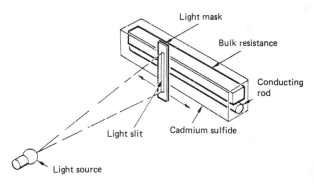

FIG. 15-96   Cadmium sulfide photo-pot sensor.

Reel Motor and Motor Drives.   In constant-tension systems using a closed-loop servo, much larger reel motors are used to provide the accelerative torque. Machines can attain full speed, or return from full speed to *stop,* in 2 s. The motors found in the more recent recorders have torque ratings of 1000 in·oz and 300 in·oz in the 1-in (2.5-cm) helical type C recorders. The motors are always dc motors, though they may be driven from a rectified ac supply. Linear dc-drive amplifiers are unsuitable because of the high power involved. SCR-controlled MDAs operating from 60-Hz alternating current have been widely used with the large quad video tape recorders, but more recently switching-type drives have come into use in the smaller 1-in tape recorders. These operate at a chop rate of 25 kHz and use careful filtering and shielding to prevent switching noise from interfering with the audio and video signals. The output signal is also filtered before being applied to the motor. Even though the motor inductance is large enough to integrate the switching waveform, the resulting magnetic field would cause interference in the audio heads.

One type of SCR control uses a motor having two field windings in a series configuration. The 60-Hz power line is full-wave rectified on each field winding, using SCRs as part of the diode bridge. To produce forward torque, a gate signal is applied to SCR1 and SCR2, and for reverse torque, SCR3 and SCR4 are fired. The gate-control logic turns the SCR on during the 60-Hz alternating cycle, and turns off at the next zero crossing. Figure 15-97 shows typical waveforms as the motor goes from forward to reverse.

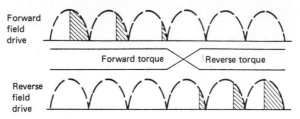

FIG. 15-97 SCR-controlled reel-motor system waveforms.

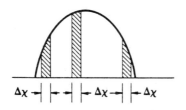

FIG. 15-98 Power as a function of SCR firing angle.

Gate control logic for the SCRs consists of a ramp generator and threshold detector to convert the incoming analog signal to a switched drive signal which controls the angle of firing of the SCR, always turning off at the power line zero crossings. A cosine corrector is also required since the output power would be a function of the sine of the firing angle. This can be seen in Fig. 15-98, where a small change in angle produces a much bigger change in power at the center of the half cycle than at the zero crossings.

Reel drive MDAs using SCRs are efficient owing to the small voltage drop across the active devices and are compact enough to mount directly on the back of the motor.

15.7.7 REEL-TO-CAPSTAN COUPLING SERVOS

by Harold V. Clark

Reel-to-capstan coupling servos are used only on machines which employ the capstan to move tape from reel to reel in all modes of operation, including high-speed shuttle. Machines which do not use the capstan in shuttle modes do not need a coupling servo, because there is sufficient tape storage in the compliance arm systems to absorb the acceleration transient from stop to play or record speed. In shuttle mode on these machines the tape is accelerated by the reel motors themselves, so the acceleration can never exceed the capabilities of the reel motors.

On a machine which uses the capstan for shuttling tape, however, high rates of acceleration may be experienced for periods up to several seconds. If the acceleration produced by the capstan is greater than the rate at which the takeup reel motor can accelerate, a loop of loose tape will be thrown, probably resulting in damage to it. It is therefore necessary to limit the capstan acceleration to a rate not greater than the reel acceleration capability in shuttle modes.

The reel-to-capstan coupling servo limits the capstan acceleration only when a tape storage sensor indicates that it is near the limit of its range due to excessive tape acceleration. In this case, a correction signal proportional to the excess displacement is subtracted from the capstan speed command, correcting the capstan speed reference signal downward to a value such that the capstan acceleration does not exceed the acceleration capabilities of the reels. The bandwidth and compensation of this feedback loop must be carefully chosen to coordinate with the capstan speed loop and the reel servo loops, because the loops are effectively in tandem and will all interact. The coupling servo loop will normally have a response intermediate between that of the reel servos and the capstan speed control servo. See Fig. 15-99.

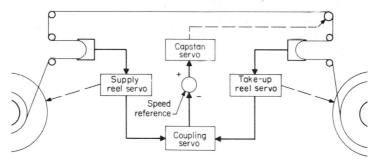

FIG. 15-99   Reel-to-capstan coupling servo block diagram.

15.7.8 QUADRUPLEX AUTOMATIC TRACKING SERVOS

by Reginald Oldershaw

Background.   Autotracking servos are used on 2-in quadruplex recorders to keep the video head tips fully registered on the recorded video tracks without the need to manually adjust the tracking control. Manual operation is still retained for editing and can be selected by means of an *auto/manual* switch.

As video is recorded by the headwheel, the control track is recorded by a stationary head placed 0.725 in (1.84 cm) ahead of the video track, the exact position being determined by means of a factory-made standard alignment tape. Since the video tracks can be only 0.005 in (0.0127 cm) wide, an error of 0.0025 in (0.0063 cm) will produce a 6-dB loss of RF and also will result in an interference pattern on the playback picture due to interference from the adjacent track.

Autotracking was used on the Ampex VR1200 and RCA TR-70 because, with the increasing use of the editing equipment, it was important to have a video tape recorder which needed no manual adjustment while in playback. Subsequently, when the Ampex VCR-25 cassette and RCA TCR-100 cartridge recorders, capable of playing spot commercials as short as ten seconds, were introduced, autotracking was mandatory, since operationally it was not possible to preview each tape and set tracking phase.

Method.   During playback, when tracking manually, it is necessary to adjust the tracking control to obtain maximum RF signal, as indicated on an oscilloscope or meter. It is normal practice to check either side of the peak reading a few times to establish the highest value. This is also necessary in the autotracking method. A low-frequency sinusoidal *dither* signal is injected into the capstan MDA along with the closed-loop error so that the video track on tape is mistracked from side to side across the video head to determine where the peak value lies. An RF-envelope detector provides the signal level resulting from the injected dither signal, and a synchronous detector using these two signals provides an off-track error signal which is coupled back into the loop to correct the capstan phase.

A system block diagram is shown in Fig. 15-100. When the *manual/auto* switch is in the *manual* position, the manual control changes the phase of the delay between the headwheel tachometer and control-track phase comparator in the normal way to move the relative phase of the recorded video tracks with respect to the playback tips on the headwheel. When the switch is set to *auto,* the tracking delay obtains an input signal from the autotracking error.

An oscillator is chosen for the dither signal which is below the audio passband and is injected at a level which does not significantly add to the audio flutter. Nevertheless, it is a flutter component, since the audio is on a longitudinal track and the dither produces

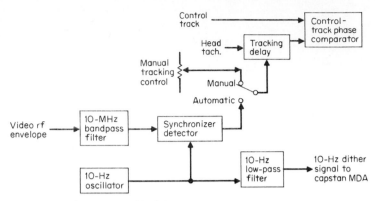

FIG. 15-100   Auto-tracking block diagram.

a longitudinal displacement. The oscillator is free running, not locked to headwheel, and so does not arithmetically add to flutter on multigeneration audio. The RF on tape is envelope-detected in the signal system, and the level is controlled by an AGC having a time constant longer than the 10-Hz dither.

The detected RF is passed through a 10-Hz bandpass filter to extract the dither components and is applied to a synchronous detector which produces an output signal proportional to the off-track error. Figure 15-101 shows the operation of the synchronous detector. When the dither signal is superimposed on the RF envelope, the return signal has a different phase on each side of the track center. At the middle of the track, a lower-amplitude, twice-frequency component is produced. On the left-hand side of track center in the diagram the return signal is in opposite phase to the inserted dither. On the right-hand side of track center the return signal is in phase.

The square-wave 10-Hz signal, used as an input to the synchronous detector, is in phase with the dither component, and is used to rectify the return signal from the 10-Hz filter. The rectified error is shown in the diagram as having a negative value on the left-hand side of track center and a positive value on the right-hand side. A long integration time, in excess of 1 s, is used to distinguish the low-level wanted signal from other low-frequency transport-flutter components.

Lock-up on Nonstandard Tapes.   When the play button is pressed, the autotracking is inhibited, and the tracking delay held to its center value, presuming the tape to have a standard control track format. The frame pulse on the control track signal is first matched to the reference frame pulse, and the 240-Hz control track is then locked to the headwheel tachometer using the control track phase comparator. Once the error signal has settled, the autotracking is *enabled,* and peak tracking is obtained.

If the control track is nonstandard and has a displacement greater than one-half track width, then it is possible during the peaking operation to be on the center of an adjacent track. When this happens, the off-tape vertical television signal is 1 μs out of phase with the reference vertical information. A separate detector referred to as the *track selector* examines the signals for this condition. It uses reference and off-tape vertical information to determine whether the correct track has been selected within a window of tolerance. If the off-tape vertical lies outside the window, an overriding bias is applied to the servo as an error signal, with sufficient amplitude to move the tape to the next track.

On some 2-in quad video tape recorders an additional feature was added by which a tape could be played back even if it had no control track at all. Alternatively, the control track could be rewritten if the control track format showed an excessive error. This was done by means of the autotracking circuits. Lock-up under these conditions was much longer than usual, but it was a means of saving a tape that had been incorrectly recorded.

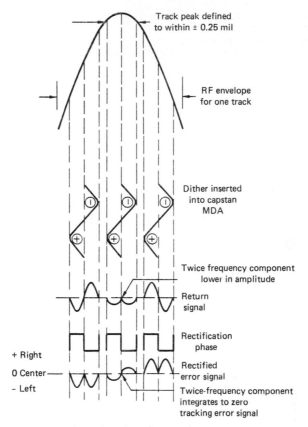

Track peak defined
to within ± 0.25 mil

RF envelope
for one track

Dither inserted
into capstan
MDA

Twice frequency component
lower in amplitude

Return
signal

Rectification
phase

+ Right

0 Center

– Left

Rectified
error signal

Twice-frequency component
integrates to zero
tracking error signal

FIG. 15-101   Operation of synchronous detector.

If the *control-track-presence detector* showed an absence of control track, then no effort
was made to use the control track frame pulse in the framing cycle of lock-up. Instead,
a frame pulse from demodulated video was decoded and compared with reference. Unlike
the frame pulse on the longitudinal control track, the demodulated pulses are disturbed
as the video head crosses tracks, but it is a converging process and after a second or two
the video tips settle on the correct track. Under these conditions, the autotracking error
signal is substituted for the control-track phase-comparator error signal, rather than
merely modulating the delay to the control-track phase comparator, and the time constant of the autotracking detector extended to several seconds. It takes approximately
five seconds for the servos to settle for a satisfactory playback, but many valuable tapes
have been recovered by this means.

15.7.9 C-FORMAT AUTOMATIC-SCAN TRACKING

by Raymond F. Ravizzo

Operational Requirements.   To obtain special-effects reproduction such as slow and
still motion, a helical recorder must transport tape at a different speed from that at
which it was recorded. More particularly, if a field-per-scan format VTR is in the stop-
motion mode, the video head must continuously scan the same recorded track. The

resulting geometric error between the stationary recorded track and the reproduce head scan is shown in Fig. 15-102.

Uninterrupted video reproduction under these conditions requires the aid of a head-tracking servo that will deflect the playback video head in a direction perpendicular to the recorded track to prevent undesirable mistracking, as shown in Fig. 15-102. Here the tracking servo must deflect the reproduce head with a sawtooth waveform at the scanning frequency. In addition, as the track angle varies owing to continuing changes of playback tape speed for various slow-motion rates, the tracking system must continuously respond to these changes by generating a complex sawtooth waveform, as shown in Fig. 15-103. A basic block diagram of a head-tracking servo is shown in Fig. 15-104. Head-tracking

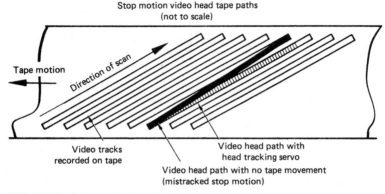

FIG. 15-102   C-format head-tracking geometry.

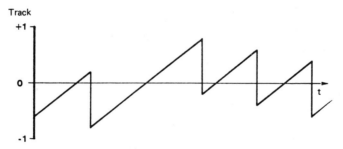

FIG. 15-103   Head-reflection waveform for one-fifth speed slow motion.

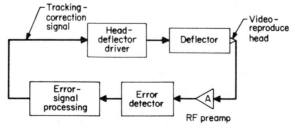

FIG. 15-104   Head-tracking servo block diagram.

errors are determined by detecting variations in the amplitude of the playback RF envelope. Any tracking variations are detected, amplified, and processed, to be used for repositioning the reproduce head to track center.

A head-deflection range of about 0.011 in (± 0.325 mm) is required to track a reproduce speed range from -1 times normal to $+3$ times normal on the 625-line type C helical-scan format. This deflection is most commonly achieved by mounting the head transducer on the free end of a thin cantilever beam made of bimorph material, as shown in Fig. 15-105.

A bimorph consists of two thin strips of piezoceramic material bonded together. With applied voltage, one portion expands and the other contracts. By anchoring the base of the beam, an effective deflection of the tip can be obtained.

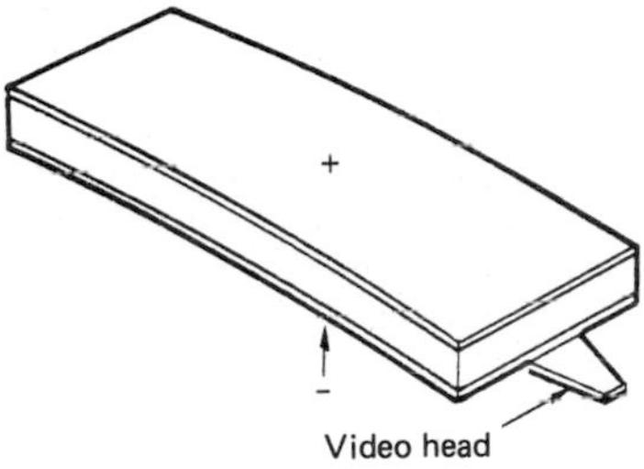

Piezoceramic material has no piezoelectric properties when first fabricated, because the electric dipoles of the granular particles are randomly oriented. The material must be polarized to align the dipoles in a *poling* operation. A deflector drive voltage of about ± 200 V is necessary to cover the required range.

Desirable characteristics of a bender-element configuration include low mass, fast response, and rigidity in directions other than the intended direction of deflection.

Another characteristic of the deflector drive is that voltages applied with a polarity opposite to the original polarization direction of the bimorph crystal must be limited to a low level, or not applied at all. Figure 15-106 shows a method of drive using two bias voltages to prevent depolarization. The arrows show the direction of original polarization with the arrow pointing from plus to minus. If the bias voltage V_B is equal to the peak value of the drive V_D, an opposite polarity drive cannot be applied to the deflector. An alternate method is to limit the amplitude of the depoling drive through the use of Zener diodes.

FIG. 15-105 Bimorph strip and video head.

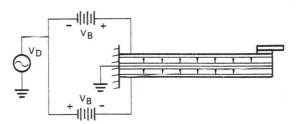

FIG. 15-106 Bimorph deflector drive using bias offsets.

At the extreme ranges of deflection, the reproduce head would have a zenith error of about one degree, as shown in Fig. 15-107. This angle reduces optimum head-to-tape contact and is significant because of the short wavelength of the video recording (<100 μin). This loss can be effectively reduced by any of several ways, as shown in Fig. 15-108.

One approach is to recurve the bimorph bender beam by cross-wiring the deflector plates at a point about one-third from the free end. The net effect is that by inverting the drive polarity to the end plates, the end segment of the beam is deflected in an opposing direction to reduce the zenith angle as shown in Fig. 15-108a. Another implementation is to use multiple bimorphs spaced from each other at the mount and at the tip by a hinged bridge. Figure 15-108*b* shows that this configuration keeps the hinge connecting the two blades parallel to the tape. As the video head tip is mounted perpendicular to the hinge, it also will remain perpendicular to the tape at all deflection positions. Since

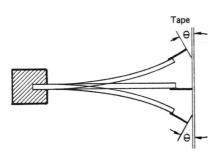

FIG. 15-107   Head-to-tape angle with bimorph deflection.

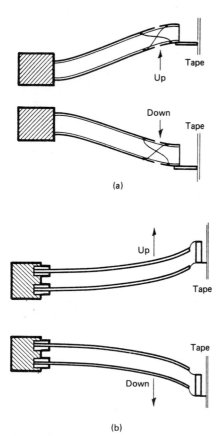

FIG. 15-108   Zenith error correction: (*a*) deflector recurving; (*b*) multiple-deflector hinging.

the deflector is in a cantilever configuration, it exhibits a natural mechanical resonant frequency. Depending on the particulars of the construction, the resonance may be at or near lower-order harmonics of the deflector sawtooth drive. Under these conditions the deflector must be damped to eliminate spurious mechanical vibrations.

By taking advantage of the fact that the piezoelectric effect is bidirectional, a small portion of the deflector is isolated and used in the generator mode for damping feedback. Any spurious vibrations of the head are picked up by the sense strip and fed back into the deflector driver input 180° out of phase for cancellation.

Other systems damp the resonance vibrations by proper placement of *dead rubber* between the moving blade and its stationary holder. The net bandwidth of such a driver and damping system is about 1 kHz.

15.7.10 TRACKING-ERROR DETECTION.   If the detected RF envelope from the reproduce video head were used as a tracking error signal, a decrease in the RF level would allow the servo to sense when mistracking was occurring. However, the servo would not be able to determine in which direction it should react to correct the error.

The method that is used in head-tracking servos to obtain both error amplitude and directional information is *dithering*. This involves intentionally introducing an error, or dither, signal by deflecting the reproduce head a small amount in each direction from track center in a sinusoidal manner in order to create a carrier amplitude modulation of a known frequency and phase. The RF signal is then passed through an envelope detector, followed by a synchronous detector, as shown in Fig. 15-109. The reference for the synchronous detector, in phase with the RF dither signal, is a square wave, usually derived from a system clock or the head-deflector sensor.

When the head is centered on track, the envelope modulation due to dither contains only twice dither frequency components. Therefore, when this signal is syn-

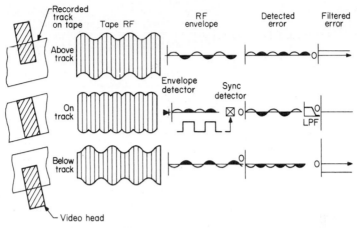

FIG. 15-109   Synchronous detection of tracking errors.

chronously detected at the dither rate, there is no output, as shown in the center of Fig. 15-109. When the reproduce head is off track center, the RF envelope will now exhibit a dither-frequency component, which is synchronously detected into a dc value. If the head is off track in the opposite direction, the fundamental component will be 180° out of phase from the previous condition. This will result in a negative dc output from the synchronous detector. The dither frequency is typically in the 500-Hz region.

An alternate method of determining tracking error is to use a computer-controlled hunting system where a known change in deflector position is made, followed by a sampling of the resultant change in RF level. Based on the results of the sample after the change, the computerized algorithm will eventually seek optimum tracking.

Since the recorded video information frequency-modulates an RF carrier, any RF amplitude change due to tracking dither can be eliminated in the limiter circuits of the demodulator. Conversely, care must be taken to ensure that any spurious AM components of the RF carrier due to video information do not reach the tracking detector. This can be accomplished either by sampling the RF envelope only during horizontal-sync time (since picture content does not affect sync amplitude) or by feeding the envelope detector at the output of the RF equalizer, but before the *straight-line network*. This is the point in the playback chain at which the amplitude versus frequency response is the flattest, i.e., the point at which the least amount of amplitude modulation is found on the FM envelope. Long-term variations of RF level are accommodated by an AGC circuit.

Error-Signal Processing.   The major function of the error-signal processing circuit is to generate the complex sawtooth waveform that represents the geometric correction necessary to keep the reproduce video head on track when reproducing at other than normal speed.

Figure 15-110 shows the required head deflection waveform used to drive the movable head in the still mode. In this mode, the video head repeatedly scans the same track, which in the type C field-per-scan format means the same television field is being played back repeatedly. As shown in Fig. 15-110, when the video head scan reaches the end of a track, the head position must be reset to its initial value to begin scanning the same field again. Advantage is taken of the format-dropout time, in which no video information is recorded for about 10 lines during vertical blanking, to reset the head deflector back to its initial position. The bandwidth of the driver/deflector system is sufficient to move the head within the 0.5-ms dropout time.

If the tape is transported slowly in the forward direction, the dc level of the deflection waveform will move down, or in the negative direction. The head-tracking servo cannot

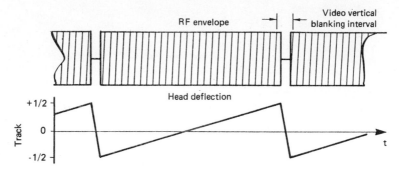

FIG. 15-110 Playback head-deflection relative to RF-envelope signal timing in *still-field* mode.

remain locked to the same field indefinitely as the tape is moved forward. What must occur is for the servo to abandon tracking the current field and lock onto the next adjacent one. This field-changing or track-changing sequence is started by simply inhibiting one of the resets in the sawtooth waveform, which allows the video head to scan onto the next track as shown in Fig. 15-111a. The decision to inhibit a reset pulse is based on the instantaneous value of the waveform at the end of the scan. Here, the logic circuitry

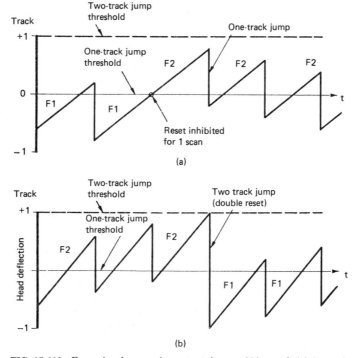

FIG. 15-111 Error-signal processing output for one-fifth speed: (*a*) forward mode; (*b*) reverse mode.

determines that if the waveform is less than zero deflection, i.e., negative, the next reset is inhibited.

Figure 15-111b shows the reset control when the tape is moved in the reverse direction. The dc level of the sawtooth waveform moves in the positive direction. Again, the servo logic must decide when to jump from field 2 to field 1. This is accompanied by determining when the instantaneous deflection at the end of scan is greater than a level of one track in the positive direction. When this occurs, a two-track negative reset command is generated.

A block diagram of one configuration of an error-processing circuit is shown in Fig. 15-112. The output of the tracking-error detector feeds the input of an operational amplifier in an integrator configuration. This allows the closed-loop tracking servo to control both the slope and the dc level of the correction waveform. A *reset* command is implemented by sampling the level of the integrator output at the end of the scan, using a scanner tachometer for basic timing, and deciding whether to generate reset pulses by triggering the track-jump pulse generator.

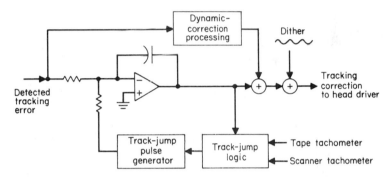

FIG. 15-112   Error-signal processing based on head-deflection information.

To maintain the head-deflection waveform favorably centered about the zero-deflection point, the reset decision levels as shown in Fig. 15-111 must be controlled by tape speed. This is accomplished by measuring the period of a tape-driven tachometer that can then be used to vary the reset levels, depending on tape speed.

The dynamic-correction processing circuit provides servo gain at the scanner rotational rate and its harmonics. Since the helical type C format recorded track is over 16 in (40 cm) long, small nonlinear tracking errors can occur. These errors will have a fundamental frequency at the scan rate, since the error pattern repeats with each head rotation. With the dynamic correction servo gain optimized at these frequencies, any residual tracking error can be corrected by the servo.

Tracking-correction output to the tracking-head deflection driver consists of a summation of the integrator output for variable-speed tracking, dynamic-correction output for nonlinear errors, and a low-level dither signal for error-direction sensing.

15.7.11 VERTICAL-SYNC PLAYBACK PROCESSING. Whenever the tracking video head jumps, either to repeat or to skip the next field, there is a predictable timing discontinuity in the reproduced video. The magnitude of the interruption is due purely to the recorded format and is related to the incremental horizontal line skew between recorded tracks. With the type C format, for 525- and 625-line television systems the horizontal-timing shift between two adjacent tracks (fields) is 260 and 309 lines, respectively. Another aspect of the type C format which must be considered is that vertical sync for a particular field is recorded after the approximately 10-line drop-out. The net result of these factors is that a track jump during the drop-out produces a timing shift

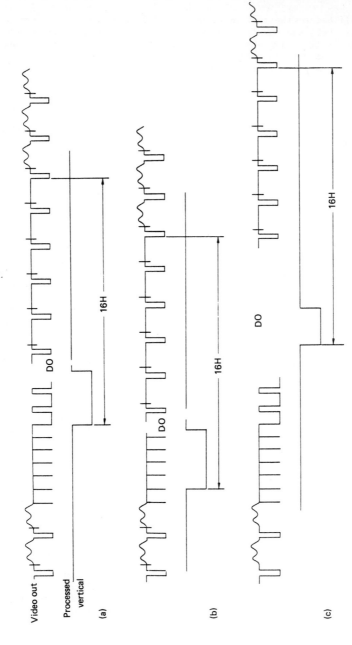

FIG. 15-113 Processed vertical-sync signal output for various playback conditions: (*a*) dropout after second serrated pulse; (*b*) dropout during broad vertical-sync pulses; (*c*) dropout after start of broad vertical-sync pulses.

between reproduced vertical sync and video of the following field. In 525-line systems a shift of $2.5H$ is produced, whereas in 625-line systems, the shift is $3.5H$. Also, the polarity of the phase shift depends on the direction of the track jump. Finally, under some reproduce conditions, there may not be any serrated vertical-sync pulses in the video output, which means that vertical sync cannot be detected by a conventional integrating circuit.

For the foregoing reasons, a vertical-sync signal must be generated that does not depend on the presence of serrated pulses and can be programmed to change by increments of 2.5 or $3.5H$ if the tracking servo system either steps or slips a track, depending on whether the playback mode is faster or slower than normal.

A timing diagram of the processed vertical-sync signal output for various playback conditions for the 525-line format is shown in Fig. 15-113. In Fig. 15-113a the normal reproduce condition is shown where the vertical-sync drop-out begins after the second serrated pulse, and the regenerated and processed vertical sync is in time with playback vertical sync. Figure 15-113b shows the *still* playback condition with the tape unfavorably parked to the extent that all vertical broad pulses are missing. Since the playback head is repeating a field by stepping back at the end of the scan, the processed sync output must be advanced from playback sync by $2.5H$ as shown, in order to be valid for the following field which begins after the vertical sync dropout. An opposite condition is shown in Fig. 15-113c, where the tape is moving at twice normal speed. Here the tracking head must step forward at the end of a scan in order to skip every other field. Processed vertical must therefore be delayed in order to be in time with the following video.

Figure 15-114 is a simplified block diagram of a processed vertical-sync generator. Input is taken from two sources, stripped sync from *demodulator-out* and the tracking-servo jump command. A horizontal phase-locked loop generates a tape-synchronous $2H$-frequency signal. This is used as a clock input to a field counter that can store 525 (or 625) $2H$ pulses per field.

As noted earlier, vertical sync may not always be present. Therefore, the first two equalizing pulses in the vertical-blanking interval are used as a signal to trigger the generation of a complete vertical-sync-interval signal. Since equalizer pulses are shorter in time duration than serrated vertical pulses, care is taken to ensure a safe margin of noise immunity. For this purpose, a *noise-immunity gate* circuit precedes the pulse detector circuit that is used to load the presettable field counter via the reset logic circuit. This is another noise protection circuit that allows only an off-tape load to reach the counter if there are eight consecutive timing disagreements when compared with the *self-load pulse*. The intention is to provide immunity against random single or multiple tape errors, yet reasonable rephasing time (eight fields) for initial synchronization.

Depending on track jump data, the counter is preset for a normal 525/625 count (no

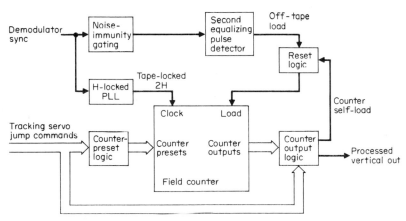

FIG. 15-114 Processed vertical-sync generator block diagram.

track jump) or modified in increments of $\pm 5\ 2H$ pulses (or $\pm 7\ 2H$ pulses in 625-line systems), as dictated by the tracking servo commands. This ensures that the counter will be in the same state at the second equalizing pulse in every field.

The counter output logic, which is also driven by jump commands, controls time shifting of the processed vertical-output signal, based on jump data.

15.7.12 MICROPROCESSOR-CONTROLLED SERVOS

by Reginald Oldershaw

System Configuration. A significant improvement in videotape recorder servos was the microprocessor. It is used in many ways and has resulted in a considerable reduction in size and cost. It provides bus compatibility with all other systems within the video tape recorder, and thereby reduces harness complexity. Field changes can often be made with only a change of erasable-program read-only memory (EPROM), and as improvements in diagnostics are made, customers can subscribe to the updates to improve their systems, all without the use of a soldering iron.

There is no doubt that system design will go even further in the future, but as early as 1982 a 1-in (2.5-cm) type C video tape recorder was introduced in which the entire servo electronics were contained on one board with many self-checking features never possible before because of the quantity and size of hardware that would have been required.

The capstan servo has been chosen as the one for which to describe the application of microprocessors, since it has more variations than the reel or scanner servos. In Sec. 15.7.5 it was shown how the digital capstan servo replaced the analog servo, and the resulting block diagram (Fig. 15-91) shows a tachometer-lock loop in which the capstan tachometer is locked in phase and frequency to a capstan reference clock. Changing the capstan clock produces a corresponding change in capstan speed or phase, as desired.

In the microprocessor-controlled servo, the microprocessor is used with a bus-compatible 16-bit-timer IC to provide the reference to the tachometer lock loop. The timer chip is loaded with a constant from the processor, runs to its maximum count, and reloads, forming a continuous output frequency. To change the frequency, a new load number is presented to the timer IC on the data bus. Each timer IC contains three or more 16-bit counters that are used to replace one-shot and ramp-generator delays, as well as clock sources. The source for the timer IC is the crystal-controlled 4-MHz clock for the processor.

Two data buses are used, one to communicate with the control system in the machine, and the other an internal data bus with which the processor can communicate with its internal devices, such as random access memory (RAM), EPROM, timer devices, and internal and external eight-bit ports. Each port or device has its own unique address, and to communicate with a particular device, an address is decoded and used as an *enable* during that particular portion of machine code.

Figure 15-115 shows the basic configuration. The EPROM stores the program. A typical size for this is 16 kbytes. No more than 1 kbyte of RAM is required. The few analog signals that exist are fed via an analog multiplexer to an analog-to-digital (A/D) converter, the values of which are routinely sampled and stored as a digital word in RAM locations to be accessed at the appropriate program step.

Scanner-tachometer and control-track signals are fed to timer ICs to generate the necessary timing delays. The processor runs through the program in a sequential manner and returns and repeats the steps over and over, but many branch decisions are encountered and the time to complete the loop will vary depending on the branch taken. For this reason, the processor cannot be used to make precision timing measurements. The method used to compare the phase difference between scanner tachometer and control track, for example, is to start a timer using the signal edge of the scanner tachometer and stop the timer using the signal edge of control track, giving a time difference as a number of 4-MHz clock cycles. When the phase difference is needed by the program, the 16-bit number is loaded into the processor in two bytes and the counter is reset, ready to make the measurement again. It is arranged that the cycle time of the program does

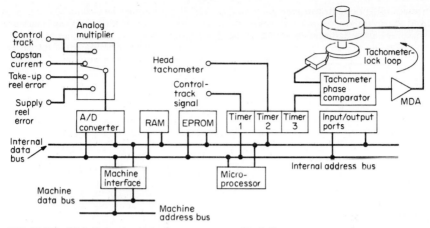

FIG. 15-115 Wide dynamic-range phase comparator block diagram.

not exceed one field, or one scanner rotation, so that the next measurement will not be missed.

System commands and status are passed over a system bus machine interface using any of the available handshaking protocols.

Arithmetic Processing and Filtering. It is desirable, when going from zero to full speed in shuttle mode, to have constant acceleration. This requires the reference clock to the capstan to be stepped in equal frequency increments. The capstan reference clock comes from a timer IC that free runs by counting the maximal count and reloading. For a 16-bit counter the maximal count M is given by

$$M = 2^{16} - 1 \tag{15-49}$$
$$= 65.535$$

The scheme is shown in Fig. 15-116 with a 4-MHz clock

$$F_{\text{out}} = \frac{1}{\text{counter period}} \tag{15-50}$$

$$= \frac{1}{\left(\dfrac{65535 - \text{load}}{4 \times 10^6}\right)}$$

$$= 4 \times \frac{10^6}{65535 - \text{load}}$$

$$\text{Load} = 65535 - \frac{4 \times 10^6}{F_{\text{out}}} \tag{15-51}$$

To change the reference clock in equal increments requires solving this simple equation, but the equation involves dividing by the variable F_{out}. This is a slow operation in the processor and would seriously affect cycle time. A table is set up in program memory to handle this case. The counter load numbers are calculated during the servo design, for example, 256 steps from zero to full speed, and the corresponding load numbers are stored sequentially in a table. In stepping through the table, the load numbers are accessed to provide a close approximation to linear acceleration.

Some calculations are done in the processor, and a good example of this is filtering.

In the analog domain the average value of a signal is obtained by means of a low-pass filter whose *corner frequency* determines the averaging time. In the processor, the same averaging can be achieved by sampling the signal and dividing by the number of samples.

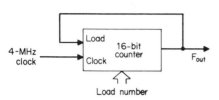

FIG. 15-116   Timer-load scheme.

The averaging time is the product of the sample period and the number of samples. To make the arithmetic run fast, it is usual to arrange for a binary division so that a series of *arithmetic shift-right* instructions is all that is needed.

To understand this, take an example of 21 to be divided by 2 or 4. The number 21 in binary is 10101, as shown in the table in Fig. 15-117. An arithmetic shift-right command yields 1010, or 10, which is 21/2 to within the least significant bit.

A second shift-right yields 101 or 5, so with two shift-right commands we have very quickly divided by 4. Similarly a shift-left instruction will multiply the number by 2. Low-pass filters can average over numbers of samples other than binary by means of simple algorithms using a *shift-and-add* technique. An average of about seven samples can be obtained using five *shifts* and one *add* instruction from the algorithm

$$\frac{1}{8} + \frac{1}{64} = \frac{9}{64} \approx \frac{1}{7.11} \tag{15-52}$$

If, however, it is wished to divide by π, then a full mathematical algorithm is required that is well within the capability of the processor, but prohibitively slow.

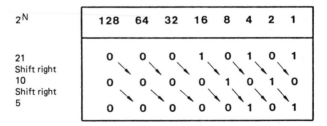

FIG. 15-117   Arithmetic shift-right operation.

A running average frequently is used in microprocessor servo filtering in which samples $S1$ and $S2$ are added together and divided by 2, to form an average. The new sample $S3$ is added to the current average and the sum again divided by 2, which provides a phase advance to the long-term average, since the last sample has a weighting factor equal to that of all previous samples.

Wide Dynamic-Range Phase Comparator.   The digital phase comparator described in Sec. 15.7.7 and used to tachometer-lock the capstan motor has a dynamic range of only one-half cycle of the high-resolution tachometer. If the reference frequency to this loop is changed abruptly, the loop will unlock, relock, and slip several tachometer lines in phase. On video tape recorders using microprocessors, the phase slip is overcome by the method shown in Fig. 15-118.

A four-bit counter is preset to its midrange when the capstan is at rest. As the reference clocks the counter up, the output of the D/A converter applies a signal to the motor, and the motor rotates. The capstan tachometer clocks the counter back down to

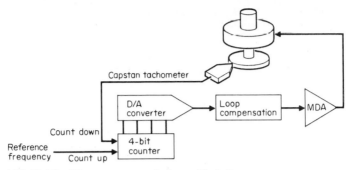

FIG. 15-118 Microprocessor-control servo block diagram.

its midrange position as a steady-state speed is reached. During acceleration, the counter may lag slightly but does not lose the count. Abrupt changes that unbalance the counter or D/A converter will slowly be clocked out by the capstan tachometer without slipping in phase, unless the step change exceeds the counter range. Eight-bit counters and D/A converters have been used, giving a dynamic range of ± 64 clock cycles before losing lock.

Measuring Parameters. Measuring tape-pack diameter in the video tape recorder is done on many models to determine the impending end of tape. In this way the video tape recorder can come to a stop without the tail end of the tape passing across the scanner tips at high speed. The method is to count the number of capstan tachometer pulses that occur for one cycle of reel tachometer and from this to determine the tape pack diameter. It is usual to average three or four measurements to be sure of the result.

Under microprocessor control, this measurement can be extended from one single dimension to as many as desired throughout the reel pack and can be used to determine the change in reel torque required to keep the tension constant across the scanner. It can also be used to modify the reel-servo gain, which is set low upon threading the transport, and modified once a pack diameter has been established.

Waveform Generation. Segments of programs are set aside for test waveforms and can be used, for example, to verify the performance of the automatic-scantracking (AST) servo or for setup when replacing an AST video head. The method of waveform generation is to store the coordinates of the waveform in a table and output the table value to a D/A converter at fixed time intervals.

If a sine wave is encoded in 256 equal period steps, the contents of the first three steps in the table would be the integer value of

$$\sin \frac{360}{256} \qquad \sin \frac{2 \times 360}{256} \qquad \sin \frac{3 \times 360}{256} \qquad \cdots \qquad (15\text{-}53)$$

The coordinate values are converted to hexadecimal, the maximum and minimum values of which are scaled to the limits of an eight-bit byte (00-FF). To display the waveform, the initial address of the table is selected and the contents at that address transferred to the D/A converter. A timer IC is used to step off the intervals, and at each one, the next table value is transferred to the D/A converter.

This method of waveform generation is obviously limited to low frequencies, because each step requires approximately 30 μs of processor time and for 256 steps gives a period of 8 ms or 125 Hz. The waveform described is quantized into 256 (eight bit) amplitude steps which is usually adequate, but waveforms with a higher resolution are possible at the expense of speed and program-memory space.

15.8 AUDIO AND AUXILIARY CHANNELS

by E. Stanley Busby

Every video recorder has at least one channel for recording program audio. Additional audio channels may be provided, either as replicas of the main channel or as communications-grade channels for nonprogram use. In addition, a control track channel carries information related to the head wheel rotation and to the video frame rate.

The physical locations of these tracks are shown in the format drawings of Sec. 15.3 (Fig. 15-49, 2-in quadruplex; Fig. 15-20, 1-in SMPTE type B; Fig. 15-51, 1-in SMPTE type C; Fig. 15-52, 1-in SMPTE type E), and details regarding the control track and its usage are discussed in Sec. 15.7.

15.8.1 DERIVATION OF STANDARD REPRODUCE CURVES. Equalization, the process of correcting for deviations from uniform frequency response, is distributed between the record and reproduce circuits. In general, losses attributable to the reproduce process are corrected in the reproduce circuit and vice versa.

The major loss during reproduce is inversely proportional to wavelength for wavelengths which are short compared with the tape-coating thickness. Assuming a recording having uniform record current with frequency, and no other losses, the system response has been found to approximate that of a simple RC low-pass circuit. The reproduce system must therefore have an inverse response, rising with frequency.

Based on measurements made on typical tape samples, a standard reproduce curve is selected and promulgated by various standards organizations to effect tape interchange.[24,25] The response at high frequencies is expressed in terms of an RC product, or time constant. The response of the reproduce system is given by Eq. (15-54). Values of R_1C_1 range from 15 to 50 μs. Thicker tape coatings and slower tape speeds require the larger values

$$\text{Gain (dB)} = 10 \log \left[1 + (2\pi f R_1 C_1)^2\right] \tag{15-54}$$

Low End Roll-off. In some systems, and in some geographic areas, it is common practice to arbitrarily boost frequencies less than about 200 Hz during record and attenuate them during replay, thus reducing interference from hum fields at power-line frequencies.

The associated reproduce inverse response is given by Eq. (15-55). Again, the response is specified in terms of an RC product. Values of 2000 s are typical for quadruplex recorders, 3180 s for helical-scan recorders, and infinity for quadruplex recorders in parts of Europe

$$\text{Gain (dB)} = 10 \log \left[1 + \frac{1}{(2\pi f R_2 C_2)^2}\right]^{-1} \tag{15-55}$$

Where R_2C_2 is finite, there is a frequency, typically between 600 and 750 Hz, at which the influences of the low- and the high-frequency circuits are equal. It is useful as a test frequency and is obtained by equating Eqs. (15-54) and (15-55) and solving for f. It is given by

$$f = \frac{1}{2\pi} \sqrt{\frac{1}{R_1 C_1 R_2 C_2}} \tag{15-56}$$

Sin (x)/x Correction. In addition, other reproduce losses must be corrected. The loss due to the reproduce gap takes the form of sin (x)/x where $x = \pi \cdot$ gap length $\cdot$ frequency/velocity. The effective gap length may be found by finding the frequency of zero response (facilitated by using a low tape speed) and applying the following equation

$$\text{Gap length} = \frac{\text{velocity}}{f_0} \tag{15-57}$$

Gap length may also be estimated by multiplying the optically measured length by 1.11.

Other losses due to the change in permeability of the core material and eddy current losses versus frequency are also corrected, but are included in both techniques of response measurements described in the following sections. The complete reproduce response is given by

$$G \text{ (dB)} = 10 \log \left\{ \frac{1 + (2\pi f R_1 C_1)^2}{1 + [1/(2\pi f R_2 C_2)^2]} \right\} \left(\frac{x}{\sin x} \right)^2 \tag{15-58}$$

The $x/\sin x$ component of the response is typically produced by allowing the inductance of the reproduce head to resonate with its associated wiring at a frequency well out of band. Figure 15-119a shows how the input resistance of the head preamplifier is used to adjust the resonant rise to equal the gap loss at the upper band limit.

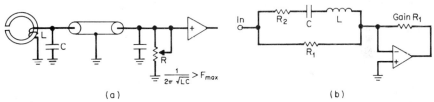

(a)   (b)

FIG. 15-119 Resonance used to correct for $(\sin x)/x$ loss: (a) R used to adjust for resonance rise to equal gap loss; (b) alternate resonance-boost circuit.

Figure 15-119b shows an alternative resonant boost circuit. Assuming a maximum resonant peak of twice that required at the upper band limit, select R_1; assume $A_1 = 0.7$, $A_2 = 0.97$, and $A_3 = 1.0$; set f_m equal to the upper frequency limit; set $x = (\pi \cdot$ gap length $\cdot f_m)$/velocity; and calculate G_1, G_2, and G_3 using

$$G_n = \frac{A_n x}{\sin (A_n x)} \tag{15-59}$$

Set $R = 1/(G_3^2 - 1)$ and find P_1 and P_2 using

$$P_n = \sqrt{\frac{(1 + R)^2 - G_n^2 R^2}{G_n^2 - 1}} \tag{15-60}$$

Then

$$R_2 = R R_1 \tag{15-61}$$

$$L = \frac{(A_1 P_1 - A_2 P_2)}{A_2^2 - A_1^2} \frac{R_1}{2\pi f_M} \tag{15-62}$$

$$C = \frac{A_2^2 - A_1^2}{A_1 A_2 (A_2 P_1 - A_1 P_2) \, 2\pi f_M R_1} \tag{15-63}$$

Need for 240- or 250-Hz Notch on Cue Track. On quadruplex recorders the auxiliary audio track is adjacent to a saturated recording whose major component is equal to the head wheel rotational rate. It is customary to put a deep notch in the response of this channel at 240 Hz (250 Hz on PAL and SECAM systems).

Means of Equalization. The most common reproduce equalizer circuit is shown in Fig. 15-120. At low frequencies the amplifier is an integrator which, with the head as a differentiator, forms a system having flat response. At higher frequencies the gain is controlled by R and the rising response of the head results in the desired response. R is usually made adjustable. If the head is treated as a current source, as shown in Fig. 15-121, it may be thought of as a source of voltage rising with frequency, in series with an

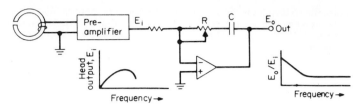

FIG. 15-120 Typical reproduce equalizer.

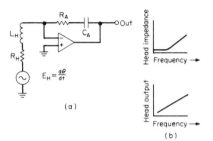

(a)

FIG. 15-121 Current-mode preamplifier: (a) circuit; (b) response.

inductor, whose impedance likewise rises with frequency. The current into the amplifier is then constant with frequency, except at low frequencies where the resistance of the head winding becomes appreciable compared with its reactance.

C_A in the amplifier feedback path compensates for the head resistance by letting $R_A C_A = L_H/R_H$. The desired overall high-frequency response is produced by subsequent RC response shaping or by use of a delay line as shown in Fig. 15-122. A good approximation to the desired $1 + (2fR_1C_1)^2$ response can be obtained by using delays of from $1/20f_{max}$ to $1/10f_{max}$ and gain A of 2 to 10, the shorter delay times requiring the higher gain. The circuit response is given by the following equation where T is delay time

$$\text{Gain (dB)} = 20 \log [1 + 2(A + A^2)(1 - \cos 2\pi fT)]^{1/2} \qquad (15\text{-}64)$$

METHOD 1. ADJUSTMENT OF RESPONSE AND METHODS OF MEASUREMENT. Either of two methods can be used to adjust the response in Method 1. An induction coil with low inductance and driven from a resistive source is brought into proximity to the reproduce head gap. It provides a magnetic field uniform with frequency. The response obtained at the reproducer output is compared with that calculated by Eq. (15-58) and adjustments made until the response is adequately close.

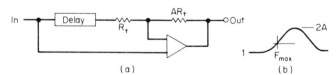

FIG. 15-122 Delay-line equalizer: (a) circuit; (b) response.

Alternatively, the test frequency source can have a shaped response which is the inverse of the standard response. In this case, the output response should include only the last term of Eq. (15-58), since $(\sin x)/x$ loss is not a factor in inductive coupling. Losses due to eddy current and permeability versus frequency are operative and will be adjusted out in this method.

Low-frequency phenomena due to the finite size of the head's pole pieces are not evident in this method.

METHOD 2. The replay of a carefully made reference tape provides the means of adjusting the head azimuth angle, the reproduce frequency response, and reproduce gain.

Reference tapes can suffer degradation of short-wavelength output with extended use. It is good operating practice to generate secondary reference tapes for routine use.

Occasionally a reference tape is furnished whose recorded track is wider than the reproduce head. At low frequencies this will cause an increase in response which must be accounted for during measurement of response and probably when setting reproduce gain.

Refer to Fig. 15-123a. The increase in response as a function of frequency is approximated by the following equation and can be pronounced at low frequencies

$$\text{Fringing gain (dB)} = 20 \log \left\{ 1 + \frac{2 - \exp(-kd_1) - \exp(-kd_2)}{2kW} \right\} \qquad (15\text{-}65)$$

where $k = \pi$ frequency/velocity and W = head width.

At frequencies suitable for setting reproduce levels, Eq. (15-65) is sufficiently accurate. At low frequencies, especially on tape having transverse particle orientation it can yield significant error. For a more exact formula see Ref. 26.

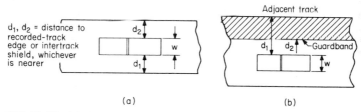

d_1, d_2 = distance to recorded-track edge or intertrack shield, whichever is nearer

(a)   (b)

FIG. 15-123 (a) Dimensions for fringing-gain calculation; (b) dimensions for reproduce crosstalk calculation.

15.8.2 RECORD EQUALIZATION.

Having standardized the reproduce frequency response, the remaining equalization needed is applied in the record circuitry. The required high-frequency boost or roll-off is usually a result of simple first-order RC response shaping, but may employ a delay line. In either case, the circuit must be capable of boosting or rolling off high-frequency components, and capable of adjustment.

The remanent magnetization as a result of the record process at high frequencies is strongly dependent on the ac bias level. The bias level is typically chosen to minimize distortion, and then the record equalization is adjusted to correct the resulting frequency response.

Low-frequency boost, when applied, typically is produced by fixed components.

15.8.3 STANDARD RECORD LEVEL.

The choice of recording level is arbitrary. If too high, then peaks of the program more frequently approach saturation of the medium and are clipped and distorted. If too low, soft passages are obscured by noise. The standard recording level is typically chosen to be 12 to 15 dB below tape saturation, approximately 8 or 9 dB below that level at which third harmonic distortion is 3 percent.

Record level is specified in terms of surface magnetization of the medium in nanowebers per meter of track width. Values of 100 to 110 nWb/m are typical of video recorders.[24,25]

Metering. Program levels are measured by two different means. The peak program meter (PPM) responds quickly to an increase in level and decays slowly, displaying, in effect, the envelope of peak amplitude. The volume unit meter, standard in the United States, is slower to respond and is more symmetric in its attack and decay.[24,25]

Operational judgments of program level using a VU meter tend to maintain signal-to-noise ratio constant and allow distortion to vary as a function of the program peak/

average power ratio. Use of a PPM tends to hold distortion constant and allow the S/N ratio to vary with program content.

15.8.4 DISTORTIONS

Odd-Order and Even-Order Distortions. The tape recording process itself produces distortion products only of an odd-order nature. Even-order-distortion products result from any of a number of causes, including:

1. Any fixed magnetic field in the vicinity of the record head, including the earth's magnetic field
2. Any direct current through the record winding
3. Even-order distortion of the bias current source
4. Any residual dc flux remaining on the medium, prior to recording, as a result of erasure
5. Faulty amplifiers

Odd-order distortion products are usually expressed as a percentage of the output amplitude at some specified record level. The absolute amplitude of odd-order products is very closely proportional to the third power of the record level. The phase of the products is such as to correspond to compression of signal peaks.

Predistortion Means. To some degree, the system may be linearized by creating, in the record circuitry, odd-order distortion products of opposite polarity to that produced in the recording process. Figure 15-124 illustrates two methods of doing this.

Intermodulation distortion, the multiplicative influence of one signal component upon another, is also reduced by this *predistortion*. First-order intermodulation products are reduced to the same degree as is odd-order harmonic distortion. Second-order effects remain unchanged, and third-order effects tend to increase, but are generally below the noise floor and insignificant.

Effect of Bias Level on Distortion. For a given gap length and coating thickness, there is a bias level which results in a minimum third-harmonic distortion. With thick coatings and optimum record gap lengths, as in audio recorders, this effect can be significant.

In video recorders, which of necessity use the same head for recording and reproduction, and thin coatings, the effect is not pronounced. Nevertheless, it is typical to set the bias level somewhat greater than that which maximizes long-wavelength output, minimizing distortion to the extent possible.

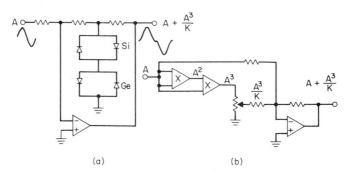

(a)   (b)

FIG. 15-124 Methods for generating odd-order harmonics.

15.8.5 CROSSTALK FROM ADJACENT CHANNELS. The influence of one recording track upon an adjacent one depends on whether they were recorded separately or together. If separately, the crosstalk of a channel into its neighbor may be approximated by the following equations referring to Fig. 15-123(b)

$$\text{Crosstalk (dB)} = 20 \log \left\{ \frac{\exp\,(-kd_2) - \exp\,(-kd_1)}{2kW} \right\} \tag{15-66}$$

where k and W are as in Eq. (15-65). Equation (15-66) assumes no intertrack shield and is grossly inadequate where wavelengths approach the width of the reproduce head structure.

In the case of simultaneous recording, the signal magnetic field penetrates the recording field of its neighbor, and some fraction of it is recorded there.

Crosstalk cancellation in simultaneous recording can be partially effected by injecting the necessary fraction of the record signal, in antiphase, into the neighboring channel.

Cancellation in the reproduce process is done by cross-coupling into a channel a portion of the reproduce signal from its neighbor, in antiphase. The cancellation is effective only in the midrange of frequencies, that is, 250 to 5000 Hz.

The most effective means of crosstalk reduction is the installation of magnetic shields between adjacent heads.

An alternate shield consists of a bar of permeable material positioned close to the reproduce gap. It must be centered on the gap and tangent to the tape path.

15.8.6 NOISE

External Sources. Anything which produces an ac magnetic field in the vicinity of a record/reproduce head contributes to the recorded track when bias current is present and is also coupled into the head during reproduce.

Reel motors generate a field whose effect may be minimized by shielding and/or adjusting the angular position of the motor frame to minimize its field at the audio head. Maximum shielding is effected by wrapping the motor with two concentric magnetic shields, the innermost having moderate permeability and capable of sustaining high fields, and the outermost having high permeability.

The headwheel motor presents a similar problem with similar solutions. The energy involved is generally smaller.

If a picture monitor is mounted in the recorder frame, the fields associated with its deflection yoke are a significant source of magnetic contamination. The horizontal scanning field is the stronger, but only its fundamental frequency is in-band. The vertical scanning field is less strong, but is rich in in-band harmonics.

Shielding of these fields at the source is difficult since the energy absorbed by the shield must be produced by the scanning circuitry. Rotation of the source is out of the question. Typically, the head assembly is encased in a magnetic shield made of a highly permeable material. The best structure is a sandwich of two or more layers of permeable ferrous material separated by layers of copper.

Internal Sources. Ideally, the noise produced by moving an unrecorded medium over the reproduce head should exceed amplifier noise by at least 10 dB. Noise with the medium at rest stems from the resistive component of the head impedance and from various noise sources in the preamplifier.

Preamplifier noise can be minimized by concentrating noise contribution at one semiconductor junction at the input and optimizing the current through it, considering the source impedance of the head at midband frequencies.

Reproduce noise will generally be higher when bias current is applied, consisting, in part, of noise concentrated about the bias frequency and stemming from the granularity of the magnetic particles in the tape. High bias frequencies reduce this effect, but involve problems in obtaining sufficient current.

The bias current must be spectrally pure to avoid the generation of in-band modulation components, which would be perceived as noise.

15.8.7 EDITING. When an audio signal is abruptly turned on or off, it has been modulated by a step function. The resulting spectrum is annoying to the ear. The purpose of editing is to attach a new recording to an old one, or replace a section of old recording with new. To avoid audible disturbances at points of commencement and cessation of recording, it is necessary to cause the old recording to be erased gradually, over a period of 5 to 100 ms, and the new recording to begin just as gradually.

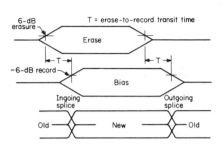

FIG. 15-125 Controlled onset and decay of bias and erase for editing.

The separation between the audio erase head and the record head requires that changes to the erase current must occur before changes at the record head. On multispeed recorders this time is a function of the selected tape speed. Figure 15-125 shows the timing relationship between the envelopes of the erase and bias signals. Some recorders use a modulation envelope resulting from RC charge and discharge, while others use a linear ramp as shown in the figure.

Spectral components introduced by adequately long modulation waveforms are usually inaudible. Clicks, pops, and thumps heard at or near the edit point usually stem from any of the following sources:

1. DC conditions which produce even-order distortion (see Sec. 15.8.4).
2. Failure to maintain symmetrical bias and erase waveforms during the ramp-up and ramp-down.
3. Very rapid turn-on and/or asymmetry of the video erase waveform that can crosstalk into an audio track, recording a low-frequency thump there.
4. Many circuits changing state at the moment of an edit. Power supply voltage changes, small voltage drops in ground wires, and interwiring crosstalk can introduce audible disturbances directly into the audio record channel.

15.8.8 SOURCES OF WOW AND FLUTTER. Perturbations from a perfectly uniform tape speed can produce frequency modulation of the reproduced audio signal. The slower modulation rates are called *wow* and the faster rates, *flutter*. The chief factor which determines tape speed is the peripheral speed of the capstan shaft. Indirect drive capstans, i.e., belt and pulley, are typically stabilized by a heavy flywheel. Elements which contribute to this source of wow and flutter are:

1. Any rotational element which is not round, or which is not rotating about its center
2. Variations in motor torque
3. Uneven races or balls in the ball bearings
4. Slipping of the tape with respect to the capstan surface

Direct-drive capstans typically are fitted with a tachometer disc which generates a frequency that is a multiple of the rotation rate. This frequency is compared with a stable reference, and the torque of the motor controlled to maintain equality of these two frequencies (see Sec. 15.7.2). Failure of the tachometer disc and the capstan to share a common center of rotation will introduce speed error.

The tape is held tensioned by the torque supplied by the supply reel motor and its servo system, if any. Perturbation in tape tension in this system will be reduced by the gain in the capstan servo loop. This gain is finite and inversely proportional to frequency. At frequencies outside the bandwidth of the capstan servo, usually in the tens of hertz, the tape may be considered to be mechanically *grounded* at the capstan.

Tape is a flexible medium. Perturbations in tape tension cause it to stretch. The resulting change in tape speed at the head is reduced by the ratio of the distance from the head-to-capstan distance divided by the distance of the disturbance to the capstan. Typical sources are:

1. The supply reel and its motor and bearings
2. Turnaround idlers and their bearings
3. The impingement of the video head on the tape
4. Out-of-round or eccentric rotation of the scanner in helical scan recorders

The difference between static and moving coefficients of friction can produce a high-frequency variation in tape speed known as *scrape flutter* or *violin-string effect*. It is very much influenced by the smoothness of the tape and the head.

Scrape flutter and other high-frequency disturbances may be reduced by the introduction into the tape path of a *scrape idler,* just upstream of the record/reproduce head, as shown in Fig. 15-126. It usually employs jeweled sleeve bearings and must have very low eccentricity.

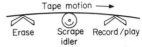

FIG. 15-126   Inertial damping of scrape flutter.

Such idlers are frequently seen on audio recorders, and occasionally on video recorders.

15.8.9 INPUT AND OUTPUT CHARACTERISTICS.
The audio input to a video recorder is typically symmetrical about ground and offers an impedance of greater than 10 kΩ. It may be made unbalanced by arbitrarily grounding one input terminal.

The output is typically balanced as well, with unbalanced output available by grounding one output terminal. The output impedance is low, less than 30 Ω, so that connecting a 600-Ω load does not unduly change the output voltage. The output is generally the secondary of a transformer. Output levels of +8 VU are typical at standard recording level with an output capability of +24 to +28 VU provided to accommodate program peaks.

15.8.10 TIME CODE.
The auxiliary audio channel is often employed to record a time code which is useful in the implementation of automated editing systems (see Sec. 15.9.2). The time code signal is recorded at fairly high levels to achieve good S/N ratio, unlike audio practice, which is to record at a low level so as to avoid crosstalk of the time code signal into adjacent audio channels.

15.9 EDITING

by E. Stanley Busby

This section deals with the ways and means of editing on a particular video recorder, or a small number of directly connected recorders. Edit controllers, which use recorders as one of many peripherals, are treated in Chap. 20.

15.9.1 DEVELOPMENT OF EDITING TECHNIQUES

Mechanical Splicing.   The first attempts at editing were performed on 2-in (5-cm) quadruplex recordings. The recorded tracks were made visible by applying to the tape a suspension of carbonyl iron in a volatile solvent. When the solvent dried, optical magnification permitted identification of the recorded tracks and location of the track containing vertical sync. By use of a simple jig, a razor cut was made in the guard band between tracks, and just after such a track. Thus any disturbance caused during playback would occur when the vertical scanning oscillator in a receiver was least susceptible to interference. After discarding the unwanted portion of the recording, the desired portion, cut in the same relative place, was joined using a thin metallic adhesive tape, spanning the width of the video tape. If any of the following conditions were not met, the splice produced a picture disturbance on playback:

1. An odd field had to be joined to an even one, and vice versa. Otherwise, a half-line time-base discrepancy would be created during playback. The horizontal phase-locked oscillators in receivers would then have to rephase, an operation extending well into the active picture area, and producing a visual effect known as *twitch*.

2. The phase of the control-track signal on the two recordings relative to the recorded tracks had to be the same. If not, mistracking could occur during playback, resulting in interference from adjacent tracks while the capstan servo reacted to the change.

3. The FM-carrier frequencies representing tip of sync could not be grossly unequal, to avoid a resulting dc step function, which could disturb video clamp circuits following the demodulator.

With the early machines, at least half of the splices were unsuccessful, since there was no clear way to identify even and odd fields. Subsequently a sharp flux reversal was superimposed on the control track recording, coincident with the beginning of an even field. This facilitated the optical location of a suitable splice point. Still later, the flux reversal was reduced in frequency to 15 Hz (525-line systems) or 6.25 Hz (625-line systems) to reflect the cyclic relation between color subcarrier frequency and horizontal and vertical rates. At the best, the yield of good splices was marginal, and in no cases could the audio, recorded about nine inches downstream from the video, be edited at the same temporal location as the video.

Electronic Editing.
Several developments led to an electronic analog of the mechanical splice:

1. A video erase head, tilted at an angle, and timed so that erasure began within an appropriate guard band

2. Circuitry to separate the control track pulse from its low-frequency component

3. Timing circuitry to start video recording about 0.6 s after the start of erasure

4. Means to alter the playback timing so that synchronism was established at the reproduce head rather than at the system output

5. Timing circuitry to control the start of audio erasure and the start of audio recording

6. Timing circuitry to control the inverse of items 3 and 5, that is, the process of coming OUT of record

Even with all the above, there remained the severe limitation that the video to be inserted into a recording must be of the same duration as that which was to be replaced. Further, the exact time of the commencement of recording was related to a human function, the pressing of a button.

To deal with the latter, devices were developed to allow rehearsal of a proposed edit, the frame-by-frame adjustment of the proposed edit point, and the actual performance of the edit. Timing of the edit was dictated by a tone burst recorded on the auxiliary audio channel. The location of the burst could be modified by an integral number of frames from a control panel, tested for suitability, and finally committed to tape. Provision was also made for the repetitive addition of a few frames of video to permit a mode of operation similar to that in film animation.

When the source of the video to be recorded was the playback of another video recorder, the relative timing of the two elements of the edit was provided by adjusting each machine's position using its tape timer as a reference, and starting them simultaneously. This method was highly susceptible to inaccuracy.

15.9.2 TIME-CODE EDITING. The need for fast and accurate conjunction of video signals from two or more sources is met by the recording of a digitally coded signal on each auxiliary audio channel. Promulgated by the Society of Motion Picture and Television Engineers, it is called the *Time and Control Code.* Each frame of a recording is uniquely defined by an 80-bit coded signal. To define the frame, 32 bits are used, expressed in hours, minutes, seconds, and frames, 32 bits are reserved for undefined purposes, and a 16-bit sequence defines the frame boundary and includes a bit sequence from which the direction of tape movement can be discerned. The channel coding is *biphase mark* in which each bit cell is delineated by a flux reversal, and a binary 1 is defined by an extra midcell transition. Since the frame rate of NTSC color television is $30 \times 1/1.001$ Hz, the indicated time falls behind real clock time at the rate of several seconds per hour. To compensate for this, two frame counts are skipped over each minute except every tenth minute. This adjustment, which is optional, is called *drop frame.* It is not necessary in 25-frame systems, whose frame rate is exact.

Vertical-Interval Time Code. The time-code signal is recoverable over a wide range of tape speeds, about one-fifth normal speed to 50 times normal. Even so, helical scan recorders offer the capability of reproducing a single field while the tape is stationary, obviating recovery of the time-code longitudinal recording. To provide frame identification in this case, vertical interval time code (VITC) may be injected into the video signal path during recording. One iteration of the code occupies one horizontal interval and is typically recorded on two nonadjacent intervals during vertical blanking.

On most helical scan recorders, VITC may be recorded using either the dedicated sync head, which records only the vertical sync interval, or the main video head, which records all other video, including much of the vertical blanking interval. If recorded on the sync channel, the advantage is obtained that time code may be recorded independently of the video, perhaps at a later time than the video. A disadvantage is that the video head, which is often able to move to adjacent tracks (see Sec. 15.7.9, AST for C-Format Machines), may not be scanning the track identified by the sync channel. If recorded on the video channel, there can be no doubt that it accurately identifies the frame, but it must be applied at the time of video recording, with no possibility of independently altering it later.

In both time codes, numbers are expressed in binary-coded decimal (BCD), using four bits per digit. Since the tens of frames and tens of hours digits never exceed two, and the tens of seconds and tens of minutes digits never exceed five, there are six bits of the 32 time-code bits that are not needed to express time. Several of these are used to convey other information:

1. In NTSC, whether drop-frame time correction is being used in the counting sequence.
2. Whether the counting sequence has been phased to reflect a particular horizontal-to-subcarrier timing relationship.
3. Whether the user bits contain coded alphabetic characters.
4. A parity bit, used to cause the total number of transitions, after coding, to be even. This is a necessary, but not sufficient requirement to edit a time-code track without loss of data during subsequent decoding.

The complete time-code specifications may be found in Refs. 27 and 28.

15.9.3 IN-MACHINE EDITORS. The advent of microprocessors greatly increased the density of logic functions available within a video recorder. Much of this logic capability is applied to editing requirements. Two modes of editing are typically offered, INSERT and ASSEMBLE. The INSERT mode implies that some previously recorded element is being totally replaced. Here, a control track recording is assumed to exist, and the recorder uses it to control the tape position during the recording, just as it would in playback. In the ASSEMBLE mode, it is assumed that a new recording is being added at the end of an old one, and a new control track is recorded, phase-coherent with the old one. In helical scan recorders, erasure of the tape into which the new video recording is to go is provided by a *flying erase head,* mounted on the rotating scanner, and located just ahead of the record head, to pave the way. Many video recorders include the ability to reproduce time code and synchronize their tape movement with an external reference. When the specified frame is approached, all the precursory actions are taken to cause a splice to occur at the specified frame number and to cease at another specified frame number. Typically, video and audio splice points are separately definable.

Communications with a video recorder that is a peripheral of an editing system are accomplished with a high-speed serial interface. Data which are time-related are transferred within a frame interval. The video recorder is expected to be able to position its tape at a specified frame, synchronize its tape movement with other recorders, begin recording at a specified frame, and cease recording at still another specified frame. Many other machine functions may be executed via the serial interface.

The serial interface may either be an external accessory to the recorder or, as is increasingly the case, be contained within the recorder's electronics.

REFERENCES

1. M. L. Williams and R. L. Comstock, "An Analytic Model of the Write Process in Digital Magnetic Recording," *AIP Conf. Proc.,* no. 5, pp. 738–742, 1971.

2. H. N. Bertram, "Long Wavelength AC Bias Recording Theory," *IEEE Trans. Magnetics,* vol. MAG-10, pp. 1039–1048, 1974.

3. H. N. Bertram, "Geometric Effects in the Magnetic Recording Process," *IEEE Trans. Magnetics,* submitted to IEEE.

4. J. C. Mallinson and H. N. Bertram, "A Theoretical and Experimental Comparison of the Longitudinal and Vertical Modes of Magnetic Recording," *IEEE Trans. Magnetics,* submitted to IEEE.

5. John G. McKnight, "Erasure of Magnetic Tape," *J. Audio Eng. Soc.,* vol. 11, no. 3, pp. 223–232, 1963.

6. H. N. Bertram, "Wavelength Response in AC Biased Recording," *IEEE Trans. Magnetics,* vol. MAG-11, no. 5, pp. 1176–1178, 1975.

7. S. Duinker and J. Guest, "Long Wavelength Response of Magnetic Reproducing Heads with Rounded Outer Edges," *Phillips Res. Rep.,* vol. 19, no. 1, 1964.

8. D. L. Lindholm, "Dependence of Reproducing Gap Null in Head Geometry," *IEEE Trans. Magnetics,* vol. MAG-11, no. 6, pp. 1962–1996, 1975.

9. J. C. Mallinson, "Gap Irregularity Effects in Tape Recording," *IEEE Trans. Magnetics,* vol. Mag-5, no. 1, p. 71, 1969.

10. E. D. Daniel, "Tape Noise in Audio Recording," *J. Audio Eng. Soc.,* vol. 20, no. 2, pp. 92–99, 1972.

11. J. C. Mallinson, "The Signal-to-Noise Ratio of a Frequency-ModulatedVideo Recorder," *EBU Review-Technical,* no. 153, 1975.

12. D. L. Lindholm, "Spacing Losses in Finite Track Width Reproducing Systems," *IEEE Trans. Magnetics,* vol. MAG-14, no. 2, pp. 55–59, 1978.

13. Soshin Chikazumi, *Physics of Magnetism,* Wiley, New York, 1964.

14. J. Smit and H. P. J. Wijn, *Ferrite,* Wiley, New York, 1959.

15. Richard M. Bozorth, *Ferromagnetism,* Van Nostrand, Princeton, N.J., 1961.

16. C. B. Pear, *Magnetic Recording in Science and Industry,* Reinhold, New York, 1967.

17. Finn Jorgensen, *The Complete Handbook of Magnetic Recording,* Tab Book, Blue Ridge Summit, Pa., 1980.

18. John D. Kraus, *Electromagnetics,* McGraw-Hill, New York, 1953.

19. C. D. Mee, *The Physics of Magnetic Recording,* Wiley, New York, 1964.

20. *ASTM Standards on Magnetic Properties,* American Society for Testing and Materials, Philadelphia, Pa., 1976.

21. E. Unger and K. Fritzsch, "Calculation of the Stray Reluctance of Gaps in Magnetic Circuits," *J. Audio Eng. Soc.,* vol. 18, no. 6, pp. 641–643, December 1970.

22. M. O. Felix and H. Walsh, "FM System of Exceptional Bandwidth," *Proc. IEEE,* vol. 112, no. 9, p. 1664, September 1965.

23. E. C. Cherry and R. S. Rivlin, "Non-linear Distortion with Particular Reference to the Theory of Frequency Modulated Waves," *Phil. Mag.,* pt. 2, vol. 33, p. 272, 1942.

24. American National Standards Institute, ANSI V98.8, 98.3, 98.17, 98.20, New York.

25. International Electrotechnical Commission, Publication 94.

26. A. Van Herk, "Side-Fringing Response of Magnetic Reproducing Heads," *AES J.,* vol. 26, no. 4, April 1978.

27. *Time-and-Control Codes for Television Tape Recordings (625 Line Systems),* EBU TECH 3097-E, Technical Centre of the EBU, Bruxelles, Belgium, April 1972.

28. *Time and Control Code for Video and Audio Tape for 525-Line/60 Field Television Systems,* ANSI V98.12M, American National Standards Institute, New York, 1981.

BIBLIOGRAPHY

Basic Principles of Magnetic Recording

Roters, Herbert C., *Electromagnetic Devices,* Wiley, New York, 1961.

Sharrock, Michael P., and D. P. Stubs, "Perpendicular Magnetic Recording Technology: A Review," *SMPTE J.,* vol. 93, pp. 1127–1133, Dec. 1984.

Head Design

Buchanan, John D., and John D. Tuttle, "A Sensitive Radiotracer Technique for Measuring Abrasivity of Magnetic Recording Tapes," *Int. J. Applied Radiation Isotapes,* vol. 19, pp. 101–121, 1968.

Camras, M., "An X-Field Micro-Gap Head for High Density Magnetic Recording," *IEEE Trans. Audio,* vol. AU-12, pp. 41–52, May–June 1964.

Carroll, J. F., Jr. and R. C. Gotham, "The Measurement of Abrasiveness of Magnetic Tape," *IEEE Trans. Magnetics,* vol. MAG-2, no. 1, pp. 6–13, March 1966.

Chikazumi, Soshin, *Physics of Magnetism,* Wiley, New York.

Eldridge, D. F., and A. Baaba, "The Effects of Track Width in Magnetic Recording," *IRE Trans. Audio,* vol. AU-9, pp. 10–15, 1961.

Gooch, Beverley R., and Edward Schiller, Magnetic Head with Protective Pockets of Glass Adjacent to the Corners of the Gap, U.S. Patent 3,818,693, May 1974.

Gooch, Beverley R., and Edward Schiller, Method of Manufacturing a Magnetic Head, U.S. Patent 3,845,550.

Ichiyama, Y., "Analytic Expressions for the Side Fringe Field of Narrow Track Heads," *IEEE Trans. Magnetics,* vol. MAG-13, no. 5, pp. 1688–1689, September 1977.

Jorgensen, Finn, *The Complete Handbook of Magnetic Recording,* Tab Books, Blue Ridge Summit, Pa., 1980.

Kolb, F. J., Jr. and R. S. Perry, "Wear of Permalloy Magnetic Heads against Striped Motion-Picture Film," *J. SMPTE,* pp. 912–919, September 1968.

Kornei, O., "Structure and Performance of Magnetic Transducer Heads," *JAES,* vol. 1, no. 3, July 1953.

Lindholm, D. A., "Magnetic Fields of Finite Track Width Heads," *IEEE Trans. Magnetics,* vol. MAG-13, no. 5, pp. 1460–1462, September 1977.

Lindholm, D. A., "Spacing Losses in Finite Track Width Reproducing Systems," *IEEE Trans. Magnetics,* vol. MAG-14, no. 2, pp. 55–59, March 1978.

McKnight, J. G., "Erasure of Magnetic Tape," *J. AES,* vol. 11, pp. 223–233, October 1963.

Mec., C. D., *The Physics of Magnetic Recording,* Wiley, New York.

Penn, C. B., *Magnetic Recording in Science and Industry,* Reinhold, New York.

Ragle, Herbert U., and Eric D. Daniel, "Head Wear," Memorex Corp., 1963.

Rittinger, Michael, "Magnetic Head Wear Investigation," *J. SMPTE,* pp. 179–183, April 1955.

van Herk, A., "Side Fringing Fields and Write and Read Crosstalk of Narrow Magnetic Recording Heads," *IEEE Trans. Magnetics,* vol. MAG-13, no. 4, pp. 1021–1028, July 1977.

van Herk, A., "Side-Fringing Response of Magnetic Reproducing Heads," *J. AES,* vol. 26, no. 4, pp. 209–211, April 1978.

Magnetic Tape—Manufacturing Process

Bate, G., "Recent Developments in Magnetic Recording Materials," *J. Appl. Phys.,* p. 2447, 1981.

Hawthorne, J. M., and C. J. Hefielinger, "Polyester Films," in N. M. Bikales (ed.), *Encyclopedia of Polymer Science and Technology,* vol. 11, Wiley, New York, 1969, p. 42.

Jorgensen, F., *The Complete Handbook of Magnetic Recording,* Tab Books, Blue Ridge Summit, Pa., 1980.

Kalil, F. (ed.), *Magnetic Tape Recording for the Eighties,* NASA References Publication 1975, April 1982.

Lueck, L. B. (ed.), *Symposium Proceedings Textbook,* Symposium on Magnetic Media Manufacturing Methods, Honolulu, May 25–27, 1983.

Nylen, P., and E. Sunderland, *Modern Surface Coatings,* Interscience Publishers Division, Wiley, London, 1965.

Perry, R. H., and A. A. Nishimura, "Magnetic Tape," in Kirk Othmer (ed.), *Encyclopedia of Chemical Technology,* 3d ed., vol. 14, Wiley, New York, 1981, pp. 732–753.

Tochihara, S., "Magnetic Coatings and Their Applications in Japan," *Prog. Organic Coatings,* vol. 10, p. 195–204, 1982.

Recording Systems and Signal Processing

Felix, M. O., and H. Walsh, "FM System of Exceptional Bandwidth," *Proc. IEEE,* vol. 112, no. 9, September 1965.

Fujiwara, Yoshio, Hitoshi Sakamoto, and Steven Sarafian, "Delta T: A New Dimension in Type-C Recording." *SMPTE J.,* vol. 93, pp. 1134–1137, Dec. 1984.

Gabor, D., *Theory of Communication,* 1946, 93, pt. III, p. 429.

Sadashige, Koichi, "Developmental Trends for Future Consumer VCRs," *SMPTE J.,* vol. 93, pp. 1138–1149, Dec. 1984.

Todorovic, A., "The MAGNUM Specialist Group and The EBU Approach to the Digital Video-Tape Recorder," *SMPTE J.,* vol. 92, pp. 568, May 1983.

Ville, J., "Theorie et applications de la notion de signal analytique,"*Cables et Transm.*, 1948, 2(1), p. 61.

Yoshida, H. and T. Eguchi, "Digital-Video Recording Based on Proposed Format from Sony," *SMPTE J.*, vol. 92, pp. 568, May,

ANSI/SMPTE Standards

Quadruplex 2-in

C98.1-1978, C98.3-1980, C98.4-1983, C98.5-1983.
V98.6-1981, V98.12-1981.
RP6-1979, RP11-1968, RP16-1982, RP36-1978, RP101-1981, RP102-1981.

Helical 1-in Type-B

C98.18M-1983, C98.19M-1983, C98.20M-1979.
RP83-1980, RP84- 1980.

Helical 1-in Type-C

V98.18M-1983, V98.19M-1983, V98.20M-1979.
RP85-1979, RP86-1979.

Helical 3/4-in Type-E

C98.21M-1980, RP87-1980.

Helical 1/2-in Types G (Beta) and H (VHS)

Type-G: V98.34M-1984, RP119-1984.
Type-H: V98.32M-1983, RP112-1983.

Video Disc Recording and Reproduction

Robert A. Castrignano

Vice President, Information Storage Systems Technology
CBS Technology Center
Stamford, Connecticut

16.1 BASIC PRINCIPLES

All video discs are information storage devices with high packing densities, which appear similar to diffraction gratings when illuminated by white light. The various systems are characterized by their playback signal detection technique, i.e., as either contact or noncontact systems. Contact systems include both grooved and nongrooved discs. In both, playback is achieved by a stylus in physical contact with the disc. The stylus may either activate a piezoelectric crystal or incorporate its own unique signal detection electrode. Noncontact systems incorporate sophisticated laser-beam technology to reproduce the information on the discs. The discs may be either reflective (opaque) or transmissive. In the case of reflective discs the signal is generated by detecting the laser beam that is reflected from the disc surface. In the transmissive system the signal is generated by detecting the laser beam that is transmitted through the disc.

16.1.1 DISC GEOMETRY. Discs may be single- or double-sided and vary in diameter from 8 to 12 in (0.2 to 0.3 m). Thickness varies from 0.006 to 0.08 in (0.15 to 2.0 mm), and visual characteristics range from a clear, thin plastic polyvinyl chloride (PVC) foil to an opaque black PVC disc similar to a conventional LP audio record.

In general an audio record contains six grooves per millimeter, whereas a video disc contains from approximately 400 to 700 grooves, or tracks, per millimeter, depending on the recording system. The minuteness of the grooves may be visualized by considering that there are 12 to 30 grooves in the width of a human hair. The groove pitch is related to the length of playing time per side. For NTSC 525-line television standards and a constant angular disc velocity of 30 r/s (one television frame per revolution), 54,000 frames must be recorded for 30 min of playing time. Information packing density varies from 280 grooves per millimeter, corresponding to a track pitch of 3.57 μm, to 740 grooves per millimeter, which corresponds to a track pitch of 1.35 μm.

A dimensional analysis of a typical one-frame-per-revolution disc, with a radius of 5.75 in, indicates a tangential length for one television line of 1,747 μm. Thus, if a modulating frequency of 8 MHz and an angular velocity of 30 r/s is assumed, it follows that there are 267,000 cycles per revolution, or 508 cycles of information depression pits per television line. This equates to a pit pitch of 3.44 μm and a pit length of 1.72 μm, assuming a 50 percent duty cycle.

Dimensional eccentricities in the order of 0.1 mm, for discs with a diameter of approximately 11.8 in (300 mm), dictate the need for sophisticated radial servo systems with a gain of 300 to 400 and a tangential, or time-base correction, servo with a gain in excess of 500.

Surface flatness must be maintained within several microns to minimize the servo and g force requirements of the signal detection system. All geometric requirements must be achieved within the nominal environmental ranges of 40 to 110°F and 5 to 90 percent relative humidity.

16.1.2 ROTATIONAL SPEEDS AND PLAYING TIME. The majority of disc systems are designed to play from the outside radius to the inside radius, similar to audio records; however, some play from the inside to outside radius. The advantage of inside-to-outside play is that the design of the start-of-play mechanism, in automatic changers, is independent of the disc diameter.

Discs are recorded at either constant angular velocity (CAV) or constant linear velocity (CLV). CAV recording results in a coherent-pattern disc with a constant number of television fields per revolution, thus permitting still-frame display in the case of two fields per revolution. CLV recording has been adapted to increase the playing time per side by increasing the packing density at the outside radius to be equal to that of the inside radius of CAV discs. The disadvantages of CLV recordings are that the signal-to-noise (S/N) performance of the entire disc is no better than that of the inside radius of CAV discs, and still-frame capability is not possible without external signal processing. In CAV recording the S/N performance improves as the radius increases because of the increase in tangential velocity and the resultant increase in pit length.

Three different rotational speeds have been adapted for use with CAV consumer disc systems. These are 450 r/min (four television frames per revolution), 900 r/min (two television frames per revolution), and 1800 r/min (one television frame per revolution). In the case of CLV recording, the speed varies from approximately 600 r/min, at the inside radius, to 1800 r/min, at the outside radius.

High-speed operation tends to produce higher-quality, less noisy (higher S/N) pictures at the expense of decreased playing time, greater disc eccentricity problems, and poorer mechanical stability of the turntable. Conversely, low-speed operation masks disc defects, simplifies the player design, and provides longer playing time.

16.1.3 PHOTOGRAPHIC SYSTEMS.

The use of film as a high-density storage medium has led to the development of several systems for video storage and reproduction. High-resolution silver halide films, on stable polyester or Mylar polyester film bases, have been used in conventional and holographic recording systems. In addition, experiments have been conducted by exposing and developing dye-sensitized material with monochromatic laser beams.

In the nonholographic systems, an analog, frequency, or pulse position-modulated video signal is used to expose a film master or to laser-machine (by an ablative process) a thin metallic coating of bismuth, tellurium, or similar low-melting-point metal on a clear-glass substrate.

Inexpensive duplication of the film master, by contact printing to silver halide, Diazo, or Kalvar films, was initially proposed by the I/O Metrics Co. and subsequently pursued by its successors, Atlantic Richfield Development Co. (ARDEV) and the McDonald Douglas Electronics Co. Albeit both AM and FM encoding techniques were considered, the AM system was adapted, although it entailed stringent system linearity requirements between the recorded and playback signals. Coherent (laser) and inexpensive noncoherent (incandescent lamp) light sources were considered for playback. Laser optical systems produced improved S/N performance at the expense of a more sophisticated optical system and increased film grain noise in the detected video signal.

Figure 16-1 illustrates the system principles adopted by I/O Metrics. An inexpensive 15-W incandescent light source is focused to a 1-μm-sized spot on the disc, thus produc-

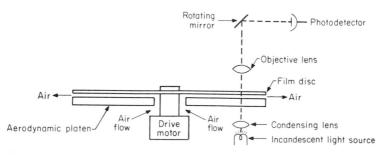

FIG. 16-1 I/O Metrics' transparent film readout system.

ing an intensity-modulated beam which is detected by a photodiode. A front-surface servo-controlled mirror provides radial tracking, and a second servo controlled the objective lens to maintain focus of the light beam. The disc flatness necessary to maintain focus within several micrometers was achieved by rotating the thin 4- to 6-mil thick, 12-in-diameter disc at 1800 r/min above an aerodynamic platen. The system was designed so that rotation at 1800 r/min results in a playing time of 14.1 min per radial inch. Analog audio recording was possible during either the back-porch or vertical retrace interval of the video signal.

More contemporary work performed by ARDEV was aimed at redesigning the audio and video formats to provide 30 s of audio per still frame of video. Ten sequential, 100:1 time-compressed audio tracks were recorded for each video track.

Concurrent with the photographic film work pursued by I/O Metrics and ARDEV, Eastman Kodak[1] experimented with producing replicas on a laser-machined master by contact printing to a reflective substrate, which was coated with a dye layer. Positive images of the master disc's apertures were formed by contact printing. Postexposure development produced a dye in the nonexposed areas of the disc, which absorbed the readout light source wavelength. Typical discs were 7 mils thick with a dye layer thickness of approximately 2 μm.

Holographic video disc experiments were conducted by Hitachi[2] in the interest of eliminating the sophisticated and expensive optical focus and tracking servos required in the optical playback systems. The 11.8-in-diameter (300-mm-diameter) disc contains 54,000 holograms, each of which redundantly stores luminance, chroma, and audio codes in an area of approximately 1 mm.

16.1.4 MECHANICAL STYLUS AND GROOVE SYSTEM. Contemporary high-density stylus and groove systems date back to June 1970 when Telefunken and Decca jointly demonstrated the TelDec disc,[3-5] a single-sided, spirally recorded hill-and-dale, floppy grooved disc. The embossed-thermoplastic disc was approximately 120 mm in diameter and 0.12 mm thick. Separate frequency-modulated video and audio carriers were recorded as hills and dales at right angles to the surface of the disc. The depth of modulation was approximately 0.3 μm. The track pitch of approximately 3.6 μm resulted in a groove density of 280 grooves per millimeter. The included angle of the groove was 130°. Two audio channels were recorded, each frequency-modulating a separate carrier, thus providing the capability of stereo or dual-language sound. Playing time was approximately 10 min. The rotational speed and recording format were designed so that playback was from outside to inside radius and one revolution of the disc corresponded to one television frame. During playback,

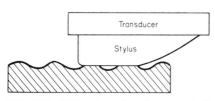

disc flatness was maintained by rotating the disc above an air cushion. A specially designed diamond stylus, illustrated in Fig. 16-2, was in pressure contact with the elastic disc, thus depressing the hills of the hill-and-dale recording. When this pressure was released as the disc rotated, the sudden change of the deformed crests in returning to their original positions caused a pulse in the pressure-sensitive trans-

FIG. 16-2 TelDec transducer/stylus and disc cross section.

ducer and thus generated an FM electrical signal which was amplified and demodulated. The stylus-in-the-groove concept obviates the need for a groove-tracking, or focus, servo.

The recorded chrominance and luminance (Y) information was coded using a line-sequential red, blue, and green (R,B,G) mixed-highs system. The low-frequency part of each sequential line contained $R - Y$, $B - Y$, or $G - Y$ color signal components, plus a high-frequency spectrum consisting of a mixed-high Y signal, as illustrated in Fig. 16-3.

The spectrum of the recorded signal is illustrated in Fig. 16-4. Superimposed on the spectrum diagram is a typical television waveform to identify the frequencies corresponding to sync tips, black level, and white level. In addition, the frequencies of 0.8 and 1.07 MHz are allocated for FM sound channels 1 and 2, respectively, each with a devia-

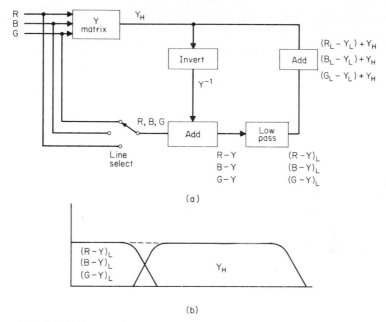

(a)

(b)

FIG. 16-3 TelDec encoding system: (a) system diagram; (b) frequency spectrum of line-sequential color and mixed-highs luminance signals.

tion of ± 50 kHz. For stereo operation, channel 1 contains the sum $(R + L)$ information and channel 2 contains the difference $(R - L)$ information.

The discs were normally enclosed in protective jackets, except when being played. To be played, the disc and its jacket were inserted into the player, where the jacket remained while the disc was automatically extracted at the beginning of play and reinserted at the end of play.

An order of magnitude in system improvement was achieved by the Matsushita Electric Co. in Japan, when they demonstrated a 9-in (22.9-cm)-diameter, 0.079-in (2-mm)-thick, double-sided, CLV disc that was capable of 1-h playing time per side. Track pitch was 2.3 μm, and playback was from the inside out. The speed at the inside was 700 r/min, corresponding to two television frames, and 300 r/min at the outside, corresponding to five television frames. The name *Visc-o-Pac* was adopted

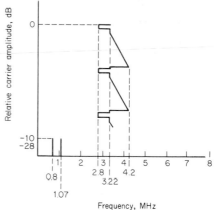

FIG. 16-4 TelDec system frequency spectrum and relative carrier amplitudes.

to describe the disc and its plastic container, the latter designed to protect the disc from damage in handling. The disc container opened automatically, in a clamshell fashion, when in the play mode inside the player.

System performance specifications for NTSC color provided a video S/N ratio of better than 45 dB, an audio S/N ratio of approximately 60 dB, a resolution in excess of 270 television lines, and two audio channels, each with a bandwidth of 20 kHz.

16.1.5 CAPACITIVE STYLUS AND GROOVE SYSTEMS. The RCA capacitive system[6,7] has been designated the *capacitive electronic disc* (CED) because the information, which is recorded on the disc, is recovered by sensing capacity changes between the hill-and-dale pit surface of the conductive V-grooved PVC disc and a discrete stylus riding in the groove. The conductive disc represents one side of the capacitor, and a metallic electrode attached to the stylus represents the other. The *stylus-to-disc* capacitance is part of a resonant circuit which is excited by an oscillator, whose frequency coincides with the upper slope of the resonance curve. See Fig. 16-5. The amplitude response to the excitation is detected with a pickup loop. The magnitude of the detected response depends on where the frequency of the oscillator signal falls on the resonance curve. As the stylus-disc capacity changes, the resonant frequency changes such that the oscillator carrier signal will be amplitude-modulated. The voltage obtained by peak detection of this carrier represents the stylus-disc capacity variations.

Initial versions of the disc, recorded in the CAV mode, were 304 mm in diameter, approximately 1.78 mm thick, and played for 30 min per side at a rotational speed of 450 r/min. In these initial versions the disc[8] consisted of a nonconductive compression-molded polyvinyl chloride–polyvinyl acetate copolymer, the surface of which contained the modulating signal information and grooves. The required surface conductivity was obtained by vacuum deposition of a 20-nm-thick metallic film, which was then covered by 20 nm of a glow-discharge-deposited organic dielectric film. See Fig. 16-6. For the first metallic-coating experiments,[9] aluminum was used; however, it proved to be too grainy. Alternatively, gold produced much better pictures but was much too expensive. Experiments in 1972 showed that copper, deposited by either evaporation or sputtering, provided smooth films with excellent adhesion to vinyl; unfortunately, it was chemically

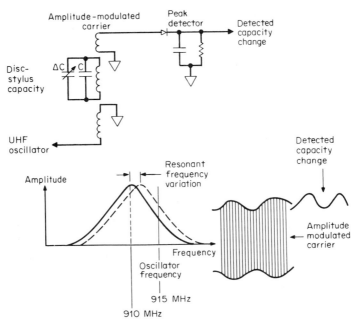

FIG. 16-5   CED RF signal detection circuit. *(RCA Corp.)*

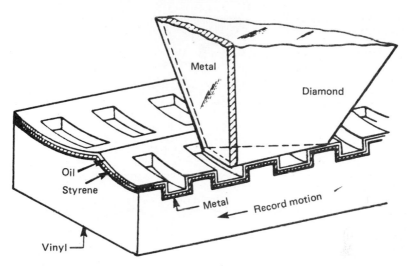

FIG. 16-6   Capacitive stylus and Inconel-coated disc. *(RCA Corp.)*

active and therefore subject to a variety of corrosion reactions. Finally in 1975, Inconel, an alloy of nickel, chromium, and iron, was adopted as an acceptable corrosion-resistant metal for the conductive layer. Adhesion problems between the vinyl and Inconel were solved by sputtering 5 nm of copper over the vinyl before the Inconel. Proper control of the oxygen content at the surface of the Inconel layer permitted good adhesion with the styrene layer. After continuing experimentation, it became apparent that metallic coatings were expensive to apply consistently and with high reliability. Consequently, the search for a conductive disc intensified. This effort culminated with the development of a conductive molding compound[8] based on Ketjenblack EC (a product of Akzo Chemie, Netherlands) and a polyvinyl chloride (PVC) homopolymer resin base. The specifications for disc warp (0.5 mm) could not be met by PVC/PVA systems but were easily met by carbon-filled PVC homopolymers. The Ketjenblack carbon loading level is fundamental to the compound performance and determines the resistivity, the melt viscosity, the physical characteristics, and the surface quality of the disc. Conductivity is achieved by the addition of carbon whose granular size is smaller than the 500-nm minimum-size requirement for information elements, and with conductive properties sufficient to achieve a resistivity of less than 5 $\Omega \cdot$ cm with 15 percent loading of the base PVC material with carbon. The conductive disc is compression molded in a 100-ton hydraulic compression press in less than 40 s and then is coated with a 30-nm-thick layer of special lubricant to extend the life of the disc and the stylus. The disc is recorded in the CAV mode, is double-sided, 11.9 in (302 mm) in diameter, and is 0.075 in (1.9 mm) thick. Luminance, chrominance, and dual-channel audio signals are recorded as frequency-modulated hills and dales, of varying width and periodicity, in a 140° V-shaped spiral groove. Playing time is 1 h per side at a rotational speed of 450 r/min, which corresponds to four television frames per revolution. The spiral groove pitch is 2.64 μm. Groove depth is approximately 0.48 μm with a delta height variation, between the hills and dales of the recorded signal, of approximately 80 nm. Figure 16-7 is a cross-sectional view of the conductive-carbon-loaded disc. The discs are enclosed in a plastic cassette to protect them from damage in handling.

The spectrum of the playback of the recorded signal is illustrated in Fig. 16-8. During recording, the signal-processing electronics separate the chroma from the composite signal and translate the interleaved chrominance subcarrier from 3.58 to 1.53 MHz. In addi-

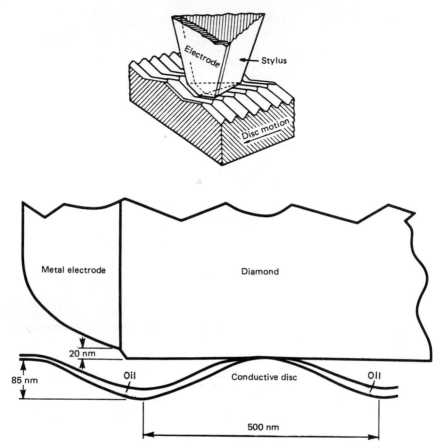

FIG. 16-7 RCA CED conductive disc. Profile of stylus and disc along a groove. Recorded video carrier is shown with its shortest wavelength. *(RCA Corp.)*

tion, two separate audio carriers are present at 0.716 and 0.905 MHz. Stereo recording is accomplished by frequency-modulating one carrier with the $L + R$ signal and the other with the $L - R$ signal. The luminance bandwidth is 3 MHz, and the chroma is 0.5 MHz. The audio bandwidth is 15 kHz, and the deviation of the sound carrier is ± 50 kHz.

During playback of a disc, velocity variations are compensated for by a unique *arm stretcher,* which advances or retracts as the stylus moves tangentially along a groove, thus generating a constant sync-pulse time base. Frequency and phase variations of the color signal, due to disc eccentricities, are subtracted out by generating a local carrier, in the player, with the same errors. A radial tracking and focus servo is unnecessary because of the grooved disc. System performance specifications are illustrated in Table 16-1.

16.1.6 CAPACITIVE STYLUS AND GROOVELESS SYSTEMS. The capacitive stylus and grooveless disc system[11] was demonstrated in September 1978 by The Victor Company of Japan, Ltd. (JVC). The disc, which has been designated very high density (VHD), is similar to the RCA capacitive grooved disc in that the information recorded

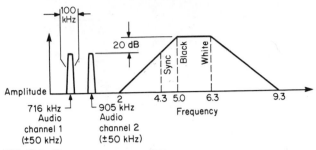

FIG. 16-8 Frequency spectrum, CED system. *(RCA Corp.)*

on the disc is recovered by sensing capacitive changes between the surface of the conductive disc and the bottom of the recorded information pits. Unlike the RCA CED disc, the VHD disc is grooveless. Luminance, chrominance, and two channels of audio information are present as rectangular depressions, recorded in a single spiral of pits, in the surface of a flat PVC conductive disc. All the pits are of the same width and depth; however, they vary in duty cycle. See Fig. 16-9. In addition to the video-audio information track, a stylus tracking signal of two different alternate low-frequency signals is recorded on either side of the video-audio track. The tracking signal pits are approximately four-tenths as deep as the video-audio pits and approximately one-tenth the frequency. As the disc plays, the stylus senses the frequency of the tracking pits and adjusts its position accordingly. Figure 16-10 shows the method of controlling the stylus. The stylus is

Table 16-1 CED (NTSC) System Parameters

Physical dimensions	
Diameter	11 in (279 mm)
Thickness	0.075 in (1.9 mm)
Center hole diameter	1.3 in (33 mm)
Rotation speed	450 r/min
Information track pitch	104 μin (2.64 μm)
Groove depth	19 μin (0.48 μm)
Signal amplitude	3.3 μin (0.085 μm)

Disc performance parameters	
Playing time	60 min/side
Luminance bandwidth	3 MHz
Chrominance bandwidth	0.5 MHz
Video signal-to-noise	>46 dB
Chrominance signal-to-noise	>40 dB
Video FM signal	4.3 to 6.3 MHz
Audio carriers	
$(L + R)$ (45.5 F_h)	715.9 kHz
$(L - R)$ (57.5 F_h)	904.7 kHz
Audio bandwidth	15 kHz
Audio deviation	± 50 kHz
Audio signal-to-noise	50 dB

Source: J. J. Brandinger, "The RCA CED VideoDisc System—An Overview," *RCA Rev.*, vol. 42, no. 3, September 1981.

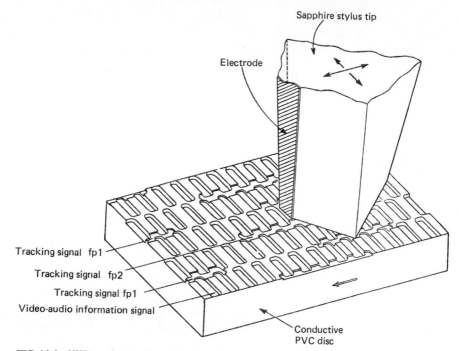

FIG. 16-9   VHD conductive disc and stylus. *(Victor Company of Japan, Ltd.)*

mounted on one end of a cantilever pickup arm and a magnet is attached to the other. Fixed coils are mounted near the magnet. A single coil is wound around, but not in contact with, the magnet, and a pair of vertical coils in phase opposition to each other are mounted on each side of the single coil. Hence the stylus can move transversely and longitudinally in response to the particular current flowing in these coils. The current is varied by the tracking error signal or by a command to move the stylus to a predetermined specific track. The stylus vertical tracking force is 40 mg. While the capacitive probe on the stylus is equal to the width of one track, the stylus itself is approximately five times the track width. Its multiple-track-width design extends its useful playing life and enables it to act as a disc surface cleaner.

The double-sided disc, recorded in the CAV mode, is 259 mm in diameter and is approximately 2 mm thick. Playing time is 1 h per side at a rotational speed of 900 r/min, which corresponds to two television frames per revolution. The spiral track pitch is 1.35 μm. The disc is enclosed in a case which protects it from dust, scratches, and mishandling. The frequency spectrum of the signal is illustrated in Fig. 16-11. Superimposed on the spectrum diagram is a typical television waveform to indicate the frequencies corresponding to sync tips, black level, and white level. The luminance bandwidth is 3.1 MHz. The chrominance bandwidth is 0.5 MHz, and the chrominance signal exists as an interleaved subcarrier translated from 3.58 to 2.56 MHz. In addition, two separate sound carriers are present at 3.43 and 3.73 MHz. Sound channel deviations are ± 75 kHz. The tracking signals are shown at 0.511 and 0.716 MHz, respectively, and the alternate frame indexing signal is shown at 0.275 MHz. System performance specifications are listed in Table 16-2.

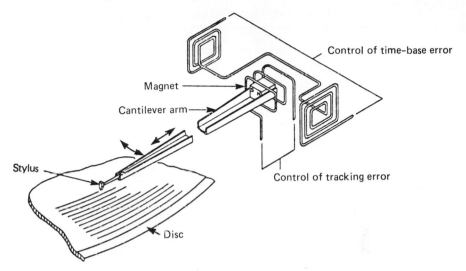

FIG. 16-10   VHD Electro tracking stylus. *(Victor Company of Japan, Ltd.)*

16.1.7 OPTICAL REFLECTIVE SYSTEMS. Optical reflective video disc systems were independently developed by the N.V. Philips Gloeilampenfabrieken Co. in the Netherlands and by MCA in the United States. As a result of a joint technical interchange agreement, the proprietary technologies of both companies were combined to produce an optical disc in which the video and audio information are recorded as rectangular depressions molded into a plastic substrate. The pits have constant width and depth; however, the length and spacings vary according to the video and audio modulating signals. See Fig. 16-12. The molded substrate is coated with a thin metallic layer to render the entire surface reflective. In addition, a second clear plastic coating is applied

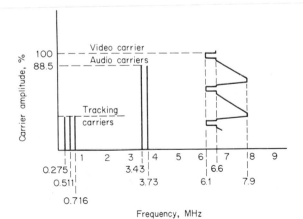

FIG. 16-11   Frequency spectrum, VHD system.

Table 16-2   VHD (NTSC) System Parameters

Physical dimensions	
Diameter	10.2 in (260 mm)
Thickness	⁓0.08 in (⁓2 mm)
Center hole diameter	1.5 in (38.2 mm)
Rotation speed	900 r/min
Information track pitch	53.2 μin (1.35 μm)
Pit depth (information signal)	11.8 μin (0.3 μm)
Pit length (information signal)	23.6 μin (0.6 μm)
Pit depth (tracking signal)	4.7 μin (0.12 μm)
Pit width (tracking signal)	0.6 μm

Disc performance parameters	
Playing time	60 min/side
Luminance bandwidth	3.1 MHz
Chrominance bandwidth	0.5 MHz
Audio bandwidth	20 kHz
Video FM signal	6.1–7.9 MHz
Audio carriers:	
Left channel (217.75 F_h)	3.426 MHz
Right channel (237.25 F_h)	3.733 MHz
Audio deviation	±75 kHz
Audio signal-to-noise	>60 dB
Tracking signal frequencies:	
F_1 (32.5 F_h)	511 kHz
F_2 (45.5 F_h)	716 kHz
Alternate frame indexing signal:	
F_3 (17.5 F_h)	275 kHz

over the metallic layer to protect the information surface. Discs are recorded in both the CAV and CLV modes. Both types are double-sided, 305 mm in diameter, and 2.5 mm thick. The single spiral track pitch is 1.6 μm. Playing time is 30 min per side at 1800 r/min (one television frame per revolution) for the CAV disc. Playing time for the CLV disc is 60 min per side. The rotational speed for the CLV discs varies from 1800 r/min at the outside radius to approximately 600 r/min at the inside radius.

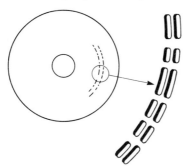

FIG. 16-12   Video/audio information pits in the surface of a reflective optical disc.

The frequency spectrum of the signal is illustrated in Fig. 16-13. A typical video waveform is superimposed on the frequency spectrum diagram to indicate the frequencies associated with the sync tip, black level, and white level of the video signal. The video information is coded such that on NTSC standards the composite signal frequency modulates a carrier at 8 MHz. Two separate audio signals frequency-modulate two discrete audio carriers at 2.3 and 2.8 MHz, respectively. Sound channel deviations are ±100 kHz. For stereo operation, channel 1 contains the left-side information and channel 2 the right-side information. A similar *color-under* system was demonstrated by the Sony Corp., in which the composite video signal was separated into its luminance and chrominance

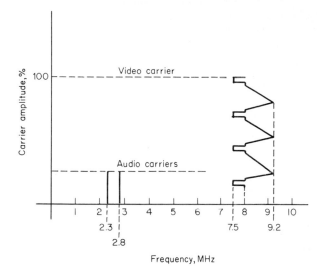

FIG. 16-13 NTSC, frequency spectrum, VLP optical system.

components before frequency-modulating two separate carriers. The luminance signal modulated a carrier from 5.5 to 7.0 MHz. The chroma signal modulated a carrier at 1.54 MHz. Two sound channels modulated frequencies at 0.401 and 0.606 MHz, respectively.

The high-frequency spectrum, typical of optical systems, dictates a system bandwidth which is defined by the following formula

$$BW = 2\pi nr \frac{2NA}{\lambda}$$

where n = speed of rotation, r/s (Hz)
r = radius, μm
λ = wavelength of readout beam, μm
NA = numerical aperture of readout objective lens

If the formula is applied to a typical optical system with an NA of 0.45, a rotational speed of 1800 r/min, an HeNe laser with a wavelength of 0.6328 μm, and a radius of 54 mm, it follows that the required cutoff frequency of the optical stylus is 10.5 MHz.

The encoding process is illustrated in Fig. 16-14. The composite video signal frequency-modulates one carrier, while two independent sound signals each frequency-modulates an associated carrier. The three carrier signals are then linearly added and fed to a limiter circuit, the output of which is a pulse-width modulated signal, as illustrated in Fig. 16-15.[12]

Playback is achieved via the detection of a reflected, low-power HeNe laser beam from the information surface of the disc. See Fig. 16-16. The light reflected from the surface of the disc is detected by a photodiode and regenerated as video and audio signals. This results from the diffraction which occurs at the information pits, causing most of the diffracted rays to fall outside the detection lens aperture; hence, less light is received when a pit passes in front of the lens than when a smooth section of the disc surface does so. The pits thus modulate the photodiode current. Maximum modulation occurs when the pit depth is one-quarter of the laser light source wavelength. In other words, the incident light reflected from the bottom of the pits is 180° out of phase with the light reflected from the disc surface, and thus results in cancellation. The optical

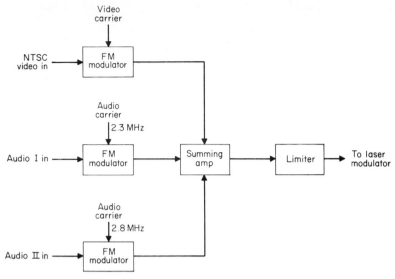

FIG. 16-14   Philips VLP carrier modulation and encoding system.

playback system[13] employed in the Philips/Magnavox player is illustrated in Fig. 16-16. The disc is scanned from below by a low-power (approximately 1-mW) HeNe laser. A unique diffraction grating is located at the exit of the laser to generate two first-order beams, which are employed in the radial tracking system. The optics[14-16] are designed so that the two subsidiary beams straddle the main scanning zero-order beam and are imaged partially on the track being read and partially on the spaces on either side of the track being scanned. See Fig. 16-17.

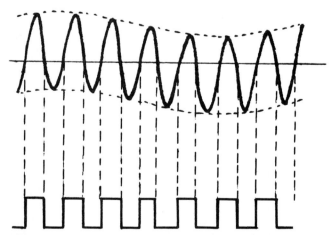

FIG. 16-15   Philips pulse-width-modulation system. (*N.V. Philips Gloeilampenfabrieken Co., Philips Technical Rev., vol. 33, no. 7, 1978.*)

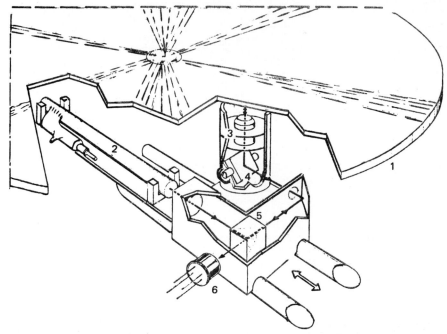

FIG. 16-16   Reflective optical disc playback system. The record 1 is scanned from below by the light from the HeNe laser 2. The objective 3 is held focused on the record by a system based on a loud-speaker mechanism. The pivoting mirror 4 ensures that the beam remains centered on the track; the mirror is operated by a rotating-coil arrangement. Incident and reflected light are separated from the prism 5. The detector 6 converts the reflected light into an electrical signal. *(N.V. Philips Gloeilampenfabrieken Co., Philips Technical Rev., vol. 33, no. 7, 1978.)*

The zero-order and the two first-order beams are each imaged onto its own detector. See Fig. 16-18. The zero-order beam is imaged onto a four-sector diode such that the sum of the four sectors represents the total FM signal while the difference, of the sum of the diagonally opposite sectors, represents the focus error signal. When the beam is perfectly focused, the image on the diodes is circular. When defocusing occurs, the cylindrical lens produces an elliptical image, whose major axis rotates as a function of being too close to or too far from the disc. See Fig. 16-19.

Each of the first-order beams is imaged on its own detector, thus generating an error-correction signal which actuates a pivoted rotating mirror to maintain radial beam tracking.

Precision turntable rotational speed and color signal time-base performance are achieved by two additional servos that are associated with the playback signal processing

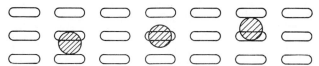

FIG. 16-17   N.V. Philips tri-beam tracking system. *(N.V. Philips Gloeilampenfabrieken Co., Philips Technical Rev., vol. 33, no. 7, 1978.)*

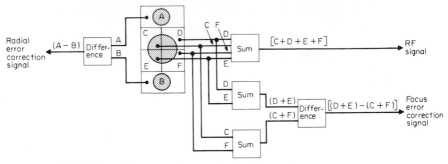

FIG. 16-18 Optical photo detectors and servo signal generation block diagram for the main RF signal and the focus and radial servos.

circuitry. The turntable is driven by a dc motor whose speed is controlled by a correction signal generated by a phase-comparison circuit which compares the time of an internal-crystal-controlled horizontal-frequency oscillator with the horizontal sync pulses from the playback video signal. High-frequency instantaneous speed errors, caused by disc

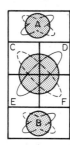

FIG. 16-19 Optical characteristics of a focused (circular) and defocused (elliptical) reflected beam.

eccentricities, are corrected by a color-burst phase detection circuit which generates an error voltage that is fed to the motor speed servo control circuit as well as to a tangentially pivoted mirror which advances or retracts the optical playback beam along a given track. System parameters for the Philips reflective optical system are listed in Table 16-3.

16.1.8 OPTICAL TRANSMISSIVE SYSTEMS. Concurrent with the development of the reflective optical system, Thomson-CSF developed a transmissive optical system. The objective was to produce consumer or institutional discs by stamping them from long, flat rolls of PVC. The disc was double-sided, 301 mm in diameter, and 0.15 mm thick. Playing time was 30 min per side (one television frame per revolution); hence, the disc had a capacity of 108,000 discrete television frames. Since the disc was transmissive, it was possible to access both sides of the disc by focusing the pickup beam on either the top or bottom, through the disc itself.

Rotational speed was 1500 r/min for the PAL and SECAM systems and 1800 r/min for the NTSC system. The track pitch was 1.56 μm for a track density of about 640 grooves per millimeter. The video plus two channels of audio as well as digital frame numbers were recorded as rectangular pits pressed into the surface of a thin plastic foil with a refractive index of 1.5. Many different approaches were investigated in which the number of pits per second represented the modulated-luminance FM carrier frequency, while the varying length, resulting from pulse-width-modulation, was a measure of the pilot frequency. The pit depth was constant as approximately one-fourth wavelength of the HeNe laser readout beam.

The spectrum of the recorded signal is illustrated in Fig. 16-20. A typical video waveform is superimposed on the frequency spectrum diagram to illustrate the frequencies corresponding to sync tips, black level, and white level. The same encoding format is used for SECAM/PAL and NTSC. A sequential chrominance signal $(R - Y)$ or $(B - Y)$ modulated an FM subcarrier which was added to the luminance signal. The technique is sim-

Table 16-3 VLP (NTSC) System Parameters

Physical dimensions	
Diameter	11.9 in (301.6 mm)
Thickness (nominal)	0.08 in (2.0 mm)
Center hole diameter	1.4 in (35 mm)
Rotational speed	1800 r/min
Information track pitch	65.8 μin (1.67 μm)
Information pit depth	~3.9 μin (~0.1 μm)
Information pit width	~23.6 μin (~0.6 μm)
Information pit length	~59 μin (~1.5 μm)

Disc performance parameters	
Playing time:	
CAV	30 min/side
CLV	60 min/side
Luminance bandwidth	3.2 MHz
Chrominance bandwidth	0.5 MHz
Audio bandwidth	20 kHz
Video FM signal deviation	7.5–9.2 MHz
Audio carriers:	
Right channel (128.5 F_h)	2.8MHz
Left channel (148.5 F_h)	2.3 MHz
Audio deviation	±100 kHz

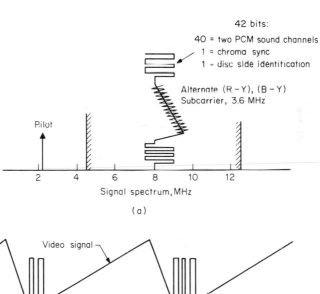

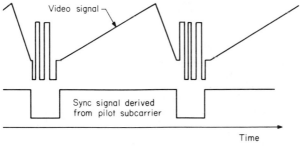

FIG. 16-20 Thomson-CSF encoding. *(Thomson-CSF.)*

ilar to that employed in the standard SECAM system, except that a lower center frequency is employed.

The horizontal sync was suppressed and replaced by a digital signal consisting of 42 bits. Forty bits were used to transmit two digital sound channels; one bit was employed for chrominance sync and one bit was used to identify the top or bottom of the transmissive disc. In addition, a pilot frequency was added to the main carrier in order to record a reference on the disc and to permit regeneration of the sync signal. The composite video signal modulated a carrier between 8.0 and 9.5 MHz.

Playback was achieved by using the laser beam as an optical stylus. Since the disc was transmissive, the differential light paths, between the areas of pits and no pits, modulated the focused laser beam as it passed through the disc before it was collected by the photodetector. The laser generated a 1-mW HeNe light beam which was imaged on the disc surface by an objective lens with a 0.45 NA. The depth of focus was on the order of 1.0 μm. The consumer prototype and the institutional version of the player focused the laser beam through aerodynamic stabilization of the thin foil and an electromagnetic lens-focusing servo. Laboratory studies have indicated that focusing can be achieved solely by aerodynamic stabilization of the disc.

Both institutional and consumer players use a caddy to protect the disc. The institutional caddy is rigid, while the consumer one is flexible so that it can be mailed.

16.2 MASTER RECORDING TECHNIQUES

16.2.1 **MECHANICAL DISC CUTTING.** Mechanical disc cutting is defined as the process wherein an electronically excited transducer drives a cutting stylus, which in turn cuts signal depressions or grooves into the surface of a *master* substrate.

The requirement for recording high-frequency video signals, in real time, dictates the ability to cut signal elements at an average rate of 5 MHz, with a constant groove pitch and depth of cut, in an ultraflat substrate. Although real-time recordings have been made successfully in the laboratory, existing production systems have been limited to one-half real time because of the problems associated with the large amount of power required to drive the cutting stylus, the friction between the stylus and the disc, and the more mundane problem of collecting and disposing of the thin filament of material that is cut from the substrate surface.

TelDec ameliorated the demanding system requirements by employing a conventional audio recording technique operating at one-twentyfifth of real time.

Initial RCA experiments† with heated audio cutterheads and lacquer-coated metal substrates led to the development of special video cutterheads using piezoelectric rather than magnetic transducers. The electromechanical recording technique that evolved offered the following advantages:

1. A high yield of defect-free masters
2. The ability to cut signals with short spatial wavelengths
3. The least demanding of environmental cleanliness in the cutting area
4. The achievement of smooth surfaces that result in high signal-to-noise ratios
5. Simple recording and processing facility requirements
6. Simultaneous readout during recording
7. The ability to make multiple copies from a single original master
8. The ability to cut the deep grooves necessary for reliable stylus tracking during playback

---

†Process details summarized and reproduced, with permission from RCA, from E. Keizer.[16a]

9. Inherent linear recording

10. The ability to cut the grooves and signal recording simultaneously

The key elements of an electromechanical recorder are illustrated in Fig. 16-21. A cutterhead is mounted on an arm with a translation mechanism that moves the cutterhead radially across the disc. A smoothly turning precision turntable is accurately locked to the signal source by a tachometer and speed servo system. On the turntable is a flat, fine-grained copper-coated aluminum disc to be cut by a sharp diamond stylus in the cutting head. Before recording, the copper surface is machined to be flat. The diamond cutting stylus has a tip-face shape that corresponds to the desired cross-sectional shape of the finished groove and cuts the groove at the same time the signal is recorded. During operation, the depth of cut is adjusted to be deeper than the desired groove depth, thus resulting in an overcut which produces a grooved disc without lands. The actual groove depth is controlled by the shape of the tip of the recording stylus and the pitch of the cutterhead radial translation lead screw. Means are provided to remove the chip during the cutting. The amplitude of the signal recorded is determined by the high-frequency motion imparted to the tip. Signals to drive the cutterhead are provided through an equalizing circuit, which is required to compensate for variations in the cutterhead's amplitude and phase response as a function of frequency.

The basic construction of present-day cutterheads is illustrated in Fig. 16-22. A piezoelectric transducer, typically made from Vernitron lead-zirconate-titanate (PZT-8) material, supports a diamond cutting tip that is bonded to it with an epoxy cement. In turn, the PZT element is bonded to the steel element with epoxy or a low-temperature metal alloy. The two, loaded by the diamond cutter, form a mechanical resonant assembly that is damped by a layer of Kapton material and Viscaloid cement used between the steel element and the support, and to some extent by losses in the bonding materials on both faces of the PZT element.

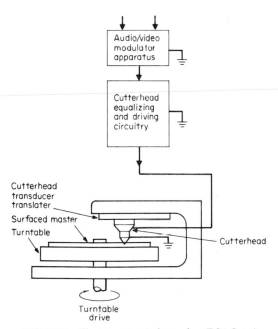

FIG. 16-21 Electromechanical recorder. (RCA Corp.)

The PZT element and the metal element have dimensions that are tapered and chosen so that the main resonance is broadened and the response is free of undesired resonance modes over a relatively wide frequency range. One of the functions of the steel element is to present, at the higher driving frequencies, a high mechanical impedance to the steel-element side of the PZT element, so that a given motion of the other side, upon which the diamond cutter is mounted, can be achieved at a lower driving voltage.

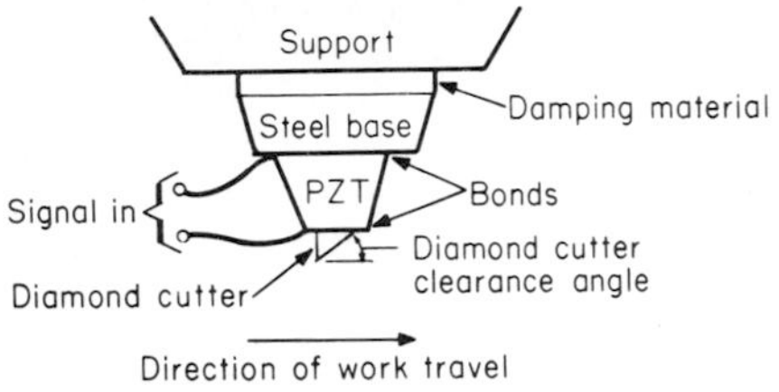

For real-time recording, the PZT element and the steel element each has a thickness of about 0.1 mm, and their widths and lengths are somewhat greater. The greatest tip-to-base dimension of the diamond is 0.1 mm. A back clearance angle of 28 to 35° is required so that the tip-to-base dimension can taper to zero. In spite of the low diamond-tip mass, the dynamic stress on the bond between it and the PZT element is in excess of 70 kg/cm^2 for approximately an 85-nm peak-to-peak motion at 2.5 MHz. Stresses in the PZT-steel bond are also high.

FIG. 16-22 RCA CED electromechanical cutterhead. *(RCA Corp.)*

In general, the cutterhead sensitivity at frequencies well below resonance is independent of the thickness of the piezoelectric elements used; i.e., the displacement is proportional to the driving voltage. Nevertheless it is desirable to limit the voltage to avoid breakdown due to heating and to excessive field intensity. Major elements of the recording system are illustrated in Fig. 16-23.[10] Signal input originates from video tape that has been carefully prepared to match the recording standards of the CED system, as illustrated in Fig. 16-8. Source signals are NTSC compatible and comply with EIA standard RS-170A. Chrominance is limited to peak levels corresponding to 75 percent saturated color bars. Program material in any format, other than 1-in type C, is transferred to this format in order to permit the necessary 2/1 reduction in speed during playback. The headwheel of the 1-in (2.5-cm) VTR plays back at full speed while the capstan

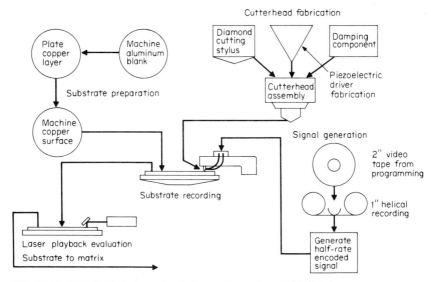

FIG. 16-23   RCA CED electromechanical recording system. *(RCA Corp.)*

advances the tape at half speed; thus each television field is played back twice. The composite video information generated by the tape recording is time-base corrected, preemphasized, and encoded into the buried subcarrier format wherein the chrominance subcarrier is translated from 3.58 to 1.53 MHz. The translated 1.53-MHz frequency was selected to be precisely the 195th harmonic of one-half the line frequency ($195f_{H/2}$) to ensure that the chrominance signal is interleaved with the luminance signal. This analog signal then is converted to digital form, with eight-bit digitizing at a 14.3-MHz sampling rate, and is stored in two-field store units from which it is read out at half rate. Ultimately the digital signal is converted back to the analog form, and the field identification signal is added in the vertical blanking interval during lines 17 and 280. The half-speed video then is encoded as an FM carrier. The half-rate audio from the tape recorder is similarly encoded on an FM carrier and added to the video FM signal to produce the composite cutterhead drive signal.

The luminance signal is preemphasized in two steps to improve the signal-to-noise ratio. The first is an RC preemphasis which boosts the signal 6 dB per octave between 249 and 995 kHz; the second is linear phase preemphasis, which boosts high frequencies by a decibel number equal to the square of the frequency in megahertz to a maximum limit of 12 dB.

The master substrate is a 0.5-in (13-mm)-thick, 14-in-(35.5-cm) diameter aluminum disc, on which an approximately 15-mil (0.038-cm) layer of fine-grained copper has been electroplated. Following plating, the copper surface is machined to be flat within 1 μm.

Immediately following recording, an optical laser playback evaluation is possible by detecting the reflected light from an HeNe laser beam that is directed at the reflective surface of the copper master disc.

16.2.2 ELECTRON-BEAM EXPOSURE.[†]

Electron-beam exposure of video disc masters was initially employed by RCA to satisfy the high-resolution requirements of recording video signals in real time. A copper-clad aluminum substrate, with a precut spiral groove, after coating with an electron-beam-sensitive resist, was mounted on a turntable in a vacuum chamber. The turntable rotated beneath an electron beam column and was moved radially, so that the electron beam tracked the groove and exposed the electron-beam-sensitized resist. The beam was blanked on and off by the video and audio signal. After exposure the resist was developed to produce a pit relief image. Recording in a pregrooved disc dictated the need for a beam-tracking servo, in addition to a focus and a radial exposure servo.

Owing to imperfect flatness of the substrate, or temperature-related dimensional changes, the distance from the final lens of the electron beam system to the surface of the substrate varied during recording. To ensure focus within ± 197 μin (5 μm), a capacitance probe was used to sense the variations and to generate a correction current for an auxiliary dynamic focusing coil. The sensor was located at the same radius as the beam, but slightly offset, so as to sense an area of the substrate surface prior to beam exposure.

The radial-exposure servo was required to ensure that the beam decreased as the recording proceeded from the outside to the inside radius. Since the substrate rotated at a constant angular velocity, the linear speed, with which its coated surface moved under the beam, decreased as the recording progressed from the outside to the inside radius. To maintain constant exposure, the beam current was reduced in proportion to the radius by using a weak nonrotating magnetic lens which was mounted immediately below the electron gun.

Application of the resist to the surface of the grooved substrate was achieved by a spinning process. The substrate was a 0.5-in (13-mm)-thick aluminum disc with a 0.02-in (0.5-mm) thickness of bright copper electroplated on it. After a brief spinning period at approximately 450 r/min, to help reduce the resist to a uniform coating thickness, the substrate was spun continuously at a lower speed in a class-100-environment clean room until the resist solvent evaporated, leaving a smooth and uniform 1.0-μm thickness of

---

[†]Process details summarized and reproduced, with permission from RCA, from E. Keizer.[16a]

resist in the groove. The treatment of the coated substrate before exposure, the type of development, and the treatment after development to prepare it for replication required controlled drying and storage in order to ensure uniform sensitivity. Development by immersion was superseded by a spray process to minimize surface staining.

A number of electron-beam-sensitive materials were screened for resolution, smoothness, and sensitivity. Included were ablative materials, dichromated gelatins, photoconducting thermoplastic materials, photochromic plastic materials, silver halide materials, and both positive and negative photoresists. Initially Shipley AZ 1350 was selected for its high resolution and smooth surface. However, the requirements for a more sensitive, 1- to 3-μC/cm^2, and high-resolution positive working electron-beam-sensitive resist, capable of imaging 0.25-μm relief patterns, led RCA to the development of a proprietary material.

A major effort was initiated to develop the electron-beam recorder into a reliable production machine. An efficient oil-driven turntable capable of over 500 r/min was developed, along with means to control its speed to synchronize accurately with the video input signal. The beam improvements required for real-time recording were achieved by using a line-source rather than a circular cross-section electron beam, and a specially designed electron-optical column that imaged the source directly onto the resist-coated substrate at a reduced size of one two hundred and fiftieth.

In the design of the column, a large electron-optical aperture was used to increase exposure density. In addition, the column was designed to reduce drift of the focus, intensity, and position of the beam, as well as to improve reproducibility of control settings. The column parameters were adjusted to provide a final beam half-angle of about 0.25 mrad, with a depth of field of approximately 5.0 μm. With a half-intensity line width of 120 nm, a final image current of approximately 100 nA/μm of line image was achieved. This was sufficient to expose the resist up to a level of approximately 3×10^{-6} C/cm^2. Figure 16-24 is a graph of the beam current per length of line image in micrometers versus line width at half-intensity level.

16.2.3 LASER-BEAM EXPOSURE.

Master recording in the Philips/MCA, Thomson, and JVC systems is accomplished by laser exposure of photoresist-coated substrates. Shipley AZ 1350J is a typical photoresist. The resist is generally spun and baked over a layer of chromium, or comparable adhesive material, which has been vacuum evaporated on a flat glass disc. The chromium layer is required as an adhesion layer between the

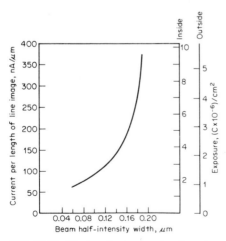

FIG. 16-24 Electron-beam recorder characteristic. (*RCA Corp.*)

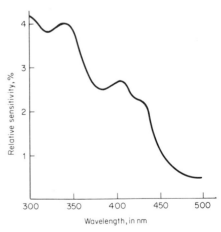

FIG. 16-25 Spectral response of Shipley AZ 1300 series positive photoresist. (*Shipley Company.*)

glass and the photoresist. Unfortunately the reflectivity of the chromium results in a reflection of the incident laser beam, which in turn generates a standing wave pattern[17] that causes nonuniform energy distribution in the photoresist, resulting in variations in pit dimensions.

Experiments[18] have indicated that for optimum pit geometry there exists a unique relationship between the exposure-sensitivity product of the photoresist and the thickness of the chromium layer.

Figure 16-25 illustrates the sensitivity of the photoresist as a function of exposing wavelength. The Shipley AZ 1350J is maximally flat in the UV region. Sensitivity at 457.9 nm is approximately 25 percent of that at 300 nm.

A simplified diagram of a laser recorder is illustrated in Fig. 16-26. An Ar-Ion laser operating at 457.9 nm is generally employed to expose the photoresist. Approximately 50 mW of power is required at the disc surface. Two optical modulators are required, one for video modulation of the laser beam and the other to vary the beam intensity as a function of radius in order to maintain a constant exposure from the maximum to minimum radius. While either acoustooptical or electrooptical modulators may be used, preference generally has been for the acoustooptical modulators because of their more stable performance over wide ambient temperature variations. Following modulation, the laser beam is focused by an objective lens, with an NA of approximately 0.7 onto the surface of the disc. The NA of the lens is chosen to generate a beam size, with resultant track width, that is approximately one-third of the beam spot of the playback system.

The exposed photoresist is developed by either a spray application or immersion in Shipley AZ developer. Developing times, at 68°F (20°C), vary from 2 min in a concentration of two parts AZ developer to one part deionized water to 3 to 4 min in a concentration of one part AZ developer to one part deionized water. Nonuniformity in the resist exposure-sensitivity parameters requires that the pit geometry be monitored during development so that the development process can be terminated when the optimum pit geometry is achieved.

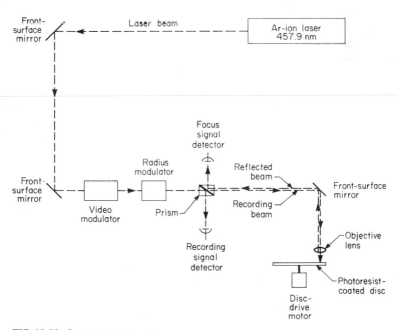

FIG. 16-26 Laser optical recorder.

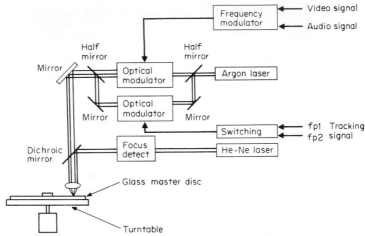

FIG. 16-27   Optical path of the VHD laser beam recorder. *(Victor Company, Japan.)*

Typically a master disc is 14 in in diameter and from 0.25 to 0.5 in thick. It is a highly polished flat glass platen with less than one micron-size defect per square millimeter of the surface. Approximately 150 nm of photoresist is coated over the chromium layer.

Optimum exposure and development results in approximately 150-nm-deep information pits, i.e., one-fourth of the wavelength of the HeNe laser beam that is used in playback.

Variations of the basic laser recording techniques, employed by each of the major video disc proponents, include different methods for implementing a radial drive servo, a focus servo, and an exposure intensity servo. Various photoresist adhesion layers were also used in an attempt to minimize the standing-wave problems that are inherent in the use of the reflective chromium. A uniquely different system was originally employed by MCA and later abandoned in favor of the one described above. The original MCA system employed a modulated laser beam to burn holes in a 20- to 30-nm-thick metal layer on a glass disc. The advantage of the process was that it permitted immediate playback, during recording, thus providing instantaneous assurance of a successful recording.

Since the thin metal layer of recorded information was not capable of producing an optical interference or scattering pattern, photoresist was applied over the metal surface and exposed with a UV light, through the holes in the metal, from the underside of the glass disc. Thus, the metal layer acted as a mask for exposing the photoresist layer. Following exposure, the disc was developed to produce a relief image in the photoresist.

The JVC recorder, illustrated in Fig. 16-27,† employs two optical modulators operating in parallel. Beam-splitting optics are used to separate the laser beam into two paths, one through the video modulator and the second through the tracking signal modulator. Hence the video information and tracking pits are recorded simultaneously.

Since the VHD video disc system employs a constant angular velocity system, the track velocity changes from the outside to inside radius. Therefore, the signal element at the innermost radius becomes shorter relative to the outer. If the spot size of the laser beam is constant through the entire disc, from outside radius to inside radius, the duty cycle of the pits becomes asymmetrical, causing an even-order harmonic distortion of the

---

†Illustration and description summarized, with permission from JVC, from T. Inoue, T. Hidaka, and V. Roberts.[11]

Table 16-4 Salient Characteristics of Metal Considered for Laser Ablation Recording Experiments

Metal	Melting point, °C	Boiling point, · °C	Approx. surface tension, dynes/ cm	Heat of fusion, cal/g	Toxicity,† Mg/m³
Bi	271.3	1560	350	12.5	X
Cd	320.9	765	550		0.1
Sn	231.88	2260	550	14.2	X
Te	449.5	989.8	175	4.27	0.1
In	156.6	2080	2550	6.8	Low
Pb	327.5	1744	450		Cumulative poison 0.1
Cr	1857	2672	1590		
Rh	1966	3737	2000		

†X indicates unknown.

playback FM signal. In order to compensate for this, the master recorder varies the spot size of the laser beam, during recording, to correspond with the radius of the disc.

All optical master-disc recording and processing operations must be conducted in a clean-room atmosphere in order to minimize the effects of dust. In addition, the undeveloped resist-coated discs must be handled under yellow light in order to avoid fogging the photoresist.

16.2.4 LASER-BEAM ABLATION OF METAL FILMS. The technique of laser-beam ablation of metal films has been investigated for its potential of creating instantly reproducible data, without any postprocessing of the disc. The technique involves imaging a laser beam on a transparent substrate which is coated with a low-melting-point metal. Table 16-4 lists the salient characteristics of several candidate metals. The most popular ones are bismuth and tellurium and alloys of tellurium, arsenide, and selenium. Tellurium-coated substrates have been widely experimented with[19] because of their high sensitivity and resolution capability. However, the toxicity and affinity to oxidation of tellurium have led to alloying it with arsenide and selenium and to hermetically sealing the alloy-coated surface in a disc sandwich configuration. Metal coating thicknesses vary from 10 to 50 nm. Since the actual ablation process involves both melting and evaporation of the metal, the selection of the optimum metal must also include consideration of its surface tension, heat of fusion, and toxicity.

Recording on tellurium (Te) or chromium (Cr) requires ventilation since the oxides of both metals are toxic when they exist in concentrations of 0.1 mg/m³.

Toxicity is not a problem with bismuth (Bi). The disadvantage in using bismuth is its marginal adherence to glass.

16.3 REPLICATION TECHNIQUES

16.3.1 MASTER PREPARATION AND TREATMENT.† The replication process of producing a consumer record from a recorded master involves a number of different processes. In all video disc systems, the process is similar to that used in producing audio records; i.e., the recorded substrate, if nonmetallic, is coated with a conductive layer

†Process details summarized and reproduced, with permission from RCA, from R. J. Ryan.[19a]

before being passivated and subjected to an electroplating process wherein hundreds of nickel stampers are produced from each recorded substrate. Figure 16-28 illustrates the electroplating fan-out matrix of metal parts where one or more nickel masters are replicated from the recorded substrate and, in turn, 5 to 10 mothers are produced from each master. In addition, 5 to 10 stampers are produced from each mother. Stampers are generally 7 to 8 mils (0.018 to 0.02 cm) thick, whereas mothers and masters are twice as thick.

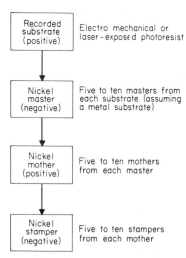

FIG. 16-28   Electroplating matrix flow.

The fan-out number of nickel masters, from a recorded substrate, is dependent on whether the recorded information is cut directly into the metal or is in the form of developed photoresist pits. If the substrate surface is photoresist, it is coated with a thin layer of metal before being electroplated; hence, only one replica is possible. If the substrate is metal, 5 to 10 nickel masters are possible. The polarity of the information surface is reversed in each electroplating stage such that the discs pressed from the stampers will have the same polarity as the recorded substrate.

The major difference in the matrix process between audio and video disc production is in the cleanliness and precise control of all matrix operating parameters.

1. Cleaner acid and passivation tanks are required.

2. Pumps and filters to remove all particles larger than 1 μm must be operated continuously.

3. Filters must be changed on a rigid schedule.

4. All processing tanks must be covered to reduce atmospheric contamination.

5. The nickel sulfamate solutions are maintained at a lower than normal level of NiCl concentration.

6. Low stress levels must be maintained by frequent carbon treatment.

After being electroplated, the parts are separated, cleaned, and critically inspected for scratches, physical damage, stains, dents, creases, shiny spots, or fingerprints.

Following inspection, the nickel stampers are prepared for use in the molding process by abrasive grinding of the back surface and trimming and coining to conform to the press mold configuration.

16.3.2 DISC-PRESSING SYSTEMS.† Four production methods have been explored actively in the video disc industry for molding video discs: compression molding, injection molding, embossing, and photopolymer casting.

1. *Compression molding* is the oldest and most widely used operational cycle and includes compressing a preformed molten disc shot of compound approximately 3 in (7.6 cm) in diameter and approximately 1 in (2.5 cm) thick, between an upper and lower mold, each of which has a nickel stamper, containing the video disc program information, clamped to the mold surface. The molten compound then is molded under controlled temperature and pressure to form a disc.

---

†Process details summarized and reproduced, with permission from RCA, from R. J. Ryan.[19a]

The mold temperature is thermally cycled from a high temperature, by the injection of steam during the compression period, to a lower temperature by the circulation of cooling water, just prior to opening the press and the removal of the disc. Following compression, the disc is transferred from the lower mold to a trimming station where the outer diameter of the disc is trimmed to remove the excess flash that has oozed out from the molds during the compression period. Typical cycle times are in the order of 40 to 45 s with a compression force of approximately 100 tons.

The shot of molten compound that is pressed into a video disc is generally a proprietary PVC compound, with plasticizer and lubricating additives, that has been specifically designed and processed under extremely clean conditions for video disc pressing. The composition typically consists of three main ingredients, i.e., a base resin homopolymer, copolymer, or a combination of both; a thermal stabilizer to retard the thermal degradation of the polymer structure; and a lubricant to control shear effects, to aid in fusion of the resin particles, and to reduce friction against the processing equipment surfaces. In addition, extremely fine-grain carbon particles are added to the mixture to make the disc conductive. Finally, the compound must exhibit three main characteristics: a uniform homogeneous composition, a good melt flow with controlled rheology and process stability, and the ability to exist as a dimensional and environmentally stable disc.

To accomplish these objectives (see Fig. 16-29), the base ingredients are first premixed in a high-intensity mixer to ensure a good distribution of all particles in the dry-blend matrix. The blend is then compounded in an extruder where a combination of external heat plus shearing action results in a uniform melt blend. The melt blend is then formed into pellets and transferred to an extruder hopper at the press, where it is plasticized through the extruder to the desired temperature and delivered as a preformed and preweighed shot of molten compound to the center of the lower mold in the press. Ideally the mold temperature should be higher than the shot melt temperature to effect good surface replication and minimal cycle times. With lower mold temperature, the cycle time must be increased to compensate for reduced melt flow.

The low disc-stress levels and excellent picture-element definition are the main attributes of the compression process. The chief disadvantages are limited stamper life, due to the movement of the metal stamper against the press mold during the

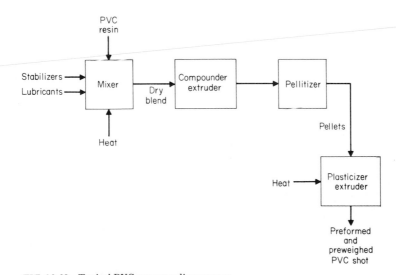

FIG. 16-29 Typical PVC compounding process.

thermal cycling, greater chance of contamination due to the exposed mold process and deflashing operations, and its limitation to the use of low-temperature materials, mainly PVC.

2. *Injection molding,* in contrast with compression molding where horizontal upper and lower molds open and close to the ambient environment during the pressing cycle, is characterized by a closed vertical mold process where hot plastic compound is injected into the mold cavity, under high pressure, after the mold halves are closed. The plastic solidifies at the mold temperature, and, after cooling, the mold is opened and the disc is removed. The advantages of injection molding include longer life of the stampers and no debris from ancillary operations like deflashing. Disadvantages include lower replication definition and higher molded-part stress. In addition, injection molding, with its increased rate and higher-temperature operation, imposes different requirements on the lubricant system; hence, selection and control of the lubricant compatibility and addition level are more critical than in compression molding.

3. The *embossing process* was adopted for producing thin, transparent optical video discs. The operation involves clamping a heated stamper against the surface of a flat section, from a long roll of PVC, for enough time to emboss the video information from the stamper onto the sheet of PVC.

4. The *photopolymer casting process* was developed to produce ultra-high-definition video disc replicas. A photopolymer liquid is poured over the surface of a metal stamper and then covered with a clear plastic disc. The clear plastic disc must be applied with sufficient care to ensure uniform distribution of the photopolymer over the stamper surface and to ensure that no air bubbles are entrapped. Ultraviolet light is used to cure the photopolymer. Typically several minutes of exposure are required, depending on the light intensity, before the disc can be separated from the stamper. The casting process is capable of reproducing the highest-definition video discs because the low temperature and pressure involved in replication result in the least physical deformation of the stamper.

16.3.3 POSTPRESSING TREATMENT OF RECORDS.

POSTPRESSING TREATMENT OF RECORDS. Postpressing treatment of video discs differs for the various disc systems.

RCA CED discs are inspected, after pressing, for any physical defects and then washed and lubricated with a silicone coating. The lubricant is required to extend playback stylus life and reduce disc wear and to prevent intimate contact between the stylus and the disc surface. The lubricant coating must be smooth, continuous, and defect-free, and must prevent adhesion of particulate matter to the stylus shoe and adhesion of dust to the disc surface. RCA found[19b] that ordinary household dust had particularly detrimental effects on video discs because of the dust's high content of organic salts. It was determined that at a relative humidity of about 80 percent, the dust particles on the surface of the disc absorbed enough moisture to cause a solution of salts to form around them. When the humidity drops, this pool of salts dries out, leaving the dust particles firmly adhered to the surface. Upon playback, these adhered particles caused either locked or skipped grooves and accompanying discontinuous playback. The optimum lubricant thickness is 20 to 25 nm. Below about 15 nm stylus wear becomes significant, and at about 30 to 40 nm the tendency for the stylus to entrap particulate matter increases. Both spin and spray processes were investigated for applying the lubricant. It was found that most of the solvent vehicles useful for spinning have an undesirable effect on the disc surface, probably because of the dissolution of some of the additives in the disc compound; hence, spray coating was adopted.

The JVC VHD disc production system does not include any significant postprocessing operations, other than the normal quality control inspection, since lubrication of the disc is accomplished via the solid lubricants that are present in the PVC compound.

Postpressing operations of reflective optical discs involve the vacuum evaporation of an aluminum reflective coating, followed by the application of a clear PVC protective coating over the aluminized surface. Double-sided discs are produced by physically cementing two individual discs, back to back, after each has been statically balanced. Typical quality control inspections include a check on flatness and the optical characteristics of the clear PVC coating.

16.3.4 QUALITY CONTROL EVALUATION. The high-definition requirements of video discs impose demanding tolerances on the various manufacturing operations. Production acceptance criteria include evaluation of both physical and electronic parameters.

Physical parameters and tolerances vary with each system. However, all 1-h-per-side PVC discs must generally satisfy the following requirements:

1. Diameter dimensional tolerance within ± 1.0 mm
2. Thickness dimensional tolerance within ± 0.2 mm
3. Eccentricity less than 0.1 mm
4. Surface flatness within 1 to 2 μm
5. Disc warp within 3 to 4 mm, depending on the system
6. Track pitch dimensional uniformity within ± 10 percent
7. Birefringence in optical discs less than 20°
8. Refractive index for optical discs 1.5 ± 0.1

Electronic parameters that must be evaluated include the following:

1. Video and audio carrier amplitudes
2. Video modulating signals that conform to FCC Rules and Regulations and EIA RS-170A Recommended Practices for NTSC standards and EBU specifications for PAL and SECAM signals
3. Playback velocity, focus, and radial tracking servo systems, to the extent required by the system, that result in playback signals meeting the requirements under no. 2
4. Video signal-to-noise ratio, unweighted, 42 dB
5. Luminance bandwidth 3.0 MHz
6. Chrominance bandwidth 0.5 MHz

All quality control evaluations must be conducted to ensure performance under a general ambient operating temperature range of approximately 16 to 35°C and a relative humidity of approximately 10 to 90 percent.

Electron microscopes and laser diffraction-measuring devices are typical of the sophisticated equipment that is required to measure both signal and groove parameters as well as for evaluating disc surfaces.

16.3.5 RECORD PACKAGING. Except for reflective optical discs, all other discs are protected from handling damage and exposure to environmental dust and dirt by a caddy. In addition to its use as a protective enclosure, the caddy provides an ideal physical format for handling, labeling, storage, and shipping.

The caddy design is such that the disc cannot be removed from the caddy and thus be subject to mishandling. Disc extraction from the caddy is accomplished automatically by inserting the loaded caddy into the player, where a release mechanism disengages the disc and permits removal of the empty caddy prior to program playback from the disc. At the end of play, the empty caddy is reinserted into the player, at which time the disc is automatically reloaded into the caddy.

The reflective optical disc, by nature of its design and construction, is an order of magnitude less sensitive to mishandling and environmental exposure; hence, it is simply jacketed in a container similar to that used for audio records.

16.4 DIRECT READ-AFTER-WRITE TECHNOLOGY

16.4.1 DIRECT READ-AFTER-WRITE CHARACTERISTICS. Direct[20,21] read-after-write (DRAW†) technology has been investigated for nonconsumer applications of high-density data storage and retrieval. Ideally a DRAW system is one in which a single laser beam is used to both record and reproduce data, with a negligible time delay between recording and playback, on a suitably sensitized substrate without any subsequent disc processing. Further, the rapid access time and high-resolution requirements of industrial data storage systems have dictated the exclusive use of optical technology for DRAW systems.

Typical characteristics for DRAW-system discs include:

1. Archival characteristics for a span of 10 to 100 years
2. Information-packing density of 10^{10} to 10^{12} bits per disc side
3. Low inherent error rate, that is, 1 per 10^{12}
4. Possible erasability
5. Optical quality substrates
6. High sensitivity at the recording wavelengths
7. Nontoxic material
8. High-definition capability, i.e., ability to record and play back pits less than 1 μm in diameter
9. Recording and playback operation under non-clean-room conditions
10. No ancillary processing of the disc
11. Substrate surface flatness of 1 to 2 μm
12. Playback laser power equal to 25 to 30 percent of the recording laser power

16.4.2 DIRECT READ-AFTER-WRITE MATERIALS.[22] The numerous materials that have been investigated for DRAW applications generally can be classified into four categories: chalcogenide, ablative, dye polymer, and silver-based thin films.

The *chalcogenides,* i.e., oxides, sulfides, selenides, and tellurides, are attractive because of their ability to change between an amorphous and crystalline state by the application of heat pulses from a laser. Bidirectional switching is accomplished by exposure to laser pulses of different peak power and duration. Film thicknesses vary from 50 to 100 nm and are either vacuum evaporated or sputtered onto glass or flexible substrates.

Ablative materials include all low-melting-point metals which may be either melted or evaporated by exposure to the heat from a laser beam (see Sec. 16.2.4). In general, archival characteristics are limited by oxidation of the recorded data surface such that increasing deterioration occurs from 2 h to 2 years. Tellurium is the most widely used metal. Recording is possible, at 10 megabits per second (Mbits/s) with approximately 8 mW of incident power at the disc surface. Unfortunately at the high ambient temperatures and humidities typical of tropical and subtropical climates, tellurium oxidizes, and the data disappear.

Dye-polymer laser-write and read (LWR) materials have been developed by the Eastman Kodak Company as a dual wavelength sensitized material. Two lasers operating at different wavelengths are used independently for recording and playback.

A *silver-based* DRAW material called *Drexon* has been developed by Drexler Technology Corp., in which very fine metal particles are dispersed in the surface of a plastic film. Heating the metal particles results in a reflective surface while also creating an absorptive area around the metal reflective particles.

---

†An N. V. Philips Gloeilampenfabrieken Co. trademark.

16.4.3 DIRECT READ-AFTER-WRITE SYSTEMS.   DRAW systems have been investigated by a number of companies,[23] including N. V. Philips, Thomson-CSF, RCA, Drexler Technology Corp., Hitachi, IBM, Exxon, Xerox, Bell Laboratories, and Wang Laboratories, for commercial data storage and retrieval applications.

The Philips system[24-26] is based on a pregrooved disc with up to 45,000 tracks per side in which each track is divided into a number of sectors. Each sector is identified with a unique address code so that the user data recorded in that sector can be readily located. The physical characteristics of the disc are illustrated in Fig. 16-30. It is configured as an *Air Sandwich*† to protect the sensitized surface from ambient degradation or contamination. Recording and playback are similar in principle to the consumer reflective optical system described in Sec. 16.1.7, except that the recording is digital and each pit represents one bit of information. Two varieties of the Air Sandwich disc have been developed by Philips to handle data rates below and above 3 Mbits/s. Both are similar in physical characteristics but differ in the sensitized material, the laser wavelength and power, and information format, and the information capacity. At data rates below 3 Mbits/s the sensitized surface consists of a layer of approximately 30 nm of a tellurium-based material. The record-playback light source is a semiconductor diode laser of gallium aluminum arsenide operating at 820 nm. The laser is capable of delivering between 10 and 15 mW of exposing power to the tellurium layer. Playback power is generally one-third to one-fourth of that used for recording. The disc contains 45,000 tracks, each of which is subdivided into 128 sectors. Up to 1000 pits which are 0.7 to 1.0 μm in size can be recorded per section, for a total of 5.76×10^9 bits per side. Rotational speed is 150 r/min. Average access time for data retrieval is 250 ms. At data rates in excess of 3 Mbits/s, that is, 5 to 10 Mbits/s, a somewhat different tellurium alloy is used instead.

The disc contains 40,000 tracks per side, with 32 sectors per track. Up to 15,200 bits which are 1 to 2 μm in size can be recorded per sector for a total nonformatted capacity of 1.95×10^{10} bits per side. Address and housekeeping bits reduce the total information capacity to 10^{10} user bits per side. The reported error rate is less than 10^{-10}. HeCd and HeNe gas lasers are used for recording and playback. Average data retrieval time is 500 ms.

Hitachi has reported development of a similar pregrooved sandwich disc with a 40-nm-thick sensitized surface of an amorphous layer of AsTeSe. Exposure is achieved by a semiconductor diode laser operating at 820 nm with a power of approximately 15 mW. The recording data rate is 1 to 2 Mbits/s with a pit size of between 31.5 and 35 μin (0.8 and 0.9 μm). Rotational speed is 240 r/min.

The RCA‡ approach to an optical digital data storage and retrieval system is based on an antireflective trilayer disc design in which three independent layers of aluminum, silicon dioxide, and titanium are successively deposited, one above the other, on a substrate. The aluminum constitutes the reflective layer, the silicon dioxide is a transparent dielectric layer, and the titanium is a strong light-absorbing layer. The thickness of the

---

†An N. V. Philips Gloeilampenfabrieken Co. trademark.

‡Illustrations and summary of description reproduced with permission from RCA from H. E. Bell, R. A Bartolini, and F. W. Spong.[27]

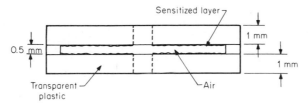

FIG. 16-30   Philips Air Sandwich.

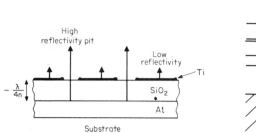

FIG. 16-31   The trilayer antireflection structure. *(RCA Corp.)*

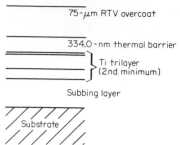

FIG. 16-32   Cross section of the structure of a
fully encapsulated second-minimum titanium
trilayer disc. *(RCA Corp.)*

dielectric and absorbing layers can be predetermined, for a given laser wavelength, to
yield a high reflectivity to absorption contrast ratio. A simplified version of the trilayer
design is illustrated in Fig. 16-31. Recording is achieved by employing an incident laser
light beam to thermally melt holes, or pits, in the titanium layer. Playback is achieved
by detecting the reflection of the laser beam from the highly reflective aluminum layer,
through the holes in the titanium layer. The low reflectivity of the titanium compared
with the high reflectivity of the aluminum results in a high contrast ratio. To minimize
the effects of surface degradation due to handling or exposure to a dusty environment, a
thin silicon rubber overcoat is applied over the titanium. In addition, a thermal barrier
of silicon dioxide is employed between the titanium and the overcoat layer to prevent
thermal degradation of the overcoat when melting the titanium at 1668°C. A cross section of a fully encapsulated disc is illustrated in Fig. 16-32.

Figure 16-33 is the schematic of an RCA optical recorder-playback system for evaluating trilayer discs. The modulated laser beam is expanded to fill the rear aperture of
the recording objective lens, which focuses the beam to expose the disc. Recent developments in diode lasers make it possible to simplify the recorder design by eliminating
the bulky gas laser and the laser modulator. Reproduction of the recorded signal is
achieved by detecting the reflected signal with a photodiode, after it has traversed the

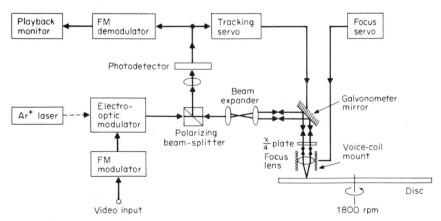

FIG. 16-33   Schematic of optical video disc recorder and playback system. *(RCA Corp.)*

reflection path through the objective lens, the one-fourth-wavelength plate, and the polarizing beam splitter. The return pass through the one-fourth-wavelength plate results in a 90° rotation in polarization, such that the polarizing beam splitter directs the return beam to the photodetector.

16.5 ERASABLE DISC TECHNOLOGY

Erasable disc technology has been pursued, in laboratories around the world, in the interest of developing a medium that includes the high-density storage characteristics of video discs and the record/erase/rerecord features of magnetic materials.

Research has been pursued in using materials with magnetooptic properties as well as materials whose properties are reversible between an amorphous and crystalline phase by the application of a laser beam.

Magnetooptic research has concentrated on the use of vertically magnetized rare-earth metals whose characteristics are such that when heated to their Curie-effect temperature by a laser beam in the presence of a local external magnetic field, a reversal in the direction of vertical magnetization occurs. Playback is achieved by detecting the change in polarization between the incident (exposing) laser beam and the reflected (readout) laser beam.

Amorphous-crystalline research has been devoted to exploiting the "change-of-state" characteristics of tellurium suboxide materials from a high-reflectivity crystalline phase to a low-reflectivity amorphous phase, by selective exposure to a laser beam. Playback is achieved by detecting the difference in reflectivity of the two states. Recording and playback have been achieved with 8 mW of incident power from a 0.83-μm-wavelength laser, and erasing has been accomplished with 10-mW of incident power from a 0.78-μm-wavelength laser.

REFERENCES

1. D. G. Howe, H. T. Thomas, and J. J. Wrobel, "Replication of Very High-Density Video Disc Master Recordings via Contact Printing," *Photographic Science and Engineering,* vol. 23, no. 6, November/December 1979.

2. Y. Tsunoda, K. Tatsuno, K. Katoaka, and Y. Takeda, "A Holographic Video Disc; an Alternate Approach to Optical Video Discs," *Applied Optics,* vol. 15, no. 6, June 1976.

3. J. C. C. Gilbert, "The Video Disc-Vision Programs on Gramaphone Records," *Wireless World,* August 1970.

4. W. Bruch, "A Color Video Disc System," *J. SMPTE,* vol. 81, April 1972.

5. G. Dickopp and H. Redlich, "Design Simplicity Cuts Costs for German Color-Video Disc System," *Electronics Magazine,* Sept. 27, 1973, p. 93.

6. "Video Disc," *RCA Rev.,* vol. 39, no. 1, March 1978.

7. H. N. Crooks, "The RCA Selectavision Video Disc System," *RCA Engineer,* vol. 26-5, March/April 1981.

8. L. P. Fox, "The Conductive Video Disc," *RCA Rev.,* vol. 39, no. 1, March 1978.

9. E. O. Keizer and D. S. McCoy, "The Evolution of the RCA 'Selectavision' Video Disc System," *RCA Rev.,* vol. 39, no. 1, March 1978.

10. J. J. Brandinger, "The RCA CED Video Disc System—An Overview," *RCA Rev.,* vol. 42, no. 3, September 1981.

11. T. Inoue, T. Hidaka, and V. Roberts, "The VHD Video Disc System," *123rd SMPTE Tech. Conf.,* Oct. 30, 1981.

12. W. Van den Bussche, A. H. Hoogendijk, and J. H. Wessels, "Signal Processing in the Philips 'VLP' System," *Philips Technical Rev.,* rev. 33, no. 7, 1973.

13. K. Compaan and P. Kramer, "The Philips 'VLP' System," *Philips Technical Rev.,* rev. 33, no. 7, 1973.

14. G. Bouwhuis and P. Burgstede, "The Optical Scanning System of the Philips 'VLP' Record Player," *Philips Technical Rev.,* rev. 33, no. 7, 1973.

15. G. Bouwhuis and J. J. M. Braat, "Video Disc Player Optics," *Applied Optics,* vol. 17, July 1, 1978.

16. J. J. M. Braat and G. Bouwhuis, "Position Sensing in the Video Disc Readout," *Applied Optics,* vol. 17, July 1, 1978.

16a. E. Keizer, "Video Disc Mastering," *RCA Rev.,* vol. 30, March 1978.

17. B. A. J. Jacobs, "Laser Beam Recording of Video Master Discs," *Applied Optics,* vol. 17, July 1, 1978.

18. R. A. Castrignano, "Video Disc Technology—A Tutorial Survey," *J. Appl. Photo. Engr.,* vol. 4, no. 1, winter, 1978.

19. D. Y. Lou, G. M. Blom, and G. C. Kenney, "Bit Oriented Optical Storage with Thin Tellurium Films," *J. Vac. Sci. Technol.,* vol.18, no. 1, January/February 1981.

19a. R. J. Ryan, "Material and Process Development for Video Disc Replication," *RCA Rev.,* vol. 39, March 1978.

19b. D. Ross, "Coatings for Video Discs," *RCA Rev.,* March 1978.

20. Di Chen and J. David Zook, "An Overview of Optical Data Storage Technology," *Proc. IEEE,* vol. 63, no. 8, August 1975.

21. R. A. Bartolini, H. A. Weakliem, and B. F. Williams, "Review and Analysis of Optical Recording Media," *Optical Engineering,* vol. 15, no. 2, March/April 1976.

22. A. W. Smith, "Injection Laser Writing on Chalcogenide Films," *Applied Optics,* vol. 13, no. 4, April 1974.

23. M. Marshall, "Optical Discs Excite Industry," *Electronics Magazine,* May 5, 1981, p. 97.

24. Ten Billion Bits Fit onto Two Sides of 12-in. Disk for Optical Data Recorder," *Elect. Mag.,* Nov. 23, 1978, p. 75.

25. Philips Unveils Optical Disks for Mass Storage," *Electronics Magazine,* Nov. 9, 1978, p. 64.

26. G. C. Kenney et al., "An Optical Disk Replaces 25 Mag Tapes," *IEEE Spectrum,* February 1979.

27. H. E. Bell, R. A. Bartolini, and F. W. Spong, "Optical Recording with the Encapsulated Titanium Trilayer," *RCA Rev.,* vol. 40, September 1979.

BIBLIOGRAPHY

Adler, R. (1974). "An Optical Video Disc Player for NTSC Receivers," *IEEE Trans Broadcast Telev. Receivers,* vol. BTR-20, pp. 230–234, August.

Adler, R. (1976). "Video Disc System Alternatives," *IEEE 17th Chicago Spring Conference Consumer Electronics,* June 8.

Ahmed, M., R. Brown, and A. Korpel (1975). "The Aerodynamic Stabilization of Video Discs," *IEEE Trans. Consum. Electron.,* vol. CE-21, pp. 131–139, May.

Bartolini, R. A., H. E. Bell, R. E. Flory, M. Lurie, and F. W. Spong (1978). "Optical Disc Systems Emerge," *IEEE Spectrum,* August.

Berg, A., R. Comier, and J. Courtney-Pratt (1974). "High-Resolution Graphics Using a He-Cd Laser to Write on Kalvar Film," *J. SMPTE,* vol. 83, pp. 588–599.

Bogels, P. W. (1976). "System Coding Parameters, Mechanics and Electro-mechanics of the Reflective Video Disc Player," *IEEE 17th Chicago Spring Conference Consumer Electronics,* June 8.

Brandinger, J. J. (1981). "The RCA CED Video Disc System. An Overview," *RCA Rev.,* vol. 42, no. 3, September.

Broadbent, K. D. (1974). "A Review of the MCA Discovision System," *J. SMPTE,* vol. 83, pp. 554–559.

Broussaud, G., E. Spitz, C. Tinet, and F. LeCarvennec (1974). "A Video Disc Optical Design," *SID Journal,* pp. 38–39.

"Cutting the VideoDisc Master," *RCA Rev.,* vol. 43, no. 1, March 1982.

Duker, H. (1975). "High Resolution Optical Video Recording in Real Time," Robert Bosch GmbH, Stuttgart, Germany, *Ninth International TV Symposium,* Montreux, Switzerland, pp. 91–92.

"Erasable Optical Disks Could Be Here in 1985," *Electronics,* July 14, 1983, p. 54.

"Fourth Layer Improves Magneto-optic Disk," *Electronics,* July 28, 1982, p. 52.

Hrbek, G. (1974). "An Experimental Optical Video Disk Playback System," *J. SMPTE,* vol. 83, pp. 580–582.

Jenssen, P., and P. Day (1973). "Control Mechanisms in the Philips 'VLP' Record Player," *Philips Tech. Rev.,* vol. 33, no. 7, pp. 190–193.

Jerome, J. A., and E. M. Kaczorowski (1974). "Film Based Video Disc System," *J. SMPTE,* vol. 83, pp. 560–563.

Kaczorowski, E. M., and J. A. Jerome (1974). "A Photo-Optical Video Disc System," *SID Journal,* pp. 36–37.

Kenney, G. C., and A. H. Hoogendijk (1974). "Signal Processing for a Video Disc System (VLP)," N. V. Philips, Eindhoven, The Netherlands, *IEEE Trans. Broadcast Telev. Receivers,* vol. BTR-20, pp. 217–229.

Korpel, A. (1974). "A Review of Video Disc Principles," *Symposium Digest of Technical Papers, SID International Symp.,* pp. 32–33.

Korpel, A. (1974). "Optical Video Disc Technology," *Proc. SPIE, Symposium no. 53,* Laser Recording and Information Handling Technology, pp. 97–100.

Laub, L. J. (1976). "Optics of Reflective Video Disc Players," *IEEE Trans. Consum. Electronics,* vol. CE-22, pp. 258–265.

"Optical Memories That Will Give Computer Storage a Big Boost," *Business Week,* May 2, 1983, p. 56.

Pritchard, D. H., J. K. Clemens, and M. D. Ross (1981). "The Principle and Quality of the Buried-Subcarrier Encoding and Decoding System and Its Application to the RCA Video Disc System," *RCA Rev.,* vol. 42, no. 3, September.

Ross, M. D., J. K. Clemens, and R. C. Palmer (1981). "The Influence of Carrier-to-Noise Ratio and Stylus Life on the RCA Video Disc System Parameters," *RCA Rev.,* vol. 42, no. 3, September.

Selectavision Video Disc, *RCA Engineer,* vol. 27, no. 1, January/February 1982.

"Video Disc," *RCA Rev.,* vol. 39, no. 1, March 1978.

"Video Disc Optics," *RCA Rev.,* vol. 39, no. 3, September 1978.

"Video Disc Player Technology," *RCA Engineer,* vol. 26, no. 9, November/December 1981.

Whitman, R. (1974). "A Transmission Mode Optical Video Disc System," *Symposium Digest of Technical Papers, SID International Symposium,* pp. 34–35.

Winslow, J. S. (1976). "Mastering and Replication of Reflective Video Discs," *IEEE 17th Chicago Spring Conference Consumer Electronics,* June 8.

Film Transmission Systems and Equipment

Anthony H. Lind

RCA Corporation (retired)
Camden, New Jersey

Kenneth G. Lisk

Eastman Kodak (retired)
Rochester, New York

with contributions by

K. Blair Benson

Engineering Consultant
Norwalk, Connecticut

Robert N. Hurst

RCA Corporation
Gibbsboro, New Jersey

17.1 SYSTEM OVERVIEW

by Anthony H. Lind

Film has been an important source of picture and sound program material for television since its beginning. Film continues to be an important medium for recording picture images and for the storage of pictures and sound. The purpose of this chapter is to provide a reference source of technical information covering the film as a medium and the related equipment for recording on film and for converting the film images and sound tracks into video and audio electrical signals for transmission through the television system.

17.1.1 TELECINE SCANNING SYSTEMS. There are three general types of film scanners in use in broadcasting and other television services at the present time.

Projector-Camera. The most widely used telecine scanner equipment in the United States is a camera system in which the optical image from the film is projected onto the photosensitive layer of beam-scanned photoconductive pickup tubes. In this case, the optical image is exposed onto the pickup tube for a substantial portion of the film frame time. This exposure accumulates charge in the photosensitive layer and, owing to the substantial exposure period, results in a relatively sensitive camera system.

Flying-Spot Scanner. This type of scanner has been in use for many years and is favored particularly in Europe, and in countries having television scanning standards of 25 frames per second. This system generates the video signal from film by scanning the film image with a very small spot of light and collecting the resulting transmitted light at a photo cell. The scanning spot is generated in conventional systems by means of a high-intensity cathode-ray tube which is scanned by an unblanked electron beam to generate a high-intensity flying spot of light. An alternative, which has been employed in some recent specialized equipment, is to utilize a laser beam to scan the film image. In this case the scanning or movement of the laser spot has been accomplished by rotating prisms that deflect the laser beam as a function of the prism drive. Flying-spot scanner systems do not involve storage but are instantaneous in their generation of the video signal.

CCD Scanner. This type of scanner uses a single line of charge-coupled elements which forms a line-scan charge-coupled device (CCD). In this case the film frame is illuminated by an incandescent light source, and the film image is imaged, by suitable optics, onto the line-scan CCD effectively one line of the picture information at a time. This system does permit a very small amount of integration (storage) as a result of one line of picture elements being exposed simultaneously.

17.1.2 TELECINE MULTIPLEX SYSTEMS. For reasons of cost effectiveness, and for floor space efficiency, it is common practice to utilize optical combining of film picture sources into a common telecine camera. Since the optical paths are changed by moving a mirror into (or out of) a given optical path (which can be done very quickly), optical switching from one projector to another can be done on the air as a normal operating practice. In an extension of the basic single-camera configuration a second telecine camera can be introduced to provide redundancy, and thus backup, in the event of a failure in one telecine camera. It also provides the capability of a second video output for preview or other transmission purposes.

In order to make physical arrangements of telecine cameras, optical multiplexers, and film projectors, it is essential that relatively long throws be designed into the optical system. This has led to the use of a relay optical system. Since the film image in the gate of the projectors is relatively small, as is the optical image on the telecine camera pickup tubes, the large image at the point of transfer minimizes picture movement due to vibration, etc., of the projector with respect to the camera. The relay lens provides proper collection and redirection of the light rays for the telecine camera optics so that illumination uniformity is maintained over the image area and, at the same time, permits a relatively large image to be formed in space which is then relayed to the telecine camera.

In telecine machines such as flying-spot scanners and CCD line scanners, it has not been practical to provide optical multiplexing as described for photoconductive-type telecine camera systems. However, it has been the practice that a given scanner transport can be adapted readily to handle different film widths. Thus, a single scanner can be used for 35- or 16-mm film, with a minimum downtime to convert the machine from one mode to the other. Some scanners also are provided with a means of scanning 2 × 2 slide transparencies.

In optically multiplexed telecine systems, which are generally referred to as *film islands,* several key parameters have been standardized in order to make possible subsequent replacement of a projector or telecine camera in an existing film island. Mechanically, the components of the multiplexed island must have a common optical axis height above the floor. Furthermore, the optics of the telecine camera and film projectors must be designed to function at a standardized field-lens image size. Also, for ease in subsequent replacement, the electrical control system of projectors and multiplexers should be compatible.

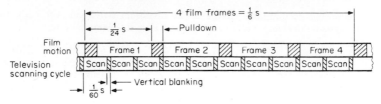

FIG. 17-1   Standard projector time-cycle.

17.2 INTERMITTENT-MOTION FILM PROJECTORS

by Anthony H. Lind

17.2.1 FRAME-RATE CONVERSION.   Motion picture film for professional (16 and 35 mm) use is universally exposed at a rate of 24 frames per second. Since television systems operate at either 30 frames per second (in 525-line systems) or 25 frames per second (in 625-line systems), there is a discrepancy between the film-taking frame rate and the television system reproduction rates. In 25-frame systems it is universally the practice to simply reproduce the 24-frame rate film at 25 frames per second. In this case the discrepancy is only 4.17 percent, but the increase in pitch of sound reproduction and reduction in running time of the film have been generally an acceptable compromise. In the case of 30-frame systems the discrepancy of six frames per second is 25 percent, and thus would be intolerable in both instances. As shown in Fig. 17-1 in the period of four film frames there are five television frames or 10 television fields. In order to avoid unacceptable flicker effects in the telecine video output, it is essential that each television field have the same exposure.

2-3 Pulldown.   This can be accomplished as shown in Figs. 17-2 and 17-3. In the first case the film-transport motion alternately holds the film in the projector gate to permit two field exposures or three field exposures to occur. The film exposure is for a very brief period that is phased to occur during the vertical blanking period of the television system. In the second case, the film pulldown time is reduced such that the period of stationary film in the gate is increased, and in this instance the exposure time can be substantially increased. With the longer application time, the exposure can occur any time during the field period, and, as shown in Fig. 17-3, the exposure occurs at the same phasing for each of the television fields. In the case of the long exposure (or application) time which approaches 50 percent of the television field period, it has been found that the film projector does not have to operate in precise synchronism with the television field rate. This makes it possible to operate the projectors from the power line source while the color television system is operating slightly nonsynchronously with the power line fre-

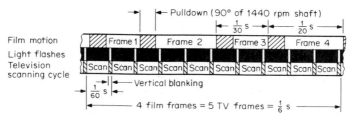

FIG. 17-2   Short-application projector time-cycle.

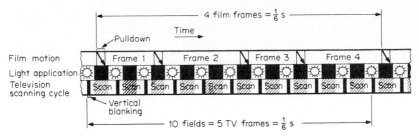

FIG. 17-3   Long-application projector time-cycle.

quency. If the exposure time becomes less than 30 percent, this nonsynchronous operation of the projector does begin to show a flicker effect.

Fast Pulldown.   An alternative approach in an intermittent projector is to cause the film pulldown to occur during the vertical blanking period, and thus position the film in the gate for full exposure during the succeeding two or three television fields. This requires extremely fast pulldown which places great stress on the film, particularly on the edges of the sprocket holes where the pulldown force is applied. Although 16-mm projectors have been constructed to operate with this approach, practical 35-mm machines have not. Because of the much simpler projector design and the much gentler handling, the long application projector design with 2-3 pulldown is quite universally used for photoconductive telecine systems.

17.2.2   PULLDOWN MECHANISMS.   In order to achieve the greatest film-positioning accuracy in the taking camera, since the film is running at a uniform rate, a claw pulldown is utilized to advance the film frame-to-frame. In addition, a registration pin is inserted into a sprocket perforation for each frame to ensure very high frame-to-frame consistency.

Claw.   One widely used technique in reproducing intermittent projectors, to advance the film, is to employ a two- or three-tooth claw which is periodically pushed horizontally toward the film so that the claw teeth engage sprocket holes and then the claw is moved vertically to advance the film in the gate.[1]† Both motions are accomplished by cam-driving the mechnisms against spring loading. Since the film motion entails a 3-2 relationship, the cams have two lobes and cause two frame advance cycles per revolution of the driving shaft. Although two and sometimes three teeth are present on the claw assembly, in general only one tooth surface engages a sprocket hole edge for driving the film. In the event of a damaged sprocket hole one of the other teeth will usually have a good sprocket hole to engage and thus provide insurance that the film will continue to advance through the projector. The two-lobe cam which drives the claw for advancing the film must be very precisely made so that the stopping point of the claw is precisely the same for each lobe. This is to ensure that there will be no spurious variation in the vertical positioning of successive film frames in the gate.

Mechanically Controlled Sprocket.   A second approach to moving the film intermittently at the gate of a projector is to use a sprocket which is rotated by an angle suitable to advance the film one frame pitch per cycle of film advance. Such a drive is shown in Fig. 17-4. The most commonly used sprocket for 16-mm film contains four teeth, while that most commonly used for 35-mm film contains 16 teeth. The sprockets are usually

†Superscript numbers refer to References at end of chapter.

rotated through the proper incremental angles by means of a Geneva drive, which is shown in Fig. 17-4. Sprocket pulldown systems are more difficult to design and fabricate than is the claw pulldown system because, in addition to requiring very precise cycle-to-cycle accuracy of the Geneva drive, the tooth-to-tooth spacing on the sprocket becomes critical with respect to vertical registration of the frames in the projector gate. However, because of the greater film mass and in an effort to minimize stress on the driving contact points of 35-mm film, the sprocket drive is universally used for 35-mm projectors.

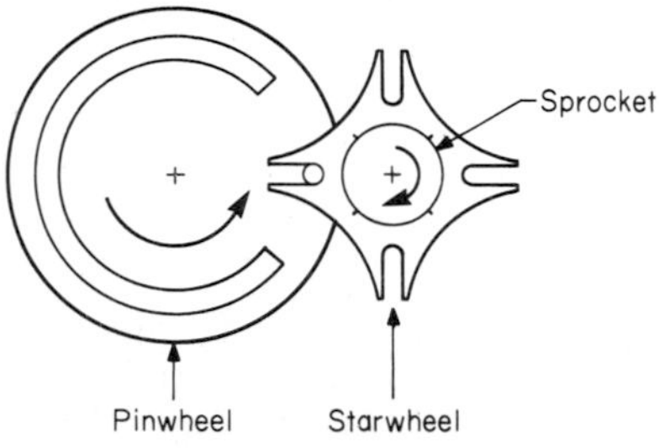

Servo-Controlled Sprocket. In the mid-1970s a sprocket pulldown utilizing a servo-controlled stepping motor for rotating the sprocket was introduced. This approach permits a fine degree of control in the stopping point of each succeeding tooth on the sprocket. This approach also lends itself readily to running the projector at widely different speeds.

The servo stepping-motor drive of sprocket pulldowns has resulted in the capability of operating the associated projector over a wide range of speeds. One projector, the RCA FR-35, can be adjusted at an incremental film rate from 0 to 48 frames per second. The servo-controlled projectors operate independently from the power line and thus can be locked to television vertical sync if so desired.

FIG. 17-4   Geneva sprocket drive.

17.3 CONTINUOUS-MOTION FILM PROJECTORS

by Anthony H. Lind

Film projectors employing continuous motion of the film with a means of optical rectification to immobilize the image at the projected image plane have been pursued throughout the life of the television industry. While there are a number of continuous-motion film transports employed in flying-spot scanner systems (which will be described in the next section), there have been very few successful continuous-motion projectors for use with storage-type camera pickup tubes. A projector which presents a totally immobilized image from frame to frame with a constant brightness as a function of time would become universal so that it could operate nonsynchronously with respect to the storage-type (photoconductive) television camera. In the United States, one continuous-motion projector has been offered for sale recently. Functionally, the film intermittent drive, the heart of the intermittent projector, is replaced by a hollow prism, and the projector optical system is suitably changed to accommodate the prism. The rotating prism assembly must be manufactured with great precision to achieve sufficiently adequate projected image stability. The prism assembly contains a sprocket around its periphery which is used to drive the assembly from the film. Steady and flicker-free images over a range of 2 to 150 frames per second for 35-mm film and 2 to 250 frames per second for 16-mm film are the specified ranges for the commercial product.

17.4 FILM-SCANNER TRANSPORTS

by Anthony H. Lind

17.4.1 FLYING-SPOT SCANNERS. The flying-spot scanner is an instantaneous (nonstorage) reading type of transducing system. The combination of the film-moving

Table 17-1　Flying-Spot Scanner Techniques

1. Moving mirror
2. Rotating polygon
3. Twin lens
4. Jump scan
5. Progressive scan and digital store

mechanism and the spot on the scanning raster of the flying-spot source tube must be coordinated so that the film image is properly scanned to generate the proper two fields of video information. One approach is to move the film intermittently at a very fast rate during the vertical blanking period so that the film is stationary in the gate of the projector during the active scan period. Transports of this type have been constructed but have proved to be very impractical because of the excessive noise plus wear and tear on the film. The only transports of this type have been designed for 16-mm film. A second and much more practical approach has been to design continuous-motion film transports of several varieties.

A number of film-motion/scanning techniques have been used with flying-spot scanners over the years. Table 17-1 lists the principal techniques that have been used. The first three techniques are no longer used in practical commercial products.

Jump Scan. In both the jump scan and the progressive scan and digital store approaches to flying-spot scanners, the optical and mechanical configurations of the transport system are considerably simplifed, when compared with earlier designs. The complicated optics of earlier systems have been replaced by more complex electronics. A typical design uses three independent dc motors, one for each film reel and one for the capstan. The capstan can be started in 100 ms since no flywheels are necessary for stabilization. Each reeling motor has its own servo to maintain film tensions, feedback being obtained from the appropriate tension roller. Both 16 and 35-mm film can be accommodated in this type of transport with only relatively simple changes in the film-loading path. In operation the film is moving at constant velocity past the gate, and each film frame is scanned twice for the 625/50 system and two or three times for the 525-line 60-field system. To do this, the scan must move on the flying-spot tube to follow the film, from which is derived the term *jump scan.* Figure 17-5 shows the vertical-scan waveforms, and Fig. 17-6 the scan rasters on the tube face for a 525/60 scanning system. Each field scan must be identical to within 0.1 percent to maintain good field-to-field registration. With pincushion correction to compensate for the flat-face tube and careful design in the scanning circuitry, it is possible to obtain 0.5 to 1.0 percent accuracy. Therefore, additional means are required to reach 0.1 percent. The main problem areas are the corners of the raster, and the errors in the four corners tend to be quite different. Therefore, independent controls or vertical and horizontal deflection in each corner are provided by means of a second scanning yoke of very low impedance which can be driven at horizontal frequencies.

In addition to special compensations which are required to ensure field-to-field registration, two other problems exist. As the film shrinks, the film velocity falls as the frame rate is fixed by the television system which is crystal controlled. Therefore as the film velocity falls, the distance moved by the film frame in one field period also decreases, which means that the distance between the two field scans must be reduced with shrinkage. This, of course, can be performed automatically by mea-

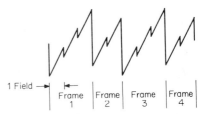

FIG. 17-5　A 525/60 jump-scan vertical waveform.

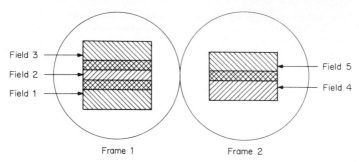

FIG. 17-6   The 525/60 jump-scan rasters on flying-spot tube.

suring the shrinkage and adjusting the vertical amplitude. Another error which requires careful control with jump scan is shading. As each of the field scans is separated on the tube face, difference in brightness between each field scan due to optics and screen burn becomes stricter. Shading correctors and a corrector to automatically compensate for screen burn are provided.

Progressive Scan and Digital Store. The advent of digital field and frame-store hardware that is both physically and economically practical has led to the most universal and mechanically simple approach of scanning each frame of film once in a progressive or sequential mode, storing all the information from the film frame in suitable digital video stores. The stored video information is then sorted and read out as required to produce the two interlaced fields of information in either the 525-line 60-field standard or 625-line 50-field standard.

If the video information were scanned from the film at a 24-frame rate, the total vertical scan would be obtained from the film motion. This would result in the flying-spot tube generating a single line of scan on the tube. Such operation, of course, would quickly cause burn and result in a very short scanning-tube life. By increasing the horizontal-scan frequency by the ratio of 4/3, the 525 lines are scanned in three-fourths of the normal period. Therefore, three-quarters of the vertical scan is contributed by a film motion, and the remaining quarter by the vertical scan on the tube. The conversion from 4/3 horizontal frequency to normal frequency takes place in the digital store-read operation which also is organized to generate a properly interlaced output signal.

Figure 17-7 shows the read-write sequence for the 525/60 system. The suffixes 1, 2, 3, etc., to the letters A, B refer to the film frame number, and the letters A, B refer to the field store. Although the total number of lines scanned in one film frame is 875, only 488 of these are stored. The first line stored is placed in field A and the second in field B, the third in field A, and so on until 488 lines have been stored. The fields stored are read out

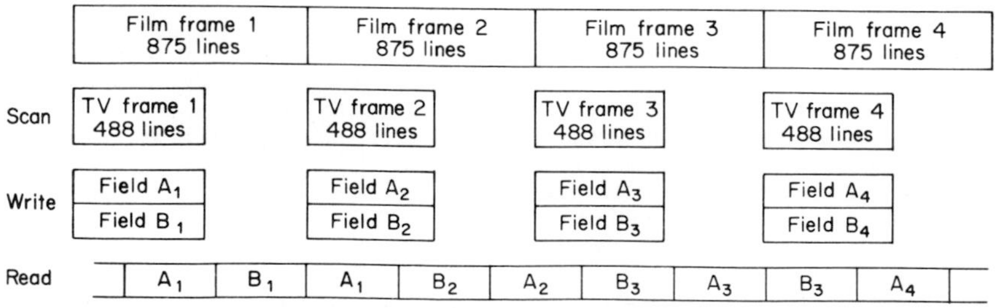

	Film frame 1 875 lines	Film frame 2 875 lines	Film frame 3 875 lines	Film frame 4 875 lines
Scan	TV frame 1 488 lines	TV frame 2 488 lines	TV frame 3 488 lines	TV frame 4 488 lines
Write	Field A_1 Field B_1	Field A_2 Field B_2	Field A_3 Field B_3	Field A_4 Field B_4
Read	A_1 B_1	A_1 B_2	A_2 B_3	A_3 B_3 A_4

FIG. 17-7   Write-read sequence for the 525/60 system.

alternately, producing a very simple sequential-to-interlace conversion. Three fields are obtained from film frame 1 in the order A1, B1, A1, and two conditions must be fulfilled for correct operation. First, A1, read, must finish later than A1, write, and in fact finishes some four lines later, and this is too small to show on this scale. Similarly, A1, read, must start before A2, write, and again there is an approximate four-line difference. The figure actually shows the time-scale for blocks of 525 lines in the write cycle and 262.5 in the read cycle, whereas, as mentioned before, only 488 lines are stored and 244 read in each field block.

17.4.2 CCD LINE SCANNER. The film transport for CCD line scan telecine is functionally very similar to that of servo-controlled flying-spot scanners. In a typical CCD-telecine transport (Marconi Communications Systems) a very important element is a frame-pulse generator that is coupled to the film by means of a film sprocket. Pulses generated by this sprocket-pulse generator are used to achieve positional control of the film frame with respect to addressing the line scan video information into the digital video store. An optical-disc signal generator is attached to the shaft which is driven by this sprocket. The optical generator produces a pair of pulses that are phased 90° apart for each film frame. These pulses are used for framing, for film counting, and also to provide the instructions for addressing and writing into the video store.

Line scanning with a continuous film transport requires horizontal deflection to be achieved with a shift register and vertical deflection to be achieved by the motion of the film. By shifting the write-in and readout addresses in the digital store and by increasing the line frequency when scanning, different film formats can be adapted by digital processing to the television format. Since the film motion is continuous, the ratio of image height on the film to the frame pitch must correspond to the vertical active picture time versus total television frame time in order that the film image fit the television raster appropriately. The proper aspect ratio is achieved by adjusting the horizontal image size to correspond with the CCD line-length dimension which is scanned during a period of 52 μs (the standard active television-line time). This condition does apply in the case of standard 16-mm film. However, it does not with the standard 35-mm format and with anamorphic formats. To scan a 35-mm film properly, one approach is to increase the line frequency to 18 kHz, which results in 602 lines being generated during the total film frame passage while only 488 lines are stored as active visible lines.

The digital-store addressing-control system makes it possible to operate the transport and the scanner system over a wide range of film speeds. For purposes of search, one transport (Bosch FDL-60) will operate over a range 2.4 to 600 f/s with 16-mm film and 2.4 to 240 f/s for 35-mm film.

17.5 TELECINE CAMERAS

by Anthony H. Lind and Robert N. Hurst

17.5.1 PHOTOCONDUCTIVE CAMERAS. The signals from a telecine camera's three photoconductive pickup tubes must be passed through several processing circuits before they are ready for encoding into the NTSC, PAL, or SECAM signals, which are suitable for transmission. Figure 17-8a is typical of the processing chains found in photoconductive telecine cameras. There are three such cascades of circuits, substantially identical; for simplicity, Fig. 17-8a shows only one of those three.

The signal from the photoconductive tube passes first into a *preamplifier,* which has the task of either providing gain for the low-level signals received from the pickup tube or altering the input time constant so that the tube and circuit capacity cannot substantially affect the circuit bandwidth. If the latter task is the designer's choice, the preamplifier is designed to have a very low input impedance and to operate as a transresistance amplifier (see Fig. 17-8b).

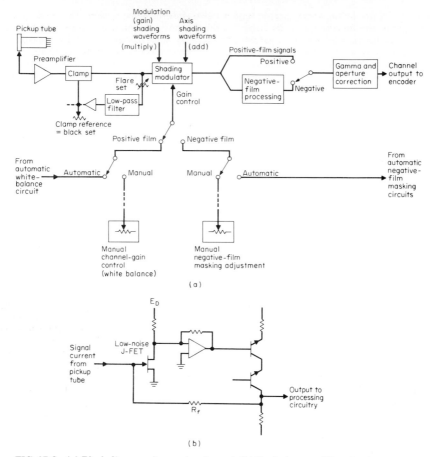

FIG. 17-8 (*a*) Block diagram of one color channel. (*b*) Typical preamplifier circuit.

The preamplifier output usually passes through a combined *clamping* and *black-level correction* circuit, which allows for the compensation of undesired backgrounds arising from dark current or bias light. One popular means for such correction consists of adding to the signal a blanking pulse of variable amplitude and polarity, thereby allowing the dc component of the active scan to be varied slightly to achieve this compensation. Alternately, the reference voltage for the clamp may be made variable, thus achieving the desired black-level correction.

Once black level has been set and clamping accomplished, the signal is passed to a *flare-correction* circuit. In this circuit, the signal is integrated (by means of a low-pass filter having a cutoff below frame rate) to determine the overall illumination of the scene. The resulting signal is fed back to the black-level control to compensate for the stray illumination arising from light reflections in the optics path and from the tube surfaces.

Shading is accomplished next, in the typical circuit cascade. Shading correction is of two types. The first, called *axis shading*, adds to the signal a combination of sawteeth and parabolas, at both horizontal and vertical rates. The second type, called *gain shading*, multiplies the signal by a different combination of horizontal and vertical sawteeth and parabolas. If both types of shading are provided, the operator is usually offered four control knobs for each type, so that independent adjustments of both axis shading and gain shading may be made. Axis shading permits the operator to correct for uneven dc conditions across the picture plane, such as might be caused by dark currents. Gain shading, on the other hand, permits correction of uneven illumination from the projector, or uneven sensitivity in the photoconductive pickup tube.

Since the multiplicative (gain) shading in a modern telecine camera is usually done through four-quadrant multipliers, it is economical to use this same multiplier as a dc-controlled, remotable *gain control*. As Fig. 17-8a shows, the lead which controls the multiplier gain is switched, with the switch position depending upon whether the telecine chain is handling positive film or negative film. For positive film, the gain control goes to a (remote) potentiometer labeled either *channel gain* or *white balance*. Since Fig. 17-8 depicts only one of the three channels, there will be three such (remote) gain controls—one each for red, blue, and green—and the operator may use these three to adjust the three channel gains to achieve equal outputs from each channel for a white signal, thus achieving white balance.

Most modern cameras will also include a servomechanism for the automatic adjustment of white balance. This possibility is indicated in Fig. 17-8a by the switch which allows the (remote) gain control to be replaced by a servo-generated direct current which can automatically adjust channel gain to achieve white balance. In most telecine cameras, this automatic action is achieved by holding the green channel gain to a fixed value and allowing the servo to vary red and blue channel gains to achieve white balance. The circuitry which detects the channel behavior and generates the red and blue control signals is called *auto-white balance* circuitry and is described below.

When negative film is being handled by the telecine chain, the switch in the multiplier gain-control lead of Fig. 17-8a is switched to the negative-film position, and the channel gain is then controlled by a potentiometer labeled *negative-film masking*. The gain controls of the three channels can then be set to different values, to compensate for the orange or yellow mask that is added to negative film to minimize crosstalk between film dyes in the printing process. Although this masking coating is of no use in television-signal processing, being present chiefly to improve positive prints made from the negative film, it is still present in all standard negative film and must be compensated for by circuitry in the telecine chain.

Initial compensation for the orange mask is usually done optically, by the addition in the optical path of cyan filters, such as one or two Kodak CC40s. Such optical filters reduce red-channel output, making it more nearly equivalent to green. The balance of the compensation is then accomplished by the individual negative-film masking potentiometers just discussed, and sometimes by additional channel gain controls in the negative-film path, which is described in the next paragraph.

By referring again to Fig. 17-8a, it can be seen that the shaded and compensated signal is then passed to a pair of paths—one for *negative-film* signals and one for *positive-film* signals. A switch at the output of the alternate paths allows the operator to choose the appropriate path for the film in use at the time. The positive path is a direct connection; the positive-film signal from the preceding circuitry is ready for use as it stands. The negative-film signal, however, must be processed further before it can be encoded for transmission.

Figure 17-9a shows the *negative-film processing path* in more detail. As shown in the figure, it is common practice to provide a gain control at the very beginning of the negative-film-processing path, to enable further correction for the orange or yellow mask of the negative film. The gain controls are usually set to unity for the red and green channels, while the blue channel is set to approximately X3 gain, to compensate further for the blue loss of the masking dyes. The result is an approximation to the same levels in each of the three channels.

Since target blanking is present in negative-film signals (even though it represents a white signal on the output picture), to clamp on this "white" pulse is feasible, using a conventional driven clamp, for example. This function is typically performed next in the circuit cascade.

The clamped signals from the three channels are shunted through a side path to a

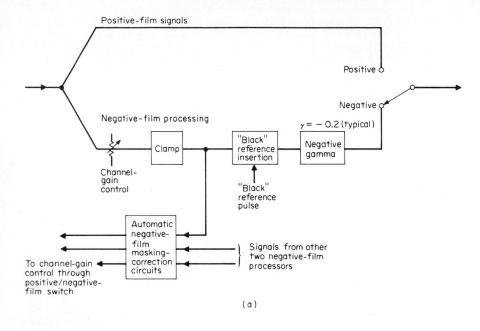

(a)

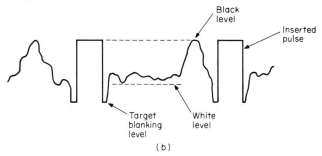

(b)

FIG. 17-9   Negative-film processing. (*a*) Simplified block diagram. (*b*) Black-level reference-pulse waveform.

circuit whose correct name is *automatic negative-film mask correction,* although the circuit is frequently called simply *auto-black set.* This circuit provides control signals to the four-quadrant multiplier (or other gain-control device) ahead of the negative-film processing channel, trimming the gains of the three channels to restore the black levels to the proper values. Such an automatic circuit makes it possible to pass intermixed negative films with different masks through the telecine chain, with automatic readjustment of the channel gains to compensate for the mask change at the splice point.

After the signals have been picked off for the automatic negative-film mask correction, it is common practice to replace the standard camera blanking or target blanking level (which is actually "white" in the negative signal) with a "black" reference pulse of known level. The resultant waveform is shown in Fig. 17-9*b*. The purpose of this modified waveform is to provide a black reference against which the auto-black set (that is, negative-film auto-mask correction) may operate.

The black-reference pulse is inserted in such a way that its absolute level is determined by the circuit constants, e.g., by forcing a transistor into cutoff so that the output level during the pulse is determined solely by precision resistors and regulated supply voltages. This pulse then can be used as the absolute-black reference, and the auto-black set circuits can operate as a servo loop which sets the video from the thinnest (blackest) portion of the film to this absolute-black reference level, or to a level which is a fixed (set-up) level difference.

When the reference-black pulse has been added to the signal, and the auto-black set servo loop has been closed around the signal, the signal is effectively clamped and may therefore have any needed *gamma correction* performed upon it. In the negative-film processing channel, normally no attempt is made to perform the usual gamma predistortion to prepare the signal for display on a (normally nonlinear) kinescope. This type of gamma circuit, which is common to both telecine cameras and live cameras, is found in a later part of the cascade, where the signal path is common to both positive-film signals and negative-film signals. The gamma circuit shown in Fig. 17-9 a (marked *negative gamma*) is designed solely to correct for the gamma characteristics of negative film, modifying its effective gamma to unity.

Since negative film typically has a gamma of about -0.5, the negative gamma circuit is designed to have a gamma of -0.2, so that the product of the film gamma and the circuit gamma is $+1.0$. This unity-gamma signal is then passed from the negative-film processing portion of the chain back into the signal path common to both negative film and positive film.

Figure 17-10 details the circuit path following the switches which choose between positive-film processing and negative-film processing. First in the cascade is an *adjustable gamma-correction circuit*. Unlike the fixed-value gamma circuits of live cameras, whose primary purpose is to predistort the signal to accommodate the inherent nonlinearity of all picture tubes, this telecine circuit combines the predistortion function with a true gamma correction function, accomplishing the latter as the operator adjusts it to compensate for the variable gamma of the film, the differences in the gammas of the various

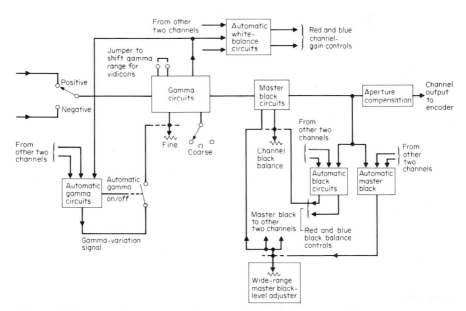

FIG. 17-10   Block diagram showing signal path and processing after positive/negative-film selector switch.

dye layers of a given film, and the differences in the gammas of various types of pickup tubes.

Fundamentally, the circuit must offer a predistortion gamma equal to the inverse of the picture-tube gamma. Since the kinescope gamma in the usual home-viewing situation is typically 2.2, the exponent of this circuit's transfer characteristic must be 1/2.2, or about 0.45. In addition, the circuit must be controllable to accommodate the use of either unity-gamma lead oxide and selenium pickup tubes or the approximately 0.6 to 0.75 gamma of vidicons. Also, the variations in the gamma of the film itself must be accommodated through operator-adjusted controls.

The result of these requirements is a gamma circuit having a variation range from about 0.2 to about 0.6, for unity-gamma pickup tubes, and (usually) a solderable or pluggable strap to shift the entire gamma range when vidicons are in use.

The fine adjustments on the gamma circuits of each channel allow the operator to change the gamma of each channel to accommodate the gamma variations of the dye layers of the film, because the characteristics of the dye layers are different for red, blue, and green, and may also vary from manufacturer to manufacturer. Independent measurements made on Eastman 5247 film showed that this film's red gamma is 0.588, while green gamma is 0.66, and blue is 0.545. While these values are derived from measurements on a single sample of film, the values are representative.

Common operating practice calls for setting the gamma controls of the three channels for gray-scale tracking on some film chosen as a standard, and accommodating other films as deviations from this standard. This operating practice is aided if the physical operating controls can be zeroed on the standard film by means of internal screwdriver adjustments on the gamma circuits, and the controls can be set for the other films as known deviations or fixed deviations established by preview.

The gamma circuits usually are followed by the *master black-level control.* In a telecine camera, this circuit is made to have a much greater range than the corresponding circuit of a live camera, since variations in film exposure may call for a sharp *pulling* of blacks in order to obtain an optimum transmitted picture. This problem occurs most often with negative film.

Following the setting of blacks, most telecine cameras provide *aperture compensation,* a function which allows the camera operator to compensate for the phaseless roll-off of the optical system and pickup-tube scanning aperture. In some cameras, both true aperture compensation (usually called *out-of-band aperture* because the circuits performing the function have a response peak above the video passband) and so-called contouring (usually called *in-band aperture*) are made available.

Although it is feasible to place aperture compensation circuits in each of the three channels (red, blue, and green), common practice places these relatively expensive circuits in the green channel only, for experience has shown that the degree of additional picture improvement obtained by duplicating these circuits three times does not justify the additional expense.

The three output signals from the three channels are usually fed to an encoder whose encoding algorithm is appropriate to the mode of transmission to be used (NTSC, PAL, SECAM, analog components, digital components).

Matrixing.   The above discussion covers only one channel of a telecine camera and extrapolates to the entire camera by suggesting that the other two channels are identical, except, perhaps, for the green-only aperture compensation. There is another aspect of the camera which may be shown only as an interaction among the three channels.

Consider what would happen if a film camera photographs a color-bar pattern from a kinescope. Assume that the color-bar generator had been perfectly set up and that the picture-tube color temperature is specially set to match the color-temperature rating of the film. The resulting film, if placed in a telecine camera, would *not* produce proper color bars at the camera output, because the picture-tube phosphors, the film dyes, and the telecine camera taking filters do not have the same spectral shapes. Consequently, the color-bar signal from the telecine camera would evidence errors in both hue and saturation.

To correct these errors, it is necessary to cause deliberate crosstalk among the three primary-color channels. Carefully calculated amounts of green signal and red signal, for example, are caused to contaminate the blue channel; similarly, green is contaminated by red and blue, and red, by blue and green. The contaminating colors may be either negative or positive, as required. This overall arrangement is usually called *linear matrixing,* but the name *linear masking* or simply *masking* is also seen in the literature. The last two names are deprecated because they can be confused with the masking process described above.

Since dye sensitivity is not a linear function, some telecine manufacturers have chosen to attempt compensation by performing *nonlinear matrixing.* In such an arrangement, the crosstalking signals are passed through logarithmic amplifiers before they are added to the neighboring channels.

Other manufacturers chose to perform corrections of this nature by circuits which operate in the domain of the color difference signals—I and Q or $R - Y$ and $B - Y$. An example of such a system is the RCA ChromaComp circuitry.

Details of Selected Circuits.   A typical preamplifier is shown in Fig. 17-8b.. Since the amplifier receives signal current at its low-impedance input and produces a signal voltage at its output, its gain is expressed in ohms and is numerically equal to the feedback resistor R_f. Since this resistor would produce the same output voltage if fed the signal current without the presence of the amplifier, the amplifier's main function is to lower the input impedance sufficiently to avoid the effects of stray capacity.

Usually, no processing is done in the preamplifier. A flat response is usually provided, and an output level comfortably high (of the order of 0.5 to 1 V) for transmission by short terminated coaxial cable from the preamplifier assembly to the next assembly or module.

Figure 17-11b shows a typical gain-shading circuit, with auxiliary input for channel gain control. The sawteeth and parabolas used to vary the gain as a function of xy posi-

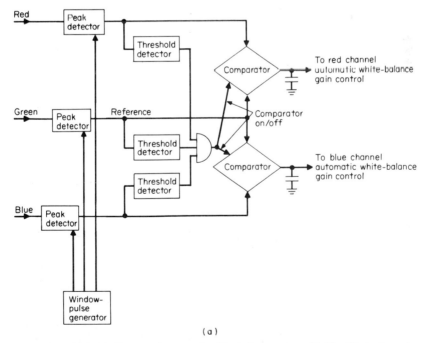

(a)

FIG. 17-11   (*a*) Block diagram of automatic white-balance system. (*b*) Simplified schematic of gain-shading circuit.

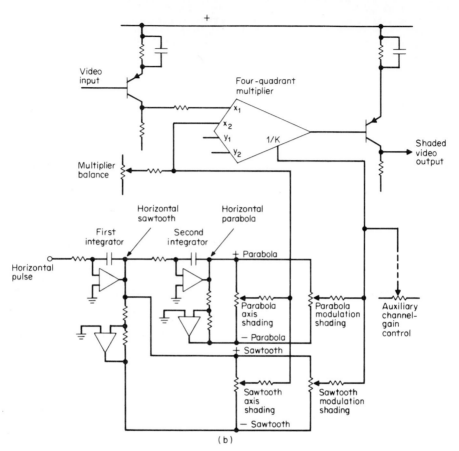

FIG. 17-11 *(Continued)*

tion on the raster are, in this figure, derived from a pulse, which is integrated once to produce a sawtooth, and once again to produce a parabola. In this drawing, only the horizontal-shading circuits are indicated; the vertical circuits are similar except for appropriate changes in time constants.

Auto-white-balance circuitry is seen in Fig. 17-11a. By taking the green channel as a reference, and by forcing the white area as seen by the red and blue channels to have outputs identical to that of the green channel, white balance is achieved. Key to the operation of the system in a telecine camera is the thresholding circuits shown. These determine when all three channels have instantaneous outputs above 80 percent, or some similar level. The AND gate determines when this condition has occurred, and the assumption (usually valid) is that such a condition signifies a white signal. The AND gate then enables the two comparators, a sample is taken, and the resulting control signals pass to the red and blue channels (Fig. 17-8a, marked "from AUTO WHITE BAL. CIR-CUIT") to control these two channel gains, to force a match to green level at that instant.

17.5.2 FLYING-SPOT SCANNERS.   In addition to continuous-motion film transports as described in Sec. 17.4.1, flying-spot scanners utilize extensive electronics to scan film horizontally, transduce the optical image to electrical signals, process the component

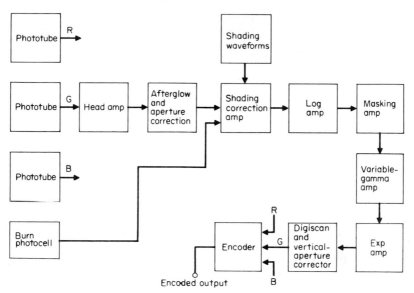

FIG. 17-12 Block diagram of video-signal channels.

color signals, and produce suitably encoded color-video output. Digital video-signal storage makes possible great flexibility in scanning and improved quality in the generated video signal.[2]

Block Diagram. A functional block diagram of the video signal channels in a flying-spot scanner is shown in Fig. 17-12. Only the green channel is shown in its entirety since the red and blue channels are identical and confining the diagram to the green channel only makes it easier to follow. The video signals are nonstandard analog signals up to the digiscan block. Within this block the analog signals are converted to digital for further processing and are delivered at its output as reconverted analog signals at the standard scanning rates for encoding in the encoder. Since basic video amplification and color encoding are functions which are common to essentially all telecine cameras, they will not be discussed further here.

Light Spot Source. The light spot for flying-spot scanning is generated by a special cathode-ray tube which utilizes a very fine grain phosphor and a high-resolution scanning beam and which operates a high voltage to generate a very high light output. The data which are summarized in Table 17-2 describe a typical flying-spot tube as is widely used in current flying-spot scanners. It should be noted that when operating at anode voltages of 20 to 30 kV, the cathode-ray tube is very likely to emit x-rays which exceed the recommended maximum permissible level of 0.5 mR/h as measured at a distance of 50 mm from any part of its external surface. Suitable shielding and operating precautions should be taken when using this equipment. Since the light is generated by electrons exciting a phosphor, there is a continuation of light output after the electrons have stopped exciting a particular area of the phosphor. This persistence or afterglow requires electronic compensation at a suitable point in the signal channel. The afterglow for a typical flying-spot tube is shown in Fig. 17-13.

A very important flying-spot tube characteristic in color-telecine scanners is the spectral characteristic of the light emitted by the flying-spot tube. A typical spectral energy distribution is shown in Fig. 17-14.

Table 17-2 Flying-Spot CRT Characteristics

General description	
Deflection	Magnetic
Focus	Magnetic
Deflection angle	42°
Useful screen dimensions	6.5 in (165 mm) diameter
Max. overall length	21.3 in (542 mm)
Screen	Aluminized
Standard phosphor	X3
Absolute maximum ratings	
Anode voltage	32 kV
Modulator voltage (negative)	200 V
Typical operating conditions and characteristics	
Heater voltage	4.0 V
Heater current	1.0 A
Anode voltage	25 kV
Modulator voltage (negative) for spot cutoff	74 to 124 V
Resolution (line width) at $I = 100\ \mu A$	4724 μin (0.12 mm)
Modulator drive voltage for 100-μA I	35 V
X3 phosphor persistence to 10%	0.15 μs approx.

Color Response. The basic color response of a typical flying-spot scanner is determined by a combination of the spectral output from the flying-spot tube, the spectral response of the photomultiplier tubes, and the optical filtering imposed ahead of the three R, G, and B photomultipliers. Figure 17-15 shows characteristics of the individual color filters that serve to divide the light ahead of the photomultipliers. The overall color response curves which combine the flying-spot tube characteristics, the photomultiplier sensitivity characteristic, and the color separation characteristics are shown in Fig. 17-

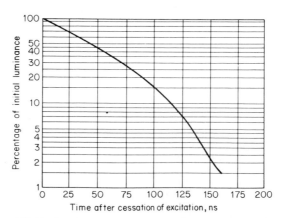

FIG. 17-13 Afterglow of a typical flying-spot CRT. (Anode voltage = 25 kV; beam current = 150μA.) *(Rank Cintel, Ltd.)*

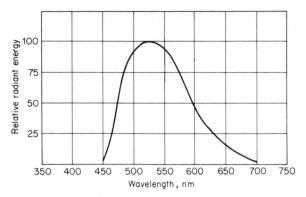

FIG. 17-14   Typical flying-spot CRT spectral output. *(Rank Cin-tel, Ltd.)*

16. Also included in this figure for reference purposes are the light output curve from the flying-spot tube and the sensitivity of the photomultiplier tubes. It is evident from the curve that the sensitivity falls rather rapidly at the blue and red ends of the spectrum. Blue suffers most in that the shorter wavelength sensitivity is determined purely by the cathode-ray tube output. In this instance the phosphor used is a single crystal composition which limits the blue response. The only way at present to widen the response

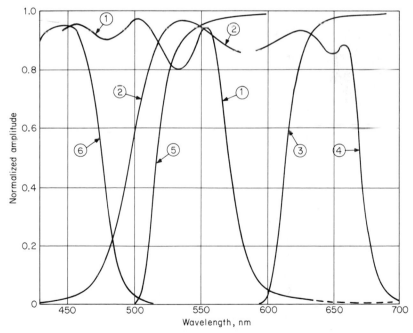

FIG. 17-15   Color-splitting color-filter curves: 1 red reflector, 2 blue reflector, 3 red trimming filter, 4 infrared rejection filter, 5 green trimming filter, 6 blue trimming filter.

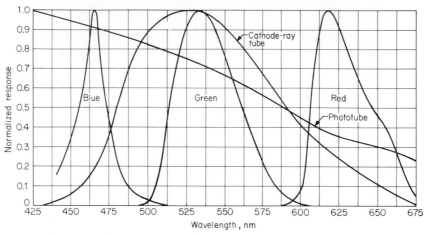

FIG. 17-16 Overall color response of Rank Cintel Mark III flying-spot scanner. *(Rank Cintel, Ltd.)*

would be to use a mixed phosphor. Such mixed phosphors produce higher screen grain, and therefore the single crystal phosphor is still considered to be a better trade-off at the expense of blue light output.

Sequential Scan. The capstan drive continuous film transport, in conjunction with digital video signal storage, has become very attractive for operation in the 525/60 television system. Since the digital video store makes it possible to manipulate the signals and, in effect, to standards-convert, it makes possible the scanning of the film frames by sequential scan and then converting the stored information into standardized, interlaced field output signals. As described in Sec. 17.4.1 and shown in Fig. 17-6, the film scan occurs at a rate higher than normal scan and is accomplished partially by the film motion and partially by limited vertical scan of the flying-spot tube to cover the total vertical image dimension in the proper vertical time period. The faster write times allow both the sequential-to-interlaced signal conversion and the 24-to-30 frame conversion to be accomplished in the read process while using only one field store memory for each luminance signal field and one field store memory for each chrominance signal field. Figure 17-17 shows a simplified block diagram of the digital store and standards conversion system. The nonstandard R, G, B analog signals, which are sequentially scanned at an accelerated line rate and at the film frame rate, are introduced at the input to the first matrix. In this matrix they are translated to $Y, R - Y,$ and $B - Y$ signals. The Y signal is analog-to-digital converted, processed in the digital vertical-aperture corrector, and passed on

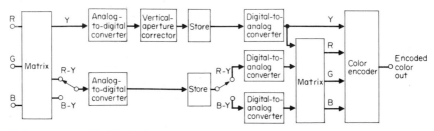

FIG. 17-17 Simplified block diagram of digital store and standards converter.

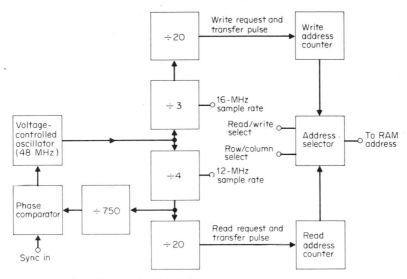

FIG. 17-18 Control system for digital video store.

to the digital store. The digital store consists of two field memories into which are written alternate lines to sort out the even and odd field signals. The $R - Y$ and $B - Y$ signals are switched at the clock rate to be combined into a single digital stream by the analog-to-digital converter and then written into the chrominance store which also consists of two field memories so that alternate lines of information can be written into the alternating field stores. The luminance and chrominance field stores are then read at $^{525}/_{60}$ sync rates. They then pass through the appropriate D/A converter stages to become analog signals which are synchronous with the local synchronizing signals. These analog video signals are then rematrixed to R, G, and B to be color encoded along with the luminance Y signal. It should be noted that the vertical aperture corrector (VAC) is located in this system at a point where the video signal is in the sequentially scanned digital form. This makes it possible for vertical aperture compensation with improved vertical resolution since spatially adjacent lines of information are available for the correction.

In Fig. 17-18 a simplified diagram of the control system for the digital video store is shown. The master oscillator operating at approximately 48 MHz is phase-loop-locked to the basic horizontal line rate through the divide-by-4, divide-by-750, and phase-comparator loop. The 48-MHz signal is divided by 3 in the upper signal path to generate a 16-MHz clock signal for the write signal sampling. This signal is further divided by 20 to form 20 parallel paths for operation of serial-to-parallel shift registers to provide slower transfer of the data into the 16-K random access memories. The lower signal path shows the 48-MHz master signal being divided by 4 to form a 12-MHz clock signal for sampling in the read cycle. Again the divide-by-20 action occurs to generate a transfer pulse related to the 20 parallel-to-serial shift registers that are used to generate the output serial bit stream at the read clock rate.

Signal Processing. In a flying-spot scanner, in addition to signal amplification, band-shaping, and color encoding, four correction-type circuits are usually provided. They are gamma, log masking, color correction, and aperture correction. The amplification, band-shaping, and encoding circuits are common to all color television cameras and are not discussed further here.

Telecine cameras, like other picture sources, incorporate gamma correctors, to compensate for the receiver characteristics which are established primarily by the picture

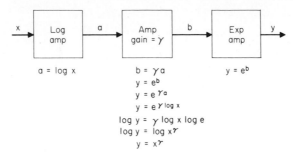

FIG. 17-19   Log-exponential gamma amplifier.

display tube. The meaning of gamma, as used here, is as the index of a power log transfer function. Thus

$$Y = X^\gamma$$

By the use of a combination of logarithmic and exponential (antilog) amplifiers, the overall gamma is varied by changing the gain of an amplifier between the log and antilog amplifiers. Figure 17-19 illustrates this principle. With this approach it is possible to achieve a range of gamma control of ± 0.25 to ± 2.0. Figure 17-20 shows the basis for a log-amplifier circuit. The base emitter diode of the log transistor provides the characteristic, and by connecting the collector to ground, the base current is reduced by transistor action and so reduces the voltage drop in the base resistance at high currents. The signal is coupled to the emitter as a current so that the voltage developed between emitter and base (ground) is the logarithm of the signal. To ensure stability, a clamp is used to stabilize the black-level potential on the emitter and, together with the oven, maintain a constant signal current at the chosen black-level potential. The exponential amplifier is the inverse of the log amplifier, the input signal being a voltage source and the output a current. The exponential transistor has a fairly low input impedance in the common-emitter mode. Therefore the driving amplifier needs to be very low impedance. The output current from the collector is converted to an output voltage by a feedback amplifier, and the two feedback clamps are used to control the stability and gain. The stability could be controlled by the white-level clamp and oven alone, but the black-level clamp

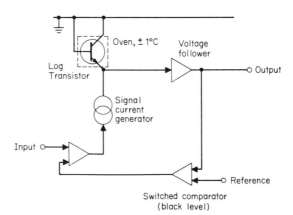

FIG. 17-20   Log-amplifier circuit.

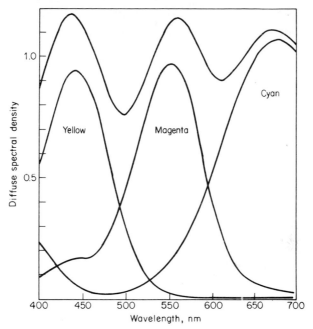

FIG. 17-21 Typical color print film spectral densities. Normalized dyes to form a visual neutral density of 1.0 for viewing illuminant of 3200 K. *(Eastman Kodak Company.)*

is added to control gain, since as the gamma is varied, the signal amplitude or pedestal will change, and the addition of the black-level clamp prevents this.

Log masking is provided to compensate for the overlapping spectral characteristics of the film dye layers. Figure 17-21 shows the spectral densities of a typical color print film. Log masking involves matrixing log signals; for example, in the red channel one may subtract fractions of log G and log B from log R, and since log R, log G, and log B represent the density of the film at particular wavelengths, the process is one of adding and/or subtracting film densities. Figure 17-22 shows an arrangement for the red channel. Since there are three resistors in each matrix, a total of nine resistors are required for one particular film. These matrixes are usually switched electronically to compensate for

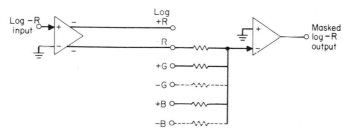

FIG. 17-22 Log-masking circuit.

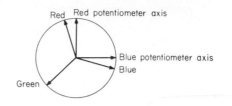

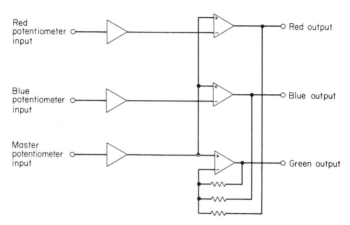

FIG. 17-23   Joystick control circuit for color correction.

different types of film. Although masking matrixes can be calculated, given information on the film characteristics and the telecine characteristics, it is usually easier to determine the matrix resistor values empirically by using a color bar slide made by the film manufacturer on the specific film to be compensated.

For color correction, the gain, gamma, and pedestal of the three color channels can be adjusted to correct for color balance errors due to the film or other causes. This results in a total of nine controls which would be difficult to operate on air. To make this operation easier, joysticks are used for differential gain, differential gamma, and differential pedestal. The master gain, gamma, and pedestal controls can be incorporated in the joystick or provided separately. In either case three potentiometers are required for each control. They are a master and two R and B controls. The R and B potentiometers operate on axis at right angles via a simple mechanical linkage from the joystick. The outputs from these potentiometers (dc-signals) are coupled to the circuit as shown in Fig. 17-23. With this arrangement if the master potentiometer is operated, all three outputs move together. If the red potentiometer is moved along the red pot axis, there will be a positive red output and a negative green output. The matrix is arranged for constant balance. Similarly this operation also applies to the blue pot axis.

To be practical in routine operations, a preprogrammed system for maintaining color balance can be employed. This preprogrammed color control is prepared by stopping the film on a scene-by-scene basis, making the desired color balance adjustments, and storing this information on a control medium, such as magnetic tape or a random access semiconductor memory.

Horizontal aperture correction is accomplished using a two-delay-line system as

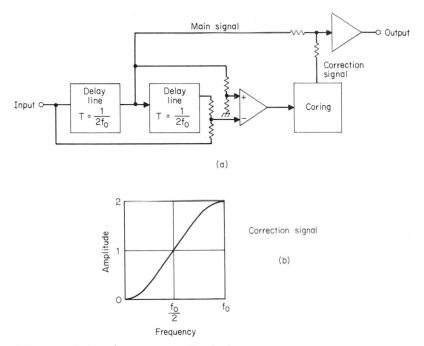

FIG. 17-24   Horizontal aperture correction circuit.

shown in Fig. 17-24a. The circuit configuration is advantageous in that it produces a
cosinusoidal amplitude—frequency response which is a good approximation to the optical losses in the film process. By separating the correction signal from the main path, the
process of coring can be carried out. *Coring* means removal of the information close to
the zero axis of the correction signal. The coring circuit is similar to a class B push-pull
amplifier where the bias is set so that the output is zero over a band in the center of the
transfer characteristic. This means that only medium and large transitions pass through
the coring circuit. Small signals, such as those generated by noise and film grain, will be
attenuated. Therefore the use of coring enables a higher level of aperture correction to
be applied without amplifying noise and film grain.

Vertical aperture correction requires two one-line (1H) delays connected in the same
configuration as shown in Fig. 17-24a. These delays are not practical in lump constant
form, and therefore they are usually accomplished by means of ultrasonic glass delay
lines. In order to achieve a video bandwidth of 5.5 MHz, an amplitude-modulated carrier
of 27 MHz is used in the glass delay line circuit. For economy of parts the vertical aperture compensation in a typical flying-spot scanner is accomplished in the luminance
channel as shown in Fig. 17-25. In this case four 1H delays are used. Three delays are
used for the R, G, and B signals while the fourth delay is used for the matrixed Y signal
that is obtained from the R, G, and B signals after one line of delay. Since the main line
or reference point for the luminance signal is at point (a), the R, G, and B signals must
also be delayed by one line in order to maintain proper Y, R, G, and B timing at the point
of color encoding. In the case of the sequential scan and digital store system, the vertical
aperture correction is accomplished in the digital Y channel since this permits an economy of parts; while it functionally is equivalent to that shown in Fig. 17-25, the hardware
including the 1H delays is substantially different from that which would be used in an
analog signal path.

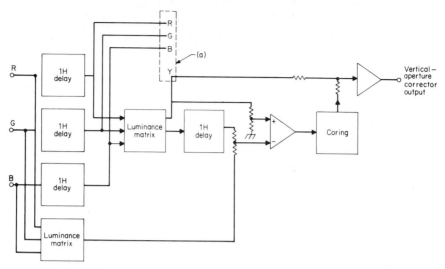

FIG. 17-25   Block diagram of flying-spot scanner vertical aperture correction.

17.5.3   CCD LINE SCANNERS.[3,4]   The combination of two developments, the line scan CCD photosensor and the digital video and frame field store, has led to the development of line-scan CCD telecines. This telecine is based on capstan-drive continuous-motion film transports. The vertical component of the film scanning is generated entirely by motion of the film. The sequentially scanned signals are subsequently stored in digital video stores and then read out in line-interlaced form. In the case of the 525/60 television system the 24-frame film rate is converted to the 60-frame television rate by means of this same digital video store system. Two commercially available CCD line-scan telecines, the Bosch FDL-60[5] and the Marconi D-3410,[6] are typical of CCD telecine scanners.

     Block Diagram.   The functional block diagram shown in Fig. 17-26 is typical of the video signal system. The *RGB* signals, following generation by the line-scan CCDs, are

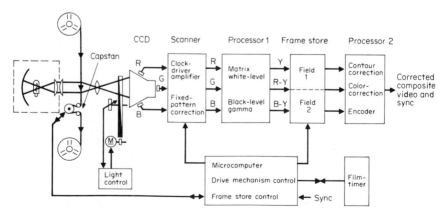

FIG. 17-26   Functional block diagram of Bosch FDL-60 video system. *(Robert Bosch GmbH, Fernseh Group.)*

amplified and processed in analog form and subsequently A/D converted for storage in the digital frame stores. As generated, these signals are line sequential RGB signals. In the readout process from the digital stores the signals are converted to line interlaced fields. After suitable matrixing to recover RGB in addition to the Y signals, they are encoded to the standard NTSC output signal. In the Marconi line-scan telecine, the RGB signals, as generated, are converted to the digital form early in the video channel with gamma correction and masking signal processing being carried out in the digital domain.

CCD Line-Scan Sensor. The CCD line-scan photosensitive sensor generally used is the Fairchild CCD133. This is a 1024-element device. The CCD consists of a single row of elements that are interdigitated with alternate elements being coupled to one of two transport shift registers. The transport shift-register outputs are brought to separate pin-outs; thus these signals must be combined off chip to provide a continuous video signal output. The clock rate for charge transfer can be as high as 20 MHz; therefore the CCD readout can be very rapid. Since the same photosensing elements are used for every line in the picture, any variation in the sensitivity among the elements will result in a vertical line pattern appearing in the output picture. In addition to the individual element sensitivity variations, there can be a difference in sensitivity between the two transport shift register signals. These variations in sensitivity must be compensated to eliminate any fixed pattern nonuniformity in the delivered video output signal.

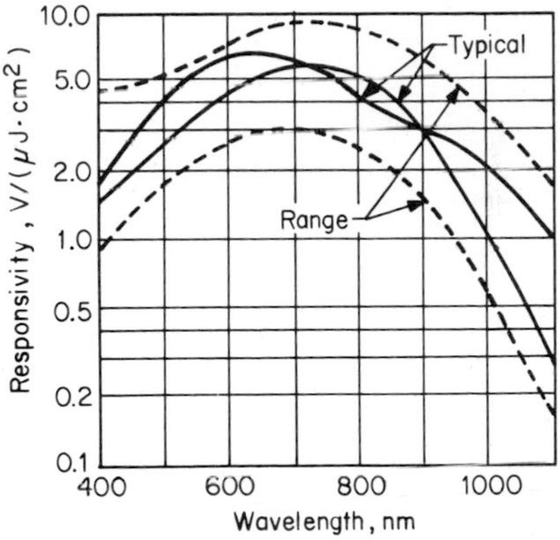

FIG. 17-27 Typical response of Fairchild CCD133. *(Fairchild Camera and Instrument Corp.)*

The typical spectral response for these devices is shown in Fig. 17-27. From the spectral curve it is evident that the sensor is low in blue sensitivity and also that it has very great infrared response. The infrared sensitivity can be suppressed by optical filtering, but the relatively low blue signal results in compromise in blue-channel signal-to-noise performance.

Sequential Scan. The capstan-driven continuous-motion film transport causes the film to be scanned by the line-scan sensors at a rate of 24 frames per second in the 525/60 system. In the time passage of the dimension of the film frame corresponding to the specified picture height, the proper number of lines must be scanned. The ratio of picture height dimension on 16-mm film, compared with the film-frame pitch, is very nearly the same as the ratio of active vertical picture time to the total vertical television frame period. However, since the film-frame rate is 24 frames per second, the scanning rate is slower than is the case for a 30 frame-per-second television rate. In this case it is 12.6 kHz rather than 15.734 kHz as in the 525/60 NTSC system. In the case of 35-mm film the line-scan rate for proper picture height scanning becomes 14.75 kHz.

In order to keep the video signals generated equal in bandwidth, the CCDs are clocked at a constant clock rate. For the case where all 1024 elements are utilized during an active picture time of 53 μs, the clock frequency becomes 19.3 MHz. In order to avoid aliasing effects, the sampled video signal is filtered to a bandwidth of approximately 6 MHz. The analog RGB signals thus generated are matrixed to Y and $R - Y$, $B - Y$ components and then digitized for distribution to the field stores. A group of three field stores is utilized for the luminance signal, and a second group of three field stores is used for the $R - Y$ and $B - Y$ chrominance signals. Figure 17-28 shows how the 24-frame

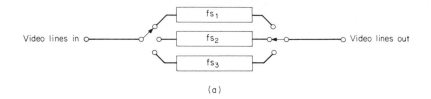

(a)

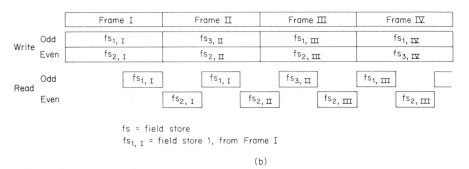

fs = field store
$fs_{1, I}$ = field store 1, from Frame I

(b)

FIG. 17-28   Diagram of 24-frame sequential scan, 60-field readout in 525/60 system.

film rate can be converted to the 60-field television rate by addressing and readout control of the video stores. By means of computer-controlled addressing, alternate lines of a given frame scan are written into two of the three field stores. For example, during film-frame I, the odd scan lines are written into field-store 1 while the even scan lines are switched to write in field-store 2. After approximately 60 percent of the frame has been scanned, readout of field-store 1 can start to generate the video lines for the first odd field. The conclusion of reading out field 1 occurs shortly after the completion of writing the last line from film-frame I. The next television field (even) is obtained from field-store 2. This field occurs during the early part of scanning of film-frame II. Since field-store 2 will be read at a rate faster than the information is written in, by scanning the film, it will be possible to begin to write in field-store 2 the information scanned from film-frame II. Also as frame II is scanned, the odd lines will be written into field-store 3. This leaves the information from frame I available in field-store 1 to be read out to form the second odd field of the television signal. This process continues as diagrammed to result in the 3-2 type of conversion which is required to generate a television 30-frame-per-second signal from the 24-frame-per-second film rate.

Colorimetry.   The spectral responses for the Bosch FDL-60, without masking or linear matrixing being employed, are as shown in Fig. 17-29. The curves are normalized to maximum response in each channel.

Signal Processing.   Owing to variations in element-to-element sensitivity in the CCD sensors and shading variations in the projection optical system, the output of the sensors will have a fixed pattern variation along the scan line. In order to compensate for this fixed-pattern noise and shading, each of the signal channels is provided with a compensating circuit following signal preamplification. The functional circuit for the Bosch telecine is shown in Fig. 17-30. The amplified video signal from the sensor is A/D converted in a parallel path and then passed through a reciprocal read-only memory. This signal is stored in a line store random access memory. It is read out line by line and D/A converted, following which it modulates the input video signal by means of a multiplier stage

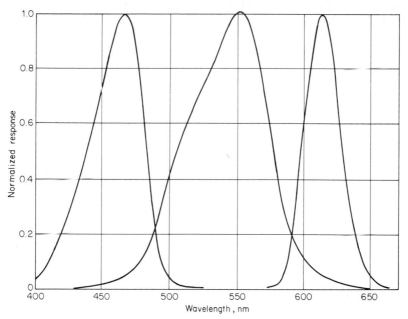

FIG. 17-29 Overall spectral characteristics of Bosch FDL-60. *(Robert Bosch GmbH, Fernseh Group.)*

to correct for the sensitivity variations. The correction signal is loaded into the memory by operating the telecine with an open gate to represent an all-white image.

In the Marconi telecine much of the signal processing is carried out in digital signal form. A block diagram of this system is shown in Fig. 17-31. The *RGB* signals from the image sensors are converted to an 11-bit-per-sample digital signal. The 11 bits are necessary to provide amplitude resolution for gamma correction, after which coding is the standard 8 bits per sample. Gamma correction and masking are carried out by logarithmic conversion, multiplication/matrixing, and exponential conversion. By utilizing extended black stretch, gamma values down to 0.15 are possible. Following gamma correction and masking, the *RGB* signals are matrixed to Y, $B - Y$, and $R - Y$, each at 8 bits per sample for subsequent sequential-to-interlace scanned conversion. This scan conversion function is described in an earlier paragraph.

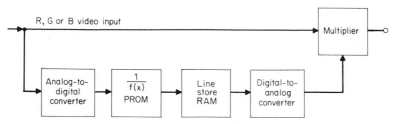

FIG. 17-30 Fixed pattern compensation circuit for FDL-60. *(Robert Bosch GmbH, Fernseh Group.)*

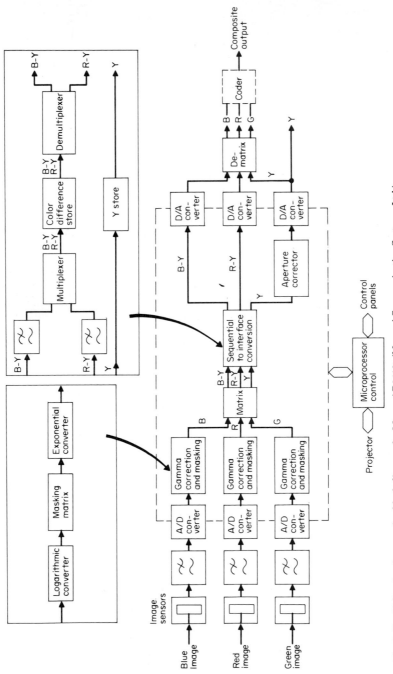

FIG. 17-31 Video signal processing block diagram for Marconi B3410. (*Marconi Communication Systems, Ltd.*)

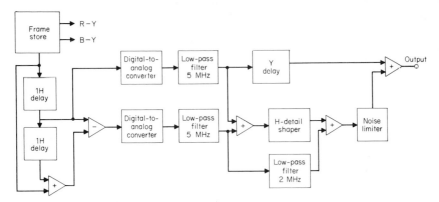

FIG. 17-32 Block diagram of FDL-60 contour corrector. *(Robert Bosch GmbH, Fernseh Group.)*

Contour Correction. Contour correction, which involves the combination of horizontal and vertical aperture compensation, can be carried out entirely in the analog domain or the digital domain. However, the Bosch telecine utilizes a combination of digital and analog circuits. A functional block diagram of this system is shown in Fig. 17-32. The Y signal is read from the field stores and passed through two 1H delays which are digital delays. The signal with one-line delay is used as the principal output Y signal component after being converted from the digital form to the analog form and then suitably low-pass filtered. The minus-one-line and plus-one-line signals are combined and then converted to the analog form and, after filtering, are combined with the principal Y component to form a wide-band vertically aperture-corrected signal which is then processed through a horizontal aperture correction circuit to generate a horizontal-rate correction signal. By means of a separate parallel path the correction signal is further low-pass filtered to a 2-MHz bandwidth to be combined with the horizontal-correction signal to form the final contour-correction signal which is combined with the main path Y signal, after a suitable delay in that path to correct any timing errors. The output from this summing function is a luminance signal which is corrected for both vertical and horizontal aperture response limitations.

In both the Bosch and Marconi telecines there are available, as remotely controllable functions, a large number of color-correction adjustments which can be used for color matching, color enhancement, and other color correction as artistically chosen changes. It is usually most convenient to preview either in a film-moving situation or still frame, make the color corrections, and store this information in a suitable memory so that the control in playback of the film is preprogrammed.

17.6 TELEVISION FILM RECORDING

by K. Blair Benson and Kenneth G. Lisk

The transfer of images from video to film has been an important and growing activity for a number of years. As more material has become available from live camera or videotape, the need for film records of this material has increased to take advantage of the economy and convenience of film distribution and display. For example, the visual images stored on film permit direct viewing by optical projection and television viewing on any transmission standard, without the need for line and frame conversion.

17.6.1 EARLY EXPERIMENTAL RECORDING

Phonograph Disc Recording. Recordings were made on ordinary phonograph discs by Baird in 1927 using a system he called *Phonovision.* At the time, Baird was experimenting with a 30-line television picture scanned at a rate of 12½ frames per second, using a scanning wheel revolving 750 times per minute. Because of the coarse line structure and low frame rate, the television signal occupied a narrow bandwidth that consequently could be transmitted over a telephone line or recorded on a phonograph disc. In the same year, a method was proposed for utilizing film at both transmitting and receiving ends. With this method, the scene to be televised would be first filmed by conventional motion-picture equipment, and the resultant film would be scanned for transmission. Of all the methods proposed in the early experimental stages of television development, for the next 30 years the only program-recording process to be of any significance, until the introduction of the Ampex videotape system, was motion-picture film.

Continuous-Motion Film Recording. In 1933, at an exhibition in Berlin, a system of theater television was demonstrated in which the television signal was recorded on film, and the film recording was then projected on a screen with a motion-picture projector. A scanning wheel was used, running at 3000 r/min, producing 180-line pictures at 25 frames per second. In 1934, a patent was granted to a German company for recording television images on film from a cathode-ray tube, and in 1935 the equipment was demonstrated at the Berlin Radio Exposition. This method produced recognizable images on the film, although a nonsynchronous, continuous-motion film-transport mechanism was required in the camera.

Variable-Density Film Recording. A novel machine was developed in England in 1934 called the *Visiogram,* in which the television video signals were recorded on motion-picture film in the form of a variable-density track, similar to sound recording.

The great interest in theater television at that time spurred development of many different types of novel apparatus, including a rapid-process system using 17.5-mm film running at 47 ft/min (14.3 m/min) available for projection in 30 s.

Cathode-Ray Tube Picture Film Recording. With the introduction of the iconoscope camera tube in 1933, interest in mechanical scanning systems and the intermediate film process rapidly waned. With the all-electronic camera, it became possible for the first time to pick up live subjects for immediate transmission without the delay and complexity of motion-picture film processing.

First attempts in the United States to film cathode-ray-tube images were made in 1938. In these early efforts, standard spring-wound 16-mm motion-picture cameras were used, operating at 16 frames per second (standard silent motion-picture rate). With the low intensity of the picture tubes available in those days, it was necessary to use the fastest type of film that could be obtained. Since the cameras were nonsynchronous with the 30-frames-per-second rate of the television scanning system, the film recordings were marred by banding, or horizontal bars, caused by uneven matching of the odd and even fields recorded in each film frame. By recording at 15 frames per second (one-half the television scanning frequency), banding was eliminated, but then only every other television frame could be recorded. Moreover, the recordings could not be played back at the standard motion-picture sound speed of 24 frames per second.

It soon became obvious that for commercial use of television film recordings, the 30-frames-per-second television picture would have to be recorded on film at 24 frames per second to conform with the already well-established standards of the motion-picture industry.

Television film recording is known by many other names, such as kinescope recording, kine-photos, teletranscriptions, tape-to-film transfer, telerecording, video printing, laser scanning, and image vision, to cite a few.

17.6.2 TIME-DELAY BROADCAST. In the earlier days of monochrome television, film copies of programs were made by photographing a cathode-ray tube display with a motion-picture film camera. This process, at that time called *kinescope* or *kine recording*, was of marginally acceptable quality. Nevertheless, since it was the only practical system available, it was used extensively for delayed broadcasting to accommodate the different time zones across the country and to program stations not connected by coaxial or microwave circuits to the network program origination centers. For the time-zone delay, the kine recordings were developed on an expedited basis in film laboratories or by real-time processors such as the Houston-Fearless Processor and Eastman Viscomat Processor at the program origination centers.

In addition to the monochrome service initiated in 1947, some very limited distribution of color programming was accomplished in the early 1950s using a trinoscope cathode-ray tube display and color film.

With the advent of the video tape recorder in 1956, the use of film for time-delay of broadcast programs was terminated. Nevertheless, some of the systems and equipment used and the problems encountered warrant description, since kinescope recording continues to be used for a variety of other purposes in broadcasting and related industries.

17.6.3 FILM-TRANSFER PROCESS. Recording of a 525-line, 30-frames-per-second television display with a conventional motion-picture camera results in an interference pattern of horizontal bars, caused by the difference between the television frame rate and the film rate of 24 frames per second. It is this frame-rate difference of 6 Hz that complicates the process of recording images on film from 30 frames per second television signals. The simplest approach is to expose the film to segments of the television frame by means of a rotating shutter in the camera with an opening angle of 72°. Figure 17-33 shows the timing cycle for this system of recording and the resultant *image-splice* problem where the frame segments must match.

In an intermittent-motion recording at 24 frames per second, the film must be advanced between exposure of motion-picture frames. Since there are only four motion-picture frames in the time interval occupied by five television frames, one-half of a television field is thrown away after each motion-picture frame is exposed. The film is advanced during that half-field interval. In terms of shutter rotation, assuming a film

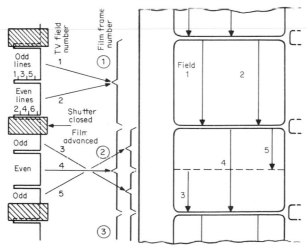

FIG. 17-33 Conversion from rate of 30 television frames per second to 24 film frames per second.

exposure interval of 360°, the television frame duration is 288°. During the first 72° of the film cycle, the film is pulled down to advance one frame. The camera shutter is closed at this time, and the actual pulldown time for the film is approximately 60° or 8.3 ms. Exposure of the film takes place during a shutter opening of 288°, or $\frac{1}{30}$ s, for the next television frame. This action is followed by the film pulldown, after which the cycle is repeated.

The sequence described requires very accurate shutter timing in order to avoid a horizontal band of varying density in the recorded picture, known as *shutter bar*. Cameras have been used with mechanical shutters, or the shuttering action has been accomplished by electronic blanking of the video signal fed to the display device.

16-mm Fast Pulldown Camera Systems.   PAL and SECAM television systems have a frame rate of 25 frames per second. By operating the recording camera at this rate of speed rather than the normal motion-picture standard of 24 frames per second, the midfield splice encountered with the NTSC and PAL-M 30 frames-per-second systems can be avoided. However, the camera must be locked and properly phased to the television frame frequency. Furthermore, the film pulldown must be substantially more rapid than that for conventional projection in order to complete the frame advance during the vertical blanking interval.

A camera design that provides these features accomplishes the rapid pulldown in 800 μs by the use of compressed air to move the film through the gate. The camera is driven from the television signal with the speed and phase synchronization established by phase comparison circuitry. Use of the camera on 30-frame-per-second systems requires operation at 24 frames per second, with the 30-to-24-frame conversion accomplished by blanking one half of every fifth television field.

Conventional Pulldown 35-mm Recording Cameras.   Although some use was made of 35-mm cameras for time-zone delay broadcasting and for very limited applications of theater projection in the early days of monochrome television, the performance left much to be desired as regards image steadiness and smoothness of motion in the picture. More recently, by the use of digital frame stores, these shortcomings have been eliminated. The digital techniques permit some useful signal manipulations, such as the ability to de-interlace the picture and scan each frame sequentially. In addition, frame stores enable color subcarrier signals to be stored and read out at rates other than broadcasting standards.

For example, blanking intervals can be compressed or expanded, active line durations can be reduced, and field blanking intervals can be lumped together to realize total freedom from the earlier restraints of vertical blanking. Thus, signal timings can be rearranged so that 10 ms of unused time interval is available in each frame for film pulldown, which makes possible the use of a conventional 35-mm motion-picture camera for recording without incurring any loss of video signal information during the pulldown time.

17.6.4 RECORDING SYSTEMS.   For a number of years, five basic systems for recording of television picture signals have been in use throughout the world. The most used film format is 16 mm, with some limited use of 35 mm, and occasional use of 8 mm for news and special applications. These systems are

1. Photography of a monochrome or color cathode-ray picture tube
2. Photography of an optically combined image from three picture tubes, each displaying a primary color
3. Exposure of film to a laser beam
4. Exposure of film to an electron beam
5. Color and component separation

Single-Tube Color Picture Recording.[7,8]   The simplest and most direct approach for television recording is to photograph on color film the image on a color picture tube, such as a shadow mask, slot mask, or a Trinitron striped mask single-gun tube. In all cases, the relationship among the film, lens, and display are closely interdependent because of the picture tube's limited light output.

Consequently, a fast lens of high quality is necessary. If a lens with an aperture of $f/1.2$ or less is used, it will provide the required speed and in addition can be stopped down to a smaller aperture for an optimum degree of resolution. This will provide a high modulation transfer function (MTF) at low spatial frequencies within the bandwidth of television systems. Because of the limitation imposed by television standards, a superior MTF at higher frequencies is of no significance, whereas good performance at lower frequencies is essential.

Display Color Balance. A second major consideration is color balance and control. A way of obtaining the correct color balance when exposing a tungsten-balanced film is the use of a Kodak Wratten Filter No. 85B and a Kodak Color Compensating Filter CC40R. The Kodak Wratten Filter No. 85B is needed because the color temperature of the television display is approximately visual match with daylight. A daylight-balanced film needs only a Kodak Color Compensating Filter CC40R. Figure 17-34 shows the reason a Kodak Color Compensating Filter CC40R is required. The energy of the red phosphor emittance is concentrated in a narrow spectral region and does not fall close to the peak of the red sensitivity of the film. By the use of these filters in front of the camera lens, the proper balance can be obtained on the film. However, the use of filters is undesirable because the light losses brought about by the filters require higher display brightness, which can degrade the picture. The preferred way to achieve correct color balance of the display tube is to eliminate the filters and simulate their effect by adjustment of video gains and screen brightness controls. This can easily be done by viewing the display in a darkened room through filters complementary to the Kodak Wratten Filter No. 85B and the Kodak Color Compensating Filter CC40R required for daylight use of the film. When recording on a tungsten-balanced film, a combination of a Kodak Wratten Filter

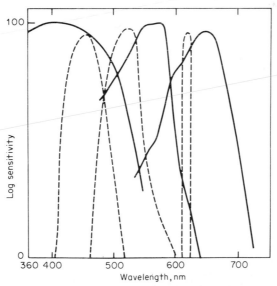

FIG. 17-34 Spectral characteristics of normalized shadow-mask phosphor emittances (dashed lines) and normalized spectral sensitivities of color-negative film. *(Eastman Kodak Company.)*

No. 80B and a Kodak Color Compensating Filter CC40C is used for viewing. When viewing the television display through this filter pack, the red, green, and blue video gains are adjusted to give a good highlight or white neutral, while the red, green, and blue screen brightness controls adjust the dark gray or black. After these adjustments are made, the actual color balance of the display, as seen without the filters, is much warmer than normal, and no correction filters are needed on recording.

Three-Tube Color Picture Recording (Trinoscope).[9]   Of all the systems used in video-to-film transfer, trinoscope-based devices are numerically in the majority. A trinoscope employs three precision cathode-ray tubes, one for each primary color. The three separate images are optically combined, registered, and photographed as a single image. The trinoscope system has been used in a number of different configurations for some time and has many desirable features. It is capable of higher resolution than a shadow-mask or Trinitron-type tube, and the individual phosphors can be selected to match the sensitivities of the recording film. Also with specially selected cathode-ray tubes, it can produce high radiant energy (brightness). Although registration of the three tubes has proved difficult in the past, the use of deflection quadrant linearity correctors has greatly minimized this problem.

Trinoscope systems have benefited on the display side by the introduction of cathode-ray tubes having rare earth phosphors in all three channels. Such phosphors are characterized by relatively high conversion efficiency, virtually zero afterglow, and a power-law shaped transfer characteristic.

Laser-Beam Recorder.[10]   The laser-beam recorder consists of three lasers for primary-color light sources. To generate a standard television raster, a galvanometer is used for vertical deflection and a multifaceted polygonal mirror for horizontal deflection. The intensity of the beams is controlled by acoustooptic light modulators.

The laser-beam recorder has two major characteristics of high resolution and a high degree of brightness. Very slow films in the ASA range of 1 to 5 can easily be exposed with this device.

Electron-Beam Recording.   Electron-beam recording (EBR) systems are fundamentally different from other methods in that the film is run in a vacuum chamber and exposed by direct electron bombardment. The first systems utilized black-and-white 16-mm film. However, 35-mm films are now being employed.

The electron-beam recorder has some distinct advantages; e.g., it is capable of very high resolution, and, in the absence of a phosphor, flare, phosphor grain, and shading are eliminated.

Electron-beam recorders also have some unique characteristics as compared with other film-transfer systems. One is that the film must be introduced to the high-vacuum recording chamber. Early methods placed the entire roll of film in the high-vacuum chamber and evacuated to the required pressure (that is, 10^{-6} torr). Later techniques involved prepumping the film in a separate chamber (roughing) and then entering the high-vacuum chamber through a vacuum feedthrough. Another characteristic is that the electron beam builds up an electrostatic charge on the film (charging effect), causing a distortion or a deflection of the beam. If beam currents are kept at reasonable levels, this problem is minimized.

Component Color-Separation Systems.   Electron-beam recording systems have been devised to sequentially record the component video signals on black-and-white film. This is done by decoding the composite video signal into its separate red, green, and blue components. These signals are then passed to a "frame switcher," delayed, and sequentially recorded on 16-mm film in a camera running at 72 frames per second to form a separation master. The printing process is carried out on a step printer that exposes each of the three separations through the correct color filter onto a single frame of color stock.

In another component system introduced in 1971 in the United States, Japan, and the United Kingdom and discarded in favor of video tape shortly thereafter, a monochrome-signal image and a coded color-signal image were recorded by an electron beam, side by side, on a 16-mm-wide black-and-white negative film. Positive prints were released in single-reel cartridges for television playback, using an automatic-threading, continuous-motion, flying-spot scanner.

A high-resolution, time-sequence separation system has been developed for use in recording video-generated special effects. Exposures with a resolution of 4000 lines are made on a color film through a three-color filter wheel between the monochrome cathode-ray tube display and the camera. The system operates frame by frame, off line (not in real time).

Recording Films.   Color-negative film stocks, primarily 16 mm, have been the major recording material in television film recording for many years. With the introduction of 35-mm cameras, use of 35-mm film will be on the rise. If only one or two copies are needed, color-reversal films are used. Laser and specially designed trinoscope systems are capable of exposing the slower print and internegative film types.

Sound.   Two different systems are used for recording sound onto motion-picture film.

Double System.   Individual sound and picture recorders are operated at constant relative speed in synchronism. This method provides the highest possible quality.

Alternatively, track recorded simultaneously on film is called *single-system* recording. This system is used most often when only a single film copy is needed. Since the film processing usually is optimized for best picture quality rather than for sound, the sound quality suffers.

17.6.5 RECORDING CHARACTERISTICS AND PROCEDURES.   Even the best television film recordings cannot compare favorably with directly shot film pictures. The inherent resolution limit of all broadcast television signals is substantially lower than that of, for example, 16-mm film, and this together with the interlaced nature of television scanning and the compromise inherent in color-encoding systems gives rise to visual effects that are foreign to the normally perceived characteristics of film pictures. Efforts in the last few years have been concentrated on the adoption of circuit techniques in signal processing intended to reduce, as far as possible, the consequence of these effects.

Image Enhancement.   The restricted bandwidth of the television signal makes the use of comb filter decoders mandatory to preserve maximum luminance resolution. Unfortunately, comb filters, by their very nature, halve the vertical chrominance resolution, and in the case of PAL where there is not a whole number of cycles of subcarrier in each scanning line, comb filtering is difficult to engineer well. The chrominance information does not sit neatly between the line-scanning multiples of luminance spectrum but has a varying position on a line-by-line basis. Much more, therefore, is demanded of the comb filter, and the problem is only solved in analog terms by further worsening the vertical chrominance resolution. The improvements in reduction of cross color and related effects normally achieved by the use of comb filter decoders are lower in PAL than in NTSC for this reason.

The aperture correction applied to the original electronic camera signals causes objectionable results on a resultant film transfer. Such correction is normally intended to compensate for the aperture losses of the shadow mask tube. However, motion-picture film does not have such aperture losses, and the applied correction appears as objectionable fringing. Modern circuit techniques permit previously applied corrections to be mostly taken out or at least reduced in visibility but in the horizontal direction only. In the vertical direction, aperture correction treats temporarily adjacent scanning lines that, because of the interlaced nature of the signal, are not spatially adjacent. In consequence, the correction tends to reduce rather than enhance the vertical definition on film, and this is further worsened by the effect of the comb filter decoder.

Some enhancement of the signal is desirable before film images are created from it.

This is not because of any system loss but to give a subjective increase in perceived sharpness when the film is viewed on a large screen. The enhancers used for this application have unique characteristics. In the horizontal direction, the peaking frequencies are at the upper limit of the signal spectrum, and vertical correction is confined to spatially adjacent scanning lines.

Scanning Interlace. Another fundamental problem arises from the interlaced nature of scanning. Since each pair of fields is captured in a single film frame, interfield differences in the signal become locked into the film frame and can give rise to frame rate modulations on film, which are not manifest in the television picture. Interlacing is also responsible for the poor edge resolution on moving objects since any moving object will be reproduced on the film with serrated edges due to the positional displacement between fields.

Noise and Grain. Noise in the television signal can also cause objectionable effects on film, particularly because it beats with film grain to produce large low-frequency components but also because the incoming noise within the frame period is locked into a single film frame so that the perceived noise on film changes at frame rate rather than at high frequency.

17.6.6 DIGITAL TECHNIQUES. The signal processing used in transfer systems is intended to reduce as far as possible the effects described above. The advent of digital techniques has made available two tools that are now indispensable in this field. The recursive filter permits a substantial reduction in incoming noise level. However, such devices are not level dependent, and it is noise in the black areas of the signal which degrades the film picture most by effectively adding a lift component that reduces contrast and gives the film pictures a "flary" look. For this reason, some form of coring that treats only the noise in the blacks is also frequently used.

Most tape-to-film transfer is undertaken from videotapes recorded to broadcast signal specifications, and it is these specifications that cause most of the problems. The near future is likely to give us digital videotape recording in component form. Such recordings offer the promise of freedom from all the effects caused by the interleaved nature of chrominance and luminance information, and this should result in a further substantial subjective improvement in the quality of transfers to film. However, electronic image capturing exists outside the field of conventional television broadcasting, and where such systems are designed with ultimate transfer to film as an objective, then the bandwidth, signal encoding, and interlace constraints can be avoided totally. All present-day high-quality transfer systems inherently are capable of handling the bandwidth needed for high-definition transmission, and since the film projector is a relatively inexpensive display device, compared with large-screen high-brightness television projectors, the need for video tape-to-film transfer may be expected to continue on a large scale for theatrical-type television program presentation.

17.7 FILMS FOR TELEVISION

by Kenneth G. Lisk

17.7.1 EXPOSURE INFORMATION

Exposure Index. The film exposure index (EI) is a measurement of film speed that can be used with an exposure meter to determine the aperture needed for specific lighting conditions. The indices reported on film data sheets by film manufacturers are based on practical picture tests but make allowance for some normal variations in equipment and film that will be used for the production. There are many variables for a single exposure. Individual cameras, lights, and meters are all different (lenses are often calibrated in T-stops). Coatings on lenses affect the amount of light that strikes the emulsion. The actual shutter speeds and f-numbers of a camera and those marked on it sometimes differ. Particular film emulsions have unique properties. Camera techniques can also affect exposure. All these variables can combine to make a real difference between the recommended exposure and the optimum exposure for specific conditions and equipment. For these reasons, it is always wise to test several combinations of camera, film, and equipment to find the exposures that produce the best results. Data sheet EI figures are applicable to meters marked for ISO or ASA† speeds and are used as a starting point for an exposure series.

For measurement exposure, there are three kinds of exposure meters: The *averaging reflection meter* and the *reflection spot meter* are most useful for daylight exposures, while the *incident light exposure meter* is designed for indoor work with incandescent illuminations.

Exposure Latitude. Exposure latitude is the range between overexposure and underexposure within which a film will still produce usable images. As the luminance ratio (the range from black to white) decreases, the exposure latitude increases. For example, on overcast days the range from darkest to lightest narrows, increasing the apparent exposure latitude. On the other hand, the exposure latitude decreases when the film is recording subjects with high luminance ratios such as black trees against a sunlit snowy field.

Lighting Contrast Ratios. When artificial light sources are used to illuminate a subject, a ratio between the relative intensity of the key light and the fill lights can be determined. First, the intensity of light is measured at the subject under both the key and fill lighting. Then the intensity of the fill light alone is measured. The ratio of the intensities of the combined key light and fill lights to the fill light alone, measured at the subject, is known as the *lighting ratio.*

Except for dramatic or special effects, the generally accepted ratio for television color photography is 2/1 or 3/1. If duplicate prints of the camera film are needed, the ratio should seldom exceed 3/1. For example, if the combined main light and fill light on a scene produce a meter reading of 6000 fc (6.48×10^4 lm/m^2) at the highlight areas and 1000 fc (1.08×10^4 lm/m^2) in the shadow areas, the ratio is 6/1. The shadow areas should be illuminated to give a reading of at least 2000 and preferably 3000 fc to bring the lighting ratio within the permissible range.

Reciprocity Characteristics. Reciprocity refers to the relationship between light intensity (illuminance) and exposure time with respect to the total amount of exposure received by the film. According to the reciprocity law, the amount of exposure H received by the film equals the illuminance E of the light striking the film multiplied by the exposure time t. In practice, any film has its maximum sensitivity at a particular exposure (i.e., normal exposure at the film's rated exposure index). This sensitivity varies with the exposure time and illumination level. This variation is called *reciprocity effect.* Within a reasonable range of illumination levels and exposure times, the film produces a good image. At extreme illumination levels or exposure times, the effective sensitivity of the film is lowered, so that predicted increases in exposure time to compensate for low illumination or increases in illumination to compensate for short exposure time fail to produce adequate exposure. This condition is called *reciprocity-law failure* because the reciprocity law fails to describe the film sensitivity at very fast and very slow exposures. The reciprocity law usually applies quite well for exposure times of ⅛ to ¹⁄₁₀₀₀ for black-and-white films. Above and below these speeds, black-and-white films are subject to reciprocity failure, but their wide exposure latitude usually compensates for the effective loss of film speed. When the law does not hold, the symptoms are underexposure and

---

†ASA (American Standards Association, now American National Standards Institute or ANSI) superseded by ISO (International Standards Organization).

Table 17-3 Conversions of Filter Factors to Exposure Increase in Stops

Filter factor	+ Stops	Filter factor	+ Stops	Filter factor	+ Stops
1.25	+1⅓	4	+2	12	+ 3⅔
1.5	+ ⅔	5	+2⅓	40	+ 5⅓
2	+1	6	+2⅔	100	+ 6⅔
2.5	+1⅓	8	+3	1000	+10
3	+1⅔	10	+3⅓		

change in contrast. For color films, the photographer must compensate for both film speed and color-balance changes because the speed change may be different for each of the three emulsion layers. However, contrast changes cannot be compensated for and contrast mismatch can occur.

Filter Factors. Since a filter absorbs part of the light that would otherwise fall on the film, the exposure must be increased when a filter is used. The filter factor is the multiple by which an exposure is increased for a specific filter with a particular film. This factor depends principally upon the absorption characteristics of the filter, the spectral sensitivity of the film emulsion, and the spectral composition of the light falling on the subject. Table 17-3 shows conversions of filter factors to exposure increase in stops.

17.7.2 SENSITOMETRY. *Sensitometry* is the science of measuring the response of photographic emulsions to light. *Image structure* refers to the properties that determine how well the film can faithfully record detail. The appearance and utility of a photographic record are closely associated with the sensitometric and image-structure characteristics of the film used to make that record. The ways in which a film is exposed, processed, and viewed affect the degree to which the film's sensitometric and image-structure potential is realized. The age of unexposed film and the conditions under which it was stored also affect the sensitivity of the emulsion. Indeed, measurements of film characteristics made by particular processors using particular equipment and those reported on data sheets may differ slightly. Still, the information on the data sheet provides a useful basis for comparing films. When cinematographers need a high degree of control over the outcome, they should have the laboratory test the film they have chosen under conditions that match as nearly as possible those expected in practice.

17.7.3 SENSITOMETRIC INFORMATION. *Transmission density D* is a measure of the light-controlling power of the silver or dye deposit in a film emulsion. In color films, the density of the cyan dye represents its controlling power to red light, that of magenta dye to green light, and that of yellow dye to blue light. Transmission density may be mathematically defined as the common logarithm (log base 10) of the ratio of the light incident on processed film (P_o) to the light transmitted by the film (P_1)

$$D = \log \frac{P_o}{P_1} \qquad (17\text{-}1)$$

The measured value of the density depends on the spectral distribution of the exposing light, the spectral absorption of the film image, and the spectral sensitivity of the receptor. When the spectral sensitivity of the receptor approximates that of the human eye, the density is called *visual density.* When it approximates that of a duplicating or print stock, the condition is called *printing density.*

For practical purposes, *transmission density* is measured in two ways. *Total diffuse density* (Fig. 17-35a) is determined by comparing all the transmitted light with the incident light perpendicular to the film plane ("normal" incidence). The receptor is placed so that all the transmitted light is collected and evaluated equally. This setup is analogous to the contact printer except that the "receptor" in the printer is film.

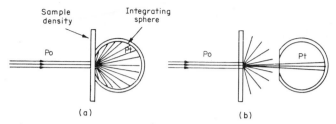

FIG. 17-35   (a) Totally diffuse density measurement. (b) Specular density measurement. (*Eastman Kodak Company.*)

Specular density (Fig. 17-35b) is determined by comparing only the transmitted light that is perpendicular ("normal") to the film plane with the "normal" incident light, analogous to optical printing or projection.

To simulate actual conditions of film use, totally diffuse density readings are routinely used when motion-picture films are to be contact-printed onto positive print stock. Specular density readings are appropriate when a film is to be optically printed or directly projected. However, totally diffuse density measurements are accepted in the trade for routine control in both contact and optical printing of color films. Totally diffuse density and specular density are almost equivalent for color films because the scattering effect of the dyes is slight, unlike the effect of silver in black-and-white emulsions.

A *characteristic curve* is a graph of the relationship between the amount of exposure given a film and its corresponding density after processing. The density values that produce the curve are measured on a film test strip that is exposed in a sensitometer under carefully controlled conditions and processed under equally controlled conditions. When a particular application requires precise information about the reactions of an emulsion to unusual light—filming action in a parking lot illuminated by sodium vapor lights, for example—the exposing light in the sensitometer can be filtered to simulate that to which the film will actually be exposed. A specially constructed step tablet consisting of a strip of film or glass containing a graduated series of neutral densities differing by a constant factor is placed on the surface of the test strip to control the amount of exposure, the exposure time being held constant. The resulting range of densities in the test strip simulates most picture-taking situations in which an object modulates the light over a wide range of illuminance, causing a range of exposures (different densities) on the film.

After processing, the graduated densities on the processed test strip are measured with a densitometer. The amount of exposure (measured in lux†) received by each step on the test strip is multiplied by the exposure time (measured in seconds) to produce exposure values in units of lux-seconds. The logarithms (base 10) of the exposure values (log H) are plotted on the horizontal scale of the graph, and the corresponding densities are plotted on the vertical scale to produce the characteristic curve. This curve is also known as the sensitometric curve, the D log H (or E) curve, or the HD (Hurter and Driffield) curve.

In Fig. 17-36 the lux-second values are shown below the log exposure values. The equivalent transmittance and opacity values are shown to the left of the density values.

The characteristic curve for a test film exposed and processed as described above is an absolute or real characteristic curve of a particular film processed in a particular manner.

Sometimes it is necessary to establish that the values produced by one densitometer are comparable with those produced by another. Status densitometry is used for this.

---

†Illumination of 1 lx (= 1 lm/m^2) is produced by 1 standard candle from a distance of 1 m. When a film is exposed for 1 s to a standard candle 1 m distant, it receives 1 lx·s of exposure.

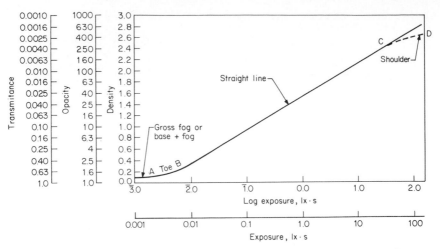

FIG. 17-36 Typical sensitometric, or Hurter and Driffield (HD), characteristic curve. *(Eastman Kodak Company.)*

Status densitometry refers to measurements made on a densitometer that conforms to a specified unfiltered spectral response. When a set of carefully matched filters is used with a densitometer, the term *status A* densitometry is used. The densities of color-positive materials (reversal, duplicating, and print) are measured by status A densitometry. When a different set of carefully matched filters is incorporated in the densitometer, the term *status M* densitometry is used. The densities of color preprint films (color negative, internegative, intermediate, low-contrast reversal original, and reversal intermediate) are measured by status M densitometry.

Representative characteristic curves are those that are typical of a product and are made by averaging the results from a number of tests made on a number of production batches of film. The curves shown in the data sheets are representative curves.

Relative characteristic curves are formed by plotting the densities of the test film against the densities of a specific uncalibrated sensitometric step scale used to produce the test film. These are commonly used in laboratories as process control tools.

Black-and-white films usually have one characteristic curve. A color film, on the other hand, has three characteristic curves, one each for the red-modulating (cyan-colored) dye layer, the green-modulating (magenta-colored) dye layer, and the blue-modulating (yellow-colored) dye layer (see Figs. 17-37 and 17-38). Because reversal films yield a positive image after processing, their characteristic curves are inverse to those of negative films.

General Curve Regions. Regardless of film type, all characteristic curves are composed of five regions: D-min, the toe, the straight-line portion, the shoulder, and D-max.

Exposures less than at *A* on negative film or greater than at *A* on reversal film will not be recorded as changes in density. This constant density area of a black-and-white film curve is called *base plus fog*. In a color film, it is termed *minimum density* or D-min.

The toe (*A* to *B*) (Fig. 17-36) is the portion of the characteristic curve where the slope (or gradient) increases gradually with constant changes in exposure (log *H*). The straight line (*B* to *C*) is the portion of the curve where the slope does not change; the density change for a given log-exposure change remains constant or linear. For optimum results, all significant picture information is placed on the straight-line portion. The shoulder (*C* to *D*) is the portion of the curve where the slope decreases. Further changes in exposure (log *H*) will produce no increase in density because the maximum density (D-max) of the film has been reached.

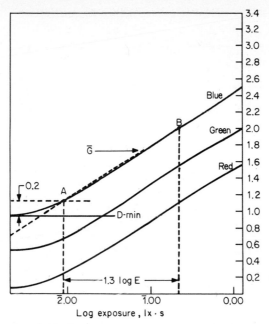

FIG. 17-37 Typical characteristic curve of negative color film. *(Eastman Kodak Company.)*

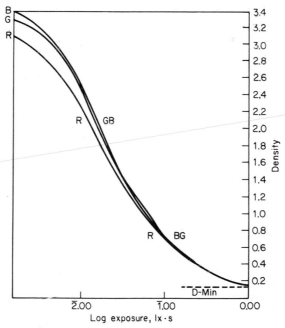

FIG. 17-38 Typical characteristic curve of positive color film. *(Eastman Kodak Company.)*

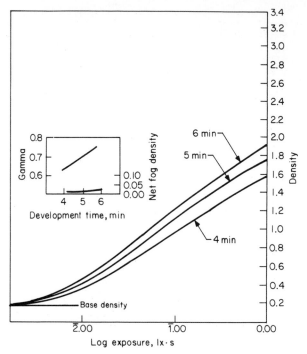

FIG. 17-39 Curves for a development-time series on a typical black-and-white negative film. *(Eastman Kodak Company.)*

Base density is the density of fixed-out (all silver removed) negative-positive film that is unexposed and undeveloped. Net densities produced by exposure and development are measured from the base density. For reversal films, the analogous term of D-min describes the area receiving total exposure and complete processing. The resulting density is that of the film base with any residual dyes.

Fog refers to the net density produced during development of negative-positive films in areas that have had no exposure. Fog caused by development may be increased with extended development time or increased developer temperatures. The type of developing agent and the pH value of the developer can also affect the degree of fog. The net fog value for a given development time is obtained by subtracting the base density from the density of the unexposed but processed film. When such values are determined for a series of development times, a time-fog curve (Fig. 17-39) showing the rate of fog growth with development can be plotted.

Curve Values.   From the characteristic curve additional values can be derived that not only illustrate properties of the film but also aid in predicting results and solving problems that may occur during picture-taking or during the developing and printing processes.

Speed describes the inherent sensitivity of an emulsion to light under specified conditions of exposure and development. The speed of a film is represented by a number derived from the film's characteristic curve.

Contrast refers to the separation of lightness and darkness (called *tones*) in a film or print and is broadly represented by the slope of the characteristic curve. Adjectives such as *flat* or *soft* and *contrasty* or *hard* are often used to describe contrast. In general, the steeper the slope of the characteristic curve, the higher the contrast. The terms *gamma* and *average gradient* refer to numerical means for indicating the contrast of the photographic image.

Gamma is the slope of the straight-line portion of the characteristic curve or the tangent of the angle ($\propto$) formed by the straight line with the horizontal. In Fig. 17-40 the tangent of the angle ($\propto$) is obtained by dividing the density increase by the log exposure change. The resulting numerical value is referred to as gamma.

Gamma does not describe contrast characteristics of the toe or the shoulder. Camera negative films record some parts of scenes, such as shadow areas, on the toe portion of the characteristic curve. Gamma does not account for this aspect of contrast.

Average gradient is the slope of the line connecting two points bordering a specified log-exposure interval on the characteristic curve. The location of the two points includes

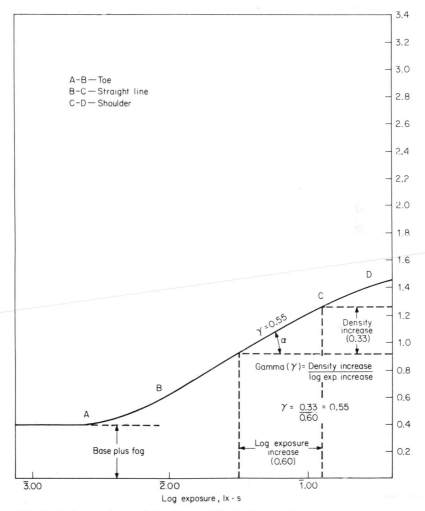

FIG. 17-40 Gamma characteristic. *(Eastman Kodak Company.)*

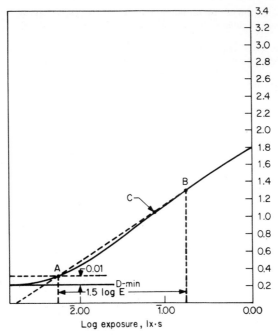

FIG. 17-41 Average gradient determination. *(Eastman Kodak Company.)*

portions of the curve beyond the straight-line portion. Thus, the average gradient can describe contrast characteristics in areas of the scene not rendered on the straight-line portion of the curve. Measurement of an average gradient extending beyond the straight-line portion is shown in Fig. 17-41.

17.7.4 IMAGE STRUCTURE.   The sharpness of image detail that a particular film type can produce cannot be measured by a single test or expressed by one number. For example, resolving-power test data give a reasonably good indication of image quality. However, because these values describe the maximum resolving power a photographic system or component is capable of, they do not indicate the capacity of the system (or component) to reproduce detail at other levels. For more complete analyses of detail quality, other evaluating methods, such as the modulation-transfer function and film granularity, are often used. An examination of the modulation-transfer curve, rms granularity, and both the high- and low-contrast resolving power listings will provide a good basis for comparison of the detail-imaging qualities of different films.

Modulation Transfer Curve.   Modulation transfer relates to the ability of a film to reproduce images of different sizes. The modulation-transfer curve describes a film's capacity to reproduce the complex spatial frequencies of detail in an object. In physical terms, the measurements evaluate the effect on the image of light diffusion within the emulsion. First, film is exposed under carefully controlled conditions to a series of special test patterns, similar to that illustrated in Fig. 17-42a. After development, the image (Fig. 17-42b) is scanned in a microdensitometer to produce trace (Fig. 17-42c). The resulting measurements show the degree of loss in image contrast at increasingly higher frequencies as the detail becomes finer. These losses in contrast are compared mathe-

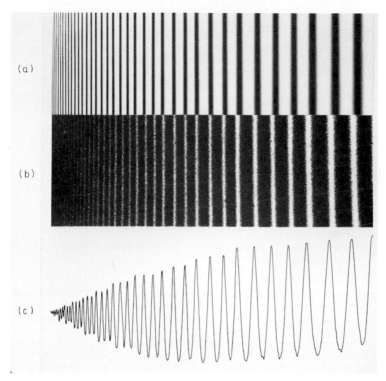

FIG. 17-42 Image (a) of a sinusoidal test object (b) recorded on a photographic emulsion and a microdensitometer tracing (c) of the image. *(Eastman Kodak Company.)*

matically with the contrast of the portion of the image unaffected by detail size. The rate of change or modulation M of each pattern can be expressed by this formula in which E represents exposure

$$M = \frac{E_{max} - E_{min}}{E_{max} + E_{min}} \qquad (17\text{-}2)$$

When the microdensitometer scans the test film, the densities of the trace are interpreted in terms of exposure, and the effective modulation of the image (M) is calculated. The modulation transfer factor is the ratio of the modulation of the developed image to the modulation of the exposing pattern (M_o), or M_i/M_o. This ratio is plotted on the vertical axis (logarithmic scale) as a percentage of response. The spatial frequency of the patterns is plotted on the horizontal axis as cycles per millimeter. Figure 17-43 shows two such curves. At lower magnifications, the test film represented by curve A appears sharper than that represented by curve B; at very high magnifications, the test film represented by curve B appears sharper.

Understanding Graininess and Granularity. The terms *graininess* and *granularity* are often confused or even used as synonyms in discussions of silver or dye-deposit distributions in photographic emulsions. The two terms refer to two distinctly different ways of evaluating the image structure. When a photographic image is viewed with sufficient magnification, the viewer experiences the visual sensation of graininess, a subjective impression of nonuniformity in an image. This nonuniformity in the image structure

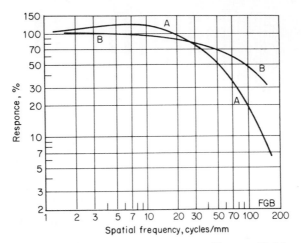

FIG. 17-43 Modulation transfer curves. *(Eastman Kodak Company.)*

can also be measured objectively with a microdensitometer. This objective evaluation measures film granularity.

Motion-picture films consist of silver halide crystals dispersed in gelatin (the emulsion) that is coated in thin layers on a support (the film base). The exposure and development of these crystals forms the photographic image, which is, at some stage, made up of discrete particles of silver. In color processes, where the silver is removed after development, the dyes form dye clouds centered on the sites of the developed silver crystals. The crystals vary in size, shape, and sensitivity, and generally are randomly distributed within the emulsion. Within an area of uniform exposure, some of the crystals will be made developable by exposure; others will not.

The location of these crystals is also random. Development usually does not change the position of a grain, and so the image of a uniformly exposed area is the result of a random distribution of either opaque silver particles (black-and-white film) or dye clouds (cloud film) separated by transparent gelatin. Although the viewer sees a granular pattern, the eye is not necessarily seeing the individual silver particles, which range from about 0.002 mm down to about a tenth of that size. At magnifications where the eye cannot distinguish individual particles, it resolves random groupings of these particles into denser and less dense areas. As magnification decreases, the observer progressively associates larger groups of spots as new units of graininess. The size of these compounded groups gets larger as the magnification decreases, but the amplitude (the difference in density between the darker and lighter areas) decreases. At still lower magnifications, the graininess disappears altogether because no granular structure can be detected visually.

17.7.5 FILM SYSTEMS

Negative Camera Films. Negative film produces an image that must be printed on another stock for final viewing. Since at least one intermediate stage is usually produced to protect the original footage, negative camera film is an efficient choice when significant editing and special effects are planned. Printing techniques for negative-positive film systems are very sophisticated and highly flexible; hence, negative film is especially appropriate for complex special effects. All negative films can go through several print "generations" without pronounced contrast buildup.

Reversal film produces a positive image after processing. With certain exceptions, reversal camera films are designed to be projected after processing. Since processing can be the only intermediate step between the camera film and the projection print, reversal film is a good choice for an absolutely accurate record without intervening duplication stages. Additional prints can be made by direct printing onto reversal print films. If more than two or three prints are needed, an internegative is usually made from flashed† and processed camera film and used to print onto positive print stocks for optimum economy and protection of the original.

Color-reversal films are balanced for projection at 5400 K, which is suitable for both television broadcast and conventional motion-picture projection. These films can be exposed at effective film speeds ranging from one-half to two times the normal exposure indices (one-half to one stop) with little loss in quality. When some loss in quality is acceptable, the effective film speed can be increased by two full lens stops.

Laboratory and Print Films. The filmmaker can maximize the effectiveness of the camera films he or she chooses by understanding the laboratory techniques through which camera film is transformed into the finished production.

While films have been categorized as laboratory films or release print film, in actual practice both are used in the laboratory and in the production of finished screen versions derived from camera originals.

While reversal camera film original can be the finished production, it is rarely used this way if prints are desired. Usually a work print is made and editing worked out and tested before the original is cut. The original material is then cut and assembled to conform to the work print and used to produce internegatives or reversal release prints. Negative film requires printing for both editing and final use. Master positives and duplicate negatives are generally produced to generate optical effects and to protect the camera original from damage during printing operations when a large number of release prints are being produced. A color reversal intermediate is a means of obtaining a duplicate negative without going through a master positive stage.

Color Balance. Color balance relates to the color of a light source that a color film is designed to record without additional filtration. All laboratory and print films are balanced for the tungsten light sources used in printers, while camera films are nominally balanced for 5500-K daylight, 3200-K tungsten, or 3400-K tungsten exposure.

When filming is done under light sources different from those recommended, filtration over the camera lens or over the light source is required. Camera film data sheets contain starting-point filter recommendations for the most common lighting sources: daylight, 3200-K tungsten, 3400-K tungsten, cool-white fluorescent, deluxe cool-white fluorescent, and Mole-Richardson H1-Arc lamps (both white-flame and yellow-flame carbons). Table 17-3 summarizes the filter recommendations from the data sheets.

17.7.6 FILM CHARACTERISTICS

Color Sensitivity and Spectral Sensitivity. The term *color sensitivity* describes the portion of the visual spectrum to which the film is sensitive (Fig. 17-44). All black-and-white camera films are panchromatic (sensitive to the entire visible spectrum).

Some films, called *orthochromatic,* are sensitive mainly to the blue-and-green portions of the visible spectrum. Films used exclusively to receive images from black-and-white materials are blue-sensitive. Some films are sensitive to blue light and ultraviolet radiation. The extended sensitivity in the ultraviolet region of the spectrum permits the film to respond to the output of cathode-ray tubes.

While color films and panchromatic black-and-white films are sensitive to all wavelengths of visible light, rarely are two films equally sensitive to all wavelengths. Spectral

---

†Reduction in contrast by an overall flash exposure before development.

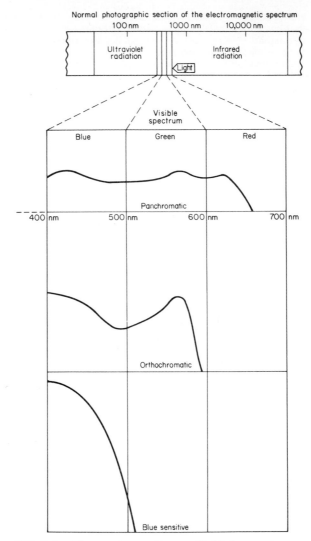

FIG. 17-44 Film sensitivities. *(Eastman Kodak Company.)*

sensitivity describes the relative sensitivity of the emulsion to the spectrum within the film's sensitivity range. The photographic emulsion has inherently the sensitivity of photosensitive silver halide crystals. These crystals are sensitive to high-energy radiation, such as x-rays, gamma rays, ultraviolet radiation, and blue-light wavelengths (blue-sensitive black-and-white films). In conventional photographic emulsions, sensitivity is limited at the short-wavelength (ultraviolet) end to about 250 nm because the gelatin used in the photographic emulsion absorbs much ultraviolet radiation. The sensitivity of an emulsion to the longer wavelengths can be extended by the addition of suitably chosen dyes. By this means, the emulsion can be made sensitive through the green region (orthochromatic black-and-white films), through the green and red regions (color and

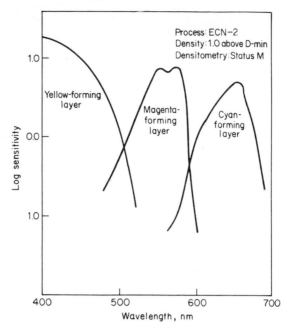

FIG. 17-45 Typical color film spectral sensitivity curves. *(Eastman Kodak Company.)*

panchromatic black-and-white films), and into the near-infrared region of the spectrum (infrared-sensitive film).

Three spectral sensitivity curves are shown for color films—one each for the red-sensitive (cyan-dye forming), the green-sensitive (magenta-dye forming), and the blue-sensitive (yellow-dye forming) emulsion layers. One curve is shown for black-and-white films. The data are derived by exposing the film to calibrated bands of radiation 10 nm wide throughout the spectrum, and the sensitivity is expressed as the reciprocal of the exposure (ergs/cm^2) required to produce a specified density. The radiation expressed in nanometers is plotted on the horizontal axis, and the logarithm of sensitivity is plotted on the vertical axis to produce a spectral sensitivity curve, as shown in Fig. 17-45.

Equivalent Neutral Density. When the amounts of the components of an image are expressed in this unit, each of the density figures tells how dense a gray that component can form. Because each emulsion layer of a color film has its own speed and contrast characteristics, equivalent neutral density (END) is derived as a standard basis for comparison of densities represented by the spectral sensitivity curve. For color films, the standard density used to specify spectral sensitivity is as follows. For reversal films, END = 1.0. For negative films, direct duplicating, and print films, END = 1.0 above D-min.

Spectral Dye-Density Curves. Processing exposed color film produces cyan, magenta, and yellow dye images in the three separate layers of the film. The spectral dye-density curves (illustrated in Fig. 17-46) indicate the total absorption by each color dye measured at a particular wavelength of light and the visual neutral density (at 1.0) of the combined layers measured at the same wavelengths.

Spectral dye-density curves for reversal and print films represent dyes normalized to form a visual neutral density of 1.0 for a specified viewing and measuring illuminant.

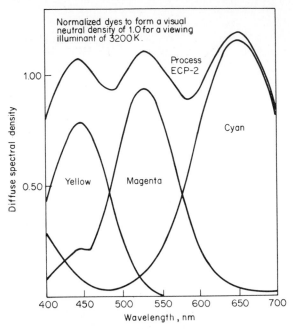

FIG. 17-46 Typical spectral dye-density curves.

Films that are generally viewed by projection are measured with light having a color temperature of 5400 K. Color-masked films have a curve that represents typical dye densities for a midscale neutral subject.

The wavelengths of light, expressed in nanometers, are plotted on the horizontal axis, and the corresponding diffuse spectral densities are plotted on the vertical axis. Ideally, a color dye should absorb only in its own region of the spectrum. All color dyes in use absorb some wavelengths in other regions of the spectrum. This unwanted absorption, which could prevent satisfactory color reproduction when the dyes are printed, is corrected in the film's manufacture.

In color negative films, some of the dye-forming couplers incorporated in the emulsion layers at the time of manufacture are colored and are evident in the D-min of the film after development. These residual couplers provide automatic masking to compensate for the effects of unwanted dye absorption when the negative is printed. This explains why negative color films look orange.

Since color reversal films and print films are usually designed for direct projection, the dye-forming couplers must be colorless. In this case, the couplers are selected to produce dyes that will, as closely as possible, absorb in only their respective regions of the spectrum. If these films are printed, they require no printing mask.

REFERENCES

1. David M. Stern, "A Broadcast Television Film-Chain Projector," *SMPTE J.*, vol. 86, no. 3, 1977, pp. 146–148.
2. John D. Millward, "Flying-Spot Scanner (Rank Cintel) on 525-line NTSC Standards," *SMPTE J.*, vol. 90, no. 9, 1981, p. 786.

3. Ian Childs and J. R. Sanders, "An Experimental Telecine Using a Line-Array CCD Sensor," *SMPTE J.*, vol. 88, no. 4, 1978, pp. 209–213.

4. Ian Childs and J. R. Sanders, "New Capabilities for a Line-Array CCD Telecine," *SMPTE J.*, vol. 92, no. 12, 1983, p. 1294.

5. Dieter Poetsch, "FDL 60 (Bosch)—An Advanced Film Scanning System," *SMPTE J.*, vol. 93, no. 3, 1984, pp. 216–227.

6. Ray Matchell, "The Marconi B3410 Line-Array Telecine," *SMPTE J.*, vol. 91, no. 11, 1982, p. 1066.

7. Kenneth G. Lisk and Charles Evans, "Film (EK) Recording, Color Television, from Shadow-Mask Picture Tube," *SMPTE. J.*, vol. 80, no. 10, 1971, pp. 801–806.

8. Kenneth G. Lisk, "A Low-Cost Color-Television Film-Recording System," *SMPTE J.*, vol. 88, no. 3, 1979, pp. 157–160.

9. Henry A. Barrett, R. Collette, K. G. Lisk, J. Sager, and R. J. Sypula, "An Experimental Trinoscope for Improved Video-to-Film Recording," *SMPTE J.*, vol. 93, no. 8, 1984, p. 746.

10. Yukio Sugiura, Y. Nojiri, and K. Okada. "HDTV Laser-Beam Recording on 35-mm Color Film and Its Application to Electro-cinematography," *SMPTE J.*, vol. 93, no. 7, 1984, pp. 642–651.

11. Kenneth G. Lisk, J. P. Pytlak, and H. A. Barrett, "New Tools for Improved Telecine Quality," *SMPTE J.*, vol. 93, no. 1, 1984, pp. 6–13.

12. Woodlief, Thomas, Jr. (ed.), *SPSE Handbook of Photographic Science and Engineering*, Wiley, New York, 1973.

13. T. H. James and G. C. Higgins, *Fundamentals of Photographic Theory*, Morgan and Morgan, Inc., Dobbs Ferry, New York, 1968.

14. R. M. Evans, W. T. Hanson, Jr., and W. L. Brewer, *Principles of Color Photography*, Wiley, New York, 1953.

15. G. Wyszecki and W. S. Stiles, *Color Science*, Wiley, New York, 1967.

16. T. H. James (ed.), *The Theory of the Photographic Process*, 4th ed., MacMillan, New York, 1977.

Digital Television

Ernest J. Tarnai

Manager, Signal Processing and Broadband Systems
Bell-Northern Research Ltd.
Ottawa, Canada

18.1 FUNDAMENTALS OF DIGITAL TELEVISION

18.1.1 WHY DIGITAL?   The two most commonly encountered signal formats in electronics are *analog,* signals which are continuous in space and/or time, and *digital,* signals which are usually binary sequences. The input and output signals of television systems, at the camera and at the receiver, respectively, are inherently analog.† Thus the question "why digital?" for television is a natural one.

Technical Considerations.   The customary answers to this paramount question are usually based on technical grounds.[1-3] The most important of these are:

- While analog signal degradations (e.g., signal-to-noise ratio, crosstalk, distortions— both linear and nonlinear) are accumulative and difficult to distinguish from the video signal (thus difficult to remove), the ability to regenerate the digital pulse train *exactly* renders the digital signals theoretically immune to impairments caused by circuit deficiencies.

- Several digital bit streams can be interleaved, i.e., *time multiplexed.* This process could expedite the transmission, storage, or processing of ancillary signals along with the associated video.

- Utilization of forward error correcting (FEC) codes will preserve the integrity of the signals to practically any desired level.

- Storage, delay, or manipulation of the signals is easier in the digital than in the analog domain.

- The integration of the digital computer into the television environment as a valuable picture source or manipulator is easier in the digital format.

- Channels characterized by bandwidth and noise can be more effectively utilized by not carrying redundant picture elements. The process of redundancy reduction is most effective in the digital domain.

Economic Considerations.   Equally compelling economic factors also justify the use of digital techniques. The major items are:

- Since the development of the transistor, the costs of digital, solid-state devices (both

---

†This statement could be questioned on quantum mechanical grounds; however, engineering approximations are often more convenient than absolute truth.

for memory and logic) are continuously plummeting (at a faster rate than their analog counterparts), resulting in lower equipment prices.

- Digital circuits do not require alignment, tuning, etc.; thus the operation and maintenance of digital hardware are usually easier and more cost effective than of their analog counterparts.

- The proliferation of computers of various sizes coupled with digital technology allows the mechanization of labor-intensive, routine-type plant operations (such as maintenance and test procedures using digital test sets) resulting in potentially substantial savings.

18.1.2 ANALOG-TO-DIGITAL (A/D) CONVERSION. Since the input and output of television systems are analog signals, the input must be represented as numeric sequences to take advantage of digital techniques and the processed numeric sequence must be mapped onto analog signals for display purposes. The total system accomplishing this is represented in block diagram form in Fig. 18-1. The "digital system" shown could be a simple component, such as a time-base corrector in an analog studio, or it may represent the whole television chain from the camera output to the input of the transmitting antenna.

The analog-to-digital (A/D) converter performs three distinct functions as illustrated on Fig. 18-2. The low-pass filter is not strictly part of the A/D converter but is required to band limit the input analog signal. This, as will be seen later, is a prerequirement of successful sampling.

FIG. 18-1 Digital television system.   FIG. 18-2 A/D converter block diagram.

Sampling. The first step in the analog-to-digital conversion process is the sampling of the input analog signal. If the input and the sampling process meet certain criteria, this will not result in any information loss. This may be shown as follows. Consider a time function $f(t)$ and its Fourier transform $F(\omega)$

$$f(t) = \frac{1}{2\pi} \int_{-\infty}^{\infty} F(\omega) e^{j\omega t}\, d\omega \leftrightarrow F(\omega) = \int_{-\infty}^{\infty} f(t) e^{-j\omega t}\, dt \qquad (18\text{-}1)$$

Taking samples of $f(t)$ at T-s intervals can be expressed mathematically as the product of $f(t)$ and a periodic impulse train defined as

$$\delta_T(t) \triangleq \sum_{n=-\infty}^{\infty} \delta(t - nT) \qquad (18\text{-}2)$$

where $\delta(t)$ is the Dirac delta function. $\delta_T(t)$ is periodic; therefore, it can be expanded in a Fourier series

$$\delta_T(t) = \sum_{n=-\infty}^{\infty} F_n \exp{(jn\omega_0 t)} \qquad \omega_0 = \frac{2\pi}{T} \qquad (18\text{-}3)$$

with Fourier coefficients

$$F_n = \frac{1}{T} \int_{-T/2}^{T/2} \delta_T(t) \exp{(-jn\omega_0 t)}\, dt = \frac{1}{T} \qquad (18\text{-}4)$$

Thus the sampled version of $f(t)$, which will be denoted by $f_s(t)$, is

$$f_s(t) = f(t)\, \delta_T(t) \qquad (18\text{-}5)$$

The Fourier transform of the sampled version of $f(t)$, $f_s(t)$ can be obtained by the convolution of $F(\omega)$ and the Fourier transform of $\delta_T(t)$

$$\mathcal{F}\{f_s(t)\} = \mathcal{F}\{f(t)\delta_T(t)\}$$

$$= \frac{1}{2\pi} F(\omega) \star \mathcal{F}\{\delta_T(t)\} \tag{18-6}$$

The operation "$\star$" denotes the convolution integral: $f(t) = \int_{-\infty}^{\infty} h(\tau)g(\tau - t)\, d\tau$. The transform of the impulse train follows from the Fourier series expansion [Eqs. (18-3 and 18-4)]

$$\mathcal{F}\{\delta_T(t)\} = \mathcal{F}\left\{\sum_{n=-\infty}^{\infty} \frac{1}{T} \exp jn\omega_0 t\right\}$$

$$= \frac{1}{T} \sum_{n=-\infty}^{\infty} \mathcal{F}\{\exp jn\omega_0 t\}$$

$$= \frac{2\pi}{T} \sum_{n=-\infty}^{\infty} \delta(\omega - n\omega_0) \triangleq \omega_0 \, \delta_{\omega_0}(\omega) \tag{18-7}$$

The explicit form of Eq. (18-6) is

$$F_s(\omega) = \frac{1}{T} \int_{-\infty}^{\infty} F(\xi) \sum_{n=-\infty}^{\infty} \delta(\omega - \xi - n\omega_0)\, d\xi$$

$$= \frac{1}{T} \sum_{n=-\infty}^{\infty} \int_{-\infty}^{\infty} F(\xi)\, \delta(\omega - \xi - n\omega_0)\, d\xi$$

$$= \frac{1}{T} \sum_{n=-\infty}^{\infty} F(\omega - n\omega_0) \tag{18-8}$$

Thus the spectrum or Fourier transform of the sampled signal is a periodic signal consisting of the original spectrum of $f(t)$ scaled by $1/T$ and shifted on the frequency domain by $n\omega_0$. The sampling process is depicted in both the time and the frequency domains in Fig. 18-3.

If the original signal does not have frequency components above $\omega_0/2$, the components of the periodic spectrum $F_s(\omega)$ will not overlap; therefore, one could, in principle, recover the scaled version of the original spectrum $F(\omega)$ by low-pass filtering. The above argument leads to one of the fundamental statements in communications, namely, *Nyquist's uniform sampling theorem.*

$f_m \rightarrow$ A signal which is band-limited to f_m Hertz is uniquely defined by its sample taken uniformly at $1/2f_m$ s apart.[4]

In practical situations the sampling process is accomplished not by impulses but by a periodic signal consisting of narrow, rectangular pulses of width τ

$$g(t) = \begin{cases} 0 & nT - \tau/2 < t < nT + \tau/2 \\ & nT + \tau/2 < t < (n + 1)T + \tau/2 \\ 1 & n = 0, \pm 1, \pm 2, \ldots \end{cases} \tag{18-9}$$

The reader can easily verify that a derivation similar to the above yields the spectrum of the signal sampled by finite width pulses as

$$\tilde{F}_s(\omega) = \frac{1}{T} \sum_{n=-\infty}^{\infty} S_a\left(\frac{n\tau\omega_0}{2}\right) F(\omega - n\omega_0) \tag{18-10}$$

where $S_a(x) = (\sin x)/x$. Thus, the sampled signal's spectrum consists of the original spectrum repeated at ω_0 intervals and scaled by a factor, $S_a(n\omega_0\tau/2)$, as shown in Fig. 18-

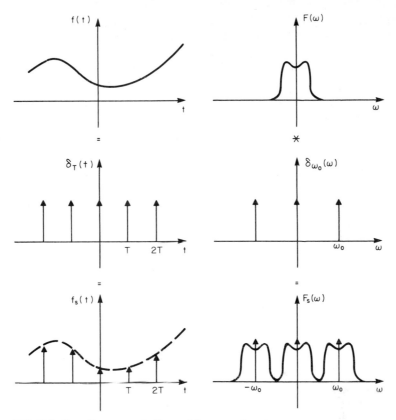

FIG. 18-3 Sampling process in time and frequency domains.

4. Thus, at least in theory, sampling with rectangular pulses preserves all the information present in the original signal.

If the input signal is not band-limited or the sampling frequency does not meet the requirement of the sampling theorem also known as the *Nyquist criteria,* the various components of the spectrum overlap as shown in Fig. 18-3 and the recovered signal will not be an exact replica of the input but will suffer distortion called *aliasing.* (An example of aliasing is the appearance of wagon wheels going backwards—a familiar annoyance to motion-picture engineers.†)

Quantization. The output of the sampler (on Fig. 18-2) is discrete in time but continuous in amplitude. Such signals are called *discrete time signals* and will be discussed with respect to digital filtering. In order to assign discrete numerical values to these samples, the quantizer maps a range of its input onto each output value. This is an irreversible process since the D/A converter cannot recreate all the input values but can assign only one analog amplitude to digital samples within the range. The impairment thus introduced is typical of digital systems and is termed *quantization error.* Figure 18-5 depicts the principle of quantization.

---

†Note that, owing to the frame repetition process, motion picture is a sampled data rather than an analog process. Similarly, "analog television" is a sampled data process by virtue of either the frame repetition or the scanning process.

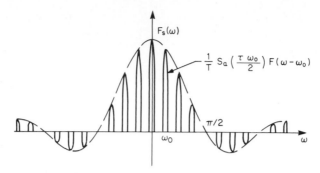

FIG. 18-4 Spectrum of a band-limited signal sampled with narrow pulses.

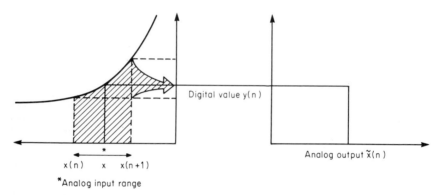

FIG. 18-5 Quantization action. All analog input values between $x(n)$ and $x(n+1)$ map onto the output value $\tilde{x}(n)$.

Analog input values between quantization levels $x(n)$ and $x(n+1)$ are mapped into a single digital word which is, in turn, decoded as the analog output $\tilde{x}(n)$. The error introduced by the process $\Delta = |\tilde{x}(n) - x|$ is the quantization error associated with an arbitrary sample x within the range $[x(n), x(n+1)]$.

The quantizer may consist of uniformly spaced quantization levels to facilitate signal processing or to reduce circuit complexity. Alternatively, it may consist of nonuniformly spaced or tapered levels to optimize signal-to-quantization-noise ratio (if coarse quantization can be accepted to reduce the bit rate).

The peak-to-peak signal-to-rms quantizing error ratio (S/Qe) for different numbers of bits per sample, n, for uniformly quantized pulse-code modulation (PCM), derived in Ref. 5, is

$$\frac{S}{Qe} \text{ (dB)} = 20 \text{ (log 2)} \, n + 10 \log 12 + 10 \log \frac{F_s}{2F_{v \text{ max}}}$$

$$= 6.02n + 10.8 + 10 \log \frac{F_s}{F_{v \text{ max}}}$$

(18-11)

where F_s is the sampling frequency and $F_{v \text{ max}}$ is the effective bandwidth of the television signal. For example, with 8-bit samples, ($n = 8$) for $F_v = 4.2$ MHz, and $F_s = 14.3$ MHz, the S/Qe equals 61.3 dB. Each bit less reduces this value by 6.02 dB. The quantitative effects of quantization error Qe depend on the number of bits used. For six or more bits per sample, Qe appears as random noise in the picture. When five or fewer bits per sample are used, the dependency of Qe on the television image becomes evident as gray step contours. This is especially visible in areas of the picture that are gradually changing in luminance or chroma level.[5]

Signal Encoding. The last step in the analog-to-digital conversion is the assignment of a binary code to the numerical value corresponding to the individual samples. In the example illustrated in Fig. 18-6 a straightforward base 2 representation of the numbers is used. In practical situations the encoding could be more complex (using multilevel codes, parity, scrambling, etc.). The actual encoding depends on the nature and error characteristics of the digital system requiring the digital signal.

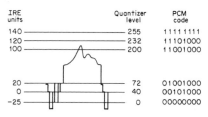

FIG. 18-6   Coding process.

18.1.3 DIGITAL-TO-ANALOG CONVERSION.

The digital to analog (D/A) conversion is quite simple in principle. The digital stream is decoded into discrete time signals corresponding to the sampled version of the desired video signal (corrupted by quantization error or other impairments). The analog signal is recovered by low-pass filtering of these samples.

In practical situations, however, the D/A conversion does not operate on a sequence $\{y(n)\}$ but rather on a continuous time function $y(t) = \Sigma y(n)h(t - nT)$. The most commonly used D/A conversions employ rectangular pulses as $h(t)$. This has the following two effects: (1) $y(t)$ contains a large amount of high-frequency energy and (2) the spectrum of $h(t)$ is a scaled version of the spectrum of the original sequence. The scaling factor is $|H(\omega)| = 2 \sin (\omega T/2)/\omega$. The first of these effects can be eliminated by a low-pass filter following the D/A conversion and the second by a predistortion filter with amplitude response $|G(\omega)| = 1/|H(\omega)|$.

18.1.4 ERRORS AND ERROR MANAGEMENT

Basic Principles. In analog systems, on one hand, impairments due to noise, linear or nonlinear distortions, etc., are more or less proportional to the length of the signal path. These contaminations, in general, are indistinguishable from the signal, hence difficult to remove. In digital systems, on the other hand, regeneration of the digital pulses, before the cumulative effects of the above impairments become catastrophic, renders the signals resistive to these deficiencies. However, other causes such as excessive attenuation, noise, interference, echoes, and circuit or memory faults can cause infrequent but serious errors in digital streams.

Digital television signals, a representation of analog video as "numerical sequences," are also susceptible to errors. These usually occur during transmission or storage of the signals.

Fortunately, the introduction of some redundancy into the digital signals allows the correction or recognition of some of these errors. When an error is merely detected, the system can either hide its effects (see Error Concealment below) or request the repeat of the information containing the error. For television the request for retransmission scheme is not acceptable because of the inherent delays involved. The process called *forward error correction (FEC)* is a very active research topic and cannot be adequately covered here. This is done extensively in the literature.[6-11] It is hoped, however, that the rest of this section will provide the reader with a rudimentary appreciation of the coding process and some of its benefits and limitations.

Before describing the two main branches of FEC, namely, *block coding* and *convolution coding,* some definitions and statements of basic principles are in order.

1. *Representation of binary sequences.* Binary sequences or strings of ones and zeros will be represented in either of the two following ways:

   n-tuple representation

   $$\mathbf{a} = (a_0, a_1, a_2, \ldots, a_{n-1}) \qquad a_i = 0 \text{ or } 1 \qquad (18\text{-}12)$$

   polynomial representation

   $$\mathbf{P}(x) = a_0 + a_1 x + a_2 x^2 + \cdots + a_{n-1} x^{n-1} \qquad (18\text{-}13)$$

   For example, the 7-tuple $\mathbf{a} = (1, 0, 1, 1, 0, 0, 1)$ which can be considered as a vector in a seven-dimensional space would have a polynomial representation as

   $$\mathbf{P}(x) = 1 + x^2 + x^3 + x^6 \qquad (18\text{-}14)$$

   because a_0, a_2, a_3, and a_6 are 1 and the rest of the coefficients are 0. The highest power of x is referred to as the *degree* of the polynomial and denoted by deg[ ].

2. *Modulo 2 addition.* The ubiquitous operation in coding theory is the modulo 2 addition of two binary numbers. This operation is defined by the *truth table* shown in Fig. 18-31 along with the standard circuit symbol for modulo 2 addition.

3. *Groups.* Groups are mathematical systems consisting of one or more elements and an operation (e.g., the binary digits 0 and 1 as elements and modulo 2 addition as the operation) such that the following requirements are met:

   - Closure—If a and b are elements of the system S and if x denotes the operation, then $a \times b$ is also an element of S.

   - Identity—The system S contains an identity element (with respect to the operation x) such that for any other element $I \times a = a \times I = a$.

   - Inverse—For every element a of the system exists, another element for which $a \times a^{-1} = a^{-1} \times a = I$.

   - Associative—If a, b, and c are in S, then $(a \times b) \times c = a \times b \times c = a \times (b \times c)$.

   If the group satisfies the following fifth requirement, it is called a *commutative* (or *abelian*) group:

   - Commutative—If a and b are in S, then $a \times b = b \times a$.

4. *Fields.* Fields are mathematical systems consisting of two or more elements and two operations (which will be denoted here by $+$ and $\times$) such that the following three conditions are met:

   - They are commutative groups under one of the operations.

   - They are commutative groups under the other operations except that the identity element (under the first operation) has no inverse (under the second).

   - The operations are distributive $a \times (b + c) = (a \times b) + (a \times c)$.

5. *Finite fields.* Finite (or Galois) fields contain a finite number of elements. A finite field containing q elements is usually denoted by GF(q). Consider the following examples:

   - The set of all integers (positive, negative, and zero) and the ordinary addition $(+)$ operation form a group since, first, the sum of all integers is also an integer, second, the number 0 is the identity element (for example, $a + 0 = 0 + a = a$), third, the "negative" of an integer is its inverse [$a + (-a) = 0$], and, obviously, fourth, the order of addition is immaterial.

   - The set of integers under ordinary addition and multiplication is *not* a field, however, since (if 1 is the identity element under $\times$) only 1 and -1 have inverses.

   - The set $\{0, 1\}$ and the modulo 2 addition and ordinary multiplication form a finite field GF(2).

6. *Irreducible polynomials.* A polynomial $\mathbf{a}(x)$ is reducible over a field F if and only if there exist two polynomials $\mathbf{b}(x)$ and $\mathbf{c}(x)$ at least of degree 1 in F such that $\mathbf{a}(x) = \mathbf{b}(x)\mathbf{c}(x)$. Otherwise $\mathbf{a}(x)$ is irreducible.

7. *Operations modulo p.* A polynomial $\mathbf{a}(x)$ (which is a result of some operation) mod-

ulo some other, irreducible, polynomial is the remainder after $\mathbf{p}(x)$ divides $\mathbf{a}(x)$, that is

$$\mathbf{a}(x) = \mathbf{q}(x)\mathbf{p}(x) + \mathbf{r}(x) \rightarrow \mathbf{a}(x) = \mathbf{r}(x) \text{ modulo } \mathbf{p}(x) \qquad (18\text{-}15)$$

Note that the $\deg[r(x)] < \deg[p(x)]$. For example, consider the two polynomials $a(x) = x^3 + 1$ and $p(x) = x^2 + 1$ over GF(2). Then,

$$x^3 + 1 = x(x^2 + 1) + (x + 1) = (x + 1) \text{ modulo } (x^2 + 1) \qquad (18\text{-}16)$$

8. *Monic polynomials.* Monic polynomials have unity as the coefficient for their highest degree term.

9. *Roots of polynomials.* α is termed the root of $\mathbf{a}(x)$ if $\mathbf{a}(\alpha) = 0$. Further, if α is a root of $\mathbf{a}(x)$, then $(x - \alpha)$ is a factor of $\mathbf{a}(x)$. [α may or may not be from the field of the coefficients of $\mathbf{a}(x)$.] The number of distinct roots will not exceed the degree of $\mathbf{a}(x)$.

10. *Order of α.* The (multiplicative) order of an element of GF(g) is defined to be the smallest integer such that $\alpha^\mu = 1$ [in GF(q)]. If $\mu = q - 1$, α is said to be a *primitive* element of GF(q). Every GF(q) contains at least one primitive element. The successive powers of a primitive element will generate all nonzero elements of GF(q).

11. *Minimal polynomial.* $\mathbf{m}_\alpha(x)$ is a minimal polynomial over GF(q) of the element α if it is the monic polynomial of the lowest degree such that $\mathbf{m}_\alpha(\alpha) = 0$.

12. *Weight of a code word $w(\mathbf{V})$.* This weight is defined as the number of nonzero elements in the code word; for example, $\mathbf{a} = (1, 0, 1, 1, 0, 1)$ has a weight $w(\mathbf{a}) = 4$.

13. *Hamming distance d.* The distance between two code words is a measure of the degree to which they are different. The most commonly used distance, the Hamming distance, is defined as the number of dissimilar bits between the two code words; e.g.,

$$\mathbf{V}_1 = (1, 0, 1, 0, 1, 1, 0)$$
$$d(\mathbf{V}_1, \mathbf{V}_2) = 5 \qquad (18\text{-}17)$$
$$+ \ \underline{\mathbf{V}_2 = (0, 1, 0, 0, 1, 0, 1)}$$
$$\mathbf{V}_3 = (1, 1, 1, 0, 0, 1, 1)$$

Note that the distance between two elements is equal to the weight of their sum

$$d(\mathbf{V}_1, \mathbf{V}_2) = w(\mathbf{V}_1 + \mathbf{V}_2) \qquad (18\text{-}18)$$

as shown (for the above example) in Fig. 18-7.

14. *Minimum distance $d_{\min}$.* This is defined as the smallest d among all possible code words. This quantity is usually accepted as a figure of merit for the code; i.e., the larger $d_{\min}$ is for a given number of redundancy (and circuit complexity), the better the code is.

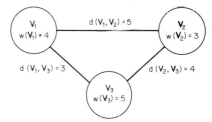

FIG. 18-7 Weights and hamming distances for code words in Eq. (18-17).

15. *The coding process.* The coding process is shown schematically in Fig. 18-8. The efficiency of the code is defined as the ratio of information digits to total number of digits of the encoded signal. All things being equal, the higher this ratio, the better the code. Two general methods are used to optimize the code, namely, *block codes* and *convolution codes,* as described below.

Block Codes. The fundamental principle of block codes is the grouping of k information digits into blocks of symbols or, in the case of binary systems, bits adding l check digits to the block (derived from the k information digits—thus fully redundant) and presenting $n = l + k$ code digits to the digital system. A great many block codes have been proposed, but to simplify the discussion, only the more important types (from a technical viewpoint) will be covered here, namely, binary *group codes.* These are schemes where all code words satisfy the group criteria stated in no. 3 above.

Linear block codes are such that each code word $\mathbf{x}$ is a linear combination of k independent vectors where k is the number of information bits. The code is denoted by (n,

FIG. 18-8   FEC coding process.

$k, t)$ where n is the dimension of $\mathbf{x}$, and $(n - k)$ is the number of redundant bits. (The parameter t denotes the number of bit errors the code can correct within the block. This number, especially when it is 1, is often omitted.) The relationship between the code words and the information block $\mathbf{u}$ can be stated in a matrix notation as

$$\mathbf{x} = G^{\mathrm{T}}\mathbf{u}^{\mathrm{T}} \tag{18-19}$$

where G is called the *generator* matrix and can have the form

$$G = [P : I_k] \tag{18-20}$$

P denotes a general $(n - k) \times k$ matrix and I_k is a $k \times k$ identity matrix. (See the example to follow.) Another matrix, called the *parity check* matrix, can be obtained from G as shown below

$$H = [I_{n-k} : P^{\mathrm{T}}] \tag{18-21}$$

The code produced by a generator matrix which has the form shown above is called a *systematic code*. The k consecutive digits of systematic codes are identical to the k information digits. Other codes generated by matrices obtained by linear operations on the columns of G have the same properties as the original systematic code. The paramount property of all valid code words is that

$$\mathbf{x}H^{\mathrm{T}} = \mathbf{0} \tag{18-22}$$

That is, the parity check matrix facilitates the decision whether an arbitrary n-tuple is a valid code word.

Consider a digital system (either transmission or processing) which receives a valid code word x at its input and produces an output signal $\hat{\mathbf{x}} = \mathbf{x} + \mathbf{e}$, where $\mathbf{e}$ is an error vector. According to the definition of H

$$\hat{\mathbf{x}}H^{\mathrm{T}} = (\mathbf{x} + \mathbf{e})H^{\mathrm{T}} = \mathbf{x}H^{\mathrm{T}} + \mathbf{e}H^{\mathrm{T}} = \mathbf{e}H^{\mathrm{T}} = \mathbf{s} \tag{18-23}$$

The vector $\mathbf{s}$ is called the *syndrome,* and obviously it is zero if $\mathbf{e}$ is a valid code word, or zero and nonzero otherwise. Thus if the coding scheme is chosen judiciously, the syndrome determines the presence of errors.

As an example, consider a (7, 4) linear block-code generator matrix

$$G = \begin{bmatrix} 1 & 0 & 0 & \vline & 1 & 0 & 0 & 0 \\ 0 & 1 & 1 & \vline & 0 & 1 & 0 & 0 \\ 1 & 1 & 1 & \vline & 0 & 0 & 1 & 0 \\ 1 & 0 & 1 & \vline & 0 & 0 & 0 & 1 \end{bmatrix} \Big\} \; k = 4 \tag{18-24}$$

$$\underset{n - k = 3}{\longleftrightarrow}$$

Valid code words are generated according to the equation $\mathbf{x}^{\mathrm{T}} = G^{\mathrm{T}}\mathbf{u}^{\mathrm{T}}$ or $\mathbf{x} = \mathbf{u}G$ where u is any arbitrary 4-tuple. Thus

$$x_0 = u_0 + u_2 + u_3$$

$$x_1 = u_0 + u_1 + u_2$$

$$x_2 = u_1 + u_2 + u_3$$

$$x_3 = u_0 \tag{18-25}$$

$$x_4 = u_1$$

$$x_5 = u_2$$

$$x_6 = u_3$$

The parity check matrix for this example is

$$H = \begin{bmatrix} 1 & 0 & 0 & | & 1 & 0 & 1 & 1 \\ 0 & 1 & 0 & | & 1 & 1 & 1 & 0 \\ 0 & 0 & 1 & | & 0 & 1 & 1 & 1 \end{bmatrix} \qquad (18\text{-}26)$$

Let the information block be the 4-tuple $\mathbf{u} = (1, 0, 0, 1)$. Then the code word generated will be $\mathbf{x} = (0, 1, 1, 1, 0, 0, 1)$ (note modulo 2 addition, $u_0 + u_3 = 1 + 1 = 0$). To verify that $\mathbf{x}$ is a code word, the multiplication $\mathbf{x}H^T$ can be performed; the result is, of course, 0. But assume that a single error occurs at the least significant bit position and the decoder receives a vector $\mathbf{x} = (1, 1, 1, 1, 0, 0, 1)$. With this vector, the syndrome $\mathbf{x}H^T$ becomes

$$\mathbf{s} = [1\,1\,1\,1\,0\,0\,1] \begin{bmatrix} 1 & 0 & 0 \\ 0 & 1 & 0 \\ 0 & 0 & 1 \\ 1 & 1 & 0 \\ 0 & 0 & 1 \\ 0 & 1 & 1 \\ 1 & 1 & 1 \\ 1 & 0 & 1 \end{bmatrix} = [1\,0\,0] \qquad (18\text{-}27)$$

which is obviously nonzero, indicating an error condition. In the following a few linear block codes will be described.

Repetition Codes. The simplest form of block coding is the repetition code which simply repeats a single information bit n times. For repetition codes $k = 1$ and the number of redundant bits is, of course, $l = n - 1$. The generator matrix and parity matrix are

$$G = [1 \ | \ 1 \ 1 \ \dots \ 1] \quad \text{and} \quad H = \begin{bmatrix} 1 & | & 1 & 0 & 0 & \dots & 0 \\ 1 & | & 0 & 1 & 0 & \dots & 0 \\ 1 & | & 0 & 0 & 1 & \dots & 0 \\ . & | & & & . & & \\ . & | & & & . & & \\ . & | & & & . & & \\ 1 & | & 0 & 0 & 0 & \dots & 1 \end{bmatrix} \qquad (18\text{-}28)$$

respectively. The syndrome $\mathbf{x}H^T$ will contain 1s in positions where the information bit differs from the check bits, i.e., where there is an error. From a transmission rate point of view, this is a very inefficient code since the information rate is only $1/n$ of the channel rate. The usefulness or value of the code lies in its extreme simplicity. This code can detect $n - 1$ errors and correct $t = (n - 1)/2$ errors per code word.

Parity Check Codes. The code words for parity check codes are obtained by adding one extra bit ($l = 1$) to the k information bits. This check bit is obtained by the modulo 2 addition of all bits in the information word (i.e., the check bit is 1 if the information word contains an odd number of 1s and 0 otherwise). The generator and parity matrices are

$$G = \begin{bmatrix} 1 & 0 & \dots & 0 & | & 1 \\ 0 & 1 & \dots & 0 & | & 1 \\ & . & . & & | & . \\ & . & . & & | & . \\ & . & . & & | & . \\ 0 & 0 & \dots & 0 & | & 1 \end{bmatrix} \quad \text{and} \quad H = [1\,1\,1.\dots1 \ | \ 1] \qquad (18\text{-}29)$$

Parity check codes are capable of detecting one single error per code word. Their strength lies in the simplicity and efficiency in which they perform the coding and decoding process.

Hamming Codes. A large class of codes which is important from historic, theoretic, and practical points are the so-called Hamming codes. These were one of the first important sets of codes described systematically (by Hamming in 1950). These can be analyzed mathematically more easily than most other codes, and they provide algorithms which are widely used in practice. These codes may be derived as follows: An (n, k) linear code (where $n = 2^l - 1$ and $k = 2^l - l - 1$) is a *Hamming code of length $2^l - 1$* if its $l \times (2^l - 1)$ parity matrix H contains $2^l - 1$ different nonzero vectors as its columns. Consider the following example.

Let the generator matrix for a $(7, 4)$ code $(k = 4, l = 3)$ be given as

$$G = \begin{bmatrix} 1 & 1 & 1 & 0 & 0 & 0 & 0 \\ 1 & 0 & 0 & 1 & 1 & 0 & 0 \\ 0 & 1 & 0 & 1 & 0 & 1 & 0 \\ 1 & 1 & 0 & 1 & 0 & 0 & 1 \end{bmatrix} \qquad (18\text{-}30)$$

(Note that this is not in the usual form since columns 3 and 4 are interchanged; however, this results only in interchanging symbols 3 and 4 in the code words. This minor depar-

Table 18-1   Hamming Code (7, 4)

000	0000000	000	1111111	000	0001111	000	1110000
001	1000000	001	0111111	001	1001111	001	0110000
010	0100000	010	1011111	010	0101111	010	1010000
011	0010000	011	1101111	011	0011111	011	1100000
100	0001000	100	1110111	100	0000111	100	1111000
101	0000100	101	1111011	101	0001011	101	1110100
110	0000010	110	1111101	110	0001101	110	1110010
111	0000001	111	1111110	111	0001110	111	1110001
000	0010110	000	1101001	000	0011001	000	1100110
001	1010110	001	0101001	001	1011001	001	0100110
010	0110110	010	1001001	010	0111001	010	1000110
011	0000110	011	1111001	011	0001001	011	1110110
100	0011110	100	1100001	100	0010001	100	1101110
101	0010010	101	1101101	101	0011101	101	1100010
110	0010100	110	1101011	110	0011011	110	1100100
111	0010111	111	1101000	111	0011000	111	1100111
000	0100101	000	1011010	000	0101010	000	1010101
001	1100101	001	0011010	001	1101010	001	0010101
010	0000101	010	1111010	010	0001010	010	1110101
011	0110101	011	1001010	011	0111010	011	1000101
100	0101101	100	1010010	100	0100010	100	1011101
101	0100001	101	1011110	101	0101110	101	1010001
110	0100111	110	1011000	110	0101000	110	1010111
111	0100100	111	1011011	111	0101011	111	1010100
000	0110011	000	1001100	000	0111100	000	1000011
001	1110011	001	0001100	001	1111100	001	0000011
010	0010011	010	1101100	010	0011100	010	1100011
011	0100011	011	1011100	011	0101100	011	1010011
100	0111011	100	1000100	100	0110100	100	1001011
101	0110111	101	1001000	101	0111000	101	1000111
110	0110001	110	1001110	110	0111110	110	1000001
111	0110010	111	1001101	111	0111101	111	1000010

ture from convention is adopted to systematize the correction process.) The parity check equation $Hx^T = 0$ for this case becomes

$$\begin{bmatrix} 0 & 0 & 0 & 1 & 1 & 1 & 1 \\ 0 & 1 & 1 & 0 & 0 & 1 & 1 \\ 1 & 0 & 1 & 0 & 1 & 0 & 1 \end{bmatrix} \mathbf{x} = \begin{bmatrix} E_0 \\ E_1 \\ E_2 \end{bmatrix} \qquad (18\text{-}31)$$

The columns of the parity matrix are the binary representation of the numbers from 1 to 7. This code contains $2^k = 16$ valid code words and $2^n = 128$ possible 7-tuples. Table 18-1 lists all possible n-tuples and their relation to valid code words. It was constructed as follows. The first row in each column contains all the 16 possible valid code words and their associated syndrome (E_0, E_1, E_2) computed by the above parity equation (which are zero by definition). Under each code word are listed the 7-tuples which are one-bit different from them and the associated syndromes.

Three important facts should be noted from this table. (1) No two valid code words have a Hamming distance less than 3, indicating that the code should correct $t = (3 - 1)/2 = 1$ error. (2) All $2^k = 16$ columns contain an n-bit code word and n words which are different from the code word in one bit position. Since $n = 2^l - 1$, the table contains

$$2^h(1 + n) = 2^k(1 + 2^l - 1) = 2^{l+k} = 2^n \qquad (18\text{-}32)$$

words, i.e., all possible 7-tuples. (3) The syndrome indicates in binary notation not only that an error occurred but also which bit position is in error. In other words, this Hamming code corrects a single error per block. It is also capable of detecting two errors per block. From a practical point of view, a significant characteristic of Hamming codes is that the hardware implementation of both the encoder and the decoder is easy and straightforward. One possible implementation for the code defined in Eqs. (18-30) and (18-31) is shown in Fig. 18-9.

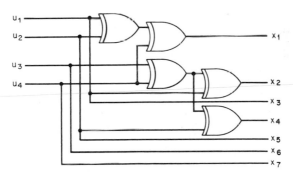

FIG. 18-9 Possible implementation of (7,4) Hamming code.

Polynomial Codes. As stated earlier, an n-tuple can be represented as a polynomial of degree $n - 1$. By use of this notation, an (n, k) code can be defined as a set of all polynomials of degree $n - 1$ or less which contain $\mathbf{g}(x)$, an $n - k$ degree polynomial as a factor. For example, a (6, 3) code can be described by the following equation:

$$\mathbf{c}(x) = (1 + x + x^3)(a_0 + a_1 x + a_2 x^2) \qquad (18\text{-}33)$$

Here $\mathbf{g}(x) = 1 + x + x^3$. The polynomial $\mathbf{c}(x)$ can be expanded as

$$c(x) = a_0 + (a_0 + a_1)x + (a_1 + a_2)x^2 + (a_0 + a_2)x^3 + a_1 x^4 + a_2 x^5 \qquad (18\text{-}34)$$

One realization of this code is shown in Fig. 18-10. This figure also shows the content of the shift registers as a function of clock pulses.

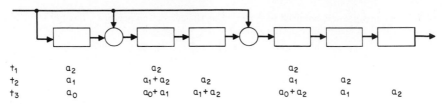

t_1	a_2	a_2		a_2		
t_2	a_1	$a_1 + a_2$	a_2	a_1	a_2	
t_3	a_0	$a_0 + a_1$	$a_1 + a_2$	$a_0 + a_2$	a_1	a_2

FIG. 18-10   Shift register coder. This circuit realizes the (6,3) code using $\mathbf{g}(x) = 1 + x + x^3$ as generator polynomial.

For certain values of n, the polynomial codes are *cyclic*; that is, a cyclic permutation of the code words is also a code word. This is shown in Fig. 18-11.

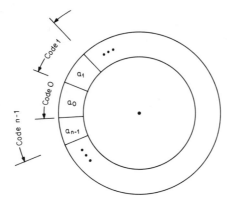

FIG. 18-11   Cyclic codes.

In general a polynomial code (n, k) is cyclic if the generator polynomial $\mathbf{g}(x)$ divides $1 + xn$ but not $1 + x^n$ where $n < \tilde{n}$. For example, our previous generating polynomial $\mathbf{g}(x) = 1 + x + x^3$ divides $1 + x^7$; that is

$$1 + x^7 = (1 + x)(1 + x + x^3)(1 + x^2 + x^3) \tag{18-35}$$

Therefore the (4, 7) code generated by $\mathbf{g}(x)$ is cyclic. The relationship between the code and the message is shown in Table 18-2.

For cyclic codes a *parity check polynomial* may be defined as

$$\mathbf{h}(x) = \frac{1 + x^n}{\mathbf{g}(x)} \tag{18-36}$$

The property of $\mathbf{h}(x)$ is that if $\mathbf{c}(x)$ is a code word, then $\mathbf{c}(x)\mathbf{h}^*(x) = 0$. [The $*$ means the reversing of the order of the polynomial; for example, $h(x) = 1 + x + x^2 + x^4 \rightarrow \mathbf{h}^*(x) = x^4 + x^2 + x + 1$. Incidentally, $(1 + x^7)/(1 + x + x^3) = 1 + x + x^2 + x^4$.]

An alternate way of writing the generator polynomial for an (n, k) code is

$$\mathbf{g}(x) = \mathrm{LCM}\{P_1(x)P_2(x) \cdots P_t(x)\} \tag{18-37}$$

where LCM stands for *least common multiplier* and $P_i(x)$ are the odd roots of $1 + x^n$; that is

$$1 + x^n = P_1(x)P_2(x) \cdots P_n(x) \tag{18-38}$$

Let us consider some important polynomial codes which are capable of correcting multiple errors.

Table 18-2   Example of Cyclic Code†

$g(x)$	Message	Code	Class
1101	0000	0000000	$c_1 = 0000000$
	1000	1101000	
	0100	0110100	
	0010	0011010	
	0001	0001101	
	1110	1000110	
	0111	0100011	
	1101	1010001	$c_2 = 1101000$
	1100	1011100	
	0110	0101110	
	0011	0010111	
	1111	1001011	
	1001	1100101	
	1010	1110010	
	1010	0111001	$c_3 = \overline{c_2}$
	1011	1111111	$c_4 = \overline{c_1}$

† This (7, 4) code is generated by the polynomial $g(x) = \{1101\}$.

Galoy Code.   Galoy discovered a (23, 11) code, which can correct for three errors. This can have either of the following two generator polynomials

$$g(x) = x^{11} + x^9 + x^6 + x^5 + x^4 + x^2 + 1 \tag{18-39}$$

or

$$g(x) = x^{11} + x^{10} + x^6 + x^5 + x^4 + x^2 + 1 \tag{18-40}$$

The Galoy code belongs to a larger class of codes, called the BCH codes, explained below.

BCH Codes.   The BCH codes (named after Bose, Chaudhuri, and Hocquenghem) are multiple-error correcting codes. Currently, these codes show great promise for video applications. They are generalized Hamming codes which can be defined in terms of roots of generator polynomials. One such definition is as follows.
$g(x)$ is a generator polynomial for a BCH code if

- It is a monic polynomial with degree $r \geq 1$ with coefficients over GF(q)
- n is the smallest integer for which $g(x)$ divides $x^n - 1$ (hence the code is cyclic)
- It is the lowest degree of polynomial to have distinct roots $\alpha^{m_0}, \alpha^{m_0+1}, \ldots, \alpha^{m_0+d_0-2}$, where m_0 is an arbitrary integer and $d_0 \geq 2(d_0)$ is a lower bond on the minimum distance of the code $d_{\min}$ and is termed the *design distance*

For binary codes, that is, for $q = 2$, the simplification of setting $m_0 = 1$ and selecting α to be a primitive element of GF(2^m) is usually adopted. In such case, the order of α, $2^m - 1$ is the length of the code word n.

The BCH codes form an infinite set. However, the most important generator polynomials are tabulated. The first few elements of such a table are given in Table 18-3.

Table 18-3 BCH Code-Generating Polynomials

n	h	t	g(x)†
7	4	1	13
15	11	1	23
	7	2	721
	5	3	2467
31	26	1	45
	21	2	3551
	16	3	107657
	11	5	5423325
	6	7	313365047
63	57	1	103
	51	2	12471
	45	3	1701317
	39	4	166623567
	36	5	1033500423
	30	6	157464165547
	24	7	17323260404441
	18	10	1363026512351725
	16	11	6331141367235453
	10	13	472622305527250155
	7	15	5231045543503271737
127	120	1	211
	113	2	41567
	106	3	11554743
	99	4	3447023271
	92	5	624730022327

n	k	t	g(x)†
255	171	11	15416214212342356077061630637
	163	12	7500415510075602551574724514601
	155	13	3757513005407665015722506464677633
	147	14	1642130173537165525304165305441011711
	139	15	461401732060175561570722730247453567445
	131	18	2157133314715101512612502774421420241 65471
	123	19	1206405224206600371721032651161412262 72506267
	115	21	6052666557210024726363640460027635255 6313472737
	107	22	2220577232206625631241730023534742017 6574750154441
	99	23	1065666725347317422274141620157433225 2411076432303431
	91	25	6750265030327444172723631724732511075 5507627072434561
	87	26	11013676341147432364352316343071720462 0672254527331172317
	79	27	6670003563765750002070344420736617462 1015326711766541342355
	71	29	2402471052064432151555417211233116320 5444250362557643221706035
	63	30	1075447505516354432531521735770700366 6111726455267613656702543301

n	k	t	g(x) (octal)†
	85	6	13070447632273
	78	7	2623000216613115
	71	9	6255010713253127753
	64	10	1206534025570773100045
	57	11	335265252505705053517721
	50	13	54446512523314012421501421
	43	14	17721772213651227521220574343
	36	15	31460746665220750447645745721735
	29	21	4031144613676706036675301411176155
	22	23	12337607040472252243545626637647043
	15	27	22057042445604554770523013762217604353
	8	31	7047264052751030651476224271567733130217
	247	1	435
	239	2	267543
	231	3	156720665
	223	4	75626641375
	215	5	23157564726421
	207	6	16176560567636227
	199	7	7633031270420722341
	191	8	2663470176115333714567
	187	9	52755313540001322236351
	179	10	22624710717340432416300455
255	55	31	731542520350110013301527530602054325414326755010557044426035473617
	47	42	2533542017062646563033041377406233175123334145446045005066024552543173
	45	43	1520205605234161131101346376423701563670024470762373033202157025051541
	37	45	5136330255067007414177447245437530420735706174323432347644354737403044003
	29	47	30257155366730714655270640123613771153422423242011741140602547574104035 65037
	21	55	1256215257060332656001773153607612103227341405653074542521153121614466513473725
	13	59	464173200505256544442657371425006600433067744547656140317467721357026134460500547
	9	63	15726025217472463201031043255351346141623672120440745451127661155477055616775116057

†To save space, the coefficients of $g(x)$ are shown in octal notation, 173 denotes 1 111 101.

Source: Reprinted from John P. Stenbit, "Table of Generators for Bose-Chaudhuri Codes," *IEEE Trans. Info. Theory*, vol. IT-10, no. 4, p. 391, October 1968. Copyright © 1964 IEEE.

18.17

Reed-Solomon (RS) codes are a subset of the BCH codes defined over $GF(q^m)$ where $m = 1$ and $q \geq 2$. Those codes have special ability to correct multiple bursts of error in a code word.

Interleaved Codes. Many digital communication channels, storage media, etc., have relatively low average error rates. However, when errors occur—because of the mechanism causing the errors or the modulation used to transmit and/or store the signal—they occur in clusters or bursts. Block codes with reasonable length and complexity can correct bursts of only moderate length. Nevertheless, by a technique called *interleaving*, practical block codes can be used to correct error bursts occurring in physical systems. Figure 18-12 shows an example of the principle of code interleaving which is as follows. Assume that we know that the digital channel we are using is via a satellite link employing 8 PSK modulation. This means that a single error in transmission destroys four consecutive bits; i.e., the error occurs in bursts of length four. Further, we wish to use a coder (shown on Fig. 18-9) producing a (7, 4, 1) Hamming code. To correct for the burst, we first read n (say 7) words into a 7×7 buffer one row at a time. Next, we read and transmit the content of the memory column by column.

At the receiver the process is reversed. Thus if a burst occurs while the fourth column is being transmitted, a single error will occur at the fourth bit of four of the code words. As we have seen earlier, the code we are using can correct this single error. In other words, the interleaving technique spread a burst of error of length n over n code words. The price paid is the cost of memory and the delay caused by buffering the signal at both ends of the channel.

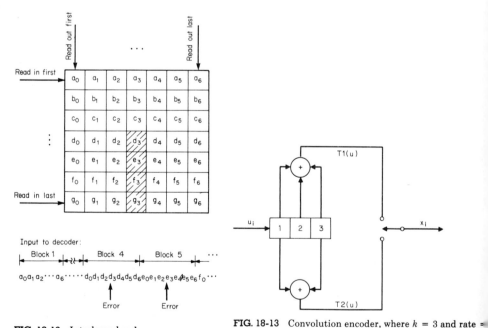

FIG. 18-12   Interleaved code.

FIG. 18-13   Convolution encoder, where $k = 3$ and rate =

Convolution Codes. The second major group of FEC codes is the so-called *convolution* or *tree* codes. These schemes are very effective when the major system requirement is energy efficiency (such as in the case of power-limited satellite communication channels). In digital video systems—where bit rates are inherently high—bandwidth efficiency is usually more important than power limitation. Thus convolution codes proba-bly could play a less significant role in video than block codes. Therefore, convolution codes will be given only a very brief coverage here. More information pertaining to con-volution codes is readily available in the literature.[7-9,11]

Figure 18-13 depicts the operation of a convolution coder. The input and output con-volution coders are semi-infinite sequences of symbols rather than finite-length blocks. An example for the coder in Fig. 18-13 is shown in Table 18-4. The input bits are clocked into a $K = 3$ stage shift register. K is referred to as the *constraint length* of the code. The content of the shift register is used to generate n (= 2) linear algebraic functions. These symbols (or bits) are outputted in time-multiplexed fashion.

Table 18-4   Example of States of Coder of Fig. 18-13

i	u_i	SR	T_1 ($G_1 = 111$)	T_2 ($G_2 = 101$)	x_i
0	1	100	1	1	(1, 1)
1	0	010	1	0	(1, 0)
2	1	101	0	0	(0, 0)
3	1	110	0	1	(0, 1)
4	0	011	0	1	(0, 1)
.	.	.	.	.	.
.	.	.	.	.	.
.	.	.	.	.	.

†Input is {1, 0, 1, 1, . . . }.

A conceptually useful way of representing this coding process is by the so-called tree diagram (see Fig. 18-14). This graph is constructed as follows. At each new information bit the graph branches either up—for a zero—or down—for a 1. The output code gen-erated is written on the various branches. For example, the code used in Table 18-4 for (10110···) is shown as a heavy line on the tree (down-up-down-down- ···). The disad-vantage of this graph is that it grows exponentially with the length of the code word.

One important characteristic of a tree diagram is that the branches following certain nodes are identical. These nodes, denoted as *states,* are shown in Fig. 18-14 as a, b, c, and d. Thus, by merging similar states or nodes, a new graph can be constructed which does not grow beyond 2^{K-1} nodes where K is the constraint length. This graph is called the *trellis* and is shown for the previous example in Fig. 18-15. This diagram contains all possible semi-infinite codes generated by the coder shown above.

Yet another way of describing a convolution code is via a *state diagram* as shown in Fig. 18-16. This depicts the transition of states caused by an information bit. The mes-sage can be recovered from a coded sequence (possibly corrupted by random errors) by locating the path through the trellis which has the highest weight or cross-correlation with the code sequence. The *path weight* is defined at each node as the accumulative sum of weights $w_i = p_{1i}r_{1i} + p_{2i}r_{2i}$ of the preceding links where p and r are parity bits associated with the ith link of the trellis and the ith segment of the code, respectively. (*Note:* Zeros are replaced by -1s.)

Consider, as an example, the correct path through the trellis for an error-free code word (11 01 00 01 11 00 · · ·). The parity bits for the path a-b-c-b-d-c-b (see Fig. 18-15) are 11 01 00 01 11 00 · · ·, obviously the same as the code word. Thus the weight of this path at the sixth set of nodes is $w = (1 \cdot 1 + -1 \cdot -1) + (-1 \cdot -1 + 1 \cdot 1) + \cdots = 12$.

Clearly, one way of decoding any sequence is by computing all $4(2^m)$ weights corre-sponding to all possible paths to each of the four nodes at $t = m$. Obviously, this process becomes impractical for larger m.

Viterbi suggested a major simplification for the decoding process by noting that for each node only two paths are possible and that only the one with the highest weight

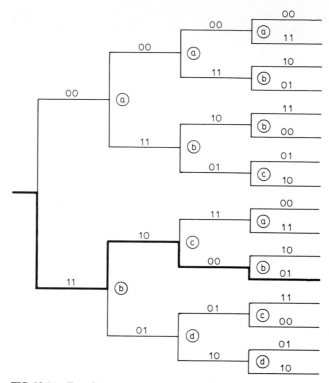

FIG. 18-14 Tree diagram of a convolution coder shown in Fig. 18-13.

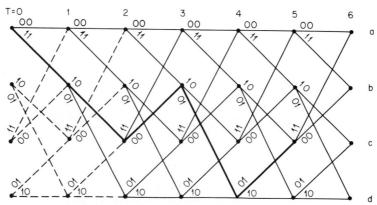

FIG. 18-15 Trellis for example in Fig. 18-13.

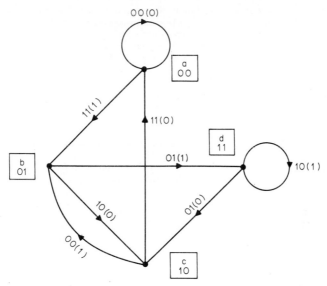

FIG. 18-16   State diagram for convolution code shown in Fig. 18-13.

needs to be retained at any node. This process is shown on Fig. 18-17. A general result is that in an error-producing channel all surviving paths will originate from a single node at $T = 5K$. Thus data decision can be made after $5K$ successive nodes where K is the code constraint length. $5K$ is called the *coding delay*.

Error Concealment.   In the previous section, various coding schemes were described which had the ability of detecting errors caused by transmission, storage, processing, etc., of digital sequences. It was also shown that certain codes can detect that a limited number of errors occurred and their exact location. These errors can be corrected by simply *complementing* (replacing a 0 by a 1 or vice versa) the bit in error. A general characteristic of FEC codes is that they can correct a certain number of errors and can detect a

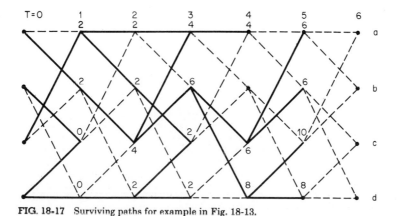

FIG. 18-17   Surviving paths for example in Fig. 18-13.

larger number of errors in a given block. Simpler codes, such as the parity check code above, can correct zero error and detect one error per block.

The way the knowledge about the existence of an error (within a block) is utilized is dependent upon the nature of the digital signals. If this signal represents digital data [such as used, for example, for electronic fund transfer (EFT)], each data bit is critical (e.g., the location of the decimal point in the dollar value of the transaction). When errors are detected in such signals, the block containing them is simply not used (e.g., in EFT the receiver requests the retransmission of the block).

In the case of video signals—which contain a large amount of redundancy—individual bits usually do not contain vital information. This allows the system designers to exploit the knowledge of errors in the signal to maintain acceptable signal quality. The technique used is known as *error concealment* and consists of hiding the error. Suppose, for example, the detector indicates that the digital word corresponding to a *pixel* (i.e., a video sample) contains an error that cannot be corrected. Simply ignoring the error could potentially degrade the picture since the probability that the most significant bit is in error is as high as ⅛. However, experimental evidence indicates that replacing a pixel by an interpolated one every once in a while leaves the video signal unimpaired.

Various interpolation techniques used in error concealment include previous pixel repetition, extrapolation using two or more previous pixels, and interpolation between preceding and following samples.

18.2 DIGITAL TELEVISION SIGNALS

Video signals are band limited (nominally to 4.2 MHz in 525-line countries and to 6.0 MHz in regions using 625 lines). Therefore, the theory discussed in Sec. 18.1 is directly applicable.

Television signals are highly structured. Consequently, quantization errors or errors due to finite precision processing can result in perceptible impairments (e.g., contouring) if proper care is not used. Another consequence of this structured nature of television is that video signals are highly redundant; i.e., a large percentage of the signal could be derived from other parts, facilitating bit rate reduction if the sampling parameters are chosen judiciously.

The digital coding process selected depends largely on the intended applications of the signals. These are discussed in some detail in the next two sections.

18.2.1 DIGITIZED VIDEO.
Today's analog television systems contain a large number of digital subsystems. The signals used by these are digitized representations of the composite NTSC, PAL, or SECAM television signals and are referred to as digitized video.[14] In this section the salient features of the digitized video signals are presented.

Three basic factors are pertinent to the digital coding process. These are:

- sampling
- number of bits per sample
- quantization

Sampling. "The task of a television system is to communicate a three-dimensional function: the time-varying scene image by the camera. Although it is necessary to sample (scan) this function and map it into a one-dimensional signal for transmission and display purposes, the basic signal remains three-dimensional."[12] Thus the analog signal, whether component or composite, is already sampled in the vertical and temporal dimension by the scanning process. Therefore to consider the spectrum of a video signal for digitization purposes as a one-dimensional phenomenon is inadequate.

Sampling of the video signal to obtain digital television signal results in a three-dimensional sampling grid. For efficient sampling the number of sampling points per unit volume should be as low as possible without introducing aliasing. Most three-dimensional sampling grids of interest can be constructed by superposition of rectangular grids. This implies that the horizontal sampling frequency is some multiple of the color subcarrier since both the line rate and the field rate are derived from this frequency. The most common sampling grids are as follows:

- *Field aligned.* This scheme consists of a sampling grid with rectangular projection both in the spatial and the horizontal-temporal direction as shown in Fig. 18-18a. For NTSC signals the sampling frequency must exceed 8.4 MHz to avoid aliasing. The lowest common multiple of f_{sc} which meets this requirement is 3.

- *Field offset.* The sampling grid for this pattern is shown in Fig. 18-18b. This scheme is more efficient from a spectrum density point of view, and therefore $2f_{sc}$ sampling for NTSC signals may be acceptable.

- *Checkerboard or line quincunx.* This sampling grid is obtained by offsetting the sampling pattern from one line to the next by half-horizontal-sampling intervals. This is shown in Fig. 18-18c.

- *Double checkerboard sampling.* This pattern has good spectral properties and was recommended for $2f_{sc}$ sampling of NTSC signals.[13] The details of this are shown in Fig. 18-18d.

Number of Bits per Sample.   Several subjective studies indicate that 8 bits per sample are adequate to produce acceptable quality video signals. This resolution, however, may not be sufficient to resolve various test or data signals coded in the vertical blanking

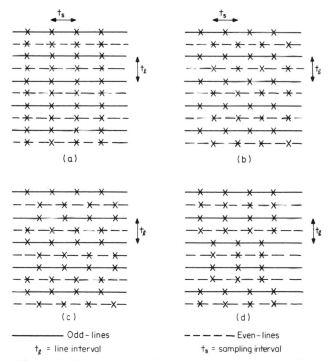

FIG. 18-18 Rectangular sampling patterns: (a) rectangular field-aligned; (b) rectangular field-offset; (c) checkerboard; (d) field-aligned, double checkerboard.

interval. If these signals should also be digitized, a quantization of 9 bits per sample is preferred to ensure the integrity of these test signals.

Quantization. Although when bandwidth reduction techniques are used (to be covered later), the quantizers employed could be nonuniform or tapered, for the primary digital coding the uniformly spaced quantization or decision levels are preferred. The resulting signal is referred to as *uniformly quantized PCM.* Although this quantization scheme is not the most efficient (from the signal to quantization noise point of view), it is popular because of the simplicity of implementation and the ease of processing the resulting signals.

Summary of Digitized Video Parameters. The most important parameters of digitized video are summarized in Table 18-5.

Table 18-5 Parameters of Digitized Television

	525-Line systems	625-Line systems
Active lines	485/s	575/s
Vertical sync	7.6% of lines	8.0% of lines
Horizontal blanking	16.0% line period	18.0% line period
Frame rate	30 Hz	25 Hz
Line frequency	15,750 Hz	15,625 Hz
Bandwidth	4.5 MHz	5.5 MHz
Active line period	53 μs	52 μs
Subcarrier	3.579545 MHz	4.4336187 MHz
	$= (227 + \frac{1}{2})$fl	$= (284 - \frac{1}{4})$fl-25
Bit rate at 3 f_{sc}	85.90908 Mbits/s	N/A
Bit rate at 4 f_{sc}	114.54544 Mbits/s	141.8758 Mbits/s
Bit rate at 3 f_{sc} active pels only	66.679192 Mbits/s	N/A
Bit rate at 4 f_{sc} active pels only	88.905589 Mbits/s	114.01139 Mbits/s

18.2.2 ALL-DIGITAL VIDEO. *All-digital video* refers to the signals used in television systems which are digital from the camera output to the transmitting antenna.[15-17] Such systems exist today, only in space transmission of video, but are planned for the future, and a large number of parameters are standardized or being negotiated by various users both nationally and internationally. For all-digital video the quantizer is presented with either the luminance and color-difference signals (Y, I, Q or Y, U, V) or the red, green, and blue signals (R, G, B) rather than the composite NTSC, SECAM, or PAL. These signals are commonly referred to as *component* video signals. The voltage levels assigned to the red, blue, and green are denoted by E_R', E_B', and E_G', respectively.

Luminance E_Y' and *color-difference* signals $E_B' - E_Y'$ and $E_R' - E_Y'$ are defined as follows:

$$E_Y' = 0.299E_R' + 0.587E_G' + 0.114E_B'$$

$$(E_R' - E_Y') = 0.701E_R' - 0.587E_G' + 0.114E_B' \qquad (18\text{-}41)$$

$$(E_B' - E_Y') = -0.299E_R' - 0.587E_G' + 0.886E_B'$$

For 1-V maximum signal levels the values of black, white, and prime colors are shown in Table 18-6. The data in this table indicate that the excursion of the color-difference signals is larger than unity (for example, $E_R' - E_y'$ can range from -0.701 to 0.701 V).

Table 18-6 Values for Normalized White, Black, and Saturated Primary and Complementary Colors

Condition	E'_R	E'_G	E'_B	E'_Y	$E'_R - E'_Y$	$E'_B - E'_Y$
White	1.0	1.0	1.0	1.0	0.0	0.0
Black	0.0	0.0	0.0	0.0	0.0	0.0
Red	1.0	0.0	0.0	0.299	0.701	−0.299
Green	0.0	1.0	0.0	0.587	−0.587	−0.587
Blue	0.0	0.0	1.0	0.114	−0.114	0.886
Yellow	1.0	1.0	0.0	0.886	0.114	−0.886
Cyan	0.0	1.0	1.0	0.701	−0.701	0.299
Magenta	1.0	0.0	1.0	0.413	0.587	0.587

To remedy the situation, the color-difference signals are renormalized as

$$C'_R = 0.713(E'_R - E_Y) = 0.701E'_R - 0.587E'_G + 0.114E'_B$$
$$C'_B = 0.564(E'_B - E_Y) = -0.299E'_R - 0.587E'_G + 0.886E'_B$$

(18-42)

Since digital video standards are to be aimed at future as well as present system needs, the CCIR recommends that an extensible family of compatible digital codes should be adopted. The members of this family will have spatially static sampling patterns with two simultaneous color-difference signals spatially cosited except when the signal represents red, green, and blue signals, in which case all three samples will be cosited. Such a sampling pattern is shown in Fig. 18-19.

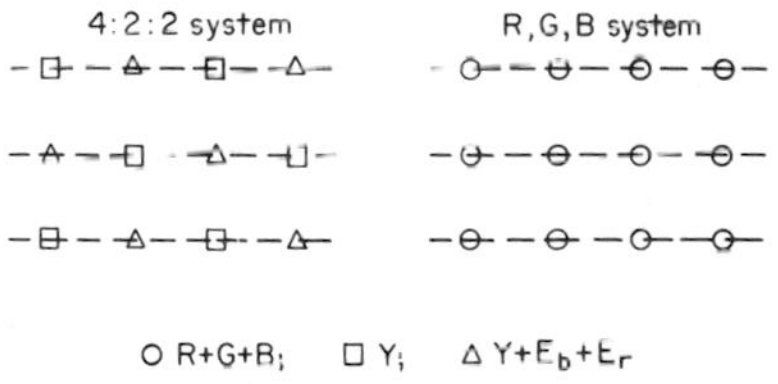

Several members of this extensible family of codes are under consideration. One member which has universal support is the so-called 4:2:2 system. The nomenclature originates from the fact that the sampling frequencies adopted for this member are "close" to $4f_{sc}$ and $2f_{sc}$ in the NTSC 525-line system. The 4 represents the sampling frequency of 13.5 MHz† and all ratios are derived from it.

FIG. 18-19 Digital video-sampling patterns.

Therefore, the 2 represents 6.75 MHz, etc. The 4 denotes the sampling rate used for the luminance and 2 refers to the sampling rates associated with the color-difference signals. The values for the encoding parameters for this member are presented in Table 18-7.

18.3 COMPONENTS OF DIGITAL TELEVISION SYSTEMS

18.3.1 FILTER THEORY. One important reason for using digital techniques in a television environment is the ease and efficiency of processing signals digitally. The theory of digital signal processing is, however, different from the conventional linear system theory. In this section, a basic introduction to digital signal processing (with emphasis on digital filter theory) is provided.

---

†In order to arrive at a compatible, international standard, 13.5 MHz was chosen, rather than the exact value for NTSC of 14.3 MHz.

Table 18-7 Encoding Parameters of Digital Television for Studios

Parameters	525-Lines 30-frame/s systems	625-Lines 25-frame/s systems
Coded signals	$Y, R - Y, B - Y$	
Sampling frequency for luminance signals	13.5 MHz	
Sampling frequency for color-difference signals	6.75 MHz	
Number of samples per digital active lines—luminance signals	720	
Number of samples per digital active line—color-difference	360	
Number of samples per total line—luminance signals	858	864
Number of samples per total line—each color-difference signal	429	432
Sampling structure	Orthogonal, line, field, and frame-locked $R - Y, B - Y$ samples cosited with odd Y samples	
Quantization	Uniformly quantized PCM for each signals	
Digital code†	The luminance Y signal is encoded in 220 quantization levels such that the binary word 16 corresponds to black and 235 to peak white. The color-difference signal will be encoded in 225 quantization levels symmetrically with respect to the 0 signal corresponding to the digital word 128.	

†*Note:* Digital words FF and 00 are reserved for timing-reference signaling.

18.3.2 DIGITAL SYSTEMS OR FILTERS. *Digital filters* (which are the building blocks of digital television) transform or map digital input signals into digital outputs. These signals represent *discrete-time* processes which usually arise as the result of sampling analog signals. Although the process of *filtering* is done using digital rather than discrete-time techniques, the latter approach is useful to facilitate an understanding of the principles involved. The effects of finite precision arithmetic, which is the basic difference between discrete-time and digital processes, are beyond the scope of this work but are well covered in excellent texts.[19,20]

Discrete-time signals are represented by sequences of numbers $\mathbf{x} = \{x(n)\}$, where $x(n)$ denotes the amplitude of the nth sample [however, for simplicity $x(n)$ is often used to refer to the sequence]. Some important sequences are defined in Table 18-8.

A digital filter maps an input sequence $x(n)$ onto an output sequence $y(n) = T[x(n)]$ as shown on Fig. 18-20. The practical and useful constraints usually imposed on $T[\ ]$ are linearity and shift invariance. These are defined as follows.

Linear Systems. If the discrete-time filter outputs are $y_1(n)$ and $y_2(n)$ for arbitrary inputs $x_1(n)$ and $x_2(n)$, respectively, then the system is linear if and only if the input $ax_1 + bx_2$ produces the output $ay_1 + by_2$.

Table 18-8 Selected Important Sequences

Definition	Name	Sequence
$\delta(n) = \begin{cases} 1 & n = 0 \\ 0 & n \neq 0 \end{cases}$	Unit impulse	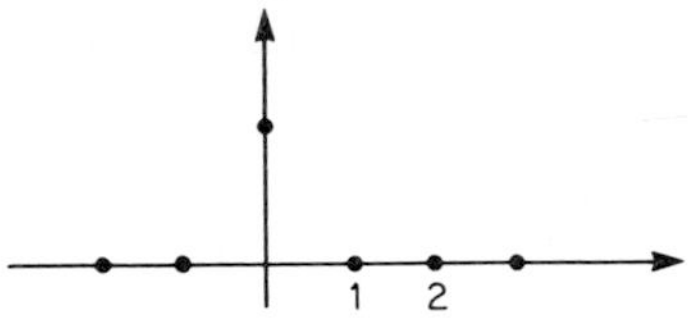
$\delta(n - m) = \begin{cases} 1 & n = m \\ 0 & n \neq m \end{cases}$	Shifted impulse	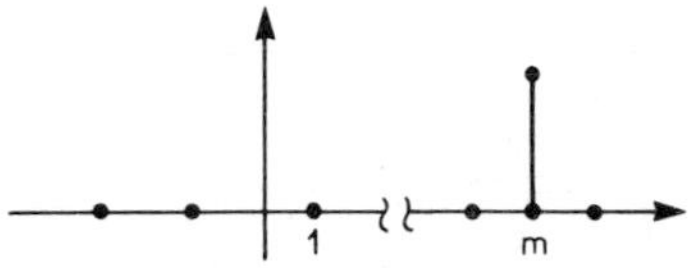
$\mu(n) = \sum_{k=0}^{\infty} \delta(n - k)$	Unit step	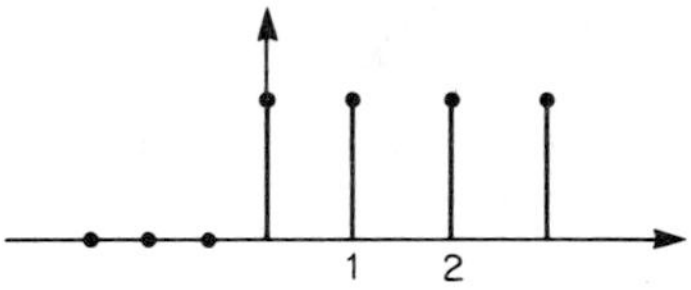
$a^{-k} = \sum_{k=0}^{\infty} a^{-k} \delta(n - k)$	Real exponential	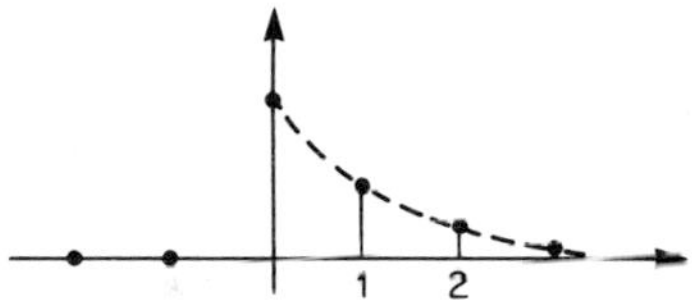
$h(n) = \cos{(2\pi n/n_0)}$	Sinusoidal	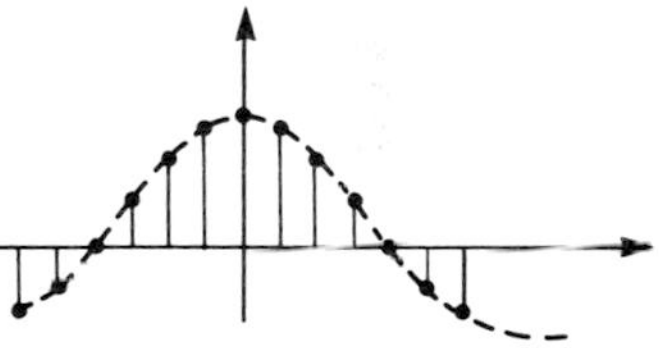

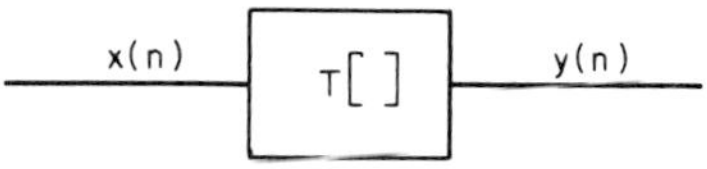

FIG. 18-20 Block diagram of a digital filter.

Shift Invariance. System with input/output relations given by $\mathbf{y} = T[x(n)]$ is shift invariant if the time displacement of the input produces similar displacement for the output; that is $y(n - k) = T[x(n - k)]$.

Two other useful concepts for discrete-time signals are *causality* and *stability*. A system is said to be causal if its output cannot anticipate the input and is stable if finite input cannot produce infinite outputs.

Unit sample response plays the same paramount role in digital systems as the impulse response does in linear system theory. It is defined as the sequence produced at the output of a linear shift invariant system as a result of a unit sample input. Mathematically

$$h(n) = T[\delta(n)] \qquad (18\text{-}43)$$

Since any arbitrary sequence can be written as a sum of weighted and shifted unit samples; that is, $\{x(n)\} = \Sigma x(m)\, \delta(n - m)$ using the linearity and shift invariance properties, the output produced by any input can be expressed in terms of unit sample response as

$$y(n) = \sum_{n=-\infty}^{\infty} x(k)h(n-k) \triangleq x(n) \star h(n) \tag{18-44}$$

The expression in Eq. 18-44 is referred to as the *convolution sum* and is analogous to the convolution integral of continuous time systems.

The linear shift invariant (LSI) system can be represented in the frequency domain by considering the input $x(n) = e^{-j\omega n}$. The output to this is [see Eq. (18-44)]

$$\begin{aligned} y(n) &= \Sigma h(k)e^{j\omega(n-k)} \\ &= e^{j\omega n}\Sigma h(k)e^{-j\omega k} \triangleq e^{j\omega n}H(e^{j\omega}) \end{aligned} \tag{18-45}$$

Since the input used above is a sampled sinusoidal (of frequency ω), $H(e^{j\omega})$ is referred to as the frequency response of the LSI system. Note that $H(e^{j\omega})$ is a periodic function of ω with period 2π. From this, Eq. (18-45) can be considered as a Fourier series expansion of $H(e^{j\omega})$ with $h(n)$ being the Fourier coefficients calculated in the usual way, i.e.

$$h(n) = \frac{1}{2\pi} \int_{-\pi}^{\pi} H(e^{j\omega})e^{j\omega n} \, d\omega \tag{18-46}$$

The Fourier coefficient pair relation in Eqs. (18-45) and (18-46) holds for any sequence; thus

$$X(e^{j\omega}) = \Sigma x(n)e^{-j\omega k}$$
$$x(n) = \frac{1}{2\pi} \int_{-\pi}^{\pi} X(e^{j\omega})e^{j\omega n} \, d\omega \tag{18-47}$$

are completely general and are referred to as the *Fourier transform* and *inverse Fourier transform* of $x(n)$ and $X(k)$, respectively. Since $x(n) = e^{jn\omega}$ produces a response $H(e^{j\omega})e^{j\omega n}$, the input in Eq. (18-44) produces the response

$$y(n) = \frac{1}{2\pi} \int_{-\pi}^{\pi} X(e^{j\omega})H(e^{j\omega})e^{j\omega} \, d\omega \tag{18-48}$$

Comparing Eqs. (18-44) and (18-48) shows that convolution in the time domain corresponds to multiplication in the frequency domain

$$Y(e^{j\omega}) = X(e^{j\omega})H(e^{j\omega}) \tag{18-49}$$

Another useful representation of sequences is the Z transform.

18.3.3 THE Z TRANSFORM.

The Z transform of a sequence $\{x(n); n = \ldots, -2, -1, 0, 1, 2, \ldots\}$ is defined as the power series in the complex variable z^{-1} whose coefficients are the members of the sequence. Explicitly

$$X(z) = \sum_{n=-\infty}^{\infty} x(n)z^{-n} \tag{18-50}$$

The Z transform in general converges in an annular space centered at the origin of the complex plain as shown in Fig. 18-21.

For causal sequences, $\{x(n); 0 < n < \infty\}$, the Z transform converges outside a circle containing all poles or singularities of x. For finite sequences the transforms converge everywhere, except on some occasions at $z = 0$ or ∞. Table 18-9 provides important examples of Z transforms.

The *inverse Z transform* $Z^{-1}[\ ]$ is defined as

$$x(n) = \frac{1}{2\pi j} \oint_C X(z)z^{n-1} \, dz \tag{18-51}$$

where the contour C encircles the origin within the region of convergence.

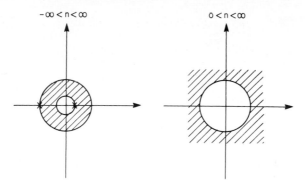

FIG. 18-21 Region of convergence for Z transforms (x denotes singularities).

Table 18-9 Some Important Z Transforms

Sequence	Z transform	Radius of convergence
$\delta(n - k)$	z^{-k}	0
$\mu(n)$	$1/(1 - z^{-1})$	1
a^n	$1/(1 - az^{-1})$	a
n	$z^{-1}/(1 - z^{-1})^2$	1

The various methods of calculation of the inverse Z transform include

residue theorem
partial fraction expansion
long division
power series expansion
table look-up

These will be demonstrated by a simple example. Consider the Z transform

$$X(z) = \frac{2z^2 - 0.5z}{z^2 - 0.5z - 0.5} \tag{18-52}$$

and evaluate its inverse transform $x(n)$.

1. *Residue theorem.* From complex variable theory, we know that

$$x(n) = \frac{1}{2\pi j} \oint_C X(z)z^{n-1} \, dz$$

$$= \text{residues of } X(z)z^{n-1} \text{ at all singularities within the contour } C \tag{18-53}$$

For Eq. (18-52), the poles are at $z = 1$ and $z = -\frac{1}{2}$; therefore

$$x(n) = X(z)z^{n-1}(z - 1)|_{z=1} + X(z)z^{n-1}(z + \frac{1}{2})|_{z=-1/2} \tag{18-54}$$

$$= 1.0 + (-\frac{1}{2})^n$$

2. *Partial fractions*

$$X(z) = \frac{z}{z-1} + \frac{z}{z+0.5}$$

$$= \frac{1}{1 - z^{-1}} + \frac{1}{1 + 0.5^{-1}} \tag{18-55}$$

which yields, after looking at Table 18-9,

$$x(n) = 1 + (-\tfrac{1}{2})^n \tag{18-56}$$

3. *Long division*

$$
\begin{array}{r}
2 + 0.5z^{-1} + 1.\,25z^{-2} + \cdots \\
1.0 - 0.5z^{-1} - 0.5z^{-2} \overline{\smash{)}\, 2 - 0.5z^{-1}} \\
\underline{2 - 1.0z^{-1} + 1.\,0z^{-2}} \\
0.5z^{-1} + 1.\,0z^{-2} \\
\underline{0.5z^{-1} - 0.\,25z^{-2} - 0.25z^{-3}}
\end{array}
$$

$$x(0) = 2 \qquad x(1) = 0.5 \qquad x(2) = 0.125 \qquad \cdots \tag{18-57}$$

The most important properties of the Z transform are summarized in Table 18-10. In addition, an important relationship exists between the Z transform and the Fourier transform of a sequence, namely, that the Z transform evaluated on the unit circle yields the Fourier transform since

$$X(z)|_{z=\exp j\omega} = X(\exp j\omega) = \sum_{n=-\infty}^{\infty} x(n) \exp(-j\omega n) \tag{18-58}$$

Table 18-10 Properties of the Z Transform

Property	Sequence	Z transform
Linearity	$ax(n) + by(n)$	$aX(z) + bY(z)$
Forward difference	$x(n+1) - x(n)$	$(z-1)X(z) - zx(0)$
Backward difference	$x(n) - x(n-1)$	$(1 - z^{-1})X(z) - x(-1)$
Shift	$x(n+k)$	$z^k X(z) - z^k \sum_{j=0}^{k-1} (j)z^{-j}$
"Frequency" scaling	$a^n x(n)$	$X(a^{-1}z)$
"Time" scaling	$nx(n)$	$z^{-1}(d/dz^{-1})X(z)$
Convolution	$y(n) = \sum_{k=0}^{n} h(n-k)x(k)$	$Y(z) = H(z)X(z)$
Multiplication	$x(n)y(n)$	$1/2\pi j \oint_c [(X(\omega)Y(z\omega^{-1})/\omega]\,d\omega$
Initial value	$f(0) = \lim_{z \to \infty} X(z)$	
Final value	$f(\infty) = \lim_{z \to 1} (1 - z^{-1})X(z)$	

18.3.4 DISCRETE FOURIER TRANSFORMS.

Sequences for which $x_p(n) = x_p(n + kN)$ for any value of k are periodic with period N. All periodic signals, including $x_p(n)$, can be expressed as a Fourier series

$$x_p(n) = \sum_{k=-\infty}^{\infty} \check{X}_p(k) \exp\left(-j \frac{2\pi}{N} nk\right) \tag{18-59}$$

Note, however, that only N of the exponential expressions for $k = 0, 1, 2, \ldots N - 1$ are different since $e^{j(2\pi/N)nk} = e^{j(2\pi/N)Nnk}$, etc. Thus the Fourier coefficients corresponding to the same exponent can be collected to yield

$$x_p(n) = \frac{1}{N} \sum_{n=0}^{N-1} X_p(k) \exp\left(j \frac{2\pi}{N} nk\right) \tag{18-60}$$

Using orthogonality properties of exponentials, $X_p(k)$ can be expressed in terms of $x_p(n)$ as

$$X_p(k) = \sum_{n=0}^{N-1} x_p(n) \exp\left(-j \frac{2\pi}{N} nk\right) \tag{18-61}$$

Equations (18-60) and (18-61) are called *discrete Fourier series* (DFS) pairs. The important properties of DFS are summarized in Table 18-11.

A finite duration sequence can be considered as one period of a periodic sequence, and its unique Fourier representation is the DFS of the periodic sequence. The nomenclature adopted to refer to this interpretation is *discrete Fourier transform (DFT)*. The properties of DFTs are shown in Table 18-12.

Table 18-11 Properties of the Discrete Fourier Series (DFS)

Property	Sequence	DFS
Linearity	$ax_p(n) + by_p(n)$	$aX_p(k) + bY_p(k)$
Shift	$x_p(n + m)$ $\times\ e^{-(2\pi/N)}x_p(n)$	$e^{j(2\pi/N)km}X_p(k)$ $\times X_p(k + 1)$
Convolution	$\sum_{m=0}^{N-1} x_p(m)y_p(n - m)$ $\times x_p(n)y_p(n)$	$X_p(k)Y_p(k)$ $\times\ (1/N)\sum_{l=0}^{N-1} X_p(l)Y_p(k - 1)$

Table 18-12 Properties of the (DFT) Discrete Fourier Transform†

Property	Sequence	DFT
Linearity	$ax(n) + by(n)$	$aX(k) + bY(k)$
Shift	$x((n + m))_N R_N(n)$ $\times\ e^{-(2\pi/N)}x(n)$	$e^{(2\pi/N)km}X(k) \times (X((k + l))_N R_N(k)$
Convolution	$\left[\sum_{m=0}^{N-1} x((m))_N y((n - m))_N\right] R_N(n)$ $\times\ x(n)y(n)$	$X(k)Y(k)$ $\times\ (1/N)\left[\sum_{k=0}^{N-1} X((K))_N Y((k - K))_N\right] R_N(k)$

†$R_N(n) - \mu(n) - \mu(n - N)$.
$x((n))_N = x(n)$ modulo N.

Digital Filter Structures. An important and easily tractable subset of LTI systems is that which can be described by *constant coefficient difference equations.* (The remainder of this section will be constrained to deal with such systems.) The general form which describes this subset is

$$\sum_{k=0}^{N} b_k y(n - k) = \sum_{r=0}^{M} a_r x(n - r) \tag{18-62}$$

$$Y(z) \sum_{k=0}^{N} b_k z^{-k} = X(z) \sum_{r=0}^{M} a_r z^{-r} \tag{18-63}$$

from which follows that the transform corresponding to the frequency response of a filter $H(z)$ is a ratio of two polynomials

$$H(z) = \frac{X(x)}{Y(z)} = \frac{\displaystyle\sum_{r=0}^{M} a_r z^{-r}}{\displaystyle\sum_{k=0}^{N} b_k z^{-k}} \tag{18-64}$$

The roots of the numerator polynomials are called the *zeros* of the filter, and the roots of the denominator are referred to as the *poles* of the system. We can assume without loss of generality that b_0 in Eq. (18-64) is 1. Then

$$y(n) = \sum_{r=0}^{M} a_r x(n - r) - \sum_{k=1}^{N} b_k y(n - k) \tag{18-65}$$

A simple realization of this filter is shown in Fig. 18-22. This structure is sometimes referred to as the first canonical form. The number of delay elements may be reduced (by as much as a factor of 2 if $N = M$) by an alternate realization called the second canonic form as shown on Fig. 18-23.

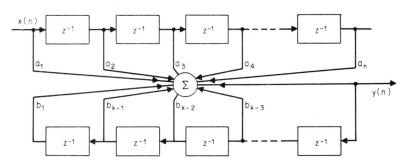

FIG. 18-22 First canonic form. z^{-1} denotes delay elements; a and b denote multiplying constants.

Another realization of the same system is possible by rewriting $H(z)$ as

$$H(z) = \frac{Y(z)}{X(z)} = a_0 \prod_{i=0}^{k} H_i(z) \tag{18-66}$$

where

$$H_i(z) = \frac{1 + a_{1i} z^{-1} + a_{2i} z^{-2}}{1 + b_{1i} z^{-1} + b_{2i} z^{-2}} \tag{18-67}$$

or

$$H_i(z) = \frac{1 + a_{1i} z^{-1}}{1 + b_{2i} z^{-1}} \tag{18-68}$$

and k equals the integer part of $(N + 1)/2$. Equation (18-67) is called a second-order section and Eq. (18-68) a first-order section. Both of these can be realized by either of the above two canonic structures with *real* coefficients. The total filter is obtained by cascading the first- and second-order sections as shown in Fig. 18-24.

One advantage of this method is that the poles and zeros can be paired in such a way as to minimize the effects of finite word length arithmetic in practical situations. By duality (or algebra) a parallel implementation of the same filter is possible as indicated in Fig. 18-25. (Note that the frequency responses H_1 in Figs. 18-24 and 18-25 are not the same.)

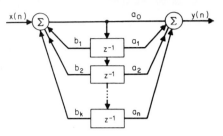

FIG. 18-23   Second canonic form.

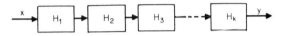

FIG. 18-24   Cascaded first- and second-order sections.

All-Zero Filters. An interesting special case of Eq. (18-64) is when the denominator is unity; i.e.

$$H(z) = \sum_{n=0}^{N-1} h(n)z^{-n} \qquad (18\text{-}69)$$

The difference equation describing this filter is

$$y(n) = \sum_{j=0}^{N-1} h(j)x(n - j) \qquad (18\text{-}70)$$

The easiest implementation of Eq. (18-70) is shown in Fig. 18-26. This is essentially a *tapped delay line,* and the common name for such a structure is *transversal filter.* Obviously other structures can also realize Eq. (18-70). These can be found, for example, in Refs. 19 and 20. **Transversal filters are a subset of a larger class of filters called FIR filters, discussed in the next section.**

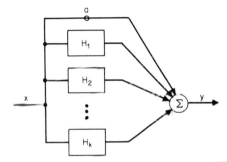

FIG. 18-25   Parallel first- and second-order sections.

18.3.5 FINITE IMPULSE RESPONSE FILTERS.

Digital filters are usually categorized into two classes, depending on the duration of their impulse response: *finite impulse response (FIR)* filters and *infinite impulse response (IIR)* filters. The impulse responses of these are $\{h_1(n)\} = \sum_{n=0}^{N-1} h_1(n)$ and $\{h_2(n)\} = \sum_{n=0}^{\infty} h_2(n)$, respectively.

FIR filters are stable and easily "implementable" with good (round-off) noise perfor-

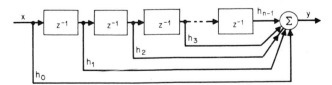

FIG. 18-26   All-zero or transversal filter.

mance but usually require large values of N for practical filters. The most attractive aspect of FIR filters is that they can be designed with exactly *linear* phase.

It can be shown that for linear-phase FIR filters, i.e., for $H(e^{j\omega}) = \pm |H(e^{j\omega})| e^{j\theta(\omega)}$ where $\theta(\omega) = \alpha = (N-1)/2$, the impulse response has the symmetry property that $h(n) = h(N-1-n)n$, $0 \le n \le N-1$ [with the center of symmetry at the midpoint of $\{h(n)\}$ whether or not this point is a sample location]. The delay is $\alpha = (N-1)$. Note that for even N, the filter delay is not an integer number of samples.

For constant group delay† filters, i.e.

$$H(\exp j\omega) = \pm |H(\exp j\omega)| \exp[j(\beta - \alpha\omega)] \tag{18-71}$$

the impulse response is antisymmetric with respect to the center of the samples; that is, $h(n) = -h(N-1-n)$, $0 < n \le N-1$. The delay is again $\alpha = (N-1)$ and $\beta = \pm \pi/2$. The amplitude responses of both constant phase and constant group delay filters, for both even and odd N, are shown in Table 18-13.

Table 18-13   Frequency Response of Linear-Phase FIR Filters

	Symmetric	Antisymmetric
Odd	$\left[\displaystyle\sum_{n=0}^{(N-1)/2} a(n) \cos (\omega n) \right] e^{-j\omega(N-1)/2}$	$j\left[\displaystyle\sum_{n=1}^{(N-1)/2} c(n) \sin (\omega n) \right] e^{-j\omega(N-1)/2}$
Even	$\left[\displaystyle\sum_{n=1}^{N/2} b(n) \cos (\omega(n-1)/2) \right] e^{-j\omega(N-1)/2}$	$j\left[\displaystyle\sum_{n=0}^{N/2-1} d(n) \sin (\omega(N/2 - n - 1)/2) \right]$ $\times e^{-j\omega(N-1)/2}$

The zeros of these filters appear in complex conjugate pairs, and each zero has another zero associated with it which is its mirror image with respect to the unit circle. Thus arbitrary zeros exist in quadruplets, zeros not on the real axes but on the unit circle are in pairs, while zeros at 0 and π are single zeros. This property is shown in Fig. 18-27.

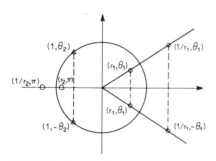

FIG. 18-27   Pole configuration of linear-phase FIR filters.

Windowing. The long-established practice of reducing the cause of Gibb's oscillation, the phenomenon caused by the time domain truncation of signals, is the use of "window" functions which smooth the discontinuities at the truncation boundaries.

A multitude of windows have been suggested by various authors. Unfortunately however, no uniform criteria exist by which an optimum window can be selected. That is, windows which yield the best result for one type of application may be inferior for others. Thus, the usefulness of various windows should match their requirements.

The windowing process may be described mathematically as

$$f(t) = s(t)w(t) \tag{18-72}$$

in the time domain, or

$$F(\omega) = S(\omega) \star W(\omega) \tag{18-73}$$

---

†*Group delay* is defined as the time derivative of the phase of $H\exp(j\omega)$.

in the frequency domain, where $w(t)$ denotes the window function of truncation, which is time limited, and $s(t)$ is the time function being truncated.

Equations (18-72) and (18-73) describe continuous systems and have to be modified to accommodate sampling, which is necessary for digital processing.

If # defines the operation

$$G\#(\omega) = \frac{1}{t_s} \sum_{k=-\infty}^{\infty} G\left(\omega - \frac{2\pi k}{t_s}\right) \qquad (18\text{-}74)$$

then

$$\tilde{F}(\omega) = F\#(\omega) + \{f(0+) - f(T-) \exp(-j\omega T)\} \qquad (18\text{-}75)$$

can be shown to be the Fourier transform of the sampled version of $f(t)$, where $t_s = T/N$ is the sampling internal and T is the truncation time $[w(t) = 0$ for $t < 0$ and $t_s > T]$. The DFT of the sequence $fn = f(nt_s)$, $n = 0, \ldots, N = 1$ is

$$F_n = \tilde{F} \frac{2n\pi}{T} \qquad (18\text{-}76)$$

Furthermore, if the usual convergence conditions are met, then

$$F\#(\omega) = [S(\omega) \star W(\omega)]\#$$
$$= S(\omega) \star W\#(\omega) \qquad (18\text{-}77)$$

If one assumes that $h(0+) = h(T-)$, then the transform pair of the sampled window function is

$$W\#(\omega) = \sum_{n=0}^{N-1} w(nt_s) \exp(-jn\omega t_s) - \frac{1}{2}\{1 - \exp(j\omega T)\}w(0+) \qquad (18\text{-}78)$$

$$w_n = w(nt_s)$$
$$= \frac{1}{N} \sum_{k=0}^{N-1} W\#\left(\frac{2\pi k}{T}\right) \exp\left(\frac{j2\pi nk}{n}\right) \qquad (18\text{-}79)$$

Thus the window function may uniquely be defined by any of the following: $w(t)$, w_n, $W(\omega)$, $W\#(\omega)$.

Let us now examine the structure of the window function in detail. By virtue of the sampling theorem, the spectrum of a time-limited signal [that is, $w(t) = 0$ for $t < 0$ and $t > T$] may be expressed as

$$W(\omega) = T \exp\left(\frac{-j\omega T}{2}\right) \sin\left(\frac{\omega T}{2}\right) \sum_{n=-\infty}^{\infty} \frac{C_n}{\omega T/2 - n\pi} \qquad (18\text{-}80)$$

where the coefficients of expansion are

$$C_n = \frac{1}{T} W\left(\frac{2\pi n}{T}\right) = \frac{W_n}{T} \qquad (18\text{-}81)$$

Practical considerations require selecting $w(t)$ in such a way that the number of terms in Eq. (18-80) is kept small; that is, $W_m = W(2\pi m/T)$ will vanish for $M < |m| < N/2$. (*Note:* This is not true for the Kaiser-Bessel window.) If we further restrict our set of functions such that $C_0 = 1$ and the other C_n's are real, then $C_n = C_{-n}$. Thus, we can define $D_n = 2C_n$ and $X = T/2$, which yields

$$W(\omega) = T \exp(-jX) \sin(X) \frac{1}{X} + \sum_{n=1}^{M} \frac{D_n X}{X^2 - n^2\pi^2} \qquad (18\text{-}82)$$

and

$$w(t) = w_T(t)\left\{1 + \sum_{n=1}^{M} D_n \cos\frac{2\pi nt}{T}\right\} \qquad (18\text{-}83)$$

where $w_T(t)$ denotes the right-angular window or gating function. Therefore, a general window function can be specified in terms of a set of coefficients $\{D_n\}$.

Classification of Window Functions.   In grouping the various window functions, we shall define three classes of windows.[21] We shall also include an additional class, namely, the Kaiser-Bessel windows.

CLASS 1.   Weighting function in this class will be denoted by

$$w_1 = w_T(t) \left\{ 1 + \sum_{n=1}^{M} D_1(n, M) \cos \frac{2\pi nt}{T} \right\} \qquad (18\text{-}84)$$

where the expansion coefficients $D_1(n, M)$ are chosen to minimize side-lobe amplitude of the spectrum of $w_1(t, M)$ for a given M. The explicit form of the coefficients is

$$D_1(n, M) = (-1)^n 2 \prod_{k=1}^{n} \frac{M + 1 - K}{M + K} \qquad (18\text{-}85)$$

Substituting Eq. (18-85) into Eq. (18-84) yields the equation of the window function in closed form as

$$w_1(t, M) = w_T(t) \frac{4^M (M!)^2}{(2M!)^2} \sin^{2M} \left(\frac{\pi t}{T} \right) \qquad (18\text{-}86)$$

Among the popular windows the *Hanning, Larson,* and *Singleton* windows are members of this class.

CLASS 2.   Functions belonging to this class are called *Taylor-Dolph-Tchebycheff* windows and will be denoted by $w_2(t, M)$. Their distinguishing characteristic is that their spectra have minimum "main-lobe width" for a given maximum side-lobe amplitude R^{-1}.

To ensure that the higher-order side lobes decrease monotonically, M has to be selected to exceed a threshold value for any given R. The relationship between $M_{\min}$ and R is empirically determined to be

$$20 \log R = 6M + 18 \qquad (18\text{-}87)$$

Using Eq. (18-87), we can eliminate the dependence of the expansion coefficients $D_2(R, M, n)$ on the side-lobe amplitude. The explicit form of those coefficients is

$$D_2(M, n) = \frac{\displaystyle\prod_{k=1}^{M} \left\{ 1 - \frac{(n/\alpha)^2}{\lambda^2 + (K - \frac{1}{2})^2} \right\}}{\displaystyle\prod_{\substack{k=1 \\ k \neq n}}^{M} \{ 1 - (n/k)^2 \}} \qquad (18\text{-}88)$$

where

$$\alpha^2 = \frac{(M + 1)^2}{\lambda^2 + (M + \frac{1}{2})^2} \qquad (18\text{-}89)$$

and

$$\lambda = -\ln [R + R^2 - 1] \qquad (18\text{-}90)$$

CLASS 3.   The set of functions $\{w_3(t, M)\}$ belonging to this class may be referred to as Rife-Vincent windows. The characteristics of $w_3(t, M)$ are that they preserve the desirable properties of the functions of the previous two classes. $w_3(t, 1)$ is set to be identical to $w_1(t, 1)$. For $M = 2$ the side lobes of $w_3(\omega)$ reach -60 dB with ω as small as possible. For $M > 2$

$$w_3(t, M) = K w_1(t, M - 2) w_2(t, 2) \qquad (18\text{-}91)$$

where the weighting factor K was chosen to ensure that $w_3(t, M)$ is in the form of Eq. (18-84).

CLASS 4.   The elements of this class are the so-called Kaiser-Bessel windows. These are the digital approximation of the prolate spheroidal wave functions which are time limited and have minimum energy outside some selected frequency interval. The explicit form of these windows is

$$w_4(n, \beta) = I_0 \left(\beta \sqrt{1 - \left(\frac{2n}{N - 1} - 1 \right)^2} \right) \Big/ I_0(\beta) \qquad 0 \leq n \leq N - 1 \quad (18\text{-}92)$$

where β is a parameter and $I_0(x)$ is the modified 0th-order Bessel function.

The expansion coefficients for the first three classes of windows are shown in Table 18-14 for order $M \leq 7$. Examples of windows (for $M = 5$) are shown in Fig. 18-28.

Table 18-14 Expansion Coefficients for Window Functions

N=	1	2	3	4	5	6	7
				Class 1			
M = 1	−1.00000						
M = 2	−1.33333	0.33333					
M = 3	−1.50000	0.60000	−0.10000				
M = 4	−1.60000	0.80000	−0.22857	0.02857			
M = 5	−1.66667	0.95238	−0.35714	0.07937	−0.00794		
M = 6	−1.71428	1.07143	−0.47619	0.14286	−0.02597	0.00216	
M = 7	−1.75000	1.16667	−0.58333	0.21212	−0.05303	0.00816	−0.00058
				Class 2			
M = 1	−0.40745						
M = 2	−0.57464	−0.02877					
M = 3	−0.70077	−0.02850	−0.00767				
M = 4	−0.80481	−0.01272	−0.00901	−0.00281			
M = 5	−0.89305	0.01323	−0.00703	−0.00342	−0.00120		
M = 6	−0.96910	0.04578	−0.00394	−0.00285	−0.00146	−0.00056	
M = 7	−1.03538	0.08249	−0.00116	−0.00189	−0.00123	−0.00067	−0.00028
				Class 3			
M = 1	1.00000						
M = 2	−1.19685	0.19685					
M = 3	−1.43596	0.49754	−0.06158				
M = 4	−1.56272	0.72545	−0.18064	0.01792			
M = 5	−1.64225	0.89658	−0.31006	0.06076	−0.00503		
M = 6	−1.69705	1.02834	−0.43310	0.11988	−0.01945	0.00138	
M = 7	−1.73718	1.13248	−0.54487	0.18726	−0.04332	0.00601	−0.00037

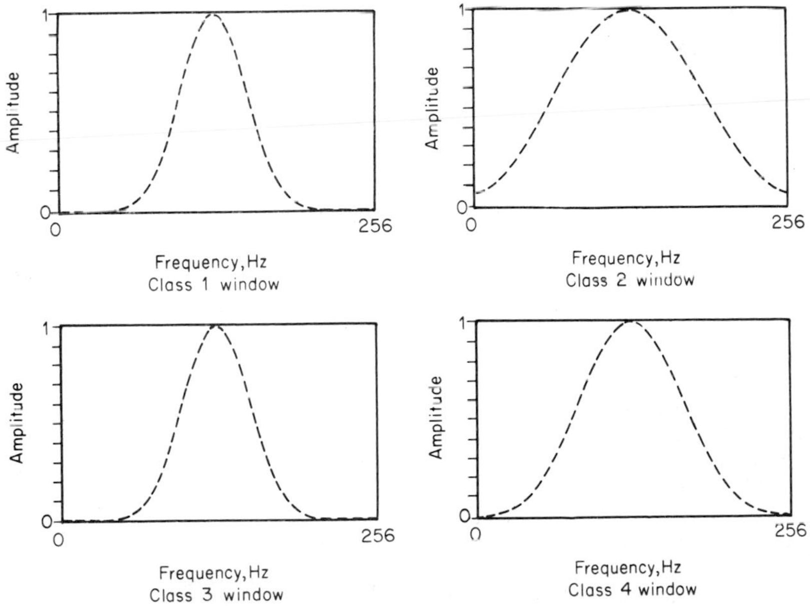

Frequency, Hz
Class 1 window

Frequency, Hz
Class 2 window

Frequency, Hz
Class 3 window

Frequency, Hz
Class 4 window

FIG. 18-28 Window functions.

Computer-Aided FIR Filter Design. An alternative way of designing FIR filters relies on computer techniques. This is theoretically more complex than applying windowing, but often results in "better" filters. Also, computer programs which are very good and easy to use are available to design digital filters (for example, the IEEE DSP package). One popular method is the so-called *frequency-sampling design.* It is based on the fact that the desired frequency response can be expressed by substituting the inverse DFT [Eq. 18-61)] into the definition of the Z transform [Eq. (18-50)] which yields

$$H(z) = \frac{1/z^{-N}}{N} \sum_{k=0}^{N-1} \frac{H(k)}{1 - \exp\,[j(2\pi/N)k]z^{-1}} \tag{18-93}$$

Letting $z = e^{j\omega}$ results in the following expression for the frequency response

$$H(\exp\,j\omega) = \frac{\exp\,[-j\omega(N-1)/2]}{N} \sum_{k=0}^{N-1} H(k)\,\frac{\sin\,[N(\omega-2\pi k/N)/2]}{\sin\,[(\omega-2\pi k/N)/2]} \tag{18-94}$$

Equation (18-94) implies that the frequency response can be approximated by N uniformly spaced points around the unit circle and using Eq. (18-94) to interpolate the values between these points.

The interpolation formula, however, can result in quite large errors between the N points, especially when the desired frequency response has large discontinuities. One way of remedying the situation is to allow a few sample points near the discontinuity to deviate from their ideal locations. Computer optimization techniques can be used to locate the position of these *unconstrained* samples to minimize the interpolation error in some sense.

Figure 18-29 illustrates the frequency-sampling method. Figure 18-29a shows the frequency samples for an ideal low-path filter, and Fig. 18-29b shows the result of the interpolation by Eq. (18-94). Figure 18-29c and d is similar, but H_1 and H_2 are not constrained to remain on the "ideal" positions. Several other computer methods are available, and for these the reader should consult the references cited.

18.3.6 INFINITE IMPULSE RESPONSE FILTERS.
By definition, the impulse response of an *IIR* filter is $\{h(n); n = 0, 1, \ldots, \infty\}$ with the constraint that $\Sigma_n|h(n)| <$

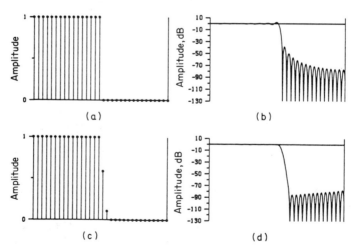

FIG. 18-29 Frequency-sampling filter design: (*a*) ideal low-pass; (*b*) amplitude response; (*c*) modified low-pass; (*d*) amplitude response.

∞ to ensure stability. The Z transform of such a filter, in the most general way, is

$$H(z) = \sum_{n=0}^{\infty} h(n)z^{-n} = \frac{\sum_{i=0}^{M} b_i z^{-i}}{1 + \sum_{i=1}^{N} a_i z^{-i}} \qquad (18\text{-}95)$$

All the important design parameters of the filter can be expressed in terms of $H(z)$ as follows:

1. Magnitude response of the filter is usually expressed in the terms of *magnitude-squared response*

$$|H(\exp j\omega)|^2 = |H(z)H(z^{-1})|_{z=\exp j\omega} \qquad (18\text{-}96)$$

2. The *phase-response* $\beta(\omega)$ can be written as

$$\beta(\exp j\omega) = \frac{1}{2j} \ln \left[\frac{H(z)}{H(z^{-1})} \right]\bigg|_{z=\exp j\omega} \qquad (18\text{-}97)$$

$$\text{since } H(z) = |H(z)|\exp\,[j\beta(z)]$$

3. The *group delay* τ which is the derivative of β can be expressed as

$$\tau(\exp j\omega) = -\,\mathcal{R}\left\{ z\frac{d}{dz}\ln H(z) \right\}\bigg|_{z=\exp j\omega} \qquad (18\text{-}98)$$

where $\mathcal{R}\{\ \}$ stands for the real part of the complex quantity.

Thus, the filter design problem can be thought of as finding a system Z transform which matches a set of design parameters. This is an analogous situation to traditional filter design where the Laplace transform of the filter $H(s)$ is matched to the design parameters. A great body of knowledge exists pertaining to analog filter design. Digital IIR filter design can utilize this knowledge by finding suitable transformation from the s plane of the Laplace transform to the z plane of the Z transform.

Digital Filter Design Based on Analog Systems.   The objectives of such transformations are:

• To preserve the frequency selective qualities of the analog filters which is accomplished by mapping the j or imaginary axis of the s plane onto the unit circle of the z plane
• To maintain stability of the system by mapping the left half s plane inside the unit circle

The most "obvious" transformation from analog to digital design is to change the differential equations describing the analog filter to constant coefficient difference equations describing digital systems. Unfortunately, such mapping of differences neither preserves the required filter characteristics nor (as a rule) results in stable systems.

A more useful method is based on digitizing the impulse response of an analog filter, i.e., approximating the analog filter impulse response $h(t)$ with a digital unit sample response $h(nT)$. Such transformation is referred to as *impulse invariant transformation*. The correspondence between the first- and second-order terms of such transformations is shown on Table 18-15. Although mapping from the s plane to the z plane using impulse invariant transformation results in stable systems with reasonable frequency characteristics, severe aliasing can result if the analog filter is not band limited to $-\pi/T < \omega < \pi/T$.

The third technique of digitization of an analog filter employs the direct mapping of the poles and zeros of the S plane onto the z plane by a simple replacement technique

$$s + a \rightarrow 1 - z^{-1}e^{-aT} \qquad (18\text{-}99)$$

Table 18-15 Impulse Invarient Transformations

$H(s)$	$H(z)$
$\dfrac{1}{s + a}$	$\dfrac{1}{1 - z^{-1}e^{-aT}}$
$\dfrac{s + a}{s^2 + 2as + a^2 + \omega^2}$	$\dfrac{1 - z^{-1}e^{-aT}\cos\omega T}{1 - 2z^{-1}e^{-aT}\cos\omega T + z^{-2}e^{-2aT}}$
$\dfrac{\omega}{s^2 + 2as + a^2 + \omega^2}$	$\dfrac{z^{-1}e^{-aT}\sin T}{1 - 2z^{-1}e^{-aT}\cos\omega T + z^{-2}e^{-2aT}}$

This method is called the *matched-Z transform*. Although this mapping is easily implemented, the resulting filter could be severely aliased or otherwise different from the analog version.

The fourth, and perhaps the most useful, mapping of the s to the z plane is the so-called bilinear transformation (a familiar concept in complex-variable theory)

$$s \rightarrow \frac{2}{T}\frac{1 - z^{-1}}{1 + z^{-1}} \tag{18-100}$$

Although bilinear transformation yields a nonlinear relationship between analog and digital frequencies, a large class of important filters (low-pass, high-pass, bandpass, and bandstop filters) can be designed by predistortion or *frequency wrapping* of the analog systems.

Frequency Transformation. Another useful digital technique which was developed for analog filters is the frequency transformation, which can be used to obtain highpass, bandpass, etc., filters from a low-pass design. For IIR filters the appropriate formulas for *frequency band transformations* are shown in Table 18-16.

Optimization Techniques. Another useful design technique for IIR filters (especially which complex design characteristics) utilizes mathematical optimization techniques such as the Fletcher-Powell algorithm. Since the required filter coefficients cannot be found by simple or explicit formulas to match design requirements, they are found by computer optimization of some error criterion—such as the minimization of the mean square error. A large number of error criteria and appropriate optimization techniques are published, and little practical purpose would be served delineating them here. A good introduction to this topic can be found in Refs. 19 and 20.

All-Pass IIR Filters. IIR filters cannot have linear phase, in general, since the requirement that $H(z) = H(z^{-1})$ would force the poles of the filter outside the unit circle.

Table 18-16 Digital Frequency Band Translations†

Low pass	$(z^{-1} - a)/(1 - az^{-1})$
High pass	$-(z^{-1} + b)/(1 + bz^{-1})$
Bandpass	$-[(k + 1)z^{-2} - 2ckz^{-1} + (k - 1)]/[(k + 1)z^{-2} - 2ckz^{-1} + (k + 1)]$
Bandstop	$[(k + 1)z^{-2} - 2cz^{-1} + (k - 1)]/[(k + 1)z^{-2} - 2cz^{-1} + (k + 1)]$

†ω_0 = original low-pass cutoff
ω_1 = desired frequency
$a = \sin[(\omega_0 - \omega_1)/2]/\sin[(\omega_0 + \omega_1)/2]$
$b = -\cos[(\omega_1 + \omega_0)/2]\cos[(\omega_1 - \omega_0)/2]$
$c = \cos[\omega_2 + \omega_1)/2]/\cos[(\omega_2 - \omega_1)/2]$
$k = \tan[(\omega_2 - \omega_1)/2]\tan(\omega_0/2)$

However, one special case of IIR systems, namely, the *all-pass filter,* is useful for equalizing phase- or group-delay characteristics. Poles and zeros of such filters, as shown in Fig. 18-30, appear in complex conjugate pairs, and they are symmetric with respect to the unit circle. Thus, the frequency response of these must be of the form

$$H(z) = \frac{[z - (1/r)e^{j\theta}][z - (1/r)e^{-j\theta}]}{(z - re^{j\theta})(z - re^{-j\theta})} \tag{18-101}$$

This function evaluated on the unit circle has constant magnitude and, therefore, can affect only the phase of the filter.

18.3.7 MULTIDIMENSIONAL FILTERS.

The digital filters discussed so far process signals which are the function of a single variable, usually time t. Images, however, are two-dimensional; i.e., they are functions of two spatial variables customarily denoted by x and y. Television signals are three-dimensional signals; that is, they are the function of two spatial variables x and y and a temporal variable t. Obviously, multidimensional signal processing requires multidimensional filters. Although the extension of filter theory of a single variable to several variables is not trivial, fortunately it is straightforward. Some two-dimensional equivalents of concepts previously covered are presented in the rest of this section. Higher-dimensional versions are similar and can be worked out by the reader.

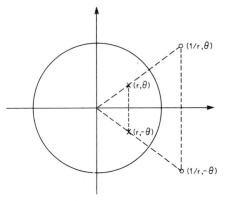

FIG. 18-30 Pole-zero configuration of all-pass filter.

Two-Dimensional LSI Systems. If the impulse† response of a two-dimensional LSI system is $h(n_1, n_2)$, then the output resulting from an arbitrary input $x(n_1, n_2)$ is

$$y(n_1, n_2) = x(n_1, n_2) * h(n_1, n_2)$$

$$= \sum_{m1=-\infty}^{\infty} \sum_{m2=-\infty}^{\infty} h(m_1, m_2) x(n_1 - m_1, n_2 - m_2) \tag{18-102}$$

$$= \sum_{m1=-\infty}^{\infty} \sum_{m2=-\infty}^{\infty} x(m_1, m_2) h(n_1 - m_1, n_2 - m_2)$$

Thus, the convolution theorem is valid for two-dimensional systems.

Two-Dimensional Discrete Fourier Transform. The two-dimensional equivalence of the discrete Fourier transform is defined as

$$x(n_1, n_2) = \frac{1}{N_1 N_2} \sum_{k1=0}^{N_1-1} \sum_{k2=0}^{N_2-1} X(k_1, k_2) \exp\left[j\left(\frac{2\pi}{N_1}\right)n_1 k_1\right] \exp\left[j\left(\frac{2\pi}{N_2}\right)n_2 k_2\right] \tag{18-103}$$

---

†Note that here the "impulse" is understood to be a two-dimensional function defined as

$$\delta(n_1, n_2) = \begin{cases} 1 & n_1 = n_2 = 0 \\ 0 & \text{otherwise} \end{cases}$$

and the inverse transform as

$$X(k_1, k_2) = \sum_{n1=0}^{N_1-1} \sum_{n2=0}^{N_2-1} x(n_1, n_2) \exp\left[-j\left(\frac{2\pi}{N_1}\right) n_1 k_1\right] \exp\left[-j\left(\frac{2\pi}{N_2}\right) n_2 k_2\right] \quad (18\text{-}104)$$

Two-Dimensional Z Transform. Similarly to the one-dimensional case, the Z transform of a two-dimensional signal may be defined as

$$X(z_1, z_2) = \sum_{n1=-\infty}^{\infty} \sum_{n2=-\infty}^{\infty} x(n_1, n_2) z_1^{-n1} z_2^{-n2} \quad (18\text{-}105)$$

The inverse-Z transform is also similar to the one-dimensional case

$$x(n_1, n_2) = \frac{1}{(2\pi j)^2} \oint_{C1} \oint_{C2} X(z_1, z_2) z_1^{n_1-1} z_2^{n_2-1} \, dz_1, \, dz_2 \quad (18\text{-}106)$$

The usefulness of the Z transform stems from the fact that (again as in the one-dimensional case) if a sequence $y(n_1, n_2)$ is the result of the two-dimensional convolution of input $x(n_1, n_2)$ and system impulse response $h(n_1, n_2)$, then its transform $Y(z_1, z_2)$ is the product of the corresponding transforms $X(z_1, z_2)$ and $H(z_1, z_2)$; that is

$$y(n_1, n_2) = x(n_1, n_2) \star h(n_1, n_2) \rightarrow Y(z_1, z_2) = X(z_1, z_2)H(z_1, z_2) \quad (18\text{-}107)$$

Several other useful properties of the Z transform carry forward from one to two dimensions.

Two-Dimensional Filters. The analytic form of both the two-dimensional difference equation characterizing digital filters and the corresponding Z transform are similar to their one-dimensional counterparts

$$y(n_1, n_2) = \sum_{i=0}^{p} \sum_{j=0}^{q} a_{ij} x(n_1 - i, n_2 - j) - \sum_{i=0}^{p} \sum_{j=0}^{q} b_{ij} y(n_1 - i, n_2 - j) \quad (18\text{-}108)$$

$$H(z_1, z_2) = \frac{\displaystyle\sum_{i=0}^{p} \sum_{j=0}^{q} a_{ij} z_1^{-i} z_2^{-j}}{\displaystyle\sum_{i=0}^{p} \sum_{j=0}^{q} b_{ij} z_1^{-i} z_2^{-j}} \quad (18\text{-}109)$$

Although these filters can be designed either as IIR or FIR filters, only the second approach is practical since the assessment of stability conditions for two-dimensional IIR filters is almost impossible.

Two-dimensional FIR filter design methods are extensions of single-dimensional techniques discussed previously. These include:

• windowing

• frequency sampling

• optimal (computer-aided) design

The paramount feature of multidimensional filters is the requirement for various delay elements. Whereas one-dimensional filters require only single delay elements, two-dimensional (video) filters will also need line delay. Three-dimensional filters utilize frame delays. Thus these higher-dimensional filters require line and frame storage (delay) which implies increased cost and complexity.

18.3.8 HARDWARE FOR DIGITAL VIDEO. Although a large number of semiconductor technologies are available, only a few are suitable for digital video applications. Because of the inherently high bit rates involved, digital video imposes very strict speed constraints on the hardware. Only those technologies which are conforming to these constraints will be reviewed briefly in the following paragraphs.

Logic Circuits. Circuit elements which are performing addition, multiplication, logic operations, etc., are commonly referred to as logic circuits. The basic building blocks of these are the *logic gates,* which are designed to perform *logic operations* such as AND, OR, and complement. The circuit symbols and the truth tables for the basic logic gates are shown in Fig. 18-31. These gates and a few simple rules of boolean algebra (shown in Table 18-17) facilitate the design of very complex circuits. An example of interconnection of simple gates to obtain a somewhat more complex circuit is shown in Fig. 18-32. A very large repertoire of logic circuits is available as off-the-shelf, specially constructed ICs. In addition, the circuit designer can customize ICs to individual needs at moderate costs.

In addition to logic circuits, the arithmetic logic unit (ALU) plays a significant role in video circuit design. The functional block diagram of such a circuit element is shown in Fig. 18-33. ALUs are especially useful when addition, subtraction, and multiplication are required.

The two semiconductor technologies which are widely used in digital video circuits are *Schottky TTL* and *emitter-coupled logic (ECL)* circuits. Both of these meet the speed requirements, and the choice between them is usually made on the basis of power consumption, cost, part availability, etc.

The electrical characteristics of the ECL devices are shown in Table 18-18.

Memory Elements. The second important type of components for digital video is the storage or memory. High-speed solid-state memory chips (developed for the computer

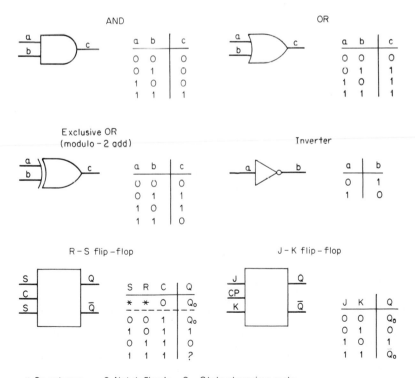

* Do not care;   ? Not defined; Q_0 State at previous cycle

FIG. 18-31 Basic logic circuits.

Table 18-17 Basic Rules of Boolean Algebra†

Rules and relations	
	$0 + 0 = 0$
	$0 \cdot 0 = 0$
	$1 + 1 = 1$
	$1 \cdot 1 = 1$
	$0 \cdot 1 = 0$
	$0 + 1 = 1$
	$\overline{0} = 1$
	$\overline{1} = 0$
	$a \cdot a = a$
	$a + a = a$
	$\overline{a} \cdot a = 0$
	$\overline{a} + a = 1$
	$0 + a = a$
	$0 \cdot a = 0$
	$1 + a = 1$
	$1 \cdot a = a$
	$a + a \cdot b = a$
	$a \cdot (a + b) = a$
	$a + a \cdot b = a + b$
Commutative law	$a + b = b + a$
	$a \cdot b = b \cdot a$
Associative law	$(a + b) + c = a + (b + c)$
	$(a \cdot b) \cdot c = a \cdot (b \cdot c)$
Distributive law	$a \cdot (b + c) = a \cdot b + a \cdot c$
DeMorgan's rules	$\overline{(a \cdot b)} = \overline{a} + \overline{b}$
	$\overline{(a + b)} = \overline{a} \cdot \overline{b}$

† $\cdot$ denotes AND, $+$ denotes OR.

industry) have been available for video application for some time. Memory chips are available in two varieties: dynamic and static. *Dynamic memories* require refreshing periodically. These are the slower of the two types but can achieve more density, i.e., more bits per IC. Also, these are the less expensive of the two. Dynamic memories are very useful for bulk storage.

Static memories are faster but not as dense as dynamic memories; i.e., more chips are required for a given storage capacity and, in addition, these chips are (bit for bit) more expensive. Static TTL random access memories (RAMs) can have access time as low as 45 ns. ECL RAMs are even faster with access time of (for a 1024×1 chip) less than 30 ns.

The size of the memory chip is an important consideration in circuit design. 1K (that is, 1024×1) chips are useful for line memories. Eight chips in parallel can store over 1 kword of information, which is more than required for one television line.

For frame memories larger ICs are more useful. Although larger-capacity chips are usually slower, these are preferred in order to reduce the number of chips, facilitate addressing (the number of address pins on a chip if finite), etc. Note that memory devices (particularly for frame stores) can be much slower than logic devices if multiplexing techniques can be employed. Some pertinent data for memory devices are shown in Table 18-19.

Truth Table

Inputs				Output
F	S	I_{0n}	I_{1n}	Z_n
H	X	X	X	L
L	H	X	L	L
L	H	X	H	H
L	L	L	X	L
L	L	H	X	H

H = high voltage level.
L = low voltage level.
X = immaterial.

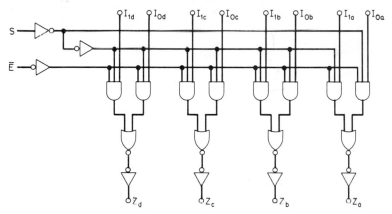

FIG. 18-32 Example of logic circuit design, quad two-input multiplexer.
Note: The '22 quad two-input digital multiplexers consist of four multiplexing circuits with common select and enable logic; each circuit contains two inputs and one output.

18.3.9 VIDEO A/D CONVERTERS.

The component of paramount importance for digital television is a practicable A/D converter. In principle, changing the analog, band-limited video signals into digital bit streams is possible (see Sec. 18.1.1). However, the inherently large bandwidth associated with television signals makes the realization of video A/D conversion difficult. Component counts, for example, are very high (e.g., a large number of comparators are required to determine the adjacent two-threshold-bracketing the sample among the $2^N - 1$ possible levels for an N-bit converter). Timing accuracy requirements are also critical. Jitter as little as 70 ps could cause errors. Both of these difficulties can be overcome, however, by utilizing very large scale integration (VLSI) techniques.

Most of the proposed VLSI implementations are very similar to the architecture shown in a simplified fashion in Fig. 18-34.[22] The technique used by this type of A/D converter is called flash conversion. Another technique suggested by some is based on succesive approximations. This second type, however, has not been a successful candidate for large-scale integration.

A reference voltage V_r is divided into n equal parts ($n = 255$ for an 8-bit converter). The input voltage V_{in} is compared with $V_r(R/255R)$, $V_r(2R/255R)$, ... The serial-to-parallel converter is presented by the 255 outputs of these comparators. The inputs

Function Table

Mode select inputs				Active low inputs and outputs		Active high inputs and outputs	
S_3	S_2	S_1	S_0	Logic (M = H)	Arithmetic (M = L) (Cn = L)	Logic (M = H)	Arithmetic (M = L) (Cn = H)
L	L	L	L	$\bar A$	A minus 1	$\bar A$	A
L	L	L	H	$\overline{AB}$	AB minus 1	$\bar A + \bar B$	A + B
L	L	H	L	$\bar A + \bar B$	$A\bar B$ minus 1	$\bar A B$	$A + \bar B$
L	L	H	H	Logic 1	minus 1	Logic 0	minus 1
L	H	L	L	$\bar A + \bar B$	A plus $(A + \bar B)$	$\bar A \bar B$	A plus $A\bar B$
L	H	L	H	$\bar B$	AB plus $(A + \bar B)$	$\bar B$	$(A + B)$ plus $A\bar B$
L	H	H	L	$\bar A \oplus \bar B$	A minus B minus 1	$A \ominus \bar B$	A minus B minus 1
L	H	H	H	$A + \bar B$	$A + \bar B$	$A\bar B$	$A\bar B$ minus 1
H	L	L	L	$\overline{AB}$	A plus $(A + B)$	$\bar A + B$	A plus AB
H	L	L	H	$A \oplus B$	A plus B	$A \oplus B$	A plus B
H	L	H	L	B	$A\bar B$ plus $(A + B)$	B	$(A + \bar B)$ plus AB
H	L	H	H	$A + B$	$A + B$	AB	AB minus 1
H	H	L	L	Logic 0	A plus A	Logic 1	A plus A
H	H	L	H	$\overline{AB}$	AB plus A	$A + \bar B$	$(A + B)$ plus A
H	H	H	L	AB	$A\bar B$ minus A	$A + B$	$(A + \bar B)$ plus A
H	H	H	H	A	A	A	A minus 1

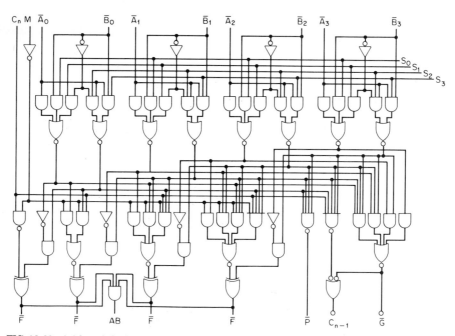

FIG. 18-33 Arithmetic logic unit (ALU).

18.46

Table 18-18 Typical Electrical Characteristics of ECL Circuits

	Min	Max
Supply voltage, V_{ee}	−4.7	−6.2
Logic 1 output, V_{oh}	−0.7	−1.05
Logic 0 output, V_{ol}	−1.62	−1.89
Logic 1 threshold, V_{oha}	−0.91	−1.065
Logic 0 threshold, V_{ola}	−1.555	−1.63
Switching times, ns	1.0	2.0
Rise time	1.4	2.2
Fall time	1.2	2.3

Table 18-19 Characteristics of Memory Devices

	Access time, ns	Number of pins
Dynamic		
4096 × 1	200–400	16–22
16384 × 1	150–300	16
65536 × 1	150–400	16
Static		
1024 × 1	30–45	16
4096 × 1	55–200	16–22

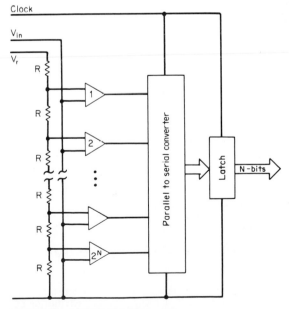

FIG. 18-34 n-Bit VLSI A/D converter.

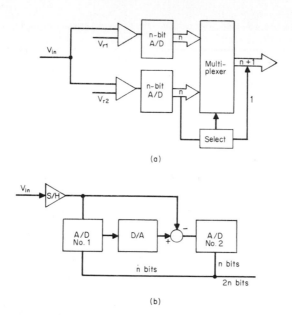

FIG. 18-35   Cascading of A/D converters.

which correspond to reference voltages less than V_{in} are high and the rest are low. The converter detects the transition from high to low inputs and produces an 8-bit word corresponding to this transition level. The latch holds this binary value until the next clock cycle.

Current technology permits VLSI video A/D converters which operate up to 50 M samples per second with an 8-bit/sample resolution or 100 M samples per second with a 6-bit/sample resolution. If higher resolution is required, a number of cascading techniques are possible, two of which are shown in Fig. 18-35. Figure 18-35a shows a parallel implementation of an $n + 1$ bit A/D converter. The analog video is simultaneously presented to two comparators to determine if it exceeds half the total voltage range. If V_{in} is less than half of the total range, A/D-1 converts the sample into an 8-bit word and the circuit sets the most significant bit to 0. If V_{in} is between half and full range of the input, A/D-2 converts $V_{in}/2$ into an 8-bit word and the most significant bit is set to 1.

More precision can be obtained by using the feed forward arrangement shown in Fig. 18-35b. Here the video signal is fed to a sample-and-hold circuit. The output of the sample and hold is digitized to n-bit resolution which is D/A converted. The difference between the D/A and V_{in} is the "quantization error" of the first A/D. This can again be digitized to n-bit resolution by a second A/D. Concatenating the two n-bit words results in a $2n$ resolution A/D. In practice, the two n-bit words are allowed to overlap resulting in an A/D resolution between n and $2n$.

VLSI implementation of D/A converters is also possible. Current state of the art is 8-bit D/A converters operating at a speed of 100 M samples per second.

18.4 TRANSMISSION OF DIGITAL TELEVISION SIGNALS

Transmission of digital television signals is required in many applications ranging from interconnection of digital hardware in the studio to the distribution of signals over large geographic areas (possibly using satellite transponders). The method of transmission largely depends on application, but in general the various alternatives can be divided into two groups, namely, bit parallel transmission (usually employed within the studio complex) and serial transmission (used for transmission of the signals over long distances). Before looking at these in some detail, the characteristics of the transmission channel will be reviewed.

18.4.1 TRANSMISSION CHANNEL.   The selection of transmission facility to deliver digital video is usually based on reach, capacity, and cost trade-offs. The actual transmission media could be copper cable, fiber cable, or radio. The salient features of these are as follow.

Copper cables can be constructed utilizing either insulated, twisted pairs or coaxial conductors. The twisted pair is the least costly of all transmission facilities and can deliver without repeaters 10 Mbits/s over distances of 0.62 mi (1 km) or up to 30 Mbits/s over distances of 164 ft (50 m) [or 984 ft (300 m) using equalization and shielding]. On the other hand, coaxial cables have higher associated terminal costs but have much higher capacity. They can handle transmission rates in the gigabit per second range over the distance approaching 1 km on a single link, without repeaters.[23]

The limiting factors for either distance or capacity for cable transmission include attenuation, crosstalk, intersymbol interference, thermal noise, impulse noise, echoes, etc. Of course, the reach of any digital transmission system can be extended by using regenerative repeaters. In fact, such repeated lines make long-distance transmission of digital signals possible. For short distances however, repeaters are avoided if possible for cost reasons.

Fiber Cables.   Higher capacities and larger reach can be obtained using fiber optic technology. Figure 18-36 shows the reach of various fiber systems as a function of capacity. Curve 1 shows capacities of systems utilizing light-emitting diode transmitters operating at 850-nm wavelength and PIN detectors. Curve 2 corresponds to laser transmitters and avalanche photodiode detectors at the same wavelength. Curve 3 shows the capacity of long-wavelength systems operating at 1300 nm. Finally, curve 4 shows the capability of systems utilizing monomode, long-wavelength technology. Although the cost of terminal equipment for fiber-based systems is usually higher than for their copper counterparts, the large bit rate capacity over long distances can make this system more attractive.

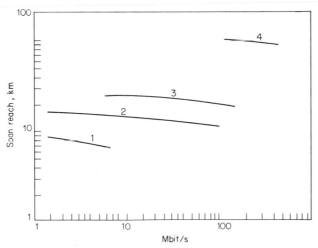

FIG. 18-36   Fiber system reach.

Radio. An alternative way of delivering high bit rate digital signals for long distances is over digital microwave radio facilities. Such radio relay systems exist in the 6-, 8-, and 12-GHz band and also on the various satellite spectrum allocations.

18.4.2 BIT-PARALLEL TRANSMISSION. Similarly to their analog counterparts, digital studios will employ a large quantity of cables for the interconnection of the various components required to produce television programs. To ensure that the digital studio operates smoothly and efficiently, the transmission scheme—or the interface standard of the various studio equipment—must be specified. The length requirement of the cables (up to 50 m without equalization, or 300 m with equalization) can be met at a reasonable cost if such standards are based on bit-parallel transmission standards as described below. The signal to be considered for this scheme is the 4:2:2 member of the family of digital codes contained in CCIR Recommendation 601 (see Table 18-7).

The digital words of this signal contain 8 bits corresponding to the 256-PCM levels used in the encoding process. Consequently, the transmission scheme suggested utilizes eight conductor pairs, each carrying a bit of the same significance of the multiplexed luminance and chrominance samples (data 0 to 7). The luminance signal is sampled at 13.5 MHz, and each of the color difference signals at 6.75 MHz. The multiplexed word rate (or bit rate of each of the eight conducting pairs) is 27 MHz. The multiplexed video signal is depicted in Fig. 18-37.

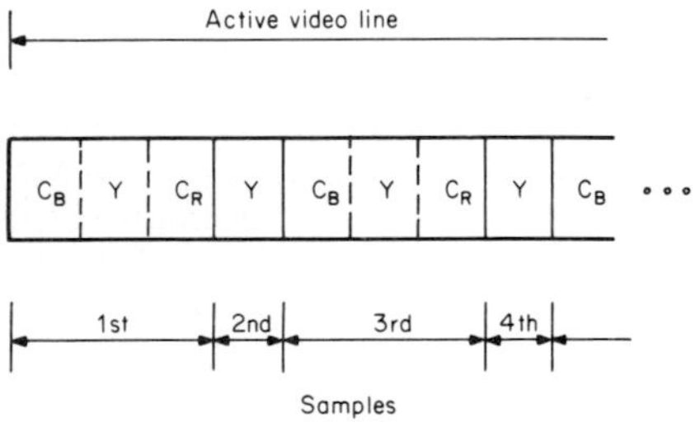

An additional pair of conductors is included in the interface to transmit the 27-MHz clock along with the data signal. All the transmitted pulses are nonreturn to zero (NRZ) binary, ECL compatible voltages.

Since the deterministic portion of the video signal (horizontal and vertical sync) need not be transmitted, the time duration of the sync signal can be utilized to transmit ancillary information as required.

Special timing reference signals of the form FF 00 00 XY are used to indicate various blanking/sync information within the video signal as shown in Fig. 18-38. The first three-word preamble, FF 00 00, is unique since neither FF (256) nor 00(0) is an allowed level to code video. (Ancillary signals are also forbidden to include these code words.) The last word, denoted XY, carries the information pertaining to the type of timing reference, as shown in Table 18-20.

FIG. 18-37 Multiplexing of luminance and color-difference words.

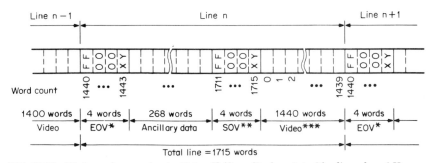

FIG. 18-38 Timing-reference signals. *Notes:* Half-amplitude point of leading edge of H sync coincides with word 1469; *, EOV → end of video; **, SOV → start of video; *** can be used for ancillary data during vertical blanking.

Table 18-20   Timing-Reference Signals

Bit position	7	6	5	4	3	2	1	0
XY	1	F	V	H	P_3	P_2	P_1	P_0

$F = 1 \rightarrow$ even field
$V = 1 \rightarrow$ vertical blanking
$H = 1 \rightarrow$ start of horizontal blanking (E V)
$P_0 - P_3$ (6, 3) Hamming code plus 1 even parity bit

(F and V can change when $H = 1$)

For 525-line systems, ancillary signals can be included anywhere in the bit stream not occupied by video or timing reference signals. For 625-line systems ancillary signals may be inserted during two specified lines of the vertical blanking period.

For housekeeping purposes, the following six-word preamble is included with the ancillary signals: 00 FF FF TT MM LL. TT denotes the data type (i.e., audio, teletex, etc.), and MM LL are used for either transmitting line numbers or data word count. The various electrical and mechanical characteristics of the interface are summarized in Table 18-21.

18.4.3 SERIAL TRANSMISSION.   Communication networks operated by various telephone companies are rapidly changing into all-digital environments. Concurrently, analog production methods in television studios are being phased out in favor of digital techniques. The result of these A/D conversions is that the necessity of transmitting video

Table 18-21   Electrical and Mechanical Characteristics of the Digital Studio Interface

Transmission cable	Nine balanced pairs—8 for parallel video signals plus 1 for clock—shielded
Equipment output impedance	$100 \perp 10 \, \Omega$
Equipment input impedance	$110 \pm 10 \, \Omega$
Signal amplitude	0.8 to 2.0 V p-p
Signal rise and fall time (20 and 80% point across 110)	5 ns
Input sensitivity	185 mV (min)
Differential delay between any two conductor pairs	8 ns
Connector	Twenty-five pin D type with side-lock mechanism
Pin assignment	1–8 data7–data0 14–21 return7–return0 11 clock 24 clock return 12,25 cable shield 9,10,22,23 not used

information digitally is imminent. For example, common carriers will have to digitize NTSC or other composite signals simply because no analog transmission channels will be available on certain routes.

In order to optimally utilize the transmission facilities, the telephone companies organize their digital networks into highly structured, multiplexed, hierarchical architectures as shown in Fig. 18-39. Any traffic entering this system must do so at the defined multiplex levels or a multiple of such level (e.g., at one or several DS-3 rates).

To ensure the efficient use of the transmission facilities, the total bit rate required for digital video must, therefore, be designed to be compatible with this digital transmission hierarchy. The first and the second levels do not appear to have sufficient capacity to carry broadcast-quality video signals. The fourth level with its high capacity could be attractive for very high-quality digital television transmission such as the 4:2:2 component standard. However, the associated cost with high-capacity transmission channels also is high. For digitized composite video, a single or multiple DS-3 stream appears to be the best alternative.

In most parts of the world, the bit rates shown in Fig. 18-39 are available to transmit information. In North America, however, the 44.736 Mbits/s ± 30 ppm includes framing bits required by various transmission and multiplexing equipment. In fact, the available bit rate for digitized television transmission is only 44.407 Mbits/s (± 30 ppm).

For digitized television, the DS-3 bit rate (or a multiple of the DS-3 rate) may have to accommodate not only the video but a number of ancillary signals. Some of these, for the NTSC system, are as follows:

1. *Associated audio.* The number of audio channels recommended by the International Radio Consultative Committee (CCIR) is two. However, some broadcasters have requirements for a third channel. Therefore, to accommodate three audio channels

	North America		Japan		Others	
	Bit rate	m_{ij}	Bit rate	m_{ij}	Bit rate	m_{ij}
DS-1	1.544		1.544		2.048	
		4		4		4
DS-2	6.132		6.132		8.448	
		7		5		4
DS-3	44.736		32.064		34.368	
		6		3		4
DS-4	274.776†		97.728		139.264	

†Not a CCITT recommendation.

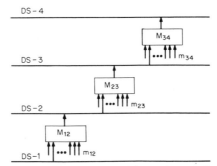

FIG. 18-39 Common-carriers' multiplex and transmission-hierarchy levels.

appears wise. The analog audio signal may be sampled at 48 kHz (to conform to the standards with the studios) or at 32 kHz, which is the accepted rate for transmission purposes. Since we are considering transmission here, this later sampling rate is preferred. Each audio sample is encoded in 14-to-11-bit, A-law companded PCM. A single parity bit is included with the 11 information bits to yield a 12-bit word. [An alternate encoding scheme is used for transmission of digitized audio signals in France and in the United Kingdom employing near instantaneous companding (NIC). These countries could, for compatibility reasons, prefer NIC for associated audio as well.]

2. *Captioning for the deaf.* Line 21 of the present NTSC signals could contain coded ASCII information which can be displayed as alphanumeric characters on the screen if the receiver is equipped with a special adaptor. The most expedient way of handling this information is to decode the analog waveform to ASCII and transmit this as binary data.

The identifications for all characters in the repertoire are similar to those recommended for videography by the CCITT:

- 26 lowercase letters
- 26 uppercase letters
- 12 accented lowercase letters
- 12 accented uppercase letters
- 10 numbers
- 9 punctuation marks and 1 space character
- 32 signs and symbols

The total number of characters is 129 (space character plus 128 symbols) requiring an 8-bit word to specify each character. The display format conforming with SMPTE recommended practice RP-27.3 is 40 characters per row. Therefore, 6 bits will be needed to specify the position of each character. Thus for 110 characters per second, the transmission bit rate will be 1540 bits/s.

3. *Videography (teletex).* Currently, videography-type digital information is transmitted on lines 10 to 21 of the field blanking interval. The repertoire of characters for videography is the same as was described above. The information structure, however, is much more complex.

The most expedient way to transmit videography information is by recovering the binary data. The videography coder's frame format contains 288 bits/line; thus the transmission rate is $11 \times 288 = 3168$ bits/frame or 95,040 bits/s for 525-line systems.

4. *Test signals.* One line of vertical-interval test signals (VITSs) is normally being transmitted on the seventeenth or eighteenth line of each field for in-service test purposes. Similarly, in line 19 of each field, a vertical-interval reference signal (VIRS) is transmitted to provide phase information. These signals are essential for monitoring the performance of analog transmission sections and therefore must be retained in a hybrid environment. Also, in order to ascertain the end-to-end analog impairment, the digital sections must transmit these signals in a "transparent" manner.

The most straightforward method of transmitting VITS or VIRS over digital facilities is using PCM coding with sufficient resolution to preserve the fine structure of these signals, for example, 4 f_{sc} sampling and 10-bit/sample PCM.

Alternatively, differential pulse-code modulation (DPCM) transmission can be used since these signals are usually slowly varying. If, for example, a 1-bit DPCM (Δmod) scheme is used, and if the signal is assumed stationary (a good approximation to the operating situation), then in 10 frames the DPCM output will converge to that of a PCM system using 10-bit resolution. The bit rate required to implement such a scheme is about 25 kbits/s (1 line/field, 820 bits/line).

5. *Color-burst information.* † Removal of the line-blanking period results in the removal of the color-burst signal. The loss of this information can result in catastrophic loss of color quality. However, the color-burst information can be transmitted at a reduced rate. Such a rate may be two color-burst signals per frame: one containing information relevant to even lines and the other to odd lines. These signals can then be stored in the receiver for the entire frame period and inserted into the newly generated horizontal sync signal interval. The resolution of color burst should be in 9 bits/sample resulting in a transmission rate requirement of

$$30 \times 9 \times 164 \times 2 = 88{,}560 \text{ bits/s}$$

Alternatively, if the sampling clock is carefully aligned with the color burst, only the field designation (even or odd) and the subcarrier phase need to be transmitted. Thus, for example, for NTSC (which is a four-field signal) a 2-bit number is sufficient to uniquely specify the state of the burst.

6. *Framing requirements.* Framing for digitized video should serve two purposes. First, it should allow the receiver to identify the significance of various bits received (functions such as video, audio, etc., and values or word boundaries). Second, it should carry synchronizing information (vertical and horizontal sync interval locations). The first of these is the usual function of framing of digital signals. The second is a convenience obtained by locking the signal frame to the field repetition rate of the video signal.

The bit-capacity requirement for framing can be estimated as

$$230 \times 60 = 1380 \text{ bits/s for field synchronization}$$
$$4 \times 525 = 2100 \text{ bits/s for line synchronization}$$

Thus the resulting total requirement is 3480 bits/s.

7. *Error-correction overhead.* On the one hand, real transmission systems can operate at just above a 10^{-4} error rate before alarm or protection equipment is activated. On the other hand, video transmission requires an effective error rate of 10^{-8} or better. Therefore, error correction is mandatory for video transmission.

Published results indicate that BCH codes with as little as 3 percent redundancy can produce a residual error rate of 10^{-8} from a nominal rate of 10^{-4}. In addition, such codes can have a burst-correcting capability of up to 26 bits. The actual implementation of this code is rather complex, and simpler but less efficient BCH or RS codes can be also considered.

8. *Miscellaneous signals.* A small percentage of the available bit rate will be used to transmit miscellaneous information such as:

• low bit rate commentary

• time code

• monitoring and alarm signals

• station identification code

• optional data channel

• order wire, etc.

The exact nature of these signals is not important provided that about 10 kbits/s capacity is assigned to carry these miscellaneous signals.

The various digital signals associated with the digitized video are summarized in Table 18-22. The functional block diagram of the digitization process is shown in Fig. 18-40.

---

†Color-burst information is not an ancillary signal in the strict sense. It is necessary only when horizontal sync is removed for bit-rate reduction purposes. This technique is discussed later.

Table 18-22   Capacity Budget for Digitized Video†

	Single DS-3	Double DS-3
Total bit rate	44407.0	88814.0
Audio	768.00	
Captioning	1.54	
Videography	95.04	
VITS, VIRS	49.20	
Color burst	88.56	
Framing	3.48	
Miscellaneous	10.00	
FEC	133.221	266.442
Video	43257.965	87531.738

†In kbits/s.

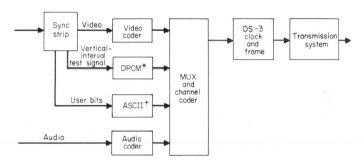

* Differential pulse-code modulation

+ American national standard code for information interchange

FIG. 18-40   Video coder block diagram.

18.4.4 BIT-RATE REDUCTION.

The transmission of video information over any facility taxes the channel because of the excessive bandwidths or bit rates associated with it. The situation is especially acute in the case of digital channels since the information capacity of digital channels is usually not as well utilized as their analog counterparts are. For example, the transmission of one analog television channel over analog, FM, microwave radio circuit occupying a 20-MHz band will displace 600 regular telephone channels. A four times subcarrier sampled, 8-bit PCM encoded video signal (114.4 Mbits/s) will require three DS-3 channels (actually, only 2.6 DS-3s, but as was stated earlier, the multiplex levels of the transmission hierarchy are well defined and cannot be violated). Thus, a digital television signal will displace 2016 telephone channels, more than three times as many as required for the analog system.

Fortunately, however, the television signals are highly redundant and (especially in the digital domain) lend themselves to bandwidth or bit-rate reduction. In fact, current state-of-art algorithms are aimed at reducing the bit rate of broadcast quality digital video signals to DS-3 rates which, in information capacity, compare favorably with analog channels. For special applications, such as video conferencing, bit-rate reduction to 1.5 Mbits/s or below is possible with slight loss in quality and increased hardware complexity.

Three distinct but complementary methods are available to reduce the bit rate of digital video signals. These will be described briefly for coding NTSC composite video. The principles outlined are also applicable to other composite or component signals. The field covered here is extremely fertile and can be given only a casual overview.

Removal of the Deterministic Portion of the Signal.   Video signals are composed of information-bearing portions and timing or synchronization components. Signals in the second category are absolutely essential to ensure that the scanning at the receiver is aligned with that of the camera. However, the long-distance transmission of these timing waveforms is not necessary since, if the locations of synchronization pulses are transmitted, the receiver terminal can insert locally generated timing reference signals. The bandwidth saving achieved by deleting sync is considerable. For vertical synchronization approximately 34 of the available 525 lines are used, as shown in Fig. 18-41. Therefore, removal of vertical synchronization interval will result in a saving of about 6.5 percent of the total capacity

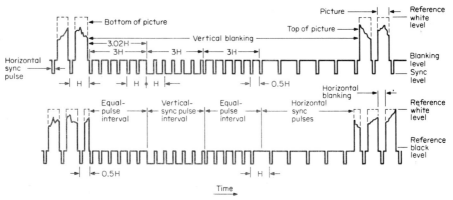

FIG. 18-41   Vertical-synchronizing interval. H is the time from start of one line to start of next line (63.5 μs).

Similarly, the horizontal synchronizing interval, 11.1 ns or 17.5 percent of the total line period of 63.5 ns, is deterministic (see Fig. 28-42). Thus, horizontal and vertical sync deletion can reduce the actual signal duration by as much as 23 percent.

Variable Word-Length Coding.   When the video information is encoded using other than PCM techniques (for example, DPCM to be discussed later), the frequency of the resultant code words usually has a very skewed distribution; i.e., certain code words—usually the ones corresponding to zero or reasonably low output values—occur much more frequently than ones representing large output values. The transmission rate can be reduced by assigning shorter code words to frequent output values and longer words to less frequent ones. This scheme is referred to as variable word-length or *entropy coding.* One possible variable word-length coding scheme is shown in Fig. 18-43.

One consequence of variable word-length coding is that the line-transmission rate is picture-dependent. That is, the distribution of the code length, hence, the transmission rate, is dependent on the information being encoded. But, since the existing transmission networks cannot accommodate variable-rate transmission, entropy coders (and decod-

	Nominal, μs	Tolerance, μs
Blanking	11.1	+0.3 −0.6
Sync	4.76	±0.32
Front porch	1.59	+0.13 −0.32
Back porch	4.76	+0.96 −0.61
Sync to burst	0.56	+0.08 −0.17
Burst	2.24	+0.27 −0
Blanking to burst	6.91	+0.08 −0.17
Sync and burst	7.56	+0.38 −0.49

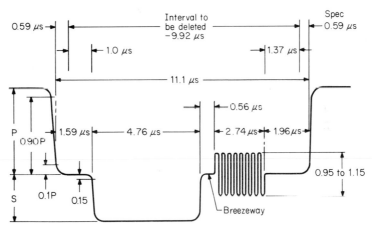

FIG. 18-42 Horizontal-synchronizing interval.

ers) must employ elastic buffers to produce uniform output rates. Of course, for any finite-length buffer, the probability of buffer overflow is also finite. Thus, variable word-length coders must be designed to handle such overflows. Various *overflow strategies* could include the reduction of sampling rate or the number of quantization level when the buffer fill reaches a certain predefined level.

Predictive Coding of Video.   Unlike the previous two techniques which resulted in information-preserving or reversible codes, this scheme utilizes the statistical properties of both the source and the receiver (including the human psychovisual system) to achieve bit-rate reduction at the expense of the loss of some information which will not significantly affect the "subjective quality" of the picture.

The various algorithms proposed or implemented to reduce the bit rate of the video signals can be divided in two classes: *time-domain coding* and *transform-domain coding*. Both of these can use either fixed parameters or adaptive techniques.

Input		Output	Entropy code	Code length
−255 to	−68	−80	100000001	9
− 67	−48	−55	10000001	8
− 47	−33	−38	1000001	7
− 32	−21	−25	100001	6
− 20	−12	−15	10001	5
− 11	− 8	− 9	1001	4
− 7	− 3	− 5	101	3
− 2	2	0	0	1
3	7	5	111	3
8	11	9	1101	4
12	20	15	11001	5
21	32	25	110001	6
33	47	38	1100001	7
48	67	55	11000001	8
68	255	80	110000001	9

FIG. 18-43　Variable word-length coding.

The literature covering this field of coding is vast. An authoritative review paper published as far back as 1972[24] cites 85 *important* references. Since then the number of publications has been growing at a phenomenal rate. Thus the coverage given here must be considered as very broadbrush indeed.

Time-Domain Coding.　These techniques have been proposed for either composite or component video signals. Component signals may be originated in this format or can be derived from composite video by various decoding techniques such as digital comb filtering.

The "heart" of most predictive techniques is the differential pulse-code modulation (DPCM) loop as shown on Fig. 18-44. The DPCM predicts the value for a video sample based on the past history of the encoded signal. The predicted value is subtracted from the actual sample, and the resultant error is quantized and transmitted. The receiver performs the same process and adds the received error signal to its prediction to obtain the correct video sample.

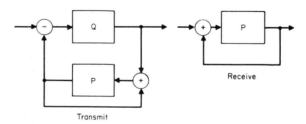

FIG. 18-44　Differential pulse-code modulation.

Among the many excellent DPCM schemes, the one given in Ref. 25 is chosen here for illustration purposes since it utilizes a number of salient features of DPCM. The algorithm employs two DPCM loops, both of which are operating on the video samples. One loop performs temporal prediction (i.e., calculates the value of the current pixel on the basis of pixels from previous frames) and the other performs spatial prediction (based on pixels within the same frame). The prediction errors of the two loops are compared (on an eight-pixel block basis), and the smaller of the two is transmitted. The rationale for this is that violent motion—which produces large temporal error—will not be perceptible by the human visual system. Similarly, details of chaotic scenes are usually not detectable by the television viewer.

For this particular algorithm the intrafield, interfield, and interframe parameters are

$$P(z) = \tfrac{3}{4}z^{-1} + z^{-2L} - \tfrac{3}{4}z^{-2L-1}$$
$$= \tfrac{3}{4}z^{-1} + z^{-F} - \tfrac{3}{4}z^{-F-1} \qquad (18\text{-}110)$$
$$= -z^{-L} + z^{-2F} + z^{-F-2L-1}$$

where L and F denote line and field delays, respectively. This coding scheme provides excellent picture quality at DS-3 rates.

Transform Coding. Experience indicates that bandwidth reduction is sometimes easier if the signal is first subjected to a linear transformation prior to prediction and quantization. This is depicted in Fig. 18-45.

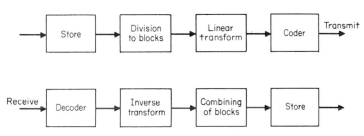

FIG. 18-45 Transform coding.

Ideally, data corresponding to an entire field of video information should be transformed. However, this is impractical owing to the large volume of data involved. Therefore, video data are subdivided into smaller blocks, and each individual block is transformed using a linear operation.

The transformation can be expressed mathematically as

$$X_{kl} = \sum_{m=1}^{M}\sum_{n=1}^{N} a_{nm}P_{mn} \qquad m = 1, \ldots, M, \quad n = 1 \ldots, N \qquad (18\text{-}111)$$

where X_{kl} = transformed variables
   P_{nm} = picture elements
   a_{nm} = coefficients of matrix corresponding to linear transformation

A number of transform coding techniques have been proposed. The most important of these are listed below.

1. *Karhunen-Loeve transform.* The transformation matrix for this method is formed from the eigenvectors and eigenvalues of the covariance matrix of the data block being transformed. The result of the transformation is totally uncorrelated data; thus the transformation is optimum from a compression point of view. However, since a great deal of calculation (and the a priori knowledge of the data statistics) are required, this method has little practical significance but can be used as a bench mark to evaluate other schemes.

2. *Discrete Fourier transform.* The fast Fourier transform makes the calculations involved in this method viable. However, the result contains highly redundant data; thus DFT is not very desirable from a data compression point of view.

3. *Discrete cosine transform.* The discrete cosine transform (DCT) is a good compromise between performance and complexity. The one-dimension DCT is defined as

$$X(k) = \begin{cases} \dfrac{2}{M} \displaystyle\sum_{m=0}^{M-1} P(m) & k = 0 \\[4mm] \dfrac{2}{M} \displaystyle\sum_{m=0}^{M-1} P(m) \cos \left(\dfrac{2m+1}{M} \right) k & 3k = 1 \cdots M - 1 \end{cases} \tag{18-112}$$

where $P(m)$ is the sequence of samples and M is the number of samples in the block. The inverse transform is given as

$$P(m) = \frac{1}{2} X(0) + \sum_{k=1}^{M} X(k) \cos \frac{k(2M+1)}{m} \tag{18-113}$$

Once the image is transformed, it can be subjected to the same types of bandwidth reduction techniques as the time domain video, for example, DPCM coding.

18.5 TYPICAL APPLICATIONS OF DIGITAL TELEVISION TECHNOLOGY

Until the "all-digital" television studio is realized, the number of digital components in the analog production plant is continuously growing. (See Sec. 21.4.5, Digital Standards for Television Studios.) Furthermore, the various types of equipment available are continuously changing as a result of rapid advances in technology. Therefore, to describe all the available digital tools television engineers have at their disposal is far beyond the scope of this work. Instead, a representative selection of digital hardware used in the production chain will be discussed in this section to provide the rationale for the preceding discussion of theoretical background.

18.5.1 DIGITAL VIDEO SOURCES

Telecine and Camera. During the infancy of digital video technology, the expectations of digital picture sources, such as production cameras or telecine, were high. With the benefit of hindsight, however, it is not surprising that the impact of digital techniques on these devices was slow and gradual in coming. The reason is simple. Analog circuits did the job well, both economically and technically. The advantages of digital techniques including reliability, performance, lower cost, etc., were realized only when the various techniques were developed for other purposes, such as for standard converters, transmission systems, time-base correctors, and synchronizers.

Perhaps the easiest way to understand why the analog components competed so successfully is to consider the telecine processing channel. The argument concerning the telecine is equally applicable to the camera processing channel.[26] Therefore the camera, although it is the most important picture source in a television production, will not be covered separately.

The simplified functional block diagram of a flying-spot telecine channel is shown on Fig. 18-46. The various functions to be performed are:[27]

- afterglow correction to reduce afterglow of the phosphor
- aperture correction to increase high-frequency content
- low-pass filter to remove out-of-band signals
- masking to correct for errors on color rendering
- gamma correction to predistort the signal to reduce nonlinear effects in the display tube

Most of these functions are easily implemented in analog form. In the digital domain, since nonlinear processes are involved, more than the usual resolution is required, that

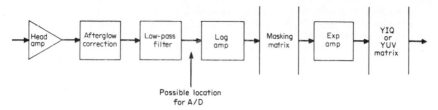

FIG. 18-46 Telecine processing chain.

is, 11 bits per sample. Also, to maintain processing accuracy, 12 to 13 bits should be retained in arithmetic operations. These constraints, plus some of the circuit complexities involved in processes such as color matrix correction, do not favor the digital approach.

Digital techniques, however, especially at the output of the source devices, can offer significant benefits in picture quality. For example, when the film is scanned sequentially using digital field stores, the required interlaced display can easily be realized. Other benefits include means of correction for element-to-element variations of sensitivity of line-array telecine sensors.

Thus, the advantages of digital output and analog head ends are obvious. To cater to both of these is possible by placing the A/D converter within the processing chains. According to current views, the optimum location of the A/D converter is as close to the head end as possible. One alternative is to perform the conversion immediately following the afterglow correction.

Computer Graphics. Digital computers are continuously gaining importance as picture sources. These significant developments are covered in Chap. 19.

Digital Video Effects. This related topic is covered in Sec. 14.6.6.

18.5.2 DIGITAL RECORDING OF TELEVISION SIGNALS. A variety of systems have been proposed for video tape and disk recording in digital formats. In the U.S. digital tape recording techniques are being reviewed by the SMPTE Working Group on Digital Television Tape Recording for the purpose of recommending specifications for industry-wide adoption under the auspices of the American National Standards Institute (ANSI). Parallel activity is being conducted by the Magnetoscope Numerique (MAGNUM) committee of the European Broadcasting Union (EBU) for consideration by the appropriate CCIR study groups. These CCIR study groups, in turn, are responsible for the formulation of recommendations for international exchange of television programs recorded on magnetic tape in a digital format. These developments, because of their complexity and state of flux, are beyond the scope of this handbook. Instead, the reader is referred to recent papers on the subject.[28-31]

Video-disk recording systems use several different combinations of analog and digital techniques, not yet standardized by industry. These are described in Chapter 16. For recent progress on system standards, the reader is referred to a SMPTE report on the video disc system.[32]

18.5.3 TIME-BASE CORRECTION. Although time-base correction concerns primarily video tape recording (Sec. 15.5.2), it is covered briefly here, because of both its historical and technical importance. On one hand, the digital time-base correctors (TBCs) are indispensable components of present-day television studios for reasons mentioned below. On the other hand, the synergism between the development of the ¾-in (1.9-cm) tape technology and the digital video hardware coupled with signal processing required for the economical realization of TBCs helped the rapid maturity of both fields of discipline.

Video Tape Recording. Time-base correctors are employed in controlling time-base errors present during playback of video tapes. These errors are elastic variations of the video signal caused by VTR perturbations, nonuniform physical and chemical properties of the tape, and the mechanical tolerances of recorder and player mechanisms. For helical-scan type equipment the uncorrected time-base error can be as large as one horizontal line period, that is, 63 μs.

These errors are usually reduced in two stages. First, the VTR is locked to a stable external sync source (either V or H, or both) to ensure that the average rate of information flow to the TBC is constant and consistent with the other sources in the plant. Second, the signal is presented to a digital TBC as shown in Fig. 18-47.[33]

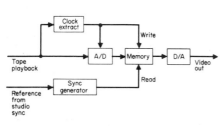

FIG. 18-47   Block diagram of digital time-base corrector.

The output signal from a VTR is digitized and stored in a memory using a "write" clock derived from the sync and color-burst information of the signal itself. Another clock signal, the so-called read clock, is generated using the studio sync and color reference. Reading the information stored in memory with this clock ensures that the signal is in perfect synchronism with other video sources; thus, the mixing of these is possible.

18.5.4 FRAME SYNCHRONIZERS.   Similarly to time base correctors (which store a video line to eliminate elastic variations of video signals), frame synchronizers also use storage techniques to reduce the visual impairments caused by the switching of nonsynchronous sources at the mixing point.

As early as 1973,[34] the necessity of digital storage of a full frame of video information to meet the quality and flexibility requirement of switching nonsynchronous inputs in production studios was well established. Frame synchronizers, consisting of full racks of vacuum-tube electronics, were used. In spite of such a complexity, quality had to be somewhat compromised in favor of reduced memory capacity; i.e., the early versions of the frame synchronizers used 3-f_{sc} sampling and 8 bits per sample quantization.

In less than a decade, steady progress in solid-state memory technology allowed the reduction of the size of frame synchronizers to a single shelf. This is over a fortyfold decrease in size. In addition to the obvious cost advantages, these new devices offer better quality (4-f_{sc} sampling and 9-bits/sample resolution) as well as extra features such as intergral or external freeze-frame capabilities.

A simplified block diagram of a frame synchronizer is shown in Fig. 18-48. The principle of operation for this circuit is identical to that of a time-base corrector except that it contains a write-inhibit feature. The purpose of this is to detect a nonsynchronous switching of sources and halt the writing into memory until the beginning of a new frame in order to eliminate disturbances at the output video.

18.5.5 SPECIAL EFFECTS.   Special-effect generators utilize frame stores to manipulate the image in various creative ways. The very important topic of special effects is covered under Video Special Effects in Sec. 14.6.6 and in Chapter 19, "Digital Video Effects."

18.5.6 STANDARDS CONVERSION.†   The inauguration of the Eurovision television network in the early 1950s created the first need by broadcasters for conversion of picture signals among the British 405-line, and the continental European 625- and 819-line, monochrome transmission standards. Prior to that time, program interchange had been handled exclusively by the universal medium of motion-picture film, either with direct photography or film recording from a black-and-white picture monitor.[35]

† From Ref. 36, by permission of Intertec Publishing Corp.

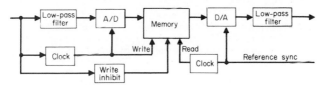

FIG. 18-48 Simplified block diagram of frame synchronizer.

However, in order to broadcast news events and programs produced from electronic television cameras and transmitted over the Eurovision network, conversion to the local standard was required. To cope with this problem, concurrently and independently broadcasters and manufacturers in Britain, France, and Germany undertook the development of the necessary equipment. In 1953 and 1954 the BBC in England, and the RTF in France, installed optical conversion systems of their own design using a picture monitor and camera, while in Germany the post office installed a similar optical system made by Fernseh to feed programs to parts of Belgium using 819-line standards.

On the other hand, program exchange with North America continued on film for several years until the advent of video tape in the late 1950s. Unlike film, as the name implies, video tape provides a picture signal that requires conversion of line and frame rates for playback on different standards. This posed a technical problem not encountered in the line-conversion equipment used in Europe, because of the different field rates of 50 and 60 Hz, and the resultant 10-Hz flicker. This was solved by a manually adjusted waveform generator, the output of which modulated the gain of a video amplifier at the flicker rate.

Although the optical converters provided signal quality acceptable for news and special events, by present-day standards the performance was less than marginal. For example, the signal-to-noise ratio was no better than 40 dB. The center and corner resolutions at 320 lines were down 2 and 4 dB, respectively. And particularly annoying was the fixed phosphor-grain pattern of the monitor screen. Thus, it was not until the introduction of all-electronic, digital systems that truly transparent conversion became possible.

Digital Conversion Developments. The first all-digital color standards converter, designed by the BBC,[36] was put into regular program service in London in 1967. Since that time, in part as a result of the great strides made in digital solid-state technology, several other organizations in England and Japan have developed and are manufacturing converters.

The Independent Broadcasting Authority (IBA) in England demonstrated its first digital intercontinental conversion equipment (DICE) to the European Broadcasting Union (EBU) Technical Committee in March 1971, after which it was used to serve the IBA's viewers in the London area, replacing the analog line-store converter.[37] This was followed by the development of a field-store converter in 1972 for the use of the Independent Television News (ITN) in London for news events received and transmitted overseas by satellite and video tape. Subsequently, Marconi was licensed to manufacture the DICE converter.

At the same time, Quantel in England introduced its DSC 4002 digital standards converter. Several of the Quantel converters, as well as the Marconi DICE, found immediate use by broadcasters and production companies. Later a new, more sophisticated BBC design, the advanced conversion equipment (ACE), was offered by McMichael, Ltd., in England, and in Japan. Oki announced its compact LT-1200 portable unit and the larger LT-1015 model intended for fixed installations.

All the various systems follow the same basic design principles. A minimum of two consecutive fields is converted from analog signals to binary digits (bits) and stored in a solid-state memory. The stored bits are read out at the rates of the converted standard, with complex interpolation among lines and fields to reduce, or avoid completely, uneven motion and fragmented reproduction of vertical and diagonal lines in the converted picture.

Interpolation. Two types of interpolation are employed in all four manufacturers' converters. Vertical or spatial interpolation is used to eliminate discontinuities in picture image lines resulting from the difference in number of scanning lines. The simplest method of vertical interpolation generates the output signal by combining picture information from the nearest two or three lines in a ratio determined by their proximity to the output line.

Field or temporal interpolation is used to reduce the uneven motion of moving objects that results from the different field-scanning rates. The process of interpolation between fields in time is similar to that of vertical interpolation between lines. The contribution from each input field to the output picture is determined by the proximity in time of the output field to the input fields.

The IBA/Marconi DICE System. Digital-storage capacity is *two fields* and the spatial interpolation is *five lines.* Field or temporal interpolation is obtained by mixing the simultaneously available outputs of the two stores in varying proportions so that the output velocity of moving objects equals the input velocity. In other words, the proportions are varied according to the timing of the input and output fields. For a 525/625-line conversion, five coefficients are used: 90, 70, 50, 30, and 10 percent. Subjectively, this method of interpolation is better in terms of movement, or so-called *judder,* than that of a conventional 24/30-frame telecine.

The MCI/Quantel converter also utilizes a two-field store and provides both spatial and temporal interpolation. An operating control is provided for adjustment of the degree of temporal interpolation or judder reduction to best favor the type of action in the picture.

The ACE of BBC and McMichael uses four fields of storage to provide an extremely smooth temporal interpolation. Spatial interpolation is over $\pm$ two lines, or a total of five lines.

Oki Electric Industry Company's converter is made in two models, one with three fields of storage, and the other with two fields. The model LT-1200 is the most compact converter available at this time, occupying only 10½ in (27 cm) of rack height. It is essentially a single-standard converter, since the appropriate decoder and coder modules must be inserted for the desired input and output standards.

Future Developments. In a little over a decade, the use of digital techniques for frame store in television equipment has become commonplace, albeit still relatively expensive. In the next decade, however, a drastic reduction in the cost and size of solid-state storage components may be expected. This will permit a wider use of standards conversion systems to a degree that the variety of standards will no longer be a restriction on program exchange among countries. In fact, it may be anticipated that the future television frame store will be manufactured on a single chip costing no more than $5, thus making a multistandard receiver technically and economically feasible.

18.5.7 DIGITAL TELEVISION RECEIVERS.

At the time of writing this section, digitization of the receiver is in the early stages of development. Nevertheless, the trends are already well established.

The key to digital receiver design is VLSI technology.[38] The digital circuitry equivalent to about 300,000 transistors, which perform the functions traditionally accomplished by analog circuits, is placed on a single semiconductor chip or a small set of semiconductor chips. The television receiver is fed an analog video signal modulating a radio-frequency carrier. To digitize it at such high frequency not only is impractical but would require an RF-channel bandwidth far in excess of that available in the VHF and UHF allocations. Therefore, the signal is demodulated by analog circuitry to a 4- or 5-MHz-wide baseband signal before presentation to the digital circuit.

The functional block diagram of the receiver is shown in Fig. 18-49. The clock driving the digital circuits is generated from the color subcarrier present on the horizontal sync of the baseband video. The audio and video signals are separated prior to the digitization process to take care of the different bandwidths and dynamic ranges of the two signals.

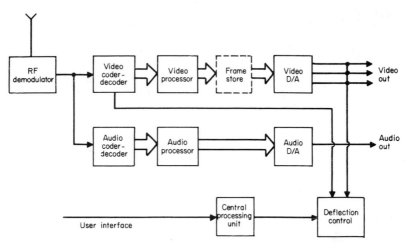

FIG. 18-49　Block diagram of digital television receiver.

The video information is digitized using three or four color-subcarrier frequencies and an effective resolution of about 8 bits per sample. The video processor performs the necessary operations of filtering, decoding, noise reduction, etc. The audio signal is sampled at about 32 kHz and digitized in 14-bit resolution. The audio processor performs the usual control functions associated with high-fidelity stereo equipment (e.g., base-treble control).

The deflection control regulates the horizontal and vertical flyback in a precise manner, thus eliminating a major source of flicker when interference is present. Frame store can be used optionally for freeze-frame display, slow motion, and pseudo-high resolution by rescanning the image with interpolated lines (possibly derived by using motion-detection techniques).

The brain of the system is the *central processing unit (CPU)*. This microcomputer-based device controls and coordinates all other circuits and provides a user interface. With this interface the user can achieve full flexibility and can display multiple signals, teletex, compute information, stretch or shrink the picture, to cite a few operations.

In summary, the use of digital circuitry in television receivers promises to provide cost-effective designs with many optional features compatible with the growing demands of the home viewer.

REFERENCES

1. David A. Howell, "A Primer on Digital Television," *J. SMPTE,* vol. 84, pp. 538–542, July 1975.

2. A. A. Goldberg, "PCM-Encoded NTSC Color Television Subjective Tests," *J. SMPTE,* vol. 82, pp. 649–654, August 1973.

3. A. A. Goldberg, "PCM NTSC Television Characteristics," *J. SMPTE,* vol. 85, pp. 141–145, March 1976.

4. H. Nyquist, "Certain Factors Affecting Telegraph Speed," *Bell System Tech. J.,* vol. 3, pp. 324–346, March 1924.

5. "Proposed Standard for Performance Measurement of A/D and D/A Converters for PCM Television Video Circuits, Project No. 746, Audio-Video Techniques Committee, IEEE Broadcast Technology Society, 1983.

6. R. W. Hamming, *Coding and Information Theory,* Prentice-Hall, Englewood Cliffs, N.J., 1980.

7. George C. Clarc, Jr. and J. Bibb Cain, *Error-Control Coding for Digital Communication,* Plenum, New York, 1981.

8. Elwyn R. Berlekamp, "The Technology of Error-Correcting Codes," *Proc. IEEE,* vol. 68, no. 5, pp. 564–593, May 1980.

9. Djimitri Wiggert, *Error-Control Coding and Applications,* Artech House, Delham, Mass., 1978.

10. R. W. Hamming, "Error Detecting and Correcting Codes," *Bell System Tech. J.,* vol. 29, pp. 147–160, 1950.

11. A. J. Viterbi, "Convolutional Codes and Their Performance in Communication Systems," *IEEE Trans. Commun.,* vol. COM-19, no. 5, October 1971.

12. E. Dudois, M. S. Sabri, and J.-Y. Ouellet, "Three-Dimensional Spectrum and Processing of Digital NTSC Color Video Signals," *J. SMPTE,* vol. 91, no. 4, pp. 372–378, April 1982.

13. J. R. Rossi, "Sub-Nyquist Encoded PCM NTSC Color Television," *J. SMPTE,* vol. 85, pp. 1–6, 1976.

14. "Digital or Mixed Analog and Digital Transmission of Television Signals," Report 646-1, CCIR Study Group CMTT, Geneva, 1982.

15. "Digital Television Transmission—General Principles," Recommendation 604, CCIR Study Group CMTT, Geneva, 1982.

16. "Digital Coding of Colour Television Signals," Report 629, CCIR Study Group 11, Geneva, 1982.

17. "The Filtering, Sampling and Multiplexing for Digital Encoding of Colour Television Signals," Report 962, CCIR Study Group 11, Geneva, 1982.

18. "Encoding Parameters of Digital Television Studios," Recommendation 601, CCIR Study Group II, Geneva, 1982.

19. Alan V. Oppenheim and Ronald W. Schafer, *Digital Signal Processing,* Prentice-Hall, Englewood Cliffs, N.J., 1975.

20. Laurence R. Rabiner and Bernard Gold, *Theory and Application of Digital Signal Processing,* Prentice-Hall, Englewood Cliffs, N.J., 1975.

21. D. C. Rief and G. A. Vincent, "Use of Discrete Fourier Transform in the Measuring of Frequencies and Levels of Tones," *Bell System Tech J.,* pp. 197–228, February 1970.

22. William K. Bucklen, "A Monolithic Video A/D Converter," *120th SMPTE Tech. Conf.,* November 1978, New York.

23. Members of the Technical Staff, *Transmission Systems for Communication,* 4th ed., Bell Telephone Laboratories, Inc., 1971.

24. Denis J. Connor, Ralph C. Brainard, and John O. Limb, "Intraframe Coding for Picture Transmission," *Proc. IEEE,* vol. 60, pp. 779–791, July 1972.

25. Y. Hatori, H. Murakami, and H. Yamamoto, "30 Mbits/s Codec for NTSC-CTV by Interfield and Intrafield Adaptive Prediction," *Proc. ICC,* 1979, Boston, pp. 23.6.1–23.6.5, June 1979.

26. J. Richard Sanders, "Digital Methods in Picture Originating Equipment—An Overview," *J. SMPTE,* vol. 89, pp. 938–942, December 1980.

27. Andrew Oliphant and Martin Weston, "A Digital Telecine Channel," *J. SMPTE,* vol. 88, pp. 474–480, July 1979.

28. J. K. R. Heitmann, "An Analytical Approach to the Standardization of Video Tape Recorders," *SMPTE J.,* vol. 91, no. 3, p. 229, March 1982.

29. H. Yoshida and Takeo Eguchi, "Digital Video Recording Based on the Proposed Format from Sony," *SMPTE J.*, vol. 92, no. 5, p. 562, May 1983.

30. Aleksandar Todorovic, "The MAGNUM Specialist Group and the EBU Approach to the Video Tape Recorder," *SMPTE J.*, vol. 92, no. 5, p. 568, May 1983.

31. J. K. R. Heitman, "Digital Video Recording: New Results in Coding and Error Protection," *SMPTE J.*, vol. 93, no. 2, p. 140, February 1984.

32. Michael J. Doyle, "The Video Disc System: A Technical Report by the SMPTE Study Group on Video Disc Recording," *SMPTE J.*, vol. 91, no. 2, p. 180, February 1982.

33. Koichi Sadashige, "Overview of Time Base Correction Techniques and Their Applications," *J. SMPTE*, vol. 85, pp. 787–791, October 1976.

34. Takehiro Yosnio and Arika Ohya, "Digital Frame Memory for Still Picture Television Receivers—PASS Encoding System and Applications," NHK Laboratory Notes, no. 235, March 1979.

35. K. B. Benson, "Television Standards Conversion Techniques," *J. SMPTE*, p. 628, August 1961.

36. K. B. Benson, "TV Standards Conversion," *Broadcast. Eng.* p. 46, October 1982.

37. J. L. E. Baldwin, et al., "Digital Video Processing—DICE," *IBA Technical Rev.*, Independent Broadcasting Authority, London, pp. 3–84, September 1976.

38. Erick J. Lerner, "Digital TV: Makers Bet on VLSI," *IEEE Spectrum.*, vol. 20, no. 2, pp. 39–43, February 1983.

Digital Video Effects

Alistair Cockburn

IBM Corporation
with contributions by

Carl Ellison

Zurich, Switzerland
Evans and Sutherland
Salt Lake City, Utah

19.1 TERMINOLOGY AND BASICS

Digital graphics comprise a large and varied set of effects for television. These techniques, being digital, are general, powerful, and repeatable. Since *digital* is almost synonymous with *computer,* the entire field of computer processing is brought to bear upon the subject, particularly digital signal theory, image processing, data base design, and systems architecture. To keep the quantity of information under control, this chapter deals with only two issues: (1) images generated by digital equipment and (2) images manipulated in space by digital equipment. The two share few terms with the other disciplines of the television industry, so that a new vocabulary has developed. The chapter is confined primarily to the identification of common algorithms and ideas, since the field of digital design is changing very rapidly and the architectures of most equipment are driven by the algorithms. The implementation in current electronic hardware is left to the reader, who will find specialized publications helpful.

19.1.1 TERMINOLOGY.   The primary terms are: *frame store, frame buffer, pixel, input sample, output sample, filtering, aliasing,* and *color table.* Other important terms are *valuator, paint systems, boundary systems, area systems,* and *run-length encoding.*

Input Sample, Output Sample.   Given a full field of video, the waveform may be sampled in time for its amplitude. According to the Whittaker-Shannon sampling theorem, the samples must occur at twice the maximum input frequency if the waveform is going to be reconstructed. The stream of input samples may then be stored and manipulated. Reversing the process, some set of discrete values *(output samples)* may be used directly in a sample-and-hold circuit or with an interpolation function to generate a new field of video (Fig. 19-1). The sampling theorem holds here also, so that it is not possible to generate a pattern of higher frequencies than twice the spacing of the output samples. The term *spacing* indicates both time and space, since a spatial distance on the television

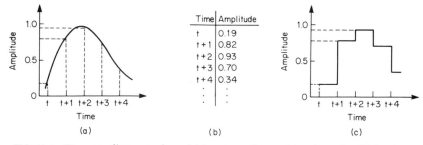

FIG. 19-1   Time-sampling a waveform: (*a*) Input waveform and input samples; (*b*) input samples stored in a frame store; (*c*) output waveform constructed from frame-store samples (sample and hold).

screen corresponds to time distance in the generation of the signal. Time and space will be used interchangeably quite often. (See Frame Buffer.)

Pixel. A unit of area on the screen. When the input samples correspond identically to the output samples, pixel is used unambiguously to mean either. Otherwise its use must be clarified, since pixel or pixel area may also indicate the area to be converted into an input sample or the area to be regenerated from an output sample.

Frame Store. A digital memory which samples the input video at a particular rate and precision, and stores the samples. The information is read out later, faster or slower, in synchronization with some other condition.

Frame Buffer. A device like a *frame store,* but with two significant differences. (1) The information stored is not the value of the input waveform, but a description of the color at that point. The description may be the percentages of *red, green,* and *blue,* or *I, Q,* and *Y,* or even just a number indicating which of a choice of colors is being used at that point. (2) The data are generally available to a computer on a random-access basis, allowing the computer to read or modify the color of any point on the screen. In many applications there is no input video at all, the computer manufacturing the appearance of the image from algorithmic definitions. The output of a frame buffer must be converted into an appropriate output waveform (RGB, NTSC, PAL, SECAM, or HDTV). When a computer generates a shape (a circle, for example), it may *sample* the space to discover which points on the screen *(output samples)* are covered by the circle. This process is a two-dimensional sampling operation, similar to the one-dimensional time sampling mentioned above (see Figs. 19-1 and 19-2).

Valuator. A system which senses the position of a pointer (generically, *pen*) and sends the position to the computer. Valuators have one or more degrees of freedom, yielding as many values to the computer. Values may indicate x position, y position, speed, height, angle (roll, pitch, or yaw), pressure (in some number of directions), etc. Principles of operation and number of degrees of freedom vary from device to device. Devices include *tablet, mouse, joystick, trackball, touchpad,* and *opto-* and *acoustic sensors.*

Paint System. A program running on a computer in conjunction with a frame buffer and valuator, simulating the effect of paint on a canvas. Using the valuator to indicate at various times brush size, color, and position, and drawing on the programmer's understanding of paint media, a paint system can simulate oil, acrylic, watercolor, airbrush,

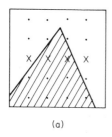

(a)

Scan line	Sample	Value
3	0	0
3	1	1
3	2	1
3	3	0
⋮	⋮	⋮

(b)

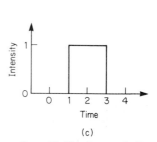

(c)

FIG. 19-2 Spatial sampling of a shape. (*a*) Triangle on a sampling grid. The dots mark the sample locations, the ×'s mark four sample points on scan-line 3. (*b*) The samples of scan-line 3 stored in a frame store. Samples 1 and 2 landed on the white triangle and have intensity 1. Samples 0 and 3 landed outside the triangle and have intensity 0. (*c*) Output waveform for scan-line 3. The two white samples represent the intersection of scan-line 3 with the triangle.

pastel, or any other two-dimensional medium to produce an image indistinguishable from a video photo of the original.

Run Length. A section of a scan line which is all the same color is referred to as a *segment*, or *run*. Most computer-generated images use a preponderance of segments six or more samples long. Describing the picture by the lengths of the runs can lead to a substantial data reduction compared with specification of each pixel.

Boundary System. A system whose primary work concerns the discovery of the outlines of objects. A boundary system manipulates the outline and allows the interior to be filled trivially or by an *area system*. Boundary systems include character generators and symbol generators.

Area System. A system whose primary work concerns the coloration within a shape. Area systems work with the samples of a portion of a frame or the entire frame. Paint systems, Squeeze/Zoom boxes, and digital frame manipulation systems are examples of area systems. Hybrid boundary-and-area systems control both boundary and coloration.

Colormap or Color Table. A table of color definitions. The entries in the table specify the color to be displayed, usually by *R, G,* and *B* components. It is used in conjunction with a frame buffer, whose pixel contents are not the actual color values but instead the index number of the color table entry defining the color (Fig. 19-3). The digital system reads the frame buffer at video speeds, uses each value to index into the colormap, and sends the color value to be converted into analog (Fig. 19-4).

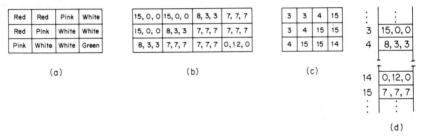

FIG. 19-3 Frame buffer with color table. (*a*) Twelve pixels are represented with their desired colors. (*b*) The colors are given numerical quantities of red, green, and blue. (*c*) Instead of storing the color quantities at each pixel, a code is given to each unique color, and that code is stored. (*d*) The color table provides a means for getting from the color code, or color number, to the color definition itself, which specifies amounts of red, green, and blue.

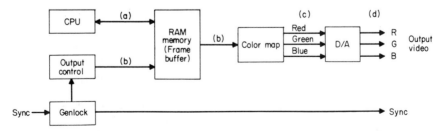

FIG. 19-4 Block diagram of frame-buffer/colormap system. (*a*) The computer writes into the frame buffer at its leisure. (*b*) The frame buffer contains color codes which are read out at video rate in order to display the picture. (*c*) The color codes are translated into red, green, and blue digital quantities. (*d*) The D/A converter generates a video waveform for each of red, green, and blue.

Filtering and Antialiasing.　Low-pass filtering a time-based signal is common. Failure to do so when appropriate leads to spurious artifacts in the output waveform called *aliases*. The shapes that digital graphics systems handle are defined with sharp boundaries and hence have very high spectral components. Failure to properly low-pass filter the image leads to stairsteps and other aliasing artifacts. Spatial aliasing artifacts include stairsteps and moiré patterns. Aliasing effects occur in time as well as space. Temporal aliasing effects include the illusion of the wagon wheel which appears to turn backward, the ragged moving stairsteps formed when panning a video camera across basketball court lines, and the jumpiness of stop-frame animation.

19.1.2　FRAME BUFFERS, LINE BUFFERS.

Given a digital memory and the concepts of *frame buffer, frame store,* and *color table,* the system designer can arrange an almost unlimited number of storage alternatives to satisfy cost, performance, and capability goals. A few of the alternatives are described in this section, sufficient for an understanding of the many types of devices available.

The variables in the system design are the amount of the screen to be buffered, the number of copies of the data, the resolution, and the precision of the data. Spatial resolution will be ignored in this discussion, since it affects the image quality and cost in a fairly straightforward manner.

Single-Buffered Frame Buffer.　Figure 19-4 shows a single buffer which stores the entire frame. Figures 19-2 and 19-3 show portions of the contents and part of the resulting waveform. Figure 19-5 shows how the two-dimensional screen can be numbered to fall naturally into a one-dimensional memory.

The characteristics of a single-buffered frame buffer are these. There must be enough memory to hold an entire frame of data. A system using 485 visible lines, 512 pixels per line, and 8 bits per pixel will require 1,986,560 bits of storage. Since there is only one buffer, the computer cannot access memory while the video circuit is reading out of it. Either the computer is constrained to operate during the blanking intervals, or the memory must be organized so that a video access reads more data than are needed, freeing the next memory access(es) for the computer.

As frame-buffer systems continue to evolve, many unusual memory organizations are constructed, allowing the computer to access not just one pixel, but two, 64, or even more pixels at a time, grouped as a row, rectangle, or other shape. The purpose of these organizations is to allow faster modification of the frame buffer. For example, a memory whose architecture allows any arbitrary rectangle to be written in one pass greatly simplifies the drawing of solid shapes.

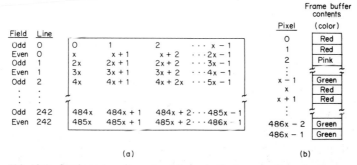

(a)　　　　　　　　　　　　　　　　(b)

FIG. 19-5　Storing a two-dimensional frame buffer in a one-dimensional memory. (*a*) The frame buffer is numbered left to right, top to bottom. In this case, there are 486 lines of *x* pixels each. (*b*) The sequential numbering serves as the address to a table (the frame buffer) which stores the data. The data shown in the illustration reveal that the picture has red on the left side of the screen and green on the right.

Double-Buffered Frame Buffer.   When there are two copies of the frame buffer, the computer can update one while the video circuit reads the other. When both finish, access is switched and the video circuit reads what was just written while a new copy is formed. Double buffering has the advantage of speed and the disadvantage of cost and space.

Line Buffer.   A small section of the screen may be buffered, or no buffer used at all (in which case the frame is generated on the fly). A common small size is one scan line. In a *line-buffered* system, the computer must generate data at video speeds, averaged over a scan-line time. Usually, line-buffered systems are double buffered, and the computer fills in the line just below the one being displayed (Fig. 19-6).

An advantage of a line-buffer system is that it can respond quickly. A disadvantage is that the computer must be able to generate the data at near-video rates.

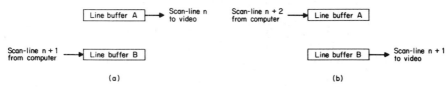

(a)   (b)

FIG. 19-6   Double-buffered line buffers. (*a*) While scan-line *n* is being displayed out of line-buffer A, scan-line *n* + 1 is being written into line-buffer B by the computer. (*b*) Scan-line *n* + 1 is then read out of line-buffer B, and the computer is free to write scan-line *n* + 2 into line-buffer A.

No-Buffer Systems.   If no buffer or a line buffer is used, a global description of the contents of the frame must be stored elsewhere. In a character generator, for example, the number of different shapes on the screen is fairly small, and all can be predefined. In this case, the definition of the letters may be stored in one place and the positioning of the letters stored in another. When the beam scans a given area of the screen, the shape definition is read out at video speeds. Such a character generator would be called a no-buffer system.

Shapes may be described by primitives other than pixels. The character generator just mentioned is one. Other primitives include run lengths, rectangles, conic sections, and triangles. In any no-buffer system, the transformation from primitive shape to scan-line data occurs at video rates.

19.1.3 COLORMAPS, GAMMA CORRECTION.   Colormaps store the translation from a color number to a color definition. They were originally used to reduce cost: It costs less to store four bits of color number at each pixel than to store 24 bits of color definition. Since most computer-generated pictures use less than 256 colors on any frame, and often less than 16, 8 bits of color number, and often 4 bits, suffice. Colormaps have two uses other than saving storage: color correction, including gamma correction, and colormap animation.

Colormap animation is a technique in which several frames of data are written into the frame buffer at one time, and by suitable camouflage are revealed in the proper sequence. Since the frame buffer contains numbers which do not represent colors directly but whose colors must be discovered via the colormap, then by redefining the meaning of the color numbers the same frame buffer information may reveal different pictures. In Fig. 19-7, a frame buffer snapshot is shown with color numbers but without color definition. A seven-frame sequence of a bouncing ball is derived in this manner:

(Frame 1) Set color numbers 0 and 2 through 9 to white, 1 to red.

(Frame 2) Change color number 1 to white, 2 to red.

(Frame 3) Change color number 2 to white, 3 and 4 to red.

(Frame 4) Change color number 3 to white, 5 and 6 to red.

(Frame 5) Change color numbers 4 and 5 to white, 7 to red.

(Frame 6) Change color numbers 6 and 7 to white, 8 to red.

(Frame 7) Change color number 8 to white, 9 to red.

On a frame-by-frame basis very little work is done, but the effect is true animation. The disadvantage of colormap animation is that if there are many overlapping areas, the number of colors used goes up steeply. The advantage is that when colormap animation is possible, very little compute power is needed for a powerful effect.

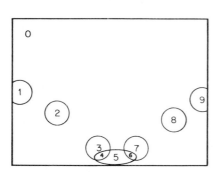

FIG. 19-7 Color table animation. By alternately changing the colors of the ball to match or contrast with the background, this single frame may be turned into a seven-frame sequence of a ball bouncing.

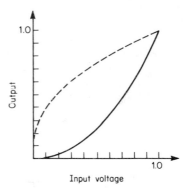

FIG. 19-8 Gamma distortion and correction curves. The solid curve shows the nonlinear effect on the input voltage at the CRT, where $V_{\text{out}} \approx V_{\text{in}}^{2.5}$. The dashed curve represents the function required to compensate for this, $V_{\text{comp}} \approx V_{\text{in}}^{0.4}$.

Gamma Correction. It is convenient to assume that color behaves linearly, that is, 0.25 red plus 0.25 red yields 0.50 red. This is not a valid assumption, since the CRT-drive electronics react exponentially with the input: $V_{\text{out}} \cong V_{\text{in}}^{2.5}$. The exponent 2.5 is referred to as *gamma* and varies, generally between 2.4 and 2.9, from one CRT to another. *Gamma correction* is the process of raising the voltage value to the 1/gamma power prior to output, so the final voltage is nearly linear: $V_{\text{out}} \cong (V_{\text{in}}^{0.4})^{2.5} \cong V_{\text{in}}$ (see Fig. 19-8). Approximations to the gamma-correction curve do quite well; other factors, like the black-threshold adjustment, mask errors in gamma. The effect of ignoring gamma, however, is quite noticeable if linearity of the color space is assumed when computing color values. When the color range is very limited or the digital system is low-cost, gamma correction is frequently ignored.

Gamma correction may be performed digitally or in analog. The analog method is to put a three- or four-stage linear approximation function after the digital-to-analog (D/A) circuit. The circuit in Fig. 19-9 shows a three-stage straight-line approximation circuit. The slope of each line segment is the gain of the amplifier, $G = R_f/R_s$. In this circuit, as the input voltage rises, the zener diodes break down in succession, lowering the gain of the amplifier and the slope of the curve.

To correct for gamma digitally, a look-up table is used, with V_{in} as the table input and $V_{\text{in}}^{0.4}$ as the table output. Colormaps may be used to provide color correction other than gamma correction. They are used to provide analog functions to compensate for phosphor or eye nonlinearities, or specific corrections for video input distortion. In some graphics systems, two colormaps are provided, one for color definition and a second for independent color correction.

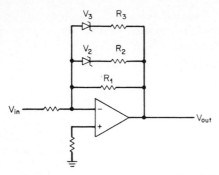

FIG. 19-9 Analog gamma correction. V_2 and V_3 are the voltages at which the zener diodes break down ($V_2 < V_3$). In the range $0 < V_{in} < V_2$, the gain is R_1/R_s; for $V_2 < V_{in} < V_3$, the gain is $(1/R_1 + 1/R_2)^{-1}/R_s = R_1 R_2/R_s(R_1 + R_2)$; for $V_{in} > V_3$, the gain is $(1/R_1 + 1/R_2 + 1/R_3)^{-1}/R_s$. This circuit provides a three-segment approximation to the gamma-correction curve in Fig. 19-8.

19.1.4 THE PIXEL-TIME CORRESPONDENCE. There is a close correspondence between space and time in video systems. A scan line may be thought of as a distance in space (height and width) or an extent in time (microseconds). The equivalence gives system designers flexibility in solving video problems. This section illustrates the difference of vocabulary between the two domains and gives some advantages of working in each.

There is always an upper limit on the number of *decision points* on a scan line, i.e., the positions at which a color transition may occur. The number of decision points is bounded ultimately by the speed of the circuitry. If there are not too many decision points, then the color at each decision point may be stored explicitly, in which case the decision points are referred to as *pixels*. A pixel's size may be defined either as a certain fraction of a scan line (for example, 512 pixels per scan line) or by its duration in time (for example, 100 ns). If there is a much larger number of decision points than actual color transitions in a typical picture, then the time of each transition may be stored instead. This latter method is equivalent to run-length encoding the scan line (see Sec. 19.2.1).

The advantage of pixel storage is that insertion of a new color transition is straightforward: the old color is overwritten with the new. However, the following problems arise:

1. Since the pixels are spread across the screen with uniform density, need for detail in one area requires provision for equal detail everywhere, resulting in excess storage in the more frequent, relatively blank parts of the screen.
2. Scan conversion, or the coloring of a run, occupies time.
3. Increasing the density of decision points increases memory costs exponentially.
4. Simple stretching and shrinking transformations along the horizontal axis require significant data movement.

In run-length encoded systems these problems are exchanged for others. The improvements are:

1. Storage is used only where detail is needed.
2. Scan conversion of a run is unnecessary.
3. Memory needs grow linearly with density of decision points.
4. Horizontal stretching and shrinking may be as easy as slowing down or speeding up the system clock.

The new problems that are encountered are:

1. Determining the amount of memory needed in the life of the system is awkward, since it is related to picture detail.
2. Insertion of new transitions requires a search through the existing transitions.
3. The hardware must be very fast to respond to short transitions. It is the speed of the decision logic that usually determines the density of the decision points.

Most systems work in the spatial domain and store the color at each point (pixel). A few character generators have been developed which operate in the time domain effectively and take advantage of the ability to change the character size horizontally or to allow great precision in aligning diagonal edges.

Presenting pixel size as the number of nanoseconds between decision points is quite common in advertising literature. In the NTSC system, where a scan line takes about 52 μs to trace, a square pixel is about 80 ns wide (646 pixels/line), and 100 ns/pixel results in about 512 pixels per line. (*Note:* The safe-title area presents some difficulty in discussing the length and/or duration of a scan line.) At the time of this writing the largest number of pixels explicitly stored is 2048 (26 ns/pixel), and the finest granularity of decision points in a transition-time system is 12 ns. The use of equivalent resolution makes comparison more difficult. *Equivalent resolution* indicates that the number of decision points is fewer, but tricks are used to enhance the image. The Vidifont V has a 48-ns pixel size, but achieves 12-ns equivalent resolution by filtering the characters off-line.

The use of filtering to increase effective resolution is described in Sec. 19.4. The growing awareness among designers of the advantages of filtering in improving image quality is evidenced by the emergence of high-quality and low-pixel count character generators, terminals, flight simulators, paint systems, and digital video optics systems.

19.2 BOUNDARY SYSTEMS

A boundary system is one whose primary work concerns the discovery of the outlines of objects. The interiors are filled either trivially or elsewhere. Shape generators and character generators are examples of boundary systems. Shapes are scaled, rotated, and deformed, and placed on the screen. In final form the shape may be solid, textured, or colored by external video, and may either be shown on a blank raster or be keyed over another image.

19.2.1 SHAPE GENERATION. The simplest shape is a rectangle. A line is a thin rectangle. The thickness of a line is dependent on the bandwidth of the display. On good *RGB* monitors, each pixel may be individually discernible, and hence the minimum thickness is one pixel. On an NTSC monitor, a colored line must have several hundred nanoseconds of thickness, or two to ten pixels, depending on pixel size.

A shape may be described in a number of ways, depending on the desired characteristics of the system. The following outline summarizes the principal ways to describe a filled shape (see Fig. 19-10):

- *Pixel enumeration:*   Bitmap
                             Run-length encoding
                             Area encoding

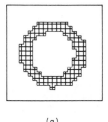 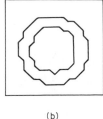 

(a)   (b)   (c)

FIG. 19-10 Small bagel shape defined by (a) pixel enumeration, (b) boundaries in screen space, and (c) boundaries in model space.

- *Boundaries in screen space:*   Vertex coordinates
                                     Adjacent pixel directions
                                     Adjacent line deltas

- *Boundaries in model space*

Pixel Enumeration.   Pixel enumeration involves identifying each pixel that is to be colored. It may be done by listing all pixels or by summarizing them in either a one- or two-dimensional fashion. Pixel enumeration is the way a picture is kept in a frame buffer.

The most direct way to encode or store a pixel enumeration of a shape is to associate one bit of storage for each pixel in its bounding rectangle, storing a 1 if the pixel is covered and 0 if not. This forms a bitmap of the shape and is the basis for bitmapped graphics. The bits are usually arranged left to right, top to bottom, and are often packed eight adjacent pixels to a byte of storage (Fig. 19-11). To display the object, each byte is read out of memory and loaded into a digital shift register at ⅛ video rate. At video rate, the data are shifted out of the shift register (Fig. 19-12). The data output from the shift register does not represent the color of the object, only its presence. A 1 coming out of the shift register enables the output of the shape's color. A 0 disables that color, enabling the output of the background, which is either another color or external video. The types of object manipulation that can be performed conveniently with bitmaps are translation (movement) and certain types of scaling. Translation of a bitmapped object is trivial, as the bitmap may be read starting at any time during the display of the raster. Reduction in size may be accomplished by periodically discarding pixels or lines. Reduction according to any fraction A/B may be done by discarding A pixels out of every B. The choice of which A pixels or lines to discard may be determined by hand or by algorithm. In all cases the picture will suffer quality degradation.

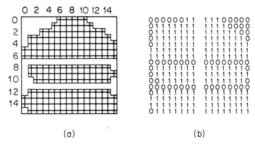

FIG. 19-11 Bitmap of a hamburger icon. (*a*) The desired icon. (*b*) 32 bytes of binary data, 1 bit per pixel, representing the image in (*a*).

A bitmap may be run-length encoded. When pixels are small or shapes are large, the savings can be significant, often on the order of 6/1. Three methods of run-length encoding are start/stop, start/run, and run-length only. In start/stop the pixel number of each transition point is stored, requiring that every number contain enough bits to store any transition point. If there are 2048 pixels per line, every transition number must be 11 bits long and is often 16 bits long (2 bytes) for convenience. In start/length the start point of each ON run is stored with the length of that run. OFF runs are not stored. Since runs can often be assumed to be less than 256 pixels long, the length may be confined to 1 byte, effecting a storage savings of 25 percent. The length/length method carries this one step further, storing only the length of each ON and OFF run, assuming for consistency that the first run on each line is always OFF.

To derive the necessary bit stream for a run-length encoded scan line, a counter is needed to count pixels on the scan line. In start/stop, the count is compared against the pixel number of the next transition point. When a match is found, the bit stream is turned from 1 to 0 or vice versa. It is left as an exercise for the reader to determine the combination of counters and comparators needed to translate the other two encoding forms into a bit stream.

Areas may be abbreviated instead of lines. Areas, usually rectangles, are marked as fully ON, fully OFF, or some of each. In one method of area encoding, only the set of fully ON rectangles is stored. In another, called *quad-tree encoding*,[2,3] the entire screen is divided into four squares (Fig. 19-13) and a value assigned to each: 0 meaning fully OFF, 1 meaning fully ON, anything else meaning some of each and requiring subdivision. If an area requires subdivision, it is quartered and marked in the same manner. This encoding is compact and yields squares which may be easily expanded. In addition, the expansion of the tree may be halted at any time, and if gray is used for the unexpanded areas, then a recognizable image may be obtained long before the expansion is complete, a valuable feature for many applications.

It is not necessary to have access to a full-frame buffer when pixel enumeration is used. Each shape on the screen may have its own definition in memory. The

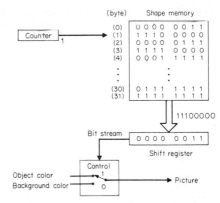

FIG. 19-12 Display a bitmap. The counter started at 0. Byte 0 of the shape memory was read out and is now in the shift register. The counter incremented to 1 and is reading out byte 1 of the shape memory. The shift register will be shifted left in sync with the video. After every eight shifts a new byte from the memory is loaded. The signal coming out of the shift register is used to control the multiplexer. If the signal is 1, the object color is sent out to the screen; if it is 0, the background is sent through.

computer keeps track of which shapes are active on any particular scan line and which line of each shape is being displayed. The manipulations that are easily handled with pixel enumerations of shapes are translation and, with certain restrictions, scaling. With run-length encoding, the shrinking of the image in the Y dimension is as described for bitmaps, above, i.e., by removing scan lines. Scaling run-length encoded images in X may require arithmetic, because each length must be multiplied by the scale factor ½, ⅔, etc. Scaling in the X dimension may also be done by changing the system-clock speed. Rota-

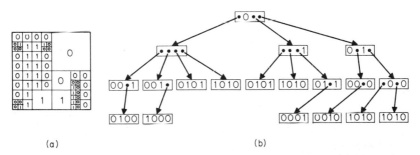

(a) (b)

FIG. 19-13 Quad-tree decomposition of a serif L: (a) shows how each square is quartered until squares are found which are completely black (0) or completely white (1); (b) the same information in a tree structure. At each node, a 0 means this quadrant is all black, 1 means this quadrant is all white, and a dot means as yet unresolved. The arrows indicate the quartering of an unresolved quadrant and point to the component quadrants. Each node is marked to correspond to the picture as (left to right in node): upper-left quadrant, upper-right, lower-left, lower-right.

tions are unwieldy and usually give slightly distorted results, although digital frame-manipulation methods may be used in a general solution (Sec. 19.3).

Boundaries in Screen Space. In order to convert data from boundaries to a bit stream, the shape must be *scan converted,* for which a line or frame buffer is needed. If the boundary definition is contrived so that all boundary changes are specified relative to the line directly above (Fig. 19-14*a*), then only a line buffer is needed and the conversion may occur a line at a time. If the boundary is specified in a closed loop (Fig. 19-14*c*), then a frame buffer or area buffer will be needed.

Once the outline is written into the memory, the computer may be instructed to fill in the interior and generate a bitmap, or the boundaries may be left alone, indicating transition points. Note that in the examples given in the figures, care was taken that every OFF-to-ON transition has a corresponding ON-to-OFF transition, even if both transitions are on the same pixel.

Boundary encoding forms generally are more complex than enumerated forms but result in a reduction of data, and often an increase in system throughput. Object placement on the screen is easily accomplished, but scaling is difficult. Some systems convert to bitmapped form before doing any manipulation. Rotation, again, is difficult.

Boundaries in Model Space. *Model space* is the space or coordinate system used to define the shape without regard to any raster. Points and shapes are defined to an arbitrary degree of precision in the manner most convenient for the application. Defining boundaries in model space is the most powerful, albeit most expensive, form of encoding. This discussion is limited to an introduction to the technique; further information can be found in the Bibliography.

Two letters defined in model space are shown in Fig. 19-15. They employ only lines and circular arcs, although any algebraic curve might be used. In practice, straight lines are the most frequently used, since they involve the least complex math. Depending on the cost of the system, model space may be constrained to closely match the intended frame buffer or may be unconstrained. In all cases, when it is time to place the object into the frame buffer, the two number systems must be matched and the pixels enumerated. To transform the model space coordinates into screen space requires one, two, or three of the transformations: scaling, rotation, and translation (Fig. 19-16).

Figure 19-16 shows the letter *p* both in model space and on the screen with both model and screen spaces marked. Note that model space is a right-handed coordinate

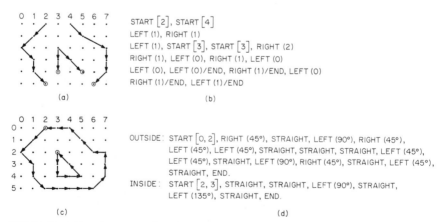

START $[2]$, START $[4]$
LEFT (1), RIGHT (1)
LEFT (1), START $[3]$, START $[3]$, RIGHT (2)
RIGHT (1), LEFT (0), RIGHT (1), LEFT (0)
LEFT (0), LEFT (0)/END, RIGHT (1)/END, LEFT (0)
RIGHT (1)/END, LEFT (1)/END

(a)   (b)

OUTSIDE: START $[0, 2]$, RIGHT (45°), STRAIGHT, LEFT (90°), RIGHT (45°),
LEFT (45°), LEFT (45°), STRAIGHT, STRAIGHT, STRAIGHT, LEFT (45°),
LEFT (45°), STRAIGHT, LEFT (90°), RIGHT (45°), STRAIGHT, LEFT (45°),
STRAIGHT, END.
INSIDE: START $[2, 3]$, STRAIGHT, STRAIGHT, LEFT (90°), STRAIGHT,
LEFT (135°), STRAIGHT, END.

(c)   (d)

FIG. 19-14 Adjacent movement boundaries in screen space. (*a*) Adjacent-line delta encoding. The downward trails always start and end in pairs. (*b*) The way to get from one scan line to the next, always moving downward. (*c*) Adjacent-pixel directions. Every boundary travels in a loop, always moving one pixel at a time. (*d*) The way to travel the loop. The first turn is relative to straight down.

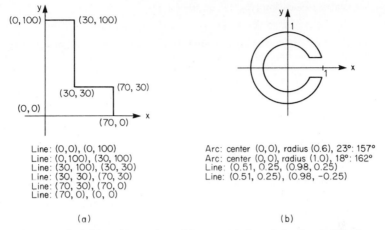

Line: (0,0), (0, 100)
Line: (0,100), (30, 100)
Line: (30, 100), (30, 30)
Line: (30, 30), (70, 30)
Line: (70, 30), (70, 0)
Line: (70, 0), (0, 0)

Arc: center (0,0), radius (0.6), 23°: 157°
Arc: center (0,0), radius (1.0), 18°: 162°
Line: (0.51, 0.25), (0.98, 0.25)
Line: (0.51, 0.25), (0.98, −0.25)

(a)   (b)

FIG. 19-15 Definition of letters in model space: (a) shape defined entirely by line segments; (b) shape defined by circular arcs and line segments.

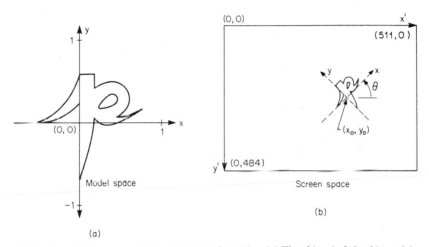

FIG. 19-16 Model-space to screen-space transformation. (a) The object is defined in model space. (b) The object is scaled, then rotated, then translated to its final position in screen space. Note that the Y axis in screen space is positive in the downward direction and that the origin is at the upper-left corner.

system (Y values increase in the upward direction), whereas screen space is a left-handed coordinate system (Y values increase in the downward direction). The p was defined, for convenience, to lie between the values 0 and 1 in model space. In this example the screen contains 485 lines of 512 pixels each. The p is first scaled up to occupy the correct number of pixels, then rotated to the desired angle, then translated to its final position. The conversion from right-handed to left-handed coordinate systems is done by negating the scale factor for the Y dimension. When rotation is not needed, the equations are derived as follows: 1 unit in model space will map to A units (pixels) in screen space horizontally, and B units vertically. After scaling, each $[X, Y]$ in the model space becomes $[X'', Y'']$

$$X'' = XA \qquad Y'' = Y(-B) \qquad (19\text{-}1)$$

The p is translated to location $[X_0, Y_0]$ on the screen

$$X' = X_0 + XA \qquad Y' = Y_0 + Y(-B) \tag{19-2}$$

Note that the vertical distance Y_0 corresponds to the number of scan lines from the top, since screen space is left-handed with the origin at the top-left corner. If the programmer had wished to describe movement from the bottom of the screen, he or she would have used $Y_0 = 484 - Y_1$, where Y_1 is the number of scan lines above the bottom. The general formulation of the multiple transformation (rotate by 0; scale by S_x in X and by S_y in Y; translate by T_x in X and T_y in Y; change to left-handed coordinate system) is given by the matrix equation

$$\begin{bmatrix} X' \\ Y' \end{bmatrix} = \begin{bmatrix} S_x \cos\theta & -\sin\theta & T_x \\ \sin\theta & -S_y \cos\theta & T_y \end{bmatrix} \begin{bmatrix} X \\ Y \\ 1 \end{bmatrix} \tag{19-3}$$

Tiling. The process of filling in a boundary is known as *tiling* or *scan conversion.*

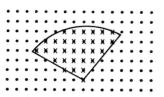

Tiling may be done either by sampling the space in the vicinity of the shape or by analytic means. The example of sampling shown in Fig. 19-17 is described as follows. Sort the boundaries into top-to-bottom, left-to-right order. Treat the scan lines as infinitely thin lines separated by Δy, and pixels as samples along the line separated by Δx. Starting on the left side of the screen with the top scan line that crosses the object, determine whether each pixel lies to the left or the right of the leftmost boundary. Turn OFF all pixels to the boundary's left. Once the boundary has been crossed, turn ON all pixels until the boundary to its right has been crossed.

FIG. 19-17 Tiling a shape by sampling. The ×'s mark which samples were found to be inside the boundaries. The bottom vertex may require special handling since it does not lie between the left and right edges.

Turn OFF all pixels between the second and third boundaries, etc. It is important to identify double crossing points such as the bottom vertex in Fig. 19-17, at which an ON and an OFF boundary meet. These points should be treated as containing two transitions.

Tiling by analytical means is done by decomposing the shape into simple atomic shapes like boxes or small triangles whose areas can be accurately computed. When it is desired to not sample the image as was done in the preceding example, but to work with areas, the pixels are treated as rectangular regions, and the intersection of each pixel region with the shape is determined (Fig. 19-18). The computation that is performed once the intersection is known depends on the application and algorithm. A common

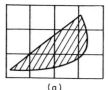

(a)

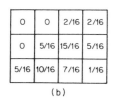

0	0	2/16	2/16
0	5/16	15/16	5/16
5/16	10/16	7/16	1/16

(b)

FIG. 19-18 Analytical tiling. (a) Pixels are shown as rectangular regions and the shape is shown superimposed. (b) Using whatever method is convenient, the fractions of each pixel covered by the shape are computed and stored. The fractions can be used for filtering, transparency, and percentage brightness calculations.

computation is to determine the area of the intersection as a fraction of the pixel area. The frame buffer will then store the percent of the pixel covered rather than just a 1 or 0. Although computationally more expensive, this method produces a vastly superior picture. If translucency is desired, the percent coverage is multiplied by the optical transmittance (between 0.0 and 1.0).

19.2.2 IMAGE PLANES. Image planes are separate frame buffers (or perhaps just area buffers) which can be linked together to assist either in pseudo-three-dimensional animation or to form a composite frame buffer larger than the original in area or depth. A common use of image planes is to allow the operator to dynamically change the size and depth of the frame buffer, e.g., change from a $512 \times 480 \times 8$ organization to a $1024 \times 480 \times 4$ organization. Having more bits per pixel is advantageous when many shades of color are desired and when the frame buffer is going to be fragmented into several shallower frame buffers as described in the next paragraph. Having a larger frame buffer area is helpful in real-time panning (see Fig. 19-19).

It is possible to treat a single-frame buffer as two completely independent frame buffers if a write-protect mask is provided. There is one bit of write-protect mask for each bit of depth in the frame buffer. When a protection bit is set to 0, the corresponding bit of any location cannot be altered; when it is a 1, the corresponding bit of any location may be altered.

The following example illustrates the use of the write-protect mask to simulate a double-buffered 1-bit frame buffer with a single-buffered 2-bit frame buffer (see also Fig. 19-20). The 2-bit frame buffer may be thought of as having a front plane and a back plane. Setting the write-protect mask to (1, 0) allows only the front plane to be altered (the most significant bit of each pixel value), and setting it to (0, 1) allows only the back plane to be modified (the least significant bit of each pixel value). Note that pictures may now consist of only two colors, since there is only 1 bit per pixel available in each plane. The protection mask is initially set to (1, 0), and the first picture drawn. The four-entry colormap is loaded with the colors black, black, red, and red in locations 0 to 3, respectively. At this point the front picture is displayed correctly on the screen regardless of the contents of the back plane. The mask is next changed to (0, 1) and a new picture is drawn.

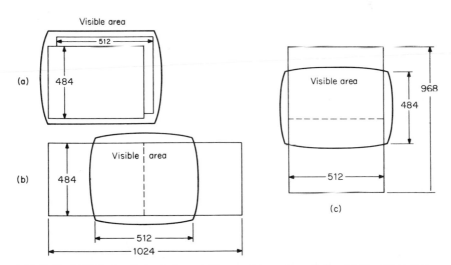

FIG. 19-19 Using image planes to dynamically reconfigure a frame buffer. (*a*) The $512 \times 484 \times 2$ memory allows four colors per pixel. All pixels are visible. Panning is not allowed. (*b*) At $1024 \times 484 \times 1$, only two colors per pixel are allowed, but horizontal panning is possible. (*c*) Vertical panning is possible at $968 \times 512 \times 1$, two colors per pixel.

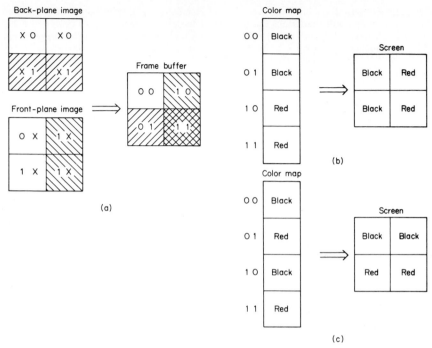

FIG. 19-20 Simulating double-buffering using write-protection masks. (*a*) Starting with a single-buffered frame buffer with two bits of storage per pixel and a protection mask, two different sets of data are written. The front image plane is loaded when the back one is protected, and the back plane is loaded when the front one is protected. (*b*) By organizing the colormap correctly, only the image in the front plane is seen. The back plane can be loaded invisibly. (*c*) Changing only two colormap locations and the write-protection mask, the back-plane image is displayed and the front plane can be reloaded invisibly.

The new data will not affect the display of the first picture. When it is time to display the second image, the colormap is loaded with black, red, black, and red, respectively, and the second picture will appear regardless of the contents of the front plane of the frame buffer. The process is continued. The single-buffered frame buffer is effectively converted into a double-buffered frame buffer with half as many colors.

Effective animation often relies on implied three-dimensionality, with objects moving behind other objects. This can be done in the most general sense by computing the shape of their overlap and eliminating the overlap from the more distant shape. It may be done in a simpler manner by drawing the farthest object first and overwriting it with closer objects. When not many levels of depth are required, hardware may be added to support each level of depth, making a pseudo-three-dimensional, or two and one-half dimensional, system.

A two and one-half dimensional *image plane* is a frame buffer or area buffer assigned a visual priority. If several of these area buffers contain objects at the same location, the added hardware will determine which plane has the highest priority and output that item (Fig. 19-21). Priority 0 usually denotes the highest priority. It can be seen from Fig. 19-21 that in addition to the object's color, the frame buffer needs to store a bit at every pixel to indicate whether an object is occupying that pixel. The same may also be done by reserving one color number to indicate "nothing present."

Commonly, the area buffers are constructed with translation registers X and Y to

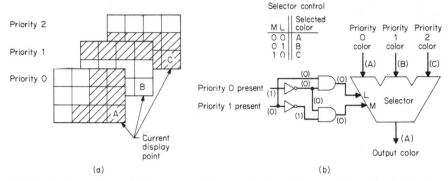

FIG. 19-21 Image planes. (a) A portion of three area buffers is shown, with the priority 0 plane having highest priority. Locations A, B, and C are in possible contention, but B has no occupant. A should block C. (b) Hardware diagram to select color output. A is selected. C will be selected only when there is no occupant at either A or B. The values in parentheses (0/1/A/B) trace the operation for the example in (a).

allow the image, or sprite, to be moved in real time. In such a case, the system will wait until the display is on scan line Y, near pixel X, before reading out the area buffer. By changing just those two data values, the sprite will move across the screen. Since the image buffers interact with each other nondestructively, the sprites can move behind each other or off the screen without being modified.

A coincidence signal may be generated to indicate when two or more objects overlap. It allows the computer to change certain parameters of the display, such as direction of motion and object shape. The coincidence signal is the logical AND of the occupancy bit from each active image plane.

19.2.3 ANIMATION. There are two kinds of animation in computer graphics systems, cel animation and colormap animation. In *cel animation,* a term from the hand-drawn film-animation industry, the image is divided into objects, or *cels.* On each update, the objects that have moved are entirely redrawn and, if necessary, recomputed. The objects that have not moved may or may not need to be redrawn, depending on the system. In *colormap animation,* all the frames are drawn into the frame buffer prior to display and only the color definitions are changed to produce the animation sequence (Fig. 19-7).

The techniques used in cel animation have been described in the preceding sections but will be summarized here. In the case of character generators, the same shapes may be used many times in the same frame and are defined in advance. They are read out as the beam scans the raster. Sufficient hardware is present to keep track of the active symbols and how much of each has already been drawn. Changing the position of a symbol is as simple as changing the time at which that section of memory is read. Changing the entire screen can be done in one vertical retrace since only the references to shapes will change, not the shape definitions themselves.

Image-plane animation, or *sprite animation,* is similar, except that when the scene changes, the object definitions themselves are more likely to change, making it more difficult to effect a total scene change in one vertical retrace. To allow object definitions to change without disturbing the display, sufficient memory is sometimes added to permit several scene definitions to be present. One scene may then be displayed while the next is being loaded (double buffering the scene memory). Systems using known, fixed object definitions do not require the recomputation of objects but have hard limits on the size and number of objects used.

When a general frame buffer and computer system are used for cel animation, no limits are imposed on the size or number of objects used, but real-time animation is difficult to achieve. Since each object is painted into a frame buffer, object interaction is destructive: The background object is overwritten with the foreground. When an object is moved, both the foreground and background objects need to be redrawn. To provide a clean changeover from one image to the next, a double-buffered frame-buffer system is required. The double buffering may be implemented either by two separate frame buffers or by one frame buffer with a write-protection mask and twice the depth needed (see Sec. 19.2.2).

Pan and zoom are two forms of animation that can be accomplished by simple frame manipulation since they do not involve interaction between objects. When panning is done across a scene that is larger than the screen, most of the image does not change. A few rows and/or columns are erased and a corresponding number of rows and/or columns are added. Panning may be implemented with a frame buffer larger than the screen so that no memory contents need be moved, but only the starting address is changed. With this method, the amount of pan is limited to the amount of memory provided. Another implementation is to let the frame buffer address space wraparound so that adding one to the address of the right side of the screen results in the address of the left side of the screen. The top and bottom wrap around similarly. Figure 19-22 illustrates the wraparound method of panning. As the top-left corner of the desired view is moved around the memory, address wraparound takes care of straightening out the picture for the display. The wrapped areas are filled in with new data in sync with the panning motion.

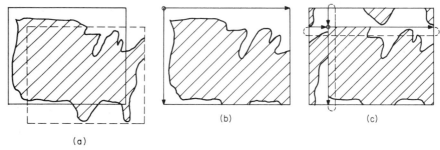

(a)   (b)   (c)

FIG. 19-22 Panning using a dynamic-origin and memory wraparound. (*a*) Solid box shows current view of map of the United States; dashed box shows next view. (*b*) Current frame-buffer contents with top-left point circled to show the start point of the readout. (*c*) Next view. Start point is moved down and to the right. Scan lines are read down to the bottom, then from the top back down to the starting scan line. Each scan line is read from the vertical starting line to the right, then from the left back to the start. (*c*) will end up on the screen looking like the dashed box in (*a*).

Zoom is similar to pan in that the contents of the frame buffer do not change and there is no object interaction. When reading the frame buffer, each pixel is read for a longer period of time than usual, and each scan line is read several times, resulting in a display showing less of the image, and enlarged pixels. This form of zoom is called *pixel replication* and has the advantages that it is easy to implement and is useful in cases where precise pixel identification is important. The disadvantage is that pixel replication magnifies the stairsteps (Fig. 19-23) and line thickness, so it is not useful if a magnification of the data base with enhanced detail is desired (required for that is a scaling of the picture primitives: line, arc, etc.).

19.2.4 EQUIPMENT. Most of the boundary-oriented digital-graphics equipment on the market is two-dimensional. Varying with price, current technology, and the designers' ingenuity, all the discussed approaches have been investigated, if not manufactured. As technology changes, new methods will be developed and many ideas previously found impractical will become attractive.

Character generators compose the largest segment of the boundary systems market. Most character generators store bitmaps of the letter shapes and read them out as

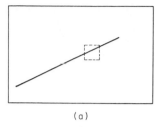
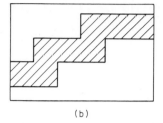

(a)   (b)

FIG. 19-23 Pixel-replication zoom. (a) Line at normal magnification. Dashed box shows area to be magnified. (b) Zoom at large magnification expands each pixel of (a) to a large box.

needed. This continues to be attractive owing to the low-memory requirement. In addition to providing a simple architecture for low-cost systems, it provides generality for those systems which use highly optimized shapes, such as those providing edge-polishing (the Vidifont V) and character rotation (Vidifont V, Compositor). Several systems use run-length encoding to achieve very high resolution. Very high resolution frame buffers have two drawbacks: The frame-buffer memory becomes expensive, and there are too many pixels to easily update in real time. An effective use of run-length encoding is done by the Bosch-Fernseh FP-2000, which performs three-dimensional manipulation on characters and symbols in real time, with an effective resolution of about 4000 pixels per line. The objects handled by the FP-2000 are defined by straight-line segments in model space. The advantage of working with line segments is that only the endpoints need be computed, and since no bitmap is produced, scan conversion is simplified. A secondary advantage of using run-length encoding is that since pixels are not used directly, the horizontal timing may be varied.

The manipulations required of character generators are primarily those of font-style selection, size and color changes, and roll and crawl. Marquee effects may be constructed on this type of equipment as well. Shape movement, roll, and crawl are done by changing the time at which the predefined shape is read out. Most systems require that a new shape definition be made for each size change. However, Quanta Corp. has overcome this limitation by providing dynamic alteration of the size on readout. One system reduces shapes by speeding up the system clock, so that a pixel "occupies" less time; another, which uses run-length encoding, multiplies each run length by an appropriate fraction, keeping track of the remainders to avoid error buildup. Both systems shrink characters vertically by selectively removing scan lines. These systems have the advantage that they produce multiple fonts with the same memory, and the disadvantages that the reduced characters are not visually optimized and all objects on the same text line need to be scaled by the same amount. The latter restriction can be overcome by a more complex design, but the former cannot.

Colors are not always defined by colormaps. In systems which need only seven or eight colors, the 3 bits stored at each pixel serve directly as the inputs to the red, green, and blue guns of the CRT. The choice of colors is fixed: red, green, blue, yellow, magenta, cyan, white, and black. External video may be used if one of the colors is eliminated.

Shape generators differ from character generators only in degree. Character generators take advantage of the fact that text shapes come from a limited library of shapes, are organized in lines, and are generally the same size and color. Shape and logo generators must be able to display multicolor logos in any position or orientation. All the methods used in character generators may be used in shape generators and vice versa.

The Bosch-Fernseh FP-2000 was the first broadcast system to use symbols defined in model space. Definition of characters with polygons had been done earlier, but only to provide a size-independent model of each character. Programs were run to construct pixel-enumerated forms of the characters in various sizes and orientations. The FP-2000 defines polygons in model space and performs three-dimensional and perspective transformations in hardware and firmware. In addition to three-dimensional manipulation of the polygons, it also maintains depth (z-axis) information to dynamically deduce which shape is in front. Sufficient hardware is provided to allow total recomputation of a fairly large number of objects on every field.

The leaders in real-time boundary-oriented digital graphics are not in television picture production but in aircraft visual simulation. At the time of this writing, Singer, General Electric, and Evans & Sutherland, to name a few companies, produce visual systems costing from hundreds of thousands to several millions of dollars. These systems provide faithful reproductions of three-dimensional shapes using polygons, curved surfaces, textured surfaces, translucent surfaces, and point lights and have lighting models for day, night, and foggy conditions. In the non-real-time field, film producers and digital special effects makers are using the most powerful computers available to assist in increasingly realistic scene generation. They use public and proprietary algorithms to produce commercials and film segments. The capabilities of these groups increases annually.

19.2.5 KEYING AND MATTING.   Chroma keying is insufficient for symbol-generator systems, where shapes need to be placed over backgrounds of all colors. The symbol generator, therefore, must output its own key signal, a 1-bit digital (two-level analog) signal which is ON whenever an object is present and OFF when nothing is present. When the key signal is ON, the upstream video signal is shut off and the internal video stream enabled. When the key is OFF, the upstream video is passed through and the internal signal is shut off. The key signal may be used internally or may be sent out for use by some other device. One of the problems with this hard on/off method is that the color subcarrier suffers a sudden phase shift, resulting in chromatic aberration along the edge.

When an object is partially present, i.e., translucent, a two-level key is insufficient. Multilevel keying is called *matting,* after the matte process used in the film industry. Although the switchers in production at the time of this writing do not accept matte signals, the principle has been demonstrated in a few digital symbol generators and in an analog, special-purpose matte system. It is likely that digital key signals of several bits will become more common with the growth of digital studio equipment. Matte signals are used for translucencies, digital chroma matting, and filtered edges.

19.3 AREA SYSTEMS

An area system is one whose primary work concerns the coloration of a given area. It does not relate to boundaries and, as a corollary, does not generate a key signal. It treats the frame or area on a pixel-by-pixel, or sample-by-sample, basis. Two systems falling into this category are paint programs and digital frame-manipulation systems.

19.3.1 PAINT PROGRAMS.   A paint program is one which allows the operator to draw a picture on the screen using some kind of pointing device. Depending on the sophistication of the system, the operations allowed may be as simple as turning ON and OFF individual pixels, or as complex as simulating dozens of art media, including oil, watercolor, and airbrush. Extreme realism is achievable depending on the artist and the system, so that a paint program output may be indistinguishable from a video photo of a live scene.

A paint system usually consists of a general-purpose computer, frame buffer with colormap, one or more monitors for output, and a valuator such as a tablet (see Fig. 19-24). The paint program runs on the computer, watching for inputs and modifying the frame buffer and color table. A discussion of paint programs generally reduces to a discussion of input devices, simulation algorithms, and the menu.

Most paint systems use input devices with two degrees of freedom, X and Y. These include touch panels, tablets, mice, trackballs, joysticks, and pressure grips. A command is selected by moving the pointer to the *menu,* an area of the screen displaying the choice of commands, and pressing a button. Drawing is accomplished by holding down a button

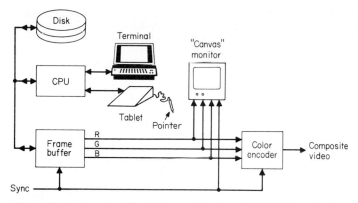

FIG. 19-24 Paint-system block diagram. Operator sits behind three devices: A terminal displaying the command options, a monitor displaying the simulated "canvas," and a tablet. The CPU runs the paint program, storing old pictures on disk and building the current picture in the frame buffer. The frame-buffer image is displayed on the monitor and may be broadcast as well.

while moving the pointer in the desired path. Often the same area on the screen is used for both menu and drawing, requiring that the program know when to display each. Multiple screens are sometimes used to reduce the confusion. Although a two-dimensional input device is acceptable, a three-dimensional device is more desirable, because it allows improved paint dynamics for brushstroke-width variation and travel through the entire color gamut. Three-dimensional variations are available for most input methods, the most notable being the pressure-sensitive stylus with tablet.

A typical paint program involves four tasks: (1) Track the pointer and catch the commands, (2) manipulate the color table if needed, (3) convert the pointer position to data for the frame buffer as required for the current command, and (4) input or output a computer file of the picture.

Tracking the pointer and echoing a cursor onto the screen require the same arithmetic operations as model space to screen space transformations (see Sec. 19.2.1). The cursor shown on the screen is not allowed to destroy the data in the frame buffer. If a hardware cursor is not available, the contents of the frame buffer at the position of the cursor may be altered nondestructively by inverting the number stored at each pixel. To restore the picture, the pixel contents are reinverted.

Commands may be selected in several ways. (1) There can be a predefined pointer area or set of pushbuttons. (2) A single button might cause the menu to be displayed at the current location. (3) A separate screen may be provided for the menu. As with the paint effects, command input style varies with the programmer: graphic icons may be used instead of text; the menu may be presented in stages instead of all at once; etc. It is accorded a matter of personal opinion and a subject of debate as to which method is "best."

The power and versatility of a paint program are largely gauged by the number of media simulated and the quality of reproduction. The following are sample effects presented in order of difficulty:

1. Set a single pixel to the selected color.
2. Form a circle of selected size where the cursor indicates.
3. Copy a section of selected size from one place on the screen to another ("cut and paste").
4. Draw a straight line of selected thickness.
5. Copy in parts of other pictures.

6. Blend the selected color into the picture at the selected location in a nontrivial manner: the method used determines the medium simulated. Watercolors subtractively mix with the background in an uneven circular pattern. Oil will replace the background and have a sharp edge. Airbrush will partially replace the background, with decreasing effect as the distance from the cursor increases.

Many media will require that pixels be treated as areas rather than as points in order to achieve proper color addition, subtraction, and transparency. The better paint programs filter the image (see Sec. 19.4) to provide exceedingly smooth edge detail. Some allow capture of external video for use as an insert. Each paint program uses methods acquired and devised by the programmer, resulting in individuality of each.

Many paint programs provide colormap animation. As the computer and frame-buffer speeds warrant, some provide limited cel or frame-to-frame animation.

19.3.2 DIGITAL FRAME MANIPULATION. Digital frame manipulation is a pure example of applying mathematics to a sample stream of known characteristics. A television picture is communicated as a single-valued function in time. For presentation to the viewer, the stream is transformed into a three-dimensional signal of X, Y, and time. The X dimension is transmitted as a continuous signal, but the Y and time dimensions are samples of the original scene. The digital frame manipulator (the digital effects processor mentioned in Sec. 14.6.6) samples the video input signal at its own sampling rate and constructs a new image, sampled in all three dimensions (Fig. 19-25). From the constructed image, a new one-dimensional signal is generated for transmission. Owing to the amount of sampling done, the mathematics described in Sec. 19.4, filtering and aliasing, is of paramount importance.

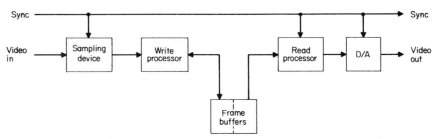

FIG. 19-25 Block diagram of digital frame-manipulation system. The sync signal acts as the master timing control for input sampling and output, but does not necessarily control the write processor. The input samples are written into the first frame buffer at locations appropriate for the effect being produced. The read processor reads the data out of the second frame buffer in order not to interfere with the writing process. The read proceeds in time with the studio sync signal.

The simplest frame manipulation involves synchronizing the frame with an external event. In this case the samples are stored and later read unmodified upon activation of a signal. The samples may be read faster or slower than written. This system requires only a frame store and not a frame buffer.

If instead of reading out the entire frame store, only every other sample is read out, the resulting picture will be half the size of the original in each dimension. By throwing away varying amounts of the frame store, a general squeeze-zoom can be implemented. For example, if the desired size is ⅗ of full scale horizontally and ¾ of full scale vertically, then two out of every five (e.g., the second and fourth) samples are discarded while reading out a scan line, and one out of every four scan lines skipped. Needless to say, owing to loss in information the quality of the image will suffer with reduction. There are, however, ways to soften the degradation.

In general, frame manipulations are accomplished by sampling the input stream, storing the results, and reading them out in a possibly nontrivial manner. The two areas of interest are types of manipulation and image quality. The effort put into each is reflected in the price of the equipment.

Some simple effects and a brief description of their implementation are given below. In each case the data are assumed to be sampled from an R-G-B signal and stored sequentially into a random access frame buffer. It is assumed that there are two frame buffers and that one is being read while the other is being written.

1. To turn the picture upside down, read it out last scan line first.

2. To rotate it 90°, read it with X and Y reversed (instead of reading top line first, left to right, read left column first, top to bottom).

3. To shrink it by A/B, skip A samples out of each B and skip A scan lines out of each B scan lines.

4. To shrink it and reposition it at (X, Y), wait until the beam of the display gets almost to (X, Y), then read it out according to the method described for shrinking it. On each scan line, wait until the beam gets almost to X horizontally before reading out that scan line.

5. To achieve pseudo-three-dimensional perspective, shrink it more on each succeeding scan line.

In order to maintain image quality when rotating, scaling, and performing nonlinear deformations, it is necessary to oversample, combine samples, or interpolate between samples. Figure 19-26 shows a rotated sampling grid which would benefit from interpolation. If an output sample lies a distance (x, y) from the input sample at $(0, 0)$, the input sampling grid is spaced X horizontally and Y vertically, and the input samples surrounding it are A, B, C, and D as shown in Fig. 19-26b, then the output sample has the value

$$\text{OUT} = \frac{x}{X}\frac{y}{Y}A + \frac{x}{X}\frac{(Y-y)}{Y}C + \frac{(X-x)}{X}\frac{y}{Y}B + \frac{(X-x)}{X}\frac{(Y-y)}{Y}D \quad (19\text{-}4)$$

This computation is done at each pixel for each of red, green, and blue.

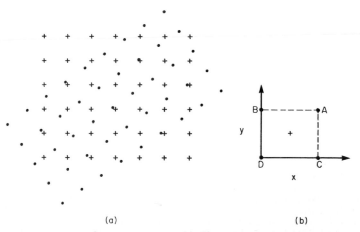

(a)   (b)

FIG. 19-26 Interpolating between samples. The output sampling grid is marked with plus signs, the input sampling grid with dots. (a) The output grid is at a 35° orientation with respect to the input-sampling grid. The mismatch suggests that interpolation is appropriate. (b) Close-up of one output sample. The colors A, B, C, and D affect the output value according to some measure of their distance from it. The text suggests bilinear interpolation.

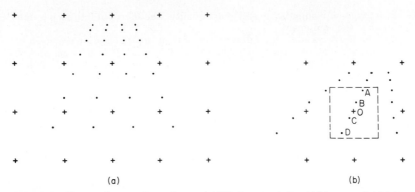

(a)   (b)

FIG. 19-27   Perspective mapping an image. (*a*) The input sample grid (shown as dots) is in receding perspective with respect to the output sampling grid (marked with plus signs). Where input samples are sparse, there may be several output samples interpolating values between them. Where input samples are dense, an output sample will combine several input samples (*b*).

A sampling grid in perspective is shown in Fig. 19-27. Depending on the quality of output desired, the value of the output for the dashed box may be determined by using just one of *A, B, C,* or *D,* by interpolating and averaging them, or by including even more distant samples in the calculation.

Assuming that interpolating and averaging are done as needed, it is possible to perform more complex frame manipulations by altering the write sequence to the frame

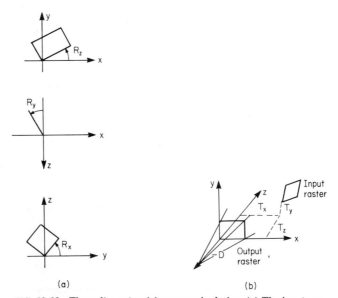

(a)   (b)

FIG. 19-28   Three-dimensional frame manipulation. (*a*) The input raster is being rotated around the *X, Y,* and *Z* axes. (*b*) A view of the scaled and rotated input raster being moved back in three-dimensional space. The output raster is located in the *Z* = 0 plane, and the eye is at distance −*D* along the *Z* axis.

buffer instead of or as well as the manner of read. The following example shows the equations that must be solved to display the input raster moving around in three-dimensional space. The input raster originally sits in the $Z = 0$ plane, with its point $(0, 0)$ touching the three-space origin, $(0, 0, 0)$. It is scaled in all three dimensions, rotated about each of the three axes, moved back some distance in X, Y, and Z, and viewed from some point on the Z axis. The output raster sits on the $Z = 0$ plane with its point $(0, 0)$ touching the three-space origin. As per Fig. 19-28, let the following symbols be defined as in the accompanying table.

	X dimension	Y dimension	Z dimension
Scaling	S_x	S_y	S_z
Rotation	R_x	R_y	R_z
Translation	T_x	T_y	T_z
Viewing distance	D		
Input point	(X_{in}, Y_{in})		
Output point	(X_{out}, Y_{out})		

Then a sample from the input raster will transform to a point on the output raster as

$$\begin{bmatrix} X_{out} \\ Y_{out} \end{bmatrix} = \begin{bmatrix} X'/W' \\ Y'/W' \end{bmatrix} \qquad (19\text{-}5a)$$

$$\begin{bmatrix} X' \\ Y' \\ Z' \\ W' \end{bmatrix} = \begin{bmatrix} S_x \cos R_z \cos R_y & -\sin R_z & -\sin R_y & T_x \\ \sin R_z & S_y \cos R_z \cos R_x & -\sin R_x & T_y \\ \sin R_y & \sin R_x & S_z \cos R_x \cos R_y & T_x T_z \\ \dfrac{\sin R_y}{D} & \dfrac{\sin R_x}{D} & \dfrac{S_z \cos R_x \cos R_y}{D} & 1 \end{bmatrix} \begin{bmatrix} X_{in} \\ Y_{in} \\ 0 \\ 1 \end{bmatrix}$$

$$(19\text{-}5b)$$

In a similar problem, the task is to make the raster take on the appearance of the parametrically defined surface $G(u, v)$, for u and v between 0 and 1. Each input sample is mapped onto the output raster according to $G(u, v)$, and some interpolation, averaging, or filtering function applied to determine its effect on neighboring pixels. The frame is read in the normal sequence: $x = 0$ to $x = (X_{max} - 1)$ for each line of $y = 0$ to $y = (Y_{max} - 1)$. For each sample, the parametric variables are $u = x/X_{max}$ and $v = y/Y_{max}$, and thus the position on the output raster is given by $G(x/X_{max}, y/Y_{max})$.

Transparency may be simulated by adding the current sample value to the contents of the destination pixel. Any function that can be computed fast enough may be used to alter the shape or color of a frame.

The set of functions that can be performed in real time is largely determined by the cost of the system, since there are always the same number of samples to compute per frame. Mathematical sophistication in the design extends the range of the system. The methods of manipulating frames mentioned above do not describe any given machine but rather provide a guide for understanding the scope of the digital frame-manipulation problem and the class of equipment which performs digital frame manipulation.

19.3.3 **HYBRID BOUNDARY-AREA SYSTEMS.** Given that every digital effects machine, whether boundary or area system, is essentially a computer using a frame buffer, it is natural to consider a system which handles both problems. This section will not consider such hybrid systems in detail, since the component parts have been examined. A hybrid system might be expected to display two-dimensional or three-dimensional shapes whose interiors are rotated and compressed pieces of external video. In short, it might do much of the postproduction work. In the case of a good hybrid system it is, in fact, meaningless to distinguish computer-generated portions of the image from live portions or shots from an art card. From the producer's point of view, the system in question will recommend the points of division.

19.4 FILTERING AND ALIASING

When an image is not filtered at all, only sampled, mistakes are made which can amount to one full pixel per scan line per object per frame. Figure 19-29 shows an object just larger than an integral number of pixels. With a very small motion, the object can change in apparent size by a full pixel, a brightness change not commensurate with the motion change. When several objects lie in close proximity, the errors combine and may form moiré patterns. The goal of filtering the image is to reduce the amount of error in displaying the picture so that the shapes and motions appear natural (Fig. 19-30).

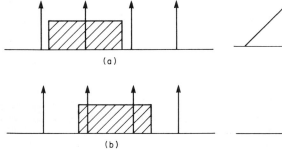

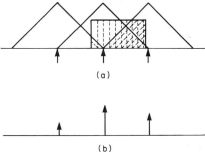

FIG. 19-29 Sampling error. A small motion of the object can result in a large change in brightness. (*a*) The object misses the side samples and has brightness = 1. (*b*) The object lands on the samples and has brightness = 2.

FIG. 19-30 The effect of filtering. A common and inexpensive filtering technique is to allow an object to affect samples some distance away according to a tent shape = (1 − distance). (*a*) The arrows show the locations of samples; the tent shapes show the region and amount of influence around each sample. The rectangle is an object affecting all three samples, according to the shading. (*b*) The resultant intensity of the three samples is shown by the length of the arrrows. With this technique, a subtle movement of the object causes a corresponding subtle change on the screen.

In a live television broadcast, the information is four-dimensional, consisting of height, width, depth, and time. The video camera divides both the time T and height Y dimensions into intervals, keeps the width X continuous, and discards the depth Z dimension. When the video signal is presented to the viewer, it is reformatted as a three-dimensional signal, continuous in X and sampled in Y and T. The information present in the scene has a certain three-dimensional frequency distribution. When the information is converted for transmission into a single-valued function of time, the conversion process is based upon bandwidth and channel-capacity assumptions about the scene, the equipment used to convey it, and the viewer. The mismatches between signal bandwidth and channel capacity at each stage in the operation cause imperfections, or artifacts, in the final image. The problems of these artifacts are evident in live broadcasts but are more severe yet in digitally generated scenes.

The Whittaker-Shannon sampling theorem describes the problem of reconstructing a signal when the channel capacity is limited. If the spectrum $A(f)$ of the signal is band-limited to F_{max} (is nonzero only for frequencies less than F_{max}), and if the signal itself, $a(x)$, is sampled more frequently than $\frac{1}{2}F_{max}$ ($F_{sample} > 2F_{max}$), then the function $a'(x)$ which results from the sampling has a frequency spectrum which is an infinite number of copies of $A(f)$, no two of which overlap. Therefore, a reconstruction filter $R(f)$ can be found which will exactly recover $A(f)$ and, therefore $a(x)$, from the samples. Figure 19-31 illustrates the theorem. If, however, the original signal $a(x)$ contains frequencies above one-half the sampling frequency, then the spectrum of $a'(x)$ is an infinite number of overlapping copies of $A(f)$, and it is not true that the original signal can always be reconstructed. The overlapping section of copies of $A(f)$ are known as *aliases*.

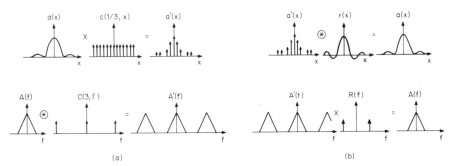

FIG. 19-31 The sampling theorem. The top graphs show the activity in the time/space domain; the lower graphs show the activity in the frequency domain. (*a*) The input signal is sampled. It is multiplied by the comb in x. The spectrum is convolved with the corresponding frequency comb, and copies of the original show up. (*b*) The sampled signal is put through a reconstruction filter: The spectrum is multiplied by the band-limited spectrum $R(f)$ to eliminate the duplicates. The samples are convolved with the filter $r(x)$ to obtain the original.

Aliasing is a problem which occurs in live broadcasts, since it is not possible to band-limit the live scene or to adequately low-pass filter it before the camera samples it. In fact, the imperfect resolution of the camera and the finite decay time of the photosensors do low-pass filter the image, but as technology improves, their frequency ranges increase, quite the opposite of what is needed to correct the aliasing problem. A typical time-based aliasing artifact is the illusion in which a wagon wheel appears to be stationary or turning backward when it is actually moving forward. A Y-dimensional aliasing artifact is the breakup of diagonal lines like guitar strings or the lines on a basketball court.

Digital graphics systems do not have any intrinsic low-pass filtering mechanisms, so that all must be provided by arithmetic computations. Without deliberate filtering, digitally generated scenes exhibit stairsteps that roll and boil as the objects move, and develop a disturbing stop-motion effect like that of conventional stop-frame animation sequences, instead of producing the more realistic motion blur.

Digital generation techniques do have one advantage over live scenes: Since all four (X, Y, Z, T) dimensions are explicitly available to the computer, the entire four-dimensional space of the image can be low-pass filtered and band-limited before any samples are taken. Digitally generated images can theoretically provide higher image quality than live shots. At the time of this writing, however, the general solution to the four-dimensional filtering problem has not been demonstrated.

19.4.1 BACKGROUND. This section presents some mathematical identities useful for understanding filtering.

The frequency spectrum of a signal $g(x)$ is given by its Fourier integral transform

$$G(f) = \int_{-\infty}^{\infty} g(x) \exp{(-2\pi jxf)} \, dx \triangleq F[g(x)] \qquad (19\text{-}6)$$

Its inverse is

$$g(x) = \int_{-\infty}^{\infty} G(f) \exp\left(+2\pi jxf\right) dx \tag{19-7}$$

The transformation is linear, that is

$$g_1(x) = k_1 g_2(x) + k_2 g_3(x) \Leftrightarrow G_1(f) = k_1 G_2(f)a + k_2 G_3(f) \tag{19-8}$$

The convolution of two functions is defined as

$$g_1(x) * g_2(x) = \int_{-\infty}^{\infty} g_1(t)g_2(x-t)\, dt = \int_{-\infty}^{\infty} g_2(t)g_1(x-t)\, dt \tag{19-9}$$

In the frequency domain convolution becomes multiplication and vice versa

$$g_3(x) = g_1(x) * g_2(x) \Leftrightarrow G_3(f) = G_1(f)G_2(f) \tag{19-10}$$

$$g_3(x) = g_1(x)g_2(x) \quad \Leftrightarrow G_3(f) = G_1(f) * G_2(f) \tag{19-11}$$

A sample is taken by multiplying the input signal with the Dirac delta function located at the sample point. The delta function $\delta(x)$ is any function such that

$$\int_{-\infty}^{x} \delta(u)\, du = \{0 \text{ for } x < 0;\ \tfrac{1}{2} \text{ for } x = 0;\ 1 \text{ for } x > 0\} \tag{19-12}$$

The delta function located at $x = X_0$ is described by $\delta(x - X_0)$. Phase-shifting a signal by P corresponds in the frequency domain to multiplying the spectrum by the spiral of radius 1: $\exp(-2\pi jPf)$. That is

$$g_2(x) \triangleq g_1(x - P)$$

$$= \int_{-\infty}^{\infty} \delta(t - P)g_1(x - t)\, dt$$

$$= g_1(x) * \delta(x - P)$$

and from Eq. (19-10)

$$G_2(f) = G_1(f)F\left[\delta(x - P)\right]$$

$$= G_1(f) \int_{-\infty}^{\infty} \delta(x - P) \exp\left(-2\pi jfx\ dx\right)$$

so finally

$$G_2(f) = (G_1(f) \exp\left(-2\pi jPf\right) \tag{19-13}$$

Sampling a stream of data every s units is equivalent to multiplying the input signal by a series of delta functions separated by distance s, that is, the comb function $c(s, x)$

$$c(s, x) \triangleq \sum_{k=-\infty}^{\infty} \delta(x - ks) \Leftrightarrow C\left(\frac{1}{s, f}\right) = F[c(s, x)]$$

$$= \frac{c(1/s, f)}{s} \tag{19-14}$$

The comb function and its transform are shown in Fig. 19-31a.

Low-pass filtering is the convolution of the input waveform (in the case of television, the image) with a curve which is band-limited in the frequency domain. This multiplies the two spectra together and results in a band-limited output signal. The output signal may then be sampled and later passed through an appropriate reconstruction filter.

Aliasing is technically the overlap of the copies of $A(f)$ which make up $A'(f)$ (refer to Fig. 19-31). A similar effect, heterodyne noise, arises when the reconstruction filter $R(f)$ is too wide and includes sections of separate copies of $A(f)$ (see Fig. 19-32b). The end result is similar, and both are grouped together as aliasing problems.

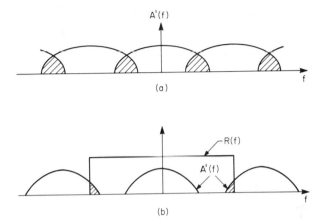

FIG. 19-32 Aliasing. (*a*) The spectra are too wide and overlap each other. In this case, no filter can sort out the overlapped areas. (*b*) The filter is too wide and brings in parts of the unwanted duplicate spectra.

The energy of aliasing is

$$E = \int_{-\infty}^{-(1/2)F \text{ sample}} [G(f)]^2 \, df + \int_{(1/2)F \text{ sample}}^{\infty} [G(f)]^2 \, df \qquad (19\text{-}15)$$

Shapes used in digital image generation are described by exact boundaries, which are based on the step function $u(x) = \{0 \text{ for } x < 0; 1 \text{ for } x \geq 0\}$. The spectrum of $u(x)$ is $U(f) = \frac{1}{2}\pi jf$, and falls off as $1/f$ but never reaches zero. Sampling a shape at frequency F_s, then, results in aliasing with energy $E = 1/n^2 F_s$. The S/N ratio is roughly proportional to the sampling frequency since the energy of aliasing decreases as the frequency of sampling increases.

Since the limits of human perception are finite, there is a sampling frequency above which the aliasing artifacts are not objectionable. Unfortunately, that frequency is quite high. The eye exhibits drop-off in detection at 50 cycles per degree of visual field and 40 to 50 Hz, with detectable residues as high as 100 Hz, and has a contrast threshold of about 2 percent regardless of frequency. The filtering techniques described in the following section produce output signals whose energy of aliasing falls off faster than $1/f$ and, hence, produce more acceptable pictures at low resolutions.

19.4.2 ONE-DIMENSIONAL FILTERING. Convolving the input signal with the perfect box filter $B(f) = \{1 \text{ for } -1/s < f < 1/s; 0 \text{ elsewhere}\}$ has three problems: (1) $b(x)$ has infinite extent, (2) $b(x)$ has negative values, and (3) it assumes a perfect reconstruction filter. These problems are overcome in practice as follows: (1) The convolution is carried for only a few samples' distance, (2) a gray wash is added to bring the negative values above zero, and (3) the output filter is approximated.

More commonly yet, the signal is convolved with an arbitrary curve which is chosen to be computationally tractable and still give adequate results. The spectrum of the curve (the filter) is nonzero above the desired cutoff frequency, producing heterodyned noise, but the energy of aliasing will nonetheless be reduced. The chosen curve, $h(x)$, generally follows three guidelines: (1) To avoid negative values, $h(x)$ is always positive. (2) To simplify computation and avoid blurring, it is nonzero only for a small region. That is, $h(x) = 0$ for $|x| > k$ where k is usually between ¾ and 1½ pixels. (3) The curve satisfies the equal-energy criterion (described below). Several acceptable curves are shown in Fig. 19-33. They are

$$c(x) = 1 \qquad\qquad \text{for } |x| < s, \text{ 0 elsewhere} \qquad\qquad (19\text{-}16)$$

$$c(x) = \left|1 - \frac{x}{s}\right| \qquad\qquad \text{for } |x| < s, \text{ 0 elsewhere} \qquad\qquad (19\text{-}17)$$

$$c(x) = \cos^2 \frac{\pi x}{2s} \qquad\qquad \text{for } |x| < s, \text{ 0 elsewhere} \qquad\qquad (19\text{-}18)$$

Even when the simplest function, Eq. 19-16, is used, the energy of aliasing falls off quite rapidly, as $1/f^2$.

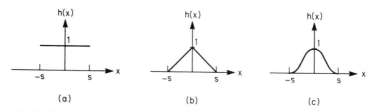

FIG. 19-33   Three ad hoc filters which satisfy the equal energy criterion. (*a*) The constant. (*b*) The tent, or pyramid. (*c*) The cosine squared. See text for the formulas.

There is a simpler but equivalent technique to that of convolving the image with a filter and sampling it. That is to center a copy of the filter around each sample point and compute the volume of intersection of each copy of the filter with the input signal (Fig. 19-30). The latter process has computational advantages and is given in detail in Ref. 1. Using this method, the image is filtered in two dimensions, *x* and *y*, as follows.

Step 1.   Prior to drawing any pictures, choose a filter size and shape. Restrict all objects to be simple shapes, like rectangles of width greater than one pixel. Compute the volume of intersection for all possible intersections of a rectangle with the filter region and put the results in a table (this table is not likely to be very large, since there are not very many ways a rectangle can intersect a filter region).

Step 2.   When drawing the picture, compute the type of intersection of the primitive object with each successive filter region. Look into the table to find the volume of the intersection. Add the volume to the value already stored at that pixel location.

Equal-Energy Criterion.   The *equal-energy criterion* (EEC) states that no matter how a given image is shifted around the sampling grid, it should generate the same total amount of light. Not all filters satisfy the equal-energy criterion. The simplest way to check that a filter satisfies the criterion is to center a filter over each sample and sum the contributions at each point. The result should be a horizontal line with no bumps ("hot spots"). Figure 19-34*a* shows a filter that satisfies the EEC, and Fig. 19-34*b* shows one that does not. Arithmetically, when filtering with the curve $h(x)$ and sampling with the comb $c(s, x)$, it can be shown that $h(x)$ satisfies the EEC when its spectrum is zero at the comb frequencies; that is, $H(f) = 0$ for all $f = k/s$, $k = 0$. The convolution of any filter with one which satisfies the EEC is a filter which satisfies the EEC. That is, since $g_3(x) = g_2(x) * g_1(x)$ transforms to $G_3(f) = G_2(f)G_1(f)$, and $g_1(x)$ satisfies the EEC, then $G_1(f) = 0$ for all $f = k/s$, $k = 0$, and $G_3(f) = 0$ for all $f = k/s$, $k = 0$. Hence $g_3(x)$ satisfies the EEC. The advantage this property provides is that the image may be filtered several times, and if any of the filters satisfy the EEC, the whole system does (Fig. 19-35).

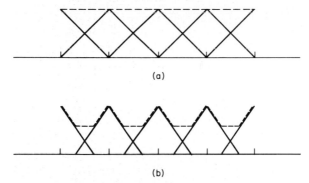

(a)

(b)

FIG. 19-34 The equal-energy criterion. (a) The filter has been replicated at each sample location. When all contributions to any point are summed, the result is a constant, and the EEC is satisfied. (b) The contributions do not sum to a constant. As objects move under this filter, they will appear brighter at some locations than at others.

(a)

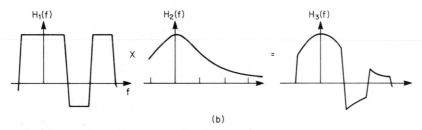

(b)

FIG. 19-35 Chaining filters. (a) A filter satisfying the equal-energy criterion is followed by one that does not. The system as a whole satisfies the criterion. (b) The (somewhat improbable) spectrum which satisfies the EEC is zero at the sampling harmonics. It is multiplied by a gaussian spectrum, which does not satisfy the EEC. The result clearly will have zeros at the sampling harmonics and, hence, will satisfy the equal-energy criterion.

19.4.3 MULTIDIMENSIONAL FILTERING. Filtering the image in three dimensions, X, Y, and T, is simply applying the one-dimensional filtering mathematics to each dimension. Time does not differ from X or Y in its need for filtering but is often ignored in practice. The two-dimensional Fourier transform is

$$G(f_x, f_y) = \int\limits_{-\infty}^{\infty}\!\!\int g(x,y) \exp\left[-2\pi j(xf_x + yf_y)\right] dy\, dx \qquad (19\text{-}19)$$

and the three-dimensional (X, Y, T) transform is

$$G(f_x, f_y, f_t) = \int\limits_{-\infty}^{\infty} \int \int g(x, y, t) \exp\left[-2\pi j(xf_x + yf_y + tf_t)\right] dt\, dy\, dx \quad (19\text{-}20)$$

The sampling function is

$$c(s_x, s_y, s_t, x, y, t) \leftrightarrow C(1/s_x, 1/s_y, 1/s_t, f_x, f_y, f_t) \quad (19\text{-}21)$$

The three-dimensional filter $h(x, y, t)$ is required to satisfy the three-dimensional equal-energy criterion: $H(f_x, f_y, f_t) = 0$ for all $f_x = k_1/s_x, f_y = k_2/s_y, f_t = k_3/s_t, k_1 + k_2 + k_3 = 0$.

A simple way to form a multidimensional filter is to assume a function separable in all dimensions, then multiply together one-dimensional functions with the desired properties to obtain it. That is, let $h(x, y, t) = h_1(x)h_2(y)h_3(t)$, where h_1, h_2, and h_3 each satisfies the equal-energy criterion and any other desired constraints. Examples of acceptable three-dimensional filters follow, where $h(x, y, t) = 0$ for $|x| > s_x$ or $|y| > s_y$ or $|t| > s_t$

$$h(x, y, t) = 1 \qquad\qquad \text{(constant in all dimensions)} \quad (19\text{-}22)$$

$$h(x, y, t) = |1 - x/s_x|\ |1 - y/s_y| \qquad \text{(constant in } t) \quad (19\text{-}23)$$

$$h(x, y, t) = \cos^2(\pi x/2s_x)\cos^2(\pi y/2s_y)\ |1 - t/s_t| \quad (19\text{-}24)$$

A set of circularly symmetric filters which satisfy the two-dimensional EEC are the first-order Bessel functions. A finite-width all-positive approximation to a circularly symmetric function is obtained by convolving together functions whose spectra are *pillboxes* (the "circ" function) of differing radii: the radii needed to be zero at each frequency sample.

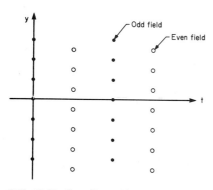

Two problems arise in filtering images in the T (time) dimension. One is that as objects move, they hide and reveal other objects. Within a single computational period, the objects effectively change shape. The simplest solution is to over-sample in T and form a straight-line approximation. The other difficulty is that for displays which are interlaced, the sampling in Y is not the same on each field (Fig. 19-36). Hence, the two-dimensional (Y, T) comb function which is used to represent the sampling grid is not $c(s_y, s_t, y, t)$, and the equations for acceptable filters should be modified accordingly. Analytical techniques are still being developed for dealing with these two problems.

FIG. 19-36 Sampling grid with interlace. The X dimension is not shown since it is not involved in interlace. In interlace, the Y-sampling comb is shifted by one half-sample distance on each field.

19.5 INPUT TECHNIQUES

19.5.1 VALUATOR INPUT. Valuator input is used for obtaining a value in a range. A valuator may be one or more dimensional, depending on the technique used and the application program reading the device. The value may be used to indicate position, speed, acceleration, time, color, angle, etc. The valuator may be used to simulate some other device; e.g., position may be used to select a command, in which case the position valuator is simulating a bank of buttons.

The techniques of generating values vary widely and include voltage, capacitance, resistance, optics, acoustics, pressure, and magnetics. It is possible to arrange a resistive or capacitive area or network so that when a point inside the area is touched, the resistance, capacitance, voltage, or current is changed. The amount of the change indicates the position of the touch, in analog form. This is converted to digital form for use by the computer. A potentiometer is a form of valuator that changes the voltage as the knob is turned. Either the voltage or the change in voltage since the last sampling may be used as the value. Voltage variation in a strain gauge may be used in a pressure-sensitive device. Time delays are used in acoustic, optical, and magnetic sensors. A sound sent by a transponder (the pointer) is detected at different times by several receivers; triangulation gives the position of the pointer. In an optical or magnetic grid, pulses are sent along each element at different times. The time at which an interruption is felt indicates the position of the pointer. The density of elements determines the accuracy of the result.

Valuators usually work in one, two, or three dimensions, although more are possible. One-dimensional devices include potentiometers and slides. Two-dimensional devices include the tablet (electrical, magnetic, acoustic, or optical), the touch panel (capacitive or optical), light pen (optical), trackball (electrical or optical), mouse (electrical or optical), and joystick (electrical or pressure). Three-dimensional devices include the tablet with a pressure-sensitive pointer, three-dimensional optical, acoustic, or mechanical sensing grids, trackball using horizontal rotation as the third input, joysticks with potentiometers mounted on the stick, and pressure-sensitive joysticks.

The simplest valuators return raw numbers. To interpret the number, the computer must express it as a percentage of the largest possible value or multiply it by a scaling factor to put it into a useful range. A simple example is the use of a potentiometer to move an object up and down on the screen. The voltage V will vary between 0 and V_{max}. The computer program wishes to place a selected object at the top of the screen when the voltage is 0 and at the bottom when it is V_{max}. The top of the screen is at $Y = 0$, the bottom at $Y = Y_{max}$. Y is computed as $Y = Y_{max} \star V/V_{max}$. The value (Y_{max}/V_{max}) will never change and so may be precomputed, simplifying the equation to $Y = V \star k$, where $k = Y_{max}/V_{max}$.

The programmer may wish to work in relative coordinates, in which the valuator value indicates the change in position since the previous sampling. Since most valuator methods provide only absolute coordinates, either the computer or some extra hardware in the valuator has to save the previous value and subtract it from the current one to find the change. This change has the same range as the absolute coordinate and is used in the same way.

Quite often the valuator exhibits jitter of several percent. When this is objectionable, the sample stream is low-pass filtered, usually by averaging the current value with the preceding eight samples, perhaps also rejecting any value which is too different from the preceding average. Two- and three-dimensional valuators are handled as several separate one-dimensional valuators.

19.5.2 CAMERA INPUT.

The signal from a video camera is a time-domain analog signal. An A/D converter samples the video stream and the computer stores the values. The precision of the A/D converter and its speed are selected by the system designer according to the type of images he or she expects the system to encounter and the size of the frame buffer. It is unlikely that the video stream will be perfect, since the camera may have an uneven response over the full field and the scene may not lie completely square to the sampling grid. For these reasons the digitized frame often is processed to reduce noise and dropouts, and to rotate or scale the image.

The methods used to rotate or scale the image are precisely those described in Sec. 19.3.2, Digital Frame Manipulation. In this case, however, there are no time constraints, so that even the most esoteric methods become viable. In order to determine the amount of scaling or rotation, registration marks are put on the card or in the scene to be shot. These can be detected and the required transformation deduced. Pincushion distortion can be corrected in the same manner.

Inadequate studio practices and postdigitization processing introduce flaws into the image which must be removed manually, an expensive procedure. Good studio practices include providing a flat shooting surface, even lighting, and consistent shooting procedures. Postdigitization processing tasks include camera compensation, contrast enhancement, and noise reduction. To identify the camera characteristics, a blank or white field is shot.

Knowing the intended intensity, the program can arrange a compensation factor at each sample point for subsequent shots. To enhance edge detail, the image is put through a differentiator, with the Laplacian at each point being subtracted from the value. The Laplacian is $d^2f/dx^2 + d^2f/dy^2$, or on a discrete grid

$$
\begin{aligned}
L(x, y) &= \{[f(x + 1, y) - f(x, y)] - [f(x, y) - f(x - 1, y)]\} \\
&\quad + \{[f(x, y + 1) - f(x, y)] - [f(x, y) - f(x, y - 1)]\} \\
&= \{f(x - 1, y) + f(x - 1, y) + f(x, y - 1) + f(x, y + 1)\} - 4f(x, y)
\end{aligned}
\tag{19-25}
$$

The new value I' at each point is

$$
\begin{aligned}
I'(x, y) &= I(x, y) - L(x, y) \\
&= 5I(x, y) - \{I(x - 1, y) + I(x + 1, y) + I(x, y - 1) + I(x, y + 1)\}
\end{aligned}
\tag{19-26}
$$

Lastly, to reduce the noise, the dynamic range of the inputs is reduced, and a floor and a ceiling are put on the input values. This requires that the input be sampled to greater precision than will ultimately be desired. If the floor is at F, the ceiling at C, and the output range is from 0 to M, then the final intensity I' may be computed from the input intensity I as (see Fig. 19-37)

$$
\begin{aligned}
I' &= 0 && \text{if } I < F \\
I' &= M \star \frac{I - F}{C - F} && \text{if } F < I < C \\
I' &= M && \text{if } I > C
\end{aligned}
\tag{19-27}
$$

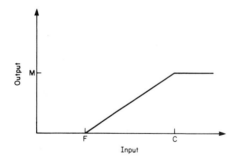

FIG. 19-37 Noise-reduction function. The input samples are obtained to greater precision than needed and then mapped into a smaller range.

19.6 DATA COMPRESSION

The quantity of data involved in digital video systems is such that data compression methods are almost mandatory. A frame buffer, $512 \times 484 \times 4$, requires 991,232 bits of storage and provides only a primitive picture. A more reasonable frame buffer, $1000 \times 484 \times 8$, requires 3,872,000 bits of information. A system generating images in real time needs to handle this much information 25 to 60 times a second. Both spatial and temporal data-reduction techniques are available and improving continually.

19.6.1 **SPATIAL DATA REDUCTION.** Section 19.2.1 dealt with shape generation in boundary systems. All the methods mentioned were devised to provide data reduction. They are summarized below.

Run-Length Encoding.   When many pixels next to each other on a line are the same color, that run may be described by its length. Describing the entire picture by run lengths may easily provide data reduction by greater than 6, and is the simplest and most common data-reduction method.

Area Encoding.   Dividing the picture into areas of the same color is less intuitive, but powerful. The method of quad trees recursively divides the picture into squares of all black, all white, or some of each.

Vertex Coordinates in Screen Space.   When all boundaries are constructed with straight lines, only the endpoints of the lines need to be stored. This results in great data reduction, but the system must include a line expander and a frame buffer, and the enclosed area must be filled in.

Screen Space: Adjacent Pixel Directions.   At each pixel, a code indicates in which of the eight possible directions the boundary proceeds. The code varies in length from 1 to 8 bits so that the most frequently used direction is assigned the 1-bit code and the least frequently used direction is assigned the 8-bit code. For example, the codes might be

0:	*turn right 90°*
10:	*turn left 90°*
110:	*straight ahead*
1110:	*turn right 45°*
11110:	*turn left 45°*
111110:	*turn right 135°*
1111110:	*turn left 135°*
11111110:	*reverse*

A great many variants on this method are possible. Each is an instance of Huffman encoding the turning direction and is not necessarily optimal for any given picture. Huffman encoding is described below. It is possible to Huffman-encode each individual picture or shape to get the optimal code. Nonetheless, this code provides excellent data reduction as it stands. To decode the picture description, the computer examines the input stream 1 bit at a time and separates out each token, or meaningful entity. The computer program checks the token against a table to discover its meaning and extends the boundary accordingly. A frame buffer is required, and, as with all boundary representations, the area must be filled in afterward.

Screen Space: Adjacent-Line Deltas.   These are similar to adjacent-pixel directions in that the picture description is a bit stream of encoded tokens which must be separated into meaningful commands. The commands in this case always proceed downward and are based upon the previous movement. Commands are also provided to start a new downward stroke and to end an old one. The added complexity and, in some cases, bit-stream length are countered by the fact that only a line buffer is needed for implementation, not a full-frame buffer. As with the *adjacent-pixel directions,* this method can be Huffman-optimized for each picture, or a general code may be used. Either run-length encoding or a one-dimensional area fill may be used to color in the segment.

Boundaries in Model Space.   The boundaries are defined in a manner independent of the sampling grid and converted on read-in. Once converted, they become boundaries in screen space (see above). The advantage lies in the generality of the description; the disadvantage in the extra step necessary to produce an image.

Huffman Encoding.   Given a sequence of numbers, Huffman encoding guarantees the fewest number of bits necessary to encode that sequence. The sequence of numbers may be run-lengths, coordinates, direction, changes in movement, etc. Huffman coding has three parts: (1) forming the code, (2) encoding the data, (3) decoding the data.

To generate the code (see Fig. 19-38), all the numbers are put into a table which also contains their frequency of occurrence. The least commonly used table entry is assigned a 1-bit code of 0; the next least commonly used item is assigned a 1-bit code of 1. The two, now decodable, are combined to form a composite entry and erased from the table. The composite entry is inserted into the table with a count which is the sum of two erased entries' counts. The process is repeated over and over, so that eventually there are only two entries in the table, probably composites of composites. The one with the lower count is given the code 0 and the other is given the code 1. As the entries are successively merged, another table (more correctly, a tree structure) is built keeping a history of the

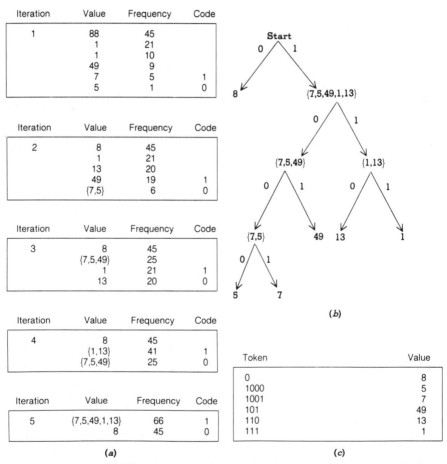

Iteration	Value	Frequency	Code
1	88	45	
	1	21	
	1	10	
	49	9	
	7	5	1
	5	1	0

Iteration	Value	Frequency	Code
2	8	45	
	1	21	
	13	20	
	49	19	1
	{7,5}	6	0

Iteration	Value	Frequency	Code
3	8	45	
	{7,5,49}	25	
	1	21	1
	13	20	0

Iteration	Value	Frequency	Code
4	8	45	
	{1,13}	41	1
	{7,5,49}	25	0

Iteration	Value	Frequency	Code
5	{7,5,49,1,13}	66	1
	8	45	0

(a)

(b)

Token	Value
0	8
1000	5
1001	7
101	49
110	13
111	1

(c)

FIG. 19-38 Huffman encoding. (a) The top table (iteration 1) shows the values to be coded and the number of times each was used. On each iteration, the two least frequently used were combined to form a composite but decodable value, and the composite entered into the table. After five iterations all the values are encoded. (b) The tree structure was built as the values were being combined. The tree shows how to encode the values for transmission and how to decode them later. (c) The final codes for each value.

decisions. When there is only one entry in the table, the decision table becomes complete and contains the encoding/decoding sequence for each data item. The decoding mechanism is simply the reverse sequence of 1s and 0s used to fold an initial value into the final composite entry. Each original datum has a unique bit sequence, and the length of the bit sequence is inversely proportional to the frequency of occurrence of the datum.

The data base, i.e., the sequence of values that describe the picture, is then encoded according to the decision table. Each value is checked against the decision table for its code. The result is a bit sequence which is meaningless until decoded. The decision table is therefore stored along with the bit sequence.

When it is desired to transmit and decode the picture description, the decision table is sent first, followed by the bit sequence. As each bit arrives, the decision table is consulted until a complete token is collected. The token is translated into the datum it represents, and the picture definition augmented. The translation of the incoming bit sequence continues until the picture is complete.

The codes used in adjacent-pixel directions and adjacent-line deltas were examples of Huffman-like encoding in which the code was determined for general use and not optimized for any particular picture.

19.6.2 TEMPORAL DATA REDUCTION. Temporal data reduction relies on the fact that images generally change fairly little on a frame-by-frame basis. The reduction scheme selected is directly related to the information the programmer has about the total set of picture changes allowable. When the movement and the objects are well defined, great data reduction is possible. When nothing is known about the frame, the problem is not one of describing the motion of objects, but of discovering similarities among large amounts of data and approximating the differences in a fairly compact form. The latter situation is the problem faced by digital television and is covered in Chap. 18.

When the motion of the objects in the scene is not well defined, occasionally it is practical to subtract one frame from another and store the areas that change by any of the spatial data reduction methods just described in Sec. 19.6.1. Such frame-to-frame coherence is used in the Dubner system, in which scan lines are run-length encoded and preserved in a list. On subsequent frames, those scan lines which have become obsolete are removed from the list and any new ones are added. The order of the lines is rearranged as necessary on each frame to produce a correct picture. The actual traffic of new scan line definitions into the system is said to be relatively low. The use of frame-to-frame coherence in digital-graphics systems is rare.

When the difference between the frames is due to moving objects, the most compact form of encoding is to describe the motion of the objects instead of the changes in the final picture. From the description of the motion, the computer recomputes the image on each frame. Most motion can be described by constant velocity or acceleration in a straight line for some number of frames. This probably is the most compact form of storage but requires that acceleration and velocity computations be made from the original sequence. An alternate method is to describe the motion as though a person were operating a valuator device and moving the object directly. On playback, the valuator data are read in real time, and the positions of the objects updated. Since valuator input is very compact, storing the valuator data is highly storage-efficient.

REFERENCES

1. Satish Gupta and Robert Sproull, "Filtering Edges for Gray-Scale Displays," *Computer Graphics,* vol. 15, no. 3, pp. 1–5, August 1981.

2. A. Klinger and C. R. Dyer, "Experiments on Picture Representation Using Regular Decomposition," *Computer Graphics and Image Processing,* vol. 5, pp. 68–105, 1976.

3. Mann-May Yau and Sargur Srihar, "A Hierarchical Data Structure for Multidimensional Digital Images," *Communications of the ACM,* vol. 27, no. 7, pp. 504–515, July 1983.

BIBLIOGRAPHY

Computer Graphics, a quarterly report of SIGGRAPH-ACM, particularly the "SIG-GRAPH Conference Proceedings" since 1978.

Computer Graphics and Image Processing (periodical).

Foley, J., and A. van Dam, *Fundamentals of Interactive Computer Graphics,* Addison-Wesley, Reading, Mass., 1982.

Newman, William, and Robert Sproull, *Principles of Interactive Computer Graphics,* McGraw-Hill, New York, 1979.

Pavlidis, Theo, *Algorithms for Graphics and Image Processing,* Computer Science Press, Rockville, Md., 1982.

Electronic Editing

Richard J. Caldwell

TVC Video, Inc.
New York, New York

20.1 EVOLUTION OF VIDEO TAPE EDITING

20.1.1 THE GROWING NEED FOR EDITING.   Following the introduction of regular television service in the United States, several problems began to emerge. The first was coping with multiple time zones unique to the United States. With the establishment of a network of stations, relaying of programs to be broadcast at a particular time could only be done on film or be retelevised live. The resolution of this problem began with the introduction of the first video tape recorder (VTR) in 1956.

The second problem was the demand for more and more hours of programming. Material was recorded to be broadcast later and to supplement live programming, then eventually to supplant it. Recording also provided the advantage of retaking a program if there was a mistake or technical failure. However, this necessarily limited the complexity of shows with multiple scene changes. This process was commonplace in the BBC even in the late 1960s. The ability to edit videotape became necessary as a cost-effective way to correct errors and polish up the flow of a program. With the advent of editing, the creativity long afforded to film editors began to emerge in the video world.

20.1.2 BASIC REQUIREMENTS FOR EDITING.   Video editing requires the ability to change over to record during the vertical interval on a predictable basis, producing a *cut* to the new material. Originally this was accomplished using a physical splice in the tape, the edit point usually chosen by listening to the audio cue. The audio was then laid off onto audio tape so that it could be relaid over the physical edit. The reason for this is the same as for optical or magnetic sound on film, which dimensionally leads the associated frame of picture. Mechanical editing reduced the reusability of the tape and increased production costs. Programs recorded in sections were thus *assembled* into a complete show.

20.1.3 TYPES OF EDITING.   With the introduction of electronic editing in the 1960s the mechanical edit was eliminated and replaced by two electronic processes. The first is assemble editing: The new material, including control track, is sequenced in recording so as to produce a continuation of the existing material. This is acceptable for *in* edits, but does not produce a clean ending of the record mode. The second process allows the insertion of new material into an original recording while preserving the integrity of the control track signal. The sequencing of audio and video erase and record current *on* and *off* times are arranged to give clean *in* and *out* edits that are undetectable. The electronic-assembly technique requires a second machine to play back the new material, or a cue to be given to a studio to pick up the new action at the edit point.

Assemble editing is used extensively for packaging commercial spots or assembling sequences machine to machine. Insert editing is the most commonly used technique in postproduction today.

20.1.4 **ELEMENTARY SYSTEMS.** The simplest system of editing basically consists of a *punch and crunch* system which goes into *edit record* on the basis of the editor's reaction time in hitting the record button on cue. There is no preview or second chance.

Cue-tone systems allow the rehearsal of an edit by recording a burst of tone on the cue track at the approximate edit point. The timing of the edit can then be reviewed and adjusted by advancing or delaying the cue-tone trigger timing to a precise point.

Another method now used more frequently is a tachometer count from either the control-track signal or the mechanical timer in contact with the tape. At the chosen edit point the count number is stored in a register and used as the trigger timing number. It can be modified by adding or subtracting frames or by a time entry from the register number to create an offset. In the case of 1-in (2.5-cm) type-C format, the edit point is selected by advancing the tape to the precise frame required and simply entering that number into the register. Thus, frame-accurate single-frame edits are extremely easy to perform. The *out* edits are computed in the same manner using a duration time added to the *in* time or an *out* time entered directly by moving the tape to a new position.

20.2 SMPTE/ANSI TIME CODE

20.2.1 **BACKGROUND.** The aforementioned system is effective but slow and labor-intensive. It also becomes limited owing to the inability to search out edit points or store edit-time data for later retrieval, and loses efficiency when more than a playback and record machine are involved.

The need for a means of accurately addressing every frame of recorded video similar to edge numbering in film was apparent. By the mid-1970s, the process of time coding recorded material as the recording was being made produced a count of all frames in real time in an increasing numerical sequence. The time information could be decoded to give a precise location of video without the need to observe action cues in the picture. For example, now it was easy to find *take 15* out of *30* by using the time code to locate the required scene. Building a program could progress much faster than heretofore.

Under the auspices of the SMPTE the time-code format was issued as a Recommended Practice (RP) and subsequently adopted by the American National Standards Institute (ANSI) as their standard ANSI V98.12-1975 for 525-line quadruplex systems. Since then SMPTE has issued RPs covering 1-in systems and vertical-interval recording of the time-code signal.

20.2.2 **BASIC FORMAT.** The time-code signal is an 80-bit biphase mark binary-coded decimal (BCD) format recorded at a rate of 2400 bits/s on every frame as data on an audio channel. The format is shown in Fig. 20-1. The channel has a wider-than-normal playback frequency response to enable accurate recovery of the timing information from the playback square wave from tape. The wide bandwidth becomes important as soon as the tape moves up in speed to *shuttle forward* or reverse at up to 20 times play speed or down to one-fifth play speed. The code contains not only data but a clock signal using the positive and negative transitions to synchronize the reader. (All 1s would double the bit rate.)

Longitudinal Time Code (LTC). This code consists of time information as hours, minutes, seconds, and frames requiring 26 bits; 32 bits of user-defined code; 16 bits of sync word; 2 bits assigned to identify color frame and drop frame, and 4 bits reserved

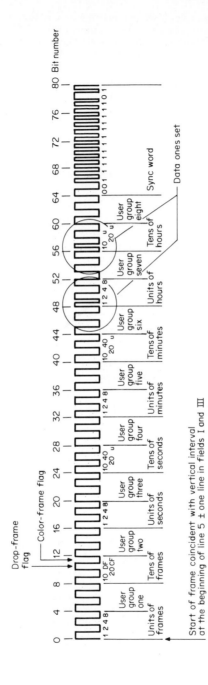

FIG. 20-1 Format of 80-bit ANSI/SMPTE time code. *Notes:* (1) User groups are set to zero if no data transmitted. (2) Time code reading: 11:00:00:00, non-drop-frame model, no color-frame identification. (3) Unassigned address bits are 27, 43, 58, and 59 to be set to zero until designated by ANSI.

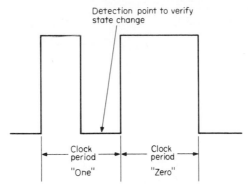

FIG. 20-2 Detection of state change. Change of state at the half clock period defines a 1. Clock edges may be positive- or negative-going. The signal is immune to 180 phase shifts. *Note*: An idealized waveform is shown for clarity of illustration.

but undefined. As shown in Fig. 20-2, a zero is defined as no change of state after half of the clock period has elapsed. A data 1 is defined as a change of state at the half point. The code data stream does not have a dc component which permits recording on a longitudinal audio track owing to the arrangement of the data bits based on the probability of bit patterns produced over 24 h of recording.

20.2.3 COLOR FRAME-RATE TIMING ERROR. The use of 29.97 Hz for color, rather than 30 Hz used for monochrome, results in an error in timing. The resultant 0.1 percent error adds 3.6 s, or 108 frames, per hour.

To correct the code so that it reflects real time, the 108 frames are dropped every hour. This is achieved by dropping frame numbers 00 and 01 every minute except the tenth, twentieth, thirtieth, fortieth, fiftieth, and sixtieth minutes, thus eliminating 108 frames. Drop-frame time code (compensated), also known as *color time* or *real time,* runs accurately 24 h, except for a 75-ms residual error. Non-drop-frame time code is favored for production work because of the ease in calculating editing cues from the contiguous numbering of frames.

The color-frame bit, used to define color framing, is not in general use, although it would be very helpful to maintain uniformity of coded tapes from different production sources.

20.2.4 VERTICAL-INTERVAL TIME CODE.₁ Vertical-interval time code (VITC) (see Fig. 20-3) is similar in bit pattern to LTC except for additional sync bits, i.e., *cyclic-redundancy check* and *field identifier,* bringing the total to 90 bits per frame. The field bit allows reading accuracy to a field instead of a frame as with LTC. Since VITC is recorded with the video on helical-scan machines, it can be read in still mode, which is its major advantage over LTC. Its major disadvantage, however, is that it cannot be rere-corded (or restriped) as with LTC, since it is an integral part of the video signal. Attempts to relay VITC using sync-track heads will result in a degraded sync-track waveform with an unstable timing relationship to the original recorded video. Thus, re-striping consists of dubbing down a generation, deleting and inserting new code. During editing, unless new code is inserted on the edit master, the source code will be transferred directly, including fades or wipes. If cuts are performed, then the code will switch to and from each playback source. This is useful when tracing back to the originals from an off-line edit session involving user data such as film edge-numbers.

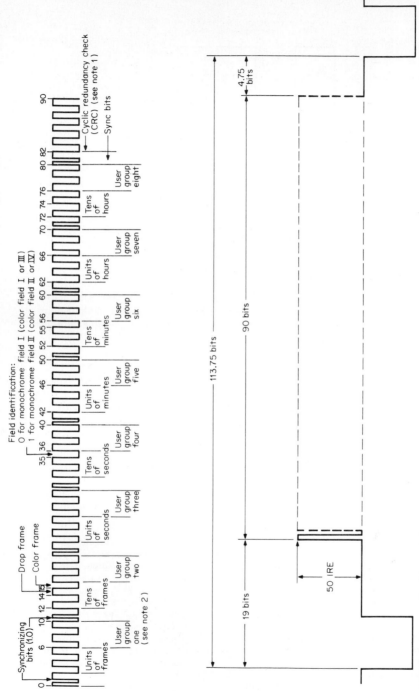

FIG. 20-3 Format of 90-bit vertical-interval time code (VITC). (*a*) Time-code format; (*b*) vertical-interval timing allocation. *Notes:* (1) CRC changes with data. Missing bits will cause an error to be detected. (2) User groups are shown encoded: 0101, 1010, 0101, 1010, 0101, 1010, 0101, 1010. (3) VTC is recorded on two nonadjacent lines in vertical interval (after line 15 if machine does not have sync track) on both fields of each frame.

Field identification:
O for monochrome field I (color field I or III)
1 for monochrome field II (color field II or IV)

Cyclic redundancy check (CRC) (see note 1)
Sync bits

Drop frame
Color frame

Synchronizing bits (1.0)

Units of frames — User group one (see note 2)
Tens of frames — User group two
Units of seconds — User group three
Tens of seconds — User group four
Units of minutes — User group five
Tens of minutes — User group six
Units of hours — User group seven
Tens of hours — User group eight

113.75 bits
90 bits
19 bits
4.75 bits
50 IRE

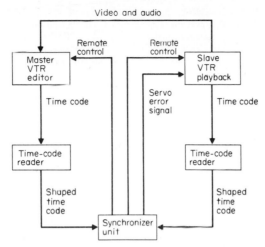

FIG. 20-4 Synchronizer-system block diagram. Some microprocessor-based systems can accommodate several devices simultaneously. Audio tape recorders (ATRs) or magnetic *dubbers* can be substituted for VTR's, permitting the system to be used for audio-sweetening operations. In a system such as the EECO type MQS-100, the time-code readers are included in the synchronizer unit. *(EECO Inc.)*

20.2.5 SYNCHRONIZERS. These are used to *slave* one machine to a master code signal (see Fig. 20-4). The master could be a generator or more usually the edit VTR machine; the slave would be the playback VTR. The slave code is compared with the master code. The time difference between the two is compared in order to generate an error signal. The error signal is used in place of the internal servo error signal to drive the slave machine capstan into synchronism. This system provides frame accuracy and repeatability. It is relatively inexpensive and allows precise edit point location using an offset value between the master and slave.

20.3 COMPUTER TECHNOLOGY

The synchronizer technique is the nucleus of small nondistributive computer systems where a central controller acts as the master and all machines are slaved, including the record machine. Interface to the central processor is usually by a bus system where devices have a set of unique addresses for commands and status similar to other peripheral devices like printers and displays.

When central controller systems first appeared, a minicomputer was used because microcomputers were not fast or powerful enough. The trend now has reversed. The minicomputer is too cumbersome compared with the microcomputers now available in amazing variety. Programmable read-only memory (PROM) is the popular and relatively inexpensive medium to store operating program software instead of core or metal-oxide semiconductor (MOS) random access memory (RAM).

20.3.1 CENTRAL PROCESSING. An example of a central controller technique is the Convergence ECS 104 editing system. The system is controlled by a 6802 microprocessor operating at 895 kHz. The operating program resides in PROM. Machine cycles are 1.12 μs; real-time interrupts occur every 15 ms. RAM of 2K bytes on a central processing unit (CPU) board is used for scratchpad calculations and temporary storage of data during interrupt processing routines. The microprocessor generates 16-bit addressing, but with the write/read line the maximum number of addresses possible, that is, 65,536, is increased to 67,584. Data are routed on an 8-bit bus which is bidirectional to and from the CPU. There are up to four interface logic boards supported by the system which control the VTR with the associated motor-drive amplifier (MDA) board. Status information is also furnished to the CPU.

20.3.2 DISTRIBUTED PROCESSING. Distributed processing delegates many repetitive tasks to microprocessor-based controllers (see Fig. 20-5). The main advantage is to free central processor time to handle more keyboard entries from the editor and to allow several events to occur at one time. For example, a nondistributive system has to sequence commands out to several devices and sequence through status coming back from them on a time-shared basis. The more devices on the system, the slower the central processor becomes in handling the many inputs.

The distributive system allows multitasking and delegation of simpler routines to smart controllers, which in turn provide control functions in accordance with the requirements of each device.

The central controller or host unit issues commands of a general nature such as "search time-code location given" and lets the local microprocessor perform the following tasks:

• control the acknowledgement of the command
• issue a fast-forward or rewind command to establish code direction
• compare the target code and then follow an algorithm to decelerate the moving tape to a predetermined park point
• issue a stop command
• listen for the stop status of the machine
• report back to the central host or supervisor that it is done

With five or six machines operating, deleting these housekeeping routines frees a great deal of processor time for listening to the operator's input via the keyboard.

An example of distributed processing editing systems is the Interactive Systems Company (ISC) Model 31. It is based on a Digital Equipment Corporation (DEC) LSI 11/2 and communicates to peripheral devices such as VTRs and switchers through serial input/output (I/O) ports. The number of ports addressable can be expanded as needed up to 16. The length of cables for serial interface interconnections is not critical. For parallel data, the length is limited to about 25 ft (7.6 m) (for TTL). The system will support either EIA RS-232 or RS-422 by changing the strapping in the communications board appropriately, and by software support of the different protocol requirements. The real-time functions such as cueing, synchronizing, and edit control are delegated to the VTR controllers. This is achieved by having a real-time clock pulse (derived from television composite sync) cause an interrupt routine to occur in the interfaces and host unit to update the internal real-time clocks synchronously. Other commands to VTRs and switchers are executed immediately and include typically RESET, STOP, and PLAY and requests for time code and status information. The model 31 uses 64K bytes of memory and supports 5¼ and 8-in dual disk drives, a printer, a display, a paper-tape reader punch, keyboard, and jogger motion control unit.

The Ampex ACE editing system has some unique features worthy of mention. Three versions of human interface are provided: TouchScreen, dedicated keyboard, and ASC II keyboard, each with a joystick. Data are distributed via RS-422 to machine interfaces and RS-232 to peripheral I/O devices. An interesting variation from the usual PROM storage of unique device-data routines is that at *power up* the executive transmits the device-file program for storage in RAM in the interface unit. The executive system consists of a single-board CPU with 256K bytes of RAM to hold the operating program. The system distributes data from the executive via an intelligent line controller and the machine interface, both of which are microprocessor controlled. The executive has four primary function blocks:

1. Receive operator input from the human interface
2. Process the input command and data and instruct the appropriate machines and interfaces to carry them out
3. Keep track of status coming back
4. Display status data and other activities via the monitor

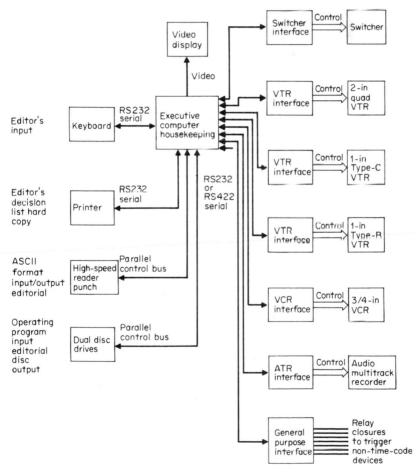

FIG. 20-5   Distributed-processing system block diagram.

As a matter of interest the TouchScreen is not actually touch sensitive. It has a matrix of infrared photo sensors and emitters arranged along the sides of the screen and at the top and bottom. The principle is the same as scanning a keyboard for an entry. The operator's finger interrupts the scanning and causes a matrix code to be sent out to the executive. The scanning can be frozen by breaking the beams or pointing to a blank space on the screen. Then the editor can move a finger from the other hand to a new location on the screen without erroneous results occurring due to the operator failing to move the finger out of the scanned area prior to changing its location.

20.4 TAPE AND SWITCHER EQUIPMENT CONTROL

Three types of equipment are controlled in editing systems. An interface is a translation device which will produce commands in a form that the machine can utilize via its remote control input, and the interpretation of the status signals back to the central system. A block diagram and a description of functions for a typical editing system are shown in Fig. 20-6.

20.4.1 VTR CONTROL AND INTERFACE. *PLAY, RECORD, FAST FORWARD, REWIND, STILL, JOG,* and *STOP* are primary functions which allow simple remote control. The machine feeds back status in the form of RECORD, PLAY, STOP, and servo lock buses, etc. These are translated to binary form and assigned to appear on particular data lines in the display memory.

The machine has its own set of ballistic factors which govern its dynamic performance, e.g., acceleration from STOP, wind-motor speeds, torque, and reel size. These vary among models and manufacturer and, to a lesser degree, among individual machines. These characteristics may be incorporated in the timing-loop routines (see Fig. 20-7). For example, if machine A is a helical format and machine B is quadruplex, B machine must be commanded to stop in advance of A to allow for the greater inertia of the tape and reel. Even though the stop command is issued to both devices almost simultaneously by the editor, A-machine interface will delay the command to allow for the slower quad, since both machines have the same preroll time programmed to allow correct parking of the tape. Mixing machines of different formats slow the editing process to the rate of the slowest machine. This is most significant in SEARCH and SYNCHRONIZE modes.

The other major function of the interface is to slew the machine servo by introducing an offset error. The purpose of this is to bring machines into synchronism by chasing a moving target window. The capstan MDA is given a *speed-up* or *slow-down* override signal to advance or retard playback of the time code, thus bringing the machine into frame-accurate synchronism during preroll with other sources. Since tape machines vary in synchronizing technique, with some using a fixed step-function for speed-change and others an analog voltage swing, the individual interface must accommodate these differences.

When synchronism has been achieved, the servo of the machine is allowed to take over and establish servo lock. If frame accuracy and servo lock are achieved, the VTR is not disturbed, since it is rigidly locked to house sync or video. This is accomplished by checking the decoded time against target time and verifying status of a servo-locked condition by the interface. When a problem arises and is not corrected by a predetermined number of frames prior to the target time, usually an *in* edit, the editor will abort and try again.

Three main problems can occur during this period:

1. A gap in the code will cause a loss of frame reference to the interface during the synchronizing period, or a distorted code could produce an erroneous number.

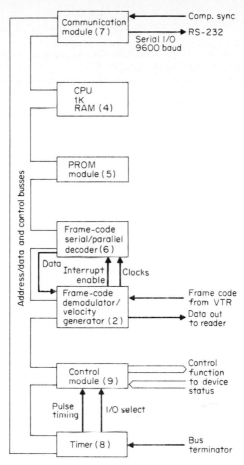

FIG. 20-6 VTR interface. *(CMX).*

Communication Module (Board 7): Selects communication lines 1 or 2, or off to disable Control Module (9) outputs for serial/parellel data communication to and from microprocessor bus; *generates* communications interrupt when receive buffer is full; *generates* real-time clock interrupts from odd-field sync; *strips* composite sync to produce 30-Hz odd-field frame pulse; *generates* initiate pulse to all modules to reset F/W (firmware) and H/W (hardware); *maintains* run light.

Central Processing Unit (Board 4): Control bus under direction of PROM program with 2-MHz clock oscillator for microprocessor unit; *scratchpad* calculations for microprocessor unit.

PROM Module (Board 5): Operating program for various device routines.

Frame-Code Serial/Parallel Decoder (Board 6): Receives 80-bit data and clock pulses to produce two, 16-bit words of hours and minutes, seconds, and frames; *detects* time-code travel direction and bit-errors; *enables* frame-code interrupt; *provides* user-bit selection. FC-error detection, and drop/non-drop FC detection.

Frame-code Demodulator Velocity Generator (Board 2): Converts 80-bit code to data and clock pulses; *generates* even-field frame pulse from sync word; *produces* 8-bit velocity word to indicate speed to microprocessor; *Programmable timer* produces an interrupt of time code of from 1 to 16 frames duration for microprocessor.

Control Module (Board 9): Outputs device commands, including tape-speed override, to VTRs and ATRs; *receives* status data from devices; *receives* fail-safe disable commands from (7) and provides play/record safety lockout.

Bus Terminator/Timer (Board 8): Provides bus pull-up and -down terminations; *decodes* status for receive or output enable; *determines* output pulse widths via timer interrupt to microprocessor; *enables* Jam-sync operation.

2. The code may not have been recorded locked to house sync, causing a drifting relationship of the replayed code with respect to the video.

3. The sync word of time code is decoded and turned into an even-field frame rate pulse in phase with a similar pulse derived from house sync on the odd field, causing it to fall outside the frame-lock window.

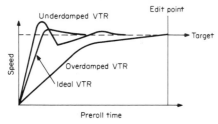

FIG. 20-7 Algorithm for synchronizing. Ideal synchronizing allows the VTR servo to reach lock as rapidly as possible. The interface can cause a tape machine to slow down more rapidly than speed up. Thus, the source usually is synchronized ahead of the cue time and then slowed down for precise synchronization. *(From CMX Patent No. 3,890,638.)*

Frame-Lock Window. A software window defines the limits of accurate frame lock. If the time-code pulse drifts out of the software-generated window, the synchronizer routine will attempt to reframe the machine servo to correct the error (see Fig. 20-8). The machine is released and the position of the pulse is checked to see if it falls in the window; if not, the error will be flagged and the system will abort. Most systems make three attempts and then halt the sequence to allow operator action. Nonsynchronous code is a problem that can be solved using an interface which will handle nonsync code at the price of frame accuracy. The usual method is to rerecord (restripe) the time code.

Color-frame timing discontinuity is identifiable by servo hunting as the color framer tries to reframe with the interface. Ideally, only color-frame edits should be made, although they may be undesirable or unacceptable, esthetically. The normal mode of operation is to disable the color framer in the record machine so that the interface positions the servo to replay the code at the correct frame timing. Because the time code is not necessarily in color-frame synchronism with the house system when it is generated, there is a 50 percent chance of it coming up out of phase when powering up. If the color-frame bit on the time code were set initially and maintained, regardless of power-up condition, then problems associated with color framing in editing and detection by the editing system could be substantially reduced. Horizontal shifts in half cycles of subcarrier on edits, whether match frames or cuts on the playback machines, would be improved when dealing with material from different sources. It would not eliminate the problem but it would make it predictable, and thus be amenable to easy correction.

20.4.2 ATR CONTROL AND INTERFACE. The control of an audio tape recorder (ATR) requires a slightly different approach. During synchronization, the ATR behaves similarly to a VTR as its servo loop is modified to change the speed of the machine.

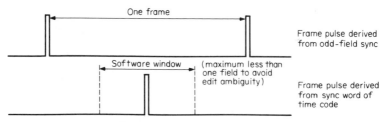

FIG. 20-8 Relationship of frame sync to real-time sync. Tracking errors on playback will displace the time-code frame pulse, possibly to the point where the software window is exceeded. This condition is encountered most frequently with cassette formats. *(From CMX Patent No. 3,890,638.)*

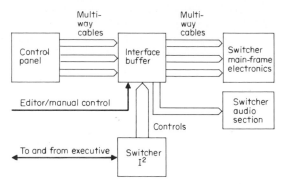

FIG. 20-9 Switcher system block diagram showing intelligent interface (I^2) between executive software and hardware. (CMX.)

However, when synchronism is achieved, the time code is used as the reference for the ATR servo loop. The machine thus remains under the full control of the interface during RECORD, EDIT, and PLAY modes. Normal practice during an original recording, when time code is being laid down, is to ensure that the machine's servo loop is not passing through the interface control circuits to avoid the introduction of random errors in the speed. The remainder of the functions and routines in the interface are very similar to the VTR. Status is reported back from the machines' indicators as before.

20.4.3 SWITCHER CONTROL AND INTERFACE.

The interface to a switcher takes two basic forms, either parallel or serial control between the interface and the switcher. Parallel control consists of a "wire per function" to operate crosspoints (usually eight), select *MIX, WIPE,* or *KEY* modes, and select patterns. A digital-to-analog converter and driver generating an analog ramp to simulate the wiper arm of a fader potentiometer is also required.

Usually there is an interface buffer (see Fig. 20-9) so that operator input via the switcher panel is possible and system control may or may not be initiated as desired. The buffer allows easier access to the control lines which run between the switcher control panels and the signal and control electronics. It also allows customizing of those crosspoints to differences in user needs.

The buffer is also used for data translation where there is no *wire-for-wire* equivalent; e.g., wires may have "binary weights" associated with them to reduce the size of control cables required. Unfortunately, the diversity of interface designs is dictated by the many different switcher designs.

The other main system is serial control via EIA RS-232C† or RS-422-A‡ Recommended Standards. The process requires a timing and translating interface rather than a controller. The interface acts as an interpreter, accepting data from the supervisor, then sending them on to the switcher for actuation. The PROM program in the interface then becomes a large *look-up* table to accomplish command translation. The data paths to and from the interface need not be at the same baud rate. The rate to the switcher may be significantly higher than to the editor. The interface is of increased importance in the system because the data to and from the interface and switcher have a higher priority than the status back to the editor. The communication in the system becomes

---

†Interface between data terminal equipment and data communication equipment employing serial binary-data interchange.

‡Electrical characteristics of balanced-voltage interface circuits.

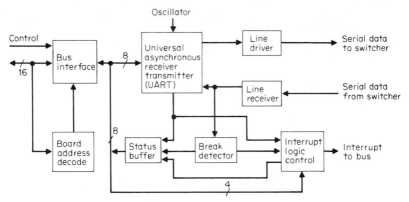

FIG. 20-10　Serial I/O module block diagram. *(CMX.)*

totally software-driven to a point that makes maintenance nearly impossible, except for replacing data path components, if no diagnostic aids are available (See Fig. 20-10). The same applies to VTRs except for devices that use a 200-kHz rotating loop of command and status, such as the Sony BVH-1100 wherein the interface holds for the 64-bit data loop to clock through before inserting a new command in the correct sequence going to the machine. Of 64 bits, 32 are command and 32 are status; also included is a sync bit which has a wider pulse to allow detection and synchronization of the interface command loop.

The SMPTE is developing a standard serial-data format compatible with all types of machines so that interfacing can be accomplished with a universal system. However, until agreement is reached and implemented in manufacture, there will continue to be as many unique interfaces in the field as there are devices. On the other hand, some manufacturers of editing systems are eliminating the interface hardware and are communicating directly with the serial port of the equipment. The interface requirements then involve software, rather than hardware. The command translation occurs in an I/O routine in a software module. Reducing the hardware increases the speed and reliability, and reduces cost. However, it does make the editing-system hardware ports dedicated to either EIA standards, RS-232C or RS-422. This is the only limitation but is not serious, since there are up to 16 ports available on most systems, which can be strapped for the system in use.

20.5 EDITING PROGRAMS

Concepts for editing-system software vary from a single-block program without any optional features to systems which operate using files for downloading interface routines, and others programmed in modular building blocks using very fast compilers to execute the program. Most systems are not designed for access by the user because the software is considered proprietary information.

20.5.1 PROGRAM ROUTINES. ISC editing software is an example of only a part of what can be included on the disk. It can have the following programs resident on the same disk: Operating system; super-edit program; utility programs for copying, deleting, formatting, and initializing other disks; system diagnostic test programs; 409 and TRACE (see Sec. 20.7.2); and user-generated custom programs. This eliminates having several disks and the need to change them when changing programs.

The operating program in the executive may contain some of or all the following routines:

- Peripheral I/O device handling
- Direct memory access for disk
- Mapped display data translator (data to ASC II for display)
- Keyboard command processor
- Channel management
- Real-time clock processor
- Communications processor
- Binary frame code processor (for interfaces)
- Edit point (target time) processor (with respect to real-time clock)
- Target duration processor (with respect to real-time clock)
- Split edit function timing processor and flags
- Record machine flag
- Preroll constant processor
- Latency (start offset processor)
- Edit-type processor (audio, video, or both)
- Transition processor (CUT, DISSOLVES, KEY)
- Delay DISSOLVE, etc. (late start on machine)
- Preview mode
- Crosspoint source selection
- Edit decision listing
- Error message and system initialization routines
- Dead man timer processor (allows for busy interface when communication requested by executive)

The interface may contain firmware with the following routine modules:

- Low core vectors and system constants
- Search routine
- Synchronize routine
- Handshaking with executive and microprocessor
- Translation device command driver for communication
- Editor timing constants
- Microprocessor stack processor
- Real-time clock processor
- Communications receive interrupt processor
- Frame code interrupt processor
- Pulse width timer processor
- Double-precision routines for frame-code calculations from BCD to binary
- Communications transmit processor

20.5.2 EDITING STYLES. Editors adapt their techniques of using a particular editing system to suit the programming they are helping the producer or director to create. The primary consideration is what kind of human interface is used. Does the system have a keyboard-driven menu system, a dedicated keyboard, a multilevel keyboard, or a joystick or other tactile controller system? Often the requirements of one editor differ from the next even on the same program material because of their distinct methods, their length of time in the business, and how they use the human interface. The more flexible the system, the more the system is adaptable to individual preference. For example, many editors like to edit open-ended, preferring to terminate a sequence after the required time and then overlap the material with the next sequence, building an unclean list. Others like to edit to the exact frame and work with a clean list all the time, particularly if they are going to assemble the program later; some like to edit record rather than previewing. The tactile controllers like Convergence *Superstick*, CMX *Gismo*, and ISC *Jogger Motion Control* bring back the lost physical contact with the tape machine caused by moving the editor into another room. To a certain extent, these devices allow manual feedback to the editor, reinforcing his or her satisfaction and sense of control over events.

Rigid or total control of the interfaced devices has advantages but creates more data input for the editor to enter when setting up a sequence. For instance, if the editing system does not control all the crosspoints of a switcher (normally the first eight on parallel-controlled switchers) and the editor wishes to select other sources to perform a programmable function, the editor has to attempt to override the system control immediately following the setting of crosspoints to obtain a controlled DISSOLVE, WIPE, or KEY transition. Some computer-assisted editing systems, such as the CMX 3400, permit manual control selection. The degree of interface is available in a number of ways, for instance, manual or automatic control of audio and/or video as with Grass Valley 1600-1L. CMX allows switching of audio control on and off on their Grass Valley Group (GVG) Model 300 CMX Interactive interface, along with the choice of which mix effects are controlled and whether video cross-point selection is manual or automatic. Ross Video switcher model 524-series CMX Interactive Interface (I^2) allows control of crosspoints and/or transition rate, and allows choice of some 11 keying modes available on the multilevel effects through a single edit-system key function. The longer the process, the more tempting it is to do the edit On the Fly and manipulate the desired transition in real time. For the most part, the editing-system control frees the editor for other manual chores, such as operating digital effects in real time or making adjustments of audio levels for proper balance. The CMX 3400 system even permits video selection and types of transitions learned by some video switchers to be captured into the CMX decision list.

20.6 DISPLAYS AND CONTROLS

20.6.1 **DISPLAYS.**   Most editing systems started with displays of the time code of each machine. Now they have blossomed into full information centers and more with the introduction of the 3400+ by CMX. The display consists of several parts: DATA, SCRATCHPAD, and EVENTS. For example, the CMX display gives the following information:

- Title or other user-defined data
- Current duration of program
- In and out times of various record and playback machines
- Durations between *in* and *out*
- Current code location
- Type of code and mode of operation
- Status on as many as seven record-playback machines at one time
- General-purpose interface status
- Switcher-memory status
- Current programmed function
- Current mode of preview
- Selected source

- Constant register
- Variable number of event lines
- Type of edit (video, audio, cut, wipe, etc.)

Maximum screen area is 2000 characters. Figure 20-11 shows the clarity of the data.

Some systems such as the Edge from CMX use multiple (approximately 36) pages depending on the mode of operation. This is done to simplify the information presented to the editor on the 5-in (12.7-cm) screen. Multiple pages tend to require multiple keystrokes to get the desired display. It is entirely a matter of personal preference and cost effectiveness.

20.6.2 KEYBOARDS. The data entry keyboard must include some fundamental functions common to most editing systems. These include MARK IN and OUT, SET IN and OUT, TRIM IN and OUT, PREVIEW, RECORD, PLAY, FAST FORWARD, REWIND, CUE, and ALL STOP. Beyond these primary motion functions, selection of machine comes next and an assortment of keys dedicated to list management, system setup, and peripheral devices like high-speed reader punch, disk drive, and printers. The keys for SET IN and OUT provide the means of entering a specific number as a start or end point. TRIM IN or OUT allows modification by a number of frames plus or minus or trimming by a constant factor. The MARK functions are used for timing *on the fly* of an event, like a sudden noise or camera cut. The introduction of VITC has allowed direct access to such points on type C format because in still-frame mode the VITC is read over and over again. The MARK IN function gives precise easy entry. With quad format a succession of previews has to be performed to zero in on the precise frame, usually with the aid of a reaction time factor. This can be set up on system initialization, in play only to reflect the speed of a particular editor.

20.6.3 CONTROL MODES. Modes provide the means to play back with time codes the picture and sound from a number of recorders and other signal sources and, more importantly, to simulate an edited sequence. Thus, the editor gains a perception of the timing and dramatic impact of the overall sequence. The simplest form of preview consists merely of a playback of all the sources involved and is very limited, creatively. For instance, an editor who is creating an action sequence which must run a particular length must meticulously preview virtually every frame of tape and tediously build up the sequence. With this initial task completed, if the sequence does not flow properly, the process must be repeated. In other words, tape editing is a linear process that does not permit the addition of more frames in one portion of a scene without the deletion of an equivalent number of frames in another portion.

To aid the editor, a variety of optional features are obtainable in commercially available editing systems and programming software, and many more custom-designed modifications and adjuncts to the standard hardware packages are in use in different editing facilities. Some of the most used modes are listed, with brief descriptions, below.

- Gismo (ganged or individual speed-motion option). Tactile control of speed in SEARCH and JOG modes, MARK functions, and transport control of all machines on a remote-control hand-held controller. The functions are duplicates of the main keyboard, except for SEARCH and MULTIPLE ROLL on a single keystroke function.
- Motion memory. Programming of slow motion or fast motion effects on a repeatable basis.
- Master-slave. The most used option that allows a number of sources to be slaved to a master machine. This is particularly useful for quad split effects or slaving an ATR to a VTR when relaying a master audio track.
- Sync roll. Provides the equivalent of real-time selection from multiple ISO (isolated) reels of tape, in the same manner as from several cameras as during taping of a live show.

CMX 340X SAMPLE EDIT DECISION LIST

			IN	OUT	DURATION	TIME-CODE
	V-ONLY				00:06:55:19	
		R-VTR	01:06:23:11			/PLA N-01:17:06:05
	DIS					
Menu area	C TO A 060	A-077	13:10:37:20	13:10:44:04	00:00:06:04	/REW N-13:09:46:11
		B-081	12:35:18:28	12:35:18:28		/STP N-12:35:19:08
		C-084	12:39:18:24	12:39:18:24		/LOS D-12:42:06:20
	SORT REC-IN	AUX				REC OFF PUNCH ON
	EVENT #020	BLACKS				CO - 13:31:25:14
System message area	RENUMBER ALL EVENTS, OK?					

	014	084	B	C		12:39:13:20	12:30:18:24	01:06:18:07	01:06:23:11
	015	081	A	C		12:35:01:06	12:35:18:28	01:06:23:11	01:06:41:03
	016	084	V	C		12:39:18:24	12:39:18:24	01:06:23:11	01:06:23:11
	016	077	V	D	060	13:10:37:20	13:10:44:04	01:06:23:11	01:06:29:25
	017	077	V	C		13:26:17:02	13:26:22:02	01:06:29:25	01:06:34:25
	017	084	V	W019	045	12:35:04:03	12:35:10:11	01:06:34:25	01:06:41:03
List area	018	081	B	C		12:37:08:02	12:37:11:02	01:06:41:03	01:06:44:03
	018	081	B	K B		12:37:11:02	12:37:18:23	01:06:44:03	01:06:51:24
	018	AX	B	K	030	00:00:00:00	00:00:06:21	01:06:44:03	01:06:50:24
	019	077	B	K B	(F)	12:19:37:19	12:19:44:12	01:06:51:26	01:06:58:19
	019	084	B	K B	090	12:53:00:19	12:53:04:12	01:06:51:26	01:06:55:19

		IN	OUT	DURATION
MENU DISPLAY:	R-VTR	01:00:00:00		
KEY				
A to B 030	A-001	02:00:00:00	02:00:15:00	00:00:15:00
00:00:03:00	B-020	04:00:00:00	04:00:11:00	00:00:11:00
	AUX			
EVENT # 001	BLACK			

EDL DISPLAY:

001	001	B	C	02:00:00:00	02:00:03:00	01:00:00:00	01:00:03:00
001	001	B	KB	02:00:03:00	02:00:15:00	01:00:03:00	01:00:15:00
001	020	B	K 030	04:00:00:00	04:00:11:00	01:00:03:00	01:00:14:00

MESSAGE AREA DISPLAY:
KEY, OK?
BACKGROUND = A, FOREGROUND = B
DELAY = 3:00 FADE TO BLACK OFF, OK?
DURATION = 30 (FRAMES)

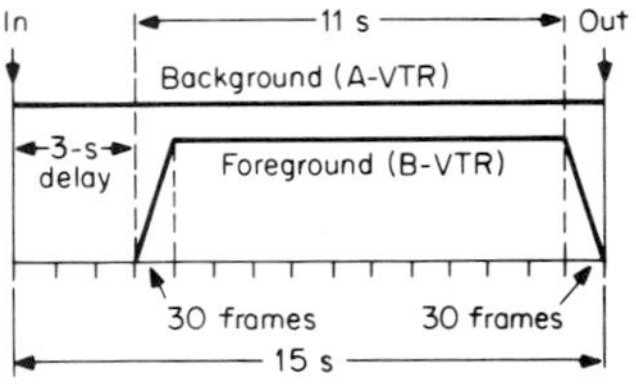

FIG. 20-11 Typical edit decision list (EDL) and display layout. *(CMX.)*

- General purpose. Interface allows the frame-accurate triggering to enable an event to occur, e.g., starting a digital-effect sequence, changing a slide, or starting a count-down generator.
- Multiple record. More than one master tape can be made at one time during the edit process.
- Switcher memory. Similar to an intelligent general-purpose interface, in that it activates a switcher automatic sequence on a particular frame by transmitting the start or run code to the switcher.

- Frame bump. Slewing of one machine with respect to another to achieve synchronism of audio or video. This is useful for relaying sweetened audio or matching ISO reels with different time codes.

- Reel summary. Used during assembly mode to provide information on the most efficient order to load up playback reels in a Mode-B (checkerboard), rather than a Mode-A (sequential) assembly. A Mode-B assembly requires a minimum number of reel changes, whereas Mode A will involve many more reel changes.

- Pre-Cue. Enables available machines not involved in the edit operation in process to pre-cue according to the time-code location data in the edit decision list (EDL).

20.6.4 PROGRAMMED EFFECTS. These may be categorized as cuts, dissolves, wipes, and keys. *Cuts* are self-explanatory and are considered as single-line events involving a single source.

Dissolves can be programmed up to several seconds (255 frames in CMX systems) in length and down to zero frames (cut). They can be either video or audio, or both, delayed from the *in* point of the edit and logged as a two-line event in the edit decision list (EDL).

Wipes are similar to *dissolves* except that the type of wipe effect must be specified. Though the list of effects is not identical from one video switcher to the next, most switchers can call up any of the 23 *standard patterns* specified by the SMPTE. Some more sophisticated switchers provide thousands of combinations (see Sec. 14.6.1).

Key involves the foreground and background and may also include a fade-to- or fade-from-black function. The key and foreground may be faded in or out together, or the key may be faded onto the already cut-to background. A key is listed as a two- or three-line event, depending upon whether a fade is used.

20.7 EDIT DECISION LISTS

20.7.1 DATA-LOGGING NOMENCLATURES. The edit decision list (EDL) is the means whereby the editor can keep a reference of decisions as laid down on the tape.[2] This is necessary even though information subsequently is lost by overrecording. The information stored by the *executive program* contains the data necessary to repeat that decision operation to a limited extent. If the decision was a simple audio and video cut function, the following data must be logged:

- record tape in and out times
- playback tape in and out times
- playback reel number
- programmed effect namely, (cut)
- whether audio 1 and audio 2 and/or video were edited

A *dissolve* involves more information as to the duration and the source direction of to and from. The second playback in and out times must also be present. Unusual conditions are noted also, such as mixed-drop and nondrop frame time codes and the code system used in the initial system setup. A wipe is handled the same way as a dissolve with the addition of a wipe code. Table 20-1 shows the dilemma of logging audio 1 and 2 and video edit functions in the EDL data.

Key functions involve three lines of data and must denote foreground, background, and fade rate. Figure 20-11 illustrates a key edit event and the expected transition.

Additional information in the form of notes can be added and tied to an event to reflect some unusual condition in that edit. *E-mem* register and GPI activations may also be stored. All information in the E-mem may be uploaded to the executive of systems such as the CMX-3400.[3]

Table 20-1   Audio-Video Edit-List Formats[†]

	Audio 1	Audio 2	Video	Audio 1 and 2	Audio 1 and video	Audio 2 and video	Audio 1 and 2 and video
1. Old single—audio	A	....	V	....	B		
2. ISC—dual audio	A1	A2	V	A	A1V	A2V	AV
3. Convergence	A1	A2	V	A12	V1	V2	V12
4. CMX 217	A	A2	V	AA	B	A2/V	AA/V
5. Proposed SMPTE	A1	A2	V	A	B1	B2	B

[†]The 409 Program can convert among these formats. They are selected with Number 2 on the terminal keyboard.

20.7.2 LIST MANAGEMENT AND PROGRAMS. List management, mentioned in Sec. 20.7.1, is the process of arranging edits in the list by number or by record-in time, depending on the method of assembly selected by the editor. Listing by number follows the order in which the edits were made, which may not correspond to the edited sequence. Sorting by record-in time builds the edits in as close a fashion as possible to the desired end product. There may be overlaps and gaps in the edit list by accident or design, and these must be noted by the editor. This kind of number crunching is a tedious procedure subject to error. A desirable practice to avoid such problems is to dump the edit list onto disc, punched paper tape, and hard-copy printout for reference *before* undertaking any alterations. Thus, if a mistake is made, the original is available. The most frequently encountered pitfall is the *move events* function, which is available with pull up on record-in times. An error here, when eliminating an edit which has an overlap, will cause the resultant gap *(hole)* to vanish and an incorrect pull-up constant to be used to correct the subsequent events. All the record in and out times will change on the list.[4]

    Sort ON/OFF. In the OFF mode the EDL will have events and numbers in the order in which they were entered. Sort ON is done by record time or event number in ascending order. Care must be taken in sort by event number because it is difficult to detect holes in a list. Moving events in this mode will cause problems in keeping track of the record in and out times. Sorting by record-in time is the best way to work with a list because holes between in and out record times are easier to spot.

    Renumber allows the editor the option of changing all or some of the event numbers.

    The *record ON/OFF* function allows the entry of predetermined edit data without the need to mount tapes.

    The *delete* function may be used to eliminate a range of events.

    The *record-start* function allows the editor to assign a new record-start time for the entire list or to create a hole in a list or close one. The most common use of this function is for auto-assembly to match the record times to a time-coded crystal-black insert tape.

    The reedit mode permits the changing of a single event that requires revision. This consists of three functions: open, insert, and close. *Close* is used only as an exit mode if no change is required. *Insert* enters the revised edit and closes the mode. At this time the editor has the option to save the edit prior to revision for later recall. An edit saved is not assembled because it is flagged with a letter.

    Two programs, called 409 and TRACE, are available which eliminate the need for laborious calculations by the editor.

    The *409 program,* created by ISC, reads data intended for automatic assembly of videotape programs and produces a cleaned-up edit list in standard format. In the process, events are reordered in the record sequence, overlaps are eliminated, and continuous segments are joined into a single event. The result is a printed event list and an EDL disk or punched tape which can be assembled in reduced time, with less chance of error.

The 409 program also allows the revision of programs by adding or deleting events from the terminal keyboard and by shifting events on the record tape. In shifting events, audio and video can be handled either separately or together. The 409 program performs the following functions in cleaning up the edit lists:

- Audio and video are treated separately, with the output including both audio and video events only if the audio and video are from the same source, synchronized with the same *in* and *out* points.

- Unused events or parts thereof which have been overwritten are deleted. To determine overlap, events which are later in the input sequence are retained and earlier ones are changed.

- Edits in the output list are in the sequence of the *record* time.

- Events are numbered with a sequential line number (on the left side of the list), and the original event number (on the right-hand side). Most 28K bytes memory systems will hold about 1200 events during the cleanup process.

- The editor may request either an automatic cleanup process, or the steps may be done separately.

- The cleaning steps available are eliminating overlaps, splitting events which have been overwritten by inserts, joining consecutive events from the same reel, and rearranging events to maximize the number of both edits.

- The editor can select (MARK) an event, either by the new sequential event number or by the original event number, and then verify the list content at that point.

- The editor can delete a selected event or can shift the record time of all events starting with a specified line number, to open or close a hole. Audio and video can be shifted separately or together.

- The output list is marked to indicate discontinuities in the record in time, in either audio or video.

- The duration of a DISSOLVE, WIPE, or KEY can be changed without reentering the entire event.

- The editor can trim the play in or out time of an event. All events later in the list will be shifted accordingly.

The TRACE *program* allows the editor to produce a *rough cut* and then to reedit that record tape into a more polished version. The reedit process may be continued until the final result is satisfactory. Such an edit process results in several generations of record tapes. The program reads in the EDL data representing one or more generations of reedits. It then computes the corresponding edits, through the several generations, back to the original source tapes. The result is an edit list representing the same material as the final tape, with all edits referenced to the source tapes. This kind of program can reduce hours of painstaking manual documentation to a few minutes at the keyboard. It makes the rough-cut and reedit technique practical and foolproof, and it allows reediting of on-line work, without concern for added dub generations, since the program can be reassembled using the original source tapes. TRACE provides more flexibility in editing off-line, since several complicated EDLs can be combined easily.

20.8 ASSEMBLY TECHNIQUES

20.8.1 **OFF-LINE.** This process is an edit session using less expensive machines like the ¾-in (1.9-cm) U-matic format. With less capital tied up in equipment, the rate per hour drops, and thus more time can be spent doing creative work on cassette than with expensive 1-in machines. The editor uses cassette duplicates of each reel, usually with *burned-in* time-code numbers for easy searching, when not using a machine equipped with a reader, such as in the director's or producer's office.

20.8.2 ON-LINE. In this editing process the program is assembled using the original master recordings. A fully equipped edit room can cost in the order of a million dollars. Thus, in order to obtain an adequate return on such a large capital investment, on-line time is very expensive and must be used with optimum efficiency. Alternatively, off-line rooms with the bare minimum of low-cost equipment can be set up on a minimal budget. Consequently, the proliferation of relatively inexpensive cassette-editing systems for off-line use has begun a trend in most full-service postproduction houses to deemphasize this editing activity and to concentrate on more lucrative business requiring extremely sophisticated and costly video and audio equipment, and specialized digital techniques.

20.8.3 AUTO-ASSEMBLY. As its name suggests, auto-assembly is a fast, minimum-staff process to put together a program. However, the speed of auto-assembly depends greatly on the kind of program and effects required. Most editors prefer to be present at the assembly, either doing the work or supervising it. There are two modes of auto-assembly. One is to lay down each edit in time sequence until it is complete (*sequential* or *mode A* assembly). The second is to build a show in a checkerboard method, going through each source tape and editing in, wherever possible, all the segments required from that tape, or editing modules of a production, not necessarily in sequence, into a contiguous recording in the proper order (mode B assembly).

Mode A is commonplace and is easier to work with than mode B. Mode B requires working from a clean list without overlaps and equipment that is precisely frame accurate so that no overlaps or holes occur in the audio and video (less likely to occur in sequential). During the sequential-assembly process, effects are usually built as the show progresses. However some editors prefer to make an *effects reel* prior to going into the off-line session. Then the editor merely slots in the cassette version of the finished effect to enable quick editing, using the effects reel as another tape source. Equipment such as VTRs play an important role in the speed of editing, whether in auto-assembly or in on-line editing. During auto-assembly the editing crew can improve efficiency, keeping ahead of the edit system by watching the assembly list. This is done by putting up the next tape required rather than waiting until the system says that it is unable to perform an event owing to an unmounted reel. The more machines available for playback, the easier and faster it becomes. Obviously it is not cost effective to have as many machines as there are tape reels involved! Two playbacks are the minimum to allow DISSOLVES and WIPES; a third machine will provide a worthwhile saving by speeding up editing and assembly about 50 percent. In other words, a 3-h assembly can be cut to 2 h. In the on-line process the third machine doubles as a record/play machine used for building effects which require more than one generation to achieve. This saves time taking down and remounting tape reels.[5]

20.9 EDITING TRENDS

There are two emerging trends in electronic editing that promise to change the direction of systems design from an emphasis on self-sufficient hardware packages to a more comprehensive and interdependent, software-oriented system approach tailored to the requirements of the editors and specific production problems.

First, an increased volume of information is being made available for the editor's use with human interfaces which can be customized at the editor's discretion. Second, editing is no longer an isolated operation. Instead, it is interlinked with a variety of other systems by increasingly sophisticated communications and control facilities.

20.9.1 SYSTEM DESIGNS. The third-generation systems, such as the Model 3400+, represents the move away from hard-wired keyboard design with predetermined and limited control functions. The keyboard can be customized by means of software to any one of eight memorized setups preferred by the editor. There are two screens for output: on one, edit data, and on the other, a field for a script or auto-assembly notes. A *touch-screen*, motion-control unit (Gismo +), and a voice controller, are available as options. The voice controller can learn any language.

The advantages are to free the editor's hands and allow more visual and audible contact during the edit process. The large data-memory system operating program is in DEC (Digital Equipment Corp.) assembly language, executed at high speed.[6]

20.9.2 SYSTEM INTEGRATION. The second trend is toward development of means to sort out the complexities of production houses and broadcasting facilities with multiple edit rooms, each controlling shared equipment. For example, five or six edit rooms may share 15 VTRs, a selection of digital-effects devices, and character generators. The problem has been solved, for the most part, by the use of a central routing switcher serving *all* technical facilities, as well as nonoperating areas such as offices and viewing rooms.

Consequently, the size of routing switchers has grown from a simple master-control studio-in/record or network-out format to complex 100-by-100 or larger matrix assemblies, handling video, stereo audio, time code, cue, intercommunications, and data.

A typical advanced system concept to solve these scheduling and facilities assignment problems is the Multi-Master SAVANT, developed by Robert Lund Associates, Inc. With this system, the operator at the editing console can (1) inquire as to the availability of support facilities and assign to this operation any that are not scheduled elsewhere, (2) program the sequence of functions, (3) recall stored data for display, and (4) program operating sequences of controls, where appropriate, on equipment assigned to the editing console.

The system, in addition, provides the nontechnical scheduling operations with a means to set up, in advance, equipment assignments to editing suites and display on CRT terminals all current and future equipment and facilities assignments.

Thus, editing systems, in addition to providing the basic creative function of program assembly, are, in effect, intelligent data-entry terminals interfaced with the master system which handles all the housekeeping chores of a facility. This places a greater burden upon the operator and consequently dictates a simplification of control and greater data memory capacity. These requirements can be achieved effectively only by an industry-wide standardization of control systems and data formats.

20.9.3 TIME-CODE RECORDING ON FILM. Until recently integration of film and video-tape recorded material in electronic editing operations has been hampered by the lack of direct compatibility between the visible edge numbers on 24-frame/s motion-picture film and the electrical time-code signal on 30-frame/s on video tape. The major problem, in addition to the difficulty in coping with the difference in frame rates between the two media, is need for a means to record a program or time-related identification on film as can be done easily on an audio track or in the vertical interval on tape.[7]

The frame-rate problem has been solved with PAL and SECAM 625-line systems by ignoring the 4 percent difference in film and video frame rates, and running film at 25 Hz. No such simple answer exists for the 525-line, 30-Hz system, however, short of shooting film at 30 frames/s. One technique frequently used is to transfer the original film negative to video cassette (VCR) for editing and then to conform the negative accordingly. This approach involves the use of costly film-to-tape and tape-to-film transfer equipment and complex facilities for conforming the film material. Consequently, it has been used with varying degrees of success.

Film/Tape Editing Methods. One approach uses software to predict and identify the relationship between film and video frames. The start edge number is known and compared with the start time-code number from the film-to-tape transfer. The required in and out times for an edit are established from the edit decision list, and the edge numbers nearest to the required edit points are extrapolated using the computer program.[8]

The end result is a list of edge numbers corresponding to the decisions made from the video tape recording.

Another method is to use visual (burn in) and vertical-interval time code with the edge numbers encoded in the user bits.

Both methods require accuracy of setup at the initial transfer and are time consuming and therefore unprofitable.

Time-Code Magnetic Recording on Film. A system has been developed by Eastman Kodak Co. by which electrical signal information, such as SMPTE time code, can be recorded magnetically on the surface of optical motion-picture film, without interfering with transmission of the visual image. It is not intended as a high-quality audio track but as a means of recording magnetically encoded data such as time code for use in editing systems, cue tracks for scene identification, magnetic edge numbers, or all these.

The Datakode recording medium consists of a layer about 8 μm (315 μin) thick of gamma-ferric oxide particles with a density of about ½₀₀ that of conventional magnetic tape. The layer is transparent, although it increases the transmission density by less than 0.15. The particle density drops the recoverable signal by 46 dB. Tests have shown that the layer is quite capable of recording SMPTE time code on 16-mm film.

The time code may be recorded during shooting or may be precoded. In either case there is a fixed relationship between time code and edge numbers prior to the film being exposed. In addition, many processes in film production from exposure to printing of release prints could be automated by the use of Datakoded film.[9]

At this time of writing, the necessary production and postproduction hardware is under development and not available for commercial applications.

REFERENCES

1. "SMPTE Recommended Practice RP-108: Vertical Interval Time and Control Code for Video Tape," *SMPTE J.* vol. 90, pp. 800–801, Sept., 1981.

2. Robert Lund, "SMPTE Committee on Video Recording and Reproduction Technology: Working Group on Editing Procedures." *SMPTE J.* vol. 91, p. 840, Sept., 1982.

3. Bruce Rayner, "The Application of Switcher-Intelligent Interfaces to Videotape Editing," *SMPTE J.* vol. 88, p. 715, Oct., 1979.

4. Arthur Schneider, "Edit List Management," *SMPTE J.* vol. 88, p. 538, Aug., 1979.

5. Douglas Thorton and Gene Simon, "VTR Modifications for Computer-assisted Editing Systems," *SMPTE J.* vol. 92, p. 646, June 1983.

6. Gene Simon and Joseph Roizen, "Novel Human Interface for Videotape Editing," *SMPTE J.* vol. 93, pp. 1147–1149, Dec., 1984.

7. Joseph E. Rooney, "Film and Videotape Editing: The Process of Conformation," *SMPTE J.* vol. 93, p. 166, Feb., 1984.

8. Arthur Schneider and Donald Kravitz, "A Videotape Editing System for Film Postproduction," *SMPTE J.* vol. 91, p. 552, June 1982.

9. Ronald E. Uhlig, "Technical Experience with Datakode Magnetic Control Surface," *SMPTE J.* vol. 93, p. 730, August 1984.

Standards and Recommended Practices

Dalton H. Pritchard

RCA Laboratories
David Sarnoff Research Center
Princeton, New Jersey

21.1 CLASSIFICATION AND DEFINITION OF STANDARDS

A definition of the term *standards,* as given in the *Random House College Dictionary,* states "something considered by an authority or by general consent as a basis of comparison; an approved model."† Within this context, television standards may be defined as the criteria which describe a specific procedure or method to be employed in implementing a television system.

Rules and regulations, in which the particular technical standards are set forth, constitute the policies to be followed and enforced by the appropriate governmental authority, such as the Federal Communications Commission (FCC) in the United States. As such, these rules and regulations, and, therefore, the technical standards contained within, are legally enforceable. By contrast, recommended practices, often referred to as standards of good engineering practice, may not necessarily be enforced by law but often form the policies of the appropriate authority when issued as constituting the rules and regulations.

The basic purpose of television standards is to ensure the autonomous nature of a system in that the parameters of the system are designed and operated on a common basis which optimizes performance under the economic and technical constraints relevant to the particular service.

Television standards may be classified in three broad categories as *Spectrum Allocation and Radiation Standards* (Sec. 21.1.1), which relate primarily to the use of the available radio frequency spectrum; *Signal Generation and Transmission Standards* (Sec. 21.1.2), that describe the specific signal parameters used to convey pictorial and aural information; and *Equipment Performance Standards* (Sec. 21.1.3), that relate to the apparatus used to generate the visual and aural signal components.

---

†Reprinted by permission from the *Random House College Dictionary*, rev. ed. Copyright © 1984 by Random House, Inc.

The following material summarizes the television engineering standards set forth by the U.S. FCC Rules and Regulations (Part 15, July 1981; Parts 73 and 74, March 1980) as well as the Recommendations of the XVth Plenary Assembly of the International Radio Consultative Committee (CCIR) held in Geneva, Switzerland, in February 1982 and published by the International Telecommunications Union (ITU).

21.1.1 SPECTRUM ALLOCATION AND RADIATION STANDARDS. Spectrum allocation and radiation standards define the manner in which the available radio frequency spectrum is to be occupied by television signal transmissions. This classification of standards has the objectives of providing maximum service, reducing interference to a minimum, and ensuring the most efficient use and occupancy of the channels available for assignment.

Some of the topics included in this classification are designated channel limits with specified radiation frequencies, grades of service, allowable radiated power levels, permissible co-channel, adjacent-channel, and separated-channel interference levels, and the geographical separation among broadcast stations required to protect the established grades of service relative to propagation characteristics. These standards also specify tolerances on carrier, subcarrier, and intercarrier frequencies, receiver intermediate frequencies, and appropriate monitoring procedures and techniques.

Related to these standards is an *allocation plan* which assigns television channels to particular geographical localities, or zones, in such a way as to ensure the widest possible choice of stations over as great an area as possible, while at the same time protecting the service from disturbances arising from undesired sources or causes of interference, including the effects arising from natural phenomena. The specific items contained within the allocation and radiation standards are summarized in Sec. 21.2.

21.1.2 SIGNAL GENERATION AND TRANSMISSION STANDARDS. The signal generation and transmission standards describe the specific nature and parameters of the television signal as it is to be transmitted within the spectrum as allocated; i.e., these standards describe the signal to be radiated by the transmitter. When the television equipment designers and operators of transmitting and receiving apparatus adhere to the signal generation and transmission standards, the quality of the information transfer is assured to be appropriate for the intended service. Satisfactory reception, free of undue interference, is ensured for all transmitters within the expected range and by any receiver designed and intended for the service. These standards also provide a basis for enforcement of rules and regulations (containing the technical standards) designed to protect the public interest in the television service.

In particular, the signal generation and transmission standards relate specific measurable quantities in the radiated signals (such as amplitude, phase, and frequency) to the visual and aural quantities undergoing transmission at any particular instant in time. The quantities may be expressed in terms of the parameters pertinent to either or both the time domain and the frequency domain. They also specify baseband values as well as the resulting radio-frequency signal values. Thus, the modulation standards define the relationships between the visual quantities (luminance and chrominance) and the visual carrier and its modulation sidebands. Likewise, the aural quantity values (sound intensity variations) relate to the aural carrier and its modulation sidebands. The variations of the visual quantities relative to time (signal waveforms in amplitude versus time) are described by the technical transmission standards that include scanning parameters and signals especially designed to provide the necessary synchronization of signal quantities and visual events. In addition, the transmission standards include those parameters that define the distribution of the radiated carriers, subcarriers, and modulation sidebands within the allocated channel. The specific items comprising the signal generation and transmission standards are summarized in Sec. 21.3.

21.1.3 EQUIPMENT PERFORMANCE STANDARDS. After the most effective use of the available radio-frequency spectrum has been defined by the allocation and radiation standards, the characteristics of the visual and aural signals to be radiated are specified and their integrity ensured by the signal generation and transmission standards. It follows, then, that equipment standards must be determined that set forth the characteristics which must be possessed by the transmitting and receiving equipment capable of generating the specified signals to ensure conformity with both the signal transmission and allocation standards.

In some instances, it becomes necessary that standards classified in the equipment category be included in the rules and regulations of the governmental authority (e.g., the FCC) in order to promulgate and enforce both the transmission and allocation standards. In addition to this category, equipment standards are also set up by appropriate professional engineering groups or trade associations. These standards serve as a basis for those included within the rules and regulations and also ensure that standards and performance can be met with realistic and safe margins with equipment capable of being manufactured within state-of-the-art technologies, or certified by the consent of the members of the industry representative organizations. Thus, these standards assure the existence of a suitable technology along with manufacturing feasibility for system functions as well as components. These *standards of good engineering practice* are recommendations based on a consensus of those skilled in the art but are not necessarily enforceable by law unless included in the governmental authority rules and regulations.

In prior years, the predominate organization in the area of television equipment standards was the Radio Electronics Television Manufacturers Association (RETMA). In more recent years, RETMA has been superseded in this role by such organizations as the Electronic Industries Association (EIA), the Society of Motion Picture and Television Engineers (SMPTE), appropriate professional groups of the Institute of Electrical and Electronics Engineers (IEEE), and others.

Details of equipment standards for television engineering as well as certain recommended practices are summarized in Sec. 21.5.

21.2 ALLOCATION AND RADIATION STANDARDS

Radio-frequency spectrum allocations for the wide variety of telecommunications services that include commercial, military, broadcast, relay, point-to-point, fixed station, mobile, and many other interests constitute a highly valuable commodity of limited supply. In order to provide an orderly and efficient use of the spectrum, there are a number of international organizations dedicated to obtaining agreement among participating national administrations and operating organizations throughout the world. These groups carry on studies of technical as well as other problems related to the interworking of the individual national telecommunications authorities to provide effective worldwide telecommunications services, frequency allocations, technical systems recommendations, and communications networks while promoting the various services for the public good and in the public interest. Accordingly, each participating national administration determines its own telecommunication rules and regulations containing the specific technical standards.

Most notable among these international activities are the World Administrative Radio Conference (WARC) and the International Telecommunications Union (ITU), under whose auspices the International Radio Consultative Committee (CCIR) operates. The latest meeting of WARC, designated as WARC-79, was held in September to December 1979 in Geneva, Switzerland.

These organizations promulgate their deliberations in the form of recommendations which are published by the ITU. The CCIR, whose recommendations pertaining to international technical standards for television services are summarized herein, functions through the actions of Study Groups and Plenary Assemblies. After each Plenary Assembly, the ITU publishes volumes which contain the recommendations of the assembly. The data contained herein are taken from the CCIR Recommendations 470-1, Report 624-2, entitled "Characteristics of Television Systems." These recommendations were the outgrowth of the latest assembly, designated as the XVth Plenary Assembly of the CCIR, held in Geneva, Switzerland, in 1982 and represents the details of the different television systems in use throughout the world in the time period between 1974 and 1982 and those continuing in use. The details of the transmission standards are given in Sec. 21.2. It should be recognized that the overall situation is of a continually shifting nature and future decisions can, and no doubt will, alter some of the details.

To facilitate the direct interchange of television programs, efforts have been made to achieve common international standards. The most notable as well as the most controversial effort of the XIth Plenary Assembly of the CCIR, held in Oslo, Norway, in 1966, was an attempt at standardization of color television systems by the contributing countries of the world. The discussions pertaining to the possibility of a universal system proved inconclusive. Therefore, the CCIR, instead of issuing a unanimous recommendation for a single system, was forced to issue a report describing the characteristics and recommendations for a variety of proposed systems. It was, therefore, left to the controlling organizations of the administrations of the individual countries to make their own choices as to which standard to adopt.

This outcome was not totally surprising since one of the primary requirements for any color television system is compatibility with a coexisting monochrome system. In many cases, the monochrome standards already existed, and, therefore, such technical factors as number of scanning lines, field rates, video bandwidths, modulation techniques, and carrier frequencies were predetermined, relevant to such factors as local power line frequencies, radio-frequency channel allocations, and pertinent telecommunications agreements which varied in many regions of the world.

The ease with which international exchange of program material is accomplished is thereby hampered, and at present is accomplished satisfactorily, with good picture quality by means of standards-conversion techniques, or *transcoders,* utilizing advanced digital signal processing techniques and carried by satellite relay in many instances, as well as on video tape.

The United States participates in the international telecommunications conferences, plenary assemblies, and study groups, and joins in specific agreements in accordance with the recommendations. The FCC constitutes the regulatory agency within the United States and its possessions. The United States is also a signatory to separate, bilateral agreements concerning television broadcast stations with the governments of Canada and Mexico. All AM, FM, and television broadcast station agreements to which the United States is a signatory are available for inspection in the Office of the Chief, Broadcast Bureau of the FCC, Washington, D.C. Copies of these agreements may be purchased from the FCC Copy Contractor, Downtown Copy Center, 1114 21st St., N.W., Washington DC 20037.

The data quoted in this chapter relative to television engineering standards are summarized from the FCC Rules and Regulations, Part 15, Radio Frequency Devices; Part 73, Radio Broadcast Services; Part 74, Experimental, Auxiliary and Special Broadcast, and Other Program Distributional Services; and Part 76, Cable TV Service.†

21.2.1 CHANNEL SPECIFICATIONS AND CARRIER FREQUENCIES IN THE UNITED STATES AND WORLDWIDE. The available radio-frequency spectrum space is assigned to two broad classes of television services that include standard broadcast and the related point-to-point transmissions. The point-to-point services include auxiliary functions such as network relays, studio-to-transmitter links, and both portable and mobile remote-pickup facilities. Certain ancillary activities such as broadcast translators and signal booster operations as well as instructional television, subscription television, satellite relay, and experimental television are also included in the specific frequency assignments and channel specifications.

---

†FCC Rules and Regulations; Part 15, July 1981; Part 73, March 1980; Part 74, March 1980; Part 76, February 1981.

The broadcast service is divided into two frequency ranges referred to as *very high frequency* (VHF) and the *ultra high frequency* (UHF) regions. Broadly speaking, the VHF region lies in a frequency range from about 40 MHz to slightly above 200 MHz, while the UHF region extends from about 470 MHz up to almost 1.0 GHz. Most of the point-to-point services are allocated in the 2- to 14-GHz region. The VHF range includes both frequency-modulation broadcasting and television broadcasting, while the UHF range contains only television channels, although there are gaps in both ranges that are assigned to other services.

The frequency bands for both FM broadcasting and television broadcasting are based upon internationally agreed upon allocation tables, referred to as the Stockholm Plan in the European broadcasting area, and a similar arrangement used in the United States and specified by the FCC. These bands are designated as Bands I, II, III, IV, and V with Bands I and II containing both television and frequency-modulation broadcast stations. Bands III, IV, and V contain only television broadcast stations. The frequencies for these bands were established at the ITU conferences, the latest having been held in Geneva, Switzerland. Table 21-1 summarizes the specific frequency limits of these bands in relation to the geographical telecommunications region designations in the world.

Within these bands, the exact channel assignments and bandwidths vary throughout the world depending upon the specifications of the particular television broadcasting system that has been adopted by the individual governmental administrations. In the United States, the frequency assignments for television comprise 68 channels, each occupying 6 MHz. The VHF bands are subdivided as follows: The five *low-band* VHF channels, designated 2 through 6, extend from 54 to 88 MHz with a gap between 72 and 76 MHz. The seven *high-band* VHF channels, numbered 7 through 13, extend contiguously from 174 to 216 MHz. The UHF band contains 55 channels, 14 through 69, that extend contiguously from 470 to 806 MHz. Originally, in Region 2, channel 1 was assigned 44 to 50 MHz, but this frequency range was transferred to other services in 1946 and is not available for television service.

Table 21-2 provides a complete list of the original 82 channel frequencies and designations in the VHF and UHF bands in the United States as per FCC Rules and Regulations, Part 73, Subpart E, Sec. 73.603, as per an allocation plan set forth in 1953. Since 1953, some exceptions have occurred as indicated in the footnotes to Table 21-2.

The channel designations and frequency band allocations in Region 1 (Europe, Eurasia, and Africa) and Region 3 (Asia and Australia) vary widely from country to country within the Zone I, II, III, IV, and V frequency bands (Table 21-1) and with respect to the particular transmission standards adopted by the individual administrations. Space dictates that the television channel designations for all the 118 countries throughout the world whose administrations have registered allocation decisions as of April 1981 cannot be included within this document. However, some typical examples that are being followed by a majority of the authorities are given in Tables 21-3 and 21-4. A later listing by country indicates the CCIR system designations for all 118 countries (Table 21-24).

Table 21-1   VHF and UHF Broadcasting Bands

Band	Region 1 (Europe, Eurasia, and Africa), MHz	Region 2 (the Americas), MHz	Region 3 (Asia and Australia), MHz
Band I (VHF)†	41 –68	54 –72 and 76–88	44 –50 and 54–68
Band II (VHF)‡	87.5–100	88 –108	87 –108
Band III (VHF)†	162 –230	174–216	170–216
Bands IV and V	470 –590	470–590	470–585
(UHF)†	610–940		610–960

†Used primarily for television broadcasting.
‡Used primarily for FM sound broadcasting.

Table 21-2 Numerical Designation of Television Channels in the United States

Channel no.	MHz	Channel no.	MHz	Channel no.	MHz	Channel no.	MHz
2	54–60	22	518–524	42	638–644	63	764–770
3	60–66	23	524–530	43	644–650	64	770–776
4	66–72	24	530–536	44	650–656	65	776–782
5‡	76–82	25	536–542	45	656–662	66	782–788
6†	82–88	26	542–548	46	662–668	67	788–794
7	174–180	27	548–554	47	668–674	68	794–800
8	180–186	28	554–560	48	674–680	69	800–806
9	186–192	29	560–566	49	680–686		
10	192–198	20	566–572	50	686–692	70¶	806–812
11	198–204	31	572–578	51	692–696	71¶	812–818
12	204–210	32	578–584	52	698–704	72¶	818–824
13	210–216	33	584–590	53	704–710	73¶	824–830
14‡	470–476	34	590–596	54	710–176	74¶	830–836
15‡	476–482	35	596–602	55	716–722	75¶	836–842
16‡	482–488	36	602–608	56	722–728	76¶	842–848
17‡	488–494	37§	608–614	57	728–734	77¶	848–854
18‡	494–500	38	614–620	58	734–740	78¶	854–860
19‡	500–506	39	620–626	59	740–746	79¶	860–866
20‡	506–512	40	626–632	60	746–752	80¶	866–872
21	519–518	41	632–638	62	758–764	81¶	872–878
						83¶	884–890

†In Alaska and Hawaii, channels 5 and 6 (76–82 MHz and 82–88 MHz) are allocated for nonbroadcast services.

‡Channels 14 through 20, previously assigned in 13 specific locations to land-mobile service, may not be reassigned to television broadcast (FCC Part 2, July 1981).

§Channel 37 (608–614 MHz) is reserved for radio astronomy.

¶Channels 70 through 83 are allocated for land-mobile service (FCC Part 2, July 1981), (FCC Parts 15, 73, 74 August 1982).

Table 21-3 Typical VHF Television Channels (Other than Region 2)

Channel	Channel limits, MHz	Video, MHz	Audio, MHz
Band I			
2	47–54	48.25	53.75
2A	48.25–55.5	49.75	55.25
3	54–61	55.25	60.75
4	61–68	62.25	67.75
Band III			
5	174–181	175.25	180.75
6	181–188	182.25	187.75
7	188–195	189.25	191.75
8	195–202	196.25	201.75
9	202–209	203.25	208.75
10	209–216	210.25	215.75
11	216–223	217.25	222.75
12	223–230	224.25	229.75

Table 21-4   Typical UHF Television Channels (Other Than Region 2)

Channel	Channel Limits, MHz	Video, MHz, G, H, I, K, L	Audio, MHz, G, H	Channel	Channel Limits, MHz	Video, MHz, G, H, I, K, L	Audio, MHz, G, H
				44	654–662	655.25	660.75
Bands IV and V				45	662–670	663.25	668.75
21	470–278	471.25	476.75	46	670–678	671.25	676.75
22	478–486	479.25	484.75	47	678–686	679.25	684.75
23	486–494	487.25	492.75	48	686–694	687.25	692.75
24	494–502	495.25	500.75	49	694–702	695.25	700.75
25	502–510	503.25	508.75	50	702–710	703.25	708.75
26	510–518	511.25	516.75	51	710–718	711.25	716.75
27	518–526	519.25	524.75	52	718–726	719.25	724.75
28	526–534	527.25	532.75	53	726–734	727.25	732.75
29	534–542	535.25	540.75	54	734–742	735.25	740.75
30	542–550	543.25	548.75	55	742–750	743.25	748.75
31	550–558	551.25	556.75	56	750–758	751.25	756.75
32	558–566	559.25	564.75	57	758–766	759.25	764.75
33	566–574	567.25	572.75	58	766–774	767.25	772.75
34	574–582	575.25	580.75	59	774–782	775.25	780.75
35	582–590	583.25	588.75	60	782–790	783.25	788.75
36	590–598	591.25	596.75	61	790–798	791.25	796.75
37	598–606	599.25	604.75	62	798–806	799.25	804.75
38	606–614	607.25	612.75	63	806–814	807.25	812.75
39	614–622	615.25	620.75	64	814–822	815.25	820.75
40	622–630	623.25	628.75	65	822–830	823.25	828.75
41	630–638	631.26	636.75	66	830–838	831.25	836.75
42	638–646	639.27	644.75	67	838–846	839.25	844.75
43	646–654	647.25	652.75	68	846–854	847.25	852.75

Most of the countries in Region 2 (29 as of April 1981) retain the same channel designations and carrier frequencies as in the United States (Table 21-2). Many (62 as of April 1981) of the countries in Region 1 utilize the designations as shown in Tables 21-3 and 21-4 with some variations in the low-frequency end and high-frequency end channel assignments for Bands I and III. A few countries in Europe, such as Bulgaria, Hungary, USSR, and Czechoslovakia, in addition to the allocations as listed in Tables 21-3 and 21-4, utilize two- or three-channel allocations out-of-band in the frequency range between about 76 and about 100 MHz. Another notable exception is the People's Republic of China that indicates 48 total channels with numbers C-1 through C-5 occupying a range between 49.75 and 91.75 MHz and channels C-6 through C-48 running contiguously from 168.25 through 795.75 MHz. The designations for Japan indicate channels 1 through 3 to be in the region between 90 and 108 MHz with channels 4 through 12 occupying frequencies between 170 and 222 MHz. The UHF bands in Japan (channels 13 through 62) occupy the frequencies between 470 and 770 MHz.

As may be seen from Table 21-2, each of the United States channels occupies 6 MHz of spectrum space; however, other systems throughout the world specify different bandwidths and different visual carrier-to-aural carrier spacings as well as different vestigial sideband frequencies.† The CCIR has compiled the specifications for all the systems in

---

†Specific details as to channel designations may be obtained from the individual administrations. However, the information herein was taken from a compilation provided by the courtesy of the RCA Frequency Bureau in a private communication.

use (13 total) and given them alphabetical letter designations. The key to understanding the CCIR designations lies in recognizing that the letters refer primarily to local monochrome television standards that, for reasons of compatibility with color techniques, determine such details as line and field rates, video channel bandwidth, and aural carrier to visual carrier frequency spacing. Further classification in terms of the particular color technique involved then adds NTSC, PAL, or SECAM as appropriate. For example, the letter M designates a 525-line/60-field, 4.2-MHz bandwidth, 4.5-MHz sound-to-picture carrier spacing monochrome system. Thus, M(NTSC) describes a color system employing the NTSC technique for introducing the chrominance information within the constraints of the above basic monochrome signal values. Likewise, M(PAL) indicates the same line and field rates and bandwidths but employing the PAL color subcarrier modulation approach. (The specifications pertaining to NTSC, PAL, and SECAM color television systems are given in Table 21-4.)

Figure 21-1 contains a chart showing the various channel bandwidths for easy comparison, drawn approximately to scale, and using the visual carrier frequency as a zero frequency reference, for the CCIR-designated systems. The detailed characteristics of the radiated signals, both monochrome and color, are given in Table 21-5 (CCIR Report 624-2, Table III).

Some general comparison statements can be made concerning the specifications of the radiated signals from the underlying monochrome systems and the existing color standards:

1. There are four different scanning standards; 405 lines/50 fields, 525 lines/60 fields, 625 lines/50 fields, and 819 lines/50 fields.

2. There are six different spacings of video-to-sound carriers, namely, 3.5, 4.5, 5.5, 6.0, 6.5, and 11.15 MHz.

3. Some systems use FM and others use AM for the sound modulation method.

4. Some systems use positive polarity (luminance proportional to voltage) modulation of the video carrier, while others, such as the United States (M/NTSC) system, use negative modulation (decreasing carrier level with increasing luminance values).

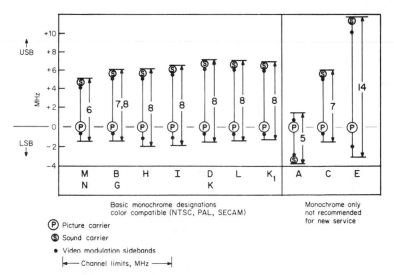

FIG. 21-1 CCIR channel designations. *(From CCIR Report 624-2, 1982, Table III.)*

Table 21-5 Basic Characteristics of the Radiated Signals for Worldwide Systems

Characteristics	CCIR system designation										
	A	M	N	C	B, G	H	I	D, K	K1	L	E
Channel bandwidth, MHz	......	6	6	7	B; 7 G; 8	8	8	8	8	8	14
Vision to sound carrier spacing, MHz	−3.5	+4.5	+4.5	+5.5	+5.5 ±0.001	+5.5	+5.9996 ±0.0005	+6.5 ±0.001	+6.5	+6.5	+11.15
Vision carrier to near-channel edge, MHz	+1.25	−1.25	−1.25	−1.25	−1.25	−1.25	−1.25	−1.25	−1.25	−1.25	±2.83
Main sideband width, MHz	3	4.2	4.2	5	5	5	5.5	6	6	6	10
Vestigial sideband width, MHz	0.75	0.75	0.75	0.75	0.75	1.25	1.25	0.75	1.25	1.25	2.0
Vestigial sideband attenuation, dB at MHz	No spec.	20(1.25) 42(3.58)	20(1.25) 42(3.5)	20(1.25) 20(3.0)	20(1.25) 20(3.0) 30(4.43)	20(1.75) 20(3.0)	20(3.0) 30(4.43)	20(1.25) 30(4.43 ± 0.1)	30(4.3) 20(2.7) 0(0.8)	30(4.3) 15(2.7) 0(0.8)	15(2.7) 0(0.8)

Visual modulation type and polarity	A5C pos.	A5C neg.	A5C neg.	A5C pos.	A5C neg.	A5C neg.	A5C neg.	A5C neg.	A5C neg.	A5C pos.	A5C pos.
Sync pulse level (% of peak carrier)	<3	100	100	<3	100	100	100	100	100	<6	<3
Blanking level	30	72.5–77.5	72.5–77.5	22.5–27.5	75 ± 2.5	72.5–77.5	76 ± 2	75 ± 2.5	75 ± 2.5	30 ± 2	30 ± 2
Setup	0	2.88–6.75	2.88–6.75	0	0–2	0–7	0	0–4.5	0–4.5	0–4.5	0–4.5
Peak white	100	10–15	10–15	100	10–12.5	10–12.5	20 ± 2	12.5	10–12.5	100–110	100
Sound modulation	A3	F3	F3	A3	F3	F3	F3	F3	F3	A3	A3
Frequency deviation, kHz	....	± 25	± 25	....	± 50	± 50	± 50	± 50	± 50		
Preemphasis, μs	....	75	75	50	50	50	50	50	50		
Visual/sound ratio	4/1	10/1 5/1	10/1 5/1	4/1	10/1	5/1 10/1	5/1	10/1 5/1	10/1	10/1	10/1
Precorrection group delay, ns	No spec.	−170	....	....	−170						

5. There are differences in the techniques for color-subcarrier encoding represented by NTSC, PAL, and SECAM, and in each case there are many differences in the details of various synchronization pulse widths, timing, and tolerance standards. It is evident that one must refer to the CCIR documents for complete information on the details of the combined monochrome/color standards. A summary of these signal generation and transmission standards is given in Secs. 21.3 and 21.4. Figure 21-2 presents a comparison, approximately to scale in the baseband domain, of the relative bandwidths, color-subcarrier frequencies, and sound-carrier relative spacing for the major color systems used in the world today.

Table 21-6 shows channel assignments in the United States to the principal communities, territories, and possessions. Channels designated by a superscript letter *b* are

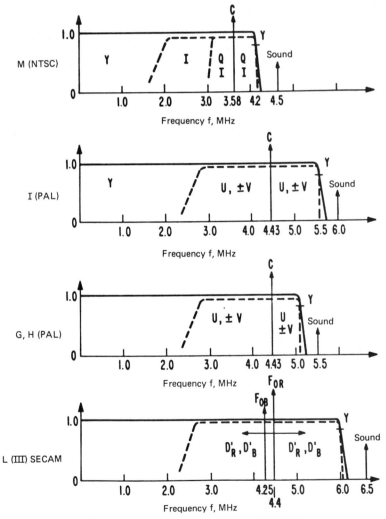

FIG. 21-2 Bandwidth comparison among NTSC, PAL, and SECAM.

Table 21-6 Channel Assignments in the United States[a]

Locality	Offset	Locality	Offset	Locality	Offset
Channel 2		TX Houston	—	OK Eufaula	0[b]
		TX Midland	+	OR Salem	+[b]
AL Andalusia	—[b]	UT Salt Lake City	—	PA Clearfield	+[b]
AK Anchorage	—	WA Spokane	—	PA Philadelphia	0
AK Fairbanks	+	WI Green Bay	+	SD Rapid City	+
AK Ketchikan	0	WY Casper	+	SD Watertown	—
AZ Flagstaff	0	PR San Juan	+	TN Chattanooga	+
AR Little Rock	—[b]			TN Memphis	—
CA Los Angeles	0	**Channel 3**		TX Bryan	0
CA San Francisco	+			TX Corpus Christi	—
CO Denver	0	AK Juneau	0[b]	TX Marfa	0
FL Daytona Beach	—	AK Seward	—	TX San Angelo	—
FL Miami	0[b]	AR Douglas	0	TX Wichita Falls	+
GA Atlanta	0	AR Phoenix	+	UT Price	+
HI Hilo (Hawaii)	0	CA Eureka	—	VT Burlington	0
HI Honolulu (Oahu)	+	CA Sacramento	0	VA Harrisonburg	—
ID Boise	0	CA Santa Barbara	—	VA Norfolk–	+
IL Chicago	—	CO Greenwood	—	Portsmouth–	
IN Terre Haute	+	Springs		Newport News	
IA Cedar Rapids	0	CO Sterling	0	WV Huntington	+
KS Great Bend	0	CT Hartford	+	WI Madison	0
LA Baton Rouge	0	FL Pensacola	—	PR Mayaguez	+
ME Bangor	—	FL Tampa	0[b]	VI Charlotte Amalie–	0[b]
MD Baltimore	+	GA Columbus	0	Christiansted	
MA Boston	+[b]	GA Savannah	+		
MI Detroit	+	HI Lihue (Kauai)	+	**Channel 4**	
MN Minneapolis–St.	—[b]	HI Wailuku (Maui)	0		
Paul		ID Idaho Falls	0	AL Dothan	0
MS Mississippi State	+[b]	ID Lewiston	—	AK Anchorage	—
MO St. Joseph	—	IL Champaign	+	AK Bethel	0[b]
MO St. Louis	0	IL Harrisburg	0	AK Fairbanks	+
MT Anaconda	+	IA Mason City	+	AK Ketchikan	0
MT Billings	0	KS Lakin	0[b]	AZ Tucson	—
NE North Platte	—	KS Wichita	—	AR Little Rock	0
NV Goldfield	—	KY Louisville	—	CA Los Angeles	0
NV Reno	0	LA Lafayette	+	CA San Francisco	—
NM Sante Fe	+	LA Shreveport	—	CO Denver	—
NY Buffalo	0	MI Escanaba	+	DC Washington	—
NY New York	0	MI Kalamazoo	—	FL Jacksonville	+
NY Utica	—	MN Duluth	0	FL Miami	0
NC Columbia	0[b]	MS Jackson	+	HI Hilo (Hawaii)	0[b]
NC Greensboro	—	MO Kirksville	—	HI Honolulu (Oahu)	—
ND Dickinson	+	MO Springfield	+	ID Boise	+[b]
ND Grand Forks	0[b]	MT Great Falls	+	IL Rock Island	+
OH Dayton	0	MT Miles City	—	IN Bloomington	0
OK Tulsa	+	NE Lexington	+[b]	IA Sioux City	—
OR Klamath Falls	—	NE Omaha	0	KS Colby	—
OR Portland	0	NV Ely	—	LA New Orleans	+
PA Pittsburgh	—	NV Las Vegas	0	MA Boston	—
SC Charleston	+	NM Gallup	0	MI Cheboygan	+
SD Seneca	—[b]	NM Portales	+[b]	MI Detroit	0
SD Vermillion	+[b]	NY Syracuse	—	MN Minneapolis–St.	0
TN Nashville	—	NC Charlotte	0	Paul	
TN Sneedville	—[b]	NC Wilmington	—	MS Columbus	—
TX Amarillo	—[b]	ND Bismarck	0[b]	MO Kansas City	0
TX Denton	0[b]	OH Cleveland	0	MO St. Louis	—

Table 21-6 Channel Assignments in the United States (*Continued*)

Locality	Offset	Locality	Offset	Locality	Offset
Channel 4		MA Boston	−	GA Thomasville	0
		MI Bay City	−	ID Nampa	0
MT Butte	0	MI Calumet	−	ID Pocatello	−
MT Hardin	+	MN Minneapolis–St. Paul	−	IN Indianapolis	0
NE Scottsbluff–Hay Springs	+	MO Kansas City	+	IA Davenport	+
NE Superior	+	MO St. Louis	−	KS Dodge City	−
NV Reno	0	MT Glendive	+	KY Paducah	+
NM Albuquerque	+	MT Great Falls	+	LA New Orleans	0
NY Buffalo	−	NE Hastings	−	ME Portland	−
NY New York	0	NE Hay Springs–Scottsbluff	+	MA New Bedford	+
NC Chapel Hill	+^{b}	NV Boulder City	+	MI Alpena	0^b
ND Valley City	−	NV Reno	0^b	MI Lansing	−
ND Williston	0^b	NM Albuquerque	+^{b}	MI Marquette	−
OH Columbus	−	NY Lake Placid	0	MN Austin	−
OK Oklahoma City	−	NY New York	+	MS Greenwood	+
OR Roseburg	+	NY Syracuse	−	MO Sedalia	0
PA Pittsburgh	+	NC Raleigh	0	MT Butte	+
SC Charleston	0	ND Bismarck	0	MT Miles City	0^b
SC Greenville	−	OH Cincinnati	−	NE Hayes Center	+
SD Pierre	0	OH Cleveland	+	NE Omaha	+
TN Nashville	+	OK Oklahoma City	0	NV Ely	+
TX Amarillo	0	OR Medford	0	NM Carlsbad	−
TX Big Spring	0	SC Charleston	+	NM Silver City	0
TX Dallas	+	SD Lead	−	NY Albany–Schenectady	0
TX El Paso	0	SD Mitchell	+	NC Wilmington	0
TX Harlingen	+	TN Memphis	+	ND Fargo	0
TX Port Arthur	−	TN Nashville	0	ND Minot	+^{b}
TX San Antonio	0	TX Fort Stockton	+	OH Columbus	+
UT Cedar City	0	TX Fort Worth	+	OK Tulsa	+
UT Salt Lake City	−	TX Lubbock	−^{b}	OR Portland	+
WA Seattle	0	TX San Antonio	0	PA Johnstown	0
WA Spokane	−	TX Weslaco	−	PA Philadelphia	−
WV Beckley	0	UT Salt Lake City	+	SD Reliance	−
WI Milwaukee	−	VA Bristol	+	TN Knoxville	0
WY Lander	0	WA Seattle	+	TX Beaumont	−
GM Agana	0^b	WV Weston	0	TX Corpus Christi	0
PR San Juan	−	WI Green Bay	+	TX San Angelo	0
		WY Cheyenne	+	TX Temple	+
Channel 5		PR Mayaguez	−	TX Texarkana	−
				TX Wichita Falls	−
AL Mobile	+	**Channel 6**		UT Vernal	0
AZ Phoenix	−			VA Richmond	+
AR Fort Smith	−	AL Birmingham	−	WA Spokane	−
CA Los Angeles	0	AZ Kingman	−	WV Bluefield	−
CA San Francisco	+	AZ Tucson	+^{b}	WI Milwaukee	0
CO Grand Junction	−	AR Mountain View	−^{b}	WI Superior	+
CO Pueblo	0	CA Eureka	−	WY Casper	+^{b}
DC Washington	−	CA Sacramento	0^b	PR San Juan	+^{b}
FL Gainsville	−^{b}	CA San Luis Obispo	+		
FL West Palm Beach	0	CO Denver	−^{b}	**Channel 7**	
GA Atlanta	−	CO Durango	+		
IL Chicago	0	FL Miami	0	AL Munford	−^{b}
IA Ames	0	FL Orlando	−	AK Anchorage	−^{b}
LA Alexandria	0	GA Augusta	+	AK Fairbanks	+
ME Bangor	+			AZ Prescott	0

Table 21-6 Channel Assignments in the United States (*Continued*)

Locality	Offset	Locality	Offset	Locality	Offset
Channel 7		WV Wheeling	0	SD Martin	$-^b$
		WI Wausau	−	TN Nashville	$+^b$
AR Little Rock	−	WY Sheridan	0	TX Boquillas	−
CA El Centro	+	PR Ponce	+	TX Dallas	0
CA Los Angeles	0			TX Houston	$-^b$
CA Redding	0	**Channel 8**		TX Laredo	0
CA San Francisco	−			TX San Angelo	+
CO Denver	0	AL Selma	−	UT Richfield	+
DC Washington	+	AK Juneau	0	VA Petersburg	0
FL Jacksonville	0^b	AZ Phoenix	$+^b$	WV Charleston	+
FL Miami	−	AR Jonesboro	−	WI LaCrosse	+
FL Panama City	+	CA Salinas–Monterey	+	WY Laramie	$+^b$
HI Wailuku (Maui)	0	CA San Diego	0	GM Agana	0
ID Boise	0	CO Grand Junction	−	VI Christiansted	+
IL Chicago	0	CO Pueblo	0^b		
IN Evansville	0	CT New Haven	+	**Channel 9**	
IA Waterloo	+	FL Tampa	−		
KS Hays	−	GA Athens	0^b	AK Fairbanks	$+^b$
KS Pittsburgh	+	GA Waycross	$-^b$	AK Ketchikan	0^b
LA Lake Charles	−	HI Lihue (Kauai)	$-^b$	AK Seward	−
ME Bangor	−	ID Idaho Falls	+	AZ Tucson	−
MA Boston	+	IL Carbondale	0^b	AR Arkadelphia	$-^b$
MI Detroit	−	IL Moline	0	CA El Centro	+
MI Traverse City	+	IN Indianapolis	−	CA Los Angeles	0
MN Alexandria	0	IA Des Moines	−	CA Redding	0^b
MS Laurel	0	KS Hutchinson	0^b	CA San Francisco	$+^b$
MO Hannibal	−	LA Monroe	+	CO Denver	−
MT Butte	$-^b$	LA New Orleans	0	DC Washington	0
NE Bassett	$-^b$	ME Lewiston	−	FL Orlando	0
NE Omaha	0	ME Presque Island	0	GA Columbus	╁
NV Winnemucca	+	MI Grand Rapids	+	GA Savannah	$-^b$
NM Albuquerque	+	MI Iron Mountain	−	HI Hilo (Hawaii)	0
NY Buffalo	+	MI Sault Sainte Marie	0	HI Honolulu (Oahu)	−
NY Carthage	−	MN Duluth	0^b	ID Caldwell	−
NY New York	0	MO Columbia	+	IL Chicago	+
NC Washington	0	MT Billings	0	IN Evansville	$+^b$
ND Dickinson	0	MT Missoula	−	IA Cedar Rapids	−
ND Jamestown	−	NE Albion	+	IA Sioux City	0
OH Dayton	+	NE McCook	−	KS Garden City	0^b
OK Lawton	+	NV Las Vegas	−	KS Randall	$-^b$
OR Corvallis	$-^b$	NV Reno	0^b	LA Baton Rouge	−
SC Charleston	$-^b$	NM Gallup	$-^b$	MI Cadillac	0
SC Spartanburg	+	NM Roswell	0	MN Bemidji	0^b
SD Rapid City	+	NY Rochester	0	MN Minneapolis–St. Paul	+
TN Jackson	+	NC High Point	−	MS Tupelo	−
TX Amarillo	0	ND Devils Lake	+	MO Kansas City	+
TX Austin	+	ND Williston	−	MO St. Louis	0^b
TX El Paso	0^b	OH Cleveland	0	MT Bozeman	0^b
TX Odessa	−	OK Elk City	+	MT Harve	+
TX Presidio	+	OK Tulsa	−	MT Kalispell	−
TX Tyler	0	OR Medford	$+^b$	NE North Platte	$+^b$
UT Salt Lake City	0^b	OR Portland	−	NV Manchester	−
VA Roanoke	−	PA Lancaster	−	NM Sante Fe	$+^b$
WA Seattle	0	SD Brookings	0^b		
WA Spokane	$+^b$				

Table 21-6 Channel Assignments in the United States (*Continued*)

Locality	Offset	Locality	Offset	Locality	Offset
Channel 9		**Channel 10**		HI Honolulu (Oahu)	$+^b$
				ID Twin Falls	0
NY New York	+	MO Springfield	0	IL Chicago	0^b
NY Syracuse	−	MT Helena	+	IA Des Moines	$+^b$
NC Charlotte	+	MT Miles City	0	KS Garden City	+
NC Greenville	−	NE Lincoln	+	KS Topeka	0^b
ND Dickinson	$−^b$	NE Scottsbluff	−	KY Louisville	+
OH Cincinnati	0	NV Elko	−	LA Houma	0
OH Steubenville	+	NV Las Vegas	$+^b$	MD Baltimore	−
OK Oklahoma City	−	NM Gallup	0	MI Alpena	0
OR Eugene	+	NM Roswell	−	MN International	0
SD Aberdeen	−	NM Silver City	+	Falls	
SD Rapid City	0^b	NY Albany–	−	MN Minneapolis–St.	−
TN Chattanooga	0	Schenectady		Paul	
TX Abilene	+	NY Rochester	+	MS Meridian	−
TX El Paso	0	ND Minot	−	MO St. Louis	−
TX Lufkin	0	OH Columbus	+	MT Billings	0^b
TX Monahana–	−	OK Ada	+	MT Havre	+
Odessa		OR Medford	+	MT Missoula	$−^b$
TX San Antonio	$−^b$	OR Portland	0^b	NE Grand Island	−
UT Ogden	$+^b$	PA Altoona	−	NH Durham	0^b
WA Seattle	0^b	PA Philadelphia	0	NM Sante Fe	−
WV Grandview	$−^b$	RI Providence	+	NY New York	+
PR Ponce	−	SC Columbia	−	NC Durham	+
		SD Pierre	$+^b$	ND Fargo	+
Channel 10		TN Knoxville	+	ND Williston	−
		TN Memphis	$+^b$	OH Toledo	−
AL Birmingham	$−^b$	TX Amarillo	0	OK Tulsa	$−^b$
AL Mobile	+	TX Corpus Christi	−	OR North Bend	0
AK Dillingham	0	TX Del Rio	0	PA Pittsburgh	0
AK Juneau	0	TX Waco	+	SD Lead	+
AZ Phoenix	−	VA Norfolk–	+	SD Lowry	$−^b$
AR El Dorado	−	Portsmouth–		SD Sioux Falls	0
CA Sacramento	0	Newport News		TN Johnson City	−
CA San Diego	0	VA Roanoke	0	TN Lexington	0^b
CO Montrose	+	WA Pullman	$−^b$	TX Fort Worth	−
FL Miami	+	WI Milwaukee	$+^b$	TX Houston	+
FL St. Petersburg	−	WY Riverton	+	TX Lubbock	0
GA Albany	0	GM Agana	0	TX Sonora	+
HI Lihue (Kauai)	0	VI Charlotte Amalie	−	UT Provo	$+^b$
HI Wailuku (Maui)	0^b			WA Tacoma	+
ID Pocatello	0^b	**Channel 11**		WI Green Bay	+
IL Quincy	−			WY Rawlins	−
IN Terre Haute	0	AK Anchorage	0	PR Caguas	−
KS Goodland	0	AK Fairbanks	+		
KS Wichita	−	AZ Tucson–Nogales	0^c	**Channel 12**	
LA Lafayette	0	AZ Yuma	−		
ME Augusta	$−^b$	AR Little Rock	+	AL Montgomery	0
ME Presque Island	$+^b$	CA Los Angeles	0	AZ Mesa	−
MI Parma	−	CA San Jose	+	CA Chico	−
MI Sault Sainte	+	CO Colorado Springs	0	CA Santa Maria	+
Marie		FL Fort Myers	+	CO Boulder	0^b
MN Appleton	$−^b$	FL Tallahassee	$−^b$	CO Lamar	−
MN Duluth	+	GA Atlanta	+	CO La Junta	−
MN Rochester	0	GA Savannah	0	DE Wilmington	0^b
MN Thief River Falls	0	HI Hilo (Hawaii)	0	FL Jacksonville	+

21.16

Table 21-6 Channel Assignments in the United States (*Continued*)

Locality	Offset	Locality	Offset	Locality	Offset
Channel 12		VI Charlotte Amalie–Christiansted	0^b	OK Oklahoma City	0^b
				OR Eugene	0
FL Panama City	0	**Channel 13**		OR La Grande	+
FL West Palm Beach	0			PA Pittsburgh	$-^b$
GA Augusta	−	AL Birmingham	−	SC Florence	+
HI Lihue (Kauai)	−	AK Anchorage	−	SD Eagle Butte	0^b
HI Wailuku (Maui)	0	AK Fairbanks	+	SD Sioux Falls	+
ID Moscow	$-^b$	AK Sitka	0	TN Memphis	+
ID Nampa	+	AZ Flagstaff	0	TX Brady	0
IL Urbana	$-^b$	AZ Tucson	−	TX Dallas	$+^b$
IA Iowa City	$+^b$	AZ Yuma	+	TX El Paso	0
KS Hutchinson	0	AR Fayetteville	$-^b$	TX Houston	−
LA New Orleans	0^b	CA Alturas	+	TX Laredo	0
LA Shreveport	0	CA Eureka	$-^b$	TX Lubbock	−
ME Orono	$-^b$	CA Los Angeles	0	VA Hampton	−
MI Flint	−	CA Stockton	+	VA Lynchburg	0
MN Mankato	0	CO Colorado Springs	0	WA Tacoma	−
MN Walker	−	FL Panama City	0	WV Huntington	+
MS Booneville	$-^b$	FL Tampa	−	WI Eau Claire	+
MS Jackson	+	GA Macon	+	WY Rock Springs	0
MO Cape Girardeau	0	HI Hilo (Hawaii)	0	PR Fajardo	+
MO Joplin	+	HI Honolulu (Oahu)	−		
MT Helena	0	ID Twin Falls	$-^b$	**Channel 14**	
NE Lincoln	$-^b$	IL Mount Vernon	+		
NE Merriman	0^b	IL Rockford	0	AZ Globe	$+^b$
NM Clovis	−	IN Indianapolis	−	AZ Kingman	$-^b$
NM Farmington	+	IA Des Moines	−	CA Bishop	$-^b$
NM Silver City	0^b	SKS Garden City	−	CA San Francisco	+
NY Binghamton	−	KS Topeka	+	CA Santa Barbara	0^d
NC New Bern	+	KY Bowling Green	0	CA Susanville	0^b
NC Winston–Salem	0	LA Monroe	0^b	CO Boulder	0
ND Bismarck	−	ME Calais	$-^b$	CO Lamar	$-^b$
ND Pembina	0	ME Portland	+	DC Washington	−
OH Cincinnati	0	MD Baltimore	+	GA Pelham	$-^b$
OK Ardmore	−	MI Grand Rapids	+	GA Rome	+
OK Cheyenne	$+^b$	MI Marquette	0^b	HI Hilo (Hawaii)	+
OR Medford	+	MN Hibbing	−	HI Honolulu (Oahu)	0
OR Portland	0	MS Biloxi	+	ID Boise	0
PA Erie	0	MO Jefferson City	0	IL Jacksonville	0^b
RI Providence	+	MT Glendive	+	IL Joliet	0^d
SD Huron	+	MT Lewiston	0	IN Evansville	−
TN Chattanooga	+	MT Missoula	−	IA Decorah	$-^b$
TX Alpine	−	NE Alliance	$-^b$	IA High Point	$-^b$
TX Beamont	−	NE Kearney	0	IH Sioux City	0
TX San Antonio	+	NV Las Vegas	−	KS Hays	0^b
TX Sweetwater	0	NV McGill	0^b	LA Monroe	−
UT Logan	−	NJ Newark	−	LA Morgan City	$+^b$
VA Richmond	−	NM Albuquerque	+	MA Worcester	0^d
WA Bellingham	+	NY Albany–Schenectady	0	MI Mount Pleasant	0^b
WV Clarksburg	+	NY Rochester	−	MN Willmar	$-^b$
WI Milwaukee	0	NC Asheville	−	MS Meridian	0^b
WI Rhinelander	+	ND Fargo	0^b	MT Billings	0
WY Sheridan	+	ND Minot	−	MT Cut Bank	$-^b$
GM Agana	0^b	OH Toledo	0	MT Dillon	$+^b$
PR Arecibo–Aguadilla	+			NV Elko	$+^b$
				NM Albuquerque	−

Table 21-6 Channel Assignments in the United States (*Continued*)

Locality	Offset	Locality	Offset	Locality	Offset
Channel 14		NH Hanover	$+^b$	NY Watertown	0^b
		NM Carlsbad	$+^b$	NC Burlington	0
NC Greenville	0	NM Farmington	$+^b$	OH Dayton	$+^b$
NC Hickory	−	NM Socorro	$−^b$	OK Guymon	0^b
ND Grand Forks	+	NM Tucumcari	0^b	OK Lawton	$−^d$
ND Minot	−	NY Oneonta	0^d	OR Eugene	+
OH Oxford	$−^b$	ND Fargo	−	OR La Grande	0^b
OK Oklahoma City	−	ND Williston	$−^b$	PA Pittsburgh	0^b
OR Brookings	$−^b$	OH Ashtabula	0^d	PA Scranton	−
SC Allendale	0^b	OK Elk City	$−^b$	RI Providence	0^d
TN Memphis	$+^b$	OK Hugo	$+^{b,d}$	SC Beaufort	$−^b$
TX Amarillo	+	OR Bend	0^b	SC Greenville	+
TX Big Spring	0^b	PA Lancaster	+	SD Aberdeen	$−^b$
TX El Paso	0	SC Florence	−	TN Jackson	+
TX Houston	0^b	SD Rapid City	−	TX Corpus Christi	0^b
TX Tyler	+	TN Knoxville	$−^b$	TX Longview	$+^d$
UT Moab	$+^b$	TX Abilene	0	UT Cedar City	$+^b$
WA Vancouver	0^b	TX Bryan	$−^b$	UT Monticello	$−^b$
WA Walla Walla	−	UT Price	0^b	UT Provo	0
WV Wheeling	0^d	VA Hampton	0^b	WA Everett	−
WI Suring	−	VA Roanoke	$+^b$	WI Manitowoc	+
WY Casper	−	WA Centralia	$+^b$	PR Mayaguez	0
PR Ponce	0	WV Parkersburg	−		
		WI Madison	0	**Channel 17**	
Channel 15		VI Christiansted	0		
				AL Tuscaloosa	0
AL Florence	0			AZ Page	0^b
AL Mobile	+	**Channel 16**		AZ Parker	$−^b$
AZ Phoenix	−			AR Batesville	0^b
CA Sacramento	0^c	AL Munford	$−^b$	CA Bakersfield	0
CA San Diego	0^b	AZ Flagstaff	0^b	CA Fort Bragg	$+^{b,d}$
CA San Luis Obispo	$+^b$	AZ Nogales	$+^b$	CO Gunnison	$−^b$
CO Leadville	$−^b$	AZ Yuma	$−^b$	FL Jacksonville	0
FL New Smyrna Beach	$+^b$	AR Little Rock	−	FL Miami	$−^b$
GA Cochran	0^b	CA Redding	0^d	GA Atlanta	−
HI Lihue (Kauai)	−	CA Santa Cruz	$−^{b,d}$	ID Burley	$+^b$
HI Wailuku (Maui)	0	CA Ventura	+	ID Weiser	0^b
ID Grangesville	$−^b$	CO Alamosa	0^b	IL Decatur	0
ID Pocatello	0	CO Craig	$+^b$	IL Rockford	−
IL Champaign	−	FL Key West	+	IA Des Moines	+
IN Fort Wayne	+	FL Marianna	$+^b$	MI Grand Rapids	0
IA Ottumwa	+	FL Tampa	0^b	MI Iron Mountain	$+^b$
KS Oakley	$−^b$	ID Sandpoint	$+^b$	MN Ely	$−^b$
KS Wichita	$+^b$	IL Olney	$−^b$	MN Minneapolis–St. Paul	0^b
KY Louisville	0^b	IL Quincy	+	MS Bude	$+^b$
LA Lafayette	0	IN South Bend	0	MO Columbia	−
MI Bad Axe	$−^{b,d}$	IA Dubuque	−	MT Missoula	−
MI Ironwood	$−^b$	KY Somerset	0	MT Wolf Point	$+^b$
MN Austin	$−^b$	MD Salisbury	+	NE Grand Island	−
MS Greenville	−	MN Fairmont	$+^b$	NV Tonopah	$+^b$
MO Lowry City	$−^b$	MS Jackson	0	NH Portsmouth	$−^d$
MO Poplar Bluff	+	MO Joplin	0	NM Clayton	0^b
MT Helena	$+^b$	MO St. Joseph	−	NY Albany–Schenectady	$+^b$
NE Omaha	0	MT Glendive	$−^b$	NY Buffalo	0
NV Winnemucca	$−^b$	MT Great Falls	0		
		NV Yerington	$+^b$		
		NM Deming	0^b		

Table 21-6 Channel Assignments in the United States (*Continued*)

Locality	Offset	Locality	Offset	Locality	Offset
Channel 17		OK Miami	−[b]	CA San Francisco	−
		OR Burns	0[b]	CA Santa Barbara	0[b,d]
NC Goldsboro	−	OR Grand Pass	+[b]	CA Yreka City	+[b]
NC Linville	0[b]	TX Austin	+[b]	CO Denver	0
ND Bismarck	−	TX Midland	0	CO Durango	−[b]
OH Canton	−	TX Wichita Falls	−	CT Waterbury	0
OK Ardmore	0[b,d]	UT Ogden	−[b]	DC Washington	+
OK Bartlesville	+	UT St. George	−[b]	FL Fort Meyers	+
OK Woodward	−[b]	WA Wenatchee	+[b]	FL Gainesville	0
OR North Bend	+[b]	WI Eau Claire	0	GA Wrens	−[b]
OR The Dalles	−[b]	WI Milwaukee	−	HI Hilo (Hawaii)	+
PA Philadelphia	−	PR San Juan	0	HI Honolulu (Oahu)	0
SD Sioux Falls	−			ID Idaho Falls	0
TN Nashville	+	**Channel 19**		IL Chicago	0[b]
TX Texarkana	−			IL Springfield	+
UT Vernal	+[b]	AL Huntsville	0	IN Indianapolis	−[b]
WY Cheyenne	0[b]	AZ Prescott	0[b]	IA Iowa City	−
VI Charlotte Amalie	0	AR Jonesboro	+[b]	LA New Orleans	−
		CA Indio	+[b,d]	MI Detroit	+
Channel 18		CA Modesto	−	MN Wadena	−[b]
		CO Glenwood Springs	+[b]	MN Worthington	0[b]
AL Dothan	0	FL Bradenton	0[b]	MO Birchtree	−[b]
AZ Holbrook	0[b]	GA Albany	−	MT Billings	+
AZ Tucson	−	IL Peoria	0	NY Utica	0[d]
AR El Dorado	−	IA Filer	−[b]	NC Lexington	0
CA Chico	0[b,d]	KY Owensboro	−	OH Athens	0[b]
CA Fresno	+[b]	LA Tallulah	0[b]	OK Enid	−
CA San Bernardino	−	MA North Adams	0	PA Williamsport	−[d]
CO Grand Junction	+[b]	MI Bay City	+[b]	TN Crossville	+[b]
CO Sterling	+[b]	MI Marquette	0	TX Houston	0
CT Hartford	−	MN St. Cloud	0	TX Sherman	−
FL Cocoa	−[b]	MS Biloxi	+[b]	UT Salt Lake City	+
GA Chatsworth	−[b]	MO Kansas City	+[b]	VT St. Johnsbury	−[b]
GA Vidalia	+[b]	NE Norfolk	+[b]	WA Tacoma	0
IL Edwardsville	−[b]	NJ New Brunswick	−[b,d]	WI Wausau	+[b]
IN Lafayette	0	NM Lovington	0[b]	WY Casper	−
IA Carroll	−[b]	NM Sante Fe	−	PR Ponce	0
IA Davenport	+	NC Jacksonville	0		
KS Salina	+	ND Ellendale	−[b]	**Channel 21**	
KS Wichita	+[b]	OH Cincinnati	+		
KY Lexington	+	OH Cleveland	0	AL Birmingham	−
LA Lake Charles	0[b]	OK Altus	−[b]	AL Mobile	+
ME Fryeburg	+[b]	OK Muskogee	0	AZ Phoenix	0
MI Jackson	+	PA Johnstown	+	CA Hanford	0
MN Hibbing	−[b]	SC Columbia	+	CO Colorado Springs	0
MS Laurel	+	TN Kingsport	0	FL Fort Pierce	−[b]
MS Oxford	0[b]	TX Nacogdoches	−	GA Brunswick	+
MO Carrollton	0[b]	TX Victoria	+	HI Lihue (Kauai)	−[b]
MT Harve	−[b]	UT Richfield	0[b]	HI Wailuku (Maui)	0
NJ Atlantic City	0[b,d]	WA Pasco	−	IL Vandalia	0[b]
NM Alamogordo	−[b]	WI La Crosse	+	IN Fort Wayne	+
NM Raton	−[b]			IA Fort Dodge	0[b]
NY Emira	+	**Channel 20**		KS Dodge City	−[b]
NY Massena	0[b]			KS Manhattan	0[b]
NC Charlotte	0	AL Montgomery	0	KY Louisville	−
OH Zanesville	−	AR Hot Springs	0[b]	KY Murray	+[b]

Table 21-6 Channel Assignments in the United States (*Continued*)

Locality	Offset	Locality	Offset	Locality	Offset
Channel 21		MO St. Joseph	0	SD Sioux Falls	0^b
		NM Las Cruces	$-^b$	TX Brownsville	0
MI Manistee	0^b	NC Raleigh	0	TX Richardson	0
MN Duluth	+	ND Devils Lake	$+^b$	TX San Antonio	$-^b$
MS Clarksdale	0^b	OH Dayton	+	VA Richmond	0^b
MO La Plata	$+^b$	OK Ada	0^b	WA Yakima	+
MO Springfield	$-^b$	OR Klamath Falls	$+^b$	WV Charleston	0
NE Albion	$+^b$	OR Salem	0	VI Charlotte Amalie	0^b
NV Las Vegas	+	PA Pittsburgh	0		
NV Reno	+	PA Scranton	−	**Channel 24**	
NH Concord	+	SD Allen	+		
NM Roswell	−	TN Cookeville	0^b	AR Fort Smith	+
NY Levittown	$-^b$	TX Galveston	0^b	CA Chico	+
NY Rochester	0^b	TX Marshall	0^b	CA Fresno	0
NC Asheville	+	UT Logan	0^b	CA San Bernardino	$-^b$
OH Youngstown	−	VT Burlington	+	CO Trinidad	0^b
OR Astoria	0^b	WA Seattle	+	CT Hartford	0^b
OR Bend	+	WA Spokane	0	FL Orlando	$-^b$
PA Harrisburg	+	WI Oshkosh	+	GA Macon	+
SC Florence	0	PR Mayaguez	0	IL Moline	$-^b$
SD Rapid City	−			IN Lafayette	0^b
TX Beaumont	0	**Channel 23**		IA Mason City	$+^b$
TX Childress	0^b			KS Wichita	−
TX Fort Worth	−	AZ Ajo	$-^b$	LA Lafayette	0^b
TX San Angelo	$+^b$	AZ Safford	$+^b$	LA Shreveport	$-^b$
VA Lynchburg	−	CA Bakersfield	−	MD Baltimore	+
WI Madison	$-^b$	CA Modesto	$+^b$	MI Ironwood	+
VI Christiansted	0^b	CO Salida	$+^b$	MN West Branch	0^b
		FL Miami	−	MN Alexandria	0^b
Channel 22		FL Pensacola	0^b	MS Meridian	−
		GA Ashburn	$+^b$	MO St. Louis	+
AZ McNary	$+^b$	IL Decatur	−	MT Butte	0
CA Blythe	$-^b$	IL Freeport	0	NE Falls City	0^b
CA Cotati	$-^b$	IN Marion	0	NY Syracuse	$+^b$
CA Los Angeles	0	IA Ames	−	ND Minot	0
CO Fort Collins	−	KY Elizabethtown	$+^b$	OH Hillsboro	$+^b$
CO La Junta	$+^b$	LA De Ridder	$-^b$	OH Toledo	−
CO Montrose	0^b	MI East Lansing	$-^b$	OR Portland	+
FL Clearwater	0	MN Minneapolis–St. Paul	+	PA Erie	0
FL Key West	+	MS Greenwood	$+^b$	TN Athens	0^b
FL Panama City	$+^b$	MO Cape Girardeau	0	TN Memphis	0
GA Roystan	$+^b$	MO Columbia	$+^b$	TX Austin	0
GA Savannah	0	MT Missoula	−	TX Del Rio	$+^b$
IL Macomb	$+^b$	NE Beatrice	$+^b$	TX Odessa	−
IN South Bend	0	NJ Camden	$+^b$	TX Wichita Falls	0^b
IN Vincennes	$-^b$	NM Albuquerque	−	UT Ogden	0
IA Waterloo	−	NY Albany–Schenectady	−	VA Danville	−
KS Phillipsburg	$-^b$	NY Buffalo	0^b	WA Anacortes	0
KY Pikeville	$-^b$	NC Morgantown	−	WV Morgantown	$-^b$
MD Annapolis	$+^b$	ND Jamestown	0^b	WI Milwaukee	+
MA Springfield	0	OH Akron	+	PR San Juan	0
MI Calumet	$-^b$	OK Tulsa	0		
MN Brainerd	0^b	PA Altoona	−	**Channel 25**	
MS Hattiesburg	0	SC Conway	$+^b$		
MO Flat River	0^b			AL Huntsville	$+^b$
MO Joplin	$-^b$			AR Pine Bluff	−

Table 21-6 Channel Assignments in the United States (*Continued*)

Locality	Offset	Locality	Offset	Locality	Offset
Channel 25		MN Bemidji	+	WI Madison	+
		MN Mankato	−^{b}	WY Cheyenne	−
CA Ridgecrest	0^b	MO Poplar Bluff	−^{b}	VI Christiansted	0
CA Watsonville	+^{b}	MT Great Falls	0		
FL Orange Park	−	NE Omaha	0^b	**Channel 28**	
GA Dawson	0^b	NY Jamestown	+		
ID Pocatello	+	NC Winston-Salem	+^{b}	AZ Douglas	0^b
IL Peoria	+	ND Bismarck	+	AR Russellville	+^{b}
IN Evansville	−	OH Springfield	+	CA Los Angeles	0^b
IA Mount Ayr	−^{b}	OK Enid	+^{b}	FL Panama City	−
IA Rock Rapids	+^{b}	TN Knoxville	−	FL Tampa	0
KS Emporia	+^{b}	TX Abilene	+^{b}	GA Columbus	0^b
KY Ashland	−^{b}	TX El Paso	+	GA Savannah	−
LA Alexandria	+^{b}	TX Houston	0	ID Preston	0^b
ME Houlton	+^{b}	TX Sherman	−^{b}	IL Elkhart	+
MD Hagerstown	−	UT Salt Lake City	−^{b}	IA Cedar Rapids	+
MA Boston	+	WI Green Bay	+	KS Sedan	0^b
MI Saginaw	−	PR Ponce	0^b	LA Natchitoches	−^{b}
MN St. Cloud	−^{b}			MD Salisbury	−^{b}
MS Biloxi	−	**Channel 27**		MA New Bedford	−
MO Jefferson City	0			MI Flint	−^{b}
NE Fallon	0^b	AZ Tucson	−^{b}	MO King City	−^{b}
NM Carlsbad	−	CA Coalinga	−^{b}	MO Rolla	0^b
NY New York	0^b	FL Sabring	0^b	NC Durham	+
NC Greenville	0^b	FL Tallahassee	+	OH Columbus	−
OH Cleveland	+^{b}	GA Draketown	−^{b}	OK Ardmore	−^{b}
OK Oklahoma City	−	HI Lihue (Kauai)	−^{b}	OR Eugene	−^{b}
SC Columbia	−	HI Wailuku (Maui)	0^b	PA Johnstown	+^{b}
TX Victoria	0	IL Marion	0	PA Wilkes-Barre	0
TX Waco	+	IL Quincy	+^{b}	TN Cookeville	+
VA Onancock	+^{b}	IL Urbana	−	TX Corpus Christi	−
WA Richland	0	IA Sioux City	−^{b}	TX Lubbock	0
WI La Crosse	0	KS Topeka	0	VT Rutland	+^{b}
		KY Lexington	−	VA Bristol	−^{b}
Channel 26		LA Baton Rouge	+^{b}	WA Seattle	+^{b}
		MA Worcester	0	WA Spokane	−
AL Florence	0	MI Cadillac	0^b	WI Colfax	−^{b}
AL Montgomery	+^{b}	MN Duluth	−	WI Sheboygan	0
AR Hot Springs	0	MS Columbus	0		
CA Brawley	0^b	MO Springfield	−	**Channel 29**	
CA San Francisco	−	NV Reno	+		
CA Tulare	+	NM Roswell	−	AL Selma	−
CO Pueblo	+	NC Canton	0^b	AR Fayetteville	+
CT New London	+	ND Grand Forks	+	CA Bakersfield	0
DC Washington	−^{b}	OH Bowling Green	+^{b}	FL Ocala	0^b
FL Daytona Beach	0	OH Youngstown	0	FL West Palm Beach	+
FL Naples	−	PA Harrisburg	−	IN Kokomo	−
GA Augusta	0	SC Sumter	−^{b}	IA Dubuque	−^{b}
HI Hilo (Hawaii)	+	TX Dallas	−	KY Paducah	0
HI Honolulu (Oahu)	0	TX Laredo	−	KY Somerset	+^{b}
ID Coeur d'Alene	+^{b}	VA Norfolk–	0	LA Lake Charles	−
IL Chicago	0	Portsmouth–		MI Traverse City	−
IN Terre Haute	−^{b}	Newport News		MN Minneapolis–St.	+
IA Burlington	−	VA Roanoke	+	Paul	
LA New Orleans	0	WA Yakima	0	MS Jackson	+^{b}
ME Portland	−^{b}	WA Wenatchee	0	MT Kalispell	−^{b}

Table 21-6 Channel Assignments in the United States (*Continued*)

Locality	Offset	Locality	Offset	Locality	Offset
Channel 29		LA Alexandria	+	IA Sibley	0[b]
		MD Hagerstown	0[b]	KS Wichita	0
NE Hastings	+[b]	MI Ann Arbor	+	LA Baton Rouge	–
NM Hobbs	+	MS Cleveland	–[b]	LA Shreveport	0
NY Albany–Schenectady	+[b]	NH Hanover	0	MN Crookston	0[b]
NY Buffalo	–	NY New York	–	MO Springfield	–
NC Wilmington	+	NY Rochester	+	NE Pawnee City	+[b]
PA Philadelphia	0	OH Newark	–[b]	NM Roswell	+[b]
PA State College	+	TX Fort Worth	–[b]	NY Utica	0
SC Greenville	0[b]	VA Kenbridge	–[b]	NC Asheville	0[b]
TX San Antonio	+	WA Richland	0[b]	OH Youngstown	0
VA Charlottesville	–	WV Williamson	+[b]	PA Harrisburg	+[b]
WA Yakima	+	WI La Crosse	0[b]	SC Florence	+[b]
WV Charleston	0			TX Dallas	+
		Channel 32		VT Burlington	–[b]
Channel 30		CA San Francisco	+[b]	VA Norfolk–Portsmouth–Newport News	0
AR El Dorado	+[b]	CA Santa Barbara	0[b]	WV Huntington	+[b]
CA Chico	–[b]	CO Pueblo	–	WI Wausau	–
CA Fresno	+	DC Washington	+[b]	WY Cheyenne	–
CA San Bernardino	0	FL Lakeland	0		
CT New Britain	+	GA Toccoa	–	**Channel 34**	
FL Fort Meyers	0[b]	HI Hilo (Hawaii)	+[b]	CA Los Angeles	0
FL Jacksonville	+	HI Honolulu (Oahu)	0	DE Dover	0[b]
GA Atlanta	0[b]	IL Chicago	0	FL Fort Pierce	0
IN Bloomington	–[b]	IA Council Bluffs	0[b]	GA Athens	0
IA Davenport	–	IA Waterloo	–[b]	IN South Bend	–[b]
KS Chanute	+[b]	KS Pratt	+[b]	IA Ames	+[b]
MN Marshall	–[b]	LA New Orleans	+[b]	KS Columbus	+[b]
MS Meridian	–	MA Greenfield	+	KS Salina	–
MO St. Louis	+	MI Sault Sainte Marie	–[b]	ME Kittery	0[b]
NY Corning	0[b]	MN St. James	+	MS Senatobia	–[b]
NC Wilson	–	MS Yazoo City	–[b]	NY Binghamton	0
OH Portsmouth	0	MT Great Falls	0[b]	NY Lake Placid	+[b]
OH Toledo	+[b]	NM Albuquerque	+[b]	NC Raleigh	–[b]
OR Portland	0[b]	NC High Point	+[b]	OH Columbus	0[b]
SC Rock Hill	+	OK McAlester	–[b]	OK Oklahoma City	–
TN Memphis	0	OR Salem	0	TX Beaumont	–[b]
TN Nashville	+	TN Jackson	+[b]	TX Lubbock	–
TX Odessa	0	TX Abilene	+	TX Texarkana	0[b]
UT Ogden	0	TX Nacogdoches	0[b]	TX Waco	+[b]
WI Milwaukee	0	WI Appleton	+	WI Fond du Lac	+
PR San Juan	0	PR Aguadilla	0[b]		
				Channel 35	
Channel 31		**Channel 33**		CA Barstow	+[b]
AL Huntsville	+	AL Tuscaloosa	0	CA Salinas–Monterey	–
AL Mobile	0[b]	AZ Phoenix	0	FL Fort Walton Beach	0
AR Harrison	+[b]	FL Miami	0	FL Orlando	+
CA Sacramento	–	FL Pensacola	+	GA Lafayette	0[b]
CO Denver	0	GA Valdosta	0[b]	IL La Salle	0
GA Albany	–	HI Wailuku (Maui)	0[b]	KY Hazard	+[b]
ID Pocatello	–	ID Idaho Falls	+[b]	KY Madisonville	–[b]
IL Peoria	+	IL DeKalb	0[b]	ME Lewiston	–
IA Centerville	–[b]	IN Fort Wayne	–		
KY Owensboro	–	IA Ottumwa	–[b]		

Table 21-6 Channel Assignments in the United States (*Continued*)

Locality	Offset	Locality	Offset	Locality	Offset
Channel 35		GA Columbus	+	NH Berlin	$-^b$
		HI Hilo (Hawaii)	$+^b$	NJ Wildwood	0
MA North Adams	0^b	HI Honolulu (Oahu)	0^b	NY Binghamton	−
MI Grand Rapids	$+^b$	IL Chicago	−	NC Fayetteville	+
MN International Falls	$+^b$	IN Terre Haute	0	PA Greensburg	+
MN Winona	$+^b$	IA Fort Madison	$+^b$	SC Anderson	0
MS Vicksburg	−	IA Spirit Lake	0^b	WV Bluefield	−
MO Bowling Green	$+^b$	KY Morehead	$+^b$	WI Superior	0
OH Lima	−	LA New Orleans	+	PR Fajardo	0^b
OK Tulsa	$-^b$	MA Boston	0	**Channel 41**	
PA Erie	+	MI Mount Clemens	+		
PA Philadelphia	$-^b$	PA Scranton	+	AL Demopolis	0^b
SC Columbia	$+^b$	SC Greenwood	0^b	CO Denver	0^b
TX Marshall	+	TX Corpus Christi	+	FL Lake City	0^b
VA Richmond	+	TX El Paso	$-^b$	GA Macon	+
WA Yakima	0	TX Tyler	0^b	IA Lansing	$+^b$
Channel 36		WI Green Bay	0^b	KY Louisville	+
		PR San Sebastian	0	LA Alexandria	+
AL Florence	$-^b$	**Channel 39**		MI Battle Creek	+
AR Little Rock	0^b			MN St. Cloud	0
CA Palm Springs	−	AL Dothan	$+^b$	MO Kansas City	−
CA San Jose	0	AL Tuscaloosa	$-^b$	NJ Paterson	−
FL Madison	$-^b$	AR Phoenix	0^b	OK Tulsa	+
GA Atlanta	0	CA Bakersfield	$-^b$	SC Georgetown	$-^b$
IA Davenport	$+^b$	CA San Diego	0	TN Johnson City	0^b
IA Red Oak	0^b	FL Miami	0	TX San Antonio	+
KS Hutchinson	+	IL Rockford	0	VT Windsor	0^b
KY Lexington	0	IN Fort Wayne	$-^b$	VA Charlottesville	$-^b$
LA New Iberia	−	KS Parsons	0^b	WV Wheeling	0^b
MD Oakland	$+^b$	LA Monroe	+	**Channel 42**	
MI Lansing	+	MO Cape Girardeau	$-^b$		
MO Jefferson City	$-^b$	NY Amsterdam	$+^b$	AL Birmingham	+
NJ Atlantic City	0^b	NC Wilmington	$-^b$	AL Mobile	0^b
NY Elmira	−	PA Allentown	0^b	CA Concord	0
NC Charlotte	0	TN Greenville	−	CA Palm Springs	0
OK Lawton	$-^b$	TN Murfreesboro	+	FL Platka	0^b
RI Providence	0^b	TX Dallas	0	FL West Palm Beach	$+^b$
SD Sioux Falls	+	TX Houston	−	KS Wichita	0^b
TX Austin	0	TX Laredo	0^b	KY Bowling Green	+
TX Odessa	$+^b$	WV Parkersburg	+	MS Natchez	$+^b$
WI Milwaukee	0^b	**Channel 40**		NE Omaha	+
WI Park Falls	$+^b$			NY Oneonta	0^b
PR Bayamon	0	AL Anniston	−	NC Charlotte	$+^b$
Channel 37^f		AZ Tucson	0	OH Portsmouth	$-^b$
		AR Fort Smith	−	OK Hugo–Paris (Texas)	+
Not assigned—reserved for radio astronomy		CA Sacramento	−	TN Nashville	0^b
		CA Santa Ana	0	TX Austin	−
Channel 38		FL Sarasota	0	VA Front Royal	0^b
AR Pine Bluff	−	FL Tallahassee	+	WA Kennewick	+
CA San Francisco	0	IN Indianapolis	0	PR Yauco	0
DE Seaford	0	IA Dubuque	−	**Channel 43**	
FL St. Petersburg	0	KY Bowling Green	+		
		MA Springfield	0	AL Louisville	$+^b$
		MS Jackson	+	AZ Coolidge	0^b
		MO St. Louis	$-^b$		

Table 21-6 Channel Assignments in the United States (*Continued*)

Locality	Offset	Locality	Offset	Locality	Offset
Channel 43		MS Columbia	0[b]	MT Joplin	0
		MS Houston	+	NE Omaha	+[b]
CA Visalia	0[b]	MO Sikeston	—	NJ Burlington	—
CT Bridgeport	—	NE Lincoln	0	NM Las Cruces	+
FL Melbourne	+	NY Albany–	0	NC Greensboro	—
IL Bloomington	0	Schenectady		OH Cincinnati	—[b]
IN Richmond	+	NC Winston-Salem	0	OK Hugo	+[b]
IA DesMoines	—[b]	OH Alliance	+[b]	TX Galveston	—
KS Topeka	0	OH Dayton	0	TX McAllen	0
ME Rumford	+[b]	OK Lawton	0	WV Keyser	+
MS Columbus	0[b]	TN Chattanooga	0[b]	PR Ponce	0
NY Syracuse	+	TX Rosenberg	0		
OH Lorain	0			**Channel 49**	
OK Oklahoma City	+	**Channel 46**			
PA York	0			CT Bridgeport	—[b]
SC Myrtle Beach	+	CA Riverside	0	IL Springfield	—
TN Knoxville	+	CA Salinas–Monterey	—	IN Muncie	0
TX Bay City	+[b]	GA Atlanta	—	IA Estherville	+[b]
VA Blacksburg	0[b]	IN South Bend	0	GA Carrollton	—[b]
WI Tomah	+	KY Lexington	0[b]	KS Topeka	0
VI Charlotte Amalie	0	ME Fort Kent	+[b]	NH Littleton	+[b]
		MI Port Huron	+	NY Buffalo	—
Channel 44		NY Binghamton	+[b]	OH Akron	+[b]
		NY Jamestown	0[b]	PA York	+
AL Gadsden	+	TX Temple	—	SC Spartanburg	0
AL Montgomery	—	VA West Point	0[b]	VA Norfolk–	—
CA San Francisco	—	WV Clarksburg	—	Portsmouth–	
FL Pensacola	0	WI Kieler	+[b]	Newport News	
FL St. Petersburg	+	PR Guayama	0	WA Vancouver	—
GA Valdosta	—			WV Charleston	—[b]
HI Honolulu (Oahu)	0[b]	**Channel 47**		WI Bloomington	0[b]
IL Chicago	0			WI Racine	+
IN Evansville	—	AL Tuscumbia	—		
IA Keokuk	+[b]	CA Fresno	0	**Channel 50**	
KS Salina	0	FL Jacksonville	—		
ME Millinocket	—[b]	GA Macon	+[b]	CA Santa Ana	—[b]
MA Boston	+[b]	IL Peoria	—	CA Santa Rosa	—
		MD Salisbury	—	DC Washington	0
MS Greenville	0	MA New Bedford	—[b]	GA Young Harris	+[b]
OH Cambridge	—[b]	MN Rochester	—	IN Gary	0[b]
OH Lima	+	MS Hattiesburg	0[b]	IA Fort Dodge	+
PA Scranton	—[b]	NJ New Brunswick	+	MI Detroit	—
SC Aiken	0[b]	NC Rocky Mount	+	MO Kansas City	—
TX Harlingen	0[b]	OH Mansfield	+[b]	NH Manchester	—
TX Waco	—	OK Tulsa	0	NJ Little Falls	+[b]
VA Danville	+	PA Altoona	0	NY Watertown	+
WV Martinsburg	0[b]	VA Norton	—[b]	WV Weirton	+[b]
PR Aguadilla	0	WA Yakima	0[b]		
		WI Madison	+	**Channel 51**	
Channel 45					
		Channel 48		CA San Diego	0
AL Montgomery	—			CA Santa Rosa	—
FL Leesburg	—[b]	AL Huntsville	—	FL Fort Lauderdale	0
FL Miami	+	CA San Jose	—	FL Ocala	—
MD Baltimore	0	GA Columbus	0[b]	GA Flintstone	—[b]
MI Saginaw	—	MA Worcester	+[b]	KY Hopkinsville	0

Table 21-6 Channel Assignments in the United States (*Continued*)

Locality	Offset	Locality	Offset	Locality	Offset
Channel 51		OH Toledo	0	MD Waldorf	$+^b$
		PA Erie	$+^b$	MI Ann Arbor	$+^b$
KY Pikeville	+	VA Lynchburg	$+^b$	NJ Asbury Park	0^c
ME Portland	0	PR Arecibo	0	NJ New Brunswick	0^b
MA Pittsfield	+			NY Glens Falls	$-^b$
MI Manistique	$+^b$	**Channel 55**		NC Concord	0^b
NE Lincoln	0			OH Youngstown	0^b
OH Newark	0	FL Leesburg	0	PR Caguas	0^b
OH Sandusky	—	IL Springfield	+		
PA Reading	0	IN Fort Wayne	0	**Channel 59**	
TX Longview	—	NY Amsterdam	0		
VA Staunton	$-^b$	NY Riverhead	+	CT New Haven	+
WI Highland	0^b	OH Akron	—	FL Jacksonville	0^b
		PA State College	$+^b$	IL Peoria	$+^b$
Channel 52		SC Rock Hill	$-^b$	IN Indianapolis	—
		TN Crossville	+	NY Utica	0^b
CA Corona	0	VA Norfolk–	$+^b$	NC Andrews	0^b
FL Cocoa	0	Portsmouth–		PA Lebanon	—
GA Carnesville	0^b	Newport News			
KY Owenton	$+^b$	WI Kenosha	—	**Channel 60**	
MD Cumberland	+				
MI Kalamazoo	$+^b$	**Channel 56**		AL Dothan	—
NH Keene	$+^b$			AL Gadsden	0
NY Ithaca	0	CA Anaheim	—	CA San Mateo	0^b
NJ Trenton	$-^b$	CA Salinas–Monterey	0^b	GA Elberton	$+^b$
OK Oklahoma City	0	FL Melbourne	0	IL Aurora	0
TN Fayetteville	$-^b$	IN Gary	+	IN Madison	$+^b$
VA Courtland	0^b	MA Boston	0	NH Manchester	+
VA Marion	$-^b$	MI Detroit	0^b	OH Toledo	—
PR Carolina	0	NC Franklin	$+^b$	PA Bethlehem	—
		OH Columbus	$-^b$	TX Harlingen	0
Channel 53		PA Hazelton	0		
		SD Lowry	0	**Channel 61**	
CA Fresno	0	VA Danville	0^b		
CT Norwich	0	VA Fairfax	$-^b$	CT Hartford	+
KY Bowling Green	$-^b$	WA Tacoma	0^b	DE Wilmington	0
MI Lansing	—			FL West Palm Beach	0
NJ Atlantic City	+	**Channel 57**		IN Muncie	$-^b$
OH Chillicothe	0			KY Ashland	+
PA Pittsburgh	+	GA Atlanta	$+^b$	MI Bay City	+
VA Fredericksburg	0^b	IA Burlington	$-^b$	NY Rochester	$+^b$
		KY Hazard	—	NC Greensboro	0
Channel 54		MA Springfield	$+^b$	OH Cleveland	0
		NY Plattsburgh	0^b	TN Chattanooga	—
AL Decatur–	0	OH Lima	$+^b$		
Huntsville		PA Altoona	$+^b$	**Channel 62**	
CA San Jose	0^b	PA Philadelphia	0		
GA Augusta	—	SC Columbia	—	AL Birmingham	$+^b$
GA Columbus	+	VA Richmond	$-^b$	CA Santa Rosa	0^b
IL Kankakee	$-^b$	WV Parkersburg	0^b	IN Hammond	+
IA Keosauqua	$+^b$	WI Janesville	+	MD Frederick	0^b
KY Covington	$+^b$			MI Detroit	0
MD Baltimore	0	**Channel 58**		MO Kansas City	+
MI Muskegon	+			NY Syracuse	+
MT Joplin	—	CA Los Angeles	$-^b$	NC Asheville	+
NY Poughkeepsie	+	CA Stockton	0	NC Fayettesville	0

21.25

Table 21-6 Channel Assignments in the United States (*Continued*)

Locality	Offset	Locality	Offset	Locality	Offset
Channel 52		IL Freeport	$-^b$	GA Toccoa	$-^b$
		IL Springfield	$+^b$	IL Danville	0
OH Steubenville	$+^b$	KY Beattyville	0	KY Louisville	$+^b$
SD Lowry	$+$	MD Cumberland	0	MA Boston	$+$
WA Tacoma	0^b	NJ Vineland	$-$	MO Kansas City	$-^b$
		NY Ithaca	$+^b$	NJ Newark	0
Channel 63		OH Defiance	$+$	OH Marion	$-$
				SD Lowry	$-$
CA Oxnard	$+$	**Channel 66**		PR Humcoa	0
FL Boca Raton	0^b				
IL Galesburg	0	AL Opelika	0	**Channel 69**	
IN Angola	0	CA Vallejo–Fairfield	0		
IN Bloomington	$+$	IL Elgin	$+^h$	FL Hollywood	0
IA Des Moines	$-$	IL Joliet	$+$	GA Atlanta	0
NY Kingston	0	MA Worcester	0	IN Indianapolis	0^b
VA Bluefield	$+^b$	MI Flint	$-$	IA Des Moines	0
VA Richmond	0	OH Springfield	0^b	MI East Lansing	$-^b$
		PA Erie	$+$	PA Allentown	0
Channel 64		VA Manassas	$+$	VA Fredericksburg	$+$
		WV Fairmont	$-$		
CA Stockton	0			**Channel 70**	
DE Seaford	$-^b$	**Channel 67**			
IL Streator	$+^b$			PR Utuado	0^b
MI Kalamazoo	0	CA Salinas–Monterey	$-$		
NC Kannapolis	$-$	HI Lihue (Kauai)	0^b	**Channel 74**	
OH Cincinnati	$-$	IN Anderson	$+$		
PA Scranton	0	MD Baltimore	$-^b$	PR San Juan	0^b
RI Providence	$+$	NY Patchogue	0		
VA Charlottesville	$+$	NC Bryson City	$-^b$	**Channel 76**	
WA Bellingham	0	NC High Point	$+$		
PR Vega Baja	0	OH Canton	0	PR Cayey	0
Channel 65		**Channel 68**		**Channel 80**	
				PR Arecibo	0
CA San Jose	0	AL Birmingham	$+$		
CT New Haven	0^b	CA Los Angeles	$-^b$		
GA Cedartown	$-^b$				

[a]See FCC Rules and Regulations, Subpart E, paragraphs 73.606 and 73.610, for certain restrictions in antenna directivity and radiated power for stations in specific United States territories and possessions.

[b]Assigned for use by noncommercial broadcast stations.

[c]Operation on this channel is subject to the conditions and terms as set out in the Report and Order in Docket No. 19075, RM 1645, released Jan. 7, 1972, FCC 72-19.

[d]Following the decision of Docket No. 19261, these channels will not be available for television use until further action by the FCC.

[e]Channel 15 will not be available for television until further action by the FCC.

[f]Channel 37 (608–614 MHz) is reserved exclusively for radio astronomy service subject to review at the first Administrative Radio Conference after Jan. 1, 1974.

[g]Channel 58 is not available for use at Asbury Park unless and until it is determined by the FCC that it is not needed for educational use at New Brunswick, N.J.

[h]Channel 66 is not available for use at Elgin unless and until it is determined by the FCC that it is not needed for use at Joliet, Ill.

Source: FCC Rules and Regulations, Subpart E, paragraph 73.606, reorganized by numerical channel number.

assigned for use by noncommercial educational broadcast stations only. A station on a channel identified by a plus (+) or minus (−) mark is required to operate with the carrier frequencies offset by 10 KHz above or below, respectively, from the nominal carrier frequencies.

Applications may be filed with the FCC to construct and operate television stations only on the channels assigned in the Table of Assignments. However, additions to the table are possible by petition to the FCC. A channel assigned to a community listed in the Table of Assignments is available upon proper application to any unlisted community that is within 15 mi (air line distance) of the listed community.

21.2.2 SERVICE COVERAGE AND INTERFERENCE PROTECTION.

For the purpose of allocation and assignment of television broadcast stations, the United States is divided into three zones. Part 73.609 of the FCC Rules and Regulations, to include the maps designated as Figs. 1 and 2 of Part 73.609 (1) and (3), describes the precise boundaries of these zones. Zone I includes the states of Illinois, Indiana, Ohio, Pennsylvania, Maryland, Delaware, New Jersey, Connecticut, Rhode Island, and Massachusetts, and parts of Wisconsin, Michigan, New York, West Virginia, Virginia, New Hampshire, Vermont, and Maine. Zone III includes Florida and the Gulf Coast states within about 175 mi (282 km) of the coast as far west as Texas and to the Mexican border and is described by arcs drawn with a 150-mi radius to the north from nine points specified by latitude and longitude coordinates. Zone II consists of that portion of the United States which is not located in either Zone I or Zone III and includes Puerto Rico, Alaska, the Hawaiian Islands, and the Virgin Islands. When the boundary of Zone I passes through a city, the city is considered to be in Zone I. When the boundary of Zone III passes through a city, the city is considered to be in Zone II. Within these zones, the individual channel assignments and designations of communities have been made in such a manner as to minimize interference and facilitate maximum coverage in the available frequency bands.

The nature and extent of the protection from interference accorded to television broadcast stations are limited to the protection which results from the minimum assignment and station separation requirements and the rules and regulations with respect to maximum power and antenna heights as set forth by the FCC. A series of reference points and a specified distance-computation technique to be used in determining the spacing between communities listed in the Table of Assignments is detailed in Part 73.611 of the Rules and Regulations. Generally, the technique considers the distance between two reference points to be the length of the hypotenuse of a right triangle, the sides of which are determined by the differences in longitudes and latitudes, of the two reference points. This method is accepted for distances up to only 220 mi (354 km).

The service provided by a television station must be protected from intereference from other television stations operating on the same channel (co-channel interference), on the immediate adjacent channels (adjacent-channel interference), as well as on other channels separated from the desired channel by one or more intervening channels. Co-channel interference is independent of receiver design, while adjacent channel interference is a function of the receiver response outside the channel to which it is tuned.

Table 21-7 Part 73.610, FCC Rules and Regulations

Interference type	Zone	Channels 2–13 VHF, mi (km)	Channels 14–83 UHF, mi (km)
Co-channel	I	170 (274)	155 (249)
	II	190 (306)	175 (282)
	III	220 (354)	205 (330)
Adjacent channel	I	60 (97)	55 (89)
	II	60	55
	III	60	55

The minimum co-channel and adjacent-channel assignments and station separations are given in Table 21-7.

Owing to the frequency spacing which exists between channels 4 and 5, between channels 6 and 7, and between channels 13 and 14 (Table 21-2), the minimum adjacent-channel separations specified in Table 21-7 are not applicable for these pairs of channels. Additional requirements for minimum separation between stations on channels 14 to 82 (UHF) are set forth in detail in Table IV of Part 73.698 of the FCC Rules and Regulations. A chart summarizing these requirements is shown in Table 21-8.

Table 21-8   Channel Separations for Channels 14–83 (UHF)

Interference type	Channel numbers and separation	Required separation, mi (km)
Co-channel	0	Zone I, 155 (249)
		Zone II, 175 (282)
		Zone III, 205 (330)
Adjacent channel	± (6 MHz)	55 (89)
Sound image	±14 (84 MHz)	60 (97)
Picture image	±15 (90 MHz)	75 (121)
Local oscillator	±7 (42 MHz)	60 (97)
IF beat	±8 (48 MHz)	20 (32)
Intermodulation	±2 through ±5	20

Source: Part 73.610, Table IV, 73.698, FCC Rules and Regulations.

A brief explanation of the basis for these regulations pertaining to various types of potential interference sources is of interest. The regulatory powers of the FCC do not extend to the matters relating to receiver intermediate frequency (IF) values. However, various industry groups, such as the EIA, in the interest of providing the best possible service to the public, have issued standards of good engineering practice that establish the television receiver IF frequencies to be 45.75 MHz for the visual carrier and 41.25 MHz for the aural carrier. This universal practice, therefore, makes it possible for the FCC to establish regulations for minimum separation of channel assignments for the adjacent channel, sound image, picture image, local oscillator, IF beat, and intermodulation types of interference (see Table 21-8).

1. *Local-oscillator radiation.* It is customary, for spurious frequency reasons, to employ a local-oscillator frequency above that of the channel frequency being received. Thus, to produce the *standard* intermediate sound carrier frequency of 41.25 MHz, the local oscillator must be 41.25 MHz higher than the radiated sound carrier frequency. This is 41 MHz higher than the upper limit of the channel and 47 MHz higher than the lower limit of the channel. As a result, the local oscillator can radiate into a channel whose *number* is 7 times higher than the channel in use (7 times 6 MHz equals 42 MHz) which lies between 41 and 47 MHz. It has been determined that a sufficient margin of protection against such radiation is afforded by separation of stations occupying channels that are seven channel numbers apart by a minimum of 60 mi (97 km).

2. *IF images and beats.* With standard IF values, the first-order images of the picture carrier fall above or below the picture carrier at intervals equal to two times the picture IF (2 times 45.75 equals 91.5 MHz), and also the sound carrier first-order images fall above and below the sound carrier at intervals of two times the sound IF (2 times 41.25 equals 82.5 MHz). Consequently, the sound image appears in channels numbered 14 above and below the desired channel, while the picture image appears in channels numbered 15 above and below the desired channel. A safe minimum separation has been adopted by the FCC of 60 mi for the sound image and 75 mi (121 km) for the picture image. Likewise, the IF beat may occur between channels separated

by eight channel numbers. A 20-mi (32-km) separation has been designated as a safe margin for this degradation.

3. *Intermodulation.* Intermodulation interference may occur when two television stations have carriers separated by a frequency interval lying within the IF response between 41 and 47 MHz. Any pair of stations separated by seven or eight channel numbers (7 times 6 equals 42 MHz and 8 times 6 equals 48 MHz) and straddling the desired channel may cause this type of interference. Protection is afforded by the FCC by requiring at least a 20-mi (32-km) separation between stations on channels from two to five above and from two to five below the channel of interest. An exception to this rule is the immediately adjacent channels which fall under a more stringent separation rule of 55 mi (89 km).

21.2.3 GRADES OF SERVICE. In the process of authorizing the operation of television stations, two field-strength contours are considered. Specified as *grade A* and *grade B,* these indicate the approximate extent of coverage over average terrain in the absence of interference from other television stations. On the other hand, the true coverage may vary greatly because the actual terrain over a specific path may be considerably different from the average terrain on which field-strength charts are based. The required field strengths in decibels above 1 μV/m (dBμ) for the grade A and grade B service as well as the local community minimum values are shown in Table 21-9.

It should be noted that in the UHF bands, the validity of the field-strength predictions is not based upon measured data for distances greater than 30 mi (48 km). Thus the curves for channels 14 to 83 should be used with an appreciation of their limitations since the actual field strength at any one given location will vary considerably. The actual extent of service will be less than indicated owing to interference from other stations, although all predictions were based upon the assumption of no interference.

In predicting the distances to the field-strength contours, the $F(50,50)$ and $F(50,10)$ field-strength charts contained in Parts 73.684 and 73.699 of the FCC Rules and Regulations should be used. The 50 percent field strength is defined as that value exceeded for 50 percent of the time. $F(50,50)$ and $F(50,10)$ charts give the estimated 50 percent field strength exceeded at 50 percent of the locations in dB above 1 μV/m for 50 and 10 percent of the time, respectively. The charts are based upon an effective power of 1 kW radiated from a half-wave dipole in free space which produces a field strength at 1 mi (1.6 km) of 103 dB above 1 μV/m. The receiving antenna is considered as having an effective height of 30 ft (9 m).

The following charts and a description of their use are given in Chap. 6, Sec. 6.1:

Figure	Channels	Field strength
6-14	2–6	$F(50, 50)$
6-15	7–13	$F(50, 50)$
6-16	14–69	$F(50, 50)$
6-17	2–6	$F(50, 10)$
6-18	7–13	$F(50, 10)$
6-19	14–69	$F(50, 10)$

Table 21-9 Grades of Television Service, FCC Regulations[†]

Channel designations	Frequency band, Hz	Grade A service	Grade B service	Local community minimum
2–6 (low VHF)	54–88	68 dBμ; 2510 μV/m	47 dBμ; 224 μV/m	74 dBμ; 5010 μV/m
7–13 (high VHF)	174–216	71 dBμ; 3550 μV/m	56 dBμ; 631 μV/m	77 dBμ; 7080 μV/m
14–83 (UHF)	470–890	74 dBμ; 5010 μV/m	64 dBμ; 1580 μV/m	80 dBμ; 10,000 μV/m

[†]dBμ = decibels above 1 μV/m.

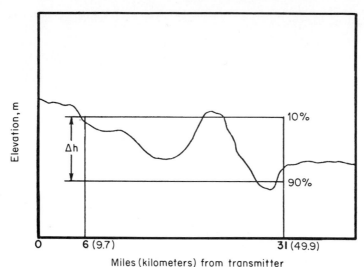

FIG. 21-3   Terrain-roughness factor Δh definition. *(FCC.)*

In predicting the grades A and B field-strength distance contours, the effective radiated power is that radiated at a vertical angle computed by the following expression

$$A_h = 0.0153 \sqrt{H}$$

where A_h = the depression angle in degrees and H = the height in feet of the transmitting antenna radiation center above the average terrain of the 2- to 10-mi (3.2- to 16-km) sector of the pertinent radial direction.

Part 73.684 of the FCC Rules and Regulations contains detailed instructions as to how the data for field-strength predictions and contours should be determined and presented as part of the applicant's documentation.

The field-strength charts (Part 73.699, Fig. 1–10C) were developed assuming a terrain roughness factor of 164 ft (50 m) which is considered representative of the average terrain in the United States. In situations where the actual terrain departs markedly from this nominal one, a terrain roughness correction factor (ΔF) may be applied.† The magnitude of this correction may be taken from a chart shown in Fig. 21-4 (the definition of the roughness factor is shown in Fig. 21-3). Alternatively, it may be computed by the following expression

$$\Delta F = C - 0.03 \, (\Delta h) \, (1 + f \, 7300)$$

where ΔF = correction factor, dB
Δh = roughness factor,
f = frequency of signal, MHz
C = factor for each of the charts as follows: 1.9 for channels 2–6; 2.5 for channels 7–13; 4.8 for channels 14–69

Part 73.685 of the FCC Rules and Regulations describes the factors relating to the choice by an applicant of the location of the transmitter and the antenna system. It defines any antenna system, other than one having a circular radiation pattern in the horizontal plane, as constituting a directional antenna. Directional antennas, to include tilting of the vertical angle, may be employed for the purpose of improving service upon

---

†All FCC rules relating to Δn and ΔF have been stayed at present (1982) pending further order.

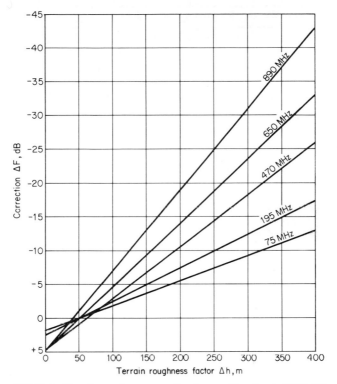

FIG. 21-4 Terrain-roughness correction for use with estimated $F(50, 50)$ and $F(50, 10)$ field-strength curves. *(FCC.)*

an appropriate showing of need. Stations operating on channels 2 to 13 cannot use directional antennas with a ratio of maximum to minimum radiation in excess of 10 dB. Channels 14 to 83 operating with peak visual power of more than 1 kW cannot utilize directional antennas with a maximum-to-minimum ratio exceeding 15 dB, while stations on channels 14 to 83 of less than 1 kW have no restrictions on antenna directivity. Detailed procedures for making field-strength measurements as well as methods for reporting such measurements are given in Part 73.686 of the Rules and Regulations. These rules are precise for the purpose of determining allocations of channel assignments and for rule making in regard to testimony presented by applicants for assignment.

21.2.4 RADIATED POWER AND ANTENNA HEIGHT. No applications for television broadcasting stations will be accepted by the FCC if they specify less than 100 W (-10 dBk) horizontally polarized visual effective radiated power in any horizontal direction.† No minimum antenna height above the average terrain is specified.

The maximum horizontally polarized visual effective radiated power (ERP) of television broadcast stations with antennas not exceeding 2000 f (610 m) above the average terrain is shown in Table 21-10.

Additional antenna-height restrictions are imposed upon stations operating in Zone I on channels 2 to 13 to not be in excess of 1000 ft above the average terrain for the power

---

†See subsection Television Broadcast Translator Stations in Sec. 21.2.5 regarding LPTV rules.

Table 21-10   Maximum Visual ERP†

Channel no.	Maximum visual ERP in dB above 1 kW (dBk)
2–6 (low VHF)	20 dBk (100 kW)
7–13 (high VHF)	25 dBk (316 kW)
14–83 (UHF)	37 dBk (5000 kW)

†*Note:* The maximum visual ERP of television stations operating on channels 14 to 83 within 250 mi (402 km) of the border between Canada and the United States may not be in excess of 30 dBk (1000 kW).

levels listed in Table 21-10. If stations on channels 2 to 13 have antennas in excess of 1000 ft (305 m) or if stations operating on channels 14 to 83 have antennas in excess of 2000 ft, the maximum effective visual radiated power shall be based on the chart for Zone I as in Fig. 21-5 (Part 73.699, Fig. 3 of the FCC Rules and Regulations).

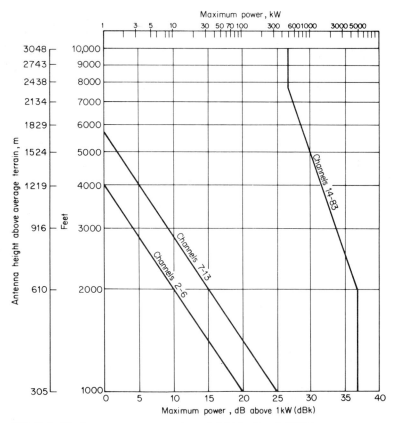

FIG. 21-5   Maximum power versus antenna height for Zone 1. *(FCC.)*

Likewise, in Zones II and III, the maximum power for stations operating with antennas exceeding 2000 ft (610 m) is determined by the chart for Zones II and III as in Fig. 21-6 (Part 73.699, Fig. 4 of the FCC Rules and Regulations).

The effective radiated power in any horizontal or vertical direction may not exceed the limits set forth in Table 21-10 and Figs. 21-5 and 21-6.

Methods for determining the operating power of a television broadcast station are given in Part 73.663 of the FCC Rules and Regulations. They are classified as either direct or indirect for both the visual and aural transmitter.

The *direct method* for the visual transmitter uses a calibrated transmission line meter (peak power reading) located at the RF output of the transmitter to include any vestigial or harmonic suppression filters. The meter must be checked at 6-month intervals and is calibrated by measuring the average power output when operating into a nonreactive dummy load, and the transmitter is modulated with a signal consisting of standard synchronizing pulses and blanking level set at 75 percent of peak amplitude as observed on a waveform monitor. Calibration is made with the transmitter operating at 80, 100, and 110 percent of the authorized power at 6-month intervals.

The direct method for the aural transmitter is essentially the same technique and procedure as employed with the visual transmitter except that the transmission line meter may be responsive to relative voltage, current, or power.

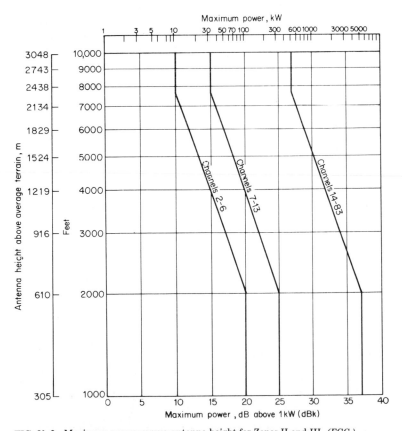

FIG. 21-6 Maximum power versus antenna height for Zones II and III. *(FCC.)*

The *indirect method* for both the visual and aural transmitters consists of applying an appropriate factor to the input power to the final radio-frequency amplifier stage of the transmitter by using the following formula

$$\text{Transmitter output power} = E_p \times I_p \times F$$

where E_p = dc input voltage of final stage
  I_p = dc input current of final stage
  F = efficiency factor

21.2.5 EXTENDED COVERAGE AND ANCILLARY BROADCAST SERVICES. A number of ancillary television broadcast services are regulated by Part 73, "Radio Broadcast Services," Part 74, "Experimental, Auxiliary, and Special Broadcast, and Other Program Distributional Services," and Part 76, "Cable Television Service," of the FCC Rules and Regulations. Technical and certain other pertinent standards relating to these services are summarized below.

Noncommercial Educational Television Broadcast Service. Part 73.621 contains the regulations pertaining to noncommercial educational broadcast stations defined as those licensed only to nonprofit educational organizations whose purpose is to serve the educational needs of the community, for the advancement of educational programs, and to furnish nonprofit and noncommercial television service.

These stations, so authorized and designated, are required to conform to all the same technical standards and requirements as commercial stations as outlined in Subpart E of Part 73 of the FCC Rules and Regulations.

Over-the-Air Subscription Television. A subscription television broadcast program is defined as a broadcast program intended to be received in intelligent form by members of the public only for a fee or charge and may be provided only upon specific authorization by the commission. The details of this service are described in Part 73, Secs. 73.641, 73.642, 73.643, and 73.644, of the Rules and Regulations.

The particular technical system to be employed for encoding and decoding must be described in complete detail and approved as meeting the technical standards required by the commission. All subscription television decoders must be leased, not sold, to subscribers. The technical system must be capable of delivering a suitable signal to the antenna terminals of receivers designed for reception of signals meeting the identical standards for color and monochrome standard broadcast facilities. The definition of *suitable signal* shall be one which complies with all the technical standards contained with Subpart E of the regulations for both visual and aural transmissions. Also, no increase in spectral energy or increase in the width of the television broadcast channel (6 MHz) is allowed.

The technical system shall be capable of producing visual and aural signal coverage and received program quality that is not significantly inferior (as judged by the commission) to that produced by normal broadcast of the same material. The decoded signal quality must be without perceptible degradation as compared with the same material broadcast in accordance with the monochrome and color standards and with no internal modifications to the subscriber's receiver. In addition, interference to reception of conventional as well as subscription television programs, co-channel and adjacent channel, monochrome and color, shall not significantly exceed that quality resulting from normal television broadcasting conducted in compliance with the technical standards outlined in the Rules and Regulations (Part 73).

Experimental Television Broadcast Stations. Experimental television broadcast stations may be licensed, upon satisfactory showing of the need thereof, for the purpose of conducting research and experimentation for the advancement of television broadcasting and may include tests of equipment, training of personnel, and experimental programs as are necessary. Any of the frequencies normally allocated to television broadcasting, as well as the various categories of the ancillary services, may be assigned by the commission for properly authorized experimentation.

Television Auxiliary Broadcast Stations. Part 74, Subpart F, Secs. 74.601 through 74.682, describe the details of the regulations pertaining to television auxiliary broadcast stations. The auxiliary services are classified as: (1) television mobile pickup, (2) television studio transmitter link, (3) television intercity relay, and (4) television translator relay. The definitions of these services are as follows:

1. *Television mobile pickup* is defined as a land-mobile station used for transmitting television program material and related communications from the scenes of events at points removed from the broadcast station such as news and sports events.

2. *Television studio transmitter link* (STL) service relates to a fixed station employed to transmit television signals from the studio to the television transmitter.

3. *Television intercity relay* station is a fixed station utilized for intercity transmission of television signals for use by the broadcast stations involved.

4. *Television translator relay* stations are fixed stations used for relaying signals from one broadcast television station to television translator stations.

Table 21-11 lists the frequencies available for assignment to television pickup, television STL, television intercity relay, and television translator relay stations.

A number of frequencies and channels listed for television broadcast assignment are also shared under specific conditions with a variety of other types of services such as: (1) incidental radiations of industrial, scientific, and medical (ISM) equipment, (2) US90 of the FCC's Table of Frequency Allocations for operations of earth-to-space transmissions in space research and earth exploration satellite services by government and nongovernment services at specific locations, (3) US111 for operations of government space-research earth stations for tracking, ranging, and telecommand purposes from specified locations, (4) US219 for operations in the Earth Explorations Satellite Service for earth-space tracking, telemetry, and telecommands from specified locations, and (5) geostationary operational environmental satellite earth stations for tracking, telemetry, and telecommands from specified locations. (See FCC Regulations, Part 74, Subpart F, Sec. 74.602, for details.)

The power output of television auxiliary broadcast stations is limited to that necessary to render satisfactory service, and the stations operating above 1000 MHz may employ any type of emission suitable for the transmission of all the visual, aural, and operational signals as are permitted under the rules of Subpart F. The transmitter frequency tolerances and spurious emissions are controlled both within the specified channel and in adjacent channels by limits indicated in Secs. 74.637 and 74.661 of Subpart F. Appropriate monitoring and inspection rules are provided in Secs. 74.662 and 74.663.

Television Broadcast Translator Stations. Subpart G of Part 74 describes the regulations governing the operation of television broadcast translator stations whose primary purpose is to provide extended coverage under special conditions for assigned broadcast stations. A television translator station is defined as a station operated to retransmit the signals of a television broadcast station without significantly altering any characteristic of the original signal other than its frequency and amplitude to provide extended coverage and service to the general public.

A station in the broadcasting service operated for the sole purpose of retransmitting the signals from a UHF translator station by amplifying and reradiating the signals received through space, without significantly altering any characteristic other than its amplitude, is termed a *translator signal booster*. Thus, television translator and booster stations provide a means whereby the signals from a television broadcast station may be retransmitted to areas in which direct reception is unsatisfactory owing to distance or intervening terrain barriers. A translator station shall be operated such that the tech-

Table 21-11 Frequencies Available†

Band A, MHz	Band B, MHz	Band D, GHz			
		Group A		Group B	
		Channel no.	Channel frequencies	Channel no.	Channel frequencies
1990–2008	6875–6900	A01	12.700–12.725	B01	12.7125–12.7375
2008–2025	6900–6925	A02	12.725–12.750	B02	12.7375–12.7625
2025–2042	6925–6950	A03	12.750–12.775	B03	12.7625–12.7875
2042–2059	6950–6975	A04	12.775–12.800	B04	12.7875–12.8125
2059–2076	6975–7000	A05	12.800–12.825	B05	12.8125–12.8375
2076–2093	7000–7025	A06	12.825–12.850	B06	12.8375–12.8625
2093–2450	7025–7050	A07	12.850–12.875	B08	12.8625–12.8875
2450–2467	7050–7075	A08	12.875–12.900	B08	12.8875–12.9125
2467–2484	7075–7100	A09	12.900–12.925	B09	12.9125–12.9375
2484–2500	7100–7125	A10	12.925–12.950	B10	12.9375–12.9625
		A11	12.950–12.975	B11	12.9625–12.9875
		A12	12.975–13.000	B12	12.9875–13.0125
		A13	13.000–13.025	B13	13.0125–13.0375
		A14	13.025–13.050	B14	13.0375–13.0625
		A15	13.050–13.075	B15	13.0625–13.0875
		A16	13.075–13.100	B16	13.0875–13.1125
		A17	13.100–13.125	B17	13.1125–13.1375
		A18	13.125–13.150	B18	13.1357–13.1625
		A19	13.150–13.175	B19	13.1625–13.1875
		A20	13.175–13.200	B20	13.1875–13.2125
		A21	13.200–13.225	B21	13.2125–13.2375
		A22	13.225–13.250		

†*Notes:* (1) For fixed stations using B and D channels, applicants are encouraged to alternate A and B channels such that adjacent RF carriers are spaced 12.5 MHz. (2) The band 13.15–13.20 GHz is reserved exclusively for assignment of television pickup and CARS pickup stations on a coequal basis within a 31-mi (50-km) radius of each of the 100 television markets delineated in Sec. 76.51. Fixed stations licensed prior to Sept. 1, 1979, may continue to operate in this range subject to periodic review. (3) The band 6.425–6.525 MHz is available for television pickup service on a shared basis with the common-carrier service. (4) The band 38.6–40.0 GHz is available for television pickup service on a shared basis with the common-carrier service.

nical characteristics of the retransmitted signals shall not deliberately be altered as to hinder reception by a conventional receiver designed to receive standard broadcast signals.

Any of the VHF channels may be assigned to a VHF translator provided that no interference is caused to the direct reception of any station operating on the same or an adjacent channel. Translators may be assigned to the UHF channels 55 through 69 provided certain output power and mileage limitations are observed as per Table 21-12.

The power output of VHF translators may not exceed 1 W visual power (10 W west of the Mississippi River) except where authorized in a community listed in the Table of Assignments. In this case, they may have power output of 100 W. UHF translators are limited to a power output of 100 W, peak visual power, except on channels listed in the Table of Assignments where the authorized peak visual power may be either 100 or 1000 W. No specifications relate to the effective radiated power resulting from the use of directional antennas.

Interference to adjacent channels is controlled by requiring any emission more than 3 MHz outside the channel be −30 dB for transmitters rated at 1 W or less and −50 dB

Table 21-12 Minimum Mileage Separation for Translator and Broadcast Stations

Interference type	Minimum separation, mi (km)
Co-channel	155 (240)
Adjacent channel	55 (89)
±2, ±3, ±4, ±5, ±8 channel numbers	20 (32)
±7, ±14 channel numbers	60 (97)
±15 channel numbers	75 (121)

Note: No minimum distance between translators operating on the same channel is specified unless obvious interference would result.

for transmitters rated at more than 1 W, with −60-dB attenuation for transmitters with power outputs of more than 100 W.

Generally, the translator equipment shall be designed such that the electrical characteristics of a standard television signal introduced into the input terminals will be maintained at the output. This includes overall frequency response, individual frequency stabilities, and relative frequency values. The details of the equipment specifications to include unattended operation, automatic control, and automatic identification techniques are contained in Secs. 74.735, 74.736, 74.737, and 74.750 of Part 74 of the FCC Rules and Regulations.

The rules governing the operation of UHF translator signal boosters are contained in Sec. 74.733 and generally are in accordance with the rules of good engineering practice. They may also be operated unattended under controlled conditions and within certain qualifications and certification.

In March 1982, the FCC ruled to allow operation of low-power television (LPTV) broadcasting service. However, the then-existing freeze on new applications will continue until the application backlog can be handled.

In effect, the new rules, in superseding the existing translator rules, will allow them to become LPTV stations simply by notification to the FCC. These translators could then originate programs as long as they meet the LPTV criteria which include having an operator on duty. The new LPTV stations will be authorized on a *secondary basis* with regard to interference from full-service stations. Full-service stations will still be protected at least to their Grade B contours.

The LPTV stations will be allowed to operate on any available VHF or UHF channel but are limited to 10 W on VHF and 1000 W on UHF. The new stations are almost entirely unregulated, with few programming rules and no ownership requirements, but still must comply with technical rules and regulations as in FCC Parts 73 and 74 where applicable.

Instructional Television Fixed Service (ITFS). In Subpart I of Part 74 of the FCC regulations, an *instructional television fixed station* (ITFS) is defined as being a station operated by an educational organization and used for the transmission of visual and aural instructional, cultural, and other types of educational material to one or more fixed receiving locations. An ITFS response station is a fixed station operated to provide communication by voice and/or data signals to an associated instructional television fixed station and must operate on assigned frequencies with a power output of no more than 250 mW.

ITFS stations are assigned specific frequencies in Groups A through G with four channels per group. The channel assignment frequencies range from 2500 MHz for the lower limit of Group A to 2686 MHz for the upper limit of Group G (see Sec. 74.902). The peak power output of an instructional television fixed station shall not exceed 10 W, and attenuation of the lower sideband is not required, assuming that interference-free operation can be obtained. See Part 74, Subpart I, Secs. 74.901 through 74.984, of the FCC Rules and Regulations for details.

Cable Television (CATV) Service. Cable television service is not a *broadcast* service in the same sense as radiated television broadcast service. However, it is very closely related in that it provides standardized television broadcast signals to the antenna terminals of a receiver designed for use with television broadcast signals and is therefore considered a critical part of the overall television services provided and regulated by rules contained within Part 76 of the FCC Rules and Regulations.† Subpart K contains the technical specifications and they are summarized herein. The original intent of CATV systems was to provide a means of advantageously receiving broadcast signals and distributing the signals to subscribers by means of a system of appropriate amplifiers and coaxial cables. Thus, service could be provided to communities in which conventional broadcast service was either poor or nonexistent. The service has expanded to include the provision of cable-delivered service in areas normally enjoying good reception to expand and enhance the total available television service by increasing the choice of programs to the subscriber.

Subpart A of Part 76 of the FCC Rules and Regulations defines CATV as being a nonbroadcast facility consisting of a set of transmission paths and associate signal generation, reception, and control equipment, under common ownership and control, that distributes or is designed to distribute to subscribers the signals of one or more television broadcast stations, but such term shall not include (1) any such facility that serves fewer than 50 subscribers, or (2) any such facility that serves or will serve only subscribers in one or more multiple-unit dwellings under common ownership, control, or management.

Unless special authorization is obtained, the channel frequencies made available by CATV systems are the same as those specified by the FCC for broadcast purposes and as listed in Table 21-2. The tolerance in the frequency of the visual carrier is ± 25 kHz, except that where subscribers employ a frequency converter to translate from one channel to another the visual carrier frequency tolerance may be ± 250 kHz from the nominal value.

The signal level to be delivered to the subscriber cable terminals, when terminated by an impedance which matches the cable impedance, shall not be less than 1 mV across 75 Ω or 2 mV across 300 Ω. The minimum level of any impedance Z may be calculated by

$$E_{\text{mV}} = \sqrt{0.0133Z}$$

Variations in the visual signal level shall not exceed 12 dB within any 24-h period and shall be within 3 dB of any other visual carrier within 6 MHz nominal frequency separation, and within 12 dB of any other channel carrier. The signal should never be so great as to cause overload of the television receiver input circuits.

The rms voltage of the aural signal shall be between 13 and 17 dB below the associated visual carrier level. The frequency response of the signal being delivered should not vary more than ± 2 dB in the range between 0.5 and 5.25 MHz relative to the level at 1.25 MHz.

Interference protection is afforded by requiring that each signal provided by cable facilities must have a ratio of the visual carrier level to any undesired co-channel signal operating on proper offset assignment (within the Grade B contour of that signal) not less than 36 dB. Also, the ratio of the visual signal level to the rms value of any coherent interfering signal, such as intermodulation products, shall not be less than 46 dB.

Isolation at the subscriber terminal shall not be less than 18 dB and shall be adequate to suppress reflections caused by open or shorted subscriber terminals anywhere else in the system.

Spurious radiations from a cable television system (CATV) shall be limited as indicated in Table 21-13.

---

†FCC Rules and Regulations, Part 76, revised as of February 1981.

Table 21-13 Radiation Limits for CATV Systems

Frequency range	Radiation limit, μV/m	Distance, ft (m)
Up to and including 54 MHz	15	100 (30.5)
Over 54 MHz up to and including 216 MHz	20	10 (3)
Over 216 MHz	15	100 (30.5)

Cable services may also include the need to transmit signals by microwave relay between a television broadcast reception point to the central cable television distribution center. The relay stations operate in the cable television relay service (Part 78 of the FCC Regulations) and utilize channel assignments in the 12.7- to 13.25-GHz band. In this service, the peak output power shall not exceed 5 W on any channel. Exceptions to this rule allow transmitted peak power values of 15, 30, or 60 W, depending upon actual frequency assignment, for frequency modulation by baseband frequency-division multiplexed standard television signals.

For local distribution, the visual signal may use vestigial sideband AM or frequency-division multiplexed FM transmission. When AM transmission is used, the visual signal relative levels will be held to within 2 dB on all channels and the aural signal level shall not be more than 7 dB below the peak power of the visual signal. The maximum power to the antenna is limited to +10 dBW in the 12.70- to 12.75-GHz band, and the isotropically radiated power shall not exceed +55 dBW.

The CATV relay stations are regulated with regard to potential interference in that spurious emissions outside the assigned channel are limited as shown in Table 21-14 and must have frequency tolerances as shown in Table 21-15.

CATV relay stations are intended to use highly directive antenna systems with the maximum beamwidth in the horizontal direction (one-half power points) not to exceed 3°. Under special conditions where need is determined, greater beamwidth or multiple

Table 21-14 CATV Relay Emission Limits

Modulation	Frequency range outside channel in % channel width	Emission limits below visual carrier, dB
1. FM and double-sideband (DSB) AM	0–50	−25
	50–150	−35
	More than 150	−43 + 10 log
2. Vestigial sideband (VSB) AM	All emissions	−50

Table 21-15 CATV Relay Service Frequency Tolerances

Transmission type	Tolerance
1. FM (frequency-division multiplexed modulation by baseband television)	0.02%
2. VSB AM	
Visual carrier	0.0005%
Aural carrier	4.5 MHz ± 1 kHz
3. FM and DSB AM	0.005%

antennas may be authorized. The FCC reserves the right to specify the direction of polarization. If the isotropically radiated power exceeds +45 dBW in band between 12.70 and 12.75 GHz, the antenna must be oriented so that the direction of maximum radiation shall be at least 1.5° away from the geostationary satellite orbit, taking into account the effects of atmospheric refraction.

21.3 SIGNAL TRANSMISSION STANDARDS

Signal transmission standards describe the specific characteristics of the broadcast television signal radiated within the allocated spectrum. These standards may be summarized as follows:

1. Definitions of fundamental functions involved in producing the radiated signal format, to include relative carrier and subcarrier frequencies and tolerances as well as modulation-sideband spectrum and radio-frequency envelope parameters

2. Transmission standards describing the salient baseband signal values relating visual psychophysical properties of luminance and chrominance values described in either the time or frequency domains

3. Synchronization and timing signal parameters, both absolute and relative

4. Specific test and monitoring signals and facilities

5. Relevant mathematical relationships describing the individual modulation signal components

The details of these signal transmission standards are contained, for United States monochrome and television broadcast services, in the FCC Rules and Regulations, Part 73, Subpart E, Secs. 73.676, 73.681, and 73.682.

21.3.1 **TECHNICAL STANDARDS DEFINITIONS.**   Section 73.681 of the FCC Regulations provides basic definitions pertaining to transmission standards. Some of the more pertinent ones are listed below:

1. *Amplitude Modulation (AM).* A system of modulation in which the envelope of the transmitted wave contains a component similar to the waveform of the baseband signal to be transmitted.

2. *Antenna Height Above Average Terrain.* The average of the antenna heights above the terrain from about 2 to 10 mi (3.2 to 16 km) from the antenna as determined for eight radial directions spaced at 45° intervals of azimuth is considered as the antenna height. Where circular or elliptical polarization is employed, the average antenna height is based upon the height of the radiation center of the antenna that produces the horizontal component of radiation.

3. *Antenna Power Gain.* The square of the ratio of the rms free space field intensity produced at 1 mi (1.6 km) in the horizontal plane, expressed in millivolts per meter for 1 kW antenna input power to 137.6 mV/m. The ratio is expressed in decibels (dB).

4. *Aspect Ratio.* The ratio of picture width to picture height as transmitted. The standard is 4:3 for 525-line NTSC and 625-line PAL and SECAM systems.

5. *Chrominance.* The colorimetric difference between any color and a reference color of equal luminance, the reference color having a specific chromaticity.

6. *Effective Radiated Power.* The product of the antenna input power and the antenna power gain expressed in kilowatts and in decibels above 1 kW (dBk). The licensed effective radiated power is based on the average antenna power gain for each direction in the horizontal plane. Where circular or elliptical polarization is employed,

the effective radiated power is applied separately to the vertical and horizontal components. For assignment purposes, only the effective radiated power for horizontal polarization is considered.

7. *Field.* A scan of the picture area once in a predetermined pattern.

8. *Frame.* Scanning all the picture area once. In the line-interlaced scanning pattern of 2/1, a frame consists of two interlaced fields.

9. *Frequency Modulation (FM).* A system of modulation where the instantaneous radio frequency varies in proportion to the instantaneous amplitude of the modulating signal, and the instantaneous radio frequency is independent of the frequency of the modulating signal.

10. *Interlaced Scanning.* A scanning pattern where successively scanned lines are spaced an integral number of line widths, and in which the adjacent lines are scanned during successive periods of the field rate.

11. *IRE Standard Scale.* A linear scale for measuring, in arbitrary IRE units, the relative amplitudes of the various components of a television signal as shown in Table 21-16.

Figure 21-7 shows the IRE level units and modulation percentage for a standard SMPTE color-bar signal (SMPTE ECR 1-1978). In practice, as specified in EIA Standard RS-189-A and illustrated in Fig. 21-44, the gray, red, green, blue, yellow, cyan, and magenta bars are composed of 75 percent level red, green, and blue signals. The peak-white bar at the bottom of the field is composed of 100 percent red,

Table 21-16   IRE Standard Scale

Level	IRE units	Modulation, %
Zero carrier	120	0
Reference white	100	12.5
Blanking	0	75
Sync peaks (max. carrier)	−40	100

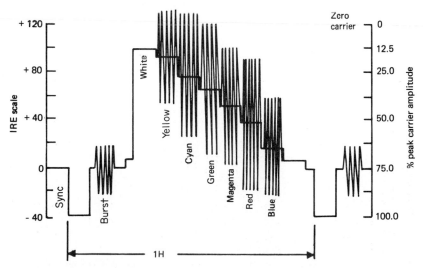

FIG. 21-7   Standard SMPTE color-bar signal.

green, and blue signals. This results in a transmitter modulation level no lower than 12.5 percent for the peak-white and yellow bars, thus avoiding differential gain and phase distortion which is inherent at near-zero carrier modulation.

12. *Luminance.* Luminance flux emitted, reflected, or transmitted per unit solid angle per unit projected area of the source (the relative light intensity of a point in the scene).

13. *Negative Transmission.* Modulation of the radio-frequency visual carrier in such a way as to cause an increase in the transmitted power with a decrease in light intensity.

14. *Polarization.* The direction of the electric field vector as radiated from the antenna.

15. *Scanning.* The process of analyzing successively, according to a predetermined method or pattern, the light values of the picture elements constituting the picture area.

16. *Vestigial Sideband Transmission.* A system of transmission wherein one of the modulation sidebands is partially attenuated at the transmitter and radiated only in part.

21.3.2 CHANNEL SPECTRUM AND CARRIER DESIGNATIONS. The basic features of the technical transmission standards for television broadcasting in the United States as specified by the FCC are summarized and listed below:

1. The width of the television broadcast channel is 6 MHz.

2. The visual carrier frequency shall be nominally 1.25 MHz above the lower boundary of the channel with a tolerance of ± 1000 Hz.

3. The aural carrier frequency shall be 4.5 MHz (± 1000 Hz) higher than the visual carrier frequency with a tolerance of ± 1000 Hz for the intercarrier spacing between the visual and aural carriers.

4. The visual amplitude characteristics of the sideband spectrum shall be in accordance with the vestigial sideband response as shown in Fig. 21-8 (Sec. 73.699, Fig. 5 of the FCC Rules and Regulations). However, the amplitude characteristics need not be

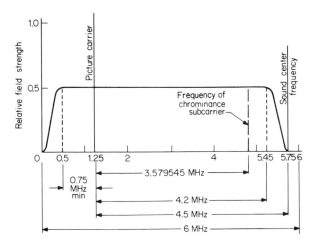

FIG. 21-8 Idealized picture transmission amplitude characteristics for all VHF and UHF channels with power outputs of greater than 1 kW. *(FCC.)*

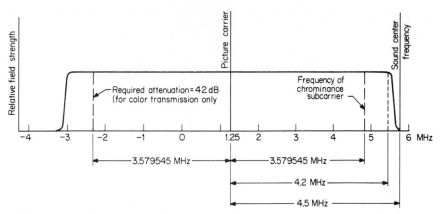

FIG. 21-9 Idealized picture transmission amplitude characteristics for all UHF channels with power output of 1 kW or less. *(FCC.)*

totally vestigial in nature for stations operating on channels 14 to 83 and having a maximum peak visual power output of 1 kW or less. In this case, the response may be in accordance with Fig. 21-9 (Sec. 73.699, Fig. 5a of the FCC Regulations).

The distribution of the carriers within the channel follows a general pattern in all cases. With the exceptions as noted in Fig. 21-9 and the British 405-line system (CCIR System A), the lower sideband of the picture signal is attenuated, resulting in a vestigial sideband format for radiation of the signal. The visual carrier is located near the vestigial end of the channel, while the aural carrier is located close to the opposite band edge. Signals carrying chrominance information exist as subcarriers within the space occupied by the luminance complete sideband and usually are placed as near the sound carrier end of the channel (high baseband frequencies) as the system modulation parameters allow. This arrangement is followed because it minimizes the potential for interference between the various carriers, both within the same channel and in the immediately adjacent channels.

The values for the channel characteristics are given in detail for all 13 CCIR-designated systems being used throughout the world in Table 21-5 (CCIR Recommendation 470-1, Report 624-2). Three of these sytems (A, C, and E) are not recommended by the CCIR for adoption in countries setting up a new television service, although they are still in operation in some countries.

In all systems, amplitude modulation (A5C) is employed for the visual signal and with negative modulation polarity in all cases except systems A, C, L, and E. (*Negative modulation* is defined as increasing scene light intensity causing a decreasing output carrier level.)

The aural carrier is FM-modulated (F3) in all systems except A, C, L, and E. The effective radiated power of the aural carrier for System M (that system designation by the CCIR which describes the television broadcast system used in the United States) shall not be less than 10 percent nor more than 20 percent of the peak radiated power of the visual transmitter. These ratios are determined by the desire to provide essentially equal coverage of the service area by both in the interest of economy and minimizing interference effects. The aural to visual carrier power ratios for all the systems used worldwide are given in Table 21-5.

21.3.3. POLARIZATION. The polarization of the electric vector of the radiated signal must be determined for stations assigned and licensed in the broadcast television service.

Generally, throughout the world, television broadcast services have employed horizontal polarization with some exceptions in the United Kingdom and France in past years.

In the United States, it is standard to employ horizontal polarization. However, circular or elliptical polarization is now permitted if desired. In this case, clockwise (right-hand) rotation, as defined in the IEEE Standard Definition 42A65-3E2, and transmission of the horizontal and vertical components in time and space quadrature shall be used. For either omnidirectional or directional antennas, the licensed effective radiated power of the vertically polarized component may not exceed the licensed effective radiated power of the horizontally polarized component.

For directional antennas, the maximum effective radiated power of the vertically polarized component shall not exceed the maximum effective radiated power of the horizontally polarized component in any specified horizontal or vertical direction.

The recent use of circular polarization is favored as a means for minimizing multipath echo (ghost) problems at the receiving antenna.

21.3.4 VISUAL CARRIER MODULATION AND SCANNING STANDARDS.

In broadcast television systems, the basic function of the image pickup apparatus is to analyze the spectral distributions of the light from the scene into tristimulus values in terms of red, green, and blue primary components on a point-by-point basis as determined by a predetermined scanning pattern. The three resulting electrical signals must then be encoded and transmitted to the receiver over a band-limited communications channel. The receiver decodes the signal and controls a three-color display device to make the *perceived* color at the receiver appear essentially the same as the *perceived* color at the scene.

In order to describe and understand the parameters of the television signal that are specified by accepted standards, the format within which these signals are presented must be determined. This format is set forth in the standards which relate to scanning techniques. Worldwide television systems employ a linear line-scan approach to analyze the visual characteristics of the scene element by element. Also, they all specify the scanning pattern, during active scanning intervals, to explore the scene progressively from left to right horizontally and from top to bottom vertically, at uniform velocity and in straight parallel paths, or lines. Interlaced scan requires that the scene be scanned consecutively by a fixed number of lines once (first field), and then scanned a second time (second field) consecutively with these lines interspersed midway between the lines of field 1. Thus, two fields constitute one complete analysis (one frame) of the scene information and will consist of an odd number of total lines. The number of lines per field and per frame is chosen by the interrelationship between a number of factors such as eye acuity at a given angle of view, channel bandwidth limitations, and the visual flicker phenomenon relating repetition rate and brightness. The field and frame rates are also chosen with regard to the above factors but, historically, were also related to the primary power line frequencies employed throughout the modern world of either 60 or 50 Hz for the field rates and resulting in 30 and 25 Hz frame rates, respectively, for 2/1 interlaced systems. Although the field rate to power line frequency relationship is no longer as critical as was once assumed, these rates are still in the acceptable range relative to flicker versus typical television display brightness values while still meeting the constraints of minimizing the required transmission channel bandwidth. The 60-field-per-second rate requires 20 percent more bandwidth than the 50-field-per-second rate for an image of the same overall resolution, but the brightness at which flicker becomes prominent is almost six times as great with the higher field scanning rate. Typically, the flicker threshold for 50-field systems occurs at a highlight brightness of approximately 30 ft·L (103 cd/m²) compared with about 170 ft·L (617 cd/m²) for 60-field systems. The standardized values for the number of lines in the television systems in use in the world today are in the range from about 400 to about 900 lines per frame. These values are intended to permit an interlaced (2/1) image to be viewed from distances of four to eight times the picture height without the line structure being objectionably visible to a *normal* observer. In the United States, the standards require 525 lines, 2/1 interlace format, with a nominal rate of 60 fields per second (59.94 for color), in a 4 by 3 aspect ratio shape factor. In 50-Hz primary power frequency areas of the world, the standards typically require 625 lines, 2/1 interlace format, with 50 fields per second (monochrome and color), in 4 by 3 aspect ratio raster format. A summary of the scanning standards for the worldwide systems is given in Table 21-17.

The specified scanning rates are carefully synchronized and interrelated by a small number of odd factors. In color systems, the horizontal and vertical rates are rigidly locked to (actually divided down from) the color subcarrier frequency.

In the United States, the FCC Rules and Regulations, Part 73.682, state that monochrome transmissions shall comply with the synchronizing waveform specifications in Fig. 7 of Sec. 73.699, reproduced here in Fig. 21-10. Likewise, the FCC regulations state

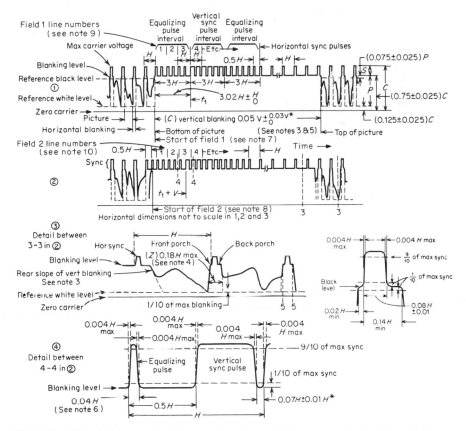

FIG. 21-10 FCC television synchronizing waveforms for monochrome transmission. *Notes:* (1) $H =$ time from start of one line to start of next line. (2) $V =$ time from start of one field to start of next field. (3) Leading and trailing edges of vertical blanking should be complete in less than $0.1H$. (4) Leading and trailing slopes of horizontal blanking must be steep enough to preserve minimum and maximum values of $(x + y)$ and (z) under all conditions of picture content. *(5) Dimensions marked with an asterisk indicate that tolerances are permitted only for long time variations and not for successive cycles. (6) Equalizing pulse area shall be between 0.45 and 0.5 of the area of a horizontal synchronizing pulse. (7) Start of field 1 is defined by a whole line between the first equalizing pulse and preceding H sync pulses. (8) Start of field 2 is defined by a half line between first equalizing pulse and preceding H sync pulses. (9) Field 1 line numbers start with first equalizing pulse in field 1. (10) Field 2 line numbers start with second equalizing pulse in field 2. (11) Refer to text for further explanations and tolerances. *(FCC.)*

Table 21-17 Basic Characteristics of Video and Synchronizing Signals†

Characteristic	CCIR system identification										
	A	M	N	C	B, G	H	I	D, K	K1	L	E
Number of lines per frame	405	525	625	625	625	625	625	625	625	625	819
Number of fields per second	50	60 (59.94)	50	50	50	50	50	50	50	50	50
Line frequency f_H, Hz, and tolerances	10,125	15,750 / 15,734 ($\pm$0.0003%)	15,625 $\pm$0.15% $\pm$0.02%	15,625 $\pm$0.02%	15,625 $\pm$0.02% ($\pm$0.0001%)	15,625 $\pm$0.02% ($\pm$0.0001%)	15,625 ($\pm$0.0001%)	15,625 $\pm$0.02% ($\pm$0.0001%)	15,625 $\pm$0.02% ($\pm$0.0001%)	15,625 $\pm$0.02% ($\pm$0.0001%)	20,475
Interlace ratio	2/1	2/1	2/1	2/1	2/1	2/1	2/1	2/1	2/1	2/1	2/1
Aspect ratio	4/3	4/3	4/3	4/3	4/3	4/3	4/3	4/3	4/3	4/3	4/3
Blanking level, IRE units	0	0	0	0	0	0	0	0	0	0	
Peak-white level	100	100	100	100	100	100	100	100	100	100	100
Sync-pulse level	−43	−40	−40	−43	−43	−43	−43	−43	−43	−43	−43
Picture-black level to blanking level (setup)	0	7.5 $\pm$2.5	7.5 $\pm$2.5	0	0	0	0	0–7	0 color 0–7 mono	0 color 0–7 mono	0–5
Nominal video bandwidth, MHz	3	4.2	4.2	5	5	5	5.5	6	6	6	10
Assumed display gamma	2.8	2.2	2.2	2.8	2.8	2.8	2.8	2.8	2.8	2.8	2.8

†*Notes:* (1) Systems A, C, and E are not recommended by CCIR for adoption by countries setting up a new television service. (2) Values of horizontal line rate tolerances in parentheses are for color television. (3) In the systems using an assumed display gamma of 2.8, an overall system gamma of 1.2 is assumed. All other systems assumed an overall transfer function of unity.

that color transmissions shall comply with the synchronizing waveforms specified in Fig. 6 of Sec. 73.699, reproduced here in Fig. 21-11.

The process of constant velocity line scanning involves a relatively short period of time to allow the scanning mechanism to return from the end of the active scan time to the start of the next scan period (retrace time). In the case of vertical scan, this is the period of time required for the scanning spot to return from the bottom of each field scan to the top of the next field scan. During this period, a small number of line scan times are utilized that do not contain picture information. In current standard systems, this retrace time is termed *vertical blanking* and occupies from 5 to 10 percent of a field time which involves from 13 to 41 lines per field, depending upon the particular system involved. The number of active scanning lines determines the actual vertical resolution of the reproduced image and ranges from 337 lines (405-line systems) to 737 lines (819-line systems).

Likewise, to allow time for the scanning spot in the horizontal direction to return from the right side to the left side of the beginning of the next line, the picture information is blanked (horizontal blanking) for a period of between 15 and 20 percent of one horizontal scan period for the various systems used worldwide.

Tables indicating the details and tolerances for both vertical and horizontal synchronizing and blanking time periods are given in Sec. 21.4, *Worldwide Transmission Standards*. In all cases, the general format of the horizontal blanking period consists of the time period devoid of picture information and containing a horizontal synchronization pulse located forward in time of the center of the blanking period. In color systems, this is followed by a *burst* of a specific number of cycles of the color subcarrier frequency, having a specific reference phase, to synchronize the color decoding process relative to the color encoding process.

The vertical blanking period contains a variety of timing pulses as follows:

1. The vertical synchronization pulse which is *serrated* at two times the horizontal scan rate provides vertical timing information as well as continuous information for maintaining horizontal synchronization throughout the vertical time period that encompasses a number of horizontal line periods.

2. Two sets of *equalizing* pulses which occur at twice horizontal rate, one group preceding and a second group following the serrated vertical synchronization pulse. The purpose of these double-frequency equalization pulses is to ensure the 2/1 interlaced scanning time relationship between the lines of Field 1 relative to the lines of Field 2. The horizontal and vertical synchronization process is continuous with time, and therefore, in the United States 525-line system, there are 262 one-half lines in each of the two interlaced fields that constitute one frame.

The modulation signal values and format for color were developed to provide reciprocal compatibility with monochrome transmission, scanning, and synchronization signals, and transmitted within the constraints of the existing channel characteristics. Thus, one signal (luminance) is chosen in all systems to occupy the wide-band portion of the modulation sideband components and to convey the brightness as well as the detail information content of the scene. A second signal, termed the *chrominance signal,* representative of the chromatic attributes of *hue* and *saturation,* is assigned less channel width in accordance with the principle that, in human vision, full three-color reproduction is not required over the entire range of resolution—commonly referred to as the *mixed-highs principle.*

Another fundamental principle involves arranging the chrominance and luminance signals within the same frequency band without generating excessive mutual interference. Recognition that the scanning process, being equivalent to sampled-data techniques, produces signal components largely concentrated in uniformly spaced groups across the channel width led to the introduction of the concept of horizontal frequency interlace (dot interlace). The color subcarrier frequency is so chosen as to be an odd multiple of one-half the horizontal line rate (in the case of NTSC) such that the phase

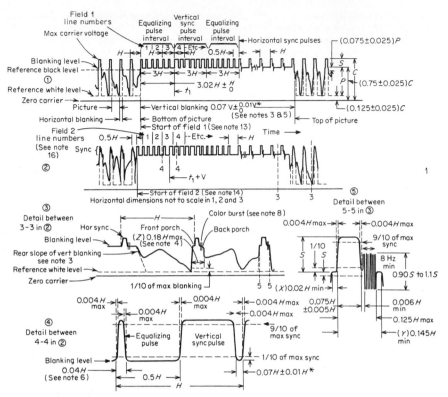

FIG. 21-11 FCC television synchronizing waveforms for color transmission. *Notes:* (1) H = time from start of one line to start of next line. (2) V = time from start of one field to start of next field. (3) Leading and trailing edges of vertical blanking should be complete in less than $0.1H$. (4) Leading and trailing slopes of horizontal blanking must be steep enough to preserve minimum and maximum values of $(x + y)$ and (z) under all conditions of picture content. *(5) Dimensions marked with an asterisk indicate that tolerances are permitted only for long time variations and not for successive cycles. (6) Equalizing pulse duration shall be between 0.45 and 0.55 of the horizontal synchronizing pulse duration. (7) Color burst follows each horizontal pulse but is omitted following the equalizing pulses and during the broad vertical pulses. (8) Color bursts to be omitted during monochrome transmission. (9) The burst frequency shall be 3.579545 MHz. The tolerance on the frequency shall be ± 10 Hz with a maximum rate of change not to exceed 0.1 Hz/s. (10) The horizontal scanning frequency shall be $\frac{2}{455}$ times the burst frequency. (11) The dimensions specified for the burst determine the times of starting and stopping the burst but not its phase. The color burst consists of amplitude modulation of a continuous sine wave. (12) Dimension P represents the peak excursion of the luminance signal from blanking level but does not include the chrominance signal. Dimension S is the synchronizing pulse amplitude above blanking level. Dimension C is the peak carrier amplitude. (13) Start of field 1 is defined by a whole line between first equalizing pulse and preceding H sync pulses. (14) Start of field 2 is defined by a half line between the first equalizing pulse and the preceding H sync pulses. (15) Field 1 line numbers start with the first equalizing pulse in field 1. (16) Field 2 line numbers start with second equalizing pulse in field 2. (17) Refer to text for further explanations and tolerances. (18) During color transmissions, the chrominance component of the picture signal may penetrate the synchronizing region and the color burst penetrate the picture region. *(FCC.)*

of the subcarrier alternates by exactly 180° on successive scanning lines. This substantially reduces the subjective visibility of the color-signal dot-pattern components.

The chrominance subcarrier frequency is 63/88 times precisely 5 MHz and equals 3.57954545 ... MHz. The tolerance is ±10 Hz, and the rate of frequency drift must not exceed 0.1 Hz/s/. The standards specify 525 scanning lines per frame, interlaced 2/1 on successive fields. Therefore, the horizontal scanning frequency is 2/455 times the chrominance subcarrier frequency. Two factors are involved in calculating the precise horizontal and vertical scanning rates relative to the color subcarrier frequency. The first factor is the odd multiple of one-half horizontal line frequency, and the other factor involves the interlace of the approximately 920-kHz beat between the unmodulated aural carrier frequency of 4.5 MHz and the color subcarrier frequency of 3.579545 MHz.

Since the picture signal energy distribution versus frequency is grouped into intervals having horizontal scan rate spacing (15.734 kHz), various prime numbers may be chosen to produce odd multiples of one-half line rate. The particular choice in the vicinity of 3.6 MHz was made for several reasons. First, the high frequency resulted in a fine interference pattern having low visibility because of small spatial dimensions. Second, this value allows about 0.5 MHz of double-sideband frequency range for the color signal sideband components, in accordance with the sound carrier being located at 4.5 MHz (±1000 Hz). The choice of the exact frequencies is

$$f_{\text{line}} = \frac{4.5 \times 10^6}{286} = 15{,}734.26 \text{ Hz}$$

$$f_{\text{field}} = \frac{f_{\text{line}}}{525/2} = 59.94 \text{ Hz}$$

$$f_{\text{sc}} = \frac{13 \times 7 \times 5}{2} \times f_{\text{line}} = 3.57954545 \dots \text{MHz}$$

These values for both horizontal and vertical scanning rates are compatible with the previously existing monochrome standards in that they fall within the tolerances allowed and represent a variation of no more than one part per thousand.

The formation of the luminance Y signal is formed by linear addition of specific proportions of the primary red, green, and blue signal values. The luminance signal should consist of values representative of the brightness sensation of the human eye. Therefore, the red, green, and blue components are adjusted in proportion to the standard luminosity response curve for the particular values of dominant wavelength represented by the individual three color primaries chosen for color television. Thus, the standard gamma-corrected luminance signal values are determined to be as follows

$$E'_Y = 0.30 \, E'_R + 0.59 \, E'_G + 0.11 \, E'_B$$

The color-picture signal corresponds to a luminance component transmitted as amplitude modulation of the visual carrier and a simultaneous pair of chrominance components, E'_I and E'_Q, transmitted as amplitude-modulation sidebands or at least two suppressed subcarrier modulators—the subcarriers having the same frequency but being in phase quadrature.

Thus, the expression for the composite signal is specified by

$$E_M = E'_Y + [E'_Q \sin (\omega t + 33°) + E'_I \cos (\omega t + 33°)]$$

where $E'_Q = 0.41(E'_B - E'_Y) + 0.48(E'_R - E'_Y)$
$E'_I = -0.27(E'_B - E'_Y) + 0.74(E'_R - E'_Y)$
$E'_Y = 0.30E'_R + 0.59E'_G + 0.11E'_B$

Expressions for E'_Q and E'_I in terms of R, G, and B are as follows

$$E'_Q = -0.522E'_G + 0.211E'_R + 0.311E'_B$$

$$E'_I = -0.274E'_G + 0.596E'_R - 0.322E'_B$$

Also, expressions for the color-difference signals of $B - Y$, $R - Y$, and $G - Y$ are shown below

$$\frac{E'_B - E'_Y}{2.03} = -0.545E'_I + 0.839E'_Q$$

$$\frac{E'_R - E'_Y}{1.141} = 0.839E'_I + 0.545E'_Q$$

$$\frac{E'_G - E'_Y}{0.703} = -0.399E'_I - 0.900E'_Q$$

Also $E'_Y = 0.589E'_G + 0.299\ E'_R + 0.114E'_B$

The following comments are pertinent to these signal value expressions:

1. E_M is the total signal voltage at a particular instant in time as applied to the modulator terminals of the transmitter.

2. E'_Y is the gamma-corrected (prime indicates the particular nonlinear exponent recommended) luminance component which is composed of a linear summation of individually gamma-corrected R', G', and B' primary signals derived from the camera.

3. E'_Q and E'_I are the amplitudes of the gamma-corrected signals representing the specific values of the orthogonal components of the chrominance signal corresponding to the narrow-band and wide-band axes, respectively. These chrominance signal values are formed by a linear matrix of the initial $\pm R'$, $\pm G'$, and $\pm B'$ signal values.

4. After individual amplitude modulation of the suppressed chrominance subcarrier by E'_I and E'_Q in two independent modulators phased in quadrature, the sum products are selected and linearly summed to form the composite, suppressed carrier chrominance signal whose instantaneous phase (relative to a zero reference) is indicative of scene *hue*, and whose instantaneous amplitude, relative to the corresponding instantaneous luminance value, is representative of the *saturation* of the scene.

5. The zero phase reference is provided in the form of a burst of eight or nine cycles of the color subcarrier frequency having a phase of $\pm 180°$ (cartesian coordinate system) and occurring at horizontal line rate immediately following each horizontal synchronizing pulse.

An additional reason for the choice of signal values in the NTSC system is that the eye is more responsive to spatial and temporal variations in luminance than it is to variations in chrominance. Therefore, the visibility of luminosity changes due to random noise and interference may be reduced by properly proportioning the relative chrominance gain and encoding angle values with respect to the luminance values. Thus, the *principle of constant luminance* is incorporated into the standards. The *IEEE Standard Dictionary of Electrical and Electronics Terms* notes that in constant-luminance transmission, the sole control of luminance is provided by the luminance signal, with no luminance provided by the chrominance signal. Noise in the chrominance band limits will, therefore, cause only chromatic variations as opposed to luminosity changes and thereby be less visible.

The choice of the I and Q color modulation components relates to the variation of color acuity characteristics of human color vision as a function of the field of view and spatial dimensions of objects in the scene. The color acuity of the eye decreases as the size of the viewed object is decreased and thereby occupies a small part of the field of view. Small objects, represented by frequencies above about 1.5 to 2.0 MHz, produce little, if any, color sensation. This principle was introduced as the use of *mixed highs* for reproducing fine detail of luminance values. Intermediate spatial dimensions (approximately 0.5 to 1.5 MHz range) are perceived satisfactorily if reproduced along a preferred orange-cyan color axis. Larger objects (0 to 0.5 MHz) require full three-color reproduction for subjectively pleasing results. Thus, the I and Q signal bandwidths are chosen accordingly, and the preferred colorimetric reproduction axis is obtained when only the

I signal exists by rotating the subcarrier modulation vector angles by 33°, maintaining a quadrature relationship between *I* and *Q* vectors.

At the encoder, therefore, the *I* signal components have a bandwidth of about 1.5 MHz and contain the preferred, high-acuity, orange-cyan color axis information. Likewise, the *Q* signal is restricted to about 0.6 MHz bandwidth and is representative of the green-purple color-axis information.

The FCC specifications for the *I* and *Q* signal response are as shown below:

Q-Channel Bandwidth
  At 400 kHz, less than 2 dB down
  At 500 kHz, less than 6 dB down
  At 600 kHz, at least 6 dB down

I-Channel Bandwidth
  At 1.3 MHz, less than 2 dB down
  At 3.6 MHz, at least 20 dB down

A vector diagram representative of the chrominance subcarrier signal is shown in Fig. 21-12 (FCC Sec. 73.699, Fig. 8).

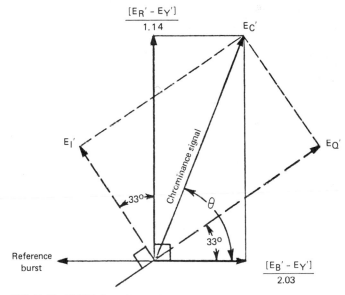

FIG. 21-12   NTSC chrominance signal vector diagram. *(FCC.)*

A pictorial representation of the *Y, I,* and *Q* signal bandwidths in both the baseband and radio-frequency domains is shown in Fig. 21-13 *a* and *b.*

In the NTSC systems adopted in the United States, the gamma-corrected voltages representing E'_R, E'_G, and E'_B are suitable for a display device having primary-color coordinates in the CIE system as listed in Table 21-18 and shown in the chromaticity diagram of Fig. 21-14.

The FCC standards specify a transfer gradient (gamma exponent) of 2.2 associated with each primary color. The *prime* notation in the signal value equations indicates values of $E_R^{1/2.2}$, $E_G^{1/2.2}$, and $E_B^{1/2.2}$, although other forms are allowed and no tolerance is assigned to the value of 2.2 at present.

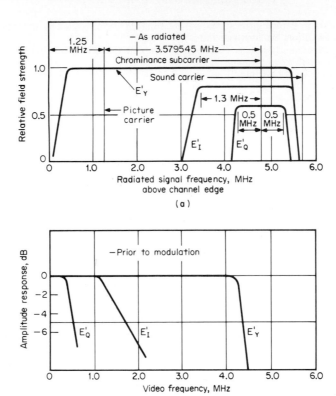

FIG. 21-13   NTSC *Y*, *I*, and *Q* relative bandwidths: (*a*) RF domain; (*b*) baseband video domain. (*D. G. Fink, Color Television Standards, McGraw-Hill, New York, 1955.*)

Table 21-18   NTSC Primary Color Coordinates

	CIE coordinates	
Color	*x*	*y*
Red	0.67	0.33
Green	0.27	0.71
Blue	0.14	0.08

Note: The color coordinates for other systems are different and are indicated in Sec. 21.4, *Worldwide Transmission Standards.*

Another colorimetric specification relates to the reference white point which is designated as being illuminant C whose CIE coordinates are $x = 0.310, y = 0.316$. Recent practice has been to adjust the studio monitor to a white value equivalent to D6500 (a practice recommended for receivers as well). This practice aids in providing colorimetric

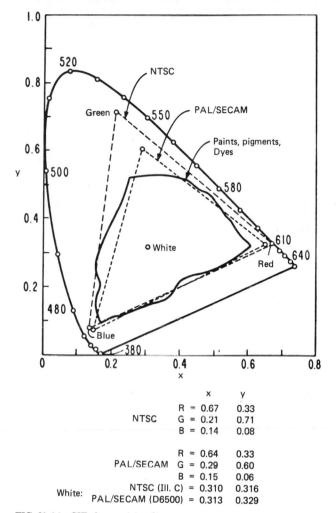

		x	y
	R =	0.67	0.33
NTSC	G =	0.21	0.71
	B =	0.14	0.08
	R =	0.64	0.33
PAL/SECAM	G =	0.29	0.60
	B =	0.15	0.06
White:	NTSC (Ill. C) =	0.310	0.316
	PAL/SECAM (D6500) =	0.313	0.329

FIG. 21-14 CIE chromaticity-diagram comparison of systems.

reproduction fidelity since the standards specify that the individual R, G, and B signal levels at the camera be adjusted to produce zero color subcarrier in the presence of *white* in the televised scene, regardless of the particular coordinates of the studio illumination. Thus, the colorimetric reproduction on a monitor adjusted for illuminant C (now D6500) will be perceived to be equivalent to that of a scene being illuminated with D6500. (The subjective difference between illuminant C and D6500 is considered negligible.)

An additional FCC specification relating to luminance and chrominance signals is that the E'_Y, E'_Q, and E'_I signal components shall match each other in time to 0.05 μs.

Specific tolerances for the radiated signal are assigned to the chrominance signal phase and amplitude values, with respect to the reference burst, when reproducing saturated primaries at 75 percent of full amplitude as follows:

Characteristic	Tolerance
Phase	$\pm 10°$
Amplitude	$\pm 20\%$

In addition, the ratio of the measured amplitude to the amplitude of the corresponding luminance value (this ratio representing saturation) shall be between the limits of 0.8 and 1.2 of the values specified for their ratios.

21.3.5 AURAL CARRIER MODULATION STANDARDS. The FCC specifies frequency modulation (F3) of the aural carrier associated with the visual carrier for television broadcasting service. Some other systems in use by other administrations in the world employ amplitude modulation for audio (see Table 21-5), but most employ FM techniques.

FM has the advantage that (1) the service area for a given signal-to-noise ratio may be provided with a relatively low transmitted carrier power, and (2) when combined with a visual AM signal, it permits the use of *intercarrier* operation of the receiver which may have economic advantages.

The maximum deviation for peak modulation is chosen as a compromise between signal-to-noise and bandwidth/stability constraints at the receiver. The United States system employs $\pm$ 25 kHz, while most of the systems in other parts of the world use $\pm$ 50 kHz (see Fig. 21-5). Both these values are less than the typical value of $\pm$ 75 kHz for FM broadcast service but are adequate to provide a service area approximately matching the service area of the associated visual signals.

As is customary in all FM modulation systems involving speech and music, the energy relationships between low- and high-frequency sound energy and signal-to-noise characteristics make it appropriate to preemphasize the high-frequency components prior to modulation and provide a complementary deemphasis at the audio detector of the receiver. The preemphasis characteristic is stated in terms of impedance versus frequency that results from a combination of inductive reactance L and pure resistance R expressed as an L/R time constant in microseconds. The deemphasis is an RC time involving capacitive reactance and a pure resistance. The United States system specifies a time constant of 75 μs, while many other systems use 50 μs. A standard preemphasis curve is specified by the FCC and is included in Sec. 21.5, *Equipment Standards* (FCC Sec. 73.699, Fig. 12). The tolerances, limits on harmonic distortion, noise, and residual amplitude modulation are also summarized in Sec. 21.5.

Multiplexing of the aural carrier may be permitted for purposes of transmitting telemetry and alerting signals from the transmitter site to the television station control point when used for remote control. The conditions for this special case are detailed in FCC Sec. 73.682, Subparagraph 23, and generally are aimed at providing the desired service without causing observable degradation of the associated visual or aural signals.

21.3.6 TEST AND MONITORING SIGNAL STANDARDS. The FCC Rules and Regulations specify certain signals that may be used for modulating the transmitter for test and monitoring purposes. These signals may coexist with the broadcast of normal picture information and are permitted to be inserted in the interval of vertical blanking beginning with line 17 and continuing through line 21 of each field.

Test signals may include signals used to supply reference modulation levels so that light intensity variations will be faithfully transmitted, and certain signals designed to check the performance of the overall transmission system or its components.

The modulation by these signals shall be confined between reference white and blanking level except in certain cases relating to chrominance subcarrier excursions. In no case shall the signals extend beyond the peak of sync, or to zero carrier level. The use of these test and cue signals shall not degrade or impair the normal program material being transmitted nor produce emission outside the normal assigned channel specifications, and no test signals are permitted to extend into horizontal blanking period.

The accompanying table describes the test signals permitted on specific lines of the vertical interval.

Line number	Signal format
(1) 17 (field 1)	Multiburst test signal† (Fig. 21-15).
(2) 17 (field 2)	Color-bar test signal† (Fig. 21-16).
(3) 18 (field 2)	Composite radiated signal† (a) Modulated stairstep, (b) $2T$ pulse,§, (c) $12.5T$ pulse, (d) White bar (Fig. 21-17).
(4) 19 (each field)	Devoted exclusively to the vertical-interval reference (VIR) signal [Fig. 21-18 (FCC Sec. 73.699, Fig. 16)].
(5) 21 (field 1), one-half of 21 (Field 2)	Program-related data signal which is related to the aural channel information; format is as per Fig. 21-19a (FCC Sec. 73.699, Fig. 17a).‡
(6) 21 (every eighth frame)	Pulse for adaptive multipath equalizer decoder; format as per Fig. 21-19b (FCC Sec. 73.699, Fig. 17b).‡
(7) 21 (field 2)	A decoder test signal representing alphanumeric characters unrelated to program material; a framing code may be inserted during the first half of line 21 with a format as shown in Fig. 21-19c (FCC Sec. 73.699, Fig. 17c).‡

†Mandatory broadcast of vertical-interval test signals (VITS) no longer required.
‡Items (5) to (7) are correct at present. However, on the basis of pending notice of rule making, they may be deleted and replaced by new information relating to *Teletext* (see Chap. 14, Sec. 10).
§T pulse = $\sin^2$ pulse with a half-amplitude duration of 125 ns ($\frac{1}{2}f_c$).

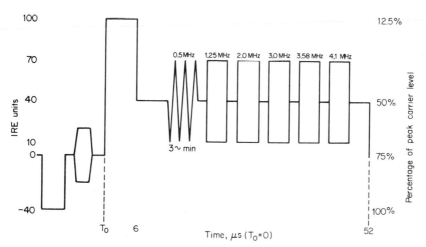

FIG. 21-15 Multiburst test signal (VITS) (field 1, line 17). *Notes:* (1) A breezeway, as shown between bursts, is recommended. Each burst equals 60 IRE units peak to peak. (2) T_0 is the nominal start of active portion of line 17, field 1. (3) Rise and fall of white box shall have rise time of not less than 0.2 μs. *(FCC.)*

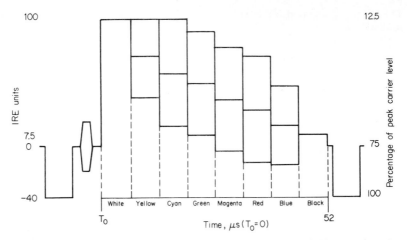

FIG. 21-16 Color-bar test signal (VITS) (field 2, line 17). *Notes:* (1) Phases and amplitudes of the colored bars are in accordance with the FCC Rules and Regulations for 100 percent saturated colors of 75 percent amplitude. (2) White flag (bar) at reference white precedes the colored bars. (3) A black bar at setup level follows the colored bars. (4) Each bar 6-μs minimum duration. (5) T_0 is the nominal start of active portion of line 17, field 2. *(FCC.)*

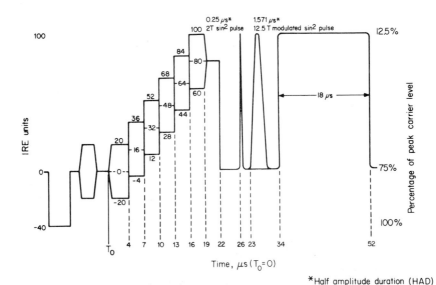

*Half amplitude duration (HAD)

FIG. 21-17 Composite radiated signal (VITS) (field 1, line 18). *Notes:* (1) Subcarrier of staircase in phase with burst. (2) Rise and decay of all luminance signals not less than 0.2 μs. Rise and decay of envelope of subcarrier component of stairstep signal shall be **approximately** sin^2-shaped and rise time 0.375 μs. (3) T_0 is the nominal start of active portion of line 18. *(FCC.)*

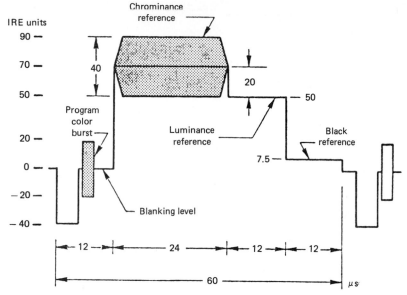

FIG. 21-18   Vertical-interval reference (VIR) signal. *Note:* The chrominance reference and the program color burst have the same phase. *(FCC.)*

The data signal shall be coded using a non-return-to-zero (NRZ) format and employs an ASCII 7-bit plus parity character code. When line 21 is not being used to transmit program-related data signals, it may be used for data signals unrelated to the program content subject to restrictions as to the nature of the information contained within the data format (see FCC Sec. 73.682, Subparagraph 22).

21.4 WORLDWIDE TRANSMISSION STANDARDS

The first compatible color television broadcast system to be placed into commercial service was developed in the United States. On Dec. 17, 1953, the FCC approved transmission standards and authorized broadcasters, as of Jan. 23, 1954, to provide regular service to the public. The decision was the culmination of the work of the National Television System Committee (NTSC) upon whose recommendation the FCC action was based. These standards have subsequently been adopted by a number of other countries as are listed later in this section. The previous existence of monochrome standards, as in all subsequent situations, provided a foundation upon which to build the innovative techniques while simultaneously imposing the requirement of compatibility.

The countries of Europe delayed the adoption of a color television system, and in the years between 1953 and 1967 a number of systems were devised that were compatible with the existing 625-line, 50-field monochrome systems. Many of the basic techniques of NTSC for incorporating the additional chromatic information within the same channel limits as the existing monochrome system also were considered in these approaches. For example, wide-band luminance and relatively narrow-band chrominance involving modulation of an appropriate color subcarrier are common elements.

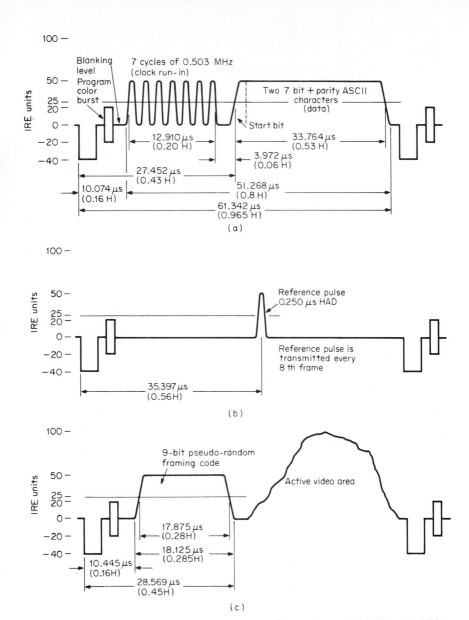

FIG. 21-19 Equipment standards, FCC Rules and Regulations, Sec. 73.699. (*a*) Line 21, field 1 data signal format. (*b*) Adaptive equalizer reference pulse. (*c*) Line 21, field 2 framing code. Horizontal dimensions not to scale. *Notes:* (1) Data 1 = 50 IRE units, data 0 = 0. (2) Data pulse rise time = 2 − *T* bar rise time. (3) Data time base = $32f_H$ (0.503496450 MHz). (4) Data bit interval = $H/32$ (1.986 μs). (5) Negative-going zero crossings of clock are coherent with data transitions. (6) Data and clock run-in coherent with *H*. (*FCC.*)

An early system that received approval was one proposed by Henri de France of the Compagnie de Télévision of Paris. Two pieces of coloring information (hue and saturation) that need to be added to the monochrome (luminance) information are transmitted as subcarrier modulation which is sequentially transmitted on alternate lines. Thus, at the receiver, a one-line memory, commonly referred to as a 1H delay element, must be employed to store one line to be concurrent with the following line. Then, a linear matrix can provide the recovered red, green, and blue signals. Of necessity, a line-switching identification technique must be incorporated. Such an approach, termed SECAM (Sequential Couleur Avec Mémoire, or sequential color with memory), was developed and officially adopted by France and the USSR, and broadcast service began in France in 1967.

The technique of a 1H delay element led to another system development, by Walter Brúch of the Telefunken Company of Germany, the phase alternation line (PAL) system. This approach was aimed at overcoming an implementation problem of NTSC that requires a high order of phase and amplitude integrity (skew symmetry) of the total signal path around the color subcarrier to prevent color quadrature distortion. Thus in PAL the line-by-line alternation of the phase of one of the color signal components results in averaging of the colorimetric distortions to the observer to that of the nominal value. The PAL system transmission standards have been adopted by numerous countries in continental Europe as well as the United Kingdom, and public broadcasting began in 1967 in Germany and the United Kingdom. Although the basic PAL system transmission standards are common, differences do exist in receiver decoding techniques, luminance bandwidth and sound-carrier-to-visual-carrier spacings, and other details in various countries.

21.4.1 NTSC, PAL, SECAM DIFFERENCES AND SIMILARITIES.

The major differences among the three main color television transmission standards of NTSC, PAL, and SECAM are in the specific modulating processes used for encoding the chrominance information. The main areas of similarity, and differences, are listed below:

1. All systems
   a. Use three primary additive colorimetric principles
   b. Use similar camera pickup and display technology
   c. Employ wide-band luminance and narrow-band chrominance
2. All are compatible with coexisting monochrome systems
   First-order differences are therefore
   a. Line and field rates
   b. Component bandwidths
   c. Frequency allocations
3. Major differences lie in color-encoding techniques
   a. NTSC: Simultaneous amplitude and phase quadrature modulation of an interlaced, suppressed subcarrier
   b. PAL: Similar to NTSC but with line alternation of one color-modulation component
   c. SECAM: Frequency modulation of line-sequential color subcarrier(s)

In all three systems, the luminance signal Y is formed in the same manner by the linear addition of specific proportions of the gamma-corrected R, G, and B signals. Signals representative of the chromatic information (hue and saturation) that relate to the difference between the luminance signal and the basic R, G, and B signals are generated by a linear matrix. This new set of signals, termed *color-difference* signals, and designated as $R - Y$, $G - Y$, and $B - Y$, are used to modulate the color subcarrier. The results are combined with luminance and the synchronizing signals and passed through a common communications channel.

The expressions for the luminance (common to all systems) and the chromaticity components are shown below:

Luminance:

$$E'_Y = 0.299E'_R + 0.587E'_G + 0.114E'_B$$

Chrominance:

$$NTSC: E'_I = 0.274E'_G + 0.596E'_R - 0.322E'_B$$

$$E'_Q = -0.522E'_G + 0.211E'_R + 0.311E'_B$$

$$B - Y = 0.493(E'_B - E'_Y) \qquad R - Y = 0.877(E'_R - E'_Y)$$

$$G - Y = 1.413(E'_G - E'_Y)$$

$$PAL: E'_U = 0.493(E'_B - E'_Y) \qquad E'_V = \pm 0.877(E'_R - E'_Y)$$

$$SECAM: D'_R = -1.9(E'_R - E'_Y) \qquad D'_B = 1.5(E'_B - E'_Y)$$

In the NTSC system, the total chroma signal expression is given by

$$C_{\text{NTSC}} = \frac{B - Y}{2.03} \sin \omega_{\text{SC}}t + \frac{R - Y}{1.14} \cos \omega_{\text{SC}}t$$

The PAL chroma signal expression is

$$C_{\text{PAL}} = \frac{U}{2.03} \sin \omega_{\text{SC}}t \pm \frac{V}{1.14} \cos \omega_{\text{SC}}t$$

where U and V are substituted for $B - Y$ and $R - Y$, respectively, and the V component is alternated 180° on a line-by-line basis.

The bandwidths of I and Q signals of the NTSC system differ in accordance with the preferred color-acuity axis (Figs. 21-16 and 21-17), whereas both U and V in PAL are equal bandwidths and approximately equal to the I signal bandwidth of NTSC.

21.4.2 PAL SYSTEM STANDARDS. Figure 21-20 is a vector diagram of the PAL quadrature modulated and line-modulating color modulation approach (CCIR Reg. 624-2, Fig. 4).

The PAL transmission system must provide the receiver with some means by which the line-switching sequence may be identified. This is accomplished by a technique known as *AB* sync, *PAL* sync, or *swinging burst* and consists of alternating the phase of the burst by ±45° at a line rate. The switched V component is representative of the switched V picture content and provides the necessary line identification sense. The fixed $(-U)$ component provides the reference for color subcarrier synchronization.

In all systems, the burst signal is eliminated during vertical synchronization time. In the line-alternation PAL system, some means must be provided for ensuring that the phase of the burst is the same for the first burst following each vertical sync pulse on a field-by-field basis. Thus, the burst reinsertion time is shifted by one line at the vertical field rate by a pulse referred to as the *meander gate*. Figure 21-21 (CCIR Rep. 624-2, Fig. 5a and b) shows this process.

In the case of the NTSC system, the visibility of the color subcarrier is minimized by causing its frequency to be an odd multiple of one-half horizontal line rate (interlaced). In the PAL system, since the V component phase is alternated 180° at a line rate, this same choice of frequency relationship would result in interlace of the U component and noninterlace of the V component. Thus, the color subcarrier is offset by one-quarter line as per the following expression

$$f_{\text{SC}} = \frac{1135}{4} f_H + \frac{1}{2} f_V$$

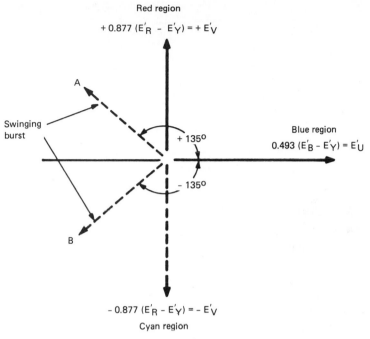

Red region

$+0.877 (E'_R - E'_Y) = +E'_V$

Blue region

$0.493 (E'_B - E'_Y) = E'_U$

Swinging burst

A

B

+ 135°

- 135°

$-0.877 (E'_R - E'_Y) = -E'_V$

Cyan region

FIG. 21-20 PAL color-modulation vector diagram. *(FCC.)*

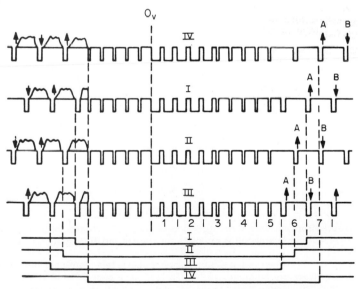

O_V

IV

I

II

III

A   B

A   B

A   B

A   B

| 1 | 2 | 3 | 4 | 5 | 6 | 7 |

I

II

III

IV

FIG. 21-21 Meander burst-blanking gate-timing diagram for systems B, G, H, and PAL/I. *(CCIR.)*

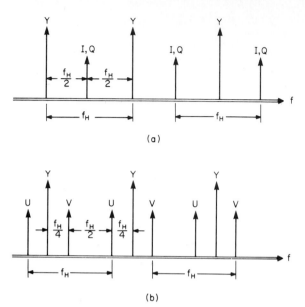

FIG. 21-22 NTSC and PAL frequency-interlace relationship. (a) NTSC ½H interlace (four fields for complete color frame). (b) PAL ¼H and ¾H offset (eight fields for complete color frame).

The 25-Hz additional offset provides motion of the dot pattern, and eight complete fields are required in PAL before a specific picture element dot position is repeated. Figure 21-22 is a diagram comparing the one-half line offset of NTSC and the one-fourth line offset of PAL.

21.4.3 SECAM SYSTEM STANDARDS. The SECAM color modulation system uses the same luminance signal values as does NTSC and PAL. Historically, the color-modulation method used for SECAM has changed through at least three versions, SECAM I, II, and III. The present system for the transmission standards, known as the *optimized* or SECAM III system, transmits the $R - Y(D_R)$ and $B - Y(D_B)$ color-difference signals as frequency modulation of two different subcarriers transmitted alternately in time sequence from one successive line to the next, with the luminance being common to every line. Since there is an odd number of total lines (625), any given line carries D_B information during one field and D_B information during the next field. The undeviated frequencies for the two subcarriers, respectively, are determined by

$$f_{OB} = 272F_H = 4.250000 \text{ MHz}$$

$$f_{OR} = 282F_H = 4.406250 \text{ MHz}$$

The accepted convention for the direction of frequency change with respect to the polarity of color-difference signal is opposite for the D_B and D_R signals, as may be seen in Fig. 21-23.

As in all frequency-modulation systems, preemphasis techniques are desirable. Two types of preemphasis are employed in SECAM. First, a conventional type of low-frequency preemphasis, as shown in Fig. 21-24, is employed with a break point reference level at 85 kHz and a maximum emphasis at about 750 kHz. The expression for this curve

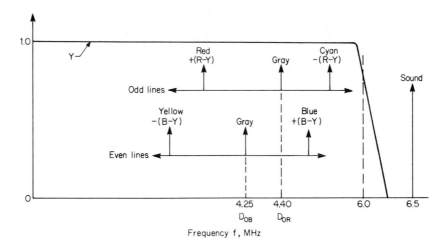

FIG. 21-23 SECAM FM color-modulation system.

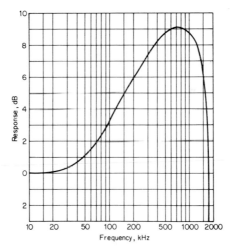

FIG. 21-24 SECAM color-signal low-frequency preemphasis. *(CCIR Rep. 624-2.)*

is given below

$$A = \left| \frac{1 + j(f/f_1)}{1 + j(f/3f_1)} \right|$$

where f = signal frequency (kHz) and f_1 = 85 kHz.

Second, another form of preemphasis is introduced where the amplitude of the subcarrier is changed as a function of the frequency deviation. The expression for this inverted *bell-shaped* characteristic is given in Table 21-19 and the characteristic is plotted in Fig. 21-25.

Table 21-19 Chrominance Subcarrier Amplitude

M/NTSC	M/PAL; B, G, H, I PAL	B, D, G, H, K, K1, L; SECAM†
$G = \sqrt{E_I'^2 + E_Q'^2}$	$G = \sqrt{E_U'^2 + E_V'^2}$	$G = M_0 \left\| \dfrac{1 + j16F}{1 + j1.26F} \right\|$

†Where $F = f/f_0 = f_0/f$, $f_0 = 4286$ kHz, f = instantaneous subcarrier frequency, and $2M_0 = 23 \pm 2.5\%$ of luminance amplitude

In order to reduce the visibility of the frequency-modulated subcarriers, the phase of the subcarriers (phase in this system carries no picture information) is reversed 180° on every third line and between each field. This coupled with the bell preemphasis, produces a degree of compatibility considered subjectively acceptable. Figure 21-26 indicates the line-switching sequence in the SECAM system.

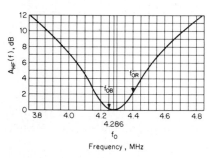

FIG. 21-25 SECAM high-frequency subcarrier preemphasis. *(CCIR Rep. 624-2.)*

As in PAL, the SECAM system must provide some means for identifying the line-switching sequence. Also, as in PAL, the specific receiver decoding process can have variations such as in SECAM V and SECAM H (not to be confused with SECAM transmission standards). At present, the CCIR Recommendations include special identification signals, termed *bottle* signals, transmitted during nine lines of vertical blanking following vertical sync, to provide the necessary line-switching sequence information.

The details of these line identification signals are shown in Fig. 21-27, with Fig. 21-28 indicating the relative position of these signals in the vertical blanking period. Also in Fig. 21-28 is an indication of the burst inserted during horizontal blanking of f_{OB}/f_{OR} that is used as a gray-level reference for the FM discriminators to establish the proper operation point at the beginning of each line. Recent practice has been to eliminate the vertical period bottle signals and to utilize the horizontal rate f_{OR}/f_{OB} burst to derive the switching synchronization information (SECAM H demodulation approach).

SECAM Line Sequential Color			
Field	Line #	Color	Subcarrier ϕ
Odd (1)	n	f_{OR}	0°
Even (2)		$n + 313$ .... f_{OB}	180°
Odd (3)	$n + 1$	f_{OB}	0°
Even (4)		$n + 314$ ...... f_{OR}	0°
Odd (5)	$n + 2$	f_{OR}	180°
Even (6)		$n + 315$ .... f_{OB}	180°
Odd (7)	$n + 3$	f_{OB}	0°
Even (8)		$n + 316$ ...... f_{OR}	180°
Odd (9)	$n + 4$	f_{OR}	0°
Even (10)		$n + 317$ .... f_{OB}	0°
Odd (11)	$n + 5$	f_{OB}	180°
Even (12)		$n + 318$ ...... f_{OR}	180°

FIG. 21-26 Color versus line-and-field timing relationship for SECAM. *Notes:* (1) Two frames (four fields) for picture completion; (2) subcarrier interlace is field-to-field and line-to-line of same color.

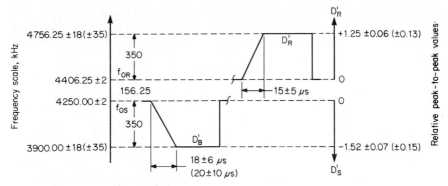

FIG. 21-27 SECAM bottle signal details. *(CCIR.)*

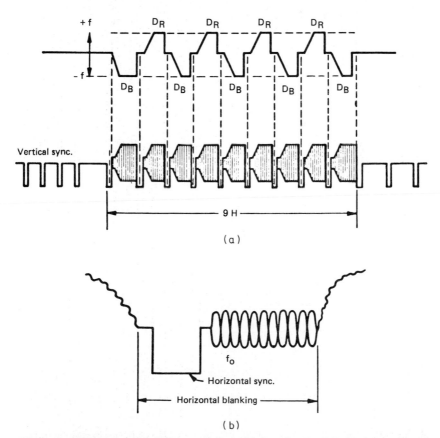

(a)

(b)

FIG. 21-28 SECAM line-identification technique: (*a*) line-identification (bottle) signal; (*b*) horizontal-blanking interval. *(CCIR.)*

21.4.4 CCIR RECOMMENDATIONS SUMMARY. Tables 21-20 to 21-23 summarize the pertinent details of the CCIR-designated transmission standards for horizontal and vertical synchronization pulses and the NTSC, PAL, and SECAM color-modulation standards in use worldwide. Reference is made to the CCIR Report 624-2 for additional details.

The general format of the field-synchronizing waveforms for the various systems are shown in Figs. 21-29 to 21-31.

It is important to note the differences in the precorrection characteristics for the receiver group delay for M/NTSC, M/PAL, B/PAL, and G/PAL systems as indicated in Fig. 21-32 (CCIR Doc. 11/1004-E, Fig. 3). Note (8) for Table III in the CCIR document provides the following additional information regarding the curves for group-delay precorrection and their use in various countries:

In the Federal Republic of Germany and the Netherlands the correction for receiver group-delay characteristics is made according to curve B in Fig. 21-32a. Tolerances are shown in the table under Fig. 21-32a. From Doc. [1966–69] it is learned that Spain uses curve A. The OIRT countries using the B/SECAM and G/SECAM systems use a nominal pre-correction of 90 ns at medium video frequencies. In Sweden, the pre-correction is 0 ± 40 ns up to 3.6 MHz. For 4.43 MHz, the correction is −170 ± 20 ns and for 5 MHz it is −350 ± 80 ns. In New Zealand the pre-correction increases linearly from 0 ± 20 ns at 0 MHz to 60 ± 50 ns at 2.25 MHz, follows curve A of Fig. 21-32a from 2.25 MHz to 4.43 MHz and then decreases linearly to −300 ± 75 ns at 5 MHz.

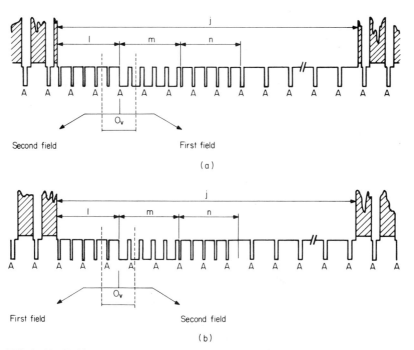

FIG. 21-29 Field-sync waveforms for all systems except M and E: (*a*) beginning of each first field; (*b*) beginning of each second field. *(CCIR.)*

Table 21-20 Details of Synchronization Pulses for Worldwide Systems

Item	A†	M	N	C	E†	B, G, H, I, D, K, K1, L
Nominal line period, μs	98.8	63.555	64	64	48, 84	64
Line blanking, μs	17.5–19	10.9 ± 0.2	10.24–11.52	11.8–12.2	9.2–9.8	12 ± 0.3
Horizontal sync pulse, μs	8–10	4.7 ± 0.1	4.22–5.76	4.8–5.2	2.4–2.6	4.7 ± 0.2
Field period, ms	20	16.6833	20	20	20	20
Field blanking (lines)	$(13\text{–}15.5)\,H + a$	$(19\text{–}21)\,H + a$	$(19\text{–}25)\,H + a$	$25\,H + a$	$33\,H + a$	$25H + a$
Duration of first equalizing pulses	None	$3H$	$3H$	$2.5H$	None	$2.5H$
Duration of second equalizing pulses	None	$3H$	$3H$	$2.5H$	None	$2.5H$
Duration of single equalizing pulse, μs	. . . .	2.3 ± 0.1	2.3–2.56	2.3–2.5	. . . .	2.35 ± 0.1
Duration of field pulse, μs	38–42	27.1	26.52–28.16	26.8–272.	19–21	27.3
Interval between field pulses, μs	11.4–7.4	4.7 ± 0.1	3.84–5.63	4.8–5.2	. . . .	4.7 ± 0.2

†Systems A and E do not use equalizing pulses.

Table 21-22 CCIR Color Systems Characteristics (II)

Item	M/NTSC		M/PAL		B, G, H, I PAL	B, D, G, H, K, K1, L, SECAM	
	x	y	x	y		x	y
Chromaticity coordinates CIE 1931	R 0.67	0.33			R	R 0.64	0.33
	G 0.21	0.71			G	G 0.29	0.60
	B 0.14	0.08			B	B 0.15	0.06
White point	I11 C 0.310 0.316				I11 D6500	I11 D6500 0.313 0.329	
Assumed gamma	2.2				2.8		
Luminance	$E'_Y = 0.299E'_R + 0.587\,E'_G + 0.114E'_B$ (prime denotes gamma-corrected signals)						
Color difference signals	$E'_I = -0.27(E'_B - E'_Y) + 0.74(E'_R - E'_Y)$ $E'_Q = 0.41(E'_B - E'_Y) + 0.48(E'_R - E'_Y)$		$E'_V = 0.493(E'_B - E'_Y)$ $E'_U = 0.877(E'_R - E'_Y)$			$D'_R = -1.9(E'_R - E'_Y)$ $D'_B = 1.5(E'_B - E'_Y)$	

Color difference signal bandwidth	dB	MHz	dB	MHz	dB	MHz	dB	MHz
	E'_I <3	@ 1.3	E'_U <2	@ 1.3	E'_U <3	@ 1.3	D'_R <3	@ 1.3
	<20	@ 3.6	>20	@ 3.6	>20	@ 4	>30	@ 3.5
	E'_Q <2	@ 0.4	E'_V <2	@ 1.3	E'_V <3	@ 1.3	D'_B <3	@ 1.3
	<6	@ 0.5	>20	@ 1.6	>20	@ 4	>30	@ 3.5
	>6	@ 0.6						

Table 21-22 CCIR Color Systems Characteristics (II) (*Continued*)

Item	M/NTSC	M/PAL	B, G, H, PAL	I/PAL	B, D, G, H, K, K1, L SECAM		
Subcarrier frequency, MHz	3.579545 ± 10	$3.575611.49 \pm 10$	$4.43618.75 \pm 5$	$4.33618.75 \pm 1$	$f_{OR} = 4.406250 \pm 2000$ $f_{OB} = 4.250000 \pm 2000$		
f_{SC} multiple of f_H	$f_{SC} = \dfrac{455}{2} f_H$	$f_{SC} = \dfrac{909}{4} f_H$	$f_{SC} = \dfrac{1135}{4} + \dfrac{1}{25} f_H$		$f_{OR} = 282 f_H$ $f_{OB} = 272 f_H$		
Chrominance sidebands or frequency deviation, KHz	$f_{SC} = \begin{cases} +620 \\ -1300 \end{cases}$	$f_{SC} = \begin{cases} +500 \\ -1300 \end{cases}$	$f_{SC} = \begin{cases} +570 \\ -1300 \end{cases}$	$f_{SC} = \begin{cases} +1070 \\ -1300 \end{cases}$	Nominal deviation		Maximum deviation
					$f_{OR} = 280 \pm 9$ $f_{OB} = 230 \pm 7$		$\begin{cases} +350 \pm 18 \\ -506 \pm 25 \end{cases}$ $\begin{cases} +506 \pm 25 \\ -350 \pm 18 \end{cases}$
Start of burst, μs	4.71–5.71 (0.38 μs after trailing edge of sync)	5.8 ± 0.1 after 0_H	5.6 ± 0.1 after 0_H				
Duration of burst, μs + cycles	2.23–3.11 8 or 9 cycles	2.52 ± 0.28 9 ± 1 cycles	2.25 ± 0.23 10 ± 1 cycles				

21.69

Table 21-23 Line-to-Line Chroma Signal Sequence

	Line (N)	Line ($N + 1$)	Line ($N + 2$)	Line ($N + 3$)
NTSC				
Chroma	I, Q	I, Q	I, Q	I, Q
Burst phase	$-(B - Y)$	$-(B - Y)$	$-(B - Y)$	$-(B - Y)$
PAL				
Chroma	$U, +V$	$U, -V$	$U, +V$	$U, -V$
Burst phase	$-U + V = +135°$	$-U - V = +225°$	$-U + V = +135°$	$-U - V = +225°$
SECAM (FM)				
Chroma	$D_R \pm 280$ KHz	$D_B \pm 230$ KHz	$D_R = \pm 280$ KHz	$D_B \pm 230$ KHz
Bursts	D_R deviation = $\begin{cases} +350 \text{ KHz} \\ -500 \text{ KHz} \end{cases}$		D_B deviation = $\begin{cases} +500 \text{ KHz} \\ -350 \text{ KHz} \end{cases}$	

Chroma switch identification lines during vertical interval†

Line no.	7	8	9	10	11	12	13	14	15
	320	321	322	323	324	325	326	327	328
Identification signals	D_R	D_B	D_R	D_B	D_R	D_B	D_R	D_B	D_R

†*Note:* Phase-reversed 180° every third line and every field.

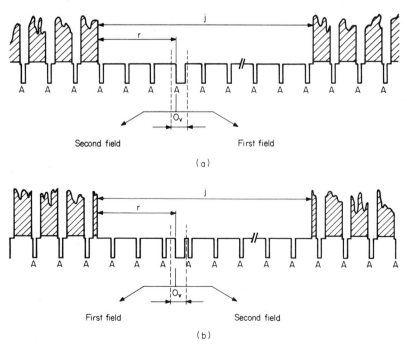

FIG. 21-30 Field-sync waveforms for system E (monochrome): (*a*) beginning of each first field; (*b*) beginning of each second field. *(CCIR.)*

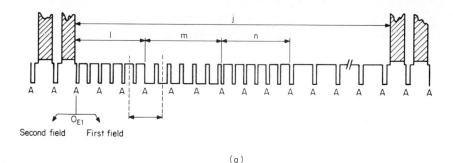

(a)

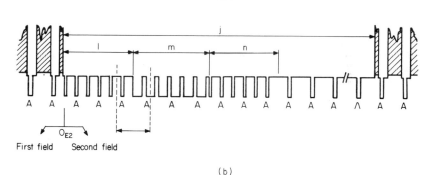

(b)

FIG. 21-31 Field-sync waveforms for system M: (a) beginning of each first field; (b) beginning of each second field. *(CCIR.)*

Table 21-24 is a list of systems in use throughout the world. The list was provided through the courtesy of the RCA Frequency Bureau in a private communication and was compiled from the following sources: *Broadcasting Yearbook, Television Factbook, World Radio Handbook, Television/Radio Age International,* and the CCIR documents. The details of such a list are subject to continual change, and this represents the situation as of February 1982.

21.4.5 DIGITAL STANDARDS FOR TELEVISION STUDIOS. The XVth Plenary Assembly of the CCIR, held in Geneva in 1982, unanimously approved (February 1982) Encoding Parameters for Digital Television for Studios, Recommendation AA/11.† These recommendations are to be used as a basis for digital coding standards for television studio usage in both 525- and 625-line areas of the world and relate to component coding based on the use of one luminance and two chrominance color-difference signals (or, if used, the red, green, and blue signals). The digital coding should allow the establishment of an extensible family of compatible digital coding standards with possible interface between the members of the family.

The family member to be used for interface between studio equipment, video recording, the transmission system, and for international program exchange should be related in the ratio of 4:2:2. A possible higher family member could be the ratio of 4:4:4.

Table 21-25 summarizes these digital coding parameters that represent an internationally agreed upon recommendation for standards.

---

†Recommendation AA/11 will be given a number when the new CCIR volumes are published by the ITU.

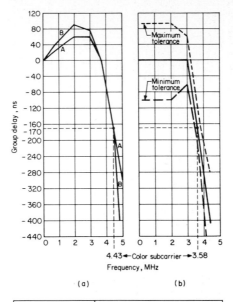

| Frequency, MHz | Nominal values and tolerances, ns | |
	Curve A	Curve B
0·25		+ 5 ± 0
1·00	+ 30 ± 50	+ 53 ± 40
2·00	+ 60 ± 50	+ 90 ± 40
3·00	+ 60 ± 50	+ 75 ± 40
3·75	0 ± 50	0 ± 40
4·43	−170 ± 35	−170 ± 40
4·80	−260 ± 75	−400 ± 90

FIG. 21-32 Group-delay precorrection curves: (*a*) B/PAL and G/PAL systems; (*b*) M/PAL and M/NTSC systems. Color subcarrier for (*a*) and (*b*) at 170-ns group delay. *(CCIR.)*

21.5 EQUIPMENT STANDARDS

The equipment standards set forth the characteristics which are possessed by the transmitting and receiving apparatus that is employed to generate the signals which will ensure conformity with both the transmission and allocation standards. Thus, the equipment and transmission standards are highly interrelated.

The various industry groups such as the Electronics Industries Association (EIA), Society of Motion Picture and Television Engineers (SMPTE), Institute of Electrical and Electronics Engineers (IEEE), and the National Association of Broadcasters (NAB), as well as others, contribute to the equipment standards and to the Recommended Practices which represent good engineering practice.

In addition, the FCC has set up procedures and rules and regulations relating to incidental radiation from television receivers and monitors and has provided for equipment certification procedures. These regulations, standards, and recommended engineering

Table 21-24 Systems† Used in Various Countries‡

Country	Band I/III (VHF)	Band IV/V (UHF)	Country	Band I/III (VHF)	Band IV/V (UHF)
Algeria	B/PAL	G§, H§/PAL	Jamaica	M	....
Angola	I/PAL§		Japan	M/NTSC	M/NTSC
Argentina	N/PAL	N/PAL	Jordan	B/PAL	....
Australia	B/PAL	....	Kenya	B/PAL	
Austria	B/PAL	G/PAL	Korea (South)	M/NTSC	M/NTSC
Bahrain	B/PAL	....	Kuwait	B/PAL	
Bangladesh	B/PAL	....	Lebanon	B/SECAM	....
Barbados	N/NTSC		Liberia	B/PAL	
Belgium	B/PAL	H/PAL	Libya	B/SECAM	....
Bermuda	M/NTSC	....	Luxembourg	C/SECAM	G/PAL
Bolivia	N/NTSC				L/SECAM
Brazil	M/PAL	M/PAL			
Brunei	B/PAL	....	Madeira	B/PAL	....
Bulgaria	D/SECAM	K/SECAM	Malagasy	K1/ SECAM	....
Canada	M/NTSC	M/NTSC			
Canary Islands	B/PAL	....	Malaysia (Fed. of)	B/PAL	
Chile	M/NTSC	M/NTSC			
China (People's Rep.)	D/PAL	K/PAL	Malta	B/PAL	H/PAL
			Martinique	K1/ SECAM	....
Colombia	M/NTSC	M/NTSC			
Costa Rica	M/NTSC	M/NTSC	Mauritius	B/SECAM	....
Cuba	M	M	Mexico	M/NTSC	M/NTSC
Cyprus	B/PAL	G,H§/PAL	Monaco	E/SECAM	G/PAL
Czechoslovakia	D/SECAM	K/SECAM			L/SECAM
Denmark	B/PAL	....	Morocco	B/SECAM	
Djibouti (Rep.)	K1/ SECAM	....	Netherlands	B/PAL	G/PAL
			Netherlands Antilles	M/NTSC	M/NTSC
Dominican Republic	M/NTSC	M/NTSC	New Caledonia	K1/ SECAM	....
Ecuador	M/NTSC	M/NTSC	New Zealand	B/PAL	
Egypt (Arab Rep.)	B/SECAM	G§, H§/ SECAM	Nicaragua	M/NTSC	M/NTSC
El Salvador	M/NTSC	M/NTSC	Niger	K1/ SECAM	....
Equatorial Guinea	B/PAL§		Nigeria	B/PAL	....
Ethiopia	B/PAL§		Norway	B/PAL	G/PAL
Finland	B/PAL	G/PAL	Oman (Sultanate of)	D/PAL	G/PAL
France	E	L/SECAM	Pakistan	B/PAL	
Gabon	K1/SECAM				
Germany (East)	B/SECAM	G/SECAM	Panama	M/NTSC	M/NTSC
Germany (West)	B/PAL	G/PAL	Paraguay	N/PAL	
Ghana	B/PAL§		Peru	M/NTSC	M/NTSC
Gibraltar	B/PAL		Philippines	M/NTSC	M/NTSC
Greece	B/SECAM		Poland	D/SECAM	K/SECAM
Guadeloupe	K1/SECAM		Portugal	B/PAL	G/PAL
Guatemala	M/NTSC	M/NTSC	Qatar	B/PAL	
Haiti	M/NTSC	M/NTSC	Reunion	K1/SECAM	
Honduras	M/NTSC	M/NTSC	Sabah/Sarawak	B/PAL	
Hong Kong	B/PAL	I/PAL	Saudi Arabia	B/SECAM	G/SECAM
Hungary	D/SECAM	K/SECAM	Sierra Leone	B/PAL	
Iceland	B/PAL		Singapore	B/PAL	
India	B		South Africa	I/PAL	I/PAL
Indonesia	B/PAL				
Iran	B/SECAM		Spain	B/PAL	G/PAL
Iraq	B/SECAM				
Ireland	A	I/PAL	St. Kitts	M/NTSC	M/NTSC
Israel	B/PAL	G/PAL	Sudan	B/PAL§	
Italy	B/PAL	G/PAL			
Ivory Coast	K1/ SECAM	....	Surinam	M/NTSC	M/NTSC
			Swaziland	B/PAL	G/PAL

Table 21-24 Systems† Used in Various Countries‡ *Continued*)

Country	Band I/III (VHF)	Band IV/V (UHF)	Country	Band I/III (VHF)	Band IV/V (UHF)
Sweden	B/PAL	G/PAL	United Arab Emirates	B/PAL	G/PAL
Switzerland	B/PAL	G/PAL	United Kingdom	A	I/PAL
Syrian Arab Rep.	B/SECAM		United States	M/NTSC	M/NTSC
Taiwan	M/NTSC	M/NTSC	Uruguay	N/PAL	N/PAL
Tanzania	B/PAL	B/PAL	USSR	D/SECAM	K/SECAM
(Zanzibar)			Venezuela	M/NTSC	M/NTSC
Thailand	B/PAL	M/PAL			
Togo	K1/SECAM		Yemen Arab Rep.	B/PAL	
Trinidad and	M/NTSC	M/NTSC	Yugoslavia	B/PAL	H/PAL
Tobago			Zaire	K1/SECAM	
Tunisia	B/SECAM		Zambia	B/PAL	
Uganda	B/PAL		Zimbabwe	B	

‡CCIR letter designations: B, D, I M, N, etc.
†As of February 1982.
§Planned.

Table 21-25 Digital Television Parameters (4:2:2) for Studios†

Parameter	525/60 Systems	625/50 Systems
Coded signals	$Y, R - Y,$ $B - Y$	$Y, R - Y,$ $B - Y$
No. of total samples Y $R - Y, B - Y$	858 429	864 432
Sampling structure	Orthogonal, line, field, and frame repetitive. $R - Y$ and $B - Y$ samples cosited with odd Y samples.	
Sampling frequency Y $R - Y, B - Y$	13.5 MHz 6.75 MHz	
Coding	Uniformly quantized PCM, 8 bits for each component signal.	
No. of samples per active line Y $R - Y, B - Y$	720 360	
Signal level vs. quantization level Y $R - Y, B - Y$	220 levels; black = level 16; peak white = level 235 224 levels centered in the range; zero signal = level 128	

†Tentative specifications for the 4:4:4 family are given in Annex I Rec. AA/11 (Mod F) involving $R, G,$ and B or $Y, R - Y, B - Y.$

practices are subject to continual revision and updating in the context of advances in the state-of-the-art technologies. This section summarizes the standards and recommendations with regard to television broadcasting practices.

21.5.1 FCC VISUAL TRANSMITTER STANDARDS.

The FCC standards in Part 73.687 set forth the equipment regulations applicable to the visual television transmitter as summarized below.

To prevent picture signal amplitude distortion, the maximum depth of modulation of the visual carrier by the peak-white signal values (negative modulation) is specified in the United States modulation standards as being 12.5 ± 2.5 percent of the peak carrier level (tip of sync waveform represents peak carrier level). The reference black level is represented by a definite carrier level independent of picture scene content. Also, the blanking level shall be transmitted at 75 ± 2.5 percent of the peak carrier level. The difference between reference black level and blanking level (setup) shall be 7.5 ± 2.5 percent of the video range between blanking level and peak-white level.

For monochrome transmission, the transmitter output shall vary in a substantially inverse logarithmic relation to the brightness of the scene, although this provision is subject to change and is not enforced at present.

An additional specification relates to the general quality of the transmitted signal with respect to hum, noise, and low-frequency response in that these effects shall not cause a peak-to-peak variation of the transmitter output of more than 5 percent of the average scanning synchronization pulse peak-signal amplitude. This rule is also subject to further consideration and is not strictly enforced.

Figure 21-33 contains representations of the visual carrier envelope as a result of

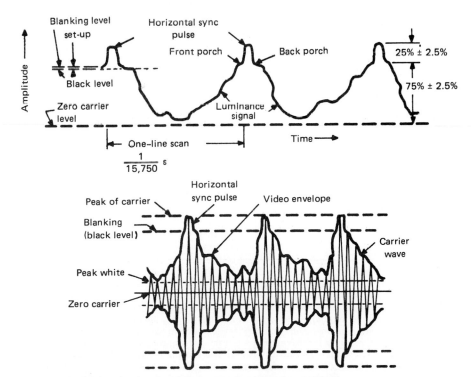

FIG. 21-33 Carrier-signal and modulation-envelope, negative modulation.

amplitude modulation by the combination of the picture information and the composite synchronization signals for the negative modulation case.

The choice between negative and positive modulation involves the interrelationships among several principal factors such as (1) the effect of impulse noise on receiver synchronization, (2) impulse noise visibility during picture time, (3) amount of peak power that may be radiated from a given transmitter tube complement, (4) the availability of a fixed reference for automatic gain control (AGC) function in the receiver, and (5) visual signal-to-noise ratio.

Positive modulation provides an advantage for synchronization in the presence of impulse noise since the downward peaks are clamped at zero carrier level for envelope detection processes. However, in this case, impulse noise causes white spots that are highly visible during picture time. Negative modulation, on the other hand, results in predominantly black impulse noise spots which are visibly less objectionable and provides improved synchronization performance during random noise since the sync pulses represent the peak carrier level and therefore the best available signal-to-noise ratio.

It is desirable to maintain best linearity for the picture content; therefore, negative modulation is best for this purpose, and the sync pulses are more tolerant to nonlinear compression of the peaks of modulation, thus allowing up to as much as 30 percent higher peak power rating to be realized with a given transmitter output amplifier.

Negative modulation also provides an advantage in AGC circuit operation in receivers since the peak level representing blanking and/or sync pulse levels can be maintained constant and thereby utilized as a stable reference for AGC circuits. In negative modulation systems, care must be taken to not overmodulate at peak luminance signal values, not only to avoid picture distortions, but also to prevent momentary interruption of the visual carrier. These periodic interruptions due to accidental overmodulation result in interruptions of the sound carrier in intercarrier receiver systems which produce undesired sound buzz in the receiver audio output. Therefore, the standards specify that the maximum modulation level for picture white shall not exceed 12.5 ± 2.5 percent of the peak carrier level. *Note:* One exception to this rule is allowed for the peaks of the color subcarrier for fully saturated primary colors in the United States color system. The frequency with which this condition occurs in typical scenes, together with the relatively small amount of signal energy involved, is not considered adequate to warrant concern. Test signals constituting 100 percent white level, fully saturated color bars would create such a problem. Therefore, 75 percent white level, fully saturated color-bar test signal format has been developed which results in the color subcarrier peak levels not exceeding the 12.5 percent modulation limit.

The various television systems used worldwide employ the vestigial sideband transmission characteristic with differences only in the details of the width and rate of attenuation of the vestigial sideband portion of the channel (see Table 21-5 and Fig. 21-11).

Corresponding to the vestigial sideband values in each of the systems, an amplitude-versus-frequency response results; when the radiated signal is demodulated with an idealized detector, the response is not flat over the modulation bandwidth. The resulting signal amplitude during the double-sideband portion of the channel is exactly twice the amplitude during the single-sideband portion of the channel. This characteristic is shown for the United States standard signal in Fig. 21-34 (FCC, Sec. 73.699, Fig. 11).

In order to equalize the amplitudes for these two portions of the signal information response, the receiver IF amplitude-versus-frequency response is designed to have an attenuation characteristic over the double-sideband region appropriate to compensate for the two-to-one relationship. This attenuation characteristic (Nyquist slope) is assumed to be in the form of a linear slope over the ± 750 kHz double-sideband region (United States standards) with the visual carrier located at the midpoint (-6 dB point) relative to the single-sideband portion of the band. Such a procedure exactly compensates the amplitude response nonsymmetry resulting from the vestigial transmission.

The television receiver industry groups follow this practice by general consent, and, of course, the FCC has no direct jurisdiction in the form of rules or regulations concerning this practice. Figure 21-35a is a drawing showing the shape of the receiver IF ampli-

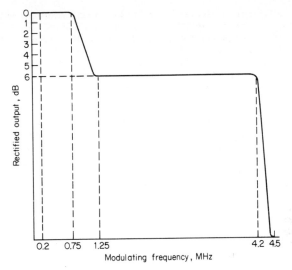

FIG. 21-34 Ideal detector output. *(FCC.)*

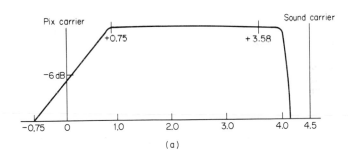

(a)

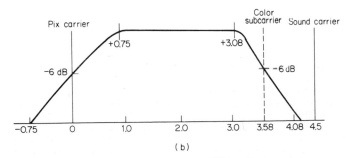

(b)

FIG. 21-35 Typical receiver characteristic of IF-amplitude response versus baseband frequency.

21.77

tude response curve for a television receiver. Modern practice, for purposes of circuit and IF filter design simplification, also provides an attenuation of the upper end of the channel such that the color subcarrier is attenuated also by 6 dB as shown in the haystack IF response in Fig. 21-35*b*.

To maintain amplitude and phase distortion of the radiated signal within acceptable limits, the amplitude and phase-versus-frequency characteristics of the signal are specified at critical frequencies relative to the response produced by an idealized detector response as per Fig. 21-16.

For monochrome transmission only, the overall attenuation characteristic of the transmitter radiated signal, as measured by a diode detector in the antenna transmission line, should not deviate below the ideal detector response, as shown in Fig. 21-16 (FCC, Fig. 11, Sec. 73.699), by amounts more than those listed in Table 21-26 (FCC, Sec. 73.687).

Table 21-26 Monochrome Transmission Relative Amplitude Response

Frequency, MHz	Decibels below ideal response, dB
0.5	2
1.25	2
2.0	3
3.0	6
3.5	12

For color transmissions, the same standards apply with the following exceptions that relate to the color subcarrier and its modulation sidebands. A sine wave at 3.58 MHz, when modulating the transmitter, should produce a radiated signal level in the idealized detector output response which is 6 ± 2 dB lower than the level produced by a sine wave at 300 kHz. (Since the idealized response indicates a reduction of 6 dB beyond 1.25 MHz due to the vestigial sideband response, this specification indicates an allowable variation from the compensated flat response of ±2 dB at 3.58 MHz relative to the response at 200 kHz.) In addition, the amplitude of the radiated signal should not vary by more than ±2 dB, relative to the value at 3.58 MHz, in the frequency range between 2.1 and 4.1 MHz. A modulating frequency of 4.18 MHz should not be more than 4 dB below the value of 3.58 MHz.

Radiation of modulation components of the lower sideband, in vestigial systems, shall be attenuated at least −30 dB at a frequency of 1.25 MHz below the level at 200 kHz. For color transmissions, the lower sideband radiation at a modulation frequency of 3.58 MHz (color subcarrier) shall be at least −42 dB below the radiated level at 200 kHz. At the opposite end of the channel, the radiated field strength in the upper sideband at 4.75 MHz and beyond shall be at least −30 dB below the radiated level at 200 kHz. These rules apply to stations whose output power is above 1 kW in both the VHF and UHF bands.

Stations operating on channels 13 to 82 (UHF) and whose output power is 1 kW or less are not regulated according to the preceding rules but, rather, must comply with the attenuation characteristic shown in Table 21-27. The response should depart from the response as shown in Fig. 21-12 (FCC, Fig. 5a, Sec. 73.699) by no more than the values shown in Table 21-27.

Similarly, as in the VHF channels, the radiation in the upper sideband at 4.75 MHz and beyond shall be at least −20 dB below the level at 200 kHz. If the existence of the lower sideband (no vestigial filter) causes interference to other stations, the rules pertaining to VHF stations (more than 1 kW power) apply.

The phase response in terms of the envelope delay (rate of change of phase) for color transmissions is specified as shown in Fig. 21-36 (FCC, Sec. 73.687, Subparagraph 6). The

Table 21-27 UHF Amplitude-Response Limits (1 kW or less)

Frequency	Decibels below ideal response, dB
0.5 MHz lower sideband	−2
0.5 MHz upper sideband	−2
1.25 MHz upper sideband	−2
2.0 MHz upper sideband	−3
3.0 MHz upper sideband	−6
3.5 MHz upper sideband	−12
3.58 MHz upper sideband (for color)	−8

envelope delay is constant up to 3.0 MHz, relative to the value between 0.05 and 0.20 MHz, and then decreases linearly to 4.18 MHz so as to have a value of −170 ns at 3.58 MHz. The tolerance increases linearly to ±0.1 μs down to 2.1 MHz and remains at ±0.1 μs down to 0.2 MHz. Also, the tolerance increases linearly to ±0.1 μs at 4.18 MHz as indicated in Fig. 21-18.

This specification is intended to approximately complement the rising value of envelope delay in the region between about 3.0 and 4.0 MHz in a typical receiver IF filter design such that both the fixed time delay of the color subcarrier and the skew-symmetry of the color modulation sidebands may be maintained within acceptable limits throughout the complete television transmission path.

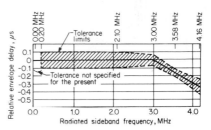

FIG. 21-36 NTSC system precorrection group-delay tolerances.

The precorrection introduced as a requirement at the transmitter permits economical receiver filter design and implementation techniques to be employed. A similar technique is incorporated in other television systems for the same reasons but involving slightly different envelope delay values, particularly in the midband frequency range. The envelope delay specifications for these systems are included in Sec. 21-4, *Worldwide Transmission Standards*.

A brief mention of the technique specified by the FCC to measure the amplitude and phase characteristics of the radiated signal is of interest. The method requires the formation of a composite signal composed of synchronizing signals to establish peak output levels. Sinusoidal signals are then added in the intervals between synchronizing pulses whose frequency may be varied over the range of interest. The sine wave zero axis is maintained at one-half of the synchronizing pulse level with the sine wave held at constant value (no greater than 75 percent of the peak signal) for all modulating frequencies to be measured. The transmitter output radiated field strength is measured at the frequencies of interest, and the level at a frequency of 200 kHz is used as a 0 dB reference level.

A frequency sweep generator signal without a synchronizing pulse pedestal, the dc level set for midrange operation, is allowed as an alternative measurement technique.

21.5.2 FCC AURAL TRANSMITTER STANDARDS.
Section 73.687, Subparagraph (b), of the FCC Rules and Regulations states that the aural transmitter must be capable of operating satisfactorily with a frequency deviation of ±25 kHz which is defined as 100 percent modulation. Additionally, it is recommended that the transmitter be technically capable of operation up to a deviation of ±40 kHz.

The total audio transmitting system from the microphone preamplifier, through all

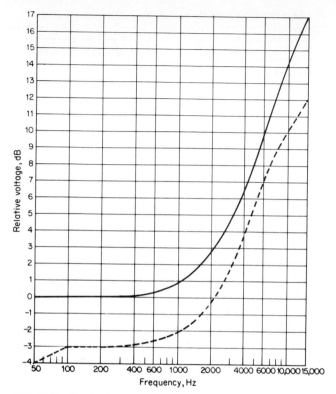

FIG. 21-37 Standard audio preemphasis curve. Time constant is 75 μs (solid line). Frequency response limits shown by use of solid and dashed lines. *(FCC.)*

the studio audio facilities (excluding equalizers for correction of microphone response), shall be capable of transmitting a band of frequencies from 15 to 15,000 Hz. A preemphasis *L, R* network having a time constant of 75 μs is utilized. The characteristic with tolerances is shown in Fig. 21-37 (FCC, Sec. 73.699, Fig. 12). The lower tolerance limit curve has break points at 100 and 7500 Hz.

The harmonic distortion at any frequency between 50 and 15,000 Hz and at modulation percentages of 25, 50 and 100 percent shall not exceed the rms values as shown below:

Modulation frequency, Hz	Distortion, %
50–100	3.5
100–7500	2.5
7500–15,000	3.0

It is further recommended that any one individual contributor to these percentages should not exceed one-half of the total for any one audio link or facility. The frequencies at which measurements should be made at 100 percent modulation include 50, 100, 400, 1000, 5000, 10,000, and 15,000 Hz. Measurements of 25 and 50 percent modulation are

Table 21-28   Receiver Incidental-Radiation Limits

Frequency	Field strength, μV/m
0.45 MHz to and including 25 MHz	†
Over 25 MHz to and including 70 MHz	32
Over 70 MHz to and including 130 MHz	50
130–174 MHz	50–150 (linear interpolation)
174–260 MHz	150
260–470 MHz	150–500 (linear interpolation)
470–1000 MHz	500‡

†The radiation limits in the range of 0.45 to 25 MHz are determined by measurement of the radio-frequency voltage between each power line and ground at the receiver utility power terminals. For television receivers, the voltage levels shall not exceed 100 μV at any frequency between 0.45 and 25 MHz. all other receivers, the voltage shall not exceed 100 μV between 0.45 and 9 MHz and 1000 μV between 10 and 25 MHz (linear interpolation).

‡For television receivers, the limit in the 470- to 1000-MHz range shall be 350 μV/m with measurements being made at the following specific frequencies: 520, 550, 600, 650, 700, 750, 800, 850, 900, and 931 MHz, respectively. The average of the 10 measurements shall not exceed 350 μV/m, with no individual measurement to exceed 750 μV/m.

made only at modulation frequencies of 50, 100, 400, 1000, and 5000 Hz. The measurement shall include audio preemphasis and deemphasis and harmonics to 30 kHz.

The aural transmitter output noise level (FM) in the band of 50 to 15,000 Hz shall be at least 55 dB below the audio-frequency level corresponding to ±25 kHz. The amplitude-modulation noise level, on the same basis, shall be 50 dB below the level represented by 100 percent modulation. The visual transmitter is inoperative during noise measurements.

When compression techniques are employed, precautions must be taken in relation to the preemphasis characteristic. The aural modulation levels are specified in detail in FCC Rules and Regulations, Sec. 73.1570.

A series of regulations applicable to both the aural and visual transmitters in the areas of carrier frequency tolerance, power output measurement, modulation percentage monitoring, video monitoring, construction practices, and safety are detailed in FCC Rules and Regulations in Sec. 73.687, Subparagraphs (c) through (k). Type approval rules and procedures are also established for modulation monitors in Secs. 73.691, 73.692, and 73.694.

Automatic means must be provided to maintain the visual carrier frequency within ±1 kHz of the authorized frequency, and automatic means shall be provided to maintain the undeviated carrier frequency 4.5 MHz above the actual visual carrier within 1 kHz.

Spurious emissions (harmonics) shall be held as low as the state of the art permits, and those in excess of 3 MHz above or below the channel edges shall be at least −60 dB below the visual carrier level.

Each television broadcast station shall be equipped with type-approved indicating instrumentation as necessary for the proper adjustment, operation, and maintenance of the transmitting system in accordance with the FCC Rules and Regulations.

21.5.3 FCC RULES AND REGULATIONS FOR RECEIVER EMISSIONS. FCC

Rules, Part 15, *Radio Frequency Devices,*† provide the requirements, technical specifications, and equipment authorization procedures for both incidental and restricted radiation devices. Subpart C, *Radio Receivers,* is of special interest to television engineers because it details the incidental radiation limits for all receivers, including FM and television broadcast units, that operate (tune) in the range between 30 and 800 MHz.

Table 21-28 lists the limits of the field strengths measured at a distance of 100 ft

---

†As of July 1981.

(30 m) or more from the receiver for receivers manufactured after Dec. 31, 1957 (see Sec. 15-72).

Figures 21-38 to 21-40 show the recommended open field test setup, the line impedance stabilization network (LISN) impedance characteristic, and the specific LISN circuit configuration, respectively.

Subpart C of Part 15 of the FCC Rules contains a number of additional regulations pertaining to television receiver operation and design that are intended to ensure the best possible performance of the overall television system in the public interest. The

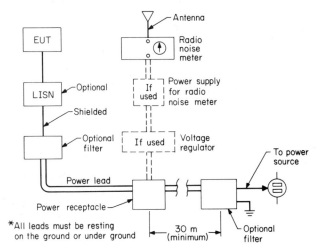

FIG. 21-38 Suggested layout for open field tests. *(FCC.)*

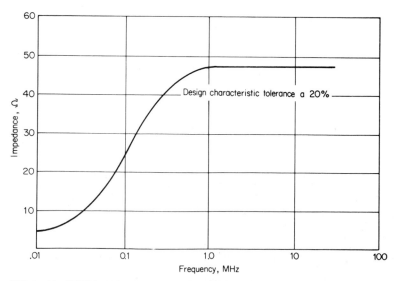

FIG. 21-39 LISN impedance characteristic. *(FCC.)*

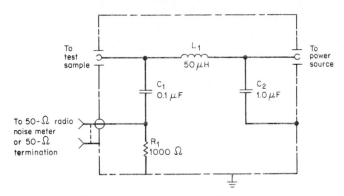

FIG. 21-40 LISN circuit configuration to provide impedance range for the 0.45 to 30-MHz frequency. *(FCC.)*

details of these items are too extensive to include herein. However, a listing by subject and section reference includes the following:

1. Section 15.65. UHF and VHF antennas as supplied by the receiver manufacturer and attached to the receiver.
2. Section 15.66. The required noise figure for the UHF channels in all-channel television receivers relative to manufacture date and certification filing date from May 1, 1946, to Oct. 1, 1984. This section also includes the data required for certification and the proof of performance for receivers certified after Oct. 1, 1979.
3. Section 15.67. The peak picture sensitivity for television receivers manufactured after April 30, 1964, relating VHF versus UHF performance.
4. Section 15.68. The rules for all-channel television receivers manufactured after July 1, 1971, regarding the equality of performance of the user interface for channel selection and tuning, tuning aids, tuning indicators, memories, detents, etc., for both VHF and UHF channel ranges.
5. Sections 15.69, 15.70, 15.71, 15.72, 15.75, and 15.77. These sections set forth the detailed rules and procedures for receiver certification, submission of tuning information, receiver identification, certification date requirements, measurement procedures, and methods for reporting of measurements.

The background and technical basis for many of these measurements and measurement techniques have originated from standardized recommendations of good engineering practice developed by industry organizations such as the EIA, IEEE, SMPTE, and NAB, and references are made to these sources where appropriate in the FCC Rules and Regulations.

Subpart H of Part 15 of the FCC Rules applies to Class I television devices. These devices are defined as being designed to operate within a radio-frequency channel allocated for television broadcasting but shall transmit its output signal to the receiving device by means of a direct connection (either wires or coaxial cable) provided by the manufacturer.

Class I devices include video tape recorders and video disc players. These devices are limited in power output to a peak envelope power of the visual carrier (as measured across a resistor equal to the impedance of the output terminals) to not exceed $346.4 \sqrt{R}$ μV. For example, if $R = 300$ Ω, the maximum rms output voltage is 6000 μV, and for $R = 75$ Ω, the maximum rms output is 3000 μV.

Sections 15.401 through 15.423 of Subpart H contain rules regarding acceptable radiation levels, certification procedures, and type approval requirements. Some of the items include:

1. At least 60-dB isolation in the receiver transfer switch.

2. At least 30-dB reduction relative to the visual carrier level of emissions more than 3 dB outside the upper and lower channel limits.

3. Built-in tuner characteristics are to be the same as those specified for a receiver.

4. The measurement methods for field strength measuring instruments operating above 50 MHz and the electrostatically shielded antenna to be used for measurements below 18 MHz along with the dipole antenna used above 30 MHz are detailed in Sec. 15.417. Either antenna may be used between 18 and 30 MHz.

5. The field strength of any radiation emanating from any part of a Class I device (with the output terminals terminated) shall not exceed 15 μV/m at a distance of $\lambda/2\,\pi$ or at 1 m, whichever is the larger distance.

6. Line-conducted radiation shall not exceed 100 μV in the range between 0.45 and 25 MHz.

The use of Class I devices that do not comply with these regulations pertaining to interference shall be discontinued until the condition causing the interference has been eliminated.

21.5.4 RECOMMENDED PRACTICES.

The preceding sections of this chapter have been concerned with the television broadcast technical standards as set forth by legally enforceable rules and regulations of a properly authorized governmental regulatory body, such as the FCC in the United States.

In addition to these rules and regulations, there exists a large body of technical information constituting recommended practices and standards that represent the consensus of opinion of those engaged in the industry. These recommendations and industry-sponsored standards represent good engineering practices that ensure practical manufacturability within that state of the art, form the basis for the generation of regulatory agency rules and regulations, and, by consent of those skilled in the art, promote the general public interest.

As a predominant source of such recommendations and standards, the EIA Engineering Standards and Publications are designed to serve the public interest through eliminating misunderstandings between manufacturers and purchasers, facilitating of interchangeability and improvement of products, and assisting purchasers in selecting appropriate products for their specific needs. It is important to note that the existence of such standards does not preclude anyone from manufacturing or selling products that do not conform to such standards or recommendations (the ones which are not legally enforceable) either domestically or internationally.

The EIA is a national trade association representing manufacturers of electronic components and systems. It consists of 10 electronic product divisions, Marketing Services, and the Engineering Department. The EIA Engineering Department provides a standardization program that has produced more than 400 standards and publications developed by about 4000 industry, governmental, and academic representatives who make up about 225 committees, study groups, and advisory councils. Thus, the technical consensus of the industry is made effective in areas of legislation, government liaison, consumer affairs, and standardization at both national and international levels. Participation and adherence to the standards so recommended are voluntary. A new international quality assessment and performance certification system for electronic components, known as the IECQ-System, was initiated as of January 1982.

The process of developing recommended standards involves standards proposals derived from the various committees and working groups made up of voluntary individuals from industry, government, and academic areas. After appropriate approval and review by the Engineering Department Executive Committee, the document is published as an EIA Recommended Standard, and copies may be purchased through the association's Standards Sales Office. Many EIA standards become National Standards through the approval of the American National Standards Institute (ANSI), some become the basis for legally enforceable FCC Rules and Regulations, and some form a basis for United States positions in international standards development through the International Electrotechnical Commission (IEC) and the International Standardization Organization (ISO).

A catalog is made available each year listing the Engineering Publications and Recommended Standards of the EIA, Joint Electron Device Engineering Council (JEDEC) of EIA, and the EIA Tube Engineering Panel Advisory Council (TEPAC). A JEDEC Solid State Products Engineering Council (SSPEC) is also quite active. Ordering details, prices, and procedures are available in the EIA catalog of EIA and JEDEC Standards and Engineering Publications. The address of the EIA Engineering Department is 2001 Eye Street, N.W., Washington, D.C. 20006. In addition, as of January 1982, the EIA Television Test Charts may be purchased directly from the vendor, Hale Color Consultants, Baltimore, Md.

In addition to the EIA, a number of other industry-sponsored organizations provide similar services and develop recommended practices and standards pertinent to their predominant areas of interst. Notable examples of these organizations are the SMPTE, the NAB, the European Broadcasting Union (EBU), the Institute of Electrical and Electronics Engineers (IEEE), and, of course, the IEC and the CCIR. Each of these organizations publishes recommended standards and practices and provides catalogs or an index listing the available publications. These recommended standards are far too lengthy and comprehensive to summarize within any one publication. Therefore, the following sections will provide a listing of excerpts from the various catalogs, by title, along with a short synopsis, that are directly pertinent to television broadcasting practices. Some examples of the most commonly used items such as certain test signal formats and test charts will be included.

It is recommended that readers secure specific documents to fulfill their particular needs directly from the appropriate organization. The material listed herein is intended to indicate the general areas covered and act as a guide in locating the required information.

21.5.5 EIA AND JEDEC STANDARDS AND PUBLICATIONS.

Tables 21-29 and 21-30 contain a list of bulletins and recommended standards applicable to broadcast television extracted from the EIA and JEDEC Standards and Publications Catalog, 1982. These publications may be purchased from the Electronic Industries Association, Engineering Department, Standards Sales office, 2001 Eye Street, N.W., Washington, D.C. 20006.

The extracts shown in Table 21-30 relate to television broadcast and display standards and represent only a small fraction of the total of about 400 EIA publications. These publications are categorized in accordance with specific topics as listed in Table 21-31.

Of the EIA publications listed, a few should be singled out as particularly relevant to television broadcasting because they either contain the standards or have formed the basis for many of the FCC Rules and Regulations. These Recommended Standards are typically presented in a format of (1) definitions, (2) minimum standards, and (3) measurement methods.

One of the more widely used publications is EIA Standard RS-170, *Electrical Performance Standards—Monochrome Television Studio Facilities,* originally issued in November 1957. EIA Subcommittee TR-4.4 on Television Studio Facilities has undertaken to revise and update RS-170, and this job has been assigned to ad hoc Subcommittee TR-4.4.1. As part of the revision, EIA Industrial Electronics Tentative Standard No. 1 entitled *Color Television Studio Picture Line Amplifier Output Drawing* (November 1977) has been issued. This updated signal waveform drawing has been considered sufficiently important to be published separately as a Tentative Standard. In March

Table 21-29 EIA Bulletins and Publications (Extracts)

Number	Title
	Consumer products engineering bulletins
CPEB1	Standard Method of Measurement of Ionizing Radiation from Television Receivers for Factory Quality Assurance
CEB2	Definition of Normal Operating Conditions for Television Receivers
CPEB3	Measurement Instrumentation for X-Radiation from Television Receivers
CPEB4-A	Standard Form for Reporting Measurements of TV and FM Broadcast Performance Receivers in Compliance with FCC Part 15 Rules
CPEB5	Performance Guidelines for Video Players
CPEB8	Susceptibility of TV Tuners to Harmonically Generated Interference
	Industrial electronics tentative standards
IETNTS1	Color Television Studio Picture Line Amplifier Output Drawing
	TEPAC engineering bulletins
TEPB21	CRT Considerations for Raster Dot Alphanumeric Presentations
TEPB22	Magnetic Deflection Yokes
	Television systems bulletins
TVSB1	EIA Recommended Practice for Use of a VIR Signal
TVSB4	EIA Recommended Practice for Horizontal Sync, Horizontal Blanking and Burst Timing in Television Broadcasting

1976, the EIA Television Systems Bulletin No. 4 entitled *EIA Recommended Practice for Horizontal Sync, Horizontal Blanking, and Burst Timing in Color Television* was prepared by the EIA Broadcast Systems Committee (BTS) and issued. The purpose of this bulletin was to minimize color signal nonuniformities that could occur with the constraints of the existing FCC Rules and Regulations. The details of timing values and tolerances of EIA Bulletin No. 4 have been incorporated in the picture line amplifier output drawing. This drawing is sufficiently comprehensive to be reproduced herein in Fig. 21-41. The notes associated with this drawing detail the pertinent waveform tolerances and measurement criteria to be adhered to in both producing and utilizing the signals, especially in studio and control facilities, and measuring the overall television system performance.

Other EIA Recommended Standards requiring special attention include RS-240, *Electrical Performance Standards for Television Broadcast Transmitters, Channels 2–6, 7–13, and 14–83* (this publication was formulated by EIA Subcommittee TR4.1 on Program Transmitters and Committee TR-4 on Television Broadcast Transmitting Equipment). In addition, RS-250-B, *Electrical Performance Standards for Television Relay Facilities* (a revision of RS-250-A), and RS-462, *Electrical Performance Standards for Television Broadcast Demodulators,* May 1979, contain many recommended practices followed throughout the television system.

Another effort of note that is indicative of the cooperation of the various industry groups is a proposed standard, still under revision, entitled *Methods of Testing Monochrome and Color Television Receivers.* This document is the result of a joint effort of the IEEE Broadcast and Television Receiver Group and the EIA Television Receiver

Table 21-30 EIA Recommended Standards (Extracts)

Number	Title and synopsis
RS-163	*RF Radiation Label:* This standard describes the label to be affixed to receivers in accordance with Part 15 of the FCC rules.
RS-170	*Electrical Performance Standards—Monochrome Television Studio Facilities:* This standard consists of definitions, minimum standards, and methods of measurement for important parameters. They are intended to apply to locally generated signals within the studio, or nearby, where control can be exercised over picture quality.
RS189A	*Encoded Color Bar Signal:* The EIA standard color-bar signal is intended to be used as a test for color monitor adjustment, color encoder adjustment, and checks of television transmission systems.
RS-240	*Electrical Performance Standards for Television Broadcast Transmitters, Channels 2–6, 7–13, and 14–83:* This standard covers definitions, minimum performance standards, and methods of measurement of both aural and visual television transmitters for monochrome and color NTSC criteria.
RS-250-B	*Electrical Performance Standards for Television Relay Facilities (ANSI/EIA RS-250-B-76):* This standard specifies the minimum electrical performance characteristics of radio relay equipment intended for transmission of NTSC color television signals from a studio to its television broadcast transmitter over long interconnection facilities, or similar applications.
RS-330	*Electrical Performance Standards for Closed Circuit Television Camera 525/ 60 Interlaced 2:1 (ANSI/EIA RS-330-68):* These standards are intended to apply to locally generated signals and cover minimum performance standards, definitions, and methods of measurement.
RS-343-A	*Electrical Performance Standard for High Resolution Monochrome Closed Circuit Television Camera:* This standard relates to equipment operating in the range of 675 to 1023 scanning lines at a field rate of 60 Hz, 2/1 interlaced, and covers definitions, performance, and measurement methods.
RS-375-A	*Electrical Performance Standards for Direct View Monochrome Closed Circuit Television Monitors 525/60 Interlaced 2:1 (ANSI/EIA RS-375-A-76):* These standards apply only to direct-view monochrome closed-circuit television monitors with a video input. They apply to video signals described in EIA RS-330.
RS-378	*Measurement of Spurious Radiation from FM and TV Broadcast Receivers in the Frequency Range of 100 to 1000 MHz—Using the EIA LAUREL Broadband Antenna:* This standard describes the potential sources of spurious radiation and sets up methods of measurement of radiated signal strength.
RS-412-A	*Electrical Performance Standards for Direct View High Resolution Monochrome Closed Circuit Television Monitors (ANSI/EIA RS-412-A-76):* These standards apply only to monochrome high-resolution monitors with a video input.
RS-413	*Recommended Test Method—Timing Error Measurements of Instrumentation Magnetic-Tape Recorder/Reproducers (ANSI/EIA RS-412-73):* This standard covers instrumentation and procedures for measurement of time-base error, intertrack time-displacement errors, composite time-base error, and pulse-to-pulse jitter in magnetic recording equipment.
RS-439	*Engineering Specifications Format for Color CCTV Camera Equipment:* This standard lists the electrical, mechanical, and environmental specifications relating to color CCTV camera equipment.
RS-462	*Electrical Performance Standards for Television Broadcast Demodulators:* This standard includes aural and visual performance criteria for various functions and operating modes which may or may not be available in a particular demodulator.
REC-109-C	*Intermediate Frequencies for Entertainment Receivers:* The standard intermediate amplifier frequencies for various types of receivers are given.
REC-141	*VHF Receiving Antenna Performance, Presentation and Measurement:* This standard relates to antenna directivity, relative gain, and VSWR to include measurement procedures.

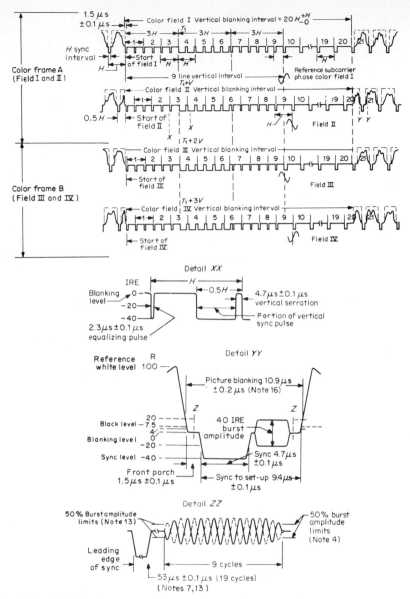

FIG. 21-41 EIA RS170A tentative standard, synchronizing waveforms. *Notes:* (1) Specifications apply to studio facilities. Common carrier, studio to transmitter, and transmitter characteristics are not included. (2) All tolerances and limits shown in this drawing permissible only for long time variations. (3) The burst frequency shall be 3.579545 MHz ± 10 Hz. (4) The horizontal scanning frequency shall be $\frac{2}{455}$ times the burst frequency [one scan period (H) = 63.556 μs]. (5) The vertical scanning frequency shall be $\frac{2}{525}$ times the horizontal scanning frequency [one scan period (V) = 16,683 μs]. (6) Start of color fields I and III is defined by a whole line between the first equalizing pulse and the preceding H sync pulse. Start of color fields II and IV is defined by a half line between the first equalizing pulse and the preceding H sync pulse. Color field I: that field with positive-going zero crossings of reference subcarrier most nearly coincident with the 50% amplitude point of the leading edges of even-numbered horizontal sync pulses. Reference subcarrier is a continuous signal with the same instantaneous phase as burst. (7) The zero crossings of reference subcarrier shall be

Table 21-31 EIA Publications Categories

Category	Subject
1	General
2	Consumer electronics
3	Communications and industrial electronics
4	Component parts (passive parts)
5	Solid-state products
6	Electron tubes
7	Government electronics
8	All

Standards Subcommittee, R-4.4, and the participating companies. This standard is intended, when completed, to replace the IEEE Standard 190, *Methods of Testing Monochrome Television Broadcast Receivers*, 1960, and describes methods for evaluating both monochrome and color receivers designed to receive signals conforming to the FCC Part 73 Rules and Regulations and specified as the NTSC system.

21.5.6 RECOMMENDED PRACTICES AND STANDARDS SPONSORED BY INDUSTRY ORGANIZATIONS

Society of Motion Picture and Television Engineers. An index to the SMPTE-sponsored American National Standards, as well as SMPTE Recommended Practices, and Engineering Guidelines of all approved standards, practices, and guidelines which are sponsored by the SMPTE, and validated during the calendar year, are published yearly in the December issue of the *Journal of the SMPTE*. Individual copies of approved standards, practices, and guidelines, and a complete set of all SMPTE-sponsored documents, may be purchased from SMPTE, 862 Scarsdale Avenue, Scarsdale, N.Y. 10583.

The 1981 Annual Index includes, for example, sections entitled *Television* and *Video Magnetic Tape Recording*. Recommended Practices are indicated as *RP* while Engineering Committee Recommendations are listed as *ECR*. *R,* followed by a date, means reaf-

---

nominally coincident with the 50% point of the leading edges of all horizontal sync pulses. When the relationship between sync and subcarrier is critical for program integration, the tolerance on this coincidence is $\pm 40°$ of reference subcarrier. (8) All rise times and fall times unless otherwise specified are to be 0.145 ± 0.02 μs measured from 10 to 90% amplitude points. All pulse widths are measured at 50% amplitude points unless otherwise specified. (9) Tolerance on sync level, reference black level (setup), and peak-to-peak burst amplitude shall be ± 2 IRE units. (10) The interval beginning with line 17 and extending through line 20 of each field may be used for test, cue, and control signals. (11) Extraneous synchronous signals during blanking intervals, including residual subcarrier, shall not exceed 1 IRE unit. Extraneous nonsynchronous signals during blanking intervals shall not exceed 0.5 IRE unit. All special-purpose signals (VITS, VIR, etc.) when added to the vertical blanking interval are excepted. Overshoot on all pulses during sync and blanking, vertical and horizontal, shall not exceed 2 IRE units. (12) Burst-envelope rise time is $0.3 + 0.2 - 0.1$ μs measured between the 10 and 90% amplitude points. Burst is not present during the nine-line vertical interval. (13) The start of burst is defined by the zero crossing (positive or negative slope) preceding the first half cycle of subcarrier that is 50% or greater of the burst amplitude. Its position is nominally 19 cycles of subcarrier from the 50% amplitude point of leading edge sync (see detail ZZ). (14) The end of burst is defined by the zero crossing (positive or negative slope) following the last half cycle of subcarrier that is 50% or greater of the burst amplitude. (15) Monochrome signals shall be in accordance with this drawing except that burst is omitted and fields III and IV are identical to fields I and II respectively. (16) Occasionally measurement of picture blanking at 20 IRE units is not possible because of scene content as verified on a picture monitor. *(Electronic Industries Association.)*

firmed as of that date. The following list gives some typical examples taken from the 1981 Annual Index† in the *Journal of the SMPTE.*

1. ECR-1978: *Alignment Color Bar Signal*
2. RP37-1969 (R 1976): *Color Temperature, Monitors*
3. RP46-1972 (R 1977): *Density, Color Films and Slides*
4. RP71-1977: *Monitors, Setting of White*
5. RP27.1-1977: *Test Patterns, Alignment*
6. PH22.95-1963 (R 1975): *Image Area, 16mm Film*
7. PH22.94-1963 (R 1975): *Image Area, 35mm Film*
8. RP 27.5-1977: *Test Patterns, Mid-Frequency Response*
9. RP93-1980: *Helical Scan Tape Recording, Code, Time and Control, Recording Requirements*
10. C98.15M-1980: *Type B 1-inch, Basic Parameters*
11. C98.18M-1979: *Type C 1-inch, Basic Parameters*
12. RP75-1974 (R 1980): *Vertical Interval Signal*
13. RP108: *Vertical Interval Time and Control Code*

European Broadcasting Union. The Technical Centre of the EBU publishes technical documents on specific topics of interest to television broadcasting engineers which have been studied by the EBU. These documents are printed in both English and French and may be purchased from the Technical Centre of the European Broadcasting Union, 32, Avenue Albert Lancaster, B-1180 Brussels, Belgium. Table 21-32 lists a few recent EBU Technical Documents pertaining to television.

International Electrotechnical Commission. A Catalogue of Publications of the IEC is published annually. Copies may be obtained through the Bureau Central de la Commission Electrotechnique Internationale, 1, rue de Varembe', Geneva, Switzerland. These publications, in a manner similar to those of the EIA, SMPTE, and EBU, provide recommended practices and standards as well as technical guidelines over a wide range of technical topics to include those subjects relating to television broadcasting, reception, and recording and playback of television material. The IEC should be contacted directly for catalogs and publications.

In addition to the examples mentioned above, a number of other industry-sponsored organizations, both in the United States and throughout the world, provide similar services relating to rules of good engineering practice. It is recommended that the reader contact the particular organization of interest directly to obtain the available catalogs and publications since it is impossible to summarize the technical content of all the material available within the space allotted.

21.5.7 CHARTS AND TEST SIGNALS. During the intervening years since the introduction of commercial broadcasting of monochrome television, and especially since the initiation of color television service in the United States in 1953, a large variety of test and monitoring techniques have evolved. A brief summary of some of the most notable and widely utilized test signals and charts is included to show the nature and scope of such procedures.

Generally, the technical as well as regulatory need to monitor and measure the performance of all segments of the total television system, both the individual components and the total signal path, has been the motivation for development of specific test signals, charts, and procedures. The requirement for both optical and electrical parameter measurements, as well as their translation interfaces, stems from the fact that the system

---

†*Note:* The 1981 Annual Index listed a total of 21 publications for television and 58 items relating to video magnetic tape recording, with about 250 standards, practices, and guidelines in all.

Table 21-32 EBU Technical Documents

Number	Title
3084	*EBU Standards for Television Tape-Recording*, 2d ed., 1975
3087	*Colour Motion Picture File Materials Especially Suited for Presentation by Colour Television*, 2d ed., 1975
3091	*Optical Viewing Conditions for Film Intended for Colour Television*, 1970
3093	*Video Player and Recorder Systems for Home Use*, 4th ed., 1975
3094	*Specifications for the Basic Signals Recommended by the EBU for the Synchronization of Television Sources*, 1971
3097	*EBU Time-and-Control Code for Television Tape Recorders (625-line TV systems)*, 2d ed., 1980
3201	*Identification of Television Transmissions in Europe*, 1974
3208	*Use of Digital Techniques in Broadcasting*, 1974
3209	*Performance Specifications of Equipment for EBU Insertion Signals (625-line systems)*, 1974
3213	*EBU Standard for Chromaticity Tolerances for Studio Monitors*
3218	*Colour Television Film-Scanners*, 2d ed., 1979
3219	*Operational Adjustments and Measurements on Transverse-Track Television Tape Machines*, 1976
3220	*Satellite Broadcasting—Design and Planning of 12 GHz Systems*, 1976
3221	*Guiding Principles for the Design of Television Waveform Monitors*, 2d ed., 1978
3222	*Analysis of the 1977 Geneva Plan for Satellite Broadcasting at 12 GHz*, 1977 (five volumes)
3229	*Operation and Maintenance of Re-Broadcast Transmitters*, 1981
3232	*Displayable Character Sets for Broadcast Teletext*, 1980
3234	*Analogue Television Transmission Tests with OTS— Synthesis of Results Prepared Jointly by the EBU and Interia Eutelsat*, 1981
Technical information sheet No. 7	*Helical-Scan Television Recording on 25.4-mm Tape*, 1981

involves the generation of pictorial information in the form of representative electrical signals, transmission of this information through a specified channel, and eventual display of the information in order to make the perceived image appear the same as the perceived original scene.

A variety of test charts of appropriate geometric formats have been developed to measure the quality of the transfer from optical information to electrical signals by the camera at one end and the picture monitor, or display, at the output end of the system. In addition, a wide variety of electrically generated test signal formats are utilized that may be injected into the system at appropriate points for both objective and subjective performance checks.

Recognition that cameras and related equipment required standardized calibration procedures led the EIA to design an initial series of five Television Test Charts. These charts were developed by the EIA TR4.4 Studio Facilities Committee and have been in wide use for over 30 years. Hale Color Consultants have consistently manufactured these high-quality charts over the years, and, as of January 1982, the EIA licensed Hale to handle the production and direct sale of these patterns. The five test charts are listed below:

1. *EIA Color Registration Chart.* This chart is used to aid the alignment and test the registration of the three-color pickup live and film color television cameras. The very fine horizontal and vertical black lines on a white background permit accurate alignment of the optical and electrical camera systems.

2. *EIA Logarithmic Reflectance Chart.* This chart has two crossed, nine-step gray scales on a 12 percent gray background, with the scale steps showing a logarithmic reflectance progression from 3 to 60 percent. Between the scales is a black level swatch for absolute black calibration. The chart is used for checking and adjusting of the camera transfer characteristic.

3. *EIA Linear Reflectance Chart.* This chart is the same as the Logarithmic Reflectance Chart except that the nine gray scale steps have a linear relationship to each other.

4. *EIA Resolution Chart.* This chart is used to measure the resolving power of the system. The chart is televised by the camera and displayed on a monitor. When adjusted according to instructions relative to the scanned area (television raster dimensions), the horizontal and vertical resolution wedges cover the range from 200 to 800 television lines. Gray scale overlays are available for the indicated sections of the chart. These gray scales are designed to provide logarithmic reflectance relationships. The chart is useful in checking scanning size, linearity, aspect ratio, focus, focus uniformity, shading, low-frequency phase shift, resolution, and transient response. Both the horizontal wedges (vertical resolution) and the vertical wedges (horizontal resolution) are calibrated in both frequency and television line-number increments. The frequency calibrations of these indicators are calculated as follows

$$f_n = A_r \eta \frac{106}{2H_a}$$

where f_n = fundamental frequency for η lines
A_r = aspect ratio = $\frac{4}{3}$
n = number of lines
H_a = active time of horizontal trace (horizontal time less the blanking time—the average blanking time is $0.16\,H$ when $H = 63.5\ \mu s$)

Thus $H_a = (63.5)(0.84) = 53.3\ \mu s$

Substitution: $f_n = \dfrac{4}{3} \dfrac{106}{(2)(53.3)}$

$$n = 0.0125\,N \text{ MHz or } n = f_n/0.0125 \text{ lines}$$

when f_n = 3.0, 4.0, 5.0, 6.0, 7.0; n = 240, 320, 400, 480, 560.
A rule of thumb, therefore, is 80 television lines per megahertz. Figure 21-42 is a reproduction of the Resolution Test Chart.

5. *EIA Linearity (Ball) Chart.* This chart is used to test geometric distortion of the camera chain. This is done by comparing, on a monitor, two superimposed patterns, one grating generated electrically and the other by viewing the television chart with the camera. The electrical frequencies required to match the chart pattern are 315 KHz for horizontal and 900 Hz for the vertical directions. The inner and outer diameters of the circles represent 1 and 2 percent of the picture height, respectively. A reduction of this pattern is shown in Fig. 21-43.

As of February 1982, three additional EIA Test Charts were made available as listed below:

1. *Multiburst Chart.* This chart consists of vertical alternate black and white bars whose repetition rates produce signal frequencies of 0.5, 1.0, 2.0, 2.5, 3.0, 3.5, and 4.0 MHz when the scanned area is adjusted according to instructions. Thus, it provides a measure of the operating modulation transfer function (MTF) (frequency response) of the camera chain and subsequent signal-processing apparatus.

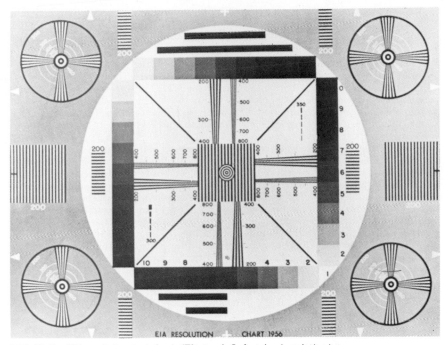

FIG. 21-42 EIA resolution test chart. *(Electronic Industries Association.)*

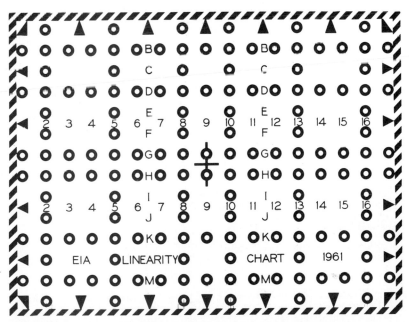

FIG. 21-43 EIA linearity (ball) test chart. *(Electronic Industries Association.)*

2. *White Pulse Chart (Window Chart).* This pattern consists of a white box against a black background to produce a full dynamic range square pulse on a waveform monitor. The chart is used to adjust image enhancement apparatus and check system rise times and transient response.

3. *Color Calibration Chart.* This chart consists of three horizontal bars on a black background. The center bar is constant at 60 percent white level. The other two bars show six saturated colors representing the hues of the NTSC primaries and their complements. On a vectorscope, these colors fall within the designated tolerance boxes when the system is properly calibrated. The color performance of two or more cameras focused on the same chart may be matched by this process.

A great variety of electronically generated equivalents of these charts as well as many other pattern formats for special test purposes are available from numerous manufacturers. These pattern generators produce idealized composite signals to be injected into the television system at the appropriate point, independent of the actual optical/electrical camera equipment. A widely used version of this type of test signal source was developed by Philips and combines almost all the EIA Test Chart configurations, including color bars, multiburst, and gray scale, etc., into one composite signal source. The reader is referred to the Philips Test Pattern Generator literature for details.

A number of test signal waveforms have been initiated over the years for purposes of testing the performance of color television processing apparatus as well as measuring the

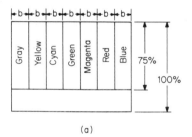

(a)

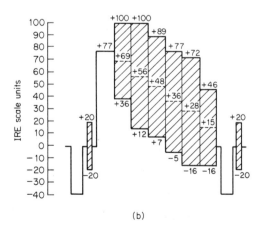

(b)

FIG. 21-44 EIA RS189A color-bar displays: (*a*) color-monitor display of gray and color bars; (*b*) waveform monitor display of reference gray and primary and complementary colors (upper part of picture display). *Notes:* (1) *b* = ⅐ active line time. (2) In (*b*) dotted lines indicate luminance values. *(Electronic Industries Association.)*

overall system performance. By far the most widely used of these test signals is the color-bar signal, in both full-field (all color bars) and split-field (additional test signals in the lower one-quarter of the scanned area) formats. The specifications for this color-bar signal are set forth in EIA Standard RS-189-A as formulated by the TR4.4 Committee on Studio Facilities. The specified standard signal has two parts. The upper three-quarters of the active scanning area in each field consists of seven equally spaced areas arranged in descending order of luminance (left to right) as follows: gray, yellow, cyan, green, magenta, red, and blue. These colored areas correspond to the three primaries of NTSC and the equal mixture complementary colors adjusted to be transmitted at 75 percent of full amplitude. The lower one-quarter of the field (split-field format) is used for transmission of special test information consisting of 100 percent white level, black level, and subcarrier signal envelope with a phase corresponding to $-I$ and a phase corresponding to $+Q$.

This test signal is extremely useful in the adjustment of transmitting encoders, the checking of signal transmission parameters such as differential phase distortion, differential amplitude distortion (at the specific frequency of the color subcarrier — 3.579545 MHz), and relative envelope delay of chrominance modulation components versus luminance components, as well as numerous other overall system parameter checks. The standardized waveforms as specified in RS-189-A are shown in Figs. 21-44 to 21-46 and the vectorscope presentation in Fig. 21-47.

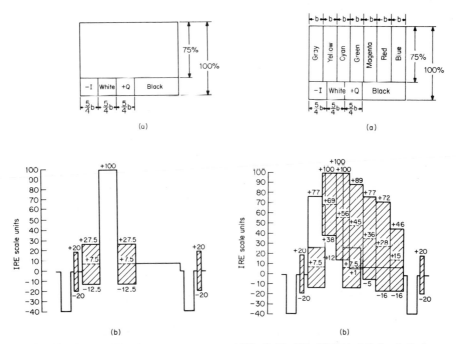

FIG. 21-45 EIA RS189A $-I$ and $+Q$ test signals and reference-white bar (lower part of picture display): (a) color monitor display of $-I$, white, $+Q$, and reference black; (b) waveform monitor display (lower part of picture display). *Note:* Dotted lines in (b) indicate luminance values. *(Electronic Industries Association.)*

FIG. 21-46 EIA RS189A full-signal displays: (a) color monitor; (b) waveform monitor. *Note:* Dotted lines in (b) indicate luminance values. *(Electronic Industries Association.)*

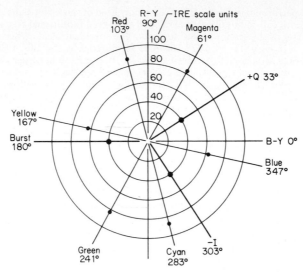

FIG. 21-47 EIA RS189A vectorscope display showing the vector and chrome-amplitude component relationships. (*Electronic Industries Association.*)

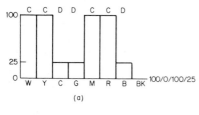

(a)

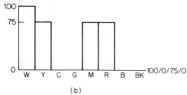

(b)

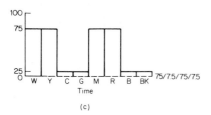

Time

(c)

FIG. 21-48 Relative amplitudes of color bars from different types of generators (red channel only): (*a*) BBC; (*b*) EBU; (*c*) monitor test-signal used in North America. (*Electronic Industries Association.*)

The SMPTE Engineering Committee Recommendation ECR 1-1978 entitled *Alignment Color Bar Test Signal for Television Picture Monitors* introduces additional test signals in the lower one quarter of the field. One strip contains the blue signal values only for all the color bars arranged in the reverse order from the upper three-quarters color-bar sequence. This is intended to provide a more accurate means of adjusting overall chroma level (saturation) and overall demodulation phase setting (hue) in the NTSC system. In addition, this recommendation includes small areas in the reference-black area of slightly *whiter than black* and slightly *blacker than black* signal levels to facilitate the adjustment of monitor visual black reference level (see SMPTE ECR 1-1978 for details).

Finally, for international purposes, the CCIR Recommendation 471 entitled *Nomenclature of Colour Bar Signals* (1970) specifies the red channel color-bar signal amplitudes for reference white, 75 percent color bar levels, and black level setup for (1) the system used by the BBC, (2) the system used by the EBU, and (3) the system used for NTSC. The red-signal waveforms are shown in Fig. 21-48.

As a final note, almost all of the Recommended Practices and Standards for handling and interconnections of television composite video signals specify a coaxial cable input and output terminating impedance of 75 Ω, with a reflected signal value of at least -30 dB, a total sync-to-peak-white signal level of 1 V, peak to peak. The *picture content polarity* is defined as that portion of the signal representing a dark area of the scene relative to that portion of the signal representing a light area to be in the negative direction. Thus, conventional signal polarity is stated as *black negative,* or in composite signal form, *sync negative* [ANSI C42.100-1972 (IEEE Std. 100-1972)]. EIA RS-250-B contains a good reference to techniques for specifying relay facilities signal-handling characteristics as well as performance checks and measurements.

BIBLIOGRAPHY

Cable Television Service, Part 76, FCC Rules and Regulations, revised as of February 1981.

"CCIR Characteristics of Systems for Monochrome and Colour Television—Recommendations and Reports," Recommendation 470-1, Report 624-1, of the XIVth Plenary Assembly of CCIR in Kyoto, Japan, 1978.

"Color Television Standards for Region 2," *IEEE Spectrum,* February 1968.

Colour Television—vol. 1: *NTSC,* vol. 2: *PAL and SECAM,* ILIFFE Books Ltd. (Wireless World), London, 1969.

EBU Technical Documents, The Technical Centre of the EBU, Brussels, Belgium.

Encoding Parameters of Digital Television for Studios, Recommendation AA/11 (Mod. F), CCIR Report (Doc. 11/1027 Rev. 1-E), Feb. 17, 1982.

Experimental, Auxiliary, and Special Broadcast, and Other Program Distributional Services, Part 74, FCC Rules and Regulations, March 1980.

Fink, D. G. (ed.), *Color Television Standards,* Selected Papers and Records of the NTSC, McGraw-Hill, New York, 1955.

Fink, D. G. (ed.), *Television Engineering Handbook,* McGraw-Hill, New York, 1957.

Frequency Allocations and Radio Treaty Matters, Part 2, General Rules and Regulations, FCC Rules and Regulations, July 1981.

Herbstreit, J. W., and H. Pouliquen, "International Standards for Color Television," *IEEE Spectrum,* March 1976.

"Index to SMPTE-Sponsored American National Standards Society-Recommended Practices, and Engineering Guidelines," *SMPTE J.,* December 1981.

International Telephone and Telegraph Corp., *Reference Data for Radio Engineers,* Sams, Indianapolis, Ind., 1975.

"LPTV Gets the FCC Go-Ahead," *Broadcasting,* vol. 102, no. 10, March 8, 1982.

"1981 Catalog of Publications," International Electrotechnical Commission, Geneva, Switzerland.

"1982 Catalog of EIA and JEDEC Standards and Engineering Publications," EIA Engineering Department, Washington, D.C., 1982.

Pritchard, D. H., "U.S. Color Television Fundamentals—A Review," *SMPTE J.,* vol. 86, pp. 819–828, November 1977.

Pritchard, D. H., and J. J. Gibson, "Worldwide Color Television Standards—Similarities and Differences," *SMPTE J.,* vol. 89, February 1980.

Radio Frequency Devices, Part 15, FCC Rules and Regulations, July 1981.

Radio Broadcast Services, Part 73, FCC Rules and Regulations, March 1980.

Reference Data and Equations

Donald G. Fink

Director Emeritus, Institute of Electrical and Electronics Engineers
Somers, New York
Editor, Electronics Engineers' Handbook and
Standard Handbook for Electrical Engineers
Fellow IEEE and SMPTE

The material in this chapter consists of constants, standard frequencies, equations, and other relationships useful in the practice of television engineering. No comprehensive compilation of mathematical and physical data has been attempted beyond those quantities (such as the table of the sine integral, Table 22-9) which are of particular interest to the television engineer and not generally available in the standard handbooks of electronics and radio engineering. An annotated bibliography of sources of mathematical and physical data appears at the end of the chapter.

22.1 RADIO-FREQUENCY DATA

22.1.1 DESIGNATION OF RADIO-FREQUENCY BANDS. See Table 22-1.

22.1.2 VELOCITY OF PROPAGATION OF RADIO WAVES IN FREE SPACE

$$c = 299.7925 \text{ m/}\mu\text{s}$$

$$= 983.57 \text{ f/}\mu\text{s}$$

$$= 327.86 \text{ yd/}\mu\text{s}$$

$$= 0.18628 \text{ statute mile/}\mu\text{s (smi/}\mu\text{s)}$$

$$= 0.16187 \text{ nautical mile/}\mu\text{s (nmi/}\mu\text{s)}$$

Table 22-1 Designation of Radio-Frequency Bands

Frequency limits	Wavelength limits	FCC designation	Exponential designation†	Uses in television
Below 30 kHz	Above 10^4 m	Very low frequency (VLF)	Up to band 4	Video and scanning frequencies
30–300 kHz	10^3–10^4 m	Low frequency (LF)	Band 5	Video and scanning frequencies
300–3000 kHz	100–1000 m	Medium frequency (MF)	Band 6	Video frequencies
3000–30,000 kHz	10–100 m	High frequency (HF)	Band 7	Video frequencies
30–300 MHz	1–10 m	Very high frequency (VHF)	Band 8	Intermediate frequencies and broadcast channels Digital video
300–3000 MHz	10–100 cm	Ultrahigh frequency (UHF)	Band 9	Broadcast, radio relay
3000–30,000 MHz	1–10 cm	Superhigh frequency (SHF)	Band 10	Radio and satellite relay
30,000–300,000 MHz	1–10 mm	Extremely high frequency (EHF)	Band 11	Experimental and satellite

†This designation is the exponent which, when applied to the base 10, gives a frequency in hertz which is approximately the geometric mean frequency of the band. Example: The frequency in band 6 is 10 to the sixth power = 1 million Hz which is the frequency within the band 0.3 to 3.0 MHz.

Reciprocal velocity of propagation

$$1/c = 0.0033356 \ \mu s/m$$
$$= 0.0010167 \ \mu s/ft$$
$$= 0.0030501 \ \mu s/yd$$
$$= 5.3682 \ \mu s/mi$$
$$= 6.1776 \ \mu s/nmi$$

22.1.3 MULTIPATH PROPAGATION. When the difference in length between one path of propagation and another is d units, the corresponding difference in propagation time is $d(1/c)$ μs where $1/c$ is expressed in the given units of length. When the difference in propagation time between one path and the other is t μs, the corresponding difference in the length of the paths is ct. Values of c and $1/c$ are given in Sec. 22.1.2.

Multipath Locus. The locus of constant path difference (constant difference in distance or constant difference in propagation time) between direct and reflected signals is

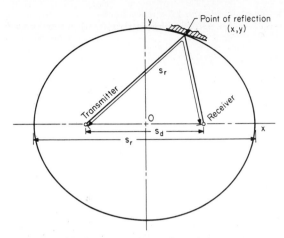

FIG. 22-1 The locus of constant difference between direct and reflected signal paths is an ellipse with transmitter and receiver locations as foci. The major axis of the ellipse is equal to the distance traversed by the reflected signal.

an ellipse with the transmitter and the receiver at its foci. The expression for the ellipse is

$$\frac{4x^2}{s_r^2} + \frac{4y^2}{s_r^2 - s_d^2} = 1 \tag{22-1}$$

where x and y = coordinates of point on ellipse (Fig. 22-1)
s_r = length of reflected path
s_d = length of direct path

To locate the source of a reflected signal, the ellipse may be constructed from s_d, the distance between the transmitter and receiver, and $s_r - s_d$, the difference in path length. The path difference may be determined from measurement of the displacement of the ghost image [Eq. (22-2)]. Computations based on a plane earth are sufficiently accurate for locating reflection points when the distances are less than 100 mi (161 km).

Image Displacement due to Multipath. The ghost image produced by a reflected signal is displaced horizontally from the primary image by

$$d = \frac{s_r - s_d}{K} \qquad \text{percent of picture width} \tag{22-2}$$

where $s_r - s_d$ = the path-length difference in feet, and the values of K are given in Table 22-2.

Table 22-2 Reflected Image Displacement Factor K

Scanning standard	Value of K
British 405-line, 25-frame	806
PAL/SECAM 625-line, 25-frame	525
NTSC 525-line, 30-frame	522
French 819-line, 25-frame	398

22.1.4 UNITED STATES CARRIER FREQUENCIES AND TOLERANCES. (See also Sec. 22.1.6.)

- *Standard IF, audio carrier:* 41.25 MHz
- *Standard IF, video carrier:* 45.75 MHz
- *Standard chrominance-subcarrier frequency:* 3.579545 MHz. Tolerance $\pm$ 0.000010 MHz. Rate of change less than 0.1 Hz/s.
- *Standard picture-carrier frequency:* The location of the picture carrier, with respect to the lower edge of the channel, for United States assignments is:
  +1.25 MHz for zero-offset stations
  +1.26 MHz for plus-offset stations
  +1.24 MHz for minus-offset stations
  The tolerance on the picture carrier is $\pm$1000 Hz. For zero-, plus-, and minus-offset assignments, see Table 21-6.
- *Standard sound carrier frequency:* The standard location of the sound carrier is 4.5 MHz higher than the picture carrier, with a tolerance of $\pm$5 kHz in monochrome transmissions and $\pm$1 kHz in color transmissions.
- *Standard intercarrier frequency:* 4.5 MHz in the NTSC standards with a tolerance of $\pm$5 kHz in monochrome transmissions and $\pm$1 kHz in color transmissions.

22.1.5 FREQUENCIES OF IMPORTANCE IN RECEIVER DESIGN

Intrachannel Beat Frequencies. The nominal values of beat frequencies within each television channel are:

- *Picture-sound beat frequency:* 4.500 MHz $\pm$ 5 kHz (monochrome), $\pm$ 1 kHz (color)
- *Picture-chrominance beat frequency:* 3.579545 MHz $\pm$ 10 Hz
- *Sound-chrominance beat frequency:* 0.920445 MHz $\pm$ 1010 Hz

Interchannel Beat Frequencies. The nominal values of beat frequencies between carriers in the desired channel and those in the upper and lower adjacent channels are given in Table 22-3. The tolerances given apply when the desired and interfering stations are offset in like fashion (both stations zero-offset, plus-offset, or minus-offset). If the stations are offset in unlike fashion, the variations in beats may be as great as $\pm$0.02 MHz.

IF Rejection Frequencies. The signal frequencies to be attenuated in picture IF amplifiers, using the standard 41.25-MHz sound and 45.75-MHz picture intermediate frequencies, are:

Upper adjacent picture	39.75 MHz
Upper adjacent sound	35.25 MHz
Upper adjacent chrominance	36.17 MHz

Table 22-3 Beats between Carriers in Adjacent Channels (megahertz)

Desired channel	Upper adjacent channel			Lower adjacent channel		
	Picture	Sound	Chrominance	Picture	Sound	Chrominance
Picture	6.00 $\pm$ 0.002	10.50 $\pm$ 0.002	9.58 $\pm$ 0.001	6.00 $\pm$ 0.002	1.50 $\pm$ 0.002	2.42 $\pm$ 0.001
Sound	1.50 $\pm$ 0.002	6.00 $\pm$ 0.002	5.08 $\pm$ 0.001	10.50 $\pm$ 0.002	6.00 $\pm$ 0.002	6.92 $\pm$ 0.001
Chrominance	2.42 $\pm$ 0.002	6.92 $\pm$ 0.002	6.00 $\pm$ 0.001	9.58 $\pm$ 0.002	5.08 $\pm$ 0.002	6.00 $\pm$ 0.001

Lower adjacent picture	51.75 MHz
Lower adjacent sound	47.25 MHz
Lower adjacent chrominance	48.17 MHz

Local Oscillator Frequencies.   The United States standard intermediate frequencies imply a local oscillator frequency located 47 MHz above the lower limit of the desired channel for zero-offset stations.

Sound Image Frequencies.   Interference to the sound system of a television receiver can occur from image responses. The principal responses of a receiver employing the standard intermediate frequencies are as follow:

- *First-order sound image* (+). The response is located above the desired sound carrier by 82.5 MHz (twice the sound intermediate frequency).
- *Second-order sound image* (−). This response is located 41.25 MHz below twice the local oscillator frequency of the desired channel.

The bandwidth of the image response in each case is equal to the effective bandwidth of the sound IF system (typically 100 kHz).

Picture Image Frequencies.   Interference to the picture system of a television receiver, due to image responses, can occur over a bandwidth equal to the effective bandwidth of the picture IF amplifier, typically 2.8 or 4.2 MHz. The 4.2-MHz bandwidth is used only in color receivers employing comb filters. When the standard intermediate frequencies are used, the principal image responses are as follows:

- *First-order picture image* (+). This response is located in the region from 82.75 to 87.25 MHz above the upper edge of the desired channel.
- *Second-order picture image* (−). This image is located in the region from 41.75 to 46.25 MHz below twice the local oscillator frequency for the desired channel.

22.1.6 UNITED STATES TERRESTRIAL BROADCAST AND CABLE TELEVISION CHANNELS.   See Tables 22-4 and 13-2.

Cable Channel Designations.   Cable television systems (including community antenna systems) in the United States have channel allocations that include the VHF and UHF channels used in terrestrial broadcasting. In addition, since the coaxial cable is protected from interference from other services, additional channels are used for the cable service. Table 13-2 lists the designations of cable channels in the United States as of 1983.

22.1.7 PRINCIPAL INTERFERING SERVICES.   Table 22-5 lists the principal services causing radio-frequency interference to television reception.

22.2 VIDEO-FREQUENCY DATA

22.2.1 VIDEO SIGNAL SPECTRA.   The spectrum of the video signal arising from the scanning process in a television camera extends from a lower limit determined by the time rate of change of the average luminance of the scene to an upper limit determined by time during which the scanning spots cross the sharpest vertical boundary in the scene as focused within the camera (Fig. 22-4). The distribution of spectrum components within these limits depends on the distribution of energy in the camera scanning spots,

Table 22-4 Frequencies Associated with Terrestrial Broadcast Television Channels in the United States†, ‡

Channel number	Frequency limits	Picture carrier	Sound carrier	Chrominance subcarrier	Local oscillator	First-order sound image (+)	Second-order sound image (−)	First-order picture image (+)	Second-order picture image (−)
2	54–60	55.25	59.75	58.829545	101.00	142.25	160.75	142.75–147.25	155.75–160.25
3	60–66	61.25	65.75	64.829545	107.00	148.25	172.75	148.75–153.25	167.75–172.25
4	66–72	67.25	71.75	70.829545	113.00	154.25	184.75	154.75–159.25	179.75–184.25
5	76–82	77.25	81.75	80.829545	123.00	164.25	204.75	164.75–169.25	199.75–204.25
6	82–88	83.25	87.75	86.829545	129.00	170.25	216.75	170.75–175.25	211.75–216.25
7	174–180	175.25	179.75	178.829545	221.00	262.25	401.75	262.75–267.25	396.75–401.25
8	180–186	181.25	185.75	184.829545	227.00	268.25	413.75	268.75–273.25	408.75–413.25
9	186–192	187.25	191.75	190.829545	233.00	274.25	425.75	274.75–279.25	420.75–425.25
10	192–198	193.25	197.75	196.829545	239.00	280.25	437.75	280.75–285.25	432.75–437.25
11	198–204	199.25	203.75	202.329545	245.00	286.25	449.75	286.75–291.25	444.75–449.25
12	204–210	205.25	209.75	208.329545	251.00	292.52	461.75	292.75–297.25	456.75–461.25
13	210–216	211.25	215.75	214.829545	257.00	298.25	473.75	298.75–303.25	468.75–473.25
14	470–476	471.25	475.75	474.829545	517.00	558.25	992.75	558.75–563.25	987.75–992.25
15	476–482	477.25	481.75	480.829545	523.00	564.25	1004.75	564.75–569.25	999.75–1004.25
16	482–488	483.25	487.75	486.829545	529.00	570.25	1016.75	570.75–575.25	1011.75–1016.25
17	488–494	489.25	493.75	492.829545	535.00	576.25	1028.75	576.75–581.25	1023.75–1028.25
18	494–500	495.25	499.75	498.829545	541.00	582.25	1040.75	582.75–587.25	1035.75–1040.25
19	500–506	501.25	505.75	504.829545	547.00	588.25	1052.75	588.75–593.25	1047.75–1052.25
20	506–512	507.25	511.75	510.829545	553.00	594.25	1064.75	594.75–599.25	1059.75–1064.25
21	512–518	513.25	517.75	516.829545	559.00	600.25	1076.75	600.75–605.25	1071.75–1076.25
22	518–524	519.25	523.75	522.829545	565.00	606.25	1088.75	606.75–611.25	1083.75–1088.25
23	524–530	525.25	529.75	528.829545	571.00	612.25	1100.75	612.75–617.25	1095.75–1100.25
24	530–536	531.25	535.75	534.829545	577.00	618.25	1112.75	618.75–623.25	1107.75–1112.25
25	536–542	537.25	541.75	540.829545	583.00	624.25	1124.75	624.75–629.25	1119.75–1124.25
26	542–548	543.25	547.75	546.829545	589.00	630.25	1136.75	630.75–635.25	1131.75–1136.25
27	548–554	549.25	553.75	552.829545	595.00	636.25	1148.75	636.75–641.25	1143.75–1148.25
28	554–560	555.25	559.75	558.829545	601.00	642.25	1160.75	642.75–647.25	1115.75–1160.25

Note: See page 22.9 for footnotes.

Table 22-4 Frequencies Associated with Terrestrial Broadcast Television Channels in the United States† ‡ (*Continued*)

Channel number	Frequency limits	Picture carrier	Sound carrier	Chrominance subcarrier	Local oscillator	First-order sound image (+)	Second-order sound image (−)	First-order picture image (+)	Second-order picture image (−)
29	560–566	561.25	565.75	564.829545	607.00	648.25	1172.75	648.75–653.25	1167.75–1172.25
30	566–572	567.25	571.75	570.829545	613.00	654.25	1184.75	654.75–659.25	1179.75–1184.25
31	572–578	573.25	577.75	576.829545	619.00	660.25	1196.75	660.75–665.25	1191.75–1196.25
32	578–584	579.25	583.75	582.829545	625.00	666.25	1208.75	666.75–671.25	1203.75–1208.25
33	584–590	585.25	589.75	588.829545	631.00	672.25	1220.75	672.75–677.25	1215.75–1220.25
34	590–596	591.25	595.75	594.829545	637.00	678.25	1232.75	678.75–683.25	1227.75–1232.25
35	596–602	597.25	601.75	600.829545	643.00	684.25	1244.75	684.75–689.25	1239.75–1244.25
36	602–608	603.25	607.75	606.829545	649.00	690.25	1256.75	690.75–695.25	1251.75–1256.25
37	608–614	609.25	613.75	612.829545	655.00	696.25	1268.75	696.75–701.25	1263.75–1268.25
38	614–620	615.25	619.75	618.829545	661.00	702.25	1280.75	702.75–707.25	1275.75–1280.25
39	620–626	621.25	625.75	624.829545	667.00	708.25	1292.75	708.75–713.25	1287.75–1292.25
40	626–632	627.25	631.75	630.829545	673.00	714.25	1304.75	714.75–719.25	1299.75–1304.25
41	632–638	633.25	637.75	636.829545	679.00	720.25	1316.75	720.75–725.25	1311.75–1316.25
42	638–644	639.25	643.75	642.829545	685.00	726.25	1328.75	726.75–731.25	1323.75–1328.25
43	644–650	645.25	649.75	648.829545	691.00	732.25	1340.75	732.75–737.25	1335.75–1340.25
44	650–656	651.25	655.75	654.829545	697.00	738.25	1352.75	738.75–743.25	1347.75–1352.25
45	656–662	357.25	661.75	660.829545	703.00	744.25	1364.75	744.75–749.25	1359.75–1364.25
46	662–668	663.25	667.75	666.829545	709.00	750.25	1376.75	750.75–755.25	1371.75–1376.25
47	668–674	669.25	673.75	672.829545	715.00	756.25	1388.75	756.75–761.25	1383.75–1388.25
48	674–680	675.25	679.75	678.829545	721.00	762.25	1400.75	762.75–767.25	1395.75–1400.25
49	680–686	681.25	685.75	684.829545	727.00	768.25	1412.75	768.75–773.25	1407.75–1412.25
50	686–692	687.25	691.75	690.829545	733.00	774.25	1424.75	774.75–779.25	1419.75–1424.25
51	692–698	693.25	697.75	696.829545	739.00	780.25	1436.75	780.75–785.25	1431.75–1436.25
52	698–704	699.25	703.75	702.829545	745.00	786.25	1448.75	786.75–791.25	1443.75–1448.25
53	704–710	705.25	709.75	708.829545	751.00	792.25	1460.75	792.75–797.25	1455.75–1460.25
54	710–716	711.25	715.75	714.829545	757.00	798.25	1472.75	798.75–803.25	1467.75–1472.25
55	716–722	717.25	721.75	720.829545	763.00	704.25	1484.75	804.75–809.25	1479.75–1484.25
56	722–728	723.25	727.75	726.829545	769.00	810.25	1496.75	810.75–815.25	1491.75–1496.25
57	728–734	729.25	733.75	732.829545	775.00	816.25	1508.75	816.75–821.25	1503.75–1508.25

Ch.	Band (MHz)								
58	734–740	735.25	739.75	738.829545	781.00	822.25	1520.75	822.75–827.25	1515.75–1520.25
59	740–746	741.25	745.75	744.829545	787.00	828.25	1532.75	828.75–833.25	1527.75–1532.25
60	746–752	747.25	751.75	750.829545	793.00	834.25	1544.75	834.75–839.25	1539.75–1544.25
61	752–758	753.25	757.75	756.829545	799.00	840.25	1556.75	840.75–845.25	1551.75–1556.25
62	758–764	759.25	763.75	762.829545	805.00	846.25	1568.75	846.75–851.25	1563.75–1568.25
63	764–770	765.25	769.75	768.829545	811.00	852.25	1580.75	852.75–857.25	1575.75–1580.25
64	770–776	771.25	775.75	774.829545	817.00	858.25	1592.75	858.75–863.25	1587.75–1592.25
65	776–782	777.25	781.75	780.829545	823.00	864.25	1604.75	864.75–869.25	1599.75–1604.25
66	782–788	783.25	787.75	786.829545	829.00	870.25	1616.75	870.75–875.25	1611.75–1616.25
67	788–794	789.25	793.75	792.829545	835.00	876.25	1628.75	876.75–881.25	1623.75–1628.25
68	794–800	795.25	799.75	798.829545	841.00	882.25	1640.75	882.75–887.25	1635.75–1640.25
69	800–806	801.25	805.75	804.829545	847.00	888.25	1652.75	888.75–893.25	1647.75–1652.25
70‡	806–812	807.25	811.75	810.829545	853.00	894.25	1664.75	894.75–899.25	1659.75–1664.25
71‡	812–818	813.25	817.75	816.829545	859.00	900.25	1676.75	900.75–905.25	1671.75–1676.25
72‡	818–824	819.25	823.75	822.829545	865.00	906.25	1688.75	906.75–911.25	1683.75–1688.25
73‡	824–830	825.25	829.75	828.829545	871.00	912.25	1700.75	912.75–917.25	1695.75–1700.25
74‡	830–836	831.25	835.75	834.829545	877.00	918.25	1712.75	918.75–923.25	1707.75–1712.25
75‡	836–842	837.25	841.75	840.829545	883.00	924.25	1724.75	924.75–929.25	1719.75–1724.25
76‡	842–848	843.25	847.75	846.829545	889.00	930.25	1736.75	930.75–935.25	1731.75–1736.25
77‡	848–854	849.25	853.75	852.829545	895.00	936.25	1748.75	936.75–941.25	1743.75–1748.25
78‡	854–860	855.25	859.75	858.829545	901.00	942.25	1760.75	942.75–947.25	1755.75–1760.25
79‡	860–866	861.25	865.75	864.829545	907.00	948.25	1772.75	948.75–953.25	1767.75–1772.25
80‡	866–872	867.25	871.75	870.829545	913.00	954.25	1784.75	954.75–959.25	1779.75–1784.25
81‡	872–878	873.25	877.75	876.829545	919.00	960.25	1796.75	960.75–965.25	1791.75–1796.25
82‡	878–884	879.25	883.75	882.829545	925.00	966.25	1808.75	966.75–971.25	1803.75–1808.25
83‡	884–890	885.25	889.75	888.829545	931.00	972.25	1820.75	972.75–977.25	1815.75–1820.25

†The frequencies listed (given in megahertz) apply to stations transmitting on the zero-offset channel assignments. For plus-offset channels add 0.01 MHz to the picture, sound, chrominance, and local oscillator frequencies. For minus-offset assignments, subtract 0.01 MHz in each case. See Table 21-6 for offset assignments to particular localities. For plus-offset assignments add 0.02 MHz to the second-order picture image frequencies; subtract 0.02 MHz for minus-offset assignments.

‡Effective September 1982, the Federal Communications Commission removed the frequencies from 806 to 890 MHz (formerly television channels 70–83 inclusive) from the television broadcast service.

22.9

Table 22-5   Services Causing Radio-Frequency Interference to Television Reception

Interfering service	Frequencies, MHz	Principal type of interference
Amateur	1.8–2.0	Transmitter harmonics
	3.5–4.0	Transmitter harmonics
	7.0–7.3	Transmitter harmonics
	10.1–10.15	Transmitter harmonics
	14.0–14.35	Transmitter harmonics
	18.07–18.17	Transmitter harmonics
	21.0–21.45	Transmitter harmonics
	24.9–25	Transmitter harmonics
	28.0–29.7	Transmitter harmonics
	50.0–54.0	Transmitter harmonics; adjacent to channel 2
	144–148	Transmitter harmonics
	220–225	Transmitter harmonics
	420–450	Image responses
	1215–1300	Second-order images $(-)$
Citizens' radio	27–29.7	Transmitter harmonics
	460–470	Adjacent to channel 14
FM radio	88–108 (200-kHz channels)	Adjacent to channel 6; transmitter harmonics
Public safety	33.42–33.98	Transmitter harmonics
	37.02–37.42	Transmitter harmonics
	37.90–37.98	Transmitter harmonics
	39.02–39.98	Transmitter harmonics
	42.02–42.94	IF pickup; harmonics
	44.62–47.66	IF pickup; harmonics
	153.77–154.43	First-order images $(+)$
	154.65–156.21	First-order images $(+)$
	158.73–159.21	First-order images $(+)$
Industrial, scientific, medical	27	Harmonics
	40.6–40.7	Harmonics

the number of lines scanned per second, the percentage of line-scan time consumed by horizontal blanking, the number of fields scanned per second, and the rates at which the luminances and chrominances of the scene change in size, position, and boundary sharpness.

To the extent that the contents and dynamic properties of the scene cannot be predicted, the spectrum limits and energy distribution are not defined. However, the spectra associated with certain static and dynamic test charts, films, and tapes may be used as the basis for television system design. Among the configurations of interest are flat fields of uniform luminance and/or chrominance and fields divided into two or more segments of different luminance by sharp vertical, horizontal, or oblique boundaries, including as particular cases the horizontal and vertical wedges of test charts and the concentric circles of zone-plate charts (Figs. 22-2 and 22-3). The reproductions of such configurations typically display diffuse boundaries and other degradations introduced by the camera scanning spots, the amplitude and phase responses of the transmission system, the receiver scanning spots, and artifacts associated with the scanning process.

The upper limit of the video spectrum actually employed in reproducing a particular image is most often determined by the amplitude-versus-frequency and phase-versus-frequency responses of the receiver in use. These responses are selected as a compromise between the image sharpness demanded by typical viewers and the deleterious effects of

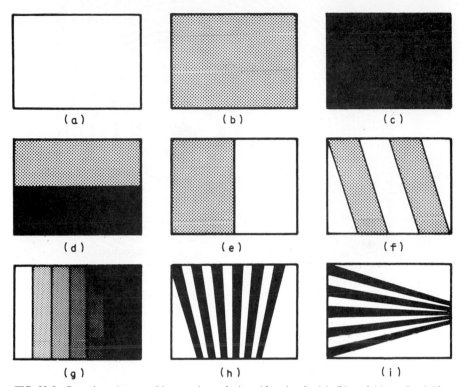

FIG. 22-2 Scanning patterns of interest in analyzing video signals. (a), (b), and (c) are flat fields useful for determining color purity and transfer gradient (gamma); (d) is a horizontal half-field pattern for measuring low-frequency performance; (e) is a vertical half field of interest in examining high-frequency transient performance; (f) displays oblique bars (see also patterns in Fig. 22-5); (g) in monochrome is a tonal wedge for determining contrast and luminance transfer characteristics; this pattern in color (bars in order, left to right: white, yellow, cyan, green, magenta, red, blue) is useful in hue measurements and adjustments; (h) and (i) are wedges for measuring horizontal and vertical resolution, respectively.

noise, interference, and incomplete separation of the luminance and chrominance signals in the receiver. Typical video spectrum limits and energy distributions are specified in Tables 22-7 and 22-9.

22.2.2 SCANNING CONSTANTS. The constants that determine the video spectrum arising from scanning a scene are given in Table 22-6.

22.2.3 MINIMUM VIDEO FREQUENCY. To reproduce a uniform value of luminance from top to bottom of an image scanned in the conventional interlaced fashion, the video signal spectrum must extend downward to include the field-scanning frequency. This frequency represents the lower limit of the spectrum arising from scanning an image whose luminance does not change. Changes in the average luminance are reproduced by extending the video spectrum to a lower frequency equal to the reciprocal of the duration of the luminance change. Since a given average luminance may persist for many minutes, the spectrum extends sensibly to zero frequency (see DC Component, Sec. 22.2.4). The technique of preserving or restoring the dc component is employed to extend the spectrum from the field frequency down to zero frequency.

FIG. 22-3 The zone-plate pattern, capable of indicating resolution at all angles, including the horizontal and vertical. Dynamic versions of the zone plate, in which the pattern changes size and position as a function of time, are useful in determining temporal as well as spatial characteristics of the imaging process.

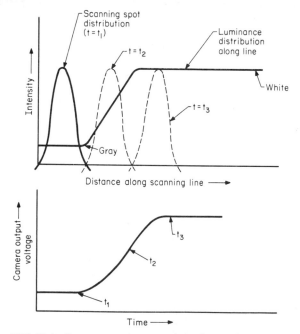

FIG. 22-4 Camera scanning spot, shown here with a gaussian distribution, passing over a luminance boundary on a scanning line. The convolution (below) of the spot and luminance distributions constitutes the corresponding camera output signal.

Table 22-6 Scanning Constants of Television Systems†

Scanning constant	NTSC 525/30 system		PAL 625/25 system	SECAM-III 625/25 system	NHK developmental 1125/30 system
	Monochrome	Color			
Nominal luminance bandwidth, MHz	4.2	4.2	5.5	6	20
Frame frequency, Hz	30	29.97	25	25	30
Frame period, ms	33.333	33.367	40	40	33.333
Active frame time, ms (maximum blanking tolerance)	30.644	30.674	36.775	36.775	32.000
Field frequency, Hz	60	59.94	50	50	60
Field period, ms	16.667	16.683	20	20	16.667
Active field time, ms (maximum blanking tolerance)	15.322	15.337	18.387	18.387	16.000
Line frequency, Hz	15,750	15,734.3	15,625	15,625	33,750
Line period, μs	63.492	63.556	64	64	29.630
Active line time, μs (maximum blanking tolerance)	52.092	52.456	51.7	51.7	24.692
Horizontal resolution factor (lines per megahertz)	78.1	78.7	77.6	77.6	29.65
Horizontal luminance resolution (lines per picture height)	328	330	427	466	593
Vertical luminance resolution (lines per picture height)	343	343	408	408	767
Picture elements per frame	150,000	151,000	232,000	254,000	758,000

†Data on NTSC, PAL, and SECAM systems based on CCIR Reports, Volume XI—Broadcasting Service (Television), International Telecommunications Union, Geneva, 1978. Data on the NHK high-definition system provided by Dr. Takashi Fujio, Deputy Director, Technical Research Laboratories, Japan Broadcasting Corporation. The latter data were those in use in the NHK-CBS HDTV demonstrations in the United States in February 1982. The vertical resolution figures are based on a number of lines equal to 71% of the active lines in the frame. The NHK system has an aspect ratio of 5/3, the other systems 4/3.

Table 22-7 Chrominance Frequencies and Resolutions

	NTSC 525/30 system		PAL 625/25 systems		SECAM-III 625/25 system	
	I signal	Q signal	E_U signal	E_V signal	D_R signal	D_B signal
Chrominance subcarrier (f_H = line-scanning frequency)	**3.579545 MHz $\pm$ 10 Hz** $= 455\, f_H/2$		**4.433618.75 MHz $\pm$ 5 Hz** $= 1135\, f_H/4 + f_H/625$		**4.406250 MHz $\pm$ 2 kHz** **4.250000 MHz $\pm$ 2 kHz**	
Nominal bandwidth	**0–1.3 MHz**	**0–0.62 MHz**	**0–0.57 MHz** 0–1.07 MHz†	**0–1.3 MHz** 0–1.3 MHz†	**0–1.3 MHz**	**0–1.3 MHz**
Maximum video frequency	**1.3 MHz**	**0.62 MHz**	**0.57 MHz** 1.07 MHz†	**1.3 MHz** 1.3 MHz†	**1.3 MHz**	**1.3 MHz**
Horizontal resolution	102 lines	49 lines	44 lines 83 lines†	101 lines 101 lines†	101 lines	101 lines
Vertical resolution	343 lines	343 lines	204 lines	204 lines	204 lines	204 lines

†Values applicable to PAL system I, whose nominal luminance bandwidth (Table 22-6) is 5.5 MHz.

Source: CCIR Recommendations and Reports Volume XI—Broadcasting Service (Television), Geneva, 1982.

Table 22-8 Spectra of Commonly Encountered Waveforms

Time function	Frequency function	Equations
Rectangular wave		$A_{avg} = A\,t_0/T$ $A_{rms} = A\sqrt{t_0/T}$ $C_n = 2A_{avg}\left(\dfrac{\sin \pi\,\dfrac{nt_0}{T}}{\pi nt_0/T}\right)$
Isosceles-triangle wave		$A_{avg} = A\,t_1/2T$ $A_{rms} = A\sqrt{2t_1/3T}$ $C_n = 2A_{avg}\left(\dfrac{\sin \pi\,\dfrac{nt_1}{T}}{\pi nt_1/T}\right)^2$
Clipped sawtooth wave		$A_{avg} = \dfrac{At_0}{2T}$ $A_{rms} = A\sqrt{\dfrac{t_0}{3T}}$ $C_n = \dfrac{4A_{avg}}{\left(\dfrac{2\pi nt_0}{T}\right)^2}\left[2\left(1-\cos\dfrac{2\pi nt_0}{T}\right)+\dfrac{2\pi nt_0}{T}\left(\dfrac{2\pi nt_0}{T}-2\sin\dfrac{2\pi nt_0}{T}\right)\right]^{1/2}$
Sawtooth wave		$A_{avg} = \dfrac{A}{2}$ $A_{rms} = \dfrac{A}{\sqrt{3}}$ $C_n = -2A_{avg}\left(\dfrac{\cos \pi n}{\pi n}\right)$

Table 22-8 Spectra of Commonly Encountered Waveforms (*Continued*)

Time function	Frequency function	Equations
Half sine wave	$F = 1/T$, $f_0 = 1/t_0$; $nF/f_0 = nt_0/T$	$A_{avg} = \dfrac{2At_0}{\pi T}$ $\qquad A_{rms} = A\sqrt{\dfrac{t_0}{2T}}$ $C_n = \dfrac{\pi}{2} A_{avg} \left[\dfrac{\sin\dfrac{\pi}{2}\left(1 - \dfrac{2nt_0}{T}\right)}{\dfrac{\pi}{2}\left(1 - \dfrac{2nt_0}{T}\right)} + \dfrac{\sin\dfrac{\pi}{2}\left(1 + \dfrac{2nt_0}{T}\right)}{\dfrac{\pi}{2}\left(1 + \dfrac{2nt_0}{T}\right)} \right]$
Full sine wave	$F = 1/T$; $f_0 = 1/t_0$; $nF/f_0 = nt_0/T$	$A_{avg} = \dfrac{A}{2}\dfrac{t_0}{T}$ $\qquad A_{rms} = \dfrac{A}{2}\sqrt{\dfrac{3t_0}{2T}}$ $C_n = A_{avg} \left[2\,\dfrac{\sin\pi\dfrac{nt_0}{T}}{\pi\dfrac{nt_0}{T}} + \dfrac{\sin\pi\left(1 - \dfrac{nt_0}{T}\right)}{\pi\left(1 - \dfrac{nt_0}{T}\right)} + \dfrac{\sin\pi\left(1 + \dfrac{nt_0}{T}\right)}{\pi\left(1 + \dfrac{nt_0}{T}\right)} \right]$
Full-wave-rectified sine wave	$F/f_0 = t_0/T = 1$	$A_{avg} = \dfrac{2}{\pi}A$ $\qquad A_{rms} = \dfrac{A}{\sqrt{2}}$ $C_n = \dfrac{\pi}{2} A_{avg} \left[\dfrac{\sin^2\dfrac{\pi}{2}(1 - 2n)}{\dfrac{\pi}{2}(1 - 2n)} + \dfrac{\sin^2\dfrac{\pi}{2}(1 + 2n)}{\dfrac{\pi}{2}(1 + 2n)} \right]$

Critically damped exponential wave

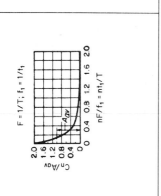

$$y(t) = \frac{A\epsilon}{t_1} t\epsilon^{-t/t_1} \qquad \text{for } T > 10t_1, \text{ where } \epsilon = 2.718$$

$$A_{\text{avg}} = \frac{A\epsilon t_1}{T} \qquad A_{\text{rms}} = \frac{A\epsilon}{2}\sqrt{\frac{t_1}{T}}$$

$$C_n = 2A_{\text{avg}} \left[\frac{1}{1 + \left(2\pi\,\dfrac{nt_1}{T}\right)^2} \right]$$

$$= 2A_{\text{avg}} \cos^2\frac{\theta_n}{2}$$

$$\frac{\theta_n}{2} = \tan^{-1}\left(2\pi\,\frac{nt_1}{T}\right)$$

F = 1/T; f₁ = 1/t₁ ... (graph axes: $F = 1/T;\ f_1 = 1/t_1$; vertical C_n/A_{av}; horizontal $nF/f_1 = nt_1/T$)

Symmetrical trapezoid wave

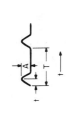

$$A_{\text{avg}} = A\,\frac{t_0 + t_1}{T} \qquad A_{\text{rms}} = A\sqrt{\frac{3t_0 + 2t_1}{3T}}$$

$$C_n = 2A_{\text{avg}} \left[\frac{\sin \pi\,\dfrac{nt_1}{T}}{\pi\,\dfrac{nt_1}{T}} \right] \left[\frac{\sin \pi\,\dfrac{n(t_0 + t_1)}{T}}{\pi\,\dfrac{n(t_0 + t_1)}{T}} \right]$$

Unsymmetrical trapezoid wave

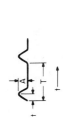

$$A_{\text{avg}} = \frac{A}{T}\left[t_0 + \frac{t_1}{2} + \frac{t_2}{2} \right] \qquad A_{\text{rms}} = A\sqrt{\frac{3t_0 + t_1 + t_2}{3T}}$$

If $t_1 \approx t_2$

$$C_n = 2A_{\text{avg}} \left[\frac{\sin \pi\,\dfrac{nt_1}{T}}{\pi\,\dfrac{nt_1}{T}} \right] \left[\frac{\sin \pi\,\dfrac{n(t_0 + t_1)}{T}}{\pi\,\dfrac{n(t_0 + t_1)}{T}} \right] \left[\frac{\sin \pi\,\dfrac{n(t_2 - t_1)}{T}}{\pi\,\dfrac{n(t_2 - t_1)}{T}} \right]$$

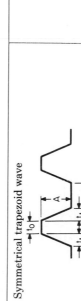

22.17

Table 22-8 Spectra of Commonly Encountered Waveforms (*Continued*)

Fractional sine wave

$$A_{\text{avg}} = \frac{A\left(\sin \pi \frac{t_0}{T} - \pi \frac{t_0}{T}\cos \pi \frac{t_0}{T}\right)}{\pi\left(1 - \cos \pi \frac{t_0}{T}\right)}$$

$$A_{\text{rms}} = \frac{A}{\left(1 - \cos \pi \frac{t_0}{T}\right)}\left[\frac{1}{2\pi}\left(\pi \frac{t_0}{T} + \frac{1}{2}\sin^2 \pi \frac{t_0}{T} - 4\cos \pi \frac{t_0}{T}\sin \pi \frac{t_0}{T} + 2\pi \frac{t_0}{T}\cos^2 \pi \frac{t_0}{T}\right)\right]$$

$$C_n = \frac{A_{\text{avg}}\,\pi \frac{t_0}{T}}{n\left(\sin \pi \frac{t_0}{T} - \pi \frac{t_0}{T}\cos \pi \frac{t_0}{T}\right)}\left[\frac{\sin \pi(n-1)\frac{t_0}{T}}{\pi(n-1)\frac{t_0}{T}} - \frac{\sin \pi(n+1)\frac{t_0}{T}}{\pi(n+1)\frac{t_0}{T}}\right]$$

Sawtooth wave

$$A_{\text{avg}} = \frac{A}{2} \qquad A_{\text{rms}} = \frac{A}{\sqrt{3}}$$

$$C_n = \frac{2A_{\text{avg}}}{\pi^2 n^2 \frac{t_1}{T}\left(1 - \frac{t_1}{T}\right)}\sin \pi \frac{t_1}{T}$$

Source: *Reference Data for Radio Engineers*, by permission.

Table 22-9 The Quantities $(\sin x)/x$ and $Si(x)$ (x in radians)

x	$(\sin x)/x$	$Si(x)$	x	$(\sin x)/x$	$Si(x)$	x	$(\sin x)/x$	$Si(x)$
0.0	1.0000	0.0000	3.5	−0.1002	1.8331	7.0	0.0939	1.4546
0.1	0.9983	0.0999	3.6	−0.1229	1.8220	7.1	0.1027	1.4644
0.2	0.9933	0.1996	3.7	−0.1432	1.8086	7.2	0.1102	1.4751
0.3	0.9851	0.2985	3.8	−0.1610	1.7933	7.3	0.1165	1.4864
0.4	0.9735	0.3965	3.9	−0.1764	1.7765	7.4	0.1214	1.4983
0.5	0.9589	0.4931	4.0	−0.1892	1.7582	7.5	0.1251	1.5107
0.6	0.9411	0.5881	4.1	−0.1996	1.7387	7.6	0.1274	1.5233
0.7	0.9203	0.6812	4.2	−0.2075	1.7184	7.7	0.1283	1.5361
0.8	0.8967	0.7721	4.3	−0.2131	1.6973	7.8	0.1280	1.5489
0.9	0.8704	0.8605	4.4	−0.2163	1.6758	7.9	0.1264	1.5617
1.0	0.8415	0.9461	4.5	−0.2172	1.6541	8.0	0.1237	1.5742
1.1	0.8102	1.0287	4.6	−0.2160	1.6325	8.1	0.1197	1.5864
1.2	0.7767	1.1081	4.7	−0.2127	1.6110	8.2	0.1147	1.5981
1.3	0.7412	1.1840	4.8	−0.2075	1.5900	8.3	0.1087	1.6093
1.4	0.7039	1.2562	4.9	−0.2005	1.5696	8.4	0.1017	1.6198
1.5	0.6650	1.3247	5.0	−0.1918	1.5499	8.5	0.0939	1.6296
1.6	0.6247	1.3892	5.1	−0.1815	1.5313	8.6	0.0854	1.6386
1.7	0.5833	1.4496	5.2	−0.1699	1.5137	8.7	0.0762	1.6467
1.8	0.5410	1.5058	5.3	−0.1570	1.4973	8.8	0.0665	1.6538
1.9	0.4981	1.5578	5.4	−0.1431	1.4823	8.9	0.0563	1.6599
2.0	0.4546	1.6054	5.5	−0.1283	1.4687	9.0	0.0458	1.6650
2.1	0.4111	1.6487	5.6	−0.1127	1.4567	9.1	0.0351	1.6691
2.2	0.3675	1.6876	5.7	−0.0966	1.4462	9.2	0.0242	1.6721
2.3	0.3242	1.7222	5.8	−0.0801	1.4374	9.3	0.0134	1.6739
2.4	0.2814	1.7525	5.9	−0.0634	1.4302	9.4	0.0026	1.6747
2.5	0.2394	1.7785	6.0	−0.0466	1.4247	9.5	−0.0079	1.6745
2.6	0.1983	1.8004	6.1	−0.0299	1.4209	9.6	−0.0182	1.6732
2.7	0.1583	1.8182	6.2	−0.0134	1.4187	9.7	−0.0280	1.6708
2.8	0.1196	1.8321	6.3	0.0027	1.4182	9.9	−0.0374	1.6676
2.9	0.0825	1.8422	6.4	0.0182	1.4192	9.9	−0.0462	1.6634
3.0	0.0470	1.8487	6.5	0.0331	1.4218	10.0	−0.0544	1.6584
3.1	0.0134	1.8517	6.6	0.0472	1.4258	10.1	−0.0619	1.6525
3.2	−0.0182	1.8514	6.7	0.0604	1.4312	10.2	−0.0686	1.6460
3.3	−0.0478	1.8481	6.8	0.0727	1.4379	10.3	−0.0745	1.6388
3.4	−0.0752	1.8419	6.9	0.0838	1.4457	10.4	−0.0796	1.6311
						∞	0.0000	1.5708

22.2.4 DC COMPONENT. It is customary in radiating the picture signal to maintain constant the carrier envelope level corresponding to black (zero luminance) irrespective of the average luminance of the scene. The corresponding level in the video signal as applied to the control electrodes of the picture tube can be held constant by conductive coupling of the amplifiers following the second detector, or by restoring the constancy of this level by peak detection or clamping. In images so reproduced, the average luminance of the image bears a direct correspondence to the average luminance of the televised scene. The time constant of the dc-restoring system may be designed to be long or short compared with the line-scanning period (Table 22-6). When peak detection of the sync pulses is used to establish the black level, the time constant is long; when a clamp circuit is used, it rapidly brings the black level of each line to the desired constant value, irrespective of prior disturbances to the black level. Since each line scan is thereby reproduced with correct average luminance, it follows that any number of line scans has the correct average value and the video spectrum is thus effectively extended to zero frequency. The peak-detection dc-restoring system is subject to overload and must be protected from strong noise pulses. DC restoration is omitted in most inexpensive black-and-white receivers, but it is essential to the correct rendition of color values and is universally employed in color television receivers.

22.2.5 MAXIMUM VIDEO FREQUENCY. Three values of maximum video frequency should be distinguished: (1) the maximum camera frequency generated by the camera, (2) the maximum modulating frequency corresponding to the fully transmitted radiated sideband, and (3) the maximum receiver video frequency present at the picture-tube control electrodes.

The *maximum camera frequency* is determined by the size and current distribution of the camera scanning spots and the time the spots take to move over the sharpest vertical boundary in the scene as focused on the scanned surfaces of the camera. The resulting transient video signal current is represented by the convolution integral of the spot distributions and the brightness distribution of the boundary, both measured along the scanning line (Fig. 22-4). The spectrum of the video signal transient includes significant amplitudes up to the frequency equal to the inverse of the time of rise of the resultant transient. Higher frequencies are also present but have a small effect on the shape of the reproduced transient after transmission of the video signal. Television cameras and scanners are usually designed to generate maximum camera frequencies of 8 to 10 MHz, that is, approximately twice the maximum modulating frequency.

The *maximum modulating frequency* is determined by the extent of the television channel reserved for the fully transmitted sideband. The channel width, in turn, is chosen to provide a value of horizontal resolution (see Sec. 22.2.6) approximately equal to the vertical resolution implicit in the scanning pattern. The maximum modulation frequencies of the principal systems are given in Table 22-6 after "Nominal Luminance Bandwidth."

The *maximum receiver video frequency* as applied to the picture tube depends on the bandwidth of the picture IF amplifier, picture second detector, and video amplifiers. The value selected is typically 2.8 MHz, a value substantially lower than the maximum modulating frequency of 4.2 MHz in the NTSC standards. Until the advent of the comb filter, it was not possible to separate the luminance signal completely from the chrominance signal in the vicinity of the 3.58-MHz color subcarrier. When strong luminance signal components are produced at or near that frequency, cross-color and cross-luminance are produced (see Sec. 22.2.8). The cutoff of the luminance signal at or about 2.8 MHz attenuates these components. It has been chosen to avoid cross effects (and also to reduce noise in the image) in receivers that do not use a comb filter. The maximum horizontal resolution given in Table 22-6 is that attainable when the full bandwidth of the luminance signal is employed.

22.2.6 HORIZONTAL RESOLUTION. The horizontal resolution factor (Table 22-6) is the proportionality factor between horizontal resolution and video frequency. When expressed in lines per megahertz, the factor is equal to three-halves the active line period in microseconds (for example, $\frac{3}{2} \times 52.5 = 78.7$ lines per megahertz for the NTSC standards). The three-halves factor represents the number of lines of horizontal resolution per hertz of the video waveform (2) divided by the aspect ratio ($\frac{4}{3}$).

22.2.7 CHROMINANCE RESOLUTION. The FCC standards specifying the chrominance signal state that the Q signal shall be attenuated, prior to modulation, less than 6 dB at 500 kHz and the I signal less than 2 dB at 1.3 MHz. Since the chrominance signals arise from the same scanning process as the luminance signal, the scanning constants given in Table 22-6 apply to chrominance resolution. See Table 22-7 for chrominance frequencies and resolutions.

22.2.8 VIDEO FREQUENCIES ARISING FROM SCANNING. The signal spectrum arising from scanning[1,2]† comprises a number of discrete components at multiples of the scanning frequencies (Table 22-6). Each spectrum component is identified by two numbers, m and n, which describe the pattern that would be produced if that component alone were present in the signal. The value of m represents the number of sinusoidal cycles of brightness measured horizontally (in the width of the picture) and n the number of cycles measured vertically (in the picture height). The 0, 0 pattern is the dc component of the signal, the 0, 1 pattern is produced by the field-scanning frequency, and the 1, 0 pattern by the line-scanning frequency. Typical patterns for various values of m and n are shown in Fig. 22-5. By combining a number of such patterns (including m and n values up to several hundred), in the appropriate amplitudes and phases, any image capable of being represented by the scanning pattern may be built up. This is a two-dimensional form of the Fourier series.

The amplitudes of the spectrum components decrease as the values of m and n increase. Since m represents the order of the harmonic of the line-scanning frequency, the corresponding amplitudes are those of the left-to-right variations in brightness. The spectra corresponding to various brightness distributions are given in Table 22-8. A typical spectrum resulting from scanning a static scene is shown in Fig. 22-6. The components of major magnitude are the dc component, the field-frequency component, and the components of the line frequency and its harmonics. Surrounding each line-frequency harmonic is a cluster of components, each separated from the next by an interval equal to the field-scanning frequency.

It is possible for the clusters surrounding adjacent line-frequency harmonics to overlap one another. This corresponds to two patterns in Fig. 22-5 situated on adjacent vertical columns, which produce the same value of video frequency when scanned. Such "intercomponent confusion" of spectral energy is fundamental to

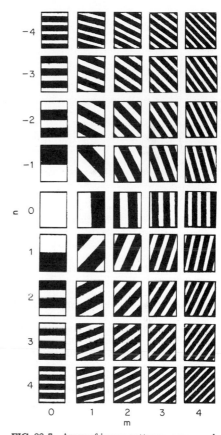

FIG. 22-5 Array of image patterns corresponding to indicated values of m and n. *(After Pierre Mertz.)*

the scanning process. Its effects are visible when a heavily striated pattern (such as that of a fabric having an accented weave) is scanned with the striations approximately parallel to the scanning lines. In the NTSC and PAL color systems, in which the luminance and chrominance signals occupy the same spectral region (one being interlaced in frequency with the other, Sec. 22.2.9), such intercomponent confusion may produce prominent color fringes. Precise filters, which sharply separate the luminance and chrominance signals ("comb filters"), can remove this effect, except in the diagonal direction.

---

†Superscript numbers refer to the References at end of chapter.

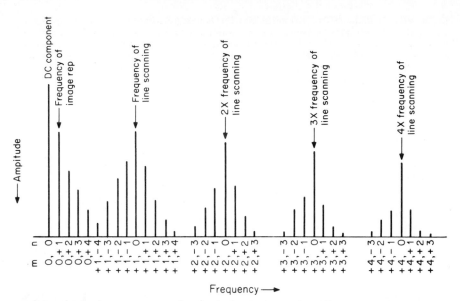

FIG. 22-6 Typical spectrum of a video signal, showing harmonics of line-scanning frequency surrounded by clusters of components separated at intervals equal to the field-scanning frequency. *(After Pierre Mertz.)*

In static and slowly moving scenes, the clusters surrounding each line-frequency harmonic are compact, seldom extending further than 1 or 2 kHz (for example, the twentieth harmonic of the field frequency) on either side of the line-harmonic frequency. The space remaining in the signal spectrum is unoccupied and may be used to accommodate the spectral components of another signal having the same structure and frequency spacing, e.g., one rigidly tied to the same scanning rates. This is the method employed in the NTSC and PAL color systems to interlace the luminance and chrominance components of the color signal. It is thus true that for scenes in which the motion is sufficiently slow for the eye to perceive the detail of the moving objects, it may be safely assumed that less than half the spectral space between line-frequency harmonics is occupied by energy of significant magnitude. It is on this principle that the NTSC and PAL compatible color television systems are based. The SECAM system uses frequency-modulated chrominance signals (see Chaps. 4 and 21), which are not frequency interlaced with the luminance signal.

22.2.9 FREQUENCY INTERLEAVING OF SPECTRUM COMPONENTS.   As noted above, the unoccupied portions of the video signal spectrum, between harmonics of the line frequency, may be filled by the harmonic clusters of another signal and both transmitted simultaneously with minimal interference. This technique is known as *frequency interleaving* (or *frequency interlace*). It is employed (1) to transmit the luminance and chrominance signals simultaneously in compatible color systems, (2) to transmit pilot signals in the operation of coaxial cable facilities, and (3) to minimize co-channel interference between stations.

The signals to be interleaved must arise from integrally related timing sources to preserve the interleaved relationship throughout the spectrum. When both signals arise from the same scanning process (as do the luminance and chrominance signals in color transmissions), the interleaved relationship can be very precisely maintained by deriving the timing of the color subcarrier and the scanning pattern from the same source. Frequency synthesizing circuits are used to produce a line-scanning frequency equal to twice an odd subharmonic of the color subcarrier frequency.

Example. In the NTSC compatible color system, the line-scanning frequency is 15,734.264 Hz and the color subcarrier frequency is equal to half the 455th multiple of that frequency, that is, 3.579545 MHz. Similarly the spacing between chrominance and sound signals is 0.920455 MHz, which is half the 117th multiple of the line-scanning frequency. For additional information on frequency interleaving, see Chap. 5.

When the interleaved signals arise from separate sources (as in offset operation of co-channel stations), such precision is not feasible, but it is possible to interleave the components less accurately with benefit. For example, in operating three or more stations in offset fashion on the same channel, it is not feasible to employ half the line-scanning frequency as the offset, since this would displace the carrier frequencies of two of the stations by the whole value of the line-scanning frequency, the most disadvantageous value.

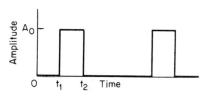

FIG. 22-7 Rectangular wave used in examining transient responses of transmission systems.

As a compromise, the FCC channel assignments are offset by 0.01 MHz, which is approximately two-thirds of the line-scanning frequency. A reduction in co-channel interference of about 16 dB results, compared with 20 dB when the offset is at the optimum value.

22.2.10 SPECTRA OF WAVEFORMS OF INTEREST IN VIDEO ENGINEERING.
Fourier analysis of repetitive waveforms gives the average and root-mean-square values of the wave and the amplitudes and phases of the harmonics. Table 22-8 presents these data for several waveforms of interest in television engineering.

22.2.11 TRANSIENT RESPONSE OF VIDEO SYSTEMS.
The transient performance of a video transmission system is revealed by the distortions suffered by a rectangular waveform (Fig. 22-7) in passing through the system. Useful information may be obtained from the quantity $(\sin x)/x$, which expresses the amplitude spectrum of the rectangular wave, and its integral $Si(x) = \int_0^x (\sin u)/u \, du$ which expresses the sum of the spectrum components, that is, the waveform after passage through a transmission system having uniform amplitude and phase characteristics up to a cutoff frequency x, beyond which the system has zero response. Figures 22-8 and 22-9 are plots of the respective

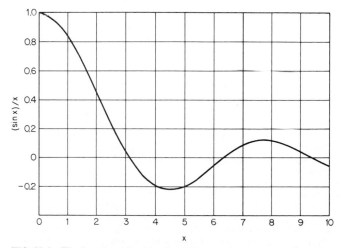

FIG. 22-8 The function $(\sin x)/x$, showing zero values at $x = \pi, 2\pi$, and 3π.

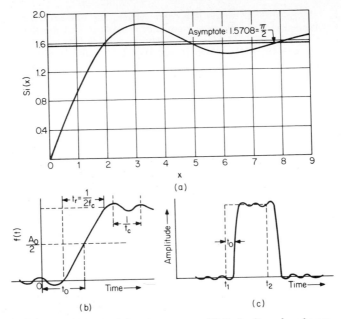

FIG. 22-9   (a) The $Si(x)$ function versus x; (b) the leading edge of a rectangular pulse after passage through a transmission system having a sharp cutoff at frequency f_c and a time delay of t_0; (c) the whole pulse at the output, shown in relation to the input pulse (dashed lines).

quantities. Table 22-9 gives the ordinates of these quantities. Figure 22-9c shows the $Si(x)$ function plotted as the output waveform in relation to the square-wave input waveform of Fig. 22-7 and identifies the cutoff and delay characteristics of the transmission system.

22.3 VISUAL DATA

22.3.1 UNITS OF VISUAL QUANTITIES. In television engineering, it is usual in texts and papers written in English to use British-American units (e.g., feet rather than meters) for visual quantities. The preferred units, used in other languages, are those of the International System of Units (Système International d'Unités, abbreviated SI). The SI units are widely used in other branches of electrical engineering and are coming into increased use in television. Table 22-10 gives the conversion factors between the SI units and the other commonly used units for visual quantities. These conversions should be used in connection with the text and illustrations that follow, which are stated in the units footlamberts, lumens per square foot (footcandles), etc., commonly used at present in television engineering.

22.3.2 VISUAL ACUITY. The ability of the eye to distinguish between small objects (and hence to resolve the details of an image) is known as visual acuity and is expressed in reciprocal minutes of the angle subtended at the eye by two objects which can just be separately identified. When the objects and background are displayed in black-and-white (as in monochrome television), at 100 percent contrast, the range of visual acuity

Table 22-10 Light Conversion Factors

*(Exact Conversions Are Shown in **Boldface** Type. Repeating Decimals Are Underlined.)*

A. Luminance units. The SI unit of luminance is the candela per square meter (cd/m²)†

	Candelas per square meter (cd/m²)	Candelas per square foot (cd/ft²)	Candelas per square inch (cd/in²)	Apostilbs (asb)	Stilbs (sb)	Lamberts (L)	Footlamberts (fL)
1 candela per square meter =	1	**0.092 903 04**	**6.451 6 × 10⁻⁴**	π = 3.141 592 65	**0.000 1**	**(0.000 1) π =** 3.141 592 65 × 10⁻⁴	0.291 863 51
1 candela per square foot =	10.763 910 4	1	**1/144 = 0.006 944 4̲4̲**	33.815 821 8	1.076 391 04 × 10⁻³	3.381 582 18 × 10⁻³	π = 3.141 592 65
1 candela per square inch =	1 550.003 1	144	1	4 869.478 4	0.155 000 31	0.486 947 84	452.389 342
1 apostilb =	1/π = 0.318 309 89	0.029 571 96	2.053 608 06 × 10⁻⁴	1	3.133 098 86 × 10⁻⁵	**0.000 1**	**0.092 903 04**
1 stilb =	**10 000**	**929.030 4**	**6.451 6**	31 415.926 5	1	π = 3.141 592 65	2 918.635
1 lambert =	10 000/π = 3 183.098 86	295.719 561	2.053 608 06	10 000	1/π = 0.318 309 89	1	**929.030 4**
1 footlambert =	3.426 259 1	1/π = 0.318 309 89	2.210 485 32 × 10⁻³	10.763 910 4	3.426 259 1 × 10⁻⁴	1.076 391 03 × 10⁻³	1

B. Illuminance units. The SI unit of illuminance is the lux (lux)‡

	Luxes (lx)	Phots (ph)	Footcandles (fc)	Lumens per square inch (lm/in²)
1 lux =	1	**0.000 1**	**0.092 903 04**	**6.451 6 × 10⁻⁴**
1 phot =	**10 000**	1	**929.030 4**	**6.451 6**
1 footcandle =	10.763 910 4	1.076 391 04 × 10⁻³	1	**1/144 = 0.006 944 4̲4̲**
1 lumen per square inch =	1 550.003 1	0.155 000 31	144	1

†*Note:* 1 nit (nt) = 1 candela per square meter (cd/m²).
1 stilb (sb) = 1 candela per square centimeter (cd/cm²).
‡*Note:* 1 lux (lux) = **1 lumen per square meter (lm/m²)**
1 phot (ph) = **1 lumen per square centimeter (lm/cm²)**
1 footcandle (fc) = **1 lumen per square foot (lm/ft²)**

Source: Fink and Beaty, *Standard Handbook for Electrical Engineers*, 11th ed., McGraw-Hill, New York, 1978.

extends from 0.2 to about 2.5 reciprocal minutes (5 to 0.4 minutes of arc, respectively). An acuity of 1 reciprocal minute is usually taken as the basis of television system design; at this value, stationary white points on two scanning lines separated by an intervening line, the remainder of the scanning pattern being dark, can be resolved at a distance of about 20 times the picture height. Adjacent scanning lines, properly interlaced, cannot ordinarily be distinguished at distances greater than six or seven times the picture height.

Visual acuity varies markedly with (1) the luminance of the background, (2) the contrast of the image, (3) the luminance of the area surrounding the image, and (4) the luminosity to which the eye is adapted. The acuity is approximately proportional to the logarithm of the background luminance, increasing from about 1 reciprocal minute at 1 ft·L (3.4 cd/m^2) background brightness and 100 percent contrast, to about 2 reciprocal minutes at 100 ft·L (340 cd/m^2). Figure 22-10 shows experimental data on visual acuity as a function of luminance, when the contrast is nearly 100 percent, using incandescent lamps as the illumination. These data apply to details viewed in the center of the field of view, i.e., by cone vision near the fovea of the retina. The acuity of rod vision, outside the foveal region, is poorer by a factor of about five times (i.e., about 0.2 reciprocal minute at 1 ft·L background luminance and 100 percent contrast).

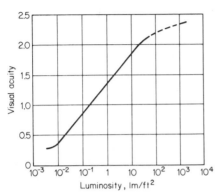

FIG. 22-10   Visual acuity as a function of the luminosity to which the eye is adapted. *(After R. J. Lythgoe.)*

Visual acuity falls off rapidly as the contrast of the image decreases. Under typical conditions (1 ft·L background luminance), the acuity increases from about 0.2 reciprocal minute at 10 percent contrast to about 1.0 reciprocal minute at 100 percent contrast, the acuity being roughly proportional to the percent contrast. Figure 22-11 shows experimental measurements of this effect.

When colored images are viewed, visual acuity depends markedly on the color. Acuity is higher for objects illuminated by monochromatic light (i.e., from a sodium lamp) than for those illuminated by a source of the same color having an extended spectrum: This improvement results from the lack of chromatic aberration in the eye. In a colored image reproduced by primary colors the acuity is highest for the green primary and lowest for the blue primary. This is explained only in part by the relative luminances of the primaries. The acuity for blue and red images is approximately two-thirds that for a white image of the same luminance (Baldwin[3,4]). For the green primary the acuity is about 90 percent that of a white image of the same luminance. When the effect of the relative contributions to luminance of the standard FCC-NTSC color primaries (approximately, green:red:blue = 6 to 3 to 1) is taken into account, the acuities for the primary images are in the approximate ratio green:red:blue = 8 to 3 to 1.

Visual acuity is also affected by glare, that is, regions in or near the field of view whose luminance is substantially greater than that of the object viewed. In television viewing acuity for an image fails rapidly if the area surrounding the image is brighter than the background luminance of the image. Acuity also decreases slightly if the image is viewed in darkness, that is, if the surround luminance is substantially lower than the average image luminance. Acuity is not adversely affected if the surround luminance has a value in the range from 0.1 to 1.0 times the average image luminance.

22.3.3 CONTRAST SENSITIVITY.

The ability of the eye to distinguish between the luminances of adjacent areas is known as contrast sensitivity; it is expressed as the ratio

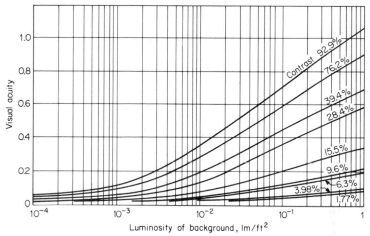

FIG. 22-11 Visual acuity as a function of luminosity and contrast. *(After J. P. Conner and R. E. Ganoung.)*

of the luminance to which the eye is adapted (usually the same as the background luminance) to the least perceptible luminance difference between the background luminance of the scene and the object luminance.

Two forms of contrast vision must be distinguished. In *rod vision*, which occurs at background luminance less than about 0.01 ft·L, the contrast sensitivity ranges from 2 to 5 (the average luminance is twice to five times the least perceptible luminance difference), increasing slowly as the background luminance increases from 0.0001 to 0.01 ft·L. In *cone vision*, which takes place above about 0.01 ft·L, the contrast sensitivity increases proportionately to background luminance from a value of 30 at 0.01 ft·L to 150 at 10 ft·L. The latter figure indicates that a luminance difference of 0.07 ft·L can be distinguished at a background luminance of 10 ft·L. Contrast sensitivity does not vary markedly with the color of the light. It is somewhat higher for blue light than for red at low background luminance, the reverse at higher luminance. Contrast sensitivity, like visual acuity, is reduced if the surround is brighter than the background. Figures 22-12 and 22-13 illustrate experimental data on contrast sensitivity.

22.3.4 FLICKER. The perceptibility of flicker varies so widely with viewing conditions† that it is difficult to describe quantitatively. In one respect, however, flicker phenomena are readily compared: When the viewing situation is constant (no change in the image or surround other than a proportional change in the luminance of all their parts, and no change in the conditions of observation), the luminance at which flicker just becomes perceptible varies logarithmically with the luminance (the Ferry-Porter law). The numerical relationship is: a positive increment in the flicker frequency of 12.6 Hz raises the luminance of flicker threshold 10 times. In television scanning the applicable flicker frequency is the field frequency, i.e., the rate at which the *area* of the image is successively illuminated. Table 22-11 gives the flicker-threshold luminance for various flicker frequencies, based on 180 ft·L for 60 fields per second.

---

†Among the factors affecting the flicker threshold are luminance of the flickering area, its color, the solid angle subtended by it at the eye, its absolute size, the luminance of the surround, the variation of luminance with time and position within the flickering area, and the adaptation and training of the observer.

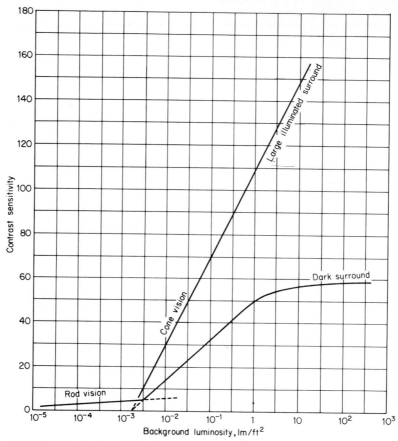

FIG. 22-12 Contrast sensitivity. *(After P. H. Moon.)*

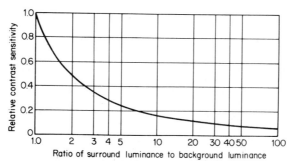

FIG. 22-13 Effect of surround luminance on contrast sensitivity. *(After P. H. Moon.)*

Table 22-11   Relative Flicker-Threshold Luminances†

Flicker frequency, Hz	System	Frames per second	Flicker-threshold luminance, ft·L
48	Home movies	16	20
48	Professional movies	24	20
50	Television scanning	25	29
60	Television scanning	30	180

†The luminances tabulated have only relative significance; they are based on a value of 180 ft·L for a flicker frequency of 60 Hz, which is typical performance under the American television standards.

22.3.5 ILLUMINATION AND LUMINANCE.   The input quantity to a television camera is the radiance of the various parts of the scene; that to the eye viewing the resulting television image is the luminance of the corresponding parts of the image.[5] If the spectral response of the camera has been adjusted to approximate that of the eye (this is the case in modern color cameras), it is appropriate to consider the camera input and the image output as the respective luminances (in common engineering parlance "brightness" is often used to mean luminance). The unit of luminance is the foot-lambert (the SI unit is the *candela per square meter,* see Table 22-10).

The luminance of a scene is inherent in those portions that are self-luminous; typical examples are a candle flame and an incandescent lamp. More commonly, the light enters the camera after reflection from a surface. The luminance, as viewed by the camera, then depends on the luminous radiation from the source in the direction of the surface, the position of the surface relative to the light source, and the reflectance of the surface (which is often a function of the angles from which it is illuminated and viewed). In view of the many variables affecting the luminance of reflecting objects, it is customary to dispense with calculations and to measure luminance with a light meter. In studio design it is necessary to know the candlepower of typical light sources and the distribution of light flux, as well as the reflectances of typical materials used in sets and costumes.

Table 22-12 gives the luminance of self-luminous sources. Table 22-13 gives the output in lumens and the efficiencies in lumens per watt of typical incandescent and fluorescent lamps.

22.3.6 LUMINANCE TRANSFER AND GAMMA CORRECTION.   When the luminances of a television image are directly proportional to the corresponding luminances of the televised scene, the system is said to be free of luminance distortion. Since such proportionality is not obtained in many of the transducers of a television system (notably the picture tube), most television systems display luminance distortion, unless the nonlinearity is specifically compensated. The basis of luminance-distortion compensation (usually called *gamma correction*) is the specific nonlinearity of the picture tube. The

Table 22-12   Luminance of Self-luminous Sources

Source	Luminance, ft·L	Source	Luminance, ft·L
Sun, at meridian	4.8×10^8	Tungsten filament	3,500,000
Sun, near horizon	1,750,000	Inside-frosted 60-W bulb	27,000
Moon	750	Fluorescent-lamp 40-W,	
Clear sky	2300	T-12 tube	1800
Overcast sky	650	High-intensity carbon arc	2×10^8
Candle flame	2900		

Table 22-13 Luminous Outputs and Efficiencies

Type and wattage of lamp	Initial lumens output	Initial lumens per watt
Incandescent:		
40-W	465	11.7
60-W	835	13.9
100-W	1,630	16.3
200-W	3,650	18.3
300-W	5,900	19.6
500-W	9,950	19.9
1,000-W	21,500	21.5
1,000-W, spot	22,500	22.5
5,000-W, studio lighting	164,000	32.7
10,000-W, studio lighting	325,000	32.7
Fluorescent:†		
15-W	600	40
20-W	860	43
30-W	1,380	46
40-W	2,100	43
100-W	4,000	40

†Hot-cathode type, 4500 K white.

light output of typical picture tubes versus control electrode voltage above cutoff is approximately a power function, the light output varying between the second and third power of the voltage. The gamma-correction systems applicable to the NTSC color television system are based on an exponent of 2.2. In the PAL and SECAM systems, the assumed exponent is 2.8. To compensate for the picture-tube characteristic, the amplitude of the luminance signal, as measured against black in the carrier envelope, is arranged to vary as the 2.2 (NTSC) or 2.8 (PAL/SECAM) root of the input luminance to the camera.

Table 22-14 Wavelengths of Various Radiations (in nanometers)

Spectrum region or line	Wavelength	Spectrum region or line	Wavelength
Ultraviolet	200–400	Peak of $\bar{x}$ distribution coefficient	600
Visible region (CIE limits)	380–780	Peak of $\bar{z}$ distribution coefficient	445
Visible region (practical		Peak of $\bar{r}$ distribution coefficient	605
limits)	400–700	Peak of $\bar{g}$ distribution coefficient	545
Violet region	400–450	Peak of $\bar{b}$ distribution coefficient	445
Blue region	450–490	Fraunhofer lines:	
Green region	490–550	A	766.1
Yellow region	550–590	B (oxygen)	687.0
Orange region	590–630	C (hydrogen)	656.28
Red region	630–700	D_1 (sodium)	589.59
Near infrared	700–3,000	D_3 (helium)	587.56
CIE red spectral primary	700	E (iron)	526.96
CIE green spectral primary	564.1	F (hydrogen)	486.14
		G (iron, calcium)	430.79
CIE blue spectral primary	435.8	H(calcium)	396.84
Peak of luminosity		K (calcium)	393.36
function ($\bar{y}$)	555		

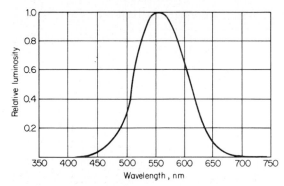

FIG. 22-14 Standard luminosity function (visibility curve) based on the CIE 1931 Standard Observer. The coordinates of this curve are given in Table 22-16, under the column labeled $\bar{y}$.

22.3.7 WAVELENGTHS OF RADIATION IN AND NEAR THE VISIBLE RANGE. The wavelengths of the various regions of the radiant-energy spectrum, spectral primary colors, and spectral lines are given in Table 22-14.

22.3.8 LUMINOSITY FUNCTION. The relative luminosities of the colors of the spectrum, that is, their relative luminance as evaluated by the average eye, is expressed as the *luminosity function*. This function, shown in Fig. 22-14, extends from 380 to 780 nm; its peak occurs at about 555 nm. The ordinates of this curve, at intervals of 10 nm, are given to four decimals in Table 22-16, under the distribution coefficient $\bar{y}$.

22.3.9 LUMINANCES OF PRIMARIES. When colors are reproduced by mixtures of primary colors, the contribution of each primary to the luminance of the mixture can be computed from their chromaticity coordinates (Table 22-17). Table 22-15 gives the relative contributions to luminance of typical primary colors in mixtures matched to illuminant C and illuminant D.

Illuminant C (average daylight) is the reference white initially specified in the NTSC recommendations to the FCC in 1953; this illuminant has a correlated color temperature

Table 22-15 Contributions to Luminance of Typical Primary Colors

Primary-color designation	Relative luminance when matching illuminant C	Relative luminance when matching illuminant D6500
CIE spectral primaries:		
Red	0.1536	
Green	0.8338	
Blue	0.0126	
	1.0000	
FCC-NTSC receiver primaries:		
Red	0.299	0.290
Green	0.587	0.606
Blue	0.114	0.104
	1.000	1.000

Table 22-16 CIE Colorimetric Data (1931 Standard Observer)

Wavelength, nm	Trichromatic coefficients		Distribution coefficients, equal-energy stimulus			Energy distributions for standard illuminants				Wavelength, nm
	r	g	$\bar{r}$	$\bar{g}$	$\bar{b}$	E_A	E_B	E_C	E_{D65}	
380	0.0272	−0.0115	0.0000	0.0000	0.0012	9.80	22.40	33.00	49.98	380
390	0.0263	−0.0114	0.0001	0.0000	0.0036	12.09	31.30	47.40	54.65	390
400	0.0247	−0.0112	0.0003	−0.0001	0.0121	14.71	41.30	63.30	82.75	400
410	0.0225	−0.0109	0.0008	−0.0004	0.0371	17.68	52.10	80.60	91.49	410
420	0.0181	−0.0094	0.0021	−0.0011	0.1154	20.99	63.20	98.10	93.43	420
430	0.0088	−0.0048	0.0022	−0.0012	0.2477	24.67	73.10	112.40	86.68	430
440	−0.0084	0.0048	−0.0026	0.0015	0.3123	28.70	80.80	121.50	104.86	440
450	−0.0390	0.0218	−0.0121	0.0068	0.3167	33.09	85.40	124.00	117.01	450
460	−0.0909	0.0517	−0.0261	0.0149	0.2982	37.81	88.30	123.10	117.81	460
470	−0.1821	0.1175	−0.0393	0.0254	0.2299	42.87	92.00	123.80	114.86	470
480	−0.3667	0.2906	−0.0494	0.0391	0.1449	48.24	95.20	123.90	115.92	480
490	−0.7150	0.6996	−0.0581	0.0569	0.0826	53.91	96.50	120.70	108.81	490
500	−1.1685	1.3905	−0.0717	0.0854	0.0478	59.86	94.20	112.10	109.35	500
510	−1.3371	1.9318	−0.0890	0.1286	0.0270	66.06	90.70	102.30	107.80	510
520	−0.9830	1.8534	−0.0926	0.1747	0.0122	72.50	89.50	96.90	104.79	520
530	−0.5159	1.4761	−0.0710	0.2032	0.0055	79.13	92.20	98.00	107.69	530
540	−0.1707	1.1628	−0.0315	0.2147	0.0015	85.95	96.90	102.10	104.41	540
550	0.0974	0.9051	0.0228	0.2118	−0.0006	92.91	101.00	105.20	104.05	550
560	0.3164	0.6881	0.0906	0.1970	−0.0013	100.00	102.80	105.30	100.00	560
570	0.4973	0.5067	0.1677	0.1709	−0.0014	107.18	102.60	102.30	96.33	570
580	0.6449	0.3579	0.2543	0.1361	−0.0011	114.44	101.00	97.80	95.79	580
590	0.7617	0.2402	0.3093	0.0975	−0.0008	121.73	99.20	93.20	88.69	590
600	0.8475	0.1537	0.3443	0.0625	−0.0005	129.04	98.00	89.70	90.01	600
610	0.9059	0.0494	0.3397	0.0356	−0.0003	136.35	98.50	88.40	89.60	610
620	0.9425	0.0580	0.2971	0.0183	−0.0002	143.62	99.70	88.10	87.70	620
630	0.9649	0.0354	0.2268	0.0083	−0.0001	150.84	101.00	88.00	83.29	630
640	0.9797	0.0205	0.1597	0.0033	0.0000	157.98	102.20	87.80	83.70	640
650	0.9888	0.0113	0.1017	0.0012	0.0000	165.03	103.90	88.20	80.03	650
660	0.9940	0.0061	0.0593	0.0004	0.0000	171.96	105.00	87.90	80.21	660
670	0.9966	0.0035	0.0315	0.0001	0.0000	178.77	104.90	86.30	82.28	670
680	0.9984	0.0016	0.0169	0.0000	0.0000	185.43	103.90	84.00	78.28	680
690	0.9996	0.0004	0.0082	0.0000	0.0000	191.93	101.60	80.20	69.72	690
700	1.0000	0.0000	0.0041	0.0000	0.0000	198.26	99.10	76.30	71.61	700
710	1.0000	0.0000	0.0021	0.0000	0.0000	204.41	96.20	72.40	74.35	710
720	1.0000	0.0000	0.0011	0.0000	0.0000	210.36	92.90	68.30	61.60	720
730	1.0000	0.0000	0.0005	0.0000	0.0000	216.12	89.40	64.40	69.89	730
740	1.0000	0.0000	0.0003	0.0000	0.0000	221.67	86.90	61.50	75.09	740
750	1.0000	0.0000	0.0001	0.0000	0.0000	227.00	85.20	59.20	63.59	750
760	1.0000	0.0000	0.0001	0.0000	0.0000	232.12	84.70	58.10	46.42	760
770	1.0000	0.0000	0.0000	0.0000	0.0000	237.01	85.40	58.20	66.81	770
780	1.0000	0.0000	0.0000	0.0000	0.0000	241.68	87.00	59.10	63.38	780

of 6770 K. Since the 1970s the reference white widely adopted for studio monitors is illuminant D6500, a slightly warmer white having a correlated color temperature of 6500 K.

It has been usual practice to match the reference white in receivers to a cooler white having a correlated color temperature of 9300 K. Recently, however, some receiver manufacturers have adopted illuminant D6500 for the reference white in receivers. See Fig. 22-17 for the location of these illuminants on the CIE chromaticity diagram.

22.3.10 CIE CHROMATICITY SYSTEM AND SPECIFICATIONS. The CIE system[6] of color specification is based on three standard monochromatic (spectral) primaries having the following wavelengths in nanometers: red, 700; green, 546.1; blue, 435.8 (Table 22-15). Any given color may be specified by three numbers, R, G, and B, called the *tristimulus values*. These are the respective amounts (lumens or watts) of the three stan-

Table 22-16 CIE Colorimetric Data (1931 Standard Observer) (*Continued*)

Wavelength, nm	Trichromatic coefficients		Distribution coefficients, equal-energy stimulus			Distribution coefficients weighted by Illuminant C			Wavelength, nm
	x	y	$\bar{x}$	$\bar{y}$	$\bar{z}$	$E_C\bar{x}$	$E_C\bar{y}$	$E_C\bar{z}$	
380	0.1741	0.0050	0.0014	0.0000	0.0065	0.0036	0.0000	0.0164	380
390	0.1738	0.0049	0.0042	0.0001	0.0201	0.0183	0.0004	0.0870	390
400	0.1733	0.0048	0.0143	0.0004	0.0679	0.0841	0.0021	0.3992	400
410	0.1726	0.0048	0.0435	0.0012	0.2074	0.3180	0.0087	1.5159	410
420	0.1714	0.0051	0.1344	0.0040	0.6456	1.2623	0.0378	6.0646	420
430	0.1689	0.0069	0.2839	0.0116	1.3856	2.9913	0.1225	14.6019	430
440	0.1644	0.0109	0.3483	0.0230	1.7471	3.9741	0.2613	19.9357	440
450	0.1566	0.0177	0.3362	0.0380	1.7721	3.9191	0.4432	20.6551	450
460	0.1440	0.0297	0.2908	0.0600	1.6692	3.3668	0.6920	19.3235	460
470	0.1241	0.0578	0.1954	0.0910	1.2876	2.2878	1.0605	15.0550	470
480	0.0913	0.1327	0.0956	0.1390	0.8130	1.1038	1.6129	9.4220	480
490	0.0454	0.2950	0.0320	0.2080	0.4652	0.3639	2.3591	5.2789	490
500	0.0082	0.5384	0.0049	0.3230	0.2720	0.0511	3.4077	2.8717	500
510	0.0139	0.7502	0.0093	0.5030	0.1582	0.0898	4.8412	1.5181	510
520	0.0743	0.8338	0.0633	0.7100	0.0782	0.5752	6.4491	0.7140	520
530	0.1547	0.8059	0.1655	0.8620	0.0422	1.5206	7.9357	0.3871	520
540	0.2296	0.7543	0.2904	0.9540	0.0203	2.7858	9.1470	0.1956	540
550	0.3016	0.6923	0.4334	0.9950	0.0087	4.2833	9.8343	0.0860	550
560	0.3731	0.6245	0.5945	0.9950	0.0039	5.8782	9.8387	0.0381	560
570	0.4441	0.5547	0.7621	0.9520	0.0021	7.3230	9.1476	0.0202	570
580	0.5125	0.4866	0.9163	0.8700	0.0017	8.4141	7.9897	0.0147	580
590	0.5752	0.4242	1.0263	0.7570	0.0011	8.9878	6.6283	0.0101	590
600	0.6270	0.3725	1.0622	0.6310	0.0008	8.9536	5.3157	0.0067	600
610	0.6658	0.3340	1.0026	0.5030	0.0003	8.3294	4.1788	0.0029	610
620	0.6915	0.3083	0.8544	0.3810	0.0002	7.0604	3.1485	0.0012	620
630	0.7079	0.2920	0.6424	0.2650	0.0000	5.3212	2.1948	0.0000	630
640	0.7190	0.2809	0.4479	0.1750	0.0000	3.6882	1.4411	0.0000	640
650	0.7260	0.2740	0.2835	0.1070	0.0000	2.3531	0.8876	0.0000	650
660	0.7300	0.2700	0.1649	0.0610	0.0000	1.3589	0.5028	0.0000	660
670	0.7320	0.2680	0.0874	0.0320	0.0000	0.7113	0.2606	0.0000	670
680	0.7334	0.2666	0.0468	0.0170	0.0000	0.3657	0.1329	0.0000	680
690	0.7344	0.2656	0.0227	0.0082	0.0000	0.1721	0.0621	0.0000	690
700	0.7347	0.2653	0.0114	0.0041	0.0000	0.0806	0.0290	0.0000	700
710	0.7347	0.2653	0.0058	0.0021	0.0000	0.0398	0.0143	0.0000	710
720	0.7347	0.2653	0.0029	0.0010	0.0000	0.0183	0.0064	0.0000	720
730	0.7347	0.2653	0.0014	0.0005	0.0000	0.0085	0.0030	0.0000	730
740	0.7347	0.2653	0.0007	0.0003	0.0000	0.0040	0.0017	0.0000	740
750	0.7347	0.2653	0.0003	0.0001	0.0000	0.0017	0.0006	0.0000	750
760	0.7347	0.2653	0.0002	0.0001	0.0000	0.0008	0.0003	0.0000	760
770	0.7347	0.2653	0.0001	0.0000	0.0000	0.0003	0.0000	0.0000	770
780	0.7347	0.2653	0.0000	0.0000	0.0000	0.0000	0.0000	0.0000	780

dard primaries needed to match the given color, each divided by the respective amount (lumens or watts) of the primary needed to match the standard reference-white light (equal-energy white, illuminant E, Fig. 22-19). To make the specification independent of the luminance of the given color, the tristimulus values are normalized to obtain the *chromaticity coordinates* (also known as *trichromatic coefficients*)

$$r = \frac{R}{R + G + B} \qquad g = \frac{G}{R + G + B} \qquad b = \frac{B}{R + G + B}$$

Since $r + g + b = 1$, only two of the chromaticity coordinates (say r and g) are needed to specify the hue and saturation of any color. These may be plotted in the *RGB chromaticity diagram* (Fig. 22-15). The values of r and g which specify the monochromatic colors of the spectrum define the spectrum locus on this diagram. The coordinates of this locus are given under r and g in Table 22-16.

Table 22-17 Chromaticity Coordinates (Trichromatic Coefficients) of Standard Illuminants, Other Light Sources, and Primaries

Designation	x	y	z
Illuminant A (2854 K)	0.4476	0.4075	0.1450
Illuminant B (4870 K)	0.3484	0.3516	0.3000
Illuminant C (6770 K)	0.3101	0.3162	0.3738
Illuminant D6500 (6500 K)	0.3127	0.3291	0.3582
Illuminant E (equal-energy white)	0.3333	0.3333	0.3333
Planckian radiator (black body):			
2000 K	0.5266	0.4133	0.0601
2848 K	0.4475	0.4075	0.1450
3500 K	0.4052	0.3907	0.2041
5000 K	0.3450	0.3516	0.3034
7500 K	0.3003	0.3103	0.3894
10,000 K	0.2806	0.2883	0.4311
Sunlight	0.336	0.350	0.314
Average daylight	0.313	0.328	0.359
North-sky daylight	0.277	0.293	0.430
Zenith sky	0.263	0.278	0.459
White-flame carbon arc	0.315	0.332	0.353
Fluorescent lamp (4500 K)	0.539	0.363	0.278
Fluorescent lamp (3500 K)	0.404	0.396	0.200
CIE standard red primary (700 nm)	0.7347	0.2653	0.0000
CIE standard green primary (546.1 nm)	0.2738	0.7174	0.0088
CIE standard blue primary (435.6 nm)	0.1666	0.0089	0.8245
NTSC primaries:			
Red	0.67	0.33	0.00
Green	0.21	0.71	0.08
Blue	0.14	0.08	0.78

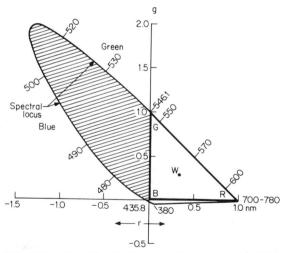

FIG. 22-15 The CIE primaries and spectral locus, plotted in terms of r and g. Note the large area involving negative values of r and small area having negative values of g. Detailed values of r and g for various wavelengths are given in Table 22-16.

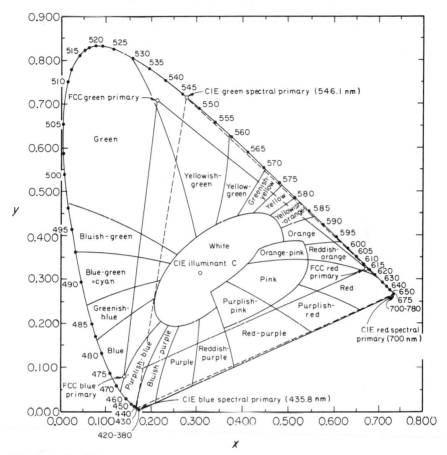

FIG. 22-16 The CIE chromaticity diagram, showing the CIE and FCC primaries, illuminant C, and the color names associated with various areas within the spectral locus. See Table 22-16 for x and y values associated with the spectral locus. [D. G. Fink and D. Christiansen (eds.), Electronics Engineers' Handbook, 2d ed., McGraw-Hill, New York, 1982.]

The color-mixture data, that is, the relative amounts of the primaries needed at each wavelength to match the equal-energy spectrum, are known as the distribution coefficients $\bar{r}$, $\bar{g}$, and $\bar{b}$. The distribution coefficients are in the same ratio as the chromaticity coordinates for any given spectral color; that is, $\bar{r}:\bar{g}:\bar{b} = r:g:b$. The $\bar{r}$, $\bar{g}$, and $\bar{b}$ values are listed in Table 22-16. Negative values of $\bar{r}$ and $\bar{g}$ occur; these imply that the primary in question is added to illuminant E to match the mixture of the remaining two primaries.

To avoid the negative values on the *RGB* chromaticity diagram, and to simplify colorimetric computations in other ways, the *RGB* diagram is linearly transformed to the *XYZ chromaticity diagram* (Figs. 22-16 to 22-18). This diagram is the accepted international standard of color specification. The quantities X, Y, and Z are tristimulus values analogous to the R, G, and B values described above; that is, they represent the amounts of three nonphysical ("fictitious") primaries required to match a given color, each divided by the respective amount needed to match illuminant E. The normalized values of these quantities (chromaticity coordinates) are $x = X/(X + Y + Z)$; $y = Y/(X + Y + Z)$; $z = Z/(X + Y + Z)$. The sum of the chromaticity coordinates of any

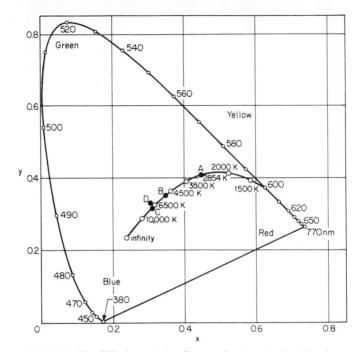

FIG. 22-17 The CIE chromaticity diagram showing the planckian locus (open circles for the color of black bodies at the indicated temperatures in absolute degrees). Also shown (black circles) are the coordinates of the standard illuminants A, B, C, and D6500.

color is unity; that is, $x + y + z = 1$. The XYZ diagram is customarily plotted in x and y; the z coordinate may be obtained from $z = 1 - (x + y)$.

The primaries on which the XYZ diagram is based lie outside the spectrum locus and hence do not correspond to physically realizable colors. They are, rather, mathematical abstractions which represent linear transformations of the RGB color-mixture data. The values of x and y specifying the spectrum locus are given in Table 22-16. The distribution coefficients $\bar{x}$ $\bar{y}$, and $\bar{z}$ are analogous to $\bar{r}$, $\bar{g}$, and $\bar{b}$; they represent the amounts of non-physical primaries required to match illuminant E at each wavelength. These values, all positive, are listed in Table 22-16. The $\bar{y}$ distribution coefficient has been chosen (by virtue of the choice of the transformation equations between RGB and XYZ) to be identical to the standard luminosity function. On this account the tristimulus value Y of any color is numerically equal to its luminance.

22.3.11 _RGB-XYZ_ TRANSFORMATION EQUATIONS. The transformations from RGB quantities to XYZ quantities and vice versa are expressed in simultaneous equations which relate the tristimulus values R, G, and B of any color in the RGB system to the corresponding tristimulus values X, Y, and Z in the XYZ system. These transformations are based on the CIE primaries matched to the equal-energy spectrum (illuminant E). The equations are:

$$X = 0.490R + 0.310G + 0.200B \tag{22-3}$$

$$Y = 0.177R + 0.813G + 0.011B \tag{22-4}$$

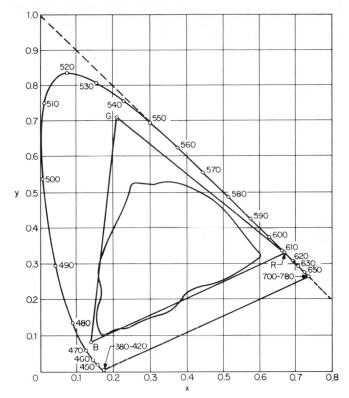

FIG. 22-18 The CIE chromaticity diagram showing the comparative color gamut of the NTSC/FCC primaries (triangle) and the approximate gamut of colored inks, pigments, and dyes. The superior rendition of color television at the extremes of red, green, and blue is evident.

$$Z = 0.000R + 0.010G + 0.990B \qquad (22\text{-}5)$$

$$R = 2.365X - 0.897Y - 0.468Z \qquad (22\text{-}6)$$

$$G = -0.515X + 1.426Y + 0.089Z \qquad (22\text{-}7)$$

$$B = 0.005X - 0.014Y + 1.009Z \qquad (22\text{-}8)$$

For the spectral colors, the particular values of the tristimulus values are the distribution coefficients $\bar{x}$, $\bar{y}$, and $\bar{z}$ and $\bar{r}$, $\bar{g}$, and $\bar{b}$, respectively. Accordingly Eqs. (22-3) to (22-8) may be used to transform one set of distribution coefficients to the other.

A different set of transformation equations applies to the NTSC-FCC standard color system since different primaries are used and the color matching is based on illuminant C.[7] The equations relating the two sets of tristimulus values in the NTSC-FCC colorimetric system are

$$X = 0.607R + 0.174G + 0.201B \qquad (22\text{-}9)$$

$$Y = 0.299R + 0.587G + 0.114B \qquad (22\text{-}10)$$

$$Z = 0.000R + 0.066G + 1.117B \qquad (22\text{-}11)$$

$$R = 1.910X - 0.533Y - 0.288Z \qquad (22\text{-}12)$$

$$G = -0.985X + 2.000Y - 0.028Z \tag{22-13}$$

$$B = 0.058X - 0.118Y + 0.896Z \tag{22-14}$$

It should be noted that the actual colorimetric measurements of tristimulus values must be performed in terms of the real primaries R, G, and B, since the nonphysical primaries X, Y, and Z cannot be realized. The corresponding tristimulus values, distribution coefficients and chromaticity coordinates in the XYZ system are then computed by the equations given in this paragraph and in Sec. 22.3.10.

22.3.12 CHROMATICITY COORDINATES OF STANDARD AND TYPICAL SOURCES. The x, y, and z chromaticity coordinates of typical illuminants, primary colors, and natural and artificial light sources are given in Table 22-17.

22.3.13 STANDARD ILLUMINANTS. The CIE set up three standard illuminants, representative of incandescent light (illuminant A), noon sunlight (illuminant B), and north-sky daylight (illuminant C). The relative energy distributions of these illuminants are given in Table 22-16 and their chromaticity coordinates in Table 22-17. Illuminant C is the reference white of the FCC-NTSC compatible color system. Accordingly, the chromaticity coordinates of an object illuminated by this white are of particular interest in color television. Table 22-16 gives the product of the illuminant C energy ordinate and the $\bar{x}$, $\bar{y}$, and $\bar{z}$ distribution coefficients under the headings $E_C\bar{x}$, $E_C\bar{y}$, and $E_C\bar{z}$. To find the chromaticity coordinates of a material illuminated by illuminant C whose reflectance R_λ is known at wavelength intervals of 10 mμ, the following relationships are employed:

$$X = \tfrac{1}{98} \sum_{\lambda=380}^{\lambda=780} R_\lambda (E_C\bar{x})_\lambda \tag{22-15}$$

$$Y = \tfrac{1}{100} \sum_{\lambda=380}^{\lambda=780} R_\lambda (E_C\bar{y})_\lambda \tag{22-16}$$

$$Z = \tfrac{1}{118} \sum_{\lambda=380}^{\lambda=780} R_\lambda (E_C\bar{z})_\lambda \tag{22-17}$$

$$x = \frac{X}{X+Y+Z} \qquad y = \frac{Y}{X+Y+Z} \qquad z = \frac{Z}{X+Y+Z} \tag{22-18}$$

In 1965, the CIE set up another standard white, illuminant D6500 (see Sec. 22.3.9), which is considered to be a closer match to average daylight than illuminant C. Since the correlated color temperatures of illuminants C and D6500 are nearly the same (6770 K and

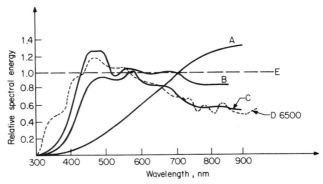

FIG. 22-19 Comparison of the spectral content of the standard illuminants A, B, C, and D6500. The horizontal dashed line is the hypothetical "equal energy" illuminant E.

6500 K, respectively), the equations above based on illuminant C can be used for most engineering purposes. Table 22-17 gives the x, y, and z coordinates of illuminant D6500. The position of illuminant D6500 on the CIE diagram is shown in Fig. 22-17. The close proximity of illuminants C and D is evident. Figure 22-19 shows the spectral distributions of illuminants A, B, C, and D6500, as well as the equal-energy white illuminant E.

The transformation equations between X, Y, and Z and R, G, and B, based on the FCC-NTSC primaries and illuminant D6500, are

$$X = 0.5879R + 0.1792G + 0.1831B \tag{22-19}$$

$$Y = 0.2896R + 0.6058G + 0.1046B \tag{22-20}$$

$$Z = 0.0000R + 0.06826G + 1.02B \tag{22-21}$$

$$R = 1.972X - 0.5497Y - 0.2975Z \tag{22-22}$$

$$G = -0.9534X + 1.936Y - 0.02741Z \tag{22-23}$$

$$B = 0.0638X - 0.1295Y + 0.9821Z \tag{22-24}$$

REFERENCES

1. P. Mertz and F. Gray, "A Theory of Scanning and Its Relation to the Characteristics of the Transmitted Signal in Telephotography and Television," *Bell System Tech. J.*, vol. 13, pp. 464–515, July 1934.

2. P. Mertz, "Television—The Scanning Process," *Proc. IRE,* vol. 29, pp. 529–537, October 1941.

3. M. W. Baldwin, Jr., "Subjective Sharpness of Additive Color Pictures," *Proc. IRE,* vol. 39, pp. 1173–1176, October 1951.

4. M. W. Baldwin, Jr., "The Subjective Sharpness of Simulated Television Images," *Proc. IRE,* vol. 28, p. 458, October 1940.

5. O. H. Schade, "Electro-Optical Characteristics of Television Systems," Pts. I–IV, *RCA Rev.,* March, June, September, and December 1948.

6. D. B. Judd, "The 1931 CIE Standard Observer and Coordinate System for Colorimetry," *J. Opt. Soc. Am.,* vol. 23, pp. 359–374, 1933.

7. W. T. Wintringham, "Color Television and Colorimetry," *Proc. IRE,* vol. 39, pp. 1135–1172, October 1951.

8. D. H. Pritchard and J. J. Gibson, "Worldwide Color TV Standards—Similarities and Differences," *J. Motion Picture Television Eng.,* vol 89, pp. 111–120, February 1980.

9. CCIR *Recommendations and Reports, CCIR, 15th Plen. Assy., Geneva, 1982,* Vol. XI, Broadcasting Service (Television), International Telecommunications Union, Geneva, 1982.

10. C. Bailey Neal, "Television Colorimetry for Receiver Engineers," *IEEE Trans. Broadcast and Television Receivers,* vol. BTR-19, no. 3, pp. 149–155, August 1973.

11. D. M. Zwick, "Television Receiver White Color: A Comparison of Picture Quality with White References of 9300 K and D6500," *IEEE Trans. Broadcast and Television Receivers,* vol. BTR-19, no. 4, November 1973.

BIBLIOGRAPHY OF PHYSICAL AND MATHEMATICAL DATA

The following list of handbooks and other sources of physical constants and mathematical data includes also references in the fields of electrical and photographic physics, electrical engineering, and electronics.

1. Donald G. Fink and H. Wayne Beaty (eds.), *Standard Handbook for Electrical Engineers,* 11th ed., McGraw-Hill, New York, 1978.

2. Donald G. Fink and Donald Christiansen (eds.), *Electronics Engineers' Handbook,* 2d ed., McGraw-Hill, New York, 1982.

3. R. G. Hudson, *The Engineer's Manual,* Wiley, New York, 1939.

4. *Handbook of Chemistry and Physics,* 62nd ed., CRC Press, Boca Raton, Fla., 1981.

5. O. W. Eshbach, *Handbook of Engineering Fundamentals,* 2d ed., Wiley, New York, 1936.

6. Keith Henney and Beverly Dudley, *Handbook of Photography,* McGraw-Hill, New York, 1939.

7. *International Critical Tables,* McGraw-Hill, New York, 1926–1933.

8. P. F. Smith and W. R. Longley, *Mathematical Tables and Formulas,* Wiley, New York, 1929.

9. Woodlief Thomas, Jr. (ed.), *SPSE Handbook for Photographic Science and Engineering,* Wiley, New York, 1973.

10. Keith Henney (ed.), *Radio Engineering Handbook,* 5th ed., McGraw-Hill, New York, 1959.

11. F. E. Terman, *Radio Engineers' Handbook,* McGraw-Hill, New York, 1943.

12. F. Langford-Smith (ed.), *Radiotron Designer's Handbook,* 4th ed., Radio Corporation of America, Harrison, N.J., 1953.

13. *Reference Data for Radio Engineers,* 5th ed., Sams, Indianapolis, 1968.

14. E. S. Allen, *Six-Place Tables,* 7th ed., McGraw-Hill, New York, 1947.

15. F. E. Fowle, *Smithsonian Physical Tables,* 9th ed., Smithsonian Institution, Washington, 1954.

16. Jahnke and Emde, *Tables of Functions,* Dover, New York, 1943.

17. H. B. Dwight, *Tables of Integrals and Other Mathematical Data,* 4th ed., Macmillan, New York, 1961.

Sources of Physical and Mathematical Data

The following mathematical functions, constants, and physical data are to be found in the references listed above. The boldface numbers refer to the corresponding volumes as listed therein.

Tables of mathematical functions

Arcs, chords—**3, 5**

Bessel functions—**13, 16, 17**

Cosine integral—**11, 16**

Error function—**4, 5, 14**

Exponential functions—**3, 4, 11, 13, 14, 15, 17**

Factorials—**4, 5, 12, 14, 15**

Hyperbolic functions (sines, cosines, tangents)—**2, 3, 4, 5, 8, 11, 12, 13, 14, 15**

Hyperbolic functions, logarithms of—**2, 4, 5, 15**

Logarithms, briggsian or common (base 10)—**3, 4, 5, 8, 11, 12, 13, 14, 15, 17**

Logarithms, napierian or natural (base 2.718282)—**3, 4, 5, 8, 11, 13, 14, 17**

Powers and roots—**3, 4, 5, 8, 14, 15**

Reciprocals—**3, 4, 5, 8, 15**

Sine integral—**11, 16**

Trigonometric functions, logarithms of (base 10)—3, 4, 5, 8, 13, 14, 17
Trigonometric functions, natural—3, 4, 5, 8, 11, 13, 14, 15, 17

Mathematical equations and formulas

Algebra—3, 4, 5, 8, 12, 13, 17
Determinants—5, 13
Differential calculus—3, 4, 8, 12, 13, 15, 17
Filter design—1, 10, 11, 12, 13
Fourier transforms—2, 11, 12, 13
Geometry—3, 4, 5, 8, 12
Hyperbolic functions—3, 4, 5, 11, 12, 13, 17
Integral calculus—3, 4, 5, 8, 11, 12, 13, 14, 15, 17
Laplace transforms—2, 13
Mathematical constants—2, 4, 5, 12, 14, 15, 16, 17
Matrices—2, 5
Mensuration—3, 4, 5, 13
Series expansions—2, 3, 4, 5, 11, 12, 13, 15, 17
Statistics—4, 5
Trigonometry—2, 3, 4, 5, 8, 11, 12, 13, 14, 17
Vector analysis—2, 3, 5, 12, 13

Properties of materials and components

Glass—2, 4, 7, 15
Insulators and dielectrics—1, 2, 5, 7, 13
Lens specifications—6
Luminescent materials—1, 2, 4, 7, 10, 12, 13
Magnetic materials—1, 2, 3, 4, 5, 7, 10, 13, 15
Optical filters—1, 6
Photographic constants—4, 6, 7, 15
Physical constants—1, 2, 4, 5, 7, 11, 13, 15
Wires and conductors—1, 2, 3, 4, 5, 7, 9, 10, 12, 13, 15

Index

1

ABOUT THE EDITOR IN CHIEF

K. BLAIR BENSON, currently an engineering consultant, has been in the forefront of many important advances in television technology. Beginning his career as an electrical engineer with General Electric, he joined the Columbia Broadcasting System Television Network as senior project engineer. From 1961 through 1966 he was responsible for the engineering design and installation of the CBS Television Network–New York Broadcast Center, a project that introduced many new techniques and equipment designs to broadcasting. He advanced to become vice president, technical development of CBS Electronics Video Recording Division. He later worked for Goldmark Communications Corporation as vice president of engineering and for Video Corp. of America as vice president of engineering and technical operations. A senior member of the Institute of Electrical and Electronics Engineers and a fellow of the Society of Motion Picture and Television Engineers, he has served on numerous engineering committees for both societies and for various terms as SMPTE Governor, television affairs vice president, and editorial vice president. He has written more than 40 scientific and technical papers on various aspects of television technology.